D1266398

Manual of Steel Construction

American Institute of Steel Construction (AISC)

Notes on the American Institute of Steel Construction (AISC) publication: ***Manual of Steel Construction***, now ***Steel Construction Manual***, with printings and revisions.

Manual of Steel Construction [Allowable Stress Design (ASD)]
Catalog main entry titles vary.

1st ed. 1927 (1928-1932 printing 2-10)
2nd ed. 1934 (1934-1936 printing 2-4)
3rd ed. 1937 (1939-1941 printing 2-4)
4th ed. 1941 (1942-1945 printing 2-5)
5th ed. 1946 (1947-162 printing 2-30)
6th ed. 1963 (1963-1964 printing with errata 1965-1967 revisions 1-4)
7th ed. 1970 (1970 printing 1973-1975 1st revision printing)

The above are accessible in PDF from the AISC website for current AISC members who are registered users. Membership> Become a Member> Student (complimentary)
https://www.aisc.org/publications/historic-steel-construction-manuals/

Lawrence Tech Library owns editions with an asterisk (*), call number, and permalink.

***6th ed. 1963** (1967 revision 4)
Manual of Steel Construction
TA 684 .A47 1967
https://ltu.on.worldcat.org/oclc/766820

***7th ed. 1970**
Manual of Steel Construction
TA 684 .A47 1970
https://ltu.on.worldcat.org/oclc/116402

***8th ed. 1980**
Manual of Steel Construction
TA 684 .A47 1980
https://ltu.on.worldcat.org/oclc/7273136

***9th ed. 1989**
Manual of Steel Construction: Allowable Stress Design
TA 684 .A47 1989
https://ltu.on.worldcat.org/oclc/20300365

Jonathan Tavarez, AISC Staff Engineer explains the “missing 10th, 11th, and 12th editions”:

There were nine ASD manuals (Edition 1-9) and three LRFD manuals (Edition 1-3). ASD and LRFD refer to two different design philosophies. The LRFD manuals were not intended to follow the ASD chronology of manuals, but to be independent documents. The 13th Edition Manual combined ASD and LRFD, and was named as such because there were a total of 12 manuals prior to this combination. Technically there is no officially-named 10th, 11th, or 12th Edition of the manual.

Manual of Steel Construction [Load & Resistance Factor Design (LRFD)]
Catalog main entry titles vary.

***1st ed. 1986**
Load & Resistance Factor Design: Manual of Steel Construction
TA 684 .M344 1986
https://ltu.on.worldcat.org/oclc/17729671

***2nd ed. 1998**
Load & Resistance Factor Design: Manual of Steel Construction
TA 684 .M344 1998 v. 1 and v. 2
https://ltu.on.worldcat.org/oclc/39955579
(Publication dates 1994, 1995-2nd ed. rev., 1998, 1998-2nd rev. of 2nd ed., 1999-metric conversion of 2nd ed. These are on assorted records owned by other libraries.)

***3rd ed. 2001**
Manual of Steel Construction. Load & Resistance Factor Design
TA 684 .M344 2001
https://ltu.on.worldcat.org/oclc/48651905

Steel Construction Manual** [combined **Allowable Stress Design (ASD)** and **Load & Resistance Factor Design (LRFD)]
Catalog main entry titles vary.

13th ed. 2005
Steel Construction Manual

***14th ed. 2011**
Steel Construction Manual
RESERVE TA 684 .A47 2011
https://ltu.on.worldcat.org/oclc/744979166

15th ed. 2017
Steel Construction Manual
Hardcover or digital edition currently for sale by AISC

Earlier editions have retention value for understanding older construction, industry changes, and research. Do not weed earlier print editions.

04/26/2022 - ssg

MANUAL
of
STEEL CONSTRUCTION

Load and Resistance Factor Design

Third Edition

ISBN 1-56424-051-7

The information presented in this publication has been prepared in accordance with recognized engineering principles and is for general information only. While it is believed to be accurate, this information should not be used or relied upon for any specific application without competent professional examination and verification of its accuracy, suitability, and applicability by a licensed professional engineer, designer, or architect. The publication of the material contained herein is not intended as a representation or warranty on the part of the American Institute of Steel Construction or of any other person named herein, that this information is suitable for any general or particular use or of freedom from infringement of any patent or patents. Anyone making use of this information assumes all liability arising from such use.

Caution must be exercised when relying upon other specifications and codes developed by other bodies and incorporated by reference herein since such material may be modified or amended from time to time subsequent to the printing of this edition. The Institute bears no responsibility for such material other than to refer to it and incorporate it by reference at the time of the initial publication of this edition.

Printed in the United States of America

First Printing: November 2001

PREFACE

This 3rd Edition *LRFD Manual of Steel Construction* is the twelfth major update of the AISC Manual of Steel Construction, which was first published in 1927. With this revision, member and connection design information has been condensed back into a single volume. It has been reorganized and reformatted to provide practical and efficient access to the information it contains, with a roadmap format to guide the user quickly to the applicable specifications, codes and standards, as well as the applicable provisions in those standards.

The following specifications, codes and standards are included in or with this Manual:

- 1999 *LRFD Specification for Structural Steel Buildings*
- 2000 *LRFD Specification for Steel Hollow Structural Sections*
- 2000 *LRFD Specification for Single-Angle Members*
- 2000 *RCSC Specification for Structural Joints Using ASTM A325 or A490 Bolts*
- 2000 *Code of Standard Practice for Steel Buildings and Bridges*
- AISC Shapes Database V3 CD

The following major improvements have been made in this revision:

- Workable gages for flange fasteners have been reintroduced.
- The revised *T*, *k* and k_1 values for W-shapes and the 0.93 wall-thickness reduction factor for HSS have been considered.
- Guidance is provided on the new OSHA safety regulations, stability bracing requirements and proper material specification.
- New information is provided on design drawing information requirements, criteria needed for connection design, mill, fabrication and erection tolerances, façade issues, temperature effects and fire protection requirements with summaries of common UL assemblies.
- Shape information has been updated to the current series.
- Coverage of round HSS has been added.
- Dimensions and properties have been added for double channels back-to-back.
- Tables of surface and box perimeter, weight/area-to-perimeter ratios and surface areas have been expanded to cover all common structural shapes.
- A new section on properly specifying materials, including shapes, plates, fasteners and other products, has been added.
- New information on corrosion protection and seismic design has been added.
- A new section has been added with design aids for tension members, including explicit consideration of net section requirements to ensure connectable member selection.
- Beam selection tables are included for selection based upon I_x, Z_x, I_y, and Z_y.
- Beam charts (ϕM_n vs. L_b) are plotted for both W-shapes and channels.
- New floor plate deflection and bending design aids have been added.
- Additional beam diagrams and formulas have been added.
- A new section has been added with design aids for W-shape beam-columns.
- New bolt length selection tables have been added.
- Bolt entering and tightening clearances have been updated.
- Bolting information has been updated for consistency with the 2000 RCSC Specification.

- Welding information, including the prequalified welded joint tables, has been updated to for consistency with AWS D1.1-2000.
- Information on prying action, Whitmore section and strength of coped beams has been updated.
- Selection tables for shear end-plate connections and single-plate connections have been improved and expanded, including single-plate connections with up to 12 rows of bolts and up to $1^1/_8$-in. diameter.
- New information and examples for flexible moment connections has been added as an update of Disque's historic "type 2 with wind" moment connection design approach.
- Previous limitations on the use of moment end-plate connections have been relaxed.
- Information on the design of anchor rods has been updated, including a new table of minimum dimensions for washers used with anchor rods.
- Composite member tables have been updated to include coverage of both 4 ksi and 5 ksi concrete.
- A cross-reference between U.S. customary and Metric shapes series has been included.

In addition, many other improvements have been made throughout this Manual.

By the AISC Committee on Manuals and Textbooks,

William A. Thornton, Chairman
Barry L. Barger, Vice Chairman

Charles J. Carter
Robert O. Disque
Marshall T. Ferrell
Lanny J. Flynn
Mark V. Holland
Bill R. Lindley II
Leonard R. Middleton
William C. Minchin
Thomas M. Murray
Charles R. Page
Davis G. Parsons II
David T. Ricker
Marc L. Sorenson
Scott T. Undershute
Gary C. Violette
Michael A. West
Heath E. Mitchell, Secretary

The Committee gratefully acknowledges the following people for their contributions to this Manual: Abbas Aminmansour, Roger L. Brockenbrough, Jennifer R. Ceccotti, Harry A. Cole, Richard A. DeVries, Guy J. Engebretson, Areti Gertos, Louis F. Geschwindner, Jr., John L. Harris III, Richard C. Kaehler, Suzanne W. Kaehler, Gerald F. Loberger, Jr., William T. Segui, Janet S. Tuegel, and Ramulu S. Vinnakota.

PART 1

DIMENSIONS AND PROPERTIES

SCOPE

The dimensions and properties for structural products commonly used in steel building design and construction are given in this part. For availability and proper material specifications for these products, as well as general specification requirements and other design considerations, see Part 2. For the design of members, see Parts 3 through 6. For the design of connections, see Parts 7 through 15. For AISC Specifications and Codes, see Part 16. For other miscellaneous information, see Part 17.

STRUCTURAL PRODUCTS

W-, M-, S- and HP-Shapes

Four types of H-shaped (or I-shaped) members are covered in this Manual:

- W-shapes, which have essentially parallel inner and outer flange surfaces.
- M-shapes, which are H-shapes that are not classified in ASTM A6 as W-, S- or HP-shapes.
- S-shapes (also known as American standard beams), which have a slope of approximately $16^2/_3$ percent (2 on 12) on the inner flange surfaces.
- HP-shapes (also known as bearing piles), which are similar to W-shapes, except their webs and flanges are of equal thickness and the depth and flange width are nominally equal for a given designation.

These shapes are designated by the mark W, M, S or HP, nominal depth (in.) and nominal weight (lbs/ft). For example, a W24×55 is a W-shape that is nominally 24-in. deep and weighs 55 lbs/ft.

The following dimensional and property information is given in this Manual for the W-, M-, S- and HP-shapes covered in ASTM A6:

- Design dimensions, detailing dimensions, and axial, strong-axis flexural and weak-axis flexural properties are given in Tables 1-1, 1-2, 1-3 and 1-4 for W-, M-, S- and HP-shapes, respectively.
- Torsional and flexural-torsional properties are given in Tables 1-25, 1-26, 1-27 and 1-28 for W-, M-, S- and HP-shapes, respectively.
- Surface areas, box perimeters and areas and W/D ratios are given in Tables 1-36, 1-37, 1-38 and 1-39 for W-, M-, S- and HP-shapes, respectively.
- SI-equivalent designations are given in Table 17-1 for W-shapes and in Table 17-2 for M-, S- and HP-shapes.

For the definitions of the tabulated variables, refer to the Nomenclature section at the back of this Manual.

Channels

Two types of channels are covered in this Manual:

- C-shapes (also known as American standard channels), which have a slope of approximately $16^2/_3$ percent (2 on 12) on the inner flange surfaces.
- MC-shapes (also known as miscellaneous channels), which have a slope other than $16^2/_3$ percent (2 on 12) on the inner flange surfaces.

These shapes are designated by the mark C or MC, nominal depth (in.) and nominal weight (lbs/ft). For example, a C12×25 is a C-shape that is nominally 12-in. deep and weighs 25 lbs/ft.

The following dimensional and property information is given in this Manual for the channels covered in ASTM A6:

- Design dimensions, detailing dimensions, and axial, strong-axis flexural and weak-axis flexural properties are given in Tables 1-5 and 1-6 for C- and MC-shapes, respectively.
- Torsional and flexural-torsional properties are given in Tables 1-29 and 1-30 for C- and MC-shapes, respectively.
- Surface areas, box perimeters and areas and W/D ratios are given in Tables 1-40 and 1-41 for C- and MC-shapes, respectively.
- SI-equivalent designations are given in Table 17-3.

For the definitions of the tabulated variables, refer to the Nomenclature section at the back of this Manual.

Angles

Angles (also known as L-shapes) have legs of equal thickness and either equal or unequal leg sizes. Angles are designated by the mark L, leg sizes (in.) and thickness (in.). For example, an L4×3×$^1/_2$ is an angle with one 4-in. leg, one 3-in. leg and $^1/_2$-in. thickness.

The following dimensional and property information is given in this Manual for the angles covered in ASTM A6:

- Design dimensions, detailing dimensions, and axial, strong-axis flexural and weak-axis flexural properties are given in Table 1-7. The effects of leg-to-leg and toe fillet radii have been considered in the determination of these section properties. For usual gages on legs of angles, see Figure 10-6.
- Torsional and flexural-torsional properties are given in Table 1-31. The effects of leg-to-leg and toe fillet radii have been considered in the determination of these properties.
- Surface areas, perimeters and W/D ratios are given in Table 1-42.
- SI-equivalent designations are given in Table 17-4.

For the definitions of the tabulated variables, refer to the Nomenclature section at the back of this Manual.

Structural Tees (WT-, MT- and ST-Shapes)

Three types of structural tees are covered in this Manual:

- WT-shapes, which are made from W-shapes.
- MT-shapes, which are made from M-shapes.
- ST-shapes, which are made from S-shapes.

These shapes are designated by the mark WT, MT or ST, nominal depth (in.) and nominal weight (lbs/ft). WT-, MT- and ST-shapes are split (sheared or flame-cut) from W-, M- and S-shapes, respectively, and have half the nominal depth and weight of that shape. For example, a WT12×27.5 is a structural tee split from a W-shape (W24×55), is nominally 12 in. deep and weighs 27.5 lbs/ft. Although off-center splitting or splitting on two lines can be obtained by special order, the resulting non-standard shape is not covered in this Manual.

The following dimensional and property information is given in this Manual for the structural tees cut from the W-, M- and S-shapes covered in ASTM A6:

- Design dimensions, detailing dimensions, and axial, strong-axis flexural and weak-axis flexural properties are given in Tables 1-8, 1-9 and 1-10 for WT-, MT- and ST-shapes, respectively.

- Torsional and flexural-torsional properties are given in Tables 1-32, 1-33 and 1-34 for WT-, MT- and ST-shapes, respectively.
- Surface areas, box perimeters and areas and W/D ratios are given in Tables 1-43, 1-44 and 1-45 for WT-, MT- and ST-shapes, respectively.
- SI-equivalent designations are given in Table 17-5 for WT-shapes and in Table 17-6 for MT- and ST-shapes.

For the definitions of the tabulated variables, refer to the Nomenclature section at the back of this Manual.

Hollow Structural Sections (HSS)

Two types of HSS are covered in this Manual:

- Rectangular (including square) HSS, which have an essentially rectangular (or square) cross-section, except for rounded corners, and uniform wall thickness, except at the weld seam(s).
- Round HSS, which have an essentially round cross-section and uniform wall thickness, except at the weld seam(s).

In both cases, ASTM A500 covers only electric-resistance-welded (ERW) HSS with a maximum periphery of 64 in. The coverage of HSS in this Manual is similarly limited.

Rectangular HSS are designated by the mark HSS, overall outside dimensions (in.) and wall thickness (in.), with all dimensions expressed as fractional numbers. For example, an HSS10×10×$^1/_2$ is nominally 10 in. by 10 in. with a $^1/_2$-in. wall thickness. Round HSS are designated by the term "HSS," nominal outside diameter (in.) and wall thickness (in.) with both dimensions expressed to three decimal places. For example, an HSS10.000×0.500 is nominally 10-in. in diameter with a $^1/_2$-in. nominal wall thickness. Some round HSS are configured to match the dimensional characteristics of steel pipe, such as an HSS5.563×0.258, which is the dimensional equivalent of a Pipe 5 Std. steel pipe.

For rectangular HSS, the outside corner radii are taken as $2t$ for ERW HSS, except that a centerline corner radius of $1.5t$ is used in all cases in the calculation of b/t and h/t ratios. Per HSS Specification Section 1.2.2, the wall thickness used in design is taken as 0.93 times the nominal wall thickness for ERW HSS. The rationale for this requirement is explained in the corresponding HSS Commentary Section 1.2.

The following dimensional and property information is given in this Manual for the HSS covered in ASTM A500, A501, A618 or A847:

- Design dimensions, detailing dimensions, and axial, strong-axis flexural and weak-axis flexural properties are given in Tables 1-11 and 1-12 for rectangular and round HSS, respectively.
- Torsional and flexural-torsional properties are given in Tables 1-11 and 1-12 for rectangular and round HSS, respectively.
- Surface areas, box perimeters and areas and A/P ratios are given in Tables 1-46 and 1-47 for rectangular and round HSS, respectively.
- SI-equivalent designations are given in Tables 17-7 and 17-8 for rectangular and round HSS, respectively.

For the definitions of the tabulated variables, refer to the Nomenclature section at the back of this Manual.

Steel Pipe

Steel pipes have an essentially round cross-section and uniform thickness, except at the weld seam(s) for welded steel pipe.

Steel pipes up to and including NPS 12 are designated by the term "Pipe," nominal diameter (in.) and weight class (Std., ×-strong, ××-strong). NPS stands for "nominal pipe size." For example, Pipe 5 Std. denotes a steel pipe with a 5-in. nominal diameter and a 0.258-in. wall thickness, which corresponds to the standard weight series. Steel pipes with wall thicknesses that do not correspond to the foregoing weight classes are designated by the term "Pipe," outside diameter (in.) and wall thickness (in.) with both expressed to three decimal places. For example, Pipe 14.000×0.375 and Pipe 5.563×0.500 are proper designations.

The following dimensional and property information is given in this Manual for the steel pipes covered in ASTM A53:

- Design dimensions, detailing dimensions, and axial, strong-axis flexural and weak-axis flexural properties are given in Table 1-13.
- Torsional and flexural-torsional properties are given in Table 1-13.
- Surface areas, box perimeters and areas and A/P ratios are given in Table 1-48.
- SI-equivalent designations are given in Table 17-9.

For the definitions of the tabulated variables, refer to the Nomenclature section at the back of this Manual.

Double Angles

Double angles (also known as 2L-shapes) are made with two angles that are interconnected through their back-to-back legs along the length of the member, either in contact for the full length or separated by spacers at the points of interconnection.

These shapes are designated by the mark 2L, the sizes and thickness of their legs (in.), and their orientation when the angle legs are not of equal size (LLBB or SLBB).[1] For example, a 2L4×3×1/2 LLBB has two angles with one 4-in. leg and one 3-in. leg and the 4-in. legs are back-to-back; a 2L4×3×1/2 SLBB is similar, except the 3-in. legs are back-to-back. In both cases, the legs are 1/2-in. thick.

The following dimensional and property information is given in this Manual for the double angles built-up from the angles covered in ASTM A6:

- Design dimensions, detailing dimensions, and axial, strong-axis flexural and weak-axis flexural properties are given in Table 1-14 for equal-leg, LLBB and SLBB angles. In each case, angle separations of zero in., 3/8 in. and 3/4 in. are covered. The effects of leg-to-leg and toe fillet radii have been considered in the determination of these section properties. For usual gages on legs of angles, see Figure 10-6.
- Torsional and flexural-torsional properties are given in Table 1-35 for equal-leg, LLBB and SLBB angles. In each case, angle separations of zero in., 3/8 in. and 3/4 in. are covered. The effects of leg-to-leg and toe fillet radii have been considered in the determination of these section properties.
- Surface areas, box perimeters and areas and W/D ratios are given in Tables 1-49, 1-50 and 1-51 for equal-leg, LLBB and SLBB angles, respectively. In each case, angle separations of zero in., 3/8 in. and 3/4 in. are covered.

---

[1] LLBB stands for long legs back-to-back. SLBB stands for short legs back-to-back. Alternatively, the orientations LLV and SLV, which stand for long legs vertical and short legs vertical, respectively, can be used.

For the definitions of the tabulated variables, refer to the Nomenclature section at the back of this Manual.

Double Channels

Double channels (also known as 2C- and 2MC-shapes) are made with two channels that are interconnected through their back-to-back webs along the length of the member, either in contact for the full length or separated by spacers at the points of interconnection.

These shapes are designated by the mark 2C or 2MC, nominal depth (in.), and nominal weight per channel (lb/ft). For example, a 2C12×25 is a double channel that consists of two channels that are each nominally 12-in. deep and each weigh 25 lbs/ft.

The following dimensional and property information is given in this Manual for the double channels built-up from the channels covered in ASTM A6:

- Design dimensions, detailing dimensions, and axial, strong-axis flexural and weak-axis flexural properties are given in Tables 1-15 and 1-16 for 2C- and 2MC-shapes, respectively. In each case, channel separations of zero, $^3/_8$ in. and $^3/_4$ in. are covered.
- Surface areas, box perimeters and areas and W/D ratios are given in Tables 1-52 and 1-53 for 2C- and 2MC-shapes, respectively. In each case, channel separations of zero, $^3/_8$ in. and $^3/_4$ in. are covered.

For the definitions of the tabulated variables, refer to the Nomenclature section at the back of this Manual.

W-Shapes with Cap Channels

W-shapes with cap channels are made with a W-shape and a channel (C- or MC-shape). The channel web is interconnected to the W-shape top flange with the flange toes down. The interconnection of the two elements must be designed for the horizontal shear q, where

$$q = \frac{VQ}{I}$$

where

q = horizontal shear, kips/in.
V = vertical shear, kips.
Q = first moment of the channel area about the neutral axis of the combined cross section, in.^{3}
I = moment of inertia of the combined cross-section, in.^{4}

The effects of other forces, such as crane horizontal and lateral forces, may also require consideration, when applicable.

The following dimensional and property information is given in this Manual for the W-shapes with cap channels built-up from the W-shapes and cap channels covered in ASTM A6:

- Design dimensions, detailing dimensions, and axial, strong-axis flexural and weak-axis flexural properties are given in Table 1-17.

For the definitions of the tabulated variables, refer to the Nomenclature section at the back of this Manual.

S-Shapes with Cap Channels

S-shapes with cap channels are made with an S-shape and a channel (C- or MC-shape). The channel web is interconnected to the S-shape top flange with the flange toes down. The interconnection of the two elements must be designed for the horizontal shear q calculated as given previously for W-shapes with cap channels.

The following dimensional and property information is given in this Manual for the S-shapes with cap channels built-up from the S-shapes and cap channels covered in ASTM A6:

- Design dimensions, detailing dimensions, and axial, strong-axis flexural and weak-axis flexural properties are given in Table 1-18.

For the definitions of the tabulated variables, refer to the Nomenclature section at the back of this Manual.

Plate Products

Weights and areas of structural plates and bars are given in Tables 1-19 through 1-21. Sheet and strip are distinguished from structural bars and plates by their dimensional characteristics as shown in Table 1-22. Weights of raised-pattern floor plates are given in Table 1-23.

The historical classification system for structural bars and plates suggests that there is only a physical difference between them based upon size (see Table 1-22) and production procedure. In raw form, flat stock has historically been classified as a bar if it is less than or equal to 8 in. wide and as a plate if it is greater than 8 in. wide. Bars are rolled between horizontal and vertical rolls and trimmed to length by shearing or flame cutting on the ends only. Plates are generally produced using one of three methods:

1. Sheared plates are rolled between horizontal rolls and trimmed to width and length by shearing or flame cutting on the edges and ends;
2. Universal mill (UM) plates are rolled between horizontal and vertical rolls and trimmed to length by shearing or flame cutting on the ends only; and,
3. Stripped plates are sheared or flame cut from wider sheared plates.

There is very little, if any, structural difference between plates and bars. Consequently, the term "plate" is becoming a universally applied term today and a PL$^1/_2 \times 4^1/_2 \times 1'$-3″, for example, might be fabricated from plate or bar stock.

For structural plates, the preferred practice is to specify thickness in $^1/_{16}$-in. increments up to $^3/_8$-in. thickness, $^1/_8$-in. increments over $^3/_8$-in. to 1-in. thickness and $^1/_4$-in. increments over 1-in. thickness. The current extreme widths for sheared and UM plates are 200 in. and 60 in., respectively. Because mill practice regarding plate widths vary, individual mills should be consulted to determine preferences.

For bars, the preferred practice is to specify width in $^1/_4$-in. increments and thickness and diameter in $^1/_8$-in. increments.

Crane Rails

Although crane rails are not listed as structural steel in Code of Standard Practice Section 2.1, this information is provided because some fabricators may choose to provide crane rails. Dimensions and properties for the crane rails shown are given in Table 1-24. For additional information or for profiles and properties of crane rails not listed, manufacturer's catalogs should be consulted. For crane-rail connections, see Part 15.

Other Structural Products

The following other structural products are covered in this Manual as indicated:

- High-strength bolts, common bolts, washers, nuts and direct tension indicators are covered in Part 7.
- Welding filler metals and fluxes are covered in Part 8.
- Forged steel structural hardware items, such as clevises, turnbuckles, sleeve nuts, recessed-pin nuts and cotter pins are covered in Part 9.
- Anchor rods and threaded rods are covered in Part 14.

STANDARD MILL PRACTICES

The production of structural products is subject to unavoidable variations relative to the theoretical dimensions and profiles, due to many factors, including roll wear, roll dressing practices and temperature effects. Such variations are limited by the dimensional and profile tolerances as summarized below.

Hot-Rolled Structural Shapes

Acceptable dimensional tolerances for hot-rolled structural shapes (W-, M-, S- and HP-shapes), channels (C- and MC-shapes) and angles are given in ASTM A6 Section 13 and summarized in Tables 1-54 through 1-58. Supplementary information can also be found in literature from structural shape producers and the Iron and Steel Society, a division of the American Institute of Mining, Metallurgical and Petroleum Engineers.

Hollow Structural Sections

Acceptable dimensional tolerances for HSS are given in ASTM A500 Section 10, A501 Section 11, A618 Section 8 or A847 Section 10, as applicable, and summarized in Tables 1-59 and 1-60, for rectangular and round HSS, respectively. Supplementary information can also be found in literature from HSS producers and the Steel Tube Institute, such as *Recommended Methods to Check Dimensional Tolerances on Hollow Structural Sections (HSS) Made to ASTM A500.*

Steel Pipe

Acceptable dimensional tolerances for steel pipes are given in ASTM A53 Section 12 and summarized in Table 1-60. Supplementary information can also be found in literature from steel pipe producers.

Plate Products

Acceptable dimensional tolerances for plate products are given in ASTM A6 Section 13 and summarized in Table 1-61. Note that plate thickness can be specified in inches or by weight per square foot, and separate tolerances apply to each method. No decimal edge thickness can be assured for plate specified by the latter method. Supplementary information, including permissible variations for sheet and strip and for other grades of steel, can also be found in literature from steel plate producers and the Iron and Steel Society, a division of the American Institute of Mining, Metallurgical and Petroleum Engineers.

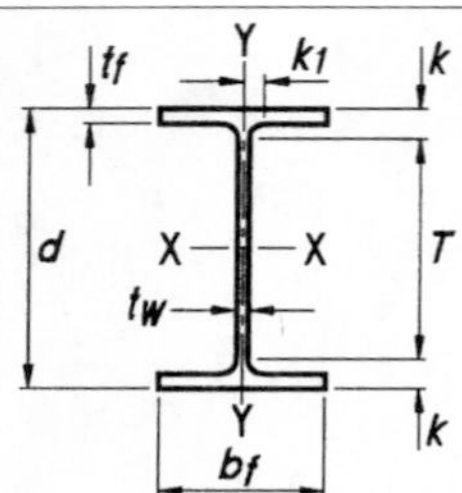

Table 1-1.
W-Shapes
Dimensions

Shape	Area, A	Depth, d		Web Thickness, t_w		Web $\frac{t_w}{2}$	Flange Width, b_f		Flange Thickness, t_f		Distance k		Distance k_1	Distance T	Workable Gage†
	in.²	in.		in.		in.	in.		in.		in.	in.	in.	in.	in.
W44×335*	98.3	44.0	44	1.02	1	1/2	16.0	16	1.77	1 3/4	2.56	2 5/8	1 5/16	38 3/4	5 1/2
×290	85.8	43.6	43 5/8	0.870	7/8	7/16	15.8	15 7/8	1.58	1 9/16	2.37	2 7/16	1 1/4	↓	↓
×262	77.2	43.3	43 1/4	0.790	13/16	7/16	15.8	15 3/4	1.42	1 7/16	2.21	2 1/4	1 3/16	↓	↓
×230	67.7	42.9	42 7/8	0.710	11/16	3/8	15.8	15 3/4	1.22	1 1/4	2.01	2 1/16	1 3/16	↓	↓
W40×593*	174	43.0	43	1.79	1 13/16	15/16	16.7	16 3/4	3.23	3 1/4	4.41	4 1/2	2 1/8	34	7 1/2
×503*	148	42.1	42	1.54	1 9/16	13/16	16.4	16 3/8	2.76	2 3/4	3.94	4	2	↓	↓
×431*	127	41.3	41 1/4	1.34	1 5/16	11/16	16.2	16 1/4	2.36	2 3/8	3.54	3 5/8	1 7/8	↓	↓
×397*	117	41.0	41	1.22	1 1/4	5/8	16.1	16 1/8	2.20	2 3/16	3.38	3 1/2	1 13/16	↓	↓
×372*	109	40.6	40 5/8	1.16	1 3/16	5/8	16.1	16	2.05	2 1/16	3.23	3 5/16	1 13/16	↓	↓
×362*	107	40.6	40 1/2	1.12	1 1/8	9/16	16.0	16	2.01	2	3.19	3 1/4	1 3/4	↓	↓
×324	95.3	40.2	40 1/8	1.00	1	1/2	15.9	15 7/8	1.81	1 13/16	2.99	3 1/16	1 11/16	↓	↓
×297	87.4	39.8	39 7/8	0.930	15/16	1/2	15.8	15 7/8	1.65	1 5/8	2.83	2 15/16	1 11/16	↓	↓
×277	81.4	39.7	39 3/4	0.830	13/16	7/16	15.8	15 7/8	1.58	1 9/16	2.76	2 7/8	1 5/8	↓	↓
×249	73.3	39.4	39 3/8	0.750	3/4	3/8	15.8	15 3/4	1.42	1 7/16	2.60	2 11/16	1 9/16	↓	↓
×215	63.4	39.0	39	0.650	5/8	5/16	15.8	15 3/4	1.22	1 1/4	2.40	2 1/2	1 9/16	↓	↓
×199	58.5	38.7	38 5/8	0.650	5/8	5/16	15.8	15 3/4	1.07	1 1/16	2.25	2 5/16	1 9/16	↓	↓
W40×392*	115	41.6	41 5/8	1.42	1 7/16	3/4	12.4	12 3/8	2.52	2 1/2	3.70	3 13/16	1 15/16	34	7 1/2
×331*	97.5	40.8	40 3/4	1.22	1 1/4	5/8	12.2	12 1/8	2.13	2 1/8	3.31	3 3/8	1 13/16	↓	↓
×327	96.0	40.8	40 3/4	1.18	1 3/16	5/8	12.1	12 1/8	2.13	2 1/8	3.31	3 3/8	1 13/16	↓	↓
×278	81.8	40.2	40 1/8	1.02	1	1/2	12.0	12	1.81	1 13/16	2.99	3 1/16	1 3/4	↓	↓
×264	77.6	40.0	40	0.960	15/16	1/2	11.9	11 7/8	1.73	1 3/4	2.91	3	1 11/16	↓	↓
×235	69.0	39.7	39 3/4	0.830	13/16	7/16	11.9	11 7/8	1.58	1 9/16	2.76	2 7/8	1 5/8	↓	↓
×211	62.0	39.4	39 3/8	0.750	3/4	3/8	11.8	11 3/4	1.42	1 7/16	2.60	2 11/16	1 9/16	↓	↓
×183	53.8	39.0	39	0.650	5/8	5/16	11.8	11 3/4	1.22	1 1/4	2.40	2 1/2	1 9/16	↓	↓
×167	49.2	38.6	38 5/8	0.650	5/8	5/16	11.8	11 3/4	1.02	1	2.21	2 5/16	1 9/16	↓	↓
×149	43.8	38.2	38 1/4	0.630	5/8	5/16	11.8	11 3/4	0.830	13/16	2.01	2 1/8	1 1/2	↓	↓
W36×798*	235	42.0	42	2.38	2 3/8	1 3/16	18.0	18	4.29	4 5/16	5.24	5 9/16	2 3/8	30 7/8	7 1/2
×650*	191	40.5	40 1/2	1.97	2	1	17.6	17 5/8	3.54	3 9/16	4.49	4 13/16	2 3/16	↓	↓
×527*	155	39.2	39 1/4	1.61	1 5/8	13/16	17.2	17 1/4	2.91	2 15/16	3.86	4 3/16	2	↓	↓
×439*	129	38.3	38 1/4	1.36	1 3/8	11/16	17.0	17	2.44	2 7/16	3.39	3 11/16	1 7/8	↓	↓
×393*	116	37.8	37 3/4	1.22	1 1/4	5/8	16.8	16 7/8	2.20	2 3/16	3.15	3 7/16	1 13/16	↓	↓
×359*	105	37.4	37 3/8	1.12	1 1/8	9/16	16.7	16 3/4	2.01	2	2.96	3 1/4	1 3/4	↓	↓
×328*	96.4	37.1	37 1/8	1.02	1	1/2	16.6	16 5/8	1.85	1 7/8	2.80	3 1/8	1 3/4	↓	↓
×300	88.3	36.7	36 3/4	0.945	15/16	1/2	16.7	16 5/8	1.68	1 11/16	2.63	2 15/16	1 11/16	↓	↓
×280	82.4	36.5	36 1/2	0.885	7/8	7/16	16.6	16 5/8	1.57	1 9/16	2.52	2 13/16	1 5/8	↓	↓
×260	76.5	36.3	36 1/4	0.840	13/16	7/16	16.6	16 1/2	1.44	1 7/16	2.39	2 11/16	1 5/8	↓	↓
×245	72.1	36.1	36 1/8	0.800	13/16	7/16	16.5	16 1/2	1.35	1 3/8	2.30	2 5/8	1 5/8	↓	↓
×230	67.6	35.9	35 7/8	0.760	3/4	3/8	16.5	16 1/2	1.26	1 1/4	2.21	2 1/2	1 9/16	↓	↓
W36×256	75.4	37.4	37 3/8	0.960	15/16	1/2	12.2	12 1/4	1.73	1 3/4	2.48	2 5/8	1 5/16	32 1/8	5 1/2
×232	68.1	37.1	37 1/8	0.870	7/8	7/16	12.1	12 1/8	1.57	1 9/16	2.32	2 7/16	1 1/4	↓	↓
×210	61.8	36.7	36 3/4	0.830	13/16	7/16	12.2	12 1/8	1.36	1 3/8	2.11	2 5/16	1 1/4	↓	↓
×194	57.0	36.5	36 1/2	0.765	3/4	3/8	12.1	12 1/8	1.26	1 1/4	2.01	2 3/16	1 3/16	↓	↓
×182	53.6	36.3	36 3/8	0.725	3/4	3/8	12.1	12 1/8	1.18	1 3/16	1.93	2 1/8	1 3/16	↓	↓
×170	50.1	36.2	36 1/8	0.680	11/16	3/8	12.0	12	1.10	1 1/8	1.85	2	1 3/16	↓	↓
×160	47.0	36.0	36	0.650	5/8	5/16	12.0	12	1.02	1	1.77	1 15/16	1 1/8	↓	↓
×150	44.2	35.9	35 7/8	0.625	5/8	5/16	12.0	12	0.940	15/16	1.69	1 7/8	1 1/8	↓	↓
×135	39.7	35.6	35 1/2	0.600	5/8	5/16	12.0	12	0.790	13/16	1.54	1 11/16	1 1/8	↓	↓

*ASTM A6 tensile group 4 or 5 shape. Special requirements may apply per LRFD Specification Section A3.1c.
†See definition of "Workable Gage" in Nomenclature section at the back of this Manual.

Table 1-1 (cont.). W-Shapes Properties

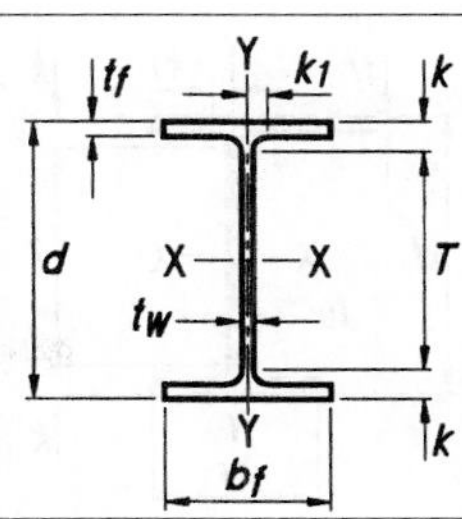

Nominal Wt.	Compact Section Criteria			X_1	$X_2 \times 10^6$	Axis X-X				Axis Y-Y			
	$\frac{b_f}{2t_f}$	$\frac{h}{t_w}$	F_y'''			I	S	r	Z	I	S	r	Z
lb/ft			ksi	ksi	(1/ksi)^2	in.^{4}	in.^{3}	in.	in.^{3}	in.^{4}	in.^{3}	in.	in.^{3}
335	4.51	38.1	44.3	2430	5110	31100	1410	17.8	1620	1200	151	3.50	236
290	5.01	44.7	32.2	2150	8170	27100	1240	17.8	1420	1050	132	3.49	206
262	5.55	49.2	26.6	1930	12300	24200	1120	17.7	1270	927	118	3.46	183
230	6.45	54.8	21.5	1690	21100	20800	971	17.5	1100	796	101	3.43	157
593	2.58	19.1	–	4760	348	50400	2340	17.0	2760	2520	302	3.80	481
503	2.97	22.2	–	4110	620	41700	1980	16.8	2320	2050	250	3.72	395
431	3.44	25.5	–	3550	1100	34800	1690	16.6	1960	1690	208	3.65	328
397	3.66	28.0	–	3310	1440	32000	1560	16.6	1800	1540	191	3.64	300
372	3.92	29.5	–	3080	1880	29600	1460	16.5	1680	1420	177	3.60	278
362	3.99	30.5	–	3040	2010	28900	1420	16.5	1640	1380	173	3.60	270
324	4.40	34.2	55.1	2720	3050	25600	1280	16.4	1460	1220	153	3.58	239
297	4.80	36.8	47.7	2500	4250	23200	1170	16.3	1330	1090	138	3.54	215
277	5.03	41.2	38.0	2360	5280	21900	1100	16.4	1250	1040	132	3.58	204
249	5.55	45.6	31.0	2130	7800	19600	993	16.3	1120	926	118	3.55	182
215	6.45	52.6	23.3	1850	13600	16700	859	16.2	964	796	101	3.54	156
199	7.39	52.6	23.3	1700	20000	14900	770	16.0	869	695	88.2	3.45	137
392	2.45	24.1	–	3910	852	29900	1440	16.1	1710	803	130	2.64	212
331	2.86	28.0	–	3360	1550	24700	1210	15.9	1430	644	106	2.57	172
327	2.85	29.0	–	3320	1620	24500	1200	16.0	1410	640	105	2.58	170
278	3.31	33.5	57.3	2860	2910	20500	1020	15.8	1190	521	87.1	2.52	140
264	3.45	35.6	50.8	2720	3500	19400	971	15.8	1130	493	82.6	2.52	132
235	3.77	41.2	38.0	2440	5200	17400	875	15.9	1010	444	74.6	2.54	118
211	4.17	45.6	31.0	2210	7680	15500	786	15.8	906	390	66.1	2.51	105
183	4.84	52.6	23.3	1920	13300	13300	683	15.7	783	336	56.9	2.50	89.6
167	5.76	52.6	23.3	1750	20700	11600	600	15.3	693	283	47.9	2.40	76.0
149	7.11	54.3	21.9	1580	33400	9800	513	15.0	598	229	38.8	2.29	62.2
798	2.10	13.2	–	6670	91.2	62600	2980	16.3	3580	4200	467	4.23	743
650	2.48	16.0	–	5560	182	48900	2420	16.0	2860	3230	367	4.11	580
527	2.96	19.6	–	4620	374	38300	1950	15.7	2280	2490	289	4.01	454
439	3.48	23.1	–	3900	721	31000	1620	15.5	1870	1990	235	3.93	367
393	3.82	25.8	–	3530	1070	27500	1450	15.4	1670	1750	208	3.90	325
359	4.16	28.1	–	3230	1490	24800	1320	15.3	1510	1570	188	3.86	292
328	4.49	30.9	–	2980	2050	22500	1210	15.3	1380	1420	171	3.84	265
300	4.96	33.3	58.0	2720	2930	20300	1110	15.2	1260	1300	156	3.83	241
280	5.29	35.6	50.9	2560	3730	18900	1030	15.1	1170	1200	144	3.81	223
260	5.75	37.5	45.8	2370	5100	17300	953	15.0	1080	1090	132	3.78	204
245	6.11	39.4	41.6	2230	6430	16100	895	15.0	1010	1010	123	3.75	190
230	6.54	41.4	37.5	2100	8190	15000	837	14.9	943	940	114	3.73	176
256	3.53	33.8	56.3	2830	2910	16800	895	14.9	1040	528	86.5	2.65	137
232	3.86	37.3	46.2	2570	4200	15000	809	14.8	936	468	77.2	2.62	122
210	4.48	39.1	42.1	2320	6560	13200	719	14.6	833	411	67.5	2.58	107
194	4.81	42.4	35.7	2140	8850	12100	664	14.6	767	375	61.9	2.56	97.7
182	5.12	44.8	32.1	2020	11200	11300	623	14.5	718	347	57.6	2.55	90.7
170	5.47	47.7	28.2	1900	14500	10500	581	14.5	668	320	53.2	2.53	83.8
160	5.88	49.9	25.8	1780	18600	9760	542	14.4	624	295	49.1	2.50	77.3
150	6.37	51.9	23.9	1680	24200	9040	504	14.3	581	270	45.1	2.47	70.9
135	7.56	54.1	22.0	1520	37900	7800	439	14.0	509	225	37.7	2.38	59.7

– in F_y''' column indicates $F_y''' > 65$ ksi.

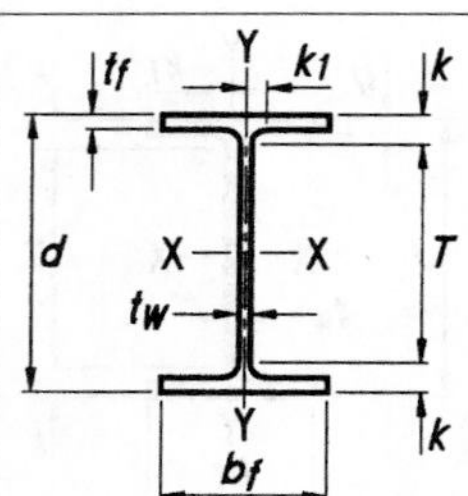

Table 1-1 (cont.). W-Shapes Dimensions

Shape	Area, A	Depth, d		Web Thickness, t_w		Web $\frac{t_w}{2}$	Flange Width, b_f		Flange Thickness, t_f		Distance k		Distance k_1	Distance T	Workable Gage†
	in.^{2}	in.		in.		in.	in.		in.		in.	in.	in.	in.	in.
W33×387*	114	36.0	36	1.26	1 1/4	5/8	16.2	16 1/4	2.28	2 1/4	3.07	3 3/16	1 7/16	29 5/8	5 1/2
×354*	104	35.6	35 1/2	1.16	1 3/16	5/8	16.1	16 1/8	2.09	2 1/16	2.88	2 15/16	1 3/8		
×318*	93.6	35.2	35 1/8	1.04	1 1/16	9/16	16.0	16	1.89	1 7/8	2.68	2 3/4	1 5/16		
×291	85.7	34.8	34 7/8	0.960	15/16	1/2	15.9	15 7/8	1.73	1 3/4	2.52	2 5/8	1 5/16		
×263	77.5	34.5	34 1/2	0.870	7/8	7/16	15.8	15 3/4	1.57	1 9/16	2.36	2 7/16	1 1/4		
×241	71.0	34.2	34 1/8	0.830	13/16	7/16	15.9	15 7/8	1.40	1 3/8	2.19	2 1/4	1 1/4		
×221	65.2	33.9	33 7/8	0.775	3/4	3/8	15.8	15 3/4	1.27	1 1/4	2.06	2 1/8	1 3/16		
×201	59.2	33.7	33 5/8	0.715	11/16	3/8	15.7	15 3/4	1.15	1 1/8	1.94	2	1 3/16		
W33×169	49.5	33.8	33 7/8	0.670	11/16	3/8	11.5	11 1/2	1.22	1 1/4	1.92	2 1/8	1 3/16	29 5/8	5 1/2
×152	44.8	33.5	33 1/2	0.635	5/8	5/16	11.6	11 5/8	1.06	1 1/16	1.76	1 15/16	1 1/8		
×141	41.6	33.3	33 1/4	0.605	5/8	5/16	11.5	11 1/2	0.960	15/16	1.66	1 13/16	1 1/8		
×130	38.3	33.1	33 1/8	0.580	9/16	5/16	11.5	11 1/2	0.855	7/8	1.56	1 3/4	1 1/8		
×118	34.7	32.9	32 7/8	0.550	9/16	5/16	11.5	11 1/2	0.740	3/4	1.44	1 5/8	1 1/8		
W30×391*	115	33.2	33 1/4	1.36	1 3/8	11/16	15.6	15 5/8	2.44	2 7/16	3.23	3 3/8	1 1/2	26 1/2	5 1/2
×357*	105	32.8	32 3/4	1.24	1 1/4	5/8	15.5	15 1/2	2.24	2 1/4	3.03	3 1/8	1 7/16		
×326*	95.8	32.4	32 3/8	1.14	1 1/8	9/16	15.4	15 3/8	2.05	2 1/16	2.84	2 15/16	1 3/8		
×292*	85.9	32.0	32	1.02	1	1/2	15.3	15 1/4	1.85	1 7/8	2.64	2 3/4	1 5/16		
×261	76.9	31.6	31 5/8	0.930	15/16	1/2	15.2	15 1/8	1.65	1 5/8	2.44	2 9/16	1 5/16		
×235	69.2	31.3	31 1/4	0.830	13/16	7/16	15.1	15	1.50	1 1/2	2.29	2 3/8	1 1/4		
×211	62.2	30.9	31	0.775	3/4	3/8	15.1	15 1/8	1.32	1 5/16	2.10	2 1/4	1 3/16		
×191	56.3	30.7	30 5/8	0.710	11/16	3/8	15.0	15	1.19	1 3/16	1.97	2 1/16	1 3/16		
×173	51.0	30.4	30 1/2	0.655	5/8	5/16	15.0	15	1.07	1 1/16	1.85	2	1 1/8		
W30×148	43.5	30.7	30 5/8	0.650	5/8	5/16	10.5	10 1/2	1.18	1 3/16	1.83	2 1/16	1 1/8	26 1/2	5 1/2
×132	38.9	30.3	30 1/4	0.615	5/8	5/16	10.5	10 1/2	1.00	1	1.65	1 7/8	1 1/8		
×124	36.5	30.2	30 1/8	0.585	9/16	5/16	10.5	10 1/2	0.930	15/16	1.58	1 13/16	1 1/8		
×116	34.2	30.0	30	0.565	9/16	5/16	10.5	10 1/2	0.850	7/8	1.50	1 3/4	1 1/8		
×108	31.7	29.8	29 7/8	0.545	9/16	5/16	10.5	10 1/2	0.760	3/4	1.41	1 11/16	1 1/8		
×99	29.1	29.7	29 5/8	0.520	1/2	1/4	10.5	10 1/2	0.670	11/16	1.32	1 9/16	1 1/16		
×90	26.4	29.5	29 1/2	0.470	1/2	1/4	10.4	10 3/8	0.610	5/8	1.26	1 1/2	1 1/16		
W27×539*	159	32.5	32 1/2	1.97	2	1	15.3	15 1/4	3.54	3 9/16	4.33	4 7/16	1 13/16	23 5/8	5 1/2
×368*	108	30.4	30 3/8	1.38	1 3/8	11/16	14.7	14 5/8	2.48	2 1/2	3.27	3 3/8	1 1/2		5 1/2
×336*	98.9	30.0	30	1.26	1 1/4	5/8	14.6	14 1/2	2.28	2 1/4	3.07	3 3/16	1 7/16		
×307*	90.4	29.6	29 5/8	1.16	1 3/16	5/8	14.4	14 1/2	2.09	2 1/16	2.88	3	1 7/16		
×281*	82.9	29.3	29 1/4	1.06	1 1/16	9/16	14.4	14 3/8	1.93	1 15/16	2.72	2 13/16	1 3/8		
×258	76.0	29.0	29	0.980	1	1/2	14.3	14 1/4	1.77	1 3/4	2.56	2 11/16	1 5/16		
×235	69.4	28.7	28 5/8	0.910	15/16	1/2	14.2	14 1/4	1.61	1 5/8	2.40	2 1/2	1 5/16		
×217	64.0	28.4	28 3/8	0.830	13/16	7/16	14.1	14 1/8	1.50	1 1/2	2.29	2 3/8	1 1/4		
×194	57.2	28.1	28 1/8	0.750	3/4	3/8	14.0	14	1.34	1 5/16	2.13	2 1/4	1 3/16		
×178	52.5	27.8	27 3/4	0.725	3/4	3/8	14.1	14 1/8	1.19	1 3/16	1.98	2 1/16	1 3/16		
×161	47.6	27.6	27 5/8	0.660	11/16	3/8	14.0	14	1.08	1 1/16	1.87	2	1 3/16		
×146	43.1	27.4	27 3/8	0.605	5/8	5/16	14.0	14	0.975	1	1.76	1 7/8	1 1/8		
W27×129	37.8	27.6	27 5/8	0.610	5/8	5/16	10.0	10	1.10	1 1/8	1.70	2	1 1/8	23 5/8	5 1/2
×114	33.5	27.3	27 1/4	0.570	9/16	5/16	10.1	10 1/8	0.930	15/16	1.53	1 13/16	1 1/8		
×102	30.0	27.1	27 1/8	0.515	1/2	1/4	10.0	10	0.830	13/16	1.43	1 3/4	1 1/16		
×94	27.7	26.9	26 7/8	0.490	1/2	1/4	9.99	10	0.745	3/4	1.34	1 5/8	1 1/16		
×84	24.8	26.7	26 3/4	0.460	7/16	1/4	9.96	10	0.640	5/8	1.24	1 9/16	1 1/16		

*ASTM A6 tensile group 4 or 5 shape. Special requirements may apply per LRFD Specification Section A3.1c.

†See definition of "Workable Gage" in Nomenclature section at the back of this Manual.

Table 1-1 (cont.). W-Shapes Properties

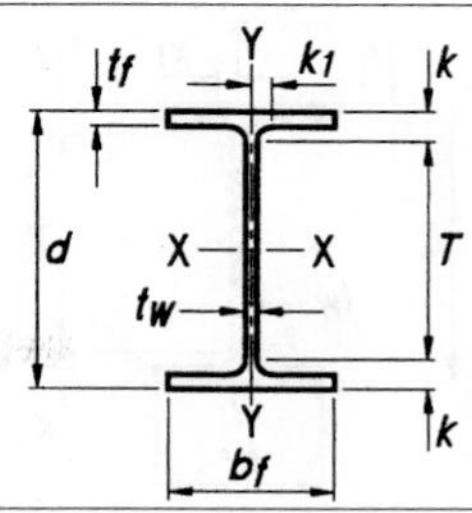

Nominal Wt.	Compact Section Criteria			X_1	$X_2 \times 10^6$	Axis X-X				Axis Y-Y			
	$\frac{b_f}{2t_f}$	$\frac{h}{t_w}$	F_y'''			I	S	r	Z	I	S	r	Z
lb/ft			ksi	ksi	(1/ksi)^2	in.^{4}	in.^{3}	in.	in.^{3}	in.^{4}	in.^{3}	in.	in.^{3}
387	3.55	23.7	–	3850	752	24300	1350	14.6	1560	1620	200	3.77	312
354	3.85	25.7	–	3530	1040	22000	1240	14.5	1420	1460	181	3.74	282
318	4.23	28.7	–	3210	1530	19500	1110	14.5	1270	1290	161	3.71	250
291	4.60	31.0	–	2930	2150	17700	1020	14.4	1160	1160	146	3.68	226
263	5.03	34.3	54.8	2680	3080	15900	919	14.3	1040	1040	131	3.66	202
241	5.66	35.9	49.9	2440	4510	14200	831	14.1	940	933	118	3.62	182
221	6.20	38.5	43.5	2250	6340	12900	759	14.1	857	840	106	3.59	164
201	6.85	41.7	37.0	2050	9180	11600	686	14.0	773	749	95.2	3.56	147
169	4.71	44.7	32.2	2160	8150	9290	549	13.7	629	310	53.9	2.50	84.4
152	5.48	47.2	28.9	1940	12900	8160	487	13.5	559	273	47.2	2.47	73.9
141	6.01	49.6	26.2	1800	17800	7450	448	13.4	514	246	42.7	2.43	66.9
130	6.73	51.7	24.1	1660	25100	6710	406	13.2	467	218	37.9	2.39	59.5
118	7.76	54.5	21.7	1510	37700	5900	359	13.0	415	187	32.6	2.32	51.3
391	3.19	19.7	–	4520	394	20700	1250	13.4	1450	1550	198	3.67	310
357	3.45	21.6	–	4170	539	18700	1140	13.3	1320	1390	179	3.64	279
326	3.75	23.4	–	3820	749	16800	1040	13.2	1190	1240	162	3.60	252
292	4.12	26.2	–	3460	1110	14900	930	13.2	1060	1100	144	3.58	223
261	4.59	28.7	–	3110	1680	13100	829	13.1	943	959	127	3.53	196
235	5.02	32.2	62.1	2830	2440	11700	748	13.0	847	855	114	3.51	175
211	5.74	34.5	54.1	2530	3840	10300	665	12.9	751	757	100	3.49	155
191	6.35	37.7	45.4	2290	5660	9200	600	12.8	675	673	89.5	3.46	138
173	7.04	40.8	38.6	2090	8270	8230	541	12.7	607	598	79.8	3.42	123
148	4.44	41.6	37.3	2310	6270	6680	436	12.4	500	227	43.3	2.28	68.0
132	5.27	43.9	33.4	2050	10500	5770	380	12.2	437	196	37.2	2.25	58.4
124	5.65	46.2	30.2	1930	13500	5360	355	12.1	408	181	34.4	2.23	54.0
116	6.17	47.8	28.2	1800	17700	4930	329	12.0	378	164	31.3	2.19	49.2
108	6.89	49.6	26.2	1680	24200	4470	299	11.9	346	146	27.9	2.15	43.9
99	7.80	51.9	23.9	1560	34100	3990	269	11.7	312	128	24.5	2.10	38.6
90	8.52	57.5	19.5	1410	49600	3610	245	11.7	283	115	22.1	2.09	34.7
539	2.15	12.1	–	7160	67.1	25600	1570	12.7	1890	2110	277	3.65	437
368	2.96	17.3	–	5120	241	16200	1060	12.2	1240	1310	179	3.48	279
336	3.19	18.9	–	4690	337	14600	972	12.1	1130	1180	162	3.45	252
307	3.46	20.6	–	4310	466	13100	887	12.0	1030	1050	146	3.41	227
281	3.72	22.5	–	3990	626	11900	814	12.0	936	953	133	3.39	206
258	4.03	24.4	–	3680	863	10800	745	11.9	852	859	120	3.36	187
235	4.41	26.2	–	3380	1210	9700	677	11.8	772	769	108	3.33	168
217	4.71	28.7	–	3130	1610	8910	627	11.8	711	704	99.8	3.32	154
194	5.24	31.8	63.6	2820	2430	7860	559	11.7	631	619	88.1	3.29	136
178	5.92	32.9	59.5	2580	3570	7020	505	11.6	570	555	78.8	3.25	122
161	6.49	36.1	49.3	2340	5150	6310	458	11.5	515	497	70.9	3.23	109
146	7.16	39.4	41.4	2130	7460	5660	414	11.5	464	443	63.5	3.20	97.7
129	4.55	39.7	40.8	2380	5420	4760	345	11.2	395	184	36.8	2.21	57.6
114	5.41	42.5	35.6	2100	9220	4080	299	11.0	343	159	31.5	2.18	49.3
102	6.03	47.1	29.1	1890	14100	3620	267	11.0	305	139	27.8	2.15	43.4
94	6.70	49.5	26.3	1740	19900	3270	243	10.9	278	124	24.8	2.12	38.8
84	7.78	52.7	23.2	1570	31100	2850	213	10.7	244	106	21.2	2.07	33.2

– in F_y''' column indicates F_y''' > 65 ksi.

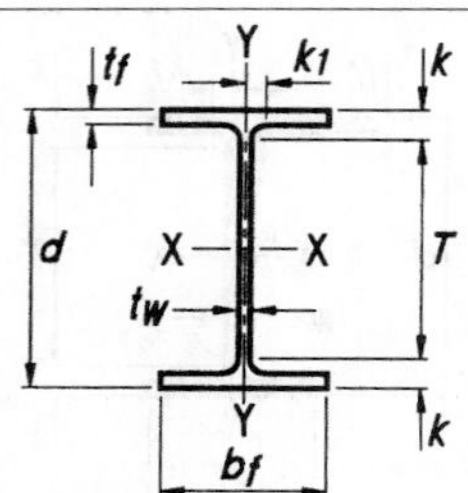

Table 1-1 (cont.). W-Shapes Dimensions

Shape	Area, A	Depth, d		Web			Flange				Distance				
				Thickness, t_w		$\frac{t_w}{2}$	Width, b_f		Thickness, t_f		k		k_1	T	Workable Gage†
	in.^{2}	in.		in.		in.	in.		in.		in.	in.	in.	in.	in.
W24×370*	109	28.0	28	1.52	1 1/2	3/4	13.7	13 5/8	2.72	2 3/4	3.22	3 5/8	1 9/16	20 3/4	5 1/2
×335*	98.4	27.5	27 1/2	1.38	1 3/8	11/16	13.5	13 1/2	2.48	2 1/2	2.98	3 3/8	1 1/2		
×306*	89.8	27.1	27 1/8	1.26	1 1/4	5/8	13.4	13 3/8	2.28	2 1/4	2.78	3 3/16	1 7/16		
×279*	82.0	26.7	26 3/4	1.16	1 3/16	5/8	13.3	13 1/4	2.09	2 1/16	2.59	3	1 7/16		
×250*	73.5	26.3	26 3/8	1.04	1 1/16	9/16	13.2	13 1/8	1.89	1 7/8	2.39	2 13/16	1 3/8		
×229	67.2	26.0	26	0.960	15/16	1/2	13.1	13 1/8	1.73	1 3/4	2.23	2 5/8	1 5/16		
×207	60.7	25.7	25 3/4	0.870	7/8	7/16	13.0	13	1.57	1 9/16	2.07	2 1/2	1 1/4		
×192	56.3	25.5	25 1/2	0.810	13/16	7/16	13.0	13	1.46	1 7/16	1.96	2 3/8	1 1/4		
×176	51.7	25.2	25 1/4	0.750	3/4	3/8	12.9	12 7/8	1.34	1 5/16	1.84	2 1/4	1 3/16		
×162	47.7	25.0	25	0.705	11/16	3/8	13.0	13	1.22	1 1/4	1.72	2 1/8	1 3/16		
×146	43.0	24.7	24 3/4	0.650	5/8	5/16	12.9	12 7/8	1.09	1 1/16	1.59	2	1 1/8		
×131	38.5	24.5	24 1/2	0.605	5/8	5/16	12.9	12 7/8	0.960	15/16	1.46	1 7/8	1 1/8		
×117	34.4	24.3	24 1/4	0.550	9/16	5/16	12.8	12 3/4	0.850	7/8	1.35	1 3/4	1 1/8		
×104	30.6	24.1	24	0.500	1/2	1/4	12.8	12 3/4	0.750	3/4	1.25	1 5/8	1 1/16		
W24×103	30.3	24.5	24 1/2	0.550	9/16	5/16	9.00	9	0.980	1	1.48	1 7/8	1 1/8	20 3/4	5 1/2
×94	27.7	24.3	24 1/4	0.515	1/2	1/4	9.07	9 1/8	0.875	7/8	1.38	1 3/4	1 1/16		
×84	24.7	24.1	24 1/8	0.470	1/2	1/4	9.02	9	0.770	3/4	1.27	1 11/16	1 1/16		
×76	22.4	23.9	23 7/8	0.440	7/16	1/4	8.99	9	0.680	11/16	1.18	1 9/16	1 1/16		
×68	20.1	23.7	23 3/4	0.415	7/16	1/4	8.97	9	0.585	9/16	1.09	1 1/2	1 1/16		
W24×62	18.3	23.7	23 3/4	0.430	7/16	1/4	7.04	7	0.590	9/16	1.19	1 1/2	1 1/16	20 3/4	3 1/2
×55	16.3	23.6	23 5/8	0.395	3/8	3/16	7.01	7	0.505	1/2	1.11	1 7/16	1	20 3/4	3 1/2
W21×201	59.2	23.0	23	0.910	15/16	1/2	12.6	12 5/8	1.63	1 5/8	2.13	2 1/2	1 5/16	18	5 1/2
×182	53.6	22.7	22 3/4	0.830	13/16	7/16	12.5	12 1/2	1.48	1 1/2	1.98	2 3/8	1 1/4		
×166	48.8	22.5	22 1/2	0.750	3/4	3/8	12.4	12 3/8	1.36	1 3/8	1.86	2 1/4	1 3/16		
×147	43.2	22.1	22	0.720	3/4	3/8	12.5	12 1/2	1.15	1 1/8	1.65	2	1 3/16		
×132	38.8	21.8	21 7/8	0.650	5/8	5/16	12.4	12 1/2	1.03	1 1/16	1.54	1 15/16	1 1/8		
×122	35.9	21.7	21 5/8	0.600	5/8	5/16	12.4	12 3/8	0.960	15/16	1.46	1 13/16	1 1/8		
×111	32.7	21.5	21 1/2	0.550	9/16	5/16	12.3	12 3/8	0.875	7/8	1.38	1 3/4	1 1/8		
×101	29.8	21.4	21 3/8	0.500	1/2	1/4	12.3	12 1/4	0.800	13/16	1.30	1 11/16	1 1/16		
W21×93	27.3	21.6	21 5/8	0.580	9/16	5/16	8.42	8 3/8	0.930	15/16	1.43	1 5/8	15/16	18 3/8	5 1/2
×83	24.3	21.4	21 3/8	0.515	1/2	1/4	8.36	8 3/8	0.835	13/16	1.34	1 1/2	7/8		
×73	21.5	21.2	21 1/4	0.455	7/16	1/4	8.30	8 1/4	0.740	3/4	1.24	1 7/16	7/8		
×68	20.0	21.1	21 1/8	0.430	7/16	1/4	8.27	8 1/4	0.685	11/16	1.19	1 3/8	7/8		
×62	18.3	21.0	21	0.400	3/8	3/16	8.24	8 1/4	0.615	5/8	1.12	1 5/16	13/16		
×55	16.2	20.8	20 3/4	0.375	3/8	3/16	8.22	8 1/4	0.522	1/2	1.02	1 3/16	13/16		
×48	14.1	20.6	20 5/8	0.350	3/8	3/16	8.14	8 1/8	0.430	7/16	0.93	1 1/8	13/16		
W21×57	16.7	21.1	21	0.405	3/8	3/16	6.56	6 1/2	0.650	5/8	1.15	1 5/16	13/16	18 3/8	3 1/2
×50	14.7	20.8	20 7/8	0.380	3/8	3/16	6.53	6 1/2	0.535	9/16	1.04	1 1/4	13/16		
×44	13.0	20.7	20 5/8	0.350	3/8	3/16	6.50	6 1/2	0.450	7/16	0.950	1 1/8	13/16		
W18×175	51.3	20.0	20	0.890	7/8	7/16	11.4	11 3/8	1.59	1 9/16	1.99	2 7/16	1 1/4	15 1/8	5 1/2
×158	46.3	19.7	19 3/4	0.810	13/16	7/16	11.3	11 1/4	1.44	1 7/16	1.84	2 3/8	1 1/4		
×143	42.1	19.5	19 1/2	0.730	3/4	3/8	11.2	11 1/4	1.32	1 5/16	1.72	2 3/16	1 3/16		
×130	38.2	19.3	19 1/4	0.670	11/16	3/8	11.2	11 1/8	1.20	1 3/16	1.60	2 1/16	1 3/16		
×119	35.1	19.0	19	0.655	5/8	5/16	11.3	11 1/4	1.06	1 1/16	1.46	1 15/16	1 3/16		
×106	31.1	18.7	18 3/4	0.590	9/16	5/16	11.2	11 1/4	0.940	15/16	1.34	1 13/16	1 1/8		
×97	28.5	18.6	18 5/8	0.535	9/16	5/16	11.1	11 1/8	0.870	7/8	1.27	1 3/4	1 1/8		
×86	25.3	18.4	18 3/8	0.480	1/2	1/4	11.1	11 1/8	0.770	3/4	1.17	1 5/8	1 1/16		
×76	22.3	18.2	18 1/4	0.425	7/16	1/4	11.0	11	0.680	11/16	1.08	1 9/16	1 1/16		

*ASTM A6 tensile group 4 or 5 shape. Special requirements may apply per LRFD Specification Section A3.1c.
†See definition of "Workable Gage" in Nomenclature section at the back of this Manual.

Table 1-1 (cont.). W-Shapes Properties

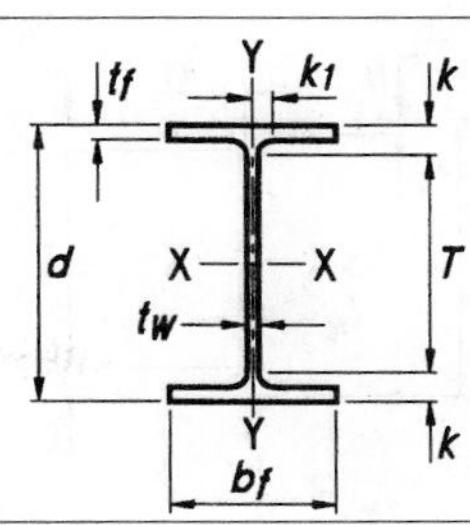

Nominal Wt.	Compact Section Criteria			X_1	$X_2 \times 10^6$	Axis X-X				Axis Y-Y			
	$\frac{b_f}{2t_f}$	$\frac{h}{t_w}$	F_y'''			I	S	r	Z	I	S	r	Z
lb/ft			ksi	ksi	(1/ksi)^2	in.^{4}	in.^{3}	in.	in.^{3}	in.^{4}	in.^{3}	in.	in.^{3}
370	2.51	14.2	–	6190	115	13400	957	11.1	1130	1160	170	3.27	267
335	2.73	15.6	–	5670	162	11900	864	11.0	1020	1030	152	3.23	238
306	2.94	17.1	–	5200	224	10700	789	10.9	922	919	137	3.20	214
279	3.18	18.6	–	4800	305	9600	718	10.8	835	823	124	3.17	193
250	3.49	20.7	–	4350	446	8490	644	10.7	744	724	110	3.14	171
229	3.79	22.5	–	4000	618	7650	588	10.7	675	651	99.4	3.11	154
207	4.14	24.8	–	3640	893	6820	531	10.6	606	578	88.8	3.08	137
192	4.43	26.6	–	3400	1170	6260	491	10.5	559	530	81.8	3.07	126
176	4.81	28.7	–	3130	1610	5680	450	10.5	511	479	74.3	3.04	115
162	5.31	30.6	–	2870	2260	5170	414	10.4	468	443	68.4	3.05	105
146	5.92	33.2	58.5	2590	3420	4580	371	10.3	418	391	60.5	3.01	93.2
131	6.70	35.6	50.7	2330	5290	4020	329	10.2	370	340	53.0	2.97	81.5
117	7.53	39.2	41.9	2090	8190	3540	291	10.1	327	297	46.5	2.94	71.4
104	8.50	43.1	34.6	1860	12900	3100	258	10.1	289	259	40.7	2.91	62.4
103	4.59	39.2	41.9	2390	5310	3000	245	9.96	280	119	26.5	1.99	41.5
94	5.18	41.9	36.7	2180	7800	2700	222	9.87	254	109	24.0	1.98	37.5
84	5.86	45.9	30.6	1950	12200	2370	196	9.79	224	94.4	20.9	1.95	32.6
76	6.61	49.0	26.8	1760	18600	2100	176	9.69	200	82.5	18.4	1.92	28.6
68	7.66	52.0	23.9	1590	29000	1830	154	9.55	177	70.4	15.7	1.87	24.5
62	5.97	49.7	26.1	1730	23800	1560	132	9.24	154	34.5	9.80	1.37	15.8
55	6.94	54.1	22.0	1570	36500	1360	115	9.13	135	29.1	8.30	1.34	13.4
201	3.86	20.6	–	4270	464	5310	461	9.47	530	542	86.1	3.02	133
182	4.22	22.6	–	3890	664	4730	417	9.40	476	483	77.2	3.00	119
166	4.57	25.0	–	3580	922	4280	380	9.36	432	435	70.0	2.99	108
147	5.44	26.1	–	3140	1590	3630	329	9.17	373	376	60.1	2.95	92.6
132	6.01	28.9	–	2840	2350	3220	295	9.12	333	333	53.5	2.93	82.3
122	6.45	31.3	–	2630	3160	2960	273	9.09	307	305	49.2	2.92	75.6
111	7.05	34.1	55.3	2400	4510	2670	249	9.05	279	274	44.5	2.90	68.2
101	7.68	37.5	45.7	2200	6400	2420	227	9.02	253	248	40.3	2.89	61.7
93	4.53	32.3	61.5	2680	3460	2070	192	8.70	221	92.9	22.1	1.84	34.7
83	5.00	36.4	48.5	2400	5250	1830	171	8.67	196	81.4	19.5	1.83	30.5
73	5.60	41.2	37.9	2140	8380	1600	151	8.64	172	70.6	17.0	1.81	26.6
68	6.04	43.6	33.8	2000	10900	1480	140	8.60	160	64.7	15.7	1.80	24.4
62	6.70	46.9	29.3	1820	15900	1330	127	8.54	144	57.5	14.0	1.77	21.7
55	7.87	50.0	25.7	1630	25800	1140	110	8.40	126	48.4	11.8	1.73	18.4
48	9.47	53.6	22.4	1450	43600	959	93	8.24	107	38.7	9.52	1.66	14.9
57	5.04	46.3	30.0	1960	13100	1170	111	8.36	129	30.6	9.35	1.35	14.8
50	6.10	49.4	26.4	1730	22600	984	94.5	8.18	110	24.9	7.64	1.30	12.2
44	7.22	53.6	22.4	1550	36600	843	81.6	8.06	95.4	20.7	6.37	1.26	10.2
175	3.58	18.0	–	4850	281	3450	344	8.20	398	391	68.8	2.76	106
158	3.92	19.8	–	4410	403	3060	310	8.12	356	347	61.4	2.74	94.8
143	4.25	22.0	–	4040	568	2750	282	8.09	322	311	55.5	2.72	85.4
130	4.65	23.9	–	3680	810	2460	256	8.03	290	278	49.9	2.70	76.7
119	5.31	24.5	–	3340	1210	2190	231	7.90	262	253	44.9	2.69	69.1
106	5.96	27.2	–	2990	1880	1910	204	7.84	230	220	39.4	2.66	60.5
97	6.41	30.0	–	2750	2580	1750	188	7.82	211	201	36.1	2.65	55.3
86	7.20	33.4	57.6	2460	4060	1530	166	7.77	186	175	31.6	2.63	48.4
76	8.11	37.8	45.2	2180	6520	1330	146	7.73	163	152	27.6	2.61	42.2

– in F_y''' column indicates $F_y''' > 65$ ksi.

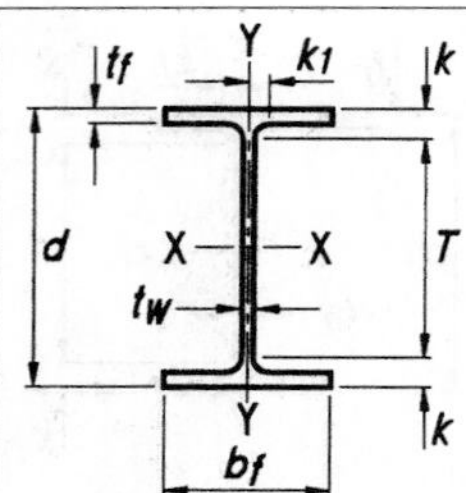

Table 1-1 (cont.). W-Shapes Dimensions

Shape	Area, A	Depth, d		Web Thickness, t_w		Web $\frac{t_w}{2}$	Flange Width, b_f		Flange Thickness, t_f		Distance k		Distance k_1	Distance T	Workable Gage†
	in.²	in.		in.		in.	in.		in.		in.	in.	in.	in.	in.
W18×71	20.8	18.5	18 1/2	0.495	1/2	1/4	7.64	7 5/8	0.810	13/16	1.21	1 1/2	7/8	15 1/2	3 1/2
×65	19.1	18.4	18 3/8	0.450	7/16	1/4	7.59	7 5/8	0.750	3/4	1.15	1 7/16	7/8	↓	↓
×60	17.6	18.2	18 1/4	0.415	7/16	1/4	7.56	7 1/2	0.695	11/16	1.10	1 3/8	13/16	↓	↓
×55	16.2	18.1	18 1/8	0.390	3/8	3/16	7.53	7 1/2	0.630	5/8	1.03	1 5/16	13/16	↓	↓
×50	14.7	18.0	18	0.355	3/8	3/16	7.50	7 1/2	0.570	9/16	0.972	1 1/4	13/16	↓	↓
W18×46	13.5	18.1	18	0.360	3/8	3/16	6.06	6	0.605	5/8	1.01	1 1/4	13/16	15 1/2	3 1/2
×40	11.8	17.9	17 7/8	0.315	5/16	3/16	6.02	6	0.525	1/2	0.927	1 3/16	13/16	↓	↓
×35	10.3	17.7	17 3/4	0.300	5/16	3/16	6.00	6	0.425	7/16	0.827	1 1/8	3/4	↓	↓
W16×100	29.7	17.0	17	0.585	9/16	5/16	10.4	10 3/8	0.985	1	1.69	1 7/8	1 1/8	13 1/4	5 1/2
×89	26.4	16.8	16 3/4	0.525	1/2	1/4	10.4	10 3/8	0.875	7/8	1.58	1 3/4	1 1/16	↓	↓
×77	22.9	16.5	16 1/2	0.455	7/16	1/4	10.3	10 1/4	0.760	3/4	1.47	1 5/8	1 1/16	↓	↓
×67	20.0	16.3	16 3/8	0.395	3/8	3/16	10.2	10 1/4	0.665	11/16	1.37	1 9/16	1	↓	↓
W16×57	16.8	16.4	16 3/8	0.430	7/16	1/4	7.12	7 1/8	0.715	11/16	1.12	1 3/8	7/8	13 5/8	3 1/2
×50	14.7	16.3	16 1/4	0.380	3/8	3/16	7.07	7 1/8	0.630	5/8	1.03	1 5/16	13/16	↓	↓
×45	13.3	16.1	16 1/8	0.345	3/8	3/16	7.04	7	0.565	9/16	0.967	1 1/4	13/16	↓	↓
×40	11.8	16.0	16	0.305	5/16	3/16	7.00	7	0.505	1/2	0.907	1 3/16	13/16	↓	↓
×36	10.6	15.9	15 7/8	0.295	5/16	3/16	6.99	7	0.430	7/16	0.832	1 1/8	3/4	↓	↓
W16×31	9.13	15.9	15 7/8	0.275	1/4	1/8	5.53	5 1/2	0.440	7/16	0.842	1 1/8	3/4	13 5/8	3 1/2
×26	7.68	15.7	15 3/4	0.250	1/4	1/8	5.50	5 1/2	0.345	3/8	0.747	1 1/16	3/4	13 5/8	3 1/2
W14×808*	237	22.8	22 7/8	3.74	3 3/4	1 7/8	18.6	18 1/2	5.12	5 1/8	5.72	6 7/16	3 1/16	10	3 – 7 1/2 – 3
×730*	215	22.4	22 3/8	3.07	3 1/16	1 9/16	17.9	17 7/8	4.91	4 15/16	5.51	6 3/16	2 3/4	↓	3 – 7 1/2 – 3
×665*	196	21.6	21 5/8	2.83	2 13/16	1 7/16	17.7	17 5/8	4.52	4 1/2	5.12	5 13/16	2 5/8	↓	↓
×605*	178	20.9	20 7/8	2.60	2 5/8	1 5/16	17.4	17 3/8	4.16	4 3/16	4.76	5 7/16	2 1/2	↓	↓
×550*	162	20.2	20 1/4	2.38	2 3/8	1 3/16	17.2	17 1/4	3.82	3 13/16	4.42	5 1/8	2 3/8	↓	↓
×500*	147	19.6	19 5/8	2.19	2 3/16	1 1/8	17.0	17	3.50	3 1/2	4.10	4 13/16	2 5/16	↓	↓
×455*	134	19.0	19	2.02	2	1	16.8	16 7/8	3.21	3 3/16	3.81	4 1/2	2 1/4	↓	↓
×426*	125	18.7	18 5/8	1.88	1 7/8	15/16	16.7	16 3/4	3.04	3 1/16	3.63	4 5/16	2 1/8	↓	↓
×398*	117	18.3	18 1/4	1.77	1 3/4	7/8	16.6	16 5/8	2.85	2 7/8	3.44	4 1/8	2 1/8	↓	↓
×370*	109	17.9	17 7/8	1.66	1 5/8	13/16	16.5	16 1/2	2.66	2 11/16	3.26	3 15/16	2 1/16	↓	↓
×342*	101	17.5	17 1/2	1.54	1 9/16	13/16	16.4	16 3/8	2.47	2 1/2	3.07	3 3/4	2	↓	↓
×311*	91.4	17.1	17 1/8	1.41	1 7/16	3/4	16.2	16 1/4	2.26	2 1/4	2.86	3 9/16	1 15/16	↓	↓
×283*	83.3	16.7	16 3/4	1.29	1 5/16	11/16	16.1	16 1/8	2.07	2 1/16	2.67	3 3/8	1 7/8	↓	↓
×257*	75.6	16.4	16 3/8	1.18	1 3/16	5/8	16.0	16	1.89	1 7/8	2.49	3 3/16	1 13/16	↓	↓
×233*	68.5	16.0	16	1.07	1 1/16	9/16	15.9	15 7/8	1.72	1 3/4	2.32	3	1 3/4	↓	↓
×211	62.0	15.7	15 3/4	0.980	1	1/2	15.8	15 3/4	1.56	1 9/16	2.16	2 7/8	1 11/16	↓	↓
×193	56.8	15.5	15 1/2	0.890	7/8	7/16	15.7	15 3/4	1.44	1 7/16	2.04	2 3/4	1 11/16	↓	↓
×176	51.8	15.2	15 1/4	0.830	13/16	7/16	15.7	15 5/8	1.31	1 5/16	1.91	2 5/8	1 5/8	↓	↓
×159	46.7	15.0	15	0.745	3/4	3/8	15.6	15 5/8	1.19	1 3/16	1.79	2 1/2	1 9/16	↓	↓
×145	42.7	14.8	14 3/4	0.680	11/16	3/8	15.5	15 1/2	1.09	1 1/16	1.69	2 3/8	1 9/16	↓	↓
W14×132	38.8	14.7	14 5/8	0.645	5/8	5/16	14.7	14 3/4	1.03	1	1.63	2 5/16	1 9/16	10	5 1/2
×120	35.3	14.5	14 1/2	0.590	9/16	5/16	14.7	14 5/8	0.940	15/16	1.54	2 1/4	1 1/2	↓	↓
×109	32.0	14.3	14 3/8	0.525	1/2	1/4	14.6	14 5/8	0.860	7/8	1.46	2 3/16	1 1/2	↓	↓
×99	29.1	14.2	14 1/8	0.485	1/2	1/4	14.6	14 5/8	0.780	3/4	1.38	2 1/16	1 7/16	↓	↓
×90	26.5	14.0	14	0.440	7/16	1/4	14.5	14 1/2	0.710	11/16	1.31	2	1 7/16	↓	↓
W14×82	24.0	14.3	14 1/4	0.510	1/2	1/4	10.1	10 1/8	0.855	7/8	1.45	1 11/16	1 1/16	10 7/8	5 1/2
×74	21.8	14.2	14 1/8	0.450	7/16	1/4	10.1	10 1/8	0.785	13/16	1.38	1 5/8	1 1/16	↓	↓
×68	20.0	14.0	14	0.415	7/16	1/4	10.0	10	0.720	3/4	1.31	1 9/16	1 1/16	↓	↓
×61	17.9	13.9	13 7/8	0.375	3/8	3/16	9.99	10	0.645	5/8	1.24	1 1/2	1	↓	↓
W14×53	15.6	13.9	13 7/8	0.370	3/8	3/16	8.06	8	0.660	11/16	1.25	1 1/2	1	10 7/8	5 1/2
×48	14.1	13.8	13 3/4	0.340	5/16	3/16	8.03	8	0.595	5/8	1.19	1 7/16	1	↓	↓
×43	12.6	13.7	13 5/8	0.305	5/16	3/16	8.00	8	0.530	1/2	1.12	1 3/8	1	↓	↓

*ASTM A6 tensile group 4 or 5 shape. Special requirements may apply per LRFD Specification Section A3.1c.
†See definition of "Workable Gage" in Nomenclature section at the back of this Manual.

Table 1-1 (cont.). W-Shapes Properties

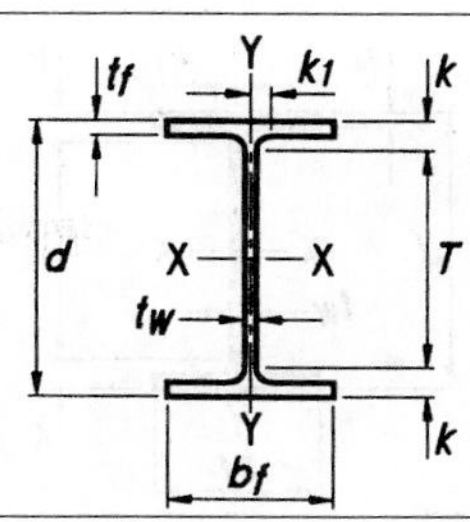

Nominal Wt.	Compact Section Criteria			X_1	$X_2 \times 10^6$	Axis X-X				Axis Y-Y			
	$\frac{b_f}{2t_f}$	$\frac{h}{t_w}$	F_y'''			I	S	r	Z	I	S	r	Z
lb/ft			ksi	ksi	(1/ksi)^2	in.^{4}	in.^{3}	in.	in.^{3}	in.^{4}	in.^{3}	in.	in.^{3}
71	4.71	32.4	61.3	2690	3290	1170	127	7.50	146	60.3	15.8	1.70	24.7
65	5.06	35.7	50.6	2470	4540	1070	117	7.49	133	54.8	14.4	1.69	22.5
60	5.44	38.7	43.1	2290	6080	984	108	7.47	123	50.1	13.3	1.68	20.6
55	5.98	41.1	38.0	2110	8540	890	98.3	7.41	112	44.9	11.9	1.67	18.5
50	6.57	45.2	31.5	1920	12400	800	88.9	7.38	101	40.1	10.7	1.65	16.6
46	5.01	44.6	32.4	2060	10100	712	78.8	7.25	90.7	22.5	7.43	1.29	11.7
40	5.73	50.9	24.8	1810	17200	612	68.4	7.21	78.4	19.1	6.35	1.27	9.95
35	7.06	53.5	22.5	1590	30800	510	57.6	7.04	66.5	15.3	5.12	1.22	8.06
100	5.29	23.2	–	3530	947	1500	177	7.10	200	186	35.7	2.50	55.0
89	5.92	25.9	–	3160	1460	1310	157	7.05	177	163	31.4	2.48	48.2
77	6.77	29.9	–	2770	2460	1120	136	7.00	152	138	26.9	2.46	41.2
67	7.70	34.4	54.5	2440	4040	970	119	6.97	132	119	23.2	2.44	35.6
57	4.98	33.0	59.1	2650	3400	758	92.2	6.72	105	43.1	12.1	1.60	18.9
50	5.61	37.4	46.1	2340	5530	659	81.0	6.68	92.0	37.2	10.5	1.59	16.3
45	6.23	41.1	38.0	2120	8280	586	72.7	6.65	82.3	32.8	9.34	1.57	14.5
40	6.93	46.5	29.7	1890	12700	518	64.7	6.63	73.0	28.9	8.25	1.57	12.7
36	8.12	48.1	27.8	1700	20400	448	56.5	6.51	64.0	24.5	7.00	1.52	10.8
31	6.28	51.6	24.2	1740	19900	375	47.2	6.41	54.0	12.4	4.49	1.17	7.03
26	7.97	56.8	20.0	1480	40300	301	38.4	6.26	44.2	9.59	3.49	1.12	5.48
808	1.81	3.05	–	18900	1.45	16000	1400	8.20	1830	5510	594	4.82	927
730	1.82	3.71	–	17500	1.90	14300	1280	8.17	1660	4720	527	4.69	816
665	1.95	4.03	–	16300	2.46	12400	1150	7.98	1480	4170	472	4.62	730
605	2.09	4.39	–	15100	3.21	10800	1040	7.80	1320	3680	423	4.55	652
550	2.25	4.79	–	14200	4.16	9430	931	7.63	1180	3250	378	4.49	583
500	2.43	5.21	–	13100	5.49	8210	838	7.48	1050	2880	339	4.43	522
455	2.62	5.66	–	12200	7.30	7190	756	7.33	936	2560	304	4.38	468
426	2.75	6.08	–	11500	8.87	6600	706	7.26	869	2360	283	4.34	434
398	2.92	6.44	–	10900	11.0	6000	656	7.16	801	2170	262	4.31	402
370	3.10	6.89	–	10300	13.9	5440	607	7.07	736	1990	241	4.27	370
342	3.31	7.41	–	9620	17.8	4900	558	6.98	672	1810	221	4.24	338
311	3.59	8.09	–	8820	24.4	4330	506	6.88	603	1610	199	4.20	304
283	3.89	8.84	–	8120	33.4	3840	459	6.79	542	1440	179	4.17	274
257	4.23	9.71	–	7460	46.1	3400	415	6.71	487	1290	161	4.13	246
233	4.62	10.7	–	6820	64.9	3010	375	6.63	436	1150	145	4.10	221
211	5.06	11.6	–	6230	91.8	2660	338	6.55	390	1030	130	4.07	198
193	5.45	12.8	–	5740	125	2400	310	6.50	355	931	119	4.05	180
176	5.97	13.7	–	5280	173	2140	281	6.43	320	838	107	4.02	163
159	6.54	15.3	–	4780	252	1900	254	6.38	287	748	96.2	4.00	146
145	7.11	16.8	–	4400	348	1710	232	6.33	260	677	87.3	3.98	133
132	7.15	17.7	–	4180	428	1530	209	6.28	234	548	74.5	3.76	113
120	7.80	19.3	–	3830	601	1380	190	6.24	212	495	67.5	3.74	102
109	8.49	21.7	–	3490	853	1240	173	6.22	192	447	61.2	3.73	92.7
99	9.34	23.5	–	3190	1220	1110	157	6.17	173	402	55.2	3.71	83.6
90	10.20	25.9	–	2900	1750	999	143	6.14	157	362	49.9	3.70	75.6
82	5.92	22.4	–	3590	849	881	123	6.05	139	148	29.3	2.48	44.8
74	6.41	25.4	–	3280	1200	795	112	6.04	126	134	26.6	2.48	40.5
68	6.97	27.5	–	3020	1660	722	103	6.01	115	121	24.2	2.46	36.9
61	7.75	30.4	–	2720	2470	640	92.1	5.98	102	107	21.5	2.45	32.8
53	6.11	30.9	–	2830	2250	541	77.8	5.89	87.1	57.7	14.3	1.92	22.0
48	6.75	33.6	57.1	2580	3250	484	70.2	5.85	78.4	51.4	12.8	1.91	19.6
43	7.54	37.4	45.9	2330	4880	428	62.6	5.82	69.6	45.2	11.3	1.89	17.3

– in F_y''' column indicates $F_y''' > 65$ ksi.

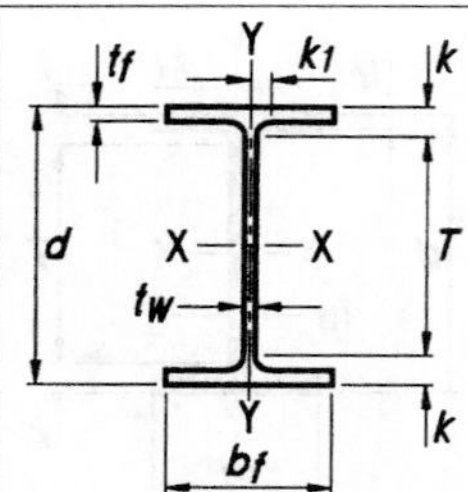

Table 1-1 (cont.). W-Shapes Dimensions

Shape	Area, A	Depth, d		Web			Flange				Distance				
				Thickness, t_w		$\frac{t_w}{2}$	Width, b_f		Thickness, t_f		k		k_1	T	Workable Gage†
	in.^{2}	in.		in.		in.	in.		in.		in.	in.	in.	in.	in.
W14×38	11.2	14.1	14 1/8	0.310	5/16	3/16	6.77	6 3/4	0.515	1/2	0.915	1 1/4	13/16	11 5/8	3 1/2
×34	10.0	14.0	14	0.285	5/16	3/16	6.75	6 3/4	0.455	7/16	0.855	1 3/16	3/4		3 1/2
×30	8.85	13.8	13 7/8	0.270	1/4	1/8	6.73	6 3/4	0.385	3/8	0.785	1 1/8	3/4		3 1/2
W14×26	7.69	13.9	13 7/8	0.255	1/4	1/8	5.03	5	0.420	7/16	0.820	1 1/8	3/4	11 5/8	2 3/4
×22	6.49	13.7	13 3/4	0.230	1/4	1/8	5.00	5	0.335	5/16	0.735	1 1/16	3/4	11 5/8	2 3/4
W12×336*	98.8	16.8	16 7/8	1.78	1 3/4	7/8	13.4	13 3/8	2.96	2 15/16	3.55	3 7/8	1 11/16	9 1/8	5 1/2
×305*	89.6	16.3	16 3/8	1.63	1 5/8	13/16	13.2	13 1/4	2.71	2 11/16	3.30	3 5/8	1 5/8		
×279*	81.9	15.9	15 7/8	1.53	1 1/2	3/4	13.1	13 1/8	2.47	2 1/2	3.07	3 3/8	1 5/8		
×252*	74.0	15.4	15 3/8	1.40	1 3/8	11/16	13.0	13	2.25	2 1/4	2.85	3 1/8	1 1/2		
×230*	67.7	15.1	15	1.29	1 5/16	11/16	12.9	12 7/8	2.07	2 1/16	2.67	2 15/16	1 1/2		
×210*	61.8	14.7	14 3/4	1.18	1 3/16	5/8	12.8	12 3/4	1.90	1 7/8	2.50	2 13/16	1 7/16		
×190	55.8	14.4	14 3/8	1.06	1 1/16	9/16	12.7	12 5/8	1.74	1 3/4	2.33	2 5/8	1 3/8		
×170	50.0	14.0	14	0.960	15/16	1/2	12.6	12 5/8	1.56	1 9/16	2.16	2 7/16	1 5/16		
×152	44.7	13.7	13 3/4	0.870	7/8	7/16	12.5	12 1/2	1.40	1 3/8	2.00	2 5/16	1 1/4		
×136	39.9	13.4	13 3/8	0.790	13/16	7/16	12.4	12 3/8	1.25	1 1/4	1.85	2 1/8	1 1/4		
×120	35.3	13.1	13 1/8	0.710	11/16	3/8	12.3	12 3/8	1.11	1 1/8	1.70	2	1 3/16		
×106	31.2	12.9	12 7/8	0.610	5/8	5/16	12.2	12 1/4	0.990	1	1.59	1 7/8	1 1/8		
×96	28.2	12.7	12 3/4	0.550	9/16	5/16	12.2	12 1/8	0.900	7/8	1.50	1 13/16	1 1/8		
×87	25.6	12.5	12 1/2	0.515	1/2	1/4	12.1	12 1/8	0.810	13/16	1.41	1 11/16	1 1/16		
×79	23.2	12.4	12 3/8	0.470	1/2	1/4	12.1	12 1/8	0.735	3/4	1.33	1 5/8	1 1/16		
×72	21.1	12.3	12 1/4	0.430	7/16	1/4	12.0	12	0.670	11/16	1.27	1 9/16	1 1/16		
×65	19.1	12.1	12 1/8	0.390	3/8	3/16	12.0	12	0.605	5/8	1.20	1 1/2	1		
W12×58	17.0	12.2	12 1/4	0.360	3/8	3/16	10.0	10	0.640	5/8	1.24	1 1/2	15/16	9 1/4	5 1/2
×53	15.6	12.1	12	0.345	3/8	3/16	9.99	10	0.575	9/16	1.17	1 3/8	15/16	9 1/4	5 1/2
W12×50	14.6	12.2	12 1/4	0.370	3/8	3/16	8.08	8 1/8	0.640	5/8	1.14	1 1/2	15/16	9 1/4	5 1/2
×45	13.1	12.1	12	0.335	5/16	3/16	8.05	8	0.575	9/16	1.08	1 3/8	15/16		
×40	11.7	11.9	12	0.295	5/16	3/16	8.01	8	0.515	1/2	1.02	1 3/8	7/8		
W12×35	10.3	12.5	12 1/2	0.300	5/16	3/16	6.56	6 1/2	0.520	1/2	0.820	1 3/16	3/4	10 1/8	3 1/2
×30	8.79	12.3	12 3/8	0.260	1/4	1/8	6.52	6 1/2	0.440	7/16	0.740	1 1/8	3/4		
×26	7.65	12.2	12 1/4	0.230	1/4	1/8	6.49	6 1/2	0.380	3/8	0.680	1 1/16	3/4		
W12×22	6.48	12.3	12 1/4	0.260	1/4	1/8	4.03	4	0.425	7/16	0.725	15/16	5/8	10 3/8	2 1/4
×19	5.57	12.2	12 1/8	0.235	1/4	1/8	4.01	4	0.350	3/8	0.650	7/8	9/16		
×16	4.71	12.0	12	0.220	1/4	1/8	3.99	4	0.265	1/4	0.565	13/16	9/16		
×14	4.16	11.9	11 7/8	0.200	3/16	1/8	3.97	4	0.225	1/4	0.525	3/4	9/16		
W10×112	32.9	11.4	11 3/8	0.755	3/4	3/8	10.4	10 3/8	1.25	1 1/4	1.75	1 15/16	1	7 1/2	5 1/2
×100	29.4	11.1	11 1/8	0.680	11/16	3/8	10.3	10 3/8	1.12	1 1/8	1.62	1 13/16	1		
×88	25.9	10.8	10 7/8	0.605	5/8	5/16	10.3	10 1/4	0.990	1	1.49	1 11/16	15/16		
×77	22.6	10.6	10 5/8	0.530	1/2	1/4	10.2	10 1/4	0.870	7/8	1.37	1 9/16	7/8		
×68	20.0	10.4	10 3/8	0.470	1/2	1/4	10.1	10 1/8	0.770	3/4	1.27	1 7/16	7/8		
×60	17.6	10.2	10 1/4	0.420	7/16	1/4	10.1	10 1/8	0.680	11/16	1.18	1 3/8	13/16		
×54	15.8	10.1	10 1/8	0.370	3/8	3/16	10.0	10	0.615	5/8	1.12	1 5/16	13/16		
×49	14.4	9.98	10	0.340	5/16	3/16	10.0	10	0.560	9/16	1.06	1 1/4	13/16		
W10×45	13.3	10.1	10 1/8	0.350	3/8	3/16	8.02	8	0.620	5/8	1.12	1 5/16	13/16	7 1/2	5 1/2
×39	11.5	9.92	9 7/8	0.315	5/16	3/16	7.99	8	0.530	1/2	1.03	1 3/16	13/16		
×33	9.71	9.73	9 3/4	0.290	5/16	3/16	7.96	8	0.435	7/16	0.935	1 1/8	3/4		
W10×30	8.84	10.5	10 1/2	0.300	5/16	3/16	5.81	5 3/4	0.510	1/2	0.810	1 1/8	11/16	8 1/4	2 3/4
×26	7.61	10.3	10 3/8	0.260	1/4	1/8	5.77	5 3/4	0.440	7/16	0.740	1 1/16	11/16		
×22	6.49	10.2	10 1/8	0.240	1/4	1/8	5.75	5 3/4	0.360	3/8	0.660	15/16	5/8		

*ASTM A6 tensile group 4 or 5 shape. Special requirements may apply per LRFD Specification Section A3.1c.
†See definition of "Workable Gage" in Nomenclature section at the back of this Manual.

Table 1-1 (cont.). W-Shapes Properties

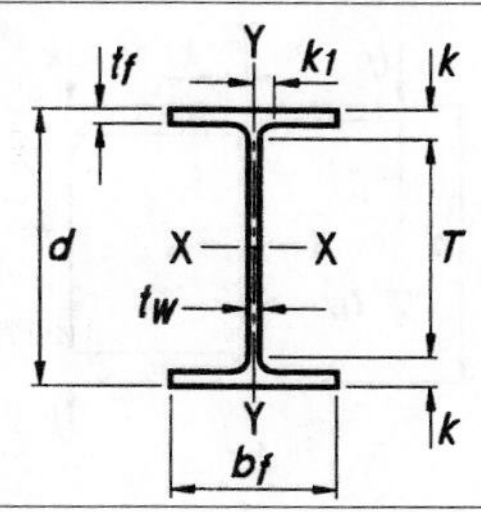

Nom-inal Wt.	Compact Section Criteria			X_1	$X_2 \times 10^6$	Axis X-X				Axis Y-Y			
	$\frac{b_f}{2t_f}$	$\frac{h}{t_w}$	F_y'''			I	S	r	Z	I	S	r	Z
lb/ft			ksi	ksi	$(1/\text{ksi})^2$	in.[4]	in.[3]	in.	in.[3]	in.[4]	in.[3]	in.	in.[3]
38	6.57	39.6	41.1	2190	6890	385	54.6	5.87	61.5	26.7	7.88	1.55	12.1
34	7.41	43.1	34.7	1970	10600	340	48.6	5.83	54.6	23.3	6.91	1.53	10.6
30	8.74	45.4	31.2	1750	17600	291	42.0	5.73	47.3	19.6	5.82	1.49	8.99
26	5.98	48.1	27.8	1880	14100	245	35.3	5.65	40.2	8.91	3.55	1.08	5.54
22	7.46	53.3	22.6	1600	27800	199	29.0	5.54	33.2	7.00	2.80	1.04	4.39
336	2.26	5.47	–	12800	6.05	4060	483	6.41	603	1190	177	3.47	274
305	2.45	5.98	–	11800	8.17	3550	435	6.29	537	1050	159	3.42	244
279	2.66	6.35	–	11000	10.8	3110	393	6.16	481	937	143	3.38	220
252	2.89	6.96	–	10100	14.7	2720	353	6.06	428	828	127	3.34	196
230	3.11	7.56	–	9390	19.7	2420	321	5.97	386	742	115	3.31	177
210	3.37	8.23	–	8670	26.6	2140	292	5.89	348	664	104	3.28	159
190	3.65	9.16	–	7940	37.0	1890	263	5.82	311	589	93.0	3.25	143
170	4.03	10.1	–	7190	54.0	1650	235	5.74	275	517	82.3	3.22	126
152	4.46	11.2	–	6510	79.3	1430	209	5.66	243	454	72.8	3.19	111
136	4.96	12.3	–	5850	119	1240	186	5.58	214	398	64.2	3.16	98.0
120	5.57	13.7	–	5240	184	1070	163	5.51	186	345	56.0	3.13	85.4
106	6.17	15.9	–	4660	285	933	145	5.47	164	301	49.3	3.11	75.1
96	6.76	17.7	–	4250	407	833	131	5.44	147	270	44.4	3.09	67.5
87	7.48	18.9	–	3880	586	740	118	5.38	132	241	39.7	3.07	60.4
79	8.22	20.7	–	3530	839	662	107	5.34	119	216	35.8	3.05	54.3
72	8.99	22.6	–	3230	1180	597	97.4	5.31	108	195	32.4	3.04	49.2
65	9.92	24.9	–	2940	1720	533	87.9	5.28	96.8	174	29.1	3.02	44.1
58	7.82	27.0	–	3070	1470	475	78.0	5.28	86.4	107	21.4	2.51	32.5
53	8.69	28.1	–	2820	2100	425	70.6	5.23	77.9	95.8	19.2	2.48	29.1
50	6.31	26.8	–	3120	1500	391	64.2	5.18	71.9	56.3	13.9	1.96	21.3
45	7.00	29.6	–	2820	2210	348	57.7	5.15	64.2	50.0	12.4	1.95	19.0
40	7.77	33.6	57.0	2530	3360	307	51.5	5.13	57.0	44.1	11.0	1.94	16.8
35	6.31	36.2	49.1	2430	4330	285	45.6	5.25	51.2	24.5	7.47	1.54	11.5
30	7.41	41.8	36.9	2080	8050	238	38.6	5.21	43.1	20.3	6.24	1.52	9.56
26	8.54	47.2	28.9	1820	13900	204	33.4	5.17	37.2	17.3	5.34	1.51	8.17
22	4.74	41.8	36.9	2170	8460	156	25.4	4.91	29.3	4.66	2.31	0.848	3.66
19	5.72	46.2	30.1	1880	15600	130	21.3	4.82	24.7	3.76	1.88	0.822	2.98
16	7.53	49.4	26.4	1630	30200	103	17.1	4.67	20.1	2.82	1.41	0.773	2.26
14	8.82	54.3	21.8	1450	48800	88.6	14.9	4.62	17.4	2.36	1.19	0.753	1.90
112	4.17	10.4	–	7080	56.7	716	126	4.66	147	236	45.3	2.68	69.2
100	4.62	11.6	–	6400	83.8	623	112	4.60	130	207	40.0	2.65	61.0
88	5.18	13.0	–	5680	132	534	98.5	4.54	113	179	34.8	2.63	53.1
77	5.86	14.8	–	5010	213	455	85.9	4.49	97.6	154	30.1	2.60	45.9
68	6.58	16.7	–	4460	334	394	75.7	4.44	85.3	134	26.4	2.59	40.1
60	7.41	18.7	–	3970	525	341	66.7	4.39	74.6	116	23.0	2.57	35.0
54	8.15	21.2	–	3580	778	303	60.0	4.37	66.6	103	20.6	2.56	31.3
49	8.93	23.1	–	3280	1090	272	54.6	4.35	60.4	93.4	18.7	2.54	28.3
45	6.47	22.5	–	3650	758	248	49.1	4.32	54.9	53.4	13.3	2.01	20.3
39	7.53	25.0	–	3190	1310	209	42.1	4.27	46.8	45.0	11.3	1.98	17.2
33	9.15	27.1	–	2720	2480	171	35.0	4.19	38.8	36.6	9.20	1.94	14.0
30	5.70	29.5	–	2900	2150	170	32.4	4.38	36.6	16.7	5.75	1.37	8.84
26	6.56	34.0	55.6	2510	3760	144	27.9	4.35	31.3	14.1	4.89	1.36	7.50
22	7.99	36.9	47.3	2150	7230	118	23.2	4.27	26.0	11.4	3.97	1.33	6.10

– in F_y''' column indicates $F_y''' > 65$ ksi.

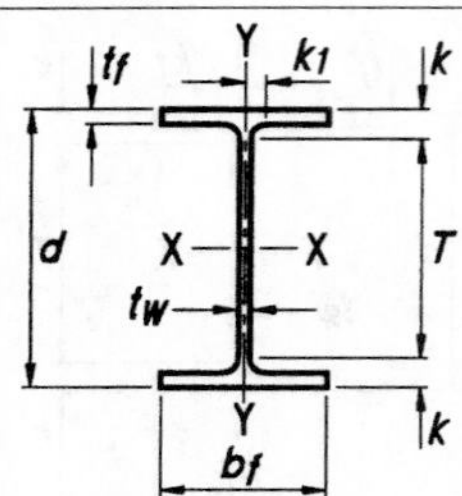

Table 1-1 (cont.). W-Shapes Dimensions

Shape	Area, A	Depth, d		Web Thickness, t_w		Web $\frac{t_w}{2}$	Flange Width, b_f		Flange Thickness, t_f		Distance k		Distance k_1	Distance T	Workable Gage†
	in.²	in.		in.		in.	in.		in.		in.	in.	in.	in.	in.
W10×19	5.62	10.2	10 1/4	0.250	1/4	1/8	4.02	4	0.395	3/8	0.695	15/16	5/8	8 3/8	2 1/4
×17	4.99	10.1	10 1/8	0.240	1/4	1/8	4.01	4	0.330	5/16	0.630	7/8	9/16	↓	↓
×15	4.41	9.99	10	0.230	1/4	1/8	4.00	4	0.270	1/4	0.570	13/16	9/16	↓	↓
×12	3.54	9.87	9 7/8	0.190	3/16	1/8	3.96	4	0.210	3/16	0.510	3/4	9/16	↓	↓
W8×67	19.7	9.00	9	0.570	9/16	5/16	8.28	8 1/4	0.935	15/16	1.33	1 5/8	15/16	5 3/4	5 1/2
×58	17.1	8.75	8 3/4	0.510	1/2	1/4	8.22	8 1/4	0.810	13/16	1.20	1 1/2	7/8	↓	↓
×48	14.1	8.50	8 1/2	0.400	3/8	3/16	8.11	8 1/8	0.685	11/16	1.08	1 3/8	13/16	↓	↓
×40	11.7	8.25	8 1/4	0.360	3/8	3/16	8.07	8 1/8	0.560	9/16	0.954	1 1/4	13/16	↓	↓
×35	10.3	8.12	8 1/8	0.310	5/16	3/16	8.02	8	0.495	1/2	0.889	1 3/16	13/16	↓	↓
×31	9.12	8.00	8	0.285	5/16	3/16	8.00	8	0.435	7/16	0.829	1 1/8	3/4	↓	↓
W8×28	8.24	8.06	8	0.285	5/16	3/16	6.54	6 1/2	0.465	7/16	0.859	15/16	5/8	6 1/8	3 1/2
×24	7.08	7.93	7 7/8	0.245	1/4	1/8	6.50	6 1/2	0.400	3/8	0.794	7/8	9/16	6 1/8	3 1/2
W8×21	6.16	8.28	8 1/4	0.250	1/4	1/8	5.27	5 1/4	0.400	3/8	0.700	7/8	9/16	6 1/2	2 3/4
×18	5.26	8.14	8 1/8	0.230	1/4	1/8	5.25	5 1/4	0.330	5/16	0.630	13/16	9/16	6 1/2	2 3/4
W8×15	4.44	8.11	8 1/8	0.245	1/4	1/8	4.01	4	0.315	5/16	0.615	13/16	9/16	6 1/2	2 1/4
×13	3.84	7.99	8	0.230	1/4	1/8	4.00	4	0.255	1/4	0.555	3/4	9/16	↓	↓
×10	2.96	7.89	7 7/8	0.170	3/16	1/8	3.94	4	0.205	3/16	0.505	11/16	1/2	↓	↓
W6×25	7.36	6.38	6 3/8	0.320	5/16	3/16	6.08	6 1/8	0.455	7/16	0.754	15/16	9/16	4 1/2	3 1/2
×20	5.89	6.20	6 1/4	0.260	1/4	1/8	6.02	6	0.365	3/8	0.664	7/8	9/16	↓	↓
×15	4.45	5.99	6	0.230	1/4	1/8	5.99	6	0.260	1/4	0.559	3/4	9/16	↓	↓
W6×16	4.74	6.28	6 1/4	0.260	1/4	1/8	4.03	4	0.405	3/8	0.655	7/8	9/16	4 1/2	2 1/4
×12	3.55	6.03	6	0.230	1/4	1/8	4.00	4	0.280	1/4	0.530	3/4	9/16	↓	↓
×9	2.68	5.90	5 7/8	0.170	3/16	1/8	3.94	4	0.215	3/16	0.465	11/16	1/2	↓	↓
×8.5	2.51	5.83	5 7/8	0.170	3/16	1/8	3.94	4	0.194	3/16	0.444	11/16	1/2	↓	↓
W5×19	5.56	5.15	5 1/8	0.270	1/4	1/8	5.03	5	0.430	7/16	0.730	13/16	7/16	3 1/2	2 3/4
×16	4.71	5.01	5	0.240	1/4	1/8	5.00	5	0.360	3/8	0.660	3/4	7/16	3 1/2	2 3/4
W4×13	3.83	4.16	4 1/8	0.280	1/4	1/8	4.06	4	0.345	3/8	0.595	3/4	1/2	2 5/8	2 1/4

*ASTM A6 tensile group 4 or 5 shape. Special requirements may apply per LRFD Specification Section A3.1c.
†See definition of "Workable Gage" in Nomenclature section at the back of this Manual.

Table 1-1 (cont.). W-Shapes Properties

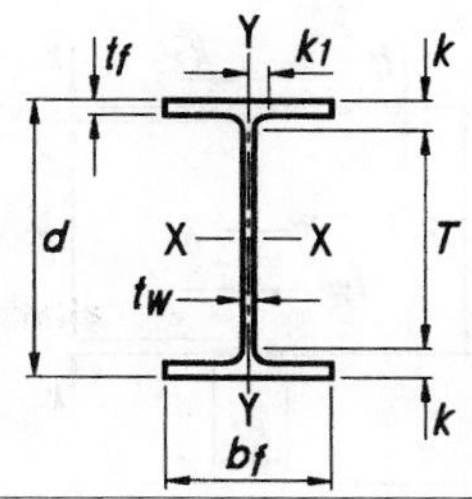

Nominal Wt.	Compact Section Criteria			X_1	$X_2 \times 10^6$	Axis X-X				Axis Y-Y			
	$\frac{b_f}{2t_f}$	$\frac{h}{t_w}$	F_y'''			I	S	r	Z	I	S	r	Z
lb/ft			ksi	ksi	(1/ksi)^2	in.^{4}	in.^{3}	in.	in.^{3}	in.^{4}	in.^{3}	in.	in.^{3}
19	5.09	35.4	51.4	2440	5030	96.3	18.8	4.14	21.6	4.29	2.14	0.874	3.35
17	6.08	36.9	47.3	2180	8220	81.9	16.2	4.05	18.7	3.56	1.78	0.845	2.80
15	7.41	38.5	43.5	1960	13300	68.9	13.8	3.95	16.0	2.89	1.45	0.810	2.30
12	9.43	46.6	29.7	1620	29500	53.8	10.9	3.90	12.6	2.18	1.10	0.785	1.74
67	4.43	11.1	–	6610	74.2	272	60.4	3.72	70.1	88.6	21.4	2.12	32.7
58	5.07	12.4	–	5810	123	228	52.0	3.65	59.8	75.1	18.3	2.10	27.9
48	5.92	15.9	–	4870	237	184	43.2	3.61	49.0	60.9	15.0	2.08	22.9
40	7.21	17.6	–	4080	474	146	35.5	3.53	39.8	49.1	12.2	2.04	18.5
35	8.10	20.5	–	3610	763	127	31.2	3.51	34.7	42.6	10.6	2.03	16.1
31	9.19	22.3	–	3220	1200	110	27.5	3.47	30.4	37.1	9.27	2.02	14.1
28	7.03	22.3	–	3470	942	98.0	24.3	3.45	27.2	21.7	6.63	1.62	10.1
24	8.12	25.9	–	3000	1650	82.7	20.9	3.42	23.1	18.3	5.63	1.61	8.57
21	6.59	27.5	–	2900	2060	75.3	18.2	3.49	20.4	9.77	3.71	1.26	5.69
18	7.95	29.9	–	2510	3800	61.9	15.2	3.43	17.0	7.97	3.04	1.23	4.66
15	6.37	28.1	–	2650	3590	48.0	11.8	3.29	13.6	3.41	1.70	0.876	2.67
13	7.84	29.9	–	2340	6170	39.6	9.91	3.21	11.4	2.73	1.37	0.843	2.15
10	9.61	40.5	39.3	1820	15800	30.8	7.81	3.22	8.87	2.09	1.06	0.841	1.66
25	6.68	15.2	–	4430	358	53.6	16.8	2.70	19.0	17.1	5.61	1.52	8.57
20	8.25	18.7	–	3600	805	41.5	13.4	2.66	15.0	13.3	4.41	1.50	6.72
15	11.5	21.2	–	2800	2270	29.3	9.77	2.56	10.8	9.32	3.11	1.45	4.75
16	4.98	19.1	–	4040	576	32.1	10.2	2.60	11.7	4.43	2.20	0.967	3.39
12	7.14	21.6	–	3100	1730	22.1	7.31	2.49	8.30	2.99	1.50	0.918	2.32
9	9.16	29.2	–	2370	4860	16.4	5.56	2.47	6.23	2.20	1.11	0.905	1.72
8.5	10.2	29.1	–	2270	6000	14.8	5.08	2.43	5.71	1.98	1.01	0.889	1.55
19	5.85	13.7	–	5200	185	26.3	10.2	2.17	11.6	9.13	3.63	1.28	5.53
16	6.94	15.4	–	4450	342	21.4	8.55	2.13	9.63	7.51	3.00	1.26	4.58
13	5.88	10.6	–	5580	152	11.3	5.46	1.72	6.28	3.86	1.90	1.00	2.92

– in F_y''' column indicates $F_y''' > 65$ ksi.

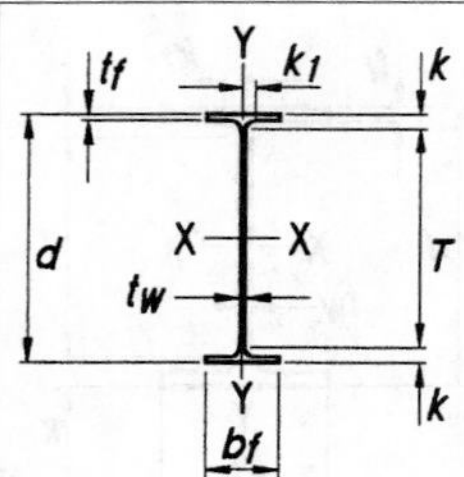

Table 1-2. M-Shapes Dimensions

Shape	Area, A	Depth, d		Web			Flange				Distance			
				Thickness, t_w		$\frac{t_w}{2}$	Width, b_f		Thickness, t_f		k	k_1	T	Workable Gage†
	in.^{2}	in.		in.		in.	in.		in.		in.	in.	in.	in.
M12×11.8	3.47	12.0	12	0.177	3/16	1/8	3.07	3 1/8	0.225	1/4	9/16	3/8	10 7/8	–
×10.8	3.18	12.0	12	0.160	3/16	1/8	3.07	3 1/8	0.210	3/16	9/16	3/8	10 7/8	–
×10	2.95	12.0	12	0.149	1/8	1/16	3.25	3 1/4	0.180	3/16	1/2	3/8	11	–
M10×9	2.65	10.0	10	0.157	3/16	1/8	2.69	2 3/4	0.206	3/16	9/16	3/8	8 7/8	–
×8	2.37	9.95	10	0.141	1/8	1/16	2.69	2 3/4	0.182	3/16	9/16	3/8	8 7/8	–
×7.5	2.22	9.99	10	0.130	1/8	1/16	2.69	2 3/4	0.173	3/16	7/16	5/16	9 1/8	–
M8×6.5	1.92	8.00	8	0.135	1/8	1/16	2.28	2 1/4	0.189	3/16	9/16	3/8	6 7/8	–
×6.2	1.82	8.00	8	0.129	1/8	1/16	2.28	2 1/4	0.177	3/16	7/16	1/4	7 1/8	–
M6×4.4	1.29	6.00	6	0.114	1/8	1/16	1.84	1 7/8	0.171	3/16	3/8	1/4	5 1/4	–
×3.7	1.09	5.92	5 7/8	0.098	1/8	1/16	2.00	2	0.129	1/8	5/16	1/4	5 1/4	–
M5×18.9*	5.56	5.00	5	0.316	5/16	3/16	5.00	5	0.416	7/16	13/16	1/2	3 3/8	2 3/4
M4×6	1.75	3.80	3 3/4	0.130	1/8	1/16	3.80	3 3/4	0.160	3/16	1/2	3/8	2 3/4	–

*This shape has tapered flanges while all other M-shapes have parallel flanges.

†See definition of "Workable Gage" in Nomenclature section at the back of this Manual.

– in Workable Gage column indicates that flange is too narrow to allow tabulation of a workable gage.

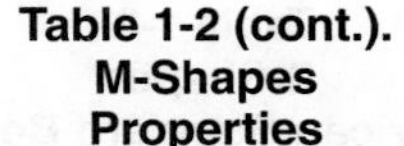

Table 1-2 (cont.). M-Shapes Properties

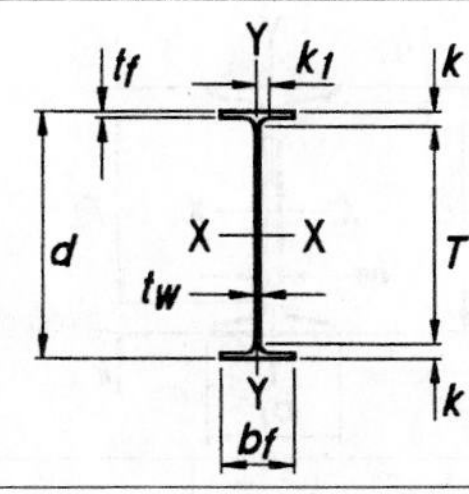

Nominal Wt.	Compact Section Criteria			X_1	$X_2 \times 10^6$	Axis X-X				Axis Y-Y			
	$\frac{b_f}{2t_f}$	$\frac{h}{t_w}$	F_y'''			I	S	r	Z	I	S	r	Z
lb/ft			ksi	ksi	(1/ksi)^2	in.^{4}	in.^{3}	in.	in.^{3}	in.^{4}	in.^{3}	in.	in.^{3}
11.8	6.81	62.5	16.5	1390	63700	72.2	12.0	4.56	14.3	1.09	0.709	0.559	1.15
10.8	7.30	69.2	13.4	1280	88000	66.7	11.1	4.58	13.2	1.01	0.661	0.564	1.07
10	9.03	74.7	11.5	1140	138000	61.7	10.3	4.57	12.2	1.03	0.636	0.592	1.02
9	6.53	58.4	18.9	1480	47100	39.0	7.79	3.83	9.22	0.672	0.500	0.503	0.809
8	7.39	65.0	15.2	1330	73200	34.6	6.95	3.82	8.20	0.593	0.441	0.500	0.711
7.5	7.77	71.0	12.8	1240	95700	33.0	6.60	3.85	7.77	0.562	0.418	0.503	0.670
6.5	6.03	53.8	22.2	1630	30800	18.5	4.63	3.11	5.43	0.376	0.329	0.443	0.529
6.2	6.44	56.5	20.2	1540	38600	17.6	4.39	3.10	5.15	0.352	0.308	0.439	0.495
4.4	5.39	47.0	29.1	1880	16100	7.23	2.41	2.36	2.80	0.180	0.195	0.372	0.311
3.7	7.75	54.7	21.5	1510	38500	5.97	2.01	2.34	2.33	0.173	0.173	0.398	0.273
18.9	6.01	11.2	–	5710	134	24.1	9.63	2.08	11.0	7.86	3.14	1.19	5.02
6	11.90	22.0	–	2900	1920	4.72	2.48	1.64	2.74	1.47	0.771	0.915	1.18

– in F_y''' column indicates $F_y''' > 65$ ksi.

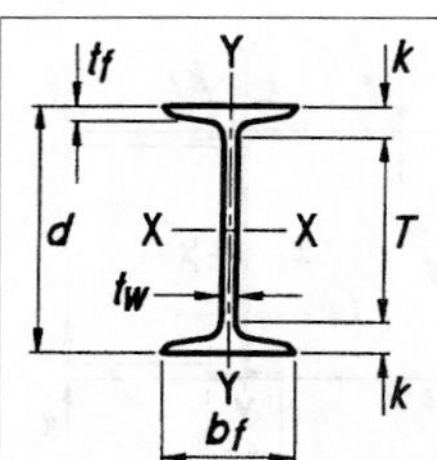

Table 1-3.
S-Shapes
(American Standard Beams)
Dimensions

Shape	Area, A	Depth, d		Web			Flange				Distance		
				Thickness, t_w		$\frac{t_w}{2}$	Width, b_f		Thickness, t_f		k	T	Workable Gage†
	in.^{2}	in.		in.		in.	in.		in.		in.	in.	in.
S24×121	35.5	24.5	24 1/2	0.800	13/16	7/16	8.05	8	1.09	1 1/16	2	20 1/2	4
×106	31.1	24.5	24 1/2	0.620	5/8	5/16	7.87	7 7/8	1.09	1 1/16	2	20 1/2	4
S24×100	29.3	24.0	24	0.745	3/4	3/8	7.25	7 1/4	0.870	7/8	1 3/4	20 1/2	4
×90	26.5	24.0	24	0.625	5/8	5/16	7.13	7 1/8	0.870	7/8	1 3/4	20 1/2	4
×80	23.5	24.0	24	0.500	1/2	1/4	7.00	7	0.870	7/8	1 3/4	20 1/2	4
S20×96	28.2	20.3	20 1/4	0.800	13/16	7/16	7.20	7 1/4	0.920	15/16	1 3/4	16 3/4	4
×86	25.3	20.3	20 1/4	0.660	11/16	3/8	7.06	7	0.920	15/16	1 3/4	16 3/4	4
S20×75	22.0	20.0	20	0.635	5/8	5/16	6.39	6 3/8	0.795	13/16	1 5/8	16 3/4	3 1/2
×66	19.4	20.0	20	0.505	1/2	1/4	6.26	6 1/4	0.795	13/16	1 5/8	16 3/4	3 1/2
S18×70	20.5	18.0	18	0.711	11/16	3/8	6.25	6 1/4	0.691	11/16	1 1/2	15	3 1/2
×54.7	16.0	18.0	18	0.461	7/16	1/4	6.00	6	0.691	11/16	1 1/2	15	3 1/2
S15×50	14.7	15.0	15	0.550	9/16	5/16	5.64	5 5/8	0.622	5/8	1 3/8	12 1/4	3 1/2
×42.9	12.6	15.0	15	0.411	7/16	1/4	5.50	5 1/2	0.622	5/8	1 3/8	12 1/4	3 1/2
S12×50	14.6	12.0	12	0.687	11/16	3/8	5.48	5 1/2	0.659	11/16	1 7/16	9 1/8	3
×40.8	11.9	12.0	12	0.462	7/16	1/4	5.25	5 1/4	0.659	11/16	1 7/16	9 1/8	3
S12×35	10.2	12.0	12	0.428	7/16	1/4	5.08	5 1/8	0.544	9/16	1 3/16	9 5/8	3
×31.8	9.31	12.0	12	0.350	3/8	3/16	5.00	5	0.544	9/16	1 3/16	9 5/8	3
S10×35	10.3	10.0	10	0.594	5/8	5/16	4.94	5	0.491	1/2	1 1/8	7 3/4	2 3/4
×25.4	7.45	10.0	10	0.311	5/16	3/16	4.66	4 5/8	0.491	1/2	1 1/8	7 3/4	2 3/4
S8×23	6.76	8.00	8	0.441	7/16	1/4	4.17	4 1/8	0.425	7/16	1	6	2 1/4
×18.4	5.4	8.00	8	0.271	1/4	1/8	4.00	4	0.425	7/16	1	6	2 1/4
S6×17.25	5.06	6.00	6	0.465	7/16	1/4	3.57	3 5/8	0.359	3/8	13/16	4 3/8	–
×12.5	3.66	6.00	6	0.232	1/4	1/8	3.33	3 3/8	0.359	3/8	13/16	4 3/8	–
S5×10	2.93	5.00	5	0.214	3/16	1/8	3.00	3	0.326	5/16	3/4	3 1/2	–
S4×9.5	2.79	4.00	4	0.326	5/16	3/16	2.80	2 3/4	0.293	5/16	3/4	2 1/2	–
×7.7	2.26	4.00	4	0.193	3/16	1/8	2.66	2 5/8	0.293	5/16	3/4	2 1/2	–
S3×7.5	2.2	3.00	3	0.349	3/8	3/16	2.51	2 1/2	0.260	1/4	5/8	1 3/4	–
×5.7	1.66	3.00	3	0.170	3/16	1/8	2.33	2 3/8	0.260	1/4	5/8	1 3/4	–

†See definition of "Workable Gage" in Nomenclature section at the back of this Manual.
– in Workable Gage column indicates that flange is too narrow to allow tabulation of a workable gage.

Table 1-3 (cont.). S-Shapes (American Standard Beams) Properties

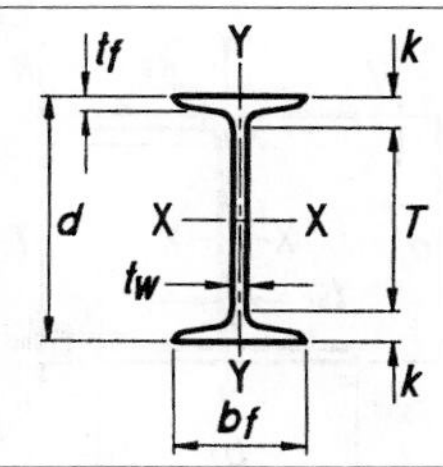

Nominal Wt.	Compact Section Criteria			X_1	$X_2 \times 10^6$	Axis X-X				Axis Y-Y			
	$\frac{b_f}{2t_f}$	$\frac{h}{t_w}$	F_y'''			I	S	r	Z	I	S	r	Z
lb/ft			ksi	ksi	(1/ksi)^2	in.^{4}	in.^{3}	in.	in.^{3}	in.^{4}	in.^{3}	in.	in.^{3}
121	3.69	25.9	–	3310	1770	3160	258	9.43	306	83.0	20.6	1.53	36.3
106	3.61	33.4	57.8	2960	2470	2940	240	9.71	279	76.8	19.5	1.57	33.4
100	4.16	27.8	–	3000	2930	2380	199	9.01	239	47.4	13.1	1.27	24.0
90	4.09	33.1	58.7	2710	4070	2250	187	9.21	222	44.7	12.5	1.30	22.4
80	4.02	41.4	37.6	2450	5460	2100	175	9.47	204	42.0	12.0	1.34	20.8
96	3.91	21.1	–	3730	1160	1670	165	7.71	198	49.9	13.9	1.33	24.9
86	3.84	25.6	–	3350	1630	1570	155	7.89	183	46.6	13.2	1.36	23.1
75	4.02	26.6	–	3140	2290	1280	128	7.62	152	29.5	9.25	1.16	16.7
66	3.93	33.5	57.4	2800	3250	1190	119	7.83	139	27.5	8.78	1.19	15.4
70	4.52	21.5	–	3560	1510	923	103	6.70	124	24.0	7.69	1.08	14.3
54.7	4.34	33.2	58.4	2750	3480	801	89.0	7.07	104	20.7	6.91	1.14	12.1
50	4.53	22.7	–	3450	1530	485	64.7	5.75	77.0	15.6	5.53	1.03	9.99
42.9	4.42	30.4	–	2970	2450	446	59.4	5.95	69.2	14.3	5.19	1.06	9.08
50	4.16	13.7	–	5030	342	303	50.6	4.55	60.9	15.6	5.69	1.03	10.3
40.8	3.98	20.6	–	3980	730	270	45.1	4.76	52.7	13.5	5.13	1.06	8.86
35	4.67	23.1	–	3440	1380	228	38.1	4.72	44.6	9.84	3.88	0.980	6.80
31.8	4.60	28.3	–	3160	1780	217	36.2	4.83	41.8	9.33	3.73	1.00	6.44
35	5.03	13.4	–	4960	374	147	29.4	3.78	35.4	8.30	3.36	0.899	6.19
25.4	4.75	25.6	–	3450	1200	123	24.6	4.07	28.3	6.73	2.89	0.950	4.99
23	4.91	14.1	–	4770	397	64.7	16.2	3.09	19.2	4.27	2.05	0.795	3.67
18.4	4.71	22.9	–	3740	845	57.5	14.4	3.26	16.5	3.69	1.84	0.827	3.18
17.25	4.97	9.67	–	6280	141	26.2	8.74	2.28	10.5	2.29	1.28	0.673	2.35
12.5	4.64	19.4	–	4260	490	22.0	7.34	2.45	8.45	1.80	1.08	0.702	1.86
10	4.61	16.8	–	4720	322	12.3	4.9	2.05	5.66	1.19	0.795	0.638	1.37
9.5	4.77	8.33	–	6850	86.9	6.76	3.38	1.56	4.04	0.887	0.635	0.564	1.13
7.7	4.54	14.1	–	5370	188	6.05	3.03	1.64	3.50	0.748	0.562	0.576	0.970
7.5	4.82	5.38	–	9160	28.1	2.91	1.94	1.15	2.35	0.578	0.461	0.513	0.821
5.7	4.48	11.0	–	6430	89.0	2.50	1.67	1.23	1.94	0.447	0.383	0.518	0.656

– in F_y''' column indicates $F_y''' > 65$ ksi.

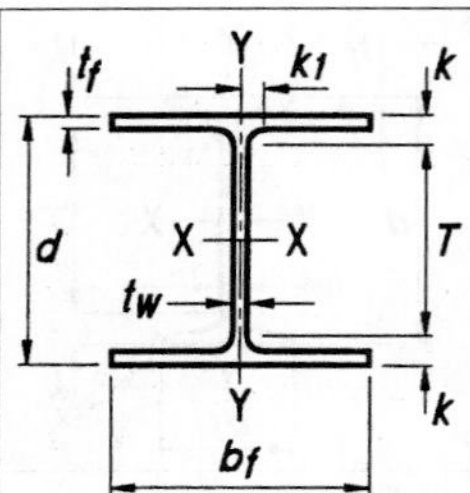

Table 1-4. HP-Shapes (Bearing Piles) Dimensions

Shape	Area, A	Depth, d		Web Thickness, t_w		Web $\frac{t_w}{2}$	Flange Width, b_f		Flange Thickness, t_f		Distance k	Distance k_1	Distance T	Distance Workable Gage[†]
	in.^{2}	in.		in.		in.	in.		in.		in.	in.	in.	in.
HP14×117	34.4	14.2	14 1/4	0.805	13/16	7/16	14.9	14 7/8	0.805	13/16	1 1/2	1 1/16	11 1/4	5 1/2
×102	30.0	14	14	0.705	11/16	3/8	14.8	14 3/4	0.705	11/16	1 3/8	1	↓	↓
×89	26.1	13.8	13 7/8	0.615	5/8	5/16	14.7	14 3/4	0.615	5/8	1 5/16	15/16	↓	↓
×73	21.4	13.6	13 5/8	0.505	1/2	1/4	14.6	14 5/8	0.505	1/2	1 3/16	7/8	↓	↓
HP12×84	24.6	12.3	12 1/4	0.685	11/16	3/8	12.3	12 1/4	0.685	11/16	1 3/8	1	9 1/2	5 1/2
×74	21.8	12.1	12 1/8	0.605	5/8	5/16	12.2	12 1/4	0.610	5/8	1 5/16	15/16	↓	↓
×63	18.4	11.9	12	0.515	1/2	1/4	12.1	12 1/8	0.515	1/2	1 1/4	7/8	↓	↓
×53	15.5	11.8	11 3/4	0.435	7/16	1/4	12.0	12	0.435	7/16	1 1/8	7/8	↓	↓
HP10×57	16.8	9.99	10	0.565	9/16	5/16	10.2	10 1/4	0.565	9/16	1 1/4	15/16	7 1/2	5 1/2
×42	12.4	9.7	9 3/4	0.415	7/16	1/4	10.1	10 1/8	0.420	7/16	1 1/8	13/16	7 1/2	5 1/2
HP8×36	10.6	8.02	8	0.445	7/16	1/4	8.15	8 1/8	0.445	7/16	1 1/8	7/8	5 3/4	5 1/2

[†] See definition of "Workable Gage" in Nomenclature section at the back of this Manual.

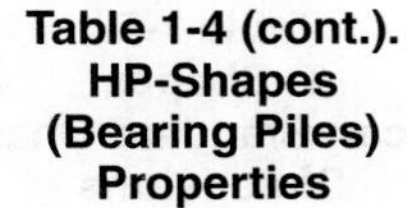

Table 1-4 (cont.). HP-Shapes (Bearing Piles) Properties

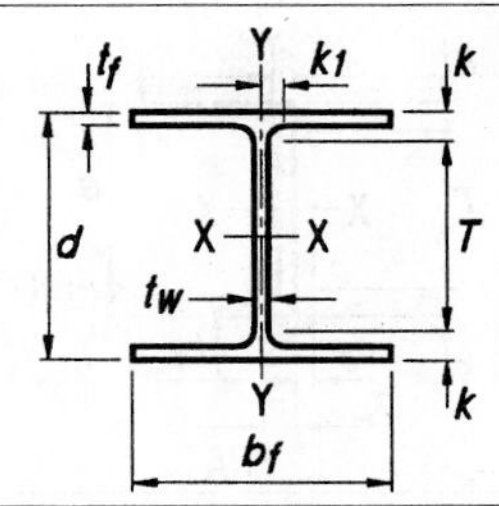

Nominal Wt.	Compact Section Criteria			X_1	$X_2 \times 10^6$	Axis X-X				Axis Y-Y			
	$\frac{b_f}{2t_f}$	$\frac{h}{t_w}$	F_y'''			I	S	r	Z	I	S	r	Z
lb/ft			ksi	ksi	(1/ksi)^2	in.^{4}	in.^{3}	in.	in.^{3}	in.^{4}	in.^{3}	in.	in.^{3}
117	9.25	14.2	–	3870	659	1220	172	5.96	194	443	59.5	3.59	91.4
102	10.5	16.2	–	3390	1090	1050	150	5.92	169	380	51.4	3.56	78.8
89	11.9	18.5	–	2960	1850	904	131	5.88	146	326	44.3	3.53	67.7
73	14.4	22.6	–	2450	3880	729	107	5.84	118	261	35.8	3.49	54.6
84	8.97	14.2	–	3860	670	650	106	5.14	120	213	34.6	2.94	53.2
74	10.0	16.1	–	3440	1050	569	93.8	5.11	105	186	30.4	2.92	46.6
63	11.8	18.9	–	2940	1940	472	79.1	5.06	88.3	153	25.3	2.88	38.7
53	13.8	22.3	–	2500	3640	393	66.7	5.03	74.0	127	21.1	2.86	32.2
57	9.05	13.9	–	3920	631	294	58.8	4.18	66.5	101	19.7	2.45	30.3
42	12.0	18.9	–	2930	1960	210	43.4	4.13	48.3	71.7	14.2	2.41	21.8
36	9.16	14.2	–	3840	685	119	29.8	3.36	33.6	40.3	9.88	1.95	15.2

– in F_y''' column indicates $F_y''' > 65$ ksi.

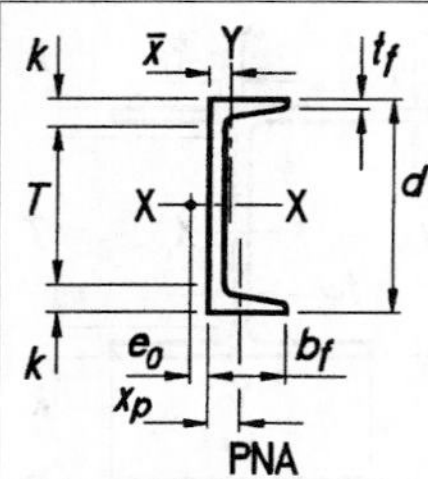

Table 1-5.
C-Shapes
(American Standard Channels)
Dimensions

Shape	Area, A	Depth, d		Web			Flange				Distance		
				Thickness, t_w		$\frac{t_w}{2}$	Width, b_f		Thickness, t_f		k	T	Workable Gage[†]
	in.^{2}	in.		in.		in.	in.		in.		in.	in.	in.
C15×50	14.7	15.0	15	0.716	11/16	3/8	3.72	3 3/4	0.650	5/8	1 7/16	12 1/8	2 1/4
×40	11.8	↓	↓	0.520	1/2	1/4	3.52	3 1/2	↓	↓	↓	↓	2
×33.9	9.95	↓	↓	0.400	3/8	3/16	3.40	3 3/8	↓	↓	↓	↓	2
C12×30	8.81	12.0	12	0.510	1/2	1/4	3.17	3 1/8	0.501	1/2	1 1/8	9 3/4	1 3/4
×25	7.34	↓	↓	0.387	3/8	3/16	3.05	3	↓	↓	↓	↓	↓
×20.7	6.08	↓	↓	0.282	5/16	3/16	2.94	3	↓	↓	↓	↓	↓
C10×30	8.81	10.0	10	0.673	11/16	3/8	3.03	3	0.436	7/16	1	8	1 3/4
×25	7.34	↓	↓	0.526	1/2	1/4	2.89	2 7/8	↓	↓	↓	↓	1 3/4
×20	5.87	↓	↓	0.379	3/8	3/16	2.74	2 3/4	↓	↓	↓	↓	1 1/2
×15.3	4.48	↓	↓	0.240	1/4	1/8	2.60	2 5/8	↓	↓	↓	↓	1 1/2
C9×20	5.87	9.00	9	0.448	7/16	1/4	2.65	2 5/8	0.413	7/16	1	7	1 1/2
×15	4.41	↓	↓	0.285	5/16	3/16	2.49	2 1/2	↓	↓	↓	↓	1 3/8
×13.4	3.94	↓	↓	0.233	1/4	1/8	2.43	2 3/8	↓	↓	↓	↓	1 3/8
C8×18.75	5.51	8.00	8	0.487	1/2	1/4	2.53	2 1/2	0.390	3/8	15/16	6 1/8	1 1/2
×13.75	4.04	↓	↓	0.303	5/16	3/16	2.34	2 3/8	↓	↓	↓	↓	1 3/8
×11.5	3.37	↓	↓	0.220	1/4	1/8	2.26	2 1/4	↓	↓	↓	↓	1 3/8
C7×14.75	4.33	7.00	7	0.419	7/16	1/4	2.30	2 1/4	0.366	3/8	7/8	5 1/4	1 1/4
×12.25	3.60	↓	↓	0.314	5/16	3/16	2.19	2 1/4	↓	↓	↓	↓	↓
×9.8	2.87	↓	↓	0.210	3/16	1/8	2.09	2 1/8	↓	↓	↓	↓	↓
C6×13	3.81	6.00	6	0.437	7/16	1/4	2.16	2 1/8	0.343	5/16	13/16	4 3/8	1 3/8
×10.5	3.08	↓	↓	0.314	5/16	3/16	2.03	2	↓	↓	↓	↓	1 1/8
×8.2	2.39	↓	↓	0.200	3/16	1/8	1.92	1 7/8	↓	↓	↓	↓	1 1/8
C5×9	2.64	5.00	5	0.325	5/16	3/16	1.89	1 7/8	0.320	5/16	3/4	3 1/2	1 1/8
×6.7	1.97	5.00	5	0.190	3/16	1/8	1.75	1 3/4	0.320	5/16	3/4	3 1/2	–
C4×7.25	2.13	4.00	4	0.321	5/16	3/16	1.72	1 3/4	0.296	5/16	3/4	2 1/2	1
×5.4	1.58	↓	↓	0.184	3/16	1/8	1.58	1 5/8	↓	↓	↓	↓	–
×4.5	1.38	↓	↓	0.125	1/8	1/16	1.58	1 5/8	↓	↓	↓	↓	–
C3×6	1.76	3.00	3	0.356	3/8	3/16	1.60	1 5/8	0.273	1/4	11/16	1 5/8	–
×5	1.47	↓	↓	0.258	1/4	1/8	1.50	1 1/2	↓	↓	↓	↓	–
×4.1	1.20	↓	↓	0.170	3/16	1/8	1.41	1 3/8	↓	↓	↓	↓	–
×3.5	1.09	↓	↓	0.132	1/8	1/16	1.37	1 3/8	↓	↓	↓	↓	–

[†] See definition of "Workable Gage" in Nomenclature section at the back of this Manual.
– in Workable Gage column indicates that flange is too narrow to allow tabulation of a workable gage.

Table 1-5 (cont.). C-Shapes (American Standard Channels) Properties

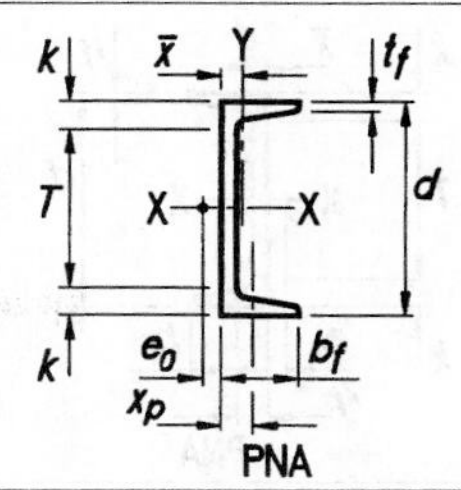

Nom-inal Wt.	Shear Center, e_o	Axis X-X				Axis Y-Y					
		I	S	r	Z	I	S	r	$\bar{x}$	Z	x_p
lb/ft	in.	in.^{4}	in.^{3}	in.	in.^{3}	in.^{4}	in.^{3}	in.	in.	in.^{3}	in.
50	0.583	404	53.8	5.24	68.5	11.0	3.77	0.865	0.799	8.14	0.490
40	0.767	348	46.5	5.45	57.5	9.17	3.34	0.883	0.778	6.84	0.392
33.9	0.896	315	42.0	5.62	50.8	8.07	3.09	0.901	0.788	6.19	0.332
30	0.618	162	27.0	4.29	33.8	5.12	2.05	0.762	0.674	4.32	0.367
25	0.746	144	24.0	4.43	29.4	4.45	1.87	0.779	0.674	3.82	0.306
20.7	0.870	129	21.5	4.61	25.6	3.86	1.72	0.797	0.698	3.47	0.253
30	0.368	103	20.7	3.42	26.7	3.93	1.65	0.668	0.649	3.78	0.441
25	0.494	91.1	18.2	3.52	23.1	3.34	1.47	0.675	0.617	3.18	0.367
20	0.636	78.9	15.8	3.66	19.4	2.80	1.31	0.690	0.606	2.70	0.294
15.3	0.796	67.3	13.5	3.87	15.9	2.27	1.15	0.711	0.634	2.34	0.224
20	0.515	60.9	13.5	3.22	16.9	2.41	1.17	0.640	0.583	2.46	0.326
15	0.681	51.0	11.3	3.40	13.6	1.91	1.01	0.659	0.586	2.04	0.245
13.4	0.742	47.8	10.6	3.49	12.6	1.75	0.954	0.666	0.601	1.94	0.219
18.75	0.431	43.9	11.0	2.82	13.9	1.97	1.01	0.598	0.565	2.17	0.344
13.75	0.604	36.1	9.02	2.99	11.0	1.52	0.848	0.613	0.554	1.73	0.252
11.5	0.697	32.5	8.14	3.11	9.63	1.31	0.775	0.623	0.572	1.57	0.211
14.75	0.441	27.2	7.78	2.51	9.75	1.37	0.772	0.561	0.532	1.63	0.309
12.25	0.538	24.2	6.92	2.60	8.46	1.16	0.696	0.568	0.525	1.42	0.257
9.8	0.647	21.2	6.07	2.72	7.19	0.957	0.617	0.578	0.541	1.26	0.205
13	0.380	17.3	5.78	2.13	7.29	1.05	0.638	0.524	0.514	1.35	0.318
10.5	0.486	15.1	5.04	2.22	6.18	0.860	0.561	0.529	0.500	1.14	0.256
8.2	0.599	13.1	4.35	2.34	5.16	0.687	0.488	0.536	0.512	0.987	0.199
9	0.427	8.89	3.56	1.83	4.39	0.624	0.444	0.486	0.478	0.913	0.264
6.7	0.552	7.48	2.99	1.95	3.55	0.470	0.372	0.489	0.484	0.757	0.215
7.25	0.386	4.58	2.29	1.47	2.84	0.425	0.337	0.447	0.459	0.695	0.266
5.4	0.501	3.85	1.92	1.56	2.29	0.312	0.277	0.444	0.457	0.565	0.231
4.5	0.587	3.65	1.83	1.63	2.12	0.289	0.265	0.457	0.493	0.531	0.321
6	0.322	2.07	1.38	1.08	1.74	0.300	0.263	0.413	0.455	0.543	0.294
5	0.392	1.85	1.23	1.12	1.52	0.241	0.228	0.405	0.439	0.464	0.245
4.1	0.461	1.65	1.10	1.17	1.32	0.191	0.196	0.398	0.437	0.399	0.262
3.5	0.493	1.57	1.04	1.20	1.24	0.169	0.182	0.394	0.443	0.364	0.296

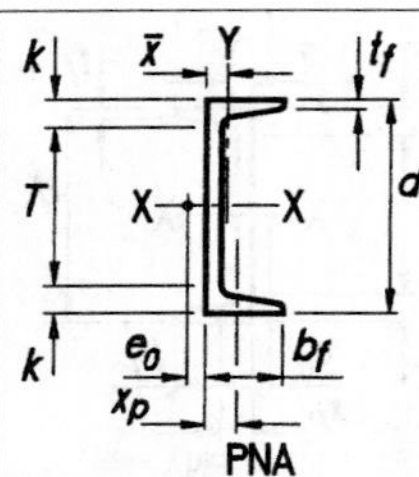

Table 1-6.
MC-Shapes
(Miscellaneous Channels)
Dimensions

Shape	Area, A	Depth, d		Web			Flange				Distance		
				Thickness, t_w		$\frac{t_w}{2}$	Width, b_f		Thickness, t_f		k	T	Workable Gage†
	in.^{2}	in.		in.		in.	in.		in.		in.	in.	in.
MC18×58	17.1	18.0	18	0.700	11/16	3/8	4.20	4 1/4	0.625	5/8	1 7/16	15 1/8	2 1/2
×51.9	15.3	↓	↓	0.600	5/8	5/16	4.10	4 1/8	↓	↓	↓	↓	↓
×45.8	13.5	↓	↓	0.500	1/2	1/4	4.00	4	↓	↓	↓	↓	↓
×42.7	12.6	↓	↓	0.450	7/16	1/4	3.95	4	↓	↓	↓	↓	↓
MC13×50	14.7	13.0	13	0.787	13/16	7/16	4.41	4 3/8	0.610	5/8	1 7/16	10 1/8	2 1/2
×40	11.8	↓	↓	0.560	9/16	5/16	4.18	4 1/8	↓	↓	↓	↓	↓
×35	10.3	↓	↓	0.447	7/16	1/4	4.07	4 1/8	↓	↓	↓	↓	↓
×31.8	9.35	↓	↓	0.375	3/8	3/16	4.00	4	↓	↓	↓	↓	↓
MC12×50	14.7	12.0	12	0.835	13/16	7/16	4.14	4 1/8	0.700	11/16	1 5/16	9 3/8	2 1/2
×45	13.2	↓	↓	0.712	11/16	3/8	4.01	4	↓	↓	↓	↓	↓
×40	11.8	↓	↓	0.590	9/16	5/16	3.89	3 7/8	↓	↓	↓	↓	↓
×35	10.3	↓	↓	0.467	7/16	1/4	3.77	3 3/4	↓	↓	↓	↓	↓
×31	9.12	↓	↓	0.370	3/8	3/16	3.67	3 5/8	↓	↓	↓	↓	2 1/4
MC12×10.6	3.10	12.0	12	0.190	3/16	1/8	1.50	1 1/2	0.309	5/16	3/4	10 1/2	–
MC10×41.1	12.1	10.0	10	0.796	13/16	7/16	4.32	4 3/8	0.575	9/16	1 5/16	7 3/8	2 1/2
×33.6	9.87	↓	↓	0.575	9/16	5/16	4.10	4 1/8	↓	↓	↓	↓	↓
×28.5	8.37	↓	↓	0.425	7/16	1/4	3.95	4	↓	↓	↓	↓	↓
MC10×25	7.35	10.0	10	0.380	3/8	3/16	3.41	3 3/8	0.575	9/16	1 5/16	7 3/8	2
×22	6.45	10.0	10	0.290	5/16	3/16	3.32	3 3/8	0.575	9/16	1 5/16	7 3/8	2
MC10×8.4	2.46	10.0	10	0.170	3/16	1/8	1.50	1 1/2	0.280	1/4	3/4	8 1/2	–
MC9×25.4	7.47	9.00	9	0.450	7/16	1/4	3.50	3 1/2	0.550	9/16	1 1/4	6 1/2	2
×23.9	7.02	9.00	9	0.400	3/8	3/16	3.45	3 1/2	0.550	9/16	1 1/4	6 1/2	2
MC8×22.8	6.7	8.00	8	0.427	7/16	1/4	3.50	3 1/2	0.525	1/2	1 3/16	5 5/8	2
×21.4	6.28	8.00	8	0.375	3/8	3/16	3.45	3 1/2	0.525	1/2	1 3/16	5 5/8	2
MC8×20	5.88	8.00	8	0.400	3/8	3/16	3.03	3	0.500	1/2	1 1/8	5 3/4	2
×18.7	5.5	8.00	8	0.353	3/8	3/16	2.98	3	0.500	1/2	1 1/8	5 3/4	2
MC8×8.5	2.5	8.00	8	0.179	3/16	1/8	1.87	1 7/8	0.311	5/16	13/16	6 3/8	1 1/8
MC7×22.7	6.67	7.00	7	0.503	1/2	1/4	3.60	3 5/8	0.500	1/2	1 1/8	4 3/4	2
×19.1	5.61	7.00	7	0.352	3/8	3/16	3.45	3 1/2	0.500	1/2	1 1/8	4 3/4	2
MC6×18	5.29	6.00	6	0.379	3/8	3/16	3.50	3 1/2	0.475	1/2	1 1/16	3 7/8	2
×15.3	4.49	6.00	6	0.340	5/16	3/16	3.50	3 1/2	0.385	3/8	7/8	4 1/4	2
MC6×16.3	4.79	6.00	6	0.375	3/8	3/16	3.00	3	0.475	1/2	1 1/16	3 7/8	1 3/4
×15.1	4.44	6.00	6	0.316	5/16	3/16	2.94	3	0.475	1/2	1 1/16	3 7/8	1 3/4
MC6×12	3.53	6.00	6	0.310	5/16	3/16	2.50	2 1/2	0.375	3/8	7/8	4 1/4	1 1/2

† See definition of "Workable Gage" in Nomenclature section at the back of this Manual.
– in Workable Gage column indicates that flange is too narrow to allow tabulation of a workable gage.

Table 1-6 (cont.). MC-Shapes (Miscellaneous Channels) Properties

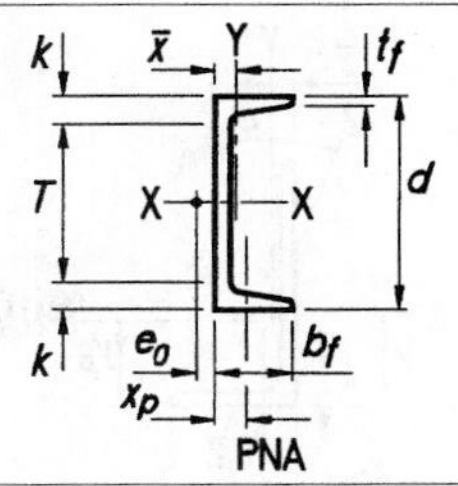

Nom-inal Wt.	Shear Center, e_o	Axis X-X				Axis Y-Y					
		I	S	r	Z	I	S	r	$\bar{x}$	Z	x_p
lb/ft	in.	in.^{4}	in.^{3}	in.	in.^{3}	in.^{4}	in.^{3}	in.	in.	in.^{3}	in.
58	0.695	675	75.0	6.29	95.4	17.6	5.28	1.02	0.862	10.7	0.474
51.9	0.797	627	69.6	6.41	87.3	16.3	5.02	1.03	0.858	9.86	0.424
45.8	0.909	578	64.2	6.55	79.2	14.9	4.77	1.05	0.866	9.14	0.374
42.7	0.969	554	61.5	6.64	75.1	14.3	4.64	1.07	0.877	8.82	0.349
50	0.815	314	48.3	4.62	60.8	16.4	4.77	1.06	0.974	10.2	0.566
40	1.03	273	41.9	4.82	51.2	13.7	4.24	1.08	0.963	8.66	0.452
35	1.16	252	38.8	4.95	46.5	12.3	3.97	1.09	0.980	8.04	0.396
31.8	1.24	239	36.7	5.05	43.4	11.4	3.79	1.10	1.00	7.69	0.36
50	0.741	269	44.9	4.28	56.5	17.4	5.64	1.09	1.05	10.9	0.613
45	0.844	252	41.9	4.36	52.0	15.8	5.31	1.09	1.04	10.1	0.551
40	0.952	234	39.0	4.46	47.7	14.2	4.98	1.10	1.04	9.31	0.49
35	1.07	216	36.1	4.59	43.2	12.6	4.65	1.11	1.05	8.63	0.429
31	1.17	202	33.7	4.71	39.7	11.3	4.37	1.11	1.08	8.15	0.425
10.6	0.284	55.3	9.2	4.22	11.6	0.378	0.307	0.349	0.269	0.635	0.129
41.1	0.864	157	31.5	3.61	39.3	15.7	4.85	1.14	1.09	9.49	0.604
33.6	1.06	139	27.8	3.75	33.7	13.1	4.35	1.15	1.09	8.28	0.494
28.5	1.21	126	25.3	3.89	30.0	11.3	3.99	1.16	1.12	7.59	0.419
25	1.03	110	22.0	3.87	26.2	7.25	2.96	0.993	0.953	5.65	0.367
22	1.12	102	20.5	3.99	23.9	6.40	2.75	0.997	0.990	5.29	0.467
8.4	0.332	31.9	6.39	3.61	7.92	0.326	0.268	0.364	0.284	0.548	0.123
25.4	0.986	87.9	19.5	3.43	23.5	7.57	2.99	1.01	0.970	5.70	0.415
23.9	1.040	84.9	18.9	3.48	22.5	7.14	2.89	1.01	0.981	5.51	0.390
22.8	1.040	63.8	15.9	3.09	19.1	7.01	2.81	1.02	1.01	5.37	0.419
21.4	1.090	61.5	15.4	3.13	18.2	6.58	2.71	1.02	1.02	5.18	0.452
20	0.843	54.4	13.6	3.04	16.4	4.42	2.02	0.867	0.840	3.86	0.367
18.7	0.889	52.4	13.1	3.09	15.6	4.15	1.95	0.868	0.849	3.72	0.344
8.5	0.542	23.3	5.82	3.05	6.95	0.624	0.431	0.500	0.428	0.875	0.156
22.7	1.01	47.4	13.5	2.67	16.4	7.24	2.83	1.04	1.04	5.38	0.477
19.1	1.15	43.1	12.3	2.77	14.5	6.06	2.55	1.04	1.08	4.85	0.579
18	1.17	29.7	9.89	2.37	11.7	5.88	2.47	1.05	1.12	4.68	0.644
15.3	1.16	25.3	8.44	2.38	9.91	4.91	2.01	1.05	1.05	3.85	0.511
16.3	0.930	26.0	8.66	2.33	10.4	3.77	1.82	0.887	0.927	3.47	0.465
15.1	0.982	24.9	8.30	2.37	9.83	3.46	1.73	0.883	0.940	3.30	0.543
12	0.725	18.7	6.24	2.30	7.47	1.85	1.03	0.724	0.704	1.97	0.294

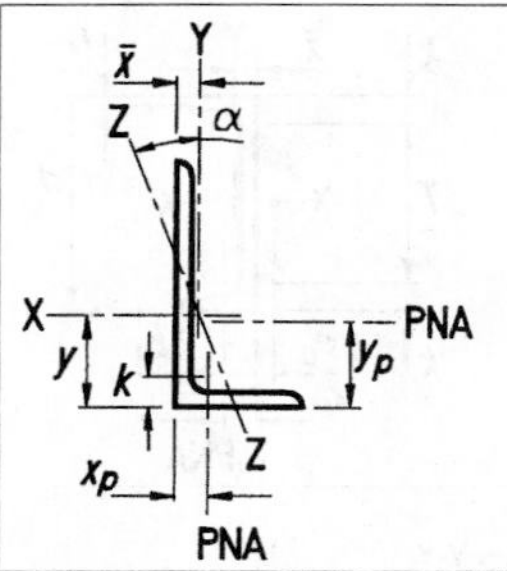

Table 1-7. Angles (L-Shapes) Properties

Shape	k	Wt.	Area, A	Axis X-X					
				I	S	r	ȳ	Z	y_p
	in.	lb/ft	in.^{2}	in.^{4}	in.^{3}	in.	in.	in.^{3}	in.
L8×8×1 1/8	1 3/4	57.2	16.8	98.1	17.5	2.41	2.40	31.6	1.05
×1	1 5/8	51.3	15.1	89.1	15.8	2.43	2.36	28.5	0.943
×7/8	1 1/2	45.3	13.3	79.7	14.0	2.45	2.31	25.3	0.832
×3/4	1 3/8	39.2	11.5	69.9	12.2	2.46	2.26	22.0	0.720
×5/8	1 1/4	33.0	9.69	59.6	10.3	2.48	2.21	18.6	0.606
×9/16	1 3/16	29.8	8.77	54.2	9.33	2.49	2.19	16.8	0.548
×1/2	1 1/8	26.7	7.84	48.8	8.36	2.49	2.17	15.1	0.490
L8×6×1	1 1/2	44.4	13.1	80.9	15.1	2.49	2.65	27.3	1.47
×7/8	1 3/8	39.3	11.5	72.4	13.4	2.50	2.60	24.3	1.41
×3/4	1 1/4	34.0	9.99	63.5	11.7	2.52	2.55	21.1	1.34
×5/8	1 1/8	28.6	8.41	54.2	9.86	2.54	2.50	17.9	1.27
×9/16	1 1/16	25.9	7.61	49.4	8.94	2.55	2.48	16.2	1.23
×1/2	1	23.2	6.80	44.4	8.01	2.55	2.46	14.6	1.20
×7/16	15/16	20.4	5.99	39.3	7.06	2.56	2.43	12.9	1.16
L8×4×1	1 1/2	37.6	11.1	69.7	14.0	2.51	3.03	24.3	2.47
×7/8	1 3/8	33.3	9.79	62.6	12.5	2.53	2.99	21.7	2.41
×3/4	1 1/4	28.9	8.49	55.0	10.9	2.55	2.94	18.9	2.34
×5/8	1 1/8	24.4	7.16	47.0	9.20	2.56	2.89	16.1	2.27
×9/16	1 1/16	22.1	6.49	42.9	8.34	2.57	2.86	14.6	2.23
×1/2	1	19.7	5.80	38.6	7.48	2.58	2.84	13.1	2.20
×7/16	15/16	17.4	5.11	34.2	6.59	2.59	2.81	11.6	2.16
L7×4×3/4	1 1/4	26.2	7.70	37.8	8.39	2.21	2.50	14.8	1.87
×5/8	1 1/8	22.1	6.50	32.4	7.12	2.23	2.45	12.5	1.80
×1/2	1	17.9	5.26	26.6	5.79	2.25	2.40	10.2	1.74
×7/16	15/16	15.8	4.63	23.6	5.11	2.26	2.38	9.0	1.70
×3/8	7/8	13.6	4.00	20.5	4.42	2.27	2.35	7.8	1.67
L6×6×1	1 1/2	37.5	11.0	35.4	8.55	1.79	1.86	15.4	0.918
×7/8	1 3/8	33.2	9.75	31.9	7.61	1.81	1.81	13.7	0.813
×3/4	1 1/4	28.8	8.46	28.1	6.64	1.82	1.77	11.9	0.705
×5/8	1 1/8	24.3	7.13	24.1	5.64	1.84	1.72	10.1	0.594
×9/16	1 1/16	22.0	6.45	22.0	5.12	1.85	1.70	9.18	0.538
×1/2	1	19.6	5.77	19.9	4.59	1.86	1.67	8.22	0.481
×7/16	15/16	17.3	5.08	17.6	4.06	1.86	1.65	7.25	0.423
×3/8	7/8	14.9	4.38	15.4	3.51	1.87	1.62	6.27	0.365
×5/16	13/16	12.5	3.67	13.0	2.95	1.88	1.60	5.26	0.306
L6×4×7/8	1 3/8	27.1	7.95	27.6	7.13	1.86	2.12	12.7	1.46
×3/4	1 1/4	23.5	6.90	24.4	6.23	1.88	2.08	11.1	1.40
×5/8	1 1/8	19.8	5.83	21.0	5.29	1.90	2.03	9.42	1.34
×9/16	1 1/16	17.9	5.27	19.2	4.81	1.91	2.01	8.57	1.31
×1/2	1	16.0	4.72	17.3	4.31	1.92	1.99	7.69	1.28
×7/16	15/16	14.1	4.15	15.4	3.81	1.92	1.97	6.79	1.26
×3/8	7/8	12.2	3.58	13.4	3.3	1.93	1.94	5.87	1.23
×5/16	13/16	10.2	2.99	11.3	2.77	1.94	1.92	4.93	1.21
L6×3 1/2×1/2	1	15.3	4.48	16.6	4.23	1.92	2.08	7.46	1.52
×3/8	7/8	11.6	3.41	12.8	3.24	1.94	2.04	5.71	1.46
×5/16	13/16	9.72	2.86	10.9	2.72	1.95	2.02	4.81	1.43
L5×5×7/8	1 3/8	27.3	8.02	17.8	5.16	1.49	1.56	9.31	0.802
×3/4	1 1/4	23.7	6.98	15.7	4.52	1.50	1.52	8.14	0.698
×5/8	1 1/8	20.1	5.90	13.6	3.85	1.52	1.47	6.93	0.590
×1/2	1	16.3	4.79	11.3	3.15	1.53	1.42	5.66	0.479
×7/16	15/16	14.4	4.22	10.0	2.78	1.54	1.40	5.00	0.422
×3/8	7/8	12.4	3.65	8.76	2.41	1.55	1.37	4.33	0.365
×5/16	13/16	10.4	3.07	7.44	2.04	1.56	1.35	3.65	0.307

Table 1-7 (cont.). Angles (L-Shapes) Properties

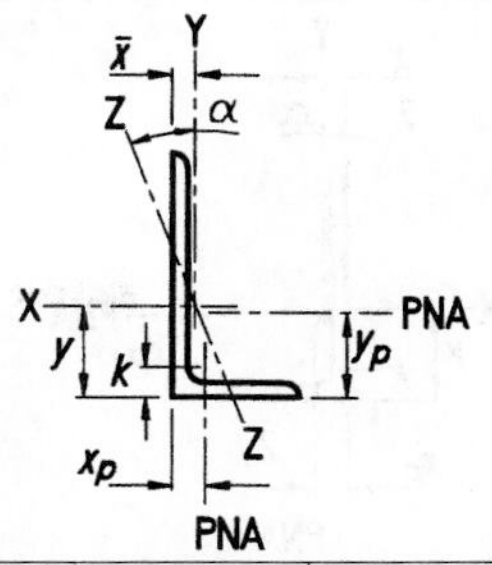

Shape	Axis Y-Y					Axis Z-Z			Q_s^*
	I	S	r	$\bar{x}$	Z	x_p	r	Tan	$F_y = 36$
	in.^{4}	in.^{3}	in.	in.	in.^{3}	in.	in.	α	ksi
L8×8×1 1/8	98.1	17.5	2.41	2.40	31.6	1.05	1.56	1.00	–
×1	89.1	15.8	2.43	2.36	28.5	0.943	1.56	1.00	–
×7/8	79.7	14.0	2.45	2.31	25.3	0.832	1.57	1.00	–
×3/4	69.9	12.2	2.46	2.26	22.0	0.720	1.57	1.00	–
×5/8	59.6	10.3	2.48	2.21	18.6	0.606	1.58	1.00	0.997
×9/16	54.2	9.33	2.49	2.19	16.8	0.548	1.58	1.00	0.959
×1/2	48.8	8.36	2.49	2.17	15.1	0.490	1.59	1.00	0.912
L8×6×1	38.8	8.92	1.72	1.65	16.2	0.816	1.28	0.542	–
×7/8	34.9	7.94	1.74	1.60	14.4	0.721	1.28	0.546	–
×3/4	30.8	6.92	1.75	1.56	12.5	0.624	1.29	0.550	–
×5/8	26.4	5.88	1.77	1.51	10.5	0.526	1.29	0.554	0.997
×9/16	24.1	5.34	1.78	1.49	9.52	0.476	1.30	0.556	0.959
×1/2	21.7	4.79	1.79	1.46	8.52	0.425	1.30	0.557	0.912
×7/16	19.3	4.23	1.80	1.44	7.50	0.374	1.31	0.559	0.850
L8×4×1	11.6	3.94	1.03	1.04	7.73	0.691	0.844	0.247	–
×7/8	10.5	3.51	1.04	0.997	6.77	0.612	0.846	0.252	–
×3/4	9.37	3.07	1.05	0.949	5.82	0.531	0.850	0.257	–
×5/8	8.11	2.62	1.06	0.902	4.86	0.448	0.856	0.262	0.997
×9/16	7.44	2.38	1.07	0.878	4.39	0.405	0.859	0.264	0.959
×1/2	6.75	2.15	1.08	0.854	3.91	0.363	0.863	0.266	0.912
×7/16	6.03	1.90	1.09	0.829	3.42	0.320	0.867	0.268	0.850
L7×4×3/4	9.00	3.01	1.08	1.00	5.60	0.550	0.855	0.324	–
×5/8	7.79	2.56	1.10	0.958	4.69	0.464	0.860	0.329	–
×1/2	6.48	2.10	1.11	0.910	3.77	0.376	0.866	0.334	0.965
×7/16	5.79	1.86	1.12	0.886	3.31	0.331	0.869	0.337	0.912
×3/8	5.06	1.61	1.12	0.861	2.84	0.286	0.873	0.339	0.84
L6×6×1	35.4	8.55	1.79	1.86	15.4	0.918	1.17	1.00	–
×7/8	31.9	7.61	1.81	1.81	13.7	0.813	1.17	1.00	–
×3/4	28.1	6.64	1.82	1.77	11.9	0.705	1.17	1.00	–
×5/8	24.1	5.64	1.84	1.72	10.1	0.594	1.17	1.00	–
×9/16	22.0	5.12	1.85	1.70	9.17	0.538	1.18	1.00	–
×1/2	19.9	4.59	1.86	1.67	8.22	0.481	1.18	1.00	–
×7/16	17.6	4.06	1.86	1.65	7.25	0.423	1.18	1.00	0.973
×3/8	15.4	3.51	1.87	1.62	6.26	0.365	1.19	1.00	0.912
×5/16	13.0	2.95	1.88	1.60	5.26	0.306	1.19	1.00	0.826
L6×4×7/8	9.70	3.37	1.10	1.12	6.25	0.663	0.855	0.422	–
×3/4	8.63	2.95	1.12	1.08	5.42	0.575	0.857	0.429	–
×5/8	7.47	2.52	1.13	1.03	4.56	0.485	0.860	0.435	–
×9/16	6.85	2.29	1.14	1.01	4.13	0.440	0.862	0.438	–
×1/2	6.21	2.06	1.15	0.986	3.69	0.393	0.865	0.441	–
×7/16	5.54	1.82	1.16	0.963	3.24	0.346	0.868	0.444	0.973
×3/8	4.84	1.58	1.16	0.940	2.79	0.298	0.871	0.447	0.912
×5/16	4.11	1.33	1.17	0.916	2.33	0.249	0.874	0.450	0.826
L6×3 1/2×1/2	4.23	1.58	0.971	0.833	2.88	0.374	0.757	0.344	–
×3/8	3.32	1.22	0.987	0.786	2.18	0.284	0.764	0.350	0.912
×5/16	2.83	1.03	0.995	0.763	1.82	0.238	0.768	0.353	0.826
L5×5×7/8	17.8	5.16	1.49	1.56	9.30	0.802	0.971	1.00	–
×3/4	15.7	4.52	1.50	1.52	8.14	0.698	0.972	1.00	–
×5/8	13.6	3.85	1.52	1.47	6.92	0.590	0.975	1.00	–
×1/2	11.3	3.15	1.53	1.42	5.66	0.479	0.980	1.00	–
×7/16	10.0	2.78	1.54	1.40	5.00	0.422	0.983	1.00	–
×3/8	8.76	2.41	1.55	1.37	4.33	0.365	0.986	1.00	0.983
×5/16	7.44	2.04	1.56	1.35	3.65	0.307	0.990	1.00	0.912

*Where no value of Q_s is shown, the angle complies with LRFD Specification Section E2.

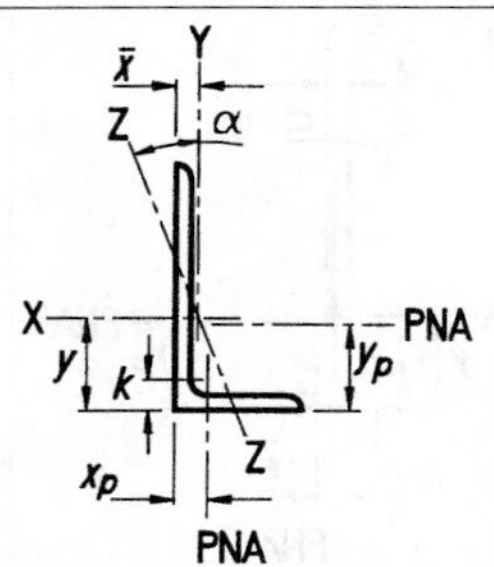

Table 1-7 (cont.). Angles (L-Shapes) Properties

Shape	k	Wt.	Area, A	Axis X-X					
				I	S	r	ȳ	Z	y_p
	in.	lb/ft	in.^{2}	in.^{4}	in.^{3}	in.	in.	in.^{3}	in.
L5×3 1/2×3/4	1 3/16	19.8	5.82	13.9	4.26	1.55	1.74	7.60	1.12
×5/8	1 1/16	16.8	4.93	12.0	3.63	1.56	1.69	6.50	1.06
×1/2	15/16	13.6	4.00	9.96	2.97	1.58	1.65	5.33	0.997
×3/8	13/16	10.4	3.05	7.75	2.28	1.59	1.60	4.09	0.933
×5/16	3/4	8.72	2.56	6.58	1.92	1.60	1.57	3.45	0.901
×1/4	11/16	7.03	2.07	5.36	1.55	1.61	1.55	2.78	0.868
L5×3×1/2	15/16	12.8	3.75	9.43	2.89	1.58	1.74	5.12	1.25
×7/16	7/8	11.3	3.31	8.41	2.56	1.59	1.72	4.53	1.21
×3/8	13/16	9.74	2.86	7.35	2.22	1.60	1.69	3.93	1.18
×5/16	3/4	8.19	2.41	6.24	1.87	1.61	1.67	3.32	1.15
×1/4	11/16	6.60	1.94	5.09	1.51	1.62	1.64	2.68	1.12
L4×4×3/4	1 1/8	18.5	5.43	7.62	2.79	1.18	1.27	5.02	0.679
×5/8	1	15.7	4.61	6.62	2.38	1.20	1.22	4.28	0.576
×1/2	7/8	12.7	3.75	5.52	1.96	1.21	1.18	3.50	0.468
×7/16	13/16	11.2	3.30	4.93	1.73	1.22	1.15	3.10	0.413
×3/8	3/4	9.72	2.86	4.32	1.5	1.23	1.13	2.69	0.357
×5/16	11/16	8.16	2.40	3.67	1.27	1.24	1.11	2.26	0.300
×1/4	5/8	6.58	1.93	3.00	1.03	1.25	1.08	1.82	0.242
L4×3 1/2×1/2	1/2	11.9	3.50	5.30	1.92	1.23	1.24	3.46	0.497
×3/8	3/8	9.10	2.68	4.15	1.48	1.25	1.20	2.66	0.433
×5/16	5/16	7.65	2.25	3.53	1.25	1.25	1.17	2.24	0.401
×1/4	1/4	6.18	1.82	2.89	1.01	1.26	1.14	1.81	0.368
L4×3×5/8	5/8	13.6	3.99	6.01	2.28	1.23	1.37	4.08	0.810
×1/2	1/2	11.1	3.25	5.02	1.87	1.24	1.32	3.36	0.747
×3/8	3/8	8.47	2.49	3.94	1.44	1.26	1.27	2.60	0.683
×5/16	5/16	7.12	2.09	3.36	1.22	1.27	1.25	2.19	0.651
×1/4	1/4	5.75	1.69	2.75	0.988	1.27	1.22	1.77	0.618
L3 1/2×3 1/2×1/2	7/8	11.1	3.27	3.63	1.48	1.05	1.05	2.66	0.466
×7/16	13/16	9.82	2.89	3.25	1.32	1.06	1.03	2.36	0.412
×3/8	3/4	8.51	2.50	2.86	1.15	1.07	1.00	2.06	0.357
×5/16	11/16	7.16	2.10	2.44	0.969	1.08	0.979	1.74	0.301
×1/4	5/8	5.79	1.70	2.00	0.787	1.09	0.954	1.41	0.243
L3 1/2×3×1/2	7/8	10.3	3.02	3.45	1.45	1.07	1.12	2.61	0.480
×7/16	13/16	9.09	2.67	3.10	1.29	1.08	1.09	2.32	0.446
×3/8	3/4	7.88	2.32	2.73	1.12	1.09	1.07	2.03	0.411
×5/16	11/16	6.65	1.95	2.33	0.951	1.09	1.05	1.72	0.375
×1/4	5/8	5.38	1.58	1.92	0.773	1.10	1.02	1.39	0.336
L3 1/2×2 1/2×1/2	7/8	9.41	2.76	3.24	1.41	1.08	1.20	2.52	0.736
×3/8	3/4	7.23	2.12	2.56	1.09	1.10	1.15	1.96	0.668
×5/16	11/16	6.10	1.79	2.20	0.925	1.11	1.13	1.67	0.633
×1/4	5/8	4.94	1.45	1.81	0.753	1.12	1.10	1.36	0.596

Table 1-7 (cont.). Angles (L-Shapes) Properties

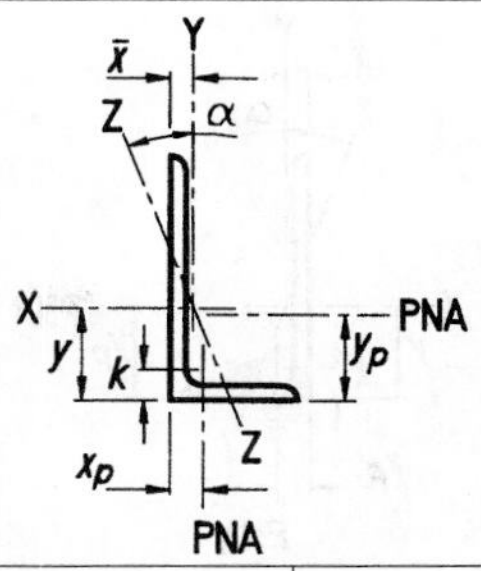

Shape	Axis Y-Y						Axis Z-Z		Q_s^*
	I	S	r	$\bar{x}$	Z	x_p	r	Tan	$F_y = 36$
	in.^{4}	in.^{3}	in.	in.	in.^{3}	in.	in.	α	ksi
L5×3 1/2×3/4	5.52	2.20	0.974	0.993	4.07	0.582	0.744	0.464	–
×5/8	4.80	1.88	0.987	0.947	3.43	0.493	0.746	0.472	–
×1/2	4.02	1.55	1.00	0.901	2.79	0.400	0.750	0.479	–
×3/8	3.15	1.19	1.02	0.854	2.12	0.305	0.755	0.486	0.983
×5/16	2.69	1.01	1.02	0.829	1.77	0.256	0.758	0.489	0.912
×1/4	2.20	0.816	1.03	0.804	1.42	0.207	0.761	0.491	0.804
L5×3×1/2	2.55	1.13	0.824	0.746	2.08	0.375	0.642	0.357	–
×7/16	2.29	1.00	0.831	0.722	1.82	0.331	0.644	0.361	–
×3/8	2.01	0.874	0.838	0.698	1.57	0.286	0.646	0.364	0.983
×5/16	1.72	0.739	0.846	0.673	1.31	0.241	0.649	0.368	0.912
×1/4	1.41	0.600	0.853	0.648	1.05	0.194	0.652	0.371	0.804
L4×4×3/4	7.62	2.79	1.18	1.27	5.01	0.679	0.774	1.00	–
×5/8	6.62	2.38	1.20	1.22	4.28	0.576	0.774	1.00	–
×1/2	5.52	1.96	1.21	1.18	3.50	0.468	0.776	1.00	–
×7/16	4.93	1.73	1.22	1.15	3.10	0.413	0.777	1.00	–
×3/8	4.32	1.50	1.23	1.13	2.68	0.357	0.779	1.00	–
×5/16	3.67	1.27	1.24	1.11	2.26	0.300	0.781	1.00	0.997
×1/4	3.00	1.03	1.25	1.08	1.82	0.242	0.783	1.00	0.912
L4×3 1/2×1/2	3.76	1.50	1.04	0.994	2.69	0.438	0.716	0.750	–
×3/8	2.96	1.16	1.05	0.947	2.06	0.334	0.719	0.755	–
×5/16	2.52	0.980	1.06	0.923	1.74	0.281	0.721	0.757	0.997
×1/4	2.07	0.794	1.07	0.897	1.40	0.227	0.723	0.759	0.912
L4×3×5/8	2.85	1.34	0.845	0.867	2.45	0.498	0.631	0.534	–
×1/2	2.40	1.10	0.858	0.822	1.99	0.407	0.633	0.543	–
×3/8	1.89	0.851	0.873	0.775	1.52	0.311	0.636	0.551	–
×5/16	1.62	0.721	0.880	0.750	1.28	0.262	0.638	0.554	0.997
×1/4	1.33	0.585	0.887	0.725	1.03	0.211	0.639	0.558	0.912
L3 1/2×3 1/2×1/2	3.63	1.48	1.05	1.05	2.66	0.466	0.679	1.00	–
×7/16	3.25	1.32	1.06	1.03	2.36	0.412	0.681	1.00	–
×3/8	2.86	1.15	1.07	1.00	2.05	0.357	0.683	1.00	–
×5/16	2.44	0.969	1.08	0.979	1.74	0.301	0.685	1.00	–
×1/4	2.00	0.787	1.09	0.954	1.41	0.243	0.688	1.00	0.965
L3 1/2×3×1/2	2.32	1.09	0.877	0.869	1.97	0.431	0.618	0.713	–
×7/16	2.09	0.971	0.885	0.846	1.75	0.382	0.620	0.717	–
×3/8	1.84	0.847	0.892	0.823	1.52	0.331	0.622	0.720	–
×5/16	1.58	0.718	0.900	0.798	1.28	0.279	0.624	0.722	–
×1/4	1.30	0.585	0.908	0.773	1.04	0.226	0.628	0.725	0.965
L3 1/2×2 1/2×1/2	1.36	0.756	0.701	0.701	1.39	0.395	0.532	0.485	–
×3/8	1.09	0.589	0.716	0.655	1.07	0.303	0.535	0.495	–
×5/16	0.937	0.501	0.723	0.632	0.900	0.256	0.538	0.500	–
×1/4	0.775	0.410	0.731	0.607	0.728	0.207	0.541	0.504	0.965

*Where no value of Q_s is shown, the angle complies with LRFD Specification Section E2.

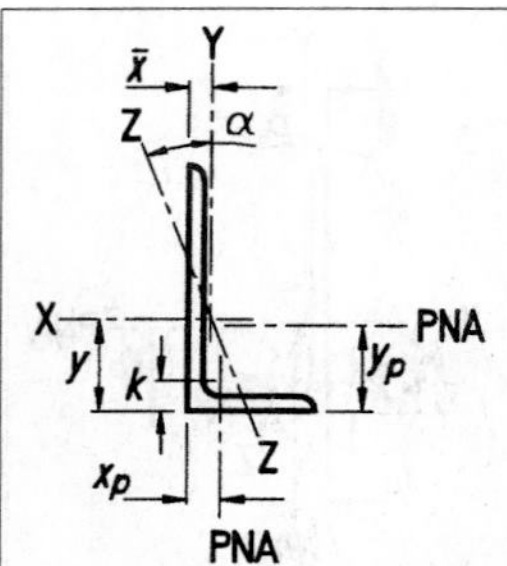

Table 1-7 (cont.). Angles (L-Shapes) Properties

Shape	k	Wt.	Area, A	Axis X-X					
				I	S	r	$\bar{y}$	Z	y_p
	in.	lb/ft	in.^{2}	in.^{4}	in.^{3}	in.	in.	in.^{3}	in.
L3×3×1/2	7/8	9.35	2.75	2.20	1.06	0.895	0.929	1.91	0.458
×7/16	13/16	8.28	2.43	1.98	0.946	0.903	0.907	1.70	0.405
×3/8	3/4	7.17	2.11	1.75	0.825	0.910	0.884	1.48	0.351
×5/16	11/16	6.04	1.78	1.50	0.699	0.918	0.860	1.26	0.296
×1/4	5/8	4.89	1.44	1.23	0.569	0.926	0.836	1.02	0.239
×3/16	9/16	3.70	1.09	0.948	0.433	0.933	0.812	0.774	0.181
L3×2 1/2×1/2	7/8	8.53	2.51	2.07	1.03	0.910	0.995	1.86	0.494
×7/16	13/16	7.56	2.22	1.87	0.921	0.917	0.972	1.66	0.462
×3/8	3/4	6.56	1.93	1.65	0.803	0.924	0.949	1.45	0.430
×5/16	11/16	5.54	1.63	1.41	0.681	0.932	0.925	1.23	0.397
×1/4	5/8	4.49	1.32	1.16	0.555	0.940	0.900	1.00	0.363
×3/16	9/16	3.41	1.00	0.899	0.423	0.947	0.874	0.761	0.328
L3×2×1/2	13/16	7.70	2.26	1.92	1.00	0.922	1.080	1.78	0.736
×3/8	11/16	5.95	1.75	1.54	0.779	0.937	1.030	1.39	0.668
×5/16	5/8	5.03	1.48	1.32	0.662	0.945	1.010	1.19	0.633
×1/4	9/16	4.09	1.20	1.09	0.541	0.953	0.980	0.969	0.596
×3/16	1/2	3.12	0.917	0.847	0.414	0.961	0.952	0.743	0.556
L2 1/2×2 1/2×1/2	3/4	7.65	2.25	1.22	0.716	0.735	0.803	1.29	0.450
×3/8	5/8	5.90	1.73	0.972	0.558	0.749	0.758	1.01	0.347
×5/16	9/16	4.98	1.46	0.837	0.474	0.756	0.735	0.853	0.293
×1/4	1/2	4.04	1.19	0.692	0.387	0.764	0.711	0.695	0.237
×3/16	7/16	3.06	0.901	0.535	0.295	0.771	0.687	0.529	0.180
L2 1/2×2×3/8	5/8	5.30	1.56	0.914	0.546	0.766	0.826	0.982	0.425
×5/16	9/16	4.49	1.32	0.790	0.465	0.774	0.803	0.839	0.391
×1/4	1/2	3.65	1.07	0.656	0.381	0.782	0.779	0.688	0.356
×3/16	7/16	2.78	0.818	0.511	0.293	0.790	0.754	0.529	0.318
L2×2×3/8	5/8	4.65	1.37	0.476	0.348	0.591	0.632	0.629	0.342
×5/16	9/16	3.94	1.16	0.414	0.298	0.598	0.609	0.537	0.290
×1/4	1/2	3.21	0.944	0.346	0.244	0.605	0.586	0.440	0.236
×3/16	7/16	2.46	0.722	0.271	0.188	0.612	0.561	0.338	0.180
×1/8	3/8	1.67	0.491	0.189	0.129	0.620	0.534	0.230	0.123

Table 1-7 (cont.). Angles (L-Shapes) Properties

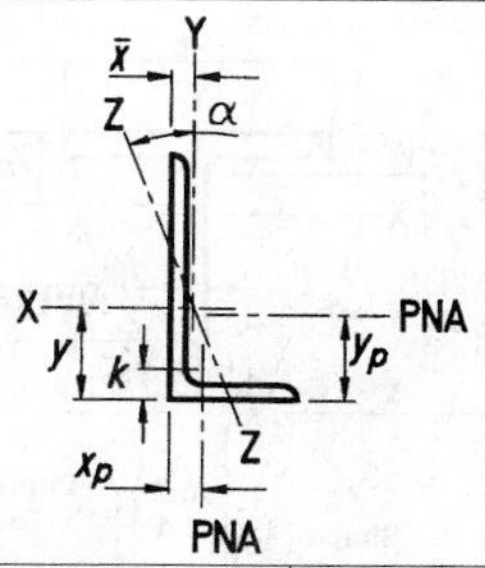

Shape	Axis Y-Y						Axis Z-Z		Q_s*
	I	S	r	$\bar{x}$	Z	x_p	r	Tan	F_y = 36
	in.⁴	in.³	in.	in.	in.³	in.	in.	α	ksi
L3×3×1/2	2.20	1.06	0.895	0.929	1.91	0.458	0.580	1.00	–
×7/16	1.98	0.946	0.903	0.907	1.70	0.405	0.580	1.00	–
×3/8	1.75	0.825	0.910	0.884	1.48	0.351	0.581	1.00	–
×5/16	1.50	0.699	0.918	0.860	1.25	0.296	0.583	1.00	–
×1/4	1.23	0.569	0.926	0.836	1.02	0.239	0.585	1.00	–
×3/16	0.948	0.433	0.933	0.812	0.774	0.181	0.586	1.00	0.912
L3×2 1/2×1/2	1.29	0.736	0.718	0.746	1.34	0.418	0.516	0.666	–
×7/16	1.17	0.656	0.724	0.724	1.19	0.370	0.516	0.671	–
×3/8	1.03	0.573	0.731	0.701	1.03	0.321	0.517	0.675	–
×5/16	0.888	0.487	0.739	0.677	0.873	0.271	0.518	0.680	–
×1/4	0.734	0.397	0.746	0.653	0.707	0.220	0.520	0.683	–
×3/16	0.568	0.303	0.753	0.627	0.536	0.167	0.521	0.687	0.912
L3×2×1/2	0.667	0.470	0.543	0.580	0.887	0.377	0.425	0.413	–
×3/8	0.539	0.368	0.555	0.535	0.679	0.291	0.426	0.426	–
×5/16	0.467	0.314	0.562	0.511	0.572	0.247	0.428	0.432	–
×1/4	0.390	0.258	0.569	0.487	0.463	0.200	0.431	0.437	–
×3/16	0.305	0.198	0.577	0.462	0.351	0.153	0.435	0.442	0.912
L2 1/2×2 1/2×1/2	1.22	0.716	0.735	0.803	1.29	0.450	0.481	1.00	–
×3/8	0.972	0.558	0.749	0.758	1.00	0.347	0.481	1.00	–
×5/16	0.837	0.474	0.756	0.735	0.853	0.293	0.481	1.00	–
×1/4	0.692	0.387	0.764	0.711	0.694	0.237	0.482	1.00	–
×3/16	0.535	0.295	0.771	0.687	0.528	0.180	0.482	1.00	0.983
L2 1/2×2×3/8	0.513	0.361	0.574	0.578	0.657	0.311	0.419	0.613	–
×5/16	0.446	0.309	0.581	0.555	0.557	0.264	0.420	0.618	–
×1/4	0.372	0.253	0.589	0.532	0.454	0.214	0.423	0.624	–
×3/16	0.292	0.195	0.597	0.508	0.347	0.164	0.426	0.628	0.983
L2×2×3/8	0.476	0.348	0.591	0.632	0.628	0.342	0.386	1.00	–
×5/16	0.414	0.298	0.598	0.609	0.536	0.290	0.386	1.00	–
×1/4	0.346	0.244	0.605	0.586	0.440	0.236	0.387	1.00	–
×3/16	0.271	0.188	0.612	0.561	0.338	0.180	0.389	1.00	–
×1/8	0.189	0.129	0.620	0.534	0.230	0.123	0.391	1.00	0.912

*Where no value of Q_s is shown, the angle complies with LRFD Specification Section E2.

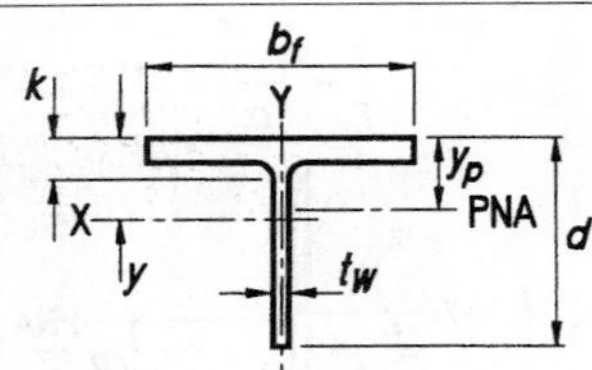

Table 1-8. WT-Shapes (Structural Tees Split from W-Shapes) Dimensions

Shape	Area, A	Depth of Tee, d		Stem: Thickness, t_w		Stem: $\frac{t_w}{2}$	Stem: Area	Flange: Width, b_f		Flange: Thickness, t_f		Distance: k		Distance: Workable Gage†
	in.²	in.		in.		in.	in.²	in.		in.		in.	in.	in.
WT22×167.5*	49.1	22.0	22	1.02	1	1/2	22.5	16.0	16	1.77	1 3/4	2.56	2 5/8	5 1/2
×145	42.9	21.8	21 3/4	0.870	7/8	7/16	19.0	15.8	15 7/8	1.58	1 9/16	2.37	2 7/16	
×131	38.6	21.7	21 5/8	0.790	13/16	7/16	17.1	15.8	15 3/4	1.42	1 7/16	2.21	2 1/4	
×115	33.8	21.5	21 1/2	0.710	11/16	3/8	15.2	15.8	15 3/4	1.22	1 1/4	2.01	2 1/16	
WT20×296.5*	87.2	21.5	21 1/2	1.79	1 13/16	15/16	38.5	16.7	16 3/4	3.23	3 1/4	4.41	4 1/2	7 1/2
×251.5*	74.0	21.0	21	1.54	1 9/16	13/16	32.4	16.4	16 3/8	2.76	2 3/4	3.94	4	
×215.5*	63.4	20.6	20 5/8	1.34	1 5/16	11/16	27.6	16.2	16 1/4	2.36	2 3/8	3.54	3 5/8	
×198.5*	58.4	20.5	20 1/2	1.22	1 1/4	5/8	25.0	16.1	16 1/8	2.2	2 3/16	3.38	3 1/2	
×186*	54.7	20.3	20 3/8	1.16	1 3/16	5/8	23.6	16.1	16	2.05	2 1/16	3.23	3 5/16	
×181*	53.3	20.3	20 1/4	1.12	1 1/8	9/16	22.7	16.0	16	2.01	2	3.19	3 1/4	
×162	47.7	20.1	20 1/8	1.00	1	1/2	20.1	15.9	15 7/8	1.81	1 13/16	2.99	3 1/16	
×148.5	43.7	19.9	19 7/8	0.930	15/16	1/2	18.5	15.8	15 7/8	1.65	1 5/8	2.83	2 15/16	
×138.5	40.7	19.8	19 7/8	0.830	13/16	7/16	16.5	15.8	15 7/8	1.58	1 9/16	2.76	2 7/8	
×124.5	36.7	19.7	19 3/4	0.750	3/4	3/8	14.8	15.8	15 3/4	1.42	1 7/16	2.60	2 11/16	
×107.5	31.7	19.5	19 1/2	0.650	5/8	5/16	12.7	15.8	15 3/4	1.22	1 1/4	2.40	2 1/2	
×99.5	29.2	19.3	19 3/8	0.650	5/8	5/16	12.6	15.8	15 3/4	1.07	1 1/16	2.25	2 5/16	
WT20×196*	57.7	20.8	20 3/4	1.42	1 7/16	3/4	29.5	12.4	12 3/8	2.52	2 1/2	3.70	3 13/16	7 1/2
×165.5*	48.7	20.4	20 3/8	1.22	1 1/4	5/8	24.9	12.2	12 1/8	2.13	2 1/8	3.31	3 3/8	
×163.5	48.0	20.4	20 3/8	1.18	1 3/16	5/8	24.1	12.1	12 1/8	2.13	2 1/8	3.31	3 3/8	
×139	40.9	20.1	20 1/8	1.02	1	1/2	20.5	12.0	12	1.81	1 13/16	2.99	3 1/16	
×132	38.8	20.0	20	0.960	15/16	1/2	19.2	11.9	11 7/8	1.73	1 3/4	2.91	3	
×117.5	34.5	19.8	19 7/8	0.830	13/16	7/16	16.5	11.9	11 7/8	1.58	1 9/16	2.76	2 7/8	
×105.5	31.0	19.7	19 5/8	0.750	3/4	3/8	14.8	11.8	11 3/4	1.42	1 7/16	2.60	2 11/16	
×91.5	26.9	19.5	19 1/2	0.650	5/8	5/16	12.7	11.8	11 3/4	1.22	1 1/4	2.40	2 1/2	
×83.5	24.6	19.3	19 1/4	0.650	5/8	5/16	12.5	11.8	11 3/4	1.02	1	2.21	2 5/16	
×74.5	21.9	19.1	19 1/8	0.630	5/8	5/16	12.0	11.8	11 3/4	0.83	13/16	2.01	2 1/8	
WT18×399*	117	21.0	21	2.38	2 3/8	1 3/16	49.9	18.0	18	4.29	4 5/16	5.24	5 9/16	7 1/2
×325*	95.5	20.2	20 1/4	1.97	2	1	39.9	17.6	17 5/8	3.54	3 9/16	4.49	4 13/16	
×263.5*	77.4	19.6	19 5/8	1.61	1 5/8	13/16	31.6	17.2	17 1/4	2.91	2 15/16	3.86	4 3/16	
×219.5*	64.5	19.1	19 1/8	1.36	1 3/8	11/16	26.0	17.0	17	2.44	2 7/16	3.39	3 11/16	
×196.5*	57.8	18.9	18 7/8	1.22	1 1/4	5/8	23.1	16.8	16 7/8	2.2	2 3/16	3.15	3 7/16	
×179.5*	52.7	18.7	18 3/4	1.12	1 1/8	9/16	20.9	16.7	16 3/4	2.01	2	2.96	3 1/4	
×164*	48.2	18.5	18 1/2	1.02	1	1/2	18.9	16.6	16 5/8	1.85	1 7/8	2.80	3 1/8	
×150	44.1	18.4	18 3/8	0.945	15/16	1/2	17.4	16.7	16 5/8	1.68	1 11/16	2.63	2 15/16	
×140	41.2	18.3	18 1/4	0.885	7/8	7/16	16.2	16.6	16 5/8	1.57	1 9/16	2.52	2 13/16	
×130	38.2	18.1	18 1/8	0.840	13/16	7/16	15.2	16.6	16 1/2	1.44	1 7/16	2.39	2 11/16	
×122.5	36.0	18.0	18	0.800	13/16	7/16	14.4	16.5	16 1/2	1.35	1 3/8	2.30	2 5/8	
×115	33.8	18.0	18	0.760	3/4	3/8	13.6	16.5	16 1/2	1.26	1 1/4	2.21	2 1/2	
WT18×128	37.7	18.7	18 3/4	0.960	15/16	1/2	18.0	12.2	12 1/4	1.73	1 3/4	2.48	2 5/8	5 1/2
×116	34.1	18.6	18 1/2	0.870	7/8	7/16	16.1	12.1	12 1/8	1.57	1 9/16	2.32	2 7/16	
×105	30.9	18.3	18 3/8	0.830	13/16	7/16	15.2	12.2	12 1/8	1.36	1 3/8	2.11	2 5/16	
×97	28.5	18.2	18 1/4	0.765	3/4	3/8	14.0	12.1	12 1/8	1.26	1 1/4	2.01	2 3/16	
×91	26.8	18.2	18 1/8	0.725	3/4	3/8	13.2	12.1	12 1/8	1.18	1 3/16	1.93	2 1/8	
×85	25.0	18.1	18 1/8	0.680	11/16	3/8	12.3	12.0	12	1.1	1 1/8	1.85	2	
×80	23.5	18.0	18	0.650	5/8	5/16	11.7	12.0	12	1.02	1	1.77	1 15/16	
×75	22.1	17.9	17 7/8	0.625	5/8	5/16	11.2	12.0	12	0.94	15/16	1.69	1 7/8	
×67.5	19.9	17.8	17 3/4	0.600	5/8	5/16	10.7	12.0	12	0.79	13/16	1.54	1 11/16	
WT16.5×193.5*	57.0	18.0	18	1.26	1 1/4	5/8	22.6	16.2	16 1/4	2.28	2 1/4	3.07	3 3/16	5 1/2
×177*	52.1	17.8	17 3/4	1.16	1 3/16	5/8	20.6	16.1	16 1/8	2.09	2 1/16	2.88	2 15/16	
×159*	46.8	17.6	17 5/8	1.04	1 1/16	9/16	18.3	16.0	16	1.89	1 7/8	2.68	2 3/4	
×145.5	42.8	17.4	17 3/8	0.960	15/16	1/2	16.7	15.9	15 7/8	1.73	1 3/4	2.52	2 5/8	
×131.5	38.7	17.3	17 1/4	0.870	7/8	7/16	15.0	15.8	15 3/4	1.57	1 9/16	2.36	2 7/16	
×120.5	35.5	17.1	17 1/8	0.830	13/16	7/16	14.2	15.9	15 7/8	1.4	1 3/8	2.19	2 1/4	
×110.5	32.6	17.0	17	0.775	3/4	3/8	13.1	15.8	15 3/4	1.27	1 1/4	2.06	2 1/8	
×100.5	29.6	16.8	16 7/8	0.715	11/16	3/8	12.0	15.7	15 3/4	1.15	1 1/8	1.94	2	

*ASTM A6 tensile group 4 or 5 shape. Special requirements may apply per LRFD Specification Section A3.1c.
†See definition of "Workable Gage" in Nomenclature section at the back of this Manual.

Table 1-8 (cont.). WT-Shapes (Structural Tees Split from W-Shapes) Properties

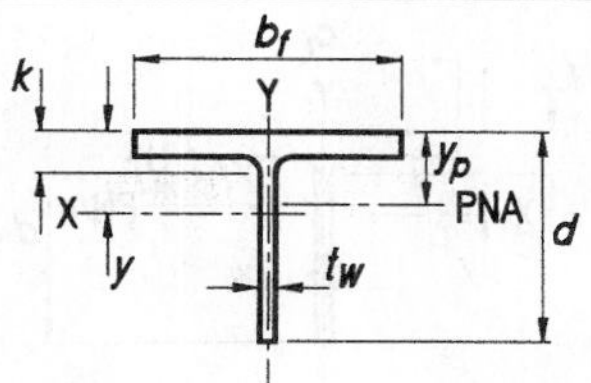

Nominal Wt.	$\frac{h}{t_w}$	Axis X-X						Axis Y-Y				Q_s**
		I	S	r	$\bar{y}$	Z	y_p	I	S	r	Z	$F_y =$ 50 ksi
lb/ft		in.^{4}	in.^{3}	in.	in.	in.^{3}	in.	in.^{4}	in.^{3}	in.	in.^{3}	
167.5	19.1	2160	131	6.63	5.51	233	1.54	600	75.3	3.50	118	0.817
145	22.3	1840	111	6.55	5.27	197	1.35	523	66.1	3.49	103	0.636
131	24.6	1650	100	6.53	5.20	177	1.23	463	58.8	3.46	91.3	0.532
115	27.4	1440	88.6	6.53	5.17	157	1.07	398	50.5	3.43	78.3	0.438
296.5	9.5	3310	209	6.16	5.66	379	2.61	1260	151	3.80	240	–
251.5	11.1	2730	175	6.07	5.39	315	2.25	1020	125	3.72	197	–
215.5	12.8	2290	148	6.01	5.18	266	1.95	843	104	3.65	164	–
198.5	14.0	2070	134	5.96	5.03	240	1.81	771	95.7	3.63	150	–
186	14.7	1930	126	5.95	4.97	225	1.70	710	88.4	3.60	139	–
181	15.3	1870	122	5.92	4.91	217	1.66	691	86.3	3.60	135	0.993
162	17.1	1650	108	5.88	4.77	192	1.50	609	76.6	3.57	119	0.893
148.5	18.4	1500	98.9	5.87	4.71	176	1.38	546	69.0	3.54	107	0.825
138.5	20.6	1360	88.6	5.78	4.50	157	1.29	522	65.9	3.58	102	0.699
124.5	22.8	1210	79.4	5.75	4.41	140	1.16	463	58.8	3.55	90.8	0.580
107.5	26.3	1030	68.0	5.71	4.28	120	1.01	398	50.5	3.54	77.8	0.445
99.5	26.3	988	66.5	5.81	4.47	117	0.929	347	44.1	3.45	68.2	0.452
196	12.0	2270	153	6.28	5.95	276	2.33	401	64.9	2.64	106	–
165.5	14.0	1880	128	6.21	5.74	231	2.00	322	52.9	2.57	85.7	–
163.5	14.5	1840	125	6.19	5.66	224	1.98	320	52.7	2.58	85.0	–
139	16.8	1540	106	6.14	5.50	190	1.71	260	43.5	2.52	69.8	0.913
132	17.8	1450	99.2	6.11	5.41	178	1.63	246	41.3	2.52	66.0	0.855
117.5	20.6	1260	85.7	6.04	5.17	153	1.45	222	37.3	2.54	59.0	0.699
105.5	22.8	1120	76.7	6.01	5.08	137	1.31	195	33.0	2.51	52.1	0.581
91.5	26.3	958	65.8	5.97	4.93	117	1.14	168	28.5	2.50	44.6	0.445
83.5	26.3	899	63.7	6.05	5.19	115	1.10	141	23.9	2.40	37.8	0.454
74.5	27.1	815	59.7	6.10	5.45	108	1.72	114	19.4	2.29	30.9	0.435
399	6.62	3930	257	5.79	5.71	478	3.26	2100	234	4.23	371	–
325	7.99	3030	203	5.64	5.26	373	2.72	1610	183	4.11	290	–
263.5	9.78	2340	159	5.50	4.87	290	2.25	1240	145	4.01	227	–
219.5	11.6	1890	130	5.41	4.60	235	1.90	996	117	3.93	183	–
196.5	12.9	1660	115	5.37	4.44	207	1.72	877	104	3.89	162	–
179.5	14.1	1500	104	5.33	4.33	187	1.58	786	94.0	3.86	146	–
164	15.4	1350	94.1	5.29	4.21	168	1.45	711	85.5	3.84	132	0.989
150	16.7	1230	86.1	5.27	4.13	153	1.33	648	77.8	3.83	120	0.925
140	17.8	1140	80.0	5.25	4.07	142	1.24	599	72.2	3.81	111	0.865
130	18.7	1060	75.1	5.26	4.05	133	1.16	545	65.8	3.77	102	0.817
122.5	19.7	995	71.0	5.26	4.03	125	1.09	507	61.4	3.75	94.8	0.768
115	20.7	934	67.0	5.25	4.01	118	1.03	470	57.0	3.73	88.0	0.714
128	16.9	1210	87.4	5.66	4.92	156	1.54	264	43.2	2.65	68.5	0.922
116	18.7	1080	78.5	5.63	4.82	140	1.40	234	38.6	2.62	60.9	0.829
105	19.6	985	73.1	5.65	4.87	131	1.27	206	33.8	2.58	53.4	0.791
97	21.2	901	67.0	5.62	4.80	120	1.18	187	30.9	2.56	48.8	0.702
91	22.4	845	63.1	5.62	4.77	113	1.11	174	28.8	2.55	45.3	0.637
85	23.9	786	58.9	5.61	4.73	105	1.04	160	26.6	2.53	41.8	0.566
80	25.0	740	55.8	5.61	4.74	100	0.980	147	24.6	2.50	38.6	0.521
75	26.0	698	53.1	5.62	4.78	95.5	0.923	135	22.5	2.47	35.4	0.486
67.5	27.1	637	49.7	5.66	4.96	90.1	1.23	113	18.9	2.38	29.8	0.456
193.5	11.8	1460	107	5.07	4.27	193	1.76	810	100	3.77	156	–
177	12.8	1320	96.8	5.03	4.15	174	1.62	729	90.6	3.74	141	–
159	14.3	1160	85.8	4.99	4.02	154	1.46	645	80.7	3.71	125	–
145.5	15.5	1060	78.3	4.96	3.93	140	1.35	581	73.1	3.68	113	0.991
131.5	17.1	943	70.2	4.93	3.83	125	1.23	517	65.5	3.65	101	0.905
120.5	18.0	872	65.8	4.96	3.84	116	1.12	466	58.8	3.62	90.8	0.867
110.5	19.2	799	60.8	4.95	3.81	107	1.03	420	53.2	3.59	82.1	0.801
100.5	20.8	725	55.5	4.95	3.77	97.8	0.940	375	47.6	3.56	73.3	0.717

**Where no value of Q_s is shown, the Tee complies with LRFD Specification Section E2.

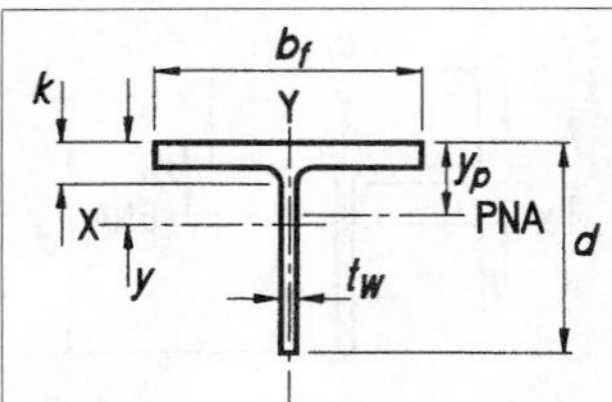

Table 1-8 (cont.). WT-Shapes (Structural Tees Split from W-Shapes) Dimensions

Shape	Area, A	Depth of Tee, d		Stem Thickness, t_w		Stem $\frac{t_w}{2}$	Stem Area	Flange Width, b_f		Flange Thickness, t_f		Distance k		Distance Workable Gage†
	in.²	in.		in.		in.	in.²	in.		in.		in.	in.	in.
WT16.5×84.5	24.8	16.9	16 7/8	0.670	11/16	3/8	11.3	11.5	11 1/2	1.22	1 1/4	1.92	2 1/8	5 1/2
×76	22.4	16.7	16 3/4	0.635	5/8	5/16	10.6	11.6	11 5/8	1.06	1 1/16	1.76	1 15/16	↓
×70.5	20.8	16.7	16 5/8	0.605	5/8	5/16	10.1	11.5	11 1/2	0.96	15/16	1.66	1 13/16	↓
×65	19.2	16.5	16 1/2	0.580	9/16	5/16	9.60	11.5	11 1/2	0.855	7/8	1.56	1 3/4	↓
×59	17.3	16.4	16 3/8	0.550	9/16	5/16	9.04	11.5	11 1/2	0.74	3/4	1.44	1 5/8	↓
WT15×195.5*	57.6	16.6	16 5/8	1.36	1 3/8	11/16	22.6	15.6	15 5/8	2.44	2 7/16	3.23	3 3/8	5 1/2
×178.5*	52.5	16.4	16 3/8	1.24	1 1/4	5/8	20.3	15.5	15 1/2	2.24	2 1/4	3.03	3 1/8	↓
×163*	47.9	16.2	16 1/4	1.14	1 1/8	9/16	18.5	15.4	15 3/8	2.05	2 1/16	2.84	2 15/16	↓
×146*	42.9	16.0	16	1.02	1	1/2	16.3	15.3	15 1/4	1.85	1 7/8	2.64	2 3/4	↓
×130.5	38.4	15.8	15 3/4	0.930	15/16	1/2	14.7	15.2	15 1/8	1.65	1 5/8	2.44	2 9/16	↓
×117.5	34.6	15.7	15 5/8	0.830	13/16	7/16	13.0	15.1	15	1.5	1 1/2	2.29	2 3/8	↓
×105.5	31.1	15.5	15 1/2	0.775	3/4	3/8	12.0	15.1	15 1/8	1.32	1 5/16	2.10	2 1/4	↓
×95.5	28.1	15.3	15 3/8	0.710	11/16	3/8	10.9	15.0	15	1.19	1 3/16	1.97	2 1/16	↓
×86.5	25.5	15.2	15 1/4	0.655	5/8	5/16	9.97	15.0	15	1.07	1 1/16	1.85	2	↓
WT15×74	21.7	15.3	15 3/8	0.650	5/8	5/16	9.97	10.5	10 1/2	1.18	1 3/16	1.83	2 1/16	5 1/2
×66	19.4	15.2	15 1/8	0.615	5/8	5/16	9.32	10.5	10 1/2	1	1	1.65	1 7/8	↓
×62	18.2	15.1	15 1/8	0.585	9/16	5/16	8.82	10.5	10 1/2	0.93	15/16	1.58	1 13/16	↓
×58	17.1	15.0	15	0.565	9/16	5/16	8.48	10.5	10 1/2	0.85	7/8	1.50	1 3/4	↓
×54	15.9	14.9	14 7/8	0.545	9/16	5/16	8.13	10.5	10 1/2	0.76	3/4	1.41	1 11/16	↓
×49.5	14.5	14.8	14 7/8	0.520	1/2	1/4	7.71	10.5	10 1/2	0.67	11/16	1.32	1 9/16	↓
×45	13.2	14.8	14 3/4	0.470	1/2	1/4	6.94	10.4	10 3/8	0.61	5/8	1.26	1 1/2	↓
WT13.5×269.5*	79.3	16.3	16 1/4	1.97	2	1	32.0	15.3	15 1/4	3.54	3 9/16	4.33	4 7/16	5 1/2
×184*	54.2	15.2	15 1/4	1.38	1 3/8	11/16	21.0	14.7	14 5/8	2.48	2 1/2	3.27	3 3/8	5 1/2
×168*	49.5	15.0	15	1.26	1 1/4	5/8	18.9	14.6	14 1/2	2.28	2 1/4	3.07	3 3/16	↓
×153.5*	45.2	14.8	14 3/4	1.16	1 3/16	5/8	17.2	14.4	14 1/2	2.09	2 1/16	2.88	3	↓
×140.5*	41.4	14.6	14 5/8	1.06	1 1/16	9/16	15.5	14.4	14 3/8	1.93	1 15/16	2.72	2 13/16	↓
×129	38.0	14.5	14 1/2	0.980	1	1/2	14.2	14.3	14 1/4	1.77	1 3/4	2.56	2 11/16	↓
×117.5	34.7	14.3	14 3/8	0.910	15/16	1/2	13.0	14.2	14 1/4	1.61	1 5/8	2.40	2 1/2	↓
×108.5	32.0	14.2	14 1/4	0.830	13/16	7/16	11.8	14.1	14 1/8	1.5	1 1/2	2.29	2 3/8	↓
×97	28.6	14.1	14	0.750	3/4	3/8	10.5	14.0	14	1.34	1 5/16	2.13	2 1/4	↓
×89	26.2	13.9	13 7/8	0.725	3/4	3/8	10.1	14.1	14 1/8	1.19	1 3/16	1.98	2 1/16	↓
×80.5	23.8	13.8	13 3/4	0.660	11/16	3/8	9.10	14.0	14	1.08	1 1/16	1.87	2	↓
×73	21.6	13.7	13 3/4	0.605	5/8	5/16	8.28	14.0	14	0.975	1	1.76	1 7/8	↓
WT13.5×64.5	18.9	13.8	13 7/8	0.610	5/8	5/16	8.43	10.0	10	1.1	1 1/8	1.70	2	5 1/2
×57	16.8	13.6	13 5/8	0.570	9/16	5/16	7.78	10.1	10 1/8	0.93	15/16	1.53	1 13/16	↓
×51	15.0	13.5	13 1/2	0.515	1/2	1/4	6.98	10.0	10	0.83	13/16	1.43	1 3/4	↓
×47	13.8	13.5	13 1/2	0.490	1/2	1/4	6.60	9.99	10	0.745	3/4	1.34	1 5/8	↓
×42	12.4	13.4	13 3/8	0.460	7/16	1/4	6.14	9.96	10	0.64	5/8	1.24	1 9/16	↓
WT12×185*	54.4	14.0	14	1.52	1 1/2	3/4	21.3	13.7	13 5/8	2.72	2 3/4	3.22	3 5/8	5 1/2
×167.5*	49.2	13.8	13 3/4	1.38	1 3/8	11/16	19.0	13.5	13 1/2	2.48	2 1/2	2.98	3 3/8	↓
×153*	44.9	13.6	13 5/8	1.26	1 1/4	5/8	17.1	13.4	13 3/8	2.28	2 1/4	2.78	3 3/16	↓
×139.5*	41.0	13.4	13 3/8	1.16	1 3/16	5/8	15.5	13.3	13 1/4	2.09	2 1/16	2.59	3	↓
×125*	36.8	13.2	13 1/8	1.04	1 1/16	9/16	13.7	13.2	13 1/8	1.89	1 7/8	2.39	2 13/16	↓
×114.5	33.6	13.0	13	0.960	15/16	1/2	12.5	13.1	13 1/8	1.73	1 3/4	2.23	2 5/8	↓
×103.5	30.4	12.9	12 7/8	0.870	7/8	7/16	11.2	13.0	13	1.57	1 9/16	2.07	2 1/2	↓
×96	28.1	12.7	12 3/4	0.810	13/16	7/16	10.3	13.0	13	1.46	1 7/16	1.96	2 3/8	↓
×88	25.8	12.6	12 5/8	0.750	3/4	3/8	9.47	12.9	12 7/8	1.34	1 5/16	1.84	2 1/4	↓
×81	23.9	12.5	12 1/2	0.705	11/16	3/8	8.81	13.0	13	1.22	1 1/4	1.72	2 1/8	↓
×73	21.5	12.4	12 3/8	0.650	5/8	5/16	8.04	12.9	12 7/8	1.09	1 1/16	1.59	2	↓
×65.5	19.3	12.2	12 1/4	0.605	5/8	5/16	7.41	12.9	12 7/8	0.96	15/16	1.46	1 7/8	↓
×58.5	17.2	12.1	12 1/8	0.550	9/16	5/16	6.67	12.8	12 3/4	0.85	7/8	1.35	1 3/4	↓
×52	15.3	12.0	12	0.500	1/2	1/4	6.02	12.8	12 3/4	0.75	3/4	1.25	1 5/8	↓

*ASTM A6 tensile group 4 or 5 shape. Special requirements may apply per LRFD Specification Section A3.1c.
†See definition of "Workable Gage" in Nomenclature section at the back of this Manual.

Table 1-8 (cont.). WT-Shapes (Structural Tees Split from W-Shapes) Properties

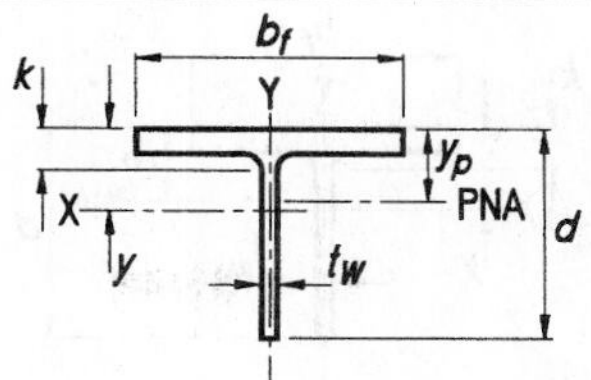

Nominal Wt.	$\frac{h}{t_w}$	Axis X-X						Axis Y-Y				Q_s**
		I	S	r	$\bar{y}$	Z	y_p	I	S	r	Z	$F_y =$ 50 ksi
lb/ft		in.^{4}	in.^{3}	in.	in.	in.^{3}	in.	in.^{4}	in.^{3}	in.	in.^{3}	
84.5	22.4	649	51.1	5.12	4.21	90.8	1.08	155	27.0	2.50	42.1	0.628
76	23.6	592	47.4	5.14	4.26	84.5	0.967	136	23.6	2.47	36.9	0.575
70.5	24.8	552	44.7	5.15	4.29	79.8	0.901	123	21.3	2.43	33.4	0.528
65	25.8	513	42.1	5.18	4.36	75.6	0.832	109	18.9	2.38	29.7	0.492
59	27.3	469	39.2	5.20	4.47	70.8	0.862	93.5	16.3	2.32	25.6	0.448
195.5	9.83	1220	96.9	4.61	4.00	177	1.85	774	99.2	3.67	155	–
178.5	10.8	1090	87.2	4.56	3.87	159	1.70	693	89.6	3.64	140	–
163	11.7	981	78.8	4.52	3.76	143	1.56	622	81.0	3.60	126	–
146	13.1	861	69.6	4.48	3.62	125	1.41	549	71.9	3.58	111	–
130.5	14.4	765	62.4	4.46	3.54	112	1.27	480	63.3	3.53	97.9	–
117.5	16.1	674	55.1	4.41	3.41	98.2	1.15	427	56.8	3.51	87.5	0.955
105.5	17.2	610	50.5	4.43	3.39	89.5	1.03	378	50.1	3.49	77.2	0.899
95.5	18.8	549	45.7	4.42	3.34	80.8	0.935	336	44.7	3.46	68.9	0.816
86.5	20.4	497	41.7	4.42	3.31	73.5	0.851	299	39.9	3.42	61.4	0.733
74	20.8	466	40.6	4.63	3.84	72.2	1.04	114	21.7	2.28	33.9	0.715
66	22.0	421	37.4	4.66	3.90	66.8	0.921	98.0	18.6	2.25	29.2	0.662
62	23.1	396	35.3	4.66	3.90	63.1	0.867	90.4	17.2	2.23	27.0	0.602
58	23.9	373	33.7	4.67	3.94	60.4	0.815	82.1	15.6	2.19	24.6	0.567
54.0	24.8	349	32.0	4.69	4.01	57.7	0.757	73.0	13.9	2.15	21.9	0.534
49.5	26.0	322	30.0	4.71	4.09	54.4	0.912	63.9	12.2	2.10	19.3	0.492
45.0	28.7	290	27.1	4.69	4.04	49.0	0.835	57.3	11.0	2.09	17.3	0.405
269.5	6.06	1530	128	4.39	4.34	242	2.60	1060	138	3.65	218	–
184.0	8.64	939	81.7	4.16	3.71	151	1.85	655	89.3	3.48	140	–
168.0	9.47	839	73.4	4.12	3.58	135	1.70	587	80.8	3.45	126	–
153.5	10.3	753	66.4	4.08	3.47	121	1.56	527	72.9	3.41	113	–
140.5	11.3	677	59.9	4.04	3.35	109	1.44	477	66.4	3.39	103	–
129.0	12.2	613	54.7	4.02	3.27	98.9	1.33	430	60.2	3.36	93.3	–
117.5	13.1	556	50.0	4.00	3.20	89.9	1.22	384	54.2	3.33	83.8	–
108.5	14.4	502	45.2	3.96	3.10	81.1	1.13	352	49.9	3.32	77.0	–
97.0	15.9	444	40.3	3.94	3.02	71.8	1.02	309	44.1	3.29	67.8	0.961
89	16.5	414	38.2	3.97	3.04	67.7	0.932	278	39.4	3.25	60.8	0.938
80.5	18.1	372	34.4	3.95	2.98	60.8	0.849	248	35.4	3.23	54.5	0.851
73	19.7	336	31.2	3.95	2.94	55.0	0.772	222	31.7	3.20	48.8	0.764
64.5	19.9	323	31.0	4.13	3.39	55.1	0.945	92.2	18.4	2.21	28.8	0.763
57	21.3	289	28.3	4.15	3.42	50.4	0.832	79.3	15.8	2.18	24.6	0.698
51	23.5	258	25.3	4.14	3.37	45.0	0.75	69.6	13.9	2.15	21.7	0.578
47	24.7	239	23.8	4.16	3.41	42.4	0.692	62.0	12.4	2.12	19.4	0.530
42	26.3	216	21.9	4.18	3.48	39.2	0.621	52.8	10.6	2.07	16.6	0.475
185	7.09	779	74.7	3.78	3.57	140	1.99	581	85.1	3.27	133	–
167.5	7.81	686	66.3	3.73	3.42	123	1.82	513	75.9	3.23	119	–
153	8.56	611	59.4	3.69	3.29	110	1.67	460	68.6	3.20	107	–
139.5	9.29	546	53.6	3.65	3.18	98.8	1.54	412	61.9	3.17	96.3	–
125	10.4	478	47.2	3.61	3.05	86.5	1.39	362	54.9	3.14	85.2	–
114.5	11.2	431	42.9	3.58	2.96	78.1	1.28	326	49.7	3.11	77.0	–
103.5	12.4	382	38.3	3.55	2.87	69.3	1.17	289	44.4	3.08	68.6	–
96	13.3	350	35.2	3.53	2.80	63.5	1.09	265	40.9	3.07	63.1	–
88	14.4	319	32.2	3.51	2.74	57.8	1.00	240	37.2	3.04	57.3	–
81	15.3	293	29.9	3.50	2.70	53.3	0.921	221	34.2	3.05	52.6	–
73	16.6	264	27.2	3.50	2.66	48.2	0.833	195	30.3	3.01	46.6	0.946
65.5	17.8	238	24.8	3.52	2.65	43.9	0.750	170	26.5	2.97	40.7	0.885
58.5	19.6	212	22.3	3.51	2.62	39.2	0.672	149	23.2	2.94	35.7	0.793
52	21.6	189	20.0	3.51	2.59	35.1	0.600	130	20.3	2.91	31.2	0.692

**Where no value of Q_s is shown, the Tee complies with LRFD Specification Section E2.

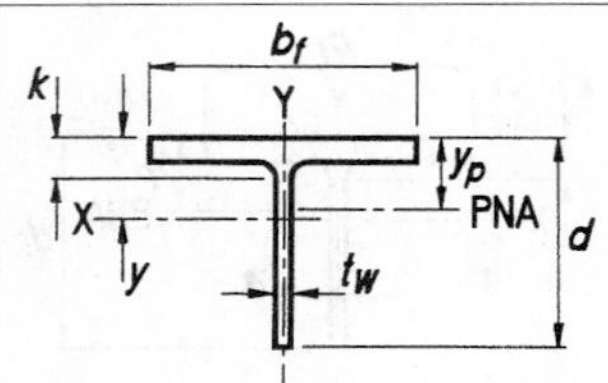

Table 1-8 (cont.). WT-Shapes (Structural Tees Split from W-Shapes) Dimensions

Shape	Area, A	Depth of Tee, d		Stem Thickness, t_w		Stem $\frac{t_w}{2}$	Stem Area	Flange Width, b_f		Flange Thickness, t_f		Distance k		Distance Workable Gage†
	in.^{2}	in.		in.		in.	in.^{2}	in.		in.		in.	in.	in.
WT12×51.5	15.1	12.3	12 1/4	0.550	9/16	5/16	6.75	9.00	9	0.98	1	1.48	1 7/8	5 1/2
×47	13.8	12.2	12 1/8	0.515	1/2	1/4	6.26	9.07	9 1/8	0.875	7/8	1.38	1 3/4	↓
×42	12.4	12.1	12	0.470	1/2	1/4	5.66	9.02	9	0.77	3/4	1.27	1 11/16	↓
×38	11.2	12.0	12	0.440	7/16	1/4	5.26	8.99	9	0.68	11/16	1.18	1 9/16	↓
×34	10.0	11.9	11 7/8	0.415	7/16	1/4	4.92	8.97	9	0.585	9/16	1.09	1 1/2	↓
WT12×31	9.16	11.9	11 7/8	0.430	7/16	1/4	5.10	7.04	7	0.59	9/16	1.19	1 1/2	3 1/2
×27.5	8.15	11.8	11 3/4	0.395	3/8	3/16	4.66	7.01	7	0.505	1/2	1.11	1 7/16	3 1/2
WT10.5×100.5	29.6	11.5	11 1/2	0.910	15/16	1/2	10.5	12.6	12 5/8	1.63	1 5/8	2.13	2 1/2	5 1/2
×91	26.8	11.4	11 3/8	0.830	13/16	7/16	9.43	12.5	12 1/2	1.48	1 1/2	1.98	2 3/8	↓
×83	24.4	11.2	11 1/4	0.750	3/4	3/8	8.43	12.4	12 3/8	1.36	1 3/8	1.86	2 1/4	↓
×73.5	21.6	11.0	11	0.720	3/4	3/8	7.94	12.5	12 1/2	1.15	1 1/8	1.65	2	↓
×66	19.4	10.9	10 7/8	0.650	5/8	5/16	7.09	12.4	12 1/2	1.03	1 1/16	1.54	1 15/16	↓
×61	17.9	10.8	10 7/8	0.600	5/8	5/16	6.50	12.4	12 3/8	0.96	15/16	1.46	1 13/16	↓
×55.5	16.3	10.8	10 3/4	0.550	9/16	5/16	5.92	12.3	12 3/8	0.875	7/8	1.38	1 3/4	↓
×50.5	14.9	10.7	10 5/8	0.500	1/2	1/4	5.34	12.3	12 1/4	0.8	13/16	1.30	1 11/16	↓
WT10.5×46.5	13.7	10.8	10 3/4	0.580	9/16	5/16	6.27	8.42	8 3/8	0.93	15/16	1.43	1 5/8	5 1/2
×41.5	12.2	10.7	10 3/4	0.515	1/2	1/4	5.52	8.36	8 3/8	0.835	13/16	1.34	1 1/2	↓
×36.5	10.7	10.6	10 5/8	0.455	7/16	1/4	4.83	8.30	8 1/4	0.74	3/4	1.24	1 7/16	↓
×34	10.0	10.6	10 5/8	0.430	7/16	1/4	4.54	8.27	8 1/4	0.685	11/16	1.19	1 3/8	↓
×31	9.13	10.5	10 1/2	0.400	3/8	3/16	4.20	8.24	8 1/4	0.615	5/8	1.12	1 5/16	↓
×27.5	8.10	10.4	10 3/8	0.375	3/8	3/16	3.90	8.22	8 1/4	0.522	1/2	1.02	1 3/16	↓
×24	7.07	10.3	10 1/4	0.350	3/8	3/16	3.61	8.14	8 1/8	0.43	7/16	0.930	1 1/8	↓
WT10.5×28.5	8.37	10.5	10 1/2	0.405	3/8	3/16	4.26	6.56	6 1/2	0.65	5/8	1.15	1 5/16	3 1/2
×25	7.36	10.4	10 3/8	0.380	3/8	3/16	3.96	6.53	6 1/2	0.535	9/16	1.04	1 1/4	↓
×22	6.49	10.3	10 3/8	0.350	3/8	3/16	3.62	6.50	6 1/2	0.45	7/16	0.950	1 1/8	↓
WT9×87.5	25.7	10.0	10	0.890	7/8	7/16	8.92	11.4	11 3/8	1.59	1 9/16	1.99	2 7/16	5 1/2
×79	23.2	9.86	9 7/8	0.810	13/16	7/16	7.99	11.3	11 1/4	1.44	1 7/16	1.84	2 3/8	↓
×71.5	21.0	9.74	9 3/4	0.730	3/4	3/8	7.11	11.2	11 1/4	1.32	1 5/16	1.72	2 3/16	↓
×65	19.1	9.63	9 5/8	0.670	11/16	3/8	6.45	11.2	11 1/8	1.2	1 3/16	1.6	2 1/16	↓
×59.5	17.5	9.48	9 1/2	0.655	5/8	5/16	6.21	11.3	11 1/4	1.06	1 1/16	1.46	1 15/16	↓
×53	15.6	9.37	9 3/8	0.590	9/16	5/16	5.53	11.2	11 1/4	0.94	15/16	1.34	1 13/16	↓
×48.5	14.3	9.30	9 1/4	0.535	9/16	5/16	4.97	11.1	11 1/8	0.87	7/8	1.27	1 3/4	↓
×43	12.7	9.20	9 1/4	0.480	1/2	1/4	4.41	11.1	11 1/8	0.77	3/4	1.17	1 5/8	↓
×38	11.2	9.11	9 1/8	0.425	7/16	1/4	3.87	11.0	11	0.68	11/16	1.08	1 9/16	↓
WT9×35.5	10.4	9.23	9 1/4	0.495	1/2	1/4	4.57	7.64	7 5/8	0.81	13/16	1.21	1 1/2	3 1/2
×32.5	9.55	9.18	9 1/8	0.450	7/16	1/4	4.13	7.59	7 5/8	0.75	3/4	1.15	1 7/16	↓
×30	8.82	9.12	9 1/8	0.415	7/16	1/4	3.78	7.56	7 1/2	0.695	11/16	1.10	1 3/8	↓
×27.5	8.10	9.06	9	0.390	3/8	3/16	3.53	7.53	7 1/2	0.63	5/8	1.03	1 5/16	↓
×25	7.33	8.99	9	0.355	3/8	3/16	3.19	7.50	7 1/2	0.57	9/16	0.972	1 1/4	↓
WT9×23	6.77	9.03	9	0.360	3/8	3/16	3.25	6.06	6	0.605	5/8	1.01	1 1/4	3 1/2
×20	5.88	8.95	9	0.315	5/16	3/16	2.82	6.02	6	0.525	1/2	0.927	1 3/16	↓
×17.5	5.15	8.85	8 7/8	0.300	5/16	3/16	2.66	6.00	6	0.425	7/16	0.827	1 1/8	↓
WT8×50	14.9	8.48	8 1/2	0.585	9/16	5/16	4.96	10.4	10 3/8	0.985	1	1.69	1 7/8	5 1/2
×44.5	13.2	8.38	8 3/8	0.525	1/2	1/4	4.40	10.4	10 3/8	0.875	7/8	1.58	1 3/4	↓
×38.5	11.5	8.26	8 1/4	0.455	7/16	1/4	3.76	10.3	10 1/4	0.76	3/4	1.47	1 5/8	↓
×33.5	9.98	8.16	8 1/8	0.395	3/8	3/16	3.23	10.2	10 1/4	0.665	11/16	1.37	1 9/16	↓

*ASTM A6 tensile group 4 or 5 shape. Special requirements may apply per LRFD Specification Section A3.1c.
†See definition of "Workable Gage" in Nomenclature section at the back of this Manual.

Table 1-8 (cont.). WT-Shapes (Structural Tees Split from W-Shapes) Properties

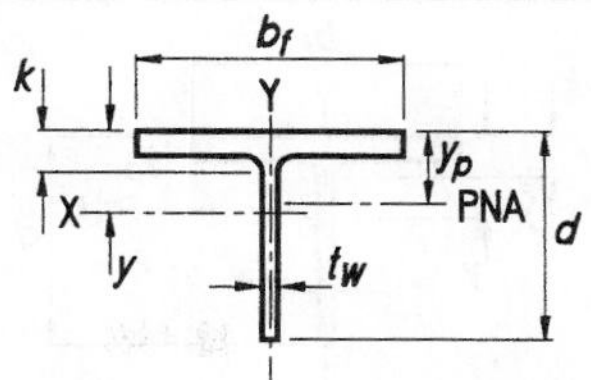

Nom-inal Wt.	$\frac{h}{t_w}$	Axis X-X						Axis Y-Y				Q_s**
		I	S	r	$\bar{y}$	Z	y_p	I	S	r	Z	$F_y =$ 50 ksi
lb/ft		in.⁴	in.³	in.	in.	in.³	in.	in.⁴	in.³	in.	in.³	
51.5	19.6	204	22.0	3.67	3.01	39.2	0.841	59.7	13.3	1.99	20.7	0.781
47	20.9	186	20.3	3.67	2.99	36.1	0.764	54.5	12.0	1.98	18.7	0.715
42	22.9	166	18.3	3.67	2.97	32.5	0.685	47.2	10.5	1.95	16.3	0.609
38	24.5	151	16.9	3.68	3.00	30.1	0.622	41.3	9.18	1.92	14.3	0.541
34	26.0	137	15.6	3.70	3.06	27.9	0.560	35.2	7.85	1.87	12.3	0.489
31	24.8	132	15.6	3.79	3.45	29.9	1.22	17.2	4.90	1.37	7.86	0.525
27.5	27.0	117	14.1	3.79	3.48	27.2	1.47	14.5	4.15	1.34	6.66	0.449
100.5	10.3	285	31.9	3.10	2.57	58.6	1.18	271	43.1	3.02	66.5	–
91	11.3	253	28.5	3.07	2.48	52.1	1.07	241	38.6	3.00	59.5	–
83	12.5	226	25.5	3.04	2.39	46.3	0.983	217	35.0	2.99	53.9	–
73.5	13.0	204	23.7	3.08	2.39	42.4	0.864	188	30.0	2.95	46.3	–
66	14.4	181	21.1	3.06	2.33	37.6	0.780	166	26.7	2.93	41.1	–
61	15.6	166	19.3	3.04	2.28	34.3	0.724	152	24.6	2.91	37.8	0.995
55.5	17.1	150	17.5	3.03	2.23	31.0	0.662	137	22.2	2.90	34.1	0.919
50.5	18.8	135	15.8	3.01	2.18	27.9	0.605	124	20.2	2.89	30.8	0.828
46.5	16.2	144	17.9	3.25	2.74	31.8	0.812	46.4	11.0	1.84	17.3	0.966
41.5	18.2	127	15.7	3.22	2.66	28.0	0.728	40.7	9.74	1.83	15.2	0.856
36.5	20.6	110	13.8	3.21	2.60	24.4	0.647	35.3	8.51	1.81	13.3	0.728
34	21.8	103	12.9	3.20	2.59	22.9	0.606	32.4	7.83	1.80	12.2	0.666
31	23.4	93.8	11.9	3.21	2.58	21.1	0.554	28.7	6.97	1.77	10.9	0.581
27.5	25.0	84.4	10.9	3.23	2.64	19.4	0.493	24.2	5.89	1.73	9.18	0.520
24	26.8	74.9	9.90	3.26	2.74	17.8	0.459	19.4	4.76	1.66	7.44	0.461
28.5	23.2	90.4	11.8	3.29	2.85	21.2	0.638	15.3	4.67	1.35	7.40	0.592
25	24.7	80.3	10.7	3.30	2.93	19.4	0.771	12.5	3.82	1.30	6.08	0.532
22	26.8	71.1	9.68	3.31	2.98	18.8	1.06	10.3	3.18	1.26	5.07	0.459
87.5	9.02	181	23.4	2.66	2.26	43.6	1.13	196	34.4	2.76	53.1	–
79	9.90	160	20.8	2.63	2.17	38.5	1.02	174	30.7	2.74	47.4	–
71.5	11.0	142	18.5	2.60	2.09	34.0	0.937	156	27.7	2.72	42.7	–
65	12.0	127	16.7	2.58	2.02	30.5	0.856	139	24.9	2.70	38.3	–
59.5	12.2	119	15.9	2.60	2.03	28.7	0.778	126	22.5	2.69	34.5	–
53	13.6	104	14.1	2.59	1.97	25.2	0.695	110	19.7	2.66	30.2	–
48.5	15.0	93.8	12.7	2.56	1.91	22.6	0.640	100	18.0	2.65	27.6	–
43	16.7	82.4	11.2	2.55	1.86	19.9	0.570	87.6	15.8	2.63	24.2	0.939
38	18.9	71.8	9.83	2.54	1.80	17.3	0.505	76.2	13.8	2.61	21.1	0.825
35.5	16.2	78.2	11.2	2.74	2.26	20.0	0.683	30.1	7.89	1.70	12.3	0.965
32.5	17.8	70.7	10.1	2.72	2.20	18.0	0.629	27.4	7.22	1.69	11.2	0.877
30	19.3	64.7	9.29	2.71	2.16	16.5	0.583	25.0	6.63	1.68	10.3	0.797
27.5	20.6	59.5	8.63	2.71	2.16	15.3	0.538	22.5	5.97	1.67	9.26	0.734
25	22.6	53.5	7.79	2.70	2.12	13.8	0.489	20.0	5.35	1.65	8.28	0.623
23	22.3	52.1	7.77	2.77	2.33	13.9	0.558	11.3	3.71	1.29	5.84	0.636
20	25.5	44.8	6.73	2.76	2.29	12.0	0.489	9.55	3.17	1.27	4.97	0.495
17.5	26.7	40.1	6.21	2.79	2.39	11.2	0.450	7.67	2.56	1.22	4.02	0.460
50	11.6	76.8	11.4	2.27	1.75	20.8	0.713	93.1	17.9	2.50	27.5	–
44.5	12.9	67.3	10.1	2.26	1.69	18.2	0.638	81.3	15.7	2.48	24.1	–
38.5	14.9	57.0	8.59	2.23	1.62	15.4	0.556	69.2	13.4	2.46	20.6	0.990
33.5	17.2	48.7	7.36	2.21	1.55	13.1	0.488	59.5	11.6	2.44	17.7	0.863

**Where no value of Q_s is shown, the Tee complies with LRFD Specification Section E2.

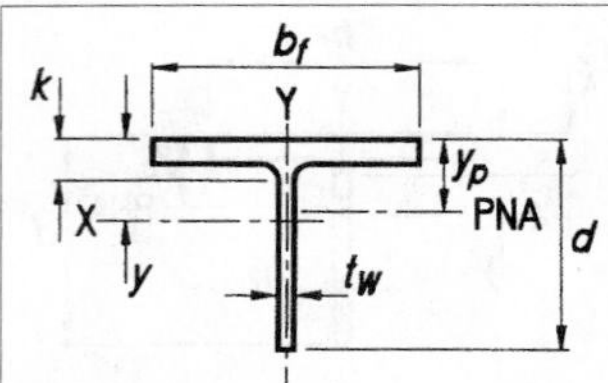

Table 1-8 (cont.). WT-Shapes (Structural Tees Split from W-Shapes) Dimensions

Shape	Area, A	Depth of Tee, d		Stem Thickness, t_w		Stem $\frac{t_w}{2}$	Stem Area	Flange Width, b_f		Flange Thickness, t_f		Distance k		Distance Workable Gage†
	in.²	in.		in.		in.	in.²	in.		in.		in.	in.	in.
WT8×28.5	8.39	8.22	8 1/4	0.430	7/16	1/4	3.53	7.12	7 1/8	0.715	11/16	1.12	1 3/8	3 1/2
×25	7.37	8.13	8 1/8	0.380	3/8	3/16	3.09	7.07	7 1/8	0.63	5/8	1.03	1 5/16	
×22.5	6.63	8.07	8 1/8	0.345	3/8	3/16	2.78	7.04	7	0.565	9/16	0.967	1 1/4	
×20	5.89	8.01	8	0.305	5/16	3/16	2.44	7.00	7	0.505	1/2	0.907	1 3/16	
×18	5.29	7.93	7 7/8	0.295	5/16	3/16	2.34	6.99	7	0.43	7/16	0.832	1 1/8	
WT8×15.5	4.56	7.94	8	0.275	1/4	1/8	2.18	5.53	5 1/2	0.44	7/16	0.842	1 1/8	3 1/2
×13	3.84	7.85	7 7/8	0.250	1/4	1/8	1.96	5.50	5 1/2	0.345	3/8	0.747	1 1/16	3 1/2
WT7×404*	119	11.4	11 3/8	3.74	3 3/4	1 7/8	42.7	18.6	18 1/2	5.12	5 1/8	5.72	6 7/16	7 1/2
×365*	107	11.2	11 1/4	3.07	3 1/16	1 9/16	34.4	17.9	17 7/8	4.91	4 15/16	5.51	6 3/16	7 1/2
×332.5*	97.8	10.8	10 7/8	2.83	2 13/16	1 7/16	30.6	17.7	17 5/8	4.52	4 1/2	5.12	5 13/16	
×302.5*	88.9	10.5	10 1/2	2.60	2 5/8	1 5/16	27.1	17.4	17 3/8	4.16	4 3/16	4.76	5 7/16	
×275*	80.9	10.1	10 1/8	2.38	2 3/8	1 3/16	24.1	17.2	17 1/4	3.82	3 13/16	4.42	5 1/8	
×250*	73.5	9.80	9 3/4	2.19	2 3/16	1 1/8	21.5	17.0	17	3.5	3 1/2	4.10	4 13/16	
×227.5*	66.9	9.51	9 1/2	2.02	2	1	19.2	16.8	16 7/8	3.21	3 3/16	3.81	4 1/2	
×213*	62.6	9.34	9 3/8	1.88	1 7/8	15/16	17.5	16.7	16 3/4	3.04	3 1/16	3.63	4 5/16	
×199*	58.5	9.15	9 1/8	1.77	1 3/4	7/8	16.2	16.6	16 5/8	2.85	2 7/8	3.44	4 1/8	
×185*	54.4	8.96	9	1.66	1 5/8	13/16	14.8	16.5	16 1/2	2.66	2 11/16	3.26	3 15/16	
×171*	50.3	8.77	8 3/4	1.54	1 9/16	13/16	13.5	16.4	16 3/8	2.47	2 1/2	3.07	3 3/4	
×155.5*	45.7	8.56	8 1/2	1.41	1 7/16	3/4	12.1	16.2	16 1/4	2.26	2 1/4	2.86	3 9/16	
×141.5*	41.6	8.37	8 3/8	1.29	1 5/16	11/16	10.8	16.1	16 1/8	2.07	2 1/16	2.67	3 3/8	
×128.5*	37.8	8.19	8 1/4	1.18	1 3/16	5/8	9.62	16.0	16	1.89	1 7/8	2.49	3 3/16	
×116.5*	34.2	8.02	8	1.07	1 1/16	9/16	8.58	15.9	15 7/8	1.72	1 3/4	2.32	3	
×105.5	31.0	7.86	7 7/8	0.980	1	1/2	7.70	15.8	15 3/4	1.56	1 9/16	2.16	2 7/8	
×96.5	28.4	7.74	7 3/4	0.890	7/8	7/16	6.89	15.7	15 3/4	1.44	1 7/16	2.04	2 3/4	
×88	25.9	7.61	7 5/8	0.830	13/16	7/16	6.32	15.7	15 5/8	1.31	1 5/16	1.91	2 5/8	
×79.5	23.4	7.49	7 1/2	0.745	3/4	3/8	5.58	15.6	15 5/8	1.19	1 3/16	1.79	2 1/2	
×72.5	21.3	7.39	7 3/8	0.680	11/16	3/8	5.03	15.5	15 1/2	1.09	1 1/16	1.69	2 3/8	
WT7×66	19.4	7.33	7 3/8	0.645	5/8	5/16	4.73	14.7	14 3/4	1.03	1	1.63	2 5/16	5 1/2
×60	17.7	7.24	7 1/4	0.590	9/16	5/16	4.27	14.7	14 5/8	0.94	15/16	1.54	2 1/4	
×54.5	16.0	7.16	7 1/8	0.525	1/2	1/4	3.76	14.6	14 5/8	0.86	7/8	1.46	2 3/16	
×49.5	14.6	7.08	7 1/8	0.485	1/2	1/4	3.43	14.6	14 5/8	0.78	3/4	1.38	2 1/16	
×45	13.2	7.01	7	0.440	7/16	1/4	3.08	14.5	14 1/2	0.71	11/16	1.31	2	
WT7×41	12.0	7.16	7 1/8	0.510	1/2	1/4	3.65	10.1	10 1/8	0.855	7/8	1.45	1 11/16	5 1/2
×37	10.9	7.09	7 1/8	0.450	7/16	1/4	3.19	10.1	10 1/8	0.785	13/16	1.38	1 5/8	
×34	9.99	7.02	7	0.415	7/16	1/4	2.91	10.0	10	0.72	3/4	1.31	1 9/16	
×30.5	8.96	6.95	7	0.375	3/8	3/16	2.60	9.99	10	0.645	5/8	1.24	1 1/2	
WT7×26.5	7.80	6.96	7	0.370	3/8	3/16	2.58	8.06	8	0.66	11/16	1.25	1 1/2	5 1/2
×24	7.07	6.90	6 7/8	0.340	5/16	3/16	2.34	8.03	8	0.595	5/8	1.19	1 7/16	
×21.5	6.31	6.83	6 7/8	0.305	5/16	3/16	2.08	8.00	8	0.53	1/2	1.12	1 3/8	
WT7×19	5.58	7.05	7	0.310	5/16	3/16	2.19	6.77	6 3/4	0.515	1/2	0.915	1 1/4	3 1/2
×17	5.00	6.99	7	0.285	5/16	3/16	1.99	6.75	6 3/4	0.455	7/16	0.855	1 3/16	3 1/2
×15	4.42	6.92	6 7/8	0.270	1/4	1/8	1.87	6.73	6 3/4	0.385	3/8	0.785	1 1/8	3 1/2
WT7×13	3.85	6.96	7	0.255	1/4	1/8	1.77	5.03	5	0.42	7/16	0.820	1 1/8	2 3/4
×11	3.25	6.87	6 7/8	0.230	1/4	1/8	1.58	5.00	5	0.335	5/16	0.735	1 1/16	2 3/4

*ASTM A6 tensile group 4 or 5 shape. Special requirements may apply per LRFD Specification Section A3.1c.

†See definition of "Workable Gage" in Nomenclature section at the back of this Manual.

Table 1-8 (cont.). WT-Shapes (Structural Tees Split from W-Shapes) Properties

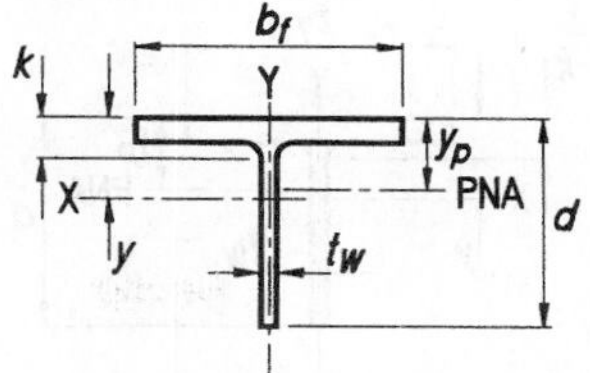

Nominal Wt.	$\frac{h}{t_w}$	Axis X-X						Axis Y-Y				Q_s**
		I	S	r	$\bar{y}$	Z	y_p	I	S	r	Z	$F_y =$ 50 ksi
lb/ft		in.^{4}	in.^{3}	in.	in.	in.^{3}	in.	in.^{4}	in.^{3}	in.	in.^{3}	
28.5	16.5	48.7	7.77	2.41	1.94	13.8	0.589	21.6	6.06	1.60	9.42	0.942
25	18.7	42.3	6.78	2.40	1.89	12.0	0.521	18.6	5.26	1.59	8.15	0.826
22.5	20.6	37.8	6.10	2.39	1.86	10.8	0.471	16.4	4.67	1.57	7.22	0.726
20	23.3	33.1	5.35	2.37	1.81	9.4	0.421	14.4	4.12	1.56	6.36	0.581
18	24.1	30.6	5.05	2.41	1.88	8.9	0.378	12.2	3.50	1.52	5.42	0.554
15.5	25.8	27.5	4.64	2.45	2.02	8.3	0.413	6.20	2.24	1.17	3.51	0.480
13	28.4	23.5	4.09	2.47	2.09	7.4	0.372	4.79	1.74	1.12	2.73	0.406
404	1.52	901	117	2.76	3.70	249	3.20	2760	297	4.82	463	–
365	1.86	739	95.4	2.62	3.47	211	3.00	2360	264	4.69	408	–
332.5	2.01	622	82.1	2.52	3.25	182	2.77	2080	236	4.62	365	–
302.5	2.20	524	70.6	2.43	3.05	157	2.55	1840	211	4.55	326	–
275	2.40	442	60.9	2.34	2.85	136	2.35	1630	189	4.49	292	–
250	2.60	375	52.7	2.26	2.67	117	2.16	1440	169	4.43	261	–
227.5	2.83	321	45.9	2.19	2.51	102	1.99	1280	152	4.38	234	–
213	3.04	287	41.4	2.14	2.40	91.7	1.88	1180	141	4.34	217	–
199	3.22	257	37.6	2.10	2.30	82.9	1.76	1090	131	4.31	201	–
185	3.45	229	33.9	2.05	2.19	74.4	1.65	994	121	4.27	185	–
171	3.70	203	30.4	2.01	2.09	66.2	1.54	903	110	4.24	169	–
155.5	4.04	176	26.7	1.96	1.97	57.7	1.41	807	99.4	4.20	152	–
141.5	4.42	153	23.5	1.92	1.86	50.4	1.29	722	89.7	4.17	137	–
128.5	4.85	133	20.7	1.88	1.75	43.9	1.18	645	80.7	4.13	123	–
116.5	5.33	116	18.2	1.84	1.65	38.2	1.08	576	72.5	4.10	110	–
105.5	5.82	102	16.2	1.81	1.57	33.4	0.980	513	65.0	4.07	98.9	–
96.5	6.41	89.8	14.4	1.78	1.49	29.4	0.903	466	59.3	4.05	90.1	–
88	6.87	80.5	13.0	1.76	1.43	26.3	0.827	419	53.5	4.02	81.3	–
79.5	7.65	70.2	11.4	1.73	1.35	22.8	0.751	374	48.1	4.00	73.0	–
72.5	8.39	62.5	10.2	1.71	1.29	20.2	0.688	338	43.7	3.98	66.2	–
66	8.84	57.8	9.57	1.73	1.29	18.6	0.658	274	37.2	3.76	56.5	–
60	9.66	51.7	8.61	1.71	1.24	16.5	0.602	247	33.7	3.74	51.2	–
54.5	10.9	45.3	7.56	1.68	1.17	14.4	0.548	223	30.6	3.73	46.3	–
49.5	11.8	40.9	6.88	1.67	1.14	12.9	0.500	201	27.6	3.71	41.8	–
45	13.0	36.5	6.16	1.66	1.09	11.5	0.456	181	25.0	3.70	37.8	–
41	11.2	41.2	7.14	1.85	1.39	13.2	0.593	74.1	14.6	2.48	22.4	–
37	12.7	36.0	6.25	1.82	1.32	11.5	0.541	66.9	13.3	2.48	20.2	–
34	13.8	32.6	5.69	1.81	1.29	10.4	0.498	60.7	12.1	2.46	18.4	–
30.5	15.2	28.9	5.07	1.80	1.25	9.15	0.448	53.7	10.7	2.45	16.4	0.972
26.5	15.4	27.6	4.94	1.88	1.38	8.87	0.484	28.8	7.15	1.92	11.0	0.957
24	16.8	24.9	4.49	1.88	1.35	8.00	0.440	25.7	6.40	1.91	9.80	0.883
21.5	18.7	21.9	3.98	1.86	1.31	7.05	0.395	22.6	5.65	1.89	8.64	0.776
19	19.8	23.3	4.22	2.04	1.54	7.45	0.412	13.3	3.94	1.55	6.07	0.758
17	21.5	20.9	3.83	2.04	1.53	6.74	0.371	11.6	3.45	1.53	5.32	0.668
15	22.7	19.0	3.55	2.07	1.58	6.25	0.329	9.79	2.91	1.49	4.49	0.609
13	24.1	17.3	3.31	2.12	1.72	5.89	0.383	4.45	1.77	1.08	2.76	0.538
11	26.7	14.8	2.91	2.14	1.76	5.20	0.325	3.50	1.40	1.04	2.19	0.448

**Where no value of Q_s is shown, the Tee complies with LRFD Specification Section E2.

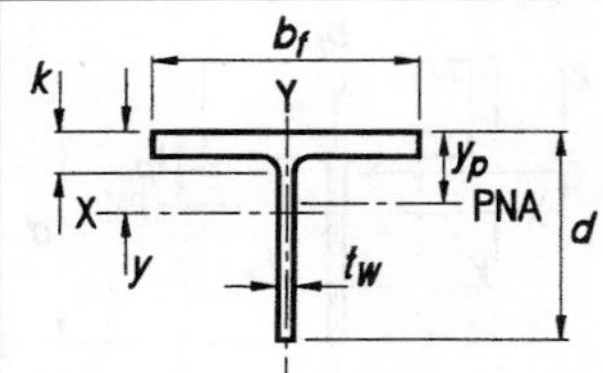

Table 1-8 (cont.). WT-Shapes (Structural Tees Split from W-Shapes) Dimensions

Shape	Area, A	Depth of Tee, d		Stem Thickness, t_w		Stem $\frac{t_w}{2}$	Stem Area	Flange Width, b_f		Flange Thickness, t_f		Distance k		Distance Workable Gage†
	in.^{2}	in.		in.		in.	in.^{2}	in.		in.		in.	in.	in.
WT6×168*	49.4	8.41	8 3/8	1.78	1 3/4	7/8	14.9	13.4	13 3/8	2.96	2 15/16	3.55	3 7/8	5 1/2
×152.5*	44.8	8.16	8 1/8	1.63	1 5/8	13/16	13.3	13.2	13 1/4	2.71	2 11/16	3.30	3 5/8	
×139.5*	41.0	7.93	7 7/8	1.53	1 1/2	3/4	12.1	13.1	13 1/8	2.47	2 1/2	3.07	3 3/8	
×126*	37.0	7.71	7 3/4	1.40	1 3/8	11/16	10.7	13.0	13	2.25	2 1/4	2.85	3 1/8	
×115*	33.9	7.53	7 1/2	1.29	1 5/16	11/16	9.67	12.9	12 7/8	2.07	2 1/16	2.67	2 15/16	
×105*	30.9	7.36	7 3/8	1.18	1 3/16	5/8	8.68	12.8	12 3/4	1.9	1 7/8	2.50	2 13/16	
×95	27.9	7.19	7 1/4	1.06	1 1/16	9/16	7.62	12.7	12 5/8	1.74	1 3/4	2.33	2 5/8	
×85	25.0	7.02	7	0.960	15/16	1/2	6.73	12.6	12 5/8	1.56	1 9/16	2.16	2 7/16	
×76	22.4	6.86	6 7/8	0.870	7/8	7/16	5.96	12.5	12 1/2	1.4	1 3/8	2.00	2 5/16	
×68	20.0	6.71	6 3/4	0.790	13/16	7/16	5.30	12.4	12 3/8	1.25	1 1/4	1.85	2 1/8	
×60	17.6	6.56	6 1/2	0.710	11/16	3/8	4.66	12.3	12 3/8	1.11	1 1/8	1.70	2	
×53	15.6	6.45	6 1/2	0.610	5/8	5/16	3.93	12.2	12 1/4	0.99	1	1.59	1 7/8	
×48	14.1	6.36	6 3/8	0.550	9/16	5/16	3.50	12.2	12 1/8	0.9	7/8	1.50	1 13/16	
×43.5	12.8	6.27	6 1/4	0.515	1/2	1/4	3.23	12.1	12 1/8	0.81	13/16	1.41	1 11/16	
×39.5	11.6	6.19	6 1/4	0.470	1/2	1/4	2.91	12.1	12 1/8	0.735	3/4	1.33	1 5/8	
×36	10.6	6.13	6 1/8	0.430	7/16	1/4	2.63	12.0	12	0.67	11/16	1.27	1 9/16	
×32.5	9.54	6.06	6	0.390	3/8	3/16	2.36	12.0	12	0.605	5/8	1.20	1 1/2	
WT6×29	8.52	6.10	6 1/8	0.360	3/8	3/16	2.19	10.0	10	0.64	5/8	1.24	1 1/2	5 1/2
×26.5	7.78	6.03	6	0.345	3/8	3/16	2.08	9.99	10	0.575	9/16	1.17	1 3/8	5 1/2
WT6×25	7.30	6.10	6 1/8	0.370	3/8	3/16	2.26	8.08	8 1/8	0.64	5/8	1.14	1 1/2	5 1/2
×22.5	6.56	6.03	6	0.335	5/16	3/16	2.02	8.05	8	0.575	9/16	1.08	1 3/8	
×20	5.84	5.97	6	0.295	5/16	3/16	1.76	8.01	8	0.515	1/2	1.02	1 3/8	
WT6×17.5	5.17	6.25	6 1/4	0.300	5/16	3/16	1.88	6.56	6 1/2	0.52	1/2	0.820	1 3/16	3 1/2
×15	4.40	6.17	6 1/8	0.260	1/4	1/8	1.60	6.52	6 1/2	0.44	7/16	0.740	1 1/8	
×13	3.82	6.11	6 1/8	0.230	1/4	1/8	1.41	6.49	6 1/2	0.38	3/8	0.680	1 1/16	
WT6×11	3.24	6.16	6 1/8	0.260	1/4	1/8	1.60	4.03	4	0.425	7/16	0.725	15/16	2 1/4
×9.5	2.79	6.08	6 1/8	0.235	1/4	1/8	1.43	4.01	4	0.35	3/8	0.650	7/8	
×8	2.36	6.00	6	0.220	1/4	1/8	1.32	3.99	4	0.265	1/4	0.565	13/16	
×7	2.08	5.96	6	0.200	3/16	1/8	1.19	3.97	4	0.225	1/4	0.525	3/4	
WT5×56	16.5	5.68	5 5/8	0.755	3/4	3/8	4.29	10.4	10 3/8	1.25	1 1/4	1.75	1 15/16	5 1/2
×50	14.7	5.55	5 1/2	0.680	11/16	3/8	3.77	10.3	10 3/8	1.12	1 1/8	1.62	1 13/16	
×44	12.9	5.42	5 3/8	0.605	5/8	5/16	3.28	10.3	10 1/4	0.99	1	1.49	1 11/16	
×38.5	11.3	5.30	5 1/4	0.530	1/2	1/4	2.81	10.2	10 1/4	0.87	7/8	1.37	1 9/16	
×34	9.99	5.20	5 1/4	0.470	1/2	1/4	2.44	10.1	10 1/8	0.77	3/4	1.27	1 7/16	
×30	8.82	5.11	5 1/8	0.420	7/16	1/4	2.15	10.1	10 1/8	0.68	11/16	1.18	1 3/8	
×27	7.91	5.05	5	0.370	3/8	3/16	1.87	10.0	10	0.615	5/8	1.12	1 5/16	
×24.5	7.21	4.99	5	0.340	5/16	3/16	1.70	10.0	10	0.56	9/16	1.06	1 1/4	
WT5×22.5	6.63	5.05	5	0.350	3/8	3/16	1.77	8.02	8	0.62	5/8	1.12	1 5/16	5 1/2
×19.5	5.73	4.96	5	0.315	5/16	3/16	1.56	7.99	8	0.53	1/2	1.03	1 3/16	
×16.5	4.85	4.87	4 7/8	0.290	5/16	3/16	1.41	7.96	8	0.435	7/16	0.935	1 1/8	
WT5×15	4.42	5.24	5 1/4	0.300	5/16	3/16	1.57	5.81	5 3/4	0.51	1/2	0.810	1 1/8	2 3/4
×13	3.81	5.17	5 1/8	0.260	1/4	1/8	1.34	5.77	5 3/4	0.44	7/16	0.740	1 1/16	
×11	3.24	5.09	5 1/8	0.240	1/4	1/8	1.22	5.75	5 3/4	0.36	3/8	0.660	15/16	
WT5×9.5	2.81	5.12	5 1/8	0.250	1/4	1/8	1.28	4.02	4	0.395	3/8	0.695	15/16	2 1/4
×8.5	2.50	5.05	5	0.240	1/4	1/8	1.21	4.01	4	0.33	5/16	0.630	7/8	
×7.5	2.21	5.00	5	0.230	1/4	1/8	1.15	4.00	4	0.27	1/4	0.570	13/16	
×6	1.77	4.93	4 7/8	0.190	3/16	1/8	0.938	3.96	4	0.21	3/16	0.510	3/4	
WT4×33.5	9.84	4.50	4 1/2	0.570	9/16	5/16	2.57	8.28	8 1/4	0.935	15/16	1.33	1 5/8	5 1/2
×29	8.54	4.38	4 3/8	0.510	1/2	1/4	2.23	8.22	8 1/4	0.81	13/16	1.20	1 1/2	
×24	7.05	4.25	4 1/4	0.400	3/8	3/16	1.70	8.11	8 1/8	0.685	11/16	1.08	1 3/8	
×20	5.87	4.13	4 1/8	0.360	3/8	3/16	1.49	8.07	8 1/8	0.56	9/16	0.954	1 1/4	
×17.5	5.14	4.06	4	0.310	5/16	3/16	1.26	8.02	8	0.495	1/2	0.889	1 3/16	
×15.5	4.56	4.00	4	0.285	5/16	3/16	1.14	8.00	8	0.435	7/16	0.829	1 1/8	

*ASTM A6 tensile group 4 or 5 shape. Special requirements may apply per LRFD Specification Section A3.1c.

†See definition of "Workable Gage" in Nomenclature section at the back of this Manual.

Table 1-8 (cont.). WT-Shapes (Structural Tees Split from W-Shapes) Properties

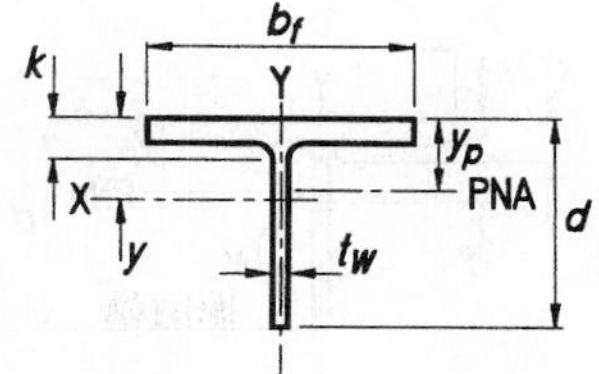

Nominal Wt.	$\frac{h}{t_w}$	Axis X-X						Axis Y-Y				Q_s**
		I	S	r	$\bar{y}$	Z	y_p	I	S	r	Z	$F_y =$ 50 ksi
lb/ft		in.[4]	in.[3]	in.	in.	in.[3]	in.	in.[4]	in.[3]	in.	in.[3]	
168	2.74	190	31.2	1.96	2.31	68.4	1.84	593	88.6	3.47	137	–
152.5	2.99	162	27.0	1.90	2.16	59.1	1.69	525	79.3	3.42	122	–
139.5	3.17	141	24.1	1.86	2.05	51.9	1.56	469	71.3	3.38	110	–
126	3.48	121	20.9	1.81	1.92	44.8	1.42	414	63.6	3.34	97.9	–
115	3.78	106	18.5	1.77	1.82	39.4	1.31	371	57.5	3.31	88.4	–
105	4.12	92.1	16.4	1.73	1.72	34.5	1.21	332	51.9	3.28	79.7	–
95	4.58	79.0	14.2	1.68	1.62	29.8	1.10	295	46.5	3.25	71.2	–
85	5.06	67.8	12.3	1.65	1.52	25.6	0.994	259	41.2	3.22	62.9	–
76	5.58	58.5	10.8	1.62	1.43	22.0	0.896	227	36.4	3.19	55.6	–
68	6.15	50.6	9.46	1.59	1.35	19.0	0.805	199	32.1	3.16	48.9	–
60	6.84	43.4	8.22	1.57	1.28	16.2	0.716	172	28.0	3.13	42.7	–
53	7.96	36.3	6.92	1.53	1.19	13.6	0.637	151	24.7	3.11	37.5	–
48	8.83	32.0	6.12	1.51	1.13	11.9	0.580	135	22.2	3.09	33.7	–
43.5	9.43	28.9	5.60	1.50	1.10	10.7	0.527	120	19.9	3.07	30.2	–
39.5	10.3	25.8	5.03	1.49	1.06	9.49	0.480	108	17.9	3.05	27.1	–
36	11.3	23.2	4.54	1.48	1.02	8.48	0.439	97.5	16.2	3.04	24.6	–
32.5	12.5	20.6	4.06	1.47	0.985	7.50	0.398	87.2	14.5	3.02	22.0	–
29	13.5	19.1	3.76	1.50	1.03	6.97	0.426	53.5	10.7	2.51	16.2	–
26.5	14.1	17.7	3.54	1.51	1.02	6.46	0.389	47.9	9.58	2.48	14.5	–
25	13.4	18.7	3.79	1.60	1.17	6.88	0.452	28.2	6.97	1.96	10.6	–
22.5	14.8	16.6	3.39	1.59	1.13	6.10	0.408	25.0	6.21	1.95	9.47	0.998
20	16.8	14.4	2.95	1.57	1.09	5.28	0.365	22.0	5.50	1.94	8.38	0.885
17.5	18.1	16.0	3.23	1.76	1.30	5.71	0.394	12.2	3.73	1.54	5.73	0.855
15	20.9	13.5	2.75	1.75	1.27	4.83	0.337	10.2	3.12	1.52	4.78	0.708
13	23.6	11.7	2.40	1.75	1.25	4.20	0.295	8.66	2.67	1.51	4.08	0.567
11	20.9	11.7	2.59	1.90	1.63	4.63	0.402	2.33	1.15	0.847	1.83	0.711
9.5	23.1	10.1	2.28	1.90	1.65	4.11	0.348	1.88	0.939	0.821	1.49	0.598
8	24.7	8.70	2.04	1.92	1.74	4.25	0.639	1.41	0.706	0.773	1.13	0.539
7	27.2	7.67	1.83	1.92	1.76	3.87	0.760	1.18	0.593	0.753	0.947	0.451
56	5.21	28.6	6.40	1.32	1.21	13.4	0.791	118	22.6	2.67	34.6	–
50	5.78	24.5	5.56	1.29	1.13	11.4	0.711	103	20.0	2.65	30.5	–
44	6.50	20.8	4.77	1.27	1.06	9.65	0.631	89.3	17.4	2.63	26.5	–
38.5	7.42	17.4	4.05	1.24	0.990	8.06	0.555	76.8	15.1	2.60	22.9	–
34	8.36	14.9	3.49	1.22	0.932	6.85	0.493	66.7	13.2	2.58	20.0	–
30	9.36	12.9	3.04	1.21	0.884	5.87	0.438	58.1	11.5	2.57	17.5	–
27	10.6	11.1	2.64	1.19	0.836	5.05	0.395	51.7	10.3	2.56	15.6	–
24.5	11.6	10.0	2.39	1.18	0.807	4.52	0.361	46.7	9.34	2.54	14.1	–
22.5	11.2	10.2	2.47	1.24	0.907	4.65	0.413	26.7	6.65	2.01	10.1	–
19.5	12.5	8.84	2.16	1.24	0.876	3.99	0.359	22.5	5.64	1.98	8.57	–
16.5	13.6	7.71	1.93	1.26	0.869	3.48	0.305	18.3	4.60	1.94	7.00	–
15	14.8	9.28	2.24	1.45	1.10	4.01	0.380	8.35	2.87	1.37	4.41	–
13	17.0	7.86	1.91	1.44	1.06	3.39	0.330	7.05	2.44	1.36	3.75	0.904
11	18.4	6.88	1.72	1.46	1.07	3.02	0.282	5.71	1.99	1.33	3.05	0.837
9.5	17.7	6.68	1.74	1.54	1.28	3.10	0.349	2.15	1.07	0.874	1.67	0.873
8.5	18.4	6.06	1.62	1.56	1.32	2.90	0.311	1.78	0.887	0.844	1.40	0.843
7.5	19.2	5.45	1.50	1.57	1.37	2.71	0.305	1.45	0.723	0.810	1.15	0.810
6	23.3	4.35	1.22	1.57	1.36	2.20	0.322	1.09	0.551	0.785	0.869	0.593
33.5	5.56	10.9	3.05	1.05	0.936	6.29	0.594	44.3	10.7	2.12	16.3	–
29	6.22	9.12	2.61	1.03	0.874	5.25	0.520	37.5	9.13	2.10	13.9	–
24	7.93	6.85	1.97	0.986	0.777	3.94	0.435	30.5	7.51	2.08	11.4	–
20	8.81	5.73	1.69	0.988	0.735	3.25	0.364	24.5	6.08	2.04	9.24	–
17.5	10.2	4.82	1.43	0.968	0.688	2.71	0.321	21.3	5.31	2.03	8.05	–
15.5	11.1	4.28	1.28	0.969	0.668	2.39	0.285	18.5	4.64	2.02	7.03	–

**Where no value of Q_s is shown, the Tee complies with LRFD Specification Section E2.

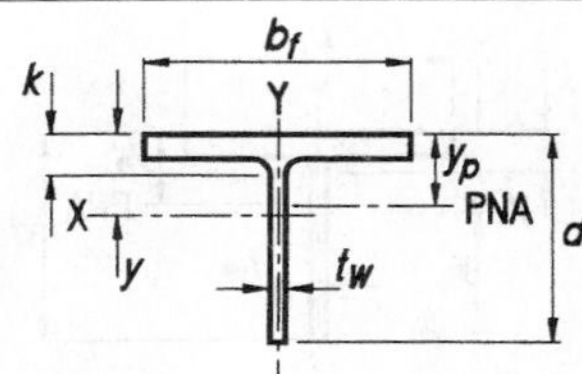

Table 1-8 (cont.). WT-Shapes (Structural Tees Split from W-Shapes) Dimensions

Shape	Area, A	Depth of Tee, d		Stem Thickness, t_w		Stem $\frac{t_w}{2}$	Stem Area	Flange Width, b_f		Flange Thickness, t_f		Distance k		Distance Workable Gage†
	in.²	in.		in.		in.	in.²	in.		in.		in.	in.	in.
WT4×14	4.12	4.03	4	0.285	5/16	3/16	1.15	6.54	6 1/2	0.465	7/16	0.859	15/16	3 1/2
×12	3.54	3.97	4	0.245	1/4	1/8	0.971	6.50	6 1/2	0.4	3/8	0.794	7/8	3 1/2
WT4×10.5	3.08	4.14	4 1/8	0.250	1/4	1/8	1.03	5.27	5 1/4	0.4	3/8	0.700	7/8	2 3/4
×9	2.63	4.07	4 1/8	0.230	1/4	1/8	0.936	5.25	5 1/4	0.33	5/16	0.630	13/16	2 3/4
WT4×7.5	2.22	4.05	4	0.245	1/4	1/8	0.993	4.01	4	0.315	5/16	0.615	13/16	2 1/4
×6.5	1.92	4.00	4	0.230	1/4	1/8	0.919	4.00	4	0.255	1/4	0.555	3/4	↓
×5	1.48	3.95	4	0.170	3/16	1/8	0.671	3.94	4	0.205	3/16	0.505	11/16	↓
WT3×12.5	3.68	3.19	3 1/4	0.320	5/16	3/16	1.02	6.08	6 1/8	0.455	7/16	0.754	15/16	3 1/2
×10	2.95	3.10	3 1/8	0.260	1/4	1/8	0.806	6.02	6	0.365	3/8	0.664	7/8	↓
×7.5	2.22	3.00	3	0.230	1/4	1/8	0.689	5.99	6	0.26	1/4	0.559	3/4	↓
WT3×8	2.37	3.14	3 1/8	0.260	1/4	1/8	0.816	4.03	4	0.405	3/8	0.655	7/8	2 1/4
×6	1.78	3.02	3	0.230	1/4	1/8	0.693	4.00	4	0.28	1/4	0.530	3/4	↓
×4.5	1.34	2.95	3	0.170	3/16	1/8	0.502	3.94	4	0.215	3/16	0.465	11/16	↓
×4.25	1.25	2.92	2 7/8	0.170	3/16	1/8	0.496	3.94	4	0.194	3/16	0.444	11/16	↓
WT2.5×9.5	2.78	2.58	2 5/8	0.270	1/4	1/8	0.695	5.03	5	0.43	7/16	0.730	13/16	2 3/4
×8	2.35	2.51	2 1/2	0.240	1/4	1/8	0.601	5.00	5	0.36	3/8	0.660	3/4	2 3/4
WT2×6.5	1.91	2.08	2 1/8	0.280	1/4	1/8	0.582	4.06	4	0.345	3/8	0.595	3/4	2 1/4

*ASTM A6 tensile group 4 or 5 shape. Special requirements may apply per LRFD Specification Section A3.1c.
†See definition of "Workable Gage" in Nomenclature section at the back of this Manual.

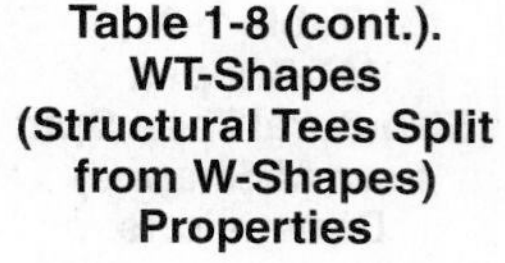

Table 1-8 (cont.). WT-Shapes (Structural Tees Split from W-Shapes) Properties

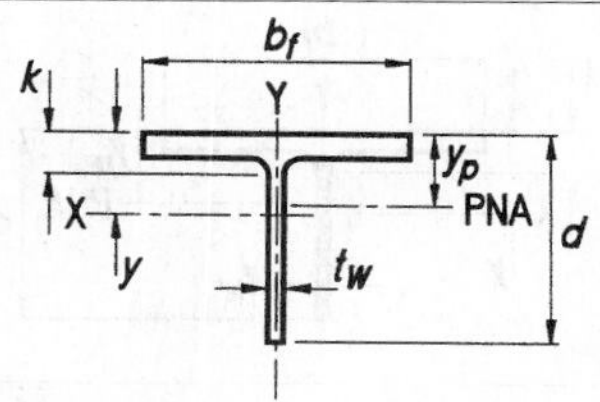

Nominal Wt.	$\frac{h}{t_w}$	Axis X-X						Axis Y-Y				Q_s**
		I	S	r	$\bar{y}$	Z	y_p	I	S	r	Z	$F_y =$ 50 ksi
lb/ft		in.^{4}	in.^{3}	in.	in.	in.^{3}	in.	in.^{4}	in.^{3}	in.	in.^{3}	
14	11.1	4.23	1.28	1.01	0.734	2.38	0.315	10.8	3.31	1.62	5.04	–
12	12.9	3.53	1.08	0.999	0.695	1.98	0.272	9.14	2.81	1.61	4.28	–
10.5	13.8	3.90	1.18	1.12	0.831	2.11	0.292	4.88	1.85	1.26	2.84	–
9	15.0	3.41	1.05	1.14	0.834	1.86	0.251	3.98	1.52	1.23	2.33	–
7.5	14.0	3.28	1.07	1.22	0.998	1.91	0.276	1.70	0.849	0.876	1.33	–
6.5	15.0	2.89	0.974	1.23	1.03	1.74	0.240	1.36	0.682	0.843	1.07	–
5	20.2	2.15	0.717	1.20	0.953	1.27	0.188	1.05	0.531	0.840	0.826	0.735
12.5	7.61	2.29	0.886	0.788	0.610	1.69	0.303	8.53	2.81	1.52	4.28	–
10	9.37	1.76	0.693	0.773	0.560	1.29	0.245	6.64	2.21	1.50	3.36	–
7.5	10.6	1.41	0.577	0.795	0.557	1.03	0.186	4.66	1.56	1.45	2.37	–
8	9.56	1.69	0.685	0.844	0.676	1.25	0.294	2.21	1.10	0.966	1.69	–
6	10.8	1.32	0.564	0.862	0.677	1.01	0.222	1.50	0.748	0.918	1.16	–
4.5	14.6	0.95	0.408	0.842	0.623	0.720	0.170	1.10	0.557	0.905	0.856	–
4.25	14.5	0.904	0.397	0.849	0.638	0.700	0.159	0.990	0.503	0.889	0.774	–
9.5	6.83	1.01	0.485	0.604	0.487	0.970	0.276	4.56	1.81	1.28	2.76	–
8	7.69	0.845	0.413	0.599	0.458	0.801	0.235	3.75	1.50	1.26	2.28	–
6.5	5.30	0.526	0.321	0.524	0.440	0.616	0.236	1.93	0.950	1.00	1.46	–

**Where no value of Q_s is shown, the Tee complies with LRFD Specification Section E2.

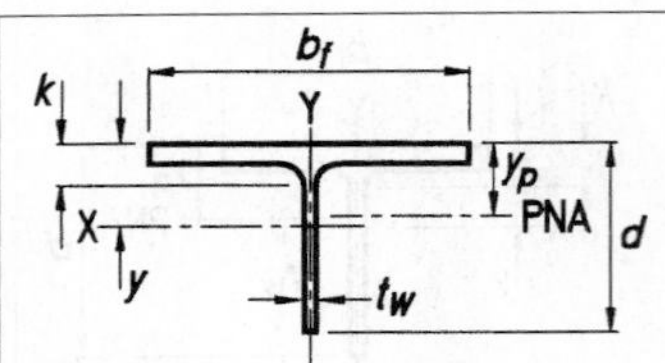

Table 1-9. MT-Shapes (Structural Tees Split from M-Shapes) Dimensions

Shape	Area, A	Depth of Tee, d		Stem				Flange				Distance	
				Thickness, t_w		$\frac{t_w}{2}$	Area	Width, b_f		Thickness, t_f		k	Workable Gage†
	in.^{2}	in.		in.		in.	in.^{2}	in.		in.		in.	in.
MT6×5.9	1.72	6.00	6	0.177	3/16	1/8	1.06	3.07	3 1/8	0.225	1/4	9/16	–
×5.4	1.58	5.99	6	0.160	3/16	1/8	0.958	3.07	3 1/8	0.210	3/16	9/16	–
×5	1.46	5.99	6	0.149	1/8	1/16	0.892	3.25	3 1/4	0.180	3/16	1/2	–
MT5×4.5	1.32	5.00	5	0.157	3/16	1/8	0.785	2.69	2 3/4	0.206	3/16	9/16	–
×4	1.17	4.97	5	0.141	1/8	1/16	0.701	2.69	2 3/4	0.182	3/16	9/16	–
×3.75	1.1	5.00	5	0.130	1/8	1/16	0.649	2.69	2 3/4	0.173	3/16	7/16	–
MT4×3.25	0.953	4.00	4	0.135	1/8	1/16	0.540	2.28	2 1/4	0.189	3/16	9/16	–
×3.1	0.904	4.00	4	0.129	1/8	1/16	0.516	2.28	2 1/4	0.177	3/16	7/16	–
MT3×2.2	0.643	3.00	3	0.114	1/8	1/16	0.342	1.84	1 7/8	0.171	3/16	3/8	–
×1.85	0.540	2.96	3	0.098	1/8	1/16	0.290	2.00	2	0.129	1/8	5/16	–
MT2.5×9.45*	2.78	2.50	2 1/2	0.316	5/16	3/16	0.790	5.00	5	0.416	7/16	13/16	2 3/4
MT2×3	0.855	1.90	1 7/8	0.130	1/8	1/16	0.247	3.80	3 3/4	0.160	3/16	1/2	–

*This shape has tapered flanges while all other MT-shapes have parallel flanges.
†See definition of "Workable Gage" in Nomenclature section at the back of this Manual.
– indicates that flange is too narrow to allow tabulation of a workable gage.

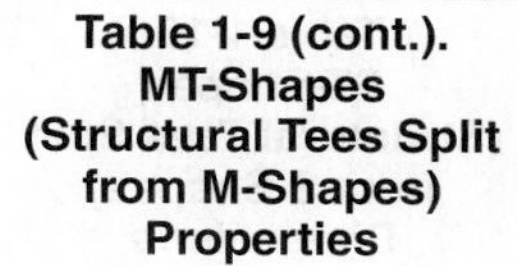

Table 1-9 (cont.). MT-Shapes (Structural Tees Split from M-Shapes) Properties

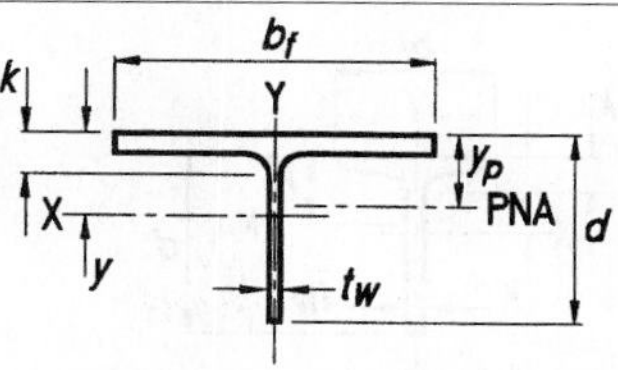

Nominal Wt.	$\frac{h}{t_w}$	Axis X-X						Axis Y-Y				Q_s**
		I	S	r	$\bar{y}$	Z	y_p	I	S	r	Z	$F_y =$ 36 ksi
lb/ft		in.⁴	in.³	in.	in.	in.³	in.	in.⁴	in.³	in.	in.³	
5.9	31.3	6.61	1.61	1.96	1.89	2.89	1.13	0.543	0.354	0.561	0.575	0.483
5.4	34.6	6.03	1.46	1.95	1.86	2.63	1.05	0.506	0.330	0.566	0.532	0.397
5	37.3	5.62	1.36	1.96	1.86	2.45	1.08	0.517	0.318	0.594	0.509	0.344
4.5	29.2	3.47	1.00	1.62	1.54	1.81	0.808	0.336	0.250	0.505	0.403	0.548
4	32.5	3.08	0.894	1.62	1.52	1.61	0.809	0.296	0.220	0.502	0.354	0.446
3.75	35.5	2.91	0.836	1.63	1.51	1.51	0.759	0.281	0.209	0.505	0.334	0.376
3.25	26.9	1.57	0.558	1.29	1.18	1.01	0.472	0.188	0.165	0.444	0.264	0.633
3.1	28.2	1.50	0.533	1.29	1.18	0.967	0.497	0.176	0.154	0.441	0.247	0.578
2.2	23.5	0.579	0.268	0.949	0.841	0.483	0.19	0.0897	0.0973	0.374	0.155	0.779
1.85	27.4	0.483	0.226	0.946	0.827	0.409	0.174	0.0863	0.0863	0.400	0.136	0.609
9.45	5.61	1.05	0.527	0.615	0.508	1.03	0.278	4.35	1.74	1.25	2.66	–
3	11.0	0.208	0.133	0.493	0.341	0.241	0.112	0.732	0.385	0.926	0.588	–

**Where no value of Q_s is shown, the angle complies with LRFD Specification Section E2.

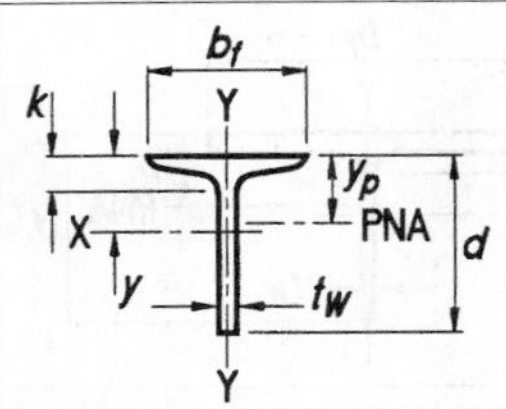

Table 1-10. ST-Shapes (Structural Tees Split from S-Shapes) Dimensions

Shape	Area, A	Depth of Tee, d		Stem				Flange				Distance	
				Thickness, t_w		$\frac{t_w}{2}$	Area	Width, b_f		Thickness, t_f		k	Workable Gage†
	in.^{2}	in.		in.		in.	in.^{2}	in.		in.		in.	in.
ST12×60.5	17.8	12.3	12 1/4	0.800	13/16	7/16	9.80	8.05	8	1.09	1 1/16	2	4
×53	15.6	12.3	12 1/4	0.620	5/8	5/16	7.60	7.87	7 7/8	1.09	1 1/16	2	4
ST12×50	14.7	12.0	12	0.745	3/4	3/8	8.94	7.25	7 1/4	0.870	7/8	1 3/4	4
×45	13.2	12.0	12	0.625	5/8	5/16	7.50	7.13	7 1/8	0.870	7/8	1 3/4	4
×40	11.7	12.0	12	0.500	1/2	1/4	6.00	7.00	7	0.870	7/8	1 3/4	4
ST10×48	14.1	10.2	10 1/8	0.800	13/16	7/16	8.12	7.20	7 1/4	0.920	15/16	1 3/4	4
×43	12.7	10.2	10 1/8	0.660	11/16	3/8	6.70	7.06	7	0.920	15/16	1 3/4	4
ST10×37.5	11.0	10.0	10	0.635	5/8	5/16	6.35	6.39	6 3/8	0.795	13/16	1 5/8	3 1/2
×33	9.69	10.0	10	0.505	1/2	1/4	5.05	6.26	6 1/4	0.795	13/16	1 5/8	3 1/2
ST9×35	10.3	9.00	9	0.711	11/16	3/8	6.40	6.25	6 1/4	0.691	11/16	1 1/2	3 1/2
×27.35	8.02	9.00	9	0.461	7/16	1/4	4.15	6.00	6	0.691	11/16	1 1/2	3 1/2
ST7.5×25	7.34	7.50	7 1/2	0.550	9/16	5/16	4.13	5.64	5 5/8	0.622	5/8	1 3/8	3 1/2
×21.45	6.30	7.50	7 1/2	0.411	7/16	1/4	3.08	5.50	5 1/2	0.622	5/8	1 3/8	3 1/2
ST6×25	7.32	6.00	6	0.687	11/16	3/8	4.12	5.48	5 1/2	0.659	11/16	1 7/16	3
×20.4	5.96	6.00	6	0.462	7/16	1/4	2.77	5.25	5 1/4	0.659	11/16	1 7/16	3
ST6×17.5	5.12	6.00	6	0.428	7/16	1/4	2.57	5.08	5 1/8	0.544	9/16	1 3/16	3
×15.9	4.65	6.00	6	0.350	3/8	3/16	2.10	5.00	5	0.544	9/16	1 3/16	3
ST5×17.5	5.14	5.00	5	0.594	5/8	5/16	2.97	4.94	5	0.491	1/2	1 1/8	2 3/4
×12.7	3.73	5.00	5	0.311	5/16	3/16	1.56	4.66	4 5/8	0.491	1/2	1 1/8	2 3/4
ST4×11.5	3.38	4.00	4	0.441	7/16	1/4	1.76	4.17	4 1/8	0.425	7/16	1	2 1/4
×9.2	2.70	4.00	4	0.271	1/4	1/8	1.08	4.00	4	0.425	7/16	1	2 1/4
ST3×8.625	2.53	3.00	3	0.465	7/16	1/4	1.40	3.57	3 5/8	0.359	3/8	13/16	–
×6.25	1.83	3.00	3	0.232	1/4	1/8	0.696	3.33	3 3/8	0.359	3/8	13/16	–
ST2.5×5	1.47	2.50	2 1/2	0.214	3/16	1/8	0.535	3.00	3	0.326	5/16	3/4	–
ST2×4.75	1.39	2.00	2	0.326	5/16	3/16	0.652	2.80	2 3/4	0.293	5/16	3/4	–
×3.85	1.13	2.00	2	0.193	3/16	1/8	0.386	2.66	2 5/8	0.293	5/16	3/4	–
ST1.5×3.75	1.10	1.50	1 1/2	0.349	3/8	3/16	0.524	2.51	2 1/2	0.260	1/4	5/8	–
×2.85	0.830	1.50	1 1/2	0.170	3/16	1/8	0.255	2.33	2 3/8	0.260	1/4	5/8	–

†See definition of "Workable Gage" in Nomenclature section at the back of this Manual.
– indicates that flange is too narrow to allow tabulation of a workable gage.

Table 1-10 (cont.). ST-Shapes (Structural Tees Split from S-Shapes) Properties

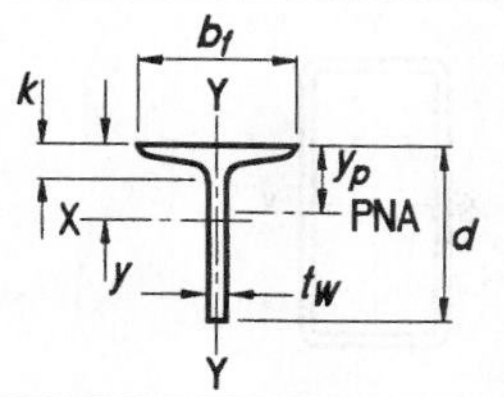

Nominal Wt.	$\frac{h}{t_w}$	Axis X-X						Axis Y-Y				Q_s**
		I	S	r	$\bar{y}$	Z	y_p	I	S	r	Z	$F_y =$ 36 ksi
lb/ft		in.^{4}	in.^{3}	in.	in.	in.^{3}	in.	in.^{4}	in.^{3}	in.	in.^{3}	
60.5	13.2	259	30.1	3.82	3.63	54.5	1.26	41.5	10.3	1.53	18.1	–
53	17.0	216	24.1	3.72	3.28	43.3	1.02	38.4	9.76	1.57	16.7	–
50	14.1	215	26.3	3.83	3.84	47.5	2.16	23.7	6.55	1.27	12.0	–
45	16.8	190	22.6	3.79	3.60	41.1	1.42	22.3	6.27	1.30	11.2	–
40	21.1	162	18.6	3.72	3.30	33.6	0.909	21.0	6.00	1.34	10.4	0.878
48	10.8	143	20.3	3.18	3.13	36.9	1.35	25.0	6.93	1.33	12.5	–
43	13.1	124	17.2	3.13	2.91	31.1	0.972	23.3	6.59	1.36	11.6	–
37.5	13.6	109	15.8	3.15	3.07	28.6	1.34	14.8	4.62	1.16	8.36	–
33	17.0	92.9	12.9	3.10	2.81	23.4	0.841	13.7	4.39	1.19	7.70	–
35	11.0	84.5	14.0	2.87	2.94	25.1	1.78	12.0	3.84	1.08	7.17	–
27.35	16.9	62.3	9.60	2.79	2.51	17.3	0.737	10.4	3.45	1.14	6.06	–
25	11.6	40.5	7.72	2.35	2.25	14.0	0.826	7.79	2.76	1.03	4.99	–
21.45	15.5	32.9	5.99	2.29	2.01	10.8	0.605	7.13	2.59	1.06	4.54	–
25	7.05	25.1	6.04	1.85	1.84	11.0	0.758	7.79	2.84	1.03	5.16	–
20.4	10.6	18.9	4.27	1.78	1.58	7.71	0.577	6.74	2.57	1.06	4.43	–
17.5	11.9	17.2	3.95	1.83	1.65	7.12	0.543	4.92	1.94	0.980	3.40	–
15.9	14.5	14.8	3.30	1.78	1.51	5.94	0.480	4.66	1.87	1.00	3.22	–
17.5	6.90	12.5	3.62	1.56	1.56	6.58	0.673	4.15	1.68	0.899	3.10	–
12.7	13.2	7.79	2.05	1.45	1.20	3.70	0.403	3.36	1.44	0.950	2.49	–
11.5	7.27	5.00	1.76	1.22	1.15	3.19	0.439	2.13	1.02	0.795	1.84	–
9.2	11.8	3.49	1.14	1.14	0.942	2.07	0.336	1.84	0.922	0.827	1.59	–
8.625	5.01	2.12	1.02	0.915	0.915	1.85	0.394	1.14	0.642	0.673	1.17	–
6.25	10.0	1.26	0.547	0.831	0.692	1.01	0.271	0.901	0.541	0.702	0.930	–
5	8.71	0.671	0.348	0.677	0.570	0.650	0.239	0.597	0.398	0.638	0.686	–
4.75	4.34	0.462	0.319	0.575	0.553	0.592	0.250	0.444	0.317	0.564	0.565	–
3.85	7.34	0.307	0.198	0.522	0.448	0.381	0.204	0.374	0.281	0.576	0.485	–
3.75	2.84	0.200	0.187	0.426	0.432	0.351	0.219	0.289	0.230	0.513	0.411	–
2.85	5.82	0.114	0.097	0.370	0.329	0.196	0.171	0.223	0.192	0.518	0.328	–

*Where no value of Q_s is shown, the angle complies with LRFD Specification Section E2.

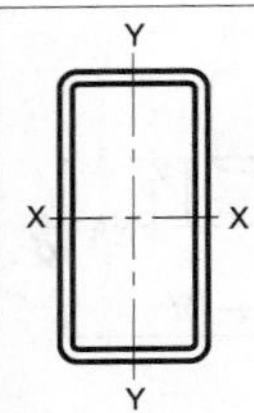

Table 1-11.
Rectangular (and Square) HSS Dimensions and Properties

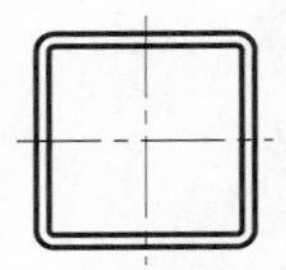

Shape	Wall Thickness, t		Nom-inal Wt.	Area, A	$\frac{b}{t}$	$\frac{h}{t}$	Axis X-X			
	nominal	design					I	S	r	Z
	in.	in.	lb/ft	in.^{2}			in.^{4}	in.^{3}	in.	in.^{3}
HSS20×12	5/8	0.581	127	35.0	17.7	31.4	1880	188	7.33	230
	1/2	0.465	103	28.3	22.8	40.0	1550	155	7.39	188
	3/8	0.349	78.4	21.5	31.4	54.3	1200	120	7.45	144
	5/16	0.291	65.8	18.1	38.2	65.7	1010	101	7.48	122
HSS20×8	5/8	0.581	110	30.3	10.8	31.4	1440	144	6.89	185
	1/2	0.465	89.6	24.6	14.2	40.0	1190	119	6.96	152
	3/8	0.349	68.2	18.7	19.9	54.3	926	92.6	7.03	117
	5/16	0.291	57.3	15.7	24.5	65.7	786	78.6	7.07	98.6
HSS20×4	1/2	0.465	75.9	20.9	5.60	40.0	838	83.8	6.33	115
	3/8	0.349	58.0	16.0	8.46	54.3	657	65.7	6.42	89.3
	5/16	0.291	48.8	13.4	10.7	65.7	560	56.0	6.46	75.6
HSS18×12	5/8	0.581	119	32.6	17.7	28.0	1450	161	6.67	196
	1/2	0.465	96.4	26.5	22.8	35.7	1200	133	6.73	161
	3/8	0.349	73.3	20.1	31.4	48.6	929	103	6.79	123
HSS18×6	5/8	0.581	93.1	25.7	7.33	28.0	923	103	6.00	135
	1/2	0.465	75.9	20.9	9.90	35.7	770	85.6	6.07	112
	3/8	0.349	58.0	16.0	14.2	48.6	602	66.9	6.15	86.4
	5/16	0.291	48.8	13.4	17.6	58.9	513	57.0	6.18	73.1
	1/4	0.233	39.4	10.8	22.8	74.3	419	46.5	6.22	59.4
HSS16×16	5/8	0.581	127	35	24.5	24.5	1370	171	6.25	200
	1/2	0.465	103	28.3	31.4	31.4	1130	141	6.31	164
	3/8	0.349	78.4	21.5	42.8	42.8	873	109	6.37	126
	5/16	0.291	65.8	18.1	52.0	52.0	739	92.3	6.39	106
HSS16×12	5/8	0.581	110	30.3	17.7	24.5	1090	136	6.00	165
	1/2	0.465	89.6	24.6	22.8	31.4	904	113	6.06	135
	3/8	0.349	68.2	18.7	31.4	42.8	702	87.7	6.12	104
	5/16	0.291	57.3	15.7	38.2	52.0	595	74.4	6.15	87.7
HSS16×8	5/8	0.581	93.1	25.7	10.8	24.5	815	102	5.64	129
	1/2	0.465	75.9	20.9	14.2	31.4	679	84.9	5.70	106
	3/8	0.349	58.0	16.0	19.9	42.8	531	66.3	5.77	82.1
	5/16	0.291	48.8	13.4	24.5	52.0	451	56.4	5.80	69.4
HSS16×4	1/2	0.465	62.3	17.2	5.60	31.4	455	56.9	5.15	77.3
	3/8	0.349	47.8	13.2	8.46	42.8	360	45.0	5.23	60.2
	5/16	0.291	40.3	11.1	10.7	52.0	308	38.5	5.27	51.1
HSS14×14	5/8	0.581	110	30.3	21.1	21.1	897	128	5.44	151
	1/2	0.465	89.6	24.6	27.1	27.1	743	106	5.49	124
	3/8	0.349	68.2	18.7	37.1	37.1	577	82.5	5.55	95.4
	5/16	0.291	57.3	15.7	45.1	45.1	490	69.9	5.58	80.5
HSS14×12	1/2	0.465	82.7	22.8	22.8	27.1	658	94.0	5.38	111
	3/8	0.349	63.1	17.3	31.4	37.1	512	73.2	5.43	85.9
HSS14×10	5/8	0.581	93.1	25.7	14.2	21.1	687	98.2	5.17	120
	1/2	0.465	75.9	20.9	18.5	27.1	573	81.8	5.23	98.8
	3/8	0.349	58.0	16.0	25.7	37.1	447	63.9	5.29	76.3
	5/16	0.291	48.8	13.4	31.4	45.1	380	54.3	5.32	64.6
	1/4	0.233	39.4	10.8	39.9	57.1	310	44.3	5.35	52.4
HSS14×6	5/8	0.581	76.1	21.0	7.33	21.1	478	68.3	4.77	88.7
	1/2	0.465	62.3	17.2	9.90	27.1	402	57.4	4.84	73.6
	3/8	0.349	47.8	13.2	14.2	37.1	317	45.3	4.91	57.3
	5/16	0.291	40.3	11.1	17.6	45.1	271	38.7	4.94	48.6
	1/4	0.233	32.6	8.96	22.8	57.1	222	31.7	4.98	39.6
	3/16	0.174	24.7	6.76	31.5	77.5	170	24.3	5.01	30.1

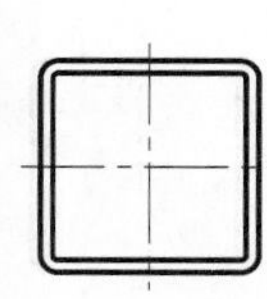

Table 1-11 (cont.). Rectangular (and Square) HSS Dimensions and Properties

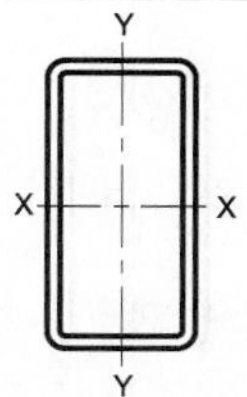

Shape	Wall Thickness, *t*		Axis Y-Y				Torsion		Surface Area Per Foot
	nominal	design	*I*	*S*	*r*	*Z*	*J*	*C*	
	in.	in.	in.^{4}	in.^{3}	in.	in.^{3}	in.^{4}	in.^{3}	ft^2
HSS20×12	5/8	0.581	851	142	4.93	162	1890	257	5.17
	1/2	0.465	705	117	4.99	132	1540	209	5.20
	3/8	0.349	547	91.1	5.04	102	1180	160	5.23
	5/16	0.291	464	77.3	5.07	85.8	997	134	5.25
HSS20×8	5/8	0.581	338	84.6	3.34	96.4	916	167	4.50
	1/2	0.465	283	70.8	3.39	79.5	757	137	4.53
	3/8	0.349	222	55.6	3.44	61.5	586	105	4.57
	5/16	0.291	189	47.4	3.47	52.0	496	88.3	4.58
HSS20×4	1/2	0.465	58.7	29.3	1.68	34.0	195	63.8	3.87
	3/8	0.349	47.6	23.8	1.73	26.8	156	49.9	3.90
	5/16	0.291	41.2	20.6	1.75	22.9	134	42.4	3.92
HSS18×12	5/8	0.581	776	129	4.87	148	1630	230	4.83
	1/2	0.465	643	107	4.93	122	1330	188	4.87
	3/8	0.349	499	83.2	4.98	93.6	1020	143	4.90
HSS18×6	5/8	0.581	158	52.7	2.48	61.0	462	109	3.83
	1/2	0.465	134	44.6	2.53	50.7	387	89.9	3.87
	3/8	0.349	106	35.5	2.58	39.5	302	69.5	3.90
	5/16	0.291	91.3	30.4	2.61	33.5	257	58.7	3.92
	1/4	0.233	75.1	25.0	2.63	27.3	210	47.7	3.93
HSS16×16	5/8	0.581	1370	171	6.25	200	2170	276	5.17
	1/2	0.465	1130	141	6.31	164	1770	224	5.20
	3/8	0.349	873	109	6.37	126	1350	171	5.23
	5/16	0.291	739	92.3	6.39	106	1140	144	5.25
HSS16×12	5/8	0.581	700	117	4.80	135	1370	204	4.50
	1/2	0.465	581	96.8	4.86	111	1120	166	4.53
	3/8	0.349	452	75.3	4.91	85.5	862	127	4.57
	5/16	0.291	384	64.0	4.94	72.2	727	107	4.58
HSS16×8	5/8	0.581	274	68.6	3.27	79.2	681	132	3.83
	1/2	0.465	230	57.6	3.32	65.5	563	108	3.87
	3/8	0.349	181	45.3	3.37	50.8	436	83.4	3.90
	5/16	0.291	155	38.7	3.40	43.0	369	70.4	3.92
HSS16×4	1/2	0.465	47.0	23.5	1.65	27.4	150	50.7	3.20
	3/8	0.349	38.3	19.1	1.71	21.7	120	39.7	3.23
	5/16	0.291	33.2	16.6	1.73	18.5	103	33.8	3.25
HSS14×14	5/8	0.581	897	128	5.44	151	1430	208	4.50
	1/2	0.465	743	106	5.49	124	1170	170	4.53
	3/8	0.349	577	82.5	5.55	95.4	900	130	4.57
	5/16	0.291	490	69.9	5.58	80.5	759	109	4.58
HSS14×12	1/2	0.465	519	86.5	4.78	100	921	145	4.20
	3/8	0.349	405	67.4	4.83	77.4	708	111	4.23
HSS14×10	5/8	0.581	407	81.5	3.98	95.1	832	146	3.83
	1/2	0.465	341	68.1	4.04	78.5	685	120	3.87
	3/8	0.349	267	53.4	4.09	60.7	528	91.8	3.90
	5/16	0.291	227	45.5	4.12	51.4	446	77.4	3.92
	1/4	0.233	186	37.2	4.14	41.8	362	62.6	3.93
HSS14×6	5/8	0.581	124	41.2	2.43	48.4	334	83.7	3.17
	1/2	0.465	105	35.1	2.48	40.4	279	69.3	3.20
	3/8	0.349	84.1	28.0	2.53	31.6	219	53.7	3.23
	5/16	0.291	72.3	24.1	2.55	26.9	186	45.5	3.25
	1/4	0.233	59.6	19.9	2.58	22	152	36.9	3.27
	3/16	0.174	45.9	15.3	2.61	16.7	116	28.0	3.28

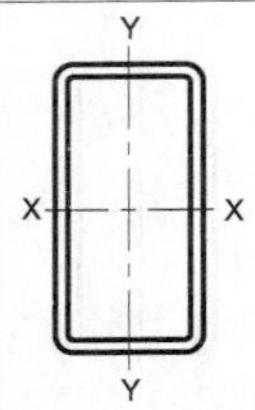

Table 1-11 (cont.). Rectangular (and Square) HSS Dimensions and Properties

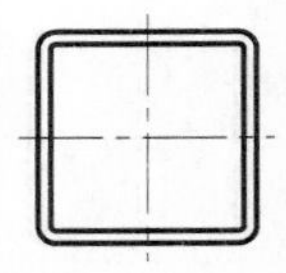

Shape	Wall Thickness, t nominal	Wall Thickness, t design	Nom-inal Wt.	Area, A	$\frac{b}{t}$	$\frac{h}{t}$	Axis X-X I	Axis X-X S	Axis X-X r	Axis X-X Z
	in.	in.	lb/ft	in.^{2}			in.^{4}	in.^{3}	in.	in.^{3}
HSS14×4	5/8	0.581	67.6	18.7	3.88	21.1	373	53.3	4.47	73.1
	1/2	0.465	55.5	15.3	5.60	27.1	317	45.3	4.55	61.0
	3/8	0.349	42.7	11.8	8.46	37.1	252	36.0	4.63	47.8
	5/16	0.291	36.0	9.92	10.7	45.1	216	30.9	4.67	40.6
	1/4	0.233	29.2	8.03	14.2	57.1	178	25.4	4.71	33.2
	3/16	0.174	22.2	6.06	20.0	77.5	137	19.5	4.74	25.3
HSS12×12	5/8	0.581	93.1	25.7	17.7	17.7	548	91.4	4.62	109
	1/2	0.465	75.9	20.9	22.8	22.8	457	76.2	4.68	89.6
	3/8	0.349	58.0	16.0	31.4	31.4	357	59.5	4.73	69.2
	5/16	0.291	48.8	13.4	38.2	38.2	304	50.7	4.76	58.6
	1/4	0.233	39.4	10.8	48.5	48.5	248	41.4	4.79	47.6
HSS12×10	1/2	0.465	69.1	19.0	18.5	22.8	395	65.9	4.56	78.8
	3/8	0.349	52.9	14.6	25.7	31.4	310	51.6	4.61	61.1
	5/16	0.291	44.6	12.2	31.4	38.2	264	44.0	4.64	51.7
	1/4	0.233	36.0	9.90	39.9	48.5	216	36.0	4.67	42.1
HSS12×8	5/8	0.581	76.1	21.0	10.8	17.7	397	66.1	4.34	82.1
	1/2	0.465	62.3	17.2	14.2	22.8	333	55.6	4.41	68.1
	3/8	0.349	47.8	13.2	19.9	31.4	262	43.7	4.47	53.0
	5/16	0.291	40.3	11.1	24.5	38.2	224	37.4	4.50	44.9
	1/4	0.233	32.6	8.96	31.3	48.5	184	30.6	4.53	36.6
	3/16	0.174	24.7	6.76	43.0	66.0	140	23.4	4.56	27.8
HSS12×6	5/8	0.581	67.6	18.7	7.33	17.7	321	53.4	4.14	68.8
	1/2	0.465	55.5	15.3	9.90	22.8	271	45.2	4.21	57.4
	3/8	0.349	42.7	11.8	14.2	31.4	215	35.9	4.28	44.8
	5/16	0.291	36.0	9.92	17.6	38.2	184	30.7	4.31	38.1
	1/4	0.233	29.2	8.03	22.8	48.5	151	25.2	4.34	31.1
	3/16	0.174	22.2	6.06	31.5	66.0	116	19.4	4.38	23.7
HSS12×4	5/8	0.581	59.1	16.4	3.88	17.7	245	40.8	3.87	55.5
	1/2	0.465	48.7	13.5	5.60	22.8	210	34.9	3.95	46.7
	3/8	0.349	37.6	10.4	8.46	31.4	168	28.0	4.02	36.7
	5/16	0.291	31.8	8.76	10.7	38.2	144	24.1	4.06	31.3
	1/4	0.233	25.8	7.10	14.2	48.5	119	19.9	4.10	25.6
	3/16	0.174	19.6	5.37	20.0	66.0	91.8	15.3	4.13	19.6
HSS12×3 1/2	3/8	0.349	36.3	10.0	7.03	31.4	156	26.0	3.94	34.7
	5/16	0.291	30.7	8.46	9.03	38.2	134	22.4	3.98	29.6
HSS12×3	5/16	0.291	29.7	8.17	7.31	38.2	124	20.7	3.90	27.9
	1/4	0.233	24.1	6.63	9.88	48.5	103	17.2	3.94	22.9
	3/16	0.174	18.3	5.02	14.2	66.0	79.6	13.3	3.98	17.5
HSS12×2	1/4	0.233	22.4	6.17	5.58	48.5	86.9	14.5	3.75	20.1
	3/16	0.174	17.1	4.67	8.49	66.0	67.4	11.2	3.80	15.5
HSS10×10	5/8	0.581	76.1	21.0	14.2	14.2	304	60.8	3.80	73.2
	1/2	0.465	62.3	17.2	18.5	18.5	256	51.2	3.86	60.7
	3/8	0.349	47.8	13.2	25.7	25.7	202	40.4	3.92	47.2
	5/16	0.291	40.3	11.1	31.4	31.4	172	34.5	3.94	40.1
	1/4	0.233	32.6	8.96	39.9	39.9	141	28.3	3.97	32.7
	3/16	0.174	24.7	6.76	54.5	54.5	108	21.6	4.00	24.8
HSS10×8	1/2	0.465	55.5	15.3	14.2	18.5	214	42.7	3.73	51.9
	3/8	0.349	42.7	11.8	19.9	25.7	169	33.9	3.79	40.5
	5/16	0.291	36.0	9.92	24.5	31.4	145	29.0	3.82	34.4
	1/4	0.233	29.2	8.03	31.3	39.9	119	23.8	3.85	28.1
	3/16	0.174	22.2	6.06	43.0	54.5	91.4	18.3	3.88	21.4

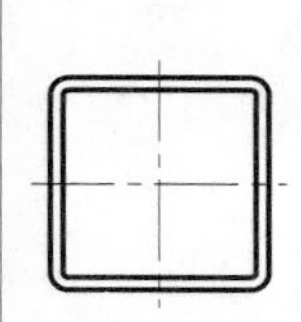

Table 1-11 (cont.). Rectangular (and Square) HSS Dimensions and Properties

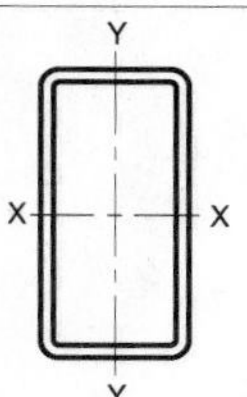

Shape	Wall Thickness, t		Axis Y-Y				Torsion		Surface Area Per Foot
	nominal	design	I	S	r	Z	J	C	
	in.	in.	in.^{4}	in.^{3}	in.	in.^{3}	in.^{4}	in.^{3}	ft^2
HSS14×4	5/8	0.581	47.2	23.6	1.59	28.5	148	52.6	2.83
	1/2	0.465	41.2	20.6	1.64	24.1	127	44.1	2.87
	3/8	0.349	33.6	16.8	1.69	19.1	102	34.6	2.90
	5/16	0.291	29.2	14.6	1.72	16.4	87.7	29.5	2.92
	1/4	0.233	24.4	12.2	1.74	13.5	72.4	24.1	2.93
	3/16	0.174	19.0	9.48	1.77	10.3	55.8	18.4	2.95
HSS12×12	5/8	0.581	548	91.4	4.62	109	885	151	3.83
	1/2	0.465	457	76.2	4.68	89.6	728	123	3.87
	3/8	0.349	357	59.5	4.73	69.2	561	94.6	3.90
	5/16	0.291	304	50.7	4.76	58.6	474	79.7	3.92
	1/4	0.233	248	41.4	4.79	47.6	384	64.5	3.93
HSS12×10	1/2	0.465	298	59.7	3.96	69.6	545	102	3.53
	3/8	0.349	234	46.9	4.01	54.0	421	78.3	3.57
	5/16	0.291	200	40.0	4.04	45.7	356	66.1	3.58
	1/4	0.233	164	32.7	4.07	37.2	289	53.5	3.60
HSS12×8	5/8	0.581	210	52.5	3.16	61.9	454	97.7	3.17
	1/2	0.465	178	44.4	3.21	51.5	377	80.4	3.20
	3/8	0.349	140	35.1	3.27	40.1	293	62.1	3.23
	5/16	0.291	120	30.1	3.29	34.1	248	52.4	3.25
	1/4	0.233	98.8	24.7	3.32	27.8	202	42.5	3.27
	3/16	0.174	75.7	18.9	3.35	21.1	153	32.2	3.28
HSS12×6	5/8	0.581	107	35.5	2.39	42.1	271	71.1	2.83
	1/2	0.465	91.1	30.4	2.44	35.2	227	59.0	2.87
	3/8	0.349	72.9	24.3	2.49	27.7	178	45.8	2.90
	5/16	0.291	62.8	20.9	2.52	23.6	152	38.8	2.92
	1/4	0.233	51.9	17.3	2.54	19.3	124	31.6	2.93
	3/16	0.174	40.0	13.3	2.57	14.7	94.6	24.0	2.95
HSS12×4	5/8	0.581	40.4	20.2	1.57	24.5	122	44.6	2.50
	1/2	0.465	35.3	17.7	1.62	20.9	105	37.5	2.53
	3/8	0.349	28.9	14.5	1.67	16.6	84.1	29.5	2.57
	5/16	0.291	25.2	12.6	1.70	14.2	72.4	25.2	2.58
	1/4	0.233	21.0	10.5	1.72	11.7	59.8	20.6	2.60
	3/16	0.174	16.4	8.20	1.75	9.00	46.1	15.7	2.62
HSS12×3 1/2	3/8	0.349	21.3	12.2	1.46	14.0	64.7	25.5	2.48
	5/16	0.291	18.6	10.6	1.48	12.1	56.0	21.8	2.50
HSS12×3	5/16	0.291	13.1	8.73	1.27	10.0	41.3	18.4	2.42
	1/4	0.233	11.1	7.38	1.29	8.28	34.5	15.1	2.43
	3/16	0.174	8.72	5.81	1.32	6.40	26.8	11.6	2.45
HSS12×2	1/4	0.233	4.41	4.41	0.845	5.08	15.1	9.64	2.27
	3/16	0.174	3.55	3.55	0.872	3.97	12.0	7.49	2.28
HSS10×10	5/8	0.581	304	60.8	3.80	73.2	498	102	3.17
	1/2	0.465	256	51.2	3.86	60.7	412	84.2	3.20
	3/8	0.349	202	40.4	3.92	47.2	320	64.8	3.23
	5/16	0.291	172	34.5	3.94	40.1	271	54.8	3.25
	1/4	0.233	141	28.3	3.97	32.7	220	44.4	3.27
	3/16	0.174	108	21.6	4.00	24.8	167	33.6	3.28
HSS10×8	1/2	0.465	151	37.8	3.14	44.5	288	66.4	2.87
	3/8	0.349	120	30.0	3.19	34.8	224	51.4	2.90
	5/16	0.291	103	25.7	3.22	29.6	190	43.5	2.92
	1/4	0.233	84.7	21.2	3.25	24.2	155	35.3	2.93
	3/16	0.174	65.1	16.3	3.28	18.4	118	26.7	2.95

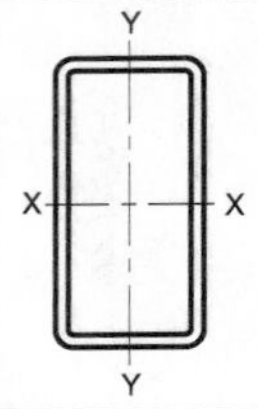

Table 1-11 (cont.). Rectangular (and Square) HSS Dimensions and Properties

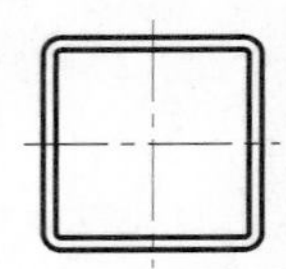

Shape	Wall Thickness, t		Nom-inal Wt.	Area, A	$\frac{b}{t}$	$\frac{h}{t}$	Axis X-X			
	nominal	design					I	S	r	Z
	in.	in.	lb/ft	in.^{2}			in.^{4}	in.^{3}	in.	in.^{3}
HSS10×6	5/8	0.581	59.1	16.4	7.33	14.2	201	40.2	3.50	51.3
	1/2	0.465	48.7	13.5	9.90	18.5	171	34.3	3.57	43.0
	3/8	0.349	37.6	10.4	14.2	25.7	137	27.4	3.63	33.8
	5/16	0.291	31.8	8.76	17.6	31.4	118	23.5	3.66	28.8
	1/4	0.233	25.8	7.10	22.8	39.9	96.9	19.4	3.69	23.6
	3/16	0.174	19.6	5.37	31.5	54.5	74.6	14.9	3.73	18.0
HSS10×5	3/8	0.349	35.1	9.67	11.3	25.7	120	24.1	3.53	30.4
	5/16	0.291	29.7	8.17	14.2	31.4	104	20.8	3.56	26.0
	1/4	0.233	24.1	6.63	18.5	39.9	85.8	17.2	3.60	21.3
	3/16	0.174	18.3	5.02	25.7	54.5	66.2	13.2	3.63	16.3
HSS10×4	5/8	0.581	50.6	14.0	3.88	14.2	149	29.9	3.26	40.3
	1/2	0.465	41.9	11.6	5.60	18.5	129	25.8	3.34	34.1
	3/8	0.349	32.5	8.97	8.46	25.7	104	20.8	3.41	27.0
	5/16	0.291	27.5	7.59	10.7	31.4	90.1	18.0	3.44	23.1
	1/4	0.233	22.4	6.17	14.2	39.9	74.7	14.9	3.48	19.0
	3/16	0.174	17.1	4.67	20.0	54.5	57.8	11.6	3.52	14.6
HSS10×3 1/2	3/16	0.174	16.4	4.50	17.1	54.5	53.6	10.7	3.45	13.7
HSS10×3	3/8	0.349	30.0	8.27	5.60	25.7	88.0	17.6	3.26	23.7
	5/16	0.291	25.4	7.01	7.31	31.4	76.3	15.3	3.30	20.3
	1/4	0.233	20.7	5.70	9.88	39.9	63.6	12.7	3.34	16.7
	3/16	0.174	15.8	4.32	14.2	54.5	49.4	9.87	3.38	12.8
	1/8	0.116	10.7	2.93	22.9	83.2	34.2	6.83	3.42	8.80
HSS10×2	3/8	0.349	27.4	7.58	2.73	25.7	71.7	14.3	3.08	20.3
	5/16	0.291	23.3	6.43	3.87	31.4	62.6	12.5	3.12	17.5
	1/4	0.233	19.0	5.24	5.58	39.9	52.5	10.5	3.17	14.4
	3/16	0.174	14.5	3.98	8.49	54.5	41.0	8.19	3.21	11.1
HSS9×7	5/8	0.581	59.1	16.4	9.05	12.5	174	38.7	3.26	48.3
	1/2	0.465	48.7	13.5	12.1	16.4	149	33.0	3.32	40.5
	3/8	0.349	37.6	10.4	17.1	22.8	119	26.4	3.38	31.8
	5/16	0.291	31.8	8.76	21.1	27.9	102	22.6	3.41	27.1
	1/4	0.233	25.8	7.10	27.0	35.6	84.1	18.7	3.44	22.2
	3/16	0.174	19.6	5.37	37.2	48.7	64.7	14.4	3.47	16.9
HSS9×5	5/8	0.581	50.6	14.0	5.61	12.5	133	29.6	3.08	38.5
	1/2	0.465	41.9	11.6	7.75	16.4	115	25.5	3.14	32.5
	3/8	0.349	32.5	8.97	11.3	22.8	92.5	20.5	3.21	25.7
	5/16	0.291	27.5	7.59	14.2	27.9	79.8	17.7	3.24	22.0
	1/4	0.233	22.4	6.17	18.5	35.6	66.1	14.7	3.27	18.1
	3/16	0.174	17.1	4.67	25.7	48.7	51.1	11.4	3.31	13.8
HSS9×3	1/2	0.465	35.1	9.74	3.45	16.4	80.8	18.0	2.88	24.6
	3/8	0.349	27.4	7.58	5.60	22.8	66.3	14.7	2.96	19.7
	5/16	0.291	23.3	6.43	7.31	27.9	57.7	12.8	3.00	16.9
	1/4	0.233	19.0	5.24	9.88	35.6	48.2	10.7	3.04	14.0
	3/16	0.174	14.5	3.98	14.2	48.7	37.6	8.35	3.07	10.8
HSS8×8	5/8	0.581	59.1	16.4	10.8	10.8	146	36.5	2.99	44.7
	1/2	0.465	48.7	13.5	14.2	14.2	125	31.2	3.04	37.5
	3/8	0.349	37.6	10.4	19.9	19.9	99.6	24.9	3.10	29.4
	5/16	0.291	31.8	8.76	24.5	24.5	85.6	21.4	3.13	25.1
	1/4	0.233	25.8	7.10	31.3	31.3	70.7	17.7	3.15	20.5
	3/16	0.174	19.6	5.37	43.0	43.0	54.4	13.6	3.18	15.7
HSS8×6	5/8	0.581	50.6	14.0	7.33	10.8	114	28.5	2.85	36.1
	1/2	0.465	41.9	11.6	9.90	14.2	98.2	24.6	2.91	30.5
	3/8	0.349	32.5	8.97	14.2	19.9	79.1	19.8	2.97	24.1
	5/16	0.291	27.5	7.59	17.6	24.5	68.3	17.1	3.00	20.6
	1/4	0.233	22.4	6.17	22.8	31.3	56.6	14.2	3.03	16.9
	3/16	0.174	17.1	4.67	31.5	43.0	43.7	10.9	3.06	13.0

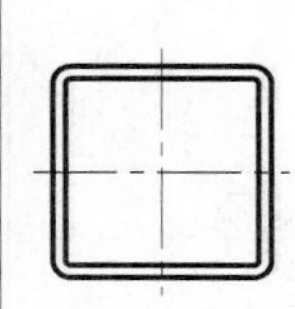

Table 1-11 (cont.). Rectangular (and Square) HSS Dimensions and Properties

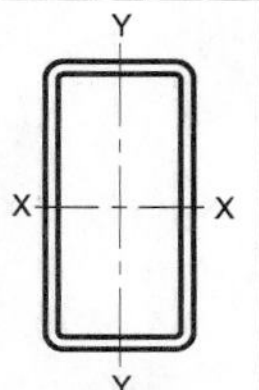

Shape	Wall Thickness, *t*		Axis Y-Y				Torsion		Surface Area Per Foot
	nominal	design	*I*	*S*	*r*	*Z*	*J*	*C*	
	in.	in.	in.^{4}	in.^{3}	in.	in.^{3}	in.^{4}	in.^{3}	ft^2
HSS10×6	5/8	0.581	89.4	29.8	2.34	35.8	209	58.6	2.50
	1/2	0.465	76.8	25.6	2.39	30.1	176	48.7	2.53
	3/8	0.349	61.8	20.6	2.44	23.7	139	37.9	2.57
	5/16	0.291	53.3	17.8	2.47	20.2	118	32.2	2.58
	1/4	0.233	44.1	14.7	2.49	16.6	96.7	26.2	2.60
	3/16	0.174	34.1	11.4	2.52	12.7	73.8	19.9	2.62
HSS10×5	3/8	0.349	40.6	16.2	2.05	18.7	100	31.2	2.40
	5/16	0.291	35.2	14.1	2.07	16.0	86.0	26.5	2.42
	1/4	0.233	29.3	11.7	2.10	13.2	70.7	21.6	2.43
	3/16	0.174	22.7	9.09	2.13	10.1	54.1	16.5	2.45
HSS10×4	5/8	0.581	33.5	16.8	1.54	20.6	95.7	36.7	2.17
	1/2	0.465	29.5	14.7	1.59	17.6	82.6	31.0	2.20
	3/8	0.349	24.3	12.1	1.64	14.0	66.5	24.4	2.23
	5/16	0.291	21.2	10.6	1.67	12.1	57.3	20.9	2.25
	1/4	0.233	17.7	8.87	1.70	9.96	47.4	17.1	2.27
	3/16	0.174	13.9	6.93	1.72	7.66	36.5	13.1	2.28
HSS10×3 1/2	3/16	0.174	10.3	5.89	1.51	6.52	28.6	11.4	2.20
HSS10×3	3/8	0.349	12.4	8.28	1.22	9.73	37.8	17.7	2.07
	5/16	0.291	11.0	7.30	1.25	8.42	33.0	15.2	2.08
	1/4	0.233	9.28	6.19	1.28	6.99	27.6	12.5	2.10
	3/16	0.174	7.33	4.89	1.30	5.41	21.5	9.64	2.12
	1/8	0.116	5.16	3.44	1.33	3.74	14.9	6.61	2.13
HSS10×2	3/8	0.349	4.70	4.70	0.787	5.76	15.9	11.0	1.90
	5/16	0.291	4.24	4.24	0.812	5.06	14.2	9.56	1.92
	1/4	0.233	3.67	3.67	0.838	4.26	12.2	7.99	1.93
	3/16	0.174	2.97	2.97	0.864	3.34	9.74	6.22	1.95
HSS9×7	5/8	0.581	117	33.5	2.68	40.5	235	62.0	2.50
	1/2	0.465	100	28.7	2.73	34.0	197	51.5	2.53
	3/8	0.349	80.4	23.0	2.78	26.7	154	40.0	2.57
	5/16	0.291	69.2	19.8	2.81	22.8	131	33.9	2.58
	1/4	0.233	57.2	16.3	2.84	18.7	107	27.6	2.60
	3/16	0.174	44.1	12.6	2.87	14.3	81.7	20.9	2.62
HSS9×5	5/8	0.581	52.0	20.8	1.92	25.3	128	42.5	2.17
	1/2	0.465	45.2	18.1	1.97	21.5	109	35.6	2.20
	3/8	0.349	36.8	14.7	2.03	17.1	86.9	27.9	2.23
	5/16	0.291	32.0	12.8	2.05	14.6	74.4	23.8	2.25
	1/4	0.233	26.6	10.6	2.08	12.0	61.2	19.4	2.27
	3/16	0.174	20.7	8.28	2.10	9.25	46.9	14.8	2.28
HSS9×3	1/2	0.465	13.2	8.81	1.17	10.8	40.0	19.7	1.87
	3/8	0.349	11.2	7.45	1.21	8.80	33.1	15.8	1.90
	5/16	0.291	9.88	6.59	1.24	7.63	28.9	13.6	1.92
	1/4	0.233	8.38	5.59	1.27	6.35	24.2	11.3	1.93
	3/16	0.174	6.64	4.42	1.29	4.92	18.9	8.66	1.95
HSS8×8	5/8	0.581	146	36.5	2.99	44.7	244	63.2	2.50
	1/2	0.465	125	31.2	3.04	37.5	204	52.4	2.53
	3/8	0.349	99.6	24.9	3.10	29.4	160	40.7	2.57
	5/16	0.291	85.6	21.4	3.13	25.1	136	34.5	2.58
	1/4	0.233	70.7	17.7	3.15	20.5	111	28.1	2.60
	3/16	0.174	54.4	13.6	3.18	15.7	84.5	21.3	2.62
HSS8×6	5/8	0.581	72.3	24.1	2.27	29.5	150	46.0	2.17
	1/2	0.465	62.5	20.8	2.32	24.9	127	38.4	2.20
	3/8	0.349	50.6	16.9	2.38	19.8	100	30.0	2.23
	5/16	0.291	43.8	14.6	2.40	16.9	85.8	25.5	2.25
	1/4	0.233	36.4	12.1	2.43	13.9	70.3	20.8	2.27
	3/16	0.174	28.2	9.39	2.46	10.7	53.7	15.8	2.28

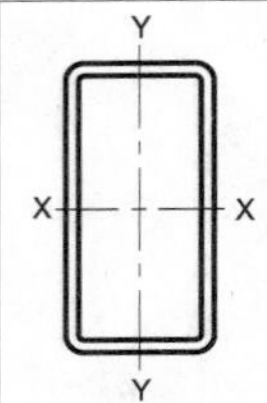

Table 1-11 (cont.). Rectangular (and Square) HSS Dimensions and Properties

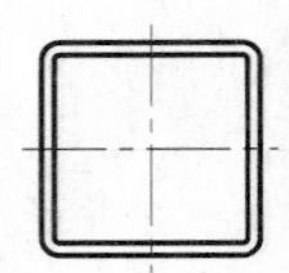

Shape	Wall Thickness, t nominal	design	Nominal Wt.	Area, A	$\frac{b}{t}$	$\frac{h}{t}$	Axis X-X I	S	r	Z
	in.	in.	lb/ft	in.^{2}			in.^{4}	in.^{3}	in.	in.^{3}
HSS8×4	5/8	0.581	42.1	11.7	3.88	10.8	82.0	20.5	2.64	27.4
	1/2	0.465	35.1	9.74	5.60	14.2	71.8	17.9	2.71	23.5
	3/8	0.349	27.4	7.58	8.46	19.9	58.7	14.7	2.78	18.8
	5/16	0.291	23.3	6.43	10.7	24.5	51.0	12.8	2.82	16.1
	1/4	0.233	19.0	5.24	14.2	31.3	42.5	10.6	2.85	13.3
	3/16	0.174	14.5	3.98	20.0	43.0	33.1	8.27	2.88	10.2
	1/8	0.116	9.85	2.70	31.5	66.0	22.9	5.73	2.92	7.02
HSS8×3	1/2	0.465	31.7	8.81	3.45	14.2	58.6	14.6	2.58	20.0
	3/8	0.349	24.9	6.88	5.60	19.9	48.5	12.1	2.65	16.1
	5/16	0.291	21.2	5.85	7.31	24.5	42.4	10.6	2.69	13.9
	1/4	0.233	17.3	4.77	9.88	31.3	35.5	8.88	2.73	11.5
	3/16	0.174	13.2	3.63	14.2	43.0	27.8	6.94	2.77	8.87
	1/8	0.116	9.00	2.46	22.9	66.0	19.3	4.83	2.80	6.11
HSS8×2	3/8	0.349	22.3	6.18	2.73	19.9	38.2	9.56	2.49	13.4
	5/16	0.291	19.0	5.26	3.87	24.5	33.7	8.43	2.53	11.6
	1/4	0.233	15.6	4.30	5.58	31.3	28.5	7.12	2.57	9.68
	3/16	0.174	12.0	3.28	8.49	43.0	22.4	5.61	2.61	7.51
	1/8	0.116	8.15	2.23	14.2	66.0	15.7	3.93	2.65	5.19
HSS7×7	5/8	0.581	50.6	14.0	9.05	9.05	93.4	26.7	2.58	33.1
	1/2	0.465	41.9	11.6	12.1	12.1	80.5	23.0	2.63	27.9
	3/8	0.349	32.5	8.97	17.1	17.1	65.0	18.6	2.69	22.1
	5/16	0.291	27.5	7.59	21.1	21.1	56.1	16.0	2.72	18.9
	1/4	0.233	22.4	6.17	27.0	27.0	46.5	13.3	2.75	15.5
	3/16	0.174	17.1	4.67	37.2	37.2	36.0	10.3	2.77	11.9
HSS7×5	5/8	0.581	42.1	11.7	5.61	9.05	69.4	19.8	2.43	25.6
	1/2	0.465	35.1	9.74	7.75	12.1	60.6	17.3	2.50	21.9
	3/8	0.349	27.4	7.58	11.3	17.1	49.5	14.1	2.56	17.5
	5/16	0.291	23.3	6.43	14.2	21.1	43.0	12.3	2.59	15.0
	1/4	0.233	19.0	5.24	18.5	27.0	35.9	10.2	2.62	12.4
	3/16	0.174	14.5	3.98	25.7	37.2	27.9	7.96	2.65	9.52
	1/8	0.116	9.85	2.70	40.1	57.3	19.3	5.52	2.68	6.53
HSS7×4	1/2	0.465	31.7	8.81	5.6	12.1	50.7	14.5	2.40	18.8
	3/8	0.349	24.9	6.88	8.46	17.1	41.8	11.9	2.46	15.1
	5/16	0.291	21.2	5.85	10.7	21.1	36.5	10.4	2.50	13.1
	1/4	0.233	17.3	4.77	14.2	27.0	30.5	8.72	2.53	10.8
	3/16	0.174	13.2	3.63	20.0	37.2	23.8	6.81	2.56	8.33
	1/8	0.116	9.00	2.46	31.5	57.3	16.6	4.73	2.59	5.73
HSS7×3	1/2	0.465	28.3	7.88	3.45	12.1	40.7	11.6	2.27	15.8
	3/8	0.349	22.3	6.18	5.60	17.1	34.1	9.73	2.35	12.8
	5/16	0.291	19.0	5.26	7.31	21.1	29.9	8.54	2.38	11.1
	1/4	0.233	15.6	4.30	9.88	27.0	25.2	7.19	2.42	9.22
	3/16	0.174	12.0	3.28	14.2	37.2	19.8	5.65	2.45	7.14
	1/8	0.116	8.15	2.23	22.9	57.3	13.8	3.95	2.49	4.93
HSS6×6	5/8	0.581	42.1	11.7	7.33	7.33	55.2	18.4	2.17	23.2
	1/2	0.465	35.1	9.74	9.90	9.90	48.3	16.1	2.23	19.8
	3/8	0.349	27.4	7.58	14.2	14.2	39.5	13.2	2.28	15.8
	5/16	0.291	23.3	6.43	17.6	17.6	34.3	11.4	2.31	13.6
	1/4	0.233	19.0	5.24	22.8	22.8	28.6	9.54	2.34	11.2
	3/16	0.174	14.5	3.98	31.5	31.5	22.3	7.42	2.37	8.63
	1/8	0.116	9.85	2.70	48.7	48.7	15.5	5.15	2.39	5.92
HSS6×5	3/8	0.349	24.9	6.88	11.3	14.2	33.9	11.3	2.22	13.8
	5/16	0.291	21.2	5.85	14.2	17.6	29.6	9.85	2.25	11.9
	1/4	0.233	17.3	4.77	18.5	22.8	24.7	8.25	2.28	9.87
	3/16	0.174	13.2	3.63	25.7	31.5	19.3	6.44	2.31	7.62

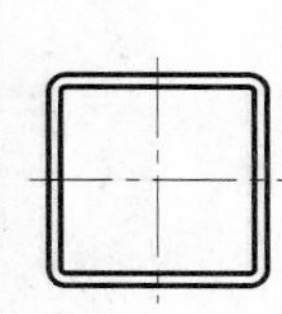

Table 1-11 (cont.). Rectangular (and Square) HSS Dimensions and Properties

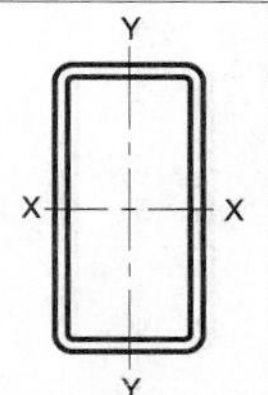

Shape	Wall Thickness, t		Axis Y-Y				Torsion		Surface Area Per Foot
	nominal	design	I	S	r	Z	J	C	
	in.	in.	in.^{4}	in.^{3}	in.	in.^{3}	in.^{4}	in.^{3}	ft^2
HSS8×4	5/8	0.581	26.6	13.3	1.51	16.6	70.3	28.7	1.83
	1/2	0.465	23.6	11.8	1.56	14.3	61.1	24.4	1.87
	3/8	0.349	19.6	9.80	1.61	11.5	49.3	19.3	1.90
	5/16	0.291	17.2	8.58	1.63	9.91	42.6	16.5	1.92
	1/4	0.233	14.4	7.21	1.66	8.20	35.3	13.6	1.93
	3/16	0.174	11.3	5.65	1.69	6.33	27.2	10.4	1.95
	1/8	0.116	7.90	3.95	1.71	4.36	18.7	7.10	1.97
HSS8×3	1/2	0.465	11.7	7.81	1.15	9.64	34.3	17.4	1.70
	3/8	0.349	9.95	6.63	1.20	7.88	28.5	14.0	1.73
	5/16	0.291	8.81	5.87	1.23	6.84	24.9	12.1	1.75
	1/4	0.233	7.49	4.99	1.25	5.70	20.8	9.97	1.77
	3/16	0.174	5.94	3.96	1.28	4.43	16.2	7.68	1.78
	1/8	0.116	4.20	2.80	1.31	3.07	11.3	5.27	1.80
HSS8×2	3/8	0.349	3.73	3.73	0.777	4.61	12.1	8.65	1.57
	5/16	0.291	3.38	3.38	0.802	4.06	10.9	7.57	1.58
	1/4	0.233	2.94	2.94	0.827	3.43	9.36	6.35	1.60
	3/16	0.174	2.39	2.39	0.853	2.70	7.48	4.95	1.62
	1/8	0.116	1.72	1.72	0.879	1.90	5.30	3.44	1.63
HSS7×7	5/8	0.581	93.4	26.7	2.58	33.1	158	47.1	2.17
	1/2	0.465	80.5	23.0	2.63	27.9	133	39.3	2.20
	3/8	0.349	65.0	18.6	2.69	22.1	105	30.7	2.23
	5/16	0.291	56.1	16.0	2.72	18.9	89.7	26.1	2.25
	1/4	0.233	46.5	13.3	2.75	15.5	73.5	21.3	2.27
	3/16	0.174	36.0	10.3	2.77	11.9	56.1	16.2	2.28
HSS7×5	5/8	0.581	40.6	16.2	1.86	20.2	88.5	32.2	1.83
	1/2	0.465	35.6	14.2	1.91	17.3	75.8	27.2	1.87
	3/8	0.349	29.3	11.7	1.97	13.8	60.6	21.4	1.90
	5/16	0.291	25.5	10.2	1.99	11.9	52.1	18.3	1.92
	1/4	0.233	21.3	8.53	2.02	9.83	42.9	15.0	1.93
	3/16	0.174	16.6	6.65	2.05	7.57	32.9	11.4	1.95
	1/8	0.116	11.6	4.63	2.07	5.20	22.5	7.79	1.97
HSS7×4	1/2	0.465	20.7	10.4	1.53	12.6	50.5	21.1	1.70
	3/8	0.349	17.3	8.63	1.58	10.2	41.0	16.8	1.73
	5/16	0.291	15.2	7.58	1.61	8.83	35.4	14.4	1.75
	1/4	0.233	12.8	6.38	1.64	7.33	29.3	11.8	1.77
	3/16	0.174	10.0	5.02	1.66	5.67	22.7	9.07	1.78
	1/8	0.116	7.03	3.51	1.69	3.91	15.6	6.20	1.80
HSS7×3	1/2	0.465	10.2	6.80	1.14	8.46	28.6	15.0	1.53
	3/8	0.349	8.71	5.81	1.19	6.95	23.9	12.1	1.57
	5/16	0.291	7.74	5.16	1.21	6.05	20.9	10.5	1.58
	1/4	0.233	6.60	4.40	1.24	5.06	17.5	8.68	1.60
	3/16	0.174	5.24	3.50	1.26	3.94	13.7	6.69	1.62
	1/8	0.116	3.71	2.48	1.29	2.73	9.48	4.60	1.63
HSS6×6	5/8	0.581	55.2	18.4	2.17	23.2	94.9	33.4	1.83
	1/2	0.465	48.3	16.1	2.23	19.8	81.1	28.1	1.87
	3/8	0.349	39.5	13.2	2.28	15.8	64.6	22.1	1.90
	5/16	0.291	34.3	11.4	2.31	13.6	55.4	18.9	1.92
	1/4	0.233	28.6	9.54	2.34	11.2	45.6	15.4	1.93
	3/16	0.174	22.3	7.42	2.37	8.63	35.0	11.8	1.95
	1/8	0.116	15.5	5.15	2.39	5.92	23.9	8.03	1.97
HSS6×5	3/8	0.349	25.5	10.2	1.92	12.2	48.1	18.2	1.73
	5/16	0.291	22.3	8.91	1.95	10.5	41.4	15.6	1.75
	1/4	0.233	18.7	7.47	1.98	8.72	34.2	12.8	1.77
	3/16	0.174	14.6	5.84	2.01	6.73	26.3	9.76	1.78

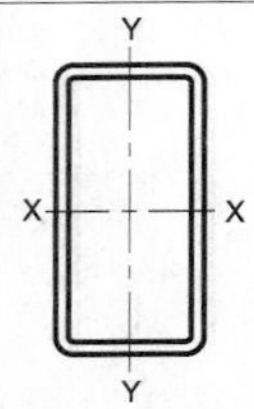

Table 1-11 (cont.). Rectangular (and Square) HSS Dimensions and Properties

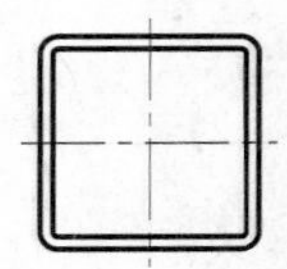

Shape	Wall Thickness, t		Nom-inal Wt.	Area, A	$\frac{b}{t}$	$\frac{h}{t}$	Axis X-X			
	nominal	design					I	S	r	Z
	in.	in.	lb/ft	in.^{2}			in.^{4}	in.^{3}	in.	in.^{3}
HSS6×4	1/2	0.465	28.3	7.88	5.60	9.90	34.0	11.3	2.08	14.6
	3/8	0.349	22.3	6.18	8.46	14.2	28.3	9.43	2.14	11.9
	5/16	0.291	19.0	5.26	10.7	17.6	24.8	8.27	2.17	10.3
	1/4	0.233	15.6	4.30	14.2	22.8	20.9	6.96	2.20	8.53
	3/16	0.174	12.0	3.28	20.0	31.5	16.4	5.46	2.23	6.60
	1/8	0.116	8.15	2.23	31.5	48.7	11.4	3.81	2.26	4.56
HSS6×3	1/2	0.465	24.9	6.95	3.45	9.90	26.8	8.95	1.97	12.1
	3/8	0.349	19.7	5.48	5.60	14.2	22.7	7.57	2.04	9.90
	5/16	0.291	16.9	4.68	7.31	17.6	20.1	6.69	2.07	8.61
	1/4	0.233	13.9	3.84	9.88	22.8	17.0	5.66	2.10	7.19
	3/16	0.174	10.7	2.93	14.2	31.5	13.4	4.47	2.14	5.59
	1/8	0.116	7.30	2.00	22.9	48.7	9.43	3.14	2.17	3.87
HSS6×2	3/8	0.349	17.2	4.78	2.73	14.2	17.1	5.71	1.89	7.93
	5/16	0.291	14.8	4.10	3.87	17.6	15.3	5.11	1.93	6.95
	1/4	0.233	12.2	3.37	5.58	22.8	13.1	4.37	1.97	5.84
	3/16	0.174	9.40	2.58	8.49	31.5	10.5	3.49	2.01	4.58
	1/8	0.116	6.45	1.77	14.2	48.7	7.42	2.47	2.05	3.19
HSS5 1/2×5 1/2	3/8	0.349	24.9	6.88	12.8	12.8	29.7	10.8	2.08	13.1
	5/16	0.291	21.2	5.85	15.9	15.9	25.9	9.43	2.11	11.3
	1/4	0.233	17.3	4.77	20.6	20.6	21.7	7.90	2.13	9.32
	3/16	0.174	13.2	3.63	28.6	28.6	17.0	6.17	2.16	7.19
	1/8	0.116	9.0	2.46	44.4	44.4	11.8	4.30	2.19	4.95
HSS5×5	1/2	0.465	28.3	7.88	7.75	7.75	26.0	10.4	1.82	13.1
	3/8	0.349	22.3	6.18	11.3	11.3	21.7	8.68	1.87	10.6
	5/16	0.291	19.0	5.26	14.2	14.2	19.0	7.62	1.90	9.16
	1/4	0.233	15.6	4.30	18.5	18.5	16.0	6.41	1.93	7.61
	3/16	0.174	12.0	3.28	25.7	25.7	12.6	5.03	1.96	5.89
	1/8	0.116	8.15	2.23	40.1	40.1	8.80	3.52	1.99	4.07
HSS5×4	1/2	0.465	24.9	6.95	5.60	7.75	21.2	8.49	1.75	10.9
	3/8	0.349	19.7	5.48	8.46	11.3	17.9	7.17	1.81	8.96
	5/16	0.291	16.9	4.68	10.7	14.2	15.8	6.32	1.84	7.79
	1/4	0.233	13.9	3.84	14.2	18.5	13.4	5.35	1.87	6.49
	3/16	0.174	10.7	2.93	20.0	25.7	10.6	4.22	1.90	5.05
HSS5×3	1/2	0.465	21.5	6.02	3.45	7.75	16.4	6.57	1.65	8.83
	3/8	0.349	17.2	4.78	5.60	11.3	14.1	5.65	1.72	7.34
	5/16	0.291	14.8	4.10	7.31	14.2	12.6	5.03	1.75	6.42
	1/4	0.233	12.2	3.37	9.88	18.5	10.7	4.29	1.78	5.38
	3/16	0.174	9.40	2.58	14.2	25.7	8.53	3.41	1.82	4.21
	1/8	0.116	6.45	1.77	22.9	40.1	6.03	2.41	1.85	2.93
HSS5×2 1/2	1/4	0.233	11.3	3.14	7.73	18.5	9.40	3.76	1.73	4.83
	3/16	0.174	8.77	2.41	11.4	25.7	7.51	3.01	1.77	3.79
	1/8	0.116	6.02	1.65	18.6	40.1	5.34	2.14	1.80	2.65
HSS5×2	3/8	0.349	14.6	4.09	2.73	11.3	10.4	4.14	1.59	5.71
	5/16	0.291	12.7	3.52	3.87	14.2	9.35	3.74	1.63	5.05
	1/4	0.233	10.5	2.91	5.58	18.5	8.08	3.23	1.67	4.27
	3/16	0.174	8.13	2.24	8.49	25.7	6.50	2.60	1.70	3.37
	1/8	0.116	5.60	1.54	14.2	40.1	4.65	1.86	1.74	2.37
HSS4 1/2×4 1/2	1/2	0.465	24.9	6.95	6.68	6.68	18.1	8.03	1.61	10.2
	3/8	0.349	19.7	5.48	9.89	9.89	15.3	6.79	1.67	8.36
	5/16	0.291	16.9	4.68	12.5	12.5	13.5	6.00	1.70	7.27
	1/4	0.233	13.9	3.84	16.3	16.3	11.4	5.08	1.73	6.06
	3/16	0.174	10.7	2.93	22.9	22.9	9.02	4.01	1.75	4.71
	1/8	0.116	7.30	2.00	35.8	35.8	6.35	2.82	1.78	3.27

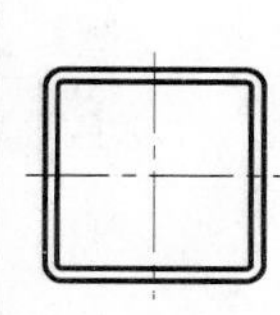

Table 1-11 (cont.). Rectangular (and Square) HSS Dimensions and Properties

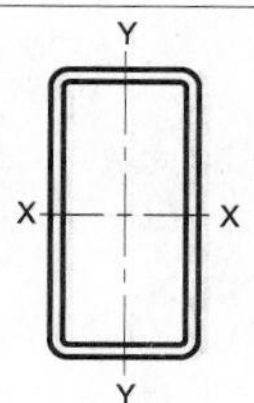

Shape	Wall Thickness, t		Axis Y-Y				Torsion		Surface Area Per Foot
	nominal	design	I	S	r	Z	J	C	
	in.	in.	in.^{4}	in.^{3}	in.	in.^{3}	in.^{4}	in.^{3}	ft^2
HSS6×4	1/2	0.465	17.8	8.89	1.50	11.0	40.3	17.8	1.53
	3/8	0.349	14.9	7.47	1.55	8.94	32.8	14.2	1.57
	5/16	0.291	13.2	6.58	1.58	7.75	28.4	12.2	1.58
	1/4	0.233	11.1	5.56	1.61	6.45	23.6	10.1	1.60
	3/16	0.174	8.76	4.38	1.63	5.00	18.2	7.74	1.62
	1/8	0.116	6.15	3.08	1.66	3.46	12.6	5.30	1.63
HSS6×3	1/2	0.465	8.69	5.79	1.12	7.28	23.1	12.7	1.37
	3/8	0.349	7.48	4.99	1.17	6.03	19.3	10.3	1.40
	5/16	0.291	6.67	4.45	1.19	5.27	16.9	8.91	1.42
	1/4	0.233	5.70	3.80	1.22	4.41	14.2	7.39	1.43
	3/16	0.174	4.55	3.03	1.25	3.45	11.1	5.71	1.45
	1/8	0.116	3.23	2.15	1.27	2.40	7.73	3.93	1.47
HSS6×2	3/8	0.349	2.77	2.77	0.760	3.46	8.42	6.35	1.23
	5/16	0.291	2.52	2.52	0.785	3.07	7.60	5.58	1.25
	1/4	0.233	2.21	2.21	0.810	2.61	6.55	4.70	1.27
	3/16	0.174	1.80	1.80	0.836	2.07	5.24	3.68	1.28
	1/8	0.116	1.31	1.31	0.861	1.46	3.72	2.57	1.30
HSS5 1/2×5 1/2	3/8	0.349	29.7	10.8	2.08	13.1	49.0	18.4	1.73
	5/16	0.291	25.9	9.43	2.11	11.3	42.2	15.7	1.75
	1/4	0.233	21.7	7.90	2.13	9.32	34.8	12.9	1.77
	3/16	0.174	17.0	6.17	2.16	7.19	26.7	9.85	1.78
	1/8	0.116	11.8	4.30	2.19	4.95	18.3	6.72	1.80
HSS5×5	1/2	0.465	26.0	10.4	1.82	13.1	44.6	18.7	1.53
	3/8	0.349	21.7	8.68	1.87	10.6	36.1	14.9	1.57
	5/16	0.291	19.0	7.62	1.90	9.16	31.2	12.8	1.58
	1/4	0.233	16.0	6.41	1.93	7.61	25.8	10.5	1.60
	3/16	0.174	12.6	5.03	1.96	5.89	19.9	8.08	1.62
	1/8	0.116	8.80	3.52	1.99	4.07	13.7	5.53	1.63
HSS5×4	1/2	0.465	14.9	7.43	1.46	9.35	30.3	14.5	1.37
	3/8	0.349	12.6	6.30	1.52	7.67	24.9	11.7	1.40
	5/16	0.291	11.1	5.57	1.54	6.67	21.7	10.1	1.42
	1/4	0.233	9.46	4.73	1.57	5.57	18.0	8.32	1.43
	3/16	0.174	7.48	3.74	1.60	4.34	14.0	6.41	1.45
HSS5×3	1/2	0.465	7.18	4.78	1.09	6.10	17.6	10.3	1.20
	3/8	0.349	6.25	4.16	1.14	5.10	14.9	8.44	1.23
	5/16	0.291	5.60	3.73	1.17	4.48	13.1	7.33	1.25
	1/4	0.233	4.81	3.21	1.19	3.77	11.0	6.10	1.27
	3/16	0.174	3.85	2.57	1.22	2.96	8.64	4.73	1.28
	1/8	0.116	2.75	1.83	1.25	2.07	6.02	3.26	1.30
HSS5×2 1/2	1/4	0.233	3.13	2.50	0.999	2.95	7.93	4.99	1.18
	3/16	0.174	2.53	2.03	1.02	2.33	6.26	3.89	1.20
	1/8	0.116	1.82	1.46	1.05	1.64	4.40	2.70	1.22
HSS5×2	3/8	0.349	2.28	2.28	0.748	2.88	6.61	5.20	1.07
	5/16	0.291	2.10	2.10	0.772	2.57	5.99	4.59	1.08
	1/4	0.233	1.84	1.84	0.797	2.20	5.17	3.88	1.10
	3/16	0.174	1.51	1.51	0.823	1.75	4.15	3.05	1.12
	1/8	0.116	1.10	1.10	0.848	1.24	2.95	2.13	1.13
HSS4 1/2×4 1/2	1/2	0.465	18.1	8.03	1.61	10.2	31.3	14.8	1.37
	3/8	0.349	15.3	6.79	1.67	8.36	25.7	11.9	1.40
	5/16	0.291	13.5	6.00	1.70	7.27	22.3	10.2	1.42
	1/4	0.233	11.4	5.08	1.73	6.06	18.5	8.44	1.43
	3/16	0.174	9.02	4.01	1.75	4.71	14.4	6.49	1.45
	1/8	0.116	6.35	2.82	1.78	3.27	9.92	4.45	1.47

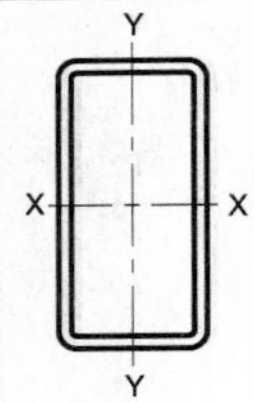

Table 1-11 (cont.). Rectangular (and Square) HSS Dimensions and Properties

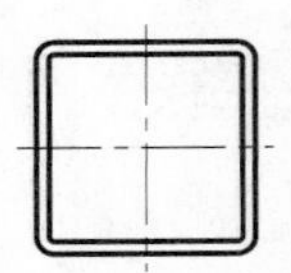

Shape	Wall Thickness, t		Nominal Wt.	Area, A	$\frac{b}{t}$	$\frac{h}{t}$	Axis X-X			
	nominal	design					I	S	r	Z
	in.	in.	lb/ft	in.^{2}			in.^{4}	in.^{3}	in.	in.^{3}
HSS4×4	1/2	0.465	21.5	6.02	5.60	5.60	11.9	5.97	1.41	7.70
	3/8	0.349	17.2	4.78	8.46	8.46	10.3	5.13	1.47	6.39
	5/16	0.291	14.8	4.10	10.7	10.7	9.14	4.57	1.49	5.59
	1/4	0.233	12.2	3.37	14.2	14.2	7.80	3.90	1.52	4.69
	3/16	0.174	9.40	2.58	20.0	20.0	6.21	3.10	1.55	3.67
	1/8	0.116	6.45	1.77	31.5	31.5	4.40	2.20	1.58	2.56
HSS4×3	3/8	0.349	14.6	4.09	5.60	8.46	7.93	3.97	1.39	5.12
	5/16	0.291	12.7	3.52	7.31	10.7	7.14	3.57	1.42	4.51
	1/4	0.233	10.5	2.91	9.88	14.2	6.15	3.07	1.45	3.81
	3/16	0.174	8.13	2.24	14.2	20.0	4.93	2.47	1.49	3.00
	1/8	0.116	5.60	1.54	22.9	31.5	3.52	1.76	1.52	2.11
HSS4×2 1/2	5/16	0.291	11.6	3.23	5.59	10.7	6.13	3.07	1.38	3.97
	1/4	0.233	9.63	2.67	7.73	14.2	5.32	2.66	1.41	3.38
	3/16	0.174	7.49	2.06	11.4	20.0	4.30	2.15	1.44	2.67
HSS4×2	3/8	0.349	12.1	3.39	2.73	8.46	5.60	2.80	1.29	3.84
	5/16	0.291	10.5	2.94	3.87	10.7	5.13	2.56	1.32	3.43
	1/4	0.233	8.78	2.44	5.58	14.2	4.49	2.25	1.36	2.94
	3/16	0.174	6.85	1.89	8.49	20.0	3.66	1.83	1.39	2.34
	1/8	0.116	4.75	1.30	14.2	31.5	2.65	1.32	1.43	1.66
HSS3 1/2×3 1/2	3/8	0.349	14.6	4.09	7.03	7.03	6.49	3.71	1.26	4.69
	5/16	0.291	12.7	3.52	9.03	9.03	5.84	3.34	1.29	4.14
	1/4	0.233	10.5	2.91	12.0	12.0	5.04	2.88	1.32	3.50
	3/16	0.174	8.13	2.24	17.1	17.1	4.05	2.31	1.35	2.76
	1/8	0.116	5.60	1.54	27.2	27.2	2.90	1.66	1.37	1.93
HSS3 1/2×2 1/2	3/8	0.349	12.1	3.39	4.16	7.03	4.75	2.72	1.18	3.59
	5/16	0.291	10.5	2.94	5.59	9.03	4.34	2.48	1.22	3.20
	1/4	0.233	8.78	2.44	7.73	12.0	3.79	2.17	1.25	2.74
	3/16	0.174	6.85	1.89	11.4	17.1	3.09	1.76	1.28	2.18
	1/8	0.116	4.75	1.30	18.6	27.2	2.23	1.28	1.31	1.54
HSS3×3	3/8	0.349	12.1	3.39	5.60	5.60	3.78	2.52	1.06	3.25
	5/16	0.291	10.5	2.94	7.31	7.31	3.45	2.30	1.08	2.90
	1/4	0.233	8.78	2.44	9.88	9.88	3.02	2.01	1.11	2.48
	3/16	0.174	6.85	1.89	14.2	14.2	2.46	1.64	1.14	1.97
	1/8	0.116	4.75	1.30	22.9	22.9	1.78	1.19	1.17	1.40
HSS3×2 1/2	5/16	0.291	9.46	2.64	5.59	7.31	2.92	1.94	1.05	2.51
	1/4	0.233	7.93	2.21	7.73	9.88	2.57	1.72	1.08	2.16
	3/16	0.174	6.21	1.71	11.4	14.2	2.11	1.41	1.11	1.73
	1/8	0.116	4.32	1.19	18.6	22.9	1.54	1.03	1.14	1.23
HSS3×2	5/16	0.291	8.40	2.35	3.87	7.31	2.38	1.59	1.01	2.11
	1/4	0.233	7.08	1.97	5.58	9.88	2.13	1.42	1.04	1.83
	3/16	0.174	5.57	1.54	8.49	14.2	1.77	1.18	1.07	1.48
	1/8	0.116	3.90	1.07	14.2	22.9	1.30	0.867	1.10	1.06
HSS3×1 1/2	1/4	0.233	6.23	1.74	3.44	9.88	1.68	1.12	0.982	1.51
	3/16	0.174	4.94	1.37	5.62	14.2	1.42	0.945	1.02	1.24
	1/8	0.116	3.47	0.956	9.93	22.9	1.06	0.706	1.05	0.895
HSS3×1	1/8	0.116	3.04	0.840	5.62	22.9	0.817	0.545	0.987	0.728
HSS2 1/2×2 1/2	5/16	0.291	8.40	2.35	5.59	5.59	1.82	1.46	0.880	1.88
	1/4	0.233	7.08	1.97	7.73	7.73	1.63	1.30	0.908	1.63
	3/16	0.174	5.57	1.54	11.4	11.4	1.35	1.08	0.937	1.32
	1/8	0.116	3.90	1.07	18.6	18.6	0.998	0.799	0.965	0.947
HSS2 1/2×1 1/2	1/4	0.233	5.38	1.51	3.44	7.73	1.03	0.822	0.826	1.11
	3/16	0.174	4.30	1.19	5.62	11.4	0.882	0.705	0.860	0.915
	1/8	0.116	3.04	0.840	9.93	18.6	0.668	0.535	0.892	0.671

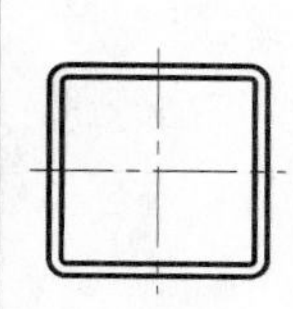

Table 1-11 (cont.). Rectangular (and Square) HSS Dimensions and Properties

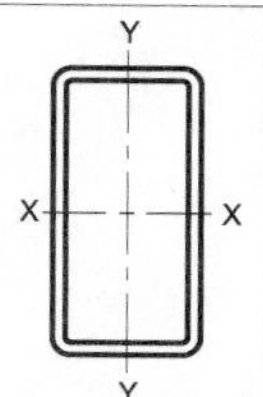

Shape	Wall Thickness, t		Axis Y-Y				Torsion		Surface Area Per Foot
	nominal	design	I	S	r	Z	J	C	
	in.	in.	in.^{4}	in.^{3}	in.	in.^{3}	in.^{4}	in.^{3}	ft^2
HSS4×4	1/2	0.465	11.9	5.97	1.41	7.70	21.0	11.2	1.20
	3/8	0.349	10.3	5.13	1.47	6.39	17.5	9.14	1.23
	5/16	0.291	9.14	4.57	1.49	5.59	15.3	7.91	1.25
	1/4	0.233	7.80	3.90	1.52	4.69	12.8	6.56	1.27
	3/16	0.174	6.21	3.10	1.55	3.67	9.96	5.07	1.28
	1/8	0.116	4.40	2.20	1.58	2.56	6.91	3.49	1.30
HSS4×3	3/8	0.349	5.01	3.34	1.11	4.18	10.6	6.59	1.07
	5/16	0.291	4.52	3.02	1.13	3.69	9.41	5.75	1.08
	1/4	0.233	3.91	2.61	1.16	3.12	7.96	4.81	1.10
	3/16	0.174	3.16	2.10	1.19	2.46	6.26	3.74	1.12
	1/8	0.116	2.27	1.51	1.21	1.73	4.38	2.59	1.13
HSS4×2 1/2	5/16	0.291	2.89	2.32	0.947	2.85	6.77	4.67	1.00
	1/4	0.233	2.53	2.02	0.973	2.43	5.78	3.93	1.02
	3/16	0.174	2.06	1.65	0.999	1.93	4.59	3.08	1.03
HSS4×2	3/8	0.349	1.80	1.80	0.729	2.31	4.83	4.04	0.900
	5/16	0.291	1.67	1.67	0.754	2.08	4.40	3.59	0.917
	1/4	0.233	1.48	1.48	0.779	1.79	3.82	3.05	0.933
	3/16	0.174	1.22	1.22	0.804	1.43	3.08	2.41	0.950
	1/8	0.116	0.898	0.898	0.830	1.02	2.20	1.69	0.967
HSS3 1/2×3 1/2	3/8	0.349	6.49	3.71	1.26	4.69	11.2	6.77	1.07
	5/16	0.291	5.84	3.34	1.29	4.14	9.89	5.90	1.08
	1/4	0.233	5.04	2.88	1.32	3.50	8.35	4.92	1.10
	3/16	0.174	4.05	2.31	1.35	2.76	6.56	3.83	1.12
	1/8	0.116	2.90	1.66	1.37	1.93	4.58	2.65	1.13
HSS3 1/2×2 1/2	3/8	0.349	2.77	2.21	0.904	2.82	6.16	4.57	0.900
	5/16	0.291	2.54	2.03	0.930	2.52	5.53	4.03	0.917
	1/4	0.233	2.23	1.78	0.956	2.16	4.75	3.40	0.933
	3/16	0.174	1.82	1.46	0.983	1.72	3.78	2.67	0.950
	1/8	0.116	1.33	1.06	1.01	1.22	2.67	1.87	0.967
HSS3×3	3/8	0.349	3.78	2.52	1.06	3.25	6.64	4.74	0.900
	5/16	0.291	3.45	2.30	1.08	2.90	5.94	4.18	0.917
	1/4	0.233	3.02	2.01	1.11	2.48	5.08	3.52	0.933
	3/16	0.174	2.46	1.64	1.14	1.97	4.03	2.76	0.950
	1/8	0.116	1.78	1.19	1.17	1.40	2.84	1.92	0.967
HSS3×2 1/2	5/16	0.291	2.18	1.74	0.908	2.20	4.34	3.39	0.833
	1/4	0.233	1.93	1.54	0.935	1.90	3.74	2.87	0.850
	3/16	0.174	1.59	1.27	0.963	1.52	3.00	2.27	0.867
	1/8	0.116	1.16	0.931	0.990	1.09	2.13	1.59	0.883
HSS3×2	5/16	0.291	1.24	1.24	0.725	1.58	2.87	2.60	0.750
	1/4	0.233	1.11	1.11	0.751	1.38	2.52	2.23	0.767
	3/16	0.174	0.932	0.932	0.778	1.12	2.05	1.78	0.784
	1/8	0.116	0.692	0.692	0.804	0.803	1.47	1.25	0.800
HSS3×1 1/2	1/4	0.233	0.543	0.725	0.559	0.911	1.44	1.58	0.683
	3/16	0.174	0.467	0.622	0.584	0.752	1.21	1.28	0.700
	1/8	0.116	0.355	0.474	0.610	0.550	0.886	0.920	0.717
HSS3×1	1/8	0.116	0.138	0.276	0.405	0.325	0.408	0.585	0.633
HSS2 1/2×2 1/2	5/16	0.291	1.82	1.46	0.880	1.88	3.20	2.74	0.750
	1/4	0.233	1.63	1.30	0.908	1.63	2.79	2.35	0.767
	3/16	0.174	1.35	1.08	0.937	1.32	2.25	1.86	0.784
	1/8	0.116	0.998	0.799	0.965	0.947	1.61	1.31	0.800
HSS2 1/2×1 1/2	1/4	0.233	0.449	0.599	0.546	0.764	1.10	1.29	0.600
	3/16	0.174	0.390	0.520	0.572	0.636	0.929	1.05	0.617
	1/8	0.116	0.300	0.399	0.597	0.469	0.687	0.759	0.633

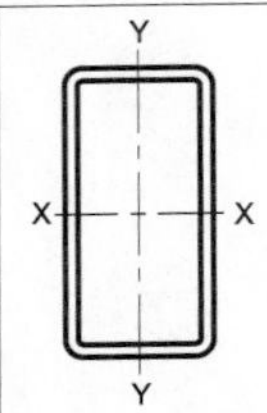

Table 1-11 (cont.). Rectangular (and Square) HSS Dimensions and Properties

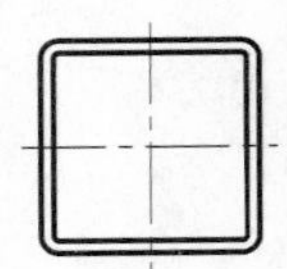

Shape	Wall Thickness, t		Nominal Wt.	Area, A	$\frac{b}{t}$	$\frac{h}{t}$	Axis X-X			
	nominal	design					I	S	r	Z
	in.	in.	lb/ft	in.^{2}			in.^{4}	in.^{3}	in.	in.^{3}
HSS2 1/4×2 1/4	1/4	0.233	6.23	1.74	6.66	6.66	1.13	1.01	0.806	1.28
	3/16	0.174	4.94	1.37	9.93	9.93	0.953	0.847	0.835	1.04
	1/8	0.116	3.47	0.956	16.4	16.4	0.712	0.633	0.863	0.755
HSS2×2	1/4	0.233	5.38	1.51	5.58	5.58	0.747	0.747	0.704	0.964
	3/16	0.174	4.30	1.19	8.49	8.49	0.641	0.641	0.733	0.797
	1/8	0.116	3.04	0.840	14.2	14.2	0.486	0.486	0.761	0.584
HSS2×1 1/2	3/16	0.174	3.66	1.02	5.62	8.49	0.495	0.495	0.697	0.639
HSS2×1	3/16	0.174	3.02	0.845	2.75	8.49	0.350	0.350	0.643	0.480
	1/8	0.116	2.19	0.608	5.62	14.2	0.280	0.280	0.679	0.366
HSS1 3/4×1 3/4	3/16	0.174	3.66	1.02	7.06	7.06	0.405	0.463	0.631	0.585
HSS1 5/8×1 5/8	3/16	0.174	3.34	0.932	6.34	6.34	0.313	0.385	0.580	0.491
	1/8	0.116	2.41	0.666	11.0	11.0	0.246	0.303	0.608	0.370
HSS1 1/2×1 1/2	3/16	0.174	3.02	0.845	5.62	5.62	0.236	0.315	0.528	0.406
	1/8	0.116	2.19	0.608	9.93	9.93	0.188	0.251	0.557	0.309
HSS1 1/4×1 1/4	3/16	0.174	2.38	0.671	4.18	4.18	0.122	0.195	0.426	0.259
	1/8	0.116	1.77	0.492	7.78	7.78	0.102	0.162	0.454	0.204

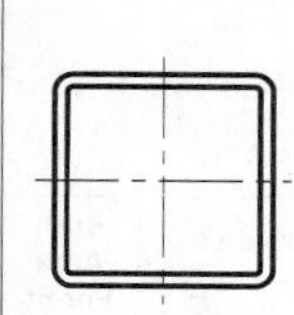

Table 1-11 (cont.). Rectangular (and Square) HSS Dimensions and Properties

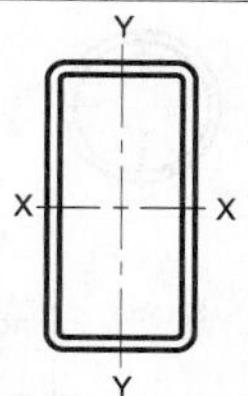

Shape	Wall Thickness, t		Axis Y-Y				Torsion		Surface Area Per Foot
	nominal	design	I	S	r	Z	J	C	
	in.	in.	in.^{4}	in.^{3}	in.	in.^{3}	in.^{4}	in.^{3}	ft^2
HSS2 1/4×2 1/4	1/4	0.233	1.13	1.01	0.806	1.28	1.96	1.85	0.683
	3/16	0.174	0.953	0.847	0.835	1.04	1.60	1.48	0.700
	1/8	0.116	0.712	0.633	0.863	0.755	1.15	1.05	0.717
HSS2×2	1/4	0.233	0.747	0.747	0.704	0.964	1.31	1.41	0.600
	3/16	0.174	0.641	0.641	0.733	0.797	1.09	1.14	0.617
	1/8	0.116	0.486	0.486	0.761	0.584	0.796	0.817	0.633
HSS2×1 1/2	3/16	0.174	0.313	0.417	0.554	0.521	0.664	0.822	0.534
HSS2×1	3/16	0.174	0.112	0.225	0.365	0.288	0.301	0.505	0.450
	1/8	0.116	0.0922	0.184	0.390	0.223	0.238	0.380	0.467
HSS1 3/4×1 3/4	3/16	0.174	0.405	0.463	0.631	0.585	0.699	0.844	0.534
HSS1 5/8×1 5/8	3/16	0.174	0.313	0.385	0.580	0.491	0.544	0.712	0.492
	1/8	0.116	0.246	0.303	0.608	0.370	0.410	0.522	0.508
HSS1 1/2×1 1/2	3/16	0.174	0.236	0.315	0.528	0.406	0.414	0.592	0.450
	1/8	0.116	0.188	0.251	0.557	0.309	0.316	0.438	0.467
HSS1 1/4×1 1/4	3/16	0.174	0.122	0.195	0.426	0.259	0.218	0.383	0.367
	1/8	0.116	0.102	0.162	0.454	0.204	0.174	0.292	0.383

Table 1-12.
Round HSS
Dimensions and Properties

Shape	Wall Thickness, t		Nom-inal Wt.	Area, A	$\frac{D}{t}$	I	S	r	Z	Torsion		Surf. Area Per Ft
	nominal	design								J	C	
	in.	in.	lb/ft	in.^{2}		in.^{4}	in.^{3}	in.	in.^{3}	in.^{4}	in.^{3}	ft^2
HSS20.000	0.500	0.465	104	28.5	43.0	1360	136	6.91	177	2720	272	5.24
	0.375	0.349	78.7	21.5	57.3	1040	104	6.95	135	2080	208	5.24
HSS18.000	0.500	0.465	93.5	25.6	38.7	985	109	6.20	143	1970	219	4.71
	0.375	0.349	70.7	19.4	51.6	754	83.8	6.24	109	1510	168	4.71
HSS16.000	0.500	0.465	82.8	22.7	34.4	685	85.7	5.49	112	1370	171	4.19
	0.438	0.407	72.9	19.9	39.3	606	75.8	5.51	99.0	1210	152	4.19
	0.375	0.349	62.6	17.2	45.8	526	65.7	5.53	85.5	1050	131	4.19
	0.312	0.291	52.3	14.4	55.0	443	55.4	5.55	71.8	886	111	4.19
HSS14.000	0.500	0.465	72.2	19.8	30.1	453	64.8	4.79	85.2	907	130	3.67
	0.375	0.349	54.6	15.0	40.1	349	49.8	4.83	65.1	698	99.7	3.67
	0.312	0.291	45.7	12.5	48.1	295	42.1	4.85	54.7	589	84.2	3.67
HSS12.750	0.500	0.465	65.5	17.9	27.4	339	53.2	4.35	70.2	678	106	3.34
	0.375	0.349	49.6	13.6	36.5	262	41.0	4.39	53.7	523	82.1	3.34
	0.250	0.233	33.4	9.16	54.7	180	28.2	4.43	36.5	359	56.3	3.34
HSS12.500	0.625	0.581	79.3	21.8	21.5	387	62.0	4.22	82.6	774	124	3.27
	0.500	0.465	64.1	17.6	26.9	319	51.0	4.26	67.4	638	102	3.27
	0.375	0.349	48.6	13.3	35.8	246	39.4	4.30	51.5	492	78.7	3.27
	0.312	0.291	40.7	11.2	43.0	208	33.3	4.32	43.4	416	66.6	3.27
	0.250	0.233	32.7	8.98	53.6	169	27.0	4.34	35.1	338	54.1	3.27
	0.188	0.174	24.7	6.74	71.8	128	20.5	4.36	26.4	256	41.0	3.27
HSS11.250	0.625	0.581	71.0	19.5	19.4	278	49.4	3.78	66.2	556	98.8	2.95
	0.500	0.465	57.5	15.8	24.2	229	40.8	3.82	54.1	459	81.6	2.95
	0.375	0.349	43.6	12.0	32.2	178	31.6	3.86	41.5	355	63.2	2.95
	0.312	0.291	36.5	10.0	38.7	151	26.8	3.88	35.0	301	53.5	2.95
	0.250	0.233	29.4	8.06	48.3	122	21.8	3.90	28.3	245	43.5	2.95
	0.188	0.174	22.2	6.05	64.7	92.9	16.5	3.92	21.3	186	33.0	2.95
HSS10.750	0.500	0.465	54.8	15.0	23.1	199	37.0	3.64	49.2	398	74.1	2.81
	0.250	0.233	28.1	7.70	46.1	106	19.8	3.72	25.8	213	39.6	2.81
HSS10.000	0.625	0.581	62.6	17.2	17.2	191	38.3	3.34	51.6	383	76.6	2.62
	0.500	0.465	50.8	13.9	21.5	159	31.7	3.38	42.3	317	63.5	2.62
	0.375	0.349	38.6	10.6	28.7	123	24.7	3.41	32.5	247	49.3	2.62
	0.312	0.291	32.3	8.88	34.4	105	20.9	3.43	27.4	209	41.9	2.62
	0.250	0.233	26.1	7.15	42.9	85.3	17.1	3.45	22.2	171	34.1	2.62
	0.188	0.174	19.7	5.37	57.5	64.8	13.0	3.47	16.8	130	25.9	2.62
HSS9.625	0.500	0.465	48.8	13.4	20.7	141	29.2	3.24	39.0	281	58.5	2.52
	0.375	0.349	37.1	10.2	27.6	110	22.8	3.28	30.0	219	45.5	2.52
	0.312	0.291	31.1	8.53	33.1	93	19.3	3.30	25.4	186	38.7	2.52
	0.250	0.233	25.1	6.87	41.3	75.9	15.8	3.32	20.6	152	31.5	2.52
	0.188	0.174	19.0	5.17	55.3	57.7	12.0	3.34	15.5	115	24.0	2.52
HSS8.750	0.500	0.465	44.1	12.1	18.8	104	23.8	2.93	32.0	208	47.6	2.29
	0.375	0.349	33.6	9.21	25.1	81.4	18.6	2.97	24.6	163	37.2	2.29
	0.312	0.291	28.1	7.73	30.1	69.3	15.8	2.99	20.8	139	31.7	2.29
	0.250	0.233	22.7	6.23	37.6	56.6	12.9	3.01	16.9	113	25.9	2.29
	0.188	0.174	17.2	4.69	50.3	43.1	9.86	3.03	12.8	86.2	19.7	2.29
HSS8.625	0.500	0.465	43.4	11.9	18.5	99.5	23.1	2.89	31.0	199	46.2	2.26
	0.375	0.349	33.1	9.07	24.7	77.8	18.0	2.93	23.9	156	36.1	2.26
	0.322	0.300	28.6	7.85	28.8	68.1	15.8	2.95	20.8	136	31.6	2.26
	0.250	0.233	22.4	6.14	37.0	54.1	12.5	2.97	16.4	108	25.1	2.26
	0.188	0.174	17.0	4.62	49.6	41.3	9.57	2.99	12.4	82.5	19.1	2.26
HSS7.625	0.125	0.116	10.0	2.74	65.7	19.3	5.06	2.66	6.54	38.6	10.1	2.00
HSS7.500	0.500	0.465	37.4	10.3	16.1	63.9	17.0	2.49	23.0	128	34.1	1.96
	0.375	0.349	28.6	7.84	21.5	50.2	13.4	2.53	17.9	100	26.8	1.96
	0.312	0.291	24.0	6.59	25.8	42.9	11.4	2.55	15.1	85.8	22.9	1.96
	0.250	0.233	19.4	5.32	32.2	35.2	9.37	2.57	12.3	70.3	18.7	1.96
	0.188	0.174	14.7	4.00	43.1	26.9	7.17	2.59	9.34	53.8	14.3	1.96

Table 1-12 (cont.). Round HSS Dimensions and Properties

Shape	Wall Thickness, t		Nom-inal Wt.	Area, A	D/t	I	S	r	Z	Torsion		Surf. Area Per Ft
	nominal	design								J	C	
	in.	in.	lb/ft	in.^{2}		in.^{4}	in.^{3}	in.	in.^{3}	in.^{4}	in.^{3}	ft^2
HSS7.000	0.500	0.465	34.7	9.55	15.1	51.2	14.6	2.32	19.9	102	29.3	1.83
	0.375	0.349	26.6	7.29	20.1	40.4	11.6	2.35	15.5	80.9	23.1	1.83
	0.312	0.291	22.3	6.13	24.1	34.6	9.88	2.37	13.1	69.1	19.8	1.83
	0.250	0.233	18.0	4.95	30.0	28.4	8.11	2.39	10.7	56.8	16.2	1.83
	0.188	0.174	13.7	3.73	40.2	21.7	6.21	2.41	8.11	43.5	12.4	1.83
	0.125	0.116	9.19	2.51	60.3	14.9	4.25	2.43	5.50	29.7	8.49	1.83
HSS6.875	0.500	0.465	34.1	9.36	14.8	48.3	14.1	2.27	19.1	96.7	28.1	1.80
	0.375	0.349	26.1	7.16	19.7	38.2	11.1	2.31	14.9	76.4	22.2	1.80
	0.312	0.291	21.9	6.02	23.6	32.7	9.51	2.33	12.6	65.4	19.0	1.80
	0.250	0.233	17.7	4.86	29.5	26.8	7.81	2.35	10.3	53.7	15.6	1.80
	0.188	0.174	13.4	3.66	39.5	20.6	5.99	2.37	7.81	41.1	12.0	1.80
HSS6.625	0.500	0.465	32.7	9.00	14.2	42.9	13.0	2.18	17.7	85.9	25.9	1.73
	0.432	0.403	28.6	7.88	16.4	38.3	11.6	2.20	15.6	76.6	23.1	1.73
	0.375	0.349	25.1	6.88	19.0	34.0	10.3	2.22	13.8	68.0	20.5	1.73
	0.312	0.291	21.1	5.79	22.8	29.1	8.79	2.24	11.7	58.2	17.6	1.73
	0.280	0.261	19.0	5.22	25.4	26.5	7.99	2.25	10.6	52.9	16.0	1.73
	0.250	0.233	17.0	4.68	28.4	23.9	7.22	2.26	9.52	47.9	14.4	1.73
	0.188	0.174	12.9	3.53	38.1	18.4	5.54	2.28	7.24	36.7	11.1	1.73
	0.125	0.116	8.69	2.37	57.1	12.6	3.79	2.30	4.92	25.1	7.59	1.73
HSS6.125	0.500	0.465	30.1	8.27	13.2	33.3	10.9	2.01	14.9	66.7	21.8	1.60
	0.375	0.349	23.1	6.33	17.6	26.5	8.66	2.05	11.7	53.0	17.3	1.60
	0.312	0.291	19.4	5.33	21.0	22.7	7.43	2.07	9.91	45.5	14.9	1.60
	0.250	0.233	15.7	4.31	26.3	18.7	6.12	2.08	8.09	37.5	12.2	1.60
	0.188	0.174	11.9	3.25	35.2	14.4	4.71	2.10	6.16	28.8	9.41	1.60
HSS6.000	0.500	0.465	29.4	8.09	12.9	31.2	10.4	1.96	14.3	62.4	20.8	1.57
	0.375	0.349	22.5	6.20	17.2	24.8	8.28	2.00	11.2	49.7	16.6	1.57
	0.312	0.291	19.0	5.22	20.6	21.3	7.11	2.02	9.49	42.6	14.2	1.57
	0.280	0.261	17.1	4.71	23.0	19.4	6.47	2.03	8.60	38.8	12.9	1.57
	0.250	0.233	15.4	4.22	25.8	17.6	5.86	2.04	7.75	35.2	11.7	1.57
	0.188	0.174	11.7	3.18	34.5	13.5	4.51	2.06	5.91	27.0	9.02	1.57
	0.125	0.116	7.85	2.14	51.7	9.28	3.09	2.08	4.02	18.6	6.19	1.57
HSS5.563	0.375	0.349	20.8	5.72	15.9	19.5	7.02	1.85	9.50	39.0	14.0	1.46
	0.258	0.241	14.6	4.03	23.1	14.3	5.14	1.88	6.83	28.6	10.3	1.46
	0.188	0.174	10.8	2.95	32.0	10.7	3.85	1.91	5.05	21.4	7.70	1.46
	0.134	0.125	7.78	2.14	44.5	7.90	2.84	1.92	3.70	15.8	5.68	1.46
HSS5.500	0.500	0.465	26.7	7.36	11.8	23.5	8.55	1.79	11.8	47.0	17.1	1.44
	0.375	0.349	20.5	5.65	15.8	18.8	6.84	1.83	9.27	37.6	13.7	1.44
	0.258	0.241	14.5	3.98	22.8	13.8	5.02	1.86	6.67	27.6	10.0	1.44
HSS5.000	0.500	0.465	24.1	6.62	10.8	17.2	6.88	1.61	9.60	34.4	13.8	1.31
	0.375	0.349	18.5	5.10	14.3	13.9	5.55	1.65	7.56	27.7	11.1	1.31
	0.312	0.291	15.6	4.30	17.2	12.0	4.79	1.67	6.46	24.0	9.58	1.31
	0.258	0.241	13.1	3.60	20.7	10.2	4.09	1.68	5.46	20.5	8.18	1.31
	0.250	0.233	12.7	3.49	21.5	9.94	3.97	1.69	5.30	19.9	7.95	1.31
	0.188	0.174	9.67	2.64	28.7	7.69	3.08	1.71	4.05	15.4	6.15	1.31
	0.125	0.116	6.51	1.78	43.1	5.31	2.12	1.73	2.77	10.6	4.25	1.31
HSS4.500	0.337	0.315	15.0	4.14	14.3	9.12	4.05	1.48	5.53	18.2	8.11	1.18
	0.237	0.221	10.8	2.97	20.4	6.82	3.03	1.51	4.05	13.6	6.06	1.18
	0.188	0.174	8.67	2.36	25.9	5.54	2.46	1.53	3.26	11.1	4.93	1.18
	0.125	0.116	5.85	1.60	38.8	3.84	1.71	1.55	2.23	7.68	3.41	1.18
HSS4.000	0.337	0.315	13.2	3.65	12.7	6.24	3.12	1.31	4.29	12.5	6.24	1.05
	0.313	0.291	12.3	3.39	13.7	5.87	2.93	1.32	4.01	11.7	5.87	1.05
	0.250	0.233	10.0	2.76	17.2	4.91	2.45	1.33	3.31	9.82	4.91	1.05
	0.237	0.221	9.53	2.62	18.1	4.70	2.35	1.34	3.16	9.40	4.70	1.05
	0.226	0.211	9.12	2.51	19.0	4.52	2.26	1.34	3.03	9.04	4.52	1.05
	0.220	0.205	8.89	2.44	19.5	4.41	2.21	1.34	2.96	8.83	4.41	1.05
	0.188	0.174	7.66	2.09	23.0	3.83	1.92	1.35	2.55	7.67	3.83	1.05
	0.125	0.116	5.18	1.42	34.5	2.67	1.34	1.37	1.75	5.34	2.67	1.05

Table 1-12 (cont.). Round HSS Dimensions and Properties

Shape	Wall Thickness, t nominal	Wall Thickness, t design	Nom-inal Wt.	Area, A	D/t	I	S	r	Z	Torsion J	Torsion C	Surf. Area Per Ft
	in.	in.	lb/ft	in.^{2}		in.^{4}	in.^{3}	in.	in.^{3}	in.^{4}	in.^{3}	ft^2
HSS3.500	0.313	0.291	10.7	2.93	12.0	3.81	2.18	1.14	3.00	7.61	4.35	0.916
	0.300	0.280	10.3	2.83	12.5	3.70	2.11	1.14	2.91	7.40	4.23	0.916
	0.250	0.233	8.69	2.39	15.0	3.21	1.83	1.16	2.49	6.41	3.66	0.916
	0.216	0.201	7.58	2.08	17.4	2.84	1.63	1.17	2.19	5.69	3.25	0.916
	0.203	0.189	7.15	1.97	18.5	2.70	1.54	1.17	2.07	5.41	3.09	0.916
	0.188	0.174	6.66	1.82	20.1	2.52	1.44	1.18	1.93	5.04	2.88	0.916
	0.125	0.116	4.51	1.23	30.2	1.77	1.01	1.20	1.33	3.53	2.02	0.916
HSS3.000	0.300	0.280	8.66	2.39	10.7	2.24	1.49	0.967	2.08	4.47	2.98	0.785
	0.250	0.233	7.35	2.03	12.9	1.95	1.30	0.982	1.79	3.90	2.60	0.785
	0.216	0.201	6.43	1.77	14.9	1.74	1.16	0.992	1.58	3.48	2.32	0.785
	0.203	0.189	6.07	1.67	15.9	1.66	1.10	1.00	1.50	3.31	2.21	0.785
	0.188	0.174	5.65	1.54	17.2	1.55	1.03	1.00	1.39	3.10	2.06	0.785
	0.152	0.142	4.63	1.27	21.1	1.30	0.870	1.01	1.16	2.61	1.74	0.785
	0.134	0.125	4.11	1.13	24.0	1.17	0.779	1.02	1.03	2.34	1.56	0.785
	0.120	0.112	3.69	1.02	26.8	1.06	0.707	1.02	0.935	2.12	1.41	0.785
HSS2.875	0.250	0.233	7.02	1.93	12.3	1.70	1.18	0.938	1.63	3.40	2.37	0.753
	0.203	0.189	5.80	1.59	15.2	1.45	1.01	0.952	1.37	2.89	2.01	0.753
	0.188	0.174	5.40	1.48	16.5	1.35	0.941	0.957	1.27	2.70	1.88	0.753
	0.125	0.116	3.67	1.01	24.8	0.958	0.667	0.976	0.884	1.92	1.33	0.753
HSS2.500	0.250	0.233	6.01	1.66	10.7	1.08	0.862	0.806	1.20	2.15	1.72	0.654
	0.188	0.174	4.65	1.27	14.4	0.865	0.692	0.825	0.943	1.73	1.38	0.654
	0.125	0.116	3.17	0.869	21.6	0.619	0.495	0.844	0.660	1.24	0.990	0.654
HSS2.375	0.250	0.233	5.68	1.57	10.2	0.910	0.766	0.762	1.07	1.82	1.53	0.622
	0.218	0.204	5.03	1.39	11.6	0.827	0.696	0.771	0.964	1.65	1.39	0.622
	0.188	0.174	4.40	1.20	13.6	0.733	0.617	0.781	0.845	1.47	1.23	0.622
	0.154	0.143	3.66	1.00	16.6	0.627	0.528	0.791	0.713	1.25	1.06	0.622
	0.125	0.116	3.01	0.823	20.5	0.527	0.443	0.800	0.592	1.05	0.887	0.622
HSS1.900	0.145	0.135	2.72	0.749	14.1	0.293	0.309	0.626	0.421	0.586	0.617	0.497
HSS1.660	0.140	0.130	2.27	0.625	12.8	0.184	0.222	0.543	0.305	0.368	0.444	0.435

Table 1-13.
Steel Pipe
Dimensions and Properties

Diameter			Wall Thickness, t	Nominal Wt.	Area, A	$\frac{D}{t}$	I	S	r	Z	J
Nominal	Outside	Inside									
in.	in.	in.	in.	lb/ft	in.^{2}		in.^{4}	in.^{3}	in.	in.^{3}	in.^{4}
Standard Weight (Std.)											
12	12.8	12.0	0.375	49.6	14.6	34.0	279	43.8	4.38	57.4	559
10	10.8	10.0	0.365	40.5	11.9	29.5	161	29.9	3.67	39.4	321
8	8.63	7.98	0.322	28.6	8.40	26.8	72.5	16.8	2.94	22.2	145
6	6.63	6.07	0.280	19.0	5.58	23.7	28.1	8.50	2.25	11.3	56.3
5	5.56	5.05	0.258	14.6	4.30	21.6	15.2	5.45	1.88	7.27	30.3
4	4.50	4.03	0.237	10.8	3.17	19.0	7.23	3.21	1.51	4.31	14.5
3 1/2	4.00	3.55	0.226	9.12	2.68	17.7	4.79	2.39	1.34	3.22	9.58
3	3.50	3.07	0.216	7.58	2.23	16.2	3.02	1.72	1.16	2.33	6.03
2 1/2	2.88	2.47	0.203	5.80	1.70	14.2	1.53	1.06	0.947	1.45	3.06
2	2.38	2.07	0.154	3.66	1.07	15.4	0.666	0.561	0.787	0.761	1.33
1 1/2	1.90	1.61	0.145	2.72	0.799	13.1	0.310	0.326	0.623	0.448	0.62
1 1/4	1.66	1.38	0.140	2.27	0.669	11.9	0.195	0.235	0.540	0.324	0.389
1	1.32	1.05	0.133	1.68	0.494	9.89	0.0873	0.133	0.421	0.187	0.175
3/4	1.05	0.824	0.113	1.13	0.333	9.29	0.0370	0.0705	0.334	0.0997	0.0741
1/2	0.840	0.622	0.109	0.852	0.250	7.71	0.0171	0.0407	0.261	0.0587	0.0342
Extra Strong (X-Strong)											
12	12.8	11.8	0.500	65.5	19.2	25.5	362	56.7	4.33	75.1	723
10	10.8	9.75	0.500	54.8	16.1	21.5	212	39.4	3.63	52.6	424
8	8.63	7.63	0.500	43.4	12.8	17.3	106	24.5	2.88	33.0	211
6	6.63	5.76	0.432	28.6	8.40	15.3	40.5	12.2	2.19	16.6	81.0
5	5.56	4.81	0.375	20.8	6.11	14.8	20.7	7.43	1.84	10.1	41.3
4	4.50	3.83	0.337	15.0	4.41	13.4	9.61	4.27	1.48	5.85	19.2
3 1/2	4.00	3.36	0.318	12.5	3.68	12.6	6.28	3.14	1.31	4.32	12.6
3	3.5	2.90	0.300	10.3	3.02	11.7	3.89	2.23	1.14	3.08	7.79
2 1/2	2.88	2.32	0.276	7.67	2.25	10.4	1.92	1.34	0.924	1.87	3.85
2	2.38	1.94	0.218	5.03	1.48	10.9	0.868	0.731	0.766	1.02	1.74
1 1/2	1.90	1.50	0.200	3.63	1.07	9.50	0.391	0.412	0.605	0.581	0.782
1 1/4	1.66	1.28	0.191	3.00	0.881	8.69	0.242	0.291	0.524	0.414	0.484
1	1.32	0.957	0.179	2.17	0.639	7.35	0.106	0.161	0.407	0.233	0.211
3/4	1.05	0.742	0.154	1.48	0.433	6.82	0.0448	0.0853	0.321	0.125	0.0896
1/2	0.840	0.546	0.147	1.09	0.320	5.71	0.0201	0.0478	0.250	0.0717	0.0402
Double-Extra Strong (XX-Strong)											
8	8.63	6.88	0.875	72.5	21.3	9.86	162	37.6	2.76	52.8	324
6	6.63	4.90	0.864	53.2	15.6	7.67	66.3	20.0	2.06	28.9	133
5	5.56	4.06	0.750	38.6	11.3	7.42	33.6	12.1	1.72	17.5	67.3
4	4.50	3.15	0.674	27.6	8.10	6.68	15.3	6.79	1.37	9.97	30.6
3	3.50	2.30	0.600	18.6	5.47	5.83	5.99	3.42	1.05	5.12	12.0
2 1/2	2.88	1.77	0.552	13.7	4.03	5.21	2.87	2.00	0.844	3.03	5.74
2	2.38	1.50	0.436	9.04	2.66	5.45	1.31	1.10	0.703	1.67	2.62

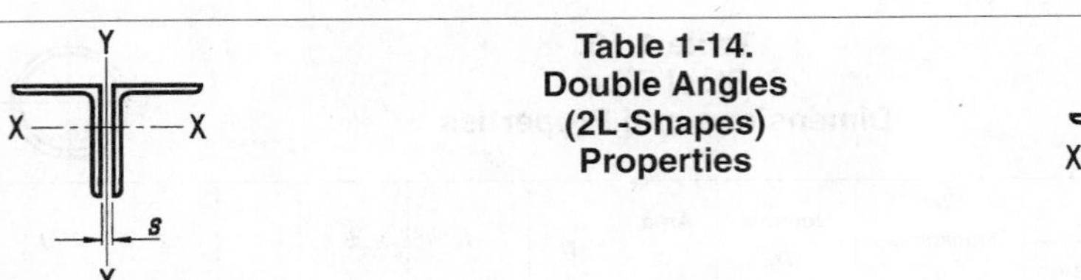

Table 1-14. Double Angles (2L-Shapes) Properties

Shape	Axis Y-Y Radii of Gyration						Q_s*			
	LLBB			SLBB			LLBB		SLBB	
	Separation *s*, in.			Separation *s*, in.			Angles in Contact	Angles Separated	Angles in Contact	Angles Separated
	0	3/8	3/4	0	3/8	3/4				
2L8×8×1 1/8	3.41	3.54	3.68	3.41	3.54	3.68	–	–	–	–
×1	3.39	3.52	3.66	3.39	3.52	3.66	–	–	–	–
×7/8	3.36	3.50	3.63	3.36	3.50	3.63	–	–	–	–
×3/4	3.34	3.47	3.61	3.34	3.47	3.61	–	–	–	–
×5/8	3.32	3.45	3.58	3.32	3.45	3.58	–	0.997	–	0.997
×9/16	3.31	3.44	3.57	3.31	3.44	3.57	–	0.959	–	0.959
×1/2	3.30	3.43	3.56	3.30	3.43	3.56	0.998	0.912	0.998	0.912
2L8×6×1	2.39	2.52	2.66	3.63	3.77	3.91	–	–	–	–
×7/8	2.37	2.50	2.63	3.61	3.75	3.89	–	–	–	–
×3/4	2.35	2.47	2.61	3.59	3.72	3.86	–	–	–	–
×5/8	2.33	2.45	2.59	3.57	3.70	3.84	–	0.997	–	0.997
×9/16	2.32	2.44	2.58	3.55	3.69	3.83	–	0.959	–	0.959
×1/2	2.31	2.43	2.56	3.54	3.68	3.81	–	0.912	0.998	0.912
×7/16	2.30	2.42	2.55	3.53	3.66	3.80	–	0.850	0.938	0.850
2L8×4×1	1.46	1.60	1.75	3.94	4.08	4.23	–	–	–	–
×7/8	1.44	1.57	1.72	3.91	4.06	4.21	–	–	–	–
×3/4	1.42	1.55	1.69	3.89	4.03	4.18	–	–	–	–
×5/8	1.39	1.52	1.66	3.86	4.00	4.15	–	0.997	–	0.997
×9/16	1.38	1.51	1.65	3.85	3.99	4.13	–	0.959	–	0.959
×1/2	1.38	1.50	1.63	3.83	3.97	4.12	–	0.912	0.998	0.912
×7/16	1.37	1.49	1.62	3.82	3.96	4.10	–	0.850	0.938	0.850
2L7×4×3/4	1.48	1.61	1.75	3.34	3.48	3.63	–	–	–	–
×5/8	1.45	1.58	1.73	3.31	3.46	3.60	–	–	–	–
×1/2	1.44	1.56	1.70	3.29	3.43	3.57	–	0.965	–	0.965
×7/16	1.43	1.55	1.68	3.28	3.42	3.56	–	0.912	0.998	0.912
×3/8	1.42	1.54	1.67	3.26	3.40	3.54	–	0.840	0.928	0.84
2L6×6×1	2.58	2.72	2.86	2.58	2.72	2.86	–	–	–	–
×7/8	2.56	2.70	2.84	2.56	2.70	2.84	–	–	–	–
×3/4	2.54	2.67	2.81	2.54	2.67	2.81	–	–	–	–
×5/8	2.52	2.65	2.79	2.52	2.65	2.79	–	–	–	–
×9/16	2.51	2.64	2.78	2.51	2.64	2.78	–	–	–	–
×1/2	2.50	2.63	2.76	2.50	2.63	2.76	–	–	–	–
×7/16	2.49	2.62	2.75	2.49	2.62	2.75	–	0.973	–	0.973
×3/8	2.48	2.60	2.74	2.48	2.60	2.74	0.998	0.912	0.998	0.912
×5/16	2.47	2.59	2.72	2.47	2.59	2.72	0.914	0.826	0.914	0.826
2L6×4×7/8	1.57	1.71	1.86	2.83	2.97	3.12	–	–	–	–
×3/4	1.55	1.69	1.83	2.8	2.95	3.09	–	–	–	–
×5/8	1.53	1.66	1.81	2.78	2.92	3.07	–	–	–	–
×9/16	1.52	1.65	1.79	2.77	2.91	3.06	–	–	–	–
×1/2	1.51	1.64	1.78	2.76	2.9	3.04	–	–	–	–
×7/16	1.50	1.63	1.77	2.75	2.89	3.03	–	0.973	–	0.973
×3/8	1.50	1.62	1.76	2.74	2.88	3.02	–	0.912	0.998	0.912
×5/16	1.49	1.61	1.74	2.73	2.87	3.01	–	0.826	0.914	0.826
2L6×3 1/2×1/2	1.28	1.41	1.55	2.84	2.98	3.12	–	–	–	–
×3/8	1.26	1.39	1.52	2.81	2.95	3.10	–	0.912	0.998	0.912
×5/16	1.25	1.38	1.51	2.80	2.94	3.08	–	0.826	0.914	0.826
2L5×5×7/8	2.16	2.30	2.44	2.16	2.30	2.44	–	–	–	–
×3/4	2.13	2.27	2.41	2.13	2.27	2.41	–	–	–	–
×5/8	2.11	2.25	2.39	2.11	2.25	2.39	–	–	–	–
×1/2	2.09	2.22	2.36	2.09	2.22	2.36	–	–	–	–
×7/16	2.08	2.21	2.35	2.08	2.21	2.35	–	–	–	–
×3/8	2.07	2.20	2.34	2.07	2.20	2.34	–	0.983	–	0.983
×5/16	2.06	2.19	2.32	2.06	2.19	2.32	0.998	0.912	0.998	0.912

*Where no value of Q_s is shown, the angle complies with LRFD Specification Section E2.

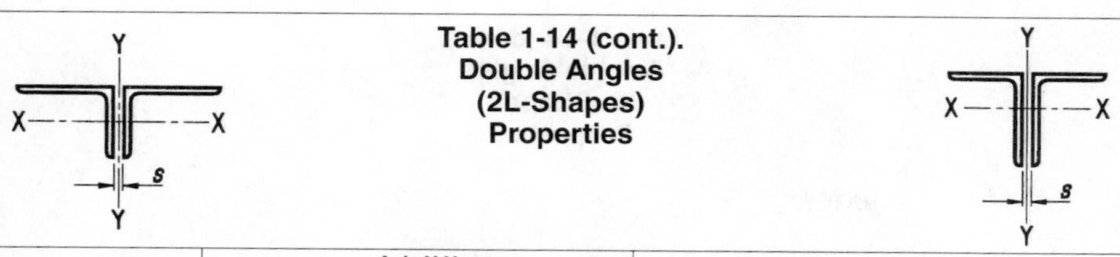

Table 1-14 (cont.). Double Angles (2L-Shapes) Properties

Shape	Axis Y-Y Radii of Gyration						Q_s*			
	LLBB			SLBB			LLBB		SLBB	
	Separation s, in.			Separation s, in.			Angles in Contact	Angles Separated	Angles in Contact	Angles Separated
	0	3/8	3/4	0	3/8	3/4				
2L5×3 1/2×3/4	1.39	1.53	1.68	2.33	2.47	2.62	–	–	–	–
×5/8	1.37	1.50	1.65	2.30	2.45	2.59	–	–	–	–
×1/2	1.35	1.48	1.62	2.28	2.42	2.57	–	–	–	–
×3/8	1.33	1.46	1.59	2.26	2.39	2.54	–	0.983	–	0.983
×5/16	1.32	1.44	1.58	2.25	2.38	2.52	–	0.912	0.998	0.912
×1/4	1.31	1.43	1.57	2.23	2.37	2.51	–	0.804	0.894	0.804
2L5×3×1/2	1.11	1.24	1.39	2.35	2.50	2.64	–	–	–	–
×7/16	1.10	1.23	1.38	2.34	2.48	2.63	–	–	–	–
×3/8	1.09	1.22	1.36	2.33	2.47	2.62	–	0.983	–	0.983
×5/16	1.08	1.21	1.35	2.32	2.46	2.60	–	0.912	0.998	0.912
×1/4	1.07	1.19	1.33	2.30	2.44	2.58	–	0.804	0.894	0.804
2L4×4×3/4	1.73	1.88	2.03	1.73	1.88	2.03	–	–	–	–
×5/8	1.71	1.85	2.00	1.71	1.85	2.00	–	–	–	–
×1/2	1.69	1.83	1.97	1.69	1.83	1.97	–	–	–	–
×7/16	1.68	1.81	1.96	1.68	1.81	1.96	–	–	–	–
×3/8	1.67	1.80	1.94	1.67	1.80	1.94	–	–	–	–
×5/16	1.66	1.79	1.93	1.66	1.79	1.93	–	0.997	–	0.997
×1/4	1.65	1.78	1.91	1.65	1.78	1.91	0.998	0.912	0.998	0.912
2L4×3 1/2×1/2	1.44	1.57	1.72	1.75	1.89	2.03	–	–	–	–
×3/8	1.42	1.55	1.69	1.73	1.86	2.00	–	–	–	–
×5/16	1.40	1.53	1.68	1.72	1.85	1.99	–	0.997	–	0.997
×1/4	1.39	1.52	1.66	1.70	1.83	1.97	–	0.912	0.998	0.912
2L4×3×5/8	1.21	1.35	1.50	1.84	1.98	2.13	–	–	–	–
×1/2	1.19	1.32	1.47	1.81	1.95	2.10	–	–	–	–
×3/8	1.17	1.30	1.44	1.79	1.93	2.07	–	–	–	–
×5/16	1.16	1.29	1.43	1.78	1.91	2.06	–	0.997	–	0.997
×1/4	1.15	1.27	1.41	1.76	1.90	2.04	–	0.912	0.998	0.912
2L3 1/2×3 1/2×1/2	1.49	1.63	1.77	1.49	1.63	1.77	–	–	–	–
×7/16	1.48	1.61	1.76	1.48	1.61	1.76	–	–	–	–
×3/8	1.47	1.60	1.74	1.47	1.60	1.74	–	–	–	–
×5/16	1.46	1.59	1.73	1.46	1.59	1.73	–	–	–	–
×1/4	1.44	1.57	1.72	1.44	1.57	1.72	–	0.965	–	0.965
2L3 1/2×3×1/2	1.23	1.37	1.52	1.55	1.69	1.84	–	–	–	–
×7/16	1.22	1.36	1.51	1.54	1.67	1.82	–	–	–	–
×3/8	1.21	1.35	1.49	1.52	1.66	1.81	–	–	–	–
×5/16	1.20	1.33	1.48	1.51	1.65	1.79	–	–	–	–
×1/4	1.19	1.32	1.46	1.50	1.63	1.78	–	0.965	–	0.965
2L3 1/2×2 1/2×1/2	0.992	1.13	1.28	1.62	1.76	1.91	–	–	–	–
×3/8	0.970	1.11	1.25	1.59	1.73	1.88	–	–	–	–
×5/16	0.960	1.09	1.24	1.58	1.72	1.87	–	–	–	–
×1/4	0.950	1.08	1.22	1.57	1.70	1.85	–	0.965	–	0.965
2L3×3×1/2	1.29	1.43	1.58	1.29	1.43	1.58	–	–	–	–
×7/16	1.28	1.42	1.57	1.28	1.42	1.57	–	–	–	–
×3/8	1.27	1.41	1.55	1.27	1.41	1.55	–	–	–	–
×5/16	1.26	1.39	1.54	1.26	1.39	1.54	–	–	–	–
×1/4	1.25	1.38	1.52	1.25	1.38	1.52	–	–	–	–
×3/16	1.24	1.37	1.51	1.24	1.37	1.51	0.998	0.912	0.998	0.912
2L3×2 1/2×1/2	1.04	1.18	1.33	1.35	1.49	1.64	–	–	–	–
×7/16	1.02	1.16	1.32	1.34	1.48	1.63	–	–	–	–
×3/8	1.01	1.15	1.30	1.32	1.46	1.61	–	–	–	–
×5/16	1.00	1.14	1.29	1.31	1.45	1.6	–	–	–	–
×1/4	0.991	1.12	1.27	1.3	1.44	1.58	–	–	–	–
×3/16	0.980	1.11	1.25	1.29	1.42	1.57	–	0.912	0.998	0.912

*Where no value of Q_s is shown, the angle complies with LRFD Specification Section E2.

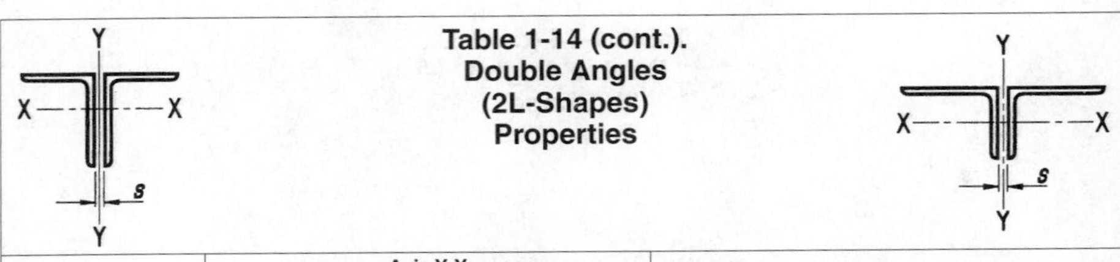

Table 1-14 (cont.). Double Angles (2L-Shapes) Properties

Shape	Axis Y-Y Radii of Gyration LLBB Separation s, in. 0	3/8	3/4	SLBB Separation s, in. 0	3/8	3/4	Q_s* LLBB Angles in Contact	LLBB Angles Separated	SLBB Angles in Contact	SLBB Angles Separated
2L3×2×1/2	0.795	0.94	1.10	1.42	1.56	1.72	–	–	–	–
×3/8	0.771	0.911	1.07	1.39	1.54	1.69	–	–	–	–
×5/16	0.760	0.897	1.05	1.38	1.52	1.67	–	–	–	–
×1/4	0.749	0.883	1.03	1.37	1.51	1.66	–	–	–	–
×3/16	0.739	0.869	1.02	1.35	1.49	1.64	–	0.912	0.998	0.912
2L2 1/2×2 1/2×1/2	1.09	1.23	1.39	1.09	1.23	1.39	–	–	–	–
×3/8	1.07	1.21	1.36	1.07	1.21	1.36	–	–	–	–
×5/16	1.05	1.19	1.34	1.05	1.19	1.34	–	–	–	–
×1/4	1.04	1.18	1.33	1.04	1.18	1.33	–	–	–	–
×3/16	1.03	1.17	1.31	1.03	1.17	1.31	–	0.983	–	0.983
2L2 1/2×2×3/8	0.815	0.957	1.11	1.13	1.27	1.42	–	–	–	–
×5/16	0.804	0.943	1.10	1.12	1.26	1.41	–	–	–	–
×1/4	0.794	0.93	1.08	1.10	1.24	1.39	–	–	–	–
×3/16	0.784	0.916	1.07	1.09	1.23	1.38	–	0.983	–	0.983
2L2×2×3/8	0.865	1.01	1.17	0.865	1.01	1.17	–	–	–	–
×5/16	0.853	0.996	1.15	0.853	0.996	1.15	–	–	–	–
×1/4	0.842	0.982	1.14	0.842	0.982	1.14	–	–	–	–
×3/16	0.831	0.967	1.12	0.831	0.967	1.12	–	–	–	–
×1/8	0.818	0.951	1.10	0.818	0.951	1.10	0.998	0.912	0.998	0.912

*Where no value of Q_s is shown, the angle complies with LRFD Specification Section E2.

Table 1-15. 2C-Shapes (Double Channels) Properties

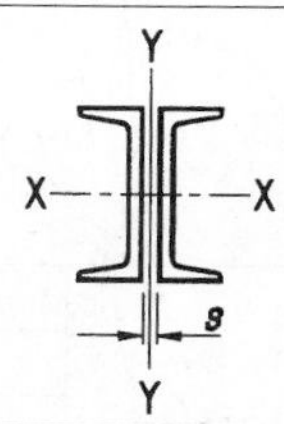

Shape	Axis Y-Y											
	Separation *s*, in.											
	0				3/8				3/4			
	I	*S*	*r*	*Z*	*I*	*S*	*r*	*Z*	*I*	*S*	*r*	*Z*
	in.^{4}	in.^{3}	in.	in.^{3}	in.^{4}	in.^{3}	in.	in.^{3}	in.^{4}	in.^{3}	in.	in.^{3}
2C15×50	40.7	11.0	1.18	30.7	50.5	12.9	1.31	36.2	62.4	15.3	1.46	41.7
×40	32.6	9.25	1.18	22.9	40.2	10.9	1.31	27.3	49.6	12.7	1.45	31.7
×33.9	28.5	8.38	1.20	19.0	35.1	9.78	1.33	22.7	43.1	11.4	1.47	26.4
2C12×30	18.2	5.75	1.02	15.1	23.3	6.94	1.15	18.4	29.6	8.36	1.30	21.7
×25	15.6	5.11	1.03	12.1	19.8	6.12	1.16	14.9	25.0	7.32	1.31	17.6
×20.7	13.6	4.64	1.06	10.0	17.2	5.51	1.19	12.3	21.7	6.55	1.34	14.6
2C10×30	15.3	5.04	0.931	15.3	20.2	6.27	1.07	18.6	26.3	7.73	1.22	21.9
×25	12.3	4.25	0.914	11.8	16.2	5.27	1.05	14.5	21.1	6.48	1.20	17.3
×20	9.91	3.62	0.918	8.84	13.0	4.44	1.05	11.0	16.9	5.43	1.20	13.2
×15.3	8.14	3.13	0.953	6.69	10.6	3.80	1.09	8.37	13.7	4.59	1.23	10.0
2C9×20	8.80	3.32	0.866	8.76	11.8	4.15	1.00	11.0	15.6	5.15	1.15	13.2
×15	6.86	2.76	0.882	6.25	9.10	3.41	1.02	7.90	12.0	4.19	1.17	9.55
×13.4	6.34	2.61	0.897	5.59	8.39	3.20	1.03	7.07	11.0	3.92	1.18	8.55
2C8×18.75	7.46	2.95	0.823	8.12	10.2	3.75	0.962	10.2	13.7	4.71	1.11	12.3
×13.75	5.51	2.35	0.826	5.49	7.47	2.95	0.962	7.00	10.0	3.68	1.11	8.52
×11.5	4.82	2.13	0.846	4.57	6.5	2.66	0.982	5.83	8.66	3.29	1.13	7.10
2C7×14.75	5.18	2.25	0.773	5.94	7.21	2.90	0.912	7.57	9.85	3.68	1.07	9.19
×12.25	4.30	1.96	0.773	4.69	5.97	2.51	0.911	6.04	8.14	3.17	1.06	7.39
×9.8	3.59	1.72	0.791	3.69	4.95	2.17	0.929	4.76	6.72	2.73	1.08	5.84
2C6×13	4.11	1.91	0.734	5.13	5.85	2.50	0.876	6.56	8.13	3.21	1.03	7.99
×10.5	3.26	1.60	0.728	3.86	4.63	2.08	0.867	5.02	6.43	2.67	1.02	6.17
×8.2	2.63	1.37	0.741	2.93	3.72	1.76	0.881	3.82	5.14	2.24	1.04	4.72
2C5×9	2.45	1.30	0.682	3.22	3.59	1.73	0.824	4.21	5.09	2.25	0.982	5.20
×6.7	1.86	1.06	0.688	2.36	2.71	1.40	0.831	3.09	3.84	1.81	0.989	3.83
2C4×7.25	1.75	1.02	0.641	2.52	2.63	1.38	0.786	3.32	3.81	1.82	0.946	4.12
×5.4	1.29	0.812	0.637	1.86	1.94	1.10	0.783	2.45	2.82	1.44	0.943	3.05
×4.5	1.25	0.789	0.673	1.95	1.86	1.05	0.820	2.47	2.66	1.36	0.981	2.98
2C3×6	1.33	0.833	0.614	2.12	2.06	1.15	0.764	2.78	3.03	1.54	0.927	3.44
×5	1.05	0.699	0.597	1.65	1.63	0.969	0.746	2.20	2.43	1.30	0.909	2.75
×4.1	0.842	0.597	0.591	1.43	1.32	0.827	0.741	1.88	1.97	1.10	0.905	2.33
×3.5	0.766	0.558	0.593	1.37	1.20	0.772	0.743	1.78	1.80	1.03	0.908	2.19

Table 1-16.
2MC-Shapes (Double Channels) Properties

Shape	Axis Y-Y, Separation s, in. 0 I				3/8 I				3/4 I			
	I	S	r	Z	I	S	r	Z	I	S	r	Z
	in.⁴	in.³	in.	in.³	in.⁴	in.³	in.	in.³	in.⁴	in.³	in.	in.³
2MC18×58	60.6	14.4	1.33	37.5	72.8	16.6	1.46	43.9	87.5	19.1	1.60	50.3
×51.9	55.0	13.4	1.34	32.7	65.9	15.4	1.47	38.4	79.0	17.6	1.61	44.1
×45.8	50.1	12.5	1.36	28.3	59.8	14.3	1.49	33.4	71.4	16.3	1.63	38.4
×42.7	47.8	12.1	1.38	26.4	57.0	13.8	1.51	31.1	67.9	15.7	1.64	35.8
2MC13×50	60.7	13.8	1.44	37.0	72.5	15.8	1.57	42.5	86.3	18.0	1.71	48.0
×40	49.1	11.7	1.45	28.0	58.4	13.4	1.58	32.4	69.4	15.2	1.72	36.8
×35	44.3	10.9	1.47	24.2	52.6	12.3	1.60	28.1	62.3	14.0	1.74	31.9
×31.8	41.5	10.4	1.49	22.1	49.2	11.7	1.62	25.6	58.2	13.3	1.76	29.1
2MC12×50	67.2	16.2	1.51	39.9	79.8	18.5	1.65	45.4	94.5	20.9	1.79	50.9
×45	60.0	15.0	1.51	34.7	71.2	17.0	1.64	39.7	84.3	19.2	1.79	44.6
×40	53.7	13.8	1.51	30.1	63.7	15.6	1.65	34.6	75.3	17.7	1.79	39.0
×35	48.0	12.8	1.53	26.1	56.9	14.4	1.66	29.9	67.2	16.2	1.81	33.8
×31	44.0	12.0	1.55	24.1	52.1	13.5	1.69	27.5	61.4	15.2	1.83	30.9
2MC12×10.6	1.21	0.804	0.441	2.07	2.05	1.21	0.575	3.23	3.33	1.78	0.733	4.40
2MC10×41.1	60.0	13.9	1.58	33.6	70.7	15.7	1.71	38.1	83.1	17.7	1.85	42.6
×33.6	49.5	12.1	1.58	26.3	58.2	13.6	1.72	30.0	68.3	15.3	1.86	33.7
×28.5	43.5	11.0	1.61	22.2	51.1	12.3	1.75	25.3	59.8	13.8	1.89	28.5
2MC10×25	27.8	8.18	1.38	16.7	33.6	9.36	1.51	19.5	40.4	10.7	1.66	22.2
×22	25.4	7.67	1.40	16.6	30.7	8.76	1.54	19.0	36.8	9.98	1.69	21.4
2MC10×8.4	1.05	0.700	0.462	1.70	1.75	1.03	0.596	2.62	2.79	1.49	0.753	3.54
2MC9×25.4	29.2	8.34	1.40	17.6	35.2	9.53	1.53	20.4	42.2	10.9	1.68	23.2
×23.9	27.8	8.05	1.41	16.5	33.4	9.19	1.54	19.1	40.1	10.5	1.69	21.8
2MC8×22.8	27.7	7.91	1.44	16.3	33.2	9.01	1.58	18.9	39.7	10.2	1.72	21.4
×21.4	26.3	7.63	1.45	16.0	31.6	8.68	1.59	18.4	37.7	9.86	1.73	20.7
2MC8×20	17.1	5.66	1.21	12.0	21.2	6.61	1.34	14.2	26.2	7.70	1.49	16.4
×18.7	16.2	5.45	1.21	11.2	20.1	6.35	1.35	13.3	24.8	7.39	1.50	15.4
2MC8×8.5	2.16	1.15	0.658	2.53	3.14	1.52	0.793	3.47	4.47	1.99	0.946	4.40
2MC7×22.7	29.0	8.06	1.47	17.1	34.7	9.16	1.61	19.6	41.3	10.4	1.76	22.1
×19.1	25.1	7.27	1.50	16.2	30.0	8.25	1.64	18.3	35.7	9.34	1.78	20.4
2MC6×18	25.0	7.13	1.54	16.2	29.8	8.07	1.68	18.2	35.3	9.11	1.83	20.1
×15.3	19.7	5.63	1.48	12.3	23.6	6.39	1.62	14.0	28.1	7.24	1.77	15.6
2MC6×16.3	15.8	5.26	1.28	11.4	19.4	6.10	1.42	13.2	23.8	7.05	1.58	15.0
×15.1	14.8	5.02	1.29	11.4	18.2	5.82	1.43	13.1	22.3	6.71	1.58	14.7
2MC6×12	7.21	2.89	1.01	6.01	9.32	3.47	1.15	7.34	11.9	4.15	1.30	8.66

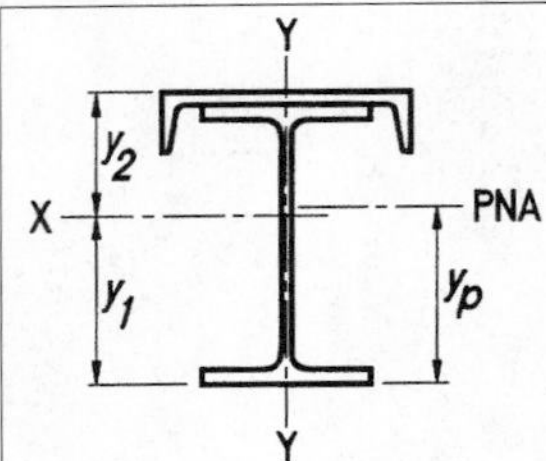

Table 1-17. W-Shapes with Cap Channels Properties

W-Shape	Channel	Total Wt.	Total Area	Axis X-X			
				I	$S_1 = \frac{I}{y_1}$	$S_2 = \frac{I}{y_2}$	r
		lb/ft	In.^{2}	in.^{4}	in.^{3}	in.^{4}	in.
W36×150	MC18×42.7	193	56.8	12000	553	831	14.6
	C15×33.9	184	54.2	11500	546	764	14.6
W33×141	MC18×42.7	184	54.1	10000	490	750	13.6
	C15×33.9	175	51.5	9580	484	689	13.6
W33×118	MC18×42.7	161	47.2	8280	400	656	13.2
	C15×33.9	152	44.6	7900	395	596	13.3
W30×116	MC18×42.7	159	46.8	6900	365	598	12.1
	C15×33.9	150	44.1	6590	360	544	12.2
W30×99	MC18×42.7	142	41.6	5830	304	533	11.8
	C15×33.9	133	39.0	5550	300	481	11.9
W27×94	C15×33.9	128	37.6	4530	268	435	11.0
W27×84	C15×33.9	118	34.7	4050	237	403	10.8
W24×84	C15×33.9	118	34.7	3340	217	367	9.82
	C12×20.7	105	30.8	3030	211	302	9.92
W24×68	C15×33.9	102	30.0	2710	173	321	9.51
	C12×20.7	88.7	26.1	2440	168	258	9.67
W21×68	C15×33.9	102	30.0	2180	156	287	8.52
	C12×20.7	88.7	26.1	1970	152	232	8.67
W21×62	C15×33.9	95.9	28.2	2000	142	272	8.41
	C12×20.7	82.7	24.3	1800	138	218	8.59
W18×50	C15×33.9	83.9	24.6	1250	100	211	7.12
	C12×20.7	70.7	20.7	1120	97.3	166	7.35
W16×36	C15×33.9	69.9	20.5	748	64.5	160	6.04
	C12×20.7	56.7	16.6	670	62.8	123	6.34
W14×30	C12×20.7	50.7	14.9	447	46.7	98.1	5.47
	C10×15.3	45.3	13.3	420	46.0	84.5	5.61
W12×26	C12×20.7	46.7	13.7	318	36.8	82.1	4.81
	C10×15.3	41.3	12.1	299	36.3	70.5	4.96

Table 1-17 (cont.). W-Shapes with Cap Channels Properties

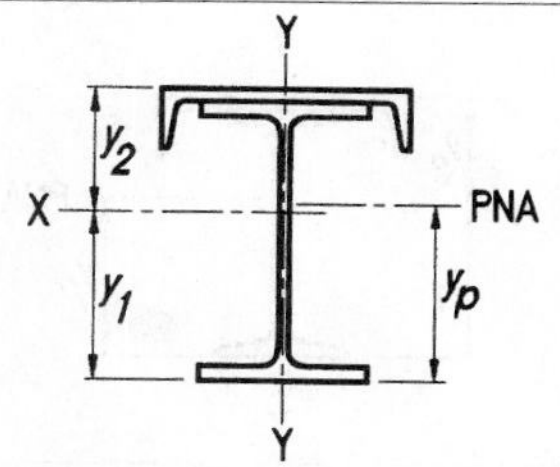

W-Shape	Channel	Axis X-X				Axis Y-Y			
		y_1	y_2	Z	y_p	I	s	r	Z
		in.	in.	in.^{3}	in.	in.^{4}	in.^{3}	in.	in.^{3}
W36×150	MC18×42.7	21.8	14.5	738	28.0	824	91.5	3.81	146
	C15×33.9	21.1	15.1	716	25.9	584	77.9	3.28	122
W33×141	MC18×42.7	20.4	13.3	652	27.0	800	88.9	3.85	142
	C15×33.9	19.8	13.9	635	24.9	561	74.8	3.30	118
W33×118	MC18×42.7	20.7	12.6	544	27.8	741	82.3	3.96	126
	C15×33.9	20.0	13.3	529	25.5	502	66.9	3.35	102
W30×116	MC18×42.7	18.9	11.5	492	26.1	718	79.8	3.92	124
	C15×33.9	18.3	12.1	480	23.8	479	63.8	3.29	100
W30×99	MC18×42.7	19.2	10.9	412	26.4	682	75.8	4.05	114
	C15×33.9	18.5	11.5	408	24.4	442	59	3.37	89.4
W27×94	C15×33.9	16.9	10.4	357	23.6	439	58.5	3.41	89.6
W27×84	C15×33.9	17.1	10.0	316	23.9	420	56.0	3.48	83.9
W24×84	C15×33.9	15.4	9.10	286	21.6	409	54.5	3.43	83.4
	C12×20.7	14.3	10.0	275	18.5	223	37.2	2.69	58.2
W24×68	C15×33.9	15.7	8.46	232	21.7	385	51.3	3.58	75.3
	C12×20.7	14.5	9.49	224	19.2	199	33.2	2.76	50.1
W21×68	C15×33.9	13.9	7.59	207	19.3	379	50.6	3.56	75.1
	C12×20.7	12.9	8.49	200	17.6	194	32.3	2.72	50.0
W21×62	C15×33.9	14.1	7.33	189	19.4	372	49.6	3.63	72.5
	C12×20.7	13.0	8.26	183	18.1	186	31.1	2.77	47.3
W18×50	C15×33.9	12.5	5.92	133	16.9	354	47.3	3.79	67.3
	C12×20.7	11.5	6.76	127	16.1	169	28.2	2.85	42.2
W16×36	C15×33.9	11.6	4.67	86.8	15.2	339	45.2	4.06	61.6
	C12×20.7	10.7	5.47	83.2	14.6	153	25.6	3.04	36.4
W14×30	C12×20.7	9.57	4.55	62.0	12.9	149	24.8	3.16	34.6
	C10×15.3	9.11	4.97	60.3	12.6	86.8	17.4	2.55	24.9
W12×26	C12×20.7	8.63	3.87	48.2	11.6	146	24.4	3.27	33.7
	C10×15.3	8.22	4.24	47.0	11.3	84.5	16.9	2.64	24.1

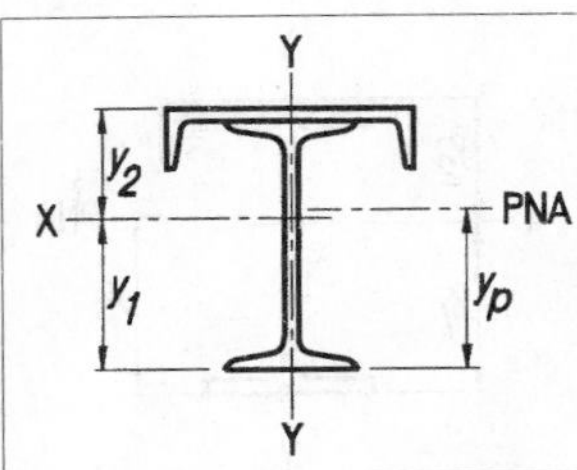

Table 1-18. S-Shapes with Cap Channels Properties

S-Shape	Channel	Total Wt.	Total Area	Axis X-X			
				I	$S_1 = \frac{I}{y_1}$	$S_2 = \frac{I}{y_2}$	r
		lb/ft	In.^{2}	in.^{4}	in.^{3}	in.^{4}	in.
S24×80	C12×20.7	101	29.5	2750	191	278	9.66
	C10×15.3	95.3	27.9	2610	188	252	9.67
S20×66	C12×20.7	86.7	25.5	1620	132	202	7.97
	C10×15.3	81.3	23.9	1530	129	181	8.00
S15×42.9	C10×15.3	58.2	17.1	615	65.7	105	6.04
	C8×11.5	54.4	16.0	583	64.7	93.9	6.04
S12×31.8	C10×15.3	47.1	13.8	314	40.2	71.2	4.77
	C8×11.5	43.3	12.7	297	39.6	63.0	4.84
S10×25.4	C10×15.3	40.7	11.9	185	27.5	52.7	3.94
	C8×11.5	36.9	10.8	175	27.1	46.3	4.02

Table 1-18 (cont.). S-Shapes with Cap Channels Properties

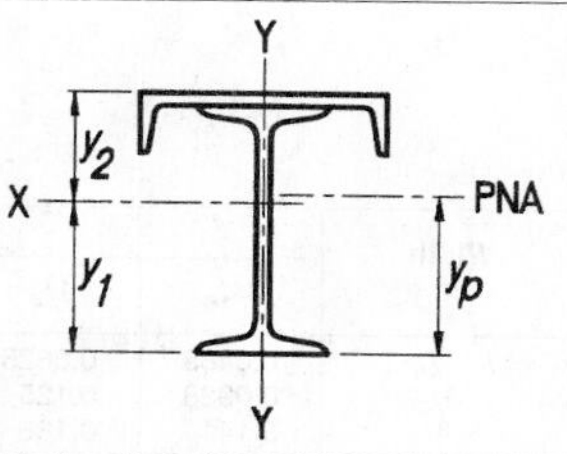

S-Shape	Channel	Axis X-X				Axis Y-Y			
		y_1	y_2	Z	y_p	I	S	r	Z
		in.	in.	in.^{3}	in.	in.^{4}	in.^{3}	in.	in.^{3}
S24×80	C12×20.7	14.4	9.90	256	18.1	171	28.5	2.41	46.4
	C10×15.3	13.9	10.4	246	16.5	109	21.8	1.98	36.8
S20×66	C12×20.7	12.3	7.99	180	16.0	156	26.1	2.48	41.0
	C10×15.3	11.8	8.44	173	14.4	94.7	18.9	1.99	31.3
S15×42.9	C10×15.3	9.37	5.87	87.6	12.8	81.5	16.3	2.18	25.0
	C8×11.5	9.01	6.21	86.5	11.6	46.8	11.7	1.71	18.7
S12×31.8	C10×15.3	7.82	4.42	54.0	10.6	76.5	15.3	2.36	22.3
	C8×11.5	7.50	4.72	52.4	10.3	41.8	10.5	1.82	16.1
S10×25.4	C10×15.3	6.73	3.51	37.2	9.03	73.9	14.8	2.49	20.9
	C8×11.5	6.45	3.77	36.1	8.82	39.2	9.81	1.90	14.6

Table 1-19. Areas of Rectangular Plates and Bars, in.^{2}

Width in.	Thickness, in.								
	3/16	1/4	5/16	3/8	1/2	5/8	3/4	7/8	1
1/4	0.0469	0.0625	0.0781	0.0938	0.125	0.156	0.188	0.219	0.250
1/2	0.0938	0.125	0.156	0.188	0.250	0.313	0.375	0.438	0.500
3/4	0.141	0.188	0.234	0.281	0.375	0.469	0.563	0.656	0.750
1	0.188	0.250	0.313	0.375	0.500	0.625	0.750	0.875	1.00
1 1/4	0.234	0.313	0.391	0.469	0.625	0.781	0.938	1.09	1.25
1 1/2	0.281	0.375	0.469	0.563	0.750	0.938	1.13	1.31	1.50
1 3/4	0.328	0.438	0.547	0.656	0.875	1.09	1.31	1.53	1.75
2	0.375	0.500	0.625	0.750	1.00	1.25	1.50	1.75	2.00
2 1/4	0.422	0.563	0.703	0.844	1.13	1.41	1.69	1.97	2.25
2 1/2	0.469	0.625	0.781	0.938	1.25	1.56	1.88	2.19	2.50
2 3/4	0.516	0.688	0.859	1.03	1.38	1.72	2.06	2.41	2.75
3	0.563	0.75	0.938	1.13	1.50	1.88	2.25	2.63	3.00
3 1/4	0.609	0.813	1.02	1.22	1.63	2.03	2.44	2.84	3.25
3 1/2	0.656	0.875	1.09	1.31	1.75	2.19	2.63	3.06	3.50
3 3/4	0.703	0.938	1.17	1.41	1.88	2.34	2.81	3.28	3.75
4	0.750	1.00	1.25	1.50	2.00	2.50	3.00	3.50	4.00
4 1/4	0.797	1.06	1.33	1.59	2.13	2.66	3.19	3.72	4.25
4 1/2	0.844	1.13	1.41	1.69	2.25	2.81	3.38	3.94	4.50
4 3/4	0.891	1.19	1.48	1.78	2.38	2.97	3.56	4.16	4.75
5	0.938	1.25	1.56	1.88	2.50	3.13	3.75	4.38	5.00
5 1/4	0.984	1.31	1.64	1.97	2.63	3.28	3.94	4.59	5.25
5 1/2	1.03	1.38	1.72	2.06	2.75	3.44	4.13	4.81	5.50
5 3/4	1.08	1.44	1.80	2.16	2.88	3.59	4.31	5.03	5.75
6	1.13	1.50	1.88	2.25	3.00	3.75	4.50	5.25	6.00
6 1/4	1.17	1.56	1.95	2.34	3.13	3.91	4.69	5.47	6.25
6 1/2	1.22	1.63	2.03	2.44	3.25	4.06	4.88	5.69	6.50
6 3/4	1.27	1.69	2.11	2.53	3.38	4.22	5.06	5.91	6.75
7	1.31	1.75	2.19	2.63	3.50	4.38	5.25	6.13	7.00
7 1/4	1.36	1.81	2.27	2.72	3.63	4.53	5.44	6.34	7.25
7 1/2	1.41	1.88	2.34	2.81	3.75	4.69	5.63	6.56	7.50
7 3/4	1.45	1.94	2.42	2.91	3.88	4.84	5.81	6.78	7.75
8	1.50	2.00	2.50	3.00	4.00	5.00	6.00	7.00	8.00
8 1/4	1.55	2.06	2.58	3.09	4.13	5.16	6.19	7.22	8.25
8 1/2	1.59	2.13	2.66	3.19	4.25	5.31	6.38	7.44	8.50
8 3/4	1.64	2.19	2.73	3.28	4.38	5.47	6.56	7.66	8.75
9	1.69	2.25	2.81	3.38	4.50	5.63	6.75	7.88	9.00
9 1/4	1.73	2.31	2.89	3.47	4.63	5.78	6.94	8.09	9.25
9 1/2	1.78	2.38	2.97	3.56	4.75	5.94	7.13	8.31	9.50
9 3/4	1.83	2.44	3.05	3.66	4.88	6.09	7.31	8.53	9.75
10	1.88	2.50	3.13	3.75	5.00	6.25	7.50	8.75	10.0
10 1/4	1.92	2.56	3.20	3.84	5.13	6.41	7.69	8.97	10.3
10 1/2	1.97	2.63	3.28	3.94	5.25	6.56	7.88	9.19	10.5
10 3/4	2.02	2.69	3.36	4.03	5.38	6.72	8.06	9.41	10.8
11	2.06	2.75	3.44	4.13	5.50	6.88	8.25	9.63	11.0
11 1/4	2.11	2.81	3.52	4.22	5.63	7.03	8.44	9.84	11.3
11 1/2	2.16	2.88	3.59	4.31	5.75	7.19	8.63	10.1	11.5
11 3/4	2.20	2.94	3.67	4.41	5.88	7.34	8.81	10.3	11.8
12	2.25	3.00	3.75	4.50	6.00	7.50	9.00	10.5	12.0

Table 1-20. Weights of Rectangular Plates and Bars, lb/ft

Width in.	Thickness, in.								
	3/16	1/4	5/16	3/8	1/2	5/8	3/4	7/8	1
1/4	0.160	0.213	0.266	0.319	0.425	0.532	0.638	0.744	0.851
1/2	0.319	0.425	0.532	0.638	0.851	1.06	1.28	1.49	1.70
3/4	0.479	0.638	0.798	0.957	1.28	1.60	1.91	2.23	2.55
1	0.638	0.851	1.06	1.28	1.70	2.13	2.55	2.98	3.40
1 1/4	0.798	1.06	1.33	1.60	2.13	2.66	3.19	3.72	4.25
1 1/2	0.957	1.28	1.60	1.91	2.55	3.19	3.83	4.47	5.10
1 3/4	1.12	1.49	1.86	2.23	2.98	3.72	4.47	5.21	5.95
2	1.28	1.70	2.13	2.55	3.4	4.25	5.10	5.95	6.81
2 1/4	1.44	1.91	2.39	2.87	3.83	4.79	5.74	6.70	7.66
2 1/2	1.60	2.13	2.66	3.19	4.25	5.32	6.38	7.44	8.51
2 3/4	1.75	2.34	2.92	3.51	4.68	5.85	7.02	8.19	9.36
3	1.91	2.55	3.19	3.83	5.10	6.38	7.66	8.93	10.2
3 1/4	2.07	2.76	3.46	4.15	5.53	6.91	8.29	9.68	11.1
3 1/2	2.23	2.98	3.72	4.47	5.95	7.44	8.93	10.4	11.9
3 3/4	2.39	3.19	3.99	4.79	6.38	7.98	9.57	11.2	12.8
4	2.55	3.40	4.25	5.10	6.81	8.51	10.2	11.9	13.6
4 1/4	2.71	3.62	4.52	5.42	7.23	9.04	10.8	12.7	14.5
4 1/2	2.87	3.83	4.79	5.74	7.66	9.57	11.5	13.4	15.3
4 3/4	3.03	4.04	5.05	6.06	8.08	10.1	12.1	14.1	16.2
5	3.19	4.25	5.32	6.38	8.51	10.6	12.8	14.9	17.0
5 1/4	3.35	4.47	5.58	6.70	8.93	11.2	13.4	15.6	17.9
5 1/2	3.51	4.68	5.85	7.02	9.36	11.7	14.0	16.4	18.7
5 3/4	3.67	4.89	6.11	7.34	9.78	12.2	14.7	17.1	19.6
6	3.83	5.10	6.38	7.66	10.2	12.8	15.3	17.9	20.4
6 1/4	3.99	5.32	6.65	7.98	10.6	13.3	16.0	18.6	21.3
6 1/2	4.15	5.53	6.91	8.29	11.1	13.8	16.6	19.4	22.1
6 3/4	4.31	5.74	7.18	8.61	11.5	14.4	17.2	20.1	23.0
7	4.47	5.95	7.44	8.93	11.9	14.9	17.9	20.8	23.8
7 1/4	4.63	6.17	7.71	9.25	12.3	15.4	18.5	21.6	24.7
7 1/2	4.79	6.38	7.98	9.57	12.8	16.0	19.1	22.3	25.5
7 3/4	4.94	6.59	8.24	9.89	13.2	16.5	19.8	23.1	26.4
8	5.10	6.81	8.51	10.2	13.6	17.0	20.4	23.8	27.2
8 1/4	5.26	7.02	8.77	10.5	14.0	17.5	21.1	24.6	28.1
8 1/2	5.42	7.23	9.04	10.8	14.5	18.1	21.7	25.3	28.9
8 3/4	5.58	7.44	9.30	11.2	14.9	18.6	22.3	26.1	29.8
9	5.74	7.66	9.57	11.5	15.3	19.1	23.0	26.8	30.6
9 1/4	5.90	7.87	9.84	11.8	15.7	19.7	23.6	27.5	31.5
9 1/2	6.06	8.08	10.1	12.1	16.2	20.2	24.2	28.3	32.3
9 3/4	6.22	8.29	10.4	12.4	16.6	20.7	24.9	29.0	33.2
10	6.38	8.51	10.6	12.8	17.0	21.3	25.5	29.8	34.0
10 1/4	6.54	8.72	10.9	13.1	17.4	21.8	26.2	30.5	34.9
10 1/2	6.70	8.93	11.2	13.4	17.9	22.3	26.8	31.3	35.7
10 3/4	6.86	9.14	11.4	13.7	18.3	22.9	27.4	32.0	36.6
11	7.02	9.36	11.7	14.0	18.7	23.4	28.1	32.8	37.4
11 1/4	7.18	9.57	12.0	14.4	19.1	23.9	28.7	33.5	38.3
11 1/2	7.34	9.78	12.2	14.7	19.6	24.5	29.3	34.2	39.1
11 3/4	7.50	10.0	12.5	15.0	20.0	25.0	30.0	35.0	40.0
12	7.66	10.2	12.8	15.3	20.4	25.5	30.6	35.7	40.8

Table 1-21. Areas and Weights of Round Bars

Size, in.	Area, in.^{2}	Wt., lb/ft	Size, in.	Area, in.^{2}	Wt., lb/ft	Size, in.	Area, in.^{2}	Wt., lb/ft	Size, in.	Area, in.^{2}	Wt., lb/ft
0	0	0	3	7.07	24.1	6	28.3	96.2	9	63.6	216
1/16	0.00307	0.0104	1/16	7.37	25.1	1/16	28.9	98.2	1/16	64.5	219
1/8	0.0123	0.0418	1/8	7.67	26.1	1/8	29.5	100	1/8	65.4	223
3/16	0.0276	0.0940	3/16	7.98	27.2	3/16	30.1	102	3/16	66.3	226
1/4	0.0491	0.167	1/4	8.30	28.2	1/4	30.7	104	1/4	67.2	229
5/16	0.0767	0.261	5/16	8.62	29.3	5/16	31.3	106	5/16	68.1	232
3/8	0.110	0.376	3/8	8.95	30.4	3/8	31.9	109	3/8	69.0	235
7/16	0.150	0.512	7/16	9.28	31.6	7/16	32.5	111	7/16	70.0	238
1/2	0.196	0.668	1/2	9.62	32.7	1/2	33.2	113	1/2	70.9	241
9/16	0.249	0.846	9/16	9.97	33.9	9/16	33.8	115	9/16	71.8	244
5/8	0.307	1.04	5/8	10.3	35.1	5/8	34.5	117	5/8	72.8	248
11/16	0.371	1.26	11/16	10.7	36.3	11/16	35.1	120	11/16	73.7	251
3/4	0.442	1.50	3/4	11.0	37.6	3/4	35.8	122	3/4	74.7	254
13/16	0.518	1.76	13/16	11.4	38.8	13/16	36.5	124	13/16	75.6	257
7/8	0.601	2.05	7/8	11.8	40.1	7/8	37.1	126	7/8	76.6	261
15/16	0.690	2.35	15/16	12.2	41.4	15/16	37.8	129	15/16	77.6	264
1	0.785	2.67	4	12.6	42.8	7	38.5	131	10	78.5	267
1/16	0.887	3.02	1/16	13.0	44.1	1/16	39.2	133	1/16	79.5	271
1/8	0.994	3.38	1/8	13.4	45.5	1/8	39.9	136	1/8	80.5	274
3/16	1.11	3.77	3/16	13.8	46.9	3/16	40.6	138	3/16	81.5	277
1/4	1.23	4.18	1/4	14.2	48.3	1/4	41.3	140	1/4	82.5	281
5/16	1.35	4.60	5/16	14.6	49.7	5/16	42.0	143	5/16	83.5	284
3/8	1.48	5.05	3/8	15.0	51.2	3/8	42.7	145	3/8	84.5	288
7/16	1.62	5.52	7/16	15.5	52.6	7/16	43.4	148	7/16	85.6	291
1/2	1.77	6.01	1/2	15.9	54.1	1/2	44.2	150	1/2	86.6	295
9/16	1.92	6.52	9/16	16.3	55.6	9/16	44.9	153	9/16	87.6	298
5/8	2.07	7.06	5/8	16.8	57.2	5/8	45.7	155	5/8	88.7	302
11/16	2.24	7.61	11/16	17.3	58.7	11/16	46.4	158	11/16	89.7	305
3/4	2.41	8.18	3/4	17.7	60.3	3/4	47.2	161	3/4	90.8	309
13/16	2.58	8.78	13/16	18.2	61.9	13/16	47.9	163	13/16	91.8	312
7/8	2.76	9.40	7/8	18.7	63.5	7/8	48.7	166	7/8	92.9	316
15/16	2.95	10.0	15/16	19.1	65.2	15/16	49.5	168	15/16	94.0	320
2	3.14	10.7	5	19.6	66.8	8	50.3	171	11	95.0	323
1/16	3.34	11.4	1/16	20.1	68.5	1/16	51.1	174	1/16	96.1	327
1/8	3.55	12.1	1/8	20.6	70.2	1/8	51.8	176	1/8	97.2	331
3/16	3.76	12.8	3/16	21.1	71.9	3/16	52.6	179	3/16	98.3	334
1/4	3.98	13.5	1/4	21.6	73.7	1/4	53.5	182	1/4	99.4	338
5/16	4.20	14.3	5/16	22.2	75.4	5/16	54.3	185	5/16	101	342
3/8	4.43	15.1	3/8	22.7	77.2	3/8	55.1	187	3/8	102	346
7/16	4.67	15.9	7/16	23.2	79.0	7/16	55.9	190	7/16	103	350
1/2	4.91	16.7	1/2	23.8	80.8	1/2	56.7	193	1/2	104	353
9/16	5.16	17.5	9/16	24.3	82.7	9/16	57.6	196	9/16	105	357
5/8	5.41	18.4	5/8	24.9	84.6	5/8	58.4	199	5/8	106	361
11/16	5.67	19.3	11/16	25.4	86.5	11/16	59.3	202	11/16	107	365
3/4	5.94	20.2	3/4	26.0	88.4	3/4	60.1	205	3/4	108	369
13/16	6.21	21.1	13/16	26.5	90.3	13/16	61.0	208	13/16	110	373
7/8	6.49	22.1	7/8	27.1	92.2	7/8	61.9	211	7/8	111	377
15/16	6.78	23.1	15/16	27.7	94.2	15/16	62.7	213	15/16	112	381
									12	113	385

Table 1-22. AISI Standard Nomenclature for Flat-Rolled Carbon Steel

Thickness, in.	Width, in. To $3^1/_2$ incl.	Over $3^1/_2$ To 6	Over 6 To 8	Over 8 To 12	Over 12 To 48	Over 48
0.2300 & thicker	Bar	Bar	Bar	Plate	Plate	Plate
0.2299 to 0.2031	Bar	Bar	Strip	Strip	Sheet	Plate
0.2030 to 0.1800	Strip	Strip	Strip	Strip	Sheet	Plate
0.1799 to 0.0449	Strip	Strip	Strip	Strip	Sheet	Sheet
0.0448 to 0.0344	Strip	Strip	Hot-rolled sheet and strip not generally produced in these widths and thicknesses			
0.0343 to 0.0255	Strip					
0.0254 & thinner						

Table 1-23. Weights of Raised-Pattern Floor Plates

Gauge No.	Wt., lb/ft^2	Nominal Thickness, in.	Wt., lb/ft^2	Nominal Thickness, in.	Wt., lb/ft^2
18	2.40	$^1/_8$	6.16	$^1/_2$	21.5
16	3.00	$^3/_{16}$	8.71	$^9/_{16}$	24.0
14	3.75	$^1/_4$	11.3	$^5/_8$	26.6
13	4.50	$^5/_{16}$	13.8	$^3/_4$	31.7
12	5.25	$^3/_8$	16.4	$^7/_8$	36.8
		$^7/_{16}$	18.9	1	41.9

Note: thickness is measured near the edge of the plate, exclusive of raised pattern.

Table 1-24.
Crane Rails
Dimensions and Properties

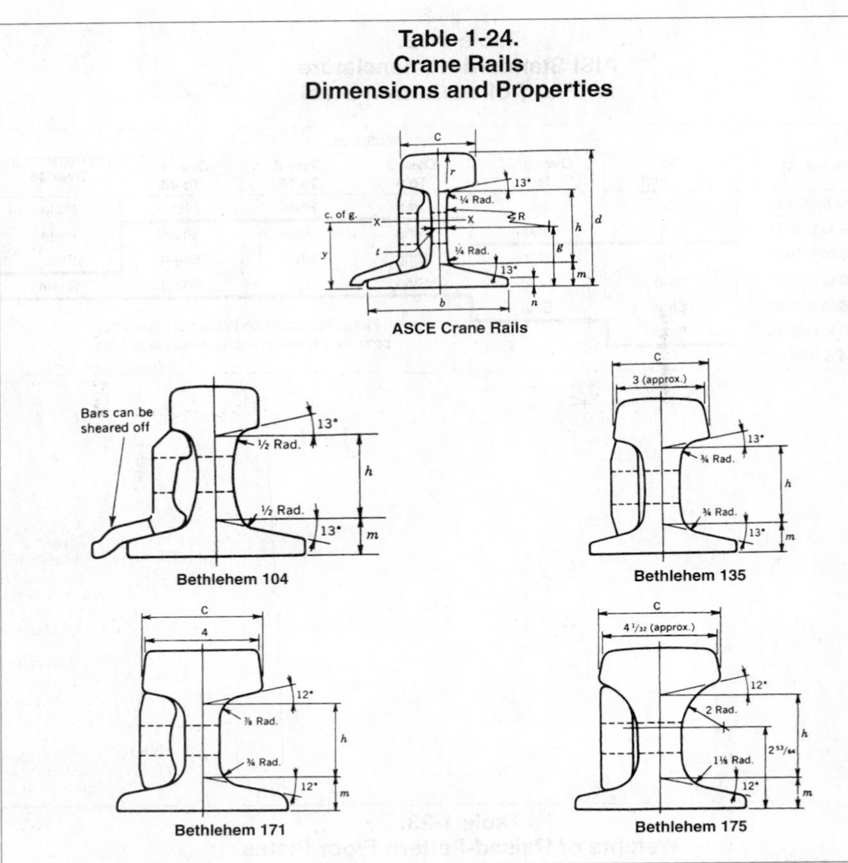

Type	Classification	Wt.	Depth, d	Gage, g	Base			Head		Web			Axis X-X				
					b	m	n	c	r	t	h	R	Area	I	S Head	S Base	y
		lb/yd	in.	in.	in.	in.	in.	in.	in.	in.	in.	in.	in.^{2}	in.^{4}	in.^{3}	in.^{3}	in.
ASCE	Light	30	3 1/8	1 25/64	3 1/8	17/32	11/64	1 11/16	12	21/64	1 23/32	12	3.00	4.10	2.55	–	–
		40	3 1/2	1 71/128	3 1/2	5/8	7/32	1 7/8	12	25/64	1 55/64	12	3.94	6.54	3.59	3.89	1.68
		50	3 7/8	1 23/32	3 7/8	11/16	1/4	2 1/8	12	7/16	2 1/16	12	4.90	10.1	5.10	–	1.88
		60	4 1/4	1 115/128	4 1/4	49/64	9/32	2 3/8	12	31/64	2 17/64	12	5.93	14.6	6.64	7.12	2.05
	–	70	4 5/8	2 3/64	4 5/8	13/16	9/32	2 7/16	12	33/64	2 15/32	12	6.81	19.7	8.19	8.87	2.22
		80	5	2 3/16	5	7/8	19/64	2 1/2	12	35/64	2 5/8	12	7.86	26.4	10.1	11.1	2.38
	Std.	85	5 3/16	2 17/64	5 3/16	57/64	19/64	2 9/16	12	9/16	2 3/4	12	8.33	30.1	11.1	12.2	2.47
		100	5 3/4	2 65/128	5 3/4	31/32	5/16	2 3/4	12	9/16	2 5/64	12	9.84	44.0	14.6	16.1	2.73
Bethlehem	Crane	104	5	2 7/16	5	1 1/16	1/2	2 1/2	12	1	2 7/16	3 1/2	10.3	29.8	10.7	13.5	2.21
		135	5 3/4	2 15/32	5 3/16	1 1/16	15/32	3 7/16	14	1 1/4	2 13/16	12	13.3	50.8	17.3	18.1	2.81
		171	6	2 5/8	6	1 1/4	5/8	4.3	Flat	1 1/4	2 3/4	Vert.	16.8	73.4	24.5	24.4	3.01
		175	6	2 21/32	6	1 9/64	1/2	4 1/4	18	1 1/2	3 7/64	Vert.	17.1	70.5	23.4	23.6	2.98

Table 1-25.
W-Shapes
Torsional Properties

Shape	Torsional Constant, J	Warping Constant, C_w	$\sqrt{\frac{EC_w}{GJ}}$	Normalized Warping Constant, W_{no}	Warping Statical Moment, S_w	Statical Moment Q_f	Statical Moment Q_w
	in.^{4}	in.^{6}	in.	in.^{2}	in.^{4}	in.^{3}	in.^{3}
W44×335*	74.4	536000	137	168	1190	279	805
×290	51.5	464000	153	166	1040	248	704
×262	37.7	407000	167	165	922	222	630
×230	24.9	346000	190	164	789	191	546
W40×593*	445	996000	76.1	166	2240	478	1370
×503*	279	791000	85.7	161	1830	403	1150
×431*	177	639000	96.7	158	1510	342	968
×397*	142	578000	103	156	1380	318	891
×372*	116	528000	109	155	1270	295	829
×362*	109	512000	110	154	1240	289	807
×324	79.4	449000	121	153	1100	259	719
×297	61.2	397000	130	151	986	235	654
×277	51.5	378000	138	151	940	225	614
×249	38.1	334000	151	149	836	202	550
×215	24.8	284000	172	149	714	174	471
×199	18.3	246000	186	148	621	151	424
W40×392*	172	306000	67.9	121	940	269	845
×331*	106	241000	76.7	118	760	225	704
×327	103	239000	77.5	117	757	225	696
×278	64.7	192000	87.6	115	622	190	586
×264	56.1	181000	91.3	114	589	182	555
×235	41.3	161000	101	113	530	166	495
×211	30.4	140000	109	112	468	148	442
×183	20.0	120000	125	111	402	129	381
×167	14.0	99800	136	111	336	107	336
×149	9.36	80000	149	110	270	86.7	288
W36×798*	1050	1490000	60.6	169	3270	631	1790
×650*	591	1100000	69.5	162	2520	510	1420
×527*	327	820000	80.6	156	1960	412	1130
×439*	193	638000	92.5	152	1570	341	931
×393*	141	554000	101	150	1390	306	829
×359*	108	492000	109	148	1240	278	751
×328*	84.1	441000	117	147	1130	254	684
×300	64.2	399000	127	146	1020	231	622
×280	52.6	366000	134	145	944	216	579
×260	41.5	330000	144	144	858	197	532
×245	34.6	305000	151	143	799	184	498
×230	28.6	282000	160	143	740	171	465
W36×256	52.9	168000	90.7	109	576	174	516
×232	39.6	148000	98.3	108	512	157	464
×210	28	128000	109	108	446	136	412
×194	22.2	116000	116	107	407	126	379
×182	18.5	107000	122	106	378	118	355
×170	15.1	98400	130	105	349	109	330
×160	12.4	90300	137	105	321	101	308
×150	10.1	82300	145	105	294	93.1	287
×135	7	68000	159	104	245	77.9	251
W33×387*	148	459000	89.6	136	1260	287	777
×354*	115	409000	95.9	135	1130	261	706
×318*	84.4	357000	105	133	1000	235	631
×291	65.1	318000	112	132	906	214	574
×263	48.7	282000	123	130	808	193	516
×241	36.2	251000	134	130	721	172	466
×221	27.8	224000	144	129	650	156	424
×201	20.8	198000	157	128	580	141	383
W33×169	17.7	82400	110	93.7	329	108	311
×152	12.4	71800	122	93.8	286	93.5	276
×141	9.7	64300	131	93.3	258	84.8	254
×130	7.37	56600	141	92.8	228	75.3	230
×118	5.3	48200	154	92.2	196	64.9	204

*Tensile group 4 or 5 shape. Special requirements may apply per LRFD Specification Section A3.1c.

Table 1-25 (cont.). W-Shapes Torsional Properties

Shape	Torsional Constant, J	Warping Constant, C_w	$\sqrt{\frac{EC_w}{GJ}}$	Normalized Warping Constant, W_{no}	Warping Statical Moment, S_w	Statical Moment Q_f	Statical Moment Q_w
	in.⁴	in.⁶	in.	in.²	in.⁴	in.³	in.³
W30×391*	173	366000	74.1	120	1140	267	721
×357*	134	325000	79.2	118	1020	244	654
×326*	103	286000	84.7	117	919	221	592
×292*	75.2	250000	92.8	115	812	199	528
×261	54.1	215000	101	114	710	176	468
×235	40.3	190000	110	112	633	159	420
×211	28.4	166000	123	112	556	140	372
×191	21	146000	134	111	494	125	334
×173	15.6	129000	146	110	439	112	300
W30×148	14.5	49400	93.9	77.3	239	85.5	247
×132	9.72	42100	106	77.3	204	72.8	216
×124	7.99	38700	112	76.9	188	67.5	202
×116	6.43	34900	118	76.5	171	61.5	187
×108	4.99	30800	127	76.1	152	54.8	170
×99	3.77	26900	136	75.7	133	48.2	154
×90	2.84	24000	148	75.2	119	43.8	139
W27×539*	496	443000	48.1	111	1490	341	942
×368*	170	255000	62.3	102	930	230	619
×336*	131	227000	66.9	101	836	210	562
×307*	101	199000	71.4	99.4	750	191	509
×281*	79.5	178000	76.2	98.2	680	175	465
×258	61.6	159000	81.8	97.1	613	160	423
×235	47	141000	88	96	548	145	383
×217	37.6	128000	93.8	95	503	134	352
×194	27.1	111000	103	93.9	442	119	312
×178	20.1	98300	113	93.7	393	106	282
×161	15.1	87300	122	92.9	352	95.6	254
×146	11.3	77200	133	92.2	314	86	229
W27×129	11.1	32400	86.9	66.4	183	68.6	195
×114	7.33	27600	98.8	66.4	155	58.2	170
×102	5.28	24000	108	65.7	137	51.8	151
×94	4.03	21200	117	65.4	122	46.3	137
×84	2.81	18000	129	64.9	103	39.6	120
W24×370*	201	185000	48.8	86.3	802	209	566
×335*	152	161000	52.4	84.6	709	188	508
×306*	117	142000	56	83.3	636	172	460
×279*	90.5	125000	59.8	82	570	156	416
×250*	66.6	108000	64.9	80.6	502	140	371
×229	51.3	96000	69.6	79.6	451	128	337
×207	38.3	84200	75.5	78.5	401	115	302
×192	30.8	76400	80.1	77.7	367	106	278
×176	23.9	68400	86.1	77	333	97.2	254
×162	18.5	62600	93.6	77	304	88.8	233
×146	13.4	54700	103	76.3	268	78.9	208
×131	9.5	47000	113	75.6	233	69.1	184
×117	6.72	40700	125	74.9	204	60.9	162
×104	4.72	35200	139	74.3	178	53.5	143
W24×103	7.07	16500	77.7	53	117	48.8	139
×94	5.26	15000	85.8	53.1	105	43.8	126
×84	3.7	12800	94.8	52.6	91.3	38.4	111
×76	2.68	11100	104	52.2	79.8	33.8	99
×68	1.87	9430	114	51.9	68	28.9	87.1
W24×62	1.77	4620	82.2	40.7	42.3	22.6	75.4
×55	1.24	3870	89.9	40.4	35.7	19.2	65.9
W21×201	40.9	62100	62.7	67.3	345	102	264
×182	30.7	54500	67.8	66.4	307	91.7	237
×166	23.6	48500	73	65.6	277	83.8	215
×147	15.4	41100	83.1	65.4	235	70.9	186
×132	11.3	36000	90.8	64.7	208	63.4	166
×122	8.98	32700	97.2	64.2	191	58.6	153
×111	6.83	29200	105	63.7	172	53.2	138
×101	5.21	26200	114	63.2	155	48.5	125

*Tensile group 4 or 5 shape. Special requirements may apply per LRFD Specification Section A3.1c.

Table 1-25 (cont.). W-Shapes Torsional Properties

Shape	Torsional Constant, J	Warping Constant, C_w	$\sqrt{\frac{EC_w}{GJ}}$	Normalized Warping Constant, W_{no}	Warping Statical Moment, S_w	Statical Moment	
						Q_f	Q_w
	in.^{4}	in.^{6}	in.	in.^{2}	in.^{4}	in.^{3}	in.^{3}
W21×93	6.03	9940	65.3	43.6	85.3	37.7	109
×83	4.34	8630	71.8	43	75	33.7	97
×73	3.02	7420	79.7	42.5	65.2	29.7	85.1
×68	2.45	6760	84.5	42.3	59.9	27.4	78.9
×62	1.83	5970	91.9	42	53.2	24.6	71.1
×55	1.24	4980	102	41.7	44.7	20.8	61.8
×48	0.803	3940	113	41.1	36	16.9	52.4
W21×57	1.77	3190	68.3	33.4	35.6	20.4	63.2
×50	1.14	2560	76.3	33.1	28.9	16.7	54
×44	0.77	2110	84.3	32.8	24	14	46.6
W18×175	33.8	33300	50.5	52.5	237	76.9	198
×158	25.2	29000	54.6	51.6	210	69	177
×143	19.2	25700	58.8	51	189	62.9	160
×130	14.5	22600	63.6	50.4	169	56.8	145
×119	10.6	20300	70.4	50.4	151	50.4	130
×106	7.48	17400	77.6	49.8	131	44.4	115
×97	5.86	15800	83.5	49.4	120	40.9	105
×86	4.1	13600	92.6	48.9	104	36	92.3
×76	2.83	11700	103	48.4	90.7	31.6	80.9
W18×71	3.49	4700	59.1	33.7	52.1	25.5	72.2
×65	2.73	4240	63.4	33.4	47.5	23.6	66.1
×60	2.17	3860	67.8	33.1	43.5	21.8	60.8
×55	1.66	3430	73.1	32.9	39	19.7	55.3
×50	1.24	3040	79.7	32.6	34.9	17.7	49.8
W18×46	1.22	1710	60.3	26.4	24.2	15	44.8
×40	0.81	1440	67.9	26.1	20.6	13	38.6
×35	0.506	1140	76.4	25.9	16.5	10.5	32.7
W16×100	8.21	11900	61.2	41.7	107	38.7	98.5
×89	5.83	10300	67.5	41.1	93.3	34.2	86.8
×77	3.86	8570	75.8	40.6	79.3	29.5	74.5
×67	2.62	7300	84.9	40.1	68.2	25.6	64.4
W16×57	2.22	2660	55.7	28	35.6	18.8	52.1
×50	1.52	2270	62.2	27.6	30.8	16.5	45.5
×45	1.11	1990	68.1	27.4	27.2	14.7	40.6
×40	0.794	1740	75.3	27.1	23.9	13.1	36
×36	0.545	1460	83.2	26.9	20.2	11.1	31.5
W16×31	0.461	739	64.4	21.3	13	8.92	26.5
×26	0.262	565	74.7	21.1	10	6.95	21.6
W14×808*	1840	433000	24.7	82.2	1950	336	916
×730*	1450	362000	25.4	78.3	1720	319	830
×665*	1120	306000	26.6	75.5	1510	287	739
×605*	869	258000	27.7	73	1320	258	659
×550*	669	219000	29.1	70.6	1160	232	587
×500*	514	187000	30.7	68.5	1020	209	523
×455*	395	160000	32.4	66.5	899	188	467
×426*	331	144000	33.6	65.3	827	176	433
×398*	273	129000	35	64.1	756	163	400
×370*	222	116000	36.8	62.9	689	150	367
×342*	178	103000	38.7	61.6	623	138	335
×311*	136	88900	41.1	60.3	553	124	301
×283*	104	77500	43.9	59.1	493	113	270
×257*	79.1	67700	47.1	57.9	438	101	242
×233*	59.5	59000	50.7	56.9	389	91.3	217
×211	44.6	51600	54.7	55.9	345	81.8	194
×193	34.8	45900	58.4	55.1	312	74.9	176
×176	26.5	40500	62.9	54.4	279	67.5	159
×159	19.7	35600	68.4	53.7	248	60.8	142
×145	15.2	31700	73.5	53	224	55.3	129

*Tensile group 4 or 5 shape. Special requirements may apply per LRFD Specification Section A3.1c.

Table 1-25 (cont.). W-Shapes Torsional Properties

Shape	Torsional Constant, J	Warping Constant, C_w	$\sqrt{\frac{EC_w}{GJ}}$	Normalized Warping Constant, W_{no}	Warping Statical Moment, S_w	Statical Moment	
						Q_f	Q_w
	in.^{4}	in.^{6}	in.	in.^{2}	in.^{4}	in.^{3}	in.^{3}
W14×132	12.3	25500	73.2	50.2	190	49.4	116
×120	9.37	22700	79.2	49.7	171	44.8	105
×109	7.12	20200	85.8	49.1	154	40.7	94.9
×99	5.37	18000	93.1	48.7	138	36.7	85.6
×90	4.06	16000	101	48.3	125	33.3	77.3
W14×82	5.07	6700	58.5	34.1	73.8	27.7	68.4
×74	3.87	6000	63.4	33.7	66.6	25.3	61.8
×68	3.01	5370	67.9	33.4	60.4	23.1	56.4
×61	2.19	4690	74.5	33.1	53.3	20.5	50.1
W14×53	1.94	2540	58.2	26.7	35.5	16.8	42.6
×48	1.45	2240	63.2	26.5	31.6	15.1	38.3
×43	1.05	1950	69.3	26.2	27.8	13.4	33.9
W14×38	0.798	1230	63.2	23	20	11.3	30.3
×34	0.569	1070	69.6	22.8	17.5	9.94	26.8
×30	0.38	887	77.7	22.6	14.7	8.37	23.2
W14×26	0.358	405	54.1	16.9	8.94	6.76	19.7
×22	0.208	314	62.6	16.8	7.02	5.36	16.1
W12×336*	243	57200	24.7	46.4	459	119	301
×305*	185	48700	26.1	45	403	107	268
×279*	143	41900	27.6	44	357	95.9	240
×252*	108	35800	29.3	42.8	313	85.9	213
×230*	83.8	31300	31.1	41.8	279	78	192
×210*	64.7	27200	33	41	249	70.6	173
×190	48.8	23500	35.3	40.1	220	63.7	155
×170	35.6	20100	38.2	39.2	192	56.5	137
×152	25.8	17200	41.5	38.4	168	50	120
×136	18.5	14700	45.4	37.7	146	44.1	106
×120	12.9	12500	50	37	126	38.5	92.3
×106	9.13	10700	55	36.4	110	34.2	81.1
×96	6.85	9410	59.7	35.9	98.2	30.9	72.8
×87	5.1	8280	64.8	35.5	87.2	27.6	65.2
×79	3.84	7320	70.3	35.2	78.1	24.8	58.7
×72	2.93	6540	76	34.9	70.3	22.5	53.1
×65	2.18	5770	82.8	34.5	62.7	20.2	47.6
W12×58	2.1	3570	66.3	28.9	46.3	17.8	42.4
×53	1.58	3160	72	28.7	41.2	15.9	38.1
W12×50	1.71	1880	53.3	23.3	30.2	14.2	35.4
×45	1.26	1650	58.2	23.1	26.7	12.7	31.5
×40	0.906	1440	64.1	22.9	23.6	11.3	27.9
W12×35	0.741	879	55.4	19.6	16.8	9.75	25.4
×30	0.457	719	63.8	19.4	13.9	8.19	21.3
×26	0.3	606	72.3	19.2	11.8	7.04	18.4
W12×22	0.293	165	38.1	12	5.13	4.76	14.4
×19	0.18	131	43.4	11.8	4.14	3.9	12.1
×16	0.103	96.9	49.4	11.7	3.09	2.93	9.81
×14	0.0704	80.6	54.4	11.6	2.59	2.48	8.5
W10×112	15.1	6030	32.2	26.3	85.7	30.5	73.2
×100	10.9	5150	35	25.8	74.7	27	64.5
×88	7.53	4340	38.6	25.3	64.2	23.5	56
×77	5.11	3640	43	24.8	54.9	20.4	48.3
×68	3.56	3110	47.5	24.4	47.6	17.9	42.2
×60	2.48	2640	52.5	24	41.2	15.7	36.8
×54	1.82	2310	57.3	23.8	36.6	14.1	32.9
×49	1.39	2070	62.1	23.6	33	12.7	29.7
W10×45	1.51	1200	45.4	19	23.6	11.3	27
×39	0.976	992	51.3	18.7	19.8	9.54	23
×33	0.583	791	59.3	18.5	16	7.75	18.9

*Tensile group 4 or 5 shape. Special requirements may apply per LRFD Specification Section A3.1c.

Table 1-25 (cont.). W-Shapes Torsional Properties

Shape	Torsional Constant, J	Warping Constant, C_w	$\sqrt{\frac{EC_w}{GJ}}$	Normalized Warping Constant, W_{no}	Warping Statical Moment, S_w	Statical Moment	
						Q_f	Q_w
	in.^{4}	in.^{6}	in.	in.^{2}	in.^{4}	in.^{3}	in.^{3}
W10×30	0.622	414	41.5	14.5	10.7	7	18.1
×26	0.402	345	47.1	14.3	9.05	5.99	15.5
×22	0.239	274	54.5	14.1	7.3	4.86	12.8
W10×19	0.233	104	34	9.89	3.93	3.67	10.6
×17	0.156	85.1	37.6	9.8	3.24	3.04	9.15
×15	0.104	68.3	41.2	9.72	2.62	2.47	7.82
×12	0.0547	50.9	49.1	9.56	1.99	1.91	6.14
W8×67	5.05	1440	27.2	16.7	32.3	14.5	34.8
×58	3.33	1180	30.3	16.3	27.2	12.4	29.7
×48	1.96	930	35	15.8	22	10.3	24.2
×40	1.12	726	41	15.5	17.5	8.3	19.7
×35	0.769	619	45.7	15.3	15.2	7.28	17.1
×31	0.536	531	50.6	15.1	13.1	6.34	15
W8×28	0.537	313	38.8	12.4	9.43	5.52	13.4
×24	0.346	259	44.1	12.2	7.94	4.71	11.3
W8×21	0.282	152	37.3	10.4	5.47	3.96	10.1
×18	0.172	122	42.8	10.3	4.44	3.23	8.37
W8×15	0.137	51.8	31.3	7.82	2.47	2.31	6.64
×13	0.0871	40.8	34.8	7.74	1.97	1.86	5.55
×10	0.0426	30.9	43.3	7.57	1.53	1.48	4.29
W6×25	0.47	150	28.8	9.01	6.23	3.88	9.39
×20	0.246	113	34.5	8.78	4.82	3.07	7.38
×15	0.105	76.5	43.4	8.58	3.34	2.15	5.32
W6×16	0.223	38.2	21.1	5.92	2.42	2.24	5.77
×12	0.0903	24.7	26.6	5.75	1.61	1.52	4.08
×9	0.0405	17.8	33.7	5.6	1.19	1.15	3.04
×8.5	0.033	15.7	35.1	5.55	1.06	1.03	2.78
W5×19	0.316	50.9	20.4	5.94	3.21	2.42	5.73
×16	0.192	40.6	23.4	5.81	2.62	1.99	4.74
W4×13	0.151	14	15.5	3.87	1.36	1.24	3.09

*Tensile group 4 or 5 shape. Special requirements may apply per LRFD Specification Section A3.1c.

Table 1-26.
M-Shapes
Torsional Properties

Shape	Torsional Constant, J	Warping Constant, C_w	$\sqrt{\frac{EC_w}{GJ}}$	Normalized Warping Constant, W_{no}	Warping Statical Moment, S_w	Statical Moment	
						Q_f	Q_w
	in.⁴	in.⁶	in.	in.²	in.⁴	in.³	in.³
M12×11.8	0.0500	37.8	44.2	9.02	1.56	1.91	7.01
×10.8	0.0393	34.9	48.0	9.01	1.45	1.79	6.45
×10	0.0292	35.8	56.3	9.58	1.40	1.65	5.96
M10×9	0.0314	16.1	36.5	6.59	0.912	1.28	4.52
×8	0.0224	14.1	40.4	6.57	0.804	1.13	4.01
×7.5	0.0187	13.5	43.3	6.60	0.767	1.09	3.79
M8×6.5	0.0184	5.74	28.4	4.45	0.480	0.792	2.66
×6.2	0.0156	5.39	29.9	4.46	0.450	0.745	2.52
M6×4.4	0.00990	1.53	20.0	2.69	0.212	0.431	1.38
×3.7	0.00530	1.45	26.6	2.90	0.187	0.355	1.14
M5×18.9*	0.340	41.3	17.7	5.73	2.98	2.28	5.53
M4×6	0.0184	4.87	26.2	3.46	0.526	0.534	1.30

*This shape has tapered flanges while all other M-shapes have parallel flanges.

Table 1-27.
S-Shapes
Torsional Properties

Shape	Torsional Constant, J	Warping Constant, C_w	$\sqrt{\frac{EC_w}{GJ}}$	Normalized Warping Constant, W_{no}	Warping Statical Moment, S_w	Statical Moment	
						Q_f	Q_w
	in.^{4}	in.^{6}	in.	in.^{2}	in.^{4}	in.^{3}	in.^{3}
S24×121	12.8	11400	48.0	47.1	103	46.2	153
×106	10.1	10500	51.9	46.1	98.8	46.2	139
S24×100	7.59	6340	46.5	41.9	66.0	32.7	119
×90	6.05	5980	50.6	41.2	63.8	32.7	110
×80	4.89	5620	54.5	40.5	61.6	32.7	101
S20×96	8.40	4690	38.0	34.9	57.8	28.5	98.3
×86	6.65	4380	41.3	34.2	55.5	28.5	91.1
S20×75	4.59	2720	39.2	30.7	38.9	21.9	75.6
×66	3.58	2540	42.8	30.0	37.3	21.9	69.1
S18×70	4.10	1800	33.7	27.0	29.2	16.6	61.9
×54.7	2.33	1550	41.5	26.0	26.9	16.6	51.8
S15×50	2.12	806	31.4	20.3	17.8	11.4	38.2
×42.9	1.54	739	35.3	19.8	16.9	11.4	34.3
S12×50	2.77	502	21.7	15.5	14.0	8.95	30.3
×40.8	1.69	434	25.8	14.9	12.9	8.95	26.2
S12×35	1.05	323	28.2	14.5	10.0	7.24	22.2
×31.8	0.878	306	30.0	14.3	9.74	7.24	20.8
S10×35	1.29	188	19.4	11.8	7.13	5.08	17.6
×25.4	0.603	152	25.6	11.1	6.34	5.08	14.0
S8×23	0.550	61.3	17.0	7.90	3.50	3.00	9.53
×18.4	0.335	52.9	20.2	7.58	3.22	3.00	8.17
S6×17.25	0.371	18.2	11.3	5.03	1.61	1.57	5.23
×12.5	0.167	14.3	14.9	4.70	1.41	1.57	4.18
S5×10	0.114	6.50	12.1	3.51	0.859	1.06	2.79
S4×9.5	0.120	3.05	8.11	2.59	0.531	0.671	1.99
×7.7	0.0732	2.57	9.53	2.47	0.481	0.671	1.73
S3×7.5	0.0896	1.08	5.60	1.72	0.280	0.385	1.16
×5.7	0.0433	0.839	7.08	1.60	0.242	0.385	0.961

Table 1-28. HP-Shapes Torsional Properties

Shape	Torsional Constant, J	Warping Constant, C_w	$\sqrt{\frac{EC_w}{GJ}}$	Normalized Warping Constant, W_{no}	Warping Statical Moment, S_w	Statical Moment	
						Q_f	Q_w
	in.^{4}	in.^{6}	in.	in.^{2}	in.^{4}	in.^{3}	in.^{3}
HP14×117	8.02	19900	80.2	49.9	149	38	96.3
×102	5.39	16800	89.9	49.2	128	33	83.3
×89	3.59	14200	101	48.5	110	28.6	71.9
×73	2.01	11200	120	47.8	88	23.3	58.3
HP12×84	4.24	7160	66.1	35.6	75	23.1	59
×74	2.98	6170	73.2	35.2	65.5	20.4	51.9
×63	1.83	4990	84	34.6	54.1	17.1	43.3
×53	1.12	4090	97.2	34.2	44.7	14.3	36.2
HP10×57	1.97	2240	54.3	24.1	34.8	12.9	32.8
×42	0.813	1540	70.1	23.4	24.7	9.41	23.7
HP8×36	0.77	578	44.1	15.4	14	6.5	16.6

Table 1-29. C-Shapes Flexural-Torsional Properties

Shape	Torsional Constant, J	Warping Constant, C_w	X_1	$X_2 \times 10^6$	Polar Radius of Gyration $\bar{r}_o^*$	Flexural Constant, H^*
	in.^{4}	in.^{6}	ksi	(1/ksi)^2	in.	No Units
C15×50	2.65	492	4640	588	5.49	0.937
×40	1.45	410	3560	1470	5.71	0.927
×33.9	1.01	358	3020	2450	5.94	0.920
C12×30	0.861	151	4080	925	4.54	0.919
×25	0.538	130	3310	1850	4.72	0.909
×20.7	0.369	112	2790	3140	4.93	0.899
C10×30	1.22	79.5	6340	186	3.63	0.921
×25	0.687	68.3	4940	458	3.76	0.912
×20	0.368	56.9	3720	1190	3.93	0.900
×15.3	0.209	45.5	2870	2670	4.19	0.884
C9×20	0.427	39.4	4700	521	3.46	0.899
×15	0.208	31.0	3390	1530	3.69	0.882
×13.4	0.168	28.2	3070	2050	3.79	0.875
C8×18.75	0.434	25.1	5630	261	3.05	0.894
×13.75	0.186	19.2	3850	947	3.26	0.874
×11.5	0.130	16.5	3260	1570	3.41	0.862
C7×14.75	0.267	13.1	5530	259	2.75	0.875
×12.25	0.161	11.2	4400	569	2.86	0.862
×9.8	0.0996	9.15	3530	1130	3.02	0.845
C6×13	0.237	7.19	6580	130	2.37	0.858
×10.5	0.128	5.91	4990	340	2.48	0.842
×8.2	0.0736	4.70	3860	762	2.65	0.824
C5×9	0.109	2.93	6030	160	2.10	0.815
×6.7	0.0549	2.22	4400	447	2.26	0.790
C4×7.25	0.0817	1.24	7290	73.1	1.75	0.767
×5.4	0.0399	0.921	5240	218	1.88	0.742
×4.5	0.0322	0.871	4610	310	2.01	0.710
C3×6	0.0725	0.462	10400	17.8	1.40	0.690
×5	0.0425	0.379	8140	42.0	1.45	0.673
×4.1	0.0269	0.307	6540	85.7	1.53	0.655
×3.5	0.0226	0.276	6040	110	1.57	0.646

*See LRFD Appendix E3.

Table 1-30.
MC-Shapes
Flexural-Torsional Properties

Shape	Torsional Constant, J	Warping Constant, C_w	X_1	$X_2 \times 10^6$	Polar Radius of Gyration $\bar{r}_o^*$	Flexural Constant, H^*
	in.[4]	in.[6]	ksi	(1/ksi)[2]	in.	No Units
MC18×58	2.81	1070	3700	1380	6.55	0.944
×51.9	2.03	985	3210	2270	6.69	0.939
×45.8	1.45	897	2760	3760	6.86	0.933
×42.7	1.23	852	2560	4750	6.96	0.930
MC13×50	2.96	558	5470	289	5.07	0.875
×40	1.55	462	4090	786	5.32	0.859
×35	1.13	412	3520	1260	5.50	0.849
×31.8	0.937	380	3230	1630	5.64	0.842
MC12×50	3.23	411	6140	146	4.76	0.859
×45	2.34	374	5310	242	4.88	0.851
×40	1.69	336	4580	402	5.00	0.841
×35	1.24	298	3960	639	5.17	0.832
×31	1.00	267	3590	856	5.33	0.822
MC12×10.6	0.0596	11.7	1870	23600	4.27	0.983
MC10×41.1	2.26	269	6650	106	4.25	0.789
×33.6	1.20	224	4960	293	4.48	0.769
×28.5	0.791	193	4070	557	4.67	0.751
MC10×25	0.638	124	3940	649	4.46	0.802
×22	0.510	110	3540	886	4.61	0.791
MC10×8.4	0.0413	7.00	2000	16400	3.67	0.972
MC9×25.4	0.691	104	4660	349	4.08	0.770
×23.9	0.599	98.0	4340	436	4.15	0.762
MC8×22.8	0.572	75.2	4930	264	3.84	0.715
×21.4	0.495	70.8	4580	332	3.91	0.709
MC8×20	0.441	47.8	4740	328	3.58	0.779
×18.7	0.38	45	4420	411	3.65	0.773
MC8×8.5	0.0587	8.21	2640	4120	3.24	0.910
MC7×22.7	0.625	58.3	6050	120	3.52	0.661
×19.1	0.407	49.3	4920	237	3.71	0.638
MC6×18	0.379	34.6	5730	128	3.46	0.562
×15.3	0.223	30	4750	279	3.41	0.579
MC6×16.3	0.336	22.1	5860	124	3.11	0.643
×15.1	0.285	20.5	5430	160	3.18	0.634
MC6×12	0.155	11.3	4750	316	2.80	0.740

*See LRFD Appendix E3.

Table 1-31.
Angles
Flexural-Torsional Properties

Shape	Torsional Constant, J	Warping Constant, C_w	Polar Radius of Gyration $\bar{r}_o^*$	Flexural Constant, H^*
	in.^{4}	in.^{6}	in.	No Units
L8×8×1 1/8	7.13	32.5	4.29	0.633
×1	5.08	23.4	4.32	0.630
×7/8	3.46	16.1	4.36	0.631
×3/4	2.21	10.4	4.39	0.631
×5/8	1.30	6.16	4.41	0.631
×9/16	0.961	4.55	4.43	0.629
×1/2	0.683	3.23	4.45	0.628
L8×6×1	4.34	16.3	3.88	–
×7/8	2.96	11.3	3.92	–
×3/4	1.90	7.28	3.95	–
×5/8	1.12	4.33	3.98	–
×9/16	0.823	3.20	3.99	–
×1/2	0.584	2.28	4.01	–
×7/16	0.396	1.55	4.02	–
L8×4×1	3.68	12.9	3.74	–
×7/8	2.51	8.89	3.78	–
×3/4	1.61	5.75	3.81	–
×5/8	0.955	3.42	3.83	–
×9/16	0.704	2.53	3.84	–
×1/2	0.501	1.80	3.86	–
×7/16	0.340	1.22	3.87	–
L7×4×3/4	1.47	3.97	3.31	–
×5/8	0.868	2.37	3.34	–
×1/2	0.456	1.25	3.37	–
×7/16	0.310	0.851	3.39	–
×3/8	0.198	0.544	3.40	–
L6×6×1	3.68	9.24	3.19	0.635
×7/8	2.51	6.41	3.21	0.635
×3/4	1.61	4.17	3.25	0.631
×5/8	0.955	2.50	3.28	0.631
×9/16	0.704	1.85	3.29	0.629
×1/2	0.501	1.32	3.30	0.631
×7/16	0.340	0.899	3.32	0.629
×3/8	0.218	0.575	3.33	0.630
×5/16	0.129	0.338	3.35	0.630
L6×4×7/8	1.93	4.04	2.83	–
×3/4	1.24	2.64	2.86	–
×5/8	0.736	1.59	2.89	–
×9/16	0.542	1.18	2.91	–
×1/2	0.385	0.843	2.92	–
×7/16	0.260	0.575	2.94	–
×3/8	0.166	0.369	2.95	–
×5/16	0.097	0.217	2.97	–
L6×3 1/2×1/2	0.364	0.779	2.89	–
×3/8	0.157	0.341	2.92	–
×5/16	0.0918	0.201	2.94	–
L5×5×7/8	2.07	3.53	2.64	0.638
×3/4	1.33	2.32	2.67	0.632
×5/8	0.792	1.40	2.70	0.633
×1/2	0.417	0.744	2.73	0.632
×7/16	0.284	0.508	2.75	0.630
×3/8	0.183	0.327	2.76	0.632
×5/16	0.108	0.193	2.78	0.630
L5×3 1/2×3/4	1.09	1.52	2.36	–
×5/8	0.651	0.918	2.39	–
×1/2	0.343	0.491	2.42	–
×3/8	0.150	0.217	2.45	–
×5/16	0.0883	0.128	2.46	–
×1/4	0.0464	0.067	2.48	–

*See LRFD Appendix E3.

Table 1-31 (cont.). Angles Flexural-Torsional Properties

Shape	Torsional Constant, J	Warping Constant, C_w	Polar Radius of Gyration $\bar{r}_o^*$	Flexural Constant, H^*
	in.^{4}	in.^{6}	in.	No Units
L5×3×1/2	0.322	0.444	2.38	–
×7/16	0.220	0.304	2.40	–
×3/8	0.141	0.196	2.41	–
×5/16	0.0832	0.116	2.42	–
×1/4	0.0438	0.0606	2.43	–
L4×4×3/4	1.02	1.12	2.10	0.637
×5/8	0.610	0.680	2.13	0.636
×1/2	0.322	0.366	2.16	0.630
×7/16	0.220	0.252	2.17	0.633
×3/8	0.141	0.162	2.19	0.629
×5/16	0.0832	0.0963	2.21	0.627
×1/4	0.0438	0.0505	2.22	0.630
L4×3 1/2×1/2	0.301	0.302	2.03	–
×3/8	0.132	0.134	2.06	–
×5/16	0.0782	0.0798	2.08	–
×1/4	0.0412	0.0419	2.08	–
L4×3×5/8	0.529	0.472	1.91	–
×1/2	0.281	0.255	1.94	–
×3/8	0.123	0.114	1.97	–
×5/16	0.0731	0.0676	1.98	–
×1/4	0.0386	0.0356	1.99	–
L3 1/2×3 1/2×1/2	0.281	0.238	1.87	0.634
×7/16	0.192	0.164	1.89	0.631
×3/8	0.123	0.106	1.90	0.634
×5/16	0.0731	0.0634	1.92	0.632
×1/4	0.0386	0.0334	1.93	0.632
L3 1/2×3×1/2	0.260	0.191	1.75	–
×7/16	0.178	0.132	1.76	–
×3/8	0.114	0.0858	1.78	–
×5/16	0.0680	0.0512	1.79	–
×1/4	0.0360	0.0270	1.81	–
L3 1/2×2 1/2×1/2	0.234	0.159	1.67	–
×3/8	0.103	0.0714	1.69	–
×5/16	0.0611	0.0426	1.71	–
×1/4	0.0322	0.0225	1.72	–
L3×3×1/2	0.230	0.144	1.59	0.635
×7/16	0.157	0.100	1.61	0.632
×3/8	0.101	0.0652	1.62	0.630
×5/16	0.0597	0.0390	1.63	0.629
×1/4	0.0313	0.0206	1.65	0.628
×3/16	0.0136	0.0090	1.66	0.627
L3×2 1/2×1/2	0.213	0.112	1.46	–
×7/16	0.146	0.0777	1.48	–
×3/8	0.0943	0.0507	1.49	–
×5/16	0.0560	0.0304	1.51	–
×1/4	0.0296	0.0161	1.52	–
×3/16	0.0130	0.0070	1.53	–
L3×2×1/2	0.192	0.0908	1.39	–
×3/8	0.0855	0.0413	1.42	–
×5/16	0.0510	0.0248	1.44	–
×1/4	0.0270	0.0132	1.45	–
×3/16	0.0119	0.0058	1.46	–

*See LRFD Appendix E3.

Table 1-31 (cont.). Angles Flexural-Torsional Properties

Shape	Torsional Constant, J	Warping Constant, C_w	Polar Radius of Gyration $\bar{r}_o^*$	Flexural Constant, H^*
	in.^{4}	in.^{6}	in.	No Units
L2 1/2×2 1/2×1/2	0.188	0.0791	1.30	0.639
×3/8	0.0833	0.0362	1.33	0.633
×5/16	0.0495	0.0218	1.35	0.632
×1/4	0.0261	0.0116	1.36	0.629
×3/16	0.0114	0.0051	1.38	0.628
L2 1/2×2×3/8	0.0746	0.0268	1.21	–
×5/16	0.0444	0.0162	1.23	–
×1/4	0.0235	0.0087	1.25	–
×3/16	0.0103	0.0038	1.26	–
L2×2×3/8	0.0658	0.0174	1.04	0.638
×5/16	0.0393	0.0106	1.06	0.635
×1/4	0.0209	0.0057	1.08	0.633
×3/16	0.0092	0.0025	1.09	0.632
×1/8	0.0029	0.0008	1.10	0.633

*See LRFD Appendix E3.

Table 1-32.
WT-Shapes
Flexural-Torsional Properties

Shape	Torsional Constant, J	Warping Constant, C_w	Polar Radius of Gyration $\bar{r}_o^{**}$	Flexural Constant, H^{**}
	in.^{4}	in.^{6}	in.	No Units
WT22×167.5*	37.1	434	8.81	0.724
×145	25.7	279	8.67	0.733
×131	18.8	204	8.65	0.731
×115	12.4	139	8.67	0.723
WT20×296.5*	221	2340	8.29	0.762
×251.5*	139	1420	8.17	0.759
×215.5*	88.2	881	8.09	0.756
×198.5*	70.6	677	8.01	0.759
×186*	58.0	559	7.99	0.756
×181*	54.2	511	7.96	0.759
×162	39.6	362	7.89	0.760
×148.5	30.5	279	7.87	0.756
×138.5	25.7	218	7.75	0.771
×124.5	19.0	158	7.70	0.769
×107.5	12.4	101	7.65	0.770
×99.5	9.12	83.5	7.82	0.747
WT20×196*	85.6	802	8.26	0.678
×165.5*	52.6	485	8.19	0.674
×163.5	51.4	449	8.13	0.681
×139	32.2	278	8.07	0.676
×132	27.9	233	8.02	0.679
×117.5	20.6	156	7.88	0.691
×105.5	15.2	113	7.84	0.689
×91.5	10.0	72.1	7.78	0.692
×83.5	6.99	62.9	8.01	0.659
×74.5	4.66	51.9	8.23	0.626
WT18×399*	523	5700	8.02	0.802
×325*	294	3010	7.80	0.800
×263.5*	163	1570	7.61	0.799
×219.5*	96.3	894	7.49	0.796
×196.5*	70.5	637	7.42	0.797
×179.5*	54.0	480	7.38	0.797
×164*	41.9	363	7.32	0.799
×150	32.0	278	7.31	0.797
×140	26.2	226	7.28	0.796
×130	20.7	181	7.29	0.791
×122.5	17.3	151	7.28	0.788
×115	14.3	125	7.28	0.784
WT18×128	26.4	205	7.45	0.704
×116	19.7	151	7.40	0.703
×105	13.9	119	7.49	0.687
×97	11.1	92.7	7.45	0.687
×91	9.2	77.6	7.45	0.685
×85	7.51	63.2	7.44	0.684
×80	6.17	53.6	7.46	0.678
×75	5.04	46.0	7.50	0.670
×67.5	3.48	37.3	7.65	0.644
WT16.5×193.5*	73.9	615	7.04	0.802
×177*	57.1	468	7.00	0.803
×159*	42.1	335	6.93	0.803
×145.5	32.5	256	6.91	0.803
×131.5	24.3	188	6.86	0.803
×120.5	18.0	146	6.90	0.793
×110.5	13.9	113	6.89	0.788
×100.5	10.4	84.9	6.88	0.784
WT16.5×84.5	8.81	55.4	6.74	0.715
×76	6.16	43.0	6.81	0.700
×70.5	4.84	35.4	6.85	0.691
×65	3.67	29.3	6.92	0.677
×59	2.64	23.4	7.02	0.659

*Tensile group 4 or 5 shape. Special requirements may apply per LRFD Specification Section A3.1c.
**See LRFD Appendix E3.

Table 1-32 (cont.). WT-Shapes Flexural-Torsional Properties

Shape	Torsional Constant, J	Warping Constant, C_w	Polar Radius of Gyration $\bar{r}_o^{**}$	Flexural Constant, H^{**}
	in.⁴	in.⁶	in.	No Units
WT15×195.5*	86.3	636	6.51	0.818
×178.5*	66.6	478	6.44	0.818
×163*	51.2	361	6.40	0.817
×146*	37.5	257	6.33	0.819
×130.5	26.9	184	6.31	0.815
×117.5	20.1	133	6.24	0.818
×105.5	14.1	96.4	6.26	0.810
×95.5	10.5	71.2	6.25	0.807
×86.5	7.78	53.0	6.24	0.802
WT15×74	7.24	37.6	6.11	0.717
×66	4.85	28.5	6.19	0.698
×62	3.98	23.9	6.21	0.694
×58	3.21	20.5	6.24	0.683
×54	2.49	17.3	6.30	0.668
×49.5	1.88	14.3	6.38	0.654
×45	1.41	10.5	6.35	0.654
WT13.5×269.5*	247	1740	6.27	0.832
×184*	84.5	532	5.96	0.828
×168*	65.4	401	5.90	0.829
×153.5*	50.5	304	5.85	0.828
×140.5*	39.6	232	5.79	0.830
×129	30.7	178	5.76	0.829
×117.5	23.4	135	5.73	0.825
×108.5	18.8	105	5.68	0.829
×97	13.5	74.3	5.64	0.826
×89	10.0	57.7	5.69	0.815
×80.5	7.53	42.7	5.66	0.814
×73	5.62	31.7	5.64	0.811
WT13.5×64.5	5.55	24.0	5.48	0.731
×57	3.65	17.5	5.54	0.715
×51	2.63	12.6	5.53	0.714
×47	2.01	10.2	5.57	0.703
×42	1.40	7.79	5.63	0.685
WT12×185*	100	553	5.47	0.837
×167.5*	75.6	405	5.40	0.837
×153*	58.4	305	5.34	0.838
×139.5*	45.1	230	5.28	0.836
×125*	33.2	165	5.22	0.837
×114.5	25.5	125	5.19	0.837
×103.5	19.1	91.3	5.14	0.835
×96	15.3	72.5	5.12	0.837
×88	11.9	55.8	5.09	0.835
×81	9.22	43.8	5.09	0.831
×73	6.70	31.9	5.08	0.827
×65.5	4.74	23.1	5.08	0.818
×58.5	3.35	16.4	5.08	0.813
×52	2.35	11.6	5.08	0.810
WT12×51.5	3.53	12.3	4.88	0.733
×47	2.62	9.57	4.89	0.728
×42	1.84	6.90	4.89	0.721
×38	1.34	5.30	4.92	0.708
×34	0.932	4.08	4.99	0.692
WT12×31	0.883	3.92	5.12	0.620
×27.5	0.615	2.93	5.15	0.607
WT10.5×100.5	20.4	85.4	4.68	0.859
×91	15.3	63.0	4.63	0.859
×83	11.8	47.3	4.59	0.861
×73.5	7.69	32.5	4.63	0.846
×66	5.62	23.4	4.60	0.844
×61	4.47	18.4	4.58	0.846
×55.5	3.40	13.8	4.56	0.845
×50.5	2.60	10.4	4.53	0.846

*Tensile group 4 or 5 shape. Special requirements may apply per LRFD Specification Section A3.1c.
**See LRFD Appendix E3.

Table 1-32 (cont.). WT-Shapes Flexural-Torsional Properties

Shape	Torsional Constant, J	Warping Constant, C_w	Polar Radius of Gyration $\bar{r}_o^{**}$	Flexural Constant, H^{**}
	in.^{4}	in.^{6}	in.	No Units
WT10.5×46.5	3.01	9.33	4.37	0.729
×41.5	2.16	6.50	4.33	0.732
×36.5	1.51	4.42	4.31	0.732
×34	1.22	3.62	4.31	0.728
×31	0.913	2.78	4.31	0.722
×27.5	0.617	2.08	4.37	0.704
×24	0.400	1.52	4.44	0.677
WT10.5×28.5	0.884	2.50	4.36	0.665
×25	0.570	1.89	4.44	0.640
×22	0.383	1.40	4.49	0.624
WT9×87.5	16.8	56.5	4.10	0.872
×79	12.5	41.2	4.06	0.872
×71.5	9.58	30.7	4.03	0.874
×65	7.23	22.8	3.99	0.873
×59.5	5.30	17.4	4.03	0.861
×53	3.73	12.1	4.00	0.859
×48.5	2.92	9.29	3.97	0.862
×43	2.04	6.42	3.94	0.860
×38	1.41	4.37	3.92	0.861
WT9×35.5	1.74	3.96	3.72	0.751
×32.5	1.36	3.01	3.69	0.755
×30	1.08	2.35	3.67	0.756
×27.5	0.830	1.84	3.68	0.749
×25	0.619	1.36	3.66	0.749
WT9×23	0.609	1.20	3.67	0.695
×20	0.404	0.788	3.65	0.691
×17.5	0.252	0.598	3.74	0.661
WT8×50	4.09	10.4	3.60	0.878
×44.5	2.91	7.19	3.58	0.878
×38.5	1.93	4.61	3.54	0.877
×33.5	1.31	3.01	3.51	0.880
WT8×28.5	1.10	1.99	3.30	0.770
×25	0.760	1.34	3.28	0.769
×22.5	0.555	0.974	3.27	0.767
×20	0.396	0.673	3.24	0.769
×18	0.272	0.516	3.30	0.745
WT8×15.5	0.230	0.366	3.26	0.695
×13	0.130	0.243	3.32	0.666
WT7×404*	897	6970	5.66	0.959
×365*	714	5250	5.48	0.966
×332.5*	555	3920	5.35	0.966
×302.5*	430	2930	5.25	0.966
×275*	331	2180	5.15	0.967
×250*	254	1620	5.05	0.967
×227.5*	196	1210	4.97	0.967
×213*	164	991	4.92	0.968
×199*	135	801	4.88	0.968
×185*	110	640	4.82	0.968
×171*	88.3	502	4.77	0.968
×155.5*	67.5	375	4.71	0.968
×141.5*	51.8	281	4.66	0.969
×128.5*	39.3	209	4.61	0.970
×116.5*	29.6	154	4.57	0.970
×105.5	22.2	113	4.52	0.969
×96.5	17.3	87.2	4.49	0.971
×88	13.2	65.2	4.46	0.970
×79.5	9.84	47.9	4.42	0.971
×72.5	7.56	36.3	4.40	0.971

*Tensile group 4 or 5 shape. Special requirements may apply per LRFD Specification Section A3.1c.
**See LRFD Appendix E3.

Table 1-32 (cont.). WT-Shapes Flexural-Torsional Properties

Shape	Torsional Constant, J	Warping Constant, C_w	Polar Radius of Gyration $\bar{r}_o^{**}$	Flexural Constant, H^{**}
	in.^{4}	in.^{6}	in.	No Units
WT7×66	6.13	26.6	4.21	0.966
×60	4.67	20.0	4.18	0.966
×54.5	3.55	15.0	4.16	0.968
×49.5	2.68	11.1	4.14	0.967
×45	2.03	8.31	4.13	0.968
WT7×41	2.53	5.63	3.25	0.912
×37	1.93	4.19	3.21	0.917
×34	1.50	3.21	3.19	0.915
×30.5	1.09	2.29	3.17	0.914
WT7×26.5	0.967	1.46	2.89	0.868
×24	0.723	1.07	2.87	0.866
×21.5	0.522	0.751	2.85	0.866
WT7×19	0.398	0.554	2.86	0.799
×17	0.284	0.400	2.86	0.793
×15	0.190	0.287	2.90	0.771
WT7×13	0.179	0.207	2.82	0.713
×11	0.104	0.134	2.86	0.690
WT6×168*	120	481	4.07	0.958
×152.5*	92.0	356	4.00	0.959
×139.5*	70.9	267	3.94	0.957
×126*	53.5	195	3.88	0.958
×115*	41.6	148	3.83	0.958
×105*	32.1	112	3.78	0.959
×95	24.3	82.1	3.74	0.960
×85	17.7	58.3	3.69	0.960
×76	12.8	41.3	3.64	0.960
×68	9.21	28.9	3.61	0.960
×60	6.42	19.7	3.57	0.959
×53	4.55	13.6	3.53	0.961
×48	3.42	10.1	3.51	0.962
×43.5	2.54	7.34	3.48	0.960
×39.5	1.91	5.43	3.47	0.960
×36	1.46	4.07	3.44	0.960
×32.5	1.09	2.97	3.43	0.960
WT6×29	1.05	2.08	3.00	0.944
×26.5	0.788	1.53	2.99	0.940
WT6×25	0.855	1.23	2.67	0.899
×22.5	0.627	0.885	2.66	0.900
×20	0.452	0.620	2.63	0.900
WT6×17.5	0.369	0.437	2.56	0.835
×15	0.228	0.267	2.55	0.830
×13	0.150	0.174	2.54	0.826
WT6×11	0.146	0.137	2.52	0.684
×9.5	0.0899	0.0934	2.54	0.663
×8	0.0511	0.0678	2.62	0.624
×7	0.0350	0.0493	2.64	0.611
WT5×56	7.50	16.9	3.04	0.963
×50	5.41	11.9	3.00	0.964
×44	3.75	8.02	2.98	0.964
×38.5	2.55	5.31	2.94	0.964
×34	1.78	3.62	2.91	0.965
×30	1.23	2.46	2.89	0.965
×27	0.909	1.78	2.87	0.966
×24.5	0.693	1.33	2.85	0.966
WT5×22.5	0.753	0.981	2.43	0.940
×19.5	0.487	0.616	2.42	0.936
×16.5	0.291	0.356	2.41	0.927

*Tensile group 4 or 5 shape. Special requirements may apply per LRFD Specification Section A3.1c.
**See LRFD Appendix E3.

Table 1-32 (cont.). WT-Shapes Flexural-Torsional Properties

Shape	Torsional Constant, J	Warping Constant, C_w	Polar Radius of Gyration $\bar{r}_o$**	Flexural Constant, H**
	in.^{4}	in.^{6}	in.	No Units
WT5×15	0.310	0.273	2.17	0.848
×13	0.201	0.173	2.15	0.847
×11	0.119	0.107	2.16	0.830
WT5×9.5	0.116	0.0796	2.08	0.729
×8.5	0.0776	0.0610	2.11	0.700
×7.5	0.0518	0.0475	2.16	0.673
×6	0.0272	0.0255	2.16	0.662
WT4×33.5	2.51	3.56	2.41	0.962
×29	1.66	2.28	2.38	0.961
×24	0.977	1.30	2.34	0.966
×20	0.558	0.715	2.31	0.961
×17.5	0.384	0.480	2.30	0.963
×15.5	0.267	0.327	2.28	0.961
WT4×14	0.268	0.230	1.97	0.935
×12	0.173	0.144	1.96	0.936
WT4×10.5	0.141	0.0916	1.80	0.877
×9	0.0855	0.0562	1.80	0.862
WT4×7.5	0.0679	0.0382	1.72	0.761
×6.5	0.0433	0.0269	1.74	0.731
×5	0.0212	0.0114	1.70	0.750
WT3×12.5	0.234	0.171	1.76	0.953
×10	0.123	0.0858	1.73	0.952
×7.5	0.0523	0.0342	1.71	0.938
WT3×8	0.111	0.0426	1.37	0.881
×6	0.0449	0.0178	1.37	0.846
×4.5	0.0202	0.00736	1.34	0.852
×4.25	0.0164	0.00616	1.34	0.837
WT2.5×9.5	0.157	0.0775	1.44	0.964
×8	0.0958	0.0453	1.43	0.962
WT2×6.5	0.0750	0.0233	1.17	0.948

*Tensile group 4 or 5 shape. Special requirements may apply per LRFD Specification Section A3.1c.
**See LRFD Appendix E3.

Table 1-33.
MT-Shapes
Flexural-Torsional Properties

Shape	Torsional Constant, J	Warping Constant, C_w	Polar Radius of Gyration $\bar{r}_o$**	Flexural Constant, H**
	in.^{4}	in.^{6}	in.	No Units
MT6×5.9	0.0249	0.0337	2.71	0.570
×5.4	0.0196	0.0250	2.69	0.574
×5	0.0145	0.0202	2.71	0.573
MT5×4.5	0.0156	0.0138	2.22	0.581
×4	0.0112	0.00989	2.22	0.586
×3.75	0.00932	0.00792	2.22	0.589
MT4×3.25	0.00917	0.00463	1.74	0.611
×3.1	0.00778	0.00403	1.75	0.611
MT3×2.2	0.00494	0.00124	1.27	0.646
×1.85	0.00265	0.000755	1.28	0.645
MT2.5×9.45*	0.156	0.0732	1.42	0.956
MT2×3	0.00919	0.00193	1.08	0.942

*This shape has tapered flanges while all other M-shapes have parallel flanges.
**See LRFD Appendix E3.

Table 1-34.
ST-Shapes
Flexural-Torsional Properties

Shape	Torsional Constant, J	Warping Constant, C_w	Polar Radius of Gyration $\bar{r}_o^*$	Flexural Constant, H^*
	in.^{4}	in.^{6}	in.	No Units
ST12×60.5	6.38	27.5	5.14	0.640
×53	5.05	15.0	4.88	0.686
ST12×50	3.76	19.5	5.28	0.584
×45	3.01	12.1	5.11	0.616
×40	2.44	6.94	4.88	0.655
ST10×48	4.16	15.0	4.36	0.625
×43	3.30	9.17	4.20	0.66
ST10×37.5	2.28	7.21	4.29	0.612
×33	1.78	4.02	4.10	0.654
ST9×35	2.02	7.03	4.01	0.581
×27.35	1.16	2.26	3.71	0.660
ST7.5×25	1.05	2.02	3.22	0.637
×21.45	0.765	0.995	3.04	0.688
ST6×25	1.36	1.97	2.60	0.662
×20.4	0.842	0.787	2.42	0.733
ST6×17.5	0.524	0.556	2.49	0.694
×15.9	0.438	0.364	2.39	0.732
ST5×17.5	0.633	0.725	2.23	0.653
×12.7	0.300	0.173	1.97	0.765
ST4×11.5	0.271	0.168	1.73	0.706
×9.2	0.167	0.0642	1.58	0.787
ST3×8.625	0.181	0.0772	1.35	0.703
×6.25	0.083	0.0197	1.20	0.818
ST2.5×5	0.0568	0.0100	1.01	0.838
ST2×4.75	0.0590	0.00995	0.904	0.798
×3.85	0.0364	0.00457	0.833	0.869
ST1.5×3.75	0.0432	0.00496	0.732	0.830
×2.85	0.0216	0.00189	0.668	0.911

*See LRFD Appendix E3.

Table 1-35. Double Angles Flexural-Torsional Properties

Shape	Long Legs Vertical						Short Legs Vertical					
	Back to Back of Angles, in.						Back to Back of Angles, in.					
	0		3/8		3/4		0		3/8		3/4	
	$\bar{r}_o^*$	H^*	$\bar{r}_o^*$	H^*	$\bar{r}_o^*$	H^*	$\bar{r}_o^*$	H^*	$\bar{r}_o^*$	H^*	$\bar{r}_o^*$	H^*
2L8×8×1 1/8	4.56	0.837	4.66	0.844	4.77	0.851	4.56	0.837	4.66	0.844	4.77	0.851
×1	4.56	0.834	4.66	0.841	4.77	0.848	4.56	0.834	4.66	0.841	4.77	0.848
×7/8	4.56	0.831	4.66	0.838	4.76	0.845	4.56	0.831	4.66	0.838	4.76	0.845
×3/4	4.56	0.829	4.66	0.836	4.76	0.843	4.56	0.829	4.66	0.836	4.76	0.843
×5/8	4.56	0.826	4.66	0.833	4.76	0.840	4.56	0.826	4.66	0.833	4.76	0.840
×9/16	4.56	0.825	4.65	0.832	4.75	0.839	4.56	0.825	4.65	0.832	4.75	0.839
×1/2	4.56	0.824	4.65	0.831	4.75	0.837	4.56	0.824	4.65	0.831	4.75	0.837
2L8×6×1	4.06	0.721	4.14	0.732	4.23	0.742	4.18	0.924	4.30	0.929	4.43	0.933
×7/8	4.07	0.718	4.14	0.728	4.23	0.739	4.17	0.922	4.29	0.926	4.42	0.930
×3/4	4.07	0.714	4.15	0.725	4.23	0.735	4.17	0.919	4.28	0.924	4.40	0.928
×5/8	4.08	0.712	4.16	0.722	4.24	0.732	4.16	0.917	4.27	0.921	4.39	0.926
×9/16	4.09	0.710	4.16	0.720	4.24	0.731	4.15	0.916	4.27	0.920	4.39	0.924
×1/2	4.09	0.709	4.16	0.719	4.24	0.729	4.15	0.915	4.26	0.919	4.38	0.923
×7/16	4.09	0.708	4.16	0.718	4.24	0.728	4.15	0.913	4.26	0.918	4.38	0.922
2L8×4×1	3.86	0.568	3.91	0.580	3.97	0.594	4.11	0.983	4.25	0.984	4.39	0.985
×7/8	3.87	0.566	3.92	0.577	3.98	0.590	4.09	0.981	4.22	0.982	4.37	0.984
×3/4	3.88	0.564	3.93	0.575	3.99	0.587	4.07	0.980	4.20	0.981	4.35	0.983
×5/8	3.89	0.562	3.94	0.573	3.99	0.585	4.05	0.979	4.18	0.980	4.32	0.981
×9/16	3.90	0.562	3.94	0.572	4.00	0.584	4.04	0.978	4.17	0.980	4.31	0.981
×1/2	3.90	0.561	3.95	0.571	4.00	0.583	4.03	0.978	4.16	0.979	4.30	0.980
×7/16	3.91	0.561	3.95	0.571	4.00	0.582	4.02	0.977	4.15	0.978	4.29	0.980
2L7×4×3/4	3.41	0.611	3.47	0.624	3.53	0.639	3.57	0.969	3.70	0.971	3.84	0.973
×5/8	3.42	0.608	3.47	0.621	3.54	0.635	3.55	0.967	3.68	0.969	3.82	0.971
×1/2	3.43	0.606	3.48	0.618	3.55	0.632	3.53	0.965	3.66	0.968	3.80	0.970
×7/16	3.43	0.605	3.49	0.617	3.55	0.630	3.53	0.964	3.66	0.967	3.79	0.969
×3/8	3.44	0.605	3.49	0.616	3.55	0.629	3.52	0.963	3.65	0.966	3.78	0.968
2L6×6×1	3.42	0.843	3.53	0.852	3.64	0.861	3.42	0.843	3.53	0.852	3.64	0.861
×7/8	3.42	0.839	3.53	0.848	3.63	0.857	3.42	0.839	3.53	0.848	3.63	0.857
×3/4	3.42	0.835	3.52	0.844	3.63	0.853	3.42	0.835	3.52	0.844	3.63	0.853
×5/8	3.42	0.831	3.52	0.840	3.62	0.849	3.42	0.831	3.52	0.840	3.62	0.849
×9/16	3.42	0.829	3.52	0.838	3.62	0.847	3.42	0.829	3.52	0.838	3.62	0.847
×1/2	3.42	0.827	3.52	0.836	3.62	0.846	3.42	0.827	3.52	0.836	3.62	0.846
×7/16	3.42	0.826	3.52	0.835	3.62	0.844	3.42	0.826	3.52	0.835	3.62	0.844
×3/8	3.42	0.824	3.51	0.833	3.61	0.842	3.42	0.824	3.51	0.833	3.61	0.842
×5/16	3.42	0.823	3.51	0.832	3.61	0.841	3.42	0.823	3.51	0.832	3.61	0.841
2L6×4×7/8	2.97	0.677	3.04	0.693	3.13	0.709	3.11	0.952	3.24	0.955	3.38	0.959
×3/4	2.98	0.672	3.05	0.687	3.13	0.704	3.10	0.949	3.23	0.953	3.36	0.956
×5/8	2.99	0.667	3.06	0.682	3.14	0.698	3.09	0.946	3.22	0.950	3.35	0.954
×9/16	2.99	0.665	3.06	0.680	3.14	0.696	3.08	0.944	3.21	0.949	3.34	0.953
×1/2	3.00	0.663	3.06	0.678	3.14	0.693	3.08	0.943	3.20	0.947	3.33	0.951
×7/16	3.00	0.661	3.07	0.676	3.14	0.691	3.08	0.941	3.20	0.946	3.33	0.950
×3/8	3.01	0.659	3.07	0.673	3.15	0.689	3.07	0.940	3.19	0.945	3.32	0.949
×5/16	3.02	0.658	3.08	0.671	3.15	0.686	3.07	0.939	3.19	0.943	3.32	0.948
2L6×3 1/2×1/2	2.95	0.613	3.01	0.628	3.08	0.644	3.05	0.964	3.18	0.966	3.32	0.969
×3/8	2.96	0.610	3.02	0.624	3.08	0.640	3.04	0.961	3.17	0.964	3.30	0.967
×5/16	2.97	0.608	3.02	0.622	3.09	0.638	3.04	0.960	3.16	0.963	3.30	0.966
2L5×5×7/8	2.85	0.845	2.96	0.856	3.07	0.866	2.85	0.845	2.96	0.856	3.07	0.866
×3/4	2.85	0.840	2.95	0.851	3.06	0.861	2.85	0.840	2.95	0.851	3.06	0.861
×5/8	2.85	0.835	2.95	0.846	3.06	0.857	2.85	0.835	2.95	0.846	3.06	0.857
×1/2	2.85	0.830	2.94	0.842	3.05	0.852	2.85	0.830	2.94	0.842	3.05	0.852
×7/16	2.85	0.828	2.94	0.839	3.05	0.850	2.85	0.828	2.94	0.839	3.05	0.850
×3/8	2.84	0.826	2.94	0.838	3.04	0.848	2.84	0.826	2.94	0.838	3.04	0.848
×5/16	2.84	0.825	2.94	0.836	3.04	0.847	2.84	0.825	2.94	0.836	3.04	0.847
2L5×3 1/2×3/4	2.49	0.699	2.57	0.717	2.66	0.736	2.60	0.943	2.73	0.949	2.86	0.953
×5/8	2.49	0.693	2.57	0.711	2.66	0.730	2.59	0.940	2.71	0.945	2.85	0.950
×1/2	2.50	0.688	2.58	0.705	2.66	0.724	2.58	0.936	2.70	0.942	2.83	0.947
×3/8	2.51	0.683	2.58	0.700	2.66	0.718	2.56	0.933	2.69	0.938	2.81	0.944
×5/16	2.51	0.682	2.58	0.698	2.66	0.716	2.56	0.931	2.68	0.937	2.81	0.942
×1/4	2.52	0.680	2.58	0.696	2.66	0.714	2.55	0.929	2.67	0.935	2.80	0.941

*See LRFD Appendix E3.

Table 1-35 (cont.). Double Angles Flexural-Torsional Properties

Shape	Long Legs Vertical						Short Legs Vertical					
	Back to Back of Angles, in.						Back to Back of Angles, in.					
	0		3/8		3/4		0		3/8		3/4	
	$\bar{r}_o^*$	H^*	$\bar{r}_o^*$	H^*	$\bar{r}_o^*$	H^*	$\bar{r}_o^*$	H^*	$\bar{r}_o^*$	H^*	$\bar{r}_o^*$	H^*
2L5×3×1/2	2.44	0.628	2.51	0.646	2.58	0.667	2.54	0.962	2.68	0.966	2.81	0.969
×7/16	2.45	0.626	2.51	0.644	2.58	0.664	2.54	0.961	2.67	0.964	2.80	0.968
×3/8	2.45	0.624	2.51	0.642	2.59	0.661	2.53	0.959	2.66	0.963	2.79	0.967
×5/16	2.46	0.623	2.52	0.640	2.59	0.659	2.52	0.958	2.65	0.962	2.78	0.965
×1/4	2.46	0.622	2.52	0.638	2.59	0.657	2.51	0.957	2.64	0.961	2.77	0.964
2L4×4×3/4	2.28	0.847	2.39	0.861	2.51	0.874	2.28	0.847	2.39	0.861	2.51	0.874
×5/8	2.28	0.841	2.39	0.854	2.50	0.868	2.28	0.841	2.39	0.854	2.50	0.868
×1/2	2.28	0.834	2.38	0.848	2.49	0.862	2.28	0.834	2.38	0.848	2.49	0.862
×7/16	2.28	0.832	2.38	0.846	2.49	0.859	2.28	0.832	2.38	0.846	2.49	0.859
×3/8	2.28	0.829	2.38	0.843	2.49	0.856	2.28	0.829	2.38	0.843	2.49	0.856
×5/16	2.28	0.826	2.37	0.840	2.48	0.854	2.28	0.826	2.37	0.840	2.48	0.854
×1/4	2.28	0.824	2.37	0.838	2.48	0.851	2.28	0.824	2.37	0.838	2.48	0.851
2L4×3 1/2×1/2	2.14	0.784	2.23	0.802	2.33	0.819	2.16	0.882	2.28	0.893	2.40	0.904
×3/8	2.14	0.778	2.23	0.795	2.33	0.813	2.16	0.876	2.27	0.888	2.39	0.899
×5/16	2.14	0.775	2.23	0.792	2.33	0.810	2.16	0.874	2.26	0.885	2.38	0.896
×1/4	2.14	0.773	2.22	0.790	2.32	0.807	2.15	0.871	2.26	0.883	2.37	0.894
2L4×3×5/8	2.02	0.728	2.11	0.750	2.21	0.773	2.10	0.930	2.22	0.938	2.36	0.945
×1/2	2.02	0.721	2.11	0.743	2.20	0.765	2.09	0.925	2.21	0.933	2.34	0.940
×3/8	2.03	0.715	2.11	0.736	2.20	0.757	2.08	0.920	2.20	0.928	2.32	0.936
×5/16	2.03	0.712	2.11	0.733	2.20	0.754	2.07	0.918	2.19	0.926	2.32	0.934
×1/4	2.03	0.710	2.11	0.730	2.20	0.751	2.06	0.915	2.18	0.924	2.31	0.932
2L3 1/2×3 1/2×1/2	1.99	0.838	2.10	0.854	2.21	0.869	1.99	0.838	2.10	0.854	2.21	0.869
×7/16	1.99	0.835	2.09	0.851	2.21	0.866	1.99	0.835	2.09	0.851	2.21	0.866
×3/8	1.99	0.832	2.09	0.848	2.20	0.863	1.99	0.832	2.09	0.848	2.20	0.863
×5/16	1.99	0.829	2.09	0.845	2.20	0.860	1.99	0.829	2.09	0.845	2.20	0.860
×1/4	1.99	0.826	2.08	0.842	2.19	0.857	1.99	0.826	2.08	0.842	2.19	0.857
2L3 1/2×3×1/2	1.85	0.780	1.94	0.801	2.05	0.822	1.88	0.892	2.00	0.904	2.13	0.915
×7/16	1.85	0.776	1.94	0.797	2.05	0.818	1.88	0.889	1.99	0.901	2.12	0.912
×3/8	1.85	0.773	1.94	0.794	2.05	0.814	1.88	0.885	1.99	0.898	2.11	0.910
×5/16	1.85	0.770	1.94	0.790	2.04	0.811	1.87	0.883	1.98	0.895	2.11	0.907
×1/4	1.85	0.767	1.94	0.787	2.04	0.807	1.87	0.880	1.98	0.893	2.10	0.905
2L3 1/2×2 1/2×1/2	1.75	0.706	1.83	0.732	1.93	0.759	1.82	0.938	1.95	0.946	2.08	0.953
×3/8	1.75	0.698	1.83	0.724	1.93	0.750	1.81	0.933	1.93	0.941	2.07	0.949
×5/16	1.76	0.695	1.83	0.720	1.92	0.746	1.80	0.930	1.92	0.939	2.06	0.947
×1/4	1.76	0.693	1.83	0.717	1.92	0.742	1.80	0.928	1.92	0.937	2.05	0.944
2L3×3×1/2	1.71	0.842	1.82	0.861	1.94	0.878	1.71	0.842	1.82	0.861	1.94	0.878
×7/16	1.71	0.838	1.82	0.857	1.94	0.874	1.71	0.838	1.82	0.857	1.94	0.874
×3/8	1.71	0.834	1.81	0.853	1.93	0.870	1.71	0.834	1.81	0.853	1.93	0.870
×5/16	1.71	0.830	1.81	0.849	1.93	0.866	1.71	0.830	1.81	0.849	1.93	0.866
×1/4	1.71	0.827	1.81	0.845	1.92	0.863	1.71	0.827	1.81	0.845	1.92	0.863
×3/16	1.71	0.823	1.80	0.842	1.91	0.859	1.71	0.823	1.80	0.842	1.91	0.859
2L3×2 1/2×1/2	1.57	0.774	1.66	0.800	1.78	0.824	1.61	0.905	1.73	0.918	1.86	0.929
×7/16	1.57	0.769	1.66	0.795	1.77	0.819	1.60	0.901	1.72	0.914	1.85	0.926
×3/8	1.57	0.764	1.66	0.790	1.77	0.815	1.60	0.897	1.72	0.911	1.85	0.923
×5/16	1.57	0.760	1.66	0.785	1.76	0.810	1.59	0.893	1.71	0.907	1.84	0.920
×1/4	1.57	0.756	1.66	0.781	1.76	0.806	1.59	0.890	1.70	0.904	1.83	0.917
×3/16	1.57	0.753	1.65	0.778	1.75	0.802	1.58	0.887	1.70	0.901	1.82	0.914
2L3×2×1/2	1.47	0.684	1.55	0.717	1.66	0.751	1.55	0.955	1.69	0.962	1.83	0.968
×3/8	1.48	0.675	1.55	0.707	1.65	0.739	1.54	0.949	1.67	0.957	1.81	0.963
×5/16	1.48	0.671	1.56	0.702	1.65	0.734	1.53	0.946	1.66	0.954	1.80	0.961
×1/4	1.48	0.668	1.56	0.698	1.65	0.730	1.52	0.944	1.65	0.952	1.79	0.959
×3/16	1.49	0.666	1.55	0.695	1.64	0.726	1.52	0.941	1.64	0.950	1.78	0.957

*See LRFD Appendix E3.

Table 1-35 (cont.). Double Angles Flexural-Torsional Properties

Shape	Long Legs Vertical						Short Legs Vertical					
	Back to Back of Angles, in.						Back to Back of Angles, in.					
	0		3/8		3/4		0		3/8		3/4	
	$\bar{r}_o^*$	H^*	$\bar{r}_o^*$	H^*	$\bar{r}_o^*$	H^*	$\bar{r}_o^*$	H^*	$\bar{r}_o^*$	H^*	$\bar{r}_o^*$	H^*
2L2 1/2×2 1/2×1/2	1.43	0.850	1.54	0.871	1.67	0.890	1.43	0.850	1.54	0.871	1.67	0.890
×3/8	1.42	0.839	1.53	0.861	1.65	0.881	1.42	0.839	1.53	0.861	1.65	0.881
×5/16	1.42	0.834	1.53	0.856	1.65	0.876	1.42	0.834	1.53	0.856	1.65	0.876
×1/4	1.42	0.829	1.52	0.852	1.64	0.872	1.42	0.829	1.52	0.852	1.64	0.872
×3/16	1.42	0.825	1.52	0.847	1.63	0.868	1.42	0.825	1.52	0.847	1.63	0.868
2L2 1/2×2×3/8	1.29	0.754	1.38	0.786	1.49	0.817	1.32	0.913	1.45	0.927	1.59	0.939
×5/16	1.29	0.748	1.38	0.781	1.49	0.812	1.32	0.909	1.44	0.923	1.58	0.936
×1/4	1.29	0.744	1.38	0.775	1.49	0.806	1.32	0.904	1.43	0.920	1.57	0.933
×3/16	1.29	0.740	1.38	0.771	1.48	0.801	1.31	0.901	1.43	0.916	1.56	0.929
2L2×2×3/8	1.14	0.847	1.25	0.874	1.38	0.897	1.14	0.847	1.25	0.874	1.38	0.897
×5/16	1.14	0.841	1.25	0.868	1.37	0.891	1.14	0.841	1.25	0.868	1.37	0.891
×1/4	1.13	0.835	1.24	0.862	1.37	0.886	1.13	0.835	1.24	0.862	1.37	0.886
×3/16	1.13	0.830	1.24	0.857	1.36	0.882	1.13	0.830	1.24	0.857	1.36	0.882
×1/8	1.13	0.826	1.23	0.853	1.35	0.877	1.13	0.826	1.23	0.853	1.35	0.877

*See LRFD Appendix E3.

Table 1-36. W-Shapes Surface and Box Perimeters, Weight-to-Perimeter Ratios, and Surface Areas

Shape	Case A			Case B			Case C			Case D		
	Perimeter	W/D Ratio	Surf. Area	Perimeter	W/D Ratio	Surf. Area	Perimeter	W/D Ratio	Surf. Area	Perimeter	W/D Ratio	Surf. Area
	in.		ft²/ft	in.		ft²/ft	in.		ft²/ft	in.		ft²/ft
W44×335	133	2.52	11.1	149	2.25	12.4	104	3.22	8.67	120	2.79	10.0
×290	132	2.20	11.0	147	1.97	12.3	103	2.82	8.58	119	2.44	9.92
×262	131	2.00	10.9	147	1.78	12.3	102	2.57	8.50	118	2.22	9.83
×230	130	1.77	10.8	146	1.58	12.2	102	2.25	8.50	117	1.97	9.75
W40×593	130	4.56	10.8	147	4.03	12.3	103	5.76	8.58	119	4.98	9.92
×503	128	3.93	10.7	144	3.49	12.0	101	4.98	8.42	117	4.30	9.75
×431	126	3.42	10.5	143	3.01	11.9	98.8	4.36	8.23	115	3.75	9.58
×397	126	3.15	10.5	142	2.80	11.8	98.1	4.05	8.18	114	3.48	9.50
×372	125	2.98	10.4	141	2.64	11.8	97.3	3.82	8.11	113	3.29	9.42
×362	125	2.90	10.4	141	2.57	11.8	97.2	3.72	8.10	113	3.20	9.42
×324	124	2.61	10.3	140	2.31	11.7	96.3	3.36	8.03	112	2.89	9.33
×297	123	2.41	10.3	139	2.14	11.6	95.4	3.11	7.95	111	2.68	9.25
×277	123	2.25	10.3	139	1.99	11.6	95.2	2.91	7.93	111	2.50	9.25
×249	123	2.02	10.3	139	1.79	11.6	94.6	2.63	7.88	110	2.26	9.17
×215	122	1.76	10.2	138	1.56	11.5	93.8	2.29	7.82	110	1.95	9.17
×199	121	1.64	10.1	137	1.45	11.4	93.2	2.14	7.77	109	1.83	9.08
W40×392	116	3.38	9.67	128	3.06	10.7	95.6	4.10	7.97	108	3.63	9.00
×331	114	2.90	9.50	126	2.63	10.5	93.8	3.53	7.82	106	3.12	8.83
×327	113	2.89	9.42	125	2.62	10.4	93.7	3.49	7.81	106	3.08	8.83
×278	112	2.48	9.33	124	2.24	10.3	92.4	3.01	7.70	104	2.67	8.67
×264	112	2.36	9.33	124	2.13	10.3	91.9	2.87	7.66	104	2.54	8.67
×235	112	2.10	9.33	124	1.90	10.3	91.3	2.57	7.61	103	2.28	8.58
×211	111	1.90	9.25	123	1.72	10.3	90.6	2.33	7.55	102	2.07	8.50
×183	110	1.66	9.17	122	1.50	10.2	89.8	2.04	7.48	102	1.79	8.50
×167	109	1.53	9.08	121	1.38	10.1	89.0	1.88	7.42	101	1.65	8.42
×149	109	1.37	9.08	121	1.23	10.1	88.2	1.69	7.35	100	1.49	8.33
W36×798	131	6.09	10.9	149	5.36	12.4	102	7.82	8.50	120	6.65	10.0
×650	128	5.08	10.7	146	4.45	12.2	98.6	6.59	8.22	116	5.60	9.67
×527	125	4.22	10.4	142	3.71	11.8	95.6	5.51	7.97	113	4.66	9.42
×439	123	3.57	10.3	140	3.14	11.7	93.6	4.69	7.80	111	3.95	9.25
×393	121	3.25	10.1	138	2.85	11.5	92.4	4.25	7.70	109	3.61	9.08
×359	121	2.97	10.1	137	2.62	11.4	91.5	3.92	7.63	108	3.32	9.00
×328	120	2.73	10.0	137	2.39	11.4	90.8	3.61	7.57	107	3.07	8.92
×300	120	2.50	10.0	136	2.21	11.3	90.1	3.33	7.51	107	2.80	8.92
×280	119	2.35	9.92	136	2.06	11.3	89.6	3.13	7.47	106	2.64	8.83
×260	119	2.18	9.92	135	1.93	11.3	89.2	2.91	7.43	106	2.45	8.83
×245	118	2.08	9.83	135	1.81	11.3	88.7	2.76	7.39	105	2.33	8.75
×230	118	1.95	9.83	134	1.72	11.2	88.3	2.60	7.36	105	2.19	8.75
W36×256	108	2.37	9.00	120	2.13	10.0	87.0	2.94	7.25	99.2	2.58	8.27
×232	108	2.15	9.00	120	1.93	10.0	86.3	2.69	7.19	98.4	2.36	8.20
×210	107	1.96	8.92	119	1.76	9.92	85.6	2.45	7.13	97.8	2.15	8.15
×194	107	1.81	8.92	119	1.63	9.92	85.1	2.28	7.09	97.2	2.00	8.10
×182	106	1.72	8.83	119	1.53	9.92	84.7	2.15	7.06	96.8	1.88	8.07
×170	106	1.60	8.83	118	1.44	9.83	84.4	2.01	7.03	96.4	1.76	8.03
×160	106	1.51	8.83	118	1.36	9.83	84.0	1.90	7.00	96.0	1.67	8.00
×150	105	1.43	8.75	117	1.28	9.75	83.8	1.79	6.98	95.8	1.57	7.98
×135	105	1.29	8.75	117	1.15	9.75	83.2	1.62	6.93	95.2	1.42	7.93
W33×387	117	3.31	9.75	133	2.91	11.1	88.2	4.39	7.35	104	3.72	8.67
×354	116	3.05	9.67	132	2.68	11.0	87.3	4.05	7.28	103	3.44	8.58
×318	115	2.77	9.58	131	2.43	10.9	86.4	3.68	7.20	102	3.12	8.50
×291	114	2.55	9.50	130	2.24	10.8	85.5	3.40	7.13	101	2.88	8.42
×263	113	2.33	9.42	129	2.04	10.8	84.8	3.10	7.07	101	2.60	8.42
×241	113	2.13	9.42	129	1.87	10.8	84.3	2.86	7.03	100	2.41	8.33
×221	112	1.97	9.33	128	1.73	10.7	83.6	2.64	6.97	99.4	2.22	8.28
×201	112	1.79	9.33	127	1.58	10.6	83.1	2.42	6.93	98.8	2.03	8.23

Case A: Shape perimeter, minus one flange surface.
Case B: Shape perimeter.
Case C: Box perimeter, minus one flange surface.
Case D: Box perimeter.

Table 1-36 (cont.). W-Shapes Surface and Box Perimeters, Weight-to-Perimeter Ratios, and Surface Areas

Shape	Case A			Case B			Case C			Case D		
	Perimeter in.	W/D Ratio	Surf. Area ft²/ft	Perimeter in.	W/D Ratio	Surf. Area ft²/ft	Perimeter in.	W/D Ratio	Surf. Area ft²/ft	Perimeter in.	W/D Ratio	Surf. Area ft²/ft
W33×169	99.6	1.70	8.30	111	1.52	9.25	79.1	2.14	6.59	90.6	1.87	7.55
×152	99.3	1.53	8.28	111	1.37	9.25	78.6	1.93	6.55	90.2	1.69	7.52
×141	98.4	1.43	8.20	110	1.28	9.17	78.1	1.81	6.51	89.6	1.57	7.47
×130	98.3	1.32	8.19	110	1.18	9.17	77.7	1.67	6.48	89.2	1.46	7.43
×118	97.8	1.21	8.15	109	1.08	9.08	77.3	1.53	6.44	88.8	1.33	7.40
W30×391	109	3.59	9.08	125	3.13	10.4	82.0	4.77	6.83	97.6	4.01	8.13
×357	108	3.31	9.00	124	2.88	10.3	81.1	4.40	6.76	96.6	3.70	8.05
×326	107	3.05	8.92	123	2.65	10.3	80.2	4.06	6.68	95.6	3.41	7.97
×292	107	2.73	8.92	122	2.39	10.2	79.3	3.68	6.61	94.6	3.09	7.88
×261	106	2.46	8.83	121	2.16	10.1	78.4	3.33	6.53	93.6	2.79	7.80
×235	105	2.24	8.75	120	1.96	10.00	77.7	3.02	6.48	92.8	2.53	7.73
×211	105	2.01	8.75	120	1.76	10.00	76.9	2.74	6.41	92.0	2.29	7.67
×191	103	1.85	8.58	118	1.62	9.83	76.4	2.50	6.37	91.4	2.09	7.62
×173	104	1.66	8.67	119	1.45	9.92	75.8	2.28	6.32	90.8	1.91	7.57
W30×148	90.3	1.64	7.53	101	1.47	8.42	71.9	2.06	5.99	82.4	1.80	6.87
×132	89.5	1.47	7.46	100	1.32	8.33	71.1	1.86	5.93	81.6	1.62	6.80
×124	89.3	1.39	7.44	99.8	1.24	8.32	70.9	1.75	5.91	81.4	1.52	6.78
×116	89.1	1.30	7.43	99.6	1.16	8.30	70.5	1.65	5.88	81.0	1.43	6.75
×108	88.9	1.21	7.41	99.4	1.09	8.28	70.1	1.54	5.84	80.6	1.34	6.72
×99	88.5	1.12	7.38	99.0	1.00	8.25	69.9	1.42	5.83	80.4	1.23	6.70
×90	88.0	1.02	7.33	98.4	0.915	8.20	69.4	1.30	5.78	79.8	1.13	6.65
W27×539	106	5.08	8.83	121	4.45	10.1	80.3	6.71	6.69	95.6	5.64	7.97
×368	101	3.64	8.42	116	3.17	9.67	75.5	4.87	6.29	90.2	4.08	7.52
×336	100	3.36	8.33	115	2.92	9.58	74.6	4.50	6.22	89.2	3.77	7.43
×307	98.8	3.11	8.23	113	2.72	9.42	73.6	4.17	6.13	88.0	3.49	7.33
×281	98.2	2.86	8.18	113	2.49	9.42	73.0	3.85	6.08	87.4	3.22	7.28
×258	97.7	2.64	8.14	112	2.30	9.33	72.3	3.57	6.03	86.6	2.98	7.22
×235	96.6	2.43	8.05	111	2.12	9.25	71.6	3.28	5.97	85.8	2.74	7.15
×217	96.0	2.26	8.00	110	1.97	9.17	70.9	3.06	5.91	85.0	2.55	7.08
×194	95.6	2.03	7.97	110	1.76	9.17	70.2	2.76	5.85	84.2	2.30	7.02
×178	95.0	1.87	7.92	109	1.63	9.08	69.7	2.55	5.81	83.8	2.12	6.98
×161	94.6	1.70	7.88	109	1.48	9.08	69.2	2.33	5.77	83.2	1.94	6.93
×146	94.3	1.55	7.86	108	1.35	9.00	68.8	2.12	5.73	82.8	1.76	6.90
W27×129	82.8	1.56	6.90	92.8	1.39	7.73	65.2	1.98	5.43	75.2	1.72	6.27
×114	82.3	1.39	6.86	92.4	1.23	7.70	64.7	1.76	5.39	74.8	1.52	6.23
×102	82.1	1.24	6.84	92.1	1.11	7.68	64.2	1.59	5.35	74.2	1.37	6.18
×94	81.5	1.15	6.79	91.5	1.03	7.63	63.8	1.47	5.32	73.8	1.27	6.15
×84	81.2	1.03	6.77	91.2	0.921	7.60	63.4	1.32	5.28	73.3	1.15	6.11
W24×370	92.9	3.98	7.74	107	3.46	8.92	69.7	5.31	5.81	83.4	4.44	6.95
×335	91.5	3.66	7.63	105	3.19	8.75	68.5	4.89	5.71	82.0	4.09	6.83
×306	90.8	3.37	7.57	104	2.94	8.67	67.6	4.53	5.63	81.0	3.78	6.75
×279	89.7	3.11	7.48	103	2.71	8.58	66.7	4.18	5.56	80.0	3.49	6.67
×250	89.0	2.81	7.42	102	2.45	8.50	65.8	3.80	5.48	79.0	3.16	6.58
×229	88.1	2.60	7.34	101	2.27	8.42	65.1	3.52	5.43	78.2	2.93	6.52
×207	87.6	2.36	7.30	101	2.05	8.42	64.4	3.21	5.37	77.4	2.67	6.45
×192	87.1	2.20	7.26	100	1.92	8.33	64.0	3.00	5.33	77.0	2.49	6.42
×176	86.5	2.03	7.21	99.4	1.77	8.28	63.3	2.78	5.28	76.2	2.31	6.35
×162	86.3	1.88	7.19	99.3	1.63	8.28	63.0	2.57	5.25	76.0	2.13	6.33
×146	85.8	1.70	7.15	98.7	1.48	8.23	62.3	2.34	5.19	75.2	1.94	6.27
×131	85.3	1.54	7.11	98.2	1.33	8.18	61.9	2.12	5.16	74.8	1.75	6.23
×117	84.5	1.38	7.04	97.3	1.20	8.11	61.4	1.91	5.12	74.2	1.58	6.18
×104	84.1	1.24	7.01	96.9	1.07	8.08	61.0	1.70	5.08	73.8	1.41	6.15
W24×103	73.5	1.40	6.13	82.5	1.25	6.88	58.0	1.78	4.83	67.0	1.54	5.58
×94	73.5	1.28	6.13	82.5	1.14	6.88	57.7	1.63	4.81	66.7	1.41	5.56
×84	73.2	1.15	6.10	82.2	1.02	6.85	57.2	1.47	4.77	66.2	1.27	5.52
×76	72.5	1.05	6.04	81.5	0.933	6.79	56.8	1.34	4.73	65.8	1.16	5.48
×68	72.2	0.942	6.02	81.2	0.837	6.77	56.4	1.21	4.70	65.3	1.04	5.44

Case A: Shape perimeter, minus one flange surface.
Case B: Shape perimeter.
Case C: Box perimeter, minus one flange surface.
Case D: Box perimeter.

Table 1-36 (cont.). W-Shapes Surface and Box Perimeters, Weight-to-Perimeter Ratios, and Surface Areas

Shape	Case A			Case B			Case C			Case D		
	Perimeter	W/D Ratio	Surf. Area	Perimeter	W/D Ratio	Surf. Area	Perimeter	W/D Ratio	Surf. Area	Perimeter	W/D Ratio	Surf. Area
	in.		ft²/ft	in.		ft²/ft	in.		ft²/ft	in.		ft²/ft
W24×62	66.4	0.934	5.53	73.5	0.844	6.13	54.4	1.14	4.53	61.5	1.01	5.13
×55	66.4	0.828	5.53	73.4	0.749	6.12	54.2	1.01	4.52	61.2	0.899	5.10
W21×201	80.5	2.50	6.71	93.1	2.16	7.76	58.6	3.43	4.88	71.2	2.82	5.93
×182	80.0	2.28	6.67	92.5	1.97	7.71	57.9	3.14	4.83	70.4	2.59	5.87
×166	79.5	2.09	6.63	91.9	1.81	7.66	57.4	2.89	4.78	69.8	2.38	5.82
×147	78.7	1.87	6.56	91.2	1.61	7.60	56.7	2.59	4.73	69.2	2.12	5.77
×132	78.5	1.68	6.54	90.9	1.45	7.58	56.0	2.36	4.67	68.4	1.93	5.70
×122	77.9	1.57	6.49	90.3	1.35	7.53	55.8	2.19	4.65	68.2	1.79	5.68
×111	77.4	1.43	6.45	89.7	1.24	7.48	55.3	2.01	4.61	67.6	1.64	5.63
×101	77.4	1.30	6.45	89.7	1.13	7.48	55.1	1.83	4.59	67.4	1.50	5.62
W21×93	66.3	1.40	5.53	74.8	1.24	6.23	51.6	1.80	4.30	60.0	1.55	5.00
×83	65.8	1.26	5.48	74.2	1.12	6.18	51.2	1.62	4.27	59.5	1.39	4.96
×73	65.5	1.11	5.46	73.8	0.989	6.15	50.7	1.44	4.23	59.0	1.24	4.92
×68	65.1	1.04	5.43	73.4	0.926	6.12	50.5	1.35	4.21	58.7	1.16	4.89
×62	65.1	0.952	5.43	73.3	0.846	6.11	50.2	1.24	4.18	58.5	1.06	4.88
×55	64.4	0.854	5.37	72.6	0.758	6.05	49.8	1.10	4.15	58.0	0.948	4.83
×48	64.0	0.750	5.33	72.1	0.666	6.01	49.3	0.974	4.11	57.5	0.835	4.79
W21×57	59.9	0.952	4.99	66.5	0.857	5.54	48.8	1.17	4.07	55.3	1.03	4.61
×50	59.7	0.838	4.98	66.3	0.754	5.53	48.1	1.04	4.01	54.7	0.914	4.56
×44	59.0	0.746	4.92	65.5	0.672	5.46	47.9	0.919	3.99	54.4	0.809	4.53
W18×175	71.1	2.46	5.93	82.5	2.12	6.88	51.4	3.40	4.28	62.8	2.79	5.23
×158	70.5	2.24	5.88	81.8	1.93	6.82	50.7	3.12	4.23	62.0	2.55	5.17
×143	69.8	2.05	5.82	81.0	1.77	6.75	50.2	2.85	4.18	61.4	2.33	5.12
×130	69.3	1.88	5.78	80.5	1.61	6.71	49.8	2.61	4.15	61.0	2.13	5.08
×119	69.2	1.72	5.77	80.5	1.48	6.71	49.3	2.41	4.11	60.6	1.96	5.05
×106	68.6	1.55	5.72	79.8	1.33	6.65	48.6	2.18	4.05	59.8	1.77	4.98
×97	68.1	1.42	5.68	79.2	1.22	6.60	48.3	2.01	4.03	59.4	1.63	4.95
×86	67.8	1.27	5.65	78.9	1.09	6.58	47.9	1.80	3.99	59.0	1.46	4.92
×76	67.3	1.13	5.61	78.3	0.971	6.53	47.4	1.60	3.95	58.4	1.30	4.87
W18×71	58.0	1.22	4.83	65.6	1.08	5.47	44.6	1.59	3.72	52.3	1.36	4.36
×65	57.6	1.13	4.80	65.2	0.997	5.43	44.4	1.46	3.70	52.0	1.25	4.33
×60	57.5	1.04	4.79	65.0	0.923	5.42	44.0	1.36	3.67	51.5	1.17	4.29
×55	57.1	0.963	4.76	64.7	0.850	5.39	43.7	1.26	3.64	51.3	1.07	4.28
×50	56.8	0.880	4.73	64.3	0.778	5.36	43.5	1.15	3.63	51.0	0.980	4.25
W18×46	52.4	0.878	4.37	58.5	0.786	4.88	42.3	1.09	3.53	48.3	0.952	4.03
×40	52.1	0.768	4.34	58.1	0.688	4.84	41.8	0.957	3.48	47.8	0.837	3.98
×35	52.1	0.672	4.34	58.1	0.602	4.84	41.4	0.845	3.45	47.4	0.738	3.95
W16×100	62.7	1.59	5.23	73.1	1.37	6.09	44.4	2.25	3.70	54.8	1.82	4.57
×89	62.4	1.43	5.20	72.8	1.22	6.07	44.0	2.02	3.67	54.4	1.64	4.53
×77	61.6	1.25	5.13	71.9	1.07	5.99	43.3	1.78	3.61	53.6	1.44	4.47
×67	61.4	1.09	5.12	71.6	0.936	5.97	42.8	1.57	3.57	53.0	1.26	4.42
W16×57	52.1	1.09	4.34	59.2	0.963	4.93	39.9	1.43	3.33	47.0	1.21	3.92
×50	52.0	0.962	4.33	59.1	0.846	4.93	39.7	1.26	3.31	46.7	1.07	3.89
×45	51.7	0.870	4.31	58.7	0.767	4.89	39.2	1.15	3.27	46.3	0.972	3.86
×40	51.3	0.780	4.28	58.3	0.686	4.86	39.0	1.03	3.25	46.0	0.870	3.83
×36	51.3	0.702	4.28	58.3	0.617	4.86	38.8	0.928	3.23	45.8	0.786	3.82
W16×31	46.9	0.661	3.91	52.4	0.592	4.37	37.3	0.831	3.11	42.9	0.723	3.58
×26	46.6	0.558	3.88	52.1	0.499	4.34	36.9	0.705	3.08	42.4	0.613	3.53

Case A: Shape perimeter, minus one flange surface.
Case B: Shape perimeter.
Case C: Box perimeter, minus one flange surface.
Case D: Box perimeter.

Table 1-36 (cont.). W-Shapes
Surface and Box Perimeters, Weight-to-Perimeter Ratios, and Surface Areas

Shape	Case A			Case B			Case C			Case D		
	Perimeter in.	W/D Ratio	Surf. Area ft²/ft	Perimeter in.	W/D Ratio	Surf. Area ft²/ft	Perimeter in.	W/D Ratio	Surf. Area ft²/ft	Perimeter in.	W/D Ratio	Surf. Area ft²/ft
W14×808	92.3	8.75	7.69	111	7.28	9.25	64.2	12.6	5.35	82.8	9.76	6.90
×730	90.4	8.08	7.53	108	6.76	9.00	62.7	11.6	5.23	80.6	9.06	6.72
×665	88.8	7.49	7.40	107	6.21	8.92	60.9	10.9	5.08	78.6	8.46	6.55
×605	86.9	6.96	7.24	104	5.82	8.67	59.2	10.2	4.93	76.6	7.90	6.38
×550	85.6	6.43	7.13	103	5.34	8.58	57.6	9.55	4.80	74.8	7.35	6.23
×500	84.0	5.95	7.00	101	4.95	8.42	56.2	8.90	4.68	73.2	6.83	6.10
×455	82.3	5.53	6.86	99.1	4.59	8.26	54.8	8.30	4.57	71.6	6.35	5.97
×426	81.8	5.21	6.82	98.5	4.32	8.21	54.1	7.87	4.51	70.8	6.02	5.90
×398	80.7	4.93	6.73	97.3	4.09	8.11	53.2	7.48	4.43	69.8	5.70	5.82
×370	79.9	4.63	6.66	96.4	3.84	8.03	52.3	7.07	4.36	68.8	5.38	5.73
×342	79.1	4.32	6.59	95.5	3.58	7.96	51.4	6.65	4.28	67.8	5.04	5.65
×311	78.1	3.98	6.51	94.3	3.30	7.86	50.4	6.17	4.20	66.6	4.67	5.55
×283	77.3	3.66	6.44	93.4	3.03	7.78	49.5	5.72	4.13	65.6	4.31	5.47
×257	76.5	3.36	6.38	92.5	2.78	7.71	48.8	5.27	4.07	64.8	3.97	5.40
×233	75.6	3.08	6.30	91.5	2.55	7.63	47.9	4.86	3.99	63.8	3.65	5.32
×211	75.2	2.81	6.27	91.0	2.32	7.58	47.2	4.47	3.93	63.0	3.35	5.25
×193	74.3	2.60	6.19	90.0	2.14	7.50	46.7	4.13	3.89	62.4	3.09	5.20
×176	74.1	2.38	6.18	89.8	1.96	7.48	46.1	3.82	3.84	61.8	2.85	5.15
×159	73.5	2.16	6.13	89.1	1.78	7.43	45.6	3.49	3.80	61.2	2.60	5.10
×145	72.7	1.99	6.06	88.2	1.64	7.35	45.1	3.22	3.76	60.6	2.39	5.05
W14×132	70.0	1.89	5.83	84.7	1.56	7.06	44.1	2.99	3.68	58.8	2.24	4.90
×120	70.1	1.71	5.84	84.8	1.42	7.07	43.7	2.75	3.64	58.4	2.05	4.87
×109	69.6	1.57	5.80	84.2	1.29	7.02	43.2	2.52	3.60	57.8	1.89	4.82
×99	69.2	1.43	5.77	83.8	1.18	6.98	43.0	2.30	3.58	57.6	1.72	4.80
×90	68.7	1.31	5.73	83.2	1.08	6.93	42.5	2.12	3.54	57.0	1.58	4.75
W14×82	56.5	1.45	4.71	66.6	1.23	5.55	38.7	2.12	3.23	48.8	1.68	4.07
×74	56.2	1.32	4.68	66.3	1.12	5.53	38.5	1.92	3.21	48.6	1.52	4.05
×68	55.7	1.22	4.64	65.7	1.04	5.48	38.0	1.79	3.17	48.0	1.42	4.00
×61	55.7	1.10	4.64	65.7	0.928	5.48	37.8	1.61	3.15	47.8	1.28	3.98
W14×53	49.8	1.06	4.15	57.9	0.915	4.83	35.9	1.48	2.99	43.9	1.21	3.66
×48	49.5	0.970	4.13	57.5	0.835	4.79	35.6	1.35	2.97	43.7	1.10	3.64
×43	49.2	0.874	4.10	57.2	0.752	4.77	35.4	1.21	2.95	43.4	0.991	3.62
W14×38	47.0	0.809	3.92	53.8	0.706	4.48	35.0	1.09	2.92	41.7	0.911	3.48
×34	46.9	0.725	3.91	53.7	0.633	4.48	34.8	0.977	2.90	41.5	0.819	3.46
×30	46.6	0.644	3.88	53.4	0.562	4.45	34.3	0.875	2.86	41.1	0.730	3.43
W14×26	41.4	0.628	3.45	46.5	0.559	3.88	32.8	0.793	2.73	37.9	0.686	3.16
×22	41.2	0.534	3.43	46.2	0.476	3.85	32.4	0.679	2.70	37.4	0.588	3.12
W12×336	69.3	4.85	5.78	82.7	4.06	6.89	47.0	7.15	3.92	60.4	5.56	5.03
×305	67.9	4.49	5.66	81.1	3.76	6.76	45.8	6.66	3.82	59.0	5.17	4.92
×279	66.6	4.19	5.55	79.7	3.50	6.64	44.9	6.21	3.74	58.0	4.81	4.83
×252	65.7	3.84	5.48	78.7	3.20	6.56	43.8	5.75	3.65	56.8	4.44	4.73
×230	64.7	3.55	5.39	77.6	2.96	6.47	43.1	5.34	3.59	56.0	4.11	4.67
×210	64.2	3.27	5.35	77.0	2.73	6.42	42.2	4.98	3.52	55.0	3.82	4.58
×190	63.4	3.00	5.28	76.1	2.50	6.34	41.5	4.58	3.46	54.2	3.51	4.52
×170	62.6	2.72	5.22	75.2	2.26	6.27	40.6	4.19	3.38	53.2	3.20	4.43
×152	62.1	2.45	5.18	74.6	2.04	6.22	39.9	3.81	3.33	52.4	2.90	4.37
×136	60.9	2.23	5.08	73.3	1.86	6.11	39.2	3.47	3.27	51.6	2.64	4.30
×120	60.4	1.99	5.03	72.7	1.65	6.06	38.5	3.12	3.21	50.8	2.36	4.23
×106	59.9	1.77	4.99	72.1	1.47	6.01	38.0	2.79	3.17	50.2	2.11	4.18
×96	59.7	1.61	4.98	71.9	1.34	5.99	37.6	2.55	3.13	49.8	1.93	4.15
×87	59.1	1.47	4.93	71.2	1.22	5.93	37.1	2.35	3.09	49.2	1.77	4.10
×79	58.8	1.34	4.90	70.9	1.11	5.91	36.9	2.14	3.08	49.0	1.61	4.08
×72	58.3	1.23	4.86	70.3	1.02	5.86	36.6	1.97	3.05	48.6	1.48	4.05
×65	58.3	1.11	4.86	70.3	0.925	5.86	36.2	1.80	3.02	48.2	1.35	4.02
W12×58	52.7	1.10	4.39	62.7	0.925	5.23	34.4	1.69	2.87	44.4	1.31	3.70
×53	52.0	1.02	4.33	62.0	0.855	5.17	34.2	1.55	2.85	44.2	1.20	3.68

Case A: Shape perimeter, minus one flange surface.
Case B: Shape perimeter.
Case C: Box perimeter, minus one flange surface.
Case D: Box perimeter.

Table 1-36 (cont.). W-Shapes Surface and Box Perimeters, Weight-to-Perimeter Ratios, and Surface Areas

Shape	Case A			Case B			Case C			Case D		
	Perimeter in.	W/D Ratio	Surf. Area ft²/ft	Perimeter in.	W/D Ratio	Surf. Area ft²/ft	Perimeter in.	W/D Ratio	Surf. Area ft²/ft	Perimeter in.	W/D Ratio	Surf. Area ft²/ft
W12×50	47.0	1.06	3.92	55.0	0.909	4.58	32.5	1.54	2.71	40.6	1.23	3.38
×45	46.2	0.974	3.85	54.3	0.829	4.53	32.3	1.39	2.69	40.3	1.12	3.36
×40	46.5	0.860	3.88	54.5	0.734	4.54	31.8	1.26	2.65	39.8	1.01	3.32
W12×35	43.2	0.810	3.60	49.8	0.703	4.15	31.6	1.11	2.63	38.1	0.919	3.18
×30	42.9	0.699	3.58	49.4	0.607	4.12	31.1	0.965	2.59	37.6	0.798	3.13
×26	42.5	0.612	3.54	49.0	0.531	4.08	30.9	0.841	2.58	37.4	0.695	3.12
W12×22	35.3	0.623	2.94	39.3	0.560	3.28	28.6	0.769	2.38	32.7	0.673	2.73
×19	35.2	0.540	2.93	39.2	0.485	3.27	28.4	0.669	2.37	32.4	0.586	2.70
×16	35.0	0.457	2.92	39.0	0.410	3.25	28.0	0.571	2.33	32.0	0.500	2.67
×14	34.6	0.405	2.88	38.6	0.363	3.22	27.8	0.504	2.32	31.7	0.442	2.64
W10×112	51.5	2.17	4.29	61.9	1.81	5.16	33.2	3.37	2.77	43.6	2.57	3.63
×100	50.7	1.97	4.23	61.0	1.64	5.08	32.5	3.08	2.71	42.8	2.34	3.57
×88	50.5	1.74	4.21	60.8	1.45	5.07	31.9	2.76	2.66	42.2	2.09	3.52
×77	49.9	1.54	4.16	60.1	1.28	5.01	31.4	2.45	2.62	41.6	1.85	3.47
×68	49.1	1.38	4.09	59.2	1.15	4.93	30.9	2.20	2.58	41.0	1.66	3.42
×60	49.1	1.22	4.09	59.2	1.01	4.93	30.5	1.97	2.54	40.6	1.48	3.38
×54	48.6	1.11	4.05	58.6	0.922	4.88	30.2	1.79	2.52	40.2	1.34	3.35
×49	48.3	1.01	4.03	58.3	0.840	4.86	30.0	1.63	2.50	40.0	1.23	3.33
W10×45	42.6	1.06	3.55	50.7	0.888	4.23	28.2	1.60	2.35	36.2	1.24	3.02
×39	42.0	0.929	3.50	50.0	0.780	4.17	27.8	1.40	2.32	35.8	1.09	2.98
×33	42.0	0.786	3.50	49.9	0.661	4.16	27.4	1.20	2.28	35.4	0.932	2.95
W10×30	37.1	0.809	3.09	42.9	0.699	3.58	26.8	1.12	2.23	32.6	0.920	2.72
×26	36.7	0.708	3.06	42.5	0.612	3.54	26.4	0.985	2.20	32.1	0.810	2.68
×22	36.3	0.606	3.03	42.1	0.523	3.51	26.2	0.840	2.18	31.9	0.690	2.66
W10×19	31.3	0.607	2.61	35.3	0.538	2.94	24.4	0.779	2.03	28.4	0.669	2.37
×17	31.3	0.543	2.61	35.3	0.482	2.94	24.2	0.702	2.02	28.2	0.603	2.35
×15	31.0	0.484	2.58	35.0	0.429	2.92	24.0	0.625	2.00	28.0	0.536	2.33
×12	30.6	0.392	2.55	34.6	0.347	2.88	23.7	0.506	1.98	27.7	0.433	2.31
W8×67	40.7	1.65	3.39	48.9	1.37	4.08	26.3	2.55	2.19	34.6	1.94	2.88
×58	40.2	1.44	3.35	48.5	1.20	4.04	25.7	2.26	2.14	33.9	1.71	2.83
×48	39.7	1.21	3.31	47.8	1.00	3.98	25.1	1.91	2.09	33.2	1.45	2.77
×40	39.0	1.03	3.25	47.1	0.849	3.93	24.6	1.63	2.05	32.6	1.23	2.72
×35	38.6	0.907	3.22	46.7	0.749	3.89	24.3	1.44	2.03	32.3	1.08	2.69
×31	38.6	0.803	3.22	46.6	0.665	3.88	24.0	1.29	2.00	32.0	0.969	2.67
W8×28	34.2	0.819	2.85	40.7	0.688	3.39	22.7	1.23	1.89	29.2	0.959	2.43
×24	34.1	0.704	2.84	40.6	0.591	3.38	22.4	1.07	1.87	28.9	0.830	2.41
W8×21	31.1	0.675	2.59	36.4	0.577	3.03	21.8	0.963	1.82	27.1	0.775	2.26
×18	30.9	0.583	2.58	36.1	0.499	3.01	21.5	0.837	1.79	26.8	0.672	2.23
W8×15	27.2	0.551	2.27	31.2	0.481	2.60	20.2	0.743	1.68	24.2	0.620	2.02
×13	26.9	0.483	2.24	30.9	0.421	2.58	20.0	0.650	1.67	24.0	0.542	2.00
×10	26.7	0.375	2.23	30.6	0.327	2.55	19.7	0.508	1.64	23.7	0.422	1.98
W6×25	29.8	0.839	2.48	35.9	0.696	2.99	18.8	1.33	1.57	24.9	1.00	2.08
×20	29.5	0.678	2.46	35.5	0.563	2.96	18.4	1.09	1.53	24.4	0.820	2.03
×15	28.8	0.521	2.40	34.8	0.431	2.90	18.0	0.833	1.50	24.0	0.625	2.00
W6×16	23.4	0.684	1.95	27.4	0.584	2.28	16.6	0.964	1.38	20.6	0.777	1.72
×12	22.8	0.526	1.90	26.8	0.448	2.23	16.1	0.745	1.34	20.1	0.597	1.68
×9	22.6	0.398	1.88	26.6	0.338	2.22	15.7	0.573	1.31	19.7	0.457	1.64
×8.5	22.7	0.374	1.89	26.6	0.320	2.22	15.6	0.545	1.30	19.5	0.436	1.63
W5×19	24.5	0.776	2.04	29.5	0.644	2.46	15.3	1.24	1.28	20.4	0.931	1.70
×16	24.1	0.664	2.01	29.1	0.550	2.43	15.0	1.07	1.25	20.0	0.800	1.67
W4×13	19.4	0.670	1.62	23.4	0.556	1.95	12.4	1.05	1.03	16.4	0.793	1.37

Case A: Shape perimeter, minus one flange surface.
Case B: Shape perimeter.
Case C: Box perimeter, minus one flange surface.
Case D: Box perimeter.

Table 1-37.
M-Shapes
Surface and Box Perimeters, Weight-to-Perimeter Ratios, and Surface Areas

Shape	Case A			Case B			Case C			Case D		
	Perimeter in.	W/D Ratio	Surf. Area ft²/ft	Perimeter in.	W/D Ratio	Surf. Area ft²/ft	Perimeter in.	W/D Ratio	Surf. Area ft²/ft	Perimeter in.	W/D Ratio	Surf. Area ft²/ft
M12×11.8	32.5	0.363	2.71	35.6	0.331	2.97	27.1	0.435	2.26	30.1	0.392	2.51
×10.8	32.5	0.332	2.71	35.6	0.303	2.97	27.1	0.399	2.26	30.1	0.359	2.51
×10	33.0	0.303	2.75	36.2	0.276	3.02	27.3	0.366	2.28	30.5	0.328	2.54
M10×9	27.4	0.328	2.28	30.1	0.299	2.51	22.7	0.396	1.89	25.4	0.354	2.12
×8	27.4	0.292	2.28	30.1	0.266	2.51	22.6	0.354	1.88	25.3	0.316	2.11
×7.5	27.4	0.274	2.28	30.1	0.249	2.51	22.7	0.330	1.89	25.4	0.295	2.12
M8×6.5	22.2	0.293	1.85	24.5	0.265	2.04	18.3	0.355	1.53	20.6	0.316	1.72
×6.2	22.4	0.277	1.87	24.7	0.251	2.06	18.3	0.339	1.53	20.6	0.301	1.72
M6×4.4	17.0	0.259	1.42	18.8	0.234	1.57	13.8	0.319	1.15	15.7	0.280	1.31
×3.7	17.2	0.215	1.43	19.2	0.193	1.60	13.8	0.268	1.15	15.8	0.234	1.32
M5×18.9	23.9	0.791	1.99	28.9	0.654	2.41	15.0	1.26	1.25	20.0	0.945	1.67
M4×6	18.2	0.330	1.52	22.0	0.273	1.83	11.4	0.526	0.950	15.2	0.395	1.27

Case A: Shape perimeter, minus one flange surface.
Case B: Shape perimeter.
Case C: Box perimeter, minus one flange surface.
Case D: Box perimeter.

Table 1-38.
S-Shapes
Surface and Box Perimeters, Weight-to-Perimeter Ratios, and Surface Areas

Shape	Case A			Case B			Case C			Case D		
	Perimeter	W/D Ratio	Surf. Area	Perimeter	W/D Ratio	Surf. Area	Perimeter	W/D Ratio	Surf. Area	Perimeter	W/D Ratio	Surf. Area
	in.		ft²/ft	in.		ft²/ft	in.		ft²/ft	in.		ft²/ft
S24×121	68.6	1.76	5.72	76.6	1.58	6.38	57.1	2.12	4.76	65.1	1.86	5.43
×106	68.4	1.55	5.70	76.3	1.39	6.36	56.9	1.86	4.74	64.7	1.64	5.39
S24×100	65.5	1.53	5.46	72.8	1.37	6.07	55.3	1.81	4.61	62.5	1.60	5.21
×90	65.4	1.38	5.45	72.5	1.24	6.04	55.1	1.63	4.59	62.3	1.44	5.19
×80	65.2	1.23	5.43	72.2	1.11	6.02	55.0	1.45	4.58	62.0	1.29	5.17
S20×96	57.9	1.66	4.83	65.1	1.47	5.43	47.8	2.01	3.98	55.0	1.75	4.58
×86	57.8	1.49	4.82	64.9	1.33	5.41	47.7	1.80	3.98	54.7	1.57	4.56
S20×75	55.4	1.35	4.62	61.8	1.21	5.15	46.4	1.62	3.87	52.8	1.42	4.40
×66	55.3	1.19	4.61	61.5	1.07	5.13	46.3	1.43	3.86	52.5	1.26	4.38
S18×70	50.9	1.38	4.24	57.2	1.22	4.77	42.3	1.65	3.53	48.5	1.44	4.04
×54.7	50.7	1.08	4.23	56.7	0.965	4.73	42.0	1.30	3.50	48.0	1.14	4.00
S15×50	43.6	1.15	3.63	49.2	1.02	4.10	35.6	1.40	2.97	41.3	1.21	3.44
×42.9	43.4	0.988	3.62	48.9	0.877	4.08	35.5	1.21	2.96	41.0	1.05	3.42
S12×50	36.9	1.36	3.08	42.4	1.18	3.53	29.5	1.69	2.46	35.0	1.43	2.92
×40.8	36.6	1.11	3.05	41.9	0.974	3.49	29.3	1.39	2.44	34.5	1.18	2.88
S12×35	36.4	0.962	3.03	41.5	0.843	3.46	29.1	1.20	2.43	34.2	1.02	2.85
×31.8	36.3	0.876	3.03	41.3	0.770	3.44	29.0	1.10	2.42	34.0	0.935	2.83
S10×35	31.7	1.10	2.64	36.7	0.954	3.06	24.9	1.41	2.08	29.9	1.17	2.49
×25.4	31.5	0.806	2.63	36.1	0.704	3.01	24.7	1.03	2.06	29.3	0.867	2.44
S8×23	26.0	0.885	2.17	30.1	0.764	2.51	20.2	1.14	1.68	24.3	0.947	2.03
×18.4	25.8	0.713	2.15	29.8	0.617	2.48	20.0	0.920	1.67	24.0	0.767	2.00
S6×17.25	20.4	0.846	1.70	24.0	0.719	2.00	15.6	1.11	1.30	19.1	0.903	1.59
×12.5	20.2	0.619	1.68	23.5	0.532	1.96	15.3	0.817	1.28	18.7	0.668	1.56
S5×10	17.3	0.578	1.44	20.3	0.493	1.69	13.0	0.769	1.08	16.0	0.625	1.33
S4×9.5	14.5	0.655	1.21	17.3	0.549	1.44	10.8	0.880	0.900	13.6	0.699	1.13
×7.7	14.4	0.535	1.20	17.1	0.450	1.43	10.7	0.720	0.892	13.3	0.579	1.11
S3×7.5	11.8	0.636	0.983	14.3	0.524	1.19	8.51	0.881	0.709	11.0	0.682	0.917
×5.7	11.6	0.491	0.967	14.0	0.407	1.17	8.33	0.684	0.694	10.7	0.533	0.892

Case A: Shape perimeter, minus one flange surface.
Case B: Shape perimeter.
Case C: Box perimeter, minus one flange surface.
Case D: Box perimeter.

Table 1-39.
HP-Shapes
Surface and Box Perimeters, Weight-to-Perimeter Ratios, and Surface Areas

Shape	Case A			Case B			Case C			Case D		
	Perimeter in.	W/D Ratio	Surf. Area ft²/ft	Perimeter in.	W/D Ratio	Surf. Area ft²/ft	Perimeter in.	W/D Ratio	Surf. Area ft²/ft	Perimeter in.	W/D Ratio	Surf. Area ft²/ft
HP14×117	70.5	1.66	5.88	85.4	1.37	7.12	43.3	2.70	3.61	58.2	2.01	4.85
×102	69.9	1.46	5.83	84.7	1.20	7.06	42.8	2.38	3.57	57.6	1.77	4.80
×89	69.7	1.28	5.81	84.4	1.05	7.03	42.3	2.10	3.53	57.0	1.56	4.75
×73	69.1	1.06	5.76	83.7	0.872	6.98	41.8	1.75	3.48	56.4	1.29	4.70
HP12×84	59.0	1.42	4.92	71.3	1.18	5.94	36.9	2.28	3.08	49.2	1.71	4.10
×74	58.7	1.26	4.89	70.9	1.04	5.91	36.4	2.03	3.03	48.6	1.52	4.05
×63	58.5	1.08	4.88	70.6	0.892	5.88	35.9	1.75	2.99	48.0	1.31	4.00
×53	57.6	0.920	4.80	69.6	0.761	5.80	35.6	1.49	2.97	47.6	1.11	3.97
HP10×57	48.4	1.18	4.03	58.6	0.973	4.88	30.2	1.89	2.52	40.4	1.41	3.37
×42	48.2	0.871	4.02	58.3	0.720	4.86	29.5	1.42	2.46	39.6	1.06	3.30
HP8×36	38.5	0.935	3.21	46.7	0.771	3.89	24.2	1.49	2.02	32.3	1.11	2.69

Case A: Shape perimeter, minus one flange surface.
Case B: Shape perimeter.
Case C: Box perimeter, minus one flange surface.
Case D: Box perimeter.

Table 1-40.
C-Shapes (American Standard Channels)
Surface and Box Perimeters, Weight-to-Perimeter Ratios, and Surface Areas

Shape	Case A			Case B			Case C			Case D		
	Perimeter	W/D Ratio	Surf. Area	Perimeter	W/D Ratio	Surf. Area	Perimeter	W/D Ratio	Surf. Area	Perimeter	W/D Ratio	Surf. Area
	in.		ft²/ft	in.		ft²/ft	in.		ft²/ft	in.		ft²/ft
C15×50	39.7	1.26	3.31	43.4	1.15	3.62	33.7	1.48	2.81	37.4	1.34	3.12
×40	39.2	1.02	3.27	42.7	0.937	3.56	33.5	1.19	2.79	37.0	1.08	3.08
×33.9	38.8	0.874	3.23	42.2	0.803	3.52	33.4	1.01	2.78	36.8	0.921	3.07
C12×30	32.3	0.929	2.69	35.5	0.845	2.96	27.2	1.10	2.27	30.3	0.990	2.53
×25	32.0	0.781	2.67	35.0	0.714	2.92	27.1	0.923	2.26	30.1	0.831	2.51
×20.7	31.7	0.653	2.64	34.6	0.598	2.88	26.9	0.770	2.24	29.9	0.692	2.49
C10×30	28.0	1.07	2.33	31.0	0.968	2.58	23.0	1.30	1.92	26.1	1.15	2.18
×25	27.6	0.906	2.30	30.5	0.820	2.54	22.9	1.09	1.91	25.8	0.969	2.15
×20	27.2	0.735	2.27	29.9	0.669	2.49	22.7	0.881	1.89	25.5	0.784	2.13
×15.3	26.8	0.571	2.23	29.4	0.520	2.45	22.6	0.677	1.88	25.2	0.607	2.10
C9×20	24.9	0.803	2.08	27.6	0.725	2.30	20.7	0.966	1.73	23.3	0.858	1.94
×15	24.5	0.612	2.04	27.0	0.556	2.25	20.5	0.732	1.71	23.0	0.652	1.92
×13.4	24.3	0.551	2.03	26.7	0.502	2.23	20.4	0.657	1.70	22.9	0.585	1.91
C8×18.75	22.6	0.830	1.88	25.1	0.747	2.09	18.5	1.01	1.54	21.1	0.889	1.76
×13.75	22.1	0.622	1.84	24.4	0.564	2.03	18.3	0.751	1.53	20.7	0.664	1.73
×11.5	21.9	0.525	1.83	24.1	0.477	2.01	18.3	0.628	1.53	20.5	0.561	1.71
C7×14.75	20.0	0.738	1.67	22.3	0.661	1.86	16.3	0.905	1.36	18.6	0.793	1.55
×12.25	19.7	0.622	1.64	21.9	0.559	1.83	16.2	0.756	1.35	18.4	0.666	1.53
×9.8	19.4	0.505	1.62	21.5	0.456	1.79	16.1	0.609	1.34	18.2	0.538	1.52
C6×13	17.6	0.739	1.47	19.8	0.657	1.65	14.2	0.915	1.18	16.3	0.798	1.36
×10.5	17.3	0.607	1.44	19.3	0.544	1.61	14.0	0.750	1.17	16.1	0.652	1.34
×8.2	17.0	0.482	1.42	18.9	0.434	1.58	13.9	0.590	1.16	15.8	0.519	1.32
C5×9	14.9	0.604	1.24	16.8	0.536	1.40	11.9	0.756	0.992	13.8	0.652	1.15
×6.7	14.5	0.462	1.21	16.3	0.411	1.36	11.8	0.568	0.983	13.5	0.496	1.13
C4×7.25	12.4	0.585	1.03	14.2	0.511	1.18	9.72	0.746	0.810	11.4	0.636	0.950
×5.4	12.1	0.446	1.01	13.6	0.397	1.13	9.58	0.564	0.798	11.2	0.482	0.933
×4.5	12.1	0.372	1.01	13.6	0.331	1.13	9.58	0.470	0.798	11.2	0.402	0.933
C3×6	10.1	0.594	0.842	11.7	0.513	0.975	7.60	0.789	0.633	9.20	0.652	0.767
×5	9.86	0.507	0.822	11.4	0.439	0.950	7.50	0.667	0.625	9.00	0.556	0.750
×4.1	9.61	0.427	0.801	11.0	0.373	0.917	7.41	0.553	0.618	8.82	0.465	0.735
×3.5	9.50	0.368	0.792	10.9	0.321	0.908	7.37	0.475	0.614	8.74	0.400	0.728

Case A: Shape perimeter, minus one flange surface.
Case B: Shape perimeter.
Case C: Box perimeter, minus one flange surface.
Case D: Box perimeter.

Table 1-41. MC-Shapes (Miscellaneous Channels) Surface and Box Perimeters, Weight-to-Perimeter Ratios, and Surface Areas

Shape	Case A			Case B			Case C			Case D		
	Perimeter	*W/D* Ratio	Surf. Area	Perimeter	*W/D* Ratio	Surf. Area	Perimeter	*W/D* Ratio	Surf. Area	Perimeter	*W/D* Ratio	Surf. Area
	in.		ft²/ft	in.		ft²/ft	in.		ft²/ft	in.		ft²/ft
MC18×58	47.0	1.23	3.92	51.2	1.13	4.27	40.2	1.44	3.35	44.4	1.31	3.70
×51.9	46.7	1.11	3.89	50.8	1.02	4.23	40.1	1.29	3.34	44.2	1.17	3.68
×45.8	46.5	0.985	3.88	50.5	0.907	4.21	40.0	1.15	3.33	44.0	1.04	3.67
×42.7	46.3	0.922	3.86	50.3	0.849	4.19	40.0	1.07	3.33	43.9	0.973	3.66
MC13×50	37.6	1.33	3.13	42.0	1.19	3.50	30.4	1.64	2.53	34.8	1.44	2.90
×40	37.0	1.08	3.08	41.1	0.973	3.43	30.2	1.32	2.52	34.4	1.16	2.87
×35	36.7	0.954	3.06	40.7	0.860	3.39	30.1	1.16	2.51	34.1	1.03	2.84
×31.8	36.5	0.871	3.04	40.5	0.785	3.38	30.0	1.06	2.50	34.0	0.935	2.83
MC12×50	35.0	1.43	2.92	39.1	1.28	3.26	28.1	1.78	2.34	32.3	1.55	2.69
×45	34.6	1.30	2.88	38.6	1.17	3.22	28.0	1.61	2.33	32.0	1.41	2.67
×40	34.3	1.17	2.86	38.2	1.05	3.18	27.9	1.43	2.33	31.8	1.26	2.65
×35	34.0	1.03	2.83	37.7	0.928	3.14	27.8	1.26	2.32	31.5	1.11	2.63
×31	33.7	0.920	2.81	37.4	0.829	3.12	27.7	1.12	2.31	31.3	0.990	2.61
MC12×10.6	27.8	0.381	2.32	29.3	0.362	2.44	25.5	0.416	2.13	27.0	0.393	2.25
MC10×41.1	31.4	1.31	2.62	35.7	1.15	2.98	24.3	1.69	2.03	28.6	1.44	2.38
×33.6	30.8	1.09	2.57	34.9	0.963	2.91	24.1	1.39	2.01	28.2	1.19	2.35
×28.5	30.4	0.938	2.53	34.3	0.831	2.86	24.0	1.19	2.00	27.9	1.02	2.33
MC10×25	28.9	0.865	2.41	32.3	0.774	2.69	23.4	1.07	1.95	26.8	0.933	2.23
×22	28.7	0.767	2.39	32.0	0.688	2.67	23.3	0.944	1.94	26.6	0.827	2.22
MC10×8.4	23.8	0.353	1.98	25.3	0.332	2.11	21.5	0.391	1.79	23.0	0.365	1.92
MC9×25.4	27.2	0.934	2.27	30.7	0.827	2.56	21.5	1.18	1.79	25.0	1.02	2.08
×23.9	27.0	0.885	2.25	30.5	0.784	2.54	21.5	1.11	1.79	24.9	0.960	2.08
MC8×22.8	25.2	0.905	2.10	28.7	0.794	2.39	19.5	1.17	1.63	23.0	0.991	1.92
×21.4	25.1	0.853	2.09	28.5	0.751	2.38	19.5	1.10	1.63	22.9	0.934	1.91
MC8×20	23.9	0.837	1.99	27.0	0.741	2.25	19.0	1.05	1.58	22.1	0.905	1.84
×18.7	23.8	0.786	1.98	26.8	0.698	2.23	19.0	0.984	1.58	22.0	0.850	1.83
MC8×8.5	20.8	0.409	1.73	22.7	0.374	1.89	17.9	0.475	1.49	19.7	0.431	1.64
MC7×22.7	23.5	0.966	1.96	27.1	0.838	2.26	17.6	1.29	1.47	21.2	1.07	1.77
×19.1	23.1	0.827	1.93	26.5	0.721	2.21	17.5	1.09	1.46	20.9	0.914	1.74
MC6×18	21.2	0.849	1.77	24.7	0.729	2.06	15.5	1.16	1.29	19.0	0.947	1.58
×15.3	21.3	0.718	1.78	24.8	0.617	2.07	15.5	0.987	1.29	19.0	0.805	1.58
MC6×16.3	19.9	0.819	1.66	22.9	0.712	1.91	15.0	1.09	1.25	18.0	0.906	1.50
×15.1	19.7	0.766	1.64	22.6	0.668	1.88	14.9	1.01	1.24	17.9	0.844	1.49
MC6×12	18.6	0.645	1.55	21.1	0.569	1.76	14.5	0.828	1.21	17.0	0.706	1.42

Case A: Shape perimeter, minus one flange surface.
Case B: Shape perimeter.
Case C: Box perimeter, minus one flange surface.
Case D: Box perimeter.

Table 1-42. Angles
Surface Perimeters, Weight-to-Perimeter Ratios, and Surface Areas

Shape	Case A-1			Case A-2			Case B		
	Perimeter, in.	W/D Ratio	Surf. Area, ft^2/ft	Perimeter, in.	W/D Ratio	Surf. Area, ft^2/ft	Perimeter, in.	W/D Ratio	Surf. Area, ft^2/ft
L8×8×1 1/8	23.7	2.41	1.98	23.7	2.41	1.98	31.7	1.80	2.64
×1	23.7	2.16	1.98	23.7	2.16	1.98	31.7	1.62	2.64
×7/8	23.7	1.91	1.98	23.7	1.91	1.98	31.7	1.43	2.64
×3/4	23.7	1.65	1.98	23.7	1.65	1.98	31.7	1.24	2.64
×5/8	23.7	1.39	1.98	23.7	1.39	1.98	31.7	1.04	2.64
×9/16	23.7	1.26	1.98	23.7	1.26	1.98	31.7	0.94	2.64
×1/2	23.7	1.13	1.98	23.7	1.13	1.98	31.7	0.842	2.64
L8×6×1	21.8	2.04	1.82	19.8	2.24	1.65	27.8	1.60	2.32
×7/8	21.8	1.80	1.82	19.8	1.98	1.65	27.8	1.41	2.32
×3/4	21.8	1.56	1.82	19.8	1.72	1.65	27.8	1.22	2.32
×5/8	21.8	1.31	1.82	19.8	1.44	1.65	27.8	1.03	2.32
×9/16	21.8	1.19	1.82	19.8	1.31	1.65	27.8	0.932	2.32
×1/2	21.8	1.06	1.82	19.8	1.17	1.65	27.8	0.835	2.32
×7/16	21.8	0.936	1.82	19.8	1.03	1.65	27.8	0.734	2.32
L8×4×1	19.8	1.90	1.65	15.8	2.38	1.32	23.8	1.58	1.98
×7/8	19.8	1.68	1.65	15.8	2.11	1.32	23.8	1.40	1.98
×3/4	19.8	1.46	1.65	15.8	1.83	1.32	23.8	1.21	1.98
×5/8	19.8	1.23	1.65	15.8	1.54	1.32	23.8	1.03	1.98
×9/16	19.8	1.12	1.65	15.8	1.40	1.32	23.8	0.929	1.98
×1/2	19.8	0.995	1.65	15.8	1.25	1.32	23.8	0.828	1.98
×7/16	19.8	0.879	1.65	15.8	1.10	1.32	23.8	0.731	1.98
L7×4×3/4	17.8	1.47	1.48	14.8	1.77	1.23	21.8	1.20	1.82
×5/8	17.8	1.24	1.48	14.8	1.49	1.23	21.8	1.01	1.82
×1/2	17.8	1.01	1.48	14.8	1.21	1.23	21.8	0.821	1.82
×7/16	17.8	0.888	1.48	14.8	1.07	1.23	21.8	0.725	1.82
×3/8	17.8	0.764	1.48	14.8	0.919	1.23	21.8	0.624	1.82
L6×6×1	17.8	2.11	1.48	17.8	2.11	1.48	23.8	1.58	1.98
×7/8	17.8	1.87	1.48	17.8	1.87	1.48	23.8	1.39	1.98
×3/4	17.8	1.62	1.48	17.8	1.62	1.48	23.8	1.21	1.98
×5/8	17.8	1.37	1.48	17.8	1.37	1.48	23.8	1.02	1.98
×9/16	17.8	1.24	1.48	17.8	1.24	1.48	23.8	0.924	1.98
×1/2	17.8	1.10	1.48	17.8	1.10	1.48	23.8	0.824	1.98
×7/16	17.8	0.972	1.48	17.8	0.972	1.48	23.8	0.727	1.98
×3/8	17.8	0.837	1.48	17.8	0.837	1.48	23.8	0.626	1.98
×5/16	17.8	0.702	1.48	17.8	0.702	1.48	23.8	0.525	1.98
L6×4×7/8	15.8	1.72	1.32	13.8	1.96	1.15	19.8	1.37	1.65
×3/4	15.8	1.49	1.32	13.8	1.70	1.15	19.8	1.19	1.65
×5/8	15.8	1.25	1.32	13.8	1.43	1.15	19.8	1.00	1.65
×9/16	15.8	1.13	1.32	13.8	1.30	1.15	19.8	0.904	1.65
×1/2	15.8	1.01	1.32	13.8	1.16	1.15	19.8	0.808	1.65
×7/16	15.8	0.892	1.32	13.8	1.02	1.15	19.8	0.712	1.65
×3/8	15.8	0.772	1.32	13.8	0.884	1.15	19.8	0.616	1.65
×5/16	15.8	0.646	1.32	13.8	0.739	1.15	19.8	0.515	1.65
L6×3 1/2×1/2	15.3	1.00	1.28	12.8	1.20	1.07	18.8	0.814	1.57
×3/8	15.3	0.758	1.28	12.8	0.906	1.07	18.8	0.617	1.57
×5/16	15.3	0.635	1.28	12.8	0.759	1.07	18.8	0.517	1.57
L5×5×7/8	14.8	1.84	1.23	14.8	1.84	1.23	19.8	1.38	1.65
×3/4	14.8	1.60	1.23	14.8	1.60	1.23	19.8	1.20	1.65
×5/8	14.8	1.36	1.23	14.8	1.36	1.23	19.8	1.02	1.65
×1/2	14.8	1.10	1.23	14.8	1.10	1.23	19.8	0.823	1.65
×7/16	14.8	0.973	1.23	14.8	0.973	1.23	19.8	0.727	1.65
×3/8	14.8	0.838	1.23	14.8	0.838	1.23	19.8	0.626	1.65
×5/16	14.8	0.703	1.23	14.8	0.703	1.23	19.8	0.525	1.65

Case A-1: Shape perimeter, minus short leg surface.
Case A-2: Shape perimeter, minus long leg surface.
Case B: Shape perimeter.

Table 1-42 (cont.). Angles
Surface Perimeters, Weight-to-Perimeter Ratios, and Surface Areas

Shape	Case A-1			Case A-2			Case B		
	Perimeter	W/D Ratio	Surf. Area	Perimeter	W/D Ratio	Surf. Area	Perimeter	W/D Ratio	Surf. Area
	in.		ft^2/ft	in.		ft^2/ft	in.		ft^2/ft
$L5\times3\frac{1}{2}\times\frac{3}{4}$	13.3	1.49	1.11	11.8	1.68	0.98	16.8	1.18	1.400
$\times\frac{5}{8}$	13.3	1.26	1.11	11.8	1.42	0.98	16.8	1.00	1.400
$\times\frac{1}{2}$	13.3	1.02	1.11	11.8	1.15	0.98	16.8	0.810	1.400
$\times\frac{3}{8}$	13.3	0.782	1.11	11.8	0.881	0.98	16.8	0.619	1.400
$\times\frac{5}{16}$	13.3	0.656	1.11	11.8	0.739	0.98	16.8	0.519	1.400
$\times\frac{1}{4}$	13.3	0.529	1.11	11.8	0.596	0.98	16.8	0.418	1.400
$L5\times3\times\frac{1}{2}$	12.8	1.00	1.07	10.8	1.19	0.90	15.8	0.810	1.320
$\times\frac{7}{16}$	12.8	0.883	1.07	10.8	1.05	0.90	15.8	0.715	1.320
$\times\frac{3}{8}$	12.8	0.761	1.07	10.8	0.902	0.90	15.8	0.616	1.320
$\times\frac{5}{16}$	12.8	0.640	1.07	10.8	0.758	0.90	15.8	0.518	1.320
$\times\frac{1}{4}$	12.8	0.516	1.07	10.8	0.611	0.90	15.8	0.418	1.320
$L4\times4\times\frac{3}{4}$	11.8	1.57	0.98	11.8	1.57	0.983	15.8	1.17	1.320
$\times\frac{5}{8}$	11.8	1.33	0.98	11.8	1.33	0.983	15.8	0.994	1.320
$\times\frac{1}{2}$	11.8	1.08	0.98	11.8	1.08	0.983	15.8	0.804	1.320
$\times\frac{7}{16}$	11.8	0.949	0.98	11.8	0.949	0.983	15.8	0.709	1.320
$\times\frac{3}{8}$	11.8	0.824	0.98	11.8	0.824	0.983	15.8	0.615	1.320
$\times\frac{5}{16}$	11.8	0.692	0.98	11.8	0.692	0.983	15.8	0.516	1.320
$\times\frac{1}{4}$	11.8	0.558	0.98	11.8	0.558	0.983	15.8	0.416	1.320
$L4\times3\frac{1}{2}\times\frac{1}{2}$	11.5	1.03	0.96	11.0	1.08	0.917	15.0	0.793	1.250
$\times\frac{3}{8}$	11.5	0.791	0.96	11.0	0.827	0.917	15.0	0.607	1.250
$\times\frac{5}{16}$	11.5	0.665	0.96	11.0	0.695	0.917	15.0	0.510	1.250
$\times\frac{1}{4}$	11.5	0.537	0.96	11.0	0.562	0.917	15.0	0.412	1.250
$L4\times3\times\frac{5}{8}$	11.0	1.24	0.92	10.0	1.36	0.833	14.0	0.971	1.170
$\times\frac{1}{2}$	11.0	1.01	0.92	10.0	1.11	0.833	14.0	0.793	1.170
$\times\frac{3}{8}$	11.0	0.770	0.92	10.0	0.847	0.833	14.0	0.605	1.170
$\times\frac{5}{16}$	11.0	0.647	0.92	10.0	0.712	0.833	14.0	0.509	1.170
$\times\frac{1}{4}$	11.0	0.523	0.92	10.0	0.575	0.833	14.0	0.411	1.170
$L3\frac{1}{2}\times3\frac{1}{2}\times\frac{1}{2}$	10.3	1.08	0.86	10.3	1.08	0.858	13.8	0.80	1.150
$\times\frac{7}{16}$	10.3	0.953	0.86	10.3	0.953	0.858	13.8	0.712	1.150
$\times\frac{3}{8}$	10.3	0.826	0.86	10.3	0.826	0.858	13.8	0.617	1.150
$\times\frac{5}{16}$	10.3	0.695	0.86	10.3	0.695	0.858	13.8	0.519	1.150
$\times\frac{1}{4}$	10.3	0.562	0.86	10.3	0.562	0.858	13.8	0.420	1.150
$L3\frac{1}{2}\times3\times\frac{1}{2}$	9.84	1.05	0.82	9.34	1.10	0.778	12.8	0.805	1.070
$\times\frac{7}{16}$	9.84	0.924	0.82	9.34	0.973	0.778	12.8	0.710	1.070
$\times\frac{3}{8}$	9.84	0.801	0.82	9.34	0.844	0.778	12.8	0.616	1.070
$\times\frac{5}{16}$	9.84	0.676	0.82	9.34	0.712	0.778	12.8	0.520	1.070
$\times\frac{1}{4}$	9.84	0.547	0.82	9.34	0.576	0.778	12.8	0.420	1.070
$L3\frac{1}{2}\times2\frac{1}{2}\times\frac{1}{2}$	9.34	1.01	0.778	8.34	1.13	0.695	11.8	0.80	0.983
$\times\frac{3}{8}$	9.34	0.774	0.778	8.34	0.867	0.695	11.8	0.613	0.983
$\times\frac{5}{16}$	9.34	0.653	0.778	8.34	0.731	0.695	11.8	0.517	0.983
$\times\frac{1}{4}$	9.34	0.529	0.778	8.34	0.592	0.695	11.8	0.419	0.983
$L3\times3\times\frac{1}{2}$	8.84	1.06	0.737	8.84	1.06	0.737	11.8	0.79	0.983
$\times\frac{7}{16}$	8.84	0.937	0.737	8.84	0.937	0.737	11.8	0.702	0.983
$\times\frac{3}{8}$	8.84	0.811	0.737	8.84	0.811	0.737	11.8	0.608	0.983
$\times\frac{5}{16}$	8.84	0.683	0.737	8.84	0.683	0.737	11.8	0.512	0.983
$\times\frac{1}{4}$	8.84	0.553	0.737	8.84	0.553	0.737	11.8	0.414	0.983
$\times\frac{3}{16}$	8.84	0.419	0.737	8.84	0.419	0.737	11.8	0.314	0.983
$L3\times2\frac{1}{2}\times\frac{1}{2}$	8.34	1.02	0.695	7.84	1.09	0.653	10.8	0.79	0.900
$\times\frac{7}{16}$	8.34	0.906	0.695	7.84	0.964	0.653	10.8	0.700	0.900
$\times\frac{3}{8}$	8.34	0.787	0.695	7.84	0.837	0.653	10.8	0.607	0.900
$\times\frac{5}{16}$	8.34	0.664	0.695	7.84	0.707	0.653	10.8	0.513	0.900
$\times\frac{1}{4}$	8.34	0.538	0.695	7.84	0.573	0.653	10.8	0.416	0.900
$\times\frac{3}{16}$	8.34	0.409	0.695	7.84	0.435	0.653	10.8	0.316	0.900

Case A-1: Shape perimeter, minus short leg surface.
Case A-2: Shape perimeter, minus long leg surface.
Case B: Shape perimeter.

Table 1-42 (cont.). Angles Surface Perimeters, Weight-to-Perimeter Ratios, and Surface Areas

Shape	Case A-1			Case A-2			Case B		
	Perimeter	W/D Ratio	Surf. Area	Perimeter	W/D Ratio	Surf. Area	Perimeter	W/D Ratio	Surf. Area
	in.		ft²/ft	in.		ft²/ft	in.		ft²/ft
L3×2×1/2	7.87	0.978	0.656	6.87	1.12	0.573	9.87	0.78	0.823
×3/8	7.87	0.756	0.656	6.87	0.866	0.573	9.87	0.603	0.823
×5/16	7.87	0.639	0.656	6.87	0.732	0.573	9.87	0.510	0.823
×1/4	7.87	0.520	0.656	6.87	0.595	0.573	9.87	0.414	0.823
×3/16	7.87	0.396	0.656	6.87	0.454	0.573	9.87	0.316	0.823
L2 1/2×2 1/2×1/2	7.39	1.04	0.616	7.39	1.04	0.616	9.89	0.77	0.824
×3/8	7.39	0.798	0.616	7.39	0.798	0.616	9.89	0.597	0.824
×5/16	7.39	0.674	0.616	7.39	0.674	0.616	9.89	0.504	0.824
×1/4	7.39	0.547	0.616	7.39	0.547	0.616	9.89	0.408	0.824
×3/16	7.39	0.414	0.616	7.39	0.414	0.616	9.89	0.309	0.824
L2 1/2×2×3/8	6.89	0.769	0.574	6.39	0.829	0.533	8.89	0.596	0.741
×5/16	6.89	0.652	0.574	6.39	0.703	0.533	8.89	0.505	0.741
×1/4	6.89	0.530	0.574	6.39	0.571	0.533	8.89	0.411	0.741
×3/16	6.89	0.403	0.574	6.39	0.435	0.533	8.89	0.313	0.741
L2×2×3/8	5.89	0.789	0.491	5.89	0.789	0.491	7.89	0.589	0.658
×5/16	5.89	0.669	0.491	5.89	0.669	0.491	7.89	0.499	0.658
×1/4	5.89	0.545	0.491	5.89	0.545	0.491	7.89	0.407	0.658
×3/16	5.89	0.418	0.491	5.89	0.418	0.491	7.89	0.312	0.658
×1/8	5.89	0.284	0.491	5.89	0.284	0.491	7.89	0.212	0.658

Case A-1: Shape perimeter, minus short leg surface.
Case A-2: Shape perimeter, minus long leg surface.
Case B: Shape perimeter.

Table 1-43.
WT-Shapes
Surface and Box Perimeters, Weight-to-Perimeter Ratios, and Surface Areas

Shape	Case A Perimeter in.	Case A W/D Ratio	Case A Surf. Area ft²/ft	Case B Perimeter in.	Case B W/D Ratio	Case B Surf. Area ft²/ft	Case C Perimeter in.	Case C W/D Ratio	Case C Surf. Area ft²/ft	Case D Perimeter in.	Case D W/D Ratio	Case D Surf. Area ft²/ft
WT22×167.5	59.3	2.82	4.94	75.3	2.22	6.28	60.0	2.79	5.00	76.0	2.20	6.33
×145	58.7	2.47	4.89	74.5	1.95	6.21	59.4	2.44	4.95	75.2	1.93	6.27
×131	58.5	2.24	4.88	74.3	1.76	6.19	59.2	2.21	4.93	75.0	1.75	6.25
×115	58.1	1.98	4.84	73.9	1.56	6.16	58.8	1.96	4.90	74.6	1.54	6.22
WT20×296.5	58.6	5.06	4.88	75.3	3.94	6.28	59.7	4.97	4.98	76.4	3.88	6.37
×251.5	57.3	4.39	4.78	73.7	3.41	6.14	58.4	4.31	4.87	74.8	3.36	6.23
×215.5	56.3	3.83	4.69	72.5	2.97	6.04	57.4	3.75	4.78	73.6	2.93	6.13
×198.5	56.0	3.54	4.67	72.1	2.75	6.01	57.1	3.48	4.76	73.2	2.71	6.10
×186	55.6	3.35	4.63	71.7	2.59	5.98	56.7	3.28	4.73	72.8	2.55	6.07
×181	55.5	3.26	4.63	71.5	2.53	5.96	56.6	3.20	4.72	72.6	2.49	6.05
×162	55.0	2.95	4.58	70.9	2.28	5.91	56.1	2.89	4.68	72.0	2.25	6.00
×148.5	54.5	2.72	4.54	70.3	2.11	5.86	55.6	2.67	4.63	71.4	2.08	5.95
×138.5	54.3	2.55	4.53	70.1	1.98	5.84	55.4	2.50	4.62	71.2	1.95	5.93
×124.5	54.1	2.30	4.51	69.9	1.78	5.83	55.2	2.26	4.60	71.0	1.75	5.92
×107.5	53.7	2.00	4.48	69.5	1.55	5.79	54.8	1.96	4.57	70.6	1.52	5.88
×99.5	53.3	1.87	4.44	69.1	1.44	5.76	54.4	1.83	4.53	70.2	1.42	5.85
WT20×196	52.9	3.71	4.41	65.3	3.00	5.44	54.0	3.63	4.50	66.4	2.95	5.53
×165.5	51.9	3.19	4.33	64.1	2.58	5.34	53.0	3.12	4.42	65.2	2.54	5.43
×163.5	51.8	3.16	4.32	63.9	2.56	5.33	52.9	3.09	4.41	65.0	2.52	5.42
×139	51.1	2.72	4.26	63.1	2.20	5.26	52.2	2.66	4.35	64.2	2.17	5.35
×132	50.8	2.60	4.23	62.7	2.11	5.23	51.9	2.54	4.33	63.8	2.07	5.32
×117.5	50.4	2.33	4.20	62.3	1.89	5.19	51.5	2.28	4.29	63.4	1.85	5.28
×105.5	50.1	2.11	4.18	61.9	1.70	5.16	51.2	2.06	4.27	63.0	1.67	5.25
×91.5	49.7	1.84	4.14	61.5	1.49	5.13	50.8	1.80	4.23	62.6	1.46	5.22
×83.5	49.3	1.69	4.11	61.1	1.37	5.09	50.4	1.66	4.20	62.2	1.34	5.18
×74.5	48.9	1.52	4.08	60.7	1.23	5.06	50.0	1.49	4.17	61.8	1.21	5.15
WT18×399	58.9	6.77	4.91	76.9	5.19	6.41	60.0	6.65	5.00	78.0	5.12	6.50
×325	56.9	5.71	4.74	74.5	4.36	6.21	58.0	5.60	4.83	75.6	4.30	6.30
×263.5	55.3	4.76	4.61	72.5	3.63	6.04	56.4	4.67	4.70	73.6	3.58	6.13
×219.5	54.1	4.06	4.51	71.1	3.09	5.93	55.2	3.98	4.60	72.2	3.04	6.02
×196.5	53.5	3.67	4.46	70.3	2.80	5.86	54.6	3.60	4.55	71.4	2.75	5.95
×179.5	53.0	3.39	4.42	69.7	2.58	5.81	54.1	3.32	4.51	70.8	2.54	5.90
×164	52.5	3.12	4.38	69.1	2.37	5.76	53.6	3.06	4.47	70.2	2.34	5.85
×150	52.4	2.86	4.37	69.1	2.17	5.76	53.5	2.80	4.46	70.2	2.14	5.85
×140	52.1	2.69	4.34	68.7	2.04	5.73	53.2	2.63	4.43	69.8	2.01	5.82
×130	51.7	2.51	4.31	68.3	1.90	5.69	52.8	2.46	4.40	69.4	1.87	5.78
×122.5	51.4	2.38	4.28	67.9	1.80	5.66	52.5	2.33	4.38	69.0	1.78	5.75
×115	51.4	2.24	4.28	67.9	1.69	5.66	52.5	2.19	4.38	69.0	1.67	5.75
WT18×128	48.8	2.62	4.07	61.0	2.10	5.08	49.6	2.58	4.13	61.8	2.07	5.15
×116	48.6	2.39	4.05	60.7	1.91	5.06	49.3	2.35	4.11	61.4	1.89	5.12
×105	48.0	2.19	4.00	60.2	1.74	5.02	48.8	2.15	4.07	61.0	1.72	5.08
×97	47.7	2.03	3.98	59.8	1.62	4.98	48.5	2.00	4.04	60.6	1.60	5.05
×91	47.7	1.91	3.98	59.8	1.52	4.98	48.5	1.88	4.04	60.6	1.50	5.05
×85	47.4	1.79	3.95	59.4	1.43	4.95	48.2	1.76	4.02	60.2	1.41	5.02
×80	47.2	1.69	3.93	59.2	1.35	4.93	48.0	1.67	4.00	60.0	1.33	5.00
×75	47.0	1.60	3.92	59.0	1.27	4.92	47.8	1.57	3.98	59.8	1.25	4.98
×67.5	46.8	1.44	3.90	58.8	1.15	4.90	47.6	1.42	3.97	59.6	1.13	4.97
WT16.5×193.5	51.4	3.76	4.28	67.6	2.86	5.63	52.2	3.71	4.35	68.4	2.83	5.70
×177	51.0	3.47	4.25	67.1	2.64	5.59	51.7	3.42	4.31	67.8	2.61	5.65
×159	50.5	3.15	4.21	66.5	2.39	5.54	51.2	3.11	4.27	67.2	2.37	5.60
×145.5	49.9	2.92	4.16	65.8	2.21	5.48	50.7	2.87	4.23	66.6	2.18	5.55
×131.5	49.7	2.65	4.14	65.5	2.01	5.46	50.4	2.61	4.20	66.2	1.99	5.52
×120.5	49.4	2.44	4.12	65.3	1.85	5.44	50.1	2.41	4.18	66.0	1.83	5.50
×110.5	49.1	2.25	4.09	64.9	1.70	5.41	49.8	2.22	4.15	65.6	1.68	5.47
×100.5	48.6	2.07	4.05	64.3	1.56	5.36	49.3	2.04	4.11	65.0	1.55	5.42

Case A: Shape perimeter, minus one flange surface.
Case B: Shape perimeter.
Case C: Box perimeter, minus one flange surface.
Case D: Box perimeter.

Table 1-43 (cont.). WT-Shapes Surface and Box Perimeters, Weight-to-Perimeter Ratios, and Surface Areas

Shape	Case A			Case B			Case C			Case D		
	Perimeter	W/D Ratio	Surf. Area	Perimeter	W/D Ratio	Surf. Area	Perimeter	W/D Ratio	Surf. Area	Perimeter	W/D Ratio	Surf. Area
	in.		ft²/ft	in.		ft²/ft	in.		ft²/ft	in.		ft²/ft
WT16.5×84.5	44.5	1.90	3.71	56.0	1.51	4.67	45.3	1.87	3.78	56.8	1.49	4.73
×76	44.2	1.72	3.68	55.8	1.36	4.65	45.0	1.69	3.75	56.6	1.34	4.72
×70.5	44.2	1.60	3.68	55.7	1.27	4.64	44.9	1.57	3.74	56.4	1.25	4.70
×65	43.7	1.49	3.64	55.2	1.18	4.60	44.5	1.46	3.71	56.0	1.16	4.67
×59	43.5	1.36	3.63	55.0	1.07	4.58	44.3	1.33	3.69	55.8	1.06	4.65
WT15×195.5	48.0	4.07	4.00	63.6	3.07	5.30	48.8	4.01	4.07	64.4	3.04	5.37
×178.5	47.5	3.76	3.96	63.0	2.83	5.25	48.3	3.70	4.03	63.8	2.80	5.32
×163	47.0	3.47	3.92	62.4	2.61	5.20	47.8	3.41	3.98	63.2	2.58	5.27
×146	46.5	3.14	3.88	61.8	2.36	5.15	47.3	3.09	3.94	62.6	2.33	5.22
×130.5	46.0	2.84	3.83	61.2	2.13	5.10	46.8	2.79	3.90	62.0	2.10	5.17
×117.5	45.7	2.57	3.81	60.8	1.93	5.07	46.5	2.53	3.88	61.6	1.91	5.13
×105.5	45.3	2.33	3.78	60.4	1.75	5.03	46.1	2.29	3.84	61.2	1.72	5.10
×95.5	44.9	2.13	3.74	59.9	1.59	4.99	45.6	2.09	3.80	60.6	1.58	5.05
×86.5	44.6	1.94	3.72	59.6	1.45	4.97	45.4	1.91	3.78	60.4	1.43	5.03
WT15×74	40.3	1.84	3.36	50.8	1.46	4.23	41.1	1.80	3.43	51.6	1.43	4.30
×66	40.1	1.65	3.34	50.6	1.30	4.22	40.9	1.61	3.41	51.4	1.28	4.28
×62	39.9	1.55	3.33	50.4	1.23	4.20	40.7	1.52	3.39	51.2	1.21	4.27
×58	39.7	1.46	3.31	50.2	1.16	4.18	40.5	1.43	3.38	51.0	1.14	4.25
×54	39.5	1.37	3.29	50.0	1.08	4.17	40.3	1.34	3.36	50.8	1.06	4.23
×49.5	39.3	1.26	3.28	49.8	0.994	4.15	40.1	1.23	3.34	50.6	0.978	4.22
×45	39.2	1.15	3.27	49.6	0.907	4.13	40.0	1.13	3.33	50.4	0.893	4.20
WT13.5×269.5	47.1	5.72	3.93	62.4	4.32	5.20	47.9	5.63	3.99	63.2	4.26	5.27
×184	44.3	4.15	3.69	59.0	3.12	4.92	45.1	4.08	3.76	59.8	3.08	4.98
×168	43.8	3.84	3.65	58.4	2.88	4.87	44.6	3.77	3.72	59.2	2.84	4.93
×153.5	43.2	3.55	3.60	57.6	2.66	4.80	44.0	3.49	3.67	58.4	2.63	4.87
×140.5	42.8	3.28	3.57	57.2	2.46	4.77	43.6	3.22	3.63	58.0	2.42	4.83
×129	42.5	3.04	3.54	56.8	2.27	4.73	43.3	2.98	3.61	57.6	2.24	4.80
×117.5	42.0	2.80	3.50	56.2	2.09	4.68	42.8	2.75	3.57	57.0	2.06	4.75
×108.5	41.7	2.60	3.48	55.8	1.94	4.65	42.5	2.55	3.54	56.6	1.92	4.72
×97	41.4	2.34	3.45	55.4	1.75	4.62	42.2	2.30	3.52	56.2	1.73	4.68
×89	41.2	2.16	3.43	55.3	1.61	4.61	41.9	2.12	3.49	56.0	1.59	4.67
×80.5	40.8	1.97	3.40	54.8	1.47	4.57	41.6	1.94	3.47	55.6	1.45	4.63
×73	40.6	1.80	3.38	54.6	1.34	4.55	41.4	1.76	3.45	55.4	1.32	4.62
WT13.5×64.5	36.8	1.75	3.07	46.8	1.38	3.90	37.6	1.72	3.13	47.6	1.36	3.97
×57	36.5	1.56	3.04	46.6	1.22	3.88	37.3	1.53	3.11	47.4	1.20	3.95
×51	36.2	1.41	3.02	46.2	1.10	3.85	37.0	1.38	3.08	47.0	1.09	3.92
×47	36.2	1.30	3.02	46.2	1.02	3.85	37.0	1.27	3.08	47.0	1.00	3.92
×42	36.0	1.17	3.00	45.9	0.915	3.83	36.8	1.14	3.07	46.7	0.899	3.89
WT12×185	40.9	4.52	3.41	54.6	3.39	4.55	41.7	4.44	3.48	55.4	3.34	4.62
×167.5	40.3	4.16	3.36	53.8	3.11	4.48	41.1	4.08	3.43	54.6	3.07	4.55
×153	39.8	3.84	3.32	53.2	2.88	4.43	40.6	3.77	3.38	54.0	2.83	4.50
×139.5	39.3	3.55	3.28	52.6	2.65	4.38	40.1	3.48	3.34	53.4	2.61	4.45
×125	38.8	3.22	3.23	52.0	2.40	4.33	39.6	3.16	3.30	52.8	2.37	4.40
×114.5	38.3	2.99	3.19	51.4	2.23	4.28	39.1	2.93	3.26	52.2	2.19	4.35
×103.5	38.0	2.72	3.17	51.0	2.03	4.25	38.8	2.67	3.23	51.8	2.00	4.32
×96	37.6	2.55	3.13	50.6	1.90	4.22	38.4	2.50	3.20	51.4	1.87	4.28
×88	37.3	2.36	3.11	50.2	1.75	4.18	38.1	2.31	3.18	51.0	1.73	4.25
×81	37.2	2.18	3.10	50.2	1.61	4.18	38.0	2.13	3.17	51.0	1.59	4.25
×73	36.9	1.98	3.08	49.8	1.47	4.15	37.7	1.94	3.14	50.6	1.44	4.22
×65.5	36.5	1.79	3.04	49.4	1.33	4.12	37.3	1.76	3.11	50.2	1.30	4.18
×58.5	36.2	1.62	3.02	49.0	1.19	4.08	37.0	1.58	3.08	49.8	1.17	4.15
×52	36.0	1.44	3.00	48.8	1.07	4.07	36.8	1.41	3.07	49.6	1.05	4.13

Case A: Shape perimeter, minus one flange surface.
Case B: Shape perimeter.
Case C: Box perimeter, minus one flange surface.
Case D: Box perimeter.

Table 1-43 (cont.). WT-Shapes Surface and Box Perimeters, Weight-to-Perimeter Ratios, and Surface Areas

Shape	Case A			Case B			Case C			Case D		
	Perimeter	*W/D* Ratio	Surf. Area	Perimeter	*W/D* Ratio	Surf. Area	Perimeter	*W/D* Ratio	Surf. Area	Perimeter	*W/D* Ratio	Surf. Area
	in.		ft²/ft	in.		ft²/ft	in.		ft²/ft	in.		ft²/ft
WT12×51.5	32.8	1.57	2.73	41.8	1.23	3.48	33.6	1.53	2.80	42.6	1.21	3.55
×47	32.7	1.44	2.73	41.8	1.12	3.48	33.5	1.40	2.79	42.5	1.11	3.54
×42	32.4	1.30	2.70	41.5	1.01	3.46	33.2	1.27	2.77	42.2	0.995	3.52
×38	32.2	1.18	2.68	41.2	0.922	3.43	33.0	1.15	2.75	42.0	0.905	3.50
×34	32.0	1.06	2.67	41.0	0.829	3.42	32.8	1.04	2.73	41.7	0.815	3.48
WT12×31	30.1	1.03	2.51	37.1	0.836	3.09	30.8	1.01	2.57	37.9	0.818	3.16
×27.5	29.8	0.923	2.48	36.8	0.747	3.07	30.6	0.899	2.55	37.6	0.731	3.13
WT10.5×100.5	34.9	2.88	2.91	47.5	2.12	3.96	35.6	2.82	2.97	48.2	2.09	4.02
×91	34.5	2.64	2.88	47.0	1.94	3.92	35.3	2.58	2.94	47.8	1.90	3.98
×83	34.0	2.44	2.83	46.4	1.79	3.87	34.8	2.39	2.90	47.2	1.76	3.93
×73.5	33.8	2.17	2.82	46.3	1.59	3.86	34.5	2.13	2.88	47.0	1.56	3.92
×66	33.4	1.98	2.78	45.8	1.44	3.82	34.2	1.93	2.85	46.6	1.42	3.88
×61	33.3	1.83	2.78	45.7	1.33	3.81	34.0	1.79	2.83	46.4	1.31	3.87
×55.5	33.1	1.68	2.76	45.4	1.22	3.78	33.9	1.64	2.83	46.2	1.20	3.85
×50.5	32.9	1.53	2.74	45.2	1.12	3.77	33.7	1.50	2.81	46.0	1.10	3.83
WT10.5×46.5	29.4	1.58	2.45	37.8	1.23	3.15	30.0	1.55	2.50	38.4	1.21	3.20
×41.5	29.2	1.42	2.43	37.5	1.11	3.13	29.8	1.39	2.48	38.1	1.09	3.18
×36.5	28.9	1.26	2.41	37.2	0.981	3.10	29.5	1.24	2.46	37.8	0.966	3.15
×34	28.9	1.18	2.41	37.1	0.916	3.09	29.5	1.15	2.46	37.7	0.902	3.14
×31	28.6	1.08	2.38	36.9	0.840	3.08	29.2	1.06	2.43	37.5	0.827	3.13
×27.5	28.4	0.968	2.37	36.7	0.749	3.06	29.0	0.948	2.42	37.2	0.739	3.10
×24	28.1	0.854	2.34	36.3	0.661	3.03	28.7	0.836	2.39	36.9	0.650	3.08
WT10.5×28.5	27.0	1.06	2.25	33.6	0.848	2.80	27.6	1.03	2.30	34.1	0.836	2.84
×25	26.7	0.936	2.23	33.2	0.753	2.77	27.3	0.916	2.28	33.9	0.737	2.83
×22	26.5	0.830	2.21	33.0	0.667	2.75	27.1	0.812	2.26	33.6	0.655	2.80
WT9×87.5	30.7	2.85	2.56	42.1	2.08	3.51	31.4	2.79	2.62	42.8	2.04	3.57
×79	30.2	2.62	2.52	41.5	1.90	3.46	31.0	2.55	2.58	42.3	1.87	3.53
×71.5	29.9	2.39	2.49	41.1	1.74	3.43	30.7	2.33	2.56	41.9	1.71	3.49
×65	29.7	2.19	2.48	40.9	1.59	3.41	30.5	2.13	2.54	41.7	1.56	3.48
×59.5	29.5	2.02	2.46	40.8	1.46	3.40	30.3	1.96	2.53	41.6	1.43	3.47
×53	29.2	1.82	2.43	40.4	1.31	3.37	29.9	1.77	2.49	41.1	1.29	3.43
×48.5	28.9	1.68	2.41	40.0	1.21	3.33	29.7	1.63	2.48	40.8	1.19	3.40
×43	28.8	1.49	2.40	39.9	1.08	3.33	29.5	1.46	2.46	40.6	1.06	3.38
×38	28.5	1.33	2.38	39.5	0.962	3.29	29.2	1.30	2.43	40.2	0.945	3.35
WT9×35.5	25.5	1.39	2.13	33.1	1.07	2.76	26.1	1.36	2.18	33.7	1.05	2.81
×32.5	25.4	1.28	2.12	32.9	0.988	2.74	26.0	1.25	2.17	33.5	0.970	2.79
×30	25.2	1.19	2.10	32.8	0.915	2.73	25.8	1.16	2.15	33.4	0.898	2.78
×27.5	25.1	1.10	2.09	32.6	0.844	2.72	25.7	1.07	2.14	33.2	0.828	2.77
×25	24.9	1.000	2.08	32.4	0.772	2.70	25.5	0.980	2.13	33.0	0.758	2.75
WT9×23	23.6	0.975	1.97	29.6	0.777	2.47	24.1	0.954	2.01	30.2	0.762	2.52
×20	23.4	0.855	1.95	29.4	0.680	2.45	23.9	0.837	1.99	29.9	0.669	2.49
×17.5	23.1	0.758	1.93	29.1	0.601	2.43	23.7	0.738	1.98	29.7	0.589	2.48
WT8×50	26.6	1.88	2.22	37.0	1.35	3.08	27.4	1.82	2.28	37.8	1.32	3.15
×44.5	26.4	1.69	2.20	36.8	1.21	3.07	27.2	1.64	2.27	37.6	1.18	3.13
×38.5	26.1	1.48	2.18	36.4	1.06	3.03	26.8	1.44	2.23	37.1	1.04	3.09
×33.5	25.7	1.30	2.14	35.9	0.933	2.99	26.5	1.26	2.21	36.7	0.913	3.06
WT8×28.5	23.0	1.24	1.92	30.1	0.947	2.51	23.6	1.21	1.97	30.7	0.928	2.56
×25	22.7	1.10	1.89	29.8	0.839	2.48	23.3	1.07	1.94	30.4	0.822	2.53
×22.5	22.6	0.996	1.88	29.6	0.760	2.47	23.2	0.970	1.93	30.2	0.745	2.52
×20	22.4	0.893	1.87	29.4	0.680	2.45	23.0	0.870	1.92	30.0	0.667	2.50
×18	22.3	0.807	1.86	29.2	0.616	2.43	22.9	0.786	1.91	29.8	0.604	2.48
WT8×15.5	20.8	0.745	1.73	26.4	0.587	2.20	21.4	0.724	1.78	26.9	0.576	2.24
×13	20.6	0.631	1.72	26.1	0.498	2.18	21.2	0.613	1.77	26.7	0.487	2.23

Case A: Shape perimeter, minus one flange surface.
Case B: Shape perimeter.
Case C: Box perimeter, minus one flange surface.
Case D: Box perimeter.

Table 1-43 (cont.). WT-Shapes Surface and Box Perimeters, Weight-to-Perimeter Ratios, and Surface Areas

Shape	Case A			Case B			Case C			Case D		
	Perimeter	W/D Ratio	Surf. Area	Perimeter	W/D Ratio	Surf. Area	Perimeter	W/D Ratio	Surf. Area	Perimeter	W/D Ratio	Surf. Area
	in.		ft²/ft	in.		ft²/ft	in.		ft²/ft	in.		ft²/ft
WT7×404	40.3	10.00	3.36	58.9	6.86	4.91	41.4	9.76	3.45	60.0	6.73	5.00
×365	39.2	9.31	3.27	57.1	6.39	4.76	40.3	9.06	3.36	58.2	6.27	4.85
×332.5	38.2	8.70	3.18	55.9	5.95	4.66	39.3	8.46	3.28	57.0	5.83	4.75
×302.5	37.3	8.11	3.11	54.7	5.53	4.56	38.4	7.88	3.20	55.8	5.42	4.65
×275	36.3	7.58	3.03	53.5	5.14	4.46	37.4	7.35	3.12	54.6	5.04	4.55
×250	35.5	7.04	2.96	52.5	4.76	4.38	36.6	6.83	3.05	53.6	4.66	4.47
×227.5	34.7	6.56	2.89	51.5	4.42	4.29	35.8	6.35	2.98	52.6	4.33	4.38
×213	34.3	6.21	2.86	51.0	4.18	4.25	35.4	6.02	2.95	52.1	4.09	4.34
×199	33.8	5.89	2.82	50.4	3.95	4.20	34.9	5.70	2.91	51.5	3.86	4.29
×185	33.3	5.56	2.78	49.8	3.71	4.15	34.4	5.38	2.87	50.9	3.63	4.24
×171	32.8	5.21	2.73	49.2	3.48	4.10	33.9	5.04	2.83	50.3	3.40	4.19
×155.5	32.2	4.83	2.68	48.4	3.21	4.03	33.3	4.67	2.78	49.5	3.14	4.13
×141.5	31.7	4.46	2.64	47.8	2.96	3.98	32.8	4.31	2.73	48.9	2.89	4.08
×128.5	31.3	4.11	2.61	47.3	2.72	3.94	32.4	3.97	2.70	48.4	2.65	4.03
×116.5	30.8	3.78	2.57	46.7	2.49	3.89	31.9	3.65	2.66	47.8	2.44	3.98
×105.5	30.4	3.47	2.53	46.2	2.28	3.85	31.5	3.35	2.63	47.3	2.23	3.94
×96.5	30.1	3.21	2.51	45.8	2.11	3.82	31.2	3.09	2.60	46.9	2.06	3.91
×88	29.8	2.95	2.48	45.5	1.93	3.79	30.9	2.85	2.58	46.6	1.89	3.88
×79.5	29.5	2.69	2.46	45.1	1.76	3.76	30.6	2.60	2.55	46.2	1.72	3.85
×72.5	29.2	2.48	2.43	44.7	1.62	3.73	30.3	2.39	2.53	45.8	1.58	3.82
WT7×66	28.3	2.33	2.36	43.0	1.53	3.58	29.4	2.24	2.45	44.1	1.50	3.68
×60	28.1	2.14	2.34	42.8	1.40	3.57	29.2	2.05	2.43	43.9	1.37	3.66
×54.5	27.8	1.96	2.32	42.4	1.29	3.53	28.9	1.89	2.41	43.5	1.25	3.63
×49.5	27.7	1.79	2.31	42.3	1.17	3.53	28.8	1.72	2.40	43.4	1.14	3.62
×45	27.4	1.64	2.28	41.9	1.07	3.49	28.5	1.58	2.38	43.0	1.05	3.58
WT7×41	23.7	1.73	1.98	33.8	1.21	2.82	24.4	1.68	2.03	34.5	1.19	2.88
×37	23.6	1.57	1.97	33.7	1.10	2.81	24.3	1.52	2.03	34.4	1.08	2.87
×34	23.3	1.46	1.94	33.3	1.02	2.78	24.0	1.42	2.00	34.0	1.00	2.83
×30.5	23.2	1.31	1.93	33.1	0.921	2.76	23.9	1.28	1.99	33.9	0.900	2.83
WT7×26.5	21.3	1.24	1.78	29.3	0.904	2.44	22.0	1.20	1.83	30.0	0.883	2.50
×24	21.1	1.14	1.76	29.1	0.825	2.43	21.8	1.10	1.82	29.9	0.803	2.49
×21.5	20.9	1.03	1.74	28.9	0.744	2.41	21.7	0.991	1.81	29.7	0.724	2.48
WT7×19	20.2	0.941	1.68	27.0	0.704	2.25	20.9	0.909	1.74	27.6	0.688	2.30
×17	20.1	0.846	1.68	26.9	0.632	2.24	20.7	0.821	1.73	27.5	0.618	2.29
×15	19.9	0.754	1.66	26.7	0.562	2.23	20.6	0.728	1.72	27.3	0.549	2.28
WT7×13	18.3	0.710	1.53	23.4	0.556	1.95	19.0	0.684	1.58	24.0	0.542	2.00
×11	18.1	0.608	1.51	23.1	0.476	1.93	18.7	0.588	1.56	23.7	0.464	1.98
WT6×168	29.4	5.71	2.45	42.8	3.93	3.57	30.2	5.56	2.52	43.6	3.85	3.63
×152.5	28.7	5.31	2.39	41.9	3.64	3.49	29.5	5.17	2.46	42.7	3.57	3.56
×139.5	28.2	4.95	2.35	41.3	3.38	3.44	29.0	4.81	2.42	42.1	3.31	3.51
×126	27.7	4.55	2.31	40.7	3.10	3.39	28.4	4.44	2.37	41.4	3.04	3.45
×115	27.2	4.23	2.27	40.1	2.87	3.34	28.0	4.11	2.33	40.9	2.81	3.41
×105	26.7	3.93	2.23	39.5	2.66	3.29	27.5	3.82	2.29	40.3	2.61	3.36
×95	26.3	3.61	2.19	39.0	2.44	3.25	27.1	3.51	2.26	39.8	2.39	3.32
×85	25.9	3.28	2.16	38.5	2.21	3.21	26.6	3.20	2.22	39.2	2.17	3.27
×76	25.4	2.99	2.12	37.9	2.01	3.16	26.2	2.90	2.18	38.7	1.96	3.23
×68	25.1	2.71	2.09	37.5	1.81	3.13	25.8	2.64	2.15	38.2	1.78	3.18
×60	24.7	2.43	2.06	37.0	1.62	3.08	25.4	2.36	2.12	37.7	1.59	3.14
×53	24.3	2.18	2.03	36.5	1.45	3.04	25.1	2.11	2.09	37.3	1.42	3.11
×48	24.1	1.99	2.01	36.3	1.32	3.03	24.9	1.93	2.08	37.1	1.29	3.09
×43.5	23.9	1.82	1.99	36.0	1.21	3.00	24.6	1.77	2.05	36.7	1.19	3.06
×39.5	23.7	1.67	1.98	35.8	1.10	2.98	24.5	1.61	2.04	36.6	1.08	3.05
×36	23.5	1.53	1.96	35.5	1.01	2.96	24.3	1.48	2.03	36.3	0.992	3.03
×32.5	23.4	1.39	1.95	35.4	0.918	2.95	24.1	1.35	2.01	36.1	0.900	3.01
WT6×29	21.5	1.35	1.79	31.5	0.921	2.63	22.2	1.31	1.85	32.2	0.901	2.68
×26.5	21.4	1.24	1.78	31.4	0.844	2.62	22.1	1.20	1.84	32.0	0.828	2.67

Case A: Shape perimeter, minus one flange surface.
Case B: Shape perimeter.
Case C: Box perimeter, minus one flange surface.
Case D: Box perimeter.

Table 1-43 (cont.). WT-Shapes Surface and Box Perimeters, Weight-to-Perimeter Ratios, and Surface Areas

Shape	Case A			Case B			Case C			Case D		
	Perimeter	W/D Ratio	Surf. Area	Perimeter	W/D Ratio	Surf. Area	Perimeter	W/D Ratio	Surf. Area	Perimeter	W/D Ratio	Surf. Area
	in.		ft²/ft	in.		ft²/ft	in.		ft²/ft	in.		ft²/ft
WT6×25	19.5	1.28	1.63	27.6	0.906	2.30	20.3	1.23	1.69	28.4	0.880	2.37
×22.5	19.4	1.16	1.62	27.5	0.818	2.29	20.1	1.12	1.68	28.2	0.798	2.35
×20	19.2	1.04	1.60	27.2	0.735	2.27	20.0	1.00	1.67	28.0	0.714	2.33
WT6×17.5	18.5	0.946	1.54	25.0	0.700	2.08	19.1	0.916	1.59	25.6	0.684	2.13
×15	18.3	0.820	1.53	24.8	0.605	2.07	18.9	0.794	1.58	25.4	0.591	2.12
×13	18.1	0.718	1.51	24.6	0.528	2.05	18.7	0.695	1.56	25.2	0.516	2.10
WT6×11	15.9	0.692	1.33	19.9	0.553	1.66	16.4	0.671	1.37	20.4	0.539	1.70
×9.5	15.7	0.605	1.31	19.7	0.482	1.64	16.2	0.586	1.35	20.2	0.470	1.68
×8	15.5	0.516	1.29	19.5	0.410	1.63	16.0	0.500	1.33	20.0	0.400	1.67
×7	15.4	0.455	1.28	19.4	0.361	1.62	15.9	0.440	1.33	19.9	0.352	1.66
WT5×56	21.2	2.64	1.77	31.6	1.77	2.63	21.8	2.57	1.82	32.2	1.74	2.68
×50	20.8	2.40	1.73	31.1	1.61	2.59	21.4	2.34	1.78	31.7	1.58	2.64
×44	20.5	2.15	1.71	30.8	1.43	2.57	21.1	2.09	1.76	31.4	1.40	2.62
×38.5	20.2	1.91	1.68	30.4	1.27	2.53	20.8	1.85	1.73	31.0	1.24	2.58
×34	19.9	1.71	1.66	30.0	1.13	2.50	20.5	1.66	1.71	30.6	1.11	2.55
×30	19.7	1.52	1.64	29.8	1.01	2.48	20.3	1.48	1.69	30.4	0.987	2.53
×27	19.5	1.38	1.63	29.5	0.915	2.46	20.1	1.34	1.68	30.1	0.897	2.51
×24.5	19.4	1.26	1.62	29.4	0.833	2.45	20.0	1.23	1.67	30.0	0.817	2.50
WT5×22.5	17.5	1.29	1.46	25.5	0.882	2.13	18.1	1.24	1.51	26.1	0.862	2.18
×19.5	17.3	1.13	1.44	25.3	0.771	2.11	17.9	1.09	1.49	25.9	0.753	2.16
×16.5	17.1	0.965	1.43	25.1	0.657	2.09	17.7	0.932	1.48	25.7	0.642	2.14
WT5×15	15.8	0.949	1.32	21.6	0.694	1.80	16.3	0.920	1.36	22.1	0.679	1.84
×13	15.6	0.833	1.30	21.3	0.610	1.78	16.1	0.807	1.34	21.9	0.594	1.83
×11	15.4	0.714	1.28	21.2	0.519	1.77	15.9	0.692	1.33	21.7	0.507	1.81
WT5×9.5	13.8	0.688	1.15	17.8	0.534	1.48	14.3	0.664	1.19	18.3	0.519	1.53
×8.5	13.6	0.625	1.13	17.7	0.480	1.48	14.1	0.603	1.18	18.1	0.470	1.51
×7.5	13.5	0.556	1.13	17.5	0.429	1.46	14.0	0.536	1.17	18.0	0.417	1.50
×6	13.4	0.448	1.12	17.3	0.347	1.44	13.8	0.435	1.15	17.8	0.337	1.48
WT4×33.5	16.7	2.01	1.39	25.0	1.34	2.08	17.3	1.94	1.44	25.6	1.31	2.13
×29	16.4	1.77	1.37	24.6	1.18	2.05	17.0	1.71	1.42	25.2	1.15	2.10
×24	16.0	1.50	1.33	24.1	0.996	2.01	16.6	1.45	1.38	24.7	0.972	2.06
×20	15.7	1.27	1.31	23.8	0.840	1.98	16.3	1.23	1.36	24.4	0.820	2.03
×17.5	15.5	1.13	1.29	23.6	0.742	1.97	16.1	1.09	1.34	24.2	0.723	2.02
×15.5	15.4	1.01	1.28	23.4	0.662	1.95	16.0	0.969	1.33	24.0	0.646	2.00
WT4×14	14.2	0.986	1.18	20.7	0.676	1.73	14.6	0.959	1.22	21.1	0.664	1.76
×12	14.0	0.857	1.17	20.5	0.585	1.71	14.4	0.833	1.20	20.9	0.574	1.74
WT4×10.5	13.1	0.802	1.09	18.4	0.571	1.53	13.6	0.772	1.13	18.8	0.559	1.57
×9	13.0	0.692	1.08	18.2	0.495	1.52	13.4	0.672	1.12	18.6	0.484	1.55
WT4×7.5	11.7	0.641	0.975	15.7	0.478	1.31	12.1	0.620	1.01	16.1	0.466	1.34
×6.5	11.6	0.560	0.967	15.6	0.417	1.30	12.0	0.542	1.00	16.0	0.406	1.33
×5	11.4	0.439	0.950	15.4	0.325	1.28	11.8	0.424	0.983	15.8	0.316	1.32
WT3×12.5	12.0	1.04	1.00	18.1	0.691	1.51	12.5	1.00	1.04	18.5	0.676	1.54
×10	11.8	0.847	0.98	17.8	0.562	1.48	12.2	0.820	1.02	18.2	0.549	1.52
×7.5	11.6	0.647	0.967	17.6	0.426	1.47	12.0	0.625	1.00	18.0	0.417	1.50
WT3×8	9.9	0.807	0.826	13.9	0.576	1.16	10.3	0.777	0.858	14.3	0.559	1.19
×6	9.64	0.622	0.803	13.6	0.441	1.13	10.00	0.600	0.833	14.0	0.429	1.17
×4.5	9.43	0.477	0.786	13.4	0.336	1.12	9.84	0.457	0.820	13.8	0.326	1.15
×4.25	9.36	0.454	0.780	13.3	0.320	1.11	9.78	0.435	0.815	13.7	0.310	1.14
WT2.5×9.5	9.86	0.963	0.822	14.9	0.638	1.24	10.2	0.931	0.850	15.2	0.625	1.27
×8	9.69	0.826	0.808	14.7	0.544	1.23	10.00	0.800	0.833	15.0	0.533	1.25
WT2×6.5	7.87	0.826	0.656	11.9	0.546	0.99	8.22	0.791	0.685	12.3	0.528	1.03

Case A: Shape perimeter, minus one flange surface.
Case B: Shape perimeter.
Case C: Box perimeter, minus one flange surface.
Case D: Box perimeter.

Table 1-44.
MT-Shapes
Surface and Box Perimeters, Weight-to-Perimeter Ratios, and Surface Areas

Shape	Case A			Case B			Case C			Case D		
	Perimeter, in.	W/D Ratio	Surf. Area, ft²/ft	Perimeter, in.	W/D Ratio	Surf. Area, ft²/ft	Perimeter, in.	W/D Ratio	Surf. Area, ft²/ft	Perimeter, in.	W/D Ratio	Surf. Area, ft²/ft
MT6×5.9	14.8	0.399	1.23	17.9	0.330	1.49	15.1	0.391	1.26	18.1	0.326	1.51
×5.4	14.7	0.367	1.23	17.8	0.303	1.48	15.1	0.358	1.26	18.1	0.298	1.51
×5	15.0	0.333	1.25	18.2	0.275	1.52	15.2	0.329	1.27	18.5	0.270	1.54
MT5×4.5	12.4	0.363	1.03	15.1	0.298	1.26	12.7	0.354	1.06	15.4	0.292	1.28
×4	12.3	0.325	1.03	15.0	0.267	1.25	12.6	0.317	1.05	15.3	0.261	1.28
×3.75	12.5	0.300	1.04	15.2	0.247	1.27	12.7	0.295	1.06	15.4	0.244	1.28
MT4×3.25	9.96	0.326	0.830	12.2	0.266	1.02	10.3	0.316	0.858	12.6	0.258	1.05
×3.1	10.1	0.307	0.842	12.3	0.252	1.03	10.3	0.301	0.858	12.6	0.246	1.05
MT3×2.2	7.66	0.287	0.638	9.50	0.232	0.792	7.84	0.281	0.653	9.68	0.227	0.807
×1.85	7.76	0.238	0.647	9.76	0.190	0.813	7.92	0.234	0.660	9.92	0.186	0.827
MT2.5×9.45	9.66	0.978	0.805	14.7	0.643	1.23	10.00	0.945	0.833	15.0	0.630	1.25
MT2×3	7.31	0.410	0.609	11.1	0.270	0.925	7.60	0.395	0.633	11.4	0.263	0.950

Case A: Shape perimeter, minus one flange surface.
Case B: Shape perimeter.
Case C: Box perimeter, minus one flange surface.
Case D: Box perimeter.

Table 1-45.
ST-Shapes
Surface and Box Perimeters, Weight-to-Perimeter Ratios, and Surface Areas

Shape	Case A			Case B			Case C			Case D		
	Perimeter (in.)	W/D Ratio	Surf. Area (ft²/ft)	Perimeter (in.)	W/D Ratio	Surf. Area (ft²/ft)	Perimeter (in.)	W/D Ratio	Surf. Area (ft²/ft)	Perimeter (in.)	W/D Ratio	Surf. Area (ft²/ft)
ST12×60.5	31.2	1.94	2.60	39.2	1.54	3.27	32.7	1.85	2.73	40.7	1.49	3.39
×53	31.0	1.71	2.58	38.8	1.37	3.23	32.5	1.63	2.71	40.3	1.32	3.36
ST12×50	29.9	1.67	2.49	37.1	1.35	3.09	31.3	1.60	2.61	38.5	1.30	3.21
×45	29.8	1.51	2.48	36.9	1.22	3.08	31.1	1.45	2.59	38.3	1.17	3.19
×40	29.6	1.35	2.47	36.6	1.09	3.05	31.0	1.29	2.58	38.0	1.05	3.17
ST10×48	26.3	1.83	2.19	33.5	1.43	2.79	27.6	1.74	2.30	34.8	1.38	2.90
×43	26.1	1.65	2.18	33.2	1.30	2.77	27.5	1.56	2.29	34.5	1.25	2.88
ST10×37.5	25.1	1.49	2.09	31.5	1.19	2.63	26.4	1.42	2.20	32.8	1.14	2.73
×33	25.0	1.32	2.08	31.3	1.05	2.61	26.3	1.25	2.19	32.5	1.02	2.71
ST9×35	23.0	1.52	1.92	29.3	1.19	2.44	24.3	1.44	2.03	30.5	1.15	2.54
×27.35	22.8	1.20	1.90	28.8	0.950	2.40	24.0	1.14	2.00	30.0	0.912	2.50
ST7.5×25	19.5	1.28	1.63	25.2	0.992	2.10	20.6	1.21	1.72	26.3	0.951	2.19
×21.45	19.4	1.11	1.62	24.9	0.861	2.08	20.5	1.05	1.71	26.0	0.825	2.17
ST6×25	16.4	1.52	1.37	21.9	1.14	1.83	17.5	1.43	1.46	23.0	1.09	1.92
×20.4	16.2	1.26	1.35	21.4	0.953	1.78	17.3	1.18	1.44	22.5	0.907	1.88
ST6×17.5	16.1	1.09	1.34	21.2	0.825	1.77	17.1	1.02	1.43	22.2	0.788	1.85
×15.9	16.0	0.994	1.33	21.0	0.757	1.75	17.0	0.935	1.42	22.0	0.723	1.83
ST5×17.5	14.0	1.25	1.17	18.9	0.926	1.58	14.9	1.17	1.24	19.9	0.879	1.66
×12.7	13.7	0.927	1.14	18.4	0.690	1.53	14.7	0.864	1.23	19.3	0.658	1.61
ST4×11.5	11.3	1.02	0.942	15.5	0.742	1.29	12.2	0.943	1.02	16.3	0.706	1.36
×9.2	11.2	0.821	0.933	15.2	0.605	1.27	12.0	0.767	1.00	16.0	0.575	1.33
ST3×8.625	8.89	0.970	0.741	12.5	0.690	1.04	9.57	0.901	0.798	13.1	0.658	1.09
×6.25	8.65	0.723	0.721	12.0	0.521	1.00	9.33	0.670	0.778	12.7	0.492	1.06
ST2.5×5	7.38	0.678	0.615	10.4	0.481	0.867	8.00	0.625	0.667	11.0	0.455	0.917
ST2×4.75	6.20	0.766	0.517	9.00	0.528	0.750	6.80	0.699	0.567	9.60	0.495	0.800
×3.85	6.06	0.635	0.505	8.72	0.442	0.727	6.66	0.578	0.555	9.32	0.413	0.777
ST1.5×3.75	5.01	0.749	0.418	7.52	0.499	0.627	5.51	0.681	0.459	8.02	0.468	0.668
×2.85	4.83	0.590	0.403	7.16	0.398	0.597	5.33	0.535	0.444	7.66	0.372	0.638

Case A: Shape perimeter, minus one flange surface.
Case B: Shape perimeter.
Case C: Box perimeter, minus one flange surface.
Case D: Box perimeter.

Table 1-46.
Rectangular (and Square) HSS
Surface and Box Perimeters, Area-to-Perimeter Ratios, and Surface Areas

Shape	Case A			Case B			Case C		
	Perimeter	A/P Ratio	Surf. Area	Perimeter	A/P Ratio	Surf. Area	Perimeter	A/P Ratio	Surf. Area
	in.		ft²/ft	in.		ft²/ft	in.		ft²/ft
HSS20×12×5/8	52.3	0.668	4.36	44.3	0.789	3.69	62.0	0.564	5.17
×1/2	52.3	0.542	4.36	44.3	0.640	3.69	62.4	0.454	5.20
×3/8	52.2	0.413	4.35	44.2	0.487	3.68	62.8	0.343	5.23
×5/16	52.2	0.346	4.35	44.2	0.409	3.68	63.0	0.287	5.25
HSS20×8×5/8	48.3	0.627	4.03	36.3	0.834	3.03	54.0	0.561	4.50
×1/2	48.3	0.510	4.03	36.3	0.679	3.03	54.4	0.453	4.53
×3/8	48.2	0.389	4.02	36.2	0.518	3.02	54.8	0.342	4.57
×5/16	48.2	0.327	4.02	36.2	0.435	3.02	55.0	0.286	4.58
HSS20×4×1/2	44.3	0.472	3.69	28.3	0.739	2.36	46.4	0.450	3.87
×3/8	44.2	0.361	3.68	28.2	0.566	2.35	46.8	0.341	3.90
×5/16	44.2	0.304	3.68	28.2	0.476	2.35	47.0	0.285	3.92
HSS18×12×5/8	48.3	0.675	4.03	42.3	0.771	3.53	58.0	0.563	4.83
×1/2	48.3	0.549	4.03	42.3	0.627	3.53	58.4	0.453	4.87
×3/8	48.2	0.418	4.02	42.2	0.477	3.52	58.8	0.342	4.90
HSS18×6×5/8	42.3	0.606	3.53	30.3	0.846	2.53	46.0	0.558	3.83
×1/2	42.3	0.494	3.53	30.3	0.691	2.53	46.4	0.450	3.87
×3/8	42.2	0.378	3.52	30.2	0.528	2.52	46.8	0.341	3.90
×5/16	42.2	0.318	3.52	30.2	0.445	2.52	47.0	0.285	3.92
×1/4	42.1	0.257	3.51	30.1	0.359	2.51	47.2	0.229	3.93
HSS16×16×5/8	48.3	0.723	4.03	48.3	0.723	4.03	62.0	0.564	5.17
×1/2	48.3	0.587	4.03	48.3	0.587	4.03	62.4	0.454	5.20
×3/8	48.2	0.447	4.02	48.2	0.447	4.02	62.8	0.343	5.23
×5/16	48.2	0.375	4.02	48.2	0.375	4.02	63.0	0.287	5.25
HSS16×12×5/8	44.3	0.684	3.69	40.3	0.752	3.36	54.0	0.561	4.50
×1/2	44.3	0.556	3.69	40.3	0.611	3.36	54.4	0.453	4.53
×3/8	44.2	0.424	3.68	40.2	0.466	3.35	54.8	0.342	4.57
×5/16	44.2	0.356	3.68	40.2	0.392	3.35	55.0	0.286	4.58
HSS16×8×5/8	40.3	0.636	3.36	32.3	0.794	2.69	46.0	0.558	3.83
×1/2	40.3	0.519	3.36	32.3	0.648	2.69	46.4	0.450	3.87
×3/8	40.2	0.397	3.35	32.2	0.495	2.68	46.8	0.341	3.90
×5/16	40.2	0.334	3.35	32.2	0.417	2.68	47.0	0.285	3.92
HSS16×4×1/2	36.3	0.474	3.03	24.3	0.708	2.03	38.4	0.447	3.20
×3/8	36.2	0.364	3.02	24.2	0.544	2.02	38.8	0.339	3.23
×5/16	36.2	0.306	3.02	24.2	0.459	2.02	39.0	0.284	3.25
HSS14×14×5/8	42.3	0.716	3.53	42.3	0.716	3.53	54.0	0.561	4.50
×1/2	42.3	0.582	3.53	42.3	0.582	3.53	54.4	0.453	4.53
×3/8	42.2	0.444	3.52	42.2	0.444	3.52	54.8	0.342	4.57
×5/16	42.2	0.373	3.52	42.2	0.373	3.52	55.0	0.286	4.58
HSS14×12×1/2	40.3	0.565	3.36	38.3	0.595	3.19	50.4	0.452	4.20
×3/8	40.2	0.432	3.35	38.2	0.454	3.18	50.8	0.341	4.23
HSS14×10×5/8	38.3	0.670	3.19	34.3	0.748	2.86	46.0	0.558	3.83
×1/2	38.3	0.546	3.19	34.3	0.610	2.86	46.4	0.450	3.87
×3/8	38.2	0.418	3.18	34.2	0.466	2.85	46.8	0.341	3.90
×5/16	38.2	0.351	3.18	34.2	0.393	2.85	47.0	0.285	3.92
×1/4	38.1	0.284	3.18	34.1	0.317	2.84	47.2	0.229	3.93
HSS14×6×5/8	34.3	0.612	2.86	26.3	0.798	2.19	38.0	0.553	3.17
×1/2	34.3	0.501	2.86	26.3	0.654	2.19	38.4	0.447	3.20
×3/8	34.2	0.385	2.85	26.2	0.502	2.18	38.8	0.339	3.23
×5/16	34.2	0.324	2.85	26.2	0.424	2.18	39.0	0.284	3.25
×1/4	34.1	0.263	2.84	26.1	0.343	2.18	39.2	0.229	3.27
×3/16	34.1	0.198	2.84	26.1	0.259	2.18	39.4	0.172	3.28

Case A: Shape perimeter, minus one short surface.
Case B: Shape perimeter, minus one long surface.
Case C: Shape perimeter.

Table 1-46 (cont.). Rectangular (and Square) HSS
Surface and Box Perimeters, Area-to-Perimeter Ratios, and Surface Areas

Shape	Case A			Case B			Case C		
	Perimeter	A/P Ratio	Surf. Area	Perimeter	A/P Ratio	Surf. Area	Perimeter	A/P Ratio	Surf. Area
	in.		ft²/ft	in.		ft²/ft	in.		ft²/ft
HSS14×4×5/8	32.3	0.578	2.69	22.3	0.837	1.86	34.0	0.550	2.83
×1/2	32.3	0.475	2.69	22.3	0.688	1.86	34.4	0.445	2.87
×3/8	32.2	0.365	2.68	22.2	0.530	1.85	34.8	0.338	2.90
×5/16	32.2	0.308	2.68	22.2	0.448	1.85	35.0	0.283	2.92
×1/4	32.1	0.250	2.68	22.1	0.363	1.84	35.2	0.228	2.93
×3/16	32.1	0.189	2.68	22.1	0.274	1.84	35.4	0.171	2.95
HSS12×12×5/8	36.3	0.707	3.03	36.3	0.707	3.03	46.0	0.558	3.83
×1/2	36.3	0.576	3.03	36.3	0.576	3.03	46.4	0.450	3.87
×3/8	36.2	0.441	3.02	36.2	0.441	3.02	46.8	0.341	3.90
×5/16	36.2	0.371	3.02	36.2	0.371	3.02	47.0	0.285	3.92
×1/4	36.1	0.300	3.01	36.1	0.300	3.01	47.2	0.229	3.93
HSS12×10×1/2	34.3	0.556	2.86	32.3	0.590	2.69	42.4	0.449	3.53
×3/8	34.2	0.426	2.85	32.2	0.452	2.68	42.8	0.340	3.57
×5/16	34.2	0.358	2.85	32.2	0.381	2.68	43.0	0.285	3.58
×1/4	34.1	0.290	2.84	32.1	0.308	2.68	43.2	0.229	3.60
HSS12×8×5/8	32.3	0.650	2.69	28.3	0.742	2.36	38.0	0.553	3.17
×1/2	32.3	0.532	2.69	28.3	0.608	2.36	38.4	0.447	3.20
×3/8	32.2	0.409	2.68	28.2	0.467	2.35	38.8	0.339	3.23
×5/16	32.2	0.345	2.68	28.2	0.394	2.35	39.0	0.284	3.25
×1/4	32.1	0.279	2.68	28.1	0.319	2.34	39.2	0.229	3.27
×3/16	32.1	0.211	2.68	28.1	0.241	2.34	39.4	0.172	3.28
HSS12×6×5/8	30.3	0.616	2.53	24.3	0.768	2.03	34.0	0.550	2.83
×1/2	30.3	0.506	2.53	24.3	0.631	2.03	34.4	0.445	2.87
×3/8	30.2	0.390	2.52	24.2	0.486	2.02	34.8	0.338	2.90
×5/16	30.2	0.329	2.52	24.2	0.410	2.02	35.0	0.283	2.92
×1/4	30.1	0.267	2.51	24.1	0.333	2.01	35.2	0.228	2.93
×3/16	30.1	0.202	2.51	24.1	0.252	2.01	35.4	0.171	2.95
HSS12×4×5/8	28.3	0.578	2.36	20.3	0.805	1.69	30.0	0.546	2.50
×1/2	28.3	0.476	2.36	20.3	0.664	1.69	30.4	0.443	2.53
×3/8	28.2	0.368	2.35	20.2	0.513	1.68	30.8	0.337	2.57
×5/16	28.2	0.311	2.35	20.2	0.434	1.68	31.0	0.282	2.58
×1/4	28.1	0.252	2.34	20.1	0.353	1.68	31.2	0.228	2.60
×3/16	28.1	0.191	2.34	20.1	0.267	1.68	31.4	0.171	2.62
HSS12×3 1/2×3/8	27.7	0.362	2.31	19.2	0.522	1.60	29.8	0.336	2.48
×5/16	27.7	0.306	2.31	19.2	0.442	1.60	30.0	0.282	2.50
HSS12×3×5/16	27.2	0.301	2.27	18.2	0.450	1.52	29.0	0.282	2.42
×1/4	27.1	0.244	2.26	18.1	0.366	1.51	29.2	0.227	2.43
×3/16	27.1	0.185	2.26	18.1	0.277	1.51	29.4	0.171	2.45
HSS12×2×1/4	26.1	0.236	2.18	16.1	0.382	1.34	27.2	0.227	2.27
×3/16	26.1	0.179	2.18	16.1	0.290	1.34	27.4	0.171	2.28
HSS10×10×5/8	30.3	0.693	2.53	30.3	0.693	2.53	38.0	0.553	3.17
×1/2	30.3	0.568	2.53	30.3	0.568	2.53	38.4	0.447	3.20
×3/8	30.2	0.436	2.52	30.2	0.436	2.52	38.8	0.339	3.23
×5/16	30.2	0.367	2.52	30.2	0.367	2.52	39.0	0.284	3.25
×1/4	30.1	0.297	2.51	30.1	0.297	2.51	39.2	0.229	3.27
×3/16	30.1	0.225	2.51	30.1	0.225	2.51	39.4	0.172	3.28
HSS10×8×1/2	28.3	0.542	2.36	26.3	0.583	2.19	34.4	0.445	2.87
×3/8	28.2	0.417	2.35	26.2	0.449	2.18	34.8	0.338	2.90
×5/16	28.2	0.352	2.35	26.2	0.379	2.18	35.0	0.283	2.92
×1/4	28.1	0.285	2.34	26.1	0.307	2.18	35.2	0.228	2.93
×3/16	28.1	0.216	2.34	26.1	0.232	2.18	35.4	0.171	2.95

Case A: Shape perimeter, minus one short surface.
Case B: Shape perimeter, minus one long surface.
Case C: Shape perimeter.

Table 1-46 (cont.). Rectangular (and Square) HSS Surface and Box Perimeters, Area-to-Perimeter Ratios, and Surface Areas

Shape	Case A Perimeter	Case A A/P Ratio	Case A Surf. Area	Case B Perimeter	Case B A/P Ratio	Case B Surf. Area	Case C Perimeter	Case C A/P Ratio	Case C Surf. Area
	in.		ft²/ft	in.		ft²/ft	in.		ft²/ft
HSS10×6×5/8	26.3	0.622	2.19	22.3	0.733	1.86	30.0	0.546	2.50
×1/2	26.3	0.512	2.19	22.3	0.605	1.86	30.4	0.443	2.53
×3/8	26.2	0.396	2.18	22.2	0.467	1.85	30.8	0.337	2.57
×5/16	26.2	0.335	2.18	22.2	0.395	1.85	31.0	0.282	2.58
×1/4	26.1	0.272	2.18	22.1	0.321	1.84	31.2	0.228	2.60
×3/16	26.1	0.206	2.18	22.1	0.243	1.84	31.4	0.171	2.62
HSS10×5×3/8	25.2	0.384	2.10	20.2	0.479	1.68	28.8	0.336	2.40
×5/16	25.2	0.325	2.10	20.2	0.405	1.68	29.0	0.282	2.42
×1/4	25.1	0.264	2.09	20.1	0.329	1.68	29.2	0.227	2.43
×3/16	25.1	0.200	2.09	20.1	0.250	1.68	29.4	0.171	2.45
HSS10×4×5/8	24.3	0.577	2.03	18.3	0.766	1.53	26.0	0.540	2.17
×1/2	24.3	0.478	2.03	18.3	0.635	1.53	26.4	0.439	2.20
×3/8	24.2	0.371	2.02	18.2	0.493	1.52	26.8	0.335	2.23
×5/16	24.2	0.314	2.02	18.2	0.418	1.52	27.0	0.281	2.25
×1/4	24.1	0.256	2.01	18.1	0.340	1.51	27.2	0.227	2.27
×3/16	24.1	0.194	2.01	18.1	0.258	1.51	27.4	0.171	2.28
HSS10×3 1/2×3/16	23.6	0.191	1.97	17.1	0.263	1.43	26.4	0.170	2.20
HSS10×3×3/8	23.2	0.357	1.93	16.2	0.511	1.35	24.8	0.334	2.07
×5/16	23.2	0.303	1.93	16.2	0.434	1.35	25.0	0.280	2.08
×1/4	23.1	0.246	1.93	16.1	0.353	1.34	25.2	0.226	2.10
×3/16	23.1	0.187	1.93	16.1	0.269	1.34	25.4	0.170	2.12
×1/8	23.1	0.127	1.93	16.1	0.182	1.34	25.6	0.114	2.13
HSS10×2×3/8	22.2	0.341	1.85	14.2	0.534	1.18	22.8	0.332	1.90
×5/16	22.2	0.290	1.85	14.2	0.454	1.18	23.0	0.279	1.92
×1/4	22.1	0.237	1.84	14.1	0.370	1.18	23.2	0.226	1.93
×3/16	22.1	0.180	1.84	14.1	0.282	1.18	23.4	0.170	1.95
HSS9×7×5/8	25.3	0.646	2.11	23.3	0.702	1.94	30.0	0.546	2.50
×1/2	25.3	0.533	2.11	23.3	0.579	1.94	30.4	0.443	2.53
×3/8	25.2	0.411	2.10	23.2	0.447	1.93	30.8	0.337	2.57
×5/16	25.2	0.348	2.10	23.2	0.378	1.93	31.0	0.282	2.58
×1/4	25.1	0.282	2.09	23.1	0.307	1.93	31.2	0.228	2.60
×3/16	25.1	0.214	2.09	23.1	0.232	1.93	31.4	0.171	2.62
HSS9×5×5/8	23.3	0.602	1.94	19.3	0.727	1.61	26.0	0.540	2.17
×1/2	23.3	0.499	1.94	19.3	0.602	1.61	26.4	0.439	2.20
×3/8	23.2	0.387	1.93	19.2	0.467	1.60	26.8	0.335	2.23
×5/16	23.2	0.328	1.93	19.2	0.396	1.60	27.0	0.281	2.25
×1/4	23.1	0.267	1.93	19.1	0.322	1.59	27.2	0.227	2.27
×3/16	23.1	0.202	1.93	19.1	0.245	1.59	27.4	0.171	2.28
HSS9×3×1/2	21.3	0.458	1.78	15.3	0.638	1.28	22.4	0.435	1.87
×3/8	21.2	0.357	1.77	15.2	0.498	1.27	22.8	0.332	1.90
×5/16	21.2	0.304	1.77	15.2	0.424	1.27	23.0	0.279	1.92
×1/4	21.1	0.248	1.76	15.1	0.346	1.26	23.2	0.226	1.93
×3/16	21.1	0.188	1.76	15.1	0.263	1.26	23.4	0.170	1.95
HSS8×8×5/8	24.3	0.673	2.03	24.3	0.673	2.03	30.0	0.546	2.50
×1/2	24.3	0.555	2.03	24.3	0.555	2.03	30.4	0.443	2.53
×3/8	24.2	0.428	2.02	24.2	0.428	2.02	30.8	0.337	2.57
×5/16	24.2	0.362	2.02	24.2	0.362	2.02	31.0	0.282	2.58
×1/4	24.1	0.294	2.01	24.1	0.294	2.01	31.2	0.228	2.60
×3/16	24.1	0.223	2.01	24.1	0.223	2.01	31.4	0.171	2.62
HSS8×6×5/8	22.3	0.629	1.86	20.3	0.691	1.69	26.0	0.540	2.17
×1/2	22.3	0.521	1.86	20.3	0.572	1.69	26.4	0.439	2.20
×3/8	22.2	0.404	1.85	20.2	0.444	1.68	26.8	0.335	2.23
×5/16	22.2	0.342	1.85	20.2	0.376	1.68	27.0	0.281	2.25
×1/4	22.1	0.279	1.84	20.1	0.306	1.68	27.2	0.227	2.27
×3/16	22.1	0.211	1.84	20.1	0.233	1.68	27.4	0.171	2.28

Case A: Shape perimeter, minus one short surface.
Case B: Shape perimeter, minus one long surface.
Case C: Shape perimeter.

Table 1-46 (cont.). Rectangular (and Square) HSS
Surface and Box Perimeters, Area-to-Perimeter Ratios, and Surface Areas

Shape	Case A Perimeter	Case A A/P Ratio	Case A Surf. Area	Case B Perimeter	Case B A/P Ratio	Case B Surf. Area	Case C Perimeter	Case C A/P Ratio	Case C Surf. Area
	in.		ft²/ft	in.		ft²/ft	in.		ft²/ft
HSS8×4×5/8	20.3	0.577	1.69	16.3	0.718	1.36	22.0	0.533	1.83
×1/2	20.3	0.481	1.69	16.3	0.599	1.36	22.4	0.435	1.87
×3/8	20.2	0.375	1.68	16.2	0.468	1.35	22.8	0.332	1.90
×5/16	20.2	0.319	1.68	16.2	0.398	1.35	23.0	0.279	1.92
×1/4	20.1	0.260	1.68	16.1	0.325	1.34	23.2	0.226	1.93
×3/16	20.1	0.198	1.68	16.1	0.247	1.34	23.4	0.170	1.95
×1/8	20.1	0.134	1.68	16.1	0.168	1.34	23.6	0.114	1.97
HSS8×3×1/2	19.3	0.457	1.61	14.3	0.618	1.19	20.4	0.432	1.70
×3/8	19.2	0.358	1.60	14.2	0.484	1.18	20.8	0.331	1.73
×5/16	19.2	0.305	1.60	14.2	0.413	1.18	21.0	0.278	1.75
×1/4	19.1	0.249	1.59	14.1	0.337	1.18	21.2	0.225	1.77
×3/16	19.1	0.190	1.59	14.1	0.257	1.18	21.4	0.170	1.78
×1/8	19.1	0.129	1.59	14.1	0.175	1.18	21.6	0.114	1.80
HSS8×2×3/8	18.2	0.340	1.52	12.2	0.507	1.02	18.8	0.329	1.57
×5/16	18.2	0.290	1.52	12.2	0.433	1.02	19.0	0.277	1.58
×1/4	18.1	0.237	1.51	12.1	0.355	1.01	19.2	0.224	1.60
×3/16	18.1	0.181	1.51	12.1	0.271	1.01	19.4	0.169	1.62
×1/8	18.1	0.124	1.51	12.1	0.185	1.01	19.6	0.114	1.63
HSS7×7×5/8	21.3	0.659	1.78	21.3	0.659	1.78	26.0	0.540	2.17
×1/2	21.3	0.545	1.78	21.3	0.545	1.78	26.4	0.439	2.20
×3/8	21.2	0.423	1.77	21.2	0.423	1.77	26.8	0.335	2.23
×5/16	21.2	0.359	1.77	21.2	0.359	1.77	27.0	0.281	2.25
×1/4	21.1	0.292	1.76	21.1	0.292	1.76	27.2	0.227	2.27
×3/16	21.1	0.221	1.76	21.1	0.221	1.76	27.4	0.171	2.28
HSS7×5×5/8	19.3	0.607	1.61	17.3	0.677	1.44	22.0	0.533	1.83
×1/2	19.3	0.506	1.61	17.3	0.564	1.44	22.4	0.435	1.87
×3/8	19.2	0.395	1.60	17.2	0.440	1.43	22.8	0.332	1.90
×5/16	19.2	0.335	1.60	17.2	0.374	1.43	23.0	0.279	1.92
×1/4	19.1	0.274	1.59	17.1	0.306	1.43	23.2	0.226	1.93
×3/16	19.1	0.208	1.59	17.1	0.233	1.43	23.4	0.170	1.95
×1/8	19.1	0.141	1.59	17.1	0.158	1.43	23.6	0.114	1.97
HSS7×4×1/2	18.3	0.482	1.53	15.3	0.577	1.28	20.4	0.432	1.70
×3/8	18.2	0.378	1.52	15.2	0.453	1.27	20.8	0.331	1.73
×5/16	18.2	0.322	1.52	15.2	0.385	1.27	21.0	0.278	1.75
×1/4	18.1	0.263	1.51	15.1	0.315	1.26	21.2	0.225	1.77
×3/16	18.1	0.201	1.51	15.1	0.240	1.26	21.4	0.170	1.78
×1/8	18.1	0.136	1.51	15.1	0.164	1.26	21.6	0.114	1.80
HSS7×3×1/2	17.3	0.456	1.44	13.3	0.594	1.11	18.4	0.428	1.53
×3/8	17.2	0.359	1.43	13.2	0.468	1.10	18.8	0.329	1.57
×5/16	17.2	0.307	1.43	13.2	0.400	1.10	19.0	0.277	1.58
×1/4	17.1	0.251	1.43	13.1	0.328	1.09	19.2	0.224	1.60
×3/16	17.1	0.192	1.43	13.1	0.250	1.09	19.4	0.169	1.62
×1/8	17.1	0.131	1.43	13.1	0.171	1.09	19.6	0.114	1.63
HSS6×6×5/8	18.3	0.640	1.53	18.3	0.640	1.53	22.0	0.533	1.83
×1/2	18.3	0.533	1.53	18.3	0.533	1.53	22.4	0.435	1.87
×3/8	18.2	0.416	1.52	18.2	0.416	1.52	22.8	0.332	1.90
×5/16	18.2	0.354	1.52	18.2	0.354	1.52	23.0	0.279	1.92
×1/4	18.1	0.289	1.51	18.1	0.289	1.51	23.2	0.226	1.93
×3/16	18.1	0.220	1.51	18.1	0.220	1.51	23.4	0.170	1.95
×1/8	18.1	0.149	1.51	18.1	0.149	1.51	23.6	0.114	1.97
HSS6×5×3/8	17.2	0.400	1.43	16.2	0.425	1.35	20.8	0.331	1.73
×5/16	17.2	0.341	1.43	16.2	0.362	1.35	21.0	0.278	1.75
×1/4	17.1	0.278	1.43	16.1	0.296	1.34	21.2	0.225	1.77
×3/16	17.1	0.212	1.43	16.1	0.225	1.34	21.4	0.170	1.78

Case A: Shape perimeter, minus one short surface.
Case B: Shape perimeter, minus one long surface.
Case C: Shape perimeter.

Table 1-46 (cont.). Rectangular (and Square) HSS Surface and Box Perimeters, Area-to-Perimeter Ratios, and Surface Areas

Shape	Case A			Case B			Case C		
	Perimeter	A/P Ratio	Surf. Area	Perimeter	A/P Ratio	Surf. Area	Perimeter	A/P Ratio	Surf. Area
	in.		ft²/ft	in.		ft²/ft	in.		ft²/ft
HSS6×4×1/2	16.3	0.484	1.36	14.3	0.552	1.19	18.4	0.428	1.53
×3/8	16.2	0.381	1.35	14.2	0.435	1.18	18.8	0.329	1.57
×5/16	16.2	0.326	1.35	14.2	0.372	1.18	19.0	0.277	1.58
×1/4	16.1	0.267	1.34	14.1	0.304	1.18	19.2	0.224	1.60
×3/16	16.1	0.204	1.34	14.1	0.233	1.18	19.4	0.169	1.62
×1/8	16.1	0.139	1.34	14.1	0.159	1.18	19.6	0.114	1.63
HSS6×3×1/2	15.3	0.455	1.28	12.3	0.567	1.03	16.4	0.424	1.37
×3/8	15.2	0.361	1.27	12.2	0.449	1.02	16.8	0.326	1.40
×5/16	15.2	0.309	1.27	12.2	0.385	1.02	17.0	0.275	1.42
×1/4	15.1	0.254	1.26	12.1	0.316	1.01	17.2	0.223	1.43
×3/16	15.1	0.194	1.26	12.1	0.242	1.01	17.4	0.169	1.45
×1/8	15.1	0.133	1.26	12.1	0.166	1.01	17.6	0.114	1.47
HSS6×2×3/8	14.2	0.337	1.18	10.2	0.469	0.850	14.8	0.323	1.23
×5/16	14.2	0.289	1.18	10.2	0.403	0.850	15.0	0.273	1.25
×1/4	14.1	0.239	1.18	10.1	0.333	0.842	15.2	0.222	1.27
×3/16	14.1	0.183	1.18	10.1	0.256	0.842	15.4	0.168	1.28
×1/8	14.1	0.126	1.18	10.1	0.176	0.842	15.6	0.113	1.30
HSS5 1/2×5 1/2×3/8	16.7	0.412	1.39	16.7	0.412	1.39	20.8	0.331	1.73
×5/16	16.7	0.351	1.39	16.7	0.351	1.39	21.0	0.278	1.75
×1/4	16.6	0.287	1.38	16.6	0.287	1.38	21.2	0.225	1.77
×3/16	16.6	0.219	1.38	16.6	0.219	1.38	21.4	0.170	1.78
×1/8	16.6	0.149	1.38	16.6	0.149	1.38	21.6	0.114	1.80
HSS5×5×1/2	15.3	0.516	1.28	15.3	0.516	1.28	18.4	0.428	1.53
×3/8	15.2	0.407	1.27	15.2	0.407	1.27	18.8	0.329	1.57
×5/16	15.2	0.347	1.27	15.2	0.347	1.27	19.0	0.277	1.58
×1/4	15.1	0.284	1.26	15.1	0.284	1.26	19.2	0.224	1.60
×3/16	15.1	0.217	1.26	15.1	0.217	1.26	19.4	0.169	1.62
×1/8	15.1	0.148	1.26	15.1	0.148	1.26	19.6	0.114	1.63
HSS5×4×1/2	14.3	0.487	1.19	13.3	0.524	1.11	16.4	0.424	1.37
×3/8	14.2	0.386	1.18	13.2	0.415	1.10	16.8	0.326	1.40
×5/16	14.2	0.330	1.18	13.2	0.356	1.10	17.0	0.275	1.42
×1/4	14.1	0.272	1.18	13.1	0.292	1.09	17.2	0.223	1.43
×3/16	14.1	0.208	1.18	13.1	0.224	1.09	17.4	0.169	1.45
HSS5×3×1/2	13.3	0.454	1.11	11.3	0.534	0.942	14.4	0.418	1.20
×3/8	13.2	0.362	1.10	11.2	0.427	0.933	14.8	0.323	1.23
×5/16	13.2	0.311	1.10	11.2	0.367	0.933	15.0	0.273	1.25
×1/4	13.1	0.257	1.09	11.1	0.303	0.925	15.2	0.222	1.27
×3/16	13.1	0.197	1.09	11.1	0.233	0.925	15.4	0.168	1.28
×1/8	13.1	0.135	1.09	11.1	0.160	0.925	15.6	0.113	1.30
HSS5×2 1/2×1/4	12.6	0.248	1.05	10.1	0.310	0.842	14.2	0.221	1.18
×3/16	12.6	0.191	1.05	10.1	0.239	0.842	14.4	0.167	1.20
×1/8	12.6	0.131	1.05	10.1	0.164	0.842	14.6	0.113	1.22
HSS5×2×3/8	12.2	0.335	1.02	9.20	0.444	0.767	12.8	0.319	1.07
×5/16	12.2	0.289	1.02	9.16	0.384	0.763	13.0	0.271	1.08
×1/4	12.1	0.239	1.01	9.13	0.318	0.761	13.2	0.220	1.10
×3/16	12.1	0.185	1.01	9.10	0.246	0.758	13.4	0.167	1.12
×1/8	12.1	0.127	1.01	9.07	0.169	0.756	13.6	0.113	1.13
HSS4 1/2×4 1/2×1/2	13.8	0.505	1.15	13.8	0.505	1.15	16.4	0.424	1.37
×3/8	13.7	0.400	1.14	13.7	0.400	1.14	16.8	0.326	1.40
×5/16	13.7	0.343	1.14	13.7	0.343	1.14	17.0	0.275	1.42
×1/4	13.6	0.281	1.13	13.6	0.281	1.13	17.2	0.223	1.43
×3/16	13.6	0.216	1.13	13.6	0.216	1.13	17.4	0.169	1.45
×1/8	13.6	0.147	1.13	13.6	0.147	1.13	17.6	0.114	1.47

Case A: Shape perimeter, minus one short surface.
Case B: Shape perimeter, minus one long surface.
Case C: Shape perimeter.

Table 1-46 (cont.). Rectangular (and Square) HSS Surface and Box Perimeters, Area-to-Perimeter Ratios, and Surface Areas

Shape	Case A			Case B			Case C		
	Perimeter	A/P Ratio	Surf. Area	Perimeter	A/P Ratio	Surf. Area	Perimeter	A/P Ratio	Surf. Area
	in.		ft²/ft	in.		ft²/ft	in.		ft²/ft
HSS4×4×1/2	12.3	0.491	1.03	12.3	0.491	1.03	14.4	0.418	1.20
×3/8	12.2	0.392	1.02	12.2	0.392	1.02	14.8	0.323	1.23
×5/16	12.2	0.337	1.02	12.2	0.337	1.02	15.0	0.273	1.25
×1/4	12.1	0.278	1.01	12.1	0.278	1.01	15.2	0.222	1.27
×3/16	12.1	0.214	1.01	12.1	0.214	1.01	15.4	0.168	1.28
×1/8	12.1	0.146	1.01	12.1	0.146	1.01	15.6	0.113	1.30
HSS4×3×3/8	11.2	0.365	0.933	10.2	0.401	0.850	12.8	0.319	1.07
×5/16	11.2	0.315	0.933	10.2	0.346	0.850	13.0	0.271	1.08
×1/4	11.1	0.261	0.925	10.1	0.287	0.842	13.2	0.220	1.10
×3/16	11.1	0.202	0.925	10.1	0.222	0.842	13.4	0.167	1.12
×1/8	11.1	0.139	0.925	10.1	0.153	0.842	13.6	0.113	1.13
HSS4×2½×5/16	10.7	0.303	0.892	9.16	0.352	0.763	12.0	0.269	1.00
×1/4	10.6	0.251	0.883	9.13	0.293	0.761	12.2	0.219	1.02
×3/16	10.6	0.195	0.883	9.10	0.227	0.758	12.4	0.166	1.03
HSS4×2×3/8	10.2	0.332	0.850	8.20	0.413	0.683	10.8	0.314	0.900
×5/16	10.2	0.289	0.850	8.16	0.359	0.680	11.0	0.267	0.917
×1/4	10.1	0.241	0.842	8.13	0.300	0.678	11.2	0.218	0.933
×3/16	10.1	0.187	0.842	8.10	0.233	0.675	11.4	0.166	0.950
×1/8	10.1	0.130	0.842	8.07	0.162	0.673	11.6	0.112	0.967
HSS3½×3½×3/8	10.7	0.382	0.892	10.7	0.382	0.892	12.8	0.319	1.07
×5/16	10.7	0.330	0.892	10.7	0.330	0.892	13.0	0.271	1.08
×1/4	10.6	0.273	0.883	10.6	0.273	0.883	13.2	0.220	1.10
×3/16	10.6	0.211	0.883	10.6	0.211	0.883	13.4	0.167	1.12
×1/8	10.6	0.145	0.883	10.6	0.145	0.883	13.6	0.113	1.13
HSS3½×2½×3/8	9.70	0.349	0.808	8.70	0.389	0.725	10.8	0.314	0.900
×5/16	9.66	0.304	0.805	8.66	0.339	0.722	11.0	0.267	0.917
×1/4	9.63	0.253	0.803	8.63	0.283	0.719	11.2	0.218	0.933
×3/16	9.60	0.197	0.800	8.60	0.220	0.717	11.4	0.166	0.950
×1/8	9.57	0.136	0.798	8.57	0.152	0.714	11.6	0.112	0.967
HSS3×3×3/8	9.20	0.368	0.767	9.20	0.368	0.767	10.8	0.314	0.900
×5/16	9.16	0.320	0.763	9.16	0.320	0.763	11.0	0.267	0.917
×1/4	9.13	0.267	0.761	9.13	0.267	0.761	11.2	0.218	0.933
×3/16	9.10	0.208	0.758	9.10	0.208	0.758	11.4	0.166	0.950
×1/8	9.07	0.144	0.756	9.07	0.144	0.756	11.6	0.112	0.967
HSS3×2½×5/16	8.66	0.305	0.722	8.16	0.324	0.680	10.0	0.264	0.833
×1/4	8.63	0.256	0.719	8.13	0.271	0.678	10.2	0.216	0.850
×3/16	8.60	0.199	0.717	8.10	0.212	0.675	10.4	0.165	0.867
×1/8	8.57	0.139	0.714	8.07	0.147	0.673	10.6	0.112	0.883
HSS3×2×5/16	8.16	0.288	0.680	7.16	0.328	0.597	9.00	0.261	0.750
×1/4	8.13	0.243	0.678	7.13	0.277	0.594	9.20	0.214	0.767
×3/16	8.10	0.190	0.675	7.10	0.217	0.592	9.40	0.164	0.783
×1/8	8.07	0.133	0.673	7.07	0.152	0.589	9.60	0.112	0.800
HSS3×1½×1/4	7.63	0.228	0.636	6.13	0.284	0.511	8.20	0.212	0.683
×3/16	7.60	0.180	0.633	6.10	0.224	0.508	8.40	0.163	0.700
×1/8	7.57	0.126	0.631	6.07	0.158	0.506	8.60	0.111	0.717
HSS3×1×1/8	7.07	0.119	0.589	5.07	0.166	0.423	7.60	0.110	0.633
HSS2½×2½×5/16	7.66	0.31	0.638	7.66	0.307	0.638	9.00	0.261	0.750
×1/4	7.63	0.259	0.636	7.63	0.259	0.636	9.20	0.214	0.767
×3/16	7.60	0.203	0.633	7.60	0.203	0.633	9.40	0.164	0.783
×1/8	7.57	0.142	0.631	7.57	0.142	0.631	9.60	0.112	0.800
HSS2½×1½×1/4	6.63	0.227	0.553	5.63	0.268	0.469	7.20	0.209	0.600
×3/16	6.60	0.181	0.550	5.60	0.213	0.467	7.40	0.161	0.617
×1/8	6.57	0.128	0.548	5.57	0.151	0.464	7.60	0.110	0.633

Case A: Shape perimeter, minus one short surface.
Case B: Shape perimeter, minus one long surface.
Case C: Shape perimeter.

Table 1-46 (cont.). Rectangular (and Square) HSS Surface and Box Perimeters, Area-to-Perimeter Ratios, and Surface Areas

Shape	Case A			Case B			Case C		
	Perimeter	*A*/*P* Ratio	Surf. Area	Perimeter	*A*/*P* Ratio	Surf. Area	Perimeter	*A*/*P* Ratio	Surf. Area
	in.		ft²/ft	in.		ft²/ft	in.		ft²/ft
HSS2¹/₄×2¹/₄×¹/₄	6.13	0.253	0.511	6.13	0.253	0.511	7.20	0.212	0.600
×³/₁₆	6.85	0.200	0.571	6.85	0.200	0.571	8.40	0.163	0.700
×¹/₈	6.82	0.140	0.568	6.82	0.140	0.568	8.60	0.111	0.717
HSS2×2×¹/₄	6.13	0.246	0.511	6.13	0.246	0.511	7.20	0.209	0.600
×³/₁₆	6.10	0.196	0.508	6.10	0.196	0.508	7.40	0.161	0.617
×¹/₈	6.07	0.138	0.506	6.07	0.138	0.506	7.60	0.110	0.633
HSS2×1¹/₂×³/₁₆	5.60	0.182	0.467	5.10	0.200	0.425	6.40	0.159	0.533
HSS2×1×³/₁₆	5.10	0.166	0.425	4.10	0.206	0.342	5.40	0.156	0.450
×¹/₈	5.07	0.120	0.423	4.07	0.149	0.339	5.60	0.108	0.467
HSS1³/₄×1³/₄×³/₁₆	5.35	0.191	0.446	5.35	0.191	0.446	6.40	0.159	0.533
HSS1⁵/₈×1⁵/₈×³/₁₆	4.97	0.187	0.414	4.97	0.187	0.414	5.90	0.158	0.492
×¹/₈	4.94	0.135	0.412	4.94	0.135	0.412	6.10	0.109	0.508
HSS1¹/₂×1¹/₂×³/₁₆	4.60	0.184	0.383	4.60	0.184	0.383	5.40	0.156	0.450
×¹/₈	4.57	0.133	0.381	4.57	0.133	0.381	5.60	0.108	0.467
HSS1¹/₄×1¹/₄×³/₁₆	3.85	0.174	0.321	3.85	0.174	0.321	4.40	0.152	0.367
×¹/₈	3.82	0.129	0.318	3.82	0.129	0.318	4.60	0.107	0.383

Case A: Shape perimeter, minus one short surface.
Case B: Shape perimeter, minus one long surface.
Case C: Shape perimeter.

Table 1-47. Round HSS Surface and Box Perimeters, Area-to-Perimeter Ratios, and Surface Areas

Shape	Case A			Case B		
	Perimeter	A/P Ratio	Surf. Area	Perimeter	A/P Ratio	Surf. Area
	in.		ft²/ft	in.		ft²/ft
HSS20.000×0.500	62.8	0.454	5.23	80.0	0.357	6.67
×0.375	62.8	0.343	5.23	80.0	0.269	6.67
HSS18.000×0.500	56.5	0.453	4.71	72.0	0.356	6.00
×0.375	56.5	0.342	4.71	72.0	0.269	6.00
HSS16.000×0.500	50.3	0.451	4.19	64.0	0.355	5.33
×0.438	50.3	0.397	4.19	64.0	0.312	5.33
×0.375	50.3	0.341	4.19	64.0	0.268	5.33
×0.312	50.3	0.286	4.19	64.0	0.224	5.33
HSS14.000×0.500	44.0	0.450	3.67	56.0	0.353	4.67
×0.375	44.0	0.340	3.67	56.0	0.267	4.67
×0.312	44.0	0.285	3.67	56.0	0.224	4.67
HSS12.750×0.500	40.1	0.448	3.34	51.0	0.352	4.25
×0.375	40.1	0.339	3.34	51.0	0.267	4.25
×0.250	40.1	0.229	3.34	51.0	0.180	4.25
HSS12.500×0.625	39.3	0.554	3.28	50.0	0.435	4.17
×0.500	39.3	0.448	3.28	50.0	0.352	4.17
×0.375	39.3	0.339	3.28	50.0	0.266	4.17
×0.312	39.3	0.284	3.28	50.0	0.223	4.17
×0.250	39.3	0.229	3.28	50.0	0.180	4.17
×0.188	39.3	0.172	3.28	50.0	0.135	4.17
HSS11.250×0.625	35.3	0.551	2.94	45.0	0.433	3.75
×0.500	35.3	0.446	2.94	45.0	0.350	3.75
×0.375	35.3	0.338	2.94	45.0	0.266	3.75
×0.312	35.3	0.283	2.94	45.0	0.223	3.75
×0.250	35.3	0.228	2.94	45.0	0.179	3.75
×0.188	35.3	0.171	2.94	45.0	0.135	3.75
HSS10.750×0.500	33.8	0.445	2.82	43.0	0.349	3.58
×0.250	33.8	0.228	2.82	43.0	0.179	3.58
HSS10.000×0.625	31.4	0.547	2.62	40.0	0.430	3.33
×0.500	31.4	0.443	2.62	40.0	0.348	3.33
×0.375	31.4	0.337	2.62	40.0	0.265	3.33
×0.312	31.4	0.283	2.62	40.0	0.222	3.33
×0.250	31.4	0.228	2.62	40.0	0.179	3.33
×0.188	31.4	0.171	2.62	40.0	0.134	3.33
HSS9.625×0.500	30.2	0.443	2.52	38.5	0.348	3.21
×0.375	30.2	0.336	2.52	38.5	0.264	3.21
×0.312	30.2	0.282	2.52	38.5	0.222	3.21
×0.250	30.2	0.227	2.52	38.5	0.179	3.21
×0.188	30.2	0.171	2.52	38.5	0.134	3.21
HSS8.750×0.500	27.5	0.440	2.29	35.0	0.346	2.92
×0.375	27.5	0.335	2.29	35.0	0.263	2.92
×0.312	27.5	0.281	2.29	35.0	0.221	2.92
×0.250	27.5	0.227	2.29	35.0	0.178	2.92
×0.188	27.5	0.171	2.29	35.0	0.134	2.92
HSS8.625×0.500	27.1	0.440	2.26	34.5	0.346	2.88
×0.375	27.1	0.335	2.26	34.5	0.263	2.88
×0.322	27.1	0.290	2.26	34.5	0.227	2.88
×0.250	27.1	0.227	2.26	34.5	0.178	2.88
×0.188	27.1	0.170	2.26	34.5	0.134	2.88
HSS7.625×0.125	24.0	0.114	2.00	30.5	0.090	2.54
HSS7.500×0.500	23.6	0.436	1.97	30.0	0.343	2.50
×0.375	23.6	0.333	1.97	30.0	0.261	2.50
×0.312	23.6	0.280	1.97	30.0	0.220	2.50
×0.250	23.6	0.226	1.97	30.0	0.177	2.50
×0.188	23.6	0.170	1.97	30.0	0.133	2.50

Case A: Shape perimeter. Case B: Box perimeter, equal to four times the depth.

Table 1-47 (cont.). Round HSS Surface and Box Perimeters, Area-to-Perimeter Ratios, and Surface Areas

Shape	Case A			Case B		
	Perimeter	A/P Ratio	Surf. Area	Perimeter	A/P Ratio	Surf. Area
	in.		ft²/ft	in.		ft²/ft
HSS7.000×0.500	22.0	0.434	1.83	28.0	0.341	2.33
×0.375	22.0	0.332	1.83	28.0	0.260	2.33
×0.312	22.0	0.279	1.83	28.0	0.219	2.33
×0.250	22.0	0.225	1.83	28.0	0.177	2.33
×0.188	22.0	0.170	1.83	28.0	0.133	2.33
×0.125	22.0	0.114	1.83	28.0	0.090	2.33
HSS6.875×0.500	21.6	0.434	1.80	27.5	0.341	2.29
×0.375	21.6	0.331	1.80	27.5	0.260	2.29
×0.312	21.6	0.279	1.80	27.5	0.219	2.29
×0.250	21.6	0.225	1.80	27.5	0.177	2.29
×0.188	21.6	0.170	1.80	27.5	0.133	2.29
HSS6.625×0.500	20.8	0.432	1.73	26.5	0.340	2.21
×0.432	20.8	0.378	1.73	26.5	0.297	2.21
×0.375	20.8	0.331	1.73	26.5	0.260	2.21
×0.312	20.8	0.278	1.73	26.5	0.219	2.21
×0.280	20.8	0.251	1.73	26.5	0.197	2.21
×0.250	20.8	0.225	1.73	26.5	0.177	2.21
×0.188	20.8	0.169	1.73	26.5	0.133	2.21
×0.125	20.8	0.114	1.73	26.5	0.0895	2.21
HSS6.125×0.500	19.2	0.430	1.60	24.5	0.337	2.04
×0.375	19.2	0.329	1.60	24.5	0.258	2.04
×0.312	19.2	0.277	1.60	24.5	0.218	2.04
×0.250	19.2	0.224	1.60	24.5	0.176	2.04
×0.188	19.2	0.169	1.60	24.5	0.133	2.04
HSS6.000×0.500	18.8	0.429	1.57	24.0	0.337	2.00
×0.375	18.8	0.329	1.57	24.0	0.258	2.00
×0.312	18.8	0.277	1.57	24.0	0.217	2.00
×0.280	18.8	0.250	1.57	24.0	0.196	2.00
×0.250	18.8	0.224	1.57	24.0	0.176	2.00
×0.188	18.8	0.169	1.57	24.0	0.133	2.00
×0.125	18.8	0.114	1.57	24.0	0.0893	2.00
HSS5.563×0.375	17.5	0.327	1.46	22.3	0.257	1.86
×0.258	17.5	0.231	1.46	22.3	0.181	1.86
×0.188	17.5	0.169	1.46	22.3	0.132	1.86
×0.134	17.5	0.122	1.46	22.3	0.0960	1.86
HSS5.500×0.500	17.3	0.426	1.44	22.0	0.334	1.83
×0.375	17.3	0.327	1.44	22.0	0.257	1.83
×0.258	17.3	0.230	1.44	22.0	0.181	1.83
HSS5.000×0.500	15.7	0.422	1.31	20.0	0.331	1.67
×0.375	15.7	0.325	1.31	20.0	0.255	1.67
×0.312	15.7	0.274	1.31	20.0	0.215	1.67
×0.258	15.7	0.229	1.31	20.0	0.180	1.67
×0.250	15.7	0.222	1.31	20.0	0.174	1.67
×0.188	15.7	0.168	1.31	20.0	0.132	1.67
×0.125	15.7	0.113	1.31	20.0	0.0890	1.67
HSS4.500×0.337	14.1	0.293	1.18	18.0	0.230	1.50
×0.237	14.1	0.210	1.18	18.0	0.165	1.50
×0.188	14.1	0.167	1.18	18.0	0.131	1.50
×0.125	14.1	0.113	1.18	18.0	0.0888	1.50
HSS4.000×0.337	12.6	0.290	1.05	16.0	0.228	1.33
×0.313	12.6	0.270	1.05	16.0	0.212	1.33
×0.250	12.6	0.219	1.05	16.0	0.172	1.33
×0.237	12.6	0.209	1.05	16.0	0.164	1.33
×0.226	12.6	0.200	1.05	16.0	0.157	1.33
×0.220	12.6	0.194	1.05	16.0	0.153	1.33
×0.188	12.6	0.166	1.05	16.0	0.131	1.33
×0.125	12.6	0.113	1.05	16.0	0.0885	1.33

Case A: Shape perimeter. Case B: Box perimeter, equal to four times the depth.

Table 1-47 (cont.).
Round HSS
Surface and Box Perimeters, Area-to-Perimeter Ratios, and Surface Areas

Shape	Case A			Case B		
	Perimeter	*A*/*P* Ratio	Surf. Area	Perimeter	*A*/*P* Ratio	Surf. Area
	in.		ft²/ft	in.		ft²/ft
HSS3.500×0.313	11.0	0.267	0.917	14.0	0.210	1.17
×0.300	11.0	0.258	0.917	14.0	0.202	1.17
×0.250	11.0	0.217	0.917	14.0	0.171	1.17
×0.216	11.0	0.189	0.917	14.0	0.149	1.17
×0.203	11.0	0.179	0.917	14.0	0.140	1.17
×0.188	11.0	0.165	0.917	14.0	0.130	1.17
×0.125	11.0	0.112	0.917	14.0	0.0881	1.17
HSS3.000×0.300	9.42	0.254	0.785	12.0	0.199	1.00
×0.250	9.42	0.215	0.785	12.0	0.169	1.00
×0.216	9.42	0.188	0.785	12.0	0.147	1.00
×0.203	9.42	0.177	0.785	12.0	0.139	1.00
×0.188	9.42	0.164	0.785	12.0	0.129	1.00
×0.152	9.42	0.135	0.785	12.0	0.106	1.00
×0.134	9.42	0.120	0.785	12.0	0.0941	1.00
×0.120	9.42	0.108	0.785	12.0	0.0847	1.00
HSS2.875×0.250	9.03	0.214	0.753	11.5	0.168	0.958
×0.203	9.03	0.177	0.753	11.5	0.139	0.958
×0.188	9.03	0.163	0.753	11.5	0.128	0.958
×0.125	9.03	0.111	0.753	11.5	0.0874	0.958
HSS2.500×0.250	7.85	0.211	0.654	10.0	0.166	0.833
×0.188	7.85	0.162	0.654	10.0	0.127	0.833
×0.125	7.85	0.111	0.654	10.0	0.0869	0.833
HSS2.375×0.250	7.46	0.210	0.622	9.50	0.165	0.792
×0.218	7.46	0.186	0.622	9.50	0.146	0.792
×0.188	7.46	0.161	0.622	9.50	0.127	0.792
×0.154	7.46	0.134	0.622	9.50	0.106	0.792
×0.125	7.46	0.110	0.622	9.50	0.0867	0.792
HSS1.900×0.145	5.97	0.125	0.498	7.60	0.0985	0.633
HSS1.660×0.140	5.22	0.120	0.435	6.64	0.0941	0.553

Case A: Shape perimeter. Case B: Box perimeter, equal to four times the depth.

Table 1-48.
Steel Pipe
Surface and Box Perimeters, Area-to-Perimeter Ratios, and Surface Areas

Shape	Case A Perimeter	Case A *A*/*P* Ratio	Case A Surf. Area	Case B Perimeter	Case B *A*/*P* Ratio	Case B Surf. Area
	in.		ft²/ft	in.		ft²/ft
Standard Weight						
12	40.1	0.364	3.34	51.0	0.286	4.25
10	33.8	0.353	2.81	43.0	0.277	3.58
8	27.1	0.310	2.26	34.5	0.243	2.88
6	20.8	0.268	1.73	26.5	0.211	2.21
5	17.5	0.246	1.46	22.3	0.193	1.85
4	14.1	0.225	1.18	18.0	0.176	1.50
3 1/2	12.6	0.213	1.05	16.0	0.167	1.33
3	11.0	0.203	0.916	14.0	0.159	1.17
2 1/2	9.03	0.189	0.753	11.5	0.148	0.958
2	7.46	0.144	0.622	9.50	0.113	0.792
1 1/2	5.97	0.134	0.497	7.60	0.105	0.633
1 1/4	5.22	0.128	0.435	6.64	0.101	0.553
1	4.13	0.120	0.344	5.26	0.0939	0.438
3/4	3.30	0.101	0.275	4.20	0.0792	0.350
1/2	2.64	0.0949	0.220	3.36	0.0745	0.280
Extra Strong						
12	40.1	0.480	3.34	51.0	0.377	4.25
10	33.8	0.477	2.81	43.0	0.374	3.58
8	27.1	0.471	2.26	34.5	0.370	2.88
6	20.8	0.404	1.73	26.5	0.317	2.21
5	17.5	0.350	1.46	22.3	0.275	1.85
4	14.1	0.312	1.18	18.0	0.245	1.50
3 1/2	12.6	0.293	1.05	16.0	0.230	1.33
3	11.0	0.274	0.916	14.0	0.215	1.17
2 1/2	9.03	0.250	0.753	11.5	0.196	0.958
2	7.46	0.198	0.622	9.50	0.156	0.792
1 1/2	5.97	0.179	0.497	7.60	0.141	0.633
1 1/4	5.22	0.169	0.435	6.64	0.133	0.553
1	4.13	0.155	0.344	5.26	0.121	0.438
3/4	3.30	0.131	0.275	4.20	0.103	0.350
1/2	2.64	0.121	0.220	3.36	0.0952	0.280
Double-Extra Strong						
8	27.1	0.786	2.26	34.5	0.618	2.88
6	20.8	0.751	1.73	26.5	0.590	2.21
5	17.5	0.649	1.46	22.3	0.510	1.85
4	14.1	0.573	1.18	18.0	0.450	1.50
3	11.0	0.497	0.916	14.0	0.390	1.17
2 1/2	9.03	0.446	0.753	11.5	0.350	0.958
2	7.46	0.356	0.622	9.50	0.280	0.792

Case A: Shape perimeter. Case B: Box perimeter, equal to four times the depth.

Table 1-49.
Equal-Leg Double Angles
Surface and Box Perimeters, Weight-to-Perimeter Ratios, and Surface Areas

Shape	Case A			Case B			Case C			Case D		
	Perimeter	W/D Ratio	Surf. Area	Perimeter	W/D Ratio	Surf. Area	Perimeter	W/D Ratio	Surf. Area	Perimeter	W/D Ratio	Surf. Area
	in.		ft²/ft	in.		ft²/ft	in.		ft²/ft	in.		ft²/ft
2L8×8×1 1/8	32.2	3.55	2.68	49.0	3.55	4.08	32.8	3.49	2.73	49.5	2.31	4.13
×1	32.2	3.19	2.68	49.0	2.09	4.08	32.8	3.13	2.73	49.5	2.07	4.13
×7/8	32.2	2.81	2.68	49.0	1.85	4.08	32.8	2.76	2.73	49.5	1.83	4.13
×3/4	32.2	2.43	2.68	49.0	1.60	4.08	32.8	2.39	2.73	49.5	1.58	4.13
×5/8	32.2	2.05	2.68	49.0	1.35	4.08	32.8	2.01	2.73	49.5	1.33	4.13
×9/16	32.2	1.85	2.68	49.0	1.22	4.08	32.8	1.82	2.73	49.5	1.20	4.13
×1/2	32.2	1.66	2.68	49.0	1.09	4.08	32.8	1.63	2.73	49.5	1.08	4.13
2L6×6×1	24.3	3.09	2.03	37.1	2.02	3.09	24.8	3.02	2.07	37.5	2.00	3.13
×7/8	24.3	2.73	2.03	37.1	1.79	3.09	24.8	2.68	2.07	37.5	1.77	3.13
×3/4	24.3	2.37	2.03	37.1	1.55	3.09	24.8	2.32	2.07	37.5	1.54	3.13
×5/8	24.3	2.00	2.03	37.1	1.31	3.09	24.8	1.96	2.07	37.5	1.30	3.13
×9/16	24.3	1.81	2.03	37.1	1.19	3.09	24.8	1.77	2.07	37.5	1.17	3.13
×1/2	24.3	1.61	2.03	37.1	1.06	3.09	24.8	1.58	2.07	37.5	1.05	3.13
×7/16	24.3	1.42	2.03	37.1	0.933	3.09	24.8	1.40	2.07	37.5	0.923	3.13
×3/8	24.3	1.23	2.03	37.1	0.803	3.09	24.8	1.20	2.07	37.5	0.795	3.13
×5/16	24.3	1.03	2.03	37.1	0.674	3.09	24.8	1.01	2.07	37.5	0.667	3.13
2L5×5×7/8	20.3	2.69	1.69	31.1	1.76	2.59	20.8	2.63	1.73	31.5	1.73	2.63
×3/4	20.3	2.33	1.69	31.1	1.52	2.59	20.8	2.28	1.73	31.5	1.50	2.63
×5/8	20.3	1.98	1.69	31.1	1.29	2.59	20.8	1.93	1.73	31.5	1.28	2.63
×1/2	20.3	1.61	1.69	31.1	1.05	2.59	20.8	1.57	1.73	31.5	1.03	2.63
×7/16	20.3	1.42	1.69	31.1	0.926	2.59	20.8	1.38	1.73	31.5	0.914	2.63
×3/8	20.3	1.22	1.69	31.1	0.797	2.59	20.8	1.19	1.73	31.5	0.787	2.63
×5/16	20.3	1.02	1.69	31.1	0.669	2.59	20.8	1.00	1.73	31.5	0.660	2.63
2L4×4×3/4	16.4	2.26	1.37	25.2	1.47	2.10	16.8	2.20	1.40	25.5	1.45	2.13
×5/8	16.4	1.91	1.37	25.2	1.25	2.10	16.8	1.87	1.40	25.5	1.23	2.13
×1/2	16.4	1.55	1.37	25.2	1.01	2.10	16.8	1.51	1.40	25.5	0.996	2.13
×7/16	16.4	1.37	1.37	25.2	0.889	2.10	16.8	1.33	1.40	25.5	0.878	2.13
×3/8	16.4	1.19	1.37	25.2	0.771	2.10	16.8	1.16	1.40	25.5	0.762	2.13
×5/16	16.4	0.995	1.37	25.2	0.648	2.10	16.8	0.971	1.40	25.5	0.640	2.13
×1/4	16.4	0.802	1.37	25.2	0.522	2.10	16.8	0.783	1.40	25.5	0.516	2.13
2L3 1/2×3 1/2×1/2	14.4	1.54	1.20	22.2	1.00	1.85	14.8	1.50	1.23	22.5	0.987	1.88
×7/16	14.4	1.36	1.20	22.2	0.885	1.85	14.8	1.33	1.23	22.5	0.873	1.88
×3/8	14.4	1.18	1.20	22.2	0.767	1.85	14.8	1.15	1.23	22.5	0.756	1.88
×5/16	14.4	0.994	1.20	22.2	0.645	1.85	14.8	0.968	1.23	22.5	0.636	1.88
×1/4	14.4	0.804	1.20	22.2	0.522	1.85	14.8	0.782	1.23	22.5	0.515	1.88
2L3×3×1/2	12.4	1.51	1.03	19.2	0.97	1.60	12.8	1.460	1.07	19.5	0.959	1.63
×7/16	12.4	1.34	1.03	19.2	0.86	1.60	12.8	1.290	1.07	19.5	0.849	1.63
×3/8	12.4	1.16	1.03	19.2	0.75	1.60	12.8	1.120	1.07	19.5	0.735	1.63
×5/16	12.4	0.974	1.03	19.2	0.629	1.60	12.8	0.944	1.07	19.5	0.619	1.63
×1/4	12.4	0.789	1.03	19.2	0.509	1.60	12.8	0.764	1.07	19.5	0.502	1.63
×3/16	12.4	0.597	1.03	19.2	0.385	1.60	12.8	0.578	1.07	19.5	0.379	1.63
2L2 1/2×2 1/2×1/2	10.5	1.46	0.875	16.3	0.94	1.360	10.8	1.420	0.90	16.5	0.927	1.38
×3/8	10.5	1.12	0.875	16.3	0.72	1.360	10.8	1.090	0.90	16.5	0.715	1.38
×5/16	10.5	0.949	0.875	16.3	0.611	1.360	10.8	0.922	0.90	16.5	0.604	1.38
×1/4	10.5	0.770	0.875	16.3	0.496	1.360	10.8	0.748	0.90	16.5	0.490	1.38
×3/16	10.5	0.583	0.875	16.3	0.375	1.360	10.8	0.567	0.90	16.5	0.371	1.38
2L2×2×3/8	8.54	1.09	0.712	13.3	0.70	1.110	8.8	1.060	0.73	13.5	0.689	1.13
×5/16	8.54	0.923	0.712	13.3	0.592	1.110	8.8	0.901	0.73	13.5	0.584	1.13
×1/4	8.54	0.752	0.712	13.3	0.483	1.110	8.8	0.734	0.73	13.5	0.476	1.13
×3/16	8.54	0.576	0.712	13.3	0.370	1.110	8.8	0.562	0.73	13.5	0.364	1.13
×1/8	8.54	0.391	0.712	13.3	0.251	1.110	8.8	0.382	0.73	13.5	0.247	1.13

Case A: Combined shape perimeter, minus two vertical leg widths, minus two horizontal leg widths, and plus a space (space = 3/4-in.).

Case B: Combined shape perimeter, minus two vertical leg widths and plus a space (space = 3/4-in.).

Case C: Combined box perimeter, equal to two horizontal leg widths plus a space (space = 3/4-in.) plus twice the depth.

Case D: Combined box perimeter, equal to four horizontal leg widths plus two spaces (space = 3/4-in.) plus twice the depth.

Table 1-50.
LLBB Double Angles
Surface and Box Perimeters, Weight-to-Perimeter Ratios, and Surface Areas

Shape	Case A			Case B			Case C			Case D		
	Perimeter	W/D Ratio	Surf. Area	Perimeter	W/D Ratio	Surf. Area	Perimeter	W/D Ratio	Surf. Area	Perimeter	W/D Ratio	Surf. Area
	in.		ft²/ft	in.		ft²/ft	in.		ft²/ft	in.		ft²/ft
2L8×6×1	28.3	3.14	2.36	41.1	2.16	3.43	28.8	3.08	2.40	41.5	2.14	3.46
×7/8	28.3	2.78	2.36	41.1	1.91	3.43	28.8	2.73	2.40	41.5	1.89	3.46
×3/4	28.3	2.40	2.36	41.1	1.65	3.43	28.8	2.36	2.40	41.5	1.64	3.46
×5/8	28.3	2.02	2.36	41.1	1.39	3.43	28.8	1.99	2.40	41.5	1.38	3.46
×9/16	28.3	1.83	2.36	41.1	1.26	3.43	28.8	1.80	2.40	41.5	1.25	3.46
×1/2	28.3	1.64	2.36	41.1	1.13	3.43	28.8	1.61	2.40	41.5	1.12	3.46
×7/16	28.3	1.44	2.36	41.1	0.993	3.43	28.8	1.42	2.40	41.5	0.983	3.46
2L8×4×1	24.3	3.09	2.03	33.1	2.27	2.76	24.8	3.03	2.07	33.5	2.24	2.79
×7/8	24.3	2.74	2.03	33.1	2.01	2.76	24.8	2.69	2.07	33.5	1.99	2.79
×3/4	24.3	2.38	2.03	33.1	1.75	2.76	24.8	2.33	2.07	33.5	1.73	2.79
×5/8	24.3	2.01	2.03	33.1	1.47	2.76	24.8	1.97	2.07	33.5	1.46	2.79
×9/16	24.3	1.82	2.03	33.1	1.34	2.76	24.8	1.78	2.07	33.5	1.32	2.79
×1/2	24.3	1.62	2.03	33.1	1.19	2.76	24.8	1.59	2.07	33.5	1.18	2.79
×7/16	24.3	1.43	2.03	33.1	1.05	2.76	24.8	1.40	2.07	33.5	1.04	2.79
2L7×4×3/4	22.3	2.35	1.86	31.1	1.68	2.59	22.8	2.30	1.90	31.5	1.66	2.63
×5/8	22.3	1.98	1.86	31.1	1.42	2.59	22.8	1.94	1.90	31.5	1.40	2.63
×1/2	22.3	1.61	1.86	31.1	1.15	2.59	22.8	1.57	1.90	31.5	1.14	2.63
×7/16	22.3	1.42	1.86	31.1	1.02	2.59	22.8	1.39	1.90	31.5	1.00	2.63
×3/8	22.3	1.22	1.86	31.1	0.875	2.59	22.8	1.19	1.90	31.5	0.863	2.63
2L6×4×7/8	20.3	2.67	1.69	29.1	1.86	2.43	20.8	2.61	1.73	29.5	1.84	2.46
×3/4	20.3	2.32	1.69	29.1	1.62	2.43	20.8	2.26	1.73	29.5	1.59	2.46
×5/8	20.3	1.95	1.69	29.1	1.36	2.43	20.8	1.90	1.73	29.5	1.34	2.46
×9/16	20.3	1.76	1.69	29.1	1.23	2.43	20.8	1.72	1.73	29.5	1.21	2.46
×1/2	20.3	1.58	1.69	29.1	1.10	2.43	20.8	1.54	1.73	29.5	1.08	2.46
×7/16	20.3	1.39	1.69	29.1	0.969	2.43	20.8	1.36	1.73	29.5	0.956	2.46
×3/8	20.3	1.20	1.69	29.1	0.838	2.43	20.8	1.17	1.73	29.5	0.827	2.46
×5/16	20.3	1.00	1.69	29.1	0.701	2.43	20.8	0.981	1.73	29.5	0.692	2.46
2L6×3 1/2×1/2	19.3	1.59	1.61	27.1	1.13	2.26	19.8	1.55	1.65	27.5	1.11	2.29
×3/8	19.3	1.20	1.61	27.1	0.856	2.26	19.8	1.17	1.65	27.5	0.844	2.29
×5/16	19.3	1.01	1.61	27.1	0.717	2.26	19.8	0.982	1.65	27.5	0.707	2.29
2L5×3 1/2×3/4	17.4	2.28	1.45	25.1	1.58	2.09	17.8	2.22	1.48	25.5	1.55	2.13
×5/8	17.4	1.93	1.45	25.1	1.34	2.09	17.8	1.89	1.48	25.5	1.32	2.13
×1/2	17.4	1.56	1.45	25.1	1.08	2.09	17.8	1.53	1.48	25.5	1.07	2.13
×3/8	17.4	1.20	1.45	25.1	0.829	2.09	17.8	1.17	1.48	25.5	0.816	2.13
×5/16	17.4	1.00	1.45	25.1	0.695	2.09	17.8	0.980	1.48	25.5	0.684	2.13
×1/4	17.4	0.808	1.45	25.1	0.560	2.09	17.8	0.790	1.48	25.5	0.551	2.13
2L5×3×1/2	16.4	1.56	1.37	23.1	1.11	1.93	16.8	1.52	1.40	23.5	1.09	1.96
×7/16	16.4	1.38	1.37	23.1	0.978	1.93	16.8	1.35	1.40	23.5	0.962	1.96
×3/8	16.4	1.19	1.37	23.1	0.843	1.93	16.8	1.16	1.40	23.5	0.829	1.96
×5/16	16.4	0.999	1.37	23.1	0.709	1.93	16.8	0.975	1.40	23.5	0.697	1.96
×1/4	16.4	0.805	1.37	23.1	0.571	1.93	16.8	0.786	1.40	23.5	0.562	1.96
2L4×3 1/2×1/2	15.8	1.51	1.32	23.5	1.01	1.96	15.8	1.51	1.32	23.5	1.01	1.96
×3/8	15.8	1.15	1.32	23.5	0.774	1.96	15.8	1.15	1.32	23.5	0.774	1.96
×5/16	15.8	0.968	1.32	23.5	0.651	1.96	15.8	0.968	1.32	23.5	0.651	1.96
×1/4	15.8	0.782	1.32	23.5	0.526	1.96	15.8	0.782	1.32	23.5	0.526	1.96
2L4×3×5/8	14.8	1.84	1.23	21.5	1.27	1.79	14.8	1.84	1.23	21.5	1.27	1.79
×1/2	14.8	1.50	1.23	21.5	1.03	1.79	14.8	1.50	1.23	21.5	1.03	1.79
×3/8	14.8	1.14	1.23	21.5	0.788	1.79	14.8	1.14	1.23	21.5	0.788	1.79
×5/16	14.8	0.962	1.23	21.5	0.662	1.79	14.8	0.962	1.23	21.5	0.662	1.79
×1/4	14.8	0.777	1.23	21.5	0.535	1.79	14.8	0.777	1.23	21.5	0.535	1.79

Case A: Combined shape perimeter, minus two vertical leg widths, minus two horizontal leg widths, and plus a space (space = 3/4-in.).

Case B: Combined shape perimeter, minus two vertical leg widths and plus a space (space = 3/4-in.).

Case C: Combined box perimeter, equal to two horizontal leg widths plus a space (space = 3/4-in.) plus twice the depth.

Case D: Combined box perimeter, equal to four horizontal leg widths plus two spaces (space = 3/4-in.) plus twice the depth.

Table 1-50 (cont.). LLBB Double Angles
Surface and Box Perimeters, Weight-to-Perimeter Ratios, and Surface Areas

Shape	Case A			Case B			Case C			Case D		
	Perimeter	W/D Ratio	Surf. Area	Perimeter	W/D Ratio	Surf. Area	Perimeter	W/D Ratio	Surf. Area	Perimeter	W/D Ratio	Surf. Area
	in.		ft²/ft	in.		ft²/ft	in.		ft²/ft	in.		ft²/ft
2L3½×3×½	13.4	1.54	1.12	20.2	1.02	1.68	13.8	1.49	1.15	20.5	1.00	1.71
×7/16	13.4	1.36	1.12	20.2	0.900	1.68	13.8	1.32	1.15	20.5	0.887	1.71
×3/8	13.4	1.18	1.12	20.2	0.780	1.68	13.8	1.14	1.15	20.5	0.769	1.71
×5/16	13.4	0.993	1.12	20.2	0.658	1.68	13.8	0.964	1.15	20.5	0.649	1.71
×1/4	13.4	0.803	1.12	20.2	0.533	1.68	13.8	0.780	1.15	20.5	0.525	1.71
2L3½×2½×½	12.4	1.52	1.03	18.2	1.03	1.52	12.8	1.47	1.07	18.5	1.02	1.54
×3/8	12.4	1.17	1.03	18.2	0.795	1.52	12.8	1.13	1.07	18.5	0.782	1.54
×5/16	12.4	0.984	1.03	18.2	0.670	1.52	12.8	0.953	1.07	18.5	0.659	1.54
×1/4	12.4	0.797	1.03	18.2	0.543	1.52	12.8	0.772	1.07	18.5	0.534	1.54
2L3×2½×½	11.4	1.50	0.950	17.2	0.992	1.43	11.8	1.45	0.983	17.5	0.975	1.46
×7/16	11.4	1.33	0.950	17.2	0.879	1.43	11.8	1.28	0.983	17.5	0.864	1.46
×3/8	11.4	1.15	0.950	17.2	0.763	1.43	11.8	1.11	0.983	17.5	0.750	1.46
×5/16	11.4	0.972	0.950	17.2	0.644	1.43	11.8	0.939	0.983	17.5	0.633	1.46
×1/4	11.4	0.788	0.950	17.2	0.522	1.43	11.8	0.761	0.983	17.5	0.513	1.46
×3/16	11.4	0.598	0.950	17.2	0.397	1.43	11.8	0.578	0.983	17.5	0.390	1.46
2L3×2×½	10.5	1.47	0.875	15.2	1.01	1.27	10.8	1.43	0.900	15.5	0.994	1.29
×3/8	10.5	1.13	0.875	15.2	0.783	1.27	10.8	1.10	0.900	15.5	0.768	1.29
×5/16	10.5	0.958	0.875	15.2	0.662	1.27	10.8	0.931	0.900	15.5	0.649	1.29
×1/4	10.5	0.779	0.875	15.2	0.538	1.27	10.8	0.757	0.900	15.5	0.528	1.29
×3/16	10.5	0.594	0.875	15.2	0.411	1.27	10.8	0.578	0.900	15.5	0.403	1.29
2L2½×2×3/8	9.54	1.11	0.800	14.3	0.741	1.19	9.75	1.09	0.813	14.5	0.731	1.21
×5/16	9.54	0.941	0.800	14.3	0.628	1.19	9.75	0.921	0.813	14.5	0.619	1.21
×1/4	9.54	0.765	0.800	14.3	0.510	1.19	9.75	0.749	0.813	14.5	0.503	1.21
×3/16	9.54	0.583	0.800	14.3	0.389	1.19	9.75	0.570	0.813	14.5	0.383	1.21

Case A: Combined shape perimeter, minus two vertical leg widths, minus two horizontal leg widths, and plus a space (space = ¾-in.).
Case B: Combined shape perimeter, minus two vertical leg widths and plus a space (space = ¾-in.).
Case C: Combined box perimeter, equal to two horizontal leg widths plus a space (space = ¾-in.) plus twice the depth.
Case D: Combined box perimeter, equal to four horizontal leg widths plus two spaces (space = ¾-in.) plus twice the depth.

Table 1-51.
SLBB Double Angles
Surface and Box Perimeters, Weight-to-Perimeter Ratios, and Surface Areas

Shape	Case A			Case B			Case C			Case D		
	Perimeter	W/D Ratio	Surf. Area	Perimeter	W/D Ratio	Surf. Area	Perimeter	W/D Ratio	Surf. Area	Perimeter	W/D Ratio	Surf. Area
	in.		ft²/ft	in.		ft²/ft	in.		ft²/ft	in.		ft²/ft
2L8×6×1	28.3	3.14	2.36	45.1	1.97	3.76	28.8	3.08	2.40	45.5	1.95	3.79
×7/8	28.3	2.78	2.36	45.1	1.74	3.76	28.8	2.73	2.40	45.5	1.73	3.79
×3/4	28.3	2.40	2.36	45.1	1.51	3.76	28.8	2.36	2.40	45.5	1.49	3.79
×5/8	28.3	2.02	2.36	45.1	1.27	3.76	28.8	1.99	2.40	45.5	1.26	3.79
×9/16	28.3	1.83	2.36	45.1	1.15	3.76	28.8	1.80	2.40	45.5	1.14	3.79
×1/2	28.3	1.64	2.36	45.1	1.03	3.76	28.8	1.61	2.40	45.5	1.02	3.79
×7/16	28.3	1.44	2.36	45.1	0.905	3.76	28.8	1.42	2.40	45.5	0.897	3.79
2L8×4×1	24.3	3.09	2.03	41.1	1.83	3.43	24.8	3.03	2.07	41.5	1.81	3.46
×7/8	24.3	2.74	2.03	41.1	1.62	3.43	24.8	2.69	2.07	41.5	1.60	3.46
×3/4	24.3	2.38	2.03	41.1	1.41	3.43	24.8	2.33	2.07	41.5	1.39	3.46
×5/8	24.3	2.01	2.03	41.1	1.19	3.43	24.8	1.97	2.07	41.5	1.18	3.46
×9/16	24.3	1.82	2.03	41.1	1.08	3.43	24.8	1.78	2.07	41.5	1.07	3.46
×1/2	24.3	1.62	2.03	41.1	0.959	3.43	24.8	1.59	2.07	41.5	0.949	3.46
×7/16	24.3	1.43	2.03	41.1	0.847	3.43	24.8	1.40	2.07	41.5	0.839	3.46
2L7×4×3/4	22.3	2.35	1.86	37.1	1.41	3.09	22.8	2.30	1.90	37.5	1.40	3.13
×5/8	22.3	1.98	1.86	37.1	1.19	3.09	22.8	1.94	1.90	37.5	1.18	3.13
×1/2	22.3	1.61	1.86	37.1	0.965	3.09	22.8	1.57	1.90	37.5	0.96	3.13
×7/16	22.3	1.42	1.86	37.1	0.852	3.09	22.8	1.39	1.90	37.5	0.84	3.13
×3/8	22.3	1.22	1.86	37.1	0.733	3.09	22.8	1.19	1.90	37.5	0.73	3.13
2L6×4×7/8	20.3	2.67	1.69	33.1	1.64	2.76	20.8	2.61	1.73	33.5	1.62	2.79
×3/4	20.3	2.32	1.69	33.1	1.42	2.76	20.8	2.26	1.73	33.5	1.40	2.79
×5/8	20.3	1.95	1.69	33.1	1.20	2.76	20.8	1.90	1.73	33.5	1.18	2.79
×9/16	20.3	1.76	1.69	33.1	1.08	2.76	20.8	1.72	1.73	33.5	1.07	2.79
×1/2	20.3	1.58	1.69	33.1	0.967	2.76	20.8	1.54	1.73	33.5	0.955	2.79
×7/16	20.3	1.39	1.69	33.1	0.852	2.76	20.8	1.36	1.73	33.5	0.842	2.79
×3/8	20.3	1.20	1.69	33.1	0.737	2.76	20.8	1.17	1.73	33.5	0.728	2.79
×5/16	20.3	1.00	1.69	33.1	0.616	2.76	20.8	0.981	1.73	33.5	0.609	2.79
2L6×3 1/2×1/2	19.3	1.59	1.61	32.1	0.953	2.68	19.8	1.55	1.65	32.5	0.942	2.71
×3/8	19.3	1.20	1.61	32.1	0.723	2.68	19.8	1.17	1.65	32.5	0.714	2.71
×5/16	19.3	1.01	1.61	32.1	0.606	2.68	19.8	0.982	1.65	32.5	0.598	2.71
2L5×3 1/2×3/4	17.4	2.28	1.45	28.1	1.41	2.34	17.8	2.22	1.48	28.5	1.39	2.38
×5/8	17.4	1.93	1.45	28.1	1.20	2.34	17.8	1.89	1.48	28.5	1.18	2.38
×1/2	17.4	1.56	1.45	28.1	0.968	2.34	17.8	1.53	1.48	28.5	0.954	2.38
×3/8	17.4	1.20	1.45	28.1	0.740	2.34	17.8	1.17	1.48	28.5	0.730	2.38
×5/16	17.4	1.00	1.45	28.1	0.621	2.34	17.8	0.980	1.48	28.5	0.612	2.38
×1/4	17.4	0.808	1.45	28.1	0.500	2.34	17.8	0.790	1.48	28.5	0.493	2.38
2L5×3×1/2	16.4	1.56	1.37	27.1	0.945	2.26	16.8	1.52	1.40	27.5	0.931	2.29
×7/16	16.4	1.38	1.37	27.1	0.834	2.26	16.8	1.35	1.40	27.5	0.822	2.29
×3/8	16.4	1.19	1.37	27.1	0.719	2.26	16.8	1.16	1.40	27.5	0.708	2.29
×5/16	16.4	0.999	1.37	27.1	0.604	2.26	16.8	0.975	1.40	27.5	0.596	2.29
×1/4	16.4	0.805	1.37	27.1	0.487	2.26	16.8	0.786	1.40	27.5	0.480	2.29
2L4×3 1/2×1/2	15.8	1.510	1.32	24.5	0.971	2.04	15.8	1.51	1.32	24.5	0.971	2.04
×3/8	15.8	1.150	1.32	24.5	0.743	2.04	15.8	1.15	1.32	24.5	0.743	2.04
×5/16	15.8	0.968	1.32	24.5	0.624	2.04	15.8	0.968	1.32	24.5	0.624	2.04
×1/4	15.8	0.782	1.32	24.5	0.504	2.04	15.8	0.782	1.32	24.5	0.504	2.04
2L4×3×5/8	14.8	1.84	1.23	23.5	1.16	1.96	14.8	1.84	1.23	23.5	1.16	1.96
×1/2	14.8	1.50	1.23	23.5	0.945	1.96	14.8	1.50	1.23	23.5	0.945	1.96
×3/8	14.8	1.14	1.23	23.5	0.721	1.96	14.8	1.14	1.23	23.5	0.721	1.96
×5/16	14.8	0.962	1.23	23.5	0.606	1.96	14.8	0.962	1.23	23.5	0.606	1.96
×1/4	14.8	0.777	1.23	23.5	0.489	1.96	14.8	0.777	1.23	23.5	0.489	1.96

Case A: Combined shape perimeter, minus two vertical leg widths, minus two horizontal leg widths, and plus a space (space = 3/4-in.).
Case B: Combined shape perimeter, minus two vertical leg widths and plus a space (space = 3/4-in.).

Case C: Combined box perimeter, equal to two horizontal leg widths plus a space (space = 3/4-in.) plus twice the depth.
Case D: Combined box perimeter, equal to four horizontal leg widths plus two spaces (space = 3/4-in.) plus twice the depth.

Table 1-51 (cont.). SLBB Double Angles Surface and Box Perimeters, Weight-to-Perimeter Ratios, and Surface Areas

Shape	Case A			Case B			Case C			Case D		
	Perimeter	W/D Ratio	Surf. Area	Perimeter	W/D Ratio	Surf. Area	Perimeter	W/D Ratio	Surf. Area	Perimeter	W/D Ratio	Surf. Area
	in.		ft²/ft	in.		ft²/ft	in.		ft²/ft	in.		ft²/ft
2L3 1/2×3×1/2	13.4	1.54	1.12	21.2	0.972	1.77	13.8	1.49	1.15	21.5	0.958	1.79
×7/16	13.4	1.36	1.12	21.2	0.858	1.77	13.8	1.32	1.15	21.5	0.846	1.79
×3/8	13.4	1.18	1.12	21.2	0.743	1.77	13.8	1.14	1.15	21.5	0.733	1.79
×5/16	13.4	0.993	1.12	21.2	0.627	1.77	13.8	0.964	1.15	21.5	0.619	1.79
×1/4	13.4	0.803	1.12	21.2	0.508	1.77	13.8	0.780	1.15	21.5	0.500	1.79
2L3 1/2×2 1/2×1/2	12.4	1.52	1.03	20.2	0.932	1.68	12.8	1.47	1.07	20.5	0.918	1.71
×3/8	12.4	1.17	1.03	20.2	0.716	1.68	12.8	1.13	1.07	20.5	0.705	1.71
×5/16	12.4	0.984	1.03	20.2	0.604	1.68	12.8	0.953	1.07	20.5	0.595	1.71
×1/4	12.4	0.797	1.03	20.2	0.489	1.68	12.8	0.772	1.07	20.5	0.482	1.71
2L3×2 1/2×1/2	11.4	1.50	0.950	18.2	0.937	1.52	11.8	1.45	0.983	18.5	0.92	1.54
×7/16	11.4	1.33	0.950	18.2	0.831	1.52	11.8	1.28	0.983	18.5	0.82	1.54
×3/8	11.4	1.15	0.950	18.2	0.721	1.52	11.8	1.11	0.983	18.5	0.71	1.54
×5/16	11.4	0.972	0.950	18.2	0.609	1.52	11.8	0.939	0.983	18.5	0.599	1.54
×1/4	11.4	0.788	0.950	18.2	0.493	1.52	11.8	0.761	0.983	18.5	0.485	1.54
×3/16	11.4	0.598	0.950	18.2	0.375	1.52	11.8	0.578	0.983	18.5	0.369	1.54
2L3×2×1/2	10.5	1.47	0.88	17.2	0.895	1.43	10.8	1.43	0.900	17.5	0.88	1.46
×3/8	10.5	1.13	0.88	17.2	0.692	1.43	10.8	1.10	0.900	17.5	0.68	1.46
×5/16	10.5	0.958	0.88	17.2	0.585	1.43	10.8	0.931	0.900	17.5	0.575	1.46
×1/4	10.5	0.779	0.88	17.2	0.476	1.43	10.8	0.757	0.900	17.5	0.467	1.46
×3/16	10.5	0.594	0.88	17.2	0.363	1.43	10.8	0.578	0.900	17.5	0.357	1.46
2L2 1/2×2×3/8	9.5	1.11	0.80	15.3	0.693	1.28	9.75	1.090	0.813	15.5	0.684	1.29
×5/16	9.5	0.941	0.80	15.3	0.587	1.28	9.75	0.921	0.813	15.5	0.579	1.29
×1/4	9.5	0.765	0.80	15.3	0.477	1.28	9.75	0.749	0.813	15.5	0.471	1.29
×3/16	9.5	0.583	0.80	15.3	0.363	1.28	9.75	0.570	0.813	15.5	0.359	1.29

Case A: Combined shape perimeter, minus two vertical leg widths, minus two horizontal leg widths, and plus a space (space = 3/4-in.).
Case B: Combined shape perimeter, minus two vertical leg widths and plus a space (space = 3/4-in.).
Case C: Combined box perimeter, equal to two horizontal leg widths plus a space (space = 3/4-in.) plus twice the depth.
Case D: Combined box perimeter, equal to four horizontal leg widths plus two spaces (space = 3/4-in.) plus twice the depth.

Table 1-52. 2C-Shapes (Double American Standard Channels) Surface and Box Perimeters, Weight-to-Perimeter Ratios, and Surface Areas

Shape	Case A			Case B			Case C			Case D		
	Perimeter	W/D Ratio	Surf. Area	Perimeter	W/D Ratio	Surf. Area	Perimeter	W/D Ratio	Surf. Area	Perimeter	W/D Ratio	Surf. Area
	in.		ft²/ft	in.		ft²/ft	in.		ft²/ft	in.		ft²/ft
2C15×50	50.2	1.99	4.18	58.4	1.71	4.87	38.2	2.62	3.18	46.4	2.16	3.87
×40	49.1	1.63	4.09	56.9	1.41	4.74	37.8	2.12	3.15	45.6	1.75	3.80
×33.9	48.4	1.40	4.03	56.0	1.21	4.67	37.6	1.80	3.13	45.1	1.50	3.76
2C12×30	41.4	1.45	3.45	48.5	1.24	4.04	31.1	1.93	2.59	38.2	1.57	3.18
×25	40.7	1.23	3.39	47.6	1.05	3.97	30.9	1.62	2.58	37.7	1.33	3.14
×20.7	40.1	1.03	3.34	46.7	0.887	3.89	30.6	1.35	2.55	37.3	1.11	3.11
2C10×30	36.7	1.63	3.06	43.5	1.38	3.63	26.8	2.24	2.23	33.6	1.79	2.80
×25	35.9	1.39	2.99	42.4	1.18	3.53	26.5	1.89	2.21	33.1	1.51	2.76
×20	35.1	1.14	2.93	41.3	0.969	3.44	26.2	1.53	2.18	32.5	1.23	2.71
×15.3	34.3	0.892	2.86	40.3	0.759	3.36	26.0	1.18	2.17	31.9	0.959	2.66
2C9×20	32.6	1.23	2.72	38.6	1.04	3.22	24.1	1.66	2.01	30.1	1.33	2.51
×15	31.7	0.946	2.64	37.4	0.802	3.12	23.7	1.27	1.98	29.5	1.02	2.46
×13.4	31.4	0.854	2.62	37.0	0.724	3.08	23.6	1.14	1.97	29.2	0.918	2.43
2C8×18.75	30.0	1.25	2.50	35.8	1.05	2.98	21.8	1.72	1.82	27.6	1.36	2.30
×13.75	28.9	0.952	2.41	34.3	0.802	2.86	21.4	1.29	1.78	26.9	1.02	2.24
×11.5	28.5	0.807	2.38	33.7	0.682	2.81	21.3	1.08	1.78	26.5	0.868	2.21
2C7×14.75	26.7	1.10	2.23	32.1	0.919	2.68	19.4	1.52	1.62	24.7	1.19	2.06
×12.25	26.1	0.939	2.18	31.3	0.783	2.61	19.1	1.28	1.59	24.3	1.01	2.03
×9.8	25.6	0.766	2.13	30.5	0.643	2.54	18.9	1.04	1.58	23.9	0.820	1.99
2C6×13	24.0	1.08	2.00	29.1	0.893	2.43	17.1	1.52	1.43	22.1	1.18	1.84
×10.5	23.3	0.901	1.94	28.1	0.747	2.34	16.8	1.25	1.40	21.6	0.972	1.80
×8.2	22.7	0.722	1.89	27.3	0.601	2.28	16.6	0.988	1.38	21.2	0.774	1.77
2C5×9	20.6	0.874	1.72	25.1	0.717	2.09	14.5	1.24	1.21	19.1	0.942	1.59
×6.7	19.8	0.677	1.65	24.1	0.556	2.01	14.3	0.937	1.19	18.5	0.724	1.54
2C4×7.25	17.6	0.824	1.47	21.8	0.665	1.82	12.2	1.19	1.02	16.4	0.884	1.37
×5.4	16.9	0.639	1.41	20.8	0.519	1.73	11.9	0.908	0.992	15.8	0.684	1.32
×4.5	16.9	0.533	1.41	20.8	0.433	1.73	11.9	0.756	0.992	15.8	0.570	1.32
2C3×6	15.0	0.800	1.25	19.0	0.632	1.58	9.95	1.21	0.829	13.9	0.863	1.16
×5	14.5	0.690	1.21	18.2	0.549	1.52	9.75	1.03	0.813	13.5	0.741	1.13
×4.1	14.0	0.586	1.17	17.5	0.469	1.46	9.57	0.857	0.798	13.1	0.626	1.09
×3.5	13.8	0.507	1.15	17.2	0.407	1.43	9.49	0.738	0.791	13.0	0.538	1.08

Case A: Combined shape perimeter, minus top flange surfaces, minus twice the depth and plus a space (space = $^3/_4$-in.).
Case B: Combined shape perimeter, minus twice the depth plus two spaces (space = $^3/_4$-in.).
Case C: Combined box perimeter, equal to two flange surfaces plus a space (space = $^3/_4$-in.) plus twice the depth.
Case D: Box perimeter, equal to four flange surfaces plus two spaces (space = $^3/_4$-in.) plus twice the depth.

Table 1-53. 2MC-Shapes (Double Miscellaneous Channels) Surface and Box Perimeters, Weight-to-Perimeter Ratios, and Surface Areas

Shape	Case A			Case B			Case C			Case D		
	Perimeter	*W/D* Ratio	Surf. Area	Perimeter	*W/D* Ratio	Surf. Area	Perimeter	*W/D* Ratio	Surf. Area	Perimeter	*W/D* Ratio	Surf. Area
	in.		ft²/ft	in.		ft²/ft	in.		ft²/ft	in.		ft²/ft
2MC18×58	58.8	1.97	4.90	67.9	1.71	5.66	45.2	2.57	3.77	54.3	2.14	4.53
×51.9	58.2	1.78	4.85	67.2	1.54	5.60	45.0	2.31	3.75	53.9	1.93	4.49
×45.8	57.7	1.59	4.81	66.5	1.38	5.54	44.8	2.04	3.73	53.5	1.71	4.46
×42.7	57.4	1.49	4.78	66.1	1.29	5.51	44.7	1.91	3.73	53.3	1.60	4.44
2MC13×50	49.9	2.00	4.16	59.5	1.68	4.96	35.6	2.81	2.97	45.1	2.22	3.76
×40	48.7	1.64	4.06	57.8	1.38	4.82	35.1	2.28	2.93	44.2	1.81	3.68
×35	48.1	1.46	4.01	57.0	1.23	4.75	34.9	2.01	2.91	43.8	1.60	3.65
×31.8	47.7	1.33	3.98	56.4	1.13	4.70	34.8	1.83	2.90	43.5	1.46	3.63
2MC12×50	46.7	2.14	3.89	55.8	1.79	4.65	33.0	3.03	2.75	42.1	2.38	3.51
×45	46.0	1.96	3.83	54.8	1.64	4.57	32.8	2.74	2.73	41.5	2.17	3.46
×40	45.3	1.77	3.78	53.9	1.48	4.49	32.5	2.46	2.71	41.1	1.95	3.43
×35	44.7	1.57	3.73	53.0	1.32	4.42	32.3	2.17	2.69	40.6	1.72	3.38
×31	44.1	1.41	3.68	52.2	1.19	4.35	32.1	1.93	2.68	40.2	1.54	3.35
2MC12×10.6	32.4	0.654	2.70	36.2	0.586	3.02	27.8	0.763	2.32	31.5	0.673	2.63
2MC10×41.1	43.6	1.89	3.63	52.9	1.55	4.41	29.4	2.80	2.45	38.8	2.12	3.23
×33.6	42.3	1.59	3.53	51.3	1.31	4.28	29.0	2.32	2.42	37.9	1.77	3.16
×28.5	41.5	1.37	3.46	50.2	1.14	4.18	28.7	1.99	2.39	37.3	1.53	3.11
2MC10×25	38.6	1.30	3.22	46.1	1.08	3.84	27.6	1.81	2.30	35.1	1.42	2.93
×22	38.1	1.15	3.18	45.4	0.969	3.78	27.4	1.61	2.28	34.8	1.26	2.90
2MC10×8.4	28.4	0.592	2.37	32.1	0.523	2.68	23.8	0.706	1.98	27.5	0.611	2.29
2MC9×25.4	37.1	1.37	3.09	44.8	1.13	3.73	25.8	1.97	2.15	33.5	1.52	2.79
×23.9	36.8	1.30	3.07	44.5	1.07	3.71	25.7	1.86	2.14	33.3	1.44	2.78
2MC8×22.8	35.1	1.30	2.93	42.9	1.06	3.58	23.8	1.92	1.98	31.5	1.45	2.63
×21.4	34.9	1.23	2.91	42.5	1.01	3.54	23.7	1.81	1.98	31.3	1.37	2.61
2MC8×20	32.6	1.23	2.72	39.4	1.02	3.28	22.8	1.75	1.90	29.6	1.35	2.47
×18.7	32.3	1.16	2.69	39.0	0.959	3.25	22.7	1.65	1.89	29.4	1.27	2.45
2MC8×8.5	26.4	0.644	2.20	30.9	0.550	2.58	20.5	0.829	1.71	25.0	0.680	2.08
2MC7×22.7	33.7	1.35	2.81	41.7	1.09	3.48	22.0	2.06	1.83	29.9	1.52	2.49
×19.1	32.9	1.16	2.74	40.6	0.941	3.38	21.7	1.76	1.81	29.3	1.30	2.44
2MC6×18	31.2	1.15	2.60	39.0	0.923	3.25	19.8	1.82	1.65	27.5	1.31	2.29
×15.3	31.4	0.975	2.62	39.1	0.783	3.26	19.8	1.55	1.65	27.5	1.11	2.29
2MC6×16.3	28.5	1.14	2.38	35.2	0.926	2.93	18.8	1.73	1.57	25.5	1.28	2.13
×15.1	28.2	1.07	2.35	34.8	0.868	2.90	18.6	1.62	1.55	25.3	1.19	2.11
2MC6×12	25.9	0.927	2.16	31.6	0.759	2.63	17.8	1.35	1.48	23.5	1.02	1.96

Case A: Combined shape perimeter, minus top flange surfaces, minus twice the depth and plus a space (space = 3/4-in.).
Case B: Combined shape perimeter, minus twice the depth plus two spaces (space = 3/4-in.).
Case C: Combined box perimeter, equal to two flange surfaces plus a space (space = 3/4-in.) plus twice the depth.
Case D: Box perimeter, equal to four flange surfaces plus two spaces (space = 3/4-in.) plus twice the depth.

Table 1-54.
W-Shapes and HP-Shapes

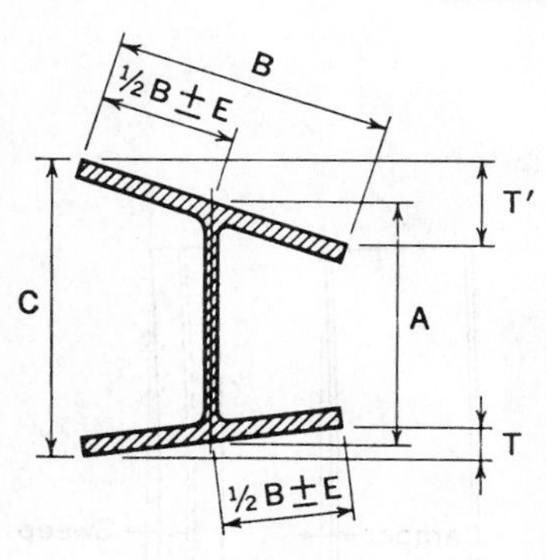

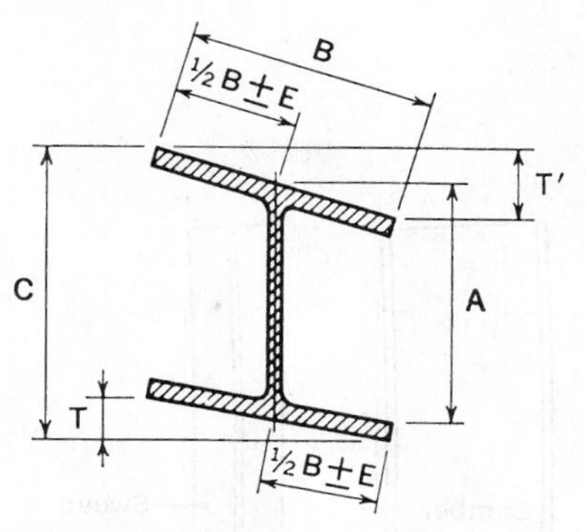

Permissible Cross-Sectional Variations

Nominal Depth, in.	*A* Depth at Web Centerline, in.		*B* Flange Width, in.		*T* + *T′* Flanges Out of Square, Max. in.	*E*[a] Web Off Center, in.	*C*, Max. Depth at any Cross-Section over Theoretical Depth, in.
	Over	Under	Over	Under			
To 12, incl.	1/8	1/8	1/4	3/16	1/4	3/16	1/4
Over 12	1/8	1/8	1/4	3/16	5/16	3/16	1/4

Permissible Variations in Length

Nominal Depth[b], in.	Variations from Specified Length for Lengths Given, in.			
	30 ft and Under		Over 30 ft	
	Over	Under	Over	Under
Beams 24 in. and under	3/8	3/8	3/8 plus 1/16 for each additional 5 ft or fraction thereof	3/8
Beams over 24 in. All columns	1/2	1/2	1/2 plus 1/16 for each additional 5 ft or fraction thereof	1/2

Mill Straightness Tolerances[c]

Sizes	Length	Permissible Variation from Straight, in.	
		Camber	Sweep
Flange width equal to or greater than 6 in.	All	$\frac{1}{8}$ in. × $\frac{\text{(total length, ft)}}{10}$	
Flange width less than 6 in.	All	$\frac{1}{8}$ in. × $\frac{\text{(total length, ft)}}{10}$	$\frac{1}{8}$ in. × $\frac{\text{(total length, ft)}}{5}$
Certain sections with a flange width approx. equal to depth & specified on order as columns[d]	45 ft and under	$\frac{1}{8}$ in. × $\frac{\text{(total length, ft)}}{10}$ with $\frac{3}{8}$ in. max.	
	Over 45 ft	$\frac{3}{8}$ in. + $\left[\frac{1}{8}\text{ in.} \times \frac{\text{(total length, ft} - 45)}{10}\right]$	

Other Permissible Rolling Variations

Area and Weight	± 2.5 percent theoretical or specified amount.
Ends Out of Square	1/64 in., per in. of depth, or of flange width if it is greater than the depth.

[a] Variation of 5/16 in. max. for sections over 426 lb/ft.
[b] For shapes specified in the order for use as bearing piles, the permitted variations are plus 5 in. and minus 0 in.
[c] The tolerances herein are taken from ASTM A6 and apply to the straightness of members received from the rolling mill, measured as illustrated in Figure 1-1. For tolerance on induced camber and sweep, see Code of Standard Practice Section 6.4.4.
[d] Applies only to W8×31and heavier, W10×49 and heavier, W12×65 and heavier, W14×90 and heavier, HP8×36, HP10×57, HP12×74 and heavier, and HP14×102 and heavier. If other sections are specified on the order as columns, the tolerance will be subject to negotiation with the manufacturer.

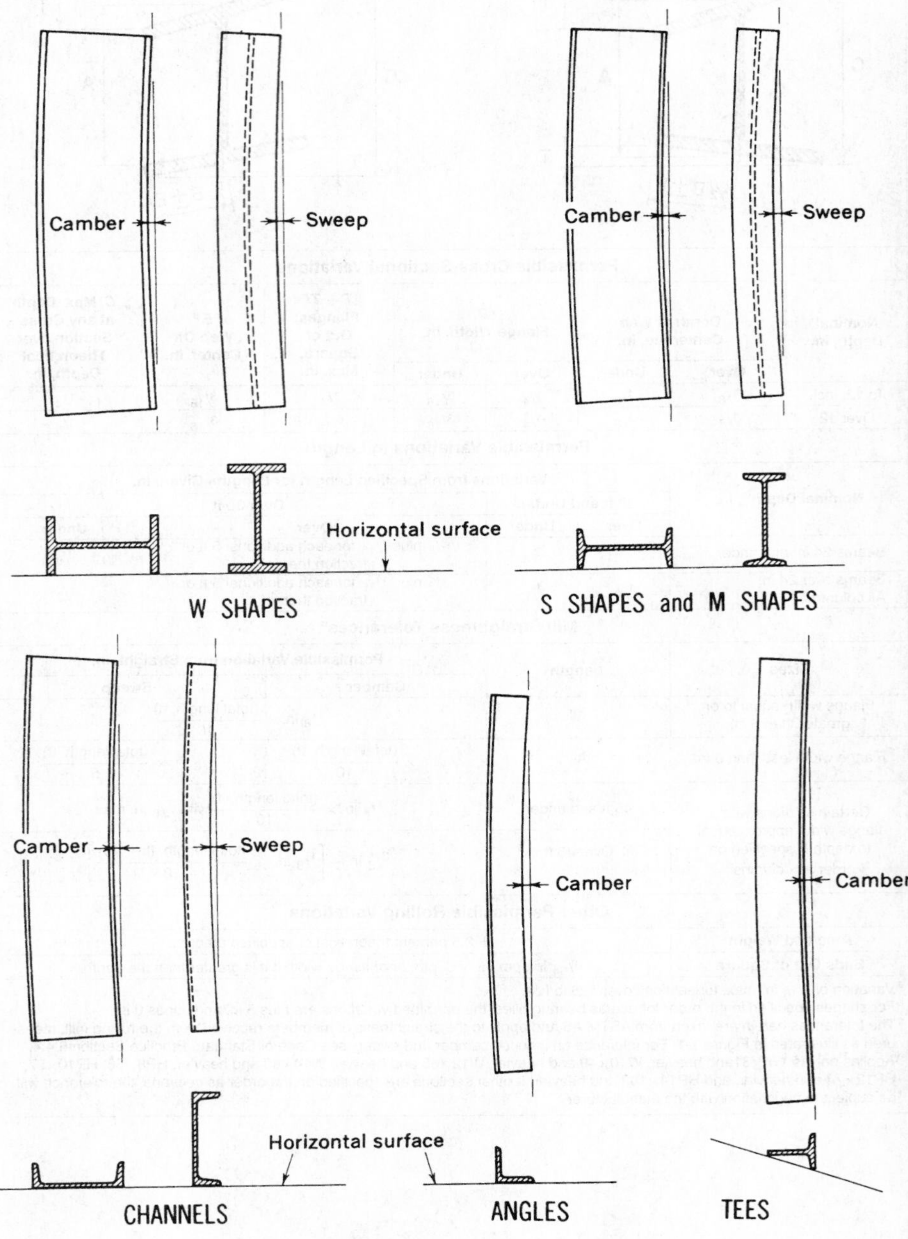

Fig. 1–1. Positions for measuring straightness.

Table 1-55.
S-Shapes, M-Shapes, and Channels

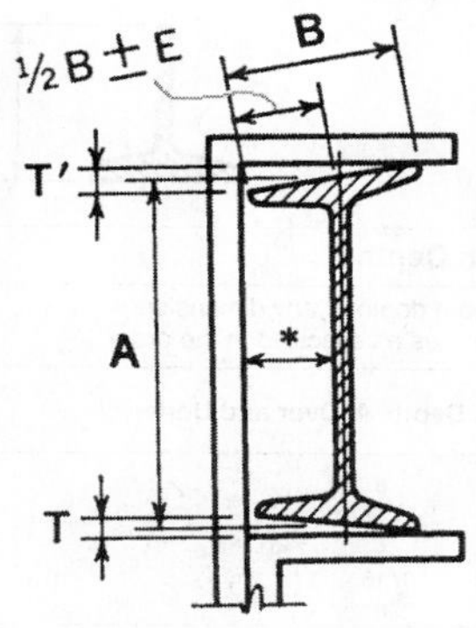

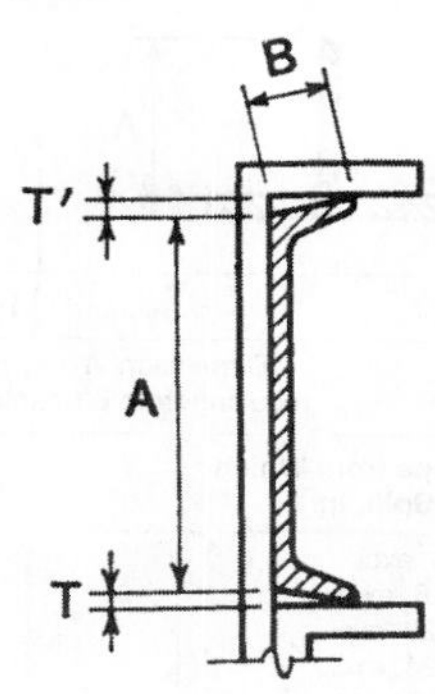

*Back of square and centerline of web to be parallel when measuring “out-of-square”

Permissible Cross-Sectional Variations

Shape	Nominal, Depth, in.	A^a Depth, in.		B Flange width, in.		$T + T'^b$ Flanges out of square, per in. of B, in.	E Web off Center, in.
		Over	Under	Over	Under		
S-shapes and M-shapes	3 to 7, incl.	$^3/_{32}$	$^1/_{16}$	$^1/_8$	$^1/_8$	$^1/_{32}$	$^3/_{16}$
	Over 7 to 14, incl.	$^1/_8$	$^3/_{32}$	$^5/_{32}$	$^5/_{32}$		
	Over 14 to 24, incl.	$^3/_{16}$	$^1/_8$	$^3/_{16}$	$^3/_{16}$		
Channels	3 to 7, incl.	$^3/_{32}$	$^1/_{16}$	$^1/_8$	$^1/_8$	$^1/_{32}$	–
	Over 7 to 14, incl.	$^1/_8$	$^3/_{32}$	$^1/_8$	$^5/_{32}$		
	Over 14	$^3/_{16}$	$^1/_8$	$^1/_8$	$^3/_{16}$		

Permissible Variations in Length

Shape	Variations Over Specified Length for Lengths Given[c], in.					
	5 to 10 ft, excl.	10 to 20 ft, excl.	20 to 30 ft, incl.	Over 30 to 40 ft, incl.	Over 40 to 65 ft, incl.	Over 65 ft
All	1	1 $^1/_2$	1 $^3/_4$	2 $^1/_4$	2 $^3/_4$	–

Mill Straightness Tolerances[d]

Chamber	$^1/_8 \text{ in.} \times \dfrac{\text{(total length, ft)}}{5}$
Sweep	Due to the extreme variations in flexibility of these shapes, permitted variations for sweep are subject to negotiation between the manufacturer and purchaser for the individual sections involved.

Other Permissible Rolling Variations

Area and Weight	± 2.5 percent theoretical or specified amount.
Ends Out of Square	S-shapes, M-shapes and channels $^1/_{64}$ in. per in. of depth.

– indicates that there is no requirement.

[a] A is measured at center line of web for beams and at back of web for channels.

[b] $T + T'$ applies when flanges of channels are toed in or out.

[c] The permitted variation under the specified length is 0 in. for all lengths. There are no requirements for lengths over 65 ft.

[d] The tolerances herein are taken from ASTM A6 and apply to the straightness of members received from the rolling mill, measured as illustrated in Figure 1-1. For tolerance on induced camber and sweep, see Code of Standard Practice Section 6.4.4.

Table 1-56.
WT-, MT-, and ST-Shapes

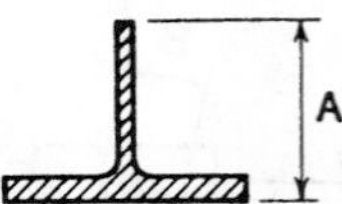

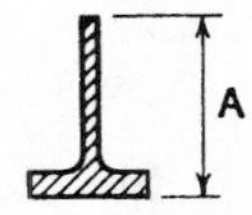

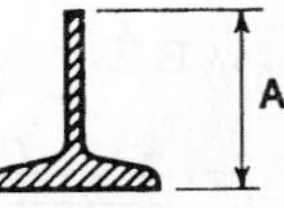

Permissible Variations in Depth

Dimension *A* may be approximately one-half beam depth or any dimension resulting from off-center splitting or splitting on two lines as specified in the order.

Depth of Shape from Which Tee is Split, in.	Variations in Depth *A*, Over and Under
To 6, excl.	$1/8$
6 to 16, excl.	$3/16$
16 to 20, excl.	$1/4$
20 to 24, excl.	$5/16$
24 and over	$3/8$

The above variations in depths of tees include the permissible variations in depth for the beams before splitting

Mill Straightness Tolerances[a]

Camber and Sweep	$1/8 \text{ in.} \times \dfrac{\text{(total length, ft)}}{5}$

Other Permissible Rolling Variations

Other permissible variations in cross section as well as permissible variations in length, area, weight, ends out-of-square, and sweep will correspond to those of the beam before splitting.

–indicates that there is no requirement.
[a]The tolerances herein are taken from ASTM A6 and apply to the straightness of members received from the rolling mill, measured as illustrated in Figure 1-1. For tolerance on induced camber and sweep, see Code of Standard Practice Section 6.4.4.

Table 1-57.
Angles, Structural Size

<table>
<tr><th colspan="5">Permissible Cross-Sectional Variations</th></tr>
<tr><th rowspan="2">Shape</th><th rowspan="2">Nominal Leg Size[a], in.</th><th colspan="2">B
Leg Size, in.</th><th rowspan="2">T
Out of Square per in. of B, in.</th></tr>
<tr><th>Over</th><th>Under</th></tr>
<tr><td rowspan="3">Angles</td><td>3 to 4, incl.</td><td>1/8</td><td>3/32</td><td rowspan="3">3/128[b]</td></tr>
<tr><td>Over 4 to 6, incl.</td><td>1/8</td><td>1/8</td></tr>
<tr><td>Over 6</td><td>3/16</td><td>1/8</td></tr>
<tr><th colspan="5">Permissible Variations in Length</th></tr>
<tr><th colspan="5">Variations Over Specified Length for Lengths Given[c], in.</th></tr>
<tr><th>5 to 10 ft, excl.</th><th>10 to 20 ft, excl.</th><th>20 to 30 ft, incl.</th><th>Over 30 to 40 ft, incl.</th><th>Over 40 to 65 ft, incl.</th></tr>
<tr><td>1</td><td>1 1/2</td><td>1 3/4</td><td>2 1/4</td><td>2 3/4</td></tr>
<tr><th colspan="5">Mill Straightness Tolerances[d]</th></tr>
<tr><td>Camber</td><td colspan="4">$^1/_8$ in. × $\frac{\text{(total length, ft)}}{5}$, applied to either leg</td></tr>
<tr><td>Sweep</td><td colspan="4">Due to the extreme variations in flexibility of these shapes, permitted variations for sweep are subject to negotiation between the manufacturer and purchaser for the individual sections involved.</td></tr>
<tr><th colspan="5">Other Permissible Rolling Variations</th></tr>
<tr><td>Area and Weight</td><td colspan="4">± 2.5 percent theoretical or specified amount.</td></tr>
<tr><td>Ends Out of Square</td><td colspan="4">3/128 in. per in. of leg length, or 1 1/2 degrees. Variations based on the longer leg of unequal angle.</td></tr>
</table>

[a]For unequal leg angles, longer leg determines classification.
[b]3/128 in. per in. = 1 1/2 degrees.
[c]The permitted variation under the specified length is 0 in. for all lengths. There are no requirements for lengths over 65 ft.
[d]The tolerances herein are taken from ASTM A6 and apply to the straightness of members received from the rolling mill, measured as illustrated in Figure 1-1. For tolerance on induced camber and sweep, see Code of Standard Practice Section 6.4.4.

Table 1-58. Angles, Bar Size[a]

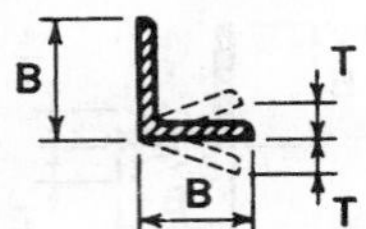

Permissible Cross-Sectional Variations

Specified Leg Size[b], in.	Variations in Thickness for Thicknesses Given, Over and Under, in.			*B* Leg Size, Over and Under, in.	*T* Out of Square per in. of *B*, in.
	3/16 and Under	Over 3/16 to 3/8, incl.	Over 3/8		
1 and Under	0.008	0.010	–	1/32	3/128[c]
Over 1 to 2, incl.	0.010	0.010	0.012	3/64	
Over 2 to 3, excl.	0.012	0.015	0.015	1/16	

Permissible Variations in Length

Section	Variations Over Specified Length for Lengths Given[d], in.				
	5 to 10 ft, excl.	10 to 20 ft, excl.	20 to 30 ft, incl.	Over 30 to 40 ft, incl.	Over 40 to 65 ft, incl.
All bar-size angles	5/8	1	1 1/2	2	2 1/2

Mill Straightness Tolerances[e]

Camber	$^1/_4$ in. in any 5 ft, or $^1/_4$ in. $\times \frac{\text{(total length, ft)}}{5}$ applied to either leg
Sweep	Due to the extreme variations in flexibility of these shapes, permitted variations for sweep are subject to negotiation between the manufacturer and purchaser for the individual sections involved.

Other Permissible Rolling Variations

Ends Out of Square	3/128 in. per in. of leg length, or 1 1/2 degrees. Variations based on the longer leg of unequal angle.

–indicates that there is no requirement.
[a] A member is "bar size" when its greatest cross-sectional dimension is less than 3 inches.
[b] For unequal angles, longer leg determines classification.
[c] 3/128 in. per in. = 1 1/2 degrees.
[d] The permitted variation under the specified length is 0 in. for all lengths. There are no requirements for lengths over 65 ft.
[e] The tolerances herein are taken from ASTM A6 and apply to the straightness of members received from the rolling mill, measured as illustrated in Figure 1-1. For tolerance on induced camber and sweep, see Code of Standard Practice Section 6.4.4.

Table 1-59. Rectangular (and Square) HSS

ASTM A500, ASTM A501, ASTM A618, and ASTM A847

<table>
<tr><td rowspan="7">Outside Dimensions</td><td colspan="4">The outside dimensions, measured across the flats at positions at least 2 in. from either end, shall not vary from the specified dimensions by more than the applicable amount given in the following table:</td></tr>
<tr><td colspan="2">Largest Outside Dimension Across Flats, in.</td><td colspan="2">Permissible Variation Over and Under Specified Dimensions[a,b], in.</td></tr>
<tr><td colspan="2">$2^1/_2$ and under</td><td colspan="2">0.020</td></tr>
<tr><td colspan="2">Over $2^1/_2$ to $3^1/_2$, incl.</td><td colspan="2">0.025</td></tr>
<tr><td colspan="2">Over $3^1/_2$ to $5^1/_2$, incl.</td><td colspan="2">0.030</td></tr>
<tr><td colspan="2">Over $5^1/_2$</td><td colspan="2">1 percent[c]</td></tr>
<tr><td colspan="4"></td></tr>
<tr><td rowspan="5">Length</td><td colspan="4">HSS are commonly produced in random lengths, in multiple lengths, and in definite cut lengths. When cut lengths are specified for HSS, the length tolerances shall be in accordance with the following table:</td></tr>
<tr><td colspan="4">Length Tolerance for Specified Cut Lengths, in.</td></tr>
<tr><td colspan="2">22 ft and under</td><td colspan="2">Over 22 to 44 ft incl.</td></tr>
<tr><td>Over</td><td>Under</td><td>Over</td><td>Under</td></tr>
<tr><td>$^1/_2$</td><td>$^1/_4$</td><td>$^3/_4$</td><td>$^1/_4$</td></tr>
<tr><td>Wall Thickness</td><td colspan="4">ASTM A500 and ASTM A847 only: The tolerance for wall thickness exclusive of the weld area shall be plus and minus 10 percent of the nominal wall thickness specified. The wall thickness is to be measured at the center of the flat.</td></tr>
<tr><td>Weight</td><td colspan="4">ASTM A501 only: The weight of HSS, as specified in ASTM A501 Tables 4, 5, and 6, shall not be less than the specified value by more than 3.5 percent.</td></tr>
<tr><td>Mass</td><td colspan="4">ASTM A618 only: The mass shall not be less than the specified value by more than 3.5 percent.</td></tr>
<tr><td>Straightness</td><td colspan="4">The permissible variation for straightness shall be $^1/_8$ in. times the number of ft of total length divided by 5.</td></tr>
<tr><td>Squareness of Sides</td><td colspan="4">Adjacent sides may deviate from 90 degrees by a tolerance of plus or minus 2 degrees maximum.</td></tr>
<tr><td>Radius of Corners</td><td colspan="4">The radius of any outside corner of the section shall not exceed 3 times the specified wall thickness[d].</td></tr>
<tr><td rowspan="9">Twist</td><td colspan="4">The tolerances for twist with respect to axial alignment of the section shall be as shown in the following table:</td></tr>
<tr><td colspan="2">Specified Dimension of Longest Side, in.</td><td colspan="2">Maximum Twist per 3 ft and in Each Additional 3 ft, in.</td></tr>
<tr><td colspan="2">$1^1/_2$ and under</td><td colspan="2">0.050</td></tr>
<tr><td colspan="2">Over $1^1/_2$ to $2^1/_2$, incl.</td><td colspan="2">0.062</td></tr>
<tr><td colspan="2">Over $2^1/_2$ to 4, incl.</td><td colspan="2">0.075</td></tr>
<tr><td colspan="2">Over 4 to 6, incl.</td><td colspan="2">0.087</td></tr>
<tr><td colspan="2">Over 6 to 8, incl.</td><td colspan="2">0.100</td></tr>
<tr><td colspan="2">Over 8</td><td colspan="2">0.112</td></tr>
<tr><td colspan="4">Twist shall be determined by holding one end of the HSS down on a flat surface plate, measuring the height that each corner on the bottom side of the tubing extends above the surface plate near the opposite ends of the HSS, and calculating the difference in the measured heights of such corners.</td></tr>
</table>

[a] The respective outside dimension tolerances include the allowances for convexity and concavity.
[b] ASTM A500 and ASTM A847 HSS only: The tolerances given are for the large flat dimension only. For HSS having a ratio of outside large to small flat dimension less than 1.5, the tolerance on the small flat dimension shall be identical to those given. For HSS having a ratio of outside large to small flat dimension in the range of 1.5 to 3.0 inclusive, the tolerance on the small flat dimension shall be 1.5 times those given. For HSS having a ratio of outside large to small flat dimension greater than 3.0, the tolerance on the small flat dimension shall be 2.0 times those given.
[c] ASTM A500 HSS only: This value is 0.1 times the large flat dimension.
[d] ASTM A501 HSS only: The radius of any outside corner must not exceed 3 times the calculated nominal wall thickness.
[e] ASTM A500, ASTM A501, and ASTM A847 HSS only: For heavier sections, it shall be permissible to use a suitable measuring device to determine twist. Twist measurements shall not be taken within 2 in. of the ends of the HSS.

<table>
<tr><th colspan="5">Table 1-60.
Round HSS and Steel Pipe</th></tr>
<tr><th colspan="5">ASTM A53</th></tr>
<tr><td>Weight</td><td colspan="4">The weight as specified in ASTM A53 Table X2.2 and Table X2.3 or as calculated from the relevant equation in ANSI/ASME B36.10M shall not vary by more than ±10 percent. Note that the weight tolerance is determined from the weights of the customary lifts of pipe as produced for shipment by the mill, divided by the number of ft of pipe in the lift. On pipe sizes over 4 in. where individual lengths may be weighed, the weight tolerance is applicable to the individual length.</td></tr>
<tr><td>Diameter</td><td colspan="4">For pipe 2 in. and over in nominal diameter, the outside diameter shall not vary more than ±1 percent from the standard specified.</td></tr>
<tr><td>Thickness</td><td colspan="4">The minimum wall thickness at any point shall not be more than 12.5 percent under the nominal wall thickness specified.</td></tr>
<tr><th colspan="5">ASTM A500 and ASTM A847</th></tr>
<tr><td>Diameter[a]</td><td colspan="4">For HSS 1.900 in. and under in nominal diameter, the outside diameter shall not vary more than ±0.5 percent, rounded to the nearest 0.005 in., from the nominal diameter specified.

For HSS 2.000 in. and over in nominal diameter, the outside diameter shall not vary more than ±0.75 percent, rounded to the nearest 0.005 in., from the nominal diameter specified.</td></tr>
<tr><td>Thickness</td><td colspan="4">The wall thickness at any point, excluding the weld seam of welded tubing, shall not be more than 10 percent under or over the nominal wall thickness specified.</td></tr>
<tr><th colspan="5">ASTM A501 and ASTM A618</th></tr>
<tr><td>Outside Dimensions</td><td colspan="4">For HSS $1^1/_2$ in. and under in nominal size, the outside diameter shall not vary more than $^1/_{64}$-in. over nor more than $^1/_{32}$-in. under the specified diameter.

For round hot-formed HSS 2 in. and over in nominal size, the outside diameter shall not vary more than ±1 percent from the specified diameter.</td></tr>
<tr><td>Weight (A501 only)</td><td colspan="4">The weight of HSS, as specified in ASTM A501 Tables 4, 5, and 6, shall not be less than the specified value by more than 3.5 percent.</td></tr>
<tr><td>Mass (A618 only)</td><td colspan="4">The mass of HSS shall not be less than the specified value by more than 3.5 percent. The mass tolerance shall be determined from individual lengths or, for HSS $4^1/_2$ in. and under in nominal size, shall be determined from masses of customary lifts produced by the mill.</td></tr>
<tr><th colspan="5">ASTM A500, ASTM A501, ASTM A618 and ASTM A847</th></tr>
<tr><td rowspan="5">Length</td><td colspan="4">HSS are commonly produced in random mill lengths, in multiple lengths, and in definite cut lengths. When cut lengths are specified for HSS, the length tolerances shall be in accordance with the following table:</td></tr>
<tr><td colspan="4">Length Tolerance for Specified Cut Lengths, in.</td></tr>
<tr><td colspan="2">22 ft and under</td><td colspan="2">Over 22 to 44 ft, incl.</td></tr>
<tr><td>Over</td><td>Under</td><td>Over</td><td>Under</td></tr>
<tr><td>$^1/_2$</td><td>$^1/_4$</td><td>$^3/_4$</td><td>$^1/_4$</td></tr>
<tr><td>Straightness</td><td colspan="4">The permissible variation for straightness of HSS shall be $^1/_8$ in. times the number of ft of total length divided by 5.</td></tr>
<tr><td colspan="5">[a]The outside diameter measurements shall be taken at least 2 in. from the end of the HSS.</td></tr>
</table>

Table 1-61.
Rectangular Sheared Plates and Universal Mill Plates

Permissible Variations in Width and Length for Sheared Plates
(1 1/2-in. and under in thickness)
Permissible Variations in Length Only for Universal Mill Plates
(2 1/2-in. and under in thickness)

Specified Dimensions, in.		Variations Over Specified Width and Length[a] for Thickness, in., and Equivalent Weights, psf, Given							
		To 3/8, excl.		3/8 to 5/8, excl.		5/8 to 1, excl.		1 to 2[b], incl.	
Length	Width	To 15.3, excl.		15.3 to 25.5, excl.		25.5 to 40.8, excl.		40.8 to 81.7, incl.	
		Width	Length	Width	Length	Width	Length	Width	Length
To 120, excl.	To 60, excl.	3/8	1/2	7/16	5/8	1/2	3/4	5/8	1
	60 to 84, excl.	7/16	5/8	1/2	11/16	5/8	7/8	3/4	1
	84 to 108, excl.	1/2	3/4	5/8	7/8	3/4	1	1	1 1/8
	108 and over	5/8	7/8	3/4	1	7/8	1 1/8	1 1/8	1 1/4
120 to 240, excl.	To 60, excl.	3/8	3/4	1/2	7/8	5/8	1	3/4	1 1/8
	60 to 84, excl.	1/2	3/4	5/8	7/8	3/4	1	7/8	1 1/4
	84 to 108, excl.	9/16	7/8	11/16	15/16	13/16	1 1/8	1	1 3/8
	108 and over	5/8	1	3/4	1 1/8	7/8	1 1/4	1 1/8	1 3/8
240 to 360, excl.	To 60, excl.	3/8	1	1/2	1 1/8	5/8	1 1/4	3/4	1 1/2
	60 to 84, excl.	1/2	1	5/8	1 1/8	3/4	1 1/4	7/8	1 1/2
	84 to 108, excl.	9/16	1	11/16	1 1/8	7/8	1 3/8	1	1 1/2
	108 and over	11/16	1 1/8	7/8	1 1/4	1	1 3/8	1 1/4	1 3/4
360 to 480, excl.	To 60, excl.	7/16	1 1/8	1/2	1 1/4	5/8	1 3/8	3/4	1 5/8
	60 to 84, excl.	1/2	1 1/4	5/8	1 3/8	3/4	1 1/2	7/8	1 5/8
	84 to 108, excl.	9/16	1 1/4	3/4	1 3/8	7/8	1 1/2	1	1 7/8
	108 and over	3/4	1 3/8	7/8	1 1/2	1	1 5/8	1 1/4	1 7/8
480 to 600, excl.	To 60, excl.	7/16	1 1/4	1/2	1 1/2	5/8	1 5/8	3/4	1 7/8
	60 to 84, excl.	1/2	1 3/8	5/8	1 1/2	3/4	1 5/8	7/8	1 7/8
	84 to 108, excl.	5/8	1 3/8	3/4	1 1/2	7/8	1 5/8	1	1 7/8
	108 and over	3/4	1 1/2	7/8	1 5/8	1	1 3/4	1 1/4	1 7/8
600 to 720, excl.	To 60, excl.	1/2	1 3/4	5/8	1 7/8	3/4	1 7/8	7/8	2 1/4
	60 to 84, excl.	5/8	1 3/4	3/4	1 7/8	7/8	1 7/8	1	2 1/4
	84 to 108, excl.	5/8	1 3/4	3/4	1 7/8	7/8	1 7/8	1 1/8	2 1/4
	108 and over	7/8	1 3/4	1	2	1 1/8	2 1/4	1 1/4	2 1/2
720 and over	To 60, excl.	9/16	2	3/4	2 1/8	7/8	2 1/4	1	2 3/4
	60 to 84, excl.	3/4	2	7/8	2 1/8	1	2 1/4	1 1/8	2 3/4
	84 to 108, excl.	3/4	2	7/8	2 1/8	1	2 1/4	1 1/4	2 3/4
	108 and over	1	2	1 1/8	2 3/8	1 1/4	2 1/2	1 3/8	3

[a] Permissible variations under specified width and length, 1/4 in. Table applies to all steels listed in ASTM A6.
[b] Permissible variations in length apply also to Universal Mill Plates up to 12 in. in width for thicknesses over 2 to 2 1/2 in., incl., except for alloy steels up to 1 3/4 in. thick.

Table 1-61 (cont.). Rectangular Sheared Plates and Universal Mill Plates

Permissible Variations from Flatness (Carbon Steel Only)

Specified Thickness, in.	Variations from Flatness for Specified Widths, in.							
	To 36, excl.	36 to 48, excl.	48 to 60, excl.	60 to 72, excl.	72 to 84, excl.	84 to 96, excl.	96 to 108, excl.	108 to 120, excl.
To 1/4, excl.	9/16	3/4	15/16	1 1/4	1 3/8	1 1/2	1 5/8	1 3/4
1/4 to 3/8, excl.	1/2	5/8	3/4	15/16	1 1/8	1 1/4	1 3/8	1 1/2
3/8 to 1/2, excl.	1/2	9/16	5/8	5/8	3/4	7/8	1	1 1/8
1/2 to 3/4, excl.	7/16	1/2	9/16	5/8	5/8	3/4	1	1
3/4 to 1, excl.	7/16	1/2	9/16	5/8	5/8	5/8	3/4	7/8
1 to 2, excl.	3/8	1/2	1/2	9/16	9/16	5/8	5/8	5/8
2 to 4, excl.	5/16	3/8	7/16	1/2	1/2	1/2	1/2	9/16
4 to 6, excl.	3/8	7/16	1/2	1/2	1/2	9/16	5/8	3/4
6 to 8, excl.	7/16	1/2	1/2	5/8	5/8	3/4	7/8	7/8

Notes:
1. The longer dimension specified is considered the length, and permissible variations in flatness along the length should not exceed the tabular amount for the specified width in plates up to 12 ft in length, or in any 12 ft for longer plates.
2. The flatness variations across the width should not exceed the tabular amount for the specified width.
3. When the longer dimension is under 36 in., the permissible variation should not exceed 1/4 in. When the longer dimension is from 36 to 72 in., inclusive, the permissible variation should not exceed 75 percent of the tabular amount for the specified width, but in no case less than 1/4 in.
4. These variations apply to plates which have a specified minimum tensile strength of not more than 60 ksi or comparable chemistry or hardness. The limits in the table are increased 50 percent for plates specified to a higher minimum tensile strength or comparable chemistry or hardness.
5. For plates 8 in. and over in thickness or 120 in. and over in width, see ASTM A6 Table 13.

Permissible Variations in Camber[a] for Carbon Steel Sheared and Gas Cut Rectangular Plates

Maximum permissible camber, in. (all thicknesses) $= 1/8 \text{ in.} \times \frac{\text{(total length, ft)}}{5}$

Permissible Variations in Camber[a] for Carbon, High-Strength Low-Alloy, and Alloy Steel Universal Mill Plates and High-Strength Low-Alloy and Alloy Steel Sheared, Special-Cut, or Gas-Cut Rectangular Plates

Dimension, in.		Camber for Thicknesses and Widths Given
Thickness	Width	
To 2, incl.	All	$1/8 \text{ in.} \times \frac{\text{(total length, ft)}}{5}$
Over 2 to 15, incl.	To 30, incl.	$3/16 \text{ in.} \times \frac{\text{(total length, ft)}}{5}$
	Over 30 to 60, incl.	$1/4 \text{ in.} \times \frac{\text{(total length, ft)}}{5}$

[a]Camber as it relates to plates is the horizontal edge curvature in the length, measured over the entire length of the plate in the flat position.

Table 1-61 (cont.). Rectangular Sheared Plates and Universal Mill Plates

Permissible Variations from Flatness (High-Strength Low-Alloy and Alloy Steel, Hot Rolled or Thermally Treated)

Specified Thickness, in.	Variations from Flatness for Specified Widths, in.							
	To 36, excl.	36 to 48, excl.	48 to 60, excl.	60 to 72, excl.	72 to 84, excl.	84 to 96, excl.	96 to 108, excl.	108 to 120, excl.
To 1/4, excl.	13/16	1 1/8	1 3/8	1 7/8	2	2 1/4	2 3/8	2 5/8
1/4 to 3/8, excl.	3/4	15/16	1 1/8	1 3/8	1 3/4	1 7/8	2	2 1/4
3/8 to 1/2, excl.	3/4	7/8	15/16	15/16	1 1/8	1 5/16	1 1/2	1 5/8
1/2 to 3/4, excl.	5/8	3/4	13/16	7/8	1	1 1/8	1 1/4	1 3/8
3/4 to 1, excl.	5/8	3/4	7/8	7/8	15/16	1	1 1/8	1 5/16
1 to 2, excl.	9/16	5/8	3/4	13/16	7/8	15/16	1	1
2 to 4, excl.	1/2	9/16	11/16	3/4	3/4	3/4	3/4	7/8
4 to 6, excl.	9/16	11/16	3/4	3/4	7/8	7/8	15/16	1 1/8
6 to 8, excl.	5/8	3/4	3/4	15/16	1	1 1/8	1 1/4	1 5/16

Notes:
1. The longer dimension specified is considered the length, and permissible variations in flatness along the length should not exceed the tabular amount for the specified width in plates up to 12 ft in length, or in any 12 ft for longer plates.
2. The flatness variations across the width should not exceed the tabular amount for the specified width.
3. When the longer dimension is under 36 in., the permissible variation should not exceed 3/8 in. When the longer dimension is from 36 to 72 in., inclusive, the permissible variation should not exceed 75 percent of the tabular amount for the specified width.
5. For plates 8 in. and over in thickness or 120 in. and over in width, see ASTM A6 Table 14.

Permissible Variations in Width for Universal Mill Plates (15 in. and Under in Thickness)

Specified Width, in.	Variations over Specified Width and Length for Thickness, in., and Equivalent Weights, psf, Given					
	To 3/8, excl.	3/8 to 5/8, excl.	5/8 to 1, excl.	1 to 2, incl.	Over 2 to 10, incl.	Over 10 to 15, incl.
Over 8 to 20, excl.	1/8	1/8	3/16	1/4	3/8	1/2
20 to 36, excl.	3/16	1/4	5/16	3/8	7/16	9/16
36 and over	5/16	3/8	7/16	1/2	9/16	5/8

Notes:
1. Permissible variation under specified width, 1/8 in.
2. Table applies to all steels listed in ASTM A6.

PART 2

GENERAL DESIGN CONSIDERATIONS

SCOPE

The specification requirements and other design considerations summarized in this Part apply in general to the design and construction of steel buildings. For seismic force resisting systems in which the seismic response modification factor R is taken greater than 3, see the Seismic Provisions, which are available from the American Institute of Steel Construction, Inc. at www.aisc.org.

APPLICABLE SPECIFICATIONS, CODES AND STANDARDS

The following specifications, codes and standards apply in the design and construction of steel buildings, per LRFD Specification Sections A1, A4, J2 and J3.1, and are covered in this Manual:

1. LRFD Specification: The 1999 AISC *Load and Resistance Factor Design Specification for Structural Steel Buildings*. This specification is included in Part 16.
2. HSS Specification: The 2000 AISC *Specification for the Design of Steel Hollow Structural Sections*. This specification is included in Part 16.
3. Single-Angle Specification: The 2000 AISC *Load and Resistance Factor Design Specification for Single-Angle Members*. This specification is included in Part 16.
4. RCSC Specification: The 2000 RCSC *Specification for Structural Joints Using ASTM A325 or A490 Bolts*. This specification is reprinted in Part 16 with the permission of the Research Council on Structural Connections.
5. AWS D1.1: *Structural Welding Code—Steel*, AWS D1.1:2000. This specification is available from the American Welding Society.
6. Code of Standard Practice: The 2000 AISC *Code of Standard Practice for Steel Buildings and Bridges*. This code is included in Part 16.
7. ASCE 7: *Minimum Design Loads for Buildings and Other Structures*, ASCE 7-98. This specification is available from the American Society of Civil Engineers.

Various other ASME, ASTM and AWS specifications and standards are also referenced in LRFD Specification Section A6.

The following specifications, codes and standards apply in the design and construction of steel buildings, per LRFD Specification Sections A1 and J10, but are not covered, except in general, in this Manual:

1. Seismic Provisions: The 1997 AISC *Seismic Provisions for Structural Steel Buildings* and 2000 *AISC Seismic Provisions Supplement No. 2*. Note: 2000 AISC *Seismic Provisions Supplement No. 2* superseded 1999 *AISC Seismic Provisions Supplement No. 1* in its entirety. These specifications are available from the American Institute of Steel Construction, Inc. at www.aisc.org.
2. Nuclear Specification: The 1994 AISC *Specification for the Design, Fabrication and Erection of Steel Safety-related Structures for Nuclear Facilities*. This specification is available from the American Institute of Steel Construction, Inc.
3. AISI Specification: The 1996 AISI *Specification for the Design of Cold-Formed Steel Structural Members* and 1999 AISI *Specification for the Design of Cold-Formed Steel Structural Members, Supplement No. 1*. These specifications are available from the American Iron and Steel Institute.
4. ACI 318: *Building Code Requirements for Structural Concrete and Commentary*, ACI 318-99. This specification is available from the American Concrete Institute.

Subject to the requirements in the applicable building code and the contract documents, the following other specifications, codes and standards may apply, but are not covered in this Manual:

1. ACI 530: *Building Code Requirements for Masonry Structures and Commentary*, ACI 530/ASCE 5/TMS402-99. This specification is available from ACI International, the American Society of Civil Engineers and the Masonry Society.
2. ASCE 3: *Standard for the Structural Design of Composite Slabs*, ANSI/ASCE 3-91. This specification is available from the American Society of Civil Engineers.
3. ASCE 8: *Specification of Cold Formed Stainless Steel Structural Members*, ANSI/ASCE 8-90. This specification is available from the American Society of Civil Engineers.
4. ASCE 9: *Standard Practice for Construction and Inspection of Composite Slabs*, ANSI/ASCE 9-91. This specification is available from the American Society of Civil Engineers.
5. ASCE 19: *Structural Applications of Steel Cables for Buildings*, ANSI/ASCE 19-96. This specification is available from the American Society of Civil Engineers.
6. CMAA 70: *Specifications for Top Running Bridge & Gantry Type Multiple Girder Electric Overhead Traveling Cranes*, CMAA 70-2000. This specification is available from the Crane Manufacturers Association of America.
7. Rack Specification: The 1997 RMI *Specification for the Design, Testing and Utilization of Industrial Steel Storage Racks*. This specification is available from the Rack Manufacturers Institute.
8. SJI Specification: The SJI *Standard Specifications for Open Web Steel Joists. K Series*, the SJI *Standard Specifications for Longspan Steel Joists, LH Series and Deep Longspan Steel Joists, DLH Series* and the SJI *Standard Specification for Joist Girders*. These specifications are available from the Steel Joist Institute.

Additionally, information on miscellaneous, architectural and ornamental metals can be found in the publications of the National Association of Architectural Metals Manufacturers (NAAMM).

Specifications, Codes and Standards for Structural Steel Buildings

Subject to the requirements in the applicable building code and the contract documents, the design, fabrication and erection of structural steel buildings is governed as indicated in LRFD Specification Sections A1 and A4 as follows:

1. ASCE 7 provides the general requirements for loads, load factors and load combinations.
2. The LRFD Specification provides the general requirements for design and construction.
3. The Code of Standard Practice provides the standard of custom and usage for the fabrication and erection of structural steel.

Additional Requirements for HSS and Steel Pipe

As indicated in LRFD Specification Section A1 and HSS Specification Section 1.1, additional requirements specific to HSS and steel pipe are provided in the HSS Specification.

Additional Requirements for Single-Angle Members

As indicated in LRFD Specification Section A1 and Single-Angle Specification Section 1, additional requirements specific to single-angle members are provided in the Single-Angle Specification.

Additional Requirements for Bolted Joints with High-Strength Bolts

As indicated in LRFD Specification Sections J3.1 and M2.5 and RCSC Specification Section 1.1, additional requirements specific to bolted joints with high-strength bolts are provided in the RCSC Specification.

Additional Requirements for Welded Joints

As indicated in LRFD Specification Section J2, additional requirements specific to welded joints are provided in AWS D1.1.

Additional Requirements for Steel-to-Concrete Anchorage

As indicated in LRFD Specification Section J10, additional requirements specific to steel-to-concrete anchorage applications are provided in ACI 318.

Additional Requirements for Seismic Applications

As indicated in LRFD Specification Section A1 and Seismic Provisions Part I, Section 1 and Part II, Section 1, when the seismic response modification factor R is taken greater than 3, additional requirements specific to high-seismic applications are provided in the Seismic Provisions. When R is taken equal to or less than 3, these additional requirements do not apply (exception: when the composite systems in Seismic Provisions Part II are used, the additional requirements in Seismic Provisions Part II apply regardless of the value of R used).

Other AISC Reference Documents

The following other AISC manuals and reference documents may be of use in the design and construction of structural steel buildings:

1. The AISC *Hollow Structural Sections (HSS) Connections Manual*. This publication covers connection design for framing with hollow structural sections (HSS) and steel pipe.
2. AISC's *Detailing for Steel Construction*. This publication covers the standard practices and recommendations for steel detailing, including the preparation of shop and erection drawings.
3. *Iron and Steel Beams: 1873 to 1952*. This publication provides a compilation of dimensions and properties of wrought iron and steel beams and columns produced in the United States through 1952, as well as a summary of historic unit-stress recommendations made by early manufacturers, ASTM tensile- and yield-strength requirements and basic AISC working-stress recommendations.
4. *A Guide to Engineering and Quality Criteria for Steel Structures: Common Questions Answered*. This publication provides a compilation of common questions and answers on a range of topics in the design and construction of structural steel buildings.

Additionally, the following AISC Design Guides are available:

1. *Column Base Plates*, AISC Design Guide No. 1 (DeWolf and Ricker, 1990).
2. *Design of Steel and Composite Buildings with Web Openings*, AISC Design Guide No. 2 (Darwin, 1990).
3. *Serviceability Design Considerations for Low-Rise Buildings*, AISC Design Guide No. 3 (Fisher and West, 1990).
4. *Extended End-Plate Moment Connections*, AISC Design Guide No. 4 (Murray, 1990). Note that coverage of extended end-plate moment connections in this design guide is based upon older criteria that are limited to use in non-seismic applications. The coverage of extended end-plate moment connection in Part 12 also includes more

recently developed criteria that are suitable for use in both wind and low-seismic applications (see also the discussion "Seismic Design").

5. *Design of Low- and Medium-Rise Steel Buildings*, AISC Design Guide No. 5 (Allison, 1991).
6. *Load and Resistance Factor Design of W-Shapes Encased in Concrete*, AISC Design Guide No. 6 (Griffis, 1997).
7. *Industrial Buildings: Roofs to Column Anchorage*, AISC Design Guide No. 7 (Fisher, 1993).
8. *Partially Restrained Composite Connections*, AISC Design Guide No. 8 (Leon, Hoffman and Staeger, 1996).
9. *Torsional Analysis of Structural Steel Members*, AISC Design Guide No. 9 (Seaburg and Carter, 1997).
10. *Erection Bracing of Low-Rise Structural Steel Frames*, AISC Design Guide No. 10 (Fisher and West, 1997).
11. *Floor Vibrations Due to Human Activity*, AISC Design Guide No. 11 (Murray, Allen and Ungar, 1997).
12. *Modification of Existing Welded Steel Moment Frames for Seismic Resistance*, AISC Design Guide No. 12 (Gross, Engelhardt, Uang, Kasai and Iwankiw, 1999).
13. *Wide-Flange Column Stiffening at Moment Connections: Wind and Seismic Applications*, AISC Design Guide No. 13(Carter, 1999).

OSHA REQUIREMENTS

OSHA Safety and Health Standards for the Construction Industry, 29 CFR 1926 Part R Safety Standards for Steel Erection must be addressed in the design, detailing, fabrication and erection of steel structures. These regulations are expected to become effective on July 18, 2001, except for requirements for slip-resistance certification of painted surfaces (see "Walking/Working Surfaces" below), which are expected to become effective on July 18, 2006.

Following is a brief summary of selected provisions. The full text of the regulations should be consulted and can be found at www.osha.gov. See also Barger and West (2001) for further information.

Columns and Column Base Plates

1. All column base plates must be designed and fabricated with a minimum of 4 anchor rods.
2. Posts (which weigh less than 300 pounds) are distinguished from columns and excluded from the 4-anchor-rod requirement.
3. Columns, column base plates and their foundations must be designed to resist a minimum eccentric gravity load of 300 lbs located 18 in. from the extreme outer face of the column in each direction at the top of the column shaft.
4. Column splices must be designed to meet the same load-resisting characteristics as columns.
5. Double connections through column webs or at beams that frame over the tops of columns[1] must be designed to have at least one installed bolt remain in place to support the first beam while the second beam is being erected. Alternatively, the fabricator must supply a seat or equivalent device with a means of positive attachment to support the first beam while the second beam is being erected.

[1] This requirement applies only at the location of the column, not at locations away from the column.

Safety Cables

1. On multi-story structures, perimeter safety cables (2 lines) are required at final interior and exterior perimeters of floors as soon as the deck is installed.
2. Perimeter columns must extend 48 in. above the finished floor (unless constructability does not allow) to allow the installation of perimeter safety cables.
3. The regulations prohibit field welding of attachments for installation of perimeter safety cables once the column has been erected.
4. Provision of some method of attaching perimeter cable is required, but responsibility is not assigned either to the fabricator or to the erector. While this will be subject to normal business arrangements between the fabricator and the erector, holes for these cables are often punched or drilled in columns by the fabricator.

Walking/Working Surfaces

1. Framed metal deck openings must have structural members configured with projecting elements turned down to allow continuous decking, except where not allowed by design constraints or constructability. The openings in the metal deck are not to be cut until the hole is needed.
2. Shear stud connectors, threaded studs, reinforcing bars and deformed anchors that will project vertically from or horizontally across the top flange of the member are not to be attached to the top flanges of beams, joists or beam attachments until after the metal decking or other walking/working surface has been installed.
3. The regulations prohibit workers from walking the top surface of painted structural members, unless the paint is certified as meeting a 0.5 slip resistance coefficient. However, the enactment of this requirement will not become effective until July 18, 2006.

Beams and Bracing

1. Solid-web members (beams) must be connected with a minimum of two bolts or their equivalent before the crane load line is released.
2. Bracing members must be connected with a minimum of one bolt or its equivalent before the crane load line is released.

Joists

1. Unless panelized, all joists 40-ft long and longer and their bearing members must have holes to allow for initial connections by bolting.
2. Bridging of joists is mandated according to manufacturer guidelines.
3. A vertical stabilizer plate to receive the joist bottom chord must be provided at columns. Minimum sizes are given and the stabilizer plate must have a hole for the attachment of guying or plumbing cables.

Controlling Contractor

1. The controlling contractor must provide adequate site access and adequate storage.
2. The controlling contractor must notify the erector of repairs or modifications to anchor rods in writing. Such modifications and repairs must be approved by the owner's designated representative for design.
3. The controlling contractor must give notice that the supporting foundations have achieved sufficient strength to allow safe steel erection.

4. The controlling contractor must either provide overhead protection or prohibit other trades from working under steel erection activities.

Fall Protection

1. Fall protection is generally required at 15 ft for workers involved in steel erection.
2. Fall protection training is required.
3. Connectors may work between 15 ft and 30 ft (or two stories, whichever is less) without being tied-off only if fall protection is provided and the connector is wearing the necessary equipment to tie-off.
4. Decking operations between 15 ft and 30 ft are subject to specific rules governing the Controlled Decking Zone (CDZ).
5. Additional training is required for workers engaged in connecting or working in the CDZ.

Lifts

1. Additional crane safety measures are required.
2. Procedures for multiple lifts (Christmas-treeing) are established.
3. Alternative site-specific erection plans are allowed in specific instances when compliance with OSHA regulations is not possible.

FUNDAMENTALS OF LIMIT-STATES DESIGN

Limit states define the boundary of structural usefulness. Strength limit states, such as flexural strength or compressive strength, relate to load carrying capability and safety. Serviceability limit states, such as deflection or drift, relate to performance under normal service conditions. Structures must be proportioned so that no applicable strength or serviceability limit state is exceeded.

Normally, several limit states will apply in the design of a structural member or connection. The controlling limit state is the one that results in the least design strength. As an example, the controlling limit state for a simple beam (exclusive of limit states that relate to the end connections) may be flexural strength, shear strength or deflection.

Loads, Load Factors and Load Combinations

Per LRFD Specification Sections A4, A5.1 and A5.3 and HSS Specification Section 1.3, the required strength (P_u, M_u, V_u, etc.) is determined for the loads, load factors and load combinations given in the applicable building code, which is usually based upon ASCE 7. The intensities of the loads below are usually specified in or taken from ASCE 7. Where applicable, the live load L should be determined with consideration of live-load reductions as permitted in the applicable building code.

D = dead load
L = live load due to occupancy
L_r = roof live load
S = snow load
R = nominal load due to initial rainwater or ice exclusive of the ponding contribution
W = wind load
E = earthquake load

The required strength is determined from the following factored combinations[2], which are based on ASCE 7:

$$1.4D \quad (1)$$
$$1.2D + 1.6L + 0.5(L_r \text{ or } S \text{ or } R) \quad (2)$$
$$1.2D + 1.6(L_r \text{ or } S \text{ or } R) + (0.5L \text{ or } 0.8W) \quad (3)$$
$$1.2D + 1.6W + 0.5L + 0.5(L_r \text{ or } S \text{ or } R) \quad (4)$$
$$1.2D \pm 1.0E + 0.5L + 0.2S \quad (5)$$
$$0.9D \pm (1.6W \text{ or } 1.0E) \quad (6)$$

Whether the loads themselves or the effects of those loads are used in these combinations, the results are the same, provided the principle of superposition is valid. This is true when deflections are small and the stress-strain behavior is nominally elastic. However, when second-order effects are significant or the behavior is inelastic, superposition is not valid and the loads should be used in these combinations.

The factored combinations above recognize that, when several transient loads act in combination, only one assumes its maximum lifetime value[3], while the other(s) are at their "arbitrary-point-in-time" (APT) values. Each combination models the total design loading condition when a different load is at its maximum. Thus, the maximum-lifetime load effect is amplified by an amount that is proportional to its relative variability and the APT load effect(s) are factored to their mean value(s).

Dead load D is present in each load combination with a load factor of 1.2, except in load combination 1, where it is the dominant (only) load effect, and load combination 6, where it is reduced for calculation of the overturning or uplift effect. The 1.2 load factor accounts for the statistical variability of the dead load. The designer must independently account for other contributions to dead load, such as the weight of additional concrete, if any, added to adjust for concrete ponding effects (Ruddy, 1986) or differing framing elevations.

When considering members subjected to gravity loads only, the number of applicable load combinations can be simplified to the following two:

$$1.4D$$
$$1.2D + 1.6L'$$

where $L' = L + (L_r \text{ or } S \text{ or } R)$. This simplification conservatively neglects the permitted reduction in load factor for the coincident effect of live load due to occupancy and controlling roof load (roof live load, snow load or rain load). For gravity floor framing, this simplification reduces to $1.2D + 1.6L$ and includes no such conservatism. Similarly, for gravity-controlled roof framing, this simplification reduces to $1.2D + 1.6(L_r \text{ or } S \text{ or } R)$ and includes no such conservatism.

Load combination 1 can only control when the dead load exceeds eight times the total live load. Designs with such loading are infrequently encountered; one possible case is an unshored composite beam during construction (i.e., when the steel beam alone resists all the construction loads). Thus, in routine gravity-load design, only the load combination $1.2D + 1.6L'$ need be considered to determine the required strength.

When wind loads or seismic loads are more critical than gravity loads, a similar approach can be taken to simplify the number of different load combinations that must be addressed, if desired.

---

[2]Exception: Per ASCE 7, the load factor on L in combinations 3, 4 and 5 shall equal 1.0 for garages, areas occupied as places of public assembly, and all areas where the live load is greater than 100 psf.

[3]Based upon a 50-year recurrence.

Nominal Strengths, Resistance Factors and Design Strengths

Per LRFD Specification Section A5.2, the design strength (ϕP_n, ϕM_n, ϕV_n, etc.) must equal or exceed the required strength (see the preceding section). The design strength is calculated as the product of the resistance factor ϕ and the nominal strength (P_n, M_n, V_n, etc.), per LRFD Specification Section A5.3. The resistance factors ϕ account for unavoidable variations in materials, design equations, fabrication and erection.

The resistance factors ϕ in the LRFD Specification are based upon research (Galambos et al., 1978) and the experience and judgment of the AISC Committee on Specifications. In general, ϕ is less than unity. The higher the variability in the test data for a given nominal strength, the lower its ϕ factor will be. Several representative ϕ factors for steel members are as follows:

$\phi = 0.90$ for limit states involving yielding
$\phi = 0.75$ for limit states involving rupture
$\phi = 0.85$ for limit states involving compression buckling

Serviceability

Serviceability requirements, per LRFD Specification Sections A5.2, A5.4 and L3, should be appropriate for the application. As indicated in LRFD Commentary Section A5.4, serviceability limit states are generally checked at the service-load level for an appropriate load or load combination based upon engineering judgment.

Some guidance is available in LRFD Commentary Section L, both in general and for specific criteria, including deflection, floor vibrations, drift, expansion/contraction, connection slip, corrosion and applications that require special cambering considerations. Additionally, the applicable building code may provide some further guidance or establish requirements. See also the serviceability discussions in Parts 3 through 6, AISC Design Guide No. 3 *Serviceability Design Considerations for Low-Rise Buildings* (Fisher and West, 1990) and AISC Design Guide No. 11 *Floor Vibrations Due to Human Activity* (Murray et al., 1997).

DESIGN BASIS

Frame Analysis, Stability and Leaning Columns

Per LRFD Specification Section A5.1, design by either elastic or plastic analysis is permitted. If design is by plastic analysis, steels with F_y greater than 65 ksi are not permitted and the special requirements in LRFD Specification Sections A5.1, B5.2, C1.1, C2.1a, C2.2a, E1.2, F1.3, H1 and I1 must be met.

In either case, second-order effects—the effects of the loads on the deflected shape of the structure and its elements—must be considered when significant, per LRFD Specification Section B4. When design is by elastic analysis, second-order effects are addressed as required in LRFD Specification Section C1.2. That is, a second-order elastic analysis or the B_1/B_2 approximate method is used. When design is by plastic analysis, second-order effects are addressed as required in LRFD Specification Section C1.1, using a second-order plastic analysis.

Per LRFD Specification Section B4, general stability is required for the structure as a whole and for each of its elements. The stability of braced frames is addressed in LRFD Specification Section C2.1. The stability of unbraced frames, including requirements regarding the destabilizing effects of leaning columns, is addressed in LRFD Specification Section C2.2.

For further information on analysis, stability, and leaning columns, see Geschwindner (2000), Salmon and Johnson (1996), Yura (1971), Lim and McNamara (1972), and LeMessurier (1977).

Stability Bracing

Per LRFD Specification Section B6, beams, girders and trusses must be restrained against rotation about their longitudinal axes at points of support. Additionally, per LRFD Specification Section C3, stability bracing must be provided consistent with that assumed in the analysis for frames (Section C3.2), columns (Section C3.3) and beams (Section C3.4). These provisions provide detailed means by which to assess the strength and stiffness required at braced points. Some guidance for special cases follows:

Approximate Method

As an alternative to the more rigorous provisions for stability bracing in LRFD Specification Section C3, the historically approximate and generally conservative procedure of designing the bracing element for a required strength equal to two percent of the factored compressive force in the braced member will normally suffice.

Simple-Span Beams

In general, adequate lateral bracing is provided to the compression flange of a simple-span beam by the connections of infill beams, joists, concrete slabs, metal deck, concrete slabs on metal deck and similar framing elements.

Beam Ends Supported on Bearing Plates

The stability of a beam end supported on a bearing plate can be provided in one of several ways (see Figure 2–1):

1. The beam end can be built into solid concrete or masonry using anchorage devices.
2. The beam top flange can be stabilized through interconnection with a floor or roof system, provided that system is itself anchored to prevent its translation relative to the beam bearing.
3. A top-flange stability connection can be provided.
4. An end-plate or transverse stiffeners located over the bearing plate extending to near the top-flange k-distance can be provided. Such stiffeners must be welded to the top of the bottom flange and to the beam web, but need not extend to or be welded to the top flange.

In each case, the beam and bearing plate must also be anchored to the support. For the design of beam bearing plates, see Part 14.

In atypical framing situations, such as when very deep beams are used, the strength and stiffness requirements in LRFD Specification Section C3 can be applied to ensure the stability of the assembly. It may also be possible to demonstrate in a limited number of cases, such as with beams with thick webs and relatively shallow depths, that the beam has been properly designed without providing the details described above. In this case, the beam and bearing plate must still be anchored to the support. In any case, it should be noted that the assembly must also meet the requirements in LRFD Specification Section K1.

Beams and Girders Framing Continuously Over Columns

Roof framing is commonly configured with cantilevered beams that frame continuously over the tops of columns to support drop-in beams between the cantilevered segments (Rongoe, 1996; CISC, 1989). It is also commonly desirable to provide an assembly in which the intersection of the beam and column can be considered a braced point for the design of both the continuous cantilevering beam and the column top. The required stability can be provided in several ways (see Figure 2–2):

1. When an infill beam frames into the continuous beam at the column top, the required stability normally can be provided by using connection element(s) for the infill beam that

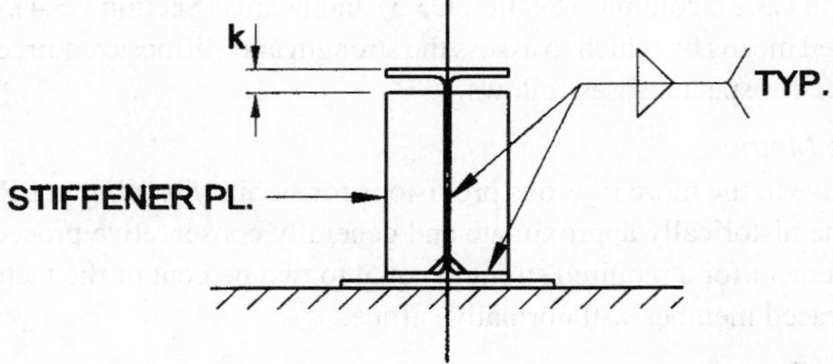

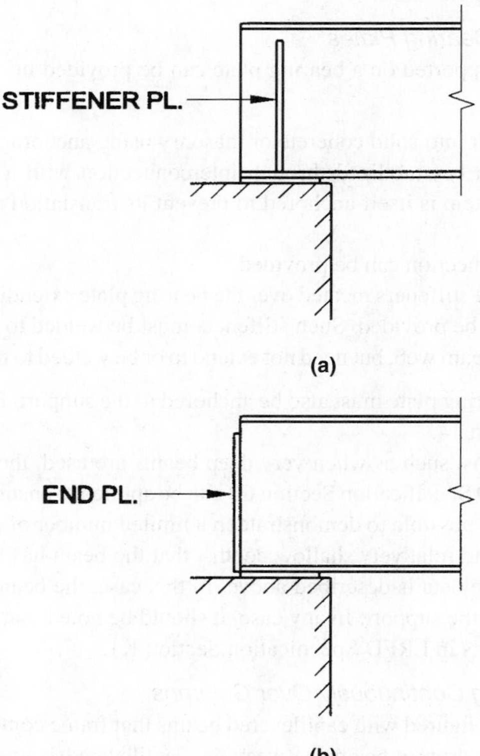

Fig. 2–1. Beam end supported on bearing plate. (a) Stability provided with transverse stiffeners. (b) Stability provided with an end plate.

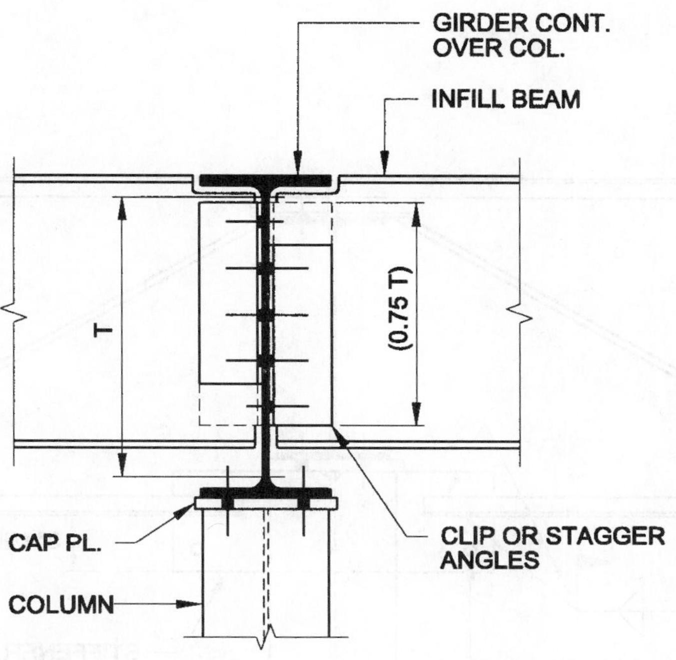

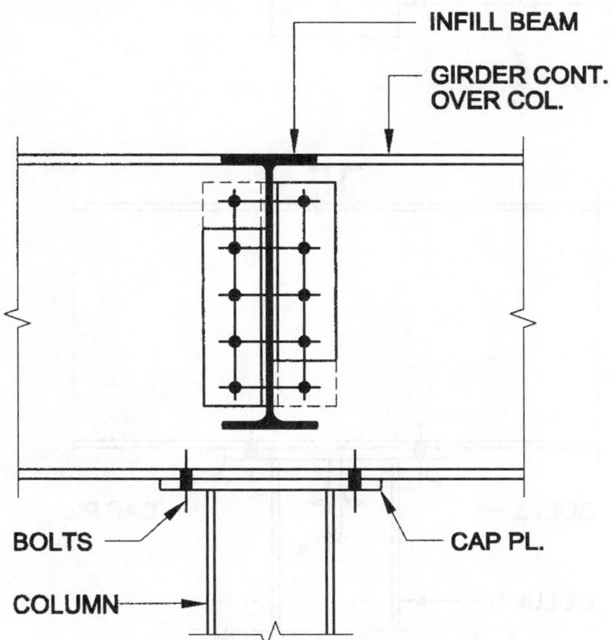

Fig. 2–2a. Beam framing continuously over column top, stability provided with connections of infill beams.

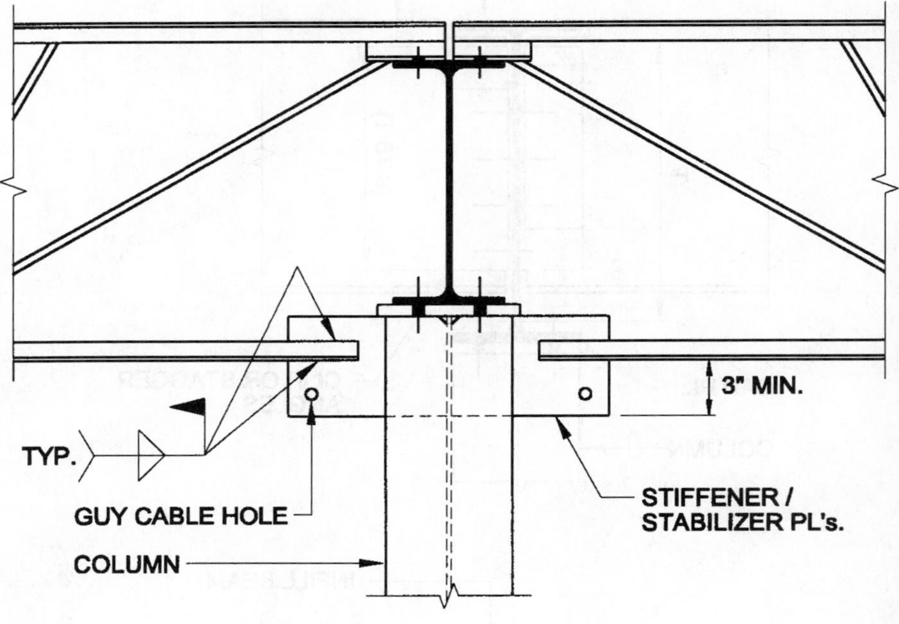

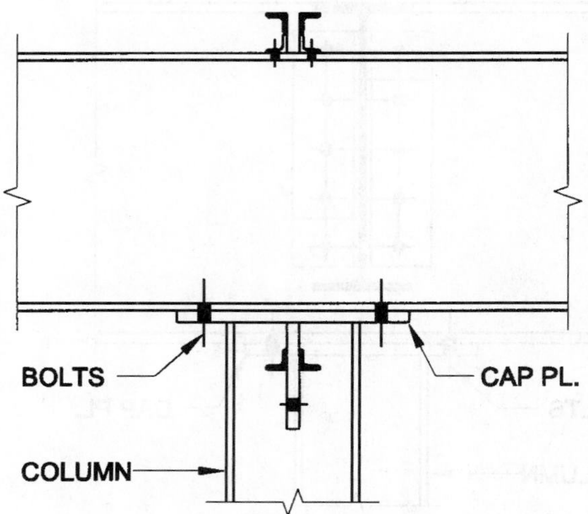

Fig. 2–2b. Beam framing continuously over column top, stability provided with welded joist-chord extensions at column top.

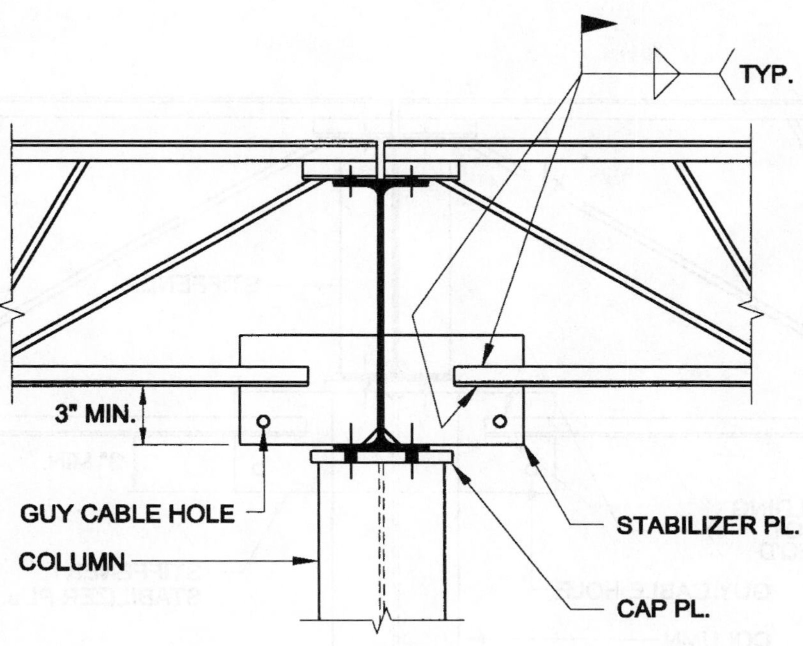

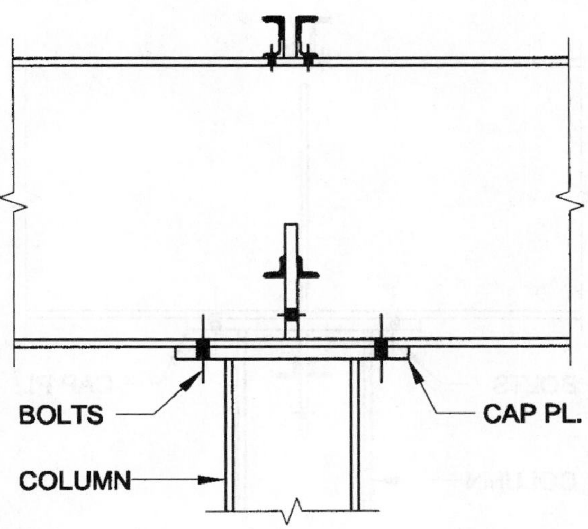

Fig. 2–2c. Beam framing continuously over column top, stability provided with welded joist-chord extensions above column top.

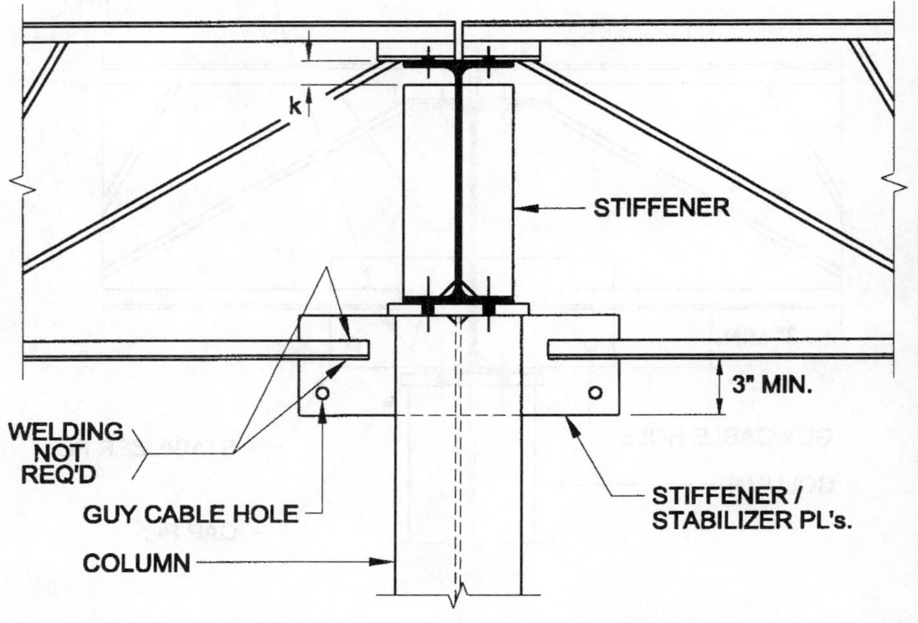

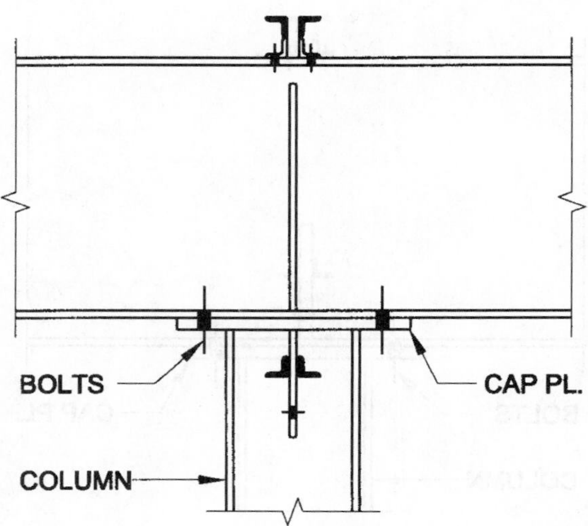

Fig. 2–2d. Beam framing continuously over column top, stability provided with transverse stiffeners, joist-chord extensions located at column top not welded.

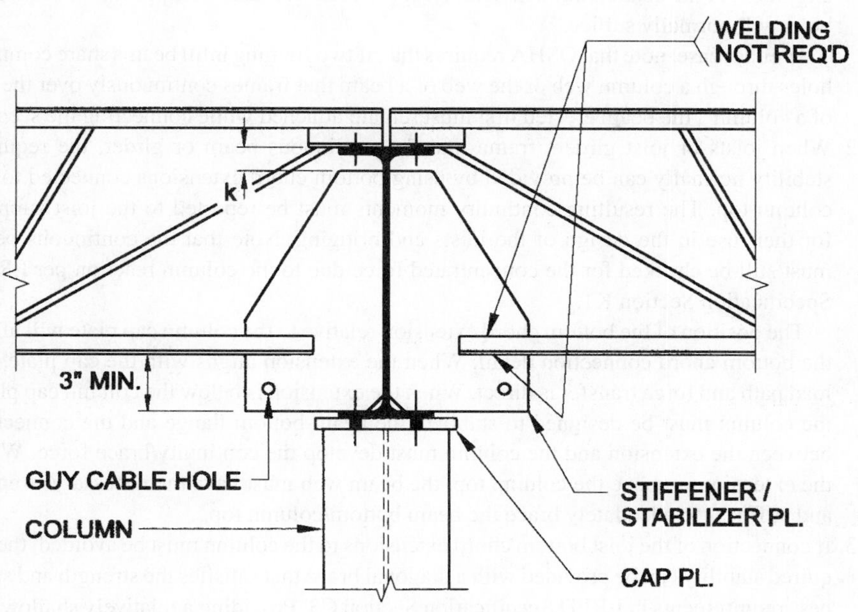

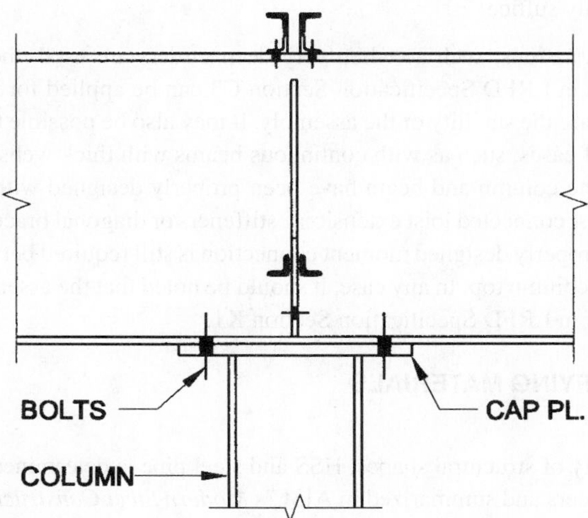

Fig. 2–2e. Beam framing continuously over column top, stability provided with stiffener plates, joist-chord extensions located above column top not welded.

cover three-quarters or more of the T-dimension of the continuous beam. Alternatively, connection elements that cover less than three-quarters of the T-dimension of the continuous beam can be used in conjunction with partial-depth stiffeners in the beam web along with a moment connection between the column top and beam bottom to maintain alignment of the beam/column assembly. A cap plate of reasonable proportions and four bolts will normally suffice.

In either case, note that OSHA requires that, if two framing infill beams share common holes through a column web or the web of a beam that frames continuously over the top of a column[4], the beam erected first must remain attached while connecting the second.

2. When joists or joist girders frame into the continuous beam or girder, the required stability normally can be provided by using bottom chord extensions connected to the column top. The resulting continuity moments must be reported to the joist supplier for their use in the design of the joists and bridging. Note that the continuous beam must still be checked for the concentrated force due to the column reaction per LRFD Specification Section K1.

   The position of the bottom chord extension relative to the column cap plate will affect the bottom chord connection detail. When the extension aligns with the cap plate, the load path and force transfer is direct. When the extension is below the column cap plate, the column must be designed to stabilize the beam bottom flange and the connection between the extension and the column must develop the continuity/brace force. When the extension is above the column top, the beam web must have the necessary strength and stiffness to adequately brace the beam bottom/column top.

3. If connection of the joist bottom chord extensions to the column must be avoided, the required stability can be provided with a diagonal brace that satisfies the strength and stiffness requirements in LRFD Specification Section C3. Providing a relatively shallow angle with respect to the horizontal can minimize gravity-load effects in the diagonal brace.

   Alternatively, the required stability can be provided with stiffeners in the beam web along with a moment connection between the column top and beam bottom to maintain alignment of the beam/column assembly. A cap plate of reasonable proportions and four bolts will normally suffice.

In atypical framing situations, such as when very deep girders are used, the strength and stiffness requirements in LRFD Specification Section C3 can be applied for both the beam and the column to ensure the stability of the assembly. It may also be possible to demonstrate in a limited number of cases, such as with continuous beams with thick webs and relatively shallow depths, that the column and beam have been properly designed without providing infill beam connections, connected joist extensions, stiffeners or diagonal braces as described above. In this case, a properly designed moment connection is still required between the beam bottom flange and the column top. In any case, it should be noted that the assembly must also meet the requirements in LRFD Specification Section K1.

PROPERLY SPECIFYING MATERIALS

Availability

The general availability of structural shapes, HSS and steel pipe is determined by an annual AISC survey of producers and summarized in AISC's *Modern Steel Construction* magazine. The availability summary for W-, M-, S- and HP-shapes, channels and angles is published in

[4]This requirement applies only at the location of the column, not at locations away from the column.

the January issue. The availability summary for HSS and steel pipe is published in the July issue. This information is also available at www.aisc.org/modernsteel.

Structural Shapes

W-Shapes

As shown in Table 2-1, the preferred material specification for W-shapes is ASTM A992 ($F_y = 50$ ksi and $F_u = 65$ ksi). The availability of W-shapes in grades other than ASTM A992 should be confirmed prior to their specification. Regardless of the material specification chosen, the cross-sectional dimensions and production tolerances for W-shapes are given in ASTM A6, a summary of which can be found in Part 1.

W-shapes with higher yield and tensile strength can be obtained by specifying:

1. ASTM A572 grade 60 or 65, which cover tensile group 1, 2 and 3 W-shapes only (see also Table 2-4); or
2. ASTM A913 grades 60, 65 or 70.

W-shapes with atmospheric corrosion resistance (weathering) characteristics can be obtained by specifying:

1. ASTM A588 grade 50; or
2. ASTM A242 grade 42, which covers tensile group 4 and 5 shapes only, A242 grade 46, which covers tensile group 3 shapes only, or A242 grade 50, which covers tensile group 1 and 2 shapes only (see also Table 2-4).

Other material specifications applicable to W-shapes include:

1. ASTM A36;
2. ASTM A529 grade 50 and 55, which cover tensile groups 1 and 2 W-shapes only (see also Table 2-4);
3. ASTM A572 grade 42 and 50; and,
4. ASTM A913 grade 50.

M-Shapes

As shown in Table 2-1, the preferred material specification for M-shapes is ASTM A36 ($F_y = 36$ ksi; $F_u = 58$ ksi), although ASTM A572 grade 50 ($F_y = 50$ ksi; $F_u = 65$ ksi) is increasingly very common. The availability of M-shapes in grades other than ASTM A36 should be confirmed prior to their specification. Regardless of the material specification chosen, the cross-sectional dimensions and production tolerances for M-shapes are given in ASTM A6, a summary of which can be found in Part 1.

M-shapes with higher yield and tensile strength can be obtained by specifying:

1. ASTM A572 grade 42, 50, 55, 60 and 65;
2. ASTM A529 grade 50 and 55; or,
3. ASTM A913 grades 50, 60, 65 or 70.

M-shapes with atmospheric corrosion resistance (weathering) characteristics can be obtained by specifying:

1. ASTM A588 grade 50; or,
2. ASTM A242 grade 50.

S-Shapes

The preceding comments for M-shapes apply equally to S-shapes.

HP-Shapes

The preceding comments for M-shapes apply equally to HP-shapes, except that:

1. A529 grade 50 and 55 and ASTM A242 grade 50 are applicable to tensile group 2 HP-shapes only (see also Table 2-4).
2. Tensile group 3 HP-shapes with atmospheric corrosion resistance (weathering) characteristics can also be obtained by specifying ASTM A242 grade 46.

Channels

The preceding comments for M-shapes apply equally to channels.

Angles

As shown in Table 2-1, the preferred material specification for angles is ASTM A36 (F_y = 36 ksi; F_u = 58 ksi). The availability of angles in grades other than ASTM A36 should be confirmed prior to their specification. Regardless of the material specification chosen, the cross-sectional dimensions and production tolerances for angles are given in ASTM A6, a summary of which can be found in Part 1.

Angles with higher yield and tensile strength can be obtained by specifying:

1. ASTM A572 grade 42, 50, 55, 60 or 65;
2. ASTM A529 grade 50 and 55, which cover tensile groups 1 and 2 angles only (see also Table 2-4); or,
3. ASTM A913 grade 50, 60, 65 or 70.

Angles with atmospheric corrosion resistance (weathering) characteristics can be obtained by specifying

1. ASTM A588 grade 50; or,
2. ASTM A242 grade 46, which covers tensile group 3 angles only, or A242 grade 50, which covers tensile group 1and 2 angles only (see also Table 2-4).

Structural Tees

Structural tees are split from W-, M- and S-shapes to make WT-, MT- and ST-shapes, respectively. For the preferred material specifications, as well as other suitable material specifications, for structural tees, refer to the preceding discussions in the sections on W-, M- or S-shapes, as appropriate.

Rectangular (and Square) HSS

As shown in Table 2-1, the preferred material specification for rectangular HSS is ASTM A500 grade B (F_y = 46 ksi; F_u = 58 ksi), although ASTM A500 grade C (F_y = 50 ksi; F_u = 62 ksi) is increasingly very common. The availability of rectangular HSS in grades other than ASTM A500 grade B should be confirmed prior to their specification. The cross-sectional dimensions and production tolerances for rectangular HSS are also given in the selected material specification, a summary of which can be found in Part 1.

Rectangular HSS with atmospheric corrosion resistance (weathering) characteristics can be obtained by specifying ASTM A847. Other material specifications applicable to rectangular HSS include:

1. ASTM A501; and,
2. ASTM A618.

Round HSS

As shown in Table 2-1, the preferred material specification for round HSS is ASTM A500 grade B ($F_y = 42$ ksi; $F_u = 58$ ksi), although ASTM A500 grade C ($F_y = 46$ ksi; $F_u = 62$ ksi) is increasingly very common. The availability of round HSS in grades other than ASTM A500 grade B should be confirmed prior to their specification. The cross-sectional dimensions and production tolerances for round HSS are also given in the selected material specification, a summary of which can be found in Part 1.

Round HSS with atmospheric corrosion resistance (weathering) characteristics can be obtained by specifying ASTM A847. Other material specifications applicable to round HSS include:

1. ASTM A501; and,
2. ASTM A618.

Steel Pipe

As shown in Table 2-1, the sole material specification for steel pipe is ASTM A53 grade B ($F_y = 35$ ksi; $F_u = 60$ ksi). The cross-sectional dimensions and production tolerances for steel pipe are also given in ASTM A53, a summary of which can be found in Part 1.

Plate Products

Structural Plates

As shown in Table 2-2, the preferred material specification for structural plates is ASTM A36 ($F_y = 36$ ksi for plate thickness is equal to or less than 8 in., $F_y = 32$ ksi otherwise; $F_u = 58$ ksi). The availability and cost effectiveness of structural plates in grades other than ASTM A36 should be confirmed prior to their specification. Note also that the availability of grades other than ASTM A36 varies through the range of thickness as shown in Table 2-2 (see also Table 2-4). Regardless of the material specification chosen, the production tolerances for structural plates are given in ASTM A6, a summary of which can be found in Part 1.

Structural plates with higher yield and tensile strength can be obtained by specifying:

1. ASTM A572 grade 42, 50, 55, 60 or 65;
2. ASTM A529 grade 50 or 55;
3. ASTM A514 grade 90 or 100; or,
4. ASTM A852.

Structural plates with atmospheric corrosion resistance (weathering) characteristics can be obtained by specifying:

1. ASTM A588 grade 42, 46 or 50; or,
2. ASTM A242 grade 42, 46 or 50.

Structural Bars

The preceding comments for structural plates apply equally to structural bars, except that neither ASTM A514 nor A852 is applicable.

Raised-Pattern Floor Plates

ASTM A786 is the standard specification for rolled steel floor plates. As floor-plate design is seldom controlled by strength considerations, ASTM A786 "commercial grade" is commonly specified. If so, per ASTM A786 Section 5.1.2, "the product will be supplied 0.33 percent maximum carbon and without specified mechanical properties." Alternatively, if a defined strength level is desired, ASTM A786 raised-pattern floor plate can be ordered to a defined

Table 2-1.
Applicable ASTM Specifications for Various Structural Shapes

Steel Type	ASTM Designation		F_y Min. Yield Stress (ksi)	F_u Tensile Stress[a] (ksi)	Applicable Shape Series									
					W	M	S	HP	C	MC	L	HSS Rect.	HSS Round	Steel Pipe
Carbon	A36		36	58–80[b]	▒	■	■	■	■	■	■			
	A53 Gr. B		35	60										■
	A500	Gr. B	42	58									■	
			46	58								■	▒	
		Gr. C	46	62									▒	
			50	62								▒		
	A501		36	58								▒	▒	
	A529[c]	Gr. 50	50	65–100	▒	▒	▒	▒	▒	▒	▒			
		Gr. 55	55	70–100	▒	▒	▒	▒	▒	▒	▒			
High-Strength Low-Alloy	A572	Gr. 42	42	60	▒	▒	▒	▒	▒	▒	▒			
		Gr. 50	50	65[d]	▒	▒	▒	▒	▒	▒	▒			
		Gr. 55	55	70	▒	▒	▒	▒	▒	▒	▒			
		Gr. 60[e]	60	75	▒	▒	▒	▒	▒	▒	▒			
		Gr. 65[e]	65	80	▒	▒	▒	▒	▒	▒	▒			
	A618[f]	Gr. I & II	50[g]	70[g]								▒	▒	
		Gr. III	50	65								▒	▒	
	A913	50	50[h]	60[h]	▒	▒	▒	▒	▒	▒	▒			
		60	60	75	▒	▒	▒	▒	▒	▒	▒			
		65	65	80	▒	▒	▒	▒	▒	▒	▒			
		70	70	90	▒	▒	▒	▒	▒	▒	▒			
	A992		50–65[i]	65[i]	■									
Corrosion Resistant High-Strength Low-Alloy	A242		42[j]	63[j]	▒									
			46[k]	67[k]	▒			▒				▒		
			50[l]	70[l]	▒	▒	▒	▒	▒	▒	▒	▒		
	A588		50	70	▒	▒	▒	▒	▒	▒	▒	▒		
	A847[f]		50	70								▒	▒	

■ = Preferred material specification.

▒ = Other applicable material specification, the availability of which should be confirmed prior to specification.

□ = Material specification does not apply.

[a] Minimum unless a range is shown.
[b] For shapes over 426 lb/ft, only the minimum of 58 ksi applies.
[c] Groups 1 and 2 shapes only. To improve weldability a maximum carbon equivalent can be specified (per ASTM Supplementary Requirement S78). If desired, maximum tensile stress of 90 ksi can be specified (per ASTM Supplementary Requirement S79).
[d] If desired, minimum tensile stress of 70 ksi can be specified (per ASTM Supplementary Requirement S81).
[e] Groups 1, 2 and 3 shapes only.
[f] ASTM A618 can also be specified as corrosion-resistant; see ASTM A618.
[g] Minimum applies for walls nominally 3/4-in. thick and under. For wall thicknesses over 3/4 in., $F_y = 46$ ksi and $F_u = 67$ ksi.
[h] If desired, maximum yield stress of 65 ksi and maximum yield-to-tensile strength ratio of 0.85 can be specified (per ASTM Supplementary Requirement S75).
[i] A maximum yield-to-tensile strength ratio of 0.85 and carbon equivalent formula are included as mandatory in ASTM A992.
[j] Groups 4 and 5 shapes only.
[k] Group 3 shapes only.
[l] Groups 1 and 2 shapes only.

Table 2-2. Applicable ASTM Specifications for Plates and Bars

Steel Type	ASTM Designation		F_y Min. Yield Stress (ksi)	F_u Tensile Stress[a] (ksi)	Plates and Bars									
					to 0.75 incl.	over 0.75 to 1.25	over 1.25 to 1.5	over 1.5 to 2 incl.	over 2 to 2.5 incl.	over 2.5 to 4 incl.	over 4 to 5 incl.	over 5 to 6 incl.	over 6 to 8 incl.	over 8
Carbon	A36		32	58–80										
			36	58–80										
	A529	Gr. 50	50	70–100		b	b	b	b					
		Gr. 55	55	70–100		b	b							
High-Strength Low-Alloy	A572	Gr. 42	42	60										
		Gr. 50	50	65										
		Gr. 55	55	70										
		Gr. 60	60	75										
		Gr. 65	65	80										
Corrosion Resistant High-Strength Low-Alloy	A242		42	63										
			46	67										
			50	70										
	A588		42	63										
			46	67										
			50	70										
Quenched and Tempered Alloy	A514[c]		90	100–130										
			100	110–130										
Quenched and Tempered Low-Alloy	A852[c]		70	90–110										

= Preferred material specification.

= Other applicable material specification, the availability of which should be confirmed prior to specification.

= Material specification does not apply.

[a] Minimum unless a range is shown.
[b] Applicable to bars only above 1-in. thickness.
[c] Available as plates only.

Table 2-3. Applicable ASTM Specifications for Various Types of Structural Fasteners

ASTM Designation		F_y Min. Yield Stress (ksi)	F_u Tensile Stress[a] (ksi)	Diameter Range (in.)	High-Strength Bolts	Common Bolts	Nuts	Washers	Direct-Tension-Indicators	Threaded Rods	Anchor Rods		
											Hooked	Headed	Threaded & Nutted
A325		–	105	over 1 to 1.5 incl.	■								
		–	120	0.5 to 1, incl.	■								
A490		–	150	0.5 to 1.5	■								
F1852		–	105	1.125	■								
		–	120	0.5 to 1, incl.	■								
A194 Gr. 2H		–	–	0.25 to 4			▒						
A563		–	–	0.25 to 4			■						
F436[b]		–	–	0.25 to 4				■					
F959		–	–	0.5 to 1.5					■				
A36		36	58–80	to 10						■	▒		▒
A193 Gr. B7		–	100	over 4 to 7						▒	▒		▒
		–	115	over 2.5 to 4						▒	▒		▒
		–	125	2.5 and under						▒	▒		▒
A307	Gr. A	–	60	0.25 to 4		■						▒	
	Gr. C	–	58–80	0.25 to 4						▒	▒		▒
A354 Gr. BD		–	140	2.5 to 4 incl.						▒	▒	▒	▒
		–	150	0.25 to 2.5, incl.						▒	▒	▒	▒
A449		–	90	1.5 to 3 incl.	c					▒	▒	▒	▒
		–	105	1.125 to 1.5, incl.	c					▒	▒	▒	▒
		–	120	0.25 to 1, incl.	c					▒	▒	▒	▒
A572	Gr. 42	42	60	to 6						▒	▒		▒
	Gr. 50	50	65	to 4						▒	▒		▒
	Gr. 55	55	70	to 2						▒	▒		▒
	Gr. 60	60	75	to 1.25						▒	▒		▒
	Gr. 65	65	80	to 1.25						▒	▒		▒
A588		42	63	Over 5 to 8, incl.						▒	▒		▒
		46	67	Over 4 to 5, incl.						▒	▒		▒
		50	70	4 and under						▒	▒		▒
A687		105	150 max.	0.625 to 3						▒	▒		▒
F1554	Gr. 36	36	58–80	0.25 to 4							■	■	■
	Gr. 55	55	75–95	0.25 to 4							▒	▒	▒
	Gr. 105	105	125–150	0.25 to 3							▒	▒	▒

■ = Preferred material specification.

▒ = Other applicable material specification, the availability of which should be confirmed prior to specification.

□ = Material specification does not apply.

– indicates that a value is not specified in the material specification.

[a] Minimum unless a range is shown or maximum (max.) is indicated.

[b] Special washer requirements may apply per RCSC Specification Table 6.1 for some steel-to-steel bolting applications and per Part 14 for anchor-rod applications.

[c] See LRFD Specification Section A3.3 for limitations on use of ASTM A449 bolts.

Table 2-4. Tensile Group Classification of Structural Shapes[a]

Shape		Group 1	Group 2	Group 3	Group 4[b]	Group 5[b]
W-Shapes	W44x	–	–	230 to 335	–	–
	W40x	–	149 to 249	264 to 331	362 to 593	–
	W36x	–	135 to 210	230 to 300	328 to 798	–
	W33x	–	118 to 169	201 to 291	318 to 387	–
	W30x	–	90 to 211	235, 261	292 to 391	–
	W27x	–	84 to 178	194 to 258	281 to 539	–
	W24x	55, 62	68 to 162	176 to 229	250 to 370	–
	W21x	44 to 57	62 to 147	166 to 201	–	–
	W18x	35 to 71	76 to 143	158, 175	–	–
	W16x	26 to 57	67 to 100	–	–	–
	W14x	22 to 53	61 to 132	145 to 211	233 to 550	605 to 808
	W12x	14 to 58	65 to 106	120 to 190	210 to 336	–
	W10x	12 to 45	49 to 112	–	–	–
	W8x	10 to 48	58, 67	–	–	–
	W6x	8.5 to 25	–	–	–	–
	W5x	16, 19	–	–	–	–
	W4x	13	–	–	–	–
M-Shapes		all	–	–	–	–
S-Shapes		to 35 lb/ft incl.	over 35 lb/ft	–	–	–
HP-Shapes		–	to 102 lb/ft incl.	over 102 lb/ft	–	–
American Standard Channels (C)		to 20.7 lb/ft incl.	over 20.7 lb/ft	–	–	–
Miscellaneous Channels (MC)		to 28.5 lb/ft incl.	over 28.5 lb/ft	–	–	–
Angles (L)		to 1/2-in. incl.	over 1/2 in. to 3/4 in. incl.	over 3/4 in.	–	–
Structural Tees (WT, MT, ST)		Structural tees cut from W-, M-, and S-shapes fall into the same group as the structural shapes from which they are cut.				

–indicates that tensile group number does not apply to that shape or shape range.
[a] This table has been adjusted from the similar table in ASTM A6 to include all shapes listed in ASTM A6 Tables A2.1 through A2.8.
[b] Special requirements may apply, per LRFD Specification Section A3.1c.

plate specification, such as ASTM A36, A572, or A588; see ASTM A786 Sections 5.1.2, Section 8 and Appendix Table X1.1.

Sheet and Strip

Sheet and strip products, which are generally thinner than structural plate and bar products (see Table 1-22), are produced to such ASTM specifications as A570, A606 or A607.

Fastening Products

Conventional Bolts

As shown in Table 2-3, the preferred material specification for conventional (heavy-hex) high-strength bolts in steel-to-steel structural connections is ASTM A325, although ASTM A490 can be specified when higher strength is desired. In either case, Type 1 (medium-carbon steel) is most commonly specified. When atmospheric corrosion resistance is desired, Type 3 can be specified.

While still formally permitted in the LRFD Specification, the use of other material specifications in steel-to-steel structural bolting applications has become quite uncommon. ASTM A307 bolts are almost as infrequently specified today as are ASTM A501 and A502 rivets.

Twist-Off-Type Tension-Control Bolt Assemblies

As shown in Table 2-3, the preferred material specification for twist-off-type tension-control bolt assemblies is ASTM F1852, which offers a strength level that is equivalent to that of ASTM A325 bolts. When a higher strength is desired, twist-off-type tension-control bolt assemblies can be obtained in a strength level that is equivalent to that of ASTM A490 bolts using the provisions for alternative-design fasteners in RCSC Specification Section 2.8. In either case, Type 1 (medium-carbon steel) is most commonly specified. When atmospheric corrosion resistance is desired, Type 3 can be specified.

Nuts

As shown in Table 2-3, the preferred material specification for heavy-hex nuts is ASTM A563. The appropriate grade and finish is specified per ASTM A563 Table X1.1 according to the bolt or threaded part with which the nut will be used. For steel-to-steel structural bolting applications, the appropriate grade and finish is summarized in RCSC Specification Section 2.4. If its availability can be confirmed prior to specification, ASTM A194 grade 2H nuts are permitted as an alternative as indicated in RCSC Specification Table 2.1.

Washers

As shown in Table 2-3, the preferred material specification for hardened steel washers is ASTM F436. This specification provides for both flat and beveled washers. While standard ASTM F436 washers are sufficient in most applications, there are several specific applications when special washers are required. The special washer requirements in RCSC Specification Section 6 apply when oversized or slotted holes are used in the outer ply of a steel-to-steel structural joint. In anchor rod and other embedment applications, hole sizes are generally larger than those for steel-to-steel structural bolting applications (see Table 14-2 for maximum anchor-rod hole sizes). Accordingly, washers used in such applications are generally larger and may require design consideration for proper force transfer, particularly when the anchorage is subject to tension. See Table 14-4 for anchor-rod washer sizes.

Compressible-Washer-Type Direct-Tension Indicators

When bolted joints are specified as pretensioned or slip-critical and the direct-tension-indicator pretensioning method is used, ASTM F959 compressible-washer-type direct-tension indicators are specified, as shown in Table 2-3. Type 325 is used with ASTM A325 high-strength bolts and type 490 is used with ASTM A490 high-strength bolts.

Anchor Rods

As shown in Table 2-3, the preferred material specification for anchor rods is ASTM F1554, which covers hooked, headed and threaded and nutted anchor rods in three strength grades: 36, 55 and 105. ASTM F1554 grade 36 is most commonly specified, although grades 55 and 105 are normally available, albeit with potentially longer lead times, when higher strength is required. ASTM F1554 grade 36 or ASTM F1554 grade 55 with weldability supplement S1 and the carbon equivalent formula in ASTM F1554 Section S1.5.2.1 can be specified to allow welded field correction should the anchor rods be placed incorrectly in the field. ASTM F1554 grades 36, 55 and 105 are essentially the anchor-rod equivalents of the generic rod specifications ASTM A36, ASTM A572 grade 55 and A193 grade B7, respectively.

Several other ASTM Specifications can also be used. For applications involving unheaded rods, ASTM A36, A193, A307, A354, A449, A572, A588 and A687 can be specified. For applications involving headed rods, ASTM A307, A354 and A449 can be specified.

Threaded Rods

As shown in Table 2-3, the preferred material specification for threaded rods, whether provided with plain or upset ends, is ASTM A36. Other material specifications that can be specified include ASTM A193, A307, A354, A449, A572, A588, and A687.

Forged Steel Structural Hardware

Forged steel structural hardware products, such as clevises, turnbuckles, eye nuts, and sleeve nuts, are occasionally used in building design and construction. These products are generally provided to AISI material specifications. AISI C-1035 material is commonly used in the manufacture of clevises and turnbuckles. AISI C-1030 material is commonly used in the manufacture of steel eye nuts and steel eye bolts. AISI C-1018 grade 2 material is commonly used in the manufacture of sleeve nuts. Other products, such as steel rod ends, steel yoke ends and pins, cotter pins and coupling nuts are commonly provided generically as "carbon steel."

The dimensional and strength characteristics of these devices are fully described in the literature provided by their manufacturer. Note that such information is usually provided as a safe working load and based upon a factor of safety as high as 5, assuming that the product will be used in rigging or similar applications subject to dynamic loading. If so, the tabular value may be overly conservative for permanent installations and similar applications subject to static loading only. If desired, the published safe working load for statically loaded applications, can be converted to a design strength by multiplying it by its factor of safety FS and the suggested resistance factor $\phi = 0.5$. That is, for design at the factored load level, the design strength is:

$$\phi R_n = 0.5 \times (\text{safe working load}) \times (FS)$$

The resulting design will be essentially equivalent to that for a factor of safety of 3 in ASD.

Filler Metal

The appropriate filler metal for structural steel is as summarized in ANSI/AWS D1.1-2000 Table 3.1 for the various combinations of base metal specification and grade and electrode specification. Weld strengths in this Manual are based upon a tensile strength level of 70 ksi.

Shear-Stud Connectors

As specified in ANSI/AWS D1.1-2000 Chapter 7 (Section 7.2.6 and Table 7.1), Type B shear stud connectors made from ASTM A108 material are used for the interconnection of steel and concrete elements in composite construction ($F_u = 60$ ksi).

Other Products

Steel Castings and Forgings

Steel castings are specified as ASTM A27 grade 65-35 or ASTM A148 grade 80-35. Steel forgings are specified as ASTM A668.

Crane Rails

Crane rails are furnished to ASTM A759, ASTM A1 and/or manufacturer's specifications and tolerances.

Most manufacturers chamfer the top and sides of the crane-rail head at the ends unless specified otherwise to reduce chipping of the running surfaces. Often, crane rails are ordered as end-hardened, which improves the resistance of the crane-rail ends to impact that occurs as the moving wheel contacts it during crane operation. Alternatively, the entire rail can be ordered as heat-treated. When maximum wheel loading or controlled cooling is needed, refer to manufacturers' catalogs. Purchase orders for crane rails should be noted "for crane service."

Light 40-lb rails are available in 30-ft lengths, 60-lb rails in 30-, 33- or 39-ft lengths, standard rails in 33- or 39-ft lengths and crane rails up to 80 ft. Consult manufacturer for availability of other lengths. Rails should be arranged so that joints on opposite sides of the crane runway will be staggered with respect to each other and with due consideration to the wheelbase of the crane. Rail joints should not occur at crane girder splices. Odd lengths that must be included to complete a run or obtain the necessary stagger, should be not less than 10-ft long. Rails are furnished with standard drilling in both standard and odd lengths unless stipulated otherwise on the order.

CONTRACT DOCUMENT INFORMATION

Design Drawings, Specifications and Other Contract Documents

Most provisions in the LRFD Specification, RCSC Specification, AWS D1.1 and the Code of Standard Practice are mandatory. Some provisions require the communication of information in the contract documents, some provisions are invoked only when specified in the contract documents, and some provisions require the approval of the owner's designated representative for design if they are to be used. Following is a summary of these provisions in the LRFD Specification, RCSC Specification and Code of Standard Practice:

Required Information

The following communication of information is required in the contract documents:

1. The type of construction, per LRFD Specification Section A2 (braced frame, fully restrained (FR) moment frame, partially restrained (PR) moment frame, simple shear framing, etc.).
2. Required drawing information, per LRFD Specification Section A7, RCSC Specification Section 1.4 (bolting products and joint type) and Code of Standard Practice Sections 3.1 and 3.1.1 through 3.1.6.
3. Drawing numbers and revision numbers, per Code of Standard Practice Section 3.5.
4. Structural system description, per Code of Standard Practice Section 7.10.1.
5. Installation schedule for non-structural steel elements in the structural system, per Code of Standard Practice Section 7.10.2.
6. Project schedule, per Code of Standard Practice Section 9.5.1.

Information Required Only When Specified

The following provisions are invoked only when specified in the contract documents:

1. Special material notch-toughness requirements, per LRFD Specification Section A3.1c.
2. Special connections requiring pretension, per LRFD Specification Section J1.11.
3. Bolted joint requirements, per LRFD Specification Section J3.1 and RCSC Specification Commentary Section 1.4.
4. Special cambering considerations, per LRFD Specification Section L1.
5. Special contours and finishing requirements for thermal cutting, per LRFD Specification Sections M2.2 and M2.3, respectively.
6. Corrosion protection requirements, if any, per LRFD Specification Sections M3.1, M3.2 and M3.5 and Code of Standard Practice Sections 6.5, 6.5.2 and 6.5.3.

7. Responsibility for field touch-up painting, if painting is specified, per LRFD Specification Section M4.6 and Code of Standard Practice Section 6.5.4.
8. Special quality assurance and inspection requirements, per LRFD Specification Sections M5 and M5.3 and Code of Standard Practice Sections 8.1.3, 8.2 and 8.3.
9. Evaluation procedures, per LRFD Specification Chapter N.
10. Fatigue requirements, if any, per LRFD Specification Appendix K3.1.
11. Modifications, if any, to the Code of Standard Practice, per Code of Standard Practice Section 1.1.
12. Submittal schedule for shop and erection drawings, per Code of Standard Practice Section 4.2.
13. Mill order timing, special mill testing and special mill tolerances, per Code of Standard Practice Section 5.1.
14. Removal of backing bars and run-off-tabs, per Code of Standard Practice Section 6.3.2.
15. Special erection mark requirements, per Code of Standard Practice Section 6.6.1.
16. Special delivery and erection sequences, per Code of Standard Practice Sections 6.7.1 and 7.1, respectively.
17. Special field splice requirements, per Code of Standard Practice Section 6.7.4.
18. Special loads to be considered during erection, per Code of Standard Practice Section 7.10.3.
19. Special safety protection treatments, per Code of Standard Practice Section 7.11.1.
20. Identification of adjustable items, per Code of Standard Practice Section 7.13.1.3.
21. Cuts, alterations and holes for other trades, per Code of Standard Practice Section 7.15.
22. Revisions to the contract, per Code of Standard Practice Section 9.3.
23. Special terms of payment, per Code of Standard Practice Section 9.6.
24. Identification of architecturally exposed structural steel, per Code of Standard Practice Section 10.

Approvals Required

The following provisions require the approval of the owner's designated representative for design if they are to be used:

1. Bolted-joint-related approvals per RCSC Commentary Section 1.4.
2. Use of electronic or other copies of the design drawings by the fabricator, per Code of Standard Practice Section 4.3.
3. Use of stock materials not conforming to specified ASTM specification, per Code of Standard Practice Section 5.2.3.
4. Correction of errors, per Code of Standard Practice Section 7.14.
5. Inspector-recommended deviations from contract documents, per Code of Standard Practice Section 8.5.6.
6. Contract price adjustment, per Code of Standard Practice Section 9.4.2.

Establishing Criteria for Connections

Code of Standard Practice Section 3.1.2 provides two methods for the establishment of connection criteria.

In the first, the complete design of all connections is shown in the structural design drawings. In this case, Code of Standard Practice Commentary Section 3.1.2 provides a summary of the information that must be included in the structural design drawings.

This method has the advantage that there is no need to provide connection loads, since the connections are completely designed in the structural design drawings. Additionally, it favors greater accuracy in the bidding process, since the connections are fully described in the contract documents.

In the second, the fabricator is allowed to select or complete the connections while preparing the shop and erection drawings, using the information provided by the owner's designated representative for design per Code of Standard Practice Section 3.1.2. In this case, Code of Standard Practice Commentary Section 3.1.2 clarifies the intention that connections that can be selected or completed by the fabricator include those for which tables appear in the contract documents or the Manual. Other connections should be shown in detail in the structural design drawings.

This method has the advantage that the fabricator's standard connections normally can be used, which often leads to project economy. However, the loads or other connection design criteria must be provided in the structural design drawings. Factored loads and reactions should be provided in the structural design drawings when LRFD has been used for the design of the frame.

In either method, the resulting shop and erection drawings must be submitted to the owner's designated representative for design for review and approval. Following is additional guidance for the communication of connection criteria to the connection designer.

Simple Shear Connections

The full force envelope should be given for each simple shear connection. Because of the potential for overestimation—and underestimation—inherent in approximate methods, it is recommended that the actual beam end reactions be indicated on the design drawings. The most effective method to communicate this information is to place a numeric value at each end of each span in the framing plans.

Although not recommended, beam end reactions can be approximated as a percentage of the tabulated uniform load from the beam uniform load tables in Manual Part 5. For example, the end reaction for a non-composite beam can often be specified as one-half the tabulated uniform load. It should be noted, however, that there are several situations in which this approach is not appropriate. For example:

1. When beams are selected for serviceability considerations or for shape repetition, the uniform load tables will often result in heavier connections than would be required by the actual design loads.
2. When beams have relatively short spans, the uniform load tables will often result in heavier connections than would be required by the actual design loads.
3. When beams support other framing beams or other concentrated loads occur on girders supporting beams, the end reactions can be higher than 50 percent of the total uniform load.
4. For composite beams, the end reactions can be higher than 50 percent of the total uniform load. The percentage requirement can be increased for this condition, but the resulting approach is still subject to the above considerations.

Moment Connections

The full force envelope should be given for each moment connection. If the owner's designated representative for design can select the governing load combination, its effect alone should be provided. Otherwise, the effects of all appropriate load combinations should be indicated. Additionally, the maximum moment imbalance should also be given for use in the check of panel-zone web shear.

Because of the potential for overestimation—and underestimation—inherent in approximate methods, it is recommended that the actual beam end reactions (moment, shear and other reactions, if any) be indicated in the structural design drawings. The most effective method to do so may be by tabulation for each joint and load combination.

Although not recommended, beam end reactions can be specified by more general criteria, such as by function of the beam strength. It should be noted, however, that there are several situations in which this approach is not appropriate. For example:

1. When beams are selected for serviceability considerations or for shape repetition, this approach will often result in heavier connections than would be required by the actual design loads.
2. When the column(s) or other members that frame at the joint could not resist the forces and moments determined from the criteria so specified, this approach will often result in heavier connections than would be required by the actual design loads.

In some cases, the structural analysis may require that the actual connections be configured to match the assumptions used in the model. For example, it may be appropriate to release weak-axis moments in a beam-column joint where only strong-axis beam moment strength is required. Such requirements should be indicated in the structural design drawings.

Truss Connections

The full force envelope should be given for each truss-member end connection. If the owner's designated representative for design can select the governing load combination for the entire truss, its effect alone should be provided. Otherwise, the effects of all appropriate load combinations should be indicated in tabular form. This approach will allow a clear understanding of all of the forces on any given joint.

Because of the potential for overestimation—and underestimation—inherent in approximate methods, it is recommended that the actual reactions at the truss member end (axial force and other reactions, if any) be indicated in the structural design drawings. It is also recommended that transfer forces, if any, be so indicated. The most effective method to do so may be by tabulation for each truss member end and load combination.

Although not recommended, truss member end reactions can be specified by more general criteria, such as by maximum member forces (tension or compression) or as a function of the member strength. It should be noted, however, that there are several situations in which such approaches are not appropriate. For example:

1. The specification of maximum member forces does not permit a check of the member forces at a joint if there are different load combinations governing the member designs at that joint. Nor does it reflect the possibility of load reversal as it may influence the design.
2. The specification of a percentage of member strength may not properly account for the interaction of forces at a joint or the transfer force through the joint. Additionally, it may not allow for a cross-check of all forces at a joint.

In either case, this approach will often result in heavier connections than would be required by the actual design loads.

Note that it is not necessary to specify a minimum connection strength as a percent of the member strength as a default. However, when trusses are shop assembled or field assembled on the ground for subsequent erection, consideration should be given to the loads that will be induced during handling, shipping and erection.

Horizontal and Vertical Bracing Connections

The recommendations for truss connections above also apply in general to bracing connections with the following additional comments.

Bracing connections may involve the interaction of gravity and lateral loads on the frame. In some cases, such as V- and inverted V-bracing (also known as Chevron bracing), gravity loads alone may govern design of the braces and their connections. Thus, clarity in the specification of loads and reactions is critical to properly consider the potential interaction of gravity and lateral loads at floors and roofs.

Strut and Tie Connections

Floor and roof members in braced bays and adjacent bays may function as struts or ties in addition to carrying gravity loads. Therefore the recommendations for simple shear connections and bracing connections above apply in combination.

TOLERANCES

The effects of mill, fabrication and erection tolerances all require consideration in the design and construction of structural steel buildings. However, the accumulation of the mill tolerances and fabrication tolerances shall not cause the erection tolerances to be exceeded, per Code of Standard Practice Section 7.12.

Mill Tolerances

Mill tolerances are those variations that could be present in the product as-delivered from the rolling mill. These tolerances are given as follows:

1. For structural shapes and plates, see ASTM A6.
2. For HSS, see ASTM A500 (or other applicable ASTM specification for HSS).
3. For steel pipe, see ASTM A53.

A summary of standard mill practices is also given in Part 1.

Fabrication Tolerances

Fabrication tolerances are generally provided in LRFD Specification Section M2 and Code of Standard Practice Section 6.4. Additional requirements that govern fabrication are as follows:

1. Compression joint fit-up, per LRFD Specification Section M4.4.
2. Roughness limits for finished surfaces, per Code of Standard Practice Section 6.2.2.
3. Straightness of projecting elements of connection materials, per Code of Standard Practice Section 6.3.1.
4. Finishing requirements at locations of removal of run-off tabs and similar devices, per Code of Standard Practice Section 6.3.2.

Erection Tolerances

Erection tolerances are generally provided in LRFD Specification Section M4 and Code of Standard Practice Section 7.13. Note that the tolerances specified therein are predicated upon the proper installation of the following items by the owner's designated representative for construction:

1. Building lines and benchmarks, per Code of Standard Practice Section 7.4.
2. Anchorage devices, per Code of Standard Practice Section 7.5.
3. Bearing devices, per Code of Standard Practice Section 7.6.
4. Grout, per Code of Standard Practice Section 7.7.

Building Façade Tolerances

The preceding mill, fabrication and erection tolerances can be maintained with standard equipment and workmanship. However, the accumulated tolerances for the structural steel and the building façade must be accounted for in the design so that the two systems can be properly mated in the field. This is normally accomplished by specifying adjustable connections in the contract documents, per Code of Standard Practice Section 7.13.1.3.

The required adjustability normally can be determined from the building façade tolerances and the accumulation of mill, fabrication and erection tolerances at the mid-span point of the spandrel beam. The actual locations of the anchor-rod group and column base, the actual slope of the columns and the actual sweep of the spandrel beam all affect the accumulation of tolerances in the structural steel at this critical location. Even if each of these is properly within the permitted envelope, significant variations will normally occur.

Figures 2–3a, 2–4a and 2–5a illustrate details that are not recommended because they do not provide for adjustment. Figures 2–3b, 2–4b and 2–5b illustrate recommended alternative details that do provide for adjustability. Note that diagonal structural and stability bracing elements have been omitted in these details to improve the clarity of presentation regarding adjustability. Also, note that all elements beyond the slab edge are normally not structural steel, per Code of Standard Practice Section 2.2, and are shown for the purposes of illustration only.

The bolted details in Figures 2–4b and 2–5b can be used to provide field adjustability with slotted holes as shown. Further adjustability can be provided in these details, if necessary, by removing the bolts and clamping the connection elements for field welding. Alternatively, when the slab edge angle or plate in Figure 2–4b is shown as field welded and identified as adjustable in the contract documents, it can be provided to within a horizontal tolerance of $\pm^3/_8$ in., per Code of Standard Practice Section 7.13.1.3. However, if the item were not shown as field welded and identified as adjustable in the contract documents, it would likely be attached in the shop or attached in the field to facilitate the concrete pour and not be suitable to provide for the necessary adjustment.

With adjustable connections specified in design and provided in fabrication, the actions taken on the job site will allow for a successful façade installation. Per the Code of Standard Practice definition of established column line (see Code of Standard Practice Glossary), proper placement of this line by the owner's designated representative for construction based upon the actual anchor-rod/column-center locations will assure that all subcontractors are working from the same information. When sufficient adjustment cannot be accommodated within the adjustable connections provided, a common solution is to allow the building façade to deviate (or drift) from the theoretical location to follow the as-built locations of the structural steel framing and concrete floor slabs. A survey of the as-built locations of these elements can be used to adjust the placement of the building façade accordingly. In this case, the adjustable connections can serve to ensure that no abrupt changes occur in the façade.

CAMBER, SWEEP AND STRAIGHTENING

Beam Camber and Sweep

Camber denotes a curve in the vertical plane. Sweep denotes a curve in the horizontal plane. Camber and sweep are provided in beams, when required, by the fabricator per Code of Standard Practice Section 6.4.4, either by cold bending or by hot bending.

Cambering and sweeping induces residual stresses similar to those that develop in rolled structural shapes as elements of the shape cool from the rolling temperature at different rates.

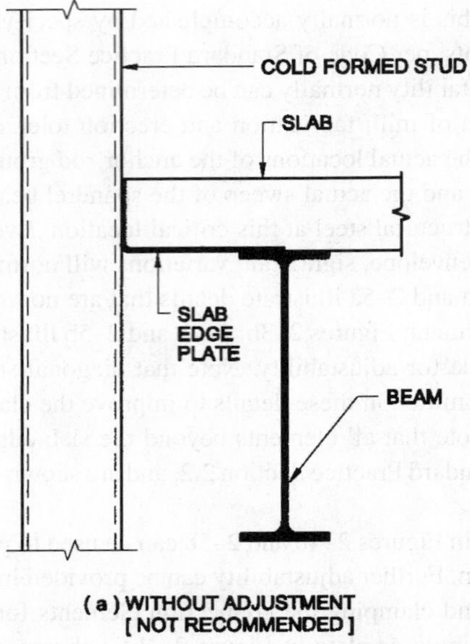

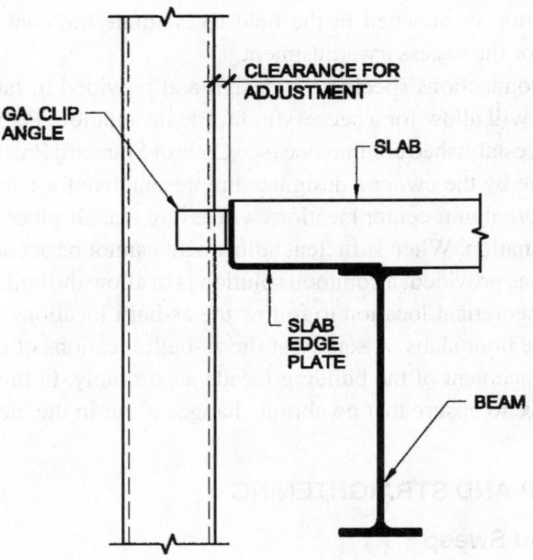

Fig. 2–3. Attaching cold-formed steel façade systems to structural steel framing.

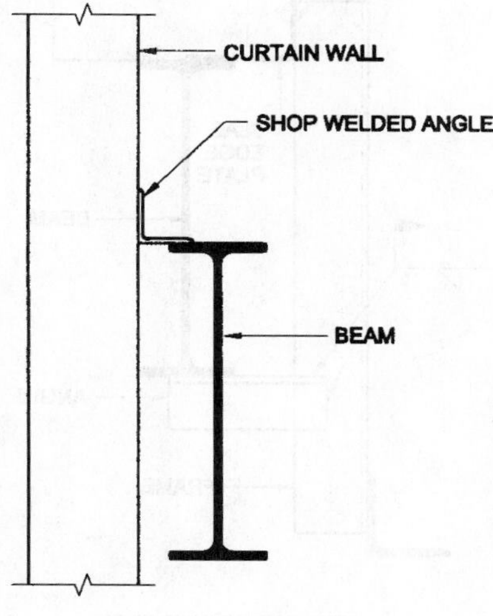

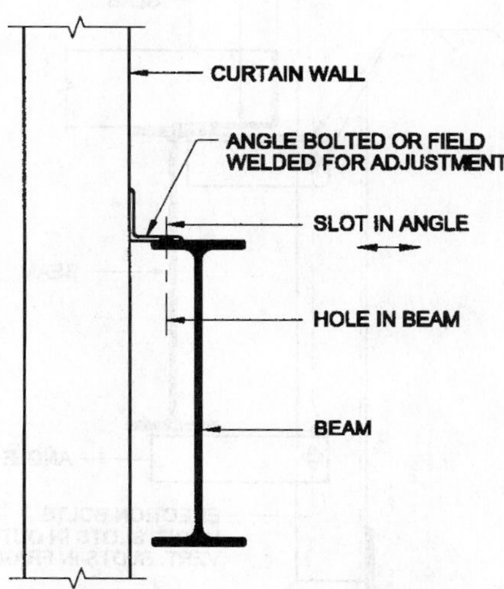

Fig. 2–4. Attaching curtain-wall façade systems to structural steel framing.

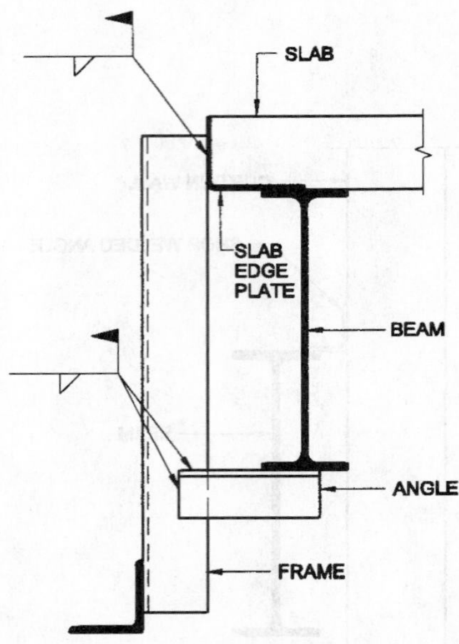

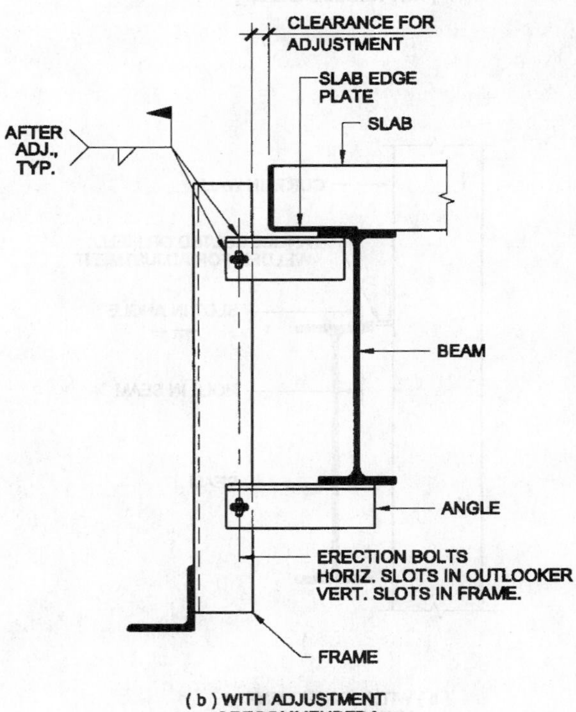

Fig. 2–5. Attaching masonry façade systems to structural steel framing.

In general, these residual stresses do not affect the design strength of structural members, since the effect of residual stresses is considered in the provisions of the LRFD Specification.

Cold Bending

The inelastic deformations required in common cold bending operations, such as for beam cambering, normally fall well short of the strain-hardening range. Specific limitations on cold-bending capabilities should be obtained from those that provide the service. However, the following general guidelines may be useful in the absence of other information:

1. The minimum radius for camber induced by cold bending in members up to a nominal depth of 30 in. is between 10 and 14 times the depth of the member. Deeper members may require a larger minimum radius.
2. Cold bending may be used to provide sweep in members to practically any radius desired.
3. A length limit of 40 to 50 ft is practical.

When curvatures and the resulting inelastic deformations are significant and corrective measures are required, the effects of cold work on the strength and ductility of the structural steels largely can be eliminated by thermal stress relief or annealing.

Hot Bending

The controlled application of heat can be used in the shop and field to provide camber or sweep. The member is rapidly heated in selected areas that tend to expand, but are restrained by the adjacent cooler areas, causing inelastic deformations in the heated areas and a change in the shape of the cooled member.

The mechanical properties of steels are largely unaffected by such heating operations, provided the maximum temperature does not exceed the temperature limitations given in LRFD Specification Section M2.1. Temperature-indicating crayons or other suitable means should be used during the heating process to ensure proper regulation of the temperature.

Heat curving induces residual stresses that are similar to those that develop in hot-rolled structural shapes as they cool from the rolling temperature because all parts of the shape do not cool at the same rate. The residual stresses from heating operations generally do not affect the design strength of structural members, since the effect of residual stresses is considered in the provisions of the LRFD Specification.

Truss Camber

Camber is provided in trusses, when required, by the fabricator per Code of Standard Practice Section 6.4.5, by geometric relocation of panel points and adjustment of member lengths based upon the camber requirements as specified in the contract documents.

Straightening

All structural shapes are straightened at the mill after rolling, either by rotary or gag straightening, to meet the aforementioned mill tolerances. Similar processes and/or the controlled application of heat can be used in the shop or field to straighten a curved or distorted member. These processes are normally applied in a manner similar to those used to induce camber and sweep and described above.

FIRE PROTECTION AND ENGINEERING

Conventional Fire-Resistant Design and Construction

Conventional fire-resistant design and construction is predicated upon the use of rated assemblies that can maintain structural stability for the duration of fire exposure, meet certain maximum temperature criteria and, in some cases, prevent the spread of fire to adjacent spaces. The performance of floor or roof construction in actual buildings when exposed to fire depends upon many factors, including member size, spacing, span, connection type, floor-slab details, fire-resistive coatings, roof material and the intensity, duration and location of the fire. Many steel structures have inherent fire resistance through a combination of the above factors and may not require additional protection from the effects of fire.

Rated Assemblies

Fire-resistive ratings for buildings are required to be identified by most model building codes used in the United States. The required ratings on the structural elements typically range from zero to four hours and are usually dependent upon such factors as the building use, occupancy, number of stories, building height, total floor area, area of each floor and building separation. Many low-rise structures are categorized by a type of construction that would require no more than a one-hour fire rating. It should also be noted that an approved sprinkler system is often permitted in lieu of a one-hour fire-rated assembly in most model building codes.

Rated assemblies are qualified through testing in accordance with ASTM E119, which specifies a standard fire for evaluating the relative fire resistance of the specific assembly tested. In effect, ASTM E119 is used to evaluate the length of time that an assembly is within certain testing limits when exposed to the standard fire. Listings of fire-resistant designs are available from a number of sources (Underwriters Laboratories, 2000; Gypsum Association, 2000). Although these listings are prescriptive in nature and deviations can be difficult to justify to the satisfaction of code officials and listing agencies, the basic heat transfer and structural principles are well defined in steel building design and construction. As a result, relatively simple analytical techniques have been developed (AISI, 1984; AISI, 1981; AISI, 1980) that enable the use a variety of different structural steel shapes in conjunction with tested assemblies. North American building code authorities specifically recognizes these analytical techniques.

Underwriters Laboratories (UL) Assemblies

A summary of UL assemblies that are commonly applicable in structural steel building design and construction is provided in Tables 2-5 through 2-9. These tables should be used in conjunction with the criteria and information contained in the UL *Fire Resistance Directory* (Underwriters Laboratories, 2000), which is also available at www.ul.com. However, the inclusion of these assemblies in this Manual should not preclude the use of other UL assemblies or any other rational approach, including the use of fire-resistive intumescent coatings, plaster finishes or mineral wool batts.

The ratings for these assemblies are given for a minimum member size that can be related to other larger member sizes as follows:

1. For W-shapes and similar members, by the ratio of the weight to the heated perimeter (W/D).
2. For HSS and steel pipe, by the ratio of the area to the heated perimeter (A/P).

Certain UL assemblies can also be used with members with smaller W/D and A/P ratios, provided certain criteria as outlined in the specific UL design are met. W/D and A/P ratios are provided in Tables 1-36 through 1-53. These values are calculated using the exact shape profiles, and can be approximated using the equations illustrated in Figure 2–6. Note that the equations for columns and braces are generally different because the heated perimeter of a beam differs from that for a column or brace.

**Table 2-5.
Roof-Ceiling Assemblies[a, b, c]**

Assembly Rating		Type of Protection System	Roof Insulation Type	Metal Deck Depth[d] (in)	U.L. Design Number
Restrained (hr)	Unrestrained (hr)				
1	3/4	Acoustical Ceiling Membrane	Rigid	1 1/2	P254
1	1		Rigid	1 1/2	P214
			Insulating Fill	9/16, 15/16, 1 5/16	P246, P255
				9/16	P261
1, 1 1/2	1, 1 1/2		Rigid	1, 1 1/2	P250
				1 1/2	P230
				1, 1 1/2, 2, 3	P225
			Insulating Fill	15/16, 1 5/16, 1 1/2	P231
1, 1 1/2, 2	1, 1 1/2, 2		Insulating Fill	9/16, 3/4, 1 1/4	P251
2	2		Rigid	1, 1 1/2	P237
1 1/2, 2	1 1/2, 2	Plaster w/ Metal Lath Membrane	Rigid	1 1/2	P404
1	1	Gypsum Wallboard Ceiling Membrane	Insulating Fill	1 5/16	P509
2	2		Rigid	1 1/2	P514
3/4, 1, 1 1/2, 2	3/4, 1, 1 1/2, 2	Spray-Applied Fire Resistive Material	Rigid	1 1/2, 3	P701
1, 1 1/2, 2	1, 1 1/2, 2		Rigid	1 1/2	P711, P740, P741
				1 1/2, 3	P714, P717, P725, P739, P819
			Insulating Fill	9/16, 15/16, 1 5/16, 1 1/2	P921
1, 1 1/2, 2	1, 1 1/2, 2		Insulating Fill	9/16, 15/16, 1 5/16, 1 1/2	P927
1, 1 1/2, 2, 3	1, 1 1/2, 2		Rigid	1 1/2, 3	P719
1, 1 1/2, 2, 3	1, 1 1/2, 2, 3		Rigid	1 1/2	P723, P733
				1 1/2, 3	P732
2	1 1/2		Rigid	1 1/2	P718

[a] The referenced assemblies are some commonly used Underwriters Laboratories (UL) assemblies used for conventional steel framed structures. For additional assemblies the reader should reference the UL Fire Resistance Directory.
[b] For additional design requirements such as beam spacing, concrete strength, density, reinforcing and clear cover, minimum metal deck gage, maximum deck span, shear connector requirements, design stress limitations, etc. see the specific referenced assembly in the Underwriters Laboratories (UL) directory.
[c] For roof designs that incorporate structural concrete slabs, D–series assemblies can be used provided that the roof insulation type, density and the appropriate D–series assembly modifications are in accordance with the UL directory.
[d] Metal deck depth for some assemblies is shown as a minimum and deeper decks may be substituted. Refer to the specific UL assembly for additional information.

Restrained and Unrestrained Ratings

One important consideration is the degree to which a standard test assembly is representative of actual building construction. Most testing of assemblies is based upon isolated floor or roof specimens in the size range of 15 ft by 18 ft. In contrast, real floor slabs and roof decks are physically, if not structurally, continuous over beams and girders with larger spans and a variety of connection details. In recognition of the practical difficulties associated with

Table 2-6.
Floor-Ceiling Assemblies[a, b]

Assembly Rating		Type of Protection System	Concrete		Metal Deck Depth (in)	U.L. Design Number
Restrained (hr)	Unrestrained (hr)		Min. Thickness above deck flutes (in)	Type		
1, 1½, 2, 3	1, 1½, 2, 3	Acoustical Ceiling Membrane	based upon required rating	NW or LW	1½, 2, 3	D216
2, 3	2, 3			NW	1½	D218
1½, 2	1½, 2	Gypsum Wallboard Ceiling Membrane	2½	NW	1½, 2, 3	D502
2	1½		2	LW	3, 4½, 6, 7½	D501
1, 1½, 2, 3	1, 1½, 2, 3	Spray-Applied Fire Resistive Material	2	NW or LW	2, 3	D743
			2½	NW or LW	9/16, 15/16, 1 5/16	D780
					1½, 2, 3	D759, D832
1, 1½, 2, 3, 4	1, 1½, 2, 3, 4		2½	NW or LW	1½, 2, 3	D739, D767, D779, D858
			3¼	LW	1½, 2, 3	D782
2	1, 1½		2½	LW	1½, 2, 3	D752
2, 3, 4	1, 1½, 2, 3		2½	NW	1½, 1⅝	D744
3, 4	1½, 2		3¼	LW	1½, 2, 3	D754
1, 1½, 2, 3	1, 1½, 2, 3	Spray-Applied Fire Resistive Material w/ Unprotected Deck	based upon required rating	NW or LW	1½, 1⅝, 2, 3	D902
					1½, 2, 3	D916, D925

[a] The referenced assemblies are some commonly used Underwriters Laboratories (UL) assemblies used for conventional steel framed structures. For additional assemblies the reader should reference the UL Fire Resistance Directory.

[b] For additional design requirements such as beam spacing, concrete strength, density, reinforcing and clear cover, minimum metal deck gage, maximum deck span, shear connector requirements, design stress limitations, etc. see the specific referenced assembly in the Underwriters Laboratories (UL) directory.

laboratory-scale testing, many building codes require that the designer assess whether the actual construction is restrained or unrestrained against thermal expansion.

Per ASTM E119, the restrained condition is applicable when the assembly is supported or surrounded by construction that is "capable of resisting substantial thermal expansion throughout the range of anticipated elevated temperatures." Otherwise, the assembly should be considered unrestrained. As shown in ASTM E119 (see Table 2-10), structural steel building construction is generally considered restrained for the purposes of resistance to thermal expansion. Analytical methods (Ioannides and Mehta, 1997) can also be used to quantify the level of thermal restraint in the structure.

Compatibility of Primer and Paint on Spray-Applied Fireproofing

The use of primer and/or paint systems under spray-applied fireproofing is strongly discouraged. Because spray-applied fireproofing has excellent adhesion to unpainted structural steel, even in the presence of rust that is not loose or flaking, mechanical anchorage devices, bonding agents or bond tests are not required to meet UL guidelines.

Table 2-7.
Beam-Only Designs—Roof[a, b]

Assembly Rating		Type of Protection System	Roof Insulation Type	Metal Deck Depth[c] (in)	U.L. Design Number
Restrained (hr)	Unrestrained (hr)				
1, $1\frac{1}{2}$, 2, 3	1, $1\frac{1}{2}$, 2, 3	Spray-Applied Fire Resistive Material	Rigid	$1\frac{1}{2}$	S715, S733
1, $1\frac{1}{2}$, 2, 3, 4	1, $1\frac{1}{2}$, 2, 3, 4		Rigid	$1\frac{1}{2}$	S701, S721, S724, S729, S734, S805
			Rigid or Insulating Fill	$1\frac{1}{2}$	S735

[a] The referenced assemblies are some commonly used Underwriters Laboratories (UL) assemblies used for conventional steel framed structures. For additional assemblies the reader should reference the UL Fire Resistance Directory.
[b] For additional design requirements such as beam spacing, concrete strength, density, reinforcing and clear cover, minimum metal deck gage, maximum deck span, shear connector requirements, design stress limitations, etc. see the specific referenced assembly in the Underwriters Laboratories (UL) directory.
[c] Metal deck depth for some assemblies is shown as a minimum and deeper decks may be substituted. Refer to the specific UL assembly for additional information.

Table 2-8.
Beam-Only Designs—Floor[a, b]

Assembly Rating		Type of Protection System	Concrete		Metal Deck Depth (in)	U.L. Design Number
Restrained (hr)	Unrestrained (hr)		Min. Thickness above Deck Flutes (in)	Type		
2	2	Gypsum Wallboard	$2\frac{1}{2}$	NW	$1\frac{1}{2}$	N501, N502
3	2		$2\frac{1}{2}$	NW	$1\frac{1}{2}$	N505
1, $1\frac{1}{2}$, 2, 3, 4	1, $1\frac{1}{2}$, 2, 3, 4	Spray-Applied Fire Resistive Material	$2\frac{1}{2}$	NW or LW	$1\frac{1}{2}$, 2, 3	N706, N734, N739, N823
					$1\frac{5}{16}$, $1\frac{1}{2}$, 2, 3	N708, N772, N782

[a] The referenced assemblies are some commonly used Underwriters Laboratories (UL) assemblies used for conventional steel framed structures. For additional assemblies the reader should reference the UL Fire Resistance Directory.
[b] For additional design requirements such as beam spacing, concrete strength, density, reinforcing and clear cover, minimum metal deck gage, maximum deck span, shear connector requirements, design stress limitations, etc. see the specific referenced assembly in the Underwriters Laboratories (UL) directory.

When steel is primed or painted, UL guidelines specify default maximum dimensions above which primed or painted steel must be wrapped with steel lath or provided with mechanical anchorage devices as follows:

1. Beams with nominal web depth greater than 16 in. or nominal flange width greater than 12 in.
2. Columns with nominal web depth or nominal flange width greater than 16 in.

For other primed or painted steel, bond tests must be performed in accordance with the ASTM E736 to ensure that adequate adhesion exists. If such testing demonstrates inadequate adhesion, a bonding agent or mechanical anchorage, such as with steel lath or other devices,

Table 2-9.
Column Assemblies[a, b]

Assembly Rating (hr)	Type of Protection	Column Types	U.L. Design Number
1, 2, 3	Gypsum Wallboard	W, HSS	X528
2		W	X516, X518, X520
3			X509, X510, X513
$^{3}/_{4}$, 1, 1$^{1}/_{2}$, 2, 3, 4	Spray-Applied Fire Resistive Material	HSS, Pipe	X771, Y707
1, 1$^{1}/_{2}$, 2, 3, 4		W	X772, X829, Y708, Y725
		W, HSS, Pipe	X790, X795
		HSS, Pipe	X827

[a] The referenced assemblies are some commonly used Underwriters Laboratories (UL) assemblies used for conventional steel framed structures. For additional assemblies the reader should reference the UL Fire Resistance Directory.
[b] For additional design requirements such as beam spacing, concrete strength, density, reinforcing and clear cover, minimum metal deck gage, maximum deck span, shear connector requirements, design stress limitations, etc. see the specific referenced assembly in the Underwriters Laboratories (UL) directory.

is required. For further information, refer to the UL *Fire Resistance Directory* (Underwriters Laboratories, 2000).

Fire Engineering

While the fire exposure in ASTM E119 is based upon a continuous fuel supply, real building fires are characterized by an exhaustible supply of fuel. Additionally, while ASTM E119 utilizes assemblies that are fully loaded, lifetime-maximum loads will rarely occur simultaneously with fire in real buildings, particularly since serviceability limit states, such as drift, deflection or vibration control many structural elements. As a result, steel structures generally have significantly more reserve strength than that accounted for in the ASTM E119 test. Alternatively, AISI (1979) provides an analytical procedure for determining maximum steel temperatures based upon realistic fire exposures for architecturally exposed exterior structural elements.

Assessment of Fire Damage

Intense fire exposure will almost certainly cause warping and twisting of members (Dill, 1960). However, it does not inevitably follow that the strength of the steel is reduced. It is almost certain that any steel that has been heated hot enough to undergo damaging grain coarsening or that has been cooled rapidly enough to harden it will be so badly distorted that it would have no consideration for re-use anyway. This leads to the general statement that steel that has been through a fire but that can be made dimensionally reusable by straightening with the methods that are available is suitable for continued service with full expectation of performance in accordance with its specified mechanical properties. Essentially then, the question is one of economics: if the steel can be straightened for less money than fabricating and installing a new piece, then that should be done.

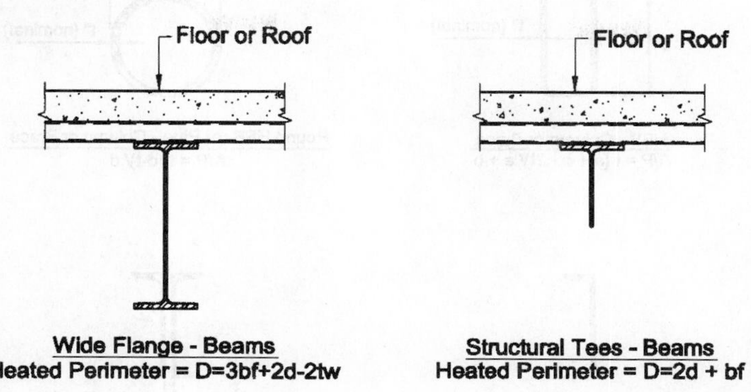

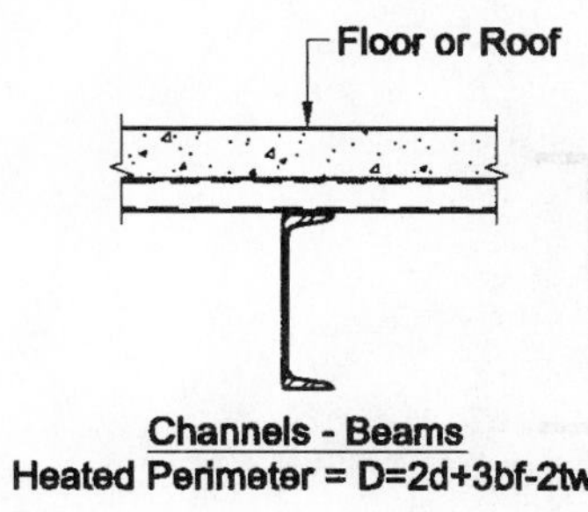

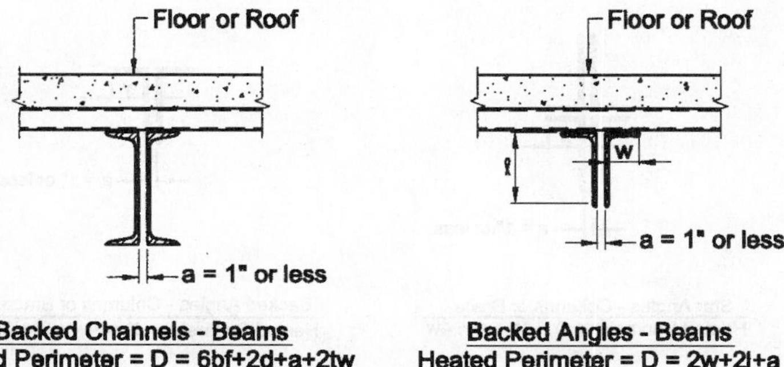

Note: For W/D ratios: W= Weight in lbs. per lineal foot
D = Heated Perimeter in inches

Fig. 2–6a. Equations for heated perimeter for beams.

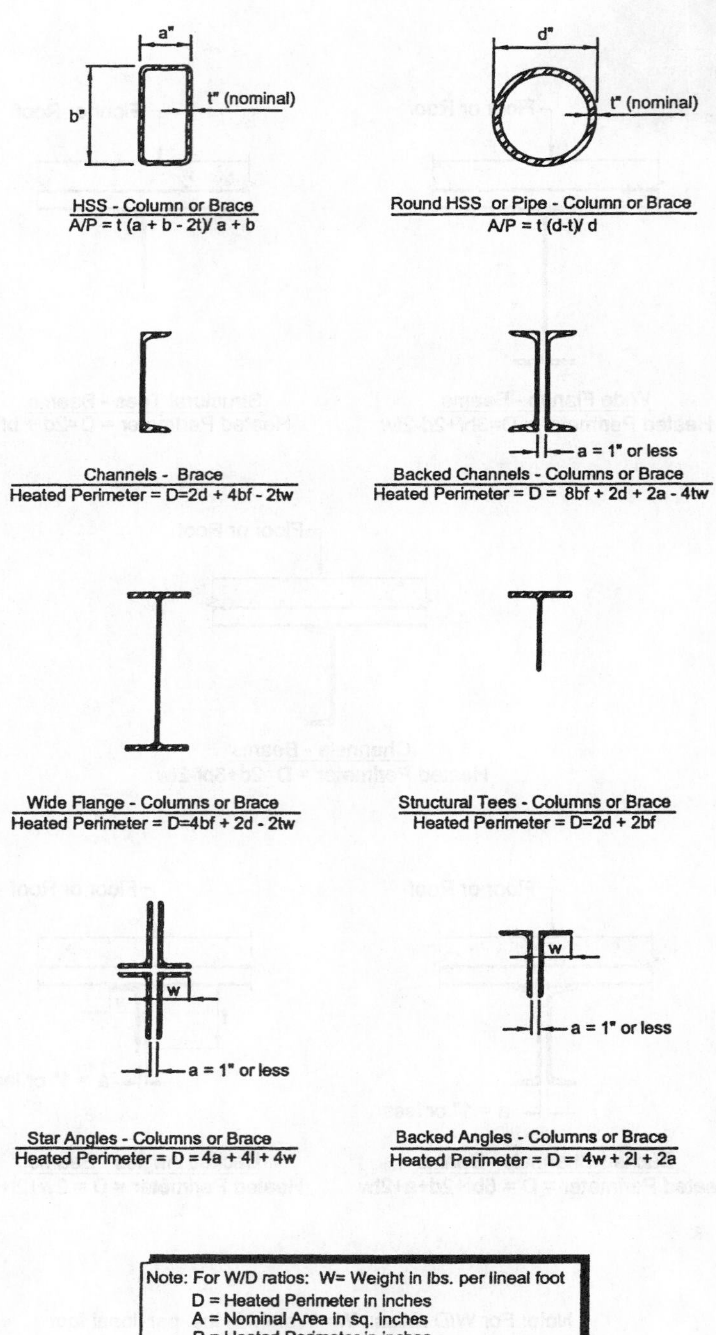

Fig. 2–6b. Equations for heated perimeter for columns and braces.

Table 2-10.
Construction Classification,
Restrained and Unrestrained

Wall Bearing	**Single-span and simply supported end spans of multiple bays**[a]	Open-web steel joists or steel beams, supporting concrete slab, precast units, or metal decking	unrestrained
		Concrete slabs, precast units, or metal decking	unrestrained
	Interior spans of multiple bays	Open-web steel joists, steel beams or metal decking, supporting continuous concrete slab	restrained
		Open-web steel joists or steel beams, supporting precast units or metal decking	unrestrained
		Cast-in-place concrete slab systems	restrained
		Precast concrete where the potential thermal expansion is resisted by adjacent construction[b]	restrained
Steel Framing		Steel beams welded, riveted, or bolted to the framing members	restrained
		All types of cast-in-place floor and roof systems (such as beam-and-slabs, flat slabs, pan joists, and waffle slabs) where the floor or roof system is secured to the framing members	restrained
		All types of prefabricated floor or roof systems where the structural members are secured to the framing members and the potential thermal expansion of the floor or roof system is resisted by the framing system or the adjoining floor or roof construction[b]	restrained
Concrete Framing		Beams securely fastened to the framing members	restrained
		All types of cast-in-place floor and roof systems (such as beam-and-slabs, flat slabs, pan joists, and waffle slabs) where the floor system is cast with the framing members	restrained
		Interior and exterior spans of precast systems with cast-in-place joints resulting in restraint equivalent to that which would exist in [concrete framing] [b(i)]	restrained
		All types of prefabricated floor or roof systems where the structural members are secured to such systems and the potential thermal expansion of the floor or roof systems is resisted by the framing system or the adjoining floor or roof construction[b]	unrestrained
Wood Construction		All types	unrestrained

[a] Floor and roof system scan be considered restrained when they are tied into walls or without tie beams, the walls being designed and detailed to resist thermal thrust from the floor or roof system.

[b] For example, resistance to potential thermal expansion is considered to be achieved when:

(i) Continuous structural concrete topping is used,

(ii) The space between the ends of precast units or between the ends of units and the vertical face of supports is filled with concrete or mortar, or

(iii) The space between the ends of precast units and the vertical faces of supports, or between the ends of solid or hollow core slab units does not exceed 0.25% of the length for normal weight concrete members of 0.1% of the length for structural light weight concrete members.

From ASTM E119-2000 Table X 3.1. Copyright ASTM. Reprinted with permission.

Heat-treated steels should not be reused unless it can be established that the steel temperature did not exceed a temperature that is 50 degrees F below the tempering temperature used in heat treatment. For example, ASTM A514 and A852 steels can be straightened and reused with no reduction in design strength if the maximum steel temperature did not exceed 1,100 degrees F. Similarly, special attention should be given to such items as high-strength bolts, which are heat treated, and cold-formed steel elements. Connections also require special attention to ensure that the stresses induced by the fire and subsequent cooling have not sheared or loosened bolts or rivets, nor cracked welds.

For further information, refer to Tide (1998) and Council on Tall Buildings and Urban Habitat (1983).

CORROSION PROTECTION

In building structures, corrosion protection is not required for steel that will be enclosed by building finish, coated with a contact-type fireproofing or in contact with concrete. When enclosed, the steel is trapped in a controlled environment and the products required for corrosion are quickly exhausted, as indicated in LRFD Commentary Section M3. A similar situation exists when steel is fireproofed or in contact with concrete. Accordingly, shop primer or paint is not required unless specified in the contract documents, per LRFD Specification Section M3.1. Per Code of Standard Practice Section 6.5, steel that is to remain unpainted need only be cleaned of heavy deposits of oil and grease by appropriate means after fabrication.

Corrosion protection is required, however, in exterior exposed applications. Likewise, steel must be protected from corrosion in aggressively corrosive applications, such as a paper processing plant, a structure with oceanfront exposure or when temperature changes can cause condensation.

When surface preparation other than the cleaning described above is required, an appropriate SSPC grade of cleaning should be specified in the contract documents. A summary of the SSPC surface preparation specifications (SSPC, 2000) is provided in Table 2-11. SSPC SP 2 is the normal grade of cleaning when cleaning is required.

For further information, refer to the publications of SSPC: The Society for Protective Coatings, the American Galvanizers Association (AGA), and the National Association of Corrosion Engineers International (NACE).

RENOVATION AND RETROFIT OF EXISTING STRUCTURES

The provisions in LRFD Specification Chapter N govern the evaluation of existing structures. Historical data on available steel grades and hot-rolled structural shapes, including dimensions and properties, is available in *Iron and Steel Beams: 1873 to 1952* (AISC, 1953). See also Ricker (1988) and Tide (1990).

THERMAL EFFECTS

Expansion and Contraction

The average coefficient of expansion ε for structural steel between 70 and 100 degrees F is 0.0000065 for each degree F. For temperatures from 100 to 1,200 degrees F, the change in length per unit length per degree F, ε, is:

$$\varepsilon = (6.1 + 0.0019t)10^{-6}$$

where t is the initial temperature in degrees F. The coefficients of expansion for other building materials can be found in Table 17-10.

Table 2-11.
Summary of Surface Preparation Specifications

SSPC Specification No.	Title	Description
SP1	Solvent Cleaning	Removal of oil, grease, dirt, soil, salts, and contaminants by cleaning with solvent, vapor, alkali, emulsion, or steam.
SP2	Hand Tool Cleaning	Removal of all loose rust, loose mill scale, and loose paint to degree specified, by hand chipping, scraping, sanding and wire brushing.
SP3	Power Tool Cleaning	Removal of all loose rust, loose mill scale, and loose paint to degree specified, by power tool chipping, descaling, sanding, wire brushing, and grinding.
SP5/NACE No. 1	Metal Blast Cleaning	Removal of all visible rust, mill scale, paint, and foreign matter by blast cleaning by wheel or nozzle (dry or wet) using sand, grit, or shot. (For very corrosive atmospheres where high cost of cleaning is warranted.)
SP6/NACE No. 3	Commercial Blast Cleaning	Blast cleaning until at least two-thirds of the surface area is free of all visible residues. (For conditions where thoroughly cleaned surface is required.)
SP7/NACE No. 4	Brush-Off Blast Cleaning	Blast cleaning of all except tightly adhering residues of mill scale, rust, and coatings, exposing numerous evenly distributed flecks of underlying metal.
SP8	Pickling	Complete removal of rust and mill scale by acid pickling, duplex pickling, or electrolytic pickling.
SP10/NACE No. 2	Near-White Blast Cleaning	Blast cleaning to nearly White Metal cleanliness, until at least 95% of the surface area is free of all visible residues. (For high humidity, chemical atmosphere, marine, or other corrosive environments.)
SP11	Power Tool Cleaning to Bare Metal	Complete removal of all rust, scale, and paint by power tools, with resultant surface profile.

Although buildings are typically constructed of flexible materials, expansion joints are often required in roofs and the supporting structure when horizontal dimensions are large. The maximum distance between expansion joints is dependent upon many variables, including ambient temperature during construction and the expected temperature range during the lifetime of the building.

Figure 2–7 (Federal Construction Council, 1974) provides guidance based on design temperature change for maximum spacing of structural expansion joints in beam-and-column-framed buildings with pinned column bases and heated interiors. The report includes data for numerous cities and gives five modification factors to be applied as appropriate:

1. If the building will be heated only and will have pinned column bases, use the maximum spacing as specified;
2. If the building will be air-conditioned as well as heated, increase the maximum spacing by 15 percent provided the environmental control system will run continuously;
3. If the building will be unheated, decrease the maximum spacing by 33 percent;
4. If the building will have fixed column bases, decrease the maximum spacing by 15 percent;
5. If the building will have substantially greater stiffness against lateral displacement in one of the plan dimensions, decrease the maximum spacing by 25 percent.

When more than one of these design conditions prevail in a building, the percentile factor to be applied is the algebraic sum of the adjustment factors of all the various applicable conditions. Most building codes include restrictions on location and maximum spacing of fire walls, which often become default locations for expansion joints.

The most effective expansion joint is a double line of columns that provides a complete and positive separation. Alternatively, low-friction sliding elements can be used. Such systems, however, are seldom totally friction-free and will induce some level of inherent restraint to movement.

Elevated-Temperature Service

For applications involving short-duration loading at elevated temperature, the variations in yield strength, tensile strength and modulus of elasticity are given in Figures 2–8a, 2–8b and 2–8c, respectively. For applications involving long-duration loading at elevated temperatures, the effects of creep must also be considered. For further information, see Brockenbrough and Merritt (1999; pp. 1.20–1.22).

Low-Temperature Service

For applications involving low-temperature service, see the discussion "Avoiding Brittle Fracture" under "Fatigue and Fracture Control."

FATIGUE AND FRACTURE CONTROL

Avoiding Brittle Fracture

By definition, brittle fracture occurs by cleavage at a stress level below the yield strength. Generally, a brittle fracture can occur when there is a sufficiently adverse combination of tensile stress, temperature, strain rate and geometrical discontinuity (notch). The exact combination of these conditions and other factors that will cause brittle fracture cannot be readily

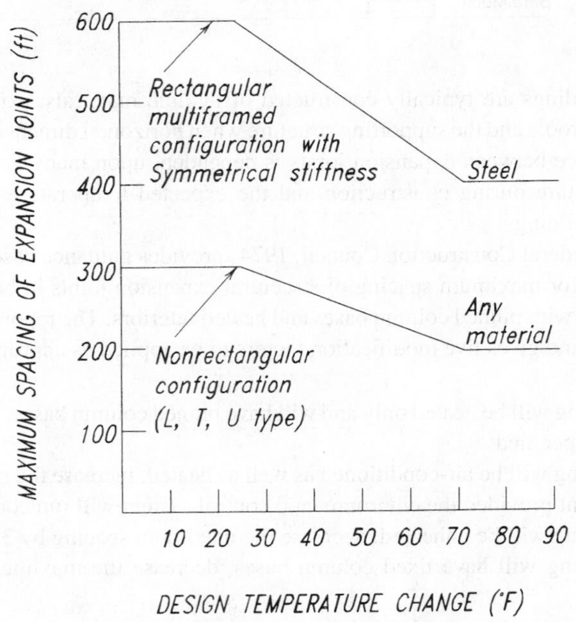

Fig. 2–7. Recommended maximum expansion-joint spacing.

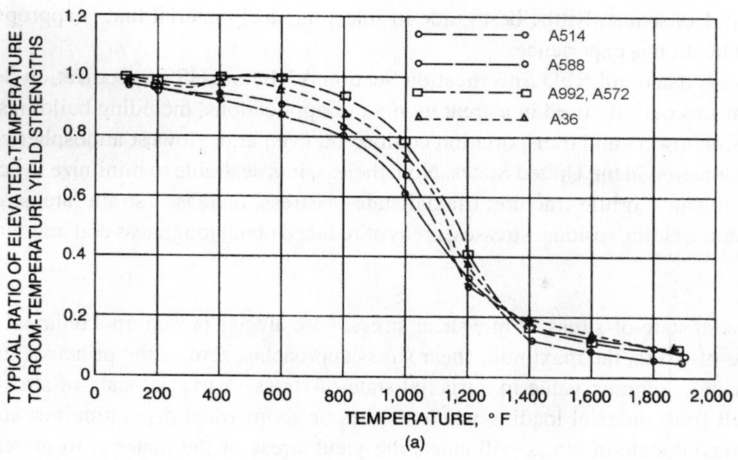

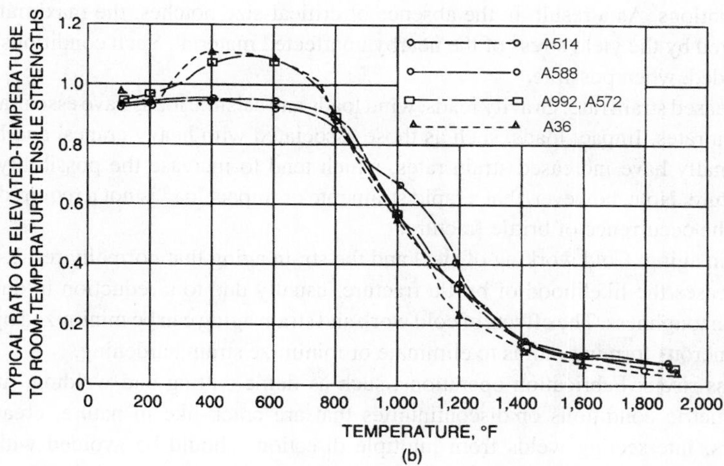

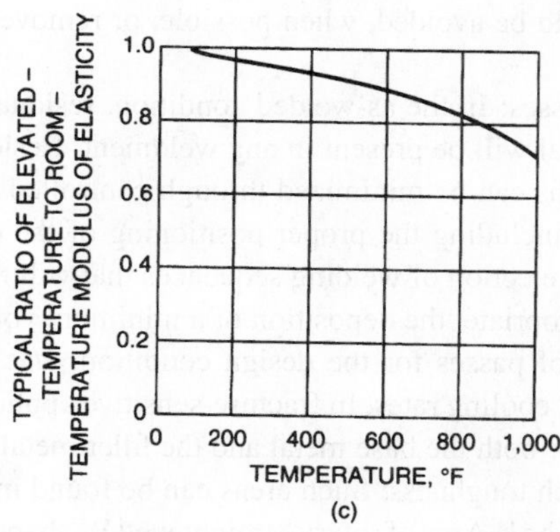

Fig. 2–8. Effect of temperature on yield strength, tensile strength and modulus of elasticity for structural steels. Reprinted with permission from R.L. Brockenbrough and B.G. Johnston, USS Steel Design Manual, R.L. Brockenbrough & Associates Inc., Pittsburgh, PA.

calculated. Consequently, the best guide in selecting steel material that is appropriate for a given application is experience.

The steels listed in LRFD Specification Section A3.1c and HSS Specification Section 1.2 have been successfully used in a great number of applications, including buildings, bridges, transmission towers and transportation equipment, even at the lowest atmospheric temperatures encountered in the United States. Nonetheless, it is desirable to minimize the conditions that tend to cause brittle fracture: triaxial state-of-stress, increased strain rate, strain aging, stress risers, welding residual stresses, areas of reduced notch toughness and low-temperature service.

1. Triaxial state-of-stress: While shear stresses are always present in a uniaxial or biaxial state-of-stress, the maximum shear stress approaches zero as the principal stresses approach a common value in a triaxial state-of-stress. A triaxial state-of-stress can also result from uniaxial loading when notches or geometrical discontinuities are present. A triaxial state-of-stress will cause the yield stress of the material to increase above its nominal value, resulting in brittle fracture by cleavage, rather than ductile shear deformations. As a result, in the absence of critical-size notches, the maximum stress is limited by the yield stress of the nearby unaffected material. Such conditions should be avoided, when possible.
2. Increased strain rate: Gravity loads, wind loads and seismic loads have essentially similar strain rates. Impact loads, such as those associated with heavy cranes, and blast loads normally have increased strain rates, which tend to increase the possibility of brittle fracture. Note, however, that a rapid strain rate or impact load is not a required condition for the occurrence of brittle fracture.
3. Strain aging: Cold working of steel and the strain aging that normally results generally increases the likelihood of brittle fracture, usually due to a reduction in ductility and notch toughness. The effects of cold work and strain aging can be minimized by selecting a generous forming radius to eliminate or minimize strain hardening.
4. Stress risers: Fabrication operations, such as flame cutting and welding, may induce geometric conditions or discontinuities that are crack-like in nature, creating stress risers. Intersecting welds from multiple directions should be avoided with properly sized weld access holes to minimize the interaction of these various stress fields. Such conditions should be avoided, when possible, or removed or repaired when they occur.
5. Welding residual stresses: In the as-welded condition, residual stresses near the yield strength of the material will be present in any weldment. Residual stresses and the accompanying distortions can be minimized through controlled welding procedures and fabrication methods, including the proper positioning of the components of the joint prior to welding, the selection of welding sequences that will minimize distortions, the use of preheat as appropriate, the deposition of a minimum volume of weld metal with a minimum number of passes for the design condition, and proper control of interpass temperatures and cooling rates. In fracture-sensitive applications, notch-toughness should be specified for both the base metal and the filler metal.
6. Areas of reduced notch toughness: Such areas can be found in the core areas of heavy shapes and plates and the k-Area of rotary-straightened W-shapes, as illustrated in LRFD Commentary Figures C–A3.1 and C–A3.2. Accordingly, LRFD Specification Section A3.1c includes special requirements for material notch toughness.

7. Low-temperature service: While steel yield strength, tensile strength, modulus of elasticity and fatigue strength increase as temperature decreases, ductility and toughness decrease. Furthermore, there is a temperature below which steel subjected to tensile stress may fracture by cleavage, with little or no plastic deformation, rather than by shear, which is usually preceded by considerable inelastic deformation. Note that cleavage and shear are used in the metallurgical sense to denote different fracture mechanisms.

When notch-toughness is important, Charpy V-notch testing can be specified to ensure a certain level of energy absorption at a given temperature, such as 15 ft-lbs at 70 degrees F. Note that the appropriate test temperature may be higher than the lowest operating temperature depending upon the rate of loading. Although it is primarily intended for bridge-related applications, the information in ASTM A709 Section S83 (including Tables S1.1, S1.2 and S1.3) may be useful in determining the proper level of notch toughness that should be specified.

In many cases, weld metal notch toughness exceeds that of the base metal. Filler metals can be selected to meet a desired minimum notch-toughness value. For each welding process, electrodes exist that have no specified notch toughness requirements. Such electrodes should not be assumed to possess any minimum notch-toughness value. When notch toughness is necessary for a given application, the desired value or an appropriate electrode should be specified in the contract documents.

For further information, refer to Fisher et al. (1998), Barsom and Rolfe (1999) and Rolfe (1977).

Avoiding Lamellar Tearing

Discussions of strength and ductility generally pertain to loadings applied in the planar direction (longitudinal or transverse orientation) of the steel plate or shape. Ductility may be significantly lower in the through-thickness direction than in the planar direction. This inherent directionality is of small consequence in many applications, but does become important in the design and fabrication of structures containing massive members with highly restrained welded joints.

The restraint against solidified weld deposit contraction inherent in some joint configurations can impose a tensile strain high enough to cause separation or tearing on planes parallel to the rolled surface of the element being joined. The incidence of this phenomenon can be reduced or eliminated through greater understanding by designers, detailers and fabricators of the inherent directionality of rolled steel, the importance of strains associated with solidified weld deposit contraction in the presence of high restraint (rather than externally applied design forces), and the need to adopt appropriate joint and welding details and procedures with proper weld metal for through-thickness connections.

Recent research (Melendrez and Dexter, 2000) demonstrates that W-shapes are not susceptible to lamellar tearing or other through-thickness failures when welded tee joints are made to the flanges at locations away from member ends. Special production practices can be specified for steel plates to enhance through-thickness ductility and assist in reducing the incidence of lamellar tearing. For further information, refer to ASTM A770. However, it must be recognized that the specification of premium-quality steel does not, of itself, eliminate the potential for lamellar tearing, or the need for careful design, detailing and fabrication of highly restrained joints. AISC (1973) provides guidelines that minimize potential problems.

See also the discussion “Avoiding Lamellar Tearing” in Part 8.

SEISMIC DESIGN

In general, nearly all building design and construction can be classified into one of three lateral-load-driven categories: wind applications, low-seismic applications and high-seismic applications.

Wind Applications and Low-Seismic Applications

Wind applications are those in which wind effects control over earthquake effects, with the seismic response modification factor R taken as equal to or less than 3 for design purposes. Low-seismic applications are those in which earthquake effects, with R taken as equal to or less than 3 for design purposes, control over wind effects. In either case, the building is designed to meet the provisions in the LRFD Specification based upon the code-specified forces distributed throughout the framing assuming a nominally elastic structural response. The resulting systems have normal levels of ductility.

The seismic response modification factor R essentially represents the ratio of the forces that would develop under the specified ground motion if the structure had an entirely linearly elastic response to the prescribed design forces (BSSC, 1997).

High-Seismic Applications

High-seismic applications are those in which R is taken greater than 3, and the building is designed to meet the provisions in both the Seismic Provisions and LRFD Specification. Note that it does not matter if wind or earthquake controls in this case. The use of R greater than 3 in the calculation of the seismic base shear requires the use of a seismically detailed system that is compatible with R even if wind effects control. High-seismic design and construction will generally cost more than wind and low-seismic design and construction.

High-seismic lateral framing systems are configured to be capable of withstanding strong ground motions as they undergo controlled ductile deformations to dissipate energy. Consider the following three examples:

1. Special Concentrically Braced Frames (SCBF)—SCBF are generally configured so that any inelasticity will occur by tension yielding and/or compression buckling in the braces. The connections of the braces to the columns and beams and between the columns and beams themselves must then be proportioned to remain nominally elastic as they undergo these deformations.
2. Eccentrically Braced Frames (EBF)—EBF are generally configured so that any inelasticity will occur by shear yielding and/or flexural yielding in the link. The beam outside the link, connections, braces and columns must then be proportioned to remain nominally elastic as they undergo these deformations.
3. Special Moment Frames (SMF)—SMF are generally configured so that any inelasticity will occur by flexural yielding in the girders near, but away from, the connection of the girders to the columns. The connections of the girders to the columns and the columns themselves must then be proportioned to remain nominally elastic as they undergo these deformations. Intermediate Moment Frames (IMF) and Ordinary Moment Frames (OMF) are also configured to provide improved seismic performance, although successively lower than that for SMF.

The code-specified base accelerations used to calculate the seismic forces are not necessarily maximums, but rather, they represent the intensity of ground motions that have been selected by the code-writing authorities as reasonable for design purposes. Accordingly, the requirements in both the Seismic Provisions and the LRFD Specification must be met so that the resulting frames can then undergo controlled deformations in a ductile, well-distributed manner.

The design provisions for high-seismic systems are also intended to result in distributed deformations throughout the frame, rather than the formation of story mechanisms, so as to increase the level of available energy dissipation and corresponding level of ground motion that can be withstood.

The member sizes in high-seismic frames will be larger than those in wind and low-seismic frames. The connections will also be much more robust so they can transmit the member-strength-driven force demands. Net sections will often require special attention so as to avoid having fracture limit states control. Special material requirements, design considerations and construction practices must be followed. For further information on the design and construction of high-seismic systems, see the Seismic Provisions, which are available from the American Institute of Steel Construction, Inc. at www.aisc.org.

PART 2 REFERENCES

ACI International, 1999a, *Building Code Requirements for Masonry Structures and Commentary*, ACI 530/ASCE 5/TMS402, ACI International, Farmington Hills, MI.

ACI International, 1999b, *Building Code Requirements for Structural Concrete and Commentary*, ACI 318, ACI International, Farmington Hills, MI.

Allison, H., 1991, AISC Design Guide No. 5 *Design of Low- and Medium-Rise Steel Buildings*, AISC, Chicago, IL.

American Institute of Steel Construction, 2000a, *Code of Standard Practice for Steel Buildings and Bridges*, AISC, Chicago, IL.

American Institute of Steel Construction, 2000b, *AISC Seismic Provisions Supplement No. 2*, AISC, Chicago, IL.

American Institute of Steel Construction 2000c, *Specification for the Design of Steel Hollow Structural Sections*, AISC, Chicago, IL.

American Institute of Steel Construction, 2000d, *Specification for Load and Resistance Factor Design of Single-Angle Members*, AISC, Chicago, IL.

American Institute of Steel Construction, 1999, *Load and Resistance Factor Design Specification for Structural Steel Buildings*, AISC, Chicago, IL.

American Institute of Steel Construction, 1997a, *A Guide to Engineering and Quality Criteria for Steel Structures: Common Questions Answered*, AISC, Chicago, IL.

American Institute of Steel Construction, 1997b, *Hollow Structural Sections (HSS) Connections Manual*, AISC, Chicago, IL.

American Institute of Steel Construction, 1997c, *Seismic Provisions for Structural Steel Buildings*, AISC, Chicago, IL.

American Institute of Steel Construction, 1994, *Specification for the Design, Fabrication and Erection of Steel Safety-related Structures for Nuclear Facilities*, AISC, Chicago, IL.

American Institute of Steel Construction, 1983, *Detailing for Steel Construction*, AISC, Chicago, IL.

American Institute of Steel Construction, 1973, "Commentary on Highly Restrained Welded Connections," *Engineering Journal*, 3rd Qtr., AISC, Chicago, IL.

American Institute of Steel Construction, 1953, *Iron and Steel Beams: 1873 to 1952*, AISC, Chicago, IL.

American Iron and Steel Institute, 1999, *Specification for the Design of Cold-Formed Steel Structural Members, Supplement No. 1*, AISI, Washington, DC.

American Iron and Steel Institute, 1996, *Specification for the Design of Cold-Formed Steel Structural Members*, AISI, Washington, DC.

American Iron and Steel Institute, 1984, *Designing Fire Protection for Steel Beams*, AISI, Washington, DC.

American Iron and Steel Institute, 1981, *Designing Fire Protection for Steel Trusses*, Second Edition, AISI, Washington, DC.

American Iron and Steel Institute, 1980, *Designing Fire Protection for Steel Columns*, Third Edition, AISI, Washington, DC.

American Iron and Steel Institute, 1979, *Fire-safe Structural Steel: A Design Guide*, AISI, Washington, DC.

American Society of Civil Engineers, 1998, *Minimum Design Loads for Buildings and Other Structures*, ASCE 7-98, ASCE, Reston, VA.

American Society of Civil Engineers, 1996, *Structural Applications of Steel Cables for Buildings*, ANSI/ASCE 19, ASCE, Reston, VA.

American Society of Civil Engineers, 1991, *Standard for the Structural Design of Composite Slabs*, ANSI/ASCE 3, ASCE, Reston, VA.

American Society of Civil Engineers, 1991, *Standard Practice for Construction and Inspection of Composite Slabs*, ANSI/ASCE 9, ASCE, Reston, VA.

American Society of Civil Engineers, 1990, *Specification of Cold Formed Stainless Steel Structural Members*, ANSI/ASCE 8, ASCE, Reston, VA.

American Welding Society, 2000, *Structural Welding Code—Steel*, AWS D1.1:2000, AWS, Miami, FL.

Barger, B.L. and M.A. West, 2001, "New OSHA Erection Rules: How They Affect Engineers, Fabricators and Contractors," *Modern Steel Construction*, May, AISC, Chicago, IL.

Barsom, J.A. and S.T. Rolfe, 1999, *Fracture and Fatigue Control in Structures: Applications of Fracture Mechanics*, Third Edition, ASTM, West Conshohocken, PA.

Bigos, J., G.W. Smith, E.F. Ball and P.J. Foehl, 1954, *Shop Paint and Painting Practice*, Proceedings of the AISC National Engineering Conference, pp. 67–87, AISC, Chicago, IL.

Brockenbrough, R.L. and B.G. Johnston, 1981, *USS Steel Design Manual*, R.L. Brockenbrough & Assoc. Inc., Pittsburgh, PA.

Brockenbrough, R.L. and F.S. Merritt, 1999, *Structural Steel Designer's Handbook*, Third Edition, McGraw-Hill, New York, NY.

Building Seismic Safety Council, 1997, National Earthquake Hazard Reduction Program (NEHRP) *Recommended Provisions For Seismic Regulations For New Buildings And Other Structures* (FEMA 302) and Commentary (FEMA 303), BSSC, Washington, DC.

Canadian Institute of Steel Construction, 1989, *Roof Framing with Cantilever (Gerber) Girders & Open Web Joists*, CISC, Willowdale, Ontario, CANADA

Carter, C.J., 1999, AISC Design Guide No. 13 *Wide-Flange Column Stiffening at Moment Connections: Wind and Seismic Applications*, AISC, Chicago, IL.

Council on Tall Buildings and Urban Habitat, 1983, *Developments in Tall Buildings*, Van Nostrand Reinhold, New York, NY.

Crane Manufacturers Association of America, 2000, *Specifications for Top Running Bridge & Gantry Type Multiple Girder Electric Overhead Traveling Cranes*, CMAA 70, CMAA, Charlotte, NC.

Darwin, D., 1990, AISC Design Guide No. 2 *Design of Steel and Composite Buildings with Web Openings*, AISC, Chicago, IL.

Deiter, G.E., Jr., 1961, *Mechanical Metallurgy*, McGraw-Hill Book Company, New York, NY.

DeWolf, J.T. and D.T. Ricker, 1990, AISC Design Guide No. 1, *Column Base Plates*, AISC, Chicago, IL.

Dill, F.H., 1960, "Structural Steel After a Fire," *Proceedings of the 1960 National Engineering Conference*, pp. 78–80, AISC, Chicago, IL.

Federal Construction Council, 1974, Technical Report No. 65 *Expansion Joints in Buildings*, National Research Council, Washington, DC.

Fisher, J.M., 1993, AISC Design Guide No. 7 *Industrial Buildings: Roofs to Column Anchorage*, AISC, Chicago, IL.

Fisher, J.M. and M.A. West, 1997, AISC Design Guide No. 10 *Erection Bracing of Low-Rise Structural Steel Frames*, AISC, Chicago, IL.

Fisher, J.M. and M.A. West, 1990, AISC Design Guide No. 3 *Serviceability Design Considerations for Low-Rise Buildings*, AISC, Chicago, IL.

Fisher, J.W., G.L. Kulak and I.F.C. Smith, 1998, *A Fatigue Primer for Structural Engineers*, NSBA/AISC, Chicago, IL.

Fisher, J.W. and A.W. Pense, 1987, "Experience with Use of Heavy W Shapes in Tension," *Engineering Journal*, Vol. 24, No. 2, (2nd Qtr.), pp. 63–77, AISC, Chicago, IL.

Geschwindner, L.F., 2000, "A Practical Look at Frame Analysis, Stability and Leaning Columns," *Proceedings of the 2000 AISC North American Steel Construction Conference*, pp. 1.1–1.29, AISC, Chicago, IL.

Griffis, L.G., 1997, AISC Design Guide No. 6 *Load and Resistance Factor Design of W-Shapes Encased in Concrete*, AISC, Chicago, IL.

Gross, J.L., M.D. Engelhardt, C.M. Uang, K. Kasai and N.R. Iwankiw, 1999, AISC Design Guide No. 12 *Modification of Existing Welded Steel Moment Frames for Seismic Resistance*, AISC, Chicago, IL.

Gypsum Association, 2000, *Fire-Resistance Design Manual GA-600*, Gypsum Association, Washington, DC.

Ioannides, S.A. and S. Mehta, 1997, "Restrained Versus Unrestrained Fire Ratings for Steel Structures—A Practical Approach," *Proceedings of the National Steel Construction Conference*, pp. 17.1–17.20, AISC, Chicago, IL.

LeMessurier, W.J., 1977, "A Practical Method of Second Order Analysis," *Engineering Journal*, Vol. 14, No. 2, (2nd Qtr.), pp. 49–67, AISC, Chicago, IL.

Leon, R.T., J.J. Hoffman and T. Staeger, 1996, AISC Design Guide No. 8 *Partially Restrained Composite Connections*, AISC, Chicago, IL.

Lightner, M.W. and R.W. Vanderbeck, 1956, "Factors Involved in Brittle Fracture," *Regional Technical Meetings*, AISI, Washington, DC.

Lim, L.C. and R.J. McNamara, 1972, "Stability of Novel Building Systems," Structural Design of Tall Steel Buildings, Volume II-16, Proceedings of the ASCE-IABSA International Conference on the Planning and Design of Tall Buildings, pp. 499–524, Lehigh University, Bethlehem, PA.

Melendrez, M.I. and R.J. Dexter, 2000, "Through-Thickness Properties of Column Flanges in Welded Moment Connections," *Journal of Structural Engineering*, Vol. 126, No. 1, pp. 24–31, ASCE, Reston, VA.

Murray, T.M., 1990, AISC Design Guide No. 4 *Extended End-Plate Moment Connections*, AISC, Chicago, IL.

Murray, T.M., D.E. Allen and E.E. Ungar, 1997, AISC Design Guide No. 11 *Floor Vibrations Due to Human Activity*, AISC, Chicago, IL.

Occupational Safety and Health Administration, 2001, *Safety and Health Standards for the Construction Industry, 29 CFR 1926 Part R Safety Standards for Steel Erection*, OSHA, Washington, DC.

Rack Manufacturers Institute, 1997, *Specification for the Design, Testing and Utilization of Industrial Steel Storage Racks*, RMI, Charlotte, NC.

Research Council on Structural Connections, 2000, *Specification for Structural Joints Using ASTM A325 or A490 Bolts*, AISC, Chicago, IL.

Ricker, D.T., 1988, "Field Welding to Existing Structures," *Engineering Journal*, Vol. 25, No. 1, (1st Qtr.), pp. 1–16, AISC, Chicago, IL.

Rolfe, S.T., 1977, "Fracture and Fatigue Control in Steel Structures," *Engineering Journal*, Vol. 14, No. 1, (1st Qtr.), pp. 2–15, AISC, Chicago, IL.

Rongoe, J., 1996, "Design Guidelines for Continuous Beams Supporting Steel Joist Roof Structures," *Proceedings of the AISC National Steel Construction Conference*, pp. 23.1–23.44, AISC, Chicago, IL.

Ruddy, J.L., 1986, "Ponding of Concrete Deck Floors," *Engineering Journal*, Vol. 23, No. 3, (3rd Qtr.), pp. 107–115, AISC, Chicago, IL.

Salmon, C.G. and J.E. Johnson, 1996 *Steel Structures: Design and Behavior, Emphasizing LRFD*, Fourth Edition, Addison-Wesley, Boston, MA.

Seaburg, P.A. and C.J. Carter, 1997, AISC Design Guide No. 9 *Torsional Analysis of Structural Steel Members*, AISC, Chicago, IL.

SSPC: The Society for Protective Coatings, 2000, *Systems and Specifications: SSPC Painting Manual, Volume II*, Eighth Edition, SSPC, Pittsburgh, PA.

Tide, R.H.R., 1998, "Integrity of Structural Steel After Exposure to Fire," *Engineering Journal*, Vol. 35, No. 1, (1st Qtr.), pp. 26–38, AISC, Chicago, IL.

Tide, R.H.R., 1990, "Reinforcing Steel Members and the Effects of Welding," *Engineering Journal*, Vol. 27, No. 4, (4th Qtr.), pp. 129–131, AISC, Chicago, IL.

Underwriters Laboratories, 2000, *Fire Resistance Directory*, UL, Northbrook, IL.

Welding Research Council, 1957, *Control of Steel Construction to Avoid Brittle Failure*, WRC, New York, NY.

Yura, J.A., 1971, "The Effective Length of Columns in Unbraced Frames," *Engineering Journal*, Vol. 8, No. 2, (2nd Qtr.), pp. 37–42, AISC, Chicago, IL.

PART 3

DESIGN OF TENSION MEMBERS

SCOPE

The specification requirements and other design considerations summarized in this Part apply to the design of members subject to axial tension. For the design of members subject to eccentric tension or combined tension and flexure, see Part 6. For tension members that are part of a seismic-force-resisting system in which the seismic response modification factor R is taken greater than 3, see the Seismic Provisions, which are available from AISC at www.aisc.org.

LOAD DETERMINATION

The required strength (factored axial tension) P_u is determined by analysis as indicated in LRFD Specification Section A5.

GROSS AREA, NET AREA AND EFFECTIVE AREA

In the determination of the design strength $\phi_t P_n$ of a tension member, the gross area A_g is needed for the tension yielding limit state and the effective area A_e is needed for the tension rupture limit state.

Gross Area

The gross area A_g is determined as specified in LRFD Specification Section B1.

Effective Area

The effective area A_e is determined by multiplying the net area A_n by the shear lag coefficient U, where A_n and U are determined as follows:

1. For tension members other than HSS, steel pipe and single angles, per LRFD Specification Section B3. Shear lag parameters for W-shapes and channels are illustrated in LRFD Commentary Figure C-B3.1.
2. For HSS and steel pipe, per HSS Specification Section 2.1. Shear lag parameters for HSS and steel pipe are illustrated in HSS Commentary Figures C2.1-1 and C2.1-2.
3. For single angles, per Single-Angle Specification Section 2. Shear lag parameters for single angles are illustrated in LRFD Commentary Figures C-B3.2 and C-B3.3.

Alternatively, LRFD Commentary Section B3 provides continued recognition of empirical U values from past editions of the AISC Specification.

TENSILE DESIGN STRENGTH

The tension yielding limit-state will control the tension-member design strength over tension rupture when the following relationship is satisfied:

$$\frac{A_e}{A_g} \geq \frac{0.9F_y}{0.75F_u}$$

Otherwise, the tension rupture limit-state will control over tension yielding.

Yielding Limit State

The tension yielding design strength $\phi_t P_n$, which must equal or exceed the required strength (factored axial tension) P_u, is determined as follows:

1. For tension members other than HSS, steel pipe and single-angles, per LRFD Specification Section D1(a) using Equation D1-1.

2. For HSS and steel pipe, per HSS Specification Section 3.1(a) using Equation 3.1-1.
3. For single angles, per Single-Angle Specification Section 2.

Rupture Limit State

The tension rupture design strength $\phi_t P_n$, which must equal or exceed the required strength (factored axial tension) P_u, is determined as follows:

1. For tension members other than HSS, steel pipe, single angles and threaded rods, per LRFD Specification Section D1(b) using Equation D1-2.
2. For HSS and steel pipe, per HSS Specification Section 3.1(b) using Equation 3.1-2.
3. For single angles, per Single-Angle Specification Section 2.
4. For threaded rods, per LRFD Specification Section J3.6.

OTHER SPECIFICATION REQUIREMENTS AND DESIGN CONSIDERATIONS

Special Requirements for Heavy Shapes and Plates

For tension members with complete-joint-penetration groove welded joints and made from heavy shapes (ASTM A6 tensile group 4 or 5 shapes) or built-up from plates exceeding 2-in. thickness, see LRFD Specification Section A3.1c.

Slenderness

Tension-member slenderness l/r should preferably be limited to a maximum of 300 per LRFD Specification Section B7, HSS Specification Section 2.3 and Single-Angle Specification Section 2. The intent of this recommendation is explained in the corresponding Commentaries.

Built-Up Tension Members

For built-up tension members, see LRFD Specification Section D2.

Pin-Connected Plates

For pin-connected tension members, see LRFD Specification Section D3.1.

Eyebars

For eyebars in tension, see LRFD Specification Section D3.2.

Concentrated Forces

If the tension member receives concentrated forces along its length, see LRFD Specification Section K1 or HSS Specification Section 8, as appropriate.

Fatigue

For tensile applications involving fatigue, see LRFD Specification Section K3.

Serviceability

Serviceability requirements, per LRFD Specification Sections A5.4 and L3, should be appropriate for the application, including an appropriate limit on the axial deformations of the tension member. See also AISC Design Guide No. 3 *Serviceability Design Considerations for Low-Rise Buildings*.

Connection Design

For tension-member connection design, see Parts 7, 8, 9, 13 and 15.

DESIGN EXAMPLES

EXAMPLE 3.1. W-shape tension member design.

Given: Determine the design strength of an ASTM A992 W8×24 with four lines of 3/4-in.-diameter bolts in standard holes, two per flange, as illustrated in Figure 3–1.

$F_y = 50$ ksi $A_g = 7.08$ in.^{2} $\bar{y} = 0.695$ (for WT4×12)
$F_u = 65$ ksi $t_f = 0.400$ in.
$r_y = 1.61$ in.

a) Assume the holes are located at the member end and the connection length is 9 in. Also, calculate at what length this tension member would cease to satisfy the slenderness limitation in LRFD Specification Section B7.

b) Assume the holes are located away from the member end so that shear-lag effects do not apply ($U = 1$).

Solution a: For tension yielding, per LRFD Specification Section D1(a),

$$\begin{aligned}\phi_t P_n &= \phi_t F_y A_g \\ &= 0.9(50\,\text{ksi})(7.08\,\text{in.}^2) \\ &= 319\ \text{kips}\end{aligned}$$

For tension rupture, per LRFD Specification Section D1(b),

$$\begin{aligned}U &= 1 - \frac{\bar{x}}{\ell} \le 0.9 \\ &= 1 - \frac{0.695\ \text{in.}}{9\ \text{in.}} \le 0.9 \\ &= 0.923 \le 0.9 \\ &= 0.9 \\ A_n &= A_g - 4(d_h + {}^1\!/_{16}\ \text{in.})t_f \\ &= 7.08\ \text{in.}^2 - 4({}^{13}\!/_{16}\ \text{in.} + {}^1\!/_{16}\ \text{in.})(0.400\ \text{in.}) \\ &= 5.68\ \text{in.}^2 \\ A_e &= UA_n \\ &= 0.9(5.68\ \text{in.}^2) \\ &= 5.11\ \text{in.}^2\end{aligned}$$

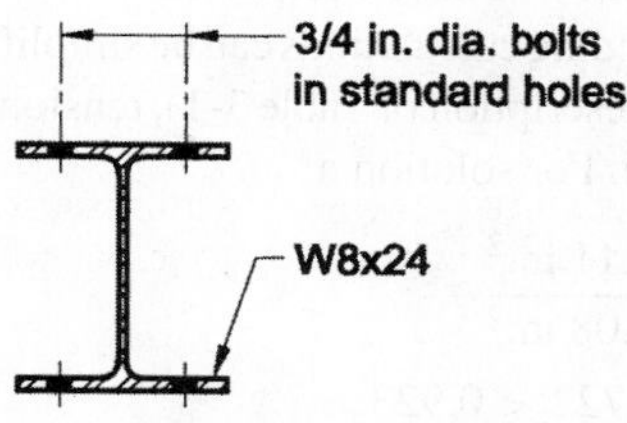

Fig. 3–1. Illustration for Example 3.1.

$$\phi_t P_n = \phi_t F_u A_e$$
$$= 0.75(65 \text{ ksi})(5.11 \text{ in.}^2)$$
$$= 249 \text{ kips}$$

Thus, the W8×24 tension member design strength is controlled by the tension rupture limit-state, where

$$\phi_t P_n = 249 \text{ kips}$$

Per LRFD Specification Section B7,

$$L_{\text{max}} = 300r$$
$$= \frac{300\,(1.61 \text{ in.})}{12 \text{ in./ft}}$$
$$= 40.3 \text{ ft}$$

Thus, the W8×24 tension member satisfies the slenderness requirements up to a 40.3-ft length.

Solution b: For tension yielding, $\phi_t P_n = 319$ kips as determined in solution a. For tension rupture, per LRFD Specification Section D1(b),

$$U = 1$$
$$A_n = 5.68 \text{ in.}^2 \text{ as determined in solution a.}$$
$$A_e = UA_n$$
$$= 1(5.68 \text{ in.}^2)$$
$$= 5.68 \text{ in.}^2$$
$$\phi_t P_n = \phi_t F_u A_e$$
$$= 0.75(65 \text{ ksi})(5.68 \text{ in.}^2)$$
$$= 277 \text{ kips}$$

Thus, the W8×24 tension member design strength is controlled by the tension rupture limit-state, where

$$\phi_t P_n = 277 \text{ kips}$$

Comments: Note that the tension rupture limit-state will often control the design strength of W-shape tension members, particularly at the end connections, where shear-lag effects must be considered. This should be considered during the member-selection process so that costly end-connection reinforcement will not be required.

The preceding calculations can be simplified using Table 3-1. If $A_e/A_g \geq 0.923$ (see description of Table 3-1), tension yielding will control over tension rupture. For solution a,

$$\frac{A_e}{A_g} = \frac{5.11 \text{ in.}^2}{7.08 \text{ in.}^2}$$
$$= 0.722 < 0.923$$

Therefore, tension rupture controls. For tension rupture, the W8×24 design strength with $A_e = 0.75A_g = 5.31$ in.^{2} is tabulated as 259 kips.

$$\phi_t P_n = 259 \text{ kips}\left(\frac{A_e}{0.75A_g}\right)$$
$$= 259 \text{ kips}\left(\frac{5.11 \text{ in.}^2}{5.31 \text{ in.}^2}\right)$$
$$= 249 \text{ kips}$$

Similarly, for solution b,

$$\frac{A_e}{A_g} = \frac{5.68 \text{ in.}^2}{7.08 \text{ in.}^2}$$
$$= 0.802 < 0.923$$

Therefore, tension rupture controls. For tension rupture, the W8×24 design strength with $A_e = 0.75A_g = 5.31$ in.^{2} is tabulated as 259 kips.

$$\phi_t P_n = 259 \text{ kips}\left(\frac{A_e}{0.75A_g}\right)$$
$$= 259 \text{ kips}\left(\frac{5.68 \text{ in.}^2}{5.31 \text{ in.}^2}\right)$$
$$= 277 \text{ kips}$$

Note that end-connection limit-states, such as block shear rupture and bolt bearing strength must also be checked.

EXAMPLE 3.2. Single-angle tension member design.

Given: Determine the design strength of an ASTM A36 L4×4×1/2 with one line of 3/4-in.-diameter bolts in standard holes, two per flange, as illustrated in Figure 3–2. Assume the connection length is 18 in. Also, calculate at what length this tension member would cease to satisfy the slenderness limitation in Single-Angle Specification Section 2.

$F_y = 36$ ksi $A_g = 3.75$ in.^{2} $r = 1.21$ in.
$F_u = 58$ ksi $\bar{y} = 1.18$

Solution: For tension yielding, per Single-Angle Specification Section 2,

$$\phi_t P_n = \phi_t F_y A_g$$
$$= 0.9(36 \text{ ksi})(3.75 \text{ in.}^2)$$
$$= 122 \text{ kips}$$

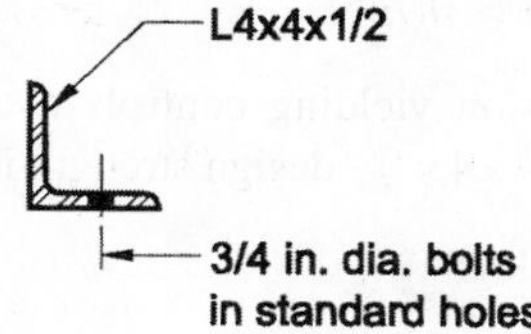

Fig. 3–2. Illustration for Example 3.2.

For tension rupture, per Single-Angle Specification Section 2,

$$U = 1 - \frac{\overline{x}}{\ell} \le 0.9$$

$$= 1 - \frac{1.18 \text{ in.}}{18 \text{ in.}} \le 0.9$$

$$= 0.934 \le 0.9$$

$$= 0.9$$

$$A_n = A_g - (d_h + {}^1/_{16} \text{ in.})t$$

$$= 3.75 \text{ in.}^2 - ({}^{13}/_{16} \text{ in.} + {}^1/_{16} \text{ in.})({}^1/_2 \text{ in.})$$

$$= 3.31 \text{ in.}^2$$

$$A_e = UA_n$$

$$= 0.9(3.31 \text{ in.}^2)$$

$$= 2.98 \text{ in.}^2$$

$$\phi_t P_n = \phi_t F_u A_e$$

$$= 0.75(58 \text{ ksi})(2.98 \text{ in.}^2)$$

$$= 130 \text{ kips}$$

Thus, the L4×4×1/2 tension member design strength is controlled by the tension yielding limit-state, where

$$\phi_t P_n = 122 \text{ kips}$$

Per Single-Angle Specification Section B2,

$$L_{\max} = 300r$$

$$= \frac{300\,(1.21 \text{ in.})}{12 \text{ in./ft}}$$

$$= 30.3 \text{ ft}$$

Thus, the L4×4×1/2 tension member satisfies the slenderness requirements up to a 30.3-ft length.

Comments: The preceding calculations can be simplified using Table 3-2. If $A_e/A_g \ge 0.745$ (see description of Table 3-2), tension yielding will control over tension rupture.

$$\frac{A_e}{A_g} = \frac{2.98 \text{ in.}^2}{3.75 \text{ in.}^2}$$

$$= 0.795 > 0.745$$

Therefore, tension yielding controls over tension rupture. For tension yielding, the L4×4×1/2 design strength is tabulated as

$$\phi_t P_n = 122 \text{ kips}$$

Note that end-connection limit-states, such as block shear rupture and bolt bearing strength must also be checked.

EXAMPLE 3.3. WT-shape tension member design.

Given: Select an ASTM A992 WT-shape tension member ($F_y = 50$ ksi, $F_u = 65$ ksi) for a required strength of 250 kips and a member length of 30 ft. Assume the end connection is fillet welded as illustrated in Figure 3–3 and has a length of 16 in.

Solution: From Table 3-3, try a WT6×20 ($A_g = 5.84$ in.^{2}, $\overline{y} = 1.09$, $r_x = 1.57$). For tension yielding, the WT6×20 design strength is tabulated as

$$\phi_t P_n = 263 \text{ kips} > 250 \text{ kips } \mathbf{o.k.}$$

For tension rupture, the WT6×20 design strength with $A_e = 0.75A_g = 4.38$ in.^{2} is tabulated as 214 kips.

$$U = 1 - \frac{\overline{x}}{\ell} \leq 0.9$$

$$= 1 - \frac{1.09 \text{ in.}}{16 \text{ in.}} \leq 0.9$$

$$= 0.932 \leq 0.9$$

$$= 0.9$$

$$A_n = A_g$$

$$= 5.84 \text{ in.}^2$$

$$A_e = UA_n$$

$$= 0.9(5.84 \text{ in.}^2)$$

$$= 5.26 \text{ in.}^2$$

$$\phi_t P_n = 214 \text{ kips} \left(\frac{A_e}{0.75A_g}\right)$$

$$= 214 \text{ kips} \left(\frac{5.26 \text{ in.}^2}{4.38 \text{ in.}^2}\right)$$

$$= 256 \text{ kips} > 250 \text{ kips } \mathbf{o.k.}$$

Per LRFD Specification Section B7,

$$L_{\max} = 300r$$

$$= \frac{300\,(1.57 \text{ in.})}{12 \text{ in./ft}}$$

$$= 39.3 \text{ ft} > 30 \text{ ft } \mathbf{o.k.}$$

Thus, the WT6×20 tension member is **o.k.**

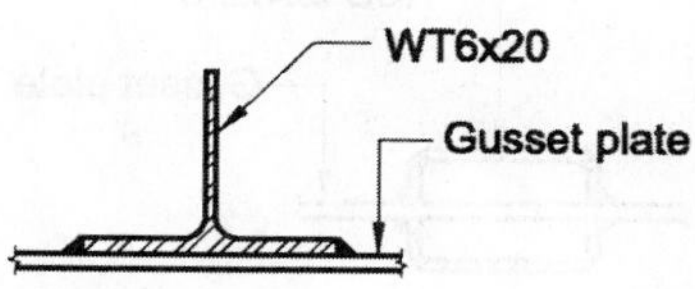

Fig. 3–3. Illustration for Example 3.3.

Comments: Note that the tension rupture limit-state will often control the design strength of W-shape tension members, particularly at the end connections, where shear-lag effects must be considered. This should be considered during the member-selection process so that costly end-connection reinforcement will not be required.

Note that end-connection limit-states, such as block shear rupture and bolt bearing strength must also be checked.

EXAMPLE 3.4. Rectangular HSS tension member design.

Given: Select an ASTM A500 rectangular HSS tension member ($F_y = 46$ ksi, $F_u = 58$ ksi) for a required strength of 225 kips and a member length of 30 ft. Assume the end connection has a length of 16 in. and is a fillet-welded $^1/_2$-in. thick single concentric gusset plate as illustrated in Figure 3–4.

Solution: From Table 3-4, try an HSS6×4×$^3/_8$ ($A_g = 6.18$ in.^{2}, $t_{des} = 0.349$, $r_y = 1.55$). For tension yielding, the HSS6×4×$^3/_8$ design strength is tabulated as

$$\phi_t P_n = 256 \text{ kips} > 225 \text{ kips } \mathbf{o.k.}$$

For tension rupture, the HSS6×4×$^3/_8$ design strength with $A_e = 0.75A_g = 4.64$ in.^{2} is tabulated as 202 kips. From HSS Specification Section 2.1,

$$\begin{aligned} \bar{x} &= \frac{B^2 + 2BH}{4\,(B + H)} \\ &= \frac{(4\text{ in.})^2 + 2\,(4\text{ in.})(6\text{ in.})}{4\,(4\text{ in.} + 6\text{ in.})} \\ &= 1.60\text{ in.} \\ U &= 1 - \frac{\bar{x}}{\ell} \le 0.9 \\ &= 1 - \frac{1.60\text{ in.}}{16\text{ in.}} \le 0.9 \\ &= 0.900 \le 0.9 \\ &= 0.9 \end{aligned}$$

Allowing for a $^1/_{16}$-in. gap in fit-up between the HSS and the gusset plate,

$$\begin{aligned} A_n &= A_g - 2(t_{gusset} + {}^1/_{16}\text{ in.})t_{HSS} \\ &= 6.18\text{ in.}^2 - 2({}^1/_2\text{ in.} + {}^1/_{16}\text{ in.})(0.349\text{ in.}) \\ &= 5.79\text{ in.}^2 \end{aligned}$$

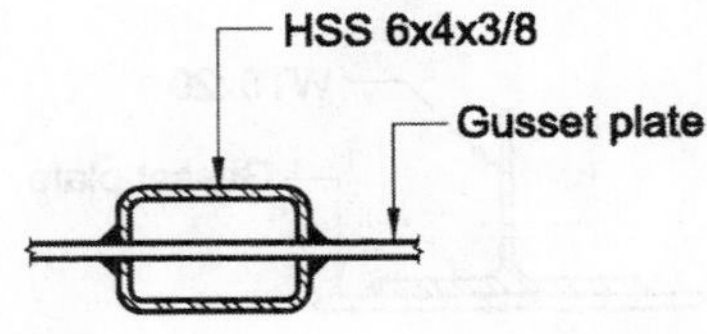

Fig. 3–4. Illustration for Example 3.4.

$$
\begin{aligned}
A_e &= UA_n \\
&= 0.9(5.79\text{ in.}^2) \\
&= 5.21\text{ in.}^2 \\
\phi_t P_n &= 202\text{ kips}\left(\frac{A_e}{0.75A_g}\right) \\
&= 202\text{ kips}\left(\frac{5.21\text{ in.}^2}{4.64\text{ in.}^2}\right) \\
&= 227\text{ kips} > 225\text{ kips } \mathbf{o.k.}
\end{aligned}
$$

Per HSS Specification Section 2.3,

$$
\begin{aligned}
L_{\max} &= 300r \\
&= \frac{300\,(1.55\text{ in.})}{12\text{ in./ft}} \\
&= 38.8\text{ ft} > 30\text{ ft } \mathbf{o.k.}
\end{aligned}
$$

Thus, the HSS6×4×3/8 tension member is **o.k.**

Comments: Note that the tension rupture limit-state will often control the design strength of HSS tension members, particularly at the end connections, where shear-lag effects must be considered. This should be considered during the member-selection process so that costly end-connection reinforcement will not be required.

Note that end-connection limit-states, such as shear rupture of the welds and HSS wall must also be checked.

EXAMPLE 3.5. Round HSS tension member design.

Given: Select an ASTM A500 round HSS tension member ($F_y = 42$ ksi, $F_u = 58$ ksi) for a required strength of 275 kips and a member length of 30 ft. Assume the end connection has a length of 16 in. and is a fillet-welded 1/2-in. thick single concentric gusset plate as illustrated in Figure 3–5.

Solution: From Table 3-5, try an HSS6.000×0.500 ($A_g = 8.09$ in.^{2}, $t_{des} = 0.465$, $r = 1.96$). For tension yielding, the HSS6.000×0.500 design strength is tabulated as

$$\phi_t P_n = 306\text{ kips} > 275\text{ kips } \mathbf{o.k.}$$

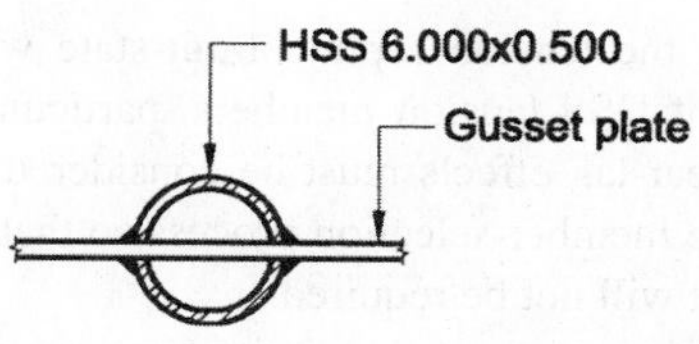

Fig. 3–5. Illustration for Examples 3.5 and 3.6.

For tension rupture, the HSS6.000×0.500 design strength with $A_e = 0.75A_g = 6.07$ in.^{2} is tabulated as 264 kips. From HSS Specification Section 2.1,

$$\begin{aligned}
\bar{x} &= \frac{D}{\pi} \\
&= \frac{6 \text{ in.}}{\pi} \\
&= 1.91 \text{ in.} \\
U &= 1 - \frac{\bar{x}}{\ell} \leq 0.9 \\
&= 1 - \frac{1.91 \text{ in.}}{16 \text{ in.}} \leq 0.9 \\
&= 0.881 \leq 0.9 \\
&= 0.881
\end{aligned}$$

Allowing for a $^1/_{16}$-in. gap in fit-up between the HSS and the gusset plate,

$$\begin{aligned}
A_n &= A_g - 2(t_{\text{gusset}} + {}^1/_{16} \text{ in.})t_{HSS} \\
&= 8.09 \text{ in.}^2 - 2({}^1/_2 \text{ in.} + {}^1/_{16} \text{ in.})(0.465 \text{ in.}) \\
&= 7.57 \text{ in.}^2 \\
A_e &= UA_n \\
&= 0.881(7.57 \text{ in.}^2) \\
&= 6.67 \text{ in.}^2 \\
\phi_t P_n &= 264 \text{ kips}\left(\frac{A_e}{0.75A_g}\right) \\
&= 264 \text{ kips}\left(\frac{6.67 \text{ in.}^2}{6.07 \text{ in.}^2}\right) \\
&= 290 \text{ kips} > 275 \text{ kips } \textbf{o.k.}
\end{aligned}$$

Per HSS Specification Section 2.3,

$$\begin{aligned}
L_{\max} &= 300r \\
&= \frac{300\,(1.96 \text{ in.})}{12 \text{ in./ft}} \\
&= 49.0 \text{ ft} > 30 \text{ ft } \textbf{o.k.}
\end{aligned}$$

Thus, the HSS6.000×0.500 tension member is **o.k.**

Comments: Note that the tension rupture limit-state will often control the design strength of HSS tension members, particularly at the end connections, where shear-lag effects must be considered. This should be considered during the member-selection process so that costly end-connection reinforcement will not be required.

Note that end-connection limit-states, such as shear rupture of the welds and HSS wall must also be checked.

EXAMPLE 3.6. Steel pipe tension member design.

Given: Repeat example 3.6 using an ASTM A53 grade B steel pipe tension member ($F_y = 35$ ksi, $F_u = 60$ ksi) for a required strength of 175 kips.

Solution: From Table 3-6, try Pipe 5 ×-strong ($OD = 5.56$ in., $A_g = 6.11$ in.^{2}, $t = 0.375$, $r = 1.84$). For tension yielding, the Pipe 5 ×-strong design strength is tabulated as

$$\phi_t P_n = 192 \text{ kips} > 175 \text{ kips } \mathbf{o.k.}$$

For tension rupture, the Pipe 5 ×-strong design strength with $A_e = 0.75A_g = 4.58$ in.^{2} is tabulated as 206 kips. From HSS Specification Section 2.1,

$$\begin{aligned}\bar{x} &= \frac{D}{\pi}\\ &= \frac{5.56 \text{ in.}}{\pi}\\ &= 1.77 \text{ in.}\end{aligned}$$

$$\begin{aligned}U &= 1 - \frac{\bar{x}}{\ell} \leq 0.9\\ &= 1 - \frac{1.77 \text{ in.}}{16 \text{ in.}} \leq 0.9\\ &= 0.889 \leq 0.9\\ &= 0.889\end{aligned}$$

Allowing for a $^1/_{16}$-in. gap in fit-up between the HSS and the gusset plate,

$$\begin{aligned}A_n &= A_g - 2(t_{\text{gusset}} + {}^1/_{16} \text{ in.})t_{\text{pipe}}\\ &= 6.11 \text{ in.}^2 - 2({}^1/_2 \text{ in.} + {}^1/_{16} \text{ in.})(0.375 \text{ in.})\\ &= 5.69 \text{ in.}^2\end{aligned}$$

$$\begin{aligned}A_e &= UA_n\\ &= 0.889(5.69 \text{ in.}^2)\\ &= 5.06 \text{ in.}^2\end{aligned}$$

$$\begin{aligned}\phi_t P_n &= 206 \text{ kips}\left(\frac{A_e}{0.75A_g}\right)\\ &= 206 \text{ kips}\left(\frac{5.06 \text{ in.}^2}{4.58 \text{ in.}^2}\right)\\ &= 228 \text{ kips} > 175 \text{ kips } \mathbf{o.k.}\end{aligned}$$

Per HSS Specification Section 2.3,

$$\begin{aligned}L_{\max} &= 300r\\ &= \frac{300\,(1.84 \text{ in.})}{12 \text{ in./ft}}\\ &= 46.0 \text{ ft} > 30 \text{ ft } \mathbf{o.k.}\end{aligned}$$

Thus, the Pipe 5 ×-strong tension member is **o.k.**

Comments: Note that end-connection limit-states, such as shear rupture of the welds and steel pipe wall must also be checked.

EXAMPLE 3.7. Double-angle tension member design.

Given: Determine the design strength of an ASTM A36 2L4×4×1/2 (3/8-in. separation) with one line of 3/4-in.-diameter bolts in standard holes as illustrated in Figure 3–6. Assume the connection length is 18 in. Also, calculate at what length and spacing of intermediate connectors this tension member would cease to satisfy the slenderness limitation in LRFD Specification Sections B7 and D2, respectively.

$F_y = 36$ ksi $A_g = 3.75$ in.^{2} (for single L4×4×1/2) $r_x = 1.21$ in.
$F_u = 58$ ksi $\overline{x} = 1.18$ in. (for single L4×4×1/2)

Solution: For tension yielding, per LRFD Specification Section D1(a),

$$\begin{aligned}\phi_t P_n &= \phi_t F_y A_g \\ &= 0.9(36\text{ ksi})(2 \times 3.75\text{ in.}^2) \\ &= 243\text{ kips}\end{aligned}$$

For tension rupture, per Single-Angle Specification Section 2,

$$\begin{aligned}U &= 1 - \frac{\overline{x}}{\ell} \leq 0.9 \\ &= 1 - \frac{1.18\text{ in.}}{18\text{ in.}} \leq 0.9 \\ &= 0.934 \leq 0.9 \\ &= 0.9\end{aligned}$$

$$\begin{aligned}A_n &= A_g - 2(d_h + {}^1\!/_{16}\text{ in.})t \\ &= 2 \times 3.75\text{ in.}^2 - 2({}^{13}\!/_{16}\text{ in.} + {}^1\!/_{16}\text{ in.})({}^1\!/_2\text{ in.}) \\ &= 6.63\text{ in.}^2\end{aligned}$$

$$\begin{aligned}A_e &= UA_n \\ &= 0.9(6.63\text{ in.}^2) \\ &= 5.97\text{ in.}^2\end{aligned}$$

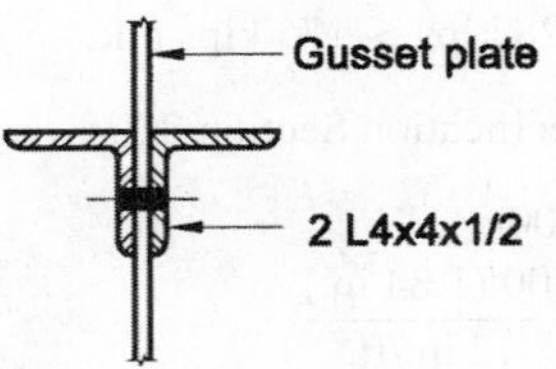

Fig. 3–6. Illustration for Example 3.7.

$$\begin{aligned}\phi_t P_n &= \phi_t F_u A_e \\ &= 0.75(58 \text{ ksi})(5.97 \text{ in.}^2) \\ &= 260 \text{ kips}\end{aligned}$$

Thus, the 2L4×4×1/2 tension member design strength is controlled by the tension yielding limit-state, where

$$\phi_t P_n = 243 \text{ kips}$$

Per LRFD Specification Section B7 for overall slenderness,

$$\begin{aligned}L_{\max} &= 300r \\ &= \frac{300\ (1.21 \text{ in.})}{12 \text{ in./ft}} \\ &= 30.3 \text{ ft}\end{aligned}$$

Thus, the 2L4×4×1/2 tension member satisfies the slenderness requirements up to a 45.8-ft length. Per LRFD Specification Section D2 for length between intermediate connectors,

$$\begin{aligned}L_{\max} &= 300r \\ &= \frac{300\ (1.21 \text{ in.})}{12 \text{ in./ft}} \\ &= 30.3 \text{ ft}\end{aligned}$$

Although this 2L4×4×1/2 tension member would satisfy the slenderness limit in LRFD Specification Section D2 with intermediate connectors spaced not more than 30.3 ft apart, a minimum of two intermediate connectors at the third points are normally provided.

Comments: The preceding calculations can be simplified using Table 3-7. If $A_e/A_g \geq$ 0.745 (see description of Table 3-7), tension yielding will control over tension rupture.

$$\begin{aligned}\frac{A_e}{A_g} &= \frac{5.97 \text{ in.}^2}{7.50 \text{ in.}^2} \\ &= 0.796 > 0.745\end{aligned}$$

Therefore, tension yielding controls over tension rupture. For tension yielding, the L4×4×1/2 design strength is tabulated as

$$\phi_t P_n = 243 \text{ kips}$$

Note that end-connection limit-states, such as block shear rupture and bolt bearing strength must also be checked.

STEEL TENSION-MEMBER SELECTION TABLES

Design strengths in axial tension for various types of tension members (see individual descriptions below) are given in Tables 3-1 through 3-7 for the limit-states of tension yielding and tension rupture. In each case, the tabulated values for design tension rupture strength $\phi_t P_n$ are based upon the assumption that $A_e = 0.75A_g$. While not a design requirement, such consideration of the effective area during the design of the member will simplify the design of its end connections, which can be difficult to configure and costly if tension members are selected based upon design tension yielding strength only.

When $A_e > 0.75A_g$, either the tabulated values for design tension rupture strength can be used conservatively or the design tension rupture strength can be calculated based upon the actual value of A_e. When $A_e < 0.75A_g$, the tabulated values of the design tension rupture strength cannot be used, but rather, must be calculated based upon the actual value of A_e.

For the design of tension-member end connections, see Parts 7, 8, 9, 13 and 15.

Table 3-1. W-Shapes

Design strengths in axial tension are given for W-shapes with $F_y = 50$ ksi and $F_u = 65$ ksi (ASTM A992). Note that tension rupture will control over tension yielding given for W-shapes with $F_y = 50$ ksi and $F_u = 65$ ksi when $A_e/A_g < 0.923$. Otherwise, tension yielding will control over tension rupture. The use of Table 3-1 is illustrated in Example 3-1.

Table 3-2. Single Angles

Design strengths in axial tension are given for single angles with $F_y = 36$ ksi and $F_u = 58$ ksi (ASTM A36). Note that tension rupture will control over tension yielding given for single angles with $F_y = 36$ ksi and $F_u = 58$ ksi when $A_e/A_g < 0.745$. Otherwise, tension yielding will control over tension rupture. The use of Table 3-2 is illustrated in Example 3-2.

Table 3-3. WT-Shapes

Design strengths in axial tension are given for WT-shapes with $F_y = 50$ ksi and $F_u = 65$ ksi (ASTM A992). Note that tension rupture will control over tension yielding given for W-shapes with $F_y = 50$ ksi and $F_u = 65$ ksi when $A_e/A_g < 0.923$. Otherwise, tension yielding will control over tension rupture. The use of Table 3-3 is illustrated in Example 3-3.

Table 3-4. Rectangular (and Square) HSS

Design strengths in axial tension are given for rectangular HSS with $F_y = 46$ ksi and $F_u = 58$ ksi (ASTM A500 grade B). Note that tension rupture will control over tension yielding given for rectangular HSS with $F_y = 46$ ksi and $F_u = 58$ ksi when $A_e/A_g < 0.952$. Otherwise, tension yielding will control over tension rupture. The use of Table 3-4 is illustrated in Example 3-4.

Table 3-5. Round HSS

Design strengths in axial tension are given for ASTM A500 round HSS with $F_y = 42$ ksi and $F_u = 58$ ksi (ASTM A500 grade B). Note that tension rupture will control over tension yielding given for round HSS with $F_y = 42$ ksi and $F_u = 58$ ksi when $A_e/A_g < 0.869$. Otherwise, tension yielding will control over tension rupture. The use of Table 3-5 is illustrated in Example 3-5.

Table 3-6. Steel Pipe

Design strengths in axial tension are given for steel pipe with $F_y = 35$ ksi and $F_u = 60$ ksi (ASTM A53 grade B). Note that tension rupture will control over tension yielding given for steel pipe with $F_y = 35$ ksi and $F_u = 60$ ksi when $A_e/A_g < 0.700$. Otherwise, tension yielding will control over tension rupture. The use of Table 3-6 is illustrated in Example 3-6.

Table 3-7. Double Angles

Design strengths in axial tension are given for double angles with $F_y = 36$ ksi and $F_u = 58$ ksi (ASTM A36). Note that tension rupture will control over tension yielding given for double angles with $F_y = 36$ ksi and $F_u = 58$ ksi when $A_e/A_g < 0.745$. Otherwise, tension yielding will control over tension rupture. The use of Table 3-7 is illustrated in Example 3-7.

$F_y = 50$ ksi
$F_u = 65$ ksi
$\phi_t P_n = 0.90 F_y A_g$ (Yielding)
$\phi_t P_n = 0.75 F_u A_e$ (Rupture)
$A_e = 0.75 A_g$

Table 3-1. W-Shapes Design Strengths in Axial Tension†

Shape	Gross Area, A_g	Effective Net Area, A_e	Design Strength, $\phi_t P_n$ Yielding	Design Strength, $\phi_t P_n$ Rupture
	in.^{2}	in.^{2}	kips	kips
W44×335	98.3	73.7	4420	3590
×290	85.8	64.4	3860	3140
×262	77.2	57.9	3470	2820
×230	67.7	50.8	3050	2480
W40×593*	174	131	7830	6390
×503*	148	111	6660	5410
×431*	127	95.3	5720	4650
×397*	117	87.8	5270	4280
×372*	109	81.8	4910	3990
×362*	107	80.3	4820	3910
×324	95.3	71.5	4290	3490
×297	87.4	65.6	3930	3200
×277	81.4	61.1	3660	2980
×249	73.3	55.0	3300	2680
×215	63.4	47.6	2850	2320
×199	58.5	43.9	2630	2140
W40×392*	115	86.3	5180	4210
×331	97.5	73.1	4390	3560
×327	96.0	72.0	4320	3510
×278	81.8	61.4	3680	2990
×264	77.6	58.2	3490	2840
×235	69.0	51.8	3110	2530
×211	62.0	46.5	2790	2270
×183	53.8	40.4	2420	1970
×167	49.2	36.9	2210	1800
×149	43.8	32.9	1970	1600
W36×798*	235	176	10600	8580
×650*	191	143	8600	6970
×527*	155	116	6980	5660
×439*	129	96.8	5810	4720
×393*	116	87.0	5220	4240
×359*	105	78.8	4730	3840
×328*	96.4	72.3	4340	3520
×300	88.3	66.2	3970	3230
×280	82.4	61.8	3710	3010
×260	76.5	57.4	3440	2800
×245	72.1	54.1	3240	2640
×230	67.6	50.7	3040	2470
W36×256	75.4	56.6	3390	2760
×232	68.1	51.1	3060	2490
×210	61.8	46.4	2780	2260
×194	57.0	42.8	2570	2090
×182	53.6	40.2	2410	1960
×170	50.1	37.6	2250	1830
×160	47.0	35.3	2120	1720
×150	44.2	33.2	1990	1620
×135	39.7	29.8	1790	1450
W33×387*	114	85.5	5130	4170
×354*	104	78.0	4680	3800
×318*	93.6	70.2	4210	3420
×291	85.7	64.3	3860	3130
×263	77.5	58.1	3490	2830
×241	71.0	53.3	3200	2600
×221	65.2	48.9	2930	2380
×201	59.2	44.4	2660	2160
W33×169	49.5	37.1	2230	1810
×152	44.8	33.6	2020	1640
×141	41.6	31.2	1870	1520
×130	38.3	28.7	1720	1400
×118	34.7	26.0	1560	1270
W30×391*	115	86.3	5180	4210
×357*	105	78.8	4730	3840
×326*	95.8	71.9	4310	3510
×292*	85.9	64.4	3870	3140
×261	76.9	57.7	3460	2810
×235	69.2	51.9	3110	2530
×211	62.2	46.7	2800	2280
×191	56.3	42.2	2530	2060
×173	51.0	38.3	2300	1870
W30×148	43.5	32.6	1960	1590
×132	38.9	29.2	1750	1420
×124	36.5	27.4	1640	1340
×116	34.2	25.7	1540	1250
×108	31.7	23.8	1430	1160
×99	29.1	21.8	1310	1060
×90	26.4	19.8	1190	965
W27×539*	159	119	7160	5800
×368*	108	81.0	4860	3950
×336*	98.9	74.2	4450	3620
×307*	90.4	67.8	4070	3310
×281*	82.9	62.2	3730	3030
×258	76.0	57.0	3420	2780
×235	69.4	52.1	3120	2540
×217	64.0	48.0	2880	2340
×194	57.2	42.9	2570	2090
×178	52.5	39.4	2360	1920
×161	47.6	35.7	2140	1740
×146	43.1	32.3	1940	1570
W27×129	37.8	28.4	1700	1380
×114	33.5	25.1	1510	1220
×102	30.0	22.5	1350	1100
×94	27.7	20.8	1250	1010
×84	24.8	18.6	1120	907

† Tension rupture on the effective net area will control over tension yielding on the gross area unless the tension member is selected so that an end connection can be configured with $A_e \geq 0.923 A_g$.
*ASTM A6 tensile group 4 or 5 shape. Special requirements may apply per LRFD Specification Section A3.1c.

Table 3-1 (cont.). W-Shapes Design Strengths in Axial Tension†

$F_y = 50$ ksi

$F_u = 65$ ksi

$\phi_t P_n = 0.90 F_y A_g$ (Yielding)

$\phi_t P_n = 0.75 F_u A_e$ (Rupture)

$A_e = 0.75 A_g$

Shape	Gross Area, A_g	Effective Net Area, A_e	Design Strength, $\phi_t P_n$	
			Yielding	Rupture
	in.^{2}	in.^{2}	kips	kips
W24×370*	109	81.8	4910	3990
×335*	98.4	73.8	4430	3600
×306*	89.8	67.4	4040	3290
×279*	82.0	61.5	3690	3000
×250*	73.5	55.1	3310	2690
×229	67.2	50.4	3020	2460
×207	60.7	45.5	2730	2220
×192	56.3	42.2	2530	2060
×176	51.7	38.8	2330	1890
×162	47.7	35.8	2150	1750
×146	43.0	32.3	1940	1570
×131	38.5	28.9	1730	1410
×117	34.4	25.8	1550	1260
×104	30.6	23.0	1380	1120
W24×103	30.3	22.7	1360	1110
×94	27.7	20.8	1250	1010
×84	24.7	18.5	1110	902
×76	22.4	16.8	1010	819
×68	20.1	15.1	905	736
W24×62	18.3	13.7	824	668
×55	16.3	12.2	734	595
W21×201	59.2	44.4	2660	2160
×182	53.6	40.2	2410	1960
×166	48.8	36.6	2200	1780
×147	43.2	32.4	1940	1580
×132	38.8	29.1	1750	1420
×122	35.9	26.9	1620	1310
×111	32.7	24.5	1470	1190
×101	29.8	22.4	1340	1090
W21×93	27.3	20.5	1230	999
×83	24.3	18.2	1090	887
×73	21.5	16.1	968	785
×68	20.0	15.0	900	731
×62	18.3	13.7	824	668
×55	16.2	12.2	729	595
×48	14.1	10.6	635	517
W21×57	16.7	12.5	752	609
×50	14.7	11.0	662	536
×44	13.0	9.75	585	475
W18×175	51.3	38.5	2310	1880
×158	46.3	34.7	2080	1690
×143	42.1	31.6	1890	1540
×130	38.2	28.7	1720	1400
×119	35.1	26.3	1580	1280
×106	31.1	23.3	1400	1140
×97	28.5	21.4	1280	1040
×86	25.3	19.0	1140	926
×76	22.3	16.7	1000	814

Shape	Gross Area, A_g	Effective Net Area, A_e	Design Strength, $\phi_t P_n$	
			Yielding	Rupture
	in.^{2}	in.^{2}	kips	kips
W18×71	20.8	15.6	936	761
×65	19.1	14.3	860	697
×60	17.6	13.2	792	644
×55	16.2	12.2	729	595
×50	14.7	11.0	662	536
W18×46	13.5	10.1	608	492
×40	11.8	8.85	531	431
×35	10.3	7.73	464	377
W16×100	29.7	22.3	1340	1090
×89	26.4	19.8	1190	965
×77	22.9	17.2	1030	839
×67	20.0	15.0	900	731
W16×57	16.8	12.6	756	614
×50	14.7	11.0	662	536
×45	13.3	9.98	599	487
×40	11.8	8.85	531	431
×36	10.6	7.95	477	388
W16×31	9.13	6.85	411	334
×26	7.68	5.76	346	281
W14×808*	237	178	10700	8680
×730*	215	161	9680	7850
×665*	196	147	8820	7170
×605*	178	134	8010	6530
×550*	162	122	7290	5950
×500*	147	110	6620	5360
×455*	134	101	6030	4920
×426*	125	93.8	5630	4570
×398*	117	87.8	5270	4280
×370*	109	81.8	4910	3990
×342*	101	75.8	4550	3700
×311*	91.4	68.6	4110	3340
×283*	83.3	62.5	3750	3050
×257*	75.6	56.7	3400	2760
×233*	68.5	51.4	3080	2510
×211	62.0	46.5	2790	2270
×193	56.8	42.6	2560	2080
×176	51.8	38.9	2330	1900
×159	46.7	35.0	2100	1710
×145	42.7	32.0	1920	1560
W14×132	38.8	29.1	1750	1420
×120	35.3	26.5	1590	1290
×109	32.0	24.0	1440	1170
×99	29.1	21.8	1310	1060
×90	26.5	19.9	1190	970

† Tension rupture on the effective net area will control over tension yielding on the gross area unless the tension member is selected so that an end connection can be configured with $A_e \geq 0.923 A_g$.

* ASTM A6 tensile group 4 or 5 shape. Special requirements may apply per LRFD Specification Section A3.1c.

$F_y = 50$ ksi
$F_u = 65$ ksi
$\phi_t P_n = 0.90 F_y A_g$ (Yielding)
$\phi_t P_n = 0.75 F_u A_e$ (Rupture)
$A_e = 0.75 A_g$

Table 3-1 (cont.). W-Shapes Design Strengths in Axial Tension†

Shape	Gross Area, A_g	Effective Net Area, A_e	Design Strength, $\phi_t P_n$	
			Yielding	Rupture
	in.²	in.²	kips	kips
W14×82	24.0	18.0	1080	878
×74	21.8	16.4	981	800
×68	20.0	15.0	900	731
×61	17.9	13.4	806	653
W14×53	15.6	11.7	702	570
×48	14.1	10.6	635	517
×43	12.6	9.45	567	461
W14×38	11.2	8.40	504	410
×34	10.0	7.50	450	366
×30	8.85	6.64	398	324
W14×26	7.69	5.77	346	281
×22	6.49	4.87	292	237
W12×336*	98.8	74.1	4450	3610
×305*	89.6	67.2	4030	3280
×279*	81.9	61.4	3690	2990
×252*	74.0	55.5	3330	2710
×230*	67.7	50.8	3050	2480
×210*	61.8	46.4	2780	2260
×190	55.8	41.9	2510	2040
×170	50.0	37.5	2250	1830
×152	44.7	33.5	2010	1630
×136	39.9	29.9	1800	1460
×120	35.3	26.5	1590	1290
×106	31.2	23.4	1400	1140
×96	28.2	21.2	1270	1030
×87	25.6	19.2	1150	936
×79	23.2	17.4	1040	848
×72	21.1	15.8	950	770
×65	19.1	14.3	860	697
W12×58	17.0	12.8	765	624
×53	15.6	11.7	702	570
W12×50	14.6	11.0	657	536
×45	13.1	9.83	590	479
×40	11.7	8.78	527	428
W12×35	10.3	7.73	464	377
×30	8.79	6.59	396	321
×26	7.65	5.74	344	280
W12×22	6.48	4.86	292	237
×19	5.57	4.18	251	204
×16	4.71	3.53	212	172
×14	4.16	3.12	187	152
W10×112	32.9	24.7	1480	1200
×100	29.4	22.1	1320	1080
×88	25.9	19.4	1170	946
×77	22.6	17.0	1020	829
×68	20.0	15.0	900	731
×60	17.6	13.2	792	644
×54	15.8	11.9	711	580
×49	14.4	10.8	648	527
W10×45	13.3	9.98	599	487
×39	11.5	8.63	518	421
×33	9.71	7.28	437	355
W10×30	8.84	6.63	398	323
×26	7.61	5.71	342	278
×22	6.49	4.87	292	237
W10×19	5.62	4.22	253	206
×17	4.99	3.74	225	182
×15	4.41	3.31	198	161
×12	3.54	2.66	159	130
W8×67	19.7	14.8	887	722
×58	17.1	12.8	770	624
×48	14.1	10.6	635	517
×40	11.7	8.78	527	428
×35	10.3	7.73	464	377
×31	9.12	6.84	410	333
W8×28	8.24	6.18	371	301
×24	7.08	5.31	319	259
W8×21	6.16	4.62	277	225
×18	5.26	3.95	237	193
W8×15	4.44	3.33	200	162
×13	3.84	2.88	173	140
×10	2.96	2.22	133	108

† Tension rupture on the effective net area will control over tension yielding on the gross area unless the tension member is selected so that an end connection can be configured with $A_e \geq 0.923 A_g$.
* ASTM A6 tensile group 4 or 5 shape. Special requirements may apply per LRFD Specification Section A3.1c.

Table 3-2. Angles (L-Shapes) Design Strengths in Axial Tension†

$F_y = 36$ ksi
$F_u = 58$ ksi
$\phi_t P_n = 0.90 F_y A_g$ (Yielding)
$\phi_t P_n = 0.75 F_u A_e$ (Rupture)
$A_e = 0.75 A_g$

Shape	Gross Area, A_g	Effective Net Area, A_e	Design Strength, $\phi_t P_n$	
			Yielding	Rupture
	in.^{2}	in.^{2}	kips	kips
L8×8×1$^1/_8$	16.8	12.6	544	548
×1	15.1	11.3	489	492
×$^7/_8$	13.3	9.98	431	434
×$^3/_4$	11.5	8.63	373	375
×$^5/_8$	9.69	7.27	314	316
×$^9/_{16}$	8.77	6.58	284	286
×$^1/_2$	7.84	5.88	254	256
L8×6×1	13.1	9.83	424	428
×$^7/_8$	11.5	8.63	373	375
×$^3/_4$	9.99	7.49	324	326
×$^5/_8$	8.41	6.31	272	274
×$^9/_{16}$	7.61	5.71	247	248
×$^1/_2$	6.80	5.10	220	222
×$^7/_{16}$	5.99	4.49	194	195
L8×4×1	11.1	8.33	360	362
×$^7/_8$	9.79	7.34	317	319
×$^3/_4$	8.49	6.37	275	277
×$^5/_8$	7.16	5.37	232	234
×$^9/_{16}$	6.49	4.87	210	212
×$^1/_2$	5.80	4.35	188	189
×$^7/_{16}$	5.11	3.83	166	167
L7×4×$^3/_4$	7.70	5.78	249	251
×$^5/_8$	6.50	4.88	211	212
×$^1/_2$	5.26	3.95	170	172
×$^7/_{16}$	4.63	3.47	150	151
×$^3/_8$	4.00	3.00	130	131
L6×6×1	11.0	8.25	356	359
×$^7/_8$	9.75	7.31	316	318
×$^3/_4$	8.46	6.35	274	276
×$^5/_8$	7.13	5.35	231	233
×$^9/_{16}$	6.45	4.84	209	211
×$^1/_2$	5.77	4.33	187	188
×$^7/_{16}$	5.08	3.81	165	166
×$^3/_8$	4.38	3.29	142	143
×$^5/_{16}$	3.67	2.75	119	120
L6×4×$^7/_8$	7.95	5.96	258	259
×$^3/_4$	6.90	5.18	224	225
×$^5/_8$	5.83	4.37	189	190
×$^9/_{16}$	5.27	3.95	171	172
×$^1/_2$	4.72	3.54	153	154
×$^7/_{16}$	4.15	3.11	134	135
×$^3/_8$	3.58	2.69	116	117
×$^5/_{16}$	2.99	2.24	96.9	97.4
L6×3$^1/_2$×$^1/_2$	4.48	3.36	145	146
×$^3/_8$	3.41	2.56	110	111
×$^5/_{16}$	2.86	2.15	92.7	93.5

Shape	Gross Area, A_g	Effective Net Area, A_e	Design Strength, $\phi_t P_n$	
			Yielding	Rupture
	in.^{2}	in.^{2}	kips	kips
L5×5×$^7/_8$	8.02	6.02	260	262
×$^3/_4$	6.98	5.24	226	228
×$^5/_8$	5.90	4.43	191	193
×$^1/_2$	4.79	3.59	155	156
×$^7/_{16}$	4.22	3.17	137	138
×$^3/_8$	3.65	2.74	118	119
×$^5/_{16}$	3.07	2.30	99.5	100
L5×3$^1/_2$×$^3/_4$	5.82	4.37	189	190
×$^5/_8$	4.93	3.70	160	161
×$^1/_2$	4.00	3.00	130	131
×$^3/_8$	3.05	2.29	98.8	99.6
×$^5/_{16}$	2.56	1.92	82.9	83.5
×$^1/_4$	2.07	1.55	67.1	67.4
L5×3×$^1/_2$	3.75	2.81	122	122
×$^7/_{16}$	3.31	2.48	107	108
×$^3/_8$	2.86	2.15	92.7	93.5
×$^5/_{16}$	2.41	1.81	78.1	78.7
×$^1/_4$	1.94	1.46	62.9	63.5
L4×4×$^3/_4$	5.43	4.07	176	177
×$^5/_8$	4.61	3.46	149	151
×$^1/_2$	3.75	2.81	122	122
×$^7/_{16}$	3.30	2.48	107	108
×$^3/_8$	2.86	2.15	92.7	93.5
×$^5/_{16}$	2.40	1.80	77.8	78.3
×$^1/_4$	1.93	1.45	62.5	63.1
L4×3$^1/_2$×$^1/_2$	3.50	2.63	113	114
×$^3/_8$	2.68	2.01	86.8	87.4
×$^5/_{16}$	2.25	1.69	72.9	73.5
×$^1/_4$	1.82	1.37	59.0	59.6
L4×3×$^5/_8$	3.99	2.99	129	130
×$^1/_2$	3.25	2.44	105	106
×$^3/_8$	2.49	1.87	80.7	81.3
×$^5/_{16}$	2.09	1.57	67.7	68.3
×$^1/_4$	1.69	1.27	54.8	55.2
L3$^1/_2$×3$^1/_2$×$^1/_2$	3.27	2.45	106	107
×$^7/_{16}$	2.89	2.17	93.6	94.4
×$^3/_8$	2.50	1.88	81.0	81.8
×$^5/_{16}$	2.10	1.58	68.0	68.7
×$^1/_4$	1.70	1.28	55.1	55.7
L3$^1/_2$×3×$^1/_2$	3.02	2.27	97.8	98.7
×$^7/_{16}$	2.67	2.00	86.5	87.0
×$^3/_8$	2.32	1.74	75.2	75.7
×$^5/_{16}$	1.95	1.46	63.2	63.5
×$^1/_4$	1.58	1.19	51.2	51.8

† Tension rupture on the effective net area will control over tension yielding on the gross area unless the tension member is selected so that an end connection can be configured with $A_e \geq 0.745 A_g$.

$F_y = 36$ ksi
$F_u = 58$ ksi
$\phi_t P_n = 0.90 F_y A_g$ (Yielding)
$\phi_t P_n = 0.75 F_u A_e$ (Rupture)
$A_e = 0.75 A_g$

Table 3-2 (cont.). Angles (L-Shapes) Design Strengths in Axial Tension†

Shape	Gross Area, A_g	Effective Net Area, A_e	Design Strength, $\phi_t P_n$	
			Yielding	Rupture
	in.^{2}	in.^{2}	kips	kips
L3$^1/_2$×2$^1/_2$×$^1/_2$	2.76	2.07	89.4	90.0
×$^3/_8$	2.12	1.59	68.7	69.2
×$^5/_{16}$	1.79	1.34	58.0	58.3
×$^1/_4$	1.45	1.09	47.0	47.4
L3×3×$^1/_2$	2.75	2.06	89.1	89.6
×$^7/_{16}$	2.43	1.82	78.7	79.2
×$^3/_8$	2.11	1.58	68.4	68.7
×$^5/_{16}$	1.78	1.34	57.7	58.3
×$^1/_4$	1.44	1.08	46.7	47.0
×$^3/_{16}$	1.09	0.818	35.3	35.6
L3×2$^1/_2$×$^1/_2$	2.51	1.88	81.3	81.8
×$^7/_{16}$	2.22	1.67	71.9	72.6
×$^3/_8$	1.93	1.45	62.5	63.1
×$^5/_{16}$	1.63	1.22	52.8	53.1
×$^1/_4$	1.32	0.990	42.8	43.1
×$^3/_{16}$	1.00	0.750	32.4	32.6
L3×2×$^1/_2$	2.26	1.70	73.2	74.0
×$^3/_8$	1.75	1.31	56.7	57.0
×$^5/_{16}$	1.48	1.11	48.0	48.3
×$^1/_4$	1.20	0.900	38.9	39.2
×$^3/_{16}$	0.917	0.688	29.7	29.9
L2$^1/_2$×2$^1/_2$×$^1/_2$	2.25	1.69	72.9	73.5
×$^3/_8$	1.73	1.30	56.1	56.6
×$^5/_{16}$	1.46	1.10	47.3	47.9
×$^1/_4$	1.19	0.893	38.6	38.8
×$^3/_{16}$	0.901	0.676	29.2	29.4
L2$^1/_2$×2×$^3/_8$	1.56	1.17	50.5	50.9
×$^5/_{16}$	1.32	0.990	42.8	43.1
×$^1/_4$	1.07	0.803	34.7	34.9
×$^3/_{16}$	0.818	0.614	26.5	26.7
L2×2×$^3/_8$	1.37	1.03	44.4	44.8
×$^5/_{16}$	1.16	0.870	37.6	37.8
×$^1/_4$	0.944	0.708	30.6	30.8
×$^3/_{16}$	0.722	0.542	23.4	23.6
×$^1/_8$	0.491	0.368	15.9	16.0

† Tension rupture on the effective net area will control over tension yielding on the gross area unless the tension member is selected so that an end connection can be configured with $A_e \geq 0.745 A_g$.

Table 3-3. WT-Shapes (Structural Tees Split from W-Shapes) Design Strengths in Axial Tension†

$F_y = 50$ ksi
$F_u = 65$ ksi
$\phi_t P_n = 0.90 F_y A_g$ (Yielding)
$\phi_t P_n = 0.75 F_u A_e$ (Rupture)
$A_e = 0.75 A_g$

Shape	Gross Area, A_g	Effective Net Area, A_e	Design Strength, $\phi_t P_n$	
			Yielding	Rupture
	in.²	in.²	kips	kips
WT22×167.5	49.1	36.8	2210	1790
×145	42.9	32.2	1930	1570
×131	38.6	29.0	1740	1410
×115	33.8	25.4	1520	1240
WT20×296.5*	87.2	65.4	3920	3190
×251.5*	74.0	55.5	3330	2710
×215.5*	63.4	47.6	2850	2320
×198.5*	58.4	43.8	2630	2140
×186*	54.7	41.0	2460	2000
×181*	53.3	40.0	2400	1950
×162	47.7	35.8	2150	1750
×148.5	43.7	32.8	1970	1600
×138.5	40.7	30.5	1830	1490
×124.5	36.7	27.5	1650	1340
×107.5	31.7	23.8	1430	1160
×99.5	29.2	21.9	1310	1070
WT20×196*	57.7	43.3	2600	2110
×165.5	48.7	36.5	2190	1780
×163.5	48.0	36.0	2160	1760
×139	40.9	30.7	1840	1500
×132	38.8	29.1	1750	1420
×117.5	34.5	25.9	1550	1260
×105.5	31.0	23.3	1400	1140
×91.5	26.9	20.2	1210	985
×83.5	24.6	18.5	1110	902
×74.5	21.9	16.4	986	800
WT18×399*	117	87.8	5270	4280
×325*	95.5	71.6	4300	3490
×263.5*	77.4	58.1	3480	2830
×219.5*	64.5	48.4	2900	2360
×196.5*	57.8	43.4	2600	2120
×179.5*	52.7	39.5	2370	1930
×164*	48.2	36.2	2170	1760
×150	44.1	33.1	1980	1610
×140	41.2	30.9	1850	1510
×130	38.2	28.7	1720	1400
×122.5	36.0	27.0	1620	1320
×115	33.8	25.4	1520	1240
WT18×128	37.7	28.3	1700	1380
×116	34.1	25.6	1530	1250
×105	30.9	23.2	1390	1130
×97	28.5	21.4	1280	1040
×91	26.8	20.1	1210	980
×85	25.0	18.8	1130	917
×80	23.5	17.6	1060	858
×75	22.1	16.6	995	809
×67.5	19.9	14.9	896	726
WT16.5×193.5*	57.0	42.8	2570	2090
×177*	52.1	39.1	2340	1910
×159*	46.8	35.1	2110	1710
×145.5	42.8	32.1	1930	1560
×131.5	38.7	29.0	1740	1410
×120.5	35.5	26.6	1600	1300
×110.5	32.6	24.5	1470	1190
×100.5	29.6	22.2	1330	1080
WT16.5×84.5	24.8	18.6	1120	907
×76	22.4	16.8	1010	819
×70.5	20.8	15.6	936	761
×65	19.2	14.4	864	702
×59	17.3	13.0	779	634
WT15×195.5*	57.6	43.2	2590	2110
×178.5*	52.5	39.4	2360	1920
×163*	47.9	35.9	2160	1750
×146*	42.9	32.2	1930	1570
×130.5	38.4	28.8	1730	1400
×117.5	34.6	26.0	1560	1270
×105.5	31.1	23.3	1400	1140
×95.5	28.1	21.1	1260	1030
×86.5	25.5	19.1	1150	931
WT15×74	21.7	16.3	977	795
×66	19.4	14.6	873	712
×62	18.2	13.7	819	668
×58	17.1	12.8	770	624
×54	15.9	11.9	716	580
×49.5	14.5	10.9	653	531
×45	13.2	9.90	594	483
WT13.5×269.5*	79.3	59.5	3570	2900
×184*	54.2	40.7	2440	1980
×168*	49.5	37.1	2230	1810
×153.5*	45.2	33.9	2030	1650
×140.5*	41.4	31.1	1860	1520
×129	38.0	28.5	1710	1390
×117.5	34.7	26.0	1560	1270
×108.5	32.0	24.0	1440	1170
×97	28.6	21.5	1290	1050
×89	26.2	19.7	1180	960
×80.5	23.8	17.9	1070	873
×73	21.6	16.2	972	790
WT13.5×64.5	18.9	14.2	851	692
×57	16.8	12.6	756	614
×51	15.0	11.3	675	551
×47	13.8	10.4	621	507
×42	12.4	9.30	558	453

† Tension rupture on the effective net area will control over tension yielding on the gross area unless the tension member is selected so that an end connection can be configured with $A_e \geq 0.923 A_g$.
*ASTM A6 tensile group 4 or 5 shape. Special requirements may apply per LRFD Specification Section A3.1c.

$F_y = 50$ ksi
$F_u = 65$ ksi
$\phi_t P_n = 0.90 F_y A_g$ (Yielding)
$\phi_t P_n = 0.75 F_u A_e$ (Rupture)
$A_e = 0.75 A_g$

Table 3-3 (cont.). WT-Shapes (Structural Tees Split from W-Shapes) Design Strengths in Axial Tension†

Shape	Gross Area, A_g	Effective Net Area, A_e	Design Strength, $\phi_t P_n$	
			Yielding	Rupture
	in.^{2}	in.^{2}	kips	kips
WT12×185*	54.4	40.8	2450	1990
×167.5*	49.2	36.9	2210	1800
×153*	44.9	33.7	2020	1640
×139.5*	41.0	30.8	1850	1500
×125*	36.8	27.6	1660	1350
×114.5	33.6	25.2	1510	1230
×103.5	30.4	22.8	1370	1110
×96	28.1	21.1	1260	1030
×88	25.8	19.4	1160	946
×81	23.9	17.9	1080	873
×73	21.5	16.1	968	785
×65.5	19.3	14.5	869	707
×58.5	17.2	12.9	774	629
×52	15.3	11.5	689	561
WT12×51.5	15.1	11.3	680	551
×47	13.8	10.4	621	507
×42	12.4	9.30	558	453
×38	11.2	8.40	504	410
×34	10.0	7.50	450	366
WT12×31	9.16	6.87	412	335
×27.5	8.15	6.11	367	298
WT10.5×100.5	29.6	22.2	1330	1080
×91	26.8	20.1	1210	980
×83	24.4	18.3	1100	892
×73.5	21.6	16.2	972	790
×66	19.4	14.6	873	712
×61	17.9	13.4	806	653
×55.5	16.3	12.2	734	595
×50.5	14.9	11.2	671	546
WT10.5×46.5	13.7	10.3	617	502
×41.5	12.2	9.15	549	446
×36.5	10.7	8.03	482	391
×34	10.0	7.50	450	366
×31	9.13	6.85	411	334
×27.5	8.10	6.08	365	296
×24	7.07	5.30	318	258
WT10.5×28.5	8.37	6.28	377	306
×25	7.36	5.52	331	269
×22	6.49	4.87	292	237
WT9×87.5	25.7	19.3	1160	941
×79	23.2	17.4	1040	848
×71.5	21.0	15.8	945	770
×65	19.1	14.3	860	697
×59.5	17.5	13.1	788	639
×53	15.6	11.7	702	570
×48.5	14.3	10.7	644	522
×43	12.7	9.53	572	465
×38	11.2	8.40	504	410
WT9×35.5	10.4	7.80	468	380
×32.5	9.55	7.16	430	349
×30	8.82	6.62	397	323
×27.5	8.10	6.08	365	296
×25	7.33	5.50	330	268
WT9×23	6.77	5.08	305	248
×20	5.88	4.41	265	215
×17.5	5.15	3.86	232	188
WT8×50	14.9	11.2	671	546
×44.5	13.2	9.90	594	483
×38.5	11.5	8.63	518	421
×33.5	9.98	7.49	449	365
WT8×28.5	8.39	6.29	378	307
×25	7.37	5.53	332	270
×22.5	6.63	4.97	298	242
×20	5.89	4.42	265	215
×18	5.29	3.97	238	194
WT8×15.5	4.56	3.42	205	167
×13	3.84	2.88	173	140
WT7×404*	119	89.3	5360	4350
×365*	107	80.3	4820	3910
×332.5*	97.8	73.4	4400	3580
×302.5*	88.9	66.7	4000	3250
×275*	80.9	60.7	3640	2960
×250*	73.5	55.1	3310	2690
×227.5*	66.9	50.2	3010	2450
×213*	62.6	47.0	2820	2290
×199*	58.5	43.9	2630	2140
×185*	54.4	40.8	2450	1990
×171*	50.3	37.7	2260	1840
×155.5*	45.7	34.3	2060	1670
×141.5*	41.6	31.2	1870	1520
×128.5*	37.8	28.4	1700	1380
×116.5*	34.2	25.7	1540	1250
×105.5	31.0	23.3	1400	1140
×96.5	28.4	21.3	1280	1040
×88	25.9	19.4	1170	946
×79.5	23.4	17.6	1050	858
×72.5	21.3	16.0	959	780
WT7×66	19.4	14.6	873	712
×60	17.7	13.3	797	648
×54.5	16.0	12.0	720	585
×49.5	14.6	11.0	657	536
×45	13.2	9.90	594	483

† Tension rupture on the effective net area will control over tension yielding on the gross area unless the tension member is selected so that an end connection can be configured with $A_e \geq 0.923 A_g$.
* ASTM A6 tensile group 4 or 5 shape. Special requirements may apply per LRFD Specification Section A3.1c.

Table 3-3 (cont.). WT-Shapes (Structural Tees Split from W-Shapes) Design Strengths in Axial Tension†

$F_y = 50$ ksi
$F_u = 65$ ksi
$\phi_t P_n = 0.90 F_y A_g$ (Yielding)
$\phi_t P_n = 0.75 F_u A_e$ (Rupture)
$A_e = 0.75 A_g$

Shape	Gross Area, A_g	Effective Net Area, A_e	Design Strength, $\phi_t P_n$	
			Yielding	Rupture
	in.^{2}	in.^{2}	kips	kips
WT7×41	12.0	9.00	540	439
×37	10.9	8.18	491	399
×34	9.99	7.49	450	365
×30.5	8.96	6.72	403	328
WT7×26.5	7.80	5.85	351	285
×24	7.07	5.30	318	258
×21.5	6.31	4.73	284	231
WT7×19	5.58	4.19	251	204
×17	5.00	3.75	225	183
×15	4.42	3.32	199	162
WT7×13	3.85	2.89	173	141
×11	3.25	2.44	146	119
WT6×168*	49.4	37.1	2220	1810
×152.5*	44.8	33.6	2020	1640
×139.5*	41.0	30.8	1850	1500
×126*	37.0	27.8	1670	1360
×115*	33.9	25.4	1530	1240
×105*	30.9	23.2	1390	1130
×95	27.9	20.9	1260	1020
×85	25.0	18.8	1130	917
×76	22.4	16.8	1010	819
×68	20.0	15.0	900	731
×60	17.6	13.2	792	644
×53	15.6	11.7	702	570
×48	14.1	10.6	635	517
×43.5	12.8	9.60	576	468
×39.5	11.6	8.70	522	424
×36	10.6	7.95	477	388
×32.5	9.54	7.16	429	349
WT6×29	8.52	6.39	383	312
×26.5	7.78	5.84	350	285
WT6×25	7.30	5.48	329	267
×22.5	6.56	4.92	295	240
×20	5.84	4.38	263	214
WT6×17.5	5.17	3.88	233	189
×15	4.40	3.30	198	161
×13	3.82	2.87	172	140
WT6×11	3.24	2.43	146	118
×9.5	2.79	2.09	126	102
×8	2.36	1.77	106	86.3
×7	2.08	1.56	93.6	76.1
WT5×56	16.5	12.4	743	605
×50	14.7	11.0	662	536
×44	12.9	9.68	581	472
×38.5	11.3	8.48	509	413
×34	9.99	7.49	450	365
×30	8.82	6.62	397	323
×27	7.91	5.93	356	289
×24.5	7.21	5.41	324	264
WT5×22.5	6.63	4.97	298	242
×19.5	5.73	4.30	258	210
×16.5	4.85	3.64	218	177
WT5×15	4.42	3.32	199	162
×13	3.81	2.86	171	139
×11	3.24	2.43	146	118
WT5×9.5	2.81	2.11	126	103
×8.5	2.50	1.88	113	91.7
×7.5	2.21	1.66	99.5	80.9
×6	1.77	1.33	79.7	64.8
WT4×33.5	9.84	7.38	443	360
×29	8.54	6.41	384	312
×24	7.05	5.29	317	258
×20	5.87	4.40	264	215
×17.5	5.14	3.86	231	188
×15.5	4.56	3.42	205	167
WT4×14	4.12	3.09	185	151
×12	3.54	2.66	159	130
WT4×10.5	3.08	2.31	139	113
×9	2.63	1.97	118	96.0
WT4×7.5	2.22	1.67	99.9	81.4
×6.5	1.92	1.44	86.4	70.2
×5	1.48	1.11	66.6	54.1

† Tension rupture on the effective net area will control over tension yielding on the gross area unless the tension member is selected so that an end connection can be configured with $A_e \geq 0.923 A_g$.
*ASTM A6 tensile group 4 or 5 shape. Special requirements may apply per LRFD Specification Section A3.1c.

$F_y = 46$ ksi
$F_u = 58$ ksi
$\phi_t P_n = 0.90 F_y A_g$ (Yielding)
$\phi_t P_n = 0.75 F_u A_e$ (Rupture)
$A_e = 0.75 A_g$

Table 3-4. Rectangular (and Square) HSS Design Strengths in Axial Tension†

Shape	Gross Area, A_g	Effective Net Area, A_e	Design Strength, $\phi_t P_n$ Yielding	Design Strength, $\phi_t P_n$ Rupture
	in.^{2}	in.^{2}	kips	kips
HSS20×12×5/8	35.0	26.3	1450	1140
×1/2	28.3	21.2	1170	922
×3/8	21.5	16.1	890	700
×5/16	18.1	13.6	749	592
HSS20×8×5/8	30.3	22.7	1250	987
×1/2	24.6	18.5	1020	805
×3/8	18.7	14.0	774	609
×5/16	15.7	11.8	650	513
HSS20×4×1/2	20.9	15.7	865	683
×3/8	16.0	12.0	662	522
×5/16	13.4	10.1	555	439
HSS18×12×5/8	32.6	24.5	1350	1070
×1/2	26.5	19.9	1100	866
×3/8	20.1	15.1	832	657
HSS18×6×5/8	25.7	19.3	1060	840
×1/2	20.9	15.7	865	683
×3/8	16.0	12.0	662	522
×5/16	13.4	10.1	555	439
×1/4	10.8	8.10	447	352
HSS16×16×5/8	35.0	26.3	1450	1140
×1/2	28.3	21.2	1170	922
×3/8	21.5	16.1	890	700
×5/16	18.1	13.6	749	592
HSS16×12×5/8	30.3	22.7	1250	987
×1/2	24.6	18.5	1020	805
×3/8	18.7	14.0	774	609
×5/16	15.7	11.8	650	513
HSS16×8×5/8	25.7	19.3	1060	840
×1/2	20.9	15.7	865	683
×3/8	16.0	12.0	662	522
×5/16	13.4	10.1	555	439
HSS16×4×1/2	17.2	12.9	712	561
×3/8	13.2	9.90	546	431
×5/16	11.1	8.33	460	362
HSS14×14×5/8	30.3	22.7	1250	987
×1/2	24.6	18.5	1020	805
×3/8	18.7	14.0	774	609
×5/16	15.7	11.8	650	513
HSS14×12×1/2	22.8	17.1	944	744
×3/8	17.3	13.0	716	566
HSS14×10×5/8	25.7	19.3	1060	840
×1/2	20.9	15.7	865	683
×3/8	16.0	12.0	662	522
×5/16	13.4	10.1	555	439
×1/4	10.8	8.10	447	352
HSS14×6×5/8	21.0	15.8	869	687
×1/2	17.2	12.9	712	561
×3/8	13.2	9.90	546	431
×5/16	11.1	8.33	460	362
×1/4	8.96	6.72	371	292
×3/16	6.76	5.07	280	221

Shape	Gross Area, A_g	Effective Net Area, A_e	Design Strength, $\phi_t P_n$ Yielding	Design Strength, $\phi_t P_n$ Rupture
	in.^{2}	in.^{2}	kips	kips
HSS14×4×5/8	18.7	14.0	774	609
×1/2	15.3	11.5	633	500
×3/8	11.8	8.85	489	385
×5/16	9.92	7.44	411	324
×1/4	8.03	6.02	332	262
×3/16	6.06	4.55	251	198
HSS12×12×5/8	25.7	19.3	1060	840
×1/2	20.9	15.7	865	683
×3/8	16.0	12.0	662	522
×5/16	13.4	10.1	555	439
×1/4	10.8	8.10	447	352
HSS12×10×1/2	19.0	14.3	787	622
×3/8	14.6	11.0	604	479
×5/16	12.2	9.15	505	398
×1/4	9.90	7.43	410	323
HSS12×8×5/8	21.0	15.8	869	687
×1/2	17.2	12.9	712	561
×3/8	13.2	9.90	546	431
×5/16	11.1	8.33	460	362
×1/4	8.96	6.72	371	292
×3/16	6.76	5.07	280	221
HSS12×6×5/8	18.7	14.0	774	609
×1/2	15.3	11.5	633	500
×3/8	11.8	8.85	489	385
×5/16	9.92	7.44	411	324
×1/4	8.03	6.02	332	262
×3/16	6.06	4.55	251	198
HSS12×4×5/8	16.4	12.3	679	535
×1/2	13.5	10.1	559	439
×3/8	10.4	7.80	431	339
×5/16	8.76	6.57	363	286
×1/4	7.10	5.33	294	232
×3/16	5.37	4.03	222	175
HSS12×3 1/2×3/8	10.0	7.50	414	326
×5/16	8.46	6.35	350	276
HSS12×3×5/16	8.17	6.13	338	267
×1/4	6.63	4.97	274	216
×3/16	5.02	3.77	208	164
HSS12×2×1/4	6.17	4.63	255	201
×3/16	4.67	3.50	193	152
HSS10×10×5/8	21.0	15.8	869	687
×1/2	17.2	12.9	712	561
×3/8	13.2	9.90	546	431
×5/16	11.1	8.33	460	362
×1/4	8.96	6.72	371	292
×3/16	6.76	5.07	280	221
HSS10×8×1/2	15.3	11.5	633	500
×3/8	11.8	8.85	489	385
×5/16	9.92	7.44	411	324
×1/4	8.03	6.02	332	262
×3/16	6.06	4.55	251	198

† Tension rupture on the effective net area will control over tension yielding on the gross area unless the tension member is selected so that an end connection can be configured with $A_e \geq 0.952 A_g$.

Table 3-4 (cont.). Rectangular (and Square) HSS Design Strengths in Axial Tension†

$F_y = 46$ ksi
$F_u = 58$ ksi
$\phi_t P_n = 0.90 F_y A_g$ (Yielding)
$\phi_t P_n = 0.75 F_u A_e$ (Rupture)
$A_e = 0.75 A_g$

Shape	Gross Area, A_g	Effective Net Area, A_e	Design Strength, $\phi_t P_n$	
			Yielding	Rupture
	in.^{2}	in.^{2}	kips	kips
HSS10×6×5/8	16.4	12.3	679	535
×1/2	13.5	10.1	559	439
×3/8	10.4	7.80	431	339
×5/16	8.76	6.57	363	286
×1/4	7.10	5.33	294	232
×3/16	5.37	4.03	222	175
HSS10×5×3/8	9.67	7.25	400	315
×5/16	8.17	6.13	338	267
×1/4	6.63	4.97	274	216
×3/16	5.02	3.77	208	164
HSS10×4×5/8	14.0	10.5	580	457
×1/2	11.6	8.70	480	378
×3/8	8.97	6.73	371	293
×5/16	7.59	5.69	314	248
×1/4	6.17	4.63	255	201
×3/16	4.67	3.50	193	152
HSS10×3 1/2×3/16	4.50	3.38	186	147
HSS10×3×3/8	8.27	6.20	342	270
×5/16	7.01	5.26	290	229
×1/4	5.70	4.28	236	186
×3/16	4.32	3.24	179	141
×1/8	2.93	2.20	121	95.7
HSS10×2×3/8	7.58	5.69	314	248
×5/16	6.43	4.82	266	210
×1/4	5.24	3.93	217	171
×3/16	3.98	2.99	165	130
HSS9×7×5/8	16.4	12.3	679	535
×1/2	13.5	10.1	559	439
×3/8	10.4	7.80	431	339
×5/16	8.76	6.57	363	286
×1/4	7.10	5.33	294	232
×3/16	5.37	4.03	222	175
HSS9×5×5/8	14.0	10.5	580	457
×1/2	11.6	8.70	480	378
×3/8	8.97	6.73	371	293
×5/16	7.59	5.69	314	248
×1/4	6.17	4.63	255	201
×3/16	4.67	3.50	193	152
HSS9×3×1/2	9.74	7.31	403	318
×3/8	7.58	5.69	314	248
×5/16	6.43	4.82	266	210
×1/4	5.24	3.93	217	171
×3/16	3.98	2.99	165	130
HSS8×8×5/8	16.4	12.3	679	535
×1/2	13.5	10.1	559	439
×3/8	10.4	7.80	431	339
×5/16	8.76	6.57	363	286
×1/4	7.10	5.33	294	232
×3/16	5.37	4.03	222	175

Shape	Gross Area, A_g	Effective Net Area, A_e	Design Strength, $\phi_t P_n$	
			Yielding	Rupture
	in.^{2}	in.^{2}	kips	kips
HSS8×6×5/8	14.0	10.5	580	457
×1/2	11.6	8.70	480	378
×3/8	8.97	6.73	371	293
×5/16	7.59	5.69	314	248
×1/4	6.17	4.63	255	201
×3/16	4.67	3.50	193	152
HSS8×4×5/8	11.7	8.78	484	382
×1/2	9.74	7.31	403	318
×3/8	7.58	5.69	314	248
×5/16	6.43	4.82	266	210
×1/4	5.24	3.93	217	171
×3/16	3.98	2.99	165	130
×1/8	2.70	2.03	112	88.3
HSS8×3×1/2	8.81	6.61	365	288
×3/8	6.88	5.16	285	224
×5/16	5.85	4.39	242	191
×1/4	4.77	3.58	197	156
×3/16	3.63	2.72	150	118
×1/8	2.46	1.85	102	80.5
HSS8×2×3/8	6.18	4.64	256	202
×5/16	5.26	3.95	218	172
×1/4	4.30	3.23	178	141
×3/16	3.28	2.46	136	107
HSS8×2×1/8	2.23	1.67	92.3	72.6
HSS7×7×5/8	14.0	10.5	580	457
×1/2	11.6	8.70	480	378
×3/8	8.97	6.73	371	293
×5/16	7.59	5.69	314	248
×1/4	6.17	4.63	255	201
×3/16	4.67	3.50	193	152
HSS7×5×5/8	11.7	8.78	484	382
×1/2	9.74	7.31	403	318
×3/8	7.58	5.69	314	248
×5/16	6.43	4.82	266	210
×1/4	5.24	3.93	217	171
×3/16	3.98	2.99	165	130
×1/8	2.70	2.03	112	88.3
HSS7×4×1/2	8.81	6.61	365	288
×3/8	6.88	5.16	285	224
×5/16	5.85	4.39	242	191
×1/4	4.77	3.58	197	156
×3/16	3.63	2.72	150	118
×1/8	2.46	1.85	102	80.5
HSS7×3×1/2	7.88	5.91	326	257
×3/8	6.18	4.64	256	202
×5/16	5.26	3.95	218	172
×1/4	4.30	3.23	178	141
×3/16	3.28	2.46	136	107
×1/8	2.23	1.67	92.3	72.6

† Tension rupture on the effective net area will control over tension yielding on the gross area unless the tension member is selected so that an end connection can be configured with $A_e \geq 0.952 A_g$.

$F_y = 46$ ksi
$F_u = 58$ ksi
$\phi_t P_n = 0.90 F_y A_g$ (Yielding)
$\phi_t P_n = 0.75 F_u A_e$ (Rupture)
$A_e = 0.75 A_g$

Table 3-4 (cont.). Rectangular (and Square) HSS Design Strengths in Axial Tension†

Shape	Gross Area, A_g	Effective Net Area, A_e	Design Strength, $\phi_t P_n$	
			Yielding	Rupture
	in.²	in.²	kips	kips
HSS6×6×5/8	11.7	8.78	484	382
×1/2	9.74	7.31	403	318
×3/8	7.58	5.69	314	248
×5/16	6.43	4.82	266	210
×1/4	5.24	3.93	217	171
×3/16	3.98	2.99	165	130
×1/8	2.70	2.03	112	88.3
HSS6×5×3/8	6.88	5.16	285	224
×5/16	5.85	4.39	242	191
×1/4	4.77	3.58	197	156
×3/16	3.63	2.72	150	118
HSS6×4×1/2	7.88	5.91	326	257
×3/8	6.18	4.64	256	202
×5/16	5.26	3.95	218	172
×1/4	4.30	3.23	178	141
×3/16	3.28	2.46	136	107
×1/8	2.23	1.67	92.3	72.6
HSS6×3×1/2	6.95	5.21	288	227
×3/8	5.48	4.11	227	179
×5/16	4.68	3.51	194	153
×1/4	3.84	2.88	159	125
×3/16	2.93	2.20	121	95.7
×1/8	2.00	1.50	82.8	65.3
HSS6×2×3/8	4.78	3.59	198	156
×5/16	4.10	3.08	170	134
×1/4	3.37	2.53	140	110
×3/16	2.58	1.94	107	84.4
×1/8	1.77	1.33	73.3	57.9
HSS5½×5½×3/8	6.88	5.16	285	224
×5/16	5.85	4.39	242	191
×1/4	4.77	3.58	197	156
×3/16	3.63	2.72	150	118
×1/8	2.46	1.85	102	80.5
HSS5×5×1/2	7.88	5.91	326	257
×3/8	6.18	4.64	256	202
×5/16	5.26	3.95	218	172
×1/4	4.30	3.23	178	141
×3/16	3.28	2.46	136	107
×1/8	2.23	1.67	92.3	72.6
HSS5×4×1/2	6.95	5.21	288	227
×3/8	5.48	4.11	227	179
×5/16	4.68	3.51	194	153
×1/4	3.84	2.88	159	125
×3/16	2.93	2.20	121	95.7
HSS5×3×1/2	6.02	4.52	249	197
×3/8	4.78	3.59	198	156
×5/16	4.10	3.08	170	134
×1/4	3.37	2.53	140	110
×3/16	2.58	1.94	107	84.4
×1/8	1.77	1.33	73.3	57.9
HSS5×2½×1/4	3.14	2.36	130	103
×3/16	2.36	1.77	97.7	77.0
×1/8	1.65	1.24	68.3	53.9
HSS5×2×3/8	4.09	3.07	169	134
×5/16	3.52	2.64	146	115
×1/4	2.91	2.18	120	94.8
×3/16	2.24	1.68	92.7	73.1
×1/8	1.54	1.16	63.8	50.5
HSS4½×4½×1/2	6.95	5.21	288	227
×3/8	5.48	4.11	227	179
×5/16	4.68	3.51	194	153
×1/4	3.84	2.88	159	125
×3/16	2.93	2.20	121	95.7
×1/8	2.00	1.50	82.8	65.3
HSS4×4×1/2	6.02	4.52	249	197
×3/8	4.78	3.59	198	156
×5/16	4.10	3.08	170	134
×1/4	3.37	2.53	140	110
×3/16	2.58	1.94	107	84.4
×1/8	1.77	1.33	73.3	57.9
HSS4×3×3/8	4.09	3.07	169	134
×5/16	3.52	2.64	146	115
×1/4	2.91	2.18	120	94.8
×3/16	2.24	1.68	92.7	73.1
×1/8	1.54	1.16	63.8	50.5
HSS4×2½×5/16	3.23	2.42	134	105
×1/4	2.67	2.00	111	87.0
×3/16	2.06	1.55	85.3	67.4
HSS4×2×3/8	3.39	2.54	140	110
×5/16	2.94	2.21	122	96.1
×1/4	2.44	1.83	101	79.6
×3/16	1.89	1.42	78.2	61.8
×1/8	1.30	0.975	53.8	42.4
HSS3½×3½×3/8	4.09	3.07	169	134
×5/16	3.52	2.64	146	115
×1/4	2.91	2.18	120	94.8
×3/16	2.24	1.68	92.7	73.1
×1/8	1.54	1.16	63.8	50.5
HSS3½×2½×3/8	3.39	2.54	140	110
×5/16	2.94	2.21	122	96.1
×1/4	2.44	1.83	101	79.6
×3/16	1.89	1.42	78.2	61.8
×1/8	1.30	0.975	53.8	42.4
HSS3×3×3/8	3.39	2.54	140	110
×5/16	2.94	2.21	122	96.1
×1/4	2.44	1.83	101	79.6
×3/16	1.89	1.42	78.2	61.8
×1/8	1.30	0.975	53.8	42.4
HSS3×2½×5/16	2.64	1.98	109	86.1
×1/4	2.21	1.66	91.5	72.2
×3/16	1.71	1.28	70.8	55.7
×1/8	1.19	0.893	49.3	38.8

† Tension rupture on the effective net area will control over tension yielding on the gross area unless the tension member is selected so that an end connection can be configured with $A_e \geq 0.952 A_g$.

Table 3-4 (cont.). Rectangular (and Square) HSS Design Strengths in Axial Tension†

$F_y = 46$ ksi
$F_u = 58$ ksi
$\phi_t P_n = 0.90 F_y A_g$ (Yielding)
$\phi_t P_n = 0.75 F_u A_e$ (Rupture)
$A_e = 0.75 A_g$

Shape	Gross Area, A_g	Effective Net Area, A_e	Design Strength, $\phi_t P_n$ Yielding	Design Strength, $\phi_t P_n$ Rupture
	in.²	in.²	kips	kips
HSS3×2×5/16	2.35	1.76	97.3	76.6
×1/4	1.97	1.48	81.6	64.4
×3/16	1.54	1.16	63.8	50.5
×1/8	1.07	0.803	44.3	34.9
HSS3×1 1/2×1/4	1.74	1.31	72.0	57.0
×3/16	1.37	1.03	56.7	44.8
×1/8	0.956	0.717	39.6	31.2
HSS3×1×1/8	0.840	0.630	34.8	27.4
HSS2 1/2×2 1/2×5/16	2.35	1.76	97.3	76.6
×1/4	1.97	1.48	81.6	64.4
×3/16	1.54	1.16	63.8	50.5
×1/8	1.07	0.803	44.3	34.9
HSS2 1/2×1 1/2×1/4	1.51	1.13	62.5	49.2
×3/16	1.19	0.893	49.3	38.8
×1/8	0.840	0.630	34.8	27.4
HSS2 1/4×2 1/4×1/4	1.74	1.31	72.0	57.0
×3/16	1.37	1.03	56.7	44.8
×1/8	0.956	0.717	39.6	31.2
HSS2×2×1/4	1.51	1.13	62.5	49.2
×3/16	1.19	0.893	49.3	38.8
×1/8	0.840	0.630	34.8	27.4
HSS2×1 1/2×3/16	1.02	0.765	42.2	33.3
HSS2×1×3/16	0.845	0.634	35.0	27.6
×1/8	0.608	0.456	25.2	19.8
HSS1 3/4×1 3/4×3/16	1.02	0.765	42.2	33.3
HSS1 5/8×1 5/8×3/16	0.932	0.699	38.6	30.4
×1/8	0.666	0.500	27.6	21.8
HSS1 1/2×1 1/2×3/16	0.845	0.634	35.0	27.6
×1/8	0.608	0.456	25.2	19.8
HSS1 1/4×1 1/4×3/16	0.671	0.503	27.8	21.9
×1/8	0.492	0.369	20.4	16.1

† Tension rupture on the effective net area will control over tension yielding on the gross area unless the tension member is selected so that an end connection can be configured with $A_e \geq 0.952A_g$.

$F_y = 42$ ksi
$F_u = 58$ ksi
$\phi_t P_n = 0.90 F_y A_g$ (Yielding)
$\phi_t P_n = 0.75 F_u A_e$ (Rupture)
$A_e = 0.75 A_g$

Table 3-5. Round HSS Design Strengths in Axial Tension†

Shape	Gross Area, A_g	Effective Net Area, A_e	Design Strength, $\phi_t P_n$	
			Yielding	Rupture
	in.^{2}	in.^{2}	kips	kips
HSS20.000×0.500	28.5	21.4	1080	931
×0.375	21.5	16.1	813	700
HSS18.000×0.500	25.6	19.2	968	835
×0.375	19.4	14.6	733	635
HSS16.000×0.500	22.7	17.0	858	740
×0.438	19.9	14.9	752	648
×0.375	17.2	12.9	650	561
×0.312	14.4	10.8	544	470
HSS14.000×0.500	19.8	14.9	748	648
×0.375	15.0	11.3	567	492
×0.312	12.5	9.38	473	408
HSS12.750×0.500	17.9	13.4	677	583
×0.375	13.6	10.2	514	444
×0.250	9.16	6.87	346	299
HSS12.500×0.625	21.8	16.4	824	713
×0.500	17.6	13.2	665	574
×0.375	13.3	9.98	503	434
×0.312	11.2	8.40	423	365
×0.250	8.98	6.74	339	293
×0.188	6.74	5.06	255	220
HSS11.250×0.625	19.5	14.6	737	635
×0.500	15.8	11.9	597	518
×0.375	12.0	9.00	454	392
×0.312	10.00	7.50	378	326
×0.250	8.06	6.05	305	263
×0.188	6.05	4.54	229	197
HSS10.750×0.500	15.0	11.3	567	492
×0.250	7.70	5.78	291	251
HSS10.000×0.625	17.2	12.9	650	561
×0.500	13.9	10.4	525	452
×0.375	10.6	7.95	401	346
×0.312	8.88	6.66	336	290
×0.250	7.15	5.36	270	233
×0.188	5.37	4.03	203	175
HSS9.625×0.500	13.4	10.1	507	439
×0.375	10.2	7.65	386	333
×0.312	8.53	6.40	322	278
×0.250	6.87	5.15	260	224
×0.188	5.17	3.88	195	169
HSS8.750×0.500	12.1	9.08	457	395
×0.375	9.21	6.91	348	301
×0.312	7.73	5.80	292	252
×0.250	6.23	4.67	235	203
×0.188	4.69	3.52	177	153
HSS8.625×0.500	11.9	8.93	450	388
×0.375	9.07	6.80	343	296
×0.322	7.85	5.89	297	256
×0.250	6.14	4.61	232	201
×0.188	4.62	3.47	175	151
HSS7.625×0.125	2.74	2.06	104	89.6
HSS7.500×0.500	10.3	7.73	389	336
×0.375	7.84	5.88	296	256
×0.312	6.59	4.94	249	215
×0.250	5.32	3.99	201	174
×0.188	4.00	3.00	151	131
HSS7.000×0.500	9.55	7.16	361	311
×0.375	7.29	5.47	276	238
×0.312	6.13	4.60	232	200
×0.250	4.95	3.71	187	161
×0.188	3.73	2.80	141	122
×0.125	2.51	1.88	94.9	81.8
HSS6.875×0.500	9.36	7.02	354	305
×0.375	7.16	5.37	271	234
×0.312	6.02	4.52	228	197
×0.250	4.86	3.65	184	159
×0.188	3.66	2.75	138	120
HSS6.625×0.500	9.00	6.75	340	294
×0.432	7.88	5.91	298	257
×0.375	6.88	5.16	260	224
×0.312	5.79	4.34	219	189
×0.280	5.22	3.92	197	171
×0.250	4.68	3.51	177	153
×0.188	3.53	2.65	133	115
×0.125	2.37	1.78	89.6	77.4
HSS6.125×0.500	8.27	6.20	313	270
×0.375	6.33	4.75	239	207
×0.312	5.33	4.00	201	174
×0.250	4.31	3.23	163	141
×0.188	3.25	2.44	123	106
HSS6.000×0.500	8.09	6.07	306	264
×0.375	6.20	4.65	234	202
×0.312	5.22	3.92	197	171
×0.280	4.71	3.53	178	154
×0.250	4.22	3.17	160	138
×0.188	3.18	2.39	120	104
×0.125	2.14	1.61	80.9	70.0
HSS5.563×0.375	5.72	4.29	216	187
×0.258	4.03	3.02	152	131
×0.188	2.95	2.21	112	96.1
×0.134	2.14	1.61	80.9	70.0
HSS5.500×0.500	7.36	5.52	278	240
×0.375	5.65	4.24	214	184
×0.258	3.98	2.99	150	130
HSS5.000×0.500	6.62	4.97	250	216
×0.375	5.10	3.83	193	167
×0.312	4.30	3.23	163	141
×0.258	3.60	2.70	136	117
×0.250	3.49	2.62	132	114
×0.188	2.64	1.98	99.8	86.1
×0.125	1.78	1.34	67.3	58.3

† Tension rupture on the effective net area will control over tension yielding on the gross area unless the tension member is selected so that an end connection can be configured with $A_e \geq 0.869 A_g$.

Table 3-5 (cont.). Round HSS Design Strengths in Axial Tension†

$F_y = 42$ ksi
$F_u = 58$ ksi
$\phi_t P_n = 0.90 F_y A_g$ (Yielding)
$\phi_t P_n = 0.75 F_u A_e$ (Rupture)
$A_e = 0.75 A_g$

Shape	Gross Area, A_g	Effective Net Area, A_e	Design Strength, $\phi_t P_n$ Yielding	Rupture	Shape	Gross Area, A_g	Effective Net Area, A_e	Design Strength, $\phi_t P_n$ Yielding	Rupture
	in.²	in.²	kips	kips		in.²	in.²	kips	kips
HSS4.500×0.337	4.14	3.11	156	135	HSS2.875×0.250	1.93	1.45	73.0	63.1
×0.237	2.97	2.23	112	97.0	×0.203	1.59	1.19	60.1	51.8
×0.188	2.36	1.77	89.2	77.0	×0.188	1.48	1.11	55.9	48.3
×0.125	1.60	1.20	60.5	52.2	×0.125	1.01	0.758	38.2	33.0
HSS4.000×0.337	3.65	2.74	138	119	HSS2.500×0.250	1.66	1.25	62.7	54.4
×0.313	3.39	2.54	128	110	×0.188	1.27	0.953	48.0	41.5
×0.250	2.76	2.07	104	90.0	×0.125	0.869	0.652	32.8	28.4
×0.237	2.62	1.97	99.0	85.7					
×0.226	2.51	1.88	94.9	81.8	HSS2.375×0.250	1.57	1.18	59.3	51.3
×0.220	2.44	1.83	92.2	79.6	×0.218	1.39	1.04	52.5	45.2
×0.188	2.09	1.57	79.0	68.3	×0.188	1.20	0.900	45.4	39.2
×0.125	1.42	1.07	53.7	46.5	×0.154	1.00	0.750	37.8	32.6
					×0.125	0.823	0.617	31.1	26.8
HSS3.500×0.313	2.93	2.20	111	95.7					
×0.300	2.83	2.12	107	92.2	HSS1.900×0.145	0.749	0.562	28.3	24.4
×0.250	2.39	1.79	90.3	77.9					
×0.216	2.08	1.56	78.6	67.9	HSS1.660×0.140	0.625	0.469	23.6	20.4
×0.203	1.97	1.48	74.5	64.4					
×0.188	1.82	1.37	68.8	59.6					
×0.125	1.23	0.923	46.5	40.2					
HSS3.000×0.300	2.39	1.79	90.3	77.9					
×0.250	2.03	1.52	76.7	66.1					
×0.216	1.77	1.33	66.9	57.9					
×0.203	1.67	1.25	63.1	54.4					
×0.188	1.54	1.16	58.2	50.5					
×0.152	1.27	0.953	48.0	41.5					
×0.134	1.13	0.848	42.7	36.9					
×0.120	1.02	0.765	38.6	33.3					

† Tension rupture on the effective net area will control over tension yielding on the gross area unless the tension member is selected so that an end connection can be configured with $A_e \geq 0.869A_g$.

$F_y = 35$ ksi
$F_u = 60$ ksi
$\phi_t P_n = 0.90 F_y A_g$ (Yielding)
$\phi_t P_n = 0.75 F_u A_e$ (Rupture)
$A_e = 0.75 A_g$

Table 3-6. Steel Pipe Design Strengths in Axial Tension†

Shape	Gross Area, A_g	Effective Net Area, A_e	Design Strength, $\phi_t P_n$		Shape	Gross Area, A_g	Effective Net Area, A_e	Design Strength, $\phi_t P_n$	
			Yielding	Rupture				Yielding	Rupture
	in.^{2}	in.^{2}	kips	kips		in.^{2}	in.^{2}	kips	kips
Pipe 12 ×-strong	19.2	14.4	605	648	Pipe 2$^1/_2$ ××-strong	4.03	3.02	127	136
Std	14.6	11.0	460	495	×-strong	2.25	1.69	70.9	76.1
					Std	1.70	1.28	53.6	57.6
Pipe 10 ×-strong	16.1	12.1	507	545					
Std	11.9	8.93	375	402	Pipe 2 ××-strong	2.66	2.00	83.8	90.0
					×-strong	1.48	1.11	46.6	50.0
Pipe 8 ××-strong	21.3	16.0	671	720	Std	1.07	0.803	33.7	36.1
×-strong	12.8	9.60	403	432					
Std	8.40	6.30	265	284	Pipe 1$^1/_2$ ×-strong	1.07	0.803	33.7	36.1
					Std	0.799	0.599	25.2	27.0
Pipe 6 ××-strong	15.6	11.7	491	527					
×-strong	8.40	6.30	265	284	Pipe 1$^1/_4$ ×-strong	0.881	0.661	27.8	29.7
Std	5.58	4.19	176	189	Std	0.669	0.502	21.1	22.6
Pipe 5 ××-strong	11.3	8.48	356	382	Pipe 1 ×-strong	0.639	0.479	20.1	21.6
×-strong	6.11	4.58	192	206	Std	0.494	0.371	15.6	16.7
Std	4.30	3.23	135	145					
					Pipe $^3/_4$ ×-strong	0.433	0.325	13.6	14.6
Pipe 4 ××-strong	8.10	6.08	255	274	Std	0.333	0.250	10.5	11.3
×-strong	4.41	3.31	139	149					
Std	3.17	2.38	99.9	107	Pipe $^1/_2$ ×-strong	0.320	0.240	10.1	10.8
					Std	0.250	0.188	7.88	8.46
Pipe 3$^1/_2$ ×-strong	3.68	2.76	116	124					
Std	2.68	2.01	84.4	90.5					
Pipe 3 ××-strong	5.47	4.10	172	185					
×-strong	3.02	2.27	95.1	102					
Std	2.23	1.67	70.2	75.2					

† Tension rupture on the effective net area will control over tension yielding on the gross area unless the tension member is selected so that an end connection can be configured with $A_e \geq 0.700 A_g$.

Table 3-7. Double Angles (2L-Shapes) Design Strengths in Axial Tension†

$F_y = 36$ ksi
$F_u = 58$ ksi
$\phi_t P_n = 0.90 F_y A_g$ (Yielding)
$\phi_t P_n = 0.75 F_u A_e$ (Rupture)
$A_e = 0.75 A_g$

Shape	Gross Area, A_g	Effective Net Area, A_e	Design Strength, $\phi_t P_n$	
			Yielding	Rupture
	in.^{2}	in.^{2}	kips	kips
2L8×8×1 1/8	33.6	25.2	1090	1100
×1	30.2	22.7	978	987
×7/8	26.6	20.0	862	870
×3/4	23.0	17.3	745	753
×5/8	19.4	14.6	629	635
×9/16	17.5	13.1	567	570
×1/2	15.7	11.8	509	513
2L8×6×1	26.2	19.7	849	857
×7/8	23.0	17.3	745	753
×3/4	20.0	15.0	648	653
×5/8	16.8	12.6	544	548
×9/16	15.2	11.4	492	496
×1/2	13.6	10.2	441	444
×7/16	12.0	9.00	389	392
2L8×4×1	22.2	16.7	719	726
×7/8	19.6	14.7	635	639
×3/4	17.0	12.8	551	557
×5/8	14.3	10.7	463	465
×9/16	13.0	9.75	421	424
×1/2	11.6	8.70	376	378
×7/16	10.2	7.65	330	333
2L7×4×3/4	15.4	11.6	499	505
×5/8	13.0	9.75	421	424
×1/2	10.5	7.88	340	343
×7/16	9.26	6.95	300	302
×3/8	8.00	6.00	259	261
2L6×6×1	22.0	16.5	713	718
×7/8	19.5	14.6	632	635
×3/4	16.9	12.7	548	552
×5/8	14.3	10.7	463	465
×9/16	12.9	9.68	418	421
×1/2	11.5	8.63	373	375
×7/16	10.2	7.65	330	333
×3/8	8.76	6.57	284	286
×5/16	7.34	5.51	238	240
2L6×4×7/8	15.9	11.9	515	518
×3/4	13.8	10.4	447	452
×5/8	11.7	8.78	379	382
×9/16	10.5	7.88	340	343
×1/2	9.44	7.08	306	308
×7/16	8.30	6.23	269	271
×3/8	7.16	5.37	232	234
×5/16	5.98	4.49	194	195
2L6×3 1/2×1/2	8.96	6.72	290	292
×3/8	6.82	5.12	221	223
×5/16	5.72	4.29	185	187

Shape	Gross Area, A_g	Effective Net Area, A_e	Design Strength, $\phi_t P_n$	
			Yielding	Rupture
	in.^{2}	in.^{2}	kips	kips
2L5×5×7/8	16.0	12.0	518	522
×3/4	14.0	10.5	454	457
×5/8	11.8	8.85	382	385
×1/2	9.58	7.19	310	313
×7/16	8.44	6.33	273	275
×3/8	7.30	5.48	237	238
×5/16	6.14	4.61	199	201
2L5×3 1/2×3/4	11.6	8.70	376	378
×5/8	9.86	7.40	319	322
×1/2	8.00	6.00	259	261
×3/8	6.10	4.58	198	199
×5/16	5.12	3.84	166	167
×1/4	4.14	3.11	134	135
2L5×3×1/2	7.50	5.63	243	245
×7/16	6.62	4.97	214	216
×3/8	5.72	4.29	185	187
×5/16	4.82	3.62	156	157
×1/4	3.88	2.91	126	127
2L4×4×3/4	10.9	8.18	353	356
×5/8	9.22	6.92	299	301
×1/2	7.50	5.63	243	245
×7/16	6.60	4.95	214	215
×3/8	5.72	4.29	185	187
×5/16	4.80	3.60	156	157
×1/4	3.86	2.90	125	126
2L4×3 1/2×1/2	7.00	5.25	227	228
×3/8	5.36	4.02	174	175
×5/16	4.50	3.38	146	147
×1/4	3.64	2.73	118	119
2L4×3×5/8	7.98	5.99	259	261
×1/2	6.50	4.88	211	212
×3/8	4.98	3.74	161	163
×5/16	4.18	3.14	135	137
×1/4	3.38	2.54	110	110
2L3 1/2×3 1/2×1/2	6.54	4.91	212	214
×7/16	5.78	4.34	187	189
×3/8	5.00	3.75	162	163
×5/16	4.20	3.15	136	137
×1/4	3.40	2.55	110	111
2L3 1/2×3×1/2	6.04	4.53	196	197
×7/16	5.34	4.01	173	174
×3/8	4.64	3.48	150	151
×5/16	3.90	2.93	126	127
×1/4	3.16	2.37	102	103

† Tension rupture on the effective net area will control over tension yielding on the gross area unless the tension member is selected so that an end connection can be configured with $A_e \geq 0.745 A_g$.

$F_y = 36$ ksi
$F_u = 58$ ksi
$\phi_t P_n = 0.90 F_y A_g$ (Yielding)
$\phi_t P_n = 0.75 F_u A_e$ (Rupture)
$A_e = 0.75 A_g$

Table 3-7 (cont.). Double Angles (2L-Shapes) Design Strengths in Axial Tension†

Shape	Gross Area, A_g	Effective Net Area, A_e	Design Strength, $\phi_t P_n$		Shape	Gross Area, A_g	Effective Net Area, A_e	Design Strength, $\phi_t P_n$	
			Yielding	Rupture				Yielding	Rupture
	in.^{2}	in.^{2}	kips	kips		in.^{2}	in.^{2}	kips	kips
2L3$^1/_2$×2$^1/_2$×$^1/_2$	5.52	4.14	179	180	2L2$^1/_2$×2$^1/_2$×$^1/_2$	4.50	3.38	146	147
×$^3/_8$	4.24	3.18	137	138	×$^3/_8$	3.46	2.60	112	113
×$^5/_{16}$	3.58	2.69	116	117	×$^5/_{16}$	2.92	2.19	94.6	95.3
×$^1/_4$	2.90	2.18	94.0	94.8	×$^1/_4$	2.38	1.79	77.1	77.9
					×$^3/_{16}$	1.80	1.35	58.3	58.7
2L3×3×$^1/_2$	5.50	4.13	178	180					
×$^7/_{16}$	4.86	3.65	157	159	2L2$^1/_2$×2×$^3/_8$	3.12	2.34	101	102
×$^3/_8$	4.22	3.17	137	138	×$^5/_{16}$	2.64	1.98	85.5	86.1
×$^5/_{16}$	3.56	2.67	115	116	×$^1/_4$	2.14	1.61	69.3	70.0
×$^1/_4$	2.88	2.16	93.3	94.0	×$^3/_{16}$	1.64	1.23	53.1	53.5
×$^3/_{16}$	2.18	1.64	70.6	71.3					
					2L2×2×$^3/_8$	2.74	2.06	88.8	89.6
2L3×2$^1/_2$×$^1/_2$	5.02	3.77	163	164	×$^5/_{16}$	2.32	1.74	75.2	75.7
×$^7/_{16}$	4.44	3.33	144	145	×$^1/_4$	1.89	1.42	61.2	61.8
×$^3/_8$	3.86	2.90	125	126	×$^3/_{16}$	1.44	1.08	46.7	47.0
×$^5/_{16}$	3.26	2.45	106	107	×$^1/_8$	0.982	0.737	31.8	32.1
×$^1/_4$	2.64	1.98	85.5	86.1					
×$^3/_{16}$	2.00	1.50	64.8	65.3					
2L3×2×$^1/_2$	4.52	3.39	146	147					
×$^3/_8$	3.50	2.63	113	114					
×$^5/_{16}$	2.96	2.22	95.9	96.6					
×$^1/_4$	2.40	1.80	77.8	78.3					
×$^3/_{16}$	1.83	1.37	59.3	59.6					

† Tension rupture on the effective net area will control over tension yielding on the gross area unless the tension member is selected so that an end connection can be configured with $A_e \geq 0.745 A_g$.

PART 4

DESIGN OF COMPRESSION MEMBERS

SCOPE

The specification requirements and other design considerations summarized in this Part apply to the design of members subject to axial compression. For the design of members subject to eccentric compression or combined axial compression and flexure, see Part 6. For compression members that are part of a seismic-force-resisting system in which the seismic response modification factor R is taken greater than 3, see the Seismic Provisions, which are available from AISC at www.aisc.org.

LOAD DETERMINATION

The required strength (factored axial compression) P_u is determined by analysis as indicated in LRFD Specification Section A5.

COMPACT, NON-COMPACT AND SLENDER-ELEMENT CROSS-SECTIONS

Determining the Width-Thickness Ratios of the Cross-Section

Steel compression members are classified on the basis of the ratios of the width-thickness ratios of the various elements of the cross-section. The width-thickness ratio λ is calculated for each element of the cross-section as follows:

1. For compression members other than HSS, steel pipe and single angles, per LRFD Specification Section B5.1.
2. For HSS and steel pipe, per HSS Specification 2.2.1.
3. For single angles, per Single-Angle Specification Section 4 (note that b/t is used in this specification, rather than λ).

See LRFD Commentary Figure C-B5.1 for selected illustrations.

Width-Thickness Limits for Compact and Non-Compact Cross-Sections

Steel compression members need not be considered slender-element cross-sections, except when λ exceeds λ_r, where λ_r is determined as follows:

1. For compression members other than HSS, steel pipe and single angles, per LRFD Specification Section B5.1 and Table B5.1.
2. For HSS and steel pipe, per HSS Specification Section 2.2.1 and Table 2.2-1.
3. For single angles, per Single-Angle Specification Section 4 with the limiting parameter given for Equation 4-3a.

In all such cases, $Q = 1$. For convenience, the width-thickness ratios in LRFD Specification Table B5.1 are given as a function of F_y in LRFD Specification Numerical Values Table 6.

Effective Width Factor for Slender-Element Cross-Sections

When λ is greater than λ_r, the cross-section is a slender-element cross-section and the reduction factor for slender compression elements (elastic local buckling effects) Q must be included in the calculation of the compressive design strength. Q is determined as follows:

Doubly Symmetric Compression Members

1. For W-shapes and similar doubly symmetric shapes, per LRFD Specification Appendix B5.3.
2. For HSS and steel pipe, per HSS Specification Section 4.2.
3. For cruciform shapes and other doubly symmetric shapes, per LRFD Specification Appendix B5.3.

Singly Symmetric Compression Members

1. For tees and double angles, per LRFD Specification Appendix B5.3.
2. For singly symmetric shapes other than tees and double angles, per LRFD Specification Appendix B5.3.

Point-Symmetric and Unsymmetric Compression Members

1. For single angles, per Single-Angle Specification Section 4.
2. For point-symmetric shapes, such as zee shapes, and other unsymmetric shapes, per LRFD Specification Appendix B5.3.

EFFECTIVE LENGTH AND COLUMN SLENDERNESS

Physical and Effective Column Lengths

The physical column length l is taken as the distance between braced points (points at which relative and/or nodal bracing is provided per LRFD Specification Section C3.3). The effective column length is then Kl, where the effective length factor K of the compression member is determined as follows:

1. For compression members other than HSS, steel pipe and single angles, per LRFD Specification Section C2.
2. For HSS and steel pipe, per HSS Specification Section 4.1.1.
3. For single angles, see Lutz (1992).

Guidance on the proper selection of a value for K is given in LRFD Commentary Section C2, including the following:

1. For columns with idealized end conditions, recommended vales of K can be determined from LRFD Specification Table C-C2.1.
2. For columns in braced frames (or steel frames that lean on shear walls or another similar structural system) and compression members in trusses, K is normally taken as unity per LRFD Specification Section C2.1, unless a smaller value can be justified by analysis. Although the alignment chart in LRFD Commentary Figure C-C2.2a (sidesway inhibited–braced frames) could be used for that purpose, it should be noted that the stability bracing provisions in LRFD Specification Section C3.3 are based upon the use of $K = 1$.
3. For columns in moment frames, the alignment charts in LRFD Commentary Figure C-C2.2b (sidesway uninhibited–moment frames) can be used.

Further information is available in the SSRC *Guide to Stability Design Criteria for Metal Structures* (Galambos, 1998). Also, for compression in beams, girders and trusses designed for flexure on the basis of simple spans, see LRFD Specification Section B8.

Stiffness Reduction Factor τ for Inelastic Buckling

Per LRFD Commentary Section C2, the stiffness reduction factor τ can be used in the determination of K for columns controlled by inelastic buckling.

Column Slenderness Parameter λ_c

The column slenderness parameter λ_c is calculated per LRFD Specification Section E2 using Equation E2-4. If desired, LRFD Specification Equation E2-4 can be rearranged and solved at a value of $\lambda_c = 1.5$ such that the dividing parameter between inelastic and elastic column buckling is $4.71\sqrt{E/F_y}$. Thus, for Kl/r equal to or less than $4.71\sqrt{E/F_y}$, inelastic column buckling will control; otherwise, elastic column buckling will control. For example, see LRFD

Commentary Section E2 and Equations C-E2-1 through C-E2-3. For convenience, Kl/r is tabulated as a function of λ_c for steel with $F_y = 50$ ksi and 36 ksi in LRFD Specification Numerical Values Table 5.

COMPRESSIVE DESIGN STRENGTH FOR COMPACT AND NON-COMPACT CROSS-SECTIONS

For compression members with elements that have width-thickness ratios λ equal to or less than λ_r, the compressive design strength ϕP_n, which must equal or exceed the required strength (factored axial compression) P_u, is determined as follows:

Doubly Symmetric Compression Members

1. For W-shapes and similar doubly symmetric shapes, per LRFD Specification Section E2 using Equation E2-1.
2. For HSS and steel pipe, per HSS Specification Section 4.2 using Equation 4.2-1.
3. For cruciform shapes and other doubly symmetric shapes, per LRFD Specification Appendix E3 using Equation A-E3-1 and F_e from Equation A-E3-5.

Singly Symmetric Compression Members

1. For tees and double angles, per LRFD Specification Section E2 using Equation E2-1 and Section E3 using Equation E3-1 and F_{crft} from Equation E3-2. As noted in LRFD Commentary Section E3, tees that conform to the limits in LRFD Commentary Table C-E3.1 need not be checked for flexural-torsional buckling.
2. For singly symmetric shapes other than tees and double angles, per LRFD Specification Section E2 using Equation E2-1 and Appendix E3 using Equation A-E3-1 and F_e from Equation A-E3-6.

Point-Symmetric and Unsymmetric Compression Members

1. For single angles, per Single-Angle Specification Section 4 using Equation 4-1.
2. For point-symmetric shapes, such as zee shapes, and other unsymmetric shapes, per LRFD Specification Appendix E3 using Equation A-E3-1 and F_e from Equation A-E3-7.

In each case, the calculation is based upon the gross area A_g, which is determined as specified in LRFD Specification Section B1.

For convenience, $\phi_c F_{cr}$ from Equations E2-2 and E2-3 is tabulated as a function of Kl/r in the numerical values section of the LRFD Specification as follows:

1. For steel with $F_y = 50$ ksi, use LRFD Specification Numerical Values Table 3-50.
2. For steel with $F_y = 36$ ksi, use LRFD Specification Numerical Values Table 3-36.
3. For steel with F_y other than 50 ksi or 36 ksi, use LRFD Specification Numerical Values Table 4.

COMPRESSIVE DESIGN STRENGTH FOR SLENDER-ELEMENT CROSS-SECTIONS

For compression members with an element or elements that have width-thickness ratios λ greater than λ_r, the compressive design strength ϕP_n, which must equal or exceed the required strength (factored axial compression) P_u, is determined as follows:

Doubly Symmetric Compression Members

1. For W-shapes and similar doubly symmetric shapes, per LRFD Specification Appendix B5.3 using Equation E3-1 and F_{cr} from Appendix B5.3d or Equations A-B5-15 and A-B5-16.

2. For HSS and steel pipe, per HSS Specification Section 4.2 using Equation 4.2-1.
3. For cruciform shapes and other doubly symmetric shapes, per LRFD Specification Appendix B5.3 and Appendix E3 using F_e from Equation A-E3-5.

Singly Symmetric Compression Members

1. For tees and double-angles, per LRFD Specification Section E2 using Equation E2-1 and Section E3 using Equation E3-1 and F_{cry} determined from LRFD Specification Appendix B5.3. As noted in LRFD Commentary Section E3, tees that conform to the limits in LRFD Commentary Table C-E3.1 need not be checked for flexural-torsional buckling.
2. For singly symmetric shapes other than tees and double angles, per LRFD Specification Appendix E3 using Equation A-E3-1 and F_e from Equation A-E3-6.

Point-Symmetric and Unsymmetric Compression Members

1. For single angles, per Single-Angle Specification Section 4 using Equation 4-1.
2. For point-symmetric shapes, such as zee shapes, and other unsymmetric shapes, per LRFD Specification Appendix E3 using Equation A-E3-1 and F_e from Equation A-E3-7.

In each case, the calculation is based upon the gross area A_g, which is determined as specified in LRFD Specification Section B1.

COMPOSITE COMPRESSION MEMBERS

For the design of reinforced-concrete-encased and concrete-filled steel compression members, see LRFD Specification Section I2. See also AISC Design Guide No. 6 *Load and Resistance Factor Design of W-Shapes Encased in Concrete* (Griffis, 1992). For further information on composite design and construction, see also Viest et al. (1997).

OTHER SPECIFICATION REQUIREMENTS AND DESIGN CONSIDERATIONS

The following other specification requirements and design considerations apply to the design of compression members.

Slenderness

Compression-member slenderness Kl/r should preferably be limited to a maximum of 200 per LRFD Specification Section B7, HSS Specification Section 2.3 and Single-Angle Specification Section 4. Note that this requirement does not apply to members that are primarily tension members, but subject to incidental compression under other load combinations.

Plastic Analysis

For design by plastic analysis, see LRFD Specification Section E1.2.

Built-Up Compression Members

For built-up compression members, see LRFD Specification Section E4.

Pin-Connected Plates

For pin-connected compression members, see LRFD Specification Section E5.

Tapered Compression Members

For tapered compression members, see LRFD Specification Appendix F3.

Serviceability

Serviceability requirements, per LRFD Specification Sections A5.4 and L3, should be appropriate for the application, including an appropriate limit on the axial deformations of the compression member. See also AISC Design Guide No. 3 *Serviceability Design Considerations for Low-Rise Buildings* (Fisher and West, 1990).

Connection Design

For compression-member connection design, see Part 13 (bracing connections and truss connections) and Part 14 (base plates, cap plates and column splices).

DESIGN EXAMPLES

EXAMPLE 4.1. W-shape compression member design (braced frame).

Given: Determine the design strength of an ASTM A992 W14×132 that is part of a braced frame.

$F_y = 50\,\text{ksi}$ $\quad A_g = 38.8\,\text{in.}^2$

$F_u = 65\,\text{ksi}$ $\quad r_x = 6.28\,\text{in.}$

$r_y = 3.76\,\text{in.}$

a) Assume the physical length $L = 30\,\text{ft}$, the ends are pinned and the column is braced at the ends only for both the X-X and Y-Y axes. Also, calculate at what length this compression member would cease to satisfy the slenderness limitation in LRFD Specification Section B7.

b) Assume the physical length $L = 30\,\text{ft}$, the ends are pinned, and the column is braced at the ends only for the X-X axis and braced at the ends and mid-height for the Y-Y axis.

Solution a: From Table 1-1, the width-thickness ratio for the flanges is

$$\lambda = \frac{b_f}{2t_f} = 7.15$$

and the width-thickness ratio for the web is

$$\lambda = \frac{h}{t_w} = 17.7$$

From LRFD Specification Section B5.1, for flange compactness

$$\begin{aligned}\lambda_r &= 0.56\sqrt{\frac{E}{F_y}}\\ &= 0.56\sqrt{\frac{29{,}000\,\text{ksi}}{50\,\text{ksi}}}\\ &= 13.5\end{aligned}$$

Since $\lambda < \lambda_r$, the flanges are not slender. From LRFD Specification Section B5.1, for web compactness

$$\begin{aligned}\lambda_r &= 1.49\sqrt{\frac{E}{F_y}}\\ &= 1.49\sqrt{\frac{29{,}000\,\text{ksi}}{50\,\text{ksi}}}\\ &= 35.9\end{aligned}$$

Since $\lambda < \lambda_r$, the web is not slender. Since neither the flanges nor the web are slender, the W14×132 can be designed per LRFD Specification Section E2.

From LRFD Specification Section C2.1 and Commentary Table C-C2.1, $K = 1.0$ for both the X-X axis and the Y-Y axis.

$$\frac{Kl_x}{r_x} = \frac{1.0(30\text{ ft} \times 12\text{ in./ft})}{6.28\text{ in.}} = 57.3$$

$$\frac{Kl_y}{r_y} = \frac{1.0(30\text{ ft} \times 12\text{ in./ft})}{3.76\text{ in.}} = 95.7$$

Since $Kl_y/r_y > Kl_x/r_x$, buckling about the Y-Y axis controls. From LRFD Specification Section E2,

$$\lambda_c = \frac{Kl}{r\pi}\sqrt{\frac{F_y}{E}} = \frac{95.7}{\pi}\sqrt{\frac{50\text{ ksi}}{29{,}000\text{ ksi}}} = 1.26$$

Since $\lambda_c < 1.5$,

$$\begin{aligned}\phi_c P_n &= \phi_c F_{cr} A_g \\ &= \phi_c \left(0.658^{\lambda_c^2}\right) F_y A_g \\ &= 0.85\left(0.658^{(1.26)^2}\right)(50\text{ ksi})\left(38.8\text{ in.}^2\right) \\ &= 844\text{ kips}\end{aligned}$$

Per LRFD Specification Section B7,

$$L_{\max} = \frac{200r}{K} = \frac{200(3.76\text{ in.})}{1.0(12\text{ in./ft})} = 62.7\text{ ft}$$

Thus, the W14×132 compression member satisfies the slenderness requirements up to a 62.7-ft length.

Solution b: As determined in solution a, neither the flanges nor the web are slender and the W14×132 can be designed per LRFD Specification Section E2.

From LRFD Specification Section C2.1 and Commentary Table C-C2.1, $K = 1.0$ for both the X-X axis and the Y-Y axis.

$$\frac{Kl_x}{r_x} = \frac{1.0(30\text{ ft} \times 12\text{ in./ft})}{6.28\text{ in.}} = 57.3$$

$$\frac{Kl_y}{r_y} = \frac{1.0\left(\dfrac{30\text{ ft}}{2} \times 12\text{ in./ft}\right)}{3.76\text{ in.}} = 47.9$$

Since $Kl_x/r_x > Kl_y/r_y$, buckling about the X-X axis controls. From LRFD Specification Section E2,

$$\begin{aligned}\lambda_c &= \frac{Kl}{r\pi}\sqrt{\frac{F_y}{E}} \\ &= \frac{57.3}{\pi}\sqrt{\frac{50\,\text{ksi}}{29{,}000\,\text{ksi}}} \\ &= 0.757\end{aligned}$$

Since $\lambda_c < 1.5$,

$$\begin{aligned}\phi_c P_n &= \phi_c F_{cr} A_g \\ &= \phi_c \left(0.658^{\lambda_c^2}\right) F_y A_g \\ &= 0.85\left(0.658^{(0.757)^2}\right)(50\,\text{ksi})\left(38.8\,\text{in.}^2\right) \\ &= 1{,}300\,\text{kips}\end{aligned}$$

Comments: The preceding calculations can be simplified using Table 4-2. For solution a, with $(KL)_y = (KL)_x = 30$ ft, buckling about the Y-Y axis controls because $r_y < r_x$ and a W14×132 has

$$\phi_c P_n = 844\,\text{kips}$$

For solution b, $(KL)_y = 15$ ft. However, $(KL)_x = 30$ ft may be more critical if $(KL)_{y\,eq} > (KL)_y$.

$$\begin{aligned}(KL)_{y\,eq} &= \frac{(KL)_x}{\dfrac{r_x}{r_y}} \\ &= \frac{1.0(30\,\text{ft})}{1.67} \\ &= 18.0\,\text{ft}\end{aligned}$$

Since $(KL)_{y\,eq} > (KL)_y$, buckling about the X-X axis controls and a W14×132 has

$$\phi_c P_n = 1{,}300\,\text{kips}$$

EXAMPLE 4.2. W-shape compression member design (moment frame).

Given: The member sizes shown for the moment frame (sidesway uninhibited) illustrated in Figure 4-1 have been determined to be adequate for lateral loads.

a) Is the ASTM A992 W14×82 adequate to support the gravity loads shown? Assume the column is continuously supported in the transverse direction (the Y-Y axis of the column).

$F_y = 50$ ksi $\quad A_g = 24.0$ in.^{2}
$F_u = 65$ ksi

b) What would the design strength be if the column base were fixed instead of pinned?

Solution a: For the column segment between the roof and the floor,

$$\frac{P_u}{A_g} = \frac{250\text{ kips}}{24.0\text{ in.}^2} = 10.4\text{ ksi}$$

From Table 4-1, $\tau = 1.00$. That is, the column buckles in the elastic region, so there is no reduction in stiffness for inelasticity.

$$G_{top} = \tau \frac{\sum\left(\frac{I}{L}\right)_c}{\sum\left(\frac{I}{L}\right)_g} = (1.00)\frac{\left(\frac{881\text{ in.}^4}{35\text{ ft}}\right)}{2\left(\frac{800\text{ in.}^4}{14\text{ ft}}\right)} = 0.220$$

$$G_{bottom} = \tau \frac{\sum\left(\frac{I}{L}\right)_c}{\sum\left(\frac{I}{L}\right)_g} = (1.00)\frac{2\left(\frac{881\text{ in.}^4}{35\text{ ft}}\right)}{2\left(\frac{1{,}360\text{ in.}^4}{14\text{ ft}}\right)} = 0.259$$

From LRFD Commentary Figure C-C2.2b, K ≈ 1.1.

$$(KL)_{y\,eq} = \frac{(KL)_x}{\frac{r_x}{r_y}} = \frac{1.1(14\text{ ft})}{2.44} = 6.31\text{ ft}$$

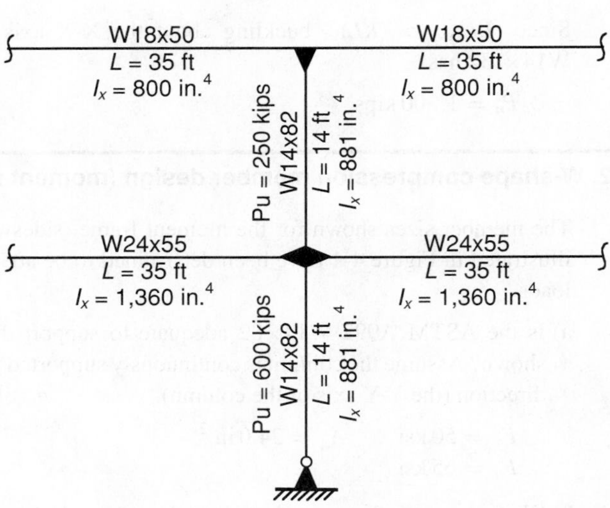

Fig. 4–1. Illustration for Example 4.2.

From Table 4-2,

$$\phi_c P_n \approx 1{,}090\,\text{kips} > 250\,\text{kips}\ \mathbf{o.k.}$$

For the column segment between the floor and the foundation,

$$\frac{P_u}{A_g} = \frac{600\,\text{kips}}{24.0\,\text{in.}^2} = 25.0\,\text{ksi}$$

From Table 4-1, $\tau = 0.85$.

$$G_{top} = \tau \frac{\Sigma\left(\frac{I}{L}\right)_c}{\Sigma\left(\frac{I}{L}\right)_g} = (0.85)\frac{2\left(\frac{881\,\text{in.}^4}{35\,\text{ft}}\right)}{2\left(\frac{1{,}360\,\text{in.}^4}{14\,\text{ft}}\right)} = 0.220$$

$$G_{bottom} = 10\,\text{(pinned end)}$$

From LRFD Commentary Figure C-C2.2b, K $\approx$ 1.7.

$$(KL)_{y\,eq} = \frac{(KL)_x}{\frac{r_x}{r_y}} = \frac{1.7(14\,\text{ft})}{2.44} = 9.75\,\text{ft}$$

From Table 4-2,

$$\phi_c P_n \approx 867\,\text{kips} > 600\,\text{kips}\ \mathbf{o.k.}$$

Thus, the W14×82 compression member is adequate.

Solution b: As determined in solution a, for the column segment between the roof and the floor,

$$\phi_c P_n \approx 1{,}090\,\text{kips}$$

As determined in solution a, for the column segment between the floor and the foundation, $G_{top} = 0.220$ and

$$G_{bottom} = 1\text{(fixed end)}$$

From LRFD Commentary Figure C-C2.2b, K $\approx$ 1.2.

$$(KL)_{y\,eq} = \frac{(KL)_x}{\frac{r_x}{r_y}} = \frac{1.2(14\,\text{ft})}{2.44} = 6.89\,\text{ft}$$

From Table 4-2,

$$\phi_c P_n \approx 940\,\text{kips}$$

EXAMPLE 4.3. Single-angle compression member design.

Given:

Determine the design strength of an ASTM A36 L4×3½×5/16 with physical length $L = 8$ ft, pinned ends and no bracing along the length of member. Also, calculate at what length this compression member would cease to satisfy the slenderness limitation in Single-Angle Specification Section 4.

$$F_y = 36\,\text{ksi} \qquad A_g = 2.25\,\text{in.}^2$$
$$F_u = 58\,\text{ksi} \qquad r_z = 0.721\,\text{in.}$$

Solution:

For L4×3½×5/16,

$$\frac{b}{t} = \frac{4\,\text{in.}}{5/16\,\text{in.}} = 12.8$$

From Single-Angle Specification Section 4,

$$\lambda_p = 0.446\sqrt{\frac{E}{F_y}} = 0.446\sqrt{\frac{29{,}000\,\text{ksi}}{36\,\text{ksi}}} = 12.7$$

$$\lambda_r = 0.910\sqrt{\frac{E}{F_y}} = 0.910\sqrt{\frac{29{,}000\,\text{ksi}}{36\,\text{ksi}}} = 25.8$$

Since $\lambda_p < b/t < \lambda_r$,

$$Q = 1.34 - 0.761\frac{b}{t}\sqrt{\frac{F_y}{E}} = 1.34 - 0.761(12.8)\sqrt{\frac{36\,\text{ksi}}{29{,}000\,\text{ksi}}} = 0.997$$

From LRFD Commentary Table C-C2.1, $K = 1.0$. From Single-Angle Specification Section 4, using $r = r_z$ (the least radius of gyration for the cross-section),

$$\lambda_c = \frac{Kl}{r\pi}\sqrt{\frac{F_y}{E}} = \frac{1.0(8\,\text{ft} \times 12\,\text{in./ft})}{(0.721\,\text{in.})\pi}\sqrt{\frac{36\,\text{ksi}}{29{,}000\,\text{ksi}}} = 1.49$$

$$\lambda_c\sqrt{Q} = 1.49\sqrt{0.997}$$
$$= 1.49$$

Since $\lambda_c < 1.5$,

$$\phi_c P_n = \phi_c F_{cr} A_g$$
$$= \phi_c Q\left(0.658^{Q\lambda_c^2}\right) F_y A_g$$
$$= 0.9(0997)\left(0.658^{(1.49)^2}\right)(36\,\text{ksi})\left(2.25\,\text{in.}^2\right)$$
$$= 28.7\,\text{kips}$$

Per Single-Angle Specification Section 4,

$$L_{\max} = \frac{200r}{K}$$
$$= \frac{200(0.721\,\text{in.})}{1.0(12\,\text{in./ft})}$$
$$= 12.0\,\text{ft}$$

Thus, the L4×3½×5/16 compression member satisfies the slenderness requirements up to a 12.0-ft length.

Comments: The preceding calculations can be simplified using Table 4-4. With $(KL)_z = 8$ ft, an L4×3½×5/16 has

$$\phi_c P_n = 28.7\,\text{kips}$$

EXAMPLE 4.4. WT-shape compression member design.

Given: Select an ASTM A992 WT-shape compression member ($F_y = 50$ ksi, $F_u = 65$ ksi) for a required strength of 100 kips, a physical length $L = 20$ ft. Assume the ends are pinned and the member is braced at the ends only for both the X-X and Y-Y.

Solution: From LRFD Commentary Table C-C2.1, $K = 1.0$. From Table 4-5, try a WT7×26.5. For $(KL)_x = 20$ ft,

$$\phi_c P_n = 102\,\text{kips} > 100\,\text{kips} \quad \textbf{o.k.}$$

For $(KL)_y = 20$ ft,

$$\phi_c P_n = 105\,\text{kips} > 100\,\text{kips} \quad \textbf{o.k.}$$

The slenderness of the WT7×26.5 satisfies the requirement in LRFD Specification Section B7, since the heavy line in the table falls below the indicated values of effective length. **o.k.**

Use WT7×26.5.

EXAMPLE 4.5. Rectangular HSS compression member design.

Given: Select a rectangular ASTM A500 grade B HSS compression member ($F_y = 46$ ksi, $F_u = 58$ ksi) for a required strength of 500 kips, a physical length $L = 20$ ft. Assume the base is fixed, the top is pinned, and the member is braced at the ends only for both the X-X and Y-Y axes.

Solution: From LRFD Commentary Table C-C2.1, $K_x = K_y = 0.8$. With $(KL)_x = (KL)_y = 16$ ft and $r_y < r_x$, buckling about the Y-Y axis controls. From Table 4-6, try an HSS12×10×1/2. For $(KL)_y = 16$ ft,

$$\phi_c P_n = 580\,\text{kips} > 500\,\text{kips} \qquad \textbf{o.k.}$$

The slenderness of the HSS12×10×1/2 satisfies the requirement in HSS Specification Section 2.3, since the heavy line in the table falls below the indicated values of effective length. **o.k.**

Use HSS12×10×1/2.

EXAMPLE 4.6. Steel pipe compression member design.

Given: Select an ASTM A53 grade B steel-pipe compression member ($F_y = 35$ ksi, $F_u = 60$ ksi) for a required strength of 200 kips, a physical length $L = 30$ ft. Assume the ends are pinned and the member is braced at the ends only for the X-X axis and at the ends and mid-height for the Y-Y axis.

Solution: From LRFD Specification Section C2.1 and Commentary Table C-C2.1, $K = 1.0$ for both the X-X axis and the Y-Y axis. With $(KL)_x > (KL)_y$ and $r_y = r_x$, buckling about the X-X axis controls. From Table 4-8, try a Pipe 10 Std. For $(KL)_y = 30$ ft,

$$\phi_c P_n = 216\,\text{kips} > 200\,\text{kips} \qquad \textbf{o.k.}$$

The slenderness of the Pipe 10 Std. satisfies the requirement in HSS Specification Section 2.3, since the heavy line in the table falls below the indicated values of effective length. **o.k.**

Use Pipe 10 Std.

EXAMPLE 4.7. Double-angle compression member design.

Given: Determine the design strength of an ASTM A36 2L4×3 1/2×5/16 LLBB (3/4-in. separation) with physical length $L = 8$ ft, pinned ends and no bracing along the length of member. Also, calculate the required number of intermediate connectors.

$F_y = 36$ ksi $\qquad r_x = 1.25$ in.

$F_u = 58$ ksi $\qquad r_z = 1.53$ in.

$r_z = 0.721$ in. (for single angle)

Solution: From LRFD Commentary Table C-C2.1, $K = 1.0$. From Table 4-10, for $(KL)_x = 8$ ft,

$$\phi_c P_n = 101\,\text{kips}$$

For buckling about the Y-Y axis, the tabulated values are for a separation of 3/8 in. To adjust to a spacing of 3/4 in, $(KL)_y$ should be multiplied by the ratio of r_y for a $^3/_8$-in. separation (1.53 in. from Table 1-14) to r_y for a $^3/_4$-in. separation (1.68 in. from Table 1-14). Thus,

$$(KL)_y = 1.0(8\text{ ft})\left(\frac{1.53\text{ in.}}{1.68\text{ in.}}\right)$$
$$= 7.29\text{ ft}$$

From Table 4-10, for $(KL)_y = 7.29$ ft,

$$\phi_c P_n \approx 102\text{ kips}$$

and at least two welded or pretensioned bolted intermediate connectors are required (a = 32 in.). Per LRFD Specification Section E4.2, the slenderness of the individual components of the built-up member based upon the distance between intermediate connectors a, must not exceed three-quarters of the controlling slenderness of the overall built-up compression member.

$$\frac{Kl_x}{r_x} = \frac{1.0(8\text{ ft} \times 12\text{ in./ft})}{1.25\text{ in.}}$$
$$= 76.8 \leftarrow \text{larger governs}$$
$$\frac{Kl_y}{r_y} = \frac{1.0(7.29\text{ ft} \times 12\text{ in./ft})}{1.53\text{ in.}}$$
$$= 57.2$$

Thus,

$$a_{max} = \frac{3}{4}\frac{\left(\frac{KL}{r}\right)_{max}(r_z)}{K}$$
$$= \frac{3}{4}\frac{(76.8\text{ in.})(0.721\text{ in.})}{1.0}$$
$$= 41.5\text{ in.} > 32\text{ in.} \quad \textbf{o.k.}$$

The slenderness of the 2L4×3 $^1/_2$×$^5/_{16}$ LLBB satisfies the requirement in LRFD Specification Section B7, since the heavy line in the table falls below the indicated values of effective length. **o.k.**

Thus the 2L4×3 $^1/_2$×$^5/_{16}$ LLBB compression member with two equally spaced welded or pretensioned bolted intermediate connectors has $\phi_c P_n = 101$ kips.

EXAMPLE 4.8. Rectangular HSS filled with 5-ksi normal-weight concrete.

Given: Determine the design strength of an ASTM A500 grade B HSS16×12×$^1/_2$ filled with 5-ksi normal-weight concrete with a physical length $L = 20$ ft. Assume the ends are pinned and the member is braced at the ends only for both the X-X and Y-Y axes.

Solution: From LRFD Commentary Table C-C2.1, $K_x = K_y = 1.0$. With $(KL)_x = (KL)_y = 20$ ft and $r_{my} < r_{mx}$, buckling about the Y-Y axis controls. From

Table 4-13, for $(KL)_y = 20$ ft, an HSS16×12×1/2 has

$$\phi_c P_n = 1{,}220 \text{ kips}$$

The slenderness of the HSS16×12×1/2 satisfies the requirement in LRFD Specification Section B7, since the heavy line in the table falls below the indicated values of effective length. **o.k.**

STEEL COMPRESSION-MEMBER SELECTION TABLES

Table 4-1. Stiffness Reduction Factor τ

When column buckling occurs in the inelastic range, the stiffness reduction factor τ can be used to adjust the value of G used in the determination of K. Values are given for steels with $F_y = 35$ ksi, 36 ksi, 42 ksi, 46 ksi and 50 ksi. The use of Table 4-1 is illustrated in Example 4.2.

Table 4-2. W-Shapes

Design strengths in axial compression are given for W-shapes with $F_y = 50$ ksi (ASTM A992). The tabulated values are given for the effective length with respect to the Y-Y axis $(KL)_y$. However, the effective length with respect to the X-X axis $(KL)_x$ must also be investigated. To determine the design strength in axial compression $\phi_c P_n$, the table should be entered at the larger of $(KL)_y$ and $(KL)_{y\ eq}$, where

$$(KL)_{y\ eq} = \frac{(KL)_x}{\dfrac{r_x}{r_y}}$$

Values of the ratio r_x/r_y and other properties useful in the design of W-shape compression members are listed at the bottom of Table 4-2. The variables P_{wo} and P_{wi} can be used in the calculation of the web local yielding design strength (LRFD Specification Equation K1-2) for the column as follows:

$$\phi R_n = P_{wo} + P_{wi} N$$

The variable P_{wb} can be used in the calculation of the web compression buckling design strength (LRFD Specification Equation K1-8) for the column as follows:

$$\phi R_n = P_{wb}$$

The variable P_{fb} can be used in the calculation of the flange local bending design strength (LRFD Specification Equation K1-1) for the column as follows:

$$\phi R_n = P_{fb}$$

The use of Table 4-2 is illustrated in Examples 4.1 and 4.2.

Table 4-3. HP-Shapes

Table 4-3 is similar to Table 4-2, except it covers HP-shapes with $F_y = 36$ ksi (ASTM A36). The use of Table 4-3 is similar to that illustrated for W-shapes in Examples 4.1 and 4.2.

Table 4-4. Single Angles

Design strengths in axial compression are given for single angles with $F_y = 36$ ksi (ASTM A36) based upon the effective length with respect to the Z-Z axis $(KL)_z$. Other properties useful in the design of single-angle compression members are listed at the bottom of Table 4-4. The use of Table 4-4 is illustrated in Example 4.3.

Table 4-5. WT-Shapes

Design strengths in axial compression are given for WT-shapes with $F_y = 50$ ksi (ASTM A992). Separate tabulated values are given for the effective lengths with respect to the X-X and Y-Y axes, $(KL)_x$ and $(KL)_y$, respectively. Other properties useful in the design of WT-shape compression members are listed at the bottom of Table 4-5. The use of Table 4-5 is illustrated in Example 4.4.

Table 4-6. Rectangular HSS

Design strengths in axial compression are given for rectangular HSS with $F_y = 46$ ksi (ASTM A500 grade B). The tabulated values are given for the effective length with respect to the Y-Y axis $(KL)_y$. However, the effective length with respect to the X-X axis $(KL)_x$ must also be investigated. To determine the design strength in axial compression $\phi_c P_n$, the table should be entered at the larger of $(KL)_y$ and $(KL)_{y\ eq}$, where $(KL)_{y\ eq}$ is determined as indicated for W-shapes in the description of Table 4-2. Values of the ratio r_x/r_y and other properties useful in the design of rectangular HSS compression members are listed at the bottom of Table 4-6. The use of Table 4-6 is illustrated in Example 4.5.

Table 4-7. Round HSS

Design strengths in axial compression are given for round HSS with $F_y = 42$ ksi (ASTM A500 grade B). To determine the design strength in axial compression $\phi_c P_n$, the table should be entered at the larger of $(KL)_y$ and $(KL)_x$. Other properties useful in the design of compression members are listed at the bottom of the column design strength tables. The use of Table 4-7 is similar to that illustrated for steel pipe in Example 4.6.

Table 4-8. Steel Pipe

Table 4-8 is similar to Table 4-7, except it covers steel pipe with $F_y = 35$ ksi (ASTM A53 grade B). The use of Table 4-8 is illustrated in Example 4.6.

Table 4-9. Equal-Leg Double Angles

Design strengths in axial compression are given for equal-leg double angles with $F_y = 36$ ksi (ASTM A36), assuming $3/8$-in. separation between the angles. These values can be used conservatively when a larger separation is provided. Alternatively, the value of $(KL)_y$ can be multiplied by the ratio of the ratio of r_y for a $3/8$-in. separation to r_y for the actual separation.

Separate tabulated values are given for the effective lengths with respect to the X-X and Y-Y axes, $(KL)_x$ and $(KL)_y$, respectively. For buckling about the X-X axis, the design strength is not affected by the number of intermediate connectors. However, for buckling about the Y-Y axis, the effects of shear deformations of the intermediate connectors must be considered. The tabulated values for $(KL)_y$ have been adjusted for the shear deformations in accordance with LRFD Specification equation E4-2, which is applicable to welded and pretensioned bolted intermediate shear connectors. The number of intermediate connectors n given in the table is selected so that the compression buckling design strength about the Y-Y axis is equal to or greater than 90 percent of that for compression buckling of the two angles as a unit. If fewer connectors or snug-tightened bolted intermediate connectors are used, the design strength must be recalculated per LRFD Specification Section E4. Per LRFD Specification Section E4.2, the slenderness of the individual components of the built-up member based upon the distance between intermediate connectors a, must not exceed three-quarters of the controlling slenderness of the overall built-up compression member.

Other properties useful in the design of double-angle compression members are listed at the bottom of Table 4-9. The use of Table 4-9 is similar to that illustrated for LLBB double angles in Example 4.7.

Table 4-10. LLBB Double Angles

Table 4-10 is the same as Table 4-9, except that it provides design strengths in axial compression for LLBB double angles. The use of Table 4-10 is illustrated in Example 4.7.

Table 4-11. SLBB Double Angles

Table 4-11 is the same as Table 4-9, except that it provides design strengths in axial compression for SLBB double angles. The use of Table 4-11 is similar to that illustrated for SLBB double angles in Example 4.7.

COMPOSITE COMPRESSION-MEMBER SELECTION TABLES

Table 4-12. Rectangular HSS Filled With 4-ksi Normal-Weight Concrete

Design strengths in axial compression are given for rectangular HSS with $F_y = 46$ ksi (ASTM A500 grade B) filled with 4-ksi normal-weight concrete. The tabulated values are given for the effective length with respect to the Y-Y axis $(KL)_y$. However, the effective length with respect to the X-X axis $(KL)_x$ must also be investigated. To determine the design strength in axial compression $\phi_c P_n$, the table should be entered at the larger of $(KL)_y$ and $(KL)_{y\,eq}$, where

$$(KL)_{y\,eq} = \frac{(KL)_x}{\dfrac{r_{mx}}{r_{my}}}$$

Values of the ratio r_{mx}/r_{my} and other properties useful in the design of composite HSS compression members are listed at the bottom of Table 4-12. The use of Table 4-12 is similar to that illustrated for rectangular HSS filled with 5-ksi normal-weight concrete in Example 4.8.

Table 4-13. Rectangular HSS Filled With 5-ksi Normal-Weight Concrete

Table 4-13 is the same as Table 4-12, except that it provides design strengths in axial compression for rectangular HSS filled with 5-ksi normal-weight concrete. The use of Table 4-13 is illustrated in Example 4.8.

Table 4-14. Round HSS Filled With 4-ksi Normal-Weight Concrete

Design strengths in axial compression are given for round HSS with $F_y = 42$ ksi (ASTM A500 grade B) filled with 4-ksi normal-weight concrete. To determine the design strength in axial compression $\phi_c P_n$, the table should be entered at the larger of $(KL)_y$ and $(KL)_x$. Other properties useful in the design of compression members are listed at the bottom of the column design strength tables. The use of Table 4-14 is similar to that illustrated for rectangular HSS filled with 5-ksi normal-weight concrete in Example 4.8.

Table 4-15. Round HSS Filled With 5-ksi Normal-Weight Concrete

Table 4-15 is the same as Table 4-14, except that it provides design strengths in axial compression for round HSS filled with 5-ksi normal-weight concrete. The use of Table 4-15 is similar to that illustrated for rectangular HSS filled with 5-ksi normal-weight concrete in Example 4.8.

Table 4-16. Steel Pipe Filled With 4-ksi Normal-Weight Concrete

Design strengths in axial compression are given for steel pipe with $F_y = 35$ ksi (ASTM A53 grade B) filled with 4-ksi normal-weight concrete. To determine the design strength in axial compression $\phi_c P_n$, the table should be entered at the larger of $(KL)_y$ and $(KL)_x$. Other properties useful in the design of compression members are listed at the bottom of the column design strength tables. The use of Table 4-16 is similar to that illustrated for rectangular HSS filled with 5-ksi normal-weight concrete in Example 4.8.

Table 4-17. Steel Pipe Filled With 5-ksi Normal-Weight Concrete

Table 4-17 is the same as Table 4-16, except that it provides design strengths in axial compression for round HSS filled with 5-ksi normal-weight concrete. The use of Table 4-17 is similar to that illustrated for rectangular HSS filled with 5-ksi normal-weight concrete in Example 4.8.

PART 4 REFERENCES

Fisher, J.M. and M.A. West, 1990, AISC Design Guide No. 3 *Serviceability Design Considerations for Low-Rise Buildings*, AISC, Chicago, IL.

Galambos, T.V., 1998, *Guide to Stability Design Criteria for Metal Structures*, Fifth Edition, John Wiley & Sons, New York, NY.

Griffis, L.G., 1997, AISC Design Guide No. 6 *Load and Resistance Factor Design of W-Shapes Encased in Concrete*, AISC, Chicago, IL.

Lutz, L.A., 1992, "Critical Slenderness of Compression Members with Effective Lengths About Nonprincipal Axes," Structural Stability Research Council, Annual Technical Session Proceedings, Pittsburgh, PA.

Viest, I.M., J.P. Colaco, R.W. Furlong, L.G. Griffis, R.T. Leon and L.A. Wylie, Jr., 1997, *Composite Construction: Design for Buildings*, ASCE and McGraw-Hill, New York, NY.

Table 4-1. Stiffness Reduction Factor τ

P_u / A_g	F_y, ksi				
	35	36	42	46	50
43	–	–	–	–	–
42	–	–	–	–	0.0319
41	–	–	–	–	0.0944
40	–	–	–	–	0.155
39	–	–	–	0.00696	0.215
38	–	–	–	0.0756	0.273
37	–	–	–	0.142	0.329
36	–	–	–	0.207	0.383
35	–	–	0.0529	0.270	0.436
34	–	–	0.127	0.331	0.486
33	–	–	0.198	0.390	0.535
32	–	–	0.267	0.447	0.582
31	–	–	0.334	0.501	0.627
30	–	0.0529	0.398	0.554	0.670
29	0.0678	0.139	0.460	0.604	0.711
28	0.155	0.221	0.519	0.651	0.749
27	0.240	0.301	0.576	0.697	0.785
26	0.321	0.377	0.629	0.739	0.819
25	0.398	0.450	0.680	0.779	0.850
24	0.472	0.519	0.727	0.816	0.879
23	0.542	0.585	0.772	0.850	0.905
22	0.608	0.646	0.813	0.882	0.929
21	0.670	0.704	0.850	0.910	0.949
20	0.727	0.757	0.884	0.934	0.966
19	0.780	0.806	0.914	0.955	0.981
18	0.828	0.850	0.941	0.973	0.991
17	0.871	0.890	0.963	0.987	0.999
16	0.909	0.924	0.980	0.996	1.00
15	0.941	0.952	0.993	1.00	↓
14	0.966	0.975	1.00	↓	↓
13	0.986	0.991	↓	↓	↓
12	0.998	1.00	↓	↓	↓
11	1.00	↓	↓	↓	↓
10	↓	↓	↓	↓	↓
9	↓	↓	↓	↓	↓
8	↓	↓	↓	↓	↓

– Indicates stiffness reduction factor is not applicable because P_u exceeds $\phi_c F_y A_g$ (column design strength for $Kl/r = 0$).

$F_y = 50$ ksi
$\phi_c P_n = 0.85 F_{cr} A_g$

Table 4-2. W-Shapes Design Strength in Axial Compression, $\phi_c P_n$, kips

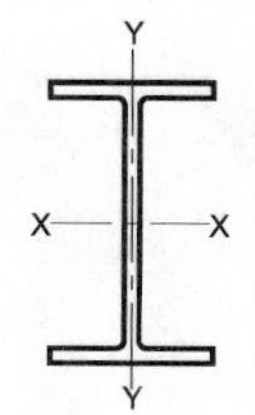

Shape	W14×										
	808*	730*	665*	605*	550*	500*	455*	426*	398*	370*	342*
Effective length KL (ft) with respect to least radius of gyration r_y											
0	10100	9140	8330	7570	6890	6250	5700	5310	4970	4630	4290
11	9540	8620	7850	7110	6460	5850	5330	4970	4640	4320	4000
12	9440	8530	7760	7030	6390	5780	5260	4900	4580	4260	3950
13	9330	8430	7660	6940	6300	5710	5190	4830	4520	4200	3890
14	9220	8320	7560	6850	6220	5620	5110	4760	4450	4140	3830
15	9100	8200	7450	6750	6120	5540	5030	4680	4380	4070	3760
16	8970	8080	7340	6640	6020	5450	4950	4600	4300	4000	3690
17	8840	7960	7220	6530	5920	5350	4860	4520	4220	3920	3620
18	8700	7820	7100	6420	5810	5250	4770	4430	4140	3840	3550
19	8550	7690	6970	6300	5700	5150	4670	4340	4050	3760	3470
20	8400	7550	6840	6170	5590	5040	4570	4250	3960	3680	3400
22	8090	7250	6560	5910	5350	4820	4370	4050	3780	3500	3230
24	7760	6940	6270	5640	5100	4590	4150	3850	3590	3320	3060
26	7410	6610	5970	5360	4840	4350	3930	3640	3390	3140	2890
28	7060	6280	5660	5080	4570	4100	3700	3430	3190	2950	2710
30	6700	5940	5340	4790	4300	3850	3480	3210	2990	2750	2530
32	6330	5600	5030	4490	4030	3610	3250	3000	2780	2560	2360
34	5970	5250	4710	4200	3760	3360	3020	2780	2580	2380	2180
36	5600	4910	4400	3910	3500	3120	2800	2570	2390	2190	2010
38	5240	4580	4090	3630	3240	2880	2580	2370	2190	2010	1840
40	4880	4250	3780	3350	2990	2650	2370	2170	2010	1840	1680
42	4530	3930	3490	3080	2740	2420	2160	1980	1830	1670	1530
44	4190	3620	3200	2820	2500	2210	1970	1800	1660	1520	1390
46	3860	3310	2930	2580	2290	2020	1800	1650	1520	1390	1270
48	3540	3040	2690	2370	2100	1860	1650	1510	1400	1280	1170
50	3260	2800	2480	2180	1940	1710	1520	1400	1290	1180	1080
Properties											
P_{wo}, kips	5350	4230	3620	3090	2630	2240	1920	1710	1520	1350	1180
P_{wi}, kips/in.	187	154	142	130	119	110	101	94.0	88.5	83.0	77.0
P_{wb}, kips	119000	66100	51700	40100	30800	23900	18800	15100	12700	10400	8320
P_{fb}, kips	7370	6780	5750	4870	4100	3450	2900	2600	2280	1990	1720
L_p, ft	17.0	16.6	16.3	16.1	15.9	15.6	15.5	15.3	15.2	15.1	15.0
L_r, ft	268	242	222	203	188	171	158	147	139	130	121
A_g, in.^{2}	237	215	196	178	162	147	134	125	117	109	101
I_x, in.^{4}	16000	14300	12400	10800	9430	8210	7190	6600	6000	5440	4900
I_y, in.^{4}	5510	4720	4170	3680	3250	2880	2560	2360	2170	1990	1810
r_y, in.	4.82	4.69	4.62	4.55	4.49	4.43	4.38	4.34	4.31	4.27	4.24
Ratio r_x/r_y	1.70	1.74	1.73	1.71	1.70	1.69	1.67	1.67	1.66	1.66	1.65
$P_{ex}(KL^2)/10^4$	458000	409000	355000	309000	270000	235000	206000	189000	172000	156000	140000
$P_{ey}(KL^2)/10^4$	158000	135000	119000	105000	93000	82400	73300	67500	62100	57000	51800

*ASTM A6 tensile group 4 or 5 shape. Special requirements may apply per LRFD Specification Section A3.1c.

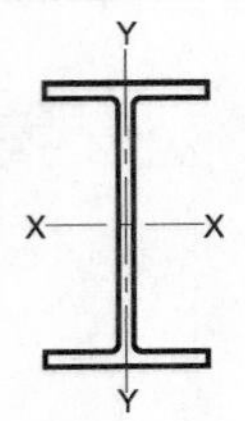

Table 4-2 (cont.). W-Shapes Design Strength in Axial Compression, $\phi_c P_n$, kips

$F_y = 50$ ksi

$\phi_c P_n = 0.85 F_{cr} A_g$

Shape		W14×									
		311*	283*	257*	233*	211	193	176	159	145	132
Effective length KL (ft) with respect to least radius of gyration r_y	0	3880	3540	3210	2910	2640	2410	2200	1980	1810	1650
	11	3610	3290	2980	2700	2440	2230	2030	1830	1670	1510
	12	3560	3240	2940	2660	2400	2200	2000	1810	1650	1480
	13	3510	3200	2890	2620	2370	2170	1970	1780	1620	1450
	14	3460	3140	2850	2570	2330	2130	1940	1740	1590	1430
	15	3400	3090	2800	2530	2280	2090	1900	1710	1560	1390
	16	3330	3030	2740	2480	2240	2050	1860	1680	1530	1360
	17	3270	2970	2690	2430	2190	2010	1820	1640	1500	1330
	18	3200	2910	2630	2380	2140	1960	1780	1600	1460	1300
	19	3130	2850	2570	2320	2090	1910	1740	1570	1430	1260
	20	3060	2780	2510	2270	2040	1870	1700	1530	1390	1220
	22	2910	2640	2380	2150	1940	1770	1610	1440	1320	1150
	24	2750	2500	2250	2030	1830	1670	1510	1360	1240	1070
	26	2590	2350	2120	1910	1710	1560	1420	1270	1160	997
	28	2430	2200	1980	1780	1600	1460	1320	1180	1080	920
	30	2270	2050	1840	1660	1490	1350	1220	1100	998	844
	32	2110	1900	1710	1530	1370	1250	1130	1010	919	769
	34	1950	1760	1570	1410	1260	1150	1040	928	842	697
	36	1790	1620	1440	1290	1160	1050	946	846	767	627
	38	1640	1480	1320	1180	1050	955	859	767	694	563
	40	1490	1340	1190	1070	951	863	775	692	626	508
	42	1350	1220	1080	967	863	783	703	628	568	461
	44	1230	1110	987	881	786	713	641	572	518	420
	46	1130	1010	903	806	719	652	586	523	474	384
	48	1040	932	829	741	660	599	538	481	435	353
	50	956	858	764	682	609	552	496	443	401	325
Properties											
P_{wo}, kips		1010	861	735	621	529	454	396	333	287	263
P_{wi}, kips/in.		70.5	64.5	59.0	53.5	49.0	44.5	41.5	37.3	34.0	32.3
P_{wb}, kips		6390	4900	3730	2780	2150	1610	1310	944	716	611
P_{fb}, kips		1440	1210	1000	832	684	583	483	398	334	298
L_p, ft		14.8	14.7	14.6	14.5	14.4	14.3	14.2	14.1	14.1	13.3
L_r, ft		110	100	91.6	83.4	76	70.1	64.5	58.9	54.7	49.6
A_g, in.^{2}		91.4	83.3	75.6	68.5	62	56.8	51.8	46.7	42.7	38.8
I_x, in.^{4}		4330	3840	3400	3010	2660	2400	2140	1900	1710	1530
I_y, in.^{4}		1610	1440	1290	1150	1030	931	838	748	677	548
r_y, in.		4.20	4.17	4.13	4.10	4.07	4.05	4.02	4.00	3.98	3.76
Ratio r_x/r_y		1.64	1.63	1.62	1.62	1.61	1.60	1.60	1.60	1.59	1.67
$P_{ex}(KL^2)/10^4$		124000	110000	97300	86200	76100	68700	61300	54400	48900	43800
$P_{ey}(KL^2)/10^4$		46100	41200	36900	32900	29500	26600	24000	21400	19400	15700

*ASTM A6 tensile group 4 or 5 shape. Special requirements may apply per LRFD Specification Section A3.1c.

$F_y = 50$ ksi
$\phi_c P_n = 0.85 F_{cr} A_g$

Table 4-2 (cont.). W-Shapes Design Strength in Axial Compression, $\phi_c P_n$, kips

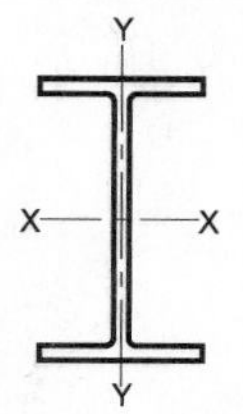

Shape	W14× 120	109	99††	90††	82	74	68	61	53	48	43††
Effective length *KL* (ft) with respect to least radius of gyration r_y											
0	1500	1360	1240	1130	1020	927	850	761	663	599	530
6	1460	1320	1200	1100	959	871	798	714	598	540	482
7	1450	1310	1190	1080	938	852	781	698	576	520	463
8	1430	1300	1180	1070	914	830	760	680	552	498	443
9	1410	1280	1160	1060	888	807	738	660	526	474	422
10	1390	1260	1150	1040	860	781	714	638	498	449	399
11	1370	1240	1130	1030	829	753	689	615	469	423	375
12	1350	1220	1110	1010	797	724	662	591	439	395	350
13	1320	1200	1090	989	764	694	633	566	409	368	325
14	1290	1170	1060	969	729	662	604	539	379	340	301
15	1270	1150	1040	947	694	630	575	513	349	313	276
16	1240	1120	1020	925	658	598	544	486	319	286	252
17	1210	1090	991	902	622	565	514	458	290	260	228
18	1180	1060	965	878	586	532	484	431	263	235	206
19	1140	1030	938	853	550	499	454	404	236	211	185
20	1110	1000	911	828	514	467	424	377	213	191	167
22	1040	943	854	776	445	405	366	325	176	157	138
24	972	879	796	723	380	345	311	276	148	132	116
26	902	815	737	670	324	294	265	236	126	113	98.7
28	832	751	679	616	279	253	229	203	109	97.2	85.1
30	762	688	621	564	243	221	199	177	94.7	84.7	74.1
32	694	627	565	512	214	194	175	155	83.2		
34	628	567	511	463	189	172	155	138			
36	565	509	458	415	169	153	138	123			
38	507	457	411	372	151	138	124	110			
40	457	412	371	336	137	124	112	99.5			
Properties											
P_{wo}, kips	227	192	167	144	185	155	136	116	116	101	85.4
P_{wi}, kips/in.	29.5	26.3	24.3	22.0	25.5	22.5	20.8	18.8	18.5	17.0	15.3
P_{wb}, kips	469	330	260	194	302	207	163	120	115	89.5	64.7
P_{fb}, kips	249	208	171	142	206	173	146	117	123	99.6	79.0
L_p, ft	13.2	13.2	13.5	15.1	8.76	8.76	8.69	8.65	6.78	6.75	6.68
L_r, ft	46.2	43.2	40.6	38.4	29.5	27.9	26.4	24.9	20.1	19.2	18.3
A_g, in.^{2}	35.3	32.0	29.1	26.5	24.0	21.8	20.0	17.9	15.6	14.1	12.6
I_x, in.^{4}	1380	1240	1110	999	881	795	722	640	541	484	428
I_y, in.^{4}	495	447	402	362	148	134	121	107	57.7	51.4	45.2
r_y, in.	3.74	3.73	3.71	3.70	2.48	2.48	2.46	2.45	1.92	1.91	1.89
Ratio r_x/r_y	1.67	1.67	1.66	1.66	2.44	2.44	2.44	2.44	3.07	3.06	3.08
$P_{ex}(KL^2)/10^4$	39500	35500	31800	28600	25200	22800	20700	18300	15500	13900	12300
$P_{ey}(KL^2)/10^4$	14200	12800	11500	10400	4240	3840	3460	3060	1650	1470	1290

††For W14×99 and W14×90, flange is non-compact. For W14×43, web may be non-compact for combined axial compression and flexure; see AISC LRFD Specification Section B5.
Note: Heavy line indicates Kl/r equal to or greater than 200.

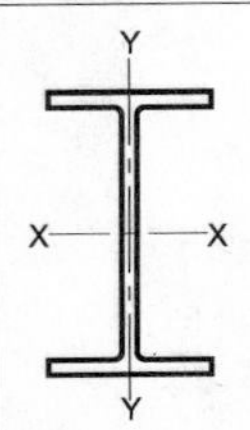

Table 4-2 (cont.). W-Shapes Design Strength in Axial Compression, $\phi_c P_n$, kips

$F_y = 50$ ksi

$\phi_c P_n = 0.85 F_{cr} A_g$

Shape	W12×										
	336	305	279	252	230	210	190	170	152	136	120
Effective length KL (ft) with respect to least radius of gyration r_y											
0	4200	3810	3480	3150	2880	2630	2370	2130	1900	1700	1500
6	4070	3690	3370	3040	2780	2540	2290	2050	1830	1630	1440
7	4020	3640	3330	3000	2740	2500	2260	2020	1810	1610	1420
8	3970	3590	3280	2960	2710	2470	2220	1990	1780	1590	1400
9	3910	3540	3230	2910	2660	2430	2190	1960	1750	1560	1380
10	3850	3480	3170	2860	2610	2380	2150	1920	1710	1530	1350
11	3780	3420	3110	2810	2560	2330	2100	1880	1680	1490	1320
12	3700	3350	3050	2750	2510	2280	2050	1840	1640	1460	1290
13	3620	3270	2980	2680	2450	2230	2000	1790	1590	1420	1250
14	3540	3190	2910	2610	2380	2170	1950	1740	1550	1380	1220
15	3450	3110	2830	2540	2320	2110	1900	1690	1510	1340	1180
16	3360	3020	2750	2470	2250	2040	1840	1640	1460	1290	1140
17	3260	2940	2670	2390	2180	1980	1780	1580	1410	1250	1100
18	3160	2840	2580	2320	2110	1910	1720	1530	1360	1210	1060
19	3060	2750	2500	2240	2030	1840	1650	1470	1310	1160	1020
20	2960	2660	2410	2160	1960	1780	1590	1420	1260	1110	976
22	2750	2460	2230	1990	1810	1640	1460	1300	1150	1020	892
24	2540	2270	2050	1830	1650	1490	1340	1180	1050	924	808
26	2330	2070	1870	1660	1500	1360	1210	1070	944	831	726
28	2120	1880	1690	1500	1350	1220	1090	959	844	742	646
30	1910	1690	1520	1340	1210	1090	967	852	749	656	569
32	1720	1510	1350	1190	1070	962	853	750	658	577	500
34	1520	1340	1200	1060	951	852	755	664	583	511	443
36	1360	1200	1070	944	848	760	674	593	520	456	395
38	1220	1080	960	847	761	682	605	532	467	409	355
40	1100	970	866	764	687	616	546	480	421	369	320
Properties											
P_{wo}, kips	1580	1340	1170	998	861	738	617	518	435	365	302
P_{wi}, kips/in.	89.0	81.5	76.5	70.0	64.5	59.0	53.0	48.0	43.5	39.5	35.5
P_{wb}, kips	15100	11600	9590	7320	5730	4400	3190	2370	1760	1320	957
P_{fb}, kips	2460	2070	1720	1420	1210	1020	852	684	551	439	347
L_p, ft	12.3	12.1	11.9	11.8	11.7	11.6	11.5	11.4	11.3	11.2	11.1
L_r, ft	131	119	110	99.7	91.9	84.2	76.6	68.9	62.1	55.7	50.0
A_g, in.^{2}	98.8	89.6	81.9	74.0	67.7	61.8	55.8	50.0	44.7	39.9	35.3
I_x, in.^{4}	4060	3550	3110	2720	2420	2140	1890	1650	1430	1240	1070
I_y, in.^{4}	1190	1050	937	828	742	664	589	517	454	398	345
r_y, in.	3.47	3.42	3.38	3.34	3.31	3.28	3.25	3.22	3.19	3.16	3.13
Ratio r_x/r_y	1.85	1.84	1.82	1.81	1.80	1.80	1.79	1.78	1.77	1.77	1.76
$P_{ex}(KL^2)/10^4$	116000	102000	89000	77900	69300	61300	54100	47200	40900	35500	30600
$P_{ey}(KL^2)/10^4$	34100	30100	26800	23700	21200	19000	16900	14800	13000	11400	9870

$F_y = 50$ ksi

$\phi_c P_n = 0.85 F_{cr} A_g$

Table 4-2 (cont.). W-Shapes Design Strength in Axial Compression, $\phi_c P_n$, kips

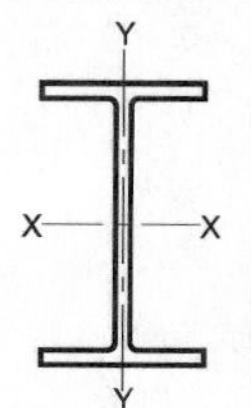

Shape	W12×										
	106	96	87	79	72	65††	58	53	50	45	40
Effective length KL (ft) with respect to least radius of gyration r_y: 0	1330	1200	1090	986	897	812	723	663	621	557	497
6	1280	1150	1050	947	861	779	680	623	562	504	450
7	1260	1140	1030	933	848	767	666	610	543	486	434
8	1240	1120	1010	917	834	754	649	594	521	466	416
9	1210	1100	994	900	818	739	631	577	497	445	396
10	1190	1070	973	880	800	723	611	559	472	422	376
11	1160	1050	950	860	781	706	590	539	445	398	354
12	1130	1020	926	838	761	687	568	518	418	374	332
13	1100	995	901	814	740	668	545	496	390	349	310
14	1070	966	874	790	717	647	521	474	363	324	287
15	1040	935	846	764	694	626	496	451	335	299	265
16	1000	904	817	738	670	604	471	428	308	274	243
17	968	871	788	711	645	581	446	404	281	250	222
18	932	838	758	683	620	558	420	381	255	227	201
19	895	805	727	655	594	535	395	357	230	204	181
20	858	771	696	627	569	512	370	334	208	185	163
22	783	703	634	570	517	464	322	290	172	152	135
24	708	635	572	514	465	417	276	247	144	128	113
26	635	569	511	459	415	372	235	210	123	109	96.5
28	565	505	453	406	367	328	202	181	106	94.1	83.2
30	497	443	397	355	321	287	176	158	92.3	82.0	72.5
32	437	390	349	312	282	252	155	139	81.2	72.1	63.7
34	387	345	309	277	250	223	137	123			
36	345	308	276	247	223	199	122	110			
38	310	276	248	221	200	179	110	98.4			
40	279	249	223	200	181	161	99.2	88.9			
Properties											
P_{wo}, kips	242	206	182	156	137	117	112	101	105	90.5	75.2
P_{wi}, kips/in.	30.5	27.5	25.8	23.5	21.5	19.5	18.0	17.3	18.5	16.8	14.8
P_{wb}, kips	609	445	365	278	213	159	125	110	133	98.6	67.4
P_{fb}, kips	276	228	185	152	126	103	115	93.0	115	93.0	74.6
L_p, ft	11.0	10.9	10.8	10.8	10.7	11.9	8.87	8.76	6.92	6.89	6.85
L_r, ft	44.9	41.4	38.4	35.7	33.6	31.7	27.0	25.6	21.5	20.3	19.2
A_g, in.^{2}	31.2	28.2	25.6	23.2	21.1	19.1	17.0	15.6	14.6	13.1	11.7
I_x, in.^{4}	933	833	740	662	597	533	475	425	391	348	307
I_y, in.^{4}	301	270	241	216	195	174	107	95.8	56.3	50.0	44.1
r_y, in.	3.11	3.09	3.07	3.05	3.04	3.02	2.51	2.48	1.96	1.95	1.94
Ratio r_x/r_y	1.76	1.76	1.75	1.75	1.75	1.75	2.10	2.11	2.64	2.64	2.64
$P_{ex}(KL^2)/10^4$	26700	23800	21200	18900	17100	15300	13600	12200	11200	9960	8790
$P_{ey}(KL^2)/10^4$	8620	7730	6900	6180	5580	4980	3060	2740	1610	1430	1260

††Flange is non-compact.

Note: Heavy line indicates Kl/r equal to or greater than 200.

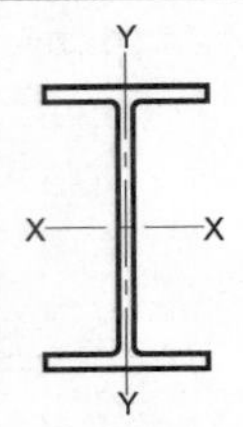

Table 4-2 (cont.). W-Shapes Design Strength in Axial Compression, $\phi_c P_n$, kips

$F_y = 50$ ksi

$\phi_c P_n = 0.85 F_{cr} A_g$

Shape	W10× 112	100	88	77	68	60	54	49	45	39	33
Effective length KL (ft) with respect to least radius of gyration r_y											
0	1400	1250	1100	961	850	748	672	612	565	489	413
6	1330	1180	1040	908	803	706	634	577	515	444	373
7	1300	1160	1020	890	787	692	621	565	497	428	360
8	1270	1140	999	869	769	675	606	551	478	412	345
9	1240	1110	973	847	749	657	590	536	458	393	329
10	1210	1080	945	822	727	638	572	520	436	374	312
11	1170	1040	916	796	703	617	553	502	412	353	294
12	1130	1010	884	768	678	595	533	484	388	332	276
13	1090	970	851	738	652	571	512	464	364	310	257
14	1050	931	817	708	625	547	490	444	339	289	238
15	1010	892	782	677	597	523	468	424	314	267	220
16	961	851	746	645	569	497	445	403	290	246	202
17	915	810	709	612	540	472	422	382	266	225	184
18	870	769	672	580	511	446	399	361	243	205	167
19	824	727	635	547	482	421	376	340	221	185	150
20	778	686	599	515	454	395	353	319	199	167	135
22	688	605	527	452	398	346	309	278	164	138	112
24	601	527	458	392	344	299	266	239	138	116	94.0
26	518	453	393	335	294	255	227	204	118	98.8	80.1
28	447	390	339	289	254	220	196	176	102	85.2	69.1
30	389	340	295	252	221	191	170	153	88.5	74.2	60.2
32	342	299	259	221	194	168	150	134	77.7	65.2	52.9
34	303	265	230	196	172	149	133	119			
36	270	236	205	175	153	133	118	106			
38	242	212	184	157	138	119	106	95.3			
40	219	191	166	141	124	108	95.9	86.0			
Properties											
P_{wo}, kips	330	275	225	182	149	124	104	90.1	98	81.1	67.8
P_{wi}, kips/in.	37.8	34.0	30.3	26.5	23.5	21.0	18.5	17.0	17.5	15.8	14.5
P_{wb}, kips	1430	1040	732	494	344	245	168	130	142	103	80.7
P_{fb}, kips	439	353	276	213	167	130	106	88.2	108	79.0	53.2
L_p, ft	9.47	9.36	9.29	9.18	9.15	9.08	9.04	8.97	7.10	6.99	6.85
L_r, ft	56.5	50.8	45.1	39.9	36.0	32.6	30.2	28.3	24.1	21.9	19.8
A_g, in.^{2}	32.9	29.4	25.9	22.6	20.0	17.6	15.8	14.4	13.3	11.5	9.71
I_x, in.^{4}	716	623	534	455	394	341	303	272	248	209	171
I_y, in.^{4}	236	207	179	154	134	116	103	93.4	53.4	45.0	36.6
r_y, in.	2.68	2.65	2.63	2.60	2.59	2.57	2.56	2.54	2.01	1.98	1.94
Ratio r_x/r_y	1.74	1.74	1.73	1.73	1.71	1.71	1.71	1.71	2.15	2.16	2.16
$P_{ex}(KL^2)/10^4$	20500	17800	15300	13000	11300	9760	8670	7790	7100	5980	4890
$P_{ey}(KL^2)/10^4$	6750	5920	5120	4410	3840	3320	2950	2670	1530	1290	1050

Note: Heavy line indicates Kl/r equal to or greater than 200.

$F_y = 36$ ksi

$\phi_c P_n = 0.85 F_{cr} A_g$

Table 4-3. HP-Shapes Design Strength in Axial Compression, $\phi_c P_n$, kips

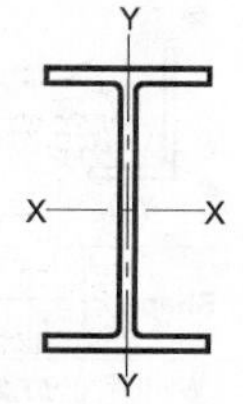

Shape	HP14×				HP12×				HP10×		HP8×
Wt/ft	117	102	89††	73††	84	74	63††	53††	57	42††	36
Effective length KL (ft) with respect to least radius of gyration r_y											
0	1050	918	799	655	753	667	563	474	514	379	324
6	1030	898	781	640	729	646	545	459	491	362	302
7	1020	891	775	635	721	639	538	453	483	356	294
8	1010	884	768	629	712	630	531	447	474	349	286
9	1000	875	760	623	701	621	523	440	464	341	276
10	993	865	752	615	690	610	514	432	453	333	266
11	980	854	742	607	677	599	504	424	441	324	255
12	967	842	732	599	663	587	494	415	429	314	243
13	953	830	721	589	649	574	482	406	415	304	232
14	938	816	709	580	634	560	471	396	401	294	219
15	922	802	696	569	618	546	458	385	387	283	207
16	905	788	683	558	601	531	446	374	372	272	195
17	888	772	670	547	584	516	432	363	357	260	182
18	870	756	656	535	567	500	419	351	341	249	170
19	851	740	641	523	548	484	405	339	326	237	158
20	832	723	626	511	530	467	391	327	310	225	146
22	792	687	595	485	492	434	362	303	279	202	123
24	750	650	563	458	454	400	333	278	248	179	104
26	707	613	529	430	416	366	304	253	219	157	88.3
28	664	574	496	402	378	332	275	229	191	136	76.2
30	620	536	462	374	342	300	247	206	166	119	66.4
32	576	498	428	346	307	268	221	183	146	104	58.3
34	533	460	395	319	273	238	196	163	129	92.3	
36	491	423	363	292	243	213	174	145	115	82.3	
38	450	387	332	267	218	191	157	130	103	73.9	
40	411	352	301	241	197	172	141	117	93.4	66.7	
Properties											
P_{wo}, kips	217	174	145	108	170	143	116	88.1	127	84.0	90.1
P_{wi}, kips/in.	29.0	25.4	22.1	18.2	24.7	21.8	18.5	15.7	20.3	14.9	16.0
P_{wb}, kips	1010	677	451	249	729	502	310	187	507	201	308
P_{fb}, kips	131	101	76.6	51.6	95.0	75.4	53.7	38.3	64.6	35.7	40.1
L_p, ft	14.9	14.8	17.2	21.4	12.2	12.2	13.9	16.7	10.2	11.9	8.12
L_r, ft	66.1	58.9	53.0	46.7	54.0	48.9	43.1	38.8	45.6	36.0	35.7
A_g, in.^{2}	34.4	30.0	26.1	21.4	24.6	21.8	18.4	15.5	16.8	12.4	10.6
I_x, in.^{4}	1220	1050	904	729	650	569	472	393	294	210	119
I_y, in.^{4}	443	380	326	261	213	186	153	127	101	71.7	40.3
r_y, in.	3.59	3.56	3.53	3.49	2.94	2.92	2.88	2.86	2.45	2.41	1.95
Ratio r_x/r_y	1.66	1.66	1.67	1.67	1.75	1.75	1.76	1.76	1.71	1.71	1.72
$P_{ex}(KL^2)/10^4$	34900	30100	25900	20900	18600	16300	13500	11200	8410	6010	3410
$P_{ey}(KL^2)/10^4$	12700	10900	9330	7470	6100	5320	4380	3630	2890	2050	1150

††Flange is non-compact.

Note: Heavy line indicates Kl/r equal to or greater than 200.

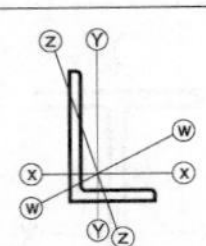

Table 4-4. Single Angles Design Strength in Axial Compression, $\phi_c P_n$, kips

$F_y = 36$ ksi

$\phi_c P_n = 0.9 F_{cr} A_g$

Shape	L8×8×							L8×6×						
	1 1/8	1	7/8	3/4	5/8†	9/16†	1/2†	1	7/8	3/4	5/8†	9/16†	1/2†	7/16†
Wt/ft	57.2	51.3	45.3	39.2	33.0	29.8	26.7	44.4	39.3	34.0	28.6	25.9	23.2	20.4
Effective length KL (ft) with respect to least radius of gyration r_z														
0	544	489	431	373	313	273	232	424	373	324	272	236	201	165
1	543	488	430	371	312	272	231	422	371	322	271	235	200	164
2	538	483	426	368	309	269	229	417	366	318	267	232	198	163
3	529	476	419	362	305	265	226	407	357	311	261	228	194	160
4	518	465	410	355	298	260	222	394	346	301	253	221	188	155
5	504	453	399	345	290	253	216	378	332	289	243	212	181	150
6	487	437	386	334	281	245	210	359	315	275	231	203	173	144
7	467	420	371	320	270	236	203	338	297	259	218	192	164	137
8	446	401	354	306	258	226	194	316	277	242	203	180	155	130
9	423	380	336	290	245	215	186	292	256	224	188	167	144	122
10	399	358	317	274	231	204	176	267	235	205	173	154	133	113
11	373	336	297	257	217	192	166	242	213	187	157	141	122	105
12	348	312	277	239	202	179	156	218	191	168	141	127	111	96.1
13	322	289	256	222	188	167	146	194	170	150	126	114	101	87.5
14	296	266	236	204	173	154	136	171	150	133	112	102	90.1	79.0
15	270	243	216	187	158	142	125	150	131	116	97.6	89.7	80.0	70.9
16	245	220	196	170	144	129	115	132	115	102	85.8	78.8	70.4	63.0
17	221	199	177	153	130	117	105	117	102	90.2	76.0	69.8	62.4	55.8
18	198	178	159	137	117	106	95.5	104	91.2	80.5	67.8	62.3	55.6	49.8
19	178	160	142	123	105	95.1	86.1	93.3	81.9	72.2	60.8	55.9	49.9	44.7
20	160	144	129	111	94.9	85.9	77.7	84.2	73.9	65.2	54.9	50.4	45.1	40.3
21	145	131	117	101	86.1	77.9	70.5	76.4	67.0	59.1	49.8	45.8	40.9	36.6
22	133	119	106	91.9	78.4	71.0	64.2							
23	121	109	97.2	84.1	71.7	64.9	58.8							
24	111	100	89.3	77.2	65.9	59.6	54.0							
25	103	92.2	82.3	71.2	60.7	55.0	49.8							
26	94.9	85.3	76.1	65.8	56.1	50.8	46.0							
27														
Properties														
A_g, in.^{2}	16.8	15.1	13.3	11.5	9.69	8.77	7.84	13.1	11.5	9.99	8.41	7.61	6.80	5.99
r_z, in.	1.56	1.56	1.57	1.57	1.58	1.58	1.59	1.28	1.28	1.29	1.29	1.30	1.30	1.31

† Section is a slender-element cross-section.

Note: Heavy line indicates Kl/r equal to or greater than 200.

$F_y = 36$ ksi
$\phi_c P_n = 0.9 F_{cr} A_g$

Table 4-4 (cont.). Single Angles Design Strength in Axial Compression, $\phi_c P_n$, kips

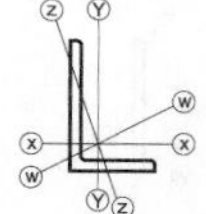

Shape		L8×4×							L7×4×				
		1	7/8	3/4	5/8†	9/16†	1/2†	7/16†	3/4	5/8	1/2†	7/16†	3/8†
Wt/ft		**37.6**	**33.3**	**28.9**	**24.4**	**22.1**	**19.7**	**17.4**	**26.2**	**22.1**	**17.9**	**15.8**	**13.6**
	0	360	317	275	231	202	171	141	249	211	164	137	109
Effective length KL (ft) with respect to least radius of gyration r_z	**1**	356	314	272	229	200	170	140	247	208	163	135	108
	2	345	304	264	222	194	165	136	239	202	158	132	105
	3	327	288	250	211	185	158	130	227	192	151	126	101
	4	303	268	233	196	172	148	123	211	179	141	118	95.3
	5	276	243	212	179	158	136	114	193	163	129	109	88.4
	6	245	217	189	160	141	123	103	172	146	116	98.4	80.6
	7	214	189	165	140	124	109	92.5	150	127	102	87.3	72.3
	8	182	161	141	120	107	94.6	81.3	128	109	88.1	76.1	63.8
	9	152	135	118	100	90.8	80.8	70.3	108	91.8	74.6	65.2	55.3
	10	124	110	96.2	82.3	75.1	67.7	59.7	88.3	75.4	61.9	54.8	47.2
	11	103	90.8	79.5	68.0	62.1	56.0	49.8	73.0	62.3	51.1	45.3	39.5
	12	86.1	76.3	66.8	57.2	52.2	47.1	41.8	61.3	52.4	43.0	38.1	33.2
	13	73.4	65.0	56.9	48.7	44.5	40.1	35.7	52.3	44.6	36.6	32.5	28.3
	14	63.3	56.1	49.1	42.0	38.3	34.6	30.7	45.1	38.5	31.6	28.0	24.4
	15												
Properties													
A_g, in.^{2}		11.1	9.79	8.49	7.16	6.49	5.80	5.11	7.70	6.50	5.26	4.63	4.00
r_z, in.		0.844	0.846	0.850	0.856	0.859	0.863	0.867	0.855	0.860	0.866	0.869	0.873

†Section is a slender-element cross-section.
Note: Heavy line indicates Kl/r equal to or greater than 200.

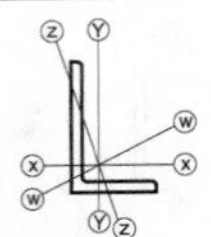

Table 4-4 (cont.). Single Angles Design Strength in Axial Compression, $\phi_c P_n$, kips

$F_y = 36$ ksi

$\phi_c P_n = 0.9 F_{cr} A_g$

Shape	L6×6×								
	1	7/8	3/4	5/8	9/16	1/2	7/16†	3/8†	5/16†
Wt/ft	**37.5**	**33.2**	**28.8**	**24.3**	**22.0**	**19.6**	**17.3**	**14.9**	**12.5**
Effective length *KL* (ft) with respect to least radius of gyration r_z									
0	356	316	274	231	209	187	160	129	98.2
1	354	314	273	230	208	186	159	129	97.8
2	349	309	268	226	204	183	157	127	96.5
3	339	301	261	220	199	178	153	124	94.4
4	326	289	251	211	192	171	147	120	91.5
5	310	275	239	201	182	163	140	115	87.9
6	292	259	225	189	172	154	132	109	83.8
7	272	241	209	176	160	143	124	102	79.1
8	250	222	192	162	147	132	114	94.7	74.0
9	228	202	175	148	134	120	104	87.1	68.6
10	205	182	158	133	121	108	94.3	79.4	63.1
11	182	162	140	118	108	96.7	84.4	71.7	57.5
12	161	142	123	104	95.4	85.4	74.7	64.1	52.0
13	140	124	108	90.6	83.3	74.5	65.4	56.7	46.5
14	121	107	92.7	78.1	71.9	64.3	56.6	49.6	41.3
15	105	93.1	80.7	68.1	62.6	56.0	49.3	43.2	36.2
16	92.3	81.8	71.0	59.8	55.0	49.2	43.3	38.0	31.8
17	81.7	72.5	62.9	53.0	48.8	43.6	38.4	33.7	28.2
18	72.9	64.6	56.1	47.3	43.5	38.9	34.2	30.0	25.2
19	65.4	58.0	50.3	42.4	39.0	34.9	30.7	27.0	22.6
20									
Properties									
A_g, in.²	11.0	9.75	8.46	7.13	6.45	5.77	5.08	4.38	3.67
r_z, in.	1.17	1.17	1.17	1.17	1.18	1.18	1.18	1.19	1.19

†Section is a slender-element cross-section.
Note: Heavy line indicates Kl/r equal to or greater than 200.

$F_y = 36$ ksi

$\phi_c P_n = 0.9 F_{cr} A_g$

Table 4-4 (cont.). Single Angles Design Strength in Axial Compression, $\phi_c P_n$, kips

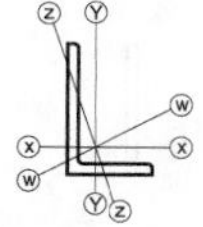

Shape		L6×4×								L6×3½×		
		7/8	3/4	5/8	9/16	1/2	7/16†	3/8†	5/16†	1/2	3/8†	5/16†
Wt/ft		27.2	23.6	19.9	18.1	16.2	14.2	12.3	10.3	15.4	11.7	9.83
Effective length KL (ft) with respect to least radius of gyration r_z	0	259	225	190	172	154	132	107	81.1	146	102	77.3
	1	256	223	188	170	152	130	106	80.4	145	100	76.5
	2	248	216	182	165	148	127	103	78.5	139	96.9	74.1
	3	235	205	173	157	140	121	98.2	75.3	130	91.3	70.3
	4	219	191	161	146	131	113	92.1	71.1	118	84.0	65.2
	5	199	174	147	133	119	103	84.9	66.1	105	75.5	59.3
	6	178	155	131	119	107	92.5	76.8	60.4	90.8	66.3	52.7
	7	155	135	115	104	93.6	81.5	68.2	54.3	76.5	56.8	45.9
	8	133	116	98.4	89.4	80.3	70.3	59.4	48.0	62.7	47.5	39.1
	9	111	97.3	82.6	75.1	67.6	59.5	50.9	41.7	50.0	38.8	32.7
	10	91.3	79.8	67.8	61.8	55.6	49.3	42.8	35.7	40.5	31.4	26.7
	11	75.5	65.9	56.1	51.0	46.0	40.7	35.4	30.0	33.5	26.0	22.0
	12	63.4	55.4	47.1	42.9	38.6	34.2	29.8	25.2	28.1	21.8	18.5
	13	54.0	47.2	40.1	36.5	32.9	29.2	25.4	21.5			
	14	46.6	40.7	34.6	31.5	28.4	25.1	21.9	18.5			
	15											
Properties												
A_g, in.^{2}		7.95	6.90	5.83	5.27	4.72	4.15	3.58	2.99	4.48	3.41	2.86
r_z, in.		0.855	0.857	0.860	0.862	0.865	0.868	0.871	0.874	0.757	0.764	0.768

Shape		L5×5×						
		7/8	3/4	5/8	1/2	7/16	3/8†	5/16†
Wt/ft		27.3	23.7	20.1	16.3	14.4	12.4	10.4
Effective length KL (ft) with respect to least radius of gyration r_z	0	260	226	191	155	137	116	90.7
	1	258	224	190	154	136	115	90.0
	2	252	219	185	150	133	113	88.1
	3	242	210	178	145	127	108	85.1
	4	228	199	168	137	121	103	81.0
	5	213	185	157	127	112	96.0	76.0
	6	195	169	143	117	103	88.2	70.3
	7	175	153	129	105	93.1	79.8	64.2
	8	155	135	115	93.6	82.8	71.2	57.7
	9	135	118	100	81.9	72.4	62.5	51.2
	10	116	101	86.1	70.5	62.4	54.0	44.8
	11	98.0	85.5	72.7	59.6	52.9	46.0	38.6
	12	82.4	71.8	61.1	50.1	44.4	38.7	32.8
	13	70.2	61.2	52.1	42.7	37.9	32.9	27.9
	14	60.5	52.8	44.9	36.8	32.6	28.4	24.1
	15	52.7	46.0	39.1	32.1	28.4	24.7	21.0
	16	46.3	40.4	34.4	28.2	25.0	21.7	18.4
	17							
Properties								
A_g, in.^{2}		8.02	6.98	5.90	4.79	4.22	3.65	3.07
r_z, in.		0.971	0.972	0.975	0.980	0.983	0.986	0.990

† Section is a slender-element cross-section.

Note: Heavy line indicates Kl/r equal to or greater than 200.

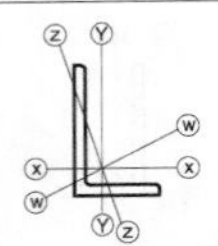

Table 4-4 (cont.). Single Angles Design Strength in Axial Compression, $\phi_c P_n$, kips

$F_y = 36$ ksi

$\phi_c P_n = 0.9 F_{cr} A_g$

Shape		L5×3½×						L5×3×				
		3/4	5/8	1/2	3/8†	5/16†	1/4†	1/2	7/16	3/8†	5/16†	1/4†
Wt/ft		**19.8**	**16.8**	**13.6**	**10.4**	**8.72**	**7.03**	**12.8**	**11.3**	**9.74**	**8.19**	**6.60**
Effective length KL (ft) with respect to least radius of gyration r_z	0	189	160	130	97.1	75.6	54.0	122	107	91.1	71.2	50.6
	1	186	158	128	95.9	74.7	53.4	119	105	89.5	70.0	49.8
	2	179	151	123	92.2	72.1	51.7	113	99.7	84.8	66.7	47.7
	3	167	141	115	86.4	67.9	49.1	103	91.0	77.6	61.4	44.4
	4	151	128	104	78.8	62.4	45.6	90.5	80.0	68.4	54.7	40.2
	5	134	114	92.5	70.1	56.0	41.5	76.7	67.9	58.3	47.2	35.3
	6	115	97.8	79.8	60.7	49.0	36.9	62.7	55.5	47.9	39.4	30.2
	7	96.4	81.9	67.0	51.2	41.9	32.2	49.3	43.8	38.0	31.9	25.0
	8	78.5	66.8	54.7	42.1	35.0	27.5	37.9	33.7	29.3	24.9	20.2
	9	62.4	53.1	43.6	33.7	28.5	23.0	29.9	26.6	23.1	19.7	16.0
	10	50.5	43.0	35.3	27.3	23.1	18.8	24.2	21.5	18.7	15.9	12.9
	11	41.8	35.6	29.2	22.5	19.1	15.5					
	12	35.1	29.9	24.5	18.9	16.0	13.1					
	13											
Properties												
A_g, in.^{2}		5.82	4.93	4.00	3.05	2.56	2.07	3.75	3.31	2.86	2.41	1.94
r_z, in.		0.744	0.746	0.750	0.755	0.758	0.761	0.642	0.644	0.646	0.649	0.652

Shape		L4×4×							L4×3½×			
		3/4	5/8	1/2	7/16	3/8	5/16†	1/4†	1/2	3/8	5/16†	1/4†
Wt/ft		**18.5**	**15.7**	**12.7**	**11.2**	**9.72**	**8.16**	**6.58**	**11.9**	**9.10**	**7.65**	**6.18**
Effective length KL (ft) with respect to least radius of gyration r_z	0	176	149	122	107	92.7	77.5	57.0	113	86.8	72.7	53.8
	1	174	147	120	106	91.5	76.6	56.4	112	85.6	71.7	53.0
	2	167	142	116	102	88.1	73.8	54.5	107	81.9	68.6	51.0
	3	157	133	108	95.5	82.8	69.4	51.5	99.3	76.1	63.8	47.7
	4	144	122	99.3	87.5	75.9	63.6	47.6	89.5	68.7	57.6	43.5
	5	128	109	88.7	78.1	67.8	56.9	43.0	78.4	60.2	50.5	38.6
	6	112	94.7	77.2	68.0	59.1	49.6	38.0	66.6	51.2	43.1	33.4
	7	94.6	80.3	65.6	57.8	50.2	42.2	32.8	54.9	42.3	35.6	28.1
	8	78.3	66.5	54.3	47.9	41.7	35.1	27.7	44.0	34.0	28.7	23.1
	9	63.0	53.5	43.7	38.6	33.6	28.4	22.9	34.8	26.8	22.7	18.4
	10	51.0	43.3	35.4	31.3	27.2	23.0	18.6	28.1	21.7	18.3	14.9
	11	42.2	35.8	29.3	25.8	22.5	19.0	15.3	23.3	18.0	15.2	12.3
	12	35.4	30.1	24.6	21.7	18.9	15.9	12.9			12.7	10.4
	13						13.6	11.0				
	14											
Properties												
A_g, in.^{2}		5.43	4.61	3.75	3.30	2.86	2.40	1.93	3.50	2.68	2.25	1.82
r_z, in.		0.774	0.774	0.776	0.777	0.779	0.781	0.783	0.716	0.719	0.721	0.723

† Section is a slender-element cross-section.

Note: Heavy line indicates Kl/r equal to or greater than 200.

$F_y = 36$ ksi

$\phi_c P_n = 0.9 F_{cr} A_g$

Table 4-4 (cont.). Single Angles Design Strength in Axial Compression, $\phi_c P_n$, kips

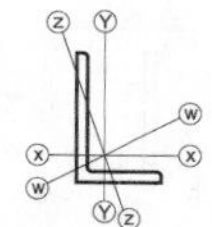

Shape		L4×3×					L3½×3½×				
		5/8	1/2	3/8	5/16†	1/4†	1/2	7/16	3/8	5/16	1/4†
Wt/ft		**13.6**	**11.1**	**8.47**	**7.12**	**5.75**	**11.1**	**9.82**	**8.51**	**7.16**	**5.79**
Effective length *KL* (ft) with respect to least radius of gyration r_z	0	129	105	80.7	67.5	49.9	106	93.6	81.0	68.0	53.2
	1	127	103	79.2	66.3	49.1	104	92.1	79.7	66.9	52.3
	2	120	97.6	74.8	62.7	46.6	99.2	87.7	75.9	63.8	50.0
	3	109	88.8	68.2	57.1	42.9	91.4	80.8	70.0	58.8	46.3
	4	95.3	77.8	59.8	50.2	38.1	81.4	72.1	62.5	52.5	41.5
	5	80.3	65.6	50.5	42.4	32.7	70.2	62.2	54.0	45.4	36.1
	6	65.1	53.3	41.1	34.6	27.1	58.6	52.0	45.1	38.0	30.5
	7	50.9	41.7	32.2	27.2	21.8	47.3	42.0	36.5	30.8	24.9
	8	38.9	31.9	24.7	20.9	16.9	37.0	32.9	28.6	24.2	19.7
	9	30.8	25.2	19.5	16.5	13.4	29.2	26.0	22.6	19.1	15.6
	10	24.9	20.4	15.8	13.3	10.8	23.7	21.0	18.3	15.5	12.6
	11						19.5	17.4	15.1	12.8	10.4
	12										
Properties											
A_g, in.^{2}		3.99	3.25	2.49	2.09	1.69	3.27	2.89	2.50	2.10	1.70
r_z, in.		0.631	0.633	0.636	0.638	0.639	0.679	0.681	0.683	0.685	0.688

Shape		L3½×3×					L3½×2½×			
		1/2	7/16	3/8	5/16	1/4†	1/2	3/8	5/16	1/4†
Wt/ft		**10.3**	**9.09**	**7.88**	**6.65**	**5.38**	**9.41**	**7.23**	**6.10**	**4.94**
Effective length *KL* (ft) with respect to least radius of gyration r_z	0	97.8	86.5	75.2	63.2	49.4	89.4	68.7	58.0	45.3
	1	95.9	84.8	73.7	62.0	48.5	87.1	66.9	56.5	44.2
	2	90.4	79.9	69.5	58.4	45.9	80.3	61.8	52.2	41.0
	3	81.8	72.4	63.0	53.0	41.8	70.3	54.1	45.8	36.2
	4	71.2	63.1	54.9	46.3	36.7	58.3	45.0	38.1	30.4
	5	59.6	52.8	46.1	38.8	31.1	45.8	35.4	30.1	24.3
	6	47.9	42.5	37.1	31.3	25.3	34.0	26.4	22.6	18.4
	7	36.9	32.9	28.7	24.3	19.9	25.0	19.4	16.6	13.6
	8	28.3	25.2	22.0	18.6	15.3	19.1	14.9	12.7	10.4
	9	22.3	19.9	17.4	14.7	12.1				8.22
	10	18.1	16.1	14.1	11.9	9.78				
	11									
Properties										
A_g, in.^{2}		3.02	2.67	2.32	1.95	1.58	2.76	2.12	1.79	1.45
r_z, in.		0.618	0.620	0.622	0.624	0.628	0.532	0.535	0.538	0.541

†Section is a slender-element cross-section.

Note: Heavy line indicates Kl/r equal to or greater than 200.

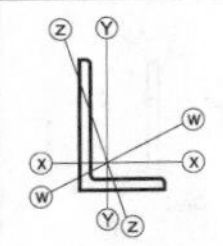

Table 4-4 (cont.). Single Angles Design Strength in Axial Compression, $\phi_c P_n$, kips

$F_y = 36$ ksi

$\phi_c P_n = 0.9 F_{cr} A_g$

Shape		L3×3×						L3×2½×					
		½	7/16	3/8	5/16	¼	3/16†	½	7/16	3/8	5/16	¼	3/16†
Wt/ft		**9.35**	**8.28**	**7.17**	**6.04**	**4.89**	**3.70**	**8.53**	**7.56**	**6.56**	**5.54**	**4.49**	**3.41**
Effective length KL (ft) with respect to least radius of gyration r_z	0	89.1	78.7	68.4	57.7	46.7	32.2	81.3	71.9	62.5	52.8	42.8	29.5
	1	87.1	77.0	66.8	56.4	45.6	31.6	79.0	69.9	60.8	51.3	41.6	28.8
	2	81.4	71.9	62.5	52.7	42.7	29.7	72.6	64.2	55.8	47.2	38.2	26.7
	3	72.7	64.3	55.9	47.2	38.2	26.9	62.9	55.7	48.4	41.0	33.2	23.5
	4	62.1	54.9	47.7	40.4	32.7	23.3	51.6	45.6	39.7	33.6	27.3	19.7
	5	50.7	44.8	39.0	33.0	26.8	19.5	39.9	35.3	30.8	26.1	21.2	15.6
	6	39.6	35.0	30.5	25.8	21.0	15.6	29.1	25.8	22.5	19.1	15.6	11.8
	7	29.6	26.2	22.8	19.4	15.8	12.0	21.4	18.9	16.5	14.0	11.4	8.69
	8	22.7	20.0	17.5	14.8	12.1	9.18	16.4	14.5	12.6	10.7	8.75	6.65
	9	17.9	15.8	13.8	11.7	9.54	7.25						
	10												
Properties													
A_g, in.²		2.75	2.43	2.11	1.78	1.44	1.09	2.51	2.22	1.93	1.63	1.32	1.00
r_z, in.		0.580	0.580	0.581	0.583	0.585	0.586	0.516	0.516	0.517	0.518	0.520	0.521

Shape		L3×2×					L2½×2½×				
		½	3/8	5/16	¼	3/16†	½	3/8	5/16	¼	3/16†
Wt/ft		**7.70**	**5.95**	**5.03**	**4.09**	**3.12**	**7.65**	**5.90**	**4.98**	**4.04**	**3.06**
Effective length KL (ft) with respect to least radius of gyration r_z	0	73.2	56.7	48.0	38.9	27.1	72.9	56.1	47.3	38.6	28.7
	1	70.2	54.4	46.0	37.3	26.1	70.6	54.2	45.8	37.3	27.8
	2	61.9	48.0	40.6	33.0	23.4	63.9	49.2	41.5	33.8	25.2
	3	50.2	38.9	33.0	26.9	19.5	54.3	41.7	35.2	28.7	21.5
	4	37.4	29.1	24.7	20.2	15.1	43.2	33.2	28.0	22.9	17.2
	5	25.6	19.9	17.0	14.0	10.9	32.1	24.7	20.9	17.1	12.9
	6	17.8	13.8	11.8	9.71	7.56	22.7	17.4	14.7	12.0	9.12
	7	13.1	10.2	8.68	7.14	5.56	16.7	12.8	10.8	8.85	6.70
	8						12.8	9.81	8.28	6.78	5.13
	9										
Properties											
A_g, in.²		2.26	1.75	1.48	1.20	0.917	2.25	1.73	1.46	1.19	0.901
r_z, in.		0.425	0.426	0.428	0.431	0.435	0.481	0.481	0.481	0.482	0.482

Shape		L2½×2×				L2×2×				
		3/8	5/16	¼	3/16†	3/8	5/16	¼	3/16	1/8†
Wt/ft		**5.30**	**4.49**	**3.65**	**2.78**	**4.65**	**3.94**	**3.21**	**2.46**	**1.67**
Effective length KL (ft) with respect to least radius of gyration r_z	0	50.5	42.8	34.7	26.1	44.4	37.6	30.6	23.4	14.5
	1	48.4	41.0	33.2	25.0	42.2	35.7	29.1	22.2	13.9
	2	42.5	36.0	29.3	22.1	36.2	30.7	25.0	19.1	12.1
	3	34.3	29.0	23.7	18.0	28.1	23.8	19.4	14.9	9.65
	4	25.3	21.5	17.6	13.5	19.7	16.7	13.6	10.5	7.04
	5	17.2	14.6	12.0	9.32	12.8	10.8	8.87	6.86	4.71
	6	11.9	10.1	8.34	6.47	8.90	7.53	6.16	4.76	3.27
	7		7.46	6.13	4.75					
	8									
Properties										
A_g, in.²		1.56	1.32	1.07	0.818	1.37	1.16	0.944	0.722	0.491
r_z, in.		0.419	0.420	0.423	0.426	0.386	0.386	0.387	0.389	0.391

†Section is a slender-element cross-section.
Note: Heavy line indicates Kl/r equal to or greater than 200.

$F_y = 50$ ksi
$\phi_c P_n = 0.85 F_{cr} A_g$

Table 4-5. WT-Shapes Design Strength in Axial Compression, $\phi_c P_n$, kips

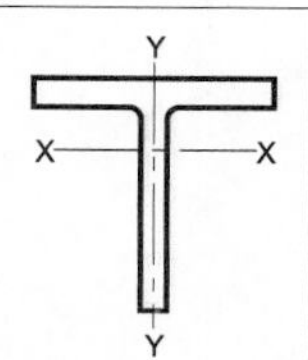

Shape		WT18× 150†	140†	130†	122.5†	115†	128†	116†	105†	97†	91†	85†	80†
Effective length *KL* (ft) with respect to indicated axis: X-X Axis	0	1730	1510	1330	1180	1030	1480	1200	1040	850	726	601	520
	10	1670	1470	1290	1140	998	1430	1170	1010	831	710	590	511
	12	1650	1440	1270	1130	986	1410	1150	1000	822	704	585	507
	14	1620	1420	1250	1110	972	1390	1140	987	812	696	579	503
	16	1580	1390	1220	1090	956	1370	1120	972	801	687	573	498
	18	1550	1360	1200	1070	939	1340	1100	955	788	677	566	492
	20	1510	1330	1170	1050	920	1310	1080	936	774	666	558	485
	22	1460	1290	1140	1020	899	1280	1050	916	759	655	549	478
	24	1420	1250	1110	993	877	1240	1030	894	743	642	539	471
	26	1370	1210	1070	964	853	1200	997	871	726	629	529	463
	28	1320	1170	1040	934	828	1160	968	847	708	614	518	454
	30	1260	1120	1000	903	802	1120	938	821	689	599	507	445
	32	1210	1080	965	871	776	1080	906	795	669	584	495	435
	34	1160	1030	926	838	748	1040	874	768	649	568	483	425
	36	1100	987	887	805	720	997	841	741	628	551	471	415
	40	989	893	807	736	663	910	773	684	585	517	444	394
Y-Y Axis	0	1730	1510	1330	1180	1030	1480	1200	1040	850	726	601	520
	10	1510	1320	1140	1010	874	1170	950	792	653	558	465	399
	12	1470	1290	1120	985	856	1110	904	756	627	538	450	387
	14	1430	1250	1080	959	836	1040	852	715	596	514	432	373
	16	1370	1210	1050	930	812	960	796	671	563	488	413	357
	18	1320	1160	1010	898	786	881	736	623	527	460	391	340
	20	1260	1110	969	863	758	799	673	572	489	430	368	321
	22	1190	1060	925	826	728	717	610	521	449	398	344	301
	24	1130	1000	879	787	696	636	547	469	409	366	319	281
	26	1060	943	831	747	662	557	485	418	369	333	293	260
	28	988	884	782	705	628	483	425	367	330	301	268	239
	30	918	825	732	662	592	423	372	323	291	269	242	218
	32	848	766	682	619	557	373	328	285	258	239	218	197
	34	780	708	632	577	521	332	292	254	230	213	194	177
	36	713	650	583	534	485	297	261	228	206	191	174	159
	40	585	540	487	452	415	241	213	186	168	156	143	130
Properties													
A, in.^{2}		44.1	41.2	38.2	36.0	33.8	37.7	34.1	30.9	28.5	26.8	25.0	23.5
r_x, in.		5.27	5.25	5.26	5.26	5.25	5.66	5.63	5.65	5.62	5.62	5.61	5.61
r_y, in.		3.83	3.81	3.77	3.75	3.73	2.65	2.62	2.58	2.56	2.55	2.53	2.50

†Section is a slender-element cross-section.

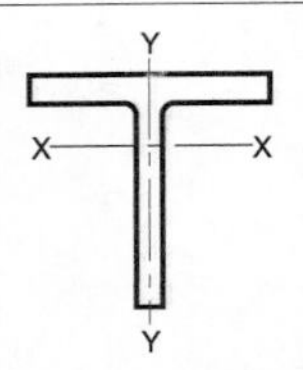

Table 4-5 (cont.). WT-Shapes Design Strength in Axial Compression, $\phi_c P_n$, kips

$F_y = 50$ ksi

$\phi_c P_n = 0.85 F_{cr} A_g$

Shape		WT18×		WT16.5×									
		75†	67.5†	193.5*	177*	159*	145.5†	131.5†	120.5†	110.5†	100.5†	84.5†	76†
Effective length KL (ft) with respect to indicated axis — X-X Axis	0	456	386	2420	2210	1990	1800	1490	1310	1110	902	662	547
	10	449	380	2330	2120	1910	1730	1430	1260	1070	875	645	535
	12	446	377	2280	2090	1870	1700	1410	1240	1060	863	638	530
	14	442	374	2240	2040	1830	1660	1380	1220	1040	849	630	523
	16	438	371	2180	1990	1780	1620	1350	1190	1020	834	621	516
	18	433	367	2120	1930	1730	1570	1310	1160	993	816	610	508
	20	428	363	2060	1870	1680	1520	1270	1130	967	797	598	499
	22	422	359	1990	1810	1620	1470	1230	1090	939	777	586	490
	24	416	354	1910	1740	1560	1410	1190	1060	910	755	572	480
	26	409	349	1840	1670	1490	1350	1140	1020	879	732	558	469
	28	402	343	1760	1600	1430	1290	1090	978	847	708	543	457
	30	395	337	1680	1520	1360	1230	1050	937	814	684	527	445
	32	387	331	1590	1450	1290	1170	996	895	780	658	511	433
	34	379	324	1510	1370	1220	1100	946	852	745	632	495	420
	36	370	318	1420	1290	1150	1040	896	809	710	605	477	407
	40	352	303	1260	1140	1010	915	795	722	640	551	442	379
Y-Y Axis	0	456	386	2420	2210	1990	1800	1490	1310	1110	902	662	547
	10	343	268	2170	1970	1750	1570	1290	1120	939	757	534	429
	12	334	261	2100	1910	1700	1520	1260	1090	915	740	512	414
	14	323	253	2030	1840	1630	1470	1220	1050	888	721	488	396
	16	310	244	1940	1760	1560	1400	1170	1010	858	699	461	377
	18	296	234	1850	1680	1490	1340	1120	970	824	674	432	355
	20	281	223	1750	1590	1410	1260	1060	924	788	648	402	332
	22	265	211	1650	1490	1320	1190	1000	875	750	620	370	309
	24	248	199	1540	1400	1240	1110	942	825	710	590	338	284
	26	231	185	1440	1300	1150	1030	880	773	668	559	306	260
	28	213	172	1330	1200	1060	953	817	720	626	527	275	235
	30	195	158	1220	1100	974	874	755	667	583	495	245	211
	32	178	144	1120	1000	888	797	693	615	540	462	216	188
	34	161	131	1010	911	805	722	633	563	498	429	192	167
	36	144	118	914	820	723	648	574	513	456	397	172	150
	40	119		742	666	588	528	468	419	377	334	140	123
Properties													
A, in.^{2}		22.1	19.9	57.0	52.1	46.8	42.8	38.7	35.5	32.6	29.6	24.8	22.4
r_x, in.		5.62	5.66	5.07	5.03	4.99	4.96	4.93	4.96	4.95	4.95	5.12	5.14
r_y, in.		2.47	2.38	3.77	3.74	3.71	3.68	3.65	3.62	3.59	3.56	2.50	2.47

*ASTM A6 tensile group 4 or 5 shape. Special requirements may apply per LRFD Specification Section A3.1c.
†Section is a slender-element cross-section.
Note: Heavy line indicates Kl/r equal to or greater than 200.

$F_y = 50$ **ksi**

$\phi_c P_n = 0.85 F_{cr} A_g$

Table 4-5 (cont.). WT-Shapes Design Strength in Axial Compression, $\phi_c P_n$, kips

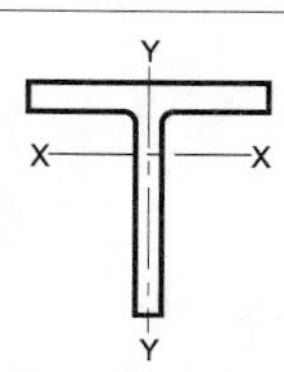

Effective length *KL* (ft) with respect to indicated axis	Shape	WT16.5×			WT15×								
		70.5†	65†	59†	195.5*	178.5*	163*	146*	130.5	117.5	105.5	95.5	86.5
X-X Axis	**0**	467	401	329	2450	2230	2040	1820	1630	1400	1190	975	794
	10	457	394	324	2330	2120	1930	1730	1550	1330	1130	933	764
	12	453	390	321	2280	2070	1890	1690	1510	1300	1110	915	750
	14	448	387	318	2220	2020	1840	1650	1470	1270	1080	894	735
	16	442	382	315	2160	1960	1780	1590	1430	1230	1050	871	718
	18	436	377	311	2080	1890	1720	1540	1370	1190	1020	845	699
	20	429	372	307	2010	1820	1660	1480	1320	1140	980	817	678
	22	422	366	303	1930	1750	1590	1410	1260	1090	941	788	656
	24	414	359	298	1840	1670	1510	1350	1200	1040	900	756	633
	26	405	352	293	1750	1580	1440	1280	1140	990	858	724	608
	28	396	345	287	1660	1500	1360	1210	1080	936	814	690	583
	30	387	337	282	1570	1410	1280	1140	1010	882	770	656	557
	32	377	329	276	1470	1330	1200	1070	949	827	725	621	530
	34	366	321	269	1380	1240	1120	994	885	773	680	586	503
	36	356	313	263	1290	1160	1040	924	822	719	636	551	476
	40	334	295	249	1110	992	893	788	700	614	549	482	422
Y-Y Axis	**0**	467	401	329	2450	2230	2040	1820	1630	1400	1190	975	794
	10	360	299	234	2210	2010	1820	1610	1420	1220	1020	830	671
	12	348	290	228	2140	1940	1760	1560	1380	1180	987	807	654
	14	335	280	220	2060	1860	1690	1500	1320	1130	951	781	635
	16	319	268	212	1960	1780	1610	1430	1260	1080	911	751	613
	18	302	254	202	1860	1690	1520	1350	1190	1030	868	718	589
	20	284	240	192	1760	1590	1430	1270	1120	970	822	683	563
	22	266	225	181	1650	1490	1340	1190	1050	909	773	646	536
	24	246	210	170	1540	1390	1250	1110	971	846	723	608	507
	26	226	194	158	1420	1280	1150	1020	895	782	672	569	477
	28	206	178	146	1310	1180	1060	938	819	718	620	529	446
	30	187	161	134	1200	1080	964	854	745	655	569	489	416
	32	167	146	121	1090	977	873	773	672	594	519	449	385
	34	149	130	109	982	881	786	694	602	534	470	410	354
	36	134	117	98.4	879	787	702	620	538	478	422	372	325
	40	109			713	639	570	504	437	389	344	304	268
Properties													
A, in.^{2}		20.8	19.2	17.3	57.6	52.5	47.9	42.9	38.4	34.6	31.1	28.1	25.5
r_x, in.		5.15	5.18	5.20	4.61	4.56	4.52	4.48	4.46	4.41	4.43	4.42	4.42
r_y, in.		2.43	2.38	2.32	3.67	3.64	3.60	3.58	3.53	3.51	3.49	3.46	3.42

*ASTM A6 tensile group 4 or 5 shape. Special requirements may apply per LRFD Specification Section A3.1c.

†Section is a slender-element cross-section.

Note: Heavy line indicates Kl/r equal to or greater than 200.

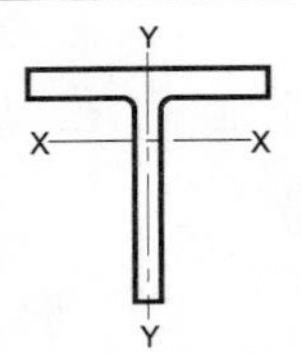

Table 4-5 (cont.). WT-Shapes Design Strength in Axial Compression, $\phi_c P_n$, kips

$F_y = 50$ ksi

$\phi_c P_n = 0.85 F_{cr} A_g$

Effective length KL (ft) with respect to indicated axis	Shape	WT15×							WT13.5×				
		74	66	62	58	54	49.5	45	129	117.5	108.5	97†	89†
X-X Axis	0	659	546	466	412	361	303	227	1620	1470	1360	1170	1040
	10	637	529	452	401	352	296	223	1510	1380	1270	1090	981
	12	627	521	446	396	348	293	221	1470	1340	1230	1060	954
	14	616	513	440	391	343	290	219	1420	1300	1190	1030	924
	16	603	503	432	384	338	286	216	1370	1250	1150	989	890
	18	588	492	424	377	332	281	213	1310	1190	1090	946	853
	20	573	480	414	369	326	276	210	1240	1130	1040	900	813
	22	556	467	404	361	319	271	207	1180	1070	983	852	771
	24	539	454	394	352	311	265	203	1110	1010	924	802	728
	26	520	439	382	342	304	259	199	1040	945	864	752	684
	28	501	424	370	332	295	252	195	969	880	803	701	639
	30	481	409	358	322	287	246	191	898	816	743	650	594
	32	460	393	345	311	278	239	186	829	752	684	599	550
	34	439	377	332	300	269	231	182	760	689	626	550	506
	36	418	360	319	289	259	224	177	694	629	570	502	464
	40	376	327	292	266	240	209	167	569	514	465	411	382
Y-Y Axis	0	659	546	466	412	361	303	227	1620	1470	1360	1170	1040
	10	514	411	350	303	256	206	157	1420	1290	1180	1010	881
	12	486	391	335	290	246	199	153	1370	1240	1140	968	849
	14	454	368	317	276	234	190	147	1310	1180	1080	925	812
	16	420	343	297	259	221	181	141	1240	1120	1030	877	771
	18	384	316	275	241	207	170	134	1160	1050	963	826	727
	20	347	287	253	223	192	159	127	1090	980	898	772	681
	22	310	259	230	203	176	146	119	1010	907	831	716	632
	24	273	230	207	183	160	134	110	925	833	764	659	583
	26	237	203	184	164	144	121	102	844	759	696	603	533
	28	206	176	161	145	127	108	93.2	765	687	629	546	484
	30	180	155	142	127	112	96.0	84.4	688	617	565	492	436
	32	159	137	125	113	99.7	85.4	75.7	613	548	502	439	389
	34	141	122	112	100	89.0	76.4	67.9	544	487	445	390	346
	36	126	109	100	90.1				486	435	398	348	310
	40								394	353	323	283	252
Properties													
A, in.^{2}		21.7	19.4	18.2	17.1	15.9	14.5	13.2	38.0	34.7	32.0	28.6	26.2
r_x, in.		4.63	4.66	4.66	4.67	4.69	4.71	4.69	4.02	4.00	3.96	3.94	3.97
r_y, in.		2.28	2.25	2.23	2.19	2.15	2.10	2.09	3.36	3.33	3.32	3.29	3.25

†Section is a slender-element cross-section.
Note: Heavy line indicates Kl/r equal to or greater than 200.

$F_y = 50$ ksi

$\phi_c P_n = 0.85 F_{cr} A_g$

Table 4-5 (cont.). WT-Shapes Design Strength in Axial Compression, $\phi_c P_n$, kips

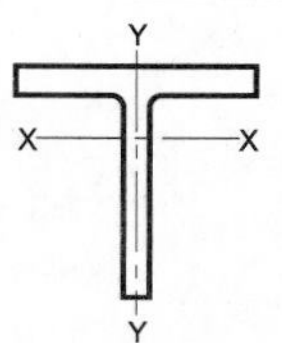

Shape		WT13.5×							WT12×				
		80.5†	73†	64.5†	57†	51†	47†	42†	185*	167.5*	153*	139.5*	125*
Effective length *KL* (ft) with respect to indicated axis — X-X Axis	0	861	701	613	498	368	311	250	2310	2090	1910	1740	1560
	10	813	666	585	478	356	301	243	2150	1940	1770	1610	1440
	12	792	651	573	469	350	297	240	2080	1880	1710	1560	1390
	14	769	634	559	458	344	292	237	2000	1800	1640	1490	1330
	16	743	615	543	447	336	286	233	1910	1720	1570	1420	1270
	18	715	593	526	434	328	280	228	1820	1640	1490	1350	1200
	20	684	571	508	420	320	273	223	1720	1540	1400	1270	1130
	22	652	546	488	405	310	266	218	1620	1450	1310	1190	1060
	24	618	521	467	390	300	258	212	1510	1350	1220	1110	982
	26	584	495	446	373	290	250	206	1400	1250	1130	1020	906
	28	549	468	424	357	279	241	200	1300	1160	1040	938	830
	30	513	441	401	339	268	233	193	1190	1060	951	856	756
	32	478	414	378	322	256	223	187	1090	963	864	776	684
	34	443	386	356	304	244	214	180	986	872	781	699	615
	36	409	360	333	287	233	205	173	889	783	699	624	548
	40	343	307	288	252	209	186	158	720	634	566	506	444
Y-Y Axis	0	861	701	613	498	368	311	250	2310	2090	1910	1740	1560
	10	725	589	473	374	281	233	180	2070	1870	1700	1540	1380
	12	701	571	443	353	268	223	173	1980	1790	1620	1470	1310
	14	673	551	410	330	252	211	165	1890	1700	1540	1400	1240
	16	643	528	375	304	236	198	156	1780	1600	1450	1310	1170
	18	609	503	338	277	218	185	147	1670	1490	1350	1220	1090
	20	574	477	301	249	199	170	136	1550	1380	1250	1130	1000
	22	537	449	264	221	180	155	125	1420	1270	1150	1040	920
	24	498	420	228	193	161	140	114	1300	1160	1050	943	835
	26	460	390	196	167	143	125	103	1180	1050	945	850	751
	28	421	360	170	145	125	110	92.1	1060	942	846	760	670
	30	383	331	148	127	109	96.8	81.3	948	838	752	673	592
	32	346	301	131	112	96.6	85.6	72.2	838	739	662	592	521
	34	309	273	116	99.6	86.0	76.3	64.4	743	655	586	525	462
	36	277	245	104	89.1				663	585	523	469	412
	40	225	200						537	474	424	380	334
Properties													
A, in.^{2}		23.8	21.6	18.9	16.8	15.0	13.8	12.4	54.4	49.2	44.9	41.0	36.8
r_x, in.		3.95	3.95	4.13	4.15	4.14	4.16	4.18	3.78	3.73	3.69	3.65	3.61
r_y, in.		3.23	3.20	2.21	2.18	2.15	2.12	2.07	3.27	3.23	3.20	3.17	3.14

*ASTM A6 tensile group 4 or 5 shape. Special requirements may apply per LRFD Specification Section A3.1c.

†Section is a slender-element cross-section.

Note: Heavy line indicates Kl/r equal to or greater than 200.

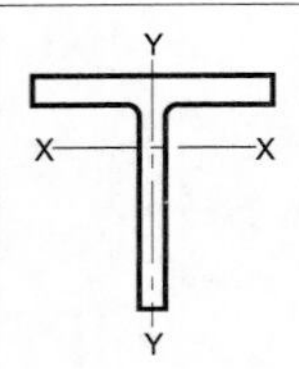

Table 4-5 (cont.). WT-Shapes Design Strength in Axial Compression, $\phi_c P_n$, kips

$F_y = 50$ ksi

$\phi_c P_n = 0.85 F_{cr} A_g$

Shape		WT12×											
		114.5	103.5	96	88	81	73†	65.5†	58.5†	52†	51.5†	47†	42†
Effective length KL (ft) with respect to indicated axis													
X-X Axis	0	1430	1290	1190	1100	1020	864	726	580	450	501	419	321
	10	1320	1190	1100	1010	932	797	673	542	424	472	397	306
	12	1270	1150	1060	970	898	769	651	526	413	459	387	300
	14	1220	1100	1010	927	858	737	626	508	401	445	376	292
	16	1160	1040	962	881	815	702	599	487	387	429	363	284
	18	1090	986	908	831	769	664	569	465	372	411	350	275
	20	1030	925	852	779	720	624	537	442	355	393	335	265
	22	960	862	793	725	670	583	504	418	338	373	320	255
	24	890	798	734	670	619	541	471	392	320	353	304	244
	26	819	734	675	615	568	499	437	367	302	332	287	233
	28	750	671	616	561	518	457	403	341	283	311	271	221
	30	682	609	558	508	469	416	369	315	264	289	254	209
	32	616	549	503	457	421	376	336	290	246	268	237	197
	34	552	491	449	407	375	338	304	265	227	247	220	185
	36	492	438	400	363	335	301	273	241	209	227	203	173
	40	399	355	324	294	271	244	221	196	175	188	171	150
Y-Y Axis	0	1430	1290	1190	1100	1020	864	726	580	450	501	419	321
	10	1250	1120	1030	936	858	723	597	472	363	372	310	239
	12	1190	1070	982	893	819	691	573	455	352	342	288	224
	14	1130	1010	929	844	775	656	545	435	339	310	264	208
	16	1060	947	871	790	727	617	515	414	324	277	237	190
	18	984	880	809	734	675	575	482	390	308	242	211	171
	20	907	811	745	675	622	532	448	365	291	209	184	152
	22	829	740	680	615	568	487	412	339	273	176	158	134
	24	752	670	615	556	514	442	376	313	254	149	134	115
	26	675	601	551	497	460	398	340	286	235	128	115	99.0
	28	601	534	490	441	409	355	305	259	216	111	99.4	85.9
	30	530	469	430	386	359	313	271	233	196	96.6	86.9	75.2
	32	466	413	379	340	316	276	239	207	178	85.1	76.6	66.4
	34	413	366	336	302	281	245	213	184	159			
	36	369	327	300	270	251	219	190	165	143			
	40	299	265	243	219	204	178	155	135	117			
Properties													
A, in.^{2}		33.6	30.4	28.1	25.8	23.9	21.5	19.3	17.2	15.3	15.1	13.8	12.4
r_x, in.		3.58	3.55	3.53	3.51	3.50	3.50	3.52	3.51	3.51	3.67	3.67	3.67
r_y, in.		3.11	3.08	3.07	3.04	3.05	3.01	2.97	2.94	2.91	1.99	1.98	1.95

†Section is a slender-element cross-section.
Note: Heavy line indicates Kl/r equal to or greater than 200.

$F_y = 50$ ksi

$\phi_c P_n = 0.85 F_{cr} A_g$

Table 4-5 (cont.). WT-Shapes Design Strength in Axial Compression, $\phi_c P_n$, kips

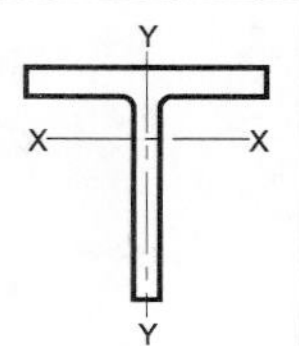

Shape		WT12×				WT10.5×							
		38†	34†	31†	27.5†	100.5	91	83	73.5	66	61†	55.5†	50.5†
Effective length KL (ft) with respect to indicated axis — X-X Axis	0	258	208	204	156	1260	1140	1040	918	825	757	637	524
	10	247	200	197	150	1130	1020	925	822	737	676	573	476
	12	242	197	193	148	1070	970	880	782	701	643	547	456
	14	237	193	190	146	1010	915	829	739	661	606	518	434
	16	231	189	185	143	950	856	775	691	618	566	486	410
	18	225	184	180	140	882	793	717	641	573	524	452	384
	20	218	179	175	136	812	729	657	589	526	481	418	357
	22	210	173	170	133	740	663	597	536	478	437	382	329
	24	202	167	164	129	669	599	538	484	431	394	347	301
	26	194	161	158	125	600	535	480	434	386	352	312	274
	28	185	155	151	120	533	474	424	385	341	311	279	247
	30	176	148	145	116	468	416	371	337	299	272	246	221
	32	167	141	138	111	412	365	326	296	263	239	217	195
	34	158	135	131	106	365	324	289	263	233	212	192	173
	36	149	128	124	102	325	289	258	234	208	189	171	154
	40	131	114	110	91.9	263	234	209	190	168	153	139	125
Y-Y Axis	0	258	208	204	156	1260	1140	1040	918	825	757	637	524
	10	190	147	123	93.0	1100	989	897	778	689	626	526	435
	12	179	140	111	85.3	1050	941	853	740	655	595	502	417
	14	167	132	98.3	76.8	986	887	803	697	617	561	476	397
	16	155	123	84.6	67.6	921	828	750	650	576	523	446	375
	18	141	113	71.0	58.1	853	766	694	601	532	484	415	352
	20	127	103	58.7	48.8	782	702	635	550	487	442	382	327
	22	113	92.0	49.2	41.1	711	637	577	498	441	401	349	301
	24	99.1	81.5			640	573	518	447	395	359	316	275
	26	85.6	71.2			571	511	462	397	351	319	283	249
	28	74.5	62.1			505	451	407	349	308	280	251	223
	30	65.3	54.5			441	394	356	305	269	244	220	199
	32	57.7				388	346	313	268	237	215	194	175
	34					344	307	277	238	210	191	172	156
	36					307	274	248	213	188	171	154	139
	40					249	222	201	173	153	139	125	113
Properties													
A, in.^{2}		11.2	10.00	9.16	8.15	29.6	26.8	24.4	21.6	19.4	17.9	16.3	14.9
r_x, in.		3.68	3.70	3.79	3.79	3.10	3.07	3.04	3.08	3.06	3.04	3.03	3.01
r_y, in.		1.92	1.87	1.37	1.34	3.02	3.00	2.99	2.95	2.93	2.91	2.90	2.89

†Section is a slender-element cross-section.

Note: Heavy line indicates Kl/r equal to or greater than 200.

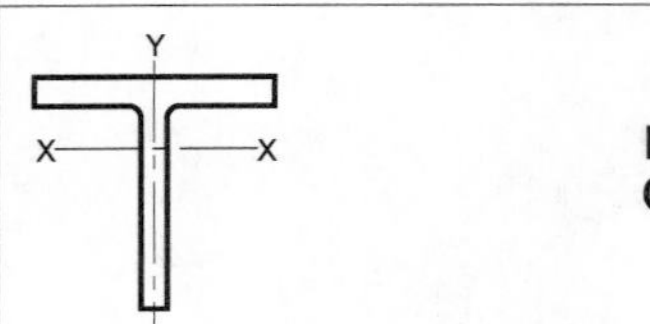

Table 4-5 (cont.). WT-Shapes Design Strength in Axial Compression, $\phi_c P_n$, kips

$F_y = 50$ ksi

$\phi_c P_n = 0.85 F_{cr} A_g$

Shape		WT10.5×										WT9×	
		46.5†	41.5†	36.5†	34†	31†	27.5†	24†	28.5†	25†	22†	87.5	79
Effective length *KL* (ft) with respect to indicated axis													
X-X Axis	**0**	562	444	331	283	225	179	139	211	166	127	1090	986
	10	511	407	307	264	212	170	132	199	158	121	941	847
	12	490	392	297	256	207	166	130	194	155	119	882	792
	14	466	374	286	247	201	162	127	188	150	116	816	732
	16	440	355	274	238	194	157	123	182	146	113	746	668
	18	412	335	260	227	186	151	119	175	141	110	674	602
	20	383	313	246	215	178	145	115	167	135	106	602	536
	22	353	291	231	203	169	139	111	159	130	102	532	472
	24	323	269	216	191	160	132	106	151	124	98.2	464	410
	26	293	247	200	178	151	126	102	143	118	94.0	399	352
	28	264	225	185	165	142	119	96.8	134	111	89.6	344	303
	30	236	203	169	153	132	112	91.8	125	105	85.1	299	264
	32	209	182	155	140	123	105	86.8	117	98.3	80.6	263	232
	34	185	162	140	128	113	97.6	81.7	108	91.8	76.0	233	206
	36	165	145	126	117	104	90.7	76.6	99.8	85.4	71.5	208	183
	40	134	117	102	94.8	87.1	77.3	66.7	83.8	73.1	62.5	168	149
Y-Y Axis	**0**	562	444	331	283	225	179	139	211	166	127	1090	986
	10	381	307	235	202	162	124	90.4	131	99.3	74.1	938	843
	12	338	276	214	186	150	117	85.5	115	88.3	66.9	884	793
	14	293	243	192	168	138	108	79.9	98.3	76.4	59.0	823	738
	16	247	210	169	150	124	98.4	73.6	81.6	64.2	50.7	758	680
	18	203	177	146	131	110	88.4	66.9	65.9	52.4	42.3	691	619
	20	166	145	124	112	96.3	78.2	59.8	54.0	43.2	35.0	623	557
	22	138	121	103	94.6	82.6	68.1	52.7	45.0			555	495
	24	117	102	87.2	80.1	70.1	58.3	45.6				489	436
	26	99.7	87.5	74.7	68.7	60.2	50.2	39.4				427	379
	28	86.2	75.7	64.7	59.5	52.2	43.7					368	327
	30	75.3	66.1	56.5	52.1							321	285
	32											282	251
	34											250	222
	36											223	198
	40											181	161
Properties													
A, in.^{2}		13.7	12.2	10.7	10.0	9.13	8.10	7.07	8.37	7.36	6.49	25.7	23.2
r_x, in.		3.25	3.22	3.21	3.20	3.21	3.23	3.26	3.29	3.30	3.31	2.66	2.63
r_y, in.		1.84	1.83	1.81	1.80	1.77	1.73	1.66	1.35	1.30	1.26	2.76	2.74

†Section is a slender-element cross-section.

Note: Heavy line indicates Kl/r equal to or greater than 200.

$F_y = 50$ ksi

$\phi_c P_n = 0.85 F_{cr} A_g$

Table 4-5 (cont.). WT-Shapes Design Strength in Axial Compression, $\phi_c P_n$, kips

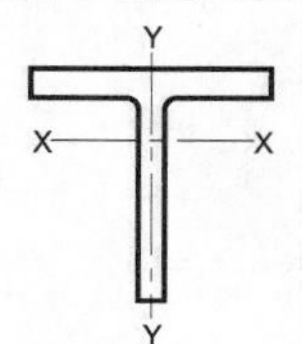

Effective length KL (ft) with respect to indicated axis	Shape	WT9×											
		71.5	65	59.5	53	48.5	43†	38†	35.5†	32.5†	30†	27.5†	25†
X-X Axis	0	893	812	744	663	608	507	393	427	356	299	253	194
	10	764	693	636	567	518	435	343	373	314	266	227	177
	12	713	646	594	529	482	407	323	351	297	253	217	170
	14	658	595	548	487	444	376	302	327	279	239	206	163
	16	599	541	499	444	403	343	278	302	259	223	193	154
	18	539	486	449	399	361	310	254	275	238	206	180	145
	20	479	431	399	354	320	276	229	248	216	189	166	135
	22	420	378	350	310	279	243	205	222	195	172	152	126
	24	364	326	303	268	241	211	181	196	173	155	138	116
	26	311	279	259	229	205	181	158	171	153	138	124	106
	28	268	240	224	198	177	156	137	148	134	122	111	95.9
	30	234	209	195	172	154	136	119	129	116	107	97.9	86.4
	32	205	184	171	151	136	119	105	113	102	93.7	86.1	77.2
	34	182	163	152	134	120	106	92.6	100	90.6	83.0	76.2	68.5
	36	162	145	135	120	107	94.4	82.6	89.3	80.8	74.1	68.0	61.1
	40	131	118	110	96.9	86.8	76.5	66.9	72.3	65.4	60.0	55.1	49.5
Y-Y Axis	0	893	812	744	663	608	507	393	427	356	299	253	194
	10	759	686	621	546	496	411	320	279	238	204	173	137
	12	714	645	584	513	467	388	304	242	209	181	155	125
	14	664	599	543	477	434	362	286	204	179	157	136	112
	16	611	551	499	438	398	334	267	167	149	133	117	97.7
	18	555	500	453	397	361	305	246	133	120	109	98.0	83.9
	20	499	449	407	356	324	275	225	109	98.3	89.3	80.4	70.5
	22	443	398	361	315	286	245	203	90.2	81.6	74.3	67.0	58.8
	24	390	349	317	276	251	216	182	76.0	68.8	62.7	56.6	49.7
	26	338	302	274	238	216	188	161	64.9	58.8	53.6	48.4	42.6
	28	292	261	237	206	187	163	140	56.1	50.8	46.3		
	30	254	228	206	179	163	142	123					
	32	224	200	182	158	144	125	108					
	34	198	178	161	140	127	111	96.1					
	36	177	158	144	125	114	99.2	85.9					
	40	143	128	117	102	92.3	80.6	69.8					
Properties													
A, in.^{2}		21.0	19.1	17.5	15.6	14.3	12.7	11.2	10.4	9.55	8.82	8.10	7.33
r_x, in.		2.60	2.58	2.60	2.59	2.56	2.55	2.54	2.74	2.72	2.71	2.71	2.70
r_y, in.		2.72	2.70	2.69	2.66	2.65	2.63	2.61	1.70	1.69	1.68	1.67	1.65

†Section is a slender-element cross-section.

Note: Heavy line indicates Kl/r equal to or greater than 200.

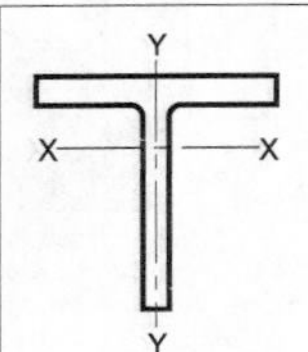

Table 4-5 (cont.). WT-Shapes Design Strength in Axial Compression, $\phi_c P_n$, kips

$F_y = 50$ ksi

$\phi_c P_n = 0.85 F_{cr} A_g$

Shape		WT9× 23†	WT9× 20†	WT9× 17.5†	WT8× 50	WT8× 44.5	WT8× 38.5†	WT8× 33.5†	WT8× 28.5†	WT8× 25†	WT8× 22.5†	WT8× 20†	WT8× 18†
Effective length KL (ft) with respect to indicated axis: X-X Axis	0	183	124	101	633	561	484	366	336	259	205	145	125
	10	168	116	94.6	516	456	392	304	283	222	179	130	113
	12	161	112	92.1	472	417	358	280	263	208	169	124	108
	14	154	108	89.1	424	375	321	254	240	192	157	117	102
	16	146	104	85.9	375	331	283	227	217	176	145	110	96.3
	18	138	99.1	82.3	327	288	245	200	193	159	133	102	90.0
	20	129	94.1	78.5	280	246	209	174	170	141	120	94.1	83.3
	22	120	88.8	74.5	235	206	175	149	147	125	107	85.9	76.6
	24	111	83.4	70.4	198	173	147	125	125	108	94.6	77.7	69.8
	26	101	77.9	66.1	168	148	125	107	107	93.0	82.8	69.7	63.2
	28	92.3	72.3	61.8	145	127	108	92.1	92.1	80.2	71.6	61.9	56.7
	30	83.4	66.8	57.5	126	111	94.1	80.2	80.2	69.9	62.3	54.5	50.4
	32	74.9	61.4	53.2	111	97.6	82.7	70.5	70.5	61.4	54.8	47.9	44.5
	34	66.6	56.1	49.0	98.4	86.4	73.3	62.5	62.5	54.4	48.5	42.4	39.4
	36	59.4	51.0	45.0	87.8	77.1	65.4	55.7	55.7	48.5	43.3	37.8	35.1
	40	48.1	41.5	37.1					45.1	39.3			28.5
Y-Y Axis	0	183	124	101	633	561	484	366	336	259	205	145	125
	10	111	80.1	61.2	522	458	390	299	214	170	138	103	84.8
	12	94.4	70.6	54.4	485	426	363	281	182	148	122	93.6	77.1
	14	77.9	60.7	47.1	446	391	333	261	151	125	105	83.1	68.8
	16	62.0	50.7	39.7	404	354	302	239	120	103	88.0	72.2	60.1
	18	49.6	41.2	32.5	361	316	269	216	95.8	82.5	71.9	61.5	51.4
	20	40.5	33.7	26.7	318	278	237	193	78.0	67.3	58.7	51.1	42.9
	22				277	241	206	171	64.7	55.9	48.8	42.6	35.8
	24				237	206	176	149	54.5	47.1	41.2	35.9	30.3
	26				202	176	150	128	46.5	40.2	35.2	30.8	
	28				175	152	130	111					
	30				152	133	113	96.6					
	32				134	117	99.8	85.0					
	34				119	103	88.5	75.4					
	36				106	92.3	79.0	67.4					
	40				85.9	74.9	64.1	54.6					
Properties													
A, in.^{2}		6.77	5.88	5.15	14.9	13.2	11.5	9.98	8.39	7.37	6.63	5.89	5.29
r_x, in.		2.77	2.76	2.79	2.27	2.26	2.23	2.21	2.41	2.40	2.39	2.37	2.41
r_y, in.		1.29	1.27	1.22	2.50	2.48	2.46	2.44	1.60	1.59	1.57	1.56	1.52

†Section is a slender-element cross-section.
Note: Heavy line indicates Kl/r equal to or greater than 200.

$F_y = 50$ ksi

$\phi_c P_n = 0.85 F_{cr} A_g$

Table 4-5 (cont.). WT-Shapes Design Strength in Axial Compression, $\phi_c P_n$, kips

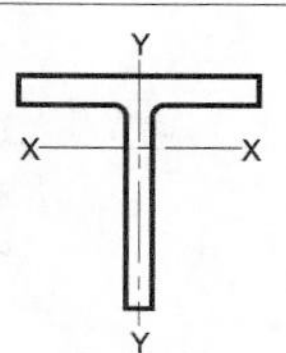

Shape		WT8×		WT7×											
		15.5†	13†	66	60	54.5	49.5	45	41	37	34	30.5†	26.5†	24†	21.5†
Effective length KL (ft) with respect to indicated axis — X-X Axis	0	93.0	66.3	825	752	680	621	561	510	463	425	370	317	265	208
	10	85.5	61.8	580	525	468	425	383	375	337	308	270	239	204	164
	12	82.4	59.9	497	448	397	360	324	327	293	267	235	210	182	148
	14	78.9	57.8	414	371	327	296	265	279	248	226	199	181	158	131
	16	75.0	55.4	335	299	261	236	211	232	205	186	165	153	135	114
	18	70.8	52.8	266	237	207	186	166	188	165	150	133	126	113	96.8
	20	66.4	50.1	215	192	167	151	135	152	134	121	108	102	92.6	80.9
	22	61.9	47.2	178	158	138	125	111	126	111	100	88.9	84.4	76.5	66.8
	24	57.3	44.3	149	133	116	105	93.6	106	92.9	84.2	74.7	70.9	64.3	56.2
	26	52.7	41.3	127	113	99.0	89.2	79.7	90.0	79.1	71.7	63.6	60.4	54.8	47.8
	28	48.1	38.3	110	97.8	85.3			77.6	68.2	61.9	54.9	52.1	47.2	41.3
	30	43.6	35.3						67.6	59.4	53.9	47.8	45.4	41.1	35.9
	32	39.3	32.3												
	34	35.1	29.5												
	36	31.3	26.7												
	40	25.3	21.7												
Y-Y Axis	0	93.0	66.3	825	752	680	621	561	510	463	425	370	317	265	208
	10	57.5	39.8	758	689	621	563	506	421	382	347	301	233	197	158
	12	49.8	35.2	734	667	601	545	490	391	355	322	281	208	177	144
	14	41.9	30.2	706	642	578	525	472	359	325	295	258	181	156	128
	16	34.1	25.2	675	614	553	502	451	324	294	267	233	154	135	112
	18	27.3	20.5	642	584	526	477	429	289	262	238	209	129	114	96.7
	20			607	551	497	450	405	254	231	209	184	105	93.9	81.7
	22			571	518	467	423	380	221	200	181	160	87.0	77.9	67.9
	24			533	484	436	395	355	188	171	154	137	73.3	65.6	57.2
	26			495	449	404	366	329	161	146	131	117	62.5	56.0	48.8
	28			457	414	373	337	303	139	126	113	101	54.0	48.3	42.2
	30			420	380	342	309	278	121	110	98.9	87.8	47.1	42.2	36.8
	32			383	346	311	281	253	106	96.5	87.0	77.3	41.4		
	34			347	314	282	255	229	94.2	85.6	77.1	68.5			
	36			312	282	253	228	205	84.1	76.4	68.8	61.2			
	40			253	229	205	185	166	68.2	61.9	55.8	49.6			
Properties															
A, in.^{2}		4.56	3.84	19.4	17.7	16.0	14.6	13.2	12.0	10.9	9.99	8.96	7.80	7.07	6.31
r_x, in.		2.45	2.47	1.73	1.71	1.68	1.67	1.66	1.85	1.82	1.81	1.80	1.88	1.88	1.86
r_y, in.		1.17	1.12	3.76	3.74	3.73	3.71	3.70	2.48	2.48	2.46	2.45	1.92	1.91	1.89

†Section is a slender-element cross-section.

Note: Heavy line indicates Kl/r equal to or greater than 200.

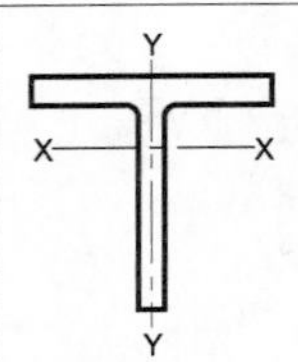

Table 4-5 (cont.). WT-Shapes Design Strength in Axial Compression, $\phi_c P_n$, kips

$F_y = 50$ ksi

$\phi_c P_n = 0.85 F_{cr} A_g$

Shape		WT7×					WT6×							
		19†	17†	15†	13†	11†	29	26.5	25	22.5†	20†	17.5†	15†	13†
Effective length *KL* (ft) with respect to indicated axis — X-X Axis	**0**	180	142	114	88.0	61.9	362	331	310	278	220	188	132	92.1
	10	148	120	98.5	77.6	55.8	227	208	206	184	151	140	104	75.7
	12	136	111	92.2	73.4	53.4	185	170	172	153	127	124	93.3	69.5
	14	123	102	85.3	68.8	50.6	145	134	139	123	105	106	82.2	62.8
	16	110	92.1	78.0	63.8	47.5	111	103	108	96.0	83.3	89.3	71.0	55.9
	18	96.6	82.1	70.4	58.5	44.3	87.7	81.1	85.5	75.8	65.8	73.3	60.2	48.9
	20	83.5	72.2	62.9	53.2	41.0	71.0	65.7	69.2	61.4	53.3	59.3	49.9	42.2
	22	71.1	62.6	55.4	47.8	37.6	58.7	54.3	57.2	50.8	44.1	49.0	41.3	35.8
	24	59.7	53.5	48.3	42.6	34.2	49.3	45.6	48.1	42.7	37.0	41.2	34.7	30.1
	26	50.9	45.6	41.5	37.6	30.8			41.0	36.4	31.6	35.1	29.5	25.6
	28	43.9	39.3	35.8	32.7	27.6						30.3	25.5	22.1
	30	38.2	34.3	31.2	28.5	24.5								
	32	33.6	30.1	27.4	25.0	21.5								
	34	29.8	26.7	24.3	22.2	19.1								
	36													
	40													
Y-Y Axis	**0**	180	142	114	88.0	61.9	362	331	310	278	220	188	132	92.1
	4	155	121	93.0	71.3	48.0	344	311	287	255	201	165	115	80.3
	6	146	115	88.8	65.6	44.8	333	301	272	242	192	154	109	76.7
	8	134	107	83.1	58.3	40.5	318	287	253	225	180	139	100	71.8
	10	120	96.8	76.1	49.8	35.5	300	271	230	205	165	123	90.2	66.0
	12	105	85.9	68.2	41.0	30.0	279	252	205	182	149	105	79.0	59.4
	14	89.4	74.3	59.7	32.3	24.4	257	231	178	158	132	86.5	67.4	52.3
	16	74.0	62.7	51.0	25.0	19.2	233	209	151	135	114	69.3	56.0	45.1
	18	59.5	51.6	42.5	19.9		208	187	126	112	96.7	55.0	45.3	38.0
	20	48.5	42.1	34.8			184	164	103	91.3	80.2	44.7	36.9	31.4
	22	40.3	35.0	29.0			160	143	85.2	75.6	66.5	37.1	30.6	26.1
	24	33.9	29.5	24.5			137	122	71.7	63.6	56.0	31.2	25.8	22.0
	26						117	104	61.1	54.3	47.8			
	28						101	89.8	52.7	46.9	41.2			
	30						87.9	78.3	46.0	40.9	36.0			
	32						77.3	68.9	40.4	35.9	31.6			
Properties														
***A*, in.²**		5.58	5.00	4.42	3.85	3.25	8.52	7.78	7.30	6.56	5.84	5.17	4.40	3.82
r_x, in.		2.04	2.04	2.07	2.12	2.14	1.50	1.51	1.60	1.59	1.57	1.76	1.75	1.75
r_y, in.		1.55	1.53	1.49	1.08	1.04	2.51	2.48	1.96	1.95	1.94	1.54	1.52	1.51

† Section is a slender-element cross-section.

Note: Heavy line indicates Kl/r equal to or greater than 200.

$F_y = 50$ ksi

$\phi_c P_n = 0.85 F_{cr} A_g$

Table 4-5 (cont.). WT-Shapes Design Strength in Axial Compression, $\phi_c P_n$, kips

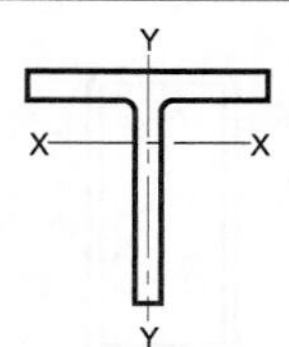

Shape		WT6×				WT5×									
		11†	9.5†	8†	7†	22.5	19.5	16.5	15	13†	11†	9.5†	8.5†	7.5†	6†
Effective length *KL* (ft) with respect to indicated axis — X-X Axis	0	97.9	70.9	54.1	39.9	282	244	206	188	146	115	104	89.6	76.1	44.6
	4	94.7	69.0	52.7	39.1	253	218	185	173	136	108	98.0	84.5	72.0	42.8
	6	90.9	66.6	51.1	38.1	220	190	162	157	124	99.3	90.7	78.5	67.2	40.7
	8	85.7	63.4	49.0	36.7	182	157	135	136	109	88.5	81.4	70.9	61.0	37.9
	10	79.6	59.6	46.3	35.0	142	123	106	114	92.5	76.2	70.8	62.2	53.8	34.6
	12	72.6	55.2	43.3	33.1	105	90.7	79.2	91.3	75.6	63.5	59.7	53.0	46.2	31.0
	14	65.2	50.4	40.0	31.0	77.1	66.6	58.2	70.3	59.5	51.3	48.8	43.8	38.6	27.2
	16	57.6	45.4	36.5	28.7	59.0	51.0	44.6	53.8	45.7	40.0	38.6	35.2	31.4	23.3
	18	50.0	40.3	32.8	26.3	46.6	40.3	35.2	42.5	36.1	31.6	30.5	27.8	24.9	19.6
	20	42.7	35.3	29.2	23.8	37.8	32.6	28.5	34.4	29.3	25.6	24.7	22.5	20.2	16.2
	22	35.8	30.5	25.7	21.4				28.4	24.2	21.1	20.4	18.6	16.7	13.4
	24	30.1	25.9	22.3	19.0				23.9	20.3	17.8	17.1	15.7	14.0	11.2
	26	25.6	22.1	19.1	16.7								13.3	11.9	9.56
	28	22.1	19.0	16.4	14.5										
	30	19.3	16.6	14.3	12.6										
	32			12.6	11.1										
Y-Y Axis	0	97.9	70.9	54.1	39.9	282	244	206	188	146	115	104	89.6	76.1	44.6
	4	71.8	50.6	34.3	24.7	266	227	188	163	126	95.8	77.0	62.2	48.4	28.1
	6	60.1	43.5	30.1	22.1	253	216	179	147	115	87.9	62.2	50.5	39.8	24.5
	8	46.3	34.8	24.5	18.7	235	201	166	126	100	77.6	45.6	37.1	29.4	19.7
	10	32.6	25.8	18.5	14.7	214	183	151	104	84.2	65.9	30.7	25.1	20.0	14.4
	12	23.1	18.4	13.3	10.9	191	163	134	81.9	67.7	53.7	21.5	17.7	14.2	10.4
	14	17.1				167	142	116	61.7	52.1	41.8	15.9	13.1		
	16					143	121	98.8	47.4	40.1	32.3				
	18					120	101	82.0	37.6	31.8	25.7				
	20					98.7	82.6	66.8	30.5	25.8	20.9				
	22					81.6	68.4	55.3	25.2	21.4	17.3				
	24					68.7	57.5	46.6							
	26					58.5	49.0	39.7							
	28					50.5	42.3	34.3							
	30					44.0	36.9	29.9							
	32					38.7	32.4	26.3							
Properties															
A, in.^{2}		3.24	2.79	2.36	2.08	6.63	5.73	4.85	4.42	3.81	3.24	2.81	2.50	2.21	1.77
r_x, in.		1.90	1.90	1.92	1.92	1.24	1.24	1.26	1.45	1.44	1.46	1.54	1.56	1.57	1.57
r_y, in.		0.847	0.821	0.773	0.753	2.01	1.98	1.94	1.37	1.36	1.33	0.874	0.844	0.810	0.785

† Section is a slender-element cross-section.

Note: Heavy line indicates Kl/r equal to or greater than 200.

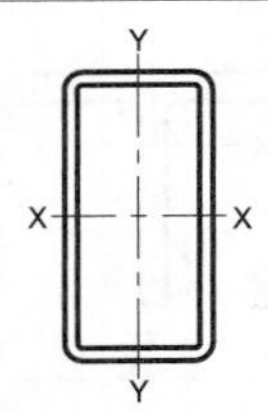

Table 4-6. Rectangular (and Square) HSS Design Strength in Axial Compression, $\phi_c P_n$, kips

$F_y = 46$ ksi

$\phi_c P_n = 0.85 F_{cr} A_g$

Shape	HSS20×12×				HSS18×12×			HSS16×16×			
	5/8††	1/2†	3/8†	5/16†	5/8	1/2†	3/8†	5/8	1/2††	3/8†	5/16†
t_{design}, in.	0.581	0.465	0.349	0.291	0.581	0.465	0.349	0.581	0.465	0.349	0.291
Wt/ft	127	103	78.4	65.8	119	96.4	73.3	127	103	78.4	65.8
Effective length *KL* (ft) with respect to least radius of gyration r_y											
0	1370	1060	723	569	1270	1030	705	1370	1110	761	579
6	1350	1050	716	564	1260	1020	698	1360	1100	756	576
7	1340	1040	713	562	1250	1010	695	1350	1090	755	575
8	1330	1040	710	560	1240	1010	692	1350	1090	753	574
9	1330	1030	707	558	1230	1000	689	1340	1080	751	572
10	1320	1020	703	555	1220	996	685	1330	1080	748	571
11	1300	1020	699	552	1210	987	681	1330	1070	746	569
12	1290	1010	695	549	1200	978	676	1320	1070	743	567
13	1280	1000	690	546	1190	969	671	1310	1060	740	565
14	1270	994	685	542	1180	958	666	1300	1060	737	563
15	1250	985	679	538	1160	947	660	1290	1050	733	560
16	1240	975	674	534	1150	936	654	1280	1040	729	558
17	1220	965	667	529	1130	923	647	1270	1030	725	555
18	1200	954	661	524	1120	911	640	1260	1020	721	552
19	1190	943	654	519	1100	897	633	1250	1010	717	549
20	1170	931	647	514	1080	883	626	1240	1000	712	546
21	1150	919	639	508	1060	869	618	1230	994	707	543
22	1130	906	632	503	1050	854	610	1210	984	702	539
23	1110	892	623	497	1030	839	601	1200	973	696	536
24	1090	878	615	491	1010	824	592	1190	962	691	532
25	1070	864	606	484	987	808	583	1170	950	685	528
26	1050	851	597	478	967	791	574	1160	939	679	524
27	1020	833	588	471	946	775	564	1140	927	672	519
28	1000	816	579	464	925	758	554	1130	914	666	515
29	979	798	569	457	904	741	544	1110	902	659	510
30	956	780	559	449	883	724	534	1090	889	652	506
32	910	743	538	434	839	689	513	1060	863	638	496
34	863	706	516	418	795	654	490	1030	835	622	485
36	816	668	494	402	751	618	467	992	807	606	474
38	770	631	471	384	707	583	444	957	779	589	463
40	723	594	447	367	663	548	420	920	750	571	450
Properties											
A_g, in.^{2}	35.0	28.3	21.5	18.1	32.6	26.5	20.1	35.0	28.3	21.5	18.1
I_x, in.^{4}	1880	1550	1200	1010	1450	1200	929	1370	1130	873	739
I_y, in.^{4}	851	705	547	464	776	643	499	1370	1130	873	739
r_x/r_y	1.49	1.48	1.48	1.48	1.37	1.37	1.36	1.00	1.00	1.00	1.00
r_y, in.	4.93	4.99	5.04	5.07	4.87	4.93	4.98	6.25	6.31	6.37	6.39

†Section is a slender-element cross-section.
††Section is a non-compact cross-section.

$F_y = 46$ ksi

$\phi_c P_n = 0.85 F_{cr} A_g$

Table 4-6 (cont.). Rectangular (and Square) HSS Design Strength in Axial Compression, $\phi_c P_n$, kips

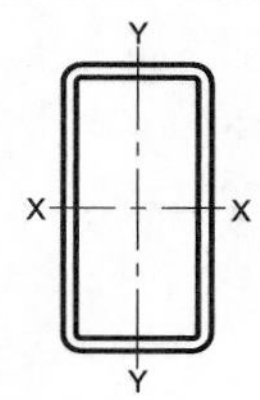

	Shape	HSS16×12×				HSS16×8×				HSS14×14×			
		5/8	1/2††	3/8†	5/16†	5/8	1/2††	3/8†	5/16†	5/8	1/2	3/8†	5/16†
	t_{design}, in.	0.581	0.465	0.349	0.291	0.581	0.465	0.349	0.291	0.581	0.465	0.349	0.291
	Wt/ft	110	89.6	68.2	57.3	93.1	75.9	58.0	48.8	110	89.6	68.2	57.3
Effective length KL (ft) with respect to least radius of gyration r_y	0	1180	962	686	538	1000	817	582	454	1180	962	710	541
	6	1170	948	679	533	973	792	568	444	1170	951	704	537
	7	1160	943	676	531	961	783	564	441	1170	947	702	535
	8	1150	937	673	529	948	772	558	437	1160	942	699	534
	9	1150	930	669	527	934	761	552	432	1150	937	696	532
	10	1140	923	665	524	918	748	545	427	1150	931	693	530
	11	1130	915	660	521	901	735	537	422	1140	925	690	527
	12	1120	907	655	519	882	720	529	416	1130	918	686	525
	13	1100	897	650	515	862	704	520	410	1120	911	682	522
	14	1090	888	644	511	841	688	510	403	1110	903	678	519
	15	1080	877	638	506	820	671	500	395	1100	895	673	516
	16	1060	866	632	502	797	653	489	388	1090	886	668	513
	17	1050	854	625	497	773	634	478	380	1080	877	662	509
	18	1030	842	618	491	749	615	466	371	1070	867	657	505
	19	1020	829	610	486	725	595	454	362	1050	856	653	501
	20	1000	816	602	480	699	575	441	353	1040	846	645	497
	21	984	803	594	474	674	555	428	344	1030	835	636	493
	22	967	789	586	468	648	534	414	334	1010	823	628	488
	23	948	774	577	461	622	513	398	323	996	811	619	483
	24	930	759	568	455	596	493	383	313	981	799	610	478
	25	911	744	558	448	570	472	367	302	966	787	601	473
	26	892	729	548	440	545	451	351	291	950	774	591	468
	27	872	713	538	433	519	431	336	280	933	761	581	462
	28	852	697	528	425	494	410	321	269	917	748	571	456
	29	832	681	517	418	469	390	305	257	900	734	561	450
	30	812	665	507	410	445	371	290	246	882	720	551	444
	32	770	632	485	393	397	332	261	222	847	692	530	431
	34	729	599	460	376	352	295	233	199	812	663	508	417
	36	687	565	434	358	314	263	208	177	775	634	486	403
	38	646	532	409	340	282	236	186	159	738	605	464	388
	40	605	499	384	322	254	213	168	143	702	575	442	373
	Properties												
	A_g, in.^{2}	30.3	24.6	18.7	15.7	25.7	20.9	16.0	13.4	30.3	24.6	18.7	15.7
	I_x, in.^{4}	1090	904	702	595	815	679	531	451	897	743	577	490
	I_y, in.^{4}	700	581	452	384	274	230	181	155	897	743	577	490
	r_x/r_y	1.25	1.25	1.25	1.24	1.72	1.72	1.71	1.71	1.00	1.00	1.00	1.00
	r_y, in.	4.80	4.86	4.91	4.94	3.27	3.32	3.37	3.40	5.44	5.49	5.55	5.58

† Section is a slender-element cross-section.

†† Section is a non-compact cross-section.

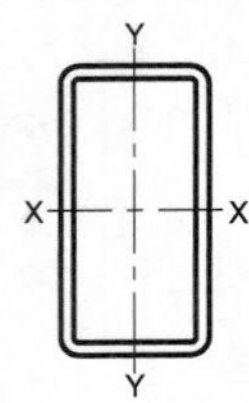

Table 4-6 (cont.). Rectangular (and Square) HSS Design Strength in Axial Compression, $\phi_c P_n$, kips

$F_y = 46$ ksi
$\phi_c P_n = 0.85 F_{cr} A_g$

Shape		HSS14×12×		HSS14×10×					HSS12×12×				
		1/2	3/8†	5/8	1/2	3/8†	5/16†	1/4†	5/8	1/2	3/8††	5/16†	1/4†
t_{design}, in.		0.465	0.349	0.581	0.465	0.349	0.291	0.233	0.581	0.465	0.349	0.291	0.233
Wt/ft		82.7	63.1	93.1	75.9	58.0	48.8	39.4	93.1	75.9	58.0	48.8	39.4
Effective length KL (ft) with respect to least radius of gyration r_y	0	891	664	1000	817	614	483	356	1000	817	626	501	358
	6	878	656	983	800	603	475	351	989	804	616	496	355
	7	873	653	975	794	599	473	350	983	800	612	494	354
	8	868	650	966	787	595	470	348	976	794	609	492	352
	9	861	646	956	779	590	466	346	969	788	604	489	351
	10	854	641	945	770	585	462	343	960	782	599	486	349
	11	847	637	933	761	579	458	341	951	775	594	483	347
	12	839	631	920	750	572	453	338	941	767	588	479	345
	13	830	626	906	739	565	448	335	931	758	581	475	342
	14	820	620	891	727	558	442	331	919	749	575	471	340
	15	810	613	876	715	549	437	328	907	740	568	467	337
	16	800	607	859	702	539	430	324	895	730	560	462	334
	17	789	600	842	688	529	424	320	881	719	552	457	331
	18	777	591	824	674	519	417	315	867	708	544	452	328
	19	765	582	806	660	508	410	310	853	697	535	446	324
	20	752	573	787	645	496	402	305	838	685	526	442	321
	21	739	563	767	629	485	394	300	823	672	517	434	317
	22	726	553	747	613	473	386	294	807	660	507	426	313
	23	712	543	727	597	461	378	288	790	647	498	418	309
	24	698	533	707	581	448	369	282	774	633	488	410	305
	25	684	522	686	564	436	361	276	757	620	477	401	300
	26	669	511	665	547	423	352	270	739	606	467	392	296
	27	654	500	643	530	410	342	264	722	592	456	384	291
	28	639	488	622	513	397	333	257	704	578	446	375	286
	29	624	477	601	496	384	323	250	686	563	435	366	281
	30	609	466	580	479	372	313	243	668	549	424	357	275
	32	578	442	537	445	346	292	229	631	520	402	338	264
	34	546	419	496	411	320	271	215	595	490	379	320	253
	36	515	395	455	379	295	250	200	558	461	357	301	241
	38	483	371	416	347	271	230	186	522	431	335	283	228
	40	452	348	377	316	248	210	171	486	403	313	264	215
Properties													
A_g, in.^{2}		22.8	17.3	25.7	20.9	16.0	13.4	10.8	25.7	20.9	16.0	13.4	10.8
I_x, in.^{4}		658	512	687	573	447	380	310	548	457	357	304	248
I_y, in.^{4}		519	405	407	341	267	227	186	548	457	357	304	248
r_x/r_y		1.13	1.12	1.30	1.29	1.29	1.29	1.29	1.00	1.00	1.00	1.00	1.00
r_y, in.		4.78	4.83	3.98	4.04	4.09	4.12	4.14	4.62	4.68	4.73	4.76	4.79

† Section is a slender-element cross-section.
†† Section is a non-compact cross-section.

$F_y = 46$ ksi

$\phi_c P_n = 0.85 F_{cr} A_g$

Table 4-6 (cont.). Rectangular (and Square) HSS Design Strength in Axial Compression, $\phi_c P_n$, kips

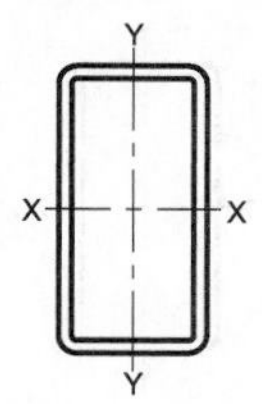

Shape	HSS12×10×				HSS12×8×					
	1/2	3/8††	5/16†	1/4†	5/8	1/2	3/8††	5/16†	1/4†	3/16†
t_{design}, in.	0.465	0.349	0.291	0.233	0.581	0.465	0.349	0.291	0.233	0.174
Wt/ft	69.1	52.9	44.6	36.0	76.1	62.3	47.8	40.3	32.6	24.7
Effective length KL (ft) with respect to least radius of gyration r_y										
0	743	571	464	344	821	673	516	422	315	209
6	727	559	456	339	793	650	500	411	307	205
7	721	554	453	337	783	642	494	407	305	204
8	714	549	450	335	772	633	487	402	302	202
9	707	544	446	333	759	623	480	397	298	201
10	698	537	442	331	745	612	471	391	294	199
11	689	531	437	328	730	600	463	385	290	196
12	680	523	432	325	714	587	453	379	286	194
13	669	516	427	322	697	574	443	371	281	191
14	658	507	421	318	679	559	432	364	276	189
15	647	499	415	314	660	544	421	355	270	186
16	634	489	408	310	641	529	409	345	264	183
17	621	480	402	306	620	513	397	335	258	179
18	608	470	394	301	600	496	385	325	252	176
19	594	459	385	297	579	479	372	314	245	172
20	580	449	376	291	557	462	359	303	238	168
21	566	438	367	286	535	444	346	292	231	163
22	551	426	358	280	513	427	333	281	224	159
23	536	415	348	274	492	409	320	270	216	154
24	520	403	339	267	470	391	306	259	208	149
25	505	392	329	261	448	374	293	248	200	144
26	489	380	319	254	426	356	280	237	192	139
27	474	368	309	247	405	339	267	226	184	134
28	458	356	300	241	384	322	254	215	176	129
29	442	344	290	233	363	305	241	204	167	124
30	426	332	280	226	343	289	228	194	159	119
32	395	308	260	212	303	256	204	174	142	108
34	364	285	240	197	269	227	181	154	127	97.2
36	334	262	221	181	240	203	161	137	113	86.7
38	304	239	202	166	215	182	145	123	101	77.8
40	276	217	184	152	194	164	131	111	91.5	70.3
Properties										
A_g, in.^{2}	19.0	14.6	12.2	9.9	21.0	17.2	13.2	11.1	8.96	6.76
I_x, in.^{4}	395	310	264	216	397	333	262	224	184	140
I_y, in.^{4}	298	234	200	164	210	178	140	120	98.8	75.7
r_x/r_y	1.15	1.15	1.15	1.15	1.37	1.37	1.37	1.37	1.36	1.36
r_y, in.	3.96	4.01	4.04	4.07	3.16	3.21	3.27	3.29	3.32	3.35

† Section is a slender-element cross-section.
†† Section is a non-compact cross-section.

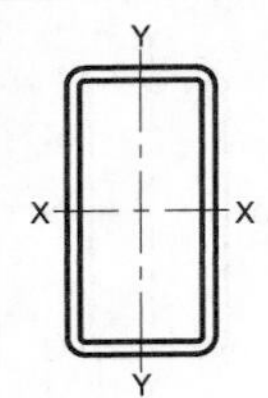

Table 4-6 (cont.). Rectangular (and Square) HSS Design Strength in Axial Compression, $\phi_c P_n$, kips

$F_y = 46$ ksi
$\phi_c P_n = 0.85 F_{cr} A_g$

Shape	HSS12×6×						HSS10×10×					
	5/8	1/2	3/8††	5/16†	1/4†	3/16†	5/8	1/2	3/8	5/16††	1/4†	3/16†
t_{design}, in.	0.581	0.465	0.349	0.291	0.233	0.174	0.581	0.465	0.349	0.291	0.233	0.174
Wt/ft	67.6	55.5	42.7	36.0	29.2	22.2	76.1	62.3	47.8	40.3	32.6	24.7
Effective length KL (ft) with respect to least radius of gyration r_y												
0	731	598	461	376	279	188	821	673	516	434	328	211
6	688	564	436	359	268	181	802	657	505	424	323	208
7	673	552	427	354	264	179	795	651	500	421	321	207
8	656	539	417	347	260	176	787	645	496	417	319	206
9	637	524	407	340	255	174	778	638	490	413	317	205
10	617	508	395	331	249	170	768	630	485	408	314	203
11	596	491	382	323	244	167	757	622	478	402	311	202
12	573	473	368	311	237	163	745	612	471	397	308	200
13	549	454	354	300	230	159	733	603	464	391	304	198
14	524	435	340	288	223	155	720	592	456	384	301	196
15	499	415	325	275	215	150	706	581	448	377	297	194
16	474	394	309	262	207	145	692	569	439	370	292	192
17	448	374	294	250	198	140	676	557	430	362	288	189
18	422	353	278	237	190	135	661	545	421	355	283	187
19	396	332	262	224	181	129	645	532	411	346	278	184
20	371	312	247	211	171	124	628	519	401	338	273	181
21	346	292	232	198	162	118	611	505	391	330	267	178
22	322	272	217	185	152	112	593	491	380	321	260	175
23	298	253	202	173	142	106	576	477	370	312	253	172
24	275	234	188	161	132	100	558	462	359	303	246	169
25	253	216	173	149	123	93.9	540	448	348	294	239	165
26	234	200	160	138	114	87.7	522	433	337	285	231	162
27	217	185	149	128	105	81.4	504	419	326	275	224	158
28	202	172	138	119	97.9	75.6	485	404	315	266	216	154
29	188	160	129	111	91.3	70.5	467	389	304	257	209	150
30	176	150	120	104	85.3	65.9	449	375	293	248	201	146
32	155	132	106	91.2	75.0	57.9	413	346	271	229	187	138
34	137	117	93.8	80.7	66.4	51.3	378	317	249	211	172	129
36	122	104	83.6	72.0	59.2	45.8	344	290	228	193	158	120
38	110	93.5	75.1	64.6	53.2	41.1	311	263	208	176	144	110
40		84.4	67.8	58.3	48.0	37.1	281	237	188	160	131	100
Properties												
A_g, in.^{2}	18.7	15.3	11.8	9.92	8.03	6.06	21.0	17.2	13.2	11.1	8.96	6.76
I_x, in.^{4}	321	271	215	184	151	116	304	256	202	172	141	108
I_y, in.^{4}	107	91.1	72.9	62.8	51.9	40.0	304	256	202	172	141	108
r_x/r_y	1.73	1.73	1.72	1.71	1.71	1.70	1.00	1.00	1.00	1.00	1.00	1.00
r_y, in.	2.39	2.44	2.49	2.52	2.54	2.57	3.80	3.86	3.92	3.94	3.97	4.00

†Section is a slender-element cross-section.
††Section is a non-compact cross-section.
Note: Heavy line indicates Kl/r equal to or greater than 200.

$F_y = 46$ ksi

$\phi_c P_n = 0.85 F_{cr} A_g$

Table 4-6 (cont.). Rectangular (and Square) HSS Design Strength in Axial Compression, $\phi_c P_n$, kips

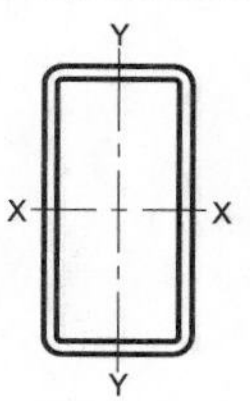

Shape		HSS10×8×					HSS10×6×					
		1/2	3/8	5/16††	1/4†	3/16†	5/8	1/2	3/8	5/16††	1/4†	3/16†
t_{design}, in.		0.465	0.349	0.291	0.233	0.174	0.581	0.465	0.349	0.291	0.233	0.174
Wt/ft		55.5	42.7	36.0	29.2	22.2	59.1	48.7	37.6	31.8	25.8	19.6
Effective length KL (ft) with respect to least radius of gyration r_y	0	598	461	388	302	200	641	528	407	343	266	180
	6	577	446	375	294	196	602	497	384	323	254	173
	7	570	440	371	291	194	588	486	375	317	250	171
	8	562	434	365	287	193	573	474	366	309	245	168
	9	552	427	360	284	191	556	460	356	301	239	165
	10	542	419	353	280	189	537	446	346	292	234	161
	11	531	411	346	275	186	518	430	334	283	227	158
	12	519	402	339	270	184	497	413	322	273	220	153
	13	507	393	331	265	181	476	396	309	262	213	149
	14	493	383	323	259	178	453	379	296	251	204	144
	15	480	372	314	253	175	431	360	282	240	195	139
	16	465	362	305	247	172	408	342	268	228	186	134
	17	450	350	296	241	168	385	323	254	216	177	129
	18	435	339	287	233	164	361	305	240	205	167	123
	19	420	327	277	225	160	339	286	226	193	158	118
	20	404	315	267	218	156	316	268	212	181	149	112
	21	388	303	257	210	152	294	250	198	170	139	106
	22	372	291	247	201	147	272	232	185	159	130	99.6
	23	356	279	237	193	142	252	215	172	148	121	93.4
	24	340	267	226	185	137	231	198	159	137	113	87.2
	25	324	254	216	177	132	213	183	147	127	104	80.8
	26	308	242	206	169	126	197	169	136	117	96.5	74.7
	27	292	231	196	161	121	183	157	126	109	89.5	69.3
	28	277	219	186	153	116	170	146	117	101	83.2	64.4
	29	262	207	177	145	110	158	136	109	94.2	77.6	60.1
	30	247	196	167	138	105	148	127	102	88.0	72.5	56.1
	32	218	174	149	123	94.2	130	112	89.6	77.3	63.7	49.3
	34	193	154	132	109	83.6	115	98.8	79.4	68.5	56.4	43.7
	36	172	137	118	97.0	74.5	103	88.2	70.8	61.1	50.3	39.0
	38	155	123	106	87.0	66.9	92.1	79.1	63.5	54.8	45.2	35.0
	40	140	111	95.2	78.5	60.4			57.3	49.5	40.8	31.6
Properties												
A_g, in.^{2}		15.3	11.8	9.92	8.03	6.06	16.4	13.5	10.4	8.76	7.10	5.37
I_x, in.^{4}		214	169	145	119	91.4	201	171	137	118	96.9	74.6
I_y, in.^{4}		151	120	103	84.7	65.1	89.4	76.8	61.8	53.3	44.1	34.1
r_x/r_y		1.19	1.19	1.19	1.18	1.18	1.50	1.49	1.49	1.48	1.48	1.48
r_y, in.		3.14	3.19	3.22	3.25	3.28	2.34	2.39	2.44	2.47	2.49	2.52

†Section is a slender-element cross-section.

††Section is a non-compact cross-section.

Note: Heavy line indicates Kl/r equal to or greater than 200.

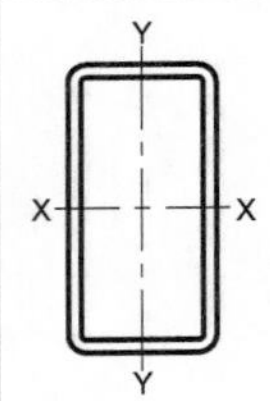

Table 4-6 (cont.). Rectangular (and Square) HSS Design Strength in Axial Compression, $\phi_c P_n$, kips

$F_y = 46$ ksi
$\phi_c P_n = 0.85 F_{cr} A_g$

Shape	HSS10×5×				HSS9×7×					
	3/8	5/16††	1/4†	3/16†	5/8	1/2	3/8	5/16	1/4†	3/16†
t_{design}, in.	0.349	0.291	0.233	0.174	0.581	0.465	0.349	0.291	0.233	0.174
Wt/ft	35.1	29.7	24.1	18.3	59.1	48.7	37.6	31.8	25.8	19.6
Effective length *KL* (ft) with respect to least radius of gyration r_y										
0	378	319	247	167	641	528	407	343	276	189
6	348	294	232	158	611	504	389	328	266	183
7	338	286	227	155	600	495	382	323	262	181
8	326	276	221	152	588	486	375	317	257	179
9	314	266	214	148	575	475	367	310	252	176
10	300	255	207	143	560	464	359	303	246	173
11	286	243	199	139	545	451	349	295	240	170
12	271	231	189	133	528	438	339	287	234	166
13	256	218	179	128	511	424	329	278	227	162
14	241	205	169	122	492	409	318	269	219	158
15	225	192	158	116	473	394	307	260	212	154
16	210	179	148	110	454	378	295	250	204	149
17	194	166	137	104	434	363	283	240	196	144
18	179	154	127	97.0	414	346	271	230	188	139
19	165	141	117	90.3	394	330	259	220	180	134
20	150	129	108	83.5	374	314	246	210	172	129
21	137	118	98.2	76.6	354	298	234	199	163	124
22	124	107	89.5	69.7	334	281	222	189	155	118
23	114	98.1	81.9	63.8	314	265	210	179	147	112
24	105	90.1	75.2	58.6	295	250	198	169	139	107
25	96.3	83.0	69.3	54.0	276	234	186	159	131	101
26	89.1	76.7	64.1	49.9	258	219	174	149	123	94.8
27	82.6	71.2	59.4	46.3	239	204	163	140	116	89.1
28	76.8	66.2	55.3	43.0	223	190	152	131	108	83.5
29	71.6	61.7	51.5	40.1	208	177	142	122	101	77.9
30	66.9	57.6	48.1	37.5	194	166	132	114	94.3	72.8
32	58.8	50.7	42.3	33.0	170	146	116	100	82.9	64.0
34	52.1	44.9	37.5	29.2	151	129	103	88.7	73.4	56.7
36					135	115	91.9	79.1	65.5	50.6
38					121	103	82.5	71.0	58.8	45.4
40					109	93.2	74.4	64.1	53.0	41.0
Properties										
A_g, in.^{2}	9.67	8.17	6.63	5.02	16.4	13.5	10.4	8.76	7.10	5.37
I_x, in.^{4}	120	104	85.8	66.2	174	149	119	102	84.1	64.7
I_y, in.^{4}	40.6	35.2	29.3	22.7	117	100	80.4	69.2	57.2	44.1
r_x/r_y	1.72	1.72	1.71	1.70	1.22	1.22	1.22	1.21	1.21	1.21
r_y, in.	2.05	2.07	2.10	2.13	2.68	2.73	2.78	2.81	2.84	2.87

†Section is a slender-element cross-section.
††Section is a non-compact cross-section.
Note: Heavy line indicates Kl/r equal to or greater than 200.

$F_y = 46$ ksi
$\phi_c P_n = 0.85 F_{cr} A_g$

Table 4-6 (cont.). Rectangular (and Square) HSS Design Strength in Axial Compression, $\phi_c P_n$, kips

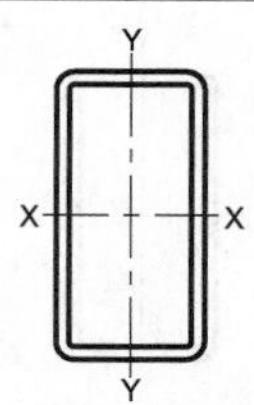

Shape		HSS9×5×						HSS8×8×					
		5/8	1/2	3/8	5/16	1/4†	3/16†	5/8	1/2	3/8	5/16	1/4††	3/16†
t_{design}, in.		0.581	0.465	0.349	0.291	0.233	0.174	0.581	0.465	0.349	0.291	0.233	0.174
Wt/ft		50.6	41.9	32.5	27.5	22.4	17.1	59.1	48.7	37.6	31.8	25.8	19.6
Effective length KL (ft) with respect to least radius of gyration r_y	0	547	454	351	297	239	163	641	528	407	343	278	190
	6	498	415	322	273	223	153	617	508	392	331	268	185
	7	481	401	313	265	216	150	608	501	387	326	265	184
	8	463	387	302	256	209	146	598	494	381	322	261	182
	9	442	371	290	246	201	142	587	485	375	316	257	180
	10	421	353	277	236	193	138	575	475	368	310	252	177
	11	398	335	264	225	184	133	562	465	360	304	247	175
	12	375	317	250	213	175	127	549	454	352	297	241	172
	13	351	297	236	201	165	122	534	442	343	290	235	169
	14	327	278	221	189	156	116	519	430	334	282	229	166
	15	303	259	207	177	146	110	503	417	324	274	223	162
	16	279	239	192	164	136	103	486	404	314	266	216	159
	17	256	220	178	152	126	96.5	469	390	304	257	209	155
	18	234	202	164	141	117	89.6	451	376	293	249	202	151
	19	212	184	150	129	108	82.6	434	362	283	240	195	147
	20	191	167	137	118	98.5	75.8	416	347	272	231	188	142
	21	173	151	124	107	89.7	69.2	398	332	261	221	180	138
	22	158	138	113	97.6	81.7	63.0	380	318	250	212	173	132
	23	145	126	104	89.3	74.8	57.7	361	303	239	203	166	126
	24	133	116	95.1	82.1	68.7	53.0	344	289	228	194	158	121
	25	122	107	87.6	75.6	63.3	48.8	326	274	217	185	151	115
	26	113	98.7	81.0	69.9	58.5	45.1	308	260	206	176	143	110
	27	105	91.5	75.1	64.8	54.3	41.9	291	246	195	167	136	104
	28	97.5	85.1	69.9	60.3	50.4	38.9	274	232	185	158	129	99.1
	29	90.9	79.3	65.1	56.2	47.0	36.3	258	219	174	149	122	93.8
	30	85.0	74.1	60.9	52.5	43.9	33.9	241	205	164	141	115	88.7
	32	74.7	65.1	53.5	46.2	38.6	29.8	212	181	145	124	102	78.6
	34				40.9	34.2	26.4	188	160	128	110	90.3	69.6
	36							168	143	114	98.1	80.5	62.1
	38							150	128	103	88.1	72.3	55.7
	40							136	116	92.6	79.5	65.2	50.3
Properties													
A_g, in.^{2}		14.0	11.6	8.97	7.59	6.17	4.67	16.4	13.5	10.4	8.76	7.10	5.37
I_x, in.^{4}		133	115	92.5	79.8	66.1	51.1	146	125	99.6	85.6	70.7	54.4
I_y, in.^{4}		52.0	45.2	36.8	32.0	26.6	20.7	146	125	99.6	85.6	70.7	54.4
r_x/r_y		1.60	1.59	1.58	1.58	1.57	1.58	1.00	1.00	1.00	1.00	1.00	1.00
r_y, in.		1.92	1.97	2.03	2.05	2.08	2.10	2.99	3.04	3.10	3.13	3.15	3.18

†Section is a slender-element cross-section.
††Section is a non-compact cross-section.
Note: Heavy line indicates Kl/r equal to or greater than 200.

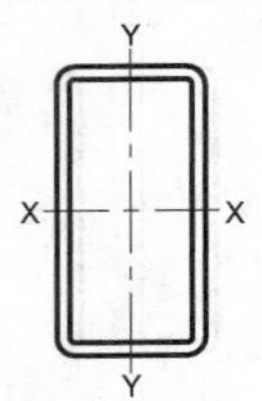

Table 4-6 (cont.). Rectangular (and Square) HSS Design Strength in Axial Compression, $\phi_c P_n$, kips

$F_y = 46$ ksi

$\phi_c P_n = 0.85 F_{cr} A_g$

Shape	HSS8×6×						HSS8×4×						
	5/8	1/2	3/8	5/16	1/4††	3/16†	5/8	1/2	3/8	5/16	1/4††	3/16†	1/8†
t_{design}, in.	0.581	0.465	0.349	0.291	0.233	0.174	0.581	0.465	0.349	0.291	0.233	0.174	0.116
Wt/ft	50.6	41.9	32.5	27.5	22.4	17.1	42.1	35.1	27.4	23.3	19.0	14.5	9.85
Effective length *KL* (ft) with respect to least radius of gyration r_y													
0	547	454	351	297	241	171	457	381	296	251	205	145	83.6
6	512	425	330	279	227	163	393	330	259	220	181	131	77.3
7	499	415	323	273	223	161	371	313	247	210	172	127	75.1
8	485	404	314	266	217	158	349	295	233	199	164	122	72.6
9	470	392	305	259	211	154	324	276	219	187	154	116	69.8
10	454	379	296	251	205	150	299	256	204	175	144	110	66.8
11	436	365	285	242	198	146	274	235	189	162	134	103	63.5
12	418	350	274	233	190	142	248	215	173	149	124	95.5	60.0
13	398	335	263	223	183	137	223	194	158	136	113	87.7	56.3
14	379	319	251	213	175	132	199	175	142	123	103	80.1	52.5
15	359	303	239	203	167	127	176	156	128	111	92.9	72.6	48.5
16	338	286	226	193	159	121	154	137	114	98.9	83.3	65.3	44.5
17	318	270	214	183	150	115	137	122	101	87.6	74.0	58.3	40.3
18	298	253	202	172	142	109	122	108	89.9	78.1	66.0	52.0	36.1
19	278	237	189	162	133	102	109	97.3	80.6	70.1	59.3	46.7	32.4
20	258	221	177	151	125	96.3	98.8	87.8	72.8	63.3	53.5	42.1	29.2
21	239	205	165	141	117	90.1	89.6	79.6	66.0	57.4	48.5	38.2	26.5
22	220	190	153	132	109	84.1	81.7	72.6	60.1	52.3	44.2	34.8	24.2
23	202	175	142	122	101	78.3	74.7	66.4	55.0	47.9	40.4	31.8	22.1
24	186	161	131	112	93.7	72.6	68.6	61.0	50.5	43.9	37.1	29.2	20.3
25	171	148	120	104	86.4	67.0	63.2	56.2	46.6	40.5	34.2	26.9	18.7
26	158	137	111	95.8	79.9	61.9		52.0	43.1	37.4	31.6	24.9	17.3
27	147	127	103	88.9	74.0	57.4				34.7	29.3	23.1	16.0
28	136	118	96.0	82.6	68.9	53.4						21.5	14.9
29	127	110	89.5	77.0	64.2	49.8							
30	119	103	83.6	72.0	60.0	46.5							
32	104	90.3	73.5	63.3	52.7	40.9							
34	92.5	80.0	65.1	56.0	46.7	36.2							
36	82.5	71.4	58.1	50.0	41.7	32.3							
38		64.1	52.1	44.9	37.4	29.0							
40				40.5	33.7	26.2							
Properties													
A_g, in.^{2}	14.0	11.6	8.97	7.59	6.17	4.67	11.7	9.74	7.58	6.43	5.24	3.98	2.70
I_x, in.^{4}	114	98.2	79.1	68.3	56.6	43.7	82.0	71.8	58.7	51.0	42.5	33.1	22.9
I_y, in.^{4}	72.3	62.5	50.6	43.8	36.4	28.2	26.6	23.6	19.6	17.2	14.4	11.3	7.90
r_x/r_y	1.26	1.25	1.25	1.25	1.25	1.24	1.75	1.74	1.73	1.73	1.72	1.70	1.71
r_y, in.	2.27	2.32	2.38	2.40	2.43	2.46	1.51	1.56	1.61	1.63	1.66	1.69	1.71

† Section is a slender-element cross-section.
†† Section is a non-compact cross-section.
Note: Heavy line indicates Kl/r equal to or greater than 200.

$F_y = 46$ ksi
$\phi_c P_n = 0.85 F_{cr} A_g$

Table 4-6 (cont.). Rectangular (and Square) HSS Design Strength in Axial Compression, $\phi_c P_n$, kips

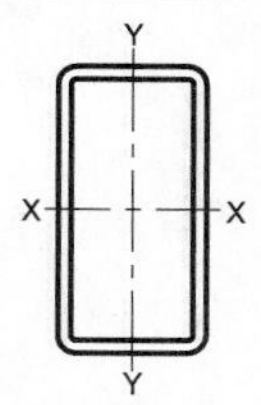

Shape	HSS7×7×						HSS7×5×						
	5/8	1/2	3/8	5/16	1/4	3/16†	5/8	1/2	3/8	5/16	1/4	3/16†	1/8†
t_{design}, in.	0.581	0.465	0.349	0.291	0.233	0.174	0.581	0.465	0.349	0.291	0.233	0.174	0.116
Wt/ft	50.6	41.9	32.5	27.5	22.4	17.1	42.1	35.1	27.4	23.3	19.0	14.5	9.85
Effective length KL (ft) with respect to least radius of gyration r_y													
0	547	454	351	297	241	177	457	381	296	251	205	153	88.8
6	519	431	334	283	230	171	414	346	271	230	188	142	84.4
7	510	423	328	278	227	169	399	334	262	223	182	139	82.7
8	499	415	322	273	222	167	382	321	253	215	176	134	81.0
9	487	405	315	267	217	164	365	307	242	206	169	129	78.7
10	473	394	307	260	212	161	346	292	231	197	162	124	76.2
11	459	383	298	253	207	157	326	276	219	187	154	118	73.5
12	444	371	289	246	201	152	306	260	207	177	146	112	70.6
13	428	358	280	238	194	148	285	243	194	166	137	105	67.5
14	412	345	270	230	188	143	264	226	182	156	129	99.1	64.2
15	395	331	260	221	181	137	244	210	169	145	120	92.6	60.8
16	377	317	249	212	174	132	223	193	156	134	112	86.3	57.3
17	359	303	238	203	167	127	204	177	144	124	103	79.9	53.8
18	342	288	227	194	159	121	185	161	132	114	94.9	73.7	50.1
19	324	274	216	185	152	116	166	146	120	104	87.0	67.7	46.4
20	306	259	205	176	145	110	150	132	109	94.3	79.2	61.9	42.7
21	288	245	194	167	137	105	136	119	98.8	85.6	71.8	56.2	38.9
22	271	230	183	157	130	99.1	124	109	90.1	78.0	65.5	51.2	35.4
23	254	216	173	148	123	93.6	113	99.5	82.4	71.3	59.9	46.8	32.4
24	237	202	162	140	115	88.2	104	91.4	75.7	65.5	55.0	43.0	29.8
25	220	189	152	131	108	83.0	96.0	84.2	69.7	60.4	50.7	39.7	27.4
26	204	176	142	122	101	77.8	88.7	77.9	64.5	55.8	46.9	36.7	25.4
27	189	163	132	114	94.8	72.7	82.3	72.2	59.8	51.8	43.5	34.0	23.5
28	176	152	123	106	88.2	67.7	76.5	67.2	55.6	48.1	40.4	31.6	21.9
29	164	141	114	98.9	82.2	63.1	71.3	62.6	51.8	44.9	37.7	29.5	20.4
30	153	132	107	92.4	76.8	59.0	66.6	58.5	48.4	41.9	35.2	27.5	19.0
32	135	116	93.9	81.3	67.5	51.8			42.6	36.8	30.9	24.2	16.7
34	119	103	83.2	72.0	59.8	45.9						21.4	14.8
36	107	91.7	74.2	64.2	53.3	41.0							
38	95.6	82.3	66.6	57.6	47.9	36.8							
40	86.3	74.3	60.1	52.0	43.2	33.2							
Properties													
A_g, in.^{2}	14.0	11.6	8.97	7.59	6.17	4.67	11.7	9.74	7.58	6.43	5.24	3.98	2.70
I_x, in.^{4}	93.4	80.5	65.0	56.1	46.5	36.0	69.4	60.6	49.5	43.0	35.9	27.9	19.3
I_y, in.^{4}	93.4	80.5	65.0	56.1	46.5	36.0	40.6	35.6	29.3	25.5	21.3	16.6	11.6
r_x/r_y	1.00	1.00	1.00	1.00	1.00	1.00	1.31	1.31	1.30	1.30	1.30	1.29	1.29
r_y, in.	2.58	2.63	2.69	2.72	2.75	2.77	1.86	1.91	1.97	1.99	2.02	2.05	2.07

† Section is a slender-element cross-section.
Note: Heavy line indicates Kl/r equal to or greater than 200.

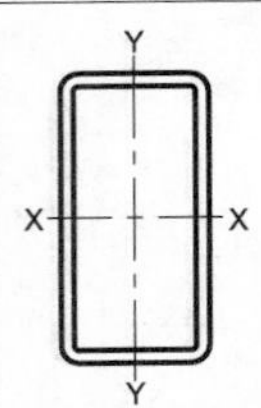

Table 4-6 (cont.). Rectangular (and Square) HSS Design Strength in Axial Compression, $\phi_c P_n$, kips

$F_y = 46$ ksi

$\phi_c P_n = 0.85 F_{cr} A_g$

Shape		HSS7×4×						HSS6×6×						
		1/2	3/8	5/16	1/4	3/16†	1/8†	5/8	1/2	3/8	5/16	1/4	3/16††	1/8†
t_{design}, in.		0.465	0.349	0.291	0.233	0.174	0.116	0.581	0.465	0.349	0.291	0.233	0.174	0.116
Wt/ft		31.7	24.9	21.2	17.3	13.2	9.00	42.1	35.1	27.4	23.3	19.0	14.5	9.85
Effective length KL (ft) with respect to least radius of gyration r_y	0	344	269	229	187	139	80.9	457	381	296	251	205	156	89.4
	6	297	234	200	164	125	74.3	425	355	277	236	192	146	86.0
	7	281	222	190	156	119	71.9	414	346	271	230	188	143	84.8
	8	264	210	180	148	113	69.3	401	336	263	224	183	139	83.4
	9	246	196	169	139	107	66.4	387	325	255	217	178	135	81.8
	10	228	182	157	130	99.9	63.2	372	313	246	210	172	131	80.0
	11	209	168	146	121	92.8	59.8	357	301	237	202	165	126	78.1
	12	190	154	134	111	85.6	56.1	340	288	227	194	159	121	76.0
	13	171	140	122	101	78.4	52.3	323	274	216	185	152	116	73.7
	14	153	126	110	92.1	71.3	48.4	306	260	206	176	145	111	71.3
	15	136	112	98.7	82.9	64.4	44.3	288	246	195	167	138	106	68.7
	16	119	99.4	87.8	74.2	57.7	40.2	270	231	184	158	130	100	65.9
	17	106	88.1	77.7	65.8	51.3	36.0	252	217	173	149	123	94.5	63.1
	18	94.3	78.5	69.3	58.7	45.7	32.1	235	203	162	140	116	89.0	60.0
	19	84.6	70.5	62.2	52.7	41.1	28.8	218	189	151	131	108	83.5	56.9
	20	76.4	63.6	56.2	47.5	37.1	26.0	201	175	141	122	101	78.1	53.6
	21	69.3	57.7	50.9	43.1	33.6	23.6	185	161	130	113	93.9	72.7	50.0
	22	63.1	52.6	46.4	39.3	30.6	21.5	169	148	120	104	87.0	67.5	46.5
	23	57.8	48.1	42.5	35.9	28.0	19.7	154	136	110	96.1	80.4	62.5	43.0
	24	53.1	44.2	39.0	33.0	25.7	18.1	142	125	101	88.3	73.8	57.5	39.7
	25	48.9	40.7	35.9	30.4	23.7	16.7	131	115	93.4	81.3	68.0	53.0	36.6
	26		37.6	33.2	28.1	21.9	15.4	121	106	86.4	75.2	62.9	49.0	33.8
	27				26.1	20.3	14.3	112	98.4	80.1	69.7	58.3	45.4	31.3
	28						13.3	104	91.5	74.5	64.8	54.2	42.2	29.1
	29							97.1	85.3	69.4	60.4	50.5	39.4	27.2
	30							90.7	79.7	64.9	56.5	47.2	36.8	25.4
	32							79.7	70.1	57.0	49.6	41.5	32.3	22.3
	34							70.6	62.1	50.5	44.0	36.8	28.7	19.8
	36							63.0	55.4	45.0	39.2	32.8	25.6	17.6
	38									40.4	35.2	29.4	22.9	15.8
	40													
Properties														
A_g, in.^{2}		8.81	6.88	5.85	4.77	3.63	2.46	11.7	9.74	7.58	6.43	5.24	3.98	2.70
I_x, in.^{4}		50.7	41.8	36.5	30.5	23.8	16.6	55.2	48.3	39.5	34.3	28.6	22.3	15.5
I_y, in.^{4}		20.7	17.3	15.2	12.8	10.0	7.03	55.2	48.3	39.5	34.3	28.6	22.3	15.5
r_x/r_y		1.57	1.56	1.55	1.54	1.54	1.53	1.00	1.00	1.00	1.00	1.00	1.00	1.00
r_y, in.		1.53	1.58	1.61	1.64	1.66	1.69	2.17	2.23	2.28	2.31	2.34	2.37	2.39

† Section is a slender-element cross-section.
†† Section is a non-compact cross-section.
Note: Heavy line indicates Kl/r equal to or greater than 200.

$F_y = 46$ ksi
$\phi_c P_n = 0.85 F_{cr} A_g$

Table 4-6 (cont.). Rectangular (and Square) HSS Design Strength in Axial Compression, $\phi_c P_n$, kips

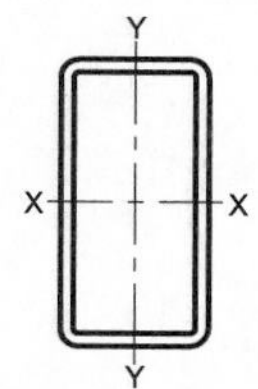

Shape		HSS6×5×				HSS6×4×					
		3/8	5/16	1/4	3/16††	1/2	3/8	5/16	1/4	3/16††	1/8†
t_{design}, in.		0.349	0.291	0.233	0.174	0.465	0.349	0.291	0.233	0.174	0.116
Wt/ft		24.9	21.2	17.3	13.2	28.3	22.3	19.0	15.6	12.0	8.15
Effective length KL (ft) with respect to least radius of gyration r_y	0	269	229	187	142	308	242	206	168	128	78.2
	1	268	228	186	142	307	241	205	168	128	78.0
	2	266	226	185	141	303	238	202	166	126	77.4
	3	263	224	182	139	296	233	199	163	124	76.4
	4	258	220	179	137	288	227	193	158	121	75.0
	5	252	215	175	134	277	218	187	153	117	73.2
	6	245	209	171	130	264	209	179	147	112	71.0
	7	237	202	165	126	250	198	170	140	107	68.5
	8	227	194	159	122	234	187	160	132	102	65.7
	9	217	186	153	117	217	174	150	124	95.5	62.6
	10	207	177	146	112	200	161	140	116	89.1	59.2
	11	196	168	138	106	183	148	129	107	82.5	55.6
	12	184	158	131	100	166	135	118	98.2	75.9	51.8
	13	173	149	123	94.6	149	122	107	89.4	69.3	47.9
	14	161	139	115	88.7	133	110	96.1	80.8	62.8	43.8
	15	149	129	107	82.8	117	97.5	85.9	72.5	56.5	39.5
	16	137	119	99.1	76.8	103	85.9	76.0	64.5	50.4	35.5
	17	126	110	91.3	71.0	90.9	76.1	67.3	57.1	44.7	31.5
	18	115	100	83.8	65.3	81.1	67.9	60.0	51.0	39.9	28.1
	19	104	91.2	76.4	59.7	72.8	60.9	53.9	45.7	35.8	25.2
	20	93.9	82.4	69.3	54.3	65.7	55.0	48.6	41.3	32.3	22.8
	21	85.2	74.7	62.8	49.3	59.6	49.9	44.1	37.4	29.3	20.6
	22	77.6	68.1	57.2	44.9	54.3	45.5	40.2	34.1	26.7	18.8
	23	71.0	62.3	52.4	41.1	49.7	41.6	36.8	31.2	24.4	17.2
	24	65.2	57.2	48.1	37.7	45.6	38.2	33.8	28.7	22.4	15.8
	25	60.1	52.7	44.3	34.8	42.0	35.2	31.1	26.4	20.7	14.6
	26	55.6	48.8	41.0	32.1			28.8	24.4	19.1	13.5
	27	51.5	45.2	38.0	29.8					17.7	12.5
	28	47.9	42.0	35.3	27.7						
	29	44.7	39.2	32.9	25.8						
	30	41.8	36.6	30.8	24.1						
Properties											
A_g, in.^{2}		6.88	5.85	4.77	3.63	7.88	6.18	5.26	4.30	3.28	2.23
I_x, in.^{4}		33.9	29.6	24.7	19.3	34.0	28.3	24.8	20.9	16.4	11.4
I_y, in.^{4}		25.5	22.3	18.7	14.6	17.8	14.9	13.2	11.1	8.76	6.15
r_x/r_y		1.16	1.15	1.15	1.15	1.39	1.38	1.37	1.37	1.37	1.36
r_y, in.		1.92	1.95	1.98	2.01	1.50	1.55	1.58	1.61	1.63	1.66

† Section is a slender-element cross-section.
†† Section is a non-compact cross-section.
Note: Heavy line indicates Kl/r equal to or greater than 200.

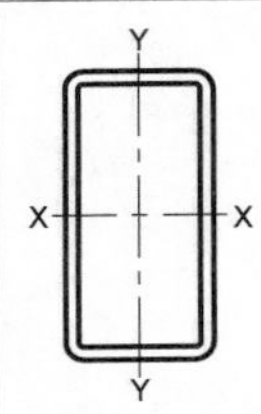

Table 4-6 (cont.). Rectangular (and Square) HSS Design Strength in Axial Compression, $\phi_c P_n$, kips

$F_y = 46$ ksi
$\phi_c P_n = 0.85 F_{cr} A_g$

Shape		HSS6×3×						HSS5½×5½×				
		1/2	3/8	5/16	1/4	3/16††	1/8†	3/8	5/16	1/4	3/16††	1/8†
t_{design}, in.		**0.465**	**0.349**	**0.291**	**0.233**	**0.174**	**0.116**	**0.349**	**0.291**	**0.233**	**0.174**	**0.116**
Wt/ft		**24.9**	**19.7**	**16.9**	**13.9**	**10.7**	**7.30**	**24.9**	**21.2**	**17.3**	**13.2**	**9.00**
Effective length KL (ft) with respect to least radius of gyration r_y	**0**	272	214	183	150	115	69.4	269	229	187	142	85.4
	1	270	213	182	149	114	69.1	268	228	186	142	85.3
	2	263	208	178	146	112	68.2	267	227	185	141	85
	3	254	201	172	142	108	66.7	264	224	183	139	84.4
	4	240	191	164	135	104	64.6	260	221	180	137	83.6
	5	224	180	154	128	98.1	62.1	254	217	177	135	82.6
	6	206	166	143	119	91.6	59.0	248	212	173	132	81.3
	7	186	151	131	109	84.6	55.4	241	206	168	128	79.9
	8	166	136	118	99.0	77.0	51.5	233	199	163	124	78.2
	9	145	121	105	88.6	69.3	47.2	224	192	157	120	76.3
	10	126	106	92.3	78.3	61.6	42.7	215	184	151	115	74.2
	11	107	91.0	80.0	68.3	54.1	37.8	205	176	144	110	71.9
	12	89.7	77.2	68.2	58.8	46.9	32.9	195	167	137	105	69.4
	13	76.4	65.8	58.1	50.1	40.1	28.3	184	158	130	99.9	66.7
	14	65.9	56.7	50.1	43.2	34.6	24.4	173	149	123	94.5	63.8
	15	57.4	49.4	43.6	37.6	30.1	21.2	163	140	115	89.0	60.8
	16	50.5	43.4	38.4	33.1	26.5	18.7	152	131	108	83.4	57.4
	17	44.7	38.5	34.0	29.3	23.5	16.5	141	122	101	77.9	53.7
	18	39.9	34.3	30.3	26.1	20.9	14.8	130	113	93.4	72.4	50.0
	19		30.8	27.2	23.5	18.8	13.2	120	104	86.3	67.1	46.4
	20				21.2	17.0	11.9	110	95.8	79.4	61.9	42.9
	21						10.8	100	87.5	72.7	56.8	39.5
	22							91.1	79.7	66.3	51.8	36.1
	23							83.4	72.9	60.6	47.4	33.0
	24							76.6	67.0	55.7	43.6	30.3
	25							70.6	61.7	51.3	40.2	28.0
	26							65.2	57.1	47.4	37.1	25.9
	27							60.5	52.9	44.0	34.4	24.0
	28							56.3	49.2	40.9	32.0	22.3
	29							52.4	45.9	38.1	29.8	20.8
	30							49.0	42.9	35.6	27.9	19.4
Properties												
A_g, in.^{2}		6.95	5.48	4.68	3.84	2.93	2.00	6.88	5.85	4.77	3.63	2.46
I_x, in.^{4}		26.8	22.7	20.1	17.0	13.4	9.43	29.7	25.9	21.7	17.0	11.8
I_y, in.^{4}		8.69	7.48	6.67	5.70	4.55	3.23	29.7	25.9	21.7	17.0	11.8
r_x/r_y		1.76	1.74	1.74	1.72	1.71	1.71	1.00	1.00	1.00	1.00	1.00
r_y, in.		1.12	1.17	1.19	1.22	1.25	1.27	2.08	2.11	2.13	2.16	2.19

†Section is a slender-element cross-section.
††Section is a non-compact cross-section.
Note: Heavy line indicates Kl/r equal to or greater than 200.

$F_y = 46$ ksi
$\phi_c P_n = 0.85 F_{cr} A_g$

Table 4-6 (cont.). Rectangular (and Square) HSS Design Strength in Axial Compression, $\phi_c P_n$, kips

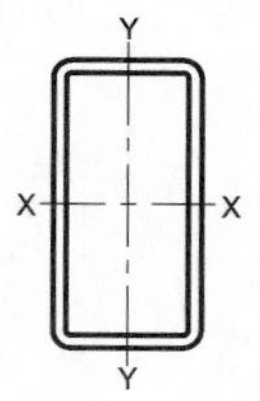

Shape	HSS5×5×						HSS5×4×				
	1/2	3/8	5/16	1/4	3/16	1/8†	1/2	3/8	5/16	1/4	3/16
t_{design}, in.	0.465	0.349	0.291	0.233	0.174	0.116	0.465	0.349	0.291	0.233	0.174
Wt/ft	28.3	22.3	19.0	15.6	12.0	8.15	24.9	19.7	16.9	13.9	10.7
Effective length *KL* (ft) with respect to least radius of gyration r_y											
0	308	242	206	168	128	81.5	272	214	183	150	115
1	307	241	205	168	128	81.4	271	213	182	150	114
2	305	239	203	166	127	80.9	267	211	180	148	113
3	300	236	201	164	125	80.2	261	206	176	145	111
4	294	231	197	161	123	79.3	253	200	171	141	108
5	286	225	192	158	120	78.0	243	193	165	136	104
6	277	219	187	153	117	76.5	231	184	158	130	100
7	267	211	180	148	113	74.7	217	174	150	124	95.2
8	256	202	173	142	109	72.6	203	164	141	117	89.9
9	243	193	165	136	105	70.3	188	153	131	109	84.3
10	230	183	157	130	99.7	67.8	173	141	122	101	78.5
11	216	173	149	123	94.5	64.9	157	129	112	93.3	72.5
12	202	162	140	116	89.2	61.3	141	117	102	85.3	66.4
13	188	151	131	108	83.7	57.7	126	105	91.8	77.3	60.4
14	174	140	122	101	78.2	54.0	112	94.2	82.2	69.5	54.6
15	160	130	112	93.7	72.7	50.3	97.6	83.4	73.0	62.0	48.9
16	146	119	103	86.4	67.3	46.6	85.7	73.3	64.2	54.8	43.4
17	132	109	94.7	79.3	61.9	43.0	76.0	64.9	56.9	48.5	38.5
18	119	98.5	86.2	72.4	56.7	39.5	67.7	57.9	50.8	43.3	34.3
19	107	88.7	77.9	65.8	51.6	36.1	60.8	52.0	45.6	38.8	30.8
20	96.7	80.1	70.3	59.3	46.7	32.7	54.9	46.9	41.1	35.1	27.8
21	87.7	72.6	63.8	53.8	42.3	29.7	49.8	42.5	37.3	31.8	25.2
22	79.9	66.2	58.1	49.0	38.6	27.0	45.4	38.8	34.0	29.0	23.0
23	73.1	60.5	53.2	44.9	35.3	24.7	41.5	35.5	31.1	26.5	21.0
24	67.1	55.6	48.8	41.2	32.4	22.7	38.1	32.6	28.6	24.3	19.3
25	61.9	51.2	45.0	38.0	29.9	20.9		30.0	26.3	22.4	17.8
26	57.2	47.4	41.6	35.1	27.6	19.4				20.7	16.4
27	53.1	43.9	38.6	32.6	25.6	17.9					
28	49.3	40.8	35.9	30.3	23.8	16.7					
29	46.0	38.1	33.5	28.2	22.2	15.6					
30	43.0	35.6	31.3	26.4	20.7	14.5					
Properties											
A_g, in.^{2}	7.88	6.18	5.26	4.30	3.28	2.23	6.95	5.48	4.68	3.84	2.93
I_x, in.^{4}	26.0	21.7	19.0	16.0	12.6	8.80	21.2	17.9	15.8	13.4	10.6
I_y, in.^{4}	26.0	21.7	19.0	16.0	12.6	8.80	14.9	12.6	11.1	9.46	7.48
r_x/r_y	1.00	1.00	1.00	1.00	1.00	1.00	1.20	1.19	1.19	1.19	1.19
r_y, in.	1.82	1.87	1.90	1.93	1.96	1.99	1.46	1.52	1.54	1.57	1.60

†Section is a slender-element cross-section.
Note: Heavy line indicates Kl/r equal to or greater than 200.

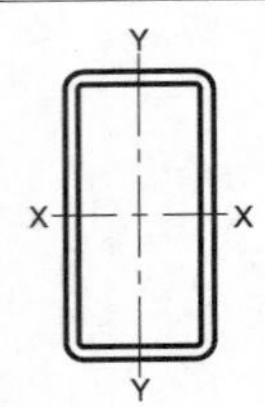

Table 4-6 (cont.). Rectangular (and Square) HSS Design Strength in Axial Compression, $\phi_c P_n$, kips

$F_y = 46$ ksi

$\phi_c P_n = 0.85 F_{cr} A_g$

Shape	HSS5×3×						HSS5×2½×		
	1/2	3/8	5/16	1/4	3/16	1/8†	1/4	3/16	1/8†
t_{design}, in.	0.465	0.349	0.291	0.233	0.174	0.116	0.233	0.174	0.116
Wt/ft	21.5	17.2	14.8	12.2	9.40	6.45	11.3	8.55	6.02
Effective length KL (ft) with respect to least radius of gyration r_y									
0	235	187	160	132	101	66.1	123	92.3	61.4
1	233	186	159	131	100	65.8	122	91.4	61.0
2	228	181	156	128	98.3	64.8	118	89.0	59.7
3	219	175	150	124	95.1	63.2	113	85.0	57.7
4	207	166	143	118	90.9	61.0	105	79.7	54.9
5	192	155	134	111	85.7	58.2	96.3	73.4	51.3
6	176	143	124	103	79.8	54.9	86.6	66.4	47.0
7	158	130	113	94.2	73.3	51.1	76.3	59.0	41.9
8	140	116	102	85.1	66.5	46.5	66.0	51.4	36.8
9	122	102	90.4	75.7	59.5	41.9	55.9	44.0	31.7
10	104	88.7	79.0	66.5	52.6	37.2	46.4	37.0	26.8
11	87.6	75.8	68.1	57.6	45.9	32.7	38.4	30.7	22.3
12	73.6	63.9	57.7	49.1	39.5	28.3	32.2	25.8	18.7
13	62.7	54.5	49.2	41.8	33.7	24.2	27.5	22.0	15.9
14	54.1	47.0	42.4	36.1	29.0	20.9	23.7	18.9	13.8
15	47.1	40.9	37.0	31.4	25.3	18.2	20.6	16.5	12.0
16	41.4	36.0	32.5	27.6	22.2	16.0	18.1	14.5	10.5
17	36.7	31.8	28.8	24.5	19.7	14.2		12.8	9.33
18	32.7	28.4	25.7	21.8	17.6	12.6			
19		25.5	23.0	19.6	15.8	11.4			
20					14.2	10.2			
21									
Properties									
A_g, in.²	6.02	4.78	4.10	3.37	2.58	1.77	3.14	2.36	1.65
I_x, in.⁴	16.4	14.1	12.6	10.7	8.53	6.03	9.40	7.37	5.34
I_y, in.⁴	7.18	6.25	5.60	4.81	3.85	2.75	3.13	2.49	1.82
r_x/r_y	1.51	1.51	1.50	1.50	1.49	1.48	1.73	1.72	1.71
r_y, in.	1.09	1.14	1.17	1.19	1.22	1.25	0.999	1.03	1.05

† Section is a slender-element cross-section.

Note: Heavy line indicates Kl/r equal to or greater than 200.

$F_y = 46$ ksi

$\phi_c P_n = 0.85 F_{cr} A_g$

Table 4-6 (cont.). Rectangular (and Square) HSS Design Strength in Axial Compression, $\phi_c P_n$, kips

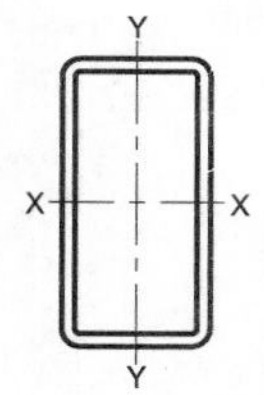

Shape	HSS4½×4½×						HSS4×4×					
	1/2	3/8	5/16	1/4	3/16	1/8†	1/2	3/8	5/16	1/4	3/16	1/8††
t_{design}, in.	0.465	0.349	0.291	0.233	0.174	0.116	0.465	0.349	0.291	0.233	0.174	0.116
Wt/ft	24.9	19.7	16.9	13.9	10.7	7.30	21.5	17.2	14.8	12.2	9.40	6.45
Effective length KL (ft) with respect to least radius of gyration r_y												
0	272	214	183	150	115	77.2	235	187	160	132	101	69.2
1	271	214	182	150	114	77.0	234	186	160	131	100	68.9
2	268	211	181	148	113	76.5	231	184	158	130	99.3	68.1
3	263	208	178	146	111	75.6	225	180	154	127	97.3	66.8
4	256	203	173	143	109	74.5	218	174	150	123	94.6	65.0
5	248	196	168	138	106	72.4	208	167	144	119	91.2	62.8
6	238	189	162	134	102	70.0	198	159	137	113	87.2	60.2
7	226	181	155	128	98.1	67.3	185	150	129	107	82.8	57.2
8	214	172	148	122	93.6	64.3	172	140	121	101	77.9	54.0
9	201	162	139	116	88.7	61.0	159	130	113	93.8	72.8	50.5
10	187	151	131	109	83.5	57.6	145	119	104	86.6	67.4	46.9
11	173	141	122	101	78.1	54.0	131	109	94.6	79.3	61.9	43.3
12	159	130	113	94.2	72.7	50.4	117	98.0	85.5	72.0	56.4	39.6
13	145	119	104	86.9	67.1	46.6	103	87.6	76.7	64.9	51.0	35.9
14	131	108	94.9	79.6	61.6	43.0	90.5	77.6	68.2	57.9	45.8	32.3
15	117	98.1	86.1	72.5	56.2	39.3	78.8	68.0	59.9	51.3	40.7	28.9
16	104	88.1	77.6	65.6	51.0	35.8	69.3	59.8	52.7	45.1	35.9	25.6
17	92.4	78.4	69.3	58.9	45.9	32.3	61.4	53.0	46.7	39.9	31.8	22.7
18	82.4	69.9	61.9	52.6	41.0	29.0	54.7	47.2	41.6	35.6	28.3	20.2
19	73.9	62.7	55.5	47.2	36.8	26.0	49.1	42.4	37.4	32.0	25.4	18.1
20	66.7	56.6	50.1	42.6	33.2	23.5	44.3	38.3	33.7	28.8	23.0	16.4
21	60.5	51.3	45.4	38.6	30.1	21.3	40.2	34.7	30.6	26.2	20.8	14.8
22	55.1	46.8	41.4	35.2	27.5	19.4	36.6	31.6	27.9	23.8	19.0	13.5
23	50.5	42.8	37.9	32.2	25.1	17.7	33.5	28.9	25.5	21.8	17.4	12.4
24	46.3	39.3	34.8	29.6	23.1	16.3		26.6	23.4	20.0	15.9	11.4
25	42.7	36.2	32.1	27.2	21.3	15.0				18.5	14.7	10.5
26	39.5	33.5	29.6	25.2	19.7	13.9						9.68
27		31.1	27.5	23.4	18.2	12.9						
28			25.6	21.7	17.0	12.0						
29					15.8	11.2						
30												
Properties												
A_g, in.^{2}	6.95	5.48	4.68	3.84	2.93	2.00	6.02	4.78	4.10	3.37	2.58	1.77
I_x, in.^{4}	18.1	15.3	13.5	11.4	9.02	6.35	11.9	10.3	9.14	7.80	6.21	4.40
I_y, in.^{4}	18.1	15.3	13.5	11.4	9.02	6.35	11.9	10.3	9.14	7.80	6.21	4.40
r_x/r_y	1.00	1.00	1.00	1.00	1.00	1.00	1.00	1.00	1.00	1.00	1.00	1.00
r_y, in.	1.61	1.67	1.70	1.73	1.75	1.78	1.41	1.47	1.49	1.52	1.55	1.58

†Section is a slender-element cross-section.

††Section is a non-compact cross-section.

Note: Heavy line indicates Kl/r equal to or greater than 200.

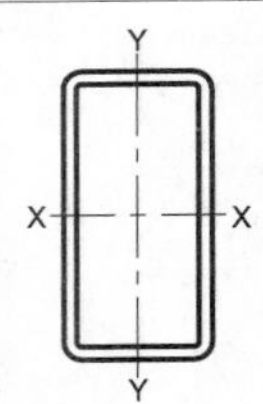

Table 4-6 (cont.). Rectangular (and Square) HSS Design Strength in Axial Compression, $\phi_c P_n$, kips

$F_y = 46$ ksi

$\phi_c P_n = 0.85 F_{cr} A_g$

Shape		HSS4×3×					HSS4×2½×			HSS4×2×				
		3/8	5/16	1/4	3/16	1/8††	5/16	1/4	3/16	3/8	5/16	1/4	3/16	1/8††
t_{design}, in.		0.349	0.291	0.233	0.174	0.116	0.291	0.233	0.174	0.349	0.291	0.233	0.174	0.116
Wt/ft		14.6	12.7	10.5	8.13	5.60	11.6	9.63	7.49	12.1	10.5	8.78	6.85	4.75
Effective length KL (ft) with respect to least radius of gyration r_y	0	160	138	114	87.6	60.2	126	104	80.5	133	115	95.4	73.9	50.8
	1	159	137	113	87.0	59.8	125	103	79.8	130	113	93.9	72.8	50.1
	2	155	134	111	85.2	58.6	121	100	77.5	123	107	89.5	69.6	48.1
	3	149	129	107	82.4	56.7	115	95.2	73.8	112	98.6	82.6	64.6	44.8
	4	141	122	101	78.5	54.2	106	88.6	69.0	99.0	87.5	73.9	58.1	40.6
	5	131	114	95.0	73.8	51.0	96.4	80.8	63.2	84.0	75.1	64.0	50.8	35.8
	6	120	105	87.8	68.5	47.5	85.6	72.2	56.8	68.8	62.2	53.7	43.1	30.6
	7	109	94.9	80.0	62.6	43.5	74.4	63.2	50.1	54.3	49.9	43.6	35.5	25.5
	8	96.7	84.7	71.8	56.5	39.4	63.3	54.2	43.3	41.7	38.7	34.3	28.3	20.7
	9	84.6	74.4	63.5	50.3	35.2	52.7	45.6	36.7	33.0	30.6	27.1	22.3	16.4
	10	72.9	64.5	55.4	44.2	31.1	42.9	37.5	30.5	26.7	24.8	21.9	18.1	13.3
	11	61.7	55.0	47.6	38.3	27.0	35.5	31.0	25.2	22.1	20.5	18.1	15.0	11.0
	12	51.9	46.2	40.3	32.6	23.2	29.8	26.0	21.2	18.5	17.2	15.2	12.6	9.21
	13	44.2	39.4	34.3	27.8	19.8	25.4	22.2	18.0				10.7	7.85
	14	38.1	34.0	29.6	24.0	17.0	21.9	19.1	15.5					
	15	33.2	29.6	25.8	20.9	14.8	19.1	16.6	13.5					
	16	29.2	26.0	22.7	18.4	13.0		14.6	11.9					
	17	25.8	23.0	20.1	16.3	11.6								
	18	23.0	20.6	17.9	14.5	10.3								
	19			16.1	13.0	9.25								
	20					8.35								
	21													
Properties														
A_g, in.^{2}		4.09	3.52	2.91	2.24	1.54	3.23	2.67	2.06	3.39	2.94	2.44	1.89	1.30
I_x, in.^{4}		7.93	7.14	6.15	4.93	3.52	6.13	5.32	4.30	5.60	5.13	4.49	3.66	2.65
I_y, in.^{4}		5.01	4.52	3.91	3.16	2.27	2.89	2.53	2.06	1.80	1.67	1.48	1.22	0.898
r_x/r_y		1.25	1.26	1.25	1.25	1.26	1.46	1.45	1.44	1.77	1.75	1.75	1.73	1.72
r_y, in.		1.11	1.13	1.16	1.19	1.21	0.947	0.973	0.999	0.729	0.754	0.779	0.804	0.830

††Section is a non-compact cross-section.
Note: Heavy line indicates Kl/r equal to or greater than 200.

$F_y = 46$ ksi

$\phi_c P_n = 0.85 F_{cr} A_g$

Table 4-6 (cont.). Rectangular (and Square) HSS Design Strength in Axial Compression, $\phi_c P_n$, kips

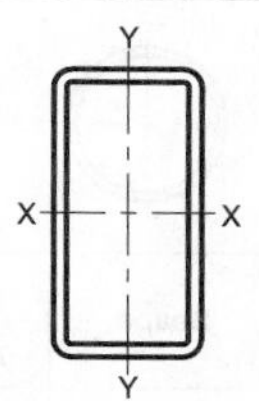

Shape	HSS3½×3½×					HSS3×3×					HSS2½×2½×			
	3/8	5/16	1/4	3/16	1/8	3/8	5/16	1/4	3/16	1/8	5/16	1/4	3/16	1/8
t_{design}, in.	**0.349**	**0.291**	**0.233**	**0.174**	**0.116**	**0.349**	**0.291**	**0.233**	**0.174**	**0.116**	**0.291**	**0.233**	**0.174**	**0.116**
Wt/ft	**14.6**	**12.7**	**10.5**	**8.13**	**5.60**	**12.1**	**10.5**	**8.78**	**6.85**	**4.75**	**8.40**	**7.08**	**5.57**	**3.90**
Effective length KL (ft) with respect to least radius of gyration r_y														
0	160	138	114	87.6	60.2	133	115	95.4	73.9	50.8	91.9	77.0	60.2	41.8
1	159	137	113	87.1	59.9	131	114	94.7	73.4	50.5	90.7	76.1	59.6	41.4
2	156	134	111	85.7	59.0	128	111	92.5	71.7	49.4	87.4	73.5	57.6	40.1
3	151	131	108	83.5	57.5	123	107	88.9	69.1	47.7	82.1	69.3	54.5	38.1
4	145	125	104	80.4	55.4	115	101	84.1	65.6	45.4	75.2	63.8	50.5	35.4
5	137	119	99	76.7	52.9	107	93.4	78.4	61.3	42.6	67.2	57.4	45.7	32.3
6	128	112	93.1	72.3	50.0	97.2	85.2	71.9	56.5	39.4	58.6	50.5	40.5	28.8
7	119	103	86.6	67.5	46.8	86.9	76.5	64.9	51.3	35.9	49.8	43.3	35.1	25.1
8	108	94.8	79.7	62.3	43.3	76.3	67.6	57.7	45.9	32.3	41.3	36.3	29.7	21.5
9	97.6	85.9	72.5	56.9	39.6	65.9	58.7	50.5	40.4	28.7	33.3	29.7	24.6	18.0
10	86.9	76.9	65.3	51.5	35.9	56.0	50.1	43.5	35.1	25.0	27.0	24.1	20.0	14.8
11	76.4	68.1	58.1	46.0	32.2	46.6	42.0	36.8	30.0	21.6	22.3	19.9	16.6	12.2
12	66.4	59.5	51.1	40.7	28.6	39.2	35.3	30.9	25.3	18.3	18.7	16.7	13.9	10.3
13	56.9	51.4	44.5	35.7	25.2	33.4	30.1	26.4	21.5	15.6	16.0	14.2	11.9	8.74
14	49.1	44.3	38.3	30.9	21.9	28.8	25.9	22.7	18.6	13.5	13.8	12.3	10.2	7.53
15	42.8	38.6	33.4	26.9	19.0	25.1	22.6	19.8	16.2	11.7		10.7	8.90	6.56
16	37.6	33.9	29.3	23.6	16.7	22.0	19.8	17.4	14.2	10.3				5.77
17	33.3	30.0	26.0	20.9	14.8	19.5	17.6	15.4	12.6	9.12				
18	29.7	26.8	23.2	18.7	13.2		15.7	13.7	11.2	8.14				
19	26.7	24.0	20.8	16.8	11.9				10.1	7.30				
20	24.1	21.7	18.8	15.1	10.7									
21	21.8	19.7	17.0	13.7	9.71									
22			15.5	12.5	8.85									
23														
Properties														
A_g, in.^{2}	4.09	3.52	2.91	2.24	1.54	3.39	2.94	2.44	1.89	1.30	2.35	1.97	1.54	1.07
I_x, in.^{4}	6.49	5.84	5.04	4.05	2.90	3.78	3.45	3.02	2.46	1.78	1.82	1.63	1.35	0.998
I_y, in.^{4}	6.49	5.84	5.04	4.05	2.90	3.78	3.45	3.02	2.46	1.78	1.82	1.63	1.35	0.998
r_x/r_y	1.00	1.00	1.00	1.00	1.00	1.00	1.00	1.00	1.00	1.00	1.00	1.00	1.00	1.00
r_y, in.	1.26	1.29	1.32	1.35	1.37	1.06	1.08	1.11	1.14	1.17	0.880	0.908	0.937	0.965

Note: Heavy line indicates Kl/r equal to or greater than 200.

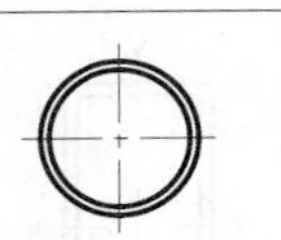

Table 4-7. Round HSS Design Strength in Axial Compression, $\phi_c P_n$, kips

$F_y = 42$ ksi

$\phi_c P_n = 0.85 F_{cr} A_g$

Shape		HSS20.000×		HSS18.000×		HSS16.000×				HSS14.000×		
		0.500	0.375	0.500	0.375	0.500	0.438	0.375	0.312	0.500	0.375	0.312
t_{design}, in.		0.465	0.349	0.465	0.349	0.465	0.407	0.349	0.291	0.465	0.349	0.291
Wt/ft		104	78.7	93.5	70.7	82.8	72.9	62.6	52.3	72.2	54.6	45.7
Effective length *KL* (ft) with respect to radius of gyration *r*	0	1020	768	914	693	810	710	614	514	707	536	446
	6	1010	763	906	687	802	703	608	509	697	528	440
	7	1010	761	904	685	799	700	605	507	694	526	438
	8	1010	759	901	683	795	697	603	505	690	523	436
	9	1000	756	897	680	791	694	600	502	685	519	433
	10	999	754	893	677	787	690	597	500	680	516	430
	11	995	751	889	674	782	686	593	497	675	511	426
	12	991	748	884	670	777	681	589	493	669	507	423
	13	986	744	879	666	771	676	585	490	662	502	419
	14	981	740	874	662	765	671	580	486	655	497	415
	15	976	737	868	658	759	665	575	482	648	492	410
	16	970	732	862	653	752	659	570	478	640	486	405
	17	964	728	855	649	744	653	565	473	632	480	400
	18	958	723	848	643	737	646	559	468	624	474	395
	19	952	718	841	638	729	640	553	463	615	467	390
	20	945	713	834	632	721	632	547	458	606	460	384
	22	930	702	818	620	703	617	534	447	587	446	372
	24	914	691	800	608	684	601	520	436	566	430	359
	25	906	685	792	601	675	592	513	430	556	423	353
	26	898	678	782	594	665	583	505	423	545	414	346
	28	880	665	763	580	644	565	489	410	522	398	332
	30	861	651	743	565	622	547	473	397	500	381	318
	31	852	644	733	557	611	537	465	390	488	372	311
	32	842	636	722	549	600	527	457	383	476	363	304
	34	821	621	700	533	577	507	440	369	453	345	289
	36	800	605	678	516	554	487	422	354	429	328	274
	37	790	597	667	507	542	477	413	347	417	319	267
	38	779	589	656	499	530	466	404	340	405	310	259
	40	756	573	632	482	507	446	387	325	381	292	245
Properties												
A_g, in.^{2}		28.5	21.5	25.6	19.4	22.7	19.9	17.2	14.4	19.8	15.0	12.5
I, in.^{4}		1360	1040	985	754	685	606	526	443	453	349	295
r, in.		6.91	6.95	6.20	6.24	5.49	5.51	5.53	5.55	4.79	4.83	4.85

$F_y = 42$ ksi
$\phi_c P_n = 0.85 F_{cr} A_g$

Table 4-7 (cont.). Round HSS Design Strength in Axial Compression, $\phi_c P_n$, kips

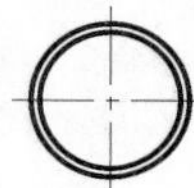

Shape		HSS12.750×			HSS12.500×						HSS11.250×		
		0.500	0.375	0.250	0.625	0.500	0.375	0.312	0.250	0.188	0.625	0.500	0.375
t_{design}, in.		0.465	0.349	0.233	0.581	0.465	0.349	0.291	0.233	0.174	0.581	0.465	0.349
Wt/ft		65.5	49.6	33.4	79.3	64.1	48.6	40.7	32.7	24.7	71.0	57.5	43.6
Effective length KL (ft) with respect to radius of gyration r	0	639	486	327	778	628	475	400	321	241	696	564	428
	6	628	478	322	764	617	467	393	315	237	681	552	419
	7	625	475	320	760	613	464	391	313	235	675	548	416
	8	620	471	318	754	609	460	388	311	234	669	543	412
	9	615	468	315	748	604	457	385	309	232	662	537	408
	10	610	464	313	741	598	453	381	306	230	654	531	404
	11	604	459	310	733	592	448	378	303	227	646	524	399
	12	597	454	306	725	586	443	373	300	225	637	517	393
	13	590	449	303	716	579	438	369	296	222	627	509	388
	14	583	444	299	706	571	432	364	292	220	617	501	381
	15	575	438	295	696	563	426	359	288	217	606	492	375
	16	567	432	291	685	555	420	354	284	214	594	483	368
	17	558	425	287	674	546	414	349	280	210	582	473	361
	18	549	418	283	663	537	407	343	275	207	570	463	353
	19	540	411	278	651	527	400	337	271	203	557	453	346
	20	530	404	273	638	517	392	331	266	200	543	443	338
	22	510	389	263	612	496	377	318	255	192	516	421	321
	24	488	373	252	585	475	360	304	245	184	487	398	304
	25	477	364	247	571	463	352	297	239	180	473	386	296
	26	466	356	241	556	452	344	290	233	176	458	374	287
	28	443	339	230	527	429	326	276	222	167	428	351	269
	30	420	321	218	498	405	309	261	210	158	399	327	251
	31	408	312	212	483	393	300	254	204	154	384	315	242
	32	396	303	206	468	381	291	246	198	149	369	303	233
	34	372	286	194	438	358	273	231	186	141	340	280	216
	36	349	268	182	409	334	255	216	174	132	312	257	198
	37	337	259	176	394	322	247	209	169	127	298	246	190
	38	325	250	171	380	311	238	202	163	123	285	235	182
	40	303	233	159	352	288	221	187	151	114	258	214	166
Properties													
A_g, in.^{2}		17.9	13.6	9.16	21.8	17.6	13.3	11.2	8.98	6.74	19.5	15.8	12.0
I, in.^{4}		339	262	180	387	319	246	208	169	128	278	229	178
r, in.		4.35	4.39	4.43	4.22	4.26	4.30	4.32	4.34	4.36	3.78	3.82	3.86

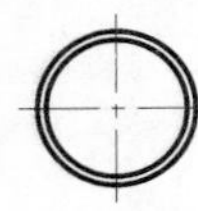

Table 4-7 (cont.). Round HSS Design Strength in Axial Compression, $\phi_c P_n$, kips

$F_y = 42$ ksi

$\phi_c P_n = 0.85 F_{cr} A_g$

Shape	HSS11.250×			HSS10.750×		HSS10.000×					
	0.312	0.250	0.188	0.500	0.250	0.625	0.500	0.375	0.312	0.250	0.188
t_{design}, in.	0.291	0.233	0.174	0.465	0.233	0.581	0.465	0.349	0.291	0.233	0.174
Wt/ft	36.5	29.4	22.2	54.8	28.1	62.6	50.8	38.6	32.3	26.1	19.7
Effective length KL (ft) with respect to radius of gyration r											
0	357	288	216	536	275	614	496	378	317	255	192
6	350	282	212	523	269	597	483	368	309	249	187
7	347	280	210	518	266	591	478	365	306	246	185
8	344	277	208	513	264	584	472	360	302	243	183
9	340	275	206	507	261	576	466	356	298	240	181
10	337	271	204	501	258	567	459	351	294	237	178
11	333	268	201	494	254	558	452	345	289	233	175
12	328	265	199	486	251	548	444	339	284	229	172
13	323	261	196	478	247	537	435	333	279	225	169
14	318	257	193	470	243	526	426	326	274	221	166
15	313	252	190	461	238	514	417	319	268	216	163
16	307	248	186	451	233	501	407	311	262	211	159
17	301	243	183	442	229	488	397	304	255	206	155
18	295	238	179	431	223	475	386	296	248	201	151
19	289	233	175	421	218	461	375	288	242	195	147
20	282	228	172	410	213	447	364	279	235	190	143
22	269	217	163	388	202	418	341	262	220	178	134
24	255	206	155	365	190	389	318	244	206	166	126
25	247	200	151	353	184	374	306	235	198	160	121
26	240	194	146	341	178	359	294	226	191	154	117
28	225	182	138	317	167	330	270	208	176	143	108
30	210	170	129	294	155	301	247	191	161	131	99.0
31	203	165	124	282	149	287	236	182	154	125	94.6
32	196	159	120	270	143	273	225	174	147	119	90.4
34	181	147	111	248	131	246	203	157	133	108	82.0
36	167	135	102	225	120	219	182	141	119	97.3	73.9
37	160	130	98.2	215	115	208	172	133	113	92.1	70.0
38	153	124	94.1	204	109	197	163	126	107	87.3	66.3
40	139	113	86.0	184	98.7	178	147	114	96.7	78.8	59.9
Properties											
A_g, in.^{2}	10.0	8.06	6.05	15.0	7.70	17.2	13.9	10.6	8.88	7.15	5.37
I, in.^{4}	151	122	92.9	199	106	191	159	123	105	85.3	64.8
r, in.	3.88	3.90	3.92	3.64	3.72	3.34	3.38	3.41	3.43	3.45	3.47

$F_y = 42$ ksi
$\phi_c P_n = 0.85 F_{cr} A_g$

Table 4-7 (cont.). Round HSS Design Strength in Axial Compression, $\phi_c P_n$, kips

Shape	HSS9.625×					HSS8.750×					HSS8.625×	
	0.500	0.375	0.312	0.250	0.188	0.500	0.375	0.312	0.250	0.188	0.500	0.375
t_{design}, in.	0.465	0.349	0.291	0.233	0.174	0.465	0.349	0.291	0.233	0.174	0.465	0.349
Wt/ft	48.8	37.1	31.1	25.1	19.0	44.1	33.6	28.1	22.7	17.2	43.4	33.1
Effective length *KL* (ft) with respect to radius of gyration *r*												
0	478	364	305	245	185	432	329	276	222	167	425	324
6	464	354	296	238	179	416	317	266	215	162	409	312
7	459	350	293	236	178	411	313	263	212	160	403	308
8	453	345	289	233	175	404	308	259	209	157	397	303
9	447	341	285	230	173	397	303	255	206	155	390	298
10	440	335	281	226	171	390	297	250	202	152	382	292
11	432	330	276	223	168	381	291	245	198	149	374	286
12	424	323	271	218	165	372	285	239	193	146	365	279
13	415	317	265	214	161	363	278	233	189	142	355	272
14	406	310	260	210	158	353	270	227	184	139	345	265
15	396	303	254	205	154	343	262	221	179	135	335	257
16	386	295	247	200	151	332	254	214	173	131	324	249
17	375	287	241	194	147	321	246	207	168	127	313	240
18	364	279	234	189	143	309	238	200	162	123	301	232
19	353	271	227	184	139	298	229	193	156	118	290	223
20	342	262	220	178	134	286	220	186	151	114	278	214
22	318	245	206	166	126	262	202	171	139	105	254	197
24	294	227	191	154	117	239	185	156	127	96.1	231	179
25	283	218	183	149	112	227	176	149	121	91.7	219	170
26	271	209	176	143	108	215	167	141	115	87.3	208	161
28	247	191	161	131	99.1	193	150	127	103	78.7	185	144
30	224	174	147	119	90.4	171	133	113	92.4	70.4	164	128
31	213	165	140	113	86.2	160	125	107	87.0	66.3	153	120
32	202	157	133	108	82.0	150	118	100	81.7	62.3	144	113
34	180	141	119	97.0	73.8	133	104	88.6	72.3	55.2	127	99.8
36	161	125	106	86.6	65.9	119	92.9	79.0	64.5	49.2	114	89.0
37	152	119	101	82.0	62.4	112	87.9	74.8	61.1	46.6	108	84.3
38	144	113	95.3	77.7	59.2	107	83.4	70.9	57.9	44.2	102	79.9
40	130	102	86.0	70.1	53.4	96.2	75.2	64.0	52.3	39.9	92.0	72.1
Properties												
A_g, in.^{2}	13.4	10.2	8.53	6.87	5.17	12.1	9.21	7.73	6.23	4.69	11.9	9.07
I, in.^{4}	141	110	93.0	75.9	57.7	104	81.4	69.3	56.6	43.1	99.5	77.8
r, in.	3.24	3.28	3.30	3.32	3.34	2.93	2.97	2.99	3.01	3.03	2.89	2.93

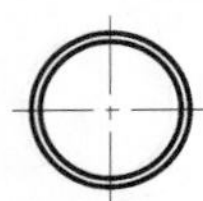

Table 4-7 (cont.). Round HSS Design Strength in Axial Compression, $\phi_c P_n$, kips

$F_y = 42$ ksi

$\phi_c P_n = 0.85 F_{cr} A_g$

Shape	HSS8.625×			HSS7.625×	HSS7.500×					HSS7.000×		
	0.322	0.250	0.188	0.125	0.500	0.375	0.312	0.250	0.188	0.500	0.375	0.312
t_{design}, in.	0.300	0.233	0.174	0.116	0.465	0.349	0.291	0.233	0.174	0.465	0.349	0.291
Wt/ft	28.6	22.4	17.0	10.0	37.4	28.6	24.0	19.4	14.7	34.7	26.6	22.3
Effective length *KL* (ft) with respect to radius of gyration *r*												
0	280	219	165	97.8	368	280	235	190	143	341	260	219
6	270	211	159	93.5	349	266	224	181	136	321	246	207
7	267	209	157	92.0	343	262	220	178	134	315	241	203
8	263	206	155	90.3	336	256	216	174	131	307	235	198
9	258	202	152	88.4	328	250	211	170	128	298	229	193
10	253	198	149	86.3	319	244	205	166	125	289	222	187
11	248	194	146	84.1	309	237	200	162	122	279	214	181
12	242	190	143	81.7	299	229	193	157	118	269	207	174
13	236	185	140	79.2	289	222	187	151	114	258	199	168
14	230	180	136	76.6	278	213	180	146	110	247	190	161
15	223	175	132	73.8	267	205	173	141	106	236	182	154
16	216	170	128	71.0	255	197	166	135	102	224	173	146
17	209	164	124	68.2	243	188	159	129	97.6	212	164	139
18	202	158	120	65.2	232	179	151	123	93.2	200	155	131
19	194	153	115	62.3	220	170	144	117	88.7	188	146	124
20	187	147	111	59.3	208	161	137	111	84.3	177	137	117
22	171	135	102	53.4	184	143	122	99.3	75.4	154	120	102
24	156	123	93.3	47.6	162	126	107	87.8	66.8	132	103	88.4
25	148	117	88.9	44.8	151	118	101	82.2	62.6	122	95.4	81.6
26	141	111	84.5	42.0	140	110	93.8	76.8	58.6	113	88.2	75.5
28	126	99.9	75.9	36.6	121	94.8	81.0	66.4	50.7	97.1	76.1	65.1
30	112	88.9	67.7	31.9	105	82.6	70.5	57.8	44.2	84.6	66.3	56.7
31	105	83.5	63.7	29.9	98.5	77.4	66.1	54.2	41.4	79.3	62.1	53.1
32	98.8	78.4	59.8	28.1	92.4	72.6	62.0	50.8	38.8	74.4	58.3	49.8
34	87.6	69.4	52.9	24.8	81.9	64.3	54.9	45.0	34.4	65.9	51.6	44.1
36	78.1	61.9	47.2	22.2	73.0	57.4	49.0	40.2	30.7	58.8	46.0	39.4
37	73.9	58.6	44.7	21.0	69.1	54.3	46.4	38.0	29.0	55.6	43.6	37.3
38	70.1	55.6	42.4	19.9	65.5	51.5	44.0	36.1	27.5	52.7	41.3	35.3
40	63.3	50.2	38.2	18.0	59.1	46.5	39.7	32.5	24.8			
Properties												
A_g, in.^{2}	7.85	6.14	4.62	2.74	10.3	7.84	6.59	5.32	4.00	9.55	7.29	6.13
I, in.^{4}	68.1	54.1	41.3	19.3	63.9	50.2	42.9	35.2	26.9	51.2	40.4	34.6
r, in.	2.95	2.97	2.99	2.66	2.49	2.53	2.55	2.57	2.59	2.32	2.35	2.37

Note: Heavy line indicates Kl/r equal to or greater than 200.

$F_y = 42$ ksi

$\phi_c P_n = 0.85 F_{cr} A_g$

Table 4-7 (cont.). Round HSS Design Strength in Axial Compression, $\phi_c P_n$, kips

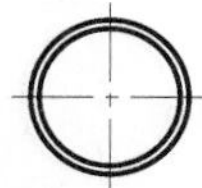

Shape		HSS7.000×			HSS6.875×					HSS6.625×			
		0.250	0.188	0.125	0.500	0.375	0.312	0.250	0.188	0.500	0.432	0.375	0.312
t_{design}, in.		0.233	0.174	0.116	0.465	0.349	0.291	0.233	0.174	0.465	0.403	0.349	0.291
Wt/ft		18.0	13.7	9.19	34.1	26.1	21.9	17.7	13.4	32.7	28.6	25.1	21.1
Effective length KL (ft) with respect to radius of gyration r	0	177	133	89.6	334	256	215	174	131	321	281	246	207
	6	167	126	84.9	314	241	203	164	123	300	263	230	194
	7	164	124	83.3	307	236	198	160	121	293	257	225	190
	8	160	121	81.4	299	230	194	157	118	285	250	219	185
	9	156	118	79.4	291	223	188	152	115	276	243	212	179
	10	151	114	77.1	281	217	183	148	112	267	234	205	173
	11	147	111	74.8	271	209	176	143	108	257	226	198	167
	12	141	107	72.2	261	201	170	138	104	246	216	190	160
	13	136	103	69.6	250	193	163	132	100	235	207	181	153
	14	130	98.8	66.8	239	185	156	127	96.0	223	197	173	146
	15	125	94.5	64.0	227	176	149	121	91.7	211	186	164	139
	16	119	90.2	61.1	215	167	142	115	87.3	200	176	155	132
	17	113	85.8	58.1	203	158	134	109	82.9	188	166	146	124
	18	107	81.3	55.2	192	149	127	103	78.4	176	156	137	117
	19	101	76.8	52.2	180	141	119	97.3	74.0	164	145	129	109
	20	95.1	72.4	49.2	168	132	112	91.4	69.6	153	135	120	102
	22	83.5	63.7	43.4	146	115	97.7	79.9	61.0	131	116	103	88.1
	24	72.4	55.4	37.8	124	98.3	84.1	69.0	52.8	110	98.1	87.2	74.7
	25	67.0	51.4	35.1	114	90.6	77.5	63.6	48.7	101	90.4	80.4	68.9
	26	62.0	47.5	32.5	106	83.7	71.6	58.8	45.1	93.7	83.6	74.3	63.7
	28	53.4	40.9	28.0	91.2	72.2	61.8	50.7	38.9	80.8	72.1	64.1	54.9
	30	46.5	35.7	24.4	79.4	62.9	53.8	44.2	33.8	70.4	62.8	55.8	47.8
	31	43.6	33.4	22.9	74.4	58.9	50.4	41.4	31.7	65.9	58.8	52.3	44.8
	32	40.9	31.3	21.4	69.8	55.3	47.3	38.8	29.7	61.9	55.2	49.1	42.0
	34	36.2	27.8	19.0	61.8	49.0	41.9	34.4	26.3	54.8	48.9	43.5	37.2
	36	32.3	24.8	16.9	55.1	43.7	37.4	30.7	23.5	48.9	43.6	38.8	33.2
	37	30.6	23.4	16.0	52.2	41.4	35.4	29.0	22.2			36.7	31.4
	38	29.0	22.2	15.2		39.2	33.5	27.5	21.1				
	40		20.1	13.7									
Properties													
A_g, in.^{2}		4.95	3.73	2.51	9.36	7.16	6.02	4.86	3.66	9.00	7.88	6.88	5.79
I, in.^{4}		28.4	21.7	14.9	48.3	38.2	32.7	26.8	20.6	42.9	38.3	34.0	29.1
r, in.		2.39	2.41	2.43	2.27	2.31	2.33	2.35	2.37	2.18	2.20	2.22	2.24

Note: Heavy line indicates Kl/r equal to or greater than 200.

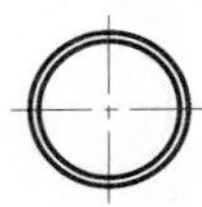

Table 4-7 (cont.). Round HSS Design Strength in Axial Compression, $\phi_c P_n$, kips

$F_y = 42$ ksi
$\phi_c P_n = 0.85 F_{cr} A_g$

Shape		HSS6.625×				HSS6.125×					HSS6.000×		
		0.280	0.250	0.188	0.125	0.500	0.375	0.312	0.250	0.188	0.500	0.375	0.312
t_{design}, in.		0.261	0.233	0.174	0.116	0.465	0.349	0.291	0.233	0.174	0.465	0.349	0.291
Wt/ft		19.0	17.0	12.9	8.69	30.1	23.1	19.4	15.7	11.9	29.4	22.5	19.0
Effective length KL (ft) with respect to radius of gyration r	0	186	167	126	84.6	295	226	190	154	116	289	221	186
	6	175	157	119	79.7	273	209	177	143	108	266	204	172
	7	171	153	116	78.0	265	204	172	139	105	258	199	168
	8	167	150	113	76.0	257	198	167	135	102	249	192	162
	9	162	145	110	73.9	247	191	161	130	98.6	240	185	156
	10	156	141	106	71.6	237	183	155	125	94.9	229	177	150
	11	151	135	103	69.1	227	175	148	120	91.0	219	169	143
	12	145	130	98.6	66.5	215	167	141	115	86.9	207	161	136
	13	139	125	94.5	63.8	204	158	134	109	82.7	196	152	129
	14	132	119	90.3	61.0	192	150	127	103	78.3	184	143	122
	15	126	113	85.9	58.1	180	141	120	97.1	73.9	172	135	114
	16	119	107	81.5	55.1	169	132	112	91.2	69.4	160	126	107
	17	112	101	77.1	52.2	157	123	105	85.2	65.0	148	117	99.6
	18	106	95.3	72.6	49.2	145	114	97.5	79.3	60.6	137	108	92.3
	19	99.2	89.4	68.2	46.3	134	106	90.3	73.6	56.3	126	99.6	85.2
	20	92.7	83.6	63.8	43.3	123	97.4	83.3	67.9	52.0	115	91.4	78.3
	22	80.0	72.3	55.3	37.7	102	81.4	69.9	57.1	43.9	95.1	75.9	65.2
	24	68.0	61.5	47.2	32.3	85.9	68.4	58.7	48.0	36.9	79.9	63.8	54.8
	25	62.6	56.7	43.5	29.7	79.2	63.1	54.1	44.2	34.0	73.7	58.8	50.5
	26	57.9	52.4	40.2	27.5	73.2	58.3	50.1	40.9	31.4	68.1	54.4	46.7
	28	49.9	45.2	34.7	23.7	63.1	50.3	43.2	35.2	27.1	58.7	46.9	40.3
	30	43.5	39.4	30.2	20.6	55.0	43.8	37.6	30.7	23.6	51.2	40.8	35.1
	31	40.7	36.9	28.3	19.3	51.5	41.0	35.2	28.7	22.1	47.9	38.2	32.8
	32	38.2	34.6	26.6	18.1	48.3	38.5	33.0	27.0	20.7	45.0	35.9	30.8
	34	33.9	30.6	23.5	16.1		34.1	29.3	23.9	18.4			
	36	30.2	27.3	21.0	14.3								
	37	28.6	25.9	19.9	13.6								
	38			18.8	12.9								
	40												
Properties													
A_g, in.^{2}		5.22	4.68	3.53	2.37	8.27	6.33	5.33	4.31	3.25	8.09	6.20	5.22
I, in.^{4}		26.5	23.9	18.4	12.6	33.3	26.5	22.7	18.7	14.4	31.2	24.8	21.3
r, in.		2.25	2.26	2.28	2.30	2.01	2.05	2.07	2.08	2.10	1.96	2.00	2.02

Note: Heavy line indicates Kl/r equal to or greater than 200.

$F_y = 42$ ksi
$\phi_c P_n = 0.85 F_{cr} A_g$

Table 4-7 (cont.). Round HSS Design Strength in Axial Compression, $\phi_c P_n$, kips

Shape	HSS6.000×				HSS5.563×				HSS5.500×		
	0.280	0.250	0.188	0.125	0.375	0.258	0.188	0.134	0.500	0.375	0.258
t_{design}, in.	0.261	0.233	0.174	0.116	0.349	0.241	0.174	0.125	0.465	0.349	0.241
Wt/ft	17.1	15.4	11.7	7.85	20.8	14.6	10.8	7.78	26.7	20.5	14.5
Effective length KL (ft) with respect to radius of gyration r											
0	168	151	114	76.4	204	144	105	76.4	263	202	142
6	156	140	105	71.0	186	131	96.5	70.1	238	183	130
7	151	136	103	69.1	180	127	93.5	67.9	230	177	125
8	147	131	99.3	67.0	173	123	90.2	65.5	220	170	121
9	141	127	95.9	64.7	166	117	86.5	62.9	210	163	116
10	136	122	92.2	62.3	158	112	82.6	60.1	199	155	110
11	130	116	88.2	59.7	149	106	78.5	57.1	188	147	104
12	123	111	84.1	56.9	141	100	74.3	54.1	177	138	98.3
13	117	105	79.8	54.1	132	94.3	69.9	50.9	165	129	92.2
14	110	99.3	75.5	51.2	123	88.1	65.5	47.7	153	120	86.1
15	104	93.4	71.0	48.2	114	81.9	61.0	44.5	141	111	79.9
16	97.1	87.4	66.6	45.3	105	75.8	56.6	41.3	130	103	73.8
17	90.4	81.5	62.2	42.3	96.8	69.8	52.3	38.2	118	94.0	67.9
18	83.9	75.7	57.8	39.4	88.4	64.0	48.0	35.1	107	85.7	62.1
19	77.5	70.0	53.5	36.5	80.3	58.3	43.9	32.1	96.8	77.7	56.5
20	71.3	64.4	49.3	33.7	72.5	52.8	39.9	29.2	87.4	70.1	51.0
21	65.2	59.0	45.3	31.0	65.8	47.9	36.2	26.5	79.2	63.6	46.3
22	59.4	53.8	41.3	28.3	59.9	43.6	32.9	24.2	72.2	57.9	42.2
23	54.4	49.2	37.8	25.9	54.8	39.9	30.1	22.1	66.1	53.0	38.6
24	49.9	45.2	34.7	23.8	50.4	36.6	27.7	20.3	60.7	48.7	35.4
25	46.0	41.6	32.0	21.9	46.4	33.8	25.5	18.7	55.9	44.9	32.6
26	42.5	38.5	29.6	20.3	42.9	31.2	23.6	17.3	51.7	41.5	30.2
27	39.4	35.7	27.4	18.8	39.8	28.9	21.9	16.0	47.9	38.5	28.0
28	36.7	33.2	25.5	17.5	37.0	26.9	20.3	14.9	44.6	35.8	26.0
29	34.2	30.9	23.8	16.3	34.5	25.1	19.0	13.9	41.5	33.3	24.3
30	32.0	28.9	22.2	15.2	32.2	23.4	17.7	13.0		31.2	22.7
31	29.9	27.1	20.8	14.3		22.0	16.6	12.2			21.2
32	28.1	25.4	19.5	13.4				11.4			
33	26.4	23.9	18.4	12.6							
34		22.5	17.3	11.9							
35											
Properties											
A_g, in.^{2}	4.71	4.22	3.18	2.14	5.72	4.03	2.95	2.14	7.36	5.65	3.98
I, in.^{4}	19.4	17.6	13.5	9.28	19.5	14.3	10.7	7.90	23.5	18.8	13.8
r, in.	2.03	2.04	2.06	2.08	1.85	1.88	1.91	1.92	1.79	1.83	1.86

Note: Heavy line indicates Kl/r equal to or greater than 200.

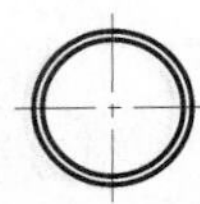

Table 4-7 (cont.). Round HSS Design Strength in Axial Compression, $\phi_c P_n$, kips

$F_y = 42$ ksi
$\phi_c P_n = 0.85 F_{cr} A_g$

Shape		HSS5.000×							HSS4.500×			
		0.500	0.375	0.312	0.258	0.250	0.188	0.125	0.337	0.237	0.188	0.125
t_{design}, in.		0.465	0.349	0.291	0.241	0.233	0.174	0.116	0.315	0.221	0.174	0.116
Wt/ft		24.1	18.5	15.6	13.1	12.7	9.67	6.51	15.0	10.8	8.67	5.85
Effective length KL (ft) with respect to radius of gyration r	0	236	182	154	129	125	94.2	63.5	148	106	84.3	57.1
	1	236	181	153	128	124	94.0	63.4	147	106	83.9	56.9
	2	233	180	152	127	123	93.1	62.8	145	104	83.0	56.3
	3	229	177	149	125	121	91.7	61.9	143	102	81.4	55.3
	4	224	173	146	122	119	89.8	60.6	139	99.6	79.3	53.9
	5	217	168	142	119	115	87.4	59.0	134	96.2	76.7	52.1
	6	209	162	137	115	111	84.5	57.1	128	92.2	73.5	50.0
	7	200	155	131	110	107	81.3	55.0	121	87.7	70.0	47.7
	8	190	148	125	105	102	77.7	52.6	114	82.7	66.2	45.1
	9	179	140	119	99.7	97.0	73.8	50.0	107	77.4	62.0	42.4
	10	168	132	112	93.9	91.4	69.6	47.3	98.7	71.9	57.7	39.5
	11	156	123	105	88.0	85.7	65.4	44.4	90.7	66.3	53.3	36.6
	12	145	114	97.2	81.8	79.8	61.0	41.5	82.6	60.7	48.9	33.6
	13	133	105	89.8	75.7	73.8	56.5	38.6	74.7	55.0	44.5	30.7
	14	121	96.3	82.5	69.5	67.9	52.1	35.6	67.0	49.6	40.2	27.8
	15	110	87.7	75.2	63.5	62.1	47.7	32.7	59.6	44.3	36.0	24.9
	16	98.7	79.3	68.2	57.6	56.4	43.5	29.8	52.5	39.2	32.0	22.2
	17	88.0	71.2	61.4	52.0	50.9	39.3	27.1	46.5	34.7	28.3	19.7
	18	78.5	63.5	54.8	46.5	45.6	35.3	24.4	41.5	31.0	25.3	17.6
	19	70.4	57.0	49.2	41.7	40.9	31.7	21.9	37.2	27.8	22.7	15.8
	20	63.6	51.4	44.4	37.6	36.9	28.6	19.7	33.6	25.1	20.5	14.2
	21	57.7	46.7	40.3	34.1	33.5	25.9	17.9	30.5	22.8	18.6	12.9
	22	52.5	42.5	36.7	31.1	30.5	23.6	16.3	27.8	20.7	16.9	11.8
	23	48.1	38.9	33.6	28.5	27.9	21.6	14.9	25.4	19.0	15.5	10.8
	24	44.1	35.7	30.8	26.1	25.6	19.9	13.7	23.3	17.4	14.2	9.89
	25	40.7	32.9	28.4	24.1	23.6	18.3	12.6		16.1	13.1	9.11
	26	37.6	30.4	26.3	22.3	21.8	16.9	11.7				
	27		28.2	24.4	20.7	20.3	15.7	10.8				
	28				19.2	18.8	14.6	10.1				
	29											
Properties												
A_g, in.^{2}		6.62	5.10	4.30	3.60	3.49	2.64	1.78	4.14	2.97	2.36	1.60
I, in.^{4}		17.2	13.9	12.0	10.2	9.94	7.69	5.31	9.12	6.82	5.54	3.84
r, in.		1.61	1.65	1.67	1.68	1.69	1.71	1.73	1.48	1.51	1.53	1.55

Note: Heavy line indicates Kl/r equal to or greater than 200.

$F_y = 42$ ksi
$\phi_c P_n = 0.85 F_{cr} A_g$

Table 4-7 (cont.). Round HSS Design Strength in Axial Compression, $\phi_c P_n$, kips

Shape	HSS4.000×							
	0.337	**0.313**	**0.250**	**0.237**	**0.226**	**0.220**	**0.188**	**0.125**
t_{design}, in.	**0.315**	**0.291**	**0.233**	**0.221**	**0.211**	**0.205**	**0.174**	**0.116**
Wt/ft	**13.2**	**12.3**	**10.0**	**9.53**	**9.12**	**8.89**	**7.66**	**5.18**
Effective length *KL* (ft) with respect to radius of gyration *r*								
0	130	121	98.5	93.5	89.6	87.1	74.6	50.7
1	130	120	98.0	93.1	89.2	86.7	74.3	50.5
2	128	119	96.6	91.7	87.9	85.4	73.2	49.7
3	124	116	94.2	89.5	85.7	83.3	71.4	48.6
4	120	112	91.0	86.4	82.8	80.5	69.0	47.0
5	115	107	87.0	82.7	79.2	77.0	66.1	45.1
6	108	101	82.3	78.3	75.0	73.0	62.7	42.8
7	101	94.4	77.1	73.5	70.4	68.4	58.8	40.2
8	93.7	87.5	71.5	68.2	65.4	63.6	54.7	37.5
9	85.8	80.2	65.7	62.8	60.1	58.5	50.4	34.6
10	77.8	72.8	59.8	57.2	54.8	53.2	45.9	31.6
11	69.8	65.5	53.8	51.5	49.4	48.0	41.5	28.7
12	62.0	58.3	48.0	46.0	44.1	42.9	37.1	25.7
13	54.5	51.3	42.3	40.7	39.0	37.9	32.9	22.9
14	47.4	44.7	36.9	35.6	34.1	33.1	28.8	20.1
15	41.2	38.9	32.2	31.0	29.7	28.9	25.1	17.6
16	36.3	34.2	28.3	27.2	26.1	25.4	22.0	15.4
17	32.1	30.3	25.0	24.1	23.1	22.5	19.5	13.7
18	28.6	27.0	22.3	21.5	20.6	20.0	17.4	12.2
19	25.7	24.2	20.0	19.3	18.5	18.0	15.6	10.9
20	23.2	21.9	18.1	17.4	16.7	16.2	14.1	9.87
21	21.0	19.8	16.4	15.8	15.1	14.7	12.8	8.95
22		18.1	14.9	14.4	13.8	13.4	11.7	8.16
23								
Properties								
A_g, in.^{2}	3.65	3.39	2.76	2.62	2.51	2.44	2.09	1.42
I, in.^{4}	6.24	5.87	4.91	4.70	4.52	4.41	3.83	2.67
r, in.	1.31	1.32	1.33	1.34	1.34	1.34	1.35	1.37

Note: Heavy line indicates Kl/r equal to or greater than 200.

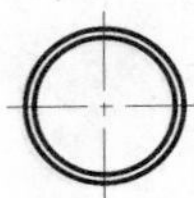

Table 4-8. Steel Pipe Design Strength in Axial Compression, $\phi_c P_n$, kips

F_y = 35 ksi

$\phi_c P_n = 0.85 F_{cr} A_g$

Shape		Pipe 12		Pipe 10		Pipe 8			Pipe 6		
		XS	Std.	XS	Std.	XXS	XS	Std.	XXS	XS	Std.
Wall Thkns., in.		0.500	0.375	0.500	0.365	0.875	0.500	0.322	0.864	0.432	0.280
Wt/ft		65.5	49.6	54.8	40.5	72.5	43.4	28.6	53.2	28.6	19.0
Effective length *KL* (ft) with respect to radius of gyration *r*	0	571	434	479	354	634	381	250	464	250	166
	6	563	428	469	347	612	369	242	436	236	158
	7	560	426	466	345	604	365	240	426	232	155
	8	557	424	462	342	596	360	237	415	226	151
	9	553	421	458	339	586	354	233	403	221	148
	10	549	418	453	335	575	348	229	390	214	144
	11	545	415	448	331	564	342	225	376	207	139
	12	540	411	442	327	551	335	221	361	200	135
	13	534	407	436	323	538	328	216	346	193	130
	14	529	403	429	318	524	320	211	330	185	125
	15	523	398	422	313	510	312	206	314	177	120
	16	517	394	415	308	495	303	201	298	169	114
	17	510	389	407	302	479	295	195	281	160	109
	18	503	384	400	297	463	286	190	264	152	104
	19	496	378	391	291	447	276	184	248	143	98.1
	20	488	372	383	284	430	267	178	232	135	92.7
	22	472	361	365	272	397	248	165	200	119	82.1
	24	455	348	347	258	363	228	153	170	103	71.8
	25	447	342	338	251	346	219	147	157	95.5	66.8
	26	438	335	328	245	329	209	140	145	88.3	61.9
	28	420	321	309	231	297	190	128	125	76.1	53.4
	30	401	307	290	216	265	171	116	109	66.3	46.5
	31	391	300	280	209	250	162	110	102	62.1	43.6
	32	382	293	270	202	235	153	104	95.8	58.3	40.9
	34	363	279	251	188	208	136	93.1	84.9	51.6	36.2
	36	343	264	232	174	186	121	83.0		46.1	32.3
	37	333	257	223	167	176	115	78.6			30.6
	38	324	249	214	161	166	109	74.5			
	40	305	235	196	148	150	98.3	67.2			
Properties											
A_g, in.^{2}		19.2	14.6	16.1	11.9	21.3	12.8	8.40	15.6	8.40	5.58
I, in.^{4}		362	279	212	161	162	106	72.5	66.3	40.5	28.1
r, in.		4.33	4.38	3.63	3.67	2.76	2.88	2.94	2.06	2.19	2.25

XS = x-strong.
XXS = xx-strong.
Note: Heavy line indicates Kl/r equal to or greater than 200.

F_y = 35 ksi

$\phi_c P_n = 0.85 F_{cr} A_g$

Table 4-8 (cont.). Steel Pipe Design Strength in Axial Compression, $\phi_c P_n$, kips

Shape		Pipe 5			Pipe 4			Pipe 3½		Pipe 3		
		XXS	XS	Std.	XXS	XS	Std.	XS	Std.	XXS	XS	Std.
Wall Thkns., in.		0.750	0.375	0.258	0.674	0.337	0.237	0.318	0.226	0.600	0.300	0.216
Wt/ft		38.6	20.8	14.6	27.6	15.0	10.8	12.5	9.12	18.6	10.3	7.58
Effective length *KL* (ft) with respect to radius of gyration *r*	0	336	182	128	241	131	94.3	109	79.7	163	89.8	66.3
	6	307	168	119	209	116	83.9	93.8	68.8	128	73.3	54.5
	7	298	163	115	199	111	80.5	88.7	65.2	117	68.0	50.7
	8	287	158	112	187	106	76.7	83.2	61.3	106	62.5	46.7
	9	275	152	108	175	99.9	72.6	77.3	57.2	94.7	56.8	42.6
	10	262	146	104	163	93.7	68.3	71.3	52.9	83.4	51.0	38.4
	11	249	140	99.4	150	87.3	63.8	65.1	48.5	72.5	45.2	34.2
	12	235	133	94.7	137	80.8	59.2	59.0	44.1	62.1	39.7	30.1
	13	221	126	89.9	124	74.3	54.6	53.0	39.8	52.9	34.4	26.3
	14	206	119	85.0	112	67.8	50.0	47.2	35.7	45.6	29.7	22.7
	15	192	111	80.0	99.6	61.5	45.6	41.6	31.7	39.7	25.8	19.8
	16	178	104	75.0	88.0	55.4	41.2	36.6	27.9	34.9	22.7	17.4
	17	164	96.9	70.0	77.9	49.5	37.1	32.4	24.7	30.9	20.1	15.4
	18	150	89.8	65.1	69.5	44.2	33.1	28.9	22.0		17.9	13.7
	19	137	82.8	60.3	62.4	39.6	29.7	25.9	19.8		16.1	12.3
	20	124	76.1	55.6	56.3	35.8	26.8	23.4	17.8			
	22	102	63.3	46.5	46.5	29.6	22.1		14.7			
	24	86.0	53.2	39.1		24.8	18.6					
	25	79.3	49.0	36.0			17.1					
	26	73.3	45.3	33.3								
	28	63.2	39.1	28.7								
	30		34.1	25.0								
	31			23.4								
	32											
Properties												
A_g, in.^{2}		11.3	6.11	4.30	8.10	4.41	3.17	3.68	2.68	5.47	3.02	2.23
I, in.^{4}		33.6	20.7	15.2	15.3	9.61	7.23	6.28	4.79	5.99	3.89	3.02
r, in.		1.72	1.84	1.88	1.37	1.48	1.51	1.31	1.34	1.05	1.14	1.16

XS = x-strong.

XXS = xx-strong.

Note: Heavy line indicates Kl/r equal to or greater than 200.

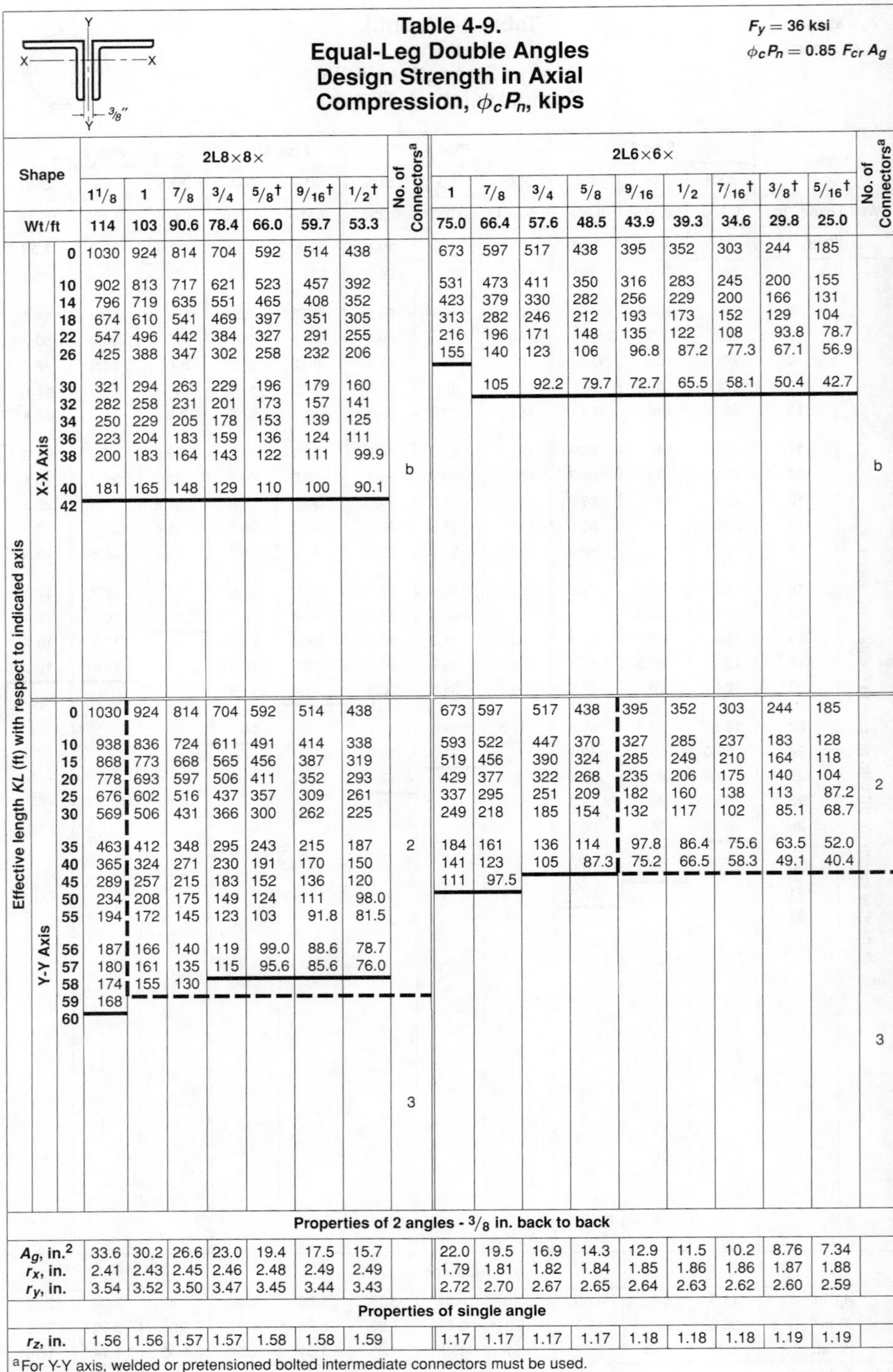

Table 4-9. Equal-Leg Double Angles Design Strength in Axial Compression, $\phi_c P_n$, kips

$F_y = 36$ ksi

$\phi_c P_n = 0.85\ F_{cr} A_g$

(Sketch: X–X and Y–Y axes, 3/8″ gap between the angles)

Shape		2L8×8×							No. of Connectors[a]	2L6×6×									No. of Connectors[a]
		1 1/8	1	7/8	3/4	5/8†	9/16†	1/2†		1	7/8	3/4	5/8	9/16	1/2	7/16†	3/8†	5/16†	
Wt/ft		114	103	90.6	78.4	66.0	59.7	53.3		75.0	66.4	57.6	48.5	43.9	39.3	34.6	29.8	25.0	
Effective length KL (ft) with respect to indicated axis — X-X Axis	0	1030	924	814	704	592	514	438		673	597	517	438	395	352	303	244	185	
	10	902	813	717	621	523	457	392		531	473	411	350	316	283	245	200	155	
	14	796	719	635	551	465	408	352		423	379	330	282	256	229	200	166	131	
	18	674	610	541	469	397	351	305		313	282	246	212	193	173	152	129	104	
	22	547	496	442	384	327	291	255		216	196	171	148	135	122	108	93.8	78.7	
	26	425	388	347	302	258	232	206		155	140	123	106	96.8	87.2	77.3	67.1	56.9	
	30	321	294	263	229	196	179	160			105	92.2	79.7	72.7	65.5	58.1	50.4	42.7	
	32	282	258	231	201	173	157	141											
	34	250	229	205	178	153	139	125											
	36	223	204	183	159	136	124	111											
	38	200	183	164	143	122	111	99.9	b										b
	40	181	165	148	129	110	100	90.1											
	42																		
Y-Y Axis	0	1030	924	814	704	592	514	438		673	597	517	438	395	352	303	244	185	
	10	938	836	724	611	491	414	338		593	522	447	370	327	285	237	183	128	
	15	868	773	668	565	456	387	319		519	456	390	324	285	249	210	164	118	
	20	778	693	597	506	411	352	293		429	377	322	268	235	206	175	140	104	2
	25	676	602	516	437	357	309	261		337	295	251	209	182	160	138	113	87.2	
	30	569	506	431	366	300	262	225		249	218	185	154	132	117	102	85.1	68.7	
	35	463	412	348	295	243	215	187	2	184	161	136	114	97.8	86.4	75.6	63.5	52.0	
	40	365	324	271	230	191	170	150		141	123	105	87.3	75.2	66.5	58.3	49.1	40.4	
	45	289	257	215	183	152	136	120		111	97.5								
	50	234	208	175	149	124	111	98.0											
	55	194	172	145	123	103	91.8	81.5											
	56	187	166	140	119	99.0	88.6	78.7											
	57	180	161	135	115	95.6	85.6	76.0											
	58	174	155	130															3
	59	168							3										
	60																		
Properties of 2 angles - 3/8 in. back to back																			
A_g, in.^{2}		33.6	30.2	26.6	23.0	19.4	17.5	15.7		22.0	19.5	16.9	14.3	12.9	11.5	10.2	8.76	7.34	
r_x, in.		2.41	2.43	2.45	2.46	2.48	2.49	2.49		1.79	1.81	1.82	1.84	1.85	1.86	1.86	1.87	1.88	
r_y, in.		3.54	3.52	3.50	3.47	3.45	3.44	3.43		2.72	2.70	2.67	2.65	2.64	2.63	2.62	2.60	2.59	
Properties of single angle																			
r_z, in.		1.56	1.56	1.57	1.57	1.58	1.58	1.59		1.17	1.17	1.17	1.17	1.18	1.18	1.18	1.19	1.19	

[a] For Y-Y axis, welded or pretensioned bolted intermediate connectors must be used.

[b] For required number of intermediate connectors, see discussion of Table 4-9.

† Section is a slender-element cross-section.

Note: Heavy line indicates Kl/r equal to or greater than 200.

$F_y = 36$ ksi

$\phi_c P_n = 0.85\ F_{cr} A_g$

Table 4-9 (cont.). Equal-Leg Double Angles Design Strength in Axial Compression, $\phi_c P_n$, kips

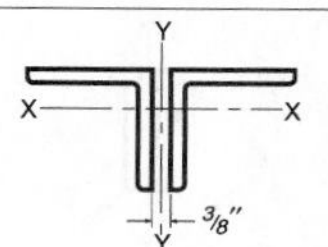

Shape		2L5×5× 7/8	3/4	5/8	1/2	7/16	3/8†	5/16†	No. of Connectors[a]	2L4×4× 3/4	5/8	1/2	7/16	3/8	5/16†	1/4†	No. of Connectors[a]
Wt/ft		54.6	47.5	40.1	32.6	28.7	24.8	20.9		37.0	31.3	25.5	22.5	19.4	16.3	13.2	
Effective length *KL* (ft) with respect to indicated axis																	
X-X Axis	0	490	428	361	293	258	219	171		334	282	229	202	175	146	108	
	2	483	423	356	289	255	217	169		326	276	224	198	171	144	106	
	4	464	406	343	278	245	209	163		306	259	211	186	161	135	101	
	6	433	379	321	261	230	196	154		274	233	190	168	146	123	92.0	
	8	393	345	293	238	210	180	143		235	201	165	146	127	107	81.3	
	10	348	306	260	212	188	161	129		194	166	137	122	106	89.6	69.3	
	12	299	264	225	184	163	140	114		152	132	109	97.1	84.9	72.2	57.1	
	14	251	221	190	155	138	120	98.0		115	100	82.9	74.4	65.3	55.8	45.4	
	16	204	181	156	128	114	99.2	82.6		87.8	76.8	63.5	56.9	50.0	42.7	35.0	
	18	162	144	125	103	91.5	80.2	68.1		69.4	60.7	50.1	45.0	39.5	33.8	27.7	
	20	132	117	101	83.1	74.1	65.0	55.3			49.1	40.6	36.4	32.0	27.3	22.4	
	22	109	96.4	83.5	68.7	61.3	53.7	45.7	b								b
	24	91.4	81.0	70.1	57.7	51.5	45.1	38.4									
	26							32.7									
Y-Y Axis	0	490	428	361	293	258	219	171		334	282	229	202	175	146	108	
	6	456	396	329	260	223	182	134		303	253	203	177	149	120	82.6	
	9	426	369	307	243	208	171	127		273	228	183	159	135	109	76.4	
	12	387	335	279	221	189	156	118		237	197	158	137	116	94.7	68.2	
	15	343	296	246	195	166	139	106		197	163	130	113	96.2	78.7	58.3	
	18	295	254	211	167	142	119	93.0		157	129	103	89.4	76.0	62.5	47.6	
	22	231	198	165	130	110	92.8	74.5	2	109	89.5	71.3	61.6	52.5	43.3	34.0	
	24	201	172	143	112	94.0	80.0	65.2		91.5	75.3	60.0	51.9	44.2	36.6	28.8	
	26	172	147	122	96.2	80.5	68.6	56.3		78.0	64.2	51.2	44.3	37.8	31.3	24.7	
	28	149	127	105	83.1	69.6	59.5	49.0		67.3	55.4	44.2	38.2	32.6	27.1	21.4	
	30	130	111	91.9	72.6	60.8	52.0	42.9		58.7	48.3	38.5	33.3	28.5			
	32	114	97.4	80.9	63.9	53.6	45.9	37.9									3
	34	101	86.3	71.7	56.7	47.5	40.7	33.7									
	36	90.1	77.0	64.0	50.6	42.5	36.4	30.2									
	38	80.9															
	40								3								
Properties of 2 angles - 3/8 in. back to back																	
A_g, in.^{2}		16.0	14.0	11.8	9.58	8.44	7.30	6.13		10.9	9.21	7.49	6.61	5.71	4.80	3.87	
r_x, in.		1.49	1.50	1.52	1.53	1.54	1.55	1.56		1.18	1.20	1.21	1.22	1.23	1.24	1.25	
r_y, in.		2.30	2.27	2.25	2.22	2.21	2.20	2.19		1.88	1.85	1.83	1.81	1.80	1.79	1.78	
Properties of single angle																	
r_z, in.		0.971	0.972	0.975	0.980	0.983	0.986	0.990		0.774	0.774	0.776	0.777	0.779	0.781	0.783	

[a]For Y-Y axis, welded or pretensioned bolted intermediate connectors must be used.

[b]For required number of intermediate connectors, see discussion of Table 4-9.

†Section is a slender-element cross-section.

Note: Heavy line indicates Kl/r equal to or greater than 200.

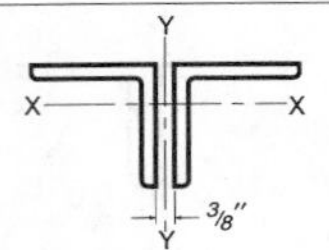

Table 4-9 (cont.). Equal-Leg Double Angles Design Strength in Axial Compression, $\phi_c P_n$, kips

$F_y = 36$ ksi

$\phi_c P_n = 0.85\ F_{cr}\ A_g$

Shape		2L3½×3½×					No. of Connectors[a]	2L3×3×						No. of Connectors[a]
		1/2	7/16	3/8	5/16	1/4†		1/2	7/16	3/8	5/16	1/4	3/16†	
Wt/ft		22.2	19.6	17.0	14.3	11.6		18.7	16.6	14.3	12.1	9.77	7.41	
Effective length *KL* (ft) with respect to indicated axis														
X-X Axis	0	200	177	153	129	101		168	149	129	109	87.8	60.8	
	1	198	175	152	128	100		167	147	128	108	87.0	60.3	
	2	194	172	149	126	98.2		162	143	124	105	84.8	58.9	
	3	188	166	144	122	95.3		155	137	119	100	81.1	56.6	
	4	179	158	138	116	91.2		145	128	112	94.1	76.2	53.5	
	5	168	149	130	110	86.3		133	118	103	86.8	70.4	49.8	
	6	156	138	121	102	80.7		120	106	92.9	78.6	63.9	45.7	
	7	143	127	111	93.7	74.5		106	94.3	82.5	69.9	56.9	41.2	
	8	129	115	100	85.0	67.9		91.8	82.0	71.9	61.1	49.9	36.6	
	9	114	102	89.5	76.1	61.2		78.2	70.0	61.5	52.4	42.9	32.0	
	10	100	89.9	78.9	67.3	54.4		65.3	58.7	51.7	44.2	36.3	27.5	
	12	74.1	66.7	58.9	50.5	41.5	b	45.3	40.8	36.0	30.8	25.3	19.5	b
	14	54.4	49.0	43.3	37.1	30.6		33.3	30.0	26.4	22.6	18.6	14.3	
	16	41.7	37.5	33.1	28.4	23.4								
	18				22.5	18.5								
	20													
Y-Y Axis	0	200	177	153	129	101		168	149	129	109	87.8	60.8	
	5	181	158	135	110	81.3		150	131	113	92.7	71.8	45.6	
	6	175	153	130	107	79.1		143	126	108	88.8	68.9	44.2	
	7	168	147	125	103	76.5		136	119	103	84.4	65.6	42.6	
	8	161	140	120	98.3	73.5		128	113	96.6	79.5	62.0	40.7	
	9	153	133	114	93.6	70.3		120	105	90.3	74.3	58.1	38.6	
	10	144	126	108	88.5	66.8		111	97.6	83.8	68.9	54.0	36.4	
	12	126	110	94.2	77.6	59.1		93.6	81.9	70.3	57.7	45.4	31.4	
	14	108	94.0	80.4	66.4	51.0		76.2	66.6	57.1	46.7	36.9	26.3	
	16	90.3	78.2	66.9	55.3	42.9		59.9	52.3	44.8	36.6	28.9	21.2	
	18	73.3	63.3	54.1	44.8	35.0		47.4	41.4	35.5	29.0	23.0	16.9	
	20	59.5	51.4	44.0	36.5	28.6	3	38.4	33.6	28.8	23.6	18.7	13.9	3
	22	49.2	42.6	36.4	30.3	23.8		31.8	27.8	23.8	19.5	15.5	11.5	
	24	41.4	35.8	30.7	25.5	20.1								
	26	35.3	30.5	26.2	21.8	17.1								
	28													
Properties of 2 angles - 3/8 in. back to back														
A_g, in.^{2}		6.53	5.77	5.00	4.21	3.41		5.50	4.86	4.22	3.55	2.87	2.18	
r_x, in.		1.05	1.06	1.07	1.08	1.09		0.895	0.903	0.910	0.918	0.926	0.933	
r_y, in.		1.63	1.61	1.60	1.59	1.57		1.43	1.42	1.41	1.39	1.38	1.37	
Properties of single angle														
r_z, in.		0.679	0.681	0.683	0.685	0.688		0.580	0.580	0.581	0.583	0.585	0.586	

[a] For Y-Y axis, welded or pretensioned bolted intermediate connectors must be used.
[b] For required number of intermediate connectors, see discussion of Table 4-9.
† Section is a slender-element cross-section.
Note: Heavy line indicates Kl/r equal to or greater than 200.

$F_y = 36$ ksi

$\phi_c P_n = 0.85\ F_{cr}\ A_g$

Table 4-9 (cont.). Equal-Leg Double Angles Design Strength in Axial Compression, $\phi_c P_n$, kips

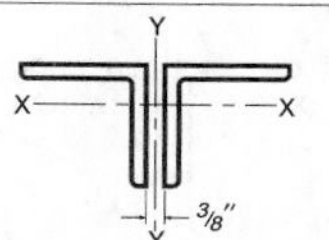

Shape		2L2½×2½×					No. of Connectors[a]	2L2×2×					No. of Connectors[a]
		1/2	3/8	5/16	1/4	3/16†		3/8	5/16	1/4	3/16	1/8†	
Wt/ft		**15.3**	**11.8**	**9.96**	**8.07**	**6.13**		**9.30**	**7.89**	**6.43**	**4.91**	**3.34**	
Effective length KL (ft) with respect to indicated axis: X-X Axis	0	138	106	89.7	72.5	54.1		83.5	71.0	57.8	44.1	27.4	
	1	136	105	88.5	71.6	53.4		81.7	69.5	56.6	43.2	26.9	
	2	130	101	85.0	68.9	51.4		76.6	65.2	53.2	40.6	25.5	
	3	121	94.0	79.6	64.5	48.3		68.7	58.7	48.0	36.7	23.3	
	4	110	85.5	72.5	58.9	44.3		59.0	50.6	41.5	31.9	20.5	
	5	97.0	75.7	64.4	52.4	39.5		48.6	41.8	34.5	26.6	17.5	
	6	83.1	65.3	55.6	45.4	34.5		38.2	33.1	27.4	21.3	14.3	
	7	69.2	54.8	46.8	38.4	29.3		28.8	25.1	20.9	16.3	11.4	
	8	56.1	44.7	38.4	31.6	24.3		22.1	19.2	16.0	12.5	8.74	
	9	44.5	35.6	30.6	25.3	19.6		17.4	15.2	12.7	9.87	6.90	
	10	36.0	28.8	24.8	20.5	15.9				10.3	7.99	5.59	
	11	29.8	23.8	20.5	16.9	13.1	b						b
	12	25.0	20.0	17.2	14.2	11.0							
	13												
Y-Y Axis	0	138	106	89.7	72.5	54.1		83.5	71.0	57.8	44.1	27.4	
	3	130	98.6	82.1	64.7	45.0		76.6	64.5	51.8	38.0	21.2	
	4	125	94.9	79.0	62.3	43.5		72.4	61.0	48.9	36.0	20.3	
	5	119	90.4	75.2	59.3	41.7		67.3	56.6	45.4	33.4	19.2	
	6	112	85.1	70.7	55.9	39.5		61.6	51.8	41.5	30.6	17.9	
	7	104	79.2	65.8	52.0	37.0		55.5	46.5	37.2	27.5	16.4	
	8	96.2	73.0	60.5	47.9	34.3		49.1	41.1	32.9	24.3	14.8	
	9	87.7	66.4	55.0	43.6	31.4		42.8	35.8	28.5	21.0	13.1	
	10	79.1	59.8	49.4	39.2	28.4		36.7	30.6	24.3	17.9	11.3	
	11	70.6	53.3	43.9	34.8	25.4		30.9	25.6	20.4	15.0	9.66	
	12	62.3	46.9	38.5	30.6	22.4		26.0	21.6	17.1	12.6	8.20	
	14	46.9	35.2	28.8	22.9	16.9	3	19.1	15.9	12.6	9.34	6.10	3
	16	36.0	27.0	22.1	17.6	13.1		14.6	12.2	9.68	7.17		
	18	28.4	21.4	17.5	13.9	10.4							
	20	23.0	17.3										
	22												
Properties of 2 angles - 3/8 in. back to back													
A_g, in.^{2}		4.50	3.47	2.93	2.37	1.80		2.73	2.32	1.89	1.44	0.982	
r_x, in.		0.735	0.749	0.756	0.764	0.771		0.591	0.598	0.605	0.612	0.620	
r_y, in.		1.23	1.21	1.19	1.18	1.17		1.01	0.996	0.982	0.967	0.951	
Properties of single angle													
r_z, in.		0.481	0.481	0.481	0.482	0.482		0.386	0.386	0.387	0.389	0.391	

[a] For Y-Y axis, welded or pretensioned bolted intermediate connectors must be used.

[b] For required number of intermediate connectors, see discussion of Table 4-9.

† Section is a slender-element cross-section.

Note: Heavy line indicates Kl/r equal to or greater than 200.

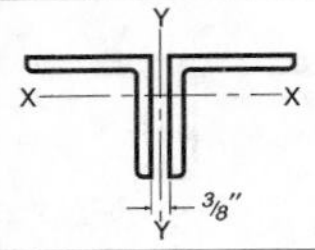

Table 4-10. LLBB Double Angles Design Strength in Axial Compression, $\phi_c P_n$, kips

$F_y = 36$ ksi

$\phi_c P_n = 0.85\ F_{cr} A_g$

Shape	2L8×6×							No. of Connectors[a]	2L8×4×							No. of Connectors[a]
	1	7/8	3/4	5/8†	9/16†	1/2†	7/16†		1	7/8	3/4	5/8†	9/16†	1/2†	7/16†	
Wt/ft	88.8	78.5	68.0	57.3	51.8	46.3	40.7		75.2	66.6	57.8	48.7	44.1	39.5	34.8	
Effective length KL (ft) with respect to indicated axis — X-X Axis																
0	799	707	612	513	446	379	312		676	600	520	436	381	323	265	
4	783	693	600	503	438	373	307		663	589	511	428	375	318	261	
6	764	677	586	491	428	365	301		648	575	499	419	367	312	256	
8	739	654	567	476	415	354	293		626	556	483	405	356	303	249	
10	707	626	543	456	399	341	283		600	533	463	389	342	292	241	
12	670	594	515	433	380	325	271		569	506	440	370	326	278	231	
14	628	557	484	407	358	308	257		534	476	414	348	307	264	220	
16	584	518	451	380	335	289	243		497	443	386	325	288	248	207	
18	537	477	416	351	311	269	227		458	409	357	300	267	231	194	
20	490	435	380	321	285	248	211		418	373	326	275	246	214	181	
22	442	393	343	291	260	227	194		378	338	296	250	224	196	167	
24	395	351	308	261	234	206	177	b	338	303	266	225	202	178	153	b
26	349	311	273	232	209	185	161		300	269	237	200	181	160	139	
28	306	273	240	205	186	165	144		263	237	209	177	161	143	125	
30	266	238	209	178	163	146	129		229	207	182	154	141	127	112	
32	234	209	184	157	143	128	114		201	182	160	136	124	112	99.0	
36	185	165	145	124	113	101	89.9		159	143	126	107	98.2	88.3	78.2	
38	166	148	130	111	101	90.7	80.7		143	129	113	96.2	88.1	79.2	70.2	
40	150	134	118	100	91.5	81.9	72.8		129	116	102	86.8	79.5	71.5	63.4	
42			107	91.0	83.0	74.3	66.1			105	92.9	78.7	72.1	64.9	57.5	
44																
Effective length KL (ft) with respect to indicated axis — Y-Y Axis																
0	799	707	612	513	446	379	312		676	600	520	436	381	323	265	
4	745	649	546	435	367	299	232		603	523	439	349	295	240	187	
6	727	633	533	426	359	293	228		566	491	413	329	279	228	179	
8	702	611	515	412	349	286	223		518	449	378	302	258	213	168	
10	671	584	493	395	336	276	217		462	400	337	270	233	193	155	
12	635	553	467	375	320	264	209		402	347	293	235	204	172	139	
14	594	518	438	353	302	251	200		340	293	247	199	174	148	122	2
16	551	481	406	328	282	236	190		280	240	203	163	144	124	104	
18	505	441	372	302	261	220	178		236	201	163	131	117	101	86.2	
20	458	400	338	275	239	203	166		192	164	133	107	95.6	83.5	71.4	
22	412	359	304	248	216	185	153		159	136	110	89.1	79.7	69.8	59.9	
24	366	319	269	220	194	167	139	2	134	115	93.0	75.3	67.5	59.2	50.9	
26	321	281	237	194	172	149	126		114	98.0						
28	279	244	205	168	150	131	113									
30	244	213	180	148	132	116	99.4									
32	215	187	158	130	116	102	88.3									
36	170	149	126	104	92.8	81.9	70.9									
38	153	134	113	93.4	83.6	73.8	64.0									
40	138	121	102	84.4	75.6	66.9	58.1									3
42	125															
44																
Properties of 2 angles - 3/8 in. back to back																
A_g, in.^{2}	26.1	23.1	20.0	16.8	15.2	13.6	12.0		22.1	19.6	17.0	14.3	13.0	11.6	10.2	
r_x, in.	2.49	2.50	2.52	2.54	2.55	2.55	2.56		2.51	2.53	2.55	2.56	2.57	2.58	2.59	
r_y, in.	2.52	2.50	2.47	2.45	2.44	2.43	2.42		1.60	1.57	1.55	1.52	1.51	1.50	1.49	
Properties of single angle																
r_z, in.	1.28	1.28	1.29	1.29	1.30	1.30	1.31		0.844	0.846	0.850	0.856	0.859	0.863	0.867	

[a] For Y-Y axis, welded or pretensioned bolted intermediate connectors must be used.
[b] For required number of intermediate connectors, see discussion of Table 4-10.
† Section is a slender-element cross-section.
Note: Heavy line indicates Kl/r equal to or greater than 200.

$F_y = 36$ ksi

$\phi_c P_n = 0.85\ F_{cr} A_g$

Table 4-10 (cont.). LLBB Double Angles Design Strength in Axial Compression, $\phi_c P_n$, kips

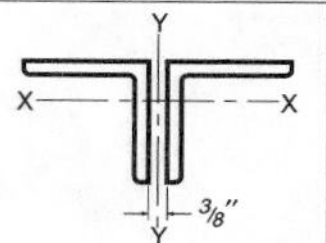

Shape		2L7×4×					No. of Connectors[a]	2L6×4×								No. of Connectors[a]
		3/4	5/8	1/2†	7/16†	3/8†		7/8	3/4	5/8	9/16	1/2	7/16†	3/8†	5/16†	
Wt/ft		52.4	44.2	35.8	31.5	27.2		54.3	47.2	39.9	36.1	32.3	28.5	24.6	20.6	
Effective length KL (ft) with respect to indicated axis — X-X Axis	0	471	398	310	258	205		490	425	358	324	291	249	201	153	
	10	403	342	268	226	182		393	343	290	263	236	204	167	129	
	12	377	319	252	213	172		357	312	264	240	216	187	154	120	
	14	348	295	234	198	161		319	279	236	215	193	168	140	110	
	16	317	269	214	183	150		279	246	208	189	171	149	125	99.9	
	18	285	243	194	167	138		241	212	180	164	148	130	110	89.2	
	20	253	216	174	150	125		204	180	153	140	127	112	95.9	78.6	
	22	222	190	154	134	113		169	150	128	117	106	94.3	82.1	68.3	
	24	193	165	135	119	101		142	126	108	98.4	89.2	79.3	69.2	58.6	
	26	165	142	117	104	89.2		121	108	91.6	83.9	76.0	67.5	58.9	49.9	
	28	142	122	100	89.5	77.9		105	92.8	79.0	72.3	65.5	58.2	50.8	43.0	
	30	124	106	87.5	77.9	67.9	b	91.1	80.9	68.8	63.0	57.1	50.7	44.3	37.5	b
	32	109	93.5	76.9	68.5	59.6							44.6	38.9	32.9	
	34	96.4	82.9	68.1	60.7	52.8										
	36	86.0	73.9	60.8	54.1	47.1										
	38															
Y-Y Axis	0	471	398	310	258	205		490	425	358	324	291	249	201	153	
	2	429	348	253	201	149		466	399	327	290	252	207	159	111	
	4	414	336	245	195	146		451	385	316	281	244	201	155	109	
	6	390	317	233	186	140		426	364	299	266	231	191	148	105	
	8	358	292	216	174	133		393	336	276	246	215	178	139	99.9	
	10	321	262	195	159	123		355	303	249	222	194	162	128	93.4	
	12	281	229	173	142	111		313	267	220	196	172	144	115	85.6	2
	14	239	195	148	124	98.7		270	230	189	169	148	125	101.0	76.8	
	16	198	161	124	105	85.3		227	193	159	142	125	106	87.1	67.5	
	18	160	130	101	86.7	72.0		196	158	130	116	102	86.9	72.9	57.9	
	20	130	106	82.8	71.4	59.6		160	134	106	94.6	83.3	71.2	60.0	48.4	
	22	108	88.2	69.0	59.7	50.0	2	132	111	87.6	78.6	69.3	59.4	50.2	40.7	
	24	91.0	74.4	58.4	50.6	42.6		111	93.8	73.9	66.3	58.5	50.2	42.5	34.7	
	26	77.8	63.6	50.0				94.9	80.0	63.1	56.6	50.0	43.0	36.5	29.8	
	28							81.9	69.1							
	30															3
Properties of 2 angles - 3/8 in. back to back																
A_g, in.²		15.4	13.0	10.5	9.27	8.00		16.0	13.9	11.7	10.6	9.50	8.36	7.22	6.05	
r_x, in.		2.21	2.23	2.25	2.26	2.27		1.86	1.88	1.89	1.90	1.91	1.92	1.93	1.94	
r_y, in.		1.61	1.58	1.56	1.55	1.54		1.71	1.68	1.66	1.65	1.64	1.62	1.61	1.60	
Properties of single angle																
r_z, in.		0.855	0.860	0.866	0.869	0.873		0.854	0.856	0.859	0.861	0.864	0.867	0.870	0.874	

[a]For Y-Y axis, welded or pretensioned bolted intermediate connectors must be used.
[b]For required number of intermediate connectors, see discussion of Table 4-10.
†Section is a slender-element cross-section.
Note: Heavy line indicates Kl/r equal to or greater than 200.

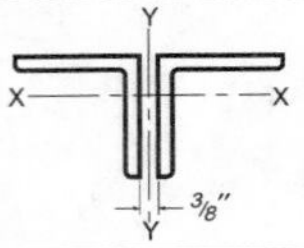

Table 4-10 (cont.). LLBB Double Angles Design Strength in Axial Compression, $\phi_c P_n$, kips

$F_y = 36$ ksi

$\phi_c P_n = 0.85\, F_{cr} A_g$

(Figure: double angles back to back, X-X and Y-Y axes, 3/8″ gap)

Shape		2L6×3½× 1/2	2L6×3½× 3/8†	2L6×3½× 5/16†	No. of Connectors[a]	2L5×3½× 3/4	2L5×3½× 5/8	2L5×3½× 1/2	2L5×3½× 3/8†	2L5×3½× 5/16†	2L5×3½× 1/4†	No. of Connectors[a]	2L5×3× 1/2	2L5×3× 7/16	2L5×3× 3/8†	2L5×3× 5/16†	2L5×3× 1/4†	No. of Connectors[a]
Wt/ft		**30.7**	**23.4**	**19.7**		**39.6**	**33.5**	**27.2**	**20.8**	**17.4**	**14.1**		**25.5**	**22.5**	**19.5**	**16.4**	**13.2**	
Effective length *KL* (ft) with respect to indicated axis — X-X Axis	0	277	192	146		355	301	245	183	143	102		230	203	172	134	95.5	
	6	257	180	138		317	269	220	165	130	93.4		206	182	155	122	87.8	
	8	243	170	131		290	247	202	152	120	87.4		189	167	143	113	82.3	
	10	225	159	124		259	221	181	137	109	80.3		170	150	129	103	75.7	
	12	206	147	115		225	192	158	120	96.8	72.4		148	132	113	91.4	68.3	
	14	185	133	105		191	164	135	103	84.1	64.1		127	113	97.4	79.5	60.6	
	16	163	119	95.4		158	136	113	86.3	71.5	55.7		106	94.0	81.8	67.8	52.7	
	18	142	105	85.2		127	110	91.4	70.5	59.6	47.4		85.7	76.5	67.1	56.6	45.0	
	20	122	91.4	75.1		103	88.8	74.1	57.1	48.6	39.7		69.4	62.0	54.3	46.2	37.7	
	22	102	78.2	65.3		85.3	73.4	61.2	47.2	40.1	32.8		57.4	51.2	44.9	38.2	31.2	
	24	85.7	65.9	56.0		71.7	61.7	51.4	39.7	33.7	27.5		48.2	43.1	37.7	32.1	26.2	
	26	73.0	56.2	47.7	b		52.5	43.8	33.8	28.7	23.5	b	41.1	36.7	32.2	27.3	22.3	b
	28	63.0	48.4	41.1														
	30	54.9	42.2	35.8														
	32	48.2	37.1	31.5														
	34																	
Y-Y Axis	0	277	192	146		355	301	245	183	143	102		230	203	172	134	95.5	
	2	237	149	105		338	282	222	154	113	73.1		205	176	143	104	67.7	
	4	227	144	102		324	271	213	149	109	71.3		193	166	135	99.6	65.2	
	6	211	135	96.9		302	252	199	140	104	68.4		175	150	123	92.0	61.3	
	8	190	124	90.3		273	228	180	127	95.6	64.3		152	131	108	81.9	55.8	
	10	165	111	82.0		240	200	158	113	85.9	59.1		126	109	90.5	70.1	49.2	2
	12	139	95.6	72.5		205	170	135	97.1	75.0	53.1	2	100	86.4	72.5	57.5	41.7	
	14	113	79.9	62.1		170	141	111	81.0	63.6	46.4		79.4	65.7	55.5	45.1	33.9	
	16	89.2	64.4	51.6		144	112	89.0	65.2	52.2	39.4		61.3	50.8	43.1	35.2	26.8	
	18	71.2	51.8	41.9		114	93.5	70.8	52.1	42.0	32.4		48.7	40.4	34.3	28.1	21.6	
	20	58.0	42.5	34.6		92.7	76.0	57.6	42.6	34.4	26.8		39.6	32.9	28.0	23.0		
	22	48.2	35.4	28.9	2	76.8	62.9	47.7	35.4	28.7	22.5							
	24					64.6	53.0	40.2	29.9	24.3								
	26																	
												3						3
Properties of 2 angles - 3/8 in. back to back																		
A_g, in.²		9.04	6.88	5.78		11.6	9.85	8.01	6.10	5.12	4.13		7.51	6.62	5.73	4.81	3.88	
r_x, in.		1.92	1.93	1.94		1.55	1.56	1.58	1.59	1.60	1.61		1.58	1.59	1.60	1.61	1.62	
r_y, in.		1.40	1.38	1.37		1.53	1.50	1.48	1.46	1.44	1.43		1.24	1.23	1.22	1.21	1.19	
Properties of single angle																		
r_z, in.		0.756	0.763	0.767		0.744	0.746	0.750	0.755	0.758	0.761		0.642	0.644	0.646	0.649	0.652	

[a] For Y-Y axis, welded or pretensioned bolted intermediate connectors must be used.
[b] For required number of intermediate connectors, see discussion of Table 4-10.
† Section is a slender-element cross-section.
Note: Heavy line indicates Kl/r equal to or greater than 200.

$F_y = 36$ ksi

$\phi_c P_n = 0.85\ F_{cr} A_g$

Table 4-10 (cont.). LLBB Double Angles Design Strength in Axial Compression, $\phi_c P_n$, kips

Axis	Shape	2L4×3½×				No. of Connectors[a]	2L4×3×					No. of Connectors[a]	2L3½×3×					No. of Connectors[a]
		½	⅜	5/16†	¼†		⅝	½	⅜	5/16†	¼†		½	7/16	⅜	5/16	¼†	
	Wt/ft	23.8	18.2	15.3	12.4		27.1	22.1	16.9	14.2	11.5		20.6	18.2	15.8	13.3	10.8	
Effective length KL (ft) with respect to indicated axis																		
X-X Axis	0	215	164	137	101		244	199	152	128	94.2		185	163	142	120	93.3	
	2	210	161	135	99.4		239	195	150	125	92.6		180	159	138	117	91.1	
	4	198	151	127	94.4		225	184	141	119	88.0		166	147	128	108	84.7	
	6	179	137	115	86.5		204	167	128	108	80.8		146	129	113	95.1	75.1	
	8	156	120	101	76.6		177	145	112	94.7	71.6		121	108	94.2	79.5	63.4	
	10	130	101	84.6	65.5		148	122	94.5	80.0	61.4		95.3	85.3	74.9	63.2	51.0	
	11	117	91.0	76.5	59.8		133	110	85.5	72.5	56.1		82.9	74.4	65.5	55.3	44.9	
	12	104	81.4	68.4	54.1		119	97.9	76.6	65.1	50.9		71.2	64.1	56.5	47.7	39.1	
	13	92.0	72.1	60.6	48.5		105	86.6	68.0	57.9	45.7		60.6	54.6	48.2	40.7	33.5	
	14	80.2	63.2	53.2	43.1		91.3	75.7	59.8	51.0	40.7		52.3	47.1	41.6	35.1	28.9	
	15	69.8	55.0	46.3	38.0		79.5	65.9	52.1	44.5	35.9		45.5	41.0	36.2	30.6	25.2	
	16	61.4	48.4	40.7	33.4	b	69.9	57.9	45.8	39.1	31.6	b	40.0	36.0	31.8	26.9	22.1	b
	18	48.5	38.2	32.2	26.4		55.2	45.8	36.2	30.9	24.9			28.5	25.2	21.2	17.5	
	20	39.3	31.0	26.0	21.3		44.7	37.1	29.3	25.0	20.2							
	22																	
Y-Y Axis	0	215	164	137	101		244	199	152	128	94.2		185	163	142	120	93.3	
	2	203	149	119	81.8		233	187	137	110	75.4		176	154	131	107	78.8	
	4	195	143	115	79.4		220	176	130	104	72.2		167	146	124	102	75.3	
	6	183	134	108	75.6		201	161	118	95.5	67.1		153	134	114	93.5	69.8	
	8	167	123	99.2	70.4		176	141	104	84.4	60.4		135	118	101	82.8	62.5	
	10	148	109	88.5	64.0		149	119	87.9	71.7	52.4	2	115	101	85.9	70.6	53.9	
	12	128	94.5	76.8	56.6	2	122	96.3	71.4	58.5	43.7		94.3	82.5	70.5	57.9	44.8	2
	14	107	79.5	64.8	48.8		100	78.5	55.5	45.7	35.0		74.6	65.2	55.7	45.7	35.8	
	16	87.6	65.0	53.0	40.9		77.3	60.5	42.8	35.3	27.3		60.5	52.8	45.0	35.3	27.9	
	18	73.2	51.9	42.4	33.2		61.2	47.9	34.0	28.1	21.8		47.9	41.8	35.7	28.0	22.2	
	20	59.4	42.2	34.6	27.2		49.6	38.9	27.7	22.9	17.8		38.9	33.9	29.0	22.8	18.1	
	22	49.2	35.0	28.7	22.7		41.0	32.2					32.2	28.1	24.0	18.9	15.0	
	24	41.4	29.5	24.2	19.2													
	26	35.3																
	28					3						3						3
Properties of 2 angles - ⅜ in. back to back																		
	A_g, in.^{2}	7.01	5.35	4.50	3.63		7.98	6.51	4.98	4.19	3.38		6.04	5.34	4.63	3.91	3.16	
	r_x, in.	1.23	1.25	1.25	1.26		1.23	1.24	1.26	1.27	1.27		1.07	1.08	1.09	1.09	1.10	
	r_y, in.	1.57	1.55	1.53	1.52		1.35	1.32	1.30	1.29	1.27		1.37	1.36	1.35	1.33	1.32	
Properties of single angle																		
	r_z, in.	0.716	0.719	0.721	0.723		0.631	0.633	0.636	0.638	0.639		0.618	0.620	0.622	0.624	0.628	

[a] For Y-Y axis, welded or pretensioned bolted intermediate connectors must be used.

[b] For required number of intermediate connectors, see discussion of Table 4-10.

† Section is a slender-element cross-section.

Note: Heavy line indicates Kl/r equal to or greater than 200.

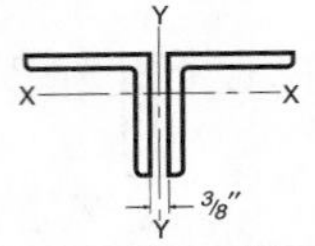

Table 4-10 (cont.). LLBB Double Angles Design Strength in Axial Compression, $\phi_c P_n$, kips

$F_y = 36$ ksi

$\phi_c P_n = 0.85\ F_{cr} A_g$

Shape		2L3½×2½× 1/2	2L3½×2½× 3/8	2L3½×2½× 5/16	2L3½×2½× 1/4†	No. of Connectors[a]	2L3×2½× 1/2	2L3×2½× 7/16	2L3×2½× 3/8	2L3×2½× 5/16	2L3×2½× 1/4	2L3×2½× 3/16†	No. of Connectors[a]
Wt/ft		**18.8**	**14.5**	**12.2**	**9.88**		**17.1**	**15.1**	**13.1**	**11.1**	**8.97**	**6.82**	
X-X Axis	**0**	169	130	110	85.6		153	136	118	99.5	80.8	55.8	
	2	165	127	107	83.7		148	131	114	96.0	78.1	54.1	
	3	160	123	104	81.3		141	125	109	91.9	74.8	52.0	
	4	153	118	99.3	78.0		132	118	102	86.5	70.4	49.3	
	5	144	111	93.9	74.0		122	108	94.6	80.0	65.2	46.0	
	6	134	104	87.8	69.4		110	98.2	85.8	72.6	59.3	42.3	
	7	123	95.7	81.0	64.3		97.9	87.3	76.4	64.8	53.1	38.2	
	8	112	87.1	73.9	59.0		85.3	76.3	66.9	56.9	46.7	34.1	
	9	100	78.3	66.6	53.4		73.0	65.5	57.5	49.0	40.3	29.9	
	10	88.3	69.5	59.2	47.8		61.4	55.2	48.6	41.6	34.3	25.8	
	12	66.4	52.8	45.2	37.0		42.7	38.4	33.9	29.0	24.0	18.5	
	14	48.8	38.9	33.3	27.5	b	31.4	28.2	24.9	21.3	17.6	13.6	b
	16	37.3	29.8	25.5	21.1								
	18	29.5	23.5	20.2	16.6								
	20												
Y-Y Axis	**0**	169	130	110	85.6		153	136	118	99.5	80.8	55.8	
	2	159	118	96.2	70.7		146	129	110	90.9	70.5	44.8	
	3	154	115	93.2	68.7		142	125	107	88.2	68.5	43.7	
	4	147	109	89.1	66.1		137	119	102	84.5	65.7	42.3	
	5	138	103	84.1	62.7		130	113	96.8	79.9	62.3	40.5	
	6	129	96.1	78.4	58.8		122	105	90.4	74.7	58.3	38.4	
	7	118	88.3	72.0	54.4	2	113	97.0	83.3	68.9	53.9	35.9	2
	8	107	79.9	65.2	49.7		103	88.2	75.8	62.7	49.1	33.2	
	9	95.5	71.4	58.2	44.7		93.7	79.2	68.1	56.4	44.1	30.4	
	10	84.2	62.9	51.2	39.6		83.8	70.3	60.3	50.0	39.2	27.4	
	12	66.2	49.0	37.9	29.8		64.8	55.9	47.8	37.7	29.5	21.4	
	14	48.8	36.3	29.4	22.2		48.2	41.4	35.4	29.3	22.9	16.1	
	16	37.5	27.9	22.7	17.2		37.0	31.8	27.2	22.5	17.6	12.4	
	18	29.7	22.1	18.0	13.7		29.2	25.2	21.5	17.9	14.0	9.91	
	20												
						3							3

Effective length KL (ft) with respect to indicated axis

Properties of 2 angles - 3/8 in. back to back

	2L3½×2½× 1/2	3/8	5/16	1/4†	2L3×2½× 1/2	7/16	3/8	5/16	1/4	3/16†
A_g, in.^{2}	5.53	4.25	3.58	2.90	5.01	4.44	3.86	3.25	2.64	2.00
r_x, in.	1.08	1.10	1.11	1.12	0.910	0.917	0.924	0.932	0.940	0.947
r_y, in.	1.13	1.11	1.09	1.08	1.18	1.16	1.15	1.14	1.12	1.11

Properties of single angle

	2L3½×2½× 1/2	3/8	5/16	1/4†	2L3×2½× 1/2	7/16	3/8	5/16	1/4	3/16†
r_z, in.	0.532	0.535	0.538	0.541	0.516	0.516	0.517	0.518	0.520	0.521

[a] For Y-Y axis, welded or pretensioned bolted intermediate connectors must be used.
[b] For required number of intermediate connectors, see discussion of Table 4-10.
† Section is a slender-element cross-section.
Note: Heavy line indicates Kl/r equal to or greater than 200.

$F_y = 36$ ksi

$\phi_c P_n = 0.85\ F_{cr} A_g$

Table 4-10 (cont.). LLBB Double Angles Design Strength in Axial Compression, $\phi_c P_n$, kips

	Shape	2L3×2×					No. of Connectors[a]	2L2½×2×				No. of Connectors[a]
		1/2	3/8	5/16	1/4	3/16†		3/8	5/16	1/4	3/16†	
	Wt/ft	15.4	11.9	10.1	8.18	6.24		10.6	8.97	7.30	5.57	
X-X Axis	0	139	107	90.6	73.4	51.0		95.2	80.8	65.5	49.3	
	2	134	103	87.6	71.0	49.5		90.4	76.8	62.3	47.0	
	3	128	99.1	83.9	68.1	47.7		84.7	72.1	58.6	44.3	
	4	120	93.3	79.1	64.3	45.3		77.4	66.0	53.7	40.7	
	5	111	86.3	73.3	59.6	42.3		68.9	58.9	48.0	36.6	
	6	101	78.5	66.7	54.4	39.0		59.8	51.2	41.9	32.1	
	7	89.5	70.2	59.8	48.8	35.4		50.5	43.5	35.7	27.5	
	8	78.3	61.6	52.6	43.0	31.6		41.6	35.9	29.6	23.0	
	9	67.3	53.2	45.5	37.4	27.8		33.4	28.9	23.9	18.7	
	10	56.8	45.2	38.8	31.9	24.2		27.0	23.4	19.4	15.2	
	11	47.2	37.6	32.4	26.7	20.6		22.3	19.4	16.0	12.5	
	12	39.6	31.6	27.2	22.4	17.4	b	18.8	16.3	13.5	10.5	b
	13	33.8	26.9	23.2	19.1	14.8				11.5	8.97	
	14	29.1	23.2	20.0	16.5	12.8						
	16					9.78						
	18											
Y-Y Axis	0	139	107	90.6	73.4	51.0		95.2	80.8	65.5	49.3	
	2	130	97.8	80.7	62.4	40.0		89.2	74.4	58.6	41.1	
	3	123	93.1	76.8	59.5	38.5		85.4	71.0	55.9	39.4	
	4	115	86.7	71.6	55.6	36.4		80.3	66.5	52.4	37.2	
	5	105	79.2	65.4	50.9	33.8		74.2	61.1	48.2	34.4	2
	6	94.5	70.8	58.5	45.7	30.8	2	67.3	55.1	43.5	31.3	
	7	83.0	62.1	51.2	40.1	27.5		60.0	48.7	38.5	27.9	
	8	71.4	53.2	43.9	34.4	24.0		52.5	42.2	33.4	24.3	
	9	63.6	46.9	36.8	28.8	20.6		45.2	35.9	28.4	20.8	
	10	53.3	38.9	31.9	24.9	17.2		38.1	31.5	24.8	18.2	
	11	44.1	32.3	26.5	20.7	14.4		31.7	26.2	20.6	15.2	
	12	37.1	27.2	22.3	17.5	12.2		26.6	22.0	17.4	12.8	
	14	27.3	20.0	16.5	12.9	9.49		19.6	16.2	12.8	9.51	
	16						3					3
Properties of 2 angles - 3/8 in. back to back												
A_g, in.^{2}		4.53	3.50	2.96	2.40	1.83		3.11	2.64	2.14	1.64	
r_x, in.		0.922	0.937	0.945	0.953	0.961		0.766	0.774	0.782	0.790	
r_y, in.		0.940	0.911	0.897	0.883	0.869		0.957	0.943	0.930	0.916	
Properties of single angle												
r_z, in.		0.425	0.426	0.428	0.431	0.435		0.419	0.420	0.423	0.426	

Effective length KL (ft) with respect to indicated axis

[a]For Y-Y axis, welded or pretensioned bolted intermediate connectors must be used.

[b]For required number of intermediate connectors, see discussion of Table 4-10.

†Section is a slender-element cross-section.

Note: Heavy line indicates Kl/r equal to or greater than 200.

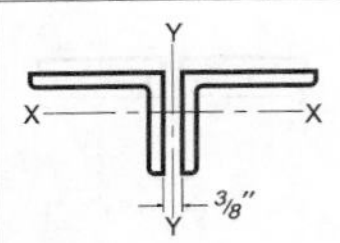

Table 4-11. SLBB Double Angles Design Strength in Axial Compression, $\phi_c P_n$, kips

$F_y = 36$ ksi

$\phi_c P_n = 0.85\ F_{cr}\ A_g$

Shape		2L8×6×							No. of Connectors[a]	2L8×4×							No. of Connectors[a]
		1	7/8	3/4	5/8†	9/16†	1/2†	7/16†		1	7/8	3/4	5/8†	9/16†	1/2†	7/16†	
Wt/ft		88.8	78.5	68.0	57.3	51.8	46.3	40.7		75.2	66.6	57.8	48.7	44.1	39.5	34.8	
Effective length KL (ft) with respect to indicated axis — X-X Axis	0	799	707	612	513	446	379	312		676	600	520	436	381	323	265	
	3	780	691	599	502	437	372	307		634	563	489	411	360	307	253	
	4	767	679	588	493	430	366	302		603	536	466	392	345	294	243	
	5	749	664	575	483	421	359	297		566	503	438	369	325	279	232	
	6	728	646	560	470	411	351	291		523	466	406	342	304	261	218	
	7	704	625	542	455	399	341	283		476	425	371	314	279	242	203	
	8	678	602	522	439	385	330	275		428	383	335	284	254	221	187	
	9	649	577	501	422	370	318	266		379	340	298	253	228	200	171	
	10	618	550	478	403	355	306	256		331	298	262	223	202	179	154	
	11	586	522	454	383	338	292	245		285	257	226	193	177	158	138	
	12	552	493	428	362	321	278	234		241	218	193	165	153	138	121	
	14	483	433	377	319	284	248	211	b	177	160	142	121	113	102	91.6	b
	16	414	372	325	276	248	218	188		136	123	108	93.0	86.1	78.3	70.1	
	18	348	314	274	235	212	189	164							61.9	55.4	
	20	286	259	227	195	178	160	141									
	22	236	214	188	161	147	133	119									
	24	199	180	158	135	124	112	100									
	26	169	153	134	115	106	95.5	85.2									
	28	146	132	116	99.5	91.0	82.4	73.5									
	30							64.0									
	32																
Y-Y Axis	0	799	707	612	513	446	379	312		676	600	520	436	381	323	265	
	4	780	686	587	481	412	341	269		669	592	512	427	372	312	252	
	6	771	678	581	476	408	338	267		663	587	507	423	368	310	250	
	8	759	668	572	469	402	334	264		654	579	501	417	364	306	248	
	10	744	654	561	460	395	329	261		643	569	492	410	358	302	244	
	12	726	638	547	449	386	322	256		630	558	482	402	351	296	240	
	16	682	600	514	423	365	306	245		598	529	457	381	333	282	230	
	20	629	553	474	390	338	286	231		559	495	427	356	312	265	218	
	24	570	501	429	354	309	262	214		515	455	393	327	288	246	203	
	28	507	445	381	315	276	237	196		468	413	356	296	262	224	187	
	32	443	389	332	275	243	210	175		418	369	317	264	234	202	169	
	36	380	333	285	236	209	182	154	3	368	325	279	232	207	179	152	6
	38	349	306	261	217	193	169	144		344	303	260	216	193	168	143	
	42	291	255	217	180	162	143	123		296	260	223	185	166	146	125	
	46	243	213	181	151	135	120	104		250	220	188	156	141	124	108	
	50	205	180	154	128	115	102	88.5		212	186	159	132	119	105	92.1	
	54	176	155	132	110	98.6	87.5	76.1		182	160	137	113	102	90.4	79.0	
	58	153	134	114	95.1	85.6	76.0	66.2		158	139	118	98.2	88.8	78.4	68.6	
	62	134	117	100						138	121	104	86.0	77.8	68.7	60.1	
	66									122	107	91.5	75.9	68.6	60.6	53.0	
	70																
Properties of 2 angles - 3/8 in. back to back																	
A_g, in.^{2}		26.1	23.1	20.0	16.8	15.2	13.6	12.0		22.1	19.6	17.0	14.3	13.0	11.6	10.2	
r_x, in.		1.72	1.74	1.75	1.77	1.78	1.79	1.80		1.03	1.04	1.05	1.06	1.07	1.08	1.09	
r_y, in.		3.77	3.75	3.72	3.70	3.69	3.68	3.66		4.08	4.06	4.03	4.00	3.99	3.97	3.96	
Properties of single angle																	
r_z, in.		1.28	1.28	1.29	1.29	1.30	1.30	1.31		0.844	0.846	0.850	0.856	0.859	0.863	0.867	

[a] For Y-Y axis, welded or pretensioned bolted intermediate connectors must be used.
[b] For required number of intermediate connectors, see discussion of Table 4-11.
† Section is a slender-element cross-section.
Note: Heavy line indicates Kl/r equal to or greater than 200.

$F_y = 36$ ksi

$\phi_c P_n = 0.85\ F_{cr} A_g$

Table 4-11 (cont.). SLBB Double Angles Design Strength in Axial Compression, $\phi_c P_n$, kips

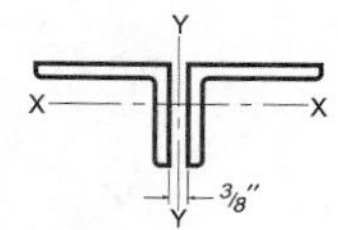

Shape		2L7×4×					No. of Connectors[a]	2L6×4×								No. of Connectors[a]
		3/4	5/8	1/2†	7/16†	3/8†		7/8	3/4	5/8	9/16	1/2	7/16†	3/8†	5/16†	
Wt/ft		**52.4**	**44.2**	**35.8**	**31.5**	**27.2**		**54.3**	**47.2**	**39.9**	**36.1**	**32.3**	**28.5**	**24.6**	**20.6**	
X-X Axis	0	471	398	310	258	205		490	425	358	324	291	249	201	153	
	6	373	317	250	212	171		391	342	289	263	236	204	167	130	
	7	343	293	232	197	160		360	316	268	244	218	189	157	122	
	8	311	266	212	182	148		328	289	245	223	200	174	145	114	
	9	278	239	192	165	136		295	261	221	202	181	158	133	106	
	10	246	213	171	149	124		262	232	198	181	162	143	121	96.8	
	11	215	186	151	133	111		229	205	175	160	144	127	108	87.9	
	12	185	161	132	117	99.0		199	178	152	140	126	111	96.1	79.1	
	13	157	138	113	102	87.2		170	153	131	121	108	96.9	84.6	70.6	
	14	136	119	97.8	87.9	75.9		146	132	113	104	93.3	83.6	73.4	62.4	
	15	118	104	85.2	76.6	66.1		127	115	98.4	90.7	81.3	72.8	64.0	54.5	
	16	104	91.0	74.9	67.3	58.1	b	112	101	86.5	79.7	71.5	64.0	56.2	47.9	b
	17	92.1	80.6	66.3	59.6	51.4		99.3	89.4	76.6	70.6	63.3	56.7	49.8	42.5	
	18	82.1	71.9	59.2	53.2	45.9		88.5	79.7	68.3	63.0	56.5	50.6	44.4	37.9	
	19										56.5	50.7	45.4	39.9	34.0	
	20															
Y-Y Axis	0	471	398	310	258	205		490	425	358	324	291	249	201	153	
	4	462	388	299	246	190		479	415	347	313	278	235	186	135	
	8	448	376	290	239	186		459	397	332	299	266	225	179	131	
	12	426	357	276	229	179		427	369	309	278	247	210	168	124	
	16	397	333	258	214	169		386	334	278	251	223	190	154	115	
	20	362	303	236	197	157		339	293	244	220	196	168	137	104	
	24	323	271	211	178	143		289	249	207	187	166	143	118	91.9	
	28	282	237	185	158	128		239	206	171	154	137	119	98.9	78.7	
	32	242	202	159	137	113		193	165	136	123	109	95.2	80.5	65.6	
	36	203	170	134	116	96.9		152	131	108	97.2	86.5	75.5	64.1	53.0	
	40	167	139	110	96.6	81.9		123	106	87.5	78.8	70.1	61.2	52.1	43.2	
	44	138	115	91.3	80.0	68.1	5	102	87.6	72.4	65.2	58.0	50.7	43.1	35.8	4
	48	116	96.7	76.8	67.3	57.3		85.7	73.7	60.8	54.8	48.8	42.6			
	52	98.7	82.5	65.5	57.4	48.9										
	56	85.1	71.1	56.5	49.6	42.2										
	60															

Effective length KL (ft) with respect to indicated axis

Properties of 2 angles - 3/8 in. back to back

	2L7×4× 3/4	5/8	1/2†	7/16†	3/8†	2L6×4× 7/8	3/4	5/8	9/16	1/2	7/16†	3/8†	5/16†
A_g, in.^{2}	15.4	13.0	10.5	9.27	8.00	16.0	13.9	11.7	10.6	9.50	8.36	7.22	6.05
r_x, in.	1.08	1.10	1.11	1.12	1.12	1.10	1.12	1.13	1.14	1.14	1.15	1.16	1.17
r_y, in.	3.48	3.46	3.43	3.42	3.40	2.96	2.94	2.91	2.90	2.89	2.88	2.86	2.85

Properties of single angle

	2L7×4× 3/4	5/8	1/2†	7/16†	3/8†	2L6×4× 7/8	3/4	5/8	9/16	1/2	7/16†	3/8†	5/16†
r_z, in.	0.855	0.860	0.866	0.869	0.873	0.854	0.856	0.859	0.861	0.864	0.867	0.870	0.874

[a] For Y-Y axis, welded or pretensioned bolted intermediate connectors must be used.
[b] For required number of intermediate connectors, see discussion of Table 4-11.
† Section is a slender-element cross-section.
Note: Heavy line indicates Kl/r equal to or greater than 200.

Table 4-11 (cont.). SLBB Double Angles Design Strength in Axial Compression, $\phi_c P_n$, kips

$F_y = 36$ ksi

$\phi_c P_n = 0.85\ F_{cr} A_g$

Shape		2L6×3½×			No. of Connectors[a]	2L5×3½×						No. of Connectors[a]	2L5×3×					No. of Connectors[a]
		1/2	3/8†	5/16†		3/4	5/8	1/2	3/8†	5/16†	1/4†		1/2	7/16	3/8†	5/16	1/4†	
Wt/ft		30.7	23.4	19.7		39.6	33.5	27.2	20.8	17.4	14.1		25.5	22.5	19.5	16.4	13.2	
X-X Axis	0	277	192	146	b	355	301	245	183	143	102	b	230	203	172	134	95.5	b
	2	268	187	142		344	292	238	178	139	99.3		220	194	165	129	92.3	
	4	243	171	132		312	266	217	163	128	92.7		192	170	145	115	83.5	
	6	207	148	116		266	228	187	142	112	82.6		154	136	118	94.7	70.6	
	8	165	122	97.1		213	183	151	116	93.3	70.3		112	100	87.4	72.3	55.8	
	10	123	94.0	77.2		160	138	115	89.6	73.5	57.2		75.6	67.7	59.6	51.0	41.3	
	11	104	80.9	67.5		135	118	97.9	77.1	63.9	50.7		62.4	56.0	49.3	42.2	34.6	
	12	87.2	68.5	58.3		113	98.7	82.4	65.3	54.8	44.4		52.5	47.0	41.4	35.4	29.0	
	13	74.3	58.4	49.8		96.5	84.1	70.2	55.6	46.7	38.4		44.7	40.1	35.3	30.2	24.8	
	14	64.0	50.4	42.9		83.2	72.5	60.6	48.0	40.3	33.1					26.0	21.3	
	15	55.8	43.9	37.4		72.5	63.2	52.7	41.8	35.1	28.9							
	16	49.0	38.6	32.9		63.7	55.5	46.4	36.7	30.8	25.4							
	17								32.5	27.3	22.5							
	18																	
Y-Y Axis	0	277	192	146	5	355	301	245	183	143	102	4	230	203	172	134	95.5	4
	6	263	179	132		336	284	229	167	127	85.3		217	190	160	123	84.1	
	8	257	175	129		324	274	221	161	123	83.4		210	184	155	119	82.1	
	10	249	170	127		310	261	210	154	118	80.9		201	176	148	115	79.6	
	12	239	165	123		292	247	198	145	112	77.8		190	166	141	109	76.6	
	14	229	158	119		273	231	185	136	106	74.3		178	156	132	103	73.0	
	16	217	151	114		253	213	171	126	98.6	70.2		165	145	123	96.5	69.0	
	20	191	135	104		210	176	141	104	83.1	60.9		138	121	103	82.0	60.0	
	24	164	117	91.5		167	140	112	82.5	67.1	50.8		111	96.6	82.5	67.0	50.2	
	28	136	99.1	78.8		127	106	84.4	62.5	51.8	40.7		85.5	74.1	63.6	52.7	40.6	
	32	110	81.6	66.1		97.3	81.4	64.7	48.0	39.8	31.6		65.5	56.8	48.8	40.5	31.6	
	36	87.4	65.4	54.0		76.9	64.4	51.2	38.0	31.6	25.1		51.8	44.9	38.6	32.1	25.1	
	38	78.5	58.8	48.6		69.0	57.8	46.0	34.1	28.4	22.6		46.5	40.3	34.6	28.8	22.6	5
	42	64.3	48.2	39.9														
	46	53.6	40.2	33.3														
	50																	
Properties of 2 angles - 3/8 in. back to back																		
A_g, in.^{2}		9.04	6.88	5.78		11.6	9.85	8.01	6.10	5.12	4.13		7.51	6.62	5.73	4.81	3.88	
r_x, in.		0.968	0.984	0.991		0.974	0.987	1.00	1.02	1.02	1.03		0.824	0.831	0.838	0.846	0.853	
r_y, in.		2.96	2.94	2.92		2.47	2.45	2.42	2.39	2.38	2.37		2.50	2.48	2.47	2.46	2.44	
Properties of single angle																		
r_z, in.		0.756	0.763	0.767		0.744	0.746	0.750	0.755	0.758	0.761		0.642	0.644	0.646	0.649	0.652	

Effective length KL (ft) with respect to indicated axis

[a]For Y-Y axis, welded or pretensioned bolted intermediate connectors must be used.

[b]For required number of intermediate connectors, see discussion of Table 4-11.

†Section is a slender-element cross-section.

Note: Heavy line indicates Kl/r equal to or greater than 200.

$F_y = 36$ ksi

$\phi_c P_n = 0.85\ F_{cr} A_g$

Table 4-11 (cont.). SLBB Double Angles Design Strength in Axial Compression, $\phi_c P_n$, kips

Shape		2L4×3½×				No. of Connectors[a]	2L4×3×					No. of Connectors[a]	2L3½×3×					No. of Connectors[a]
		1/2	3/8	5/16†	1/4†		5/8	1/2	3/8	5/16†	1/4†		1/2	7/16	3/8	5/16	1/4†	
Wt/ft		23.8	18.2	15.3	12.4		27.1	22.1	16.9	14.2	11.5		20.6	18.2	15.8	13.3	10.8	
Effective length KL (ft) with respect to indicated axis — X-X Axis	0	215	164	137	101		244	199	152	128	94.2		185	163	142	120	93.3	
	2	209	159	134	98.8		234	191	146	123	91.0		178	157	136	115	90.1	
	4	192	147	123	91.9		206	169	130	109	81.9		158	140	122	103	81.0	
	6	167	128	108	81.4		167	137	107	90.0	68.7		130	115	101	85.4	67.8	
	8	137	105	89.3	68.8		124	103	80.6	68.4	53.7		98.4	88.0	77.0	65.7	52.9	
	10	106	82.3	70.1	55.4		84.4	71.0	56.2	48.1	39.2		68.8	62.0	54.6	46.9	38.4	
	11	91.9	71.2	60.8	48.8		69.8	58.7	46.5	39.7	32.6		56.9	51.2	45.1	38.8	31.9	
	12	78.0	60.7	52.0	42.5		58.6	49.3	39.1	33.4	27.4		47.8	43.0	37.9	32.6	26.8	
	13	66.5	51.7	44.3	36.4		50.0	42.0	33.3	28.4	23.3		40.7	36.7	32.3	27.8	22.8	
	14	57.3	44.6	38.2	31.4		43.1	36.2	28.7	24.5	20.1		35.1	31.6	27.8	23.9	19.7	
	15	49.9	38.8	33.3	27.4											20.9	17.2	
	16	43.9	34.1	29.3	24.1	b						b						b
	17	38.9	30.2	25.9	21.3													
	18																	
Y-Y Axis	0	215	164	137	101		244	199	152	128	94.2		185	163	142	120	93.3	
	6	193	144	117	82.4		225	182	137	113	80.8		164	144	124	103	77.4	
	8	182	135	110	78.4		213	172	130	106	76.9		152	133	114	94.9	72.0	
	10	168	125	102	73.5		198	160	120	98.7	72.1		137	120	103	85.8	65.5	
	12	152	113	93.0	67.7		181	146	110	90.0	66.5		121	106	91.0	75.7	58.2	
	14	135	101	82.9	61.3		163	131	98.7	80.6	60.3		105	91.3	78.3	65.2	50.4	
	16	118	88.0	72.6	54.4		144	116	87.1	71.0	53.8	3	88.3	76.8	65.9	54.8	42.7	
	18	102	75.4	62.3	47.5		126	101	75.5	61.3	47.2		72.8	63.1	54.0	44.9	35.2	
	20	85.8	63.4	52.5	40.6		108	85.9	64.4	52.0	40.7		59.1	51.2	43.9	36.5	28.7	
	22	71.2	52.6	43.6	34.1		90.7	71.9	53.9	43.4	34.4		48.9	42.3	36.3	30.2	23.8	
	24	59.9	44.3	36.7	28.8		76.3	60.5	45.4	36.5	29.0		41.1	35.6	30.5	25.5	20.1	
	26	51.1	37.8	31.4	24.7	3	65.0	51.6	38.7	31.2	24.8		35.0	30.4	26.1	21.7	17.1	3
	28	44.1	32.7	27.1	21.3		56.1	44.5	33.4	26.9	21.4		30.2					
	30	38.4	28.5	23.7	18.6		48.9	38.8	29.1	23.5	18.7							
	32						43.0	34.1	25.6									
	34											4						
Properties of 2 angles - 3/8 in. back to back																		
A_g, in.^{2}		7.01	5.35	4.50	3.63		7.98	6.51	4.98	4.19	3.38		6.04	5.34	4.63	3.91	3.16	
r_x, in.		1.04	1.05	1.06	1.07		0.845	0.858	0.873	0.880	0.887		0.877	0.885	0.892	0.900	0.908	
r_y, in.		1.89	1.86	1.85	1.83		1.98	1.95	1.93	1.91	1.90		1.69	1.67	1.66	1.65	1.63	
Properties of single angle																		
r_z, in.		0.716	0.719	0.721	0.723		0.631	0.633	0.636	0.638	0.639		0.618	0.620	0.622	0.624	0.628	

[a] For Y-Y axis, welded or pretensioned bolted intermediate connectors must be used.

[b] For required number of intermediate connectors, see discussion of Table 4-11.

† Section is a slender-element cross-section.

Note: Heavy line indicates Kl/r equal to or greater than 200.

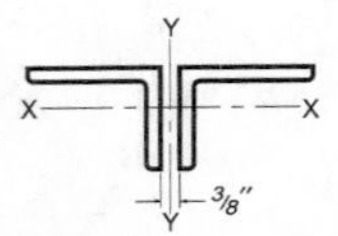

Table 4-11 (cont.). SLBB Double Angles Design Strength in Axial Compression, $\phi_c P_n$, kips

$F_y = 36$ ksi

$\phi_c P_n = 0.85\ F_{cr} A_g$

Shape		2L3½×2½×				No. of Connectors[a]	2L3×2½×						No. of Connectors[a]
		1/2	3/8	5/16	1/4†		1/2	7/16	3/8	5/16	1/4	3/16†	
Wt/ft		**18.8**	**14.5**	**12.2**	**9.88**		**17.1**	**15.1**	**13.1**	**11.1**	**8.97**	**6.82**	
X-X Axis	0	169	130	110	85.6		153	136	118	99.5	80.8	55.8	
	2	159	123	103	81.1		145	128	112	94.1	76.5	53.1	
	3	147	114	96.1	75.7		134	119	104	87.8	71.5	50.0	
	4	132	103	86.9	68.8		121	108	94.1	79.6	65.0	45.9	
	5	115	89.9	76.2	60.8		106	94.6	82.8	70.3	57.5	41.1	
	6	97.1	76.4	65.0	52.3		90.3	80.7	70.9	60.3	49.5	36.0	
	7	79.5	63.0	53.8	43.8		74.6	66.9	58.9	50.4	41.4	30.7	
	8	62.9	50.4	43.3	35.7		59.8	53.8	47.6	40.9	33.8	25.6	
	9	49.7	39.9	34.2	28.3		47.2	42.6	37.7	32.5	26.9	20.7	
	10	40.3	32.3	27.7	23.0		38.3	34.5	30.6	26.3	21.8	16.8	
	12			19.3	15.9	b		23.9	21.2	18.3	15.1	11.7	b
	13												
Y-Y Axis	0	169	130	110	85.6		153	136	118	99.5	80.8	55.8	
	2	166	126	105	80.1		150	132	114	94.8	75.0	48.6	
	4	161	122	102	77.8		143	126	109	90.7	71.9	47.0	
	6	153	116	96.9	74.1		133	117	101	84.2	66.8	44.2	
	8	142	108	90.1	69.2		120	106	91.0	75.8	60.3	40.5	
	10	130	98.4	82.0	63.3		105	92.6	79.5	66.2	52.8	36.1	
	12	116	87.7	73.1	56.7		89.3	78.6	67.3	56.0	44.7	31.2	
	14	102	76.5	63.8	49.7		73.8	64.8	55.3	46.0	36.7	26.2	
	16	87.0	65.4	54.5	42.6		59.0	51.8	43.9	36.5	29.2	21.2	
	18	73.1	54.6	45.5	35.8		46.7	40.9	34.7	28.9	23.1	16.9	
	20	60.0	44.7	37.2	29.4		37.8	33.2	28.2	23.4	18.8	13.8	
	22	49.6	37.0	30.8	24.3		31.3	27.4	23.3	19.4	15.5	11.4	
	24	41.7	31.1	25.9	20.5	4	26.3	23.1	19.6	16.3	13.1		3
	26	35.6	26.5	22.1	17.5								
	28	30.7	22.9	19.1	15.1								
	30												
Properties of 2 angles - 3/8 in. back to back													
A_g, in.^{2}		5.53	4.25	3.58	2.90		5.01	4.44	3.86	3.25	2.64	2.00	
r_x, in.		0.701	0.716	0.723	0.731		0.718	0.724	0.731	0.739	0.746	0.753	
r_y, in.		1.76	1.73	1.72	1.70		1.49	1.48	1.46	1.45	1.44	1.42	
Properties of single angle													
r_z, in.		0.532	0.535	0.538	0.541		0.516	0.516	0.517	0.518	0.520	0.521	

Effective length KL (ft) with respect to indicated axis

[a] For Y-Y axis, welded or pretensioned bolted intermediate connectors must be used.
[b] For required number of intermediate connectors, see discussion of Table 4-11.
† Section is a slender-element cross-section.
Note: Heavy line indicates Kl/r equal to or greater than 200.

$F_y = 36$ ksi
$\phi_c P_n = 0.85\ F_{cr} A_g$

Table 4-11 (cont.). SLBB Double Angles Design Strength in Axial Compression, $\phi_c P_n$, kips

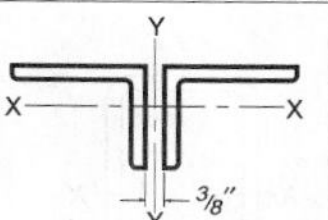

Shape		2L3×2×					No. of Connectors[a]	2L2½×2×				No. of Connectors[a]
		1/2	3/8	5/16	1/4	3/16†		3/8	5/16	1/4	3/16†	
Wt/ft		15.4	11.9	10.1	8.18	6.24		10.6	8.97	7.30	5.57	
Effective length KL (ft) with respect to indicated axis — X-X Axis	0	139	107	90.6	73.4	51.0		95.2	80.8	65.5	49.3	
	1	135	104	88.4	71.7	50.0		93.0	79.0	64.1	48.3	
	2	125	97.1	82.3	66.9	47.0		86.8	73.8	60.0	45.3	
	3	110	85.8	73.0	59.5	42.3		77.4	66.0	53.8	40.8	
	4	91.9	72.2	61.7	50.5	36.6		65.9	56.4	46.2	35.3	
	5	72.9	57.9	49.7	40.9	30.4		53.5	46.1	37.9	29.2	
	6	54.9	44.2	38.2	31.6	24.2		41.6	36.0	29.8	23.2	
	7	40.4	32.6	28.3	23.5	18.4		31.0	26.9	22.4	17.7	
	8	30.9	25.0	21.6	18.0	14.1		23.7	20.6	17.2	13.5	
	9	24.4	19.7	17.1	14.2	11.1		18.7	16.3	13.6	10.7	
	10						b					b
Y-Y Axis	0	139	107	90.6	73.4	51.0		95.2	80.8	65.5	49.3	
	4	131	101	84.5	67.6	45.7		87.0	73.3	58.4	42.5	
	6	123	94.1	78.9	63.2	43.1		78.8	66.4	52.7	38.6	
	8	112	85.7	71.8	57.5	39.7		68.7	57.8	45.6	33.6	
	10	99.3	75.9	63.5	50.9	35.5		57.5	48.3	37.9	28.1	
	12	85.9	65.5	54.6	43.8	31.0		46.3	38.8	30.1	22.5	3
	13	79.1	60.3	50.2	40.2	28.7		40.9	34.3	26.4	19.8	
	14	72.3	55.0	45.7	36.6	26.4		35.7	29.9	22.9	17.2	
	15	65.7	49.9	41.4	33.2	24.1		31.1	26.1	19.9	15.0	
	17	53.2	40.2	33.2	26.5	19.6		24.3	20.3	15.6	11.7	
	18	47.4	35.9	29.6	23.7	17.5		21.6	18.1	13.9	10.5	
	20	38.4	29.1	24.0	19.2	14.2		17.5	14.7	11.3	8.50	
	22	31.8	24.0	19.8	15.9	11.8	4					
	24	26.7	20.2	16.7	13.4	9.92						
	26	22.7										
	28											4
Properties of 2 angles - 3/8 in. back to back												
A_g, in.^{2}		4.53	3.50	2.96	2.40	1.83		3.11	2.64	2.14	1.64	
r_x, in.		0.543	0.555	0.562	0.569	0.577		0.574	0.581	0.589	0.597	
r_y, in.		1.56	1.54	1.52	1.51	1.49		1.27	1.26	1.24	1.23	
Properties of single angle												
r_z, in.		0.425	0.426	0.428	0.431	0.435		0.419	0.420	0.423	0.426	

[a]For Y-Y axis, welded or pretensioned bolted intermediate connectors must be used.
[b]For required number of intermediate connectors, see discussion of Table 4-11.
†Section is a slender-element cross-section.
Note: Heavy line indicates Kl/r equal to or greater than 200.

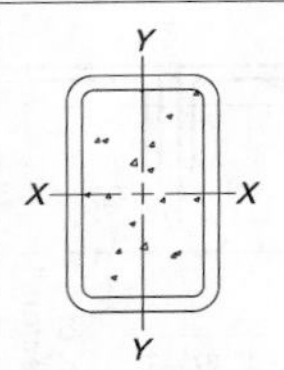

Table 4-12. Composite Rectangular (and Square) HSS Design Strength in Axial Compression, $\phi_c P_n$, kips

$F_y = 46$ ksi
$f'_c = 4$ ksi
$\phi_c = 0.85$

Shape		HSS20×12×		HSS18×12×		HSS16×16×			HSS16×12×		
		5/8	1/2	5/8	1/2	5/8	1/2	3/8	5/8	1/2	3/8
t_{design}, in.		0.581	0.465	0.581	0.465	0.581	0.465	0.349	0.581	0.465	0.349
Wt/ft		127	103	119	96.4	127	103	78.4	110	89.6	68.2
Effective length KL (ft) with respect to least radius of gyration, r_{my}	0	1870	1640	1720	1510	1910	1690	1460	1570	1380	1180
	6	1840	1610	1690	1490	1890	1670	1440	1540	1360	1160
	7	1830	1610	1680	1480	1890	1660	1440	1540	1350	1150
	8	1810	1590	1670	1470	1880	1660	1430	1520	1340	1150
	9	1800	1580	1650	1460	1870	1650	1430	1510	1330	1140
	10	1790	1570	1640	1440	1860	1640	1420	1500	1320	1130
	11	1770	1560	1630	1430	1850	1630	1410	1490	1300	1120
	12	1750	1540	1610	1420	1840	1620	1400	1470	1290	1100
	13	1730	1520	1590	1400	1830	1610	1390	1450	1280	1090
	14	1710	1500	1570	1380	1810	1600	1380	1430	1260	1080
	15	1690	1490	1550	1370	1800	1580	1370	1420	1240	1060
	16	1670	1470	1530	1350	1780	1570	1360	1400	1230	1050
	17	1640	1440	1510	1330	1770	1560	1340	1370	1210	1030
	18	1620	1420	1480	1310	1750	1540	1330	1350	1190	1020
	19	1590	1400	1460	1290	1730	1530	1320	1330	1170	998
	20	1570	1380	1430	1260	1710	1510	1300	1310	1150	980
	21	1540	1350	1410	1240	1690	1490	1290	1280	1130	962
	22	1510	1330	1380	1220	1670	1480	1270	1260	1100	943
	23	1480	1300	1350	1190	1650	1460	1260	1230	1080	923
	24	1450	1270	1330	1170	1630	1440	1240	1200	1060	904
	25	1420	1250	1300	1140	1610	1420	1220	1180	1030	883
	26	1390	1220	1270	1120	1590	1400	1210	1150	1010	863
	27	1350	1190	1240	1090	1560	1380	1190	1120	987	842
	28	1320	1160	1210	1060	1540	1360	1170	1100	962	821
	29	1290	1130	1180	1040	1520	1340	1150	1070	937	799
	30	1260	1110	1150	1010	1490	1310	1130	1040	912	778
	34	1120	988	1020	901	1390	1230	1050	923	811	691
	38	989	870	898	792	1290	1130	973	808	711	604
	42	860	756	778	686	1180	1040	890	698	614	521
	46	736	648	664	586	1070	940	806	592	522	442
	50	624	549	562	496	961	845	724	501	442	374
Properties											
r_{my}, in.		4.93	4.99	4.87	4.93	6.25	6.31	6.37	4.80	4.86	4.91
r_{mx}/r_{my}		1.49	1.48	1.37	1.37	1.00	1.00	1.00	1.25	1.25	1.25
$\phi_b M_{nx}$, kip-ft		794	649	676	555	690	566	435	569	466	359
$\phi_b M_{ny}$, kip-ft		559	455	511	421	690	566	435	466	383	295
$P_{ex}(K_x L_x)^2/10^4$		462	405	353	310	343	301	257	263	230	195
$P_{ey}(K_y L_y)^2/10^4$		209	184	189	166	343	301	257	169	148	126

$F_y = 46$ ksi
$f'_c = 4$ ksi
$\phi_c = 0.85$

Table 4-12 (cont.). Composite Rectangular (and Square) HSS Design Strength in Axial Compression, $\phi_c P_n$, kips

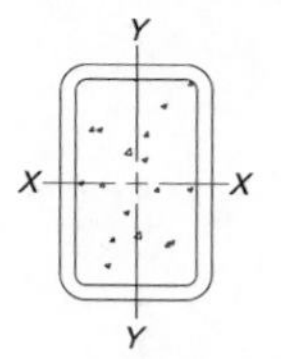

Shape	HSS16×8×			HSS14×14×			HSS14×12×		HSS14×10×		
	5/8	1/2	3/8	5/8	1/2	3/8	1/2	3/8	5/8	1/2	3/8
t_{design}, in.	0.581	0.465	0.349	0.581	0.465	0.349	0.465	0.349	0.581	0.465	0.349
Wt/ft	93.1	75.9	58.0	110	89.6	68.2	82.7	63.1	93.1	75.9	58.0
Effective length *KL* (ft) with respect to least radius of gyration, r_{my}											
0	1230	1070	905	1580	1390	1190	1250	1060	1270	1100	940
6	1190	1030	875	1560	1370	1180	1230	1050	1240	1080	918
7	1180	1020	864	1550	1370	1170	1220	1040	1230	1070	910
8	1160	1010	852	1550	1360	1160	1210	1030	1210	1060	901
9	1140	990	838	1540	1350	1160	1200	1020	1200	1050	892
10	1120	972	823	1530	1340	1150	1190	1010	1190	1030	880
11	1100	953	807	1510	1330	1140	1180	1000	1170	1020	868
12	1070	932	789	1500	1320	1130	1170	993	1150	1010	855
13	1050	910	770	1490	1310	1120	1150	982	1130	989	842
14	1020	887	751	1470	1300	1110	1140	969	1110	972	827
15	991	863	730	1460	1280	1100	1120	956	1090	954	811
16	962	837	709	1440	1270	1090	1110	942	1070	935	795
17	932	811	687	1430	1250	1070	1090	927	1050	915	778
18	900	785	664	1410	1240	1060	1070	912	1020	894	760
19	869	757	641	1390	1220	1040	1050	896	999	873	742
20	837	730	618	1370	1200	1030	1030	880	973	851	723
21	804	701	594	1350	1190	1010	1010	863	947	828	704
22	771	673	570	1330	1170	999	995	845	921	806	685
23	738	645	546	1310	1150	983	974	827	894	782	665
24	706	616	522	1290	1130	966	952	809	867	759	644
25	673	588	498	1260	1110	949	931	790	839	735	624
26	640	560	475	1240	1090	932	909	771	812	711	604
27	608	532	451	1220	1070	914	886	752	784	686	583
28	577	505	428	1190	1050	896	864	733	756	662	562
29	546	478	405	1170	1030	878	841	713	728	638	542
30	516	452	383	1150	1010	859	818	693	700	614	521
34	404	355	301	1040	917	783	725	614	591	520	441
38	324	284	241	942	826	705	633	535	489	430	365
42	265	233	197	840	736	628	544	460	400	352	298
46	221	194	164	741	649	552	460	388	334	294	249
50	187	164	139	645	565	480	389	328	282	249	211
Properties											
r_{my}, in.	3.27	3.32	3.37	5.44	5.49	5.55	4.78	4.83	3.98	4.04	4.09
r_{mx}/r_{my}	1.72	1.72	1.71	1.00	1.00	1.00	1.13	1.12	1.30	1.29	1.29
$\phi_b M_{nx}$, kip-ft	445	366	283	521	428	329	383	296	414	341	263
$\phi_b M_{ny}$, kip-ft	273	226	175	521	428	329	345	267	328	271	209
$P_{ex}(K_x L_x)^2/10^4$	186	162	136	217	190	162	165	140	160	140	118
$P_{ey}(K_y L_y)^2/10^4$	62.5	54.9	46.5	217	190	162	130	111	94.6	83.3	70.5

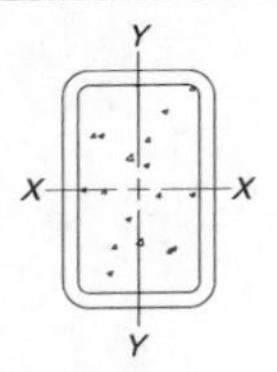

Table 4-12 (cont.). Composite Rectangular (and Square) HSS Design Strength in Axial Compression, $\phi_c P_n$, kips

$F_y = 46$ ksi
$f'_c = 4$ ksi
$\phi_c = 0.85$

Shape		HSS12×12×				HSS12×10×			HSS12×8×			
		5/8	1/2	3/8	5/16	1/2	3/8	5/16	5/8	1/2	3/8	5/16
t_{design}, in.		0.581	0.465	0.349	0.291	0.465	0.349	0.291	0.581	0.465	0.349	0.291
Wt/ft		93.1	75.9	58.0	48.8	69.1	52.9	44.6	76.1	62.3	47.8	40.3
Effective length KL (ft) with respect to least radius of gyration, r_{my}	0	1280	1120	952	864	983	836	754	983	854	720	649
	6	1260	1100	935	849	960	816	736	947	824	694	626
	7	1250	1090	929	843	951	808	730	935	813	685	617
	8	1240	1080	922	837	942	800	722	920	801	675	608
	9	1230	1070	915	830	931	791	714	904	787	663	598
	10	1220	1060	906	822	919	781	705	887	772	651	586
	11	1200	1050	897	814	906	770	695	868	756	637	574
	12	1190	1040	887	805	892	758	684	848	738	623	561
	13	1180	1030	876	795	877	745	673	826	720	607	547
	14	1160	1010	864	784	861	732	660	803	700	591	532
	15	1140	1000	852	773	845	718	647	780	680	574	517
	16	1130	985	839	761	827	703	634	755	659	556	501
	17	1110	970	826	749	809	687	620	730	637	538	484
	18	1090	953	812	736	790	671	605	704	615	520	468
	19	1070	936	797	723	771	654	590	678	592	501	450
	20	1050	918	782	709	751	637	575	651	569	481	433
	21	1030	900	766	694	730	620	559	624	546	462	415
	22	1010	882	750	680	709	602	543	597	523	442	398
	23	986	862	734	665	688	584	526	570	499	423	380
	24	964	843	717	650	667	566	510	543	476	403	362
	25	941	823	700	634	645	547	493	517	453	384	345
	26	918	803	683	618	623	529	476	490	430	365	327
	27	894	782	665	602	601	510	459	464	408	346	310
	28	870	762	647	586	579	491	442	439	385	327	294
	29	846	741	629	570	557	473	426	414	364	309	277
	30	822	720	611	553	536	454	409	389	343	291	261
	34	725	635	539	487	451	382	343	303	267	227	204
	38	630	552	468	423	370	314	282	243	214	182	163
	42	538	472	400	361	303	257	231	199	175	149	133
	46	452	397	336	303	253	214	192	166	146	124	111
	50	383	336	284	256	214	181	163	140	124	105	94.2
Properties												
r_{my}, in.		4.62	4.68	4.73	4.76	3.96	4.01	4.04	3.16	3.21	3.27	3.29
r_{mx}/r_{my}		1.00	1.00	1.00	1.00	1.15	1.15	1.15	1.37	1.37	1.37	1.37
$\phi_b M_{nx}$, kip-ft		376	309	239	202	272	211	178	283	235	183	155
$\phi_b M_{ny}$, kip-ft		376	309	239	202	240	186	158	214	178	138	118
$P_{ex}(K_x L_x)^2/10^4$		128	113	95.1	86.0	95.1	80.2	72.4	89.0	77.9	65.5	58.9
$P_{ey}(K_y L_y)^2/10^4$		128	113	95.1	86.0	71.7	60.6	54.8	47.1	41.6	35.0	31.5

$F_y = 46$ ksi
$f'_c = 4$ ksi
$\phi_c = 0.85$

Table 4-12 (cont.). Composite Rectangular (and Square) HSS Design Strength in Axial Compression, $\phi_c P_n$, kips

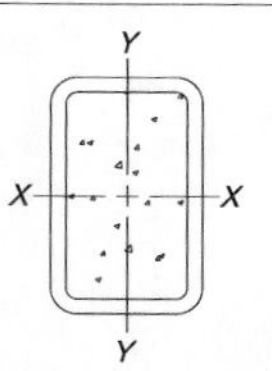

Shape	HSS12×6×				HSS10×10×				
	5/8	1/2	3/8	5/16	5/8	1/2	3/8	5/16	1/4
t_{design}, in.	0.581	0.465	0.349	0.291	0.581	0.465	0.349	0.291	0.233
Wt/ft	67.6	55.5	42.7	36.0	76.1	62.3	47.8	40.3	32.6
Effective length KL (ft) with respect to least radius of gyration, r_{my}									
0	837	722	604	540	994	866	731	660	588
6	785	678	568	508	969	844	713	644	574
7	767	663	555	497	960	837	706	638	568
8	747	646	541	485	950	828	699	631	562
9	725	627	526	471	939	818	691	624	555
10	701	607	509	456	926	807	682	615	548
11	675	585	491	440	912	795	672	606	540
12	648	562	472	423	897	782	661	596	531
13	620	539	453	406	881	769	649	586	521
14	591	514	432	388	865	754	637	575	511
15	562	489	412	369	847	739	624	563	501
16	532	463	390	350	828	723	611	551	490
17	501	438	369	331	809	706	597	538	479
18	471	412	348	312	789	689	582	525	467
19	441	386	327	293	768	671	567	511	455
20	412	361	306	275	747	653	552	497	442
21	383	336	285	256	726	635	536	483	429
22	355	312	265	238	704	616	520	469	416
23	327	289	245	221	682	596	504	454	403
24	301	266	226	203	659	577	488	439	390
25	277	245	208	188	636	557	471	424	376
26	256	226	193	173	614	538	455	409	363
27	238	210	179	161	591	518	438	394	349
28	221	195	166	150	568	498	422	379	336
29	206	182	155	139	545	479	405	364	322
30	192	170	145	130	523	459	389	349	309
34	150	132	113	101	436	383	325	291	257
38	120	106	90.1	81.2	354	312	265	237	209
42				66.4	290	256	217	194	171
46					242	213	181	162	143
50					204	180	153	137	121
Properties									
r_{my}, in.	2.39	2.44	2.49	2.52	3.80	3.86	3.92	3.94	3.97
r_{mx}/r_{my}	1.73	1.73	1.72	1.71	1.00	1.00	1.00	1.00	1.00
$\phi_b M_{nx}$, kip-ft	237	198	155	131	253	209	163	138	113
$\phi_b M_{ny}$, kip-ft	145	121	95.6	81.4	253	209	163	138	113
$P_{ex}(K_x L_x)^2/10^4$	69.8	61.1	51.3	45.9	68.7	60.4	51.1	45.8	40.4
$P_{ey}(K_y L_y)^2/10^4$	23.3	20.5	17.4	15.7	68.7	60.4	51.1	45.8	40.4

Note: Heavy line indicates Kl/r equal to or greater than 200.

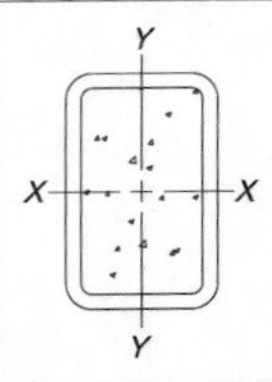

Table 4-12 (cont.). Composite Rectangular (and Square) HSS Design Strength in Axial Compression, $\phi_c P_n$, kips

$F_y = 46$ ksi
$f'_c = 4$ ksi
$\phi_c = 0.85$

Shape		HSS10×8×				HSS10×6×					HSS10×5×		
		1/2	3/8	5/16	1/4	5/8	1/2	3/8	5/16	1/4	3/8	5/16	1/4
t_{design}, in.		0.465	0.349	0.291	0.233	0.581	0.465	0.349	0.291	0.233	0.349	0.291	0.233
Wt/ft		55.5	42.7	36.0	29.2	59.1	48.7	37.6	31.8	25.8	35.1	29.7	24.1
Effective length KL (ft) with respect to least radius of gyration, r_{my}	0	745	627	563	500	726	627	522	467	411	469	418	366
	6	717	603	542	481	679	588	490	438	386	429	383	335
	7	707	595	535	475	663	574	479	428	377	415	371	325
	8	696	586	527	467	645	559	467	417	367	400	357	313
	9	684	576	517	459	625	542	453	405	356	384	343	301
	10	670	564	507	450	604	524	438	392	345	366	327	287
	11	655	552	496	440	581	505	422	378	332	348	310	272
	12	640	539	484	429	557	484	405	363	319	328	293	258
	13	623	525	472	418	532	463	388	347	305	309	276	242
	14	606	510	459	406	506	441	370	331	291	289	258	227
	15	587	495	445	394	479	419	351	315	276	269	240	211
	16	568	479	431	382	453	396	333	298	262	249	223	196
	17	549	463	416	369	426	373	314	281	247	230	205	181
	18	529	446	401	355	399	351	295	265	232	210	188	166
	19	509	429	386	342	373	328	276	248	218	192	172	152
	20	488	412	370	328	347	306	258	232	203	174	156	137
	21	468	395	355	314	322	284	240	216	189	158	141	125
	22	447	377	339	300	297	263	222	200	175	144	129	114
	23	426	360	324	286	273	242	206	185	162	132	118	104
	24	406	343	308	272	251	223	189	170	149	121	108	95.5
	25	385	325	293	259	231	205	174	157	137	111	99.7	88.0
	26	365	309	277	245	214	190	161	145	127	103	92.2	81.3
	27	345	292	262	232	198	176	149	134	117	95.5	85.5	75.4
	28	326	276	248	219	184	164	139	125	109	88.8	79.5	70.1
	29	307	260	233	206	172	153	129	116	102	82.8	74.1	65.4
	30	288	244	219	193	161	143	121	109	95.1	77.4	69.2	61.1
	32	253	214	193	170	141	125	106	95.6	83.6	68.0	60.9	53.7
	34	224	190	171	151	125	111	94.1	84.7	74.1	60.2	53.9	47.6
	36	200	169	152	134	111	99.0	83.9	75.5	66.1			
	38	179	152	137	121	100	88.8	75.3	67.8	59.3			
	40	162	137	123	109			68.0	61.2	53.5			
Properties													
r_{my}, in.		3.14	3.19	3.22	3.25	2.34	2.39	2.44	2.47	2.49	2.05	2.07	2.10
r_{mx}/r_{my}		1.19	1.19	1.19	1.18	1.50	1.49	1.49	1.48	1.48	1.72	1.72	1.71
$\phi_b M_{nx}$, kip-ft		179	140	119	96.9	177	148	117	99.4	81.4	105	89.7	73.5
$\phi_b M_{ny}$, kip-ft		154	120	102	83.5	124	104	81.8	69.7	57.3	64.5	55.2	45.5
$P_{ex}(K_x L_x)^2/10^4$		49.3	41.4	37.3	32.8	43.4	38.2	32.3	29.0	25.3	27.6	24.8	21.6
$P_{ey}(K_y L_y)^2/10^4$		34.8	29.4	26.5	23.3	19.3	17.1	14.6	13.1	11.5	9.33	8.41	7.39

Note: Heavy line indicates Kl/r equal to or greater than 200.

$F_y = 46$ ksi
$f'_c = 4$ ksi
$\phi_c = 0.85$

Table 4-12 (cont.). Composite Rectangular (and Square) HSS Design Strength in Axial Compression, $\phi_c P_n$, kips

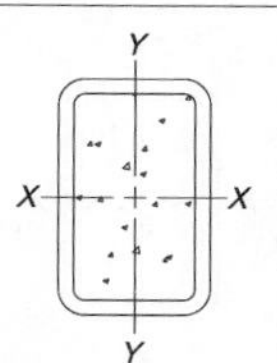

Shape		HSS9×7×					HSS9×5×				
		5/8	1/2	3/8	5/16	1/4	5/8	1/2	3/8	5/16	1/4
t_{design}, in.		0.581	0.465	0.349	0.291	0.233	0.581	0.465	0.349	0.291	0.233
Wt/ft		59.1	48.7	37.6	31.8	25.8	50.6	41.9	32.5	27.5	22.4
Effective length KL (ft) with respect to least radius of gyration, r_{my}	0	735	636	531	475	420	602	520	431	384	337
	6	698	605	505	453	400	546	474	394	351	308
	7	685	594	497	445	393	527	458	381	340	298
	8	671	582	486	436	385	506	440	367	327	287
	9	655	569	475	426	376	483	421	352	314	275
	10	638	554	463	415	366	459	401	335	299	263
	11	619	538	450	403	356	434	379	318	284	249
	12	599	521	436	391	345	407	357	300	268	235
	13	579	503	422	378	333	381	335	232	252	221
	14	557	485	406	364	321	354	312	263	235	207
	15	535	466	391	350	309	327	289	245	219	192
	16	512	446	374	336	296	301	267	226	202	178
	17	489	427	358	321	283	275	245	208	186	164
	18	465	406	341	306	270	250	223	191	171	151
	19	441	386	324	291	257	226	202	174	156	137
	20	418	366	308	276	243	204	183	157	141	124
	21	394	346	291	261	230	185	166	143	128	113
	22	371	326	274	246	217	168	151	130	116	103
	23	348	306	258	232	204	154	138	119	107	94.0
	24	326	287	242	217	192	141	127	109	97.8	86.4
	25	304	268	226	203	179	130	117	101	90.2	79.6
	26	283	250	211	190	167	120	108	93.1	83.4	73.6
	27	262	232	196	176	155	112	100	86.3	77.3	68.2
	28	244	216	182	164	144	104	93.2	80.3	71.9	63.4
	29	227	201	170	153	135	96.8	86.9	74.8	67.0	59.1
	30	212	188	159	143	126	90.5	81.2	69.9	62.6	55.3
	32	187	165	139	125	110	79.5	71.4	61.5	55.0	48.6
	34	165	146	124	111	97.9				48.8	43.0
	36	147	130	110	99.1	87.3					
	38	132	117	98.9	88.9	78.3					
	40	119	106	89.3	80.3	70.7					
Properties											
r_{my}, in.		2.68	2.73	2.78	2.81	2.84	1.92	1.97	2.03	2.05	2.08
r_{mx}/r_{my}		1.22	1.22	1.22	1.21	1.21	1.60	1.59	1.58	1.58	1.57
$\phi_b M_{nx}$, kip-ft		167	140	110	93.5	76.6	133	112	88.7	75.9	62.4
$\phi_b M_{ny}$, kip-ft		140	117	92.1	78.7	64.5	87.3	74.2	59.0	50.4	41.4
$P_{ex}(K_xL_x)^2/10^4$		37.9	33.6	28.4	25.4	22.3	28.2	25.1	21.1	18.9	16.5
$P_{ey}(K_yL_y)^2/10^4$		25.5	22.5	19.2	17.2	15.2	11.0	9.85	8.41	7.58	6.65

Note: Heavy line indicates Kl/r equal to or greater than 200.

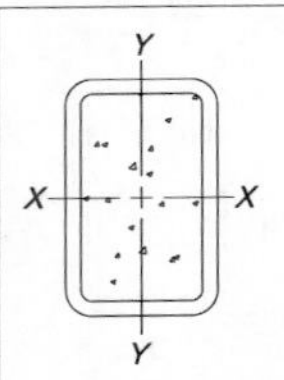

Table 4-12 (cont.). Composite Rectangular (and Square) HSS Design Strength in Axial Compression, $\phi_c P_n$, kips

$F_y = 46$ ksi
$f'_c = 4$ ksi
$\phi_c = 0.85$

Shape		HSS8×8×						HSS8×6×					
		5/8	1/2	3/8	5/16	1/4	3/16	5/8	1/2	3/8	5/16	1/4	3/16
t_{design}, in.		**0.581**	**0.465**	**0.349**	**0.291**	**0.233**	**0.174**	**0.581**	**0.465**	**0.349**	**0.291**	**0.233**	**0.174**
Wt/ft		**59.1**	**48.7**	**37.6**	**31.8**	**25.8**	**19.6**	**50.6**	**41.9**	**32.5**	**27.5**	**22.4**	**17.1**
Effective length KL (ft) with respect to least radius of gyration, r_{my}	**0**	737	639	534	478	422	364	611	529	440	393	345	295
	6	708	614	513	460	406	350	569	494	412	368	323	276
	7	698	605	506	453	400	345	555	482	402	359	316	269
	8	686	595	498	446	394	339	539	469	391	349	307	262
	9	673	584	488	438	386	333	522	454	379	339	298	254
	10	658	571	478	429	378	326	503	438	366	327	287	245
	11	643	558	467	419	369	318	482	421	352	315	277	236
	12	626	544	456	408	360	310	461	403	337	302	265	226
	13	609	529	443	397	350	301	439	384	322	288	253	216
	14	590	513	430	386	340	292	417	365	307	274	241	205
	15	571	497	417	374	329	283	394	345	291	260	229	195
	16	551	480	403	361	318	273	371	326	275	246	216	184
	17	531	463	389	348	307	263	348	306	258	231	203	173
	18	510	445	374	335	295	253	325	286	242	217	191	162
	19	490	427	359	322	283	243	302	267	226	202	178	151
	20	468	409	344	308	271	232	280	248	211	188	166	141
	21	447	391	329	295	259	222	258	229	195	175	154	131
	22	426	372	314	281	247	212	238	212	180	161	142	121
	23	405	354	298	268	235	201	217	194	166	148	131	111
	24	384	336	283	254	223	191	200	178	152	136	120	102
	25	363	318	269	241	211	181	184	164	140	126	111	93.8
	26	342	300	254	228	200	170	170	152	130	116	102	86.8
	27	322	283	240	215	188	161	158	141	120	108	94.9	80.4
	28	303	266	225	202	177	151	147	131	112	100	88.2	74.8
	29	284	250	212	190	166	142	137	122	104	93.3	82.2	69.7
	30	265	234	198	178	155	132	128	114	97.5	87.2	76.8	65.2
	32	233	205	174	156	137	116	112	100	85.7	76.6	67.5	57.3
	34	206	182	154	138	121	103	99.4	88.7	75.9	67.9	59.8	50.7
	36	184	162	138	123	108	91.8	88.7	79.1	67.7	60.6	53.4	45.3
	38	165	146	123	111	96.9	82.4		71.0	60.8	54.3	47.9	40.6
	40	149	131	111	100	87.4	74.4				49.0	43.2	36.7
Properties													
r_{my}, in.		2.99	3.04	3.10	3.13	3.15	3.18	2.27	2.32	2.38	2.40	2.43	2.46
r_{mx}/r_{my}		1.00	1.00	1.00	1.00	1.00	1.00	1.26	1.25	1.25	1.25	1.25	1.24
$\phi_b M_{nx}$, kip-ft		154	129	101	86.6	70.7	54.2	125	105	83.1	71.1	58.3	44.9
$\phi_b M_{ny}$, kip-ft		154	129	101	86.6	70.7	54.2	102	85.9	68.3	58.3	48.0	36.9
$P_{ex}(K_x L_x)^2/10^4$		31.9	28.2	23.8	21.4	18.8	16.0	24.4	21.6	18.3	16.4	14.4	12.2
$P_{ey}(K_y L_y)^2/10^4$		31.9	28.2	23.8	21.4	18.8	16.0	15.5	13.8	11.7	10.5	9.27	7.85

Note: Heavy line indicates Kl/r equal to or greater than 200.

$F_y = 46$ ksi
$f'_c = 4$ ksi
$\phi_c = 0.85$

Table 4-12 (cont.). Composite Rectangular (and Square) HSS Design Strength in Axial Compression, $\phi_c P_n$, kips

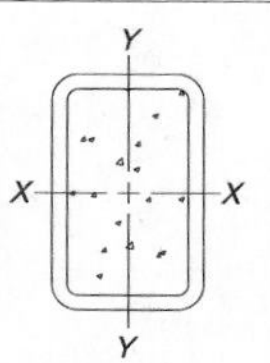

Shape	HSS8×4×						HSS7×7×					
	5/8	1/2	3/8	5/16	1/4	3/16	5/8	1/2	3/8	5/16	1/4	3/16
t_{design}, in.	0.581	0.465	0.349	0.291	0.233	0.174	0.581	0.465	0.349	0.291	0.233	0.174
Wt/ft	42.1	35.1	27.4	23.3	19.0	14.5	50.6	41.9	32.5	27.5	22.4	17.1
Effective length KL (ft) with respect to least radius of gyration, r_{my}												
0	488	421	347	308	268	226	614	532	443	396	348	298
6	417	363	301	268	234	197	581	504	420	376	331	282
7	394	344	286	255	222	187	570	495	412	369	325	277
8	370	323	270	240	210	177	557	484	404	361	318	271
9	343	301	252	225	197	166	543	472	394	353	310	265
10	316	279	234	209	183	154	528	459	383	343	302	257
11	289	256	216	192	169	142	511	445	372	333	293	250
12	261	233	197	176	154	130	494	430	360	322	283	241
13	234	210	178	160	140	119	475	414	347	311	273	233
14	208	188	160	144	127	107	456	398	334	299	263	224
15	183	167	143	128	113	95.9	437	382	320	287	252	214
16	161	147	126	113	100	85.1	417	364	306	274	241	205
17	143	130	112	100	88.9	75.4	396	347	292	262	230	195
18	127	116	99.9	89.6	79.3	67.2	376	330	277	249	219	186
19	114	104	89.7	80.4	71.2	60.3	356	312	263	236	207	176
20	103	93.8	80.9	72.6	64.2	54.5	335	295	249	223	196	166
21	93.5	85.1	73.4	65.8	58.3	49.4	315	277	234	210	185	156
22	85.2	77.5	66.9	60.0	53.1	45.0	295	260	220	198	174	147
23	78.0	70.9	61.2	54.9	48.6	41.2	276	244	206	185	163	138
24	71.6	65.1	56.2	50.4	44.6	37.8	257	227	193	173	152	128
25	66.0	60.0	51.8	46.5	41.1	34.8	238	211	180	162	142	120
26		55.5	47.9	43.0	38.0	32.2	220	196	167	150	132	111
27				39.8	35.2	29.9	204	181	154	139	122	103
28						27.8	190	169	144	129	114	95.5
29							177	157	134	120	106	89.1
30							166	147	125	113	99.0	83.2
32							145	129	110	99.0	87.0	73.2
34							129	114	97.4	87.7	77.1	64.8
36							115	102	86.9	78.2	68.8	57.8
38							103	91.6	78.0	70.2	61.7	51.9
40							93.1	82.7	70.4	63.3	55.7	46.8
Properties												
r_{my}, in.	1.51	1.56	1.61	1.63	1.66	1.69	2.58	2.63	2.69	2.72	2.75	2.77
r_{mx}/r_{my}	1.75	1.74	1.73	1.73	1.72	1.70	1.00	1.00	1.00	1.00	1.00	1.00
$\phi_b M_{nx}$, kip-ft	94.5	81.1	64.9	55.5	45.9	35.2	114	96.3	76.2	65.2	53.5	41.1
$\phi_b M_{ny}$, kip-ft	57.3	49.3	39.7	34.2	28.3	21.8	114	96.3	76.2	65.2	53.5	41.1
$P_{ex}(K_xL_x)^2/10^4$	17.0	15.2	13.0	11.6	10.1	8.51	20.0	17.8	15.1	13.6	11.9	10.1
$P_{ey}(K_yL_y)^2/10^4$	5.52	5.01	4.33	3.92	3.44	2.90	20.0	17.8	15.1	13.6	11.9	10.1

Note: Heavy line indicates Kl/r equal to or greater than 200.

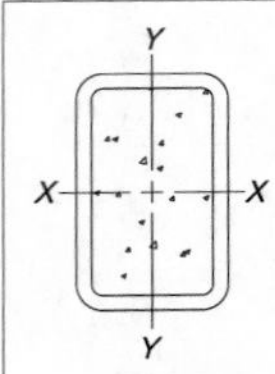

Table 4-12 (cont.). Composite Rectangular (and Square) HSS Design Strength in Axial Compression, $\phi_c P_n$, kips

$F_y = 46$ ksi
$f'_c = 4$ ksi
$\phi_c = 0.85$

Shape		HSS7×5×						HSS7×4×				
		5/8	1/2	3/8	5/16	1/4	3/16	1/2	3/8	5/16	1/4	3/16
t_{design}, in.		0.581	0.465	0.349	0.291	0.233	0.174	0.465	0.349	0.291	0.233	0.174
Wt/ft		42.1	35.1	27.4	23.3	19.0	14.5	31.7	24.9	21.2	17.3	13.2
Effective length KL (ft) with respect to least radius of gyration, r_{my}	0	497	429	356	317	277	234	378	312	277	241	202
	6	448	389	323	288	252	213	324	270	240	209	176
	7	431	375	313	278	243	206	307	256	228	199	167
	8	413	360	300	268	234	198	288	241	215	188	157
	9	393	343	287	256	224	190	268	224	201	175	147
	10	372	326	273	243	213	180	247	208	186	163	137
	11	351	307	258	230	202	171	226	191	171	150	126
	12	328	288	243	217	190	161	204	174	156	137	115
	13	305	269	227	203	178	151	184	157	142	124	104
	14	282	250	212	189	166	140	164	140	127	112	93.8
	15	260	231	196	175	154	130	144	125	113	99.9	83.7
	16	238	212	181	161	142	120	127	110	99.9	88.3	74.0
	17	216	193	165	148	130	110	112	97.3	88.5	78.3	65.5
	18	195	175	151	135	119	101	100	86.8	78.9	69.8	58.4
	19	175	158	137	122	108	91.3	90.1	77.9	70.9	62.6	52.5
	20	158	143	123	110	97.3	82.4	81.3	70.3	63.9	56.5	47.3
	21	144	129	112	100	88.2	74.7	73.7	63.8	58.0	51.3	42.9
	22	131	118	102	91.2	80.4	68.1	67.2	58.1	52.8	46.7	39.1
	23	120	108	93.2	83.4	73.6	62.3	61.5	53.1	48.4	42.8	35.8
	24	110	99.0	85.6	76.6	67.6	57.2	56.4	48.8	44.4	39.3	32.9
	25	101	91.2	78.9	70.6	62.3	52.7	52.0	45.0	40.9	36.2	30.3
	26	93.7	84.4	72.9	65.3	57.6	48.7		41.6	37.8	33.5	28.0
	27	86.9	78.2	67.6	60.5	53.4	45.2				31.0	26.0
	28	80.8	72.7	62.9	56.3	49.6	42.0					
	29	75.3	67.8	58.6	52.5	46.3	39.2					
	30	70.4	63.4	54.8	49.0	43.2	36.6					
	32			48.1	43.1	38.0	32.2					
	34						28.5					
	36											
	38											
	40											
Properties												
r_{my}, in.		1.86	1.91	1.97	1.99	2.02	2.05	1.53	1.58	1.61	1.64	1.66
r_{mx}/r_{my}		1.31	1.31	1.30	1.30	1.30	1.29	1.57	1.56	1.55	1.54	1.54
$\phi_b M_{nx}$, kip-ft		88.3	75.6	60.4	51.8	42.8	32.8	64.9	52.1	45.2	37.3	28.7
$\phi_b M_{ny}$, kip-ft		69.7	59.7	47.6	41.1	33.9	26.1	43.5	35.2	30.5	25.3	19.6
$P_{ex}(K_x L_x)^2/10^4$		14.6	13.0	11.1	10.00	8.77	7.37	10.7	9.18	8.26	7.21	6.04
$P_{ey}(K_y L_y)^2/10^4$		8.52	7.66	6.59	5.93	5.20	4.39	4.38	3.80	3.44	3.03	2.54

Note: Heavy line indicates Kl/r equal to or greater than 200.

$F_y = 46$ ksi
$f'_c = 4$ ksi
$\phi_c = 0.85$

Table 4-12 (cont.). Composite Rectangular (and Square) HSS Design Strength in Axial Compression, $\phi_c P_n$, kips

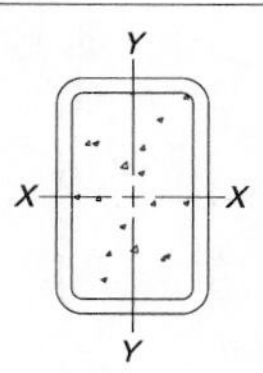

Shape		HSS6×6×						HSS6×5×			
		5/8	1/2	3/8	5/16	1/4	3/16	3/8	5/16	1/4	3/16
t_{design}, in.		0.581	0.465	0.349	0.291	0.233	0.174	0.349	0.291	0.233	0.174
Wt/ft		42.1	35.1	27.4	23.3	19.0	14.5	24.9	21.2	17.3	13.2
Effective length KL (ft) with respect to least radius of gyration, r_{my}	0	500	432	359	320	280	237	318	283	247	208
	6	463	402	334	298	261	221	288	257	224	189
	7	450	391	326	290	254	215	278	248	216	182
	8	436	380	316	282	247	209	266	238	207	175
	9	421	367	305	273	239	202	254	227	198	167
	10	404	353	294	263	230	195	241	215	188	159
	11	387	338	282	252	221	187	227	203	178	150
	12	368	323	270	241	211	179	213	191	167	141
	13	349	307	257	230	201	170	199	178	156	132
	14	330	290	243	218	191	162	185	166	145	123
	15	310	274	230	206	180	153	170	153	134	113
	16	290	257	216	194	170	144	156	141	124	104
	17	271	240	202	181	159	135	143	129	113	95.5
	18	251	224	189	169	149	126	129	117	103	86.9
	19	232	208	175	158	138	117	116	105	92.9	78.5
	20	214	192	162	146	128	108	105	95.2	83.8	70.8
	21	196	176	150	135	118	100	95.3	86.3	76.0	64.3
	22	179	161	137	124	109	91.8	86.9	78.6	69.3	58.6
	23	164	148	126	113	99.5	84.0	79.5	72.0	63.4	53.6
	24	150	136	115	104	91.3	77.2	73.0	66.1	58.2	49.2
	25	138	125	106	95.8	84.2	71.1	67.3	60.9	53.6	45.3
	26	128	116	98.2	88.5	77.8	65.7	62.2	56.3	49.6	41.9
	27	119	107	91.1	82.1	72.2	61.0	57.7	52.2	46.0	38.9
	28	110	99.6	84.7	76.3	67.1	56.7	53.6	48.6	42.8	36.1
	29	103	92.9	78.9	71.2	62.6	52.8	50.0	45.3	39.9	33.7
	30	96.1	86.8	73.8	66.5	58.5	49.4	46.7	42.3	37.2	31.5
	32	84.5	76.3	64.8	58.4	51.4	43.4	41.1	37.2	32.7	27.7
	34	74.9	67.6	57.4	51.8	45.5	38.4				
	36	66.8	60.3	51.2	46.2	40.6	34.3				
	38			46.0	41.4	36.4	30.8				
	40										
Properties											
r_{my}, in.		2.17	2.23	2.28	2.31	2.34	2.37	1.92	1.95	1.98	2.01
r_{mx}/r_{my}		1.00	1.00	1.00	1.00	1.00	1.00	1.16	1.15	1.15	1.15
$\phi_b M_{nx}$, kip-ft		80.0	68.3	54.5	46.9	38.6	29.8	47.6	41.1	34.1	26.3
$\phi_b M_{ny}$, kip-ft		80.0	68.3	54.5	46.9	38.6	29.8	42.1	36.2	30.1	23.2
$P_{ex}(K_x L_x)^2/10^4$		11.6	10.4	8.93	8.03	7.04	5.95	7.54	6.79	5.94	5.00
$P_{ey}(K_y L_y)^2/10^4$		11.6	10.4	8.93	8.03	7.04	5.95	5.67	5.12	4.50	3.78

Note: Heavy line indicates Kl/r equal to or greater than 200.

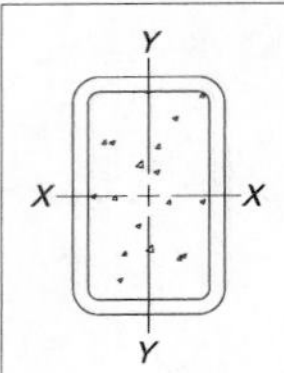

Table 4-12 (cont.). Composite Rectangular (and Square) HSS Design Strength in Axial Compression, $\phi_c P_n$, kips

$F_y = 46$ ksi
$f'_c = 4$ ksi
$\phi_c = 0.85$

Shape		HSS6×4×					HSS6×3×				
		1/2	3/8	5/16	1/4	3/16	1/2	3/8	5/16	1/4	3/16
t_{design}, in.		0.465	0.349	0.291	0.233	0.174	0.465	0.349	0.291	0.233	0.174
Wt/ft		28.3	22.3	19.0	15.6	12.0	24.9	19.7	16.9	13.9	10.7
Effective length KL (ft) with respect to least radius of gyration, r_{my}	0	336	277	246	214	179	287	237	209	181	150
	6	286	238	212	185	155	216	182	162	141	118
	7	270	226	201	175	147	195	165	147	129	108
	8	253	212	189	165	138	174	148	132	116	97.4
	9	234	197	176	154	129	152	131	117	103	86.9
	10	215	182	163	143	120	131	114	102	90.6	76.4
	11	196	166	150	131	110	111	97.7	88.1	78.3	66.3
	12	177	151	136	119	100	93.1	82.5	74.6	66.7	56.6
	13	159	136	123	108	90.4	79.3	70.3	63.6	56.8	48.3
	14	141	121	110	96.8	81.1	68.4	60.6	54.8	49.0	41.6
	15	124	107	97.5	86.1	72.1	59.6	52.8	47.8	42.7	36.3
	16	109	94.2	85.7	75.9	63.5	52.3	46.4	42.0	37.5	31.9
	17	96.2	83.5	75.9	67.2	56.2	46.4	41.1	37.2	33.2	28.2
	18	85.8	74.4	67.7	59.9	50.2	41.4	36.6	33.2	29.6	25.2
	19	77.0	66.8	60.8	53.8	45.0		32.9	29.8	26.6	22.6
	20	69.5	60.3	54.9	48.6	40.6				24.0	20.4
	21	63.0	54.7	49.8	44.0	36.9					
	22	57.4	49.8	45.3	40.1	33.6					
	23	52.6	45.6	41.5	36.7	30.7					
	24	48.3	41.9	38.1	33.7	28.2					
	25	44.5	38.6	35.1	31.1	26.0					
	26			32.5	28.7	24.0					
	27					22.3					
	28										
	29										
	30										
	32										
	34										
	36										
	38										
	40										
Properties											
r_{my}, in.		1.50	1.55	1.58	1.61	1.63	1.12	1.17	1.19	1.22	1.25
r_{mx}/r_{my}		1.39	1.38	1.37	1.37	1.37	1.76	1.74	1.74	1.72	1.71
$\phi_b M_{nx}$, kip-ft		50.4	41.1	35.5	29.4	22.8	41.7	34.2	29.7	24.8	19.3
$\phi_b M_{ny}$, kip-ft		38.0	30.8	26.7	22.3	17.3	25.1	20.8	18.2	15.2	11.9
$P_{ex}(K_x L_x)^2/10^4$		7.15	6.17	5.56	4.89	4.10	5.53	4.82	4.37	3.83	3.20
$P_{ey}(K_y L_y)^2/10^4$		3.74	3.25	2.96	2.59	2.19	1.79	1.59	1.45	1.28	1.09

Note: Heavy line indicates Kl/r equal to or greater than 200.

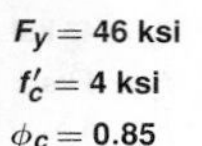

$F_y = 46$ ksi
$f'_c = 4$ ksi
$\phi_c = 0.85$

Table 4-12 (cont.). Composite Rectangular (and Square) HSS Design Strength in Axial Compression, $\phi_c P_n$, kips

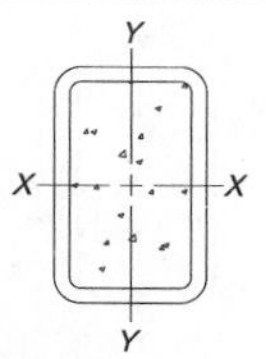

Shape		HSS5½×5½×				HSS5×5×					
		3/8	5/16	1/4	3/16	1/2	3/8	5/16	1/4	3/16	1/8
t_{design}, in.		0.349	0.291	0.233	0.174	0.465	0.349	0.291	0.233	0.174	0.116
Wt/ft		24.9	21.2	17.3	13.2	28.3	22.3	19.0	15.6	12.0	8.15
Effective length KL (ft) with respect to least radius of gyration, r_{my}	0	319	284	247	209	339	280	249	216	182	147
	6	293	261	227	192	304	252	225	196	165	132
	7	284	253	221	186	292	243	216	189	159	128
	8	274	244	213	180	279	233	207	181	152	122
	9	263	235	205	173	265	221	198	172	145	117
	10	252	225	196	165	250	210	187	163	137	110
	11	239	214	186	157	235	197	176	154	130	104
	12	227	203	177	149	219	184	165	144	121	97.4
	13	214	191	167	141	203	171	154	134	113	90.7
	14	200	180	156	132	187	158	142	125	105	84.0
	15	187	168	146	123	172	146	131	115	96.7	77.3
	16	174	156	136	115	156	133	120	105	88.7	70.8
	17	161	145	126	106	141	121	109	95.8	80.8	64.4
	18	148	133	116	98.0	127	109	98.7	86.8	73.2	58.3
	19	136	122	106	89.9	114	97.9	88.6	78.1	65.9	52.4
	20	124	112	97.2	82.0	103	88.4	80.0	70.4	59.4	47.3
	21	112	101	88.2	74.4	93.3	80.2	72.6	63.9	53.9	42.9
	22	102	92.2	80.3	67.8	85.1	73.1	66.1	58.2	49.1	39.1
	23	93.4	84.4	73.5	62.0	77.8	66.8	60.5	53.3	44.9	35.7
	24	85.8	77.5	67.5	57.0	71.5	61.4	55.5	48.9	41.3	32.8
	25	79.1	71.4	62.2	52.5	65.9	56.6	51.2	45.1	38.0	30.3
	26	73.1	66.0	57.5	48.5	60.9	52.3	47.3	41.7	35.2	28.0
	27	67.8	61.2	53.3	45.0	56.5	48.5	43.9	38.7	32.6	25.9
	28	63.0	56.9	49.6	41.8	52.5	45.1	40.8	35.9	30.3	24.1
	29	58.8	53.1	46.2	39.0	48.9	42.0	38.0	33.5	28.3	22.5
	30	54.9	49.6	43.2	36.5	45.7	39.3	35.5	31.3	26.4	21.0
	32	48.3	43.6	38.0	32.0				27.5	23.2	18.5
	34	42.8	38.6	33.6	28.4						
	36				25.3						
	38										
	40										
Properties											
r_{my}, in.		2.08	2.11	2.13	2.16	1.82	1.87	1.90	1.93	1.96	1.99
r_{mx}/r_{my}		1.00	1.00	1.00	1.00	1.00	1.00	1.00	1.00	1.00	1.00
$\phi_b M_{nx}$, kip-ft		45.2	39.0	32.2	24.8	45.2	36.6	31.6	26.3	20.3	14.0
$\phi_b M_{ny}$, kip-ft		45.2	39.0	32.2	24.8	45.2	36.6	31.6	26.3	20.3	14.0
$P_{ex}(K_x L_x)^2/10^4$		6.62	5.96	5.23	4.42	5.50	4.76	4.29	3.78	3.19	2.53
$P_{ey}(K_y L_y)^2/10^4$		6.62	5.96	5.23	4.42	5.50	4.76	4.29	3.78	3.19	2.53

Note: Heavy line indicates Kl/r equal to or greater than 200.

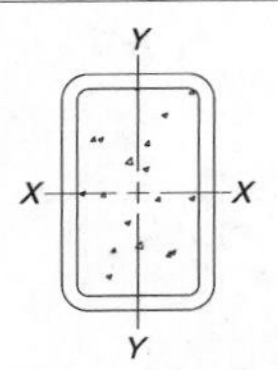

Table 4-12 (cont.). Composite Rectangular (and Square) HSS Design Strength in Axial Compression, $\phi_c P_n$, kips

$F_y = 46$ ksi
$f'_c = 4$ ksi
$\phi_c = 0.85$

Shape	HSS5×4×					HSS5×3×					
	1/2	3/8	5/16	1/4	3/16	1/2	3/8	5/16	1/4	3/16	1/8
t_{design}, in.	0.465	0.349	0.291	0.233	0.174	0.465	0.349	0.291	0.233	0.174	0.116
Wt/ft	24.9	19.7	16.9	13.9	10.7	21.5	17.2	14.8	12.2	9.40	6.45
Effective length *KL* (ft) with respect to least radius of gyration, r_{my}											
0	293	242	215	187	156	248	205	181	157	130	103
1	292	241	214	186	155	246	203	180	155	129	102
2	288	238	212	184	153	240	199	176	152	126	99.7
3	281	233	207	180	150	230	191	170	147	122	96.3
4	272	226	201	175	146	217	181	161	139	116	91.7
5	261	217	193	168	141	201	169	151	131	109	86.1
6	248	207	184	160	134	184	155	139	121	101	79.7
7	233	196	174	152	127	165	140	126	110	91.9	72.8
8	217	183	163	143	119	146	125	113	98.5	82.6	65.6
9	201	170	152	133	111	126	110	99.8	87.1	73.3	58.2
10	184	157	140	122	103	108	94.9	86.8	75.9	64.1	51.0
11	167	143	128	112	94.2	90.6	80.8	74.3	65.2	55.3	44.0
12	150	129	116	102	85.5	76.1	67.9	62.7	55.1	46.9	37.4
13	133	116	104	91.5	77.1	64.8	57.8	53.4	47.0	40.0	31.9
14	118	103	92.5	81.7	68.9	55.9	49.9	46.1	40.5	34.5	27.5
15	103	90.5	81.5	72.2	61.0	48.7	43.4	40.1	35.3	30.0	23.9
16	90.1	79.6	71.6	63.5	53.6	42.8	38.2	35.3	31.0	26.4	21.0
17	79.8	70.5	63.4	56.2	47.5	37.9	33.8	31.2	27.5	23.4	18.6
18	71.2	62.9	56.6	50.2	42.4	33.8	30.2	27.9	24.5	20.8	16.6
19	63.9	56.4	50.8	45.0	38.0		27.1	25.0	22.0	18.7	14.9
20	57.7	50.9	45.8	40.6	34.3					16.9	13.5
21	52.3	46.2	41.6	36.8	31.1						
22	47.7	42.1	37.9	33.6	28.4						
23	43.6	38.5	34.7	30.7	26.0						
24	40.1	35.4	31.8	28.2	23.8						
25		32.6	29.3	26.0	22.0						
26				24.0	20.3						
27											
28											
29											
30											
Properties											
r_{my}, in.	1.46	1.52	1.54	1.57	1.60	1.09	1.14	1.17	1.19	1.22	1.25
r_{mx}/r_{my}	1.20	1.19	1.19	1.19	1.19	1.51	1.51	1.50	1.50	1.49	1.48
$\phi_b M_{nx}$, kip-ft	37.6	30.9	26.9	22.4	17.4	30.5	25.3	22.1	18.6	14.5	10.1
$\phi_b M_{ny}$, kip-ft	32.3	26.5	23.0	19.2	15.0	21.0	17.6	15.5	13.0	10.2	7.14
$P_{ex}(K_x L_x)^2/10^4$	4.43	3.86	3.50	3.09	2.60	3.37	2.98	2.72	2.39	2.01	1.58
$P_{ey}(K_y L_y)^2/10^4$	3.11	2.72	2.46	2.18	1.84	1.48	1.32	1.21	1.07	0.908	0.718

Note: Heavy line indicates Kl/r equal to or greater than 200.

$F_y = 46$ ksi
$f'_c = 4$ ksi
$\phi_c = 0.85$

Table 4-12 (cont.). Composite Rectangular (and Square) HSS Design Strength in Axial Compression, $\phi_c P_n$, kips

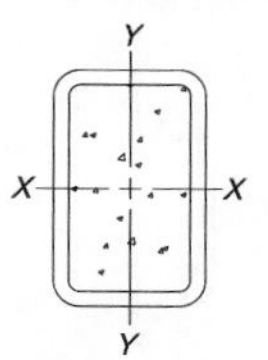

Shape	HSS5×2½×			HSS4½×4½×					
	¼	3/16	⅛	½	⅜	5/16	¼	3/16	⅛
t_{design}, in.	**0.233**	**0.174**	**0.116**	**0.465**	**0.349**	**0.291**	**0.233**	**0.174**	**0.116**
Wt/ft	**11.3**	**8.55**	**6.02**	**24.9**	**19.7**	**16.9**	**13.9**	**10.7**	**7.30**
Effective length KL (ft) with respect to least radius of gyration, r_{my}									
0	142	115	91.3	294	243	216	187	157	125
1	140	114	90.4	293	242	215	187	156	125
2	136	111	87.8	289	240	213	185	154	124
3	129	105	83.6	284	235	209	182	152	121
4	120	98.3	78.1	276	230	204	177	148	119
5	110	90.0	71.5	267	222	198	172	144	115
6	98.2	80.8	64.2	256	213	190	165	138	110
7	86.0	71.1	56.5	243	204	182	158	132	106
8	73.9	61.4	48.8	230	193	172	150	125	100
9	62.1	52.0	41.3	215	181	162	141	118	94.3
10	51.1	43.0	34.3	200	169	152	132	111	88.2
11	42.2	35.6	28.3	185	157	141	123	103	81.9
12	35.5	29.9	23.8	169	144	130	114	94.8	75.6
13	30.2	25.5	20.3	154	132	119	104	86.9	69.2
14	26.1	22.0	17.5	139	120	108	94.8	79.1	62.9
15	22.7	19.1	15.2	124	108	97.4	85.7	71.4	56.8
16	20.0	16.8	13.4	110	96.2	87.2	77.0	64.1	50.9
17		14.9	11.9	97.3	85.3	77.5	68.5	57.0	45.3
18				86.8	76.0	69.1	61.1	50.9	40.4
19				77.9	68.3	62.0	54.8	45.6	36.2
20				70.3	61.6	56.0	49.5	41.2	32.7
21				63.7	55.9	50.8	44.9	37.4	29.7
22				58.1	50.9	46.3	40.9	34.0	27.0
23				53.1	46.6	42.3	37.4	31.2	24.7
24				48.8	42.8	38.9	34.3	28.6	22.7
25				45.0	39.4	35.8	31.7	26.4	20.9
26				41.6	36.4	33.1	29.3	24.4	19.4
27					33.8	30.7	27.1	22.6	17.9
28						28.6	25.2	21.0	16.7
29								19.6	15.6
30									
Properties									
r_{my}, in.	0.999	1.03	1.05	1.61	1.67	1.70	1.73	1.75	1.78
r_{mx}/r_{my}	1.73	1.72	1.71	1.00	1.00	1.00	1.00	1.00	1.00
$\phi_b M_{nx}$, kip-ft	16.7	12.8	9.14	35.2	28.8	25.1	20.9	16.2	11.3
$\phi_b M_{ny}$, kip-ft	10.2	7.90	5.66	35.2	28.8	25.1	20.9	16.2	11.3
$P_{ex}(K_x L_x)^2/10^4$	2.06	1.70	1.35	3.79	3.31	3.00	2.63	2.22	1.76
$P_{ey}(K_y L_y)^2/10^4$	0.685	0.574	0.460	3.79	3.31	3.00	2.63	2.22	1.76

Note: Heavy line indicates Kl/r equal to or greater than 200.

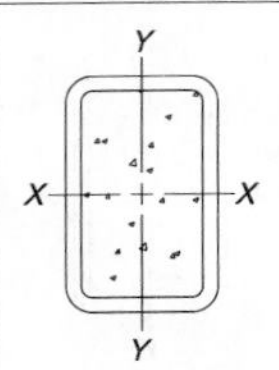

Table 4-12 (cont.). Composite Rectangular (and Square) HSS Design Strength in Axial Compression, $\phi_c P_n$, kips

$F_y = 46$ ksi
$f'_c = 4$ ksi
$\phi_c = 0.85$

Shape		HSS4×4×						HSS4×3×				
		1/2	3/8	5/16	1/4	3/16	1/8	3/8	5/16	1/4	3/16	1/8
t_{design}, in.		0.465	0.349	0.291	0.233	0.174	0.116	0.349	0.291	0.233	0.174	0.116
Wt/ft		21.5	17.2	14.8	12.2	9.40	6.45	14.6	12.7	10.5	8.13	5.60
Effective length KL (ft) with respect to least radius of gyration, r_{my}	0	251	208	184	159	133	105	173	153	133	110	86.3
	1	249	207	183	159	132	105	172	152	132	109	85.6
	2	246	204	181	157	130	104	168	149	129	107	83.7
	3	240	199	177	153	128	101	161	143	124	103	80.7
	4	231	193	171	148	124	98.3	152	135	117	97.7	76.6
	5	221	185	164	143	119	94.5	141	126	110	91.4	71.7
	6	209	176	156	136	113	90.0	129	116	101	84.3	66.1
	7	196	165	147	128	107	85.0	117	104	91.5	76.6	60.0
	8	182	154	138	120	100	79.5	103	92.7	81.7	68.6	53.7
	9	167	143	127	111	92.9	73.8	90.0	81.1	71.8	60.5	47.4
	10	152	130	117	102	85.5	67.9	77.2	69.9	62.2	52.6	41.2
	11	137	118	106	92.9	77.9	61.9	65.0	59.2	53.0	45.1	35.3
	12	122	106	95.4	83.9	70.4	55.9	54.6	49.7	44.6	38.0	29.7
	13	108	94.7	85.1	75.0	63.1	50.1	46.6	42.4	38.0	32.4	25.3
	14	94.3	83.5	75.3	66.5	56.0	44.5	40.1	36.5	32.8	28.0	21.8
	15	82.1	72.9	65.8	58.3	49.2	39.0	35.0	31.8	28.6	24.4	19.0
	16	72.2	64.1	57.8	51.2	43.3	34.3	30.7	28.0	25.1	21.4	16.7
	17	63.9	56.8	51.2	45.4	38.3	30.4	27.2	24.8	22.2	19.0	14.8
	18	57.0	50.6	45.7	40.5	34.2	27.1	24.3	22.1	19.8	16.9	13.2
	19	51.2	45.5	41.0	36.3	30.7	24.3			17.8	15.2	11.9
	20	46.2	41.0	37.0	32.8	27.7	22.0					10.7
	21	41.9	37.2	33.6	29.7	25.1	19.9					
	22	38.2	33.9	30.6	27.1	22.9	18.1					
	23	34.9	31.0	28.0	24.8	20.9	16.6					
	24		28.5	25.7	22.8	19.2	15.2					
	25				21.0	17.7	14.1					
	26						13.0					
	27											
	28											
	29											
	30											
Properties												
r_{my}, in.		1.41	1.47	1.49	1.52	1.55	1.58	1.11	1.13	1.16	1.19	1.21
r_{mx}/r_{my}		1.00	1.00	1.00	1.00	1.00	1.00	1.25	1.26	1.25	1.25	1.26
$\phi_b M_{nx}$, kip-ft		26.6	22.0	19.3	16.2	12.7	8.83	17.7	15.6	13.1	10.4	7.28
$\phi_b M_{ny}$, kip-ft		26.6	22.0	19.3	16.2	12.7	8.83	14.4	12.7	10.8	8.49	5.97
$P_{ex}(K_x L_x)^2/10^4$		2.46	2.20	1.99	1.76	1.49	1.17	1.66	1.53	1.35	1.14	0.897
$P_{ey}(K_y L_y)^2/10^4$		2.46	2.20	1.99	1.76	1.49	1.17	1.05	0.966	0.861	0.732	0.578

Note: Heavy line indicates Kl/r equal to or greater than 200.

$F_y = 46$ ksi
$f'_c = 4$ ksi
$\phi_c = 0.85$

Table 4-12 (cont.). Composite Rectangular (and Square) HSS Design Strength in Axial Compression, $\phi_c P_n$, kips

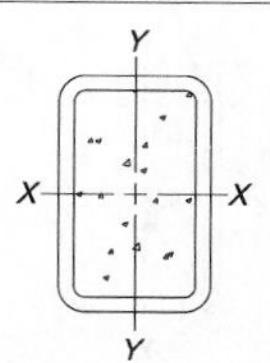

Shape	HSS4×2½×			HSS4×2×					HSS3½×3½×				
	5/16	1/4	3/16	3/8	5/16	1/4	3/16	1/8	3/8	5/16	1/4	3/16	1/8
t_{design}, in.	**0.291**	**0.233**	**0.174**	**0.349**	**0.291**	**0.233**	**0.174**	**0.116**	**0.349**	**0.291**	**0.233**	**0.174**	**0.116**
Wt/ft	**11.6**	**9.63**	**7.49**	**12.1**	**10.5**	**8.78**	**6.85**	**4.75**	**14.6**	**12.7**	**10.5**	**8.13**	**5.60**
Effective length *KL* (ft) with respect to least radius of gyration, r_{my}													
0	138	119	98.0	138	123	105	86.6	66.6	174	154	133	111	87.0
1	136	118	97.0	136	120	104	85.3	65.6	173	153	133	110	86.5
2	132	114	94.1	128	114	98.6	81.3	62.7	170	150	130	108	85.0
3	125	108	89.3	117	105	90.8	75.1	58.1	164	146	126	105	82.6
4	115	100	83.1	103	92.8	80.9	67.3	52.2	157	140	121	101	79.3
5	104	91.0	75.7	87.0	79.3	69.7	58.4	45.4	149	133	115	95.8	75.3
6	92.4	80.9	67.5	71.0	65.5	58.2	49.1	38.4	139	124	108	90.0	70.6
7	80.0	70.4	59.0	55.9	52.2	46.9	40.0	31.4	128	115	100	83.5	65.5
8	67.7	60.0	50.6	42.9	40.4	36.6	31.5	24.9	116	105	91.6	76.6	60.1
9	56.0	50.0	42.4	33.9	31.9	28.9	24.9	19.7	105	94.5	83.0	69.5	54.5
10	45.5	40.8	34.8	27.4	25.8	23.4	20.1	16.0	92.8	84.2	74.2	62.3	48.8
11	37.6	33.7	28.7	22.7	21.4	19.4	16.6	13.2	81.3	74.2	65.6	55.3	43.2
12	31.6	28.3	24.1	19.1	17.9	16.3	14.0	11.1	70.4	64.5	57.4	48.5	37.8
13	26.9	24.2	20.6				11.9	9.44	60.2	55.4	49.4	41.9	32.7
14	23.2	20.8	17.7						51.9	47.7	42.6	36.1	28.2
15	20.2	18.1	15.5						45.2	41.6	37.1	31.5	24.5
16		15.9	13.6						39.7	36.6	32.6	27.7	21.6
17									35.2	32.4	28.9	24.5	19.1
18									31.4	28.9	25.8	21.9	17.0
19									28.2	25.9	23.1	19.6	15.3
20									25.4	23.4	20.9	17.7	13.8
21									23.1	21.2	18.9	16.1	12.5
22											17.3	14.6	11.4
23													
24													
25													
26													
27													
28													
29													
30													
Properties													
r_{my}, in.	0.947	0.973	0.999	0.729	0.754	0.779	0.804	0.830	1.26	1.29	1.32	1.35	1.37
r_{mx}/r_{my}	1.46	1.45	1.44	1.77	1.75	1.75	1.73	1.72	1.00	1.00	1.00	1.00	1.00
$\phi_b M_{nx}$, kip-ft	13.7	11.7	9.21	13.2	11.8	10.1	8.07	5.73	16.2	14.3	12.1	9.52	6.66
$\phi_b M_{ny}$, kip-ft	9.83	8.38	6.66	7.97	7.18	6.18	4.93	3.52	16.2	14.3	12.1	9.52	6.66
$P_{ex}(K_x L_x)^2/10^4$	1.29	1.15	0.976	1.14	1.06	0.953	0.809	0.634	1.36	1.25	1.11	0.943	0.743
$P_{ey}(K_y L_y)^2/10^4$	0.609	0.548	0.467	0.368	0.346	0.314	0.270	0.215	1.36	1.25	1.11	0.943	0.743

Note: Heavy line indicates Kl/r equal to or greater than 200.

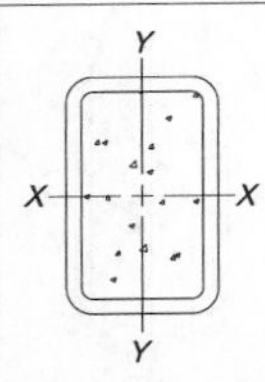

Table 4-12 (cont.). Composite Rectangular (and Square) HSS Design Strength in Axial Compression, $\phi_c P_n$, kips

$F_y = 46$ ksi
$f'_c = 4$ ksi
$\phi_c = 0.85$

Shape	HSS3×3×					HSS2½×2½×			
	3/8	5/16	1/4	3/16	1/8	5/16	1/4	3/16	1/8
t_{design}, in.	**0.349**	**0.291**	**0.233**	**0.174**	**0.116**	**0.291**	**0.233**	**0.174**	**0.116**
Wt/ft	**12.1**	**10.5**	**8.78**	**6.85**	**4.75**	**8.40**	**7.08**	**5.57**	**3.90**
Effective length *KL* (ft) with respect to least radius of gyration, r_{my}									
0	141	125	108	89.5	69.5	97.8	84.6	69.9	53.9
1	140	124	107	88.8	69.0	96.6	83.6	69.1	53.3
2	136	121	105	86.7	67.4	92.9	80.6	66.7	51.6
3	130	116	101	83.3	64.8	87.2	75.9	63.0	48.7
4	123	109	94.9	78.8	61.4	79.7	69.7	58.1	45.0
5	113	101	88.2	73.4	57.2	71.1	62.5	52.3	40.7
6	103	92.2	80.6	67.3	52.5	61.7	54.7	46.0	35.9
7	91.6	82.5	72.4	60.7	47.4	52.3	46.7	39.6	31.0
8	80.3	72.6	64.0	53.9	42.2	43.2	38.9	33.3	26.2
9	69.1	62.7	55.7	47.1	36.9	34.7	31.6	27.3	21.6
10	58.5	53.3	47.7	40.5	31.9	28.1	25.6	22.1	17.5
11	48.6	44.5	40.0	34.2	27.0	23.2	21.2	18.3	14.5
12	40.8	37.4	33.6	28.8	22.7	19.5	17.8	15.4	12.2
13	34.8	31.9	28.7	24.5	19.3	16.6	15.2	13.1	10.4
14	30.0	27.5	24.7	21.1	16.7	14.3	13.1	11.3	8.95
15	26.1	23.9	21.5	18.4	14.5		11.4	9.84	7.80
16	23.0	21.0	18.9	16.2	12.8				6.85
17	20.3	18.6	16.8	14.3	11.3				
18		16.6	15.0	12.8	10.1				
19				11.5	9.06				
20									
21									
22									
23									
24									
25									
26									
27									
28									
29									
30									
Properties									
r_{my}, in.	1.06	1.08	1.11	1.14	1.17	0.880	0.908	0.937	0.965
r_{mx}/r_{my}	1.00	1.00	1.00	1.00	1.00	1.00	1.00	1.00	1.00
$\phi_b M_{nx}$, kip-ft	11.2	10.0	8.56	6.80	4.83	6.49	5.62	4.55	3.27
$\phi_b M_{ny}$, kip-ft	11.2	10.0	8.56	6.80	4.83	6.49	5.62	4.55	3.27
$P_{ex}(K_x L_x)^2/10^4$	0.783	0.727	0.653	0.556	0.439	0.377	0.345	0.296	0.236
$P_{ey}(K_y L_y)^2/10^4$	0.783	0.727	0.653	0.556	0.439	0.377	0.345	0.296	0.236

Note: Heavy line indicates Kl/r equal to or greater than 200.

$F_y = 46$ ksi
$f'_c = 5$ ksi
$\phi_c = 0.85$

Table 4-13. Composite Rectangular (and Square) HSS Design Strength in Axial Compression, $\phi_c P_n$, kips

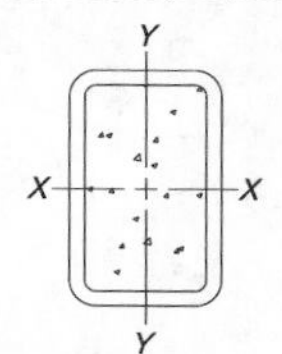

Shape	HSS20×12×		HSS18×12×		HSS16×16×			HSS16×12×		
	5/8	1/2	5/8	1/2	5/8	1/2	3/8	5/8	1/2	3/8
t_{design}, in.	0.581	0.465	0.581	0.465	0.581	0.465	0.349	0.581	0.465	0.349
Wt/ft	127	103	119	96.4	127	103	78.4	110	89.6	68.2
Effective length *KL* (ft) with respect to least radius of gyration, r_{my}										
0	1990	1770	1830	1630	2050	1830	1610	1670	1480	1290
6	1960	1740	1800	1600	2030	1810	1600	1640	1460	1270
7	1950	1730	1790	1590	2020	1810	1590	1630	1450	1260
8	1930	1720	1770	1580	2010	1800	1580	1620	1440	1250
9	1920	1710	1760	1570	2000	1790	1570	1600	1430	1240
10	1900	1690	1740	1560	1990	1780	1570	1590	1410	1230
11	1880	1680	1730	1540	1980	1770	1560	1570	1400	1220
12	1860	1660	1710	1520	1960	1760	1540	1560	1380	1200
13	1840	1640	1690	1510	1950	1740	1530	1540	1370	1190
14	1820	1620	1670	1490	1940	1730	1520	1520	1350	1170
15	1800	1600	1640	1470	1920	1720	1510	1500	1330	1160
16	1770	1580	1620	1450	1900	1700	1490	1470	1310	1140
17	1740	1550	1600	1420	1880	1680	1480	1450	1290	1120
18	1720	1530	1570	1400	1870	1670	1460	1430	1270	1100
19	1690	1500	1540	1380	1850	1650	1450	1400	1250	1080
20	1660	1470	1520	1350	1830	1630	1430	1380	1220	1060
21	1630	1450	1490	1320	1800	1610	1410	1350	1200	1040
22	1590	1420	1460	1300	1780	1590	1400	1320	1170	1020
23	1560	1390	1430	1270	1760	1570	1380	1290	1150	997
24	1530	1360	1400	1240	1740	1550	1360	1270	1120	974
25	1490	1330	1360	1220	1710	1530	1340	1240	1100	951
26	1460	1300	1330	1190	1690	1500	1320	1210	1070	927
27	1420	1270	1300	1160	1660	1480	1300	1180	1040	903
28	1390	1240	1270	1130	1630	1460	1280	1150	1020	879
29	1350	1200	1230	1100	1610	1430	1250	1110	990	855
30	1320	1170	1200	1070	1580	1410	1230	1080	962	831
34	1170	1040	1060	947	1470	1310	1140	959	851	733
38	1030	911	930	827	1350	1200	1050	836	741	636
42	887	786	801	711	1230	1100	953	717	635	544
46	753	667	678	602	1110	989	858	605	535	457
50	638	564	574	509	997	884	765	512	453	387
Properties										
r_{my}, in.	4.93	4.99	4.87	4.93	6.25	6.31	6.37	4.80	4.86	4.91
r_{mx}/r_{my}	1.49	1.48	1.37	1.37	1.00	1.00	1.00	1.25	1.25	1.25
$\phi_b M_{nx}$, kip-ft	794	649	676	555	690	566	435	569	466	359
$\phi_b M_{ny}$, kip-ft	559	455	511	421	690	566	435	466	383	295
$P_{ex}(K_x L_x)^2/10^4$	473	416	361	318	351	310	266	268	236	202
$P_{ey}(K_y L_y)^2/10^4$	214	189	193	170	351	310	266	172	152	130

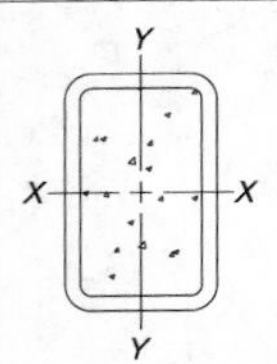

Table 4-13 (cont.). Composite Rectangular (and Square) HSS Design Strength in Axial Compression, $\phi_c P_n$, kips

$F_y = 46$ ksi
$f'_c = 5$ ksi
$\phi_c = 0.85$

Shape		HSS16×8×			HSS14×14×			HSS14×12×		HSS14×10×		
		5/8	1/2	3/8	5/8	1/2	3/8	1/2	3/8	5/8	1/2	3/8
t_{design}, in.		0.581	0.465	0.349	0.581	0.465	0.349	0.465	0.349	0.581	0.465	0.349
Wt/ft		93.1	75.9	58.0	110	89.6	68.2	82.7	63.1	93.1	75.9	58.0
Effective length KL (ft) with respect to least radius of gyration, r_{my}	0	1290	1130	975	1680	1500	1310	1340	1160	1330	1180	1020
	6	1240	1090	941	1660	1480	1290	1320	1140	1300	1150	994
	7	1230	1080	929	1650	1470	1280	1310	1130	1290	1140	985
	8	1210	1060	915	1640	1460	1270	1300	1120	1280	1130	975
	9	1190	1050	899	1630	1450	1270	1290	1110	1260	1110	963
	10	1170	1030	882	1620	1440	1260	1270	1100	1240	1100	951
	11	1140	1000	864	1610	1430	1250	1260	1090	1230	1080	937
	12	1120	982	844	1590	1420	1240	1250	1080	1210	1070	922
	13	1090	958	823	1580	1400	1220	1230	1070	1190	1050	907
	14	1060	933	801	1560	1390	1210	1220	1050	1170	1030	890
	15	1030	906	778	1550	1370	1200	1200	1040	1140	1010	872
	16	999	879	755	1530	1360	1180	1180	1020	1120	989	854
	17	966	850	730	1510	1340	1170	1160	1000	1090	967	835
	18	933	821	705	1490	1320	1150	1140	986	1070	944	815
	19	899	791	679	1470	1310	1140	1120	968	1040	920	794
	20	865	761	653	1450	1290	1120	1100	949	1010	896	773
	21	830	731	627	1430	1270	1100	1080	930	987	872	751
	22	795	700	600	1400	1250	1080	1050	910	958	847	729
	23	760	670	574	1380	1230	1070	1030	890	929	821	707
	24	726	639	547	1360	1200	1050	1010	869	900	795	685
	25	691	609	521	1330	1180	1030	984	848	870	769	662
	26	657	579	495	1310	1160	1010	960	826	841	743	639
	27	623	549	470	1280	1140	987	935	805	811	717	616
	28	590	520	444	1250	1110	966	910	783	781	690	593
	29	557	491	420	1230	1090	945	885	760	751	664	570
	30	525	463	396	1200	1070	924	859	738	722	638	547
	34	410	362	309	1090	967	837	757	649	606	536	459
	38	328	290	247	981	868	749	657	562	497	440	375
	42	269	237	202	870	769	662	561	478	407	360	307
	46	224	198	169	763	673	578	471	401	339	300	256
	50	190	167	143	661	582	498	399	339	287	254	217
Properties												
r_{my}, in.		3.27	3.32	3.37	5.44	5.49	5.55	4.78	4.83	3.98	4.04	4.09
r_{mx}/r_{my}		1.72	1.72	1.71	1.00	1.00	1.00	1.13	1.12	1.30	1.29	1.29
$\phi_b M_{nx}$, kip-ft		445	366	283	521	428	329	383	296	414	341	263
$\phi_b M_{ny}$, kip-ft		273	226	175	521	428	329	345	267	328	271	209
$P_{ex}(K_x L_x)^2/10^4$		189	165	140	222	196	167	169	144	162	143	121
$P_{ey}(K_y L_y)^2/10^4$		63.4	56.0	47.7	222	196	167	133	114	96.3	85.2	72.5

$F_y = 46$ ksi
$f'_c = 5$ ksi
$\phi_c = 0.85$

Table 4-13 (cont.). Composite Rectangular (and Square) HSS Design Strength in Axial Compression, $\phi_c P_n$, kips

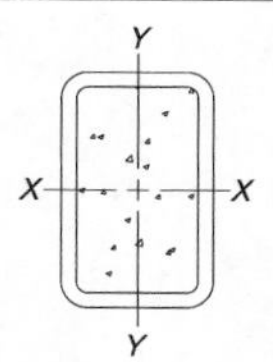

Shape		HSS12×12×				HSS12×10×			HSS12×8×			
		5/8	1/2	3/8	5/16	1/2	3/8	5/16	5/8	1/2	3/8	5/16
t_{design}, in.		0.581	0.465	0.349	0.291	0.465	0.349	0.291	0.581	0.465	0.349	0.291
Wt/ft		93.1	75.9	58.0	48.8	69.1	52.9	44.6	76.1	62.3	47.8	40.3
Effective length KL (ft) with respect to least radius of gyration, r_{my}	0	1350	1190	1030	949	1040	902	824	1020	900	770	703
	6	1320	1170	1010	931	1020	879	803	985	866	742	676
	7	1310	1160	1010	925	1010	871	795	972	855	732	667
	8	1300	1150	999	918	998	862	787	956	841	720	656
	9	1290	1140	991	910	986	851	777	939	826	708	645
	10	1280	1130	981	900	973	840	767	921	810	694	632
	11	1270	1120	970	891	959	827	755	900	792	679	618
	12	1250	1110	959	880	943	814	743	879	774	662	603
	13	1240	1090	947	868	927	800	729	856	753	645	587
	14	1220	1080	933	856	909	784	715	832	732	627	570
	15	1200	1060	919	843	891	768	701	807	710	608	553
	16	1180	1050	905	830	872	752	685	781	688	589	535
	17	1160	1030	889	815	852	734	669	754	664	569	516
	18	1140	1010	873	800	831	716	652	727	640	548	498
	19	1120	991	857	785	810	698	635	699	616	527	478
	20	1100	972	840	769	788	679	618	670	591	506	459
	21	1080	952	822	753	766	659	600	642	566	485	439
	22	1050	931	804	736	743	639	581	613	541	463	420
	23	1030	910	785	719	720	619	563	585	516	442	400
	24	1010	888	767	701	696	599	544	557	491	421	381
	25	981	867	747	683	673	578	525	529	467	400	361
	26	956	844	728	665	649	558	506	501	442	379	342
	27	930	822	708	647	626	537	487	473	418	358	323
	28	905	799	688	628	602	516	469	447	395	338	305
	29	879	776	668	610	578	496	450	421	372	318	287
	30	853	753	648	591	555	475	431	395	349	299	269
	34	749	661	568	517	463	396	358	307	272	233	210
	38	647	571	489	444	378	322	291	246	218	186	168
	42	550	485	414	375	309	264	238	201	178	153	137
	46	460	406	346	313	258	220	199	168	149	127	114
	50	390	344	293	265	218	186	168	142	126	108	96.9
Properties												
r_{my}, in.		4.62	4.68	4.73	4.76	3.96	4.01	4.04	3.16	3.21	3.27	3.29
r_{mx}/r_{my}		1.00	1.00	1.00	1.00	1.15	1.15	1.15	1.37	1.37	1.37	1.37
$\phi_b M_{nx}$, kip-ft		376	309	239	202	272	211	178	283	235	183	155
$\phi_b M_{ny}$, kip-ft		376	309	239	202	240	186	158	214	178	138	118
$P_{ex}(K_xL_x)^2/10^4$		131	115	97.9	89.0	97.0	82.4	74.7	90.2	79.2	67.0	60.6
$P_{ey}(K_yL_y)^2/10^4$		131	115	97.9	89.0	73.2	62.2	56.6	47.7	42.4	35.8	32.5

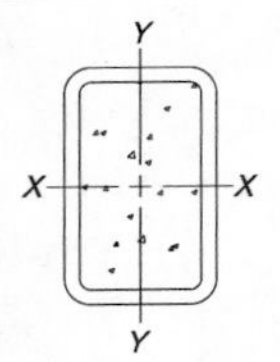

Table 4-13 (cont.). Composite Rectangular (and Square) HSS Design Strength in Axial Compression, $\phi_c P_n$, kips

$F_y = 46$ ksi
$f'_c = 5$ ksi
$\phi_c = 0.85$

Shape		HSS12×6×				HSS10×10×				
		5/8	1/2	3/8	5/16	5/8	1/2	3/8	5/16	1/4
t_{design}, in.		0.581	0.465	0.349	0.291	0.581	0.465	0.349	0.291	0.233
Wt/ft		67.6	55.5	42.7	36.0	76.1	62.3	47.8	40.3	32.6
Effective length KL (ft) with respect to least radius of gyration, r_{my}	0	863	752	639	578	1040	914	785	717	648
	6	809	706	600	542	1010	891	764	698	631
	7	790	690	586	530	1000	882	757	691	624
	8	769	671	571	516	990	873	749	684	617
	9	746	651	554	501	978	862	740	675	609
	10	720	630	536	484	964	850	729	665	601
	11	694	607	516	467	949	837	718	655	591
	12	665	582	496	448	934	823	706	644	581
	13	636	557	474	429	917	808	693	632	570
	14	605	531	452	409	899	792	680	619	558
	15	574	504	429	388	880	776	665	606	546
	16	543	477	407	368	860	758	650	592	533
	17	512	450	383	347	839	740	635	578	520
	18	480	423	360	326	818	722	619	563	507
	19	449	396	338	305	796	702	602	548	492
	20	418	369	315	285	773	683	585	532	478
	21	388	343	293	265	751	662	568	516	463
	22	359	318	272	246	727	642	550	499	449
	23	331	294	251	227	704	621	532	483	433
	24	304	269	230	208	680	600	514	466	418
	25	280	248	212	192	656	579	496	450	403
	26	259	230	196	178	632	558	477	433	388
	27	240	213	182	165	607	537	459	416	372
	28	223	198	169	153	583	516	441	399	357
	29	208	185	158	143	559	495	423	383	342
	30	194	172	147	133	536	474	405	366	327
	34	151	134	115	104	444	393	335	302	269
	38	121	108	91.9	83.1	359	318	271	244	217
	42				68.0	294	260	222	200	177
	46					245	217	185	167	148
	50					207	184	157	141	125
Properties										
r_{my}, in.		2.39	2.44	2.49	2.52	3.80	3.86	3.92	3.94	3.97
r_{mx}/r_{my}		1.73	1.73	1.72	1.71	1.00	1.00	1.00	1.00	1.00
$\phi_b M_{nx}$, kip-ft		237	198	155	131	253	209	163	138	113
$\phi_b M_{ny}$, kip-ft		145	121	95.6	81.4	253	209	163	138	113
$P_{ex}(K_x L_x)^2/10^4$		70.5	62.0	52.3	47.0	69.7	61.5	52.3	47.2	41.9
$P_{ey}(K_y L_y)^2/10^4$		23.5	20.8	17.7	16.1	69.7	61.5	52.3	47.2	41.9

Note: Heavy line indicates Kl/r equal to or greater than 200.

$F_y = 46$ ksi
$f'_c = 5$ ksi
$\phi_c = 0.85$

Table 4-13 (cont.). Composite Rectangular (and Square) HSS Design Strength in Axial Compression, $\phi_c P_n$, kips

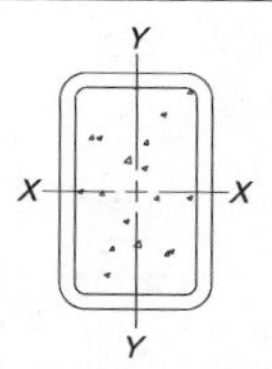

Shape		HSS10×8×				HSS10×6×					HSS10×5×		
		1/2	3/8	5/16	1/4	5/8	1/2	3/8	5/16	1/4	3/8	5/16	1/4
t_{design}, in.		0.465	0.349	0.291	0.233	0.581	0.465	0.349	0.291	0.233	0.349	0.291	0.233
Wt/ft		55.5	42.7	36.0	29.2	59.1	48.7	37.6	31.8	25.8	35.1	29.7	24.1
Effective length KL (ft) with respect to least radius of gyration, r_{my}	0	781	668	607	546	747	652	551	498	444	491	443	393
	6	751	642	583	525	698	610	516	466	416	448	404	358
	7	741	633	575	517	681	596	504	455	406	434	391	347
	8	729	623	566	509	662	580	490	443	395	417	376	334
	9	715	611	555	499	641	562	475	429	382	400	360	319
	10	701	599	544	489	619	542	459	415	369	381	343	304
	11	685	585	531	477	595	522	442	399	355	361	325	288
	12	668	571	518	465	570	500	424	383	340	340	306	272
	13	650	555	504	452	544	478	405	366	325	319	287	255
	14	631	539	489	439	517	454	385	348	309	298	268	238
	15	611	522	474	425	489	431	365	330	293	277	249	221
	16	591	505	458	411	462	407	345	312	277	256	230	204
	17	570	487	442	396	434	383	325	294	260	235	211	188
	18	549	469	425	381	406	359	305	276	244	215	193	171
	19	527	450	408	365	379	335	285	258	228	196	176	156
	20	505	431	391	350	352	312	265	240	212	177	159	141
	21	483	413	374	334	326	289	246	223	196	160	144	128
	22	461	394	356	318	301	267	228	206	181	146	131	116
	23	439	375	339	303	276	246	209	189	166	134	120	107
	24	417	356	322	287	253	226	192	174	153	123	110	97.9
	25	396	337	305	272	233	208	177	160	141	113	102	90.2
	26	374	319	289	257	216	192	164	148	130	105	94.0	83.4
	27	353	301	272	242	200	178	152	137	121	97.0	87.2	77.3
	28	333	284	256	228	186	166	141	128	112	90.2	81.1	71.9
	29	313	266	241	214	173	154	132	119	105	84.1	75.6	67.0
	30	293	249	225	200	162	144	123	111	97.8	78.6	70.6	62.6
	32	257	219	198	176	142	127	108	97.8	86.0	69.1	62.1	55.0
	34	228	194	175	156	126	112	95.8	86.6	76.2	61.2	55.0	48.8
	36	203	173	156	139	113	100	85.4	77.3	67.9			
	38	182	155	140	125	101	90.0	76.7	69.3	61.0			
	40	165	140	127	112			69.2	62.6	55.0			
Properties													
r_{my}, in.		3.14	3.19	3.22	3.25	2.34	2.39	2.44	2.47	2.49	2.05	2.07	2.10
r_{mx}/r_{my}		1.19	1.19	1.19	1.18	1.50	1.49	1.49	1.48	1.48	1.72	1.72	1.71
$\phi_b M_{nx}$, kip-ft		179	140	119	96.9	177	148	117	99.4	81.4	105	89.7	73.5
$\phi_b M_{ny}$, kip-ft		154	120	102	83.5	124	104	81.8	69.7	57.3	64.5	55.2	45.5
$P_{ex}(K_x L_x)^2/10^4$		50.1	42.4	38.3	33.8	43.8	38.6	32.9	29.7	26.0	28.0	25.3	22.2
$P_{ey}(K_y L_y)^2/10^4$		35.4	30.1	27.2	24.1	19.5	17.4	14.8	13.4	11.8	9.48	8.57	7.58

Note: Heavy line indicates Kl/r equal to or greater than 200.

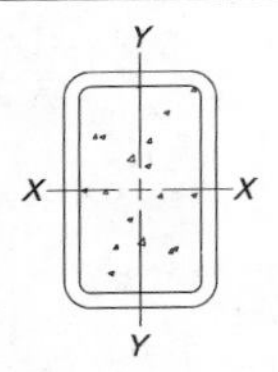

Table 4-13 (cont.). Composite Rectangular (and Square) HSS Design Strength in Axial Compression, $\phi_c P_n$, kips

$F_y = 46$ ksi
$f'_c = 5$ ksi
$\phi_c = 0.85$

Shape		HSS9×7×					HSS9×5×				
		5/8	1/2	3/8	5/16	1/4	5/8	1/2	3/8	5/16	1/4
t_{design}, in.		0.581	0.465	0.349	0.291	0.233	0.581	0.465	0.349	0.291	0.233
Wt/ft		59.1	48.7	37.6	31.8	25.8	50.6	41.9	32.5	27.5	22.4
Effective length KL (ft) with respect to least radius of gyration, r_{my}	0	758	663	562	509	455	616	537	451	406	360
	6	720	630	534	483	432	558	488	411	370	328
	7	706	618	524	474	424	538	471	397	358	317
	8	691	605	513	464	415	516	453	382	344	305
	9	674	591	501	453	405	493	433	366	329	292
	10	656	575	488	441	394	467	411	348	313	278
	11	637	558	473	428	383	441	389	330	297	263
	12	616	540	458	415	370	414	366	311	280	248
	13	594	521	442	400	357	387	342	291	262	232
	14	571	502	426	385	344	359	318	271	244	217
	15	548	482	409	370	330	331	295	252	227	201
	16	524	461	391	354	315	304	271	232	209	185
	17	500	440	373	337	301	278	248	213	192	170
	18	475	418	355	321	286	252	226	195	175	155
	19	450	397	337	305	271	227	205	177	159	141
	20	426	376	319	288	256	205	185	160	144	127
	21	401	354	301	272	242	186	167	145	130	116
	22	377	333	283	256	227	169	153	132	119	105
	23	354	313	266	240	213	155	140	121	109	96.3
	24	330	292	248	224	199	142	128	111	99.7	88.4
	25	308	273	232	209	186	131	118	102	91.9	81.5
	26	286	253	215	194	172	121	109	94.5	85.0	75.4
	27	265	235	200	180	160	113	101	87.7	78.8	69.9
	28	246	219	186	168	149	105	94.2	81.5	73.3	65.0
	29	230	204	173	156	138	97.5	87.8	76.0	68.3	60.6
	30	214	190	162	146	129	91.2	82.1	71.0	63.8	56.6
	32	189	167	142	128	114	80.1	72.1	62.4	56.1	49.7
	34	167	148	126	114	101				49.7	44.1
	36	149	132	112	101	89.9					
	38	134	119	101	91.0	80.7					
	40	121	107	91.0	82.2	72.8					
Properties											
r_{my}, in.		2.68	2.73	2.78	2.81	2.84	1.92	1.97	2.03	2.05	2.08
r_{mx}/r_{my}		1.22	1.22	1.22	1.21	1.21	1.60	1.59	1.58	1.58	1.57
$\phi_b M_{nx}$, kip-ft		167	140	110	93.5	76.6	133	112	88.7	75.9	62.4
$\phi_b M_{ny}$, kip-ft		140	117	92.1	78.7	64.5	87.3	74.2	59.0	50.4	41.4
$P_{ex}(K_x L_x)^2/10^4$		38.3	34.0	28.9	26.0	22.9	28.4	25.3	21.5	19.3	16.9
$P_{ey}(K_y L_y)^2/10^4$		25.7	22.8	19.5	17.6	15.6	11.1	9.95	8.53	7.73	6.81

Note: Heavy line indicates Kl/r equal to or greater than 200.

$F_y = 46$ ksi
$f'_c = 5$ ksi
$\phi_c = 0.85$

Table 4-13 (cont.). Composite Rectangular (and Square) HSS Design Strength in Axial Compression, $\phi_c P_n$, kips

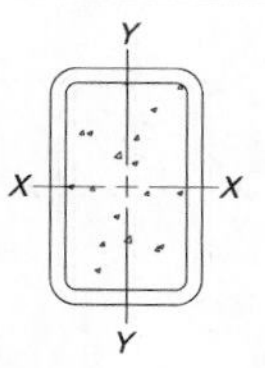

Shape		HSS8×8×						HSS8×6×					
		5/8	1/2	3/8	5/16	1/4	3/16	5/8	1/2	3/8	5/16	1/4	3/16
t_{design}, in.		0.581	0.465	0.349	0.291	0.233	0.174	0.581	0.465	0.349	0.291	0.233	0.174
Wt/ft		59.1	48.7	37.6	31.8	25.8	19.6	50.6	41.9	32.5	27.5	22.4	17.1
Effective length KL (ft) with respect to least radius of gyration, r_{my}	0	761	666	566	512	459	403	627	548	462	417	371	323
	6	730	640	543	492	440	386	583	511	431	389	346	301
	7	719	630	535	484	433	380	569	498	421	380	338	293
	8	707	619	526	476	426	373	552	484	409	369	328	285
	9	693	607	516	467	417	366	534	468	396	357	318	276
	10	678	594	505	457	408	358	514	451	382	345	306	266
	11	662	580	493	446	398	349	493	433	367	331	294	255
	12	644	565	480	434	388	339	471	414	351	317	282	244
	13	626	549	466	422	376	329	448	395	335	302	268	232
	14	607	532	452	409	365	319	425	375	318	287	255	220
	15	587	515	437	396	353	308	401	354	301	271	241	208
	16	566	497	422	382	340	297	377	333	284	256	227	196
	17	545	478	407	368	327	285	353	313	267	240	213	184
	18	523	460	391	353	314	273	329	292	250	225	199	171
	19	501	440	375	339	301	262	306	272	233	209	186	159
	20	479	421	358	324	287	250	283	252	216	194	172	148
	21	457	402	342	309	274	238	261	233	200	180	159	136
	22	434	382	325	294	261	226	239	214	184	165	147	125
	23	412	363	309	279	247	214	219	196	169	151	134	115
	24	390	344	293	265	234	202	201	180	155	139	123	105
	25	369	325	277	250	221	191	185	166	143	128	114	97.0
	26	348	307	261	236	208	179	171	153	132	118	105	89.7
	27	327	289	246	222	196	168	159	142	122	110	97.3	83.2
	28	307	271	231	208	183	158	148	132	114	102	90.5	77.3
	29	287	253	216	195	171	147	138	123	106	95.2	84.4	72.1
	30	268	237	202	182	160	137	129	115	99.1	89.0	78.8	67.4
	32	235	208	178	160	141	121	113	101	87.1	78.2	69.3	59.2
	34	208	184	157	142	125	107	100	89.7	77.2	69.3	61.4	52.4
	36	186	164	140	126	111	95.4	89.4	80.0	68.8	61.8	54.7	46.8
	38	167	148	126	114	99.8	85.6		71.8	61.8	55.5	49.1	42.0
	40	151	133	114	102	90.0	77.2				50.1	44.3	37.9
Properties													
r_{my}, in.		2.99	3.04	3.10	3.13	3.15	3.18	2.27	2.32	2.38	2.40	2.43	2.46
r_{mx}/r_{my}		1.00	1.00	1.00	1.00	1.00	1.00	1.26	1.25	1.25	1.25	1.25	1.24
$\phi_b M_{nx}$, kip-ft		154	129	101	86.6	70.7	54.2	125	105	83.1	71.1	58.3	44.9
$\phi_b M_{ny}$, kip-ft		154	129	101	86.6	70.7	54.2	102	85.9	68.3	58.3	48.0	36.9
$P_{ex}(K_x L_x)^2/10^4$		32.2	28.6	24.3	21.9	19.4	16.6	24.6	21.9	18.6	16.8	14.8	12.6
$P_{ey}(K_y L_y)^2/10^4$		32.2	28.6	24.3	21.9	19.4	16.6	15.6	13.9	11.9	10.8	9.51	8.12

Note: Heavy line indicates Kl/r equal to or greater than 200.

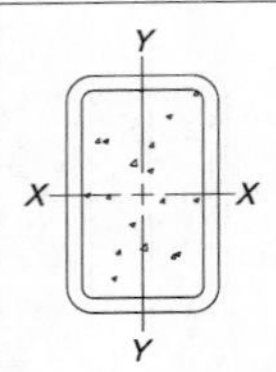

Table 4-13 (cont.). Composite Rectangular (and Square) HSS Design Strength in Axial Compression, $\phi_c P_n$, kips

$F_y = 46$ ksi
$f'_c = 5$ ksi
$\phi_c = 0.85$

Shape		HSS8×4×						HSS7×7×					
		5/8	1/2	3/8	5/16	1/4	3/16	5/8	1/2	3/8	5/16	1/4	3/16
t_{design}, in.		0.581	0.465	0.349	0.291	0.233	0.174	0.581	0.465	0.349	0.291	0.233	0.174
Wt/ft		42.1	35.1	27.4	23.3	19.0	14.5	50.6	41.9	32.5	27.5	22.4	17.1
Effective length KL (ft) with respect to least radius of gyration, r_{my}	0	496	431	360	322	284	243	630	551	466	421	375	327
	6	423	371	311	279	246	211	596	522	441	399	355	309
	7	400	351	295	265	234	200	584	512	433	391	348	303
	8	374	330	278	249	220	188	571	501	423	382	340	296
	9	347	307	260	233	206	176	556	488	413	373	332	288
	10	320	284	240	216	191	163	540	474	401	362	323	280
	11	291	260	221	198	175	150	523	459	389	351	312	271
	12	263	236	201	181	160	137	505	444	376	339	302	261
	13	236	212	182	164	145	124	486	427	362	327	291	251
	14	210	190	163	147	130	111	466	410	348	314	279	241
	15	184	168	145	131	116	99.1	446	392	333	301	267	231
	16	162	148	128	115	102	87.4	425	374	318	287	255	220
	17	143	131	113	102	90.7	77.4	404	356	302	273	243	209
	18	128	117	101	91.0	80.9	69.0	383	338	287	259	230	198
	19	115	105	90.7	81.7	72.6	61.9	361	319	272	245	218	187
	20	104	94.5	81.9	73.7	65.5	55.9	340	301	256	232	205	176
	21	94.0	85.7	74.3	66.8	59.4	50.7	320	283	241	218	193	165
	22	85.6	78.1	67.7	60.9	54.1	46.2	299	265	226	204	181	154
	23	78.4	71.5	61.9	55.7	49.5	42.3	279	248	211	191	169	144
	24	72.0	65.6	56.9	51.2	45.5	38.8	260	231	197	178	158	134
	25	66.3	60.5	52.4	47.2	41.9	35.8	240	214	183	166	146	124
	26		55.9	48.4	43.6	38.8	33.1	222	198	169	153	135	115
	27				40.4	35.9	30.7	206	184	157	142	125	106
	28						28.5	192	171	146	132	117	98.8
	29							179	159	136	123	109	92.1
	30							167	149	127	115	102	86.1
	32							147	131	112	101	89.3	75.7
	34							130	116	99.1	89.5	79.1	67.0
	36							116	103	88.4	79.8	70.6	59.8
	38							104	92.7	79.3	71.7	63.3	53.7
	40							93.9	83.6	71.6	64.7	57.2	48.4
Properties													
r_{my}, in.		1.51	1.56	1.61	1.63	1.66	1.69	2.58	2.63	2.69	2.72	2.75	2.77
r_{mx}/r_{my}		1.75	1.74	1.73	1.73	1.72	1.70	1.00	1.00	1.00	1.00	1.00	1.00
$\phi_b M_{nx}$, kip-ft		94.5	81.1	64.9	55.5	45.9	35.2	114	96.3	76.2	65.2	53.5	41.1
$\phi_b M_{ny}$, kip-ft		57.3	49.3	39.7	34.2	28.3	21.8	114	96.3	76.2	65.2	53.5	41.1
$P_{ex}(K_xL_x)^2/10^4$		17.1	15.4	13.1	11.8	10.3	8.74	20.2	18.0	15.4	13.9	12.2	10.4
$P_{ey}(K_yL_y)^2/10^4$		5.54	5.05	4.38	3.98	3.51	2.98	20.2	18.0	15.4	13.9	12.2	10.4

Note: Heavy line indicates Kl/r equal to or greater than 200.

$F_y = 46$ ksi
$f'_c = 5$ ksi
$\phi_c = 0.85$

Table 4-13 (cont.). Composite Rectangular (and Square) HSS Design Strength in Axial Compression, $\phi_c P_n$, kips

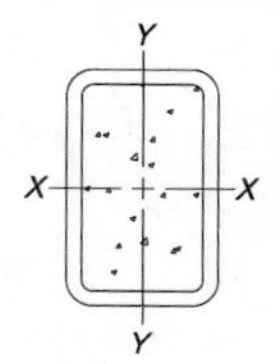

Shape	HSS7×5×						HSS7×4×				
	5/8	1/2	3/8	5/16	1/4	3/16	1/2	3/8	5/16	1/4	3/16
t_{design}, in.	0.581	0.465	0.349	0.291	0.233	0.174	0.465	0.349	0.291	0.233	0.174
Wt/ft	42.1	35.1	27.4	23.3	19.0	14.5	31.7	24.9	21.2	17.3	13.2
Effective length KL (ft) with respect to least radius of gyration, r_{my}											
0	507	442	371	333	295	254	387	323	289	254	218
6	456	399	336	302	267	230	331	278	250	220	188
7	439	385	324	292	258	222	313	263	237	209	178
8	420	369	311	280	248	213	293	247	223	196	167
9	400	352	297	267	236	203	272	230	208	183	156
10	378	333	282	254	225	193	251	213	192	170	144
11	356	314	267	240	212	182	229	195	176	156	132
12	333	294	250	225	199	171	207	177	161	142	120
13	309	274	234	210	186	160	186	160	145	128	109
14	286	254	217	195	173	148	165	143	130	115	97.2
15	263	234	201	180	160	137	146	126	115	102	86.3
16	240	215	185	166	147	126	128	111	101	90.0	75.9
17	218	196	169	152	134	115	113	98.4	89.8	79.7	67.2
18	197	177	153	138	122	104	101	87.8	80.1	71.1	59.9
19	177	159	138	124	110	93.9	90.7	78.8	71.9	63.8	53.8
20	159	144	125	112	99.4	84.8	81.8	71.1	64.9	57.6	48.6
21	144	130	113	102	90.2	76.9	74.2	64.5	58.8	52.2	44.0
22	132	119	103	92.7	82.2	70.1	67.6	58.7	53.6	47.6	40.1
23	120	109	94.4	84.9	75.2	64.1	61.9	53.7	49.0	43.6	36.7
24	111	99.9	86.7	77.9	69.0	58.9	56.8	49.4	45.0	40.0	33.7
25	102	92.1	79.9	71.8	63.6	54.3	52.4	45.5	41.5	36.9	31.1
26	94.3	85.1	73.9	66.4	58.8	50.2		42.1	38.4	34.1	28.7
27	87.4	78.9	68.5	61.6	54.6	46.5				31.6	26.6
28	81.3	73.4	63.7	57.3	50.7	43.3					
29	75.8	68.4	59.4	53.4	47.3	40.3					
30	70.8	63.9	55.5	49.9	44.2	37.7					
32			48.8	43.8	38.8	33.1					
34						29.3					
36											
38											
40											
Properties											
r_{my}, in.	1.86	1.91	1.97	1.99	2.02	2.05	1.53	1.58	1.61	1.64	1.66
r_{mx}/r_{my}	1.31	1.31	1.30	1.30	1.30	1.29	1.57	1.56	1.55	1.54	1.54
$\phi_b M_{nx}$, kip-ft	88.3	75.6	60.4	51.8	42.8	32.8	64.9	52.1	45.2	37.3	28.7
$\phi_b M_{ny}$, kip-ft	69.7	59.7	47.6	41.1	33.9	26.1	43.5	35.2	30.5	25.3	19.6
$P_{ex}(K_x L_x)^2/10^4$	14.7	13.2	11.3	10.2	8.96	7.59	10.8	9.28	8.38	7.35	6.20
$P_{ey}(K_y L_y)^2/10^4$	8.57	7.73	6.68	6.03	5.32	4.51	4.41	3.84	3.49	3.08	2.60

Note: Heavy line indicates Kl/r equal to or greater than 200.

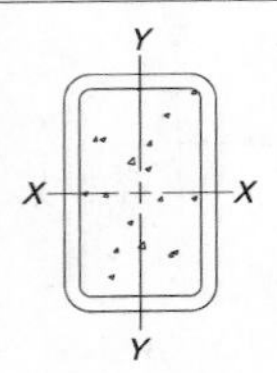

Table 4-13 (cont.). Composite Rectangular (and Square) HSS Design Strength in Axial Compression, $\phi_c P_n$, kips

$F_y = 46$ ksi
$f'_c = 5$ ksi
$\phi_c = 0.85$

Shape	HSS6×6×						HSS6×5×			
	5/8	1/2	3/8	5/16	1/4	3/16	3/8	5/16	1/4	3/16
t_{design}, in.	0.581	0.465	0.349	0.291	0.233	0.174	0.349	0.291	0.233	0.174
Wt/ft	42.1	35.1	27.4	23.3	19.0	14.5	24.9	21.2	17.3	13.2
Effective length KL (ft) with respect to least radius of gyration, r_{my}										
0	510	445	374	337	298	258	330	297	262	225
6	472	413	348	313	277	239	298	268	236	203
7	459	402	339	305	270	233	287	258	228	196
8	445	390	329	296	262	226	275	248	219	187
9	429	376	317	286	253	218	262	236	208	179
10	411	362	305	275	243	210	248	224	198	169
11	393	346	292	263	233	201	234	211	186	160
12	374	330	279	251	223	191	219	198	175	149
13	355	314	265	239	211	182	204	184	163	139
14	335	297	251	226	200	172	189	171	151	129
15	314	279	237	213	189	162	174	157	139	119
16	294	262	222	200	177	152	159	144	128	109
17	274	245	208	187	166	142	145	131	116	99.1
18	254	227	193	174	154	132	131	119	105	89.8
19	235	211	179	162	143	122	118	107	94.8	80.7
20	216	194	165	150	132	113	106	96.7	85.5	72.8
21	198	178	152	138	122	104	96.5	87.7	77.6	66.0
22	180	163	139	126	111	94.6	88.0	79.9	70.7	60.2
23	165	149	127	115	102	86.5	80.5	73.1	64.7	55.0
24	151	137	117	106	93.4	79.5	73.9	67.1	59.4	50.6
25	139	126	108	97.5	86.1	73.2	68.1	61.9	54.7	46.6
26	129	117	99.6	90.1	79.6	67.7	63.0	57.2	50.6	43.1
27	119	108	92.4	83.6	73.8	62.8	58.4	53.0	46.9	39.9
28	111	101	85.9	77.7	68.6	58.4	54.3	49.3	43.6	37.1
29	104	93.7	80.1	72.4	64.0	54.4	50.6	46.0	40.7	34.6
30	96.8	87.6	74.8	67.7	59.8	50.9	47.3	43.0	38.0	32.4
32	85.1	77.0	65.8	59.5	52.5	44.7	41.6	37.8	33.4	28.4
34	75.4	68.2	58.2	52.7	46.5	39.6				
36	67.2	60.8	52.0	47.0	41.5	35.3				
38			46.6	42.2	37.3	31.7				
40										
Properties										
r_{my}, in.	2.17	2.23	2.28	2.31	2.34	2.37	1.92	1.95	1.98	2.01
r_{mx}/r_{my}	1.00	1.00	1.00	1.00	1.00	1.00	1.16	1.15	1.15	1.15
$\phi_b M_{nx}$, kip-ft	80.0	68.3	54.5	46.9	38.6	29.8	47.6	41.1	34.1	26.3
$\phi_b M_{ny}$, kip-ft	80.0	68.3	54.5	46.9	38.6	29.8	42.1	36.2	30.1	23.2
$P_{ex}(K_x L_x)^2/10^4$	11.7	10.5	9.05	8.17	7.20	6.13	7.63	6.90	6.06	5.14
$P_{ey}(K_y L_y)^2/10^4$	11.7	10.5	9.05	8.17	7.20	6.13	5.74	5.20	4.59	3.89

Note: Heavy line indicates Kl/r equal to or greater than 200.

$F_y = 46$ ksi
$f'_c = 5$ ksi
$\phi_c = 0.85$

Table 4-13 (cont.). Composite Rectangular (and Square) HSS Design Strength in Axial Compression, $\phi_c P_n$, kips

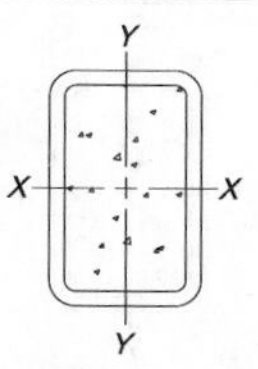

Shape	HSS6×4×					HSS6×3×				
	1/2	3/8	5/16	1/4	3/16	1/2	3/8	5/16	1/4	3/16
t_{design}, in.	0.465	0.349	0.291	0.233	0.174	0.465	0.349	0.291	0.233	0.174
Wt/ft	28.3	22.3	19.0	15.6	12.0	24.9	19.7	16.9	13.9	10.7
Effective length *KL* (ft) with respect to least radius of gyration, r_{my}										
0	343	286	256	225	192	291	242	216	189	159
1	341	285	255	224	191	289	241	214	187	158
2	336	281	252	221	189	282	235	210	183	155
3	329	275	247	217	185	271	227	202	177	149
4	319	267	239	210	179	256	215	192	168	142
5	306	257	230	203	173	239	201	180	158	133
6	291	245	220	194	165	219	185	166	146	124
7	275	232	208	183	156	197	168	151	133	113
8	257	217	195	172	146	175	151	135	120	102
9	238	202	182	160	136	153	133	120	106	90.1
10	219	186	168	148	126	132	115	104	92.6	78.8
11	199	170	154	136	115	111	98.6	89.2	79.8	68.0
12	179	154	139	123	104	93.5	83.1	75.4	67.6	57.8
13	160	138	125	111	93.9	79.6	70.8	64.2	57.6	49.2
14	142	123	112	99.2	83.7	68.7	61.0	55.4	49.7	42.4
15	124	108	98.8	87.9	74.0	59.8	53.2	48.2	43.3	37.0
16	109	95.2	86.9	77.2	65.0	52.6	46.7	42.4	38.0	32.5
17	96.8	84.3	77.0	68.4	57.6	46.6	41.4	37.6	33.7	28.8
18	86.4	75.2	68.6	61.0	51.4	41.5	36.9	33.5	30.0	25.7
19	77.5	67.5	61.6	54.8	46.1		33.1	30.1	27.0	23.0
20	70.0	60.9	55.6	49.4	41.6				24.3	20.8
21	63.5	55.3	50.4	44.8	37.7					
22	57.8	50.4	46.0	40.8	34.4					
23	52.9	46.1	42.0	37.4	31.5					
24	48.6	42.3	38.6	34.3	28.9					
25	44.8	39.0	35.6	31.6	26.6					
26			32.9	29.2	24.6					
27					22.8					
28										
29										
30										
Properties										
r_{my}, in.	1.50	1.55	1.58	1.61	1.63	1.12	1.17	1.19	1.22	1.25
r_{mx}/r_{my}	1.39	1.38	1.37	1.37	1.37	1.76	1.74	1.74	1.72	1.71
$\phi_b M_{nx}$, kip-ft	50.4	41.1	35.5	29.4	22.8	41.7	34.2	29.7	24.8	19.3
$\phi_b M_{ny}$, kip-ft	38.0	30.8	26.7	22.3	17.3	25.1	20.8	18.2	15.2	11.9
$P_{ex}(K_x L_x)^2/10^4$	7.20	6.23	5.63	4.97	4.20	5.55	4.86	4.42	3.88	3.27
$P_{ey}(K_y L_y)^2/10^4$	3.77	3.28	3.00	2.64	2.24	1.80	1.60	1.47	1.30	1.11

Note: Heavy line indicates Kl/r equal to or greater than 200.

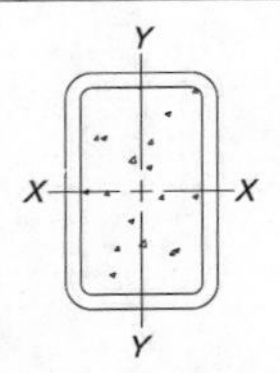

Table 4-13 (cont.). Composite Rectangular (and Square) HSS Design Strength in Axial Compression, $\phi_c P_n$, kips

$F_y = 46$ ksi
$f'_c = 5$ ksi
$\phi_c = 0.85$

Shape		HSS5½×5½×				HSS5×5×					
		3/8	5/16	1/4	3/16	1/2	3/8	5/16	1/4	3/16	1/8
t_{design}, in.		**0.349**	**0.291**	**0.233**	**0.174**	**0.465**	**0.349**	**0.291**	**0.233**	**0.174**	**0.116**
Wt/ft		**24.9**	**21.2**	**17.3**	**13.2**	**28.3**	**22.3**	**19.0**	**15.6**	**12.0**	**8.15**
Effective length KL (ft) with respect to least radius of gyration, r_{my}	**0**	331	298	263	226	346	290	260	229	196	162
	6	303	273	241	207	310	260	234	206	176	145
	7	294	264	233	200	298	251	225	198	169	139
	8	284	255	225	193	285	240	215	190	162	133
	9	272	245	216	185	270	228	205	180	154	127
	10	260	234	206	177	255	215	194	171	146	119
	11	247	222	196	168	239	202	182	160	137	112
	12	233	210	185	158	223	189	170	150	128	105
	13	220	198	174	149	206	175	158	139	119	96.9
	14	206	186	163	139	190	162	146	129	110	89.3
	15	192	173	152	130	174	148	134	118	101	81.8
	16	178	161	141	120	158	135	122	108	92.0	74.5
	17	164	148	130	111	143	123	111	98.2	83.5	67.4
	18	151	136	120	102	128	110	100	88.6	75.2	60.5
	19	138	125	109	93.0	115	99.0	89.9	79.5	67.5	54.3
	20	125	113	99.2	84.3	104	89.4	81.1	71.8	60.9	49.0
	21	113	103	90.0	76.5	94.0	81.1	73.6	65.1	55.3	44.4
	22	103	93.7	82.0	69.7	85.7	73.9	67.0	59.3	50.4	40.5
	23	94.6	85.8	75.0	63.7	78.4	67.6	61.3	54.3	46.1	37.1
	24	86.9	78.8	68.9	58.5	72.0	62.1	56.3	49.8	42.3	34.0
	25	80.1	72.6	63.5	54.0	66.3	57.2	51.9	45.9	39.0	31.4
	26	74.0	67.1	58.7	49.9	61.3	52.9	48.0	42.5	36.1	29.0
	27	68.7	62.2	54.4	46.3	56.9	49.0	44.5	39.4	33.4	26.9
	28	63.8	57.9	50.6	43.0	52.9	45.6	41.4	36.6	31.1	25.0
	29	59.5	53.9	47.2	40.1	49.3	42.5	38.6	34.1	29.0	23.3
	30	55.6	50.4	44.1	37.5	46.1	39.7	36.1	31.9	27.1	21.8
	32	48.9	44.3	38.8	32.9				28.0	23.8	19.1
	34	43.3	39.2	34.3	29.2						
	36				26.0						
	38										
	40										
Properties											
r_{my}, in.		2.08	2.11	2.13	2.16	1.82	1.87	1.90	1.93	1.96	1.99
r_{mx}/r_{my}		1.00	1.00	1.00	1.00	1.00	1.00	1.00	1.00	1.00	1.00
$\phi_b M_{nx}$, kip-ft		45.2	39.0	32.2	24.8	45.2	36.6	31.6	26.3	20.3	14.0
$\phi_b M_{ny}$, kip-ft		45.2	39.0	32.2	24.8	45.2	36.6	31.6	26.3	20.3	14.0
$P_{ex}(K_x L_x)^2/10^4$		6.70	6.05	5.34	4.54	5.54	4.82	4.36	3.85	3.27	2.62
$P_{ey}(K_y L_y)^2/10^4$		6.70	6.05	5.34	4.54	5.54	4.82	4.36	3.85	3.27	2.62

Note: Heavy line indicates Kl/r equal to or greater than 200.

$F_y = 46$ ksi
$f'_c = 5$ ksi
$\phi_c = 0.85$

Table 4-13 (cont.). Composite Rectangular (and Square) HSS Design Strength in Axial Compression, $\phi_c P_n$, kips

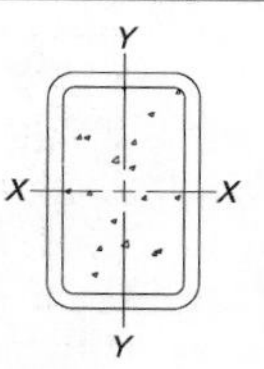

Shape		HSS5×4×					HSS5×3×					
		1/2	3/8	5/16	1/4	3/16	1/2	3/8	5/16	1/4	3/16	1/8
t_{design}, in.		0.465	0.349	0.291	0.233	0.174	0.465	0.349	0.291	0.233	0.174	0.116
Wt/ft		24.9	19.7	16.9	13.9	10.7	21.5	17.2	14.8	12.2	9.40	6.45
Effective length KL (ft) with respect to least radius of gyration, r_{my}	0	298	250	223	196	166	251	209	187	163	137	111
	1	297	248	222	195	166	249	208	185	162	136	110
	2	293	245	219	192	163	243	203	181	158	133	108
	3	286	240	215	188	160	233	195	174	152	128	104
	4	277	232	208	183	155	219	185	166	145	122	98.6
	5	265	223	200	176	149	203	172	155	135	114	92.3
	6	252	213	190	167	142	186	158	142	125	105	85.1
	7	237	201	180	158	134	166	143	129	113	95.8	77.4
	8	221	188	168	148	126	147	127	115	101	85.8	69.3
	9	204	174	156	138	117	127	111	102	89.2	75.8	61.1
	10	186	160	144	127	108	109	95.9	88.0	77.4	66.0	53.2
	11	169	146	131	116	98.3	90.9	81.4	75.2	66.2	56.6	45.6
	12	151	131	118	105	88.9	76.4	68.4	63.3	55.8	47.8	38.5
	13	134	118	106	93.8	79.7	65.1	58.2	53.9	47.6	40.7	32.8
	14	118	104	93.9	83.4	70.9	56.1	50.2	46.5	41.0	35.1	28.3
	15	103	91.4	82.5	73.4	62.4	48.9	43.7	40.5	35.7	30.6	24.6
	16	90.7	80.3	72.5	64.5	54.8	43.0	38.5	35.6	31.4	26.9	21.6
	17	80.3	71.1	64.2	57.1	48.6	38.1	34.1	31.5	27.8	23.8	19.2
	18	71.6	63.5	57.3	51.0	43.3	34.0	30.4	28.1	24.8	21.2	17.1
	19	64.3	57.0	51.4	45.7	38.9		27.3	25.2	22.3	19.1	15.3
	20	58.0	51.4	46.4	41.3	35.1					17.2	13.8
	21	52.6	46.6	42.1	37.4	31.8						
	22	48.0	42.5	38.3	34.1	29.0						
	23	43.9	38.9	35.1	31.2	26.5						
	24	40.3	35.7	32.2	28.7	24.4						
	25		32.9	29.7	26.4	22.5						
	26				24.4	20.8						
	27											
	28											
	29											
	30											
Properties												
r_{my}, in.		1.46	1.52	1.54	1.57	1.60	1.09	1.14	1.17	1.19	1.22	1.25
r_{mx}/r_{my}		1.20	1.19	1.19	1.19	1.19	1.51	1.51	1.50	1.50	1.49	1.48
$\phi_b M_{nx}$, kip-ft		37.6	30.9	26.9	22.4	17.4	30.5	25.3	22.1	18.6	14.5	10.1
$\phi_b M_{ny}$, kip-ft		32.3	26.5	23.0	19.2	15.0	21.0	17.6	15.5	13.0	10.2	7.14
$P_{ex}(K_x L_x)^2/10^4$		4.46	3.90	3.54	3.14	2.66	3.38	3.00	2.74	2.42	2.05	1.62
$P_{ey}(K_y L_y)^2/10^4$		3.13	2.74	2.49	2.21	1.88	1.48	1.33	1.22	1.09	0.925	0.739

Note: Heavy line indicates Kl/r equal to or greater than 200.

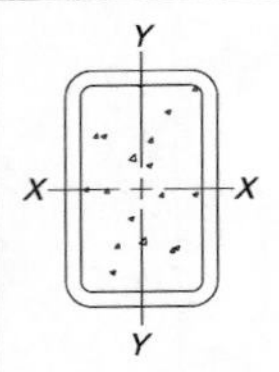

Table 4-13 (cont.). Composite Rectangular (and Square) HSS Design Strength in Axial Compression, $\phi_c P_n$, kips

$F_y = 46$ ksi
$f'_c = 5$ ksi
$\phi_c = 0.85$

Shape		HSS5×2½×			HSS4½×4½×					
		1/4	3/16	1/8	1/2	3/8	5/16	1/4	3/16	1/8
t_{design}, in.		0.233	0.174	0.116	0.465	0.349	0.291	0.233	0.174	0.116
Wt/ft		11.3	8.55	6.02	24.9	19.7	16.9	13.9	10.7	7.30
Effective length KL (ft) with respect to least radius of gyration, r_{my}	0	146	121	98.0	299	250	224	197	167	137
	1	145	119	97.0	298	250	223	196	167	137
	2	140	116	94.1	295	247	221	194	165	135
	3	133	110	89.4	289	242	217	190	162	133
	4	124	103	83.2	281	236	212	186	158	129
	5	113	93.7	75.9	272	228	205	180	153	125
	6	101	83.8	67.8	260	219	197	173	147	120
	7	87.9	73.5	59.3	247	209	188	165	140	114
	8	75.2	63.1	50.9	233	198	178	156	132	108
	9	63.0	53.1	42.7	219	186	167	147	124	101
	10	51.6	43.8	35.1	203	173	156	137	116	94.5
	11	42.7	36.2	29.0	187	160	144	127	108	87.4
	12	35.9	30.4	24.4	171	147	133	117	99.0	80.2
	13	30.6	25.9	20.8	155	134	121	107	90.4	73.1
	14	26.4	22.3	17.9	140	121	110	97.3	81.9	66.1
	15	23.0	19.4	15.6	125	109	99.0	87.7	73.7	59.3
	16	20.2	17.1	13.7	110	97.2	88.5	78.5	65.8	52.8
	17		15.1	12.2	97.8	86.1	78.4	69.6	58.3	46.8
	18				87.3	76.8	70.0	62.1	52.0	41.7
	19				78.3	68.9	62.8	55.7	46.7	37.4
	20				70.7	62.2	56.7	50.3	42.1	33.8
	21				64.1	56.4	51.4	45.6	38.2	30.7
	22				58.4	51.4	46.8	41.5	34.8	27.9
	23				53.4	47.0	42.8	38.0	31.9	25.6
	24				49.1	43.2	39.4	34.9	29.3	23.5
	25				45.2	39.8	36.3	32.2	27.0	21.6
	26				41.8	36.8	33.5	29.7	24.9	20.0
	27					34.1	31.1	27.6	23.1	18.5
	28						28.9	25.6	21.5	17.2
	29								20.0	16.1
	30									
Properties										
r_{my}, in.		0.999	1.03	1.05	1.61	1.67	1.70	1.73	1.75	1.78
r_{mx}/r_{my}		1.73	1.72	1.71	1.00	1.00	1.00	1.00	1.00	1.00
$\phi_b M_{nx}$, kip-ft		16.7	12.8	9.14	35.2	28.8	25.1	20.9	16.2	11.3
$\phi_b M_{ny}$, kip-ft		10.2	7.90	5.66	35.2	28.8	25.1	20.9	16.2	11.3
$P_{ex}(K_x L_x)^2/10^4$		2.08	1.73	1.38	3.81	3.34	3.04	2.68	2.27	1.82
$P_{ey}(K_y L_y)^2/10^4$		0.692	0.584	0.471	3.81	3.34	3.04	2.68	2.27	1.82

Note: Heavy line indicates Kl/r equal to or greater than 200.

$F_y = 46$ ksi
$f'_c = 5$ ksi
$\phi_c = 0.85$

Table 4-13 (cont.). Composite Rectangular (and Square) HSS Design Strength in Axial Compression, $\phi_c P_n$, kips

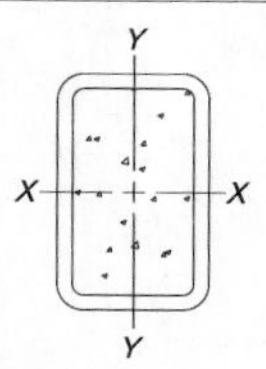

Shape	HSS4×4×						HSS4×3×				
	1/2	3/8	5/16	1/4	3/16	1/8	3/8	5/16	1/4	3/16	1/8
t_{design}, in.	0.465	0.349	0.291	0.233	0.174	0.116	0.349	0.291	0.233	0.174	0.116
Wt/ft	21.5	17.2	14.8	12.2	9.40	6.45	14.6	12.7	10.5	8.13	5.60
Effective length *KL* (ft) with respect to least radius of gyration, r_{my}											
0	254	213	190	166	141	115	176	157	137	115	92.8
1	253	212	189	166	140	114	175	156	136	115	92.0
2	249	209	187	163	138	112	171	152	133	112	89.9
3	243	204	182	160	135	110	164	146	128	108	86.5
4	235	197	177	155	131	106	155	138	121	102	81.9
5	224	189	169	148	125	102	144	129	113	95.4	76.4
6	212	180	161	141	119	96.9	131	118	104	87.8	70.2
7	199	169	151	133	112	91.2	118	106	93.9	79.5	63.5
8	184	157	141	124	105	85.1	104	94.3	83.6	70.9	56.5
9	169	145	130	115	97.1	78.6	90.9	82.3	73.3	62.3	49.5
10	154	133	119	105	89.1	72.0	77.8	70.7	63.2	54.0	42.7
11	138	120	108	95.5	80.9	65.3	65.4	59.6	53.7	46.0	36.3
12	123	108	97.1	85.9	72.8	58.7	55.0	50.1	45.1	38.7	30.5
13	109	95.8	86.4	76.6	65.0	52.2	46.8	42.7	38.5	33.0	26.0
14	94.7	84.3	76.2	67.7	57.4	46.1	40.4	36.8	33.2	28.4	22.4
15	82.5	73.5	66.5	59.1	50.2	40.2	35.2	32.1	28.9	24.8	19.5
16	72.5	64.6	58.4	52.0	44.1	35.3	30.9	28.2	25.4	21.8	17.2
17	64.2	57.2	51.8	46.0	39.1	31.3	27.4	25.0	22.5	19.3	15.2
18	57.3	51.0	46.2	41.1	34.9	27.9	24.4	22.3	20.1	17.2	13.6
19	51.4	45.8	41.4	36.9	31.3	25.1			18.0	15.4	12.2
20	46.4	41.3	37.4	33.3	28.2	22.6					11.0
21	42.1	37.5	33.9	30.2	25.6	20.5					
22	38.4	34.2	30.9	27.5	23.3	18.7					
23	35.1	31.3	28.3	25.1	21.4	17.1					
24		28.7	26.0	23.1	19.6	15.7					
25				21.3	18.1	14.5					
26						13.4					
27											
28											
29											
30											
Properties											
r_{my}, in.	1.41	1.47	1.49	1.52	1.55	1.58	1.11	1.13	1.16	1.19	1.21
r_{mx}/r_{my}	1.00	1.00	1.00	1.00	1.00	1.00	1.25	1.26	1.25	1.25	1.26
$\phi_b M_{nx}$, kip-ft	26.6	22.0	19.3	16.2	12.7	8.83	17.7	15.6	13.1	10.4	7.28
$\phi_b M_{ny}$, kip-ft	26.6	22.0	19.3	16.2	12.7	8.83	14.4	12.7	10.8	8.49	5.97
$P_{ex}(K_x L_x)^2/10^4$	2.48	2.21	2.01	1.79	1.52	1.21	1.67	1.54	1.37	1.16	0.920
$P_{ey}(K_y L_y)^2/10^4$	2.48	2.21	2.01	1.79	1.52	1.21	1.06	0.974	0.871	0.744	0.593

Note: Heavy line indicates Kl/r equal to or greater than 200.

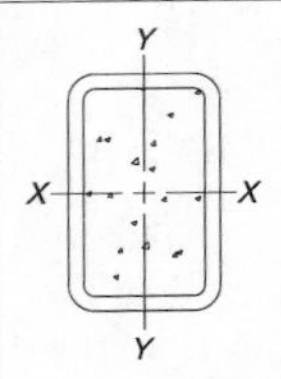

Table 4-13 (cont.). Composite Rectangular (and Square) HSS Design Strength in Axial Compression, $\phi_c P_n$, kips

$F_y = 46$ ksi
$f'_c = 5$ ksi
$\phi_c = 0.85$

Shape	HSS4×2½×			HSS4×2×					HSS3½×3½×				
	5/16	1/4	3/16	3/8	5/16	1/4	3/16	1/8	3/8	5/16	1/4	3/16	1/8
t_{design}, in.	**0.291**	**0.233**	**0.174**	**0.349**	**0.291**	**0.233**	**0.174**	**0.116**	**0.349**	**0.291**	**0.233**	**0.174**	**0.116**
Wt/ft	**11.6**	**9.63**	**7.49**	**12.1**	**10.5**	**8.78**	**6.85**	**4.75**	**14.6**	**12.7**	**10.5**	**8.13**	**5.60**
Effective length KL (ft) with respect to least radius of gyration, r_{my}													
0	141	122	102	140	124	108	89.8	70.6	177	158	138	116	93.7
1	139	121	101	137	122	106	88.4	69.5	176	157	137	116	93.1
2	135	117	98.1	130	116	101	84.2	66.2	173	154	135	114	91.4
3	127	111	93.1	118	106	92.7	77.6	61.2	167	150	131	110	88.7
4	118	103	86.4	104	93.9	82.5	69.3	54.7	160	143	126	106	85.0
5	106	93.2	78.5	87.7	80.2	70.9	59.9	47.4	151	136	119	100	80.5
6	93.9	82.7	69.8	71.4	66.1	59.0	50.2	39.8	141	127	111	94.0	75.3
7	81.1	71.7	60.8	56.1	52.5	47.4	40.6	32.4	130	117	103	87.0	69.6
8	68.4	60.9	51.8	43.0	40.6	36.9	31.8	25.4	118	107	94.1	79.6	63.5
9	56.5	50.6	43.2	34.0	32.1	29.1	25.2	20.1	106	96.1	85.0	72.0	57.3
10	45.8	41.2	35.3	27.5	26.0	23.6	20.4	16.3	93.8	85.5	75.8	64.3	51.1
11	37.9	34.1	29.2	22.7	21.5	19.5	16.8	13.5	82.1	75.2	66.9	56.8	45.0
12	31.8	28.6	24.5	19.1	18.0	16.4	14.1	11.3	70.9	65.2	58.2	49.6	39.1
13	27.1	24.4	20.9				12.1	9.63	60.5	55.9	50.0	42.6	33.5
14	23.4	21.0	18.0						52.2	48.2	43.1	36.8	28.9
15	20.4	18.3	15.7						45.5	42.0	37.6	32.0	25.2
16		16.1	13.8						40.0	36.9	33.0	28.1	22.1
17									35.4	32.7	29.2	24.9	19.6
18									31.6	29.1	26.1	22.2	17.5
19									28.3	26.1	23.4	20.0	15.7
20									25.6	23.6	21.1	18.0	14.2
21									23.2	21.4	19.2	16.3	12.9
22											17.5	14.9	11.7
23													
24													
25													
26													
27													
28													
29													
30													
Properties													
r_{my}, in.	0.947	0.973	0.999	0.729	0.754	0.779	0.804	0.830	1.26	1.29	1.32	1.35	1.37
r_{mx}/r_{my}	1.46	1.45	1.44	1.77	1.75	1.75	1.73	1.72	1.00	1.00	1.00	1.00	1.00
$\phi_b M_{nx}$, kip-ft	13.7	11.7	9.21	13.2	11.8	10.1	8.07	5.73	16.2	14.3	12.1	9.52	6.66
$\phi_b M_{ny}$, kip-ft	9.83	8.38	6.66	7.97	7.18	6.18	4.93	3.52	16.2	14.3	12.1	9.52	6.66
$P_{ex}(K_x L_x)^2/10^4$	1.30	1.16	0.990	1.15	1.07	0.960	0.819	0.646	1.37	1.26	1.13	0.959	0.763
$P_{ey}(K_y L_y)^2/10^4$	0.613	0.553	0.474	0.369	0.348	0.317	0.273	0.219	1.37	1.26	1.13	0.959	0.763

Note: Heavy line indicates Kl/r equal to or greater than 200.

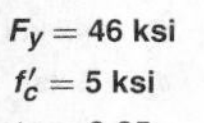

$F_y = 46$ ksi
$f'_c = 5$ ksi
$\phi_c = 0.85$

Table 4-13 (cont.). Composite Rectangular (and Square) HSS Design Strength in Axial Compression, $\phi_c P_n$, kips

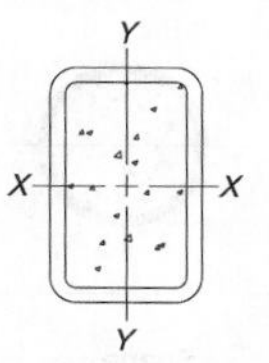

Shape		HSS3×3×					HSS2$^1/_2$×2$^1/_2$×			
		$^3/_8$	$^5/_{16}$	$^1/_4$	$^3/_{16}$	$^1/_8$	$^5/_{16}$	$^1/_4$	$^3/_{16}$	$^1/_8$
t_{design}, in.		0.349	0.291	0.233	0.174	0.116	0.291	0.233	0.174	0.116
Wt/ft		12.1	10.5	8.78	6.85	4.75	8.40	7.08	5.57	3.90
Effective length KL (ft) with respect to least radius of gyration, r_{my}	0	143	128	111	93.4	74.2	99.3	86.5	72.3	57.0
	1	142	127	111	92.7	73.6	98.0	85.5	71.5	56.3
	2	138	124	108	90.4	71.8	94.3	82.4	69.0	54.4
	3	132	118	103	86.8	69.0	88.4	77.4	65.0	51.3
	4	124	111	97.5	82.0	65.1	80.8	71.1	59.8	47.2
	5	115	103	90.4	76.2	60.5	71.9	63.6	53.8	42.5
	6	104	93.7	82.5	69.6	55.4	62.4	55.5	47.2	37.4
	7	92.6	83.7	74.0	62.6	49.8	52.7	47.3	40.4	32.1
	8	81.0	73.5	65.3	55.4	44.1	43.5	39.3	33.8	26.9
	9	69.7	63.4	56.6	48.2	38.4	34.9	31.8	27.6	22.1
	10	58.8	53.8	48.3	41.3	32.9	28.2	25.8	22.4	17.9
	11	48.8	44.8	40.4	34.7	27.6	23.3	21.3	18.5	14.8
	12	41.0	37.6	34.0	29.2	23.2	19.6	17.9	15.5	12.4
	13	35.0	32.1	28.9	24.9	19.8	16.7	15.3	13.2	10.6
	14	30.1	27.7	25.0	21.4	17.1	14.4	13.2	11.4	9.12
	15	26.3	24.1	21.7	18.7	14.9		11.5	9.95	7.94
	16	23.1	21.2	19.1	16.4	13.1				6.98
	17	20.4	18.8	16.9	14.5	11.6				
	18		16.7	15.1	13.0	10.3				
	19				11.6	9.26				
	20									
	21									
	22									
	23									
	24									
	25									
	26									
	27									
	28									
	29									
	30									
Properties										
r_{my}, in.		1.06	1.08	1.11	1.14	1.17	0.880	0.908	0.937	0.965
r_{my}, in.		1.00	1.00	1.00	1.00	1.00	1.00	1.00	1.00	1.00
$\phi_b M_{nx}$, kip-ft		11.2	10.00	8.56	6.80	4.83	6.49	5.62	4.55	3.27
$\phi_b M_{ny}$, kip-ft		11.2	10.00	8.56	6.80	4.83	6.49	5.62	4.55	3.27
$P_{ex}(K_xL_x)^2/10^4$		0.787	0.731	0.659	0.564	0.449	0.379	0.347	0.300	0.240
$P_{ey}(K_yL_y)^2/10^4$		0.787	0.731	0.659	0.564	0.449	0.379	0.347	0.300	0.240

Note: Heavy line indicates Kl/r equal to or greater than 200.

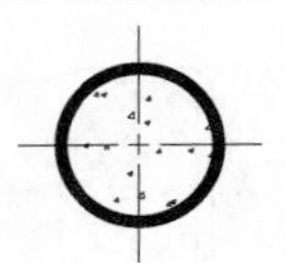

Table 4-14. Composite Round HSS Design Strength in Axial Compression, $\phi_c P_n$, kips

$F_y = 42$ ksi
$f'_c = 4$ ksi
$\phi_c = 0.85$

Shape		HSS20.000x		HSS18.000x		HSS16.000x				HSS14.000x		
		0.500	0.375	0.500	0.375	0.500	0.438	0.375	0.312	0.500	0.375	0.312
t_{design}, in.		0.465	0.349	0.465	0.349	0.465	0.407	0.349	0.291	0.465	0.349	0.291
Wt/ft		104	78.7	93.5	70.7	82.8	72.9	62.6	52.3	72.2	54.6	45.7
Effective length KL (ft) with respect to radius of gyration r	0	1840	1610	1580	1370	1330	1230	1150	1050	1090	937	855
	6	1830	1600	1560	1360	1310	1220	1130	1040	1080	922	841
	7	1820	1600	1550	1350	1300	1210	1130	1030	1070	916	836
	8	1820	1590	1550	1350	1300	1210	1120	1030	1060	910	830
	9	1810	1580	1540	1340	1290	1200	1110	1020	1060	903	823
	10	1800	1580	1530	1330	1280	1190	1110	1020	1050	895	816
	11	1790	1570	1520	1330	1270	1180	1100	1010	1040	886	808
	12	1780	1560	1510	1320	1260	1170	1090	1000	1030	877	799
	13	1770	1550	1500	1310	1250	1160	1080	991	1010	867	790
	14	1760	1540	1490	1300	1240	1150	1070	981	1000	856	780
	15	1750	1530	1480	1290	1230	1140	1060	971	989	845	770
	16	1740	1520	1470	1280	1210	1130	1050	960	975	833	759
	17	1730	1510	1450	1260	1200	1110	1030	949	961	821	747
	18	1710	1500	1440	1250	1180	1100	1020	937	946	808	735
	19	1700	1480	1430	1240	1170	1090	1010	924	930	794	723
	20	1680	1470	1410	1220	1150	1070	993	911	914	780	710
	22	1650	1440	1380	1200	1120	1040	964	884	880	750	682
	24	1620	1410	1340	1160	1080	1010	933	855	845	719	654
	25	1600	1400	1320	1150	1070	990	917	840	826	703	639
	26	1580	1380	1310	1130	1050	973	900	825	807	687	624
	28	1540	1340	1270	1100	1010	937	866	793	769	654	593
	30	1500	1310	1230	1060	969	899	831	760	730	620	562
	31	1480	1290	1210	1040	949	880	813	744	710	603	546
	32	1460	1270	1190	1030	928	861	795	727	690	586	530
	34	1420	1230	1140	988	886	822	759	693	651	551	499
	36	1370	1190	1100	949	844	782	722	659	611	517	467
	37	1350	1170	1080	930	823	762	703	642	591	500	452
	38	1330	1150	1060	910	802	743	685	624	571	483	436
	40	1280	1110	1010	871	759	703	648	590	533	450	405
Properties												
r_m, in.^{2}		6.91	6.95	6.20	6.24	5.49	5.51	5.53	5.55	4.79	4.83	4.85
$\phi_b M_n$, kip-ft		558	425	450	343	353	312	269	226	268	205	172
$P_e(KL)^2/10^4$		401	342	280	237	188	173	158	143	119	100	90.6

$F_y = 42$ ksi
$f'_c = 4$ ksi
$\phi_c = 0.85$

Table 4-14 (cont.). Composite Round HSS Design Strength in Axial Compression, $\phi_c P_n$, kips

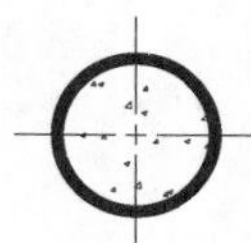

Shape		HSS12.750x			HSS12.500x						HSS11.250x		
		0.500	0.375	0.250	0.625	0.500	0.375	0.312	0.250	0.188	0.625	0.500	0.375
t_{design}, in.		0.465	0.349	0.233	0.581	0.465	0.349	0.291	0.233	0.174	0.581	0.465	0.349
Wt/ft		65.5	49.6	33.4	79.3	64.1	48.6	40.7	32.7	24.7	71.0	57.5	43.6
Effective length KL (ft) with respect to radius of gyration r	0	956	815	670	1070	932	791	722	649	576	927	806	681
	6	938	799	656	1050	913	775	707	636	563	904	786	664
	7	931	794	651	1040	907	769	702	631	559	896	779	658
	8	924	787	646	1030	899	763	696	625	554	887	771	651
	9	915	780	639	1020	891	755	689	619	548	877	762	644
	10	906	772	633	1010	881	747	682	612	542	865	752	635
	11	896	763	625	1000	871	738	673	605	535	853	741	626
	12	885	753	617	987	860	729	664	596	527	840	729	616
	13	873	743	608	974	848	718	655	588	520	825	717	605
	14	860	732	599	959	835	707	645	578	511	810	704	594
	15	847	721	589	944	821	696	634	568	502	794	690	582
	16	833	708	579	927	807	684	623	558	493	777	675	569
	17	818	696	568	911	792	671	611	547	483	760	660	557
	18	803	682	557	893	777	657	599	536	473	742	644	543
	19	787	669	545	875	761	644	586	525	462	723	628	529
	20	771	655	533	856	745	630	573	513	451	704	611	515
	22	736	625	508	817	710	600	546	488	429	664	577	486
	24	701	594	483	776	675	569	517	462	405	624	541	455
	25	682	578	469	755	656	554	503	449	394	603	523	440
	26	664	563	456	734	638	538	488	436	382	582	505	425
	28	626	530	429	691	600	506	459	409	357	540	469	394
	30	588	497	401	647	562	473	429	382	333	499	433	363
	31	569	481	388	626	543	457	414	368	321	478	415	348
	32	550	465	374	604	525	441	399	355	309	458	398	333
	34	512	432	347	561	487	409	370	328	285	418	363	304
	36	475	400	320	519	450	378	341	302	262	380	330	275
	37	457	385	307	498	432	362	327	289	250	361	313	261
	38	439	369	295	478	414	347	313	277	239	342	297	248
	40	403	339	270	438	380	317	286	252	217	309	268	224
Properties													
r_m, in.^{2}		4.35	4.39	4.43	4.22	4.26	4.30	4.32	4.34	4.36	3.78	3.82	3.86
$\phi_b M_n$, kip-ft		221	169	115	260	212	162	137	111	83.2	209	170	131
$P_e(KL)^2/10^4$		87.3	73.1	58.1	94.1	81.6	68.3	61.2	54.1	46.5	66.2	57.1	47.8

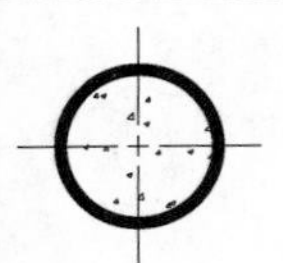

Table 4-14 (cont.). Composite Round HSS Design Strength in Axial Compression, $\phi_c P_n$, kips

$F_y = 42$ ksi
$f'_c = 4$ ksi
$\phi c = 0.85$

Shape		HSS11.250x			HSS10.750x		HSS10.000x					
		0.312	0.250	0.188	0.500	0.250	0.625	0.500	0.375	0.312	0.250	0.188
t_{design}, in.		0.291	0.233	0.174	0.465	0.233	0.581	0.465	0.349	0.291	0.233	0.174
Wt/ft		36.5	29.4	22.2	54.8	28.1	62.6	50.8	38.6	32.3	26.1	19.7
Effective length KL (ft) with respect to radius of gyration r	0	615	552	486	754	515	791	683	575	518	462	403
	6	600	538	473	734	501	767	662	557	502	447	390
	7	594	533	468	727	495	758	655	550	496	442	385
	8	588	527	463	719	490	748	646	543	490	436	380
	9	581	520	457	710	483	737	637	535	482	429	374
	10	573	513	451	699	476	725	626	526	474	422	367
	11	565	506	444	688	468	712	615	517	465	414	360
	12	556	497	437	677	460	698	603	506	456	405	353
	13	546	488	428	664	451	683	590	495	446	396	344
	14	536	479	420	650	441	667	576	483	435	386	336
	15	525	469	411	636	431	650	562	471	424	376	327
	16	513	459	402	622	421	633	547	459	413	366	318
	17	502	448	392	606	410	615	531	445	401	355	308
	18	489	437	382	590	399	597	515	432	388	344	298
	19	477	425	372	574	387	578	499	418	376	333	288
	20	464	413	361	557	376	558	482	404	363	321	278
	22	437	389	339	523	352	519	448	375	337	298	257
	24	409	364	317	488	327	479	414	346	310	274	236
	25	395	352	305	470	315	459	396	331	297	262	225
	26	381	339	294	452	302	439	379	316	284	250	215
	28	353	313	271	417	278	399	345	288	258	227	194
	30	325	288	249	382	253	361	312	260	232	204	174
	31	312	276	238	365	242	342	296	246	220	193	165
	32	298	264	227	348	230	324	280	233	208	182	155
	34	271	240	206	315	207	288	249	207	185	162	137
	36	246	217	185	282	185	257	222	184	165	144	122
	37	233	205	175	267	175	243	210	175	156	136	116
	38	221	194	166	254	166	231	199	166	148	129	110
	40	199	175	150	229	150	208	180	149	133	117	99.2
Properties												
r_m, in.^{2}		3.88	3.90	3.92	3.64	3.72	3.34	3.38	3.41	3.43	3.45	3.47
$\phi_b M_n$, kip-ft		110	89.1	67.1	155	81.3	163	133	102	86.3	69.9	52.9
$P_e(KL)^2/10^4$		42.9	37.5	32.2	49.2	32.0	44.5	38.7	32.0	28.8	25.1	21.3

F_y = 42 ksi
f'_c = 4 ksi
ϕ_c = 0.85

Table 4-14 (cont.). Composite Round HSS Design Strength in Axial Compression, $\phi_c P_n$, kips

Shape		HSS9.625x					HSS8.750x					HSS8.625x	
		0.500	0.375	0.312	0.250	0.188	0.500	0.375	0.312	0.250	0.188	0.500	0.375
t_{design}, in.		0.465	0.349	0.291	0.233	0.174	0.465	0.349	0.291	0.233	0.174	0.465	0.349
Wt/ft		48.8	37.1	31.1	25.1	19.0	44.1	33.6	28.1	22.7	17.2	43.4	33.1
Effective length KL (ft) with respect to radius of gyration r	0	650	545	490	436	380	571	476	427	378	328	559	466
	6	628	527	474	421	366	548	457	410	363	314	536	447
	7	621	520	468	415	362	540	450	404	357	309	528	440
	8	612	513	461	409	356	531	442	397	351	303	519	433
	9	602	505	454	403	350	520	434	389	344	297	509	424
	10	591	496	445	395	344	509	424	381	336	291	497	415
	11	580	486	436	387	337	497	414	372	328	283	485	405
	12	567	475	427	379	329	484	404	362	319	276	472	394
	13	554	464	417	369	321	471	392	352	310	267	459	382
	14	540	452	406	360	312	456	380	341	301	259	444	370
	15	526	440	395	350	303	441	368	330	291	250	430	358
	16	510	427	383	339	294	426	355	318	280	241	414	345
	17	495	414	371	329	284	410	342	306	270	232	398	332
	18	479	401	359	318	275	394	328	294	259	222	382	319
	19	462	387	347	306	265	378	315	282	248	212	366	305
	20	446	373	334	295	255	361	301	269	237	203	350	291
	22	412	344	308	272	234	328	273	244	215	183	317	264
	24	377	315	282	248	213	296	246	220	193	164	285	237
	25	360	301	269	237	203	279	233	208	182	155	269	224
	26	343	287	256	225	193	264	219	196	171	145	253	211
	28	310	259	231	203	173	233	194	173	151	128	223	186
	30	278	232	207	181	154	204	169	151	132	111	194	162
	31	263	219	195	171	145	191	159	141	123	104	182	152
	32	247	206	183	160	136	179	149	133	116	97.8	171	142
	34	219	182	162	142	120	159	132	117	102	86.6	151	126
	36	195	163	145	127	107	141	118	105	91.4	77.3	135	112
	37	185	154	137	120	102	134	111	99.2	86.5	73.1	128	106
	38	175	146	130	114	96.4	127	106	94.1	82.1	69.3	121	101
	40	158	132	117	103	87.0	115	95.3	84.9	74.1	62.6	109	91.0
Properties													
r_m, in.^{2}		3.24	3.28	3.30	3.32	3.34	2.93	2.97	2.99	3.01	3.03	2.89	2.93
$\phi_b M_n$, kip-ft		123	94.5	80.0	64.9	48.8	101	77.5	65.5	53.2	40.3	97.7	75.3
$P_e(KL)^2/10^4$		34.0	28.3	25.2	22.1	18.7	24.6	20.5	18.3	15.9	13.4	23.5	19.5

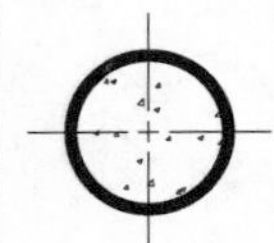

Table 4-14 (cont.). Composite Round HSS Design Strength in Axial Compression, $\phi_c P_n$, kips

$F_y = 42$ ksi
$f'_c = 4$ ksi
$\phi_c = 0.85$

Shape	HSS8.625x			HSS7.625x	HSS7.500x					HSS7.000x		
	0.322	0.250	0.188	0.125	0.500	0.375	0.312	0.250	0.188	0.500	0.375	0.312
t_{design}, in.	0.300	0.233	0.174	0.116	0.465	0.349	0.291	0.233	0.174	0.465	0.349	0.291
Wt/ft	28.6	22.4	17.0	10.0	37.4	28.6	24.0	19.4	14.7	34.7	26.6	22.3
Effective length *KL* (ft) with respect to radius of gyration *r*												
0	426	370	320	222	466	385	344	302	259	425	350	312
6	409	355	307	209	440	364	325	286	244	398	329	293
7	402	349	302	205	431	357	319	280	239	389	321	286
8	395	343	296	200	421	348	311	273	234	379	313	279
9	388	336	290	195	410	339	303	266	227	367	303	270
10	379	328	283	189	398	330	294	258	220	355	293	261
11	370	320	276	182	386	319	285	250	213	342	282	251
12	360	311	268	176	372	308	275	241	205	328	271	241
13	349	302	260	169	358	296	264	232	197	314	259	231
14	338	293	252	162	343	284	253	222	189	299	247	220
15	327	283	243	154	328	271	242	212	180	284	234	209
16	315	272	234	147	312	259	231	202	171	269	222	197
17	303	262	224	139	297	246	219	192	163	253	209	186
18	291	251	215	131	281	233	207	181	154	238	196	175
19	278	240	205	124	265	220	196	171	145	223	184	164
20	266	229	196	116	250	207	184	161	136	208	171	153
22	241	207	177	101	219	181	162	141	119	179	147	131
24	216	185	158	87.5	190	157	140	122	102	152	125	111
25	204	175	148	80.6	175	146	129	113	94.4	140	115	102
26	192	164	139	74.6	162	135	120	104	87.3	129	106	94.7
28	169	144	122	64.3	140	116	103	89.8	75.2	111	91.8	81.6
30	147	126	106	56.0	122	101	89.9	78.2	65.5	97.0	79.9	71.1
31	138	118	99.4	52.4	114	94.6	84.2	73.2	61.4	90.8	74.9	66.6
32	130	111	93.3	49.2	107	88.8	79.0	68.7	57.6	85.2	70.3	62.5
34	115	97.9	82.6	43.6	94.8	78.7	70.0	60.9	51.0	75.5	62.2	55.4
36	102	87.3	73.7	38.9	84.6	70.2	62.5	54.3	45.5	67.3	55.5	49.4
37	96.9	82.7	69.8	36.8	80.1	66.4	59.1	51.4	43.1	63.8	52.6	46.7
38	91.9	78.4	66.2	34.9	75.9	63.0	56.0	48.7	40.9	60.4	49.8	44.3
40	82.9	70.7	59.7	31.5	68.5	56.8	50.6	44.0	36.9			
Properties												
r_m, in.^{2}	2.95	2.97	2.99	2.66	2.49	2.53	2.55	2.57	2.59	2.32	2.35	2.37
$\phi_b M_n$, kip-ft	65.5	51.7	39.1	20.6	72.5	56.4	47.6	38.7	29.4	62.7	48.8	41.3
$P_e(KL)^2/10^4$	17.7	15.2	12.8	6.73	14.7	12.2	10.9	9.46	7.93	11.7	9.69	8.63

Note: Heavy line indicates Kl/r equal to or greater than 200.

$F_y = 42$ ksi
$f'_c = 4$ ksi
$\phi_c = 0.85$

Table 4-14 (cont.). Composite Round HSS Design Strength in Axial Compression, $\phi_c P_n$, kips

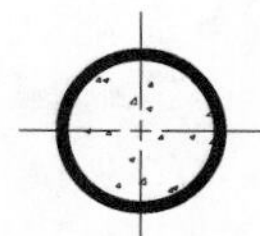

Shape	HSS7.000x			HSS6.875x					HSS6.625x			
	0.250	0.188	0.125	0.500	0.375	0.312	0.250	0.188	0.500	0.432	0.375	0.312
t_{design}, in.	0.233	0.174	0.116	0.465	0.349	0.291	0.233	0.174	0.465	0.403	0.349	0.291
Wt/ft	18.0	13.7	9.19	34.1	26.1	21.9	17.7	13.4	32.7	28.6	25.1	21.1
Effective length *KL* (ft) with respect to radius of gyration *r*												
0	274	234	194	414	342	305	267	227	395	358	325	290
6	256	219	181	388	320	285	249	212	367	333	303	270
7	250	213	176	378	313	278	243	207	358	325	295	263
8	244	208	171	368	304	271	237	201	347	315	286	255
9	236	201	166	356	295	262	229	195	335	305	277	246
10	228	194	160	344	285	253	221	188	323	293	267	237
11	220	187	154	331	274	244	213	181	310	281	256	228
12	211	179	147	317	262	234	204	173	296	268	244	217
13	202	171	140	302	250	223	195	165	281	255	232	207
14	192	163	133	288	238	212	185	157	266	242	220	196
15	182	154	126	273	226	201	175	148	251	228	208	185
16	172	146	119	257	213	190	165	140	236	215	195	174
17	162	137	111	242	201	179	156	131	221	201	183	163
18	152	129	104	227	188	167	146	123	206	187	171	152
19	142	120	97.0	212	176	156	136	114	191	174	158	141
20	133	112	90.0	197	163	145	126	106	177	161	147	130
22	114	95.7	76.6	168	140	124	108	90.5	149	136	124	110
24	96.5	80.7	64.4	142	118	105	91.1	76.2	125	114	104	92.6
25	88.9	74.4	59.4	131	109	96.8	84.0	70.2	115	105	95.9	85.3
26	82.2	68.8	54.9	121	101	89.5	77.6	64.9	107	97.2	88.7	78.9
28	70.9	59.3	47.3	104	86.8	77.1	66.9	56.0	91.9	83.8	76.5	68.0
30	61.7	51.7	41.2	90.7	75.6	67.2	58.3	48.7	80.0	73.0	66.6	59.2
31	57.8	48.4	38.6	85.0	70.8	62.9	54.6	45.7	74.9	68.4	62.4	55.5
32	54.3	45.4	36.3	79.8	66.4	59.1	51.3	42.8	70.3	64.2	58.5	52.1
34	48.1	40.2	32.1	70.7	58.8	52.3	45.4	38.0	62.3	56.8	51.9	46.1
36	42.9	35.9	28.6	63.0	52.5	46.7	40.5	33.9	55.6	50.7	46.3	41.1
37	40.6	34.0	27.1	59.7	49.7	44.2	38.3	32.0			43.8	38.9
38	38.5	32.2	25.7		47.1	41.9	36.3	30.4				
40		29.1	23.2									
Properties												
r_m, in.^{2}	2.39	2.41	2.43	2.27	2.31	2.33	2.35	2.37	2.18	2.20	2.22	2.24
$\phi_b M_n$, kip-ft	33.7	25.5	17.3	60.2	46.9	39.7	32.4	24.6	55.8	49.1	43.5	36.9
$P_e(KL)^2/10^4$	7.49	6.25	5.01	11.0	9.12	8.12	7.03	5.90	9.69	8.85	8.06	7.16

Note: Heavy line indicates Kl/r equal to or greater than 200.

Table 4-14 (cont.). Composite Round HSS Design Strength in Axial Compression, $\phi_c P_n$, kips

$F_y = 42$ ksi
$f'_c = 4$ ksi
$\phi_c = 0.85$

Shape	HSS6.625x				HSS6.125x					HSS6.000x		
	0.280	0.250	0.188	0.125	0.500	0.375	0.312	0.250	0.188	0.500	0.375	0.312
t_{design}, in.	0.261	0.233	0.174	0.116	0.465	0.349	0.291	0.233	0.174	0.465	0.349	0.291
Wt/ft	19.0	17.0	12.9	8.69	30.1	23.1	19.4	15.7	11.9	29.4	22.5	19.0
Effective length KL (ft) with respect to radius of gyration r												
0	271	253	215	177	356	293	260	227	192	347	285	253
6	252	236	200	164	328	269	239	208	176	318	261	232
7	246	229	195	160	318	261	232	202	171	308	253	225
8	238	223	189	155	307	252	224	195	165	296	244	217
9	230	215	183	149	295	243	216	187	158	284	234	208
10	222	207	176	143	282	232	206	179	151	271	224	198
11	213	199	168	137	268	221	196	170	144	257	212	189
12	203	190	160	131	254	210	186	162	136	243	201	178
13	193	180	152	124	240	198	176	152	128	229	189	168
14	183	171	144	117	225	186	165	143	120	214	177	157
15	173	161	136	110	210	174	154	134	112	199	165	147
16	162	151	128	103	195	162	144	124	104	184	153	136
17	152	142	119	96.1	181	150	133	115	96.2	170	141	125
18	142	132	111	89.2	166	138	123	106	88.5	156	130	115
19	132	123	103	82.5	152	127	113	97.0	81.0	142	119	105
20	122	113	95.1	75.9	139	116	103	88.5	73.7	129	108	95.7
22	103	95.6	79.9	63.4	115	95.8	85.2	73.1	60.9	107	88.9	79.1
24	86.3	80.3	67.1	53.3	96.6	80.5	71.6	61.5	51.2	89.6	74.7	66.4
25	79.6	74.0	61.9	49.1	89.0	74.2	66.0	56.6	47.2	82.5	68.9	61.2
26	73.6	68.5	57.2	45.4	82.3	68.6	61.0	52.4	43.6	76.3	63.7	56.6
28	63.4	59.0	49.3	39.2	70.9	59.1	52.6	45.1	37.6	65.8	54.9	48.8
30	55.2	51.4	43.0	34.1	61.8	51.5	45.8	39.3	32.8	57.3	47.8	42.5
31	51.7	48.2	40.2	31.9	57.9	48.2	42.9	36.8	30.7	53.7	44.8	39.8
32	48.6	45.2	37.8	30.0	54.3	45.3	40.3	34.6	28.8	50.4	42.0	37.4
34	43.0	40.0	33.5	26.6		40.1	35.7	30.6	25.5			
36	38.4	35.7	29.8	23.7								
37	36.3	33.8	28.2	22.4								
38			26.8	21.3								
40												
Properties												
r_m, in.^{2}	2.25	2.26	2.28	2.30	2.01	2.05	2.07	2.08	2.10	1.96	2.00	2.02
$\phi_b M_n$, kip-ft	33.4	30.0	22.8	15.5	46.9	36.9	31.2	25.5	19.4	45.0	35.3	29.9
$P_e(KL)^2/10^4$	6.69	6.21	5.20	4.14	7.44	6.19	5.50	4.76	3.97	6.95	5.77	5.13

Note: Heavy line indicates Kl/r equal to or greater than 200.

$F_y = 42$ ksi
$f'_c = 4$ ksi
$\phi_c = 0.85$

Table 4-14 (cont.). Composite Round HSS Design Strength in Axial Compression, $\phi_c P_n$, kips

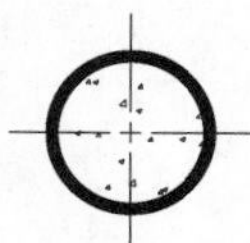

Shape	HSS6.000x				HSS5.563x				HSS5.500x		
	0.280	0.250	0.188	0.125	0.375	0.258	0.188	0.134	0.500	0.375	0.258
t_{design}, in.	0.261	0.233	0.174	0.116	0.349	0.241	0.174	0.125	0.465	0.349	0.241
Wt/ft	17.1	15.4	11.7	7.85	20.8	14.6	10.8	7.78	26.7	20.5	14.5
Effective length *KL* (ft) with respect to radius of gyration *r*: 0	236	220	186	152	258	202	167	140	310	254	199
6	216	202	170	139	233	183	151	126	279	229	180
7	210	195	165	134	225	176	145	122	269	221	173
8	202	188	159	129	215	169	139	116	257	211	166
9	194	181	152	123	205	161	133	111	244	201	158
10	185	173	145	118	194	152	126	105	231	190	149
11	176	164	138	111	183	144	118	98.4	217	179	140
12	166	155	130	105	172	135	111	91.9	203	168	131
13	157	146	122	98.6	160	125	103	85.4	189	156	122
14	147	137	115	92.0	148	116	95.5	78.9	174	144	113
15	137	127	107	85.4	137	107	87.9	72.4	160	133	104
16	127	118	98.7	78.9	125	98.0	80.5	66.1	146	121	95.0
17	117	109	91.0	72.5	114	89.2	73.3	60.0	133	110	86.4
18	108	100	83.4	66.3	103	80.8	66.3	54.1	119	99.9	78.1
19	98.3	91.4	76.1	60.3	92.9	72.6	59.6	48.5	107	89.6	70.0
20	89.2	82.9	69.0	54.5	83.9	65.5	53.8	43.8	96.7	80.9	63.2
21	80.9	75.2	62.6	49.4	76.1	59.5	48.8	39.7	87.7	73.4	57.3
22	73.7	68.5	57.0	45.0	69.3	54.2	44.4	36.2	79.9	66.9	52.2
23	67.5	62.7	52.2	41.2	63.4	49.6	40.7	33.1	73.1	61.2	47.8
24	62.0	57.6	47.9	37.8	58.2	45.5	37.3	30.4	67.2	56.2	43.9
25	57.1	53.1	44.2	34.9	53.7	42.0	34.4	28.0	61.9	51.8	40.5
26	52.8	49.1	40.8	32.2	49.6	38.8	31.8	25.9	57.2	47.9	37.4
27	49.0	45.5	37.9	29.9	46.0	36.0	29.5	24.0	53.1	44.4	34.7
28	45.5	42.3	35.2	27.8	42.8	33.4	27.4	22.3	49.4	41.3	32.3
29	42.4	39.4	32.8	25.9	39.9	31.2	25.6	20.8	46.0	38.5	30.1
30	39.7	36.9	30.7	24.2	37.3	29.1	23.9	19.5		36.0	28.1
31	37.1	34.5	28.7	22.7		27.3	22.4	18.2			26.3
32	34.9	32.4	26.9	21.3				17.1			
33	32.8	30.5	25.3	20.0							
34		28.7	23.9	18.8							
35											
Properties											
r_m, in.^{2}	2.03	2.04	2.06	2.08	1.85	1.88	1.91	1.92	1.79	1.83	1.86
$\phi_b M_n$, kip-ft	27.1	24.4	18.6	12.7	29.9	21.5	15.9	11.7	37.2	29.2	21.0
$P_e(KL)^2/10^4$	4.79	4.46	3.70	2.93	4.48	3.53	2.87	2.35	5.17	4.31	3.40

Note: Heavy line indicates Kl/r equal to or greater than 200.

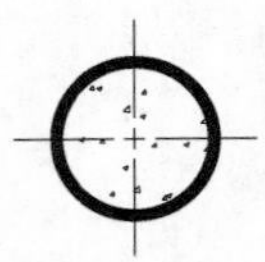

Table 4-14 (cont.). Composite Round HSS Design Strength in Axial Compression, $\phi_c P_n$, kips

$F_y = 42$ ksi
$f'_c = 4$ ksi
$\phi_c = 0.85$

Shape		HSS5.000x							HSS4.500x			
		0.500	0.375	0.312	0.258	0.250	0.188	0.125	0.337	0.237	0.188	0.125
t_{design}, in.		0.465	0.349	0.291	0.241	0.233	0.174	0.116	0.315	0.221	0.174	0.116
Wt/ft		24.1	18.5	15.6	13.1	12.7	9.67	6.51	15.0	10.8	8.67	5.85
Effective length KL (ft) with respect to radius of gyration r	0	274	224	198	175	171	143	115	182	143	123	98.5
	1	273	223	197	174	171	143	115	181	143	123	98.0
	2	270	221	195	172	169	141	113	179	141	121	96.7
	3	265	217	192	169	166	139	111	175	138	119	94.6
	4	259	212	187	165	162	136	109	170	134	115	91.7
	5	250	205	181	160	157	131	105	163	129	111	88.1
	6	241	197	174	154	151	126	101	155	123	106	83.9
	7	229	189	167	147	144	121	96.5	147	116	99.8	79.2
	8	217	179	158	140	137	115	91.4	137	109	93.5	74.1
	9	204	169	149	132	129	108	86.0	128	101	86.9	68.7
	10	191	158	140	123	121	101	80.3	117	92.9	80.0	63.2
	11	177	146	130	114	112	93.7	74.4	107	84.9	73.0	57.6
	12	163	135	120	105	104	86.5	68.5	96.9	76.8	66.1	52.0
	13	149	124	110	96.6	95.1	79.2	62.6	86.9	68.9	59.3	46.5
	14	135	112	99.9	87.9	86.6	72.0	56.7	77.2	61.3	52.8	41.3
	15	122	102	90.3	79.4	78.3	65.1	51.1	67.9	53.9	46.4	36.2
	16	109	91.1	81.0	71.2	70.3	58.3	45.7	59.7	47.4	40.8	31.8
	17	96.3	81.0	72.0	63.3	62.5	51.8	40.5	52.9	42.0	36.2	28.2
	18	85.9	72.2	64.3	56.4	55.7	46.2	36.1	47.1	37.5	32.2	25.1
	19	77.1	64.8	57.7	50.7	50.0	41.5	32.4	42.3	33.6	28.9	22.6
	20	69.6	58.5	52.1	45.7	45.1	37.5	29.3	38.2	30.3	26.1	20.4
	21	63.1	53.1	47.2	41.5	41.0	34.0	26.5	34.6	27.5	23.7	18.5
	22	57.5	48.3	43.0	37.8	37.3	31.0	24.2	31.6	25.1	21.6	16.8
	23	52.6	44.2	39.4	34.6	34.1	28.3	22.1	28.9	22.9	19.8	15.4
	24	48.3	40.6	36.1	31.7	31.4	26.0	20.3	26.5	21.1	18.1	14.1
	25	44.5	37.4	33.3	29.3	28.9	24.0	18.7		19.4	16.7	13.0
	26	41.2	34.6	30.8	27.0	26.7	22.2	17.3				
	27		32.1	28.6	25.1	24.8	20.6	16.1				
	28				23.3	23.0	19.1	14.9				
	29											
Properties												
r_m, in.^{2}		1.61	1.65	1.67	1.68	1.69	1.71	1.73	1.48	1.51	1.53	1.55
$\phi_b M_n$, kip-ft		30.2	23.8	20.3	17.2	16.7	12.8	8.73	17.4	12.8	10.3	7.02
$P_e(KL)^2/10^4$		3.74	3.14	2.79	2.46	2.42	2.00	1.57	2.06	1.64	1.41	1.09

Note: Heavy line indicates Kl/r equal to or greater than 200.

$F_y = 42$ ksi
$f'_c = 4$ ksi
$\phi_c = 0.85$

Table 4-14 (cont.). Composite Round HSS Design Strength in Axial Compression, $\phi_c P_n$, kips

Shape		HSS4.000x							
		0.337	0.313	0.250	0.237	0.226	0.220	0.188	0.125
t_{design}, in.		0.315	0.291	0.233	0.221	0.211	0.205	0.174	0.116
Wt/ft		13.2	12.3	10.0	9.53	9.12	8.89	7.66	5.18
Effective length *KL* (ft) with respect to radius of gyration *r*	0	156	148	127	122	119	116	105	82.9
	1	155	147	126	122	118	116	104	82.4
	2	153	144	124	120	116	114	103	81.1
	3	149	140	121	116	113	111	99.8	78.8
	4	143	135	116	112	109	107	96.1	75.8
	5	136	129	111	107	104	101	91.4	72.1
	6	128	121	104	101	97.5	95.5	86.1	67.8
	7	119	113	96.9	93.6	90.8	89.0	80.1	63.0
	8	110	104	89.2	86.3	83.6	81.9	73.8	58.0
	9	99.8	94.7	81.3	78.7	76.2	74.6	67.2	52.7
	10	89.8	85.3	73.2	70.9	68.7	67.2	60.5	47.4
	11	80.0	76.1	65.2	63.3	61.2	59.9	53.9	42.1
	12	70.5	67.1	57.5	55.8	54.0	52.8	47.5	37.1
	13	61.4	58.5	50.1	48.7	47.1	46.1	41.4	32.2
	14	52.9	50.5	43.2	42.1	40.6	39.7	35.7	27.8
	15	46.1	44.0	37.7	36.6	35.4	34.6	31.1	24.2
	16	40.5	38.6	33.1	32.2	31.1	30.4	27.4	21.3
	17	35.9	34.2	29.3	28.5	27.6	27.0	24.2	18.8
	18	32.0	30.5	26.1	25.4	24.6	24.0	21.6	16.8
	19	28.7	27.4	23.5	22.8	22.1	21.6	19.4	15.1
	20	25.9	24.7	21.2	20.6	19.9	19.5	17.5	13.6
	21	23.5	22.4	19.2	18.7	18.1	17.7	15.9	12.3
	22		20.4	17.5	17.0	16.5	16.1	14.5	11.2
	23								
Properties									
r_m, in.^{2}		1.31	1.32	1.33	1.34	1.34	1.34	1.35	1.37
$\phi_b M_n$, kip-ft		13.5	12.6	10.4	9.95	9.54	9.32	8.03	5.51
$P_e(KL)^2/10^4$		1.39	1.32	1.14	1.11	1.07	1.05	0.945	0.731

Note: Heavy line indicates *Kl*/*r* equal to or greater than 200.

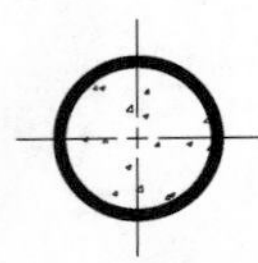

Table 4-15. Composite Round HSS Design Strength in Axial Compression, $\phi_c P_n$, kips

$F_y = 42$ ksi
$f'_c = 5$ ksi
$\phi_c = 0.85$

Shape		HSS20.000×		HSS18.000×		HSS16.000×				HSS14.000×		
		0.500	0.375	0.500	0.375	0.500	0.438	0.375	0.312	0.500	0.375	0.312
t_{design}, in.		0.465	0.349	0.465	0.349	0.465	0.407	0.349	0.291	0.465	0.349	0.291
Wt/ft		104	78.7	93.5	70.7	82.8	72.9	62.6	52.3	72.2	54.6	45.7
Effective length KL (ft) with respect to radius of gyration r	0	2050	1820	1740	1540	1450	1360	1280	1190	1190	1040	957
	6	2030	1810	1720	1520	1440	1350	1260	1170	1170	1020	940
	7	2030	1800	1720	1520	1430	1340	1250	1170	1160	1010	934
	8	2020	1800	1710	1510	1420	1330	1250	1160	1160	1010	927
	9	2010	1790	1700	1500	1410	1320	1240	1150	1150	997	919
	10	2000	1780	1690	1500	1400	1310	1230	1140	1140	988	910
	11	1990	1770	1680	1490	1390	1300	1220	1130	1120	978	901
	12	1980	1760	1670	1480	1380	1290	1210	1120	1110	967	891
	13	1970	1750	1660	1460	1370	1280	1200	1110	1100	955	879
	14	1950	1740	1640	1450	1350	1270	1190	1100	1090	942	868
	15	1940	1720	1630	1440	1340	1250	1170	1090	1070	929	855
	16	1930	1710	1610	1430	1320	1240	1160	1080	1050	915	842
	17	1910	1700	1600	1410	1310	1220	1140	1060	1040	900	828
	18	1890	1680	1580	1400	1290	1210	1130	1050	1020	885	814
	19	1880	1670	1560	1380	1270	1190	1110	1030	1000	869	799
	20	1860	1650	1550	1360	1250	1170	1100	1020	985	853	783
	22	1820	1620	1510	1330	1220	1140	1060	984	946	818	751
	24	1780	1580	1470	1290	1180	1100	1030	949	906	782	717
	25	1760	1560	1450	1270	1150	1080	1010	931	885	764	700
	26	1740	1540	1430	1250	1130	1060	987	912	864	745	682
	28	1690	1500	1380	1210	1090	1020	947	875	820	706	646
	30	1650	1460	1330	1170	1040	973	906	836	776	667	610
	31	1620	1430	1310	1150	1020	951	885	816	754	648	591
	32	1600	1410	1290	1130	996	929	864	796	732	628	573
	34	1550	1360	1240	1080	948	884	822	756	687	589	536
	36	1500	1320	1190	1040	900	839	779	716	643	550	500
	37	1470	1290	1160	1010	876	816	758	696	621	530	482
	38	1440	1270	1140	991	852	794	736	676	599	511	464
	40	1390	1220	1080	945	805	748	694	636	556	474	429
Properties												
r_m, in.^{2}		6.91	6.95	6.20	6.24	5.49	5.51	5.53	5.55	4.79	4.83	4.85
$\phi_b M_n$, kip-ft		558	425	450	343	353	312	269	226	268	205	172
$P_e(KL)^2/10^4$		416	358	290	248	194	180	165	150	123	104	94.4

$F_y = 42$ ksi
$f'_c = 5$ ksi
$\phi_c = 0.85$

Table 4-15 (cont.). Composite Round HSS Design Strength in Axial Compression, $\phi_c P_n$, kips

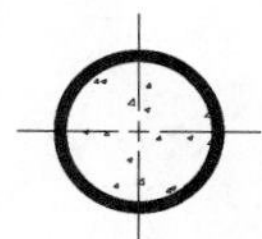

Shape		HSS12.750×			HSS12.500×						HSS11.250×		
		0.500	0.375	0.250	0.625	0.500	0.375	0.312	0.250	0.188	0.625	0.500	0.375
t_{design}, in.		0.465	0.349	0.233	0.581	0.465	0.349	0.291	0.233	0.174	0.581	0.465	0.349
Wt/ft		65.5	49.6	33.4	79.3	64.1	48.6	40.7	32.7	24.7	71.0	57.5	43.6
Effective length KL (ft) with respect to radius of gyration r	0	1040	898	755	1140	1010	870	803	731	660	985	866	744
	6	1010	879	739	1120	987	851	785	715	644	960	844	725
	7	1010	872	733	1110	979	845	779	709	639	951	836	718
	8	999	865	726	1100	971	837	772	702	632	941	827	710
	9	989	856	719	1090	961	828	763	695	625	929	817	701
	10	978	847	710	1080	950	819	754	686	617	917	806	691
	11	967	836	701	1060	938	808	745	677	609	903	793	680
	12	954	825	691	1050	926	797	734	667	600	888	780	669
	13	941	813	681	1040	912	785	723	657	590	872	766	657
	14	926	800	670	1020	898	772	711	646	579	856	751	644
	15	911	787	658	1000	882	759	698	634	568	838	736	630
	16	895	773	645	984	866	745	685	621	557	820	720	616
	17	878	758	632	965	850	730	671	608	545	800	703	601
	18	861	743	619	946	832	715	657	595	532	781	685	585
	19	843	727	605	926	814	699	642	581	519	760	667	570
	20	825	711	591	905	796	682	627	567	506	739	648	553
	22	786	676	561	862	757	648	595	537	479	696	610	520
	24	746	641	530	816	717	613	562	507	450	651	571	486
	25	725	623	515	793	696	595	545	491	436	629	551	468
	26	705	605	499	770	676	577	528	476	422	606	531	451
	28	663	568	467	723	634	540	494	444	393	561	491	416
	30	620	530	435	676	592	504	460	412	364	516	451	382
	31	599	512	419	652	571	485	443	397	349	494	432	365
	32	578	493	403	629	550	467	426	381	335	472	413	349
	34	536	457	372	582	509	431	392	350	307	430	375	316
	36	495	421	341	536	468	396	360	320	280	389	339	285
	37	475	403	326	514	448	379	344	306	267	368	321	269
	38	455	386	311	492	429	362	328	292	253	349	304	255
	40	416	352	282	449	391	329	297	263	229	315	274	231
Properties													
r_m, in.^{2}		4.35	4.39	4.43	4.22	4.26	4.30	4.32	4.34	4.36	3.78	3.82	3.86
$\phi_b M_n$, kip-ft		221	169	115	260	212	162	137	111	83.2	209	170	131
$P_e(KL)^2/10^4$		89.6	75.6	60.7	96.1	83.8	70.6	63.5	56.5	49.0	67.4	58.5	49.3

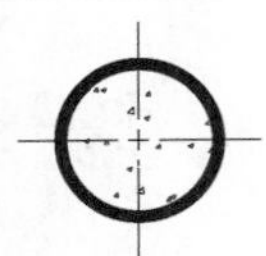

Table 4-15 (cont.). Composite Round HSS Design Strength in Axial Compression, $\phi_c P_n$, kips

$F_y = 42$ ksi
$f'_c = 5$ ksi
$\phi_c = 0.85$

Shape		HSS11.250×			HSS10.750×		HSS10.000×					
		0.312	0.250	0.188	0.500	0.250	0.625	0.500	0.375	0.312	0.250	0.188
t_{design}, in.		0.291	0.233	0.174	0.465	0.233	0.581	0.465	0.349	0.291	0.233	0.174
Wt/ft		36.5	29.4	22.2	54.8	28.1	62.6	50.8	38.6	32.3	26.1	19.7
Effective length KL (ft) with respect to radius of gyration r	0	680	618	553	809	575	836	730	624	569	513	456
	6	662	601	537	786	558	809	706	603	550	496	440
	7	655	595	532	778	552	799	698	596	543	489	434
	8	648	588	525	769	545	789	688	588	535	482	428
	9	640	580	518	759	537	777	678	578	527	474	420
	10	630	572	510	748	528	763	666	568	517	466	412
	11	621	562	502	735	519	749	654	557	507	456	404
	12	610	552	493	722	509	734	640	545	496	446	394
	13	598	542	483	708	498	717	626	533	484	435	385
	14	586	531	472	693	487	700	610	520	472	424	374
	15	574	519	462	677	475	682	595	506	459	412	364
	16	560	507	450	661	463	663	578	491	446	400	352
	17	547	494	438	643	450	643	561	477	432	388	341
	18	532	481	426	626	437	623	543	461	418	375	329
	19	518	467	414	608	424	603	525	446	404	361	317
	20	503	453	401	589	410	582	507	430	389	348	305
	22	472	424	375	551	382	539	470	397	359	321	280
	24	440	395	348	512	353	496	432	365	329	293	255
	25	424	381	334	493	339	475	413	348	314	280	243
	26	408	366	321	473	325	453	394	332	300	266	231
	28	376	336	294	435	296	411	357	300	270	240	207
	30	345	307	268	396	269	370	322	270	242	214	184
	31	329	293	255	378	255	350	304	255	229	202	173
	32	314	279	243	359	242	331	287	240	215	189	162
	34	284	252	218	324	216	293	255	213	190	168	144
	36	255	226	195	289	193	261	227	190	170	150	128
	37	241	214	184	274	182	248	215	179	161	142	121
	38	229	203	175	259	173	235	204	170	152	134	115
	40	207	183	158	234	156	212	184	154	138	121	104
Properties												
r_m, in.^{2}		3.88	3.90	3.92	3.64	3.72	3.34	3.38	3.41	3.43	3.45	3.47
$\phi_b M_n$, kip-ft		110	89.1	67.1	155	81.3	163	133	102	86.3	69.9	52.9
$P_e(KL)^2/10^4$		44.5	39.0	33.8	50.3	33.3	45.3	39.5	32.9	29.7	26.1	22.3

$F_y = 42$ ksi
$f'_c = 5$ ksi
$\phi_c = 0.85$

Table 4-15 (cont.). Composite Round HSS Design Strength in Axial Compression, $\phi_c P_n$, kips

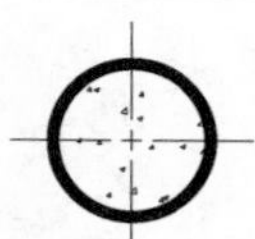

Shape	HSS9.625×					HSS8.750×					HSS8.625×	
	0.500	0.375	0.312	0.250	0.188	0.500	0.375	0.312	0.250	0.188	0.500	0.375
t_{design}, in.	0.465	0.349	0.291	0.233	0.174	0.465	0.349	0.291	0.233	0.174	0.465	0.349
Wt/ft	48.8	37.1	31.1	25.1	19.0	44.1	33.6	28.1	22.7	17.2	43.4	33.1
Effective length *KL* (ft) with respect to radius of gyration *r*												
0	693	590	537	483	429	605	513	465	417	368	593	502
6	669	569	517	465	412	580	491	445	399	351	567	480
7	660	562	510	459	407	571	483	438	392	345	558	473
8	650	553	503	452	400	561	475	430	385	339	548	464
9	640	544	494	444	393	550	465	421	377	331	537	454
10	628	534	485	435	385	538	455	412	368	323	525	444
11	615	523	474	426	376	524	443	401	359	314	511	432
12	601	511	463	416	367	510	431	390	348	305	497	420
13	587	498	452	405	357	495	418	378	338	296	482	407
14	571	485	439	394	347	480	405	366	327	285	467	394
15	555	471	427	382	336	463	391	353	315	275	450	380
16	538	457	413	370	325	447	377	340	303	264	434	366
17	521	442	400	358	314	429	362	327	291	253	417	351
18	503	427	386	345	302	412	347	313	278	242	399	336
19	485	411	372	332	290	394	332	299	266	231	382	321
20	467	395	357	318	278	376	317	285	253	219	364	306
22	430	363	328	292	254	340	286	258	228	197	328	276
24	393	331	298	265	230	305	256	230	203	175	293	246
25	374	316	284	252	218	288	241	217	191	164	276	232
26	356	300	270	239	206	271	227	204	179	153	260	218
28	320	269	242	213	184	238	199	178	156	133	227	190
30	285	240	215	189	162	208	174	155	136	116	198	166
31	269	225	201	177	152	194	163	145	128	109	186	155
32	252	211	189	166	142	182	153	136	120	102	174	146
34	223	187	167	147	126	162	135	121	106	90.3	154	129
36	199	167	149	131	112	144	121	108	94.6	80.6	138	115
37	189	158	141	124	106	136	114	102	89.6	76.3	130	109
38	179	150	134	118	101	129	108	96.8	84.9	72.3	123	103
40	161	135	121	106	91.0	117	97.6	87.4	76.6	65.3	111	93.2
Properties												
r_m, in.^{2}	3.24	3.28	3.30	3.32	3.34	2.93	2.97	2.99	3.01	3.03	2.89	2.93
$\phi_b M_n$, kip-ft	123	94.5	80.0	64.9	48.8	101	77.5	65.5	53.2	40.3	97.7	75.3
$P_e(KL)^2/10^4$	34.7	29.1	26.0	22.9	19.5	25.1	21.0	18.8	16.5	14.0	23.9	20.0

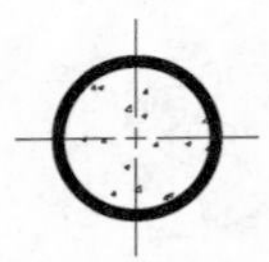

Table 4-15 (cont.). Composite Round HSS Design Strength in Axial Compression, $\phi_c P_n$, kips

$F_y = 42$ ksi
$f'_c = 5$ ksi
$\phi_c = 0.85$

Shape		HSS8.625×			HSS7.625×	HSS7.500×					HSS7.000×		
		0.322	0.250	0.188	0.125	0.500	0.375	0.312	0.250	0.188	0.500	0.375	0.312
t_{design}, in.		0.300	0.233	0.174	0.116	0.465	0.349	0.291	0.233	0.174	0.465	0.349	0.291
Wt/ft		28.6	22.4	17.0	10.0	37.4	28.6	24.0	19.4	14.7	34.7	26.6	22.3
Effective length KL (ft) with respect to radius of gyration r	0	463	408	359	253	490	411	371	330	288	445	373	336
	6	443	390	343	237	462	388	350	311	271	417	349	314
	7	436	383	337	232	453	380	342	304	265	407	340	306
	8	427	376	330	226	442	371	334	297	258	396	331	298
	9	418	368	323	219	430	360	325	288	250	384	321	288
	10	409	359	315	212	417	350	315	279	242	370	309	278
	11	398	350	306	205	403	338	304	270	234	356	298	267
	12	387	339	297	196	388	325	293	260	225	341	285	256
	13	375	329	287	188	373	312	281	249	215	326	272	244
	14	363	318	277	179	357	299	269	238	205	310	259	232
	15	350	306	267	170	341	285	256	227	195	294	245	220
	16	336	294	256	161	324	271	244	215	185	278	231	207
	17	323	282	245	152	307	257	231	204	175	261	218	195
	18	309	270	234	143	290	243	218	192	165	245	204	183
	19	295	257	223	134	274	229	205	181	154	229	190	170
	20	281	245	212	125	257	215	192	169	144	213	177	158
	22	253	220	189	108	224	187	168	147	125	182	151	135
	24	226	195	168	91.9	193	161	144	126	106	154	127	114
	25	212	183	157	84.7	178	149	133	116	98.0	142	117	105
	26	199	172	147	78.3	165	137	123	107	90.6	131	109	96.9
	28	174	149	127	67.5	142	119	106	92.5	78.1	113	93.6	83.6
	30	151	130	111	58.8	124	103	92.2	80.6	68.1	98.4	81.6	72.8
	31	142	122	104	55.1	116	96.7	86.4	75.5	63.7	92.2	76.4	68.2
	32	133	114	97.2	51.7	109	90.7	81.0	70.8	59.8	86.5	71.7	64.0
	34	118	101	86.1	45.8	96.4	80.4	71.8	62.8	53.0	76.6	63.5	56.7
	36	105	90.3	76.8	40.9	85.9	71.7	64.0	56.0	47.3	68.4	56.6	50.6
	37	99.6	85.5	72.7	38.7	81.4	67.9	60.6	53.0	44.7	64.7	53.6	47.9
	38	94.4	81.1	69.0	36.7	77.1	64.3	57.5	50.2	42.4	61.3	50.8	45.4
	40	85.2	73.2	62.2	33.1	69.6	58.1	51.9	45.3	38.3			

Properties

r_m, in.^{2}	2.95	2.97	2.99	2.66	2.49	2.53	2.55	2.57	2.59	2.32	2.35	2.37
$\phi_b M_n$, kip-ft	65.5	51.7	39.1	20.6	72.5	56.4	47.6	38.7	29.4	62.7	48.8	41.3
$P_e(KL)^2/10^4$	18.2	15.7	13.4	7.07	15.0	12.5	11.1	9.75	8.24	11.8	9.88	8.83

Note: Heavy line indicates Kl/r equal to or greater than 200.

$F_y = 42$ ksi
$f'_c = 5$ ksi
$\phi_c = 0.85$

Table 4-15 (cont.). Composite Round HSS Design Strength in Axial Compression, $\phi_c P_n$, kips

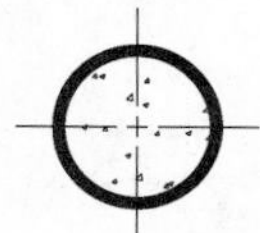

Shape	HSS7.000×			HSS6.875×					HSS6.625×			
	0.250	**0.188**	**0.125**	**0.500**	**0.375**	**0.312**	**0.250**	**0.188**	**0.500**	**0.432**	**0.375**	**0.312**
t_{design}, in.	**0.233**	**0.174**	**0.116**	**0.465**	**0.349**	**0.291**	**0.233**	**0.174**	**0.465**	**0.403**	**0.349**	**0.291**
Wt/ft	**18.0**	**13.7**	**9.19**	**34.1**	**26.1**	**21.9**	**17.7**	**13.4**	**32.7**	**28.6**	**25.1**	**21.1**
Effective length *KL* (ft) with respect to radius of gyration *r*												
0	298	259	220	434	364	327	290	252	413	377	345	310
6	278	241	204	405	339	305	270	234	384	350	320	288
7	271	235	198	395	331	298	263	228	373	341	312	280
8	264	228	192	384	322	289	256	221	362	330	302	271
9	255	221	186	372	311	280	247	213	349	319	292	262
10	246	212	178	358	300	270	238	205	336	307	281	252
11	236	204	171	344	288	259	228	197	321	294	269	241
12	226	195	163	329	276	247	218	188	306	280	256	230
13	216	185	155	314	263	236	208	178	291	266	243	218
14	205	176	146	298	249	224	197	169	275	251	230	206
15	194	166	138	282	236	211	186	159	259	237	216	194
16	183	156	129	265	222	199	175	150	243	222	203	182
17	171	146	121	249	209	187	164	140	227	207	190	170
18	160	137	112	233	195	175	153	130	211	193	176	158
19	149	127	104	217	182	162	142	121	195	178	163	146
20	139	118	95.9	201	168	151	132	112	180	165	151	135
22	118	99.6	80.4	171	143	128	112	93.9	151	138	126	113
24	99.3	83.7	67.6	144	120	107	93.7	78.9	127	116	106	94.7
25	91.5	77.1	62.3	133	111	99.0	86.4	72.7	117	107	97.7	87.2
26	84.6	71.3	57.6	123	103	91.6	79.9	67.2	108	98.8	90.4	80.7
28	72.9	61.5	49.6	106	88.5	79.0	68.9	58.0	93.2	85.2	77.9	69.6
30	63.5	53.6	43.2	92.1	77.1	68.8	60.0	50.5	81.1	74.2	67.9	60.6
31	59.5	50.2	40.5	86.2	72.2	64.4	56.2	47.3	76.0	69.5	63.6	56.7
32	55.8	47.1	38.0	80.9	67.7	60.4	52.7	44.4	71.3	65.2	59.7	53.3
34	49.5	41.7	33.7	71.7	60.0	53.5	46.7	39.3	63.2	57.8	52.8	47.2
36	44.1	37.2	30.0	63.9	53.5	47.8	41.7	35.1	56.4	51.5	47.1	42.1
37	41.8	35.2	28.4	60.5	50.7	45.2	39.4	33.2			44.6	39.8
38	39.6	33.4	26.9		48.0	42.9	37.4	31.5				
40		30.1	24.3									
Properties												
r_m, in.^{2}	2.39	2.41	2.43	2.27	2.31	2.33	2.35	2.37	2.18	2.20	2.22	2.24
$\phi_b M_n$, kip-ft	33.7	25.5	17.3	60.2	46.9	39.7	32.4	24.6	55.8	49.1	43.5	36.9
$P_e(KL)^2/10^4$	7.70	6.48	5.25	11.1	9.30	8.31	7.23	6.11	9.83	9.00	8.22	7.33

Note: Heavy line indicates Kl/r equal to or greater than 200.

Table 4-15 (cont.). Composite Round HSS Design Strength in Axial Compression, $\phi_c P_n$, kips

$F_y = 42$ ksi
$f'_c = 5$ ksi
$\phi_c = 0.85$

Shape		HSS6.625×				HSS6.125×					HSS6.000×		
		0.280	0.250	0.188	0.125	0.500	0.375	0.312	0.250	0.188	0.500	0.375	0.312
t_{design}, in.		0.261	0.233	0.174	0.116	0.465	0.349	0.291	0.233	0.174	0.465	0.349	0.291
Wt/ft		19.0	17.0	12.9	8.69	30.1	23.1	19.4	15.7	11.9	29.4	22.5	19.0
Effective length *KL* (ft) with respect to radius of gyration *r*	0	292	275	238	201	372	310	277	245	211	362	301	270
	6	271	255	220	185	341	284	254	224	192	330	275	246
	7	263	248	214	179	330	275	246	217	186	319	266	238
	8	255	240	207	173	318	265	238	209	179	307	256	229
	9	246	231	199	167	306	255	228	200	172	294	245	220
	10	237	222	191	159	292	243	218	191	163	281	234	209
	11	226	213	183	152	277	231	207	181	155	266	222	198
	12	216	202	174	144	262	219	196	171	146	251	209	187
	13	205	192	164	136	247	206	184	161	137	235	197	176
	14	193	181	155	128	231	193	173	151	128	220	184	164
	15	182	170	146	120	216	180	161	140	119	204	171	152
	16	170	160	136	112	200	167	149	130	110	189	158	141
	17	159	149	127	103	185	154	138	120	101	174	145	130
	18	148	138	117	95.5	170	142	127	110	92.6	159	133	119
	19	137	128	108	87.7	155	130	116	100	84.3	144	121	108
	20	126	118	99.4	80.2	141	118	105	90.8	76.2	131	109	97.7
	22	105	98.3	82.7	66.4	116	97.5	87.0	75.0	62.9	108	90.5	80.7
	24	88.5	82.6	69.5	55.8	97.8	81.9	73.1	63.0	52.9	90.7	76.0	67.8
	25	81.6	76.1	64.0	51.4	90.1	75.5	67.3	58.1	48.7	83.6	70.1	62.5
	26	75.4	70.4	59.2	47.5	83.3	69.8	62.3	53.7	45.1	77.3	64.8	57.8
	28	65.0	60.7	51.1	41.0	71.9	60.2	53.7	46.3	38.9	66.6	55.9	49.8
	30	56.6	52.8	44.5	35.7	62.6	52.4	46.8	40.3	33.8	58.0	48.7	43.4
	31	53.0	49.5	41.6	33.4	58.6	49.1	43.8	37.8	31.7	54.4	45.6	40.7
	32	49.8	46.4	39.1	31.4	55.0	46.1	41.1	35.5	29.7	51.0	42.8	38.1
	34	44.1	41.1	34.6	27.8		40.8	36.4	31.4	26.3			
	36	39.3	36.7	30.9	24.8								
	37	37.2	34.7	29.2	23.5								
	38			27.7	22.2								
	40												
Properties													
r_m, in.^{2}		2.25	2.26	2.28	2.30	2.01	2.05	2.07	2.08	2.10	1.96	2.00	2.02
$\phi_b M_n$, kip-ft		33.4	30.0	22.8	15.5	46.9	36.9	31.2	25.5	19.4	45.0	35.3	29.9
$P_e(KL)^2/10^4$		6.86	6.38	5.38	4.33	7.53	6.30	5.61	4.89	4.11	7.03	5.87	5.24

Note: Heavy line indicates *Kl*/*r* equal to or greater than 200.

$F_y = 42$ ksi
$f'_c = 5$ ksi
$\phi_c = 0.85$

Table 4-15 (cont.). Composite Round HSS Design Strength in Axial Compression, $\phi_c P_n$, kips

Shape	HSS6.000×				HSS5.563×				HSS5.500×		
	0.280	0.250	0.188	0.125	0.375	0.258	0.188	0.134	0.500	0.375	0.258
t_{design}, in.	0.261	0.233	0.174	0.116	0.349	0.241	0.174	0.125	0.465	0.349	0.241
Wt/ft	17.1	15.4	11.7	7.85	20.8	14.6	10.8	7.78	26.7	20.5	14.5
Effective length KL (ft) with respect to radius of gyration r											
0	253	238	204	171	271	217	182	156	322	267	214
6	231	217	186	155	244	195	164	140	289	240	191
7	224	210	179	149	235	188	157	134	278	231	184
8	215	202	173	143	225	180	150	128	265	221	176
9	206	193	165	137	214	171	143	121	252	210	167
10	196	184	157	130	203	161	135	114	238	198	158
11	186	174	149	122	191	152	127	107	223	186	148
12	176	164	140	115	178	142	118	99.4	208	174	138
13	165	154	131	107	166	131	109	91.8	193	161	128
14	154	144	122	99.5	153	121	101	84.3	178	149	118
15	143	133	113	91.9	141	111	92.4	76.9	163	137	108
16	132	123	104	84.3	128	101	84.2	69.8	149	124	98.3
17	121	113	95.5	77.0	117	92.0	76.2	62.9	135	113	88.9
18	111	104	87.1	69.9	105	82.8	68.4	56.2	121	101	79.8
19	101	94.1	78.9	63.0	94.4	74.3	61.4	50.4	108	91.1	71.6
20	91.3	85.0	71.2	56.8	85.2	67.1	55.4	45.5	97.8	82.2	64.7
21	82.8	77.1	64.6	51.6	77.3	60.8	50.3	41.3	88.7	74.5	58.6
22	75.4	70.3	58.9	47.0	70.4	55.4	45.8	37.6	80.9	67.9	53.4
23	69.0	64.3	53.9	43.0	64.4	50.7	41.9	34.4	74.0	62.1	48.9
24	63.4	59.0	49.5	39.5	59.2	46.6	38.5	31.6	67.9	57.1	44.9
25	58.4	54.4	45.6	36.4	54.5	42.9	35.5	29.1	62.6	52.6	41.4
26	54.0	50.3	42.1	33.6	50.4	39.7	32.8	26.9	57.9	48.6	38.3
27	50.1	46.7	39.1	31.2	46.8	36.8	30.4	25.0	53.7	45.1	35.5
28	46.6	43.4	36.3	29.0	43.5	34.2	28.3	23.2	49.9	41.9	33.0
29	43.4	40.4	33.9	27.0	40.5	31.9	26.4	21.7	46.5	39.1	30.8
30	40.6	37.8	31.7	25.3	37.9	29.8	24.6	20.2		36.5	28.7
31	38.0	35.4	29.6	23.7		27.9	23.1	18.9			26.9
32	35.7	33.2	27.8	22.2				17.8			
33	33.5	31.2	26.2	20.9							
34		29.4	24.6	19.7							
35											
Properties											
r_m, in.^{2}	2.03	2.04	2.06	2.08	1.85	1.88	1.91	1.92	1.79	1.83	1.86
$\phi_b M_n$, kip-ft	27.1	24.4	18.6	12.7	29.9	21.5	15.9	11.7	37.2	29.2	21.0
$P_e(KL)^2/10^4$	4.89	4.57	3.82	3.06	4.55	3.61	2.96	2.45	5.23	4.38	3.48

Note: Heavy line indicates Kl/r equal to or greater than 200.

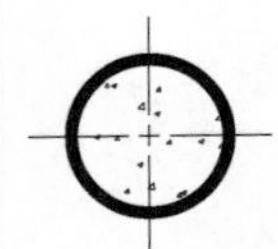

Table 4-15 (cont.). Composite Round HSS Design Strength in Axial Compression, $\phi_c P_n$, kips

$F_y = 42$ ksi
$f'_c = 5$ ksi
$\phi_c = 0.85$

Shape	HSS5.000×							HSS4.500×			
	0.500	0.375	0.312	0.258	0.250	0.188	0.125	0.337	0.237	0.188	0.125
t_{design}, in.	0.465	0.349	0.291	0.241	0.233	0.174	0.116	0.315	0.221	0.174	0.116
Wt/ft	24.1	18.5	15.6	13.1	12.7	9.67	6.51	15.0	10.8	8.67	5.85
Effective length *KL* (ft) with respect to radius of gyration *r*											
0	283	235	209	186	183	156	128	190	153	133	109
1	282	234	208	186	182	155	128	189	152	133	108
2	279	231	206	184	180	153	126	187	150	131	107
3	274	227	202	180	177	151	124	183	147	128	104
4	267	221	197	176	173	147	120	177	142	124	101
5	258	214	191	170	167	142	116	170	136	119	96.7
6	248	206	183	163	160	136	111	162	130	113	91.8
7	236	196	175	156	153	130	106	153	122	107	86.3
8	224	186	166	147	145	123	100	143	114	99.5	80.4
9	210	175	156	139	136	115	93.6	132	106	92.0	74.2
10	196	163	145	129	127	107	87.0	121	97.1	84.4	67.8
11	181	151	135	120	118	99.4	80.2	110	88.3	76.7	61.4
12	166	139	124	110	108	91.2	73.4	99.4	79.6	69.0	55.0
13	152	127	113	100	98.9	83.2	66.6	88.8	71.1	61.6	48.9
14	137	115	103	90.9	89.6	75.2	60.0	78.6	62.9	54.5	43.0
15	123	104	92.5	81.7	80.7	67.6	53.7	68.9	55.1	47.6	37.5
16	110	92.7	82.7	72.9	72.0	60.2	47.5	60.5	48.4	41.9	33.0
17	97.3	82.1	73.3	64.6	63.8	53.3	42.1	53.6	42.9	37.1	29.2
18	86.8	73.2	65.4	57.6	56.9	47.5	37.5	47.8	38.2	33.1	26.0
19	77.9	65.7	58.7	51.7	51.1	42.7	33.7	42.9	34.3	29.7	23.4
20	70.3	59.3	53.0	46.7	46.1	38.5	30.4	38.7	31.0	26.8	21.1
21	63.8	53.8	48.0	42.3	41.8	34.9	27.6	35.1	28.1	24.3	19.1
22	58.1	49.0	43.8	38.6	38.1	31.8	25.1	32.0	25.6	22.1	17.4
23	53.2	44.9	40.0	35.3	34.9	29.1	23.0	29.3	23.4	20.3	16.0
24	48.8	41.2	36.8	32.4	32.0	26.7	21.1	26.9	21.5	18.6	14.6
25	45.0	38.0	33.9	29.9	29.5	24.6	19.5		19.8	17.1	13.5
26	41.6	35.1	31.3	27.6	27.3	22.8	18.0				
27		32.6	29.1	25.6	25.3	21.1	16.7				
28				23.8	23.5	19.6	15.5				
29											
Properties											
r_m, in.^{2}	1.61	1.65	1.67	1.68	1.69	1.71	1.73	1.48	1.51	1.53	1.55
$\phi_b M_n$, kip-ft	30.2	23.8	20.3	17.2	16.7	12.8	8.73	17.4	12.8	10.3	7.02
$P_e(KL)^2/10^4$	3.78	3.19	2.84	2.51	2.47	2.06	1.63	2.09	1.67	1.44	1.13

Note: Heavy line indicates Kl/r equal to or greater than 200.

$F_y = 42$ ksi
$f'_c = 5$ ksi
$\phi_c = 0.85$

Table 4-15 (cont.). Composite Round HSS Design Strength in Axial Compression, $\phi_c P_n$, kips

Shape		HSS4.000×							
		0.337	0.313	0.250	0.237	0.226	0.220	0.188	0.125
t_{design}, in.		0.315	0.291	0.233	0.221	0.211	0.205	0.174	0.116
Wt/ft		13.2	12.3	10.0	9.53	9.12	8.89	7.66	5.18
Effective length KL (ft) with respect to radius of gyration r	0	163	154	134	129	126	124	112	91.0
	1	162	153	133	129	125	123	112	90.4
	2	159	151	131	127	123	121	110	88.8
	3	154	147	127	123	120	117	107	86.2
	4	148	141	122	118	115	113	103	82.7
	5	141	134	116	112	109	107	97.4	78.4
	6	132	126	109	106	103	101	91.4	73.4
	7	123	117	101	98.1	95.3	93.5	84.8	68.0
	8	113	107	93.0	90.1	87.5	85.8	77.8	62.2
	9	103	97.6	84.4	81.9	79.4	77.8	70.5	56.2
	10	92.1	87.7	75.7	73.5	71.3	69.8	63.2	50.2
	11	81.7	77.9	67.2	65.3	63.2	61.9	56.0	44.3
	12	71.7	68.4	58.9	57.3	55.5	54.3	49.1	38.7
	13	62.1	59.3	51.0	49.7	48.0	47.0	42.4	33.2
	14	53.6	51.2	44.0	42.8	41.4	40.5	36.6	28.7
	15	46.7	44.6	38.3	37.3	36.1	35.3	31.9	25.0
	16	41.0	39.2	33.7	32.8	31.7	31.0	28.0	21.9
	17	36.3	34.7	29.8	29.1	28.1	27.5	24.8	19.4
	18	32.4	30.9	26.6	25.9	25.1	24.5	22.1	17.3
	19	29.1	27.8	23.9	23.3	22.5	22.0	19.9	15.6
	20	26.3	25.1	21.5	21.0	20.3	19.9	17.9	14.0
	21	23.8	22.7	19.5	19.0	18.4	18.0	16.3	12.7
	22		20.7	17.8	17.3	16.8	16.4	14.8	11.6
	23								
Properties									
r_m, in.^{2}		1.31	1.32	1.33	1.34	1.34	1.34	1.35	1.37
$\phi_b M_n$, kip-ft		13.5	12.6	10.4	9.95	9.54	9.32	8.03	5.51
$P_e(KL)^2/10^4$		1.40	1.34	1.16	1.13	1.09	1.07	0.967	0.755

Note: Heavy line indicates Kl/r equal to or greater than 200.

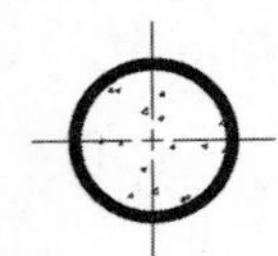

Table 4-16. Composite Steel Pipe Design Strength in Axial Compression, $\phi_c P_n$, kips

$F_y = 35$ ksi
$f'_c = 4$ ksi
$\phi_c = 0.85$

Shape		Pipe 12		Pipe 10		Pipe 8			Pipe 6		
		XS	Std.	XS	Std.	XXS	XS	Std.	XXS	XS	Std.
Wall Thkns., in		0.500	0.375	0.500	0.365	0.875	0.500	0.322	0.864	0.432	0.280
Wt/ft		65.5	49.6	54.8	40.5	72.5	43.4	28.6	53.2	28.6	19.0
Effective length *KL* (ft) with respect to least radius of gyration, r_{my}	0	888	764	697	584	741	513	395	519	325	250
	6	872	751	681	570	714	494	380	486	306	234
	7	867	746	675	565	704	488	375	474	299	229
	8	861	740	668	559	693	480	369	461	291	223
	9	854	734	661	553	681	472	363	447	283	217
	10	846	727	652	546	668	463	355	432	273	210
	11	838	720	643	538	653	453	348	416	264	202
	12	828	712	634	530	638	443	339	398	253	194
	13	819	703	623	521	621	432	331	381	242	186
	14	808	694	612	511	604	420	322	362	231	177
	15	797	684	600	501	586	408	312	343	220	168
	16	785	674	588	491	567	395	302	324	208	159
	17	773	663	575	480	548	382	292	305	197	150
	18	760	652	562	468	528	368	281	286	185	142
	19	747	640	549	457	508	355	271	268	174	133
	20	733	628	535	445	488	341	260	249	162	124
	21	719	615	520	432	468	327	249	231	151	115
	22	704	602	506	420	447	313	238	214	140	107
	23	689	589	491	407	427	299	227	196	130	98.8
	24	673	576	476	394	406	285	216	180	119	90.8
	25	658	562	460	381	386	271	205	166	110	83.7
	26	642	548	445	368	366	257	195	154	102	77.4
	27	626	534	430	355	346	244	184	142	94.1	71.8
	28	610	520	414	342	327	230	174	132	87.5	66.7
	29	593	506	399	329	308	217	164	123	81.6	62.2
	30	577	491	384	316	290	205	154	115	76.3	58.1
	32	543	462	353	290	255	180	135	101	67.0	51.1
	34	510	433	324	265	225	159	120	89.8	59.4	45.3
	36	477	404	295	241	201	142	107		53.0	40.4
	38	444	376	267	218	180	128	95.9			
	40	412	348	241	196	163	115	86.6			
Properties											
r_m, in.		4.33	4.38	3.63	3.67	2.76	2.88	2.94	2.06	2.19	2.25
$\phi_b M_n$, kip-ft		197	151	138	103	139	86.6	58.3	75.9	43.6	29.7
$P_e(KL)^2/10^4$		91.7	76.3	51.7	42.3	34.9	24.7	18.5	13.9	9.26	6.98

XS = x-strong.
XXS = xx-strong.
Note: Heavy line indicates Kl/r equal to or greater than 200.

$F_y = 35$ ksi
$f'_c = 4$ ksi
$\phi_c = 0.85$

Table 4-16 (cont.). Composite Steel Pipe Design Strength in Axial Compression, $\phi_c P_n$, kips

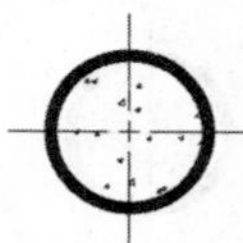

Shape	Pipe 5			Pipe 4			Pipe 3½		Pipe 3		
	XXS	XS	Std.	XXS	XS	Std.	XS	Std.	XXS	XS	Std.
Wall Thkns., in	0.750	0.375	0.258	0.674	0.337	0.237	0.318	0.226	0.600	0.300	0.216
Wt/ft	38.6	20.8	14.6	27.6	15.0	10.8	12.5	9.12	18.6	10.3	7.58
Effective length KL (ft) with respect to least radius of gyration, r_{my}											
0	374	234	186	264	164	131	135	108	175	109	87.7
6	340	214	170	227	144	114	114	91.3	136	87.1	70.0
7	329	208	164	216	137	109	107	85.9	124	80.3	64.6
8	316	200	158	203	129	103	99.7	80.0	112	73.2	58.8
9	302	192	152	189	121	96.6	92.0	73.8	99.7	65.8	52.9
10	287	183	145	175	113	90.0	84.1	67.5	87.4	58.5	47.0
11	272	174	138	160	104	83.1	76.1	61.1	75.6	51.3	41.2
12	256	165	130	146	95.9	76.2	68.2	54.8	64.3	44.5	35.7
13	240	155	122	132	87.3	69.4	60.6	48.7	54.8	38.0	30.5
14	223	145	114	118	78.9	62.7	53.3	42.8	47.3	32.8	26.3
15	207	135	106	105	70.8	56.2	46.4	37.3	41.2	28.6	22.9
16	191	125	98.6	92.1	62.9	49.9	40.8	32.8	36.2	25.1	20.1
17	175	115	90.9	81.6	55.7	44.2	36.1	29.1	32.0	22.2	17.8
18	160	106	83.3	72.8	49.7	39.4	32.2	25.9		19.8	15.9
19	145	96.6	76.0	65.3	44.6	35.4	28.9	23.3		17.8	14.3
20	131	87.6	68.9	58.9	40.3	32.0	26.1	21.0			
21	119	79.5	62.5	53.4	36.5	29.0	23.7	19.0			
22	108	72.4	56.9	48.7	33.3	26.4		17.3			
23	98.8	66.2	52.1		30.5	24.2					
24	90.8	60.8	47.8		28.0	22.2					
25	83.6	56.1	44.1			20.5					
26	77.3	51.8	40.8								
27	71.7	48.1	37.8								
28	66.7	44.7	35.2								
29		41.7	32.8								
30		38.9	30.6								
32											
34											
36											
38											
40											
Properties											
r_m, in.	1.72	1.84	1.88	1.37	1.48	1.51	1.31	1.34	1.05	1.14	1.16
$\phi_b M_n$, kip-ft	45.9	26.5	19.1	26.2	15.4	11.3	11.3	8.45	13.4	8.09	6.12
$P_e(KL)^2/10^4$	7.05	4.70	3.70	3.18	2.15	1.72	1.39	1.12	1.23	0.855	0.696

XS = x-strong.
XXS = xx-strong.
Note: Heavy line indicates Kl/r equal to or greater than 200.

Table 4-17. Composite Steel Pipe Design Strength in Axial Compression, $\phi_c P_n$, kips

$F_y = 35$ ksi
$f'_c = 5$ ksi
$\phi_c = 0.85$

Shape		Pipe 12		Pipe 10		Pipe 8			Pipe 6		
		XS	Std.	XS	Std.	XXS	XS	Std.	XXS	XS	Std.
Wall Thkns., in		0.500	0.375	0.500	0.365	0.875	0.500	0.322	0.864	0.432	0.280
Wt/ft		65.5	49.6	54.8	40.5	72.5	43.4	28.6	53.2	28.6	19.0
Effective length KL (ft) with respect to least radius of gyration, r_{my}	0	967	846	752	642	768	546	431	532	344	271
	6	949	831	733	625	739	525	414	498	323	253
	7	943	825	726	619	729	518	408	486	315	247
	8	936	818	719	613	717	510	401	472	307	240
	9	928	811	710	605	704	501	394	458	297	233
	10	919	803	701	597	690	491	385	442	287	225
	11	909	794	691	588	675	480	377	425	276	216
	12	898	785	680	578	658	468	367	407	265	207
	13	887	774	668	568	641	456	357	388	253	198
	14	875	763	655	556	623	443	346	369	241	188
	15	862	752	642	545	603	429	335	350	229	178
	16	849	740	628	533	584	415	324	330	216	168
	17	834	727	614	520	563	401	312	310	204	158
	18	820	714	599	507	543	386	300	290	191	148
	19	804	700	584	493	522	371	288	271	179	139
	20	789	686	568	480	500	356	276	252	167	129
	21	772	671	552	465	479	341	264	233	155	119
	22	756	656	535	451	457	325	251	215	143	110
	23	739	641	519	436	436	310	239	198	132	101
	24	721	625	502	422	414	295	227	181	121	93.0
	25	703	609	485	407	393	280	215	167	112	85.7
	26	685	593	468	392	372	265	203	155	103	79.2
	27	667	576	451	377	352	251	191	143	95.6	73.5
	28	649	560	434	362	331	236	180	133	88.9	68.3
	29	630	543	417	348	312	222	169	124	82.9	63.7
	30	611	527	400	333	292	208	158	116	77.4	59.5
	32	574	493	367	304	257	183	139	102	68.1	52.3
	34	537	460	334	276	228	162	123	90.4	60.3	46.3
	36	500	428	303	249	203	145	110		53.8	41.3
	38	464	395	273	224	182	130	98.4			
	40	428	364	246	202	164	117	88.8			
Properties											
r_m, in.		4.33	4.38	3.63	3.67	2.76	2.88	2.94	2.06	2.19	2.25
$\phi_b M_n$, kip-ft		197	151	138	103	139	86.6	58.3	75.9	43.6	29.7
$P_e(KL)^2/10^4$		94.0	78.8	52.8	43.5	35.2	25.1	19.0	14.0	9.40	7.15

XS = x-strong.
XXS = xx-strong.
Note: Heavy line indicates Kl/r equal to or greater than 200.

$F_y = 35$ ksi
$f'_c = 5$ ksi
$\phi_c = 0.85$

Table 4-17 (cont.). Composite Steel Pipe Design Strength in Axial Compression, $\phi_c P_n$, kips

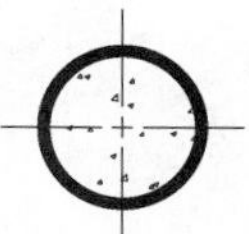

Shape	Pipe 5			Pipe 4			Pipe 3½		Pipe 3		
	XXS	XS	Std.	XXS	XS	Std.	XS	Std.	XXS	XS	Std.
Wall Thkns., in	0.750	0.375	0.258	0.674	0.337	0.237	0.318	0.226	0.600	0.300	0.216
Wt/ft	38.6	20.8	14.6	27.6	15.0	10.8	12.5	9.12	18.6	10.3	7.58
Effective length KL (ft) with respect to least radius of gyration, r_{my}											
0	383	247	200	269	173	140	142	115	178	114	93.0
6	348	226	182	232	150	122	119	96.6	138	90.2	73.6
7	336	218	176	219	143	116	111	90.5	126	83.0	67.6
8	323	210	169	206	135	109	103	84.1	113	75.4	61.3
9	308	201	162	192	126	102	95.1	77.3	101	67.6	54.9
10	293	192	154	177	117	94.5	86.6	70.3	88.1	59.8	48.5
11	277	182	146	163	108	87.0	78.1	63.4	76.1	52.3	42.3
12	261	171	137	148	98.7	79.4	69.8	56.5	64.6	45.1	36.4
13	244	161	129	133	89.6	71.9	61.7	49.9	55.0	38.5	31.0
14	227	150	120	119	80.7	64.6	53.9	43.6	47.5	33.2	26.7
15	210	139	111	105	72.1	57.6	47.0	38.0	41.3	28.9	23.3
16	193	129	103	92.6	63.8	50.9	41.3	33.4	36.3	25.4	20.5
17	177	118	94.1	82.0	56.5	45.1	36.6	29.6	32.2	22.5	18.1
18	161	108	85.9	73.1	50.4	40.2	32.6	26.4		20.1	16.2
19	146	98.4	78.0	65.6	45.2	36.1	29.3	23.7		18.0	14.5
20	131	88.9	70.4	59.2	40.8	32.6	26.4	21.4			
21	119	80.6	63.8	53.7	37.0	29.5	24.0	19.4			
22	109	73.5	58.2	49.0	33.7	26.9		17.7			
23	99.4	67.2	53.2		30.9	24.6					
24	91.3	61.7	48.9		28.3	22.6					
25	84.2	56.9	45.0			20.8					
26	77.8	52.6	41.6								
27	72.1	48.8	38.6								
28	67.1	45.4	35.9								
29		42.3	33.5								
30		39.5	31.3								
32											
34											
36											
38											
40											
Properties											
r_m, in.	1.72	1.84	1.88	1.37	1.48	1.51	1.31	1.34	1.05	1.14	1.16
$\phi_b M_n$, kip-ft	45.9	26.5	19.1	26.2	15.4	11.3	11.3	8.45	13.4	8.09	6.12
$P_e(KL)^2/10^4$	7.09	4.77	3.78	3.20	2.18	1.75	1.41	1.14	1.24	0.864	0.707

XS = x-strong.
XXS = xx-strong.
Note: Heavy line indicates Kl/r equal to or greater than 200.

PART 5

DESIGN OF FLEXURAL MEMBERS

BEAM

SCOPE

The specification requirements and other design considerations summarized in this Part apply to the design of flexural members, except plate girders, subject to uniaxial flexure without axial forces and torsion. For the design of plate girders, see LRFD Specification Chapter G. For the design of members subject to biaxial flexure and/or flexure in combination with axial tension or compression and/or torsion, see Part 6. For flexural members that are part of a seismic-force-resisting system in which the seismic response modification factor R is taken greater than 3, see the Seismic Provisions, which are available from AISC at www.aisc.org.

LOAD DETERMINATION

The controlling required strengths (factored shear force) V_u and (factored bending moment) M_u are determined by analysis as indicated in LRFD Specification Section A5.

GROSS AREA, NET AREA AND EFFECTIVE AREA

For Flexure

Per LRFD Specification Section B10, flexural design properties are based upon the full cross-section with no reduction for bolt holes when LRFD Specification Equation B10-1 is satisfied. Otherwise, the flexural design properties are based upon an effective tension flange area as given in LRFD Specification Equation B10-3. Special provisions apply to hybrid girders per LRFD Specification Section B10.

For Shear

For shear, the area is determined per LRFD Specification Section F2.1.

COMPACT, NON-COMPACT AND SLENDER-ELEMENT CROSS-SECTIONS (LOCAL BUCKLING)

Determining the Width-Thickness Ratios of the Cross-Section

Flexural members are classified for flexure on the basis of the width-thickness ratios of the various elements of the cross-section. The width-thickness ratio λ is calculated for each element of the cross-section as follows:

1. For flexural members other than HSS, steel pipe and single angles, per LRFD Specification Section B5.1.
2. For HSS and steel pipe, per HSS Specification Section 2.2.1.
3. For single angles, see Single-Angle Specification Section 5.

For convenience, the width-thickness ratios in LRFD Specification Table B5.1 are given as a function of F_y in LRFD Specification Numerical Values Table 6.

Width-Thickness Limits for Compact and Non-Compact Cross-Sections

Flexural members are considered compact (the plastic moment can be reached without local buckling) when λ is equal to or less than λ_p and the flange(s) are continuously connected to the web(s). Flexural members are considered non-compact (local buckling will occur, but only after initial yielding) when λ exceeds λ_p but is equal to or less than λ_r. Flexural members are considered slender-element cross-sections (local buckling will occur prior to yielding) when λ exceeds λ_r. The values of λ_p and λ_r are determined as follows:

1. For W-shapes, other similar doubly symmetric rolled and built-up shapes, and channels, per LRFD Specification Section B5.1 and Table B5.1.
2. For singly symmetric I-shapes, per LRFD Specification Section B5.1, Table B5.1 and Appendix B5.1.
3. For HSS and steel pipe, per HSS Specification 2.2.1 and Table 2.2-1.
4. For single angles, see Single-Angle Specification Section 5.
5. For other flexural members, per LRFD Specification Section B5.1.

See also LRFD Specification Table A-F1.1

UNBRACED LENGTH (LATERAL-TORSIONAL BUCKLING)

Classification of Spans for Flexure

Flexural members bent about their strong axis are classified on the basis of the length L_b between braced points. Braced points are points at which support against lateral-torsional buckling is provided per LRFD Specification Section C3.4. Flexural members are considered braced when L_b is equal to or less than L_p and unbraced when L_b exceeds L_p. An unbraced beam with L_b equal to or less than L_r is subject to inelastic lateral-torsional buckling. An unbraced beam with L_b greater than L_r is subject to elastic lateral-torsional buckling. The values of L_p and L_r are determined as follows:

1. For flexural members other than HSS, steel pipe and single angles, per LRFD Specification Section F1.2.
2. For HSS and steel pipe, lateral-torsional buckling does not apply per HSS Specification Section 5.1.
3. For single angles, see Single-Angle Specification Section 5.1.3.

Lateral-torsional buckling does not apply to flexural members bent about their weak axis per LRFD Specification Section F1.2.

Consideration of Moment Gradient

The moment gradient between braced points should be considered in the determination of the design strength for the limit state of lateral-torsional buckling using the beam bending coefficient C_b. In the case of a uniform moment between braced points causing single curvature of the member, $C_b = 1$. This represents the worst case, and C_b can be conservatively taken as unity for use with the maximum moment between braced points in all designs per LRFD Specification Section F1.2a. However, when desired, a non-uniform moment gradient between braced points can be considered using C_b calculated as given in LRFD Specification Equation F1-3. Exceptions are provided as follows:

1. As an alternative, when the moment diagram between braced points is a straight line between the moments at the braced points, C_b can be calculated as given in LRFD Commentary Equation C-F1-1.
2. For cantilevered members where the free end is unbraced, C_b must be taken as unity per LRFD Specification Section F1.2a.
3. For tees with the stem in compression, C_b should be taken as unity per LRFD Commentary Section 1.2c.

FLEXURAL DESIGN STRENGTH FOR STRONG-AXIS BENDING

The general flexural design strength is illustrated in Figure 5-1.

Requirements for Braced, Compact Members

For flexural members that are braced ($L_b \leq L_p$) and compact ($\lambda \leq \lambda_p$), the flexural design strength ϕM_n, which must equal or exceed the required strength (factored bending moment)

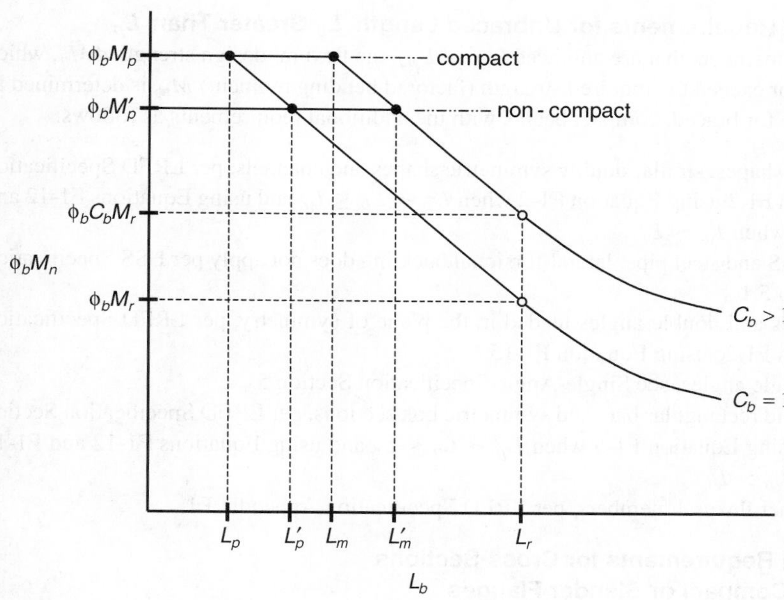

$$\phi_b M_p = 0.9 F_y Z_x \qquad \phi_b M_r = 0.9 F_L S_x$$

$$\phi_b M'_p = \phi_b M_p - (\phi_b M_p - \phi_b M_r)\frac{(\lambda - \lambda_p)}{(\lambda_r - \lambda_p)}$$

$$L_m = L_p + (L_r - L_p)\frac{\phi_b M_p (C_b - 1)}{C_b(\phi_b M_p - \phi_b M_r)} \text{ when } L_m \leq L_r \qquad L_p = 1.76 r_y \sqrt{\frac{E}{F_y}}$$

$$= \frac{\phi_b C_b S_x r_y X_1}{\phi_b M_p}\sqrt{1 + \sqrt{1 + X_2\left(\frac{\phi_b M_p}{\phi_b C_b S_x}\right)^2}} \text{ when } L_m > L_r \qquad L_r = \frac{r_y X_1}{F_L}\sqrt{1 + \sqrt{1 + X_2 F_L^2}}$$

$$L'_m = L'_p + (L_r - L_p)\frac{\phi_b M'_p (C_b - 1)}{C_b(\phi_b M_p - \phi_b M_r)} \text{ when } L'_m \leq L_r \qquad L'_p = L_p + (L_r - L_p)\frac{(\phi_b M_p - \phi_b M'_p)}{(\phi_b M_p - \phi_b M_r)}$$

$$= L_m \text{ when } L'_m > L_r$$

Fig. 5–1. General Flexural design strength of beams.

M_u, is determined as follows:

1. For W-shapes, similar doubly symmetric shapes and channels, per LRFD Specification Section F1.1 using Equation F1-1. Note that for all W-, M-, S- and HP-shapes and channels listed in Part 1, $F_y Z_x$ is equal to or less than $1.5 F_y S_x$.
2. For rectangular and square HSS, per HSS Specification Section 5.1 using Equation 5.1-4.
3. For round HSS and steel pipe, per HSS Specification Section 5.1 using Equation 5.1-1.
4. For single angles, see Single-Angle Specification Section 5.
5. For other flexural members, per LRFD Specification Appendix F1.

Additional Requirements for Unbraced Length L_b Greater Than L_p

For flexural members that are unbraced ($L_b > L_p$), the flexural design strength ϕM_n, which must equal or exceed the required strength (factored bending moment) M_u, is determined as given before for braced, compact beams with the additional requirements as follows:

1. For W-shapes, similar doubly symmetric shapes and channels, per LRFD Specification Section F1.2 using Equation F1-2 when $L_p < L_b \leq L_r$ and using Equations F1-12 and F1-13 when $L_b > L_r$.
2. For HSS and steel pipe, lateral-torsional buckling does not apply per HSS Specification Section 5.1.
3. For tees and double angles loaded in the plane of symmetry, per LRFD Specification Section F1.2c using Equation F1-15.
4. For single angles, see Single-Angle Specification Section 5.
5. For solid rectangular bars and symmetric box sections, per LRFD Specification Section F1.2 using Equation F1-5 when $L_p < L_b \leq L_r$ and using Equations F1-12 and F1-14 when $L_b > L_r$.
6. For other flexural members, per LRFD Specification Appendix F1.

Additional Requirements for Cross-Sections with Non-Compact or Slender Flanges

For flexural members that have flange width-thickness ratios such that ($\lambda > \lambda_p$), the flexural design strength ϕM_n, which must equal or exceed the required strength (factored bending moment) M_u, is determined as given before for braced, compact beams with the additional requirements as follows:

1. For W-shapes, similar doubly symmetric shapes and channels, per LRFD Specification Appendix F1 using Equation A-F1-3 when $\lambda_p < \lambda \leq \lambda_r$ and using Equation A-F1-4 when $\lambda > \lambda_r$.
2. For rectangular and square HSS, per HSS Specification Section 5.1 using Equation 5.1-5 when $\lambda_p < \lambda \leq \lambda_r$ and Equation 5.1-6 when $\lambda > \lambda_r$.
3. For round HSS and steel pipe, per HSS Specification Section 5.1 using Equation 5.1-2 when $\lambda_p < \lambda \leq \lambda_r$ and Equation 5.1-3 when $\lambda > \lambda_r$.
4. For single angles, see Single-Angle Specification Section 5.
5. For other flexural members, per LRFD Specification Appendix F1.

Additional Requirements for Cross-Sections with Non-Compact or Slender Webs

For flexural members that have web width-thickness ratios such that ($\lambda > \lambda_p$), the flexural design strength ϕM_n, which must equal or exceed the required strength (factored bending moment) M_u, is determined as given before for braced, compact beams with the additional requirements as follows:

1. For W-shapes, similar doubly symmetric shapes and channels, per LRFD Specification Appendix F1 using Equation A-F1-3 when $\lambda_p < \lambda \leq \lambda_r$ and per LRFD Specification Appendix G when $\lambda > \lambda_r$.
2. For rectangular and square HSS, per HSS Specification Section 5.1 using Equation 5.1-5 when $\lambda_p < \lambda \leq \lambda_r$ and Equation 5.1-6 when $\lambda > \lambda_r$.
3. For round HSS and steel pipe, per HSS Specification Section 5.1 using Equation 5.1-2 when $\lambda_p < \lambda \leq \lambda_r$ and Equation 5.1-3 when $\lambda > \lambda_r$.
4. For single angles, see Single-Angle Specification Section 5.
5. For other flexural members, per LRFD Specification Appendix F1.

FLEXURAL DESIGN STRENGTH FOR WEAK-AXIS BENDING

The design of flexural members subject to weak-axis bending is identical to that for strong-axis bending, except that web local buckling does not apply to H-shapes and lateral-torsional buckling does not apply per LRFD Specification Section F1.2.

SHEAR DESIGN STRENGTH

For flexural members, the shear design strength ϕV_n, which must equal or exceed the required strength (factored shear force) V_u, is determined as follows:

1. For W-shapes, similar doubly symmetric shapes, singly symmetric I-shapes and channels with shear in the plane of the web, per LRFD Specification Section F2.
2. For W-shapes, similar doubly symmetric shapes, singly symmetric I-shapes and channels with shear in the weak axis, per LRFD Specification Section H2.
3. For HSS and steel pipe, per HSS Specification Section 5.2.
4. For single angles, per Single-Angle Specification Section 3.
5. For unsymmetric and other shapes, per LRFD Specification Section H2.

STEEL W-SHAPE BEAMS WITH COMPOSITE SLABS

The following pertains to W-shapes with composite concrete slabs in regions of positive moment. For composite flexural members in regions of negative moment, see LRFD Specification Sections I3.2 and I5.2. For concrete-encased beams, see LRFD Specification Section I3.3. For further information on composite design and construction, see also Viest et al. (1997).

Concrete Slab Effective Width

The effective width of a concrete slab acting compositely with a steel beam is determined per LRFD Specification Section I3.1.

Shear Stud Connectors

Material, placement and spacing requirements for shear stud connectors are given in LRFD Specification Sections I5.1 and I5.6. The design shear strength Q_n of one shear stud connector is determined per LRFD Specification Section I5. See also LRFD Specification Numerical Values Table 10. Note that special reduction factors may apply to shear stud connectors in formed steel deck, per LRFD Specification Section I3.5.

If channel shear connectors are used, see LRFD Specification Section I5.4. If other types of shear connectors are used, see LRFD Specification Section I6.

Flexural Design Strength for Positive Moment

The flexural design strength of a composite beam subject to positive moment is determined per LRFD Specification Section I3.2 assuming a uniform compressive stress of $0.85 f_c'$ and zero tensile strength in the concrete and a uniform stress of F_y in the tension area (and compression area, if any) in the steel section. The position of the plastic neutral axis (PNA) can then be determined by static equilibrium.

Per LRFD Specification Section I5.5, enough shear stud connectors must be provided between a point of maximum moment and the nearest point of zero moment to transfer the horizontal shear force V_h between the steel beam and concrete slab, where V_h is determined per LRFD Specification Section I5.2. The corresponding number of shear stud connectors n required is determined per LRFD Specification Section I5.5. For partial-composite design, the design shear strength of the shear stud connectors ΣQ_n controls the design flexural strength of the composite flexural member.

Shored and Unshored Construction

The flexural design strength $\phi_b M_n$ is identical for both shored and unshored construction. However, in unshored construction, the steel beam must meet the requirements in LRFD Specification Section I3.4. Additionally, issues such as lateral support during construction and construction-load deflection may require consideration in unshored construction.

Shear Design Strength

Per LRFD Specification Section I3.6, the shear design strength for composite beams is determined as illustrated previously for steel beams.

OTHER SPECIFICATION REQUIREMENTS AND DESIGN CONSIDERATIONS

The following other specification requirements and design considerations apply to the design of flexural members.

Special Requirements for Heavy Shapes and Plates

For beams with complete-joint-penetration groove welded joints and made from heavy shapes (ASTM A6 tensile group 4 or 5 shapes) or built up from plates exceeding 2-in. thickness, see LRFD Specification Section A3.1c.

Plastic Analysis

For design by plastic analysis, see LRFD Specification Section F1.3.

Web Openings

For beams with web openings, see LRFD Specification Section F4. Also, see AISC Design Guide No. 2 *Steel and Composite Beams with Web Openings* (Darwin, 1990).

Plate Girders

For plate girders, see LRFD Specification Chapter G. See also LRFD Specification Numerical Values Tables 8-36, 8-50, 9-36 and 9-50. See also Zahn (1987).

Tapered Members

For tapered members, see LRFD Specification Appendix F3.

Concentrated Forces

If the flexural member receives concentrated forces along its length, see LRFD Specification Section K1 or HSS Specification Section 8, as appropriate.

Ponding

For flexural members that are part of a roof system subject to ponding, see LRFD Specification Section K2.

Fatigue

For flexural applications involving fatigue, see LRFD Specification Section K3.

Torsion

For beams subject to torsion, see AISC Design Guide No. 9 *Torsional Analysis of Structural Steel Members* (Seaburg and Carter, 1997).

Serviceability

Serviceability requirements, per LRFD Specification Section L3, should be appropriate for the application, including an appropriate limit on the deflection of the flexural member, as well as the vibration characteristics of the system of which the flexural member is a part. See also AISC Design Guide No. 3 *Serviceability Design Considerations for Low-Rise Buildings* (Fisher and West, 1990), AISC Design Guide No. 5 *Low- and Medium-Rise Steel Buildings* (Allison, 1991) and AISC Design Guide No. 11 *Floor Vibrations Due to Human Activity* (Murray, Allen and Ungar, 1997).

The maximum vertical deflection Δ, in., can be calculated using the equations given in Tables 5-16 and 5-17. Alternatively, for common cases of simple-span beams and I-shaped members and channels, the equation can be used.

$$\Delta = ML^2/(C_1 I_x)$$

where

M = maximum service-load moment, kip-ft
L = span length, ft
I_x = moment of inertia, in.^{4}
C_1 = loading constant (see Figure 5-2)

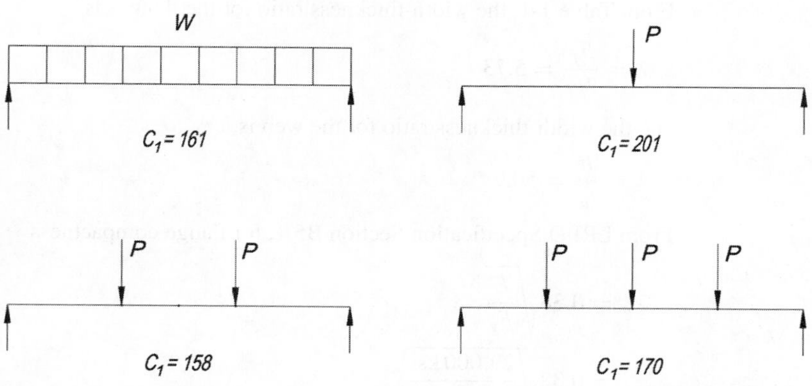

Fig. 5–2. Loading constants for use in determining simple-beam deflections.

Connection Design

For flexural-member connection design, see Part 10 (simple shear connections), Part 11 (flexible wind connections) and Part 12 (fully restrained moment connections).

DESIGN EXAMPLES

EXAMPLE 5.1. W-shape flexural member design (design strength of compact shape in strong-axis bending).

Given:

Determine the design strength of an ASTM A992 W18×40 bent about its strong-axis on a 35-ft span with a uniformly distributed load.

$F_y = 50\,\text{ksi}$ $\quad Z_x = 78.4\,\text{in.}^3$ $\quad S_x = 68.4\,\text{in.}^3$

$F_u = 65\,\text{ksi}$ $\quad I_x = 612\,\text{in.}^4$ $\quad r_y = 1.27\,\text{in.}$

$X_1 = 1{,}810\,\text{ksi}$ $\quad X_2 = 0.0172\,\text{ksi}^{-2}$

$d = 17.9\,\text{in.}$ $\quad t_w = 0.315\,\text{in.}$

a) Assume the beam is braced such that $L_b = 2\,\text{ft}$. Also, calculate what the deflection would be for a uniformly distributed 1 kip/ft service load.
b) Assume the beam is braced at the end and third points ($L_b = 11.7\,\text{ft}$).
c) Assume the beam is braced at the end points only ($L_b = 35\,\text{ft}$).

Solution a:

From Table 5-17, Diagram 1, the maximum deflection Δ_{max} occurs at mid-span and can be calculated as:

$$\begin{aligned}\Delta_{max} &= \frac{5wl^4}{384EI}\\ &= \frac{5(1\,\text{kip/ft})(35\,\text{ft})^4 \times (12\,\text{in./ft})^3}{384\,(29{,}000\,\text{ksi})\left(612\,\text{in.}^4\right)}\\ &= 1.90\,\text{in.}\end{aligned}$$

Note that this deflection must be compared with an appropriate deflection limit, which is dependent upon the application. Deflection will often be more critical than strength in beam design.

From Table 1-1, the width-thickness ratio for the flanges is

$$\lambda = \frac{b_f}{2t_f} = 5.73$$

and the width-thickness ratio for the web is

$$\lambda = \frac{h}{t_w} = 50.9$$

From LRFD Specification Section B5.1, for flange compactness

$$\begin{aligned}\lambda_p &= 0.38\sqrt{\frac{E}{F_y}}\\ &= 0.38\sqrt{\frac{29{,}000\,\text{ksi}}{50\,\text{ksi}}}\\ &= 9.15\end{aligned}$$

Since $\lambda < \lambda_p$, the flanges are compact. From LRFD Specification Section B5.1, for web compactness

$$\lambda_p = 3.76\sqrt{\frac{E}{F_y}}$$
$$= 3.76\sqrt{\frac{29{,}000\,\text{ksi}}{50\,\text{ksi}}}$$
$$= 90.6$$

Since $\lambda < \lambda_p$, the web is compact. Since both the flanges and the web are compact, the W18×40 is compact.

From LRFD Specification Section F1.2a,

$$L_p = 1.76 r_y \sqrt{\frac{E}{F_y}}$$
$$= \frac{1.76\,(1.27\,\text{in.})}{12\,\text{in./ft}}\sqrt{\frac{29{,}000\,\text{ksi}}{50\,\text{ksi}}}$$
$$= 4.49\,\text{ft}$$

Since $L_b < L_p$, the beam is braced.

For a beam that is both braced and compact, the flexural yielding limit-state controls the flexural design strength. From LRFD Specification Section F1.1, the flexural yielding design strength $\phi_b M_n$ is

$$\phi_b M_n = \phi_b F_y Z_x$$
$$= \frac{0.9(50\,\text{ksi})\left(78.4\,\text{in.}^3\right)}{12\,\text{in./ft}}$$
$$= 294\,\text{kip-ft}$$

From LRFD Specification Section F2, the shear yielding design strength $\phi_v V_n$ is

$$\phi_v V_n = \phi_v F_y d t_w$$
$$= 0.9(50\,\text{ksi})(17.9\,\text{in.})(0.315\,\text{in.})$$
$$= 254\,\text{kips}$$

Solution b: As determined in solution a, the beam is compact.

As determined in solution a, $L_p = 4.49$ ft. From LRFD Specification Section F1.2a,

$$L_r = \frac{r_y X_1}{F_L}\sqrt{1+\sqrt{1+X_2 F_L^2}}$$
$$= \frac{(1.27\,\text{in.})(1{,}810\,\text{ksi})}{(50\,\text{ksi}-10\,\text{ksi})(12\,\text{in./ft})}\sqrt{1+\sqrt{1+\left(\frac{0.0172}{\text{ksi}^2}\right)(50\,\text{ksi}-10\,\text{ksi})^2}}$$
$$= 12.0\,\text{ft}$$

For a beam that is compact, but unbraced with $L_p < L_b \le L_r$, either the flexural yielding limit-state or the inelastic lateral-torsional buckling limit-state controls the flexural design strength. For lateral-torsional buckling, the moment gradient effect can be calculated using LRFD Specification Equation F1-3, where

$$C_b = \frac{12.5M_{\text{max}}}{2.5M_{\text{max}} + 3M_A + 4M_B + 3M_C}$$

For the center-span beam segment, using relative values determined from the moment diagram for a uniformly loaded simple-span beam ($M_{\text{max}} =$ 1.00, $M_A = 0.972$, $M_B = 1.00$, and $M_C = 0.972$)

$$\begin{aligned} C_b &= \frac{12.5(1.00)}{2.5(1.00) + 3(0.972) + 4(1.00) + 3(0.972)} \\ &= 1.01 \end{aligned}$$

For the end-span beam segments, using relative values determined from the moment diagram for a uniformly loaded simple-span beam ($M_{\text{max}} =$ 0.889, $M_A = 0.306$, $M_B = 0.556$, and $M_C = 0.750$)

$$\begin{aligned} C_b &= \frac{12.5(0.889)}{2.5(0.889) + 3(0.306) + 4(0.556) + 3(0.750)} \\ &= 1.46 \end{aligned}$$

Thus, the center span with $C_b = 1.01$ is more critical.

As determined in solution a, $\phi_b M_p = 294$ kip-ft. From LRFD Specification Section F1.2a,

$$\begin{aligned} \phi_b M_r &= \phi_b F_L S_x \\ &= \frac{0.9(50\,\text{ksi} - 10\,\text{ksi})\left(68.4\,\text{in.}^3\right)}{12\,\text{in./ft}} \\ &= 205\,\text{kip-ft} \end{aligned}$$

and the flexural design strength $\phi_b M_n$ is

$$\begin{aligned} \phi_b M_n &= C_b\left[\phi_b M_p - (\phi_b M_p - \phi_b M_r)\left(\frac{L_b - L_p}{L_r - L_p}\right)\right] \le \phi_b M_p \\ &= (1.01)\left[(294\,\text{kip-ft}) - (294\,\text{kip-ft} - 250\,\text{kip-ft})\left(\frac{11.7\,\text{ft} - 4.48\,\text{ft}}{12.0\,\text{ft} - 4.48\,\text{ft}}\right)\right] \\ &\le 294\,\text{kip-ft} \\ &= 211\,\text{kip-ft} \le 294\,\text{kip-ft} \\ &= 211\,\text{kip-ft} \end{aligned}$$

As determined in solution a, the shear yielding design strength $\phi_v V_n$ is

$$\phi_v V_n = 254\,\text{kips}$$

Solution c: As determined in solution a, the beam is compact.

As determined in solution b, $L_r = 12.0$ ft

For a beam that is compact, but unbraced with $L_b > L_r$, the elastic lateral-torsional buckling limit-state controls the flexural design strength.

The moment gradient effect can be calculated using LRFD Specification Equation F1-3, where

$$C_b = \frac{12.5M_{max}}{2.5M_{max} + 3M_A + 4M_B + 3M_C}$$

Using relative values determined from the moment diagram for a uniformly loaded simple-span beam ($M_{max} = 1.00$, $M_A = 0.500$, $M_B = 1.00$, and $M_C = 0.500$)

$$C_b = \frac{12.5(1.00)}{2.5(1.00) + 3(0.750) + 4(1.00) + 3(0.750)}$$
$$= 1.14$$

From LRFD Specification Section F1.2b, the flexural design strength $\phi_b M_n$ is

$$\phi_b M_n = \phi_b \frac{C_b S_x X_1 \sqrt{2}}{\frac{L_b}{r_y}} \sqrt{1 + \frac{X_1^2 X_2}{2\left(\frac{L_b}{r_y}\right)^2}}$$

$$= 0.9 \frac{(1.14)\,(68.4\,\text{in.}^3)\,(1{,}810\,\text{ksi})\sqrt{2}}{\frac{(35\,\text{ft} \times 12\,\text{in./ft})}{(1.27\,\text{in.})}} \sqrt{1 + \frac{(1{,}810\,\text{ksi})^2\left(\frac{0.0172}{\text{ksi}^2}\right)}{2\left(\frac{(35\,\text{ft} \times 12\,\text{in./ft})}{(1.27\,\text{in.})}\right)^2}}$$

$$= 50.8\,\text{kip-ft}$$

As determined in solution a, the shear yielding design strength $\phi_v V_n$ is

$$\phi_v V_n = 254\,\text{kips}$$

Comments: The preceding calculations can be simplified using Tables 5-1, 5-3, 5-4 and 5-5. For solution a, from Table 5-3, a W18×40 has $L_p = 4.49$ ft. Thus, for $L_b = 2$ ft,

$$\phi_b M_n = 294\,\text{kip-ft}$$

and

$$\phi_v V_n = 254\,\text{kips}$$

For solution b, from Table 5-3, a W18×40 has $L_p = 4.49$ ft, $\phi_b M_{px} = 294$ kip-ft, $L_r = 12.0$ ft, $\phi_b M_{rx} = 205$ kip-ft and $BF = 11.7$ kips. From Table 5-1, $C_b = 1.46$ for the end span and 1.01 for the center span. Thus, for $L_b = 11.7$ ft,

$$\phi_b M_n = C_b[\phi_b M_{px} - BF(L_b - L_p)] \leq \phi_b M_{px}$$
$$= (1.01)\big[(294\,\text{kip-ft}) - 11.7\,\text{kips}\,(11.7\,\text{ft} - 4.48\,\text{ft})\big] \leq 294\,\text{kip-ft}$$
$$= 211\,\text{kip-ft} \leq 294\,\text{kip-ft}$$
$$= 211\,\text{kip-ft}$$

and

$$\phi_v V_n = 254\,\text{kips}$$

EXAMPLE 5.2. W-shape flexural member design (design strength of non-compact shape in strong-axis bending).

Given:

Determine the design strength of an ASTM A992 W21×48 bent about its strong axis on a 40-ft span with two equal concentrated loads acting at the third points along the beam.

$F_y = 50\,\text{ksi}$ $\quad Z_x = 107\,\text{in.}^3$ $\quad S_x = 93.0\,\text{in.}^3$

$F_u = 65\,\text{ksi}$ $\quad I_x = 959\,\text{in.}^4$ $\quad r_y = 1.66\,\text{in.}$

$X_1 = 1{,}450\,\text{ksi}$ $\quad X_2 = 0.0436\,\text{ksi}^{-2}$

$d = 20.6\,\text{in.}$ $\quad t_w = 0.350\,\text{in.}$

a) Assume the beam is braced such that $L_b = 5\,\text{ft}$. Also, calculate what the deflection would be if each concentrated load were a 10 kip service load.

b) Assume the beam is braced at the end and third points ($L_b = 13.3\,\text{ft}$).

c) Assume the beam is braced at the end points only ($L_b = 40\,\text{ft}$).

Solution a:

From Table 5-17, Diagram 9, with $x = l/3$, the maximum deflection Δ_{max} occurs at mid-span and can be calculated as:

$$\begin{aligned}\Delta_{max} &= \frac{23Pl^3}{648EI} \\ &= \frac{23\,(10\,\text{kips})(40\,\text{ft})^3 \times (12\,\text{in./ft})^3}{648\,(29{,}000\,\text{ksi})\left(959\,\text{in.}^4\right)} \\ &= 1.41\,\text{in.}\end{aligned}$$

Note that this deflection must be compared with an appropriate deflection limit, which is dependent upon the application. Deflection will often be more critical than strength in beam design.

From Table 1-1, the width-thickness ratio for the flanges is

$$\lambda = \frac{b_f}{2t_f} = 9.47$$

and the width-thickness ratio for the web is

$$\lambda = \frac{h}{t_w} = 53.6$$

From LRFD Specification Section B5.1, for flange compactness

$$\begin{aligned}\lambda_p &= 0.38\sqrt{\frac{E}{F_y}} \\ &= 0.38\sqrt{\frac{29{,}000\,\text{ksi}}{50\,\text{ksi}}} \\ &= 9.15\end{aligned}$$

$$\begin{aligned}\lambda_r &= 0.83\sqrt{\frac{E}{F_L}} \\ &= 0.83\sqrt{\frac{29{,}000\,\text{ksi}}{50\,\text{ksi} - 10\,\text{ksi}}} \\ &= 22.3\end{aligned}$$

Since $\lambda_p < \lambda \le \lambda_r$, the flanges are non-compact. From LRFD Specification Section B5.1, for web compactness

$$\lambda_p = 3.76\sqrt{\frac{E}{F_y}}$$
$$= 3.76\sqrt{\frac{29{,}000\text{ ksi}}{50\text{ ksi}}}$$
$$= 90.6$$

Since $\lambda < \lambda_p$, the web is compact. However, since the flanges are non-compact, the W21×48 must be designed as a non-compact shape.

Assuming that the beam is braced, with non-compact flanges, the flange local buckling limit-state will control the flexural design strength. From LRFD Specification Appendix F1, the flange local buckling flexural design strength $\phi_b M_n$ is adjusted from $\phi_b M_p$ to $\phi_b M'_p$ to account for the non-compactness,

$$\phi_b M_p = \phi_b F_y Z_x$$
$$= \frac{0.9(50\text{ ksi})\left(107\text{ in.}^3\right)}{12\text{ in./ft}}$$
$$= 401\text{ kip-ft}$$

$$\phi_b M_r = \phi_b F_L S_x$$
$$= \frac{0.9(50\text{ ksi} - 10\text{ ksi})\left(93.0\text{ in.}^3\right)}{12\text{ in./ft}}$$
$$= 279\text{ kip-ft}$$

$$\phi_b M'_p = \phi_b M_p - (\phi_b M_p - \phi_b M_r)\left(\frac{\lambda - \lambda_p}{\lambda_r - \lambda_p}\right)$$
$$= (401\text{ kip-ft}) - (401\text{ kip-ft} - 279\text{ kip-ft})\left(\frac{9.47 - 9.15}{22.3 - 9.15}\right)$$
$$= 398\text{ kip-ft}$$

From LRFD Specification Appendix F1, adjusting L_p to L'_p to account for the non-compactness,

$$L_p = 1.76 r_y\sqrt{\frac{E}{F_y}}$$
$$= \frac{1.76\,(1.66\text{ in.})}{12\text{ in./ft}}\sqrt{\frac{29{,}000\text{ ksi}}{50\text{ ksi}}}$$
$$= 5.86\text{ ft}$$

$$L_r = \frac{r_y X_1}{F_L}\sqrt{1 + \sqrt{1 + X_2 F_L^2}}$$
$$= \frac{(1.66\text{ in.})(1{,}450\text{ ksi})}{(50\text{ ksi} - 10\text{ ksi})(12\text{ in./ft})}\sqrt{1 + \sqrt{1 + \left(\frac{0.0436}{\text{ksi}^2}\right)(50\text{ ksi} - 10\text{ ksi})^2}}$$
$$= 15.4\text{ ft}$$

$$L'_p = L_p - (L_p - L_r)\left(\frac{\phi_b M_p - \phi_b M'_p}{\phi_b M_p - \phi_b M_r}\right)$$
$$= \left[(5.86\text{ ft}) - (5.86\text{ ft} - 15.4\text{ ft})\left(\frac{401\text{ kip-ft} - 398\text{ kip-ft}}{401\text{ kip-ft} - 279\text{ kip-ft}}\right)\right]$$
$$= 6.09\text{ ft}$$

Since $L_b < L'_p$, the beam is braced as assumed.

From LRFD Specification Section F2, the shear yielding design strength $\phi_v V_n$ is

$$\phi_v V_n = \phi_v F_y d t_w$$
$$= 0.9(50\text{ ksi})(20.6\text{ in.})(0.350\text{ in.})$$
$$= 324\text{ kips}$$

Solution b: As determined in solution a, the beam has non-compact flanges.

As determined in solution a, $L'_p = 6.09$ ft and $L_r = 15.4$ ft.

For a beam that has non-compact flanges with $L'_p < L_b \leq L_r$, either the flange-local buckling limit-state or the inelastic lateral-torsional buckling limit-state controls the flexural design strength. As determined in solution a, the flexural design strength for flange local buckling is

$$\phi_b M_n = 398\text{ kip-ft}$$

For inelastic lateral-torsional buckling, the moment gradient effect can be calculated using LRFD Specification Equation F1-3, where

$$C_b = \frac{12.5M_{\max}}{2.5M_{\max} + 3M_A + 4M_B + 3M_C}$$

For the center-span beam segment, the moment is constant and $C_b = 1$. For the end-span beam segments, using relative values determined from the moment diagram for a simple-span beam with equal concentrated loads at the third points ($M_{\max} = 1.00$, $M_A = 0.250$, $M_B = 0.500$, and $M_C = 0.750$)

$$C_b = \frac{12.5(1.00)}{2.5(1.00) + 3(0.250) + 4(0.500) + 3(0.750)}$$
$$= 1.67$$

Thus, the center span with $C_b = 1$ is more critical.

As determined in solution a, $\phi_b M_p = 401$ kip-ft, $L_p = 5.86$ ft, $\phi_b M_r = 279$ kip-ft, $L_r = 15.4$ ft, $\phi_b M'_p = 398$ kip-ft and $L'_p = 6.09$ ft. From LRFD Specification Appendix F1, the flexural design strength $\phi_b M_n$ is

$$\phi_b M_n = C_b\left[\phi_b M_p - (\phi_b M_p - \phi_b M_r)\left(\frac{L_b - L_p}{L_r - L_p}\right)\right] \leq \phi_b M'_p$$
$$= (1)\left[(401\text{ kip-ft}) - (401\text{ kip-ft} - 279\text{ kip-ft})\left(\frac{13.3\text{ ft} - 5.86\text{ ft}}{15.4\text{ ft} - 5.86\text{ ft}}\right)\right]$$

$$\le 398\text{ kip-ft}$$
$$= 306\text{ kip-ft} \le 398\text{ kip-ft}$$
$$= 306\text{ kip-ft}$$

As determined in solution a, the shear yielding design strength $\phi_v V_n$ is

$$\phi_v V_n = 324\text{ kips}$$

Solution c: As determined in solution a, the beam has non-compact flanges.

As determined in solution a, $L'_p = 6.09$ ft and $L_r = 15.4$ ft.

For a beam that has non-compact flanges with $L_b > L_r$, either the flange-local buckling limit-state or the elastic lateral-torsional buckling limit-state controls the flexural design strength. As determined in solution a, the flexural design strength for flange local buckling is

$$\phi_b M_n = 398\text{ kip-ft}$$

For elastic lateral-torsional buckling, the moment gradient effect can be calculated using LRFD Specification Equation F1-3, where

$$C_b = \frac{12.5M_{\max}}{2.5M_{\max} + 3M_A + 4M_B + 3M_C}$$

Using relative values determined from the moment diagram for a simple-span beam with equal concentrated loads at the third points ($M_{\max} = 1.00$, $M_A = 0.750$, $M_B = 1.00$, and $M_C = 0.750$)

$$C_b = \frac{12.5(1.00)}{2.5(1.00) + 3(0.750) + 4(1.00) + 3(0.750)}$$
$$= 1.14$$

As determined in solution a, $\phi_b M_p = 401$ kip-ft, $L_p = 5.86$ ft, $\phi_b M_r = 279$ kip-ft, $L_r = 15.4$ ft, $\phi_b M'_p = 398$ kip-ft and $L'_p = 6.09$ ft. Since $C_b > 1$, the value of L'_m is needed. Assuming $L'_m \le L_r$,

$$L'_m = L'_p - (L_p - L_r)\frac{\phi_b M'_p(C_b - 1)}{C_b(\phi_b M_p - \phi_b M_r)}$$
$$= (6.09\text{ ft}) - (5.86\text{ ft} - 15.4\text{ ft})\frac{(398\text{ kip-ft})(1.14 - 1)}{1.14(401\text{ kip-ft} - 279\text{ kip-ft})}$$
$$= 9.91\text{ ft}$$

Since $L'_m < L_r$ as assumed, $L'_m = 9.91$ ft.

From LRFD Specification Section F1.2b and Appendix F1, the flexural design strength $\phi_b M_n$ is

$$\phi_b M_n = \phi_b \frac{C_b S_x X_1 \sqrt{2}}{\left(\frac{L_b}{r_y}\right)}\sqrt{1 + \frac{X_1^2 X_2}{2\left(\frac{L_b}{r_y}\right)^2}} \le \phi_b M'_p$$

$$= 0.9\frac{(1.14)\left(93.0\,\text{in.}^3\right)(1{,}450\,\text{ksi})\sqrt{2}}{\dfrac{(40\,\text{ft}\times 12\,\text{in./ft})}{(1.66\,\text{in.})}}\sqrt{1+\frac{(1{,}450\,\text{ksi})^2\left(\dfrac{0.0436}{\text{ksi}^2}\right)}{2\left(\dfrac{(40\,\text{ft}\times 12\,\text{in./ft})}{(1.66\,\text{in.})}\right)^2}}$$

$$\leq 398\,\text{kip-ft}$$
$$= 70.2\,\text{kip-ft}$$

As determined in solution a, the shear yielding design strength $\phi_v V_n$ is

$$\phi_v V_n = 254\,\text{kips}$$

Comments: The preceding calculations can be simplified using Tables 5-1, 5-3, 5-4 and 5-5. For solution a, from Table 5-3, a W21×48 is indicated as non-compact and provides $\phi_b M_n = 401$ kip-ft, $L'_p = 6.09$ ft, $\phi_b M_r = 279$ kip-ft and $L_r = 15.4$ ft. These values can then be used as illustrated in solution a to determine

$$\phi_b M'_n = 398\,\text{kip-ft}$$

Also, $\phi_v V_n = 254$ kips is given.

For solutions b and c, the values tabulated in Tables 5-1 and 5-3 can be used in a similar fashion as illustrated in solutions b and c.

EXAMPLE 5.3. W-shape flexural member design (selection by moment of inertia for strong-axis bending).

Given: Select an ASTM A992 W-shape flexural member ($F_y = 50$ ksi, $F_u = 65$ ksi) for a required flexural strength M_u of 250 kip-ft, a required shear strength V_u of 40 kips, and a deflection limit of 1 in. For the deflection calculations, assume the load is a uniformly distributed service load of 2 kips/ft and the length of the simple span is 30 ft. For the strength calculations, assume the beam is braced.

Solution: From Table 5-17, Diagram 1, the maximum deflection Δ_{max} occurs at mid-span and can be calculated as:

$$\Delta_{max} = \frac{5wl^4}{384EI}$$

Rearranging and substituting $\Delta_{max} = 1$ in.,

$$I_{min} = \frac{5(2\,\text{kips/ft})(30\,\text{ft})^4\times(12\,\text{in./ft})^3}{384\,(29{,}000\,\text{ksi})(1\,\text{in.})}$$
$$= 1{,}260\,\text{in.}^4$$

From Table 5-2, a W24×55 has

$$I_x = 1{,}360\,\text{in.}^4 > 1{,}260\,\text{in.}^4\ \textbf{o.k.}$$

Because the W24×55 is braced (given) and compact, $\phi_b M_n = \phi_b M_{px}$.

From Table 5-2,

$$\phi_b M_{px} = 398\,\text{kip-ft} > 250\,\text{kip-ft}\ \textbf{o.k.}$$
$$\phi_v V_n = 419\,\text{kips} > 40\,\text{kips}\ \textbf{o.k.}$$

Thus, the W24×55 flexural member is **o.k.**

Comments: Note that end connection limit states, such as block shear rupture and bolt bearing strength must also be checked.

EXAMPLE 5.4. W-shape flexural member design (selection using plots of $\phi_b M_p$ vs. L_b for strong-axis bending).

Given: Select an ASTM A992 W-shape flexural member ($F_y = 50$ ksi, $F_u = 65$ ksi) for a required flexural strength M_u of 150 kip-ft, a required shear strength V_u of 20 kips, and a deflection limit of 1 in. For the deflection calculations, assume the load is a uniformly distributed service load of 2 kips/ft and the length of the simple span is 20 ft. For the strength calculations, assume the beam is braced at the ends and midpoint only ($L_b = 10$ ft).

Solution: From Table 5-5, for an unbraced length $M_u = 150$ kip-ft and $L_b = 10$ ft, a W16×31 with $C_b = 1$ has $\phi_b M_{nx} \approx 150$ kip-ft. Since $C_b > 1$, the actual flexural strength will be higher, so the W16×31 is **o.k.** for flexural design strength.

From Table 5-17, Diagram 1, the maximum deflection Δ_{max} occurs at mid-span and can be calculated as:

$$\begin{aligned}\Delta_{max} &= \frac{5wl^4}{384EI}\\ &= \frac{5(2\,\text{kips/ft})(20\,\text{ft})^4 \times (12\,\text{in./ft})^3}{384\,(29{,}000\,\text{ksi})\,(375\,\text{in.}^4)}\\ &= 0.662\,\text{in.} < 1\,\text{in.}\ \textbf{o.k.}\end{aligned}$$

From Table 5-2,

$$\phi_v V_n = 197\,\text{kips} > 20\,\text{kips}\ \textbf{o.k.}$$

Thus, the W16×31 flexural member is **o.k.**

Comments: Note that end connection limit states, such as block shear rupture and bolt bearing strength must also be checked.

EXAMPLE 5.5. W-shape flexural member design (determination of maximum end reaction for strong-axis bending).

Given: For an ASTM A992 W21×62, estimate the beam end reaction based upon the maximum simple-span end reaction assuming the beam is subject to its maximum factored uniform load, is braced, and spans

a) 40 ft.
b) 30 ft.
c) 8 ft.

In each case, compare the result to the actual end reaction assuming a 10-ft tributary width and a 100-psf factored uniform load.

Solution a: From Table 5-4, the tabulated value of the maximum factored uniform load for a braced, simple-span ASTM A992 W21×62 on a 40-ft span is 108 kips. The corresponding estimated maximum end reaction is

$$R_u = \frac{108\text{ kips}}{2}$$
$$= 54\text{ kips}$$

The actual end reaction resulting from a 100-psf factored uniform load with a 10-ft tributary width is

$$R_u = \frac{100\text{ psf}(40\text{ ft} \times 10\text{ ft})}{2(1000\text{ lb/kip})}$$
$$= 20\text{ kips}$$

Solution b: From Table 5-4, the tabulated value of the maximum factored uniform load for a braced, simple-span ASTM A992 W21×62 on a 30-ft span is 144 kips. The corresponding estimated maximum end reaction is

$$R_u = \frac{144\text{ kips}}{2}$$
$$= 72\text{ kips}$$

The actual end reaction resulting from a 100 psf factored uniform load with a 10-ft tributary width is

$$R_u = \frac{100\text{ psf}(30\text{ ft} \times 10\text{ ft})}{2(1000\text{ lb/kip})}$$
$$= 15\text{ kips}$$

Solution c: From Table 5-4, the tabulated value of the maximum factored uniform load for a braced, simple-span ASTM A992 W21×62 on an 8-ft span is 454 kips. Note that this value is controlled by the shear yielding design strength of the beam web, since it appears above the heavy line in the table. The corresponding estimated maximum end reaction is

$$R_u = \frac{454\text{ kips}}{2}$$
$$= 227\text{ kips}$$

The actual end reaction resulting from a 100-psf factored uniform load with a 10-ft tributary width is

$$R_u = \frac{100\text{ psf}(8\text{ ft} \times 10\text{ ft})}{2(1000\text{ lb/kip})}$$
$$= 4\text{ kips}$$

Comments: In solutions a, b and c, the estimated end reactions calculated based upon the braced, uniformly loaded member size and its span length are significantly larger than the end reactions calculated based upon the actual loading. Such conservative estimates of the beam end reactions should be avoided as they can be particularly problematic when selecting end connections for these members.

EXAMPLE 5.6. Composite W-shape flexural member design (unshored construction).

Given:

Select an ASTM A992 W-shape flexural member ($F_y = 50\,\text{ksi}$, $F_u = 65\,\text{ksi}$) for a uniformly distributed service dead load of 90 psf and service live load of 2.5 kips/ft. The beam span is 40 ft and the beam spacing is 10 ft. The slab is $4^1/_2$ in. normal-weight concrete ($f'_c = 4\,\text{ksi}$, $145\,\text{lb/ft}^3$) supported by a 3-in. deep composite metal deck. Assume that the stud reduction factor is 1.0 and that unshored construction is specified. Also, determine the service dead and live load deflections.

Solution:

The total service and factored loads are

Load Type	Service Load	Load Factor	Factored Load
Dead	0.9 kips/ft	1.2	1.1 kips/ft
Live	2.5 kips/ft	1.6	4.0 kips/ft
Total	3.4 kips/ft		5.1 kips/ft

The resulting flexural required strengths for the total factored load are

$$M_u = \frac{wL^2}{8} = \frac{(5.1\,\text{kips/ft})(40\,\text{ft})^2}{8} = 1{,}020\,\text{kip-ft}$$

$$V_u = \frac{wL}{2} = \frac{(5.1\,\text{kips/ft})(40\,\text{ft})}{2} = 102\,\text{kips}$$

Assuming $a = 2$ in.,

$$Y2 = Y_{\text{con}} - \frac{a}{2} = 7.5\,\text{in.} - \frac{2\,\text{in.}}{2} = 6.5\,\text{in.}$$

From Table 5-14, with $F_y = 50$ ksi and $Y2 = 6.5$ in., a W21×62, W24×55, and W24×62 are possible sizes. Try a W24×55. For $Y1 = 0.0$ in., $Q_n = 810$ kips and $\phi_b M_n = 1{,}050$ kip-ft.

The assumed value of $Y2$ must be checked. Per LRFD Specification Section I3.1, the effective width of the concrete flange is

$$b \le \begin{cases} 2 \times L/8 = 2 \times 40\,\text{ft}/8 = 10\,\text{ft} \\ 10\,\text{ft spacing} \end{cases} = 120\,\text{in.}$$

$$a_{req} = \frac{\Sigma Q_n}{0.85 f_c' b}$$
$$= \frac{810\,\text{kips}}{0.85(4\,\text{ksi})(120\,\text{in.})}$$
$$= 1.99\,\text{in.}$$
$$Y2 = Y_{con} - \frac{a}{2}$$
$$= 7.5\,\text{in.} - \frac{1.99\,\text{in.}}{2}$$
$$= 6.5\,\text{in.} \therefore \text{ assumed value of } Y2 \text{ is } \textbf{o.k.}$$

From Table 5-13, $Q_n = 26.1$ kips/stud and the number of shear stud connectors required is

$$\frac{2\Sigma Q_n}{Q_n} = \frac{2(810\,\text{kips})}{26.1\,\text{kips/stud}}$$
$$= 62.1 \rightarrow 64\text{ studs}$$

With unshored construction, the strength of the steel beam during construction must be checked. Assuming a construction live load of 20 psf will be assumed. For the 1.4D load case,

Load Type	Service Load	Load Factor	Factored Load
Dead	0.9 kips/ft	1.4	1.3 kips/ft

For the 1.2D + 1.6L load case,

Load Type	Service Load	Load Factor	Factored Load
Dead	0.9 kips/ft	1.2	1.1 kips/ft
Live	0.2 kips/ft	1.6	0.3 kips/ft
Total	1.1 kips/ft		1.4 kips/ft

The 1.2D + 1.6L load case is more critical for strength and the resulting flexural required strengths for the total factored load are

$$M_u = \frac{wL^2}{8}$$
$$= \frac{(1.4\,\text{kips/ft})(40\,\text{ft})^2}{8}$$
$$= 280\,\text{kip-ft}$$
$$V_u = \frac{wL}{2}$$
$$= \frac{(1.4\,\text{kips/ft})(40\,\text{ft})}{2}$$
$$= 28\,\text{kips}$$

From Table 5-3, assuming adequate lateral support is provided by the attachment of the steel deck to the compression flange, a W24×55 has

$$\phi_b M_n = 506\,\text{kip-ft} > 280\,\text{kip-ft}\ \textbf{o.k.}$$

and

$$\phi_v V_n = 252 \text{ kips} > 28 \text{ kips } \mathbf{o.k.}$$

From Table 5-15, a W24×55 with $F_y = 50$ ksi, $Y2 = 6.5$ in. and $Y1 = 0$ in. has a lower bound moment of inertia $I_{LB} = 4.060$ in.^{4} From Table 5-16, Diagram 1, the maximum deflection Δ_{max} occurs at mid-span and can be calculated as:

$$\Delta_{max} = \frac{5wl^4}{384EI}$$

$$= \frac{5(2.5 \text{ kip/ft})(40 \text{ ft})^4 \times (12 \text{ in./ft})^3}{384\,(29{,}000 \text{ ksi})\left(4{,}060 \text{ in.}^4\right)}$$

$$= 1.22 \text{ in.}$$

Note that this deflection must be compared with an appropriate deflection limit, which is dependent upon the application. Deflection will often be more critical than strength in beam design.

The construction dead-load deflection can be calculated as:

$$\Delta_{max} = \frac{5(0.9 \text{ kip/ft})(40 \text{ ft})^4 \times (12 \text{ in./ft})^3}{384\,(29{,}000 \text{ ksi})\left(1{,}360 \text{ in.}^4\right)}$$

$$= 1.31 \text{ in.}$$

Note: If necessary or desired, a beam camber of 1 in. or 1¹/₄ in. can be specified to offset the construction dead-load deflection.

For shear design strength

$$\phi V_n = \phi 0.6 F_{yw} A_w$$

$$= (0.9)(0.6)(50 \text{ ksi})(23.6 \text{ in.})(0.395 \text{ in.})$$

$$= 252 \text{ kips} > 102 \text{ kips } \mathbf{o.k.}$$

EXAMPLE 5.7. Composite W-shape flexural member design (shored construction).

Given: Select an ASTM A992 W-shape flexural member ($F_y = 50$ ksi, $F_u = 65$ ksi) for a uniformly distributed service dead load of 0.9 kips/ft and service live load of 1.3 kips/ft. The beam span is 30 ft and the beam spacing 10 ft. The slab is 3¹/₄-in. light-weight concrete ($f'_c = 3.5$ ksi, 115 lb/ft^3) supported by a 3-in. deep composite metal deck with an average rib width of 6 in. The ribs are oriented perpendicular to the beam. Shored construction is specified. Also, determine the number of ³/₄-in.-diameter shear stud connectors required and the service live load deflection.

Solution: The total service and factored loads are

Load Type	Service Load	Load Factor	Factored Load
Dead	0.9 kips/ft	1.2	1.1 kips/ft
Live	1.3 kips/ft	1.6	2.1 kips/ft
Total	2.2 kips/ft		3.2 kips/ft

The resulting flexural required strengths for the total factored load are

$$\begin{aligned} M_u &= \frac{wL^2}{8} \\ &= \frac{(3.2\,\text{kips/ft})(30\,\text{ft})^2}{8} \\ &= 360\,\text{kip-ft} \\ V_u &= \frac{wL}{2} \\ &= \frac{(3.2\,\text{kips/ft})(30\,\text{ft})}{2} \\ &= 48.0\,\text{kips} \end{aligned}$$

A trial beam size can be selected such that

$$\text{Beam weight} = 3.4\left[\frac{M_u(12\,\text{in./ft})}{\phi F_y\left(\dfrac{d}{2} + Y_{\text{con}} - \dfrac{a}{2}\right)}\right]$$

where

$$\begin{aligned} Y_{\text{con}} &= 3 + 3.25 = 6.25\,\text{in.} \\ a/2 &= 1\,\text{in. (estimate)} \\ \phi &= 0.85 \\ F_y &= 50\,\text{ksi} \end{aligned}$$

d	$\dfrac{M_u(12)(3.4)}{\phi F_y}$	$d/2$	$(Y_{con} - a/2)$	Beam Weight
16	346	8	5.25	26
18	346	9	5.25	24

From the results above, try a W16×26 and let

$$\begin{aligned} \Sigma Q_n &= F_y A_s \\ &= (50\,\text{ksi})(7.68\,\text{in.}^2) \\ &= 384\,\text{kips} \end{aligned}$$

Per LRFD Specification Section I3.1, the effective width of the concrete flange is

$$b \le \begin{cases} 2 \times L/8 = 2 \times 30\,\text{ft}/8 = 7.5\,\text{ft} = 90\,\text{in. (governs)} \\ 10\,\text{ft spacing} \end{cases}$$

$$\begin{aligned} a_{\text{req}} &= \frac{\Sigma Q_n}{0.85 f'_c b} \\ &= \frac{384\,\text{kips}}{0.85(3.5\,\text{ksi})(90\,\text{in.})} \\ &= 1.43\,\text{in.} \\ Y2 &= Y_{\text{con}} - \frac{a}{2} \\ &= 6.25\,\text{in.} - \frac{1.43\,\text{in.}}{2} \\ &= 5.53\,\text{in.} \end{aligned}$$

By interpolation from Table 5-14, a W16×26 with $Y1 = 0$ and $Y2 =$ 5.53 in. has

$$\phi_b M_n = 363\,\text{kip-ft} + \frac{5.53\,\text{in.} - 5.5\,\text{in.}}{6\,\text{in.} - 5.5\,\text{in.}}(377\,\text{kip-ft} - 363\,\text{kip-ft})$$
$$= 364\,\text{kip-ft} > 360\,\text{kip-ft}\ \textbf{o.k.}$$

Per LRFD Specification Equation I3-1, the stud reduction factor, assuming $N_r = 2$ and $H_s = 5.5$ in., is

$$\frac{0.85}{\sqrt{N_r}}(W_r/h_r)(H_s/h_r - 1.0) \le 1.0$$

$$\frac{0.85}{\sqrt{2}}\left(\frac{6\,\text{in.}}{3\,\text{in.}}\right)\left(\frac{5.5\,\text{in.}}{3\,\text{in.}} - 1.0\right) \le 1.0$$

Thus, the stud reduction factor is 1.0. Note that H_s cannot exceed $(h_r + 3)$ in computations, although actual length may be greater.

From Table 5-13, $Q_n = 19.8$ kips/stud and the number of shear stud connectors required is

$$\frac{2\Sigma Q_n}{Q_n} = \frac{2(384\,\text{kips})}{19.8\,\text{kips/stud}}$$
$$= 38.8 \rightarrow 40\,\text{studs}$$

From Table 5-15, a W16×26 with $F_y = 50$ ksi, $Y2 = 5.5$ in. and $Y1 = 0$ in. has a lower bound moment of inertia $I_{LB} = 985\,\text{in.}^4$ From Table 5-16, Diagram 1, the maximum deflection Δ_{max} occurs at mid-span and can be calculated as:

$$\Delta_{max} = \frac{5wl^4}{384EI}$$
$$= \frac{5(1.3\,\text{kip/ft})(30\,\text{ft})^4 \times (12\,\text{in./ft})^3}{384\,(29{,}000\,\text{ksi})\left(985\,\text{in.}^4\right)}$$
$$= 0.829\,\text{in.}$$

Note that this deflection must be compared with an appropriate deflection limit, which is dependent upon the application. Deflection will often be more critical than strength in beam design.

For shear design strength

$$\phi V_n = \phi 0.6 F_{yw} A_w$$
$$= (0.9)(0.6)(50\,\text{ksi})(15.7\,\text{in.})(0.250\,\text{in.})$$
$$= 106\,\text{kips} > 48.0\,\text{kips}\ \textbf{o.k.}$$

EXAMPLE 5.8. Composite W-shape flexural member design (determination of maximum end reaction for strong-axis bending).

Given: An ASTM A992 W21×44 ($F_y = 50$ ksi, $F_u = 65$ ksi) steel girder spans 30 feet and supports intermediate beams at the third points. A total of fifty 3/4-in. diameter headed studs are applied to the beam as follows: 24 between each support and the beams at the one-third points, and two between the intermediate beams. The slab consists of 3 1/4-in. light-weight

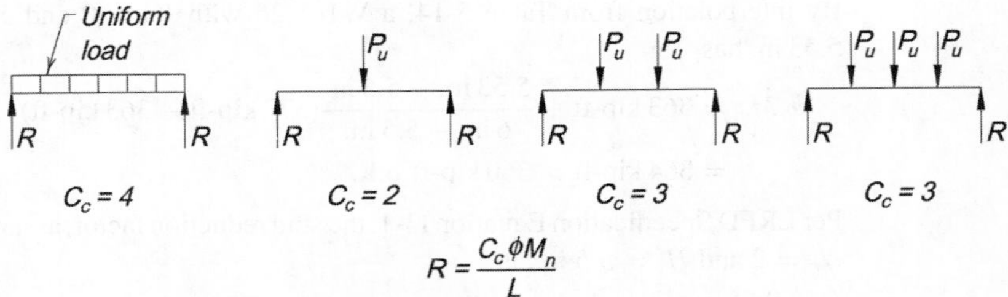

Fig. 5–3. Coefficients for use in determining composite simple-beam and reactions.

concrete (115 pcf) with a specified design strength of 3.5 ksi over a 3-in.-deep composite metal deck with an average rib width of 6 in. The ribs are oriented parallel to the beam centerline. Determine the design beam reactions.

Solution: For shear stud connectors in a single row the spacing between the support and first intermediate beam is 10(12)/24 = 5.0 in. which is greater than the specified minimum of 6 diameters per LRFD Specification Section I5.6. Since $w_r/h_r = 6/3 = 2$ is greater than 1.5, the stud reduction factor is 1, per LRFD Specification I3.5c. Therefore, from Table 5-13, the stud shear strength is:

$$\Sigma Q_n = nQ_n = (24 \text{ studs})(19.8 \text{ kips/stud}) = 475 \text{ kips}$$

For $\Sigma Q_n = 475$ kips, the required effective concrete flange thickness can be calculated to be:

$$\begin{aligned} a &= \frac{475 \text{ kips}}{0.85(3.5 \text{ ksi})(7.5 \text{ ft})(12 \text{ in./ft})} \\ &= 1.77 \text{ in.} \\ Y2 &= Y_{\text{con}} - \frac{a}{2} \\ &= 6.25 \text{ in.} - \frac{1.77 \text{ in.}}{2} \\ &= 5.36 \text{ in.} \end{aligned}$$

From Table 5-14, for a W21×44 with $F_y = 50$ ksi, $\Sigma Q_n = 475$ kips places the PNA at $Y1 = 0.27$ in. For $Y2 = 5.36$ in. and $Y1 = 0.27$ in., $\phi_b M_n = 655$ kip-ft and the end reaction is

$$\begin{aligned} R_u &= C_c \phi M_n / L \\ &= 3(655 \text{ kip-ft})/30 \text{ ft} \\ &= 65.5 \text{ kips} \end{aligned}$$

where C_c is a coefficient from Figure 5-3. Note that the beam weight was neglected in this example.

STEEL BEAM SELECTION TABLES

Table 5-1. Beam Bending Coefficient C_b

Values of the beam bending coefficient C_b are given for various loading conditions on simple-span beams in Table 5-1.

Table 5-2. W-Shapes – Selection by I_x

W-shapes are sorted in descending order by strong-axis moment of inertia I_x and then grouped in ascending order by weight with the lightest W-shape in each range in bold. Strong-axis design strengths in flexure and shear are given for W-shapes with $F_y = 50$ ksi (ASTM A992). C_b is taken as unity.

When $L_b \leq L_p$, the strong-axis flexural design strength can be checked using the tabulated values of Z_x or $\phi_b M_{px}$. When $L_p < L_b \leq L_r$, the tabulated value of the bending factor BF, can be used to linearly interpolate between $(\phi_b M_{px}, L_p)$ and $(\phi_b M_{rx}, L_r)$ to determine the strong-axis flexural design strength $\phi_b M_{nx}$ as follows:

$$\phi_b M_{nx} = C_b[\phi_b M_{px} - BF(L_b - L_p)] \leq \phi_b M_{px}$$

When $L_b > L_r$, see Table 5-5.

For non-compact W-shapes, the tabulated value of L_p has been adjusted to L'_p to account for the non-compactness. The value of $\phi_b M_{px}$ must be adjusted to $\phi_b M'_{px}$ as illustrated in Figure 5-1.

The strong-axis shear design strength can be checked using the tabulated value of $\phi_v V_n$. Values are also tabulated for Z_y, I_y and $\phi_b M_{py}$ for reference.

The use of Table 5-2 is illustrated in Examples 5.3 and 5.4.

Table 5-3. W-Shapes – Selection by Z_x

W-shapes are sorted in descending order by strong-axis plastic section modulus Z_x and then grouped in ascending order by weight with the lightest W-shape in each range in bold. Strong-axis design strengths in flexure and shear are given for W-shapes with $F_y = 50$ ksi (ASTM A992). C_b is taken as unity.

When $L_b \leq L_p$, the flexural member can be selected on the basis of the tabulated values of Z_x or $\phi_b M_{px}$. When $L_p < L_b \leq L_r$, the tabulated value of the bending factor BF, can be used to linearly interpolate between $(\phi_b M_{px}, L_p)$ and $(\phi_b M_{rx}, L_r)$ to determine the strong-axis flexural design strength $\phi_b M_{nx}$ as illustrated previously for Table 5-2. When $L_b > L_r$, see Table 5-5.

For non-compact W-shapes, the tabulated value of L_p has been adjusted to L'_p to account for the non-compactness. The value of $\phi_b M_{px}$ must be adjusted to $\phi_b M'_{px}$ as illustrated in Figure 5-1.

The strong-axis shear design strength and deflection can be checked using the tabulated values of $\phi_v V_n$ and I_x, respectively. Values are also tabulated for I_y, Z_y and $\phi_b M_{py}$ for reference.

The use of Table 5-3 is illustrated in Examples 5.1 and 5.2.

Table 5-4. W-Shapes – Maximum Total Factored Uniform Load

Maximum total factored uniform loads on braced simple-span beams bent about the strong axis are given for W-shapes with $F_y = 50$ ksi (ASTM A992). The tabulated values include the beam weight, which should be deducted when calculating the maximum factored uniform load the beam will support. Above the heavy horizontal line in the tables, the maximum total factored uniform load is limited by the strong-axis shear design strength. When the flexural design strength controls, the tabulated values can also be used for braced simple-span beams with equal concentrated loads spaced as shown in Table 5-16 if the concentrated loads are first converted to an equivalent uniform load.

When $L_b \leq L_p$, the flexural member can be selected on the basis of the tabulated values of maximum total factored uniform load. For non-compact W-shapes, the tabulated values have been adjusted to account for the non-compactness. The uniform load constant $\phi_b W_c$ for braced simple-span beams bent about the strong axis is also tabulated. For compact shapes

$$\phi_b W_c = 2\phi_b F_y Z_x/3$$

This constant can be divided by the span length L (ft) to calculate the maximum total factored uniform load for a braced simple-span beam bent about the strong axis. For non-compact shapes, this constant can be determined from LRFD Specification Appendix F1 using Equation A-F1-3.

When $L_b > L_p$, the tabulated values for maximum total factored uniform load in Table 5-4 cannot be used. See Tables 5-2, 5-3, or 5-5.

The strong-axis shear design strength and deflection can be checked using the tabulated values of $\phi_v V_n$ and I_x, respectively. Values are also tabulated for Z_x, Z_y and I_y for reference.

The use of Table 5-4 is illustrated in Example 5.5.

Table 5-5. W-Shapes – Plots of $\phi_b M_{nx}$ vs. L_b

The strong-axis flexural design strength $\phi_b M_{nx}$ is plotted as a function of unbraced length L_b for W-shapes with $F_y = 50$ ksi (ASTM A992). The tabulated values include the beam weight, which should be deducted when calculating the maximum factored uniform load the beam will support. C_b is taken as unity.

When the plotted curve is solid, the W-shape for that curve is the lightest cross-section for a given combination of flexural design strength and unbraced length. When the plotted curve is dashed, a lighter W-shape than that for the plotted curve exists. The plotted curves are arbitrarily terminated at a span-to-depth ratio of 30.

The strong-axis shear design strength and deflection must be checked independently.

The use of Table 5-5 is illustrated in Example 5.4.

Table 5-6. W-Shapes – Selection by I_y

W-shapes are sorted in descending order by weak-axis moment of inertia I_y and then grouped in ascending order by weight with the lightest W-shape in each range in bold. Weak-axis design strengths in flexure are given for W-shapes with $F_y = 50$ ksi (ASTM A992). C_b is taken as unity.

The weak-axis flexural design strength can be checked using the tabulated values of $\phi_b M_{py}$. For non-compact W-shapes, the tabulated values must be adjusted to account for the non-compactness per LRFD Specification Appendix F1.

The weak-axis shear design strength must be checked independently. Values are also tabulated for I_x, Z_x, $\phi_b M_{px}$, L_p, $\phi_b M_{rx}$, L_r, BF and $\phi_v V_n$ for reference.

Table 5-7. W-Shapes – Selection by Z_y

W-shapes are sorted in descending order by weak-axis plastic section modulus Z_y and then grouped in ascending order by weight with the lightest W-shape in each range in bold. Weak-axis design strengths in flexure shear are given for W-shapes with $F_y = 50$ ksi (ASTM A992). C_b is taken as unity.

The weak-axis flexural design strength can be checked using the tabulated values of $\phi_b M_{py}$. For non-compact W-shapes, the tabulated values must be adjusted to account for the non-compactness per LRFD Specification Appendix F1.

The weak-axis deflection can be checked using the tabulated values of I_x. The weak-axis shear design strength must be checked independently. Values are also tabulated for Z_x, I_x, $\phi_b M_{px}$, L_p, $\phi_b M_{rx}$, L_r, BF and $\phi_v V_n$ for reference.

Table 5-8. S-Shapes – Maximum Factored Uniform Load

The preceding comments for Table 5-4 apply equally to Table 5-8, except Table 5-8 is for S-shapes with $F_y = 36$ ksi (ASTM A36).

Table 5-9. C-Shapes – Maximum Factored Uniform Load

The preceding comments for Table 5-4 apply equally to Table 5-9, except Table 5-9 is for C-shapes with $F_y = 36$ ksi (ASTM A36).

Table 5-10. MC-Shapes — Maximum Factored Uniform Load

The preceding comments for Table 5-4 apply equally to Table 5-10, except Table 5-10 is for MC-shapes with $F_y = 36$ ksi (ASTM A36).

Table 5-11. C- and MC-Shapes – Plots of $\phi_b M_{nx}$ vs. L_b

The preceding comments for Table 5-5 apply equally to Table 5-11, except Table 5-11 is for C- and MC-shapes with $F_y = 36$ ksi (ASTM A36).

Table 5-12. Floor Plates

The recommended maximum uniformly distributed loads are given for floor plates with thickness and span as indicated. The calculated values are based upon simple-span bending between supports. The top half of the table is for deflection-controlled applications, where the maximum total service load corresponding to a maximum deflection of $L/100$ is indicated. The bottom half of the table is for flexural-strength-controlled cases, where the maximum total factored load corresponding to a maximum bending stress of 24 ksi is indicated.

COMPOSITE BEAM SELECTION TABLES

Table 5-13. Shear Stud Connector Strength

The nominal shear strengths of 3/4-in. headed shear stud connectors embedded in concrete slabs are given in Table 5-13. Note that special reduction factors may apply to shear stud connectors in formed steel deck, per LRFD Specification Section I3.5. The use of Table 5-13 is illustrated in Examples 5.6 and 5.7.

Table 5-14. Composite W-Shapes

Values are given for the design flexural strength $\phi_b M_n$ for W-shapes with $F_y = 50$ ksi (ASTM A992). The values tabulated are independent of the concrete flange properties, the strength evaluation of which is left to the design engineer.

In the preparation of these tables, it is recognized that the location of the plastic neutral axis (PNA) is uniquely determined by the horizontal shear force ΣQ_n at the interface between the steel section and the concrete slab. With the knowledge of the location of the PNA and the distance to the centroid of the concrete flange force ΣQ_n, the design flexural strengths $\phi_b M_n$ for the rolled section can be computed.

Flexural design strengths are tabulated for PNA locations at the top and intermediate quarter points through the thickness of the steel beam top flange. In addition, PNA locations are computed at the point where ΣQ_n equals $0.25F_y A_s$, and the point where ΣQ_n is the average of $0.25F_y A_s$ and the value of ΣQ_n when the PNA is at the bottom of the top flange. $\Sigma Q_n = 0.25F_y A_s$ is suggested as a minimum value based upon engineering judgment.

Table 5-14 can be used by selecting a valid value of ΣQ_n per LRFD Specification Section I5.2. With the effective width of the concrete flange b determined per LRFD Specification

Section I3.1, the appropriate value of the distance from concrete flange force to beam top flange $Y2$ can be determined as

$$Y2 = Y_{con} - \frac{a}{2}$$

where

Y_{con} = distance from top of steel beam to top of concrete, in.

$$a = \frac{\Sigma Q_n}{0.85 f_c' b}$$

and the flexural design strength $\phi_b M_n$ can then be determined from Table 5-14. Values for the distance from the PNA to the beam top flange $Y1$ are also tabulated for convenience. The parameters $Y1$ and $Y2$ are illustrated in Figure 5-4.

Table 5-14 can also be used to determine the beam end reactions for symmetrically loaded composite W-shapes. When the properties of the composite concrete flange have been computed, ΣQ_n can be taken as the smaller of nQ_n, $F_y A_s$, or $0.85 f_c' A_c$. With $Y2$ taken as the distance from the top of the steel beam to the top of the concrete slab less $[\Sigma Q_n/(0.85 f_c' b)]/2$, the value of $\phi_b M_n$ can be selected from Table 5-14 and the beam end reaction R_u can be determined as:

$$R_u = C_c \phi_b M_n / L$$

where

C_c = coefficient from Figure 5-3
$\phi_b M_n$ = composite beam flexural design strength, kip-ft
L = span length, ft

When the properties of the composite concrete flange have not been computed, a conservative value for ΣQ_n can be taken as the smaller of $F_s A_s$ or nQ_n, where n is the number of shear stud connectors between the beam end and the point of maximum moment. In this case, $Y2$ is equal to the distance from the top of the steel beam to the top of the concrete slab.

The use of Table 5-14 is illustrated in Examples 5.6, 5.7 and 5.8.

Table 5-15. Lower-Bound Elastic Moments of Inertia

The lower-bound elastic moments of inertia of a composite beam can be used to calculate the deflection. If calculated deflections using the lower bound moment of inertia are acceptable, a complete elastic analysis of the composite section can be avoided. The lower-bound elastic moment of inertia is based upon the area of the beam and an equivalent concrete area equal to $\Sigma Q_n/F_y$ as illustrated in Figure 5-5. The analysis includes only the horizontal shear force transferred by the shear connectors supplied, thus neglecting the contribution of the concrete flange not considered in the plastic distribution of forces.

The value for the lower bound moment of inertia can be calculated as illustrated in LRFD Commentary Section I3.2. The lower bound moment of inertia, therefore, is the moment of inertia of the cross-section at the factored load. This is smaller than the corresponding moment of inertia at the service load where deflection is calculated.

The use of Table 5-15 is illustrated in Examples 5.6, 5.7 and 5.8.

BEAM DIAGRAMS AND FORMULAS

Table 5-16. Concentrated Load Equivalents

Concentrated load equivalents are given in Table 5-16 for beams with various support conditions and loading characteristics.

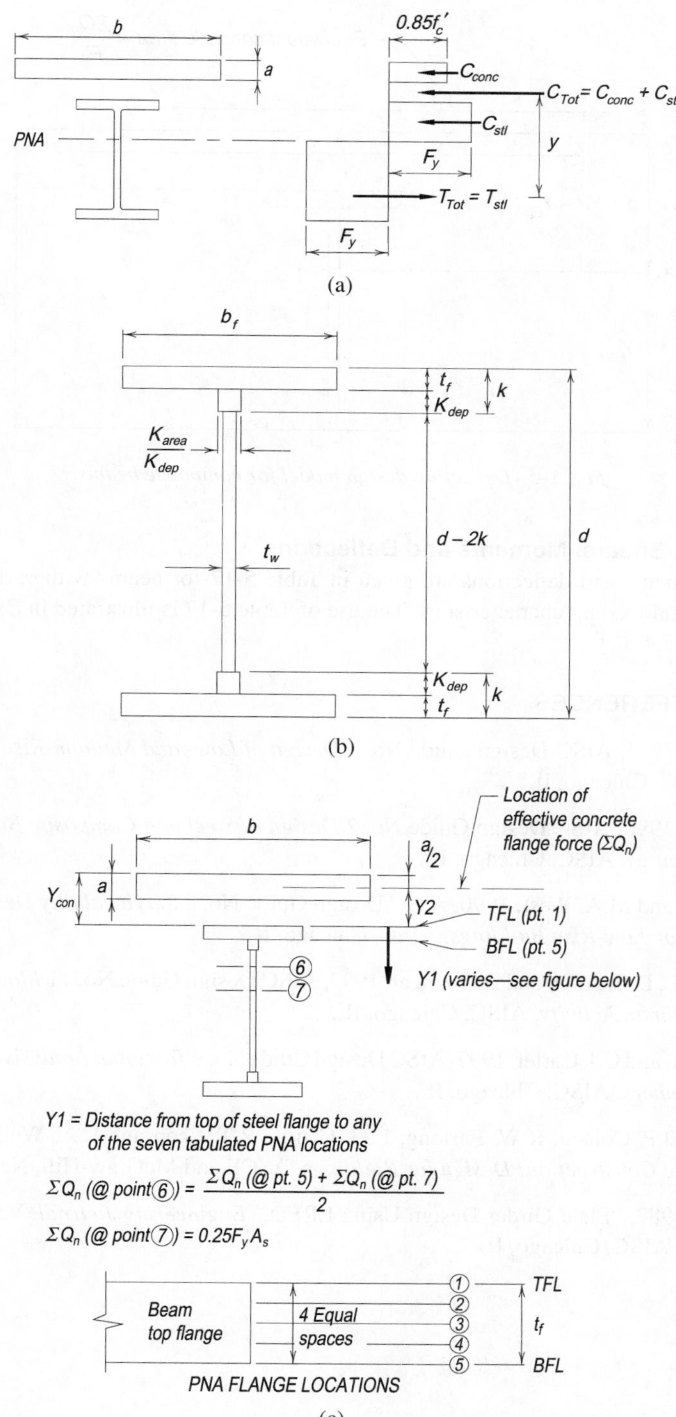

Fig. 5–4. Strength design models for composite beams.

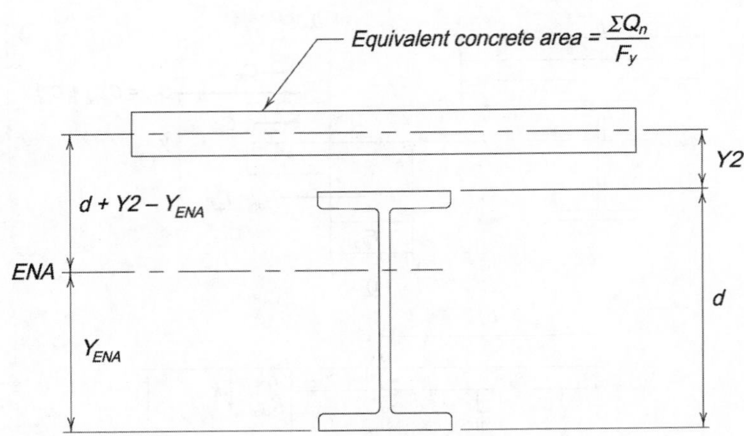

Fig. 5–5. Deflection design model for composite beams.

Table 5-17. Shears, Moments and Deflections

Shears, moments and deflections are given in Table 5-17 for beams with various support conditions and loading characteristics. The use of Table 5-17 is illustrated in Examples 5.1, 5.2, 5.3 and 5.4.

PART 5 REFERENCES

Allison, H., 1991, AISC Design Guide No. 5 *Design of Low- and Medium-Rise Steel Buildings*, AISC, Chicago, IL.

Darwin, D., 1990, AISC Design Guide No. 2 *Design of Steel and Composite Buildings with Web Openings*, AISC, Chicago, IL.

Fisher, J.M. and M.A. West, 1990, AISC Design Guide No. 3 *Serviceability Design Considerations for Low-Rise Buildings*, AISC, Chicago, IL.

Murray, T.M., D.E. Allen and E.E. Ungar, 1997, AISC Design Guide No. 11 *Floor Vibrations Due to Human Activity*, AISC, Chicago, IL.

Seaburg, P.A. and C.J. Carter, 1997, AISC Design Guide No. 9 *Torsional Analysis of Structural Steel Members*, AISC, Chicago, IL.

Viest, I.M., J.P. Colaco, R.W. Furlong, L.G. Griffis, R.T. Leon and L.A. Wylie, Jr., 1997, *Composite Construction: Design for Buildings*, ASCE and McGraw-Hill, New York, NY.

Zahn, C.J. 1987, "Plate-Girder Design Using LRFD," *Engineering Journal*, Vol. 24, No. 1, (1st Qtr.), AISC, Chicago, IL.

Table 5-1.
Values of C_b for Simply Supported Beams

Load	Lateral Bracing Along Span	C_b
Single concentrated load at midspan	None	1.32
	At load points	1.67, 1.67
Two concentrated loads	None	1.14
	At load points	1.67, 1.00, 1.67
Three concentrated loads	None	1.14
	At load points	1.67, 1.11, 1.11, 1.67
Uniformly distributed load	None	1.14
	At centerline	1.30, 1.30

I_x

Table 5-2. W-Shapes Selection by I_x

$F_y = 50$ ksi
$\phi_b = 0.90$
$\phi_v = 0.90$

Shape	X-X Axis								Y-Y Axis		
	I_x	Z_x	$\phi_b M_{px}$	$\phi_b M_{rx}$	L_p	L_r	BF	$\phi_v V_n$	I_y	Z_y	$\phi_b M_{py}$
	in.^{4}	in.^{3}	kip-ft	kip-ft	ft	ft	kips	kips	in.^{4}	in.^{3}	kip-ft
W36×798*	**62600**	**3580**	**13400**	**8940**	**14.9**	**84.6**	**64.4**	**2700**	**4200**	**743**	**2630**
W40×593*	**50400**	**2760**	**10400**	**7020**	**13.4**	**56.5**	**77.3**	**2080**	**2520**	**481**	**1700**
W36×650*	48900	2860	10700	7260	14.5	69.6	62.9	2150	3230	580	2060
W40×503*	**41700**	**2320**	**8700**	**5940**	**13.1**	**49.4**	**76.0**	**1750**	**2050**	**395**	**1410**
W36×527*	38300	2280	8550	5850	14.2	58.1	61.5	1700	2490	454	1630
W40×431*	**34800**	**1960.0**	**7350**	**5070**	**12.9**	**44.1**	**73.2**	**1490**	**1690**	**328**	**1170**
W40×397*	**32000**	**1800**	**6750**	**4680**	**12.9**	**42.1**	**70.7**	**1350**	**1540**	**300**	**1070**
W44×335	**31100**	**1620**	**6080**	**4230**	**12.4**	**35.5**	**79.6**	**1210**	**1200**	**236**	**849**
W36×439*	31000	1870	7010	4860	13.9	50.1	59.4	1410	1990	367	1320
W40×392*	29900	1710	6410	4320	9.33	34.3	83.9	1590	803	212	731
W40×372*	29600	1680.0	6300	4380	12.7	40.1	70.2	1270	1420	278	996
W40×362*	28900	1640	6150	4260	12.7	39.9	69.5	1230	1380	270	973
W36×393*	27500	1670	6260	4350	13.8	46.7	58.1	1250	1750	325	1170
W44×290	**27100**	**1420**	**5330**	**3720**	**12.3**	**34.0**	**74.0**	**1020**	**1050**	**206**	**743**
W40×324	25600	1460	5480	3840	12.6	37.6	65.6	1090	1220	239	861
W27×539*	25600	1890	7090	4710	12.9	78.0	36.5	1730	2110	437	1560
W36×359*	24800	1510	5660	3960	13.6	43.8	56.5	1130	1570	292	1060
W40×331	24700	1430	5360	3630	9.08	30.5	80.9	1340	644	172	596
W40×327	24500	1410	5290	3600	9.11	30.3	79.5	1300	640	170	591
W33×387*	24300	1560	5850	4050	13.3	47.7	52.4	1220	1620	312	1130
W44×262	**24200**	**1270**	**4760**	**3360**	**12.2**	**32.8**	**68.3**	**924**	**927**	**183**	**664**
W40×297	23200	1330	4990	3510	12.5	35.9	63.1	999	1090	215	776
W36×328*	22500	1380	5180	3630	13.6	41.8	54.8	1020	1420	265	962
W33×354*	22000	1420	5330	3720	13.2	44.6	51.1	1110	1460	282	1020
W40×277	21900	1250	4690	3300	12.6	35.5	60.8	890	1040	204	743
W44×230	**20800**	**1100**	**4130**	**2910**	**12.1**	**31.8**	**61.7**	**822**	**796**	**157**	**568**
W30×391*	20700	1450	5440	3750	13.0	52.1	43.1	1220	1550	310	1110
W40×278	20500	1190	4460	3060	8.90	27.6	75.1	1110	521	140	490
W36×300	20300	1260	4730	3330	13.5	39.9	53.0	936	1300	241	878
W40×249	19600	1120	4200	2980	12.5	34.1	56.7	798	926	182	664
W33×318*	19500	1270	4760	3330	13.1	41.9	49.8	988	1290	250	906
W40×264	19400	1130	4240	2910	8.90	27.0	73.2	1040	493	132	465
W36×280	18900	1170	4390	3090	13.5	38.7	51.3	872	1200	223	810
W30×357*	18700	1320.0	4950	3420	12.9	48.6	42.8	1100	1390	279	1010
W33×291	17700	1160	4350	3060	13	39.6	48.5	902	1160	226	821
W40×235	17400	1010	3790	2630	8.97	26.0	68.2	890	444	118	420
W36×260	17300	1080	4050	2860	13.4	37.4	49.5	823	1090	204	743
W36×256	16800	1040	3900	2690	9.36	28.7	63.0	969	528	137	487
W30×326*	16800	1190	4460	3120	12.7	45.2	41.3	997	1240	252	911

*ASTM A6 tensile group 4 or 5 shape. Special requirements may apply per LRFD Specification Section A3.1c.

$F_y = 50$ ksi
$\phi_b = 0.90$
$\phi_v = 0.90$

Table 5-2 (cont.). W-Shapes Selection by I_x

I_x

Shape	X-X Axis								Y-Y Axis		
	I_x	Z_x	$\phi_b M_{px}$	$\phi_b M_{rx}$	L_p	L_r	BF	$\phi_v V_n$	I_y	Z_y	$\phi_b M_{py}$
	in.^{4}	in.^{3}	kip-ft	kip-ft	ft	ft	kips	kips	in.^{4}	in.^{3}	kip-ft
W40×215	16700	964.0	3620	2580	12.5	32.8	51.2	684	796	156	568
W27×368*	16200	1240	4650	3180	12.3	54.7	34.6	1130	1310	279	1010
W36×245	16100	1010	3790	2690	13.2	36.5	47.5	780	1010	190	692
W14×808*	16000	1830	6860	4200	17.0	268	10.6	2300	5510	927	3340
W33×263	15900	1040	3900	2760	12.9	37.8	45.9	810	1040	202	737
W40×211	15500	906	3400	2360	8.87	24.9	64.8	798	390	105	372
W36×230	15000	943	3540	2510	13.2	35.6	45.6	737	940	176	641
W36×232	15000	936	3510	2430	9.25	27.3	60.1	871	468	122	434
W40×199	14900	869	3260	2310	12.2	31.7	48.5	679	695	137	496
W30×292*	14900	1060	3980	2790	12.6	42.1	40.2	881	1100	223	810
W27×336*	14600	1130	4240	2920	12.2	50.4	34.5	1020	1180	252	911
W14×730*	14300	1660	6230	3840	16.6	241	10.6	1860	4720	816	2960
W33×241	14200	940	3530	2490	12.8	36.2	44.0	766	933	182	664
W24×370*	13400	1130	4240	2870	11.6	61.0	27.7	1150	1160	267	956
W40×183	13300	783	2940	2050	8.83	23.9	58.7	684	336	89.6	320
W36×210	13200	833	3120	2160	9.11	26.1	57.0	822	411	107	380
W30×261	13100	943	3540	2490	12.5	39.1	39.3	793	959	196	714
W27×307*	13100	1030	3860	2660	12.0	46.7	34.7	927	1050	227	821
W33×221	12900	857	3210	2280	12.7	35.0	42.0	709	840	164	596
W14×665*	12400	1480	5550	3450	16.3	222	10.2	1650	4170	730	2660
W36×194	12100	767	2880	1990	9.04	25.3	54.4	754	375	97.7	348
W27×281*	11900	936	3510	2440	12.0	43.8	33.5	839	953	206	748
W24×335*	11900	1020	3830	2590	11.4	55.5	27.9	1020	1030	238	855
W30×235	11700	847	3180	2240	12.4	37.1	37.8	701	855	175	641
W40×167	11600	693	2600	1800	8.48	22.9	55.4	677	283	76.0	269
W33×201	11600	773	2900	2060	12.6	33.8	39.6	651	749	147	536
W36×182	11300	718	2690	1870	9.01	24.8	52.0	711	347	90.7	324
W27×258	10800	852	3200	2240	11.9	41.0	32.9	767	859	187	675
W14×605*	10800	1320	4950	3120	16.1	203	9.79	1470	3680	652	2380
W24×306*	10700	922	3460	2370	11.3	51.0	27.5	922	919	214	771
W36×170	10500	668	2510	1740	8.94	24.3	49.6	665	320	83.8	299
W30×211	10300	751	2820	2000	12.3	35.2	35.8	647	757	155	563
W40×149	9800	598	2240	1540	8.09	21.8	51.2	650	229	62.2	218
W36×160	9760	624	2340	1630	8.83	23.8	47.8	632	295	77.3	276
W27×235	9700	772	2900	2030	11.8	38.6	32.2	705	769	168	608
W24×279*	9600	835	3130	2150	11.2	47.3	27.1	836	823	193	698
W14×550*	9430	1180	4430	2790	15.9	187	9.51	1300	3250	583	2130
W33×169	9290	629	2360	1650	8.83	24.5	45.4	611	310	84.4	303
W30×191	9200	675	2530	1800	12.2	33.8	33.9	589	673	138	503
W36×150	9040	581	2180	1510	8.72	23.3	45.6	606	270	70.9	254
W27×217	8910	711	2670	1880	11.7	36.8	31.3	636	704	154	561
W24×250*	8490	744	2790	1930	11.1	43.2	26.7	739	724	171	619
W30×173	8230	607	2280	1620	12.1	32.5	32.0	538	598	123	449
W14×500*	8210	1050	3940	2510	15.6	172	9.13	1160	2880	522	1910
W33×152	8160	559	2100	1460	8.72	23.7	42.4	574	273	73.9	266
W27×194	7860	631	2370	1680	11.6	34.6	29.9	569	619	136	496

*ASTM A6 tensile group 4 or 5 shape. Special requirements may apply per LRFD Specification Section A3.1c.

I_x

Table 5-2 (cont.). W-Shapes Selection by I_x

$F_y = 50$ ksi
$\phi_b = 0.90$
$\phi_v = 0.90$

Shape	X-X Axis								Y-Y Axis		
	I_x	Z_x	$\phi_b M_{px}$	$\phi_b M_{rx}$	L_p	L_r	BF	$\phi_v V_n$	I_y	Z_y	$\phi_b M_{py}$
	in.^{4}	in.^{3}	kip-ft	kip-ft	ft	ft	kips	kips	in.^{4}	in.^{3}	kip-ft
W36×135	**7800**	**509**	**1910**	**1320**	**8.41**	**22.4**	**42.2**	**577**	**225**	**59.7**	**212**
W24×229	7650	675	2530	1760	11.0	40.2	26.3	674	651	154	559
W33×141	7450	514	1930	1340	8.58	23.0	40.4	544	246	66.9	240
W14×455*	7190	936	3510	2270	15.5	157	8.75	1040	2560	468	1710
W27×178	7020	570	2140	1520	11.5	33.0	28.9	544	555	122	443
W24×207	6820	606	2270	1590	10.9	37.3	25.7	604	578	137	500
W33×130	**6710**	**467**	**1750**	**1220**	**8.44**	**22.5**	**38.0**	**518**	**218**	**59.5**	**213**
W30×148	6680	500	1880	1310	8.05	22.8	38.5	539	227	68.0	244
W14×426*	6600	869	3260	2120	15.3	148	8.61	949	2360	434	1590
W27×161	6310	515	1930	1370	11.4	31.7	27.5	492	497	109	399
W24×192	6260	559	2100	1470	10.8	35.6	25.1	558	530	126	460
W14×398*	6000	801	3000	1970	15.2	139	8.38	875	2170	402	1470
W33×118	**5900**	**415**	**1560**	**1080**	**8.19**	**21.7**	**35.4**	**489**	**187**	**51.3**	**183**
W30×132	5770	437	1640	1140	7.95	21.9	35.7	503	196	58.4	209
W24×176	5680	511	1920	1350	10.7	33.7	24.7	510	479	115	418
W27×146	5660	464	1740	1240	11.3	30.5	25.9	448	443	97.7	357
W14×370*	5440	736	2760	1820	15.1	129	8.21	802	1990	370	1360
W30×124	5360	408	1530	1070	7.88	21.4	34.3	477	181	54.0	194
W21×201	5310	530	1990	1380	10.7	41.0	19.9	565	542	133	484
W24×162	5170	468	1760	1240	10.8	32.4	23.7	476	443	105	385
W30×116	**4930**	**378**	**1420**	**987**	**7.74**	**20.9**	**32.8**	**458**	**164**	**49.2**	**176**
W14×342*	4900	672	2520	1670	15.0	121	8.01	728	1810	338	1240
W27×129	4760	395	1480	1040	7.81	22.2	31.0	455	184	57.6	207
W21×182	4740	477	1790	1250	10.6	38.1	19.5	509	483	119	434
W24×146	4580	418	1570	1110	10.6	30.6	22.8	433	391	93.2	340
W30×108	**4470**	**346**	**1300**	**897**	**7.59**	**20.4**	**31.3**	**439**	**146**	**43.9**	**157**
W14×311*	4330	603	2260	1520	14.8	110	7.84	651	1610	304	1120
W21×166	4280	432	1620	1140	10.5	35.7	19.0	456	435	108	394
W27×114	4080	343	1290	897	7.70	21.2	28.8	420	159	49.3	177
W12×336*	4060	603	2260	1450	12.3	131	6.81	807	1190	274	996
W24×131	4020	370	1390	987	10.5	29.1	21.6	400	340	81.5	298
W30×99	**3990**	**312**	**1170**	**807**	**7.42**	**19.8**	**29.2**	**417**	**128**	**38.6**	**138**
W14×283*	3840	542	2030	1380	14.7	100	7.65	582	1440	274	1010
W21×147	3640	374	1400	990	10.4	32.8	18.5	430	376	92.7	338
W27×102	3620	305	1140	801	7.59	20.4	26.7	377	139	43.4	156

*ASTM A6 tensile group 4 or 5 shape. Special requirements may apply per LRFD Specification Section A3.1c.

$F_y = 50$ ksi
$\phi_b = 0.90$
$\phi_v = 0.90$

Table 5-2 (cont.). W-Shapes Selection by I_x

I_x

Shape	X-X Axis								Y-Y Axis		
	I_x	Z_x	$\phi_b M_{px}$	$\phi_b M_{rx}$	L_p	L_r	BF	$\phi_v V_n$	I_y	Z_y	$\phi_b M_{py}$
	in.^{4}	in.^{3}	kip-ft	kip-ft	ft	ft	kips	kips	in.^{4}	in.^{3}	kip-ft
W30×90	**3610**	**283**	**1060**	**735**	**7.38**	**19.4**	**27.1**	**374**	**115**	**34.7**	**124**
W12×305*	3550	537	2010	1310	12.1	120	6.59	717	1050	244	894
W24×117	3540	327	1230	873	10.4	27.9	20.1	361	297	71.4	262
W18×175	3450	398	1490	1030	9.75	41.4	14.6	481	391	106	387
W14×257*	3400	487	1830	1250	14.6	91.6	7.55	523	1290	246	906
W27×94	3270	278	1040	729	7.49	19.9	25.2	356	124	38.8	140
W21×132	3230	334	1250	888	10.3	31.0	17.7	383	333	82.3	301
W12×279*	3110	481	1800	1180	11.9	110	6.37	657	937	220	804
W24×104	3100	289	1080	774	10.3	26.9	18.6	325	259	62.4	229
W18×158	3060	356	1340	930	9.68	38.0	14.3	431	347	94.8	345
W14×233*	3010	436	1640	1130	14.5	83.4	7.40	462	1150	221	816
W24×103	3000	280	1050	735	7.03	20.0	24.2	364	119	41.5	149
W21×122	2970	308	1160	822	10.3	29.7	17.2	352	305	75.6	277
W27×84	**2850**	**244**	**915**	**639**	**7.31**	**19.3**	**23.0**	**332**	**106**	**33.2**	**119**
W18×143	2750	322	1210	846	9.61	35.3	14.1	384	311	85.4	312
W12×252*	2720	428	1610	1060	11.8	100	6.19	582	828	196	714
W24×94	2700	254	953	666	6.99	19.4	23.1	338	109	37.5	135
W21×111	2680	279	1050	747	10.2	28.6	16.3	319	274	68.2	250
W14×211	2660	390	1460	1010	14.4	76.0	7.28	415	1030	198	731
W18×130	2460	290	1090	768	9.54	32.8	13.7	349	278	76.7	281
W21×101	2430	253	949	681	10.2	27.5	15.4	289	248	61.7	227
W12×230*	2420	386	1450	963	11.7	92.0	6.04	526	742	177	647
W14×193	2400	355	1330	930	14.3	70.1	7.19	372	931	180	669
W24×84	**2370**	**224**	**840**	**588**	**6.89**	**18.6**	**21.5**	**306**	**94.4**	**32.6**	**118**
W18×119	2190	262	983	693	9.50	30.9	13.5	336	253	69.1	253
W14×176	2140	320	1200	843	14.2	64.5	7.09	341	838	163	602
W12×210*	2140	348	1310	876	11.6	84.2	5.91	468	664	159	585
W24×76	**2100**	**200**	**750**	**528**	**6.78**	**18.0**	**19.8**	**284**	**82.5**	**28.6**	**103**
W21×93	2070	221	829	576	6.50	19.3	19.7	338	92.9	34.7	124
W18×106	1910	230	863	612	9.40	28.7	13.0	298	220	60.5	222
W14×159	1900	287	1080	762	14.1	58.9	7.02	302	748	146	541
W12×190	1890	311	1170	789	11.5	76.6	5.79	412	589	143	523
W24×68	**1830**	**177**	**664**	**462**	**6.61**	**17.4**	**18.6**	**266**	**70.4**	**24.5**	**88.3**
W21×83	1830	196	735	513	6.46	18.5	18.5	298	81.4	30.5	110
W18×97	1750	211	791	564	9.36	27.5	12.6	269	201	55.3	203
W14×145	1710	260	975	696	14.1	54.7	6.87	272	677	133	491
W12×170	1650	275	1030	705	11.4	68.9	5.67	363	517	126	463
W21×73	1600	172	645	453	6.39	17.6	17.1	260	70.6	26.6	95.6
W24×62	**1560**	**154**	**578**	**396**	**4.84**	**13.3**	**21.6**	**275**	**34.5**	**15.8**	**55.1**
W18×86	1530	186	698	498	9.29	26.0	11.9	238	175	48.4	178
W14×132	1530	234	878	627	13.3	49.7	6.88	256	548	113	419
W16×100	1490	199	746	528	8.87	29.5	10.6	269	186	54.9	201
W21×68	1480	160	600	420	6.36	17.3	16.5	245	64.7	24.4	88.3
W12×152	1430	243	911	627	11.3	62.1	5.59	322	454	111	410
W14×120	1380	212	795	570	13.2	46.3	6.81	231	495	102	380

*ASTM A6 tensile group 4 or 5 shape. Special requirements may apply per LRFD Specification Section A3.1c.

I_x

Table 5-2 (cont.). W-Shapes Selection by I_x

$F_y = 50$ ksi
$\phi_b = 0.90$
$\phi_v = 0.90$

Shape	X-X Axis								Y-Y Axis		
	I_x	Z_x	$\phi_b M_{px}$	$\phi_b M_{rx}$	L_p	L_r	BF	$\phi_v V_n$	I_y	Z_y	$\phi_b M_{py}$
	in.^{4}	in.^{3}	kip-ft	kip-ft	ft	ft	kips	kips	in.^{4}	in.^{3}	kip-ft
W24×55	**1360**	**135**	**506**	**345**	**4.73**	**12.9**	**19.8**	**252**	**29.1**	**13.4**	**46.7**
W21×62	1330	144	540	381	6.25	16.7	15.2	227	57.5	21.7	78.8
W18×76	1330	163	611	438	9.22	24.8	11.1	209	152	42.2	155
W16×89	1310	176	660	468	8.80	27.5	10.3	238	163	48.2	177
W14×109	1240	192	720	519	13.2	43.2	6.69	203	447	92.7	344
W12×136	1240	214	803	558	11.2	55.7	5.49	286	398	98.0	361
W21×57	1170	129	484	333	4.77	13.2	17.8	231	30.7	14.9	52.6
W18×71	1170	146	548	381	6.00	17.8	14.1	247	60.3	24.7	88.9
W21×55	**1140**	**126**	**473**	**330**	**6.11**	**16.1**	**14.3**	**211**	**48.4**	**18.4**	**66.4**
W16×77	1120	151	566	405	8.69	25.3	9.68	203	138	41.2	151
W14×99††	1110	173	649	471	13.5	40.6	6.56	186	402	83.6	311
W18×65	1070	133	499	351	5.97	17.1	13.3	224	54.8	22.5	81.0
W12×120	1070	186	698	489	11.1	50.0	5.36	251	345	85.4	315
W14×90††	999	157	589	429	15.1	38.4	6.85	166	362	75.6	281
W21×50	**989**	**111**	**416**	**285**	**4.59**	**12.5**	**16.5**	**213**	**24.9**	**12.2**	**43.0**
W18×60	984	123	461	324	5.93	16.6	12.9	204	50.1	20.6	74.8
W16×67	963	131	491	354	8.65	23.8	9.04	174	119	35.5	131
W21×48††	**959**	**107**	**401**	**279**	**6.09**	**15.4**	**13.2**	**195**	**38.7**	**14.9**	**53.5**
W12×106	933	164	615	435	11.0	44.9	5.31	212	301	75.1	277
W18×55	890	112	420	295	5.90	16.1	12.2	191	44.9	18.5	66.9
W14×82	882	139	521	369	8.76	29.6	7.30	197	148	44.8	165
W21×44	**847**	**95.8**	**359**	**246**	**4.45**	**12.0**	**15.0**	**196**	**20.7**	**10.2**	**35.8**
W12×96	833	147	551	393	10.9	41.3	5.20	189	270	67.5	250
W18×50	800	101	379	267	5.83	15.6	11.5	173	40.1	16.6	60.2
W14×74	796	126	473	336	8.76	27.9	7.12	173	134	40.5	150
W16×57	758	105	394	277	5.65	16.6	10.7	190	43.1	18.9	68.1
W12×87	740	132	495	354	10.8	38.4	5.13	174	241	60.4	223
W14×68	722	115	431	309	8.69	26.4	6.91	157	121	36.9	136
W10×112	716	147	551	378	9.47	56.5	3.68	232	236	69.2	255
W18×46	712	90.7	340	236	4.56	12.6	12.9	176	22.5	11.7	41.8
W12×79	662	119	446	321	10.8	35.7	5.03	157	216	54.3	201
W16×50	659	92.0	345	243	5.62	15.7	10.1	167	37.2	16.3	59.1
W14×61	640	102	383	277	8.65	25.0	6.50	141	107	32.8	121
W10×100	623	130	488	336	9.36	50.8	3.66	204	207	61.0	225
W18×40	**612**	**78.4**	**294**	**205**	**4.49**	**12.0**	**11.7**	**152**	**19.1**	**9.95**	**35.7**
W12×72	597	108	405	292	10.7	33.6	4.93	143	195	49.2	182
W16×45	586	82.3	309	218	5.55	15.1	9.45	150	32.8	14.5	52.5
W14×53	541	87.1	327	233	6.78	20.1	7.01	139	57.7	22.0	80.4
W10×88	534	113	424	296	9.29	45.1	3.58	176	179	53.1	196
W12×65††	533	96.8	363	264	11.9	31.7	5.01	127	174	44.1	164
W16×40	**518**	**73.0**	**274**	**194**	**5.55**	**14.7**	**8.71**	**132**	**28.9**	**12.7**	**46.4**

††Indicates flange is non-compact.

$F_y = 50$ ksi
$\phi_b = 0.90$
$\phi_v = 0.90$

Table 5-2 (cont.). W-Shapes Selection by I_x

I_x

Shape	X-X Axis								Y-Y Axis		
	I_x	Z_x	$\phi_b M_{px}$	$\phi_b M_{rx}$	L_p	L_r	BF	$\phi_v V_n$	I_y	Z_y	$\phi_b M_{py}$
	in.⁴	in.³	kip-ft	kip-ft	ft	ft	kips	kips	in.⁴	in.³	kip-ft
W18×35	**510**	**66.5**	**249**	**173**	**4.31**	**11.5**	**10.7**	**143**	**15.3**	**8.06**	**28.8**
W14×48	485	78.4	294	211	6.75	19.2	6.70	127	51.4	19.6	72.0
W12×58	475	86.4	324	234	8.87	27.0	4.97	119	107	32.5	120
W10×77	455	97.6	366	258	9.18	39.9	3.53	152	154	45.9	169
W16×36	448	64.0	240	170	5.37	14.1	8.11	127	24.5	10.8	39.4
W14×43	428	69.6	261	188	6.68	18.2	6.31	113	45.2	17.3	63.6
W12×53	425	77.9	292	212	8.76	25.6	4.78	113	95.8	29.1	108
W10×68	394	85.3	320	227	9.15	36.0	3.45	132	134	40.1	149
W12×50	391	71.9	270	193	6.92	21.5	5.30	122	56.3	21.3	78.2
W14×38	383	61.1	229	163	5.47	14.9	7.05	118	26.7	12.1	44.3
W16×31	**375**	**54.0**	**203**	**142**	**4.13**	**11.0**	**8.86**	**118**	**12.4**	**7.03**	**25.3**
W12×45	348	64.2	241	173	6.89	20.3	5.06	109	50.0	19.0	69.8
W10×60	341	74.6	280	200	9.08	32.6	3.39	116	116	35.0	129
W14×34	337	54.2	203	145	5.40	14.3	6.58	108	23.3	10.6	38.9
W12×40	307	57.0	214	155	6.85	19.2	4.79	94.8	44.1	16.8	61.9
W10×54	303	66.6	250	180	9.04	30.2	3.30	101	103	31.3	116
W16×26	**301**	**44.2**	**166**	**115**	**3.96**	**10.4**	**7.83**	**106**	**9.59**	**5.48**	**19.6**
W14×30	288	46.9	176	125	5.26	13.7	6.05	101	19.6	8.97	32.7
W12×35	285	51.2	192	137	5.44	15.2	5.65	101	24.5	11.5	42.0
W10×49	272	60.4	227	164	8.97	28.3	3.24	91.6	93.4	28.3	105
W10×45	248	54.9	206	147	7.10	24.1	3.44	95.4	53.4	20.3	74.8
W14×26	**243**	**39.9**	**150**	**105**	**3.81**	**10.2**	**7.01**	**95.7**	**8.90**	**5.53**	**19.9**
W12×30	238	43.1	162	116	5.37	14.3	5.12	86.3	20.3	9.56	35.1
W10×39	209	46.8	176	126	6.99	21.8	3.32	84.4	45.0	17.2	63.6
W12×26	**204**	**37.2**	**140**	**100**	**5.33**	**13.8**	**4.66**	**75.8**	**17.3**	**8.17**	**30.0**
W14×22	**197**	**32.8**	**123**	**85.8**	**3.67**	**9.65**	**6.22**	**85.1**	**7.00**	**4.37**	**15.8**
W10×33	171	38.8	146	105	6.85	19.8	3.14	76.2	36.6	14.0	51.8
W10×30	170	36.6	137	97.2	4.84	14.6	4.11	85.0	16.7	8.84	32.3
W12×22	**156**	**29.3**	**110**	**76.2**	**3.00**	**8.42**	**6.21**	**86.3**	**4.66**	**3.66**	**13.0**
W10×26	144	31.3	117	83.7	4.80	13.6	3.84	72.3	14.1	7.50	27.5
W12×19	**130**	**24.7**	**92.6**	**63.9**	**2.90**	**7.95**	**5.69**	**77.4**	**3.76**	**2.98**	**10.6**
W10×22	118	26.0	97.5	69.6	4.70	12.7	3.49	66.1	11.4	6.10	22.3
W12×16	**103**	**20.1**	**75.4**	**51.3**	**2.73**	**7.44**	**5.11**	**71.3**	**2.82**	**2.26**	**7.93**
W10×19	96.2	21.6	81.0	56.4	3.09	8.88	4.24	68.8	4.29	3.35	12.0
W12×14	**88.5**	**17.4**	**65.2**	**44.7**	**2.66**	**7.17**	**4.55**	**64.3**	**2.36**	**1.90**	**6.69**
W10×17	81.9	18.7	70.1	48.6	2.98	8.38	3.99	65.4	3.56	2.80	10.0
W10×15	68.9	16.0	60.0	41.4	2.86	7.93	3.67	62.0	2.89	2.30	8.16
W10×12[††]	**53.8**	**12.6**	**47.3**	**32.7**	**2.87**	**7.45**	**3.18**	**50.6**	**2.18**	**1.74**	**6.19**

[††] Indicates flange is non-compact.

Z_x

Table 5-3. W-Shapes Selection by Z_x

$F_y = 50$ ksi
$\phi_b = 0.90$
$\phi_v = 0.90$

Shape	X-X Axis								Y-Y Axis		
	Z_x	I_x	$\phi_b M_{px}$	$\phi_b M_{rx}$	L_p	L_r	BF	$\phi_v V_n$	Z_y	I_y	$\phi_b M_{py}$
	in.³	in.⁴	kip-ft	kip-ft	ft	ft	kips	kips	in.³	in.⁴	kip-ft
W36×798*	**3580**	**62600**	**13400**	**8940**	**14.9**	**84.6**	**64.4**	**2700**	**743**	**4200**	**2630**
W36×650*	**2860**	**48900**	**10700**	**7260**	**14.5**	**69.6**	**62.9**	**2150**	**580**	**3230**	**2060**
W40×593*	**2760**	**50400**	**10400**	**7020**	**13.4**	**56.5**	**77.3**	**2080**	**481**	**2520**	**1700**
W40×503*	**2320**	**41700**	**8700**	**5940**	**13.1**	**49.4**	**76.0**	**1750**	**395**	**2050**	**1410**
W36×527*	2280	38300	8550	5850	14.2	58.1	61.5	1700	454	2490	1630
W40×431*	**1960**	**34800**	**7350**	**5070**	**12.9**	**44.1**	**73.2**	**1490**	**328**	**1690**	**1170**
W27×539*	1890	25600	7090	4710	12.9	78.0	36.5	1730	437	2110	1560
W36×439*	1870	31000	7010	4860	13.9	50.1	59.4	1410	367	1990	1320
W14×808*	1830	16000	6860	4200	17.0	268	10.6	2300	927	5510	3340
W40×397*	**1800**	**32000**	**6750**	**4680**	**12.9**	**42.1**	**70.7**	**1350**	**300**	**1540**	**1070**
W40×392*	**1710**	**29900**	**6410**	**4320**	**9.33**	**34.3**	**83.9**	**1590**	**212**	**803**	**731**
W40×372*	**1680**	**29600**	**6300**	**4380**	**12.7**	**40.1**	**70.2**	**1270**	**278**	**1420**	**996**
W36×393*	1670	27500	6260	4350	13.8	46.7	58.1	1250	325	1750	1170
W14×730*	1660	14300	6230	3840	16.6	241	10.6	1860	816	4720	2960
W40×362*	**1640**	**28900**	**6150**	**4260**	**12.7**	**39.9**	**69.5**	**1230**	**270**	**1380**	**973**
W44×335	**1620**	**31100**	**6080**	**4230**	**12.4**	**35.5**	**79.6**	**1210**	**236**	**1200**	**849**
W33×387*	1560	24300	5850	4050	13.3	47.7	52.4	1220	312	1620	1130
W36×359*	1510	24800	5660	3960	13.6	43.8	56.5	1130	292	1570	1060
W14×665*	1480	12400	5550	3450	16.3	222	10.2	1650	730	4170	2660
W40×324	**1460**	**25600**	**5480**	**3840**	**12.6**	**37.6**	**65.6**	**1090**	**239**	**1220**	**861**
W30×391*	1450	20700	5440	3750	13.0	52.1	43.1	1220	310	1550	1110
W40×331	1430	24700	5360	3630	9.08	30.5	80.9	1340	172	644	596
W44×290	**1420**	**27100**	**5330**	**3720**	**12.3**	**34.0**	**74.0**	**1020**	**206**	**1050**	**743**
W33×354*	1420	22000	5330	3720	13.2	44.6	51.1	1110	282	1460	1020
W40×327	1410	24500	5290	3600	9.11	30.3	79.5	1300	170	640	591
W36×328*	1380	22500	5180	3630	13.6	41.8	54.8	1020	265	1420	962
W40×297	1330	23200	4990	3510	12.5	35.9	63.1	999	215	1090	776
W30×357*	1320	18700	4950	3420	12.9	48.6	42.8	1100	279	1390	1010
W14×605*	1320	10800	4950	3120	16.1	203	9.79	1470	652	3680	2380

*ASTM A6 tensile group 4 or 5 shape. Special requirements may apply per LRFD Specification Section A3.1c.

$F_y = 50$ ksi
$\phi_b = 0.90$
$\phi_v = 0.90$

Table 5-3 (cont.). W-Shapes Selection by Z_x

Z_x

Shape	X-X Axis								Y-Y Axis		
	Z_x	I_x	$\phi_b M_{px}$	$\phi_b M_{rx}$	L_p	L_r	BF	$\phi_v V_n$	Z_y	I_y	$\phi_b M_{py}$
	in.³	in.⁴	kip-ft	kip-ft	ft	ft	kips	kips	in.³	in.⁴	kip-ft
W44×262	**1270**	**24200**	**4760**	**3360**	**12.2**	**32.8**	**68.3**	**924**	**183**	**927**	**664**
W33×318*	1270	19500	4760	3330	13.1	41.9	49.8	988	250	1290	906
W36×300	1260	20300	4730	3330	13.5	39.9	53.0	936	241	1300	878
W40×277	1250	21900	4690	3300	12.6	35.5	60.8	890	204	1040	743
W27×368*	1240	16200	4650	3180	12.3	54.7	34.6	1130	279	1310	1010
W40×278	1190	20500	4460	3060	8.90	27.6	75.1	1110	140	521	490
W30×326*	1190	16800	4460	3120	12.7	45.2	41.3	997	252	1240	911
W14×550*	1180	9430	4430	2790	15.9	187	9.51	1300	583	3250	2130
W36×280	1170	18900	4390	3090	13.5	38.7	51.3	872	223	1200	810
W33×291	1160	17700	4350	3060	13.0	39.6	48.5	902	226	1160	821
W40×264	1130	19400	4240	2910	8.90	27.0	73.2	1040	132	493	465
W27×336*	1130	14600	4240	2920	12.2	50.4	34.5	1020	252	1180	911
W24×370*	1130	13400	4240	2870	11.6	61.0	27.7	1150	267	1160	956
W40×249	**1120**	**19600**	**4200**	**2980**	**12.5**	**34.1**	**56.7**	**798**	**182**	**926**	**664**
W44×230	**1100**	**20800**	**4130**	**2910**	**12.1**	**31.8**	**61.7**	**822**	**157**	**796**	**568**
W36×260	1080	17300	4050	2860	13.4	37.4	49.5	823	204	1090	743
W30×292*	1060	14900	3980	2790	12.6	42.1	40.2	881	223	1100	810
W14×500*	1050	8210	3940	2510	15.6	172	9.13	1160	522	2880	1910
W36×256	1040	16800	3900	2690	9.36	28.7	63.0	969	137	528	487
W33×263	1040	15900	3900	2760	12.9	37.8	45.9	810	202	1040	737
W27×307*	1030	13100	3860	2660	12.0	46.7	34.7	927	227	1050	821
W24×335*	1020	11900	3830	2590	11.4	55.5	27.9	1020	238	1030	855
W40×235	1010	17400	3790	2630	8.97	26.0	68.2	890	118	444	420
W36×245	1010	16100	3790	2690	13.2	36.5	47.5	780	190	1010	692
W40×215	**964**	**16700**	**3620**	**2580**	**12.5**	**32.8**	**51.2**	**684**	**156**	**796**	**568**
W36×230	943	15000	3540	2510	13.2	35.6	45.6	737	176	940	641
W30×261	943	13100	3540	2490	12.5	39.1	39.3	793	196	959	714
W33×241	940	14200	3530	2490	12.8	36.2	44.0	766	182	933	664
W36×232	936	15000	3510	2430	9.25	27.3	60.1	871	122	468	434
W27×281*	936	11900	3510	2440	12.0	43.8	33.5	839	206	953	748
W14×455*	936	7190	3510	2270	15.5	157	8.75	1040	468	2560	1710
W24×306*	922	10700	3460	2370	11.3	51.0	27.5	922	214	919	771
W40×211	**906**	**15500**	**3400**	**2360**	**8.87**	**24.9**	**64.8**	**798**	**105**	**390**	**372**
W40×199	**869**	**14900**	**3260**	**2310**	**12.2**	**31.7**	**48.5**	**679**	**137**	**695**	**496**
W14×426*	869	6600	3260	2120	15.3	148	8.61	949	434	2360	1590
W33×221	857	12900	3210	2280	12.7	35.0	42.0	709	164	840	596
W27×258	852	10800	3200	2240	11.9	41.0	32.9	767	187	859	675
W30×235	847	11700	3180	2240	12.4	37.1	37.8	701	175	855	641
W24×279*	835	9600	3130	2150	11.2	47.3	27.1	836	193	823	698
W36×210	833	13200	3120	2160	9.11	26.1	57.0	822	107	411	380
W14×398*	801	6000	3000	1970	15.2	139	8.38	875	402	2170	1470

*ASTM A6 tensile group 4 or 5 shape. Special requirements may apply per LRFD Specification Section A3.1c.

Z_x

Table 5-3 (cont.). W-Shapes Selection by Z_x

$F_y = 50$ ksi
$\phi_b = 0.90$
$\phi_v = 0.90$

Shape	X-X Axis								Y-Y Axis		
	Z_x	I_x	$\phi_b M_{px}$	$\phi_b M_{rx}$	L_p	L_r	BF	$\phi_v V_n$	Z_y	I_y	$\phi_b M_{py}$
	in.^{3}	in.^{4}	kip-ft	kip-ft	ft	ft	kips	kips	in.^{3}	in.^{4}	kip-ft
W40×183	783	13300	2940	2050	8.83	23.9	58.7	684	89.6	336	320
W33×201	773	11600	2900	2060	12.6	33.8	39.6	651	147	749	536
W27×235	772	9700	2900	2030	11.8	38.6	32.2	705	168	769	608
W36×194	767	12100	2880	1990	9.04	25.3	54.4	754	97.7	375	348
W30×211	751	10300	2820	2000	12.3	35.2	35.8	647	155	757	563
W24×250*	744	8490	2790	1930	11.1	43.2	26.7	739	171	724	619
W14×370*	736	5440	2760	1820	15.1	129	8.21	802	370	1990	1360
W36×182	718	11300	2690	1870	9.01	24.8	52.0	711	90.7	347	324
W27×217	711	8910	2670	1880	11.7	36.8	31.3	636	154	704	561
W40×167	693	11600	2600	1800	8.48	22.9	55.4	677	76.0	283	269
W30×191	675	9200	2530	1800	12.2	33.8	33.9	589	138	673	503
W24×229	675	7650	2530	1760	11.0	40.2	26.3	674	154	651	559
W14×342*	672	4900	2520	1670	15.0	121	8.01	728	338	1810	1240
W36×170	668	10500	2510	1740	8.94	24.3	49.6	665	83.8	320	299
W27×194	631	7860	2370	1680	11.6	34.6	29.9	569	136	619	496
W33×169	629	9290	2360	1650	8.83	24.5	45.4	611	84.4	310	303
W36×160	624	9760	2340	1630	8.83	23.8	47.8	632	77.3	295	276
W30×173	607	8230	2280	1620	12.1	32.5	32.0	538	123	598	449
W24×207	606	6820	2270	1590	10.9	37.3	25.7	604	137	578	500
W14×311*	603	4330	2260	1520	14.8	110	7.84	651	304	1610	1120
W12×336*	603	4060	2260	1450	12.3	131	6.81	807	274	1190	996
W40×149	598	9800	2240	1540	8.09	21.8	51.2	650	62.2	229	218
W36×150	581	9040	2180	1510	8.72	23.3	45.6	606	70.9	270	254
W27×178	570	7020	2140	1520	11.5	33.0	28.9	544	122	555	443
W33×152	559	8160	2100	1460	8.72	23.7	42.4	574	73.9	273	266
W24×192	559	6260	2100	1470	10.8	35.6	25.1	558	126	530	460
W14×283*	542	3840	2030	1380	14.7	100	7.65	582	274	1440	1010
W12×305*	537	3550	2010	1310	12.1	120	6.59	717	244	1050	894
W21×201	530	5310	1990	1380	10.7	41.0	19.9	565	133	542	484
W27×161	515	6310	1930	1370	11.4	31.7	27.5	492	109	497	399
W33×141	514	7450	1930	1340	8.58	23.0	40.4	544	66.9	246	240
W24×176	511	5680	1920	1350	10.7	33.7	24.7	510	115	479	418
W36×135	509	7800	1910	1320	8.41	22.4	42.2	577	59.7	225	212
W30×148	500	6680	1880	1310	8.05	22.8	38.5	539	68.0	227	244
W14×257*	487	3400	1830	1250	14.6	91.6	7.55	523	246	1290	906
W12×279*	481	3110	1800	1180	11.9	110	6.37	657	220	937	804
W21×182	477	4740	1790	1250	10.6	38.1	19.5	509	119	483	434
W24×162	468	5170	1760	1240	10.8	32.4	23.7	476	105	443	385

*ASTM A6 tensile group 4 or 5 shape. Special requirements may apply per LRFD Specification Section A3.1c.

$F_y = 50$ ksi
$\phi_b = 0.90$
$\phi_v = 0.90$

Table 5-3 (cont.). W-Shapes Selection by Z_x

Z_x

Shape	X-X Axis								Y-Y Axis		
	Z_x	I_x	$\phi_b M_{px}$	$\phi_b M_{rx}$	L_p	L_r	BF	$\phi_v V_n$	Z_y	I_y	$\phi_b M_{py}$
	in.^{3}	in.^{4}	kip-ft	kip-ft	ft	ft	kips	kips	in.^{3}	in.^{4}	kip-ft
W33×130	**467**	**6710**	**1750**	**1220**	**8.44**	**22.5**	**38.0**	**518**	**59.5**	**218**	**213**
W27×146	464	5660	1740	1240	11.3	30.5	25.9	448	97.7	443	357
W30×132	437	5770	1640	1140	7.95	21.9	35.7	503	58.4	196	209
W14×233*	436	3010	1640	1130	14.5	83.4	7.40	462	221	1150	816
W21×166	432	4280	1620	1140	10.5	35.7	19.0	456	108	435	394
W12×252*	428	2720	1610	1060	11.8	100	6.19	582	196	828	714
W24×146	418	4580	1570	1110	10.6	30.6	22.8	433	93.2	391	340
W33×118	**415**	**5900**	**1560**	**1080**	**8.19**	**21.7**	**35.4**	**489**	**51.3**	**187**	**183**
W30×124	408	5360	1530	1070	7.88	21.4	34.3	477	54.0	181	194
W18×175	398	3450	1490	1030	9.75	41.4	14.6	481	106	391	387
W27×129	395	4760	1480	1040	7.81	22.2	31.0	455	57.6	184	207
W14×211	390	2660	1460	1010	14.4	76.0	7.28	415	198	1030	731
W12×230*	386	2420	1450	963	11.7	92.0	6.04	526	177	742	647
W30×116	**378**	**4930**	**1420**	**987**	**7.74**	**20.9**	**32.8**	**458**	**49.2**	**164**	**176**
W21×147	374	3640	1400	990	10.4	32.8	18.5	430	92.7	376	338
W24×131	370	4020	1390	987	10.5	29.1	21.6	400	81.5	340	298
W18×158	356	3060	1340	930	9.68	38.0	14.3	431	94.8	347	345
W14×193	355	2400	1330	930	14.3	70.1	7.19	372	180	931	669
W12×210*	348	2140	1310	876	11.6	84.2	5.91	468	159	664	585
W30×108	**346**	**4470**	**1300**	**897**	**7.59**	**20.4**	**31.3**	**439**	**43.9**	**146**	**157**
W27×114	343	4080	1290	897	7.70	21.2	28.8	420	49.3	159	177
W21×132	334	3230	1250	888	10.3	31.0	17.7	383	82.3	333	301
W24×117	327	3540	1230	873	10.4	27.9	20.1	361	71.4	297	262
W18×143	322	2750	1210	846	9.61	35.3	14.1	384	85.4	311	312
W14×176	320	2140	1200	843	14.2	64.5	7.09	341	163	838	602
W30×99	**312**	**3990**	**1170**	**807**	**7.42**	**19.8**	**29.2**	**417**	**38.6**	**128**	**138**
W12×190	311	1890	1170	789	11.5	76.6	5.79	412	143	589	523
W21×122	308	2970	1160	822	10.3	29.7	17.2	352	75.6	305	277
W27×102	305	3620	1140	801	7.59	20.4	26.7	377	43.4	139	156
W18×130	290	2460	1090	768	9.54	32.8	13.7	349	76.7	278	281
W24×104	289	3100	1080	774	10.3	26.9	18.6	325	62.4	259	229
W14×159	287	1900	1080	762	14.1	58.9	7.02	302	146	748	541
W30×90	**283**	**3610**	**1060**	**735**	**7.38**	**19.4**	**27.1**	**374**	**34.7**	**115**	**124**
W24×103	280	3000	1050	735	7.03	20.0	24.2	364	41.5	119	149
W21×111	279	2680	1050	747	10.2	28.6	16.3	319	68.2	274	250
W27×94	278	3270	1040	729	7.49	19.9	25.2	356	38.8	124	140
W12×170	275	1650	1030	705	11.4	68.9	5.67	363	126	517	463
W18×119	262	2190	983	693	9.50	30.9	13.5	336	69.1	253	253
W14×145	260	1710	975	696	14.1	54.7	6.87	272	133	677	491
W24×94	254	2700	953	666	6.99	19.4	23.1	338	37.5	109	135
W21×101	253	2430	949	681	10.2	27.5	15.4	289	61.7	248	227

*ASTM A6 tensile group 4 or 5 shape. Special requirements may apply per LRFD Specification Section A3.1c.

Z_x

Table 5-3 (cont.). W-Shapes Selection by Z_x

$F_y = 50$ ksi
$\phi_b = 0.90$
$\phi_v = 0.90$

Shape	X-X Axis								Y-Y Axis		
	Z_x	I_x	$\phi_b M_{px}$	$\phi_b M_{rx}$	L_p	L_r	BF	$\phi_v V_n$	Z_y	I_y	$\phi_b M_{py}$
	in.³	in.⁴	kip-ft	kip-ft	ft	ft	kips	kips	in.³	in.⁴	kip-ft
W27×84	**244**	**2850**	**915**	**639**	**7.31**	**19.3**	**23.0**	**332**	**33.2**	**106**	**119**
W12×152	243	1430	911	627	11.3	62.1	5.59	322	111	454	410
W14×132	234	1530	878	627	13.3	49.7	6.88	256	113	548	419
W18×106	230	1910	863	612	9.40	28.7	13.0	298	60.5	220	222
W24×84	**224**	**2370**	**840**	**588**	**6.89**	**18.6**	**21.5**	**306**	**32.6**	**94.4**	**118**
W21×93	221	2070	829	576	6.50	19.3	19.7	338	34.7	92.9	124
W12×136	214	1240	803	558	11.2	55.7	5.49	286	98.0	398	361
W14×120	212	1380	795	570	13.2	46.3	6.81	231	102	495	380
W18×97	211	1750	791	564	9.36	27.5	12.6	269	55.3	201	203
W24×76	**200**	**2100**	**750**	**528**	**6.78**	**18.0**	**19.8**	**284**	**28.6**	**82.5**	**103**
W16×100	199	1490	746	528	8.87	29.5	10.6	269	54.9	186	201
W21×83	196	1830	735	513	6.46	18.5	18.5	298	30.5	81.4	110
W14×109	192	1240	720	519	13.2	43.2	6.69	203	92.7	447	344
W18×86	186	1530	698	498	9.29	26.0	11.9	238	48.4	175	178
W12×120	186	1070	698	489	11.1	50.0	5.36	251	85.4	345	315
W24×68	**177**	**1830**	**664**	**462**	**6.61**	**17.4**	**18.6**	**266**	**24.5**	**70.4**	**88.3**
W16×89	176	1310	660	468	8.80	27.5	10.3	238	48.2	163	177
W14×99††	173	1110	649	471	13.5	40.6	6.56	186	83.6	402	311
W21×73	172	1600	645	453	6.39	17.6	17.1	260	26.6	70.6	95.6
W12×106	164	933	615	435	11.0	44.9	5.31	212	75.1	301	277
W18×76	163	1330	611	438	9.22	24.8	11.1	209	42.2	152	155
W21×68	**160**	**1480**	**600**	**420**	**6.36**	**17.3**	**16.5**	**245**	**24.4**	**64.7**	**88.3**
W14×90††	157	999	589	429	15.1	38.4	6.85	166	75.6	362	281
W24×62	**154**	**1560**	**578**	**396**	**4.84**	**13.3**	**21.6**	**275**	**15.8**	**34.5**	**55.1**
W16×77	151	1120	566	405	8.69	25.3	9.68	203	41.2	138	151
W12×96	147	833	551	393	10.9	41.3	5.20	189	67.5	270	250
W10×112	147	716	551	378	9.47	56.5	3.68	232	69.2	236	255
W18×71	146	1170	548	381	6.00	17.8	14.1	247	24.7	60.3	88.9
W21×62	**144**	**1330**	**540**	**381**	**6.25**	**16.7**	**15.2**	**227**	**21.7**	**57.5**	**78.8**
W14×82	139	882	521	369	8.76	29.6	7.30	197	44.8	148	165
W24×55	**135**	**1360**	**506**	**345**	**4.73**	**12.9**	**19.8**	**252**	**13.4**	**29.1**	**46.7**
W18×65	133	1070	499	351	5.97	17.1	13.3	224	22.5	54.8	81.0
W12×87	132	740	495	354	10.8	38.4	5.13	174	60.4	241	223
W16×67	131	963	491	354	8.65	23.8	9.04	174	35.5	119	131
W10×100	130	623	488	336	9.36	50.8	3.66	204	61.0	207	225
W21×57	129	1170	484	333	4.77	13.2	17.8	231	14.9	30.7	52.6

†† Indicates flange is non-compact.

$F_y = 50$ ksi
$\phi_b = 0.90$
$\phi_v = 0.90$

Table 5-3 (cont.). W-Shapes Selection by Z_x

Z_x

Shape	X-X Axis								Y-Y Axis		
	Z_x	I_x	$\phi_b M_{px}$	$\phi_b M_{rx}$	L_p	L_r	BF	$\phi_v V_n$	Z_y	I_y	$\phi_b M_{py}$
	in.^{3}	in.^{4}	kip-ft	kip-ft	ft	ft	kips	kips	in.^{3}	in.^{4}	kip-ft
W21×55	**126**	**1140**	**473**	**330**	**6.11**	**16.1**	**14.3**	**211**	**18.4**	**48.4**	**66.4**
W14×74	126	796	473	336	8.76	27.9	7.12	173	40.5	134	150
W18×60	123	984	461	324	5.93	16.6	12.9	204	20.6	50.1	74.8
W12×79	119	662	446	321	10.8	35.7	5.03	157	54.3	216	201
W14×68	115	722	431	309	8.69	26.4	6.91	157	36.9	121	136
W10×88	113	534	424	296	9.29	45.1	3.58	176	53.1	179	196
W18×55	**112**	**890**	**420**	**295**	**5.90**	**16.1**	**12.2**	**191**	**18.5**	**44.9**	**66.9**
W21×50	**111**	**989**	**416**	**285**	**4.59**	**12.5**	**16.5**	**213**	**12.2**	**24.9**	**43.0**
W12×72	108	597	405	292	10.7	33.6	4.93	143	49.2	195	182
W21×48[††]	**107**	**959**	**401**	**279**	**6.09**	**15.4**	**13.2**	**195**	**14.9**	**38.7**	**53.5**
W16×57	105	758	394	277	5.65	16.6	10.7	190	18.9	43.1	68.1
W14×61	102	640	383	277	8.65	25.0	6.50	141	32.8	107	121
W18×50	101	800	379	267	5.83	15.6	11.5	173	16.6	40.1	60.2
W10×77	97.6	455	366	258	9.18	39.9	3.53	152	45.9	154	169
W12×65[††]	96.8	533	363	264	11.9	31.7	5.01	127	44.1	174	164
W21×44	**95.8**	**847**	**359**	**246**	**4.45**	**12.0**	**15.0**	**196**	**10.2**	**20.7**	**35.8**
W16×50	92.0	659	345	243	5.62	15.7	10.1	167	16.3	37.2	59.1
W18×46	90.7	712	340	236	4.56	12.6	12.9	176	11.7	22.5	41.8
W14×53	87.1	541	327	233	6.78	20.1	7.01	139	22.0	57.7	80.4
W12×58	86.4	475	324	234	8.87	27.0	4.97	119	32.5	107	120
W10×68	85.3	394	320	227	9.15	36.0	3.45	132	40.1	134	149
W16×45	82.3	586	309	218	5.55	15.1	9.45	150	14.5	32.8	52.5
W18×40	**78.4**	**612**	**294**	**205**	**4.49**	**12.0**	**11.7**	**152**	**9.95**	**19.1**	**35.7**
W14×48	78.4	485	294	211	6.75	19.2	6.70	127	19.6	51.4	72.0
W12×53	77.9	425	292	212	8.76	25.6	4.78	113	29.1	95.8	108
W10×60	74.6	341	280	200	9.08	32.6	3.39	116	35.0	116	129
W16×40	**73.0**	**518**	**274**	**194**	**5.55**	**14.7**	**8.71**	**132**	**12.7**	**28.9**	**46.4**
W12×50	71.9	391	270	193	6.92	21.5	5.30	122	21.3	56.3	78.2
W14×43	69.6	428	261	188	6.68	18.2	6.31	113	17.3	45.2	63.6
W10×54	66.6	303	250	180	9.04	30.2	3.30	101	31.3	103	116
W18×35	**66.5**	**510**	**249**	**173**	**4.31**	**11.5**	**10.7**	**143**	**8.06**	**15.3**	**28.8**
W12×45	64.2	348	241	173	6.89	20.3	5.06	109	19.0	50.0	69.8
W16×36	64.0	448	240	170	5.37	14.1	8.11	127	10.8	24.5	39.4
W14×38	61.1	383	229	163	5.47	14.9	7.05	118	12.1	26.7	44.3
W10×49	60.4	272	227	164	8.97	28.3	3.24	91.6	28.3	93.4	105
W12×40	57.0	307	214	155	6.85	19.2	4.79	94.8	16.8	44.1	61.9
W10×45	54.9	248	206	147	7.10	24.1	3.44	95.4	20.3	53.4	74.8
W14×34	**54.2**	**337**	**203**	**145**	**5.40**	**14.3**	**6.58**	**108**	**10.6**	**23.3**	**38.9**

[††] Indicates flange is non-compact.

Z_x

Table 5-3 (cont.). W-Shapes Selection by Z_x

$F_y = 50$ ksi
$\phi_b = 0.90$
$\phi_v = 0.90$

Shape	X-X Axis								Y-Y Axis		
	Z_x	I_x	$\phi_b M_{px}$	$\phi_b M_{rx}$	L_p	L_r	BF	$\phi_v V_n$	Z_y	I_y	$\phi_b M_{py}$
	in.^{3}	in.^{4}	kip-ft	kip-ft	ft	ft	kips	kips	in.^{3}	in.^{4}	kip-ft
W16×31	**54.0**	**375**	**203**	**142**	**4.13**	**11.0**	**8.86**	**118**	**7.03**	**12.4**	**25.3**
W12×35	51.2	285	192	137	5.44	15.2	5.65	101	11.5	24.5	42.0
W14×30	**46.9**	**288**	**176**	**125**	**5.26**	**13.7**	**6.05**	**101**	**8.97**	**19.6**	**32.7**
W10×39	46.8	209	176	126	6.99	21.8	3.32	84.4	17.2	45.0	63.6
W16×26	**44.2**	**301**	**166**	**115**	**3.96**	**10.4**	**7.83**	**106**	**5.48**	**9.59**	**19.6**
W12×30	43.1	238	162	116	5.37	14.3	5.12	86.3	9.56	20.3	35.1
W14×26	**39.9**	**243**	**150**	**105**	**3.81**	**10.2**	**7.01**	**95.7**	**5.53**	**8.90**	**19.9**
W10×33	38.8	171	146	105	6.85	19.8	3.14	76.2	14.0	36.6	51.8
W12×26	**37.2**	**204**	**140**	**100**	**5.33**	**13.8**	**4.66**	**75.8**	**8.17**	**17.3**	**30.0**
W10×30	36.6	170	137	97.2	4.84	14.6	4.11	85.0	8.84	16.7	32.3
W14×22	**32.8**	**197**	**123**	**85.8**	**3.67**	**9.65**	**6.22**	**85.1**	**4.37**	**7.00**	**15.8**
W10×26	31.3	144	117	83.7	4.80	13.6	3.84	72.3	7.50	14.1	27.5
W12×22	**29.3**	**156**	**110**	**76.2**	**3.00**	**8.42**	**6.21**	**86.3**	**3.66**	**4.66**	**13.0**
W10×22	**26.0**	**118**	**97.5**	**69.6**	**4.70**	**12.7**	**3.49**	**66.1**	**6.10**	**11.4**	**22.3**
W12×19	**24.7**	**130**	**92.6**	**63.9**	**2.90**	**7.95**	**5.69**	**77.4**	**2.98**	**3.76**	**10.6**
W10×19	**21.6**	**96.2**	**81.0**	**56.4**	**3.09**	**8.88**	**4.24**	**68.8**	**3.35**	**4.29**	**12.0**
W12×16	**20.1**	**103**	**75.4**	**51.3**	**2.73**	**7.44**	**5.11**	**71.3**	**2.26**	**2.82**	**7.93**
W10×17	18.7	81.9	70.1	48.6	2.98	8.38	3.99	65.4	2.80	3.56	10.0
W12×14	**17.4**	**88.5**	**65.2**	**44.7**	**2.66**	**7.17**	**4.55**	**64.3**	**1.90**	**2.36**	**6.69**
W10×15	16.0	68.9	60.0	41.4	2.86	7.93	3.67	62.0	2.30	2.89	8.16
W10×12[††]	**12.6**	**53.8**	**47.3**	**32.7**	**2.87**	**7.45**	**3.18**	**50.6**	**1.74**	**2.18**	**6.19**

[††] Indicates flange is non-compact.

$F_y = 50$ ksi
$\phi_b = 0.90$
$\phi_v = 0.90$

Table 5-4. W-Shapes Maximum Total Factored Uniform Load[a]

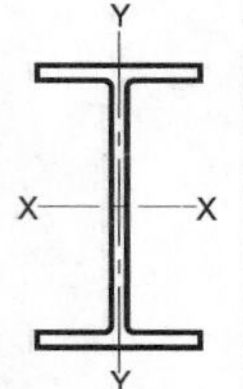

Shape		W44×				W40×						
		335*	290	262	230	593*	503*	431*	397*	372*	362*	324
Span, ft	19					4160	3500	2990		2540		
	20	2420	2050	1850	1640	4140	3480	2940	2700	2520	2460	2170
	21	2310	2030	1810	1570	3940	3310	2800	2570	2400	2340	2090
	22	2210	1940	1730	1500	3760	3160	2670	2450	2290	2240	1990
	23	2110	1850	1660	1430	3600	3030	2560	2350	2190	2140	1900
	24	2030	1780	1590	1380	3450	2900	2450	2250	2100	2050	1830
	25	1940	1700	1520	1320	3310	2780	2350	2160	2020	1970	1750
	26	1870	1640	1470	1270	3180	2680	2260	2080	1940	1890	1680
	27	1800	1580	1410	1220	3070	2580	2180	2000	1870	1820	1620
	28	1740	1520	1360	1180	2960	2490	2100	1930	1800	1760	1560
	29	1680	1470	1310	1140	2860	2400	2030	1860	1740	1700	1510
	30	1620	1420	1270	1100	2760	2320	1960	1800	1680	1640	1460
	32	1520	1330	1190	1030	2590	2180	1840	1690	1580	1540	1370
	34	1430	1250	1120	971	2440	2050	1730	1590	1480	1450	1290
	36	1350	1180	1060	917	2300	1930	1630	1500	1400	1370	1220
	38	1280	1120	1000	868	2180	1830	1550	1420	1330	1290	1150
	40	1220	1070	953	825	2070	1740	1470	1350	1260	1230	1100
	42	1160	1010	907	786	1970	1660	1400	1290	1200	1170	1040
	44	1100	968	866	750	1880	1580	1340	1230	1150	1120	995
	46	1060	926	828	717	1800	1510	1280	1170	1100	1070	952
	48	1010	888	794	688	1730	1450	1230	1130	1050	1030	913
	50	972	852	762	660	1660	1390	1180	1080	1010	984	876
	52	935	819	733	635	1590	1340	1130	1040	969	946	842
	54	900	789	706	611	1530	1290	1090	1000	933	911	811
	56	868	761	680	589	1480	1240	1050	964	900	879	782
	58	838	734	657	569	1430	1200	1010	931	869	848	755
	60	810	710	635	550	1380	1160	980	900	840	820	730
	62	784	687	615	532	1340	1120	948	871	813	794	706
	64	759	666	595	516	1290	1090	919	844	788	769	684
	66	736	645	577	500	1250	1050	891	818	764	745	664
	68	715	626	560	485	1220	1020	865	794	741	724	644
	70	694	609	544	471	1180	994	840	771	720	703	626
	72	675	592	529	458	1150	967	817	750	700	683	608
Beam Properties												
Z_x, in.^{3}		1620	1420	1270	1100	2760	2320	1960	1800	1680	1640	1460
$\phi_b W_c$, kip-ft		48600	42600	38100	33000	82800	69600	58800	54000	50400	49200	43800
$\phi_v V_n$, kips		1210	1020	924	822	2080	1750	1490	1350	1270	1230	1090
BF, kips		79.6	74.0	68.3	61.7	77.3	76.0	73.2	70.7	70.2	69.5	65.6
L_r, ft		35.5	34.0	32.8	31.8	56.5	49.4	44.1	42.1	40.1	39.9	37.6
L_p, ft		12.4	12.3	12.2	12.1	13.4	13.1	12.9	12.9	12.7	12.7	12.6
$\phi_b M_r$, kip-ft		4230	3720	3360	2910	7020	5940	5070	4680	4380	4260	3840
$\phi_b M_p$, kip-ft		6080	5330	4760	4130	10400	8700	7350	6750	6300	6150	5480

*ASTM A6 tensile group 4 or 5 shape. Special requirements may apply per LRFD Specification Section A3.1c.
[a]For beams laterally unsupported, see Table 5-5.
Note: Design strength tabulated above heavy line is limited by design shear strength.

Table 5-4 (cont.). W-Shapes Maximum Total Factored Uniform Load[a]

$F_y = 50$ ksi
$\phi_b = 0.90$
$\phi_v = 0.90$

Shape	W40× 297	277	249	215	199	392*	331*	327	278	264	235
Span, ft											
15							2690				
16						3190	2680	2600	2210	2070	
17						3020	2520	2490	2100	1990	1780
18						2850	2380	2350	1980	1880	1680
19					1360	2700	2260	2230	1880	1780	1590
20	2000				1300	2570	2150	2120	1790	1700	1520
21	1900	1780	1600	1370	1240	2440	2040	2010	1700	1610	1440
22	1810	1700	1530	1310	1190	2330	1950	1920	1620	1540	1380
23	1730	1630	1460	1260	1130	2230	1870	1840	1550	1470	1320
24	1660	1560	1400	1210	1090	2140	1790	1760	1490	1410	1260
25	1600	1500	1340	1160	1040	2050	1720	1690	1430	1360	1210
26	1530	1440	1290	1110	1000	1970	1650	1630	1370	1300	1170
27	1480	1390	1240	1070	966	1900	1590	1570	1320	1260	1120
28	1430	1340	1200	1030	931	1830	1530	1510	1280	1210	1080
29	1380	1290	1160	997	899	1770	1480	1460	1230	1170	1040
30	1330	1250	1120	964	869	1710	1430	1410	1190	1130	1010
32	1250	1170	1050	904	815	1600	1340	1320	1120	1060	947
34	1170	1100	988	851	767	1510	1260	1240	1050	997	891
36	1110	1040	933	803	724	1430	1190	1180	992	942	842
38	1050	987	884	761	686	1350	1130	1110	939	892	797
40	998	938	840	723	652	1280	1070	1060	893	848	758
42	950	893	800	689	621	1220	1020	1010	850	807	721
44	907	852	764	657	593	1170	975	961	811	770	689
46	867	815	730	629	567	1120	933	920	776	737	659
48	831	781	700	603	543	1070	894	881	744	706	631
50	798	750	672	578	521	1030	858	846	714	678	606
52	767	721	646	556	501	987	825	813	687	652	583
54	739	694	622	536	483	950	794	783	661	628	561
56	713	670	600	516	466	916	766	755	638	605	541
58	688	647	579	499	449	884	740	729	616	584	522
60	665	625	560	482	435	855	715	705	595	565	505
62	644	605	542	466	420	827	692	682	576	547	489
64	623	586	525	452	407	802	670	661	558	530	473
66	605	568	509	438	395	777	650	641	541	514	459
68	587	551	494	425	383	754	631	622	525	499	446
70	570	536	480	413	372	733	613	604	510	484	433
72	554	521	467	402	362	713	596	588	496	471	421
Beam Properties											
Z_x, in.^{3}	1330	1250	1120	964	869	1710	1430	1410	1190	1130	1010
$\phi_b W_c$, kip-ft	39900	37500	33600	28900	26100	51300	42900	42300	35700	33900	30300
$\phi_v V_n$, kips	999	890	798	684	679	1590	1340	1300	1110	1040	890
BF, kips	63.1	60.8	56.7	51.2	48.5	83.9	80.9	79.5	75.1	73.2	68.2
L_r, ft	35.9	35.5	34.1	32.8	31.7	34.3	30.5	30.3	27.6	27.0	26.0
L_p, ft	12.5	12.6	12.5	12.5	12.2	9.33	9.08	9.11	8.90	8.90	8.97
$\phi_b M_r$, kip-ft	3510	3300	2980	2580	2310	4320	3630	3600	3060	2910	2630
$\phi_b M_p$, kip-ft	4990	4690	4200	3620	3260	6410	5360	5290	4460	4240	3790

*ASTM A6 tensile group 4 or 5 shape. Special requirements may apply per LRFD Specification Section A3.1c.
[a]For beams laterally unsupported, see Table 5-5.
Note: Design strength tabulated above heavy line is limited by design shear strength.

$F_y = 50$ ksi
$\phi_b = 0.90$
$\phi_v = 0.90$

Table 5-4 (cont.). W-Shapes Maximum Total Factored Uniform Load[a]

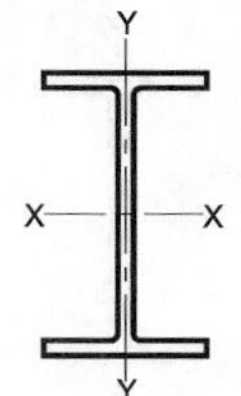

Shape	W40×				W36×						
Span, ft	211	183	167	149	798*	650*	527*	439*	393*	359*	328*
13				1300							
14				1280							
15			1350	1200							
16			1300	1120							
17	1600	1370	1220	1060							
18	1510	1310	1160	997							
19	1430	1240	1090	944	5400	4310					
20	1360	1170	1040	897	5370	4290	3410	2810	2490	2260	2040
21	1290	1120	990	854	5110	4090	3260	2670	2390	2160	1970
22	1240	1070	945	815	4880	3900	3110	2550	2280	2060	1880
23	1180	1020	904	780	4670	3730	2970	2440	2180	1970	1800
24	1130	979	866	748	4480	3580	2850	2340	2090	1890	1730
25	1090	940	832	718	4300	3430	2740	2240	2000	1810	1660
26	1050	903	800	690	4130	3300	2630	2160	1930	1740	1590
27	1010	870	770	664	3980	3180	2530	2080	1860	1680	1530
28	971	839	743	641	3840	3060	2440	2000	1790	1620	1480
29	937	810	717	619	3700	2960	2360	1930	1730	1560	1430
30	906	783	693	598	3580	2860	2280	1870	1670	1510	1380
32	849	734	650	561	3360	2680	2140	1750	1570	1420	1290
34	799	691	611	528	3160	2520	2010	1650	1470	1330	1220
36	755	653	578	498	2980	2380	1900	1560	1390	1260	1150
38	715	618	547	472	2830	2260	1800	1480	1320	1190	1090
40	680	587	520	449	2690	2150	1710	1400	1250	1130	1040
42	647	559	495	427	2560	2040	1630	1340	1190	1080	986
44	618	534	473	408	2440	1950	1550	1280	1140	1030	941
46	591	511	452	390	2330	1870	1490	1220	1090	985	900
48	566	489	433	374	2240	1790	1430	1170	1040	944	863
50	544	470	416	359	2150	1720	1370	1120	1000	906	828
52	523	452	400	345	2070	1650	1320	1080	963	871	796
54	503	435	385	332	1990	1590	1270	1040	928	839	767
56	485	419	371	320	1920	1530	1220	1000	895	809	739
58	469	405	358	309	1850	1480	1180	967	864	781	714
60	453	392	347	299	1790	1430	1140	935	835	755	690
62	438	379	335	289	1730	1380	1100	905	808	731	668
64	425	367	325	280	1680	1340	1070	877	783	708	647
66	412	356	315	272	1630	1300	1040	850	759	686	627
68	400	345	306	264	1580	1260	1010	825	737	666	609
70	388	336	297	256	1530	1230	977	801	716	647	591
72	378	326	289	249	1490	1190	950	779	696	629	575
Beam Properties											
Z_x, in.^{3}	906	783	693	598	3580	2860	2280	1870	1670	1510	1380
$\phi_b W_c$, kip-ft	27200	23500	20800	17900	107000	85800	68400	56100	50100	45300	41400
$\phi_v V_n$, kips	798	684	677	650	2700	2150	1700	1410	1250	1130	1020
BF, kips	64.8	58.7	55.4	51.2	64.4	62.9	61.5	59.4	58.1	56.5	54.8
L_r, ft	24.9	23.9	22.9	21.8	84.6	69.6	58.1	50.1	46.7	43.8	41.8
L_p, ft	8.87	8.83	8.48	8.09	14.9	14.5	14.2	13.9	13.8	13.6	13.6
$\phi_b M_r$, kip-ft	2360	2050	1800	1540	8940	7260	5850	4860	4350	3960	3630
$\phi_b M_p$, kip-ft	3400	2940	2600	2240	13400	10700	8550	7010	6260	5660	5180

*ASTM A6 tensile group 4 or 5 shape. Special requirements may apply per LRFD Specification Section A3.1c.
[a]For beams laterally unsupported, see Table 5-5.
Note: Design strength tabulated above heavy line is limited by design shear strength.

Table 5-4 (cont.). W-Shapes Maximum Total Factored Uniform Load[a]

$F_y = 50$ ksi
$\phi_b = 0.90$
$\phi_v = 0.90$

Shape		W36×										
		300	280	260	245	230	256	232	210	194	182	170
Span, ft	15								1640	1510	1420	1330
	16						1940	1740	1560	1440	1350	1250
	17						1840	1650	1470	1350	1270	1180
	18						1730	1560	1390	1280	1200	1110
	19			1650	1560	1470	1640	1480	1320	1210	1130	1050
	20	1870	1740	1620	1520	1410	1560	1400	1250	1150	1080	1000
	21	1800	1670	1540	1440	1350	1490	1340	1190	1100	1030	954
	22	1720	1600	1470	1380	1290	1420	1280	1140	1050	979	911
	23	1640	1530	1410	1320	1230	1360	1220	1090	1000	937	871
	24	1580	1460	1350	1260	1180	1300	1170	1040	959	898	835
	25	1510	1400	1300	1210	1130	1250	1120	1000	920	862	802
	26	1450	1350	1250	1170	1090	1200	1080	961	885	828	771
	27	1400	1300	1200	1120	1050	1160	1040	926	852	798	742
	28	1350	1250	1160	1080	1010	1110	1000	893	822	769	716
	29	1300	1210	1120	1040	976	1080	968	862	793	743	691
	30	1260	1170	1080	1010	943	1040	936	833	767	718	668
	32	1180	1100	1010	947	884	975	878	781	719	673	626
	34	1110	1030	953	891	832	918	826	735	677	634	589
	36	1050	975	900	842	786	867	780	694	639	598	557
	38	995	924	853	797	744	821	739	658	606	567	527
	40	945	878	810	758	707	780	702	625	575	539	501
	42	900	836	771	721	674	743	669	595	548	513	477
	44	859	798	736	689	643	709	638	568	523	490	455
	46	822	763	704	659	615	678	610	543	500	468	436
	48	788	731	675	631	589	650	585	521	479	449	418
	50	756	702	648	606	566	624	562	500	460	431	401
	52	727	675	623	583	544	600	540	481	443	414	385
	54	700	650	600	561	524	578	520	463	426	399	371
	56	675	627	579	541	505	557	501	446	411	385	358
	58	652	605	559	522	488	538	484	431	397	371	346
	60	630	585	540	505	472	520	468	417	384	359	334
	62	610	566	523	489	456	503	453	403	371	347	323
	64	591	548	506	473	442	488	439	390	360	337	313
	66	573	532	491	459	429	473	425	379	349	326	304
	68	556	516	476	446	416	459	413	368	338	317	295
	70	540	501	463	433	404	446	401	357	329	308	286
	72	525	488	450	421	393	433	390	347	320	299	278
Beam Properties												
Z_x, in.^{3}		1260	1170	1080	1010	943	1040	936	833	767	718	668
$\phi_b W_c$, kip-ft		37800	35100	32400	30300	28300	31200	28100	25000	23000	21500	20000
$\phi_v V_n$, kips		936	872	823	780	737	969	871	822	754	711	665
BF, kips		53.0	51.3	49.5	47.5	45.6	63.0	60.1	57.0	54.4	52.0	49.6
L_r, ft		39.9	38.7	37.4	36.5	35.6	28.7	27.3	26.1	25.3	24.8	24.3
L_p, ft		13.5	13.5	13.4	13.2	13.2	9.36	9.25	9.11	9.04	9.01	8.94
$\phi_b M_r$, kip-ft		3330	3090	2860	2690	2510	2690	2430	2160	1990	1870	1740
$\phi_b M_p$, kip-ft		4730	4390	4050	3790	3540	3900	3510	3120	2880	2690	2510

[a] For beams laterally unsupported, see Table 5-5.

Note: Design strength tabulated above heavy line is limited by design shear strength.

$F_y = 50$ ksi
$\phi_b = 0.90$
$\phi_v = 0.90$

Table 5-4 (cont.). W-Shapes Maximum Total Factored Uniform Load[a]

Shape	W36×			W33×							
	160	150	135	387*	354*	318*	291	263	241	221	201
Span, ft											
13			1150								
14	1260	1210	1090								
15	1250	1160	1020								
16	1170	1090	954								
17	1100	1030	898								1300
18	1040	968	848						1530	1420	1290
19	985	917	804	2450	2230	1980	1800	1620	1480	1350	1220
20	936	872	764	2340	2130	1910	1740	1560	1410	1290	1160
21	891	830	727	2230	2030	1810	1660	1490	1340	1220	1100
22	851	792	694	2130	1940	1730	1580	1420	1280	1170	1050
23	814	758	664	2030	1850	1660	1510	1360	1230	1120	1010
24	780	726	636	1950	1780	1590	1450	1300	1180	1070	966
25	749	697	611	1870	1700	1520	1390	1250	1130	1030	928
26	720	670	587	1800	1640	1470	1340	1200	1080	989	892
27	693	646	566	1730	1580	1410	1290	1160	1040	952	859
28	669	623	545	1670	1520	1360	1240	1110	1010	918	828
29	646	601	527	1610	1470	1310	1200	1080	972	887	800
30	624	581	509	1560	1420	1270	1160	1040	940	857	773
32	585	545	477	1460	1330	1190	1090	975	881	803	725
34	551	513	449	1380	1250	1120	1020	918	829	756	682
36	520	484	424	1300	1180	1060	967	867	783	714	644
38	493	459	402	1230	1120	1000	916	821	742	677	610
40	468	436	382	1170	1070	953	870	780	705	643	580
42	446	415	364	1110	1010	907	829	743	671	612	552
44	425	396	347	1060	968	866	791	709	641	584	527
46	407	379	332	1020	926	828	757	678	613	559	504
48	390	363	318	975	888	794	725	650	588	536	483
50	374	349	305	936	852	762	696	624	564	514	464
52	360	335	294	900	819	733	669	600	542	494	446
54	347	323	283	867	789	706	644	578	522	476	429
56	334	311	273	836	761	680	621	557	504	459	414
58	323	301	263	807	734	657	600	538	486	443	400
60	312	291	255	780	710	635	580	520	470	429	387
62	302	281	246	755	687	615	561	503	455	415	374
64	293	272	239	731	666	595	544	488	441	402	362
66	284	264	231	709	645	577	527	473	427	390	351
68	275	256	225	688	626	560	512	459	415	378	341
70	267	249	218	669	609	544	497	446	403	367	331
72	260	242	212	650	592	529	483	433	392	357	322
Beam Properties											
Z_x, in.^{3}	624	581	509	1560	1420	1270	1160	1040	940	857	773
$\phi_b W_c$, kip-ft	18700	17400	15300	46800	42600	38100	34800	31200	28200	25700	23200
$\phi_v V_n$, kips	632	606	577	1220	1110	988	902	810	766	709	651
BF, kips	47.8	45.6	42.2	52.4	51.1	49.8	48.5	45.9	44.0	42.0	39.6
L_r, ft	23.8	23.3	22.4	47.7	44.6	41.9	39.6	37.8	36.2	35.0	33.8
L_p, ft	8.83	8.72	8.41	13.3	13.2	13.1	13.0	12.9	12.8	12.7	12.6
$\phi_b M_r$, kip-ft	1630	1510	1320	4050	3720	3330	3060	2760	2490	2280	2060
$\phi_b M_p$, kip-ft	2340	2180	1910	5850	5330	4760	4350	3900	3530	3210	2900

*ASTM A6 tensile group 4 or 5 shape. Special requirements may apply per LRFD Specification Section A3.1c.
[a]For beams laterally unsupported, see Table 5-5.
Note: Design strength tabulated above heavy line is limited by design shear strength.

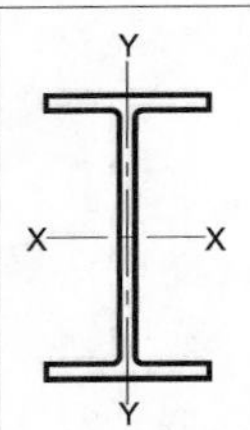

Table 5-4 (cont.). W-Shapes Maximum Total Factored Uniform Load[a]

$F_y = 50$ ksi
$\phi_b = 0.90$
$\phi_v = 0.90$

Shape		W33×					W30×					
		169	152	141	130	118	391*	357*	326*	292*	261	235
Span, ft	12					977						
	13				1040	958						
	14		1150	1090	1000	889						
	15	1220	1120	1030	934	830						
	16	1180	1050	964	876	778						
	17	1110	986	907	824	732	2440		1990		1590	
	18	1050	932	857	778	692	2420	2200	1980	1760	1570	1400
	19	993	883	812	737	655	2290	2080	1880	1670	1490	1340
	20	944	839	771	701	623	2180	1980	1790	1590	1410	1270
	22	858	762	701	637	566	1980	1800	1620	1450	1290	1160
	24	786	699	643	584	519	1810	1650	1490	1330	1180	1060
	26	726	645	593	539	479	1670	1520	1370	1220	1090	977
	28	674	599	551	500	445	1550	1410	1280	1140	1010	908
	30	629	559	514	467	415	1450	1320	1190	1060	943	847
	32	590	524	482	438	389	1360	1240	1120	994	884	794
	34	555	493	454	412	366	1280	1160	1050	935	832	747
	36	524	466	428	389	346	1210	1100	992	883	786	706
	38	497	441	406	369	328	1140	1040	939	837	744	669
	40	472	419	386	350	311	1090	990	893	795	707	635
	42	449	399	367	334	296	1040	943	850	757	674	605
	44	429	381	350	318	283	989	900	811	723	643	578
	46	410	365	335	305	271	946	861	776	691	615	552
	48	393	349	321	292	259	906	825	744	663	589	529
	50	377	335	308	280	249	870	792	714	636	566	508
	52	363	323	297	269	239	837	762	687	612	544	489
	54	349	311	286	259	231	806	733	661	589	524	471
	56	337	299	275	250	222	777	707	638	568	505	454
	58	325	289	266	242	215	750	683	616	548	488	438
	60	315	280	257	234	208	725	660	595	530	472	424
	62	304	270	249	226	201	702	639	576	513	456	410
	64	295	262	241	219	195	680	619	558	497	442	397
	66	286	254	234	212	189	659	600	541	482	429	385
	68	278	247	227	206	183	640	582	525	468	416	374
	70	270	240	220	200	178	621	566	510	454	404	363
	72	262	233	214	195	173	604	550	496	442	393	353
Beam Properties												
Z_x, in.^{3}		629	559	514	467	415	1450	1320	1190	1060	943	847
$\phi_b W_c$, kip-ft		18900	16800	15400	14000	12500	43500	39600	35700	31800	28300	25400
$\phi_v V_n$, kips		611	574	544	518	489	1220	1100	997	881	793	701
BF, kips		45.4	42.4	40.4	38.0	35.4	43.1	42.8	41.3	40.2	39.3	37.8
L_r, ft		24.5	23.7	23.0	22.5	21.7	52.1	48.6	45.2	42.1	39.1	37.1
L_p, ft		8.83	8.72	8.58	8.44	8.19	13.0	12.9	12.7	12.6	12.5	12.4
$\phi_b M_r$, kip-ft		1650	1460	1340	1220	1080	3750	3420	3120	2790	2490	2240
$\phi_b M_p$, kip-ft		2360	2100	1930	1750	1560	5440	4950	4460	3980	3540	3180

*ASTM A6 tensile group 4 or 5 shape. Special requirements may apply per LRFD Specification Section A3.1c.
[a] For beams laterally unsupported, see Table 5-5.
Note: Design strength tabulated above heavy line is limited by design shear strength.

$F_y = 50$ ksi
$\phi_b = 0.90$
$\phi_v = 0.90$

Table 5-4 (cont.). W-Shapes Maximum Total Factored Uniform Load[a]

Shape	W30× 211	191	173	148	132	124	116	108	99	90
Span, ft										
11								877	834	749
12						954	915	865	780	708
13				1080	1010	942	872	798	720	653
14				1070	936	874	810	741	669	606
15				1000	874	816	756	692	624	566
16			1080	938	819	765	709	649	585	531
17	1290	1180	1070	882	771	720	667	611	551	499
18	1250	1130	1010	833	728	680	630	577	520	472
19	1190	1070	958	789	690	644	597	546	493	447
20	1130	1010	911	750	656	612	567	519	468	425
22	1020	920	828	682	596	556	515	472	425	386
24	939	844	759	625	546	510	473	433	390	354
26	867	779	700	577	504	471	436	399	360	327
28	805	723	650	536	468	437	405	371	334	303
30	751	675	607	500	437	408	378	346	312	283
32	704	633	569	469	410	383	354	324	293	265
34	663	596	536	441	386	360	334	305	275	250
36	626	563	506	417	364	340	315	288	260	236
38	593	533	479	395	345	322	298	273	246	223
40	563	506	455	375	328	306	284	260	234	212
42	536	482	434	357	312	291	270	247	223	202
44	512	460	414	341	298	278	258	236	213	193
46	490	440	396	326	285	266	247	226	203	185
48	469	422	379	313	273	255	236	216	195	177
50	451	405	364	300	262	245	227	208	187	170
52	433	389	350	288	252	235	218	200	180	163
54	417	375	337	278	243	227	210	192	173	157
56	402	362	325	268	234	219	203	185	167	152
58	388	349	314	259	226	211	196	179	161	146
60	376	338	304	250	219	204	189	173	156	142
62	363	327	294	242	211	197	183	167	151	137
64	352	316	285	234	205	191	177	162	146	133
66	341	307	276	227	199	185	172	157	142	129
68	331	298	268	221	193	180	167	153	138	125
70	322	289	260	214	187	175	162	148	134	121
72	313	281	253	208	182	170	158	144	130	118
Beam Properties										
Z_x, in.^{3}	751	675	607	500	437	408	378	346	312	283
$\phi_b W_c$, kip-ft	22500	20300	18200	15000	13100	12200	11300	10400	9360	8490
$\phi_v V_n$, kips	647	589	538	539	503	477	458	439	417	374
BF, kips	35.8	33.9	32.0	38.5	35.7	34.3	32.8	31.3	29.2	27.1
L_r, ft	35.2	33.8	32.5	22.8	21.9	21.4	20.9	20.4	19.8	19.4
L_p, ft	12.3	12.2	12.1	8.05	7.95	7.88	7.74	7.59	7.42	7.38
$\phi_b M_r$, kip-ft	2000	1800	1620	1310	1140	1070	987	897	807	735
$\phi_b M_p$, kip-ft	2820	2530	2280	1880	1640	1530	1420	1300	1170	1060

[a]For beams laterally unsupported, see Table 5-5.
Note: Design strength tabulated above heavy line is limited by design shear strength.

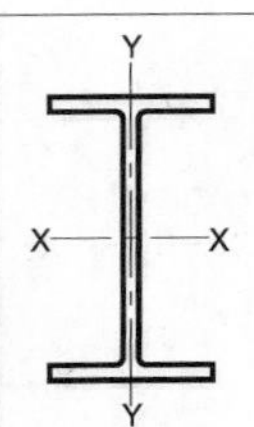

Table 5-4 (cont.). W-Shapes Maximum Total Factored Uniform Load[a]

$F_y = 50$ ksi
$\phi_b = 0.90$
$\phi_v = 0.90$

Shape		W27×										
		539*	368*	336*	307*	281*	258	235	217	194	178	161
Span, ft	15										1090	984
	16	3460	2270	2040	1850	1680	1530	1410	1270	1140	1070	966
	17	3340	2190	1990	1820	1650	1500	1360	1250	1110	1010	909
	18	3150	2070	1880	1720	1560	1420	1290	1190	1050	950	858
	19	2980	1960	1780	1630	1480	1350	1220	1120	996	900	813
	20	2840	1860	1700	1550	1400	1280	1160	1070	947	855	773
	21	2700	1770	1610	1470	1340	1220	1100	1020	901	814	736
	22	2580	1690	1540	1400	1280	1160	1050	970	860	777	702
	23	2470	1620	1470	1340	1220	1110	1010	927	823	743	672
	24	2360	1550	1410	1290	1170	1070	965	889	789	713	644
	25	2270	1490	1360	1240	1120	1020	926	853	757	684	618
	26	2180	1430	1300	1190	1080	983	891	820	728	658	594
	27	2100	1380	1260	1140	1040	947	858	790	701	633	572
	28	2030	1330	1210	1100	1000	913	827	762	676	611	552
	29	1960	1280	1170	1070	968	881	799	736	653	590	533
	30	1890	1240	1130	1030	936	852	772	711	631	570	515
	32	1770	1160	1060	966	878	799	724	667	592	534	483
	34	1670	1090	997	909	826	752	681	627	557	503	454
	36	1580	1030	942	858	780	710	643	593	526	475	429
	38	1490	979	892	813	739	673	609	561	498	450	407
	40	1420	930	848	773	702	639	579	533	473	428	386
	42	1350	886	807	736	669	609	551	508	451	407	368
	44	1290	845	770	702	638	581	526	485	430	389	351
	46	1230	809	737	672	610	556	503	464	412	372	336
	48	1180	775	706	644	585	533	483	444	394	356	322
	50	1130	744	678	618	562	511	463	427	379	342	309
	52	1090	715	652	594	540	492	445	410	364	329	297
	54	1050	689	628	572	520	473	429	395	351	317	286
	56	1010	664	605	552	501	456	414	381	338	305	276
	58	978	641	584	533	484	441	399	368	326	295	266
	60	945	620	565	515	468	426	386	356	316	285	258
	62	915	600	547	498	453	412	374	344	305	276	249
	64	886	581	530	483	439	399	362	333	296	267	241
	66	859	564	514	468	425	387	351	323	287	259	234
	68	834	547	499	454	413	376	341	314	278	251	227
	70	810	531	484	441	401	365	331	305	270		
	72	788	517	471	429	390	355					
Beam Properties												
Z_x, in.^{3}		1890	1240	1130	1030	936	852	772	711	631	570	515
$\phi_b W_c$, kip-ft		56700	37200	33900	30900	28100	25600	23200	21300	18900	17100	15500
$\phi_v V_n$, kips		1730	1130	1020	927	839	767	705	636	569	544	492
BF, kips		36.5	34.6	34.5	34.7	33.5	32.9	32.2	31.3	29.9	28.9	27.5
L_r, ft		78.0	54.7	50.4	46.7	43.8	41.0	38.6	36.8	34.6	33.0	31.7
L_p, ft		12.9	12.3	12.2	12.0	12.0	11.9	11.8	11.7	11.6	11.5	11.4
$\phi_b M_r$, kip-ft		4710	3180	2920	2660	2440	2240	2030	1880	1680	1520	1370
$\phi_b M_p$, kip-ft		7090	4650	4240	3860	3510	3200	2900	2670	2370	2140	1930

*ASTM A6 tensile group 4 or 5 shape. Special requirements may apply per LRFD Specification Section A3.1c.
[a] For beams laterally unsupported, see Table 5-5.
Note: Design strength tabulated above heavy line is limited by design shear strength.

$F_y = 50$ ksi
$\phi_b = 0.90$
$\phi_v = 0.90$

Table 5-4 (cont.). W-Shapes Maximum Total Factored Uniform Load[a]

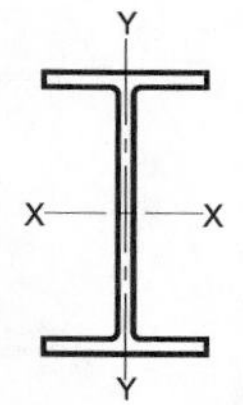

Shape	W27×						W24×				
Span, ft	146	129	114	102	94	84	370*	335*	306*	279*	250*
11					712	663					
12			840	754	695	610					
13		909	792	704	642	563					
14		846	735	654	596	523	2300	2050			
15	895	790	686	610	556	488	2260	2040	1840	1670	1480
16	870	741	643	572	521	458	2120	1910	1730	1570	1400
17	819	697	605	538	491	431	1990	1800	1630	1470	1310
18	773	658	572	508	463	407	1880	1700	1540	1390	1240
19	733	624	542	482	439	385	1780	1610	1460	1320	1170
20	696	593	515	458	417	366	1700	1530	1380	1250	1120
21	663	564	490	436	397	349	1610	1460	1320	1190	1060
22	633	539	468	416	379	333	1540	1390	1260	1140	1010
23	605	515	447	398	363	318	1470	1330	1200	1090	970
24	580	494	429	381	348	305	1410	1280	1150	1040	930
25	557	474	412	366	334	293	1360	1220	1110	1000	893
26	535	456	396	352	321	282	1300	1180	1060	963	858
27	516	439	381	339	309	271	1260	1130	1020	928	827
28	497	423	368	327	298	261	1210	1090	988	895	797
29	480	409	355	316	288	252	1170	1060	954	864	770
30	464	395	343	305	278	244	1130	1020	922	835	744
32	435	370	322	286	261	229	1060	956	864	783	698
34	409	349	303	269	245	215	997	900	814	737	656
36	387	329	286	254	232	203	942	850	768	696	620
38	366	312	271	241	219	193	892	805	728	659	587
40	348	296	257	229	209	183	848	765	692	626	558
42	331	282	245	218	199	174	807	729	659	596	531
44	316	269	234	208	190	166	770	695	629	569	507
46	303	258	224	199	181	159	737	665	601	545	485
48	290	247	214	191	174	153	706	638	576	522	465
50	278	237	206	183	167	146	678	612	553	501	446
52	268	228	198	176	160	141	652	588	532	482	429
54	258	219	191	169	154	136	628	567	512	464	413
56	249	212	184	163	149	131	605	546	494	447	399
58	240	204	177	158	144	126	584	528	477	432	385
60	232	198	172	153	139	122	565	510	461	418	372
62	225	191	166	148	135	118	547	494	446	404	360
64	218	185	161	143	130	114	530	478	432	391	349
66	211	180	156	139	126	111	514	464	419	380	
68	205	174	151				499	450			
Beam Properties											
Z_x, in.^{3}	464	395	343	305	278	244	1130	1020	922	835	744
$\phi_b W_c$, kip-ft	13900	11900	10300	9150	8340	7320	33900	30600	27700	25100	22300
$\phi_v V_n$, kips	448	455	420	377	356	332	1150	1020	922	836	739
BF, kips	25.9	31.0	28.8	26.7	25.2	23.0	27.7	27.9	27.5	27.1	26.7
L_r, ft	30.5	22.2	21.2	20.4	19.9	19.3	61.0	55.5	51.0	47.3	43.2
L_p, ft	11.3	7.81	7.70	7.59	7.49	7.31	11.6	11.4	11.3	11.2	11.1
$\phi_b M_r$, kip-ft	1240	1040	897	801	729	639	2870	2590	2370	2150	1930
$\phi_b M_p$, kip-ft	1740	1480	1290	1140	1040	915	4240	3830	3460	3130	2790

*ASTM A6 tensile group 4 or 5 shape. Special requirements may apply per LRFD Specification Section A3.1c.
[a] For beams laterally unsupported, see Table 5-5.
Note: Design strength tabulated above heavy line is limited by design shear strength.

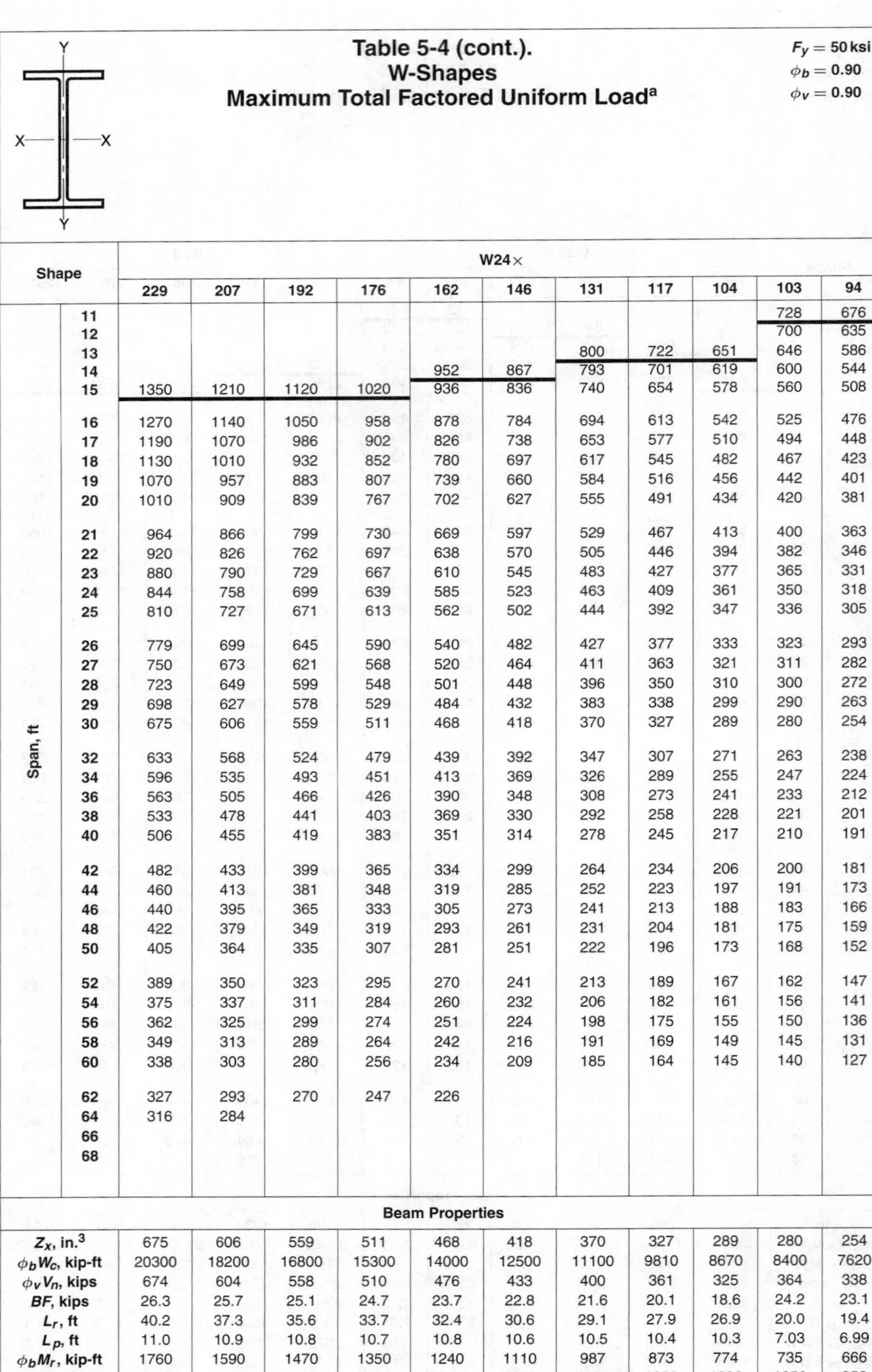

Table 5-4 (cont.). W-Shapes Maximum Total Factored Uniform Load[a]

$F_y = 50$ ksi
$\phi_b = 0.90$
$\phi_v = 0.90$

Shape	W24× 229	207	192	176	162	146	131	117	104	103	94
Span, ft											
11										728	676
12										700	635
13							800	722	651	646	586
14					952	867	793	701	619	600	544
15	1350	1210	1120	1020	936	836	740	654	578	560	508
16	1270	1140	1050	958	878	784	694	613	542	525	476
17	1190	1070	986	902	826	738	653	577	510	494	448
18	1130	1010	932	852	780	697	617	545	482	467	423
19	1070	957	883	807	739	660	584	516	456	442	401
20	1010	909	839	767	702	627	555	491	434	420	381
21	964	866	799	730	669	597	529	467	413	400	363
22	920	826	762	697	638	570	505	446	394	382	346
23	880	790	729	667	610	545	483	427	377	365	331
24	844	758	699	639	585	523	463	409	361	350	318
25	810	727	671	613	562	502	444	392	347	336	305
26	779	699	645	590	540	482	427	377	333	323	293
27	750	673	621	568	520	464	411	363	321	311	282
28	723	649	599	548	501	448	396	350	310	300	272
29	698	627	578	529	484	432	383	338	299	290	263
30	675	606	559	511	468	418	370	327	289	280	254
32	633	568	524	479	439	392	347	307	271	263	238
34	596	535	493	451	413	369	326	289	255	247	224
36	563	505	466	426	390	348	308	273	241	233	212
38	533	478	441	403	369	330	292	258	228	221	201
40	506	455	419	383	351	314	278	245	217	210	191
42	482	433	399	365	334	299	264	234	206	200	181
44	460	413	381	348	319	285	252	223	197	191	173
46	440	395	365	333	305	273	241	213	188	183	166
48	422	379	349	319	293	261	231	204	181	175	159
50	405	364	335	307	281	251	222	196	173	168	152
52	389	350	323	295	270	241	213	189	167	162	147
54	375	337	311	284	260	232	206	182	161	156	141
56	362	325	299	274	251	224	198	175	155	150	136
58	349	313	289	264	242	216	191	169	149	145	131
60	338	303	280	256	234	209	185	164	145	140	127
62	327	293	270	247	226						
64	316	284									
66											
68											
Beam Properties											
Z_x, in.^{3}	675	606	559	511	468	418	370	327	289	280	254
$\phi_b W_c$, kip-ft	20300	18200	16800	15300	14000	12500	11100	9810	8670	8400	7620
$\phi_v V_n$, kips	674	604	558	510	476	433	400	361	325	364	338
BF, kips	26.3	25.7	25.1	24.7	23.7	22.8	21.6	20.1	18.6	24.2	23.1
L_r, ft	40.2	37.3	35.6	33.7	32.4	30.6	29.1	27.9	26.9	20.0	19.4
L_p, ft	11.0	10.9	10.8	10.7	10.8	10.6	10.5	10.4	10.3	7.03	6.99
$\phi_b M_r$, kip-ft	1760	1590	1470	1350	1240	1110	987	873	774	735	666
$\phi_b M_p$, kip-ft	2530	2270	2100	1920	1760	1570	1390	1230	1080	1050	953

[a]For beams laterally unsupported, see Table 5-5.

Note: Design strength tabulated above heavy line is limited by design shear strength.

$F_y = 50$ ksi
$\phi_b = 0.90$
$\phi_v = 0.90$

Table 5-4 (cont.).
W-Shapes
Maximum Total Factored Uniform Load[a]

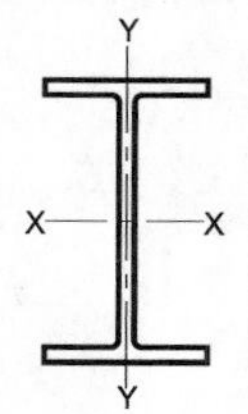

Shape	W24×					W21×					
	84	76	68	62	55	201	182	166	147	132	122
Span, ft											
8				550	503						
9				513	450						
10	612	568	531	462	405						
11	611	545	483	420	368						
12	560	500	443	385	338						
13	517	462	408	355	312				859	765	703
14	480	429	379	330	289	1130	1020	911	801	716	660
15	448	400	354	308	270	1060	954	864	748	668	616
16	420	375	332	289	253	994	894	810	701	626	578
17	395	353	312	272	238	935	842	762	660	589	544
18	373	333	295	257	225	883	795	720	623	557	513
19	354	316	279	243	213	837	753	682	591	527	486
20	336	300	266	231	203	795	716	648	561	501	462
21	320	286	253	220	193	757	681	617	534	477	440
22	305	273	241	210	184	723	650	589	510	455	420
23	292	261	231	201	176	691	622	563	488	436	402
24	280	250	221	193	169	663	596	540	468	418	385
25	269	240	212	185	162	636	572	518	449	401	370
26	258	231	204	178	156	612	550	498	432	385	355
27	249	222	197	171	150	589	530	480	416	371	342
28	240	214	190	165	145	568	511	463	401	358	330
29	232	207	183	159	140	548	493	447	387	346	319
30	224	200	177	154	135	530	477	432	374	334	308
32	210	188	166	144	127	497	447	405	351	313	289
34	198	176	156	136	119	468	421	381	330	295	272
36	187	167	148	128	113	442	398	360	312	278	257
38	177	158	140	122	107	418	377	341	295	264	243
40	168	150	133	116	101	398	358	324	281	251	231
42	160	143	126	110	96.4	379	341	309	267	239	220
44	153	136	121	105	92.0	361	325	295	255	228	210
46	146	130	115	100	88.0	346	311	282	244	218	201
48	140	125	111	96.3	84.4	331	298	270	234	209	193
50	134	120	106	92.4	81.0	318	286	259	224	200	185
52	129	115	102	88.8	77.9	306	275	249	216	193	178
54	124	111	98.3	85.6	75.0	294	265	240	208	186	171
56	120	107	94.8	82.5	72.3	284	256	231			
58	116	103	91.6	79.7	69.8						
60	112										
Beam Properties											
Z_x, in.^{3}	224	200	177	154	135	530	477	432	374	334	308
$\phi_b W_c$, kip-ft	6720	6000	5310	4620	4050	15900	14300	13000	11200	10000	9240
$\phi_v V_n$, kips	306	284	266	275	252	565	509	456	430	383	352
BF, kips	21.5	19.8	18.6	21.6	19.8	19.9	19.5	19.0	18.5	17.7	17.2
L_r, ft	18.6	18.0	17.4	13.3	12.9	41.0	38.1	35.7	32.8	31.0	29.7
L_p, ft	6.89	6.78	6.61	4.84	4.73	10.7	10.6	10.5	10.4	10.3	10.3
$\phi_b M_r$, kip-ft	588	528	462	396	345	1380	1250	1140	990	888	822
$\phi_b M_p$, kip-ft	840	750	664	578	506	1990	1790	1620	1400	1250	1160

[a]For beams laterally unsupported, see Table 5-5.
Note: Design strength tabulated above heavy line is limited by design shear strength.

Table 5-4 (cont.). W-Shapes Maximum Total Factored Uniform Load[a]

$F_y = 50$ ksi
$\phi_b = 0.90$
$\phi_v = 0.90$

Shape	W21×											
Span, ft	111	101	93	83	73	68	62	55	48††	57	50	44
7											427	391
8								421	389	461	416	359
9			677	595	521	490	454	420	354	430	370	319
10			663	588	516	480	432	378	319	387	333	287
11			603	535	469	436	393	344	290	352	303	261
12			553	490	430	400	360	315	266	323	278	240
13	639	578	510	452	397	369	332	291	245	298	256	221
14	598	542	474	420	369	343	309	270	228	276	238	205
15	558	506	442	392	344	320	288	252	212	258	222	192
16	523	474	414	368	323	300	270	236	199	242	208	180
17	492	446	390	346	304	282	254	222	187	228	196	169
18	465	422	368	327	287	267	240	210	177	215	185	160
19	441	399	349	309	272	253	227	199	168	204	175	151
20	419	380	332	294	258	240	216	189	159	194	167	144
21	399	361	316	280	246	229	206	180	152	184	159	137
22	380	345	301	267	235	218	196	172	145	176	151	131
23	364	330	288	256	224	209	188	164	139	168	145	125
24	349	316	276	245	215	200	180	158	133	161	139	120
25	335	304	265	235	206	192	173	151	127	155	133	115
26	322	292	255	226	198	185	166	145	123	149	128	111
27	310	281	246	218	191	178	160	140	118	143	123	106
28	299	271	237	210	184	171	154	135	114	138	119	103
29	289	262	229	203	178	166	149	130	110	133	115	99.1
30	279	253	221	196	172	160	144	126	106	129	111	95.8
32	262	237	207	184	161	150	135	118	99.6	121	104	89.8
34	246	223	195	173	152	141	127	111	93.7	114	97.9	84.5
36	233	211	184	163	143	133	120	105	88.5	108	92.5	79.8
38	220	200	174	155	136	126	114	99.5	83.9	102	87.6	75.6
40	209	190	166	147	129	120	108	94.5	79.7	96.8	83.3	71.9
42	199	181	158	140	123	114	103	90.0	75.9	92.1	79.3	68.4
44	190	173	151	134	117	109	98.2	85.9	72.4	88.0	75.7	65.3
46	182	165	144	128	112	104	93.9	82.2	69.3	84.1	72.4	62.5
48	174	158	138	123	108	100	90.0	78.8	66.4	80.6	69.4	59.9
50	167	152	133	118	103	96.0	86.4	75.6	63.7	77.4	66.6	57.5
52	161	146	128	113	99.2	92.3	83.1	72.7		74.4	64.0	
54			123									
Beam Properties												
Z_x, in.^{3}	279	253	221	196	172	160	144	126	107	129	111	95.8
$\phi_b W_c$, kip-ft	8370	7590	6630	5880	5160	4800	4320	3780	3190	3870	3330	2870
$\phi_v V_n$, kips	319	289	338	298	260	245	227	211	195	231	213	196
BF, kips	16.3	15.4	19.7	18.5	17.1	16.5	15.2	14.3	13.2	17.8	16.5	15.0
L_r, ft	28.6	27.5	19.3	18.5	17.6	17.3	16.7	16.1	15.4	13.2	12.5	12.0
L_p, ft	10.2	10.2	6.50	6.46	6.39	6.36	6.25	6.11	6.09	4.77	4.59	4.45
$\phi_b M_r$, kip-ft	747	681	576	513	453	420	381	330	279	333	285	246
$\phi_b M_p$, kip-ft	1050	949	829	735	645	600	540	473	401	484	416	359

[a] For beams laterally unsupported, see Table 5-5.
†† Indicates flange is noncompact.
Note: Design strength tabulated above heavy line is limited by design shear strength.

$F_y = 50$ ksi
$\phi_b = 0.90$
$\phi_v = 0.90$

Table 5-4 (cont.).
W-Shapes
Maximum Total Factored Uniform Load[a]

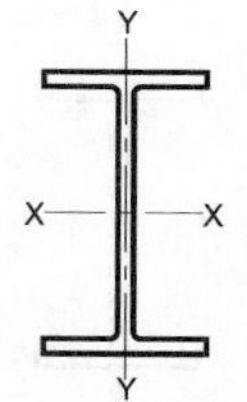

Shape		W18×										
		175	158	143	130	119	106	97	86	76	71	65
Span, ft	8										495	447
	9										487	443
	10										438	399
	11					672	596	537	477	418	398	363
	12	961	862	769	698	655	575	528	465	408	365	333
	13	918	822	743	669	605	531	487	429	376	337	307
	14	853	763	690	621	561	493	452	399	349	313	285
	15	796	712	644	580	524	460	422	372	326	292	266
	16	746	668	604	544	491	431	396	349	306	274	249
	17	702	628	568	512	462	406	372	328	288	258	235
	18	663	593	537	483	437	383	352	310	272	243	222
	19	628	562	508	458	414	363	333	294	257	231	210
	20	597	534	483	435	393	345	317	279	245	219	200
	21	569	509	460	414	374	329	301	266	233	209	190
	22	543	485	439	395	357	314	288	254	222	199	181
	23	519	464	420	378	342	300	275	243	213	190	173
	24	498	445	403	363	328	288	264	233	204	183	166
	25	478	427	386	348	314	276	253	223	196	175	160
	26	459	411	372	335	302	265	243	215	188	168	153
	27	442	396	358	322	291	256	234	207	181	162	148
	28	426	381	345	311	281	246	226	199	175	156	143
	29	412	368	333	300	271	238	218	192	169	151	138
	30	398	356	322	290	262	230	211	186	163	146	133
	31	385	345	312	281	254	223	204	180	158	141	129
	32	373	334	302	272	246	216	198	174	153	137	125
	33	362	324	293	264	238	209	192	169	148	133	121
	34	351	314	284	256	231	203	186	164	144	129	117
	35	341	305	276	249	225	197	181	159	140	125	114
	36	332	297	268	242	218	192	176	155	136	122	111
	37	323	289	261	235	212	186	171	151	132	118	108
	38	314	281	254	229	207	182	167	147	129	115	105
	39	306	274	248	223	202	177	162	143	125	112	102
	40	299	267	242	218	197	173	158	140	122	110	99.8
	42	284	254	230	207	187	164	151	133	116	104	95.0
	44	271	243	220	198	179	157	144	127	111	99.5	90.7
	46	260	232	210	189	171	150	138	121		95.2	86.7
Beam Properties												
Z_x, in.^{3}		398	356	322	290	262	230	211	186	163	146	133
$\phi_b W_c$, kip-ft		11900	10700	9660	8700	7860	6900	6330	5580	4890	4380	3990
$\phi_v V_n$, kips		481	431	384	349	336	298	269	238	209	247	224
BF, kips		14.6	14.3	14.1	13.7	13.5	13.0	12.6	11.9	11.1	14.1	13.3
L_r, ft		41.4	38.0	35.3	32.8	30.9	28.7	27.5	26.0	24.8	17.8	17.1
L_p, ft		9.75	9.68	9.61	9.54	9.50	9.40	9.36	9.29	9.22	6.00	5.97
$\phi_b M_r$, kip-ft		1030	930	846	768	693	612	564	498	438	381	351
$\phi_b M_p$, kip-ft		1490	1340	1210	1090	983	863	791	698	611	548	499

[a] For beams laterally unsupported, see Table 5-5.
Note: Design strength tabulated above heavy line is limited by design shear strength.

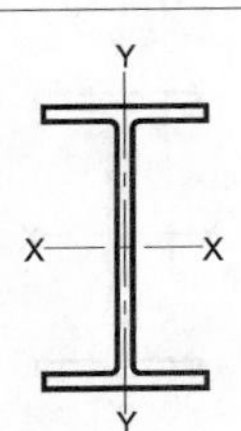

Table 5-4 (cont.). W-Shapes Maximum Total Factored Uniform Load[a]

$F_y = 50$ ksi
$\phi_b = 0.90$
$\phi_v = 0.90$

Shape	W18×						W16×				
Span, ft	60	55	50	46	40	35	100	89	77	67	57
6						287					
7				352	304	285					
8		381	345	340	294	249					381
9	408	373	337	302	261	222					350
10	369	336	303	272	235	200					315
11	335	305	275	247	214	181	537	476	405	348	286
12	308	280	253	227	196	166	498	440	378	328	263
13	284	258	233	209	181	153	459	406	348	302	242
14	264	240	216	194	168	143	426	377	324	281	225
15	246	224	202	181	157	133	398	352	302	262	210
16	231	210	189	170	147	125	373	330	283	246	197
17	217	198	178	160	138	117	351	311	266	231	185
18	205	187	168	151	131	111	332	293	252	218	175
19	194	177	159	143	124	105	314	278	238	207	166
20	185	168	152	136	118	99.8	299	264	227	197	158
21	176	160	144	130	112	95.0	284	251	216	187	150
22	168	153	138	124	107	90.7	271	240	206	179	143
23	160	146	132	118	102	86.7	260	230	197	171	137
24	154	140	126	113	98.0	83.1	249	220	189	164	131
25	148	134	121	109	94.1	79.8	239	211	181	157	126
26	142	129	117	105	90.5	76.7	230	203	174	151	121
27	137	124	112	101	87.1	73.9	221	196	168	146	117
28	132	120	108	97.2	84.0	71.3	213	189	162	140	113
29	127	116	104	93.8	81.1	68.8	206	182	156	136	109
30	123	112	101	90.7	78.4	66.5	199	176	151	131	105
31	119	108	97.7	87.8	75.9	64.4	193	170	146	127	102
32	115	105	94.7	85.0	73.5	62.3	187	165	142	123	98.4
33	112	102	91.8	82.5	71.3	60.5	181	160	137	119	95.5
34	109	98.8	89.1	80.0	69.2	58.7	176	155	133	116	92.6
35	105	96.0	86.6	77.7	67.2	57.0	171	151	129	112	90.0
36	103	93.3	84.2	75.6	65.3	55.4	166	147	126	109	87.5
37	99.7	90.8	81.9	73.5	63.6	53.9	161	143	122	106	85.1
38	97.1	88.4	79.7	71.6	61.9	52.5	157	139	119	103	82.9
39	94.6	86.2	77.7	69.8	60.3	51.2	153	135	116	101	80.8
40	92.3	84.0	75.8	68.0	58.8	49.9	149	132	113	98.3	78.8
42	87.9	80.0	72.1	64.8	56.0	47.5	142	126			
44	83.9	76.4	68.9	61.8	53.5	45.3					
46											
Beam Properties											
Z_x, in.^{3}	123	112	101	90.7	78.4	66.5	199	176	151	131	105
$\phi_b W_c$, kip-ft	3690	3360	3030	2720	2350	2000	5970	5280	4530	3930	3150
$\phi_v V_n$, kips	204	191	173	176	152	143	269	238	203	174	190
BF, kips	12.9	12.2	11.5	12.9	11.7	10.7	10.6	10.3	9.68	9.04	10.7
L_r, ft	16.6	16.1	15.6	12.6	12.0	11.5	29.5	27.5	25.3	23.8	16.6
L_p, ft	5.93	5.90	5.83	4.56	4.49	4.31	8.87	8.80	8.69	8.65	5.65
$\phi_b M_r$, kip-ft	324	295	267	236	205	173	528	468	405	354	277
$\phi_b M_p$, kip-ft	461	420	379	340	294	249	746	660	566	491	394

[a] For beams laterally unsupported, see Table 5-5.

Note: Design strength tabulated above heavy line is limited by design shear strength.

$F_y = 50$ ksi
$\phi_b = 0.90$
$\phi_v = 0.90$

Table 5-4 (cont.). W-Shapes Maximum Total Factored Uniform Load[a]

Shape	W16×						W14×				
Span, ft	50	45	40	36	31	26	808*	730*	665*	605*	550*
6					236	212					
7				253	231	189					
8	334	300	264	240	203	166					
9	307	274	243	213	180	147					
10	276	247	219	192	162	133					
11	251	224	199	175	147	121	4600				
12	230	206	183	160	135	111	4580				
13	212	190	168	148	125	102	4220	3710	3300	2930	2600
14	197	176	156	137	116	94.7	3920	3560	3170	2830	2530
15	184	165	146	128	108	88.4	3660	3320	2960	2640	2360
16	173	154	137	120	101	82.9	3430	3110	2780	2480	2210
17	162	145	129	113	95.3	78.0	3230	2930	2610	2330	2080
18	153	137	122	107	90.0	73.7	3050	2770	2470	2200	1970
19	145	130	115	101	85.3	69.8	2890	2620	2340	2080	1860
20	138	123	110	96.0	81.0	66.3	2750	2490	2220	1980	1770
21	131	118	104	91.4	77.1	63.1	2610	2370	2110	1890	1690
22	125	112	99.5	87.3	73.6	60.3	2500	2260	2020	1800	1610
23	120	107	95.2	83.5	70.4	57.7	2390	2170	1930	1720	1540
24	115	103	91.3	80.0	67.5	55.3	2290	2080	1850	1650	1480
25	110	98.8	87.6	76.8	64.8	53.0	2200	1990	1780	1580	1420
26	106	95.0	84.2	73.8	62.3	51.0	2110	1920	1710	1520	1360
27	102	91.4	81.1	71.1	60.0	49.1	2030	1840	1640	1470	1310
28	98.6	88.2	78.2	68.6	57.9	47.4	1960	1780	1590	1410	1260
29	95.2	85.1	75.5	66.2	55.9	45.7	1890	1720	1530	1370	1220
30	92.0	82.3	73.0	64.0	54.0	44.2	1830	1660	1480	1320	1180
31	89.0	79.6	70.6	61.9	52.3	42.8	1770	1610	1430	1280	1140
32	86.3	77.2	68.4	60.0	50.6	41.4	1720	1560	1390	1240	1110
33	83.6	74.8	66.4	58.2	49.1	40.2	1660	1510	1350	1200	1070
34	81.2	72.6	64.4	56.5	47.6	39.0	1610	1460	1310	1160	1040
35	78.9	70.5	62.6	54.9	46.3	37.9	1570	1420	1270	1130	1010
36	76.7	68.6	60.8	53.3	45.0	36.8	1530	1380	1230	1100	983
37	74.6	66.7	59.2	51.9	43.8	35.8	1480	1350	1200	1070	957
38	72.6	65.0	57.6	50.5	42.6	34.9	1440	1310	1170	1040	932
39	70.8	63.3	56.2	49.2	41.5	34.0	1410	1280	1140	1020	908
40	69.0	61.7	54.8				1370	1250	1110	990	885
42							1310	1190	1060	943	843
44							1250	1130	1010	900	805
46							1190	1080	965	861	770
Beam Properties											
Z_x, in.^{3}	92.0	82.3	73.0	64.0	54.0	44.2	1830	1660	1480	1320	1180
$\phi_b W_c$, kip-ft	2760	2470	2190	1920	1620	1330	54900	49800	44400	39600	35400
$\phi_v V_n$, kips	167	150	132	127	118	106	2300	1860	1650	1470	1300
BF, kips	10.1	9.45	8.71	8.11	8.86	7.83	10.6	10.6	10.2	9.79	9.51
L_r, ft	15.7	15.1	14.7	14.1	11.0	10.4	268	241	222	203	187
L_p, ft	5.62	5.55	5.55	5.37	4.13	3.96	17.0	16.6	16.3	16.1	15.9
$\phi_b M_r$, kip-ft	243	218	194	170	142	115	4200	3840	3450	3120	2790
$\phi_b M_p$, kip-ft	345	309	274	240	203	166	6860	6230	5550	4950	4430

*ASTM A6 tensile group 4 or 5 shape. Special requirements may apply per LRFD Specification Section A3.1c.
[a] For beams laterally unsupported, see Table 5-5.
Note: Design strength tabulated above heavy line is limited by design shear strength.

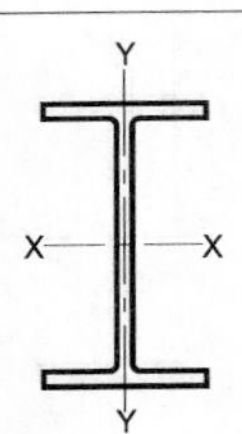

Table 5-4 (cont.). W-Shapes Maximum Total Factored Uniform Load[a]

$F_y = 50$ ksi
$\phi_b = 0.90$
$\phi_v = 0.90$

Shape		W14×										
		500*	455*	426*	398*	370*	342*	311*	283*	257*	233*	211
Span, ft	13	2320	2070	1900	1750	1600	1460	1300		1050		
	14	2250	2010	1860	1720	1580	1440	1290	1160	1040	924	831
	15	2100	1870	1740	1600	1470	1340	1210	1080	974	872	780
	16	1970	1760	1630	1500	1380	1260	1130	1020	913	818	731
	17	1850	1650	1530	1410	1300	1190	1060	956	859	769	688
	18	1750	1560	1450	1340	1230	1120	1010	903	812	727	650
	19	1660	1480	1370	1260	1160	1060	952	856	769	688	616
	20	1580	1400	1300	1200	1100	1010	905	813	731	654	585
	21	1500	1340	1240	1140	1050	960	861	774	696	623	557
	22	1430	1280	1190	1090	1000	916	822	739	664	595	532
	23	1370	1220	1130	1040	960	877	787	707	635	569	509
	24	1310	1170	1090	1000	920	840	754	678	609	545	488
	25	1260	1120	1040	961	883	806	724	650	584	523	468
	26	1210	1080	1000	924	849	775	696	625	562	503	450
	27	1170	1040	966	890	818	747	670	602	541	484	433
	28	1130	1000	931	858	789	720	646	581	522	467	418
	29	1090	968	899	829	761	695	624	561	504	451	403
	30	1050	936	869	801	736	672	603	542	487	436	390
	31	1020	906	841	775	712	650	584	525	471	422	377
	32	984	878	815	751	690	630	565	508	457	409	366
	33	955	851	790	728	669	611	548	493	443	396	355
	34	926	826	767	707	649	593	532	478	430	385	344
	35	900	802	745	687	631	576	517	465	417	374	334
	36	875	780	724	668	613	560	503	452	406	363	325
	37	851	759	705	649	597	545	489	439	395	354	316
	38	829	739	686	632	581	531	476	428	384	344	308
	39	808	720	668	616	566	517	464	417	375	335	300
	40	788	702	652	601	552	504	452	407	365	327	
	42	750	669	621	572	526	480	431				
	44	716	638	593	546	502						
	46	685	610	567								
Beam Properties												
Z_x, in.^{3}		1050	936	869	801	736	672	603	542	487	436	390
$\phi_b W_c$, kip-ft		31500	28100	26100	24000	22100	20200	18100	16300	14600	13100	11700
$\phi_v V_n$, kips		1160	1040	949	875	802	728	651	582	523	462	415
BF, kips		9.13	8.75	8.61	8.38	8.21	8.01	7.84	7.65	7.55	7.40	7.28
L_r, ft		172	157	148	139	129	121	110	100	91.6	83.4	76.0
L_p, ft		15.6	15.5	15.3	15.2	15.1	15.0	14.8	14.7	14.6	14.5	14.4
$\phi_b M_r$, kip-ft		2510	2270	2120	1970	1820	1670	1520	1380	1250	1130	1010
$\phi_b M_p$, kip-ft		3940	3510	3260	3000	2760	2520	2260	2030	1830	1640	1460

*ASTM A6 tensile group 4 or 5 shape. Special requirements may apply per LRFD Specification Section A3.1c.

[a] For beams laterally unsupported, see Table 5-5.

Note: Design strength tabulated above heavy line is limited by design shear strength.

$F_y = 50$ ksi
$\phi_b = 0.90$
$\phi_v = 0.90$

Table 5-4 (cont.). W-Shapes Maximum Total Factored Uniform Load[a]

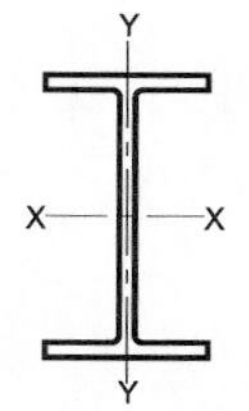

Shape		W14×										
		193	176	159	145	132	120	109	99††	90††	82	74
Span, ft	10										394	345
	11										379	344
	12										348	315
	13					512	462		372	333	321	291
	14	745	681	603	543	501	454	405	369	329	298	270
	15	710	640	574	520	468	424	384	345	307	278	252
	16	666	600	538	488	439	398	360	323	288	261	236
	17	626	565	506	459	413	374	339	304	271	245	222
	18	592	533	478	433	390	353	320	287	256	232	210
	19	561	505	453	411	369	335	303	272	242	219	199
	20	533	480	431	390	351	318	288	258	230	209	189
	21	507	457	410	371	334	303	274	246	219	199	180
	22	484	436	391	355	319	289	262	235	209	190	172
	23	463	417	374	339	305	277	250	225	200	181	164
	24	444	400	359	325	293	265	240	215	192	174	158
	25	426	384	344	312	281	254	230	207	184	167	151
	26	410	369	331	300	270	245	222	199	177	160	145
	27	394	356	319	289	260	236	213	191	171	154	140
	28	380	343	308	279	251	227	206	185	165	149	135
	29	367	331	297	269	242	219	199	178	159	144	130
	30	355	320	287	260	234	212	192	172	154	139	126
	31	344	310	278	252	226	205	186	167	149	135	122
	32	333	300	269	244	219	199	180	161	144	130	118
	33	323	291	261	236	213	193	175	157	140	126	115
	34	313	282	253	229	206	187	169	152	136	123	111
	35	304	274	246	223	201	182	165	148	132	119	108
	36	296	267	239	217	195	177					
	37	288	259	233	211							
	38	280	253									
Beam Properties												
Z_x, in.^{3}		355	320	287	260	234	212	192	173	157	139	126
$\phi_b W_c$, kip-ft		10700	9600	8610	7800	7020	6360	5760	5170	4610	4170	3780
$\phi_v V_n$, kips		372	341	302	272	256	231	203	186	166	197	173
BF, kips		7.19	7.09	7.02	6.87	6.88	6.81	6.69	6.56	6.85	7.30	7.12
L_r, ft		70.1	64.5	58.9	54.7	49.7	46.3	43.2	40.6	38.4	29.6	27.9
L_p, ft		14.3	14.2	14.1	14.1	13.3	13.2	13.2	13.5	15.1	8.76	8.76
$\phi_b M_r$, kip-ft		930	843	762	696	627	570	519	471	429	369	336
$\phi_b M_p$, kip-ft		1330	1200	1080	975	878	795	720	649	589	521	473

[a] For beams laterally unsupported, see Table 5-5.
Note: Design strength tabulated above heavy line is limited by design shear strength.
†† Indicates flange is non-compact.

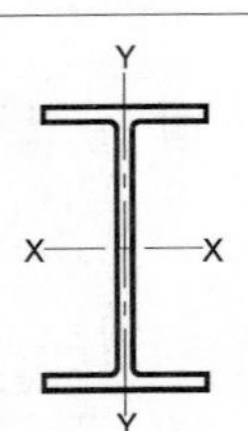

Table 5-4 (cont.). W-Shapes Maximum Total Factored Uniform Load[a]

$F_y = 50$ ksi
$\phi_b = 0.90$
$\phi_v = 0.90$

Shape		W14×									
		68	61	53	48	43	38	34	30	26	22
Span, ft	5										170
	6									191	164
	7						236	215	201	171	141
	8						229	203	176	150	123
	9			278	253	226	204	181	156	133	109
	10		281	261	235	209	183	163	141	120	98.4
	11	314	278	238	214	190	167	148	128	109	89.5
	12	288	255	218	196	174	153	136	117	99.7	82.0
	13	265	235	201	181	161	141	125	108	92.1	75.7
	14	246	219	187	168	149	131	116	101	85.5	70.3
	15	230	204	174	157	139	122	108	93.8	79.8	65.6
	16	216	191	163	147	130	115	102	87.9	74.8	61.5
	17	203	180	154	138	123	108	95.6	82.8	70.4	57.9
	18	192	170	145	131	116	102	90.3	78.2	66.5	54.7
	19	182	161	138	124	110	96.5	85.6	74.1	63.0	51.8
	20	173	153	131	118	104	91.7	81.3	70.4	59.8	49.2
	21	164	146	124	112	99.4	87.3	77.4	67.0	57.0	46.9
	22	157	139	119	107	94.9	83.3	73.9	64.0	54.4	44.7
	23	150	133	114	102	90.8	79.7	70.7	61.2	52.0	42.8
	24	144	128	109	98.0	87.0	76.4	67.8	58.6	49.9	41.0
	25	138	122	105	94.1	83.5	73.3	65.0	56.3	47.9	39.4
	26	133	118	101	90.5	80.3	70.5	62.5	54.1	46.0	37.8
	27	128	113	96.8	87.1	77.3	67.9	60.2	52.1	44.3	36.4
	28	123	109	93.3	84.0	74.6	65.5	58.1	50.2	42.7	35.1
	29	119	106	90.1	81.1	72.0	63.2	56.1	48.5	41.3	33.9
	30	115	102	87.1	78.4	69.6	61.1	54.2	46.9	39.9	32.8
	31	111	98.7	84.3	75.9	67.4	59.1	52.5	45.4	38.6	31.7
	32	108	95.6	81.7	73.5	65.2	57.3	50.8	44.0	37.4	30.8
	33	105	92.7	79.2	71.3	63.3	55.5	49.3	42.6	36.3	29.8
	34	101	90.0	76.9	69.2	61.4	53.9	47.8	41.4	35.2	28.9
	35	98.6					52.4	46.5			
Beam Properties											
Z_x, in.^{3}		115	102	87.1	78.4	69.6	61.1	54.2	46.9	39.9	32.8
$\phi_b W_c$, kip-ft		3450	3060	2610	2350	2090	1830	1630	1410	1200	984
$\phi_v V_n$, kips		157	141	139	127	113	118	108	101	95.7	85.1
BF, kips		6.91	6.50	7.01	6.70	6.31	7.05	6.58	6.05	7.01	6.22
L_r, ft		26.4	25.0	20.1	19.2	18.2	14.9	14.3	13.7	10.2	9.65
L_p, ft		8.69	8.65	6.78	6.75	6.68	5.47	5.40	5.26	3.81	3.67
$\phi_b M_r$, kip-ft		309	277	233	211	188	163	145	125	105	85.8
$\phi_b M_p$, kip-ft		431	383	327	294	261	229	203	176	150	123

[a] For beams laterally unsupported, see Table 5-5.
Note: Design strength tabulated above heavy line is limited by design shear strength.

$F_y = 50$ ksi
$\phi_b = 0.90$
$\phi_v = 0.90$

Table 5-4 (cont.). W-Shapes Maximum Total Factored Uniform Load[a]

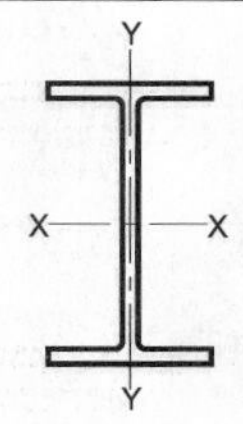

Shape		W12×										
		336*	305*	279*	252*	230*	210*	190	170	152	136	120
Span, ft	11	1610	1430	1310	1160	1050	937	824	726	644	572	502
	12	1510	1340	1200	1070	965	870	778	688	608	535	465
	13	1390	1240	1110	988	891	803	718	635	561	494	429
	14	1290	1150	1030	917	827	746	666	589	521	459	399
	15	1210	1070	962	856	772	696	622	550	486	428	372
	16	1130	1010	902	803	724	653	583	516	456	401	349
	17	1060	948	849	755	681	614	549	485	429	378	328
	18	1010	895	802	713	643	580	518	458	405	357	310
	19	952	848	759	676	609	549	491	434	384	338	294
	20	905	806	722	642	579	522	467	413	365	321	279
	21	861	767	687	611	551	497	444	393	347	306	266
	22	822	732	656	584	526	475	424	375	331	292	254
	23	787	700	627	558	503	454	406	359	317	279	243
	24	754	671	601	535	483	435	389	344	304	268	233
	25	724	644	577	514	463	418	373	330	292	257	223
	26	696	620	555	494	445	402	359	317	280	247	215
	27	670	597	534	476	429	387	346	306	270	238	207
	28	646	575	515	459	414	373	333	295	260	229	199
	29	624	556	498	443	399	360	322	284	251	221	192
	30	603	537	481	428	386	348	311	275	243	214	186
	31	584	520	465	414	374	337	301	266	235	207	180
	32	565	503	451	401	362	326	292	258	228	201	174
	33	548	488	437	389	351	316	283	250	221	195	
	34	532	474	424	378	341	307	274	243	214		
	35	517	460	412	367	331	298	267	236			
	36	503	448	401	357	322	290	259				
	37	489	435	390	347	313						
	38	476	424	380	338							
	39	464	413	370								
	40	452	403									
	41	441										
	42	431										
Beam Properties												
Z_x, in.^{3}		603	537	481	428	386	348	311	275	243	214	186
$\phi_b W_c$, kip-ft		18100	16100	14400	12800	11600	10400	9330	8250	7290	6420	5580
$\phi_v V_n$, kips		807	717	657	582	526	468	412	363	322	286	251
BF, kips		6.81	6.59	6.37	6.19	6.04	5.91	5.79	5.67	5.59	5.49	5.36
L_r, ft		131	120	110	100	92.0	84.2	76.6	68.9	62.1	55.7	50.0
L_p, ft		12.3	12.1	11.9	11.8	11.7	11.6	11.5	11.4	11.3	11.2	11.1
$\phi_b M_r$, kip-ft		1450	1310	1180	1060	963	876	789	705	627	558	489
$\phi_b M_p$, kip-ft		2260	2010	1800	1610	1450	1310	1170	1030	911	803	698

*ASTM A6 tensile group 4 or 5 shape. Special requirements may apply per LRFD Specification Section A3.1c.
[a]For beams laterally unsupported, see Table 5-5.
Note: Design strength tabulated above heavy line is limited by design shear strength.

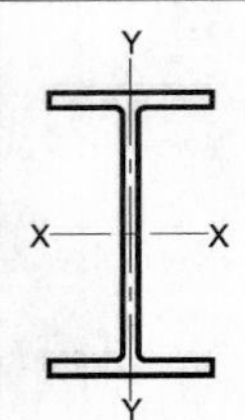

Table 5-4 (cont.). W-Shapes Maximum Total Factored Uniform Load[a]

$F_y = 50$ ksi
$\phi_b = 0.90$
$\phi_v = 0.90$

Shape		W12×											
		106	96	87	79	72	65[††]	58	53	50	45	40	35
Span, ft	7												203
	8									244	219		192
	9									240	214	190	171
	10							237	225	216	193	171	154
	11	425	377	348	315	286	255	236	212	196	175	155	140
	12	410	368	330	298	270	238	216	195	180	161	143	128
	13	378	339	305	275	249	220	199	180	166	148	132	118
	14	351	315	283	255	231	204	185	167	154	138	122	110
	15	328	294	264	238	216	191	173	156	144	128	114	102
	16	308	276	248	223	203	179	162	146	135	120	107	96.0
	17	289	259	233	210	191	168	152	137	127	113	101	90.4
	18	273	245	220	198	180	159	144	130	120	107	95.0	85.3
	19	259	232	208	188	171	150	136	123	114	101	90.0	80.8
	20	246	221	198	179	162	143	130	117	108	96.3	85.5	76.8
	21	234	210	189	170	154	136	123	111	103	91.7	81.4	73.1
	22	224	200	180	162	147	130	118	106	98.0	87.5	77.7	69.8
	23	214	192	172	155	141	124	113	102	93.8	83.7	74.3	66.8
	24	205	184	165	149	135	119	108	97.4	89.9	80.3	71.3	64.0
	25	197	176	158	143	130	114	104	93.5	86.3	77.0	68.4	61.4
	26	189	170	152	137	125	110	99.7	89.9	83.0	74.1	65.8	59.1
	27	182	163	147	132	120	106	96.0	86.6	79.9	71.3	63.3	56.9
	28	176	158	141	128	116	102	92.6	83.5	77.0	68.8	61.1	54.9
	29	170	152	137	123	112	98.5	89.4	80.6	74.4	66.4	59.0	53.0
	30	164	147	132	119	108	95.3	86.4	77.9	71.9	64.2		51.2
	31	159	142	128	115								49.5
	32	154											
Beam Properties													
Z_x, in.^{3}		164	147	132	119	108	96.8	86.4	77.9	71.9	64.2	57.0	51.2
$\phi_b W_c$, kip-ft		4920	4410	3960	3570	3240	2860	2590	2340	2160	1930	1710	1540
$\phi_v V_n$, kips		212	189	174	157	143	127	119	113	122	109	94.8	101
BF, kips		5.31	5.20	5.13	5.03	4.93	5.01	4.97	4.78	5.30	5.06	4.79	5.65
L_r, ft		44.9	41.3	38.4	35.7	33.6	31.7	27.0	25.6	21.5	20.3	19.2	15.2
L_p, ft		11.0	10.9	10.8	10.8	10.7	11.9	8.87	8.76	6.92	6.89	6.85	5.44
$\phi_b M_r$, kip-ft		435	393	354	321	292	264	234	212	193	173	155	137
$\phi_b M_p$, kip-ft		615	551	495	446	405	363	324	292	270	241	214	192

[a] For beams laterally unsupported, see Table 5-5.
Note: Design strength tabulated above heavy line is limited by design shear strength.
[††] Indicates flange is non-compact.

$F_y = 50$ ksi
$\phi_b = 0.90$
$\phi_v = 0.90$

Table 5-4 (cont.). W-Shapes Maximum Total Factored Uniform Load[a]

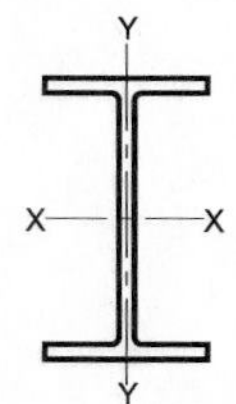

Shape	W12×						W10×					
Span, ft	30	26	22	19	16	14	112	100	88	77	68	60
4				155	143	129						
5			173	148	121	104						
6			147	124	101	87.0						
7	173	152	126	106	86.1	74.6						
8	162	140	110	92.6	75.4	65.2						
9	144	124	97.7	82.3	67.0	58.0	465	408	353	303	264	231
10	129	112	87.9	74.1	60.3	52.2	441	390	339	293	256	224
11	118	101	79.9	67.4	54.8	47.5	401	355	308	266	233	203
12	108	93.0	73.3	61.8	50.3	43.5	368	325	283	244	213	187
13	99.5	85.8	67.6	57.0	46.4	40.2	339	300	261	225	197	172
14	92.4	79.7	62.8	52.9	43.1	37.3	315	279	242	209	183	160
15	86.2	74.4	58.6	49.4	40.2	34.8	294	260	226	195	171	149
16	80.8	69.8	54.9	46.3	37.7	32.6	276	244	212	183	160	140
17	76.1	65.6	51.7	43.6	35.5	30.7	259	229	199	172	151	132
18	71.8	62.0	48.8	41.2	33.5	29.0	245	217	188	163	142	124
19	68.1	58.7	46.3	39.0	31.7	27.5	232	205	178	154	135	118
20	64.7	55.8	44.0	37.1	30.2	26.1	221	195	170	146	128	112
21	61.6	53.1	41.9	35.3	28.7	24.9	210	186	161	139	122	107
22	58.8	50.7	40.0	33.7	27.4	23.7	200	177	154	133	116	102
23	56.2	48.5	38.2	32.2	26.2	22.7	192	170	147	127	111	97.3
24	53.9	46.5	36.6	30.9	25.1	21.7	184	163	141	122	107	93.3
25	51.7	44.6	35.2	29.6	24.1	20.9	176	156	136	117	102	89.5
26	49.7	42.9	33.8	28.5	23.2	20.1	170	150	130	113	98.4	
27	47.9	41.3	32.6	27.4	22.3	19.3	163	144	126			
28	46.2	39.9	31.4	26.5	21.5	18.6	158					
29	44.6	38.5	30.3	25.6	20.8	18.0						
30	43.1	37.2	29.3	24.7	20.1							
31												
Beam Properties												
Z_x, in.^{3}	43.1	37.2	29.3	24.7	20.1	17.4	147	130	113	97.6	85.3	74.6
$\phi_b W_c$, kip-ft	1290	1120	879	741	603	522	4410	3900	3390	2930	2560	2240
$\phi_v V_n$, kips	86.3	75.8	86.3	77.4	71.3	64.3	232	204	176	152	132	116
BF, kips	5.12	4.66	6.21	5.69	5.11	4.55	3.68	3.66	3.58	3.53	3.45	3.39
L_r, ft	14.3	13.8	8.42	7.95	7.44	7.17	56.5	50.8	45.1	39.9	36.0	32.6
L_p, ft	5.37	5.33	3.00	2.90	2.73	2.66	9.47	9.36	9.29	9.18	9.15	9.08
$\phi_b M_r$, kip-ft	116	100	76.2	63.9	51.3	44.7	378	336	296	258	227	200
$\phi_b M_p$, kip-ft	162	140	110	92.6	75.4	65.2	551	488	424	366	320	280

[a]For beams laterally unsupported, see Table 5-5.
Note: Design strength tabulated above heavy line is limited by design shear strength.

Table 5-4 (cont.). W-Shapes Maximum Total Factored Uniform Load[a]

$F_y = 50$ ksi
$\phi_b = 0.90$
$\phi_v = 0.90$

Shape		W10×											
Wt./ft		54	49	45	39	33	30	26	22	19	17	15	12[††]
Span, ft	3											124	101
	4									138	131	120	93.9
	5								132	130	112	96.0	75.1
	6						170	145	130	108	93.5	80.0	62.6
	7					152	157	134	111	92.6	80.1	68.6	53.7
	8			191	169	146	137	117	97.5	81.0	70.1	60.0	46.9
	9	202	183	183	156	129	122	104	86.7	72.0	62.3	53.3	41.7
	10	200	181	165	140	116	110	93.9	78.0	64.8	56.1	48.0	37.6
	11	182	165	150	128	106	99.8	85.4	70.9	58.9	51.0	43.6	34.1
	12	167	151	137	117	97.0	91.5	78.3	65.0	54.0	46.7	40.0	31.3
	13	154	139	127	108	89.5	84.5	72.2	60.0	49.8	43.2	36.9	28.9
	14	143	129	118	100	83.1	78.4	67.1	55.7	46.3	40.1	34.3	26.8
	15	133	121	110	93.6	77.6	73.2	62.6	52.0	43.2	37.4	32.0	25.0
	16	125	113	103	87.8	72.8	68.6	58.7	48.8	40.5	35.1	30.0	23.5
	17	118	107	96.9	82.6	68.5	64.6	55.2	45.9	38.1	33.0	28.2	22.1
	18	111	101	91.5	78.0	64.7	61.0	52.2	43.3	36.0	31.2	26.7	20.9
	19	105	95.4	86.7	73.9	61.3	57.8	49.4	41.1	34.1	29.5	25.3	19.8
	20	99.9	90.6	82.4	70.2	58.2	54.9	47.0	39.0	32.4	28.1	24.0	18.8
	21	95.1	86.3	78.4	66.9	55.4	52.3	44.7	37.1	30.9	26.7	22.9	17.9
	22	90.8	82.4	74.9	63.8	52.9	49.9	42.7	35.5	29.5	25.5	21.8	17.1
	23	86.9	78.8	71.6	61.0	50.6	47.7	40.8	33.9	28.2	24.4	20.9	16.3
	24	83.3	75.5	68.6	58.5	48.5	45.8	39.1	32.5	27.0	23.4	20.0	15.6
	25	79.9		65.9			43.9	37.6	31.2	25.9	22.4		
	26						42.2						
Beam Properties													
Z_x, in.^{3}		66.6	60.4	54.9	46.8	38.8	36.6	31.3	26.0	21.6	18.7	16.0	12.6
$\phi_b W_c$, kip-ft		2000	1810	1650	1400	1160	1100	939	780	648	561	480	376
$\phi_v V_n$, kips		101	91.6	95.4	84.4	76.2	85.0	72.3	66.1	68.8	65.4	62.0	50.6
BF, kips		3.30	3.24	3.44	3.32	3.14	4.11	3.84	3.49	4.24	3.99	3.67	3.18
L_r, ft		30.2	28.3	24.1	21.8	19.8	14.6	13.6	12.7	8.88	8.38	7.93	7.45
L_p, ft		9.04	8.97	7.10	6.99	6.85	4.84	4.80	4.70	3.09	2.98	2.86	2.87
$\phi_b M_r$, kip-ft		180	164	147	126	105	97.2	83.7	69.6	56.4	48.6	41.4	32.7
$\phi_b M_p$, kip-ft		250	227	206	176	146	137	117	97.5	81.0	70.1	60.0	47.3

[a] For beams laterally unsupported, see Table 5-5.
Note: Design strength tabulated above heavy line is limited by design shear strength.
[††] Indicates flange is non-compact.

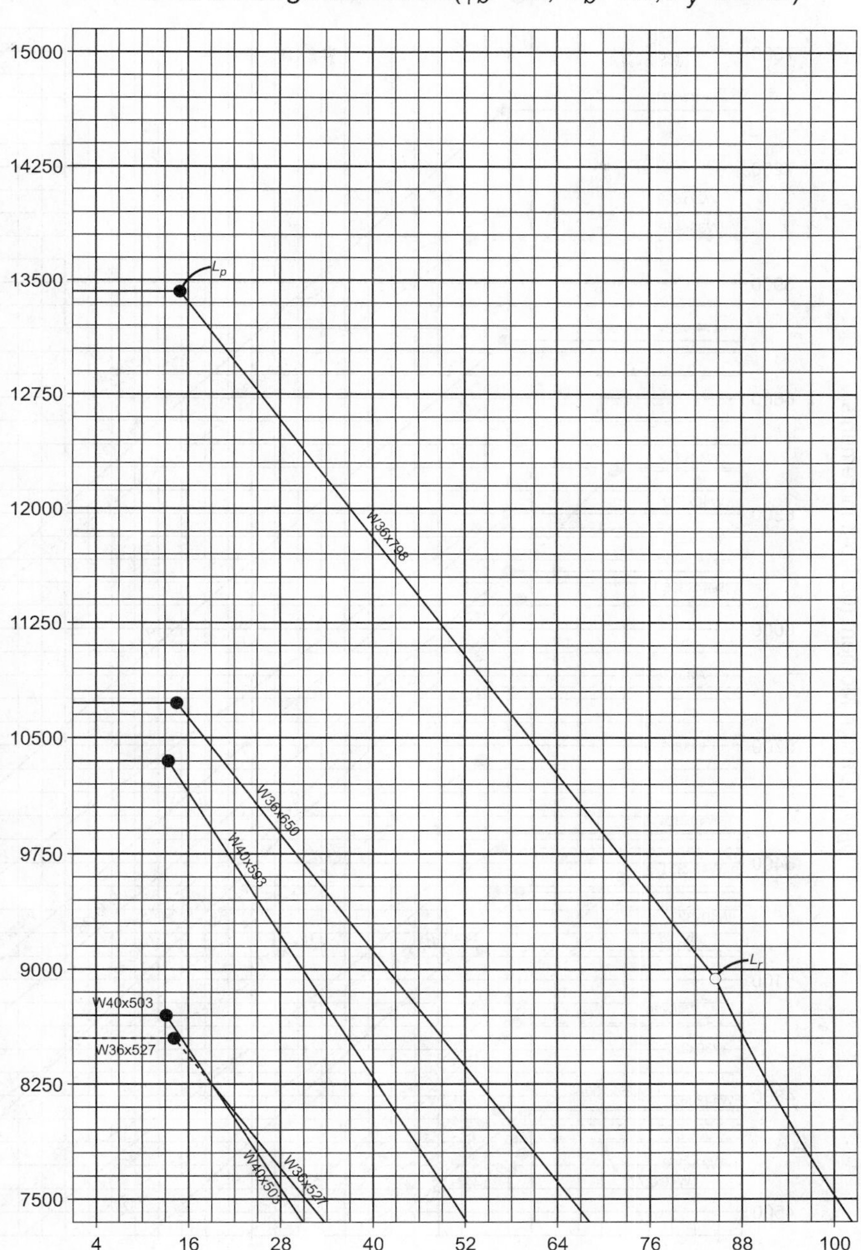
Beam Design Moments (ϕ_b=0.9, C_b=1.0, F_y=50 ksi)
$\phi_b M_n$, Design Moment (150 kip-ft increments)
15000
14250
13500
12750
12000
11250
10500
9750
9000
8250
7500
L_p
L_r
W36x798
W36x650
W40x593
W40x503
W36x527
4
16
28
40
52
64
76
88
100
L_b, Unbraced Length (3 ft increments)

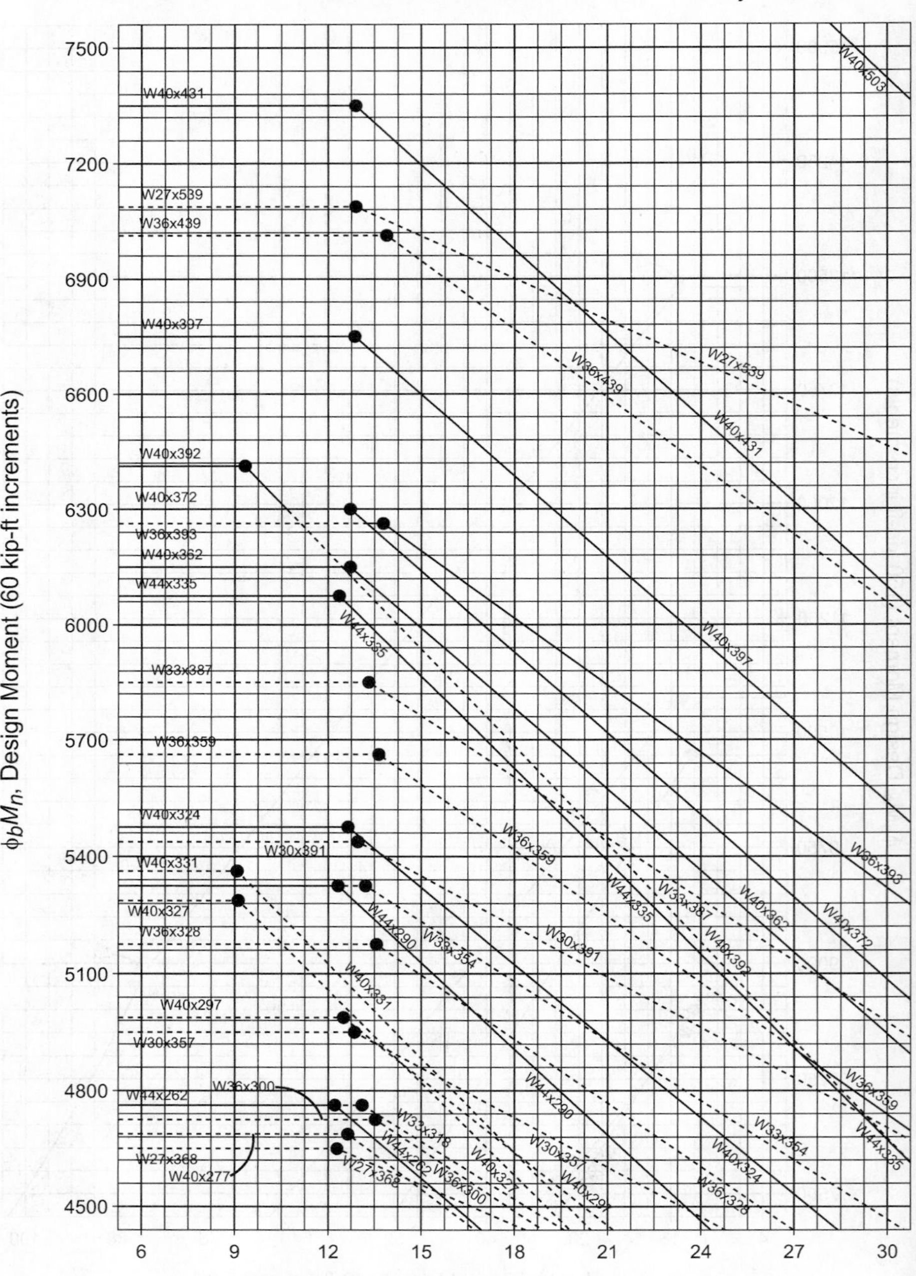
Beam Design Moments (ϕ_b=0.9, C_b=1.0, F_y=50 ksi)
$\phi_b M_n$, Design Moment (60 kip-ft increments)
L_b, Unbraced Length (0.75 ft increments)
7500
7200
6900
6600
6300
6000
5700
5400
5100
4800
4500
6
9
12
15
18
21
24
27
30
W40x503
W40x431
W27x539
W36x439
W40x397
W40x392
W40x372
W36x393
W40x362
W44x335
W33x387
W36x359
W40x324
W30x391
W40x331
W40x327
W36x328
W40x297
W30x357
W36x300
W44x262
W27x368
W40x277
W44x290
W33x354
W33x318
W40x327
W30x357
W40x297
W36x328
W40x324
W33x364
W44x335
W36x359
W40x392
W40x362
W40x372
W36x393
W33x387
W44x335

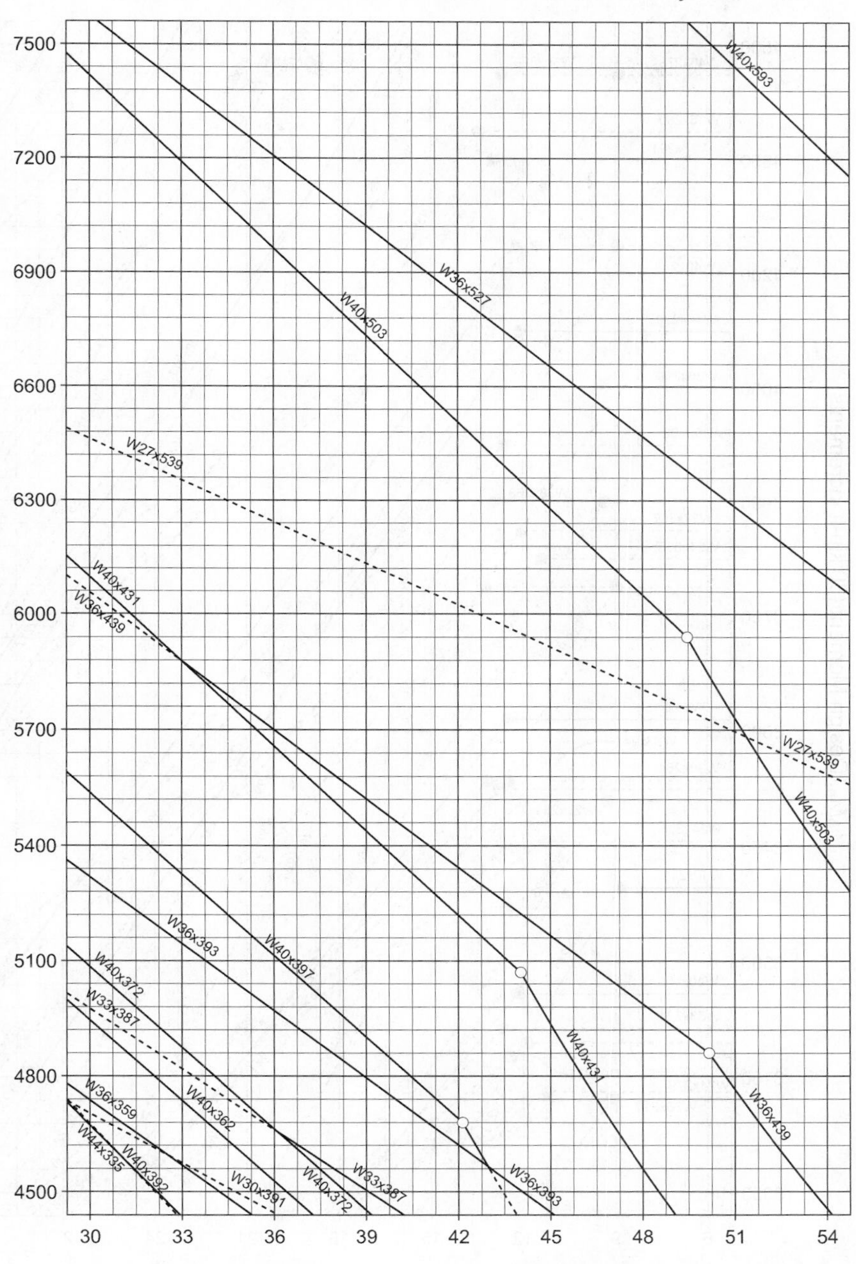

Beam Design Moments (ϕ_b=0.9, C_b=1.0, F_y=50 ksi)
$\phi_b M_n$, Design Moment (60 kip-ft increments)
L_b, Unbraced Length (0.75 ft increments)
7500
7200
6900
6600
6300
6000
5700
5400
5100
4800
4500
30
33
36
39
42
45
48
51
54
W40x593
W36x527
W40x503
W27x539
W40x431
W36x439
W36x393
W40x397
W40x372
W33x387
W36x359
W40x362
W44x335
W40x392
W30x391
W33x387
W36x393

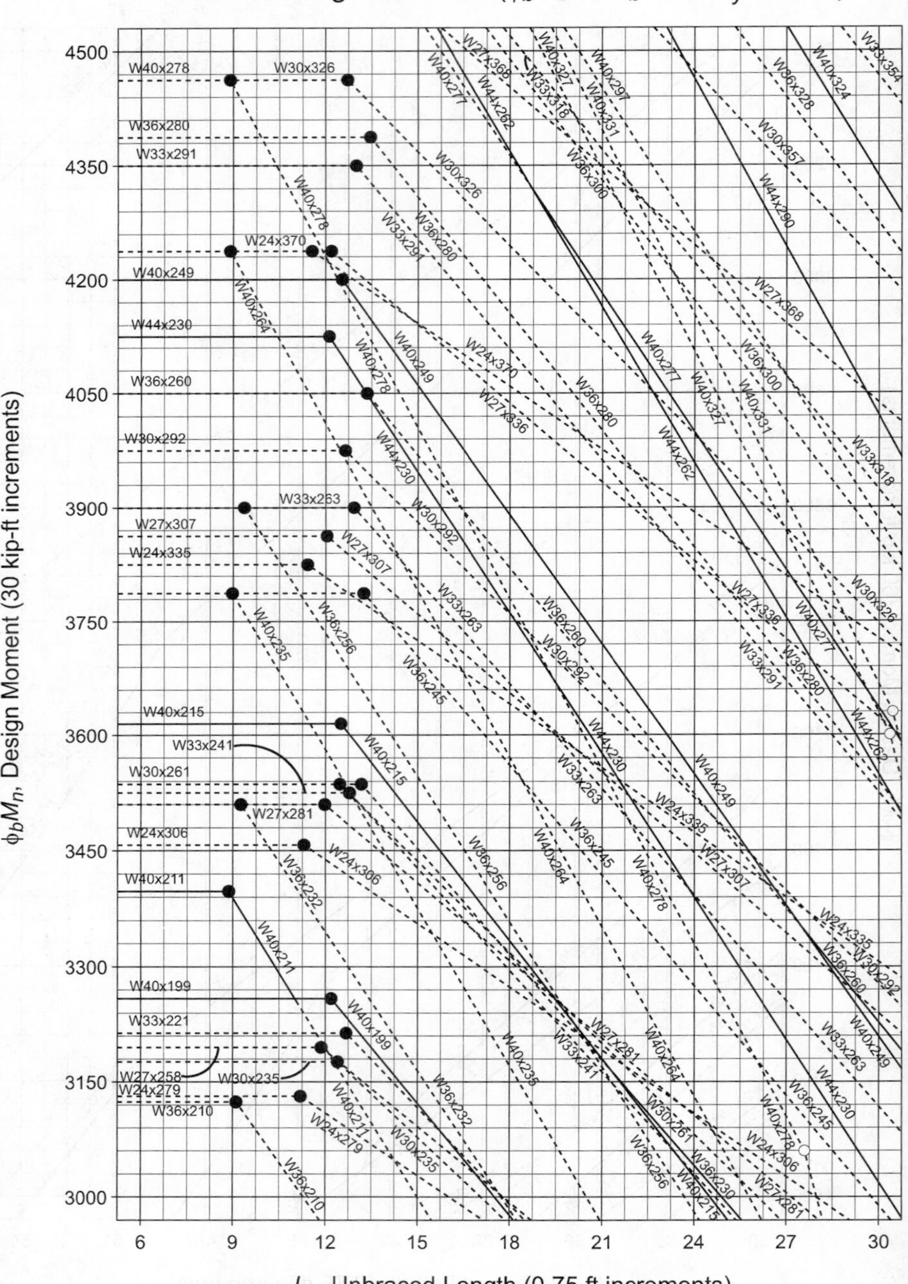

Beam Design Moments (ϕb=0.9, Cb=1.0, Fy=50 ksi)
ϕbMn, Design Moment (30 kip-ft increments)
Lb, Unbraced Length (0.75 ft increments)

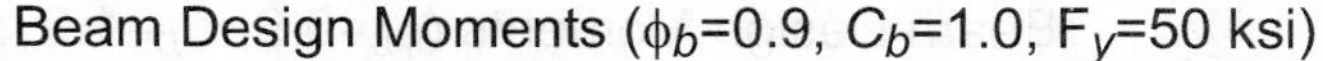
Beam Design Moments (ϕ_b=0.9, C_b=1.0, F_y=50 ksi)
$\phi_b M_n$, Design Moment (30 kip-ft increments)
4500
4350
4200
4050
3900
3750
3600
3450
3300
3150
3000
30
33
36
39
42
45
48
51
54
L_b, Unbraced Length (0.75 ft increments)
W36x439
W40x431
W36x393
W33x387
W40x397
W30x391
W40x362
W36x359
W40x372
W33x354
W44x335
W40x392
W40x324
W36x328
W30x357
W27x368
W44x290
W33x318
W40x297
W36x300
W30x326
W24x370
W40x327
W27x336
W40x331
W40x277
W36x280
W33x291
W44x262
W24x335
W30x292
W27x307
W36x260
W40x249
W33x263

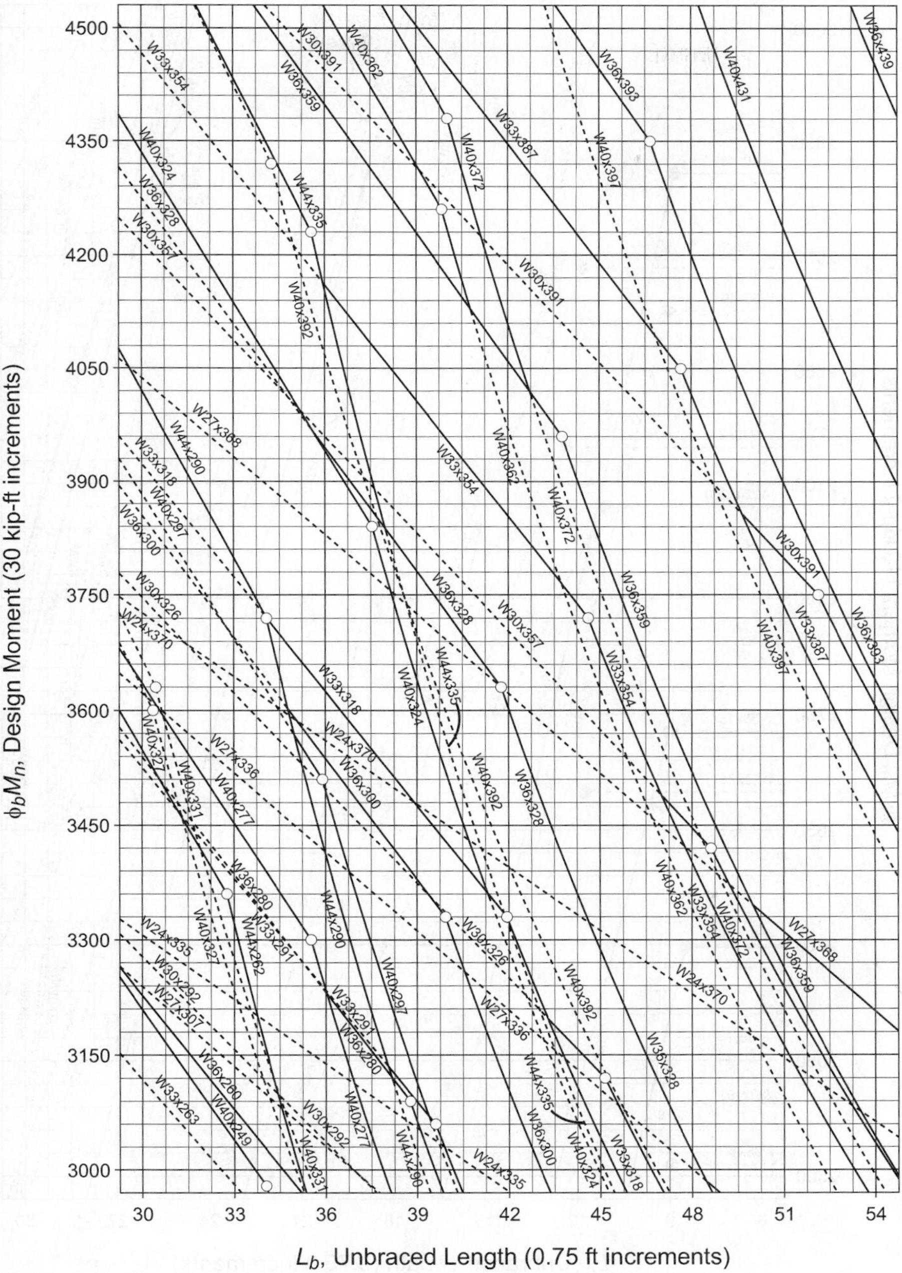

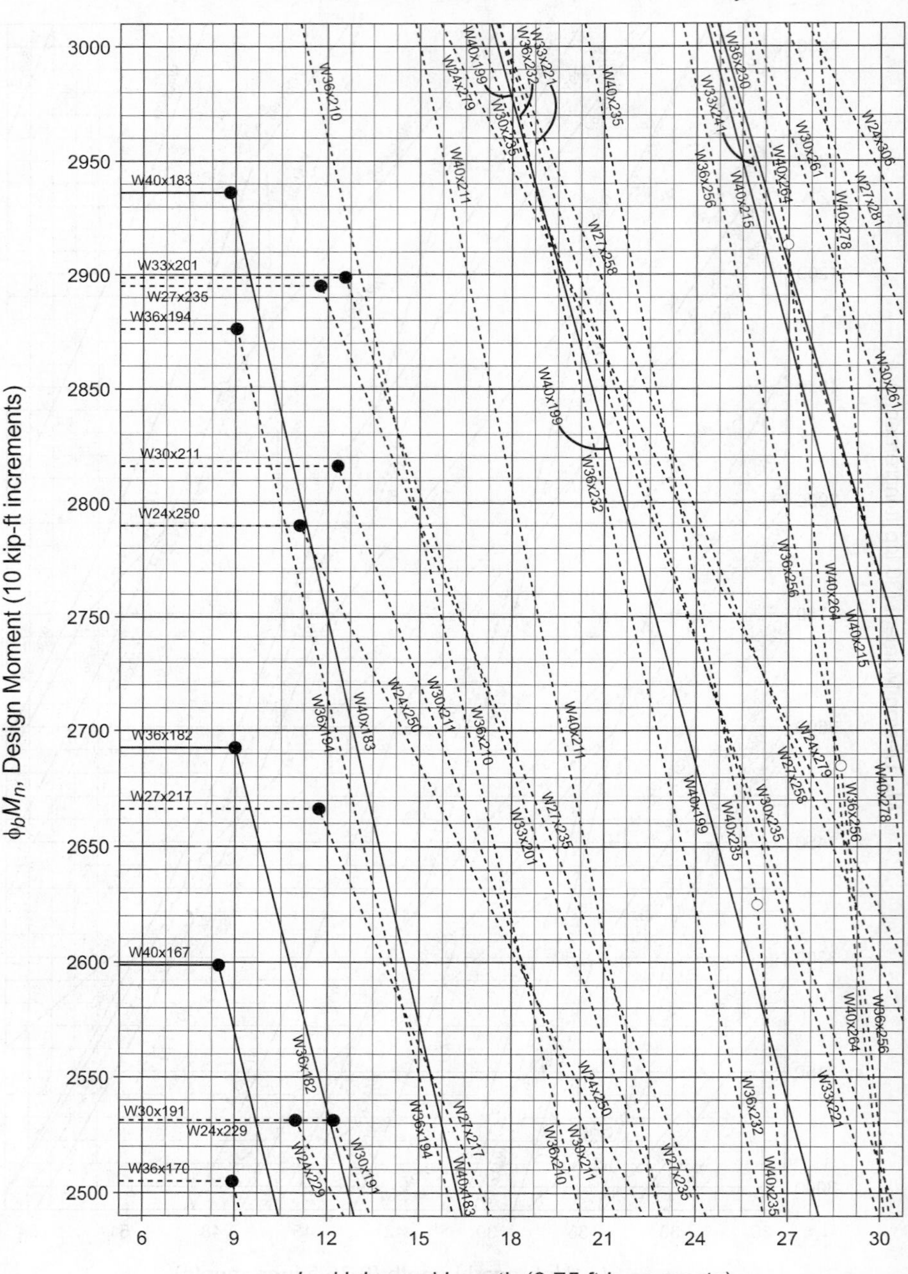
Beam Design Moments (ϕ_b=0.9, C_b=1.0, F_y=50 ksi)
$\phi_b M_n$, Design Moment (10 kip-ft increments)
L_b, Unbraced Length (0.75 ft increments)
3000
2950
2900
2850
2800
2750
2700
2650
2600
2550
2500
6
9
12
15
18
21
24
27
30
W40x183
W33x201
W27x235
W36x194
W30x211
W24x250
W36x182
W27x217
W40x167
W30x191
W24x229
W36x170
W36x210
W24x279
W40x199
W36x232
W33x221
W30x235
W40x211
W40x235
W27x258
W36x230
W33x241
W36x256
W40x215
W40x264
W30x261
W24x306
W27x281
W40x278
W36x194
W40x183
W24x250
W30x211
W36x210
W40x211
W27x235
W33x201
W40x199
W40x235
W30x235
W24x279
W27x258
W36x256
W36x182
W36x194
W27x217
W40x183
W24x229
W30x191
W24x250
W36x210
W30x211
W27x235
W36x232
W40x235
W33x221
W40x264
W36x256

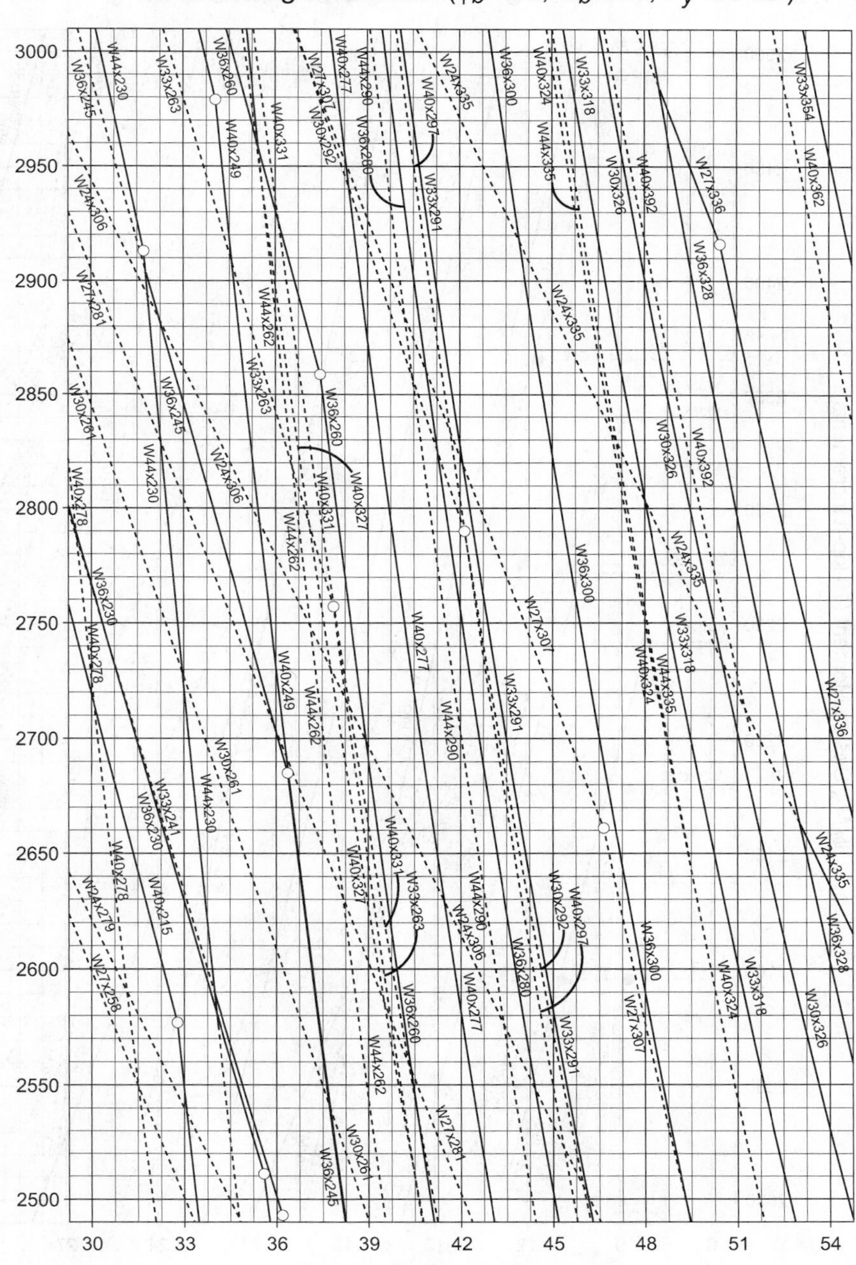
Beam Design Moments (ϕ_b=0.9, C_b=1.0, F_y=50 ksi)
$\phi_b M_n$, Design Moment (10 kip-ft increments)
L_b, Unbraced Length (0.75 ft increments)
3000
2950
2900
2850
2800
2750
2700
2650
2600
2550
2500
30
33
36
39
42
45
48
51
54
W36x245
W44x230
W33x263
W36x260
W40x249
W40x331
W27x307
W30x292
W40x277
W44x290
W36x280
W40x297
W24x335
W33x291
W36x300
W40x324
W44x335
W33x318
W30x326
W40x392
W27x336
W36x328
W33x354
W40x362
W24x306
W27x281
W30x261
W40x278
W36x230
W40x215
W33x241
W24x279
W27x258
W44x262
W33x263
W40x327
W27x281

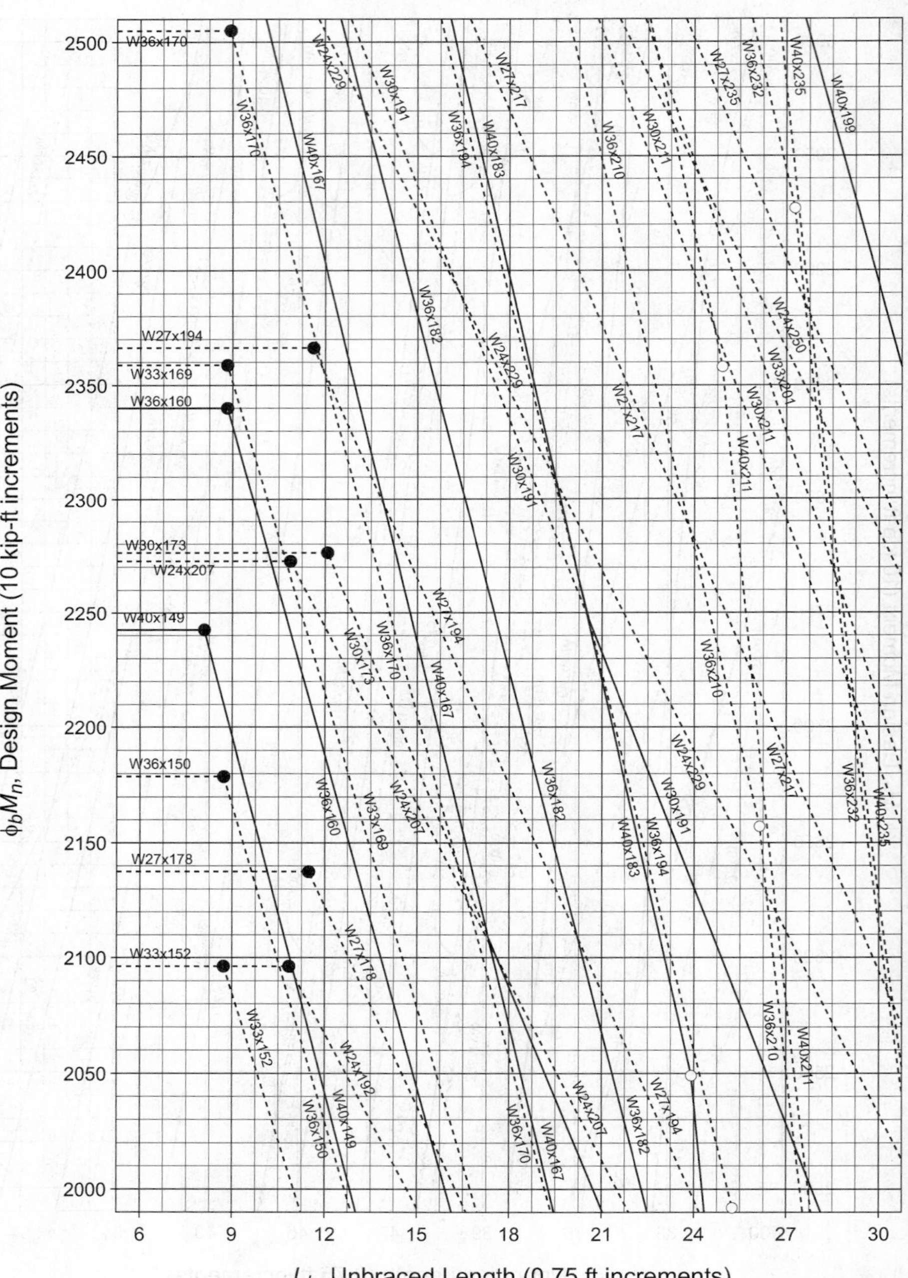
Beam Design Moments (ϕ_b=0.9, C_b=1.0, F_y=50 ksi)
$\phi_b M_n$, Design Moment (10 kip-ft increments)
L_b, Unbraced Length (0.75 ft increments)
2500
2450
2400
2350
2300
2250
2200
2150
2100
2050
2000
6
9
12
15
18
21
24
27
30
W36x170
W27x194
W33x169
W36x160
W30x173
W24x207
W40x149
W36x150
W27x178
W33x152
W24x229
W30x191
W27x217
W36x194
W40x183
W36x210
W30x211
W27x235
W36x232
W40x235
W40x199
W40x167
W36x182
W24x250
W33x201
W40x211
W24x192
W27x178

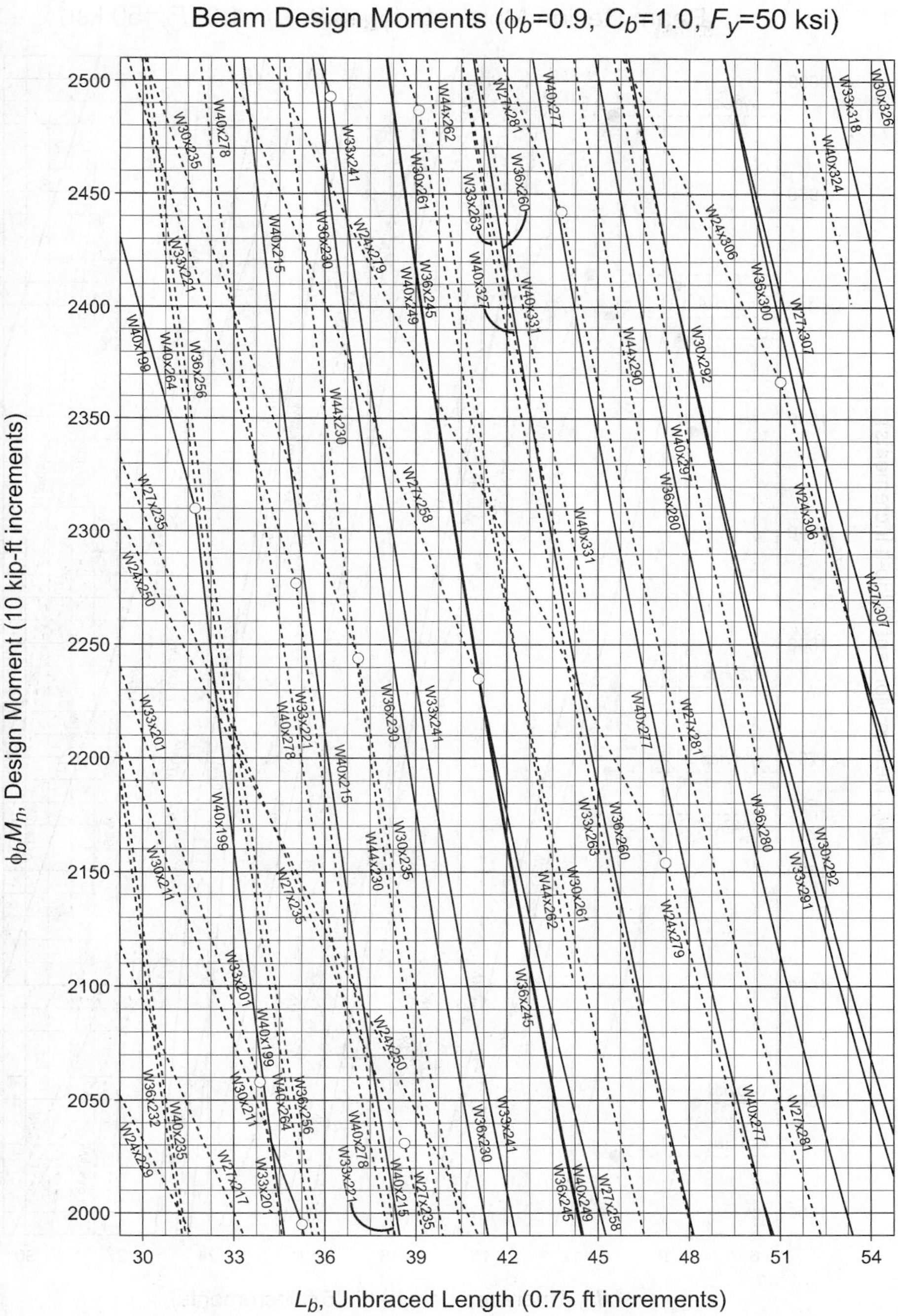

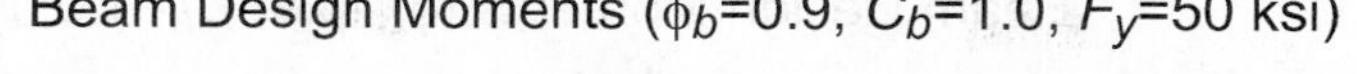

$\phi_b M_n$, Design Moment (10 kip-ft increments)

2500, 2450, 2400, 2350, 2300, 2250, 2200, 2150, 2100, 2050, 2000

30, 33, 36, 39, 42, 45, 48, 51, 54

L_b, Unbraced Length (0.75 ft increments)

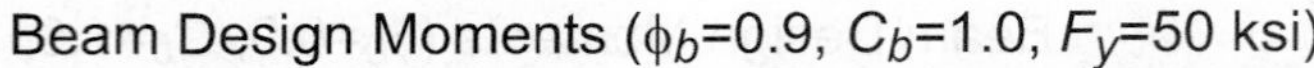

$\phi_b M_n$, Design Moment (8 kip-ft increments)

2000
1960
1920
1880
1840
1800
1760
1720
1680
1640
1600

6 9 12 15 18 21 24 27 30

L_b, Unbraced Length (0.75 ft increments)

W21x201
W27x161
W24x176
W36x135
W30x148
W21x182
W24x162
W27x146
W30x132
W21x166
W33x152
W24x192
W36x160
W36x182
W30x173
W36x194
W40x183
W24x207
W36x210
W40x211
W30x191
W27x178
W27x194
W33x141
W27x161
W21x201
W40x167
W36x170
W24x176
W40x149
W36x150
W33x169
W24x192
W36x135
W33x130
W36x135
W33x152
W36x160
W40x167
W27x178
W36x170
W36x182
W40x183
W30x173
W36x194
W36x210
W33x141
W21x182
W24x176
W27x161
W21x201
W30x148
W36x135
W24x162
W40x149
W36x150
W33x169
W36x182
W40x183
W24x192

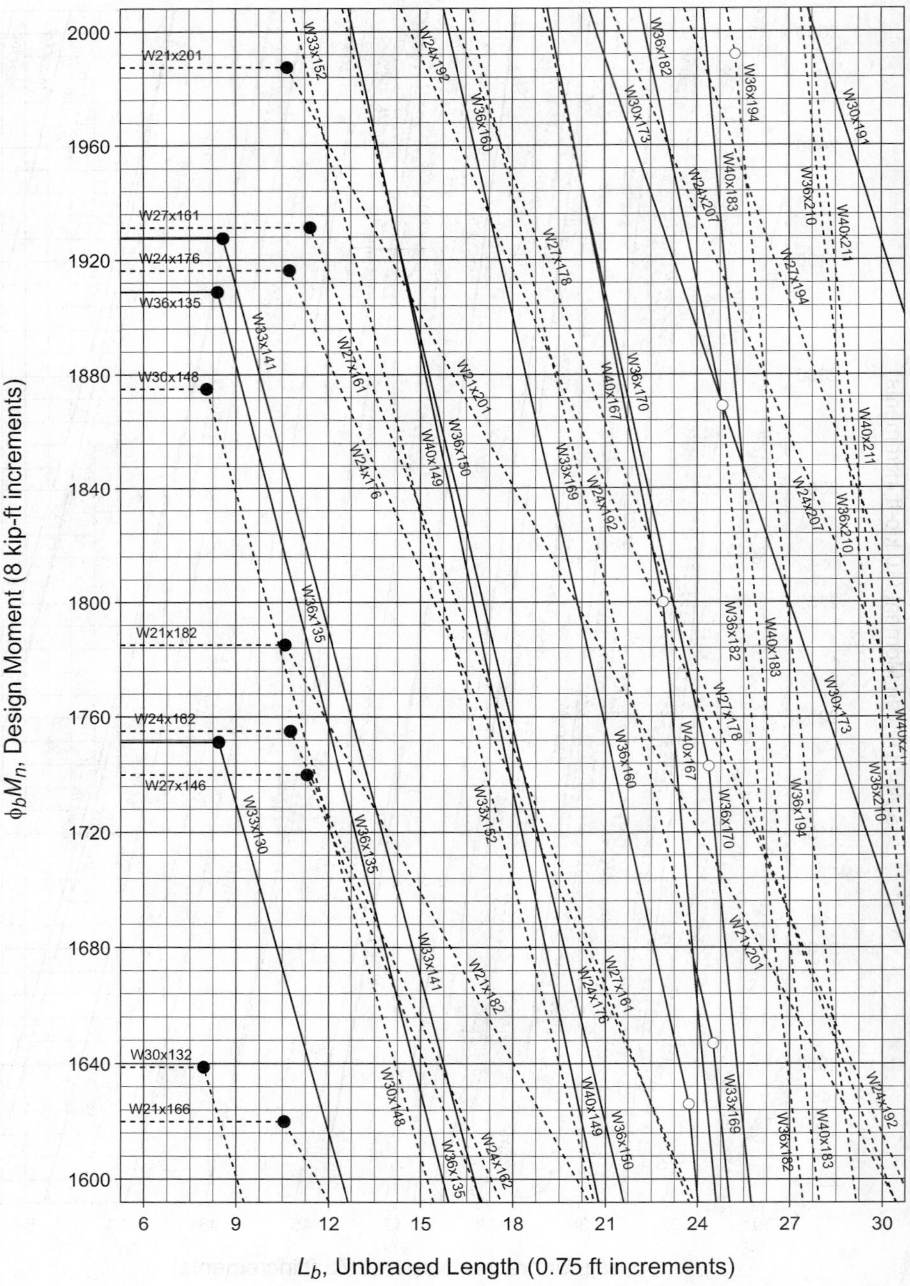

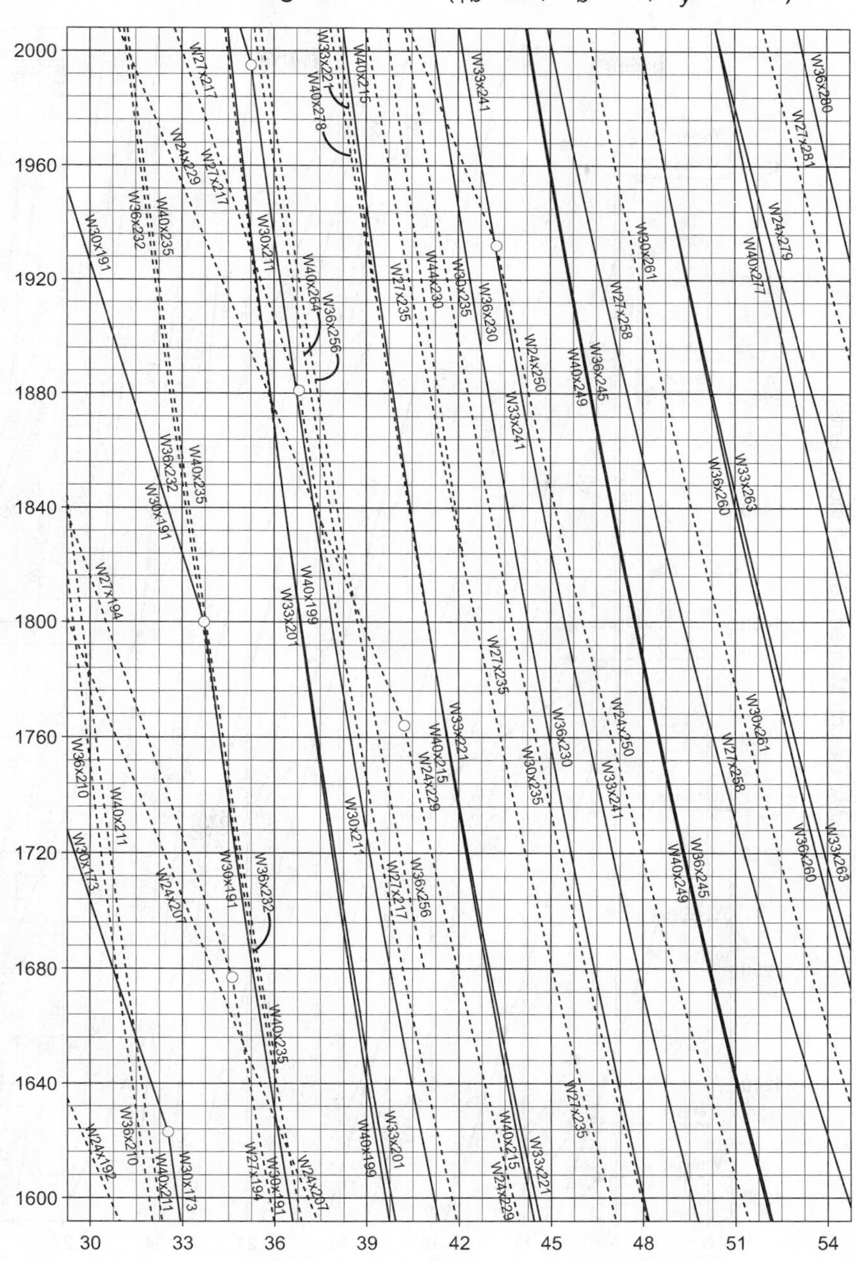
Beam Design Moments (ϕ_b=0.9, C_b=1.0, F_y=50 ksi)
$\phi_b M_n$, Design Moment (8 kip-ft increments)
L_b, Unbraced Length (0.75 ft increments)
2000
1960
1920
1880
1840
1800
1760
1720
1680
1640
1600
30
33
36
39
42
45
48
51
54
W36x280
W27x281
W24x279
W40x277
W30x261
W27x258
W36x245
W40x249
W24x250
W33x241
W36x230
W30x235
W44x230
W27x235
W36x256
W40x264
W30x211
W40x215
W33x221
W40x278
W27x217
W24x229
W40x235
W36x232
W30x191
W36x260
W33x263
W40x199
W33x201
W27x194
W36x210
W40x211
W30x173
W24x207
W36x282
W24x192
W30x191
W24x207

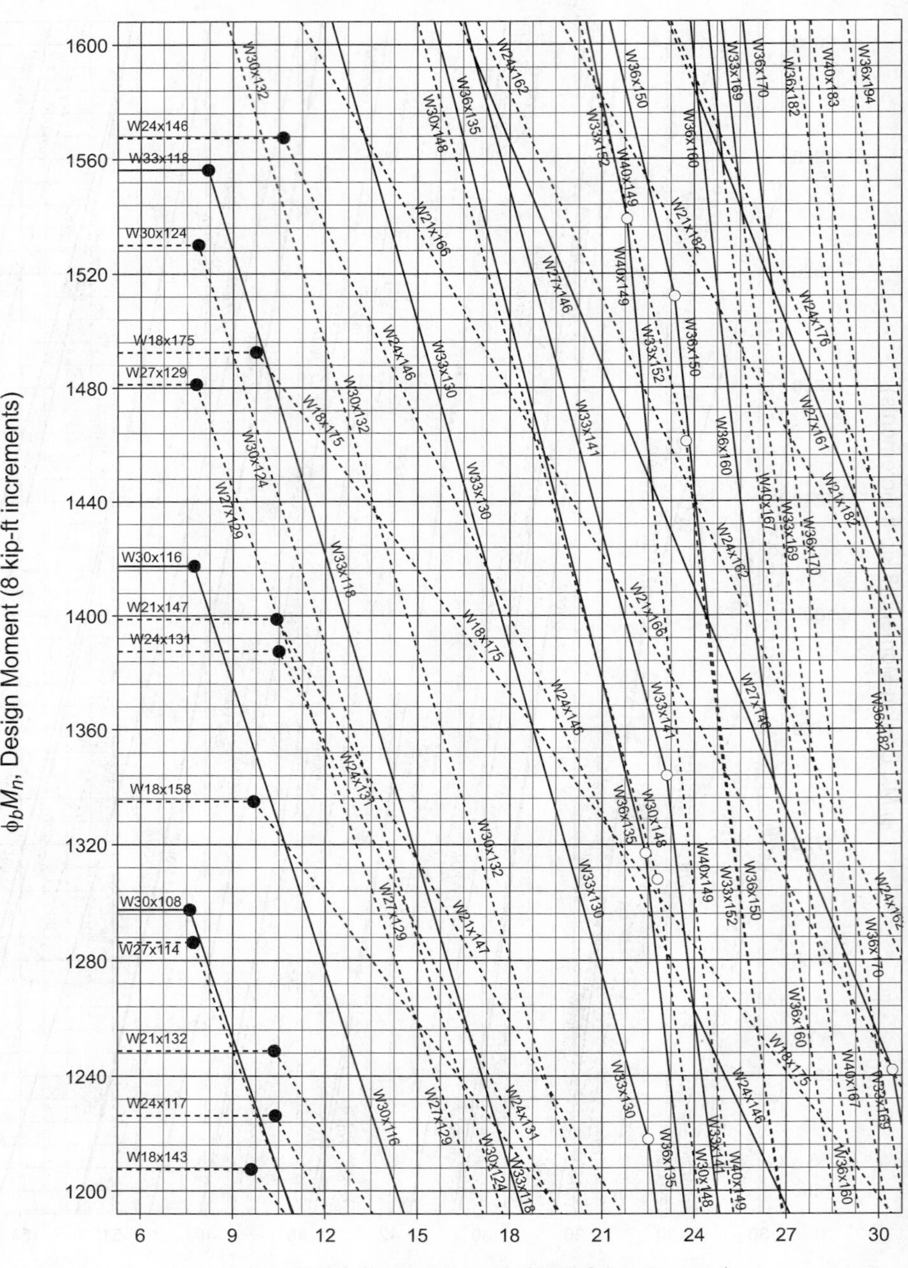

Beam Design Moments (ϕ_b=0.9, C_b=1.0, F_y=50 ksi)
$\phi_b M_n$, Design Moment (8 kip-ft increments)
L_b, Unbraced Length (0.75 ft increments)

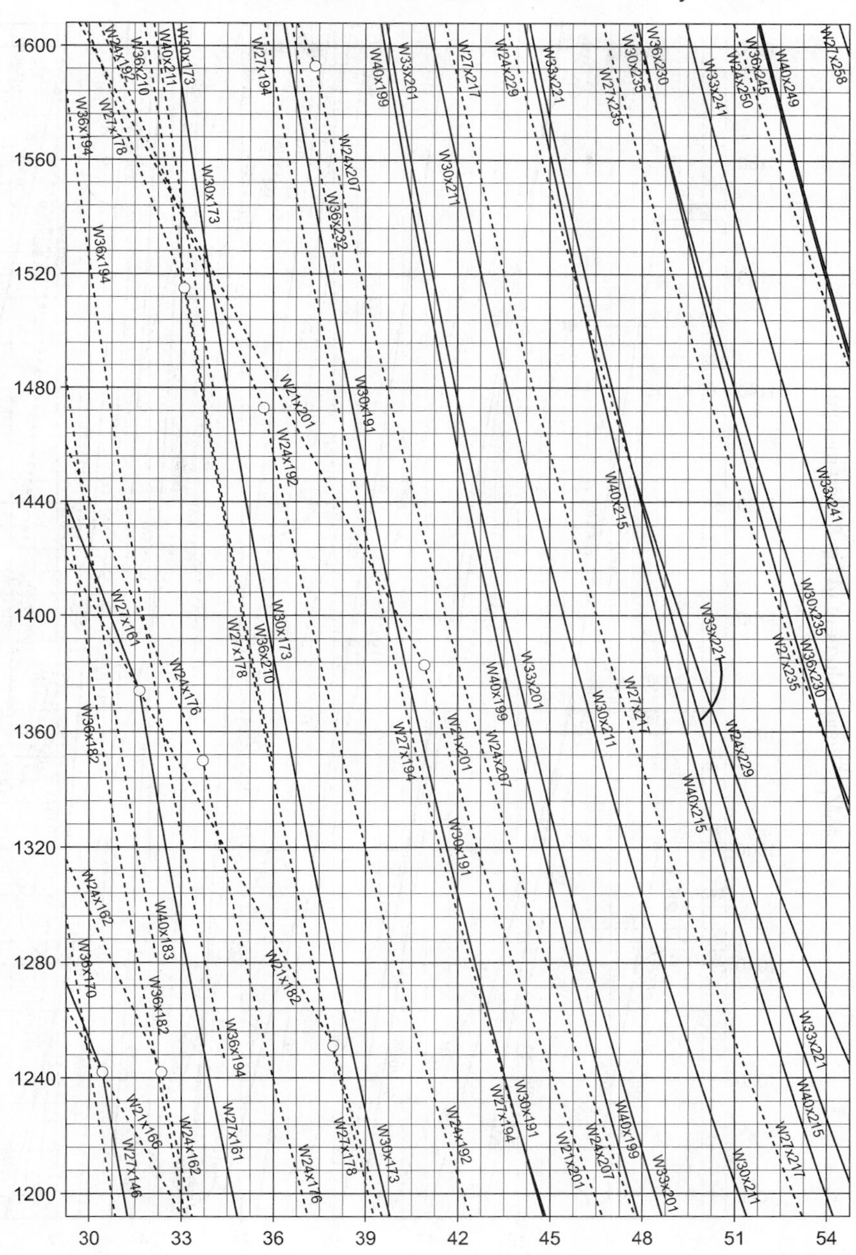
Beam Design Moments (ϕb=0.9, Cb=1.0, Fy=50 ksi)
ϕbMn, Design Moment (8 kip-ft increments)
Lb, Unbraced Length (0.75 ft increments)
1600
1560
1520
1480
1440
1400
1360
1320
1280
1240
1200
30
33
36
39
42
45
48
51
54

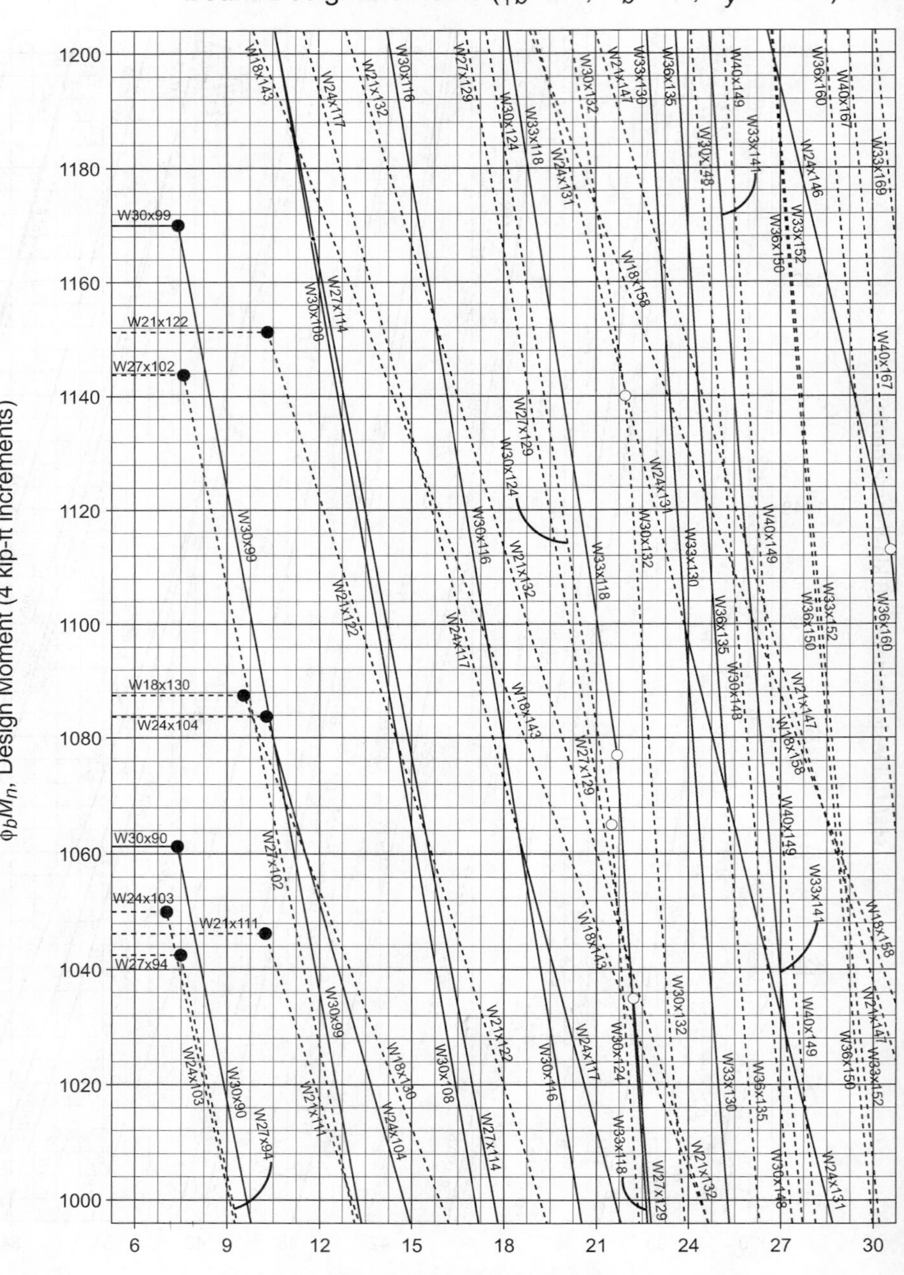
Beam Design Moments (ϕ_b=0.9, C_b=1.0, F_y=50 ksi)
$\phi_b M_n$, Design Moment (4 kip-ft increments)
L_b, Unbraced Length (0.75 ft increments)
W30x99
W21x122
W27x102
W18x130
W24x104
W30x90
W24x103
W21x111
W27x94
W18x143
W24x117
W21x132
W30x116
W27x129
W30x124
W33x118
W24x131
W30x132
W21x147
W33x130
W36x135
W30x148
W40x149
W33x141
W36x160
W40x167
W24x146
W33x169
W33x152
W36x150
W18x158
W30x108
W27x114

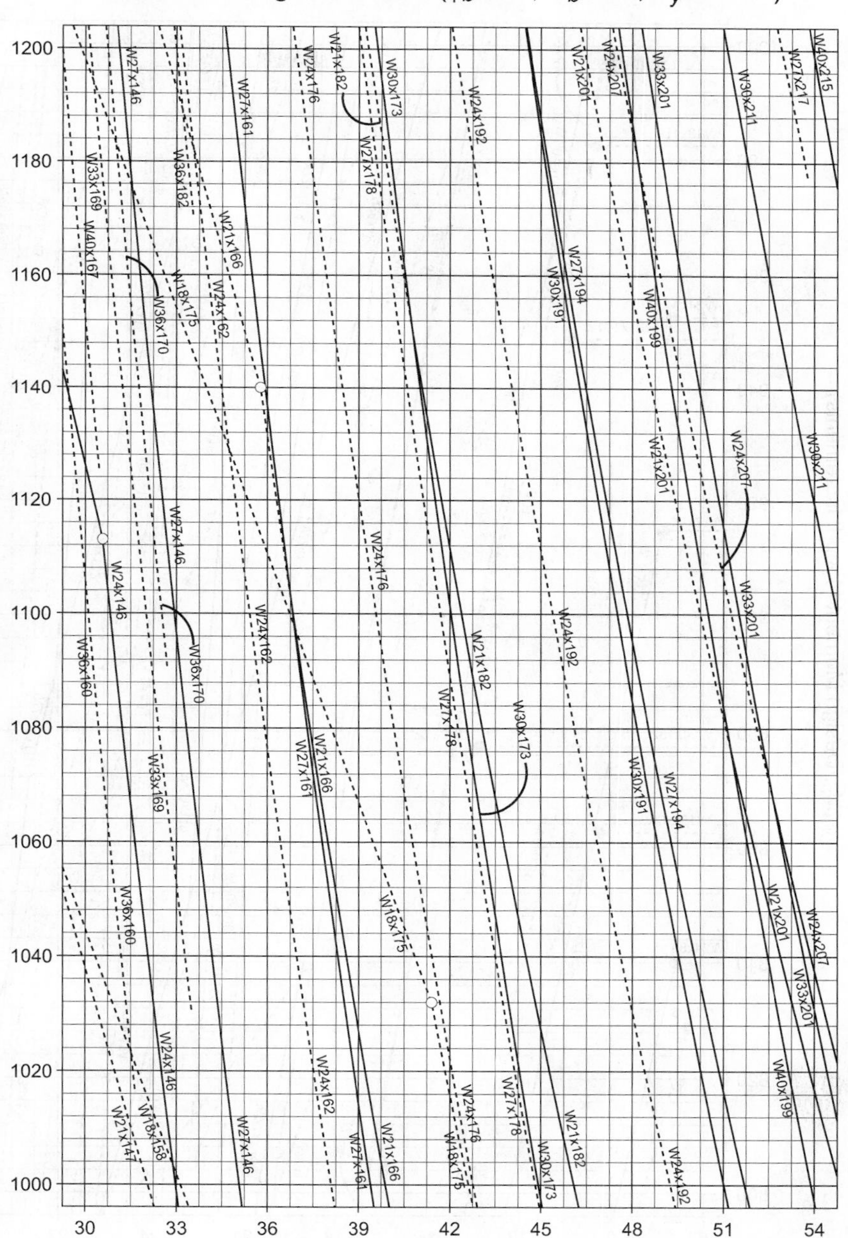
Beam Design Moments (ϕ_b=0.9, C_b=1.0, F_y=50 ksi)
$\phi_b M_n$, Design Moment (4 kip-ft increments)
L_b, Unbraced Length (0.75 ft increments)
1000
1020
1040
1060
1080
1100
1120
1140
1160
1180
1200
30
33
36
39
42
45
48
51
54
W27x146
W27x161
W24x176
W21x182
W30x173
W27x178
W24x192
W21x201
W24x207
W33x201
W30x211
W27x217
W40x215
W33x169
W36x182
W40x167
W21x166
W18x175
W24x162
W36x170
W27x194
W30x191
W40x199
W21x201
W24x207
W30x211
W24x146
W36x160
W33x201
W24x192
W21x182
W27x178
W30x173
W21x166
W27x161
W33x169
W18x175
W36x160
W24x146
W21x147
W18x158
W27x146
W24x162
W27x161
W21x166
W24x176
W18x175
W27x178
W30x173
W21x182
W24x192
W30x191
W27x194
W21x201
W24x207
W33x201
W40x199

Beam Design Moments (ϕ_b=0.9, C_b=1.0, F_y=50 ksi)

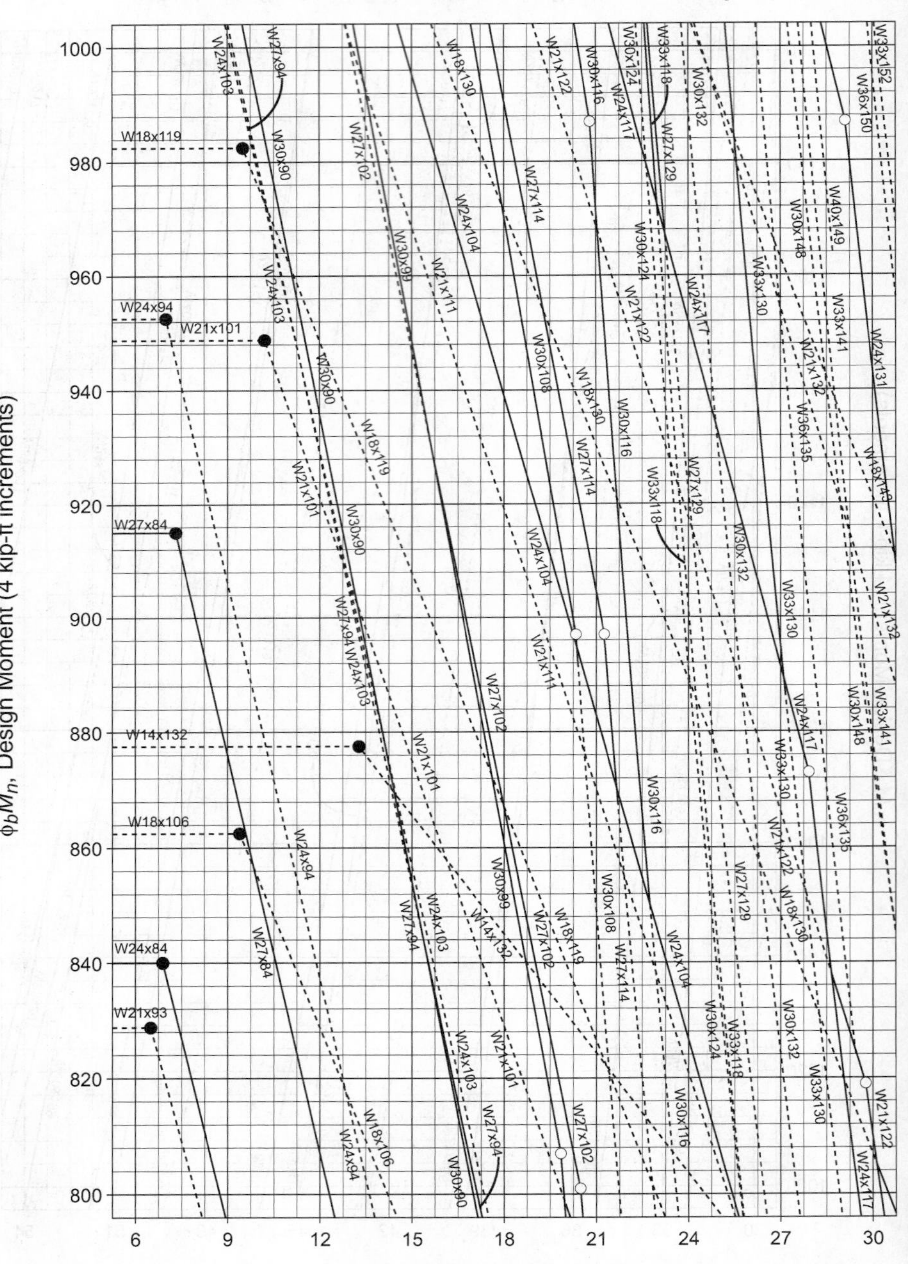

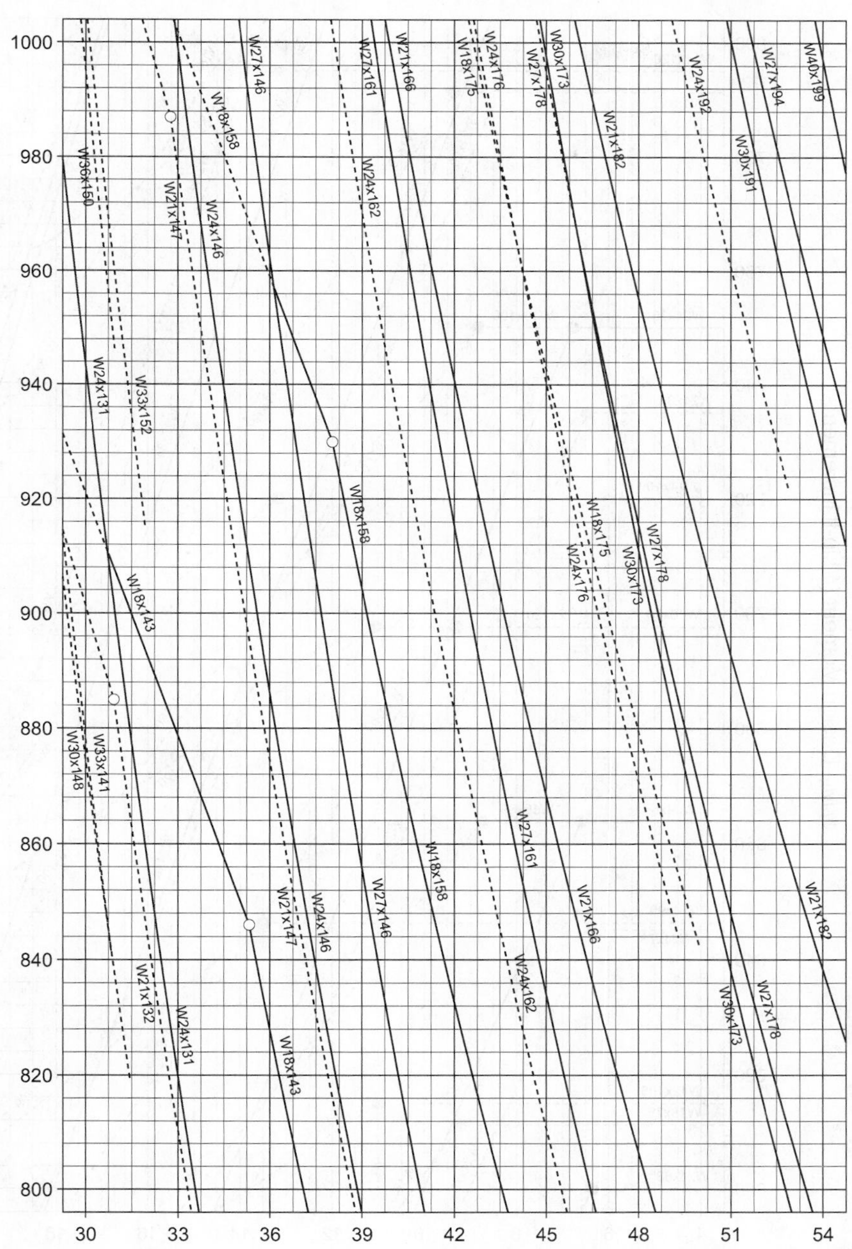
Beam Design Moments (ϕb=0.9, Cb=1.0, Fy=50 ksi)
ϕbMn, Design Moment (4 kip-ft increments)
Lb, Unbraced Length (0.75 ft increments)
1000
980
960
940
920
900
880
860
840
820
800
30
33
36
39
42
45
48
51
54
W27x146
W27x161
W21x166
W18x175
W24x176
W27x178
W30x173
W24x192
W27x194
W40x199
W18x158
W21x182
W30x191
W36x150
W21x147
W24x146
W24x162
W24x131
W33x152
W18x175
W24x176
W30x173
W27x178
W18x143
W30x148
W33x141
W27x161
W18x158
W27x146
W21x147
W24x146
W21x166
W21x182
W24x162
W21x132
W24x131
W18x143
W30x173
W27x178

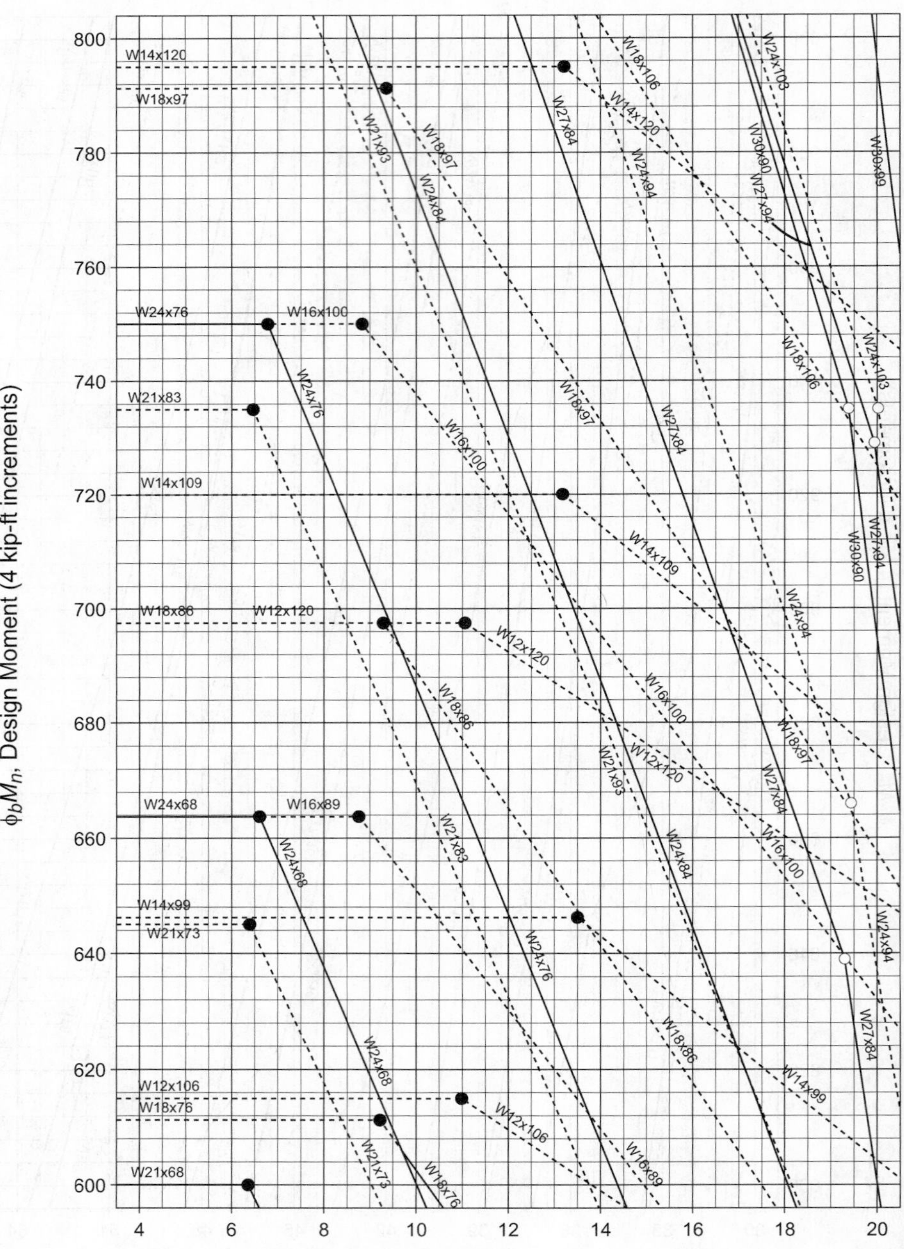
Beam Design Moments (ϕ_b=0.9, C_b=1.0, F_y=50 ksi)
$\phi_b M_n$, Design Moment (4 kip-ft increments)
L_b, Unbraced Length (0.5 ft increments)
800
780
760
740
720
700
680
660
640
620
600
4
6
8
10
12
14
16
18
20
W14x120
W18x97
W24x76
W16x100
W21x83
W14x109
W18x86
W12x120
W24x68
W16x89
W14x99
W21x73
W12x106
W18x76
W21x68
W21x93
W18x97
W24x84
W27x84
W18x106
W14x120
W24x94
W24x103
W30x90
W27x94
W30x99
W16x100
W14x109
W12x120
W18x86
W21x93
W21x83
W24x76
W18x86
W14x99
W12x106
W16x89
W24x68
W21x73
W18x76

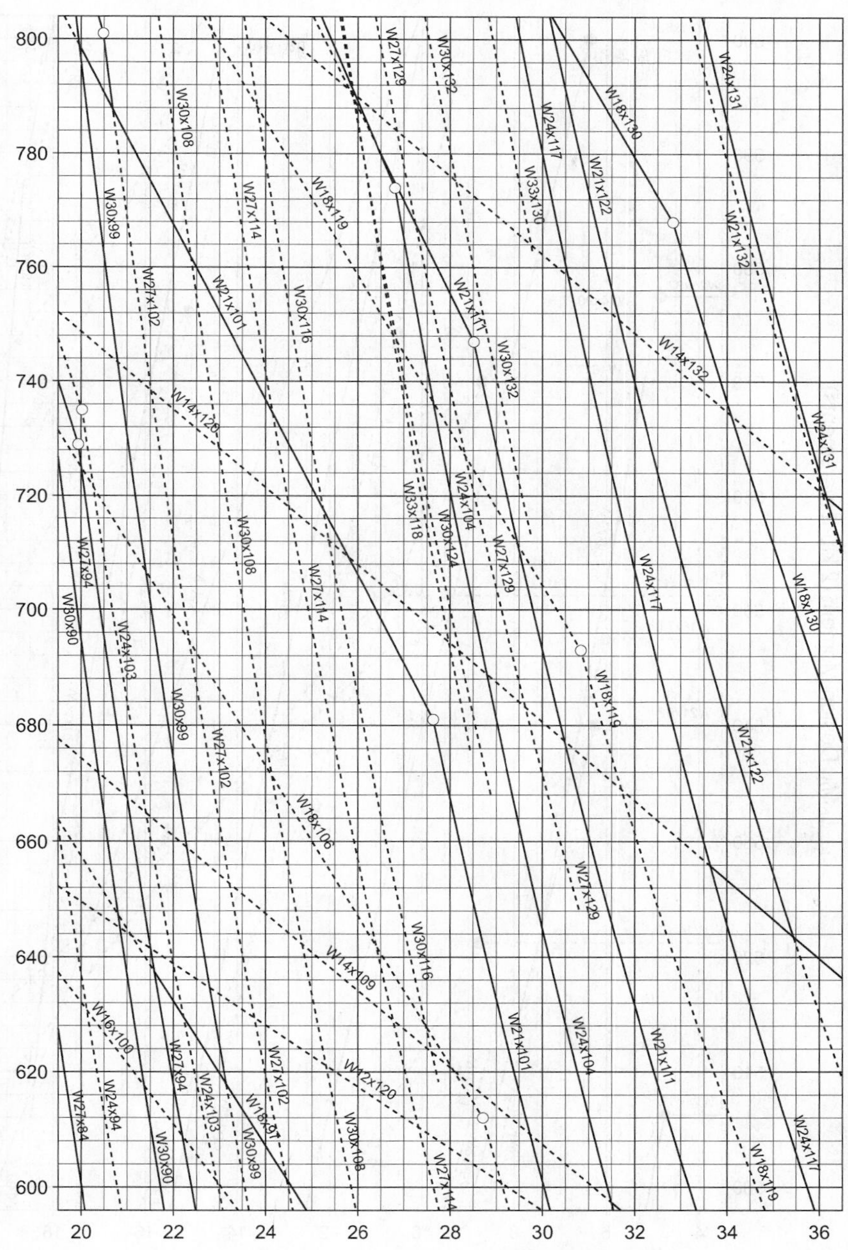
Beam Design Moments (ϕ_b=0.9, C_b=1.0, F_y=50 ksi)
$\phi_b M_n$, Design Moment (4 kip-ft increments)
L_b, Unbraced Length (0.5 ft increments)
600
620
640
660
680
700
720
740
760
780
800
20
22
24
26
28
30
32
34
36
W27x129
W30x132
W24x131
W18x130
W24x117
W21x122
W30x108
W33x130
W30x99
W27x114
W18x119
W21x132
W27x102
W21x101
W30x116
W21x111
W14x132
W30x132
W14x120
W24x104
W33x118
W30x124
W27x129
W30x108
W24x117
W18x130
W27x94
W27x114
W30x90
W24x103
W18x119
W30x99
W27x102
W21x122
W18x106
W27x129
W30x116
W14x109
W16x100
W21x101
W24x104
W21x111
W27x94
W27x102
W12x120
W24x103
W27x84
W24x94
W18x91
W24x117
W30x108
W30x90
W30x99
W27x114
W18x119

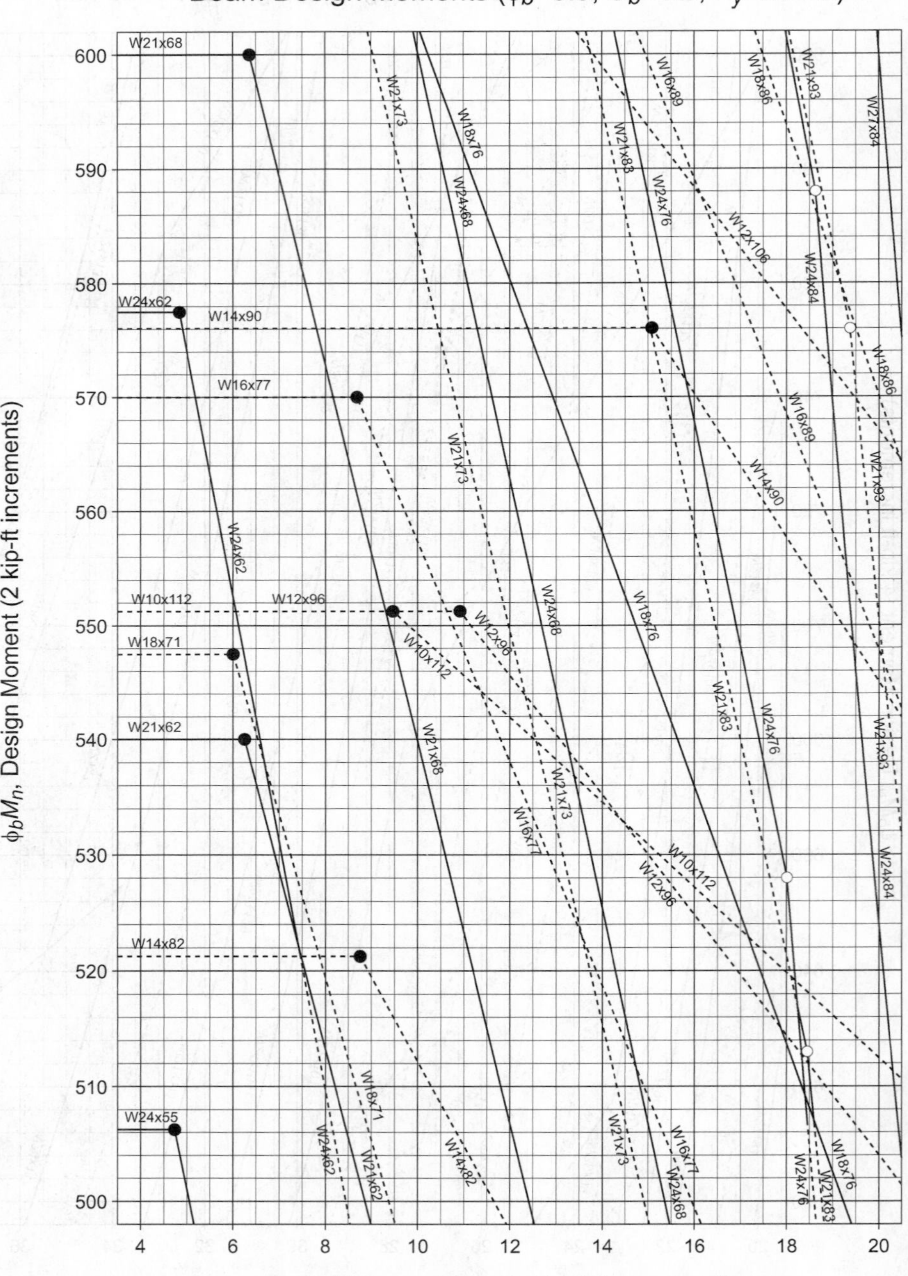
Beam Design Moments (ϕ_b=0.9, C_b=1.0, F_y=50 ksi)
$\phi_b M_n$, Design Moment (2 kip-ft increments)
L_b, Unbraced Length (0.5 ft increments)
W21x68
W24x62
W14x90
W16x77
W10x112
W12x96
W18x71
W21x62
W14x82
W24x55
W21x73
W18x76
W24x68
W21x83
W16x89
W24x76
W18x86
W21x93
W27x84
W12x106
W24x84
W14x90
W10x112
W12x96
W21x73
W16x77
W18x71
W24x62
W21x62
W14x82
W24x76
W21x83
W18x76
W18x86
W21x93
W24x84
600
590
580
570
560
550
540
530
520
510
500
4
6
8
10
12
14
16
18
20

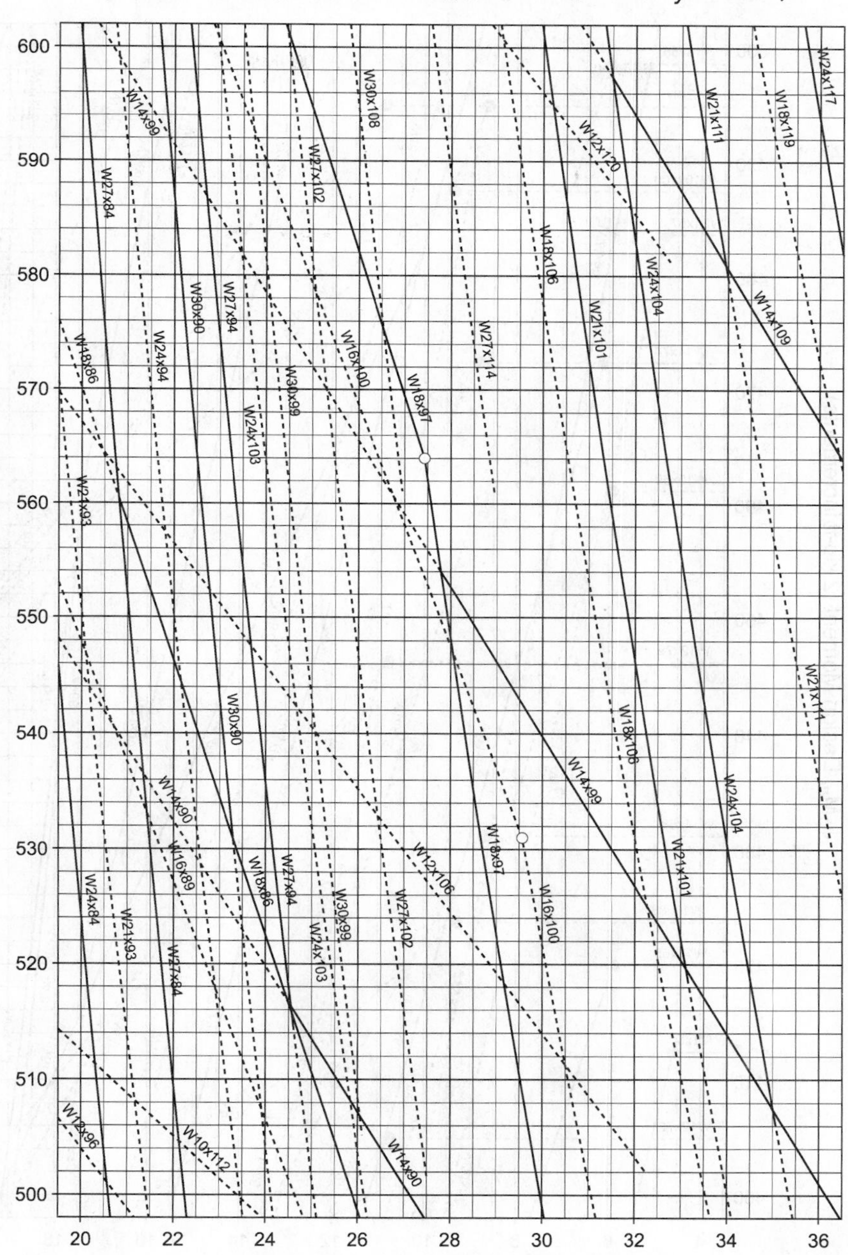

Beam Design Moments (ϕ_b=0.9, C_b=1.0, F_y=50 ksi)
$\phi_b M_n$, Design Moment (2 kip-ft increments)
L_b, Unbraced Length (0.5 ft increments)
W14x99
W27x84
W30x108
W27x102
W21x111
W18x119
W24x117
W12x120
W18x106
W24x104
W14x109
W30x90
W27x94
W24x94
W18x86
W16x100
W18x97
W27x114
W21x101
W30x99
W24x103
W21x93
W24x84
W14x90
W16x89
W12x106
W10x112
W12x96

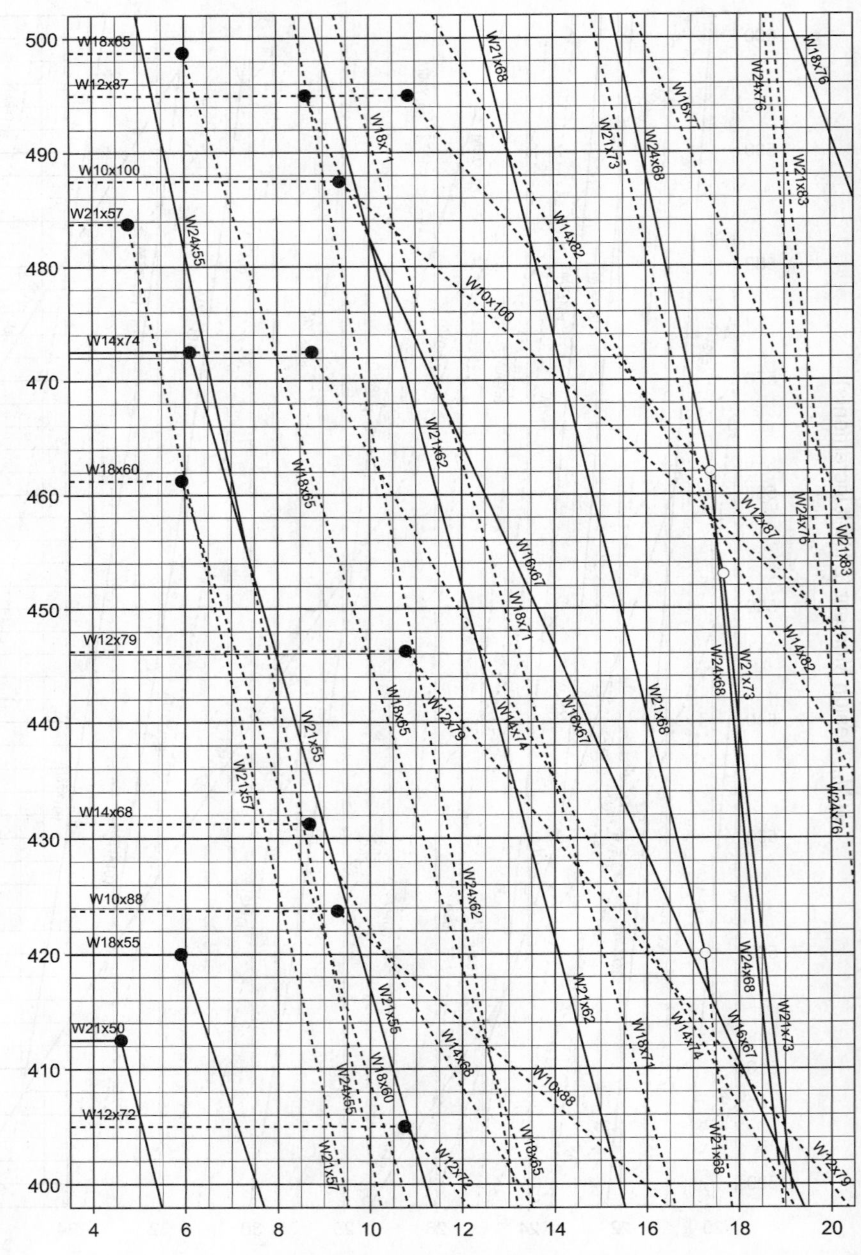

Beam Design Moments (ϕ_b=0.9, C_b=1.0, F_y=50 ksi)
$\phi_b M_n$, Design Moment (2 kip-ft increments)
L_b, Unbraced Length (0.5 ft increments)
W18x65
W12x87
W10x100
W21x57
W14x74
W18x60
W12x79
W14x68
W10x88
W18x55
W21x50
W12x72
W24x55
W18x71
W21x68
W21x73
W24x68
W16x77
W24x76
W18x76
W21x83
W14x82
W10x100
W21x62
W18x65
W16x67
W12x87
W24x76
W21x55
W21x57
W18x65
W12x79
W14x74
W24x62
W21x55
W24x55
W18x60
W14x68
W10x88
W18x65
W12x72
W21x62
W18x71
W14x74
W16x67
W21x73
W21x68
W12x79

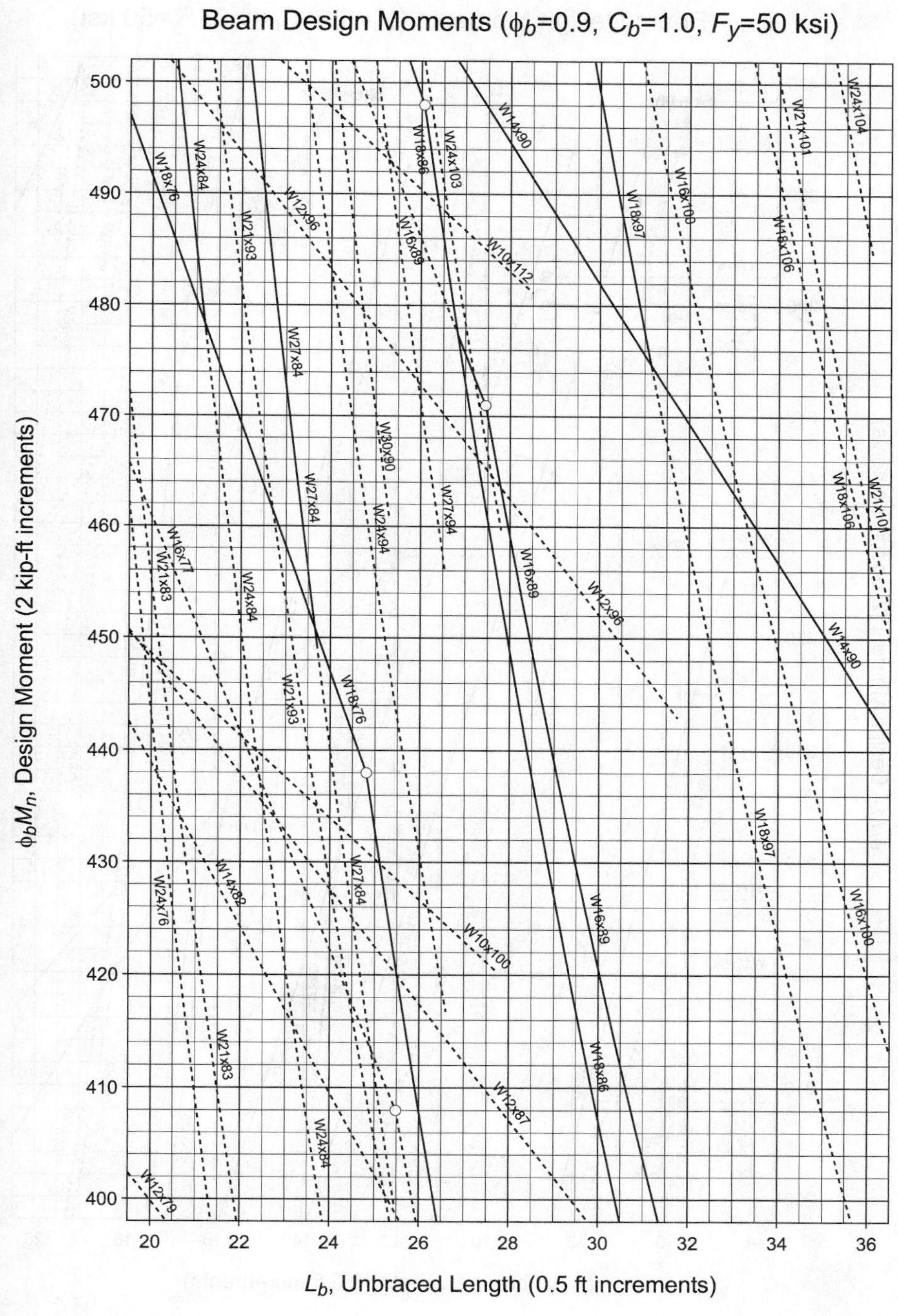

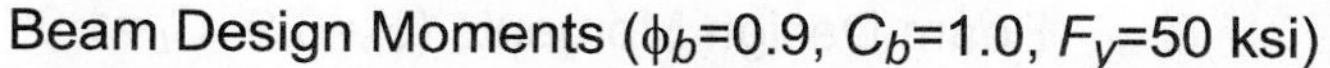
Beam Design Moments (ϕ_b=0.9, C_b=1.0, F_y=50 ksi)
$\phi_b M_n$, Design Moment (2 kip-ft increments)
L_b, Unbraced Length (0.5 ft increments)
500
490
480
470
460
450
440
430
420
410
400
20
22
24
26
28
30
32
34
36
W18x76
W24x84
W21x93
W12x96
W27x84
W18x86
W24x103
W16x89
W10x112
W14x90
W18x97
W16x100
W21x101
W18x106
W24x104
W30x90
W24x94
W27x94
W16x77
W21x83
W24x76
W14x82
W10x100
W12x87
W12x79
W18x97
W16x100
W21x101
W18x106
W14x90
W16x89
W18x86
W24x84
W21x93
W27x84
W18x76
W24x84
W21x83

Beam Design Moments (ϕ_b=0.9, C_b=1.0, F_y=50 ksi)

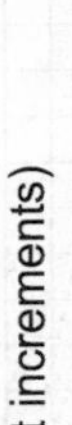

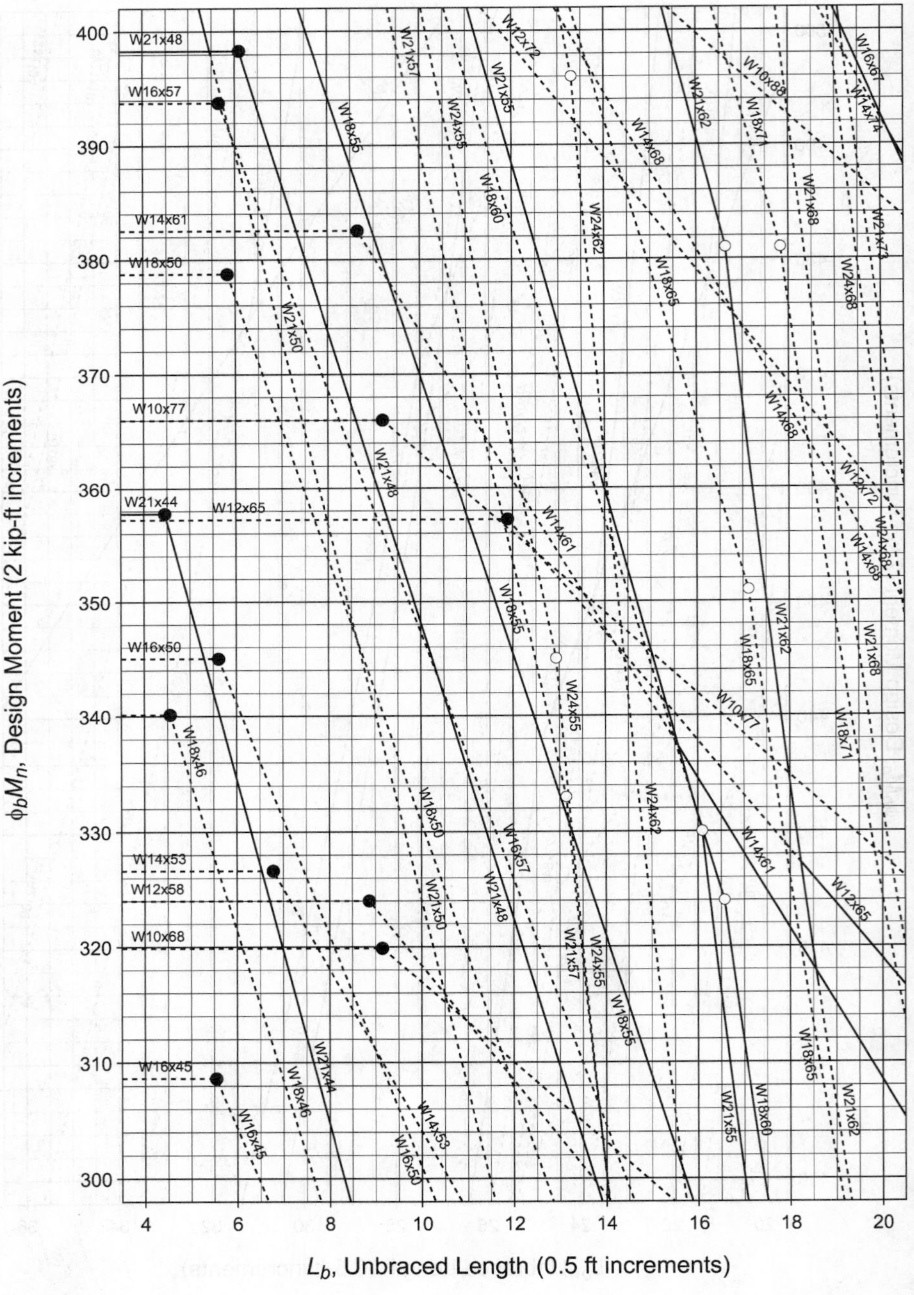

$\phi_b M_n$, Design Moment (2 kip-ft increments)

400
390
380
370
360
350
340
330
320
310
300

W21x48
W16x57
W14x61
W18x50
W10x77
W21x44
W12x65
W16x50
W14x53
W12x58
W10x68
W16x45
W18x55
W21x57
W24x55
W21x55
W12x72
W18x60
W24x62
W14x68
W18x65
W21x62
W10x88
W18x71
W21x68
W16x67
W14x74
W21x73
W24x68
W21x50
W14x61
W18x46
W18x50
W16x57
W21x48
W24x55
W21x57
W18x55
W10x77
W14x61
W12x65
W21x44
W18x46
W16x45
W14x53
W16x50
W21x55
W18x60
W18x65
W21x62
W14x68
W12x72
W14x68
W24x68
W21x68
W18x71
W24x62
W21x50

4 6 8 10 12 14 16 18 20

L_b, Unbraced Length (0.5 ft increments)

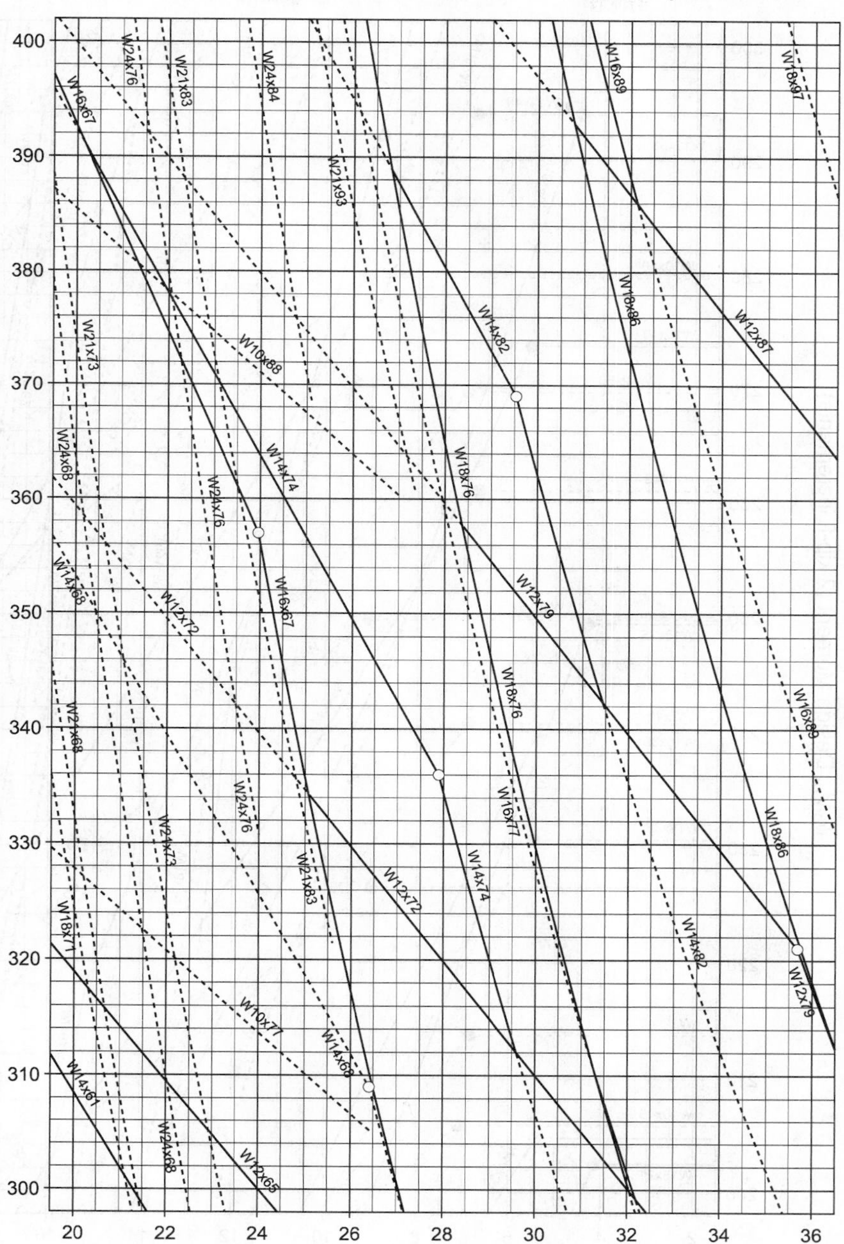

Beam Design Moments (ϕ_b=0.9, C_b=1.0, F_y=50 ksi)
$\phi_b M_n$, Design Moment (2 kip-ft increments)
L_b, Unbraced Length (0.5 ft increments)
W16x67
W24x76
W21x83
W24x84
W21x93
W16x89
W18x97
W21x73
W10x88
W14x82
W18x86
W12x87
W24x68
W14x74
W18x76
W24x76
W16x67
W14x68
W12x72
W12x79
W18x76
W16x89
W21x68
W24x76
W21x73
W21x83
W12x72
W14x74
W16x77
W18x86
W18x71
W14x82
W12x79
W10x77
W14x68
W14x61
W24x68
W12x65

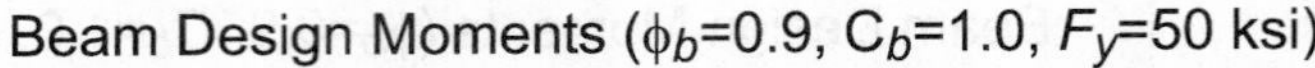

Beam Design Moments (ϕ_b=0.9, C_b=1.0, F_y=50 ksi)

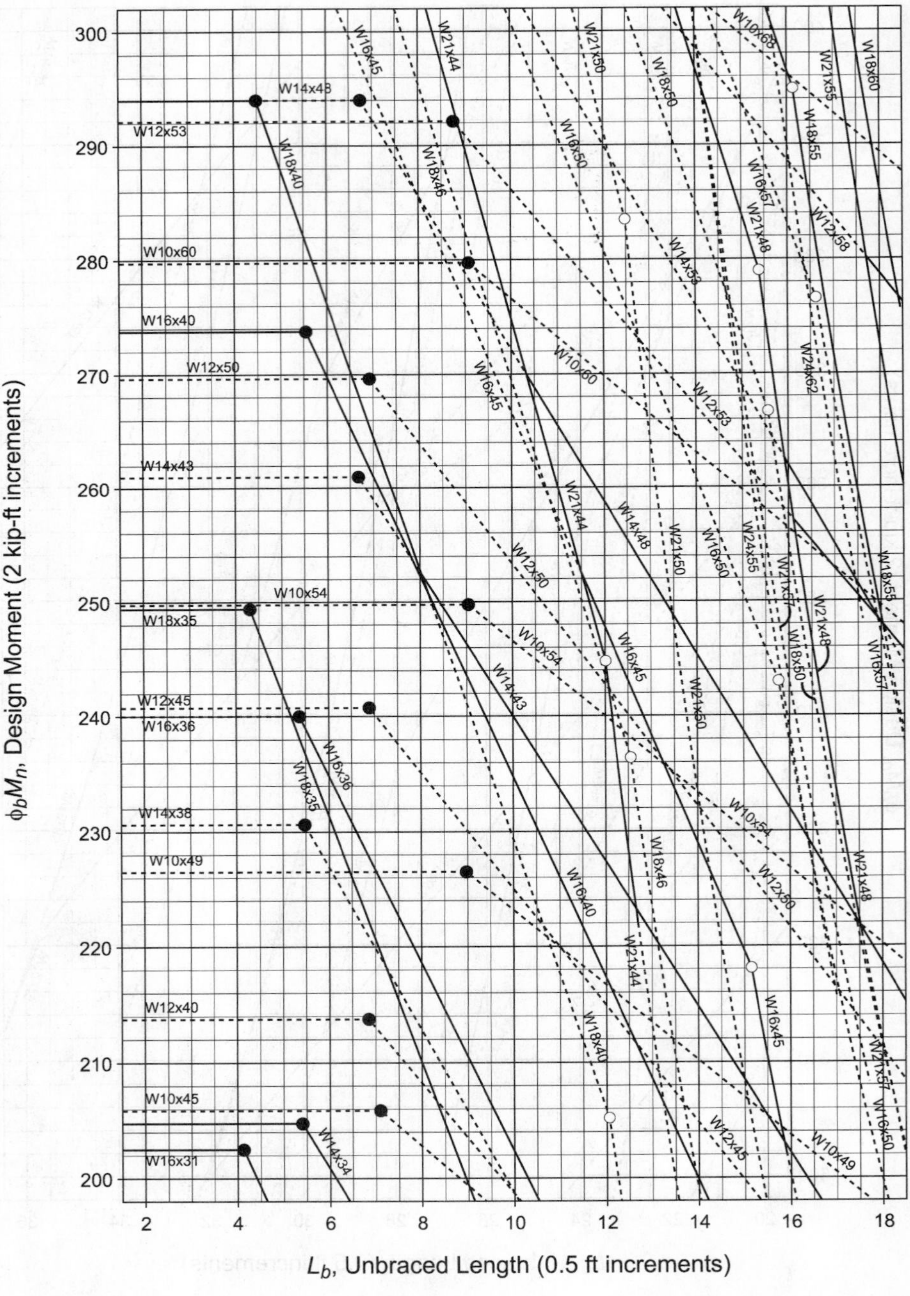

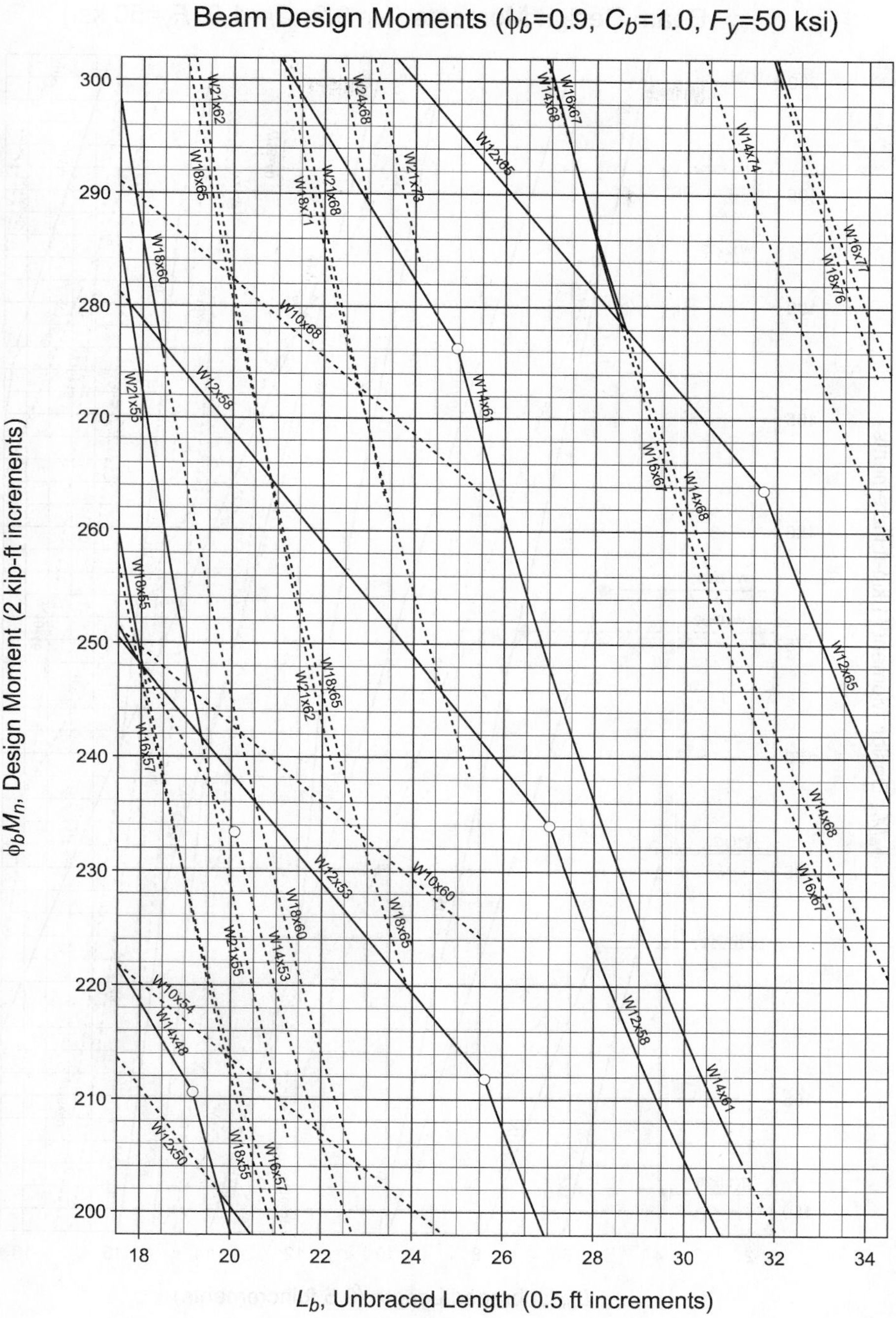

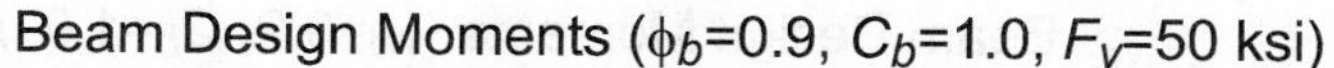
Beam Design Moments (ϕ_b=0.9, C_b=1.0, F_y=50 ksi)
$\phi_b M_n$, Design Moment (2 kip-ft increments)
L_b, Unbraced Length (0.5 ft increments)
300
290
280
270
260
250
240
230
220
210
200
18
20
22
24
26
28
30
32
34
W21x62
W18x65
W24x68
W21x68
W18x71
W21x73
W14x68
W16x67
W12x65
W14x74
W16x77
W18x76
W18x60
W10x68
W21x55
W12x58
W14x61
W16x67
W14x68
W18x65
W18x65
W21x62
W16x57
W12x65
W14x68
W16x67
W10x60
W12x53
W18x60
W18x65
W21x55
W14x53
W10x54
W14x48
W12x58
W14x61
W12x50
W18x55
W16x57

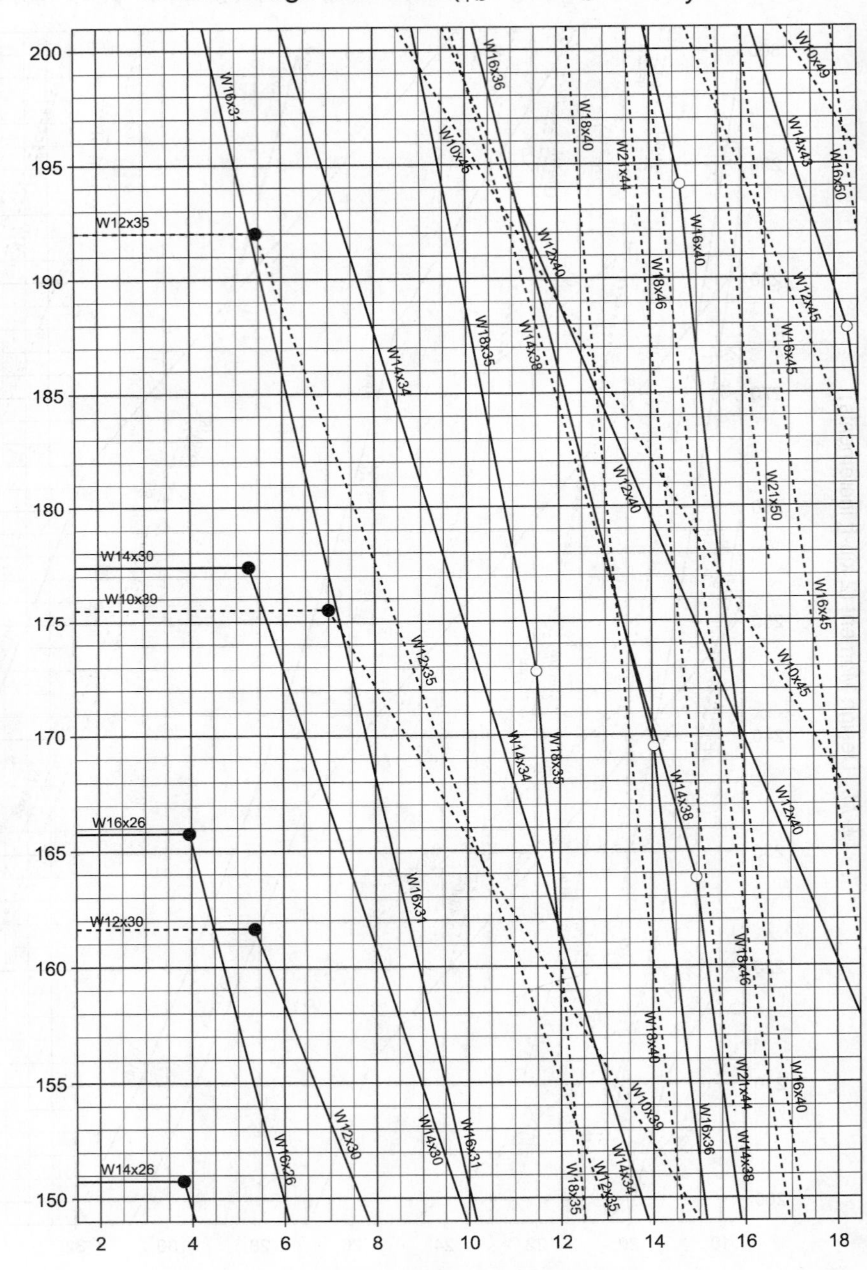
Beam Design Moments (ϕb=0.9, Cb=1.0, Fy=50 ksi)
ϕbMn, Design Moment (1 kip-ft increments)
Lb, Unbraced Length (0.5 ft increments)
150
155
160
165
170
175
180
185
190
195
200
2
4
6
8
10
12
14
16
18
W12x35
W14x30
W10x39
W16x26
W12x30
W14x26
W16x31
W14x34
W10x46
W16x36
W18x40
W21x44
W10x49
W14x43
W16x50
W12x40
W18x46
W16x40
W18x35
W14x38
W12x45
W16x45
W21x50
W10x45
W12x35
W14x30
W16x26
W12x30
W21x44
W16x36
W14x38
W18x40
W10x39

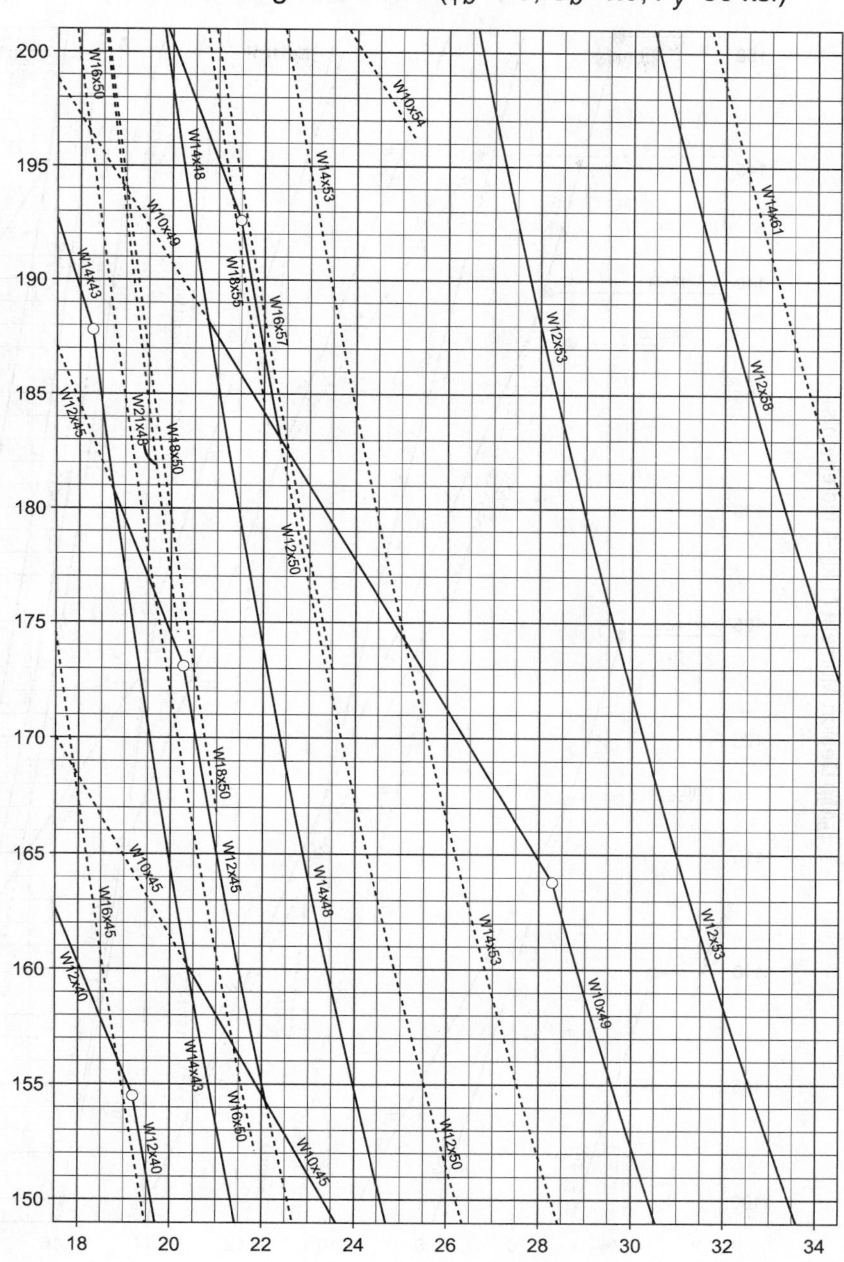
Beam Design Moments (φb=0.9, Cb=1.0, Fy=50 ksi)
φbMn, Design Moment (1 kip-ft increments)
Lb, Unbraced Length (0.5 ft increments)
200
195
190
185
180
175
170
165
160
155
150
18
20
22
24
26
28
30
32
34
W16x50
W10x54
W14x48
W14x53
W10x49
W14x61
W14x43
W18x55
W16x57
W12x53
W12x58
W12x45
W21x48
W18x50
W12x50
W18x50
W10x45
W12x45
W14x48
W14x53
W12x53
W16x45
W12x40
W10x49
W14x43
W16x50
W12x40
W10x45
W12x50

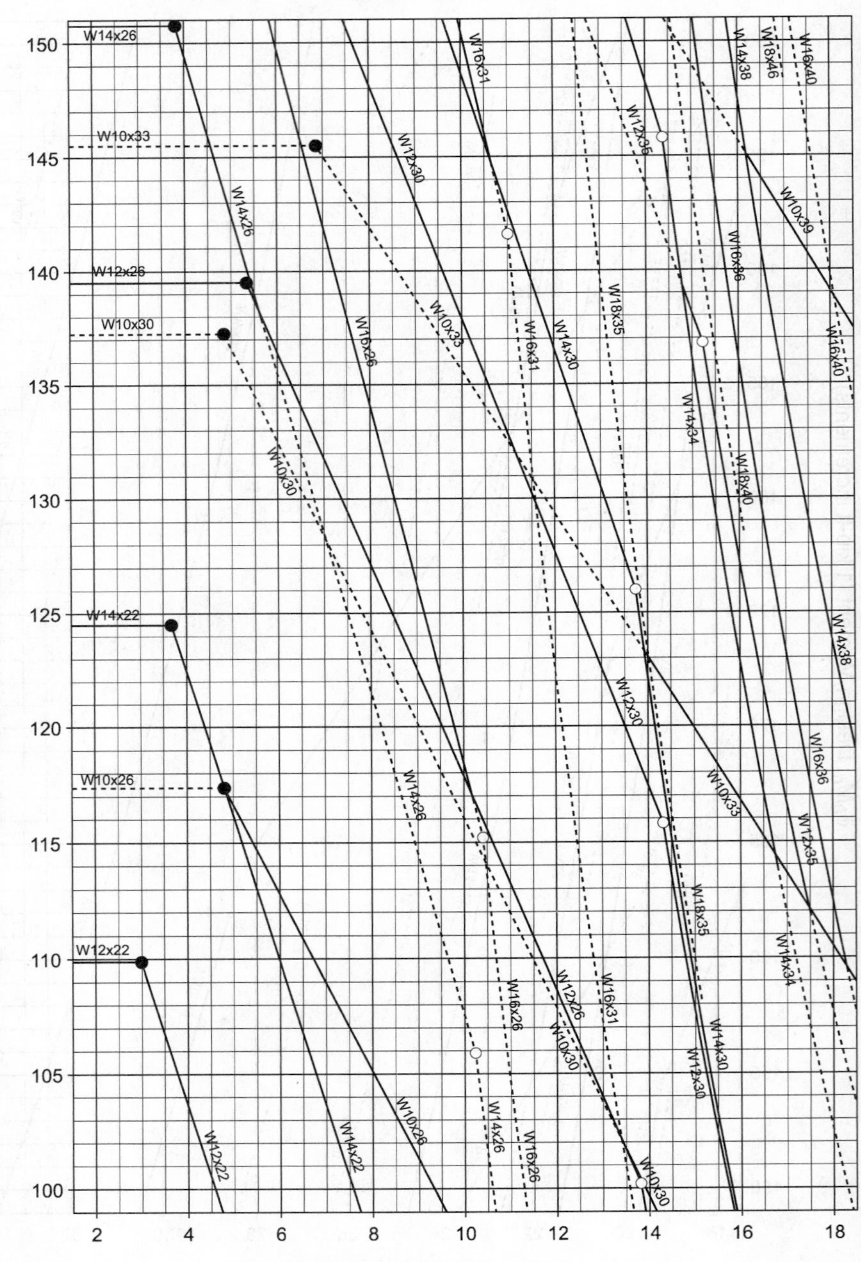
Beam Design Moments (ϕ_b=0.9, C_b=1.0, F_y=50 ksi)
$\phi_b M_n$, Design Moment (1 kip-ft increments)
L_b, Unbraced Length (0.5 ft increments)
W14x26
W10x33
W12x26
W10x30
W14x22
W10x26
W12x22
W16x31
W12x30
W14x26
W12x35
W14x38
W18x46
W16x40
W10x39
W16x36
W16x26
W10x33
W16x31
W14x30
W18x35
W14x34
W18x40
W10x30
W12x30
W14x38
W16x36
W10x33
W12x35
W14x26
W18x35
W14x34
W16x26
W12x26
W16x31
W10x30
W14x30
W12x30
W12x22
W14x22
W10x26
W14x26
W16x26
W10x30
150
145
140
135
130
125
120
115
110
105
100
2
4
6
8
10
12
14
16
18

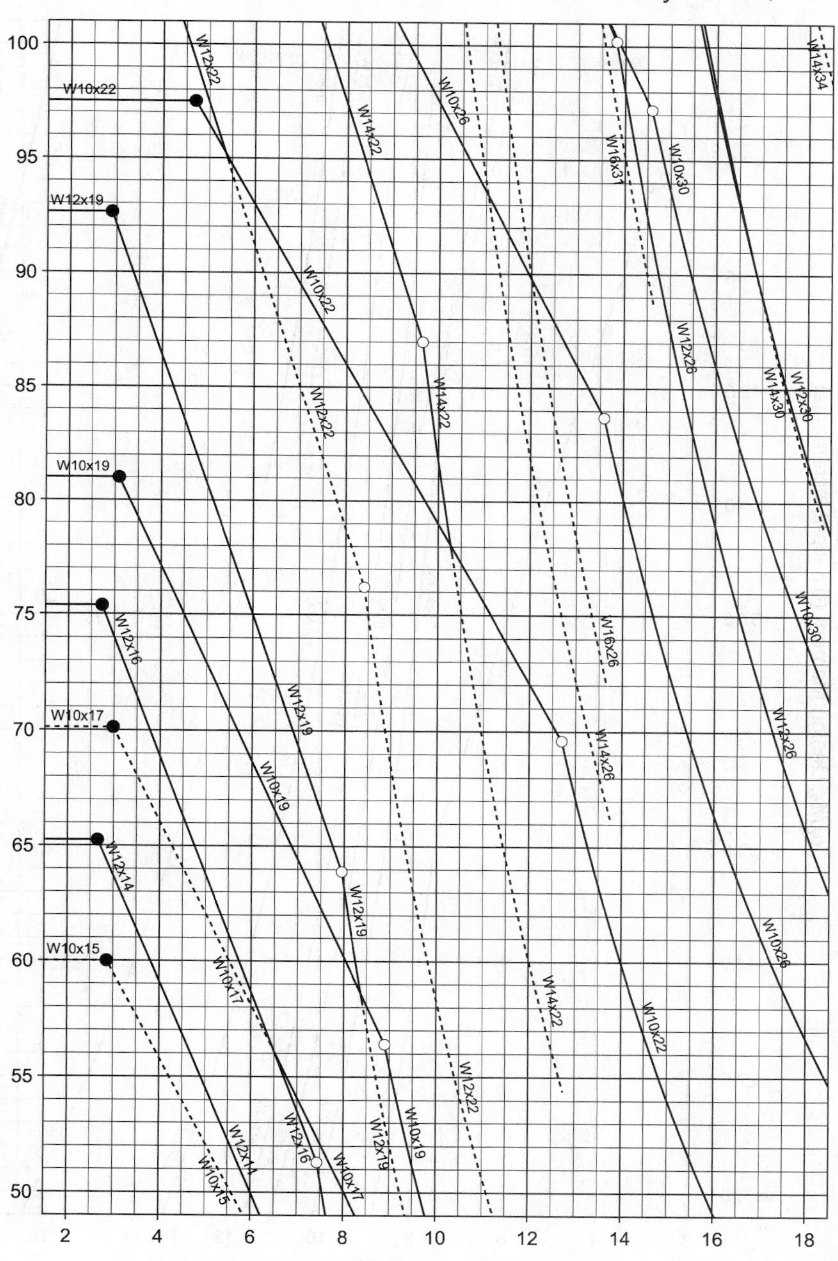
Beam Design Moments (ϕ_b=0.9, C_b=1.0, F_y=50 ksi)
$\phi_b M_n$, Design Moment (1 kip-ft increments)
L_b, Unbraced Length (0.5 ft increments)
100
95
90
85
80
75
70
65
60
55
50
2
4
6
8
10
12
14
16
18
W10x22
W12x19
W10x19
W10x17
W10x15
W12x22
W14x22
W10x26
W16x31
W10x30
W14x34
W12x16
W12x14
W12x26
W14x30
W12x30
W16x26
W14x26
W10x17
W12x14
W10x15

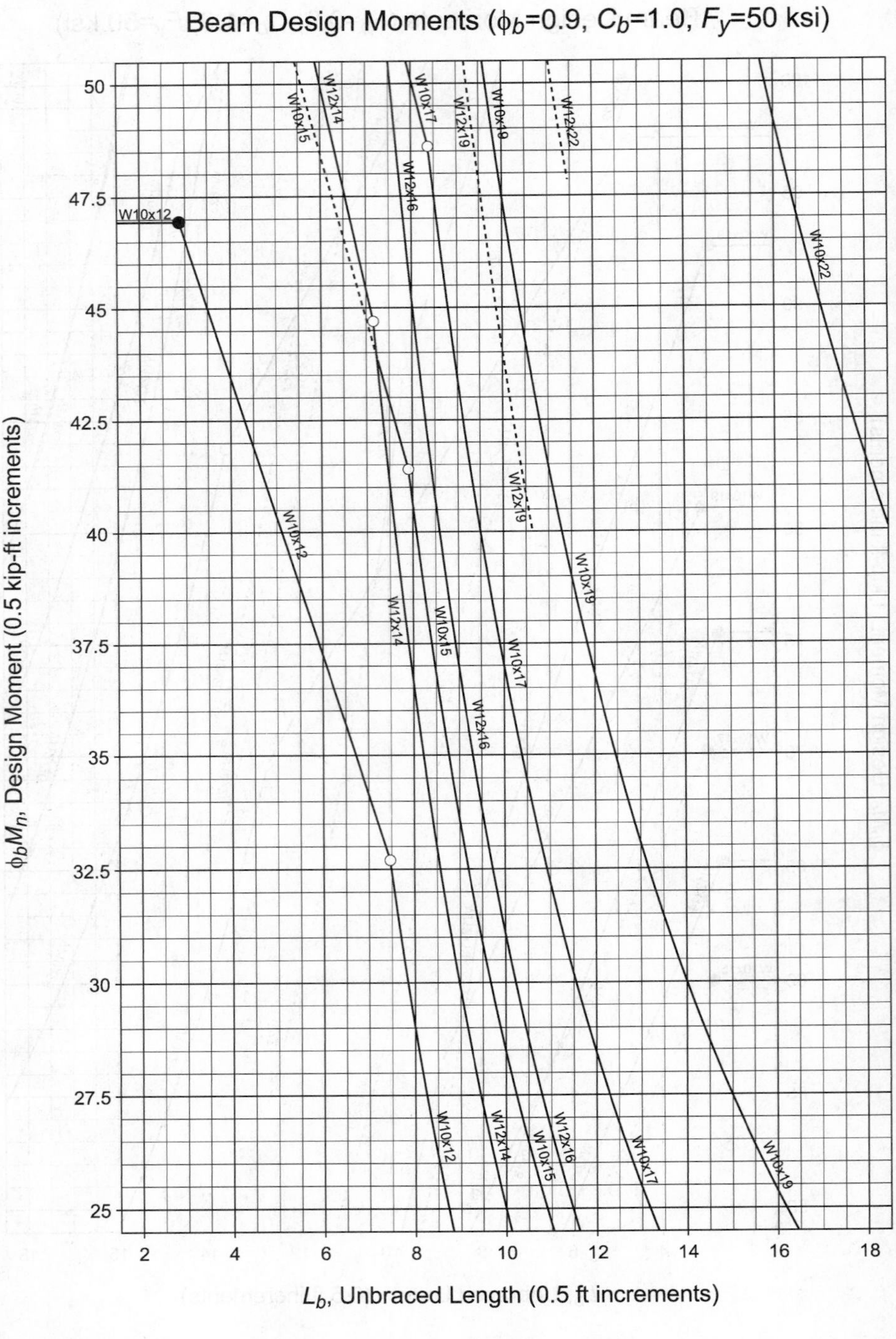

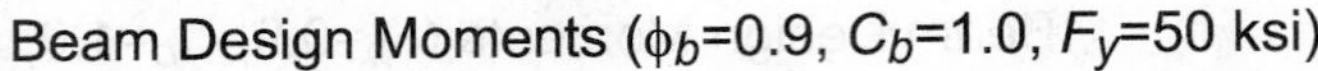

Beam Design Moments (ϕ_b=0.9, C_b=1.0, F_y=50 ksi)
$\phi_b M_n$, Design Moment (0.5 kip-ft increments)
L_b, Unbraced Length (0.5 ft increments)
50
47.5
45
42.5
40
37.5
35
32.5
30
27.5
25
2
4
6
8
10
12
14
16
18
W10x12
W10x15
W12x14
W10x17
W12x19
W10x19
W12x22
W12x16
W10x22
W10x12
W12x19
W10x19
W12x14
W10x15
W10x17
W12x16
W10x12
W12x14
W10x15
W12x16
W10x17
W10x19

$F_y = 50$ ksi
$\phi_b = 0.90$
$\phi_v = 0.90$

Table 5-6. W-Shapes Selection by I_y

I_y

Shape	Y-Y Axis			X-X Axis							
	I_y	Z_y	$\phi_b M_{py}$	I_x	Z_x	$\phi_b M_{px}$	$\phi_b M_{rx}$	L_p	L_r	BF	$\phi_v V_n$
	in.^{4}	in.^{3}	kip-ft	in.^{4}	in.^{3}	kip-ft	kip-ft	ft	ft	kips	kips
W14×808*	**5510**	**927**	**3340**	**16000**	**1830**	**6860**	**4200**	**17.0**	**268**	**10.6**	**2300**
W14×730*	**4720**	**816**	**2960**	**14300**	**1660**	**6230**	**3840**	**16.6**	**241**	**10.6**	**1860**
W36×798*	4200	743	2630	62600	3580	13400	8940	14.9	84.6	64.4	2700
W14×665*	**4170**	**730**	**2660**	**12400**	**1480**	**5550**	**3450**	**16.3**	**222**	**10.2**	**1650**
W14×605*	**3680**	**652**	**2380**	**10800**	**1320**	**4950**	**3120**	**16.1**	**203**	**9.79**	**1470**
W14×550*	**3250**	**583**	**2130**	**9430**	**1180**	**4430**	**2790**	**15.9**	**187**	**9.51**	**1300**
W36×650*	3230	580	2060	48900	2860	10700	7260	14.5	69.6	62.9	2150
W14×500*	**2880**	**522**	**1910**	**8210**	**1050**	**3940**	**2510**	**15.6**	**172**	**9.13**	**1160**
W14×455*	**2560**	**468**	**1710**	**7190**	**936**	**3510**	**2270**	**15.5**	**157**	**8.75**	**1040**
W40×593*	2520	481	1700	50400	2760	10400	7020	13.4	56.5	77.3	2080
W36×527*	2490	454	1630	38300	2280	8550	5850	14.2	58.1	61.5	1700
W14×426*	**2360**	**434**	**1590**	**6600**	**869**	**3260**	**2120**	**15.3**	**148**	**8.61**	**949**
W14×398*	**2170**	**402**	**1470**	**6000**	**801**	**3000**	**1970**	**15.2**	**139**	**8.38**	**875**
W27×539*	2110	437	1560	25600	1890	7090	4710	12.9	78.0	36.5	1730
W40×503*	2050	395	1410	41700	2320	8700	5940	13.1	49.4	76.0	1750
W36×439*	1990	367	1320	31000	1870	7010	4860	13.9	50.1	59.4	1410
W14×370*	**1990**	**370**	**1360**	**5440**	**736**	**2760**	**1820**	**15.1**	**129**	**8.21**	**802**
W14×342*	**1810**	**338**	**1240**	**4900**	**672**	**2520**	**1670**	**15.0**	**121**	**8.01**	**728**
W36×393*	1750	325	1170	27500	1670	6260	4350	13.8	46.7	58.1	1250
W40×431*	1690	328	1170	34800	1960	7350	5070	12.9	44.1	73.2	1490
W33×387*	1620	312	1130	24300	1560	5850	4050	13.3	47.7	52.4	1220
W14×311*	**1610**	**304**	**1120**	**4330**	**603**	**2260**	**1520**	**14.8**	**110**	**7.84**	**651**
W36×359*	1570	292	1060	24800	1510	5660	3960	13.6	43.8	56.5	1130
W30×391*	1550	310	1110	20700	1450	5440	3750	13.0	52.1	43.1	1220
W40×397*	1540	300	1070	32000	1800	6750	4680	12.9	42.1	70.7	1350
W33×354*	1460	282	1020	22000	1420	5330	3720	13.2	44.6	51.1	1110
W14×283*	**1440**	**274**	**1010**	**3840**	**542**	**2030**	**1380**	**14.7**	**100**	**7.65**	**582**
W40×372*	1420	278	996	29600	1680	6300	4380	12.7	40.1	70.2	1270
W36×328*	1420	265	962	22500	1380	5180	3630	13.6	41.8	54.8	1020
W30×357*	1390	279	1010	18700	1320	4950	3420	12.9	48.6	42.8	1100
W40×362*	1380	270	973	28900	1640	6150	4260	12.7	39.9	69.5	1230
W27×368*	1310	279	1010	16200	1240	4650	3180	12.3	54.7	34.6	1130
W36×300	1300	241	878	20300	1260	4730	3330	13.5	39.9	53.0	936
W33×318*	1290	250	906	19500	1270	4760	3330	13.1	41.9	49.8	988

*ASTM A6 tensile group 4 or 5 shape. Special requirements may apply per LRFD Specification Section A3.1c.

I_y Table 5-6 (cont.). W-Shapes Selection by I_y

$F_y = 50$ ksi
$\phi_b = 0.90$
$\phi_v = 0.90$

Shape	Y-Y Axis			X-X Axis							
	I_y	Z_y	$\phi_b M_{py}$	I_x	Z_x	$\phi_b M_{px}$	$\phi_b M_{rx}$	L_p	L_r	BF	$\phi_v V_n$
	in.^{4}	in.^{3}	kip-ft	in.^{4}	in.^{3}	kip-ft	kip-ft	ft	ft	kips	kips
W14×257*	**1290**	**246**	**906**	**3400**	**487**	**1830**	**1250**	**14.6**	**91.6**	**7.55**	**523**
W30×326*	1240	252	911	16800	1190	4460	3120	12.7	45.2	41.3	997
W40×324	1220	239	861	25600	1460	5480	3840	12.6	37.6	65.6	1090
W44×335	1200	236	849	31100	1620	6080	4230	12.4	35.5	79.6	1210
W36×280	1200	223	810	18900	1170	4390	3090	13.5	38.7	51.3	872
W12×336*	1190	274	996	4060	603	2260	1450	12.3	131	6.81	807
W27×336*	1180	252	911	14600	1130	4240	2920	12.2	50.4	34.5	1020
W33×291	1160	226	821	17700	1160	4350	3060	13.0	39.6	48.5	902
W24×370*	1160	267	956	13400	1130	4240	2870	11.6	61.0	27.7	1150
W14×233*	**1150**	**221**	**816**	**3010**	**436**	**1640**	**1130**	**14.5**	**83.4**	**7.40**	**462**
W30×292*	1100	223	810	14900	1060	3980	2790	12.6	42.1	40.2	881
W40×297	1090	215	776	23200	1330	4990	3510	12.5	35.9	63.1	999
W36×260	1090	204	743	17300	1080	4050	2860	13.4	37.4	49.5	823
W44×290	1050	206	743	27100	1420	5330	3720	12.3	34.0	74.0	1020
W27×307*	1050	227	821	13100	1030	3860	2660	12.0	46.7	34.7	927
W12×305*	1050	244	894	3550	537	2010	1310	12.1	120	6.59	717
W40×277	1040	204	743	21900	1250	4690	3300	12.6	35.5	60.8	890
W33×263	1040	202	737	15900	1040	3900	2760	12.9	37.8	45.9	810
W24×335*	1030	238	855	11900	1020	3830	2590	11.4	55.5	27.9	1020
W14×211	**1030**	**198**	**731**	**2660**	**390**	**1460**	**1010**	**14.4**	**76.0**	**7.28**	**415**
W36×245	1010	190	692	16100	1010	3790	2690	13.2	36.5	47.5	780
W30×261	959	196	714	13100	943	3540	2490	12.5	39.1	39.3	793
W27×281*	953	206	748	11900	936	3510	2440	12.0	43.8	33.5	839
W36×230	940	176	641	15000	943	3540	2510	13.2	35.6	45.6	737
W12×279*	937	220	804	3110	481	1800	1180	11.9	110	6.37	657
W33×241	933	182	664	14200	940	3530	2490	12.8	36.2	44.0	766
W14×193	**931**	**180**	**669**	**2400**	**355**	**1330**	**930**	**14.3**	**70.1**	**7.19**	**372**
W44×262	927	183	664	24200	1270	4760	3360	12.2	32.8	68.3	924
W40×249	926	182	664	19600	1120	4200	2980	12.5	34.1	56.7	798
W24×306*	919	214	771	10700	922	3460	2370	11.3	51.0	27.5	922
W27×258	859	187	675	10800	852	3200	2240	11.9	41.0	32.9	767
W30×235	855	175	641	11700	847	3180	2240	12.4	37.1	37.8	701
W33×221	840	164	596	12900	857	3210	2280	12.7	35.0	42.0	709
W14×176	**838**	**163**	**602**	**2140**	**320**	**1200**	**843**	**14.2**	**64.5**	**7.09**	**341**
W12×252*	828	196	714	2720	428	1610	1060	11.8	100	6.19	582
W24×279*	823	193	698	9600	835	3130	2150	11.2	47.3	27.1	836
W40×392*	803	212	731	29900	1710	6410	4320	9.33	34.3	83.9	1590
W44×230	796	157	568	20800	1100	4130	2910	12.1	31.8	61.7	822
W40×215	796	156	568	16700	964	3620	2580	12.5	32.8	51.2	684
W27×235	769	168	608	9700	772	2900	2030	11.8	38.6	32.2	705
W30×211	757	155	563	10300	751	2820	2000	12.3	35.2	35.8	647
W33×201	749	147	536	11600	773	2900	2060	12.6	33.8	39.6	651

*ASTM A6 tensile group 4 or 5 shape. Special requirements may apply per LRFD Specification Section A3.1c.

$F_y = 50$ ksi
$\phi_b = 0.90$
$\phi_v = 0.90$

Table 5-6 (cont.). W-Shapes Selection by I_y

I_y

Shape	Y-Y Axis			X-X Axis							
	I_y	Z_y	$\phi_b M_{py}$	I_x	Z_x	$\phi_b M_{px}$	$\phi_b M_{rx}$	L_p	L_r	BF	$\phi_v V_n$
	in.^{4}	in.^{3}	kip-ft	in.^{4}	in.^{3}	kip-ft	kip-ft	ft	ft	kips	kips
W14×159	**748**	**146**	**541**	**1900**	**287**	**1080**	**762**	**14.1**	**58.9**	**7.02**	**302**
W12×230*	742	177	647	2420	386	1450	963	11.7	92.0	6.04	526
W24×250*	724	171	619	8490	744	2790	1930	11.1	43.2	26.7	739
W27×217	704	154	561	8910	711	2670	1880	11.7	36.8	31.3	636
W40×199	695	137	496	14900	869	3260	2310	12.2	31.7	48.5	679
W14×145	**677**	**133**	**491**	**1710**	**260**	**975**	**696**	**14.1**	**54.7**	**6.87**	**272**
W30×191	673	138	503	9200	675	2530	1800	12.2	33.8	33.9	589
W12×210*	664	159	585	2140	348	1310	876	11.6	84.2	5.91	468
W24×229	651	154	559	7650	675	2530	1760	11.0	40.2	26.3	674
W40×331	644	172	596	24700	1430	5360	3630	9.08	30.5	80.9	1340
W40×327	640	170	591	24500	1410	5290	3600	9.11	30.3	79.5	1300
W27×194	619	136	496	7860	631	2370	1680	11.6	34.6	29.9	569
W30×173	598	123	449	8230	607	2280	1620	12.1	32.5	32.0	538
W12×190	589	143	523	1890	311	1170	789	11.5	76.6	5.79	412
W24×207	578	137	500	6820	606	2270	1590	10.9	37.3	25.7	604
W27×178	555	122	443	7020	570	2140	1520	11.5	33.0	28.9	544
W14×132	**548**	**113**	**419**	**1530**	**234**	**878**	**627**	**13.3**	**49.7**	**6.88**	**256**
W21×201	542	133	484	5310	530	1990	1380	10.7	41.0	19.9	565
W24×192	530	126	460	6260	559	2100	1470	10.8	35.6	25.1	558
W36×256	528	137	487	16800	1040	3900	2690	9.36	28.7	63.0	969
W40×278	521	140	490	20500	1190	4460	3060	8.90	27.6	75.1	1110
W12×170	517	126	463	1650	275	1030	705	11.4	68.9	5.67	363
W27×161	497	109	399	6310	515	1930	1370	11.4	31.7	27.5	492
W14×120	**495**	**102**	**380**	**1380**	**212**	**795**	**570**	**13.2**	**46.3**	**6.81**	**231**
W40×264	493	132	465	19400	1130	4240	2910	8.90	27.0	73.2	1040
W21×182	483	119	434	4740	477	1790	1250	10.6	38.1	19.5	509
W24×176	479	115	418	5680	511	1920	1350	10.7	33.7	24.7	510
W36×232	468	122	434	15000	936	3510	2430	9.25	27.3	60.1	871
W12×152	454	111	410	1430	243	911	627	11.3	62.1	5.59	322
W14×109	**447**	**92.7**	**344**	**1240**	**192**	**720**	**519**	**13.2**	**43.2**	**6.69**	**203**
W40×235	444	118	420	17400	1010	3790	2630	8.97	26.0	68.2	890
W27×146	443	97.7	357	5660	464	1740	1240	11.3	30.5	25.9	448
W24×162	443	105	385	5170	468	1760	1240	10.8	32.4	23.7	476
W21×166	435	108	394	4280	432	1620	1140	10.5	35.7	19.0	456
W36×210	411	107	380	13200	833	3120	2160	9.11	26.1	57.0	822
W14×99††	**402**	**83.6**	**311**	**1110**	**173**	**649**	**471**	**13.5**	**40.6**	**6.56**	**186**
W12×136	398	98.0	361	1240	214	803	558	11.2	55.7	5.49	286
W24×146	391	93.2	340	4580	418	1570	1110	10.6	30.6	22.8	433
W18×175	391	106	387	3450	398	1490	1030	9.75	41.4	14.6	481
W40×211	390	105	372	15500	906	3400	2360	8.87	24.9	64.8	798
W21×147	376	92.7	338	3640	374	1400	990	10.4	32.8	18.5	430
W36×194	375	97.7	348	12100	767	2880	1990	9.04	25.3	54.4	754

*ASTM A6 tensile group 4 or 5 shape. Special requirements may apply per LRFD Specification Section A3.1c.
††Indicates flange is non-compact.

I_y Table 5-6 (cont.). W-Shapes Selection by I_y

$F_y = 50$ ksi
$\phi_b = 0.90$
$\phi_v = 0.90$

Shape	Y-Y Axis			X-X Axis							
	I_y	Z_y	$\phi_b M_{py}$	I_x	Z_x	$\phi_b M_{px}$	$\phi_b M_{rx}$	L_p	L_r	BF	$\phi_v V_n$
	in.⁴	in.³	kip-ft	in.⁴	in.³	kip-ft	kip-ft	ft	ft	kips	kips
W14×90††	**362**	**75.6**	**281**	**999**	**157**	**589**	**429**	**15.1**	**38.4**	**6.85**	**166**
W36×182	347	90.7	324	11300	718	2690	1870	9.01	24.8	52.0	711
W18×158	347	94.8	345	3060	356	1340	930	9.68	38.0	14.3	431
W12×120	345	85.4	315	1070	186	698	489	11.1	50.0	5.36	251
W24×131	340	81.5	298	4020	370	1390	987	10.5	29.1	21.6	400
W40×183	336	89.6	320	13300	783	2940	2050	8.83	23.9	58.7	684
W21×132	333	82.3	301	3230	334	1250	888	10.3	31.0	17.7	383
W36×170	320	83.8	299	10500	668	2510	1740	8.94	24.3	49.6	665
W18×143	311	85.4	312	2750	322	1210	846	9.61	35.3	14.1	384
W33×169	310	84.4	303	9290	629	2360	1650	8.83	24.5	45.4	611
W21×122	305	75.6	277	2970	308	1160	822	10.3	29.7	17.2	352
W12×106	301	75.1	277	933	164	615	435	11.0	44.9	5.31	212
W24×117	297	71.4	262	3540	327	1230	873	10.4	27.9	20.1	361
W36×160	295	77.3	276	9760	624	2340	1630	8.83	23.8	47.8	632
W40×167	283	76.0	269	11600	693	2600	1800	8.48	22.9	55.4	677
W18×130	278	76.7	281	2460	290	1090	768	9.54	32.8	13.7	349
W21×111	274	68.2	250	2680	279	1050	747	10.2	28.6	16.3	319
W33×152	273	73.9	266	8160	559	2100	1460	8.72	23.7	42.4	574
W36×150	270	70.9	254	9040	581	2180	1510	8.72	23.3	45.6	606
W12×96	270	67.5	250	833	147	551	393	10.9	41.3	5.20	189
W24×104	259	62.4	229	3100	289	1080	774	10.3	26.9	18.6	325
W18×119	253	69.1	253	2190	262	983	693	9.50	30.9	13.5	336
W21×101	248	61.7	227	2430	253	949	681	10.2	27.5	15.4	289
W33×141	246	66.9	240	7450	514	1930	1340	8.58	23.0	40.4	544
W12×87	**241**	**60.4**	**223**	**740**	**132**	**495**	**354**	**10.8**	**38.4**	**5.13**	**174**
W10×112	236	69.2	255	716	147	551	378	9.47	56.5	3.68	232
W40×149	229	62.2	218	9800	598	2240	1540	8.09	21.8	51.2	650
W30×148	227	68.0	244	6680	500	1880	1310	8.05	22.8	38.5	539
W36×135	225	59.7	212	7800	509	1910	1320	8.41	22.4	42.2	577
W18×106	220	60.5	222	1910	230	863	612	9.40	28.7	13.0	298
W33×130	218	59.5	213	6710	467	1750	1220	8.44	22.5	38.0	518
W12×79	**216**	**54.3**	**201**	**662**	**119**	**446**	**321**	**10.8**	**35.7**	**5.03**	**157**
W10×100	207	61.0	225	623	130	488	336	9.36	50.8	3.66	204
W18×97	201	55.3	203	1750	211	791	564	9.36	27.5	12.6	269
W30×132	196	58.4	209	5770	437	1640	1140	7.95	21.9	35.7	503
W12×72	**195**	**49.2**	**182**	**597**	**108**	**405**	**292**	**10.7**	**33.6**	**4.93**	**143**
W33×118	187	51.3	183	5900	415	1560	1080	8.19	21.7	35.4	489
W16×100	186	54.9	201	1490	199	746	528	8.87	29.5	10.6	269
W27×129	184	57.6	207	4760	395	1480	1040	7.81	22.2	31.0	455
W30×124	181	54.0	194	5360	408	1530	1070	7.88	21.4	34.3	477
W10×88	179	53.1	196	534	113	424	296	9.29	45.1	3.58	176
W18×86	175	48.4	178	1530	186	698	498	9.29	26.0	11.9	238

†† Indicates flange is non-compact.

$F_y = 50$ ksi
$\phi_b = 0.90$
$\phi_v = 0.90$

Table 5-6 (cont.). W-Shapes Selection by I_y

I_y

Shape	Y-Y Axis			X-X Axis							
	I_y	Z_y	$\phi_b M_{py}$	I_x	Z_x	$\phi_b M_{px}$	$\phi_b M_{rx}$	L_p	L_r	BF	$\phi_v V_n$
	in.^{4}	in.^{3}	kip-ft	in.^{4}	in.^{3}	kip-ft	kip-ft	ft	ft	kips	kips
W12×65††	**174**	**44.1**	**164**	**533**	**96.8**	**363**	**264**	**11.9**	**31.7**	**5.01**	**127**
W30×116	164	49.2	176	4930	378	1420	987	7.74	20.9	32.8	458
W16×89	163	48.2	177	1310	176	660	468	8.80	27.5	10.3	238
W27×114	159	49.3	177	4080	343	1290	897	7.70	21.2	28.8	420
W10×77	154	45.9	169	455	97.6	366	258	9.18	39.9	3.53	152
W18×76	152	42.2	155	1330	163	611	438	9.22	24.8	11.1	209
W14×82	148	44.8	165	882	139	521	369	8.76	29.6	7.30	197
W30×108	146	43.9	157	4470	346	1300	897	7.59	20.4	31.3	439
W27×102	139	43.4	156	3620	305	1140	801	7.59	20.4	26.7	377
W16×77	138	41.2	151	1120	151	566	405	8.69	25.3	9.68	203
W14×74	134	40.5	150	796	126	473	336	8.76	27.9	7.12	173
W10×68	134	40.1	149	394	85.3	320	227	9.15	36.0	3.45	132
W30×99	128	38.6	138	3990	312	1170	807	7.42	19.8	29.2	417
W27×94	124	38.8	140	3270	278	1040	729	7.49	19.9	25.2	356
W14×68	121	36.9	136	722	115	431	309	8.69	26.4	6.91	157
W24×103	119	41.5	149	3000	280	1050	735	7.03	20.0	24.2	364
W16×67	119	35.5	131	963	131	491	354	8.65	23.8	9.04	174
W10×60	**116**	**35.0**	**129**	**341**	**74.6**	**280**	**200**	**9.08**	**32.6**	**3.39**	**116**
W30×90	115	34.7	124	3610	283	1060	735	7.38	19.4	27.1	374
W24×94	109	37.5	135	2700	254	953	666	6.99	19.4	23.1	338
W14×61	107	32.8	121	640	102	383	277	8.65	25.0	6.50	141
W12×58	**107**	**32.5**	**120**	**475**	**86.4**	**324**	**234**	**8.87**	**27.0**	**4.97**	**119**
W27×84	106	33.2	119	2850	244	915	639	7.31	19.3	23.0	332
W10×54	**103**	**31.3**	**116**	**303**	**66.6**	**250**	**180**	**9.04**	**30.2**	**3.30**	**101**
W12×53	**95.8**	**29.1**	**108**	**425**	**77.9**	**292**	**212**	**8.76**	**25.6**	**4.78**	**113**
W24×84	94.4	32.6	118	2370	224	840	588	6.89	18.6	21.5	306
W10×49	**93.4**	**28.3**	**105**	**272**	**60.4**	**227**	**164**	**8.97**	**28.3**	**3.24**	**91.6**
W21×93	92.9	34.7	124	2070	221	829	576	6.50	19.3	19.7	338
W24×76	82.5	28.6	103	2100	200	750	528	6.78	18.0	19.8	284
W21×83	81.4	30.5	110	1830	196	735	513	6.46	18.5	18.5	298
W21×73	70.6	26.6	95.6	1600	172	645	453	6.39	17.6	17.1	260
W24×68	70.4	24.5	88.3	1830	177	664	462	6.61	17.4	18.6	266
W21×68	64.7	24.4	88.3	1480	160	600	420	6.36	17.3	16.5	245
W18×71	60.3	24.7	88.9	1170	146	548	381	6.00	17.8	14.1	247
W14×53	57.7	22.0	80.4	541	87.1	327	233	6.78	20.1	7.01	139
W21×62	57.5	21.7	78.8	1330	144	540	381	6.25	16.7	15.2	227
W12×50	56.3	21.3	78.2	391	71.9	270	193	6.92	21.5	5.30	122
W18×65	54.8	22.5	81.0	1070	133	499	351	5.97	17.1	13.3	224
W10×45	**53.4**	**20.3**	**74.8**	**248**	**54.9**	**206**	**147**	**7.10**	**24.1**	**3.44**	**95.4**
W14×48	51.4	19.6	72.0	485	78.4	294	211	6.75	19.2	6.70	127
W18×60	50.1	20.6	74.8	984	123	461	324	5.93	16.6	12.9	204

†† Indicates flange is non-compact.

I_y

Table 5-6 (cont.). W-Shapes Selection by I_y

$F_y = 50$ ksi
$\phi_b = 0.90$
$\phi_v = 0.90$

Shape	Y-Y Axis			X-X Axis							
	I_y	Z_y	$\phi_b M_{py}$	I_x	Z_x	$\phi_b M_{px}$	$\phi_b M_{rx}$	L_p	L_r	BF	$\phi_v V_n$
	in.^{4}	in.^{3}	kip-ft	in.^{4}	in.^{3}	kip-ft	kip-ft	ft	ft	kips	kips
W12×45	**50.0**	**19.0**	**69.8**	**348**	**64.2**	**241**	**173**	**6.89**	**20.3**	**5.06**	**109**
W21×55	48.4	18.4	66.4	1140	126	473	330	6.11	16.1	14.3	211
W14×43	**45.2**	**17.3**	**63.6**	**428**	**69.6**	**261**	**188**	**6.68**	**18.2**	**6.31**	**113**
W10×39	**45.0**	**17.2**	**63.6**	**209**	**46.8**	**176**	**126**	**6.99**	**21.8**	**3.32**	**84.4**
W18×55	44.9	18.5	66.9	890	112	420	295	5.90	16.1	12.2	191
W12×40	44.1	16.8	61.9	307	57.0	214	155	6.85	19.2	4.79	94.8
W16×57	43.1	18.9	68.1	758	105	394	277	5.65	16.6	10.7	190
W18×50	40.1	16.6	60.2	800	101	379	267	5.83	15.6	11.5	173
W21×48[††]	38.7	14.9	53.5	959	107	401	279	6.09	15.4	13.2	195
W16×50	37.2	16.3	59.1	659	92.0	345	243	5.62	15.7	10.1	167
W10×33	**36.6**	**14.0**	**51.8**	**171**	**38.8**	**146**	**105**	**6.85**	**19.8**	**3.14**	**76.2**
W24×62	34.5	15.8	55.1	1560	154	578	396	4.84	13.3	21.6	275
W16×45	32.8	14.5	52.5	586	82.3	309	218	5.55	15.1	9.45	150
W21×57	30.7	14.9	52.6	1170	129	484	333	4.77	13.2	17.8	231
W24×55	29.1	13.4	46.7	1360	135	506	345	4.73	12.9	19.8	252
W16×40	28.9	12.7	46.4	518	73.0	274	194	5.55	14.7	8.71	132
W14×38	26.7	12.1	44.3	383	61.1	229	163	5.47	14.9	7.05	118
W21×50	24.9	12.2	43.0	989	111	416	285	4.59	12.5	16.5	213
W16×36	24.5	10.8	39.4	448	64.0	240	170	5.37	14.1	8.11	127
W12×35	24.5	11.5	42.0	285	51.2	192	137	5.44	15.2	5.65	101
W14×34	23.3	10.6	38.9	337	54.2	203	145	5.40	14.3	6.58	108
W21×44	20.7	10.2	35.8	847	95.8	359	246	4.45	12.0	15.0	196
W12×30	**20.3**	**9.56**	**35.1**	**238**	**43.1**	**162**	**116**	**5.37**	**14.3**	**5.12**	**86.3**
W14×30	19.6	8.97	32.7	288	46.9	176	125	5.26	13.7	6.05	101
W18×40	19.1	9.95	35.7	612	78.4	294	205	4.49	12.0	11.7	152
W12×26	**17.3**	**8.17**	**30.0**	**204**	**37.2**	**140**	**100**	**5.33**	**13.8**	**4.66**	**75.8**
W10×30	16.7	8.84	32.3	170	36.6	137	97.2	4.84	14.6	4.11	85.0
W18×35	15.3	8.06	28.8	510	66.5	249	173	4.31	11.5	10.7	143
W10×26	**14.1**	**7.50**	**27.5**	**144**	**31.3**	**117**	**83.7**	**4.80**	**13.6**	**3.84**	**72.3**
W16×31	12.4	7.03	25.3	375	54.0	203	142	4.13	11.0	8.86	118
W10×22	**11.4**	**6.10**	**22.3**	**118**	**26.0**	**97.5**	**69.6**	**4.70**	**12.7**	**3.49**	**66.1**
W16×26	9.59	5.48	19.6	301	44.2	166	115	3.96	10.4	7.83	106
W14×26	8.90	5.53	19.9	243	39.9	150	105	3.81	10.2	7.01	95.7
W14×22	**7.00**	**4.37**	**15.8**	**197**	**32.8**	**123**	**85.8**	**3.67**	**9.65**	**6.22**	**85.1**
W12×22	**4.66**	**3.66**	**13.0**	**156**	**29.3**	**110**	**76.2**	**3.00**	**8.42**	**6.21**	**86.3**
W10×19	**4.29**	**3.35**	**12.0**	**96.2**	**21.6**	**81.0**	**56.4**	**3.09**	**8.88**	**4.24**	**68.8**
W12×19	**3.76**	**2.98**	**10.6**	**130**	**24.7**	**92.6**	**63.9**	**2.90**	**7.95**	**5.69**	**77.4**

[††] Indicates flange is non-compact.

$F_y = 50$ ksi
$\phi_b = 0.90$
$\phi_v = 0.90$

Table 5-6 (cont.). W-Shapes Selection by I_y

I_y

Shape	Y-Y Axis			X-X Axis							
	I_y	Z_y	$\phi_b M_{py}$	I_x	Z_x	$\phi_b M_{px}$	$\phi_b M_{rx}$	L_p	L_r	BF	$\phi_v V_n$
	in.^{4}	in.^{3}	kip-ft	in.^{4}	in.^{3}	kip-ft	kip-ft	ft	ft	kips	kips
W10×17	**3.56**	**2.80**	**10.0**	**81.9**	**18.7**	**70.1**	**48.6**	**2.98**	**8.38**	**3.99**	**65.4**
W10×15	**2.89**	**2.30**	**8.16**	**68.9**	**16.0**	**60.0**	**41.4**	**2.86**	**7.93**	**3.67**	**62.0**
W12×16	2.82	2.26	7.93	103	20.1	75.4	51.3	2.73	7.44	5.11	71.3
W12×14	**2.36**	**1.90**	**6.69**	**88.5**	**17.4**	**65.2**	**44.7**	**2.66**	**7.17**	**4.55**	**64.3**
W10×12††	**2.18**	**1.74**	**6.19**	**53.8**	**12.6**	**47.3**	**32.7**	**2.87**	**7.45**	**3.18**	**50.6**

††Indicates flange is non-compact.

Z_y

Table 5-7. W-Shapes Selection by Z_y

$F_y = 50$ ksi
$\phi_b = 0.90$
$\phi_v = 0.90$

Shape	Y-Y Axis			X-X Axis							
	Z_y	I_y	$\phi_b M_{py}$	Z_x	I_x	$\phi_b M_{px}$	$\phi_b M_{rx}$	L_p	L_r	BF	$\phi_v V_n$
	in.^{3}	in.^{4}	kip-ft	in.^{3}	in.^{4}	kip-ft	kip-ft	ft	ft	kips	kips
W14×808*	**927**	**5510**	**3340**	**1830**	**16000**	**6860**	**4200**	**17.0**	**268**	**10.6**	**2300**
W14×730*	**816**	**4720**	**2960**	**1660**	**14300**	**6230**	**3840**	**16.6**	**241**	**10.6**	**1860**
W36×798*	743	4200	2630	3580	62600	13400	8940	14.9	84.6	64.4	2700
W14×665*	**730**	**4170**	**2660**	**1480**	**12400**	**5550**	**3450**	**16.3**	**222**	**10.2**	**1650**
W14×605*	**652**	**3680**	**2380**	**1320**	**10800**	**4950**	**3120**	**16.1**	**203**	**9.79**	**1470**
W14×550*	**583**	**3250**	**2130**	**1180**	**9430**	**4430**	**2790**	**15.9**	**187**	**9.51**	**1300**
W36×650*	580	3230	2060	2860	48900	10700	7260	14.5	69.6	62.9	2150
W14×500*	**522**	**2880**	**1910**	**1050**	**8210**	**3940**	**2510**	**15.6**	**172**	**9.13**	**1160**
W40×593*	481	2520	1700	2760	50400	10400	7020	13.4	56.5	77.3	2080
W14×455*	**468**	**2560**	**1710**	**936**	**7190**	**3510**	**2270**	**15.5**	**157**	**8.75**	**1040**
W36×527*	454	2490	1630	2280	38300	8550	5850	14.2	58.1	61.5	1700
W27×539*	437	2110	1560	1890	25600	7090	4710	12.9	78.0	36.5	1730
W14×426*	**434**	**2360**	**1590**	**869**	**6600**	**3260**	**2120**	**15.3**	**148**	**8.61**	**949**
W14×398*	**402**	**2170**	**1470**	**801**	**6000**	**3000**	**1970**	**15.2**	**139**	**8.38**	**875**
W40×503*	395	2050	1410	2320	41700	8700	5940	13.1	49.4	76.0	1750
W14×370*	**370**	**1990**	**1360**	**736**	**5440**	**2760**	**1820**	**15.1**	**129**	**8.21**	**802**
W36×439*	367	1990	1320	1870	31000	7010	4860	13.9	50.1	59.4	1410
W14×342*	**338**	**1810**	**1240**	**672**	**4900**	**2520**	**1670**	**15.0**	**121**	**8.01**	**728**
W40×431*	328	1690	1170	1960	34800	7350	5070	12.9	44.1	73.2	1490
W36×393*	325	1750	1170	1670	27500	6260	4350	13.8	46.7	58.1	1250
W30×391*	310	1550	1110	1450	20700	5440	3750	13.0	52.1	43.1	1220
W33×387*	312	1620	1130	1560	24300	5850	4050	13.3	47.7	52.4	1220
W14×311*	**304**	**1610**	**1120**	**603**	**4330**	**2260**	**1520**	**14.8**	**110**	**7.84**	**651**
W40×397*	300	1540	1070	1800	32000	6750	4680	12.9	42.1	70.7	1350
W36×359*	292	1570	1060	1510	24800	5660	3960	13.6	43.8	56.5	1130
W33×354*	282	1460	1020	1420	22000	5330	3720	13.2	44.6	51.1	1110
W27×368*	279	1310	1010	1240	16200	4650	3180	12.3	54.7	34.6	1130
W30×357*	279	1390	1010	1320	18700	4950	3420	12.9	48.6	42.8	1100
W40×372*	278	1420	996	1680	29600	6300	4380	12.7	40.1	70.2	1270

*ASTM A6 tensile group 4 or 5 shape. Special requirements may apply per LRFD Specification Section A3.1c.

$F_y = 50$ ksi
$\phi_b = 0.90$
$\phi_v = 0.90$

Table 5-7 (cont.). W-Shapes Selection by Z_y

Z_y

Shape	Y-Y Axis			X-X Axis							
	Z_y	I_y	$\phi_b M_{py}$	Z_x	I_x	$\phi_b M_{px}$	$\phi_b M_{rx}$	L_p	L_r	BF	$\phi_v V_n$
	in.³	in.⁴	kip-ft	in.³	in.⁴	kip-ft	kip-ft	ft	ft	kips	kips
W14×283*	**274**	**1440**	**1010**	**542**	**3840**	**2030**	**1380**	**14.7**	**100**	**7.65**	**582**
W12×336*	274	1190	996	603	4060	2260	1450	12.3	131	6.81	807
W40×362*	270	1380	973	1640	28900	6150	4260	12.7	39.9	69.5	1230
W24×370*	267	1160	956	1130	13400	4240	2870	11.6	61.0	27.7	1150
W36×328*	265	1420	962	1380	22500	5180	3630	13.6	41.8	54.8	1020
W30×326*	252	1240	911	1190	16800	4460	3120	12.7	45.2	41.3	997
W27×336*	252	1180	911	1130	14600	4240	2920	12.2	50.4	34.5	1020
W33×318*	250	1290	906	1270	19500	4760	3330	13.1	41.9	49.8	988
W14×257*	**246**	**1290**	**906**	**487**	**3400**	**1830**	**1250**	**14.6**	**91.6**	**7.55**	**523**
W12×305*	244	1050	894	537	3550	2010	1310	12.1	120	6.59	717
W36×300	241	1300	878	1260	20300	4730	3330	13.5	39.9	53.0	936
W40×324	239	1220	861	1460	25600	5480	3840	12.6	37.6	65.6	1090
W24×335*	238	1030	855	1020	11900	3830	2590	11.4	55.5	27.9	1020
W44×335	236	1200	849	1620	31100	6080	4230	12.4	35.5	79.6	1210
W27×307*	227	1050	821	1030	13100	3860	2660	12.0	46.7	34.7	927
W33×291	226	1160	821	1160	17700	4350	3060	13.0	39.6	48.5	902
W36×280	223	1200	810	1170	18900	4390	3090	13.5	38.7	51.3	872
W30×292*	223	1100	810	1060	14900	3980	2790	12.6	42.1	40.2	881
W14×233*	**221**	**1150**	**816**	**436**	**3010**	**1640**	**1130**	**14.5**	**83.4**	**7.40**	**462**
W12×279*	220	937	804	481	3110	1800	1180	11.9	110	6.37	657
W40×297	215	1090	776	1330	23200	4990	3510	12.5	35.9	63.1	999
W24×306*	214	919	771	922	10700	3460	2370	11.3	51.0	27.5	922
W40×392*	212	803	731	1710	29900	6410	4320	9.33	34.3	83.9	1590
W44×290	206	1050	743	1420	27100	5330	3720	12.3	34.0	74.0	1020
W27×281*	206	953	748	936	11900	3510	2440	12.0	43.8	33.5	839
W40×277	204	1040	743	1250	21900	4690	3300	12.6	35.5	60.8	890
W36×260	204	1090	743	1080	17300	4050	2860	13.4	37.4	49.5	823
W33×263	202	1040	737	1040	15900	3900	2760	12.9	37.8	45.9	810
W14×211	**198**	**1030**	**731**	**390**	**2660**	**1460**	**1010**	**14.4**	**76.0**	**7.28**	**415**
W30×261	196	959	714	943	13100	3540	2490	12.5	39.1	39.3	793
W12×252*	196	828	714	428	2720	1610	1060	11.8	100	6.19	582
W24×279*	193	823	698	835	9600	3130	2150	11.2	47.3	27.1	836
W36×245	190	1010	692	1010	16100	3790	2690	13.2	36.5	47.5	780
W27×258	187	859	675	852	10800	3200	2240	11.9	41.0	32.9	767
W44×262	183	927	664	1270	24200	4760	3360	12.2	32.8	68.3	924
W40×249	182	926	664	1120	19600	4200	2980	12.5	34.1	56.7	798
W33×241	182	933	664	940	14200	3530	2490	12.8	36.2	44.0	766

*ASTM A6 tensile group 4 or 5 shape. Special requirements may apply per LRFD Specification Section A3.1c.

Z_y

Table 5-7 (cont.). W-Shapes Selection by Z_y

$F_y = 50$ ksi
$\phi_b = 0.90$
$\phi_v = 0.90$

Shape	Y-Y Axis			X-X Axis							
	Z_y	I_y	$\phi_b M_{py}$	Z_x	I_x	$\phi_b M_{px}$	$\phi_b M_{rx}$	L_p	L_r	BF	$\phi_v V_n$
	in.^{3}	in.^{4}	kip-ft	in.^{3}	in.^{4}	kip-ft	kip-ft	ft	ft	kips	kips
W14×193	**180**	**931**	**669**	**355**	**2400**	**1330**	**930**	**14.3**	**70.1**	**7.19**	**372**
W12×230*	177	742	647	386	2420	1450	963	11.7	92.0	6.04	526
W36×230	176	940	641	943	15000	3540	2510	13.2	35.6	45.6	737
W30×235	175	855	641	847	11700	3180	2240	12.4	37.1	37.8	701
W40×331	172	644	596	1430	24700	5360	3630	9.08	30.5	80.9	1340
W24×250*	171	724	619	744	8490	2790	1930	11.1	43.2	26.7	739
W40×327	170	640	591	1410	24500	5290	3600	9.11	30.3	79.5	1300
W27×235	168	769	608	772	9700	2900	2030	11.8	38.6	32.2	705
W33×221	164	840	596	857	12900	3210	2280	12.7	35.0	42.0	709
W14×176	**163**	**838**	**602**	**320**	**2140**	**1200**	**843**	**14.2**	**64.5**	**7.09**	**341**
W12×210*	159	664	585	348	2140	1310	876	11.6	84.2	5.91	468
W44×230	157	796	568	1100	20800	4130	2910	12.1	31.8	61.7	822
W40×215	156	796	568	964	16700	3620	2580	12.5	32.8	51.2	684
W30×211	155	757	563	751	10300	2820	2000	12.3	35.2	35.8	647
W27×217	154	704	561	711	8910	2670	1880	11.7	36.8	31.3	636
W24×229	154	651	559	675	7650	2530	1760	11.0	40.2	26.3	674
W33×201	147	749	536	773	11600	2900	2060	12.6	33.8	39.6	651
W14×159	**146**	**748**	**541**	**287**	**1900**	**1080**	**762**	**14.1**	**58.9**	**7.02**	**302**
W12×190	143	589	523	311	1890	1170	789	11.5	76.6	5.79	412
W40×278	140	521	490	1190	20500	4460	3060	8.90	27.6	75.1	1110
W30×191	138	673	503	675	9200	2530	1800	12.2	33.8	33.9	589
W40×199	137	695	496	869	14900	3260	2310	12.2	31.7	48.5	679
W36×256	137	528	487	1040	16800	3900	2690	9.36	28.7	63.0	969
W24×207	137	578	500	606	6820	2270	1590	10.9	37.3	25.7	604
W27×194	136	619	496	631	7860	2370	1680	11.6	34.6	29.9	569
W21×201	133	542	484	530	5310	1990	1380	10.7	41.0	19.9	565
W14×145	**133**	**677**	**491**	**260**	**1710**	**975**	**696**	**14.1**	**54.7**	**6.87**	**272**
W40×264	132	493	465	1130	19400	4240	2910	8.90	27.0	73.2	1040
W24×192	126	530	460	559	6260	2100	1470	10.8	35.6	25.1	558
W12×170	126	517	463	275	1650	1030	705	11.4	68.9	5.67	363
W30×173	123	598	449	607	8230	2280	1620	12.1	32.5	32.0	538
W36×232	122	468	434	936	15000	3510	2430	9.25	27.3	60.1	871
W27×178	122	555	443	570	7020	2140	1520	11.5	33.0	28.9	544
W21×182	119	483	434	477	4740	1790	1250	10.6	38.1	19.5	509
W40×235	118	444	420	1010	17400	3790	2630	8.97	26.0	68.2	890
W24×176	115	479	418	511	5680	1920	1350	10.7	33.7	24.7	510
W14×132	**113**	**548**	**419**	**234**	**1530**	**878**	**627**	**13.3**	**49.7**	**6.88**	**256**
W12×152	111	454	410	243	1430	911	627	11.3	62.1	5.59	322
W27×161	109	497	399	515	6310	1930	1370	11.4	31.7	27.5	492
W21×166	108	435	394	432	4280	1620	1140	10.5	35.7	19.0	456
W36×210	107	411	380	833	13200	3120	2160	9.11	26.1	57.0	822
W18×175	106	391	387	398	3450	1490	1030	9.75	41.4	14.6	481
W40×211	105	390	372	906	15500	3400	2360	8.87	24.9	64.8	798
W24×162	105	443	385	468	5170	1760	1240	10.8	32.4	23.7	476

*ASTM A6 tensile group 4 or 5 shape. Special requirements may apply per LRFD Specification Section A3.1c.

$F_y = 50$ ksi
$\phi_b = 0.90$
$\phi_v = 0.90$

Table 5-7 (cont.). W-Shapes Selection by Z_y

Z_y

Shape	Y-Y Axis			X-X Axis							
	Z_y	I_y	$\phi_b M_{py}$	Z_x	I_x	$\phi_b M_{px}$	$\phi_b M_{rx}$	L_p	L_r	BF	$\phi_v V_n$
	in.^{3}	in.^{4}	kip-ft	in.^{3}	in.^{4}	kip-ft	kip-ft	ft	ft	kips	kips
W14×120	**102**	**495**	**380**	**212**	**1380**	**795**	**570**	**13.2**	**46.3**	**6.81**	**231**
W12×136	98.0	398	361	214	1240	803	558	11.2	55.7	5.49	286
W36×194	97.7	375	348	767	12100	2880	1990	9.04	25.3	54.4	754
W27×146	97.7	443	357	464	5660	1740	1240	11.3	30.5	25.9	448
W18×158	94.8	347	345	356	3060	1340	930	9.68	38.0	14.3	431
W24×146	93.2	391	340	418	4580	1570	1110	10.6	30.6	22.8	433
W21×147	92.7	376	338	374	3640	1400	990	10.4	32.8	18.5	430
W14×109	**92.7**	**447**	**344**	**192**	**1240**	**720**	**519**	**13.2**	**43.2**	**6.69**	**203**
W36×182	90.7	347	324	718	11300	2690	1870	9.01	24.8	52.0	711
W40×183	89.6	336	320	783	13300	2940	2050	8.83	23.9	58.7	684
W18×143	85.4	311	312	322	2750	1210	846	9.61	35.3	14.1	384
W12×120	85.4	345	315	186	1070	698	489	11.1	50.0	5.36	251
W33×169	84.4	310	303	629	9290	2360	1650	8.83	24.5	45.4	611
W36×170	83.8	320	299	668	10500	2510	1740	8.94	24.3	49.6	665
W14×99††	**83.6**	**402**	**311**	**173**	**1110**	**649**	**471**	**13.5**	**40.6**	**6.56**	**186**
W21×132	82.3	333	301	334	3230	1250	888	10.3	31.0	17.7	383
W24×131	81.5	340	298	370	4020	1390	987	10.5	29.1	21.6	400
W36×160	77.3	295	276	624	9760	2340	1630	8.83	23.8	47.8	632
W18×130	76.7	278	281	290	2460	1090	768	9.54	32.8	13.7	349
W40×167	76.0	283	269	693	11600	2600	1800	8.48	22.9	55.4	677
W21×122	75.6	305	277	308	2970	1160	822	10.3	29.7	17.2	352
W14×90††	**75.6**	**362**	**281**	**157**	**999**	**589**	**429**	**15.1**	**38.4**	**6.85**	**166**
W12×106	75.1	301	277	164	933	615	435	11.0	44.9	5.31	212
W33×152	73.9	273	266	559	8160	2100	1460	8.72	23.7	42.4	574
W24×117	71.4	297	262	327	3540	1230	873	10.4	27.9	20.1	361
W36×150	70.9	270	254	581	9040	2180	1510	8.72	23.3	45.6	606
W10×112	69.2	236	255	147	716	551	378	9.47	56.5	3.68	232
W18×119	69.1	253	253	262	2190	983	693	9.50	30.9	13.5	336
W21×111	68.2	274	250	279	2680	1050	747	10.2	28.6	16.3	319
W30×148	68.0	227	244	500	6680	1880	1310	8.05	22.8	38.5	539
W12×96	67.5	270	250	147	833	551	393	10.9	41.3	5.20	189
W33×141	66.9	246	240	514	7450	1930	1340	8.58	23.0	40.4	544
W24×104	62.4	259	229	289	3100	1080	774	10.3	26.9	18.6	325
W40×149	62.2	229	218	598	9800	2240	1540	8.09	21.8	51.2	650
W21×101	61.7	248	227	253	2430	949	681	10.2	27.5	15.4	289
W10×100	61.0	207	225	130	623	488	336	9.36	50.8	3.66	204
W18×106	60.5	220	222	230	1910	863	612	9.40	28.7	13.0	298
W12×87	**60.4**	**241**	**223**	**132**	**740**	**495**	**354**	**10.8**	**38.4**	**5.13**	**174**
W36×135	59.7	225	212	509	7800	1910	1320	8.41	22.4	42.2	577
W33×130	59.5	218	213	467	6710	1750	1220	8.44	22.5	38.0	518
W30×132	58.4	196	209	437	5770	1640	1140	7.95	21.9	35.7	503
W27×129	57.6	184	207	395	4760	1480	1040	7.81	22.2	31.0	455
W18×97	55.3	201	203	211	1750	791	564	9.36	27.5	12.6	269
W16×100	54.9	186	201	199	1490	746	528	8.87	29.5	10.6	269

††Indicates flange is non-compact.

Z_y

Table 5-7 (cont.). W-Shapes Selection by Z_y

$F_y = 50$ ksi
$\phi_b = 0.90$
$\phi_v = 0.90$

Shape	Y-Y Axis			X-X Axis							
	Z_y	I_y	$\phi_b M_{py}$	Z_x	I_x	$\phi_b M_{px}$	$\phi_b M_{rx}$	L_p	L_r	BF	$\phi_v V_n$
	in.^{3}	in.^{4}	kip-ft	in.^{3}	in.^{4}	kip-ft	kip-ft	ft	ft	kips	kips
W12×79	**54.3**	**216**	**201**	**119**	**662**	**446**	**321**	**10.8**	**35.7**	**5.03**	**157**
W30×124	54.0	181	194	408	5360	1530	1070	7.88	21.4	34.3	477
W10×88	53.1	179	196	113	534	424	296	9.29	45.1	3.58	176
W33×118	51.3	187	183	415	5900	1560	1080	8.19	21.7	35.4	489
W27×114	49.3	159	177	343	4080	1290	897	7.70	21.2	28.8	420
W30×116	49.2	164	176	378	4930	1420	987	7.74	20.9	32.8	458
W12×72	**49.2**	**195**	**182**	**108**	**597**	**405**	**292**	**10.7**	**33.6**	**4.93**	**143**
W18×86	48.4	175	178	186	1530	698	498	9.29	26.0	11.9	238
W16×89	48.2	163	177	176	1310	660	468	8.80	27.5	10.3	238
W10×77	45.9	154	169	97.6	455	366	258	9.18	39.9	3.53	152
W14×82	44.8	148	165	139	882	521	369	8.76	29.6	7.30	197
W12×65[††]	**44.1**	**174**	**164**	**96.8**	**533**	**363**	**264**	**11.9**	**31.7**	**5.01**	**127**
W30×108	43.9	146	157	346	4470	1300	897	7.59	20.4	31.3	439
W27×102	43.4	139	156	305	3620	1140	801	7.59	20.4	26.7	377
W18×76	42.2	152	155	163	1330	611	438	9.22	24.8	11.1	209
W24×103	41.5	119	149	280	3000	1050	735	7.03	20.0	24.2	364
W16×77	41.2	138	151	151	1120	566	405	8.69	25.3	9.68	203
W14×74	40.5	134	150	126	796	473	336	8.76	27.9	7.12	173
W10×68	40.1	134	149	85.3	394	320	227	9.15	36.0	3.45	132
W27×94	38.8	124	140	278	3270	1040	729	7.49	19.9	25.2	356
W30×99	38.6	128	138	312	3990	1170	807	7.42	19.8	29.2	417
W24×94	37.5	109	135	254	2700	953	666	6.99	19.4	23.1	338
W14×68	36.9	121	136	115	722	431	309	8.69	26.4	6.91	157
W16×67	35.5	119	131	131	963	491	354	8.65	23.8	9.04	174
W10×60	**35.0**	**116**	**129**	**74.6**	**341**	**280**	**200**	**9.08**	**32.6**	**3.39**	**116**
W30×90	34.7	115	124	283	3610	1060	735	7.38	19.4	27.1	374
W21×93	34.7	92.9	124	221	2070	829	576	6.50	19.3	19.7	338
W27×84	33.2	106	119	244	2850	915	639	7.31	19.3	23.0	332
W14×61	32.8	107	121	102	640	383	277	8.65	25.0	6.50	141
W24×84	32.6	94.4	118	224	2370	840	588	6.89	18.6	21.5	306
W12×58	**32.5**	**107**	**120**	**86.4**	**475**	**324**	**234**	**8.87**	**27.0**	**4.97**	**119**
W10×54	**31.3**	**103**	**116**	**66.6**	**303**	**250**	**180**	**9.04**	**30.2**	**3.30**	**101**
W21×83	30.5	81.4	110	196	1830	735	513	6.46	18.5	18.5	298
W12×53	**29.1**	**95.8**	**108**	**77.9**	**425**	**292**	**212**	**8.76**	**25.6**	**4.78**	**113**
W24×76	28.6	82.5	103	200	2100	750	528	6.78	18.0	19.8	284

[††] Indicates flange is non-compact.

$F_y = 50$ ksi
$\phi_b = 0.90$
$\phi_v = 0.90$

Table 5-7 (cont.). W-Shapes Selection by Z_y

Z_y

Shape	Y-Y Axis			X-X Axis							
	Z_y	I_y	$\phi_b M_{py}$	Z_x	I_x	$\phi_b M_{px}$	$\phi_b M_{rx}$	L_p	L_r	BF	$\phi_v V_n$
	in.^{3}	in.^{4}	kip-ft	in.^{3}	in.^{4}	kip-ft	kip-ft	ft	ft	kips	kips
W10×49	**28.3**	**93.4**	**105**	**60.4**	**272**	**227**	**164**	**8.97**	**28.3**	**3.24**	**91.6**
W21×73	26.6	70.6	95.6	172	1600	645	453	6.39	17.6	17.1	260
W18×71	24.7	60.3	88.9	146	1170	548	381	6.00	17.8	14.1	247
W24×68	24.5	70.4	88.3	177	1830	664	462	6.61	17.4	18.6	266
W21×68	24.4	64.7	88.3	160	1480	600	420	6.36	17.3	16.5	245
W18×65	22.5	54.8	81.0	133	1070	499	351	5.97	17.1	13.3	224
W14×53	22.0	57.7	80.4	87.1	541	327	233	6.78	20.1	7.01	139
W21×62	21.7	57.5	78.8	144	1330	540	381	6.25	16.7	15.2	227
W12×50	21.3	56.3	78.2	71.9	391	270	193	6.92	21.5	5.30	122
W18×60	20.6	50.1	74.8	123	984	461	324	5.93	16.6	12.9	204
W10×45	**20.3**	**53.4**	**74.8**	**54.9**	**248**	**206**	**147**	**7.10**	**24.1**	**3.44**	**95.4**
W14×48	19.6	51.4	72.0	78.4	485	294	211	6.75	19.2	6.70	127
W12×45	**19.0**	**50.0**	**69.8**	**64.2**	**348**	**241**	**173**	**6.89**	**20.3**	**5.06**	**109**
W16×57	18.9	43.1	68.1	105	758	394	277	5.65	16.6	10.7	190
W18×55	18.5	44.9	66.9	112	890	420	295	5.90	16.1	12.2	191
W21×55	18.4	48.4	66.4	126	1140	473	330	6.11	16.1	14.3	211
W14×43	17.3	45.2	63.6	69.6	428	261	188	6.68	18.2	6.31	113
W10×39	**17.2**	**45.0**	**63.6**	**46.8**	**209**	**176**	**126**	**6.99**	**21.8**	**3.32**	**84.4**
W12×40	16.8	44.1	61.9	57.0	307	214	155	6.90	19.2	4.80	94.8
W18×50	16.6	40.1	60.2	101	800	379	267	5.83	15.6	11.5	173
W16×50	16.3	37.2	59.1	92.0	659	345	243	5.62	15.7	10.1	167
W24×62	15.8	34.5	55.1	154	1560	578	396	4.84	13.3	21.6	275
W21×48††	14.9	38.7	53.5	107	959	401	279	6.09	15.4	13.2	195
W21×57	14.9	30.7	52.6	129	1170	484	333	4.77	13.2	17.8	231
W16×45	14.5	32.8	52.5	82.3	586	309	218	5.55	15.1	9.45	150
W10×33	**14.0**	**36.6**	**51.8**	**38.8**	**171**	**146**	**105**	**6.85**	**19.8**	**3.14**	**76.2**
W24×55	13.4	29.1	46.7	135	1360	506	345	4.73	12.9	19.8	252
W16×40	12.7	28.9	46.4	73.0	518	274	194	5.55	14.7	8.71	132
W21×50	12.2	24.9	43.0	111	989	416	285	4.59	12.5	16.5	213
W14×38	12.1	26.7	44.3	61.1	383	229	163	5.47	14.9	7.05	118
W18×46	11.7	22.5	41.8	90.7	712	340	236	4.56	12.6	12.9	176
W12×35	11.5	24.5	42.0	51.2	285	192	137	5.44	15.2	5.65	101
W16×36	10.8	24.5	39.4	64.0	448	240	170	5.37	14.1	8.11	127
W14×34	10.6	23.3	38.9	54.2	337	203	145	5.40	14.3	6.58	108
W21×44	10.2	20.7	35.8	95.8	847	359	246	4.45	12.0	15.0	196
W18×40	10.0	19.1	35.7	78.4	612	294	205	4.49	12.0	11.7	152
W12×30	**9.56**	**20.3**	**35.1**	**43.1**	**238**	**162**	**116**	**5.37**	**14.3**	**5.12**	**86.3**
W14×30	**8.97**	**19.6**	**32.7**	**46.9**	**288**	**176**	**125**	**5.26**	**13.7**	**6.05**	**101**
W10×30	**8.84**	**16.7**	**32.3**	**36.6**	**170**	**137**	**97.2**	**4.84**	**14.6**	**4.11**	**85.0**

†† Indicates flange is non-compact.

Z_y

Table 5-7 (cont.). W-Shapes Selection by Z_y

$F_y = 50$ ksi
$\phi_b = 0.90$
$\phi_v = 0.90$

Shape	Y-Y Axis			X-X Axis							
	Z_y	I_y	$\phi_b M_{py}$	Z_x	I_x	$\phi_b M_{px}$	$\phi_b M_{rx}$	L_p	L_r	BF	$\phi_v V_n$
	in.^{3}	in.^{4}	kip-ft	in.^{3}	in.^{4}	kip-ft	kip-ft	ft	ft	kips	kips
W12×26	**8.17**	**17.3**	**30.0**	**37.2**	**204**	**140**	**100**	**5.33**	**13.8**	**4.66**	**75.8**
W18×35	8.06	15.3	28.8	66.5	510	249	173	4.31	11.5	10.7	143
W10×26	**7.50**	**14.1**	**27.5**	**31.3**	**144**	**117**	**83.7**	**4.80**	**13.6**	**3.84**	**72.3**
W16×31	7.03	12.4	25.3	54.0	375	203	142	4.13	11.0	8.9	118
W10×22	**6.10**	**11.4**	**22.3**	**26.0**	**118**	**97.5**	**69.6**	**4.70**	**12.7**	**3.49**	**66.1**
W14×26	5.53	8.90	19.9	39.9	243	150	105	3.81	10.2	7.01	95.7
W16×26	5.48	9.59	19.6	44.2	301	166	115	3.96	10.4	7.83	106
W14×22	**4.37**	**7.00**	**15.8**	**32.8**	**197**	**123**	**85.8**	**3.67**	**9.65**	**6.22**	**85.1**
W12×22	**3.66**	**4.66**	**13.0**	**29.3**	**156**	**110**	**76.2**	**3.00**	**8.42**	**6.21**	**86.3**
W10×19	**3.35**	**4.29**	**12.0**	**21.6**	**96.2**	**81.0**	**56.4**	**3.09**	**8.88**	**4.24**	**68.8**
W12×19	**2.98**	**3.76**	**10.6**	**24.7**	**130**	**92.6**	**63.9**	**2.90**	**7.95**	**5.69**	**77.4**
W10×17	**2.80**	**3.56**	**10.0**	**18.7**	**81.9**	**70.1**	**48.6**	**2.98**	**8.38**	**3.99**	**65.4**
W10×15	**2.30**	**2.89**	**8.16**	**16.0**	**68.9**	**60.0**	**41.4**	**2.86**	**7.93**	**3.67**	**62.0**
W12×16	2.26	2.82	7.93	20.1	103	75.4	51.3	2.73	7.44	5.11	71.3
W12×14	**1.90**	**2.36**	**6.69**	**17.4**	**88.5**	**65.2**	**44.7**	**2.66**	**7.17**	**4.55**	**64.3**
W10×12††	**1.74**	**2.18**	**6.19**	**12.60**	**53.8**	**47.3**	**32.7**	**2.87**	**7.45**	**3.18**	**50.6**

†† Indicates flange is non-compact.

$F_y = 36$ ksi
$\phi_b = 0.90$
$\phi_v = 0.90$

Table 5-8. S-Shapes Maximum Total Factored Uniform Load, kips[a]

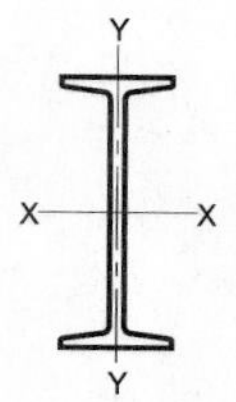

Shape	S24×					S20×				S18×	
Span, ft	121	106	100	90	80	96	86	75	66	70	54.7
5										498	
6						631		494		446	323
7			695			611	521	469	393	383	321
8	762		645	583		535	494	410	375	335	281
9	734		574	533	467	475	439	365	334	298	250
10	661	591	516	480	441	428	395	328	300	268	225
11	601	548	469	436	401	389	359	298	273	243	204
12	551	502	430	400	367	356	329	274	250	223	187
13	508	464	397	369	339	329	304	253	231	206	173
14	472	430	369	343	315	305	282	235	214	191	160
15	441	402	344	320	294	285	264	219	200	179	150
16	413	377	323	300	275	267	247	205	188	167	140
17	389	354	304	282	259	252	233	193	177	158	132
18	367	335	287	266	245	238	220	182	167	149	125
19	348	317	272	252	232	225	208	173	158	141	118
20	330	301	258	240	220	214	198	164	150	134	112
21	315	287	246	228	210	204	188	156	143	128	107
22	300	274	235	218	200	194	180	149	136	122	102
23	287	262	224	208	192	186	172	143	131	116	97.7
24	275	251	215	200	184	178	165	137	125	112	93.6
25	264	241	206	192	176	171	158	131	120	107	89.9
26	254	232	199	184	169	164	152	126	115	103	86.4
27	245	223	191	178	163	158	146	122	111	99.2	83.2
28	236	215	184	171	157	153	141	117	107	95.7	80.2
29	228	208	178	165	152	147	136	113	104	92.4	77.5
30	220	201	172	160	147	143	132	109	100	89.3	74.9
32	207	188	161	150	138	134	124	103	93.8	83.7	70.2
34	194	177	152	141	130	126	116	96.6	88.3	78.8	66.1
36	184	167	143	133	122	119	110	91.2	83.4	74.4	62.4
38	174	159	136	126	116	113	104	86.4	79.0	70.5	59.1
40	165	151	129	120	110	107	98.8	82.1	75.1	67.0	56.2
42	157	143	123	114	105	102	94.1	78.2	71.5	63.8	53.5
44	150	137	117	109	100	97.2	89.8	74.6	68.2	60.9	51.1
46	144	131	112	104	95.8	93.0	85.9	71.4	65.3		
48	138	126	108	99.9	91.8	89.1	82.4	68.4	62.6		
50	132	121	103	95.9	88.1	85.5	79.1	65.7	60.0		
52	127	116	99.3	92.2	84.7						
54	122	112	95.6	88.8	81.6						
56	118	108	92.2	85.6	78.7						
Beam Properties											
Z_x, in.^{3}	306	279	239	222	204	198	183	152	139	124	104
$\phi_b W_c$, kip-ft	6610	6030	5160	4800	4410	4280	3950	3280	3000	2680	2250
$\phi_v V_n$, kips	381	295	348	292	233	316	260	247	196	249	161
BF, kips	16.8	16.2	17.3	16.8	15.9	11.3	11.2	11.5	11.1	9.11	8.83
L_r, ft	25.6	24.1	20.2	19.4	18.7	24.3	22.9	18.8	17.9	19.2	16.9
L_p, ft	6.37	6.54	5.29	5.41	5.58	5.54	5.66	4.83	4.95	4.50	4.75
$\phi_b M_r$, kip-ft	503	468	388	365	341	322	302	250	232	201	174
$\phi_b M_p$, kip-ft	826	753	645	599	551	535	494	410	375	335	281

[a]Beam must be laterally supported if Table 5-8 is used.
Note: Design strength tabulated above heavy line is limited by design shear strength.

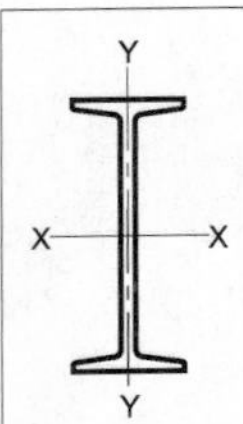

Table 5-8 (cont.). S-Shapes Maximum Total Factored Uniform Load, kips[a]

$F_y = 36$ ksi
$\phi_b = 0.90$
$\phi_v = 0.90$

Shape		S15×		S12×				S10×		S8×	
		50	42.9	50	40.8	35	31.8	35	25.4	23	18.4
Span, ft	3							231		137	
	4			321		200		191		104	84.3
	5	321		263	216	193	163	153	121	82.9	71.3
	6	277	240	219	190	161	150	127	102	69.1	59.4
	7	238	214	188	163	138	129	109	87.3	59.2	50.9
	8	208	187	164	142	120	113	95.6	76.4	51.8	44.6
	9	185	166	146	126	107	100	85.0	67.9	46.1	39.6
	10	166	149	132	114	96.3	90.3	76.5	61.1	41.5	35.6
	11	151	136	120	103	87.6	82.1	69.5	55.6	37.7	32.4
	12	139	125	110	94.9	80.3	75.2	63.7	50.9	34.6	29.7
	13	128	115	101	87.6	74.1	69.5	58.8	47.0	31.9	27.4
	14	119	107	94.0	81.3	68.8	64.5	54.6	43.7	29.6	25.5
	15	111	99.6	87.7	75.9	64.2	60.2	51.0	40.8	27.6	23.8
	16	104	93.4	82.2	71.1	60.2	56.4	47.8	38.2	25.9	22.3
	17	97.8	87.9	77.4	67.0	56.7	53.1	45.0	36.0	24.4	21.0
	18	92.4	83.0	73.1	63.2	53.5	50.2	42.5	34.0	23.0	19.8
	19	87.5	78.7	69.2	59.9	50.7	47.5	40.2	32.2	21.8	18.8
	20	83.2	74.7	65.8	56.9	48.2	45.1	38.2	30.6	20.7	17.8
	21	79.2	71.2	62.6	54.2	45.9	43.0	36.4	29.1		
	22	75.6	67.9	59.8	51.7	43.8	41.0	34.8	27.8		
	23	72.3	65.0	57.2	49.5	41.9	39.3	33.2	26.6		
	24	69.3	62.3	54.8	47.4	40.1	37.6	31.9	25.5		
	25	66.5	59.8	52.6	45.5	38.5	36.1	30.6	24.5		
	26	64.0	57.5	50.6	43.8	37.1	34.7				
	27	61.6	55.4	48.7	42.2	35.7	33.4				
	28	59.4	53.4	47.0	40.7	34.4	32.2				
	29	57.4	51.5	45.4	39.3	33.2	31.1				
	30	55.4	49.8	43.8	37.9	32.1	30.1				
	32	52.0	46.7								
	34	48.9	44.0								
	36	46.2	41.5								
	38										
	40										
	42										
	44										
Beam Properties											
Z_x, in.^{3}		77.0	69.2	60.9	52.7	44.6	41.8	35.4	28.3	19.2	16.5
$\phi_b W_c$, kip-ft		1660	1490	1320	1140	963	903	765	611	415	356
$\phi_v V_n$, kips		160	120	160	108	99.8	81.6	115	60.5	68.6	42.1
BF, kips		6.06	5.94	3.31	3.45	3.66	3.58	2.24	2.34	1.41	1.44
L_r, ft		17.8	16.4	24.1	20.2	16.7	16.0	20.8	16.1	17.7	14.9
L_p, ft		4.29	4.41	4.29	4.41	4.08	4.16	3.74	3.95	3.31	3.44
$\phi_b M_r$, kip-ft		126	116	98.7	87.9	74.3	70.6	57.3	48.0	31.6	28.1
$\phi_b M_p$, kip-ft		208	187	164	142	120	113	95.6	76.4	51.8	44.6

[a] Beam must be laterally supported if Table 5-8 is used.
Note: Design strength tabulated above heavy line is limited by design shear strength.

$F_y = 36$ ksi
$\phi_b = 0.90$
$\phi_v = 0.90$

Table 5-8 (cont.). S-Shapes Maximum Total Factored Uniform Load, kips[a]

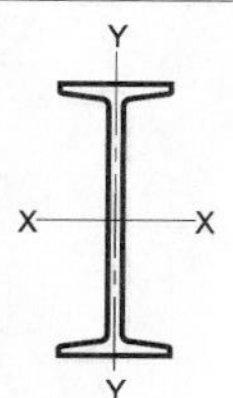

Shape	S6×		S5×	S4×		S3×	
	17.25	12.5	10	9.5	7.7	7.5	5.7
Span, ft 2	108		41.6	43.6	30.0	25.4	19.8
3	75.6	54.1	40.8	29.1	25.2	16.9	14.0
4	56.7	45.6	30.6	21.8	18.9	12.7	10.5
5	45.4	36.5	24.5	17.5	15.1	10.2	8.38
6	37.8	30.4	20.4	14.5	12.6	8.46	6.98
7	32.4	26.1	17.5	12.5	10.8	7.25	5.99
8	28.4	22.8	15.3	10.9	9.45		
9	25.2	20.3	13.6	9.70	8.40		
10	22.7	18.3	12.2	8.73	7.56		
11	20.6	16.6	11.1				
12	18.9	15.2	10.2				
13	17.4	14.0					
14	16.2	13.0					
15	15.1	12.2					
16							
17							
18							
19							
20							
Beam Properties							
Z_x, in.^{3}	10.5	8.45	5.66	4.04	3.50	2.35	1.94
$\phi_b W_c$, kip-ft	227	183	122	87.3	75.6	50.8	41.9
$\phi_v V_n$, kips	54.2	27.1	20.8	25.3	15.0	20.4	9.91
BF, kips	0.683	0.762	0.505	0.282	0.299	0.133	0.152
L_r, ft	19.4	14.1	14.0	17.6	14.2	21.4	15.2
L_p, ft	2.80	2.92	2.66	2.35	2.40	2.14	2.16
$\phi_b M_r$, kip-ft	17.0	14.3	9.56	6.59	5.91	3.78	3.26
$\phi_b M_p$, kip-ft	28.4	22.8	15.3	10.9	9.45	6.35	5.24

[a] Beam must be laterally supported if Table 5-8 is used.
Note: Design strength tabulated above heavy line is limited by design shear strength.

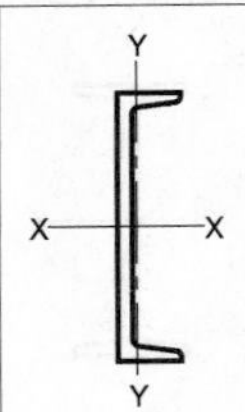

Table 5-9.
C-Shapes
Maximum Total Factored Uniform Load, kips[a]

$F_y = 36$ ksi
$\phi_b = 0.90$
$\phi_v = 0.90$

Shape		C15×			C12×			C10×			
		50	40	33.9	30	25	20.7	30	25	20	15.3
Span, ft	2							262	205	147	
	3	418			238	181		192	166	140	93.3
	4	370	303	233	183	159	132	144	125	105	85.9
	5	296	248	219	146	127	111	115	99.8	83.8	68.7
	6	247	207	183	122	106.0	92.2	96.1	83.2	69.8	57.2
	7	211	177	157	104	90.7	79.0	82.4	71.3	59.9	49.1
	8	185	155	137	91.3	79.4	69.1	72.1	62.4	52.4	42.9
	9	164	138	122	81.1	70.6	61.4	64.1	55.4	46.6	38.2
	10	148	124	110	73.0	63.5	55.3	57.7	49.9	41.9	34.3
	11	135	113	99.8	66.4	57.7	50.3	52.4	45.4	38.1	31.2
	12	123	104	91.4	60.8	52.9	46.1	48.1	41.6	34.9	28.6
	13	114	95.5	84.4	56.2	48.8	42.5	44.4	38.4	32.2	26.4
	14	106	88.7	78.4	52.1	45.4	39.5	41.2	35.6	29.9	24.5
	15	98.6	82.8	73.2	48.7	42.3	36.9	38.4	33.3	27.9	22.9
	16	92.5	77.6	68.6	45.6	39.7	34.6	36.0	31.2	26.2	21.5
	17	87.0	73.1	64.5	42.9	37.4	32.5	33.9	29.4	24.6	20.2
	18	82.2	69.0	61.0	40.6	35.3	30.7	32.0	27.7	23.3	19.1
	19	77.9	65.4	57.8	38.4	33.4	29.1	30.4	26.3	22.1	18.1
	20	74.0	62.1	54.9	36.5	31.8	27.6	28.8	24.9	21.0	17.2
	21	70.5	59.1	52.3	34.8	30.2	26.3	27.5	23.8	20.0	16.4
	22	67.3	56.5	49.9	33.2	28.9	25.1	26.2	22.7	19.0	15.6
	23	64.3	54.0	47.7	31.7	27.6	24.0	25.1	21.7	18.2	14.9
	24	61.7	51.8	45.7	30.4	26.5	23.0	24.0	20.8	17.5	14.3
	25	59.2	49.7	43.9	29.2	25.4	22.1	23.1	20.0	16.8	13.7
	26	56.9	47.8	42.2	28.1	24.4	21.3				
	27	54.8	46.0	40.6	27.0	23.5	20.5				
	28	52.8	44.4	39.2	26.1	22.7	19.7				
	29	51.0	42.8	37.8	25.2	21.9	19.1				
	30	49.3	41.4	36.6	24.3	21.2	18.4				
	32	46.2	38.8	34.3							
	34	43.5	36.5	32.3							
	36	41.1	34.5	30.5							
	38										
	40										
	42										
	44										
Beam Properties											
Z_x, in.^{3}		68.5	57.5	50.8	33.8	29.4	25.6	26.7	23.1	19.4	15.9
$\phi_b W_c$, kip-ft		1480	1240	1100	730	635	553	577	499	419	343
$\phi_v V_n$, kips		209	152	117	119	90.3	65.8	131	102	73.7	46.7
BF, kips		5.19	5.39	5.31	3.25	3.31	3.19	1.90	2.09	2.21	2.15
L_r, ft		19.0	15.7	14.1	15.0	13.1	11.9	19.5	15.6	12.6	10.7
L_p, ft		3.60	3.68	3.75	3.17	3.24	3.32	2.78	2.81	2.87	2.96
$\phi_b M_r$, kip-ft		105	90.7	81.9	52.7	46.8	41.9	40.4	35.5	30.8	26.3
$\phi_b M_p$, kip-ft		185	155	137	91.3	79.4	69.1	72.1	62.4	52.4	42.9

[a] Beam must be laterally supported if Table 5-9 is used.

Note: Design strength tabulated above heavy line is limited by design shear strength.

$F_y = 36$ ksi
$\phi_b = 0.90$
$\phi_v = 0.90$

Table 5-9 (cont.). C-Shapes Maximum Total Factored Uniform Load, kips[a]

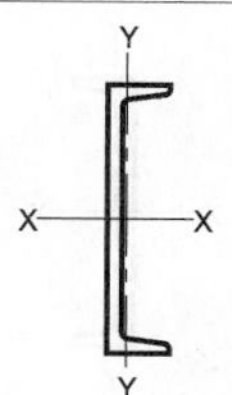

Shape		C9×			C8×			C7×		
		20	15	13.4	18.75	13.75	11.5	14.75	12.25	9.8
Span, ft	2	157	99.7		150	94.2		105	85.5	57.2
	3	122	97.9	81.5	100	79.2	68.4	70.2	60.9	51.8
	4	91.3	73.4	68.0	75.1	59.4	52.0	52.7	45.7	38.8
	5	73.0	58.8	54.4	60.0	47.5	41.6	42.1	36.5	31.1
	6	60.8	49.0	45.4	50.0	39.6	34.7	35.1	30.5	25.9
	7	52.1	42.0	38.9	42.9	33.9	29.7	30.1	26.1	22.2
	8	45.6	36.7	34.0	37.5	29.7	26.0	26.3	22.8	19.4
	9	40.6	32.6	30.2	33.4	26.4	23.1	23.4	20.3	17.3
	10	36.5	29.4	27.2	30.0	23.8	20.8	21.1	18.3	15.5
	11	33.2	26.7	24.7	27.3	21.6	18.9	19.1	16.6	14.1
	12	30.4	24.5	22.7	25.0	19.8	17.3	17.6	15.2	12.9
	13	28.1	22.6	20.9	23.1	18.3	16.0	16.2	14.1	11.9
	14	26.1	21.0	19.4	21.4	17.0	14.9	15.0	13.1	11.1
	15	24.3	19.6	18.1	20.0	15.8	13.9	14.0	12.2	10.4
	16	22.8	18.4	17.0	18.8	14.9	13.0	13.2	11.4	9.71
	17	21.5	17.3	16.0	17.7	14.0	12.2	12.4	10.7	9.14
	18	20.3	16.3	15.1	16.7	13.2	11.6			
	19	19.2	15.5	14.3	15.8	12.5	10.9			
	20	18.3	14.7	13.6	15.0	11.9	10.4			
	21	17.4	14.0	13.0						
	22	16.6	13.4	12.4						
	23									
	24									
	25									
	26									
	27									
	28									
	29									
	30									
	32									
	34									
	36									
	38									
	40									
	42									
	44									
Beam Properties										
Z_x, in.^{3}		16.9	13.6	12.6	13.9	11.0	9.63	9.75	8.46	7.19
$\phi_b W_c$, kip-ft		365	294	272	300	238	208	211	183	155
$\phi_v V_n$, kips		78.4	49.9	40.8	75.7	47.1	34.2	57.0	42.7	28.6
BF, kips		1.68	1.74	1.74	1.23	1.37	1.34	0.927	0.988	1.00
L_r, ft		14.2	11.2	10.5	15.6	11.4	10.1	14.4	11.8	9.97
L_p, ft		2.66	2.74	2.77	2.49	2.55	2.59	2.34	2.36	2.41
$\phi_b M_r$, kip-ft		26.3	22.0	20.7	21.5	17.6	15.9	15.2	13.5	11.8
$\phi_b M_p$, kip-ft		45.6	36.7	34.0	37.5	29.7	26.0	26.3	22.8	19.4

[a] Beam must be laterally supported if Table 5-9 is used.
Note: Design strength tabulated above heavy line is limited by design shear strength.

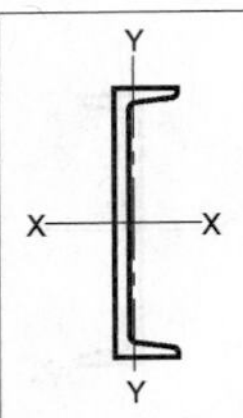

Table 5-9 (cont.). C-Shapes Maximum Total Factored Uniform Load, kips[a]

$F_y = 36$ ksi
$\phi_b = 0.90$
$\phi_v = 0.90$

Shape	C6×			C5×		C4×			C3×			
	13	10.5	8.2	9	6.7	7.25	5.4	4.5	6	5	4.1	3.5
Span, ft 2	78.7	66.7	46.7	47.4	36.9	30.7	24.7	19.4	18.8	16.4	14.3	13.4
3	52.5	44.5	37.2	31.6	25.6	20.4	16.5	15.3	12.5	10.9	9.50	8.93
4	39.4	33.4	27.9	23.7	19.2	15.3	12.4	11.4	9.40	8.21	7.13	6.70
5	31.5	26.7	22.3	19.0	15.3	12.3	9.89	9.16	7.52	6.57	5.70	5.36
6	26.2	22.2	18.6	15.8	12.8	10.2	8.24	7.63	6.26	5.47	4.75	4.46
7	22.5	19.1	15.9	13.5	11.0	8.76	7.07	6.54	5.37	4.69	4.07	3.83
8	19.7	16.7	13.9	11.9	9.58	7.67	6.18	5.72				
9	17.5	14.8	12.4	10.5	8.52	6.82	5.50	5.09				
10	15.7	13.3	11.1	9.48	7.67	6.13	4.95	4.58				
11	14.3	12.1	10.1	8.62	6.97							
12	13.1	11.1	9.29	7.90	6.39							
13	12.1	10.3	8.57									
14	11.2	9.53	7.96									
15	10.5	8.90	7.43									
16												
17												
18												
19												
20												
21												
22												
23												
24												
25												
26												
27												
28												
29												
30												
32												
34												
36												
38												
40												
42												
44												
Beam Properties												
Z_x, in.^{3}	7.29	6.18	5.16	4.39	3.55	2.84	2.29	2.12	1.74	1.52	1.32	1.24
$\phi_b W_c$, kip-ft	157	133	111	94.8	76.7	61.3	49.5	45.8	37.6	32.8	28.5	26.8
$\phi_v V_n$, kips	51.0	36.6	23.3	31.6	18.5	25.0	14.3	9.72	20.8	15.0	9.91	7.70
BF, kips	0.618	0.680	0.710	0.429	0.466	0.246	0.275	0.273	0.113	0.128	0.139	0.143
L_r, ft	15.8	12.3	9.90	13.5	10.1	14.9	10.7	9.79	19.4	15.0	11.9	10.9
L_p, ft	2.18	2.20	2.23	2.02	2.04	1.86	1.85	1.90	1.72	1.69	1.66	1.64
$\phi_b M_r$, kip-ft	11.3	9.83	8.48	6.94	5.83	4.47	3.74	3.57	2.69	2.40	2.15	2.03
$\phi_b M_p$, kip-ft	19.7	16.7	13.9	11.9	9.58	7.67	6.18	5.72	4.70	4.10	3.56	3.35

[a] Beam must be laterally supported if Table 5-9 is used.

Note: Design strength tabulated above heavy line is limited by design shear strength.

$F_y = 36$ ksi
$\phi_b = 0.90$
$\phi_v = 0.90$

Table 5-10.
MC-Shapes
Maximum Total Factored Uniform Load, kips[a]

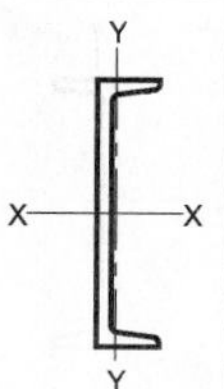

Shape	MC18×				MC13×				MC12×		
	58	51.9	45.8	42.7	50	40	35	31.8	50	45	40
Span, ft 2											
3					398	283			390	332	275
4	490	420	350		328	276	226	190	305	281	258
5	412	377	342	315	263	221	201	187	244	225	206
6	343	314	285	270	219	184	167	156	203	187	172
7	294	269	244	232	188	158	143	134	174	160	147
8	258	236	214	203	164	138	126	117	153	140	129
9	229	210	190	180	146	123	112	104	136	125	114
10	206	189	171	162	131	111	100	93.7	122	112	103
11	187	171	156	147	119	101.0	91.3	85.2	111	102	93.7
12	172	157	143	135	109	92.2	83.7	78.1	102	93.6	85.9
13	159	145	132	125	101	85.1	77.3	72.1	93.9	86.4	79.3
14	147	135	122	116	93.8	79.0	71.7	67.0	87.2	80.2	73.6
15	137	126	114	108	87.6	73.7	67.0	62.5	81.4	74.9	68.7
16	129	118	107	101.0	82.1	69.1	62.8	58.6	76.3	70.2	64.4
17	121	111	101	95.4	77.3	65.1	59.1	55.1	71.8	66.1	60.6
18	114	105	95	90.1	73.0	61.4	55.8	52.1	67.8	62.4	57.2
19	108	99.2	90.0	85.4	69.1	58.2	52.9	49.3	64.2	59.1	54.2
20	103	94.3	85.5	81.1	65.7	55.3	50.2	46.9	61.0	56.2	51.5
21	98.1	89.8	81.5	77.2	62.5	52.7	47.8	44.6	58.1	53.5	49.1
22	93.7	85.7	77.8	73.7	59.7	50.3	45.7	42.6	55.5	51.1	46.8
23	89.6	82.0	74.4	70.5	57.1	48.1	43.7	40.8	53.1	48.8	44.8
24	85.9	78.6	71.3	67.6	54.7	46.1	41.9	39.1	50.9	46.8	42.9
25	82.4	75.4	68.4	64.9	52.5	44.2	40.2	37.5	48.8	44.9	41.2
26	79.3	72.5	65.8	62.4	50.5	42.5	38.6	36.1	46.9	43.2	39.6
27	76.3	69.8	63.4	60.1	48.6	41.0	37.2	34.7	45.2	41.6	38.2
28	73.6	67.3	61.1	57.9	46.9	39.5	35.9	33.5	43.6	40.1	36.8
29	71.1	65.0	59.0	55.9	45.3	38.1	34.6	32.3	42.1	38.7	35.5
30	68.7	62.9	57.0	54.1	43.8	36.9	33.5	31.2	40.7	37.4	34.3
32	64.4	58.9	53.5	50.7	42.4	35.7	32.4	30.2			
34	60.6	55.5	50.3	47.7	41.0	34.6	31.4	29.3			
36	57.2	52.4	47.5	45.1							
38	54.2	49.6	45.0	42.7							
40	51.5	47.1	42.8	40.6							
42	49.1	44.9	40.7	38.6							
44	46.8	42.9	38.9	36.9							
Beam Properties											
Z_x, in.^{3}	95.4	87.3	79.2	75.1	60.8	51.2	46.5	43.4	56.5	52.0	47.7
$\phi_b W_c$, kip-ft	2060	1890	1710	1620	1310	1110	1000	937	1220	1120	1030
$\phi_v V_n$, kips	245	210	175	157	199	142	113	94.8	195	166	138
BF, kips	7.70	7.84	7.78	7.60	3.12	3.39	3.47	3.45	2.48	2.64	2.78
L_r, ft	18.7	17.0	15.8	15.4	26.9	21.2	18.9	17.8	30.7	26.7	23.6
L_p, ft	4.25	4.29	4.37	4.45	4.41	4.50	4.54	4.58	4.54	4.54	4.58
$\phi_b M_r$, kip-ft	146	136	125	120	94.2	81.7	75.7	71.6	87.6	81.7	76.1
$\phi_b M_p$, kip-ft	258	236	214	203	164	138	126	117	153	140	129

[a] Beam must be laterally supported if Table 5-10 is used.
Note: Design strength tabulated above heavy line is limited by design shear strength.

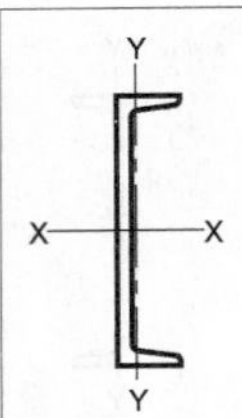

Table 5-10 (cont.). MC-Shapes Maximum Total Factored Uniform Load, kips[a]

$F_y = 36$ ksi
$\phi_b = 0.90$
$\phi_v = 0.90$

Shape	MC12×			MC10×						MC9×	
	35	31	10.6	41.1	33.6	28.5	25	22	8.4	25.4	23.9
Span, ft											
2			88.6	309					66.1		
3			83.5	283	224	165	148		57.0	157	140
4	218	173	62.6	212	182	162	141	113	42.8	127	122
5	187	172	50.1	170	146	130	113	103	34.2	102	97.2
6	156	143	41.8	141	121	108	94.3	86.0	28.5	84.6	81.0
7	133	123	35.8	121	104	92.6	80.8	73.7	24.4	72.5	69.4
8	117	107	31.3	106	91.0	81.0	70.7	64.5	21.4	63.5	60.8
9	104	95.3	27.8	94.3	80.9	72.0	62.9	57.4	19.0	56.4	54.0
10	93.3	85.8	25.1	84.9	72.8	64.8	56.6	51.6	17.1	50.8	48.6
11	84.8	78.0	22.8	77.2	66.2	58.9	51.4	46.9	15.6	46.1	44.2
12	77.8	71.5	20.9	70.7	60.7	54.0	47.2	43.0	14.3	42.3	40.5
13	71.8	66.0	19.3	65.3	56.0	49.8	43.5	39.7	13.2	39.0	37.4
14	66.7	61.3	17.9	60.6	52.0	46.3	40.4	36.9	12.2	36.3	34.7
15	62.2	57.2	16.7	56.6	48.5	43.2	37.7	34.4	11.4	33.8	32.4
16	58.3	53.6	15.7	53.1	45.5	40.5	35.4	32.3	10.7	31.7	30.4
17	54.9	50.4	14.7	49.9	42.8	38.1	33.3	30.4	10.1	29.9	28.6
18	51.8	47.6	13.9	47.2	40.4	36.0	31.4	28.7	9.50	28.2	27.0
19	49.1	45.1	13.2	44.7	38.3	34.1	29.8	27.2	9.00	26.7	25.6
20	46.7	42.9	12.5	42.4	36.4	32.4	28.3	25.8	8.55	25.4	24.3
21	44.4	40.8	11.9	40.4	34.7	30.9	26.9	24.6	8.15	24.2	23.1
22	42.4	39.0	11.4	38.6	33.1	29.5	25.7	23.5	7.78	23.1	22.1
23	40.6	37.3	10.9	36.9	31.6	28.2	24.6	22.4	7.44		
24	38.9	35.7	10.4	35.4	30.3	27.0	23.6	21.5	7.13		
25	37.3	34.3	10.0	34.0	29.1	25.9	22.6	20.6	6.84		
26	35.9	33.0	9.64								
27	34.6	31.8	9.28								
28	33.3	30.6	8.95								
29	32.2	29.6	8.64								
30	31.1	28.6	8.35								
32											
34											
36											
38											
40											
42											
44											
Beam Properties											
Z_x, in.^{3}	43.2	39.7	11.6	39.3	33.7	30.0	26.2	23.9	7.92	23.5	22.5
$\phi_b W_c$, kip-ft	933	858	251	849	728	648	566	516	171	508	486
$\phi_v V_n$, kips	109	86.3	44.3	155	112	82.6	73.9	56.4	33.0	78.7	70.0
BF, kips	2.84	2.85	4.08	1.49	1.70	1.81	1.92	1.91	2.61	1.43	1.46
L_r, ft	20.9	19.2	4.72	34.6	26.4	22.3	18.6	17.0	4.93	21.9	20.6
L_p, ft	4.62	4.62	1.45	4.75	4.79	4.83	4.13	4.15	1.52	4.20	4.20
$\phi_b M_r$, kip-ft	70.4	65.7	18.0	61.4	54.2	49.3	42.9	40.0	12.5	38.0	36.9
$\phi_b M_p$, kip-ft	117	107	31.3	106	91.0	81.0	70.7	64.5	21.4	63.5	60.8

[a] Beam must be laterally supported if Table 5-10 is used.

Note: Design strength tabulated above heavy line is limited by design shear strength.

$F_y = 36$ ksi
$\phi_b = 0.90$
$\phi_v = 0.90$

Table 5-10 (cont.). MC-Shapes Maximum Total Factored Uniform Load, kips[a]

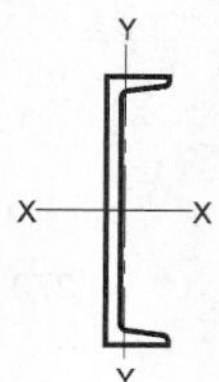

Shape	MC8×					MC7×		MC6×				
	22.8	21.4	20	18.7	8.5	22.7	19.1	18	15.3	16.3	15.1	12
Span, ft												
2			124		55.7	137		88.4	79.3	87.5	73.7	72.3
3	133	117	118	110	50.0	118	95.8	84.2	71.4	74.9	70.8	53.8
4	103	98.3	88.6	84.2	37.5	88.6	78.3	63.2	53.5	56.2	53.1	40.3
5	82.5	78.6	70.8	67.4	30.0	70.8	62.6	50.5	42.8	44.9	42.5	32.3
6	68.8	65.5	59.0	56.2	25.0	59.0	52.2	42.1	35.7	37.4	35.4	26.9
7	58.9	56.2	50.6	48.1	21.4	50.6	44.7	36.1	30.6	32.1	30.3	23.1
8	51.6	49.1	44.3	42.1	18.8	44.3	39.1	31.6	26.8	28.1	26.5	20.2
9	45.8	43.7	39.4	37.4	16.7	39.4	34.8	28.1	23.8	25.0	23.6	17.9
10	41.3	39.3	35.4	33.7	15.0	35.4	31.3	25.3	21.4	22.5	21.2	16.1
11	37.5	35.7	32.2	30.6	13.6	32.2	28.5	23.0	19.5	20.4	19.3	14.7
12	34.4	32.8	29.5	28.1	12.5	29.5	26.1	21.1	17.8	18.7	17.7	13.4
13	31.7	30.2	27.2	25.9	11.5	27.2	24.1	19.4	16.5	17.3	16.3	12.4
14	29.5	28.1	25.3	24.1	10.7	25.3	22.4	18.1	15.3	16.0	15.2	11.5
15	27.5	26.2	23.6	22.5	10.0	23.6	20.9	16.8	14.3	15.0	14.2	10.8
16	25.8	24.6	22.1	21.1	9.38	22.1	19.6					
17	24.3	23.1	20.8	19.8	8.83	20.8	18.4					
18	22.9	21.8	19.7	18.7	8.34							
19	21.7	20.7	18.6	17.7	7.90							
20	20.6	19.7	17.7	16.8	7.51							
21												
22												
23												
24												
25												
26												
27												
28												
29												
30												
32												
34												
36												
38												
40												
42												
44												
Beam Properties												
Z_x, in.^{3}	19.1	18.2	16.4	15.6	6.95	16.4	14.5	11.7	9.91	10.4	9.83	7.47
$\phi_b W_c$, kip-ft	413	393	354	337	150	354	313	253	214	225	212	161
$\phi_v V_n$, kips	66.4	58.3	62.2	54.9	27.8	68.4	47.9	44.2	39.7	43.7	36.9	36.2
BF, kips	1.08	1.09	1.15	1.16	1.44	0.733	0.786	0.530	0.550	0.556	0.565	0.618
L_r, ft	23.3	21.7	19.1	17.9	7.25	28.8	23.6	27.6	23.1	23.8	22.0	16.0
L_p, ft	4.25	4.25	3.61	3.61	2.08	4.33	4.33	4.37	4.37	3.69	3.68	3.01
$\phi_b M_r$, kip-ft	31.0	30.0	26.5	25.5	11.3	26.3	24.0	19.3	16.5	16.9	16.2	12.2
$\phi_b M_p$, kip-ft	51.6	49.1	44.3	42.1	18.8	44.3	39.1	31.6	26.8	28.1	26.5	20.2

[a] Beam must be laterally supported if Table 5-10 is used.
Note: Design strength tabulated above heavy line is limited by design shear strength.

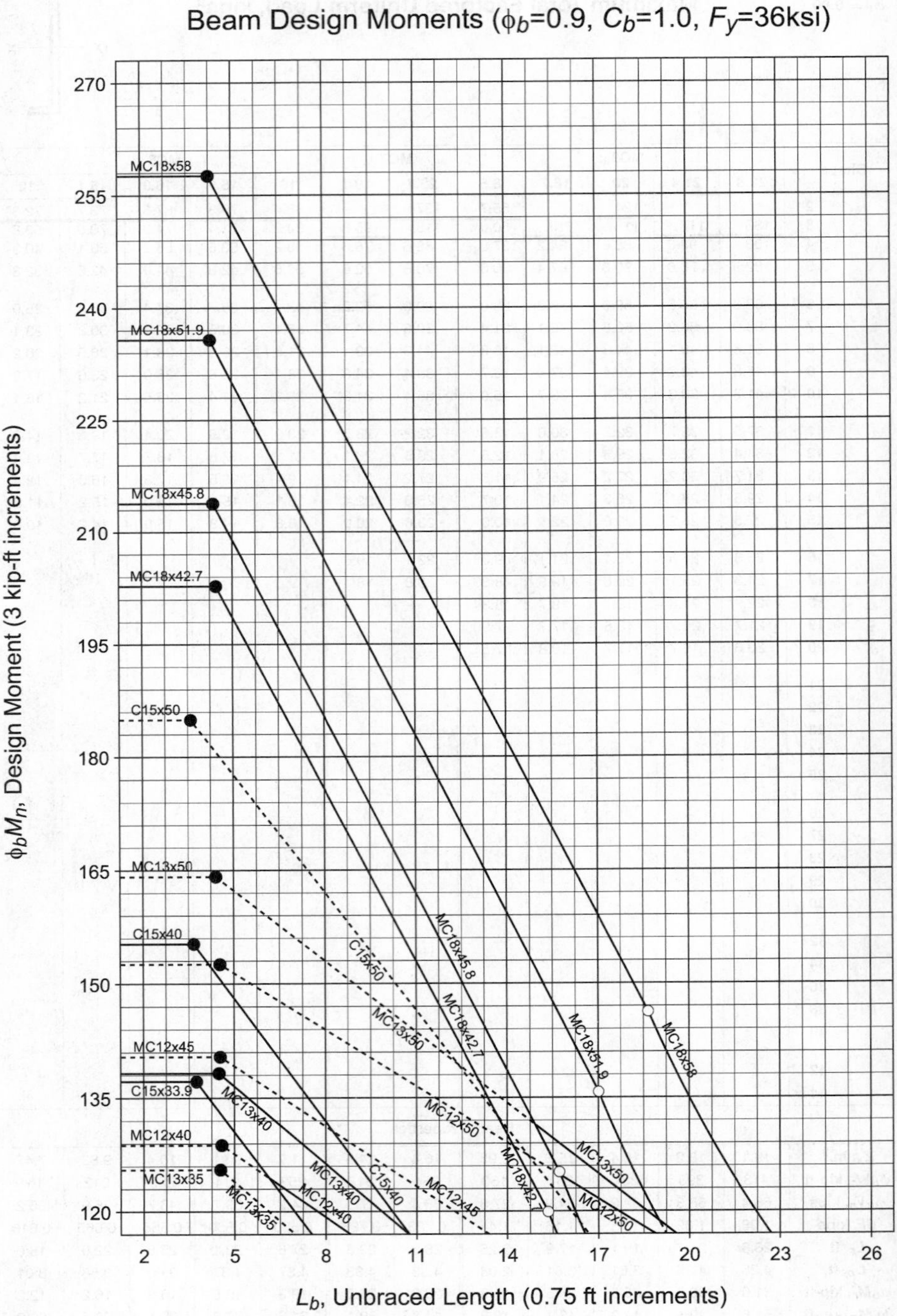

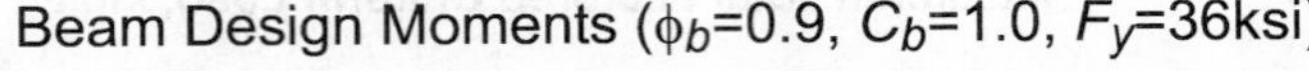

$\phi_b M_n$, Design Moment (3 kip-ft increments)

270
255
240
225
210
195
180
165
150
135
120

MC18x58
MC18x51.9
MC18x45.8
MC18x42.7
C15x50
MC13x50
C15x40
MC12x45
C15x33.9
MC12x40
MC13x35
MC13x40
MC12x50
MC12x40
MC13x40
C15x40
MC12x45
MC13x35
MC18x42.7
MC13x50
MC12x50
MC18x45.8
C15x50
MC18x42.7
MC13x50
MC18x51.9
MC18x58

2 5 8 11 14 17 20 23 26

L_b, Unbraced Length (0.75 ft increments)

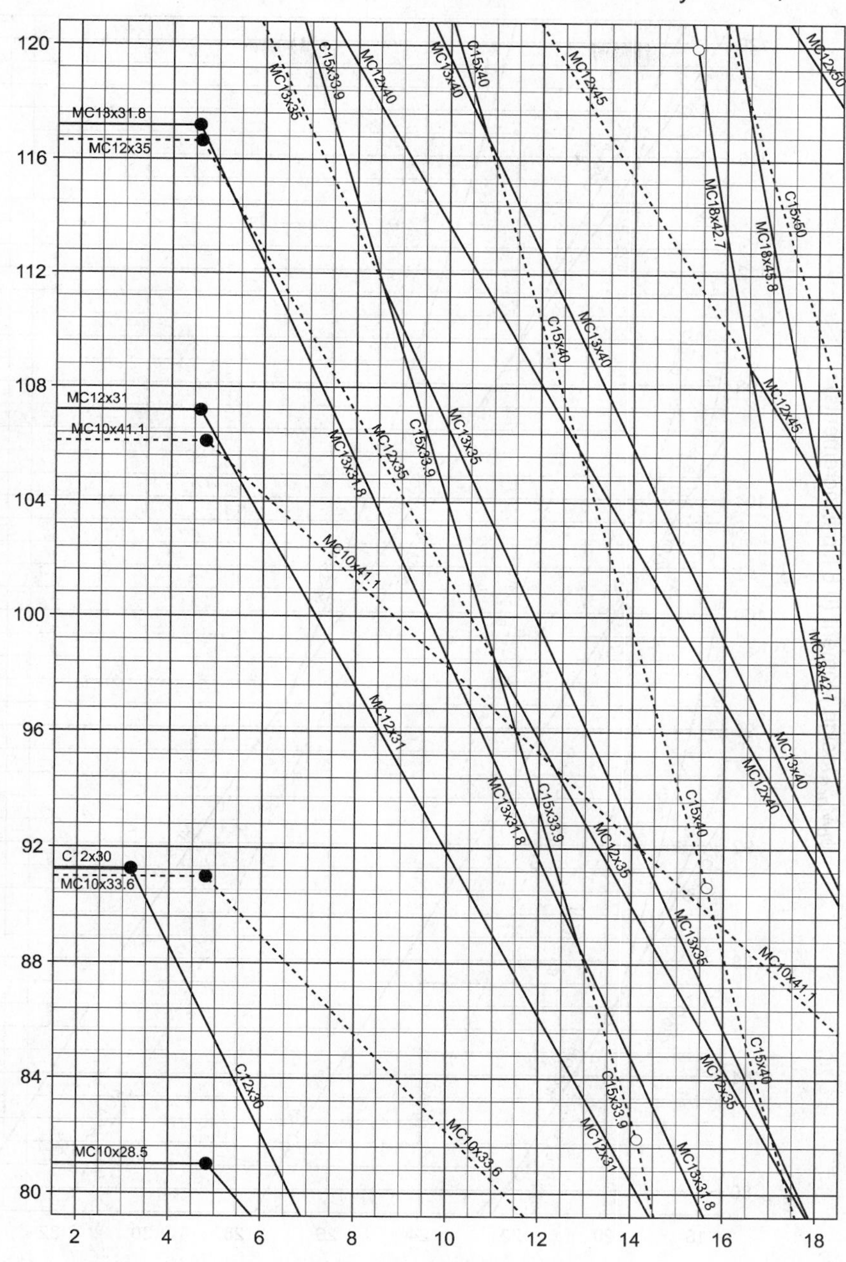
Beam Design Moments (ϕ_b=0.9, C_b=1.0, F_y=36ksi)
$\phi_b M_n$, Design Moment (0.8 kip-ft increments)
L_b, Unbraced Length (0.5 ft increments)
MC13x31.8
MC12x35
MC12x31
MC10x41.1
C12x30
MC10x33.6
MC10x28.5
MC13x35
C15x33.9
MC12x40
MC13x40
C15x40
MC12x45
MC12x50
MC18x42.7
MC18x45.8
C15x50

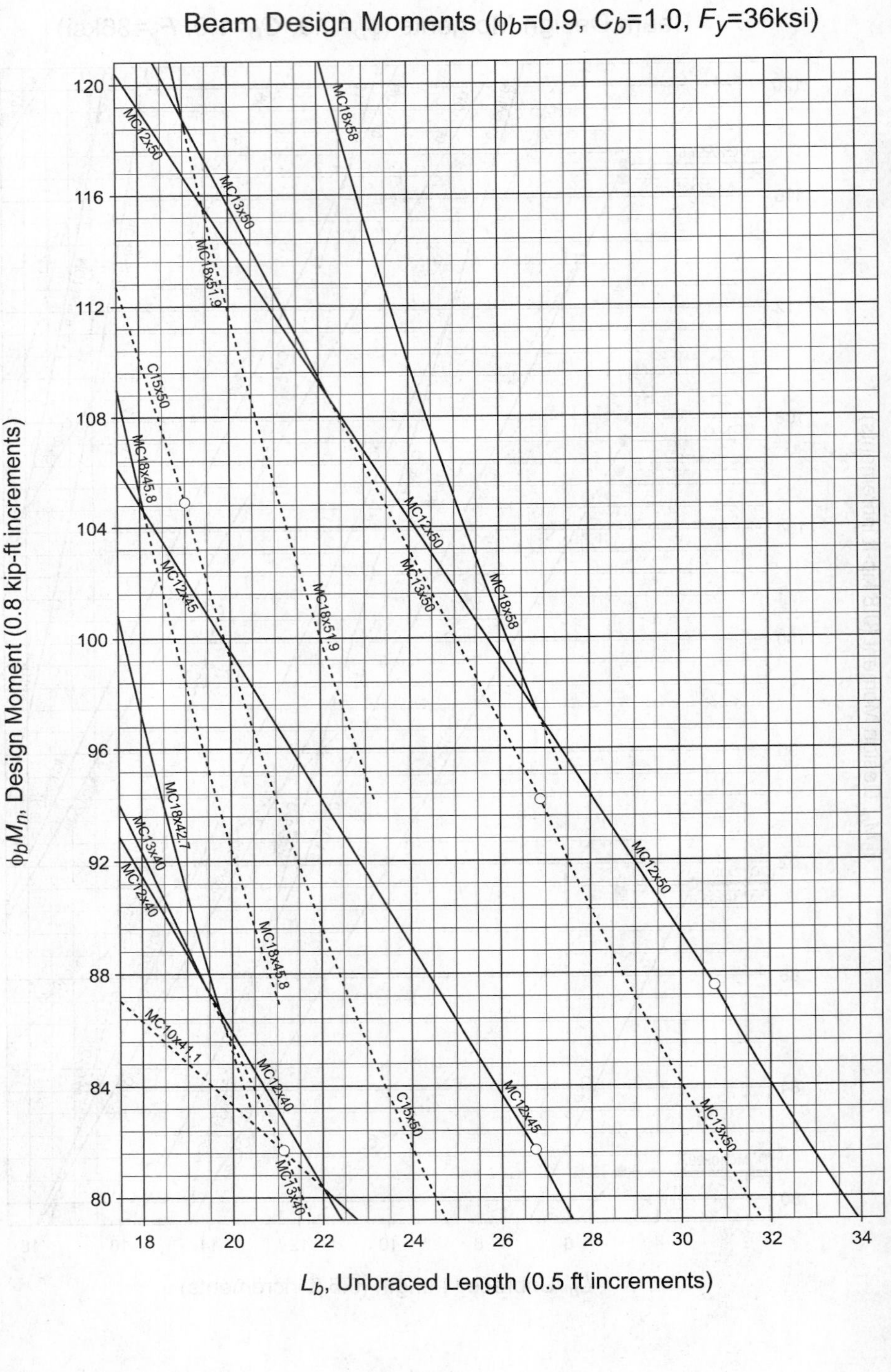

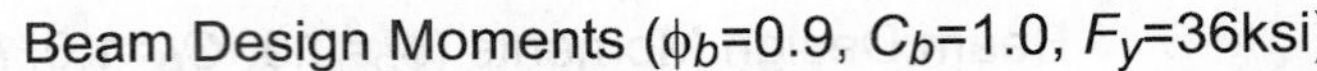
Beam Design Moments (ϕ_b=0.9, C_b=1.0, F_y=36ksi)

$\phi_b M_n$, Design Moment (0.8 kip-ft increments)
120
116
112
108
104
100
96
92
88
84
80
18
20
22
24
26
28
30
32
34
L_b, Unbraced Length (0.5 ft increments)
MC12x50
MC13x50
MC18x58
MC18x51.9
C15x50
MC18x45.8
MC12x45
MC18x42.7
MC13x40
MC12x40
MC10x41.1

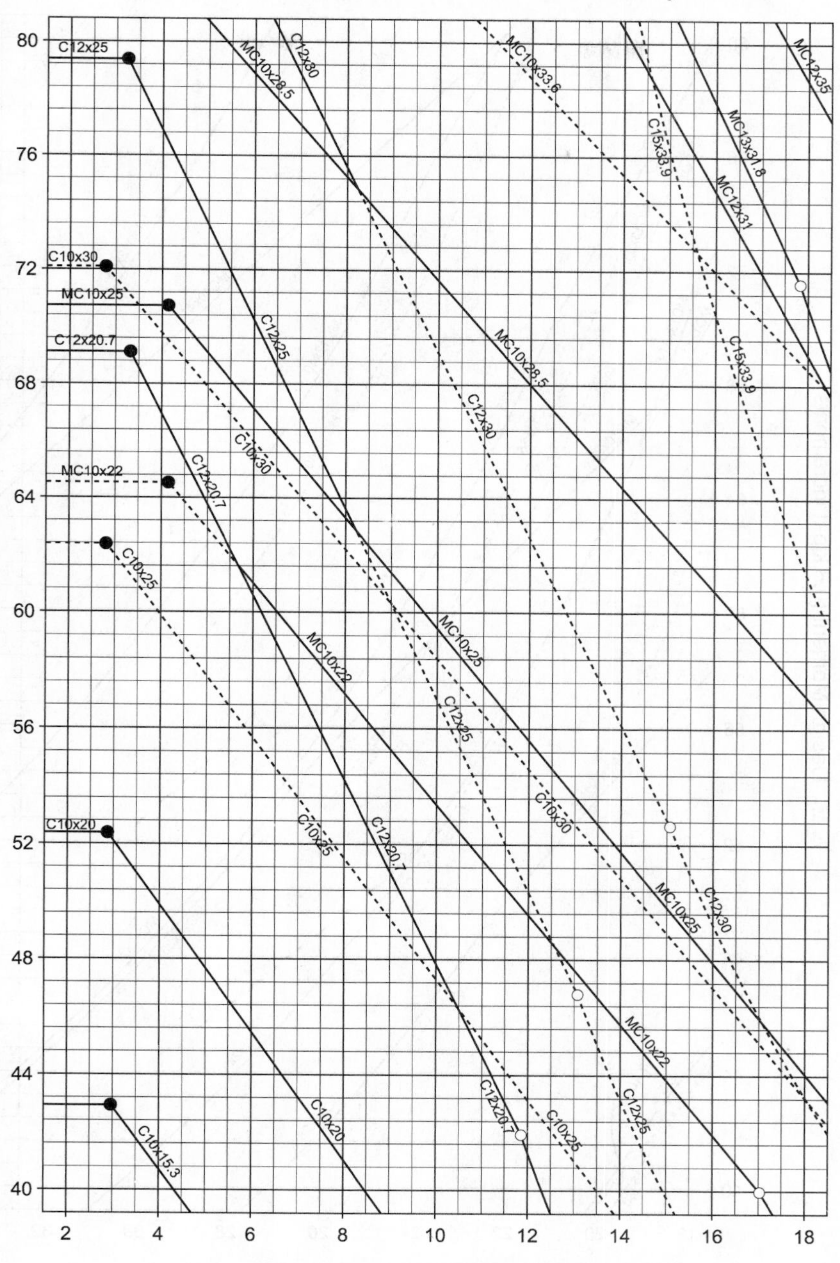
Beam Design Moments (ϕb=0.9, Cb=1.0, Fy=36ksi)
ϕbMn, Design Moment (0.8 kip-ft increments)
Lb, Unbraced Length (0.5 ft increments)
C12x25
C10x30
MC10x25
C12x20.7
MC10x22
C10x20
C10x25
C10x15.3
MC10x28.5
C12x30
MC10x33.6
C15x33.9
MC13x31.8
MC12x31
MC12x35
80
76
72
68
64
60
56
52
48
44
40
2
4
6
8
10
12
14
16
18

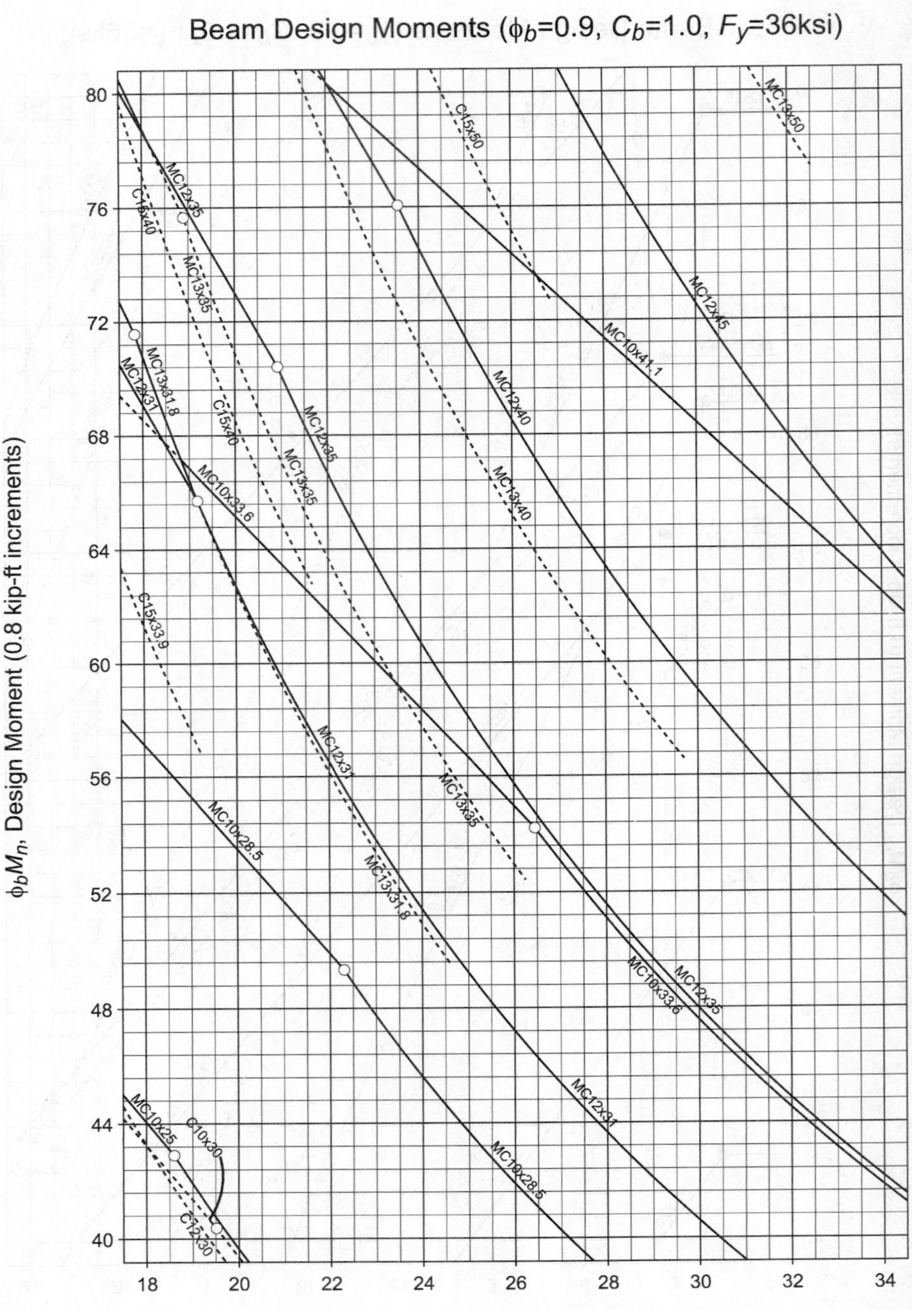

Beam Design Moments (ϕ_b=0.9, C_b=1.0, F_y=36ksi)
$\phi_b M_n$, Design Moment (0.8 kip-ft increments)
L_b, Unbraced Length (0.5 ft increments)
80
76
72
68
64
60
56
52
48
44
40
18
20
22
24
26
28
30
32
34
MC13x50
C15x50
MC12x45
MC10x41.1
MC12x40
MC13x40
MC12x35
C15x40
MC13x35
MC13x31.8
MC12x31
MC10x33.6
C15x33.9
MC10x28.5
MC10x25
C10x30
C12x30

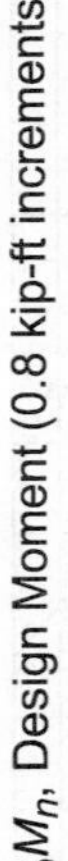

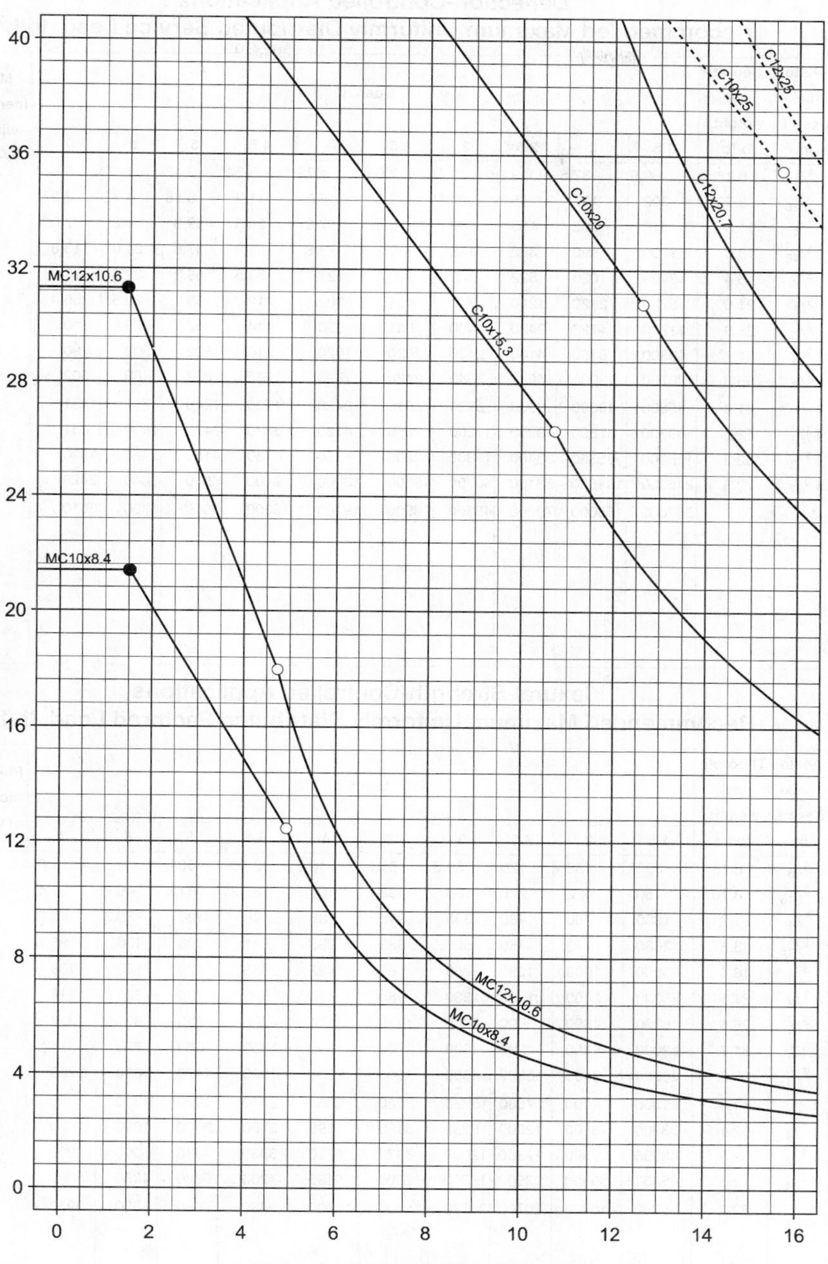
Beam Design Moments (ϕb=0.9, Cb=1.0, Fy=36ksi)
C12x25
C10x25
C12x20.7
C10x20
C10x15.3
MC12x10.6
MC10x8.4
MC12x10.6
MC10x8.4
ϕbMn, Design Moment (0.8 kip-ft increments)
Lb, Unbraced Length (0.5 ft increments)
0
2
4
6
8
10
12
14
16
0
4
8
12
16
20
24
28
32
36
40

Table 5-12.
Floor Plate

Deflection-Controlled Applications
Recommended Maximum Uniformly Distributed Service Load, lb/ft²

Plate thickness t, in.	Theoretical weight, lb/ft²	Span, ft										Moment of inertia per ft of width, in.⁴/ft
		1.5	2	2.5	3	3.5	4	4.5	5	6	7	
1/8	6.15	89.5	37.8	19.3	11.2	7.05	4.72	3.32	2.42	1.40	0.881	0.00195
3/16	8.70	302	127	65.3	37.8	23.8	15.9	11.2	8.16	4.72	2.97	0.00659
1/4	11.3	716	302	155	89.5	56.4	37.8	26.5	19.3	11.2	7.05	0.0156
5/16	13.8	1400	590	302	175	110	73.8	51.8	37.8	21.9	13.8	0.0305
3/8	16.4	2420	1020	522	302	190	127	89.5	65.3	37.8	23.8	0.0527
1/2	21.5	5730	2420	1240	716	451	302	212	155	89.5	56.4	0.125
5/8	26.6	11200	4720	2420	1400	881	590	414	302	175	110	0.244
3/4	31.7	19300	8160	4180	2420	1520	1020	716	522	302	190	0.422
7/8	36.8	30700	13000	6630	3840	2420	1620	1140	829	480	302	0.670
1	41.9	45800	19300	9900	5730	3610	2420	1700	1240	716	451	1.00
1 1/4	52.1	89500	37800	19300	11200	7050	4720	3320	2420	1400	881	1.95
1 1/2	62.3	155000	65300	33400	19300	12200	8160	5730	4180	2420	1520	3.38
1 3/4	72.5	246000	104000	53100	30700	19300	13000	9100	6630	3840	2420	5.36
2	82.7	367000	155000	79200	45800	28900	19300	13600	9900	5730	3610	8.00

Flexural-Strength-Controlled Applications
Recommended Maximum Uniformly Distributed Factored Load, lb/ft²

Plate thickness t, in.	Theoretical weight, lb/ft²	Span, ft										Plastic section modulus per ft of width, in.³/ft
		1.5	2	2.5	3	3.5	4	4.5	5	6	7	
1/8	6.15	333	188	120	83.3	61.2	46.9	37.0	30.0	20.8	15.3	0.0469
3/16	8.70	750	422	270	188	138	105	83.3	67.5	46.9	34.4	0.105
1/4	11.3	1330	750	480	333	245	188	148	120	83.3	61.2	0.188
5/16	13.8	2080	1170	750	521	383	293	231	188	130	95.7	0.293
3/8	16.4	3000	1690	1080	750	551	422	333	270	188	138	0.422
1/2	21.5	5330	3000	1920	1330	980	750	593	480	333	245	0.750
5/8	26.6	8330	4690	3000	2080	1530	1170	926	750	521	383	1.17
3/4	31.7	12000	6750	4320	3000	2200	1690	1330	1080	750	551	1.69
7/8	36.8	16300	9190	5880	4080	3000	2300	1820	1470	1020	750	2.30
1	41.9	21300	12000	7680	5330	3920	3000	2370	1920	1330	980	3.00
1 1/4	52.1	33300	18800	12000	8330	6120	4690	3700	3000	2080	1530	4.69
1 1/2	62.3	48000	27000	17300	12000	8820	6750	5330	4320	3000	2200	6.75
1 3/4	72.5	65300	36800	23500	16300	12000	9190	7260	5880	4080	3000	9.19
2	82.7	85300	48000	30700	21300	15700	12000	9480	7680	5330	3920	12.0

Table 5-13.
Shear Stud Connectors
Nominal Shear Strength Q_n, kips[a]

Specified Compressive Strength of Concrete f'_c, ksi	Light-Weight Concrete (115 lb/ft³)				Normal-Weight Concrete (145 lb/ft³)			
	Nominal Shear Stud Connector Diameter, in.				Nominal Shear Stud Connector Diameter, in.			
	$^1/_2$	$^5/_8$	$^3/_4$	$^7/_8$	$^1/_2$	$^5/_8$	$^3/_4$	$^7/_8$
3	7.86	12.3	17.7	24.1	9.35	14.6	21.0	28.6
3.5	8.82	13.8	19.8	27.0	10.5	16.4	23.6	32.1
4	9.75	15.2	21.9	29.9	11.6	18.1	26.1	35.5
4.5	10.7	16.6	24.0	32.6	11.8	18.4	26.5	36.1
5	11.5	18.0	25.9	35.3	11.8	18.4	26.5	36.1
Minimum Stud Length, in.	**2**	**2 $^1/_2$**	**3**	**3 $^1/_2$**	**2**	**2 $^1/_2$**	**3**	**3 $^1/_2$**

[a] Applicable only to concrete made with ASTM C33 aggregates.

Table 5-14. Composite W-Shapes Design Strength in Flexure $\phi_b M_n$, kip-ft

$F_y = 50$ ksi
$\phi_b = 0.90$
$\phi_c = 0.85$

Shape	$\phi_b M_p$	PNA[c]	$Y1$[a]	ΣQ_n	$Y2$[b], in.										
	kip-ft		in.	kips	2	2.5	3	3.5	4	4.5	5	5.5	6	6.5	7
W40×297	4990	TFL	0.000	4370	6780	6930	7090	7240	7400	7550	7710	7860	8020	8170	8330
		2	0.413	3720	6680	6810	6940	7070	7200	7340	7470	7600	7730	7860	7990
		3	0.825	3070	6560	6660	6770	6880	6990	7100	7210	7320	7430	7530	7640
		4	1.24	2410	6420	6500	6590	6670	6760	6840	6930	7010	7100	7190	7270
		BFL	1.65	1760	6260	6320	6380	6440	6510	6570	6630	6690	6760	6820	6880
		6	4.55	1430	6140	6200	6250	6300	6350	6400	6450	6500	6550	6600	6650
		7	8.15	1090	5950	5980	6020	6060	6100	6140	6180	6220	6250	6290	6330
W40×278	4460	TFL	0.000	4090	6400	6550	6690	6840	6980	7130	7270	7420	7560	7710	7850
		2	0.453	3550	6320	6440	6570	6690	6820	6950	7070	7200	7320	7450	7570
		3	0.905	3000	6210	6320	6430	6530	6640	6750	6850	6960	7060	7170	7280
		4	1.36	2460	6090	6180	6270	6350	6440	6530	6620	6700	6790	6880	6960
		BFL	1.81	1920	5960	6020	6090	6160	6230	6300	6360	6430	6500	6570	6640
		6	5.69	1470	5780	5830	5890	5940	5990	6040	6090	6150	6200	6250	6300
		7	10.1	1020	5470	5500	5540	5580	5610	5650	5680	5720	5760	5790	5830
W40×277	4690	TFL	0.000	4070	6300	6440	6590	6730	6880	7020	7160	7310	7450	7600	7740
		2	0.395	3450	6200	6320	6450	6570	6690	6810	6930	7060	7180	7300	7420
		3	0.790	2820	6090	6190	6290	6390	6490	6590	6690	6790	6890	6990	7090
		4	1.19	2200	5960	6030	6110	6190	6270	6340	6420	6500	6580	6660	6730
		BFL	1.58	1570	5810	5860	5920	5970	6030	6080	6140	6200	6250	6310	6360
		6	4.24	1300	5720	5760	5810	5850	5900	5950	5990	6040	6080	6130	6180
		7	7.59	1020	5560	5600	5630	5670	5700	5740	5780	5810	5850	5890	5920
W40×264	4240	TFL	0.000	3880	6050	6180	6320	6460	6600	6730	6870	7010	7150	7280	7420
		2	0.433	3370	5970	6080	6200	6320	6440	6560	6680	6800	6920	7040	7160
		3	0.865	2850	5870	5970	6070	6170	6270	6370	6470	6580	6680	6780	6880
		4	1.30	2340	5760	5840	5920	6010	6090	6170	6250	6340	6420	6500	6580
		BFL	1.73	1820	5630	5690	5760	5820	5890	5950	6020	6080	6140	6210	6270
		6	5.46	1400	5470	5520	5570	5620	5670	5720	5770	5820	5870	5920	5970
		7	9.90	970	5180	5210	5250	5280	5320	5350	5380	5420	5450	5490	5520
W40×249	4200	TFL	0.000	3670	5630	5760	5890	6020	6150	6280	6410	6540	6670	6800	6930
		2	0.355	3100	5550	5660	5770	5880	5990	6100	6210	6320	6430	6540	6650
		3	0.710	2540	5450	5540	5630	5720	5810	5900	5990	6080	6170	6260	6350
		4	1.07	1980	5330	5400	5470	5540	5610	5680	5750	5820	5890	5960	6030
		BFL	1.42	1420	5200	5250	5300	5350	5400	5450	5500	5550	5610	5660	5710
		6	4.12	1170	5120	5160	5210	5250	5290	5330	5370	5410	5450	5500	5540
		7	7.48	916	4980	5020	5050	5080	5110	5150	5180	5210	5240	5280	5310
W40×235	3790	TFL	0.000	3450	5340	5460	5580	5710	5830	5950	6070	6190	6320	6440	6560
		2	0.395	2980	5270	5370	5480	5580	5690	5790	5900	6010	6110	6220	6320
		3	0.790	2510	5180	5270	5360	5450	5540	5620	5710	5800	5890	5980	6070
		4	1.19	2040	5080	5150	5230	5300	5370	5440	5510	5590	5660	5730	5800
		BFL	1.58	1570	4970	5020	5080	5130	5190	5250	5300	5360	5410	5470	5520
		6	5.20	1220	4840	4880	4930	4970	5010	5060	5100	5140	5190	5230	5270
		7	9.46	863	4610	4640	4670	4700	4730	4760	4790	4820	4850	4880	4910

[a] $Y1$ = distance from top of the steel beam to plastic neutral axis.
[b] $Y2$ = distance from top of the steel beam to concrete flange force.
[c] See Figure 5-4c for PNA locations.

$F_y = 50$ ksi
$\phi_b = 0.90$
$\phi_c = 0.85$

Table 5-14 (cont.). Composite W-Shapes Design Strength in Flexure $\phi_b M_n$, kip-ft

Shape	$\phi_b M_p$	PNA[c]	$Y1$[a]	ΣQ_n	$Y2$[b], in.										
	kip-ft		in.	kips	2	2.5	3	3.5	4	4.5	5	5.5	6	6.5	7
W40×215	3620	TFL	0.000	3170	4830	4940	5050	5160	5280	5390	5500	5610	5730	5840	5950
		2	0.305	2690	4750	4850	4940	5040	5140	5230	5330	5420	5520	5610	5710
		3	0.610	2210	4670	4750	4830	4900	4980	5060	5140	5220	5300	5370	5450
		4	0.915	1720	4580	4640	4700	4760	4820	4880	4940	5000	5060	5120	5190
		BFL	1.22	1240	4470	4520	4560	4600	4650	4690	4730	4780	4820	4870	4910
		6	3.85	1020	4410	4440	4480	4510	4550	4590	4620	4660	4690	4730	4770
		7	7.31	793	4280	4310	4340	4370	4400	4420	4450	4480	4510	4540	4560
W40×211	3400	TFL	0.000	3100	4760	4870	4980	5090	5200	5310	5420	5530	5640	5750	5860
		2	0.355	2680	4700	4800	4890	4990	5080	5180	5270	5370	5460	5550	5650
		3	0.710	2260	4630	4710	4790	4870	4950	5030	5110	5190	5270	5350	5430
		4	1.07	1840	4540	4600	4670	4740	4800	4870	4930	5000	5060	5130	5190
		BFL	1.42	1420	4440	4490	4540	4590	4640	4690	4750	4800	4850	4900	4950
		6	5.04	1100	4330	4370	4410	4450	4490	4530	4560	4600	4640	4680	4720
		7	9.37	775	4120	4150	4170	4200	4230	4260	4280	4310	4340	4370	4390
W40×199	3260	TFL	0.000	2930	4420	4530	4630	4730	4840	4940	5050	5150	5250	5360	5460
		2	0.268	2500	4360	4450	4540	4630	4710	4800	4890	4980	5070	5160	5250
		3	0.535	2080	4290	4360	4440	4510	4580	4660	4730	4800	4880	4950	5020
		4	0.803	1660	4210	4270	4330	4380	4440	4500	4560	4620	4680	4740	4800
		BFL	1.07	1230	4120	4160	4210	4250	4290	4340	4380	4430	4470	4510	4560
		6	4.23	983	4040	4080	4110	4150	4180	4220	4250	4290	4320	4360	4390
		7	8.10	731	3900	3920	3950	3980	4000	4030	4050	4080	4100	4130	4160
W40×183	2940	TFL	0.000	2690	4100	4190	4290	4380	4480	4570	4670	4760	4860	4950	5050
		2	0.305	2330	4040	4120	4210	4290	4370	4450	4540	4620	4700	4780	4870
		3	0.610	1970	3980	4050	4120	4190	4260	4330	4400	4470	4540	4610	4680
		4	0.915	1610	3910	3970	4020	4080	4140	4190	4250	4310	4360	4420	4480
		BFL	1.22	1250	3830	3870	3920	3960	4010	4050	4100	4140	4180	4230	4270
		6	4.71	961	3740	3770	3810	3840	3870	3910	3940	3980	4010	4040	4080
		7	9.15	673	3550	3580	3600	3630	3650	3670	3700	3720	3750	3770	3790
W40×167	2600	TFL	0.000	2460	3710	3800	3890	3970	4060	4150	4230	4320	4410	4500	4580
		2	0.255	2160	3670	3740	3820	3900	3970	4050	4130	4200	4280	4350	4430
		3	0.510	1860	3620	3680	3750	3810	3880	3940	4010	4080	4140	4210	4270
		4	0.765	1560	3560	3610	3670	3720	3780	3840	3890	3950	4000	4060	4110
		BFL	1.02	1260	3500	3540	3590	3630	3680	3720	3760	3810	3850	3900	3940
		6	4.90	936	3390	3430	3460	3490	3530	3560	3590	3630	3660	3690	3730
		7	9.84	615	3180	3200	3230	3250	3270	3290	3310	3330	3360	3380	3400
W40×149	2240	TFL	0.000	2190	3270	3350	3430	3510	3580	3660	3740	3820	3890	3970	4050
		2	0.208	1950	3240	3310	3370	3440	3510	3580	3650	3720	3790	3860	3930
		3	0.415	1700	3200	3260	3320	3380	3440	3500	3560	3620	3680	3740	3800
		4	0.623	1460	3150	3200	3260	3310	3360	3410	3460	3510	3570	3620	3670
		BFL	0.830	1210	3110	3150	3190	3230	3280	3320	3360	3410	3450	3490	3530
		6	5.15	879	3000	3030	3060	3090	3120	3150	3180	3220	3250	3280	3310
		7	10.4	548	2770	2790	2810	2830	2850	2870	2890	2900	2920	2940	2960

[a] $Y1$ = distance from top of the steel beam to plastic neutral axis.
[b] $Y2$ = distance from top of the steel beam to concrete flange force.
[c] See Figure 5-4c for PNA locations.

Table 5-14 (cont.). Composite W-Shapes Design Strength in Flexure $\phi_b M_n$, kip-ft

$F_y = 50$ ksi
$\phi_b = 0.90$
$\phi_c = 0.85$

Shape	$\phi_b M_p$	PNA[c]	$Y1$[a]	ΣQ_n	$Y2$[b], in.										
	kip-ft		in.	kips	2	2.5	3	3.5	4	4.5	5	5.5	6	6.5	7
W36×300	4730	TFL	0.000	4420	6360	6520	6680	6830	6990	7150	7300	7460	7610	7770	7930
		2	0.420	3710	6250	6390	6520	6650	6780	6910	7040	7170	7310	7440	7570
		3	0.840	3010	6120	6230	6340	6440	6550	6660	6760	6870	6980	7080	7190
		4	1.26	2310	5970	6050	6140	6220	6300	6380	6460	6540	6630	6710	6790
		BFL	1.68	1610	5800	5860	5910	5970	6030	6080	6140	6200	6260	6310	6370
		6	3.99	1360	5720	5770	5810	5860	5910	5960	6010	6050	6100	6150	6200
		7	6.67	1100	5590	5620	5660	5700	5740	5780	5820	5860	5900	5940	5980
W36×280	4390	TFL	0.000	4120	5910	6060	6200	6350	6490	6640	6790	6930	7080	7220	7370
		2	0.393	3470	5810	5930	6050	6180	6300	6420	6550	6670	6790	6910	7040
		3	0.785	2820	5690	5790	5890	5990	6090	6190	6290	6390	6490	6590	6690
		4	1.18	2170	5550	5630	5700	5780	5860	5940	6010	6090	6170	6240	6320
		BFL	1.57	1510	5400	5450	5500	5560	5610	5660	5720	5770	5820	5880	5930
		6	3.88	1270	5320	5360	5410	5450	5500	5540	5590	5630	5680	5720	5770
		7	6.61	1030	5190	5230	5270	5300	5340	5380	5410	5450	5490	5520	5560
W36×260	4050	TFL	0.000	3830	5460	5590	5730	5870	6000	6140	6270	6410	6540	6680	6810
		2	0.360	3230	5370	5480	5600	5710	5820	5940	6050	6170	6280	6400	6510
		3	0.720	2630	5260	5350	5450	5540	5630	5730	5820	5910	6000	6100	6190
		4	1.08	2030	5140	5210	5280	5350	5420	5500	5570	5640	5710	5780	5860
		BFL	1.44	1430	5000	5050	5100	5150	5200	5250	5300	5350	5400	5460	5510
		6	3.92	1200	4920	4970	5010	5050	5090	5140	5180	5220	5260	5310	5350
		7	6.77	956	4800	4830	4870	4900	4930	4970	5000	5040	5070	5100	5140
W36×256	3900	TFL	0.000	3770	5530	5660	5790	5930	6060	6200	6330	6460	6600	6730	6860
		2	0.433	3240	5440	5560	5670	5790	5900	6020	6130	6250	6360	6480	6590
		3	0.865	2710	5350	5440	5540	5630	5730	5830	5920	6020	6110	6210	6310
		4	1.30	2190	5230	5310	5390	5460	5540	5620	5700	5770	5850	5930	6010
		BFL	1.73	1660	5100	5160	5220	5280	5330	5390	5450	5510	5570	5630	5690
		6	5.15	1300	4970	5010	5060	5100	5150	5200	5240	5290	5330	5380	5430
		7	8.88	943	4740	4770	4800	4840	4870	4900	4940	4970	5000	5040	5070
W36×245	3790	TFL	0.000	3610	5120	5250	5380	5500	5630	5760	5890	6010	6140	6270	6400
		2	0.338	3050	5030	5140	5250	5360	5470	5570	5680	5790	5900	6010	6110
		3	0.675	2490	4940	5020	5110	5200	5290	5380	5460	5550	5640	5730	5820
		4	1.01	1930	4820	4890	4960	5030	5100	5170	5230	5300	5370	5440	5510
		BFL	1.35	1380	4700	4750	4800	4840	4890	4940	4990	5040	5090	5140	5190
		6	3.81	1140	4620	4670	4710	4750	4790	4830	4870	4910	4950	4990	5030
		7	6.78	901	4500	4530	4570	4600	4630	4660	4690	4730	4760	4790	4820
W36×232	3510	TFL	0.000	3410	4960	5080	5200	5320	5440	5560	5680	5800	5920	6040	6160
		2	0.393	2930	4880	4990	5090	5190	5300	5400	5510	5610	5710	5820	5920
		3	0.785	2460	4800	4880	4970	5060	5140	5230	5320	5400	5490	5580	5670
		4	1.18	1980	4700	4770	4840	4910	4980	5050	5120	5190	5260	5330	5400
		BFL	1.57	1510	4580	4640	4690	4740	4800	4850	4900	4960	5010	5060	5120
		6	5.01	1180	4460	4510	4550	4590	4630	4670	4710	4760	4800	4840	4880
		7	8.77	851	4260	4290	4320	4350	4380	4410	4440	4470	4500	4530	4560

[a] $Y1$ = distance from top of the steel beam to plastic neutral axis.
[b] $Y2$ = distance from top of the steel beam to concrete flange force.
[c] See Figure 5-4c for PNA locations.

$F_y = 50$ ksi
$\phi_b = 0.90$
$\phi_c = 0.85$

Table 5-14 (cont.). Composite W-Shapes Design Strength in Flexure $\phi_b M_n$, kip-ft

Shape	$\phi_b M_p$	PNA[c]	$Y1$[a]	ΣQ_n	$Y2$[b], in.										
	kip-ft		in.	kips	2	2.5	3	3.5	4	4.5	5	5.5	6	6.5	7
W36×230	3540	TFL	0.000	3380	4780	4900	5020	5140	5260	5370	5490	5610	5730	5850	5970
		2	0.315	2860	4700	4800	4900	5000	5100	5200	5300	5410	5510	5610	5710
		3	0.630	2340	4610	4690	4770	4850	4940	5020	5100	5190	5270	5350	5430
		4	0.945	1820	4500	4570	4630	4700	4760	4830	4890	4950	5020	5080	5150
		BFL	1.26	1300	4390	4440	4480	4530	4570	4620	4670	4710	4760	4800	4850
		6	3.83	1070	4320	4360	4400	4430	4470	4510	4550	4580	4620	4660	4700
		7	6.83	845	4200	4230	4260	4290	4320	4350	4380	4410	4440	4470	4500
W36×210	3120	TFL	0.000	3090	4450	4560	4670	4780	4890	5000	5110	5220	5330	5440	5550
		2	0.340	2680	4390	4490	4580	4670	4770	4860	4960	5050	5150	5240	5340
		3	0.680	2260	4320	4400	4480	4560	4640	4720	4800	4880	4960	5040	5120
		4	1.02	1850	4230	4300	4360	4430	4490	4560	4630	4690	4760	4820	4890
		BFL	1.36	1430	4140	4190	4240	4290	4340	4390	4440	4490	4540	4600	4650
		6	5.08	1100	4020	4060	4100	4140	4180	4220	4250	4290	4330	4370	4410
		7	9.04	773	3810	3840	3860	3890	3920	3950	3970	4000	4030	4060	4080
W36×194	2880	TFL	0.000	2850	4090	4190	4290	4390	4490	4590	4690	4790	4900	5000	5100
		2	0.315	2470	4030	4120	4200	4290	4380	4470	4550	4640	4730	4820	4900
		3	0.630	2090	3960	4040	4110	4180	4260	4330	4410	4480	4550	4630	4700
		4	0.945	1710	3890	3950	4010	4070	4130	4190	4250	4310	4370	4430	4490
		BFL	1.26	1330	3800	3850	3900	3950	3990	4040	4090	4130	4180	4230	4270
		6	4.93	1020	3700	3730	3770	3810	3840	3880	3910	3950	3990	4020	4060
		7	8.94	713	3500	3530	3550	3580	3600	3630	3650	3680	3710	3730	3760
W36×182	2690	TFL	0.000	2680	3830	3920	4010	4110	4200	4300	4390	4490	4580	4680	4770
		2	0.295	2320	3770	3850	3940	4020	4100	4180	4260	4350	4430	4510	4590
		3	0.590	1970	3710	3780	3850	3920	3990	4060	4130	4200	4270	4340	4410
		4	0.885	1610	3640	3700	3750	3810	3870	3920	3980	4040	4100	4150	4210
		BFL	1.18	1250	3560	3610	3650	3700	3740	3780	3830	3870	3920	3960	4010
		6	4.89	961	3460	3500	3530	3560	3600	3630	3670	3700	3740	3770	3800
		7	8.91	670	3280	3300	3330	3350	3370	3400	3420	3450	3470	3490	3520
W36×170	2510	TFL	0.000	2510	3570	3660	3740	3830	3920	4010	4100	4190	4280	4360	4450
		2	0.275	2180	3520	3590	3670	3750	3820	3900	3980	4060	4130	4210	4290
		3	0.550	1850	3460	3530	3590	3660	3720	3790	3850	3920	3980	4050	4110
		4	0.825	1520	3400	3450	3500	3560	3610	3670	3720	3770	3830	3880	3930
		BFL	1.10	1190	3330	3370	3410	3450	3500	3540	3580	3620	3660	3710	3750
		6	4.78	906	3230	3270	3300	3330	3360	3390	3430	3460	3490	3520	3560
		7	8.89	626	3060	3080	3100	3130	3150	3170	3190	3210	3240	3260	3280
W36×160	2340	TFL	0.000	2350	3330	3410	3500	3580	3660	3750	3830	3910	4000	4080	4160
		2	0.255	2040	3280	3360	3430	3500	3570	3640	3720	3790	3860	3930	4010
		3	0.510	1740	3230	3290	3350	3420	3480	3540	3600	3660	3720	3790	3850
		4	0.765	1430	3170	3220	3280	3330	3380	3430	3480	3530	3580	3630	3680
		BFL	1.02	1130	3110	3150	3190	3230	3270	3310	3350	3390	3430	3470	3510
		6	4.82	857	3020	3050	3080	3110	3140	3170	3200	3230	3260	3290	3320
		7	8.96	588	2850	2870	2890	2910	2940	2960	2980	3000	3020	3040	3060

[a] $Y1$ = distance from top of the steel beam to plastic neutral axis.
[b] $Y2$ = distance from top of the steel beam to concrete flange force.
[c] See Figure 5-4c for PNA locations.

Table 5-14 (cont.). Composite W-Shapes Design Strength in Flexure $\phi_b M_n$, kip-ft

$F_y = 50$ ksi
$\phi_b = 0.90$
$\phi_c = 0.85$

Shape	$\phi_b M_p$	PNA[c]	$Y1^a$	ΣQ_n	$Y2^b$, in.										
	kip-ft		in.	kips	2	2.5	3	3.5	4	4.5	5	5.5	6	6.5	7
W36×150	2180	TFL	0.000	2210	3120	3200	3280	3360	3440	3510	3590	3670	3750	3830	3910
		2	0.235	1930	3080	3150	3220	3290	3350	3420	3490	3560	3630	3700	3760
		3	0.470	1650	3030	3090	3150	3210	3270	3330	3380	3440	3500	3560	3620
		4	0.705	1360	2980	3030	3080	3130	3170	3220	3270	3320	3370	3420	3460
		BFL	0.940	1080	2930	2960	3000	3040	3080	3120	3160	3190	3230	3270	3310
		6	4.87	817	2840	2870	2890	2920	2950	2980	3010	3040	3070	3100	3130
		7	9.11	553	2670	2690	2710	2730	2750	2770	2790	2810	2820	2840	2860
W36×135	1910	TFL	0.000	1990	2780	2850	2920	2990	3070	3140	3210	3280	3350	3420	3490
		2	0.198	1750	2750	2810	2870	2930	3000	3060	3120	3180	3240	3310	3370
		3	0.395	1510	2710	2760	2820	2870	2920	2980	3030	3080	3140	3190	3250
		4	0.593	1270	2670	2710	2760	2800	2850	2890	2940	2980	3030	3070	3120
		BFL	0.790	1040	2620	2660	2700	2730	2770	2810	2840	2880	2920	2950	2990
		6	5.02	767	2530	2560	2590	2610	2640	2670	2700	2720	2750	2780	2800
		7	9.53	496	2350	2370	2390	2410	2420	2440	2460	2480	2490	2510	2530
W33×221	3210	TFL	0.000	3260	4380	4490	4610	4720	4840	4950	5070	5180	5300	5410	5530
		2	0.318	2760	4300	4400	4490	4590	4690	4790	4890	4980	5080	5180	5280
		3	0.635	2260	4210	4290	4370	4450	4530	4610	4690	4770	4850	4930	5010
		4	0.953	1760	4110	4170	4240	4300	4360	4420	4490	4550	4610	4670	4730
		BFL	1.27	1250	4000	4050	4090	4130	4180	4220	4270	4310	4360	4400	4440
		6	3.61	1030	3940	3970	4010	4050	4080	4120	4150	4190	4230	4260	4300
		7	6.43	815	3830	3860	3880	3910	3940	3970	4000	4030	4060	4090	4120
W33×201	2900	TFL	0.000	2960	3950	4060	4160	4270	4370	4480	4580	4690	4790	4900	5000
		2	0.288	2510	3880	3970	4060	4150	4240	4330	4420	4510	4590	4680	4770
		3	0.575	2060	3810	3880	3950	4020	4100	4170	4240	4320	4390	4460	4530
		4	0.863	1610	3720	3780	3830	3890	3950	4000	4060	4120	4170	4230	4290
		BFL	1.15	1150	3620	3660	3700	3740	3790	3830	3870	3910	3950	3990	4030
		6	3.60	947	3560	3600	3630	3660	3700	3730	3760	3800	3830	3860	3900
		7	6.50	740	3460	3480	3510	3540	3560	3590	3620	3640	3670	3690	3720
W33×169	2360	TFL	0.000	2480	3310	3400	3490	3580	3660	3750	3840	3930	4010	4100	4190
		2	0.305	2120	3260	3340	3410	3490	3560	3640	3710	3790	3860	3940	4010
		3	0.610	1770	3200	3260	3320	3390	3450	3510	3580	3640	3700	3760	3830
		4	0.915	1420	3130	3180	3230	3280	3330	3380	3430	3480	3530	3580	3630
		BFL	1.22	1070	3050	3090	3130	3170	3210	3240	3280	3320	3360	3400	3430
		6	4.28	845	2980	3010	3040	3070	3100	3130	3160	3190	3220	3250	3280
		7	7.66	619	2850	2870	2900	2920	2940	2960	2980	3010	3030	3050	3070
W33×152	2100	TFL	0.000	2240	2980	3050	3130	3210	3290	3370	3450	3530	3610	3690	3770
		2	0.265	1930	2930	3000	3070	3130	3200	3270	3340	3410	3480	3540	3610
		3	0.530	1630	2880	2930	2990	3050	3110	3160	3220	3280	3340	3390	3450
		4	0.795	1320	2820	2870	2910	2960	3010	3050	3100	3150	3190	3240	3290
		BFL	1.06	1010	2750	2790	2830	2860	2900	2930	2970	3010	3040	3080	3110
		6	4.38	785	2680	2710	2740	2760	2790	2820	2850	2880	2900	2930	2960
		7	7.93	560	2550	2570	2590	2610	2630	2650	2670	2690	2710	2730	2750

[a] $Y1$ = distance from top of the steel beam to plastic neutral axis.
[b] $Y2$ = distance from top of the steel beam to concrete flange force.
[c] See Figure 5-4c for PNA locations.

$F_y = 50$ ksi
$\phi_b = 0.90$
$\phi_c = 0.85$

Table 5-14 (cont.). Composite W-Shapes Design Strength in Flexure $\phi_b M_n$, kip-ft

Shape	$\phi_b M_p$	PNA[c]	$Y1$[a]	ΣQ_n	$Y2$[b], in.										
	kip-ft		in.	kips	2	2.5	3	3.5	4	4.5	5	5.5	6	6.5	7
W33×141	1930	TFL	0.000	2080	2750	2820	2900	2970	3040	3120	3190	3260	3340	3410	3480
		2	0.240	1800	2710	2770	2830	2900	2960	3030	3090	3150	3220	3280	3350
		3	0.480	1530	2660	2710	2770	2820	2880	2930	2980	3040	3090	3150	3200
		4	0.720	1250	2610	2650	2700	2740	2790	2830	2880	2920	2960	3010	3050
		BFL	0.960	976	2550	2590	2620	2660	2690	2730	2760	2800	2830	2860	2900
		6	4.29	748	2480	2510	2540	2560	2590	2610	2640	2670	2690	2720	2750
		7	8.05	520	2350	2370	2390	2410	2420	2440	2460	2480	2500	2520	2540
W33×130	1750	TFL	0.000	1920	2520	2580	2650	2720	2790	2860	2920	2990	3060	3130	3190
		2	0.214	1670	2480	2540	2600	2660	2720	2780	2830	2890	2950	3010	3070
		3	0.428	1420	2440	2490	2540	2590	2640	2690	2740	2790	2840	2890	2940
		4	0.641	1180	2400	2440	2480	2520	2560	2600	2650	2690	2730	2770	2810
		BFL	0.855	932	2350	2380	2410	2450	2480	2510	2550	2580	2610	2640	2680
		6	4.39	705	2280	2300	2330	2350	2380	2400	2430	2450	2480	2500	2530
		7	8.30	479	2140	2160	2180	2190	2210	2230	2240	2260	2280	2290	2310
W33×118	1560	TFL	0.000	1740	2270	2330	2390	2450	2510	2570	2640	2700	2760	2820	2880
		2	0.185	1520	2240	2290	2340	2400	2450	2510	2560	2610	2670	2720	2770
		3	0.370	1310	2200	2250	2290	2340	2390	2430	2480	2530	2570	2620	2670
		4	0.555	1100	2160	2200	2240	2280	2320	2360	2400	2440	2480	2510	2550
		BFL	0.740	884	2120	2160	2190	2220	2250	2280	2310	2340	2380	2410	2440
		6	4.47	659	2050	2080	2100	2120	2150	2170	2190	2220	2240	2260	2290
		7	8.56	434	1920	1930	1950	1960	1980	2000	2010	2030	2040	2060	2070
W30×116	1420	TFL	0.000	1710	2060	2120	2180	2240	2300	2360	2420	2480	2540	2600	2660
		2	0.213	1490	2030	2080	2130	2180	2240	2290	2340	2390	2450	2500	2550
		3	0.425	1260	1990	2030	2080	2120	2170	2210	2260	2300	2350	2390	2440
		4	0.638	1040	1950	1990	2020	2060	2100	2130	2170	2210	2240	2280	2320
		BFL	0.850	818	1910	1930	1960	1990	2020	2050	2080	2110	2140	2170	2200
		6	3.98	623	1850	1870	1890	1910	1940	1960	1980	2000	2020	2050	2070
		7	7.43	428	1740	1760	1770	1790	1800	1820	1830	1850	1860	1880	1890
W30×108	1300	TFL	0.000	1590	1900	1950	2010	2070	2120	2180	2230	2290	2350	2400	2460
		2	0.190	1390	1870	1920	1970	2020	2060	2110	2160	2210	2260	2310	2360
		3	0.380	1190	1840	1880	1920	1960	2000	2050	2090	2130	2170	2210	2260
		4	0.570	987	1800	1840	1870	1910	1940	1980	2010	2050	2080	2120	2150
		BFL	0.760	787	1760	1790	1820	1850	1870	1900	1930	1960	1990	2010	2040
		6	4.04	592	1700	1730	1750	1770	1790	1810	1830	1850	1870	1890	1910
		7	7.63	396	1600	1610	1620	1640	1650	1670	1680	1690	1710	1720	1740
W30×99	1170	TFL	0.000	1460	1740	1790	1840	1890	1940	1990	2050	2100	2150	2200	2250
		2	0.168	1280	1710	1760	1800	1850	1890	1940	1980	2030	2070	2120	2160
		3	0.335	1100	1680	1720	1760	1800	1840	1880	1920	1960	1990	2030	2070
		4	0.503	927	1650	1690	1720	1750	1780	1820	1850	1880	1920	1950	1980
		BFL	0.670	752	1620	1650	1670	1700	1730	1750	1780	1810	1830	1860	1890
		6	4.13	558	1560	1580	1600	1620	1640	1660	1680	1700	1720	1740	1760
		7	7.85	364	1450	1470	1480	1490	1500	1520	1530	1540	1560	1570	1580

[a] $Y1$ = distance from top of the steel beam to plastic neutral axis.
[b] $Y2$ = distance from top of the steel beam to concrete flange force.
[c] See Figure 5-4c for PNA locations.

Table 5-14 (cont.). Composite W-Shapes Design Strength in Flexure $\phi_b M_n$, kip-ft

$F_y = 50$ ksi
$\phi_b = 0.90$
$\phi_c = 0.85$

Shape	$\phi_b M_p$	PNA[c]	Y1[a]	ΣQ_n	Y2[b], in.										
	kip-ft		in.	kips	2	2.5	3	3.5	4	4.5	5	5.5	6	6.5	7
W30×90	1060	TFL	0.000	1320	1570	1610	1660	1710	1750	1800	1850	1890	1940	1990	2030
		2	0.153	1160	1540	1580	1630	1670	1710	1750	1790	1830	1870	1910	1950
		3	0.305	1000	1520	1550	1590	1620	1660	1700	1730	1770	1800	1840	1870
		4	0.458	844	1490	1520	1550	1580	1610	1640	1670	1700	1730	1760	1790
		BFL	0.610	686	1460	1490	1510	1540	1560	1580	1610	1630	1660	1680	1710
		6	3.95	508	1410	1430	1450	1470	1480	1500	1520	1540	1560	1570	1590
		7	7.73	330	1310	1320	1340	1350	1360	1370	1380	1390	1410	1420	1430
W27×102	1140	TFL	0.000	1500	1650	1710	1760	1810	1860	1920	1970	2020	2080	2130	2180
		2	0.208	1290	1620	1670	1710	1760	1800	1850	1900	1940	1990	2030	2080
		3	0.415	1090	1590	1630	1660	1700	1740	1780	1820	1860	1900	1930	1970
		4	0.623	878	1550	1580	1610	1640	1670	1710	1740	1770	1800	1830	1860
		BFL	0.830	670	1510	1530	1560	1580	1610	1630	1650	1680	1700	1720	1750
		6	3.40	523	1470	1490	1510	1520	1540	1560	1580	1600	1620	1640	1650
		7	6.27	375	1400	1410	1420	1440	1450	1460	1480	1490	1500	1520	1530
W27×94	1040	TFL	0.000	1390	1520	1560	1610	1660	1710	1760	1810	1860	1910	1960	2010
		2	0.186	1200	1490	1530	1570	1620	1660	1700	1740	1790	1830	1870	1910
		3	0.373	1010	1460	1490	1530	1570	1600	1640	1670	1710	1740	1780	1820
		4	0.559	827	1430	1450	1480	1510	1540	1570	1600	1630	1660	1690	1720
		BFL	0.745	641	1390	1410	1440	1460	1480	1500	1530	1550	1570	1590	1620
		6	3.38	493	1350	1370	1390	1400	1420	1440	1450	1470	1490	1510	1520
		7	6.38	346	1280	1290	1300	1320	1330	1340	1350	1360	1380	1390	1400
W27×84	915	TFL	0.000	1240	1350	1390	1440	1480	1520	1570	1610	1660	1700	1740	1790
		2	0.160	1080	1320	1360	1400	1440	1480	1520	1550	1590	1630	1670	1710
		3	0.320	921	1300	1330	1360	1400	1430	1460	1500	1530	1560	1590	1630
		4	0.480	762	1270	1300	1330	1350	1380	1410	1430	1460	1490	1520	1540
		BFL	0.640	603	1240	1260	1290	1310	1330	1350	1370	1390	1410	1440	1460
		6	3.43	456	1200	1220	1240	1250	1270	1280	1300	1320	1330	1350	1370
		7	6.61	310	1130	1140	1150	1160	1170	1190	1200	1210	1220	1230	1240
W24×76	750	TFL	0.000	1120	1110	1150	1190	1230	1270	1310	1340	1380	1420	1460	1500
		2	0.170	967	1080	1120	1150	1190	1220	1260	1290	1320	1360	1390	1430
		3	0.340	814	1060	1090	1120	1150	1180	1200	1230	1260	1290	1320	1350
		4	0.510	662	1030	1060	1080	1100	1130	1150	1170	1200	1220	1240	1270
		BFL	0.680	509	1010	1020	1040	1060	1080	1100	1110	1130	1150	1170	1190
		6	2.99	394	975	989	1000	1020	1030	1050	1060	1070	1090	1100	1120
		7	5.59	280	924	934	944	954	964	974	984	994	1000	1010	1020
W24×68	664	TFL	0.000	1010	986	1020	1060	1090	1130	1160	1200	1240	1270	1310	1340
		2	0.146	874	967	998	1030	1060	1090	1120	1150	1180	1210	1250	1280
		3	0.293	743	946	972	999	1020	1050	1080	1100	1130	1160	1180	1210
		4	0.439	611	924	946	967	989	1010	1030	1050	1080	1100	1120	1140
		BFL	0.585	480	901	918	935	952	969	986	1000	1020	1040	1050	1070
		6	3.04	366	871	884	897	910	923	936	949	962	975	988	1000
		7	5.80	251	819	828	837	846	854	863	872	881	890	899	908

[a] $Y1$ = distance from top of the steel beam to plastic neutral axis.
[b] $Y2$ = distance from top of the steel beam to concrete flange force.
[c] See Figure 5-4c for PNA locations.

$F_y = 50$ ksi
$\phi_b = 0.90$
$\phi_c = 0.85$

Table 5-14 (cont.). Composite W-Shapes Design Strength in Flexure $\phi_b M_n$, kip-ft

Shape	$\phi_b M_p$	PNA[c]	$Y1$[a]	ΣQ_n	$Y2$[b], in.										
	kip-ft		in.	kips	2	2.5	3	3.5	4	4.5	5	5.5	6	6.5	7
W24×62	578	TFL	0.000	915	898	930	962	995	1030	1060	1090	1120	1160	1190	1220
		2	0.148	811	882	911	940	969	997	1030	1050	1080	1110	1140	1170
		3	0.295	707	866	891	916	941	966	991	1020	1040	1070	1090	1120
		4	0.443	603	849	870	891	913	934	955	977	998	1020	1040	1060
		BFL	0.590	500	830	848	866	883	901	919	936	954	972	989	1010
		6	3.38	364	794	806	819	832	845	858	871	884	897	910	923
		7	6.53	229	727	735	743	751	759	767	776	784	792	800	808
W24×55	506	TFL	0.000	815	797	826	854	883	912	941	970	999	1030	1060	1090
		2	0.126	726	784	809	835	861	887	912	938	964	990	1020	1040
		3	0.253	638	770	793	815	838	860	883	906	928	951	973	996
		4	0.379	549	755	775	794	814	833	853	872	892	911	931	950
		BFL	0.505	461	740	757	773	789	805	822	838	854	871	887	903
		6	3.39	332	706	718	729	741	753	765	776	788	800	812	823
		7	6.64	204	642	649	657	664	671	678	685	693	700	707	714
W21×62	540	TFL	0.000	915	810	843	875	907	940	972	1000	1040	1070	1100	1130
		2	0.154	788	792	819	847	875	903	931	959	987	1010	1040	1070
		3	0.308	662	772	795	818	842	865	889	912	936	959	982	1010
		4	0.461	535	750	769	788	807	826	845	864	883	902	921	940
		BFL	0.615	408	727	742	756	771	785	800	814	828	843	857	872
		6	2.54	318	706	717	728	739	751	762	773	784	796	807	818
		7	4.78	229	670	678	686	694	702	710	718	726	734	743	751
W21×57	484	TFL	0.000	835	742	772	801	831	861	890	920	949	979	1010	1040
		2	0.163	728	727	752	778	804	830	856	881	907	933	959	985
		3	0.325	622	710	732	754	776	798	820	842	864	886	908	930
		4	0.488	515	691	710	728	746	764	783	801	819	837	856	874
		BFL	0.650	409	672	687	701	716	730	744	759	773	788	802	817
		6	2.93	309	646	657	668	679	690	701	712	723	733	744	755
		7	5.40	209	602	610	617	625	632	639	647	654	662	669	676
W21×55	473	TFL	0.000	810	711	740	769	798	826	855	884	912	941	970	998
		2	0.131	703	696	721	746	770	795	820	845	870	895	920	945
		3	0.261	595	679	700	721	742	763	784	806	827	848	869	890
		4	0.392	488	661	679	696	713	731	748	765	782	800	817	834
		BFL	0.522	381	643	656	670	683	697	710	724	737	751	764	778
		6	2.62	292	621	631	642	652	662	673	683	693	704	714	724
		7	5.00	203	584	592	599	606	613	620	627	635	642	649	656
W21×50	413	TFL	0.000	735	646	672	698	724	750	776	802	828	854	880	906
		2	0.134	648	633	656	679	702	725	747	770	793	816	839	862
		3	0.268	560	619	639	659	679	699	718	738	758	778	798	818
		4	0.40	473	605	621	638	655	672	688	705	722	739	755	772
		BFL	0.535	386	589	603	617	630	644	658	671	685	699	712	726
		6	2.91	285	564	574	584	594	604	614	624	634	645	655	665
		7	5.56	184	519	526	532	539	545	552	558	565	571	578	585

[a] $Y1$ = distance from top of the steel beam to plastic neutral axis.
[b] $Y2$ = distance from top of the steel beam to concrete flange force.
[c] See Figure 5-4c for PNA locations.

Table 5-14 (cont.). Composite W-Shapes Design Strength in Flexure $\phi_b M_n$, kip-ft

$F_y = 50$ ksi
$\phi_b = 0.90$
$\phi_c = 0.85$

Shape	$\phi_b M_p$	PNA[c]	$Y1$[a]	ΣQ_n	$Y2$[b], in.											
	kip-ft		in.	kips	2	2.5	3	3.5	4	4.5	5	5.5	6	6.5	7
W21×48	401	TFL	0.000	705	614	639	664	689	714	739	764	789	814	839	864
		2	0.108	617	601	623	645	667	689	711	733	755	776	798	820
		3	0.215	530	588	607	626	644	663	682	701	719	738	757	776
		4	0.323	442	574	590	605	621	637	652	668	684	699	715	731
		BFL	0.430	355	559	572	584	597	610	622	635	647	660	672	685
		6	2.71	266	538	547	556	566	575	585	594	604	613	622	632
		7	5.26	176	500	506	512	518	525	531	537	543	550	556	562
W21×44	358	TFL	0.000	650	569	592	615	638	661	684	707	730	753	776	799
		2	0.113	577	558	578	599	619	640	660	681	701	721	742	762
		3	0.225	504	547	565	582	600	618	636	654	672	689	707	725
		4	0.338	431	535	550	565	581	596	611	626	642	657	672	687
		BFL	0.450	358	523	535	548	561	573	586	599	611	624	637	649
		6	2.92	260	498	507	517	526	535	544	553	563	572	581	590
		7	5.71	163	455	460	466	472	478	483	489	495	501	506	512
W18×60	461	TFL	0.000	880	692	723	754	785	817	848	879	910	941	972	1000
		2	0.174	749	672	699	726	752	779	805	832	858	885	911	938
		3	0.348	617	651	673	695	717	739	761	783	804	826	848	870
		4	0.521	486	629	646	663	680	698	715	732	749	766	784	801
		BFL	0.695	355	605	617	630	642	655	667	680	693	705	718	730
		6	2.18	287	589	599	609	619	629	639	650	660	670	680	690
		7	3.80	220	565	573	580	588	596	604	612	619	627	635	643
W18×55	420	TFL	0.000	810	634	663	691	720	749	777	806	835	863	892	921
		2	0.158	691	617	641	665	690	714	739	763	788	812	837	861
		3	0.315	573	598	618	638	659	679	699	719	740	760	780	801
		4	0.473	454	578	594	610	626	642	658	674	690	706	722	738
		BFL	0.630	336	556	568	580	592	604	616	628	639	651	663	675
		6	2.15	269	541	550	560	569	579	588	598	607	617	626	636
		7	3.86	203	517	524	531	539	546	553	560	567	575	582	589
W18×50	379	TFL	0.000	735	573	599	625	651	677	703	729	755	781	807	833
		2	0.143	628	557	579	601	624	646	668	690	713	735	757	779
		3	0.285	521	540	559	577	596	614	633	651	669	688	706	725
		4	0.428	414	522	537	552	566	581	596	610	625	640	654	669
		BFL	0.570	308	504	514	525	536	547	558	569	580	591	602	613
		6	2.08	246	489	498	507	516	524	533	542	550	559	568	577
		7	3.82	184	468	474	481	487	494	500	507	513	520	526	533
W18×46	340	TFL	0.000	675	528	552	576	600	624	648	672	696	720	743	767
		2	0.151	583	515	536	556	577	597	618	639	659	680	701	721
		3	0.303	492	500	518	535	553	570	587	605	622	640	657	675
		4	0.454	400	485	499	513	527	542	556	570	584	598	612	627
		BFL	0.605	308	469	479	490	501	512	523	534	545	556	567	578
		6	2.42	239	452	460	469	477	485	494	502	511	519	528	536
		7	4.36	169	425	431	437	443	449	455	461	467	473	479	485

[a] $Y1$ = distance from top of the steel beam to plastic neutral axis.
[b] $Y2$ = distance from top of the steel beam to concrete flange force.
[c] See Figure 5-4c for PNA locations.

$F_y = 50$ ksi
$\phi_b = 0.90$
$\phi_c = 0.85$

Table 5-14 (cont.). Composite W-Shapes Design Strength in Flexure $\phi_b M_n$, kip-ft

Shape	$\phi_b M_p$	PNA[c]	$Y1$[a]	ΣQ_n	$Y2$[b], in.										
	kip-ft		in.	kips	2	2.5	3	3.5	4	4.5	5	5.5	6	6.5	7
W18×40	294	TFL	0.000	590	458	479	499	520	541	562	583	604	625	646	667
		2	0.131	511	446	464	482	500	518	537	555	573	591	609	627
		3	0.263	432	434	449	464	480	495	510	526	541	556	571	587
		4	0.394	353	421	433	446	458	471	483	496	508	521	533	546
		BFL	0.525	274	407	417	426	436	446	455	465	475	485	494	504
		6	2.26	211	392	400	407	415	422	430	437	445	452	460	467
		7	4.27	148	369	374	379	385	390	395	400	405	411	416	421
W18×35	249	TFL	0.000	515	396	414	432	451	469	487	505	523	542	560	578
		2	0.106	451	387	403	418	434	450	466	482	498	514	530	546
		3	0.213	388	377	391	404	418	432	445	459	473	487	500	514
		4	0.319	324	367	378	389	401	412	424	435	447	458	470	481
		BFL	0.425	260	356	365	374	383	393	402	411	420	430	439	448
		6	2.37	194	341	347	354	361	368	375	382	389	396	402	409
		7	4.56	129	315	320	324	329	333	338	343	347	352	356	361
W16×36	240	TFL	0.000	530	374	392	411	430	449	467	486	505	524	542	561
		2	0.108	455	363	379	395	411	427	443	459	475	491	508	524
		3	0.215	380	351	365	378	391	405	418	432	445	459	472	486
		4	0.323	305	339	350	361	371	382	393	404	415	425	436	447
		BFL	0.430	229	326	334	343	351	359	367	375	383	391	399	408
		6	1.82	181	316	322	329	335	342	348	355	361	367	374	380
		7	3.46	133	300	305	310	314	319	324	328	333	338	343	347
W16×31	203	TFL	0.000	457	322	338	354	370	386	403	419	435	451	467	483
		2	0.110	396	313	327	341	355	369	383	397	411	425	439	453
		3	0.220	335	304	315	327	339	351	363	375	387	398	410	422
		4	0.330	274	294	303	313	323	333	342	352	362	371	381	391
		BFL	0.440	213	283	291	299	306	314	321	329	336	344	351	359
		6	2.00	164	273	278	284	290	296	302	307	313	319	325	331
		7	3.80	114	255	259	263	268	272	276	280	284	288	292	296
W16×26	166	TFL	0.000	384	268	282	295	309	322	336	350	363	377	390	404
		2	0.086	337	261	273	285	297	309	321	333	345	356	368	380
		3	0.173	289	254	264	274	285	295	305	315	326	336	346	356
		4	0.259	242	246	255	264	272	281	289	298	306	315	324	332
		BFL	0.345	194	239	246	252	259	266	273	280	287	294	301	307
		6	2.05	145	228	233	238	243	249	254	259	264	269	274	279
		7	4.01	96	211	214	217	221	224	228	231	234	238	241	245
W14×38	231	TFL	0.000	560	359	379	399	418	438	458	478	498	518	537	557
		2	0.129	473	346	363	380	396	413	430	447	463	480	497	514
		3	0.258	386	333	346	360	374	387	401	415	428	442	456	469
		4	0.386	299	318	329	340	350	361	371	382	392	403	414	424
		BFL	0.515	211	303	311	318	326	333	341	348	356	363	371	378
		6	1.38	176	296	302	309	315	321	327	334	340	346	352	358
		7	2.53	140	286	291	296	301	306	311	316	321	326	331	336

[a] $Y1$ = distance from top of the steel beam to plastic neutral axis.
[b] $Y2$ = distance from top of the steel beam to concrete flange force.
[c] See Figure 5-4c for PNA locations.

Table 5-14 (cont.). Composite W-Shapes Design Strength in Flexure $\phi_b M_n$, kip-ft

$F_y = 50$ ksi
$\phi_b = 0.90$
$\phi_c = 0.85$

Shape	$\phi_b M_p$	PNA[c]	$Y1$[a]	ΣQ_n	$Y2$[b], in.										
	kip-ft		in.	kips	2	2.5	3	3.5	4	4.5	5	5.5	6	6.5	7
W14×34	205	TFL	0.000	500	319	336	354	372	390	407	425	443	460	478	496
		2	0.114	423	308	323	338	353	368	383	397	412	427	442	457
		3	0.228	346	296	308	320	333	345	357	369	382	394	406	418
		4	0.341	270	283	293	302	312	322	331	341	350	360	369	379
		BFL	0.455	193	270	277	284	291	298	304	311	318	325	332	339
		6	1.42	159	264	269	275	280	286	292	297	303	309	314	320
		7	2.61	125	254	258	263	267	272	276	280	285	289	294	298
W14×30	177	TFL	0.000	443	279	295	310	326	342	357	373	389	404	420	436
		2	0.096	378	270	283	296	310	323	336	350	363	377	390	403
		3	0.193	313	260	271	282	293	304	315	326	337	348	359	371
		4	0.289	248	249	258	267	276	285	293	302	311	320	329	337
		BFL	0.385	183	239	245	252	258	265	271	278	284	291	297	304
		6	1.46	147	231	237	242	247	252	258	263	268	273	278	284
		7	2.80	111	221	225	229	233	237	241	244	248	252	256	260
W14×26	151	TFL	0.000	385	244	257	271	285	298	312	325	339	353	366	380
		2	0.105	332	236	248	260	271	283	295	307	318	330	342	354
		3	0.210	279	228	238	248	258	268	277	287	297	307	317	327
		4	0.315	226	220	228	236	244	252	260	268	276	284	292	300
		BFL	0.420	173	211	217	223	229	235	241	247	254	260	266	272
		6	1.67	135	203	207	212	217	222	227	231	236	241	246	250
		7	3.18	96.1	191	194	197	201	204	208	211	214	218	221	225
W14×22	125	TFL	0.000	325	203	215	226	238	249	261	272	284	295	307	318
		2	0.084	283	197	207	217	227	237	247	257	267	277	287	297
		3	0.168	241	191	200	208	217	225	234	242	251	259	268	276
		4	0.251	199	185	192	199	206	213	220	227	234	241	248	255
		BFL	0.335	157	178	183	189	194	200	206	211	217	222	228	233
		6	1.67	119	170	174	178	183	187	191	195	200	204	208	212
		7	3.32	81.1	158	161	164	167	169	172	175	178	181	184	187
W12×30	162	TFL	0.000	440	254	269	285	300	316	332	347	363	378	394	409
		2	0.110	368	243	256	269	282	295	308	321	334	347	361	374
		3	0.220	296	232	243	253	264	274	285	295	306	316	327	337
		4	0.330	224	221	229	237	245	252	260	268	276	284	292	300
		BFL	0.440	153	209	214	219	225	230	236	241	247	252	257	263
		6	1.10	131	205	209	214	218	223	228	232	237	242	246	251
		7	1.92	110	199	203	207	211	215	219	223	227	230	234	238
W12×26	140	TFL	0.000	383	219	233	247	260	274	287	301	314	328	341	355
		2	0.095	321	211	222	233	245	256	267	279	290	301	313	324
		3	0.190	259	201	210	220	229	238	247	256	265	275	284	293
		4	0.285	198	191	198	205	212	219	226	233	240	247	254	261
		BFL	0.380	136	181	186	191	196	200	205	210	215	220	225	229
		6	1.07	116	177	182	186	190	194	198	202	206	210	214	218
		7	1.94	95.6	172	176	179	183	186	189	193	196	200	203	206

[a] $Y1$ = distance from top of the steel beam to plastic neutral axis.
[b] $Y2$ = distance from top of the steel beam to concrete flange force.
[c] See Figure 5-4c for PNA locations.

$F_y = 50$ ksi
$\phi_b = 0.90$
$\phi_c = 0.85$

Table 5-14 (cont.). Composite W-Shapes Design Strength in Flexure $\phi_b M_n$, kip-ft

Shape	$\phi_b M_p$	PNA[c]	$Y1$[a]	ΣQ_n	$Y2$[b], in.										
	kip-ft		in.	kips	2	2.5	3	3.5	4	4.5	5	5.5	6	6.5	7
W12×22	110	TFL	0.000	324	187	199	210	221	233	244	256	267	279	290	302
		2	0.106	281	181	191	201	211	221	231	241	251	260	270	280
		3	0.213	238	174	183	191	200	208	216	225	233	242	250	259
		4	0.319	196	167	174	181	188	195	202	209	216	223	230	237
		BFL	0.425	153	160	166	171	176	182	187	193	198	204	209	214
		6	1.66	117	153	157	161	165	169	173	177	182	186	190	194
		7	3.03	81.0	142	145	147	150	153	156	159	162	165	168	170
W12×19	92.6	TFL	0.000	279	160	170	180	189	199	209	219	229	239	249	258
		2	0.088	243	155	163	172	181	189	198	206	215	224	232	241
		3	0.175	208	149	157	164	172	179	186	194	201	208	216	223
		4	0.263	173	144	150	156	162	168	175	181	187	193	199	205
		BFL	0.350	138	138	143	148	153	158	163	168	172	177	182	187
		6	1.68	104	131	135	138	142	146	149	153	157	160	164	168
		7	3.14	69.6	120	123	125	128	130	133	135	138	140	142	145
W12×16	75.4	TFL	0.000	236	133	142	150	158	167	175	183	192	200	209	217
		2	0.066	209	130	137	144	152	159	167	174	181	189	196	204
		3	0.133	183	126	132	139	145	152	158	165	171	177	184	190
		4	0.199	156	122	127	133	138	144	149	155	160	166	171	177
		BFL	0.265	130	117	122	127	131	136	140	145	150	154	159	163
		6	1.71	94.3	110	114	117	120	124	127	130	134	137	140	144
		7	3.32	58.9	98.9	101	103	105	107	109	111	113	116	118	120
W12×14	65.3	TFL	0.000	208	117	124	132	139	147	154	161	169	176	183	191
		2	0.056	186	114	121	127	134	140	147	153	160	167	173	180
		3	0.113	163	111	116	122	128	134	140	145	151	157	163	168
		4	0.169	141	107	112	117	122	127	132	137	142	147	152	157
		BFL	0.225	119	104	108	112	116	121	125	129	133	137	142	146
		6	1.68	85.3	97	100	103	106	109	112	115	118	121	124	127
		7	3.35	52.0	86.4	88.2	90	91.9	93.7	95.6	97.4	99.2	101	103	105
W10×26	117	TFL	0.000	381	193	206	220	233	247	260	274	287	301	314	327
		2	0.110	317	183	195	206	217	228	240	251	262	273	285	296
		3	0.220	254	174	183	192	201	210	219	228	237	246	255	264
		4	0.330	190	164	170	177	184	190	197	204	211	217	224	231
		BFL	0.440	127	153	157	162	166	171	175	180	184	189	193	198
		6	0.886	111	150	154	158	162	166	170	173	177	181	185	189
		7	1.49	95.1	146	150	153	156	160	163	167	170	173	177	180
W10×22	97.5	TFL	0.000	325	163	175	186	198	209	221	232	244	255	267	278
		2	0.090	273	156	165	175	185	194	204	214	223	233	243	252
		3	0.180	221	148	156	164	171	179	187	195	203	210	218	226
		4	0.270	169	140	146	152	158	164	170	176	182	188	194	200
		BFL	0.360	118	131	135	140	144	148	152	156	160	165	169	173
		6	0.962	99.3	128	131	135	138	142	145	149	153	156	160	163
		7	1.72	81.1	124	126	129	132	135	138	141	144	147	149	152

[a] $Y1$ = distance from top of the steel beam to plastic neutral axis.
[b] $Y2$ = distance from top of the steel beam to concrete flange force.
[c] See Figure 5-4c for PNA locations.

Table 5-14 (cont.). Composite W-Shapes Design Strength in Flexure $\phi_b M_n$, kip-ft

$F_y = 50$ ksi
$\phi_b = 0.90$
$\phi_c = 0.85$

Shape	$\phi_b M_p$	PNA[c]	$Y1$[a]	ΣQ_n	$Y2$[b], in.										
	kip-ft		in.	kips	2	2.5	3	3.5	4	4.5	5	5.5	6	6.5	7
W10×19	81	TFL	0.000	281	141	151	161	171	181	191	201	211	221	231	241
		2	0.099	241	136	144	153	161	170	178	187	195	204	212	221
		3	0.198	202	130	137	144	151	158	165	172	180	187	194	201
		4	0.296	162	123	129	135	140	146	152	158	163	169	175	181
		BFL	0.395	122	117	121	125	130	134	138	143	147	151	156	160
		6	1.25	96.2	112	115	118	122	125	129	132	135	139	142	146
		7	2.29	70.3	105	107	110	112	115	117	120	122	125	127	130
W10×17	70.1	TFL	0.000	250	125	133	142	151	160	169	178	186	195	204	213
		2	0.083	216	120	127	135	143	150	158	166	173	181	189	196
		3	0.165	183	115	121	128	134	141	147	154	160	167	173	180
		4	0.248	150	110	115	120	126	131	136	142	147	152	158	163
		BFL	0.330	117	104	108	113	117	121	125	129	133	137	142	146
		6	1.31	89.8	99.0	102	105	109	112	115	118	121	124	128	131
		7	2.45	62.4	91.5	93.7	95.9	98.1	100	103	105	107	109	111	114
W10×15	60	TFL	0.000	221	109	117	125	133	140	148	156	164	172	180	187
		2	0.068	194	105	112	119	126	133	140	146	153	160	167	174
		3	0.135	167	101	107	113	119	125	131	137	143	149	154	160
		4	0.203	140	97.2	102	107	112	117	122	127	132	137	142	147
		BFL	0.270	113	92.9	96.9	101	105	109	113	117	121	125	129	133
		6	1.35	83.8	87.4	90.3	93.3	96.3	99.3	102	105	108	111	114	117
		7	2.60	55.1	79.3	81.2	83.2	85.1	87.1	89.0	91	92.9	94.9	96.9	98.8
W10×12	47.3	TFL	0.000	177	86.9	93.2	99.5	106	112	118	125	131	137	143	150
		2	0.0525	156	84.0	89.5	95.0	101	106	112	117	123	128	134	139
		3	0.105	135	80.9	85.7	90.5	95.3	100	105	110	114	119	124	129
		4	0.158	115	77.8	81.8	85.9	90.0	94.0	98.1	102	106	110	114	118
		BFL	0.210	93.8	74.5	77.9	81.2	84.5	87.8	91.2	94.5	97.8	101	104	108
		6	1.30	69.0	69.9	72.3	74.8	77.2	79.7	82.1	84.6	87.0	89.4	91.9	94.3
		7	2.61	44.3	62.9	64.5	66.1	67.6	69.2	70.8	72.3	73.9	75.5	77.0	78.6

[a] $Y1$ = distance from top of the steel beam to plastic neutral axis.
[b] $Y2$ = distance from top of the steel beam to concrete flange force.
[c] See Figure 5-4c for PNA locations.

Table 5-15.
Lower Bound Elastic Moment of Inertia I_{LB} for Plastic Composite Sections

Shape[d]	PNA[c]	$Y1^a$	ΣQ_n	$Y2^b$, in.										
		in.	kips	2	2.5	3	3.5	4	4.5	5	5.5	6	6.5	7
W40×297	TFL	0.000	4370	44200	45100	46100	47100	48200	49200	50300	51400	52500	53700	54800
(23200)	2	0.413	3720	42500	43400	44300	45200	46200	47100	48100	49100	50200	51200	52300
	3	0.825	3070	40500	41300	42100	42900	43800	44700	45600	46500	47400	48300	49300
	4	1.24	2410	38100	38800	39500	40200	40900	41700	42500	43200	44000	44900	45700
	BFL	1.65	1760	35200	35800	36400	36900	37500	38100	38800	39400	40000	40700	41400
	6	4.55	1430	33500	34000	34500	35000	35500	36000	36600	37100	37700	38200	38800
	7	8.15	1090	31600	32000	32300	32800	33200	33600	34000	34500	34900	35400	35800
W40×278	TFL	0.000	4090	40500	41400	42300	43300	44300	45300	46300	47300	48400	49400	50500
(20500)	2	0.453	3550	39100	39900	40800	41700	42600	43500	44400	45400	46400	47400	48400
	3	0.905	3000	37400	38200	39000	39800	40600	41400	42300	43200	44100	45000	45900
	4	1.36	2460	35500	36200	36900	37600	38300	39100	39900	40600	41400	42200	43100
	BFL	1.81	1920	33300	33800	34400	35100	35700	36300	37000	37600	38300	39000	39700
	6	5.69	1470	31100	31500	32000	32500	33100	33600	34100	34700	35200	35800	36400
	7	10.1	1020	28500	28800	29200	29600	30000	30400	30800	31200	31600	32100	32500
W40×277	TFL	0.000	4070	41300	42200	43200	44100	45100	46000	47000	48100	49100	50200	51200
(21900)	2	0.395	3450	39700	40600	41400	42300	43100	44000	45000	45900	46900	47800	48800
	3	0.790	2820	37800	38500	39300	40100	40900	41700	42500	43300	44200	45000	45900
	4	1.19	2200	35500	36200	36800	37500	38100	38800	39500	40300	41000	41700	42500
	BFL	1.58	1570	32700	33200	33700	34300	34800	35300	35900	36500	37000	37600	38200
	6	4.24	1300	31300	31700	32200	32600	33100	33600	34100	34600	35100	35600	36100
	7	7.59	1020	29700	30000	30400	30800	31200	31600	32000	32400	32800	33200	33700
W40×264	TFL	0.000	3880	38200	39000	39900	40800	41700	42700	43700	44600	45600	46600	47700
(19400)	2	0.433	3370	36900	37700	38500	39300	40200	41100	41900	42900	43800	44700	45700
	3	0.865	2850	35300	36000	36800	37500	38300	39100	39900	40800	41600	42500	43400
	4	1.30	2340	33500	34200	34800	35500	36200	36900	37600	38400	39100	39900	40700
	BFL	1.730	1820	31400	31900	32500	33100	33700	34300	34900	35500	36100	36800	37500
	6	5.46	1400	29400	29800	30300	30800	31300	31800	32300	32800	33300	33800	34400
	7	9.90	970	26900	27300	27600	28000	28300	28700	29100	29500	29900	30300	30700
W40×249	TFL	0.000	3670	36900	37700	38500	39300	40200	41100	42000	42900	43800	44800	45700
(19600)	2	0.355	3100	35400	36200	36900	37700	38500	39300	40100	40900	41800	42700	43500
	3	0.710	2540	33700	34400	35100	35700	36500	37200	37900	38700	39400	40200	41000
	4	1.07	1980	31700	32300	32800	33400	34000	34700	35300	35900	36600	37200	37900
	BFL	1.42	1420	29200	29700	30100	30600	31100	31600	32100	32600	33100	33700	34200
	6	4.12	1170	28000	28300	28700	29100	29600	30000	30400	30900	31300	31800	32200
	7	7.48	916	26500	26800	27200	27500	27800	28200	28500	28900	29300	29700	30000
W40×235	TFL	0.000	3450	33900	34600	35400	36200	37000	37900	38700	39600	40500	41400	42300
(17400)	2	0.395	2980	32700	33400	34100	34800	35600	36400	37100	37900	38800	39600	40500
	3	0.790	2510	31300	31900	32600	33200	33900	34600	35300	36100	36800	37600	38300
	4	1.19	2040	29600	30200	30800	31400	32000	32600	33200	33900	34500	35200	35900
	BFL	1.58	1570	27700	28200	28700	29200	29700	30200	30700	31300	31800	32400	33000
	6	5.20	1220	26000	26400	26800	27200	27700	28100	28500	29000	29400	29900	30400
	7	9.46	863	24000	24300	24600	24900	25300	25600	25900	26300	26600	27000	27400

[a] $Y1$ = distance from top of the steel beam to plastic neutral axis.
[b] $Y1$ = distance from top of the steel beam to concrete flange force.
[c] See Figure 5-4c for PNA locations.
[d] Value in parentheses is I_x (in.^{4}) of non-composite steel shape.

I_{LB}

Table 5-15 (cont.). Lower Bound Elastic Moment of Inertia I_{LB} for Plastic Composite Sections

Shape[d]	PNA[c]	$Y1$[a]	ΣQ_n	$Y2$[b], in.										
		in.	kips	2	2.5	3	3.5	4	4.5	5	5.5	6	6.5	7
W40×215	TFL	0.000	3170	31400	32000	32700	33500	34200	35000	35700	36500	37300	38100	39000
(16700)	2	0.305	2690	30200	30800	31400	32100	32800	33500	34200	34900	35600	36400	37100
	3	0.610	2210	28700	29300	29900	30500	31100	31700	32300	33000	33600	34300	35000
	4	0.915	1720	27000	27500	28000	28500	29000	29500	30100	30600	31200	31800	32400
	BFL	1.22	1240	24900	25300	25700	26100	26500	27000	27400	27800	28300	28800	29200
	6	3.85	1020	23800	24200	24500	24900	25200	25600	26000	26300	26700	27100	27500
	7	7.31	793	22600	22800	23100	23400	23700	24000	24300	24600	24900	25300	25600
W40×211	TFL	0.000	3100	30100	30800	31500	32200	32900	33700	34400	35200	36000	36800	37600
(15500)	2	0.355	2680	29000	29700	30300	31000	31600	32300	33000	33800	34500	35200	36000
	3	0.710	2260	27800	28400	29000	29600	30200	30800	31400	32100	32800	33400	34100
	4	1.07	1840	26400	26900	27400	27900	28500	29000	29600	30200	30800	31400	32000
	BFL	1.420	1420	24700	25100	25500	26000	26400	26900	27400	27900	28400	28900	29400
	6	5.04	1100	23100	23500	23900	24200	24600	25000	25400	25800	26200	26600	27100
	7	9.37	775	21300	21600	21900	22200	22500	22800	23100	23400	23700	24000	24300
W40×199	TFL	0.000	2930	28200	28900	29500	30200	30900	31600	32300	33000	33700	34500	35200
(14900)	2	0.268	2500	27200	27800	28400	29000	29600	30200	30900	31500	32200	32900	33600
	3	0.535	2080	26000	26500	27000	27600	28200	28700	29300	29900	30500	31100	31800
	4	0.803	1660	24600	25000	25500	26000	26400	26900	27500	28000	28500	29100	29600
	BFL	1.07	1230	22800	23200	23600	23900	24300	24800	25200	25600	26000	26500	26900
	6	4.23	983	21600	21900	22300	22600	22900	23300	23600	24000	24400	24700	25100
	7	8.10	731	20200	20500	20700	21000	21300	21600	21800	22100	22400	22700	23000
W40×183	TFL	0.000	2690	25700	26300	26900	27500	28200	28800	29400	30100	30800	31500	32200
(13300)	2	0.305	2330	24800	25400	25900	26500	27100	27700	28300	28900	29500	30200	30800
	3	0.610	1970	23800	24300	24800	25300	25900	26400	27000	27500	28100	28700	29300
	4	0.915	1610	22600	23000	23500	24000	24400	24900	25400	25900	26400	26900	27400
	BFL	1.22	1250	21200	21600	21900	22300	22700	23100	23500	24000	24400	24800	25300
	6	4.71	961	19800	20200	20500	20800	21100	21500	21800	22200	22500	22900	23200
	7	9.15	673	18300	18500	18800	19000	19200	19500	19800	20000	20300	20600	20900
W40×167	TFL	0.000	2460	22800	23300	23800	24400	25000	25500	26100	26700	27300	28000	28600
(11600)	2	0.255	2160	22000	22500	23000	23600	24100	24600	25200	25700	26300	26900	27500
	3	0.510	1860	21200	21700	22100	22600	23100	23600	24100	24600	25200	25700	26300
	4	0.765	1560	20300	20700	21100	21500	22000	22400	22900	23300	23800	24300	24800
	BFL	1.02	1260	19200	19500	19900	20300	20600	21000	21400	21800	22300	22700	23100
	6	4.90	936	17800	18000	18300	18600	19000	19300	19600	19900	20300	20600	21000
	7	9.84	615	16100	16300	16500	16700	16900	17200	17400	17700	17900	18100	18400
W40×149	TFL	0.000	2190	19600	20000	20500	21000	21500	22000	22500	23100	23600	24200	24700
(9800)	2	0.208	1950	19000	19400	19900	20300	20800	21300	21800	22300	22800	23300	23900
	3	0.415	1700	18300	18700	19100	19600	20000	20500	20900	21400	21900	22300	22800
	4	0.623	1460	17600	18000	18400	18700	19100	19600	20000	20400	20800	21300	21700
	BFL	0.830	1210	16700	17100	17400	17800	18100	18500	18900	19200	19600	20000	20400
	6	5.15	879	15400	15700	15900	16200	16500	16800	17100	17400	17700	18000	18300
	7	10.4	548	13700	13900	14100	14300	14500	14700	14900	15100	15300	15500	15800

[a] $Y1$ = distance from top of the steel beam to plastic neutral axis.
[b] $Y1$ = distance from top of the steel beam to concrete flange force.
[c] See Figure 5-4c for PNA locations.
[d] Value in parentheses is I_x (in.^{4}) of non-composite steel shape.

Table 5-15 (cont.). Lower Bound Elastic Moment of Inertia I_{LB} for Plastic Composite Sections

Shape[d]	PNA[c]	$Y1$[a]	ΣQ_n	$Y2$[b], in.										
		in.	kips	2	2.5	3	3.5	4	4.5	5	5.5	6	6.5	7
W36×300	TFL	0.000	4420	38600	39500	40400	41400	42400	43400	44400	45400	46500	47600	48700
(20300)	2	0.420	3710	37000	37800	38700	39500	40400	41400	42300	43200	44200	45200	46200
	3	0.840	3010	35100	35900	36600	37400	38200	39000	39800	40700	41500	42400	43300
	4	1.26	2310	32900	33500	34100	34800	35500	36100	36800	37600	38300	39000	39800
	BFL	1.68	1610	30100	30600	31100	31600	32100	32600	33200	33700	34300	34900	35500
	6	3.99	1360	28900	29300	29800	30200	30700	31200	31600	32100	32600	33100	33700
	7	6.67	1100	27600	28000	28300	28700	29100	29500	29900	30300	30700	31200	31600
W36×280	TFL	0.000	4120	35800	36600	37500	38400	39300	40200	41200	42100	43100	44100	45200
(18900)	2	0.393	3470	34300	35100	35900	36700	37500	38400	39300	40100	41100	42000	42900
	3	0.785	2820	32600	33300	34000	34700	35500	36200	37000	37800	38600	39400	40200
	4	1.18	2170	30600	31100	31700	32300	33000	33600	34300	34900	35600	36300	37000
	BFL	1.57	1510	28000	28400	28900	29400	29800	30300	30800	31400	31900	32400	33000
	6	3.88	1270	26900	27300	27700	28100	28500	28900	29400	29900	30300	30800	31300
	7	6.61	1030	25700	26000	26300	26700	27100	27400	27800	28200	28600	29000	29400
W36×260	TFL	0.000	3830	32800	33600	34400	35200	36100	36900	37800	38700	39600	40600	41500
(17300)	2	0.360	3230	31500	32200	33000	33700	34500	35300	36100	36900	37700	38600	39500
	3	0.720	2630	30000	30600	31200	31900	32600	33300	34000	34700	35500	36200	37000
	4	1.08	2030	28100	28600	29200	29700	30300	30900	31500	32100	32800	33400	34100
	BFL	1.44	1430	25800	26200	26600	27100	27500	28000	28500	28900	29400	29900	30500
	6	3.92	1200	24700	25100	25500	25900	26300	26700	27100	27500	28000	28400	28900
	7	6.77	956	23500	23800	24100	24500	24800	25100	25500	25900	26200	26600	27000
W36×256	TFL	0.000	3770	33000	33700	34600	35400	36200	37100	38000	38900	39800	40700	41700
(16800)	2	0.433	3240	31700	32500	33200	34000	34800	35600	36400	37200	38100	38900	39800
	3	0.865	2710	30300	31000	31600	32300	33000	33800	34500	35300	36000	36800	37600
	4	1.30	2190	28700	29300	29800	30500	31100	31700	32400	33000	33700	34400	35100
	BFL	1.730	1660	26700	27200	27700	28200	28700	29200	29700	30300	30900	31400	32000
	6	5.15	1300	25100	25500	25900	26300	26800	27200	27700	28100	28600	29100	29600
	7	8.88	943	23300	23600	23900	24200	24600	24900	25300	25600	26000	26400	26800
W36×245	TFL	0.000	3610	30600	31300	32100	32900	33600	34400	35300	36100	37000	37800	38700
(16100)	2	0.338	3050	29400	30100	30700	31400	32200	32900	33700	34400	35200	36000	36800
	3	0.675	2490	27900	28500	29200	29800	30400	31100	31700	32400	33100	33900	34600
	4	1.01	1930	26200	26700	27200	27800	28300	28900	29500	30000	30600	31300	31900
	BFL	1.35	1380	24100	24500	24900	25400	25800	26200	26700	27200	27600	28100	28600
	6	3.81	1140	23100	23400	23800	24100	24500	24900	25300	25700	26100	26500	27000
	7	6.78	901	21900	22200	22500	22800	23100	23400	23800	24100	24400	24800	25100
W36×232	TFL	0.000	3410	29400	30100	30800	31600	32300	33100	33900	34700	35500	36400	37200
(15000)	2	0.393	2930	28300	29000	29600	30300	31000	31700	32500	33200	34000	34800	35600
	3	0.785	2460	27100	27700	28300	28900	29500	30200	30800	31500	32200	32900	33600
	4	1.18	1980	25600	26100	26600	27200	27700	28300	28900	29500	30100	30700	31300
	BFL	1.57	1510	23800	24300	24700	25200	25600	26100	26600	27100	27600	28100	28700
	6	5.01	1180	22400	22800	23100	23500	23900	24300	24700	25100	25600	26000	26400
	7	8.77	851	20800	21000	21300	21600	21900	22200	22600	22900	23200	23500	23900

[a] $Y1$ = distance from top of the steel beam to plastic neutral axis.
[b] $Y1$ = distance from top of the steel beam to concrete flange force.
[c] See Figure 5-4c for PNA locations.
[d] Value in parentheses is I_x (in.^{4}) of non-composite steel shape.

Table 5-15 (cont.). Lower Bound Elastic Moment of Inertia I_{LB} for Plastic Composite Sections

Shape[d]	PNA[c]	$Y1$[a]	ΣQ_n	$Y2$[b], in.										
		in.	kips	2	2.5	3	3.5	4	4.5	5	5.5	6	6.5	7
W36×230	TFL	0.000	3380	28500	29100	29800	30600	31300	32000	32800	33600	34400	35200	36000
(15000)	2	0.315	2860	27300	28000	28600	29300	29900	30600	31300	32000	32800	33500	34300
	3	0.630	2340	26000	26600	27100	27700	28300	28900	29600	30200	30900	31500	32200
	4	0.945	1820	24400	24900	25400	25900	26400	26900	27500	28000	28600	29100	29700
	BFL	1.26	1300	22500	22900	23200	23600	24000	24500	24900	25300	25800	26200	26700
	6	3.83	1070	21500	21800	22100	22500	22800	23200	23600	23900	24300	24700	25100
	7	6.83	845	20400	20700	20900	21200	21500	21800	22100	22400	22800	23100	23400
W36×210	TFL	0.000	3090	26000	26600	27300	28000	28600	29300	30000	30800	31500	32300	33100
(13200)	2	0.340	2680	25100	25700	26300	26900	27500	28200	28900	29500	30200	30900	31600
	3	0.680	2260	24000	24500	25100	25700	26200	26800	27400	28000	28700	29300	30000
	4	1.02	1850	22800	23300	23700	24200	24800	25300	25800	26400	26900	27500	28100
	BFL	1.36	1430	21300	21700	22100	22500	23000	23400	23900	24300	24800	25300	25800
	6	5.08	1100	19900	20300	20600	20900	21300	21700	22000	22400	22800	23200	23600
	7	9.04	773	18300	18600	18800	19100	19400	19700	19900	20200	20500	20800	21100
W36×194	TFL	0.000	2850	23800	24400	25000	25600	26200	26900	27500	28200	28900	29600	30300
(12100)	2	0.315	2470	23000	23500	24100	24600	25200	25800	26400	27000	27700	28300	29000
	3	0.630	2090	22000	22500	23000	23500	24000	24600	25100	25700	26300	26900	27500
	4	0.945	1710	20900	21300	21800	22200	22700	23200	23700	24200	24700	25200	25700
	BFL	1.260	1330	19500	19900	20300	20700	21100	21500	21900	22300	22800	23200	23700
	6	4.93	1020	18300	18600	18900	19200	19500	19900	20200	20600	20900	21300	21700
	7	8.94	713	16800	17000	17300	17500	17700	18000	18300	18500	18800	19100	19400
W36×182	TFL	0.000	2680	22200	22700	23300	23900	24400	25000	25700	26300	26900	27600	28300
(11300)	2	0.295	2320	21400	21900	22400	23000	23500	24100	24600	25200	25800	26400	27000
	3	0.590	1970	20500	21000	21500	21900	22400	22900	23500	24000	24500	25100	25700
	4	0.885	1610	19500	19900	20300	20700	21200	21600	22100	22600	23000	23500	24000
	BFL	1.18	1250	18200	18600	18900	19300	19700	20000	20400	20800	21200	21700	22100
	6	4.89	961	17000	17300	17600	17900	18200	18600	18900	19200	19600	19900	20200
	7	8.91	670	15700	15900	16100	16300	16600	16800	17000	17300	17600	17800	18100
W36×170	TFL	0.000	2510	20600	21100	21700	22200	22700	23300	23900	24500	25100	25700	26300
(10500)	2	0.275	2180	19900	20400	20900	21400	21900	22400	22900	23500	24000	24600	25200
	3	0.550	1850	19100	19500	20000	20400	20900	21400	21900	22400	22900	23400	23900
	4	0.825	1520	18100	18500	18900	19300	19700	20200	20600	21000	21500	21900	22400
	BFL	1.10	1190	17000	17300	17700	18000	18400	18700	19100	19500	19900	20300	20700
	6	4.78	906	15900	16100	16400	16700	17000	17300	17600	17900	18200	18600	18900
	7	8.89	626	14500	14800	15000	15200	15400	15600	15800	16100	16300	16600	16800
W36×160	TFL	0.000	2350	19200	19600	20100	20600	21100	21700	22200	22700	23300	23900	24400
(9760)	2	0.255	2040	18500	18900	19400	19900	20300	20800	21300	21800	22300	22900	23400
	3	0.510	1740	17800	18200	18600	19000	19400	19900	20300	20800	21300	21800	22300
	4	0.765	1430	16900	17200	17600	18000	18400	18800	19200	19600	20000	20400	20900
	BFL	1.02	1130	15900	16200	16500	16800	17100	17500	17800	18200	18600	18900	19300
	6	4.82	857	14800	15000	15300	15600	15800	16100	16400	16700	17000	17300	17600
	7	8.96	588	13500	13700	13900	14100	14300	14500	14700	15000	15200	15400	15600

[a] $Y1$ = distance from top of the steel beam to plastic neutral axis.
[b] $Y1$ = distance from top of the steel beam to concrete flange force.
[c] See Figure 5-4c for PNA locations.
[d] Value in parentheses is I_x (in.^{4}) of non-composite steel shape.

Table 5-15 (cont.). Lower Bound Elastic Moment of Inertia I_{LB} for Plastic Composite Sections

Shape[d]	PNA[c]	$Y1$[a]	ΣQ_n	$Y2$[b], in.										
		in.	kips	2	2.5	3	3.5	4	4.5	5	5.5	6	6.5	7
W36×150	TFL	0.000	2210	17800	18300	18700	19200	19700	20200	20700	21200	21700	22300	22800
(9040)	2	0.235	1930	17200	17700	18100	18500	19000	19400	19900	20400	20900	21400	21900
	3	0.470	1650	16600	16900	17300	17700	18100	18600	19000	19400	19900	20300	20800
	4	0.705	1360	15700	16100	16400	16800	17200	17500	17900	18300	18700	19100	19500
	BFL	0.94	1080	14800	15100	15400	15700	16000	16400	16700	17000	17400	17700	18100
	6	4.87	817	13800	14000	14300	14500	14800	15100	15300	15600	15900	16200	16500
	7	9.11	553	12600	12700	12900	13100	13300	13500	13700	13900	14100	14300	14500
W36×135	TFL	0.000	1990	15600	16000	16400	16800	17200	17700	18100	18600	19100	19500	20000
(7800)	2	0.198	1750	15100	15500	15800	16200	16600	17100	17500	17900	18300	18800	19200
	3	0.395	1510	14500	14900	15200	15600	16000	16300	16700	17100	17500	17900	18300
	4	0.593	1270	13900	14200	14500	14800	15200	15500	15900	16200	16600	16900	17300
	BFL	0.790	1040	13200	13400	13700	14000	14300	14600	14900	15200	15500	15900	16200
	6	5.02	767	12100	12400	12600	12800	13100	13300	13600	13800	14100	14300	14600
	7	9.53	496	10900	11100	11200	11400	11600	11700	11900	12100	12300	12500	12700
W33×221	TFL	0.000	3260	24600	25200	25900	26500	27200	27900	28600	29300	30100	30800	31600
(12900)	2	0.318	2760	23600	24200	24800	25400	26000	26700	27300	28000	28600	29300	30000
	3	0.635	2260	22500	23000	23500	24100	24600	25200	25800	26400	27000	27600	28200
	4	0.953	1760	21100	21500	22000	22500	22900	23400	23900	24400	24900	25500	26000
	BFL	1.27	1250	19400	19700	20100	20500	20800	21200	21600	22000	22400	22800	23300
	6	3.61	1030	18500	18800	19100	19400	19800	20100	20400	20800	21100	21500	21900
	7	6.43	815	17600	17800	18100	18400	18600	18900	19200	19500	19800	20100	20400
W33×201	TFL	0.000	2960	22100	22700	23300	23900	24500	25100	25700	26400	27100	27700	28400
(11600)	2	0.288	2510	21300	21800	22300	22800	23400	24000	24600	25200	25800	26400	27100
	3	0.575	2060	20200	20700	21200	21700	22200	22700	23200	23700	24300	24800	25400
	4	0.863	1610	19000	19400	19800	20200	20700	21100	21600	22000	22500	23000	23500
	BFL	1.15	1150	17500	17800	18100	18500	18800	19200	19500	19900	20200	20600	21000
	6	3.60	947	16700	17000	17300	17500	17800	18100	18500	18800	19100	19400	19800
	7	6.50	740	15800	16000	16300	16500	16700	17000	17300	17500	17800	18100	18300
W33×169	TFL	0.000	2480	18100	18600	19100	19600	20100	20600	21200	21700	22300	22900	23400
(9290)	2	0.305	2120	17400	17900	18300	18800	19300	19700	20200	20700	21300	21800	22300
	3	0.610	1770	16700	17100	17500	17900	18300	18700	19200	19600	20100	20600	21100
	4	0.915	1420	15700	16100	16400	16800	17200	17600	17900	18300	18800	19200	19600
	BFL	1.22	1070	14600	14900	15200	15500	15800	16100	16500	16800	17100	17500	17800
	6	4.28	845	13800	14000	14300	14500	14800	15100	15300	15600	15900	16200	16500
	7	7.66	619	12800	13000	13200	13400	13600	13800	14000	14300	14500	14700	14900
W33×152	TFL	0.000	2240	16000	16500	16900	17300	17800	18300	18800	19200	19800	20300	20800
(8160)	2	0.265	1930	15400	15800	16200	16700	17100	17500	18000	18400	18900	19400	19900
	3	0.530	1630	14800	15200	15500	15900	16300	16700	17100	17500	17900	18400	18800
	4	0.795	1320	14000	14300	14600	15000	15300	15700	16000	16400	16800	17100	17500
	BFL	1.06	1010	13100	13300	13600	13900	14200	14400	14700	15100	15400	15700	16000
	6	4.38	785	12200	12500	12700	12900	13200	13400	13700	13900	14200	14400	14700
	7	7.93	560	11300	11500	11700	11800	12000	12200	12400	12600	12800	13000	13200

[a] $Y1$ = distance from top of the steel beam to plastic neutral axis.
[b] $Y1$ = distance from top of the steel beam to concrete flange force.
[c] See Figure 5-4c for PNA locations.
[d] Value in parentheses is I_x (in.^{4}) of non-composite steel shape.

Table 5-15 (cont.). Lower Bound Elastic Moment of Inertia I_{LB} for Plastic Composite Sections

Shape[d]	PNA[c]	$Y1$[a]	ΣQ_n	$Y2$[b], in.										
		in.	kips	2	2.5	3	3.5	4	4.5	5	5.5	6	6.5	7
W33×141	TFL	0.000	2080	14700	15100	15500	15900	16300	16800	17200	17700	18100	18600	19100
(7450)	2	0.240	1800	14200	14500	14900	15300	15700	16100	16500	16900	17400	17800	18200
	3	0.480	1530	13600	13900	14300	14600	15000	15300	15700	16100	16500	16900	17300
	4	0.720	1250	12900	13200	13500	13800	14100	14400	14800	15100	15500	15800	16200
	BFL	0.960	976	12100	12300	12600	12800	13100	13400	13700	14000	14300	14600	14900
	6	4.29	748	11300	11500	11700	11900	12100	12400	12600	12800	13100	13300	13600
	7	8.05	520	10300	10500	10700	10800	11000	11200	11300	11500	11700	11900	12100
W33×130	TFL	0.000	1920	13300	13700	14000	14400	14800	15200	15600	16000	16500	16900	17300
(6710)	2	0.214	1670	12800	13200	13500	13900	14200	14600	15000	15400	15800	16200	16600
	3	0.428	1420	12300	12600	12900	13300	13600	13900	14300	14600	15000	15400	15800
	4	0.641	1180	11700	12000	12300	12600	12900	13200	13500	13800	14100	14500	14800
	BFL	0.855	932	11000	11300	11500	11800	12000	12300	12500	12800	13100	13400	13700
	6	4.39	705	10300	10500	10600	10900	11100	11300	11500	11700	12000	12200	12400
	7	8.30	479	9350	9490	9640	9790	9950	10100	10300	10400	10600	10800	11000
W33×118	TFL	0.000	1740	11800	12100	12500	12800	13200	13500	13900	14300	14700	15100	15500
(5900)	2	0.185	1520	11400	11700	12000	12300	12700	13000	13400	13700	14100	14400	14800
	3	0.370	1310	11000	11300	11500	11800	12100	12500	12800	13100	13400	13800	14100
	4	0.555	1100	10500	10700	11000	11300	11500	11800	12100	12400	12700	13000	13300
	BFL	0.740	884	9890	10100	10300	10600	10800	11000	11300	11500	11800	12100	12300
	6	4.47	659	9150	9330	9510	9700	9890	10100	10300	10500	10700	10900	11200
	7	8.56	434	8260	8390	8530	8660	8800	8950	9090	9250	9400	9560	9720
W30×116	TFL	0.000	1710	9870	10200	10500	10800	11100	11400	11800	12100	12500	12800	13200
(4930)	2	0.213	1490	9530	9810	10100	10400	10700	11000	11300	11600	12000	12300	12600
	3	0.425	1260	9120	9370	9630	9900	10200	10400	10700	11000	11300	11600	12000
	4	0.638	1040	8670	8890	9120	9360	9600	9850	10100	10400	10600	10900	11200
	BFL	0.850	818	8130	8320	8520	8720	8920	9140	9360	9580	9810	10000	10300
	6	3.98	623	7570	7730	7890	8060	8230	8400	8580	8770	8960	9150	9350
	7	7.43	428	6910	7030	7150	7270	7400	7530	7670	7810	7950	8090	8240
W30×108	TFL	0.000	1590	9000	9280	9560	9840	10100	10400	10800	11100	11400	11700	12100
(4470)	2	0.190	1390	8700	8950	9220	9480	9760	10000	10300	10600	10900	11300	11600
	3	0.380	1190	8350	8590	8830	9070	9330	9590	9850	10100	10400	10700	11000
	4	0.570	987	7940	8150	8370	8590	8820	9050	9290	9530	9780	10000	10300
	BFL	0.760	787	7470	7650	7840	8030	8230	8430	8640	8850	9060	9290	9510
	6	4.04	592	6930	7080	7230	7390	7550	7710	7880	8060	8240	8420	8600
	7	7.63	396	6280	6390	6500	6620	6730	6850	6980	7110	7240	7370	7510
W30×99	TFL	0.000	1460	8130	8380	8630	8900	9170	9450	9730	10000	10300	10600	10900
(3990)	2	0.168	1280	7860	8090	8330	8580	8830	9090	9360	9630	9910	10200	10500
	3	0.335	1100	7550	7760	7980	8210	8440	8680	8930	9180	9440	9700	9970
	4	0.503	927	7210	7400	7600	7800	8010	8230	8450	8680	8910	9150	9400
	BFL	0.670	752	6810	6970	7150	7330	7510	7700	7900	8100	8300	8510	8720
	6	4.13	558	6280	6420	6560	6710	6860	7010	7170	7330	7500	7670	7840
	7	7.85	364	5640	5740	5850	5950	6060	6170	6280	6400	6520	6640	6770

[a] $Y1$ = distance from top of the steel beam to plastic neutral axis.

[b] $Y1$ = distance from top of the steel beam to concrete flange force.

[c] See Figure 5-4c for PNA locations.

[d] Value in parentheses is I_x (in.^{4}) of non-composite steel shape.

Table 5-15 (cont.). Lower Bound Elastic Moment of Inertia I_{LB} for Plastic Composite Sections

Shape[d]	PNA[c]	Y1[a]	ΣQn	Y2[b], in.												
		in.	kips	2	2.5	3	3.5	4	4.5	5	5.5	6	6.5	7
W30×90	TFL	0.000	1320	7310	7540	7770	8010	8250	8500	8760	9020	9290	9570	9850
(3610)	2	0.153	1160	7070	7280	7500	7720	7950	8190	8430	8670	8930	9190	9450
	3	0.305	1000	6800	7000	7200	7400	7610	7830	8050	8280	8510	8750	8990
	4	0.458	844	6500	6670	6850	7040	7230	7430	7630	7830	8040	8260	8480
	BFL	0.610	686	6140	6300	6450	6620	6780	6960	7130	7310	7500	7690	7880
	6	3.95	508	5670	5790	5920	6050	6190	6330	6470	6620	6770	6920	7080
	7	7.73	330	5090	5180	5270	5370	5470	5570	5670	5780	5880	5990	6110
W27×102	TFL	0.000	1500	7250	7480	7730	7980	8240	8510	8780	9060	9350	9650	9950
(3620)	2	0.208	1290	6970	7190	7420	7650	7890	8140	8390	8650	8920	9200	9480
	3	0.415	1090	6670	6870	7080	7290	7510	7730	7960	8200	8450	8700	8950
	4	0.623	878	6300	6470	6650	6840	7030	7230	7430	7640	7850	8070	8300
	BFL	0.830	670	5860	6010	6160	6310	6470	6640	6810	6980	7160	7340	7530
	6	3.40	523	5500	5620	5740	5870	6010	6150	6290	6430	6580	6740	6900
	7	6.27	375	5070	5170	5260	5360	5470	5570	5680	5800	5910	6030	6150
W27×94	TFL	0.000	1390	6580	6800	7020	7260	7490	7740	7990	8250	8520	8790	9070
(3270)	2	0.186	1200	6340	6540	6750	6960	7190	7410	7650	7890	8130	8390	8650
	3	0.373	1010	6060	6240	6430	6630	6830	7030	7250	7460	7690	7920	8160
	4	0.559	827	5740	5900	6070	6250	6420	6610	6800	6990	7190	7390	7600
	BFL	0.745	641	5360	5500	5640	5790	5940	6090	6250	6420	6590	6760	6940
	6	3.38	493	5010	5120	5240	5360	5480	5610	5750	5880	6020	6160	6310
	7	6.38	346	4590	4680	4770	4860	4960	5050	5150	5260	5360	5470	5590
W27×84	TFL	0.000	1240	5770	5970	6160	6370	6580	6800	7030	7260	7490	7740	7990
(2850)	2	0.160	1080	5570	5750	5940	6130	6330	6530	6740	6950	7170	7400	7630
	3	0.320	921	5340	5510	5680	5850	6030	6220	6410	6610	6810	7010	7230
	4	0.480	762	5070	5220	5370	5530	5690	5860	6030	6200	6380	6570	6760
	BFL	0.640	603	4760	4890	5020	5150	5290	5440	5580	5730	5890	6050	6210
	6	3.43	456	4420	4530	4630	4740	4860	4970	5100	5220	5350	5480	5610
	7	6.61	310	4020	4100	4180	4260	4340	4430	4520	4610	4710	4800	4900
W24×76	TFL	0.000	1120	4280	4440	4600	4770	4950	5130	5320	5510	5710	5910	6120
(2100)	2	0.170	967	4120	4270	4420	4580	4740	4910	5080	5260	5440	5630	5830
	3	0.340	814	3930	4070	4210	4350	4500	4650	4810	4970	5140	5310	5490
	4	0.510	662	3720	3840	3960	4090	4220	4350	4490	4630	4780	4930	5090
	BFL	0.680	509	3460	3560	3660	3770	3880	3990	4110	4230	4360	4480	4610
	6	2.99	394	3230	3320	3400	3490	3580	3680	3770	3880	3980	4080	4190
	7	5.59	280	2970	3040	3100	3170	3240	3310	3390	3460	3540	3630	3710
W24×68	TFL	0.000	1010	3760	3900	4050	4200	4360	4520	4690	4860	5040	5220	5410
(1830)	2	0.146	874	3620	3760	3890	4030	4180	4330	4480	4640	4810	4980	5150
	3	0.293	743	3470	3590	3710	3840	3980	4110	4260	4400	4550	4710	4870
	4	0.439	611	3290	3390	3510	3620	3740	3860	3990	4120	4250	4390	4530
	BFL	0.585	480	3080	3170	3260	3360	3460	3570	3670	3790	3900	4020	4140
	6	3.04	366	2860	2930	3010	3090	3180	3260	3350	3450	3540	3640	3740
	7	5.80	251	2600	2660	2720	2780	2840	2900	2970	3040	3110	3180	3260

[a] Y1 = distance from top of the steel beam to plastic neutral axis.
[b] Y1 = distance from top of the steel beam to concrete flange force.
[c] See Figure 5-4c for PNA locations.
[d] Value in parentheses is I_x (in.^{4}) of non-composite steel shape.

Table 5-15 (cont.). Lower Bound Elastic Moment of Inertia I_{LB} for Plastic Composite Sections

Shape[d]	PNA[c]	$Y1$[a]	ΣQ_n	$Y2$[b], in.										
		in.	kips	2	2.5	3	3.5	4	4.5	5	5.5	6	6.5	7
W24×62	TFL	0.000	915	3320	3440	3580	3720	3860	4010	4160	4310	4480	4640	4810
(1560)	2	0.148	811	3210	3330	3460	3590	3720	3860	4000	4150	4300	4460	4620
	3	0.295	707	3090	3200	3320	3440	3560	3690	3820	3960	4100	4250	4390
	4	0.443	603	2950	3060	3160	3270	3390	3500	3620	3750	3880	4010	4140
	BFL	0.590	500	2800	2890	2990	3080	3180	3290	3400	3510	3620	3740	3860
	6	3.38	364	2560	2630	2710	2790	2870	2950	3040	3130	3220	3310	3410
	7	6.53	229	2260	2310	2370	2420	2480	2540	2600	2660	2730	2790	2860
W24×55	TFL	0.000	815	2910	3030	3150	3270	3390	3530	3660	3800	3940	4090	4240
(1360)	2	0.126	726	2820	2930	3040	3160	3280	3400	3530	3660	3790	3930	4070
	3	0.253	638	2720	2820	2930	3040	3150	3260	3380	3500	3630	3760	3890
	4	0.379	549	2610	2700	2800	2900	3000	3100	3210	3320	3440	3560	3680
	BFL	0.505	461	2480	2560	2650	2740	2830	2920	3020	3120	3230	3330	3440
	6	3.39	332	2260	2320	2390	2460	2540	2610	2690	2770	2850	2940	3030
	7	6.64	204	1980	2030	2070	2120	2170	2230	2280	2340	2390	2450	2510
W21×62	TFL	0.000	915	2760	2880	3000	3120	3250	3390	3530	3670	3820	3970	4130
(1330)	2	0.154	788	2650	2760	2870	2990	3110	3240	3360	3500	3640	3780	3920
	3	0.308	662	2530	2630	2730	2840	2950	3060	3180	3300	3420	3550	3680
	4	0.461	535	2390	2470	2560	2650	2750	2850	2950	3060	3170	3280	3400
	BFL	0.615	408	2210	2280	2360	2440	2520	2600	2690	2770	2870	2960	3060
	6	2.54	318	2070	2130	2190	2260	2320	2390	2460	2540	2610	2690	2780
	7	4.78	229	1900	1950	2000	2050	2100	2150	2210	2270	2330	2390	2450
W21×57	TFL	0.000	835	2490	2590	2700	2820	2940	3060	3190	3320	3460	3600	3740
(1170)	2	0.163	728	2400	2490	2600	2710	2820	2930	3050	3170	3300	3430	3570
	3	0.325	622	2290	2380	2480	2580	2680	2780	2890	3010	3120	3240	3370
	4	0.488	515	2170	2250	2340	2430	2520	2610	2710	2810	2910	3020	3130
	BFL	0.650	409	2030	2110	2180	2250	2330	2410	2500	2580	2670	2770	2860
	6	2.93	309	1880	1940	2000	2060	2120	2190	2260	2330	2410	2480	2560
	7	5.40	209	1700	1740	1780	1830	1880	1930	1980	2030	2090	2140	2200
W21×55	TFL	0.000	810	2390	2490	2590	2710	2820	2940	3060	3190	3320	3450	3590
(1140)	2	0.131	703	2300	2390	2490	2590	2700	2810	2930	3040	3160	3290	3420
	3	0.261	595	2190	2280	2370	2470	2560	2660	2770	2870	2990	3100	3220
	4	0.392	488	2080	2150	2230	2320	2400	2490	2580	2680	2780	2880	2980
	BFL	0.522	381	1940	2000	2070	2140	2210	2290	2370	2450	2530	2620	2710
	6	2.62	292	1800	1850	1910	1970	2030	2090	2160	2230	2290	2370	2440
	7	5.00	203	1640	1680	1720	1770	1810	1860	1910	1960	2010	2070	2120
W21×50	TFL	0.000	735	2110	2210	2300	2400	2510	2620	2730	2840	2960	3080	3210
(984)	2	0.134	648	2040	2130	2220	2310	2410	2510	2620	2730	2840	2950	3070
	3	0.268	560	1960	2040	2130	2210	2300	2400	2490	2590	2690	2800	2910
	4	0.401	473	1870	1940	2020	2100	2180	2260	2350	2440	2530	2630	2730
	BFL	0.535	386	1760	1830	1890	1960	2030	2110	2180	2260	2350	2430	2520
	6	2.91	285	1620	1670	1720	1780	1840	1900	1960	2020	2090	2160	2230
	7	5.56	184	1440	1470	1510	1550	1590	1640	1680	1730	1780	1820	1880

[a] $Y1$ = distance from top of the steel beam to plastic neutral axis.

[b] $Y1$ = distance from top of the steel beam to concrete flange force.

[c] See Figure 5-4c for PNA locations.

[d] Value in parentheses is I_x (in.^{4}) of non-composite steel shape.

Table 5-15 (cont.). Lower Bound Elastic Moment of Inertia I_{LB} for Plastic Composite Sections

Shape[d]	PNA[c]	$Y1$[a]	ΣQ_n	$Y2$[b], in.										
		in.	kips	2	2.5	3	3.5	4	4.5	5	5.5	6	6.5	7
W21×48	TFL	0.000	705	2030	2110	2210	2300	2400	2500	2610	2720	2830	2950	3070
(959)	2	0.108	617	1950	2040	2120	2210	2300	2400	2500	2600	2710	2820	2930
	3	0.215	530	1870	1950	2030	2110	2200	2280	2380	2470	2570	2670	2770
	4	0.323	442	1780	1850	1920	1990	2070	2150	2230	2320	2400	2490	2590
	BFL	0.430	355	1670	1730	1790	1860	1920	1990	2060	2140	2210	2290	2370
	6	2.71	266	1540	1590	1640	1690	1750	1810	1860	1920	1990	2050	2120
	7	5.26	176	1390	1420	1460	1500	1540	1580	1620	1660	1710	1750	1800
W21×44	TFL	0.000	650	1830	1920	2000	2090	2180	2280	2370	2480	2580	2690	2800
(843)	2	0.113	577	1780	1850	1930	2020	2100	2190	2280	2380	2480	2580	2680
	3	0.225	504	1710	1780	1850	1930	2010	2100	2180	2270	2360	2460	2550
	4	0.338	431	1630	1700	1770	1840	1910	1990	2060	2150	2230	2310	2400
	BFL	0.450	358	1550	1610	1670	1730	1790	1860	1930	2000	2080	2150	2230
	6	2.92	260	1410	1460	1500	1560	1610	1660	1720	1780	1840	1900	1960
	7	5.71	163	1240	1270	1310	1340	1380	1420	1460	1500	1540	1580	1630
W18×60	TFL	0.000	880	2070	2170	2270	2380	2490	2610	2730	2860	2990	3130	3270
(984)	2	0.174	749	1980	2070	2170	2270	2370	2480	2590	2710	2830	2950	3080
	3	0.348	617	1880	1960	2050	2140	2230	2330	2430	2530	2640	2750	2860
	4	0.521	486	1760	1830	1900	1980	2060	2140	2230	2320	2410	2510	2610
	BFL	0.695	355	1610	1660	1720	1790	1850	1920	1990	2060	2140	2220	2300
	6	2.18	287	1520	1570	1620	1670	1730	1780	1840	1910	1970	2040	2110
	7	3.80	220	1420	1460	1500	1540	1590	1640	1680	1730	1790	1840	1900
W18×55	TFL	0.000	810	1880	1970	2070	2170	2270	2380	2490	2600	2720	2850	2980
(890)	2	0.158	691	1800	1880	1970	2060	2160	2260	2360	2470	2580	2690	2810
	3	0.315	573	1710	1790	1860	1950	2030	2120	2210	2310	2410	2510	2620
	4	0.473	454	1600	1670	1730	1810	1880	1960	2040	2120	2210	2300	2390
	BFL	0.630	336	1470	1520	1580	1640	1700	1760	1830	1900	1970	2040	2110
	6	2.15	269	1380	1430	1480	1530	1580	1630	1690	1750	1800	1870	1930
	7	3.86	203	1290	1320	1360	1400	1440	1490	1530	1580	1630	1670	1730
W18×50	TFL	0.000	735	1690	1770	1860	1950	2040	2140	2240	2350	2450	2570	2680
(800)	2	0.143	628	1620	1700	1780	1860	1940	2030	2130	2220	2320	2430	2530
	3	0.285	521	1540	1610	1680	1750	1830	1910	2000	2080	2170	2260	2360
	4	0.428	414	1440	1500	1560	1630	1700	1770	1840	1910	1990	2070	2160
	BFL	0.570	308	1330	1370	1430	1480	1530	1590	1650	1710	1780	1840	1910
	6	2.08	246	1250	1290	1330	1380	1420	1470	1520	1580	1630	1690	1740
	7	3.82	184	1160	1190	1220	1260	1300	1340	1380	1420	1460	1510	1550
W18×46	TFL	0.000	675	1540	1610	1690	1780	1860	1950	2040	2140	2240	2340	2450
(712)	2	0.151	583	1480	1550	1620	1700	1780	1860	1950	2040	2130	2220	2320
	3	0.303	492	1410	1470	1540	1610	1680	1760	1840	1920	2000	2090	2180
	4	0.454	400	1330	1380	1440	1500	1570	1630	1700	1780	1850	1930	2010
	BFL	0.605	308	1230	1280	1330	1380	1430	1490	1550	1610	1670	1730	1800
	6	2.42	239	1140	1180	1220	1270	1310	1360	1410	1460	1510	1570	1620
	7	4.36	169	1040	1070	1100	1140	1170	1210	1250	1280	1320	1370	1410

[a] $Y1$ = distance from top of the steel beam to plastic neutral axis.
[b] $Y1$ = distance from top of the steel beam to concrete flange force.
[c] See Figure 5-4c for PNA locations.
[d] Value in parentheses is I_x (in.^{4}) of non-composite steel shape.

Table 5-15 (cont.). Lower Bound Elastic Moment of Inertia I_{LB} for Plastic Composite Sections

Shape[d]	PNA[c]	$Y1$[a]	ΣQ_n	$Y2$[b], in.										
		in.	kips	2	2.5	3	3.5	4	4.5	5	5.5	6	6.5	7
W18×40	TFL	0.000	590	1320	1390	1450	1530	1600	1680	1760	1840	1930	2020	2110
(612)	2	0.131	511	1270	1330	1390	1460	1530	1600	1680	1760	1840	1920	2010
	3	0.263	432	1210	1270	1320	1390	1450	1510	1580	1650	1730	1800	1880
	4	0.394	353	1140	1190	1240	1300	1350	1410	1470	1530	1600	1670	1740
	BFL	0.525	274	1060	1100	1150	1190	1240	1290	1340	1390	1450	1510	1560
	6	2.26	211	985	1020	1060	1090	1130	1170	1220	1260	1310	1350	1400
	7	4.27	148	896	922	950	979	1010	1040	1070	1110	1140	1180	1210
W18×35	TFL	0.000	515	1120	1170	1230	1300	1360	1430	1500	1570	1650	1720	1800
(510)	2	0.106	451	1080	1130	1190	1240	1300	1370	1430	1500	1570	1640	1720
	3	0.213	388	1030	1080	1130	1190	1240	1300	1360	1420	1490	1550	1620
	4	0.319	324	978	1020	1070	1120	1170	1220	1270	1330	1390	1450	1510
	BFL	0.425	260	917	955	995	1040	1080	1130	1170	1220	1270	1320	1380
	6	2.37	194	842	873	906	940	975	1010	1050	1090	1130	1170	1220
	7	4.56	129	753	776	800	825	851	878	906	935	965	996	1030
W16×36	TFL	0.000	530	973	1030	1080	1140	1200	1270	1340	1410	1480	1550	1630
(448)	2	0.108	455	933	983	1040	1090	1150	1210	1270	1330	1400	1470	1540
	3	0.215	380	886	931	979	1030	1080	1130	1190	1250	1310	1370	1440
	4	0.323	305	831	871	912	956	1000	1050	1100	1150	1200	1260	1310
	BFL	0.430	229	765	797	831	867	905	944	984	1030	1070	1120	1160
	6	1.82	181	715	743	772	802	833	866	901	936	973	1010	1050
	7	3.46	133	659	680	703	727	752	778	805	833	862	892	923
W16×31	TFL	0.000	457	827	874	923	974	1030	1080	1140	1200	1260	1330	1400
(375)	2	0.110	396	795	838	884	931	981	1030	1090	1140	1200	1260	1320
	3	0.220	335	758	797	838	882	927	974	1020	1070	1130	1180	1240
	4	0.330	274	714	749	786	824	864	906	949	995	1040	1090	1140
	BFL	0.440	213	663	692	723	756	790	825	862	900	940	982	1020
	6	2.00	164	614	639	664	691	720	749	780	812	845	879	914
	7	3.80	114	556	574	594	614	636	658	681	705	730	756	783
W16×26	TFL	0.000	384	674	712	753	796	840	887	935	985	1040	1090	1150
(301)	2	0.0863	337	649	686	724	763	805	849	894	941	990	1040	1090
	3	0.173	289	621	654	689	726	764	804	846	889	934	980	1030
	4	0.259	242	589	619	651	683	718	754	791	830	871	912	956
	BFL	0.345	194	551	577	604	633	663	694	727	760	795	832	869
	6	2.05	145	505	527	549	572	597	622	649	676	705	734	765
	7	4.01	96	450	466	482	499	517	535	555	575	596	617	640
W14×38	TFL	0.000	560	844	896	951	1010	1070	1130	1200	1270	1340	1410	1490
(385)	2	0.129	473	805	853	903	956	1010	1070	1130	1190	1260	1330	1400
	3	0.258	386	759	802	847	894	943	995	1050	1100	1160	1220	1290
	4	0.386	299	704	741	779	819	861	905	951	999	1050	1100	1150
	BFL	0.515	211	636	665	695	726	759	794	830	868	907	948	990
	6	1.38	176	604	629	656	683	712	742	774	807	841	877	914
	7	2.53	140	568	589	611	634	659	684	710	738	766	796	827

[a] $Y1$ = distance from top of the steel beam to plastic neutral axis.
[b] $Y1$ = distance from top of the steel beam to concrete flange force.
[c] See Figure 5-4c for PNA locations.
[d] Value in parentheses is I_x (in.^{4}) of non-composite steel shape.

Table 5-15 (cont.). Lower Bound Elastic Moment of Inertia I_{LB} for Plastic Composite Sections

Shape[d]	PNA[c]	$Y1$[a]	ΣQ_n	$Y2$[b], in.										
		in.	kips	2	2.5	3	3.5	4	4.5	5	5.5	6	6.5	7
W14×34	TFL	0.000	500	745	791	840	891	945	1000	1060	1120	1190	1250	1320
(340)	2	0.114	423	711	754	798	845	895	946	1000	1060	1110	1180	1240
	3	0.228	346	671	709	749	791	835	881	929	979	1030	1090	1140
	4	0.341	270	624	656	691	727	764	804	845	888	933	979	1030
	BFL	0.455	193	566	591	618	647	677	708	741	775	811	848	886
	6	1.42	159	535	558	581	606	632	659	687	717	748	780	813
	7	2.61	125	502	521	540	561	582	605	628	653	678	705	732
W14×30	TFL	0.000	443	642	682	725	770	817	866	918	972	1030	1090	1150
(291)	2	0.0963	378	614	651	691	732	775	821	868	918	969	1020	1080
	3	0.193	313	581	615	650	688	727	767	810	855	901	949	999
	4	0.289	248	543	572	603	635	669	704	741	780	820	862	905
	BFL	0.385	183	496	520	545	571	599	627	658	689	722	756	791
	6	1.46	147	466	486	507	530	553	578	604	630	658	687	717
	7	2.80	111	432	448	465	483	502	522	542	564	586	610	634
W14×26	TFL	0.000	385	553	589	626	665	706	749	794	841	890	941	994
(245)	2	0.105	332	530	563	598	634	672	712	754	797	843	890	938
	3	0.210	279	504	534	565	598	633	669	707	746	787	830	874
	4	0.315	226	473	499	527	556	586	618	652	686	722	760	799
	BFL	0.420	173	436	458	481	506	531	558	586	615	645	677	709
	6	1.67	135	405	423	443	463	485	507	530	555	580	607	634
	7	3.18	96.1	368	382	397	413	429	447	465	483	503	523	544
W14×22	TFL	0.000	325	453	483	514	547	581	617	655	694	735	778	822
(199)	2	0.0838	283	436	463	492	523	555	588	624	660	698	738	779
	3	0.168	241	416	441	467	495	525	555	587	621	656	692	730
	4	0.251	199	392	415	438	463	489	517	545	575	606	639	672
	BFL	0.335	157	365	384	404	426	448	472	496	522	548	576	605
	6	1.67	119	335	351	368	386	404	423	444	465	487	509	533
	7	3.32	81.1	301	312	325	338	352	366	381	397	413	430	448
W12×30	TFL	0.000	440	530	567	606	648	691	737	785	835	887	942	998
(238)	2	0.110	368	504	538	573	611	651	692	736	782	829	879	931
	3	0.220	296	473	503	534	567	602	639	678	718	760	804	850
	4	0.330	224	435	460	486	514	544	575	607	641	676	713	751
	BFL	0.440	153	389	408	428	449	472	495	520	546	573	601	631
	6	1.10	131	372	389	407	426	446	467	489	512	536	561	587
	7	1.92	110	355	370	385	402	419	438	457	477	498	520	542
W12×26	TFL	0.000	383	455	487	521	557	594	634	676	719	764	812	861
(204)	2	0.095	321	433	462	493	526	560	596	634	674	715	758	803
	3	0.190	259	407	432	460	489	519	551	585	620	656	694	734
	4	0.285	198	375	397	420	444	470	497	525	555	586	618	652
	BFL	0.380	136	336	352	370	389	409	429	451	474	498	523	548
	6	1.07	116	321	336	351	368	386	404	423	444	465	487	509
	7	1.94	95.6	304	317	331	345	360	376	392	410	428	447	467

[a] $Y1$ = distance from top of the steel beam to plastic neutral axis.
[b] $Y1$ = distance from top of the steel beam to concrete flange force.
[c] See Figure 5-4c for PNA locations.
[d] Value in parentheses is I_x (in.^{4}) of non-composite steel shape.

Table 5-15 (cont.). Lower Bound Elastic Moment of Inertia I_{LB} for Plastic Composite Sections

Shape[d]	PNA[c]	$Y1$[a]	ΣQ_n	$Y2$[b], in.										
		in.	kips	2	2.5	3	3.5	4	4.5	5	5.5	6	6.5	7
W12×22	TFL	0.000	324	371	398	427	458	490	523	559	596	634	674	716
(156)	2	0.106	281	356	381	408	436	466	497	530	564	600	638	676
	3	0.213	238	338	361	386	412	439	467	497	528	561	595	631
	4	0.319	196	318	339	360	383	408	433	460	487	517	547	578
	BFL	0.425	153	294	312	330	350	370	392	414	438	463	489	515
	6	1.66	117	270	285	300	316	333	351	370	389	410	431	453
	7	3.03	81	242	253	265	277	290	303	317	332	347	363	380
W12×19	TFL	0.000	279	313	336	361	387	414	443	473	505	538	573	608
(130)	2	0.0875	243	300	322	345	369	395	422	450	479	510	542	575
	3	0.175	208	286	306	327	349	373	398	423	450	479	508	539
	4	0.263	173	270	288	307	327	348	370	393	417	442	469	496
	BFL	0.350	138	251	266	283	300	318	337	357	378	400	423	447
	6	1.68	104	229	242	255	270	284	300	317	334	352	370	390
	7	3.14	69.6	203	212	222	233	244	255	267	280	293	307	321
W12×16	TFL	0.000	236	254	273	294	316	339	363	388	415	442	471	501
(103)	2	0.0663	209	245	263	282	303	324	347	371	396	422	449	477
	3	0.133	183	235	252	270	289	309	330	352	375	400	425	451
	4	0.199	156	223	239	255	272	291	310	330	351	373	396	420
	BFL	0.265	130	210	224	239	254	271	288	306	325	344	365	386
	6	1.71	94.3	189	200	212	225	238	251	266	281	297	313	331
	7	3.32	58.9	163	171	179	188	197	207	217	228	239	250	262
W12×14	TFL	0.000	208	220	237	255	274	295	316	338	361	386	411	437
(88.6)	2	0.0563	186	213	229	246	264	283	303	324	346	369	393	418
	3	0.113	163	204	219	235	252	270	288	308	328	350	372	395
	4	0.169	141	195	209	223	239	255	272	290	309	329	349	370
	BFL	0.225	119	184	197	210	224	238	254	270	287	305	323	342
	6	1.68	85.3	165	175	186	197	208	221	234	247	261	276	291
	7	3.35	52	141	148	155	163	171	179	188	198	207	218	228
W10×26	TFL	0.000	381	339	367	397	429	463	499	536	576	617	661	706
(144)	2	0.110	317	321	346	374	403	434	466	500	536	574	613	655
	3	0.220	254	300	322	346	372	399	428	458	490	523	557	594
	4	0.330	190	274	292	312	334	356	380	405	431	459	488	518
	BFL	0.440	127	241	255	270	286	303	321	340	360	381	402	425
	6	0.886	111	232	245	258	273	288	304	321	339	358	377	398
	7	1.49	95.1	222	233	245	258	271	286	301	317	333	351	369
W10×22	TFL	0.000	325	282	306	331	358	387	417	449	483	518	555	593
(118)	2	0.090	273	267	289	313	337	364	391	420	451	483	517	552
	3	0.180	221	251	270	291	312	336	360	386	413	442	472	503
	4	0.270	169	230	246	264	282	302	323	345	368	392	417	443
	BFL	0.360	118	205	218	232	246	261	277	295	312	331	351	371
	6	0.962	99.3	195	206	218	230	244	258	273	289	305	323	341
	7	1.72	81.1	183	193	203	214	225	238	250	264	278	293	308

[a] $Y1$ = distance from top of the steel beam to plastic neutral axis.

[b] $Y1$ = distance from top of the steel beam to concrete flange force.

[c] See Figure 5-4c for PNA locations.

[d] Value in parentheses is I_x (in.^{4}) of non-composite steel shape.

Table 5-15 (cont.). Lower Bound Elastic Moment of Inertia I_{LB} for Plastic Composite Sections

Shape[d]	PNA[c]	Y1[a]	ΣQn	Y2[b], in.											
		in.	kips	2	2.5	3	3.5	4	4.5	5	5.5	6	6.5	7	
W10×19	TFL	0.000	281	238	259	281	304	329	355	383	412	443	474	508	
(96.3)	2	0.0988	241	227	246	267	288	311	335	361	388	416	445	476	
	3	0.198	202	215	232	251	270	291	313	336	360	386	413	440	
	4	0.296	162	200	215	231	248	266	286	306	327	350	373	397	
	BFL	0.395	122	182	195	208	222	237	253	270	287	306	325	345	
	6	1.25	96.2	169	179	190	202	215	228	243	257	273	289	306	
	7	2.29	70.3	153	161	170	179	189	200	211	223	235	248	261	
W10×17	TFL	0.000	250	206	224	244	264	286	310	334	360	387	415	445	
(81.9)	2	0.0825	216	197	214	232	251	272	293	316	340	365	391	418	
	3	0.165	183	187	202	219	236	255	274	295	317	340	364	388	
	4	0.248	150	175	189	203	219	235	253	271	290	311	332	354	
	BFL	0.330	117	161	173	185	198	212	227	243	259	276	294	313	
	6	1.31	89.8	148	157	167	178	190	202	215	229	243	258	274	
	7	2.45	62.4	132	139	147	155	164	173	183	193	204	215	227	
W10×15	TFL	0.000	221	177	193	210	228	248	268	289	312	336	361	387	
(68.9)	2	0.0675	194	170	185	201	218	236	255	275	296	318	342	366	
	3	0.135	167	162	176	190	206	223	240	259	278	299	320	342	
	4	0.203	140	153	165	178	192	207	223	240	258	276	295	315	
	BFL	0.270	113	142	153	164	177	190	204	218	233	250	266	284	
	6	1.35	83.8	128	137	147	157	167	178	190	203	216	229	244	
	7	2.60	55.1	112	118	125	133	140	148	157	166	175	185	196	
W10×12	TFL	0.000	177	139	152	165	180	195	211	229	247	265	285	306	
(53.8)	2	0.0525	156	134	145	158	172	186	201	217	234	252	271	290	
	3	0.105	135	127	138	150	163	176	190	205	221	237	254	272	
	4	0.158	115	121	131	142	153	165	178	191	206	221	236	252	
	BFL	0.210	93.8	113	122	131	141	152	163	175	187	200	214	228	
	6	1.30	69.0	102	109	116	124	133	142	152	162	173	184	195	
	7	2.61	44.3	87.9	93.0	98.4	104	110	117	124	131	139	146	155	

[a] $Y1$ = distance from top of the steel beam to plastic neutral axis.
[b] $Y1$ = distance from top of the steel beam to concrete flange force.
[c] See Figure 5-4c for PNA locations.
[d] Value in parentheses is I_x (in.^{4}) of non-composite steel shape.

Table 5-16
Concentrated Load Equivalents

n	Loading	Coeff.	Simple Beam 	Beam Fixed One End, Supported at Other 	Beam Fixed Both Ends 
∞	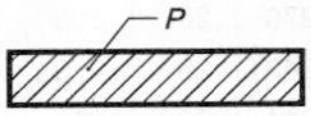	a	0.125	0.070	0.042
		b	—	0.125	0.083
		c	0.500	0.375	—
		d	—	0.625	0.500
		e	0.013	0.005	0.003
		f	1.000	1.000	0.667
		g	1.000	0.415	0.300
2	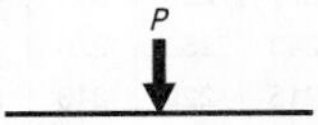	a	0.250	0.156	0.125
		b	—	0.188	0.125
		c	0.500	0.313	—
		d	—	0.688	0.500
		e	0.021	0.009	0.005
		f	2.000	1.500	1.000
		g	0.800	0.477	0.400
3	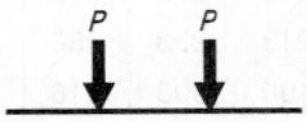	a	0.333	0.222	0.111
		b	—	0.333	0.222
		c	1.000	0.667	—
		d	—	1.333	1.000
		e	0.036	0.015	0.008
		f	2.667	2.667	1.778
		g	1.022	0.438	0.333
4	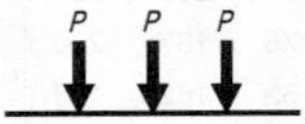	a	0.500	0.266	0.188
		b	—	0.469	0.313
		c	1.500	1.031	—
		d	—	1.969	1.500
		e	0.050	0.021	0.010
		f	4.000	3.750	2.500
		g	0.950	0.428	0.320
5	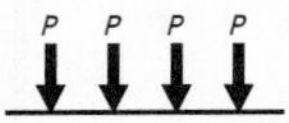	a	0.600	0.360	0.200
		b	—	0.600	0.400
		c	2.000	1.400	—
		d	—	2.600	2.000
		e	0.063	0.027	0.013
		f	4.800	4.800	3.200
		g	1.008	0.424	0.312

Maximum positive moment (kip-ft): aPL
Maximum negative moment (kip-ft): bPL
Pinned end reaction (kips): cP
Fixed end reaction (kips): dP
Maximum deflection (in.): ePl^3 / EI
Equivalent simple span uniform load (kips): fP
Deflection coefficient for equivalent simple span uniform load: g
Number of equal load spaces: n
Span of beam (ft): L
Span of beam (in.): l

Table 5-17.
Shears, Moments, and Deflections

The formulas given below are frequently required in structural designing. They are included herein for the convenience of those engineers who have infrequent use for such formulas and hence may find reference necessary. Variation from the standard nomenclature on page 4-187 is noted.

BEAMS

Flexural stress at extreme fiber:

$f = Mc/I = M/S$

Flexural stress at any fiber:

$f = My/I$ $y =$ distance from neutral axis to fiber

Average vertical shear (for maximum see below):

$v = V/A = V/dt$ (for beams and girders)

Horizontal shearing stress at any section A-A:

$v = VQ/Ib$ $Q =$ statical moment about the neutral axis of that portion of the cross section lying outside of section A-A
$b =$ width at section A-A

(Intensity of vertical shear is equal to that of horizontal shear acting normal to it at the same point and both are usually a maximum at mid-height of beam.)

Shear and deflection at any point:

$EI\dfrac{d^2y}{dx^2} = M$ x and y are abscissa and ordinate respectively of a point on the neutral axis, referred to axes of rectangular coordinates through a selected point of support.

(First integration gives slopes; second integration gives deflections. Constants of integration must be determined.)

CONTINUOUS BEAMS (the theorem of three moments)

Uniform load:

$$M_a\frac{l_1}{I_1} + 2M_b\left(\frac{l_1}{I_1} + \frac{l_2}{I_2}\right) + M_c\frac{l_2}{I_2} = -{}^1/_4\left(\frac{w_1 l_1^3}{I_1} + \frac{w_2 l_2^3}{I_2}\right)$$

Concentrated loads:

$$M_a\frac{l_1}{I_1} + 2M_b\left(\frac{l_1}{I_1} + \frac{l_2}{I_2}\right) + M_c\frac{l_2}{I_2} = -\frac{P_1 a_1 b_1}{I_1}\left(1 + \frac{a_1}{l_1}\right) - \frac{P_2 a_2 b_2}{I_2}\left(1 + \frac{b_2}{l_2}\right)$$

Considering any two consecutive spans in any continuous structure:

M_a, M_b, M_c = moments at left, center, and right supports respectively, of any pair of adjacent spans
l_1 and l_2 = length of left and right spans, respectively, of the pair
I_1 and I_2 = moment of inertia of left and right spans, respectively
w_1 and w_2 = load per unit of length on left and right spans, respectively
P_1 and P_2 = concentrated loads on left and right spans, respectively
a_1 and a_2 = distance of concentrated loads from left support, in left and right spans, respectively
b_1 and b_2 = distance of concentrated loads from right support, in left and right spans,respectively

The above equations are for beam with moment of inertia constant in each span but differing in different spans, continuous over three or more supports. By writing such an equation for each successive pair of spans and introducing the known values (usually zero) of end moments, all other moments can be found.

Table 5-17 (cont.). Shears, Moments, and Deflections

1. SIMPLE BEAM—UNIFORMLY DISTRIBUTED LOAD

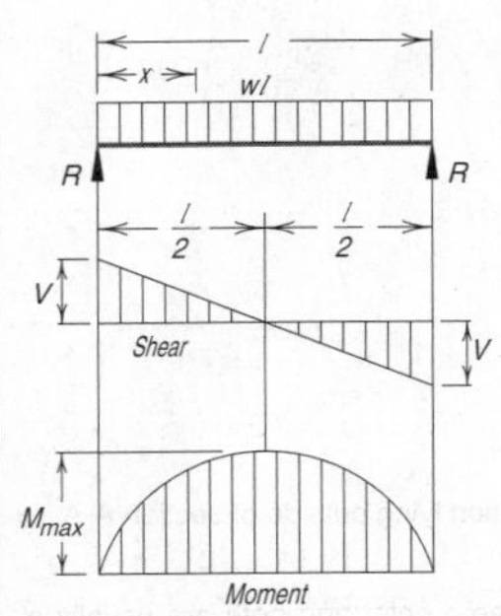

Total Equiv. Uniform Load .... $= wl$

$R = V$ .... $= \frac{wl}{2}$

V_x .... $= w\left(\frac{l}{2} - x\right)$

M_{max} (at center) .... $= \frac{wl^2}{8}$

M_x .... $= \frac{wx}{2}(l - x)$

Δ_{max} (at center) .... $= \frac{5wl^4}{384EI}$

Δ_x .... $= \frac{wx}{24EI}(l^3 - 2lx^2 + x^3)$

2. SIMPLE BEAM—LOAD INCREASING UNIFORMLY TO ONE END

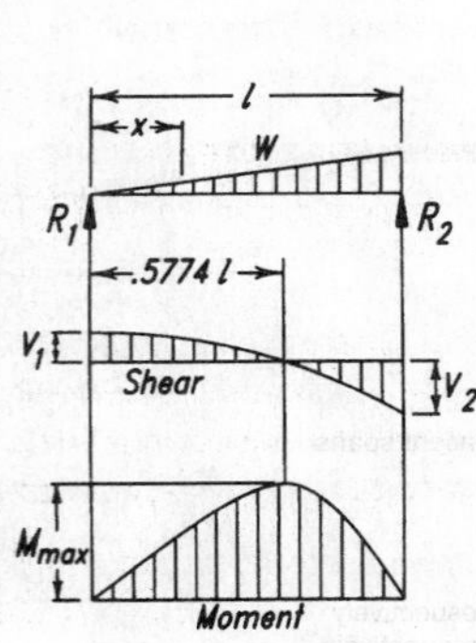

Total Equiv. Uniform Load .... $= \frac{16W}{9\sqrt{3}} = 1.03W$

$R_1 = V_1$ .... $= \frac{W}{3}$

$R_2 = V_2 = V_{max}$ .... $= \frac{2W}{3}$

V_x .... $= \frac{W}{3} - \frac{Wx^2}{l^2}$

M_{max} (at $x = \frac{l}{\sqrt{3}} = .577l$) .... $= \frac{2Wl}{9\sqrt{3}} = .128Wl$

M_x .... $= \frac{Wx}{3l^2}(l^2 - x^2)$

Δ_{max} (at $x = l\sqrt{1 - \sqrt{\frac{8}{15}}} = .519l$) .... $= 0.130 = \frac{Wl^3}{EI}$

Δ_x .... $= \frac{Wx}{180EIl^2}(3x^4 - 10l^2x^2 + 7l^4)$

3. SIMPLE BEAM—LOAD INCREASING UNIFORMLY TO CENTER

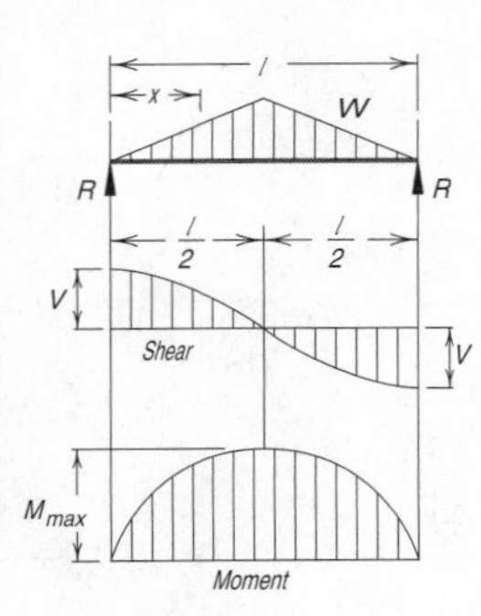

Total Equiv. Uniform Load .... $= \frac{4W}{3}$

$R = V$ .... $= \frac{W}{2}$

V_x (when $x < \frac{l}{2}$) .... $= \frac{W}{2l^2}(l^2 - 4x^2)$

M_{max} (at center) .... $= \frac{Wl}{6}$

M_x (when $x < \frac{l}{2}$) .... $= Wx\left(\frac{1}{2} - \frac{2x^2}{3l^2}\right)$

Δ_{max} (at center) .... $= \frac{Wl^3}{60EI}$

Δ_x (when $x < \frac{l}{2}$) .... $= \frac{Wx}{480EIl^2}(5l^2 - 4x^2)^2$

Table 5-17 (cont.). Shears, Moments, and Deflections

4. SIMPLE BEAM—UNIFORMLY LOAD PARTIALLY DISTRIBUTED

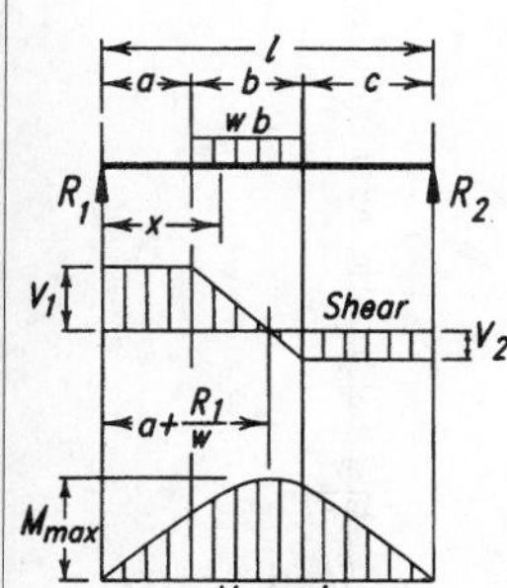

$R_1 = V_1$ (max. when $a < c$) .... $= \frac{wb}{2l}(2c + b)$

$R_2 = V_2$ (max. when $a > c$) .... $= \frac{wb}{2l}(2a + b)$

V_x (When $x > a$ and $< (a + b)$) ...... $= R_1 - w(x - a)$

M_{max} $\left(\text{at } x = a + \frac{R_1}{w}\right)$ .... $= R_1\left(a + \frac{R_1}{2w}\right)$

M_x (when $x < a$) .... $= R_1 x$

M_x (when $x > a$ and $< (a + b)$) ...... $= R_1 x - \frac{w}{2}(x - a)^2$

M_x (when $x > (a + b)$) .... $= R_2(l - x)$

5. SIMPLE BEAM—UNIFORM LOAD PARTIALLY DISTRIBUTED AT ONE END

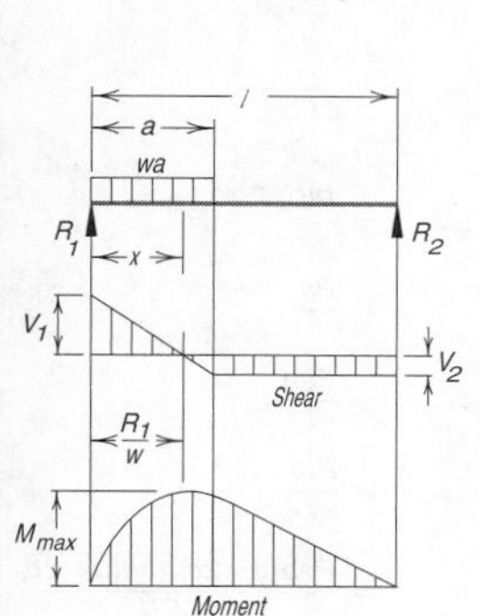

$R_1 = V_1$ max .... $= \frac{wa}{2l}(2l - a)$

$R_2 = V_2$ .... $= \frac{wa^2}{2l}$

V_x (when $x < a$) .... $= R_1 - wx$

M_{max} $\left(\text{at } x = \frac{R_1}{w}\right)$ .... $= \frac{R_1^2}{2w}$

M_x (when $x < a$) .... $= R_1 x - \frac{wx^2}{2}$

M_x (when $x > a$) .... $= R_2(l - x)$

Δ_x (when $x < a$) .... $= \frac{wx}{24EIl}(a^2(2l - a)^2 - 2ax^2(2l - a) + lx^3)$

Δ_x (when $x > a$) .... $= \frac{wa^2(l - x)}{24EIl}(4xl - 2x^2 - a^2)$

6. SIMPLE BEAM—UNIFORM LOAD PARTIALLY DISTRIBUTED AT EACH END

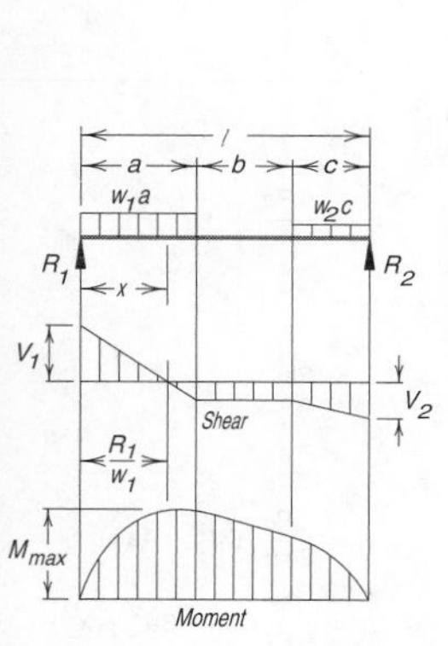

$R_1 = V_1$ .... $= \frac{w_1 a(2l - a) + w_2 c^2}{2l}$

$R_2 = V_2$ .... $= \frac{w_2 c(2l - c) + w_1 a^2}{2l}$

V_x (when $x < a$) .... $= R_1 - w_1 x$

V_x (when $a < x < (a + b)$) .... $= R_1 - w_1 a$

V_x (when $x > (a + b)$) .... $= R_2 - w_2(l - x)$

M_{max} $\left(\text{at } x = \frac{R_1}{w_1} \text{ when } R_1 < w_1 a\right)$ ... $= \frac{R_1^2}{2w_1}$

M_{max} $\left(\text{at } x = l - \frac{R_2}{w_2} \text{ when } R_2 < w_2 c\right)$ $= \frac{R_2^2}{2w_2}$

M_x (when $x < a$) .... $= R_1 x - \frac{w_1 x^2}{2}$

M_x (when $a < x < (a + b)$) .... $= R_1 x - \frac{w_1 a}{2}(2x - a)$

M_x (when $x > (a + b)$) .... $= R_2(l - x) - \frac{w_2(l - x)^2}{2}$

Table 5-17 (cont.). Shears, Moments, and Deflections

7. SIMPLE BEAM—CONCENTRATED LOAD AT CENTER

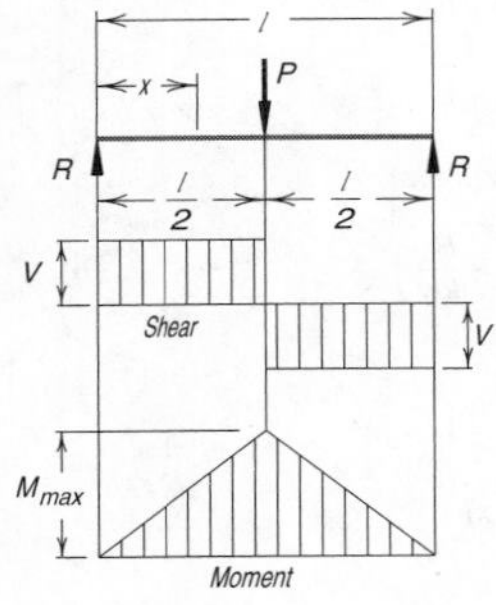

Total Equiv. Uniform Load .... $= 2P$

$R = V$ .... $= \frac{P}{2}$

M_{max} (at point of load) .... $= \frac{Pl}{4}$

M_x $\left(\text{when } x < \frac{l}{2}\right)$ .... $= \frac{Px}{2}$

Δ_{max} (at point of load) .... $= \frac{Pl^3}{48EI}$

Δ_x $\left(\text{when } x < \frac{l}{2}\right)$ .... $= \frac{Px}{48EI}(3l^2 - 4x^2)$

8. SIMPLE BEAM—CONCENTRATED LOAD AT ANY POINT

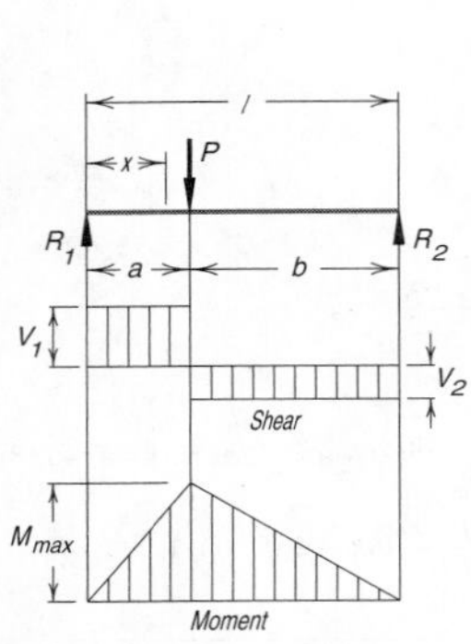

Total Equiv. Uniform Load .... $= \frac{8Pab}{l^2}$

$R_1 = V_1$ ($= V_{max}$ when $a < b$) .... $= \frac{Pb}{l}$

$R_2 = V_2$ ($= V_{max}$ when $a > b$) .... $= \frac{Pa}{l}$

M_{max} (at point of load) .... $= \frac{Pab}{l}$

M_x (when $x < a$) .... $= \frac{Pbx}{l}$

Δ_{max} $\left(\text{at } x = \sqrt{\frac{a(a+2b)}{3}} \text{ when } a > b\right)$ .... $= \frac{Pab(a+2b)\sqrt{3a(a+2b)}}{27EIl}$

Δ_a (at point of load) .... $= \frac{Pa^2b^2}{3EIl}$

Δ_x (when $x < a$) .... $= \frac{Pbx}{6EIl}(l^2 - b^2 - x^2)$

9. SIMPLE BEAM—TWO EQUAL CONCENTRATED LOADS SYMMETRICALLY PLACED

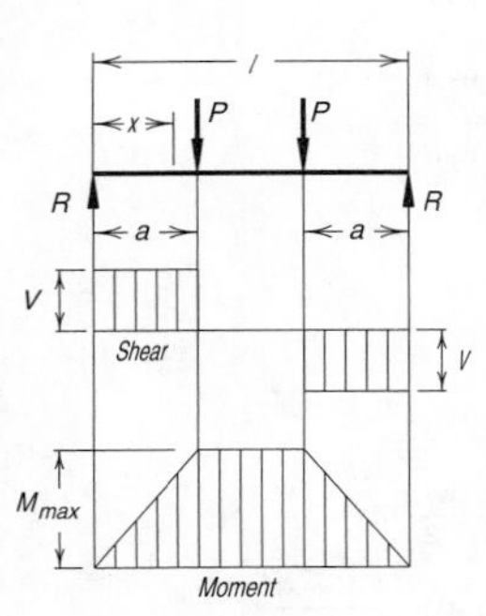

Total Equiv. Uniform Load .... $= \frac{8Pa}{l}$

$R = V$ .... $= P$

M_{max} (between loads) .... $= Pa$

M_x (when $x < a$) .... $= Px$

Δ_{max} (at center) .... $= \frac{Pa}{24EI}(3l^2 - 4a^2)$

Δ_x (when $x < a$) .... $= \frac{Px}{6EI}(3la - 3a^2 - x^2)$

Δ_x (when $a < x < (l - a)$) .... $= \frac{Pa}{6EI}(3lx - 3x^2 - a^2)$

Table 5-17 (cont.). Shears, Moments, and Deflections

10. SIMPLE BEAM—TWO EQUAL CONCENTRATED LOADS UNSYMMETRICALLY PLACED

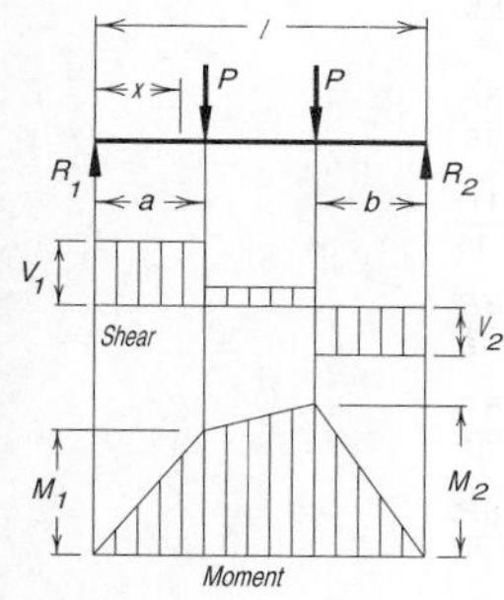

$R_1 = V_1$ (= V_{max} when $a < b$) .... $= \frac{P}{l}(l - a + b)$

$R_2 = V_2$ (= V_{max} when $a > b$) .... $= \frac{P}{l}(l - b + a)$

V_x (when $a < x < (l - b)$) .... $= \frac{P}{l}(b - a)$

M_1 (= M_{max} when $a > b$) .... $= R_1 a$

M_2 (= M_{max} when $a < b$) .... $= R_2 b$

M_x (when $x < a$) .... $= R_1 x$

M_x (when $a < x < (l - b)$) .... $= R_1 x - P(x - a)$

11. SIMPLE BEAM—TWO UNEQUAL CONCENTRATED LOADS UNSYMMETRICALLY PLACED

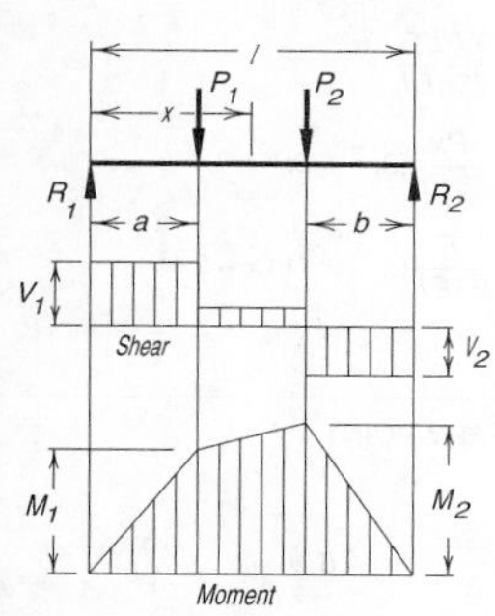

$R_1 = V_1$ .... $= \frac{P_1(l - a) + P_2 b}{l}$

$R_2 = V_2$ .... $= \frac{P_1 a + P_2(l - b)}{l}$

V_x (when $a < x < (l - b)$) .... $= R_1 - P_1$

M_1 (= M_{max} when $R_1 < P_1$) .... $= R_1 a$

M_2 (= M_{max} when $R_2 < P_2$) .... $= R_2 b$

M_x (when $x < a$) .... $= R_1 x$

M_x (when $a < x < (l - b)$) .... $= R_1 x - P(x - a)$

12. BEAM FIXED AT ONE END, SUPPORTED AT OTHER—UNIFORMLY DISTRIBUTED LOAD

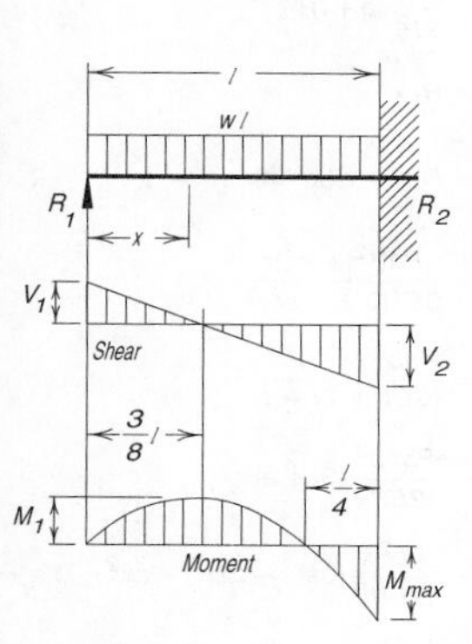

Total Equiv. Uniform Load .... $= wl$

$R_1 = V_1$ .... $= \frac{3wl}{8}$

$R_2 = V_{2\,max}$ .... $= \frac{5wl}{8}$

V_x .... $= R_1 - wx$

M_{max} .... $= \frac{wl^2}{8}$

M_x $\left(\text{at } x = \frac{3}{8}l\right)$ .... $= \frac{9}{128}wl^2$

M_x .... $= R_1 x - \frac{wx^2}{2}$

Δ_{max} $\left(\text{at } x = \frac{l}{16}(1 + \sqrt{33}) = 0.4215l\right)$ ... $= \frac{wl^4}{185EI}$

Δ_x .... $= \frac{wx}{48EI}(l^3 - 3lx^2 + 2x^3)$

Table 5-17 (cont.). Shears, Moments and Deflections

13. BEAM FIXED AT ONE END, SUPPORTED AT OTHER—CONCENTRATED LOAD AT CENTER

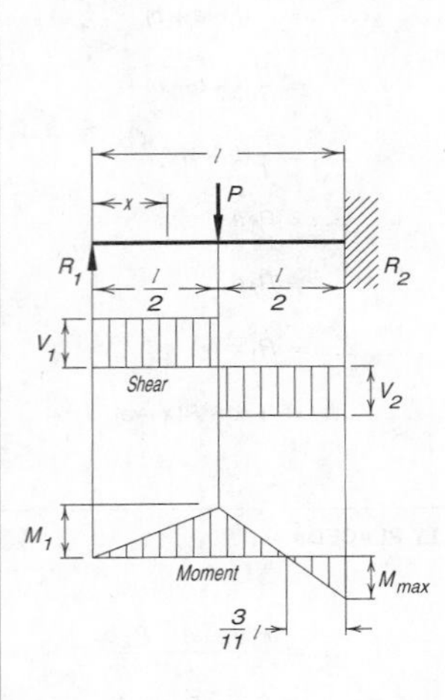

Total Equiv. Uniform Load .... $= \frac{3P}{2}$

$R_1 = V_1$ .... $= \frac{5P}{15}$

$R_2 = V_{2\ max}$ .... $= \frac{11P}{16}$

M_{max} (at fixed end) .... $= \frac{3Pl}{16}$

M_1 (at point of load) .... $= \frac{5Pl}{32}$

$M_x \left(\text{when } x < \frac{l}{2}\right)$ .... $= \frac{5Px}{16}$

$M_x \left(\text{when } x > \frac{l}{2}\right)$ .... $= P\left(\frac{l}{2} - \frac{11x}{16}\right)$

$\Delta_{max} \left(\text{at } x = l\sqrt{\frac{1}{5}} = 0.447l\right)$ .... $= \frac{Pl^3}{48EI\sqrt{5}} = 0.00932\frac{Pl^3}{EI}$

Δ_x (at point of load) .... $= \frac{7PL^3}{768EI}$

$\Delta_x \left(\text{when } x < \frac{l}{2}\right)$ .... $= \frac{Px}{96EI}(3l^2 - 5x^2)$

$\Delta_x \left(\text{when } x > \frac{l}{2}\right)$ .... $= \frac{P}{96EI}(x - l)^2(11x - 2l)$

14. BEAM FIXED AT ONE END, SUPPORTED AT OTHER—CONCENTRATED LOAD AT ANY POINT

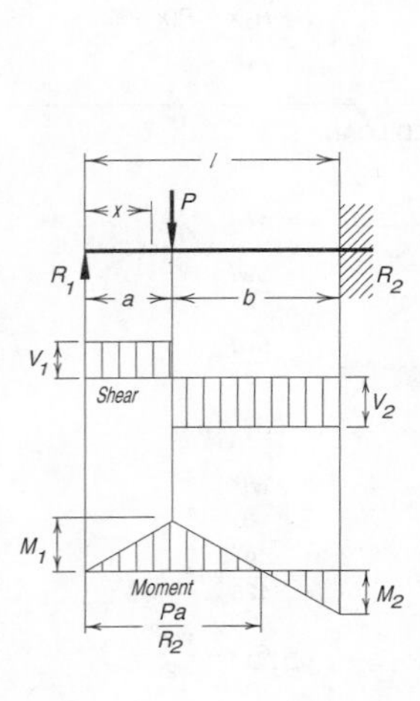

$R_1 = V_1$ .... $= \frac{Pb^2}{2l^3}(a + 2l)$

$R_2 = V_2$ .... $= \frac{Pa}{2l^3}(3l^2 - a^2)$

M (at point of load) .... $= R_1 a$

M_2 (at fixed end) .... $= \frac{Pab}{2l^2}(a + l)$

M_x (when $x < a$) .... $= R_1 x$

M_x (when $x > a$) .... $= R_1 x - P(x - a)$

$\Delta_{max} \left(\text{when } a < 0.414l \text{ at } x = l\frac{(l^2 + a^2)}{(3l^2 - a^2)}\right) = \frac{Pa(l^2 + a^2)^3}{3EI(3l^2 - a^2)^2}$

$\Delta_{max} \left(\text{when } a > 0.414l \text{ at } x = l\sqrt{\frac{a}{2l + a}}\right)$ .. $= \frac{Pab^2}{6EI}\sqrt{\frac{a}{2l + a}}$

Δ_a (at point of load) .... $= \frac{Pb^2b^3}{12EIl^3}(3l + a)$

Δ (when $x < a$) .... $= \frac{Pb^2x}{12EIl^3}(3al^2 - 2lx^2 - ax^2)$

Δ_x (when $x > a$) .... $= \frac{Pa}{12EIl^2}(l - x)^2(3l^2x - a^2x - 2a^2l)$

Table 5-17 (cont.). Shears, Moments and Deflections

15. BEAM FIXED AT BOTH ENDS—UNIFORMLY DISTRIBUTED LOADS

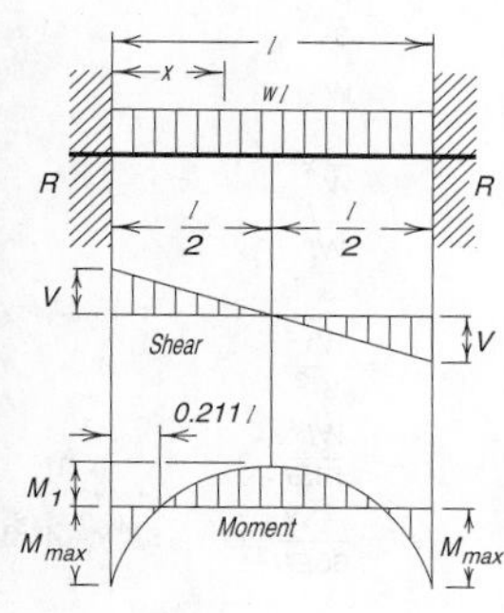

Total Equiv. Uniform Load .... $= \frac{2wl}{3}$

$R = V$ .... $= \frac{wl}{2}$

V_x .... $= w\left(\frac{l}{2} - x\right)$

M_{max} (at ends) .... $= \frac{wl^2}{12}$

M_1 (at center) .... $= \frac{wl^2}{24}$

M_x .... $= \frac{w}{12}(6lx - l^2 - 6x^2)$

Δ_{max} (at center) .... $= \frac{wl^4}{384EI}$

Δ_x .... $= \frac{wx^2}{24EI}(l - x)^2$

16. BEAM FIXED AT BOTH ENDS—CONCENTRATED LOAD AT CENTER

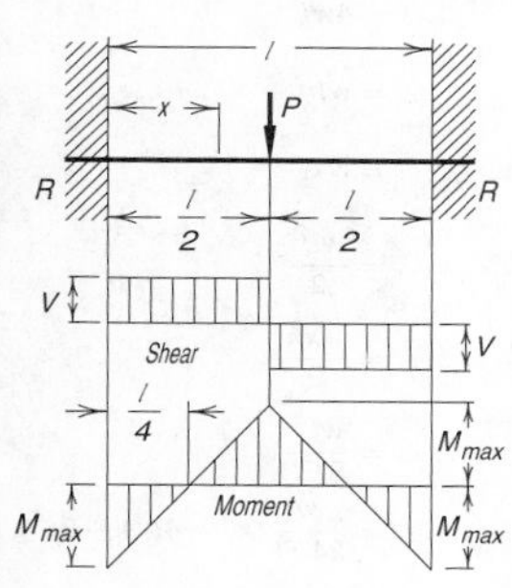

Total Equiv. Uniform Load .... $= P$

$R = V$ .... $= \frac{P}{2}$

M_{max} (at center and ends) .... $= \frac{Pl}{8}$

M_x $\left(\text{when } x < \frac{l}{2}\right)$ .... $= \frac{P}{8}(4x - l)$

Δ_{max} (at center) .... $= \frac{Pl^3}{192EI}$

Δ_x $\left(\text{when } x < \frac{l}{2}\right)$ .... $= \frac{Px^2}{48EI}(3l - 4x)$

17. BEAM FIXED AT BOTH ENDS—CONCENTRATED LOAD AT ANY POINT

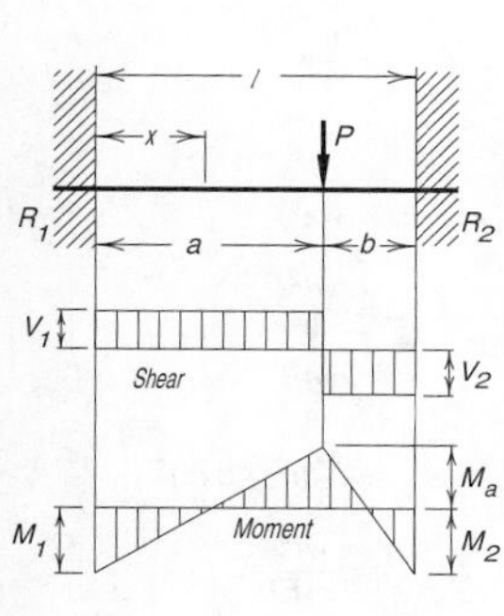

$R_1 = V_1 (= V_{max}$ when $a < b)$ .... $= \frac{Pb^2}{l^3}(3a + b)$

$R_2 = V_2 (= V_{max}$ when $a > b)$ .... $= \frac{Pa^2}{l^3}(a + 3b)$

M_1 $(= M_{max}$ when $a < b)$ .... $= \frac{Pab^2}{l^2}$

M_2 $(= M_{max}$ when $a > b)$ .... $= \frac{Pa^2b}{l^2}$

M_a (at point of load) .... $= \frac{2Pa^2b^2}{l^3}$

M_x (when $x < a$) .... $= R_1x - \frac{Pab^2}{l^2}$

Δ_{max} $\left(\text{when } a > b \text{ at } x = \frac{2al}{3a + b}\right)$ .... $= \frac{2Pa^3b^2}{3EI(3a + b)^2}$

Δ_a (at point of load) .... $= \frac{Pa^3b^3}{3EIl^3}$

Δ_x (when $x < a$) .... $= \frac{Pb^2x^2}{6EIl^3}(3al - 3ax - bx)$

Table 5-17 (cont.). Shears, Moments and Deflections

18. CANTILEVERED BEAM—LOAD INCREASING UNIFORMLY TO FIXED END

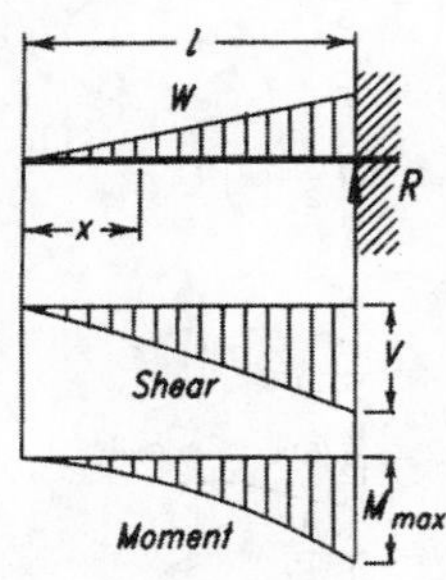

Total Equiv. Uniform Load .... $= \frac{8}{3}W$

$R = V$ .... $= W$

V_x .... $= W\frac{x^2}{l^2}$

M_{max} (at fixed end) .... $= \frac{Wl}{3}$

M_x .... $= \frac{Wx^3}{3l^2}$

Δ_{max} (at free end) .... $= \frac{Wl^3}{15EI}$

Δ_x .... $= \frac{W}{60EIl^2}(x^5 - 5l^4x + 4l^5)$

19. CANTILEVERED BEAM—UNIFORMLY DISTRIBUTED LOAD

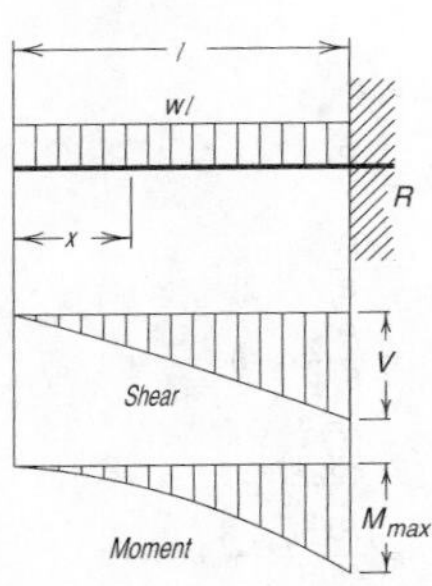

Total Equiv. Uniform Load .... $= 4wl$

$R = V$ .... $= wl$

V_x .... $= wx$

M_{max} (at fixed end) .... $= \frac{wl^2}{2}$

M_x .... $= \frac{wx^2}{2}$

Δ_{max} (at free end) .... $= \frac{wl^4}{8EI}$

Δ_x .... $= \frac{w}{24EI}(x^4 - 4l^3x + 3l^4)$

20. BEAM FIXED AT ONE END, FREE TO DEFLECT VERTICALLY BUT NOT ROTATE AT OTHER—UNIFORMLY DISTRIBUTED LOAD

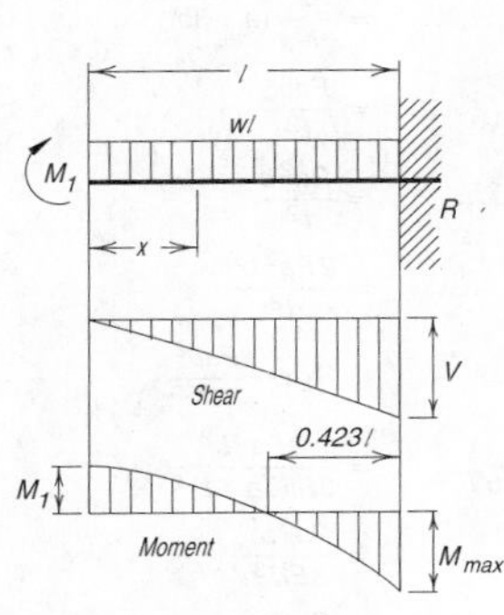

Total Equiv. Uniform Load .... $= \frac{8}{3}wl$

$R = V$ .... $= wl$

V_x .... $= wx$

M_1 (at deflected end) .... $= \frac{wl^2}{6}$

M_{max} (at fixed end) .... $= \frac{wl^2}{3}$

M_x .... $= \frac{w}{6}(l^2 - 3x^2)$

Δ_{max} (at deflected end) .... $= \frac{wl^4}{24EI}$

Δ_x .... $= \frac{w(l^2 - x^2)^2}{24EI}$

Table 5-17 (cont.). Shears, Moments and Deflections

21. CANTILEVERED BEAM—CONCENTRATED LOAD AT ANY POINT

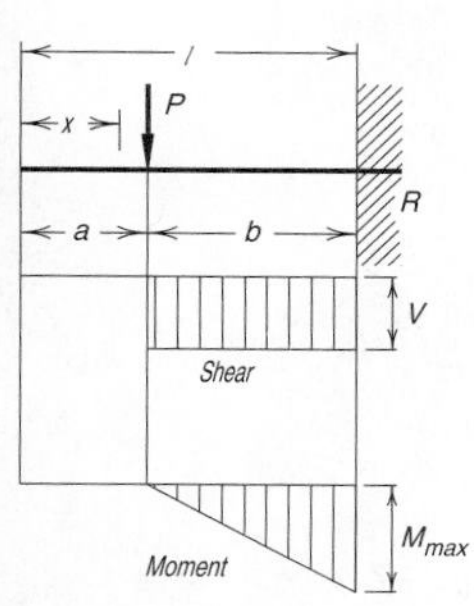

Total Equiv. Uniform Load .... $= \frac{8Pb}{l}$

$R = V$ .... $= P$

M_{max} (at fixed end) .... $= Pb$

M_x (when $x > a$) .... $= P(x - a)$

Δ_{max} (at free end) .... $= \frac{Pb^2}{6EI}(3l - b)$

Δ_a (at point of load) .... $= \frac{Pb^3}{3EI}$

Δ_x (when $x < a$) .... $= \frac{Pb^2}{6EI}(3l - 3x - b)$

Δ_x (when $x > a$) .... $= \frac{P(l - x)^2}{6EI}(3b - l + x)$

22. CANTILEVERED BEAM—CONCENTRATED LOAD AT FREE END

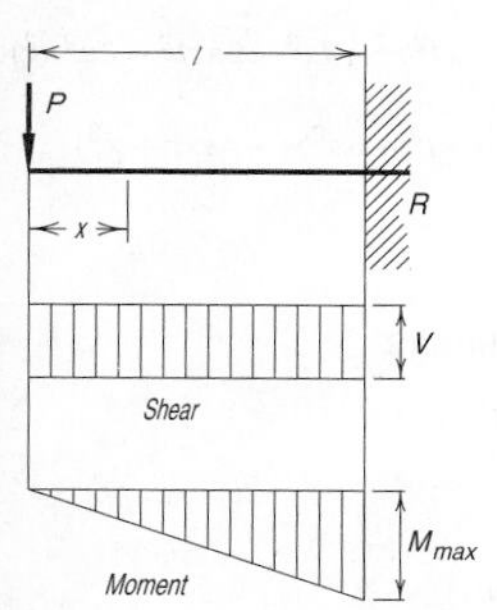

Total Equiv. Uniform Load .... $= 8P$

$R = V$ .... $= P$

M_{max} (at fixed end) .... $= Pl$

M_x .... $= Px$

Δ_{max} (at free end) .... $= \frac{Pl^3}{3EI}$

Δ_x .... $= \frac{P}{6EI}(2l^3 - 3l^2x + x^3)$

23. BEAM FIXED AT ONE END, FREE TO DEFLECT VERTICALLY BUT NOT ROTATE AT OTHER—CONCENTRATED LOAD AT DEFLECTED END

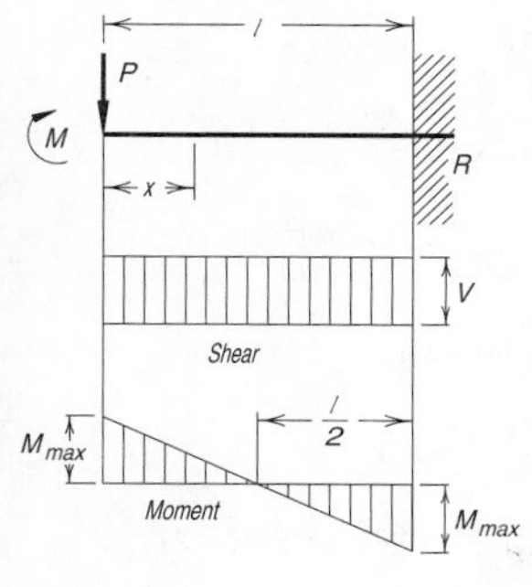

Total Equiv. Uniform Load .... $= 4P$

$R = V$ .... $= P$

M_{max} (at both ends) .... $= \frac{Pl}{2}$

M_x .... $= P\left(\frac{l}{2} - x\right)$

Δ_{max} (at deflected end) .... $= \frac{Pl^3}{12EI}$

Δ_x .... $= \frac{P(l - x)^2}{12EI}(l + 2x)$

Table 5-17 (cont.). Shears, Moments and Deflections

24. BEAM OVERHANGING ONE SUPPORT—UNIFORMLY DISTRIBUTED LOAD

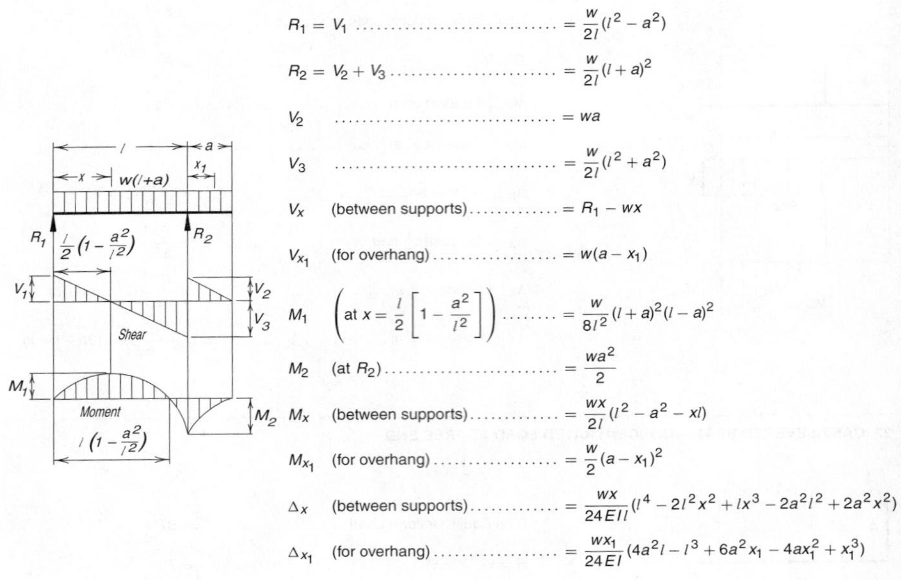

$R_1 = V_1$ .... $= \frac{w}{2l}(l^2 - a^2)$

$R_2 = V_2 + V_3$ .... $= \frac{w}{2l}(l + a)^2$

V_2 .... $= wa$

V_3 .... $= \frac{w}{2l}(l^2 + a^2)$

V_x (between supports) .... $= R_1 - wx$

V_{x_1} (for overhang) .... $= w(a - x_1)$

M_1 $\left(\text{at } x = \frac{l}{2}\left[1 - \frac{a^2}{l^2}\right]\right)$ .... $= \frac{w}{8l^2}(l + a)^2(l - a)^2$

M_2 (at R_2) .... $= \frac{wa^2}{2}$

M_x (between supports) .... $= \frac{wx}{2l}(l^2 - a^2 - xl)$

M_{x_1} (for overhang) .... $= \frac{w}{2}(a - x_1)^2$

Δ_x (between supports) .... $= \frac{wx}{24EIl}(l^4 - 2l^2x^2 + lx^3 - 2a^2l^2 + 2a^2x^2)$

Δ_{x_1} (for overhang) .... $= \frac{wx_1}{24EI}(4a^2l - l^3 + 6a^2x_1 - 4ax_1^2 + x_1^3)$

25. BEAM OVERHANGING ONE SUPPORT—UNIFORMLY DISTRIBUTED LOAD ON OVERHANG

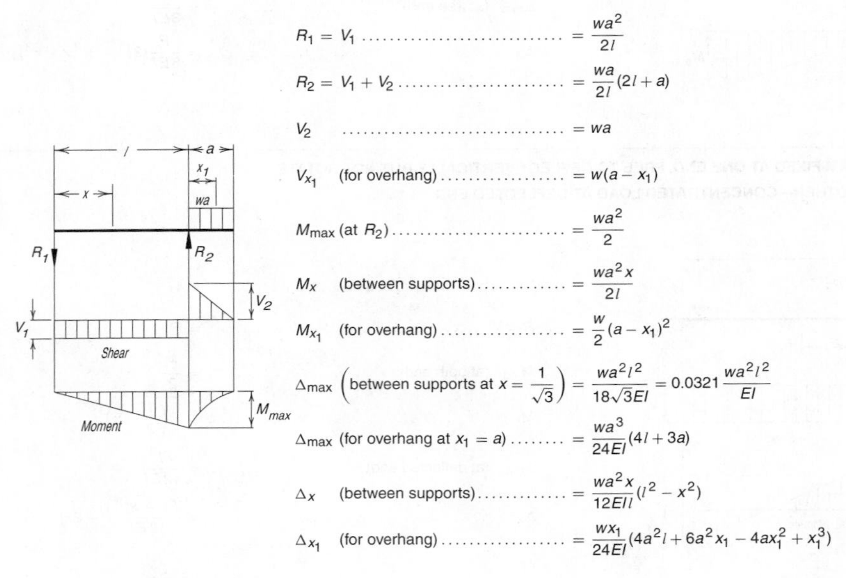

$R_1 = V_1$ .... $= \frac{wa^2}{2l}$

$R_2 = V_1 + V_2$ .... $= \frac{wa}{2l}(2l + a)$

V_2 .... $= wa$

V_{x_1} (for overhang) .... $= w(a - x_1)$

M_{max} (at R_2) .... $= \frac{wa^2}{2}$

M_x (between supports) .... $= \frac{wa^2x}{2l}$

M_{x_1} (for overhang) .... $= \frac{w}{2}(a - x_1)^2$

Δ_{max} $\left(\text{between supports at } x = \frac{1}{\sqrt{3}}\right)$ $= \frac{wa^2l^2}{18\sqrt{3}EI} = 0.0321\frac{wa^2l^2}{EI}$

Δ_{max} (for overhang at $x_1 = a$) .... $= \frac{wa^3}{24EI}(4l + 3a)$

Δ_x (between supports) .... $= \frac{wa^2x}{12EIl}(l^2 - x^2)$

Δ_{x_1} (for overhang) .... $= \frac{wx_1}{24EI}(4a^2l + 6a^2x_1 - 4ax_1^2 + x_1^3)$

Table 5-17 (cont.). Shears, Moments and Deflections

26. BEAM OVERHANGING ONE SUPPORT—CONCENTRATED LOAD AT END OF OVERHANG

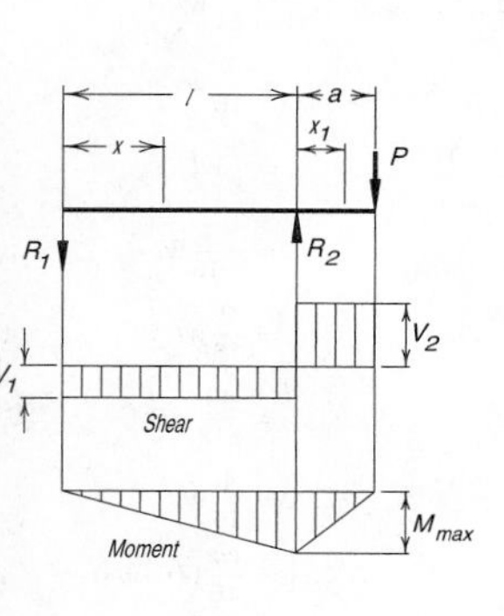

$R_1 = V_1$ .... $= \dfrac{Pa}{l}$

$R_2 = V_1 + V_2$ .... $= \dfrac{P}{l}(l + a)$

V_2 .... $= P$

M_{max} (at R_2) .... $= Pa$

M_x (between supports) .... $= \dfrac{Pax}{l}$

M_{x_1} (for overhang) .... $= P(a - x_1)$

Δ_{max} $\left(\text{between supports at } x = \dfrac{l}{\sqrt{3}}\right)$ .... $= \dfrac{Pal^2}{9\sqrt{3}EI} = 0.0642\dfrac{Pal^2}{EI}$

Δ_{max} (for overhang at $x_1 = a$) .... $= \dfrac{Pa^2}{3EI}(l + a)$

Δ_x (between supports) .... $= \dfrac{Pax}{6EIl}(l^2 - x^2)$

Δ_{x_1} (for overhang) .... $= \dfrac{Px_1}{6EI}(2al + 3ax_1 - x_1^2)$

27. BEAM OVERHANGING ONE SUPPORT—UNIFORMLY DISTRIBUTED LOAD BETWEEN SUPPORTS

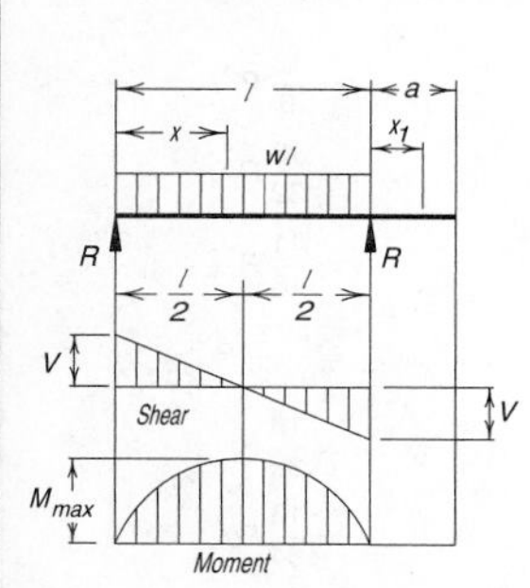

Total Equiv. Uniform Load .... $= wl$

$R = V$ .... $= \dfrac{wl}{2}$

V_x .... $= w\left(\dfrac{l}{2} - x\right)$

M_{max} (at center) .... $= \dfrac{wl^2}{8}$

M_x .... $= \dfrac{wx}{2}(l - x)$

Δ_{max} (at center) .... $= \dfrac{5wl^4}{384EI}$

Δ_x .... $= \dfrac{wx}{24EI}(l^3 - 2lx^2 + x^3)$

Δ_{x_1} .... $= \dfrac{wl^3x_1}{24EI}$

28. BEAM OVERHANGING ONE SUPPORT—CONCENTRATED LOAD AT ANY POINT BETWEEN SUPPORTS

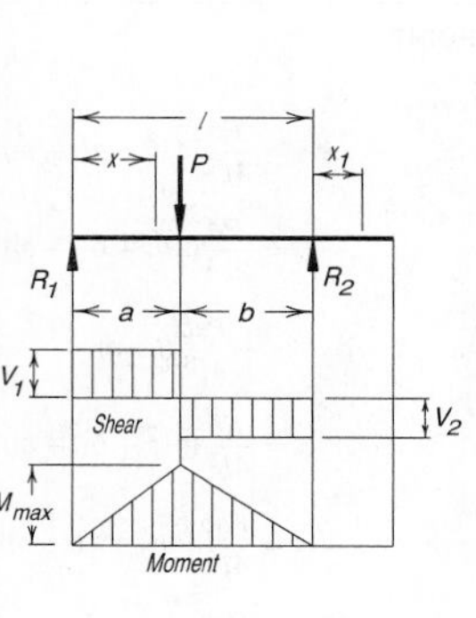

Total Equiv. Uniform Load .... $= \dfrac{8Pab}{l^2}$

$R_1 = V_1$ (= V_{max} when $a < b$) .... $= \dfrac{Pb}{l}$

$R_2 = V_2$ (= V_{max} when $a > b$) .... $= \dfrac{Pa}{l}$

M_{max} (at point of load) .... $= \dfrac{Pab}{l}$

M_x (when $x < a$) .... $= \dfrac{Pbx}{l}$

Δ_{max} $\left(\text{at } x = \sqrt{\dfrac{a(a+2b)}{3}} \text{ when } a > b\right)$ .... $= \dfrac{Pab(a+2b)\sqrt{3a(a+2b)}}{27EIl}$

Δ_a (at point of load) .... $= \dfrac{Pa^2b^2}{3EIl}$

Δ_x (when $x < a$) .... $= \dfrac{Pbx}{6EIl}(l^2 - b^2 - x^2)$

Δ_x (when $x > a$) .... $= \dfrac{Pa(l-x)}{6EIl}(2lx - x^2 - a^2)$

Δ_{x_1} .... $= \dfrac{Pabx_1}{6EIl}(l + a)$

Table 5-17 (cont.). Shears, Moments and Deflections

29. CONTINUOUS BEAM—TWO EQUAL SPANS—UNIFORM LOAD ON ONE SPAN

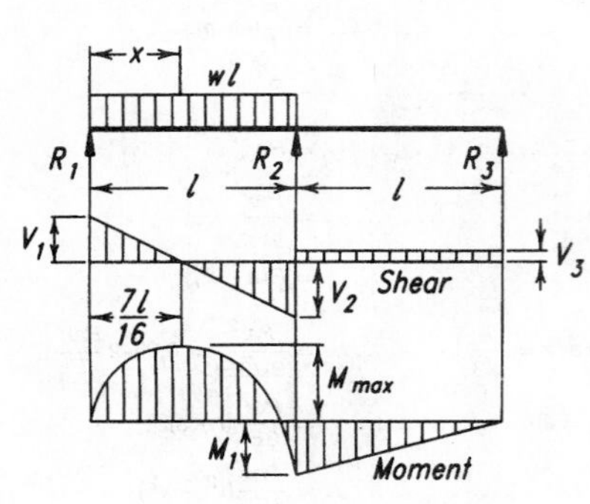

Total Equiv. Uniform Load .... $= \frac{49}{64} wl$

$R_1 = V_1$ .... $= \frac{7}{16} wl$

$R_2 = V_2 + V_3$ .... $= \frac{5}{8} wl$

$R_3 = V_3$ .... $= -\frac{1}{16} wl$

V_2 .... $= \frac{9}{16} wl$

$M_{max} \left(\text{at } x = \frac{7}{16} l\right)$ .... $= \frac{49}{512} wl^2$

M_1 (at support R_2) .... $= \frac{1}{16} wl^2$

M_x (when $x < l$) .... $= \frac{wx}{16}(7l - 8x)$

Δ_{max} (at 0.472 l from R_1) .... $= 0.0092 wl^4/EI$

30. CONTINUOUS BEAM—TWO EQUAL SPANS—CONCENTRATED LOAD AT CENTER OF ONE SPAN

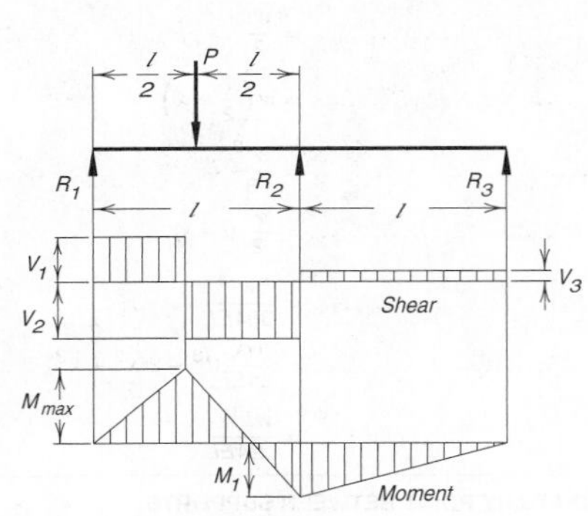

Total Equiv. Uniform Load .... $= \frac{13}{8} P$

$R_1 = V_1$ .... $= \frac{13}{32} P$

$R_2 = V_2 + V_3$ .... $= \frac{11}{16} P$

$R_3 = V_3$ .... $= -\frac{3}{32} P$

V_2 .... $= \frac{19}{32} P$

M_{max} (at point of load) .... $= \frac{13}{64} Pl$

M_1 (at support R_2) .... $= \frac{3}{32} Pl$

Δ_{max} (at 0.480 l from R_1) .... $= 0.015 Pl^3/EI$

31. CONTINUOUS BEAM—TWO EQUAL SPANS—CONCENTRATED LOAD AT ANY POINT

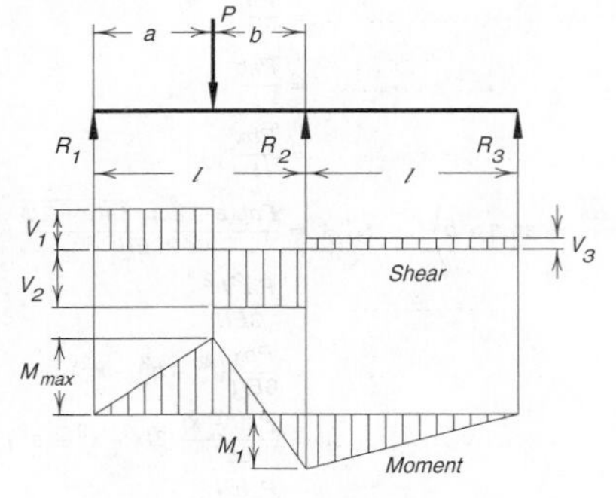

$R_1 = V_1$ .... $= \frac{Pb}{4l^3}(4l^2 - a(l + a))$

$R_2 = V_2 + V_3$ .... $= \frac{Pa}{2l^3}(2l^2 + b(l + a))$

$R_3 = V_3$ .... $= -\frac{Pab}{4l^3}(l + a)$

V_2 .... $= \frac{Pa}{4l^3}(4l^2 + b(l + a))$

M_{max} (at point of load) .... $= \frac{Pab}{4l^3}(4l^2 - a(l + a))$

M_1 (at support R_2) .... $= \frac{Pab}{4l^2}(l + a)$

Table 5-17 (cont.).
Shears, Moments and Deflections

32. BEAM—UNIFORMLY DISTRIBUTED LOAD AND VARIABLE END MOMENTS

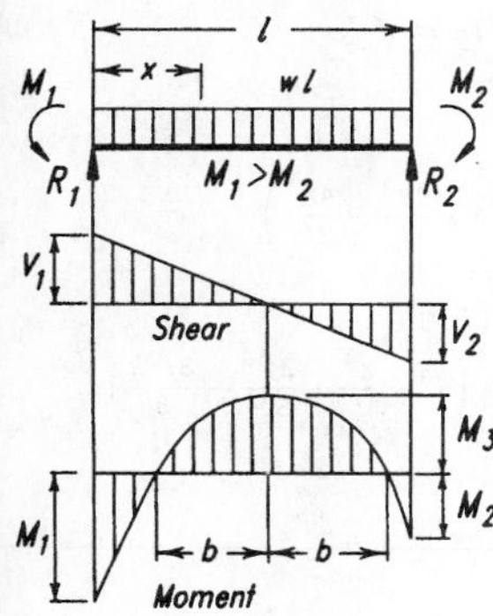

$$R_1 = V_1 \text{ .... } = \frac{wl}{2} + \frac{M_1 - M_2}{l}$$

$$R_2 = V_2 \text{ .... } = \frac{wl}{2} - \frac{M_1 - M_2}{l}$$

$$V_x \text{ .... } = w\left(\frac{l}{2} - x\right) + \frac{M_1 - M_2}{l}$$

$$M_3 \left(\text{at } x = \frac{l}{2} + \frac{M_1 - M_2}{wl}\right) = \frac{wl^2}{8} - \frac{M_1 + M_2}{2} + \frac{(M_1 - M_2)^2}{2wl^2}$$

$$M_x \text{ .... } = \frac{wx}{2}(l - x) + \left(\frac{M_1 - M_2}{l}\right)x - M_1$$

$$b \text{(to locate inflection points)} = \sqrt{\frac{l^2}{4} - \left(\frac{M_1 + M_2}{w}\right) + \left(\frac{M_1 - M_2}{wl}\right)^2}$$

$$\Delta_x = \frac{wx}{24EI}\left[x^3 - \left(2l + \frac{4M_1}{wl} - \frac{4M_2}{wl}\right)x^2 + \frac{12M_1}{w}x + l^3 - \frac{8M_1 l}{w} - \frac{4M_2 l}{w}\right]$$

33. BEAM—CONCENTRATED LOAD AT CENTER AND VARIABLE END MOMENTS

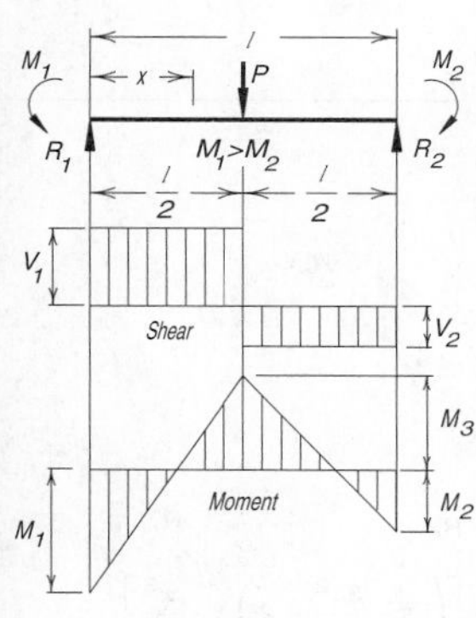

$$R_1 = V_1 \text{ .... } = \frac{P}{2} + \frac{M_1 - M_2}{l}$$

$$R_2 = V_2 \text{ .... } = \frac{P}{2} - \frac{M_1 - M_2}{l}$$

$$M_3 \text{(at center)} \text{ .... } = \frac{Pl}{4} - \frac{M_1 + M_2}{2}$$

$$M_x \left(\text{when } x < \frac{l}{2}\right) \text{ .... } = \left(\frac{P}{2} + \frac{M_1 - M_2}{l}\right)x - M_1$$

$$M_x \left(\text{when } x > \frac{l}{2}\right) \text{ .... } = \frac{P}{2}(l - x) + \frac{(M_1 - M_2)x}{l} - M_1$$

$$\Delta_x \left(\text{when } x < \frac{l}{2}\right) = \frac{Px}{48EI}\left(3l^2 - 4x^2 - \frac{8(l - x)}{Pl}[M_1(2l - x) + M_2(l + x)]\right)$$

Table 5-17 (cont.). Shears, Moments and Deflections

34. SIMPLE BEAM—LOAD INCREASING UNIFORMLY FROM CENTER

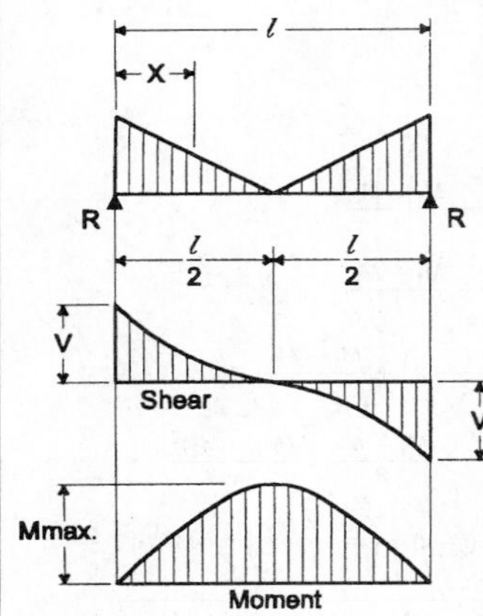

Total Equiv. Uniform Load .... $= \frac{2W}{3}$

$R = V$ .... $= \frac{W}{2}$

V_x $\left(\text{when } x < \frac{l}{2}\right)$ .... $= \frac{W}{2}\left(\frac{l-2x}{l}\right)^2$

M_{max} (at center) .... $= \frac{Wl}{12}$

M_x $\left(\text{when } x < \frac{l}{2}\right)$ .... $= \frac{W}{2}\left(x - \frac{2x^2}{l} + \frac{4x^3}{3l^2}\right)$

Δ_{max} (at center) .... $= \frac{3\,Wl^3}{320\,EI}$

Δ_x $\left(\text{when } x < \frac{l}{2}\right)$ .... $= \frac{W}{12\,EI}\left(x^3 - \frac{x^4}{l} + \frac{2\,x^6}{5\,l^2} - \frac{3\,l^2 x}{8}\right)$

35. SIMPLE BEAM—CONCENTRATED MOMENT AT END

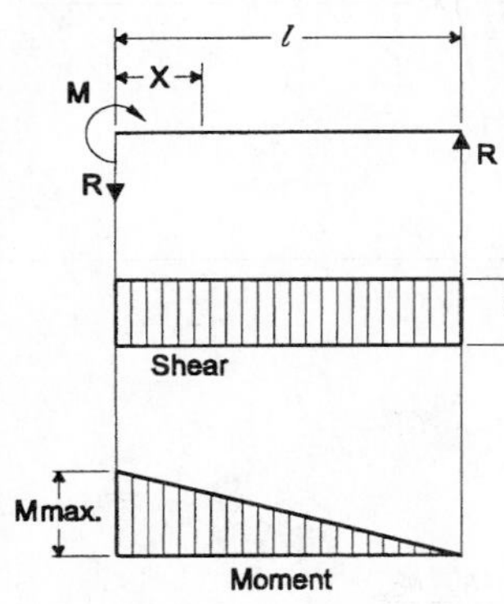

Total Equiv. Uniform Load .... $= \frac{8M}{l}$

$R = V$ .... $= \frac{M}{l}$

M_{max} .... $= M$

M_x .... $= M\left(1 - \frac{x}{l}\right)$

Δ_{max} (at $x = 0.442l$) .... $= 0.0042\,\frac{Ml^2}{EI}$

Δ_x .... $= \frac{M}{6\,EI}\left(3x^2 - \frac{x^3}{l} - 2\,lx\right)$

36. SIMPLE BEAM—CONCENTRATED MOMENT AT ANY POINT

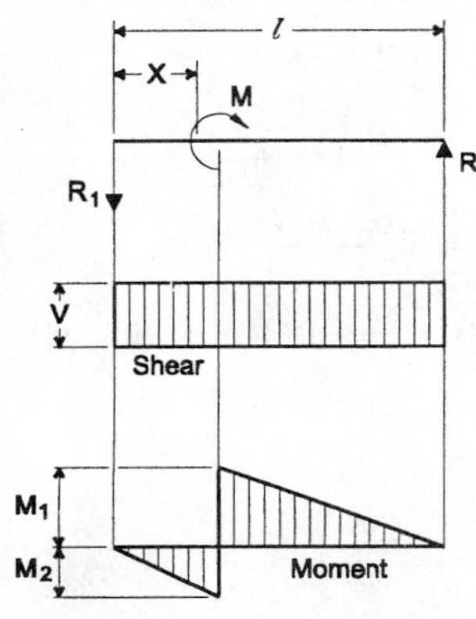

Total Equiv. Uniform Load .... $= \frac{8M}{l}$

$R = V$ .... $= \frac{M}{l}$

M_x (when $x < a$)............. $= Rx$

M_x (when $x > a$)............. $= M + Rx$

Δ_x (when $x < a$)............. $= \frac{M}{6\,EI}\left[\left(6a - \frac{3a^2}{l} - 2l\right)x - \frac{x^3}{l}\right]$

Δ_x .... $= \frac{M}{6\,EI}\left[3(a^2 + x^2) - \frac{x^3}{l} - \left(2l + \frac{3a^2}{l}\right)x\right]$

Table 5-17 (cont.). Shears, Moments and Deflections

37. CONTINUOUS BEAM—THREE EQUAL SPANS—ONE END SPAN UNLOADED

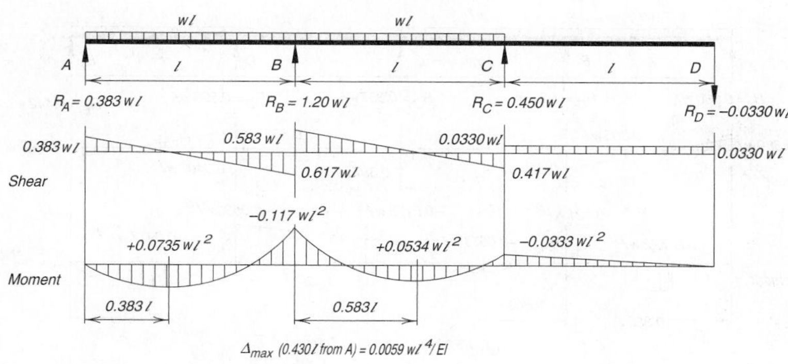

38. CONTINUOUS BEAM—THREE EQUAL SPANS—END SPANS LOADED

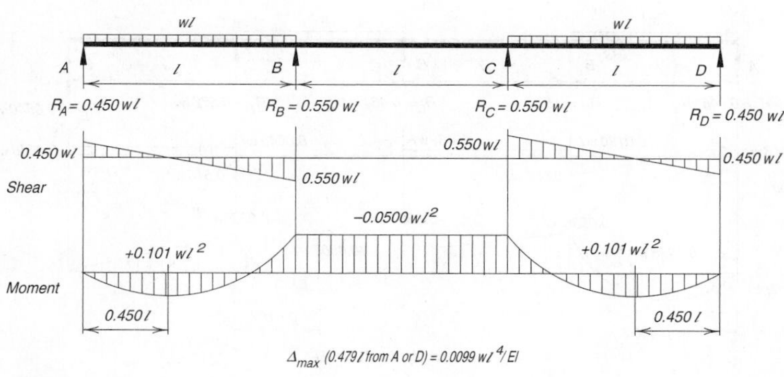

39. CONTINUOUS BEAM—THREE EQUAL SPANS—ALL SPANS LOADED

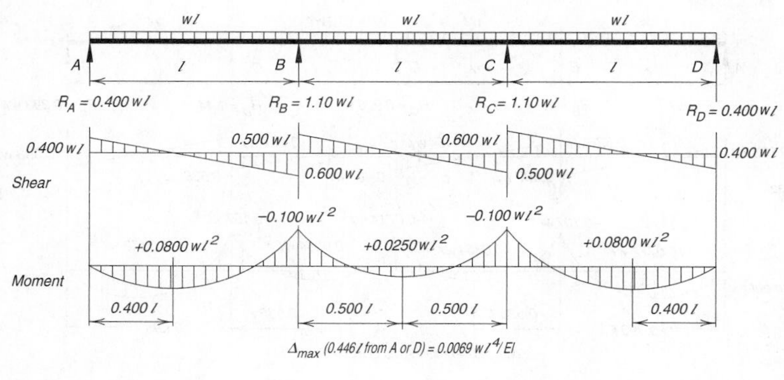

Table 5-17 (cont.). Shears, Moments and Deflections

40. CONTINUOUS BEAM—FOUR EQUAL SPANS—THIRD SPAN UNLOADED

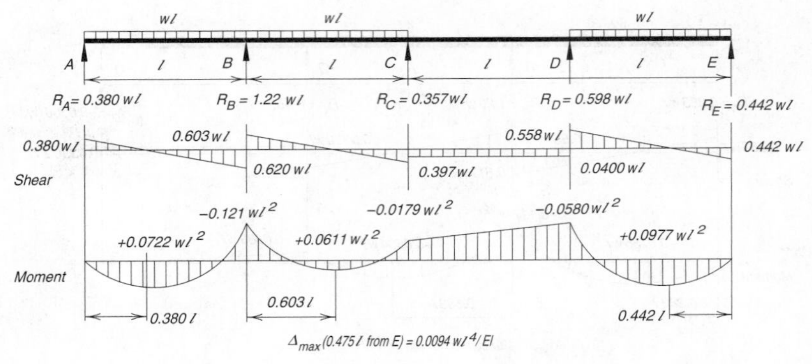

Δ_{max} (0.475 l from E) = 0.0094 wl^4/EI

41. CONTINUOUS BEAM—FOUR EQUAL SPANS—LOAD FIRST AND THIRD SPANS

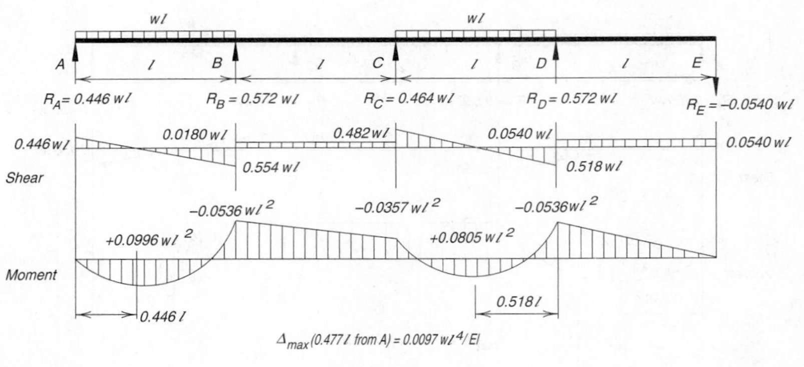

Δ_{max} (0.477 l from A) = 0.0097 wl^4/EI

42. CONTINUOUS BEAM—FOUR EQUAL SPANS—ALL SPANS LOADED

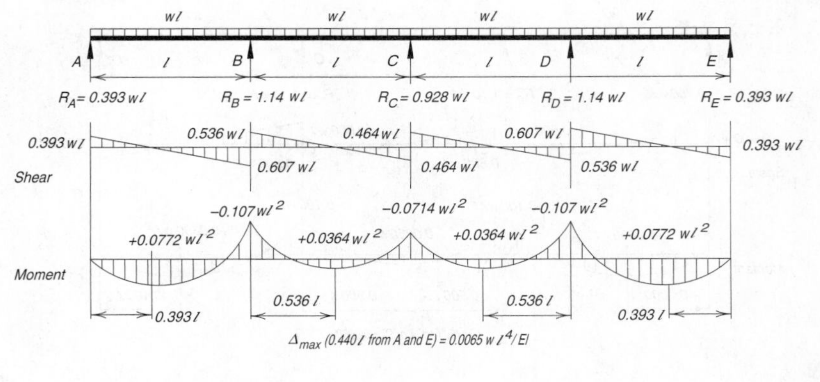

Δ_{max} (0.440 l from A and E) = 0.0065 wl^4/EI

Table 5-17 (cont.).
Shears, Moments and Deflections

43. SIMPLE BEAM—ONE CONCENTRATED MOVING LOAD

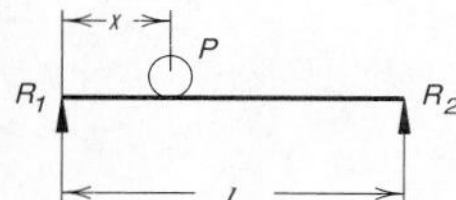

$$R_{1\,max} = V_{1\,max}(\text{at } x = 0) \ldots\ldots = P$$

$$M_{max}\left(\text{at point of load, when } x = \frac{l}{2}\right) \ldots\ldots = \frac{Pl}{4}$$

44. SIMPLE BEAM—TWO EQUAL CONCENTRATED MOVING LOADS

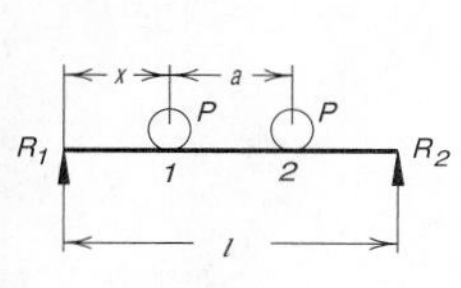

$$R_{1\,max} = V_{1\,max}(\text{at } x = 0) \ldots\ldots = P\left(2 - \frac{a}{l}\right)$$

$$M_{max}\begin{cases}\left[\begin{array}{l}\text{when } a < (2-\sqrt{2})l = 0.586l \\ \text{under load 1 at } x = \frac{1}{2}\left(l - \frac{a}{2}\right)\end{array}\right] \ldots\ldots = \frac{P}{2l}\left(l - \frac{a}{2}\right)^2 \\ \left[\begin{array}{l}\text{when } a > (2-\sqrt{2})l = 0.586l \\ \text{with one load at center of span (Case 43)}\end{array}\right] \ldots\ldots = \frac{Pl}{4}\end{cases}$$

45. SIMPLE BEAM—TWO UNEQUAL CONCENTRATED MOVING LOADS

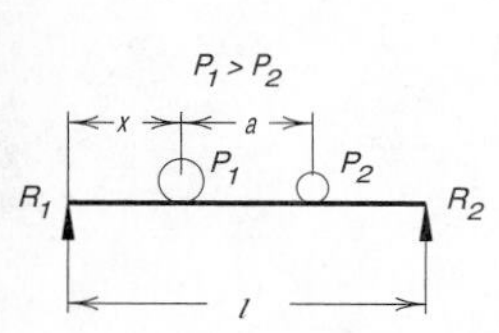

$$R_{1\,max} = V_{1\,max}(\text{at } x = 0) \ldots\ldots = P_1 + P_2\frac{l-a}{l}$$

$$M_{max}\begin{cases}\left[\text{under } P_1\text{, at } x = \frac{1}{2}\left(l - \frac{P_2 a}{P_1 + P_2}\right)\right] \ldots\ldots = (P_1 + P_2)\frac{x^2}{l} \\ \left[\begin{array}{l}M_{max} \text{ may occur with larger} \\ \text{load at center of span and other} \\ \text{load off span (Case 43)}\end{array}\right] \ldots\ldots = \frac{P_1 l}{4}\end{cases}$$

GENERAL RULES FOR SIMPLE BEAMS CARRYING MOVING CONCENTRATED LOADS

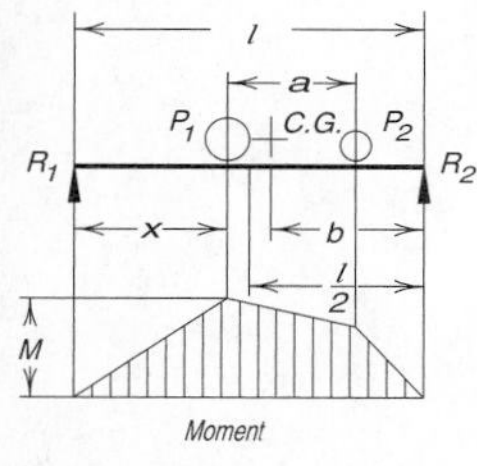

The maximum shear due to moving concentrated loads occurs at one support when one of the loads is at that support. With several moving loads, the location that will produce maximum shear must be determined by trial.

The maximum bending moment produced by moving concentrated loads occurs under one of the loads when that load is as far from one support as the center of gravity of all the moving loads on the beam is from the other support.

In the accompanying diagram, the maximum bending moment occurs under load P_1 when $x = b$. It should also be noted that this condition occurs when the center-line of the span is midway between the center of gravity of loads and the nearest concentrated load.

PART 6

DESIGN OF MEMBERS SUBJECT TO COMBINED LOADING

SCOPE

The specification requirements and other design considerations summarized in this Part apply to the design of members subject to combined loading. For the design of members subject to axial tension only, see Part 3. For the design of members subject to axial compression only, see Part 4. For the design of members subject to uniaxial flexure only, see Part 5. For members that are part of a seismic-force-resisting system in which the seismic response modification factor R is taken greater than 3, see the Seismic Provisions, which are available from AISC at www.aisc.org.

LOAD DETERMINATION

The controlling combination(s) of required strengths (factored axial force) P_u, (factored shear force) V_u, [factored bending moment(s)] M_u and/or (factored torsion) T_u are determined by analysis as indicated in LRFD Specification Section A5. When compressive axial forces occur in combination with moment(s), second-order effects must be considered per LRFD Specification Section C1. Additionally, the destabilizing effects of leaning columns must be considered per LRFD Specification Section C2.2.

COMPACT, NON-COMPACT AND SLENDER-ELEMENT CROSS-SECTIONS

Based upon the types of load transmitted by the member, the discussions of width-thickness ratios (Part 4 for compression members and Part 5 for flexural members) apply to the design of members subject to combined loading, with the exception of the web width-thickness criteria as follows. The web of a member subject to combined axial compression and flexure has different width-thickness limitations in LRFD Specification Table B5.1. As illustrated in Figure 6-1, the limiting values transition from that for a flexural member at $P_u/\phi_b P_y = 0$ to that for a compression member at $P_u/\phi_b P_y = 1$.

MEMBERS SUBJECT TO COMBINED AXIAL TENSION AND FLEXURE

The interaction of the combined effects of the required strengths [factored axial tension and factored bending moment(s)] must satisfy the unity check as follows:

1. For doubly symmetric and singly symmetric members other than HSS and steel pipe, per LRFD Specification Section H1.1.
2. For HSS and steel pipe, per HSS Specification Section 7.1.
3. For single angles, per Single-Angle Specification Section 6.2.
4. For unsymmetric and other members, per LRFD Specification Section H2.

MEMBERS SUBJECT TO COMBINED AXIAL COMPRESSION AND FLEXURE

The interaction of the combined effects of the required strengths [factored axial compression and factored bending moment(s)] must satisfy the unity check as follows:

1. For doubly symmetric and singly symmetric members other than HSS and steel pipe, per LRFD Specification Section H1.2.
2. For HSS and steel pipe, per HSS Specification Section 7.1.
3. For single angles, per Single-Angle Specification Section 6.1.
4. For unsymmetric and other members, per LRFD Specification Section H2.

MEMBERS SUBJECT TO COMBINED TORSION, FLEXURE, SHEAR AND/OR AXIAL FORCE

The interaction of the combined effects of the required strengths [factored torsion, factored bending moment(s), factored shear force and/or factored axial force] must satisfy the requirements as follows:

1. For members other than HSS and steel pipe, per LRFD Specification Section H2.
2. For HSS and steel pipe, per HSS Specification Sections 6 and 7.2.

See also AISC Design Guide No. 9 *Torsional Analysis of Structural Steel Members.*

COMPOSITE MEMBERS SUBJECT TO COMBINED AXIAL COMPRESSION AND FLEXURE

For the design of composite members subject to combined axial compression and flexure, see LRFD Specification Section I4.

OTHER SPECIFICATION REQUIREMENTS AND DESIGN CONSIDERATIONS

Based upon the types of load transmitted by the member, the summaries of other specification requirements and design considerations given in Part 3 for tension members, Part 4 for compression members and Part 5 for flexural members apply to the design of members subject to combined loading, with the exception of the web width-thickness criteria, as discussed previously.

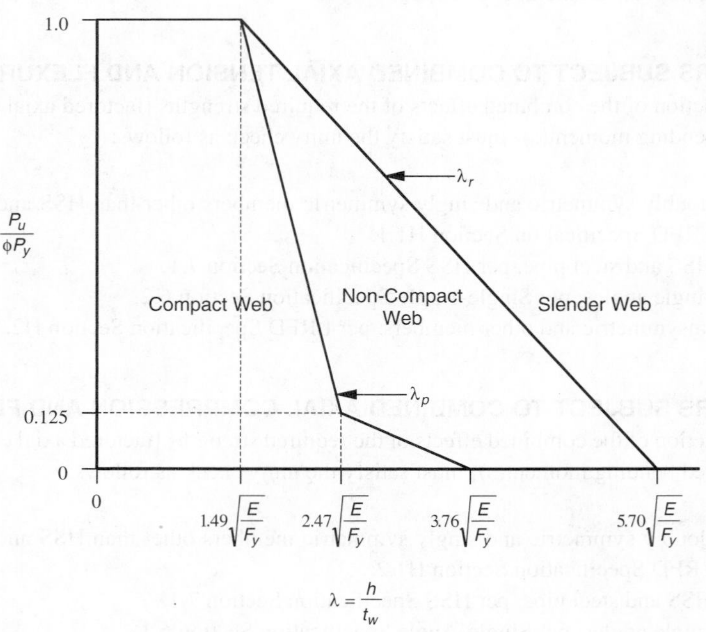

Fig. 6–1. Effect of axial load on web slenderness.

DESIGN EXAMPLES

EXAMPLE 6.1. W-shape subject to combined axial tension and flexure.

Given:

Check the adequacy of an ASTM A992 W10×22 subject to axial tension and flexure as follows:

$$F_y = 50\,\text{ksi} \qquad A_g = 6.49\,\text{in.}^2$$
$$F_u = 65\,\text{ksi} \qquad Z_y = 6.10\,\text{in.}^3$$

a) $P_u = 55\,\text{kips}$, $M_{uy} = 20$ kip-ft, $M_{ux} = 0$.
b) $P_u = 140\,\text{kips}$, $M_{ux} = 55$ kip-ft, $M_{uy} = 0$, $L_b = 4\,\text{ft}$, $C_b = 1.0$.

Solution a:

From LRFD Specification Section D1,

$$\begin{aligned}\phi_t P_n &= 0.9 F_y A_g \\ &= 0.9(50\,\text{ksi})(6.49\,\text{in.}^2) \\ &= 292\,\text{kips}\end{aligned}$$

Note: for convenience, it is assumed that tension yielding controls over tension rupture. If this is not the case, the above design strength should be based upon the tension rupture design strength.

From LRFD Specification Section H1.1,

$$\begin{aligned}\frac{P_u}{\phi_t P_n} &= \frac{55\,\text{kips}}{292\,\text{kips}} \\ &= 0.188 < 0.2\end{aligned}$$

Therefore, LRFD Specification Equation H1-1b governs. With $M_{ux} = 0$, this equation becomes:

$$\frac{P_u}{2\phi_t P_n} + \frac{M_{uy}}{\phi_b M_{ny}} \le 1.0$$

From LRFD Specification Section F1.1

$$\begin{aligned}\phi_b M_{ny} &= 0.9 F_y Z_y \\ &= \frac{0.9(50\,\text{ksi})\left(6.10\,\text{in.}^3\right)}{12\,\text{in./ft}} \\ &= 22.9\,\text{kip-ft}\end{aligned}$$

Thus,

$$\begin{aligned}\frac{P_u}{2\phi_t P_n} + \frac{M_{uy}}{\phi_b M_{ny}} &= \frac{0.188}{2} + \frac{20\,\text{kip-ft}}{22.9\,\text{kip-ft}} \\ &= 0.0940 + 0.873 \\ &= 0.967 < 1.0\ \textbf{o.k.}\end{aligned}$$

Solution b:

As determined in solution a, $\phi_t P_n = 292$ kips. From LRFD Specification Section H1.1,

$$\begin{aligned}\frac{P_u}{\phi_t P_n} &= \frac{140\,\text{kips}}{292\,\text{kips}} \\ &= 0.479 > 0.2\end{aligned}$$

Therefore, LRFD Specification Equation H1-1a governs. With $M_{uy} = 0$, this equation becomes:

$$\frac{P_u}{\phi_t P_n} + \frac{8}{9}\frac{M_{ux}}{\phi_b M_{nx}} \le 1.0$$

From Table 5-3, a W10×22 is compact, has $L_p = 4.70$ ft $> L_b = 4$ ft (braced) and, $\phi_b M_{nx} = 97.5$ kip-ft. Thus,

$$\begin{aligned}\frac{P_u}{\phi_t P_n} + \frac{8}{9}\frac{M_{ux}}{\phi_b M_{nx}} &= 0.479 + \frac{8}{9}\frac{55\text{ kip-ft}}{97.5\text{ kip-ft}}\\ &= 0.479 + 0.501\\ &= 0.980 < 1.0 \textbf{ o.k.}\end{aligned}$$

EXAMPLE 6.2. W-Shape subject to combined axial compression and flexure (braced frame).

Given: Check the adequacy of an ASTM A992 W14×176 with $L_x = L_y = 14.0$ ft in a symmetric braced frame subject to the loading $P_u = 1{,}400$ kips, $M_{ux} = 200$ kip-ft, $M_{uy} = 70$ kip-ft. Assume reverse-curvature bending with equal end moments about both axes and no loads along the member.

$F_y = 50$ ksi	$A = 51.8$ in.^{2}	$r_x = 6.43$ in.
$F_u = 65$ ksi	$Z_x = 320$ in.^{3}	$r_y = 4.02$ in.
	$Z_y = 163$ in.^{3}	

Solution: For a braced frame, $K = 1.0$ and $K_x L_x = K_y L_y = 14.0$ ft. From Table 4-2,

$$\phi_c P_n = 1{,}940 \text{ kips}$$

From LRFD Specification Section H1.2,

$$\begin{aligned}\frac{P_u}{\phi_c P_n} &= \frac{1{,}400 \text{ kips}}{1{,}940 \text{ kips}}\\ &= 0.722 > 0.2\end{aligned}$$

Therefore, LRFD Specification Equation H1-1a governs.

From LRFD Specification Section C1.2, for a braced frame, $M_{lt} = 0$. From LRFD Specification Equation C1-1

$$M_{ux} = B_{1x} M_{ntx}, \quad \text{where } M_{ntx} = 200 \text{ kip-ft; and}$$
$$M_{uy} = B_{1y} M_{nty}, \quad \text{where } M_{nty} = 70 \text{ kip-ft}$$

From LRFD Specification Equations C1-2 and C1-3:

$$B_1 = \frac{C_m}{(1 - P_u/P_{e1})} \geq 1$$

where in this case (a braced frame with no transverse loading),

$$C_m = 0.6 - 0.4(M_1/M_2)$$

For reverse curvature bending and equal end moments:

$$\begin{aligned} M_1/M_2 &= +1.0 \\ C_m &= 0.6 - 0.4(1.0) = 0.2 \\ P_{e1} &= \frac{\pi^2 E}{(KL)^2} \end{aligned}$$

From Table 4-2,

$$\begin{aligned} P_{e1x} &= 61{,}300\,\text{kip-in.}^2 \times 10^4/(14.0\,\text{ft} \times 12\,\text{in./ft})^2 \\ &= 21{,}700\,\text{kips} \end{aligned}$$

$$\begin{aligned} P_{e1y} &= 24{,}000\,\text{kip-in.}^2 \times 10^4/(14.0\,\text{ft} \times 12\,\text{in./ft})^2 \\ &= 8{,}500\,\text{kips} \end{aligned}$$

Thus,

$$\begin{aligned} B_{1x} &= \frac{C_{mx}}{(1 - P_u/P_{e1x})} \geq 1 \\ &= \frac{0.2}{(1 - 1{,}400\,\text{kips}/21{,}700\,\text{kips})} \geq 1 \\ &= 0.214 \geq 1 \\ &= 1 \end{aligned}$$

$$\begin{aligned} B_{1y} &= \frac{C_{my}}{(1 - P_u/P_{e1y})} \geq 1 \\ &= \frac{0.2}{(1 - 1{,}400\,\text{kips}/8{,}500\,\text{kips})} \geq 1 \\ &= 0.239 \geq 1 \\ &= 1 \end{aligned}$$

$$\begin{aligned} M_{ux} &= 1.0 \times 200\,\text{kip-ft} \\ M_{uy} &= 1.0 \times 70\,\text{kip-ft} \end{aligned}$$

From LRFD Specification Equation F1-4,

$$\begin{aligned} L_p &= 1.76 r_y \sqrt{\frac{E}{F_{yf}}} \\ &= 1.76(4.02\,\text{in.})\sqrt{\frac{29{,}000\,\text{ksi}}{50\,\text{ksi}}}(1\,\text{ft}/12\,\text{in.}) \\ &= 14.2\,\text{ft} \end{aligned}$$

Since $L_b = 14.0\,\text{ft} < L_p = 14.2\,\text{ft}$, $\phi_b M_{nx} = \phi_b M_{px} = \phi_b F_y Z_x$

$$\begin{aligned}\phi_b M_{nx} &= \phi_b F_y Z_x \\ &= \frac{0.9 \times 50\,\text{ksi} \times 320\,\text{in.}^3}{12\,\text{in./ft}} \\ &= 1{,}200\,\text{kip-ft} \\ \phi_b M_{ny} &= \phi_b F_y Z_y \\ &= \frac{0.9 \times 50\,\text{ksi} \times 163\,\text{in.}^3}{12\,\text{in./ft}} \\ &= 611\,\text{kip-ft}\end{aligned}$$

From LRFD Specification Equation H1-1a,

$$\begin{aligned}\frac{1{,}400\,\text{kips}}{1{,}940\,\text{kips}} + \frac{8}{9}\left(\frac{200\,\text{kip-ft}}{1{,}200\,\text{kip-ft}} + \frac{70\,\text{kip-ft}}{611\,\text{kip-ft}}\right) &= 0.722 + {}^{8}/_{9}\,(0.167 + 0.115) \\ &= 0.722 + 0.251 \\ &= 0.973 < 1.0\ \textbf{o.k.}\end{aligned}$$

Therefore, the W14×176 is **o.k.**

EXAMPLE 6.3. W-shape subject to combined axial compression and flexure (moment frame).

Given:

Repeat Example 6.2 for an ASTM A992 W14×176 with $L_x = L_y =$ 14.0 ft in a symmetric moment frame subject to the loading $P_u =$ 1,400 kips, $M_{ux} = 300$ kip-ft, $M_{uy} = 0$ kip-ft. Assume a drift index $\Delta_{oh}/L \leq 0.0025$ (or $^{1}/_{400}$) and that

$$\begin{aligned}\Sigma P_u &= 24{,}000\,\text{kips} \\ \Sigma H &= 800\,\text{kips}\end{aligned}$$

Solution:

As determined in Example 6.2, $\phi_c P_n = 1{,}940$ kips.

From LRFD Specification Section H1.2,

$$\begin{aligned}\frac{P_u}{\phi_c P_n} &= \frac{1{,}400\,\text{kips}}{1{,}940\,\text{kips}} \\ &= 0.722 > 0.2\end{aligned}$$

Therefore, LRFD Specification Equation H1-1a governs.

Because $M_{ntx} = M_{nty} = M_{lty} = 0$ and only $M_{ltx} \neq 0$, $M_{ux} = B_2 M_{ltx}$ and $M_{uy} = 0$.

$$M_{ltx} = 300\,\text{kip-ft}$$

From LRFD Specification Equation C1-4,

$$B_2 = \frac{1}{1 - \frac{\Sigma P_u}{\Sigma H}\left(\frac{\Delta_{oh}}{L}\right)}$$
$$= \frac{1}{1 - \frac{24{,}000\text{ kips}}{800\text{ kips}}(0.0025)}$$
$$= 1.08$$
$$M_{ux} = 1.08 \times 300\text{ kip-ft}$$
$$= 324\text{ kip-ft}$$

As determined in Example 6.2, $\phi_b M_{nx} = 1{,}200$ kip-ft.

From LRFD Specification Equation H1-1a:

$$\frac{1{,}400\text{ kips}}{1{,}940\text{ kips}} + \frac{8}{9}\frac{324\text{ kip-ft}}{1{,}200\text{ kip-ft}} = 0.722 + {}^8\!/_9\,(0.270)$$
$$= 0.962 < 1.0 \textbf{ o.k.}$$

Therefore the W14×176 is **o.k.**

EXAMPLE 6.4. W-shape member design (braced frame).

Given:

Select the lightest ASTM A992 W14 ($F_y = 50$ ksi, $F_u = 65$ ksi) that is adequate for the following combination of axial compression and flexure:

$$P_u = 400\text{ kips} \quad M_{ntx} = 250\text{ kip-ft} \quad M_{nty} = 80\text{ kip-ft}$$
$$M_{ltx} = 0\text{ (braced frame)} \quad M_{lty} = 0\text{ (braced frame)}$$

Assume $KL_x = KL_y = L_b = 14$ ft and $B_{1x} = B_{1y} = 1.0$

Solution:

From the information given,

$$M_{ux} = B_{1x} M_{ntx}$$
$$= 1.0 \times 250\text{ kip-ft}$$
$$= 250\text{ kip-ft}$$
$$M_{uy} = B_{1y} M_{nty}$$
$$= 1.0 \times 80\text{ kip-ft}$$
$$= 80\text{ kip-ft}$$

From Table 6-2, try W14×99, which has:

$$b = 0.939 \times 10^{-3}\text{ (kips)}^{-1}$$
$$m = 1.38 \times 10^{-3}\text{ (kip-ft)}^{-1}$$
$$n = 2.88 \times 10^{-3}\text{ (kip-ft)}^{-1}$$

and

$$bP_u = 0.939 \times 10^{-3}\text{ (kips)}^{-1} \times 400\text{ kips}$$
$$= 0.376$$
$$mM_{ux} = 1.38 \times 10^{-3}\text{ (kip-ft)}^{-1} \times 250\text{ kip-ft}$$
$$= 0.345$$
$$nM_{uy} = 2.88 \times 10^{-3}\text{ (kip-ft)}^{-1} \times 80\text{ kip-ft}$$
$$= 0.230$$

From Table 6-2, $\phi_c P_n = 1/b = 1{,}060$ kips and

$$\frac{P_u}{\phi_c P_n} = \frac{400\text{ kips}}{1{,}060\text{ kips}} = 0.377 > 0.2$$

Since $P_u/\phi_c P_n > 0.2$,

$$bP_u + mM_{ux} + nM_{uy} = 0.376 + 0.345 + 0.230 = 0.951 < 1.0 \textbf{ o.k.}$$

From Table 6-2, try W14×90, which has:

$$b = 1.03 \times 10^{-3}\text{ (kips)}^{-1}$$
$$m = 1.54 \times 10^{-3}\text{ (kip-ft)}^{-1}$$
$$n = 3.25 \times 10^{-3}\text{ (kip-ft)}^{-1}$$

and

$$bP_u = 1.03 \times 10^{-3}\text{ (kips)}^{-1} \times 400\text{ kips} = 0.412$$
$$mM_{ux} = 1.54 \times 10^{-3}\text{ (kip-ft)}^{-1} \times 250\text{ kip-ft} = 0.385$$
$$nM_{uy} = 3.25 \times 10^{-3}\text{ (kip-ft)}^{-1} \times 80\text{ kip-ft} = 0.260$$

From Table 6-2, $\phi_c P_n = 1/b = 969$ kips and

$$\frac{P_u}{\phi_c P_n} = \frac{400\text{ kips}}{969\text{ kips}} = 0.413 > 0.2$$

Since $P_u/\phi_c P_n > 0.2$,

$$bP_u + mM_{ux} + nM_{uy} = 0.412 + 0.385 + 0.260 = 1.06 > 1.0 \textbf{ n.g.}$$

Therefore, a W14×99 is the lightest W14 that will be adequate.

Comments: If desired, the median values of b, m and n in Table 6-1 can be used to assist in making the trial selection of a shape. For further design examples illustrating the use of the design approach reflected in Tables 6-1 and 6-2, refer to Aminmansour (2000).

STEEL BEAM-COLUMN SELECTION TABLES

Table 6-1. Median Values of *b*, *m* and *n* for Beam-Columns

To assist in making a trial selection of a W-shape to resist combined axial compression and flexure, median values of b, m and n have been tabulated in Table 6-1 based upon nominal depth and steel with $F_y = 50$ ksi (ASTM A992). These variables are defined as given in the discussion below of Table 6-2.

When axial effects dominate, it is recommended that the initial value for m be taken from Table 6-1 to determine a corresponding estimate of the required b. When X-X axis flexural effects dominate, it is recommended that initial value for b taken from Table 6-1 to determine a corresponding estimate of the required m. If Y-Y axis flexural effects are present in either case, the tabulated value of n can also be taken from Table 6-1.

For further information, see Aminmansour (2000).

Table 6-2. W-Shapes: Values of *b*, *m* and *n* for Beam-Columns

The adequacy of W-shapes with $F_y = 50$ ksi (ASTM A992) subject to combined axial compression and flexure can be determined using the values of b, m and n tabulated in Table 6-2. These variables are defined as follows:

$$b = \frac{1}{\phi_c P_n}, \text{ (kips)}^{-1}$$

$$m = \frac{8}{9\phi_b M_{nx}}, \text{ (kip-ft)}^{-1}$$

$$n = \frac{8}{9\phi_b M_{ny}}, \text{ (kip-ft)}^{-1}$$

The proper value of b is selected for the larger of the unbraced length for compression buckling about the Y-Y axis $(KL)_y$ and the unbraced length for compression buckling about the X-X axis $(KL)_{y\,eq}$, as described in Part 4. The tabulated values can be used directly if the cross-section is not a slender-element cross-section (see Figure 6-1).

The proper value of m is selected for the unbraced length L_b, as described in Part 5. Values of m have been included for $L_b \leq L_r$. Because unbraced length is not a factor in weak-axis bending, a single value of n applies for any given W-shape.

When $P_u/\phi_c P_n \geq 0.2$, the tabulated values of b, m and n can be used as follows to solve LRFD Specification Equation H1-1a:

$$bP_u + mM_{ux} + nM_{uy} \leq 1.0$$

When $P_u/\phi_c P_n < 0.2$, the tabulated values of b, m and n can be used as follows to solve LRFD Specification Equation H1-1b:

$$^1/_2\, bP_u + {}^9/_8(mM_{ux} + nM_{uy}) \leq 1.0$$

For further information, see Aminmansour (2000).

PART 6 REFERENCE

Aminmansour, A., 2000, "A New Approach for Design of Steel Beam-Columns," *Engineering Journal*, Vol. 37, No. 2, (2nd Qtr.), pp. 41–72, AISC, Chicago, IL.

Table 6-1. W-Shapes Median Values of b^a, m^b, and n^b for Beam-Columns

$F_y = 50$ ksi
$\phi_c = 0.85$
$\phi_b = 0.90$

Shape	W44		W40		W36		W33		W30		W27	
Effective length KL (ft) with respect to least radius of gyration r_y or unbraced length L_b (ft) for X-X axis bending	b $\times 10^3$	m $\times 10^3$	b $\times 10^3$	m $\times 10^3$	b $\times 10^3$	m $\times 10^3$	b $\times 10^3$	m $\times 10^3$	b $\times 10^3$	m $\times 10^3$	b $\times 10^3$	m $\times 10^3$
0	0.310	0.177	0.423	0.260	0.334	0.235	0.368	0.277	0.450	0.371	0.411	0.376
6	–	–	–	–	–	–	–	–	–	–	–	–
7	–	–	–	–	–	–	–	–	–	–	–	–
8	–	–	–	–	–	–	–	–	–	–	–	–
9	–	–	–	–	–	–	–	–	–	–	–	–
10	–	–	–	–	–	–	–	–	–	–	–	–
11	0.339	0.177	0.461	0.260	0.362	0.235	0.401	0.277	0.494	0.371	0.463	0.376
12	0.345	0.177	0.469	0.260	0.367	0.235	0.408	0.277	0.504	0.371	0.473	0.377
13	0.351	0.179	0.477	0.262	0.374	0.235	0.414	0.278	0.515	0.376	0.485	0.382
14	0.358	0.182	0.487	0.266	0.380	0.237	0.424	0.281	0.527	0.381	0.498	0.387
15	0.366	0.184	0.497	0.270	0.388	0.240	0.434	0.285	0.539	0.386	0.512	0.392
16	0.375	0.187	0.508	0.274	0.395	0.243	0.445	0.289	0.554	0.392	0.528	0.398
17	0.384	0.189	0.521	0.279	0.405	0.246	0.457	0.293	0.569	0.397	0.545	0.403
18	0.394	0.192	0.535	0.283	0.416	0.250	0.470	0.297	0.587	0.404	0.564	0.409
19	0.406	0.196	0.550	0.288	0.428	0.253	0.485	0.301	0.607	0.410	0.584	0.414
20	0.419	0.199	0.566	0.292	0.440	0.256	0.500	0.306	0.628	0.416	0.607	0.420
22	0.447	0.205	0.603	0.302	0.469	0.264	0.536	0.315	0.677	0.429	0.659	0.432
24	0.481	0.212	0.648	0.313	0.502	0.271	0.578	0.324	0.735	0.443	0.720	0.445
26	0.523	0.220	0.701	0.324	0.541	0.280	0.627	0.335	0.803	0.458	0.794	0.459
28	0.574	0.227	0.765	0.336	0.587	0.288	0.685	0.346	0.884	0.475	0.882	0.474
30	0.635	0.236	0.842	0.349	0.640	0.297	0.753	0.357	0.981	0.492	0.987	0.490
32	0.708	0.245	0.936	0.365	0.702	0.307	0.833	0.370	1.10	0.510	1.12	0.506
34	0.796	0.260	1.051	0.399	0.775	0.318	0.928	0.383	1.24	0.544	1.26	0.524
36	0.892	0.284	1.18	0.439	0.863	0.329	1.04	0.407	1.39	0.593	1.41	0.559
38	0.992	0.310	1.315	0.479	0.961	0.354	1.16	0.441	1.54	0.643	1.57	0.601
40	1.10	0.335	1.46	0.520	1.07	0.382	1.29	0.476	1.71	0.694	1.74	0.643
42	1.22	0.361	1.61	0.561	1.17	0.411	1.42	0.511	1.89	0.745	1.92	0.685
44	1.33	0.386	1.76	0.603	1.29	0.440	1.55	0.547	2.07	0.795	2.11	0.728
46	1.46	0.412	1.93	0.645	1.41	0.469	1.70	0.582	2.26	0.846	2.31	0.770
48	1.59	0.439	2.10	0.688	1.53	0.498	1.85	0.618	2.46	0.898	2.51	0.812
50	1.72	0.465	2.27	0.730	1.66	0.527	2.01	0.654	2.67	0.948	2.73	0.853
$n \times 10^3$	1.27		1.68		1.28		1.49		1.88		1.79	

[a] b has the units (kips)$^{-1}$.
[b] m and n have the units (kip-ft)$^{-1}$.
– indicates that values for this effective length are not given in Table 6-2.

$F_y = 50$ ksi
$\phi_c = 0.85$
$\phi_b = 0.90$

Table 6-1 (cont.). W-Shapes Median Values of b^a, m^b, and n^b for Beam-Columns

Shape	W24		W21		W18		W16		W14		W12		W10	
Effective length KL (ft) with respect to least radius of gyration r_y or unbraced length L_b (ft) for X-X axis bending	b $\times 10^3$	m $\times 10^3$	b $\times 10^3$	m $\times 10^3$	b $\times 10^3$	m $\times 10^3$	b $\times 10^3$	m $\times 10^3$	b $\times 10^3$	m $\times 10^3$	b $\times 10^3$	m $\times 10^3$	b $\times 10^3$	m $\times 10^3$
0	0.547	0.567	0.919	1.14	1.07	1.45	1.62	2.58	0.504	0.826	1.01	1.99	1.91	4.69
6	–	–	1.02	1.14	1.12	1.45	1.86	2.61	–	–	1.06	1.99	2.10	4.69
7	–	–	1.07	1.16	1.14	1.45	1.96	2.68	–	–	1.07	1.99	2.17	4.70
8	–	–	1.12	1.19	1.17	1.45	2.09	2.77	–	–	1.09	1.99	2.26	4.77
9	–	–	1.18	1.22	1.20	1.45	2.24	2.86	–	–	1.11	1.99	2.36	4.86
10	–	–	1.26	1.25	1.23	1.48	2.43	2.95	–	–	1.14	1.99	2.49	4.96
11	0.630	0.570	1.34	1.29	1.27	1.50	2.65	3.06	0.546	0.826	1.16	2.00	2.63	5.05
12	0.647	0.579	1.44	1.32	1.32	1.53	2.92	3.16	0.554	0.826	1.19	2.02	2.79	5.15
13	0.666	0.587	1.56	1.36	1.37	1.56	3.24	3.28	0.563	0.826	1.23	2.04	2.99	5.25
14	0.687	0.596	1.69	1.40	1.43	1.59	3.62	3.41	0.573	0.826	1.27	2.07	3.21	5.36
15	0.711	0.606	1.85	1.44	1.49	1.62	4.09	3.55	0.584	0.831	1.31	2.09	3.46	5.48
16	0.737	0.615	2.04	1.49	1.57	1.66	4.65	3.74	0.596	0.836	1.36	2.12	3.76	5.59
17	0.766	0.625	2.26	1.54	1.65	1.69	5.25	4.09	0.609	0.842	1.41	2.14	4.11	5.71
18	0.797	0.635	2.53	1.59	1.74	1.73	5.88	4.45	0.624	0.847	1.46	2.17	4.50	5.84
19	0.832	0.646	2.82	1.67	1.84	1.77	6.56	4.81	0.639	0.853	1.53	2.20	4.97	5.97
20	0.871	0.656	3.12	1.78	1.96	1.81	7.26	5.17	0.656	0.859	1.59	2.22	5.51	6.11
22	0.960	0.679	3.77	2.05	2.23	1.89	8.79	5.90	0.693	0.871	1.75	2.28	6.67	6.43
24	1.07	0.704	4.50	2.31	2.57	1.99	10.5	6.62	0.736	0.883	1.95	2.34	7.93	6.97
26	1.20	0.730	5.28	2.57	3.00	2.18	12.3	7.35	0.786	0.895	2.18	2.40	9.30	7.67
28	1.36	0.759	6.11	2.83	3.48	2.44	–	–	0.844	0.908	2.46	2.47	10.78	8.37
30	1.56	0.789	7.02	3.09	4.00	2.70	–	–	0.911	0.921	2.81	2.54	12.4	9.10
32	1.77	0.852	–	–	4.55	2.96	–	–	0.988	0.935	3.20	2.62	14.1	9.79
34	2.00	0.928	–	–	5.14	3.22	–	–	1.08	0.949	3.62	2.70	–	–
36	2.25	1.00	–	–	5.76	3.48	–	–	1.18	0.963	4.05	2.80	–	–
38	2.50	1.08	–	–	6.42	3.75	–	–	1.30	0.978	4.52	2.99	–	–
40	2.77	1.15	–	–	7.11	4.01	–	–	1.45	0.994	5.00	3.17	–	–
42	3.06	1.23	–	–	–	–	–	–	1.59	1.01	–	–	–	–
44	3.35	1.31	–	–	–	–	–	–	1.75	1.03	–	–	–	–
46	3.67	1.38	–	–	–	–	–	–	1.91	1.04	–	–	–	–
48	3.99	1.46	–	–	–	–	–	–	2.08	1.06	–	–	–	–
50	4.33	1.53	–	–	–	–	–	–	2.26	1.08	–	–	–	–
$n \times 10^3$	2.61		7.63		5.73		15.0		1.64		4.41		13.0	

[a] b has the units (kips)$^{-1}$.
[b] m and n have the units (kip-ft)$^{-1}$.
– indicates that values for this effective length are not given in Table 6-2.

Table 6-2.
W-Shapes
Values of b^a, m^b, and n^b for Beam-Columns

$F_y = 50$ ksi
$\phi_c = 0.85$
$\phi_b = 0.90$

Shape	W44×								W40×			
	335		290		262		230		593*		503*	
	$b \times 10^3$	$m \times 10^3$	$b \times 10^3$	$m \times 10^3$	$b \times 10^3$	$m \times 10^3$	$b \times 10^3$	$m \times 10^3$	$b \times 10^3$	$m \times 10^3$	$b \times 10^3$	$m \times 10^3$
Effective length *KL* (ft) with respect to least radius of gyration r_y or unbraced length L_b (ft) for X-X axis bending												
0	0.244†	0.146	0.290†	0.167	0.330†	0.187	0.388†	0.215	0.135	0.0859	0.159	0.102
11	0.267†	0.146	0.317†	0.167	0.361†	0.187	0.423†	0.215	0.148	0.0859	0.174	0.102
12	0.271	0.146	0.322†	0.167	0.367†	0.187	0.430†	0.215	0.150	0.0859	0.177	0.102
13	0.277	0.148	0.328†	0.168	0.374†	0.189	0.438†	0.218	0.153	0.0859	0.181	0.102
14	0.283	0.150	0.335†	0.171	0.381†	0.192	0.446†	0.222	0.156	0.0863	0.185	0.103
15	0.290	0.152	0.342†	0.173	0.389†	0.194	0.456†	0.225	0.159	0.0869	0.189	0.104
16	0.298	0.154	0.351†	0.176	0.398†	0.197	0.466†	0.229	0.163	0.0876	0.193	0.105
17	0.307	0.156	0.359†	0.178	0.409†	0.200	0.478†	0.233	0.167	0.0882	0.198	0.106
18	0.316	0.158	0.369†	0.181	0.419†	0.203	0.491†	0.236	0.171	0.0889	0.203	0.107
19	0.326	0.160	0.380†	0.184	0.432†	0.207	0.505†	0.240	0.176	0.0896	0.209	0.108
20	0.338	0.163	0.392†	0.187	0.445†	0.210	0.520†	0.244	0.181	0.0903	0.216	0.109
22	0.363	0.167	0.418†	0.193	0.475†	0.217	0.555†	0.253	0.192	0.0918	0.230	0.111
24	0.393	0.173	0.451	0.199	0.511†	0.225	0.596†	0.262	0.206	0.0933	0.246	0.113
26	0.428	0.178	0.492	0.206	0.554†	0.233	0.646†	0.272	0.221	0.0948	0.266	0.115
28	0.470	0.184	0.540	0.213	0.607	0.241	0.706†	0.283	0.240	0.0964	0.289	0.117
30	0.519	0.190	0.597	0.221	0.673	0.250	0.779†	0.294	0.261	0.0980	0.315	0.120
32	0.577	0.197	0.665	0.230	0.750	0.260	0.869	0.310	0.285	0.0997	0.347	0.122
34	0.648	0.204	0.747	0.239	0.844	0.281	0.980	0.343	0.314	0.101	0.383	0.125
36	0.726	0.214	0.837	0.260	0.946	0.308	1.10	0.378	0.348	0.103	0.427	0.128
38	0.809	0.232	0.933	0.283	1.05	0.336	1.22	0.413	0.388	0.105	0.476	0.131
40	0.897	0.250	1.03	0.306	1.17	0.363	1.36	0.449	0.430	0.107	0.527	0.133
42	0.989	0.268	1.14	0.329	1.29	0.392	1.49	0.485	0.474	0.109	0.581	0.137
44	1.09	0.286	1.25	0.352	1.41	0.420	1.64	0.522	0.520	0.111	0.638	0.140
46	1.19	0.304	1.37	0.375	1.55	0.449	1.79	0.559	0.568	0.114	0.697	0.143
48	1.29	0.323	1.49	0.399	1.68	0.478	1.95	0.596	0.619	0.116	0.759	0.147
50	1.40	0.341	1.61	0.422	1.83	0.507	2.12	0.634	0.672	0.118	0.824	0.152
Properties												
L_p, ft	12.4		12.3		12.2		12.1		13.4		13.1	
$\phi_b M_p$, kip-ft	6080		5330		4760		4130		10400		8700	
$n \times 10^3$	1.05		1.20		1.34		1.56		0.523		0.632	

[a] b has the units (kips)$^{-1}$.
[b] m and n have the units (kip-ft)$^{-1}$.
† Web may be slender for combined axial and bending stress.
* ASTM A6 tensile group 4 or 5 shape. Special requirements may apply per LRFD Specification Section A3.1c.

F_y = 50 ksi
ϕ_c = 0.85
ϕ_b = 0.90

Table 6-2 (cont.). W-Shapes Values of b^a, m^b, and n^b for Beam-Columns

Shape	W40×											
	431*		397*		372*		362*		324		297	
Effective length *KL* (ft) with respect to least radius of gyration r_y or unbraced length L_b (ft) for X-X axis bending	*b* ×10³	*m* ×10³	*b* ×10³	*m* ×10³	*b* ×10³	*m* ×10³	*b* ×10³	*m* ×10³	*b* ×10³	*m* ×10³	*b* ×10³	*m* ×10³
0	0.185	0.121	0.201	0.132	0.216	0.141	0.220	0.145	0.247	0.162	0.272†	0.178
11	0.204	0.121	0.221	0.132	0.238	0.141	0.243	0.145	0.273	0.162	0.298	0.178
12	0.208	0.121	0.225	0.132	0.243	0.141	0.247	0.145	0.278	0.162	0.304	0.178
13	0.212	0.121	0.230	0.132	0.248	0.142	0.252	0.145	0.284	0.163	0.310	0.179
14	0.216	0.122	0.235	0.133	0.253	0.143	0.258	0.147	0.290	0.165	0.317	0.182
15	0.221	0.124	0.240	0.135	0.259	0.145	0.264	0.148	0.297	0.167	0.325	0.184
16	0.227	0.125	0.246	0.136	0.266	0.146	0.271	0.150	0.305	0.169	0.334	0.186
17	0.233	0.126	0.253	0.138	0.273	0.148	0.278	0.152	0.313	0.171	0.343	0.189
18	0.239	0.127	0.260	0.139	0.281	0.150	0.286	0.154	0.322	0.173	0.353	0.192
19	0.246	0.129	0.268	0.141	0.289	0.152	0.295	0.156	0.332	0.176	0.365	0.194
20	0.254	0.130	0.276	0.142	0.299	0.154	0.304	0.158	0.343	0.178	0.377	0.197
22	0.272	0.133	0.295	0.146	0.320	0.157	0.326	0.162	0.367	0.183	0.404	0.203
24	0.292	0.136	0.318	0.149	0.345	0.161	0.351	0.166	0.396	0.188	0.437	0.209
26	0.316	0.139	0.344	0.153	0.374	0.166	0.381	0.170	0.430	0.193	0.475	0.215
28	0.344	0.142	0.375	0.157	0.408	0.170	0.416	0.175	0.470	0.199	0.520	0.222
30	0.377	0.146	0.411	0.161	0.448	0.175	0.457	0.180	0.517	0.205	0.573	0.229
32	0.416	0.149	0.454	0.165	0.496	0.180	0.505	0.185	0.573	0.211	0.636	0.237
34	0.462	0.153	0.504	0.169	0.552	0.185	0.562	0.190	0.639	0.218	0.712	0.245
36	0.517	0.157	0.564	0.174	0.619	0.191	0.631	0.196	0.716	0.226	0.799	0.254
38	0.576	0.161	0.629	0.179	0.690	0.197	0.703	0.203	0.798	0.235	0.890	0.275
40	0.638	0.166	0.697	0.184	0.764	0.203	0.779	0.210	0.884	0.253	0.986	0.296
42	0.704	0.170	0.768	0.190	0.843	0.216	0.859	0.224	0.975	0.271	1.09	0.317
44	0.772	0.175	0.843	0.201	0.925	0.230	0.942	0.238	1.07	0.288	1.19	0.338
46	0.844	0.185	0.921	0.212	1.01	0.243	1.03	0.252	1.17	0.306	1.30	0.359
48	0.919	0.195	1.00	0.224	1.10	0.257	1.12	0.266	1.27	0.323	1.42	0.380
50	0.997	0.205	1.09	0.236	1.19	0.270	1.22	0.280	1.38	0.341	1.54	0.402

Properties

	431*	397*	372*	362*	324	297
L_p, ft	12.9	12.9	12.7	12.7	12.6	12.5
$\phi_b M_p$, kip-ft	7350	6750	6300	6150	5480	4990
$n \times 10^3$	0.760	0.827	0.893	0.913	1.03	1.15

[a] *b* has the units (kips)$^{-1}$.
[b] *m* and *n* have the units (kip-ft)$^{-1}$.
† Web may be slender for combined axial and bending stress.
* ASTM A6 tensile group 4 or 5 shape. Special requirements may apply per LRFD Specification Section A3.1c.

Table 6-2 (cont.). W-Shapes Values of b^a, m^b, and n^b for Beam-Columns

$F_y = 50$ ksi
$\phi_c = 0.85$
$\phi_b = 0.90$

Shape	W40×											
	277		249		215		199		392*		331	
	$b \times 10^3$	$m \times 10^3$	$b \times 10^3$	$m \times 10^3$	$b \times 10^3$	$m \times 10^3$	$b \times 10^3$	$m \times 10^3$	$b \times 10^3$	$m \times 10^3$	$b \times 10^3$	$m \times 10^3$
Effective length *KL* (ft) with respect to least radius of gyration r_y or unbraced length L_b (ft) for X-X axis bending												
0	0.299†	0.190	0.339†	0.212	0.405†	0.246	0.441†	0.273	0.205	0.139	0.241	0.166
11	0.326†	0.190	0.370†	0.212	0.440†	0.246	0.482†	0.273	0.246	0.142	0.293	0.171
12	0.331†	0.190	0.376†	0.212	0.447†	0.246	0.490†	0.273	0.254	0.144	0.304	0.173
13	0.337†	0.191	0.383†	0.213	0.455†	0.248	0.499†	0.276	0.264	0.146	0.316	0.176
14	0.344†	0.193	0.391†	0.216	0.464†	0.251	0.509†	0.280	0.275	0.148	0.330	0.179
15	0.352†	0.196	0.399†	0.219	0.474†	0.255	0.520†	0.285	0.287	0.150	0.345	0.182
16	0.360†	0.198	0.408†	0.222	0.484†	0.259	0.532†	0.289	0.301	0.152	0.363	0.185
17	0.369†	0.201	0.418†	0.225	0.496†	0.263	0.545†	0.294	0.317	0.154	0.383	0.188
18	0.378†	0.204	0.429†	0.229	0.509†	0.267	0.560†	0.299	0.334	0.156	0.404	0.192
19	0.389	0.207	0.441†	0.232	0.523†	0.271	0.576†	0.304	0.353	0.159	0.429	0.195
20	0.402	0.210	0.454†	0.235	0.538†	0.275	0.594†	0.309	0.374	0.161	0.457	0.199
22	0.430	0.216	0.484†	0.243	0.572†	0.284	0.634†	0.319	0.425	0.166	0.522	0.206
24	0.464	0.222	0.519	0.250	0.614†	0.294	0.682†	0.331	0.488	0.172	0.604	0.214
26	0.504	0.229	0.565	0.259	0.663†	0.304	0.739†	0.343	0.569	0.177	0.708	0.223
28	0.550	0.237	0.618	0.268	0.721†	0.315	0.808†	0.357	0.660	0.183	0.822	0.232
30	0.605	0.245	0.681	0.277	0.791†	0.327	0.892	0.371	0.758	0.190	0.943	0.242
32	0.670	0.253	0.755	0.287	0.877	0.340	0.995	0.390	0.862	0.197	1.07	0.261
34	0.748	0.262	0.845	0.298	0.982	0.366	1.12	0.432	0.973	0.205	1.21	0.282
36	0.838	0.275	0.947	0.325	1.10	0.402	1.26	0.475	1.09	0.218	1.36	0.303
38	0.934	0.298	1.06	0.354	1.23	0.438	1.40	0.519	1.22	0.233	1.51	0.324
40	1.04	0.321	1.17	0.382	1.36	0.475	1.55	0.564	1.35	0.247	1.68	0.344
42	1.14	0.344	1.29	0.411	1.50	0.512	1.71	0.610	1.49	0.262	1.85	0.365
44	1.25	0.368	1.41	0.439	1.64	0.550	1.88	0.656	1.63	0.276		
46	1.37	0.391	1.55	0.468	1.80	0.588	2.05	0.702				
48	1.49	0.415	1.68	0.497	1.96	0.626	2.23	0.749				
50	1.62	0.438	1.83	0.526	2.12	0.664	2.42	0.796				
Properties												
L_p, ft	12.6		12.5		12.5		12.2		9.33		9.08	
$\phi_b M_p$, kip-ft	4690		4200		3620		3260		6410		5360	
$n \times 10^3$	1.20		1.34		1.56		1.79		1.22		1.49	

[a] *b* has the units $(\text{kips})^{-1}$.
[b] *m* and *n* have the units $(\text{kip-ft})^{-1}$.
† Web may be slender for combined axial and bending stress.
* ASTM A6 tensile group 4 or 5 shape. Special requirements may apply per LRFD Specification Section A3.1c.
Note: Heavy line indicates Kl/r equal to or greater than 200.

$F_y = 50$ ksi
$\phi_c = 0.85$
$\phi_b = 0.90$

Table 6-2 (cont.). W-Shapes Values of b^a, m^b, and n^b for Beam-Columns

Shape	W40×											
	327		278		264		235		211		183	
	b ×10³	m ×10³	b ×10³	m ×10³	b ×10³	m ×10³	b ×10³	m ×10³	b ×10³	m ×10³	b ×10³	m ×10³
Effective length *KL* (ft) with respect to least radius of gyration r_y or unbraced length L_b (ft) for X-X axis bending												
0	0.245	0.168	0.288	0.199	0.303	0.210	0.354†	0.235	0.405†	0.262	0.483†	0.303
11	0.297	0.173	0.352	0.206	0.371	0.218	0.420†	0.244	0.480†	0.273	0.570†	0.316
12	0.308	0.176	0.365	0.210	0.385	0.222	0.434†	0.248	0.496†	0.278	0.589†	0.323
13	0.320	0.179	0.381	0.214	0.401	0.226	0.449†	0.253	0.514†	0.284	0.610†	0.330
14	0.334	0.181	0.398	0.218	0.420	0.230	0.470	0.258	0.535†	0.290	0.634†	0.338
15	0.350	0.184	0.418	0.222	0.440	0.234	0.492	0.263	0.559†	0.296	0.662†	0.345
16	0.367	0.188	0.440	0.226	0.464	0.239	0.518	0.269	0.586†	0.303	0.692†	0.353
17	0.387	0.191	0.464	0.231	0.490	0.244	0.547	0.274	0.615	0.310	0.727†	0.362
18	0.409	0.194	0.492	0.235	0.519	0.249	0.579	0.280	0.652	0.317	0.767†	0.371
19	0.434	0.197	0.523	0.240	0.552	0.254	0.615	0.286	0.694	0.324	0.812†	0.380
20	0.461	0.201	0.558	0.245	0.589	0.260	0.655	0.293	0.741	0.332	0.863†	0.390
22	0.527	0.208	0.642	0.255	0.676	0.271	0.751	0.307	0.852	0.349	0.988	0.411
24	0.610	0.217	0.748	0.267	0.789	0.284	0.873	0.322	0.995	0.368	1.16	0.436
26	0.714	0.225	0.878	0.280	0.926	0.298	1.02	0.339	1.17	0.403	1.36	0.497
28	0.828	0.235	1.02	0.296	1.07	0.322	1.19	0.378	1.35	0.451	1.57	0.559
30	0.951	0.245	1.17	0.326	1.23	0.355	1.36	0.418	1.56	0.500	1.81	0.623
32	1.08	0.264	1.33	0.356	1.40	0.387	1.55	0.458	1.77	0.550	2.06	0.687
34	1.22	0.286	1.50	0.386	1.58	0.420	1.75	0.498	2.00	0.600	2.32	0.753
36	1.37	0.307	1.68	0.415	1.77	0.453	1.96	0.539	2.24	0.650	2.60	0.818
38	1.53	0.328	1.88	0.445	1.98	0.486	2.19	0.579	2.50	0.700	2.90	0.884
40	1.69	0.349	2.08	0.475	2.19	0.519	2.43	0.620	2.76	0.750	3.21	0.951
42	1.86	0.370	2.29	0.504	2.42	0.552	2.67	0.660				
44												
46												
48												
50												
Properties												
L_p, ft	9.11		8.90		8.90		8.97		8.87		8.83	
$\phi_b M_p$, kip-ft	5290		4460		4240		3790		3400		2940	
$n \times 10^3$	1.50		1.81		1.91		2.12		2.39		2.78	

[a] b has the units (kips)$^{-1}$.
[b] m and n have the units (kip-ft)$^{-1}$.
† Web may be slender for combined axial and bending stress.
Note: Heavy line indicates Kl/r equal to or greater than 200.

Table 6-2 (cont.).
W-Shapes
Values of b^a, m^b, and n^b for Beam-Columns

$F_y = 50$ ksi
$\phi_c = 0.85$
$\phi_b = 0.90$

Shape	W40×				W36×							
	167		149		798*		650*		527*		439*	
Effective length KL (ft) with respect to least radius of gyration r_y or unbraced length L_b (ft) for X-X axis bending	b ×10³	m ×10³	b ×10³	m ×10³	b ×10³	m ×10³	b ×10³	m ×10³	b ×10³	m ×10³	b ×10³	m ×10³
0	0.533†	0.342	0.610†	0.396	0.100	0.0662	0.123	0.0829	0.152	0.104	0.182	0.127
11	0.636†	0.361	0.736†	0.425	0.108	0.0662	0.133	0.0829	0.164	0.104	0.198	0.127
12	0.658†	0.370	0.764†	0.435	0.109	0.0662	0.135	0.0829	0.167	0.104	0.201	0.127
13	0.684†	0.379	0.796†	0.446	0.111	0.0662	0.137	0.0829	0.170	0.104	0.205	0.127
14	0.713†	0.388	0.833†	0.458	0.112	0.0662	0.139	0.0829	0.173	0.104	0.208	0.127
15	0.746†	0.397	0.875†	0.471	0.114	0.0662	0.142	0.0831	0.176	0.105	0.213	0.128
16	0.784†	0.407	0.923†	0.484	0.116	0.0665	0.145	0.0836	0.180	0.105	0.217	0.129
17	0.827†	0.418	0.979†	0.498	0.119	0.0669	0.148	0.0841	0.183	0.106	0.222	0.130
18	0.876†	0.429	1.04†	0.512	0.121	0.0672	0.151	0.0846	0.188	0.107	0.227	0.131
19	0.932†	0.441	1.12†	0.528	0.124	0.0675	0.154	0.0851	0.192	0.108	0.233	0.132
20	0.996†	0.454	1.20†	0.545	0.127	0.0679	0.158	0.0856	0.197	0.109	0.240	0.134
22	1.16	0.481	1.42	0.586	0.133	0.0685	0.167	0.0867	0.208	0.110	0.254	0.136
24	1.37	0.535	1.69	0.682	0.141	0.0692	0.176	0.0878	0.221	0.112	0.270	0.139
26	1.61	0.612	1.99	0.782	0.149	0.0699	0.188	0.0889	0.236	0.114	0.289	0.141
28	1.87	0.691	2.30	0.885	0.159	0.0706	0.201	0.0900	0.254	0.115	0.311	0.144
30	2.14	0.771	2.64	0.992	0.170	0.0714	0.216	0.0912	0.274	0.117	0.337	0.147
32	2.44	0.854	3.01	1.10	0.183	0.0721	0.233	0.0924	0.297	0.119	0.367	0.150
34	2.75	0.937	3.40	1.21	0.198	0.0729	0.253	0.0936	0.324	0.121	0.401	0.153
36	3.09	1.02	3.81	1.32	0.215	0.0737	0.276	0.0948	0.355	0.123	0.441	0.156
38	3.44	1.11	4.24	1.44	0.234	0.0744	0.303	0.0961	0.391	0.125	0.489	0.159
40	3.81	1.19			0.257	0.0753	0.335	0.0974	0.433	0.128	0.542	0.163
42					0.283	0.0761	0.369	0.0988	0.478	0.130	0.598	0.166
44					0.311	0.0769	0.405	0.100	0.524	0.132	0.656	0.170
46					0.340	0.0778	0.443	0.102	0.573	0.135	0.717	0.174
48					0.370	0.0787	0.482	0.103	0.624	0.137	0.780	0.178
50					0.401	0.0796	0.523	0.105	0.677	0.140	0.847	0.183

Properties

	W40×167	W40×149	W36×798*	W36×650*	W36×527*	W36×439*
L_p, ft	8.48	8.09	14.9	14.5	14.2	13.9
$\phi_b M_p$, kip-ft	2600	2240	13400	10700	8550	7010
$n \times 10^3$	3.30	4.07	0.338	0.431	0.547	0.672

[a] b has the units (kips)$^{-1}$.
[b] m and n have the units (kip-ft)$^{-1}$.
† Web may be slender for combined axial and bending stress.
* ASTM A6 tensile group 4 or 5 shape. Special requirements may apply per LRFD Specification Section A3.1c.
Note: Heavy line indicates Kl/r equal to or greater than 200.

$F_y = 50$ ksi
$\phi_c = 0.85$
$\phi_b = 0.90$

Table 6-2 (cont.).
W-Shapes
Values of b^a, m^b, and n^b for Beam-Columns

Shape		W36×											
		393*		359*		328*		300		280		260	
		$b \times 10^3$	$m \times 10^3$	$b \times 10^3$	$m \times 10^3$	$b \times 10^3$	$m \times 10^3$	$b \times 10^3$	$m \times 10^3$	$b \times 10^3$	$m \times 10^3$	$b \times 10^3$	$m \times 10^3$
Effective length KL (ft) with respect to least radius of gyration r_y or unbraced length L_b (ft) for X-X axis bending	0	0.203	0.142	0.224	0.157	0.244	0.172	0.266	0.188	0.286	0.203	0.312†	0.219
	11	0.221	0.142	0.244	0.157	0.266	0.172	0.291	0.188	0.312	0.203	0.336	0.219
	12	0.224	0.142	0.248	0.157	0.271	0.172	0.295	0.188	0.317	0.203	0.342	0.219
	13	0.228	0.142	0.253	0.157	0.275	0.172	0.301	0.188	0.323	0.203	0.348	0.219
	14	0.232	0.142	0.257	0.158	0.281	0.173	0.307	0.189	0.329	0.204	0.355	0.221
	15	0.237	0.144	0.263	0.159	0.287	0.174	0.313	0.191	0.336	0.206	0.363	0.224
	16	0.242	0.145	0.269	0.161	0.293	0.176	0.320	0.193	0.344	0.209	0.371	0.227
	17	0.248	0.146	0.275	0.162	0.300	0.178	0.328	0.196	0.352	0.211	0.381	0.230
	18	0.254	0.148	0.282	0.164	0.308	0.180	0.336	0.198	0.361	0.214	0.391	0.233
	19	0.260	0.149	0.289	0.166	0.316	0.182	0.345	0.200	0.371	0.217	0.401	0.236
	20	0.268	0.151	0.297	0.168	0.325	0.184	0.355	0.203	0.382	0.219	0.413	0.239
	22	0.284	0.154	0.315	0.171	0.345	0.189	0.377	0.208	0.406	0.225	0.439	0.245
	24	0.302	0.157	0.337	0.175	0.368	0.193	0.403	0.213	0.434	0.231	0.470	0.252
	26	0.324	0.160	0.361	0.179	0.396	0.198	0.433	0.219	0.466	0.237	0.506	0.260
	28	0.349	0.164	0.390	0.183	0.427	0.203	0.468	0.224	0.504	0.244	0.548	0.267
	30	0.378	0.167	0.423	0.188	0.464	0.208	0.508	0.231	0.549	0.251	0.597	0.275
	32	0.412	0.171	0.462	0.192	0.507	0.213	0.556	0.237	0.600	0.259	0.654	0.284
	34	0.452	0.175	0.507	0.197	0.557	0.219	0.611	0.244	0.660	0.267	0.721	0.293
	36	0.497	0.179	0.560	0.202	0.616	0.225	0.676	0.251	0.731	0.275	0.800	0.303
	38	0.552	0.183	0.623	0.207	0.686	0.232	0.752	0.259	0.815	0.284	0.892	0.318
	40	0.612	0.188	0.690	0.213	0.760	0.239	0.834	0.268	0.903	0.301	0.988	0.343
	42	0.675	0.192	0.761	0.219	0.838	0.247	0.919	0.286	0.995	0.323	1.09	0.368
	44	0.741	0.197	0.835	0.226	0.919	0.263	1.01	0.305	1.09	0.344	1.20	0.394
	46	0.809	0.203	0.913	0.239	1.00	0.278	1.10	0.324	1.19	0.366	1.31	0.419
	48	0.881	0.212	0.994	0.253	1.09	0.294	1.20	0.343	1.30	0.388	1.42	0.445
	50	0.956	0.222	1.08	0.266	1.19	0.310	1.30	0.362	1.41	0.410	1.54	0.470

Properties

	393*	359*	328*	300	280	260
L_p, ft	13.8	13.6	13.6	13.5	13.5	13.4
$\phi_b M_p$, kip-ft	6260	5660	5180	4730	4390	4050
$n \times 10^3$	0.760	0.841	0.924	1.01	1.10	1.20

[a] b has the units (kips)$^{-1}$.
[b] m and n have the units (kip-ft)$^{-1}$.
† Web may be slender for combined axial and bending stress.
*ASTM A6 tensile group 4 or 5 shape. Special requirements may apply per LRFD Specification Section A3.1c.

Table 6-2 (cont.). W-Shapes Values of b^a, m^b, and n^b for Beam-Columns

$F_y = 50$ ksi
$\phi_c = 0.85$
$\phi_b = 0.90$

Shape	W36× 245		230		256		232		210		194	
	$b \times 10^3$	$m \times 10^3$	$b \times 10^3$	$m \times 10^3$	$b \times 10^3$	$m \times 10^3$	$b \times 10^3$	$m \times 10^3$	$b \times 10^3$	$m \times 10^3$	$b \times 10^3$	$m \times 10^3$
0	0.334†	0.235	0.360†	0.251	0.312	0.228	0.350†	0.253	0.391†	0.285	0.433†	0.309
11	0.362†	0.235	0.390†	0.251	0.374	0.234	0.416	0.261	0.461	0.295	0.511†	0.321
12	0.367†	0.235	0.396†	0.251	0.387	0.238	0.431	0.266	0.478	0.300	0.528†	0.327
13	0.374†	0.235	0.403†	0.251	0.402	0.242	0.448	0.271	0.497	0.306	0.547†	0.334
14	0.380†	0.237	0.410†	0.254	0.419	0.246	0.467	0.276	0.519	0.312	0.568†	0.341
15	0.388†	0.240	0.418†	0.257	0.437	0.251	0.488	0.281	0.543	0.319	0.593	0.348
16	0.395	0.243	0.427†	0.261	0.458	0.255	0.512	0.286	0.571	0.325	0.623	0.356
17	0.405	0.246	0.437†	0.264	0.481	0.260	0.538	0.292	0.601	0.332	0.657	0.364
18	0.416	0.250	0.448†	0.268	0.507	0.265	0.568	0.298	0.636	0.339	0.695	0.372
19	0.428	0.253	0.459†	0.272	0.536	0.270	0.601	0.304	0.674	0.347	0.737	0.381
20	0.440	0.256	0.471	0.276	0.568	0.275	0.638	0.310	0.717	0.355	0.785	0.390
22	0.469	0.264	0.502	0.284	0.645	0.286	0.726	0.324	0.819	0.372	0.898	0.410
24	0.502	0.271	0.538	0.292	0.740	0.298	0.836	0.339	0.947	0.390	1.04	0.431
26	0.541	0.280	0.581	0.301	0.862	0.311	0.976	0.355	1.11	0.411	1.22	0.467
28	0.587	0.288	0.630	0.311	0.999	0.326	1.13	0.381	1.29	0.458	1.42	0.525
30	0.640	0.297	0.688	0.321	1.15	0.352	1.30	0.421	1.48	0.508	1.63	0.583
32	0.702	0.307	0.755	0.332	1.31	0.384	1.48	0.461	1.68	0.558	1.85	0.642
34	0.775	0.318	0.835	0.344	1.47	0.417	1.67	0.501	1.90	0.608	2.09	0.701
36	0.863	0.329	0.930	0.361	1.65	0.450	1.87	0.541	2.13	0.659	2.34	0.761
38	0.961	0.354	1.04	0.392	1.84	0.482	2.08	0.581	2.37	0.710	2.61	0.821
40	1.07	0.382	1.15	0.424	2.04	0.515	2.31	0.621	2.63	0.761	2.89	0.881
42	1.17	0.411	1.27	0.457	2.25	0.547	2.55	0.661	2.89	0.812	3.19	0.941
44	1.29	0.440	1.39	0.489	2.47	0.580						
46	1.41	0.469	1.52	0.522								
48	1.53	0.498	1.65	0.555								
50	1.66	0.527	1.79	0.589								

Effective length KL (ft) with respect to least radius of gyration r_y or unbraced length L_b (ft) for X-X axis bending

Properties

	245	230	256	232	210	194
L_p, ft	13.2	13.2	9.36	9.25	9.11	9.04
$\phi_b M_p$, kip-ft	3790	3540	3900	3510	3120	2880
$n \times 10^3$	1.28	1.39	1.83	2.05	2.34	2.55

[a] b has the units (kips)$^{-1}$.
[b] m and n have the units (kip-ft)$^{-1}$.
† Web may be slender for combined axial and bending stress.
Note: Heavy line indicates Kl/r equal to or greater than 200.

$F_y = 50$ ksi
$\phi_c = 0.85$
$\phi_b = 0.90$

Table 6-2 (cont.). W-Shapes Values of b^a, m^b, and n^b for Beam-Columns

Shape	W36×										W33×	
	182		170		160		150		135		387*	
	b ×10³	m ×10³	b ×10³	m ×10³	b ×10³	m ×10³	b ×10³	m ×10³	b ×10³	m ×10³	b ×10³	m ×10³
Effective length *KL* (ft) with respect to least radius of gyration r_y or unbraced length L_b (ft) for X-X axis bending												
0	0.468†	0.330	0.509†	0.355	0.550†	0.380	0.592†	0.408	0.672†	0.466	0.206	0.152
11	0.551†	0.343	0.599†	0.370	0.648†	0.398	0.699†	0.428	0.800†	0.494	0.226	0.152
12	0.569†	0.350	0.619†	0.378	0.670†	0.406	0.723†	0.438	0.828†	0.506	0.230	0.152
13	0.589†	0.358	0.641†	0.386	0.694†	0.415	0.749†	0.448	0.860†	0.518	0.234	0.152
14	0.612†	0.365	0.666†	0.394	0.721†	0.425	0.779†	0.459	0.896†	0.531	0.239	0.153
15	0.638†	0.373	0.694†	0.403	0.752†	0.435	0.813†	0.470	0.938†	0.545	0.244	0.154
16	0.668†	0.382	0.726†	0.412	0.787†	0.445	0.851†	0.481	0.985†	0.560	0.249	0.156
17	0.701	0.391	0.763†	0.422	0.827†	0.456	0.895†	0.493	1.04†	0.575	0.256	0.157
18	0.742	0.400	0.804†	0.432	0.873†	0.468	0.944†	0.506	1.10†	0.591	0.262	0.159
19	0.788	0.409	0.850	0.443	0.924†	0.480	1.00†	0.520	1.17†	0.608	0.270	0.160
20	0.839	0.419	0.907	0.454	0.982	0.493	1.07†	0.534	1.25†	0.626	0.278	0.162
22	0.961	0.441	1.04	0.478	1.13	0.520	1.23	0.565	1.46	0.666	0.295	0.165
24	1.12	0.465	1.21	0.505	1.32	0.558	1.44	0.616	1.73	0.761	0.316	0.168
26	1.31	0.513	1.42	0.568	1.55	0.637	1.69	0.706	2.03	0.874	0.341	0.171
28	1.52	0.577	1.65	0.640	1.80	0.720	1.96	0.798	2.35	0.992	0.369	0.175
30	1.74	0.642	1.89	0.713	2.07	0.804	2.25	0.893	2.70	1.11	0.402	0.179
32	1.98	0.708	2.16	0.788	2.35	0.890	2.56	0.991	3.07	1.24	0.441	0.182
34	2.24	0.775	2.43	0.864	2.66	0.977	2.89	1.09	3.47	1.36	0.486	0.187
36	2.51	0.842	2.73	0.941	2.98	1.06	3.24	1.19	3.89	1.49	0.540	0.191
38	2.80	0.909	3.04	1.02	3.32	1.15	3.61	1.29	4.33	1.63	0.601	0.195
40	3.10	0.977	3.37	1.10	3.68	1.24	4.00	1.39			0.666	0.200
42	3.42	1.05	3.71	1.17							0.735	0.204
44											0.806	0.210
46											0.881	0.215
48											0.960	0.221
50											1.04	0.232
Properties												
L_p, ft	9.01		8.94		8.83		8.72		8.41		13.3	
$\phi_b M_p$, kip-ft	2690		2510		2340		2180		1910		5850	
$n \times 10^3$	2.74		2.97		3.22		3.50		4.19		0.790	

[a] b has the units (kips)$^{-1}$.
[b] m and n have the units (kip-ft)$^{-1}$.
† Web may be slender for combined axial and bending stress.
* ASTM A6 tensile group 4 or 5 shape. Special requirements may apply per LRFD Specification Section A3.1c.
Note: Heavy line indicates Kl/r equal to or greater than 200.

Table 6-2 (cont.). W-Shapes Values of b^a, m^b, and n^b for Beam-Columns

$F_y = 50$ ksi
$\phi_c = 0.85$
$\phi_b = 0.90$

Shape	W33×											
	354*		318*		291		263		241		221	
	$b \times 10^3$	$m \times 10^3$	$b \times 10^3$	$m \times 10^3$	$b \times 10^3$	$m \times 10^3$	$b \times 10^3$	$m \times 10^3$	$b \times 10^3$	$m \times 10^3$	$b \times 10^3$	$m \times 10^3$
Effective length *KL* (ft) with respect to least radius of gyration r_y or unbraced length L_b (ft) for X-X axis bending												
0	0.226	0.167	0.251	0.187	0.275	0.204	0.304	0.228	0.331†	0.252	0.368†	0.277
11	0.248	0.167	0.276	0.187	0.302	0.204	0.334	0.228	0.365	0.252	0.401†	0.277
12	0.252	0.167	0.281	0.187	0.307	0.204	0.340	0.228	0.372	0.252	0.408†	0.277
13	0.257	0.167	0.286	0.187	0.313	0.204	0.347	0.228	0.380	0.253	0.414†	0.278
14	0.262	0.168	0.292	0.188	0.320	0.207	0.354	0.231	0.388	0.256	0.424	0.281
15	0.268	0.170	0.299	0.190	0.327	0.209	0.362	0.234	0.397	0.259	0.434	0.285
16	0.274	0.172	0.306	0.192	0.335	0.211	0.371	0.236	0.407	0.263	0.445	0.289
17	0.281	0.173	0.314	0.195	0.344	0.214	0.381	0.239	0.418	0.266	0.457	0.293
18	0.289	0.175	0.322	0.197	0.353	0.216	0.392	0.242	0.430	0.270	0.470	0.297
19	0.297	0.177	0.331	0.199	0.364	0.219	0.403	0.245	0.443	0.273	0.485	0.301
20	0.306	0.179	0.341	0.201	0.375	0.222	0.416	0.249	0.457	0.277	0.500	0.306
22	0.326	0.182	0.364	0.206	0.400	0.227	0.444	0.255	0.489	0.285	0.536	0.315
24	0.349	0.186	0.391	0.211	0.430	0.233	0.477	0.262	0.526	0.293	0.578	0.324
26	0.376	0.190	0.422	0.216	0.464	0.239	0.516	0.269	0.570	0.302	0.627	0.335
28	0.408	0.195	0.458	0.221	0.505	0.245	0.562	0.277	0.622	0.311	0.685	0.346
30	0.445	0.199	0.500	0.227	0.553	0.252	0.616	0.285	0.683	0.321	0.753	0.357
32	0.489	0.204	0.550	0.232	0.609	0.259	0.679	0.294	0.755	0.332	0.833	0.370
34	0.540	0.209	0.609	0.239	0.674	0.267	0.753	0.303	0.839	0.343	0.928	0.383
36	0.601	0.214	0.679	0.245	0.754	0.275	0.843	0.313	0.940	0.355	1.04	0.407
38	0.670	0.219	0.756	0.252	0.840	0.283	0.939	0.324	1.05	0.383	1.16	0.441
40	0.742	0.225	0.838	0.259	0.930	0.295	1.04	0.348	1.16	0.413	1.29	0.476
42	0.818	0.231	0.924	0.267	1.03	0.314	1.15	0.372	1.28	0.442	1.42	0.511
44	0.898	0.237	1.01	0.284	1.13	0.334	1.26	0.397	1.40	0.472	1.55	0.547
46	0.982	0.248	1.11	0.301	1.23	0.354	1.38	0.421	1.53	0.502	1.70	0.582
48	1.07	0.261	1.21	0.317	1.34	0.374	1.50	0.445	1.67	0.532	1.85	0.618
50	1.16	0.275	1.31	0.333	1.45	0.394	1.63	0.470	1.81	0.561	2.01	0.654
Properties												
L_p, ft	13.2		13.1		13.0		12.9		12.8		12.7	
$\phi_b M_p$, kip-ft	5330		4760		4350		3900		3530		3210	
$n \times 10^3$	0.873		0.982		1.08		1.21		1.34		1.49	

[a] *b* has the units (kips)$^{-1}$.
[b] *m* and *n* have the units (kip-ft)$^{-1}$.
† Web may be slender for combined axial and bending stress.
* ASTM A6 tensile group 4 or 5 shape. Special requirements may apply per LRFD Specification Section A3.1c.

$F_y = 50$ ksi
$\phi_c = 0.85$
$\phi_b = 0.90$

Table 6-2 (cont.). W-Shapes Values of b^a, m^b, and n^b for Beam-Columns

Shape	W33× 201		169		152		141		130		118	
Effective length KL (ft) with respect to least radius of gyration r_y or unbraced length L_b (ft) for X-X axis bending	$b \times 10^3$	$m \times 10^3$	$b \times 10^3$	$m \times 10^3$	$b \times 10^3$	$m \times 10^3$	$b \times 10^3$	$m \times 10^3$	$b \times 10^3$	$m \times 10^3$	$b \times 10^3$	$m \times 10^3$
0	0.412†	0.307	0.504†	0.377	0.566†	0.424	0.619†	0.461	0.682†	0.508	0.768†	0.571
11	0.450†	0.307	0.599†	0.393	0.673†	0.444	0.738†	0.486	0.815†	0.537	0.924†	0.610
12	0.458†	0.307	0.619†	0.401	0.696†	0.454	0.764†	0.497	0.844†	0.550	0.958†	0.625
13	0.466†	0.308	0.643†	0.410	0.723†	0.464	0.793†	0.508	0.878†	0.563	0.997†	0.642
14	0.475†	0.313	0.669†	0.418	0.753†	0.475	0.827†	0.520	0.915†	0.577	1.04†	0.658
15	0.486†	0.317	0.700†	0.428	0.787†	0.486	0.865†	0.532	0.959†	0.592	1.09†	0.676
16	0.497†	0.322	0.734†	0.437	0.826†	0.497	0.908†	0.545	1.01†	0.607	1.15†	0.695
17	0.509†	0.326	0.773	0.447	0.870†	0.509	0.958†	0.559	1.06†	0.623	1.22†	0.715
18	0.523†	0.331	0.820	0.458	0.919	0.522	1.01†	0.574	1.13†	0.640	1.30†	0.736
19	0.536†	0.336	0.873	0.469	0.979	0.535	1.08†	0.589	1.20†	0.658	1.38†	0.758
20	0.554	0.341	0.933	0.480	1.05	0.549	1.15	0.605	1.28	0.677	1.49†	0.782
22	0.594	0.352	1.07	0.505	1.21	0.579	1.34	0.640	1.50	0.718	1.75	0.846
24	0.641	0.363	1.26	0.532	1.42	0.620	1.58	0.705	1.78	0.815	2.08	0.986
26	0.697	0.375	1.47	0.592	1.67	0.706	1.86	0.805	2.09	0.933	2.44	1.13
28	0.762	0.388	1.71	0.663	1.94	0.794	2.15	0.907	2.42	1.05	2.83	1.28
30	0.839	0.402	1.96	0.735	2.22	0.883	2.47	1.01	2.78	1.18	3.25	1.44
32	0.931	0.417	2.23	0.808	2.53	0.974	2.81	1.12	3.16	1.31	3.70	1.60
34	1.04	0.435	2.52	0.881	2.85	1.07	3.18	1.23	3.57	1.44	4.18	1.76
36	1.17	0.476	2.83	0.955	3.20	1.16	3.56	1.34	4.00	1.57	4.68	1.92
38	1.30	0.517	3.15	1.03	3.57	1.25	3.97	1.45	4.45	1.70	5.22	2.09
40	1.44	0.560	3.49	1.10	3.95	1.34	4.40	1.56				
42	1.59	0.602										
44	1.74	0.645										
46	1.90	0.688										
48	2.07	0.732										
50	2.25	0.775										

Properties

	201	169	152	141	130	118
L_p, ft	12.6	8.83	8.72	8.58	8.44	8.19
$\phi_b M_p$, kip-ft	2900	2360	2100	1930	1750	1560
$n \times 10^3$	1.66	2.93	3.35	3.70	4.17	4.85

[a] *b* has the units (kips)$^{-1}$.
[b] *m* and *n* have the units (kip-ft)$^{-1}$.
† Web may be slender for combined axial and bending stress.
Note: Heavy line indicates Kl/r equal to or greater than 200.

Table 6-2 (cont.). W-Shapes Values of b^a, m^b, and n^b for Beam-Columns

$F_y = 50$ ksi
$\phi_c = 0.85$
$\phi_b = 0.90$

Shape	W30×											
	391*		357*		326*		292*		261		235	
	$b \times 10^3$	$m \times 10^3$	$b \times 10^3$	$m \times 10^3$	$b \times 10^3$	$m \times 10^3$	$b \times 10^3$	$m \times 10^3$	$b \times 10^3$	$m \times 10^3$	$b \times 10^3$	$m \times 10^3$
0	0.205	0.163	0.224	0.180	0.246	0.199	0.274	0.224	0.306	0.251	0.340	0.280
11	0.225	0.163	0.247	0.180	0.271	0.199	0.303	0.224	0.339	0.251	0.377	0.280
12	0.229	0.163	0.251	0.180	0.276	0.199	0.308	0.224	0.346	0.251	0.385	0.280
13	0.234	0.164	0.256	0.180	0.282	0.200	0.315	0.224	0.353	0.253	0.393	0.282
14	0.238	0.165	0.262	0.181	0.288	0.202	0.322	0.227	0.361	0.256	0.402	0.285
15	0.244	0.166	0.268	0.183	0.295	0.204	0.330	0.229	0.370	0.259	0.412	0.289
16	0.250	0.168	0.275	0.185	0.302	0.205	0.338	0.231	0.380	0.262	0.423	0.292
17	0.256	0.169	0.282	0.186	0.311	0.207	0.347	0.234	0.391	0.265	0.435	0.296
18	0.264	0.170	0.290	0.188	0.320	0.209	0.357	0.236	0.402	0.268	0.448	0.300
19	0.271	0.172	0.299	0.190	0.329	0.212	0.368	0.239	0.415	0.271	0.463	0.304
20	0.280	0.173	0.308	0.191	0.340	0.214	0.380	0.242	0.429	0.274	0.479	0.308
22	0.299	0.176	0.329	0.195	0.364	0.218	0.408	0.247	0.461	0.281	0.514	0.316
24	0.321	0.179	0.354	0.199	0.392	0.222	0.440	0.253	0.498	0.288	0.556	0.325
26	0.347	0.182	0.383	0.203	0.425	0.227	0.477	0.259	0.542	0.296	0.606	0.334
28	0.378	0.186	0.418	0.207	0.464	0.232	0.522	0.265	0.593	0.304	0.664	0.344
30	0.413	0.189	0.458	0.211	0.510	0.237	0.574	0.271	0.655	0.312	0.734	0.354
32	0.456	0.192	0.506	0.215	0.564	0.243	0.635	0.278	0.727	0.321	0.816	0.365
34	0.505	0.196	0.562	0.220	0.628	0.248	0.709	0.285	0.814	0.331	0.915	0.376
36	0.565	0.200	0.629	0.224	0.704	0.254	0.794	0.293	0.913	0.341	1.03	0.389
38	0.629	0.204	0.701	0.229	0.785	0.260	0.885	0.301	1.02	0.351	1.14	0.409
40	0.697	0.208	0.776	0.235	0.870	0.267	0.981	0.309	1.13	0.368	1.27	0.439
42	0.769	0.212	0.856	0.240	0.959	0.273	1.08	0.318	1.24	0.392	1.40	0.468
44	0.844	0.217	0.939	0.246	1.05	0.281	1.19	0.336	1.36	0.416	1.53	0.498
46	0.922	0.221	1.03	0.252	1.15	0.291	1.30	0.355	1.49	0.440	1.68	0.527
48	1.00	0.226	1.12	0.258	1.25	0.306	1.41	0.374	1.62	0.464	1.82	0.557
50	1.09	0.231	1.21	0.268	1.36	0.321	1.53	0.393	1.76	0.488	1.98	0.586

Effective length KL (ft) with respect to least radius of gyration r_y or unbraced length L_b (ft) for X-X axis bending

Properties

	391*	357*	326*	292*	261	235
L_p, ft	13.0	12.9	12.7	12.6	12.5	12.4
$\phi_b M_p$, kip-ft	5440	4950	4460	3980	3540	3180
$n \times 10^3$	0.798	0.883	0.975	1.10	1.24	1.39

[a] b has the units (kips)$^{-1}$.
[b] m and n have the units (kip-ft)$^{-1}$.
*ASTM A6 tensile group 4 or 5 shape. Special requirements may apply per LRFD Specification Section A3.1c.

$F_y = 50$ ksi
$\phi_c = 0.85$
$\phi_b = 0.90$

Table 6-2 (cont.). W-Shapes
Values of b^a, m^b, and n^b for Beam-Columns

Effective length KL (ft) with respect to least radius of gyration r_y or unbraced length L_b (ft) for X-X axis bending

Shape	W30× 211		191		173		148		132		124	
	$b \times 10^3$	$m \times 10^3$	$b \times 10^3$	$m \times 10^3$	$b \times 10^3$	$m \times 10^3$	$b \times 10^3$	$m \times 10^3$	$b \times 10^3$	$m \times 10^3$	$b \times 10^3$	$m \times 10^3$
0	0.378	0.316	0.424†	0.351	0.476†	0.391	0.563†	0.474	0.640†	0.542	0.692†	0.581
11	0.420	0.316	0.465	0.351	0.523†	0.391	0.697†	0.505	0.792†	0.581	0.857†	0.624
12	0.428	0.316	0.474	0.351	0.533†	0.391	0.726†	0.516	0.827†	0.595	0.894†	0.640
13	0.438	0.318	0.485	0.355	0.544†	0.396	0.762	0.528	0.866†	0.609	0.937†	0.656
14	0.448	0.322	0.497	0.360	0.556†	0.401	0.805	0.540	0.909	0.625	0.986†	0.673
15	0.460	0.327	0.509	0.365	0.569†	0.407	0.853	0.553	0.966	0.641	1.04†	0.691
16	0.472	0.331	0.523	0.370	0.584†	0.413	0.908	0.566	1.03	0.657	1.11	0.709
17	0.486	0.336	0.539	0.375	0.598	0.419	0.971	0.581	1.10	0.675	1.19	0.729
18	0.501	0.340	0.556	0.381	0.618	0.426	1.04	0.595	1.19	0.694	1.28	0.750
19	0.517	0.345	0.574	0.386	0.639	0.433	1.12	0.611	1.28	0.714	1.38	0.773
20	0.535	0.350	0.594	0.392	0.661	0.439	1.22	0.628	1.39	0.735	1.50	0.796
22	0.575	0.360	0.640	0.404	0.713	0.454	1.44	0.664	1.66	0.782	1.80	0.866
24	0.622	0.371	0.694	0.417	0.775	0.469	1.72	0.734	1.97	0.899	2.14	0.998
26	0.679	0.382	0.757	0.431	0.848	0.485	2.02	0.828	2.32	1.02	2.51	1.13
28	0.745	0.394	0.833	0.446	0.934	0.503	2.34	0.922	2.69	1.14	2.92	1.27
30	0.824	0.407	0.922	0.461	1.04	0.522	2.69	1.02	3.08	1.26	3.35	1.41
32	0.917	0.421	1.03	0.478	1.16	0.542	3.06	1.11	3.51	1.38	3.81	1.55
34	1.03	0.436	1.16	0.500	1.31	0.587	3.45	1.21	3.96	1.51	4.30	1.69
36	1.15	0.459	1.30	0.545	1.47	0.641	3.87	1.30	4.44	1.63	4.82	1.83
38	1.29	0.496	1.45	0.590	1.63	0.696						
40	1.43	0.533	1.60	0.636	1.81	0.751						
42	1.57	0.571	1.77	0.682	2.00	0.807						
44	1.72	0.608	1.94	0.727	2.19	0.863						
46	1.89	0.645	2.12	0.773	2.39	0.919						
48	2.05	0.682	2.31	0.819	2.61	0.976						
50	2.23	0.720	2.50	0.865	2.83	1.03						

Properties

	211	191	173	148	132	124
L_p, ft	12.3	12.2	12.1	8.05	7.95	7.88
$\phi_b M_p$, kip-ft	2820	2530	2280	1880	1640	1530
$n \times 10^3$	1.58	1.77	1.98	3.65	4.25	4.59

[a] b has the units (kips)$^{-1}$.
[b] m and n have the units (kip-ft)$^{-1}$.
† Web may be slender for combined axial and bending stress.
Note: Heavy line indicates Kl/r of 200.

Table 6-2 (cont.). W-Shapes Values of b^a, m^b, and n^b for Beam-Columns

$F_y = 50$ ksi
$\phi_c = 0.85$
$\phi_b = 0.90$

Shape	W30× 116		W30× 108		W30× 99		W30× 90		W27× 539*		W27× 368*	
	$b \times 10^3$	$m \times 10^3$	$b \times 10^3$	$m \times 10^3$	$b \times 10^3$	$m \times 10^3$	$b \times 10^3$	$m \times 10^3$	$b \times 10^3$	$m \times 10^3$	$b \times 10^3$	$m \times 10^3$
Effective length KL (ft) with respect to least radius of gyration r_y or unbraced length L_b (ft) for X-X axis bending												
0	0.747†	0.627	0.817†	0.685	0.904†	0.760	1.03†	0.838	0.148	0.125	0.218	0.191
11	0.930†	0.679	1.02†	0.747	1.14†	0.834	1.29†	0.923	0.163	0.125	0.242	0.191
12	0.971†	0.696	1.07†	0.767	1.19†	0.858	1.34†	0.950	0.166	0.125	0.247	0.191
13	1.02†	0.715	1.12†	0.788	1.25†	0.883	1.41†	0.978	0.169	0.125	0.252	0.192
14	1.07†	0.734	1.18†	0.811	1.32†	0.909	1.49†	1.01	0.173	0.126	0.258	0.194
15	1.14†	0.754	1.25†	0.835	1.41†	0.937	1.58†	1.04	0.177	0.127	0.265	0.195
16	1.21†	0.776	1.34†	0.860	1.50†	0.967	1.69†	1.07	0.181	0.127	0.272	0.197
17	1.30	0.799	1.43	0.887	1.62†	0.999	1.81†	1.11	0.186	0.128	0.280	0.198
18	1.40	0.824	1.55	0.916	1.75	1.03	1.96†	1.15	0.191	0.129	0.289	0.200
19	1.52	0.850	1.69	0.947	1.91	1.07	2.13†	1.19	0.197	0.129	0.298	0.201
20	1.66	0.877	1.85	0.979	2.10	1.12	2.34	1.28	0.203	0.130	0.308	0.203
22	1.99	0.989	2.23	1.13	2.55	1.32	2.83	1.52	0.217	0.132	0.332	0.206
24	2.37	1.14	2.65	1.31	3.03	1.53	3.37	1.76	0.233	0.133	0.359	0.209
26	2.78	1.30	3.11	1.50	3.56	1.75	3.96	2.02	0.252	0.134	0.392	0.213
28	3.23	1.46	3.61	1.68	4.12	1.97	4.59	2.29	0.275	0.136	0.431	0.216
30	3.70	1.62	4.15	1.88	4.73	2.20	5.27	2.56	0.301	0.138	0.476	0.220
32	4.21	1.78	4.72	2.07	5.39	2.44	5.99	2.84	0.332	0.139	0.531	0.224
34	4.76	1.95	5.32	2.27	6.08	2.68	6.77	3.13	0.369	0.141	0.597	0.228
36	5.33	2.12							0.413	0.142	0.669	0.232
38									0.460	0.144	0.745	0.236
40									0.510	0.146	0.826	0.241
42									0.562	0.148	0.910	0.245
44									0.617	0.149	0.999	0.250
46									0.674	0.151	1.09	0.255
48									0.734	0.153	1.19	0.260
50									0.797	0.155	1.29	0.266
Properties												
L_p, ft	7.74		7.59		7.42		7.38		12.9		12.3	
$\phi_b M_p$, kip-ft	1420		1300		1170		1060		7090		4650	
$n \times 10^3$	5.05		5.66		6.45		7.15		0.570		0.883	

[a] b has the units (kips)$^{-1}$.
[b] m and n have the units (kip-ft)$^{-1}$.
† Web may be slender for combined axial and bending stress.
* ASTM A6 tensile group 4 or 5 shape. Special requirements may apply per LRFD Specification Section A3.1c.
Note: Heavy line indicates Kl/r equal to or greater than 200.

$F_y = 50$ ksi
$\phi_c = 0.85$
$\phi_b = 0.90$

Table 6-2 (cont.). W-Shapes Values of b^a, m^b, and n^b for Beam-Columns

Shape	W27×											
	336*		307*		281*		258		235		217	
	$b \times 10^3$	$m \times 10^3$	$b \times 10^3$	$m \times 10^3$	$b \times 10^3$	$m \times 10^3$	$b \times 10^3$	$m \times 10^3$	$b \times 10^3$	$m \times 10^3$	$b \times 10^3$	$m \times 10^3$
Effective length KL (ft) with respect to least radius of gyration r_y or unbraced length L_b (ft) for X-X axis bending												
0	0.238	0.210	0.260	0.230	0.284	0.253	0.310	0.278	0.339	0.307	0.368	0.333
11	0.265	0.210	0.290	0.230	0.317	0.253	0.347	0.278	0.380	0.307	0.413	0.333
12	0.270	0.210	0.297	0.230	0.324	0.253	0.354	0.279	0.389	0.308	0.422	0.334
13	0.276	0.211	0.303	0.232	0.331	0.256	0.362	0.281	0.398	0.311	0.432	0.338
14	0.283	0.213	0.311	0.234	0.340	0.258	0.372	0.284	0.408	0.315	0.443	0.343
15	0.290	0.215	0.319	0.236	0.349	0.261	0.382	0.287	0.420	0.318	0.456	0.347
16	0.298	0.216	0.328	0.239	0.359	0.263	0.393	0.291	0.432	0.322	0.469	0.351
17	0.307	0.218	0.338	0.241	0.370	0.266	0.405	0.294	0.446	0.326	0.485	0.355
18	0.317	0.220	0.349	0.243	0.382	0.269	0.419	0.297	0.461	0.330	0.501	0.360
19	0.327	0.222	0.361	0.245	0.395	0.271	0.434	0.300	0.478	0.334	0.519	0.365
20	0.339	0.224	0.374	0.248	0.409	0.274	0.450	0.304	0.496	0.338	0.539	0.369
22	0.365	0.228	0.403	0.253	0.442	0.280	0.486	0.311	0.537	0.346	0.584	0.379
24	0.396	0.232	0.438	0.258	0.481	0.286	0.530	0.318	0.586	0.355	0.637	0.390
26	0.433	0.236	0.480	0.263	0.527	0.292	0.582	0.326	0.644	0.365	0.701	0.401
28	0.476	0.241	0.529	0.269	0.582	0.299	0.643	0.334	0.714	0.375	0.777	0.412
30	0.527	0.245	0.588	0.274	0.647	0.306	0.717	0.342	0.797	0.385	0.869	0.424
32	0.589	0.250	0.658	0.280	0.725	0.313	0.805	0.351	0.898	0.396	0.980	0.438
34	0.663	0.255	0.742	0.287	0.819	0.321	0.909	0.360	1.01	0.408	1.11	0.451
36	0.743	0.260	0.832	0.293	0.918	0.329	1.02	0.370	1.14	0.420	1.24	0.466
38	0.828	0.266	0.927	0.300	1.02	0.337	1.14	0.380	1.27	0.433	1.38	0.493
40	0.917	0.271	1.03	0.307	1.13	0.346	1.26	0.392	1.40	0.457	1.53	0.526
42	1.01	0.277	1.13	0.315	1.25	0.355	1.39	0.409	1.55	0.486	1.69	0.560
44	1.11	0.283	1.24	0.323	1.37	0.366	1.52	0.432	1.70	0.514	1.85	0.593
46	1.21	0.290	1.36	0.331	1.50	0.385	1.66	0.455	1.86	0.543	2.02	0.626
48	1.32	0.296	1.48	0.345	1.63	0.405	1.81	0.479	2.02	0.571	2.20	0.659
50	1.43	0.303	1.61	0.361	1.77	0.424	1.97	0.502	2.19	0.599	2.39	0.693

Properties

	336*	307*	281*	258	235	217
L_p, ft	12.2	12.0	12.0	11.9	11.8	11.7
$\phi_b M_p$, kip-ft	4240	3860	3510	3200	2900	2670
$n \times 10^3$	0.975	1.08	1.19	1.32	1.46	1.58

[a] b has the units (kips)$^{-1}$.
[b] m and n have the units (kip-ft)$^{-1}$.
*ASTM A6 tensile group 4 or 5 shape. Special requirements may apply per LRFD Specification Section A3.1c.

Table 6-2 (cont.). W-Shapes Values of b^a, m^b, and n^b for Beam-Columns

$F_y = 50$ ksi
$\phi_c = 0.85$
$\phi_b = 0.90$

Shape	W27× 194		178		161		146		129		114	
Effective length *KL* (ft) with respect to least radius of gyration r_y or unbraced length L_b (ft) for X-X axis bending	*b* ×10³	*m* ×10³	*b* ×10³	*m* ×10³	*b* ×10³	*m* ×10³	*b* ×10³	*m* ×10³	*b* ×10³	*m* ×10³	*b* ×10³	*m* ×10³
0	0.411	0.376	0.448	0.416	0.496†	0.460	0.559†	0.511	0.640†	0.600	0.736†	0.691
11	0.463	0.376	0.506	0.416	0.559	0.460	0.624†	0.511	0.808	0.643	0.927†	0.746
12	0.473	0.377	0.517	0.419	0.572	0.464	0.637†	0.516	0.849	0.658	0.971†	0.764
13	0.485	0.382	0.530	0.425	0.586	0.471	0.652†	0.524	0.896	0.673	1.02	0.784
14	0.498	0.387	0.545	0.430	0.602	0.478	0.668	0.532	0.950	0.689	1.08	0.804
15	0.512	0.392	0.561	0.437	0.620	0.485	0.688	0.541	1.01	0.706	1.16	0.826
16	0.528	0.398	0.578	0.443	0.640	0.493	0.710	0.549	1.08	0.724	1.24	0.848
17	0.545	0.403	0.598	0.449	0.662	0.500	0.735	0.558	1.16	0.743	1.33	0.872
18	0.564	0.409	0.619	0.456	0.686	0.508	0.762	0.568	1.25	0.763	1.44	0.897
19	0.584	0.414	0.642	0.463	0.712	0.516	0.791	0.577	1.36	0.783	1.56	0.924
20	0.607	0.420	0.668	0.470	0.740	0.525	0.824	0.587	1.47	0.805	1.70	0.953
22	0.659	0.432	0.726	0.485	0.806	0.542	0.898	0.608	1.77	0.853	2.05	1.05
24	0.720	0.445	0.796	0.500	0.884	0.561	0.987	0.631	2.11	0.964	2.44	1.20
26	0.794	0.459	0.879	0.517	0.978	0.581	1.09	0.655	2.47	1.08	2.87	1.35
28	0.882	0.474	0.979	0.535	1.09	0.603	1.22	0.681	2.87	1.20	3.32	1.51
30	0.987	0.490	1.10	0.554	1.23	0.626	1.38	0.709	3.29	1.32	3.82	1.67
32	1.12	0.506	1.25	0.575	1.39	0.658	1.57	0.773	3.74	1.44	4.34	1.83
34	1.26	0.524	1.41	0.610	1.57	0.720	1.77	0.848	4.23	1.56	4.90	1.99
36	1.41	0.559	1.58	0.661	1.76	0.782	1.98	0.924	4.74	1.69	5.49	2.15
38	1.57	0.601	1.76	0.713	1.96	0.845	2.21	1.00				
40	1.74	0.643	1.95	0.764	2.17	0.908	2.45	1.08				
42	1.92	0.685	2.15	0.816	2.40	0.971	2.70	1.15				
44	2.11	0.728	2.36	0.868	2.63	1.03	2.96	1.23				
46	2.31	0.770	2.58	0.919	2.88	1.10	3.24	1.31				
48	2.51	0.812	2.80	0.971	3.13	1.16	3.52	1.39				
50	2.73	0.853	3.04	1.02	3.40	1.22	3.82	1.46				

Properties

	194	178	161	146	129	114
L_p, ft	11.6	11.5	11.4	11.3	7.81	7.70
$\phi_b M_p$, kip-ft	2370	2140	1930	1740	1480	1290
$n \times 10^3$	1.79	2.01	2.23	2.49	4.29	5.02

[a] *b* has the units (kips)$^{-1}$.
[b] *m* and *n* have the units (kip-ft)$^{-1}$.
† Web may be slender for combined axial and bending stress.
Note: Heavy line indicates Kl/r equal to or greater than 200.

$F_y = 50$ ksi
$\phi_c = 0.85$
$\phi_b = 0.90$

Table 6-2 (cont.). W-Shapes Values of b^a, m^b, and n^b for Beam-Columns

Shape	W27×						W24×					
	102		94		84		370*		335*		306*	
	b ×10³	m ×10³	b ×10³	m ×10³	b ×10³	m ×10³	b ×10³	m ×10³	b ×10³	m ×10³	b ×10³	m ×10³
0	0.844†	0.777	0.928†	0.853	1.06†	0.971	0.216	0.210	0.239	0.232	0.262	0.257
11	1.06†	0.844	1.17†	0.932	1.34†	1.07	0.243	0.210	0.270	0.232	0.297	0.257
12	1.11†	0.866	1.23†	0.957	1.41†	1.10	0.249	0.210	0.277	0.233	0.304	0.259
13	1.17†	0.889	1.29†	0.984	1.49†	1.13	0.255	0.212	0.284	0.235	0.312	0.261
14	1.24†	0.913	1.37†	1.01	1.57†	1.17	0.262	0.213	0.291	0.237	0.321	0.263
15	1.31†	0.939	1.45†	1.04	1.68†	1.20	0.269	0.215	0.300	0.239	0.330	0.265
16	1.41	0.966	1.55†	1.07	1.80†	1.24	0.278	0.216	0.310	0.240	0.341	0.267
17	1.51	0.995	1.67	1.11	1.94†	1.28	0.287	0.218	0.320	0.242	0.353	0.269
18	1.64	1.03	1.81	1.14	2.10	1.33	0.297	0.219	0.332	0.244	0.366	0.272
19	1.78	1.06	1.98	1.18	2.30	1.38	0.308	0.220	0.344	0.246	0.380	0.274
20	1.95	1.09	2.17	1.23	2.54	1.48	0.320	0.222	0.358	0.248	0.395	0.276
22	2.36	1.25	2.62	1.44	3.07	1.74	0.348	0.225	0.390	0.252	0.431	0.281
24	2.80	1.43	3.12	1.66	3.66	2.02	0.381	0.228	0.428	0.256	0.474	0.286
26	3.29	1.63	3.66	1.89	4.29	2.30	0.420	0.232	0.473	0.260	0.525	0.291
28	3.82	1.82	4.25	2.12	4.98	2.60	0.467	0.235	0.528	0.264	0.587	0.296
30	4.38	2.02	4.88	2.35	5.72	2.89	0.524	0.238	0.593	0.269	0.661	0.302
32	4.98	2.22	5.55	2.59	6.50	3.20	0.593	0.242	0.673	0.273	0.752	0.308
34	5.63	2.42	6.27	2.83	7.34	3.50	0.669	0.246	0.760	0.278	0.848	0.314
36							0.750	0.250	0.852	0.283	0.951	0.320
38							0.836	0.254	0.949	0.288	1.06	0.326
40							0.926	0.258	1.05	0.294	1.17	0.333
42							1.02	0.262	1.16	0.299	1.29	0.340
44							1.12	0.266	1.27	0.305	1.42	0.347
46							1.23	0.271	1.39	0.311	1.55	0.355
48							1.33	0.275	1.51	0.317	1.69	0.363
50							1.45	0.280	1.64	0.323	1.83	0.371

Row labels: Effective length KL (ft) with respect to least radius of gyration r_y or unbraced length L_b (ft) for X-X axis bending

Properties

	W27×102	W27×94	W27×84	W24×370*	W24×335*	W24×306*
L_p, ft	7.59	7.49	7.31	11.6	11.4	11.3
$\phi_b M_p$, kip-ft	1140	1040	915	4240	3830	3460
$n \times 10^3$	5.68	6.37	7.45	0.930	1.04	1.15

[a] b has the units (kips)$^{-1}$.
[b] m and n have the units (kip-ft)$^{-1}$.
† Web may be slender for combined axial and bending stress.
*ASTM A6 tensile group 4 or 5 shape. Special requirements may apply per LRFD Specification Section A3.1c.
Note: Heavy line indicates Kl/r equal to or greater than 200.

Table 6-2 (cont.). W-Shapes Values of b^a, m^b, and n^b for Beam-Columns

$F_y = 50$ ksi
$\phi_c = 0.85$
$\phi_b = 0.90$

Shape	W24×											
	279*		250*		229		207		192		176	
	$b \times 10^3$	$m \times 10^3$	$b \times 10^3$	$m \times 10^3$	$b \times 10^3$	$m \times 10^3$	$b \times 10^3$	$m \times 10^3$	$b \times 10^3$	$m \times 10^3$	$b \times 10^3$	$m \times 10^3$
0	0.287	0.284	0.320	0.319	0.350	0.351	0.388	0.391	0.418	0.424	0.455	0.464
11	0.326	0.284	0.364	0.319	0.399	0.351	0.443	0.392	0.478	0.425	0.522	0.465
12	0.334	0.286	0.373	0.321	0.410	0.355	0.455	0.396	0.491	0.430	0.536	0.472
13	0.343	0.288	0.383	0.325	0.421	0.359	0.468	0.401	0.505	0.435	0.552	0.478
14	0.352	0.291	0.395	0.328	0.433	0.362	0.482	0.405	0.520	0.441	0.569	0.484
15	0.363	0.294	0.407	0.331	0.447	0.366	0.498	0.410	0.537	0.446	0.588	0.491
16	0.375	0.296	0.421	0.334	0.463	0.370	0.515	0.415	0.556	0.452	0.609	0.498
17	0.388	0.299	0.436	0.338	0.480	0.375	0.534	0.420	0.577	0.458	0.633	0.505
18	0.403	0.302	0.452	0.341	0.498	0.379	0.555	0.425	0.600	0.464	0.658	0.512
19	0.419	0.304	0.471	0.345	0.519	0.383	0.579	0.431	0.626	0.470	0.687	0.519
20	0.436	0.307	0.491	0.348	0.541	0.387	0.604	0.436	0.653	0.476	0.718	0.527
22	0.476	0.313	0.537	0.356	0.593	0.396	0.663	0.447	0.718	0.489	0.790	0.542
24	0.525	0.319	0.592	0.363	0.655	0.406	0.735	0.459	0.795	0.503	0.877	0.559
26	0.583	0.326	0.659	0.372	0.731	0.416	0.821	0.472	0.889	0.518	0.983	0.577
28	0.652	0.332	0.739	0.380	0.822	0.426	0.925	0.485	1.00	0.534	1.11	0.596
30	0.737	0.339	0.838	0.389	0.935	0.437	1.05	0.499	1.14	0.550	1.27	0.617
32	0.839	0.346	0.954	0.398	1.06	0.449	1.20	0.514	1.30	0.568	1.45	0.639
34	0.947	0.354	1.08	0.408	1.20	0.461	1.35	0.529	1.47	0.587	1.63	0.666
36	1.06	0.362	1.21	0.418	1.35	0.474	1.52	0.546	1.65	0.610	1.83	0.717
38	1.18	0.370	1.34	0.429	1.50	0.488	1.69	0.570	1.84	0.652	2.04	0.768
40	1.31	0.378	1.49	0.440	1.66	0.502	1.88	0.606	2.04	0.695	2.26	0.819
42	1.44	0.387	1.64	0.452	1.83	0.530	2.07	0.642	2.24	0.737	2.49	0.870
44	1.59	0.397	1.80	0.469	2.01	0.559	2.27	0.678	2.46	0.779	2.73	0.920
46	1.73	0.406	1.97	0.493	2.20	0.588	2.48	0.714	2.69	0.821	2.99	0.971
48	1.89	0.420	2.15	0.517	2.39	0.617	2.70	0.750	2.93	0.863	3.25	1.02
50	2.05	0.439	2.33	0.541	2.60	0.646	2.93	0.786	3.18	0.904	3.53	1.07

Row labels: Effective length KL (ft) with respect to least radius of gyration r_y or unbraced length L_b (ft) for X-X axis bending

Properties

	279*	250*	229	207	192	176
L_p, ft	11.2	11.1	11.0	10.9	10.8	10.7
$\phi_b M_p$, kip-ft	3130	2790	2530	2270	2100	1920
$n \times 10^3$	1.27	1.44	1.59	1.78	1.93	2.13

[a] b has the units $(\text{kips})^{-1}$.
[b] m and n have the units $(\text{kip-ft})^{-1}$.
*ASTM A6 tensile group 4 or 5 shape. Special requirements may apply per LRFD Specification Section A3.1c.

$F_y = 50$ ksi
$\phi_c = 0.85$
$\phi_b = 0.90$

Table 6-2 (cont.). W-Shapes
Values of b^a, m^b, and n^b for Beam-Columns

Shape	W24×											
	162		146		131		117		104		103	
Effective length KL (ft) with respect to least radius of gyration r_y or unbraced length L_b (ft) for X-X axis bending	$b \times 10^3$	$m \times 10^3$	$b \times 10^3$	$m \times 10^3$	$b \times 10^3$	$m \times 10^3$	$b \times 10^3$	$m \times 10^3$	$b \times 10^3$	$m \times 10^3$	$b \times 10^3$	$m \times 10^3$
0	0.493	0.506	0.547	0.567	0.611	0.641	0.699†	0.725	0.803†	0.820	0.796†	0.847
11	0.566	0.508	0.630	0.570	0.706	0.646	0.797†	0.732	0.915†	0.831	1.07	0.932
12	0.581	0.515	0.647	0.579	0.726	0.656	0.815†	0.745	0.938†	0.845	1.14	0.956
13	0.597	0.522	0.666	0.587	0.748	0.667	0.840	0.757	0.965†	0.861	1.22	0.982
14	0.616	0.530	0.687	0.596	0.772	0.678	0.868	0.771	0.994†	0.877	1.31	1.01
15	0.636	0.537	0.711	0.606	0.799	0.689	0.900	0.784	1.03†	0.893	1.41	1.04
16	0.659	0.545	0.737	0.615	0.830	0.700	0.934	0.799	1.06†	0.910	1.53	1.07
17	0.684	0.553	0.766	0.625	0.863	0.713	0.973	0.813	1.10†	0.928	1.67	1.10
18	0.712	0.561	0.797	0.635	0.900	0.725	1.01	0.829	1.15	0.947	1.84	1.13
19	0.742	0.570	0.832	0.646	0.940	0.738	1.06	0.844	1.20	0.966	2.03	1.17
20	0.776	0.579	0.871	0.656	0.985	0.751	1.11	0.861	1.26	0.986	2.25	1.21
22	0.853	0.597	0.960	0.679	1.09	0.780	1.23	0.896	1.40	1.03	2.72	1.39
24	0.947	0.617	1.07	0.704	1.22	0.810	1.38	0.934	1.57	1.08	3.24	1.58
26	1.06	0.638	1.20	0.730	1.37	0.843	1.56	0.975	1.78	1.13	3.80	1.76
28	1.20	0.661	1.36	0.759	1.56	0.879	1.78	1.02	2.04	1.23	4.41	1.95
30	1.37	0.685	1.56	0.789	1.79	0.942	2.04	1.14	2.34	1.38	5.06	2.14
32	1.56	0.711	1.77	0.852	2.04	1.04	2.32	1.26	2.67	1.53	5.76	2.33
34	1.76	0.765	2.00	0.928	2.30	1.13	2.62	1.38	3.01	1.68		
36	1.97	0.826	2.25	1.00	2.58	1.23	2.94	1.50	3.38	1.83		
38	2.20	0.886	2.50	1.08	2.87	1.32	3.28	1.62	3.76	1.98		
40	2.43	0.946	2.77	1.15	3.18	1.42	3.63	1.74	4.17	2.14		
42	2.68	1.01	3.06	1.23	3.51	1.52	4.00	1.86	4.59	2.29		
44	2.94	1.07	3.35	1.31	3.85	1.61	4.39	1.98	5.04	2.45		
46	3.22	1.13	3.67	1.38	4.21	1.71	4.80	2.11	5.51	2.60		
48	3.50	1.19	3.99	1.46	4.58	1.80	5.23	2.23	6.00	2.76		
50	3.80	1.25	4.33	1.53								

Properties

	162	146	131	117	104	103
L_p, ft	10.8	10.6	10.5	10.4	10.3	7.03
$\phi_b M_p$, kip-ft	1760	1570	1390	1230	1080	1050
$n \times 10^3$	2.31	2.61	2.98	3.40	3.88	5.96

[a] b has the units (kips)$^{-1}$.
[b] m and n have the units (kip-ft)$^{-1}$.
† Web may be slender for combined axial and bending stress.
Note: Heavy line indicates Kl/r equal to or greater than 200.

Table 6-2 (cont.). W-Shapes Values of b^a, m^b, and n^b for Beam-Columns

$F_y = 50$ ksi
$\phi_c = 0.85$
$\phi_b = 0.90$

Shape	W24× 94		84		76		68		62		55	
Effective length *KL* (ft) with respect to least radius of gyration r_y or unbraced length L_b (ft) for X-X axis bending	$b \times 10^3$	$m \times 10^3$	$b \times 10^3$	$m \times 10^3$	$b \times 10^3$	$m \times 10^3$	$b \times 10^3$	$m \times 10^3$	$b \times 10^3$	$m \times 10^3$	$b \times 10^3$	$m \times 10^3$
0	0.886†	0.933	1.02†	1.06	1.14†	1.19	1.30†	1.34	1.43†	1.54	1.65†	1.76
6	0.962†	0.933	1.11†	1.06	1.24†	1.19	1.42†	1.34	1.67†	1.61	1.93†	1.85
7	0.991†	0.933	1.14†	1.06	1.28†	1.19	1.46†	1.35	1.77†	1.67	2.06†	1.93
8	1.03†	0.956	1.18†	1.09	1.33†	1.22	1.52†	1.39	1.90†	1.74	2.21†	2.01
9	1.07†	0.981	1.23†	1.12	1.38†	1.26	1.58†	1.44	2.07†	1.82	2.40†	2.10
10	1.12†	1.01	1.29†	1.15	1.45†	1.30	1.66†	1.48	2.27†	1.90	2.64†	2.21
11	1.18	1.03	1.35†	1.18	1.53†	1.33	1.75†	1.53	2.53	2.00	2.96†	2.32
12	1.25	1.06	1.43†	1.22	1.62†	1.37	1.86†	1.58	2.88	2.10	3.36	2.45
13	1.34	1.09	1.53†	1.25	1.72†	1.42	1.98†	1.63	3.32	2.21	3.90	2.60
14	1.44	1.12	1.64	1.29	1.85†	1.46	2.14†	1.69	3.85	2.45	4.52	2.96
15	1.55	1.16	1.78	1.33	2.00	1.51	2.32†	1.75	4.42	2.75	5.19	3.34
16	1.69	1.19	1.94	1.38	2.18	1.57	2.53	1.82	5.03	3.07	5.90	3.72
17	1.85	1.23	2.12	1.43	2.40	1.62	2.79	1.89	5.68	3.38	6.66	4.12
18	2.03	1.27	2.34	1.48	2.65	1.68	3.11	2.04	6.37	3.71	7.47	4.53
19	2.24	1.31	2.59	1.56	2.95	1.84	3.47	2.24	7.09	4.04	8.32	4.94
20	2.49	1.40	2.87	1.70	3.27	2.01	3.84	2.45	7.86	4.37	9.22	5.36
21	2.74	1.51	3.17	1.83	3.60	2.17	4.23	2.65	8.67	4.71	10.2	5.79
22	3.01	1.62	3.48	1.97	3.96	2.34	4.65	2.87	9.51	5.05	11.2	6.22
23	3.29	1.73	3.80	2.11	4.32	2.51	5.08	3.08				
24	3.58	1.84	4.14	2.25	4.71	2.69	5.53	3.30				
25	3.88	1.95	4.49	2.40	5.11	2.86	6.00	3.53				
26	4.20	2.06	4.86	2.54	5.53	3.04	6.49	3.75				
27	4.53	2.18	5.24	2.69	5.96	3.22	7.00	3.98				
28	4.87	2.29	5.63	2.83	6.41	3.40	7.53	4.21				
29	5.23	2.41	6.04	2.98	6.87	3.58	8.08	4.44				
30	5.59	2.52	6.47	3.12	7.36	3.76	8.64	4.67				
Properties												
L_p, ft	6.99		6.89		6.78		6.61		4.84		4.73	
$\phi_b M_p$, kip-ft	953		840		750		664		578		506	
$n \times 10^3$	6.58		7.56		8.59		10.1		16.1		19.0	

[a] *b* has the units (kips)$^{-1}$.
[b] *m* and *n* have the units (kip-ft)$^{-1}$.
† Web may be slender for combined axial and bending stress.
Note: Heavy line indicates Kl/r equal to or greater than 200.

$F_y = 50$ ksi
$\phi_c = 0.85$
$\phi_b = 0.90$

Table 6-2 (cont.). W-Shapes Values of b^a, m^b, and n^b for Beam-Columns

Shape	W21×											
	201		182		166		147		132		122	
	$b \times 10^3$	$m \times 10^3$	$b \times 10^3$	$m \times 10^3$	$b \times 10^3$	$m \times 10^3$	$b \times 10^3$	$m \times 10^3$	$b \times 10^3$	$m \times 10^3$	$b \times 10^3$	$m \times 10^3$
Effective length KL (ft) with respect to least radius of gyration r_y or unbraced length L_b (ft) for X-X axis bending												
0	0.397	0.447	0.439	0.498	0.482	0.549	0.545	0.635	0.606	0.712	0.655	0.772
6	0.414	0.447	0.458	0.498	0.503	0.549	0.569	0.635	0.634	0.712	0.685	0.772
7	0.421	0.447	0.465	0.498	0.511	0.549	0.578	0.635	0.644	0.712	0.696	0.772
8	0.428	0.447	0.473	0.498	0.520	0.549	0.589	0.635	0.656	0.712	0.709	0.772
9	0.436	0.447	0.483	0.498	0.530	0.549	0.601	0.635	0.670	0.712	0.724	0.772
10	0.446	0.447	0.493	0.498	0.542	0.549	0.615	0.635	0.686	0.712	0.742	0.772
11	0.457	0.449	0.506	0.500	0.556	0.552	0.631	0.640	0.703	0.718	0.761	0.780
12	0.469	0.453	0.520	0.506	0.571	0.558	0.648	0.649	0.724	0.729	0.783	0.792
13	0.483	0.458	0.535	0.511	0.588	0.565	0.668	0.658	0.746	0.740	0.808	0.804
14	0.498	0.463	0.552	0.517	0.607	0.572	0.690	0.667	0.771	0.751	0.835	0.817
15	0.515	0.468	0.571	0.523	0.628	0.579	0.715	0.676	0.799	0.762	0.865	0.830
16	0.534	0.473	0.592	0.529	0.652	0.586	0.742	0.686	0.830	0.774	0.899	0.843
17	0.555	0.478	0.616	0.535	0.678	0.594	0.773	0.696	0.864	0.786	0.936	0.857
18	0.578	0.483	0.641	0.542	0.706	0.601	0.806	0.706	0.902	0.798	0.978	0.871
19	0.603	0.488	0.670	0.548	0.738	0.609	0.843	0.716	0.944	0.811	1.02	0.886
20	0.631	0.494	0.701	0.555	0.772	0.617	0.884	0.727	0.990	0.824	1.07	0.902
22	0.695	0.505	0.773	0.569	0.853	0.634	0.978	0.750	1.10	0.852	1.19	0.934
24	0.773	0.516	0.861	0.584	0.950	0.652	1.09	0.774	1.23	0.882	1.33	0.969
26	0.867	0.529	0.968	0.599	1.07	0.670	1.23	0.800	1.39	0.914	1.51	1.01
28	0.983	0.542	1.10	0.615	1.21	0.690	1.41	0.827	1.59	0.949	1.73	1.05
30	1.13	0.555	1.26	0.632	1.39	0.711	1.62	0.856	1.82	0.986	1.98	1.10
32	1.28	0.569	1.43	0.650	1.58	0.734	1.84	0.888	2.07	1.05	2.26	1.20
34	1.45	0.584	1.62	0.669	1.79	0.757	2.08	0.945	2.34	1.14	2.55	1.31
36	1.62	0.600	1.81	0.690	2.00	0.786	2.33	1.02	2.63	1.23	2.86	1.41
38	1.81	0.617	2.02	0.712	2.23	0.839	2.59	1.09	2.93	1.32	3.18	1.51
40	2.00	0.634	2.24	0.756	2.48	0.892	2.87	1.16	3.24	1.41	3.53	1.62

Properties

	201	182	166	147	132	122
L_p, ft	10.7	10.6	10.6	10.4	10.3	10.3
$\phi_b M_p$, kip-ft	1990	1790	1620	1400	1250	1150
$n \times 10^3$	1.84	2.05	2.26	2.63	2.95	3.21

[a] b has the units (kips)$^{-1}$.
[b] m and n have the units (kip-ft)$^{-1}$.

Table 6-2 (cont.). W-Shapes Values of b^a, m^b, and n^b for Beam-Columns

$F_y = 50$ ksi
$\phi_c = 0.85$
$\phi_b = 0.90$

Shape	W21×											
	111		101		93		83		73		68	
Effective length KL (ft) with respect to least radius of gyration r_y or unbraced length L_b (ft) for X-X axis bending	$b \times 10^3$	$m \times 10^3$	$b \times 10^3$	$m \times 10^3$	$b \times 10^3$	$m \times 10^3$	$b \times 10^3$	$m \times 10^3$	$b \times 10^3$	$m \times 10^3$	$b \times 10^3$	$m \times 10^3$
0	0.720	0.850	0.799†	0.937	0.862	1.07	0.975†	1.21	1.14†	1.38	1.24†	1.48
6	0.753	0.850	0.832†	0.937	0.964	1.07	1.08	1.21	1.25†	1.38	1.37†	1.48
7	0.765	0.850	0.844†	0.937	1.00	1.09	1.13	1.23	1.30†	1.40	1.42†	1.51
8	0.780	0.850	0.859†	0.937	1.05	1.11	1.18	1.26	1.36†	1.44	1.48†	1.55
9	0.796	0.850	0.874	0.937	1.11	1.14	1.25	1.29	1.43†	1.48	1.55†	1.60
10	0.816	0.850	0.896	0.937	1.18	1.17	1.33	1.33	1.51	1.52	1.64†	1.65
11	0.837	0.860	0.920	0.949	1.26	1.20	1.42	1.37	1.61	1.57	1.74	1.70
12	0.862	0.874	0.947	0.965	1.35	1.23	1.52	1.41	1.74	1.62	1.88	1.75
13	0.889	0.888	0.977	0.981	1.46	1.27	1.65	1.45	1.88	1.67	2.04	1.81
14	0.920	0.903	1.01	0.998	1.59	1.30	1.79	1.49	2.05	1.72	2.22	1.88
15	0.954	0.918	1.05	1.02	1.74	1.34	1.96	1.54	2.26	1.78	2.44	1.94
16	0.991	0.934	1.09	1.03	1.91	1.38	2.17	1.59	2.49	1.85	2.70	2.02
17	1.03	0.950	1.14	1.05	2.12	1.43	2.40	1.65	2.77	1.91	3.01	2.10
18	1.08	0.967	1.19	1.07	2.37	1.47	2.69	1.71	3.10	2.02	3.37	2.27
19	1.13	0.985	1.24	1.09	2.64	1.52	2.99	1.81	3.46	2.20	3.76	2.47
20	1.19	1.00	1.31	1.11	2.92	1.61	3.32	1.95	3.83	2.38	4.17	2.67
22	1.32	1.04	1.45	1.16	3.53	1.85	4.01	2.24	4.64	2.74	5.04	3.09
24	1.48	1.08	1.63	1.21	4.21	2.08	4.78	2.53	5.52	3.11	6.00	3.52
26	1.68	1.13	1.85	1.26	4.94	2.31	5.61	2.82	6.48	3.48	7.04	3.95
28	1.92	1.18	2.13	1.33	5.72	2.54	6.50	3.12	7.51	3.86	8.17	4.38
30	2.21	1.28	2.44	1.48	6.57	2.77	7.46	3.41	8.62	4.23	9.37	4.82
32	2.51	1.41	2.78	1.63								
34	2.84	1.53	3.13	1.78								
36	3.18	1.66	3.51	1.93								
38	3.54	1.79	3.92	2.08								
40	3.93	1.91	4.34	2.23								

Properties

	111	101	93	83	73	68
L_p, ft	10.2	10.2	6.50	6.46	6.39	6.36
$\phi_b M_p$, kip-ft	1050	949	829	735	645	600
$n \times 10^3$	3.55	3.92	7.15	8.10	9.30	10.1

[a] b has the units (kips)$^{-1}$.
[b] m and n have the units (kip-ft)$^{-1}$.
† Web may be slender for combined axial and bending stress.
Note: Heavy line indicates Kl/r equal to or greater than 200.

$F_y = 50$ ksi
$\phi_c = 0.85$
$\phi_b = 0.90$

Table 6-2 (cont.). W-Shapes Values of b^a, m^b, and n^b for Beam-Columns

Shape	W21× 62		55		48		57		50		44	
Effective length KL (ft) with respect to least radius of gyration r_y or unbraced length L_b (ft) for X-X axis bending	b ×10³	m ×10³	b ×10³	m ×10³	b ×10³	m ×10³	b ×10³	m ×10³	b ×10³	m ×10³	b ×10³	m ×10³
0	1.38†	1.65	1.59†	1.88	1.88†	2.23	1.52†	1.84	1.77†	2.15	2.05†	2.48
6	1.53†	1.65	1.76†	1.88	2.09†	2.23	1.80†	1.93	2.11†	2.28	2.47†	2.66
7	1.58†	1.68	1.83†	1.93	2.17†	2.28	1.92†	2.00	2.26†	2.38	2.65†	2.78
8	1.65†	1.73	1.91†	2.00	2.28†	2.36	2.07†	2.09	2.45†	2.49	2.88†	2.92
9	1.74†	1.79	2.01†	2.06	2.40†	2.45	2.26†	2.18	2.69†	2.61	3.17†	3.07
10	1.84†	1.84	2.13†	2.13	2.55†	2.54	2.51	2.28	2.99†	2.74	3.55†	3.23
11	1.96†	1.90	2.27†	2.21	2.73†	2.64	2.83	2.39	3.40	2.89	4.04†	3.42
12	2.10†	1.97	2.44†	2.29	2.95†	2.75	3.24	2.51	3.93	3.05	4.71	3.63
13	2.27	2.04	2.65†	2.38	3.22†	2.86	3.75	2.65	4.59	3.36	5.53	4.17
14	2.48	2.11	2.89	2.47	3.54†	2.99	4.35	2.96	5.32	3.81	6.41	4.74
15	2.74	2.19	3.21	2.57	3.94	3.13	4.99	3.30	6.11	4.27	7.36	5.33
16	3.04	2.28	3.57	2.68	4.45	3.42	5.68	3.66	6.95	4.74	8.37	5.94
17	3.40	2.42	4.02	2.96	5.02	3.80	6.41	4.01	7.85	5.22	9.45	6.56
18	3.81	2.66	4.51	3.26	5.63	4.20	7.18	4.38	8.80	5.71	10.6	7.19
19	4.25	2.90	5.03	3.57	6.27	4.61	8.01	4.74	9.81	6.20	11.8	7.83
20	4.71	3.15	5.57	3.89	6.95	5.03	8.87	5.11	10.9	6.70	13.1	8.49
21	5.19	3.40	6.14	4.21	7.66	5.46	9.78	5.48	12.0	7.21	14.4	9.15
22	5.70	3.66	6.74	4.54	8.41	5.89	10.7	5.85				
23	6.23	3.92	7.36	4.87	9.19	6.34						
24	6.78	4.18	8.02	5.20	10.0	6.79						
25	7.36	4.44	8.70	5.54	10.9	7.24						
26	7.96	4.70	9.41	5.88	11.7	7.71						
27	8.58	4.97	10.1	6.23	12.7	8.17						
28	9.23	5.23	10.9	6.57								
29	9.90	5.50										
30												

Properties

	62	55	48	57	50	44
L_p, ft	6.25	6.11	6.09	4.77	4.59	4.45
$\phi_b M_p$, kip-ft	540	473	401	484	413	358
$n \times 10^3$	11.3	13.4	16.7	16.9	20.7	24.8

[a] b has the units $(\text{kips})^{-1}$.
[b] m and n have the units $(\text{kip-ft})^{-1}$.
† Web may be slender for combined axial and bending stress.
Note: Heavy line indicates Kl/r equal to or greater than 200.

Table 6-2 (cont.). W-Shapes Values of b^a, m^b, and n^b for Beam-Columns

$F_y = 50$ ksi
$\phi_c = 0.85$
$\phi_b = 0.90$

Shape	W18×											
	175		158		143		130		119		106	
Effective length KL (ft) with respect to least radius of gyration r_y or unbraced length L_b (ft) for X-X axis bending	$b \times 10^3$	$m \times 10^3$	$b \times 10^3$	$m \times 10^3$	$b \times 10^3$	$m \times 10^3$	$b \times 10^3$	$m \times 10^3$	$b \times 10^3$	$m \times 10^3$	$b \times 10^3$	$m \times 10^3$
0	0.459	0.596	0.508	0.666	0.559	0.736	0.616	0.817	0.670	0.905	0.757	1.03
6	0.482	0.596	0.535	0.666	0.588	0.736	0.649	0.817	0.706	0.905	0.798	1.03
7	0.491	0.596	0.544	0.666	0.599	0.736	0.661	0.817	0.720	0.905	0.814	1.03
8	0.501	0.596	0.556	0.666	0.612	0.736	0.676	0.817	0.736	0.905	0.832	1.03
9	0.513	0.596	0.569	0.666	0.627	0.736	0.692	0.817	0.754	0.905	0.853	1.03
10	0.527	0.597	0.585	0.668	0.644	0.740	0.712	0.822	0.775	0.911	0.878	1.04
11	0.542	0.603	0.602	0.675	0.664	0.748	0.734	0.833	0.799	0.924	0.906	1.06
12	0.560	0.609	0.622	0.683	0.686	0.757	0.758	0.844	0.827	0.937	0.937	1.07
13	0.579	0.615	0.644	0.690	0.711	0.766	0.786	0.855	0.857	0.951	0.973	1.09
14	0.601	0.621	0.669	0.698	0.739	0.776	0.818	0.866	0.892	0.965	1.01	1.11
15	0.626	0.628	0.697	0.706	0.770	0.785	0.852	0.878	0.930	0.979	1.06	1.13
16	0.653	0.634	0.728	0.714	0.805	0.795	0.892	0.890	0.973	0.994	1.11	1.14
17	0.684	0.641	0.762	0.722	0.843	0.805	0.935	0.902	1.02	1.01	1.16	1.16
18	0.718	0.648	0.801	0.731	0.886	0.816	0.984	0.915	1.07	1.03	1.23	1.18
19	0.755	0.655	0.843	0.740	0.934	0.826	1.04	0.928	1.13	1.04	1.29	1.20
20	0.797	0.662	0.891	0.748	0.988	0.837	1.10	0.942	1.20	1.06	1.37	1.23
22	0.895	0.676	1.00	0.767	1.11	0.860	1.24	0.970	1.36	1.09	1.55	1.27
24	1.02	0.692	1.14	0.786	1.27	0.884	1.42	1.00	1.55	1.13	1.78	1.32
26	1.17	0.708	1.31	0.807	1.46	0.910	1.64	1.03	1.80	1.17	2.07	1.37
28	1.35	0.724	1.52	0.828	1.70	0.937	1.90	1.07	2.08	1.22	2.40	1.43
30	1.55	0.742	1.75	0.851	1.95	0.965	2.18	1.10	2.39	1.26	2.76	1.54
32	1.77	0.761	1.99	0.875	2.22	0.996	2.48	1.14	2.72	1.34	3.14	1.67
34	2.00	0.780	2.24	0.900	2.50	1.03	2.80	1.21	3.07	1.45	3.55	1.81
36	2.24	0.800	2.52	0.927	2.81	1.07	3.14	1.29	3.44	1.55	3.97	1.94
38	2.49	0.822	2.80	0.955	3.13	1.14	3.50	1.38	3.84	1.66	4.43	2.08
40	2.76	0.845	3.11	1.01	3.47	1.21	3.88	1.46	4.25	1.76	4.91	2.21

Properties

	175	158	143	130	119	106
L_p, ft	9.75	9.68	9.61	9.54	9.50	9.40
$\phi_b M_p$, kip-ft	1490	1340	1210	1090	983	863
$n \times 10^3$	2.30	2.57	2.85	3.17	3.52	4.01

[a] b has the units (kips)$^{-1}$.
[b] m and n have the units (kip-ft)$^{-1}$.

$F_y = 50$ ksi
$\phi_c = 0.85$
$\phi_b = 0.90$

Table 6-2 (cont.). W-Shapes Values of b^a, m^b, and n^b for Beam-Columns

Shape	W18× 97		86		76		71		65		60	
	b ×10³	m ×10³	b ×10³	m ×10³	b ×10³	m ×10³	b ×10³	m ×10³	b ×10³	m ×10³	b ×10³	m ×10³
Effective length KL (ft) with respect to least radius of gyration r_y or unbraced length L_b (ft) for X-X axis bending												
0	0.826	1.12	0.930	1.27	1.07†	1.45	1.13	1.62	1.23	1.78	1.37†	1.93
6	0.871	1.12	0.982	1.27	1.12†	1.45	1.29	1.62	1.41	1.78	1.54†	1.93
7	0.889	1.12	1.00	1.27	1.14†	1.45	1.35	1.67	1.48	1.83	1.61	1.99
8	0.909	1.12	1.03	1.27	1.17†	1.45	1.43	1.71	1.56	1.88	1.70	2.05
9	0.932	1.12	1.05	1.27	1.20	1.45	1.52	1.76	1.66	1.94	1.81	2.11
10	0.959	1.13	1.08	1.29	1.23	1.48	1.63	1.81	1.78	2.00	1.94	2.17
11	0.990	1.15	1.12	1.31	1.27	1.50	1.76	1.86	1.92	2.06	2.10	2.25
12	1.02	1.17	1.16	1.34	1.32	1.53	1.91	1.92	2.09	2.12	2.29	2.32
13	1.06	1.19	1.20	1.36	1.37	1.56	2.09	1.98	2.30	2.19	2.51	2.40
14	1.11	1.21	1.25	1.39	1.43	1.59	2.31	2.04	2.54	2.27	2.78	2.49
15	1.16	1.23	1.31	1.41	1.49	1.62	2.57	2.11	2.82	2.34	3.09	2.58
16	1.21	1.26	1.37	1.44	1.57	1.66	2.87	2.19	3.17	2.43	3.48	2.68
17	1.27	1.28	1.44	1.47	1.65	1.69	3.24	2.26	3.58	2.52	3.93	2.85
18	1.34	1.30	1.52	1.50	1.74	1.73	3.64	2.36	4.01	2.73	4.40	3.11
19	1.42	1.33	1.61	1.53	1.84	1.77	4.05	2.55	4.47	2.95	4.90	3.37
20	1.50	1.35	1.71	1.56	1.96	1.81	4.49	2.74	4.95	3.17	5.43	3.63
22	1.71	1.41	1.94	1.63	2.23	1.89	5.43	3.12	5.99	3.62	6.58	4.15
24	1.96	1.46	2.23	1.70	2.57	1.99	6.47	3.49	7.13	4.07	7.83	4.68
26	2.28	1.53	2.61	1.78	3.00	2.18	7.59	3.87	8.36	4.51	9.18	5.20
28	2.64	1.62	3.02	1.98	3.48	2.44	8.80	4.24	9.70	4.96	10.7	5.73
30	3.03	1.78	3.47	2.18	4.00	2.70						
32	3.45	1.94	3.95	2.39	4.55	2.96						
34	3.90	2.10	4.46	2.59	5.14	3.22						
36	4.37	2.26	5.00	2.79	5.76	3.48						
38	4.87	2.42	5.57	3.00	6.42	3.75						
40	5.40	2.58	6.17	3.20	7.11	4.01						
Properties												
L_p, ft	9.36		9.29		9.22		6.00		5.97		5.93	
$\phi_b M_p$, kip-ft	791		698		611		548		499		461	
$n \times 10^3$	4.38		5.00		5.73		10.0		11.0		11.9	

[a] b has the units (kips)$^{-1}$.
[b] m and n have the units (kip-ft)$^{-1}$.
† Web may be slender for combined axial and bending stress.
Note: Heavy line indicates Kl/r equal to or greater than 200.

Table 6-2 (cont.). W-Shapes Values of b^a, m^b, and n^b for Beam-Columns

$F_y = 50$ ksi
$\phi_c = 0.85$
$\phi_b = 0.90$

Shape	W18×									
	55		50		46		40		35	
	b ×10³	m ×10³	b ×10³	m ×10³	b ×10³	m ×10³	b ×10³	m ×10³	b ×10³	m ×10³
Effective length KL (ft) with respect to least radius of gyration r_y or unbraced length L_b (ft) for X-X axis bending										
0	1.51†	2.12	1.70†	2.35	1.85†	2.61	2.19†	3.02	2.57†	3.56
6	1.69†	2.12	1.91†	2.36	2.24†	2.77	2.66†	3.22	3.14†	3.84
7	1.77†	2.19	1.99†	2.43	2.41†	2.88	2.85†	3.36	3.39†	4.03
8	1.86†	2.25	2.10†	2.51	2.63†	3.01	3.11†	3.52	3.71†	4.23
9	1.97	2.33	2.22†	2.60	2.91	3.15	3.43†	3.69	4.12†	4.46
10	2.12	2.40	2.38†	2.69	3.28	3.30	3.85†	3.88	4.66†	4.71
11	2.29	2.49	2.56	2.78	3.75	3.46	4.39	4.08	5.38	4.99
12	2.50	2.57	2.79	2.89	4.33	3.65	5.10	4.32	6.34	5.54
13	2.75	2.67	3.08	3.00	5.08	3.98	5.99	4.91	7.44	6.36
14	3.04	2.77	3.42	3.12	5.89	4.47	6.95	5.54	8.63	7.21
15	3.40	2.88	3.82	3.25	6.76	4.97	7.98	6.19	9.91	8.08
16	3.82	3.00	4.32	3.48	7.69	5.48	9.08	6.85	11.3	8.98
17	4.32	3.27	4.87	3.84	8.68	5.99	10.2	7.52	12.7	9.89
18	4.84	3.58	5.46	4.21	9.73	6.51	11.5	8.19	14.3	10.8
19	5.39	3.88	6.09	4.58	10.8	7.03	12.8	8.88	15.9	11.8
20	5.98	4.19	6.75	4.95	12.0	7.55	14.2	9.56	17.6	12.7
21	6.59	4.50	7.44	5.33	13.2	8.07	15.6	10.3		
22	7.23	4.81	8.16	5.71						
23	7.90	5.13	8.92	6.09						
24	8.60	5.44	9.71	6.48						
25	9.34	5.76	10.5	6.86						
26	10.1	6.07	11.4	7.25						
27	10.9	6.39	12.3	7.64						
28										
Properties										
L_p, ft	5.90		5.83		4.56		4.49		4.31	
$\phi_b M_p$, kip-ft	420		379		340		294		249	
$n \times 10^3$	13.3		14.8		21.3		24.9		30.9	

[a] b has the units (kips)$^{-1}$.
[b] m and n have the units (kip-ft)$^{-1}$.
† Web may be slender for combined axial and bending stress.
Note: Heavy line indicates Kl/r equal to or greater than 200.

$F_y = 50$ ksi
$\phi_c = 0.85$
$\phi_b = 0.90$

Table 6-2 (cont.). W-Shapes Values of b^a, m^b, and n^b for Beam-Columns

Shape	W16×											
	100		89		77		67		57		50	
	$b \times 10^3$	$m \times 10^3$	$b \times 10^3$	$m \times 10^3$	$b \times 10^3$	$m \times 10^3$	$b \times 10^3$	$m \times 10^3$	$b \times 10^3$	$m \times 10^3$	$b \times 10^3$	$m \times 10^3$
Effective length *KL* (ft) with respect to least radius of gyration r_y or unbraced length L_b (ft) for X-X axis bending												
0	0.792	1.19	0.891	1.34	1.03	1.56	1.18	1.80	1.40	2.26	1.62†	2.58
6	0.842	1.19	0.948	1.34	1.09	1.56	1.25	1.80	1.62	2.28	1.86	2.61
7	0.860	1.19	0.969	1.34	1.12	1.56	1.28	1.80	1.71	2.34	1.96	2.68
8	0.882	1.19	0.994	1.34	1.15	1.56	1.32	1.80	1.82	2.41	2.09	2.77
9	0.908	1.19	1.02	1.34	1.18	1.57	1.36	1.81	1.95	2.48	2.24	2.86
10	0.938	1.21	1.06	1.37	1.22	1.59	1.40	1.84	2.11	2.56	2.43	2.95
11	0.971	1.22	1.10	1.39	1.27	1.62	1.46	1.88	2.30	2.64	2.65	3.06
12	1.01	1.24	1.14	1.41	1.32	1.65	1.52	1.91	2.53	2.73	2.92	3.16
13	1.05	1.26	1.19	1.43	1.38	1.68	1.59	1.95	2.81	2.82	3.24	3.28
14	1.10	1.28	1.25	1.46	1.45	1.71	1.66	1.99	3.14	2.92	3.62	3.41
15	1.16	1.30	1.31	1.48	1.52	1.75	1.75	2.03	3.53	3.02	4.09	3.55
16	1.22	1.32	1.38	1.51	1.60	1.78	1.85	2.07	4.02	3.14	4.65	3.74
17	1.29	1.34	1.46	1.54	1.70	1.81	1.96	2.12	4.54	3.32	5.25	4.09
18	1.37	1.36	1.55	1.56	1.81	1.85	2.09	2.16	5.08	3.60	5.88	4.45
19	1.46	1.38	1.65	1.59	1.93	1.89	2.23	2.21	5.67	3.88	6.56	4.81
20	1.55	1.41	1.77	1.62	2.06	1.93	2.39	2.26	6.28	4.16	7.26	5.17
22	1.79	1.46	2.04	1.69	2.39	2.01	2.77	2.37	7.60	4.72	8.79	5.90
24	2.09	1.51	2.39	1.75	2.81	2.10	3.26	2.50	9.04	5.28	10.5	6.62
26	2.46	1.56	2.81	1.83	3.29	2.24	3.83	2.80	10.6	5.83	12.3	7.35
28	2.85	1.62	3.26	1.94	3.82	2.48	4.44	3.11				
30	3.27	1.70	3.74	2.11	4.38	2.71	5.10	3.42				
32	3.72	1.84	4.26	2.29	4.99	2.95	5.80	3.73				
34	4.20	1.98	4.81	2.47	5.63	3.18	6.55	4.04				
36	4.71	2.12	5.39	2.64	6.31	3.42	7.35	4.34				
38	5.25	2.25	6.00	2.82	7.03	3.65	8.18	4.65				
40	5.82	2.39	6.65	2.99	7.79	3.89	9.07	4.96				

Properties

	100	89	77	67	57	50
L_p, ft	8.83	8.76	8.69	8.62	5.65	5.62
$\phi_b M_p$, kip-ft	750	664	570	495	394	345
$n \times 10^3$	4.43	5.03	5.87	6.81	13.1	15.0

[a] *b* has the units $(\text{kips})^{-1}$.
[b] *m* and *n* have the units $(\text{kip-ft})^{-1}$.
† Web may be slender for combined axial and bending stress.
Note: Heavy line indicates Kl/r equal to or greater than 200.

Table 6-2 (cont.). W-Shapes Values of b^a, m^b, and n^b for Beam-Columns

$F_y = 50$ ksi
$\phi_c = 0.85$
$\phi_b = 0.90$

Shape	W16× 45		40		36		31		26	
	$b \times 10^3$	$m \times 10^3$	$b \times 10^3$	$m \times 10^3$	$b \times 10^3$	$m \times 10^3$	$b \times 10^3$	$m \times 10^3$	$b \times 10^3$	$m \times 10^3$
0	1.83†	2.88	2.12†	3.25	2.39†	3.70	2.84†	4.39	3.50†	5.36
6	2.09†	2.92	2.42†	3.29	2.74†	3.78	3.56†	4.78	4.43†	5.94
7	2.20†	3.01	2.54†	3.41	2.88†	3.92	3.88†	5.02	4.85†	6.26
8	2.33	3.11	2.68†	3.52	3.06†	4.07	4.30†	5.28	5.40†	6.63
9	2.50	3.22	2.86†	3.65	3.28†	4.22	4.84†	5.58	6.13†	7.04
10	2.71	3.33	3.08†	3.79	3.54†	4.39	5.56	5.91	7.11†	7.51
11	2.97	3.46	3.34†	3.93	3.87†	4.58	6.54	6.28	8.48	8.50
12	3.27	3.59	3.69	4.09	4.28	4.78	7.78	7.26	10.1	9.90
13	3.64	3.73	4.10	4.26	4.79	5.00	9.13	8.29	11.8	11.3
14	4.09	3.88	4.61	4.45	5.42	5.23	10.6	9.34	13.7	12.9
15	4.63	4.05	5.22	4.76	6.20	5.87	12.2	10.4	15.8	14.4
16	5.27	4.44	5.94	5.28	7.05	6.53	13.8	11.5	17.9	16.0
17	5.95	4.87	6.71	5.82	7.96	7.21	15.6	12.6	20.2	17.6
18	6.67	5.31	7.52	6.36	8.93	7.91	17.5	13.7	22.7	19.2
19	7.43	5.76	8.38	6.91	9.95	8.61	19.5	14.8		
20	8.23	6.20	9.28	7.46	11.0	9.33				
21	9.08	6.65	10.2	8.02	12.2	10.0				
22	9.96	7.10	11.2	8.58	13.3	10.8				
23	10.9	7.55	12.3	9.15	14.6	11.5				
24	11.9	8.00	13.4	9.72	15.9	12.3				
25	12.9	8.46	14.5	10.3	17.2	13.0				
26	13.9	8.91	15.7	10.9						
27										

Row labels: Effective length KL (ft) with respect to least radius of gyration r_y or unbraced length L_b (ft) for X-X axis bending

Properties

	45	40	36	31	26
L_p, ft	5.55	5.55	5.37	4.13	3.96
$\phi_b M_p$, kip-ft	309	274	240	203	166
$n \times 10^3$	16.9	19.2	22.6	35.2	45.3

[a] b has the units (kips)$^{-1}$.
[b] m and n have the units (kip-ft)$^{-1}$.
† Web may be slender for combined axial and bending stress.
Note: Heavy line indicates Kl/r equal to or greater than 200.

$F_y = 50$ ksi
$\phi_c = 0.85$
$\phi_b = 0.90$

Table 6-2 (cont.). W-Shapes Values of b^a, m^b, and n^b for Beam-Columns

Shape	W14×											
	808*		730*		665*		605*		550*		500*	
Effective length KL (ft) with respect to least radius of gyration r_y or unbraced length L_b (ft) for X-X axis bending	b ×10³	m ×10³	b ×10³	m ×10³	b ×10³	m ×10³	b ×10³	m ×10³	b ×10³	m ×10³	b ×10³	m ×10³
0	0.0993	0.130	0.109	0.143	0.120	0.160	0.132	0.180	0.145	0.201	0.160	0.226
11	0.105	0.130	0.116	0.143	0.127	0.160	0.141	0.180	0.155	0.201	0.171	0.226
12	0.106	0.130	0.117	0.143	0.129	0.160	0.142	0.180	0.157	0.201	0.173	0.226
13	0.107	0.130	0.119	0.143	0.130	0.160	0.144	0.180	0.159	0.201	0.175	0.226
14	0.109	0.130	0.120	0.143	0.132	0.160	0.146	0.180	0.161	0.201	0.178	0.226
15	0.110	0.130	0.122	0.143	0.134	0.160	0.148	0.180	0.163	0.201	0.181	0.226
16	0.111	0.130	0.124	0.143	0.136	0.160	0.151	0.180	0.166	0.201	0.184	0.226
17	0.113	0.130	0.126	0.143	0.138	0.160	0.153	0.180	0.169	0.201	0.187	0.226
18	0.115	0.130	0.128	0.143	0.141	0.161	0.156	0.180	0.172	0.202	0.190	0.227
19	0.117	0.130	0.130	0.143	0.143	0.161	0.159	0.181	0.175	0.202	0.194	0.228
20	0.119	0.130	0.133	0.144	0.146	0.161	0.162	0.181	0.179	0.203	0.198	0.228
22	0.124	0.131	0.138	0.144	0.152	0.162	0.169	0.182	0.187	0.204	0.208	0.229
24	0.129	0.131	0.144	0.145	0.159	0.162	0.177	0.182	0.196	0.204	0.218	0.230
26	0.135	0.131	0.151	0.145	0.168	0.163	0.186	0.183	0.207	0.205	0.230	0.231
28	0.142	0.132	0.159	0.146	0.177	0.164	0.197	0.184	0.219	0.206	0.244	0.232
30	0.149	0.132	0.168	0.146	0.187	0.164	0.209	0.185	0.232	0.207	0.259	0.234
32	0.158	0.133	0.179	0.147	0.199	0.165	0.223	0.185	0.248	0.208	0.277	0.235
34	0.168	0.133	0.190	0.147	0.212	0.166	0.238	0.186	0.266	0.209	0.298	0.236
36	0.179	0.133	0.204	0.148	0.228	0.166	0.256	0.187	0.286	0.210	0.321	0.237
38	0.191	0.134	0.218	0.148	0.245	0.167	0.276	0.188	0.309	0.211	0.347	0.238
40	0.205	0.134	0.235	0.149	0.264	0.167	0.298	0.189	0.335	0.212	0.378	0.239
42	0.221	0.135	0.255	0.149	0.287	0.168	0.324	0.189	0.365	0.213	0.413	0.240
44	0.239	0.135	0.276	0.150	0.312	0.169	0.355	0.190	0.400	0.214	0.453	0.242
46	0.259	0.136	0.302	0.150	0.341	0.169	0.388	0.191	0.437	0.215	0.495	0.243
48	0.282	0.136	0.329	0.151	0.372	0.170	0.422	0.192	0.476	0.216	0.539	0.244
50	0.306	0.136	0.357	0.151	0.403	0.171	0.458	0.193	0.517	0.217	0.585	0.245

Properties

	808*	730*	665*	605*	550*	500*
L_p, ft	17.0	16.6	16.3	16.1	15.9	15.6
$\phi_b M_p$, kip-ft	6860	6230	5550	4950	4430	3940
$n \times 10^3$	0.266	0.300	0.335	0.374	0.418	0.466

[a] b has the units (kips)$^{-1}$.
[b] m and n have the units (kip-ft)$^{-1}$.
*ASTM A6 tensile group 4 or 5 shape. Special requirements may apply per LRFD Specification Section A3.1c.

Table 6-2 (cont.). W-Shapes Values of b^a, m^b, and n^b for Beam-Columns

$F_y = 50$ ksi
$\phi_c = 0.85$
$\phi_b = 0.90$

Shape	W14×											
	455*		426*		398*		370*		342*		311*	
	b ×10³	m ×10³	b ×10³	m ×10³	b ×10³	m ×10³	b ×10³	m ×10³	b ×10³	m ×10³	b ×10³	m ×10³
0	0.176	0.253	0.188	0.273	0.201	0.296	0.216	0.322	0.233	0.353	0.257	0.393
11	0.188	0.253	0.201	0.273	0.215	0.296	0.231	0.322	0.250	0.353	0.277	0.393
12	0.190	0.253	0.204	0.273	0.218	0.296	0.235	0.322	0.253	0.353	0.281	0.393
13	0.193	0.253	0.207	0.273	0.221	0.296	0.238	0.322	0.257	0.353	0.285	0.393
14	0.196	0.253	0.210	0.273	0.225	0.296	0.242	0.322	0.261	0.353	0.289	0.393
15	0.199	0.253	0.213	0.273	0.228	0.296	0.246	0.322	0.266	0.353	0.294	0.393
16	0.202	0.254	0.217	0.273	0.233	0.297	0.250	0.323	0.271	0.354	0.300	0.395
17	0.206	0.254	0.221	0.274	0.237	0.297	0.255	0.324	0.276	0.355	0.306	0.396
18	0.210	0.255	0.226	0.275	0.242	0.298	0.260	0.325	0.282	0.356	0.312	0.397
19	0.214	0.255	0.230	0.275	0.247	0.299	0.266	0.326	0.288	0.357	0.319	0.399
20	0.219	0.256	0.235	0.276	0.252	0.300	0.272	0.327	0.294	0.358	0.327	0.400
22	0.229	0.257	0.247	0.278	0.265	0.302	0.285	0.329	0.309	0.361	0.344	0.403
24	0.241	0.259	0.260	0.279	0.279	0.303	0.301	0.331	0.326	0.363	0.363	0.406
26	0.254	0.260	0.275	0.281	0.295	0.305	0.319	0.333	0.346	0.366	0.385	0.409
28	0.270	0.261	0.292	0.282	0.314	0.307	0.339	0.335	0.369	0.368	0.411	0.412
30	0.288	0.263	0.311	0.284	0.335	0.309	0.363	0.337	0.395	0.370	0.441	0.415
32	0.308	0.264	0.334	0.285	0.359	0.310	0.390	0.339	0.424	0.373	0.474	0.418
34	0.331	0.265	0.359	0.287	0.387	0.312	0.421	0.341	0.458	0.375	0.513	0.421
36	0.358	0.267	0.388	0.289	0.419	0.314	0.456	0.343	0.498	0.378	0.558	0.424
38	0.388	0.268	0.422	0.290	0.456	0.316	0.497	0.346	0.543	0.381	0.610	0.427
40	0.423	0.270	0.460	0.292	0.498	0.318	0.544	0.348	0.595	0.383	0.670	0.431
42	0.463	0.271	0.506	0.294	0.548	0.320	0.599	0.350	0.656	0.386	0.738	0.434
44	0.508	0.273	0.555	0.295	0.601	0.322	0.657	0.352	0.720	0.389	0.810	0.437
46	0.556	0.274	0.607	0.297	0.657	0.324	0.719	0.355	0.787	0.391	0.886	0.441
48	0.605	0.276	0.660	0.299	0.715	0.326	0.782	0.357	0.856	0.394	0.964	0.444
50	0.656	0.277	0.717	0.300	0.776	0.328	0.849	0.359	0.929	0.397	1.05	0.448

Effective length KL (ft) with respect to least radius of gyration r_y or unbraced length L_b (ft) for X-X axis bending

Properties

	455*	426*	398*	370*	342*	311*
L_p, ft	15.5	15.3	15.2	15.1	15.0	14.8
$\phi_b M_p$, kip-ft	3510	3260	3000	2760	2520	2260
$n \times 10^3$	0.520	0.558	0.603	0.656	0.715	0.794

[a] b has the units $(\text{kips})^{-1}$.
[b] m and n have the units $(\text{kip-ft})^{-1}$.
*ASTM A6 tensile group 4 or 5 shape. Special requirements may apply per LRFD Specification Section A3.1c.

$F_y = 50$ ksi
$\phi_c = 0.85$
$\phi_b = 0.90$

Table 6-2 (cont.). W-Shapes Values of b^a, m^b, and n^b for Beam-Columns

Shape	W14×											
	283*		257*		233*		211		193		176	
	$b \times 10^3$	$m \times 10^3$	$b \times 10^3$	$m \times 10^3$	$b \times 10^3$	$m \times 10^3$	$b \times 10^3$	$m \times 10^3$	$b \times 10^3$	$m \times 10^3$	$b \times 10^3$	$m \times 10^3$
Effective length KL (ft) with respect to least radius of gyration r_y or unbraced length L_b (ft) for X-X axis bending												
0	0.282	0.437	0.311	0.487	0.343	0.544	0.380	0.608	0.414	0.668	0.454	0.741
11	0.304	0.437	0.335	0.487	0.371	0.544	0.410	0.608	0.448	0.668	0.491	0.741
12	0.308	0.437	0.340	0.487	0.376	0.544	0.416	0.608	0.454	0.668	0.499	0.741
13	0.313	0.437	0.345	0.487	0.382	0.544	0.423	0.608	0.462	0.668	0.507	0.741
14	0.318	0.437	0.351	0.487	0.388	0.544	0.430	0.608	0.470	0.668	0.516	0.741
15	0.324	0.438	0.358	0.488	0.395	0.545	0.438	0.610	0.479	0.670	0.526	0.744
16	0.330	0.439	0.365	0.490	0.403	0.547	0.447	0.613	0.488	0.674	0.537	0.749
17	0.336	0.441	0.372	0.492	0.412	0.550	0.456	0.616	0.499	0.678	0.548	0.753
18	0.344	0.443	0.380	0.494	0.421	0.552	0.466	0.619	0.510	0.681	0.561	0.758
19	0.351	0.444	0.389	0.496	0.431	0.555	0.477	0.622	0.522	0.685	0.575	0.762
20	0.360	0.446	0.398	0.498	0.441	0.558	0.489	0.625	0.536	0.689	0.589	0.767
22	0.379	0.450	0.420	0.502	0.465	0.563	0.516	0.632	0.565	0.697	0.623	0.777
24	0.400	0.453	0.444	0.506	0.493	0.568	0.547	0.638	0.600	0.705	0.661	0.786
26	0.425	0.457	0.472	0.511	0.525	0.574	0.583	0.645	0.639	0.713	0.706	0.796
28	0.454	0.460	0.505	0.515	0.561	0.579	0.625	0.652	0.685	0.721	0.757	0.807
30	0.487	0.464	0.542	0.520	0.604	0.585	0.672	0.659	0.738	0.730	0.816	0.817
32	0.525	0.468	0.586	0.524	0.652	0.590	0.728	0.666	0.799	0.738	0.885	0.828
34	0.569	0.472	0.635	0.529	0.709	0.596	0.791	0.674	0.870	0.747	0.965	0.839
36	0.619	0.475	0.693	0.534	0.773	0.602	0.865	0.681	0.952	0.756	1.06	0.850
38	0.677	0.479	0.759	0.539	0.849	0.608	0.950	0.689	1.05	0.766	1.16	0.862
40	0.746	0.483	0.837	0.544	0.938	0.615	1.05	0.697	1.16	0.775	1.29	0.874
42	0.822	0.487	0.923	0.549	1.03	0.621	1.16	0.705	1.28	0.785	1.42	0.886
44	0.902	0.491	1.01	0.554	1.13	0.627	1.27	0.713	1.40	0.795	1.56	0.899
46	0.986	0.496	1.11	0.559	1.24	0.634	1.39	0.721	1.53	0.806	1.71	0.912
48	1.07	0.500	1.21	0.565	1.35	0.641	1.51	0.730	1.67	0.816	1.86	0.926
50	1.16	0.504	1.31	0.570	1.47	0.648	1.64	0.739	1.81	0.827	2.02	0.940
Properties												
L_p, ft	14.7		14.6		14.5		14.4		14.3		14.2	
$\phi_b M_p$, kip-ft	2030		1830		1640		1460		1330		1200	
$n \times 10^3$	0.883		0.982		1.09		1.22		1.33		1.48	

[a] b has the units (kips)$^{-1}$.
[b] m and n have the units (kip-ft)$^{-1}$.
*ASTM A6 tensile group 4 or 5 shape. Special requirements may apply per LRFD Specification Section A3.1c.

Table 6-2 (cont.). W-Shapes Values of b^a, m^b, and n^b for Beam-Columns

$F_y = 50$ ksi
$\phi_c = 0.85$
$\phi_b = 0.90$

Shape	W14× 159		145		132		120		109		99	
Effective length KL (ft) with respect to least radius of gyration r_y or unbraced length L_b (ft) for X-X axis bending	$b \times 10^3$	$m \times 10^3$	$b \times 10^3$	$m \times 10^3$	$b \times 10^3$	$m \times 10^3$	$b \times 10^3$	$m \times 10^3$	$b \times 10^3$	$m \times 10^3$	$b \times 10^3$	$m \times 10^3$
0	0.504	0.826	0.551	0.912	0.606	1.01	0.667	1.12	0.735	1.23	0.809	1.38
11	0.546	0.826	0.597	0.912	0.664	1.01	0.730	1.12	0.806	1.23	0.887	1.38
12	0.554	0.826	0.606	0.912	0.675	1.01	0.743	1.12	0.820	1.23	0.903	1.38
13	0.563	0.826	0.617	0.912	0.688	1.01	0.757	1.12	0.836	1.23	0.920	1.38
14	0.573	0.826	0.628	0.912	0.702	1.02	0.773	1.13	0.853	1.24	0.939	1.38
15	0.584	0.831	0.640	0.918	0.717	1.03	0.790	1.14	0.872	1.26	0.960	1.39
16	0.596	0.836	0.653	0.924	0.734	1.04	0.808	1.15	0.892	1.27	0.983	1.41
17	0.609	0.842	0.668	0.931	0.752	1.04	0.829	1.16	0.915	1.28	1.01	1.42
18	0.624	0.847	0.683	0.938	0.772	1.05	0.851	1.17	0.940	1.29	1.04	1.44
19	0.639	0.853	0.700	0.945	0.793	1.06	0.875	1.18	0.966	1.31	1.07	1.45
20	0.656	0.859	0.719	0.951	0.817	1.07	0.901	1.19	0.995	1.32	1.10	1.47
22	0.693	0.871	0.760	0.966	0.870	1.09	0.960	1.21	1.06	1.34	1.17	1.50
24	0.736	0.883	0.808	0.980	0.931	1.11	1.03	1.23	1.14	1.37	1.26	1.53
26	0.786	0.895	0.864	0.995	1.00	1.13	1.11	1.26	1.23	1.40	1.36	1.57
28	0.844	0.908	0.928	1.01	1.09	1.15	1.20	1.28	1.33	1.43	1.47	1.61
30	0.911	0.921	1.00	1.03	1.19	1.17	1.31	1.31	1.45	1.46	1.61	1.64
32	0.988	0.935	1.09	1.04	1.30	1.19	1.44	1.33	1.60	1.50	1.77	1.68
34	1.08	0.949	1.19	1.06	1.43	1.21	1.59	1.36	1.76	1.53	1.96	1.73
36	1.18	0.963	1.30	1.08	1.59	1.23	1.77	1.39	1.96	1.57	2.18	1.77
38	1.30	0.978	1.44	1.10	1.78	1.26	1.97	1.42	2.19	1.60	2.43	1.82
40	1.45	0.994	1.60	1.12	1.97	1.28	2.19	1.45	2.43	1.64	2.70	1.87
42	1.59	1.01	1.76	1.13	2.17	1.31	2.41	1.48	2.67	1.69	2.97	1.97
44	1.75	1.03	1.93	1.16	2.38	1.33	2.65	1.52	2.93	1.75	3.26	2.08
46	1.91	1.04	2.11	1.18	2.60	1.36	2.89	1.56	3.21	1.85	3.57	2.20
48	2.08	1.06	2.30	1.20	2.83	1.39	3.15	1.63	3.49	1.94	3.88	2.32
50	2.26	1.08	2.49	1.22	3.08	1.43	3.42	1.71	3.79	2.04	4.21	2.43
Properties												
L_p, ft	14.1		14.1		13.3		13.2		13.2		13.5	
$\phi_b M_p$, kip-ft	1080		975		878		795		720		649	
$n \times 10^3$	1.64		1.81		2.12		2.34		2.58		2.88	

[a] b has the units (kips)$^{-1}$.
[b] m and n have the units (kip-ft)$^{-1}$.

$F_y = 50$ ksi
$\phi_c = 0.85$
$\phi_b = 0.90$

Table 6-2 (cont.). W-Shapes Values of b^a, m^b, and n^b for Beam-Columns

Shape	W14×											
	90		82		74		68		61		53	
	$b \times 10^3$	$m \times 10^3$	$b \times 10^3$	$m \times 10^3$	$b \times 10^3$	$m \times 10^3$	$b \times 10^3$	$m \times 10^3$	$b \times 10^3$	$m \times 10^3$	$b \times 10^3$	$m \times 10^3$
Effective length KL (ft) with respect to least radius of gyration r_y or unbraced length L_b (ft) for X-X axis bending												
0	0.888	1.54	0.980	1.71	1.08	1.88	1.18	2.06	1.31	2.32	1.51	2.72
6	0.913	1.54	1.04	1.71	1.15	1.88	1.25	2.06	1.40	2.32	1.67	2.72
7	0.922	1.54	1.07	1.71	1.17	1.88	1.28	2.06	1.43	2.32	1.73	2.73
8	0.933	1.54	1.09	1.71	1.20	1.88	1.32	2.06	1.47	2.32	1.81	2.79
9	0.945	1.54	1.13	1.71	1.24	1.89	1.35	2.07	1.52	2.34	1.90	2.86
10	0.959	1.54	1.16	1.74	1.28	1.92	1.40	2.11	1.57	2.38	2.01	2.92
11	0.974	1.54	1.21	1.76	1.33	1.95	1.45	2.14	1.63	2.42	2.13	2.99
12	0.992	1.54	1.25	1.79	1.38	1.98	1.51	2.18	1.69	2.46	2.28	3.06
13	1.01	1.54	1.31	1.81	1.44	2.01	1.58	2.21	1.77	2.51	2.44	3.14
14	1.03	1.54	1.37	1.84	1.51	2.04	1.65	2.25	1.85	2.56	2.64	3.22
15	1.06	1.54	1.44	1.87	1.59	2.08	1.74	2.29	1.95	2.61	2.87	3.30
16	1.08	1.54	1.52	1.90	1.67	2.11	1.84	2.33	2.06	2.66	3.13	3.39
17	1.11	1.54	1.61	1.93	1.77	2.15	1.95	2.38	2.18	2.71	3.44	3.49
18	1.14	1.56	1.71	1.96	1.88	2.19	2.07	2.42	2.32	2.76	3.81	3.58
19	1.17	1.58	1.82	1.99	2.00	2.23	2.20	2.47	2.48	2.82	4.24	3.69
20	1.21	1.60	1.94	2.03	2.14	2.27	2.36	2.52	2.65	2.88	4.69	3.80
22	1.29	1.64	2.25	2.10	2.47	2.35	2.73	2.62	3.07	3.01	5.68	4.31
24	1.38	1.68	2.63	2.17	2.90	2.44	3.21	2.73	3.62	3.15	6.76	4.82
26	1.49	1.73	3.09	2.25	3.40	2.54	3.77	2.85	4.25	3.41	7.93	5.34
28	1.62	1.78	3.58	2.34	3.95	2.66	4.37	3.10	4.92	3.76	9.20	5.86
30	1.77	1.83	4.12	2.45	4.53	2.90	5.02	3.39	5.65	4.12	10.6	6.37
32	1.95	1.88	4.68	2.65	5.15	3.14	5.71	3.67	6.43	4.48	12.0	6.88
34	2.16	1.94	5.29	2.85	5.82	3.37	6.45	3.96	7.26	4.83		
36	2.41	2.00	5.93	3.04	6.52	3.61	7.23	4.24	8.14	5.19		
38	2.69	2.06	6.60	3.24	7.27	3.84	8.05	4.52	9.07	5.54		
40	2.98	2.19	7.32	3.43	8.05	4.08	8.92	4.80	10.1	5.89		

Properties

	90	82	74	68	61	53
L_p, ft	15.1	8.8	8.8	8.7	8.7	6.8
$\phi_b M_p$, kip-ft	589	521	473	431	383	327
$n \times 10^3$	3.25	5.39	5.94	6.53	7.35	11.1

[a] b has the units (kips)$^{-1}$.
[b] m and n have the units (kip-ft)$^{-1}$.
Note: Heavy line indicates Kl/r equal to or greater than 200.

Table 6-2 (cont.). W-Shapes Values of b^a, m^b, and n^b for Beam-Columns

$F_y = 50$ ksi
$\phi_c = 0.85$
$\phi_b = 0.90$

Shape: W14×	48		43		38		34		30		26		22	
Effective length KL (ft) with respect to least radius of gyration r_y or unbraced length L_b (ft) for X-X axis bending	$b \times 10^3$	$m \times 10^3$	$b \times 10^3$	$m \times 10^3$	$b \times 10^3$	$m \times 10^3$	$b \times 10^3$	$m \times 10^3$	$b \times 10^3$	$m \times 10^3$	$b \times 10^3$	$m \times 10^3$	$b \times 10^3$	$m \times 10^3$
0	1.67	3.02	1.89†	3.41	2.15†	3.85	2.46†	4.34	2.82†	5.01	3.30†	5.90	4.04†	7.14
6	1.85	3.02	2.08	3.41	2.48†	3.92	2.83†	4.43	3.26†	5.14	4.35†	6.56	5.37†	8.09
7	1.92	3.04	2.16	3.43	2.60	4.04	2.98†	4.58	3.44†	5.33	4.82†	6.92	6.00†	8.58
8	2.01	3.11	2.26	3.52	2.78	4.18	3.17†	4.74	3.66†	5.53	5.46†	7.32	6.83†	9.13
9	2.11	3.19	2.37	3.61	3.00	4.32	3.39†	4.91	3.94†	5.74	6.36	7.76	7.98†	9.76
10	2.23	3.27	2.51	3.70	3.26	4.47	3.69	5.09	4.27	5.98	7.55	8.27	9.61	10.9
11	2.37	3.35	2.67	3.80	3.57	4.64	4.05	5.29	4.72	6.23	9.10	9.45	11.6	12.8
12	2.53	3.44	2.85	3.91	3.95	4.82	4.50	5.51	5.26	6.51	10.8	10.9	13.8	14.8
13	2.72	3.53	3.07	4.02	4.41	5.01	5.03	5.74	5.93	6.81	12.7	12.3	16.2	16.9
14	2.94	3.62	3.33	4.14	4.96	5.21	5.68	6.00	6.74	7.28	14.7	13.8	18.8	19.0
15	3.19	3.72	3.62	4.26	5.64	5.46	6.49	6.53	7.73	8.16	16.9	15.3	21.6	21.1
16	3.49	3.83	3.97	4.40	6.42	6.02	7.38	7.23	8.79	9.07	19.3	16.8	24.6	23.3
17	3.84	3.95	4.38	4.54	7.25	6.60	8.33	7.95	9.93	10.0	21.7	18.3	27.8	25.5
18	4.25	4.07	4.86	4.69	8.13	7.18	9.34	8.67	11.1	10.9	24.4	19.8		
19	4.74	4.20	5.41	5.01	9.05	7.76	10.4	9.40	12.4	11.9				
20	5.25	4.48	6.00	5.41	10.0	8.35	11.5	10.1	13.7	12.9				
21	5.79	4.79	6.61	5.80	11.1	8.94	12.7	10.9	15.1	13.8				
22	6.35	5.11	7.26	6.19	12.1	9.53	14.0	11.6	16.6	14.8				
23	6.94	5.43	7.93	6.59	13.3	10.1	15.3	12.4	18.2	15.8				
24	7.56	5.74	8.64	6.99	14.4	10.7	16.6	13.1	19.8	16.8				
25	8.20	6.06	9.37	7.38	15.7	11.3	18.0	13.9						
26	8.87	6.38	10.1	7.78										
27	9.57	6.69	10.9	8.18										
28	10.3	7.01	11.8	8.57										
29	11.0	7.32	12.6	8.97										
30	11.8	7.64	13.5	9.37										

Properties

	48	43	38	34	30	26	22
L_p, ft	6.75	6.68	5.47	5.40	5.26	3.81	3.67
$\phi_b M_p$, kip-ft	294	261	231	205	177	151	125
$n \times 10^3$	12.3	14.0	20.1	22.9	27.2	44.5	56.4

[a] b has the units $(\text{kips})^{-1}$.
[b] m and n have the units $(\text{kip-ft})^{-1}$.
† Web may be slender for combined axial and bending stress.
Note: Heavy line indicates Kl/r equal to or greater than 200.

$F_y = 50$ ksi
$\phi_c = 0.85$
$\phi_b = 0.90$

Table 6-2 (cont.). W-Shapes Values of b^a, m^b, and n^b for Beam-Columns

Shape	W12×											
	336*		305*		279*		252*		230*		210*	
Effective length KL (ft) with respect to least radius of gyration r_y or unbraced length L_b (ft) for X-X axis bending	$b \times 10^3$	$m \times 10^3$	$b \times 10^3$	$m \times 10^3$	$b \times 10^3$	$m \times 10^3$	$b \times 10^3$	$m \times 10^3$	$b \times 10^3$	$m \times 10^3$	$b \times 10^3$	$m \times 10^3$
0	0.238	0.393	0.263	0.441	0.287	0.493	0.318	0.554	0.348	0.614	0.381	0.681
6	0.246	0.393	0.271	0.441	0.297	0.493	0.329	0.554	0.360	0.614	0.394	0.681
7	0.249	0.393	0.274	0.441	0.301	0.493	0.333	0.554	0.364	0.614	0.399	0.681
8	0.252	0.393	0.278	0.441	0.305	0.493	0.338	0.554	0.370	0.614	0.405	0.681
9	0.256	0.393	0.282	0.441	0.310	0.493	0.343	0.554	0.376	0.614	0.412	0.681
10	0.260	0.393	0.287	0.441	0.315	0.493	0.349	0.554	0.383	0.614	0.420	0.681
11	0.265	0.393	0.293	0.441	0.321	0.493	0.356	0.554	0.390	0.614	0.429	0.681
12	0.270	0.393	0.299	0.441	0.328	0.493	0.364	0.554	0.399	0.615	0.438	0.682
13	0.276	0.394	0.306	0.443	0.336	0.495	0.373	0.556	0.409	0.617	0.449	0.686
14	0.283	0.395	0.313	0.444	0.344	0.496	0.383	0.559	0.420	0.620	0.461	0.689
15	0.290	0.396	0.322	0.446	0.353	0.498	0.393	0.561	0.431	0.623	0.475	0.692
16	0.298	0.398	0.331	0.447	0.364	0.500	0.405	0.563	0.444	0.625	0.489	0.695
17	0.307	0.399	0.341	0.449	0.375	0.502	0.418	0.565	0.459	0.628	0.505	0.698
18	0.316	0.400	0.352	0.450	0.387	0.504	0.432	0.567	0.475	0.631	0.523	0.702
19	0.327	0.401	0.363	0.452	0.401	0.505	0.447	0.570	0.492	0.633	0.542	0.705
20	0.338	0.403	0.376	0.453	0.415	0.507	0.464	0.572	0.510	0.636	0.563	0.708
22	0.364	0.405	0.406	0.456	0.449	0.511	0.502	0.577	0.553	0.642	0.611	0.715
24	0.394	0.408	0.441	0.459	0.489	0.515	0.548	0.581	0.605	0.647	0.669	0.722
26	0.430	0.410	0.483	0.463	0.536	0.519	0.602	0.586	0.666	0.653	0.738	0.729
28	0.473	0.413	0.532	0.466	0.592	0.523	0.666	0.591	0.738	0.659	0.820	0.736
30	0.523	0.415	0.590	0.469	0.658	0.526	0.744	0.596	0.825	0.665	0.919	0.743
32	0.583	0.418	0.660	0.472	0.739	0.530	0.837	0.601	0.932	0.671	1.04	0.750
34	0.656	0.421	0.744	0.476	0.834	0.535	0.945	0.606	1.05	0.677	1.17	0.758
36	0.735	0.424	0.835	0.479	0.935	0.539	1.06	0.611	1.18	0.683	1.32	0.766
38	0.819	0.426	0.930	0.483	1.04	0.543	1.18	0.616	1.31	0.690	1.47	0.774
40	0.908	0.429	1.03	0.486	1.15	0.547	1.31	0.622	1.46	0.696	1.62	0.782

Properties

	336*	305*	279*	252*	230*	210*
L_p, ft	12.3	12.1	11.9	11.8	11.7	11.6
$\phi_b M_p$, kip-ft	2260	2010	1800	1610	1450	1310
$n \times 10^3$	0.893	0.994	1.11	1.24	1.37	1.52

[a] b has the units (kips)$^{-1}$.
[b] m and n have the units (kip-ft)$^{-1}$.
*ASTM A6 tensile group 4 or 5 shape. Special requirements may apply per LRFD Specification Section A3.1c.

Table 6-2 (cont.). W-Shapes Values of b^a, m^b, and n^b for Beam-Columns

$F_y = 50$ ksi
$\phi_c = 0.85$
$\phi_b = 0.90$

Shape	W12×											
	190		170		152		136		120		106	
	b ×10³	m ×10³	b ×10³	m ×10³	b ×10³	m ×10³	b ×10³	m ×10³	b ×10³	m ×10³	b ×10³	m ×10³
0	0.422	0.762	0.471	0.862	0.526	0.975	0.590	1.11	0.667	1.27	0.754	1.45
6	0.437	0.762	0.488	0.862	0.546	0.975	0.613	1.11	0.693	1.27	0.784	1.45
7	0.443	0.762	0.495	0.862	0.554	0.975	0.621	1.11	0.703	1.27	0.795	1.45
8	0.449	0.762	0.502	0.862	0.562	0.975	0.631	1.11	0.714	1.27	0.809	1.45
9	0.457	0.762	0.511	0.862	0.572	0.975	0.642	1.11	0.727	1.27	0.824	1.45
10	0.466	0.762	0.521	0.862	0.584	0.975	0.655	1.11	0.742	1.27	0.841	1.45
11	0.476	0.762	0.532	0.862	0.597	0.975	0.670	1.11	0.759	1.27	0.860	1.45
12	0.487	0.764	0.545	0.865	0.611	0.980	0.686	1.11	0.778	1.28	0.882	1.46
13	0.499	0.768	0.559	0.870	0.627	0.986	0.705	1.12	0.799	1.29	0.906	1.47
14	0.513	0.772	0.574	0.875	0.645	0.992	0.725	1.13	0.823	1.30	0.934	1.48
15	0.528	0.776	0.591	0.879	0.664	0.998	0.748	1.14	0.849	1.31	0.963	1.50
16	0.544	0.780	0.610	0.884	0.686	1.00	0.772	1.15	0.878	1.32	0.997	1.51
17	0.562	0.784	0.631	0.889	0.710	1.01	0.800	1.15	0.909	1.34	1.03	1.52
18	0.582	0.788	0.654	0.895	0.736	1.02	0.830	1.16	0.944	1.35	1.07	1.54
19	0.604	0.792	0.679	0.900	0.765	1.02	0.863	1.17	0.982	1.36	1.12	1.55
20	0.628	0.796	0.706	0.905	0.796	1.03	0.899	1.18	1.02	1.37	1.17	1.57
22	0.683	0.804	0.769	0.915	0.869	1.04	0.982	1.20	1.12	1.39	1.28	1.60
24	0.749	0.813	0.845	0.926	0.955	1.06	1.08	1.21	1.24	1.42	1.41	1.63
26	0.827	0.821	0.935	0.937	1.06	1.07	1.20	1.23	1.38	1.44	1.57	1.66
28	0.921	0.830	1.04	0.949	1.18	1.09	1.35	1.25	1.55	1.47	1.77	1.69
30	1.03	0.839	1.17	0.960	1.34	1.10	1.52	1.27	1.76	1.49	2.01	1.73
32	1.17	0.849	1.33	0.972	1.52	1.12	1.73	1.29	2.00	1.52	2.29	1.77
34	1.32	0.858	1.50	0.984	1.72	1.13	1.96	1.31	2.26	1.55	2.59	1.80
36	1.48	0.868	1.69	0.997	1.92	1.15	2.20	1.33	2.53	1.58	2.90	1.84
38	1.65	0.878	1.88	1.01	2.14	1.17	2.45	1.36	2.82	1.61	3.23	1.89
40	1.83	0.888	2.08	1.02	2.37	1.18	2.71	1.38	3.12	1.64	3.58	1.93

Effective length KL (ft) with respect to least radius of gyration r_y or unbraced length L_b (ft) for X-X axis bending

Properties

	190	170	152	136	120	106
L_p, ft	11.5	11.4	11.3	11.2	11.1	11.0
$\phi_b M_p$, kip-ft	1170	1030	911	803	698	615
$n \times 10^3$	1.70	1.92	2.17	2.46	2.82	3.21

[a] b has the units $(\text{kips})^{-1}$.
[b] m and n have the units $(\text{kip-ft})^{-1}$.

$F_y = 50$ ksi
$\phi_c = 0.85$
$\phi_b = 0.90$

Table 6-2 (cont.). W-Shapes Values of b^a, m^b, and n^b for Beam-Columns

Shape	W12×											
	96		87		79		72		65		58	
Effective length KL (ft) with respect to least radius of gyration r_y or unbraced length L_b (ft) for X-X axis bending	b ×10³	m ×10³	b ×10³	m ×10³	b ×10³	m ×10³	b ×10³	m ×10³	b ×10³	m ×10³	b ×10³	m ×10³
0	0.834	1.61	0.919	1.80	1.01	1.99	1.12	2.19	1.23	2.49	1.38	2.74
6	0.868	1.61	0.957	1.80	1.06	1.99	1.16	2.19	1.28	2.49	1.47	2.74
7	0.881	1.61	0.971	1.80	1.07	1.99	1.18	2.19	1.30	2.49	1.50	2.74
8	0.895	1.61	0.987	1.80	1.09	1.99	1.20	2.19	1.33	2.49	1.54	2.74
9	0.912	1.61	1.01	1.80	1.11	1.99	1.22	2.19	1.35	2.49	1.58	2.75
10	0.932	1.61	1.03	1.80	1.14	1.99	1.25	2.19	1.38	2.49	1.64	2.79
11	0.953	1.61	1.05	1.80	1.16	2.00	1.28	2.20	1.42	2.49	1.69	2.84
12	0.978	1.63	1.08	1.82	1.19	2.02	1.31	2.23	1.45	2.49	1.76	2.88
13	1.01	1.64	1.11	1.84	1.23	2.04	1.35	2.26	1.50	2.49	1.84	2.93
14	1.04	1.66	1.14	1.86	1.27	2.07	1.39	2.29	1.54	2.52	1.92	2.98
15	1.07	1.68	1.18	1.88	1.31	2.09	1.44	2.31	1.60	2.56	2.02	3.03
16	1.11	1.69	1.22	1.90	1.36	2.12	1.49	2.35	1.66	2.60	2.12	3.08
17	1.15	1.71	1.27	1.92	1.41	2.14	1.55	2.38	1.72	2.63	2.24	3.13
18	1.19	1.73	1.32	1.94	1.46	2.17	1.61	2.41	1.79	2.67	2.38	3.19
19	1.24	1.75	1.38	1.96	1.53	2.20	1.68	2.44	1.87	2.72	2.53	3.25
20	1.30	1.76	1.44	1.98	1.59	2.22	1.76	2.47	1.95	2.76	2.70	3.31
22	1.42	1.80	1.58	2.03	1.75	2.28	1.94	2.54	2.15	2.85	3.11	3.43
24	1.57	1.84	1.75	2.08	1.95	2.34	2.15	2.62	2.40	2.94	3.63	3.57
26	1.76	1.88	1.96	2.13	2.18	2.40	2.41	2.70	2.69	3.04	4.26	3.72
28	1.98	1.92	2.21	2.18	2.46	2.47	2.72	2.78	3.05	3.15	4.94	3.98
30	2.26	1.97	2.52	2.24	2.81	2.54	3.12	2.87	3.49	3.27	5.67	4.34
32	2.57	2.01	2.86	2.30	3.20	2.62	3.54	2.96	3.97	3.41	6.45	4.70
34	2.90	2.06	3.23	2.36	3.62	2.70	4.00	3.09	4.48	3.69	7.28	5.06
36	3.25	2.11	3.63	2.43	4.05	2.80	4.49	3.31	5.02	3.97	8.17	5.42
38	3.62	2.17	4.04	2.50	4.52	2.99	5.00	3.54	5.59	4.25	9.10	5.78
40	4.01	2.22	4.48	2.63	5.00	3.17	5.54	3.77	6.20	4.53	10.1	6.14

Properties

	96	87	79	72	65	58
L_p, ft	10.9	10.8	10.8	10.7	11.9	8.87
$\phi_b M_p$, kip-ft	551	495	446	405	363	324
$n \times 10^3$	3.56	3.98	4.41	4.88	5.54	7.38

[a] b has the units (kips)$^{-1}$.
[b] m and n have the units (kip-ft)$^{-1}$.

Table 6-2 (cont.). W-Shapes Values of b^a, m^b, and n^b for Beam-Columns

$F_y = 50$ ksi
$\phi_c = 0.85$
$\phi_b = 0.90$

Shape	W12×											
	53		50		45		40		35		30	
	$b \times 10^3$	$m \times 10^3$	$b \times 10^3$	$m \times 10^3$	$b \times 10^3$	$m \times 10^3$	$b \times 10^3$	$m \times 10^3$	$b \times 10^3$	$m \times 10^3$	$b \times 10^3$	$m \times 10^3$
Effective length *KL* (ft) with respect to least radius of gyration r_y or unbraced length L_b (ft) for X-X axis bending												
0	1.51	3.04	1.61	3.30	1.80	3.69	2.01	4.16	2.30†	4.63	2.77†	5.50
6	1.60	3.04	1.78	3.30	1.98	3.69	2.22	4.16	2.68	4.71	3.20†	5.61
7	1.64	3.04	1.84	3.30	2.06	3.70	2.31	4.17	2.84	4.85	3.38†	5.80
8	1.68	3.04	1.92	3.37	2.14	3.78	2.41	4.27	3.04	5.01	3.60†	6.00
9	1.73	3.05	2.01	3.44	2.25	3.86	2.52	4.37	3.27	5.17	3.87	6.21
10	1.79	3.11	2.12	3.51	2.37	3.95	2.66	4.47	3.56	5.35	4.22	6.45
11	1.86	3.16	2.25	3.58	2.51	4.04	2.82	4.59	3.91	5.53	4.65	6.69
12	1.93	3.21	2.39	3.66	2.68	4.14	3.01	4.70	4.33	5.73	5.16	6.96
13	2.01	3.27	2.56	3.74	2.87	4.24	3.23	4.82	4.84	5.95	5.78	7.25
14	2.11	3.33	2.76	3.83	3.09	4.34	3.48	4.95	5.45	6.18	6.54	7.57
15	2.22	3.39	2.99	3.92	3.35	4.45	3.77	5.09	6.22	6.44	7.48	8.26
16	2.34	3.45	3.25	4.01	3.65	4.57	4.12	5.23	7.07	6.98	8.51	9.12
17	2.47	3.52	3.56	4.11	4.00	4.69	4.51	5.39	7.98	7.62	9.60	9.99
18	2.63	3.58	3.92	4.21	4.41	4.82	4.98	5.55	8.95	8.26	10.8	10.9
19	2.80	3.65	4.34	4.32	4.89	4.95	5.53	5.72	9.97	8.90	12.0	11.8
20	2.99	3.73	4.81	4.43	5.42	5.10	6.13	6.10	11.1	9.54	13.3	12.6
22	3.45	3.88	5.82	4.76	6.56	5.73	7.42	6.96	13.4	10.8	16.1	14.4
24	4.05	4.05	6.93	5.31	7.80	6.42	8.83	7.83	15.9	12.1	19.1	16.2
26	4.76	4.28	8.13	5.87	9.16	7.11	10.4	8.69				
28	5.51	4.73	9.43	6.42	10.6	7.80	12.0	9.56				
30	6.33	5.17	10.8	6.97	12.2	8.48	13.8	10.4				
32	7.20	5.61	12.3	7.51	13.9	9.16	15.7	11.3				
34	8.13	6.05										
36	9.12	6.49										
38	10.2	6.93										
40	11.3	7.37										
Properties												
L_p, ft	8.76		6.92		6.89		6.85		5.44		5.37	
$\phi_b M_p$, kip-ft	292		270		241		214		192		162	
$n \times 10^3$	8.23		11.4		12.7		14.4		21.2		25.3	

[a] b has the units $(\text{kips})^{-1}$.
[b] m and n have the units $(\text{kip-ft})^{-1}$.
† Web may be slender for combined axial and bending stress.
Note: Heavy line indicates Kl/r equal to or greater than 200.

$F_y = 50$ ksi
$\phi_c = 0.85$
$\phi_b = 0.90$

Table 6-2 (cont.). W-Shapes Values of b^a, m^b, and n^b for Beam-Columns

Shape	W12×									
	26		22		19		16		14	
	$b \times 10^3$	$m \times 10^3$	$b \times 10^3$	$m \times 10^3$	$b \times 10^3$	$m \times 10^3$	$b \times 10^3$	$m \times 10^3$	$b \times 10^3$	$m \times 10^3$
Effective length KL (ft) with respect to least radius of gyration r_y or unbraced length L_b (ft) for X-X axis bending										
0	3.26†	6.37	3.80†	8.09	4.55†	9.60	5.53†	11.8	6.47†	13.6
2	3.31†	6.37	3.99†	8.09	4.79†	9.60	5.84†	11.8	6.84†	13.6
3	3.38†	6.37	4.24†	8.09	5.10†	9.65	6.26†	12.0	7.34†	14.0
4	3.48†	6.37	4.64†	8.58	5.59†	10.3	6.93†	12.9	8.13†	15.0
5	3.60†	6.37	5.24	9.13	6.32†	11.0	7.93†	13.9	9.33†	16.3
6	3.76†	6.52	6.15	9.75	7.40	11.9	9.44†	15.2	11.2 †	17.8
7	3.97†	6.75	7.44	10.5	9.06	12.8	11.8	16.6	14.1	19.6
8	4.22†	6.99	9.27	11.3	11.5	14.1	15.3	19.7	18.3	24.3
9	4.53†	7.26	11.7	13.0	14.5	17.1	19.4	24.0	23.2	29.8
10	4.91†	7.54	14.5	15.2	17.9	20.2	24.0	28.6	28.6	35.6
11	5.38	7.85	17.5	17.5	21.7	23.4	29.0	33.3	34.6	41.6
12	5.98	8.18	20.9	19.9	25.8	26.6	34.5	38.1	41.2	47.9
13	6.71	8.55	24.5	22.2	30.3	29.9				
14	7.60	9.07	28.4	24.5						
15	8.71	10.2								
16	9.91	11.3								
17	11.2	12.4								
18	12.5	13.5								
19	14.0	14.7								
20	15.5	15.8								
21	17.1	17.0								
22	18.7	18.2								
23	20.5	19.4								
24	22.3	20.6								
25	24.2	21.8								
26										

Properties

	26	22	19	16	14
L_p, ft	5.33	3.00	2.90	2.73	2.66
$\phi_b M_p$, kip-ft	140	110	92.6	75.4	65.3
$n \times 10^3$	29.6	68.4	84.1	112	133

[a] b has the units $(\text{kips})^{-1}$.
[b] m and n have the units $(\text{kip-ft})^{-1}$.
† Web may be slender for combined axial and bending stress.
Note: Heavy line indicates Kl/r equal to or greater than 200.

Table 6-2 (cont.). W-Shapes Values of b^a, m^b, and n^b for Beam-Columns

$F_y = 50$ ksi
$\phi_c = 0.85$
$\phi_b = 0.90$

Shape	W10× 112		100		88		77		68		60	
	b ×10³	*m* ×10³	*b* ×10³	*m* ×10³	*b* ×10³	*m* ×10³	*b* ×10³	*m* ×10³	*b* ×10³	*m* ×10³	*b* ×10³	*m* ×10³
0	0.715	1.61	0.800	1.82	0.908	2.10	1.04	2.43	1.18	2.78	1.34	3.18
6	0.754	1.61	0.845	1.82	0.960	2.10	1.10	2.43	1.24	2.78	1.42	3.18
7	0.768	1.61	0.861	1.82	0.979	2.10	1.12	2.43	1.27	2.78	1.45	3.18
8	0.786	1.61	0.881	1.82	1.00	2.10	1.15	2.43	1.30	2.78	1.48	3.18
9	0.805	1.61	0.904	1.82	1.03	2.10	1.18	2.43	1.34	2.78	1.52	3.18
10	0.828	1.62	0.930	1.83	1.06	2.11	1.22	2.45	1.38	2.80	1.57	3.21
11	0.854	1.63	0.960	1.85	1.09	2.13	1.26	2.47	1.42	2.84	1.62	3.25
12	0.883	1.64	0.993	1.86	1.13	2.15	1.30	2.50	1.47	2.87	1.68	3.29
13	0.916	1.65	1.03	1.87	1.17	2.17	1.35	2.52	1.53	2.90	1.75	3.34
14	0.953	1.66	1.07	1.89	1.22	2.18	1.41	2.55	1.60	2.93	1.83	3.38
15	0.995	1.67	1.12	1.90	1.28	2.20	1.48	2.57	1.67	2.97	1.91	3.42
16	1.04	1.69	1.17	1.92	1.34	2.22	1.55	2.60	1.76	3.00	2.01	3.47
17	1.09	1.70	1.23	1.93	1.41	2.24	1.63	2.63	1.85	3.04	2.12	3.51
18	1.15	1.71	1.30	1.95	1.49	2.26	1.72	2.65	1.96	3.07	2.24	3.56
19	1.21	1.72	1.38	1.97	1.57	2.29	1.83	2.68	2.07	3.11	2.38	3.61
20	1.29	1.73	1.46	1.98	1.67	2.31	1.94	2.71	2.20	3.15	2.53	3.66
22	1.45	1.76	1.65	2.01	1.90	2.35	2.21	2.77	2.51	3.23	2.89	3.77
24	1.66	1.79	1.90	2.05	2.18	2.40	2.55	2.83	2.91	3.31	3.35	3.88
26	1.93	1.81	2.21	2.08	2.55	2.44	2.99	2.90	3.40	3.40	3.92	3.99
28	2.24	1.84	2.56	2.12	2.95	2.49	3.46	2.97	3.94	3.49	4.55	4.12
30	2.57	1.87	2.94	2.16	3.39	2.54	3.98	3.04	4.53	3.59	5.23	4.25
32	2.92	1.90	3.35	2.20	3.86	2.60	4.52	3.11	5.15	3.69	5.95	4.40
34	3.30	1.93	3.78	2.24	4.36	2.65	5.11	3.19	5.82	3.80	6.71	4.66
36	3.70	1.96	4.24	2.28	4.88	2.71	5.73	3.28	6.52	3.91	7.52	4.97
38	4.12	1.99	4.72	2.32	5.44	2.77	6.38	3.36	7.26	4.15	8.38	5.28
40	4.57	2.03	5.23	2.37	6.03	2.83	7.07	3.46	8.05	4.39	9.29	5.59

Effective length KL (ft) with respect to least radius of gyration r_y or unbraced length L_b (ft) for X-X axis bending

Properties

	112	100	88	77	68	60
L_p, ft	9.47	9.36	9.29	9.18	9.15	9.08
$\phi_b M_p$, kip-ft	551	488	424	366	320	280
$n \times 10^3$	3.49	3.95	4.54	5.25	5.99	6.87

[a] *b* has the units $(\text{kips})^{-1}$.
[b] *m* and *n* have the units $(\text{kip-ft})^{-1}$.

$F_y = 50$ ksi
$\phi_c = 0.85$
$\phi_b = 0.90$

Table 6-2 (cont.). W-Shapes Values of b^a, m^b, and n^b for Beam-Columns

Shape		W10× 54		49		45		39		33		30	
		$b \times 10^3$	$m \times 10^3$	$b \times 10^3$	$m \times 10^3$	$b \times 10^3$	$m \times 10^3$	$b \times 10^3$	$m \times 10^3$	$b \times 10^3$	$m \times 10^3$	$b \times 10^3$	$m \times 10^3$
Effective length KL (ft) with respect to least radius of gyration r_y or unbraced length L_b (ft) for X-X axis bending	0	1.49	3.56	1.63	3.92	1.77	4.32	2.05	5.06	2.42	6.11	2.66	6.48
	6	1.58	3.56	1.73	3.92	1.94	4.32	2.25	5.06	2.68	6.11	3.26	6.71
	7	1.61	3.56	1.77	3.92	2.01	4.32	2.33	5.07	2.78	6.13	3.50	6.92
	8	1.65	3.56	1.81	3.92	2.09	4.38	2.43	5.16	2.90	6.26	3.81	7.15
	9	1.70	3.56	1.86	3.93	2.18	4.46	2.54	5.26	3.04	6.41	4.19	7.40
	10	1.75	3.60	1.92	3.98	2.30	4.54	2.68	5.37	3.21	6.55	4.66	7.66
	11	1.81	3.65	1.99	4.04	2.42	4.62	2.83	5.48	3.40	6.71	5.25	7.94
	12	1.88	3.70	2.07	4.10	2.57	4.70	3.01	5.59	3.63	6.87	5.97	8.24
	13	1.95	3.76	2.15	4.16	2.75	4.79	3.22	5.71	3.89	7.04	6.87	8.57
	14	2.04	3.81	2.25	4.23	2.95	4.88	3.46	5.84	4.19	7.22	7.97	8.92
	15	2.14	3.86	2.36	4.30	3.18	4.98	3.74	5.97	4.55	7.41	9.15	9.49
	16	2.25	3.92	2.48	4.36	3.45	5.07	4.07	6.10	4.96	7.61	10.4	10.3
	17	2.37	3.98	2.62	4.43	3.76	5.17	4.45	6.24	5.44	7.82	11.8	11.2
	18	2.51	4.04	2.77	4.51	4.12	5.28	4.88	6.39	6.00	8.04	13.2	12.0
	19	2.66	4.10	2.95	4.58	4.53	5.39	5.40	6.55	6.67	8.28	14.7	12.9
	20	2.83	4.16	3.14	4.66	5.02	5.51	5.99	6.71	7.39	8.61	16.3	13.7
	22	3.24	4.29	3.60	4.83	6.08	5.75	7.25	7.10	8.94	9.80	19.7	15.4
	24	3.76	4.44	4.18	5.00	7.23	6.02	8.62	7.92	10.6	11.0		
	26	4.41	4.59	4.91	5.19	8.49	6.60	10.1	8.73	12.5	12.2		
	28	5.11	4.75	5.70	5.40	9.85	7.19	11.7	9.54	14.5	13.3		
	30	5.87	4.92	6.54	5.84	11.3	7.79	13.5	10.4	16.6	14.5		
	32	6.67	5.30	7.44	6.31	12.9	8.37	15.3	11.2	18.9	15.7		
	34	7.53	5.69	8.40	6.79								
	36	8.45	6.07	9.42	7.26								
	38	9.41	6.46	10.5	7.73								
	40	10.4	6.85	11.6	8.20								
Properties													
L_p, ft		9.04		8.97		7.10		6.99		6.85		4.84	
$\phi_b M_p$, kip-ft		250		227		206		176		146		137	
$n \times 10^3$		7.67		8.45		11.9		14.0		17.2		27.5	

[a] b has the units (kips)$^{-1}$.
[b] m and n have the units (kip-ft)$^{-1}$.
Note: Heavy line indicates Kl/r equal to or greater than 200.

Table 6-2 (cont.). W-Shapes Values of b^a, m^b, and n^b for Beam-Columns

$F_y = 50$ ksi
$\phi_c = 0.85$
$\phi_b = 0.90$

Shape	W10×											
	26		22		19		17		15		12	
	$b \times 10^3$	$m \times 10^3$	$b \times 10^3$	$m \times 10^3$	$b \times 10^3$	$m \times 10^3$	$b \times 10^3$	$m \times 10^3$	$b \times 10^3$	$m \times 10^3$	$b \times 10^3$	$m \times 10^3$
Effective length KL (ft) with respect to least radius of gyration r_y or unbraced length L_b (ft) for X-X axis bending												
0	3.09	7.57	3.66†	9.12	4.19	11.0	4.77†	12.7	5.46†	14.8	7.20†	18.9
2	3.16	7.57	3.74†	9.12	4.42	11.0	5.00	12.7	5.77†	14.8	7.60†	18.9
3	3.25	7.57	3.83†	9.12	4.74	11.0	5.38	12.7	6.16	14.9	8.15†	19.0
4	3.39	7.57	3.99	9.12	5.22	11.5	5.97	13.5	6.90	15.9	9.01†	20.4
5	3.56	7.62	4.21	9.22	5.91	12.2	6.82	14.3	7.97	17.1	10.3†	21.9
6	3.80	7.88	4.49	9.56	6.88	12.9	8.02	15.3	9.51	18.4	12.3	23.8
7	4.09	8.16	4.85	9.93	8.23	13.8	9.71	16.4	11.7	19.9	15.4	26.0
8	4.45	8.46	5.31	10.3	10.1	14.8	12.1	17.7	14.9	21.9	19.8	30.6
9	4.90	8.78	5.87	10.8	12.7	16.0	15.3	20.5	18.9	26.5	25.1	37.4
10	5.46	9.12	6.57	11.2	15.7	18.7	18.9	24.0	23.3	31.2	30.9	44.5
11	6.16	9.50	7.45	11.8	19.0	21.4	22.9	27.6	28.2	36.0	37.4	51.8
12	7.02	9.90	8.54	12.3	22.6	24.2	27.3	31.3	33.6	41.0	44.6	59.4
13	8.10	10.3	9.94	13.2	26.6	26.9	32.0	34.9	39.4	45.9	52.3	67.0
14	9.40	11.1	11.5	14.8	30.8	29.6	37.1	38.6				
15	10.8	12.2	13.2	16.4								
16	12.3	13.4	15.1	18.1								
17	13.9	14.5	17.0	19.7								
18	15.5	15.7	19.0	21.3								
19	17.3	16.8	21.2	23.0								
20	19.2	18.0	23.5	24.6								
21	21.1	19.1	25.9	26.3								
22	23.2	20.3	28.5	27.9								
23												
Properties												
L_p, ft	4.80		4.70		3.09		2.98		2.86		2.87	
$\phi_b M_p$, kip-ft	117		97.5		81.0		70.1		60.0		47.3	
$n \times 10^3$	32.3		39.8		73.8		88.8		109		145	

[a] b has the units (kips)$^{-1}$.
[b] m and n have the units (kip-ft)$^{-1}$.
† Web may be slender for combined axial and bending stress.
Note: Heavy line indicates Kl/r equal to or greater than 200.

PART 7

DESIGN CONSIDERATIONS FOR BOLTS

BOLTS

SCOPE

The specification requirements and other design considerations summarized in this Part apply to the design of bolts in steel-to-steel structural connections. For the design of steel-to-concrete anchorage, see Part 14. For the design of connection elements, see Part 9. For the design of simple shear, moment, bracing and other connections, see Parts 10 through 15. For the design of bolted joints for HSS and steel pipe, see the AISC *Hollow Structural Sections Connections Manual*. For bolted joints that are part of a seismic-force-resisting system in which the seismic response modification factor R is taken greater than 3, see the Seismic Provisions, which are available from AISC at www.aisc.org.

LOAD DETERMINATION

The required strength for bolted joints is determined using force-transfer models such as those described in Parts 10 through 15 to distribute the member end reactions, which are determined by analysis as indicated in LRFD Specification Section A5.

GENERAL REQUIREMENTS FOR BOLTED JOINTS

Fastener Components

The applicable material specifications for fastener components are as given in Part 2. The materials and storage requirements fastener components are as given in LRFD Specification Section A3.3 and RCSC Specification Section 2. The compatibility of ASTM A563 nuts and F436 washers with ASTM A325, F1852 and A490 bolts is as given in RCSC Specification Table 2.1. These products are given identifying marks as illustrated in RCSC Specification Figure C-2.1. Alternative-design fasteners, including twist-off-type tension-control bolt assemblies with a strength level matching that of ASTM A490 bolts, and alternative washer-type indicating devices are permitted, subject to the requirements in RCSC Specification Sections 2.8 and 2.6.2, respectively.

Mixing grades of fasteners raises inventory and quality control issues associated with the use of multiple fastener grades. When both ASTM A325 and A490 bolts are used on a project, different diameters can be specified for each to help ensure that the ASTM A490 bolts are installed in the proper location.

Regardless of the bolt type selected, the normal sizes of $^3/_4$-in., $^7/_8$-in., 1-in. and 1 $^1/_8$-in. diameter are usually preferred. Diameters above 1 in. require special consideration for availability as well as installation, when pretensioned installation is required. Special equipment may be required to pretension large-diameter ASTM A490 bolts.

Proper Selection of Bolt Length

Per RCSC Specification Section 2.3.2, adequate thread engagement is developed when the end of the bolt is at least flush with or projects beyond the face of the nut. To provide for this, the ordered length of ASTM A325, F1852 and A490 bolts should be calculated as the grip (see Figure 7-1) plus the nominal thickness of washers and/or direct-tension indicators, if used, plus the allowance from Table 7-1, with the total rounded to the next higher increment of $^1/_4$ in. up to 5-in. length and the next higher $^1/_2$ in. over 5-in. length. Note that bolts longer than five inches are generally available only in $^1/_2$-in. increments, except by special arrangement with the manufacturer or vendor. While longer lengths may be ordered, an 8-in. length is generally the maximum stock length available. The foregoing process of bolt selection is tabulated in Table 7-2. See Carter (1996) for further information.

Washer Requirements

Requirements for the use of ASTM F436 washers and/or plate washers are given in RCSC Specification Section 6.

Bolted Parts

The requirements for connected plies, faying surfaces, bolt holes and burrs are given in AISC Specification Sections J3.2 and M2.5 and RCSC Specification Section 3. Spacing and edge distance requirements are given in LRFD Specification Sections J3.3, J3.4 and J3.5.

PROPER SPECIFICATION OF JOINT TYPE

When ASTM A325, F1852 or A490 high-strength bolts are to be used, the joint type must be specified as snug-tightened, pretensioned or slip-critical, per RCSC Specification Section 4.

Snug-Tightened Joints

Snug-tightened joints simplify design, installation and inspection and should be specified whenever pretensioned joints and slip-critical joints are not required. The applicability is summarized and design requirements, installation requirements and inspection requirements are stipulated for snug-tightened joints per RCSC Specification Section 4.1. Faying surfaces in snug-tightened joints must meet the requirements in RCSC Specification Sections 3.2 and 3.2.1, but not those for slip-critical joints in RCSC Specification Section 3.2.2. Note that there is generally no need to limit the actual level of pretension provided in snug-tightened joints, per RCSC Specification Section 9.1.

Pretensioned Joints

When pretension is required, but slip-resistance is not of concern, a pretensioned joint should be specified. The applicability is summarized and design requirements, installation

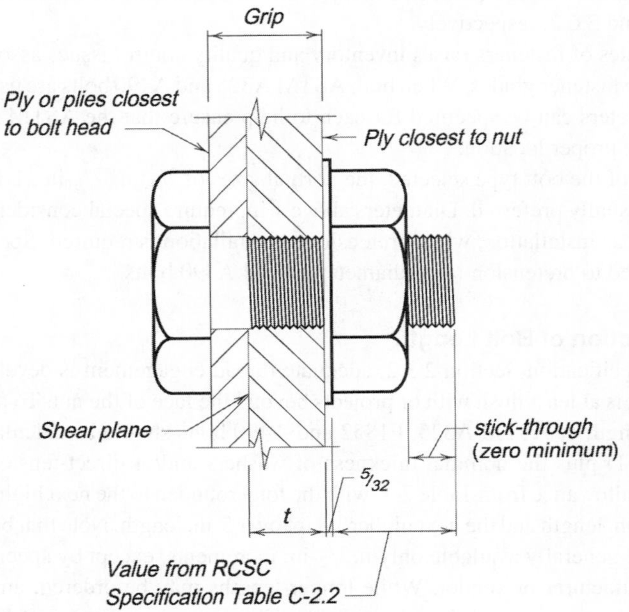

Fig. 7–1. Grip and other parameters for bolt length selection.

requirements and inspection requirements are stipulated for pretensioned joints per RCSC Specification Section 4.2. Additionally, pretensioned joints are required by default in some cases per LRFD Specification Section J1.11. Faying surfaces in pretensioned joints must meet the requirements in RCSC Specification Sections 3.2 and 3.2.1, but not those for slip-critical joints in RCSC Specification Section 3.2.2.

Slip-Critical Joints

The applicability is summarized and design requirements, installation requirements and inspection requirements are stipulated for slip-critical joints per RCSC Specification Section 4.3. Faying surfaces in slip-critical joints must meet the requirements in RCSC Specification Sections 3.2 and 3.2.2. Note that the surfaces under the bolt head, washer and/or nut are not faying surfaces.

Subject to the requirements in RCSC Specification Section 4.3, slip-critical joints are rarely required in building design. Slip-critical joints are appreciably more expensive because of the associated costs of faying-surface preparation. When slip resistance is required and the steel is to be painted, the fabricator should be consulted to determine the most economical approach to providing the necessary slip resistance. Special paint systems that are rated for slip-resistance can be specified. Alternatively, a normal paint system can be used with the faying surfaces masked.

DESIGN REQUIREMENTS

The design requirements for bolted joints in RCSC Specification Section 5 apply as they are stipulated by joint type in RCSC Specification Section 4. Essentially similar design requirements are also found in the LRFD Specification as follows. In each case, the design strength determined in accordance with these provisions must equal or exceed the required strength.

Shear

Design shear strength is determined as given in RCSC Specification Section 5.1 and LRFD Specification Section J3.6, with consideration of the presence of fillers and/or shims, per RCSC Specification Section 5.1 and LRFD Specification Section J6. When the length of a bolted joint measured parallel to the line of force exceeds 50 in., a 20 percent strength reduction may be applicable, per LRFD Specification Table J3.2 footnote e.

Tension

Design tensile strength is determined as given in RCSC Specification Section 5.1 and LRFD Specification Section J3.6, with consideration of the effects of prying action, if any. Prying action is a phenomenon (in bolted construction only) whereby the deformation of a fitting under a tensile force increases the tensile force in the bolt. While the effect of prying action is relevant to the design of the bolts, it is primarily a function of the strength and stiffness of the connection elements. Prying action is addressed in Part 9.

Combined Shear and Tension

Design strength for combined shear and tension is determined as given in RCSC Specification Section 5.2 and LRFD Specification Section J3.7.

Bearing Strength at Bolt Holes

Design bearing strength at bolt holes is determined as given in RCSC Specification Section 5.3 and LRFD Specification Section J3.10.

Slip Resistance

Slip-resistance can be checked at the factored-load level, as given in RCSC Specification Section 5.4.1 and LRFD Specification Sections J3.8a and J3.9a, or at the service-load level, as given in RCSC Specification Section 5.4.2 and LRFD Specification Appendices J3.8b and J3.9b. See RCSC Commentary Section 5.4 for guidance. Note, however, that the design criterion in either case is the same: the prevention of slip in the service-load range.

ECCENTRICALLY LOADED BOLT GROUPS

Eccentricity in the Plane of the Faying Surface

When eccentricity occurs in the plane of the faying surface, the bolts must be designed to resist the combined effect of the direct shear P_u and the additional shear from the induced moment $P_u e$. Two analysis methods for this type of eccentricity are the instantaneous center of rotation method and the elastic method.

The instantaneous center of rotation method is more accurate, but generally requires the use of tabulated values or an iterative solution. The elastic method is simplified, but may be excessively conservative because it neglects the ductility of the bolt group and the potential for load redistribution.

Instantaneous Center of Rotation Method

Eccentricity produces both a rotation and a translation of one connection element with respect to the other. The combined effect of this rotation and translation is equivalent to a rotation about a point defined as the instantaneous center of rotation (IC) as illustrated in Figure 7-2a. The location of the IC depends upon the geometry of the bolt group as well as the direction and point of application of the load.

The load-deformation relationship for one bolt is illustrated in Figure 7-3, where

$$R = R_{ult}(1 - e^{-10\Delta})^{0.55}$$

where

R = nominal shear strength of one bolt at a deformation Δ, kips.
R_{ult} = ultimate shear strength of one bolt, kips.
Δ = total deformation, including shear, bearing and bending deformation in the bolt and bearing deformation of the connection elements, in.
e = 2.718 . . ., base of the natural logarithm.

The nominal shear strength of the bolt most remote from the IC can be determined by applying a maximum deformation Δ_{max} to that bolt. The load-deformation relationship is based upon data obtained experimentally for $^3/_4$-in.-diameter ASTM A325 bolts, where $R_{ult} = 74$ kips, and $\Delta_{max} = 0.34$ in.

The nominal shear strengths of the other bolts in the joint can be determined by applying a deformation Δ that varies linearly with distance from the IC. The nominal shear strength of the bolt group is, then, the sum of the individual strengths of all bolts.

The individual resistance of each bolt is assumed to act on a line perpendicular to a ray passing through the IC and the centroid of that bolt as illustrated in Figure 7-2b. If the correct location of the IC has been selected, the three equations of in-plane static equilibrium ($\Sigma F_x = 0$, $\Sigma F_y = 0$ and $\Sigma M = 0$) will be satisfied.

For further information, see Crawford and Kulak (1968).

Elastic Method

For a force applied as illustrated in Figure 7-4, the eccentric force P_u is resolved into a direct shear P_u acting through the center of gravity (CG) of the bolt group and a moment $P_u e$ where e is the eccentricity. Each bolt is then assumed to resist an equal share of the direct shear P_u, and a share of the eccentric moment $P_u e$ proportional to its distance from the CG. The resultant vectorial sum of these forces r_u is the required strength for the bolt.

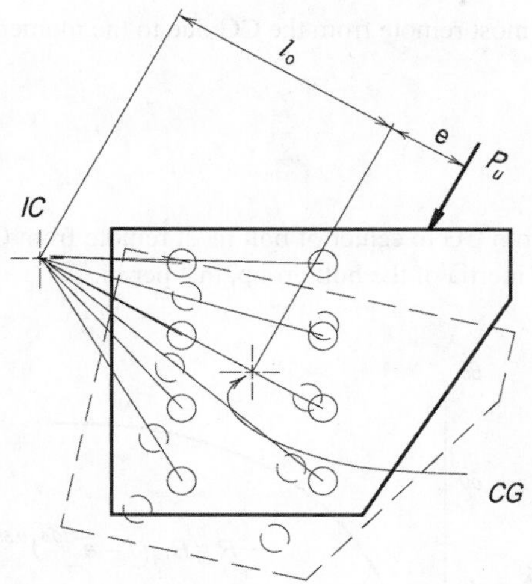

(a) Instantaneous center of rotation (IC)

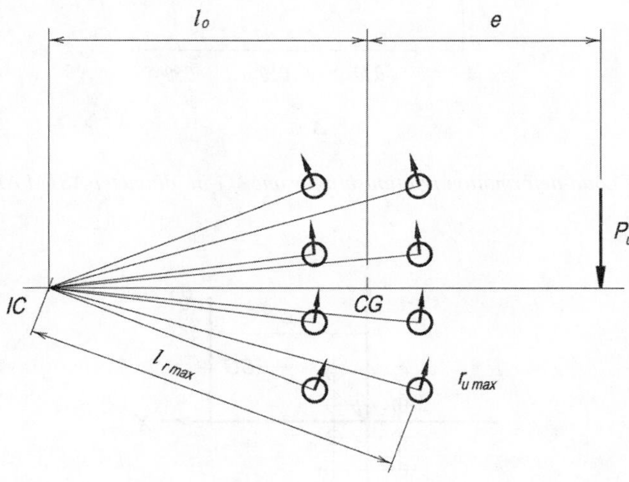

(b) Forces on bolts in group

Fig. 7–2. Illustration for instantaneous center of rotation method.

The shear per bolt due to the concentric force P_u is r_p, where

$$r_p = \frac{P_u}{n}$$

where n is the number of bolts. To determine the resultant forces on each bolt, r_p must be resolved into horizontal component r_{px} and vertical component r_{py}, where

$$r_{px} = r_p \sin\theta \quad \text{and} \quad r_{py} = r_p \cos\theta$$

The shear on the bolt most remote from the CG due to the moment $P_u e$ is r_m, where

$$r_m = \frac{P_u ec}{I_p}$$

where,

c = radial distance from CG to center of bolt most remote from CG, in.
I_p = polar moment of inertia of the bolt group, in.^{4} per in.^{2}

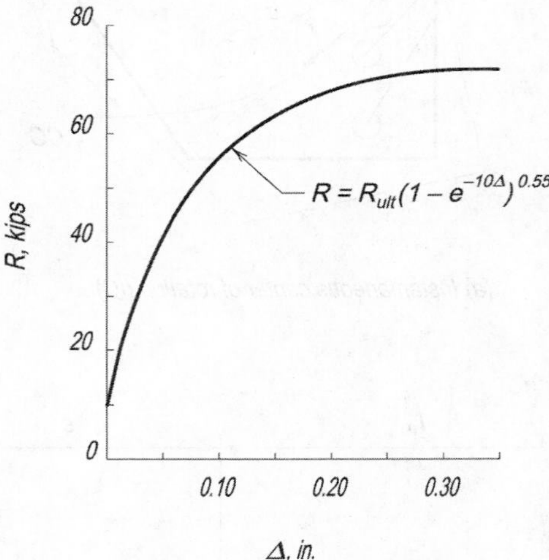

Fig. 7–3. Load-deformation relationship for one 3/4-in.-diameter ASTM A325 bolt.

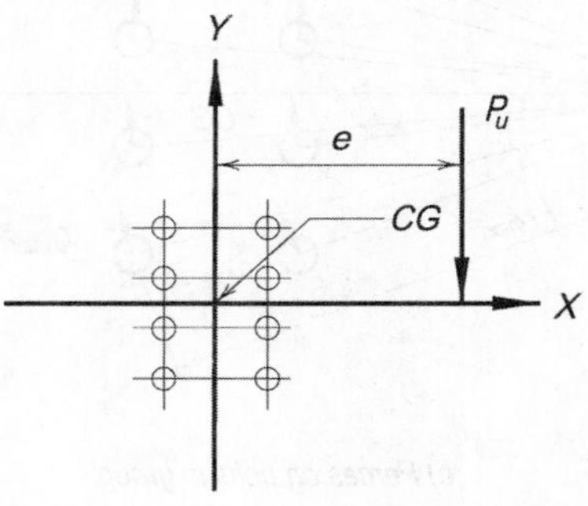

Fig. 7–4. Illustration for elastic method.

To determine the resultant force on the most highly stressed bolt, r_m must be resolved into horizontal component r_{mx} and vertical component r_{my}, where

$$r_{mx} = \frac{P_u e c_y}{I_p} \quad \text{and} \quad r_{my} = \frac{P_u e c_x}{I_p}$$

In the above equation, c_x and c_y are the horizontal and vertical components of the diagonal distance c. Thus, the required strength per bolt is r_u, where

$$r_u = \sqrt{(r_{px} + r_{mx})^2 + (r_{py} + r_{my})^2}$$

For further information, see Higgins (1971).

Eccentricity Normal to the Plane of the Faying Surface

Eccentricity normal to the plane of the faying surface produces tension above and compression below the neutral axis as illustrated in Figure 7-5 for a bracket connection. The eccentric force P_u is resolved into a direct shear P_u acting at the faying surface of the joint and a moment normal to the plane of the faying surface $P_u e$ where e is the eccentricity. Each bolt is then assumed to resist an equal share of the concentric force P_u, and the moment is resisted by tension in the bolts above the neutral axis and compression below the neutral axis. Two design approaches for this type of eccentricity are Case I, in which the neutral axis is not taken at the center of gravity (CG), and Case II, in which the neutral axis is taken at the CG.

Case I—Neutral Axis Not at Center of Gravity

The shear per bolt due to the concentric force P_u is r_{uv}, where

$$r_{uv} = \frac{P_u}{n}$$

where n is the number of bolts in the connection.

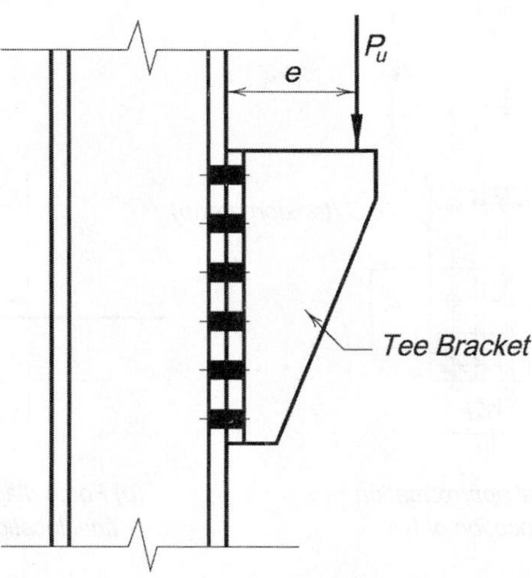

Fig. 7–5. Tee bracket subject to eccentric loading normal to the plane of the faying surface.

A trial position for the neutral axis can be selected at one-sixth of the total bracket depth, measured upward from the bottom (line X-X in Figure 7-6a). To provide for reasonable proportions and to account for the bending stiffness of the connection elements, the effective width of the compression block W_{eff} should be taken as:

$$W_{eff} = 8t_f \leq b_f$$

where

t_f = lesser connection element thickness, in.
b_f = connection element width, in.

This effective width is valid for bracket flanges made from W-shapes, S-shapes, welded plates and angles. Where the bracket flange thickness is not constant, the average flange thickness should be used.

The assumed location of the neutral axis can be evaluated by checking static equilibrium assuming an elastic stress distribution. Equating the moment of the bolt area above the neutral axis with the moment of the compression block area below the neutral axis,

$$\Sigma A_b \times y = W_{eff} \times d \times \frac{d}{2}$$

In the above equation,

ΣA_b = sum of the areas of all bolts above the neutral axis, in.^{2}
y = distance from line X-X to CG of the bolt group above neutral axis, in.
d = depth of compression block, in.

The value of d may then be adjusted until a reasonable equality exists.

Once the neutral axis has been located, the tensile force per bolt r_{ut}, as illustrated in Figure 7-6b may be determined as:

$$r_{ut} = \frac{P_u ec}{I_x} \times A_b$$

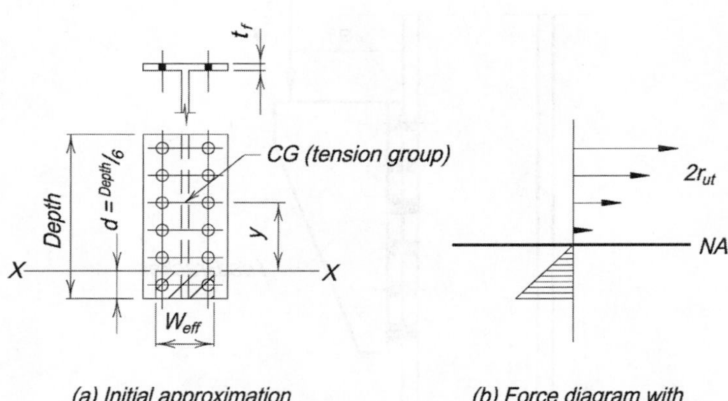

Fig. 7–6. Location of neutral axis (NA) for out-of-plane eccentric loading using Case I.

where

c = distance from neutral axis to most remote bolt in group, in.
I_x = combined moment of inertia of bolt group and compression block about neutral axis, in.^{4}

Bolts above the neutral axis are subjected to the shear force r_{uv}, the tensile force r_{ut}, and the effect of prying action; bolts below the neutral axis are subjected to the shear force r_{uv} only.

Case II—Neutral Axis at Center of Gravity

This method provides a more direct, but also a more conservative result. As for Case I, the shear force per bolt due to the concentric force P_u is r_{uv}, where

$$r_{uv} = \frac{P_u}{n}$$

and n is the number of bolts in the connection.

The neutral axis is assumed to be located at the CG of the bolt group as illustrated in Figure 7-7. The bolts above the neutral axis are in tension and the bolts below the neutral axis are said to be in "compression." To obtain a more accurate result, a plastic stress distribution is assumed; this assumption is justified because this method is still more conservative than Case I. Accordingly, the tensile force r_{ut} in each bolt above the neutral axis due to the moment P_{ue} is:

$$r_{uv} = \frac{P_u e}{n' d_m}$$

where

n' = number of bolts above the neutral axis
d_m = moment arm between resultant tensile force and resultant compressive force, in.

Bolts above the neutral axis are subjected to the shear force r_{uv}, the tensile force r_{ut}, and the effect of prying action; bolts below the neutral axis are subjected to the shear force r_{uv} only.

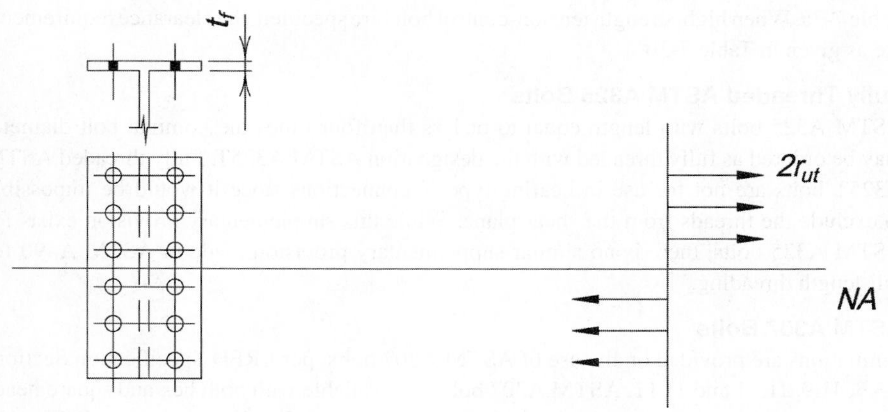

Fig. 7–7. Location of neutral axis (NA) for out-of-plane eccentric loading using Case II.

OTHER SPECIFICATION REQUIREMENTS AND DESIGN CONSIDERATIONS

The following other specification requirements and design considerations apply to the design of bolted joints:

Placement of Bolt Groups

For the required placement of bolt groups at the ends of axially loaded members, see LRFD Specification Section J1.8.

Bolts in Combination with Welds or Rivets

For bolts used in combination with welds or rivets, see LRFD Specification Section J1.9 or J1.10, respectively.

Galvanizing High-Strength Bolts

Galvanizing of high-strength bolts is permitted as follows:

1. By the hot-dip or mechanical process for ASTM A325 Type 1 high-strength bolts, per ASTM A325 Section 4.3.
2. By the mechanical process only for ASTM F1852 twist-off-type tension-control bolt assemblies, per ASTM F1852 Section 6.3.
3. By the hot-dip or mechanical process for ASTM A449 bolts, per ASTM A449 Section 5.1.

ASTM A490 bolts are not permitted to be galvanized, per ASTM A490 Section 5.4. See also RCSC Specification Commentary Section 2.3.

Reuse of Bolts

The reuse of high-strength bolts is limited, per RCSC Specification Section 2.3.3. See also Bowman and Betancourt (1991).

Fatigue Applications

For applications involving fatigue, see RCSC Specification Sections 4.2, 4.3 and 5.5 and LRFD Specification Appendix K3.

Entering and Tightening Clearances

Clearances must be provided for the entering and tightening of the bolts with an impact wrench. The clearance requirements for conventional high-strength bolts are as given in Table 7-3a. When high-strength tension-control bolts are specified, the clearance requirements are as given in Table 7-3b.

Fully Threaded ASTM A325 Bolts

ASTM A325 bolts with length equal to or less than four times the nominal bolt diameter may be ordered as fully threaded with the designation ASTM A325T. Fully threaded ASTM A325T bolts are not for use in bearing-type X connections since it would be impossible to exclude the threads from the shear plane. While this supplementary provision exists for ASTM A325 bolts, there is no similar supplementary provision made in ASTM A490 for full-length threading.

ASTM A307 Bolts

Limitations are provided on the use of ASTM A307 bolts, per LRFD Specification Sections A3.3, J1.9, J1.11 and J3.11. ASTM A307 bolts are available with both hex and square heads in diameters from $^{1}/_{4}$ in. to 4 in. in grade A for general applications and grade B for cast-iron-flanged piping joints. ASTM A563 Grade A nuts are recommended for use with ASTM A307 bolts. Other suitable grades are listed in ASTM A563 Table X1.1.

ASTM A449 Bolts

Limitations are provided on the use of ASTM A449 bolts, per LRFD Specification Sections A3.3 and J3.1.

DESIGN EXAMPLES

EXAMPLE 7.1. Eccentrically loaded bolt groups (IC method).

Given: Refer to Figure 7-8. Determine the largest eccentric force P_u for which the design shear strength of the bolts in the connection is adequate using the instantaneous center of rotation method. Use $^7/_8$-in. diameter A325-N bolts, $\phi r_n = 21.6$ kips/bolt.

a. Assume the load is vertical as illustrated in Figure 7-8 ($\theta = 0°$)

b. Assume the load acts at an angle of 75° with respect to vertical ($\theta = 75°$)

Solution a: From Table 7-19 with $\theta = 0°$, with $s = 3$ in., $e = 16$ in., and $n = 6$:

$$C = 3.55$$

Design Shear Strength

$$\phi R_n = C \times \phi r_n$$
$$= 3.55 \times 21.6 \text{ kips/bolt}$$
$$= 76.7 \text{ kips}$$

Thus, P_u must be less than or equal to 76.7 kips.

Comment: Note that this eccentricity has effectively reduced the shear strength of this bolt group by about 70 percent when compared with the concentrically loaded case.

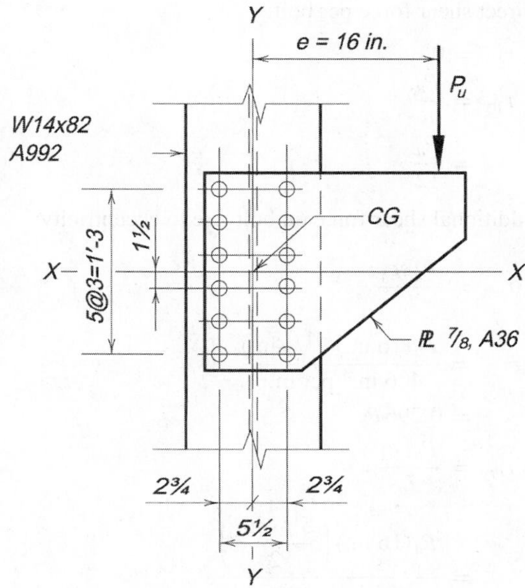

Fig. 7–8. Illustration for Examples 7.1 and 7.2.

Solution b: From Table 7-19 with $\theta = 75°$, $s = 3$ in., $e = 16$ in., and $n = 6$:

$$C = 7.90$$

Design shear strength

$$\begin{aligned}\phi R_n &= C \times \phi r_n \\ &= 7.90 \times 21.6 \text{ kips/bolt} \\ &= 171 \text{ kips}\end{aligned}$$

Thus, P_u must be less than or equal to 171 kips.

Comment: In Solution b, the horizontal component of the design strength is

$$\begin{aligned}\phi R_n \sin 75° &= (171 \text{ kips})(0.966) \\ &= 165 \text{ kips}\end{aligned}$$

and the vertical component of the design strength is

$$\begin{aligned}\phi R_n \cos 75° &= (171 \text{ kips})(0.259) \\ &= 44.3 \text{ kips}\end{aligned}$$

EXAMPLE 7.2. Eccentrically loaded bolt groups (elastic method).

Given: Refer to Example 7-1a. Recalculate the largest eccentric force P_u for which the design shear strength of the bolts in the connection is adequate using the elastic method. Compare the result with that of Example 7-1. Use $^7/_8$-in. diameter A325-N bolts, $\phi r_n = 21.6$ kips.

$$I_p = 406 \text{ in.}^4 \text{ per in.}^2$$

Solution: Direct shear force per bolt:

$$\begin{aligned}r_{px} &= 0 \\ r_{py} &= \frac{P_u}{n} \\ &= \frac{P_u}{12}\end{aligned}$$

Additional shear force on bolt due to eccentricity:

$$\begin{aligned}r_{mx} &= \frac{P_u e c_y}{I_p} \\ &= \frac{P_u(16 \text{ in.})(7^1/_2\text{-in.})}{406 \text{ in.}^4 \text{ per in.}^2} \\ &= 0.296 P_u \\ r_{my} &= \frac{P_u e c_x}{I_p} \\ &= \frac{P_u(16 \text{ in.})\left(\dfrac{5^1/_2\text{-in.}}{2}\right)}{406 \text{ in.}^4 \text{ per in.}^2} \\ &= 0.108\ P_u\end{aligned}$$

Resultant shear force:

$$
\begin{aligned}
r_u &= \sqrt{(r_{px} + r_{mx})^2 + (r_{py} + r_{my})^2} \\
&= \sqrt{(0 + 0.296P_u)^2 + \left(\frac{P_u}{12} + 0.108P_u\right)^2} \\
&= 0.352P_u
\end{aligned}
$$

Since r_u must be less than or equal to ϕr_n,

$$
\begin{aligned}
P_u &\le \frac{\phi r_n}{0.352} \\
&\le \frac{21.6 \text{ kips}}{0.352} \\
&\le 61.3 \text{ kips}
\end{aligned}
$$

This 20 percent reduction in the strength predicted by the instantaneous center of rotation method in Example 7-1a is indicative of the conservatism of the elastic method.

DESIGN TABLES

Table 7-1. Dimensions of High-Strength Fasteners

Dimensions of ASTM A325 and A490 bolts, A563 nuts, and F436 washers are given and illustrated in Table 7-1.

Table 7-2. Bolt Length Selection

To provide for adequate thread engagement, the ordered length of ASTM A325, F1852 and A490 bolts must be properly selected. Bolt lengths are provided in Table 7-2 for the indicated combinations of bolt diameter, grip and number of washers used under the nut. See Carter (1996) for further information.

Table 7-3. Entering and Tightening Clearances

Clearance is required for entering and tightening bolts with an impact wrench. The required clearances are given for conventional high-strength bolts and twist-off-type tension-control bolt assemblies in Tables 7-3a and 7-3b, respectively.

Table 7-4. Threading Dimensions for High-Strength and Non-High-Strength Bolts

Data regarding the characteristics of the threading dimensions of high-strength and non-high-strength bolts is provided in Table 7-4.

Table 7-5. Weights of High-Strength Fasteners

Weights of conventional ASTM A325 and A490 bolts, A563 nuts, and F436 washers are given in Table 7-5. For dimensions and weights of tension-control ASTM A325 and A490 bolts, refer to manufacturers' literature or IFI. For dimensions and weights of ASTM A449 bolts, refer to Table 7-6.

Table 7-6. Dimensions of Non-High-Strength Bolts and Nuts

Typical non-high-strength bolt head and nut dimensions are given in Table 7-6. Thread lengths listed in this table may be calculated for non-high-strength bolts as $2d_b + 1/4$-in. for bolts up to six inches long and $2d_b + 1/2$-in. for bolts over six inches long, where d_b is the bolt diameter.

Note that these thread lengths are longer than those given previously for high-strength bolts in Table 7-1. Threading dimensions are given in Table 7-4.

Tables 7-7, 7-8 and 7-9. Weights of Non-High-Strength Fasteners

Weights of non-high-strength bolts are given in Tables 7-7, 7-8, and 7-9.

Tables 7-10 and 7-11. Design Shear Strength of Bolts

The design bolt shear strengths of various bolts are summarized in Table 7-10; design bolt shear strengths of vertical rows of n bolts are summarized in Table 7-11.

Tables 7-12 and 7-13. Design Bearing Strength at Bolt Holes

The design bearing strength at bolt holes is tabulated for various spacings and edge distances in Tables 7-12 and 7-13, respectively. Note that these tables may be applied to bolts with countersunk heads, by subtracting one-half the depth of the countersink from the material thickness t. As illustrated in Figure 7-9, this is equivalent to subtracting $d_b/4$ from the material thickness t.

Table 7-14. Design Tensile Strength of Bolts

The design bolt tensile strengths of various bolts are summarized in Table 7-14.

Tables 7-15 and 7-16. Design Resistance to Slip

The design slip resistances of various bolts are summarized in Tables 7-15 and 7-16. In Table 7-15, the design resistance to slip is tabulated for comparison to the factored load. In Table 7-16, the design resistance to slip is tabulated for comparison to the service load. In either case, the design criterion is the prevention of slip in the service-load range.

Tables 7-17 through 7-24. Coefficients *C* for Eccentrically Loaded Bolt Groups

Tables 7-17 through 7-24 employ the instantaneous center of rotation method for the bolt patterns and eccentric conditions indicated and inclined loads at 0°, 15°, 30°, 45°, 60°, and 75°. The tabulated non-dimensional coefficient C is obtained by dividing the required eccentric force P_u by R_{ult}.

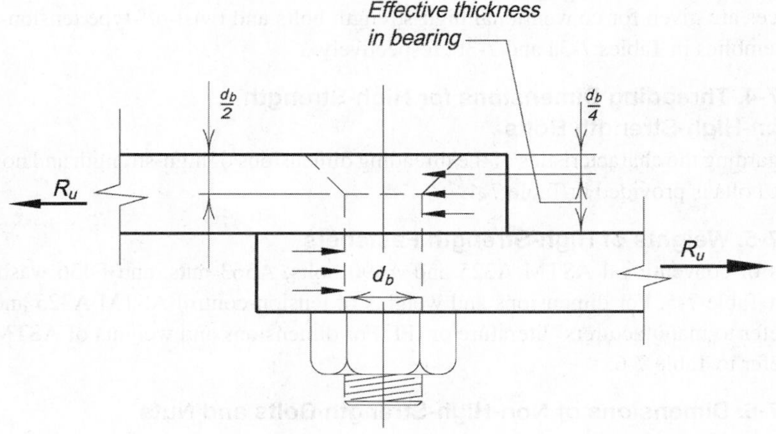

Fig. 7–9. Effective bearing thickness for bolts with countersunk heads.

For any of the bolt group geometries shown, the design strength of the eccentrically loaded bolt group is ϕR_n, where

$$\phi R_n = C \times \phi r_n$$

In the above equation, ϕr_n is the least design strength of one bolt determined from the limit states of bolt shear strength, bearing strength at bolt holes, and slip resistance (if the connection is to be slip-critical). The design strength ϕR_n must be greater than or equal to the required strength P_u. Thus, by dividing P_u by ϕr_n, the minimum coefficient C is obtained, and a bolt group can be selected for which the coefficient is of that magnitude or greater.

These tables may be used with any bolt diameter and are conservative when used with ASTM A490 bolts. Linear interpolation within a given table between adjacent values of e_x is permitted. Design strengths determined with these tables provide a reliability equivalent to that for bolts in joints less than 50 in. long subject to shear produced by a concentric load in either bearing-type or slip-critical connections. Although this procedure is based on connections which may experience slip under load, both load tests and analytical studies indicate that it may be conservatively extended to slip-critical connections (Kulak, 1975).

A convergence criterion of one percent was employed for the tabulated iterative solutions. Straight-line interpolation between values for loads at different angles may be significantly unconservative. Either a direct analysis should be performed or the values for the next lower angle increment in the tables should be used for design. For bolt group patterns not treated in these tables, a direct analysis is required if the instantaneous center of rotation method is to be used.

PART 7 REFERENCES

Bowman, M.D. and M. Betancourt, 1991, "Reuse of A325 and A490 High-Strength Bolts," *Engineering Journal*, Vol. 28, No. 3, (3rd Qtr.), pp. 110–118, AISC, Chicago, IL.

Carter, C.J., 1996, "Specifying Bolt Length for High-Strength Bolts," *Engineering Journal*, Vol. 33, No. 2, (2nd Qtr.), pp. 43–53, AISC, Chicago, IL.

Crawford, S.F. and G.L. Kulak, 1968, "Behavior of Eccentrically Loaded Bolted Connections," *Studies in Structural Engineering*, (No. 4), Department of Civil Engineering, Nova Scotia Technical College, Halifax, Nova Scotia.

Higgins, T.R., 1971, "Treatment of Eccentrically Loaded Connections in the AISC Manual," *Engineering Journal*, Vol. 8, No. 2, (April), pp. 52–54, AISC, Chicago, IL.

Kulak, G.L., 1975, Eccentrically Loaded Slip-Resistant Connections," *Engineering Journal*, Vol. 12, No. 2, (2nd Qtr.), pp. 52–55, AISC, Chicago, IL.

Table 7-1.
Dimensions of High-Strength Fasteners, in.

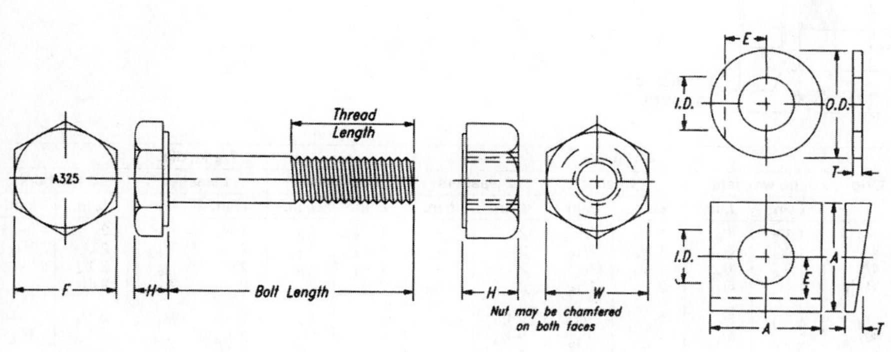

	Measurement		Nominal Bolt Diameter d_b, in.								
			1/2	5/8	3/4	7/8	1	1 1/8	1 1/4	1 3/8	1 1/2
A325 and A490 Bolts[a]	Width Across Flats, *F*		7/8	1 1/16	1 1/4	1 7/16	1 5/8	1 13/16	2	2 3/16	2 3/8
	Height, *H*		5/16	25/64	15/32	35/64	39/64	11/16	25/32	27/32	15/16
	Thread Length		1	1 1/4	1 3/8	1 1/2	1 3/4	2	2	2 1/4	2 1/4
	Bolt Length[f] = Grip +− >		11/16	7/8	1	1 1/8	1 1/4	1 1/2	1 5/8	1 3/4	1 7/8
A563 Nuts[b]	Width Across Flats, *W*		7/8	1 1/16	1 1/4	1 7/16	1 5/8	1 13/16	2	2 3/16	2 3/8
	Height, *H*		31/64	39/64	47/64	55/64	63/64	1 7/64	1 7/32	1 11/32	1 15/32
F436 Circular Washers[c]	Nom. Outside Diameter, *OD*		1 1/16	1 5/16	1 15/32	1 3/4	2	2 1/4	2 1/2	2 3/4	3
	Diameter, *ID*		17/32	11/16	13/16	15/16	1 1/8	1 1/4	1 3/8	1 1/2	1 5/8
	Thckns., *T*	Min.	0.097	0.122	0.122	0.136	0.136	0.136	0.136	0.136	0.136
		Max.	0.177	0.177	0.177	0.177	0.177	0.177	0.177	0.177	0.177
	Min. Edge Distance, *E*[d]		7/16	9/16	21/32	25/32	7/8	1	1 3/32	1 7/32	1 5/16
F436 Square or Rect. Washers[c,e]	Dimension, *A*		1 3/4	1 3/4	1 3/4	1 3/4	1 3/4	2 1/4	2 1/4	2 1/4	2 1/4
	Mean Thckns., *T*		5/16	5/16	5/16	5/16	5/16	5/16	5/16	5/16	5/16
	Taper in Thickness		2:12	2:12	2:12	2:12	2:12	2:12	2:12	2:12	2:12
	Min. Edge Distance, *E*[d]		7/16	9/16	21/32	25/32	7/8	1	1 3/32	1 7/32	1 5/16

[a]Tolerances as specified in ASTM A325 and A490.
[b]Tolerances as specified in ASTM A563.
[c]ASTM F436 Washer Tolerances, in.:

Nominal Outside Diameter	−1/32; +1/32
Nominal Diameter of Hole	−0; +1/32
Flatness; max. deviation from straight-edge placed on cut side shall not exceed	0.010
Concentricity: center of hole to outside diameter (full indicator runout)	0.030
Burr shall not project above immediately adjacent washer surface more than	0.010

[d]For clipped washers only.
[e]For use with American standard beams (S) and channel (C).
[f]Tabular value does not include thickness of washer(s).

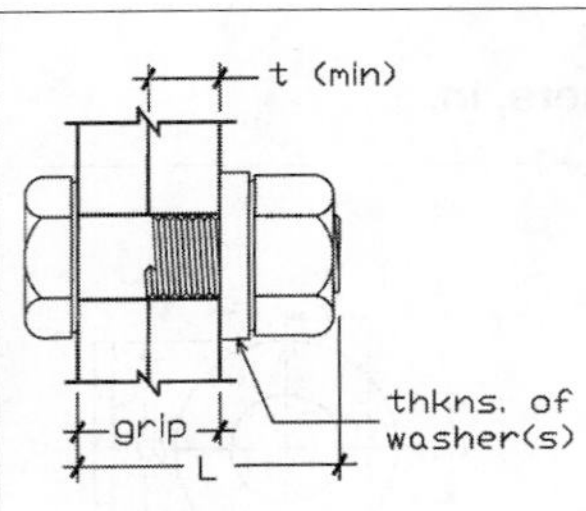

Table 7-2.
Bolt Length Selection

Grip	3/4 in. dia.						7/8 in. dia.					
	no washers		1 washer		2 washers		no washers		1 washer		2 washers	
	L, in.	*t*, in.	*L*, in.	*t*, in.	*L*, in.	*t*, in.	*L*, in.	*t*, in.	*L*, in.	*t*, in.	*L*, in.	*t*, in.
1/2	1 1/2	3/8	1 3/4	1/8	2	–	1 3/4	1/4	1 3/4	1/4	2	–
9/16	1 3/4	3/16	1 3/4	3/16	2	–	1 3/4	5/16	2	1/16	2	1/16
5/8	1 3/4	1/4	1 3/4	1/4	2	–	1 3/4	3/8	2	1/8	2 1/4	–
11/16	1 3/4	5/16	2	1/16	2	1/16	2	3/16	2	3/16	2 1/4	–
3/4	1 3/4	3/8	2	1/8	2 1/4	–	2	1/4	2	1/4	2 1/4	–
13/16	2	3/16	2	3/16	2 1/4	–	2	5/16	2 1/4	1/16	2 1/4	1/16
7/8	2	1/4	2	1/4	2 1/4	–	2	3/8	2 1/4	1/8	2 1/2	–
15/16	2	5/16	2 1/4	1/16	2 1/4	1/16	2 1/4	3/16	2 1/4	3/16	2 1/2	–
1	2	3/8	2 1/4	1/8	2 1/2	–	2 1/4	1/4	2 1/4	1/4	2 1/2	–
1 1/16	2 1/4	3/16	2 1/4	3/16	2 1/2	–	2 1/4	5/16	2 1/2	1/16	2 1/2	1/16
1 1/8	2 1/4	1/4	2 1/4	1/4	2 1/2	–	2 1/4	3/8	2 1/2	1/8	2 3/4	–
1 3/16	2 1/4	5/16	2 1/2	1/16	2 1/2	1/16	2 1/2	3/16	2 1/2	3/16	2 3/4	–
1 1/4	2 1/4	3/8	2 1/2	1/8	2 3/4	–	2 1/2	1/4	2 1/2	1/4	2 3/4	–
1 5/16	2 1/2	3/16	2 1/2	3/16	2 3/4	–	2 1/2	5/16	2 3/4	1/16	2 3/4	1/16
1 3/8	2 1/2	1/4	2 1/2	1/4	2 3/4	–	2 1/2	3/8	2 3/4	1/8	3	–
1 7/16	2 1/2	5/16	2 3/4	1/16	2 3/4	1/16	2 3/4	3/16	2 3/4	3/16	3	–
1 1/2	2 1/2	3/8	2 3/4	1/8	3	–	2 3/4	1/4	2 3/4	1/4	3	–
1 9/16	2 3/4	3/16	2 3/4	3/16	3	–	2 3/4	5/16	3	1/16	3	1/16
1 5/8	2 3/4	1/4	2 3/4	1/4	3	–	2 3/4	3/8	3	1/8	3 1/4	–
1 11/16	2 3/4	5/16	3	1/16	3	1/16	3	3/16	3	3/16	3 1/4	–
1 3/4	2 3/4	3/8	3	1/8	3 1/4	–	3	1/4	3	1/4	3 1/4	–
1 13/16	3	3/16	3	3/16	3 1/4	–	3	5/16	3 1/4	1/16	3 1/4	1/16
1 7/8	3	1/4	3	1/4	3 1/4	–	3	3/8	3 1/4	1/8	3 1/2	–
1 15/16	3	5/16	3 1/4	1/16	3 1/4	1/16	3 1/4	3/16	3 1/4	3/16	3 1/2	–
2	3	3/8	3 1/4	1/8	3 1/2	–	3 1/4	1/4	3 1/4	1/4	3 1/2	–
2 1/16	3 1/4	3/16	3 1/4	3/16	3 1/2	–	3 1/4	5/16	3 1/2	1/16	3 1/2	1/16
2 1/8	3 1/4	1/4	3 1/4	1/4	3 1/2	–	3 1/4	3/8	3 1/2	1/8	3 3/4	–
2 3/16	3 1/4	5/16	3 1/2	1/16	3 1/2	1/16	3 1/2	3/16	3 1/2	3/16	3 3/4	–
2 1/4	3 1/4	3/8	3 1/2	1/8	3 3/4	–	3 1/2	1/4	3 1/2	1/4	3 3/4	–
2 5/16	3 1/2	3/16	3 1/2	3/16	3 3/4	–	3 1/2	5/16	3 3/4	1/16	3 3/4	1/16
2 3/8	3 1/2	1/4	3 1/2	1/4	3 3/4	–	3 1/2	3/8	3 3/4	1/8	4	–
2 7/16	3 1/2	5/16	3 3/4	1/16	3 3/4	1/16	3 3/4	3/16	3 3/4	3/16	4	–
2 1/2	3 1/2	3/8	3 3/4	1/8	4	–	3 3/4	1/4	3 3/4	1/4	4	–
2 9/16	3 3/4	3/16	3 3/4	3/16	4	–	3 3/4	5/16	4	1/16	4	1/16
2 5/8	3 3/4	1/4	3 3/4	1/4	4	–	3 3/4	3/8	4	1/8	4 1/4	–
2 11/16	3 3/4	5/16	4	1/16	4	1/16	4	3/16	4	3/16	4 1/4	–
2 3/4	3 3/4	3/8	4	1/8	4 1/4	–	4	1/4	4	1/4	4 1/4	–
2 13/16	4	3/16	4	3/16	4 1/4	–	4	5/16	4 1/4	1/16	4 1/4	1/16
2 7/8	4	1/4	4	1/4	4 1/4	–	4	3/8	4 1/4	1/8	4 1/2	–
2 15/16	4	5/16	4 1/4	1/16	4 1/4	1/16	4 1/4	3/16	4 1/4	3/16	4 1/2	–
3	4	3/8	4 1/4	1/8	4 1/2	–	4 1/4	1/4	4 1/4	1/4	4 1/2	–
3 1/16	4 1/4	3/16	4 1/4	3/16	4 1/2	–	4 1/4	5/16	4 1/2	1/16	4 1/2	1/16
3 1/8	4 1/4	1/4	4 1/4	1/4	4 1/2	–	4 1/4	3/8	4 1/2	1/8	4 3/4	–
3 3/16	4 1/4	5/16	4 1/2	1/16	4 1/2	1/16	4 1/2	3/16	4 1/2	3/16	4 3/4	–

Notes:
1. Minimum bolt length calculated per RCSC Specification Table C-2.2.
2. Below 5-in. bolt length, minimum length increment is 1/4 in.; otherwise, 1/2 in..
3. *t* is the minimum thickness of ply closest to nut required to exclude threads from shear plane.
4. Washers assumed to be located under nut. When washer(s) are under bolt head, increase *t* by thickness of washer(s).
5. Shaded cells indicate longer bolt and additional washers may be required to prevent nut from jamming on the thread run-out.
6. – indicates that threads are excluded regardless of ply thickness.

Table 7-2 (cont.). Bolt Length Selection

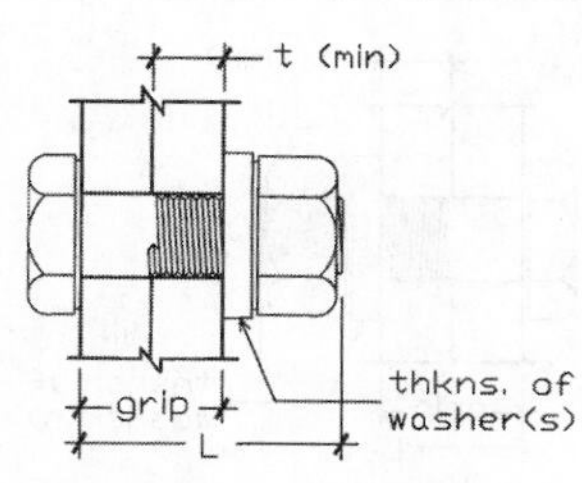

Grip	1 in. dia.						1 1/8 in. dia.					
	no washers		1 washer		2 washers		no washers		1 washer		2 washers	
	L, in.	*t*, in.	*L*, in	*t*, in.	*L*, in.	*t*, in	*L*, in.	*t*, in.	*L*, in.	*t*, in.	*L*, in.	*t*, in.
1/2	1 3/4	1/2	2	1/4	2 1/4	–	2	1/2	2 1/4	1/4	2 1/2	–
9/16	2	5/16	2	5/16	2 1/4	1/16	2 1/4	5/16	2 1/4	5/16	2 1/2	1/16
5/8	2	3/8	2	3/8	2 1/4	1/8	2 1/4	3/8	2 1/4	3/8	2 1/2	1/8
11/16	2	7/16	2 1/4	3/16	2 1/4	3/16	2 1/4	7/16	2 1/2	3/16	2 1/2	3/16
3/4	2	1/2	2 1/4	1/4	2 1/2	–	2 1/4	1/2	2 1/2	1/4	2 3/4	–
13/16	2 1/4	5/16	2 1/4	5/16	2 1/2	1/16	2 1/2	5/16	2 1/2	5/16	2 3/4	1/16
7/8	2 1/4	3/8	2 1/4	3/8	2 1/2	1/8	2 1/2	3/8	2 1/2	3/8	2 3/4	1/8
15/16	2 1/4	7/16	2 1/2	3/16	2 1/2	3/16	2 1/2	7/16	2 3/4	3/16	2 3/4	3/16
1	2 1/4	1/2	2 1/2	1/4	2 3/4	–	2 1/2	1/2	2 3/4	1/4	3	–
1 1/16	2 1/2	5/16	2 1/2	5/16	2 3/4	1/16	2 3/4	5/16	2 3/4	5/16	3	1/16
1 1/8	2 1/2	3/8	2 1/2	3/8	2 3/4	1/8	2 3/4	3/8	2 3/4	3/8	3	1/8
1 3/16	2 1/2	7/16	2 3/4	3/16	2 3/4	3/16	2 3/4	7/16	3	3/16	3	3/16
1 1/4	2 1/2	1/2	2 3/4	1/4	3	–	2 3/4	1/2	3	1/4	3 1/4	–
1 5/16	2 3/4	5/16	2 3/4	5/16	3	1/16	3	5/16	3	5/16	3 1/4	1/16
1 3/8	2 3/4	3/8	2 3/4	3/8	3	1/8	3	3/8	3	3/8	3 1/4	1/8
1 7/16	2 3/4	7/16	3	3/16	3	3/16	3	7/16	3 1/4	3/16	3 1/4	3/16
1 1/2	2 3/4	1/2	3	1/4	3 1/4	–	3	1/2	3 1/4	1/4	3 1/2	–
1 9/16	3	5/16	3	5/16	3 1/4	1/16	3 1/4	5/16	3 1/4	5/16	3 1/2	1/16
1 5/8	3	3/8	3	3/8	3 1/4	1/8	3 1/4	3/8	3 1/4	3/8	3 1/2	1/8
1 11/16	3	7/16	3 1/4	3/16	3 1/4	3/16	3 1/4	7/16	3 1/2	3/16	3 1/2	3/16
1 3/4	3	1/2	3 1/4	1/4	3 1/2	–	3 1/4	1/2	3 1/2	1/4	3 3/4	–
1 13/16	3 1/4	5/16	3 1/4	5/16	3 1/2	1/16	3 1/2	5/16	3 1/2	5/16	3 3/4	1/16
1 7/8	3 1/4	3/8	3 1/4	3/8	3 1/2	1/8	3 1/2	3/8	3 1/2	3/8	3 3/4	1/8
1 15/16	3 1/4	7/16	3 1/2	3/16	3 1/2	3/16	3 1/2	7/16	3 3/4	3/16	3 3/4	3/16
2	3 1/4	1/2	3 1/2	1/4	3 3/4	–	3 1/2	1/2	3 3/4	1/4	4	–
2 1/16	3 1/2	5/16	3 1/2	5/16	3 3/4	1/16	3 3/4	5/16	3 3/4	5/16	4	1/16
2 1/8	3 1/2	3/8	3 1/2	3/8	3 3/4	1/8	3 3/4	3/8	3 3/4	3/8	4	1/8
2 3/16	3 1/2	7/16	3 3/4	3/16	3 3/4	3/16	3 3/4	7/16	4	3/16	4	3/16
2 1/4	3 1/2	1/2	3 3/4	1/4	4	–	3 3/4	1/2	4	1/4	4 1/4	–
2 5/16	3 3/4	5/16	3 3/4	5/16	4	1/16	4	5/16	4	5/16	4 1/4	1/16
2 3/8	3 3/4	3/8	3 3/4	3/8	4	1/8	4	3/8	4	3/8	4 1/4	1/8
2 7/16	3 3/4	7/16	4	3/16	4	3/16	4	7/16	4 1/4	3/16	4 1/4	3/16
2 1/2	3 3/4	1/2	4	1/4	4 1/4	–	4	1/2	4 1/4	1/4	4 1/2	–
2 9/16	4	5/16	4	5/16	4 1/4	1/16	4 1/4	5/16	4 1/4	5/16	4 1/2	1/16
2 5/8	4	3/8	4	3/8	4 1/4	1/8	4 1/4	3/8	4 1/4	3/8	4 1/2	1/8
2 11/16	4	7/16	4 1/4	3/16	4 1/4	3/16	4 1/4	7/16	4 1/2	3/16	4 1/2	3/16
2 3/4	4	1/2	4 1/4	1/4	4 1/2	–	4 1/4	1/2	4 1/2	1/4	4 3/4	–
2 13/16	4 1/4	5/16	4 1/4	5/16	4 1/2	1/16	4 1/2	5/16	4 1/2	5/16	4 3/4	1/16
2 7/8	4 1/4	3/8	4 1/4	3/8	4 1/2	1/8	4 1/2	3/8	4 1/2	3/8	4 3/4	1/8
2 15/16	4 1/4	7/16	4 1/2	3/16	4 1/2	3/16	4 1/2	7/16	4 3/4	3/16	4 3/4	3/16
3	4 1/4	1/2	4 1/2	1/4	4 3/4	–	4 1/2	1/2	4 3/4	1/4	5	–
3 1/16	4 1/2	5/16	4 1/2	5/16	4 3/4	1/16	4 3/4	5/16	4 3/4	5/16	5	1/16
3 1/8	4 1/2	3/8	4 1/2	3/8	4 3/4	1/8	4 3/4	3/8	4 3/4	3/8	5	1/8
3 3/16	4 1/2	7/16	4 3/4	3/16	4 3/4	3/16	4 3/4	7/16	5	3/16	5	3/16

Notes:
1. Minimum bolt length calculated per RCSC Specification Table C-2.2.
2. Below 5-in. bolt length, minimum length increment is 1/4 in.; otherwise, 1/2 in..
3. *t* is the minimum thickness of ply closest to nut required to exclude threads from shear plane.
4. Washers assumed to be located under nut. When washer(s) are under bolt head, increase *t* by thickness of washer(s).
5. Shaded cells indicate longer bolt and additional washers may be required to prevent nut from jamming on the thread run-out.
6. – indicates that threads are excluded regardless of ply thickness.

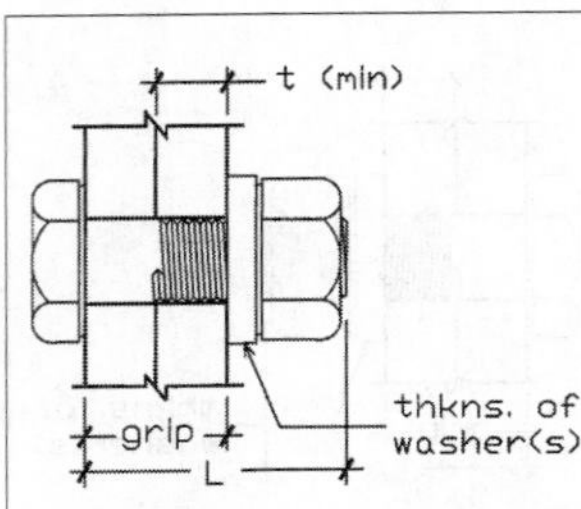

Table 7-2 (cont.). Bolt Length Selection

Grip	3/4 in. dia.						7/8 in. dia.					
	no washers		1 washer		2 washers		no washers		1 washer		2 washers	
	L, in.	*t*, in.	*L*, in.	*t*, in.	*L*, in.	*t*, in.	*L*, in.	*t*, in.	*L*, in.	*t*, in.	*L*, in.	*t*, in.
3 1/4	4 1/4	3/8	4 1/2	1/8	4 3/4	–	4 1/2	1/4	4 1/2	1/4	4 3/4	–
3 5/16	4 1/2	3/16	4 1/2	3/16	4 3/4	–	4 1/2	5/16	4 3/4	1/16	4 3/4	1/16
3 3/8	4 1/2	1/4	4 1/2	1/4	4 3/4	–	4 1/2	3/8	4 3/4	1/8	5	–
3 7/16	4 1/2	5/16	4 3/4	1/16	4 3/4	1/16	4 3/4	3/16	4 3/4	3/16	5	–
3 1/2	4 1/2	3/8	4 3/4	1/8	5	–	4 3/4	1/4	4 3/4	1/4	5	–
3 9/16	4 3/4	3/16	4 3/4	3/16	5	–	4 3/4	5/16	5	1/16	5	1/16
3 5/8	4 3/4	1/4	4 3/4	1/4	5	–	4 3/4	3/8	5	1/8	5 1/2	–
3 11/16	4 3/4	5/16	5	1/16	5	1/16	5	3/16	5	3/16	5 1/2	–
3 3/4	4 3/4	3/8	5	1/8	5 1/2	–	5	1/4	5	1/4	5 1/2	–
3 13/16	5	3/16	5	3/16	5 1/2	–	5	5/16	5 1/2	–	5 1/2	–
3 7/8	5	1/4	5	1/4	5 1/2	–	5	3/8	5 1/2	–	5 1/2	–
3 15/16	5	5/16	5 1/2	–	5 1/2	–	5 1/2	–	5 1/2	–	5 1/2	–
4	5	3/8	5 1/2	–	5 1/2	–	5 1/2	–	5 1/2	–	5 1/2	–
4 1/16	5 1/2	–	5 1/2	–	5 1/2	–	5 1/2	1/16	5 1/2	1/16	5 1/2	1/16
4 1/8	5 1/2	–	5 1/2	–	5 1/2	–	5 1/2	1/8	5 1/2	1/8	6	–
4 3/16	5 1/2	1/16	5 1/2	1/16	5 1/2	1/16	5 1/2	3/16	5 1/2	3/16	6	–
4 1/4	5 1/2	1/8	5 1/2	1/8	6	–	5 1/2	1/4	5 1/2	1/4	6	–
4 5/16	5 1/2	3/16	5 1/2	3/16	6	–	5 1/2	5/16	6	–	6	–
4 3/8	5 1/2	1/4	5 1/2	1/4	6	–	5 1/2	3/8	6	–	6	–
4 7/16	5 1/2	5/16	6	–	6	–	6	–	6	–	6	–
4 1/2	5 1/2	3/8	6	–	6	–	6	–	6	–	6	–
4 9/16	6	–	6	–	6	–	6	1/16	6	1/16	6	1/16
4 5/8	6	–	6	–	6	–	6	1/8	6	1/8	6 1/2	–
4 11/16	6	1/16	6	1/16	6	1/16	6	3/16	6	3/16	6 1/2	–
4 3/4	6	1/8	6	1/8	6 1/2	–	6	1/4	6	1/4	6 1/2	–
4 13/16	6	3/16	6	3/16	6 1/2	–	6	5/16	6 1/2	–	6 1/2	–
4 7/8	6	1/4	6	1/4	6 1/2	–	6	3/8	6 1/2	–	6 1/2	–
4 15/16	6	5/16	6 1/2	–	6 1/2	–	6 1/2	–	6 1/2	–	6 1/2	–
5	6	3/8	6 1/2	–	6 1/2	–	6 1/2	–	6 1/2	–	6 1/2	–
5 1/16	6 1/2	–	6 1/2	–	6 1/2	–	6 1/2	1/16	6 1/2	1/16	6 1/2	1/16
5 1/8	6 1/2	–	6 1/2	–	6 1/2	–	6 1/2	1/8	6 1/2	1/8	7	–
5 3/16	6 1/2	1/16	6 1/2	1/16	6 1/2	1/16	6 1/2	3/16	6 1/2	3/16	7	–
5 1/4	6 1/2	1/8	6 1/2	1/8	7	–	6 1/2	1/4	6 1/2	1/4	7	–
5 5/16	6 1/2	3/16	6 1/2	3/16	7	–	6 1/2	5/16	7	–	7	–
5 3/8	6 1/2	1/4	6 1/2	1/4	7	–	6 1/2	3/8	7	–	7	–
5 7/16	6 1/2	5/16	7	–	7	–	7	–	7	–	7	–
5 1/2	6 1/2	3/8	7	–	7	–	7	–	7	–	7	–
5 9/16	7	–	7	–	7	–	7	1/16	7	1/16	7	1/16
5 5/8	7	–	7	–	7	–	7	1/8	7	1/8	7 1/2	–
5 11/16	7	1/16	7	1/16	7	1/16	7	3/16	7	3/16	7 1/2	–
5 3/4	7	1/8	7	1/8	7 1/2	–	7	1/4	7	1/4	7 1/2	–
5 13/16	7	3/16	7	3/16	7 1/2	–	7	5/16	7 1/2	–	7 1/2	–
5 7/8	7	1/4	7	1/4	7 1/2	–	7	3/8	7 1/2	–	7 1/2	–
5 15/16	7	5/16	7 1/2	–	7 1/2	–	7 1/2	–	7 1/2	–	7 1/2	–

Notes:
1. Minimum bolt length calculated per RCSC Specification Table C-2.2.
2. Below 5-in. bolt length, minimum length increment is 1/4 in.; otherwise, 1/2 in..
3. *t* is the minimum thickness of ply closest to nut required to exclude threads from shear plane.
4. Washers assumed to be located under nut. When washer(s) are under bolt head, increase *t* by thickness of washer(s).
5. Shaded cells indicate longer bolt and additional washers may be required to prevent nut from jamming on the thread run-out.
6. – indicates that threads are excluded regardless of ply thickness.

Table 7-2 (cont.). Bolt Length Selection

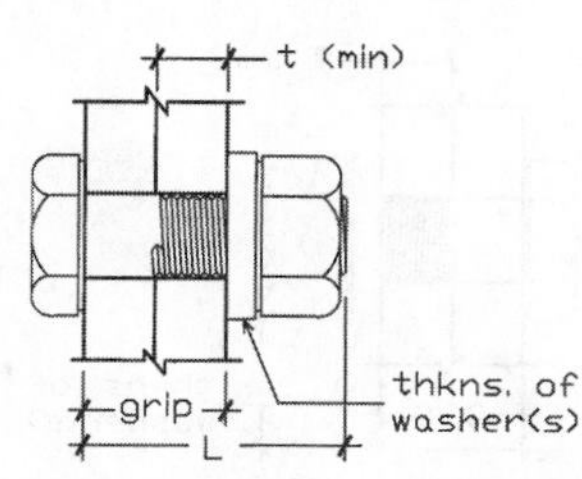

Grip	1 in. dia.						1 1/8 in. dia.					
	no washers		1 washer		2 washers		no washers		1 washer		2 washers	
	L, in.	*t*, in.	*L*, in	*t*, in.	*L*, in.	*t*, in	*L*, in.	*t*, in.	*L*, in.	*t*, in.	*L*, in.	*t*, in.
3 1/4	4 1/2	1/2	4 3/4	1/4	5	–	4 3/4	1/2	5	1/4	5 1/2	–
3 5/16	4 3/4	5/16	4 3/4	5/16	5	1/16	5	5/16	5	5/16	5 1/2	–
3 3/8	4 3/4	3/8	4 3/4	3/8	5	1/8	5	3/8	5	3/8	5 1/2	–
3 7/16	4 3/4	7/16	5	3/16	5	3/16	5	7/16	5 1/2	–	5 1/2	–
3 1/2	4 3/4	1/2	5	1/4	5 1/2	–	5	1/2	5 1/2	–	5 1/2	–
3 9/16	5	5/16	5	5/16	5 1/2	–	5 1/2	1/16	5 1/2	1/16	5 1/2	1/16
3 5/8	5	3/8	5	3/8	5 1/2	–	5 1/2	1/8	5 1/2	1/8	5 1/2	1/8
3 11/16	5	7/16	5 1/2	–	5 1/2	–	5 1/2	3/16	5 1/2	3/16	5 1/2	3/16
3 3/4	5	1/2	5 1/2	–	5 1/2	–	5 1/2	1/4	5 1/2	1/4	6	–
3 13/16	5 1/2	1/16	5 1/2	1/16	5 1/2	1/16	5 1/2	5/16	5 1/2	5/16	6	–
3 7/8	5 1/2	1/8	5 1/2	1/8	5 1/2	1/8	5 1/2	3/8	5 1/2	3/8	6	–
3 15/16	5 1/2	3/16	5 1/2	3/16	5 1/2	3/16	5 1/2	7/16	6	–	6	–
4	5 1/2	1/4	5 1/2	1/4	6	–	5 1/2	1/2	6	–	6	–
4 1/16	5 1/2	5/16	5 1/2	5/16	6	–	6	1/16	6	1/16	6	1/16
4 1/8	5 1/2	3/8	5 1/2	3/8	6	–	6	1/8	6	1/8	6	1/8
4 3/16	5 1/2	7/16	6	–	6	–	6	3/16	6	3/16	6	3/16
4 1/4	5 1/2	1/2	6	–	6	–	6	1/4	6	1/4	6 1/2	–
4 5/16	6	1/16	6	1/16	6	1/16	6	5/16	6	5/16	6 1/2	–
4 3/8	6	1/8	6	1/8	6	1/8	6	3/8	6	3/8	6 1/2	–
4 7/16	6	3/16	6	3/16	6	3/16	6	7/16	6 1/2	–	6 1/2	–
4 1/2	6	1/4	6	1/4	6 1/2	–	6	1/2	6 1/2	–	6 1/2	–
4 9/16	6	5/16	6	5/16	6 1/2	–	6 1/2	1/16	6 1/2	1/16	6 1/2	1/16
4 5/8	6	3/8	6	3/8	6 1/2	–	6 1/2	1/8	6 1/2	1/8	6 1/2	1/8
4 11/16	6	7/16	6 1/2	–	6 1/2	–	6 1/2	3/16	6 1/2	3/16	6 1/2	3/16
4 3/4	6	1/2	6 1/2	–	6 1/2	–	6 1/2	1/4	6 1/2	1/4	7	–
4 13/16	6 1/2	1/16	6 1/2	1/16	6 1/2	1/16	6 1/2	5/16	6 1/2	5/16	7	–
4 7/8	6 1/2	1/8	6 1/2	1/8	6 1/2	1/8	6 1/2	3/8	6 1/2	3/8	7	–
4 15/16	6 1/2	3/16	6 1/2	3/16	6 1/2	3/16	6 1/2	7/16	7	–	7	–
5	6 1/2	1/4	6 1/2	1/4	7	–	6 1/2	1/2	7	–	7	–
5 1/16	6 1/2	5/16	6 1/2	5/16	7	–	7	1/16	7	1/16	7	1/16
5 1/8	6 1/2	3/8	6 1/2	3/8	7	–	7	1/8	7	1/8	7	1/8
5 3/16	6 1/2	7/16	7	–	7	–	7	3/16	7	3/16	7	3/16
5 1/4	6 1/2	1/2	7	–	7	–	7	1/4	7	1/4	7 1/2	–
5 5/16	7	1/16	7	1/16	7	1/16	7	5/16	7	5/16	7 1/2	–
5 3/8	7	1/8	7	1/8	7	1/8	7	3/8	7	3/8	7 1/2	–
5 7/16	7	3/16	7	3/16	7	3/16	7	7/16	7 1/2	–	7 1/2	–
5 1/2	7	1/4	7	1/4	7 1/2	–	7	1/2	7 1/2	–	7 1/2	–
5 9/16	7	5/16	7	5/16	7 1/2	–	7 1/2	1/16	7 1/2	1/16	7 1/2	1/16
5 5/8	7	3/8	7	3/8	7 1/2	–	7 1/2	1/8	7 1/2	1/8	7 1/2	1/8
5 11/16	7	7/16	7 1/2	–	7 1/2	–	7 1/2	3/16	7 1/2	3/16	7 1/2	3/16
5 3/4	7	1/2	7 1/2	–	7 1/2	–	7 1/2	1/4	7 1/2	1/4	8	–
5 13/16	7 1/2	1/16	7 1/2	1/16	7 1/2	1/16	7 1/2	5/16	7 1/2	5/16	8	–
5 7/8	7 1/2	1/8	7 1/2	1/8	7 1/2	1/8	7 1/2	3/8	7 1/2	3/8	8	–
5 15/16	7 1/2	3/16	7 1/2	3/16	7 1/2	3/16	7 1/2	7/16	8	–	8	–

Notes:
1. Minimum bolt length calculated per RCSC Specification Table C-2.2.
2. Below 5-in. bolt length, minimum length increment is 1/4 in.; otherwise, 1/2 in..
3. *t* is the minimum thickness of ply closest to nut required to exclude threads from shear plane.
4. Washers assumed to be located under nut. When washer(s) are under bolt head, increase *t* by thickness of washer(s).
5. Shaded cells indicate longer bolt and additional washers may be required to prevent nut from jamming on the thread run-out.
6. – indicates that threads are excluded regardless of ply thickness.

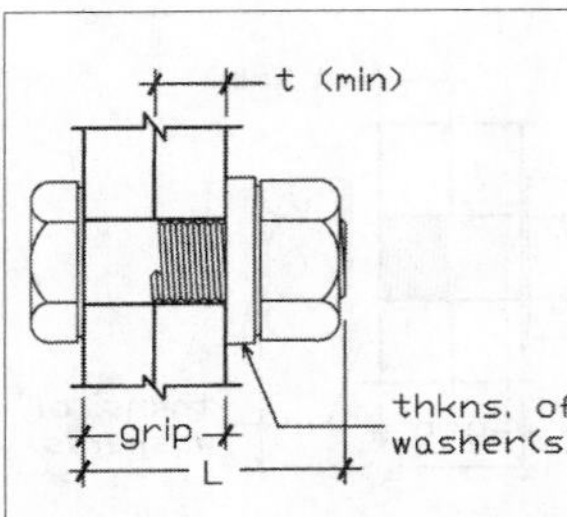

Table 7-2 (cont.). Bolt Length Selection

Grip	3/4 in. dia.						7/8 in. dia.					
	no washers		1 washer		2 washers		no washers		1 washer		2 washers	
	L, in.	*t*, in.	*L*, in.	*t*, in.	*L*, in.	*t*, in.	*L*, in.	*t*, in.	*L*, in.	*t*, in.	*L*, in.	*t*, in.
6	7	3/8	7 1/2	–	7 1/2	–	7 1/2	–	7 1/2	–	7 1/2	–
6 1/16	7 1/2	–	7 1/2	–	7 1/2	–	7 1/2	1/16	7 1/2	1/16	7 1/2	1/16
6 1/8	7 1/2	–	7 1/2	–	7 1/2	–	7 1/2	1/8	7 1/2	1/8	8	–
6 3/16	7 1/2	1/16	7 1/2	1/16	7 1/2	1/16	7 1/2	3/16	7 1/2	3/16	8	–
6 1/4	7 1/2	1/8	7 1/2	1/8	8	–	7 1/2	1/4	7 1/2	1/4	8	–
6 5/16	7 1/2	3/16	7 1/2	3/16	8	–	7 1/2	5/16	8	–	8	–
6 3/8	7 1/2	1/4	7 1/2	1/4	8	–	7 1/2	3/8	8	–	8	–
6 7/16	7 1/2	5/16	8	–	8	–	8	–	8	–	8	–
6 1/2	7 1/2	3/8	8	–	8	–	8	–	8	–	8	–
6 9/16	8	–	8	–	8	–	8	1/16	8	1/16	8	1/16
6 5/8	8	–	8	–	8	–	8	1/8	8	1/8		
6 11/16	8	1/16	8	1/16	8	1/16	8	3/16	8	3/16		
6 3/4	8	1/8	8	1/8			8	1/4	8	1/4		
6 13/16	8	3/16	8	3/16			8	5/16				
6 7/8	8	1/4	8	1/4			8	3/8				
6 15/16	8	5/16										
7	8	3/8										

Note: bolt lengths exceeding 8″ are not generally available except by special order

Notes:
1. Minimum bolt length calculated per RCSC Specification Table C-2.2.
2. Below 5-in. bolt length, minimum length increment is 1/4 in.; otherwise, 1/2 in..
3. *t* is the minimum thickness of ply closest to nut required to exclude threads from shear plane.
4. Washers assumed to be located under nut. When washer(s) are under bolt head, increase *t* by thickness of washer(s).
5. Shaded cells indicate longer bolt and additional washers may be required to prevent nut from jamming on the thread run-out.
6. – indicates that threads are excluded regardless of ply thickness.

Table 7-2 (cont.). Bolt Length Selection

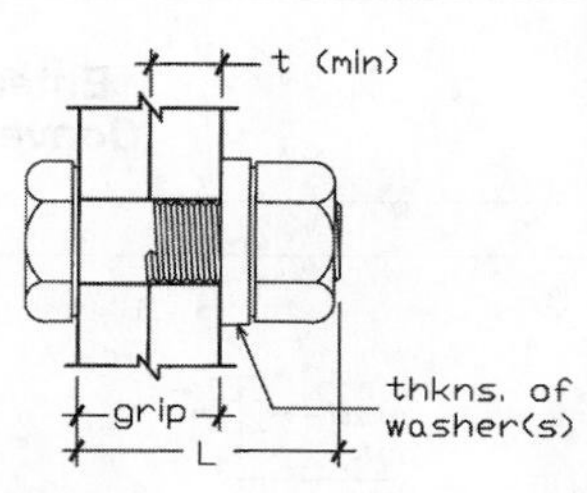

Grip	1 in. dia.						1 1/8 in. dia.					
	no washers		1 washer		2 washers		no washers		1 washer		2 washers	
	L, in.	*t*, in.	*L*, in	*t*, in.	*L*, in.	*t*, in	*L*, in.	*t*, in.	*L*, in.	*t*, in.	*L*, in.	*t*, in.
6	7 1/2	1/4	7 1/2	1/4	8	–	7 1/2	1/2	8	–	8	–
6 1/16	7 1/2	5/16	7 1/2	5/16	8	–	8	1/16	8	1/16	8	1/16
6 1/8	7 1/2	3/8	7 1/2	3/8	8	–	8	1/8	8	1/8	8	1/8
6 3/16	7 1/2	7/16	8	–	8	–	8	3/16	8	3/16	8	3/16
6 1/4	7 1/2	1/2	8	–	8	–	8	1/4	8	1/4		
6 5/16	8	1/16	8	1/16	8	1/16	8	5/16	8	5/16		
6 3/8	8	1/8	8	1/8	8	1/8	8	3/8	8	3/8		
6 7/16	8	3/16	8	3/16	8	3/16	8	7/16				
6 1/2	8	1/4	8	1/4			8	1/2				
6 9/16	8	5/16	8	5/16								
6 5/8	8	3/8	8	3/8								
6 11/16	8	7/16										
6 3/4	8	1/2										

Note: bolt lengths exceeding 8″ are not generally available except by special order

Notes:
1. Minimum bolt length calculated per RCSC Specification Table C-2.2.
2. Below 5-in. bolt length, minimum length increment is 1/4 in.; otherwise, 1/2 in..
3. *t* is the minimum thickness of ply closest to nut required to exclude threads from shear plane.
4. Washers assumed to be located under nut. When washer(s) are under bolt head, increase *t* by thickness of washer(s).
5. Shaded cells indicate longer bolt and additional washers may be required to prevent nut from jamming on the thread run-out.
6. – indicates that threads are excluded regardless of ply thickness.

Table 7-3a.
Entering and Tightening Clearance, in.
Conventional ASTM A325 and A490 Bolts

Aligned Bolts

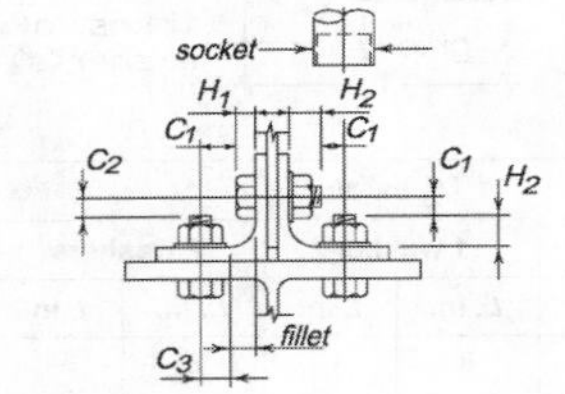

Nominal Bolt Dia., in.	Socket Dia., in.	H_1	H_2	C_1	C_2	C_3	
						Circular	Clipped
5/8	1 3/4	25/64	1 1/4	1	11/16	11/16	9/16
3/4	2 1/4	15/32	1 3/8	1 1/4	3/4	3/4	11/16
7/8	2 1/2	35/64	1 1/2	1 3/8	7/8	7/8	13/16
1	2 5/8	39/64	1 5/8	1 7/16	15/16	1	7/8
1 1/8	2 7/8	11/16	1 7/8	1 9/16	1 1/16	1 1/8	1
1 1/4	3 1/8	25/32	2	1 11/16	1 1/8	1 1/4	1 1/8
1 3/8	3 1/4	27/32	2 1/8	1 3/4	1 1/4	1 3/8	1 1/4
1 1/2	3 1/2	15/16	2 1/4	1 7/8	1 5/16	1 1/2	1 5/16

Staggered Bolts

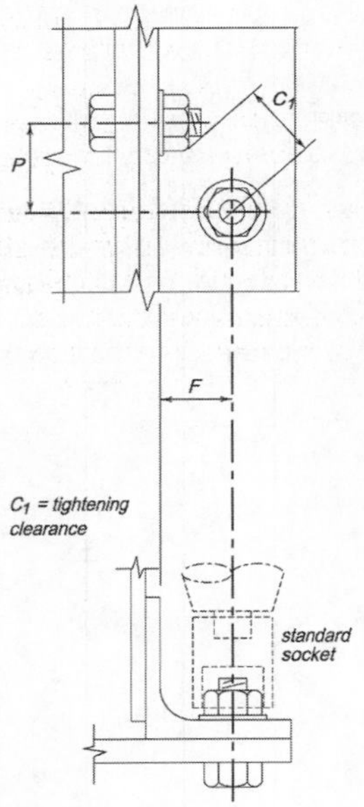

	Stagger *P*, in.							
	Nominal Bolt Diameter d_b, in.							
F	5/8	3/4	7/8	1	1 1/8	1 1/4	1 3/8	1 1/2
1	1 5/8							
1 1/8	1 1/2							
1 1/4	1 1/2	1 15/16						
1 3/8	1 7/16	1 7/8	2 3/16					
1 1/2	1 1/4	1 13/16	2 1/8	2 5/16				
1 5/8	1 1/4	1 3/4	2 1/16	2 5/16	2 9/16			
1 3/4	1 3/16	1 11/16	2	2 1/4	2 9/16	2 13/16	3	
1 7/8	1 1/8	1 9/16	1 15/16	2 3/16	2 1/2	2 3/4	3	3 3/4
2	1	1 1/2	1 13/16	2 1/8	2 7/16	2 3/4	2 15/16	3 1/4
2 1/8	13/16	1 3/8	1 11/16	2	2 3/8	2 11/16	2 15/16	3 3/16
2 1/4		1 1/4	1 9/16	1 7/8	2 1/4	2 5/8	2 7/8	3 3/16
2 3/8		1 1/8	1 1/2	1 3/4	2 1/8	2 1/2	2 13/16	3 1/8
2 1/2		7/8	1 3/8	1 5/8	2	2 7/16	2 3/4	3 1/16
2 5/8			1 3/16	1 1/2	1 15/16	2 5/16	2 7/8	3
2 3/4			15/16	1 3/8	1 7/8	2 1/8	2 1/2	2 7/8
2 7/8				1 3/16	1 3/4	2 1/16	2 3/8	2 13/16
3				7/8	1 5/8	2	2 1/4	2 11/16
3 1/8					1 1/2	1 7/8	2 1/8	2 1/2
3 1/4					1 1/4	1 3/4	2	2 3/8
3 3/8					15/16	1 5/8	1 15/16	2 1/4
3 1/2						1 3/8	1 3/4	2 1/8
3 5/8						1 1/16	1 9/16	2
3 3/4							1 5/16	1 7/8
3 7/8								1 11/16
4								1 3/8

Notes:
H_1 = height of head, in.
H_2 = maximum shank extension, * in.
C_1 = clearance for tighening, in.
C_2 = clearance for entering, in.
C_3 = clearance for fillet, * in.
P = bolt stagger, in.
F = clearance for tightening staggered bolts, in.
*Based on one standard hardened washer.

Table 7-3b.
Entering and Tightening Clearance, in.
Tension Control ASTM F1852 and A490 Bolts

Aligned Bolts

Tools	Nominal Bolt Dia., in.	H_1	H_2	C_1	C_2	C_3 Circular	C_3 Clipped
Large Tools	**4 1/4-in. Diameter Critical**						
	3/4	1/2	1 3/8	1 7/8	7/8	3/4	–
	7/8	9/16	1 1/2	1 7/8	1	7/8	–
	1	5/8	1 3/4	1 7/8	1 1/8	1	–
	2 3/4-in. Diameter Critical						
	3/4	1/2	1 3/8	1 3/8	7/8	3/4	–
	7/8	9/16	1 1/2	1 3/8	1	7/8	–
	1	5/8	1 3/4	1 3/8	1 1/8	1	–
Small Tools	**3 1/8-in. Diameter Critical**						
	5/8	7/16	1 1/4	1 5/8	13/16	11/16	–
	3/4	1/2	1 3/8	1 5/8	7/8	3/4	–
	7/8	9/16	1 1/2	1 5/8	1	7/8	–
	2 1/4-in. Diameter Critical						
	5/8	7/16	1 1/4	1 1/8	13/16	11/16	–
	3/4	1/2	1 3/8	1 1/8	7/8	3/4	–
	7/8	9/16	1 1/2	1 1/8	1	7/8	–

Staggered Bolts

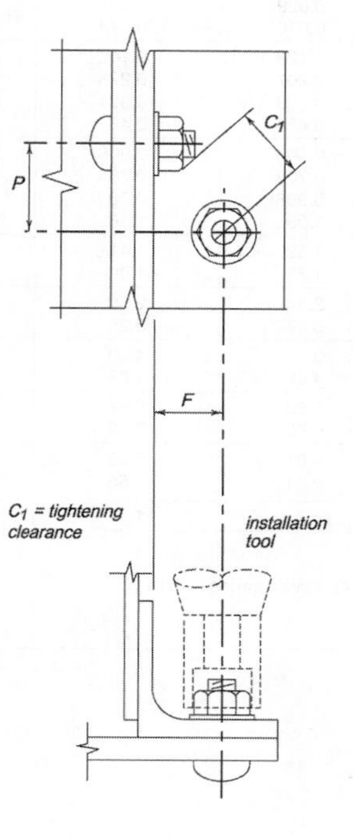

F	Stagger P, in. — Nominal Bolt Diameter d_b, in. 5/8	3/4	7/8	1
1 1/4	1 13/16			
1 3/8	1 3/4	2 1/16	2 1/4	2 7/16
1 1/2	1 11/16	2	2 3/16	2 3/8
1 5/8	1 9/16	1 7/8	2 1/16	2 1/4
1 3/4	1 1/2	1 13/16	2	2 3/16
1 7/8	1 7/16	1 3/4	1 7/8	2 1/8
2	1 5/16	1 5/8	1 3/4	2
2 1/8	1 1/4	1 9/16	1 11/16	1 15/16
2 1/4	1 3/16	1 1/2	1 9/16	1 7/8
2 3/8	1 1/8	1 3/8	1 1/2	1 3/4
2 1/2	1	1 5/16	1 3/8	1 11/16
2 5/8		1 3/16	1 5/16	1 9/16
2 3/4		1 1/8	1 3/16	1 1/2
2 7/8			1 1/8	1 3/8
3				1 5/16
3 3/8				1 5/16

Notes:
H_1 = height of head, in.
H_2 = maximum shank extension, *in.
C_1 = clearance for tighening, in.
C_2 = clearance for entering, in.
C_3 = clearance for fillet, *in.
P = bolt stagger, in.
F = clearance for tightening staggered bolts, in.
*Based on one standard hardened washer.

Table 7-4. Threading Dimensions for High-Strength and Non-High-Strength Bolts

SCREW THREADS

Unified Standard Series-UNC/UNRC and 4UN/4UNR

ANSI B1.1

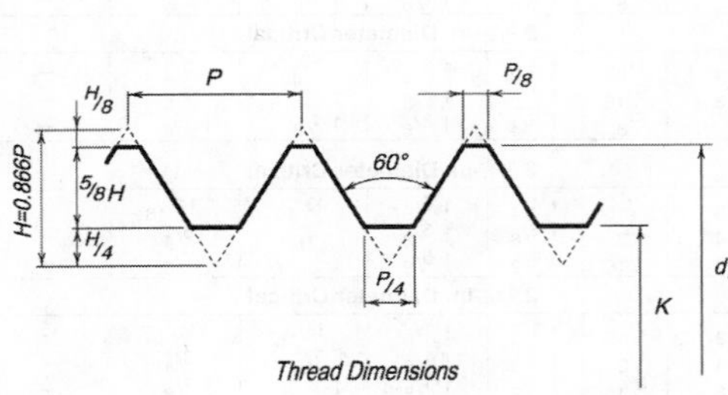

Thread Dimensions

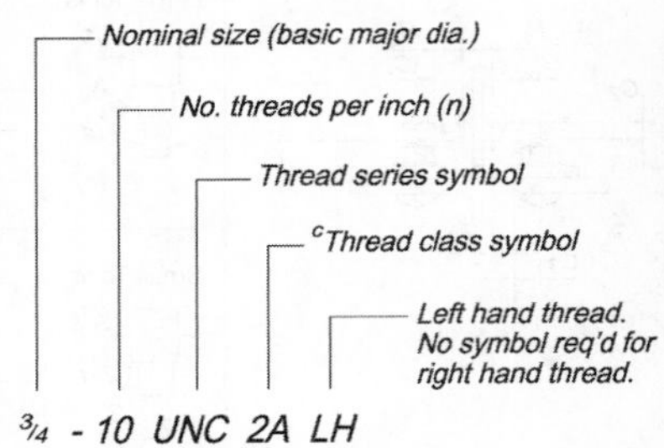

Standard Designations

Diameter		Area			Threads per Inch, n[b]
Bolt Diameter d_b, in.	Min. Root K, in.	Gross Bolt Area, in.^{2}	Min. Root Area, in.^{2}	Net Tensile Area, in.^{2}[a]	
1/4	0.189	0.049	0.029	0.032	20
3/8	0.298	0.110	0.070	0.078	16
1/2	0.406	0.196	0.129	0.142	13
5/8	0.514	0.307	0.207	0.226	11
3/4	0.627	0.442	0.309	0.334	10
7/8	0.739	0.601	0.429	0.462	9
1	0.847	0.785	0.563	0.606	8
1 1/8	0.950	0.994	0.709	0.763	7
1 1/4	1.08	1.23	0.908	0.969	7
1 3/8	1.17	1.49	1.08	1.16	6
1 1/2	1.30	1.77	1.32	1.41	6
1 3/4	1.51	2.41	1.78	1.90	5
2	1.73	3.14	2.34	2.50	4.5
2 1/4	1.98	3.98	3.07	3.25	4.5
2 1/2	2.19	4.91	3.78	4.00	4
2 3/4	2.44	5.94	4.69	4.93	4
3	2.69	7.07	5.70	5.97	4
3 1/4	2.94	8.30	6.80	7.10	4
3 1/2	3.19	9.62	8.01	8.33	4
3 3/4	3.44	11.0	9.31	9.66	4
4	3.69	12.6	10.7	11.1	4

[a] Net tensile area $= 0.7854 \times \left(d_b - \frac{0.9743}{n}\right)^2$

[b] For diameters listed, thread series is UNC (coarse). For larger diameters, thread series is 4UN.

[c] 2A denotes Class 2A fit applicable to external threads; 2B denotes corresponding Class 2B fit for internal threads.

Table 7-5. Weights of High-Strength Fasteners, pounds per 100 count

	Bolt Length, in.	Nominal Bolt Diameter d_b, in.								
		1/2	5/8	3/4	7/8	1	1 1/8	1 1/4	1 3/8	1 1/2
100 Conventional A325 or A490 Bolts with A563 Nuts	1	16.5	29.4	47.0	–	–	–	–	–	–
	1 1/4	17.8	31.1	49.6	74.4	104	–	–	–	–
	1 1/2	19.2	33.1	52.2	78.0	109	148	197	–	–
	1 3/4	20.5	35.3	55.3	81.9	114	154	205	261	333
	2	21.9	37.4	58.4	86.1	119	160	212	270	344
	2 1/4	23.3	39.8	61.6	90.3	124	167	220	279	355
	2 1/2	24.7	41.7	64.7	94.6	130	174	229	290	366
	2 3/4	26.1	43.9	67.8	98.8	135	181	237	300	379
	3	27.4	46.1	70.9	103	141	188	246	310	391
	3 1/4	28.8	48.2	74.0	107	146	195	255	321	403
	3 1/2	30.2	50.4	77.1	111	151	202	263	332	416
	3 3/4	31.6	52.5	80.2	116	157	209	272	342	428
	4	33.0	54.7	83.3	120	162	216	280	353	441
	4 1/4	34.3	56.9	86.4	124	168	223	289	363	453
	4 1/2	35.7	59.0	89.5	128	173	230	298	374	465
	4 3/4	37.1	61.2	92.7	133	179	237	306	384	478
	5	38.5	63.3	95.8	137	184	244	315	395	490
	5 1/4	39.9	65.5	98.9	141	190	251	324	405	503
	5 1/2	41.2	67.7	102	146	196	258	332	416	515
	5 3/4	42.6	69.8	105	150	201	265	341	426	527
	6	44.0	71.9	108	154	207	272	349	437	540
	6 1/4	–	74.1	111	158	212	279	358	447	552
	6 1/2	–	76.3	114	163	218	286	367	458	565
	6 3/4	–	78.5	118	167	223	293	375	468	577
	7	–	80.6	121	171	229	300	384	479	589
	7 1/4	–	82.8	124	175	234	307	392	489	602
	7 1/2	–	84.9	127	179	240	314	401	500	614
	7 3/4	–	87.1	130	183	246	321	410	510	626
	8	–	89.2	133	187	251	328	418	521	639
	8 1/4	–	–	–	192	257	335	427	531	651
	8 1/2	–	–	–	196	262	342	435	542	664
	8 3/4	–	–	–	–	–	–	444	552	676
	9	–	–	–	–	–	–	453	563	689
	Per inch add'tl. add	5.5	8.6	12.4	16.9	22.1	28.0	34.4	42.5	49.7
100 F436 Circular Washers		2.1	3.6	4.8	7.0	9.4	11.3	13.8	16.8	20.0
100 F436 Square Washers		23.1	22.4	21.0	20.2	19.2	34.0	31.6	31.2	32.9

This table conforms to weight standards adopted by the Industrial Fasteners Institute (IFI), updated for washer weights.

Table 7-6.
Dimensions of Non-High-Strength Bolts and Nuts, in.

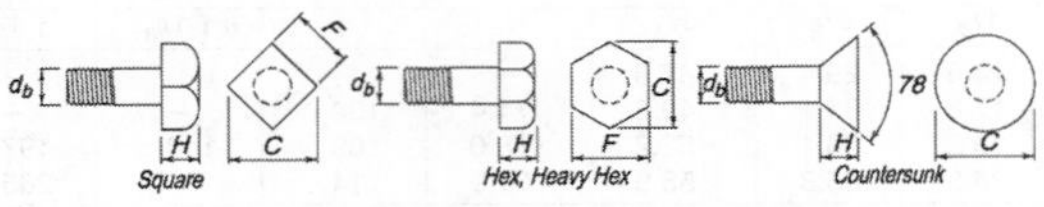

	Bolts Dia. d_b, in.	Square F, in.	Square C, in.	Square H, in.	Hex F, in.	Hex C, in.	Hex H, in.	Heavy Hex F, in.	Heavy Hex C, in.	Heavy Hex H, in.	Countersunk C, in.	Countersunk H, in.	Min. Thrd. Length, in. L ≤ 6 in.	Min. Thrd. Length, in. L > 6 in.
Bolts	1/4	3/8	1/2	3/16	7/16	1/2	3/16	–	–	–	1/2	1/8	3/4	1
	3/8	9/16	13/16	1/4	9/16	5/8	1/4	–	–	–	11/16	3/16	1	1 1/4
	1/2	3/4	1 1/16	5/16	3/4	7/8	3/8	7/8	1	3/8	7/8	1/4	1 1/4	1 1/2
	5/8	15/16	1 5/16	7/16	15/16	1 1/16	7/16	1 1/16	1 1/4	1 1/8	1 1/8	5/16	1 1/2	1 3/4
	3/4	1 1/8	1 9/16	1/2	1 1/8	1 5/16	1/2	1 1/4	1 7/16	1/2	1 3/8	3/8	1 3/4	2
	7/8	1 5/16	1 7/8	5/8	1 5/16	1 1/2	9/16	1 7/16	1 11/16	9/16	1 9/16	7/16	2	2 1/4
	1	1 1/2	2 1/8	11/16	1 1/2	1 3/4	11/16	1 5/8	1 7/8	11/16	1 13/16	1/2	2 1/4	2 1/2
	1 1/8	1 11/16	2 3/8	3/4	1 11/16	1 15/16	3/4	1 13/16	2 1/16	3/4	2 1/16	9/16	2 1/2	2 3/4
	1 1/4	1 7/8	2 5/8	7/8	1 7/8	2 3/16	7/8	2	2 5/16	7/8	2 1/4	5/8	2 3/4	3
	1 3/8	2 1/16	2 15/16	15/16	2 1/16	2 3/8	15/16	2 3/16	2 1/2	15/16	2 1/2	11/16	3	3 1/4
	1 1/2	2 1/4	3 3/16	1	2 1/4	2 5/8	1	2 3/8	2 3/4	1	2 11/16	3/4	3 1/4	3 1/2
	1 3/4	–	–	–	2 5/8	3	1 3/16	2 3/4	3 3/16	1 3/16	–	–	3 3/4	4
	2	–	–	–	3	3 7/16	1 3/8	3 1/8	3 5/8	1 3/8	–	–	4 1/4	4 1/2
	2 1/4	–	–	–	3 3/8	3 7/8	1 1/2	3 1/2	4 1/16	1 1/2	–	–	4 3/4	5
	2 1/2	–	–	–	3 3/4	4 5/16	1 11/16	3 7/8	4 1/2	1 11/16	–	–	5 1/4	5 1/2
	2 3/4	–	–	–	4 1/8	4 3/4	1 13/16	4 1/4	4 15/16	1 13/16	–	–	5 3/4	6
	3	–	–	–	4 1/2	5 3/16	2	4 5/8	5 5/16	2	–	–	6	6 1/2
	3 1/4	–	–	–	4 7/8	5 5/8	2 3/16	–	–	–	–	–	6	7
	3 1/2	–	–	–	5 1/4	6 1/16	2 5/16	–	–	–	–	–	6	7 1/2
	3 3/4	–	–	–	5 5/8	6 1/2	2 1/2	–	–	–	–	–	6	8
	4	–	–	–	6	6 15/16	2 11/16	–	–	–	–	–	6	8 1/2

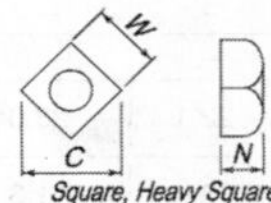

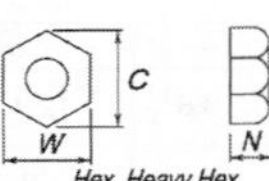

	Nut Size, in.	Square W, in.	Square C, in.	Square N, in.	Hex W, in.	Hex C, in.	Hex N, in.	Heavy Square W, in.	Heavy Square C, in.	Heavy Square N, in.	Heavy Hex W, in.	Heavy Hex C, in.	Heavy Hex N, in.
Nuts	1/4	7/16	5/8	1/4	7/16	1/2	3/16	1/2	11/16	1/4	1/2	9/16	1/4
	3/8	5/8	7/8	5/16	9/16	5/8	1/4	11/16	1	3/8	11/16	13/16	3/8
	1/2	4/5	1 1/8	7/16	3/4	7/8	3/8	7/8	1 1/4	1/2	7/8	1	1/2
	5/8	1	1 7/16	9/16	15/16	1 1/16	7/16	1 1/16	1 1/2	5/8	1 1/16	1 1/4	5/8
	3/4	1 1/8	1 9/16	11/16	1 1/8	1 5/16	1/2	1 1/4	1 3/4	3/4	1 1/4	1 7/16	3/4
	7/8	1 5/16	1 7/8	3/4	1 5/16	1 1/2	9/16	1 7/16	2 1/16	7/8	1 7/16	1 11/16	7/8
	1	1 1/2	2 1/8	7/8	1 1/2	1 3/4	11/16	1 5/8	2 5/16	1	1 5/8	1 7/8	1
	1 1/8	1 11/16	2 3/8	1	1 11/16	1 15/16	3/4	1 13/16	2 9/16	1 1/8	1 13/16	2 1/16	1 1/8
	1 1/4	1 7/8	2 5/8	1 1/8	1 7/8	2 3/16	7/8	2	2 13/16	1 1/4	2	2 5/16	1 1/4
	1 3/8	2 1/16	2 15/16	1 1/4	2 1/16	2 3/8	15/16	2 3/16	3 1/8	1 3/8	2 3/16	2 1/2	1 3/8
	1 1/2	2 1/4	3 3/16	1 5/16	2 1/4	2 5/8	1	2 3/8	3 3/8	1 1/2	2 3/8	2 3/4	1 1/2
	1 3/4	–	–	–	–	–	–	–	–	–	2 3/4	3 3/16	1 3/4
	2	–	–	–	–	–	–	–	–	–	3 1/8	3 5/8	2
	2 1/4	–	–	–	–	–	–	–	–	–	3 1/2	4 1/16	2 3/16
	2 1/2	–	–	–	–	–	–	–	–	–	3 7/8	4 1/2	2 7/16
	2 3/4	–	–	–	–	–	–	–	–	–	4 1/4	4 15/16	2 11/16
	3	–	–	–	–	–	–	–	–	–	4 5/8	5 5/16	2 15/16
	3 1/4	–	–	–	–	–	–	–	–	–	5	5 3/4	3 3/16
	3 1/2	–	–	–	–	–	–	–	–	–	5 3/8	6 3/16	3 7/16
	3 3/4	–	–	–	–	–	–	–	–	–	5 3/4	6 5/8	3 11/16
	4	–	–	–	–	–	–	–	–	–	6 1/8	7 1/16	3 15/16

Notes:
For high-strength bolt and nut dimensions, refer to Table 7-1.
Square, hex and heavy hex bolt dimensions, rounded to nearest 1/16-in., are in accordance with ANSI B18.2.1.
Countersunk bolt dimensions, rounded to the nearest 1/16-in., are in accordance with ANSI 18.5.
Minimum thread length = $2d_b$ + 1/4-in. for bolts up to 6-in. long, and $2d_b$ + 1/2-in. for bolts longer than 6-in..

Table 7-7. Weights of Non-High-Strength Fasteners, pounds

Bolt Length, in. (100 Square Bolts with Hexagonal Nuts)	Nominal Bolt Diameter d_b, in.								
	1/4	3/8	1/2	5/8	3/4	7/8	1	1 1/8	1 1/4
1	2.38	6.11	13.0	24.1	38.9	–	–	–	–
1 1/4	2.71	6.71	14.0	25.8	41.5	–	–	–	–
1 1/2	3.05	7.47	15.1	27.6	44.0	67.3	95.1	–	–
1 3/4	3.39	8.23	16.5	29.3	46.5	70.8	99.7	–	–
2	3.73	8.99	17.8	31.4	49.1	74.4	104	143	–
2 1/4	4.06	9.75	19.1	33.5	52.1	77.9	109	149	–
2 1/2	4.40	10.5	20.5	35.6	55.1	82.0	114	155	206
2 3/4	4.74	11.3	21.8	37.7	58.2	86.1	119	161	213
3	5.07	12.0	23.2	39.8	61.2	90.2	124	168	221
3 1/4	5.41	12.8	24.5	41.9	64.2	94.4	129	174	229
3 1/2	5.75	13.5	25.9	44.0	67.2	98.5	135	181	237
3 3/4	6.09	14.3	27.2	46.1	70.2	103	140	188	246
4	6.42	15.1	28.6	48.2	73.3	107	145	195	254
4 1/4	6.76	15.8	29.9	50.3	76.3	111	151	202	262
4 1/2	7.10	16.6	31.3	52.3	79.3	115	156	208	271
4 3/4	7.43	17.3	32.6	54.4	82.3	119	162	215	279
5	7.77	18.1	33.9	56.5	85.3	123	167	222	288
5 1/4	8.11	18.9	35.3	58.6	88.4	127	172	229	296
5 1/2	8.44	19.6	36.6	60.7	91.4	131	178	236	304
5 3/4	8.78	20.4	38.0	62.8	94.4	136	183	242	313
6	9.12	21.1	39.3	64.9	97.4	140	188	249	321
6 1/4	9.37	21.7	40.4	66.7	100	143	193	255	329
6 1/2	9.71	22.5	41.8	68.7	103	147	198	262	337
6 3/4	10.1	23.3	43.1	70.8	106	151	204	269	345
7	10.4	24.0	44.4	72.9	109	156	209	275	354
7 1/4	10.7	24.8	45.8	75.0	112	160	214	282	362
7 1/2	11.0	25.5	47.1	77.1	115	164	220	289	371
7 3/4	11.4	26.3	48.5	79.2	118	168	225	296	379
8	11.7	27.0	49.8	81.3	121	172	231	303	387
8 1/2	–	28.6	52.5	85.5	127	180	241	316	404
9	–	30.1	55.2	89.7	133	189	252	330	421
9 1/2	–	31.6	57.9	93.9	139	197	263	343	438
10	–	66.1	60.6	98.1	145	205	274	357	454
10 1/2	–	34.6	63.3	102	151	213	284	371	471
11	–	36.2	66.0	106	157	221	295	384	488
11 1/5	–	37.7	68.7	110	163	230	306	398	505
12	–	39.2	71.3	115	170	238	316	411	522
12 1/2	–	–	74.0	119	176	246	327	425	538
13	–	–	76.7	123	182	254	338	439	556
13 1/2	–	–	79.4	127	188	263	349	452	572
14	–	–	82.1	131	194	271	359	466	589
14 1/2	–	–	84.8	135	200	279	370	479	605
15	–	–	87.5	140	206	287	381	493	622
15 1/2	–	–	90.2	144	212	296	392	507	639
16	–	–	92.9	148	218	304	402	520	656
Per inch add'tl.add	1.3	3.0	5.4	8.4	12.1	16.5	21.4	27.2	33.6

Notes:
For weight of high-strength fasteners, see Table 8-3.
This table conforms to weight standards adopted by the Industrial Fasteners Institute (IFI).
*Square bolt per ANSI B 18.2.1, hexagonal nut per ANSI B18.2.2. For other non-high-strength fasteners, refer to Tables 7-8 and 7-9.

Table 7-8. Weight Adjustments for Combinations of Non-High-Strength Fasteners Other than Tabulated in Table 7-7

Combinations of 100		Add or Subtr.	Nominal Bolt Diameter d_b, in.								
			1/4	3/8	1/2	5/8	3/4	7/8	1	1 1/8	1 1/4
Square Bolts With	Square Nuts	+	0.1	1.0	2.0	3.4	3.5	5.5	8.0	12.2	16.3
	Heavy Square Nuts	+	0.6	2.1	4.1	7.0	11.6	17.2	23.2	32.1	41.2
	Heavy Hex Nuts	+	0.4	1.5	2.8	4.6	7.6	10.7	14.2	18.9	24.3
100 Square Bolts with Hexagonal Nuts*	Square Nuts	+	0.1	0.6	1.1	1.4	0.2	0.5	−0.2	−0.1	−1.7
	Hex Nuts	−	0.0	0.4	0.9	2.0	3.3	5.0	8.2	12.3	18.0
	Heavy Square Nuts	+	0.6	1.7	3.2	5.0	8.3	12.2	15.0	19.8	23.2
	Heavy Hex Nuts	+	0.4	1.1	1.9	2.6	4.3	5.7	6.0	6.6	6.3
100 Hex Bolts	Heavy Square Nuts	+	–	–	4.7	7.3	11.3	16.5	20.7	27.0	33.6
	Heavy Hex Nuts	+	–	–	3.4	4.9	7.3	10.0	11.7	13.8	16.7

Notes:
For weights of high-strength fasteners, see Table 7-5.
This table conforms to weight standards adopted by the Industrial Fasteners Institute (IFI).
*Add or subtract value in this table to or from the value in Table 7-7.

Table 7-9. Weights of Non-High-Strength Bolts of Diameter Greater Than 1 1/4 in., pounds

Weight of 100 Each		Nominal Bolt Diameter d_b, in.											
		1 3/8	1 1/2	1 3/4	2	2 1/4	2 1/2	2 3/4	3	3 1/4	3 1/2	3 3/4	4
Heads of:	Square Bolts	105	130	–	–	–	–	–	–	–	–	–	–
	Hex Bolts	84.0	112	178	259	369	508	680	900	1120	1390	1730	2130
	Heavy Hex Bolts	95.0	124	195	280	397	541	720	950	–	–	–	–
One Linear Inch, Unthreaded Shank		42.0	50.0	68.2	89.0	113	139	168	200	235	272	313	356
One Linear Inch, Threaded Shank		35.0	42.5	57.4	75.5	97.4	120	147	178	210	246	284	325
Square Nuts		94.5	122	–	–	–	–	–	–	–	–	–	–
Heavy Square Nuts		125	161	–	–	–	–	–	–	–	–	–	–
Heavy Hex Nuts		102	131	204	299	419	564	738	950	1190	1530	1810	2180

Notes:
For weights of high-strength fasteners, see Table 7-5.
This table conforms to weight standards adopted by the Industrial Fasteners Institute (IFI).

$\phi = 0.75$

Table 7-10. Design Shear Strength of One Bolt, kips

ASTM Desig.	Thread Cond.	ϕF_v (ksi)	Loading	Nominal Bolt Diameter d_b, in.							
				$^5/_8$	$^3/_4$	$^7/_8$	1	$1^1/_8$	$1^1/_4$	$1^3/_8$	$1^1/_2$
				Nominal Bolt Area, in.^{2}							
				0.307	0.442	0.601	0.785	0.994	1.23	1.48	1.77
A325 F1852	N	36.0	S	11.0	15.9	21.6	28.3	35.8	44.2	53.5	63.6
			D	22.1	31.8	43.3	56.5	71.6	88.4	107	127
	X	45.0	S	13.8	19.9	27.1	35.3	44.7	55.2	66.8	79.5
			D	27.6	39.8	54.1	70.7	89.5	110	134	159
A490	N	45.0	S	13.8	19.9	27.1	35.3	44.7	55.2	66.8	79.5
			D	27.6	39.8	54.1	70.7	89.5	110	134	159
	X	56.3	S	17.3	24.9	33.8	44.2	55.9	69.0	83.5	99.4
			D	34.5	49.7	67.6	88.4	112	138	167	199
A307	–	18.0	S	5.52	7.95	10.8	14.1	17.9	22.1	26.7	31.8
			D	11.0	15.9	21.6	28.3	35.8	44.2	53.5	63.6

N = Threads included in shear plane.
X = Threads excluded from shear plane.
S = Single shear.
D = Double shear.

$\phi = 0.75$

Table 7-11. Design Shear Strength of *n* Bolts, kips

n	ASTM A325 & F1852											
	N						X					
	$^3/_4$		$^7/_8$		1		$^3/_4$		$^7/_8$		1	
	S	D	S	D	S	D	S	D	S	D	S	D
12	191	382	260	520	339	679	239	477	325	649	424	848
11	175	350	238	476	311	622	219	437	298	595	389	778
10	159	318	216	433	283	565	199	398	271	541	353	707
9	143	286	195	390	254	509	179	358	244	487	318	636
8	127	254	173	346	226	452	159	318	216	433	283	565
7	111	223	152	303	198	396	139	278	189	379	247	495
6	95.4	191	130	260	170	339	119	239	162	325	212	424
5	79.5	159	108	216	141	283	99.4	199	135	271	177	353
4	63.6	127	86.6	173	113	226	79.5	159	108	216	141	283
3	47.7	95.4	64.9	130	84.8	170	59.6	119	81.2	162	106	212
2	31.8	63.6	43.3	86.6	56.5	113	39.8	79.5	54.1	108	70.7	141
1	15.9	31.8	21.6	43.3	28.3	56.5	19.9	39.8	27.1	54.1	35.3	70.7
n	ASTM A490											
	N						X					
	$^3/_4$		$^7/_8$		1		$^3/_4$		$^7/_8$		1	
	S	D	S	D	S	D	S	D	S	D	S	D
12	239	477	325	649	424	848	298	596	406	812	530	1060
11	219	437	298	595	389	778	273	547	372	744	486	972
10	199	398	271	541	353	707	249	497	338	676	442	884
9	179	358	244	487	318	636	224	447	304	609	398	795
8	159	318	216	433	283	565	199	398	271	541	353	707
7	139	278	189	379	247	495	174	348	237	474	309	619
6	119	239	162	325	212	424	149	298	203	406	265	530
5	99.4	199	135	271	177	353	124	249	169	338	221	442
4	79.5	159	108	216	141	283	99.4	199	135	271	177	353
3	59.6	119	81.2	162	106	212	74.6	149	101	203	133	265
2	39.8	79.5	54.1	108	70.7	141	49.7	99.4	67.6	135	88.4	177
1	19.9	39.8	27.1	54.1	35.3	70.7	24.9	49.7	33.8	67.6	44.2	88.4

N = Threads included in shear plane.
X = Threads excluded from shear plane.
S = Single shear.
D = Double shear.

Table 7-12. Design Bearing Strength at Bolt Holes for Various Bolt Spacings[a] kips/in. thickness

Hole Type	Bolt Spacing s, in.	F_u, ksi	Nominal Bolt Diameter d_b, in. 5/8	3/4	7/8	1	1 1/8	1 1/4	1 3/8	1 1/2
STD SSLT	$2^2/_3\ d_b$	58	51.1	62	72.9	83.7	94.6	105	116	127
		65	57.3	69.5	81.7	93.8	106	118	130	143
	3 in.	58	65.2	78.3	91.3	101	94.6	–	–	–
		65	73.1	87.7	102	113	106	–	–	–
SSLP	$2^2/_3\ d_b$	58	41.3	52.2	63.1	70.7	78.3	89.2	100	111
		65	46.3	58.5	70.7	79.2	87.8	99.9	112	124
	3 in.	58	65.2	78.3	91.3	88.1	78.3	–	–	–
		65	73.1	87.7	102	98.7	87.7	–	–	–
OVS	$2^2/_3\ d_b$	58	44.6	55.5	66.3	74	81.6	92.4	103	114
		65	50.0	62.2	74.3	82.9	91.4	104	116	128
	3 in.	58	65.2	78.3	91.3	91.3	81.6	–	–	–
		65	73.1	87.7	102	102	91.4	–	–	–
LSLP	$2^2/_3\ d_b$	58	5.44	6.53	7.61	8.70	9.79	10.9	12.0	13.1
		65	6.09	7.31	8.53	9.75	11.0	12.2	13.4	14.6
	3 in.	58	65.2	58.7	42.4	26.1	9.79	–	–	–
		65	73.1	65.8	47.5	29.2	11.0	–	–	–
LSLT	$2^2/_3\ d_b$	58	42.6	51.7	60.7	69.8	78.8	87.9	97.0	106
		65	47.7	57.9	68.0	78.2	88.4	98.5	109	119
	3 in.	58	54.4	65.2	76.1	84.3	78.8	–	–	–
		65	60.9	73.1	85.3	94.5	88.4	–	–	–
STD, SSLT, SSLP, OVS, LSLP	$s \geq s_{full}$	58	65.2	78.3	91.3	104	117	130	144	157
		65	73.1	87.7	102	117	132	146	161	175
LSLT	$s \geq s_{full}$	58	54.4	65.2	76.1	87.0	97.9	109	120	130
		65	60.9	73.1	85.3	97.5	110	122	134	146
Spacing for full bearing strength s_{full}[b], in.		STD, SSLT, LSLT	1 15/16	2 5/16	2 11/16	3 1/16	3 7/16	3 13/16	4 3/16	4 9/16
		OVS	2 1/16	2 7/16	2 13/16	3 1/4	3 11/16	4 1/16	4 7/16	4 13/16
		SSLP	2 1/8	2 1/2	2 7/8	3 5/16	3 3/4	4 1/8	4 1/2	4 7/8
		LSLP	2 13/16	3 3/8	3 15/16	4 1/2	5 1/16	5 5/8	6 3/16	6 3/4
Minimum Spacing[b] $= 2^2/_3\ d_b$, in.			1 2/3	2	2 1/3	2 2/3	3	3 1/3	3 2/3	4

LSLP = Long-Slotted Hole oriented parallel to the line of force.
LSLT = Long-Slotted Hole oriented transverse to the line of force.
OVS = Oversized Hole.
SSLP = Short-Slotted Hole oriented parallel to the line of force.
SSLT = Short-Slotted Hole oriented transverse to the line of force.
STD = Standard Hole.

–indicates spacing less than minimum spacing required per LRFD Specification Section J3.3.

[a] Spacing indicated is from the center of the hole or slot to the center of the adjacent hole or slot in the line of force. Hole deformation is considered. When hole deformation is not a consideration, see LRFD Specification Section J3.10.

[b] Decimal value has been rounded to the nearest sixteenth of an inch.

Table 7-13.
Design Bearing Strength at Bolt Holes for Various Edge Distances[a]
kips/in. thickness

Hole Type	Edge Distance L_e, in.	F_u, ksi	Nominal Bolt Diameter d_b, in. 5/8	3/4	7/8	1	1 1/8	1 1/4	1 3/8	1 1/2
STD SSLT	1 1/4	58	47.3	44.0	40.8	37.5	34.3	31.0	27.7	24.5
		65	53.0	49.4	45.7	42.0	38.4	34.7	31.1	27.4
	2	58	65.2	78.3	79.9	76.7	73.4	70.1	66.9	63.6
		65	73.1	87.7	89.6	85.9	82.3	78.6	75.0	71.3
SSLP	1 1/4	58	42.4	39.2	35.9	31.0	26.1	22.8	19.6	16.3
		65	47.5	43.9	40.2	34.7	29.2	25.6	21.9	18.3
	2	58	65.2	78.3	75.0	70.1	65.2	62.0	58.7	55.5
		65	73.1	87.7	84.1	78.6	73.1	69.5	65.8	62.2
OVS	1 1/4	58	44.0	40.8	37.5	32.6	27.7	24.5	21.2	17.9
		65	49.4	45.7	42.0	36.6	31.1	27.4	23.8	20.1
	2	58	65.2	78.3	76.7	71.8	66.9	63.6	60.4	57.1
		65	73.1	87.7	85.9	80.4	75.0	71.3	67.6	64.0
LSLP	1 1/4	58	24.5	16.3	8.16	–	–	–	–	–
		65	27.4	18.3	9.14	–	–	–	–	–
	2	58	63.6	55.5	47.3	39.2	31.0	22.8	14.7	6.5
		65	71.3	62.2	53.0	43.9	34.7	25.6	16.5	7.3
LSLT	1 1/4	58	39.4	36.7	34.0	31.3	28.5	25.8	23.1	20.4
		65	44.2	41.1	38.1	35.0	32.0	28.9	25.9	22.9
	2	58	54.4	65.2	66.6	63.9	61.2	58.5	55.7	53.0
		65	60.9	73.1	74.6	71.6	68.6	65.5	62.5	59.4
STD, SSLT, SSLP, OVS, LSLP	$L_e \geq L_{e\,full}$	58	65.2	78.3	91.3	104	117	130	144	157
		65	73.1	87.7	102	117	132	146	161	175
LSLT	$L_e \geq L_{e\,full}$	58	54.4	65.2	76.1	87.0	97.9	109	120	130
		65	60.9	73.1	85.3	97.5	110	122	134	146
Edge distance for full bearing strength $L_{e\,full}$[b], in.		STD, SSLT, LSLT	1 15/16	2 5/16	2 11/16	3 1/16	3 7/16	3 13/16	4 3/16	4 9/16
		OVS	2 1/16	2 7/16	2 13/16	3 1/4	3 11/16	4 1/16	4 7/16	4 13/16
		SSLP	2 1/8	2 1/2	2 7/8	3 5/16	3 3/4	4 1/8	4 1/2	4 7/8
		LSLP	2 13/16	3 3/8	3 15/16	4 1/2	5 1/16	5 5/8	6 3/16	6 3/4

LSLP = Long-Slotted Hole oriented parallel to the line of force.
LSLT = Long-Slotted Hole oriented transverse to the line of force.
OVS = Oversized Hole.
SSLP = Short-Slotted Hole oriented parallel to the line of force.
SSLT = Short-Slotted Hole oriented transverse to the line of force.
STD = Standard Hole.

–indicates edge distance is inadequate for this hole size.
[a] Edge distance indicated is from the center of the hole or slot to the edge of the material in the line of force. Hole deformation is considered. When hole deformation is not a consideration, see LRFD Specification Section J3.10.
[b] Decimal value has been rounded to the nearest sixteenth of an inch.

$\phi = 0.75$

Table 7-14.
Design Tensile Strength of Bolts, kips

ASTM Desig.	ϕF_t (ksi)	Nominal Bolt Diameter d_b, in. 5/8	3/4	7/8	1	1 1/8	1 1/4	1 3/8	1 1/2
		Nominal Bolt Area, in.^{2}							
		0.307	**0.442**	**0.601**	**0.785**	**0.994**	**1.23**	**1.48**	**1.77**
A325 & F1852	**67.5**	20.7	29.8	40.6	53.0	67.1	82.8	100	119
A490	**84.8**	26.0	37.4	51.0	66.6	84.2	104	126	150
A307	**33.8**	10.4	14.9	20.3	26.5	33.5	41.4	50.1	59.6

$\phi = 1.0$ STD
$\phi = 0.85$ OVS & SSL
$\phi = 0.7$ LSLT
$\phi = 0.6$ LSLP

Table 7-15. Slip-Critical Connections Design Resistance to Shear at Service Loads Using Factored Loads, ϕR_n, kips[1] (Class A Faying Surface, $\mu = 0.33$)[2]

ASTM Desig.	Hole Type	Loading	Nominal Bolt Diameter d_b, in.							
			5/8	3/4	7/8	1	1 1/8	1 1/4	1 3/8	1 1/2
			Minimum ASTM A325/F1852 Bolt Pretension, kips							
			19	28	39	51	56	71	85	103
A325 F1852	STD	S	7.09	10.4	14.5	19.0	20.9	26.5	31.7	38.4
		D	14.2	20.9	29.1	38.0	41.8	53.0	63.4	76.8
	OVS SSL	S	6.02	8.88	12.4	16.2	17.8	22.5	26.9	32.6
		D	12.0	17.8	24.7	32.3	35.5	45.0	53.9	65.3
	LSLT	S	4.96	7.31	10.2	13.3	14.6	18.5	22.2	26.9
		D	9.92	14.6	20.4	26.6	29.2	37.1	44.4	53.8
	LSLP	S	4.25	6.26	8.73	11.4	12.5	15.9	19.0	23.0
		D	8.50	12.5	17.5	22.8	25.1	31.8	38.0	46.1
			Minimum ASTM A490 Bolt Pretension, kips							
			24	35	49	64	80	102	121	148
A490	STD	S	8.95	13.1	18.3	23.9	29.8	38.0	45.1	55.2
		D	17.9	26.1	36.5	47.7	59.7	76.1	90.2	110
	OVS SSL	S	7.61	11.1	15.5	20.3	25.4	32.3	38.4	46.9
		D	15.2	22.2	31.1	40.6	50.7	64.7	76.7	93.8
	LSLP	S	5.37	7.83	11.0	14.3	17.9	22.8	27.1	33.1
		D	10.7	15.7	21.9	28.6	35.8	45.6	54.1	66.2
	LSLT	S	6.26	9.14	12.8	16.7	20.9	26.6	31.6	38.6
		D	12.5	18.3	25.6	33.4	41.8	53.3	63.2	77.3

STD = Standard Hole.
OVS = Oversized Hole.
SSL = Short-Slotted Hole.
LSLP = Long-Slotted Hole oriented parallel to line of force.
LSLT = Long-Slotted Hole oriented transverse to line of force.
S = Single Shear.
D = Double Shear.

[1] For design slip resistance using service loads, refer to Table 7-16.
[2] For Class B faying surfaces, multiply the tabled design resistance by 1.52.
For Class C faying surfaces, multiply the tabled design resistance by 1.06.

$\phi = 1.0$ STD
$\phi = 0.85$ OVS & SSL
$\phi = 0.7$ LSLT
$\phi = 0.6$ LSLP

Table 7-16. Slip-Critical Connections Design Resistance to Shear at Service Loads Using Service Loads, ϕR_n, kips[1] (Class A Faying Surface, $\mu = 0.33$)[2]

ASTM Desig.	Hole Type	Loading	Nominal Bolt Diameter d_b, in.							
			5/8	3/4	7/8	1	1 1/8	1 1/4	1 3/8	1 1/2
			Nominal Bolt Area, in.[2]							
			0.307	**0.442**	**0.601**	**0.785**	**0.994**	**1.23**	**1.48**	**1.77**
A325 F1852	STD	S	5.22	7.51	10.2	13.4	16.9	20.9	25.2	30.0
		D	10.4	15.0	20.4	26.7	33.8	41.7	50.5	60.1
	OVS SSL	S	4.60	6.63	9.02	11.8	14.9	18.4	22.3	26.5
		D	9.20	13.3	18.0	23.6	29.8	36.8	44.5	53.0
	LSLT	S	3.68	5.30	7.22	9.42	11.9	14.7	17.8	21.2
		D	7.36	10.6	14.4	18.8	23.9	29.5	35.6	42.4
	LSLP	S	3.07	4.42	6.01	7.85	9.94	12.3	14.8	17.7
		D	6.14	8.84	12.0	15.7	19.9	24.5	29.7	35.3
A490	STD	S	6.44	9.28	12.6	16.5	20.9	25.8	31.2	37.1
		D	12.9	18.6	25.3	33.0	41.7	51.5	62.4	74.2
	OVS SSL	S	5.52	7.95	10.8	14.1	17.9	22.1	26.7	31.8
		D	11.0	15.9	21.6	28.3	35.8	44.2	53.5	63.6
	LSLT	S	4.60	6.63	9.02	11.8	14.9	18.4	22.3	26.5
		D	9.20	13.3	18.0	23.6	29.8	36.8	44.5	53.0
	LSLP	S	3.99	5.74	7.82	10.2	12.9	16.0	19.3	23.0
		D	7.98	11.5	15.6	20.4	25.8	31.9	38.6	45.9

STD = Standard Hole.
OVS = Oversized Hole.
SSL = Short-Slotted Hole.
LSLP = Long-Slotted Hole parallel to line of force.
LSLT = Long-Slotted Hole transverse to line of force.
S = Single Shear.
D = Double Shear.

[1] For design slip resistance using loads, refer to Table 7-15.
[2] For Class B faying surfaces, multiply the tabled design resistance by 1.52.
For Class C faying surfaces, multiply the tabled design resistance by 1.06.

Table 7-17. Coefficients C for Eccentrically Loaded Bolt Groups Angle = 0°

$$C_{req} = \frac{P_u}{\phi r_n} \quad \text{or} \quad \phi R_n = C \times \phi r_n$$

where

P_u = factored force, kips
ϕr_n = design strength per bolt, kips
ϕR_n = design strength per bolt group, kips
e = eccentricity of P_u with respect to centroid of bolt group, in. (not tabulated, may be determined by geometry)
e_x = horizontal component of e, in.
s = bolt spacing, in.
C = coefficient tabulated below

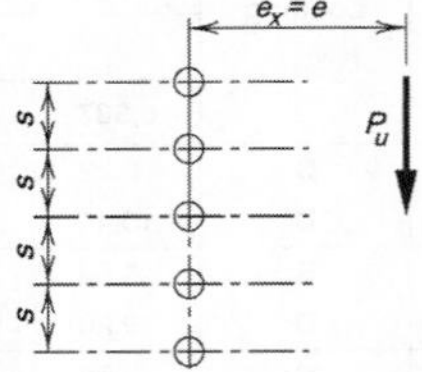

s, in.	e_x, in.	Number of Bolts in One Vertical Row, n										
		2	3	4	5	6	7	8	9	10	11	12
3	2	1.18	2.23	3.32	4.39	5.45	6.48	7.51	8.52	9.53	10.5	11.5
	3	0.88	1.75	2.81	3.90	4.98	6.06	7.12	8.17	9.21	10.2	11.3
	4	0.69	1.40	2.36	3.40	4.47	5.56	6.64	7.72	8.78	9.84	10.9
	5	0.56	1.15	2.01	2.96	3.98	5.05	6.13	7.22	8.30	9.38	10.4
	6	0.48	0.97	1.73	2.59	3.55	4.57	5.63	6.70	7.79	8.87	9.96
	7	0.41	0.83	1.51	2.28	3.17	4.13	5.15	6.20	7.28	8.36	9.44
	8	0.36	0.73	1.34	2.04	2.85	3.75	4.72	5.73	6.78	7.85	8.93
	9	0.32	0.65	1.21	1.83	2.59	3.42	4.34	5.31	6.32	7.36	8.42
	10	0.29	0.59	1.09	1.66	2.36	3.14	4.00	4.92	5.89	6.90	7.94
	12	0.24	0.49	0.92	1.40	2.00	2.68	3.44	4.27	5.15	6.09	7.06
	14	0.21	0.42	0.79	1.21	1.74	2.33	3.01	3.75	4.55	5.41	6.31
	16	0.18	0.37	0.70	1.06	1.53	2.06	2.67	3.33	4.06	4.85	5.68
	18	0.16	0.33	0.62	0.95	1.37	1.84	2.39	3.00	3.66	4.38	5.15
	20	0.15	0.29	0.56	0.85	1.24	1.67	2.16	2.72	3.33	3.99	4.70
	24	0.12	0.25	0.47	0.71	1.03	1.40	1.82	2.29	2.81	3.37	3.99
	28	0.11	0.21	0.40	0.61	0.89	1.20	1.57	1.97	2.42	2.92	3.45
	32	0.09	0.18	0.35	0.54	0.78	1.05	1.37	1.73	2.13	2.57	3.04
	36	0.08	0.16	0.31	0.48	0.69	0.94	1.22	1.54	1.90	2.29	2.72
6	2	1.63	2.71	3.75	4.77	5.77	6.77	7.76	8.75	9.74	10.7	11.7
	3	1.39	2.48	3.56	4.60	5.63	6.65	7.65	8.66	9.66	10.7	11.6
	4	1.18	2.23	3.32	4.39	5.45	6.48	7.51	8.52	9.53	10.5	11.5
	5	1.01	1.98	3.07	4.15	5.23	6.28	7.33	8.36	9.38	10.4	11.4
	6	0.88	1.75	2.81	3.90	4.98	6.06	7.12	8.17	9.21	10.2	11.3
	7	0.77	1.56	2.58	3.64	4.73	5.81	6.89	7.95	9.00	10.1	11.1
	8	0.69	1.40	2.36	3.40	4.47	5.56	6.64	7.72	8.78	9.84	10.9
	9	0.62	1.26	2.17	3.17	4.22	5.30	6.39	7.47	8.55	9.61	10.7
	10	0.56	1.15	2.01	2.96	3.98	5.05	6.13	7.22	8.30	9.38	10.4
	12	0.48	0.97	1.73	2.59	3.55	4.57	5.63	6.70	7.79	8.87	9.96
	14	0.41	0.83	1.51	2.28	3.17	4.13	5.15	6.20	7.28	8.36	9.44
	16	0.36	0.73	1.34	2.04	2.85	3.75	4.72	5.73	6.78	7.85	8.93
	18	0.32	0.65	1.21	1.83	2.59	3.42	4.34	5.31	6.32	7.36	8.42
	20	0.29	0.59	1.09	1.66	2.36	3.14	4.00	4.92	5.89	6.90	7.94
	24	0.24	0.49	0.92	1.40	2.00	2.68	3.44	4.27	5.15	6.09	7.06
	28	0.21	0.42	0.79	1.21	1.74	2.33	3.01	3.75	4.55	5.41	6.31
	32	0.18	0.37	0.70	1.06	1.53	2.06	2.67	3.33	4.06	4.85	5.68
	36	0.16	0.33	0.62	0.95	1.37	1.84	2.39	3.00	3.66	4.38	5.15

Table 7-17 (cont.).
Coefficients *C* for Eccentrically Loaded Bolt Groups
Angle = 15°

$$C_{req} = \frac{P_u}{\phi r_n} \quad \text{or} \quad \phi R_n = C \times \phi r_n$$

where

P_u = factored force, kips
ϕr_n = design strength per bolt, kips
ϕR_n = design strength per bolt group, kips
e = eccentricity of P_u with respect to centroid of bolt group, in. (not tabulated, may be determined by geometry)
e_x = horizontal component of e, in.
s = bolt spacing, in.
C = coefficient tabulated below

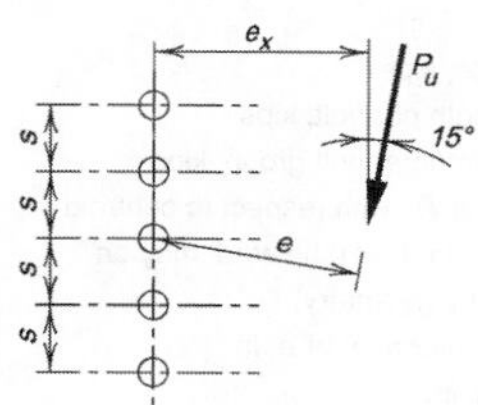

s, in.	e_x, in.	Number of Bolts in One Vertical Row, *n*										
		2	3	4	5	6	7	8	9	10	11	12
	2	1.15	2.20	3.28	4.34	5.39	6.42	7.45	8.46	9.47	10.5	11.5
	3	0.86	1.76	2.78	3.85	4.92	5.98	7.03	8.08	9.11	10.1	11.2
	4	0.67	1.42	2.35	3.36	4.41	5.48	6.55	7.61	8.67	9.72	10.8
	5	0.55	1.17	2.00	2.94	3.94	4.98	6.04	7.11	8.18	9.24	10.3
	6	0.47	0.99	1.73	2.58	3.52	4.52	5.55	6.61	7.67	8.74	9.81
	7	0.41	0.86	1.52	2.30	3.16	4.11	5.10	6.13	7.18	8.24	9.30
	8	0.36	0.75	1.35	2.06	2.86	3.74	4.69	5.68	6.70	7.74	8.80
	9	0.32	0.67	1.22	1.86	2.60	3.43	4.32	5.27	6.26	7.28	8.31
	10	0.29	0.61	1.10	1.69	2.38	3.16	4.00	4.90	5.85	6.84	7.85
3	12	0.24	0.51	0.93	1.43	2.03	2.71	3.46	4.28	5.15	6.06	7.01
	14	0.21	0.43	0.81	1.24	1.76	2.37	3.04	3.78	4.57	5.41	6.30
	16	0.19	0.38	0.71	1.09	1.56	2.10	2.70	3.37	4.09	4.87	5.69
	18	0.17	0.34	0.63	0.97	1.39	1.88	2.43	3.04	3.70	4.42	5.18
	20	0.15	0.3	0.57	0.88	1.26	1.70	2.20	2.76	3.37	4.03	4.74
	24	0.12	0.25	0.48	0.73	1.06	1.43	1.86	2.33	2.86	3.43	4.04
	28	0.11	0.22	0.41	0.63	0.91	1.23	1.60	2.02	2.47	2.97	3.51
	32	0.09	0.19	0.36	0.55	0.80	1.08	1.41	1.77	2.18	2.62	3.10
	36	0.08	0.17	0.32	0.49	0.71	0.96	1.26	1.58	1.95	2.34	2.78
	2	1.61	2.69	3.72	4.74	5.74	6.74	7.73	8.73	9.71	10.7	11.7
	3	1.36	2.45	3.52	4.56	5.59	6.60	7.61	8.61	9.61	10.6	11.6
	4	1.15	2.20	3.28	4.34	5.39	6.42	7.45	8.46	9.47	10.5	11.5
	5	0.98	1.96	3.03	4.10	5.16	6.21	7.25	8.28	9.30	10.3	11.3
	6	0.86	1.76	2.78	3.85	4.92	5.98	7.03	8.08	9.11	10.1	11.2
	7	0.75	1.57	2.55	3.60	4.66	5.73	6.80	7.85	8.90	9.94	11.0
	8	0.67	1.42	2.35	3.36	4.41	5.48	6.55	7.61	8.67	9.72	10.8
	9	0.61	1.29	2.16	3.14	4.17	5.23	6.30	7.36	8.43	9.49	10.5
	10	0.55	1.17	2.00	2.94	3.94	4.98	6.04	7.11	8.18	9.24	10.3
6	12	0.47	0.99	1.73	2.58	3.52	4.52	5.55	6.61	7.67	8.74	9.81
	14	0.41	0.86	1.52	2.3	3.16	4.11	5.10	6.13	7.18	8.24	9.30
	16	0.36	0.75	1.35	2.06	2.86	3.74	4.69	5.68	6.70	7.74	8.80
	18	0.32	0.67	1.22	1.86	2.60	3.43	4.32	5.27	6.26	7.28	8.31
	20	0.29	0.61	1.10	1.69	2.38	3.16	4.00	4.90	5.85	6.84	7.85
	24	0.24	0.51	0.93	1.43	2.03	2.71	3.46	4.28	5.15	6.06	7.01
	28	0.21	0.43	0.81	1.24	1.76	2.37	3.04	3.78	4.57	5.41	6.30
	32	0.19	0.38	0.71	1.09	1.56	2.10	2.70	3.37	4.09	4.87	5.69
	36	0.17	0.34	0.63	0.97	1.39	1.88	2.43	3.04	3.70	4.42	5.18

Table 7-17 (cont.). Coefficients C for Eccentrically Loaded Bolt Groups Angle = 30°

$$C_{req} = \frac{P_u}{\phi r_n} \quad \text{or} \quad \phi R_n = C \times \phi r_n$$

where

P_u = factored force, kips
ϕr_n = design strength per bolt, kips
ϕR_n = design strength per bolt group, kips
e = eccentricity of P_u with respect to centroid of bolt group, in. (not tabulated, may be determined by geometry)
e_x = horizontal component of e, in.
s = bolt spacing, in.
C = coefficient tabulated below

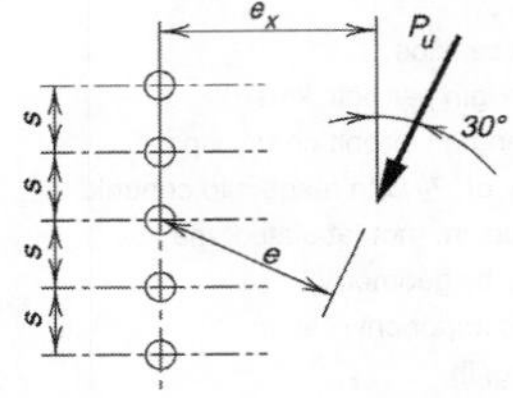

s, in.	e_x, in.	Number of Bolts in One Vertical Row, n										
		2	3	4	5	6	7	8	9	10	11	12
	2	1.14	2.20	3.25	4.30	5.33	6.36	7.38	8.39	9.40	10.4	11.4
	3	0.86	1.80	2.79	3.83	4.87	5.92	6.96	7.99	9.02	10.0	11.1
	4	0.69	1.50	2.40	3.39	4.41	5.45	6.49	7.53	8.57	9.61	10.6
	5	0.57	1.27	2.08	3.00	3.98	4.99	6.02	7.06	8.11	9.15	10.2
	6	0.49	1.09	1.82	2.68	3.60	4.57	5.58	6.60	7.64	8.68	9.72
	7	0.43	0.95	1.61	2.40	3.27	4.20	5.17	6.17	7.18	8.21	9.25
	8	0.38	0.83	1.44	2.17	2.98	3.86	4.79	5.76	6.75	7.77	8.79
	9	0.34	0.75	1.30	1.98	2.74	3.57	4.46	5.39	6.35	7.34	8.35
	10	0.31	0.67	1.19	1.82	2.52	3.31	4.15	5.05	5.98	6.95	7.93
3	12	0.26	0.56	1.01	1.55	2.17	2.87	3.64	4.46	5.33	6.24	7.17
	14	0.23	0.48	0.87	1.35	1.90	2.53	3.23	3.98	4.78	5.63	6.51
	16	0.20	0.42	0.77	1.20	1.69	2.26	2.89	3.58	4.33	5.11	5.94
	18	0.18	0.38	0.69	1.07	1.52	2.04	2.62	3.25	3.94	4.67	5.45
	20	0.16	0.34	0.62	0.97	1.37	1.85	2.38	2.97	3.61	4.30	5.02
	24	0.14	0.28	0.52	0.81	1.16	1.57	2.02	2.53	3.09	3.69	4.33
	28	0.12	0.24	0.45	0.70	1.00	1.36	1.75	2.20	2.69	3.22	3.79
	32	0.10	0.21	0.40	0.61	0.88	1.19	1.54	1.94	2.38	2.85	3.37
	36	0.09	0.19	0.35	0.55	0.78	1.07	1.38	1.74	2.13	2.56	3.03
	2	1.59	2.66	3.69	4.70	5.71	6.70	7.70	8.69	9.68	10.7	11.7
	3	1.34	2.43	3.48	4.52	5.54	6.55	7.55	8.56	9.55	10.6	11.5
	4	1.14	2.20	3.25	4.30	5.33	6.36	7.38	8.39	9.40	10.4	11.4
	5	0.98	1.99	3.02	4.06	5.11	6.14	7.17	8.20	9.22	10.2	11.2
	6	0.86	1.80	2.79	3.83	4.87	5.92	6.96	7.99	9.02	10.0	11.1
	7	0.77	1.64	2.59	3.60	4.64	5.68	6.73	7.77	8.80	9.83	10.9
	8	0.69	1.50	2.40	3.39	4.41	5.45	6.49	7.53	8.57	9.61	10.6
	9	0.63	1.37	2.23	3.19	4.19	5.22	6.26	7.30	8.34	9.38	10.4
	10	0.57	1.27	2.08	3.00	3.98	4.99	6.02	7.06	8.11	9.15	10.2
6	12	0.49	1.09	1.82	2.68	3.60	4.57	5.58	6.60	7.64	8.68	9.72
	14	0.43	0.95	1.61	2.40	3.27	4.20	5.17	6.17	7.18	8.21	9.25
	16	0.38	0.83	1.44	2.17	2.98	3.86	4.79	5.76	6.75	7.77	8.79
	18	0.34	0.75	1.30	1.98	2.74	3.57	4.46	5.39	6.35	7.34	8.35
	20	0.31	0.67	1.19	1.82	2.52	3.31	4.15	5.05	5.98	6.95	7.93
	24	0.26	0.56	1.01	1.55	2.17	2.87	3.64	4.46	5.33	6.24	7.17
	28	0.23	0.48	0.87	1.35	1.90	2.53	3.23	3.98	4.78	5.63	6.51
	32	0.20	0.42	0.77	1.20	1.69	2.26	2.89	3.58	4.33	5.11	5.94
	36	0.18	0.38	0.69	1.07	1.52	2.04	2.62	3.25	3.94	4.67	5.45

Table 7-17 (cont.). Coefficients C for Eccentrically Loaded Bolt Groups Angle = 45°

$$C_{req} = \frac{P_u}{\phi r_n} \quad \text{or} \quad \phi R_n = C \times \phi r_n$$

where

P_u = factored force, kips
ϕr_n = design strength per bolt, kips
ϕR_n = design strength per bolt group, kips
e = eccentricity of P_u with respect to centroid of bolt group, in. (not tabulated, may be determined by geometry)
e_x = horizontal component of e, in.
s = bolt spacing, in.
C = coefficient tabulated below

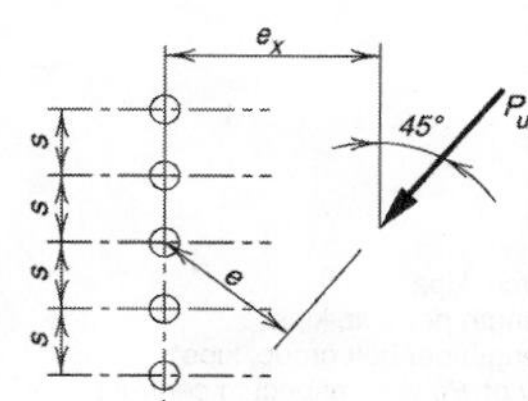

s, in.	e_x, in.	Number of Bolts in One Vertical Row, n										
		2	3	4	5	6	7	8	9	10	11	12
3	2	1.17	2.23	3.26	4.28	5.29	6.30	7.31	8.32	9.32	10.3	11.3
	3	0.92	1.89	2.87	3.87	4.88	5.90	6.91	7.93	8.94	9.95	11.0
	4	0.75	1.63	2.54	3.5	4.49	5.49	6.51	7.52	8.53	9.55	10.6
	5	0.64	1.42	2.25	3.17	4.13	5.11	6.11	7.11	8.12	9.14	10.2
	6	0.55	1.25	2.01	2.88	3.8	4.76	5.73	6.73	7.73	8.73	9.74
	7	0.49	1.11	1.81	2.63	3.51	4.43	5.38	6.36	7.34	8.34	9.34
	8	0.44	0.99	1.64	2.41	3.25	4.14	5.06	6.01	6.98	7.96	8.96
	9	0.40	0.90	1.49	2.22	3.02	3.87	4.77	5.69	6.64	7.61	8.58
	10	0.36	0.81	1.37	2.06	2.82	3.63	4.50	5.39	6.32	7.27	8.23
	12	0.31	0.68	1.17	1.79	2.47	3.22	4.02	4.87	5.74	6.65	7.58
	14	0.27	0.59	1.03	1.58	2.20	2.88	3.62	4.41	5.24	6.11	6.99
	16	0.24	0.52	0.91	1.41	1.97	2.60	3.29	4.03	4.81	5.63	6.48
	18	0.21	0.46	0.82	1.27	1.78	2.36	3.00	3.70	4.43	5.21	6.02
	20	0.19	0.41	0.74	1.16	1.62	2.16	2.76	3.41	4.1	4.84	5.61
	24	0.16	0.35	0.63	0.98	1.38	1.85	2.37	2.94	3.56	4.22	4.92
	28	0.14	0.30	0.54	0.85	1.19	1.61	2.08	2.58	3.14	3.73	4.37
	32	0.12	0.26	0.48	0.75	1.05	1.43	1.84	2.30	2.80	3.34	3.92
	36	0.11	0.23	0.43	0.67	0.94	1.28	1.65	2.07	2.53	3.02	3.55
6	2	1.57	2.64	3.66	4.67	5.67	6.66	7.66	8.65	9.64	10.6	11.6
	3	1.35	2.43	3.46	4.48	5.49	6.49	7.50	8.49	9.49	10.5	11.5
	4	1.17	2.23	3.26	4.28	5.29	6.30	7.31	8.32	9.32	10.3	11.3
	5	1.03	2.05	3.06	4.07	5.09	6.10	7.12	8.13	9.13	10.1	11.1
	6	0.92	1.89	2.87	3.87	4.88	5.90	6.91	7.93	8.94	9.95	11.0
	7	0.83	1.75	2.70	3.68	4.68	5.69	6.71	7.72	8.74	9.75	10.8
	8	0.75	1.63	2.54	3.50	4.49	5.49	6.51	7.52	8.53	9.55	10.6
	9	0.69	1.52	2.39	3.33	4.30	5.30	6.30	7.31	8.33	9.34	10.4
	10	0.64	1.42	2.25	3.17	4.13	5.11	6.11	7.11	8.12	9.14	10.2
	12	0.55	1.25	2.01	2.88	3.80	4.76	5.73	6.73	7.73	8.73	9.74
	14	0.49	1.11	1.81	2.63	3.51	4.43	5.38	6.36	7.34	8.34	9.34
	16	0.44	0.99	1.64	2.41	3.25	4.14	5.06	6.01	6.98	7.96	8.96
	18	0.40	0.9	1.49	2.22	3.02	3.87	4.77	5.69	6.64	7.61	8.58
	20	0.36	0.81	1.37	2.06	2.82	3.63	4.50	5.39	6.32	7.27	8.23
	24	0.31	0.68	1.17	1.79	2.47	3.22	4.02	4.87	5.74	6.65	7.58
	28	0.27	0.59	1.03	1.58	2.20	2.88	3.62	4.41	5.24	6.11	6.99
	32	0.24	0.52	0.91	1.41	1.97	2.60	3.29	4.03	4.81	5.63	6.48
	36	0.21	0.46	0.82	1.27	1.78	2.36	3.00	3.70	4.43	5.21	6.02

Table 7-17 (cont.). Coefficients C for Eccentrically Loaded Bolt Groups Angle = 60°

$$C_{req} = \frac{P_u}{\phi r_n} \quad \text{or} \quad \phi R_n = C \times \phi r_n$$

where

P_u = factored force, kips
ϕr_n = design strength per bolt, kips
ϕR_n = design strength per bolt group, kips
e = eccentricity of P_u with respect to centroid of bolt group, in. (not tabulated, may be determined by geometry)
e_x = horizontal component of e, in.
s = bolt spacing, in.
C = coefficient tabulated below

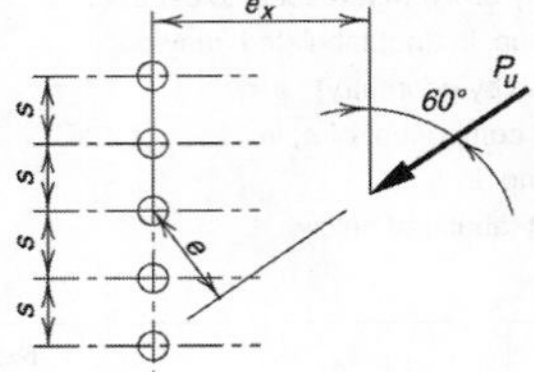

s, in.	e_x, in.	Number of Bolts in One Vertical Row, n										
		2	3	4	5	6	7	8	9	10	11	12
	2	1.27	2.32	3.32	4.31	5.30	6.30	7.29	8.27	9.27	10.3	11.3
	3	1.05	2.05	3.02	4.00	4.98	5.97	6.96	7.94	8.94	9.93	10.9
	4	0.89	1.83	2.77	3.72	4.69	5.66	6.64	7.62	8.61	9.60	10.6
	5	0.77	1.65	2.54	3.47	4.41	5.37	6.34	7.32	8.29	9.28	10.3
	6	0.68	1.49	2.34	3.24	4.16	5.10	6.06	7.02	7.99	8.97	9.95
	7	0.61	1.37	2.17	3.03	3.93	4.85	5.79	6.74	7.71	8.67	9.64
	8	0.56	1.26	2.01	2.83	3.71	4.61	5.54	6.48	7.43	8.39	9.35
	9	0.51	1.16	1.87	2.66	3.51	4.39	5.30	6.23	7.17	8.12	9.07
	10	0.47	1.07	1.74	2.50	3.32	4.19	5.08	5.99	6.92	7.86	8.81
3	12	0.40	0.93	1.52	2.22	3.00	3.82	4.67	5.55	6.45	7.37	8.30
	14	0.35	0.81	1.35	2.00	2.73	3.50	4.32	5.16	6.03	6.92	7.83
	16	0.32	0.72	1.21	1.81	2.49	3.23	4.00	4.81	5.65	6.51	7.40
	18	0.29	0.65	1.09	1.66	2.30	2.98	3.72	4.50	5.31	6.14	7.00
	20	0.26	0.58	1.00	1.53	2.12	2.77	3.47	4.21	4.99	5.80	6.63
	24	0.22	0.49	0.85	1.32	1.84	2.41	3.05	3.73	4.45	5.21	5.99
	28	0.19	0.42	0.74	1.15	1.61	2.13	2.71	3.34	4.00	4.70	5.44
	32	0.17	0.37	0.65	1.02	1.43	1.91	2.44	3.02	3.63	4.28	4.97
	36	0.15	0.33	0.59	0.92	1.29	1.72	2.21	2.74	3.31	3.92	4.57
	2	1.60	2.65	3.65	4.64	5.64	6.63	7.62	8.61	9.60	10.6	11.6
	3	1.42	2.48	3.48	4.48	5.47	6.46	7.45	8.44	9.44	10.4	11.4
	4	1.27	2.32	3.32	4.31	5.30	6.30	7.29	8.27	9.27	10.3	11.3
	5	1.15	2.18	3.17	4.15	5.14	6.13	7.12	8.11	9.10	10.1	11.1
	6	1.05	2.05	3.02	4.00	4.98	5.97	6.96	7.94	8.94	9.93	10.9
	7	0.96	1.93	2.89	3.86	4.83	5.81	6.80	7.78	8.77	9.76	10.8
	8	0.89	1.83	2.77	3.72	4.69	5.66	6.64	7.62	8.61	9.60	10.6
	9	0.83	1.73	2.65	3.59	4.55	5.51	6.49	7.47	8.45	9.43	10.4
	10	0.77	1.65	2.54	3.47	4.41	5.37	6.34	7.32	8.29	9.28	10.3
6	12	0.68	1.49	2.34	3.24	4.16	5.10	6.06	7.02	7.99	8.97	9.95
	14	0.61	1.37	2.17	3.03	3.93	4.85	5.79	6.74	7.71	8.67	9.64
	16	0.56	1.26	2.01	2.83	3.71	4.61	5.54	6.48	7.43	8.39	9.35
	18	0.51	1.16	1.87	2.66	3.51	4.39	5.30	6.23	7.17	8.12	9.07
	20	0.47	1.07	1.74	2.50	3.32	4.19	5.08	5.99	6.92	7.86	8.81
	24	0.40	0.93	1.52	2.22	3.00	3.82	4.67	5.55	6.45	7.37	8.30
	28	0.35	0.81	1.35	2.00	2.73	3.50	4.32	5.16	6.03	6.92	7.83
	32	0.32	0.72	1.21	1.81	2.49	3.23	4.00	4.81	5.65	6.51	7.40
	36	0.29	0.65	1.09	1.66	2.30	2.98	3.72	4.50	5.31	6.14	7.00

Table 7-17 (cont.). Coefficients *C* for Eccentrically Loaded Bolt Groups Angle = 75°

$$C_{req} = \frac{P_u}{\phi r_n} \quad \text{or} \quad \phi R_n = C \times \phi r_n$$

where

P_u = factored force, kips
ϕr_n = design strength per bolt, kips
ϕR_n = design strength per bolt group, kips
e = eccentricity of P_u with respect to centroid of bolt group, in. (not tabulated, may be determined by geometry)
e_x = horizontal component of e, in.
s = bolt spacing, in.
C = coefficient tabulated below

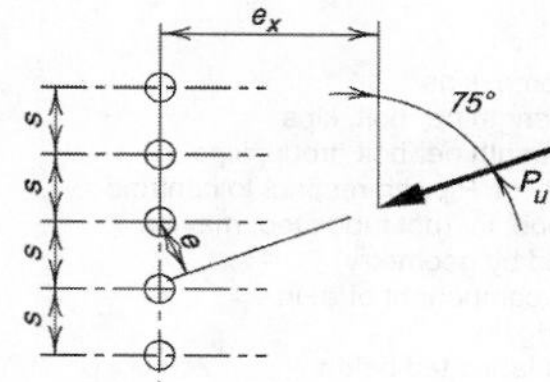

s, in.	e_x, in.	Number of Bolts in One Vertical Row, *n*										
		2	3	4	5	6	7	8	9	10	11	12
	2	1.49	2.51	3.49	4.46	5.44	6.42	7.40	8.38	9.36	10.3	11.3
	3	1.32	2.33	3.30	4.27	5.24	6.21	7.18	8.15	9.13	10.1	11.1
	4	1.18	2.18	3.14	4.09	5.05	6.01	6.98	7.95	8.92	9.89	10.9
	5	1.07	2.04	2.99	3.93	4.88	5.84	6.79	7.75	8.72	9.68	10.7
	6	0.98	1.92	2.85	3.79	4.73	5.67	6.62	7.57	8.53	9.49	10.5
	7	0.90	1.82	2.73	3.65	4.58	5.52	6.46	7.40	8.36	9.31	10.3
	8	0.84	1.72	2.62	3.52	4.44	5.37	6.30	7.24	8.19	9.14	10.1
	9	0.78	1.63	2.51	3.40	4.31	5.23	6.16	7.09	8.03	8.97	9.92
	10	0.73	1.55	2.41	3.29	4.19	5.10	6.02	6.94	7.88	8.81	9.76
3	12	0.65	1.41	2.23	3.08	3.95	4.84	5.75	6.66	7.59	8.51	9.45
	14	0.58	1.30	2.06	2.88	3.73	4.60	5.50	6.40	7.31	8.23	9.16
	16	0.53	1.20	1.92	2.70	3.52	4.38	5.26	6.15	7.05	7.96	8.88
	18	0.48	1.11	1.78	2.53	3.33	4.17	5.03	5.91	6.80	7.70	8.61
	20	0.44	1.03	1.66	2.38	3.16	3.97	4.82	5.69	6.56	7.45	8.35
	24	0.38	0.89	1.46	2.12	2.85	3.63	4.44	5.27	6.13	6.99	7.87
	28	0.34	0.79	1.29	1.90	2.59	3.33	4.11	4.91	5.73	6.57	7.43
	32	0.30	0.70	1.16	1.73	2.38	3.08	3.81	4.58	5.37	6.19	7.02
	36	0.27	0.62	1.05	1.58	2.19	2.85	3.55	4.28	5.05	5.84	6.65
	2	1.71	2.72	3.70	4.69	5.67	6.66	7.64	8.79	9.78	10.8	11.7
	3	1.60	2.61	3.59	4.57	5.55	6.53	7.52	8.50	9.48	10.5	11.5
	4	1.49	2.51	3.49	4.46	5.44	6.42	7.40	8.38	9.36	10.3	11.3
	5	1.40	2.42	3.39	4.37	5.34	6.31	7.29	8.26	9.24	10.2	11.2
	6	1.32	2.33	3.30	4.27	5.24	6.21	7.18	8.15	9.13	10.1	11.1
	7	1.25	2.25	3.22	4.18	5.14	6.11	7.07	8.05	9.01	10.0	11.0
	8	1.18	2.18	3.14	4.09	5.05	6.01	6.98	7.95	8.92	9.89	10.9
	9	1.13	2.11	3.06	4.01	4.97	5.92	6.88	7.85	8.81	9.78	10.8
	10	1.07	2.04	2.99	3.93	4.88	5.84	6.79	7.75	8.72	9.68	10.7
6	12	0.98	1.92	2.85	3.79	4.73	5.67	6.62	7.57	8.53	9.49	10.5
	14	0.90	1.82	2.73	3.65	4.58	5.52	6.46	7.40	8.36	9.31	10.3
	16	0.84	1.72	2.62	3.52	4.44	5.37	6.30	7.24	8.19	9.14	10.1
	18	0.78	1.63	2.51	3.40	4.31	5.23	6.16	7.09	8.03	8.97	9.92
	20	0.73	1.55	2.41	3.29	4.19	5.10	6.02	6.94	7.88	8.81	9.76
	24	0.65	1.41	2.23	3.08	3.95	4.84	5.75	6.66	7.59	8.51	9.45
	28	0.58	1.30	2.06	2.88	3.73	4.60	5.50	6.40	7.31	8.23	9.16
	32	0.53	1.20	1.92	2.70	3.52	4.38	5.26	6.15	7.05	7.96	8.88
	36	0.48	1.11	1.78	2.53	3.33	4.17	5.03	5.91	6.80	7.70	8.61

Table 7-18.
Coefficients *C* for Eccentrically Loaded Bolt Groups
Angle = 0°

$$C_{req} = \frac{P_u}{\phi r_n} \quad \text{or} \quad \phi R_n = C \times \phi r_n$$

where

P_u = factored force, kips
ϕr_n = design strength per bolt, kips
ϕR_n = design strength per bolt group, kips
e = eccentricity of P_u with respect to centroid of bolt group, in. (not tabulated, may be determined by geometry)
e_x = horizontal component of e, in.
s = bolt spacing, in.
C = coefficient tabulated below

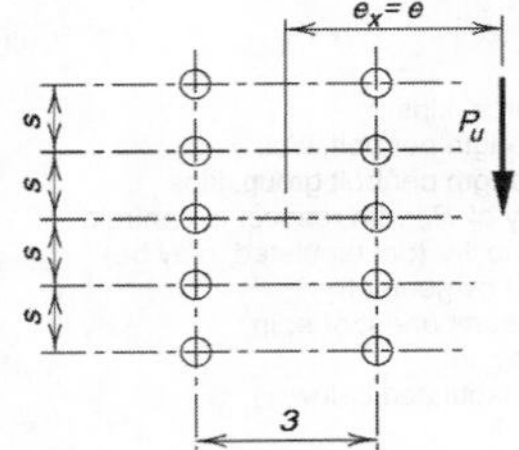

s, in.	e_x, in.	Number of Bolts in One Vertical Row, n											
		1	2	3	4	5	6	7	8	9	10	11	12
	2	0.84	2.54	4.48	6.59	8.72	10.8	12.9	15.0	17.0	19.0	21.0	23.0
	3	0.65	2.03	3.68	5.67	7.77	9.91	12.1	14.2	16.3	18.3	20.4	22.5
	4	0.54	1.67	3.06	4.86	6.84	8.93	11.1	13.2	15.4	17.5	19.6	21.7
	5	0.45	1.42	2.59	4.21	6.01	8.00	10.1	12.2	14.4	16.5	18.7	20.8
	6	0.39	1.22	2.25	3.69	5.32	7.17	9.16	11.2	13.4	15.5	17.7	19.8
	7	0.35	1.08	1.99	3.27	4.74	6.46	8.33	10.3	12.4	14.5	16.7	18.8
	8	0.31	0.96	1.78	2.93	4.27	5.86	7.60	9.50	11.5	13.6	15.7	17.8
	9	0.28	0.86	1.60	2.65	3.87	5.34	6.97	8.75	10.7	12.7	14.7	16.8
	10	0.26	0.78	1.46	2.42	3.53	4.90	6.42	8.10	9.91	11.8	13.8	15.9
3	12	0.22	0.66	1.24	2.06	3.01	4.19	5.51	7.01	8.63	10.4	12.2	14.2
	14	0.19	0.57	1.08	1.78	2.62	3.66	4.82	6.15	7.61	9.19	10.9	12.7
	16	0.17	0.51	0.95	1.57	2.32	3.24	4.27	5.47	6.79	8.23	9.78	11.4
	18	0.15	0.45	0.85	1.41	2.07	2.90	3.83	4.92	6.11	7.43	8.85	10.4
	20	0.14	0.41	0.77	1.27	1.88	2.63	3.48	4.47	5.55	6.76	8.07	9.48
	24	0.12	0.34	0.65	1.07	1.58	2.21	2.93	3.77	4.69	5.72	6.85	8.06
	28	0.10	0.29	0.56	0.92	1.36	1.90	2.53	3.25	4.05	4.95	5.93	7.00
	32	0.09	0.26	0.49	0.80	1.19	1.67	2.22	2.86	3.57	4.36	5.23	6.18
	36	0.08	0.23	0.43	0.72	1.06	1.49	1.98	2.55	3.18	3.90	4.67	5.52
	2	0.84	3.24	5.39	7.47	9.51	11.5	13.5	15.5	17.5	19.5	21.5	23.4
	3	0.65	2.79	4.93	7.08	9.17	11.2	13.3	15.3	17.3	19.3	21.3	23.3
	4	0.54	2.41	4.44	6.60	8.75	10.9	12.9	15.0	17.0	19.1	21.1	23.1
	5	0.45	2.10	3.97	6.11	8.27	10.4	12.5	14.6	16.7	18.7	20.8	22.8
	6	0.39	1.85	3.55	5.62	7.77	9.93	12.1	14.2	16.3	18.4	20.4	22.5
	7	0.35	1.64	3.18	5.17	7.27	9.43	11.6	13.7	15.9	18.0	20.1	22.1
	8	0.31	1.47	2.87	4.75	6.79	8.92	11.1	13.3	15.4	17.5	19.6	21.7
	9	0.28	1.34	2.61	4.39	6.34	8.43	10.6	12.7	14.9	17.1	19.2	21.3
	10	0.26	1.22	2.39	4.06	5.92	7.96	10.1	12.2	14.4	16.6	18.7	20.9
6	12	0.22	1.04	2.04	3.52	5.20	7.10	9.12	11.2	13.4	15.5	17.7	19.9
	14	0.19	0.90	1.77	3.09	4.61	6.36	8.27	10.3	12.4	14.5	16.7	18.9
	16	0.17	0.80	1.57	2.75	4.12	5.74	7.52	9.44	11.5	13.5	15.7	17.8
	18	0.15	0.71	1.41	2.48	3.72	5.21	6.87	8.68	10.6	12.6	14.7	16.8
	20	0.14	0.64	1.28	2.25	3.38	4.77	6.31	8.02	9.85	11.8	13.8	15.9
	24	0.12	0.54	1.07	1.90	2.86	4.06	5.40	6.91	8.55	10.3	12.2	14.1
	28	0.10	0.46	0.93	1.64	2.47	3.52	4.70	6.05	7.52	9.12	10.8	12.6
	32	0.09	0.41	0.81	1.44	2.18	3.11	4.16	5.37	6.69	8.15	9.71	11.4
	36	0.08	0.36	0.73	1.29	1.94	2.78	3.72	4.81	6.02	7.34	8.78	10.3

Table 7-18 (cont.). Coefficients *C* for Eccentrically Loaded Bolt Groups Angle = 15°

$$C_{req} = \frac{P_u}{\phi r_n} \quad \text{or} \quad \phi R_n = C \times \phi r_n$$

where

P_u = factored force, kips
ϕr_n = design strength per bolt, kips
ϕR_n = design strength per bolt group, kips
e = eccentricity of P_u with respect to centroid of bolt group, in. (not tabulated, may be determined by geometry)
e_x = horizontal component of e, in.
s = bolt spacing, in.
C = coefficient tabulated below

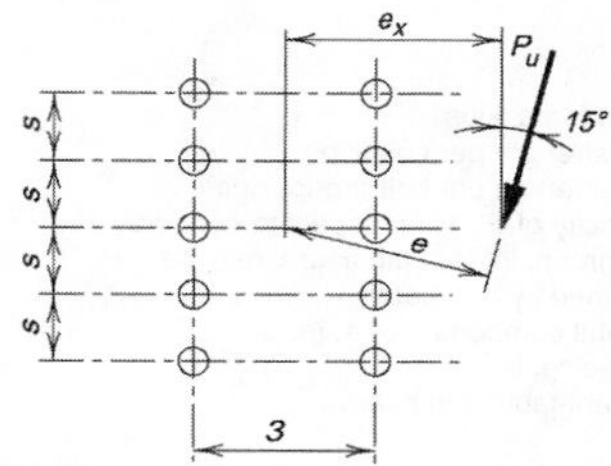

s, in.	e_x, in.	Number of Bolts in One Vertical Row, *n*											
		1	2	3	4	5	6	7	8	9	10	11	12
	2	0.87	2.54	4.47	6.54	8.63	10.7	12.8	14.8	16.9	18.9	20.9	22.9
	3	0.68	2.04	3.71	5.63	7.69	9.80	11.9	14.0	16.1	18.2	20.2	22.3
	4	0.55	1.69	3.11	4.85	6.79	8.84	10.9	13.1	15.2	17.3	19.4	21.5
	5	0.47	1.44	2.66	4.21	6.00	7.94	9.98	12.1	14.2	16.3	18.4	20.5
	6	0.41	1.25	2.31	3.70	5.34	7.15	9.09	11.1	13.2	15.3	17.4	19.6
	7	0.36	1.10	2.04	3.29	4.79	6.46	8.30	10.2	12.3	14.3	16.4	18.6
	8	0.32	0.98	1.83	2.96	4.32	5.87	7.60	9.45	11.4	13.4	15.5	17.6
	9	0.29	0.88	1.65	2.68	3.94	5.37	6.99	8.74	10.6	12.6	14.6	16.6
	10	0.27	0.81	1.51	2.45	3.61	4.93	6.45	8.11	9.88	11.8	13.7	15.7
3	12	0.23	0.68	1.28	2.09	3.08	4.24	5.58	7.05	8.66	10.4	12.2	14.1
	14	0.20	0.59	1.11	1.82	2.69	3.71	4.90	6.21	7.67	9.23	10.9	12.7
	16	0.17	0.52	0.98	1.61	2.38	3.29	4.36	5.54	6.86	8.29	9.83	11.5
	18	0.16	0.47	0.88	1.44	2.13	2.96	3.92	4.99	6.20	7.51	8.93	10.4
	20	0.14	0.42	0.79	1.31	1.93	2.68	3.56	4.54	5.65	6.85	8.17	9.57
	24	0.12	0.35	0.67	1.10	1.62	2.26	3.00	3.84	4.79	5.82	6.96	8.17
	28	0.10	0.30	0.57	0.94	1.40	1.95	2.60	3.32	4.15	5.05	6.05	7.12
	32	0.09	0.27	0.50	0.83	1.23	1.72	2.28	2.93	3.66	4.46	5.34	6.29
	36	0.08	0.24	0.45	0.74	1.10	1.53	2.04	2.61	3.27	3.98	4.78	5.64
	2	0.87	3.21	5.35	7.42	9.45	11.5	13.5	15.5	17.4	19.4	21.4	23.4
	3	0.68	2.76	4.88	7.00	9.09	11.1	13.2	15.2	17.2	19.2	21.2	23.2
	4	0.55	2.38	4.40	6.53	8.65	10.7	12.8	14.9	16.9	18.9	20.9	22.9
	5	0.47	2.07	3.96	6.04	8.17	10.3	12.4	14.5	16.5	18.6	20.6	22.6
	6	0.41	1.83	3.56	5.56	7.67	9.80	11.9	14.0	16.1	18.2	20.3	22.3
	7	0.36	1.63	3.22	5.12	7.19	9.30	11.4	13.6	15.7	17.8	19.9	21.9
	8	0.32	1.47	2.92	4.73	6.72	8.81	10.9	13.1	15.2	17.3	19.4	21.5
	9	0.29	1.34	2.66	4.37	6.29	8.33	10.4	12.6	14.7	16.8	18.9	21.0
	10	0.27	1.23	2.45	4.05	5.90	7.88	9.95	12.1	14.2	16.3	18.5	20.6
6	12	0.23	1.05	2.09	3.53	5.21	7.06	9.04	11.1	13.2	15.3	17.5	19.6
	14	0.2	0.91	1.83	3.11	4.64	6.35	8.22	10.2	12.2	14.3	16.5	18.6
	16	0.17	0.81	1.62	2.78	4.17	5.75	7.51	9.38	11.4	13.4	15.5	17.6
	18	0.16	0.72	1.45	2.50	3.77	5.24	6.88	8.66	10.5	12.5	14.5	16.6
	20	0.14	0.66	1.32	2.28	3.45	4.80	6.34	8.02	9.82	11.7	13.7	15.7
	24	0.12	0.55	1.11	1.93	2.93	4.10	5.46	6.95	8.57	10.3	12.1	14.0
	28	0.10	0.48	0.96	1.67	2.54	3.57	4.78	6.11	7.58	9.15	10.8	12.6
	32	0.09	0.42	0.84	1.47	2.24	3.16	4.24	5.44	6.77	8.21	9.75	11.4
	36	0.08	0.37	0.75	1.32	2.00	2.83	3.80	4.89	6.10	7.42	8.85	10.4

Table 7-18 (cont.). Coefficients *C* for Eccentrically Loaded Bolt Groups Angle = 30°

$$C_{req} = \frac{P_u}{\phi r_n} \quad \text{or} \quad \phi R_n = C \times \phi r_n$$

where

P_u = factored force, kips
ϕr_n = design strength per bolt, kips
ϕR_n = design strength per bolt group, kips
e = eccentricity of P_u with respect to centroid of bolt group, in. (not tabulated, may be determined by geometry)
e_x = horizontal component of e, in.
s = bolt spacing, in.
C = coefficient tabulated below

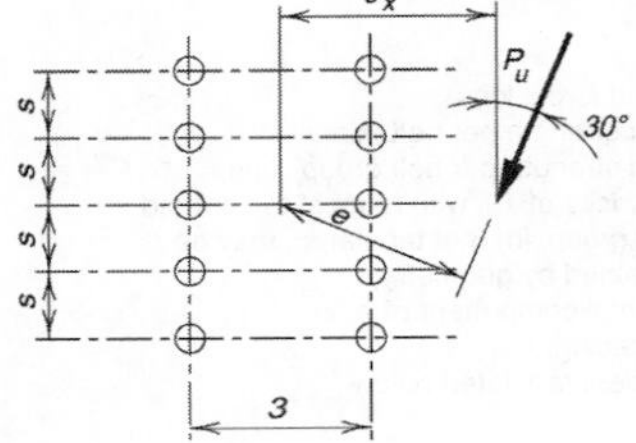

s, in.	e_x, in.	Number of Bolts in One Vertical Row, *n*											
		1	2	3	4	5	6	7	8	9	10	11	12
3	2	0.97	2.60	4.52	6.54	8.59	10.6	12.7	14.7	16.7	18.8	20.8	22.8
	3	0.75	2.12	3.83	5.71	7.71	9.75	11.8	13.9	15.9	18.0	20.0	22.1
	4	0.62	1.78	3.29	4.99	6.88	8.87	10.9	13.0	15.1	17.1	19.2	21.3
	5	0.52	1.53	2.85	4.39	6.16	8.06	10.0	12.1	14.1	16.2	18.3	20.4
	6	0.45	1.34	2.51	3.89	5.54	7.33	9.23	11.2	13.2	15.3	17.3	19.4
	7	0.40	1.19	2.23	3.48	5.01	6.70	8.51	10.4	12.4	14.4	16.4	18.5
	8	0.36	1.07	2.00	3.15	4.57	6.14	7.86	9.68	11.6	13.6	15.6	17.6
	9	0.32	0.97	1.81	2.87	4.19	5.66	7.28	9.02	10.9	12.8	14.7	16.7
	10	0.30	0.88	1.66	2.64	3.87	5.24	6.77	8.43	10.2	12.0	13.9	15.9
	12	0.25	0.75	1.41	2.27	3.34	4.54	5.92	7.43	9.04	10.8	12.5	14.4
	14	0.22	0.65	1.23	1.98	2.93	3.99	5.24	6.61	8.09	9.67	11.4	13.1
	16	0.19	0.58	1.08	1.76	2.60	3.56	4.69	5.94	7.30	8.77	10.3	12.0
	18	0.17	0.52	0.97	1.58	2.34	3.21	4.24	5.38	6.64	8.0	9.45	11.0
	20	0.16	0.47	0.88	1.43	2.12	2.92	3.87	4.92	6.08	7.3	8.70	10.1
	24	0.13	0.39	0.74	1.21	1.79	2.48	3.29	4.18	5.19	6.3	7.48	8.75
	28	0.12	0.34	0.64	1.04	1.55	2.14	2.85	3.63	4.52	5.5	6.54	7.68
	32	0.10	0.30	0.56	0.92	1.36	1.89	2.51	3.21	4.00	4.9	5.81	6.83
	36	0.09	0.26	0.50	0.82	1.21	1.69	2.25	2.87	3.59	4.4	5.22	6.15
6	2	0.97	3.20	5.31	7.37	9.39	11.4	13.4	15.4	17.4	19.4	21.3	23.3
	3	0.75	2.75	4.86	6.95	9.01	11.1	13.1	15.1	17.1	19.1	21.1	23.1
	4	0.62	2.39	4.42	6.49	8.57	10.6	12.7	14.7	16.8	18.8	20.8	22.8
	5	0.52	2.10	4.02	6.04	8.11	10.2	12.3	14.3	16.4	18.4	20.4	22.5
	6	0.45	1.87	3.67	5.61	7.66	9.73	11.8	13.9	16.0	18.0	20.1	22.1
	7	0.40	1.69	3.36	5.21	7.21	9.27	11.4	13.4	15.5	17.6	19.6	21.7
	8	0.36	1.53	3.08	4.84	6.79	8.82	10.9	13.0	15.1	17.1	19.2	21.3
	9	0.32	1.40	2.84	4.51	6.40	8.39	10.4	12.5	14.6	16.7	18.7	20.8
	10	0.30	1.29	2.63	4.21	6.04	7.98	9.99	12.0	14.1	16.2	18.3	20.4
	12	0.25	1.12	2.28	3.70	5.39	7.23	9.16	11.2	13.2	15.3	17.3	19.4
	14	0.22	0.98	2.00	3.29	4.86	6.57	8.41	10.3	12.3	14.4	16.4	18.5
	16	0.19	0.87	1.78	2.95	4.40	6.01	7.75	9.6	11.5	13.5	15.5	17.6
	18	0.17	0.79	1.60	2.68	4.02	5.52	7.17	8.9	10.8	12.7	14.7	16.7
	20	0.16	0.71	1.45	2.45	3.70	5.09	6.65	8.3	10.1	12.0	13.9	15.9
	24	0.13	0.60	1.23	2.08	3.17	4.39	5.79	7.3	8.95	10.7	12.5	14.4
	28	0.12	0.52	1.06	1.82	2.77	3.85	5.11	6.5	7.99	9.59	11.3	13.0
	32	0.10	0.46	0.93	1.61	2.45	3.42	4.56	5.8	7.20	8.68	10.3	11.9
	36	0.09	0.41	0.83	1.44	2.20	3.08	4.12	5.3	6.53	7.91	9.37	10.9

Table 7-18 (cont.). Coefficients C for Eccentrically Loaded Bolt Groups Angle = 45°

$C_{req} = \dfrac{P_u}{\phi r_n}$ or $\phi R_n = C \times \phi r_n$

where

P_u = factored force, kips
ϕr_n = design strength per bolt, kips
ϕR_n = design strength per bolt group, kips
e = eccentricity of P_u with respect to centroid of bolt group, in. (not tabulated, may be determined by geometry)
e_x = horizontal component of e, in.
s = bolt spacing, in.
C = coefficient tabulated below

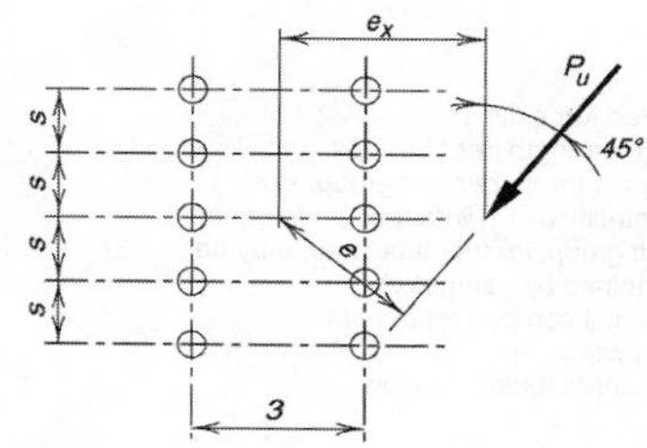

s, in.	ex, in.	Number of Bolts in One Vertical Row, n											
		1	2	3	4	5	6	7	8	9	10	11	12
	2	1.17	2.79	4.67	6.62	8.61	10.6	12.6	14.6	16.6	18.6	20.6	22.6
	3	0.92	2.32	4.06	5.92	7.86	9.83	11.8	13.9	15.9	17.9	19.9	21.9
	4	0.75	1.99	3.57	5.31	7.16	9.09	11.1	13.1	15.1	17.1	19.1	21.1
	5	0.64	1.74	3.17	4.78	6.53	8.39	10.3	12.3	14.3	16.3	18.3	20.3
	6	0.55	1.54	2.84	4.33	5.98	7.76	9.63	11.6	13.5	15.5	17.5	19.5
	7	0.49	1.38	2.57	3.93	5.49	7.20	9.00	10.9	12.8	14.8	16.7	18.7
	8	0.44	1.25	2.33	3.60	5.06	6.70	8.43	10.3	12.1	14.0	16.0	18.0
	9	0.40	1.14	2.13	3.31	4.69	6.25	7.91	9.67	11.5	13.4	15.3	17.2
	10	0.36	1.05	1.96	3.06	4.36	5.85	7.44	9.14	10.9	12.7	14.6	16.5
3	12	0.31	0.90	1.68	2.65	3.83	5.17	6.63	8.20	9.86	11.6	13.4	15.2
	14	0.27	0.78	1.47	2.33	3.40	4.61	5.95	7.41	8.97	10.6	12.3	14.1
	16	0.24	0.69	1.31	2.08	3.05	4.16	5.38	6.74	8.20	9.75	11.4	13.1
	18	0.21	0.62	1.17	1.88	2.76	3.77	4.91	6.18	7.55	9.00	10.5	12.1
	20	0.19	0.56	1.06	1.71	2.52	3.45	4.51	5.69	6.97	8.34	9.80	11.3
	24	0.16	0.48	0.90	1.45	2.14	2.94	3.87	4.91	6.04	7.26	8.57	9.95
	28	0.14	0.41	0.77	1.26	1.86	2.56	3.38	4.30	5.30	6.41	7.59	8.85
	32	0.12	0.36	0.68	1.11	1.64	2.27	3.00	3.82	4.73	5.73	6.80	7.94
	36	0.11	0.32	0.61	0.99	1.47	2.03	2.70	3.44	4.26	5.17	6.15	7.20
	2	1.17	3.24	5.30	7.32	9.33	11.3	13.3	15.3	17.3	19.3	21.3	23.2
	3	0.92	2.84	4.90	6.93	8.96	11.0	13.0	15.0	17.0	19.0	21.0	23.0
	4	0.75	2.51	4.52	6.53	8.56	10.6	12.6	14.6	16.6	18.6	20.6	22.6
	5	0.64	2.24	4.17	6.15	8.15	10.2	12.2	14.2	16.2	18.3	20.3	22.3
	6	0.55	2.03	3.86	5.78	7.76	9.77	11.8	13.8	15.8	17.9	19.9	21.9
	7	0.49	1.85	3.59	5.45	7.39	9.38	11.4	13.4	15.4	17.5	19.5	21.5
	8	0.44	1.70	3.35	5.13	7.03	9.00	11.0	13.0	15.0	17.1	19.1	21.1
	9	0.40	1.57	3.13	4.85	6.70	8.63	10.6	12.6	14.6	16.7	18.7	20.7
	10	0.36	1.46	2.94	4.58	6.38	8.28	10.2	12.2	14.2	16.3	18.3	20.3
6	12	0.31	1.28	2.60	4.11	5.81	7.64	9.54	11.5	13.5	15.5	17.5	19.5
	14	0.27	1.13	2.32	3.71	5.31	7.06	8.89	10.8	12.7	14.7	16.7	18.7
	16	0.24	1.01	2.09	3.36	4.88	6.55	8.31	10.2	12.0	14.0	15.9	17.9
	18	0.21	0.92	1.90	3.07	4.50	6.09	7.78	9.56	11.4	13.3	15.2	17.2
	20	0.19	0.84	1.73	2.83	4.18	5.69	7.31	9.02	10.8	12.7	14.6	16.5
	24	0.16	0.72	1.47	2.43	3.64	5.00	6.48	8.08	9.76	11.5	13.3	15.2
	28	0.14	0.62	1.28	2.13	3.22	4.45	5.80	7.28	8.86	10.5	12.2	14.0
	32	0.12	0.55	1.13	1.90	2.88	3.99	5.24	6.62	8.09	9.65	11.3	13.0
	36	0.11	0.49	1.01	1.71	2.61	3.62	4.77	6.05	7.43	8.90	10.4	12.1

Table 7-18 (cont.).
Coefficients *C* for Eccentrically Loaded Bolt Groups
Angle = 60°

$C_{req} = \dfrac{P_u}{\phi r_n}$ or $\phi R_n = C \times \phi r_n$

where

P_u = factored force, kips
ϕr_n = design strength per bolt, kips
ϕR_n = design strength per bolt group, kips
e = eccentricity of P_u with respect to centroid of bolt group, in. (not tabulated, may be determined by geometry)
e_x = horizontal component of e, in.
s = bolt spacing, in.
C = coefficient tabulated below

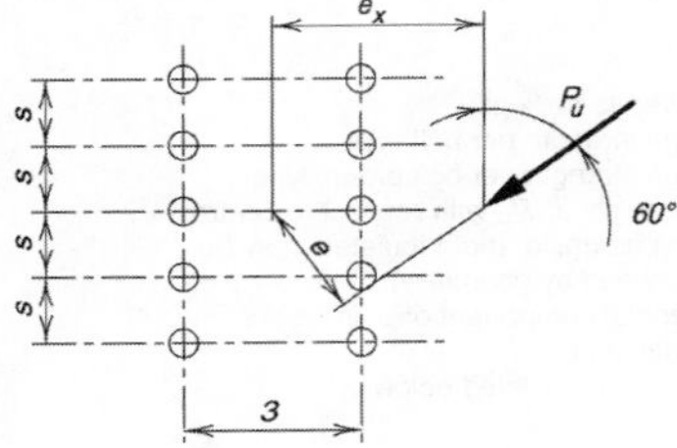

s, in.	e_x, in.	Number of Bolts in One Vertical Row, *n*											
		1	2	3	4	5	6	7	8	9	10	11	12
	2	1.51	3.17	4.97	6.85	8.77	10.7	12.7	14.6	16.6	18.6	20.6	22.5
	3	1.24	2.76	4.47	6.30	8.19	10.1	12.0	14.0	16.0	17.9	19.9	21.9
	4	1.04	2.43	4.04	5.81	7.65	9.53	11.5	13.4	15.3	17.3	19.3	21.2
	5	0.89	2.16	3.70	5.39	7.17	9.01	10.9	12.8	14.7	16.7	18.6	20.6
	6	0.77	1.95	3.40	5.01	6.73	8.52	10.4	12.3	14.2	16.1	18.0	20.0
	7	0.68	1.77	3.13	4.67	6.33	8.07	9.88	11.7	13.6	15.5	17.4	19.4
	8	0.61	1.62	2.90	4.37	5.96	7.65	9.42	11.2	13.1	15.0	16.9	18.8
	9	0.56	1.49	2.70	4.09	5.62	7.26	8.98	10.8	12.6	14.5	16.3	18.2
	10	0.51	1.38	2.52	3.84	5.31	6.89	8.58	10.3	12.1	14.0	15.8	17.7
3	12	0.43	1.20	2.21	3.40	4.76	6.25	7.85	9.53	11.3	13.0	14.9	16.7
	14	0.38	1.06	1.96	3.05	4.30	5.71	7.23	8.83	10.5	12.2	14.0	15.8
	16	0.34	0.95	1.76	2.75	3.92	5.24	6.68	8.20	9.79	11.5	13.2	14.9
	18	0.30	0.85	1.60	2.51	3.59	4.84	6.19	7.64	9.16	10.8	12.4	14.1
	20	0.27	0.78	1.46	2.30	3.32	4.48	5.76	7.14	8.60	10.1	11.7	13.4
	24	0.23	0.66	1.24	1.97	2.87	3.90	5.04	6.29	7.64	9.06	10.6	12.1
	28	0.20	0.57	1.07	1.72	2.52	3.44	4.47	5.61	6.85	8.17	9.55	11.0
	32	0.18	0.50	0.95	1.52	2.24	3.07	4.01	5.06	6.20	7.41	8.70	10.1
	36	0.16	0.45	0.85	1.37	2.02	2.77	3.63	4.59	5.65	6.77	7.98	9.26
	2	1.51	3.39	5.36	7.33	9.31	11.3	13.3	15.2	17.2	19.2	21.2	23.2
	3	1.24	3.08	5.04	7.01	8.98	11.0	12.9	14.9	16.9	18.9	20.9	22.8
	4	1.04	2.80	4.73	6.69	8.66	10.6	12.6	14.6	16.6	18.6	20.5	22.5
	5	0.89	2.57	4.45	6.39	8.35	10.3	12.3	14.3	16.2	18.2	20.2	22.2
	6	0.77	2.37	4.20	6.11	8.05	10.0	12.0	13.9	15.9	17.9	19.9	21.8
	7	0.68	2.19	3.98	5.85	7.76	9.70	11.7	13.6	15.6	17.6	19.5	21.5
	8	0.61	2.04	3.77	5.61	7.49	9.41	11.4	13.3	15.3	17.2	19.2	21.2
	9	0.56	1.91	3.59	5.38	7.24	9.13	11.1	13.0	15.0	16.9	18.9	20.9
	10	0.51	1.80	3.42	5.17	7.00	8.87	10.8	12.7	14.7	16.6	18.6	20.5
6	12	0.43	1.60	3.11	4.78	6.54	8.37	10.2	12.1	14.1	16.0	18.0	19.9
	14	0.38	1.44	2.85	4.43	6.13	7.91	9.74	11.6	13.5	15.4	17.4	19.3
	16	0.34	1.31	2.63	4.12	5.74	7.48	9.27	11.1	13.0	14.9	16.8	18.7
	18	0.30	1.20	2.43	3.84	5.40	7.08	8.84	10.7	12.5	14.4	16.3	18.2
	20	0.27	1.10	2.26	3.58	5.08	6.71	8.43	10.2	12.0	13.9	15.7	17.6
	24	0.23	0.95	1.97	3.15	4.53	6.06	7.69	9.39	11.2	12.9	14.8	16.6
	28	0.20	0.84	1.73	2.80	4.08	5.52	7.06	8.68	10.4	12.1	13.9	15.7
	32	0.18	0.74	1.54	2.52	3.71	5.05	6.51	8.05	9.66	11.3	13.1	14.8
	36	0.16	0.67	1.39	2.28	3.39	4.65	6.02	7.49	9.03	10.7	12.3	14.0

Table 7-18 (cont.). Coefficients C for Eccentrically Loaded Bolt Groups Angle = 75°

$$C_{req} = \frac{P_u}{\phi r_n} \quad \text{or} \quad \phi R_n = C \times \phi r_n$$

where

P_u = factored force, kips
ϕr_n = design strength per bolt, kips
ϕR_n = design strength per bolt group, kips
e = eccentricity of P_u with respect to centroid of bolt group, in. (not tabulated, may be determined by geometry)
e_x = horizontal component of e, in.
s = bolt spacing, in.
C = coefficient tabulated below

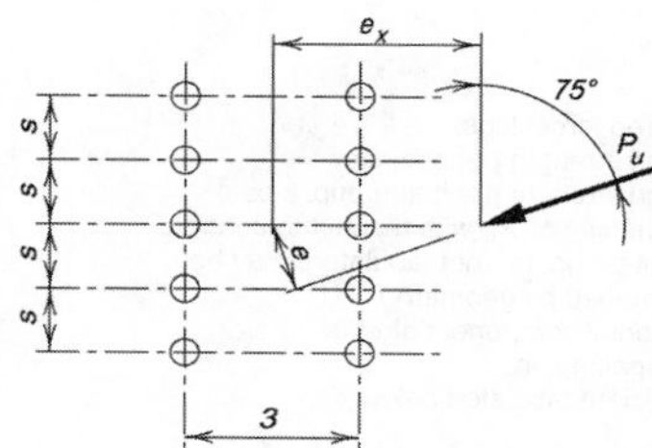

s, in.	e_x, in.	Number of Bolts in One Vertical Row, n											
		1	2	3	4	5	6	7	8	9	10	11	12
3	2	1.84	3.63	5.44	7.29	9.17	11.1	13.0	14.9	16.9	18.8	20.8	22.7
	3	1.71	3.41	5.17	6.97	8.82	10.7	12.6	14.5	16.4	18.4	20.3	22.3
	4	1.57	3.19	4.90	6.67	8.50	10.4	12.2	14.1	16.0	18.0	19.9	21.8
	5	1.44	2.98	4.65	6.39	8.19	10.0	11.9	13.8	15.7	17.6	19.5	21.4
	6	1.31	2.79	4.41	6.12	7.90	9.71	11.6	13.4	15.3	17.2	19.1	21.0
	7	1.20	2.61	4.19	5.88	7.62	9.42	11.3	13.1	15.0	16.9	18.8	20.7
	8	1.10	2.45	3.99	5.65	7.37	9.14	11.0	12.8	14.7	16.5	18.4	20.3
	9	1.01	2.31	3.81	5.43	7.14	8.89	10.7	12.5	14.3	16.2	18.1	20.0
	10	0.93	2.18	3.63	5.23	6.91	8.65	10.4	12.2	14.1	15.9	17.8	19.6
	12	0.81	1.95	3.33	4.86	6.49	8.19	9.94	11.7	13.5	15.3	17.2	19.0
	14	0.71	1.77	3.06	4.53	6.11	7.76	9.47	11.2	13.0	14.8	16.6	18.4
	16	0.63	1.61	2.83	4.23	5.75	7.36	9.03	10.8	12.5	14.3	16.1	17.9
	18	0.57	1.48	2.63	3.96	5.42	6.98	8.61	10.3	12.0	13.8	15.6	17.4
	20	0.52	1.36	2.45	3.72	5.12	6.63	8.23	9.88	11.6	13.3	15.1	16.9
	24	0.44	1.18	2.15	3.30	4.60	6.02	7.53	9.12	10.8	12.4	14.2	15.9
	28	0.38	1.04	1.91	2.95	4.16	5.49	6.93	8.45	10.0	11.7	13.3	15.0
	32	0.34	0.92	1.71	2.67	3.78	5.04	6.41	7.86	9.37	10.9	12.6	14.2
	36	0.30	0.83	1.55	2.43	3.47	4.65	5.94	7.32	8.78	10.3	11.9	13.5
6	2	1.84	3.66	5.55	7.48	9.42	11.4	13.3	15.3	17.6	19.6	21.5	23.5
	3	1.71	3.49	5.36	7.27	9.20	11.2	13.1	15.1	17.0	19.0	21.0	22.9
	4	1.57	3.32	5.18	7.08	9.00	10.9	12.9	14.8	16.8	18.7	20.7	22.7
	5	1.44	3.16	5.01	6.89	8.81	10.7	12.7	14.6	16.6	18.5	20.5	22.4
	6	1.31	3.02	4.84	6.72	8.62	10.5	12.5	14.4	16.3	18.3	20.2	22.2
	7	1.20	2.88	4.69	6.55	8.44	10.4	12.3	14.2	16.1	18.1	20.0	22.0
	8	1.10	2.75	4.54	6.39	8.27	10.2	12.1	14.0	15.9	17.9	19.8	21.8
	9	1.01	2.63	4.40	6.24	8.11	10.0	11.9	13.8	15.7	17.7	19.6	21.5
	10	0.93	2.52	4.27	6.09	7.95	9.83	11.7	13.6	15.6	17.5	19.4	21.3
	12	0.81	2.32	4.03	5.82	7.66	9.52	11.4	13.3	15.2	17.1	19.0	20.9
	14	0.71	2.15	3.82	5.57	7.38	9.22	11.1	13.0	14.9	16.7	18.7	20.6
	16	0.63	2.00	3.62	5.35	7.13	8.95	10.8	12.7	14.5	16.4	18.3	20.2
	18	0.57	1.87	3.44	5.14	6.90	8.69	10.5	12.4	14.2	16.1	18.0	19.9
	20	0.52	1.75	3.28	4.94	6.67	8.45	10.3	12.1	13.9	15.8	17.7	19.5
	24	0.44	1.55	2.98	4.57	6.24	7.98	9.75	11.6	13.4	15.2	17.1	18.9
	28	0.38	1.40	2.74	4.24	5.85	7.54	9.28	11.1	12.9	14.7	16.5	18.3
	32	0.34	1.27	2.52	3.95	5.49	7.13	8.83	10.6	12.4	14.1	16.0	17.8
	36	0.30	1.16	2.33	3.68	5.16	6.75	8.41	10.1	11.9	13.7	15.4	17.3

Table 7-19.
Coefficients *C* for Eccentrically Loaded Bolt Groups
Angle = 0°

$$C_{req} = \frac{P_u}{\phi r_n} \quad \text{or} \quad \phi R_n = C \times \phi r_n$$

where

P_u = factored force, kips
ϕr_n = design strength per bolt, kips
ϕR_n = design strength per bolt group, kips
e = eccentricity of P_u with respect to centroid of bolt group, in. (not tabulated, may be determined by geometry)
e_x = horizontal component of e, in.
s = bolt spacing, in.
C = coefficient tabulated below

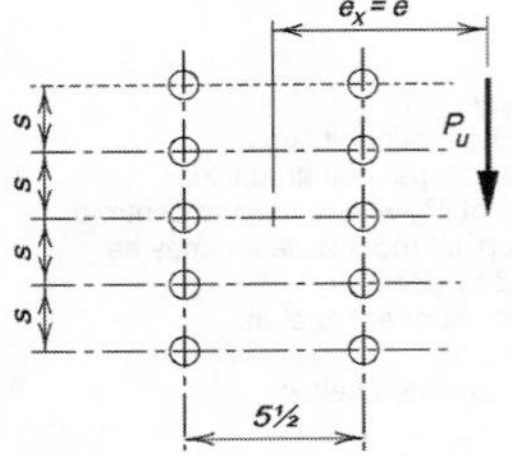

s, in.	e_x, in.	Number of Bolts in One Vertical Row, *n*											
		1	2	3	4	5	6	7	8	9	10	11	12
3	2	1.14	2.75	4.59	6.61	8.69	10.8	12.9	14.9	17.0	19.0	21.0	23.0
	3	0.94	2.32	3.92	5.80	7.82	9.90	12.0	14.1	16.2	18.3	20.4	22.4
	4	0.80	1.99	3.39	5.10	6.98	9.00	11.1	13.2	15.3	17.4	19.6	21.7
	5	0.70	1.74	2.96	4.51	6.24	8.15	10.2	12.3	14.4	16.5	18.6	20.8
	6	0.62	1.54	2.62	4.03	5.60	7.39	9.30	11.3	13.4	15.5	17.7	19.8
	7	0.55	1.38	2.36	3.63	5.07	6.72	8.53	10.5	12.5	14.6	16.7	18.8
	8	0.50	1.25	2.14	3.30	4.61	6.15	7.84	9.67	11.6	13.6	15.7	17.8
	9	0.46	1.14	1.96	3.01	4.22	5.66	7.23	8.97	10.8	12.8	14.8	16.9
	10	0.42	1.04	1.80	2.78	3.89	5.23	6.70	8.34	10.1	12.0	13.9	15.9
	12	0.37	0.90	1.55	2.39	3.36	4.53	5.82	7.28	8.87	10.6	12.4	14.3
	14	0.32	0.79	1.36	2.10	2.96	3.99	5.13	6.44	7.87	9.42	11.1	12.8
	16	0.29	0.70	1.21	1.87	2.64	3.55	4.58	5.76	7.05	8.47	9.99	11.6
	18	0.26	0.63	1.09	1.68	2.37	3.20	4.14	5.21	6.38	7.68	9.08	10.6
	20	0.24	0.57	0.99	1.53	2.16	2.91	3.77	4.75	5.82	7.02	8.30	9.69
	24	0.20	0.48	0.84	1.29	1.83	2.46	3.19	4.03	4.94	5.97	7.07	8.28
	28	0.18	0.42	0.73	1.11	1.58	2.13	2.77	3.49	4.29	5.19	6.15	7.21
	32	0.16	0.37	0.64	0.98	1.39	1.88	2.44	3.08	3.79	4.58	5.44	6.38
	36	0.14	0.33	0.57	0.88	1.24	1.68	2.18	2.75	3.39	4.10	4.87	5.72
6	2	1.14	3.25	5.37	7.45	9.49	11.5	13.5	15.5	17.5	19.5	21.4	23.4
	3	0.94	2.86	4.93	7.05	9.14	11.2	13.2	15.3	17.3	19.3	21.3	23.3
	4	0.80	2.52	4.47	6.59	8.72	10.8	12.9	15.0	17.0	19.0	21.0	23.1
	5	0.70	2.24	4.04	6.12	8.25	10.4	12.5	14.6	16.7	18.7	20.8	22.8
	6	0.62	2.00	3.65	5.66	7.77	9.91	12.1	14.2	16.3	18.4	20.4	22.5
	7	0.55	1.80	3.31	5.23	7.29	9.42	11.6	13.7	15.8	17.9	20.0	22.1
	8	0.50	1.64	3.02	4.84	6.83	8.93	11.1	13.2	15.4	17.5	19.6	21.7
	9	0.46	1.50	2.77	4.49	6.39	8.45	10.6	12.7	14.9	17.0	19.2	21.3
	10	0.42	1.38	2.56	4.18	5.99	7.99	10.1	12.2	14.4	16.5	18.7	20.8
	12	0.37	1.19	2.21	3.65	5.29	7.16	9.15	11.2	13.4	15.5	17.7	19.8
	14	0.32	1.04	1.95	3.24	4.72	6.44	8.32	10.3	12.4	14.5	16.7	18.8
	16	0.29	0.93	1.74	2.90	4.24	5.83	7.59	9.48	11.5	13.6	15.7	17.8
	18	0.26	0.84	1.57	2.62	3.84	5.31	6.95	8.74	10.7	12.6	14.7	16.8
	20	0.24	0.76	1.43	2.39	3.50	4.87	6.39	8.08	9.89	11.8	13.8	15.9
	24	0.20	0.64	1.21	2.02	2.98	4.16	5.49	6.99	8.61	10.4	12.2	14.1
	28	0.18	0.55	1.05	1.76	2.59	3.63	4.80	6.13	7.59	9.18	10.9	12.7
	32	0.16	0.49	0.93	1.55	2.29	3.21	4.25	5.45	6.77	8.21	9.76	11.4
	36	0.14	0.43	0.83	1.38	2.05	2.88	3.81	4.90	6.09	7.41	8.83	10.4

Table 7-19 (cont.).
Coefficients *C* for Eccentrically Loaded Bolt Groups
Angle = 15°

$C_{req} = \dfrac{P_u}{\phi r_n}$ or $\phi R_n = C \times \phi r_n$

where

P_u = factored force, kips
ϕr_n = design strength per bolt, kips
ϕR_n = design strength per bolt group, kips
e = eccentricity of P_u with respect to centroid of bolt group, in. (not tabulated, may be determined by geometry)
e_x = horizontal component of e, in.
s = bolt spacing, in.
C = coefficient tabulated below

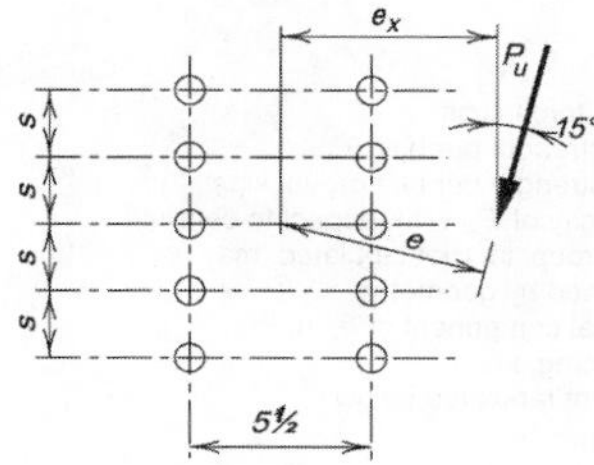

s, in.	e_x, in.	Number of Bolts in One Vertical Row, n											
		1	2	3	4	5	6	7	8	9	10	11	12
3	2	1.18	2.78	4.61	6.59	8.64	10.7	12.8	14.8	16.8	18.9	20.9	22.9
	3	0.97	2.34	3.97	5.80	7.78	9.83	11.9	14.0	16.1	18.1	20.2	22.2
	4	0.83	2.02	3.45	5.11	6.97	8.94	11.0	13.1	15.2	17.3	19.3	21.4
	5	0.72	1.77	3.03	4.54	6.26	8.12	10.1	12.1	14.2	16.3	18.4	20.5
	6	0.64	1.57	2.70	4.06	5.65	7.39	9.27	11.2	13.3	15.4	17.5	19.6
	7	0.57	1.41	2.43	3.66	5.13	6.74	8.52	10.4	12.4	14.4	16.5	18.6
	8	0.52	1.28	2.20	3.34	4.68	6.18	7.86	9.65	11.6	13.5	15.6	17.6
	9	0.48	1.17	2.01	3.06	4.30	5.70	7.27	8.97	10.8	12.7	14.7	16.7
	10	0.44	1.07	1.85	2.82	3.98	5.27	6.76	8.36	10.1	11.9	13.8	15.8
	12	0.38	0.93	1.60	2.44	3.44	4.58	5.90	7.34	8.91	10.6	12.4	14.2
	14	0.33	0.81	1.40	2.15	3.03	4.05	5.22	6.51	7.94	9.47	11.1	12.8
	16	0.30	0.72	1.25	1.91	2.70	3.62	4.68	5.84	7.14	8.54	10.1	11.7
	18	0.27	0.65	1.13	1.72	2.44	3.27	4.23	5.28	6.48	7.77	9.16	10.7
	20	0.25	0.59	1.02	1.57	2.22	2.98	3.86	4.83	5.93	7.11	8.40	9.78
	24	0.21	0.50	0.87	1.33	1.88	2.53	3.27	4.11	5.05	6.07	7.19	8.39
	28	0.18	0.43	0.75	1.15	1.63	2.19	2.84	3.57	4.39	5.29	6.28	7.33
	32	0.16	0.38	0.66	1.01	1.43	1.93	2.50	3.15	3.88	4.68	5.56	6.50
	36	0.14	0.34	0.59	0.90	1.28	1.73	2.24	2.82	3.48	4.19	4.99	5.84
6	2	1.18	3.24	5.34	7.40	9.43	11.5	13.5	15.4	17.4	19.4	21.4	23.4
	3	0.97	2.85	4.90	6.99	9.07	11.1	13.2	15.2	17.2	19.2	21.2	23.2
	4	0.83	2.51	4.45	6.53	8.63	10.7	12.8	14.8	16.9	18.9	20.9	22.9
	5	0.72	2.23	4.05	6.07	8.16	10.3	12.4	14.5	16.5	18.6	20.6	22.6
	6	0.64	2.00	3.68	5.62	7.69	9.80	11.9	14.0	16.1	18.2	20.2	22.3
	7	0.57	1.81	3.36	5.20	7.22	9.31	11.4	13.5	15.7	17.7	19.8	21.9
	8	0.52	1.65	3.08	4.82	6.78	8.83	10.9	13.1	15.2	17.3	19.4	21.5
	9	0.48	1.52	2.83	4.48	6.36	8.37	10.5	12.6	14.7	16.8	18.9	21.0
	10	0.44	1.40	2.62	4.18	5.98	7.93	9.97	12.1	14.2	16.3	18.4	20.6
	12	0.38	1.21	2.27	3.66	5.31	7.13	9.08	11.1	13.2	15.3	17.4	19.6
	14	0.33	1.07	2.00	3.25	4.76	6.44	8.28	10.2	12.3	14.3	16.4	18.6
	16	0.30	0.95	1.79	2.92	4.29	5.85	7.58	9.43	11.4	13.4	15.5	17.6
	18	0.27	0.86	1.62	2.65	3.90	5.34	6.97	8.72	10.6	12.5	14.6	16.6
	20	0.25	0.78	1.47	2.42	3.58	4.91	6.43	8.09	9.87	11.7	13.7	15.7
	24	0.21	0.66	1.25	2.06	3.05	4.21	5.55	7.03	8.64	10.4	12.2	14.1
	28	0.18	0.57	1.08	1.79	2.66	3.68	4.87	6.19	7.65	9.22	10.9	12.6
	32	0.16	0.50	0.95	1.58	2.35	3.26	4.33	5.52	6.84	8.27	9.81	11.4
	36	0.14	0.45	0.85	1.42	2.11	2.93	3.90	4.97	6.18	7.49	8.91	10.4

Table 7-19 (cont.). Coefficients C for Eccentrically Loaded Bolt Groups Angle = 30°

$$C_{req} = \frac{P_u}{\phi r_n} \quad \text{or} \quad \phi R_n = C \times \phi r_n$$

where

P_u = factored force, kips
ϕr_n = design strength per bolt, kips
ϕR_n = design strength per bolt group, kips
e = eccentricity of P_u with respect to centroid of bolt group, in. (not tabulated, may be determined by geometry)
e_x = horizontal component of e, in.
s = bolt spacing, in.
C = coefficient tabulated below

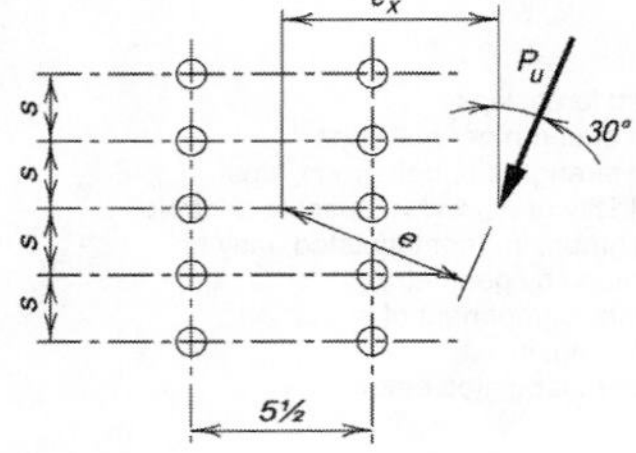

s, in.	e_x, in.	Number of Bolts in One Vertical Row, n											
		1	2	3	4	5	6	7	8	9	10	11	12
3	2	1.30	2.90	4.72	6.66	8.65	10.7	12.7	14.7	16.7	18.7	20.8	22.8
	3	1.08	2.47	4.13	5.94	7.86	9.85	11.9	13.9	16.0	18.0	20.0	22.1
	4	0.92	2.14	3.64	5.30	7.12	9.04	11.0	13.0	15.1	17.1	19.2	21.2
	5	0.80	1.89	3.24	4.76	6.46	8.29	10.2	12.2	14.2	16.3	18.3	20.4
	6	0.71	1.69	2.91	4.29	5.88	7.61	9.45	11.4	13.4	15.4	17.4	19.5
	7	0.64	1.53	2.63	3.90	5.38	7.01	8.76	10.6	12.5	14.5	16.5	18.6
	8	0.58	1.39	2.40	3.57	4.95	6.49	8.14	9.92	11.8	13.7	15.7	17.7
	9	0.53	1.28	2.20	3.29	4.58	6.02	7.59	9.29	11.1	12.9	14.9	16.8
	10	0.49	1.18	2.03	3.04	4.26	5.61	7.09	8.72	10.4	12.2	14.1	16.0
	12	0.42	1.02	1.76	2.65	3.72	4.92	6.25	7.73	9.31	11.0	12.8	14.6
	14	0.37	0.90	1.55	2.34	3.29	4.37	5.58	6.93	8.38	9.93	11.6	13.3
	16	0.33	0.80	1.38	2.09	2.95	3.92	5.03	6.26	7.59	9.03	10.6	12.2
	18	0.30	0.72	1.25	1.89	2.67	3.55	4.57	5.70	6.93	8.27	9.70	11.2
	20	0.27	0.66	1.13	1.73	2.43	3.25	4.19	5.23	6.36	7.62	8.95	10.4
	24	0.23	0.56	0.96	1.46	2.07	2.77	3.57	4.47	5.47	6.56	7.73	8.99
	28	0.20	0.48	0.83	1.27	1.79	2.41	3.11	3.90	4.78	5.75	6.78	7.91
	32	0.18	0.43	0.73	1.12	1.58	2.13	2.76	3.46	4.25	5.11	6.04	7.06
	36	0.16	0.38	0.66	1.00	1.42	1.91	2.47	3.10	3.81	4.59	5.44	6.36
6	2	1.30	3.27	5.33	7.36	9.38	11.4	13.4	15.4	17.4	19.3	21.3	23.3
	3	1.08	2.89	4.91	6.96	9.01	11.0	13.1	15.1	17.1	19.1	21.1	23.1
	4	0.92	2.56	4.50	6.53	8.58	10.6	12.7	14.7	16.8	18.8	20.8	22.8
	5	0.80	2.29	4.13	6.10	8.14	10.2	12.3	14.3	16.4	18.4	20.4	22.5
	6	0.71	2.08	3.80	5.69	7.70	9.75	11.8	13.9	15.9	18.0	20.0	22.1
	7	0.64	1.89	3.51	5.31	7.27	9.30	11.4	13.4	15.5	17.6	19.6	21.7
	8	0.58	1.74	3.25	4.96	6.86	8.86	10.9	13.0	15.0	17.1	19.2	21.3
	9	0.53	1.61	3.02	4.64	6.49	8.44	10.5	12.5	14.6	16.7	18.7	20.8
	10	0.49	1.49	2.81	4.35	6.13	8.04	10.0	12.1	14.1	16.2	18.3	20.4
	12	0.42	1.30	2.47	3.85	5.51	7.31	9.22	11.2	13.2	15.3	17.3	19.4
	14	0.37	1.15	2.19	3.44	4.98	6.67	8.49	10.4	12.4	14.4	16.4	18.5
	16	0.33	1.03	1.96	3.11	4.54	6.12	7.83	9.66	11.6	13.5	15.6	17.6
	18	0.30	0.93	1.78	2.83	4.16	5.63	7.26	9.00	10.8	12.8	14.7	16.7
	20	0.27	0.85	1.62	2.60	3.83	5.21	6.74	8.41	10.2	12.0	13.9	15.9
	24	0.23	0.72	1.38	2.23	3.30	4.51	5.89	7.40	9.02	10.7	12.5	14.4
	28	0.20	0.63	1.20	1.95	2.89	3.96	5.21	6.59	8.07	9.66	11.3	13.1
	32	0.18	0.55	1.06	1.73	2.57	3.53	4.67	5.92	7.28	8.75	10.3	12.0
	36	0.16	0.50	0.95	1.55	2.31	3.18	4.22	5.36	6.61	7.98	9.43	11.0

Table 7-19 (cont.). Coefficients *C* for Eccentrically Loaded Bolt Groups Angle = 45°

$C_{req} = \frac{P_u}{\phi r_n}$ or $\phi R_n = C \times \phi r_n$

where

P_u = factored force, kips
ϕr_n = design strength per bolt, kips
ϕR_n = design strength per bolt group, kips
e = eccentricity of P_u with respect to centroid of bolt group, in. (not tabulated, may be determined by geometry)
e_x = horizontal component of e, in.
s = bolt spacing, in.
C = coefficient tabulated below

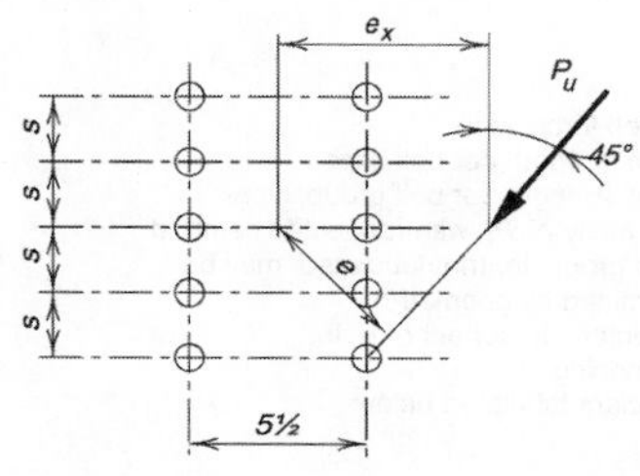

s, in.	e_x, in.	Number of Bolts in One Vertical Row, *n*											
		1	2	3	4	5	6	7	8	9	10	11	12
	2	1.53	3.18	4.96	6.84	8.77	10.7	12.7	14.7	16.7	18.7	20.7	22.6
	3	1.30	2.76	4.42	6.22	8.09	10.0	12.0	14.0	15.9	17.9	19.9	21.9
	4	1.11	2.43	3.97	5.67	7.46	9.32	11.2	13.2	15.2	17.2	19.2	21.2
	5	0.98	2.17	3.60	5.19	6.89	8.68	10.6	12.5	14.4	16.4	18.4	20.4
	6	0.87	1.95	3.28	4.77	6.37	8.09	9.90	11.8	13.7	15.6	17.6	19.6
	7	0.78	1.78	3.01	4.40	5.91	7.56	9.31	11.1	13.0	14.9	16.9	18.8
	8	0.71	1.63	2.77	4.07	5.50	7.07	8.76	10.5	12.4	14.2	16.2	18.1
	9	0.65	1.50	2.57	3.78	5.13	6.64	8.26	9.97	11.8	13.6	15.5	17.4
	10	0.60	1.39	2.39	3.52	4.81	6.25	7.81	9.45	11.2	13.0	14.8	16.7
3	12	0.52	1.22	2.08	3.09	4.26	5.58	7.01	8.54	10.2	11.9	13.6	15.4
	14	0.45	1.08	1.85	2.75	3.82	5.02	6.34	7.76	9.28	10.9	12.6	14.3
	16	0.41	0.96	1.65	2.48	3.45	4.55	5.77	7.09	8.53	10.1	11.6	13.3
	18	0.37	0.87	1.50	2.25	3.14	4.16	5.29	6.53	7.87	9.30	10.8	12.4
	20	0.33	0.79	1.37	2.06	2.88	3.82	4.87	6.04	7.30	8.65	10.1	11.6
	24	0.28	0.68	1.16	1.76	2.47	3.28	4.21	5.23	6.35	7.55	8.85	10.2
	28	0.25	0.59	1.01	1.53	2.15	2.87	3.69	4.61	5.61	6.69	7.87	9.11
	32	0.22	0.52	0.89	1.35	1.91	2.55	3.29	4.11	5.01	6.00	7.07	8.20
	36	0.20	0.46	0.80	1.21	1.71	2.29	2.96	3.70	4.53	5.43	6.40	7.44
	2	1.53	3.39	5.36	7.35	9.35	11.3	13.3	15.3	17.3	19.3	21.3	23.2
	3	1.30	3.04	4.99	6.98	8.98	11.0	13.0	15.0	17.0	19.0	21.0	22.9
	4	1.11	2.74	4.64	6.60	8.60	10.6	12.6	14.6	16.6	18.6	20.6	22.6
	5	0.98	2.49	4.31	6.24	8.21	10.2	12.2	14.2	16.3	18.3	20.3	22.3
	6	0.87	2.28	4.02	5.89	7.84	9.82	11.8	13.8	15.9	17.9	19.9	21.9
	7	0.78	2.10	3.76	5.57	7.48	9.44	11.4	13.4	15.5	17.5	19.5	21.5
	8	0.71	1.94	3.53	5.28	7.13	9.07	11.0	13.0	15.1	17.1	19.1	21.1
	9	0.65	1.81	3.32	5.00	6.81	8.71	10.7	12.7	14.7	16.7	18.7	20.7
	10	0.60	1.69	3.13	4.74	6.50	8.37	10.3	12.3	14.3	16.3	18.3	20.3
6	12	0.52	1.50	2.80	4.29	5.94	7.74	9.61	11.5	13.5	15.5	17.5	19.5
	14	0.45	1.34	2.52	3.89	5.45	7.17	8.98	10.9	12.8	14.7	16.7	18.7
	16	0.41	1.21	2.29	3.55	5.02	6.67	8.41	10.2	12.1	14.0	16.0	17.9
	18	0.37	1.10	2.09	3.26	4.65	6.22	7.89	9.65	11.5	13.4	15.3	17.2
	20	0.33	1.01	1.92	3.01	4.33	5.82	7.42	9.11	10.9	12.7	14.6	16.5
	24	0.28	0.86	1.64	2.61	3.79	5.13	6.60	8.17	9.84	11.6	13.4	15.2
	28	0.25	0.75	1.44	2.30	3.36	4.58	5.92	7.38	8.95	10.6	12.3	14.1
	32	0.22	0.67	1.27	2.05	3.02	4.12	5.35	6.72	8.18	9.73	11.4	13.0
	36	0.20	0.60	1.14	1.85	2.73	3.74	4.88	6.15	7.52	8.98	10.5	12.1

Table 7-19 (cont.). Coefficients *C* for Eccentrically Loaded Bolt Groups Angle = 60°

$$C_{req} = \frac{P_u}{\phi r_n} \quad \text{or} \quad \phi R_n = C \times \phi r_n$$

where

P_u = factored force, kips
ϕr_n = design strength per bolt, kips
ϕR_n = design strength per bolt group, kips
e = eccentricity of P_u with respect to centroid of bolt group, in. (not tabulated, may be determined by geometry)
e_x = horizontal component of e, in.
s = bolt spacing, in.
C = coefficient tabulated below

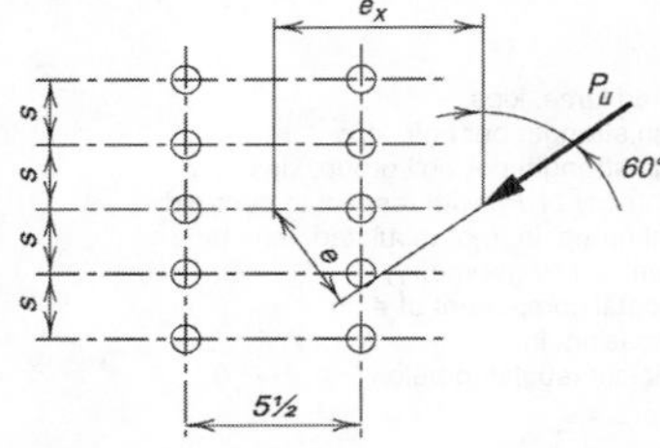

s, in.	e_x, in.	Number of Bolts in One Vertical Row, *n*											
		1	2	3	4	5	6	7	8	9	10	11	12
	2	1.78	3.55	5.34	7.17	9.04	10.9	12.9	14.8	16.7	18.7	20.6	22.6
	3	1.62	3.26	4.95	6.71	8.53	10.4	12.3	14.2	16.1	18.1	20.0	22.0
	4	1.45	2.97	4.57	6.27	8.04	9.86	11.7	13.6	15.5	17.5	19.4	21.4
	5	1.31	2.71	4.23	5.86	7.58	9.36	11.2	13.1	15.0	16.9	18.8	20.7
	6	1.18	2.48	3.93	5.50	7.16	8.90	10.7	12.5	14.4	16.3	18.2	20.1
	7	1.07	2.28	3.66	5.18	6.79	8.48	10.2	12.0	13.9	15.7	17.6	19.5
	8	0.98	2.11	3.43	4.88	6.45	8.09	9.80	11.6	13.4	15.2	17.1	19.0
	9	0.90	1.97	3.22	4.61	6.12	7.72	9.39	11.1	12.9	14.7	16.6	18.4
	10	0.83	1.84	3.03	4.37	5.82	7.37	9.00	10.7	12.5	14.2	16.1	17.9
3	12	0.72	1.62	2.70	3.93	5.28	6.73	8.28	9.91	11.6	13.4	15.1	16.9
	14	0.64	1.45	2.43	3.56	4.81	6.19	7.66	9.22	10.9	12.5	14.3	16.0
	16	0.57	1.31	2.21	3.24	4.42	5.71	7.11	8.60	10.2	11.8	13.5	15.2
	18	0.52	1.19	2.02	2.98	4.07	5.29	6.63	8.05	9.55	11.1	12.7	14.4
	20	0.47	1.09	1.85	2.75	3.77	4.93	6.19	7.55	8.98	10.5	12.1	13.7
	24	0.40	0.93	1.59	2.37	3.28	4.32	5.46	6.69	8.01	9.41	10.9	12.4
	28	0.35	0.82	1.39	2.08	2.90	3.83	4.86	5.99	7.21	8.51	9.88	11.3
	32	0.31	0.72	1.24	1.86	2.59	3.43	4.37	5.41	6.54	7.75	9.02	10.4
	36	0.28	0.65	1.11	1.67	2.34	3.11	3.97	4.93	5.98	7.10	8.29	9.55
	2	1.78	3.59	5.48	7.41	9.36	11.3	13.3	15.3	17.2	19.2	21.2	23.2
	3	1.62	3.35	5.20	7.12	9.06	11.0	13.0	15.0	16.9	18.9	20.9	22.9
	4	1.45	3.11	4.93	6.82	8.75	10.7	12.7	14.6	16.6	18.6	20.6	22.5
	5	1.31	2.89	4.66	6.53	8.45	10.4	12.3	14.3	16.3	18.2	20.2	22.2
	6	1.18	2.70	4.42	6.26	8.16	10.1	12.0	14.0	15.9	17.9	19.9	21.9
	7	1.07	2.52	4.19	6.01	7.88	9.79	11.7	13.7	15.6	17.6	19.6	21.5
	8	0.98	2.36	3.99	5.77	7.62	9.51	11.4	13.4	15.3	17.3	19.2	21.2
	9	0.90	2.23	3.81	5.55	7.37	9.24	11.1	13.1	15.0	17.0	18.9	20.9
	10	0.83	2.10	3.64	5.35	7.13	8.98	10.9	12.8	14.7	16.7	18.6	20.6
6	12	0.72	1.89	3.34	4.97	6.70	8.49	10.3	12.2	14.1	16.1	18.0	19.9
	14	0.64	1.71	3.08	4.63	6.29	8.04	9.85	11.7	13.6	15.5	17.4	19.3
	16	0.57	1.57	2.85	4.32	5.92	7.62	9.39	11.2	13.1	15.0	16.9	18.8
	18	0.52	1.44	2.65	4.04	5.58	7.22	8.95	10.7	12.6	14.4	16.3	18.2
	20	0.47	1.33	2.47	3.79	5.26	6.86	8.55	10.3	12.1	13.9	15.8	17.7
	24	0.40	1.16	2.17	3.36	4.71	6.21	7.82	9.50	11.2	13.0	14.8	16.7
	28	0.35	1.02	1.92	3.00	4.26	5.67	7.19	8.80	10.5	12.2	14.0	15.8
	32	0.31	0.91	1.72	2.71	3.88	5.20	6.64	8.17	9.77	11.4	13.1	14.9
	36	0.28	0.82	1.56	2.46	3.55	4.80	6.16	7.61	9.14	10.7	12.4	14.1

Table 7-19 (cont.).
Coefficients C for Eccentrically Loaded Bolt Groups
Angle = 75°

$C_{req} = \frac{P_u}{\phi r_n}$ or $\phi R_n = C \times \phi r_n$

where

P_u = factored force, kips
ϕr_n = design strength per bolt, kips
ϕR_n = design strength per bolt group, kips
e = eccentricity of P_u with respect to centroid of bolt group, in. (not tabulated, may be determined by geometry)
e_x = horizontal component of e, in.
s = bolt spacing, in.
C = coefficient tabulated below

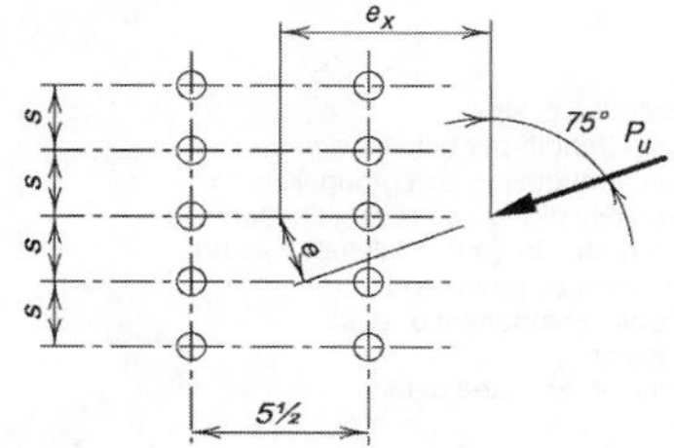

s, in.	e_x, in.	Number of Bolts in One Vertical Row, n											
		1	2	3	4	5	6	7	8	9	10	11	12
3	2	1.92	3.82	5.70	7.57	9.45	11.3	13.2	15.2	17.1	19.0	20.9	22.9
	3	1.87	3.72	5.54	7.36	9.19	11.1	12.9	14.8	16.7	18.6	20.5	22.5
	4	1.82	3.60	5.37	7.14	8.94	10.8	12.6	14.5	16.3	18.2	20.1	22.1
	5	1.75	3.47	5.18	6.92	8.68	10.5	12.3	14.1	16.0	17.9	19.8	21.7
	6	1.68	3.33	5.00	6.69	8.42	10.2	12.0	13.8	15.7	17.5	19.4	21.3
	7	1.60	3.19	4.81	6.47	8.17	9.92	11.7	13.5	15.3	17.2	19.1	20.9
	8	1.52	3.06	4.63	6.26	7.93	9.66	11.4	13.2	15.0	16.9	18.7	20.6
	9	1.45	2.93	4.46	6.05	7.70	9.41	11.2	12.9	14.7	16.5	18.4	20.3
	10	1.38	2.80	4.29	5.85	7.48	9.16	10.9	12.6	14.4	16.2	18.1	19.9
	12	1.25	2.57	3.98	5.48	7.07	8.71	10.4	12.1	13.9	15.7	17.5	19.3
	14	1.13	2.36	3.70	5.15	6.69	8.29	9.96	11.7	13.4	15.2	16.9	18.7
	16	1.03	2.18	3.45	4.85	6.34	7.90	9.53	11.2	12.9	14.7	16.4	18.2
	18	0.95	2.02	3.23	4.57	6.01	7.54	9.13	10.8	12.5	14.2	15.9	17.7
	20	0.87	1.88	3.03	4.32	5.71	7.19	8.75	10.4	12.0	13.7	15.4	17.2
	24	0.75	1.65	2.69	3.87	5.17	6.57	8.05	9.60	11.2	12.9	14.5	16.2
	28	0.66	1.46	2.42	3.50	4.71	6.03	7.44	8.93	10.5	12.1	13.7	15.4
	32	0.59	1.31	2.18	3.19	4.32	5.56	6.90	8.32	9.81	11.4	12.9	14.6
	36	0.53	1.19	1.99	2.92	3.98	5.15	6.42	7.78	9.21	10.7	12.2	13.8
6	2	1.92	3.80	5.69	7.59	9.51	11.5	13.4	15.4	17.6	19.6	21.5	23.5
	3	1.87	3.70	5.55	7.42	9.32	11.2	13.2	15.1	17.1	19.0	21.0	23.0
	4	1.82	3.59	5.40	7.25	9.14	11.1	13.0	14.9	16.9	18.8	20.8	22.7
	5	1.75	3.48	5.26	7.09	8.96	10.9	12.8	14.7	16.6	18.6	20.5	22.5
	6	1.68	3.36	5.11	6.93	8.78	10.7	12.6	14.5	16.4	18.4	20.3	22.2
	7	1.60	3.24	4.97	6.77	8.62	10.5	12.4	14.3	16.2	18.1	20.1	22.0
	8	1.52	3.13	4.84	6.62	8.45	10.3	12.2	14.1	16.0	17.9	19.9	21.8
	9	1.45	3.02	4.71	6.47	8.29	10.2	12.0	13.9	15.8	17.7	19.7	21.6
	10	1.38	2.91	4.58	6.33	8.14	9.98	11.9	13.7	15.6	17.6	19.5	21.4
	12	1.25	2.72	4.34	6.07	7.85	9.67	11.5	13.4	15.3	17.2	19.1	21.0
	14	1.13	2.54	4.13	5.82	7.57	9.38	11.2	13.1	15.0	16.8	18.7	20.6
	16	1.03	2.38	3.92	5.59	7.32	9.10	10.9	12.8	14.6	16.5	18.4	20.3
	18	0.95	2.24	3.74	5.38	7.09	8.85	10.7	12.5	14.3	16.2	18.1	19.9
	20	0.87	2.11	3.57	5.17	6.87	8.61	10.4	12.2	14.0	15.9	17.7	19.6
	24	0.75	1.88	3.27	4.80	6.44	8.15	9.90	11.7	13.5	15.3	17.1	19.0
	28	0.66	1.70	3.00	4.47	6.06	7.72	9.43	11.2	13.0	14.8	16.6	18.4
	32	0.59	1.55	2.77	4.17	5.70	7.31	8.99	10.7	12.5	14.3	16.1	17.9
	36	0.53	1.42	2.57	3.90	5.37	6.93	8.57	10.3	12.0	13.8	15.5	17.3

Table 7-20.
Coefficients C for Eccentrically Loaded Bolt Groups
Angle = 0°

$$C_{req} = \frac{P_u}{\phi r_n} \quad \text{or} \quad \phi R_n = C \times \phi r_n$$

where

P_u = factored force, kips
ϕr_n = design strength per bolt, kips
ϕR_n = design strength per bolt group, kips
e = eccentricity of P_u with respect to centroid of bolt group, in. (not tabulated, may be determined by geometry)
e_x = horizontal component of e, in.
s = bolt spacing, in.
C = coefficient tabulated below

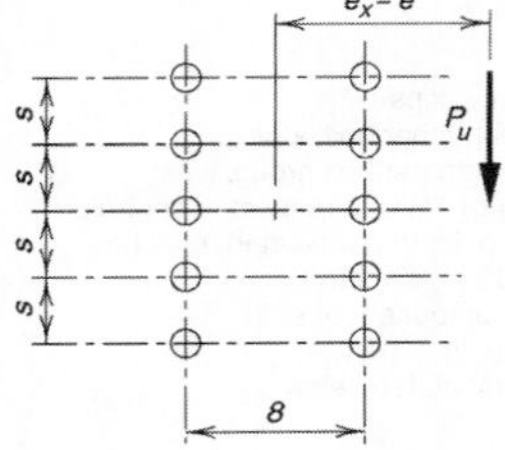

s, in.	e_x, in.	Number of Bolts in One Vertical Row, n											
		1	2	3	4	5	6	7	8	9	10	11	12
	2	1.31	2.91	4.71	6.66	8.69	10.8	12.8	14.9	16.9	18.9	21.0	23.0
	3	1.12	2.54	4.14	5.95	7.90	9.93	12.0	14.1	16.2	18.2	20.3	22.4
	4	0.98	2.24	3.66	5.33	7.15	9.10	11.1	13.2	15.3	17.4	19.5	21.6
	5	0.87	1.99	3.27	4.80	6.48	8.33	10.3	12.3	14.4	16.5	18.6	20.7
	6	0.79	1.80	2.95	4.35	5.90	7.63	9.49	11.5	13.5	15.6	17.7	19.8
	7	0.71	1.63	2.68	3.97	5.40	7.02	8.77	10.7	12.6	14.6	16.7	18.8
	8	0.65	1.49	2.46	3.65	4.97	6.48	8.13	9.91	11.8	13.8	15.8	17.9
	9	0.60	1.38	2.27	3.37	4.59	6.01	7.55	9.24	11.1	13.0	14.9	17.0
	10	0.56	1.28	2.11	3.13	4.27	5.59	7.04	8.64	10.4	12.2	14.1	16.1
3	12	0.49	1.11	1.84	2.73	3.73	4.90	6.19	7.63	9.18	10.9	12.6	14.5
	14	0.44	0.99	1.64	2.42	3.31	4.36	5.50	6.80	8.20	9.73	11.4	13.1
	16	0.39	0.89	1.47	2.17	2.98	3.91	4.95	6.13	7.40	8.80	10.3	11.9
	18	0.36	0.80	1.33	1.97	2.70	3.55	4.50	5.57	6.73	8.02	9.39	10.9
	20	0.33	0.73	1.22	1.80	2.47	3.25	4.12	5.10	6.17	7.35	8.62	9.99
	24	0.28	0.63	1.04	1.53	2.10	2.77	3.51	4.35	5.28	6.30	7.39	8.59
	28	0.25	0.55	0.91	1.33	1.83	2.41	3.06	3.79	4.60	5.50	6.46	7.51
	32	0.22	0.48	0.80	1.18	1.62	2.13	2.71	3.36	4.08	4.87	5.73	6.67
	36	0.20	0.43	0.72	1.06	1.45	1.91	2.43	3.01	3.66	4.37	5.15	5.99
	2	1.31	3.28	5.35	7.42	9.47	11.5	13.5	15.5	17.5	19.5	21.4	23.4
	3	1.12	2.93	4.94	7.03	9.12	11.2	13.2	15.3	17.3	19.3	21.3	23.3
	4	0.98	2.63	4.52	6.59	8.70	10.8	12.9	14.9	17.0	19.0	21.0	23.0
	5	0.87	2.37	4.13	6.15	8.25	10.4	12.5	14.6	16.6	18.7	20.7	22.8
	6	0.79	2.15	3.78	5.72	7.78	9.90	12.0	14.1	16.2	18.3	20.4	22.4
	7	0.71	1.97	3.47	5.32	7.33	9.43	11.6	13.7	15.8	17.9	20.0	22.1
	8	0.65	1.81	3.19	4.95	6.89	8.95	11.1	13.2	15.4	17.5	19.6	21.7
	9	0.60	1.67	2.95	4.62	6.48	8.49	10.6	12.7	14.9	17.0	19.1	21.3
	10	0.56	1.55	2.75	4.33	6.10	8.05	10.1	12.2	14.4	16.5	18.7	20.8
6	12	0.49	1.35	2.40	3.82	5.43	7.25	9.21	11.3	13.4	15.5	17.7	19.8
	14	0.44	1.20	2.14	3.41	4.86	6.56	8.40	10.4	12.4	14.5	16.7	18.8
	16	0.39	1.08	1.92	3.07	4.40	5.96	7.69	9.56	11.5	13.6	15.7	17.8
	18	0.36	0.97	1.75	2.79	4.00	5.46	7.06	8.83	10.7	12.7	14.7	16.8
	20	0.33	0.89	1.60	2.56	3.67	5.02	6.52	8.18	9.97	11.9	13.9	15.9
	24	0.28	0.76	1.37	2.18	3.14	4.32	5.62	7.11	8.71	10.4	12.3	14.2
	28	0.25	0.66	1.19	1.90	2.75	3.78	4.93	6.26	7.70	9.27	11.0	12.7
	32	0.22	0.58	1.05	1.68	2.44	3.35	4.38	5.58	6.88	8.31	9.85	11.5
	36	0.20	0.52	0.95	1.51	2.19	3.01	3.94	5.02	6.21	7.52	8.93	10.4

Table 7-20 (cont.). Coefficients C for Eccentrically Loaded Bolt Groups Angle = 15°

$$C_{req} = \frac{P_u}{\phi r_n} \quad \text{or} \quad \phi R_n = C \times \phi r_n$$

where

P_u = factored force, kips
ϕr_n = design strength per bolt, kips
ϕR_n = design strength per bolt group, kips
e = eccentricity of P_u with respect to centroid of bolt group, in. (not tabulated, may be determined by geometry)
e_x = horizontal component of e, in.
s = bolt spacing, in.
C = coefficient tabulated below

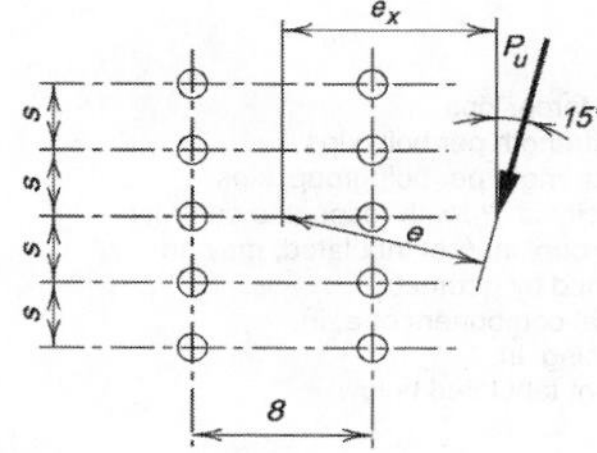

s, in.	e_x, in.	Number of Bolts in One Vertical Row, n											
		1	2	3	4	5	6	7	8	9	10	11	12
3	2	1.35	2.96	4.75	6.67	8.67	10.7	12.7	14.8	16.8	18.8	20.9	22.9
	3	1.16	2.58	4.20	5.98	7.90	9.89	11.9	14.0	16.0	18.1	20.2	22.2
	4	1.02	2.28	3.73	5.37	7.17	9.08	11.1	13.1	15.2	17.3	19.3	21.4
	5	0.90	2.03	3.35	4.85	6.53	8.34	10.3	12.2	14.3	16.3	18.4	20.5
	6	0.81	1.84	3.03	4.40	5.96	7.66	9.48	11.4	13.4	15.4	17.5	19.6
	7	0.74	1.67	2.76	4.02	5.48	7.06	8.79	10.6	12.6	14.5	16.6	18.6
	8	0.68	1.53	2.53	3.70	5.05	6.53	8.17	9.91	11.8	13.7	15.7	17.7
	9	0.63	1.42	2.34	3.43	4.68	6.07	7.61	9.27	11.0	12.9	14.8	16.8
	10	0.58	1.31	2.17	3.19	4.36	5.66	7.12	8.69	10.4	12.2	14.0	16.0
	12	0.51	1.15	1.90	2.79	3.82	4.97	6.28	7.69	9.23	10.9	12.6	14.4
	14	0.45	1.02	1.69	2.48	3.40	4.43	5.61	6.88	8.29	9.79	11.4	13.1
	16	0.41	0.91	1.51	2.23	3.05	3.99	5.05	6.21	7.50	8.88	10.4	11.9
	18	0.37	0.83	1.37	2.02	2.77	3.63	4.60	5.66	6.84	8.11	9.48	11.0
	20	0.34	0.76	1.26	1.85	2.54	3.32	4.21	5.19	6.28	7.45	8.73	10.1
	24	0.29	0.65	1.07	1.58	2.16	2.84	3.60	4.45	5.39	6.40	7.52	8.71
	28	0.25	0.56	0.93	1.37	1.89	2.47	3.14	3.88	4.71	5.61	6.59	7.64
	32	0.23	0.50	0.83	1.22	1.67	2.19	2.78	3.44	4.18	4.98	5.86	6.80
	36	0.20	0.45	0.74	1.09	1.50	1.96	2.49	3.09	3.75	4.47	5.27	6.12
6	2	1.35	3.29	5.33	7.39	9.42	11.4	13.4	15.4	17.4	19.4	21.4	23.4
	3	1.16	2.94	4.93	6.99	9.05	11.1	13.1	15.2	17.2	19.2	21.2	23.2
	4	1.02	2.64	4.52	6.55	8.63	10.7	12.8	14.8	16.9	18.9	20.9	22.9
	5	0.90	2.38	4.15	6.12	8.18	10.3	12.4	14.4	16.5	18.5	20.6	22.6
	6	0.81	2.17	3.82	5.70	7.72	9.80	11.9	14.0	16.1	18.2	20.2	22.3
	7	0.74	1.99	3.52	5.31	7.28	9.33	11.4	13.5	15.6	17.7	19.8	21.9
	8	0.68	1.83	3.25	4.95	6.86	8.87	11.0	13.1	15.2	17.3	19.4	21.5
	9	0.63	1.69	3.02	4.63	6.46	8.43	10.5	12.6	14.7	16.8	18.9	21.0
	10	0.58	1.58	2.81	4.34	6.10	8.00	10.0	12.1	14.2	16.3	18.4	20.5
	12	0.51	1.38	2.47	3.84	5.45	7.23	9.15	11.2	13.2	15.3	17.4	19.6
	14	0.45	1.23	2.20	3.44	4.91	6.56	8.38	10.3	12.3	14.4	16.5	18.6
	16	0.41	1.10	1.98	3.11	4.46	5.99	7.69	9.52	11.5	13.5	15.5	17.6
	18	0.37	1.00	1.80	2.83	4.08	5.49	7.09	8.82	10.7	12.6	14.6	16.6
	20	0.34	0.92	1.65	2.60	3.75	5.06	6.56	8.20	9.96	11.8	13.8	15.7
	24	0.29	0.78	1.41	2.23	3.22	4.36	5.70	7.15	8.74	10.4	12.2	14.1
	28	0.25	0.68	1.23	1.95	2.82	3.83	5.02	6.32	7.76	9.31	11.0	12.7
	32	0.23	0.60	1.09	1.73	2.50	3.41	4.47	5.64	6.96	8.38	9.90	11.5
	36	0.20	0.54	0.97	1.55	2.25	3.07	4.03	5.09	6.30	7.60	9.01	10.5

Table 7-20 (cont.). Coefficients *C* for Eccentrically Loaded Bolt Groups Angle = 30°

$$C_{req} = \frac{P_u}{\phi r_n} \quad \text{or} \quad \phi R_n = C \times \phi r_n$$

where

P_u = factored force, kips
ϕr_n = design strength per bolt, kips
ϕR_n = design strength per bolt group, kips
e = eccentricity of P_u with respect to centroid of bolt group, in. (not tabulated, may be determined by geometry)
e_x = horizontal component of e, in.
s = bolt spacing, in.
C = coefficient tabulated below

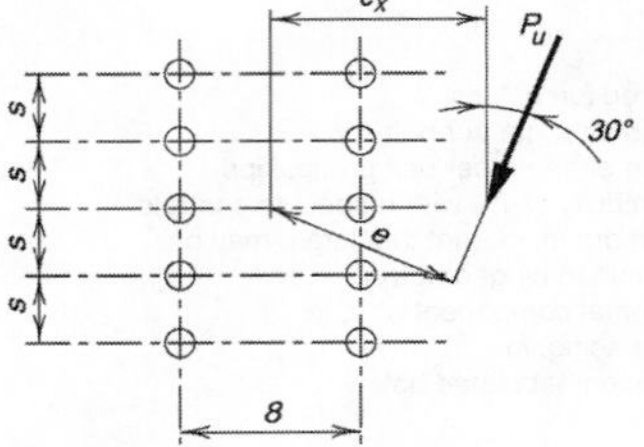

s, in.	e_x, in.	Number of Bolts in One Vertical Row, *n*											
		1	2	3	4	5	6	7	8	9	10	11	12
	2	1.49	3.12	4.91	6.80	8.75	10.7	12.7	14.7	16.7	18.7	20.8	22.7
	3	1.29	2.74	4.39	6.16	8.04	9.98	12.0	14.0	16.0	18.0	20.0	22.1
	4	1.13	2.43	3.95	5.60	7.37	9.24	11.2	13.2	15.2	17.2	19.2	21.3
	5	1.00	2.18	3.58	5.10	6.77	8.55	10.4	12.4	14.3	16.3	18.4	20.4
	6	0.90	1.98	3.26	4.67	6.23	7.93	9.72	11.6	13.5	15.5	17.5	19.5
	7	0.82	1.81	2.99	4.30	5.76	7.37	9.08	10.9	12.8	14.7	16.7	18.7
	8	0.75	1.67	2.76	3.97	5.35	6.87	8.49	10.2	12.0	13.9	15.9	17.8
	9	0.70	1.55	2.56	3.69	4.98	6.42	7.96	9.62	11.4	13.2	15.1	17.0
	10	0.65	1.44	2.38	3.44	4.66	6.02	7.49	9.07	10.8	12.5	14.4	16.2
3	12	0.57	1.26	2.09	3.03	4.13	5.34	6.66	8.12	9.67	11.3	13.0	14.8
	14	0.50	1.12	1.86	2.71	3.69	4.78	5.99	7.33	8.75	10.3	11.9	13.6
	16	0.45	1.01	1.67	2.44	3.33	4.33	5.44	6.66	7.98	9.39	10.9	12.5
	18	0.41	0.92	1.52	2.22	3.03	3.95	4.97	6.10	7.32	8.64	10.1	11.5
	20	0.38	0.84	1.39	2.03	2.78	3.62	4.57	5.62	6.75	7.98	9.30	10.7
	24	0.32	0.72	1.19	1.74	2.38	3.11	3.93	4.84	5.83	6.92	8.08	9.32
	28	0.28	0.63	1.04	1.52	2.08	2.72	3.44	4.24	5.13	6.09	7.12	8.24
	32	0.25	0.56	0.92	1.35	1.84	2.41	3.06	3.77	4.57	5.43	6.36	7.37
	36	0.23	0.50	0.83	1.21	1.66	2.17	2.75	3.40	4.11	4.89	5.74	6.66
	2	1.49	3.36	5.36	7.37	9.38	11.4	13.4	15.4	17.4	19.3	21.3	23.3
	3	1.29	3.02	4.97	6.99	9.01	11.0	13.1	15.1	17.1	19.1	21.1	23.1
	4	1.13	2.73	4.60	6.58	8.61	10.7	12.7	14.7	16.7	18.8	20.8	22.8
	5	1.00	2.48	4.26	6.18	8.18	10.2	12.3	14.3	16.4	18.4	20.4	22.4
	6	0.90	2.27	3.96	5.80	7.76	9.79	11.8	13.9	15.9	18.0	20.0	22.1
	7	0.82	2.09	3.68	5.44	7.36	9.35	11.4	13.5	15.5	17.6	19.6	21.7
	8	0.75	1.93	3.43	5.11	6.97	8.93	11.0	13.0	15.1	17.1	19.2	21.2
	9	0.70	1.80	3.21	4.81	6.61	8.53	10.5	12.6	14.6	16.7	18.7	20.8
	10	0.65	1.68	3.01	4.53	6.27	8.14	10.1	12.1	14.2	16.2	18.3	20.4
6	12	0.57	1.49	2.67	4.05	5.67	7.43	9.31	11.3	13.3	15.3	17.4	19.4
	14	0.50	1.33	2.39	3.65	5.15	6.81	8.60	10.5	12.4	14.4	16.5	18.5
	16	0.45	1.20	2.16	3.31	4.71	6.27	7.96	9.76	11.7	13.6	15.6	17.6
	18	0.41	1.09	1.97	3.03	4.34	5.79	7.39	9.12	10.9	12.8	14.8	16.8
	20	0.38	1.00	1.81	2.80	4.01	5.37	6.89	8.53	10.3	12.1	14.0	15.9
	24	0.32	0.86	1.55	2.41	3.48	4.68	6.04	7.53	9.14	10.8	12.6	14.5
	28	0.28	0.75	1.35	2.12	3.06	4.13	5.36	6.72	8.19	9.76	11.4	13.2
	32	0.25	0.67	1.20	1.89	2.73	3.69	4.81	6.05	7.40	8.86	10.4	12.0
	36	0.23	0.60	1.08	1.70	2.46	3.34	4.36	5.50	6.74	8.09	9.53	11.1

Table 7-20 (cont.). Coefficients *C* for Eccentrically Loaded Bolt Groups Angle = 45°

$C_{req} = \dfrac{P_u}{\phi r_n}$ or $\phi R_n = C \times \phi r_n$

where

P_u = factored force, kips
ϕr_n = design strength per bolt, kips
ϕR_n = design strength per bolt group, kips
e = eccentricity of P_u with respect to centroid of bolt group, in. (not tabulated, may be determined by geometry)
e_x = horizontal component of e, in.
s = bolt spacing, in.
C = coefficient tabulated below

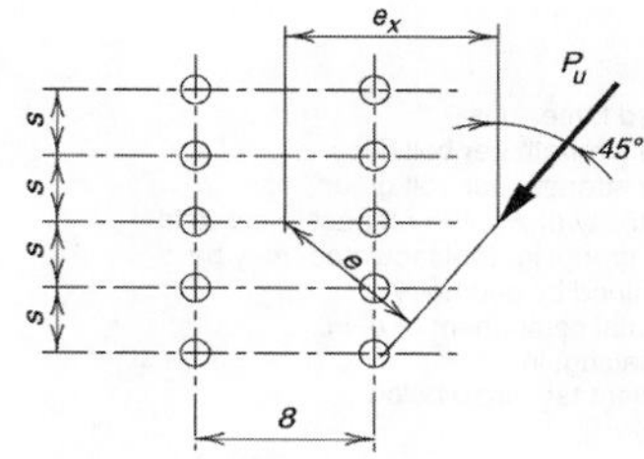

s, in.	e_x, in.	Number of Bolts in One Vertical Row, n											
		1	2	3	4	5	6	7	8	9	10	11	12
	2	1.70	3.43	5.22	7.06	8.95	10.9	12.8	14.8	16.8	18.7	20.7	22.7
	3	1.51	3.09	4.76	6.52	8.35	10.2	12.2	14.1	16.1	18.0	20.0	22.0
	4	1.35	2.78	4.34	6.01	7.78	9.60	11.5	13.4	15.3	17.3	19.3	21.3
	5	1.21	2.52	3.97	5.57	7.25	9.01	10.8	12.7	14.6	16.6	18.5	20.5
	6	1.10	2.30	3.67	5.17	6.78	8.47	10.2	12.1	13.9	15.9	17.8	19.8
	7	1.00	2.12	3.40	4.82	6.35	7.97	9.67	11.5	13.3	15.2	17.1	19.0
	8	0.92	1.96	3.17	4.51	5.96	7.51	9.15	10.9	12.7	14.5	16.4	18.3
	9	0.85	1.82	2.96	4.23	5.60	7.08	8.68	10.4	12.1	13.9	15.7	17.6
	10	0.79	1.70	2.78	3.97	5.28	6.70	8.24	9.86	11.5	13.3	15.1	17.0
3	12	0.69	1.50	2.46	3.54	4.73	6.04	7.46	8.97	10.6	12.2	14.0	15.7
	14	0.61	1.34	2.21	3.18	4.27	5.48	6.80	8.21	9.70	11.3	12.9	14.6
	16	0.55	1.21	2.00	2.88	3.89	5.01	6.23	7.54	8.95	10.4	12.0	13.6
	18	0.50	1.11	1.82	2.64	3.56	4.60	5.74	6.97	8.30	9.71	11.2	12.7
	20	0.46	1.02	1.67	2.42	3.29	4.25	5.31	6.47	7.73	9.06	10.5	11.9
	24	0.40	0.87	1.43	2.09	2.84	3.68	4.62	5.65	6.77	7.96	9.23	10.6
	28	0.35	0.76	1.26	1.83	2.49	3.24	4.07	5.00	6.00	7.08	8.24	9.47
	32	0.31	0.68	1.12	1.63	2.22	2.89	3.64	4.47	5.38	6.37	7.43	8.56
	36	0.28	0.61	1.00	1.46	2.00	2.60	3.29	4.04	4.87	5.78	6.75	7.79
	2	1.70	3.52	5.44	7.40	9.37	11.4	13.3	15.3	17.3	19.3	21.3	23.2
	3	1.51	3.23	5.11	7.06	9.03	11.0	13.0	15.0	17.0	19.0	21.0	22.9
	4	1.35	2.96	4.79	6.70	8.67	10.7	12.7	14.6	16.6	18.6	20.6	22.6
	5	1.21	2.72	4.48	6.36	8.30	10.3	12.3	14.3	16.3	18.3	20.3	22.3
	6	1.10	2.51	4.20	6.03	7.94	9.90	11.9	13.9	15.9	17.9	19.9	21.9
	7	1.00	2.33	3.96	5.73	7.60	9.53	11.5	13.5	15.5	17.5	19.5	21.5
	8	0.92	2.18	3.73	5.45	7.27	9.17	11.1	13.1	15.1	17.1	19.1	21.1
	9	0.85	2.04	3.53	5.19	6.96	8.83	10.8	12.7	14.7	16.7	18.7	20.7
	10	0.79	1.92	3.35	4.94	6.67	8.50	10.4	12.4	14.3	16.3	18.3	20.3
6	12	0.69	1.71	3.02	4.50	6.13	7.88	9.73	11.6	13.6	15.5	17.5	19.5
	14	0.61	1.55	2.75	4.12	5.65	7.33	9.11	11.0	12.9	14.8	16.8	18.8
	16	0.55	1.41	2.51	3.78	5.22	6.83	8.55	10.3	12.2	14.1	16.0	18.0
	18	0.50	1.29	2.31	3.49	4.85	6.39	8.04	9.77	11.6	13.4	15.3	17.3
	20	0.46	1.19	2.13	3.24	4.53	6.00	7.57	9.25	11.0	12.8	14.7	16.6
	24	0.40	1.03	1.84	2.82	3.99	5.32	6.76	8.32	9.97	11.7	13.5	15.3
	28	0.35	0.90	1.62	2.50	3.56	4.76	6.09	7.53	9.08	10.7	12.4	14.2
	32	0.31	0.80	1.44	2.24	3.20	4.30	5.52	6.86	8.32	9.85	11.5	13.1
	36	0.28	0.72	1.30	2.02	2.90	3.92	5.04	6.30	7.66	9.10	10.6	12.2

Table 7-20 (cont.). Coefficients *C* for Eccentrically Loaded Bolt Groups Angle = 60°

$$C_{req} = \frac{P_u}{\phi r_n} \quad \text{or} \quad \phi R_n = C \times \phi r_n$$

where

P_u = factored force, kips
ϕr_n = design strength per bolt, kips
ϕR_n = design strength per bolt group, kips
e = eccentricity of P_u with respect to centroid of bolt group, in. (not tabulated, may be determined by geometry)
e_x = horizontal component of e, in.
s = bolt spacing, in.
C = coefficient tabulated below

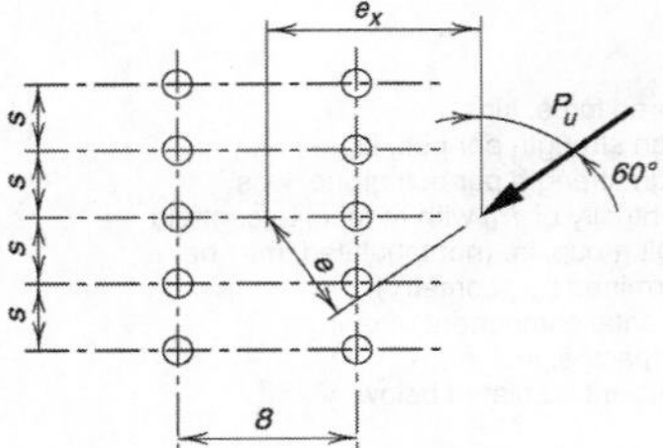

s, in.	e_x, in.	Number of Bolts in One Vertical Row, n											
		1	2	3	4	5	6	7	8	9	10	11	12
3	2	1.86	3.71	5.56	7.41	9.28	11.2	13.1	15.0	16.9	18.8	20.8	22.7
	3	1.77	3.52	5.29	7.07	8.88	10.7	12.6	14.5	16.4	18.3	20.2	22.1
	4	1.66	3.31	4.99	6.70	8.45	10.3	12.1	13.9	15.8	17.7	19.6	21.6
	5	1.54	3.10	4.70	6.34	8.04	9.79	11.6	13.4	15.3	17.1	19.0	21.0
	6	1.43	2.90	4.41	6.00	7.64	9.35	11.1	12.9	14.7	16.6	18.5	20.4
	7	1.33	2.71	4.15	5.68	7.27	8.94	10.7	12.4	14.2	16.1	17.9	19.8
	8	1.24	2.54	3.92	5.39	6.94	8.56	10.3	12.0	13.8	15.6	17.4	19.3
	9	1.16	2.38	3.70	5.12	6.63	8.22	9.86	11.6	13.3	15.1	16.9	18.7
	10	1.08	2.24	3.51	4.88	6.34	7.89	9.49	11.2	12.9	14.6	16.4	18.2
	12	0.96	2.00	3.17	4.44	5.82	7.28	8.81	10.4	12.1	13.8	15.5	17.3
	14	0.86	1.81	2.88	4.07	5.36	6.73	8.19	9.72	11.3	13.0	14.7	16.4
	16	0.77	1.64	2.64	3.74	4.95	6.25	7.64	9.11	10.7	12.2	13.9	15.6
	18	0.70	1.51	2.43	3.46	4.59	5.83	7.15	8.56	10.0	11.6	13.2	14.8
	20	0.65	1.39	2.25	3.21	4.28	5.45	6.71	8.06	9.48	11.0	12.5	14.1
	24	0.56	1.20	1.95	2.80	3.76	4.81	5.96	7.19	8.50	9.88	11.3	12.8
	28	0.49	1.06	1.72	2.48	3.34	4.29	5.34	6.47	7.68	8.97	10.3	11.7
	32	0.43	0.94	1.54	2.22	3.00	3.87	4.83	5.87	6.99	8.19	9.46	10.8
	36	0.39	0.85	1.39	2.01	2.72	3.52	4.40	5.36	6.41	7.53	8.71	9.96
6	2	1.86	3.72	5.59	7.50	9.43	11.4	13.3	15.3	17.3	19.2	21.2	23.2
	3	1.77	3.55	5.37	7.25	9.16	11.1	13.0	15.0	17.0	18.9	20.9	22.9
	4	1.66	3.36	5.14	6.98	8.88	10.8	12.7	14.7	16.7	18.6	20.6	22.6
	5	1.54	3.17	4.90	6.72	8.59	10.5	12.4	14.4	16.3	18.3	20.3	22.2
	6	1.43	2.99	4.67	6.46	8.31	10.2	12.1	14.1	16.0	18.0	19.9	21.9
	7	1.33	2.82	4.46	6.21	8.05	9.92	11.8	13.8	15.7	17.7	19.6	21.6
	8	1.24	2.67	4.26	5.98	7.79	9.65	11.5	13.5	15.4	17.3	19.3	21.3
	9	1.16	2.52	4.08	5.76	7.55	9.39	11.3	13.2	15.1	17.0	19.0	20.9
	10	1.08	2.40	3.91	5.56	7.32	9.14	11.0	12.9	14.8	16.7	18.7	20.6
	12	0.96	2.17	3.61	5.20	6.90	8.66	10.5	12.4	14.2	16.1	18.1	20.0
	14	0.86	1.98	3.35	4.87	6.51	8.23	10.0	11.8	13.7	15.6	17.5	19.4
	16	0.77	1.82	3.11	4.57	6.15	7.81	9.56	11.4	13.2	15.1	16.9	18.9
	18	0.70	1.69	2.91	4.30	5.81	7.43	9.13	10.9	12.7	14.5	16.4	18.3
	20	0.65	1.57	2.72	4.05	5.50	7.07	8.73	10.5	12.2	14.1	15.9	17.8
	24	0.56	1.37	2.41	3.61	4.96	6.43	8.00	9.67	11.4	13.2	15.0	16.8
	28	0.49	1.22	2.15	3.25	4.49	5.88	7.38	8.97	10.6	12.3	14.1	15.9
	32	0.43	1.09	1.94	2.94	4.10	5.41	6.83	8.34	9.92	11.6	13.3	15.0
	36	0.39	0.99	1.76	2.69	3.77	5.00	6.35	7.78	9.30	10.9	12.5	14.2

Table 7-20 (cont.). Coefficients *C* for Eccentrically Loaded Bolt Groups Angle = 75°

$C_{req} = \dfrac{P_u}{\phi r_n}$ or $\phi R_n = C \times \phi r_n$

where

P_u = factored force, kips
ϕr_n = design strength per bolt, kips
ϕR_n = design strength per bolt group, kips
e = eccentricity of P_u with respect to centroid of bolt group, in. (not tabulated, may be determined by geometry)
e_x = horizontal component of e, in.
s = bolt spacing, in.
C = coefficient tabulated below

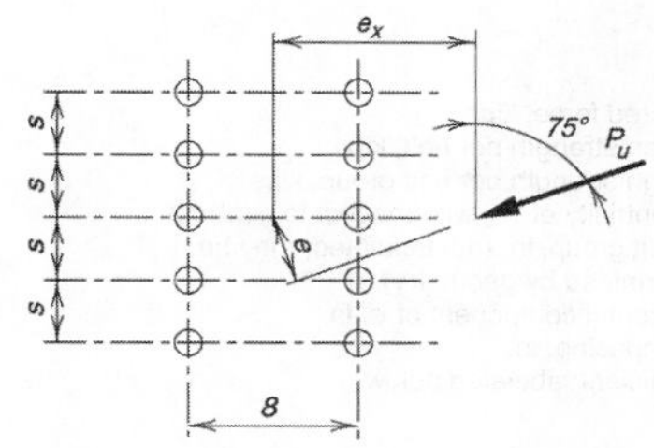

s, in.	e_x, in.	Number of Bolts in One Vertical Row, *n*											
		1	2	3	4	5	6	7	8	9	10	11	12
	2	1.94	3.87	5.79	7.70	9.61	11.5	13.4	15.3	17.3	19.2	21.1	23.0
	3	1.92	3.82	5.70	7.58	9.45	11.3	13.2	15.1	17.0	18.9	20.8	22.7
	4	1.89	3.75	5.60	7.43	9.26	11.1	12.9	14.8	16.7	18.5	20.4	22.3
	5	1.85	3.67	5.48	7.28	9.07	10.9	12.7	14.5	16.4	18.2	20.1	22.0
	6	1.81	3.59	5.35	7.11	8.87	10.6	12.4	14.2	16.1	17.9	19.8	21.6
	7	1.76	3.50	5.22	6.94	8.67	10.4	12.2	14.0	15.8	17.6	19.4	21.3
	8	1.71	3.40	5.08	6.76	8.46	10.2	11.9	13.7	15.5	17.3	19.1	21.0
	9	1.66	3.30	4.94	6.59	8.26	9.96	11.7	13.4	15.2	17.0	18.8	20.6
	10	1.61	3.20	4.80	6.42	8.06	9.73	11.4	13.2	14.9	16.7	18.5	20.3
3	12	1.51	3.01	4.53	6.08	7.67	9.30	11.0	12.7	14.4	16.2	17.9	19.7
	14	1.41	2.82	4.27	5.76	7.31	8.90	10.5	12.2	13.9	15.6	17.4	19.2
	16	1.31	2.65	4.03	5.47	6.96	8.52	10.1	11.8	13.4	15.2	16.9	18.6
	18	1.23	2.48	3.80	5.19	6.64	8.16	9.73	11.3	13.0	14.7	16.4	18.1
	20	1.15	2.34	3.60	4.93	6.34	7.82	9.36	10.9	12.6	14.2	15.9	17.7
	24	1.01	2.08	3.23	4.48	5.80	7.20	8.67	10.2	11.8	13.4	15.0	16.7
	28	0.90	1.87	2.93	4.08	5.33	6.65	8.06	9.52	11.0	12.6	14.2	15.9
	32	0.81	1.69	2.67	3.75	4.91	6.17	7.51	8.91	10.4	11.9	13.5	15.1
	36	0.73	1.54	2.45	3.45	4.55	5.74	7.01	8.36	9.77	11.2	12.8	14.3
	2	1.94	3.86	5.77	7.68	9.60	11.5	13.5	15.4	17.6	19.6	21.5	23.5
	3	1.92	3.80	5.68	7.55	9.45	11.4	13.3	15.2	17.2	19.1	21.1	23.0
	4	1.89	3.74	5.57	7.42	9.29	11.2	13.1	15.0	16.9	18.9	20.8	22.8
	5	1.85	3.66	5.46	7.29	9.14	11.0	12.9	14.8	16.7	18.7	20.6	22.6
	6	1.81	3.58	5.35	7.15	8.98	10.8	12.7	14.6	16.5	18.5	20.4	22.3
	7	1.76	3.49	5.23	7.01	8.83	10.7	12.5	14.4	16.3	18.3	20.2	22.1
	8	1.71	3.40	5.12	6.88	8.68	10.5	12.4	14.3	16.2	18.1	20.0	21.9
	9	1.66	3.31	5.00	6.74	8.53	10.4	12.2	14.1	16.0	17.9	19.8	21.7
	10	1.61	3.22	4.89	6.61	8.38	10.2	12.0	13.9	15.8	17.7	19.6	21.5
6	12	1.51	3.05	4.67	6.36	8.10	9.89	11.7	13.6	15.4	17.3	19.2	21.1
	14	1.41	2.88	4.46	6.12	7.84	9.61	11.4	13.3	15.1	17.0	18.9	20.8
	16	1.31	2.73	4.26	5.89	7.59	9.33	11.1	12.9	14.8	16.6	18.5	20.4
	18	1.23	2.58	4.08	5.68	7.35	9.08	10.8	12.7	14.5	16.3	18.2	20.1
	20	1.15	2.45	3.90	5.47	7.13	8.84	10.6	12.4	14.2	16.0	17.9	19.7
	24	1.01	2.21	3.59	5.10	6.71	8.38	10.1	11.9	13.6	15.5	17.3	19.1
	28	0.90	2.01	3.32	4.77	6.32	7.96	9.65	11.4	13.1	14.9	16.7	18.5
	32	0.81	1.84	3.08	4.47	5.97	7.56	9.21	10.9	12.7	14.4	16.2	18.0
	36	0.73	1.70	2.87	4.19	5.64	7.19	8.80	10.5	12.2	13.9	15.7	17.5

Table 7-21.
Coefficients *C* for Eccentrically Loaded Bolt Groups
Angle = 0°

$C_{req} = \dfrac{P_u}{\phi r_n}$ or $\phi R_n = C \times \phi r_n$

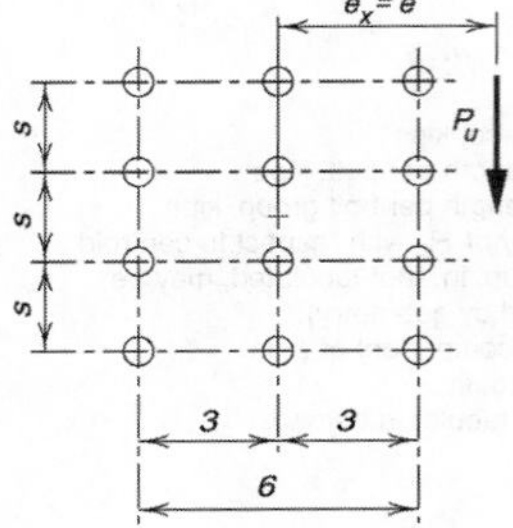

where

P_u = factored force, kips
ϕr_n = design strength per bolt, kips
ϕR_n = design strength per bolt group, kips
e = eccentricity of P_u with respect to centroid of bolt group, in. (not tabulated, may be determined by geometry)
e_x = horizontal component of e, in.
s = bolt spacing, in.
C = coefficient tabulated below

s, in.	e_x, in.	Number of Bolts in One Vertical Row, *n*											
		1	2	3	4	5	6	7	8	9	10	11	12
	2	1.71	4.07	6.81	9.86	13.0	16.1	19.3	22.3	25.4	28.5	31.5	34.5
	3	1.42	3.40	5.79	8.61	11.7	14.8	18.0	21.1	24.3	27.4	30.5	33.6
	4	1.21	2.90	4.97	7.53	10.4	13.4	16.6	19.8	23.0	26.1	29.3	32.5
	5	1.05	2.51	4.34	6.64	9.24	12.1	15.2	18.3	21.5	24.7	27.9	31.1
	6	0.92	2.21	3.85	5.91	8.27	11.0	13.9	16.9	20.0	23.2	26.4	29.7
	7	0.81	1.96	3.44	5.31	7.46	9.95	12.7	15.6	18.6	21.8	25.0	28.2
	8	0.72	1.76	3.11	4.80	6.78	9.09	11.6	14.4	17.3	20.4	23.5	26.7
	9	0.64	1.60	2.83	4.38	6.20	8.34	10.7	13.3	16.1	19.1	22.1	25.2
	10	0.58	1.46	2.59	4.02	5.71	7.70	9.91	12.4	15.0	17.9	20.8	23.8
3	12	0.49	1.24	2.21	3.44	4.91	6.65	8.59	10.8	13.2	15.7	18.5	21.3
	14	0.42	1.08	1.92	3.00	4.30	5.83	7.57	9.53	11.7	14.0	16.5	19.2
	16	0.37	0.95	1.70	2.66	3.82	5.19	6.75	8.51	10.5	12.6	14.9	17.3
	18	0.33	0.85	1.52	2.39	3.43	4.67	6.08	7.68	9.45	11.4	13.5	15.8
	20	0.29	0.77	1.37	2.16	3.11	4.24	5.53	6.99	8.61	10.4	12.3	14.4
	24	0.24	0.64	1.15	1.82	2.62	3.57	4.67	5.92	7.30	8.84	10.5	12.3
	28	0.21	0.55	0.99	1.57	2.26	3.08	4.04	5.12	6.33	7.67	9.13	10.7
	32	0.18	0.49	0.87	1.38	1.98	2.71	3.55	4.51	5.58	6.77	8.06	9.47
	36	0.16	0.43	0.77	1.23	1.77	2.42	3.17	4.03	4.99	6.05	7.21	8.48
	2	1.71	4.85	8.04	11.2	14.2	17.3	20.3	23.2	26.2	29.2	32.2	35.1
	3	1.42	4.24	7.36	10.6	13.7	16.8	19.9	22.9	25.9	28.9	31.9	34.9
	4	1.21	3.72	6.66	9.86	13.1	16.2	19.4	22.4	25.5	28.5	31.6	34.6
	5	1.05	3.29	6.00	9.14	12.4	15.6	18.7	21.9	25.0	28.1	31.1	34.2
	6	0.92	2.93	5.41	8.44	11.6	14.9	18.1	21.2	24.4	27.5	30.6	33.7
	7	0.81	2.63	4.90	7.79	10.9	14.1	17.3	20.6	23.7	26.9	30.0	33.2
	8	0.72	2.38	4.46	7.20	10.2	13.4	16.6	19.8	23.0	26.2	29.4	32.6
	9	0.64	2.17	4.09	6.67	9.54	12.6	15.8	19.1	22.3	25.5	28.7	31.9
	10	0.58	2.00	3.78	6.20	8.94	12.0	15.1	18.3	21.6	24.8	28.0	31.2
6	12	0.49	1.71	3.27	5.41	7.88	10.7	13.7	16.8	20.0	23.3	26.5	29.8
	14	0.42	1.49	2.87	4.78	7.01	9.61	12.4	15.4	18.6	21.8	25.0	28.2
	16	0.37	1.32	2.55	4.28	6.29	8.69	11.3	14.2	17.2	20.3	23.5	26.7
	18	0.33	1.19	2.30	3.86	5.70	7.91	10.4	13.1	15.9	18.9	22.0	25.2
	20	0.29	1.08	2.09	3.51	5.20	7.25	9.54	12.1	14.8	17.7	20.7	23.8
	24	0.24	0.91	1.76	2.97	4.42	6.19	8.19	10.4	12.9	15.5	18.3	21.2
	28	0.21	0.78	1.52	2.57	3.84	5.39	7.14	9.15	11.4	13.7	16.3	19.0
	32	0.18	0.69	1.33	2.27	3.39	4.77	6.33	8.13	10.1	12.3	14.6	17.1
	36	0.16	0.61	1.19	2.03	3.03	4.27	5.67	7.30	9.10	11.1	13.2	15.5

Table 7-21 (cont.). Coefficients C for Eccentrically Loaded Bolt Groups Angle = 15°

$$C_{req} = \frac{P_u}{\phi r_n} \quad \text{or} \quad \phi R_n = C \times \phi r_n$$

where

P_u = factored force, kips
ϕr_n = design strength per bolt, kips
ϕR_n = design strength per bolt group, kips
e = eccentricity of P_u with respect to centroid of bolt group, in. (not tabulated, may be determined by geometry)
e_x = horizontal component of e, in.
s = bolt spacing, in.
C = coefficient tabulated below

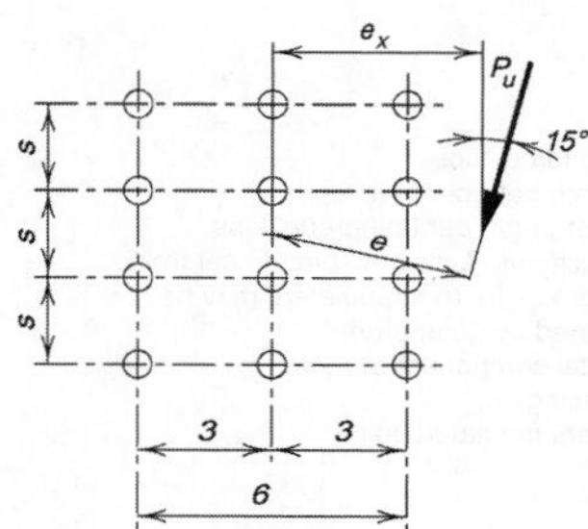

s, in.	e_x, in.	Number of Bolts in One Vertical Row, n											
		1	2	3	4	5	6	7	8	9	10	11	12
	2	1.77	4.10	6.84	9.82	12.9	16.0	19.1	22.2	25.2	28.3	31.3	34.3
	3	1.47	3.45	5.86	8.61	11.6	14.7	17.8	20.9	24.1	27.2	30.3	33.3
	4	1.25	2.95	5.07	7.55	10.4	13.3	16.4	19.5	22.7	25.8	29.0	32.1
	5	1.08	2.57	4.44	6.67	9.26	12.1	15.1	18.1	21.3	24.4	27.6	30.7
	6	0.94	2.26	3.93	5.96	8.33	11.0	13.8	16.8	19.8	23.0	26.1	29.3
	7	0.83	2.01	3.52	5.37	7.55	9.97	12.7	15.5	18.5	21.5	24.7	27.8
	8	0.74	1.81	3.18	4.87	6.88	9.13	11.7	14.4	17.2	20.2	23.2	26.4
	9	0.66	1.64	2.90	4.45	6.31	8.40	10.8	13.3	16.1	18.9	21.9	25.0
	10	0.60	1.50	2.65	4.10	5.81	7.77	9.99	12.4	15.0	17.8	20.7	23.6
3	12	0.50	1.28	2.27	3.52	5.01	6.74	8.71	10.9	13.2	15.8	18.4	21.2
	14	0.43	1.11	1.98	3.08	4.40	5.93	7.69	9.62	11.8	14.1	16.5	19.1
	16	0.38	0.98	1.75	2.73	3.91	5.29	6.87	8.62	10.6	12.7	15.0	17.4
	18	0.34	0.88	1.57	2.45	3.52	4.77	6.20	7.80	9.59	11.5	13.6	15.9
	20	0.30	0.79	1.42	2.22	3.19	4.33	5.65	7.12	8.76	10.5	12.5	14.6
	24	0.25	0.67	1.19	1.87	2.69	3.66	4.78	6.04	7.45	8.99	10.7	12.5
	28	0.22	0.57	1.02	1.61	2.32	3.17	4.14	5.24	6.47	7.82	9.31	10.9
	32	0.19	0.50	0.90	1.42	2.04	2.79	3.65	4.62	5.72	6.92	8.24	9.66
	36	0.17	0.45	0.80	1.26	1.82	2.49	3.26	4.13	5.11	6.20	7.38	8.66
	2	1.77	4.83	7.98	11.1	14.1	17.2	20.2	23.2	26.1	29.1	32.1	35.0
	3	1.47	4.22	7.31	10.5	13.6	16.7	19.7	22.8	25.8	28.8	31.8	34.8
	4	1.25	3.71	6.64	9.77	12.9	16.1	19.2	22.3	25.3	28.3	31.4	34.4
	5	1.08	3.28	6.01	9.06	12.2	15.4	18.5	21.7	24.8	27.8	30.9	33.9
	6	0.94	2.94	5.45	8.38	11.5	14.7	17.8	21.0	24.1	27.2	30.3	33.4
	7	0.83	2.65	4.97	7.75	10.8	13.9	17.1	20.3	23.5	26.6	29.7	32.8
	8	0.74	2.40	4.55	7.17	10.1	13.2	16.4	19.6	22.7	25.9	29.1	32.2
	9	0.66	2.20	4.18	6.66	9.49	12.5	15.6	18.8	22.0	25.2	28.4	31.5
	10	0.60	2.02	3.86	6.20	8.92	11.9	14.9	18.1	21.3	24.5	27.6	30.8
6	12	0.50	1.74	3.34	5.43	7.91	10.6	13.6	16.6	19.8	23.0	26.1	29.3
	14	0.43	1.52	2.94	4.82	7.07	9.60	12.4	15.3	18.4	21.5	24.6	27.8
	16	0.38	1.35	2.62	4.32	6.38	8.71	11.3	14.1	17.0	20.1	23.2	26.3
	18	0.34	1.22	2.36	3.91	5.79	7.95	10.4	13.0	15.8	18.8	21.8	24.9
	20	0.30	1.10	2.14	3.57	5.30	7.31	9.60	12.1	14.8	17.6	20.5	23.5
	24	0.25	0.93	1.81	3.03	4.52	6.26	8.28	10.5	12.9	15.5	18.2	21.1
	28	0.22	0.80	1.56	2.63	3.93	5.47	7.26	9.24	11.4	13.8	16.3	18.9
	32	0.19	0.71	1.37	2.32	3.47	4.85	6.45	8.23	10.2	12.4	14.7	17.1
	36	0.17	0.63	1.23	2.08	3.11	4.35	5.80	7.41	9.23	11.2	13.3	15.6

Table 7-21 (cont.). Coefficients *C* for Eccentrically Loaded Bolt Groups Angle = 30°

$$C_{req} = \frac{P_u}{\phi r_n} \quad \text{or} \quad \phi R_n = C \times \phi r_n$$

where

P_u = factored force, kips
ϕr_n = design strength per bolt, kips
ϕR_n = design strength per bolt group, kips
e = eccentricity of P_u with respect to centroid of bolt group, in. (not tabulated, may be determined by geometry)
e_x = horizontal component of e, in.
s = bolt spacing, in.
C = coefficient tabulated below

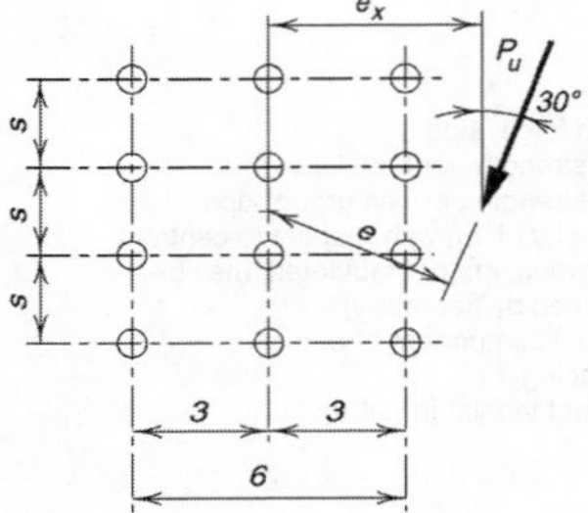

s, in.	e_x, in.	Number of Bolts in One Vertical Row, *n*											
		1	2	3	4	5	6	7	8	9	10	11	12
	2	1.94	4.26	6.99	9.90	12.9	16.0	19.0	22.0	25.1	28.1	31.1	34.1
	3	1.61	3.63	6.09	8.80	11.7	14.7	17.7	20.8	23.9	27.0	30.0	33.1
	4	1.37	3.15	5.35	7.83	10.6	13.5	16.5	19.5	22.6	25.7	28.7	31.8
	5	1.19	2.77	4.74	7.00	9.54	12.3	15.2	18.2	21.2	24.3	27.4	30.5
	6	1.04	2.45	4.23	6.30	8.67	11.3	14.1	17.0	19.9	23.0	26.0	29.1
	7	0.92	2.19	3.81	5.71	7.92	10.4	13.0	15.8	18.7	21.7	24.7	27.8
	8	0.82	1.98	3.45	5.22	7.27	9.58	12.1	14.8	17.6	20.5	23.4	26.4
	9	0.74	1.80	3.16	4.79	6.71	8.88	11.2	13.8	16.5	19.3	22.2	25.2
	10	0.67	1.65	2.90	4.42	6.22	8.26	10.5	12.9	15.5	18.2	21.1	24.0
3	12	0.56	1.41	2.49	3.82	5.41	7.22	9.23	11.5	13.8	16.4	19.0	21.8
	14	0.48	1.23	2.18	3.36	4.78	6.40	8.22	10.3	12.4	14.8	17.2	19.8
	16	0.42	1.08	1.93	2.99	4.26	5.73	7.40	9.25	11.3	13.4	15.7	18.2
	18	0.38	0.97	1.73	2.69	3.85	5.18	6.71	8.41	10.3	12.3	14.4	16.7
	20	0.34	0.88	1.57	2.44	3.50	4.73	6.14	7.70	9.42	11.3	13.3	15.4
	24	0.28	0.74	1.32	2.06	2.96	4.01	5.22	6.58	8.08	9.72	11.5	13.4
	28	0.24	0.64	1.14	1.78	2.56	3.48	4.54	5.73	7.05	8.51	10.1	11.8
	32	0.21	0.56	1.00	1.57	2.26	3.07	4.01	5.07	6.25	7.55	8.96	10.5
	36	0.19	0.50	0.89	1.40	2.02	2.75	3.59	4.54	5.61	6.78	8.06	9.44
	2	1.94	4.86	7.96	11.0	14.1	17.1	20.1	23.1	26.0	29.0	32.0	35.0
	3	1.61	4.27	7.32	10.4	13.5	16.6	19.6	22.6	25.6	28.6	31.6	34.6
	4	1.37	3.78	6.70	9.75	12.9	15.9	19.0	22.1	25.1	28.1	31.1	34.2
	5	1.19	3.39	6.14	9.10	12.2	15.3	18.4	21.5	24.5	27.6	30.6	33.7
	6	1.04	3.06	5.64	8.48	11.5	14.6	17.7	20.8	23.9	27.0	30.1	33.1
	7	0.92	2.78	5.19	7.91	10.9	13.9	17.0	20.1	23.2	26.3	29.4	32.5
	8	0.82	2.54	4.80	7.38	10.3	13.3	16.3	19.4	22.6	25.7	28.8	31.9
	9	0.74	2.34	4.45	6.90	9.67	12.6	15.7	18.7	21.9	25.0	28.1	31.2
	10	0.67	2.16	4.14	6.46	9.14	12.0	15.0	18.1	21.2	24.3	27.4	30.5
6	12	0.56	1.87	3.61	5.71	8.20	10.9	13.8	16.8	19.8	22.9	26.0	29.1
	14	0.48	1.65	3.20	5.10	7.41	9.95	12.7	15.6	18.5	21.5	24.6	27.7
	16	0.42	1.47	2.86	4.60	6.74	9.12	11.7	14.5	17.3	20.3	23.3	26.4
	18	0.38	1.33	2.58	4.19	6.17	8.39	10.8	13.5	16.2	19.1	22.0	25.0
	20	0.34	1.21	2.35	3.84	5.68	7.75	10.1	12.6	15.2	18.0	20.9	23.8
	24	0.28	1.02	2.00	3.29	4.89	6.71	8.78	11.1	13.5	16.1	18.8	21.6
	28	0.24	0.88	1.73	2.86	4.28	5.90	7.77	9.83	12.1	14.5	17.0	19.6
	32	0.21	0.78	1.52	2.54	3.80	5.25	6.95	8.83	10.9	13.1	15.4	17.9
	36	0.19	0.70	1.36	2.27	3.41	4.73	6.28	8.00	9.88	11.9	14.1	16.4

Table 7-21 (cont.). Coefficients C for Eccentrically Loaded Bolt Groups Angle = 45°

$$C_{req} = \frac{P_u}{\phi r_n} \quad \text{or} \quad \phi R_n = C \times \phi r_n$$

where

P_u = factored force, kips
ϕr_n = design strength per bolt, kips
ϕR_n = design strength per bolt group, kips
e = eccentricity of P_u with respect to centroid of bolt group, in. (not tabulated, may be determined by geometry)
e_x = horizontal component of e, in.
s = bolt spacing, in.
C = coefficient tabulated below

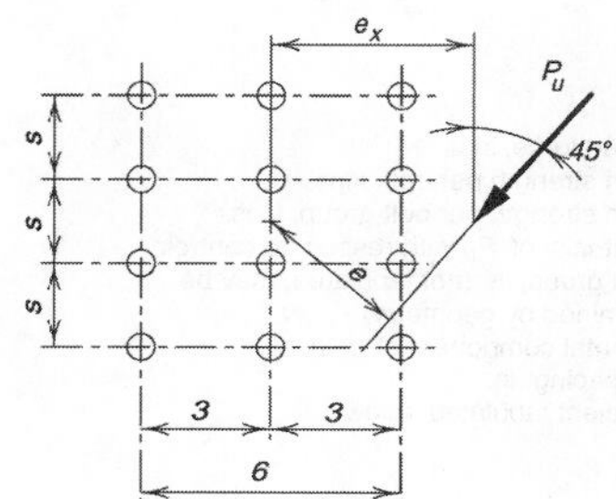

s, in.	e_x, in.	1	2	3	4	5	6	7	8	9	10	11	12
		Number of Bolts in One Vertical Row, n											
	2	2.23	4.67	7.33	10.2	13.1	16.0	19.0	22.0	25.0	28.0	31.0	33.9
	3	1.89	4.06	6.50	9.19	12.0	14.9	17.9	20.9	23.9	26.9	29.9	32.9
	4	1.63	3.57	5.84	8.36	11.1	13.9	16.8	19.7	22.7	25.7	28.7	31.7
	5	1.42	3.17	5.27	7.63	10.2	12.9	15.7	18.6	21.5	24.5	27.5	30.5
	6	1.25	2.84	4.78	6.99	9.40	12.0	14.7	17.6	20.4	23.4	26.3	29.3
	7	1.11	2.57	4.36	6.42	8.70	11.2	13.8	16.6	19.4	22.3	25.2	28.2
	8	0.99	2.33	3.99	5.92	8.09	10.5	13.0	15.7	18.4	21.2	24.1	27.0
	9	0.90	2.13	3.68	5.49	7.54	9.80	12.2	14.8	17.5	20.3	23.1	26.0
	10	0.81	1.96	3.40	5.10	7.05	9.21	11.6	14.0	16.6	19.3	22.1	24.9
3	12	0.68	1.68	2.95	4.46	6.22	8.19	10.4	12.7	15.1	17.7	20.3	23.0
	14	0.59	1.47	2.59	3.95	5.55	7.35	9.34	11.5	13.8	16.2	18.7	21.3
	16	0.52	1.31	2.31	3.54	4.99	6.65	8.49	10.5	12.7	14.9	17.3	19.8
	18	0.46	1.17	2.08	3.20	4.54	6.06	7.77	9.64	11.7	13.8	16.1	18.5
	20	0.41	1.06	1.89	2.92	4.15	5.56	7.15	8.90	10.8	12.8	15.0	17.2
	24	0.35	0.90	1.60	2.48	3.54	4.76	6.15	7.70	9.39	11.2	13.1	15.2
	28	0.30	0.77	1.38	2.15	3.08	4.16	5.39	6.77	8.28	9.91	11.7	13.5
	32	0.26	0.68	1.22	1.90	2.72	3.68	4.79	6.03	7.39	8.87	10.5	12.2
	36	0.23	0.61	1.08	1.69	2.44	3.30	4.30	5.42	6.66	8.02	9.49	11.1
	2	2.23	5.02	8.01	11.0	14.0	17.0	20.0	23.0	25.9	28.9	31.9	34.8
	3	1.89	4.50	7.44	10.4	13.5	16.5	19.5	22.5	25.5	28.4	31.4	34.4
	4	1.63	4.05	6.89	9.86	12.9	15.9	18.9	21.9	24.9	27.9	30.9	33.9
	5	1.42	3.68	6.40	9.30	12.3	15.3	18.3	21.3	24.4	27.4	30.4	33.4
	6	1.25	3.36	5.96	8.78	11.7	14.7	17.7	20.7	23.8	26.8	29.8	32.8
	7	1.11	3.09	5.57	8.29	11.2	14.1	17.1	20.1	23.2	26.2	29.2	32.3
	8	0.99	2.86	5.22	7.84	10.6	13.6	16.5	19.5	22.6	25.6	28.6	31.7
	9	0.90	2.65	4.90	7.43	10.2	13.0	16.0	19.0	22.0	25.0	28.0	31.1
	10	0.81	2.47	4.61	7.04	9.69	12.5	15.4	18.4	21.4	24.4	27.4	30.4
6	12	0.68	2.16	4.11	6.35	8.85	11.6	14.4	17.3	20.2	23.2	26.2	29.2
	14	0.59	1.92	3.69	5.76	8.11	10.7	13.4	16.2	19.1	22.1	25.0	28.0
	16	0.52	1.72	3.34	5.25	7.47	9.94	12.6	15.3	18.1	21.0	23.9	26.9
	18	0.46	1.56	3.04	4.82	6.91	9.26	11.8	14.4	17.2	20.0	22.9	25.8
	20	0.41	1.43	2.79	4.44	6.43	8.66	11.1	13.6	16.3	19.0	21.9	24.7
	24	0.35	1.22	2.38	3.84	5.62	7.64	9.84	12.2	14.7	17.3	20.0	22.8
	28	0.30	1.06	2.08	3.37	4.98	6.81	8.82	11.0	13.4	15.8	18.4	21.1
	32	0.26	0.94	1.84	3.00	4.46	6.12	7.97	10.0	12.2	14.6	17.0	19.5
	36	0.23	0.84	1.65	2.71	4.04	5.56	7.27	9.18	11.2	13.4	15.7	18.1

Table 7-21 (cont.).
Coefficients C for Eccentrically Loaded Bolt Groups
Angle = 60°

$$C_{req} = \frac{P_u}{\phi r_n} \quad \text{or} \quad \phi R_n = C \times \phi r_n$$

where

P_u = factored force, kips
ϕr_n = design strength per bolt, kips
ϕR_n = design strength per bolt group, kips
e = eccentricity of P_u with respect to centroid of bolt group, in. (not tabulated, may be determined by geometry)
e_x = horizontal component of e, in.
s = bolt spacing, in.
C = coefficient tabulated below

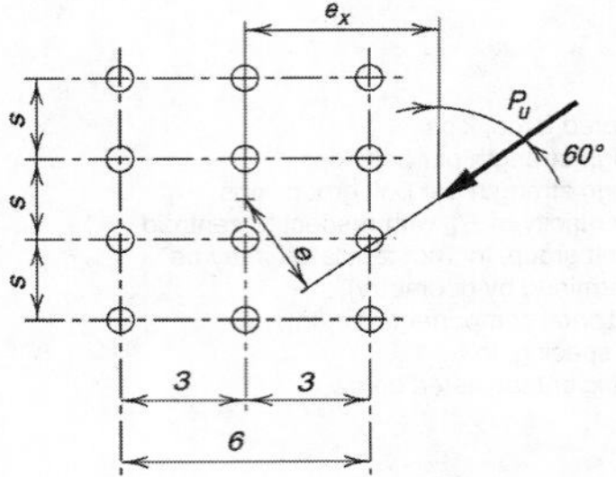

s, in.	e_x, in.	Number of Bolts in One Vertical Row, n											
		1	2	3	4	5	6	7	8	9	10	11	12
	2	2.59	5.21	7.88	10.6	13.4	16.3	19.2	22.1	25.0	28.0	30.9	33.9
	3	2.32	4.73	7.27	9.91	12.7	15.5	18.3	21.2	24.1	27.0	30.0	32.9
	4	2.07	4.29	6.69	9.23	11.9	14.6	17.5	20.3	23.2	26.1	29.0	32.0
	5	1.84	3.90	6.18	8.63	11.2	13.9	16.6	19.5	22.3	25.2	28.1	31.0
	6	1.65	3.56	5.73	8.08	10.6	13.2	15.9	18.7	21.5	24.3	27.2	30.1
	7	1.49	3.27	5.32	7.59	10.0	12.6	15.2	17.9	20.7	23.5	26.3	29.2
	8	1.35	3.01	4.95	7.13	9.48	12.0	14.5	17.2	19.9	22.7	25.5	28.4
	9	1.23	2.78	4.63	6.71	8.98	11.4	13.9	16.5	19.2	22.0	24.7	27.6
	10	1.12	2.58	4.34	6.33	8.52	10.9	13.3	15.9	18.5	21.2	24.0	26.8
3	12	0.95	2.25	3.84	5.67	7.70	9.91	12.3	14.7	17.3	19.9	22.6	25.3
	14	0.83	1.98	3.43	5.11	7.00	9.08	11.3	13.7	16.1	18.7	21.3	23.9
	16	0.73	1.77	3.09	4.64	6.40	8.36	10.5	12.7	15.1	17.5	20.1	22.6
	18	0.65	1.60	2.81	4.24	5.89	7.73	9.74	11.9	14.2	16.5	19.0	21.5
	20	0.59	1.46	2.57	3.9	5.44	7.19	9.09	11.1	13.3	15.6	17.9	20.4
	24	0.49	1.24	2.20	3.35	4.72	6.27	7.99	9.85	11.9	14.0	16.2	18.5
	28	0.42	1.07	1.91	2.93	4.15	5.55	7.10	8.81	10.7	12.6	14.7	16.8
	32	0.37	0.95	1.69	2.60	3.70	4.97	6.38	7.95	9.65	11.5	13.4	15.4
	36	0.33	0.85	1.51	2.34	3.34	4.49	5.79	7.23	8.81	10.5	12.3	14.2
	2	2.59	5.32	8.17	11.1	14.0	17.0	19.9	22.9	25.8	28.8	31.8	34.7
	3	2.32	4.94	7.73	10.6	13.5	16.5	19.4	22.4	25.4	28.3	31.3	34.3
	4	2.07	4.57	7.31	10.2	13.1	16.0	19.0	21.9	24.9	27.8	30.8	33.8
	5	1.84	4.25	6.91	9.73	12.6	15.5	18.5	21.4	24.4	27.4	30.3	33.3
	6	1.65	3.95	6.55	9.32	12.2	15.1	18.0	20.9	23.9	26.9	29.8	32.8
	7	1.49	3.69	6.22	8.94	11.8	14.6	17.5	20.5	23.4	26.4	29.3	32.3
	8	1.35	3.46	5.92	8.58	11.4	14.2	17.1	20.0	22.9	25.9	28.8	31.8
	9	1.23	3.25	5.64	8.25	11.0	13.8	16.7	19.6	22.5	25.4	28.4	31.3
	10	1.12	3.06	5.39	7.94	10.6	13.4	16.3	19.1	22.0	24.9	27.9	30.8
6	12	0.95	2.73	4.92	7.37	9.97	12.7	15.5	18.3	21.2	24.1	27.0	29.9
	14	0.83	2.46	4.52	6.85	9.36	12.0	14.7	17.5	20.3	23.2	26.1	29.0
	16	0.73	2.23	4.18	6.39	8.80	11.4	14.0	16.8	19.6	22.4	25.3	28.1
	18	0.65	2.04	3.87	5.97	8.28	10.8	13.4	16.1	18.8	21.6	24.4	27.3
	20	0.59	1.88	3.60	5.59	7.81	10.2	12.8	15.4	18.1	20.9	23.7	26.5
	24	0.49	1.63	3.15	4.94	6.99	9.25	11.7	14.2	16.8	19.5	22.2	25.0
	28	0.42	1.43	2.79	4.41	6.31	8.44	10.7	13.1	15.7	18.2	20.9	23.6
	32	0.37	1.27	2.49	3.97	5.74	7.74	9.90	12.2	14.6	17.1	19.7	22.3
	36	0.33	1.15	2.25	3.61	5.26	7.13	9.17	11.4	13.7	16.1	18.6	21.1

Table 7-21 (cont.). Coefficients *C* for Eccentrically Loaded Bolt Groups Angle = 75°

$$C_{req} = \frac{P_u}{\phi r_n} \quad \text{or} \quad \phi R_n = C \times \phi r_n$$

where

P_u = factored force, kips
ϕr_n = design strength per bolt, kips
ϕR_n = design strength per bolt group, kips
e = eccentricity of P_u with respect to centroid of bolt group, in. (not tabulated, may be determined by geometry)
e_x = horizontal component of e, in.
s = bolt spacing, in.
C = coefficient tabulated below

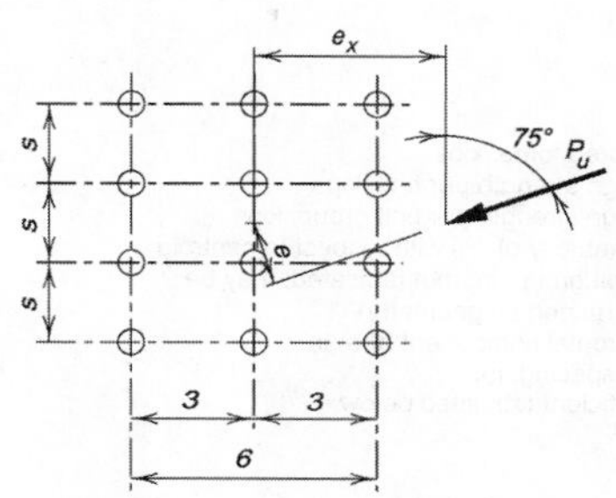

s, in.	e_x, in.	Number of Bolts in One Vertical Row, n											
		1	2	3	4	5	6	7	8	9	10	11	12
	2	2.86	5.68	8.47	11.3	14.1	16.9	19.8	22.6	25.5	28.4	31.3	34.2
	3	2.77	5.49	8.19	10.9	13.7	16.4	19.2	22.1	24.9	27.8	30.7	33.6
	4	2.66	5.27	7.89	10.5	13.2	16.0	18.8	21.6	24.4	27.2	30.1	33.0
	5	2.53	5.04	7.58	10.2	12.8	15.5	18.3	21.0	23.9	26.7	29.5	32.4
	6	2.40	4.81	7.27	9.81	12.4	15.1	17.8	20.6	23.3	26.2	29.0	31.8
	7	2.26	4.57	6.97	9.47	12.0	14.7	17.4	20.1	22.9	25.6	28.4	31.3
	8	2.13	4.35	6.69	9.13	11.7	14.3	16.9	19.6	22.4	25.1	27.9	30.7
	9	2.00	4.13	6.41	8.82	11.3	13.9	16.5	19.2	21.9	24.7	27.4	30.2
	10	1.89	3.93	6.15	8.51	11.0	13.5	16.1	18.8	21.5	24.2	27.0	29.8
3	12	1.67	3.57	5.67	7.95	10.4	12.9	15.4	18.0	20.7	23.4	26.1	28.8
	14	1.49	3.25	5.25	7.44	9.77	12.2	14.7	17.3	19.9	22.6	25.3	28.0
	16	1.34	2.97	4.87	6.98	9.23	11.6	14.1	16.6	19.2	21.8	24.5	27.2
	18	1.21	2.73	4.54	6.56	8.74	11.1	13.5	16.0	18.5	21.1	23.7	26.4
	20	1.10	2.53	4.24	6.18	8.28	10.5	12.9	15.3	17.8	20.4	23.0	25.6
	24	0.93	2.19	3.75	5.52	7.48	9.59	11.8	14.2	16.6	19.1	21.6	24.2
	28	0.80	1.93	3.34	4.97	6.79	8.78	10.9	13.2	15.5	17.9	20.4	22.9
	32	0.71	1.72	3.01	4.51	6.20	8.08	10.1	12.3	14.5	16.8	19.2	21.7
	36	0.63	1.55	2.74	4.12	5.70	7.47	9.40	11.5	13.6	15.9	18.2	20.6
	2	2.86	5.66	8.48	11.3	14.2	17.1	20.1	23.0	26.4	29.3	32.3	35.2
	3	2.77	5.49	8.25	11.1	13.9	16.8	19.7	22.7	25.6	28.5	31.5	34.4
	4	2.66	5.30	8.02	10.8	13.6	16.5	19.4	22.3	25.2	28.2	31.1	34.0
	5	2.53	5.10	7.79	10.6	13.4	16.2	19.1	22.0	24.9	27.8	30.8	33.7
	6	2.40	4.91	7.56	10.3	13.1	15.9	18.8	21.7	24.6	27.5	30.4	33.3
	7	2.26	4.72	7.34	10.1	12.9	15.7	18.5	21.4	24.3	27.2	30.1	33.0
	8	2.13	4.54	7.14	9.83	12.6	15.4	18.3	21.1	24.0	26.9	29.8	32.7
	9	2.00	4.37	6.94	9.61	12.4	15.2	18.0	20.8	23.7	26.6	29.5	32.4
	10	1.89	4.21	6.75	9.40	12.1	14.9	17.7	20.6	23.4	26.3	29.2	32.1
6	12	1.67	3.90	6.39	9.00	11.7	14.4	17.2	20.0	22.9	25.7	28.6	31.5
	14	1.49	3.63	6.06	8.63	11.3	14.0	16.8	19.6	22.4	25.2	28.1	30.9
	16	1.34	3.39	5.75	8.29	10.9	13.6	16.3	19.1	21.9	24.7	27.5	30.4
	18	1.21	3.17	5.47	7.96	10.6	13.2	15.9	18.7	21.4	24.2	27.0	29.9
	20	1.10	2.98	5.22	7.66	10.2	12.9	15.5	18.2	21.0	23.8	26.6	29.4
	24	0.93	2.65	4.76	7.10	9.57	12.2	14.8	17.5	20.2	22.9	25.7	28.5
	28	0.80	2.38	4.37	6.60	8.99	11.5	14.1	16.7	19.4	22.1	24.8	27.6
	32	0.71	2.16	4.03	6.15	8.45	10.9	13.4	16.0	18.7	21.3	24.0	26.8
	36	0.63	1.97	3.73	5.75	7.96	10.3	12.8	15.3	17.9	20.6	23.3	26.0

Table 7-22.
Coefficients C for Eccentrically Loaded Bolt Groups
Angle = 0°

$$C_{req} = \frac{P_u}{\phi r_n} \quad \text{or} \quad \phi R_n = C \times \phi r_n$$

where

P_u = factored force, kips
ϕr_n = design strength per bolt, kips
ϕR_n = design strength per bolt group, kips
e = eccentricity of P_u with respect to centroid of bolt group, in. (not tabulated, may be determined by geometry)
e_x = horizontal component of e, in.
s = bolt spacing, in.
C = coefficient tabulated below

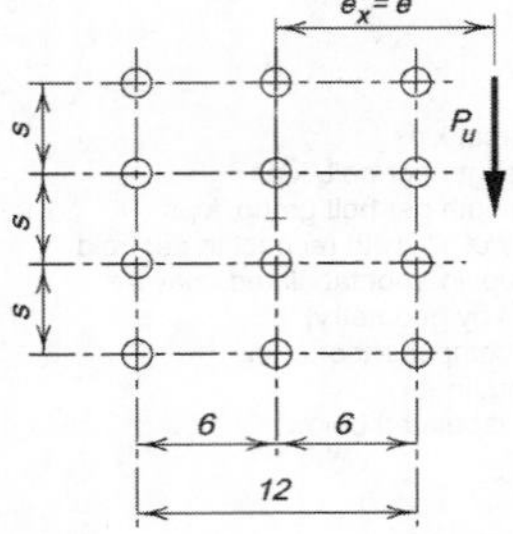

s, in.	e_x, in.	Number of Bolts in One Vertical Row, n											
		1	2	3	4	5	6	7	8	9	10	11	12
	2	2.15	4.55	7.17	10.0	13.0	16.0	19.1	22.2	25.3	28.3	31.4	34.4
	3	1.91	4.06	6.43	9.06	11.9	14.9	17.9	21.0	24.1	27.2	30.3	33.4
	4	1.71	3.65	5.80	8.23	10.9	13.7	16.7	19.8	22.9	26.0	29.1	32.3
	5	1.55	3.31	5.27	7.51	9.97	12.7	15.5	18.5	21.5	24.7	27.8	31.0
	6	1.42	3.02	4.82	6.88	9.16	11.7	14.4	17.3	20.3	23.3	26.4	29.6
	7	1.31	2.77	4.44	6.34	8.46	10.8	13.4	16.1	19.0	22.0	25.1	28.2
	8	1.21	2.56	4.10	5.87	7.85	10.1	12.5	15.1	17.9	20.7	23.7	26.8
	9	1.12	2.38	3.81	5.46	7.31	9.39	11.7	14.1	16.8	19.6	22.5	25.5
	10	1.05	2.21	3.55	5.09	6.84	8.79	10.9	13.3	15.8	18.5	21.3	24.2
3	12	0.92	1.94	3.12	4.48	6.03	7.78	9.70	11.8	14.1	16.6	19.1	21.9
	14	0.81	1.72	2.77	3.99	5.38	6.95	8.69	10.6	12.7	14.9	17.3	19.9
	16	0.72	1.53	2.48	3.58	4.84	6.27	7.85	9.60	11.5	13.6	15.8	18.1
	18	0.64	1.38	2.25	3.25	4.40	5.70	7.15	8.75	10.5	12.4	14.4	16.6
	20	0.58	1.26	2.05	2.96	4.02	5.21	6.55	8.03	9.65	11.4	13.3	15.3
	24	0.49	1.06	1.73	2.52	3.42	4.45	5.60	6.88	8.29	9.82	11.5	13.2
	28	0.42	0.92	1.50	2.19	2.97	3.87	4.88	6.00	7.24	8.59	10.1	11.6
	32	0.37	0.81	1.32	1.93	2.63	3.42	4.32	5.32	6.42	7.62	8.93	10.3
	36	0.33	0.72	1.18	1.72	2.35	3.06	3.87	4.77	5.76	6.84	8.02	9.29
	2	2.15	4.94	7.98	11.1	14.2	17.2	20.2	23.2	26.2	29.2	32.1	35.1
	3	1.91	4.48	7.39	10.5	13.6	16.7	19.8	22.8	25.8	28.9	31.9	34.8
	4	1.71	4.07	6.81	9.86	13.0	16.1	19.3	22.3	25.4	28.5	31.5	34.5
	5	1.55	3.71	6.27	9.22	12.3	15.5	18.6	21.8	24.9	28.0	31.0	34.1
	6	1.42	3.40	5.79	8.61	11.7	14.8	18.0	21.1	24.3	27.4	30.5	33.6
	7	1.31	3.13	5.35	8.05	11.0	14.1	17.3	20.5	23.6	26.8	29.9	33.1
	8	1.21	2.90	4.97	7.53	10.4	13.4	16.6	19.8	23.0	26.1	29.3	32.5
	9	1.12	2.69	4.64	7.07	9.78	12.8	15.9	19.0	22.2	25.4	28.6	31.8
	10	1.05	2.51	4.34	6.64	9.24	12.1	15.2	18.3	21.5	24.7	27.9	31.1
6	12	0.92	2.21	3.85	5.91	8.27	11.0	13.9	16.9	20.0	23.2	26.4	29.7
	14	0.81	1.96	3.44	5.31	7.46	9.95	12.7	15.6	18.6	21.8	25.0	28.2
	16	0.72	1.76	3.11	4.80	6.78	9.09	11.6	14.4	17.3	20.4	23.5	26.7
	18	0.64	1.60	2.83	4.38	6.20	8.34	10.7	13.3	16.1	19.1	22.1	25.2
	20	0.58	1.46	2.59	4.02	5.71	7.70	9.91	12.4	15.0	17.9	20.8	23.8
	24	0.49	1.24	2.21	3.44	4.91	6.65	8.59	10.8	13.2	15.7	18.5	21.3
	28	0.42	1.08	1.92	3.00	4.30	5.83	7.57	9.53	11.7	14.0	16.5	19.2
	32	0.37	0.95	1.70	2.66	3.82	5.19	6.75	8.51	10.5	12.6	14.9	17.3
	36	0.33	0.85	1.52	2.39	3.43	4.67	6.08	7.68	9.45	11.4	13.5	15.8

Table 7-22 (cont.). Coefficients C for Eccentrically Loaded Bolt Groups Angle = 15°

$C_{req} = \dfrac{P_u}{\phi r_n}$ or $\phi R_n = C \times \phi r_n$

where

P_u = factored force, kips
ϕr_n = design strength per bolt, kips
ϕR_n = design strength per bolt group, kips
e = eccentricity of P_u with respect to centroid of bolt group, in. (not tabulated, may be determined by geometry)
e_x = horizontal component of e, in.
s = bolt spacing, in.
C = coefficient tabulated below

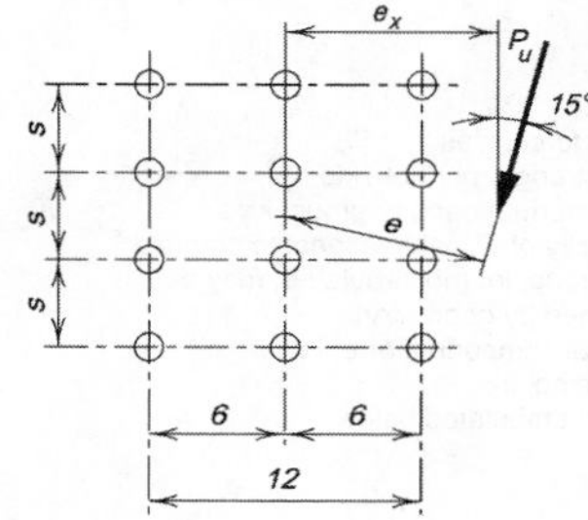

s, in.	e_x, in.	Number of Bolts in One Vertical Row, n											
		1	2	3	4	5	6	7	8	9	10	11	12
	2	2.22	4.62	7.25	10.1	13.0	16.0	19.0	22.1	25.1	28.2	31.2	34.2
	3	1.97	4.13	6.53	9.13	11.9	14.9	17.9	20.9	24.0	27.1	30.1	33.2
	4	1.77	3.72	5.91	8.31	10.9	13.7	16.7	19.7	22.7	25.8	28.9	32.0
	5	1.61	3.38	5.39	7.60	10.1	12.7	15.5	18.4	21.4	24.5	27.6	30.7
	6	1.47	3.10	4.93	6.98	9.28	11.8	14.4	17.2	20.2	23.2	26.2	29.3
	7	1.35	2.85	4.54	6.45	8.59	10.9	13.5	16.1	19.0	21.9	24.9	27.9
	8	1.25	2.63	4.21	5.98	7.98	10.2	12.6	15.1	17.8	20.7	23.6	26.6
	9	1.16	2.44	3.91	5.57	7.45	9.51	11.8	14.2	16.8	19.5	22.4	25.3
	10	1.08	2.28	3.65	5.21	6.97	8.92	11.1	13.4	15.9	18.5	21.2	24.1
3	12	0.94	2.00	3.20	4.59	6.16	7.91	9.84	11.9	14.2	16.6	19.2	21.9
	14	0.83	1.77	2.85	4.09	5.50	7.08	8.84	10.8	12.8	15.0	17.4	19.9
	16	0.74	1.58	2.56	3.68	4.96	6.40	8.00	9.75	11.7	13.7	15.9	18.2
	18	0.66	1.43	2.31	3.34	4.51	5.83	7.30	8.91	10.7	12.6	14.6	16.8
	20	0.60	1.30	2.11	3.05	4.13	5.34	6.70	8.19	9.82	11.6	13.5	15.5
	24	0.50	1.10	1.79	2.59	3.52	4.56	5.74	7.03	8.45	10.0	11.7	13.4
	28	0.43	0.95	1.55	2.25	3.06	3.98	5.01	6.15	7.40	8.77	10.2	11.8
	32	0.38	0.84	1.37	1.99	2.70	3.52	4.43	5.45	6.57	7.79	9.12	10.5
	36	0.34	0.75	1.22	1.78	2.42	3.15	3.98	4.89	5.90	7.01	8.20	9.49
	2	2.22	4.97	7.97	11.0	14.1	17.1	20.1	23.1	26.1	29.1	32.1	35.0
	3	1.97	4.50	7.40	10.5	13.5	16.6	19.7	22.7	25.7	28.7	31.7	34.7
	4	1.77	4.10	6.84	9.82	12.9	16.0	19.1	22.2	25.2	28.3	31.3	34.3
	5	1.61	3.75	6.32	9.20	12.3	15.4	18.5	21.6	24.7	27.8	30.8	33.9
	6	1.47	3.45	5.86	8.61	11.6	14.7	17.8	20.9	24.1	27.2	30.3	33.3
	7	1.35	3.18	5.44	8.06	11.0	14.0	17.1	20.3	23.4	26.5	29.6	32.7
	8	1.25	2.95	5.07	7.55	10.4	13.3	16.4	19.5	22.7	25.8	29.0	32.1
	9	1.16	2.75	4.73	7.09	9.78	12.7	15.7	18.8	22.0	25.1	28.3	31.4
	10	1.08	2.57	4.44	6.67	9.26	12.1	15.1	18.1	21.3	24.4	27.6	30.7
6	12	0.94	2.26	3.93	5.96	8.33	11.0	13.8	16.8	19.8	23.0	26.1	29.3
	14	0.83	2.01	3.52	5.37	7.55	9.97	12.7	15.5	18.5	21.5	24.7	27.8
	16	0.74	1.81	3.18	4.87	6.88	9.13	11.7	14.4	17.2	20.2	23.2	26.4
	18	0.66	1.64	2.90	4.45	6.31	8.40	10.8	13.3	16.1	18.9	21.9	25.0
	20	0.60	1.50	2.65	4.10	5.81	7.77	9.99	12.4	15.0	17.8	20.7	23.6
	24	0.50	1.28	2.27	3.52	5.01	6.74	8.71	10.9	13.2	15.8	18.4	21.2
	28	0.43	1.11	1.98	3.08	4.40	5.93	7.69	9.62	11.8	14.1	16.5	19.1
	32	0.38	0.98	1.75	2.73	3.91	5.29	6.87	8.62	10.6	12.7	15.0	17.4
	36	0.34	0.88	1.57	2.45	3.52	4.77	6.20	7.80	9.59	11.5	13.6	15.9

Table 7-22 (cont.). Coefficients *C* for Eccentrically Loaded Bolt Groups Angle = 30°

$$C_{req} = \frac{P_u}{\phi r_n} \quad \text{or} \quad \phi R_n = C \times \phi r_n$$

where

P_u = factored force, kips
ϕr_n = design strength per bolt, kips
ϕR_n = design strength per bolt group, kips
e = eccentricity of P_u with respect to centroid of bolt group, in. (not tabulated, may be determined by geometry)
e_x = horizontal component of e, in.
s = bolt spacing, in.
C = coefficient tabulated below

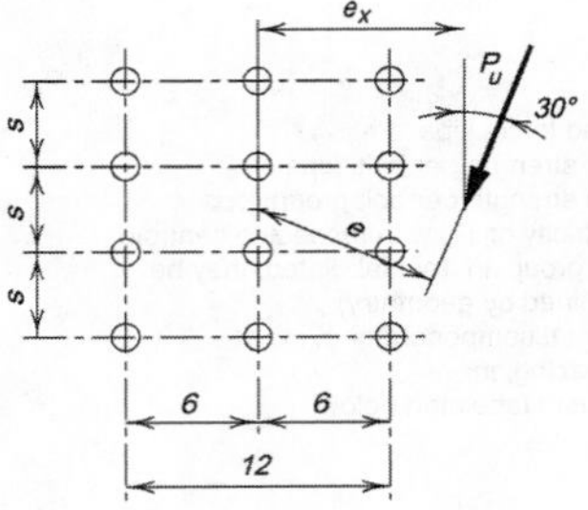

s, in.	e_x, in.	Number of Bolts in One Vertical Row, n											
		1	2	3	4	5	6	7	8	9	10	11	12
	2	2.40	4.89	7.53	10.3	13.2	16.1	19.1	22.1	25.1	28.1	31.1	34.1
	3	2.15	4.40	6.84	9.45	12.2	15.1	18.0	21.0	24.0	27.0	30.0	33.0
	4	1.94	3.99	6.24	8.69	11.3	14.0	16.9	19.8	22.8	25.8	28.8	31.9
	5	1.76	3.65	5.74	8.02	10.5	13.1	15.8	18.7	21.6	24.6	27.6	30.6
	6	1.61	3.35	5.29	7.42	9.72	12.2	14.8	17.6	20.4	23.4	26.3	29.3
	7	1.49	3.10	4.90	6.89	9.06	11.4	13.9	16.6	19.3	22.2	25.1	28.1
	8	1.37	2.87	4.55	6.42	8.47	10.7	13.1	15.6	18.3	21.1	23.9	26.9
	9	1.28	2.67	4.24	6.00	7.94	10.1	12.4	14.8	17.4	20.0	22.8	25.7
	10	1.19	2.49	3.97	5.63	7.47	9.49	11.7	14.0	16.5	19.1	21.8	24.6
3	12	1.04	2.19	3.50	4.98	6.64	8.48	10.5	12.6	14.9	17.3	19.9	22.5
	14	0.92	1.95	3.12	4.46	5.97	7.64	9.46	11.4	13.6	15.8	18.2	20.7
	16	0.82	1.75	2.81	4.03	5.40	6.93	8.61	10.4	12.4	14.5	16.7	19.1
	18	0.74	1.58	2.55	3.66	4.92	6.33	7.89	9.59	11.4	13.4	15.5	17.7
	20	0.67	1.44	2.33	3.35	4.52	5.82	7.27	8.85	10.6	12.4	14.4	16.4
	24	0.56	1.22	1.98	2.86	3.87	5.00	6.26	7.65	9.16	10.8	12.5	14.4
	28	0.48	1.06	1.72	2.49	3.37	4.37	5.48	6.71	8.06	9.51	11.1	12.8
	32	0.42	0.93	1.52	2.20	2.99	3.88	4.87	5.97	7.18	8.49	9.91	11.4
	36	0.38	0.83	1.36	1.97	2.68	3.48	4.38	5.38	6.47	7.66	8.95	10.3
	2	2.40	5.11	8.05	11.1	14.1	17.1	20.1	23.0	26.0	29.0	32.0	34.9
	3	2.15	4.66	7.51	10.5	13.5	16.5	19.6	22.6	25.6	28.6	31.6	34.6
	4	1.94	4.26	6.99	9.90	12.9	16.0	19.0	22.0	25.1	28.1	31.1	34.1
	5	1.76	3.92	6.52	9.34	12.3	15.3	18.4	21.5	24.5	27.6	30.6	33.6
	6	1.61	3.63	6.09	8.80	11.7	14.7	17.7	20.8	23.9	27.0	30.0	33.1
	7	1.49	3.38	5.70	8.30	11.1	14.1	17.1	20.2	23.2	26.3	29.4	32.5
	8	1.37	3.15	5.35	7.83	10.6	13.5	16.5	19.5	22.6	25.7	28.7	31.8
	9	1.28	2.95	5.03	7.40	10.0	12.9	15.8	18.8	21.9	25.0	28.1	31.2
	10	1.19	2.77	4.74	7.00	9.54	12.3	15.2	18.2	21.2	24.3	27.4	30.5
6	12	1.04	2.45	4.23	6.30	8.67	11.3	14.1	17.0	19.9	23.0	26.0	29.1
	14	0.92	2.19	3.81	5.71	7.92	10.4	13.0	15.8	18.7	21.7	24.7	27.8
	16	0.82	1.98	3.45	5.22	7.27	9.58	12.1	14.8	17.6	20.5	23.4	26.4
	18	0.74	1.80	3.16	4.79	6.71	8.88	11.2	13.8	16.5	19.3	22.2	25.2
	20	0.67	1.65	2.90	4.42	6.22	8.26	10.5	12.9	15.5	18.2	21.1	24.0
	24	0.56	1.41	2.49	3.82	5.41	7.22	9.23	11.5	13.8	16.4	19.0	21.8
	28	0.48	1.23	2.18	3.36	4.78	6.40	8.22	10.3	12.4	14.8	17.2	19.8
	32	0.42	1.08	1.93	2.99	4.26	5.73	7.40	9.25	11.3	13.4	15.7	18.2
	36	0.38	0.97	1.73	2.69	3.85	5.18	6.71	8.41	10.3	12.3	14.4	16.7

Table 7-22 (cont.). Coefficients *C* for Eccentrically Loaded Bolt Groups Angle = 45°

$$C_{req} = \frac{P_u}{\phi r_n} \quad \text{or} \quad \phi R_n = C \times \phi r_n$$

where

P_u = factored force, kips
ϕr_n = design strength per bolt, kips
ϕR_n = design strength per bolt group, kips
e = eccentricity of P_u with respect to centroid of bolt group, in. (not tabulated, may be determined by geometry)
e_x = horizontal component of e, in.
s = bolt spacing, in.
C = coefficient tabulated below

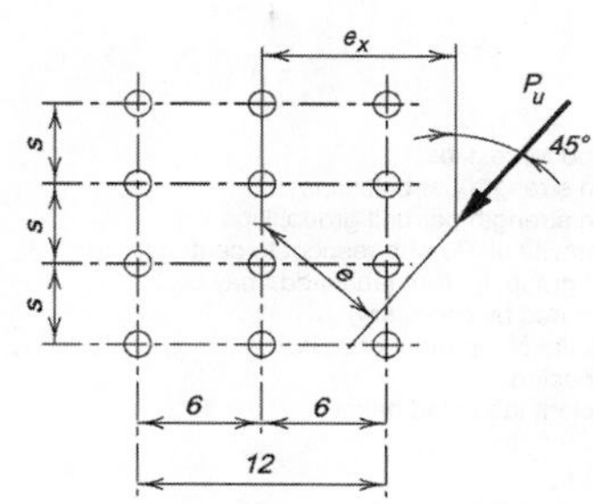

s, in.	*e*$_x$, in.	Number of Bolts in One Vertical Row, *n*											
		1	2	3	4	5	6	7	8	9	10	11	12
	2	2.64	5.30	8.01	10.8	13.6	16.4	19.3	22.3	25.2	28.1	31.1	34.0
	3	2.43	4.90	7.44	10.1	12.8	15.6	18.4	21.3	24.2	27.1	30.1	33.1
	4	2.23	4.52	6.89	9.38	12.0	14.7	17.5	20.3	23.2	26.1	29.0	32.0
	5	2.05	4.17	6.40	8.75	11.2	13.9	16.6	19.3	22.2	25.0	27.9	30.9
	6	1.89	3.86	5.96	8.20	10.6	13.1	15.7	18.4	21.2	24.0	26.9	29.8
	7	1.75	3.59	5.57	7.70	9.99	12.4	14.9	17.5	20.2	23.0	25.8	28.7
	8	1.63	3.35	5.22	7.25	9.43	11.7	14.2	16.7	19.3	22.1	24.8	27.7
	9	1.52	3.13	4.90	6.83	8.91	11.1	13.5	15.9	18.5	21.2	23.9	26.7
	10	1.42	2.94	4.61	6.45	8.44	10.6	12.8	15.2	17.7	20.3	23.0	25.7
3	12	1.25	2.60	4.11	5.78	7.60	9.58	11.7	14.0	16.3	18.8	21.3	23.9
	14	1.11	2.32	3.69	5.21	6.90	8.73	10.7	12.8	15.0	17.4	19.8	22.3
	16	0.99	2.09	3.34	4.74	6.29	8.00	9.85	11.8	13.9	16.1	18.5	20.9
	18	0.90	1.90	3.04	4.33	5.77	7.36	9.10	11.0	12.9	15.0	17.3	19.5
	20	0.81	1.73	2.79	3.98	5.33	6.81	8.44	10.2	12.1	14.1	16.2	18.4
	24	0.68	1.47	2.38	3.42	4.60	5.91	7.35	8.91	10.6	12.4	14.3	16.3
	28	0.59	1.28	2.08	2.99	4.03	5.20	6.49	7.90	9.42	11.1	12.8	14.6
	32	0.52	1.13	1.84	2.65	3.59	4.63	5.80	7.07	8.46	9.95	11.6	13.3
	36	0.46	1.01	1.65	2.38	3.23	4.17	5.23	6.40	7.67	9.04	10.5	12.1
	2	2.64	5.38	8.22	11.1	14.1	17.0	20.0	23.0	25.9	28.9	31.9	34.8
	3	2.43	5.02	7.78	10.7	13.6	16.6	19.5	22.5	25.5	28.5	31.4	34.4
	4	2.23	4.67	7.33	10.2	13.1	16.0	19.0	22.0	25.0	28.0	31.0	33.9
	5	2.05	4.34	6.90	9.66	12.5	15.5	18.4	21.4	24.4	27.4	30.4	33.4
	6	1.89	4.06	6.50	9.19	12.0	14.9	17.9	20.9	23.9	26.9	29.9	32.9
	7	1.75	3.80	6.16	8.76	11.5	14.4	17.3	20.3	23.3	26.3	29.3	32.3
	8	1.63	3.57	5.84	8.36	11.1	13.9	16.8	19.7	22.7	25.7	28.7	31.7
	9	1.52	3.36	5.54	7.99	10.6	13.4	16.2	19.2	22.1	25.1	28.1	31.1
	10	1.42	3.17	5.27	7.63	10.2	12.9	15.7	18.6	21.5	24.5	27.5	30.5
6	12	1.25	2.84	4.78	6.99	9.40	12.0	14.7	17.6	20.4	23.4	26.3	29.3
	14	1.11	2.57	4.36	6.42	8.70	11.2	13.8	16.6	19.4	22.3	25.2	28.2
	16	0.99	2.33	3.99	5.92	8.09	10.5	13.0	15.7	18.4	21.2	24.1	27.0
	18	0.90	2.13	3.68	5.49	7.54	9.80	12.2	14.8	17.5	20.3	23.1	26.0
	20	0.81	1.96	3.40	5.10	7.05	9.21	11.6	14.0	16.6	19.3	22.1	24.9
	24	0.68	1.68	2.95	4.46	6.22	8.19	10.4	12.7	15.1	17.7	20.3	23.0
	28	0.59	1.47	2.59	3.95	5.55	7.35	9.34	11.5	13.8	16.2	18.7	21.3
	32	0.52	1.31	2.31	3.54	4.99	6.65	8.49	10.5	12.7	14.9	17.3	19.8
	36	0.46	1.17	2.08	3.20	4.54	6.06	7.77	9.64	11.7	13.8	16.1	18.5

Table 7-22 (cont.). Coefficients C for Eccentrically Loaded Bolt Groups Angle = 60°

$C_{req} = \dfrac{P_u}{\phi r_n}$ or $\phi R_n = C \times \phi r_n$

where

P_u = factored force, kips
ϕr_n = design strength per bolt, kips
ϕR_n = design strength per bolt group, kips
e = eccentricity of P_u with respect to centroid of bolt group, in. (not tabulated, may be determined by geometry)
e_x = horizontal component of e, in.
s = bolt spacing, in.
C = coefficient tabulated below

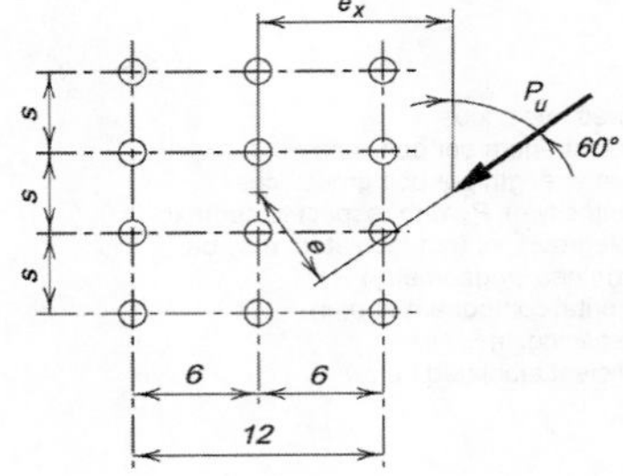

s, in.	e_x, in.	Number of Bolts in One Vertical Row, n											
		1	2	3	4	5	6	7	8	9	10	11	12
3	2	2.83	5.64	8.45	11.3	14.1	16.9	19.8	22.6	25.5	28.4	31.3	34.2
	3	2.72	5.43	8.13	10.8	13.6	16.3	19.1	21.9	24.8	27.6	30.5	33.4
	4	2.59	5.18	7.77	10.4	13.0	15.7	18.5	21.2	24.0	26.8	29.7	32.5
	5	2.46	4.92	7.40	9.92	12.5	15.1	17.8	20.5	23.2	26.0	28.9	31.7
	6	2.32	4.66	7.03	9.46	12.0	14.5	17.1	19.8	22.5	25.2	28.0	30.8
	7	2.19	4.41	6.68	9.02	11.4	13.9	16.5	19.1	21.8	24.5	27.2	30.0
	8	2.07	4.17	6.35	8.61	11.0	13.4	15.9	18.4	21.1	23.7	26.5	29.2
	9	1.95	3.95	6.04	8.22	10.5	12.9	15.3	17.8	20.4	23.0	25.7	28.5
	10	1.84	3.74	5.75	7.86	10.1	12.4	14.8	17.3	19.8	22.4	25.0	27.7
	12	1.65	3.38	5.22	7.19	9.28	11.5	13.8	16.2	18.6	21.1	23.7	26.3
	14	1.49	3.06	4.76	6.61	8.58	10.7	12.9	15.2	17.5	20.0	22.5	25.0
	16	1.35	2.79	4.37	6.09	7.95	9.93	12.0	14.2	16.5	18.9	21.3	23.8
	18	1.23	2.55	4.02	5.64	7.39	9.28	11.3	13.4	15.6	17.9	20.3	22.7
	20	1.12	2.35	3.72	5.24	6.90	8.69	10.6	12.6	14.8	17.0	19.3	21.7
	24	0.95	2.02	3.22	4.57	6.06	7.68	9.43	11.3	13.3	15.4	17.5	19.8
	28	0.83	1.76	2.84	4.04	5.39	6.86	8.47	10.2	12.0	14.0	16.0	18.1
	32	0.73	1.56	2.53	3.61	4.84	6.19	7.66	9.26	11.0	12.8	14.7	16.7
	36	0.65	1.40	2.27	3.26	4.38	5.62	6.98	8.46	10.1	11.7	13.5	15.4
6	2	2.83	5.64	8.47	11.3	14.2	17.1	20.0	23.0	25.9	28.9	31.8	34.8
	3	2.72	5.44	8.19	11.0	13.8	16.7	19.6	22.6	25.5	28.4	31.4	34.3
	4	2.59	5.21	7.88	10.6	13.4	16.3	19.2	22.1	25.0	28.0	30.9	33.9
	5	2.46	4.97	7.57	10.3	13.1	15.9	18.8	21.7	24.6	27.5	30.4	33.4
	6	2.32	4.73	7.27	9.91	12.7	15.5	18.3	21.2	24.1	27.0	30.0	32.9
	7	2.19	4.51	6.97	9.56	12.3	15.0	17.9	20.8	23.7	26.6	29.5	32.4
	8	2.07	4.29	6.69	9.23	11.9	14.6	17.5	20.3	23.2	26.1	29.0	32.0
	9	1.95	4.09	6.43	8.92	11.5	14.3	17.0	19.9	22.8	25.6	28.6	31.5
	10	1.84	3.90	6.18	8.63	11.2	13.9	16.6	19.5	22.3	25.2	28.1	31.0
	12	1.65	3.56	5.73	8.08	10.6	13.2	15.9	18.7	21.5	24.3	27.2	30.1
	14	1.49	3.27	5.32	7.59	10.0	12.6	15.2	17.9	20.7	23.5	26.3	29.2
	16	1.35	3.01	4.95	7.13	9.48	12.0	14.5	17.2	19.9	22.7	25.5	28.4
	18	1.23	2.78	4.63	6.71	8.98	11.4	13.9	16.5	19.2	22.0	24.7	27.6
	20	1.12	2.58	4.34	6.33	8.52	10.9	13.3	15.9	18.5	21.2	24.0	26.8
	24	0.95	2.25	3.84	5.67	7.70	9.91	12.3	14.7	17.3	19.9	22.6	25.3
	28	0.83	1.98	3.43	5.11	7.00	9.08	11.3	13.7	16.1	18.7	21.3	23.9
	32	0.73	1.77	3.09	4.64	6.40	8.36	10.5	12.7	15.1	17.5	20.1	22.6
	36	0.65	1.60	2.81	4.24	5.89	7.73	9.74	11.9	14.2	16.5	19.0	21.5

Table 7-22 (cont.). Coefficients C for Eccentrically Loaded Bolt Groups Angle = 75°

$$C_{req} = \frac{P_u}{\phi r_n} \quad \text{or} \quad \phi R_n = C \times \phi r_n$$

where

P_u = factored force, kips
ϕr_n = design strength per bolt, kips
ϕR_n = design strength per bolt group, kips
e = eccentricity of P_u with respect to centroid of bolt group, in. (not tabulated, may be determined by geometry)
e_x = horizontal component of e, in.
s = bolt spacing, in.
C = coefficient tabulated below

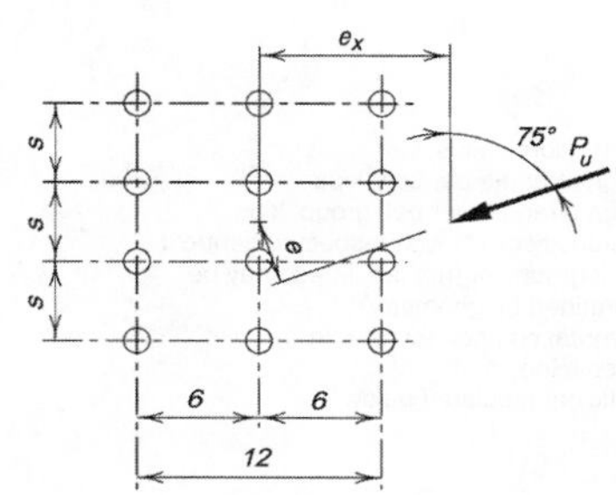

s, in.	e_x, in.	Number of Bolts in One Vertical Row, n											
		1	2	3	4	5	6	7	8	9	10	11	12
	2	2.92	5.83	8.73	11.6	14.5	17.4	20.3	23.1	26.0	28.9	31.8	34.7
	3	2.89	5.77	8.63	11.5	14.3	17.2	20.0	22.8	25.7	28.5	31.4	34.2
	4	2.86	5.70	8.51	11.3	14.1	16.9	19.7	22.5	25.3	28.1	30.9	33.7
	5	2.82	5.61	8.38	11.1	13.9	16.6	19.4	22.1	24.9	27.7	30.5	33.3
	6	2.77	5.51	8.23	10.9	13.6	16.3	19.0	21.8	24.5	27.2	30.0	32.8
	7	2.72	5.40	8.06	10.7	13.4	16.0	18.7	21.4	24.1	26.8	29.6	32.3
	8	2.66	5.29	7.89	10.5	13.1	15.7	18.3	21.0	23.7	26.4	29.1	31.9
	9	2.60	5.16	7.71	10.3	12.8	15.4	18.0	20.6	23.3	26.0	28.7	31.4
	10	2.53	5.04	7.53	10.0	12.6	15.1	17.7	20.3	22.9	25.6	28.3	31.0
3	12	2.40	4.78	7.16	9.57	12.0	14.5	17.0	19.6	22.1	24.8	27.4	30.1
	14	2.26	4.52	6.80	9.12	11.5	13.9	16.4	18.9	21.4	24.0	26.6	29.3
	16	2.13	4.27	6.45	8.68	11.0	13.3	15.8	18.2	20.7	23.3	25.9	28.5
	18	2.00	4.03	6.12	8.27	10.5	12.8	15.2	17.6	20.1	22.6	25.1	27.7
	20	1.89	3.81	5.80	7.88	10.1	12.3	14.6	17.0	19.4	21.9	24.4	27.0
	24	1.67	3.41	5.24	7.18	9.22	11.4	13.6	15.9	18.2	20.7	23.1	25.6
	28	1.49	3.06	4.75	6.56	8.49	10.5	12.6	14.9	17.1	19.5	21.9	24.3
	32	1.34	2.77	4.33	6.02	7.84	9.77	11.8	13.9	16.1	18.4	20.7	23.1
	36	1.21	2.52	3.97	5.56	7.27	9.10	11.1	13.1	15.2	17.4	19.7	22.0
	2	2.92	5.82	8.71	11.6	14.5	17.4	20.3	23.5	26.4	29.3	32.3	35.2
	3	2.89	5.76	8.60	11.4	14.3	17.1	20.0	22.9	25.8	28.7	31.7	34.6
	4	2.86	5.68	8.47	11.3	14.1	16.9	19.8	22.6	25.5	28.4	31.3	34.2
	5	2.82	5.59	8.34	11.1	13.9	16.7	19.5	22.4	25.2	28.1	31.0	33.9
	6	2.77	5.49	8.19	10.9	13.7	16.4	19.2	22.1	24.9	27.8	30.7	33.6
	7	2.72	5.39	8.04	10.7	13.4	16.2	19.0	21.8	24.6	27.5	30.4	33.3
	8	2.66	5.27	7.89	10.5	13.2	16.0	18.8	21.6	24.4	27.2	30.1	33.0
	9	2.60	5.16	7.74	10.4	13.0	15.8	18.5	21.3	24.1	27.0	29.8	32.7
	10	2.53	5.04	7.58	10.2	12.8	15.5	18.3	21.0	23.9	26.7	29.5	32.4
6	12	2.40	4.81	7.27	9.81	12.4	15.1	17.8	20.6	23.3	26.2	29.0	31.8
	14	2.26	4.57	6.97	9.47	12.0	14.7	17.4	20.1	22.9	25.6	28.4	31.3
	16	2.13	4.35	6.69	9.13	11.7	14.3	16.9	19.6	22.4	25.1	27.9	30.7
	18	2.00	4.13	6.41	8.82	11.3	13.9	16.5	19.2	21.9	24.7	27.4	30.2
	20	1.89	3.93	6.15	8.51	11.0	13.5	16.1	18.8	21.5	24.2	27.0	29.8
	24	1.67	3.57	5.67	7.95	10.4	12.9	15.4	18.0	20.7	23.4	26.1	28.8
	28	1.49	3.25	5.25	7.44	9.77	12.2	14.7	17.3	19.9	22.6	25.3	28.0
	32	1.34	2.97	4.87	6.98	9.23	11.6	14.1	16.6	19.2	21.8	24.5	27.2
	36	1.21	2.73	4.54	6.56	8.74	11.1	13.5	16.0	18.5	21.1	23.7	26.4

Table 7-23.
Coefficients *C* for Eccentrically Loaded Bolt Groups
Angle = 0°

$$C_{req} = \frac{P_u}{\phi r_n} \quad \text{or} \quad \phi R_n = C \times \phi r_n$$

where

P_u = factored force, kips
ϕr_n = design strength per bolt, kips
ϕR_n = design strength per bolt group, kips
e = eccentricity of P_u with respect to centroid of bolt group, in. (not tabulated, may be determined by geometry)
e_x = horizontal component of e, in.
s = bolt spacing, in.
C = coefficient tabulated below

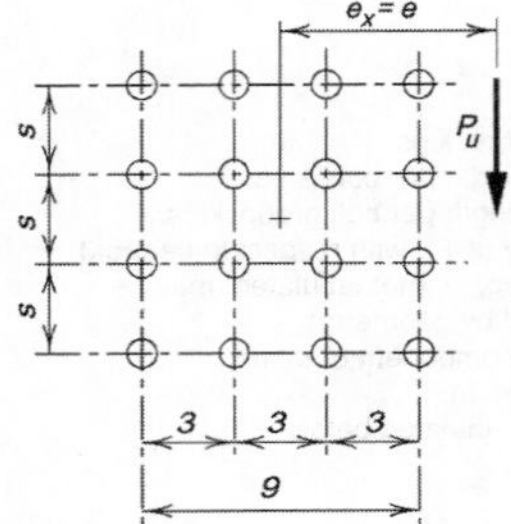

s, in.	e_x, in.	Number of Bolts in One Vertical Row, n											
		1	2	3	4	5	6	7	8	9	10	11	12
	2	2.60	5.70	9.24	13.2	17.3	21.4	25.6	29.7	33.8	37.8	41.9	45.9
	3	2.23	4.92	8.05	11.7	15.6	19.7	23.9	28.1	32.3	36.4	40.6	44.7
	4	1.94	4.30	7.09	10.4	14.0	18.0	22.1	26.3	30.5	34.7	38.9	43.1
	5	1.69	3.79	6.30	9.29	12.6	16.4	20.3	24.4	28.6	32.9	37.1	41.4
	6	1.49	3.37	5.65	8.37	11.5	14.9	18.7	22.6	26.7	30.9	35.2	39.4
	7	1.32	3.03	5.10	7.59	10.4	13.7	17.2	21.0	24.9	29.0	33.2	37.5
	8	1.18	2.74	4.63	6.92	9.56	12.6	15.9	19.5	23.3	27.3	31.4	35.5
	9	1.07	2.50	4.24	6.35	8.81	11.6	14.7	18.1	21.7	25.6	29.6	33.7
	10	0.98	2.29	3.89	5.86	8.15	10.8	13.7	16.9	20.3	24.0	27.9	31.9
3	12	0.83	1.96	3.34	5.06	7.06	9.37	12.0	14.8	17.9	21.3	24.9	28.6
	14	0.73	1.72	2.92	4.44	6.21	8.27	10.6	13.2	16.0	19.1	22.3	25.8
	16	0.65	1.52	2.59	3.95	5.54	7.39	9.48	11.8	14.4	17.2	20.2	23.4
	18	0.58	1.37	2.33	3.55	4.99	6.67	8.57	10.7	13.1	15.6	18.4	21.4
	20	0.53	1.24	2.11	3.23	4.53	6.07	7.81	9.77	11.9	14.3	16.9	19.6
	24	0.44	1.04	1.78	2.72	3.83	5.14	6.62	8.30	10.2	12.2	14.4	16.8
	28	0.38	0.90	1.54	2.35	3.31	4.45	5.73	7.20	8.82	10.6	12.6	14.7
	32	0.34	0.79	1.36	2.07	2.91	3.92	5.05	6.35	7.79	9.38	11.1	13.0
	36	0.30	0.71	1.21	1.85	2.60	3.50	4.51	5.68	6.96	8.39	9.95	11.6
	2	2.60	6.48	10.7	14.8	18.9	23.0	27.0	31.0	34.9	38.9	42.9	46.8
	3	2.23	5.75	9.79	14.0	18.2	22.3	26.4	30.5	34.5	38.5	42.5	46.5
	4	1.94	5.12	8.91	13.1	17.4	21.6	25.7	29.9	33.9	38.0	42.0	46.1
	5	1.69	4.58	8.10	12.2	16.4	20.7	24.9	29.1	33.2	37.4	41.4	45.5
	6	1.49	4.13	7.37	11.3	15.5	19.7	24.0	28.3	32.5	36.6	40.8	44.9
	7	1.32	3.74	6.74	10.5	14.5	18.8	23.1	27.3	31.6	35.8	40.0	44.1
	8	1.18	3.41	6.20	9.73	13.6	17.8	22.1	26.4	30.6	34.9	39.1	43.3
	9	1.07	3.13	5.73	9.05	12.8	16.9	21.1	25.4	29.7	34.0	38.2	42.5
	10	0.98	2.89	5.31	8.45	12.0	16.0	20.1	24.4	28.7	33.0	37.3	41.5
6	12	0.83	2.50	4.63	7.43	10.7	14.3	18.3	22.4	26.7	31.0	35.3	39.6
	14	0.73	2.19	4.09	6.60	9.53	12.9	16.7	20.6	24.7	29.0	33.3	37.6
	16	0.65	1.95	3.65	5.93	8.59	11.7	15.2	19.0	22.9	27.1	31.3	35.5
	18	0.58	1.76	3.29	5.37	7.81	10.7	14.0	17.5	21.3	25.3	29.4	33.6
	20	0.53	1.60	2.99	4.90	7.15	9.85	12.9	16.2	19.8	23.6	27.6	31.7
	24	0.44	1.35	2.53	4.16	6.10	8.44	11.1	14.0	17.3	20.8	24.4	28.3
	28	0.38	1.17	2.19	3.61	5.31	7.37	9.69	12.3	15.2	18.4	21.8	25.3
	32	0.34	1.03	1.93	3.19	4.69	6.53	8.61	11.0	13.6	16.5	19.6	22.9
	36	0.30	0.92	1.72	2.85	4.20	5.85	7.73	9.89	12.3	14.9	17.7	20.8

Table 7-23 (cont.).
Coefficients C for Eccentrically Loaded Bolt Groups
Angle = 15°

$$C_{req} = \frac{P_u}{\phi r_n} \quad \text{or} \quad \phi R_n = C \times \phi r_n$$

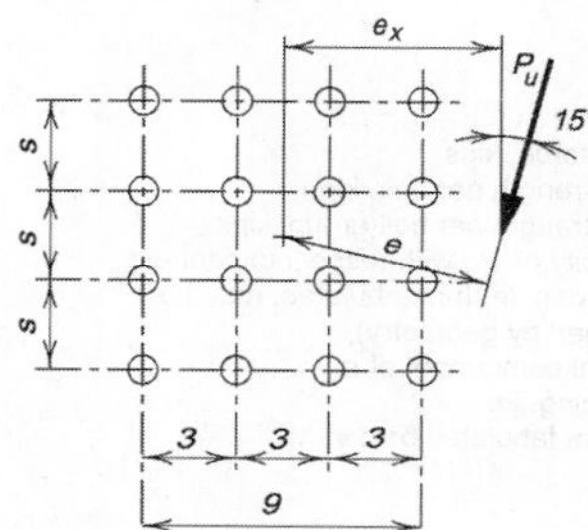

where

P_u = factored force, kips
ϕr_n = design strength per bolt, kips
ϕR_n = design strength per bolt group, kips
e = eccentricity of P_u with respect to centroid of bolt group, in. (not tabulated, may be determined by geometry)
e_x = horizontal component of e, in.
s = bolt spacing, in.
C = coefficient tabulated below

s, in.	e_x, in.	Number of Bolts in One Vertical Row, *n*											
		1	2	3	4	5	6	7	8	9	10	11	12
	2	2.68	5.77	9.31	13.2	17.2	21.3	25.4	29.5	33.6	37.6	41.7	45.7
	3	2.30	5.00	8.17	11.7	15.6	19.6	23.7	27.8	32.0	36.1	40.2	44.3
	4	1.99	4.38	7.22	10.4	14.1	17.9	21.9	26.0	30.2	34.4	38.5	42.7
	5	1.74	3.88	6.43	9.37	12.7	16.4	20.2	24.2	28.3	32.5	36.7	40.9
	6	1.53	3.45	5.77	8.47	11.6	15.0	18.6	22.5	26.5	30.6	34.8	39.0
	7	1.36	3.10	5.21	7.71	10.6	13.7	17.2	20.9	24.8	28.8	32.9	37.1
	8	1.22	2.81	4.74	7.05	9.70	12.7	15.9	19.5	23.2	27.1	31.1	35.2
	9	1.11	2.57	4.34	6.48	8.95	11.7	14.8	18.1	21.7	25.5	29.4	33.4
	10	1.01	2.36	4.00	5.98	8.29	10.9	13.8	17.0	20.4	24.0	27.7	31.6
3	12	0.86	2.02	3.44	5.18	7.21	9.52	12.1	15.0	18.1	21.4	24.9	28.5
	14	0.75	1.77	3.01	4.55	6.36	8.43	10.8	13.3	16.1	19.2	22.4	25.8
	16	0.67	1.57	2.68	4.05	5.67	7.54	9.66	12.0	14.6	17.3	20.3	23.5
	18	0.60	1.41	2.40	3.65	5.12	6.81	8.74	10.9	13.3	15.8	18.6	21.5
	20	0.54	1.28	2.18	3.32	4.66	6.21	7.98	9.95	12.1	14.5	17.1	19.8
	24	0.46	1.08	1.84	2.80	3.94	5.26	6.78	8.47	10.4	12.4	14.6	17.0
	28	0.40	0.93	1.59	2.43	3.41	4.56	5.89	7.37	9.02	10.8	12.8	14.9
	32	0.35	0.82	1.40	2.14	3.00	4.03	5.19	6.51	7.98	9.59	11.3	13.2
	36	0.31	0.73	1.25	1.91	2.68	3.60	4.65	5.83	7.15	8.59	10.2	11.9
	2	2.68	6.48	10.6	14.7	18.8	22.9	26.9	30.9	34.8	38.8	42.8	46.7
	3	2.30	5.75	9.75	13.9	18.1	22.2	26.3	30.3	34.3	38.3	42.3	46.3
	4	1.99	5.13	8.91	13.0	17.2	21.4	25.5	29.6	33.7	37.7	41.8	45.8
	5	1.74	4.61	8.14	12.1	16.3	20.5	24.7	28.8	33.0	37.1	41.1	45.2
	6	1.53	4.17	7.45	11.2	15.3	19.5	23.7	27.9	32.1	36.3	40.4	44.5
	7	1.36	3.79	6.84	10.4	14.4	18.6	22.8	27.0	31.2	35.4	39.6	43.7
	8	1.22	3.46	6.30	9.71	13.6	17.6	21.8	26.0	30.3	34.5	38.7	42.9
	9	1.11	3.19	5.83	9.05	12.8	16.7	20.9	25.1	29.3	33.5	37.8	42.0
	10	1.01	2.94	5.42	8.47	12.0	15.9	19.9	24.1	28.3	32.6	36.8	41.0
6	12	0.86	2.55	4.73	7.47	10.7	14.3	18.2	22.2	26.4	30.6	34.8	39.1
	14	0.75	2.24	4.18	6.66	9.62	12.9	16.6	20.5	24.5	28.6	32.8	37.1
	16	0.67	2.00	3.74	6.00	8.71	11.8	15.2	18.9	22.8	26.8	30.9	35.1
	18	0.60	1.80	3.38	5.45	7.94	10.8	14.0	17.5	21.2	25.1	29.1	33.2
	20	0.54	1.64	3.08	4.98	7.28	9.92	13.0	16.2	19.8	23.5	27.4	31.4
	24	0.46	1.39	2.60	4.25	6.23	8.54	11.2	14.1	17.3	20.8	24.4	28.1
	28	0.40	1.20	2.26	3.69	5.43	7.48	9.85	12.5	15.4	18.5	21.8	25.3
	32	0.35	1.06	1.99	3.26	4.81	6.65	8.77	11.1	13.8	16.6	19.7	22.9
	36	0.31	0.94	1.78	2.92	4.31	5.97	7.89	10.0	12.5	15.1	17.9	20.9

Table 7-23 (cont.). Coefficients C for Eccentrically Loaded Bolt Groups Angle = 30°

$$C_{req} = \frac{P_u}{\phi r_n} \quad \text{or} \quad \phi R_n = C \times \phi r_n$$

where

P_u = factored force, kips
ϕr_n = design strength per bolt, kips
ϕR_n = design strength per bolt group, kips
e = eccentricity of P_u with respect to centroid of bolt group, in. (not tabulated, may be determined by geometry)
e_x = horizontal component of e, in.
s = bolt spacing, in.
C = coefficient tabulated below

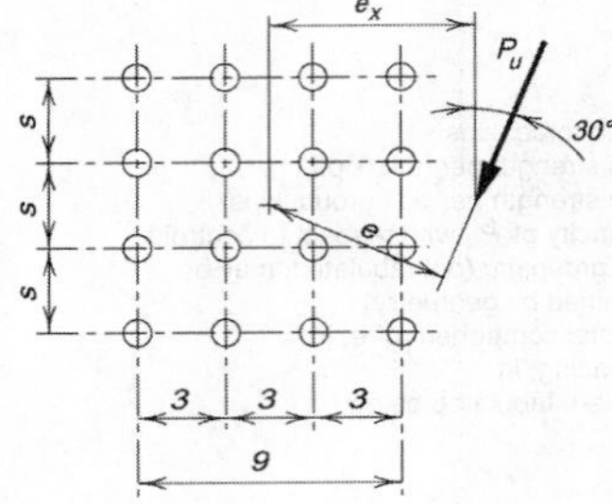

s, in.	e_x, in.	Number of Bolts in One Vertical Row, n											
		1	2	3	4	5	6	7	8	9	10	11	12
	2	2.90	6.06	9.59	13.4	17.3	21.3	25.3	29.4	33.4	37.4	41.4	45.4
	3	2.50	5.31	8.52	12.1	15.8	19.7	23.7	27.8	31.8	35.9	40.0	44.0
	4	2.18	4.70	7.62	10.9	14.4	18.2	22.1	26.1	30.1	34.2	38.3	42.4
	5	1.91	4.18	6.85	9.86	13.2	16.8	20.5	24.4	28.4	32.5	36.6	40.7
	6	1.69	3.75	6.19	8.98	12.1	15.5	19.1	22.9	26.8	30.7	34.8	38.9
	7	1.51	3.38	5.63	8.21	11.1	14.3	17.8	21.4	25.2	29.1	33.1	37.1
	8	1.36	3.07	5.14	7.55	10.3	13.3	16.6	20.0	23.7	27.5	31.4	35.4
	9	1.23	2.81	4.73	6.97	9.54	12.4	15.5	18.8	22.3	26.0	29.8	33.7
	10	1.13	2.59	4.37	6.46	8.88	11.6	14.5	17.7	21.1	24.7	28.3	32.2
3	12	0.96	2.23	3.78	5.62	7.78	10.2	12.9	15.8	18.9	22.2	25.7	29.3
	14	0.84	1.95	3.32	4.96	6.90	9.08	11.5	14.2	17.1	20.1	23.4	26.8
	16	0.74	1.73	2.96	4.43	6.19	8.17	10.4	12.9	15.5	18.4	21.4	24.6
	18	0.67	1.56	2.66	4.00	5.60	7.41	9.46	11.7	14.2	16.8	19.7	22.7
	20	0.61	1.42	2.42	3.65	5.11	6.77	8.67	10.8	13.1	15.5	18.2	21.0
	24	0.51	1.20	2.04	3.09	4.34	5.77	7.41	9.22	11.2	13.4	15.7	18.2
	28	0.44	1.03	1.77	2.68	3.77	5.01	6.46	8.05	9.83	11.8	13.9	16.1
	32	0.39	0.91	1.56	2.36	3.32	4.43	5.71	7.14	8.72	10.5	12.3	14.4
	36	0.35	0.81	1.39	2.11	2.97	3.97	5.12	6.40	7.84	9.41	11.1	13.0
	2	2.90	6.59	10.6	14.7	18.7	22.7	26.7	30.7	34.7	38.7	42.6	46.6
	3	2.50	5.88	9.83	13.9	18.0	22.0	26.1	30.1	34.1	38.1	42.1	46.1
	4	2.18	5.30	9.05	13.0	17.1	21.2	25.3	29.4	33.5	37.5	41.5	45.5
	5	1.91	4.81	8.35	12.2	16.3	20.4	24.5	28.6	32.7	36.8	40.8	44.9
	6	1.69	4.38	7.72	11.4	15.4	19.5	23.6	27.7	31.8	35.9	40.0	44.1
	7	1.51	4.01	7.15	10.7	14.6	18.6	22.7	26.8	31.0	35.1	39.2	43.3
	8	1.36	3.69	6.64	10.0	13.8	17.7	21.8	25.9	30.0	34.2	38.3	42.4
	9	1.23	3.41	6.19	9.41	13.0	16.9	20.9	25.0	29.1	33.3	37.4	41.6
	10	1.13	3.16	5.79	8.85	12.4	16.1	20.1	24.1	28.2	32.4	36.5	40.6
6	12	0.96	2.76	5.09	7.88	11.1	14.7	18.5	22.4	26.4	30.5	34.6	38.8
	14	0.84	2.44	4.54	7.08	10.1	13.4	17.0	20.8	24.7	28.8	32.8	36.9
	16	0.74	2.18	4.08	6.41	9.21	12.3	15.7	19.4	23.2	27.1	31.1	35.1
	18	0.67	1.97	3.70	5.85	8.45	11.4	14.6	18.1	21.7	25.5	29.4	33.4
	20	0.61	1.80	3.38	5.37	7.80	10.5	13.6	16.9	20.4	24.1	27.9	31.8
	24	0.51	1.53	2.87	4.61	6.74	9.16	11.9	14.9	18.1	21.5	25.1	28.8
	28	0.44	1.32	2.49	4.02	5.91	8.07	10.5	13.3	16.2	19.4	22.7	26.2
	32	0.39	1.17	2.20	3.57	5.26	7.20	9.45	11.9	14.6	17.6	20.7	23.9
	36	0.35	1.05	1.97	3.21	4.73	6.49	8.55	10.8	13.3	16.0	18.9	22.0

Table 7-23 (cont.). Coefficients *C* for Eccentrically Loaded Bolt Groups Angle = 45°

$C_{req} = \dfrac{P_u}{\phi r_n}$ or $\phi R_n = C \times \phi r_n$

where

P_u = factored force, kips
ϕr_n = design strength per bolt, kips
ϕR_n = design strength per bolt group, kips
e = eccentricity of P_u with respect to centroid of bolt group, in. (not tabulated, may be determined by geometry)
e_x = horizontal component of e, in.
s = bolt spacing, in.
C = coefficient tabulated below

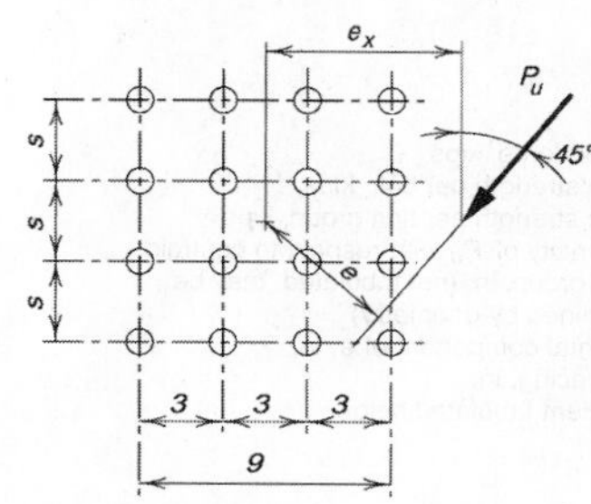

s, in.	e_x, in.	Number of Bolts in One Vertical Row, *n*											
		1	2	3	4	5	6	7	8	9	10	11	12
	2	3.26	6.62	10.2	13.9	17.7	21.5	25.5	29.4	33.4	37.3	41.3	45.3
	3	2.87	5.92	9.19	12.7	16.4	20.2	24.0	28.0	31.9	35.9	39.9	43.9
	4	2.54	5.31	8.36	11.7	15.2	18.8	22.6	26.5	30.4	34.4	38.4	42.4
	5	2.25	4.78	7.63	10.8	14.1	17.6	21.3	25.1	29.0	32.9	36.8	40.8
	6	2.01	4.33	6.99	9.94	13.1	16.5	20.1	23.8	27.5	31.4	35.3	39.3
	7	1.81	3.93	6.42	9.20	12.2	15.5	18.9	22.5	26.2	30.0	33.8	37.7
	8	1.64	3.60	5.92	8.55	11.4	14.6	17.9	21.3	24.9	28.6	32.4	36.3
	9	1.49	3.31	5.49	7.96	10.7	13.7	16.9	20.3	23.8	27.4	31.1	34.9
	10	1.37	3.06	5.10	7.44	10.1	12.9	16.0	19.2	22.7	26.2	29.8	33.6
3	12	1.17	2.65	4.46	6.55	8.93	11.6	14.4	17.5	20.7	24.0	27.5	31.1
	14	1.03	2.33	3.95	5.83	8.00	10.4	13.1	15.9	18.9	22.1	25.4	28.8
	16	0.91	2.08	3.54	5.24	7.23	9.47	11.9	14.6	17.4	20.4	23.6	26.8
	18	0.82	1.88	3.20	4.75	6.59	8.66	10.9	13.4	16.1	18.9	21.9	25.0
	20	0.74	1.71	2.92	4.35	6.04	7.96	10.1	12.4	15.0	17.6	20.5	23.5
	24	0.63	1.45	2.48	3.71	5.18	6.84	8.71	10.8	13.0	15.4	18.0	20.7
	28	0.54	1.26	2.15	3.23	4.52	5.99	7.65	9.50	11.5	13.7	16.0	18.5
	32	0.48	1.11	1.90	2.86	4.00	5.31	6.81	8.48	10.3	12.3	14.4	16.7
	36	0.43	0.99	1.69	2.56	3.59	4.77	6.13	7.64	9.30	11.1	13.1	15.2
	2	3.26	6.89	10.8	14.7	18.7	22.7	26.6	30.6	34.6	38.5	42.5	46.5
	3	2.87	6.28	10.1	14.0	18.0	22.0	26.0	30.0	33.9	37.9	41.9	45.9
	4	2.54	5.74	9.38	13.3	17.2	21.2	25.2	29.2	33.2	37.2	41.2	45.2
	5	2.25	5.27	8.75	12.6	16.5	20.4	24.5	28.5	32.5	36.5	40.5	44.5
	6	2.01	4.85	8.20	11.9	15.7	19.7	23.7	27.7	31.7	35.7	39.7	43.8
	7	1.81	4.49	7.70	11.3	15.0	18.9	22.9	26.9	30.9	34.9	39.0	43.0
	8	1.64	4.16	7.25	10.7	14.4	18.2	22.1	26.1	30.1	34.1	38.2	42.2
	9	1.49	3.87	6.83	10.2	13.7	17.5	21.4	25.3	29.3	33.3	37.4	41.4
	10	1.37	3.62	6.45	9.65	13.1	16.8	20.7	24.6	28.5	32.5	36.6	40.6
6	12	1.17	3.19	5.78	8.75	12.0	15.6	19.3	23.1	27.0	31.0	35.0	39.0
	14	1.03	2.84	5.21	7.97	11.1	14.5	18.1	21.8	25.6	29.5	33.4	37.4
	16	0.91	2.56	4.74	7.30	10.2	13.5	16.9	20.5	24.3	28.1	32.0	35.9
	18	0.82	2.33	4.33	6.72	9.48	12.6	15.9	19.4	23.0	26.7	30.6	34.4
	20	0.74	2.13	3.98	6.21	8.83	11.8	15.0	18.3	21.8	25.5	29.2	33.1
	24	0.63	1.82	3.42	5.38	7.74	10.4	13.3	16.5	19.8	23.2	26.8	30.5
	28	0.54	1.59	2.99	4.74	6.87	9.30	12.0	14.9	18.0	21.3	24.7	28.2
	32	0.48	1.41	2.65	4.22	6.17	8.38	10.8	13.6	16.5	19.5	22.8	26.1
	36	0.43	1.26	2.38	3.81	5.59	7.62	9.89	12.4	15.2	18.0	21.1	24.3

Table 7-23 (cont.). Coefficients *C* for Eccentrically Loaded Bolt Groups Angle = 60°

$$C_{req} = \frac{P_u}{\phi r_n} \quad \text{or} \quad \phi R_n = C \times \phi r_n$$

where

P_u = factored force, kips
ϕr_n = design strength per bolt, kips
ϕR_n = design strength per bolt group, kips
e = eccentricity of P_u with respect to centroid of bolt group, in. (not tabulated, may be determined by geometry)
e_x = horizontal component of e, in.
s = bolt spacing, in.
C = coefficient tabulated below

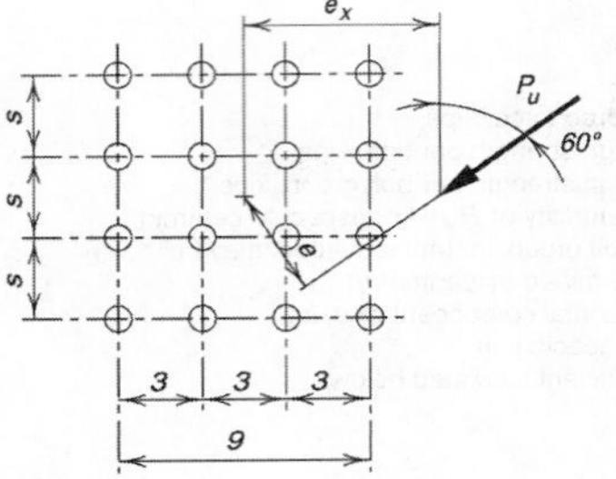

s, in.	e_x, in.	Number of Bolts in One Vertical Row, *n*											
		1	2	3	4	5	6	7	8	9	10	11	12
	2	3.63	7.25	10.9	14.6	18.3	22.1	25.9	29.7	33.6	37.5	41.4	45.3
	3	3.38	6.77	10.2	13.8	17.4	21.1	24.8	28.6	32.4	36.3	40.2	44.1
	4	3.10	6.27	9.55	13.0	16.5	20.1	23.7	27.5	31.3	35.1	38.9	42.8
	5	2.84	5.80	8.92	12.2	15.6	19.1	22.7	26.4	30.1	33.9	37.8	41.6
	6	2.60	5.36	8.33	11.5	14.8	18.2	21.7	25.4	29.1	32.8	36.6	40.4
	7	2.38	4.96	7.79	10.8	14.1	17.4	20.9	24.4	28.0	31.7	35.5	39.3
	8	2.19	4.60	7.30	10.2	13.4	16.6	20.0	23.5	27.1	30.7	34.4	38.2
	9	2.02	4.28	6.85	9.68	12.7	15.9	19.2	22.6	26.1	29.7	33.4	37.1
	10	1.87	3.99	6.45	9.17	12.1	15.2	18.4	21.8	25.3	28.8	32.4	36.1
3	12	1.62	3.51	5.75	8.27	11.0	13.9	17.0	20.3	23.6	27.0	30.6	34.1
	14	1.43	3.12	5.18	7.50	10.1	12.8	15.8	18.9	22.1	25.4	28.9	32.4
	16	1.27	2.81	4.70	6.85	9.23	11.9	14.7	17.6	20.7	24.0	27.3	30.7
	18	1.15	2.56	4.29	6.28	8.52	11.0	13.7	16.5	19.5	22.6	25.8	29.1
	20	1.04	2.34	3.95	5.80	7.89	10.2	12.8	15.5	18.4	21.4	24.5	27.7
	24	0.88	2.00	3.39	5.01	6.87	8.98	11.3	13.8	16.4	19.2	22.1	25.2
	28	0.76	1.74	2.96	4.39	6.07	7.97	10.1	12.3	14.8	17.4	20.1	23.0
	32	0.67	1.54	2.63	3.91	5.43	7.15	9.06	11.2	13.4	15.8	18.4	21.1
	36	0.60	1.38	2.36	3.52	4.91	6.48	8.22	10.2	12.3	14.5	16.9	19.4
	2	3.63	7.29	11.1	14.9	18.8	22.7	26.6	30.5	34.5	38.4	42.4	46.3
	3	3.38	6.88	10.6	14.3	18.2	22.1	26.0	29.9	33.9	37.8	41.8	45.7
	4	3.10	6.46	10.0	13.8	17.6	21.5	25.4	29.3	33.2	37.2	41.1	45.1
	5	2.84	6.06	9.55	13.2	17.0	20.9	24.7	28.7	32.6	36.5	40.4	44.4
	6	2.60	5.69	9.09	12.7	16.4	20.3	24.1	28.0	31.9	35.9	39.8	43.8
	7	2.38	5.34	8.66	12.2	15.9	19.7	23.5	27.4	31.3	35.2	39.2	43.1
	8	2.19	5.03	8.27	11.7	15.4	19.1	22.9	26.8	30.7	34.6	38.5	42.4
	9	2.02	4.74	7.90	11.3	14.9	18.6	22.4	26.2	30.1	34.0	37.9	41.8
	10	1.87	4.47	7.55	10.9	14.4	18.1	21.8	25.6	29.5	33.4	37.3	41.2
6	12	1.62	4.01	6.93	10.1	13.6	17.1	20.8	24.5	28.3	32.2	36.0	39.9
	14	1.43	3.63	6.38	9.46	12.8	16.2	19.8	23.5	27.3	31.0	34.9	38.7
	16	1.27	3.31	5.91	8.84	12.0	15.4	18.9	22.5	26.2	30.0	33.8	37.6
	18	1.15	3.04	5.49	8.28	11.3	14.6	18.0	21.6	25.2	28.9	32.7	36.5
	20	1.04	2.81	5.12	7.77	10.7	13.9	17.2	20.7	24.3	28.0	31.7	35.4
	24	0.88	2.44	4.49	6.90	9.62	12.6	15.8	19.1	22.6	26.1	29.8	33.4
	28	0.76	2.15	3.99	6.18	8.70	11.5	14.5	17.7	21.1	24.5	28.0	31.6
	32	0.67	1.91	3.58	5.58	7.93	10.6	13.4	16.5	19.7	23.0	26.4	29.9
	36	0.60	1.73	3.24	5.08	7.27	9.76	12.5	15.4	18.4	21.6	24.9	28.3

Table 7-23 (cont.).
Coefficients C for Eccentrically Loaded Bolt Groups
Angle = 75°

$$C_{req} = \frac{P_u}{\phi r_n} \quad \text{or} \quad \phi R_n = C \times \phi r_n$$

where

P_u = factored force, kips
ϕr_n = design strength per bolt, kips
ϕR_n = design strength per bolt group, kips
e = eccentricity of P_u with respect to centroid of bolt group, in. (not tabulated, may be determined by geometry)
e_x = horizontal component of e, in.
s = bolt spacing, in.
C = coefficient tabulated below

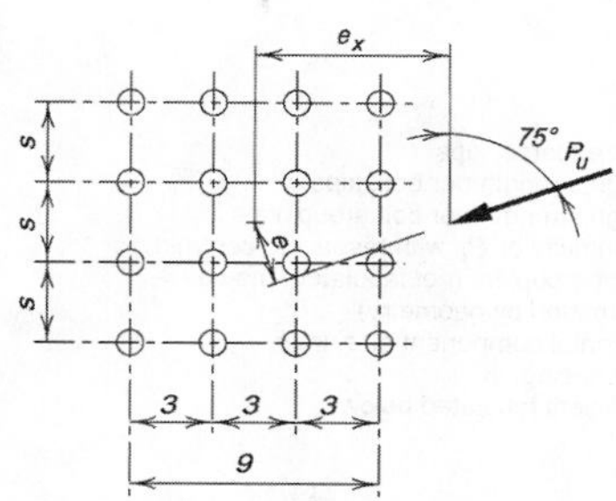

s, in.	e_x, in.	Number of Bolts in One Vertical Row, n											
		1	2	3	4	5	6	7	8	9	10	11	12
	2	3.86	7.69	11.5	15.3	19.1	22.9	26.7	30.5	34.3	38.2	42.0	45.9
	3	3.79	7.53	11.2	14.9	18.6	22.4	26.1	29.9	33.6	37.4	41.3	45.1
	4	3.70	7.34	11.0	14.6	18.2	21.8	25.5	29.2	33.0	36.7	40.5	44.3
	5	3.59	7.13	10.6	14.2	17.7	21.3	24.9	28.6	32.3	36.1	39.8	43.6
	6	3.47	6.89	10.3	13.8	17.2	20.8	24.4	28.0	31.7	35.4	39.1	42.9
	7	3.34	6.65	9.98	13.4	16.8	20.3	23.8	27.4	31.1	34.7	38.4	42.2
	8	3.20	6.40	9.64	12.9	16.3	19.8	23.3	26.8	30.4	34.1	37.8	41.5
	9	3.07	6.16	9.31	12.6	15.9	19.3	22.8	26.3	29.9	33.5	37.1	40.8
	10	2.94	5.91	8.98	12.2	15.4	18.8	22.2	25.7	29.3	32.9	36.5	40.2
3	12	2.68	5.45	8.36	11.4	14.6	17.9	21.3	24.7	28.2	31.8	35.4	39.0
	14	2.45	5.03	7.79	10.7	13.8	17.1	20.4	23.8	27.2	30.7	34.3	37.9
	16	2.24	4.65	7.28	10.1	13.1	16.3	19.5	22.9	26.3	29.7	33.2	36.8
	18	2.06	4.31	6.81	9.55	12.5	15.5	18.7	22.0	25.4	28.8	32.2	35.8
	20	1.90	4.01	6.40	9.03	11.9	14.8	18.0	21.2	24.5	27.9	31.3	34.8
	24	1.63	3.51	5.69	8.13	10.8	13.6	16.6	19.7	22.8	26.1	29.5	32.9
	28	1.43	3.11	5.11	7.36	9.83	12.5	15.3	18.3	21.4	24.6	27.8	31.1
	32	1.27	2.79	4.62	6.71	9.02	11.5	14.2	17.1	20.0	23.1	26.3	29.5
	36	1.14	2.53	4.22	6.15	8.31	10.7	13.3	16.0	18.8	21.8	24.9	28.0
	2	3.86	7.67	11.5	15.3	19.1	23.0	26.9	30.8	35.2	39.1	43.0	47.0
	3	3.79	7.51	11.2	15.0	18.8	22.6	26.4	30.3	34.2	38.1	42.1	46.0
	4	3.70	7.32	11.0	14.7	18.4	22.2	26.0	29.9	33.8	37.7	41.6	45.5
	5	3.59	7.12	10.7	14.4	18.1	21.8	25.6	29.5	33.3	37.2	41.1	45.0
	6	3.47	6.92	10.4	14.1	17.7	21.5	25.3	29.1	32.9	36.8	40.7	44.6
	7	3.34	6.70	10.2	13.8	17.4	21.1	24.9	28.7	32.5	36.4	40.2	44.1
	8	3.20	6.49	9.92	13.5	17.1	20.8	24.5	28.3	32.1	36.0	39.8	43.7
	9	3.07	6.28	9.66	13.2	16.8	20.5	24.2	28.0	31.8	35.6	39.4	43.3
	10	2.94	6.08	9.42	12.9	16.5	20.2	23.9	27.6	31.4	35.2	39.0	42.9
6	12	2.68	5.69	8.95	12.4	15.9	19.5	23.2	26.9	30.7	34.5	38.3	42.1
	14	2.45	5.33	8.51	11.9	15.4	19.0	22.6	26.3	30.0	33.8	37.6	41.4
	16	2.24	4.99	8.10	11.4	14.9	18.4	22.0	25.7	29.4	33.1	36.9	40.7
	18	2.06	4.69	7.72	11.0	14.4	17.9	21.5	25.1	28.8	32.5	36.2	40.0
	20	1.90	4.42	7.36	10.6	13.9	17.4	21.0	24.6	28.2	31.9	35.6	39.3
	24	1.63	3.95	6.74	9.83	13.1	16.5	20.0	23.5	27.1	30.7	34.4	38.1
	28	1.43	3.57	6.21	9.16	12.3	15.6	19.0	22.5	26.1	29.7	33.3	36.9
	32	1.27	3.25	5.74	8.56	11.6	14.8	18.2	21.6	25.1	28.6	32.2	35.9
	36	1.14	2.98	5.33	8.02	11.0	14.1	17.3	20.7	24.1	27.6	31.2	34.8

Table 7-24.
Coefficients *C* for Eccentrically Loaded Bolt Groups
Angle = 0°

$$C_{req} = \frac{P_u}{\phi r_n} \quad \text{or} \quad \phi R_n = C \times \phi r_n$$

where

P_u = factored force, kips
ϕr_n = design strength per bolt, kips
ϕR_n = design strength per bolt group, kips
e = eccentricity of P_u with respect to centroid of bolt group, in. (not tabulated, may be determined by geometry)
e_x = horizontal component of e, in.
s = bolt spacing, in.
C = coefficient tabulated below

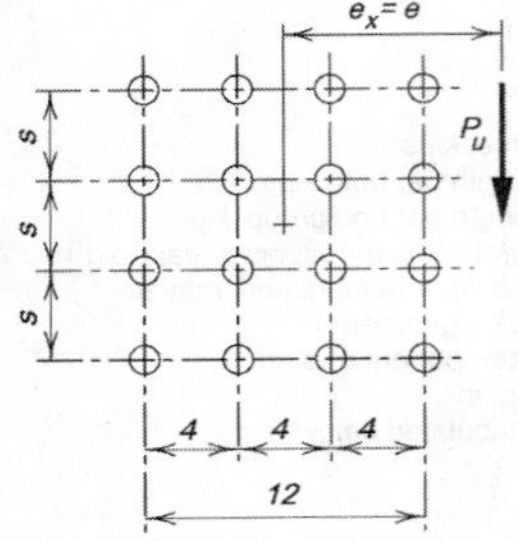

s, in.	*e*ₓ, in.	Number of Bolts in One Vertical Row, *n*											
		1	2	3	4	5	6	7	8	9	10	11	12
	2	2.82	5.98	9.46	13.3	17.3	21.3	25.5	29.6	33.7	37.7	41.8	45.8
	3	2.50	5.31	8.43	12.0	15.7	19.7	23.8	28.0	32.2	36.3	40.4	44.6
	4	2.23	4.74	7.58	10.8	14.3	18.2	22.2	26.3	30.4	34.6	38.8	43.0
	5	2.01	4.27	6.86	9.82	13.1	16.7	20.5	24.5	28.6	32.8	37.0	41.3
	6	1.81	3.86	6.24	8.96	12.0	15.4	19.0	22.9	26.9	31.0	35.2	39.4
	7	1.64	3.52	5.70	8.22	11.1	14.2	17.6	21.3	25.2	29.2	33.3	37.5
	8	1.49	3.22	5.24	7.57	10.2	13.2	16.4	19.9	23.6	27.5	31.5	35.6
	9	1.36	2.96	4.83	7.01	9.48	12.3	15.3	18.6	22.1	25.9	29.8	33.8
	10	1.25	2.73	4.47	6.51	8.83	11.4	14.3	17.5	20.8	24.4	28.2	32.1
3	12	1.07	2.37	3.89	5.68	7.74	10.1	12.6	15.5	18.5	21.8	25.3	29.0
	14	0.94	2.08	3.42	5.02	6.86	8.95	11.3	13.8	16.6	19.6	22.8	26.2
	16	0.83	1.86	3.05	4.49	6.15	8.04	10.2	12.5	15.0	17.8	20.7	23.9
	18	0.75	1.67	2.75	4.06	5.56	7.29	9.22	11.4	13.7	16.3	19.0	21.9
	20	0.68	1.52	2.50	3.70	5.07	6.65	8.43	10.4	12.6	14.9	17.5	20.2
	24	0.58	1.29	2.12	3.14	4.30	5.66	7.18	8.88	10.8	12.8	15.0	17.4
	28	0.50	1.12	1.84	2.72	3.73	4.92	6.24	7.73	9.37	11.2	13.1	15.2
	32	0.44	0.98	1.62	2.40	3.30	4.34	5.51	6.84	8.29	9.90	11.6	13.5
	36	0.40	0.88	1.45	2.15	2.95	3.89	4.94	6.13	7.43	8.88	10.4	12.1
	2	2.82	6.54	10.6	14.8	18.9	22.9	26.9	30.9	34.9	38.9	42.8	46.8
	3	2.50	5.90	9.81	14.0	18.1	22.3	26.4	30.4	34.5	38.5	42.5	46.5
	4	2.23	5.33	9.01	13.1	17.3	21.5	25.7	29.8	33.9	37.9	42.0	46.0
	5	2.01	4.84	8.27	12.2	16.4	20.6	24.8	29.0	33.2	37.3	41.4	45.5
	6	1.81	4.42	7.60	11.4	15.5	19.7	24.0	28.2	32.4	36.6	40.7	44.8
	7	1.64	4.05	7.02	10.6	14.6	18.8	23.0	27.3	31.5	35.7	39.9	44.1
	8	1.49	3.73	6.51	9.94	13.7	17.8	22.0	26.3	30.6	34.8	39.1	43.3
	9	1.36	3.45	6.06	9.30	13.0	16.9	21.1	25.3	29.6	33.9	38.2	42.4
	10	1.25	3.20	5.66	8.72	12.2	16.1	20.2	24.4	28.6	32.9	37.2	41.5
6	12	1.07	2.80	4.98	7.73	10.9	14.5	18.4	22.5	26.7	30.9	35.2	39.5
	14	0.94	2.47	4.43	6.92	9.81	13.2	16.8	20.7	24.8	29.0	33.2	37.5
	16	0.83	2.21	3.98	6.25	8.90	12.0	15.4	19.1	23.0	27.1	31.3	35.5
	18	0.75	2.00	3.60	5.68	8.13	11.0	14.2	17.7	21.4	25.3	29.4	33.6
	20	0.68	1.82	3.29	5.21	7.47	10.1	13.1	16.4	20.0	23.7	27.7	31.7
	24	0.58	1.55	2.79	4.45	6.40	8.72	11.3	14.3	17.5	20.9	24.5	28.3
	28	0.50	1.34	2.42	3.87	5.59	7.64	9.96	12.6	15.5	18.6	21.9	25.5
	32	0.44	1.18	2.14	3.43	4.95	6.79	8.87	11.2	13.8	16.7	19.7	23.0
	36	0.40	1.06	1.92	3.07	4.44	6.10	7.98	10.1	12.5	15.1	17.9	20.9

Table 7-24 (cont.). Coefficients C for Eccentrically Loaded Bolt Groups Angle = 15°

$$C_{req} = \frac{P_u}{\phi r_n} \quad \text{or} \quad \phi R_n = C \times \phi r_n$$

where

P_u = factored force, kips
ϕr_n = design strength per bolt, kips
ϕR_n = design strength per bolt group, kips
e = eccentricity of P_u with respect to centroid of bolt group, in. (not tabulated, may be determined by geometry)
e_x = horizontal component of e, in.
s = bolt spacing, in.
C = coefficient tabulated below

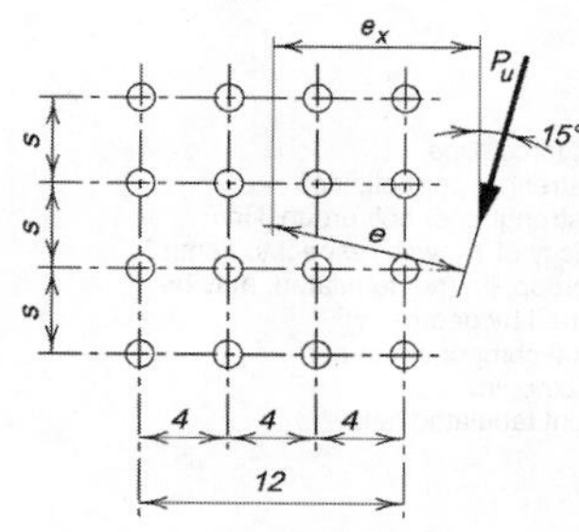

s, in.	e_x, in.	Number of Bolts in One Vertical Row, n											
		1	2	3	4	5	6	7	8	9	10	11	12
	2	2.91	6.06	9.56	13.3	17.2	21.3	25.3	29.4	33.5	37.5	41.6	45.6
	3	2.57	5.40	8.57	12.0	15.8	19.7	23.7	27.8	31.9	36.1	40.2	44.3
	4	2.30	4.84	7.72	10.9	14.4	18.2	22.1	26.1	30.2	34.3	38.5	42.6
	5	2.06	4.37	6.99	9.93	13.2	16.7	20.5	24.4	28.5	32.6	36.7	40.9
	6	1.86	3.96	6.37	9.09	12.1	15.5	19.0	22.8	26.7	30.8	34.9	39.0
	7	1.69	3.61	5.83	8.36	11.2	14.3	17.7	21.3	25.1	29.0	33.1	37.2
	8	1.53	3.31	5.36	7.72	10.4	13.3	16.5	19.9	23.6	27.4	31.3	35.3
	9	1.40	3.04	4.95	7.15	9.64	12.4	15.4	18.7	22.2	25.8	29.7	33.6
	10	1.29	2.81	4.59	6.65	9.0	11.6	14.5	17.6	20.9	24.4	28.1	31.9
3	12	1.11	2.44	4.00	5.82	7.9	10.2	12.8	15.6	18.7	21.9	25.3	28.9
	14	0.97	2.15	3.52	5.15	7.0	9.12	11.5	14.0	16.8	19.8	22.9	26.3
	16	0.86	1.92	3.15	4.61	6.3	8.21	10.3	12.7	15.2	18.0	20.9	24.0
	18	0.78	1.73	2.84	4.17	5.7	7.45	9.41	11.6	13.9	16.5	19.2	22.1
	20	0.71	1.57	2.59	3.80	5.2	6.81	8.61	10.6	12.8	15.2	17.7	20.4
	24	0.60	1.33	2.19	3.23	4.4	5.80	7.36	9.07	11.0	13.0	15.3	17.6
	28	0.52	1.15	1.90	2.80	3.9	5.05	6.41	7.91	9.59	11.4	13.4	15.5
	32	0.46	1.02	1.68	2.48	3.4	4.46	5.67	7.01	8.50	10.1	11.9	13.8
	36	0.41	0.91	1.50	2.22	3.0	4.00	5.08	6.29	7.63	9.09	10.7	12.4
	2	2.91	6.57	10.6	14.7	18.8	22.8	26.8	30.8	34.8	38.8	42.7	46.7
	3	2.57	5.93	9.81	13.9	18.0	22.1	26.2	30.3	34.3	38.3	42.3	46.3
	4	2.30	5.37	9.04	13.0	17.2	21.3	25.5	29.6	33.6	37.7	41.7	45.8
	5	2.06	4.89	8.33	12.2	16.3	20.5	24.6	28.8	32.9	37.0	41.1	45.1
	6	1.86	4.48	7.70	11.4	15.4	19.5	23.7	27.9	32.1	36.2	40.3	44.4
	7	1.69	4.12	7.13	10.6	14.5	18.6	22.8	27.0	31.2	35.4	39.5	43.7
	8	1.53	3.80	6.62	9.95	13.7	17.7	21.8	26.0	30.2	34.4	38.6	42.8
	9	1.40	3.52	6.17	9.32	12.9	16.8	20.9	25.1	29.3	33.5	37.7	41.9
	10	1.29	3.27	5.77	8.76	12.2	16.0	20.0	24.1	28.3	32.5	36.8	41.0
6	12	1.11	2.86	5.09	7.80	11.0	14.5	18.3	22.3	26.4	30.6	34.8	39.0
	14	0.97	2.54	4.53	7.00	9.92	13.2	16.8	20.6	24.6	28.7	32.8	37.1
	16	0.86	2.27	4.08	6.34	9.02	12.0	15.4	19.0	22.9	26.9	30.9	35.1
	18	0.78	2.06	3.70	5.78	8.26	11.1	14.2	17.7	21.3	25.2	29.1	33.2
	20	0.71	1.88	3.38	5.30	7.60	10.2	13.2	16.4	19.9	23.6	27.5	31.4
	24	0.60	1.59	2.88	4.54	6.54	8.84	11.5	14.4	17.5	20.9	24.5	28.2
	28	0.52	1.38	2.50	3.96	5.72	7.77	10.1	12.7	15.6	18.7	22.0	25.4
	32	0.46	1.22	2.21	3.51	5.08	6.92	9.03	11.4	14.0	16.8	19.9	23.1
	36	0.41	1.09	1.98	3.15	4.56	6.23	8.15	10.3	12.7	15.3	18.1	21.1

Table 7-24 (cont.). Coefficients C for Eccentrically Loaded Bolt Groups Angle = 30°

$$C_{req} = \frac{P_u}{\phi r_n} \quad \text{or} \quad \phi R_n = C \times \phi r_n$$

where

P_u = factored force, kips
ϕr_n = design strength per bolt, kips
ϕR_n = design strength per bolt group, kips
e = eccentricity of P_u with respect to centroid of bolt group, in. (not tabulated, may be determined by geometry)
e_x = horizontal component of e, in.
s = bolt spacing, in.
C = coefficient tabulated below

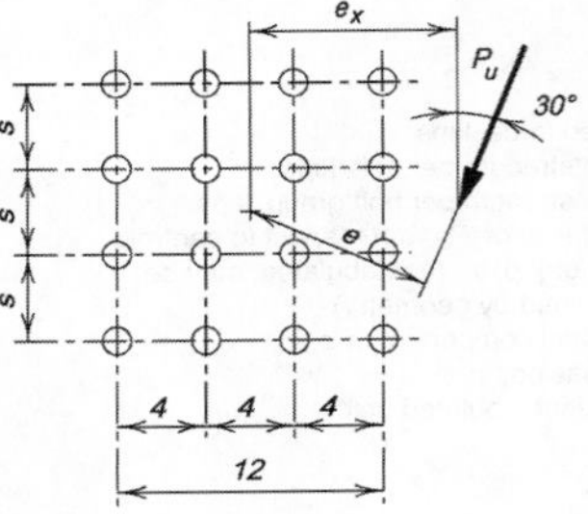

s, in.	e_x, in.	Number of Bolts in One Vertical Row, n											
		1	2	3	4	5	6	7	8	9	10	11	12
	2	3.14	6.41	9.91	13.6	17.5	21.4	25.4	29.4	33.4	37.4	41.4	45.4
	3	2.79	5.75	8.95	12.4	16.1	20.0	23.9	27.9	31.9	35.9	40.0	44.0
	4	2.50	5.19	8.16	11.4	14.9	18.5	22.4	26.3	30.3	34.3	38.4	42.4
	5	2.25	4.71	7.45	10.5	13.7	17.2	20.9	24.7	28.6	32.6	36.7	40.7
	6	2.04	4.29	6.83	9.65	12.7	16.0	19.6	23.3	27.1	31.0	35.0	39.0
	7	1.85	3.93	6.28	8.92	11.8	15.0	18.3	21.9	25.6	29.4	33.3	37.3
	8	1.69	3.61	5.80	8.27	11.0	14.0	17.2	20.6	24.2	27.9	31.7	35.6
	9	1.55	3.33	5.38	7.70	10.3	13.1	16.2	19.4	22.9	26.5	30.2	34.0
	10	1.43	3.08	5.00	7.19	9.64	12.3	15.3	18.4	21.7	25.2	28.8	32.5
3	12	1.23	2.68	4.37	6.32	8.52	11.0	13.6	16.5	19.6	22.8	26.2	29.8
	14	1.08	2.36	3.88	5.62	7.61	9.83	12.3	14.9	17.8	20.8	24.0	27.3
	16	0.96	2.11	3.47	5.05	6.86	8.89	11.1	13.6	16.2	19.0	22.0	25.2
	18	0.87	1.91	3.14	4.57	6.24	8.10	10.2	12.4	14.9	17.5	20.3	23.3
	20	0.79	1.74	2.86	4.18	5.71	7.43	9.35	11.5	13.8	16.2	18.9	21.6
	24	0.67	1.48	2.43	3.56	4.88	6.36	8.03	9.87	11.9	14.1	16.4	18.9
	28	0.58	1.28	2.11	3.10	4.25	5.55	7.02	8.65	10.4	12.4	14.5	16.7
	32	0.51	1.13	1.87	2.74	3.76	4.92	6.23	7.69	9.29	11.0	12.9	14.9
	36	0.46	1.01	1.67	2.45	3.37	4.41	5.60	6.91	8.36	9.95	11.7	13.5
	2	3.14	6.75	10.7	14.7	18.7	22.7	26.7	30.7	34.7	38.6	42.6	46.6
	3	2.79	6.12	9.94	13.9	18.0	22.0	26.1	30.1	34.1	38.1	42.1	46.1
	4	2.50	5.58	9.23	13.1	17.2	21.2	25.3	29.4	33.4	37.5	41.5	45.5
	5	2.25	5.13	8.58	12.4	16.3	20.4	24.5	28.6	32.7	36.7	40.8	44.8
	6	2.04	4.73	8.00	11.6	15.5	19.5	23.6	27.7	31.8	35.9	40.0	44.1
	7	1.85	4.38	7.47	10.9	14.7	18.7	22.7	26.8	31.0	35.1	39.2	43.3
	8	1.69	4.06	6.98	10.3	14.0	17.9	21.9	25.9	30.1	34.2	38.3	42.4
	9	1.55	3.78	6.55	9.72	13.3	17.1	21.0	25.1	29.2	33.3	37.4	41.5
	10	1.43	3.53	6.15	9.18	12.6	16.3	20.2	24.2	28.3	32.4	36.5	40.6
6	12	1.23	3.10	5.47	8.25	11.4	14.9	18.6	22.5	26.5	30.6	34.7	38.8
	14	1.08	2.76	4.90	7.46	10.4	13.7	17.2	21.0	24.9	28.8	32.9	37.0
	16	0.96	2.48	4.43	6.79	9.55	12.6	16.0	19.6	23.3	27.2	31.2	35.2
	18	0.87	2.25	4.04	6.22	8.79	11.7	14.9	18.3	21.9	25.7	29.5	33.5
	20	0.79	2.06	3.70	5.72	8.14	10.9	13.9	17.1	20.6	24.2	28.0	31.9
	24	0.67	1.76	3.17	4.93	7.06	9.48	12.2	15.2	18.3	21.7	25.3	28.9
	28	0.58	1.53	2.76	4.32	6.22	8.38	10.8	13.5	16.5	19.6	22.9	26.3
	32	0.51	1.35	2.45	3.84	5.54	7.50	9.73	12.2	14.9	17.8	20.9	24.1
	36	0.46	1.21	2.19	3.46	5.00	6.77	8.82	11.1	13.6	16.3	19.1	22.2

Table 7-24 (cont.). Coefficients C for Eccentrically Loaded Bolt Groups Angle = 45°

$$C_{req} = \frac{P_u}{\phi r_n} \quad \text{or} \quad \phi R_n = C \times \phi r_n$$

where

P_u = factored force, kips
ϕr_n = design strength per bolt, kips
ϕR_n = design strength per bolt group, kips
e = eccentricity of P_u with respect to centroid of bolt group, in. (not tabulated, may be determined by geometry)
e_x = horizontal component of e, in.
s = bolt spacing, in.
C = coefficient tabulated below

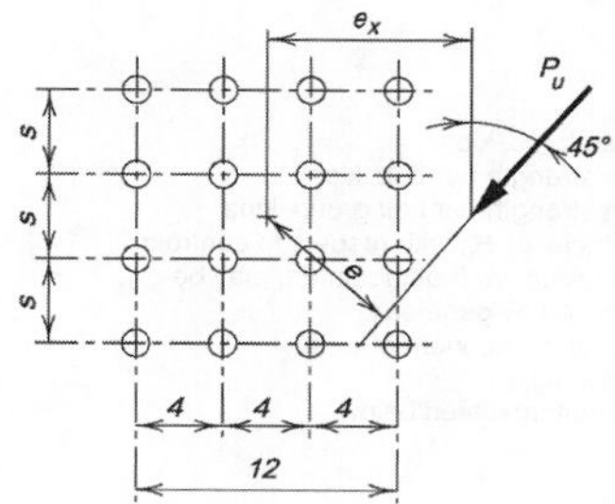

s, in.	e_x, in.	Number of Bolts in One Vertical Row, n											
		1	2	3	4	5	6	7	8	9	10	11	12
	2	3.46	6.96	10.5	14.2	18.0	21.8	25.7	29.6	33.5	37.4	41.4	45.3
	3	3.15	6.38	9.73	13.2	16.8	20.6	24.4	28.2	32.1	36.1	40.0	44.0
	4	2.87	5.84	8.97	12.3	15.7	19.3	23.1	26.9	30.7	34.6	38.6	42.5
	5	2.61	5.36	8.30	11.4	14.7	18.2	21.8	25.5	29.3	33.2	37.1	41.0
	6	2.39	4.93	7.69	10.7	13.9	17.2	20.7	24.3	28.0	31.8	35.6	39.5
	7	2.19	4.55	7.15	9.98	13.0	16.2	19.6	23.1	26.7	30.4	34.2	38.1
	8	2.01	4.21	6.66	9.34	12.2	15.3	18.6	22.0	25.5	29.2	32.9	36.7
	9	1.86	3.90	6.21	8.76	11.5	14.5	17.7	21.0	24.4	27.9	31.6	35.3
	10	1.72	3.63	5.82	8.24	10.9	13.8	16.8	20.0	23.3	26.8	30.4	34.0
3	12	1.49	3.18	5.14	7.33	9.76	12.4	15.2	18.3	21.4	24.7	28.1	31.6
	14	1.32	2.82	4.59	6.58	8.81	11.3	13.9	16.7	19.7	22.8	26.1	29.5
	16	1.17	2.53	4.14	5.95	8.00	10.3	12.7	15.4	18.2	21.2	24.3	27.5
	18	1.06	2.29	3.76	5.43	7.32	9.44	11.7	14.2	16.9	19.7	22.7	25.7
	20	0.96	2.10	3.44	4.98	6.74	8.71	10.9	13.2	15.7	18.4	21.2	24.2
	24	0.82	1.79	2.94	4.26	5.81	7.53	9.43	11.5	13.8	16.2	18.7	21.4
	28	0.71	1.56	2.56	3.73	5.09	6.61	8.31	10.2	12.2	14.4	16.7	19.2
	32	0.63	1.38	2.26	3.31	4.52	5.89	7.42	9.11	11.0	12.9	15.1	17.3
	36	0.56	1.23	2.03	2.97	4.06	5.30	6.69	8.23	9.91	11.7	13.7	15.8
	2	3.46	7.09	10.9	14.8	18.7	22.7	26.7	30.6	34.6	38.5	42.5	46.5
	3	3.15	6.58	10.3	14.1	18.1	22.0	26.0	30.0	33.9	37.9	41.9	45.9
	4	2.87	6.09	9.65	13.4	17.3	21.3	25.3	29.3	33.3	37.3	41.2	45.2
	5	2.61	5.66	9.07	12.8	16.6	20.6	24.5	28.5	32.5	36.5	40.5	44.5
	6	2.39	5.26	8.54	12.1	15.9	19.8	23.8	27.8	31.8	35.8	39.8	43.8
	7	2.19	4.91	8.07	11.6	15.3	19.1	23.0	27.0	31.0	35.0	39.0	43.0
	8	2.01	4.59	7.63	11.0	14.6	18.4	22.3	26.2	30.2	34.2	38.2	42.2
	9	1.86	4.30	7.23	10.5	14.0	17.7	21.5	25.5	29.4	33.4	37.4	41.4
	10	1.72	4.04	6.85	10.0	13.4	17.1	20.8	24.7	28.6	32.6	36.6	40.6
6	12	1.49	3.59	6.19	9.14	12.4	15.9	19.5	23.3	27.2	31.1	35.1	39.1
	14	1.32	3.22	5.62	8.38	11.4	14.8	18.3	22.0	25.8	29.6	33.5	37.5
	16	1.17	2.91	5.13	7.71	10.6	13.8	17.2	20.8	24.4	28.2	32.1	36.0
	18	1.06	2.66	4.71	7.12	9.87	12.9	16.2	19.6	23.2	26.9	30.7	34.6
	20	0.96	2.44	4.35	6.61	9.22	12.1	15.3	18.6	22.1	25.7	29.4	33.2
	24	0.82	2.10	3.76	5.76	8.11	10.8	13.7	16.7	20.0	23.4	27.0	30.6
	28	0.71	1.83	3.30	5.08	7.22	9.64	12.3	15.2	18.3	21.5	24.9	28.4
	32	0.63	1.63	2.94	4.54	6.50	8.71	11.2	13.9	16.7	19.8	23.0	26.3
	36	0.56	1.46	2.64	4.11	5.90	7.93	10.2	12.7	15.4	18.3	21.3	24.5

Table 7-24 (cont.). Coefficients *C* for Eccentrically Loaded Bolt Groups Angle = 60°

$$C_{req} = \frac{P_u}{\phi r_n} \quad \text{or} \quad \phi R_n = C \times \phi r_n$$

where

P_u = factored force, kips
ϕr_n = design strength per bolt, kips
ϕR_n = design strength per bolt group, kips
e = eccentricity of P_u with respect to centroid of bolt group, in. (not tabulated, may be determined by geometry)
e_x = horizontal component of e, in.
s = bolt spacing, in.
C = coefficient tabulated below

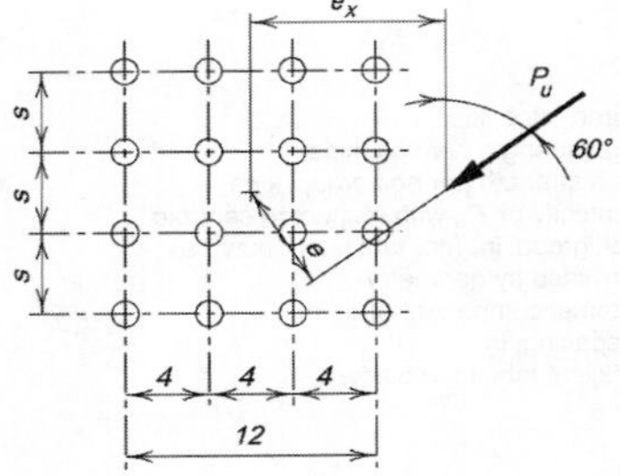

s, in.	e_x, in.	Number of Bolts in One Vertical Row, *n*											
		1	2	3	4	5	6	7	8	9	10	11	12
	2	3.74	7.46	11.2	14.9	18.6	22.4	26.2	30.0	33.9	37.7	41.6	45.5
	3	3.57	7.12	10.7	14.3	17.9	21.6	25.3	29.0	32.8	36.7	40.5	44.4
	4	3.38	6.75	10.2	13.6	17.1	20.7	24.3	28.0	31.8	35.6	39.4	43.2
	5	3.17	6.36	9.61	12.9	16.4	19.8	23.4	27.0	30.7	34.5	38.2	42.0
	6	2.97	5.99	9.09	12.3	15.6	19.0	22.5	26.1	29.7	33.4	37.1	40.9
	7	2.78	5.63	8.59	11.7	14.9	18.2	21.6	25.1	28.7	32.3	36.0	39.8
	8	2.60	5.29	8.13	11.1	14.2	17.5	20.8	24.3	27.8	31.4	35.0	38.7
	9	2.44	4.98	7.69	10.6	13.6	16.8	20.1	23.4	26.9	30.4	34.0	37.7
	10	2.28	4.69	7.28	10.1	13.0	16.1	19.3	22.7	26.1	29.5	33.1	36.7
3	12	2.02	4.18	6.56	9.16	11.9	14.9	18.0	21.2	24.5	27.8	31.3	34.8
	14	1.80	3.76	5.95	8.38	11.0	13.8	16.7	19.8	23.0	26.3	29.6	33.1
	16	1.62	3.40	5.43	7.70	10.2	12.8	15.6	18.6	21.6	24.8	28.1	31.4
	18	1.47	3.10	4.99	7.11	9.42	11.9	14.6	17.4	20.4	23.5	26.7	29.9
	20	1.34	2.85	4.61	6.59	8.76	11.1	13.7	16.4	19.3	22.2	25.3	28.5
	24	1.15	2.45	3.99	5.73	7.67	9.82	12.2	14.6	17.3	20.1	23.0	26.0
	28	1.00	2.15	3.51	5.06	6.80	8.76	10.9	13.2	15.6	18.2	20.9	23.8
	32	0.88	1.91	3.13	4.52	6.11	7.89	9.83	11.9	14.2	16.6	19.2	21.8
	36	0.79	1.72	2.81	4.08	5.53	7.16	8.95	10.9	13.0	15.3	17.7	20.2
	2	3.74	7.47	11.2	15.0	18.9	22.8	26.7	30.6	34.5	38.5	42.4	46.4
	3	3.57	7.16	10.8	14.6	18.4	22.2	26.1	30.0	33.9	37.9	41.8	45.8
	4	3.38	6.82	10.4	14.1	17.8	21.7	25.5	29.4	33.3	37.3	41.2	45.1
	5	3.17	6.47	9.94	13.6	17.3	21.1	24.9	28.8	32.7	36.6	40.5	44.5
	6	2.97	6.14	9.52	13.1	16.7	20.5	24.3	28.2	32.1	36.0	39.9	43.8
	7	2.78	5.82	9.11	12.6	16.2	19.9	23.7	27.6	31.5	35.3	39.3	43.2
	8	2.60	5.52	8.73	12.1	15.7	19.4	23.2	27.0	30.8	34.7	38.6	42.5
	9	2.44	5.24	8.37	11.7	15.2	18.9	22.6	26.4	30.2	34.1	38.0	41.9
	10	2.28	4.98	8.03	11.3	14.8	18.4	22.1	25.8	29.7	33.5	37.4	41.3
6	12	2.02	4.51	7.41	10.6	14.0	17.5	21.1	24.8	28.5	32.3	36.2	40.1
	14	1.80	4.10	6.86	9.91	13.2	16.6	20.1	23.8	27.5	31.2	35.0	38.9
	16	1.62	3.76	6.37	9.29	12.4	15.8	19.2	22.8	26.5	30.2	33.9	37.7
	18	1.47	3.46	5.94	8.74	11.8	15.0	18.4	21.9	25.5	29.2	32.9	36.6
	20	1.34	3.21	5.56	8.23	11.2	14.3	17.6	21.0	24.6	28.2	31.9	35.6
	24	1.15	2.79	4.91	7.34	10.1	13.0	16.2	19.5	22.9	26.4	30.0	33.6
	28	1.00	2.47	4.38	6.61	9.13	11.9	14.9	18.1	21.4	24.7	28.2	31.8
	32	0.88	2.21	3.95	5.99	8.33	11.0	13.8	16.8	20.0	23.2	26.6	30.1
	36	0.79	2.00	3.58	5.46	7.65	10.1	12.8	15.7	18.7	21.9	25.1	28.5

Table 7-24 (cont.). Coefficients *C* for Eccentrically Loaded Bolt Groups Angle = 75°

$C_{req} = \dfrac{P_u}{\phi r_n}$ or $\phi R_n = C \times \phi r_n$

where

P_u = factored force, kips
ϕr_n = design strength per bolt, kips
ϕR_n = design strength per bolt group, kips
e = eccentricity of P_u with respect to centroid of bolt group, in. (not tabulated, may be determined by geometry)
e_x = horizontal component of e, in.
s = bolt spacing, in.
C = coefficient tabulated below

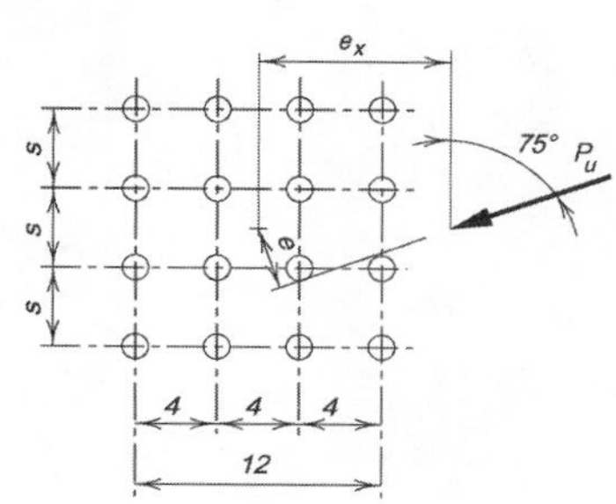

s, in.	e_x, in.	Number of Bolts in One Vertical Row, *n*											
		1	2	3	4	5	6	7	8	9	10	11	12
	2	3.89	7.75	11.6	15.5	19.3	23.1	26.9	30.8	34.6	38.5	42.3	46.2
	3	3.84	7.66	11.5	15.2	19.0	22.7	26.5	30.3	34.1	37.9	41.7	45.5
	4	3.79	7.54	11.3	15.0	18.7	22.4	26.1	29.8	33.5	37.3	41.0	44.8
	5	3.72	7.40	11.1	14.7	18.3	21.9	25.6	29.3	32.9	36.7	40.4	44.1
	6	3.65	7.25	10.8	14.4	17.9	21.5	25.1	28.7	32.4	36.1	39.8	43.5
	7	3.56	7.08	10.6	14.1	17.6	21.1	24.6	28.2	31.8	35.5	39.1	42.8
	8	3.47	6.90	10.3	13.7	17.2	20.6	24.1	27.7	31.3	34.9	38.5	42.2
	9	3.37	6.71	10.0	13.4	16.8	20.2	23.7	27.2	30.7	34.3	37.9	41.6
	10	3.27	6.52	9.77	13.1	16.4	19.8	23.2	26.7	30.2	33.7	37.3	41.0
3	12	3.07	6.14	9.23	12.4	15.6	18.9	22.3	25.7	29.1	32.6	36.2	39.8
	14	2.87	5.76	8.71	11.8	14.9	18.1	21.4	24.7	28.1	31.6	35.1	38.7
	16	2.68	5.40	8.22	11.1	14.2	17.3	20.5	23.8	27.2	30.6	34.1	37.6
	18	2.50	5.07	7.76	10.6	13.5	16.6	19.7	23.0	26.3	29.7	33.1	36.6
	20	2.34	4.76	7.33	10.0	12.9	15.9	19.0	22.2	25.5	28.8	32.2	35.6
	24	2.06	4.23	6.57	9.10	11.8	14.7	17.6	20.7	23.9	27.1	30.4	33.8
	28	1.82	3.78	5.94	8.30	10.9	13.5	16.4	19.3	22.4	25.5	28.7	32.0
	32	1.63	3.41	5.41	7.61	10.0	12.6	15.3	18.1	21.0	24.1	27.2	30.4
	36	1.48	3.11	4.95	7.01	9.26	11.7	14.3	17.0	19.8	22.8	25.8	28.9
	2	3.89	7.74	11.6	15.4	19.3	23.1	27.0	30.9	35.2	39.1	43.0	47.0
	3	3.84	7.64	11.4	15.2	19.0	22.8	26.6	30.5	34.4	38.3	42.2	46.1
	4	3.79	7.52	11.2	14.9	18.7	22.5	26.3	30.1	34.0	37.8	41.7	45.6
	5	3.72	7.38	11.0	14.7	18.4	22.1	25.9	29.7	33.6	37.4	41.3	45.2
	6	3.65	7.23	10.8	14.4	18.1	21.8	25.6	29.3	33.2	37.0	40.8	44.7
	7	3.56	7.07	10.6	14.2	17.8	21.5	25.2	29.0	32.8	36.6	40.4	44.3
	8	3.47	6.90	10.4	13.9	17.5	21.2	24.9	28.6	32.4	36.2	40.0	43.9
	9	3.37	6.73	10.1	13.6	17.2	20.8	24.5	28.3	32.0	35.8	39.6	43.5
	10	3.27	6.56	9.92	13.4	16.9	20.5	24.2	27.9	31.7	35.5	39.3	43.1
6	12	3.07	6.21	9.48	12.9	16.4	19.9	23.6	27.3	31.0	34.7	38.5	42.3
	14	2.87	5.88	9.07	12.4	15.9	19.4	23.0	26.6	30.3	34.1	37.8	41.6
	16	2.68	5.57	8.67	11.9	15.4	18.8	22.4	26.0	29.7	33.4	37.1	40.9
	18	2.50	5.27	8.29	11.5	14.9	18.3	21.9	25.5	29.1	32.8	36.5	40.2
	20	2.34	4.99	7.94	11.1	14.4	17.8	21.3	24.9	28.5	32.2	35.8	39.6
	24	2.06	4.50	7.29	10.3	13.6	16.9	20.4	23.9	27.4	31.0	34.7	38.3
	28	1.82	4.08	6.73	9.67	12.8	16.1	19.4	22.9	26.4	30.0	33.6	37.2
	32	1.63	3.73	6.25	9.06	12.1	15.3	18.6	22.0	25.4	29.0	32.5	36.1
	36	1.48	3.43	5.82	8.51	11.4	14.5	17.8	21.1	24.5	28.0	31.5	35.1

PART 8

DESIGN CONSIDERATIONS FOR WELDS

SCOPE

The specification requirements and other design considerations summarized in this Part apply to the design of welded joints. For the design of connection elements, see Part 9. For the design of connections, see Parts 10 through 15. For the design of welded joints for HSS and steel pipe, see the AISC *Hollow Structural Sections Connections Manual*. For welded joints that are part of a seismic-force-resisting system in which the seismic response modification factor R is taken greater than 3, see the Seismic Provisions, which are available from AISC at www.aisc.org.

LOAD DETERMINATION

The required strength(s) for welded joints are determined using force-transfer models such as those described in Parts 10 through 15 to distribute the member end reactions, which are determined by analysis as indicated in LRFD Specification Section A5.

GENERAL REQUIREMENTS FOR WELDED JOINTS

The requirements for welded construction are given in AISC Specification Section M2.4, which references AWS D1.1, except as modified in LRFD Specification Section J2. For further information, see also Blodgett et al. (1997).

A general summary of ASTM requirements for material properties that affect weldability of steel is provided in Table 8-1.

Filler Metals

Filler metal requirements are given in LRFD Specification Sections A3.5, J2.6 and J2.7, based upon the matching of the filler metal and base metal per LRFD Specification Table J2.5 and AWS D1.1 Table 3.1. Filler metal notch-toughness requirements are given in LRFD Specification Section J2.6. When low-hydrogen consumables are required, drying ovens are necessary near points of use in the shop.

Shielded Metal Arc Welding (SMAW)

SMAW or "stick" consumables are manufactured to meet the requirements in AWS A5.1 or AWS A5.5. A condensed excerpt of AWS specifications for SMAW electrodes is given in Table 8-2.

Submerged Arc Welding (SAW)

SAW consumables are manufactured to meet the requirements in AWS A5.17 or AWS A5.23. Separate flux provides the shielding of the weld metal during deposition and subsequent solidification, and must be kept within the acceptable limits for moisture content. The SAW process (automatic or semi-automatic) is particularly suitable for long straight and circumferential welds, but the work must be positioned to allow for near flat or horizontal welding.

Gas Metal Arc Welding (GMAW)

GMAW consumables are manufactured to meet the requirements in AWS 5.18 or AWS 5.28. Gases, such as pure argon, argon and oxygen, argon and carbon dioxide, or pure carbon dioxide provide the shielding of the weld metal during deposition and subsequent solidification.

Flux-Cored Arc Welding (FCAW)

FCAW consumables are manufactured to meet the requirements in A5.20 or AWS 5.29. Flux contained inside the hollow-tube-shaped consumable provides the shielding of the weld metal during deposition and subsequent solidification, and must be kept within the acceptable limits for moisture content. Some FCAW consumables are also gas-shielded. The FCAW process can be used with semi-automatic equipment in any position with the appropriate electrode.

Electroslag Welding (ESW) and Electrogas Welding (EGW)

ESW and EGW consumables are manufactured to meet the requirements in A5.25. Note that AWS prohibits the use of these welding processes on quenched and tempered steels.

Thermal Cutting

Oxygen-fuel gas cutting can be used to cut almost any commercially available plate thickness. If the plate being cut contains large discontinuities or non-metallic inclusions, turbulence may be created in the cutting stream, resulting in notches or gouges in the edge of the cut. Plasma-arc cutting is much faster and less susceptible to the effects of discontinuities or non-metallic inclusions, but leaves a slight taper in the cut as it descends and can be used only up to about $1^1/_2$-in. thickness. Within the depth limits given in the LRFD Specification and AWS D1.1, it is usually better practice to remove and fair-in notches or gouges by grinding than to weld repair and grind. In either case, however, a smooth transition should be provided.

Air-Arc Gouging

In this method, a carbon arc is used to melt a nugget-shaped area of the base metal, which is blown away with a jet of compressed air. Air-arc gouging can be used to remove weld defects, gouge the weld root to sound weld metal, form a U groove on one side of a square butt joint, and for similar operations. Because the arc quench is similar in both air-arc gouging and welding, any pre-heat required for welding should also be used for air-arc gouging.

Inspection

The owner's designated representative for design must specify in the contract documents which type of welding inspection is required as well as the extent and application of each type. Visual testing (VT) is the most commonly specified welding inspection procedure. Other, more stringent methods can add significant cost to the project and should be specified only when necessary.

In the absence of instruction, AWS D1.1 Section 6.6.5 states that the fabricator or erector is responsible only for those weld discontinuities found with VT. If additional inspection more stringent than VT is later required, the owner is normally responsible for the cost of weld repairs other than those identified by VT.

Visual Testing (VT)

VT provides the most economical approach to weld inspection. It is particularly well suited for inspecting single-pass welds, but is limited in that only surface imperfections can be detected. Proper VT includes a check of the joint for accuracy and cleanliness before welding, proper observation during the welding process and post-weld inspection. Acceptance criteria are specified AWS D1.1.

Penetrant Testing (PT)

As illustrated in Figure 8-1, PT can be used to detect tight cracks as long as they are open to the surface. Like VT, however, only surface cracks are detectable. Furthermore, deep weld ripples and scratches may give a false indication when PT is used.

Magnetic-Particle Testing (MT)

As illustrated in Figure 8-2, MT can be used to detect surface cracks. Additionally, the powder may be picked up and preserved with clear tape, providing accurate and detailed records of inspection results. However, relatively smooth surfaces are required for MT and, while cleanup is easy, demagnetization, when necessary, may not be.

Ultrasonic Testing (UT)

As illustrated in Figure 8-3, UT can be used to detect favorably oriented, flat discontinuities smaller than $^1/_{64}$ in. in carbon and low-alloy structural steels. However, welds with extremely coarse grain structure, such as electroslag welds, are difficult to inspect. Also, certain joint geometries can limit the use of UT and it is difficult to inspect members less than $^5/_{16}$-in. thick because there is a "dead area" at the origin of the sound wave. For more information, see Krautkramer (1977) and Institute of Welding (1972).

Radiographic Testing (RT)

RT can be used to detect porosity, slag, voids, cracks, irregularities, and lack of fusion and the film negative provides a permanent record of the inspection. However, defects smaller than about $1^1/_2$ percent of the thickness and defects not parallel to the beam may not register. RT of closed, inaccessible pipe joints is difficult to obtain and interpret and should be discouraged. Additionally, when the particle beam must penetrate varying thicknesses, as at fillets and tee or corner joints, RT is not readily interpreted and the resulting inspection may be less consistent. When this is the case, other inspection methods should be used. Other limitations of RT are that the required exposure time increases with material thickness and there is a worker hazard

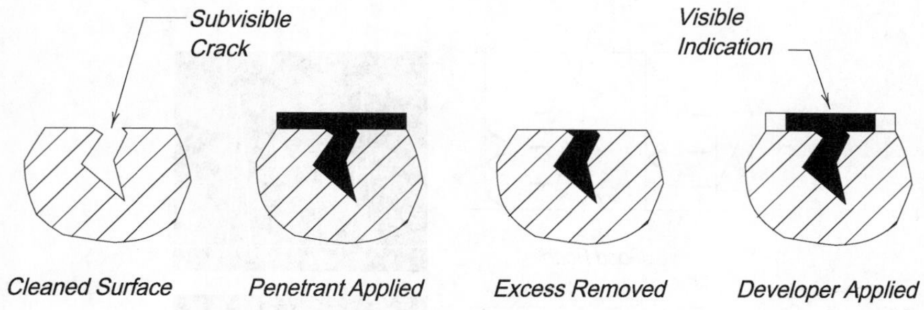

Fig. 8–1. Schematic illustration of penetrant testing (PT).

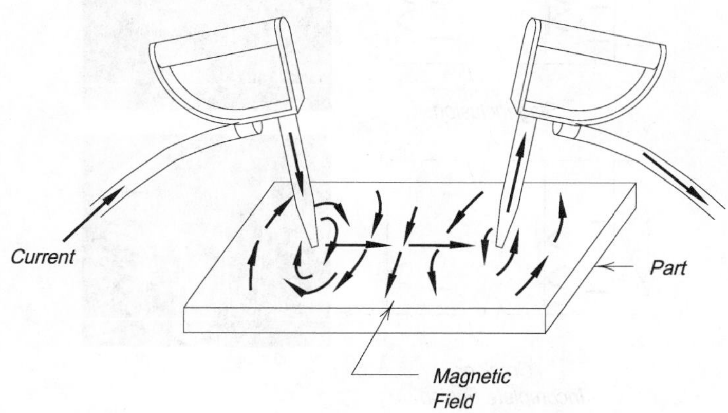

Fig. 8–2. Schematic illustration of magnetic particle testing (MT).

due to the radiation used in the method. The precautions for avoiding these hazards and the equipment and film costs make this method the most expensive inspection method.

PROPER SPECIFICATION OF JOINT TYPE

Selection of Weld Type

Fillet welds are normally more economical than groove welds and generally should be used in applications for which groove welds are not required. Additionally, fillet welds around the insides of holes or slots require less weld metal than plug or slot welds of the same size, even though the diameters of holes and widths of slots for fillet welds must be larger to accommodate the necessary tilt of the electrode.

When groove welds are required, bevel and V-groove welds, which can be flame cut, are usually more economical than J- and U-groove welds, which must be air-arc gouged or planed. Also, double-bevel, double-V, double-J and double-U welds are typically more economical than welds of the same type with single-sided preparation because they use less weld metal, particularly as the thickness of the connection element(s) being welded increases. The symmetry also results in less rotational distortion strain. However, in thinner connection elements, the savings in weld-metal volume may not offset the additional cost of double edge preparation, weld-root cleaning and repositioning. As a general rule of thumb, double-sided joint preparation is normally less expensive than single-sided preparation above 1-in. thickness.

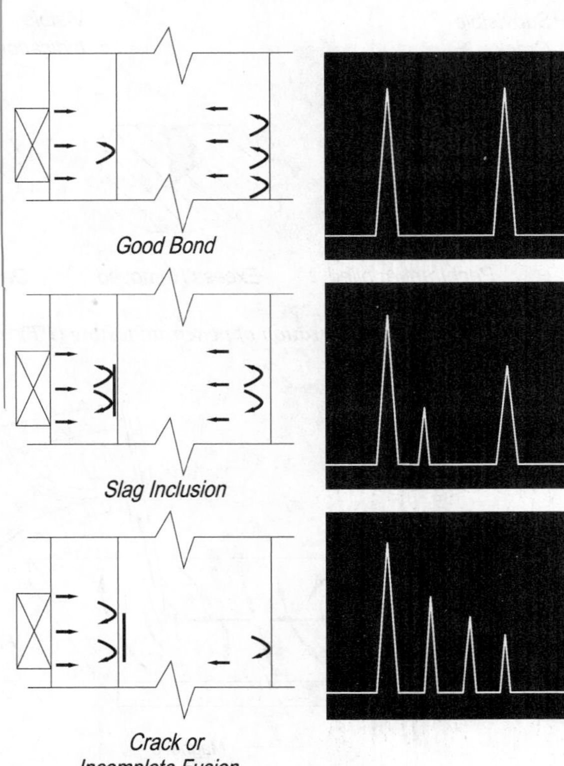

Fig. 8–3. Variations in UT reflections caused by defects at the boundary.

DESIGN REQUIREMENTS

The following summary is based upon the requirements in the LRFD Specification. The requirements in AWS D1.1 are also applicable, except as modified in LRFD Specification Section J2.

Effective Area

The effective areas of welded joints are determined as follows.

Groove Welds

For complete-joint-penetration groove welds, partial-joint-penetration groove welds and flare groove welds, the effective area is as given in LRFD Specification J2.1a subject to the limitations on effective throat thickness in LRFD Specification Section J2.1b.

In each of the prequalified welded joint details for partial-joint-penetration groove welds, an effective throat thickness E is established based upon the material thickness, weld-preparation size, or depth S. As shown in Figure 8-4, E is less than S. The required effective throat thickness E should be specified in the design drawings. Subsequently, the groove depth S and geometry that will provide for this required effective throat thickness should be shown in the shop drawings; E may also be shown to eliminate confusion.

Fillet Welds

For fillet welds, the effective area is as given in LRFD Specification Section J2.2a subject to the limitations in LRFD Specification Section J2.2b on:

1. Minimum and maximum weld size;
2. Minimum effective length;
3. Maximum effective length for end-loaded fillet welds;
4. Use of intermittent fillet welds; and,
5. Minimum length of fillet-welded lap joints.

Requirements are also given in LRFD Specification Section J2.2b for fillet weld terminations and use of fillet welds in holes or slots.

Per AWS D1.1, the size of the root opening, if any, for root openings equal to or greater than $^{1}/_{16}$ in., but not exceeding $^{3}/_{16}$ in., is added to the fillet-weld leg size required for strength.

Plug and Slot Welds

For plug and slot welds, the effective area is as given in LRFD Specification Section J2.3a subject to the limitations in LRFD Specification Section J2.3b.

Design Strength

The design strength of a welded joint is determined in accordance with LRFD Specification Section J2.4 and Table J2.5 and LRFD Specification Appendix J2.4. The calculation of the

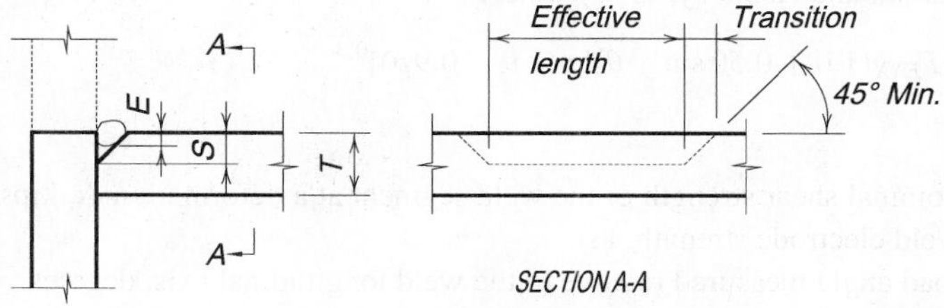

Fig. 8–4. Partial-joint-penetration (PJP) groove weld nomenclature.

design strength of a longitudinally loaded fillet weld can be simplified from that given in LRFD Specification Table J2.5. For a fillet weld with $F_{EXX} = 70$ ksi, length l, in., and a leg size of D sixteenths as follows:

$$\begin{aligned}\phi R_n &= 0.75 \times 0.6 F_{EXX} \times \frac{\sqrt{2}}{2} \times \frac{D}{16} \times l \\ &= 1.392\, Dl\end{aligned}$$

When the fillet weld is not longitudinally loaded, the provisions in LRFD Specification Appendix J2.4 may be used to take advantage of the increased strength due to load angle. The maximum strength increase will be for a transversely loaded fillet weld, which is 50 percent stronger than the same fillet weld longitudinally loaded.

Welds that are oversized waste weld metal and labor time, resulting in a significant and unnecessary increase in the cost of the connection. Thus, it is important to use the proper weld size required for strength or the minimum weld size from LRFD Specification Table J2.4. Fillet welds sizes up to $^5/_{16}$ in. may be deposited in a single pass when deposited in the flat or horizontal position. Larger-size welds must be deposited in multiple passes, which will require appreciably more labor. Also, when weld metal is deposited in the flat position, it can be deposited more quickly since gravity does not adversely affect the deposit—on the order of four times faster than deposition rates in the vertical and overhead positions.

ECCENTRICALLY LOADED WELD GROUPS

Eccentricity in the Plane of the Faying Surface

Eccentricity in the plane of the faying surface produces additional shear. The welds must be designed to resist the combined effect of the direct shear P_u and the additional shear from the induced moment $P_u e$. Two methods of analysis for this type of eccentricity are the instantaneous center of rotation method and the elastic method.

The instantaneous center of rotation method is more accurate, but generally requires the use of tabulated values or an iterative solution. The elastic method is simplified, but may be excessively conservative because it neglects the ductility of the weld group and the potential for load redistribution.

Instantaneous Center of Rotation Method

Eccentricity produces both a rotation and a translation of one connection element with respect to the other. The combined effect of this rotation and translation is equivalent to a rotation about a point defined as the instantaneous center of rotation (IC) as illustrated in Figure 8-5a. The location of the IC depends upon the geometry of the weld group as well as the direction and point of application of the load.

The load-deformation relationship for a unit-length segment of the weld is illustrated in Figure 8-6 (Lesik and Kennedy, 1990), where

$$R = 0.60 F_{EXX}(1.0 + 0.50 \sin^{1.5}\theta)[p(1.9 - 0.9p)]^{0.3}$$

where

R = nominal shear strength of the weld segment at a deformation Δ, kips.
F_{EXX} = weld electrode strength, ksi.
θ = load angle measured relative to the weld longitudinal axis, degrees.
p = ratio of element deformation to its deformation at maximum stress.

Unlike the load-deformation relationship for bolts, the strength and deformation of welds are dependent upon the angle θ that the resultant elemental force makes with the axis of the weld element. Load-deformation curves in Figure 8-6 for values of $\theta = 0°, 30°, 45°, 60°, 75°$, and $90°$ are shown relative to $R_o = 0.6F_{EXX}$. For further information, see LRFD Specification Appendix J2.4 and its Commentary.

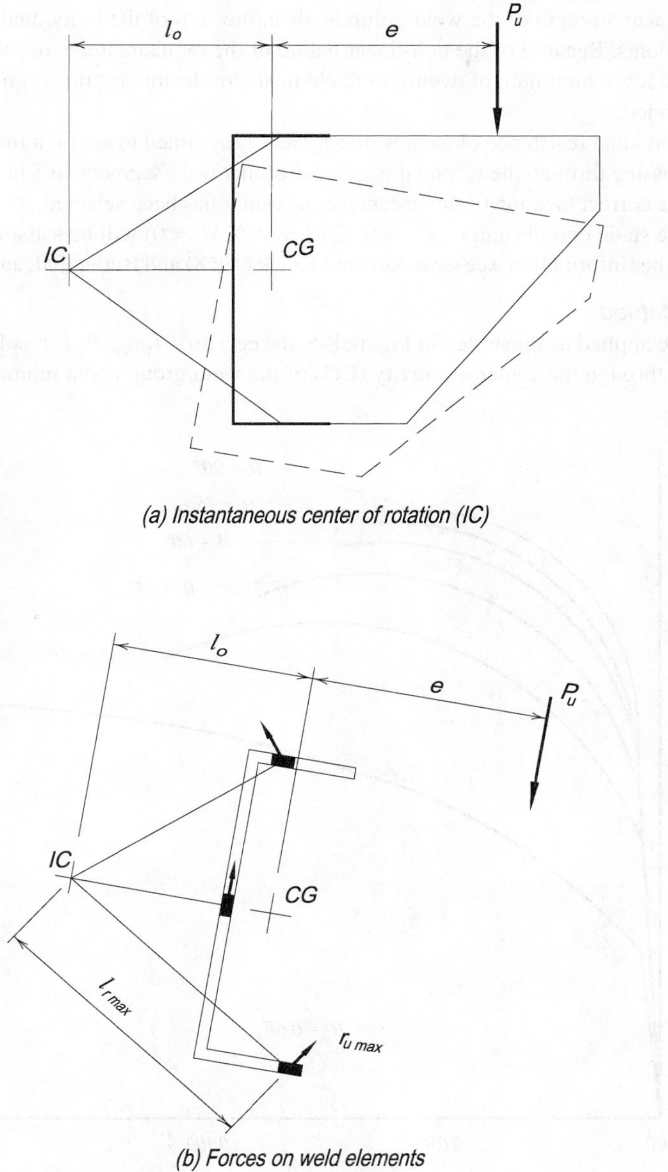

Fig. 8–5. Instantaneous center of rotation method.

The nominal shear strength of the weld group is governed by Δ_{max} of the weld segment that first reaches its limit, where

$$\Delta_{max} = 1.087w(\theta + 6)^{-0.65} \leq 0.17w$$

where w is the weld leg size, in.

The nominal shear strengths of the other unit-length weld segments in the joint can be determined by applying a deformation Δ that varies linearly with distance from the IC. The nominal shear strength of the weld group is, then, the sum of the individual strengths of all weld segments. Because of the non-linear nature of the requisite iterative solution, for sufficient accuracy, a minimum of twenty weld elements for the longest line segment is generally recommended.

The individual resistance of each weld segment is assumed to act on a line perpendicular to a ray passing through the IC and the centroid of that weld segment, as illustrated in Figure 8-5b. If the correct location of the instantaneous center has been selected, the three equations of in-plane static equilibrium ($\Sigma F_x = 0$, $\Sigma F_y = 0$, $\Sigma M = 0$) will be satisfied.

For further information, see Crawford and Kulak (1968) and Butler, Pal, and Kulak (1972).

Elastic Method

For a force applied as illustrated in Figure 8-5, the eccentric force P_u is resolved into a force P_u acting through the center of gravity (CG) of the weld group and a moment $P_u e$ where e

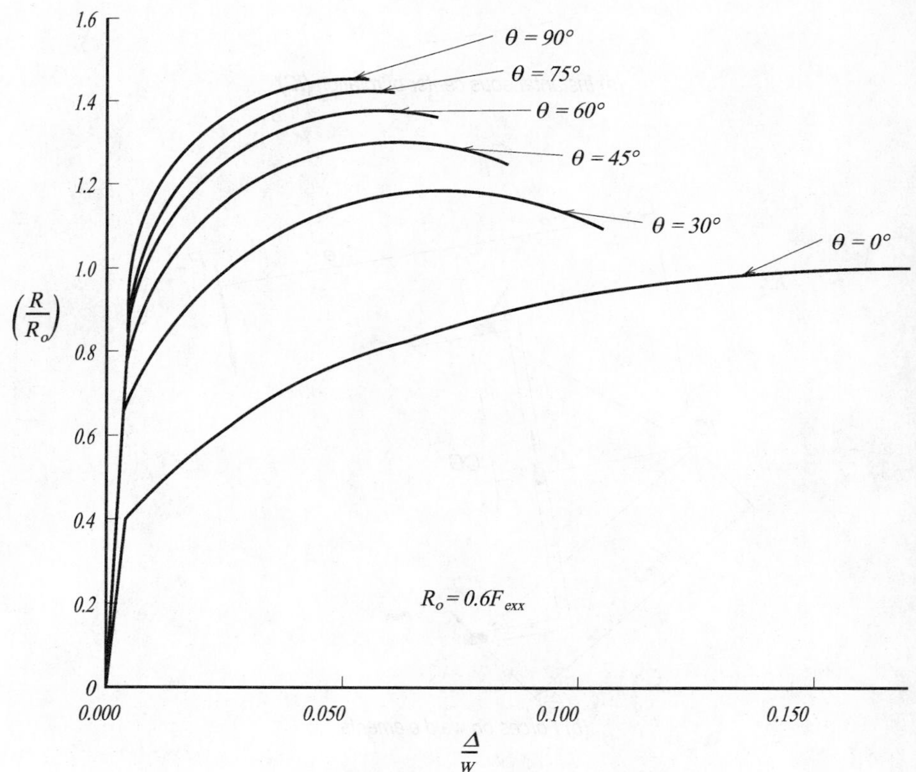

Fig. 8–6. Fillet-weld strength as a function of load angle, θ.

is the eccentricity. Each weld element is then assumed to resist an equal share of the direct shear P_u, and a share of the eccentric moment $P_u e$ proportional to its distance from the CG. The resultant vectorial sum of these forces r_u is the required strength for the weld.

The shear per linear inch of weld due to the concentric force P_u is r_p,

where

$$r_p = \frac{P_u}{l}$$

and l is the total length of the weld in the weld group. To determine the resultant shear per linear inch of weld, r_p must be resolved into horizontal component r_{px} and vertical component r_{py}, where

$$r_{px} = r_p \sin\theta \quad \text{and} \quad r_{py} = r_p \cos\theta$$

The shear per linear inch of weld due to the moment $P_u e$ is r_m, where

$$r_m = \frac{P_u ec}{I_p}$$

where,

c = radial distance from CG to point in weld group most remote from CG, in.

I_p = polar moment of inertia of the weld group, in.^{4} per in.^{2} ($I_p = I_x + I_y$). Refer to Figure 8-7. For section moduli and torsional constants of various welds treated as line elements, refer to Table 5 (page 7.4-7) of Blodgett (1966).

To determine the resultant force on the most highly stressed weld element, r_m must be resolved into horizontal component r_{mx} and vertical component r_{my}, where

$$r_{mx} = \frac{P_u ec_y}{I_p} \quad \text{and} \quad r_{my} = \frac{P_u ec_x}{I_p}$$

In the above equations, c_x and c_y are the horizontal and vertical components of the radial distance c. Thus, the resultant force is r_u, where

$$r_u = \sqrt{(r_{px} + r_{mx})^2 + (r_{py} + r_{my})^2}$$

For further information, see Higgins (1971).

Eccentricity Normal to the Plane of the Faying Surface

Eccentricity normal to the plane of the faying surface produces tension above and compression below the neutral axis as illustrated in Figure 8-8 for a bracket connection. The eccentric force P_u is resolved into a direct shear P_u acting at the faying surface of the joint and a moment normal to the plane of the faying surface $P_u e$ where e is the eccentricity. Each unit-length segment of weld is then assumed to resist an equal share of the concentric force P_u, and the moment is resisted by tension in the welds above the neutral axis and compression below the neutral axis.

In contrast to bolts, where the interaction of shear and tension must be considered, for welds, shear and tension can be combined vectorially into a resultant shear. Thus, the solution of a weld loaded eccentrically normal to the plane of the faying surface is similar to that discussed previously for welds loaded eccentrically in the plane of the faying surface.

OTHER SPECIFICATION REQUIREMENTS AND DESIGN CONSIDERATIONS

The following other specification requirements and design considerations apply to the design of tension members.

Special Requirements for Heavy Shapes and Plates

For complete-joint-penetration groove welded joints in heavy shapes (ASTM A6 tensile group 4 or 5 shapes) and plates exceeding 2-in. thickness, see LRFD Specification Section A3.1c.

Placement of Weld Groups

For the required placement of weld groups at the ends of axially loaded members, see LRFD Specification Section J1.8.

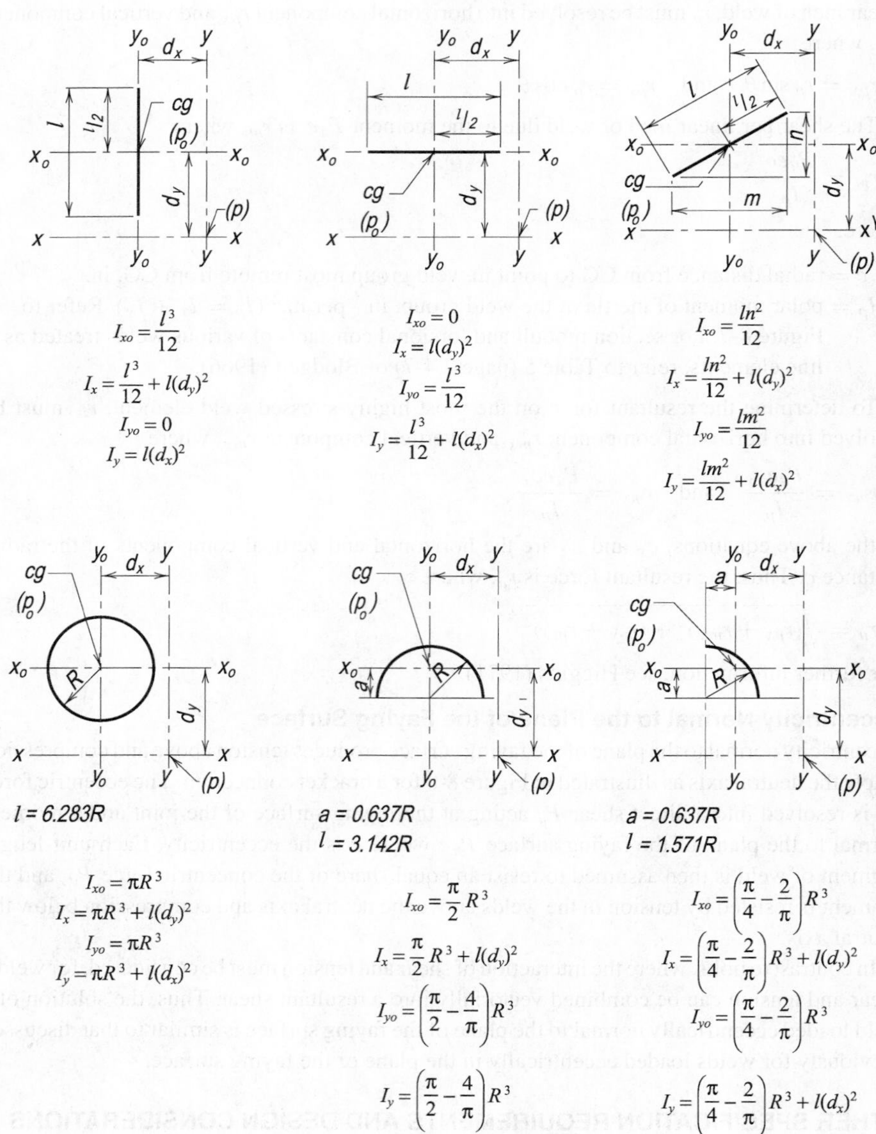

Fig. 8–7. Moments of inertia of various weld segments.

Welds in Combination with Bolts or Rivets

For welds used in combination with bolts or rivets, see LRFD Specification Section J1.9 or J1.10, respectively.

Fatigue

For applications involving fatigue, see LRFD Specification Appendix K3.

One-Sided Fillet Welds

When lateral deformation is not otherwise prevented, a severe notch can result at locations of one-sided welds. For the fillet-welded joint illustrated in Figure 8-9, the unwelded side has no

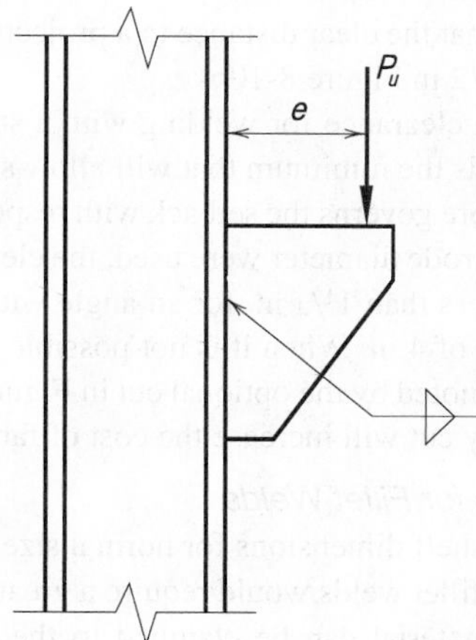

Fig. 8–8. Welds subject to eccentricity normal to the plane of the faying surface.

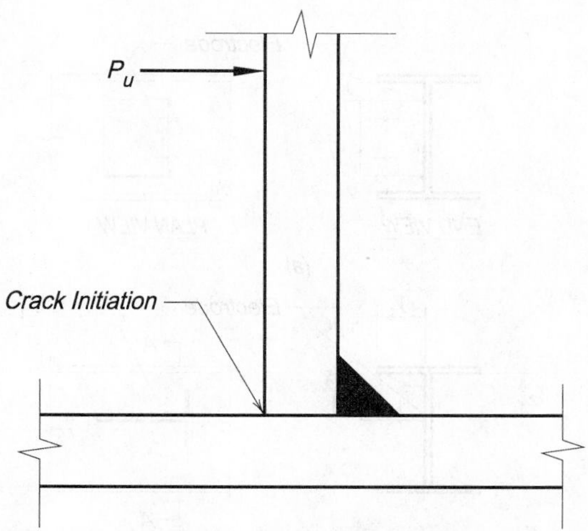

Fig. 8–9. Notch effect at one-sided weld.

strength in tension and a notch may form from the unwelded side. Using one fillet weld on each side will eliminate this condition. This is also true with partial-joint penetration groove welds.

Welding Considerations and Appurtenances

Clearance Requirements

Clearances are required to allow the welder to make proper welds. Ample room must be provided so that the welder or welding operator may manipulate the electrode and observe the weld as it is being deposited.

In the SMAW process, the preferred position of the electrode when welding in the horizontal position is in a plane forming 30° with the vertical side of the fillet weld being made. However, this angle, shown as angle x in Figure 8-10, may be varied somewhat to avoid contact with some projecting part of the work. A simple rule to provide adequate clearance for the electrode in horizontal fillet welding is that the clear distance to a projecting element should be at least one-half its height; distance $y/2$ in Figure 8-10b.

A special case of minimum clearance for welding with a straight electrode is illustrated in Figure 8-11. The 20° angle is the minimum that will allow satisfactory welding along the bottom of the angle and therefore governs the setback with respect to the end of the beam. If a $^1/_2$-in. setback and $^3/_8$-in. electrode diameter were used, the clearance between the angle and the beam flange could be no less than $1^1/_4$ in. for an angle with a leg dimension w of 3 in., nor less than $1^5/_8$ in. with a w of 4 in. When it is not possible to provide this clearance, the end of the angle may be cut as noted by the optional cut in Figure 8-11 to allow the necessary angle. However, this secondary cut will increase the cost of fabricating the connection.

Minimum Shelf Dimensions for Fillet Welds

The recommended minimum shelf dimensions for normal size SMAW fillet welds are summarized in Figure 8-12. SAW fillet welds would require a greater shelf dimension to contain the flux, although auxiliary material can be clamped to the member to provide for this. The dimension b illustrated in Figure 8-13 must be sufficient to accommodate the combined dimensional variations of the angle length, cope depth, beam depth and weld size.

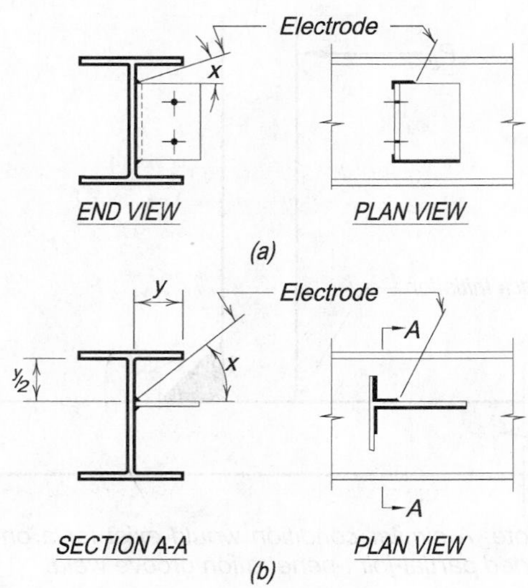

Fig. 8–10. Clearances for SMAW welding.

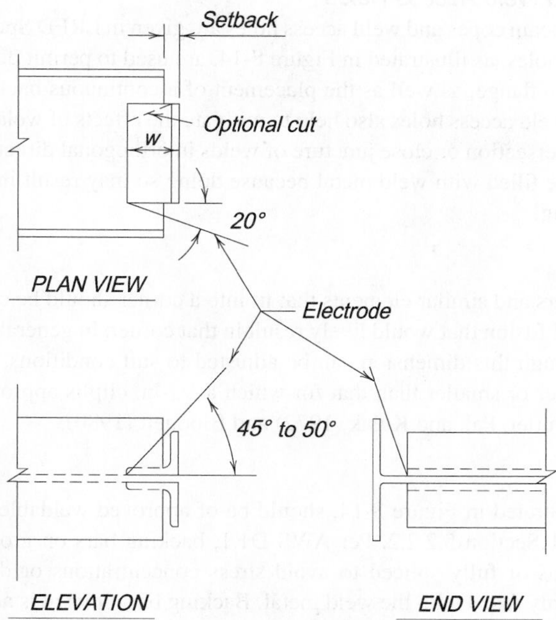

Fig. 8–11. Clearances for SMAW welding.

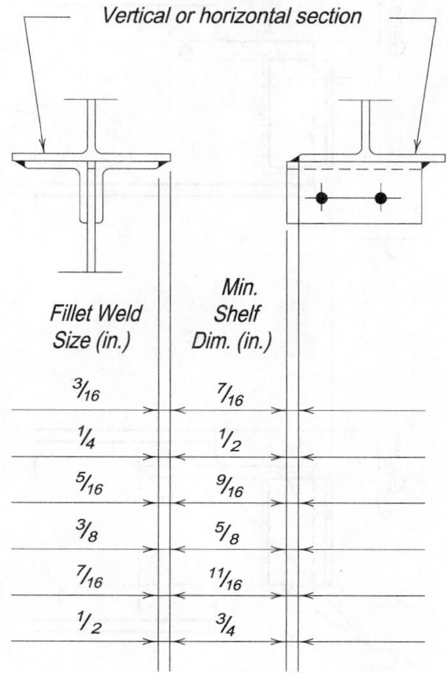

Fig. 8–12. Recommended minimum shelf dimensions for SMAW fillet welds.

Beam Copes and Weld Access Holes

Requirements for beam copes and weld access holes are given in LRFD Specification Section J1.6. Weld access holes, as illustrated in Figure 8-14, are used to permit down-hand welding to the beam bottom flange, as well as the placement of a continuous backing bar under the beam top flange. Weld access holes also help to mitigate the effects of weld shrinkage strains and prevent the intersection or close juncture of welds in orthogonal directions. Weld access holes should not be filled with weld metal because doing so may result in a state of triaxial stress under loading.

Corner Clips

Corners of stiffeners and similar elements that fit into a corner should be clipped generously to avoid the lack of fusion that would likely result in that corner. In general, a $^3/_4$-in. clip will be adequate, although this dimension can be adjusted to suit conditions, such as when the fillet radius is larger or smaller than that for which a $^3/_4$-in. clip is appropriate. For further information, see Butler, Pal, and Kulak (1972) and Blodgett (1980).

Backing Bars

Backing bars, illustrated in Figure 8-14, should be of approved weldable material as specified in AWS D1.1 Section 5.2.2.2. Per AWS D1.1, backing bars on groove-welded joints must be continuous or fully spliced to avoid stress concentrations or discontinuities and should be thoroughly fused with the weld metal. Backing bar removal is addressed in LRFD Specification Section J2.6 and AWS D1.1.

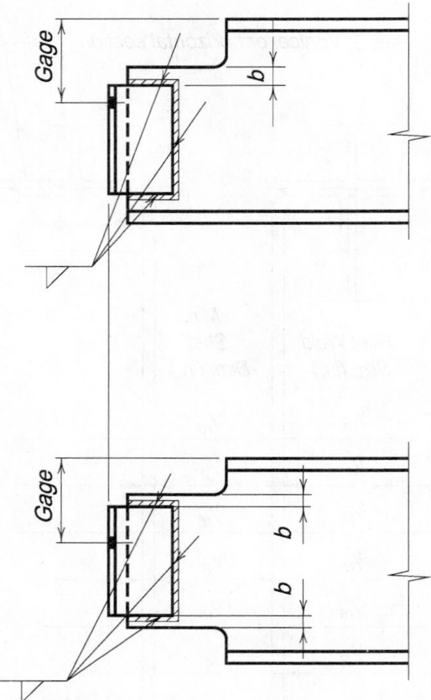

Fig. 8–13. Illustration of shelf dimensions for fillet welding.

Spacer Bars

Spacer bars, illustrated in Figure 8-14, must be of the same material specification as the base metal, per AWS D1.1 Section 5.2.2.3. This can create a procurement problem since small tonnage requirements may make them difficult to obtain in the specified ASTM designation.

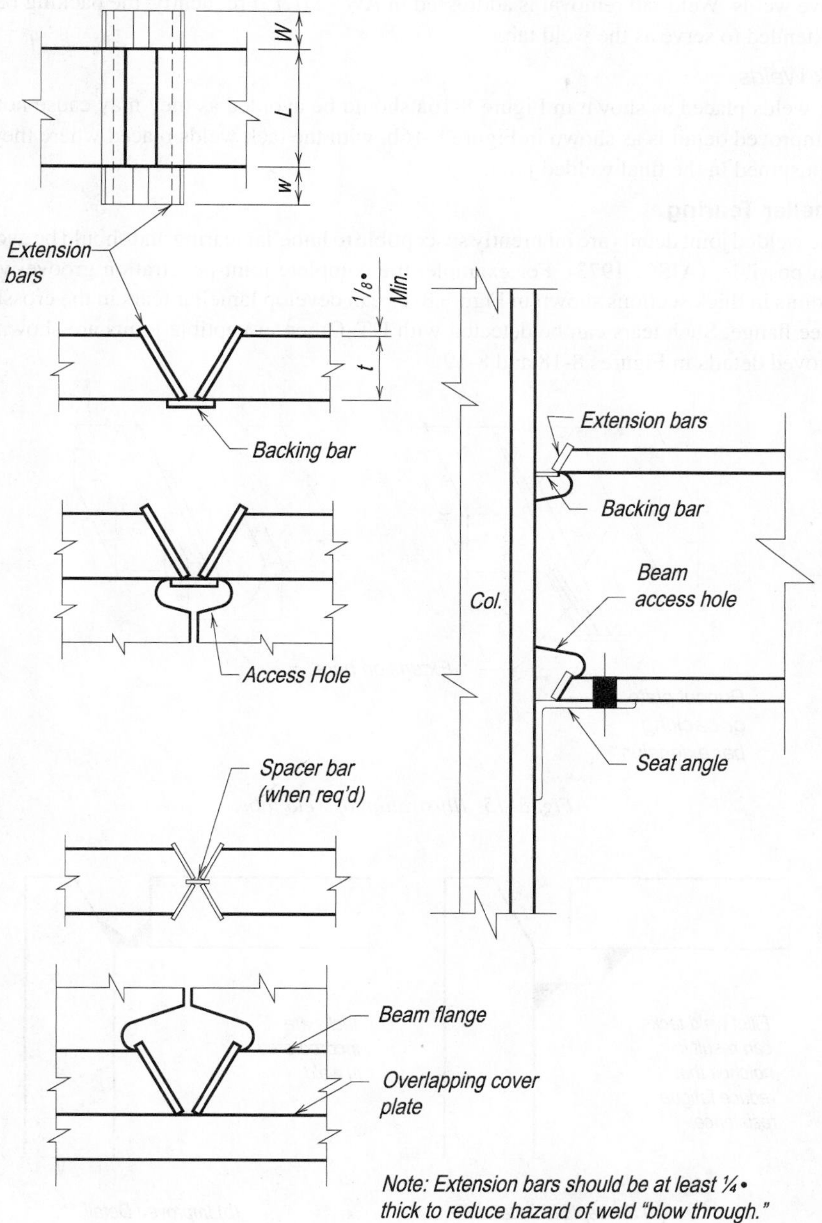

Fig. 8–14. Illustration of backing bars, spacer bars, weld tabs and other fittings for welding.

Weld Tabs

To obtain a fully welded cross section, the termination at either end of the joint must be of sound weld metal. Weld tabs, illustrated in Figure 8-14, should be of approved weldable material as specified in AWS D1.1 Section 5.2.2.1. Various configurations of weld tabs are illustrated in Figure 8-15, including flat-type weld tabs, which are normally used with bevel and V-groove welds, and contour-type weld tabs, which are normally used with J- and U-groove welds. Weld-tab removal is addressed in AWS D1.1. Frequently, the backing bar can be extended to serve as the weld tab.

Tack Welds

Tack welds placed as shown in Figure 8-16a should be avoided as they may cause notches. An improved detail is as shown in Figure 8-16b, with the tack welds placed where they will be consumed in the final welded joint.

Lamellar Tearing

Some welded joint details are inherently susceptible to lamellar tearing and should be avoided, when possible. (AISC, 1973). For example, the complete-joint-penetration groove-welded tee joints in thick sections shown in Figure 8-17 can develop lamellar tears in the crossbar of the tee flange. Such tears can be detected with UT. Other susceptible joints are shown with improved details in Figures 8-18 and 8-19.

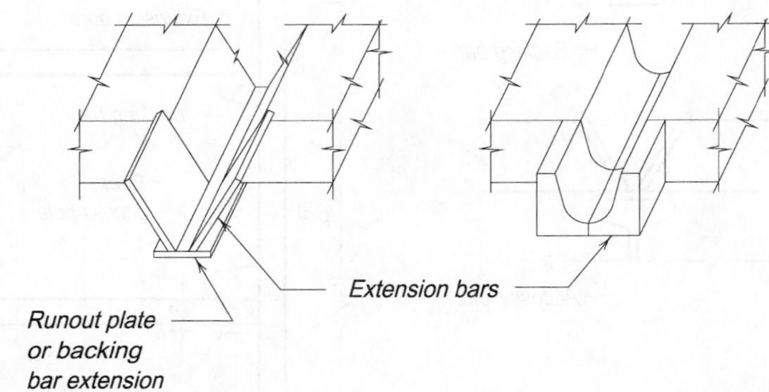

Fig. 8–15. Illustration of weld tabs.

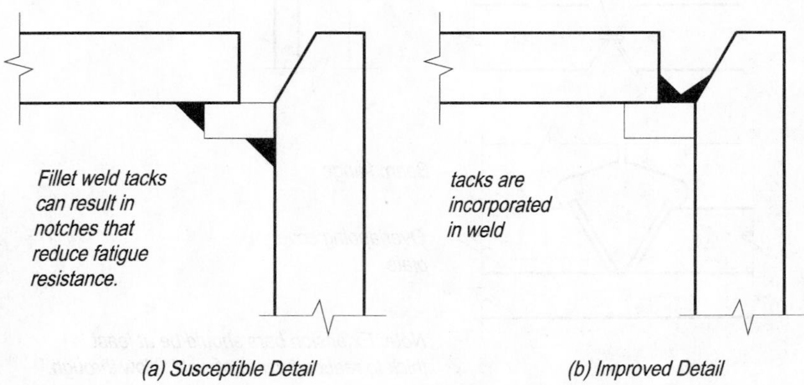

Fig. 8–16. Backing bar tack welding.

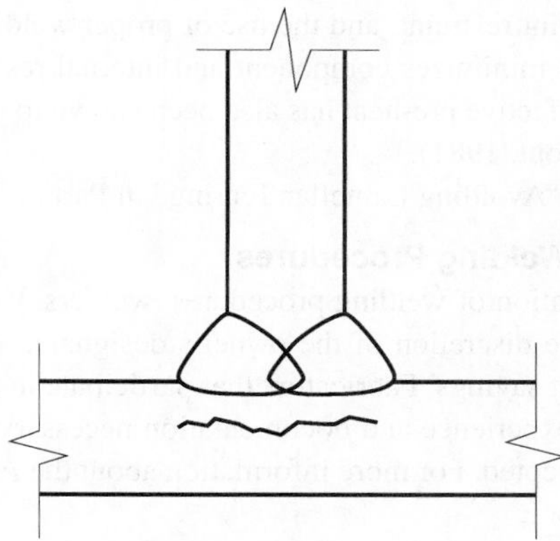

Fig. 8–17. Lamellar tearing due to weld shrinkage in large welds with thick material under high restraint.

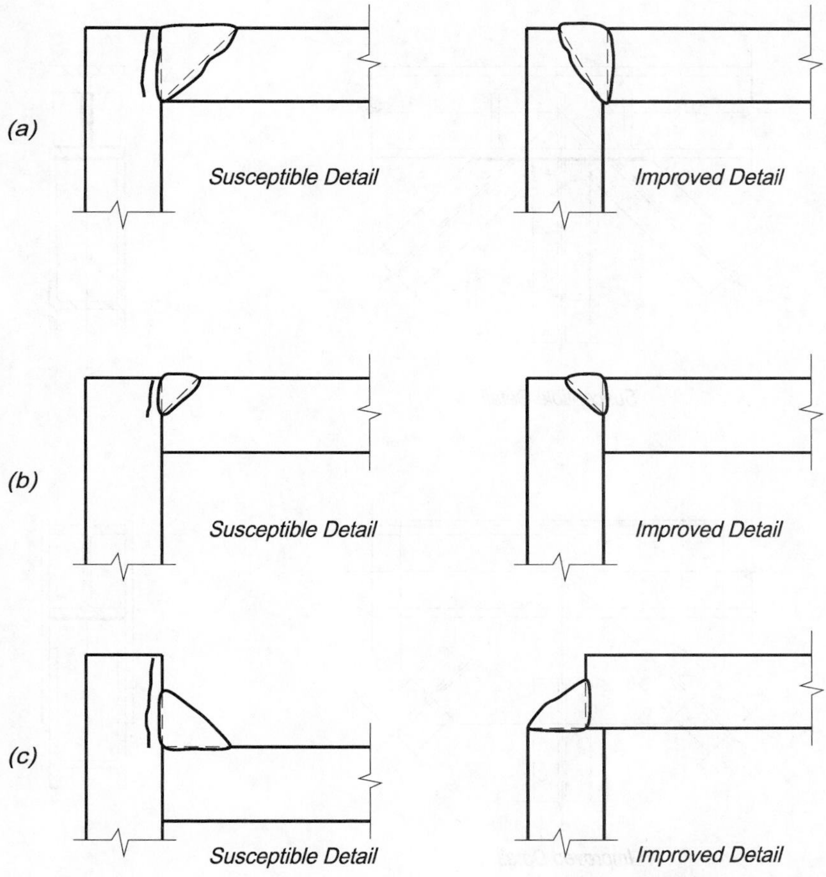

(a)

Susceptible Detail

Improved Detail

(b)

Susceptible Detail

Improved Detail

(c)

Susceptible Detail

Improved Detail

Fig. 8–18. Susceptible and improved details to reduce the incidence of lamellar tearing.

Lamellar tearing can be minimized through good joint design, including minimization of weld-metal volume and joint restraint, and the use of proper welding procedures, including the use of a sequence that minimizes component and internal restraint. Welding with low-hydrogen processes and effective pre-heat has also been shown to minimize lamellar tearing (Kaufmann, Pense, and Stout, 1981).

See also the discussion "Avoiding Lamellar Tearing" in Part 2.

Prior Qualification of Welding Procedures

Evidence of prior qualification of welding procedures, welders, welding operators, or tackers may be accepted at the discretion of the owner's designated representative for design, resulting in significant cost savings. Fabricators that participate in the AISC Quality Certification Program have the experience and documentation necessary to assure that such prior qualifications could be accepted. For more information about the AISC Quality Certification Program, visit www.aisc.org.

Painting Welded Connections

Paint is normally omitted in areas to be field welded, per LRFD Specification Section M3.5. Note that this requirement does not generally apply to shop-assembled connections, because painting is normally done after the welds are made. When required, the small paint-free areas

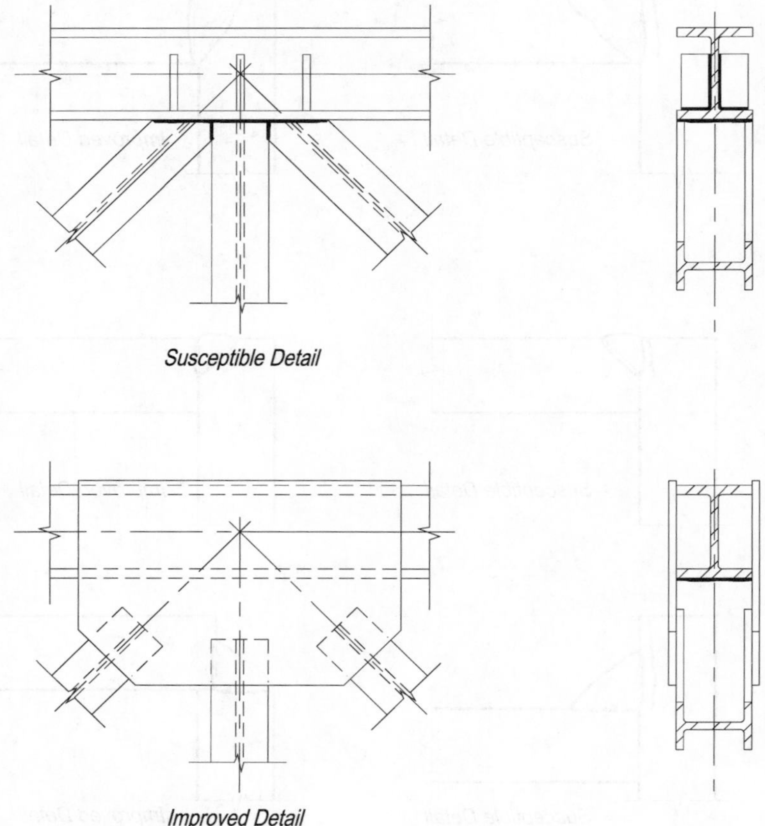

Fig. 8–19. Susceptible and improved details to avoid intersecting welds with high restraint.

can generally be identified with a general note (e.g., "no paint on OSL of connection angles," where OSL stands for outstanding leg).

DESIGN EXAMPLES

EXAMPLE 8.1. Eccentrically loaded weld groups (IC method).

Given: Refer to Figure 8-20. Determine the largest eccentric force P_u for which the design shear strength of the welds in the connection is adequate using the instantaneous center of rotation method. Use $^3/_8$-in. fillet weld and 70 ksi electrode weld size.

a. Assume the load is vertical as illustrated in Figure 8-20 ($\theta = 0°$)
b. Assume the load acts at an angle of 75° with respect to vertical ($\theta = 75°$)

Solution a: From Figure 8-20,

$$l = 10\,\text{in.}$$
$$kl = 5\,\text{in.}$$
$$k = 0.5$$

From Table 8-9 with $\theta = 0°$, $x = 0.125$

$$xl + al = 10\,\text{in.}$$
$$0.125(10\,\text{in.}) + a(10\,\text{in.}) = 10\,\text{in.}$$
$$a = 0.875$$

By interpolation from Table 8-9 with $\theta = 0°$,

$$C = 1.41$$

Design shear strength

$$\begin{aligned}\phi R_n &= CC_1Dl \\ &= 1.41(1.0)(6\,\text{sixteenths})(10\,\text{in.}) \\ &= 84.6\,\text{kips}\end{aligned}$$

Thus, P_u must be less than or equal to 84.6 kips.

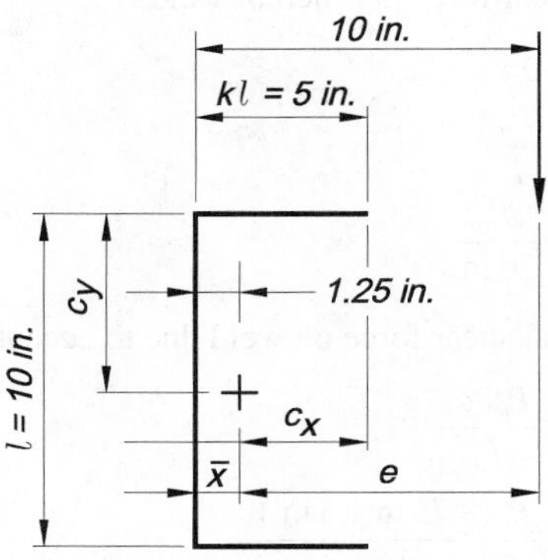

Fig. 8–20. Illustration for Examples 8.1 and 8.2.

Comment: Note that this eccentricity has effectively reduced the shear strength of this weld group by 60 percent when compared with the concentrically loaded case.

Solution b: As determined in solution a,

$$k = 0.5$$
$$a = 0.875$$

By interpolation from Table 8-9 with $\theta = 75°$,

$$C = 2.59$$

Design shear strength

$$\begin{aligned}\phi R_n &= CC_1Dl \\ &= 2.59(1.0)(6\text{ sixteenths})(10\text{ in.}) \\ &= 155\text{ kips}\end{aligned}$$

Thus, P_u must be less than or equal to 155 kips.

Comment: In Solution B, the horizontal component of the design strength is

$$\begin{aligned}\phi R_n \sin 75° &= (155\text{ kips})(0.966) \\ &= 150\text{ kips}\end{aligned}$$

and the vertical component of the design strength is

$$\begin{aligned}\phi R_n \cos 75° &= (155\text{ kips})(0.259) \\ &= 40.1\text{ kips}\end{aligned}$$

EXAMPLE 8.2. Eccentrically loaded weld groups (elastic method).

Given: Refer to Example 8-1a. Recalculate the largest eccentric force P_u for which the design shear strength of the welds in the connection is adequate using the elastic method. Compare the result with that of Example 8-1a. Use $^3/_8$-in. weld size, E70XX electrodes

$$I_p = 385\text{ in.}^4\text{ per in.}^2$$

Solution: Direct shear force per inch of weld

$$\begin{aligned}r_{px} &= 0 \\ r_{py} &= \frac{P_u}{l} \\ &= \frac{P_u}{20\text{ in.}}\end{aligned}$$

Additional shear force on weld due to eccentricity

$$\begin{aligned}r_{mx} &= \frac{P_u e c_y}{I_p} \\ &= \frac{P_u(8.75\text{ in.})(5\text{ in.})}{385\text{ in.}^4\text{ per in.}^2} \\ &= 0.114\ P_u\end{aligned}$$

$$r_{my} = \frac{P_u e c_x}{I_p}$$
$$= \frac{P_u(8.75\text{ in.})(3.75\text{ in.})}{385\text{ in.}^4\text{ per in.}^2}$$
$$= 0.0852\,P_u$$

Resultant shear force per inch of weld

$$r_u\sqrt{(r_{px}+r_{mx})^2+(r_{py}+r_{my})^2}$$
$$= \sqrt{(0+0.114P_u)^2+\left(\frac{P_u}{20}+0.0852P_u\right)^2}$$
$$= 0.177P_u$$

Since r_u must be less than or equal to ϕr_n,

$$P_u \le \frac{\phi r_n}{0.177}$$
$$\le \frac{1.392D}{0.177}$$
$$\le \frac{1.392(6\text{ sixteenths})}{0.177}$$
$$\le 47.2\text{ kips}$$

This is a 44 percent reduction in the strength predicted by the instantaneous center of rotation method in Example 8-1a.

DESIGN TABLES

Table 8-1. ASTM Requirements for Properties Affecting Weldability of Steels

A general summary of ASTM requirements for material properties that affect weldability of steel is provided in Table 8-1.

Table 8-2. Condensed AWS Specifications for SMAW Electrodes

A condensed excerpt of AWS specifications for SMAW electrodes is given in Table 8-2.

Table 8-3. Prequalified Welded Joints

The prequalified welded joints details given in AWS D1.1 and Table 8-3 provide joint geometries, such as root openings, angles, and clearances (see Figures 8-21 and 8-22) that will permit the deposition of sound weld material. Prequalified welded joints are not, in themselves, adequate consideration of welded design details and the other provisions in AWS D1.1 must be satisfied as they are referenced in LRFD Specification Section J2.2. The design and detailing for successful welded construction requires consideration of factors which include, but are not limited to, the magnitude, type, and distribution of forces to be transmitted, access, restraint against weld shrinkage, thickness of connected materials, residual stress, and distortion.

The designations such as B-L1a, B-U2, and B-P3 are those used in AWS D1.1. Note that lowercase letters, e.g., a, b, c, etc., are often used to differentiate between joints that would otherwise have the same joint designation. These prequalified welded joints are limited to those made by the SMAW, SAW, GMAW (except short circuit transfer), and FCAW procedures. Small deviations from dimensions, angles of grooves, and variation in depth of groove joints are permissible within the tolerances given.

In general, all fillet welds are prequalified, provided they conform to the requirements in AWS D1.1. Groove welds are classified using the conventions indicated in the tables. Welded joints other than those prequalified by AWS may be qualified, provided they are tested and qualified in accordance with AWS D1.1.

Table 8-4. Electrode Strength Coefficient C_1

Electrode strength coefficients C_1 that can be used to adjust tabular values for electrodes other than E70XX are given in Table 8-4. Note that this coefficient includes an additional reduction factor of 0.90 for E80 and E90 electrodes and 0.85 for E100 and E110; this accounts for the uncertainty of extrapolation to these higher-strength electrodes.

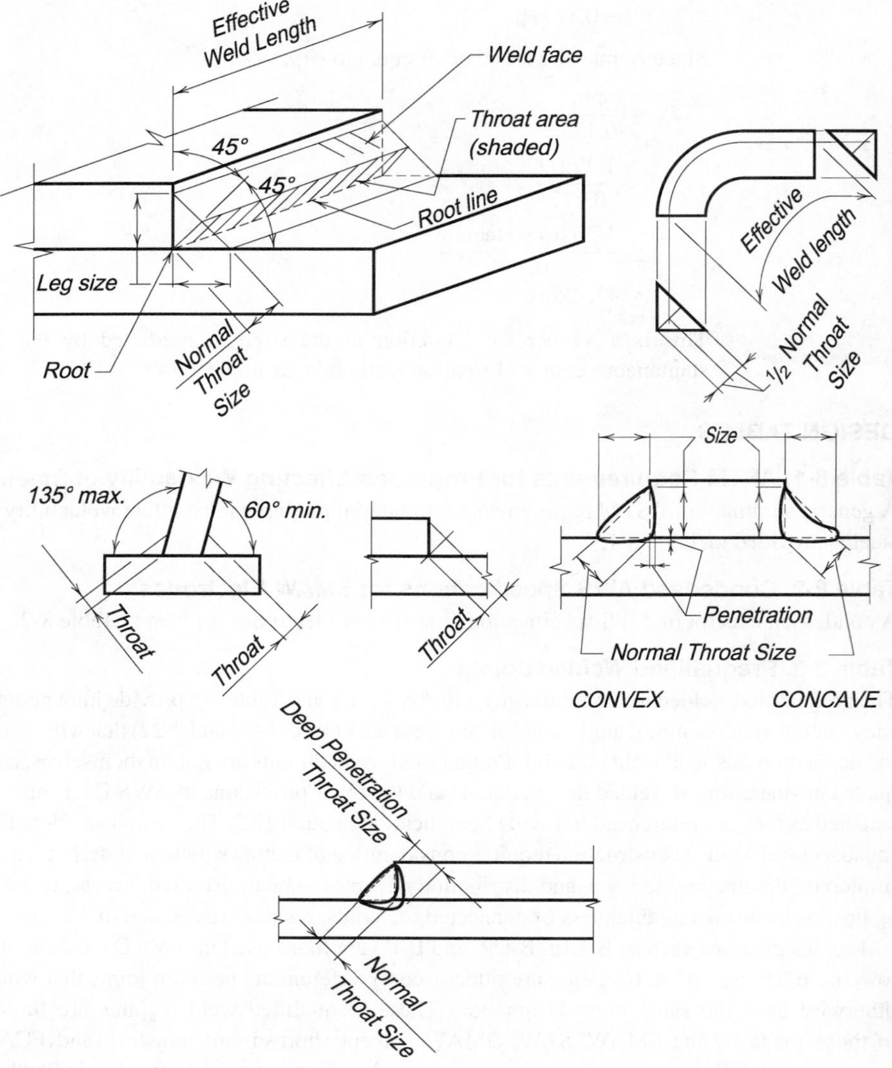

Fig. 8–21. Fillet weld nomenclature.

Tables 8-5 Through 8-12. Coefficients C for Eccentrically Loaded Weld Groups

Tables 8-5 through 8-12 employ the instantaneous center of rotation method in accordance with LRFD Specification Appendix J2.4 for the weld patterns and eccentric conditions

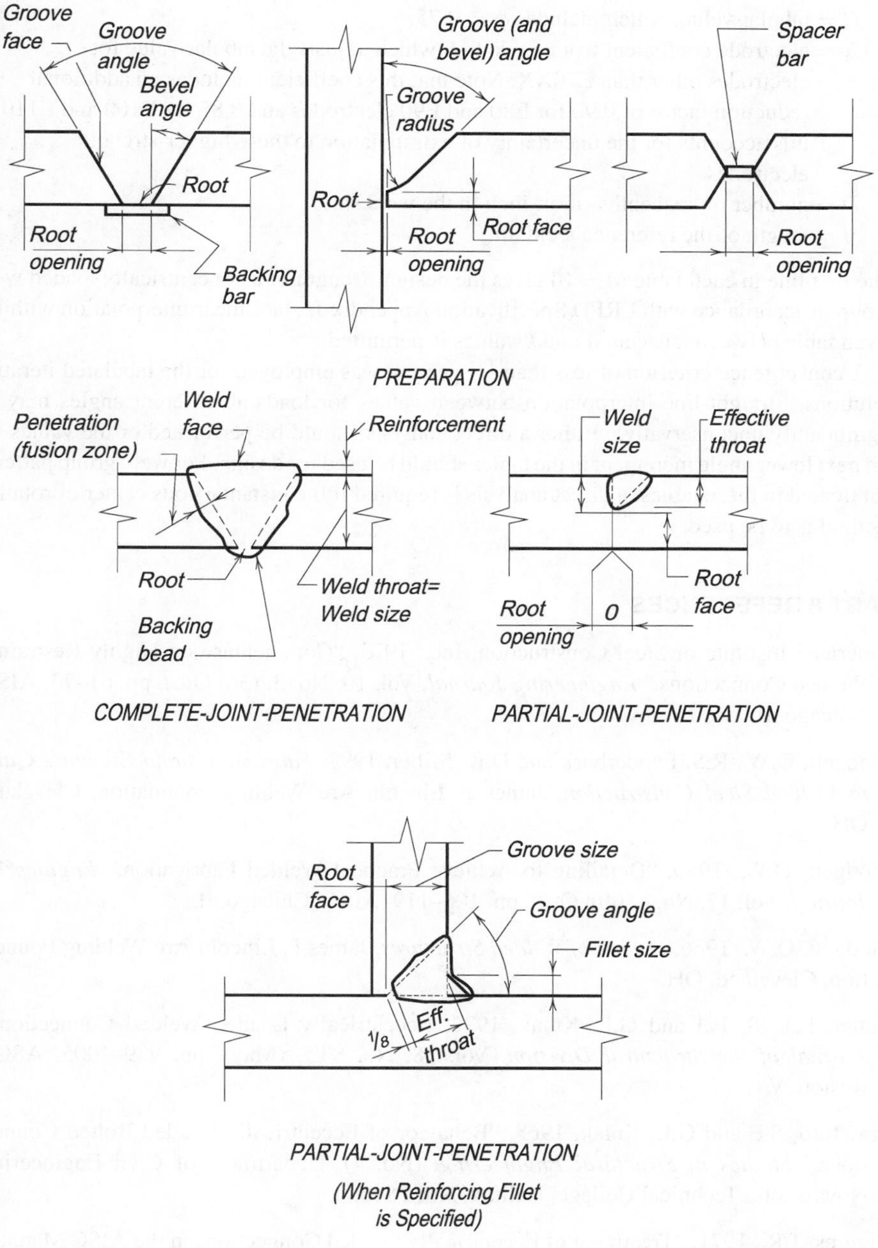

PREPARATION

COMPLETE-JOINT-PENETRATION

PARTIAL-JOINT-PENETRATION

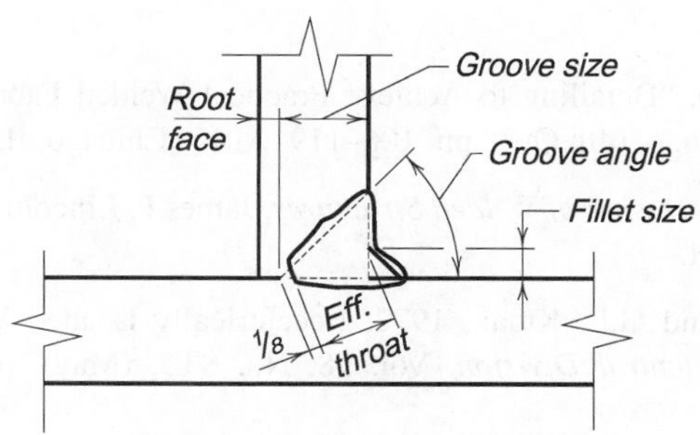

PARTIAL-JOINT-PENETRATION (When Reinforcing Fillet is Specified)

Fig. 8–22. Groove weld nomenclature.

indicated and inclined loads at 0°, 15°, 30°, 45°, 60°, and 75°. For any of the weld group geometries shown, the design strength of the eccentrically loaded weld group is ϕR_n, where

$$\phi R_n = CC_1 Dl$$

where

C = tabular value, which includes $\phi = 0.75$.
C_1 = electrode coefficient from Table 8-4, which adjusts the tabular value for electrodes other than E70XX. Note that this coefficient includes an additional reduction factor of 0.90 for E80 and E90 electrodes and 0.85 for E100 and E110; this accounts for the uncertainty of extrapolation to these higher-strength electrodes.
D = number of sixteenths-of-an-inch in the weld size
l = length of the reference weld, in.

The first line in each table ($a = 0$) gives the design strength of a concentrically loaded weld group in accordance with LRFD Specification Appendix J2.4a. Linear interpolation within a given table between adjacent a and k values is permitted.

A convergence criterion of less than $^1/_2$ percent was employed for the tabulated iterative solutions. Straight-line interpolation between values for loads at different angles may be significantly unconservative. Either a direct analysis should be performed or the values for the next lower angle increment in the tables should be used for design. For weld group patterns not treated in these tables, a direct analysis is required if the instantaneous center of rotation method is to be used.

PART 8 REFERENCES

American Institute of Steel Construction, Inc., 1973, "Commentary on Highly Restrained Welded Connections," *Engineering Journal*, Vol. 10, No. 3, (3rd Qtr.), pp. 61–73, AISC, Chicago, IL.

Blodgett, O.W., R.S. Funderburk and D.K. Miller, 1997, *Fabricator's and Erector's Guide to Welded Steel Construction*, James F. Lincoln Arc Welding Foundation, Cleveland, OH.

Blodgett, O.W., 1980, "Detailing to Achieve Practical Welded Fabrication," *Engineering Journal*, Vol. 17, No. 4, (4th Qtr.), pp. 106–119, AISC, Chicago, IL.

Blodgett, O.W., 1966, *Design of Welded Structures*, James F. Lincoln Arc Welding Foundation, Cleveland, OH.

Butler, L.J., S. Pal and G.L. Kulak, 1972, Eccentrically Loaded Welded Connections," *Journal of the Structural Division*, Vol. 98, No. ST5, (May), pp. 989–1005, ASCE, Reston, VA.

Crawford, S.F and G.L. Kulak, 1968, "Behavior of Eccentrically Loaded Bolted Connections," *Studies in Structural Engineering*, (No. 4), Department of Civil Engineering, Nova Scotia Technical College, Halifax, Nova Scotia.

Higgins, T.R., 1971, "Treatment of Eccentrically Loaded Connections in the AISC Manual," *Engineering Journal*, Vol. 8, No. 2, (April), pp. 52–54, AISC, Chicago, IL.

Institute of Welding, 1972, *Procedures and Recommendations for the Ultrasonic Testing of Butt Welds*," London, England.

Kaufmann, J.A., A.W. Pense and R.D. Stout, 1981, "An Evaluation of Factors Significant to Lamellar Tearing," *Welding Journal Research Supplement*, Vol. 60, No. 3, (March), AWS, Miami, FL.

Krautkramer, J., 1977, *Ultrasonic Testing of Materials*, 2nd Edition, Springer-Verlag, Berlin, West Germany.

Lesik, D.F. and D.J.L. Kennedy, 1990, "Ultimate Strength of Fillet-Welded Connections Loaded in Plane," *Canadian Journal of Civil Engineering*, Vol. 17, No. 1, National Research Council of Canada, Ottawa, Canada.

Table 8-1.
ASTM Requirements for Properties Affecting Weldability of Steels

ASTM Specification	Products Covered	Max. Carbon Content, % by Weight (heat analysis)	Max. Tensile Strength, ksi	Grain Size	Max. Thickness, in.
A36	Shapes	0.26	80[b]	–[c]	none
	Plates	0.25–0.29[a]			
	Bars	0.26–0.29[a]			
A53 Grade B	Steel Pipe	0.3	none	–	2.344, 24 dia.
A242	Shapes, Plates, Bars	0.15	none	–	4
A500	HSS	Gr. A, B, D: 0.26 Gr. C: 0.23	none	–	$^5/_8$
A501	HSS	0.26	none	–	1
A514	Plates–quenched and tempered	varies among 13 grades	130	killed, fine grain	[d]
A529	Shapes	0.27	100	–	Groups 1 & 2
	Plates				1
	Bars				Gr 50: 2.5 Gr. 55: 1.5
A570 Gr. 36, 50	Sheet, Strip	0.25	Gr. 36 Type 2: 80	–	0.229
A572	Shapes, Plates, Bars, Sheet Piling	Varies among 5 grades, 0.21–0.26	none	–[c]	Gr. 42: 6[e] Gr. 50: 4[e] Gr. 55: 2[e] Gr. 60, 65: 1.25[f,g]
A588	Shapes, Plates, Bars	Gr. A: 0.19 Gr. B: 0.20 Gr. C: 0.15 Gr. K: 0.17	none	fine grain	F_y = 42: 8[e] F_y = 46: 5[e] F_y = 50: 4[e]
A606	Sheet, Strip	0.22	none	–	none
A607	Sheet, Strip	Varies among 6 grades, Class 1: 0.22–0.26, Class 2: 0.15	none	–	none
A618	HSS	Gr. Ia: 0.15	none	–	Gr Ia, Ib, II: $1^1/_2$
		Gr. Ib: 0.20			
		Gr. II: 0.22			
		Gr. III: 0.23			
A852	Plates	0.19	110	fine grain	4
A992	W-Shapes	0.23	none	–[c]	none

[a] Varies by thickness.
[b] For W-shapes over 426 lb/ft, the 80 ksi maximum tensile strength does not apply.
[c] Supplemental Requirements can specify killed fine grain.
[d] Varies among 13 Grades, $1^1/_4$-6
[e] Applies to plates and bars only.
[f] Applies to plates and bars only. Shapes in Groups 1, 2, and 3 are available.
[g] No max. thickness in Zees or Rolled Tees; Sheet Piling not available in Gr. 65.

Table 8-2. Condensed AWS Specifications for SMAW Electrodes

Electrode		Grades	Min. Tensile Strength, ksi	Criteria for Composition of Deposited Weld Metal	Impact Test Criteria		Criteria for Radiographic Soundness
Type	AWS Spec.				Charpy V-Notch Test	Weld Metal Condition	
Carbon Steel	A5.1	60	60	Not Stipulated	Required for some grades only	As-welded	Stipulated for all but E6022
		70	70	Stipulated			
Low Alloy	A5.5	70	75[a]	Stipulated (all grades)	Required for some grades only	Some as-welded. Some post-weld heat-treated	Stipulated for all grades
		80	80				
		90	90				
		100	100				
		110	110				
		120	120				

[a] E7015-B2L, E7016-B2L, and E7018-B2L have a min. tensile strength of 75 ksi. All other 70 grade electrodes have a min. tensile strength of 70 ksi.

Note: A particular production welding condition may be more severe than the test conditions specified for the above.

Table 8-3. Prequalified Welded Joints

Symbols for Joint Types

B	butt joint	BC	butt or corner joint
C	corner joint	TC	T- or corner joint
T	T-joint	BTC	butt, T-, or corner joint

Symbols for Base Metal Thickness and Penetration

L	limited thickness, complete-joint-penetration
U	unlimited thickness, complete-joint-penetration
P	partial-joint-penetration

Symbols for Weld Types

1	square-groove	6	single-U-groove
2	single-V-groove	7	double-U-groove
3	double-V-groove	8	single-J-groove
4	single-bevel-groove	9	double-J-groove
5	double-bevel-groove	10	flare-bevel-groove

Symbols for Welding Processes if not Shielded Metal Arc Welding (SMAW):

S	submerged arc welding (SAW)
G	gas metal arc welding (GMAW)
F	flux cored arc welding (FCAW)

Symbols for Welding Positions

F	flat
H	horizontal
V	vertical
OH	overhead

Symbols for Joint Designation

The lower case letters (e.g., a, b, c, d, etc.) are used to differentiate between joints that would otherwise have the same joint designation.

Symbols for Dimensions

R	Root opening
α, β	Groove angles
f	Root face
r	J- or U-groove radius
S, S_1, S_2	PJP groove weld depth of groove
E, E_1, E_2	PJP groove weld sizes corresponding to S, S_1, S_2, respectively

Notes to Prequalified Welded Joints

A	Not prequalified for gas metal arc welding (GMAW) using short circuiting transfer nor GTAW. Refer to AWS D1.1 Annex A.
B	Joint is welded from one side only.
Br	Cyclic load application limits these joints to the horizontal welding position. Refer to AWS D1.1 Section 2.27.5.
C	Backgouge root to sound metal before welding second side.
D	SMAW joints may be used for prequalified GMAW (except GMAW-S) and FCAW.
E	Minimum effective throat thickness (E) as shown in LRFD Specification Table J2.3; S as specified on drawings.
J	If fillet welds are used in buildings to reinforce groove welds in corner and T-joints, they shall be equal to $1/4$ T_1, but need not exceed $3/8$-in. Groove welds in corner and T-joints of cyclically loaded structures shall be reinforced with fillet welds equal to $1/4$ T_1, but need not exceed $3/8$-in.
M	Double-groove welds may have grooves of unequal depth, but the depth of the shallower groove shall be no less than one-fourth of the thickness of the thinner part joined.
Mp	Double-groove welds may have grooves of unequal depth, provided these conform to the limitations of Note E. Also, the effective throat thickness (E) applies individually to each groove.
N	The orientation of the two members in the joints may vary from 135° to 180° for butt joints, or 45° to 135° for corner joints, or 45° to 90° for T-joints.
V	For corner joints, the ouside groove preparation may be in either or both members, provided the basic groove configuration is not changed and adequate edge distance is maintained to support the welding operations without excessive edge melting.
Z	Effective throat thickness (E) is based on joints welded flush.

The information contained in Table 8-3 has been adapted from AWS D1.1:2000 with the permission of the American Welding Society (AWS).

Table 8-3 (cont.). Prequalified Welded Joints

Basic Weld Symbols

Back	Fillet	Plug or Slot	Groove or Butt: Square	V	Bevel	U	J	Flare V	Flare Bevel

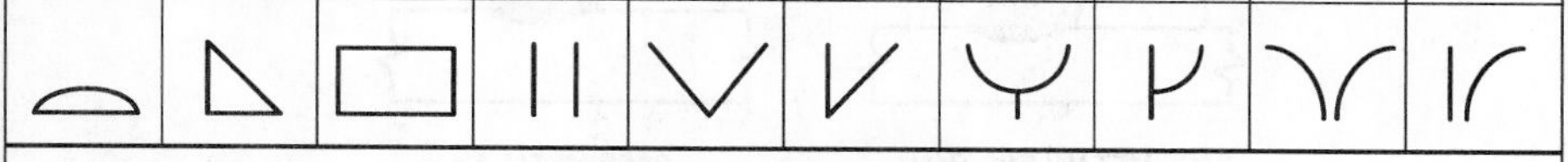

Supplementary Weld Symbols

Backing	Spacer	Weld All Around	Field Weld	Contour: Flush	Contour: Convex	For other basic and supplementary weld symbols, see AWS A2.4

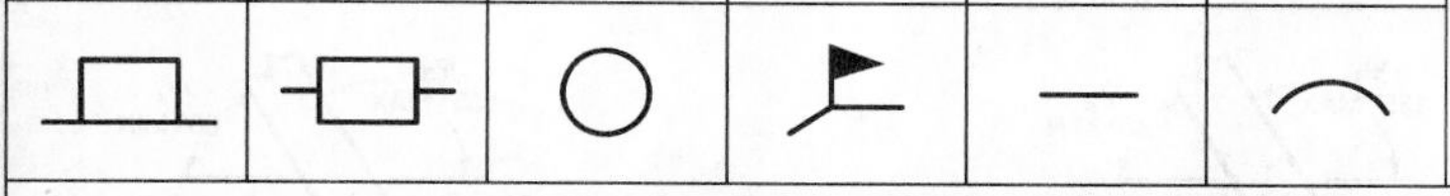

Standard Location of Elements of a Welding Symbol

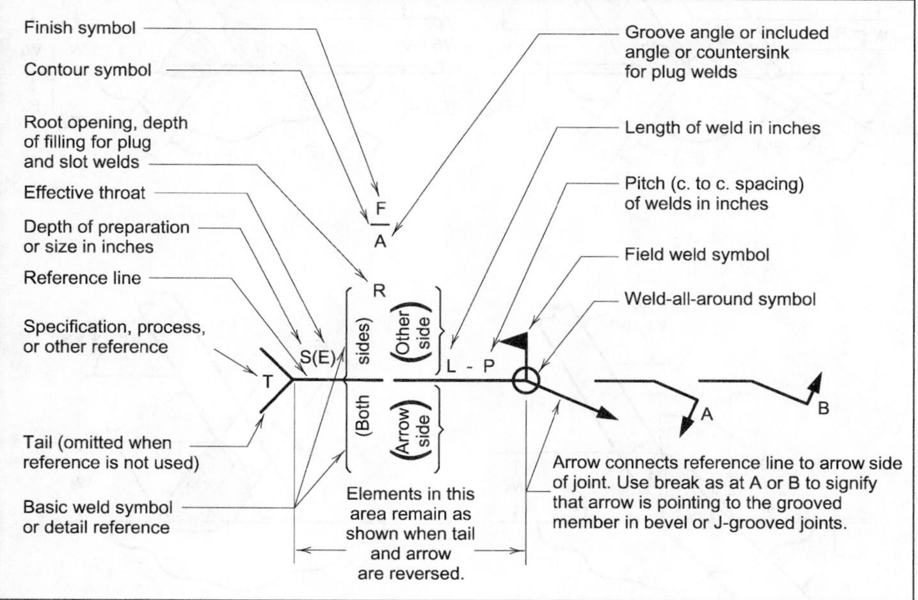

Finish symbol
Contour symbol
Root opening, depth of filling for plug and slot welds
Effective throat
Depth of preparation or size in inches
Reference line
Specification, process, or other reference
Tail (omitted when reference is not used)
Basic weld symbol or detail reference
Groove angle or included angle or countersink for plug welds
Length of weld in inches
Pitch (c. to c. spacing) of welds in inches
Field weld symbol
Weld-all-around symbol
F
A
R
S(E)
T
L - P
(Both sides)
(Other side)
(Arrow side)
A
B
Elements in this area remain as shown when tail and arrow are reversed.
Arrow connects reference line to arrow side of joint. Use break as at A or B to signify that arrow is pointing to the grooved member in bevel or J-grooved joints.

Note:
Size, weld symbol, length of weld, and spacing must read in that order, from left to right, along the reference line. Neither orientation of reference nor location of the arrow alters this rule.

The perpendicular leg of fillet, bevel, J, flare bevel weld symbols must be at left.

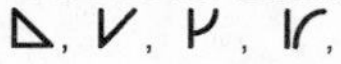

Arrow and other side welds are of the same size unless otherwise shown. Dimensions of fillet welds must be shown on both the arrow side and the other side symbol.

The point of the field weld symbol must point toward the tail.

Symbols apply between abrupt changes in direction of welding unless governed by the "all around" symbol or otherwise dimensioned.

These symbols do not explicitly provide for the case that frequently occurs in structural work, where duplicate material (such as stiffeners) occurs on the far side of a web or gusset plate. The fabricating industry has adopted this convention: that when the billing of the detail material discloses the existence of a member on the far side as well as on the near side, the welding shown for the near side shall be duplicated on the far side.

Table 8-3 (cont.). Prequalified Welded Joints Fillet Welds

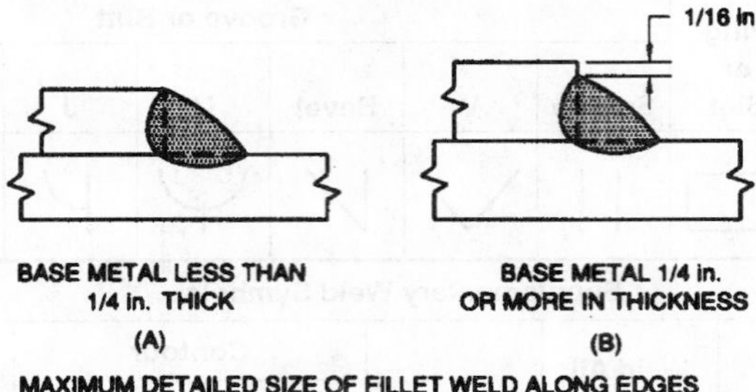

MAXIMUM DETAILED SIZE OF FILLET WELD ALONG EDGES

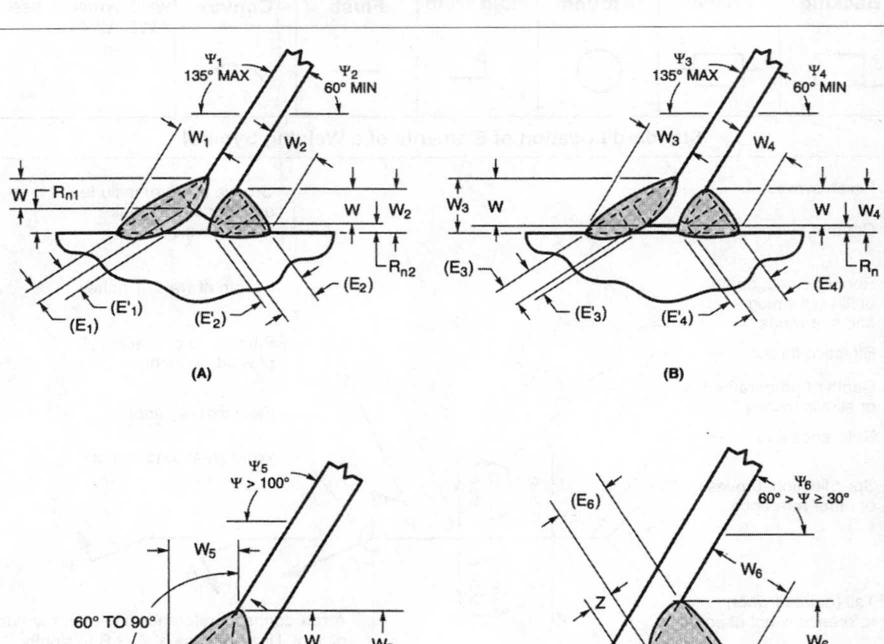

Notes:

1. (E_n), (E'_n) = Effective throat thicknesses dependant on magnitude of root opening (R_n). Refer to AWS D1.1 Section 5.22.1. Subscript n represents 1, 2, 3, 4, or 5.
2. t = thickness of thinner part.
3. Not prequalified for gas metal arc welding (GMAW) using short circuit transfer nor GTAW. Refer to AWS D1.1 Annex A for GMAW-S.
4. Figure D. Apply Z loss dimension of AWS D1.1 Table 2.2 to determine effective throat thickness.
5. Figure D. Not prequalified for angles under 30°. For welder qualifications see AWS D1.1 Table 4.8.
6. Angles under 60° are permissible, however, if the weld is considered to be a partial-joint-penetration groove weld.

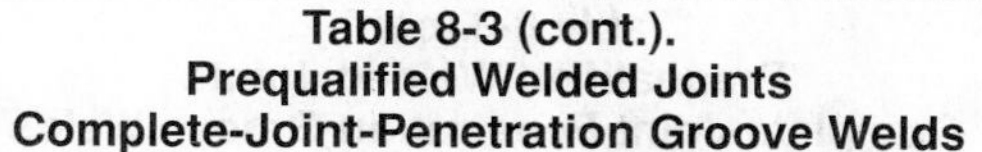

Table 8-3 (cont.). Prequalified Welded Joints Complete-Joint-Penetration Groove Welds

Square-groove weld (1)
Butt joint (B)
Corner joint (C)

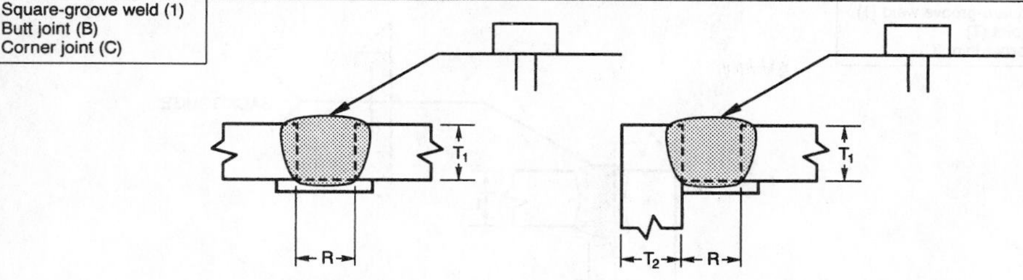

Welding Process	Joint Designation	Base Metal Thickness (U = unlimited)		Groove Preparation			Permitted Welding Positions	Gas Shielding for FCAW	Notes
					Tolerances				
		T_1	T_2	Root Opening	As Detailed (see 3.13.1)	As Fit-Up (see 3.13.1)			
SMAW	B-L1a	1/4 max	—	$R = T_1$	+1/16, –0	+1/4, –1/16	All	—	D, N
SMAW	C-L1a	1/4 max	U	$R = T_1$	+1/16, –0	+1/4, –1/16	All	—	D, N
FCAW GMAW	B-L1a-GF	3/8 max	—	$R = T_1$	+1/16, –0	+1/4, –1/16	All	Not required	A, N

Square-groove weld (1)
Butt joint (B)

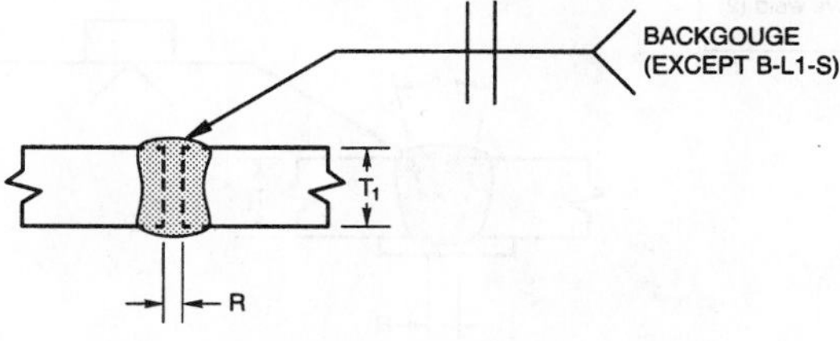

Welding Process	Joint Designation	Base Metal Thickness (U = unlimited)		Groove Preparation			Permitted Welding Positions	Gas Shielding for FCAW	Notes
					Tolerances				
		T_1	T_2	Root Opening	As Detailed (see 3.13.1)	As Fit-Up (see 3.13.1)			
SMAW	B-L1b	1/4 max	—	$R = \frac{T_1}{2}$	+1/16, –0	+1/16, –1/8	All	—	C, D, N
GMAW FCAW	B-L1b-GF	3/8 max	—	R = 0 to 1/8	+1/16, –0	+1/16, –1/8	All	Not required	A, C, N
SAW	B-L1-S	3/8 max	—	R = 0	±0	+1/16, –0	F	—	N
SAW	B-L1a-S	5/8 max	—	R = 0	±0	+1/16, –0	F	—	C, N

Table 8-3 (cont.). Prequalified Welded Joints Complete-Joint-Penetration Groove Welds

Square-groove weld (1)
T-joint (T)
Corner joint (C)

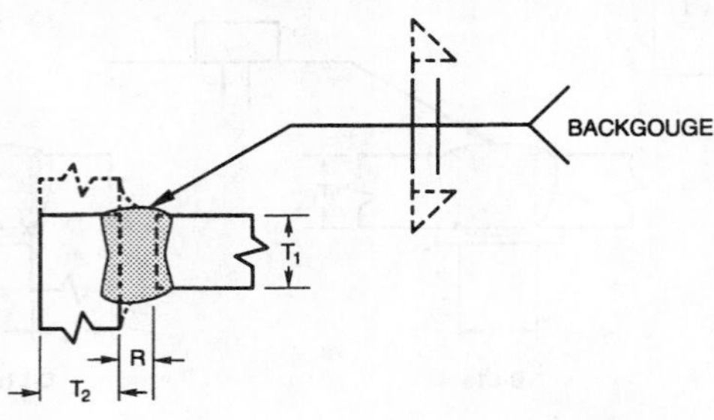

Welding Process	Joint Designation	Base Metal Thickness (U = unlimited) T_1	T_2	Groove Preparation: Root Opening	Tolerances: As Detailed (see 3.13.1)	Tolerances: As Fit-Up (see 3.13.1)	Permitted Welding Positions	Gas Shielding for FCAW	Notes
SMAW	TC-L1b	1/4 max	U	$R = \frac{T_1}{2}$	+1/16, –0	+1/16, –1/8	All	—	C, D, J
GMAW FCAW	TC-L1-GF	3/8 max	U	R = 0 to 1/8	+1/16, –0	+1/16, –1/8	All	Not required	A, C, J
SAW	TC-L1-S	3/8 max	U	R = 0	±0	+1/16, –0	F	—	C, J

Single-V-groove weld (2)
Butt joint (B)

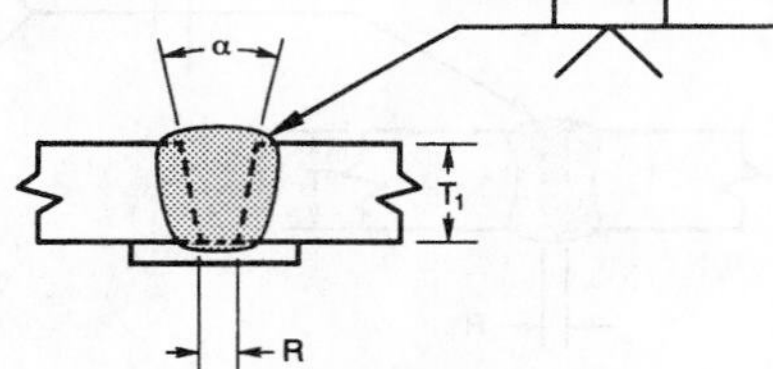

Tolerances: As Detailed (see 3.13.1)	As Fit-Up (see 3.13.1)
R = +1/16, –0	+1/4, –1/16
α = +10°, –0°	+10°, –5°

Welding Process	Joint Designation	Base Metal Thickness (U = unlimited) T_1	T_2	Groove Preparation: Root Opening	Groove Angle	Permitted Welding Positions	Gas Shielding for FCAW	Notes
SMAW	B-U2a	U	—	R = 1/4	α = 45°	All	—	D, N
				R = 3/8	α = 30°	F, V, OH	—	D, N
				R = 1/2	α = 20°	F, V, OH	—	D, N
GMAW FCAW	B-U2a-GF	U	—	R = 3/16	α = 30°	F, V, OH	Required	A, N
				R = 3/8	α = 30°	F, V, OH	Not req.	A, N
				R = 1/4	α = 45°	F, V, OH	Not req.	A, N
SAW	B-L2a-S	2 max	—	R = 1/4	α = 30°	F	—	N
SAW	B-U2-S	U	—	R = 5/8	α = 20°	F	—	N

Table 8-3 (cont.).
Prequalified Welded Joints
Complete-Joint-Penetration Groove Welds

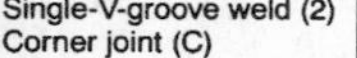
Single-V-groove weld (2)
Corner joint (C)

Tolerances	
As Detailed (see 3.13.1)	As Fit-Up (see 3.13.1)
R = +1/16, −0	+1/4, −1/16
α = +10°, −0°	+10°, −5°

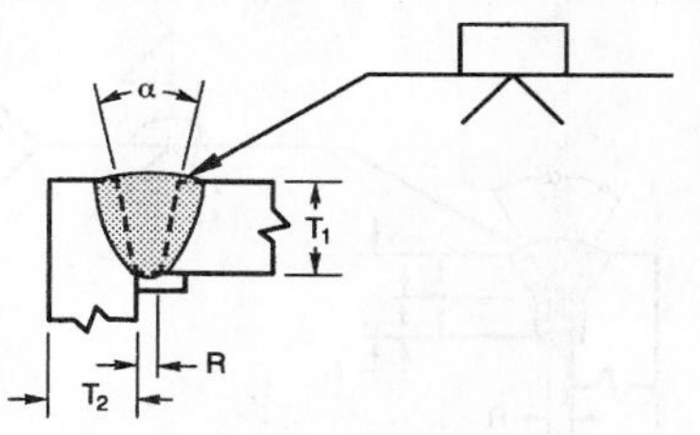

Welding Process	Joint Designation	Base Metal Thickness (U = unlimited) T_1	T_2	Groove Preparation Root Opening	Groove Angle	Permitted Welding Positions	Gas Shielding for FCAW	Notes
SMAW	C-U2a	U	U	R = 1/4	α = 45°	All	—	D, N
				R = 3/8	α = 30°	F, V, OH	—	D, N
				R = 1/2	α = 20°	F, V, OH	—	D, N
GMAW FCAW	C-U2a-GF	U	U	R = 3/16	α = 30°	F, V, OH	Required	A
				R = 3/8	α = 30°	F, V, OH	Not req.	A, N
				R = 1/4	α = 45°	F, V, OH	Not req.	A, N
SAW	C-L2a-S	2 max	U	R = 1/4	α = 30°	F	—	N
SAW	C-U2-S	U	U	R = 5/8	α = 20°	F	—	N

Single-V-groove weld (2)
Butt joint (B)

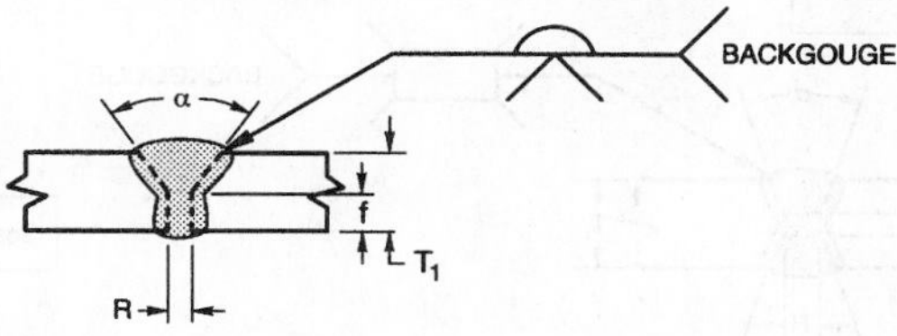

Welding Process	Joint Designation	Base Metal Thickness (U = unlimited) T_1	T_2	Groove Preparation: Root Opening Root Face Groove Angle	Tolerances: As Detailed (see 3.13.1)	Tolerances: As Fit-Up (see 3.13.1)	Permitted Welding Positions	Gas Shielding for FCAW	Notes
SMAW	B-U2	U	—	R = 0 to 1/8 f = 0 to 1/8 α = 60°	+1/16, −0 +1/16, −0 +10°, −0°	+1/16, −1/8 Not limited +10°, −5°	All	—	C, D, N
GMAW FCAW	B-U2-GF	U	—	R = 0 to 1/8 f = 0 to 1/8 α = 60°	+1/16, −0 +1/16, −0 +10°, −0°	+1/16, −1/8 Not limited +10°, −5°	All	Not required	A, C, N
SAW	B-L2c-S	Over 1/2 to 1	—	R = 0 f = 1/4 max α = 60°	R = ±0 f = +0, −f α = +10°, −0°	+1/16, −0 ±1/16 +10°, −5°	F	—	C, N
		Over 1 to 1-1/2	—	R = 0 f = 1/2 max α = 60°					
		Over 1-1/2 to 2	—	R = 0 f = 5/8 max α = 60°					

Table 8-3 (cont.). Prequalified Welded Joints Complete-Joint-Penetration Groove Welds

Single-V-groove weld (2)
Corner joint (C)

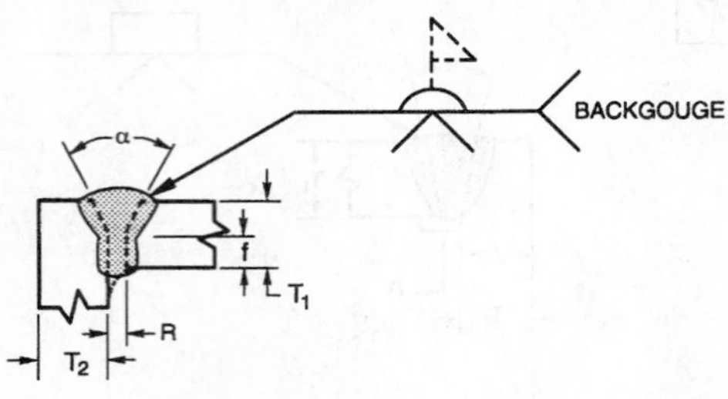

Welding Process	Joint Designation	Base Metal Thickness (U = unlimited) T_1	T_2	Groove Preparation: Root Opening, Root Face, Groove Angle	Tolerances: As Detailed (see 3.13.1)	Tolerances: As Fit-Up (see 3.13.1)	Permitted Welding Positions	Gas Shielding for FCAW	Notes
SMAW	C-U2	U	U	R = 0 to 1/8 f = 0 to 1/8 α = 60°	+1/16, −0 +1/16, −0 +10°, −0°	+1/16, −1/8 Not limited +10°, −5°	All	—	C, D, J, N
GMAW FCAW	C-U2-GF	U	U	R = 0 to 1/8 f = 0 to 1/8 α = 60°	+1/16, −0 +1/16, −0 +10°, −0°	+1/16, −1/8 Not limited +10°, −5°	All	Not required	A, C, J, N
SAW	C-U2b-S	U	U	R = 0 to 1/8 f = 1/4 max α = 60°	±0 +0, −1/4 +10°, −0°	+1/16, −0 ±1/16 +10°, −5°	F	—	C, J, N

Double-V-groove weld (3)
Butt joint (B)

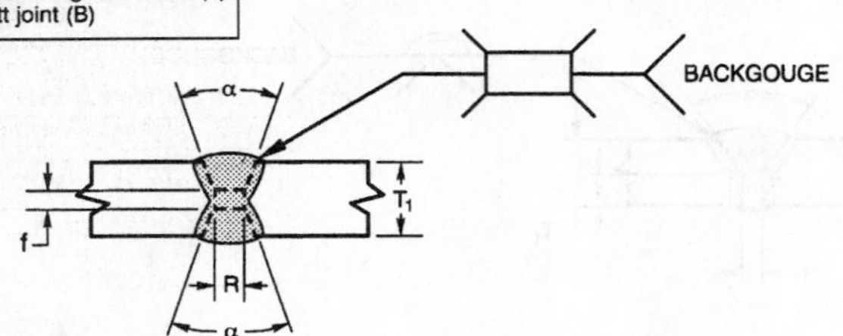

		Tolerances: As Detailed (see 3.13.1)	Tolerances: As Fit-Up (see 3.13.1)
		R = ±0	+1/4, −0
		f = ±0	+1/16, −0
		α = +10°, −0°	+10°, −5°
Spacer	SAW	±0	+1/16, −0
Spacer	SMAW	±0	+1/8, −0

Welding Process	Joint Designation	Base Metal Thickness (U = unlimited) T_1	T_2	Groove Preparation: Root Opening	Root Face	Groove Angle	Permitted Welding Positions	Gas Shielding for FCAW	Notes
SMAW	B-U3a	U Spacer = 1/8 × R	—	R = 1/4	f = 0 to 1/8	α = 45°	All	—	C, D, M, N
				R = 3/8	f = 0 to 1/8	α = 30°	F, V, OH	—	
				R = 1/2	f = 0 to 1/8	α = 20°	F, V, OH	—	
SAW	B-U3a-S	U Spacer = 1/4 × R	—	R = 5/8	f = 0 to 1/4	α = 20°	F	—	C, M, N

Table 8-3 (cont.). Prequalified Welded Joints Complete-Joint-Penetration Groove Welds

Double-V-groove weld (3)
Butt joint (B)

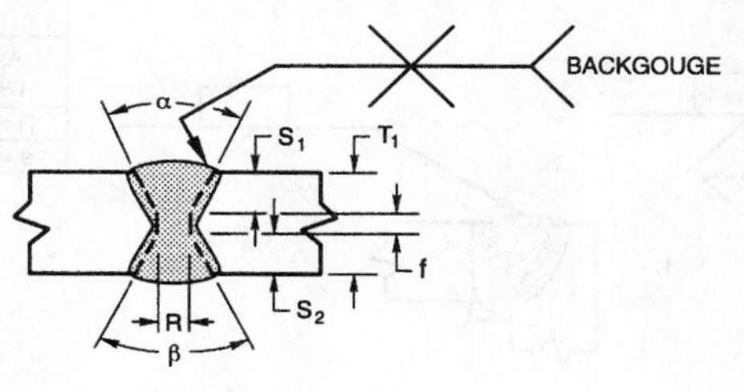

For B-U3c-S only		
T_1		S_1
Over	to	
2	2-1/2	1-3/8
2-1/2	3	1-3/4
3	3-5/8	2-1/8
3-5/8	4	2-3/8
4	4-3/4	2-3/4
4-3/4	5-1/2	3-1/4
5-1/2	6-1/4	3-3/4
For T_1 > 6-1/4 or $T_1 \le 2$ $S_1 = 2/3\ (T_1 - 1/4)$		

Welding Process	Joint Designation	Base Metal Thickness (U = unlimited) T_1	T_2	Groove Preparation: Root Opening / Root Face / Groove Angle	Tolerances: As Detailed (see 3.13.1)	Tolerances: As Fit-Up (see 3.13.1)	Permitted Welding Positions	Gas Shielding for FCAW	Notes
SMAW	B-U3b	U	—	R = 0 to 1/8 f = 0 to 1/8 α = β = 60°	+1/16, –0 +1/16, –0 +10°, –0°	+1/16, –1/8 Not limited +10°, –5°	All	—	C, D, M, N
GMAW FCAW	B-U3-GF						All	Not required	A, C, M, N
SAW	B-U3c-S	U	—	R = 0 f = 1/4 min α = β = 60°	+1/16, –0 +1/4, –0 +10°, –0°	+1/16, –0 +1/4, –0 +10°, –5°	F	—	C, M, N
				To find S_1 see table above: $S_2 = T_1 - (S_1 + f)$					

Single-bevel-groove weld (4)
Butt joint (B)

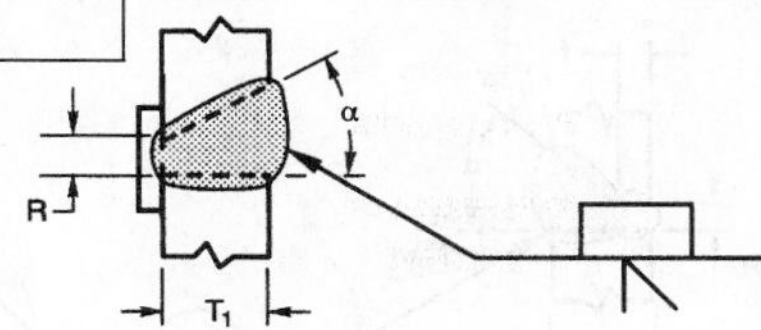

Tolerances	
As Detailed (see 3.13.1)	As Fit-Up (see 3.13.1)
R = +1/16, –0	+1/4, –1/16
α = +10°, –0°	+10°, –5°

Welding Process	Joint Designation	Base Metal Thickness (U = unlimited) T_1	T_2	Groove Preparation: Root Opening	Groove Preparation: Groove Angle	Permitted Welding Positions	Gas Shielding for FCAW	Notes
SMAW	B-U4a	U	—	R = 1/4	α = 45°	All	—	Br, D, N
				R = 3/8	α = 30°	All	—	Br, D, N
GMAW FCAW	B-U4a-GF	U	—	R = 3/16	α = 30°	All	Required	A, Br, N
				R = 1/4	α = 45°	All	Not req.	A, Br, N
				R = 3/8	α = 30°	F	Not req.	A, Br, N
SAW	B-U4a-S	U	U	R = 3/8	α = 30°	F	—	Br, N
				R = 1/4	α = 45°			

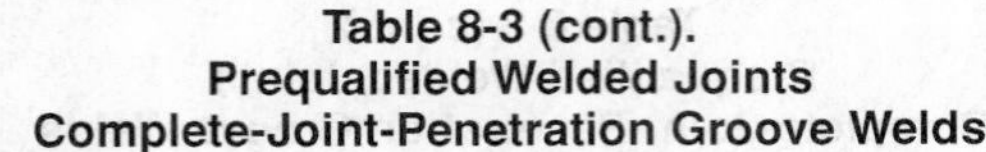

Table 8-3 (cont.). Prequalified Welded Joints Complete-Joint-Penetration Groove Welds

Single-bevel-groove weld (4)
T-joint (T)
Corner joint (C)

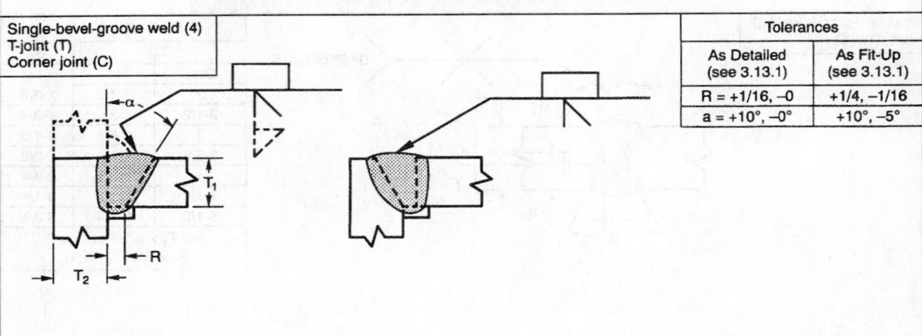

Tolerances	
As Detailed (see 3.13.1)	As Fit-Up (see 3.13.1)
R = +1/16, –0	+1/4, –1/16
a = +10°, –0°	+10°, –5°

Welding Process	Joint Designation	Base Metal Thickness (U = unlimited) T_1	T_2	Groove Preparation Root Opening	Groove Angle	Permitted Welding Positions	Gas Shielding for FCAW	Notes
SMAW	TC-U4a	U	U	R = 1/4	α = 45°	All	—	D, J, N, V
				R = 3/8	α = 30°	F, V, OH	—	D, J, N, V
GMAW FCAW	TC-U4a-GF	U	U	R = 3/16	α = 30°	All	Required	A, J, N, V
				R = 3/8	α = 30°	F	Not req.	A, J, N, V
				R = 1/4	α = 45°	All	Not req.	A, J, N, V
SAW	TC-U4a-S	U	U	R = 3/8	α = 30°	F	—	J, N, V
				R = 1/4	α = 45°			

Single-bevel-groove weld (4)
Butt joint (B)

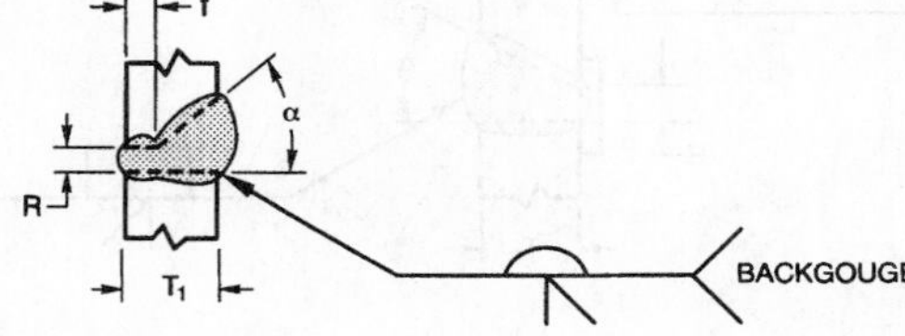

Welding Process	Joint Designation	Base Metal Thickness (U = unlimited) T_1	T_2	Groove Preparation: Root Opening Root Face Groove Angle	Tolerances: As Detailed (see 3.13.1)	As Fit-Up (see 3.13.1)	Permitted Welding Positions	Gas Shielding for FCAW	Notes
SMAW	B-U4b	U	—	R = 0 to 1/8	+1/16, –0	+1/16, –1/8	All	—	Br, C, D, N
GMAW FCAW	B-U4b-GF	U	—	f = 0 to 1/8 α = 45°	+1/16, –0 +10°, –0°	Not limited 10°, –5°	All	Not required	A, Br, C, N
SAW	B-U4b-S	U	U	R = 0 f = 1/4 max α = 60°	±0 +0, –1/8 +10°, –0°	+1/4, –0 ±1/16 10°, –5°	F	—	Br, C, N

Table 8-3 (cont.). Prequalified Welded Joints Complete-Joint-Penetration Groove Welds

Single-bevel-groove weld (4)
T-joint (T)
Corner joint (C)

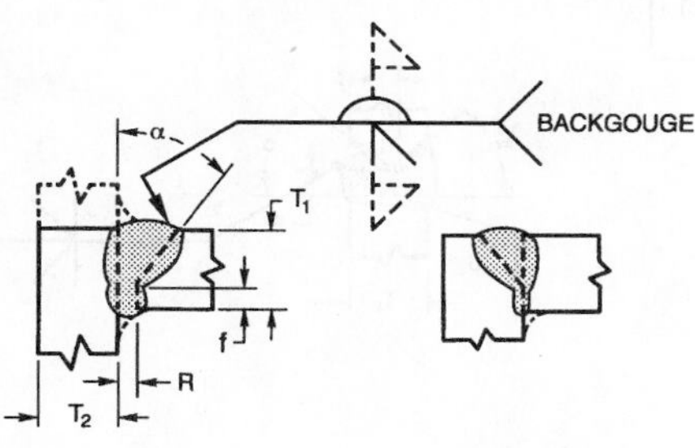

Welding Process	Joint Designation	Base Metal Thickness (U = unlimited) T_1	T_2	Groove Preparation: Root Opening / Root Face / Groove Angle	Tolerances: As Detailed (see 3.13.1)	Tolerances: As Fit-Up (see 3.13.1)	Permitted Welding Positions	Gas Shielding for FCAW	Notes
SMAW	TC-U4b	U	U	R = 0 to 1/8 f = 0 to 1/8 α = 45°	+1/16, –0 +1/16, –0 +10°, –0°	+1/16, –1/8 Not limited 10°, –5°	All	—	C, D, J, N, V
GMAW FCAW	TC-U4b-GF	U	U				All	Not required	A, C, J, N, V
SAW	TC-U4b-S	U	U	R = 0 f = 1/4 max α = 60°	±0 +0, –1/8 +10°, –0°	+1/4, –0 ±1/16 10°, –5°	F	—	C, J, N, V

Double-bevel-groove weld (5)
Butt joint (B)
T-joint (T)
Corner joint (C)

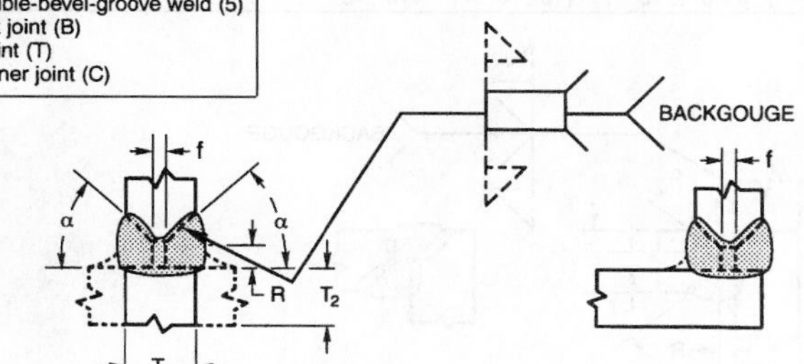

	Tolerances: As Detailed (see 3.13.1)	Tolerances: As Fit-Up (see 3.13.1)
	R = ±0	+1/4, –0
	f = +1/16, –0	±1/16
	α = +10°, –0°	+10°, –5°
Spacer	+1/16, –0	+1/8, –0

Welding Process	Joint Designation	Base Metal Thickness (U = unlimited) T_1	T_2	Groove Preparation: Root Opening	Root Face	Groove Angle	Permitted Welding Positions	Gas Shielding for FCAW	Notes
SMAW	B-U5b	U Spacer = 1/8 × R	U	R = 1/4	f = 0 to 1/8	α = 45°	All	—	Br, C, D, M, N
	TC-U5a	U Spacer = 1/4 × R	U	R = 1/4	f = 0 to 1/8	α = 45°	All	—	C, D, J, M, N, V
				R = 3/8	f = 0 to 1/8	α = 30°	F, OH	—	C, D, J, M, N, V

Table 8-3 (cont.). Prequalified Welded Joints Complete-Joint-Penetration Groove Welds

Double-bevel-groove weld (5)
Butt joint (B)

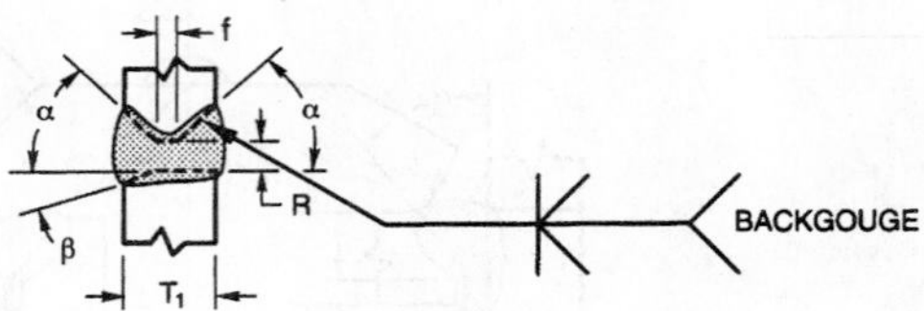

Welding Process	Joint Designation	Base Metal Thickness (U = unlimited)		Groove Preparation			Permitted Welding Positions	Gas Shielding for FCAW	Notes
					Tolerances				
		T_1	T_2	Root Opening Root Face Groove Angle	As Detailed (see 3.13.1)	As Fit-Up (see 3.13.1)			
SMAW	B-U5a	U	—	R = 0 to 1/8 f = 0 to 1/8 α = 45° β = 0° to 15°	+1/16, −0 +1/16, −0 $\alpha + \beta^{+10°}_{-0°}$	+1/16, −1/8 Not limited $\alpha + \beta^{+10°}_{-5°}$	All	—	Br, C, D, M, N
GMAW FCAW	B-U5-GF	U	—	R = 0 to 1/8 f = 0 to 1/8 α = 45° β = 0° to 15°	+1/16, −0 +1/16, −0 α + β = +10°, −0°	+1/16, −1/8 Not limited α + β = +10°, −5°	All	Not required	A, Br, C, M, N

Double-bevel-groove weld (5)
T-joint (T)
Corner joint (C)

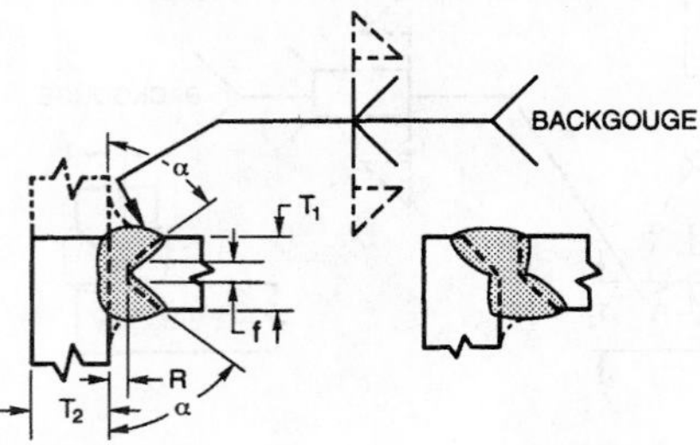

Welding Process	Joint Designation	Base Metal Thickness (U = unlimited)		Groove Preparation			Permitted Welding Positions	Gas Shielding for FCAW	Notes
					Tolerances				
		T_1	T_2	Root Opening Root Face Groove Angle	As Detailed (see 3.13.1)	As Fit-Up (see 3.13.1)			
SMAW	TC-U5b	U	U	R = 0 to 1/8 f = 0 to 1/8 α = 45°	+1/16, −0 +1/16, −0 +10°, −0°	+1/16, −1/8 Not limited +10°, −5°	All	—	C, D, J, M, N, V
GMAW FCAW	TC-U5-GF	U	U	R = 0 to 1/8 f = 0 to 1/8 α = 45°	+1/16, −0 +1/16, −0 +10°, −0°	+1/16, −1/8 Not limited +10°, −5°	All	Not required	A, C, J, M, N, V
SAW	TC-U5-S	U	U	R = 0 f = 3/16 max α = 60°	±0 +0, −3/16 +10°, −0°	+1/16, −0 ±1/16 +10°, −5°	F	—	C, J, M, N, V

Table 8-3 (cont.). Prequalified Welded Joints Complete-Joint-Penetration Groove Welds

Single-U-groove weld (6)
Butt joint (B)
Corner joint (C)

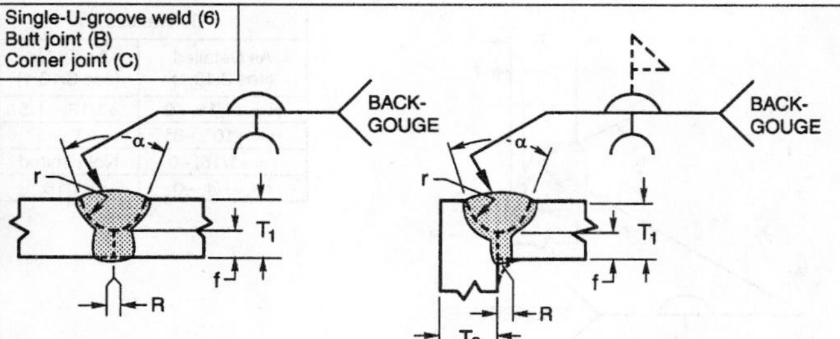

Tolerances	
As Detailed (see 3.13.1)	As Fit-Up (see 3.13.1)
R = +1/16, –0	+1/16, –1/8
α = +10°, –0°	+10°, –5°
f = ±1/16	Not Limited
r = +1/8, –0	+1/8, –0

Welding Process	Joint Designation	Base Metal Thickness (U = unlimited) T_1	T_2	Groove Preparation: Root Opening	Groove Angle	Root Face	Groove Radius	Permitted Welding Positions	Gas Shielding for FCAW	Notes
SMAW	B-U6	U	U	R = 0 to 1/8	α = 45°	f = 1/8	r = 1/4	All	—	C, D, N
				R = 0 to 1/8	α = 20°	f = 1/8	r = 1/4	F, OH	—	C, D, N
	C-U6	U	U	R = 0 to 1/8	α = 45°	f = 1/8	r = 1/4	All	—	C, D, J, N
				R = 0 to 1/8	α = 20°	f = 1/8	r = 1/4	F, OH	—	C, D, J, N
GMAW FCAW	B-U6-GF	U	U	R = 0 to 1/8	α = 20°	f = 1/8	r = 1/4	All	Not req.	A, C, N
	C-U6-GF	U	U	R = 0 to 1/8	α = 20°	f = 1/8	r = 1/4	All	Not req.	A, C, J, N

Double-U-groove weld (7)
Butt joint (B)

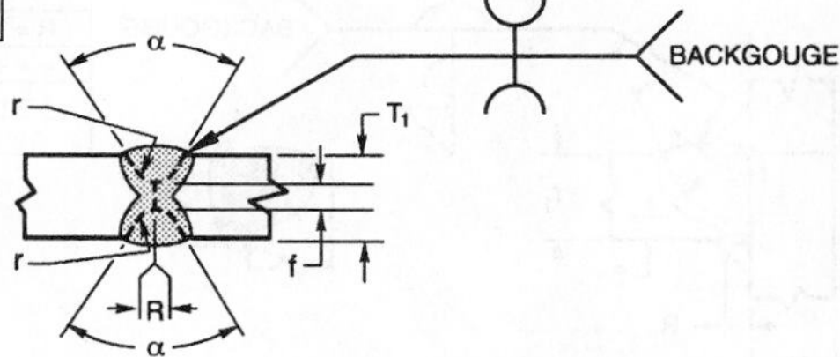

Tolerances	
As Detailed (see 3.13.1)	As Fit-Up (see 3.13.1)
For B-U7 and B-U7-GF	
R = +1/16, –0	+1/16, –1/8
α = +10°, –0°	+10°, –5°
f = ±1/16, –0	Not Limited
r = +1/4, –0	±1/16
For B-U7-S	
R = ±0	+1/16, –0
f = +0, –1/4	±1/16

Welding Process	Joint Designation	Base Metal Thickness (U = unlimited) T_1	T_2	Groove Preparation: Root Opening	Groove Angle	Root Face	Groove Radius	Permitted Welding Positions	Gas Shielding for FCAW	Notes
SMAW	B-U7	U	—	R = 0 to 1/8	α = 45°	f = 1/8	r = 1/4	All	—	C, D, M, N
				R = 0 to 1/8	α = 20°	f = 1/8	r = 1/4	F, OH	—	C, D, M, N
GMAW FCAW	B-U7-GF	U	—	R = 0 to 1/8	α = 20°	f = 1/8	r = 1/4	All	Not required	A, C, N, M
SAW	B-U7-S	U	—	R = 0	α = 20°	f = 1/4 max	r = 1/4	F	—	C, M, N

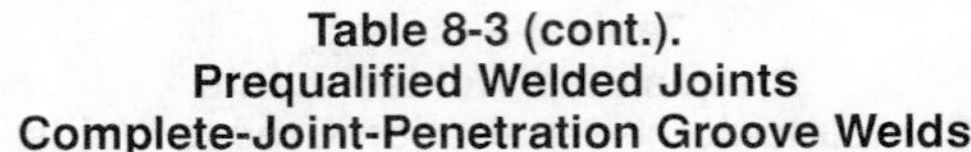

Table 8-3 (cont.). Prequalified Welded Joints Complete-Joint-Penetration Groove Welds

Single-J-groove weld (8)
Butt joint (B)

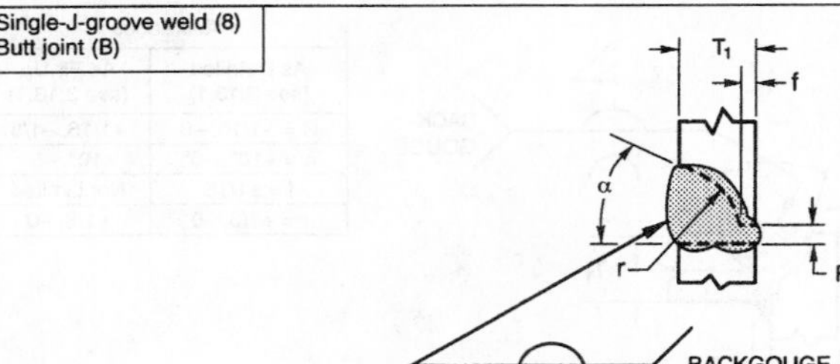

Tolerances	
As Detailed (see 3.13.1)	As Fit-Up (see 3.13.1)
R = +1/16, −0	+1/16, −1/8
α = +10°, −0°	+10°, −5°
f = +1/16, −0	Not Limited
r = +1/4, −0	±1/16

Welding Process	Joint Designation	Base Metal Thickness (U = unlimited)		Groove Preparation				Permitted Welding Positions	Gas Shielding for FCAW	Notes
		T_1	T_2	Root Opening	Groove Angle	Root Face	Groove Radius			
SMAW	B-U8	U	—	R = 0 to 1/8	α = 45°	f = 1/8	r = 3/8	All	—	Br, C, D, N
GMAW FCAW	B-U8-GF	U	—	R = 0 to 1/8	α = 30°	f = 1/8	r = 3/8	All	Not req.	A, Br, C, N
SAW	B-U8-S	U	U	R = 0 f = 1/4 max α = 45°	±0 +0, −1/8 +10°, −0°	+1/4, −0 ±1/16 +10°, −5°	r = 3/8	F	—	Br, C, N

Single-J-groove weld (8)
T-joint (T)
Corner joint (C)

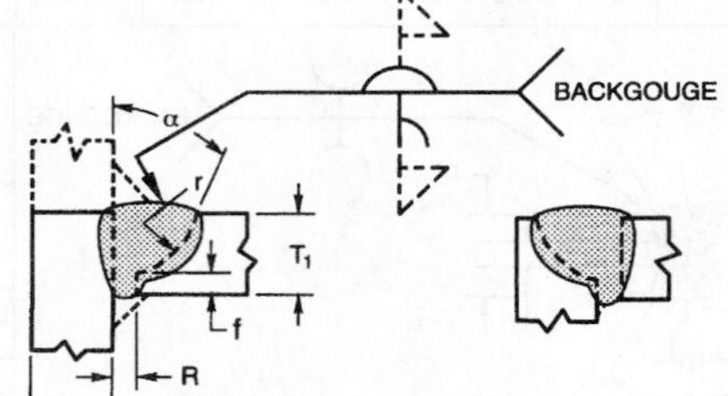

Tolerances	
As Detailed (see 3.13.1)	As Fit-Up (see 3.13.1)
R = +1/16, −0	+1/16, −1/8
α = +10°, −0°	+10°, −5°
f = +1/16, −0	Not Limited
r = +1/4, −0	±1/16

Welding Process	Joint Designation	Base Metal Thickness (U = unlimited)		Groove Preparation				Permitted Welding Positions	Gas Shielding for FCAW	Notes
		T_1	T_2	Root Opening	Groove Angle	Root Face	Groove Radius			
SMAW	TC-U8a	U	U	R = 0 to 1/8	α = 45°	f = 1/8	r = 3/8	All	—	C, D, J, N, V
				R = 0 to 1/8	α = 30°	f = 1/8	r = 3/8	F, OH	—	C, D, J, N, V
GMAW FCAW	TC-U8a-GF	U	U	R = 0 to 1/8	α = 30°	f = 1/8	r = 3/8	All	Not required	A, C, J, N, V
SAW	TC-U8a-S	U	U	R = 0 f = 1/4 max α = 45°	±0 +0, −1/8 +10°, −0°	+1/4, −0 ±1/16 +10°, −5°	r = 3/8	F	—	C, J, R, V

Table 8-3 (cont.). Prequalified Welded Joints Complete-Joint-Penetration Groove Welds

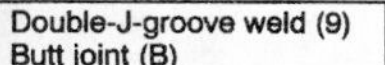
Double-J-groove weld (9)
Butt joint (B)

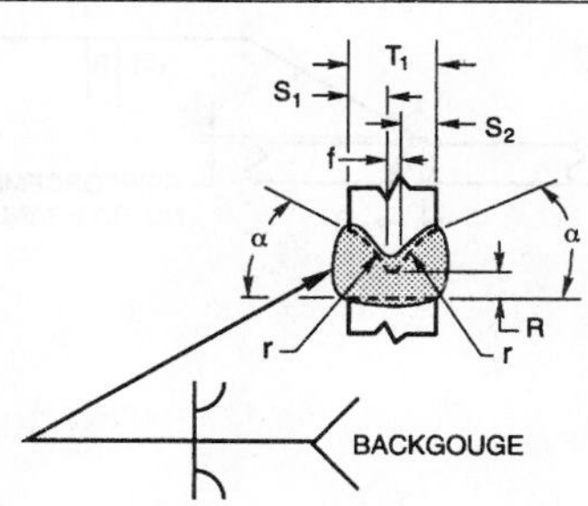

Tolerances	
As Detailed (see 3.13.1)	As Fit-Up (see 3.13.1)
R = +1/16, –0	+1/16, –1/8
α = +10°, –0°	+10°, –5°
f = +1/16, –0	Not Limited
r = +1/8, –0	±1/16

Welding Process	Joint Designation	Base Metal Thickness (U = unlimited)		Groove Preparation				Permitted Welding Positions	Gas Shielding for FCAW	Notes
		T_1	T_2	Root Opening	Groove Angle	Root Face	Groove Radius			
SMAW	B-U9	U	—	R = 0 to 1/8	α = 45°	f = 1/8	r = 3/8	All	—	Br, C, D, M, N
GMAW FCAW	B-U9-GF	U	—	R = 0 to 1/8	α = 30°	f = 1/8	r = 3/8	All	Not required	A, Br, C, M, N

Double-J-groove weld (9)
T-joint (T)
Corner joint (C)

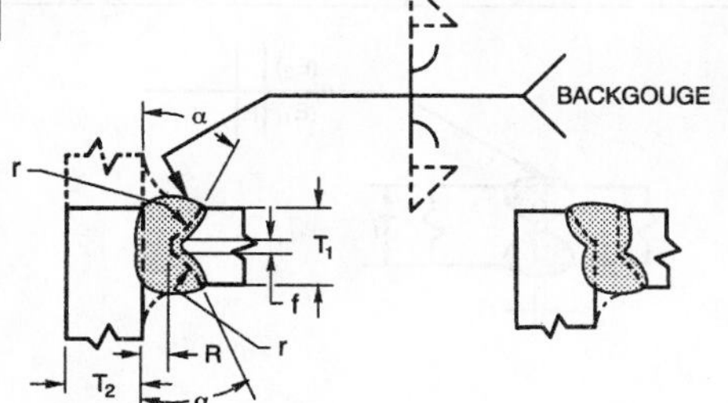

Tolerances	
As Detailed (see 3.13.1)	As Fit-Up (see 3.13.1)
R = +1/16, –0	+1/16, –1/8
α = +10°, –0°	+10°, –5°
f = +1/16, –0	Not Limited
r = 1/8, –0	±1/16

Welding Process	Joint Designation	Base Metal Thickness (U = unlimited)		Groove Preparation				Permitted Welding Positions	Gas Shielding for FCAW	Notes
		T_1	T_2	Root Opening	Groove Angle	Root Face	Groove Radius			
SMAW	TC-U9a	U	U	R = 0 to 1/8	α = 45°	f = 1/8	r = 3/8	All	—	C, D, J, M, N, V
				R = 0 to 1/8	α = 30°	f = 1/8	r = 3/8	F, OH	—	C, D, J, M, V
GMAW FCAW	TC-U9a-GF	U	U	R = 0 to 1/8	α = 30°	f = 1/8	r = 3/8	All	Not required	A, C, J, M, N, V

Table 8-3 (cont.). Prequalified Welded Joints Partial-Joint-Penetration Groove Welds

Square-groove weld (1)
Butt joint (B)

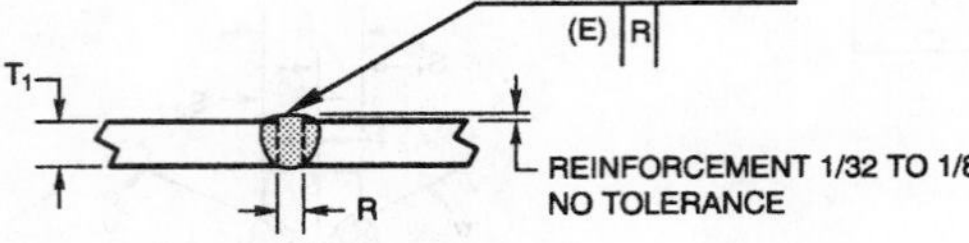

Welding Process	Joint Designation	Base Metal Thickness (U = unlimited) T_1	T_2	Groove Preparation: Root Opening	Tolerances: As Detailed (see 3.12.3)	Tolerances: As Fit-Up (see 3.12.3)	Permitted Welding Positions	Weld Size (E)	Notes
SMAW	B-P1a	1/8	—	R = 0 to 1/16	+1/16, –0	±1/16	All	T_1 – 1/32	B, D
	B-P1c	1/4 max	—	$R = \frac{T_1}{2}$ min	+1/16, –0	±1/16	All	$\frac{T_1}{2}$	B, D

Square-groove weld (1)
Butt joint (B)

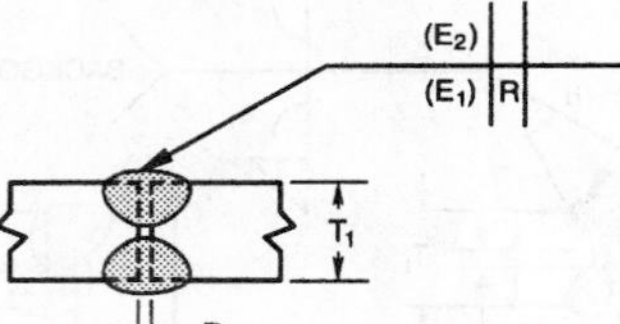

$E_1 + E_2$ MUST NOT EXCEED $\frac{3T_1}{4}$

Welding Process	Joint Designation	Base Metal Thickness (U = unlimited) T_1	T_2	Groove Preparation: Root Opening	Tolerances: As Detailed (see 3.12.3)	Tolerances: As Fit-Up (see 3.12.3)	Permitted Welding Positions	Total Weld Size ($E_1 + E_2$)	Notes
SMAW	B-P1b	1/4 max	—	$R = \frac{T_1}{2}$	+1/16, –0	±1/16	All	$\frac{3T_1}{4}$	D

Table 8-3 (cont.).
Prequalified Welded Joints
Partial-Joint-Penetration Groove Welds

Single-V-groove weld (2)
Butt joint (B)
Corner joint (C)

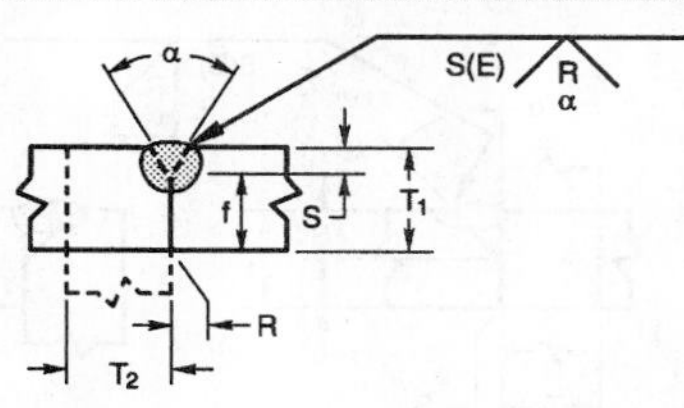

Welding Process	Joint Designation	Base Metal Thickness (U = unlimited)		Groove Preparation			Permitted Welding Positions	Weld Size (E)	Notes
		T_1	T_2	Root Opening Root Face Groove Angle	Tolerances: As Detailed (see 3.12.3)	Tolerances: As Fit-Up (see 3.12.3)			
SMAW	BC-P2	1/4 min	U	R = 0 f = 1/32 min α = 60°	0, +1/16 +U, –0 +10°, –0°	+1/8, –1/16 ±1/16 +10°, –5°	All	S	B, D, E, N
GMAW FCAW	BC-P2-GF	1/4 min	U	R = 0 f = 1/8 min α = 60°	0, +1/16 +U, –0 +10°, –0°	+1/8, –1/16 ±1/16 +10°, –5°	All	S	A, B, E, N
SAW	BC-P2-S	7/16 min	U	R = 0 f = 1/4 min α = 60°	±0 +U, –0 +10°, –0°	+1/16, –0 ±1/16 +10°, –5°	F	S	B, E, N

Double-V-groove weld (3)
Butt joint (B)

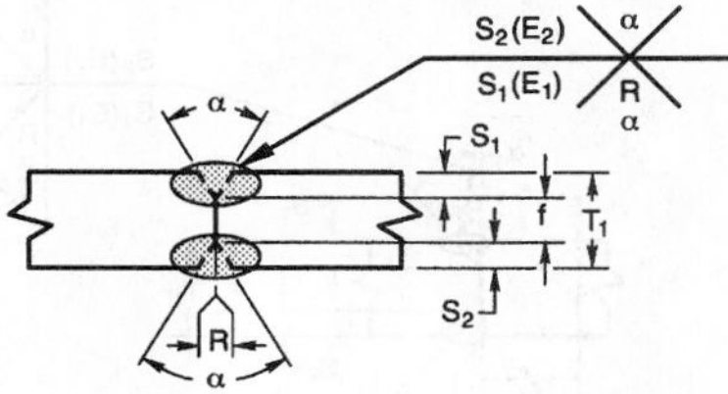

Welding Process	Joint Designation	Base Metal Thickness (U = unlimited)		Groove Preparation			Permitted Welding Positions	Total Weld Size ($E_1 + E_2$)	Notes
		T_1	T_2	Root Opening Root Face Groove Angle	Tolerances: As Detailed (see 3.12.3)	Tolerances: As Fit-Up (see 3.12.3)			
SMAW	B-P3	1/2 min	—	R = 0 f = 1/8 min α = 60°	+1/16, –0 +U, –0 +10°, –0°	+1/8, –1/16 ±1/16 +10°, –5°	All	$S_1 + S_2$	D, E, Mp, N
GMAW FCAW	B-P3-GF	1/2 min	—	R = 0 f = 1/8 min α = 60°	+1/16, –0 +U, –0 +10°, –0°	+1/8, –1/16 ±1/16 +10°, –5°	All	$S_1 + S_2$	A, E, Mp, N
SAW	B-P3-S	3/4 min	—	R = 0 f = 1/4 min α = 60°	±0 +U, –0 +10°, –0°	+1/16, –0 ±1/16 +10°, –5°	F	$S_1 + S_2$	E, Mp, N

Table 8-3 (cont.).
Prequalified Welded Joints
Partial-Joint-Penetration Groove Welds

Single-bevel-groove weld (4)
Butt joint (B)
T-joint (T)
Corner joint (C)

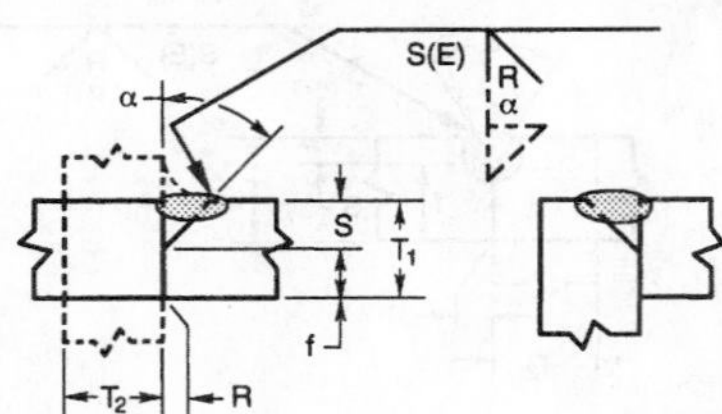

Welding Process	Joint Designation	Base Metal Thickness (U = unlimited) T_1	T_2	Groove Preparation: Root Opening Root Face Groove Angle	Tolerances: As Detailed (see 3.12.3)	As Fit-Up (see 3.12.3)	Permitted Welding Positions	Weld Size (E)	Notes
SMAW	BTC-P4	U	U	R = 0 f = 1/8 min α = 45°	+1/16, –0 unlimited +10°, –0°	+1/8, –1/16 ±1/16 +10°, –5°	All	S–1/8	B, D, E, J, N, V
GMAW FCAW	BTC-P4-GF	1/4 min	U	R = 0 f = 1/8 min α = 45°	+1/16, –0 unlimited +10°, –0°	+1/8, –1/16 ±1/16 +10°, –5°	F, H V, OH	S S–1/8	A, B, E, J, N, V
SAW	TC-P4-S	7/16 min	U	R = 0 f = 1/4 min α = 60°	±0 +U, –0 +10°, –0°	+1/16, –0 ±1/16 +10°, –5°	F	S	B, E, J, N, V

Double-bevel-groove weld (5)
Butt joint (B)
T-joint (T)
Corner joint (C)

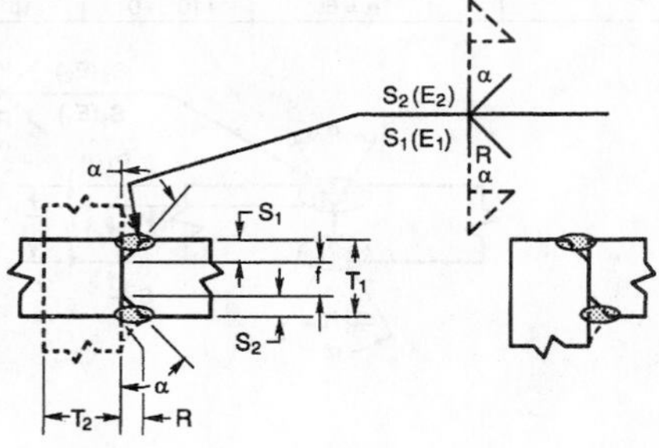

Welding Process	Joint Designation	Base Metal Thickness (U = unlimited) T_1	T_2	Groove Preparation: Root Opening Root Face Groove Angle	Tolerances: As Detailed (see 3.12.3)	As Fit-Up (see 3.12.3)	Permitted Welding Positions	Total Weld Size ($E_1 + E_2$)	Notes
SMAW	BTC-P5	5/16 min	U	R = 0 f = 1/8 min α = 45°	+1/16, –0 unlimited +10°, –0°	+1/8, –1/16 ±1/16 +10°, –5°	All	$S_1 + S_2$ –1/4	D, E, J, Mp, N, V
GMAW FCAW	BTC-P5-GF	1/2 min	U	R = 0 f = 1/8 min α = 45°	+1/16, –0 unlimited +10°, –0°	+1/8, –1/16 ±1/16 +10°, –5°	F, H V, OH	$S_1 + S_2$ $S_1 + S_2$ –1/4	A, E, J, Mp, N, V
SAW	TC-P5-S	3/4 min	U	R = 0 f = 1/4 min α = 60°	±0 +U, –0 +10°, –0°	+1/16, –0 ±1/16 +10°, –5°	F	$S_1 + S_2$	E, J, Mp, N, V

Table 8-3 (cont.).
Prequalified Welded Joints
Partial-Joint-Penetration Groove Welds

Single-U-groove weld (6)
Butt joint (B)
Corner joint (C)

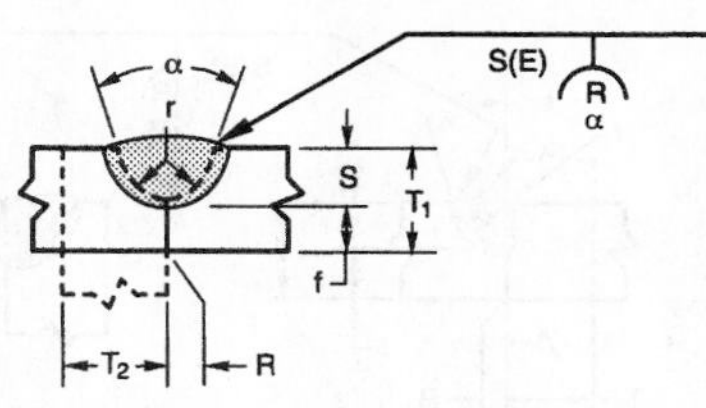

Welding Process	Joint Designation	Base Metal Thickness (U = unlimited) T_1	T_2	Groove Preparation: Root Opening, Root Face, Groove Radius, Groove Angle	Tolerances: As Detailed (see 3.12.3)	Tolerances: As Fit-Up (see 3.12.3)	Permitted Welding Positions	Weld Size (E)	Notes
SMAW	BC-P6	1/4 min	U	R = 0 f = 1/32 min r = 1/4 α = 45°	+1/16, –0 +U, –0 +1/4, –0 +10°, –0°	+1/8, –1/16 ±1/16 ±1/16 +10°, –5°	All	S	B, D, E, N
GMAW FCAW	BC-P6-GF	1/4 min	U	R = 0 f = 1/8 min r = 1/4 α = 20°	+1/16, –0 +U, –0 +1/4, –0 +10°, –0°	+1/8, –1/16 ±1/16 ±1/16 +10°, –5°	All	S	A, B, E, N
SAW	BC-P6-S	7/16 min	U	R = 0 f = 1/4 min r = 1/4 α = 20°	±0 +U, –0 +1/4, –0 +10°, –0°	+1/16, –0 ±1/16 ±1/16 +10°, –5°	F	S	B, E, N

Double-U-groove weld (7)
Butt joint (B)

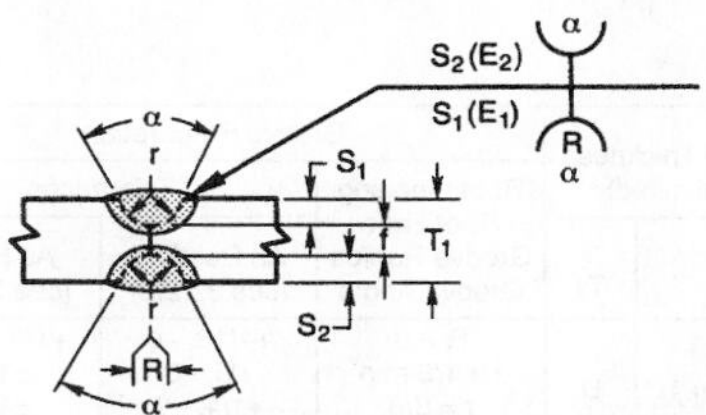

Welding Process	Joint Designation	Base Metal Thickness (U = unlimited) T_1	T_2	Groove Preparation: Root Opening, Root Face, Groove Radius, Groove Angle	Tolerances: As Detailed (see 3.12.3)	Tolerances: As Fit-Up (see 3.12.3)	Permitted Welding Positions	Total Weld Size ($E_1 + E_2$)	Notes
SMAW	B-P7	1/2 min	—	R = 0 f = 1/8 min r = 1/4 α = 45°	+1/16, –0 +U, –0 +1/4, –0 +10°, –0°	+1/8, –1/16 ±1/16 ±1/16 +10°, –5°	All	$S_1 + S_2$	D, E, Mp, N
GMAW FCAW	B-P7-GF	1/2 min	—	R = 0 f = 1/8 min r = 1/4 α = 20°	+1/16, –0 +U, –0 +1/4, –0 +10°, –0°	+1/8, –1/16 ±1/16 ±1/16 +10°, –5°	All	$S_1 + S_2$	A, E, Mp, N
SAW	B-P7-S	3/4 min	—	R = 0 f = 1/4 min r = 1/4 α = 20°	±0 +U, –0 +1/4, –0 +10°, –0°	+1/16, –0 ±1/16 ±1/16 +10°, –5°	F	$S_1 + S_2$	E, Mp, N

Table 8-3 (cont.). Prequalified Welded Joints Partial-Joint-Penetration Groove Welds

Single-J-groove weld (8)
Butt joint (B)
T-joint (T)
Corner joint (C)

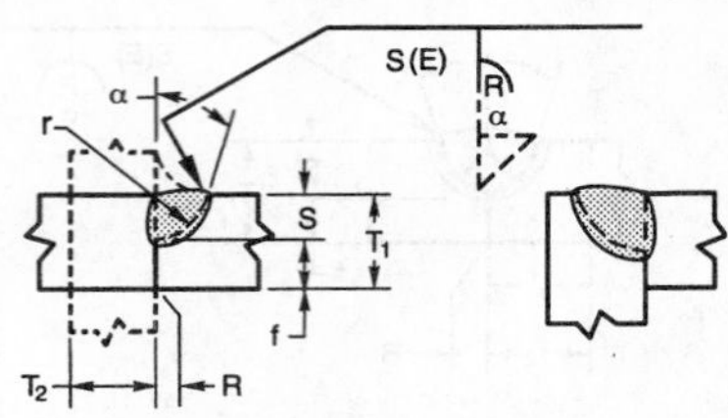

Welding Process	Joint Designation	Base Metal Thickness (U = unlimited) T_1	T_2	Groove Preparation: Root Opening Root Face Groove Radius Groove Angle	Tolerances: As Detailed (see 3.12.3)	Tolerances: As Fit-Up (see 3.12.3)	Permitted Welding Positions	Weld Size (E)	Notes
SMAW	TC-P8*	1/4 min	U	R = 0 f = 1/8 min r = 3/8 α = 45°	+1/16, −0 +U, −0 +1/4, −0 +10°, −0°	+1/8, −1/16 ±1/16 ±1/16 +10°, −5°	All	S	D, E, J, N, V
SMAW	BC-P8**	1/4 min	U	R = 0 f = 1/8 min r = 3/8 α = 30°	+1/16, −0 +U, −0 +1/4, −0 +10°, −0°	+1/8, −1/16 ±1/16 ±1/16 +10°, −5°	All	S	D, E, J, N, V
GMAW FCAW	TC-P8-GF*	1/4 min	U	R = 0 f = 1/8 min r = 3/8 α = 45°	+1/16, −0 +U, −0 +1/4, −0 +10°, −0°	+1/8, −1/16 ±1/16 ±1/16 +10°, −5°	All	S	A, E, J, N, V
GMAW FCAW	BC-P8-GF**	1/4 min	U	R = 0 f = 1/8 min r = 3/8 α = 30°	+1/16, −0 +U, −0 +1/4, −0 +10°, −0°	+1/8, −1/16 ±1/16 ±1/16 +10°, −5°	All	S	A, E, J, N, V
SAW	TC-P8-S*	7/16 min	U	R = 0 f = 1/4 min r = 1/2 α = 45°	±0 +U, −0 +1/4, −0 +10°, −0°	+1/16, −0 ±1/16 ±1/16 +10°, −5°	F	S	E, J, N, V
SAW	C-P8-S**	7/16 min	U	R = 0 f = 1/4 min r = 1/2 α = 20°	±0 +U, −0 +1/4, −0 +10°, −0°	+1/16, −0 ±1/16 ±1/16 +10°, −5°	F	S	E, J, N, V

*Applies to inside corner joints.
**Applies to outside corner joints.

Table 8-3 (cont.). Prequalified Welded Joints Complete-Joint-Penetration Groove Welds

Double-J-groove weld (9)
Butt joint (B)
T-joint (T)
Corner joint (C)

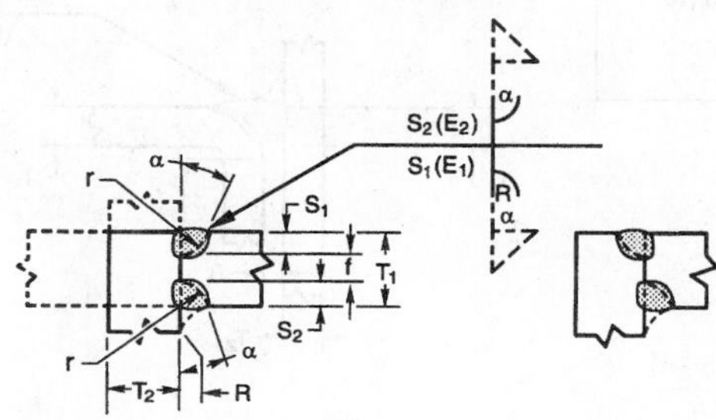

Welding Process	Joint Designation	Base Metal Thickness (U = unlimited) T_1	T_2	Groove Preparation: Root Opening, Root Face, Groove Radius, Groove Angle	Tolerances: As Detailed (see 3.12.3)	Tolerances: As Fit-Up (see 3.12.3)	Permitted Welding Positions	Weld Size (E)	Notes
SMAW	TC-P8*	1/4 min	U	R = 0 f = 1/8 min r = 3/8 α = 45°	+1/16, –0 +U, –0 +1/4, –0 +10°, –0°	+1/8, –1/16 ±1/16 ±1/16 +10°, –5°	All	S	D, E, J, N, V
SMAW	BC-P8**	1/4 min	U	R = 0 f = 1/8 min r = 3/8 α = 30°	+1/16, –0 +U, –0 +1/4, –0 +10°, –0°	+1/8, –1/16 ±1/16 ±1/16 +10°, –5°	All	S	D, E, J, N, V
GMAW FCAW	TC-P8-GF*	1/4 min	U	R = 0 f = 1/8 min r = 3/8 α = 45°	+1/16, –0 +U, –0 +1/4, –0 +10°, –0°	+1/8, –1/16 ±1/16 ±1/16 +10°, –5°	All	S	A, E, J, N, V
GMAW FCAW	BC-P8-GF**	1/4 min	U	R = 0 f = 1/8 min r = 3/8 α = 30°	+1/16, –0 +U, –0 +1/4, –0 +10°, –0°	+1/8, –1/16 ±1/16 ±1/16 +10°, –5°	All	S	A, E, J, N, V
SAW	TC-P8-S*	7/16 min	U	R = 0 f = 1/4 min r = 1/2 α = 45°	±0 +U, –0 +1/4, –0 +10°, –0°	+1/16, –0 ±1/16 ±1/16 +10°, –5°	F	S	E, J, N, V
SAW	C-P8-S**	7/16 min	U	R = 0 f = 1/4 min r = 1/2 α = 20°	±0 +U, –0 +1/4, –0 +10°, –0°	+1/16, –0 ±1/16 ±1/16 +10°, –5°	F	S	E, J, N, V

*Applies to inside corner joints.
**Applies to outside corner joints.

Table 8-3 (cont.).
Prequalified Welded Joints
Flare-Bevel Groove Welds

Flare-bevel-groove weld (10)
Butt joint (B)
T-joint (T)
Corner joint (C)

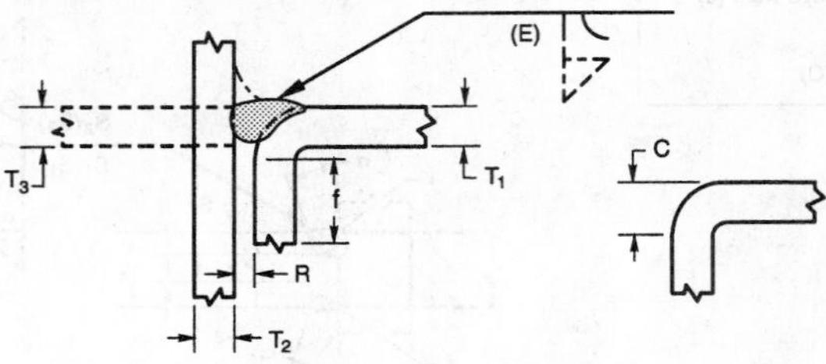

Welding Process	Joint Designation	Base Metal Thickness (U = unlimited)			Groove Preparation			Permitted Welding Positions	Weld Size (E)	Notes
					Root Opening Root Face Bend Radius*	Tolerances				
		T_1	T_2	T_3		As Detailed (see 3.12.3)	As Fit-Up (see 3.12.3)			
SMAW	BTC-P10	3/16 min	U	T_1 min	R = 0 f = 3/16 min $C = \frac{3T_1}{2}$ min	+1/16, –0 +U, –0 –0, +Not-Limited	+1/8, –1/16 +U, –1/16 –0, +Not-Limited	All	$\frac{5T_1}{8}$	D, J, N, Z
GMAW FCAW	BTC-P10-GF	3/16 min	U	T_1 min	R = 0 f = 3/16 min $C = \frac{3T_1}{2}$ min	+1/16, –0 +U, –0 –0, +Not-Limited	+1/8, –1/16 +U, –1/16 –0, +Not-Limited	All	$\frac{5T_1}{8}$	A, J, N, Z
SAW	T-P10-S	1/2 min	1/2 min	N/A	R = 0 f = 1/2 min $C = \frac{3T_1}{2}$ min	±0 +U, –0 –0, +Not-Limited	+1/16, –0 +U, –1/16 –0, +Not-Limited	F	$\frac{5T_1}{8}$	J, N, Z

*For cold formed (A500) rectangular tubes, C dimension is not limited. See the following:

Effective Weld Size of Flare-Bevel-Groove Welded Joints. Tests have been performed on cold formed ASTM A 500 material exhibiting a "c" dimension as small as T_1 with a nominal radius of 2t. As the radius increases, the "c" dimension also increases. The corner curvature may not be a quadrant of a circle tangent to the sides. The corner dimension, "c," may be less than the radius of the corner.

Table 8-4.
Electrode Strength Coefficient C_1

Electrode	F_{EXX} (ksi)	C_1
E60	60	0.857
E70	70	1.00
E80	80	1.03
E90	90	1.16
E100	100	1.21
E110	110	1.34

Table 8-5.
Coefficients *C* for Eccentrically Loaded Weld Groups
Angle = 0°

$\phi R_n = CC_1 Dl \qquad C_{min} = \frac{P_u}{C_1 Dl} \qquad D_{min} = \frac{P_u}{CC_1 l} \qquad l_{min} = \frac{P_u}{CC_1 D}$

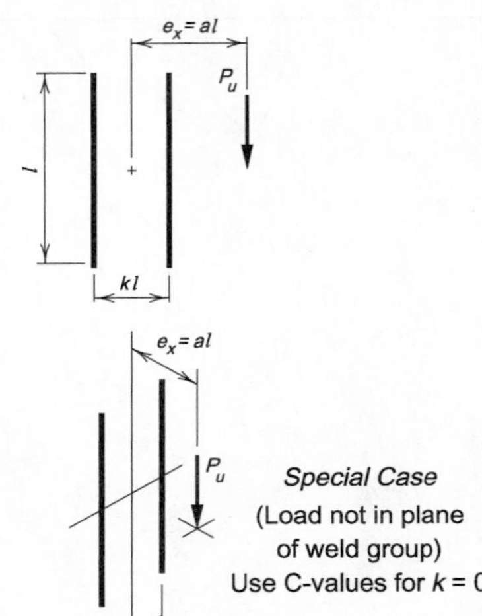

where

P_u = factored force, kips
D = number of sixteenths-of-an-inch in the fillet weld size
l = characteristic length of weld group, in.
$a = e_x/l$
e_x = horizontal component of eccentricity of P_u with respect to centroid of weld group, in.
C = coefficient tabulated below, which includes $\phi = 0.75$
C_1 = electrode strength coefficient from Table 8-4 (1.0 for E70XX electrodes)

a	*k*															
	0	**0.1**	**0.2**	**0.3**	**0.4**	**0.5**	**0.6**	**0.7**	**0.8**	**0.9**	**1.0**	**1.2**	**1.4**	**1.6**	**1.8**	**2.0**
0.00	2.78	2.78	2.78	2.78	2.78	2.78	2.78	2.78	2.78	2.78	2.78	2.78	2.78	2.78	2.78	2.78
0.10	2.79	2.79	2.79	2.79	2.78	2.77	2.75	2.74	2.73	2.71	2.70	2.67	2.64	2.61	2.59	2.79
0.15	2.75	2.75	2.74	2.73	2.71	2.70	2.69	2.67	2.66	2.64	2.63	2.60	2.58	2.55	2.53	2.50
0.20	2.64	2.63	2.63	2.62	2.60	2.59	2.58	2.57	2.56	2.55	2.54	2.52	2.50	2.48	2.46	2.44
0.25	2.48	2.48	2.48	2.47	2.47	2.46	2.46	2.45	2.45	2.44	2.44	2.43	2.41	2.40	2.39	2.38
0.30	2.32	2.32	2.32	2.32	2.33	2.33	2.33	2.33	2.33	2.33	2.33	2.33	2.33	2.32	2.32	2.31
0.40	2.00	2.00	2.01	2.03	2.05	2.07	2.08	2.10	2.11	2.12	2.14	2.15	2.16	2.17	2.18	2.18
0.50	1.72	1.72	1.74	1.77	1.80	1.83	1.86	1.89	1.91	1.93	1.95	1.99	2.01	2.03	2.05	2.06
0.60	1.50	1.50	1.52	1.55	1.59	1.63	1.67	1.71	1.74	1.77	1.79	1.84	1.87	1.90	1.92	1.94
0.70	1.32	1.32	1.34	1.38	1.42	1.47	1.51	1.55	1.59	1.62	1.65	1.71	1.75	1.79	1.81	1.84
0.80	1.17	1.18	1.20	1.24	1.28	1.33	1.38	1.42	1.46	1.50	1.53	1.59	1.64	1.68	1.71	1.74
0.90	1.05	1.06	1.08	1.12	1.17	1.22	1.27	1.31	1.35	1.39	1.43	1.49	1.54	1.59	1.62	1.66
1.00	0.957	0.963	0.986	1.02	1.07	1.12	1.17	1.21	1.26	1.29	1.33	1.40	1.45	1.50	1.54	1.58
1.20	0.806	0.812	0.835	0.872	0.916	0.963	1.01	1.06	1.10	1.14	1.17	1.24	1.30	1.35	1.40	1.44
1.40	0.695	0.701	0.724	0.758	0.799	0.844	0.889	0.932	0.973	1.01	1.05	1.12	1.18	1.23	1.28	1.32
1.60	0.611	0.616	0.638	0.670	0.708	0.750	0.792	0.833	0.873	0.911	0.947	1.01	1.07	1.13	1.17	1.22
1.80	0.544	0.550	0.570	0.600	0.635	0.674	0.714	0.753	0.791	0.828	0.863	0.928	0.987	1.04	1.09	1.13
2.00	0.491	0.496	0.515	0.542	0.576	0.612	0.650	0.687	0.723	0.758	0.792	0.855	0.912	0.964	1.01	1.05
2.20	0.447	0.452	0.470	0.495	0.526	0.560	0.596	0.631	0.665	0.699	0.731	0.792	0.848	0.899	0.945	0.988
2.40	0.410	0.415	0.431	0.455	0.484	0.516	0.550	0.583	0.616	0.648	0.679	0.738	0.792	0.842	0.887	0.929
2.60	0.379	0.384	0.399	0.421	0.448	0.478	0.510	0.542	0.573	0.604	0.634	0.691	0.743	0.791	0.836	0.877
2.80	0.352	0.357	0.371	0.392	0.417	0.446	0.476	0.506	0.536	0.565	0.594	0.649	0.699	0.746	0.790	0.830
3.00	0.329	0.333	0.347	0.366	0.390	0.417	0.446	0.474	0.503	0.531	0.559	0.611	0.661	0.706	0.748	0.788

Table 8-5 (cont.). Coefficients *C* for Eccentrically Loaded Weld Groups Angle = 15°

$\phi R_n = CC_1 Dl \qquad C_{min} = \dfrac{P_u}{C_1 Dl} \qquad D_{min} = \dfrac{P_u}{CC_1 l} \qquad l_{min} = \dfrac{P_u}{CC_1 D}$

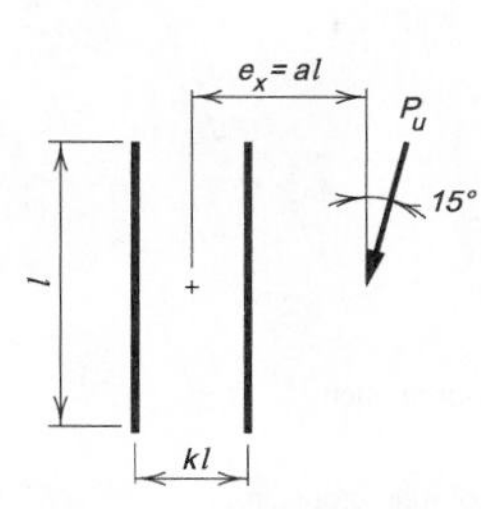

where

P_u = factored force, kips
D = number of sixteenths-of-an-inch in the fillet weld size
l = characteristic length of weld group, in.
$a = e_x/l$
e_x = horizontal component of eccentricity of P_u with respect to centroid of weld group, in.
C = coefficient tabulated below, which includes $\phi = 0.75$
C_1 = electrode strength coefficient from Table 8-4 (1.0 for E70XX electrodes)

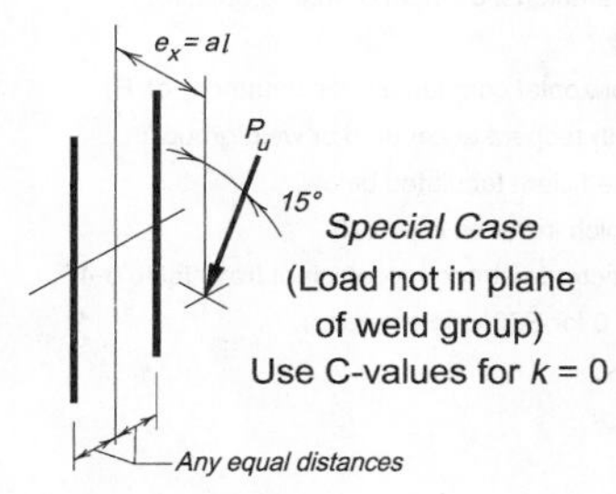

a	k: 0	0.1	0.2	0.3	0.4	0.5	0.6	0.7	0.8	0.9	1.0	1.2	1.4	1.6	1.8	2.0
0.00	2.97	2.97	2.97	2.97	2.97	2.97	2.97	2.97	2.97	2.97	2.97	2.97	2.97	2.97	2.97	2.97
0.10	2.84	2.84	2.84	2.83	2.82	2.82	2.81	2.80	2.80	2.79	2.78	2.77	2.75	2.74	2.73	2.72
0.15	2.76	2.76	2.75	2.75	2.74	2.73	2.72	2.72	2.71	2.70	2.70	2.69	2.68	2.67	2.66	2.65
0.20	2.63	2.63	2.63	2.62	2.62	2.62	2.61	2.61	2.61	2.61	2.60	2.60	2.59	2.59	2.58	2.58
0.25	2.48	2.48	2.48	2.48	2.49	2.49	2.49	2.49	2.49	2.50	2.50	2.50	2.51	2.51	2.51	2.51
0.30	2.32	2.32	2.32	2.33	2.34	2.35	2.36	2.37	2.38	2.39	2.39	2.41	2.42	2.43	2.43	2.43
0.40	2.01	2.01	2.02	2.04	2.06	2.09	2.12	2.14	2.16	2.18	2.19	2.22	2.25	2.27	2.28	2.30
0.50	1.74	1.74	1.76	1.78	1.82	1.86	1.89	1.93	1.96	1.99	2.01	2.05	2.09	2.12	2.15	2.17
0.60	1.52	1.52	1.54	1.57	1.62	1.66	1.70	1.75	1.78	1.82	1.85	1.90	1.95	1.99	2.02	2.05
0.70	1.34	1.35	1.37	1.40	1.45	1.50	1.54	1.59	1.63	1.67	1.70	1.77	1.82	1.87	1.90	1.94
0.80	1.20	1.20	1.22	1.26	1.31	1.36	1.41	1.46	1.50	1.54	1.58	1.65	1.71	1.76	1.80	1.84
0.90	1.08	1.08	1.11	1.15	1.19	1.24	1.30	1.34	1.39	1.43	1.47	1.54	1.60	1.66	1.70	1.74
1.00	0.979	0.985	1.01	1.05	1.09	1.15	1.20	1.25	1.29	1.33	1.37	1.45	1.51	1.57	1.62	1.66
1.20	0.826	0.832	0.856	0.893	0.938	0.987	1.04	1.09	1.13	1.17	1.21	1.29	1.35	1.41	1.46	1.51
1.40	0.714	0.719	0.743	0.778	0.820	0.866	0.913	0.960	1.00	1.05	1.09	1.16	1.23	1.28	1.34	1.39
1.60	0.628	0.633	0.656	0.688	0.727	0.770	0.815	0.859	0.901	0.941	0.980	1.05	1.12	1.18	1.23	1.28
1.80	0.560	0.566	0.587	0.617	0.653	0.693	0.735	0.777	0.817	0.855	0.893	0.963	1.03	1.09	1.14	1.19
2.00	0.506	0.511	0.530	0.558	0.592	0.630	0.669	0.708	0.746	0.783	0.819	0.887	0.949	1.01	1.06	1.11
2.20	0.461	0.466	0.484	0.510	0.541	0.577	0.613	0.650	0.687	0.722	0.757	0.822	0.882	0.938	0.989	1.04
2.40	0.423	0.428	0.445	0.469	0.499	0.532	0.566	0.601	0.636	0.670	0.703	0.765	0.824	0.878	0.928	0.974
2.60	0.391	0.396	0.412	0.434	0.462	0.493	0.526	0.559	0.591	0.624	0.656	0.716	0.772	0.825	0.873	0.918
2.80	0.363	0.368	0.383	0.404	0.430	0.460	0.491	0.522	0.553	0.584	0.614	0.672	0.727	0.778	0.825	0.869
3.00	0.339	0.344	0.358	0.378	0.403	0.430	0.460	0.489	0.519	0.549	0.578	0.634	0.686	0.736	0.781	0.824

Table 8-5 (cont.).
Coefficients *C* for Eccentrically Loaded Weld Groups
Angle = 30°

$\phi R_n = CC_1 Dl$ $\quad C_{min} = \dfrac{P_u}{C_1 Dl}$ $\quad D_{min} = \dfrac{P_u}{CC_1 l}$ $\quad l_{min} = \dfrac{P_u}{CC_1 D}$

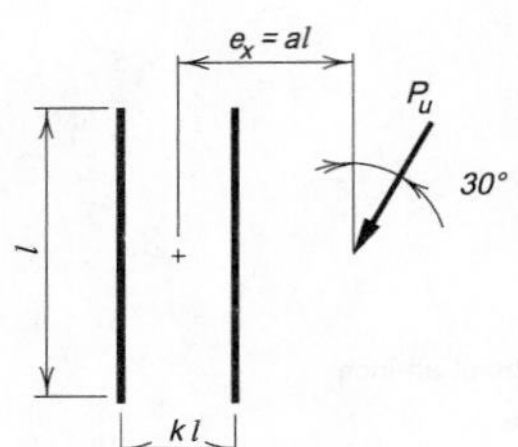

where

P_u = factored force, kips
D = number of sixteenths-of-an-inch in the fillet weld size
l = characteristic length of weld group, in.
$a = e_x/l$
e_x = horizontal component of eccentricity of P_u with respect to centroid of weld group, in.
C = coefficient tabulated below, which includes $\phi = 0.75$
C_1 = electrode strength coefficient from Table 8-4 (1.0 for E70XX electrodes)

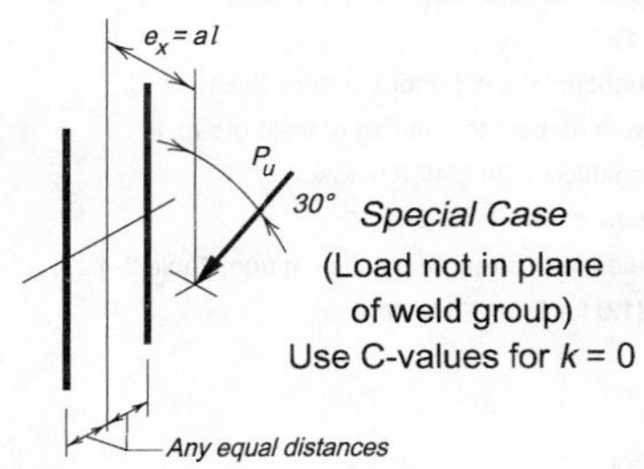

a	*k*															
	0	**0.1**	**0.2**	**0.3**	**0.4**	**0.5**	**0.6**	**0.7**	**0.8**	**0.9**	**1.0**	**1.2**	**1.4**	**1.6**	**1.8**	**2.0**
0.00	3.28	3.28	3.28	3.28	3.28	3.28	3.28	3.28	3.28	3.28	3.28	3.28	3.28	3.28	3.28	3.28
0.10	3.03	3.03	3.04	3.04	3.04	3.05	3.05	3.06	3.06	3.06	3.07	3.07	3.07	3.06	3.06	3.05
0.15	2.87	2.87	2.87	2.87	2.88	2.88	2.88	2.89	2.90	2.90	2.91	2.92	2.94	2.94	2.95	2.95
0.20	2.72	2.73	2.73	2.73	2.74	2.74	2.75	2.76	2.76	2.77	2.78	2.79	2.81	2.82	2.83	2.84
0.25	2.57	2.57	2.57	2.58	2.59	2.61	2.62	2.63	2.64	2.66	2.67	2.68	2.70	2.72	2.73	2.74
0.30	2.41	2.41	2.42	2.43	2.45	2.47	2.49	2.51	2.53	2.54	2.56	2.59	2.61	2.63	2.64	2.66
0.40	2.11	2.11	2.12	2.14	2.17	2.20	2.24	2.28	2.30	2.33	2.35	2.40	2.43	2.46	2.49	2.51
0.50	1.84	1.85	1.86	1.89	1.93	1.98	2.02	2.06	2.10	2.13	2.17	2.22	2.27	2.31	2.34	2.37
0.60	1.62	1.63	1.65	1.68	1.73	1.78	1.83	1.88	1.92	1.96	2.00	2.07	2.12	2.17	2.21	2.25
0.70	1.44	1.45	1.47	1.51	1.56	1.61	1.66	1.72	1.77	1.81	1.85	1.92	1.99	2.04	2.09	2.13
0.80	1.30	1.30	1.33	1.37	1.42	1.47	1.52	1.58	1.63	1.68	1.72	1.80	1.87	1.93	1.98	2.02
0.90	1.17	1.18	1.20	1.24	1.29	1.35	1.41	1.46	1.51	1.56	1.61	1.69	1.76	1.82	1.88	1.92
1.00	1.07	1.08	1.10	1.14	1.19	1.25	1.30	1.36	1.41	1.46	1.51	1.59	1.66	1.73	1.78	1.84
1.20	0.907	0.913	0.939	0.979	1.03	1.08	1.14	1.19	1.24	1.29	1.33	1.42	1.49	1.56	1.62	1.68
1.40	0.786	0.792	0.817	0.855	0.901	0.951	1.00	1.05	1.10	1.15	1.20	1.28	1.35	1.42	1.49	1.54
1.60	0.693	0.699	0.723	0.759	0.801	0.848	0.897	0.946	0.993	1.04	1.08	1.16	1.24	1.31	1.37	1.42
1.80	0.619	0.625	0.648	0.681	0.721	0.765	0.811	0.857	0.901	0.945	0.987	1.07	1.14	1.21	1.27	1.32
2.00	0.559	0.565	0.587	0.617	0.655	0.696	0.740	0.783	0.825	0.866	0.907	0.984	1.05	1.12	1.18	1.24
2.20	0.510	0.516	0.536	0.564	0.599	0.638	0.679	0.720	0.761	0.800	0.838	0.912	0.981	1.04	1.10	1.16
2.40	0.469	0.474	0.493	0.520	0.552	0.589	0.628	0.667	0.705	0.743	0.779	0.850	0.917	0.978	1.04	1.09
2.60	0.433	0.439	0.456	0.481	0.512	0.546	0.583	0.620	0.657	0.693	0.728	0.795	0.860	0.920	0.975	1.03
2.80	0.403	0.408	0.424	0.448	0.477	0.509	0.544	0.579	0.614	0.649	0.683	0.748	0.809	0.867	0.922	0.972
3.00	0.376	0.382	0.397	0.419	0.446	0.477	0.510	0.543	0.577	0.610	0.642	0.705	0.764	0.821	0.873	0.923

Table 8-5 (cont.). Coefficients *C* for Eccentrically Loaded Weld Groups Angle = 45°

$\phi R_n = CC_1 Dl \qquad C_{min} = \frac{P_u}{C_1 Dl} \qquad D_{min} = \frac{P_u}{CC_1 l} \qquad l_{min} = \frac{P_u}{CC_1 D}$

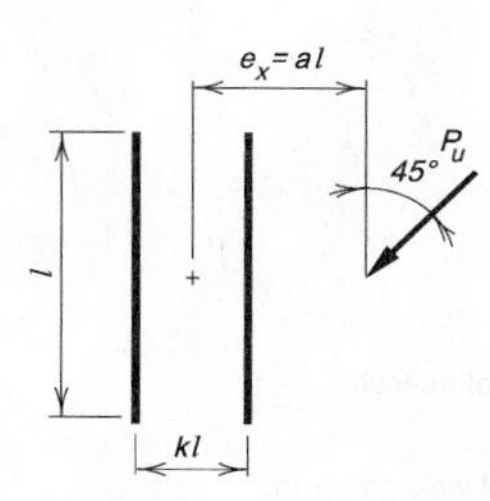

where

P_u = factored force, kips

D = number of sixteenths-of-an-inch in the fillet weld size

l = characteristic length of weld group, in.

$a = e_x / l$

e_x = horizontal component of eccentricity of P_u with respect to centroid of weld group, in.

C = coefficient tabulated below, which includes $\phi = 0.75$

C_1 = electrode strength coefficient from Table 8-4 (1.0 for E70XX electrodes)

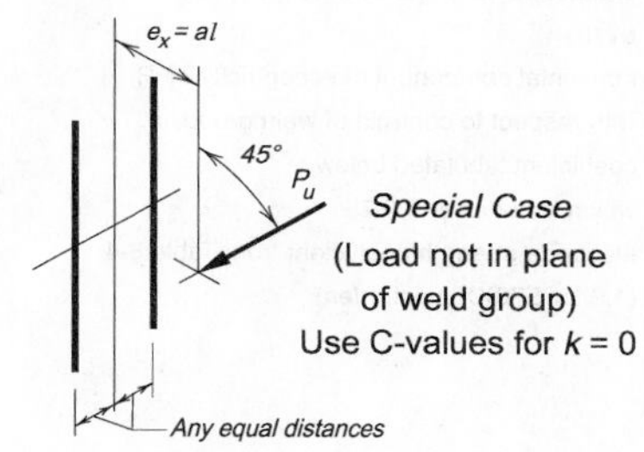

a	k															
	0	0.1	0.2	0.3	0.4	0.5	0.6	0.7	0.8	0.9	1.0	1.2	1.4	1.6	1.8	2.0
0.00	3.61	3.61	3.61	3.61	3.61	3.61	3.61	3.61	3.61	3.61	3.61	3.61	3.61	3.61	3.61	3.61
0.10	3.37	3.37	3.38	3.38	3.40	3.42	3.43	3.44	3.46	3.47	3.48	3.50	3.51	3.52	3.52	3.52
0.15	3.13	3.13	3.15	3.17	3.20	3.23	3.25	3.28	3.30	3.33	3.35	3.38	3.41	3.43	3.45	3.46
0.20	2.94	2.94	2.95	2.97	2.99	3.03	3.06	3.10	3.13	3.17	3.20	3.25	3.29	3.33	3.35	3.38
0.25	2.77	2.77	2.78	2.80	2.83	2.86	2.89	2.93	2.97	3.01	3.04	3.11	3.16	3.21	3.25	3.28
0.30	2.61	2.61	2.63	2.65	2.68	2.71	2.75	2.79	2.83	2.86	2.90	2.97	3.04	3.09	3.14	3.18
0.40	2.32	2.32	2.34	2.37	2.41	2.45	2.50	2.54	2.59	2.63	2.66	2.73	2.80	2.87	2.92	2.97
0.50	2.06	2.07	2.09	2.12	2.17	2.22	2.27	2.33	2.38	2.42	2.47	2.54	2.61	2.68	2.74	2.79
0.60	1.84	1.85	1.87	1.91	1.96	2.02	2.08	2.14	2.19	2.25	2.30	2.38	2.45	2.52	2.58	2.63
0.70	1.66	1.66	1.69	1.73	1.79	1.85	1.91	1.97	2.03	2.09	2.14	2.23	2.31	2.38	2.44	2.50
0.80	1.50	1.51	1.54	1.58	1.64	1.70	1.76	1.83	1.89	1.95	2.00	2.10	2.18	2.26	2.32	2.38
0.90	1.37	1.38	1.40	1.45	1.51	1.57	1.64	1.70	1.76	1.82	1.88	1.98	2.07	2.14	2.21	2.27
1.00	1.26	1.26	1.29	1.34	1.40	1.46	1.53	1.59	1.65	1.71	1.77	1.87	1.96	2.04	2.11	2.17
1.20	1.08	1.08	1.11	1.16	1.21	1.28	1.34	1.40	1.46	1.52	1.58	1.68	1.77	1.85	1.93	2.00
1.40	0.938	0.946	0.975	1.02	1.07	1.13	1.19	1.25	1.31	1.37	1.42	1.52	1.62	1.70	1.77	1.84
1.60	0.831	0.838	0.866	0.908	0.958	1.01	1.07	1.13	1.19	1.24	1.29	1.39	1.48	1.56	1.64	1.71
1.80	0.745	0.752	0.779	0.817	0.864	0.917	0.972	1.03	1.08	1.13	1.19	1.28	1.37	1.45	1.52	1.59
2.00	0.675	0.682	0.707	0.743	0.787	0.836	0.888	0.941	0.992	1.04	1.09	1.18	1.27	1.35	1.42	1.49
2.20	0.617	0.624	0.647	0.681	0.722	0.768	0.818	0.868	0.917	0.964	1.01	1.10	1.18	1.26	1.33	1.40
2.40	0.568	0.574	0.596	0.628	0.667	0.710	0.756	0.804	0.851	0.897	0.941	1.03	1.11	1.18	1.25	1.32
2.60	0.526	0.532	0.553	0.583	0.619	0.660	0.703	0.749	0.794	0.838	0.881	0.963	1.04	1.12	1.18	1.25
2.80	0.489	0.496	0.515	0.544	0.578	0.617	0.658	0.701	0.743	0.786	0.827	0.906	0.982	1.05	1.12	1.18
3.00	0.458	0.464	0.482	0.509	0.542	0.579	0.618	0.658	0.699	0.739	0.779	0.856	0.929	1.00	1.06	1.12

Table 8-5 (cont.). Coefficients C for Eccentrically Loaded Weld Groups Angle = 60°

$$\phi R_n = CC_1 Dl \qquad C_{min} = \frac{P_u}{C_1 Dl} \qquad D_{min} = \frac{P_u}{CC_1 l} \qquad l_{min} = \frac{P_u}{CC_1 D}$$

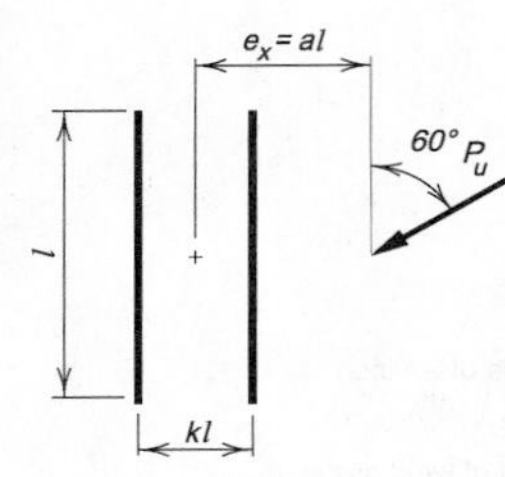

where

P_u = factored force, kips

D = number of sixteenths-of-an-inch in the fillet weld size

l = characteristic length of weld group, in.

$a = e_x/l$

e_x = horizontal component of eccentricity of P_u with respect to centroid of weld group, in.

C = coefficient tabulated below, which includes $\phi = 0.75$

C_1 = electrode strength coefficient from Table 8-4 (1.0 for E70XX electrodes)

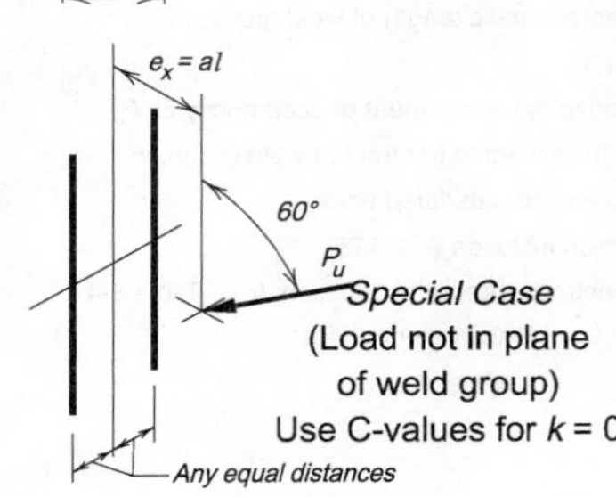

a	k															
	0	0.1	0.2	0.3	0.4	0.5	0.6	0.7	0.8	0.9	1.0	1.2	1.4	1.6	1.8	2.0
0.00	3.91	3.91	3.91	3.91	3.91	3.91	3.91	3.91	3.91	3.91	3.91	3.91	3.91	3.91	3.91	3.91
0.10	3.65	3.66	3.68	3.71	3.74	3.78	3.80	3.83	3.84	3.85	3.86	3.86	3.87	3.86	3.86	3.86
0.15	3.46	3.47	3.49	3.53	3.58	3.63	3.68	3.73	3.76	3.79	3.80	3.83	3.84	3.85	3.85	3.86
0.20	3.27	3.28	3.30	3.35	3.40	3.47	3.54	3.60	3.65	3.69	3.73	3.78	3.80	3.82	3.83	3.84
0.25	3.10	3.10	3.13	3.17	3.23	3.30	3.38	3.46	3.53	3.58	3.63	3.70	3.75	3.78	3.80	3.82
0.30	2.95	2.95	2.98	3.02	3.07	3.14	3.22	3.31	3.39	3.46	3.53	3.62	3.68	3.73	3.76	3.78
0.40	2.68	2.69	2.71	2.76	2.81	2.88	2.95	3.03	3.12	3.21	3.29	3.43	3.53	3.61	3.66	3.70
0.50	2.44	2.45	2.48	2.53	2.59	2.66	2.73	2.80	2.88	2.96	3.05	3.22	3.36	3.46	3.54	3.60
0.60	2.24	2.24	2.27	2.32	2.39	2.46	2.54	2.62	2.69	2.77	2.84	3.01	3.17	3.29	3.39	3.47
0.70	2.05	2.06	2.09	2.15	2.21	2.29	2.37	2.45	2.53	2.60	2.67	2.82	2.98	3.12	3.24	3.34
0.80	1.89	1.90	1.93	1.99	2.06	2.14	2.22	2.30	2.38	2.45	2.52	2.66	2.81	2.95	3.08	3.20
0.90	1.75	1.76	1.79	1.85	1.92	2.00	2.08	2.16	2.24	2.32	2.39	2.53	2.65	2.80	2.93	3.05
1.00	1.62	1.63	1.67	1.73	1.80	1.88	1.96	2.04	2.12	2.20	2.27	2.41	2.53	2.66	2.79	2.91
1.20	1.42	1.43	1.46	1.52	1.59	1.67	1.75	1.83	1.91	1.98	2.06	2.19	2.32	2.43	2.54	2.66
1.40	1.25	1.26	1.30	1.36	1.42	1.50	1.57	1.65	1.73	1.81	1.88	2.01	2.13	2.25	2.35	2.45
1.60	1.12	1.13	1.17	1.22	1.28	1.35	1.43	1.50	1.58	1.65	1.72	1.85	1.98	2.09	2.19	2.29
1.80	1.01	1.02	1.06	1.11	1.17	1.24	1.31	1.38	1.45	1.52	1.59	1.72	1.84	1.95	2.05	2.14
2.00	0.922	0.932	0.964	1.01	1.07	1.13	1.20	1.27	1.34	1.41	1.47	1.60	1.72	1.82	1.92	2.02
2.20	0.847	0.856	0.887	0.932	0.986	1.05	1.11	1.18	1.25	1.31	1.37	1.49	1.61	1.71	1.81	1.90
2.40	0.782	0.791	0.820	0.863	0.914	0.971	1.03	1.10	1.16	1.22	1.28	1.40	1.51	1.62	1.71	1.80
2.60	0.726	0.734	0.762	0.803	0.852	0.907	0.964	1.03	1.09	1.15	1.21	1.32	1.43	1.53	1.62	1.71
2.80	0.677	0.686	0.712	0.750	0.797	0.849	0.905	0.963	1.02	1.08	1.14	1.24	1.35	1.45	1.54	1.62
3.00	0.635	0.643	0.668	0.704	0.749	0.799	0.852	0.906	0.963	1.02	1.07	1.18	1.28	1.37	1.46	1.55

Table 8-5 (cont.). Coefficients *C* for Eccentrically Loaded Weld Groups Angle = 75°

$\phi R_n = CC_1 Dl \qquad C_{min} = \dfrac{P_u}{C_1 Dl} \qquad D_{min} = \dfrac{P_u}{CC_1 l} \qquad l_{min} = \dfrac{P_u}{CC_1 D}$

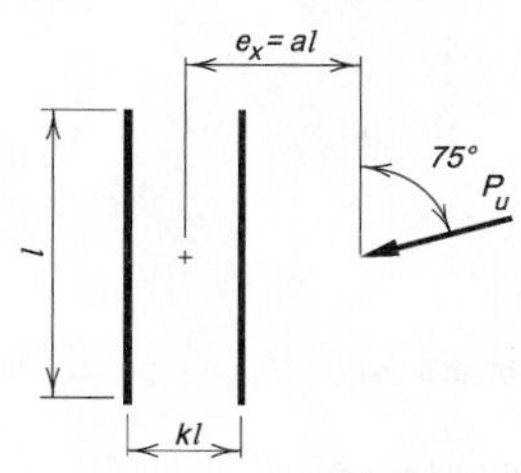

where

P_u = factored force, kips
D = number of sixteenths-of-an-inch in the fillet weld size
l = characteristic length of weld group, in.
$a = e_x / l$
e_x = horizontal component of eccentricity of P_u with respect to centroid of weld group, in.
C = coefficient tabulated below, which includes $\phi = 0.75$
C_1 = electrode strength coefficient from Table 8-4 (1.0 for E70XX electrodes)

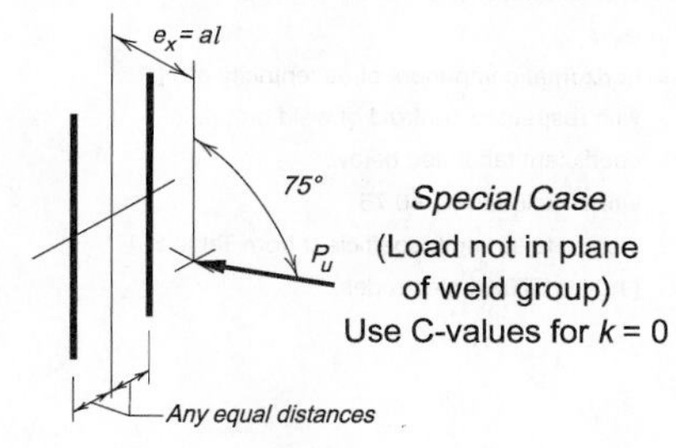

a	k = 0	0.1	0.2	0.3	0.4	0.5	0.6	0.7	0.8	0.9	1.0	1.2	1.4	1.6	1.8	2.0
0.00	4.11	4.11	4.11	4.11	4.11	4.11	4.11	4.11	4.11	4.11	4.11	4.11	4.11	4.11	4.11	4.11
0.10	3.88	3.90	3.95	4.00	4.04	4.07	4.08	4.09	4.09	4.09	4.09	4.09	4.09	4.09	4.09	4.09
0.15	3.76	3.77	3.83	3.90	3.96	4.01	4.04	4.06	4.08	4.08	4.09	4.09	4.09	4.09	4.09	4.09
0.20	3.64	3.65	3.71	3.79	3.88	3.94	3.99	4.03	4.05	4.06	4.07	4.08	4.09	4.09	4.09	4.09
0.25	3.53	3.54	3.60	3.69	3.78	3.87	3.93	3.98	4.01	4.03	4.05	4.07	4.08	4.08	4.09	4.09
0.30	3.43	3.44	3.49	3.58	3.69	3.78	3.86	3.92	3.97	4.00	4.02	4.05	4.07	4.07	4.08	4.08
0.40	3.24	3.25	3.29	3.38	3.50	3.62	3.72	3.80	3.86	3.91	3.95	4.00	4.03	4.05	4.06	4.07
0.50	3.07	3.08	3.12	3.20	3.32	3.45	3.57	3.67	3.75	3.82	3.87	3.94	3.99	4.02	4.04	4.05
0.60	2.91	2.92	2.97	3.05	3.15	3.29	3.42	3.54	3.63	3.71	3.78	3.88	3.94	3.98	4.01	4.03
0.70	2.77	2.78	2.82	2.90	3.00	3.13	3.27	3.40	3.51	3.60	3.68	3.80	3.88	3.93	3.97	4.00
0.80	2.63	2.64	2.69	2.77	2.87	2.99	3.13	3.26	3.39	3.49	3.58	3.72	3.81	3.88	3.93	3.96
0.90	2.50	2.52	2.57	2.64	2.74	2.86	3.00	3.13	3.26	3.38	3.48	3.64	3.75	3.83	3.88	3.93
1.00	2.38	2.40	2.45	2.53	2.63	2.74	2.87	3.01	3.14	3.26	3.37	3.55	3.68	3.77	3.83	3.88
1.20	2.17	2.18	2.24	2.32	2.41	2.52	2.64	2.78	2.91	3.04	3.16	3.37	3.52	3.64	3.73	3.80
1.40	1.99	2.00	2.05	2.13	2.23	2.33	2.45	2.57	2.71	2.84	2.96	3.18	3.36	3.50	3.61	3.69
1.60	1.83	1.84	1.89	1.97	2.06	2.17	2.28	2.39	2.52	2.65	2.78	3.01	3.20	3.36	3.49	3.59
1.80	1.69	1.70	1.75	1.83	1.92	2.02	2.13	2.24	2.35	2.48	2.61	2.84	3.05	3.22	3.36	3.48
2.00	1.57	1.58	1.63	1.70	1.79	1.89	1.99	2.10	2.21	2.33	2.45	2.68	2.89	3.08	3.23	3.36
2.20	1.46	1.48	1.52	1.59	1.68	1.77	1.87	1.97	2.08	2.19	2.30	2.54	2.75	2.94	3.10	3.24
2.40	1.37	1.38	1.43	1.49	1.57	1.67	1.76	1.86	1.97	2.07	2.18	2.40	2.61	2.81	2.98	3.13
2.60	1.28	1.30	1.34	1.40	1.48	1.57	1.66	1.76	1.86	1.96	2.07	2.28	2.49	2.68	2.85	3.01
2.80	1.21	1.22	1.27	1.33	1.40	1.49	1.58	1.67	1.77	1.86	1.96	2.16	2.37	2.56	2.73	2.89
3.00	1.14	1.15	1.20	1.26	1.33	1.41	1.49	1.59	1.68	1.77	1.87	2.06	2.26	2.45	2.62	2.78

Table 8-6. Coefficients C for Eccentrically Loaded Weld Groups Angle = 0°

$\phi R_n = CC_1 Dl \qquad C_{min} = \frac{P_u}{C_1 Dl} \qquad D_{min} = \frac{P_u}{CC_1 l} \qquad l_{min} = \frac{P_u}{CC_1 D}$

where

P_u = factored force, kips

D = number of sixteenths-of-an-inch in the fillet weld size

l = characteristic length of weld group, in.

$a = e_x/l$

e_x = horizontal component of eccentricity of P_u with respect to centroid of weld group, in.

C = coefficient tabulated below, which includes $\phi = 0.75$

C_1 = electrode strength coefficient from Table 8-4 (1.0 for E70XX electrodes)

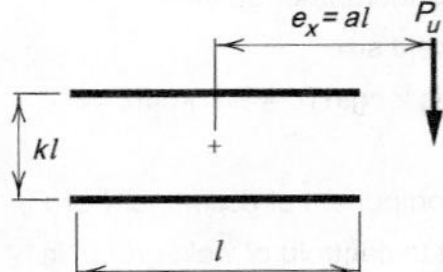

a	k															
	0	0.1	0.2	0.3	0.4	0.5	0.6	0.7	0.8	0.9	1.0	1.2	1.4	1.6	1.8	2.0
0.00	4.18	4.18	4.18	4.18	4.18	4.18	4.18	4.18	4.18	4.18	4.18	4.18	4.18	4.18	4.18	4.18
0.10	3.24	3.27	3.36	3.48	3.61	3.73	3.83	3.91	3.97	4.01	4.05	4.09	4.12	4.13	4.14	4.15
0.15	2.92	2.95	3.03	3.15	3.29	3.43	3.56	3.68	3.77	3.84	3.90	3.98	4.04	4.07	4.09	4.11
0.20	2.65	2.68	2.75	2.85	2.99	3.15	3.30	3.43	3.55	3.65	3.73	3.85	3.93	3.99	4.03	4.06
0.25	2.41	2.44	2.50	2.60	2.73	2.88	3.04	3.19	3.33	3.44	3.54	3.70	3.81	3.89	3.95	3.99
0.30	2.20	2.23	2.29	2.39	2.50	2.64	2.81	2.96	3.11	3.24	3.36	3.54	3.68	3.79	3.86	3.92
0.40	1.86	1.88	1.95	2.03	2.14	2.26	2.39	2.55	2.71	2.86	2.99	3.22	3.40	3.55	3.66	3.74
0.50	1.60	1.62	1.68	1.76	1.85	1.96	2.08	2.21	2.36	2.51	2.66	2.91	3.12	3.29	3.43	3.55
0.60	1.40	1.42	1.47	1.54	1.63	1.73	1.84	1.96	2.08	2.22	2.36	2.63	2.86	3.05	3.21	3.34
0.70	1.24	1.26	1.30	1.37	1.45	1.54	1.64	1.75	1.86	1.98	2.11	2.38	2.61	2.81	2.99	3.14
0.80	1.11	1.13	1.17	1.23	1.30	1.39	1.48	1.58	1.68	1.79	1.90	2.15	2.39	2.60	2.78	2.95
0.90	1.00	1.02	1.06	1.11	1.18	1.26	1.35	1.44	1.53	1.63	1.73	1.96	2.19	2.40	2.59	2.76
1.00	0.914	0.929	0.965	1.02	1.08	1.15	1.23	1.31	1.40	1.49	1.59	1.79	2.02	2.22	2.41	2.59
1.20	0.777	0.789	0.821	0.866	0.920	0.984	1.05	1.13	1.20	1.28	1.36	1.53	1.72	1.92	2.11	2.28
1.40	0.674	0.685	0.713	0.753	0.802	0.856	0.918	0.982	1.05	1.12	1.19	1.34	1.50	1.68	1.85	2.02
1.60	0.594	0.604	0.629	0.665	0.709	0.759	0.812	0.871	0.931	0.993	1.06	1.19	1.33	1.48	1.64	1.80
1.80	0.531	0.541	0.563	0.595	0.635	0.680	0.729	0.780	0.836	0.892	0.950	1.07	1.20	1.33	1.47	1.62
2.00	0.481	0.489	0.509	0.538	0.574	0.616	0.661	0.708	0.757	0.810	0.862	0.971	1.09	1.21	1.33	1.47
2.20	0.439	0.446	0.465	0.492	0.524	0.562	0.604	0.647	0.693	0.740	0.789	0.890	0.994	1.10	1.22	1.34
2.40	0.404	0.410	0.427	0.452	0.483	0.517	0.555	0.596	0.638	0.681	0.727	0.820	0.917	1.02	1.12	1.23
2.60	0.374	0.379	0.396	0.419	0.447	0.479	0.514	0.552	0.592	0.632	0.674	0.760	0.850	0.941	1.04	1.14
2.80	0.348	0.353	0.368	0.390	0.416	0.446	0.479	0.514	0.551	0.589	0.628	0.709	0.793	0.878	0.966	1.06
3.00	0.325	0.330	0.344	0.364	0.389	0.417	0.448	0.481	0.516	0.552	0.588	0.664	0.742	0.822	0.904	0.989

Table 8-6 (cont.).
Coefficients C for Eccentrically Loaded Weld Groups
Angle = 15°

$\phi R_n = CC_1 Dl$ $C_{min} = \dfrac{P_u}{C_1 Dl}$ $D_{min} = \dfrac{P_u}{CC_1 l}$ $l_{min} = \dfrac{P_u}{CC_1 D}$

where

P_u = factored force, kips
D = number of sixteenths-of-an-inch in the fillet weld size
l = characteristic length of weld group, in.
$a = e_x/l$
e_x = horizontal component of eccentricity of P_u with respect to centroid of weld group, in.
C = coefficient tabulated below, which includes $\phi = 0.75$
C_1 = electrode strength coefficient from Table 8-4 (1.0 for E70XX electrodes)

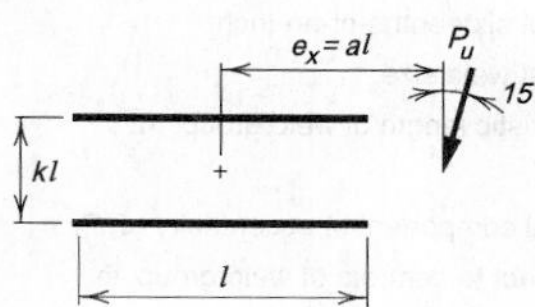

a	k															
	0	0.1	0.2	0.3	0.4	0.5	0.6	0.7	0.8	0.9	1.0	1.2	1.4	1.6	1.8	2.0
0.00	4.11	4.11	4.11	4.11	4.11	4.11	4.11	4.11	4.11	4.11	4.11	4.11	4.11	4.11	4.11	4.11
0.10	3.29	3.30	3.34	3.43	3.55	3.66	3.76	3.83	3.89	3.94	3.97	4.02	4.04	4.06	4.07	4.07
0.15	2.97	2.99	3.03	3.11	3.22	3.35	3.48	3.59	3.68	3.76	3.82	3.90	3.96	4.00	4.02	4.04
0.20	2.70	2.71	2.76	2.84	2.94	3.07	3.20	3.34	3.45	3.55	3.64	3.76	3.85	3.91	3.95	3.98
0.25	2.46	2.48	2.53	2.61	2.71	2.82	2.95	3.09	3.22	3.34	3.44	3.61	3.72	3.81	3.87	3.91
0.30	2.25	2.27	2.32	2.40	2.50	2.61	2.73	2.87	3.00	3.13	3.25	3.44	3.58	3.69	3.77	3.83
0.40	1.91	1.93	1.98	2.05	2.15	2.25	2.37	2.49	2.62	2.75	2.87	3.10	3.29	3.44	3.56	3.65
0.50	1.65	1.66	1.71	1.79	1.87	1.97	2.08	2.19	2.30	2.42	2.55	2.79	2.99	3.17	3.32	3.44
0.60	1.44	1.46	1.50	1.57	1.65	1.75	1.85	1.95	2.06	2.17	2.28	2.51	2.72	2.91	3.08	3.22
0.70	1.28	1.29	1.34	1.40	1.48	1.56	1.66	1.76	1.85	1.96	2.06	2.27	2.48	2.67	2.85	3.00
0.80	1.15	1.16	1.20	1.26	1.33	1.41	1.50	1.59	1.69	1.78	1.88	2.07	2.27	2.45	2.63	2.79
0.90	1.04	1.05	1.09	1.14	1.21	1.29	1.37	1.45	1.54	1.63	1.72	1.90	2.08	2.26	2.43	2.59
1.00	0.945	0.958	0.993	1.05	1.11	1.18	1.26	1.34	1.42	1.50	1.59	1.75	1.92	2.10	2.26	2.42
1.20	0.803	0.814	0.846	0.892	0.946	1.01	1.08	1.15	1.22	1.30	1.37	1.52	1.67	1.82	1.97	2.12
1.40	0.697	0.707	0.735	0.776	0.825	0.881	0.942	1.01	1.07	1.14	1.21	1.34	1.47	1.61	1.74	1.87
1.60	0.615	0.624	0.649	0.686	0.731	0.781	0.835	0.894	0.954	1.01	1.08	1.20	1.32	1.44	1.56	1.68
1.80	0.550	0.559	0.581	0.614	0.655	0.701	0.751	0.802	0.858	0.913	0.968	1.08	1.19	1.30	1.41	1.52
2.00	0.497	0.505	0.526	0.556	0.593	0.635	0.681	0.729	0.778	0.830	0.881	0.985	1.09	1.19	1.29	1.39
2.20	0.454	0.461	0.480	0.508	0.542	0.580	0.623	0.667	0.713	0.760	0.808	0.904	1.00	1.10	1.19	1.28
2.40	0.418	0.424	0.442	0.467	0.499	0.534	0.573	0.615	0.657	0.700	0.745	0.834	0.924	1.01	1.10	1.19
2.60	0.387	0.392	0.409	0.433	0.462	0.495	0.531	0.570	0.609	0.650	0.691	0.775	0.859	0.942	1.03	1.11
2.80	0.360	0.365	0.381	0.403	0.430	0.461	0.494	0.530	0.568	0.606	0.644	0.724	0.802	0.881	0.959	1.03
3.00	0.336	0.341	0.356	0.377	0.402	0.431	0.463	0.496	0.532	0.568	0.604	0.678	0.753	0.827	0.900	0.972

Table 8-6 (cont.).
Coefficients *C* for Eccentrically Loaded Weld Groups
Angle = 30°

$\phi R_n = CC_1 Dl \qquad C_{min} = \dfrac{P_u}{C_1 Dl} \qquad D_{min} = \dfrac{P_u}{CC_1 l} \qquad l_{min} = \dfrac{P_u}{CC_1 D}$

where

P_u = factored force, kips

D = number of sixteenths-of-an-inch in the fillet weld size

l = characteristic length of weld group, in.

$a = e_x/l$

e_x = horizontal component of eccentricity of P_u with respect to centroid of weld group, in.

C = coefficient tabulated below, which includes $\phi = 0.75$

C_1 = electrode strength coefficient from Table 8-4 (1.0 for E70XX electrodes)

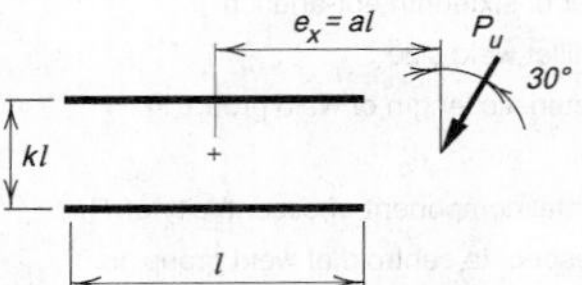

a	k															
	0	0.1	0.2	0.3	0.4	0.5	0.6	0.7	0.8	0.9	1.0	1.2	1.4	1.6	1.8	2.0
0.00	3.91	3.91	3.91	3.91	3.91	3.91	3.91	3.91	3.91	3.91	3.91	3.91	3.91	3.91	3.91	3.91
0.10	3.37	3.38	3.40	3.45	3.50	3.56	3.62	3.67	3.71	3.75	3.77	3.81	3.83	3.84	3.85	3.85
0.15	3.07	3.07	3.10	3.14	3.20	3.27	3.35	3.43	3.50	3.56	3.61	3.69	3.74	3.77	3.79	3.81
0.20	2.82	2.83	2.85	2.89	2.94	3.01	3.09	3.17	3.26	3.35	3.42	3.54	3.62	3.68	3.72	3.75
0.25	2.60	2.61	2.63	2.68	2.73	2.80	2.87	2.94	3.03	3.12	3.21	3.36	3.48	3.56	3.62	3.67
0.30	2.40	2.41	2.44	2.49	2.55	2.62	2.69	2.76	2.84	2.92	3.01	3.18	3.32	3.43	3.51	3.57
0.40	2.06	2.07	2.10	2.16	2.22	2.30	2.38	2.46	2.54	2.61	2.69	2.83	2.99	3.14	3.25	3.35
0.50	1.80	1.80	1.84	1.89	1.97	2.04	2.13	2.21	2.29	2.36	2.44	2.57	2.70	2.85	2.98	3.10
0.60	1.58	1.59	1.62	1.68	1.75	1.83	1.91	2.00	2.07	2.15	2.23	2.36	2.49	2.61	2.74	2.86
0.70	1.41	1.42	1.45	1.51	1.58	1.66	1.74	1.82	1.90	1.97	2.05	2.18	2.30	2.42	2.53	2.64
0.80	1.26	1.27	1.31	1.37	1.43	1.51	1.59	1.66	1.74	1.82	1.89	2.03	2.15	2.26	2.36	2.46
0.90	1.15	1.15	1.19	1.25	1.31	1.38	1.46	1.53	1.61	1.68	1.75	1.88	2.01	2.12	2.22	2.32
1.00	1.05	1.06	1.09	1.14	1.20	1.27	1.35	1.42	1.49	1.57	1.63	1.76	1.88	1.99	2.09	2.19
1.20	0.891	0.900	0.932	0.979	1.03	1.10	1.17	1.23	1.30	1.37	1.43	1.56	1.67	1.78	1.88	1.97
1.40	0.774	0.783	0.812	0.855	0.906	0.963	1.02	1.09	1.15	1.21	1.27	1.39	1.50	1.60	1.70	1.79
1.60	0.684	0.692	0.719	0.757	0.805	0.857	0.913	0.972	1.03	1.09	1.14	1.26	1.36	1.46	1.55	1.64
1.80	0.612	0.620	0.644	0.679	0.723	0.771	0.823	0.876	0.931	0.985	1.04	1.14	1.24	1.33	1.42	1.51
2.00	0.554	0.562	0.584	0.616	0.655	0.700	0.749	0.799	0.848	0.899	0.949	1.05	1.14	1.23	1.31	1.39
2.20	0.506	0.513	0.534	0.563	0.600	0.641	0.686	0.733	0.779	0.826	0.874	0.965	1.05	1.14	1.22	1.30
2.40	0.465	0.472	0.491	0.519	0.553	0.591	0.633	0.676	0.720	0.764	0.808	0.896	0.979	1.06	1.14	1.21
2.60	0.431	0.437	0.455	0.481	0.512	0.548	0.587	0.628	0.669	0.711	0.752	0.835	0.915	0.992	1.07	1.14
2.80	0.401	0.406	0.423	0.448	0.477	0.511	0.547	0.585	0.625	0.664	0.704	0.782	0.858	0.931	1.00	1.07
3.00	0.375	0.380	0.396	0.419	0.447	0.478	0.513	0.548	0.586	0.623	0.661	0.734	0.808	0.878	0.946	1.01

Table 8-6 (cont.). Coefficients C for Eccentrically Loaded Weld Groups Angle = 45°

$\phi R_n = CC_1 Dl$ $C_{min} = \frac{P_u}{C_1 Dl}$ $D_{min} = \frac{P_u}{CC_1 l}$ $l_{min} = \frac{P_u}{CC_1 D}$

where

P_u = factored force, kips

D = number of sixteenths-of-an-inch in the fillet weld size

l = characteristic length of weld group, in.

$a = e_x/l$

e_x = horizontal component of eccentricity of P_u with respect to centroid of weld group, in.

C = coefficient tabulated below, which includes $\phi = 0.75$

C_1 = electrode strength coefficient from Table 8-4 (1.0 for E70XX electrodes)

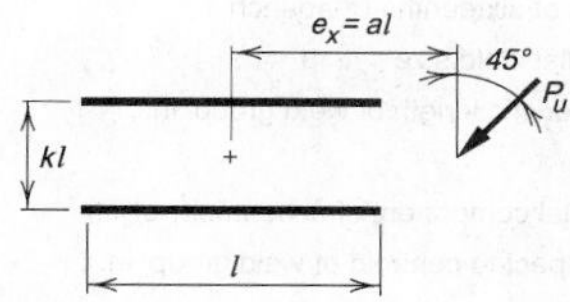

a	k															
	0	0.1	0.2	0.3	0.4	0.5	0.6	0.7	0.8	0.9	1.0	1.2	1.4	1.6	1.8	2.0
0.00	3.61	3.61	3.61	3.61	3.61	3.61	3.61	3.61	3.61	3.61	3.61	3.61	3.61	3.61	3.61	3.61
0.10	3.37	3.37	3.38	3.38	3.40	3.42	3.43	3.44	3.46	3.47	3.48	3.50	3.51	3.52	3.52	3.52
0.15	3.13	3.13	3.15	3.17	3.20	3.23	3.25	3.28	3.30	3.33	3.35	3.38	3.41	3.43	3.45	3.47
0.20	2.94	2.94	2.95	2.97	2.99	3.03	3.06	3.10	3.13	3.17	3.20	3.25	3.29	3.33	3.35	3.38
0.25	2.77	2.77	2.78	2.80	2.83	2.86	2.89	2.93	2.97	3.01	3.04	3.11	3.16	3.21	3.25	3.28
0.30	2.61	2.61	2.63	2.65	2.68	2.71	2.75	2.79	2.83	2.86	2.90	2.97	3.04	3.09	3.14	3.18
0.40	2.32	2.32	2.34	2.37	2.41	2.45	2.50	2.54	2.59	2.63	2.66	2.73	2.80	2.87	2.92	2.97
0.50	2.06	2.07	2.09	2.12	2.17	2.22	2.27	2.33	2.38	2.42	2.47	2.54	2.61	2.68	2.74	2.79
0.60	1.84	1.85	1.87	1.91	1.96	2.02	2.08	2.14	2.19	2.25	2.30	2.38	2.45	2.52	2.58	2.63
0.70	1.66	1.66	1.69	1.73	1.79	1.85	1.91	1.97	2.03	2.09	2.14	2.23	2.31	2.38	2.44	2.50
0.80	1.50	1.51	1.54	1.58	1.64	1.70	1.76	1.83	1.89	1.95	2.00	2.10	2.18	2.26	2.32	2.38
0.90	1.37	1.38	1.40	1.45	1.51	1.57	1.64	1.70	1.76	1.82	1.88	1.98	2.07	2.14	2.21	2.27
1.00	1.26	1.26	1.29	1.34	1.40	1.46	1.53	1.59	1.65	1.71	1.77	1.87	1.96	2.04	2.11	2.17
1.20	1.08	1.08	1.11	1.16	1.21	1.28	1.34	1.40	1.46	1.52	1.58	1.68	1.77	1.85	1.93	2.00
1.40	0.938	0.946	0.975	1.02	1.07	1.13	1.19	1.25	1.31	1.37	1.42	1.52	1.62	1.70	1.77	1.84
1.60	0.831	0.838	0.866	0.908	0.958	1.01	1.07	1.13	1.19	1.24	1.29	1.39	1.48	1.56	1.64	1.71
1.80	0.745	0.752	0.779	0.817	0.864	0.917	0.972	1.03	1.08	1.13	1.19	1.28	1.37	1.45	1.52	1.59
2.00	0.675	0.682	0.707	0.743	0.787	0.836	0.888	0.941	0.992	1.04	1.09	1.18	1.27	1.35	1.42	1.49
2.20	0.617	0.624	0.647	0.681	0.722	0.768	0.818	0.868	0.917	0.964	1.01	1.10	1.18	1.26	1.33	1.40
2.40	0.568	0.574	0.596	0.628	0.667	0.710	0.756	0.804	0.851	0.897	0.941	1.03	1.11	1.18	1.25	1.32
2.60	0.526	0.532	0.553	0.583	0.619	0.660	0.703	0.749	0.794	0.838	0.881	0.963	1.04	1.12	1.18	1.25
2.80	0.489	0.496	0.515	0.544	0.578	0.617	0.658	0.701	0.743	0.786	0.827	0.906	0.982	1.05	1.12	1.18
3.00	0.458	0.464	0.482	0.509	0.542	0.579	0.618	0.658	0.699	0.739	0.779	0.856	0.929	1.00	1.06	1.12

Table 8-6 (cont.). Coefficients C for Eccentrically Loaded Weld Groups Angle = 60°

$\phi R_n = CC_1 Dl \qquad C_{min} = \dfrac{P_u}{C_1 Dl} \qquad D_{min} = \dfrac{P_u}{CC_1 l} \qquad l_{min} = \dfrac{P_u}{CC_1 D}$

where

P_u = factored force, kips

D = number of sixteenths-of-an-inch in the fillet weld size

l = characteristic length of weld group, in.

$a = e_x/l$

e_x = horizontal component of eccentricity of P_u with respect to centroid of weld group, in.

C = coefficient tabulated below, which includes $\phi = 0.75$

C_1 = electrode strength coefficient from Table 8-4 (1.0 for E70XX electrodes)

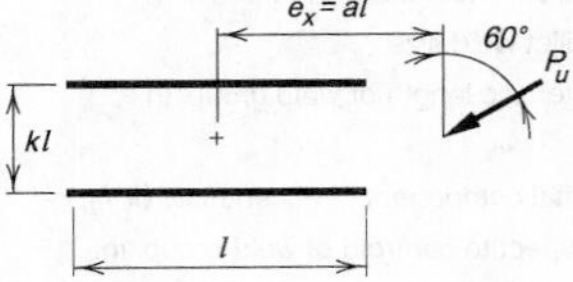

a	k															
	0	0.1	0.2	0.3	0.4	0.5	0.6	0.7	0.8	0.9	1.0	1.2	1.4	1.6	1.8	2.0
0.00	3.28	3.28	3.28	3.28	3.28	3.28	3.28	3.28	3.28	3.28	3.28	3.28	3.28	3.28	3.28	3.28
0.10	3.20	3.19	3.19	3.19	3.19	3.19	3.19	3.19	3.18	3.18	3.18	3.17	3.16	3.15	3.14	3.13
0.15	3.10	3.09	3.10	3.10	3.10	3.10	3.10	3.10	3.10	3.10	3.10	3.10	3.10	3.09	3.09	3.08
0.20	2.98	2.98	2.98	2.98	2.98	2.99	2.99	3.00	3.01	3.01	3.02	3.02	3.03	3.03	3.02	3.02
0.25	2.89	2.89	2.89	2.89	2.89	2.90	2.90	2.91	2.91	2.92	2.92	2.94	2.95	2.96	2.96	2.96
0.30	2.80	2.80	2.80	2.81	2.81	2.82	2.82	2.83	2.83	2.84	2.85	2.86	2.87	2.89	2.89	2.90
0.40	2.63	2.63	2.63	2.64	2.65	2.66	2.67	2.68	2.69	2.70	2.71	2.73	2.74	2.76	2.77	2.78
0.50	2.44	2.44	2.45	2.47	2.48	2.50	2.52	2.54	2.55	2.57	2.58	2.61	2.63	2.65	2.66	2.68
0.60	2.27	2.27	2.28	2.29	2.32	2.35	2.37	2.40	2.42	2.44	2.46	2.49	2.52	2.55	2.57	2.59
0.70	2.09	2.10	2.11	2.13	2.16	2.19	2.23	2.27	2.29	2.32	2.34	2.39	2.42	2.45	2.48	2.50
0.80	1.94	1.94	1.96	1.98	2.02	2.06	2.10	2.14	2.18	2.21	2.23	2.29	2.33	2.37	2.40	2.42
0.90	1.80	1.80	1.82	1.85	1.89	1.93	1.98	2.02	2.06	2.10	2.13	2.19	2.24	2.28	2.32	2.34
1.00	1.67	1.67	1.69	1.73	1.77	1.82	1.87	1.92	1.96	2.00	2.04	2.10	2.15	2.20	2.24	2.27
1.20	1.46	1.46	1.48	1.52	1.57	1.62	1.67	1.73	1.78	1.82	1.86	1.93	2.00	2.05	2.10	2.14
1.40	1.28	1.29	1.31	1.35	1.40	1.46	1.51	1.57	1.62	1.67	1.71	1.79	1.86	1.92	1.97	2.01
1.60	1.15	1.15	1.18	1.22	1.27	1.32	1.38	1.44	1.49	1.54	1.58	1.66	1.73	1.80	1.85	1.90
1.80	1.03	1.04	1.06	1.11	1.16	1.21	1.27	1.32	1.37	1.42	1.47	1.55	1.63	1.69	1.75	1.80
2.00	0.940	0.946	0.971	1.01	1.06	1.11	1.17	1.22	1.28	1.32	1.37	1.45	1.53	1.60	1.66	1.71
2.20	0.861	0.867	0.892	0.932	0.979	1.03	1.09	1.14	1.19	1.24	1.28	1.37	1.44	1.51	1.57	1.63
2.40	0.794	0.800	0.825	0.863	0.909	0.959	1.01	1.06	1.11	1.16	1.21	1.29	1.36	1.43	1.49	1.55
2.60	0.736	0.743	0.767	0.804	0.848	0.896	0.946	1.00	1.05	1.09	1.14	1.22	1.29	1.36	1.42	1.48
2.80	0.686	0.693	0.716	0.752	0.794	0.841	0.889	0.937	0.985	1.03	1.07	1.16	1.23	1.30	1.36	1.42
3.00	0.643	0.649	0.672	0.706	0.746	0.792	0.838	0.885	0.931	0.975	1.02	1.10	1.17	1.24	1.30	1.36

Table 8-6 (cont.). Coefficients C for Eccentrically Loaded Weld Groups Angle = 75°

$\phi R_n = CC_1 Dl$ $\quad C_{min} = \dfrac{P_u}{C_1 Dl}$ $\quad D_{min} = \dfrac{P_u}{CC_1 l}$ $\quad l_{min} = \dfrac{P_u}{CC_1 D}$

where

P_u = factored force, kips
D = number of sixteenths-of-an-inch in the fillet weld size
l = characteristic length of weld group, in.
$a = e_x/l$
e_x = horizontal component of eccentricity of P_u with respect to centroid of weld group, in.
C = coefficient tabulated below, which includes $\phi = 0.75$
C_1 = electrode strength coefficient from Table 8-4 (1.0 for E70XX electrodes)

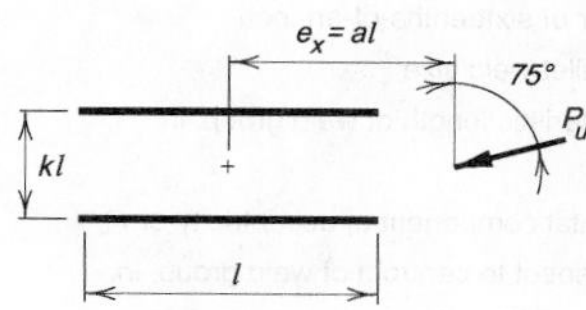

a	k															
	0	0.1	0.2	0.3	0.4	0.5	0.6	0.7	0.8	0.9	1.0	1.2	1.4	1.6	1.8	2.0
0.00	2.97	2.97	2.97	2.97	2.97	2.97	2.97	2.97	2.97	2.97	2.97	2.97	2.97	2.97	2.97	2.97
0.10	2.86	2.86	2.87	2.87	2.88	2.88	2.88	2.89	2.89	2.89	2.89	2.88	2.87	2.85	2.84	2.55
0.15	2.89	2.89	2.89	2.89	2.89	2.89	2.89	2.89	2.88	2.88	2.87	2.86	2.85	2.83	2.82	2.81
0.20	2.88	2.88	2.88	2.88	2.88	2.87	2.87	2.86	2.86	2.85	2.85	2.84	2.82	2.81	2.80	2.78
0.25	2.87	2.87	2.87	2.87	2.86	2.86	2.85	2.85	2.84	2.84	2.83	2.81	2.80	2.79	2.78	2.76
0.30	2.86	2.86	2.86	2.85	2.85	2.84	2.84	2.83	2.82	2.82	2.81	2.80	2.78	2.77	2.76	2.74
0.40	2.84	2.83	2.83	2.82	2.82	2.81	2.80	2.79	2.79	2.78	2.77	2.76	2.74	2.73	2.72	2.71
0.50	2.79	2.79	2.78	2.77	2.77	2.76	2.76	2.75	2.74	2.73	2.73	2.71	2.70	2.69	2.68	2.67
0.60	2.74	2.73	2.73	2.72	2.72	2.71	2.70	2.70	2.69	2.68	2.68	2.67	2.66	2.65	2.64	2.63
0.70	2.66	2.66	2.66	2.66	2.65	2.65	2.64	2.64	2.64	2.63	2.63	2.62	2.61	2.61	2.60	2.59
0.80	2.59	2.59	2.59	2.58	2.58	2.58	2.58	2.58	2.58	2.58	2.58	2.57	2.57	2.56	2.56	2.56
0.90	2.51	2.51	2.51	2.51	2.51	2.51	2.51	2.51	2.52	2.52	2.52	2.52	2.52	2.52	2.52	2.52
1.00	2.42	2.42	2.43	2.43	2.43	2.44	2.44	2.45	2.45	2.46	2.46	2.47	2.48	2.48	2.48	2.48
1.20	2.25	2.25	2.25	2.27	2.28	2.30	2.31	2.32	2.33	2.34	2.35	2.37	2.38	2.39	2.40	2.40
1.40	2.08	2.08	2.09	2.11	2.13	2.15	2.18	2.19	2.21	2.23	2.24	2.27	2.29	2.31	2.32	2.33
1.60	1.92	1.93	1.94	1.96	1.99	2.02	2.05	2.08	2.10	2.12	2.14	2.17	2.20	2.22	2.24	2.26
1.80	1.78	1.78	1.80	1.83	1.86	1.89	1.93	1.96	1.99	2.02	2.04	2.08	2.12	2.15	2.17	2.19
2.00	1.65	1.66	1.67	1.70	1.74	1.78	1.82	1.86	1.89	1.92	1.95	2.00	2.04	2.07	2.10	2.12
2.20	1.54	1.54	1.56	1.60	1.64	1.68	1.72	1.76	1.80	1.83	1.86	1.92	1.96	2.00	2.03	2.06
2.40	1.44	1.44	1.46	1.50	1.54	1.59	1.63	1.68	1.71	1.75	1.78	1.84	1.89	1.93	1.97	2.00
2.60	1.35	1.35	1.37	1.41	1.45	1.50	1.55	1.60	1.64	1.67	1.71	1.77	1.82	1.87	1.91	1.94
2.80	1.26	1.27	1.29	1.33	1.37	1.42	1.47	1.52	1.56	1.60	1.64	1.71	1.76	1.81	1.85	1.88
3.00	1.19	1.20	1.22	1.26	1.30	1.35	1.40	1.45	1.50	1.54	1.58	1.64	1.70	1.75	1.79	1.83

Table 8-7. Coefficients C for Eccentrically Loaded Weld Groups Angle = 0°

$\phi R_n = CC_1 Dl$ $\quad C_{min} = \dfrac{P_u}{C_1 Dl}$ $\quad D_{min} = \dfrac{P_u}{CC_1 l}$ $\quad l_{min} = \dfrac{P_u}{CC_1 D}$

where

P_u = factored force, kips

D = number of sixteenths-of-an-inch in the fillet weld size

l = characteristic length of weld group, in.

$a = e_x / l$

e_x = horizontal component of eccentricity of P_u with respect to centroid of weld group, in.

C = coefficient tabulated below, which includes $\phi = 0.75$

C_1 = electrode strength coefficient from Table 8-4 (1.0 for E70XX electrodes)

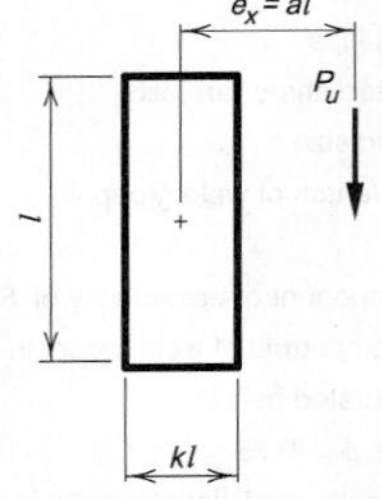

a	k															
	0	0.1	0.2	0.3	0.4	0.5	0.6	0.7	0.8	0.9	1.0	1.2	1.4	1.6	1.8	2.0
0.00	2.78	3.20	3.62	4.04	4.45	4.87	5.29	5.71	6.12	6.54	6.96	7.80	8.63	9.47	10.3	11.1
0.10	2.79	3.07	3.42	3.78	4.15	4.53	4.91	5.30	5.69	6.08	6.47	7.25	8.03	8.82	9.61	10.4
0.15	2.75	3.05	3.37	3.71	4.06	4.42	4.78	5.15	5.52	5.89	6.27	7.02	7.78	8.54	9.31	10.1
0.20	2.64	2.95	3.25	3.57	3.91	4.25	4.59	4.94	5.30	5.66	6.02	6.75	7.49	8.23	8.98	9.74
0.25	2.48	2.79	3.10	3.40	3.72	4.04	4.38	4.71	5.06	5.40	5.75	6.46	7.18	7.91	8.65	9.39
0.30	2.32	2.61	2.92	3.22	3.52	3.83	4.15	4.47	4.80	5.14	5.48	6.17	6.88	7.59	8.32	9.05
0.40	2.00	2.26	2.54	2.83	3.12	3.41	3.71	4.01	4.32	4.64	4.96	5.62	6.30	6.99	7.70	8.42
0.50	1.72	1.95	2.20	2.47	2.75	3.03	3.31	3.59	3.89	4.19	4.50	5.13	5.78	6.46	7.14	7.84
0.60	1.50	1.70	1.92	2.17	2.44	2.70	2.97	3.24	3.52	3.80	4.09	4.70	5.33	5.98	6.64	7.32
0.70	1.32	1.50	1.70	1.93	2.17	2.43	2.68	2.93	3.20	3.47	3.75	4.33	4.93	5.56	6.20	6.86
0.80	1.17	1.33	1.52	1.73	1.95	2.20	2.43	2.67	2.92	3.18	3.45	4.00	4.58	5.18	5.80	6.44
0.90	1.05	1.20	1.37	1.56	1.77	2.00	2.23	2.45	2.69	2.93	3.18	3.71	4.26	4.84	5.44	6.05
1.00	0.957	1.09	1.24	1.42	1.62	1.83	2.05	2.26	2.49	2.72	2.96	3.45	3.98	4.53	5.10	5.69
1.20	0.806	0.916	1.05	1.21	1.38	1.57	1.76	1.95	2.15	2.36	2.57	3.02	3.50	4.00	4.53	5.08
1.40	0.695	0.790	0.908	1.05	1.20	1.37	1.54	1.72	1.89	2.08	2.27	2.68	3.12	3.58	4.06	4.57
1.60	0.611	0.694	0.800	0.923	1.06	1.21	1.37	1.53	1.69	1.86	2.03	2.40	2.80	3.23	3.67	4.15
1.80	0.544	0.619	0.714	0.825	0.950	1.09	1.23	1.37	1.52	1.68	1.84	2.18	2.54	2.93	3.35	3.79
2.00	0.491	0.558	0.645	0.746	0.860	0.984	1.12	1.25	1.38	1.53	1.67	1.99	2.33	2.69	3.08	3.49
2.20	0.447	0.509	0.588	0.680	0.785	0.898	1.02	1.14	1.27	1.40	1.54	1.83	2.14	2.48	2.84	3.22
2.40	0.410	0.467	0.540	0.625	0.721	0.827	0.939	1.05	1.17	1.29	1.42	1.69	1.99	2.30	2.64	3.00
2.60	0.379	0.431	0.499	0.578	0.667	0.765	0.869	0.977	1.09	1.20	1.32	1.57	1.85	2.15	2.46	2.80
2.80	0.352	0.401	0.464	0.538	0.621	0.712	0.809	0.911	1.01	1.12	1.23	1.47	1.73	2.01	2.31	2.63
3.00	0.329	0.375	0.434	0.503	0.580	0.666	0.757	0.853	0.949	1.05	1.16	1.38	1.62	1.89	2.17	2.47

Table 8-7 (cont.). Coefficients *C* for Eccentrically Loaded Weld Groups Angle = 15°

$\phi R_n = CC_1 Dl$ $C_{min} = \frac{P_u}{C_1 Dl}$ $D_{min} = \frac{P_u}{CC_1 l}$ $l_{min} = \frac{P_u}{CC_1 D}$

where

P_u = factored force, kips
D = number of sixteenths-of-an-inch in the fillet weld size
l = characteristic length of weld group, in.
$a = e_x/l$
e_x = horizontal component of eccentricity of P_u with respect to centroid of weld group, in.
C = coefficient tabulated below, which includes $\phi = 0.75$
C_1 = electrode strength coefficient from Table 8-4 (1.0 for E70XX electrodes)

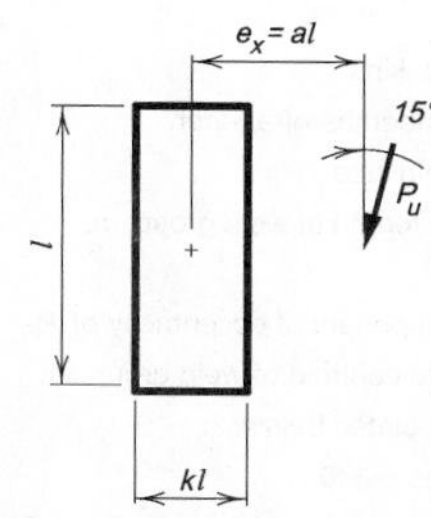

a	*k*															
	0	0.1	0.2	0.3	0.4	0.5	0.6	0.7	0.8	0.9	1.0	1.2	1.4	1.6	1.8	2.0
0.00	2.97	3.38	3.79	4.20	4.61	5.02	5.43	5.84	6.25	6.66	7.07	7.89	8.71	9.54	10.4	11.2
0.10	2.84	3.16	3.52	3.89	4.28	4.66	5.05	5.43	5.83	6.22	6.62	7.41	8.20	8.99	9.79	10.6
0.15	2.76	3.10	3.44	3.79	4.14	4.51	4.87	5.24	5.61	5.98	6.35	7.11	7.86	8.63	9.41	10.2
0.20	2.63	2.96	3.30	3.64	3.98	4.32	4.67	5.02	5.37	5.73	6.09	6.82	7.55	8.30	9.05	9.81
0.25	2.48	2.79	3.12	3.45	3.78	4.11	4.45	4.78	5.13	5.47	5.82	6.53	7.25	7.98	8.72	9.46
0.30	2.32	2.61	2.92	3.24	3.57	3.89	4.22	4.54	4.88	5.21	5.55	6.24	6.95	7.66	8.39	9.13
0.40	2.01	2.26	2.54	2.84	3.15	3.46	3.77	4.08	4.39	4.71	5.04	5.70	6.38	7.07	7.78	8.50
0.50	1.74	1.96	2.21	2.48	2.77	3.07	3.37	3.66	3.96	4.27	4.58	5.21	5.87	6.54	7.23	7.93
0.60	1.52	1.72	1.94	2.19	2.46	2.74	3.03	3.30	3.59	3.88	4.18	4.79	5.42	6.07	6.74	7.43
0.70	1.34	1.52	1.72	1.95	2.19	2.46	2.73	3.00	3.27	3.55	3.83	4.42	5.03	5.66	6.31	6.97
0.80	1.20	1.36	1.54	1.75	1.98	2.22	2.48	2.74	3.00	3.26	3.53	4.09	4.67	5.28	5.91	6.55
0.90	1.08	1.22	1.39	1.58	1.80	2.03	2.27	2.52	2.76	3.01	3.26	3.80	4.36	4.94	5.54	6.16
1.00	0.979	1.11	1.27	1.45	1.65	1.87	2.09	2.32	2.55	2.79	3.03	3.54	4.07	4.63	5.21	5.81
1.20	0.826	0.938	1.07	1.23	1.41	1.60	1.81	2.01	2.21	2.43	2.65	3.10	3.59	4.10	4.64	5.19
1.40	0.714	0.810	0.930	1.07	1.23	1.40	1.58	1.77	1.95	2.14	2.34	2.76	3.20	3.67	4.16	4.68
1.60	0.628	0.713	0.820	0.946	1.09	1.24	1.41	1.57	1.74	1.91	2.09	2.48	2.89	3.32	3.78	4.26
1.80	0.560	0.636	0.733	0.846	0.974	1.11	1.26	1.42	1.57	1.73	1.89	2.24	2.62	3.02	3.45	3.89
2.00	0.506	0.575	0.663	0.766	0.882	1.01	1.15	1.29	1.43	1.57	1.73	2.05	2.40	2.77	3.17	3.59
2.20	0.461	0.524	0.605	0.699	0.805	0.922	1.05	1.18	1.31	1.45	1.59	1.89	2.21	2.56	2.93	3.32
2.40	0.423	0.481	0.556	0.643	0.741	0.849	0.965	1.09	1.21	1.33	1.47	1.75	2.05	2.38	2.72	3.09
2.60	0.391	0.445	0.514	0.595	0.686	0.786	0.894	1.01	1.12	1.24	1.36	1.63	1.91	2.22	2.54	2.89
2.80	0.363	0.414	0.478	0.554	0.638	0.732	0.832	0.938	1.05	1.16	1.27	1.52	1.79	2.07	2.38	2.71
3.00	0.339	0.386	0.447	0.518	0.597	0.684	0.779	0.878	0.981	1.09	1.19	1.43	1.68	1.95	2.24	2.55

Table 8-7 (cont.).
Coefficients C for Eccentrically Loaded Weld Groups
Angle = 30°

$\phi R_n = CC_1 Dl \qquad C_{min} = \dfrac{P_u}{C_1 Dl} \qquad D_{min} = \dfrac{P_u}{CC_1 l} \qquad l_{min} = \dfrac{P_u}{CC_1 D}$

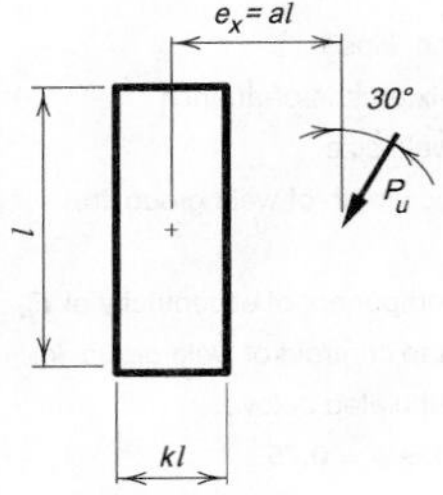

where

P_u = factored force, kips
D = number of sixteenths-of-an-inch in the fillet weld size
l = characteristic length of weld group, in.
$a = e_x/l$
e_x = horizontal component of eccentricity of P_u with respect to centroid of weld group, in.
C = coefficient tabulated below, which includes $\phi = 0.75$
C_1 = electrode strength coefficient from Table 8-4 (1.0 for E70XX electrodes)

a	k															
	0	0.1	0.2	0.3	0.4	0.5	0.6	0.7	0.8	0.9	1.0	1.2	1.4	1.6	1.8	2.0
0.00	3.28	3.67	4.06	4.45	4.84	5.23	5.62	6.01	6.40	6.79	7.18	7.96	8.74	9.53	10.3	11.1
0.10	3.03	3.45	3.85	4.24	4.62	5.00	5.38	5.76	6.14	6.52	6.90	7.67	8.44	9.22	10.0	10.8
0.15	2.87	3.25	3.63	4.02	4.40	4.77	5.15	5.52	5.89	6.26	6.64	7.39	8.15	8.93	9.70	10.5
0.20	2.72	3.07	3.43	3.80	4.17	4.53	4.89	5.25	5.61	5.97	6.34	7.09	7.84	8.60	9.37	10.2
0.25	2.57	2.88	3.22	3.57	3.93	4.29	4.65	4.99	5.34	5.69	6.05	6.77	7.52	8.27	9.04	9.81
0.30	2.41	2.70	3.02	3.35	3.70	4.05	4.40	4.75	5.09	5.43	5.77	6.48	7.21	7.95	8.71	9.48
0.40	2.11	2.36	2.64	2.95	3.27	3.60	3.94	4.28	4.61	4.94	5.27	5.95	6.65	7.36	8.09	8.85
0.50	1.84	2.07	2.32	2.60	2.90	3.21	3.53	3.86	4.19	4.50	4.83	5.48	6.16	6.85	7.56	8.28
0.60	1.62	1.83	2.06	2.31	2.59	2.88	3.18	3.50	3.82	4.12	4.43	5.07	5.73	6.40	7.09	7.79
0.70	1.44	1.63	1.84	2.07	2.33	2.60	2.89	3.19	3.50	3.79	4.09	4.70	5.33	5.99	6.66	7.35
0.80	1.30	1.46	1.65	1.87	2.11	2.37	2.64	2.93	3.22	3.50	3.79	4.38	4.98	5.61	6.26	6.93
0.90	1.17	1.32	1.50	1.70	1.93	2.18	2.43	2.70	2.98	3.25	3.52	4.08	4.67	5.28	5.90	6.55
1.00	1.07	1.21	1.37	1.56	1.78	2.01	2.25	2.51	2.77	3.02	3.28	3.82	4.38	4.97	5.57	6.20
1.20	0.907	1.03	1.17	1.34	1.53	1.73	1.95	2.18	2.41	2.65	2.88	3.37	3.89	4.44	5.00	5.59
1.40	0.786	0.890	1.02	1.17	1.34	1.52	1.72	1.93	2.14	2.34	2.56	3.01	3.49	3.99	4.52	5.07
1.60	0.693	0.785	0.901	1.04	1.19	1.36	1.53	1.72	1.91	2.10	2.30	2.72	3.16	3.62	4.12	4.63
1.80	0.619	0.703	0.808	0.931	1.07	1.22	1.38	1.55	1.73	1.90	2.08	2.47	2.88	3.31	3.77	4.25
2.00	0.559	0.635	0.731	0.844	0.970	1.11	1.26	1.41	1.57	1.74	1.91	2.26	2.64	3.05	3.48	3.93
2.20	0.510	0.579	0.668	0.771	0.887	1.01	1.15	1.30	1.45	1.60	1.75	2.08	2.44	2.82	3.22	3.64
2.40	0.469	0.533	0.614	0.710	0.818	0.935	1.06	1.20	1.34	1.48	1.62	1.93	2.27	2.62	3.00	3.4
2.60	0.433	0.493	0.569	0.658	0.758	0.867	0.984	1.11	1.24	1.37	1.51	1.80	2.11	2.45	2.80	3.18
2.80	0.403	0.458	0.529	0.613	0.706	0.808	0.918	1.03	1.16	1.28	1.41	1.68	1.98	2.29	2.63	2.99
3.00	0.376	0.428	0.495	0.573	0.661	0.757	0.860	0.969	1.08	1.20	1.32	1.58	1.86	2.16	2.48	2.82

Table 8-7 (cont.). Coefficients C for Eccentrically Loaded Weld Groups Angle = 45°

$\phi R_n = CC_1 Dl \qquad C_{min} = \frac{P_u}{C_1 Dl} \qquad D_{min} = \frac{P_u}{CC_1 l} \qquad l_{min} = \frac{P_u}{CC_1 D}$

where

P_u = factored force, kips

D = number of sixteenths-of-an-inch in the fillet weld size

l = characteristic length of weld group, in.

$a = e_x/l$

e_x = horizontal component of eccentricity of P_u with respect to centroid of weld group, in.

C = coefficient tabulated below, which includes $\phi = 0.75$

C_1 = electrode strength coefficient from Table 8-4 (1.0 for E70XX electrodes)

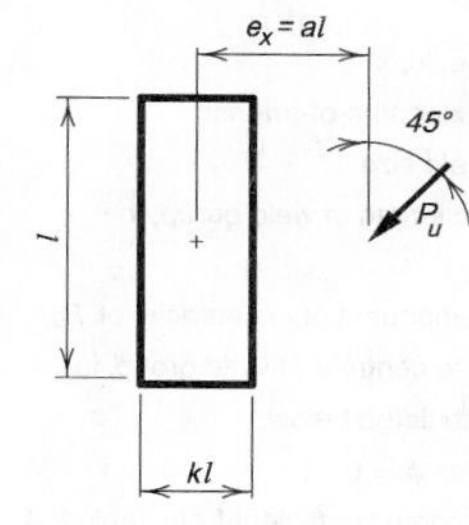

a	k															
	0	0.1	0.2	0.3	0.4	0.5	0.6	0.7	0.8	0.9	1.0	1.2	1.4	1.6	1.8	2.0
0.00	3.61	3.97	4.33	4.70	5.06	5.42	5.78	6.14	6.50	6.86	7.22	7.95	8.67	9.39	10.1	10.8
0.10	3.37	3.74	4.11	4.48	4.84	5.21	5.58	5.94	6.31	6.68	7.04	7.78	8.52	9.25	9.99	10.7
0.15	3.13	3.52	3.89	4.26	4.62	4.99	5.36	5.74	6.11	6.49	6.86	7.61	8.36	9.11	9.85	10.6
0.20	2.94	3.29	3.65	4.02	4.38	4.75	5.12	5.50	5.88	6.26	6.64	7.39	8.15	8.91	9.66	10.4
0.25	2.77	3.10	3.44	3.79	4.14	4.51	4.87	5.25	5.63	6.01	6.40	7.16	7.92	8.68	9.45	10.2
0.30	2.61	2.91	3.24	3.57	3.92	4.27	4.63	5.00	5.38	5.76	6.15	6.91	7.67	8.44	9.21	9.98
0.40	2.32	2.59	2.87	3.18	3.51	3.85	4.20	4.55	4.92	5.29	5.67	6.42	7.18	7.95	8.72	9.50
0.50	2.06	2.30	2.57	2.86	3.16	3.49	3.82	4.17	4.52	4.89	5.26	5.97	6.71	7.47	8.24	9.01
0.60	1.84	2.06	2.30	2.58	2.87	3.17	3.50	3.83	4.18	4.53	4.89	5.57	6.28	7.02	7.78	8.54
0.70	1.66	1.86	2.08	2.34	2.61	2.91	3.22	3.54	3.88	4.22	4.56	5.22	5.90	6.62	7.35	8.10
0.80	1.50	1.69	1.90	2.13	2.40	2.68	2.98	3.29	3.61	3.93	4.27	4.91	5.56	6.25	6.97	7.70
0.90	1.37	1.54	1.74	1.96	2.21	2.48	2.76	3.06	3.37	3.68	4.00	4.62	5.25	5.91	6.61	7.32
1.00	1.26	1.41	1.60	1.81	2.05	2.30	2.58	2.86	3.15	3.45	3.76	4.35	4.97	5.61	6.28	6.97
1.20	1.08	1.21	1.38	1.57	1.78	2.01	2.26	2.52	2.78	3.06	3.34	3.90	4.48	5.07	5.70	6.35
1.40	0.938	1.06	1.21	1.38	1.57	1.78	2.01	2.24	2.49	2.74	3.00	3.52	4.06	4.62	5.21	5.81
1.60	0.831	0.939	1.07	1.23	1.41	1.60	1.80	2.02	2.24	2.48	2.71	3.20	3.70	4.23	4.78	5.36
1.80	0.745	0.843	0.966	1.11	1.27	1.45	1.63	1.83	2.04	2.25	2.48	2.93	3.40	3.90	4.42	4.96
2.00	0.675	0.764	0.877	1.01	1.16	1.32	1.49	1.67	1.87	2.06	2.27	2.69	3.14	3.60	4.09	4.61
2.20	0.617	0.699	0.804	0.925	1.06	1.21	1.37	1.54	1.72	1.90	2.10	2.49	2.91	3.35	3.81	4.30
2.40	0.568	0.644	0.741	0.854	0.98	1.12	1.27	1.43	1.59	1.77	1.95	2.32	2.71	3.13	3.57	4.03
2.60	0.526	0.597	0.687	0.792	0.911	1.04	1.18	1.33	1.48	1.65	1.82	2.16	2.54	2.93	3.35	3.78
2.80	0.489	0.556	0.641	0.739	0.850	0.972	1.10	1.24	1.39	1.54	1.70	2.03	2.38	2.75	3.15	3.57
3.00	0.458	0.520	0.600	0.693	0.796	0.911	1.03	1.17	1.30	1.45	1.60	1.91	2.24	2.60	2.97	3.37

Table 8-7 (cont.). Coefficients C for Eccentrically Loaded Weld Groups Angle = 60°

$\phi R_n = CC_1 Dl$ $\quad C_{min} = \dfrac{P_u}{C_1 Dl}$ $\quad D_{min} = \dfrac{P_u}{CC_1 l}$ $\quad l_{min} = \dfrac{P_u}{CC_1 D}$

where

P_u = factored force, kips
D = number of sixteenths-of-an-inch in the fillet weld size
l = characteristic length of weld group, in.
$a = e_x/l$
e_x = horizontal component of eccentricity of P_u with respect to centroid of weld group, in.
C = coefficient tabulated below, which includes $\phi = 0.75$
C_1 = electrode strength coefficient from Table 8-4 (1.0 for E70XX electrodes)

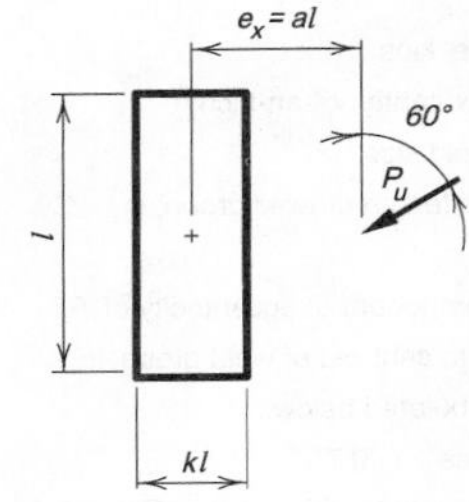

a	k: 0	0.1	0.2	0.3	0.4	0.5	0.6	0.7	0.8	0.9	1.0	1.2	1.4	1.6	1.8	2.0
0.00	3.91	4.23	4.56	4.89	5.22	5.54	5.87	6.20	6.53	6.85	7.18	7.84	8.49	9.15	9.80	10.5
0.10	3.65	3.97	4.30	4.64	4.99	5.35	5.70	6.05	6.40	6.74	7.08	7.75	8.42	9.08	9.74	10.4
0.15	3.46	3.78	4.11	4.45	4.80	5.17	5.53	5.90	6.26	6.61	6.97	7.66	8.34	9.02	9.69	10.4
0.20	3.27	3.60	3.92	4.26	4.61	4.97	5.35	5.72	6.09	6.46	6.82	7.54	8.24	8.93	9.61	10.3
0.25	3.10	3.42	3.74	4.07	4.41	4.78	5.15	5.53	5.92	6.29	6.67	7.40	8.12	8.83	9.52	10.2
0.30	2.95	3.25	3.57	3.89	4.23	4.59	4.97	5.35	5.74	6.12	6.50	7.25	7.98	8.70	9.41	10.1
0.40	2.68	2.96	3.26	3.57	3.90	4.24	4.61	4.99	5.38	5.77	6.16	6.93	7.69	8.43	9.16	9.87
0.50	2.44	2.70	2.98	3.29	3.61	3.95	4.31	4.68	5.06	5.45	5.84	6.61	7.37	8.12	8.87	9.60
0.60	2.24	2.47	2.74	3.04	3.35	3.69	4.04	4.40	4.77	5.15	5.54	6.31	7.08	7.83	8.58	9.31
0.70	2.05	2.28	2.53	2.81	3.12	3.45	3.79	4.14	4.50	4.87	5.25	6.02	6.79	7.55	8.30	9.04
0.80	1.89	2.10	2.34	2.62	2.92	3.23	3.56	3.91	4.26	4.62	4.99	5.75	6.51	7.27	8.02	8.76
0.90	1.75	1.95	2.18	2.44	2.73	3.04	3.36	3.69	4.03	4.38	4.74	5.49	6.24	6.99	7.74	8.48
1.00	1.62	1.81	2.03	2.29	2.56	2.86	3.17	3.49	3.82	4.17	4.52	5.24	5.98	6.72	7.46	8.20
1.20	1.42	1.59	1.79	2.02	2.28	2.55	2.84	3.15	3.46	3.78	4.11	4.80	5.50	6.21	6.93	7.66
1.40	1.25	1.41	1.59	1.81	2.04	2.30	2.57	2.85	3.14	3.45	3.76	4.41	5.07	5.75	6.44	7.16
1.60	1.12	1.26	1.43	1.63	1.85	2.09	2.34	2.60	2.88	3.16	3.46	4.08	4.69	5.34	6.01	6.70
1.80	1.01	1.14	1.30	1.48	1.69	1.91	2.14	2.39	2.65	2.92	3.20	3.78	4.36	4.97	5.61	6.27
2.00	0.922	1.04	1.19	1.36	1.55	1.75	1.97	2.20	2.45	2.70	2.97	3.52	4.08	4.65	5.26	5.90
2.20	0.847	0.956	1.09	1.25	1.43	1.62	1.83	2.04	2.27	2.51	2.76	3.28	3.82	4.37	4.94	5.55
2.40	0.782	0.884	1.01	1.16	1.33	1.51	1.70	1.91	2.12	2.35	2.58	3.08	3.59	4.11	4.66	5.24
2.60	0.726	0.821	0.943	1.08	1.24	1.41	1.59	1.78	1.99	2.20	2.42	2.89	3.38	3.88	4.40	4.96
2.80	0.677	0.767	0.881	1.01	1.16	1.32	1.49	1.67	1.87	2.07	2.28	2.73	3.20	3.68	4.18	4.70
3.00	0.635	0.719	0.827	0.952	1.09	1.24	1.40	1.58	1.76	1.95	2.15	2.58	3.03	3.49	3.97	4.47

Table 8-7 (cont.).
Coefficients *C* for Eccentrically Loaded Weld Groups
Angle = 75°

$\phi R_n = CC_1 Dl \qquad C_{min} = \frac{P_u}{C_1 Dl} \qquad D_{min} = \frac{P_u}{CC_1 l} \qquad l_{min} = \frac{P_u}{CC_1 D}$

where

P_u = factored force, kips
D = number of sixteenths-of-an-inch in the fillet weld size
l = characteristic length of weld group, in.
$a = e_x/l$
e_x = horizontal component of eccentricity of P_u with respect to centroid of weld group, in.
C = coefficient tabulated below, which includes $\phi = 0.75$
C_1 = electrode strength coefficient from Table 8-4 (1.0 for E70XX electrodes)

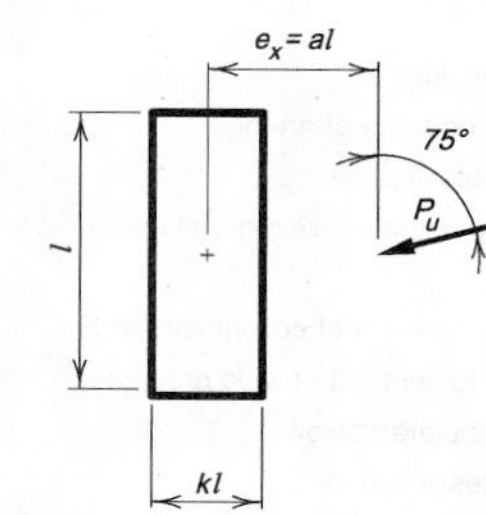

a	k															
	0	0.1	0.2	0.3	0.4	0.5	0.6	0.7	0.8	0.9	1.0	1.2	1.4	1.6	1.8	2.0
0.00	4.11	4.40	4.70	5.00	5.29	5.59	5.89	6.18	6.48	6.78	7.07	7.67	8.26	8.85	9.45	10.0
0.10	3.88	4.17	4.49	4.81	5.14	5.45	5.76	6.07	6.36	6.66	6.96	7.53	8.11	8.69	9.27	9.84
0.15	3.76	4.04	4.36	4.69	5.03	5.36	5.69	6.00	6.31	6.61	6.91	7.50	8.09	8.67	9.25	9.83
0.20	3.64	3.92	4.23	4.57	4.92	5.26	5.60	5.92	6.24	6.55	6.86	7.46	8.05	8.64	9.23	9.81
0.25	3.53	3.80	4.11	4.45	4.81	5.16	5.50	5.84	6.16	6.48	6.80	7.41	8.01	8.60	9.19	9.78
0.30	3.43	3.70	4.00	4.34	4.70	5.05	5.40	5.75	6.08	6.41	6.72	7.35	7.96	8.56	9.16	9.75
0.40	3.24	3.51	3.80	4.14	4.49	4.86	5.22	5.57	5.91	6.25	6.57	7.22	7.85	8.46	9.07	9.67
0.50	3.07	3.34	3.63	3.95	4.31	4.67	5.04	5.40	5.75	6.10	6.43	7.08	7.71	8.34	8.96	9.57
0.60	2.91	3.17	3.46	3.78	4.13	4.49	4.86	5.23	5.60	5.95	6.30	6.96	7.60	8.23	8.84	9.46
0.70	2.77	3.02	3.31	3.63	3.97	4.33	4.70	5.07	5.44	5.80	6.16	6.84	7.50	8.13	8.75	9.36
0.80	2.63	2.87	3.16	3.48	3.81	4.17	4.53	4.91	5.28	5.65	6.01	6.72	7.39	8.04	8.66	9.27
0.90	2.50	2.74	3.02	3.34	3.67	4.02	4.38	4.75	5.13	5.50	5.87	6.58	7.27	7.93	8.57	9.20
1.00	2.38	2.62	2.89	3.20	3.53	3.87	4.23	4.60	4.98	5.35	5.72	6.45	7.15	7.83	8.48	9.11
1.20	2.17	2.39	2.66	2.96	3.28	3.61	3.96	4.32	4.69	5.06	5.44	6.18	6.90	7.60	8.28	8.93
1.40	1.99	2.20	2.45	2.74	3.05	3.38	3.71	4.06	4.42	4.79	5.16	5.91	6.64	7.36	8.06	8.74
1.60	1.83	2.03	2.27	2.54	2.84	3.16	3.49	3.82	4.17	4.53	4.90	5.64	6.39	7.12	7.83	8.53
1.80	1.69	1.88	2.11	2.37	2.66	2.96	3.28	3.61	3.95	4.29	4.65	5.39	6.13	6.87	7.59	8.30
2.00	1.57	1.75	1.97	2.22	2.49	2.79	3.10	3.41	3.74	4.08	4.43	5.15	5.88	6.62	7.34	8.07
2.20	1.46	1.63	1.84	2.08	2.34	2.63	2.92	3.23	3.55	3.88	4.22	4.92	5.64	6.37	7.10	7.83
2.40	1.37	1.53	1.73	1.96	2.21	2.48	2.77	3.07	3.37	3.69	4.02	4.71	5.42	6.14	6.87	7.59
2.60	1.28	1.44	1.63	1.85	2.09	2.35	2.62	2.91	3.21	3.52	3.84	4.51	5.20	5.91	6.63	7.36
2.80	1.21	1.36	1.54	1.75	1.98	2.23	2.50	2.78	3.07	3.36	3.67	4.32	5.00	5.70	6.41	7.13
3.00	1.14	1.28	1.46	1.66	1.88	2.12	2.38	2.65	2.92	3.22	3.52	4.14	4.80	5.49	6.19	6.91

Table 8-8.
Coefficients *C* for Eccentrically Loaded Weld Groups
Angle = 0°

$\phi R_n = CC_1 Dl \qquad C_{min} = \dfrac{P_u}{C_1 Dl} \qquad D_{min} = \dfrac{P_u}{CC_1 l} \qquad l_{min} = \dfrac{P_u}{CC_1 D}$

where

P_u = factored force, kips

D = number of sixteenths-of-an-inch in the fillet weld size

l = characteristic length of weld group, in.

$a = e_x/l$

e_x = horizontal component of eccentricity of P_u with respect to centroid of weld group, in.

C = coefficient tabulated below, which includes $\phi = 0.75$

C_1 = electrode strength coefficient from Table 8-4 (1.0 for E70XX electrodes)

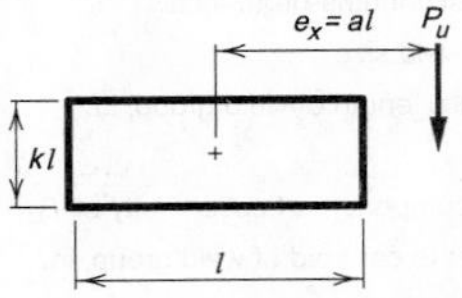

a	*k*															
	0	**0.1**	**0.2**	**0.3**	**0.4**	**0.5**	**0.6**	**0.7**	**0.8**	**0.9**	**1.0**	**1.2**	**1.4**	**1.6**	**1.8**	**2.0**
0.00	4.18	4.45	4.73	5.01	5.29	5.57	5.85	6.12	6.40	6.68	6.96	7.52	8.07	8.63	9.19	9.74
0.10	3.24	3.51	3.81	4.15	4.51	4.87	5.22	5.55	5.87	6.17	6.47	7.03	7.56	8.08	8.59	9.09
0.15	2.92	3.18	3.49	3.81	4.16	4.53	4.89	5.25	5.60	5.94	6.27	6.89	7.47	8.03	8.58	9.11
0.20	2.65	2.89	3.19	3.52	3.85	4.21	4.58	4.95	5.31	5.67	6.02	6.69	7.32	7.92	8.50	9.06
0.25	2.41	2.64	2.93	3.25	3.58	3.92	4.28	4.65	5.02	5.39	5.75	6.45	7.12	7.76	8.37	8.96
0.30	2.20	2.43	2.70	3.00	3.33	3.66	4.01	4.37	4.74	5.11	5.48	6.20	6.90	7.57	8.21	8.83
0.40	1.86	2.07	2.31	2.59	2.90	3.22	3.54	3.88	4.23	4.60	4.96	5.70	6.43	7.14	7.83	8.49
0.50	1.60	1.79	2.01	2.27	2.55	2.85	3.16	3.48	3.80	4.15	4.50	5.22	5.95	6.68	7.39	8.09
0.60	1.40	1.57	1.77	2.01	2.26	2.54	2.83	3.13	3.44	3.76	4.09	4.79	5.50	6.22	6.95	7.66
0.70	1.24	1.39	1.58	1.79	2.03	2.28	2.56	2.84	3.13	3.44	3.75	4.40	5.09	5.80	6.51	7.23
0.80	1.11	1.25	1.42	1.62	1.84	2.07	2.32	2.59	2.87	3.15	3.45	4.07	4.72	5.41	6.11	6.82
0.90	1.00	1.13	1.29	1.48	1.68	1.89	2.13	2.38	2.64	2.91	3.18	3.77	4.39	5.05	5.73	6.42
1.00	0.914	1.03	1.18	1.35	1.54	1.74	1.96	2.19	2.44	2.69	2.96	3.51	4.10	4.73	5.38	6.06
1.20	0.777	0.880	1.01	1.16	1.32	1.50	1.69	1.89	2.11	2.34	2.57	3.08	3.61	4.18	4.78	5.40
1.40	0.674	0.764	0.879	1.01	1.16	1.31	1.48	1.67	1.86	2.06	2.27	2.73	3.21	3.73	4.28	4.86
1.60	0.594	0.675	0.777	0.895	1.03	1.17	1.32	1.48	1.66	1.84	2.03	2.44	2.89	3.36	3.87	4.39
1.80	0.531	0.604	0.696	0.803	0.921	1.05	1.19	1.34	1.49	1.66	1.84	2.21	2.62	3.06	3.52	4.00
2.00	0.481	0.547	0.630	0.728	0.836	0.953	1.08	1.22	1.36	1.51	1.67	2.02	2.39	2.80	3.23	3.67
2.20	0.439	0.499	0.576	0.665	0.765	0.873	0.990	1.11	1.25	1.39	1.54	1.86	2.20	2.58	2.97	3.39
2.40	0.404	0.459	0.530	0.612	0.705	0.805	0.913	1.03	1.15	1.28	1.42	1.72	2.04	2.39	2.76	3.14
2.60	0.374	0.425	0.491	0.567	0.653	0.747	0.847	0.955	1.07	1.19	1.32	1.60	1.90	2.22	2.57	2.93
2.80	0.348	0.396	0.457	0.529	0.608	0.696	0.790	0.890	1.00	1.11	1.23	1.49	1.77	2.08	2.40	2.74
3.00	0.325	0.370	0.428	0.495	0.569	0.651	0.740	0.834	0.935	1.04	1.16	1.40	1.66	1.95	2.25	2.57

Table 8-8 (cont.).
Coefficients *C* for Eccentrically Loaded Weld Groups
Angle = 15°

$\phi R_n = CC_1 Dl \qquad C_{min} = \frac{P_u}{C_1 Dl} \qquad D_{min} = \frac{P_u}{CC_1 l} \qquad l_{min} = \frac{P_u}{CC_1 D}$

where

P_u = factored force, kips
D = number of sixteenths-of-an-inch in the fillet weld size
l = characteristic length of weld group, in.
$a = e_x/l$
e_x = horizontal component of eccentricity of P_u with respect to centroid of weld group, in.
C = coefficient tabulated below, which includes $\phi = 0.75$
C_1 = electrode strength coefficient from Table 8-4 (1.0 for E70XX electrodes)

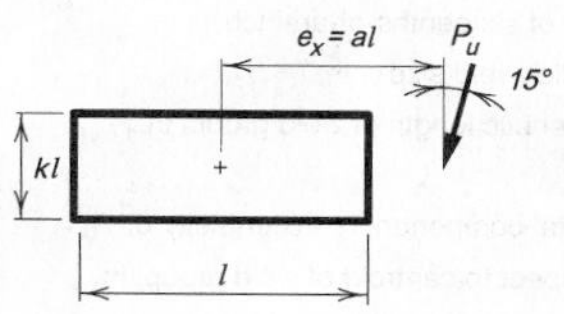

a	k: 0	0.1	0.2	0.3	0.4	0.5	0.6	0.7	0.8	0.9	1.0	1.2	1.4	1.6	1.8	2.0
0.00	4.11	4.40	4.70	5.00	5.29	5.59	5.89	6.18	6.48	6.78	7.07	7.67	8.26	8.85	9.45	10.0
0.10	3.29	3.56	3.85	4.19	4.55	4.91	5.26	5.61	5.95	6.29	6.62	7.26	7.88	8.49	9.09	9.69
0.15	2.97	3.24	3.53	3.85	4.20	4.56	4.93	5.30	5.66	6.01	6.35	7.01	7.65	8.27	8.89	9.51
0.20	2.70	2.95	3.24	3.56	3.89	4.25	4.62	4.99	5.37	5.73	6.09	6.78	7.45	8.08	8.71	9.31
0.25	2.46	2.70	2.98	3.29	3.63	3.97	4.33	4.70	5.08	5.45	5.82	6.54	7.23	7.90	8.54	9.17
0.30	2.25	2.48	2.75	3.05	3.38	3.71	4.06	4.43	4.80	5.17	5.55	6.29	7.01	7.70	8.36	9.01
0.40	1.91	2.12	2.36	2.65	2.95	3.27	3.61	3.95	4.30	4.67	5.04	5.78	6.52	7.25	7.95	8.64
0.50	1.65	1.83	2.06	2.32	2.60	2.90	3.22	3.54	3.88	4.22	4.58	5.31	6.04	6.78	7.51	8.23
0.60	1.44	1.61	1.82	2.06	2.32	2.60	2.89	3.20	3.52	3.84	4.18	4.88	5.60	6.33	7.06	7.78
0.70	1.28	1.43	1.62	1.84	2.08	2.34	2.62	2.91	3.21	3.51	3.83	4.49	5.19	5.90	6.62	7.34
0.80	1.15	1.29	1.46	1.67	1.89	2.13	2.38	2.65	2.93	3.23	3.53	4.15	4.82	5.50	6.21	6.92
0.90	1.04	1.17	1.33	1.52	1.72	1.95	2.19	2.44	2.70	2.98	3.26	3.86	4.49	5.15	5.83	6.52
1.00	0.945	1.07	1.22	1.39	1.59	1.79	2.02	2.25	2.50	2.76	3.03	3.60	4.19	4.83	5.48	6.15
1.20	0.803	0.908	1.04	1.19	1.36	1.54	1.74	1.95	2.17	2.40	2.65	3.16	3.70	4.27	4.86	5.48
1.40	0.697	0.789	0.907	1.04	1.19	1.35	1.53	1.72	1.91	2.12	2.34	2.80	3.30	3.82	4.35	4.92
1.60	0.615	0.698	0.803	0.924	1.06	1.21	1.36	1.53	1.71	1.90	2.09	2.51	2.97	3.45	3.93	4.45
1.80	0.550	0.625	0.719	0.829	0.951	1.08	1.23	1.38	1.54	1.71	1.89	2.28	2.69	3.13	3.58	4.06
2.00	0.497	0.565	0.651	0.752	0.863	0.984	1.12	1.26	1.40	1.56	1.73	2.08	2.47	2.87	3.29	3.73
2.20	0.454	0.516	0.595	0.687	0.790	0.902	1.02	1.15	1.29	1.43	1.59	1.91	2.27	2.65	3.04	3.45
2.40	0.418	0.475	0.548	0.633	0.728	0.832	0.943	1.06	1.19	1.32	1.47	1.77	2.10	2.45	2.82	3.20
2.60	0.387	0.440	0.508	0.586	0.675	0.772	0.875	0.987	1.11	1.23	1.36	1.65	1.96	2.28	2.63	2.99
2.80	0.360	0.409	0.473	0.546	0.629	0.719	0.817	0.921	1.03	1.15	1.27	1.54	1.83	2.14	2.46	2.80
3.00	0.336	0.383	0.442	0.511	0.589	0.674	0.765	0.863	0.967	1.08	1.19	1.44	1.72	2.01	2.31	2.63

Table 8-8 (cont.). Coefficients *C* for Eccentrically Loaded Weld Groups Angle = 30°

$\phi R_n = CC_1 Dl \qquad C_{min} = \dfrac{P_u}{C_1 Dl} \qquad D_{min} = \dfrac{P_u}{CC_1 l} \qquad l_{min} = \dfrac{P_u}{CC_1 D}$

where

P_u = factored force, kips
D = number of sixteenths-of-an-inch in the fillet weld size
l = characteristic length of weld group, in.
$a = e_x/l$
e_x = horizontal component of eccentricity of P_u with respect to centroid of weld group, in.
C = coefficient tabulated below, which includes $\phi = 0.75$
C_1 = electrode strength coefficient from Table 8-4 (1.0 for E70XX electrodes)

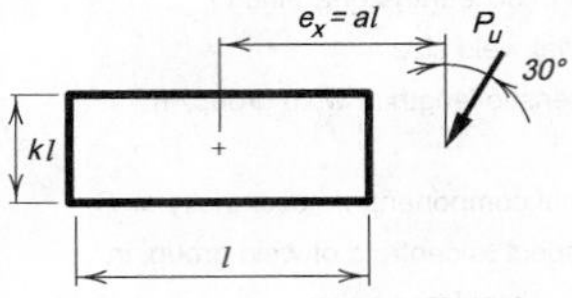

a	k: 0	0.1	0.2	0.3	0.4	0.5	0.6	0.7	0.8	0.9	1.0	1.2	1.4	1.6	1.8	2.0
0.00	3.91	4.23	4.56	4.89	5.22	5.54	5.87	6.20	6.53	6.85	7.18	7.84	8.49	9.15	9.80	10.5
0.10	3.37	3.70	4.02	4.36	4.71	5.08	5.45	5.81	6.18	6.55	6.90	7.61	8.30	8.98	9.65	10.3
0.15	3.07	3.38	3.71	4.03	4.38	4.74	5.12	5.50	5.88	6.26	6.64	7.37	8.10	8.80	9.50	10.2
0.20	2.82	3.11	3.42	3.74	4.07	4.42	4.80	5.18	5.57	5.96	6.34	7.10	7.85	8.58	9.30	10.0
0.25	2.60	2.87	3.16	3.47	3.80	4.14	4.51	4.88	5.27	5.66	6.05	6.82	7.58	8.33	9.06	9.79
0.30	2.40	2.65	2.93	3.24	3.56	3.89	4.25	4.62	5.00	5.39	5.77	6.55	7.31	8.06	8.81	9.55
0.40	2.06	2.29	2.55	2.83	3.14	3.46	3.81	4.16	4.52	4.89	5.27	6.04	6.81	7.57	8.32	9.06
0.50	1.80	2.00	2.23	2.50	2.79	3.10	3.42	3.76	4.11	4.46	4.83	5.57	6.33	7.09	7.84	8.58
0.60	1.58	1.76	1.98	2.23	2.50	2.79	3.10	3.42	3.75	4.09	4.43	5.15	5.88	6.61	7.35	8.09
0.70	1.41	1.57	1.77	2.01	2.26	2.54	2.82	3.13	3.44	3.76	4.09	4.77	5.47	6.18	6.90	7.63
0.80	1.26	1.42	1.60	1.82	2.06	2.31	2.59	2.87	3.17	3.47	3.79	4.44	5.10	5.78	6.48	7.19
0.90	1.15	1.29	1.46	1.66	1.89	2.13	2.38	2.65	2.93	3.22	3.52	4.14	4.77	5.42	6.09	6.79
1.00	1.05	1.18	1.34	1.53	1.74	1.96	2.20	2.46	2.72	3.00	3.28	3.87	4.47	5.09	5.74	6.41
1.20	0.891	1.01	1.15	1.32	1.50	1.70	1.91	2.14	2.38	2.62	2.88	3.42	3.97	4.54	5.13	5.76
1.40	0.774	0.875	1.00	1.15	1.32	1.49	1.69	1.89	2.10	2.33	2.56	3.05	3.56	4.08	4.63	5.20
1.60	0.684	0.774	0.889	1.02	1.17	1.33	1.51	1.69	1.88	2.09	2.30	2.75	3.22	3.70	4.21	4.74
1.80	0.612	0.694	0.798	0.919	1.05	1.20	1.36	1.53	1.70	1.89	2.08	2.50	2.94	3.38	3.86	4.35
2.00	0.554	0.629	0.723	0.834	0.957	1.09	1.24	1.39	1.55	1.73	1.91	2.29	2.69	3.11	3.55	4.01
2.20	0.506	0.574	0.662	0.763	0.877	1.00	1.14	1.28	1.43	1.59	1.75	2.11	2.48	2.87	3.29	3.72
2.40	0.465	0.528	0.609	0.703	0.809	0.924	1.05	1.18	1.32	1.47	1.62	1.95	2.30	2.67	3.06	3.47
2.60	0.431	0.489	0.565	0.652	0.750	0.857	0.972	1.10	1.23	1.36	1.51	1.82	2.15	2.49	2.86	3.24
2.80	0.401	0.456	0.526	0.608	0.699	0.800	0.908	1.02	1.15	1.27	1.41	1.70	2.01	2.34	2.68	3.05
3.00	0.375	0.426	0.492	0.569	0.654	0.749	0.851	0.959	1.07	1.20	1.32	1.60	1.89	2.20	2.52	2.87

Table 8-8 (cont.).
Coefficients *C* for Eccentrically Loaded Weld Groups
Angle = 45°

$\phi R_n = CC_1 Dl \qquad C_{min} = \dfrac{P_u}{C_1 Dl} \qquad D_{min} = \dfrac{P_u}{CC_1 l} \qquad l_{min} = \dfrac{P_u}{CC_1 D}$

where

P_u = factored force, kips
D = number of sixteenths-of-an-inch in the fillet weld size
l = characteristic length of weld group, in.
$a = e_x/l$
e_x = horizontal component of eccentricity of P_u with respect to centroid of weld group, in.
C = coefficient tabulated below, which includes $\phi = 0.75$
C_1 = electrode strength coefficient from Table 8-4 (1.0 for E70XX electrodes)

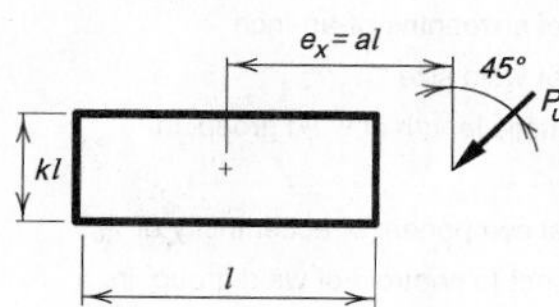

a	k = 0	0.1	0.2	0.3	0.4	0.5	0.6	0.7	0.8	0.9	1.0	1.2	1.4	1.6	1.8	2.0
0.00	3.61	3.97	4.33	4.70	5.06	5.42	5.78	6.14	6.50	6.86	7.22	7.95	8.67	9.39	10.1	10.8
0.10	3.37	3.74	4.11	4.48	4.84	5.21	5.58	5.94	6.31	6.68	7.04	7.78	8.52	9.25	9.98	10.7
0.15	3.13	3.52	3.89	4.26	4.62	4.99	5.36	5.74	6.11	6.49	6.86	7.61	8.36	9.11	9.85	10.6
0.20	2.94	3.29	3.65	4.02	4.38	4.75	5.12	5.50	5.88	6.26	6.64	7.39	8.15	8.91	9.67	10.4
0.25	2.77	3.10	3.44	3.79	4.14	4.51	4.87	5.25	5.63	6.01	6.40	7.16	7.92	8.68	9.45	10.2
0.30	2.61	2.91	3.24	3.57	3.92	4.27	4.63	5.00	5.38	5.76	6.15	6.91	7.67	8.44	9.21	9.98
0.40	2.32	2.59	2.87	3.18	3.51	3.85	4.20	4.55	4.92	5.29	5.67	6.42	7.18	7.95	8.72	9.50
0.50	2.06	2.30	2.57	2.86	3.16	3.49	3.82	4.17	4.52	4.89	5.26	5.97	6.71	7.47	8.24	9.01
0.60	1.84	2.06	2.30	2.58	2.87	3.17	3.50	3.83	4.18	4.53	4.89	5.57	6.28	7.02	7.78	8.54
0.70	1.66	1.86	2.08	2.34	2.61	2.91	3.22	3.54	3.88	4.22	4.56	5.22	5.90	6.62	7.35	8.10
0.80	1.50	1.69	1.90	2.13	2.40	2.68	2.98	3.29	3.61	3.93	4.27	4.91	5.56	6.25	6.97	7.70
0.90	1.37	1.54	1.74	1.96	2.21	2.48	2.76	3.06	3.37	3.68	4.00	4.62	5.25	5.91	6.61	7.32
1.00	1.26	1.41	1.60	1.81	2.05	2.30	2.58	2.86	3.15	3.45	3.76	4.35	4.97	5.61	6.28	6.97
1.20	1.08	1.21	1.38	1.57	1.78	2.01	2.26	2.52	2.78	3.06	3.34	3.90	4.48	5.07	5.70	6.35
1.40	0.938	1.06	1.21	1.38	1.57	1.78	2.01	2.24	2.49	2.74	3.00	3.52	4.06	4.62	5.21	5.81
1.60	0.831	0.939	1.07	1.23	1.41	1.60	1.80	2.02	2.24	2.48	2.71	3.20	3.70	4.23	4.78	5.36
1.80	0.745	0.843	0.966	1.11	1.27	1.45	1.63	1.83	2.04	2.25	2.48	2.93	3.40	3.90	4.42	4.96
2.00	0.675	0.764	0.877	1.01	1.16	1.32	1.49	1.67	1.87	2.06	2.27	2.69	3.14	3.60	4.09	4.61
2.20	0.617	0.699	0.804	0.925	1.06	1.21	1.37	1.54	1.72	1.90	2.10	2.49	2.91	3.35	3.81	4.30
2.40	0.568	0.644	0.741	0.854	0.981	1.12	1.27	1.43	1.59	1.77	1.95	2.32	2.71	3.13	3.57	4.03
2.60	0.526	0.597	0.687	0.792	0.911	1.04	1.18	1.33	1.48	1.65	1.82	2.16	2.54	2.93	3.35	3.78
2.80	0.489	0.556	0.641	0.739	0.850	0.972	1.10	1.24	1.39	1.54	1.70	2.03	2.38	2.75	3.15	3.57
3.00	0.458	0.520	0.600	0.693	0.796	0.911	1.03	1.17	1.30	1.45	1.60	1.91	2.24	2.60	2.97	3.37

Table 8-8 (cont.). Coefficients C for Eccentrically Loaded Weld Groups Angle = 60°

$\phi R_n = CC_1 Dl$ $\quad C_{min} = \dfrac{P_u}{C_1 Dl}$ $\quad D_{min} = \dfrac{P_u}{CC_1 l}$ $\quad l_{min} = \dfrac{P_u}{CC_1 D}$

where

P_u = factored force, kips

D = number of sixteenths-of-an-inch in the fillet weld size

l = characteristic length of weld group, in.

$a = e_x/l$

e_x = horizontal component of eccentricity of P_u with respect to centroid of weld group, in.

C = coefficient tabulated below, which includes $\phi = 0.75$

C_1 = electrode strength coefficient from Table 8-4 (1.0 for E70XX electrodes)

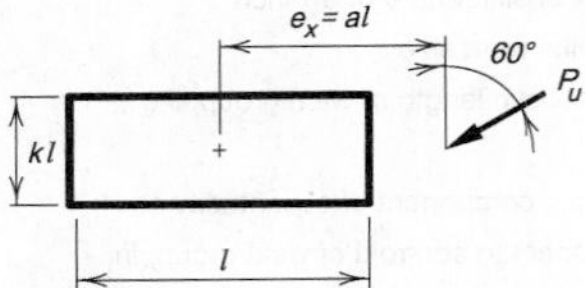

a	k: 0	0.1	0.2	0.3	0.4	0.5	0.6	0.7	0.8	0.9	1.0	1.2	1.4	1.6	1.8	2.0
0.00	3.28	3.67	4.06	4.45	4.84	5.23	5.62	6.01	6.40	6.79	7.18	7.96	8.74	9.53	10.3	11.1
0.10	3.20	3.59	3.98	4.37	4.76	5.14	5.53	5.92	6.30	6.69	7.08	7.86	8.63	9.41	10.2	11.0
0.15	3.10	3.50	3.90	4.29	4.67	5.05	5.43	5.82	6.20	6.58	6.96	7.74	8.51	9.29	10.1	10.9
0.20	2.98	3.38	3.79	4.17	4.56	4.94	5.31	5.69	6.07	6.45	6.83	7.59	8.36	9.13	9.91	10.7
0.25	2.89	3.27	3.66	4.04	4.43	4.80	5.17	5.55	5.92	6.29	6.67	7.43	8.19	8.96	9.74	10.5
0.30	2.80	3.16	3.54	3.92	4.29	4.66	5.02	5.39	5.76	6.13	6.50	7.25	8.01	8.77	9.55	10.3
0.40	2.63	2.96	3.30	3.66	4.02	4.38	4.74	5.09	5.44	5.80	6.16	6.89	7.64	8.40	9.17	9.94
0.50	2.44	2.74	3.06	3.40	3.75	4.10	4.46	4.80	5.14	5.49	5.84	6.54	7.28	8.02	8.78	9.56
0.60	2.27	2.54	2.84	3.16	3.49	3.83	4.18	4.53	4.86	5.20	5.54	6.23	6.93	7.66	8.42	9.18
0.70	2.09	2.35	2.63	2.93	3.25	3.58	3.92	4.26	4.59	4.92	5.25	5.93	6.63	7.34	8.07	8.82
0.80	1.94	2.18	2.44	2.73	3.03	3.35	3.68	4.01	4.34	4.66	4.99	5.65	6.34	7.04	7.75	8.49
0.90	1.80	2.02	2.27	2.54	2.83	3.14	3.46	3.78	4.11	4.43	4.74	5.40	6.07	6.76	7.46	8.18
1.00	1.67	1.88	2.11	2.37	2.65	2.95	3.26	3.58	3.90	4.21	4.52	5.16	5.82	6.50	7.19	7.90
1.20	1.46	1.64	1.85	2.09	2.35	2.62	2.91	3.21	3.52	3.82	4.11	4.73	5.36	6.02	6.69	7.38
1.40	1.28	1.45	1.64	1.86	2.10	2.35	2.63	2.91	3.20	3.48	3.76	4.35	4.96	5.58	6.23	6.89
1.60	1.15	1.30	1.47	1.67	1.89	2.13	2.39	2.66	2.93	3.19	3.46	4.01	4.60	5.20	5.82	6.47
1.80	1.03	1.17	1.33	1.51	1.72	1.95	2.19	2.44	2.69	2.94	3.20	3.72	4.28	4.85	5.45	6.07
2.00	0.940	1.06	1.21	1.39	1.58	1.79	2.01	2.25	2.49	2.72	2.97	3.47	3.99	4.55	5.12	5.72
2.20	0.861	0.974	1.11	1.27	1.46	1.65	1.86	2.09	2.31	2.53	2.76	3.24	3.74	4.27	4.82	5.40
2.40	0.794	0.899	1.03	1.18	1.35	1.54	1.73	1.94	2.15	2.36	2.58	3.03	3.52	4.02	4.55	5.11
2.60	0.736	0.834	0.956	1.10	1.26	1.43	1.62	1.82	2.02	2.22	2.42	2.86	3.32	3.80	4.31	4.84
2.80	0.686	0.778	0.893	1.03	1.18	1.34	1.52	1.70	1.89	2.08	2.28	2.69	3.13	3.60	4.08	4.59
3.00	0.643	0.729	0.837	0.964	1.11	1.26	1.43	1.61	1.79	1.97	2.15	2.55	2.97	3.41	3.88	4.37

Table 8-8 (cont.). Coefficients C for Eccentrically Loaded Weld Groups Angle = 75°

$\phi R_n = CC_1 Dl$ $\quad C_{min} = \frac{P_u}{C_1 Dl}$ $\quad D_{min} = \frac{P_u}{CC_1 l}$ $\quad l_{min} = \frac{P_u}{CC_1 D}$

where

P_u = factored force, kips

D = number of sixteenths-of-an-inch in the fillet weld size

l = characteristic length of weld group, in.

$a = e_x/l$

e_x = horizontal component of eccentricity of P_u with respect to centroid of weld group, in.

C = coefficient tabulated below, which includes $\phi = 0.75$

C_1 = electrode strength coefficient from Table 8-4 (1.0 for E70XX electrodes)

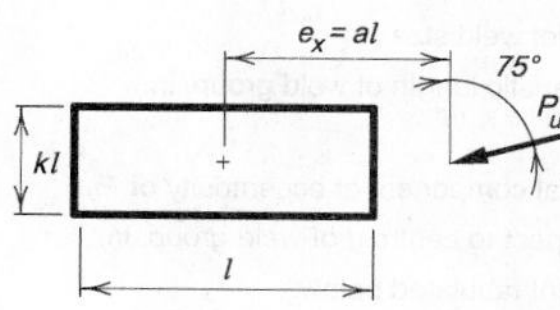

a	k = 0	0.1	0.2	0.3	0.4	0.5	0.6	0.7	0.8	0.9	1.0	1.2	1.4	1.6	1.8	2.0
0.00	2.97	3.38	3.79	4.20	4.61	5.02	5.43	5.84	6.25	6.66	7.07	7.89	8.71	9.54	10.4	11.2
0.10	2.86	3.27	3.67	4.08	4.49	4.90	5.31	5.72	6.14	6.55	6.96	7.78	8.60	9.42	10.2	11.1
0.15	2.89	3.24	3.65	4.06	4.47	4.87	5.28	5.69	6.10	6.50	6.91	7.72	8.54	9.36	10.2	11.0
0.20	2.88	3.19	3.61	4.02	4.43	4.83	5.24	5.64	6.05	6.45	6.86	7.67	8.48	9.29	10.1	10.9
0.25	2.87	3.17	3.57	3.98	4.38	4.79	5.19	5.59	5.99	6.39	6.80	7.60	8.40	9.21	10.0	10.8
0.30	2.86	3.17	3.54	3.93	4.33	4.73	5.13	5.53	5.93	6.33	6.73	7.52	8.32	9.12	9.92	10.7
0.40	2.84	3.16	3.51	3.88	4.26	4.64	5.02	5.41	5.79	6.18	6.57	7.36	8.15	8.94	9.73	10.5
0.50	2.79	3.13	3.47	3.83	4.19	4.56	4.93	5.30	5.68	6.05	6.43	7.20	7.96	8.75	9.53	10.3
0.60	2.74	3.07	3.42	3.76	4.11	4.47	4.83	5.19	5.56	5.93	6.30	7.04	7.80	8.56	9.33	10.1
0.70	2.66	3.00	3.34	3.68	4.02	4.37	4.72	5.07	5.43	5.79	6.16	6.89	7.63	8.38	9.14	9.90
0.80	2.59	2.91	3.25	3.59	3.93	4.26	4.61	4.95	5.30	5.66	6.01	6.74	7.47	8.21	8.96	9.71
0.90	2.51	2.82	3.15	3.49	3.82	4.15	4.49	4.83	5.17	5.52	5.87	6.58	7.30	8.04	8.78	9.52
1.00	2.42	2.73	3.05	3.38	3.71	4.03	4.37	4.70	5.03	5.38	5.72	6.42	7.14	7.86	8.60	9.34
1.20	2.25	2.53	2.84	3.15	3.48	3.80	4.12	4.44	4.77	5.10	5.44	6.12	6.82	7.53	8.26	8.99
1.40	2.08	2.35	2.63	2.94	3.25	3.57	3.88	4.19	4.51	4.83	5.16	5.83	6.52	7.22	7.93	8.65
1.60	1.92	2.17	2.44	2.73	3.04	3.35	3.65	3.95	4.26	4.58	4.90	5.55	6.23	6.92	7.62	8.33
1.80	1.78	2.01	2.26	2.54	2.84	3.14	3.43	3.73	4.04	4.34	4.65	5.30	5.96	6.63	7.33	8.03
2.00	1.65	1.87	2.11	2.37	2.65	2.95	3.24	3.53	3.82	4.12	4.43	5.05	5.70	6.37	7.05	7.75
2.20	1.54	1.74	1.97	2.22	2.49	2.77	3.06	3.34	3.63	3.92	4.22	4.83	5.47	6.12	6.79	7.48
2.40	1.44	1.63	1.84	2.08	2.34	2.61	2.89	3.17	3.44	3.73	4.02	4.62	5.24	5.89	6.55	7.23
2.60	1.35	1.52	1.73	1.95	2.20	2.47	2.74	3.01	3.28	3.56	3.84	4.43	5.04	5.67	6.32	6.98
2.80	1.26	1.43	1.63	1.84	2.08	2.34	2.60	2.86	3.13	3.40	3.67	4.25	4.84	5.46	6.10	6.75
3.00	1.19	1.35	1.53	1.74	1.97	2.22	2.48	2.73	2.99	3.25	3.52	4.07	4.66	5.26	5.89	6.53

Table 8-9.
Coefficients *C* for Eccentrically Loaded Weld Groups
Angle = 0°

$\phi R_n = CC_1 Dl \qquad C_{min} = \dfrac{P_u}{C_1 Dl} \qquad D_{min} = \dfrac{P_u}{CC_1 l} \qquad l_{min} = \dfrac{P_u}{CC_1 D}$

where

P_u = factored force, kips
D = number of sixteenths-of-an-inch in the fillet weld size
l = characteristic length of weld group, in.
$a = e_x/l$
e_x = horizontal component of eccentricity of P_u with respect to centroid of weld group, in.
C = coefficient tabulated below, which includes $\phi = 0.75$
C_1 = electrode strength coefficient from Table 8-4 (1.0 for E70XX electrodes)

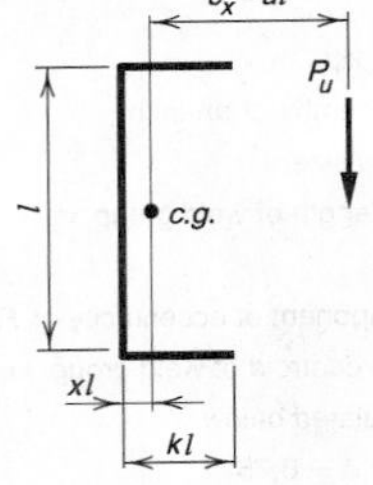

a	k: 0	0.1	0.2	0.3	0.4	0.5	0.6	0.7	0.8	0.9	1.0	1.2	1.4	1.6	1.8	2.0
0.00	1.39	1.81	2.28	2.65	3.06	3.48	3.90	4.32	4.73	5.15	5.57	6.40	7.24	8.07	8.91	9.74
0.10	1.40	1.71	2.09	2.48	2.88	3.28	3.69	4.10	4.51	4.92	5.33	6.16	6.99	7.82	8.65	9.48
0.15	1.37	1.69	2.05	2.43	2.81	3.20	3.60	4.00	4.40	4.80	5.21	6.02	6.84	7.65	8.47	9.28
0.20	1.32	1.63	1.98	2.33	2.70	3.08	3.46	3.84	4.23	4.62	5.01	5.80	6.58	7.38	8.17	8.97
0.25	1.24	1.56	1.88	2.22	2.57	2.93	3.29	3.65	4.03	4.40	4.77	5.53	6.30	7.07	7.84	8.62
0.30	1.16	1.46	1.77	2.09	2.42	2.76	3.10	3.45	3.81	4.16	4.53	5.26	6.00	6.75	7.51	8.27
0.40	1.00	1.27	1.55	1.84	2.13	2.43	2.74	3.06	3.38	3.71	4.04	4.73	5.43	6.14	6.87	7.61
0.50	0.860	1.09	1.35	1.61	1.87	2.14	2.41	2.70	3.00	3.30	3.61	4.25	4.92	5.60	6.30	7.01
0.60	0.748	0.952	1.17	1.41	1.65	1.89	2.14	2.40	2.67	2.95	3.24	3.84	4.47	5.13	5.80	6.49
0.70	0.659	0.838	1.04	1.25	1.46	1.68	1.91	2.15	2.40	2.66	2.93	3.50	4.09	4.72	5.36	6.03
0.80	0.586	0.746	0.922	1.11	1.31	1.51	1.72	1.94	2.17	2.42	2.67	3.20	3.77	4.36	4.98	5.62
0.90	0.527	0.671	0.829	1.00	1.18	1.37	1.56	1.77	1.98	2.21	2.44	2.95	3.48	4.05	4.64	5.25
1.00	0.478	0.609	0.752	0.909	1.08	1.25	1.43	1.62	1.82	2.03	2.25	2.73	3.23	3.77	4.33	4.92
1.20	0.403	0.512	0.633	0.766	0.910	1.06	1.22	1.39	1.56	1.75	1.94	2.36	2.81	3.29	3.80	4.34
1.40	0.348	0.441	0.546	0.661	0.787	0.922	1.06	1.21	1.36	1.53	1.70	2.08	2.48	2.92	3.38	3.86
1.60	0.305	0.387	0.479	0.581	0.692	0.813	0.938	1.07	1.21	1.36	1.51	1.85	2.21	2.61	3.03	3.48
1.80	0.272	0.345	0.427	0.518	0.618	0.727	0.840	0.958	1.09	1.22	1.36	1.66	1.99	2.35	2.74	3.15
2.00	0.245	0.311	0.385	0.467	0.558	0.657	0.760	0.868	0.983	1.10	1.23	1.51	1.81	2.14	2.50	2.88
2.20	0.223	0.283	0.350	0.425	0.508	0.599	0.694	0.793	0.897	1.01	1.13	1.38	1.66	1.97	2.30	2.65
2.40	0.205	0.260	0.321	0.390	0.467	0.551	0.639	0.729	0.826	0.929	1.04	1.27	1.53	1.82	2.12	2.46
2.60	0.189	0.240	0.297	0.360	0.431	0.509	0.591	0.675	0.765	0.860	0.961	1.18	1.42	1.69	1.98	2.29
2.80	0.176	0.223	0.276	0.335	0.401	0.474	0.550	0.628	0.712	0.801	0.895	1.10	1.33	1.57	1.85	2.14
3.00	0.164	0.208	0.257	0.313	0.375	0.443	0.514	0.588	0.666	0.749	0.838	1.03	1.24	1.48	1.73	2.01
x	0.000	0.008	0.029	0.056	0.089	0.125	0.164	0.204	0.246	0.289	0.333	0.424	0.516	0.610	0.704	0.800

Table 8-9 (cont.). Coefficients *C* for Eccentrically Loaded Weld Groups Angle = 15°

$\phi R_n = CC_1 Dl$ $\quad C_{min} = \dfrac{P_u}{C_1 Dl}$ $\quad D_{min} = \dfrac{P_u}{CC_1 l}$ $\quad l_{min} = \dfrac{P_u}{CC_1 D}$

where

P_u = factored force, kips

D = number of sixteenths-of-an-inch in the fillet weld size

l = characteristic length of weld group, in.

$a = e_x/l$

e_x = horizontal component of eccentricity of P_u with respect to centroid of weld group, in.

C = coefficient tabulated below, which includes $\phi = 0.75$

C_1 = electrode strength coefficient from Table 8-4 (1.0 for E70XX electrodes)

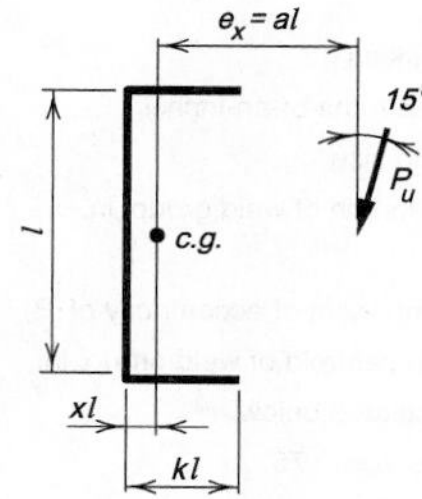

a	*k* = 0	0.1	0.2	0.3	0.4	0.5	0.6	0.7	0.8	0.9	1.0	1.2	1.4	1.6	1.8	2.0
0.00	1.48	1.89	2.31	2.72	3.13	3.54	3.95	4.36	4.77	5.18	5.59	6.41	7.23	8.05	8.87	9.69
0.10	1.42	1.77	2.15	2.55	2.96	3.38	3.79	4.20	4.61	5.02	5.43	6.25	7.08	7.90	8.73	9.55
0.15	1.38	1.73	2.09	2.47	2.86	3.25	3.64	4.04	4.44	4.84	5.24	6.05	6.85	7.66	8.46	9.28
0.20	1.32	1.66	2.01	2.37	2.74	3.11	3.49	3.87	4.25	4.64	5.02	5.80	6.57	7.36	8.15	8.94
0.25	1.24	1.56	1.91	2.25	2.60	2.96	3.32	3.68	4.05	4.41	4.79	5.53	6.29	7.06	7.82	8.60
0.30	1.16	1.46	1.79	2.12	2.45	2.79	3.13	3.48	3.83	4.19	4.54	5.27	6.01	6.75	7.51	8.27
0.40	1.00	1.27	1.56	1.85	2.16	2.46	2.77	3.09	3.41	3.74	4.07	4.76	5.46	6.17	6.90	7.63
0.50	0.869	1.10	1.35	1.62	1.89	2.17	2.45	2.74	3.04	3.34	3.65	4.30	4.96	5.64	6.34	7.06
0.60	0.759	0.961	1.18	1.42	1.67	1.92	2.18	2.44	2.72	3.00	3.29	3.90	4.53	5.18	5.86	6.55
0.70	0.670	0.849	1.05	1.26	1.48	1.72	1.95	2.20	2.45	2.71	2.98	3.55	4.15	4.78	5.43	6.10
0.80	0.598	0.758	0.934	1.12	1.33	1.54	1.76	1.99	2.22	2.47	2.72	3.26	3.83	4.43	5.06	5.70
0.90	0.539	0.683	0.842	1.02	1.20	1.40	1.60	1.81	2.03	2.26	2.50	3.01	3.55	4.12	4.71	5.33
1.00	0.490	0.621	0.766	0.924	1.10	1.28	1.47	1.66	1.87	2.08	2.31	2.79	3.30	3.84	4.41	5.00
1.20	0.413	0.524	0.646	0.781	0.928	1.09	1.25	1.42	1.61	1.80	1.99	2.42	2.88	3.36	3.88	4.42
1.40	0.357	0.452	0.558	0.675	0.804	0.943	1.09	1.24	1.40	1.57	1.75	2.13	2.54	2.99	3.45	3.95
1.60	0.314	0.398	0.491	0.595	0.708	0.833	0.967	1.10	1.25	1.40	1.56	1.90	2.27	2.67	3.11	3.56
1.80	0.280	0.355	0.438	0.531	0.633	0.746	0.867	0.988	1.12	1.26	1.40	1.71	2.05	2.42	2.81	3.24
2.00	0.253	0.320	0.395	0.479	0.572	0.675	0.784	0.896	1.01	1.14	1.27	1.55	1.87	2.20	2.57	2.96
2.20	0.230	0.291	0.360	0.437	0.522	0.616	0.717	0.819	0.926	1.04	1.16	1.43	1.71	2.03	2.37	2.73
2.40	0.211	0.267	0.330	0.401	0.479	0.566	0.659	0.753	0.853	0.959	1.07	1.31	1.58	1.87	2.19	2.53
2.60	0.195	0.247	0.305	0.370	0.444	0.524	0.610	0.697	0.790	0.889	0.993	1.22	1.47	1.74	2.04	2.36
2.80	0.182	0.230	0.284	0.344	0.412	0.487	0.568	0.649	0.736	0.827	0.925	1.14	1.37	1.62	1.90	2.20
3.00	0.170	0.214	0.265	0.321	0.386	0.456	0.531	0.607	0.688	0.774	0.866	1.06	1.28	1.52	1.79	2.07
x	0.000	0.008	0.029	0.056	0.089	0.125	0.164	0.204	0.246	0.289	0.333	0.424	0.516	0.610	0.704	0.800

Table 8-9 (cont.). Coefficients C for Eccentrically Loaded Weld Groups Angle = 30°

$\phi R_n = CC_1 Dl$ $C_{min} = \frac{P_u}{C_1 Dl}$ $D_{min} = \frac{P_u}{CC_1 l}$ $l_{min} = \frac{P_u}{CC_1 D}$

where

P_u = factored force, kips
D = number of sixteenths-of-an-inch in the fillet weld size
l = characteristic length of weld group, in.
$a = e_x/l$
e_x = horizontal component of eccentricity of P_u with respect to centroid of weld group, in.
C = coefficient tabulated below, which includes $\phi = 0.75$
C_1 = electrode strength coefficient from Table 8-4 (1.0 for E70XX electrodes)

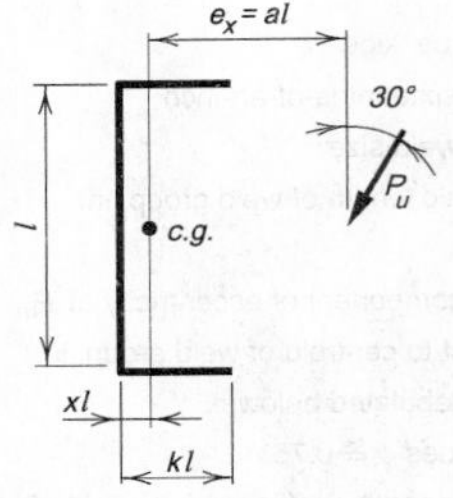

a	k = 0	0.1	0.2	0.3	0.4	0.5	0.6	0.7	0.8	0.9	1.0	1.2	1.4	1.6	1.8	2.0
0.00	1.64	2.03	2.42	2.81	3.20	3.59	3.98	4.37	4.76	5.15	5.54	6.33	7.11	7.89	8.67	9.45
0.10	1.52	1.93	2.32	2.72	3.11	3.50	3.89	4.28	4.68	5.07	5.46	6.24	7.03	7.81	8.60	9.39
0.15	1.44	1.82	2.21	2.60	2.99	3.37	3.75	4.14	4.52	4.91	5.29	6.07	6.84	7.62	8.41	9.19
0.20	1.36	1.72	2.09	2.47	2.84	3.21	3.58	3.95	4.33	4.70	5.08	5.84	6.60	7.38	8.15	8.93
0.25	1.28	1.61	1.97	2.33	2.69	3.04	3.40	3.76	4.11	4.48	4.84	5.59	6.34	7.10	7.87	8.65
0.30	1.20	1.51	1.84	2.18	2.53	2.88	3.22	3.56	3.91	4.26	4.62	5.33	6.07	6.82	7.58	8.35
0.40	1.05	1.32	1.61	1.91	2.23	2.55	2.87	3.19	3.52	3.85	4.18	4.86	5.57	6.28	7.02	7.77
0.50	0.921	1.16	1.41	1.68	1.96	2.26	2.56	2.86	3.16	3.47	3.79	4.44	5.11	5.80	6.51	7.24
0.60	0.812	1.02	1.25	1.49	1.74	2.01	2.29	2.57	2.86	3.15	3.45	4.07	4.71	5.38	6.06	6.77
0.70	0.722	0.908	1.11	1.33	1.56	1.81	2.07	2.33	2.60	2.87	3.15	3.74	4.36	5.00	5.67	6.35
0.80	0.647	0.816	1.00	1.20	1.41	1.64	1.88	2.12	2.37	2.63	2.90	3.46	4.05	4.66	5.30	5.96
0.90	0.586	0.739	0.905	1.09	1.28	1.49	1.71	1.95	2.18	2.42	2.68	3.21	3.77	4.36	4.97	5.60
1.00	0.535	0.674	0.827	0.994	1.18	1.37	1.58	1.80	2.02	2.25	2.48	2.99	3.52	4.08	4.67	5.28
1.20	0.454	0.572	0.703	0.847	1.00	1.17	1.36	1.55	1.75	1.95	2.16	2.61	3.10	3.61	4.15	4.71
1.40	0.393	0.496	0.610	0.736	0.874	1.02	1.19	1.36	1.53	1.72	1.91	2.32	2.75	3.22	3.72	4.24
1.60	0.347	0.437	0.538	0.650	0.773	0.909	1.06	1.21	1.37	1.53	1.70	2.07	2.47	2.90	3.36	3.85
1.80	0.310	0.391	0.481	0.582	0.693	0.816	0.949	1.09	1.23	1.38	1.54	1.87	2.24	2.64	3.06	3.52
2.00	0.280	0.353	0.435	0.526	0.627	0.740	0.861	0.988	1.12	1.26	1.40	1.71	2.05	2.41	2.81	3.23
2.20	0.255	0.322	0.397	0.481	0.573	0.677	0.789	0.904	1.02	1.15	1.28	1.57	1.88	2.22	2.59	2.98
2.40	0.234	0.296	0.365	0.442	0.528	0.623	0.727	0.833	0.944	1.06	1.18	1.45	1.74	2.06	2.40	2.77
2.60	0.217	0.273	0.337	0.409	0.489	0.577	0.674	0.772	0.875	0.983	1.10	1.34	1.62	1.92	2.24	2.59
2.80	0.201	0.254	0.314	0.381	0.455	0.538	0.628	0.719	0.815	0.916	1.02	1.26	1.51	1.79	2.10	2.43
3.00	0.188	0.238	0.293	0.356	0.426	0.504	0.588	0.673	0.763	0.858	0.959	1.18	1.42	1.68	1.97	2.28
x	0.000	0.008	0.029	0.056	0.089	0.125	0.164	0.204	0.246	0.289	0.333	0.424	0.516	0.610	0.704	0.800

Table 8-9 (cont.). Coefficients C for Eccentrically Loaded Weld Groups Angle = 45°

$\phi R_n = CC_1 Dl$ $C_{min} = \frac{P_u}{C_1 Dl}$ $D_{min} = \frac{P_u}{CC_1 l}$ $l_{min} = \frac{P_u}{CC_1 D}$

where

P_u = factored force, kips

D = number of sixteenths-of-an-inch in the fillet weld size

l = characteristic length of weld group, in.

$a = e_x / l$

e_x = horizontal component of eccentricity of P_u with respect to centroid of weld group, in.

C = coefficient tabulated below, which includes $\phi = 0.75$

C_1 = electrode strength coefficient from Table 8-4 (1.0 for E70XX electrodes)

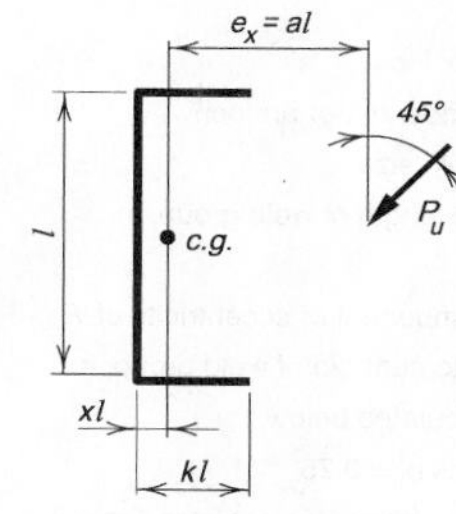

a	k															
	0	0.1	0.2	0.3	0.4	0.5	0.6	0.7	0.8	0.9	1.0	1.2	1.4	1.6	1.8	2.0
0.00	1.81	2.17	2.53	2.89	3.25	3.61	3.97	4.33	4.70	5.06	5.42	6.14	6.86	7.58	8.31	9.03
0.10	1.68	2.06	2.43	2.80	3.17	3.55	3.92	4.29	4.66	5.03	5.40	6.14	6.88	7.61	8.34	9.06
0.15	1.57	1.95	2.32	2.68	3.05	3.42	3.80	4.17	4.54	4.92	5.29	6.03	6.77	7.51	8.25	8.99
0.20	1.47	1.83	2.19	2.55	2.91	3.28	3.64	4.02	4.39	4.76	5.13	5.88	6.62	7.37	8.12	8.86
0.25	1.39	1.72	2.06	2.41	2.76	3.12	3.48	3.85	4.22	4.58	4.95	5.69	6.44	7.19	7.94	8.69
0.30	1.31	1.62	1.94	2.27	2.61	2.96	3.31	3.67	4.03	4.39	4.75	5.49	6.23	6.99	7.75	8.50
0.40	1.16	1.43	1.72	2.03	2.34	2.66	2.99	3.33	3.68	4.03	4.37	5.08	5.82	6.57	7.32	8.09
0.50	1.03	1.28	1.54	1.81	2.10	2.40	2.71	3.03	3.36	3.70	4.03	4.70	5.41	6.15	6.90	7.66
0.60	0.921	1.14	1.38	1.63	1.90	2.18	2.47	2.77	3.08	3.41	3.72	4.37	5.04	5.76	6.49	7.23
0.70	0.829	1.03	1.25	1.48	1.73	1.99	2.26	2.54	2.84	3.15	3.46	4.08	4.72	5.40	6.11	6.84
0.80	0.751	0.935	1.13	1.35	1.58	1.82	2.08	2.35	2.63	2.93	3.22	3.81	4.43	5.08	5.77	6.47
0.90	0.685	0.854	1.04	1.24	1.45	1.68	1.92	2.18	2.45	2.73	3.00	3.57	4.17	4.79	5.45	6.14
1.00	0.629	0.785	0.956	1.14	1.34	1.56	1.78	2.03	2.29	2.55	2.81	3.35	3.93	4.53	5.16	5.83
1.20	0.538	0.674	0.822	0.985	1.16	1.35	1.56	1.78	2.01	2.25	2.49	2.98	3.51	4.07	4.66	5.28
1.40	0.469	0.589	0.720	0.864	1.02	1.19	1.38	1.58	1.79	2.00	2.22	2.67	3.16	3.68	4.23	4.81
1.60	0.416	0.522	0.639	0.769	0.911	1.07	1.24	1.42	1.61	1.80	2.00	2.42	2.87	3.35	3.87	4.41
1.80	0.373	0.468	0.574	0.692	0.821	0.964	1.12	1.29	1.46	1.63	1.82	2.20	2.62	3.07	3.56	4.06
2.00	0.338	0.424	0.521	0.628	0.746	0.879	1.02	1.17	1.33	1.49	1.66	2.02	2.41	2.84	3.29	3.76
2.20	0.308	0.388	0.477	0.575	0.685	0.806	0.939	1.08	1.22	1.37	1.53	1.86	2.23	2.63	3.05	3.50
2.40	0.284	0.357	0.439	0.531	0.632	0.745	0.868	1.00	1.13	1.27	1.42	1.73	2.07	2.44	2.84	3.27
2.60	0.263	0.331	0.407	0.492	0.587	0.692	0.807	0.928	1.05	1.18	1.32	1.61	1.93	2.28	2.66	3.06
2.80	0.245	0.308	0.379	0.458	0.548	0.646	0.753	0.867	0.983	1.10	1.23	1.51	1.81	2.14	2.50	2.88
3.00	0.229	0.288	0.355	0.429	0.513	0.606	0.707	0.814	0.922	1.04	1.16	1.42	1.70	2.02	2.36	2.72
x	0.000	0.008	0.029	0.056	0.089	0.125	0.164	0.204	0.246	0.289	0.333	0.424	0.516	0.610	0.704	0.800

Table 8-9 (cont.). Coefficients C for Eccentrically Loaded Weld Groups Angle = 60°

$\phi R_n = CC_1 Dl$ $\quad C_{min} = \dfrac{P_u}{C_1 Dl}$ $\quad D_{min} = \dfrac{P_u}{CC_1 l}$ $\quad l_{min} = \dfrac{P_u}{CC_1 D}$

where

P_u = factored force, kips
D = number of sixteenths-of-an-inch in the fillet weld size
l = characteristic length of weld group, in.
$a = e_x/l$
e_x = horizontal component of eccentricity of P_u with respect to centroid of weld group, in.
C = coefficient tabulated below, which includes $\phi = 0.75$
C_1 = electrode strength coefficient from Table 8-4 (1.0 for E70XX electrodes)

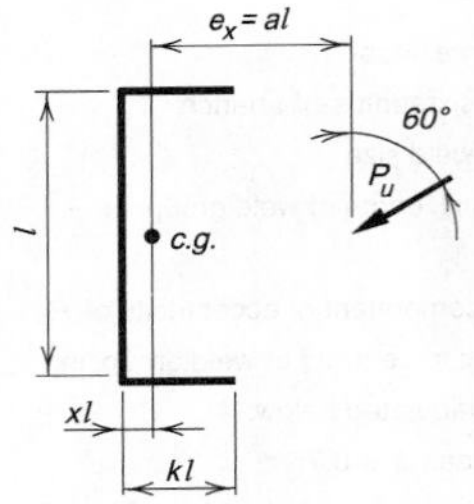

a	k = 0	0.1	0.2	0.3	0.4	0.5	0.6	0.7	0.8	0.9	1.0	1.2	1.4	1.6	1.8	2.0
0.00	1.95	2.28	2.61	2.94	3.26	3.59	3.92	4.25	4.57	4.90	5.23	5.88	6.54	7.19	7.85	8.51
0.10	1.82	2.15	2.48	2.82	3.16	3.51	3.86	4.21	4.56	4.90	5.24	5.92	6.59	7.25	7.90	8.56
0.15	1.73	2.05	2.38	2.71	3.06	3.41	3.76	4.12	4.47	4.83	5.18	5.87	6.55	7.23	7.89	8.55
0.20	1.63	1.96	2.28	2.61	2.94	3.29	3.65	4.01	4.37	4.73	5.09	5.80	6.49	7.18	7.85	8.52
0.25	1.55	1.87	2.18	2.50	2.83	3.17	3.53	3.89	4.25	4.62	4.98	5.70	6.41	7.11	7.79	8.47
0.30	1.47	1.78	2.09	2.40	2.72	3.05	3.40	3.76	4.13	4.50	4.87	5.60	6.32	7.02	7.72	8.40
0.40	1.34	1.62	1.91	2.20	2.51	2.83	3.16	3.52	3.88	4.24	4.61	5.36	6.09	6.82	7.53	8.23
0.50	1.22	1.48	1.75	2.03	2.32	2.63	2.95	3.28	3.63	4.00	4.36	5.10	5.84	6.57	7.30	8.01
0.60	1.12	1.36	1.61	1.88	2.15	2.44	2.75	3.07	3.41	3.76	4.13	4.86	5.59	6.32	7.03	7.75
0.70	1.03	1.25	1.49	1.74	2.00	2.28	2.58	2.89	3.22	3.56	3.91	4.62	5.33	6.05	6.78	7.50
0.80	0.945	1.16	1.38	1.62	1.87	2.14	2.42	2.72	3.04	3.36	3.70	4.39	5.09	5.80	6.52	7.24
0.90	0.874	1.07	1.28	1.51	1.75	2.01	2.28	2.57	2.87	3.18	3.51	4.17	4.85	5.55	6.26	6.98
1.00	0.812	1.00	1.20	1.41	1.64	1.89	2.15	2.43	2.72	3.02	3.34	3.97	4.63	5.32	6.02	6.73
1.20	0.709	0.875	1.06	1.25	1.46	1.69	1.93	2.19	2.46	2.74	3.03	3.62	4.23	4.89	5.56	6.26
1.40	0.626	0.776	0.939	1.12	1.31	1.52	1.75	1.98	2.23	2.49	2.76	3.31	3.89	4.50	5.15	5.82
1.60	0.560	0.696	0.845	1.01	1.18	1.38	1.59	1.81	2.04	2.28	2.54	3.05	3.59	4.17	4.79	5.43
1.80	0.506	0.630	0.766	0.916	1.08	1.26	1.46	1.66	1.88	2.11	2.34	2.82	3.33	3.88	4.46	5.07
2.00	0.461	0.575	0.701	0.839	0.993	1.16	1.34	1.53	1.74	1.95	2.17	2.62	3.11	3.62	4.17	4.75
2.20	0.423	0.528	0.645	0.774	0.917	1.07	1.24	1.42	1.61	1.81	2.02	2.44	2.90	3.39	3.91	4.47
2.40	0.391	0.489	0.597	0.718	0.851	1.00	1.16	1.33	1.51	1.69	1.89	2.29	2.72	3.19	3.68	4.21
2.60	0.363	0.454	0.556	0.669	0.794	0.932	1.08	1.24	1.41	1.59	1.77	2.15	2.56	3.01	3.48	3.98
2.80	0.339	0.424	0.519	0.625	0.744	0.875	1.02	1.17	1.33	1.49	1.66	2.02	2.42	2.84	3.29	3.77
3.00	0.317	0.398	0.488	0.587	0.700	0.824	0.956	1.10	1.25	1.41	1.57	1.91	2.29	2.69	3.12	3.58
x	0.000	0.008	0.029	0.056	0.089	0.125	0.164	0.204	0.246	0.289	0.333	0.424	0.516	0.610	0.704	0.800

Table 8-9 (cont.).
Coefficients *C* for Eccentrically Loaded Weld Groups
Angle = 75°

$\phi R_n = CC_1 Dl$ $\quad C_{min} = \dfrac{P_u}{C_1 Dl}$ $\quad D_{min} = \dfrac{P_u}{CC_1 l}$ $\quad l_{min} = \dfrac{P_u}{CC_1 D}$

where

P_u = factored force, kips

D = number of sixteenths-of-an-inch in the fillet weld size

l = characteristic length of weld group, in.

$a = e_x / l$

e_x = horizontal component of eccentricity of P_u with respect to centroid of weld group, in.

C = coefficient tabulated below, which includes $\phi = 0.75$

C_1 = electrode strength coefficient from Table 8-4 (1.0 for E70XX electrodes)

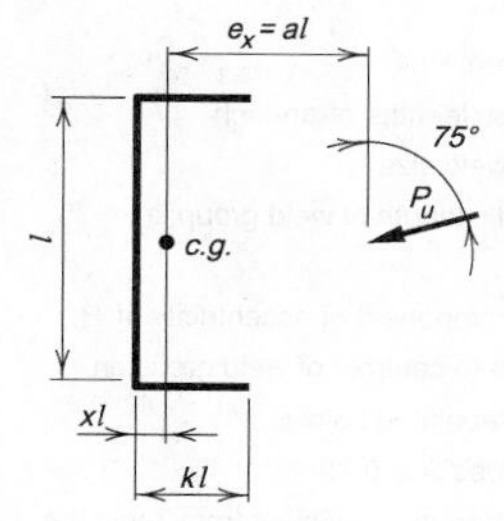

a	*k*															
	0	0.1	0.2	0.3	0.4	0.5	0.6	0.7	0.8	0.9	1.0	1.2	1.4	1.6	1.8	2.0
0.00	2.05	2.35	2.65	2.94	3.24	3.54	3.83	4.13	4.43	4.72	5.02	5.61	6.21	6.80	7.39	7.99
0.10	1.94	2.22	2.51	2.81	3.13	3.44	3.75	4.05	4.35	4.65	4.94	5.53	6.11	6.68	7.26	7.83
0.15	1.88	2.15	2.44	2.75	3.07	3.39	3.70	4.01	4.32	4.63	4.93	5.52	6.10	6.68	7.26	7.84
0.20	1.82	2.09	2.38	2.69	3.01	3.33	3.65	3.97	4.29	4.60	4.90	5.50	6.09	6.67	7.25	7.83
0.25	1.76	2.04	2.32	2.63	2.95	3.27	3.60	3.92	4.24	4.56	4.87	5.47	6.07	6.66	7.24	7.82
0.30	1.71	1.99	2.27	2.57	2.89	3.21	3.54	3.87	4.19	4.51	4.83	5.44	6.05	6.64	7.23	7.81
0.40	1.62	1.89	2.16	2.46	2.77	3.09	3.42	3.76	4.09	4.41	4.73	5.36	5.98	6.58	7.18	7.77
0.50	1.53	1.80	2.07	2.35	2.65	2.98	3.31	3.64	3.98	4.31	4.64	5.27	5.90	6.51	7.11	7.71
0.60	1.46	1.71	1.97	2.25	2.55	2.86	3.19	3.53	3.87	4.21	4.54	5.19	5.83	6.44	7.04	7.64
0.70	1.38	1.63	1.89	2.16	2.45	2.76	3.08	3.42	3.76	4.10	4.44	5.11	5.75	6.38	6.99	7.59
0.80	1.31	1.56	1.81	2.07	2.35	2.66	2.98	3.31	3.65	3.99	4.34	5.01	5.67	6.31	6.93	7.54
0.90	1.25	1.49	1.73	1.99	2.27	2.56	2.88	3.21	3.54	3.89	4.23	4.92	5.59	6.24	6.87	7.48
1.00	1.19	1.42	1.66	1.91	2.18	2.47	2.78	3.10	3.44	3.78	4.13	4.82	5.49	6.16	6.80	7.43
1.20	1.09	1.30	1.53	1.77	2.03	2.31	2.60	2.91	3.24	3.57	3.92	4.61	5.30	5.98	6.64	7.29
1.40	0.994	1.20	1.41	1.65	1.90	2.16	2.44	2.74	3.05	3.38	3.71	4.40	5.09	5.78	6.46	7.13
1.60	0.914	1.11	1.31	1.53	1.77	2.03	2.30	2.58	2.88	3.20	3.52	4.20	4.89	5.58	6.27	6.94
1.80	0.845	1.03	1.22	1.43	1.66	1.91	2.16	2.44	2.73	3.03	3.35	4.00	4.68	5.38	6.07	6.76
2.00	0.784	0.956	1.14	1.34	1.56	1.80	2.04	2.31	2.58	2.88	3.18	3.82	4.49	5.18	5.86	6.55
2.20	0.730	0.894	1.07	1.26	1.47	1.70	1.94	2.19	2.45	2.73	3.03	3.65	4.30	4.97	5.65	6.33
2.40	0.683	0.838	1.01	1.19	1.39	1.61	1.84	2.08	2.33	2.60	2.89	3.49	4.12	4.77	5.43	6.11
2.60	0.641	0.788	0.949	1.13	1.32	1.53	1.75	1.98	2.22	2.48	2.76	3.34	3.95	4.58	5.23	5.90
2.80	0.604	0.744	0.897	1.07	1.25	1.45	1.66	1.89	2.12	2.37	2.63	3.20	3.78	4.39	5.04	5.70
3.00	0.570	0.704	0.851	1.01	1.19	1.38	1.59	1.80	2.03	2.27	2.52	3.07	3.63	4.23	4.85	5.50
x	0.000	0.008	0.029	0.056	0.089	0.125	0.164	0.204	0.246	0.289	0.333	0.424	0.516	0.610	0.704	0.800

Table 8-10.
Coefficients *C* for Eccentrically Loaded Weld Groups
Angle = 0°

$\phi R_n = CC_1 Dl \qquad C_{min} = \dfrac{P_u}{C_1 Dl} \qquad D_{min} = \dfrac{P_u}{CC_1 l} \qquad l_{min} = \dfrac{P_u}{CC_1 D}$

where

P_u = factored force, kips
D = number of sixteenths-of-an-inch in the fillet weld size
l = characteristic length of weld group, in.
$a = e_x/l$
e_x = horizontal component of eccentricity of P_u with respect to centroid of weld group, in.
C = coefficient tabulated below, which includes $\phi = 0.75$
C_1 = electrode strength coefficient from Table 8-4 (1.0 for E70XX electrodes)

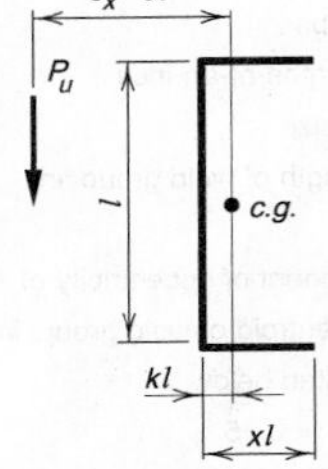

a	*k*															
	0	**0.1**	**0.2**	**0.3**	**0.4**	**0.5**	**0.6**	**0.7**	**0.8**	**0.9**	**1.0**	**1.2**	**1.4**	**1.6**	**1.8**	**2.0**
0.00	1.39	1.81	2.23	2.65	3.06	3.48	3.90	4.32	4.73	5.15	5.57	6.40	7.24	8.07	8.91	9.74
0.10	1.40	1.72	2.10	2.48	2.86	3.24	3.62	4.00	4.38	4.76	5.13	5.88	6.63	7.37	8.12	8.87
0.15	1.38	1.70	2.05	2.41	2.77	3.13	3.50	3.86	4.22	4.58	4.94	5.66	6.38	7.11	7.84	8.57
0.20	1.32	1.63	1.96	2.31	2.65	2.99	3.34	3.68	4.03	4.38	4.72	5.42	6.12	6.83	7.55	8.27
0.25	1.24	1.55	1.86	2.18	2.51	2.84	3.17	3.50	3.83	4.16	4.50	5.17	5.86	6.55	7.25	7.96
0.30	1.16	1.45	1.75	2.05	2.36	2.68	2.99	3.31	3.63	3.95	4.27	4.93	5.60	6.27	6.96	7.66
0.40	1.00	1.26	1.52	1.79	2.08	2.36	2.65	2.94	3.24	3.54	3.84	4.46	5.09	5.74	6.41	7.08
0.50	0.860	1.08	1.31	1.56	1.81	2.07	2.34	2.60	2.88	3.15	3.44	4.02	4.63	5.25	5.89	6.55
0.60	0.748	0.942	1.14	1.35	1.58	1.83	2.07	2.32	2.58	2.84	3.11	3.66	4.23	4.83	5.45	6.08
0.70	0.659	0.828	1.00	1.20	1.41	1.63	1.86	2.10	2.34	2.58	2.83	3.35	3.90	4.47	5.06	5.67
0.80	0.586	0.735	0.895	1.07	1.27	1.47	1.69	1.91	2.13	2.36	2.60	3.09	3.61	4.15	4.72	5.30
0.90	0.527	0.661	0.807	0.971	1.15	1.34	1.54	1.75	1.96	2.18	2.40	2.86	3.35	3.87	4.41	4.97
1.00	0.478	0.599	0.734	0.885	1.05	1.23	1.41	1.61	1.81	2.01	2.22	2.66	3.13	3.62	4.14	4.68
1.20	0.403	0.505	0.621	0.751	0.893	1.05	1.21	1.38	1.56	1.75	1.94	2.33	2.75	3.20	3.68	4.17
1.40	0.348	0.436	0.537	0.651	0.776	0.912	1.06	1.21	1.37	1.54	1.71	2.07	2.45	2.86	3.30	3.75
1.60	0.305	0.383	0.472	0.574	0.685	0.806	0.936	1.07	1.22	1.37	1.53	1.86	2.21	2.58	2.98	3.41
1.80	0.272	0.341	0.422	0.512	0.613	0.722	0.839	0.963	1.10	1.23	1.38	1.68	2.00	2.35	2.72	3.11
2.00	0.245	0.308	0.381	0.463	0.554	0.654	0.760	0.873	0.993	1.12	1.25	1.53	1.84	2.15	2.50	2.87
2.20	0.223	0.280	0.347	0.422	0.506	0.597	0.694	0.798	0.909	1.02	1.15	1.41	1.69	1.99	2.31	2.65
2.40	0.205	0.257	0.318	0.388	0.465	0.548	0.639	0.735	0.836	0.942	1.06	1.30	1.56	1.85	2.15	2.47
2.60	0.189	0.238	0.294	0.358	0.430	0.508	0.592	0.681	0.773	0.872	0.976	1.20	1.46	1.72	2.00	2.30
2.80	0.176	0.221	0.274	0.333	0.400	0.472	0.550	0.633	0.720	0.811	0.908	1.12	1.36	1.61	1.88	2.16
3.00	0.164	0.207	0.256	0.311	0.374	0.442	0.515	0.593	0.673	0.758	0.850	1.05	1.27	1.51	1.76	2.03
x	0.000	0.008	0.029	0.056	0.089	0.125	0.164	0.204	0.246	0.289	0.333	0.424	0.516	0.610	0.704	0.800

Table 8-10 (cont.). Coefficients *C* for Eccentrically Loaded Weld Groups Angle = 15°

$\phi R_n = CC_1 Dl \qquad C_{min} = \dfrac{P_u}{C_1 Dl} \qquad D_{min} = \dfrac{P_u}{CC_1 l} \qquad l_{min} = \dfrac{P_u}{CC_1 D}$

where

P_u = factored force, kips
D = number of sixteenths-of-an-inch in the fillet weld size
l = characteristic length of weld group, in.
$a = e_x/l$
e_x = horizontal component of eccentricity of P_u with respect to centroid of weld group, in.
C = coefficient tabulated below, which includes $\phi = 0.75$
C_1 = electrode strength coefficient from Table 8-4 (1.0 for E70XX electrodes)

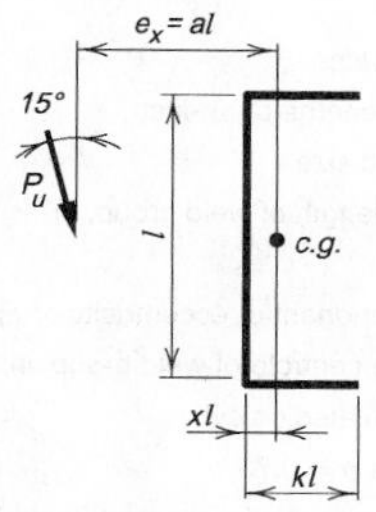

a	k															
	0	0.1	0.2	0.3	0.4	0.5	0.6	0.7	0.8	0.9	1.0	1.2	1.4	1.6	1.8	2.0
0.00	1.48	1.89	2.31	2.72	3.13	3.54	3.95	4.36	4.77	5.18	5.59	6.41	7.23	8.05	8.87	9.69
0.10	1.42	1.77	2.15	2.53	2.91	3.29	3.66	4.04	4.41	4.78	5.14	5.89	6.63	7.38	8.14	8.90
0.15	1.38	1.73	2.09	2.45	2.80	3.16	3.52	3.87	4.22	4.58	4.93	5.64	6.36	7.08	7.81	8.54
0.20	1.32	1.65	1.99	2.33	2.67	3.01	3.35	3.69	4.03	4.36	4.70	5.38	6.08	6.78	7.49	8.21
0.25	1.24	1.55	1.87	2.20	2.53	2.85	3.17	3.50	3.82	4.15	4.47	5.13	5.80	6.49	7.18	7.89
0.30	1.16	1.45	1.75	2.06	2.37	2.68	3.00	3.31	3.62	3.93	4.25	4.89	5.54	6.21	6.89	7.58
0.40	1.00	1.25	1.51	1.78	2.06	2.35	2.64	2.94	3.23	3.53	3.82	4.43	5.05	5.69	6.35	7.03
0.50	0.869	1.09	1.31	1.55	1.80	2.05	2.32	2.60	2.88	3.17	3.45	4.03	4.63	5.25	5.89	6.54
0.60	0.759	0.950	1.14	1.35	1.58	1.81	2.06	2.33	2.59	2.87	3.14	3.69	4.27	4.86	5.47	6.11
0.70	0.670	0.838	1.01	1.20	1.41	1.62	1.86	2.10	2.35	2.61	2.87	3.40	3.95	4.52	5.10	5.71
0.80	0.598	0.747	0.905	1.08	1.27	1.47	1.68	1.91	2.15	2.39	2.64	3.14	3.66	4.21	4.77	5.36
0.90	0.539	0.672	0.818	0.980	1.15	1.34	1.54	1.75	1.97	2.20	2.44	2.92	3.41	3.94	4.48	5.04
1.00	0.490	0.611	0.746	0.896	1.06	1.23	1.42	1.61	1.82	2.04	2.26	2.72	3.19	3.69	4.21	4.75
1.20	0.413	0.516	0.633	0.763	0.905	1.06	1.22	1.39	1.58	1.77	1.97	2.39	2.82	3.27	3.75	4.25
1.40	0.357	0.446	0.548	0.663	0.789	0.925	1.07	1.22	1.39	1.56	1.74	2.12	2.52	2.93	3.38	3.84
1.60	0.314	0.393	0.484	0.586	0.698	0.820	0.951	1.09	1.24	1.39	1.56	1.91	2.27	2.65	3.06	3.49
1.80	0.280	0.351	0.432	0.524	0.626	0.737	0.854	0.981	1.11	1.26	1.41	1.73	2.06	2.42	2.80	3.19
2.00	0.253	0.317	0.391	0.474	0.567	0.668	0.776	0.891	1.01	1.14	1.28	1.58	1.89	2.22	2.57	2.95
2.20	0.230	0.289	0.356	0.433	0.518	0.611	0.710	0.816	0.929	1.05	1.18	1.45	1.74	2.05	2.38	2.73
2.40	0.211	0.265	0.327	0.398	0.476	0.562	0.654	0.753	0.857	0.967	1.09	1.34	1.61	1.90	2.21	2.54
2.60	0.195	0.245	0.303	0.368	0.441	0.521	0.606	0.698	0.795	0.898	1.01	1.24	1.50	1.77	2.06	2.37
2.80	0.182	0.228	0.282	0.343	0.410	0.485	0.565	0.650	0.741	0.837	0.937	1.16	1.40	1.66	1.93	2.23
3.00	0.170	0.213	0.263	0.320	0.384	0.453	0.528	0.609	0.694	0.782	0.877	1.08	1.31	1.56	1.82	2.10
x	0.000	0.008	0.029	0.056	0.089	0.125	0.164	0.204	0.246	0.289	0.333	0.424	0.516	0.610	0.704	0.800

Table 8-10 (cont.). Coefficients C for Eccentrically Loaded Weld Groups Angle = 30°

$\phi R_n = CC_1 Dl \qquad C_{min} = \dfrac{P_u}{C_1 Dl} \qquad D_{min} = \dfrac{P_u}{CC_1 l} \qquad l_{min} = \dfrac{P_u}{CC_1 D}$

where

P_u = factored force, kips

D = number of sixteenths-of-an-inch in the fillet weld size

l = characteristic length of weld group, in.

$a = e_x / l$

e_x = horizontal component of eccentricity of P_u with respect to centroid of weld group, in.

C = coefficient tabulated below, which includes $\phi = 0.75$

C_1 = electrode strength coefficient from Table 8-4 (1.0 for E70XX electrodes)

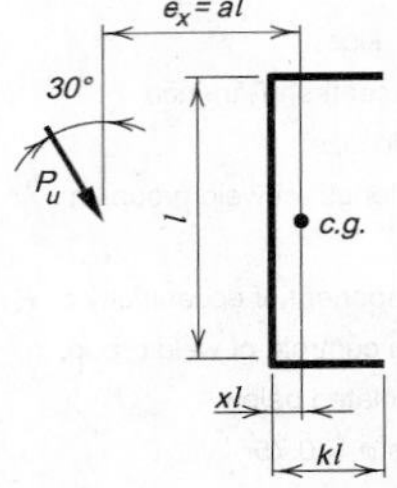

a	*k*															
	0	**0.1**	**0.2**	**0.3**	**0.4**	**0.5**	**0.6**	**0.7**	**0.8**	**0.9**	**1.0**	**1.2**	**1.4**	**1.6**	**1.8**	**2.0**
0.00	1.64	2.03	2.42	2.81	3.20	3.59	3.98	4.37	4.76	5.15	5.54	6.33	7.11	7.89	8.67	9.45
0.10	1.52	1.92	2.30	2.66	3.02	3.38	3.74	4.10	4.46	4.83	5.19	5.93	6.67	7.43	8.19	8.96
0.15	1.44	1.81	2.17	2.53	2.87	3.21	3.55	3.90	4.24	4.59	4.94	5.66	6.39	7.13	7.88	8.64
0.20	1.36	1.70	2.04	2.37	2.70	3.02	3.35	3.67	4.00	4.34	4.68	5.37	6.09	6.82	7.56	8.32
0.25	1.28	1.60	1.91	2.22	2.53	2.83	3.14	3.45	3.77	4.09	4.42	5.10	5.79	6.51	7.24	8.00
0.30	1.20	1.49	1.79	2.08	2.37	2.66	2.96	3.25	3.56	3.89	4.21	4.86	5.53	6.23	6.95	7.68
0.40	1.05	1.30	1.56	1.82	2.08	2.35	2.63	2.92	3.22	3.52	3.84	4.46	5.10	5.77	6.45	7.15
0.50	0.921	1.14	1.36	1.59	1.83	2.09	2.35	2.63	2.92	3.21	3.51	4.12	4.73	5.37	6.03	6.70
0.60	0.812	1.01	1.20	1.41	1.63	1.87	2.12	2.39	2.66	2.94	3.23	3.82	4.41	5.01	5.64	6.30
0.70	0.722	0.895	1.07	1.26	1.47	1.69	1.93	2.17	2.43	2.70	2.98	3.55	4.11	4.69	5.30	5.93
0.80	0.647	0.803	0.966	1.15	1.34	1.54	1.76	2.00	2.24	2.49	2.75	3.30	3.85	4.41	4.99	5.60
0.90	0.586	0.726	0.878	1.05	1.23	1.42	1.62	1.84	2.07	2.31	2.56	3.08	3.61	4.15	4.71	5.30
1.00	0.535	0.663	0.805	0.960	1.13	1.31	1.50	1.71	1.92	2.15	2.39	2.88	3.39	3.91	4.45	5.02
1.20	0.454	0.563	0.687	0.825	0.973	1.13	1.30	1.48	1.68	1.88	2.10	2.55	3.02	3.50	4.01	4.53
1.40	0.393	0.489	0.599	0.721	0.855	1.00	1.15	1.31	1.49	1.67	1.86	2.28	2.72	3.16	3.63	4.12
1.60	0.347	0.432	0.530	0.639	0.760	0.890	1.03	1.18	1.33	1.50	1.68	2.06	2.46	2.88	3.31	3.77
1.80	0.310	0.386	0.475	0.574	0.684	0.802	0.928	1.06	1.21	1.36	1.52	1.87	2.25	2.63	3.04	3.47
2.00	0.280	0.349	0.430	0.521	0.620	0.729	0.846	0.969	1.10	1.24	1.39	1.71	2.06	2.42	2.80	3.21
2.20	0.255	0.319	0.393	0.476	0.568	0.668	0.776	0.890	1.01	1.14	1.28	1.58	1.91	2.25	2.60	2.98
2.40	0.234	0.293	0.361	0.439	0.524	0.616	0.716	0.823	0.936	1.06	1.19	1.46	1.77	2.09	2.42	2.78
2.60	0.217	0.271	0.335	0.406	0.486	0.572	0.665	0.765	0.870	0.983	1.10	1.36	1.65	1.95	2.27	2.60
2.80	0.201	0.252	0.311	0.378	0.453	0.534	0.621	0.714	0.813	0.919	1.03	1.28	1.55	1.83	2.13	2.45
3.00	0.188	0.236	0.291	0.354	0.424	0.500	0.582	0.669	0.763	0.862	0.967	1.20	1.45	1.72	2.00	2.31
x	0.000	0.008	0.029	0.056	0.089	0.125	0.164	0.204	0.246	0.289	0.333	0.424	0.516	0.610	0.704	0.800

Table 8-10 (cont.). Coefficients *C* for Eccentrically Loaded Weld Groups Angle = 45°

$\phi R_n = CC_1 Dl$ $C_{min} = \frac{P_u}{C_1 Dl}$ $D_{min} = \frac{P_u}{CC_1 l}$ $l_{min} = \frac{P_u}{CC_1 D}$

where

P_u = factored force, kips
D = number of sixteenths-of-an-inch in the fillet weld size
l = characteristic length of weld group, in.
$a = e_x/l$
e_x = horizontal component of eccentricity of P_u with respect to centroid of weld group, in.
C = coefficient tabulated below, which includes $\phi = 0.75$
C_1 = electrode strength coefficient from Table 8-4 (1.0 for E70XX electrodes)

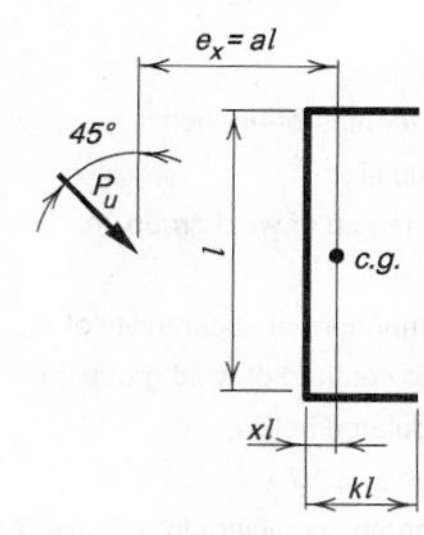

a	k															
	0	0.1	0.2	0.3	0.4	0.5	0.6	0.7	0.8	0.9	1.0	1.2	1.4	1.6	1.8	2.0
0.00	1.81	2.17	2.53	2.89	3.25	3.61	3.97	4.33	4.70	5.06	5.42	6.14	6.86	7.58	8.31	9.03
0.10	1.68	2.04	2.38	2.71	3.04	3.37	3.71	4.06	4.41	4.77	5.12	5.84	6.56	7.30	8.03	8.76
0.15	1.57	1.93	2.25	2.56	2.86	3.18	3.51	3.85	4.19	4.55	4.91	5.64	6.37	7.10	7.85	8.59
0.20	1.47	1.81	2.12	2.41	2.69	2.99	3.30	3.64	3.98	4.33	4.69	5.41	6.14	6.88	7.63	8.38
0.25	1.39	1.70	1.99	2.27	2.53	2.82	3.12	3.44	3.77	4.12	4.47	5.18	5.92	6.66	7.40	8.15
0.30	1.31	1.60	1.88	2.14	2.40	2.68	2.97	3.28	3.60	3.94	4.28	4.98	5.70	6.42	7.17	7.92
0.40	1.16	1.41	1.67	1.91	2.16	2.43	2.71	3.01	3.31	3.63	3.96	4.64	5.33	6.03	6.74	7.46
0.50	1.03	1.26	1.48	1.71	1.96	2.22	2.49	2.77	3.06	3.37	3.68	4.34	5.02	5.70	6.38	7.09
0.60	0.921	1.13	1.33	1.55	1.78	2.03	2.29	2.55	2.84	3.13	3.43	4.06	4.72	5.38	6.06	6.75
0.70	0.829	1.02	1.20	1.41	1.63	1.86	2.11	2.37	2.63	2.91	3.21	3.82	4.45	5.09	5.75	6.42
0.80	0.751	0.920	1.10	1.29	1.50	1.72	1.95	2.20	2.46	2.72	3.00	3.59	4.21	4.82	5.46	6.12
0.90	0.685	0.840	1.01	1.19	1.38	1.59	1.81	2.05	2.30	2.55	2.82	3.38	3.98	4.58	5.20	5.83
1.00	0.629	0.772	0.930	1.10	1.28	1.48	1.69	1.92	2.15	2.40	2.65	3.19	3.77	4.35	4.95	5.57
1.20	0.538	0.664	0.804	0.957	1.12	1.30	1.49	1.69	1.91	2.13	2.37	2.87	3.40	3.95	4.50	5.09
1.40	0.469	0.581	0.706	0.845	1.00	1.16	1.33	1.51	1.71	1.91	2.13	2.59	3.09	3.60	4.12	4.67
1.60	0.416	0.515	0.629	0.754	0.892	1.04	1.20	1.36	1.54	1.73	1.93	2.36	2.82	3.31	3.80	4.31
1.80	0.373	0.463	0.566	0.681	0.807	0.943	1.09	1.24	1.41	1.58	1.77	2.17	2.60	3.05	3.52	4.00
2.00	0.338	0.420	0.515	0.621	0.736	0.862	1.00	1.14	1.29	1.45	1.62	2.00	2.40	2.83	3.27	3.73
2.20	0.308	0.384	0.471	0.569	0.677	0.793	0.918	1.05	1.19	1.34	1.50	1.85	2.23	2.64	3.05	3.48
2.40	0.284	0.354	0.435	0.526	0.626	0.734	0.850	0.974	1.11	1.25	1.40	1.72	2.08	2.46	2.85	3.27
2.60	0.263	0.328	0.403	0.488	0.582	0.683	0.792	0.909	1.03	1.16	1.31	1.61	1.95	2.31	2.68	3.07
2.80	0.245	0.306	0.376	0.455	0.543	0.639	0.741	0.851	0.966	1.09	1.22	1.51	1.83	2.17	2.53	2.90
3.00	0.229	0.286	0.352	0.427	0.510	0.600	0.697	0.799	0.910	1.03	1.15	1.43	1.73	2.05	2.39	2.74
x	0.000	0.008	0.029	0.056	0.089	0.125	0.164	0.204	0.246	0.289	0.333	0.424	0.516	0.610	0.704	0.800

Table 8-10 (cont.). Coefficients *C* for Eccentrically Loaded Weld Groups Angle = 60°

$\phi R_n = CC_1 Dl$ $C_{min} = \frac{P_u}{C_1 Dl}$ $D_{min} = \frac{P_u}{CC_1 l}$ $l_{min} = \frac{P_u}{CC_1 D}$

where

P_u = factored force, kips
D = number of sixteenths-of-an-inch in the fillet weld size
l = characteristic length of weld group, in.
$a = e_x/l$
e_x = horizontal component of eccentricity of P_u with respect to centroid of weld group, in.
C = coefficient tabulated below, which includes $\phi = 0.75$
C_1 = electrode strength coefficient from Table 8-4 (1.0 for E70XX electrodes)

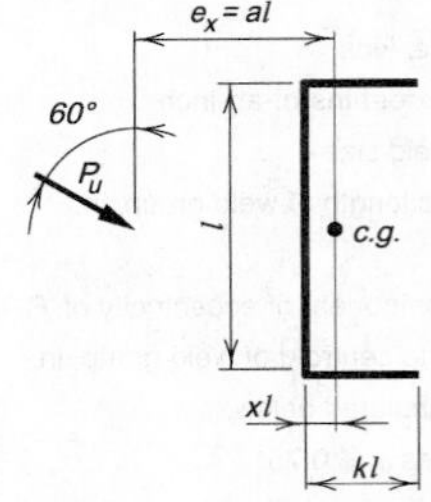

a	k = 0	0.1	0.2	0.3	0.4	0.5	0.6	0.7	0.8	0.9	1.0	1.2	1.4	1.6	1.8	2.0
0.00	1.95	2.28	2.61	2.94	3.26	3.59	3.92	4.25	4.57	4.90	5.23	5.88	6.54	7.19	7.85	8.51
0.10	1.83	2.13	2.42	2.72	3.03	3.36	3.69	4.03	4.37	4.71	5.05	5.73	6.41	7.08	7.74	8.40
0.15	1.73	2.03	2.30	2.58	2.88	3.19	3.52	3.86	4.20	4.55	4.89	5.59	6.27	6.96	7.63	8.31
0.20	1.63	1.93	2.19	2.45	2.74	3.04	3.36	3.69	4.03	4.38	4.73	5.43	6.13	6.82	7.51	8.19
0.25	1.55	1.84	2.09	2.34	2.62	2.92	3.23	3.55	3.88	4.22	4.56	5.26	5.97	6.67	7.37	8.06
0.30	1.47	1.76	2.00	2.25	2.52	2.81	3.12	3.43	3.76	4.09	4.42	5.11	5.80	6.51	7.22	7.93
0.40	1.34	1.60	1.84	2.08	2.34	2.62	2.91	3.22	3.54	3.87	4.20	4.87	5.54	6.23	6.91	7.61
0.50	1.22	1.46	1.69	1.93	2.18	2.45	2.73	3.04	3.35	3.67	3.99	4.66	5.33	6.01	6.69	7.37
0.60	1.12	1.34	1.56	1.79	2.04	2.30	2.57	2.86	3.16	3.48	3.80	4.46	5.14	5.81	6.49	7.16
0.70	1.03	1.23	1.44	1.67	1.91	2.16	2.42	2.70	3.00	3.30	3.62	4.27	4.94	5.62	6.30	6.98
0.80	0.945	1.14	1.34	1.55	1.79	2.03	2.29	2.56	2.84	3.14	3.45	4.08	4.75	5.42	6.10	6.79
0.90	0.874	1.06	1.25	1.45	1.68	1.91	2.16	2.42	2.70	2.98	3.28	3.91	4.56	5.24	5.91	6.60
1.00	0.812	0.983	1.17	1.37	1.58	1.81	2.05	2.30	2.57	2.84	3.13	3.75	4.39	5.05	5.73	6.41
1.20	0.709	0.862	1.03	1.21	1.41	1.62	1.84	2.08	2.33	2.59	2.86	3.44	4.06	4.70	5.36	6.03
1.40	0.626	0.766	0.922	1.09	1.27	1.46	1.67	1.89	2.13	2.37	2.63	3.18	3.76	4.38	5.02	5.66
1.60	0.560	0.688	0.831	0.988	1.16	1.34	1.53	1.74	1.95	2.18	2.43	2.94	3.50	4.08	4.70	5.32
1.80	0.506	0.623	0.756	0.901	1.06	1.23	1.41	1.60	1.80	2.02	2.25	2.73	3.26	3.82	4.41	5.00
2.00	0.461	0.569	0.692	0.828	0.975	1.13	1.30	1.48	1.67	1.88	2.09	2.55	3.05	3.58	4.14	4.71
2.20	0.423	0.524	0.638	0.765	0.902	1.05	1.21	1.38	1.56	1.75	1.95	2.39	2.86	3.37	3.90	4.45
2.40	0.391	0.484	0.591	0.710	0.840	0.979	1.13	1.29	1.46	1.64	1.83	2.24	2.69	3.17	3.69	4.21
2.60	0.363	0.451	0.551	0.662	0.785	0.916	1.06	1.21	1.37	1.54	1.72	2.11	2.54	3.00	3.49	3.99
2.80	0.339	0.421	0.515	0.620	0.736	0.861	0.994	1.14	1.29	1.45	1.62	2.00	2.40	2.85	3.31	3.79
3.00	0.317	0.395	0.484	0.584	0.693	0.812	0.939	1.07	1.22	1.37	1.53	1.89	2.28	2.70	3.15	3.60
x	0.000	0.008	0.029	0.056	0.089	0.125	0.164	0.204	0.246	0.289	0.333	0.424	0.516	0.610	0.704	0.800

Table 8-10 (cont.).
Coefficients C for Eccentrically Loaded Weld Groups
Angle = 75°

$$\phi R_n = CC_1 Dl \qquad C_{min} = \frac{P_u}{C_1 Dl} \qquad D_{min} = \frac{P_u}{CC_1 l} \qquad l_{min} = \frac{P_u}{CC_1 D}$$

where

P_u = factored force, kips

D = number of sixteenths-of-an-inch in the fillet weld size

l = characteristic length of weld group, in.

$a = e_x/l$

e_x = horizontal component of eccentricity of P_u with respect to centroid of weld group, in.

C = coefficient tabulated below, which includes $\phi = 0.75$

C_1 = electrode strength coefficient from Table 8-4 (1.0 for E70XX electrodes)

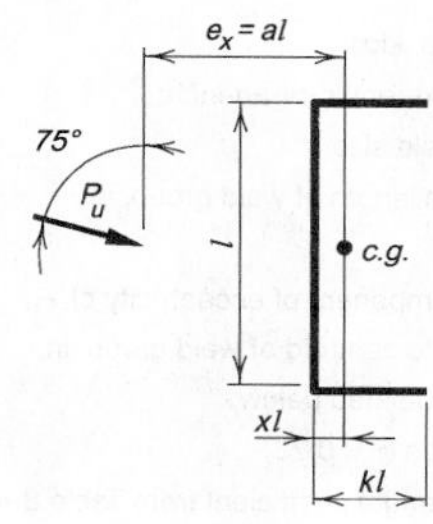

a	k															
	0	0.1	0.2	0.3	0.4	0.5	0.6	0.7	0.8	0.9	1.0	1.2	1.4	1.6	1.8	2.0
0.00	2.05	2.35	2.65	2.94	3.24	3.54	3.83	4.13	4.43	4.72	5.02	5.61	6.21	6.80	7.39	7.99
0.10	1.94	2.21	2.48	2.77	3.06	3.37	3.68	3.99	4.31	4.60	4.88	5.46	6.04	6.61	7.19	7.77
0.15	1.88	2.14	2.40	2.67	2.96	3.26	3.58	3.89	4.20	4.51	4.82	5.42	6.00	6.58	7.16	7.74
0.20	1.82	2.07	2.32	2.59	2.88	3.18	3.48	3.78	4.10	4.42	4.73	5.35	5.96	6.54	7.12	7.70
0.25	1.76	2.01	2.26	2.53	2.82	3.11	3.41	3.71	4.02	4.32	4.63	5.26	5.88	6.49	7.08	7.67
0.30	1.71	1.95	2.20	2.47	2.76	3.05	3.36	3.66	3.96	4.26	4.56	5.17	5.79	6.42	7.03	7.62
0.40	1.62	1.86	2.10	2.36	2.65	2.95	3.25	3.56	3.86	4.16	4.46	5.06	5.66	6.26	6.88	7.49
0.50	1.53	1.77	2.01	2.27	2.55	2.84	3.15	3.46	3.77	4.07	4.38	4.98	5.58	6.17	6.76	7.35
0.60	1.46	1.69	1.93	2.18	2.45	2.74	3.05	3.36	3.67	3.98	4.29	4.91	5.51	6.10	6.69	7.28
0.70	1.38	1.61	1.85	2.09	2.36	2.65	2.95	3.26	3.57	3.89	4.21	4.83	5.44	6.05	6.64	7.22
0.80	1.31	1.54	1.77	2.01	2.28	2.56	2.85	3.16	3.48	3.79	4.11	4.75	5.37	5.98	6.58	7.17
0.90	1.25	1.47	1.70	1.94	2.19	2.47	2.76	3.07	3.38	3.70	4.02	4.66	5.29	5.92	6.52	7.13
1.00	1.19	1.40	1.63	1.87	2.12	2.39	2.68	2.98	3.29	3.60	3.93	4.57	5.21	5.84	6.46	7.07
1.20	1.09	1.29	1.50	1.73	1.98	2.24	2.51	2.80	3.11	3.42	3.74	4.39	5.04	5.69	6.32	6.95
1.40	0.994	1.18	1.39	1.61	1.85	2.10	2.36	2.64	2.94	3.24	3.56	4.21	4.86	5.52	6.17	6.80
1.60	0.914	1.10	1.29	1.50	1.73	1.97	2.23	2.50	2.78	3.08	3.39	4.02	4.68	5.34	6.00	6.65
1.80	0.845	1.02	1.21	1.41	1.63	1.86	2.10	2.36	2.64	2.93	3.23	3.85	4.50	5.16	5.82	6.49
2.00	0.784	0.947	1.13	1.32	1.53	1.75	1.99	2.24	2.50	2.78	3.07	3.68	4.33	4.98	5.65	6.31
2.20	0.730	0.886	1.06	1.25	1.45	1.66	1.89	2.13	2.38	2.65	2.93	3.53	4.16	4.81	5.47	6.13
2.40	0.683	0.832	1.00	1.18	1.37	1.57	1.79	2.03	2.27	2.53	2.80	3.38	3.99	4.64	5.29	5.96
2.60	0.641	0.782	0.940	1.11	1.30	1.49	1.71	1.93	2.17	2.42	2.68	3.24	3.84	4.47	5.12	5.78
2.80	0.604	0.738	0.890	1.06	1.23	1.42	1.62	1.84	2.07	2.31	2.57	3.11	3.70	4.31	4.95	5.61
3.00	0.570	0.699	0.844	1.00	1.17	1.36	1.55	1.76	1.98	2.22	2.46	2.99	3.56	4.16	4.79	5.44
x	0.000	0.008	0.029	0.056	0.089	0.125	0.164	0.204	0.246	0.289	0.333	0.424	0.516	0.610	0.704	0.800

Table 8-11.
Coefficients *C* for Eccentrically Loaded Weld Groups
Angle = 0°

$\phi R_n = CC_1 Dl \qquad C_{min} = \frac{P_u}{C_1 Dl} \qquad D_{min} = \frac{P_u}{CC_1 l} \qquad l_{min} = \frac{P_u}{CC_1 D}$

where

P_u = factored force, kips
D = number of sixteenths-of-an-inch in the fillet weld size
l = characteristic length of weld group, in.
$a = e_x/l$
e_x = horizontal component of eccentricity of P_u with respect to centroid of weld group, in.
C = coefficient tabulated below, which includes ϕ = 0.75
C_1 = electrode strength coefficient from Table 8-4 (1.0 for E70XX electrodes)

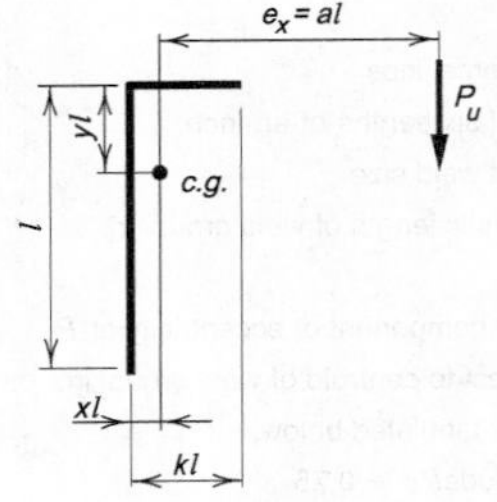

a	k = 0	0.1	0.2	0.3	0.4	0.5	0.6	0.7	0.8	0.9	1.0	1.2	1.4	1.6	1.8	2.0
0.00	1.39	1.60	1.81	2.02	2.23	2.44	2.65	2.85	3.06	3.27	3.48	3.90	4.32	4.73	5.15	5.57
0.10	1.40	1.53	1.71	1.90	2.09	2.29	2.48	2.68	2.88	3.08	3.29	3.70	4.11	4.52	4.94	5.35
0.15	1.38	1.52	1.69	1.87	2.05	2.24	2.43	2.62	2.82	3.01	3.21	3.61	4.01	4.42	4.83	5.26
0.20	1.32	1.47	1.63	1.80	1.98	2.15	2.33	2.52	2.70	2.89	3.08	3.46	3.85	4.25	4.65	5.05
0.25	1.24	1.39	1.55	1.71	1.88	2.04	2.21	2.39	2.57	2.74	2.93	3.30	3.68	4.06	4.46	4.85
0.30	1.16	1.30	1.45	1.61	1.77	1.92	2.08	2.25	2.42	2.59	2.77	3.13	3.50	3.88	4.26	4.65
0.40	1.00	1.12	1.25	1.39	1.54	1.68	1.82	1.97	2.13	2.29	2.45	2.79	3.15	3.51	3.89	4.27
0.50	0.860	0.965	1.08	1.20	1.32	1.46	1.60	1.73	1.87	2.02	2.17	2.50	2.84	3.19	3.55	3.92
0.60	0.748	0.840	0.935	1.04	1.15	1.27	1.40	1.53	1.66	1.80	1.94	2.24	2.57	2.90	3.26	3.62
0.70	0.659	0.739	0.822	0.913	1.01	1.12	1.24	1.36	1.48	1.61	1.74	2.03	2.34	2.66	3.00	3.35
0.80	0.586	0.658	0.732	0.813	0.901	1.00	1.11	1.22	1.34	1.45	1.58	1.85	2.14	2.45	2.77	3.11
0.90	0.527	0.591	0.658	0.731	0.811	0.900	1.00	1.11	1.21	1.32	1.44	1.69	1.97	2.26	2.57	2.90
1.00	0.478	0.536	0.597	0.663	0.736	0.818	0.909	1.01	1.11	1.21	1.32	1.56	1.82	2.10	2.40	2.71
1.20	0.403	0.452	0.503	0.558	0.620	0.690	0.769	0.856	0.947	1.04	1.13	1.35	1.58	1.83	2.10	2.38
1.40	0.348	0.389	0.433	0.481	0.535	0.596	0.665	0.743	0.824	0.904	0.990	1.18	1.39	1.62	1.86	2.12
1.60	0.305	0.342	0.381	0.423	0.470	0.525	0.586	0.655	0.728	0.800	0.877	1.05	1.24	1.44	1.67	1.91
1.80	0.272	0.305	0.339	0.377	0.419	0.468	0.524	0.585	0.652	0.717	0.787	0.941	1.11	1.30	1.51	1.73
2.00	0.245	0.275	0.306	0.340	0.378	0.423	0.473	0.529	0.590	0.649	0.713	0.854	1.01	1.19	1.38	1.58
2.20	0.223	0.250	0.278	0.309	0.344	0.385	0.431	0.483	0.539	0.593	0.652	0.780	0.927	1.09	1.26	1.45
2.40	0.205	0.229	0.255	0.284	0.316	0.354	0.396	0.443	0.495	0.546	0.600	0.719	0.855	1.00	1.17	1.34
2.60	0.189	0.212	0.236	0.262	0.292	0.327	0.366	0.410	0.458	0.505	0.555	0.667	0.792	0.933	1.09	1.25
2.80	0.176	0.197	0.219	0.244	0.271	0.304	0.340	0.381	0.426	0.470	0.517	0.621	0.739	0.870	1.01	1.17
3.00	0.164	0.184	0.204	0.228	0.254	0.284	0.318	0.356	0.398	0.440	0.483	0.581	0.692	0.815	0.950	1.10
x	0.000	0.005	0.017	0.035	0.057	0.083	0.113	0.144	0.178	0.213	0.250	0.327	0.408	0.492	0.579	0.667
y	0.500	0.455	0.417	0.385	0.357	0.333	0.313	0.294	0.278	0.263	0.250	0.227	0.208	0.192	0.179	0.167

Table 8-11 (cont.). Coefficients C for Eccentrically Loaded Weld Groups Angle = 15°

$\phi R_n = CC_1 Dl$ $\quad C_{min} = \dfrac{P_u}{C_1 Dl}$ $\quad D_{min} = \dfrac{P_u}{CC_1 l}$ $\quad l_{min} = \dfrac{P_u}{CC_1 D}$

where

P_u = factored force, kips

D = number of sixteenths-of-an-inch in the fillet weld size

l = characteristic length of weld group, in.

$a = e_x/l$

e_x = horizontal component of eccentricity of P_u with respect to centroid of weld group, in.

C = coefficient tabulated below, which includes $\phi = 0.75$

C_1 = electrode strength coefficient from Table 8-4 (1.0 for E70XX electrodes)

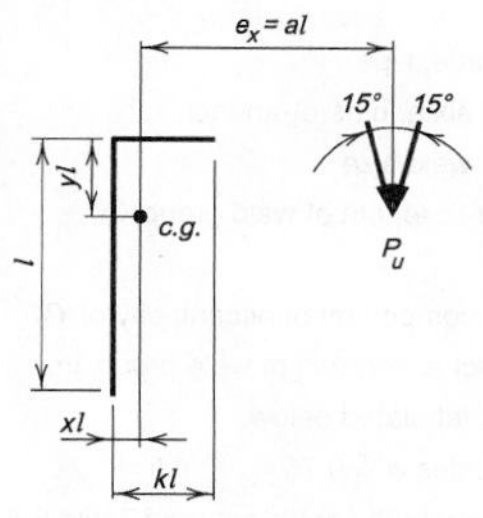

a	k: 0	0.1	0.2	0.3	0.4	0.5	0.6	0.7	0.8	0.9	1.0	1.2	1.4	1.6	1.8	2.0
0.00	1.48	1.69	1.89	2.10	2.31	2.51	2.72	2.92	3.13	3.33	3.54	3.95	4.36	4.77	5.18	5.59
0.10	1.42	1.56	1.73	1.90	2.09	2.28	2.47	2.67	2.88	3.09	3.30	3.73	4.17	4.59	5.02	5.45
0.15	1.38	1.53	1.69	1.85	2.03	2.20	2.39	2.57	2.76	2.95	3.15	3.54	3.95	4.37	4.80	5.23
0.20	1.32	1.48	1.63	1.78	1.94	2.11	2.28	2.46	2.64	2.82	3.00	3.38	3.77	4.16	4.57	4.97
0.25	1.24	1.39	1.55	1.69	1.85	2.00	2.16	2.33	2.50	2.67	2.85	3.21	3.59	3.97	4.36	4.76
0.30	1.16	1.31	1.46	1.60	1.74	1.89	2.04	2.20	2.36	2.53	2.70	3.05	3.42	3.79	4.18	4.57
0.40	1.00	1.13	1.27	1.41	1.53	1.66	1.80	1.94	2.09	2.25	2.41	2.74	3.09	3.45	3.83	4.21
0.50	0.869	0.978	1.09	1.22	1.34	1.46	1.59	1.72	1.85	2.00	2.15	2.47	2.80	3.15	3.51	3.88
0.60	0.759	0.854	0.953	1.06	1.18	1.29	1.41	1.53	1.65	1.79	1.93	2.23	2.55	2.89	3.24	3.60
0.70	0.670	0.753	0.840	0.936	1.04	1.15	1.26	1.37	1.48	1.61	1.74	2.03	2.33	2.65	2.99	3.34
0.80	0.598	0.672	0.749	0.834	0.927	1.03	1.13	1.23	1.34	1.46	1.58	1.85	2.14	2.45	2.78	3.12
0.90	0.539	0.605	0.675	0.750	0.834	0.925	1.02	1.12	1.23	1.33	1.45	1.71	1.98	2.28	2.59	2.91
1.00	0.490	0.550	0.613	0.681	0.758	0.842	0.933	1.03	1.12	1.23	1.34	1.58	1.84	2.12	2.42	2.73
1.20	0.413	0.464	0.517	0.574	0.639	0.712	0.791	0.878	0.964	1.06	1.15	1.36	1.60	1.86	2.13	2.41
1.40	0.357	0.401	0.446	0.496	0.552	0.616	0.686	0.764	0.841	0.921	1.01	1.20	1.41	1.64	1.89	2.15
1.60	0.314	0.352	0.392	0.436	0.485	0.542	0.605	0.675	0.745	0.818	0.897	1.07	1.26	1.47	1.70	1.94
1.80	0.280	0.314	0.350	0.389	0.433	0.484	0.541	0.604	0.668	0.735	0.806	0.962	1.14	1.33	1.54	1.77
2.00	0.253	0.283	0.315	0.351	0.391	0.437	0.489	0.546	0.606	0.666	0.731	0.875	1.04	1.21	1.41	1.62
2.20	0.230	0.258	0.287	0.319	0.356	0.398	0.446	0.499	0.554	0.609	0.669	0.801	0.950	1.11	1.29	1.49
2.40	0.211	0.237	0.263	0.293	0.327	0.366	0.410	0.458	0.510	0.561	0.616	0.738	0.877	1.03	1.20	1.38
2.60	0.195	0.219	0.243	0.271	0.302	0.338	0.379	0.424	0.472	0.520	0.571	0.685	0.814	0.957	1.12	1.28
2.80	0.182	0.203	0.226	0.252	0.281	0.314	0.352	0.394	0.439	0.484	0.532	0.639	0.759	0.894	1.04	1.20
3.00	0.170	0.190	0.211	0.235	0.262	0.294	0.329	0.368	0.411	0.453	0.498	0.598	0.711	0.838	0.977	1.13
x	0.000	0.005	0.017	0.035	0.057	0.083	0.113	0.144	0.178	0.213	0.250	0.327	0.408	0.492	0.579	0.667
y	0.500	0.455	0.417	0.385	0.357	0.333	0.313	0.294	0.278	0.263	0.250	0.227	0.208	0.192	0.179	0.167

Table 8-11 (cont.).
Coefficients *C* for Eccentrically Loaded Weld Groups
Angle = 30°

$\phi R_n = CC_1 Dl$ $\quad C_{min} = \frac{P_u}{C_1 Dl}$ $\quad D_{min} = \frac{P_u}{CC_1 l}$ $\quad l_{min} = \frac{P_u}{CC_1 D}$

where

P_u = factored force, kips
D = number of sixteenths-of-an-inch in the fillet weld size
l = characteristic length of weld group, in.
$a = e_x / l$
e_x = horizontal component of eccentricity of P_u with respect to centroid of weld group, in.
C = coefficient tabulated below, which includes $\phi = 0.75$
C_1 = electrode strength coefficient from Table 8-4 (1.0 for E70XX electrodes)

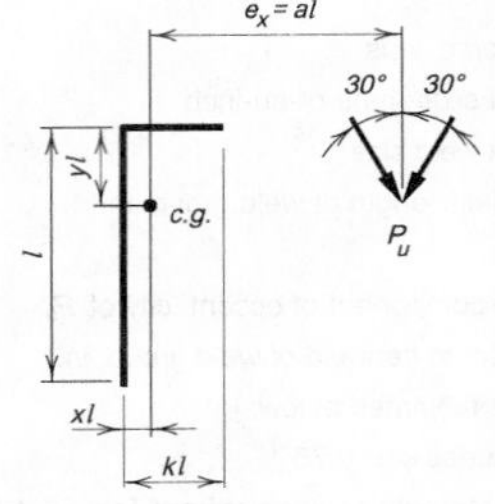

a	k = 0	0.1	0.2	0.3	0.4	0.5	0.6	0.7	0.8	0.9	1.0	1.2	1.4	1.6	1.8	2.0
0.00	1.64	1.83	2.03	2.22	2.42	2.61	2.81	3.01	3.20	3.40	3.59	3.98	4.37	4.76	5.15	5.54
0.10	1.52	1.68	1.85	2.03	2.21	2.39	2.58	2.77	2.97	3.16	3.36	3.77	4.18	4.59	5.00	5.41
0.15	1.44	1.60	1.76	1.91	2.08	2.25	2.42	2.60	2.78	2.97	3.16	3.55	3.95	4.37	4.79	5.20
0.20	1.36	1.51	1.67	1.83	1.98	2.14	2.30	2.47	2.64	2.82	3.00	3.37	3.76	4.16	4.57	4.98
0.25	1.28	1.43	1.58	1.73	1.88	2.03	2.18	2.34	2.51	2.68	2.85	3.21	3.59	3.98	4.37	4.78
0.30	1.20	1.34	1.49	1.63	1.78	1.92	2.06	2.22	2.37	2.54	2.71	3.06	3.43	3.81	4.20	4.60
0.40	1.05	1.18	1.31	1.44	1.58	1.71	1.84	1.98	2.13	2.28	2.44	2.78	3.13	3.50	3.88	4.27
0.50	0.921	1.03	1.15	1.27	1.40	1.52	1.64	1.77	1.91	2.06	2.21	2.53	2.87	3.23	3.60	3.98
0.60	0.812	0.910	1.02	1.13	1.25	1.36	1.47	1.59	1.72	1.86	2.00	2.31	2.64	2.98	3.34	3.71
0.70	0.722	0.811	0.908	1.01	1.12	1.22	1.33	1.44	1.56	1.69	1.83	2.12	2.43	2.77	3.12	3.48
0.80	0.647	0.729	0.815	0.908	1.01	1.11	1.21	1.31	1.43	1.55	1.67	1.95	2.25	2.58	2.91	3.26
0.90	0.586	0.660	0.736	0.820	0.912	1.01	1.10	1.20	1.31	1.42	1.54	1.81	2.10	2.41	2.73	3.07
1.00	0.535	0.601	0.671	0.747	0.831	0.920	1.01	1.11	1.21	1.31	1.43	1.68	1.96	2.25	2.56	2.89
1.20	0.454	0.510	0.568	0.632	0.704	0.783	0.868	0.952	1.04	1.14	1.24	1.47	1.72	1.99	2.28	2.58
1.40	0.393	0.441	0.492	0.547	0.610	0.680	0.756	0.835	0.916	1.00	1.10	1.30	1.53	1.77	2.04	2.32
1.60	0.347	0.389	0.433	0.482	0.537	0.600	0.668	0.742	0.814	0.893	0.978	1.16	1.37	1.60	1.84	2.10
1.80	0.310	0.347	0.387	0.430	0.480	0.536	0.599	0.666	0.733	0.805	0.882	1.05	1.24	1.45	1.67	1.92
2.00	0.280	0.314	0.349	0.389	0.433	0.485	0.542	0.604	0.666	0.732	0.802	0.958	1.13	1.33	1.54	1.76
2.20	0.255	0.286	0.318	0.354	0.395	0.442	0.495	0.552	0.610	0.670	0.735	0.880	1.04	1.22	1.42	1.63
2.40	0.234	0.263	0.292	0.325	0.363	0.406	0.455	0.508	0.562	0.618	0.679	0.813	0.963	1.13	1.31	1.51
2.60	0.217	0.243	0.270	0.301	0.335	0.376	0.421	0.471	0.521	0.573	0.630	0.754	0.896	1.05	1.22	1.41
2.80	0.201	0.226	0.251	0.280	0.312	0.349	0.392	0.438	0.486	0.535	0.588	0.704	0.836	0.984	1.15	1.32
3.00	0.188	0.211	0.235	0.261	0.292	0.327	0.366	0.410	0.455	0.501	0.550	0.660	0.784	0.924	1.08	1.24
x	0.000	0.005	0.017	0.035	0.057	0.083	0.113	0.144	0.178	0.213	0.250	0.327	0.408	0.492	0.579	0.667
y	0.500	0.455	0.417	0.385	0.357	0.333	0.313	0.294	0.278	0.263	0.250	0.227	0.208	0.192	0.179	0.167

Table 8-11 (cont.).
Coefficients C for Eccentrically Loaded Weld Groups
Angle = 45°

$\phi R_n = CC_1 Dl$ $\quad C_{min} = \dfrac{P_u}{C_1 Dl}$ $\quad D_{min} = \dfrac{P_u}{CC_1 l}$ $\quad l_{min} = \dfrac{P_u}{CC_1 D}$

where

P_u = factored force, kips
D = number of sixteenths-of-an-inch in the fillet weld size
l = characteristic length of weld group, in.
$a = e_x / l$
e_x = horizontal component of eccentricity of P_u with respect to centroid of weld group, in.
C = coefficient tabulated below, which includes $\phi = 0.75$
C_1 = electrode strength coefficient from Table 8-4 (1.0 for E70XX electrodes)

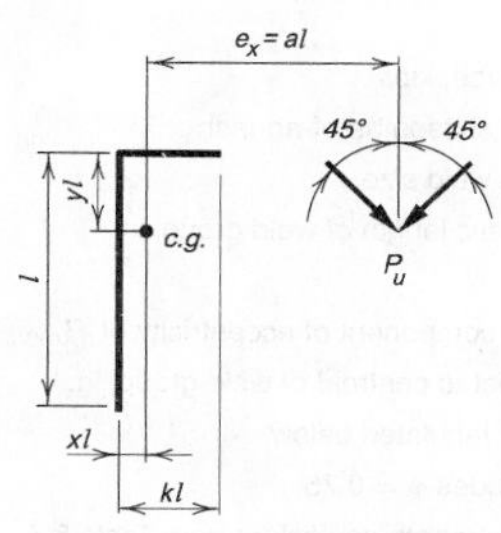

a	k: 0	0.1	0.2	0.3	0.4	0.5	0.6	0.7	0.8	0.9	1.0	1.2	1.4	1.6	1.8	2.0
0.00	1.81	1.99	2.17	2.35	2.53	2.71	2.89	3.07	3.25	3.43	3.61	3.97	4.33	4.70	5.06	5.42
0.10	1.68	1.83	1.99	2.14	2.30	2.47	2.64	2.82	3.00	3.18	3.37	3.76	4.15	4.55	4.95	5.35
0.15	1.57	1.71	1.86	2.01	2.17	2.33	2.50	2.67	2.84	3.02	3.21	3.59	3.99	4.39	4.80	5.21
0.20	1.47	1.61	1.75	1.90	2.06	2.22	2.37	2.54	2.71	2.88	3.06	3.43	3.83	4.23	4.64	5.06
0.25	1.39	1.52	1.65	1.80	1.96	2.11	2.26	2.42	2.58	2.75	2.93	3.29	3.68	4.07	4.48	4.90
0.30	1.31	1.43	1.56	1.71	1.86	2.01	2.15	2.31	2.46	2.63	2.80	3.16	3.54	3.93	4.33	4.75
0.40	1.16	1.27	1.40	1.53	1.67	1.81	1.95	2.10	2.25	2.40	2.57	2.92	3.28	3.66	4.06	4.47
0.50	1.03	1.14	1.25	1.38	1.51	1.64	1.77	1.91	2.05	2.20	2.36	2.70	3.05	3.43	3.81	4.21
0.60	0.921	1.02	1.13	1.24	1.36	1.49	1.62	1.74	1.88	2.02	2.18	2.50	2.85	3.21	3.59	3.98
0.70	0.829	0.919	1.02	1.13	1.24	1.36	1.48	1.60	1.73	1.87	2.01	2.32	2.66	3.02	3.39	3.77
0.80	0.751	0.835	0.928	1.03	1.14	1.25	1.36	1.47	1.60	1.73	1.87	2.17	2.49	2.84	3.20	3.58
0.90	0.685	0.764	0.849	0.943	1.05	1.15	1.26	1.36	1.48	1.61	1.74	2.03	2.34	2.68	3.03	3.39
1.00	0.629	0.702	0.782	0.870	0.966	1.07	1.16	1.27	1.38	1.50	1.63	1.90	2.21	2.53	2.87	3.22
1.20	0.538	0.603	0.674	0.751	0.836	0.926	1.01	1.11	1.21	1.32	1.43	1.69	1.97	2.27	2.59	2.92
1.40	0.469	0.527	0.589	0.655	0.730	0.811	0.894	0.979	1.07	1.17	1.28	1.51	1.77	2.05	2.34	2.65
1.60	0.416	0.467	0.521	0.580	0.646	0.721	0.799	0.877	0.961	1.05	1.15	1.36	1.60	1.86	2.14	2.43
1.80	0.373	0.419	0.466	0.519	0.579	0.647	0.720	0.793	0.869	0.953	1.04	1.24	1.46	1.70	1.96	2.23
2.00	0.338	0.379	0.422	0.470	0.524	0.587	0.654	0.722	0.794	0.871	0.954	1.14	1.34	1.56	1.80	2.06
2.20	0.308	0.346	0.385	0.429	0.479	0.536	0.599	0.663	0.730	0.801	0.878	1.05	1.24	1.45	1.67	1.92
2.40	0.284	0.318	0.355	0.395	0.441	0.493	0.552	0.613	0.675	0.741	0.812	0.971	1.15	1.34	1.56	1.78
2.60	0.263	0.295	0.328	0.365	0.408	0.457	0.511	0.570	0.627	0.689	0.756	0.904	1.07	1.26	1.46	1.67
2.80	0.245	0.274	0.305	0.340	0.380	0.425	0.476	0.532	0.585	0.643	0.707	0.845	1.00	1.18	1.37	1.57
3.00	0.229	0.256	0.286	0.318	0.355	0.398	0.446	0.498	0.549	0.604	0.663	0.794	0.942	1.11	1.29	1.48
x	0.000	0.005	0.017	0.035	0.057	0.083	0.113	0.144	0.178	0.213	0.250	0.327	0.408	0.492	0.579	0.667
y	0.500	0.455	0.417	0.385	0.357	0.333	0.313	0.294	0.278	0.263	0.250	0.227	0.208	0.192	0.179	0.167

Table 8-11 (cont.). Coefficients *C* for Eccentrically Loaded Weld Groups Angle = 60°

$\phi R_n = CC_1 Dl \qquad C_{min} = \frac{P_u}{C_1 Dl} \qquad D_{min} = \frac{P_u}{CC_1 l} \qquad l_{min} = \frac{P_u}{CC_1 D}$

where

P_u = factored force, kips

D = number of sixteenths-of-an-inch in the fillet weld size

l = characteristic length of weld group, in.

$a = e_x/l$

e_x = horizontal component of eccentricity of P_u with respect to centroid of weld group, in.

C = coefficient tabulated below, which includes $\phi = 0.75$

C_1 = electrode strength coefficient from Table 8-4 (1.0 for E70XX electrodes)

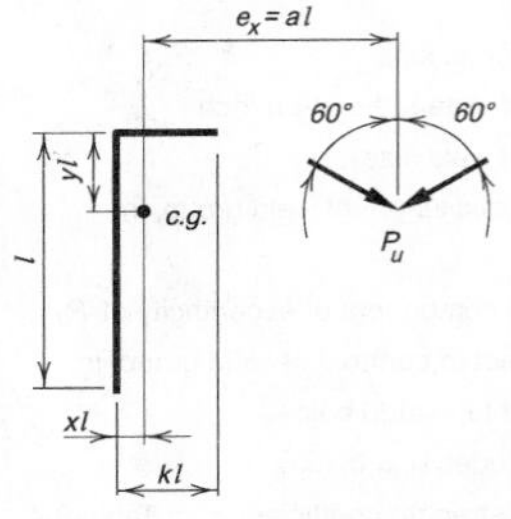

a	k = 0	0.1	0.2	0.3	0.4	0.5	0.6	0.7	0.8	0.9	1.0	1.2	1.4	1.6	1.8	2.0
0.00	1.95	2.12	2.28	2.44	2.61	2.77	2.94	3.10	3.26	3.43	3.59	3.92	4.25	4.57	4.90	5.23
0.10	1.83	1.94	2.07	2.20	2.35	2.51	2.68	2.85	3.03	3.21	3.40	3.78	4.17	4.55	4.92	5.27
0.15	1.73	1.84	1.96	2.10	2.25	2.40	2.57	2.74	2.92	3.10	3.28	3.67	4.07	4.47	4.86	5.23
0.20	1.63	1.74	1.87	2.00	2.15	2.31	2.47	2.64	2.81	2.99	3.18	3.56	3.96	4.37	4.78	5.16
0.25	1.55	1.66	1.78	1.92	2.06	2.22	2.38	2.55	2.72	2.89	3.07	3.45	3.85	4.27	4.68	5.08
0.30	1.47	1.58	1.70	1.83	1.98	2.13	2.29	2.46	2.63	2.80	2.98	3.35	3.75	4.16	4.58	5.00
0.40	1.34	1.44	1.56	1.69	1.83	1.98	2.13	2.29	2.45	2.62	2.80	3.16	3.55	3.96	4.38	4.81
0.50	1.22	1.32	1.43	1.56	1.69	1.83	1.99	2.14	2.30	2.46	2.63	2.99	3.37	3.77	4.19	4.61
0.60	1.12	1.21	1.32	1.44	1.57	1.71	1.85	2.00	2.15	2.31	2.48	2.83	3.21	3.60	4.01	4.43
0.70	1.03	1.12	1.22	1.33	1.46	1.59	1.73	1.88	2.02	2.18	2.34	2.68	3.05	3.44	3.85	4.26
0.80	0.945	1.03	1.13	1.24	1.36	1.49	1.63	1.76	1.90	2.05	2.21	2.55	2.91	3.29	3.69	4.09
0.90	0.874	0.958	1.05	1.16	1.27	1.40	1.53	1.66	1.79	1.94	2.09	2.42	2.77	3.15	3.53	3.93
1.00	0.812	0.893	0.983	1.08	1.19	1.31	1.44	1.56	1.69	1.83	1.98	2.30	2.65	3.01	3.39	3.79
1.20	0.709	0.783	0.865	0.957	1.06	1.17	1.29	1.40	1.52	1.65	1.79	2.09	2.42	2.76	3.13	3.51
1.40	0.626	0.695	0.771	0.855	0.948	1.05	1.16	1.26	1.38	1.50	1.63	1.91	2.22	2.55	2.89	3.25
1.60	0.560	0.623	0.693	0.771	0.857	0.952	1.05	1.15	1.25	1.36	1.49	1.75	2.04	2.35	2.68	3.03
1.80	0.506	0.564	0.629	0.701	0.781	0.868	0.957	1.05	1.15	1.25	1.37	1.61	1.89	2.18	2.50	2.83
2.00	0.461	0.515	0.575	0.642	0.716	0.798	0.880	0.964	1.06	1.16	1.26	1.49	1.75	2.03	2.33	2.64
2.20	0.423	0.473	0.530	0.590	0.659	0.735	0.813	0.891	0.978	1.07	1.17	1.39	1.63	1.90	2.18	2.48
2.40	0.391	0.438	0.489	0.545	0.608	0.680	0.755	0.829	0.910	1.00	1.09	1.30	1.53	1.78	2.05	2.33
2.60	0.363	0.407	0.454	0.506	0.565	0.632	0.704	0.774	0.850	0.932	1.02	1.22	1.43	1.67	1.93	2.20
2.80	0.339	0.380	0.424	0.472	0.527	0.590	0.658	0.726	0.797	0.875	0.958	1.14	1.35	1.57	1.82	2.08
3.00	0.317	0.356	0.397	0.442	0.494	0.553	0.618	0.683	0.750	0.824	0.903	1.08	1.27	1.49	1.72	1.97
x	0.000	0.005	0.017	0.035	0.057	0.083	0.113	0.144	0.178	0.213	0.250	0.327	0.408	0.492	0.579	0.667
y	0.500	0.455	0.417	0.385	0.357	0.333	0.313	0.294	0.278	0.263	0.250	0.227	0.208	0.192	0.179	0.167

Table 8-11 (cont.).
Coefficients C for Eccentrically Loaded Weld Groups
Angle = 75°

$\phi R_n = CC_1 Dl \qquad C_{min} = \frac{P_u}{C_1 Dl} \qquad D_{min} = \frac{P_u}{CC_1 l} \qquad l_{min} = \frac{P_u}{CC_1 D}$

where

P_u = factored force, kips
D = number of sixteenths-of-an-inch in the fillet weld size
l = characteristic length of weld group, in.
$a = e_x/l$
e_x = horizontal component of eccentricity of P_u with respect to centroid of weld group, in.
C = coefficient tabulated below, which includes $\phi = 0.75$
C_1 = electrode strength coefficient from Table 8-4 (1.0 for E70XX electrodes)

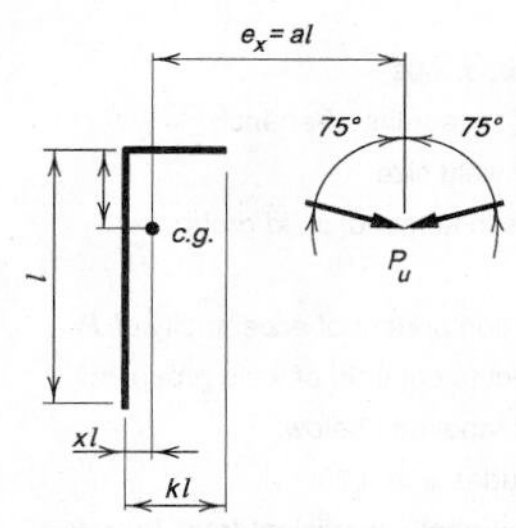

a	k: 0	0.1	0.2	0.3	0.4	0.5	0.6	0.7	0.8	0.9	1.0	1.2	1.4	1.6	1.8	2.0
0.00	2.05	2.20	2.35	2.50	2.65	2.79	2.94	3.09	3.24	3.39	3.54	3.83	4.13	4.43	4.72	5.02
0.10	1.94	2.01	2.11	2.23	2.37	2.52	2.69	2.87	3.05	3.24	3.44	3.80	4.09	4.38	4.67	4.96
0.15	1.88	1.95	2.05	2.17	2.31	2.47	2.63	2.81	3.00	3.19	3.39	3.77	4.09	4.38	4.67	4.96
0.20	1.82	1.90	2.00	2.12	2.26	2.41	2.58	2.76	2.95	3.14	3.34	3.73	4.09	4.39	4.68	4.96
0.25	1.76	1.84	1.95	2.07	2.21	2.36	2.53	2.71	2.89	3.09	3.29	3.69	4.06	4.39	4.68	4.97
0.30	1.71	1.79	1.90	2.02	2.16	2.31	2.48	2.66	2.84	3.04	3.24	3.65	4.03	4.38	4.68	4.97
0.40	1.62	1.70	1.81	1.93	2.07	2.22	2.39	2.56	2.75	2.94	3.14	3.54	3.95	4.32	4.66	4.96
0.50	1.53	1.62	1.72	1.84	1.98	2.13	2.30	2.47	2.66	2.85	3.04	3.44	3.85	4.25	4.61	4.94
0.60	1.46	1.54	1.64	1.76	1.90	2.05	2.21	2.38	2.57	2.76	2.95	3.34	3.75	4.16	4.54	4.90
0.70	1.38	1.47	1.57	1.69	1.82	1.97	2.13	2.30	2.48	2.67	2.86	3.25	3.65	4.07	4.46	4.84
0.80	1.31	1.40	1.50	1.62	1.75	1.90	2.06	2.22	2.40	2.59	2.77	3.16	3.56	3.97	4.38	4.76
0.90	1.25	1.34	1.44	1.55	1.68	1.83	1.98	2.15	2.33	2.51	2.69	3.07	3.47	3.88	4.29	4.68
1.00	1.19	1.28	1.38	1.49	1.62	1.76	1.91	2.08	2.25	2.43	2.60	2.98	3.38	3.79	4.20	4.60
1.20	1.09	1.17	1.27	1.38	1.50	1.64	1.79	1.95	2.11	2.28	2.45	2.82	3.21	3.61	4.02	4.43
1.40	0.994	1.07	1.17	1.28	1.40	1.53	1.67	1.83	1.98	2.14	2.31	2.67	3.04	3.44	3.85	4.26
1.60	0.914	0.992	1.08	1.19	1.30	1.43	1.57	1.72	1.86	2.02	2.18	2.53	2.89	3.28	3.68	4.10
1.80	0.845	0.920	1.01	1.11	1.22	1.34	1.47	1.62	1.76	1.90	2.06	2.39	2.75	3.13	3.52	3.93
2.00	0.784	0.857	0.941	1.04	1.14	1.26	1.39	1.52	1.66	1.80	1.95	2.27	2.62	2.99	3.37	3.77
2.20	0.730	0.801	0.881	0.973	1.08	1.19	1.31	1.44	1.57	1.70	1.85	2.16	2.50	2.86	3.23	3.62
2.40	0.683	0.751	0.828	0.916	1.01	1.12	1.24	1.36	1.49	1.62	1.76	2.06	2.39	2.73	3.10	3.48
2.60	0.641	0.706	0.781	0.865	0.960	1.07	1.18	1.29	1.41	1.54	1.67	1.96	2.28	2.62	2.97	3.35
2.80	0.604	0.666	0.738	0.819	0.910	1.01	1.12	1.23	1.34	1.46	1.59	1.87	2.18	2.51	2.86	3.22
3.00	0.570	0.631	0.700	0.777	0.863	0.961	1.07	1.17	1.28	1.39	1.52	1.79	2.09	2.41	2.74	3.10
x	0.000	0.005	0.017	0.035	0.057	0.083	0.113	0.144	0.178	0.213	0.250	0.327	0.408	0.492	0.579	0.667
y	0.500	0.455	0.417	0.385	0.357	0.333	0.313	0.294	0.278	0.263	0.250	0.227	0.208	0.192	0.179	0.167

Table 8-12.
Coefficients C for Eccentrically Loaded Weld Groups
Angle = 0°

$\phi R_n = CC_1 Dl$ $C_{min} = \dfrac{P_u}{C_1 Dl}$ $D_{min} = \dfrac{P_u}{CC_1 l}$ $l_{min} = \dfrac{P_u}{CC_1 D}$

where

P_u = factored force, kips
D = number of sixteenths-of-an-inch in the fillet weld size
l = characteristic length of weld group, in.
$a = e_x/l$
e_x = horizontal component of eccentricity of P_u with respect to centroid of weld group, in.
C = coefficient tabulated below, which includes $\phi = 0.75$
C_1 = electrode strength coefficient from Table 8-4 (1.0 for E70XX electrodes)

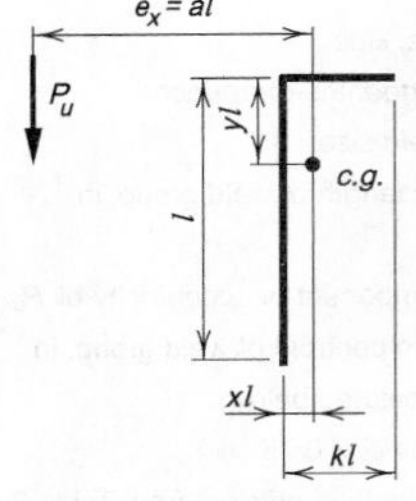

a	k															
	0	0.1	0.2	0.3	0.4	0.5	0.6	0.7	0.8	0.9	1.0	1.2	1.4	1.6	1.8	2.0
0.00	1.39	1.60	1.81	2.02	2.23	2.44	2.65	2.85	3.06	3.27	3.48	3.90	4.32	4.73	5.15	5.57
0.10	1.40	1.55	1.74	1.93	2.12	2.31	2.49	2.66	2.83	2.99	3.14	3.45	3.76	4.08	4.42	4.76
0.15	1.38	1.53	1.71	1.88	2.06	2.22	2.38	2.54	2.69	2.83	2.98	3.28	3.59	3.91	4.24	4.58
0.20	1.32	1.47	1.63	1.79	1.94	2.09	2.23	2.37	2.52	2.67	2.81	3.12	3.42	3.75	4.08	4.41
0.25	1.24	1.38	1.52	1.66	1.80	1.94	2.07	2.21	2.35	2.51	2.66	2.96	3.27	3.59	3.91	4.25
0.30	1.16	1.29	1.42	1.54	1.67	1.79	1.92	2.06	2.20	2.35	2.51	2.82	3.12	3.44	3.76	4.10
0.40	1.00	1.11	1.22	1.32	1.42	1.54	1.66	1.80	1.94	2.09	2.24	2.55	2.85	3.16	3.48	3.81
0.50	0.860	0.958	1.05	1.14	1.24	1.34	1.46	1.58	1.72	1.86	2.01	2.31	2.61	2.91	3.22	3.54
0.60	0.748	0.833	0.913	1.00	1.09	1.18	1.29	1.41	1.54	1.67	1.81	2.10	2.40	2.69	2.99	3.31
0.70	0.659	0.733	0.805	0.884	0.967	1.06	1.15	1.26	1.38	1.51	1.64	1.92	2.21	2.50	2.79	3.09
0.80	0.586	0.652	0.719	0.791	0.868	0.951	1.04	1.14	1.25	1.37	1.50	1.77	2.05	2.32	2.61	2.90
0.90	0.527	0.586	0.648	0.715	0.786	0.864	0.949	1.04	1.15	1.26	1.38	1.63	1.90	2.17	2.44	2.73
1.00	0.478	0.532	0.589	0.651	0.717	0.791	0.870	0.957	1.05	1.16	1.27	1.51	1.77	2.03	2.29	2.57
1.20	0.403	0.448	0.497	0.551	0.609	0.674	0.744	0.820	0.904	1.00	1.10	1.31	1.55	1.80	2.04	2.30
1.40	0.348	0.387	0.430	0.476	0.528	0.586	0.648	0.716	0.790	0.874	0.963	1.16	1.37	1.60	1.83	2.07
1.60	0.305	0.340	0.378	0.419	0.466	0.518	0.573	0.634	0.702	0.776	0.856	1.03	1.23	1.44	1.66	1.88
1.80	0.272	0.303	0.337	0.374	0.416	0.463	0.514	0.569	0.630	0.697	0.771	0.932	1.11	1.31	1.51	1.72
2.00	0.245	0.273	0.304	0.338	0.376	0.419	0.465	0.516	0.572	0.633	0.700	0.848	1.02	1.20	1.39	1.58
2.20	0.223	0.249	0.277	0.308	0.343	0.382	0.425	0.472	0.523	0.579	0.641	0.778	0.932	1.10	1.28	1.47
2.40	0.205	0.228	0.254	0.283	0.315	0.351	0.391	0.434	0.482	0.534	0.591	0.718	0.861	1.02	1.18	1.36
2.60	0.189	0.211	0.235	0.261	0.291	0.325	0.362	0.402	0.446	0.495	0.548	0.667	0.800	0.945	1.10	1.27
2.80	0.176	0.196	0.218	0.243	0.271	0.302	0.337	0.375	0.416	0.461	0.511	0.622	0.747	0.882	1.03	1.18
3.00	0.164	0.183	0.204	0.227	0.253	0.282	0.315	0.350	0.389	0.431	0.478	0.582	0.700	0.825	0.962	1.11
x	0.000	0.005	0.017	0.035	0.057	0.083	0.113	0.144	0.178	0.213	0.250	0.327	0.408	0.492	0.579	0.667
y	0.500	0.455	0.417	0.385	0.357	0.333	0.313	0.294	0.278	0.263	0.250	0.227	0.208	0.192	0.179	0.167

Table 8-12 (cont.). Coefficients *C* for Eccentrically Loaded Weld Groups Angle = 15°

$\phi R_n = CC_1 Dl \qquad C_{min} = \frac{P_u}{C_1 Dl} \qquad D_{min} = \frac{P_u}{CC_1 l} \qquad l_{min} = \frac{P_u}{CC_1 D}$

where

P_u = factored force, kips

D = number of sixteenths-of-an-inch in the fillet weld size

l = characteristic length of weld group, in.

$a = e_x/l$

e_x = horizontal component of eccentricity of P_u with respect to centroid of weld group, in.

C = coefficient tabulated below, which includes $\phi = 0.75$

C_1 = electrode strength coefficient from Table 8-4 (1.0 for E70XX electrodes)

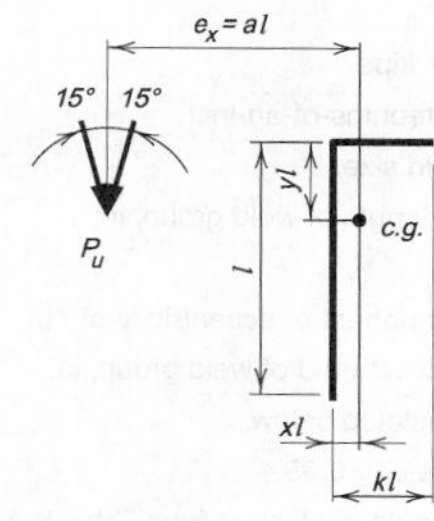

a	k: 0	0.1	0.2	0.3	0.4	0.5	0.6	0.7	0.8	0.9	1.0	1.2	1.4	1.6	1.8	2.0
0.00	1.48	1.69	1.89	2.10	2.31	2.51	2.72	2.92	3.13	3.33	3.54	3.95	4.36	4.77	5.18	5.59
0.10	1.42	1.57	1.74	1.92	2.09	2.27	2.44	2.61	2.78	2.95	3.12	3.45	3.78	4.12	4.46	4.81
0.15	1.38	1.54	1.70	1.86	2.02	2.19	2.35	2.51	2.67	2.83	2.98	3.30	3.62	3.95	4.29	4.64
0.20	1.32	1.47	1.63	1.78	1.94	2.09	2.22	2.36	2.51	2.66	2.82	3.13	3.45	3.78	4.12	4.46
0.25	1.24	1.39	1.54	1.68	1.81	1.94	2.07	2.21	2.35	2.50	2.66	2.98	3.30	3.63	3.96	4.30
0.30	1.16	1.30	1.43	1.55	1.68	1.80	1.93	2.06	2.21	2.36	2.51	2.83	3.15	3.48	3.81	4.15
0.40	1.00	1.12	1.23	1.33	1.44	1.55	1.68	1.81	1.95	2.10	2.25	2.57	2.88	3.20	3.53	3.87
0.50	0.869	0.970	1.06	1.15	1.25	1.36	1.48	1.60	1.74	1.88	2.02	2.33	2.64	2.95	3.28	3.61
0.60	0.759	0.847	0.926	1.01	1.10	1.20	1.31	1.43	1.56	1.69	1.83	2.12	2.43	2.74	3.05	3.37
0.70	0.670	0.746	0.820	0.899	0.983	1.07	1.18	1.29	1.41	1.53	1.67	1.95	2.25	2.54	2.84	3.15
0.80	0.598	0.666	0.734	0.807	0.885	0.969	1.06	1.17	1.28	1.40	1.52	1.79	2.08	2.37	2.66	2.96
0.90	0.539	0.600	0.663	0.730	0.803	0.881	0.968	1.06	1.17	1.28	1.40	1.66	1.93	2.21	2.49	2.78
1.00	0.490	0.545	0.604	0.666	0.734	0.807	0.888	0.979	1.08	1.18	1.30	1.54	1.80	2.08	2.34	2.63
1.20	0.413	0.461	0.511	0.565	0.625	0.690	0.761	0.840	0.927	1.02	1.12	1.34	1.58	1.84	2.09	2.35
1.40	0.357	0.398	0.442	0.490	0.543	0.601	0.664	0.734	0.812	0.897	0.988	1.19	1.40	1.64	1.88	2.12
1.60	0.314	0.350	0.389	0.432	0.479	0.532	0.589	0.651	0.721	0.798	0.879	1.06	1.26	1.48	1.70	1.93
1.80	0.280	0.312	0.347	0.386	0.429	0.476	0.528	0.585	0.649	0.718	0.793	0.958	1.14	1.34	1.55	1.77
2.00	0.253	0.282	0.313	0.348	0.388	0.431	0.479	0.531	0.589	0.652	0.721	0.873	1.04	1.23	1.43	1.63
2.20	0.230	0.257	0.286	0.318	0.354	0.393	0.438	0.486	0.538	0.597	0.660	0.801	0.958	1.13	1.32	1.51
2.40	0.211	0.236	0.262	0.292	0.325	0.362	0.403	0.447	0.496	0.550	0.609	0.740	0.886	1.05	1.22	1.40
2.60	0.195	0.218	0.242	0.270	0.301	0.335	0.373	0.415	0.460	0.510	0.565	0.687	0.824	0.975	1.13	1.31
2.80	0.182	0.203	0.225	0.251	0.280	0.312	0.348	0.386	0.428	0.476	0.527	0.641	0.770	0.910	1.06	1.22
3.00	0.170	0.189	0.211	0.234	0.261	0.291	0.325	0.361	0.401	0.445	0.494	0.601	0.722	0.852	0.993	1.15
x	0.000	0.005	0.017	0.035	0.057	0.083	0.113	0.144	0.178	0.213	0.250	0.327	0.408	0.492	0.579	0.667
y	0.500	0.455	0.417	0.385	0.357	0.333	0.313	0.294	0.278	0.263	0.250	0.227	0.208	0.192	0.179	0.167

Table 8-12 (cont.).
Coefficients *C* for Eccentrically Loaded Weld Groups
Angle = 30°

$\phi R_n = CC_1 Dl \qquad C_{min} = \dfrac{P_u}{C_1 Dl} \qquad D_{min} = \dfrac{P_u}{CC_1 l} \qquad l_{min} = \dfrac{P_u}{CC_1 D}$

where

P_u = factored force, kips

D = number of sixteenths-of-an-inch in the fillet weld size

l = characteristic length of weld group, in.

$a = e_x/l$

e_x = horizontal component of eccentricity of P_u with respect to centroid of weld group, in.

C = coefficient tabulated below, which includes $\phi = 0.75$

C_1 = electrode strength coefficient from Table 8-4 (1.0 for E70XX electrodes)

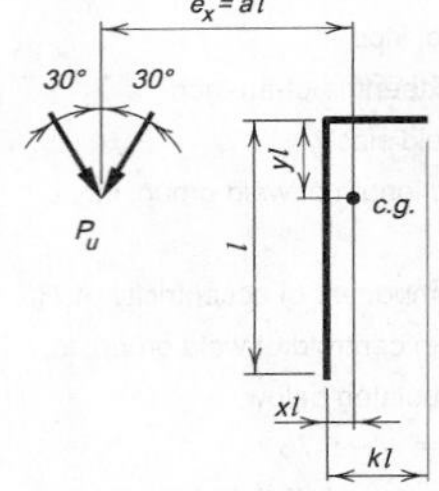

a	*k*															
	0	**0.1**	**0.2**	**0.3**	**0.4**	**0.5**	**0.6**	**0.7**	**0.8**	**0.9**	**1.0**	**1.2**	**1.4**	**1.6**	**1.8**	**2.0**
0.00	1.64	1.83	2.03	2.22	2.42	2.61	2.81	3.01	3.20	3.40	3.59	3.98	4.37	4.76	5.15	5.54
0.10	1.52	1.68	1.85	2.03	2.20	2.38	2.55	2.73	2.90	3.06	3.23	3.59	3.95	4.32	4.70	5.08
0.15	1.44	1.59	1.75	1.91	2.07	2.23	2.40	2.56	2.72	2.88	3.05	3.41	3.77	4.13	4.51	4.89
0.20	1.36	1.51	1.66	1.81	1.96	2.12	2.26	2.41	2.56	2.72	2.89	3.24	3.59	3.94	4.31	4.68
0.25	1.28	1.42	1.56	1.71	1.85	1.99	2.12	2.26	2.41	2.57	2.74	3.09	3.43	3.77	4.13	4.49
0.30	1.20	1.34	1.47	1.60	1.74	1.87	1.99	2.13	2.28	2.43	2.60	2.95	3.28	3.62	3.97	4.32
0.40	1.05	1.17	1.29	1.40	1.51	1.64	1.76	1.90	2.04	2.19	2.35	2.68	3.03	3.35	3.69	4.04
0.50	0.921	1.03	1.13	1.22	1.33	1.45	1.57	1.70	1.83	1.98	2.13	2.45	2.80	3.13	3.46	3.80
0.60	0.812	0.904	0.993	1.08	1.18	1.29	1.41	1.53	1.66	1.80	1.94	2.25	2.59	2.93	3.25	3.58
0.70	0.722	0.804	0.884	0.968	1.06	1.16	1.27	1.38	1.51	1.64	1.78	2.08	2.40	2.73	3.06	3.37
0.80	0.647	0.722	0.796	0.874	0.957	1.05	1.15	1.26	1.38	1.51	1.64	1.92	2.23	2.55	2.87	3.19
0.90	0.586	0.654	0.722	0.794	0.872	0.957	1.05	1.16	1.27	1.39	1.51	1.78	2.08	2.39	2.70	3.02
1.00	0.535	0.596	0.659	0.727	0.799	0.879	0.968	1.07	1.17	1.29	1.40	1.66	1.94	2.24	2.55	2.85
1.20	0.454	0.506	0.560	0.620	0.685	0.755	0.833	0.920	1.02	1.12	1.22	1.46	1.72	1.99	2.28	2.56
1.40	0.393	0.438	0.487	0.540	0.597	0.660	0.729	0.806	0.891	0.984	1.08	1.29	1.53	1.79	2.05	2.32
1.60	0.347	0.386	0.430	0.477	0.529	0.585	0.648	0.717	0.794	0.878	0.967	1.16	1.38	1.61	1.87	2.12
1.80	0.310	0.345	0.384	0.427	0.474	0.526	0.583	0.645	0.715	0.791	0.874	1.05	1.25	1.47	1.70	1.94
2.00	0.280	0.312	0.347	0.386	0.429	0.477	0.529	0.586	0.650	0.720	0.796	0.961	1.15	1.35	1.57	1.79
2.20	0.255	0.285	0.317	0.352	0.392	0.436	0.484	0.537	0.595	0.660	0.730	0.884	1.06	1.24	1.45	1.66
2.40	0.234	0.262	0.291	0.324	0.360	0.401	0.446	0.495	0.549	0.609	0.674	0.817	0.978	1.15	1.34	1.55
2.60	0.217	0.242	0.269	0.299	0.334	0.372	0.413	0.459	0.510	0.566	0.626	0.760	0.910	1.08	1.25	1.44
2.80	0.201	0.225	0.250	0.278	0.311	0.346	0.385	0.428	0.475	0.527	0.584	0.710	0.851	1.01	1.17	1.35
3.00	0.188	0.210	0.234	0.260	0.290	0.324	0.360	0.401	0.445	0.494	0.547	0.666	0.799	0.944	1.10	1.27
x	0.000	0.005	0.017	0.035	0.057	0.083	0.113	0.144	0.178	0.213	0.250	0.327	0.408	0.492	0.579	0.667
y	0.500	0.455	0.417	0.385	0.357	0.333	0.313	0.294	0.278	0.263	0.250	0.227	0.208	0.192	0.179	0.167

Table 8-12 (cont.). Coefficients *C* for Eccentrically Loaded Weld Groups Angle = 45°

$\phi R_n = CC_1 Dl \qquad C_{min} = \dfrac{P_u}{C_1 Dl} \qquad D_{min} = \dfrac{P_u}{CC_1 l} \qquad l_{min} = \dfrac{P_u}{CC_1 D}$

where

P_u = factored force, kips
D = number of sixteenths-of-an-inch in the fillet weld size
l = characteristic length of weld group, in.
$a = e_x / l$
e_x = horizontal component of eccentricity of P_u with respect to centroid of weld group, in.
C = coefficient tabulated below, which includes $\phi = 0.75$
C_1 = electrode strength coefficient from Table 8-4 (1.0 for E70XX electrodes)

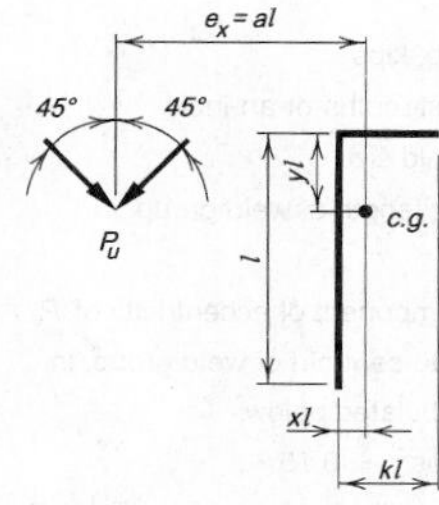

a	k															
	0	0.1	0.2	0.3	0.4	0.5	0.6	0.7	0.8	0.9	1.0	1.2	1.4	1.6	1.8	2.0
0.00	1.81	1.99	2.17	2.35	2.53	2.71	2.89	3.07	3.25	3.43	3.61	3.97	4.33	4.70	5.06	5.42
0.10	1.68	1.83	1.99	2.15	2.32	2.50	2.67	2.85	3.02	3.19	3.37	3.74	4.10	4.48	4.85	5.21
0.15	1.57	1.71	1.86	2.02	2.18	2.36	2.53	2.69	2.86	3.03	3.21	3.58	3.95	4.33	4.71	5.08
0.20	1.47	1.60	1.74	1.88	2.04	2.21	2.38	2.54	2.71	2.88	3.06	3.42	3.80	4.18	4.56	4.94
0.25	1.39	1.51	1.64	1.77	1.92	2.07	2.24	2.41	2.58	2.75	2.93	3.29	3.66	4.04	4.42	4.80
0.30	1.31	1.43	1.55	1.67	1.81	1.95	2.11	2.28	2.45	2.62	2.80	3.16	3.54	3.91	4.29	4.66
0.40	1.16	1.27	1.38	1.49	1.63	1.77	1.92	2.07	2.23	2.39	2.57	2.92	3.27	3.64	4.03	4.41
0.50	1.03	1.13	1.23	1.35	1.48	1.61	1.74	1.88	2.03	2.19	2.36	2.72	3.06	3.40	3.76	4.14
0.60	0.921	1.01	1.11	1.22	1.34	1.46	1.58	1.72	1.86	2.01	2.18	2.52	2.87	3.21	3.55	3.91
0.70	0.829	0.911	1.00	1.11	1.21	1.33	1.45	1.58	1.71	1.86	2.01	2.35	2.70	3.04	3.37	3.71
0.80	0.751	0.828	0.915	1.01	1.11	1.22	1.33	1.45	1.58	1.72	1.87	2.19	2.53	2.88	3.21	3.54
0.90	0.685	0.757	0.839	0.927	1.02	1.12	1.22	1.34	1.47	1.60	1.74	2.05	2.38	2.73	3.05	3.38
1.00	0.629	0.696	0.773	0.854	0.938	1.03	1.14	1.25	1.36	1.49	1.63	1.92	2.24	2.58	2.91	3.23
1.20	0.538	0.598	0.666	0.735	0.810	0.892	0.985	1.09	1.20	1.31	1.43	1.70	2.00	2.31	2.64	2.96
1.40	0.469	0.523	0.582	0.644	0.712	0.786	0.868	0.960	1.06	1.16	1.28	1.52	1.80	2.09	2.40	2.71
1.60	0.416	0.464	0.516	0.572	0.633	0.700	0.774	0.857	0.948	1.05	1.15	1.38	1.63	1.90	2.19	2.50
1.80	0.373	0.416	0.463	0.513	0.570	0.631	0.699	0.775	0.858	0.947	1.04	1.25	1.49	1.74	2.02	2.31
2.00	0.338	0.377	0.419	0.466	0.518	0.574	0.636	0.705	0.782	0.865	0.954	1.15	1.37	1.60	1.86	2.14
2.20	0.308	0.344	0.383	0.426	0.474	0.526	0.583	0.647	0.718	0.795	0.878	1.06	1.26	1.49	1.73	1.99
2.40	0.284	0.317	0.353	0.392	0.436	0.485	0.538	0.598	0.664	0.736	0.812	0.983	1.17	1.38	1.61	1.85
2.60	0.263	0.293	0.327	0.363	0.405	0.450	0.500	0.555	0.617	0.684	0.756	0.916	1.09	1.29	1.50	1.73
2.80	0.245	0.273	0.304	0.338	0.377	0.420	0.467	0.518	0.576	0.638	0.707	0.857	1.03	1.21	1.41	1.62
3.00	0.229	0.256	0.285	0.317	0.353	0.393	0.437	0.486	0.540	0.599	0.663	0.805	0.964	1.14	1.33	1.53
x	0.000	0.005	0.017	0.035	0.057	0.083	0.113	0.144	0.178	0.213	0.250	0.327	0.408	0.492	0.579	0.667
y	0.500	0.455	0.417	0.385	0.357	0.333	0.313	0.294	0.278	0.263	0.250	0.227	0.208	0.192	0.179	0.167

Table 8-12 (cont.). Coefficients *C* for Eccentrically Loaded Weld Groups Angle = 60°

$\phi R_n = CC_1 Dl \qquad C_{min} = \dfrac{P_u}{C_1 Dl} \qquad D_{min} = \dfrac{P_u}{CC_1 l} \qquad l_{min} = \dfrac{P_u}{CC_1 D}$

where

P_u = factored force, kips

D = number of sixteenths-of-an-inch in the fillet weld size

l = characteristic length of weld group, in.

$a = e_x/l$

e_x = horizontal component of eccentricity of P_u with respect to centroid of weld group, in.

C = coefficient tabulated below, which includes $\phi = 0.75$

C_1 = electrode strength coefficient from Table 8-4 (1.0 for E70XX electrodes)

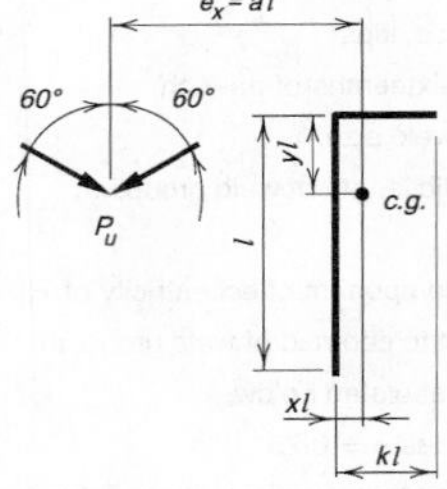

a	*k*															
	0	**0.1**	**0.2**	**0.3**	**0.4**	**0.5**	**0.6**	**0.7**	**0.8**	**0.9**	**1.0**	**1.2**	**1.4**	**1.6**	**1.8**	**2.0**
0.00	1.95	2.12	2.28	2.44	2.61	2.77	2.94	3.10	3.26	3.43	3.59	3.92	4.25	4.57	4.90	5.23
0.10	1.83	1.94	2.07	2.21	2.36	2.53	2.70	2.88	3.06	3.24	3.41	3.76	4.11	4.45	4.78	5.12
0.15	1.73	1.84	1.96	2.09	2.24	2.40	2.57	2.75	2.94	3.13	3.31	3.67	4.03	4.38	4.72	5.06
0.20	1.63	1.74	1.86	1.98	2.12	2.28	2.45	2.63	2.82	3.00	3.20	3.57	3.94	4.30	4.65	4.99
0.25	1.55	1.65	1.76	1.88	2.02	2.18	2.35	2.53	2.72	2.90	3.08	3.46	3.84	4.21	4.56	4.91
0.30	1.47	1.58	1.68	1.80	1.94	2.09	2.26	2.44	2.62	2.81	2.99	3.36	3.73	4.11	4.48	4.83
0.40	1.34	1.44	1.54	1.66	1.79	1.94	2.10	2.27	2.45	2.64	2.83	3.20	3.56	3.93	4.29	4.65
0.50	1.22	1.31	1.41	1.53	1.67	1.81	1.97	2.13	2.30	2.48	2.66	3.04	3.42	3.78	4.14	4.49
0.60	1.12	1.20	1.30	1.42	1.55	1.69	1.85	2.01	2.17	2.35	2.52	2.89	3.27	3.64	4.00	4.36
0.70	1.03	1.11	1.20	1.32	1.45	1.59	1.74	1.90	2.06	2.23	2.40	2.75	3.12	3.50	3.87	4.23
0.80	0.945	1.02	1.12	1.23	1.35	1.49	1.64	1.79	1.95	2.11	2.28	2.63	2.99	3.36	3.73	4.10
0.90	0.874	0.950	1.04	1.15	1.27	1.40	1.54	1.68	1.83	1.99	2.16	2.51	2.87	3.23	3.60	3.96
1.00	0.812	0.886	0.973	1.08	1.19	1.32	1.45	1.58	1.73	1.88	2.05	2.40	2.75	3.10	3.46	3.83
1.20	0.709	0.777	0.858	0.952	1.06	1.17	1.28	1.41	1.54	1.68	1.84	2.17	2.52	2.87	3.21	3.56
1.40	0.626	0.690	0.765	0.852	0.945	1.04	1.15	1.26	1.39	1.52	1.66	1.97	2.31	2.66	3.00	3.34
1.60	0.560	0.619	0.689	0.769	0.850	0.938	1.04	1.14	1.26	1.38	1.52	1.81	2.12	2.46	2.80	3.13
1.80	0.506	0.561	0.626	0.697	0.771	0.853	0.943	1.04	1.15	1.27	1.39	1.66	1.96	2.28	2.61	2.94
2.00	0.461	0.512	0.573	0.636	0.705	0.780	0.864	0.957	1.06	1.17	1.28	1.53	1.82	2.12	2.44	2.77
2.20	0.423	0.471	0.527	0.585	0.649	0.719	0.797	0.883	0.977	1.08	1.19	1.43	1.69	1.98	2.28	2.60
2.40	0.391	0.436	0.487	0.541	0.600	0.666	0.738	0.819	0.907	1.00	1.10	1.33	1.58	1.85	2.14	2.44
2.60	0.363	0.405	0.452	0.502	0.558	0.620	0.687	0.763	0.847	0.937	1.03	1.25	1.48	1.74	2.01	2.30
2.80	0.339	0.379	0.422	0.469	0.521	0.580	0.644	0.715	0.793	0.878	0.969	1.17	1.39	1.63	1.90	2.17
3.00	0.317	0.355	0.395	0.440	0.489	0.545	0.605	0.672	0.746	0.826	0.913	1.10	1.31	1.54	1.79	2.06
x	0.000	0.005	0.017	0.035	0.057	0.083	0.113	0.144	0.178	0.213	0.250	0.327	0.408	0.492	0.579	0.667
y	0.500	0.455	0.417	0.385	0.357	0.333	0.313	0.294	0.278	0.263	0.250	0.227	0.208	0.192	0.179	0.167

Table 8-12 (cont.). Coefficients C for Eccentrically Loaded Weld Groups Angle = 75°

$\phi R_n = CC_1 Dl$ $\quad C_{min} = \dfrac{P_u}{C_1 Dl}$ $\quad D_{min} = \dfrac{P_u}{CC_1 l}$ $\quad l_{min} = \dfrac{P_u}{CC_1 D}$

where

P_u = factored force, kips

D = number of sixteenths-of-an-inch in the fillet weld size

l = characteristic length of weld group, in.

$a = e_x/l$

e_x = horizontal component of eccentricity of P_u with respect to centroid of weld group, in.

C = coefficient tabulated below, which includes $\phi = 0.75$

C_1 = electrode strength coefficient from Table 8-4 (1.0 for E70XX electrodes)

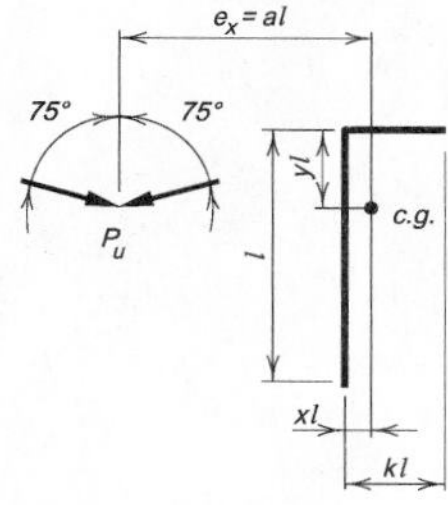

a	k															
	0	0.1	0.2	0.3	0.4	0.5	0.6	0.7	0.8	0.9	1.0	1.2	1.4	1.6	1.8	2.0
0.00	2.05	2.20	2.35	2.50	2.65	2.79	2.94	3.09	3.24	3.39	3.54	3.83	4.13	4.43	4.72	5.02
0.10	1.94	2.01	2.08	2.20	2.34	2.49	2.65	2.81	2.97	3.13	3.28	3.59	3.93	4.25	4.57	4.88
0.15	1.88	1.94	2.03	2.15	2.29	2.44	2.61	2.77	2.94	3.10	3.25	3.56	3.87	4.20	4.52	4.84
0.20	1.82	1.89	1.97	2.09	2.24	2.39	2.56	2.73	2.90	3.07	3.23	3.53	3.84	4.15	4.48	4.80
0.25	1.76	1.83	1.92	2.04	2.19	2.35	2.52	2.69	2.86	3.03	3.20	3.51	3.81	4.11	4.42	4.75
0.30	1.71	1.78	1.87	2.00	2.14	2.30	2.47	2.65	2.82	3.00	3.17	3.49	3.79	4.09	4.39	4.70
0.40	1.62	1.69	1.79	1.91	2.05	2.21	2.38	2.56	2.74	2.92	3.10	3.43	3.75	4.05	4.35	4.65
0.50	1.53	1.61	1.70	1.83	1.97	2.12	2.29	2.47	2.66	2.84	3.03	3.38	3.70	4.02	4.32	4.61
0.60	1.46	1.53	1.63	1.75	1.89	2.05	2.21	2.39	2.57	2.76	2.95	3.31	3.65	3.97	4.28	4.58
0.70	1.38	1.46	1.56	1.68	1.82	1.97	2.14	2.31	2.49	2.68	2.87	3.25	3.60	3.93	4.24	4.55
0.80	1.31	1.39	1.49	1.61	1.75	1.90	2.06	2.24	2.41	2.60	2.79	3.17	3.54	3.88	4.20	4.51
0.90	1.25	1.33	1.43	1.54	1.68	1.83	1.99	2.16	2.34	2.52	2.71	3.10	3.47	3.82	4.16	4.47
1.00	1.19	1.27	1.37	1.48	1.62	1.77	1.93	2.10	2.27	2.45	2.64	3.03	3.41	3.76	4.10	4.43
1.20	1.09	1.16	1.26	1.37	1.50	1.64	1.80	1.97	2.14	2.31	2.50	2.87	3.26	3.64	3.99	4.34
1.40	0.994	1.07	1.16	1.27	1.40	1.54	1.69	1.85	2.01	2.19	2.36	2.73	3.12	3.50	3.87	4.23
1.60	0.914	0.987	1.08	1.18	1.30	1.44	1.58	1.74	1.90	2.07	2.24	2.59	2.97	3.36	3.74	4.11
1.80	0.845	0.915	1.00	1.11	1.22	1.35	1.49	1.64	1.80	1.95	2.11	2.45	2.82	3.21	3.61	3.99
2.00	0.784	0.852	0.937	1.04	1.15	1.27	1.40	1.55	1.70	1.84	2.00	2.33	2.69	3.06	3.46	3.85
2.20	0.730	0.797	0.878	0.972	1.08	1.20	1.33	1.46	1.60	1.74	1.89	2.21	2.56	2.92	3.31	3.71
2.40	0.683	0.748	0.825	0.915	1.02	1.13	1.26	1.39	1.52	1.65	1.80	2.10	2.44	2.80	3.17	3.56
2.60	0.641	0.704	0.778	0.865	0.963	1.07	1.19	1.32	1.44	1.57	1.71	2.00	2.33	2.68	3.05	3.42
2.80	0.604	0.664	0.736	0.819	0.913	1.02	1.13	1.25	1.37	1.49	1.62	1.91	2.23	2.57	2.93	3.30
3.00	0.570	0.628	0.698	0.777	0.867	0.966	1.07	1.18	1.30	1.42	1.55	1.82	2.13	2.46	2.81	3.18
x	0.000	0.005	0.017	0.035	0.057	0.083	0.113	0.144	0.178	0.213	0.250	0.327	0.408	0.492	0.579	0.667
y	0.500	0.455	0.417	0.385	0.357	0.333	0.313	0.294	0.278	0.263	0.250	0.227	0.208	0.192	0.179	0.167

PART 9

DESIGN OF CONNECTION ELEMENTS

CONN.

SCOPE

The specification requirements and other design considerations summarized in this Part apply to the design of connection elements (angles, plates, tees, gussets, etc.) used to transfer load from one structural member to another, as well as the affected elements of the connected members (beam webs, beam flanges, column webs, column flanges, etc.). For design considerations for bolted and welded joints, see Parts 7 and 8, respectively. For the design of connections, see Parts 10 through 15. For the design of connection elements for HSS and steel pipe, see the AISC *Hollow Structural Sections Connections Manual*. For connection elements that are part of a seismic-force-resisting system in which the seismic response modification factor R is taken greater than 3, see the Seismic Provisions, which are available from AISC at www.aisc.org.

LOAD DETERMINATION

The required strength(s) for connection elements are determined using force-transfer models such as those described in Parts 10 through 15 to distribute the member end reactions, which are determined by analysis as indicated in LRFD Specification Section A5.

GROSS AREA, NET AREA AND WHITMORE SECTION

In the determination of the design strength of connection elements, the gross area A_g is of interest for yielding limit states and the net area A_n is of interest for rupture limit states. In either case, the Whitmore section may limit the effective width to less than the overall dimension of a connection element.

Gross Area

The gross area A_g is determined as specified in LRFD Specification Section B1, subject to the limitations given below for the Whitmore section.

Net Area

The net area A_n is determined as specified in LRFD Specification Section B2, except as limited in LRFD Specification Section J5.2(b) and subject to the limitations given below for the Whitmore section. The reduction in area for bolt holes can be determined using Table 9-1.

Whitmore Section (Effective Width)

When connection elements are large in comparison to the bolted or welded joints within them, the Whitmore section may limit the gross and net areas of the connection element to less than the full area (Whitmore, 1952). As illustrated in Figure 9-1, the width of the Whitmore section is determined at the end of the joint by spreading the force from the start of the joint 30 degrees to each side in the connection element along the line of force. The Whitmore section may spread across the joint between connection elements (see Figure 13-1), but cannot spread beyond an unconnected edge.

YIELDING LIMIT STATES

Connection Elements Subject to Shear

The shear yielding design strength ϕR_n, which must equal or exceed the required strength (factored shear) R_u, is determined in accordance with LRFD Specification Section J5.3 using Equation J5-3.

Connection Elements Subject to Tension

The tension yielding design strength ϕR_n, which must equal or exceed the required strength (factored tension) R_u, is determined in accordance with LRFD Specification Section J5.2(a) using Equation J5-1.

Connection Elements Subject to Flexure

The flexural yielding design strength $\phi_b M_n$, which must equal or exceed the required strength (factored moment) M_u, is determined in accordance with LRFD Specification Section F1.1 using Equation F1-1. When connection elements are long enough that lateral-torsional buckling and/or local buckling could control, see LRFD Specification Section F1.2, LRFD Specification Appendix F1 and Cheng et al. (1984).

In some cases, such as those illustrated below (see also Parts 10 through 15), the elastic section modulus S is used conservatively instead of the plastic section modulus Z in the calculation of the flexural yielding design strength.

Coped Beams

For a beam coped at the top flange and/or bottom flange, the flexural yielding design strength, which must equal or exceed the required strength (factored moment) M_u, is $\phi_b M_n$ where

$$\phi_b = 0.90$$
$$M_n = F_y S_{net}$$

where

$M_u = R_u e$, kip-in.
$R_u =$ factored beam end reaction, kips
$e =$ distance from the face of the cope to the point of inflection of the beam, in. It is usually assumed that the point of inflection is located at the face of the supporting member and e is as shown in Figure 9-2. However, depending upon the connection type and stiffness and support condition, the point of inflection may move away from the face of the supporting member; when this is the case, a lesser value of e may be justified. The choice of e shown in Figure 9-2 will be conservative.
$F_y =$ specified minimum yield strength, ksi
$S_{net} =$ net elastic section modulus, in.^{3} Values of S_{net} are tabulated in Table 9-2.

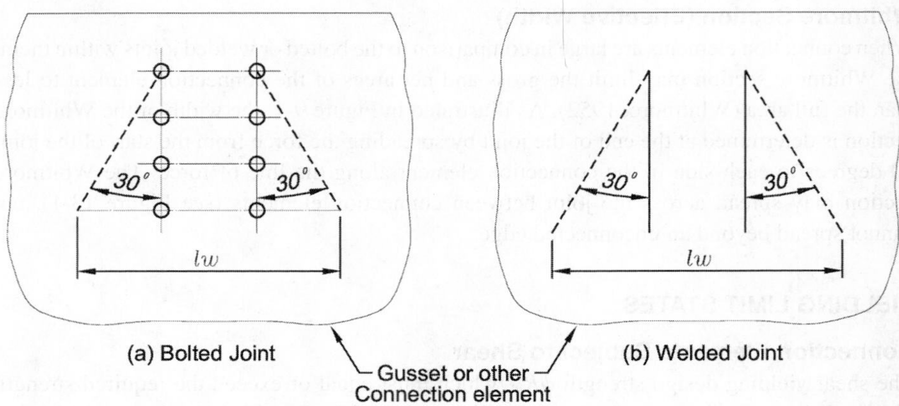

Fig. 9–1. Illustration of the width of the Whitmore section.

RUPTURE LIMIT STATES

Connection Elements Subject to Shear

The shear rupture design strength ϕR_n, which must equal or exceed the required strength (factored shear) R_u, is determined in accordance with LRFD Specification Section J4.1 using Equation J4-1.

Connection Elements Subject to Block Shear Rupture

The block shear rupture design strength ϕR_n, which must equal or exceed the required strength (factored shear) R_u, is determined in accordance with LRFD Specification Section J4.3 using Equations J4-3a and J4-3b. The values tabulated in Tables 9-3 and 9-4 can be used to calculate the block shear rupture design strength.

Connection Elements Subject to Tension

The tension rupture design strength ϕR_n, which must equal or exceed the required strength (factored tension) R_u, is determined in accordance with LRFD Specification Section J4.2 and J5.2(b) using Equation J5-2.

Connection Elements Subject to Flexure

The flexural rupture design strength ϕM_n, which must equal or exceed the required strength (factored moment) M_u, can be determined as

$$\phi_b = 0.75$$
$$M_n = F_u Z_{net}$$

Beams and Girders with Bolt Holes in the Tension Flange

For beams and rolled girders with bolt holes in the tension flange, see LRFD Specification Section B10.

Coped Beams

For a beam coped at the top flange and/or bottom flange, the flexural rupture design strength, which must equal or exceed the required strength (factored moment) M_u, is ϕM_n, where

$$\phi = 0.75$$
$$M_n = F_u S_{net}$$

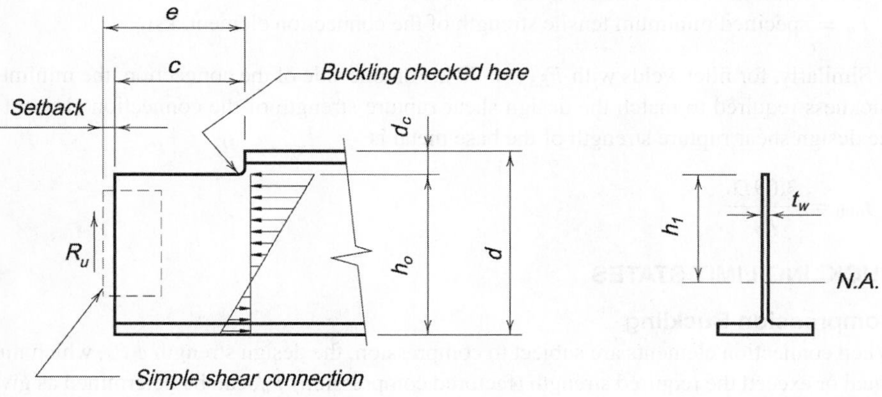

Fig. 9–2. Local buckling of beam web coped at top flange only.

where

$M_u = R_u e$, kip-in.
R_u = factored beam end reaction, kips
e = distance from the face of the cope to the point of inflection of the beam, in. It is usually assumed that the point of inflection is located at the face of the supporting member and e is as shown in Figure 9-2. However, depending upon the connection type and stiffness and support condition, the point of inflection may move away from the face of the supporting member; when this is the case, a lesser value of e may be justified. The choice of e shown in Figure 9-2 will be conservative.
F_u = specified minimum tensile strength, ksi
S_{net} = net elastic section modulus, in.^{3} Values of S_{net} are tabulated in Table 9-2.

Connection Element Rupture Strength at Welds

In many cases, the load path from a weld to the connection element is such that the strength of the connection element can be evaluated directly. However, in some cases, the design strength of the connection element is not directly calculable. For example, while the strength of the beam-web welds for a double-angle connection can be directly calculated, the strength of the beam web at this weld cannot. In cases such as these, it is often convenient to calculate the minimum base-metal thickness that will match the design shear rupture strength of the base metal to the design shear rupture strength of the weld(s).

For fillet welds with $F_{EXX} = 70$ ksi on both sides of the connection element, the minimum thickness required to match the design shear rupture strength of the connection element to the design shear rupture strength of the base metal is

$$t_{min} = \frac{0.75 \times 0.6F_{EXX} \times \dfrac{\sqrt{2}}{2} \times \dfrac{D}{16} \times 2}{0.75 \times 0.6F_u}$$
$$= \frac{6.19D}{F_u}$$

where

D = number of sixteenths of an in. in the weld size.
F_u = specified minimum tensile strength of the connection element, ksi.

Similarly, for fillet welds with $F_{EXX} = 70$ ksi on one side of the connection, the minimum thickness required to match the design shear rupture strength of the connection element to the design shear rupture strength of the base metal is

$$t_{min} = \frac{3.09D}{F_u}$$

BUCKLING LIMIT STATES

Compression Buckling

When connection elements are subject to compression, the design strength ϕP_n, which must equal or exceed the required strength (factored compression) P_u, can be determined as given in Part 4.

Lateral-Torsional Buckling

Most connection elements are short enough that lateral-torsional buckling will not limit the design strength for flexure. However, when connection elements are long enough that lateral-torsional buckling could control, see Part 5, LRFD Specification Section F1.2, LRFD Specification Appendix F1, and Cheng et al. (1984).

Local Buckling

Most connection elements are thick enough that local buckling will not limit the design strength for flexure. When connection elements are thin enough that local buckling could control, see LRFD Specification Appendix F1 and Cheng et al. (1984).

Coped Beams

For beam ends with short copes no greater than the length of the connection angle(s), plate, or tee, local web buckling will generally not occur. For a beam coped at the top flange or both the top and bottom flanges, the flexural local buckling design strength ϕM_n, which must equal or exceed the required strength (factored moment) M_u, is

$$\phi M_n = \phi F_{cr} S_{net}$$

where

$M_u = R_u e$, kip-in.
$R_u =$ factored beam end reaction, kips
$e =$ distance from the face of the cope to the point of inflection of the beam, in. It is usually assumed that the point of inflection is located at the face of the supporting member and e is as shown in Figure 9-2. However, depending upon the connection type and stiffness and support condition, the point of inflection may move away from the face of the supporting member; when this is the case, a lesser value of e may be justified. The choice of e shown in Figure 9-2 will be conservative.
$\phi F_{cr} =$ design buckling stress, determined as given below, ksi.
$S_{net} =$ net elastic section modulus, in.^{3} Values of S_{net} are tabulated in Table 9-2.

When a beam is coped at the top flange only, the design buckling stress ϕF_{cr} is based upon the classical plate buckling formula with a K-factor corresponding to the condition with three edges simply supported and one free edge. An additional factor f is applied to generally account for stress concentration at the cope and to correlate the solutions with experimental results (Cheng et al., 1984). The design buckling stress ϕF_{cr} for a beam coped at the top flange only when $c \leq 2d$ and $d_c \leq d/2$ (see Figure 9-2) is:

$$\phi F_{cr} = \phi \frac{\pi^2 E}{12(1-\nu^2)} \left(\frac{t_w}{h_o}\right)^2 fk$$

$$= 23{,}590 \left(\frac{t_w}{h_o}\right)^2 fk$$

where

$E = 29{,}000$ ksi, modulus of elasticity of steel
$\nu = 0.3$, Poisson's ratio

f = plate buckling model adjustment factor

$$= \frac{2c}{d} \quad \text{when} \quad \frac{c}{d} \leq 1.0$$

$$= 1 + \frac{c}{d} \quad \text{when} \quad \frac{c}{d} > 1.0$$

k = plate buckling coefficient

$$= 2.2\left(\frac{h_o}{c}\right)^{1.65} \quad \text{when} \quad \frac{h_o}{c} \leq 1.0$$

$$= \frac{2.2h_o}{c} \quad \text{when} \quad \frac{h_o}{c} > 1.0$$

$h_o = d - d_c$, reduced beam depth, in. Note that, for convenience, the dimension h_0 as illustrated in Figure 9-2 is used in these calculations instead of the more correct dimension h_1 to eliminate the detailed calculation required to locate the neutral axis of the coped beam. Alternatively, the dimension h_1 may be substituted for h_0 in the local buckling calculations.

c = cope length as illustrated in Figure 9-2, in.

d = beam depth, in.

d_c = cope depth as illustrated in Figure 9-2, in.

When a beam is coped at both flanges, the design buckling stress ϕF_{cr} is based upon a web buckling model with an adjustment factor f_d (Cheng et al., 1984). The design buckling stress ϕF_{cr} for a beam coped at both flanges when $c \leq 2d$ and $d_c \leq 0.2d$ (see Figure 9-3) is:

$$\phi F_{cr} = \phi 0.62\pi E \frac{t_w^{\,2}}{ch_o} f_d$$

$$= 50{,}840 \frac{t_w^{\,2}}{ch_o} f_d$$

where

$$f_d = 3.5 - 7.5\left(\frac{d_c}{d}\right)$$

d_c = cope depth at the compression flange, in.

and all other variables are as defined previously.

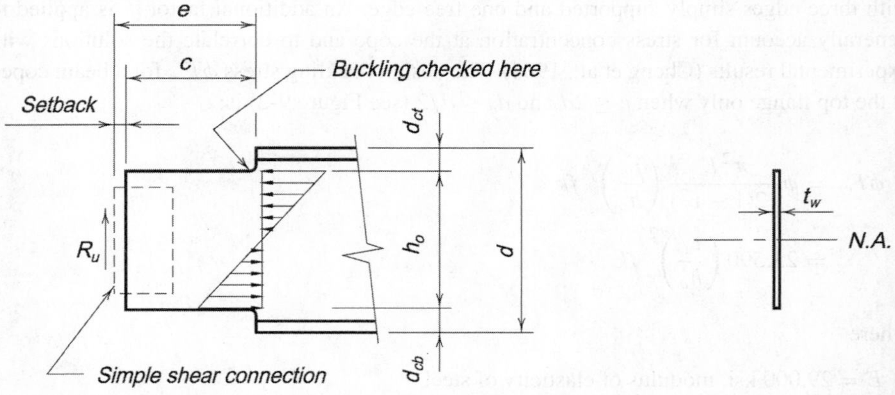

Fig. 9–3. Local buckling of beam web coped at both flanges.

When a beam is coped at both flanges and $d_c > 0.2d$, a conservative procedure also based upon the aforementioned classical plate buckling equation can be used. Including both elastic and inelastic buckling, the design buckling stress ϕF_{cr} is

$$\phi F_{cr} = 0.9 F_y Q$$

where

$Q = 1$ for $\lambda \leq 0.7$
$= (1.34 - 0.486\lambda)$ for $0.7 < \lambda \leq 1.41$
$= (1.30/\lambda^2)$ for $\lambda > 1.41$

$\lambda = \dfrac{\sqrt{F_y}}{167} \dfrac{1}{K} \dfrac{h_o}{2t_w}$

K = plate buckling coefficient, which depends upon the plate aspect ratio. If the length is denoted by c and the half depth of the beam web remaining is denoted by $h_o/2$, then K is related to the aspect ratio $2c/h_o$ as given in the following table:

$2c/h_o$	K
0.25	16
0.3	13
0.4	10
0.5	6
0.6	4.5
0.75	2.5
1	1.3
1.5	0.8
2	0.6
3	0.5
≥4	0.425

When $2c/h_o$ is equal to or greater than 4, K has the constant value of 0.425 and this approach can be reduced to that for single angles given in LRFD Specification Section B5.3a(a).

BEARING LIMIT STATES

Bearing Strength at Bolt Holes

For design bearing strength at bolt holes, see Part 7.

Steel-on-Steel Bearing Strength (Other Than at Bolt Holes)

Bearing strength for applications other than at bolt holes is determined as given in LRFD Specification Section J8. The fabrication and erection requirements in LRFD Specification Sections M2.6, M2.8, and M4.4 are applicable to connection elements that transfer load by contact bearing on steel.

Bearing Strength on Concrete or Masonry

The bearing strength of concrete or masonry is determined as given in LRFD Specification Section J9. The fabrication and erection requirements in LRFD Specification Sections M2.8 and M4.1 are applicable to connection elements that transfer load by contact bearing on concrete or masonry.

OTHER SPECIFICATION REQUIREMENTS AND DESIGN CONSIDERATIONS

The following other specification requirements and design considerations apply to the design of connection elements:

Prying Action

Prying action is a phenomenon (in bolted construction only and for tensile bolt forces only) whereby the deformation of a connection element under a tensile force increases the tensile force in the bolt above that due to the direct tensile force alone. Proper design for prying action includes the selection of bolt diameter and fitting thickness such that there is sufficient stiffness and strength in the connection element and strength in the bolt.

The following procedures for designing and analyzing a tension connection for prying action are based upon the use of factored loads. For treatment of prying action for fatigue and other situations in which service loads are of interest, see Thornton (1985), Astaneh (1985) and AISC (1989).

Consider the tee or angle used in a hanger connection as shown in Figure 9-4. The thickness required to eliminate prying action t_{min} can be determined as:

$$t_{min} = \sqrt{\frac{4.44 r_{ut} b'}{p F_y}}$$

where

r_{ut} = required strength (factored tension) per bolt, kips.

$b' = \left(b - \frac{d_b}{2}\right)$

b = for a tee-type connection element, the distance from bolt centerline to the face of the tee stem, in.; for an angle-type connection element, the distance from bolt centerline to centerline of angle leg, in.

d_b = bolt diameter, in.

p = tributary length per pair of bolts for tee (perpendicular to the plane of the page in Figure 9-4), which should preferably not exceed the gage between the pair of bolts, g.

F_y = specified minimum yield strength of connection element, in.

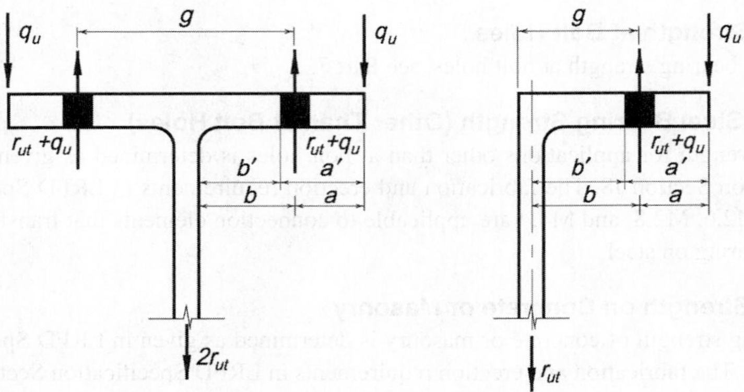

Fig. 9–4. Illustration of variables in prying action calculations.

In this solution, the additional force in the bolt due to prying action q_u is essentially zero as long as the bolt force does not exceed r_{ut}. When the resulting fitting thickness is reasonable, no further check of prying action is necessary.

Alternatively, it is usually possible to determine a lesser required thickness by designing the connection element and bolted joint for the actual effects of prying action with q_u greater than zero. To do so, the preliminary fitting thickness t can be selected based upon flexural yielding such that:

$$r_{ut} \le \frac{F_y t^2 p}{2.22b}$$

Table 15-1 can be used to select the preliminary fitting thickness. Subsequently, the thickness required to ensure an acceptable combination of fitting strength and stiffness and bolt strength t_{min} can be determined as:

$$t_{min} = \sqrt{\frac{4.44 r_{ut} b'}{p F_y (1 + \delta \alpha')}}$$

where

$\delta = 1 - \dfrac{d'}{p}$ ratio of the net area at the bolt line to the gross area at the face of the stem

$\alpha' = 1.0$ if $\beta \ge 1$

$\quad = $ the lesser of 1 and $\dfrac{1}{\delta}\left(\dfrac{\beta}{1-\beta}\right)$ if $\beta < 1$

$d' =$ width of the hole along the length of the fitting, in.

$$\beta = \frac{1}{\rho}\left(\frac{\phi r_n}{r_{ut}} - 1\right)$$

$$\rho = \frac{b'}{a'}$$

$$a' = \left(a + \frac{d_h}{2}\right) \le \left(1.25b + \frac{d_h}{2}\right)$$

$a =$ distance from the bolt centerline to the edge of the fitting, in.

If $t_{min} \le t$, the preliminary fitting thickness is satisfactory. Otherwise, a fitting with a thicker flange, or a change in geometry (i.e., b and p) is required.

Although it is not necessary to do so, if desired, the factored prying force per bolt q_u can be determined as:

$$q_u = \phi r_n \left[\delta \alpha \rho \left(\frac{t}{t_c}\right)^2\right]$$

where

$$\alpha = \frac{1}{\delta}\left[\frac{r_{ut}}{\phi r_n}\left(\frac{t_c}{t}\right)^2 - 1\right] \ge 0$$

$$t_c = \sqrt{\frac{4.44 \phi r_n b'}{p F_y}}$$

$\quad =$ flange or angle thickness required to develop the design strength of the bolt ϕr_n with no prying action, in.

The total factored force per bolt including the effects of prying action can then be determined as $r_u + q_u$.

Alternatively, when the fitting geometry is known, the design tensile strength per bolt ϕr_n, including the effects of prying action, can be determined as Q times the bolt design tensile strength ϕr_n determined per LRFD Specification Section J3.6.That is,

$$\phi r_n = \phi F_t A_b Q$$

where

$Q = 1$ if $\alpha' < 0$, which means that the fitting has sufficient strength and stiffness to develop the full bolt design tensile strength.

$= r_{ut} \leq \phi r_n \left(\frac{t}{t_c}\right)^2 (1 + \delta\alpha')$ if $0 \leq \alpha' \leq 1$, which means that fitting has sufficient strength, but insufficient stiffness to develop the full bolt design tensile strength.

$= r_{ut} \leq \phi r_n \left(\frac{t}{t_c}\right)^2 (1 + \delta)$ if $\alpha' > 1$, which means that the fitting has insufficient strength and stiffness to develop the full bolt design tensile strength.

$\alpha' = \frac{1}{\delta(1+\rho)}\left[\left(\frac{t_c}{t}\right)^2 - 1\right]$ = value of α that either maximizes the bolt design tensile strength for a given thickness or minimizes the thickness required for a given bolt design tensile strength.

Rotational Ductility

The simple shear connections described in Part 10 provide for the rotational ductility required per LRFD Specification Sections A2.2 and J1.2. as follows:

1. For double-angle, shear end-plate, single-angle, single-plate and tee shear connections, the geometry and thickness of the connection elements attached to the support (angle legs, plate or tee flange) are configured so that flexing of those connection elements accommodates the simple-beam end rotation.
2. For unstiffened and stiffened seated connections, the geometry and thickness of the top or side stability angle is configured so that flexing of that connection element accommodates the simple-beam end rotation.
3. For single-plate connections, the geometry and thickness of the plate are configured so that the plate will yield, the bolt group will rotate and/or or the bolt holes will elongate at failure prior to the failure of the welds or bolts.

For each of the simple-shear connections in Part 10, except tee shear connections, prescriptive guidance is provided in Part 10 to ensure adequate rotational ductility. Rotational ductility can be ensured for tee shear connections as follows. Note that this approach can also be used to demonstrate adequate rotational ductility in other simple shear connections that flex to accommodate the simple-beam end rotation, but with configurations that differ from those prescribed in Part 10.

When the connection elements are welded to the support and bolted to the supported beam, the 70 ksi weld size w must be such that

$$w_{\min} = 0.0158 \frac{F_y {t_f}^2}{b}\left(\frac{b^2}{L^2} + 2\right)$$

but need not exceed $^3/_4\, t_s$, where

t_f = thickness of the tee flange, in.
t_s = thickness of the tee stem, in.
b = flexible width in connection element as illustrated in Figure 9-5, in.
L = depth of connection element as illustrated in Figure 9-5, in.

For a tee bolted to the support and bolted or welded to the supported beam, the minimum diameter for bolts through the tee flange for ductility must be such that

$$d_{b\,\min} = 0.613\, t_f \sqrt{\frac{F_y}{b}\left(\frac{b^2}{L^2} + 2\right)}$$

but need not exceed $0.69\sqrt{t_s}$. Additionally, to provide for rotational ductility when the tee stem is bolted to the supported beam, the maximum tee stem thickness should be such that

$$t_{s\,\max} = \frac{d_b}{2} + {}^1/_{16} \text{ in.}$$

When the tee stem is welded to the supported beam, there is no perceived ductility problem for this weld.

Concentrated Forces

If the connection element delivers a concentrated force to a member or other connection element, see LRFD Specification Section K1 or HSS Specification Section 8, as appropriate. See also AISC Design Guide No. 13 *Wide-Flange Column Stiffening at Moment Connections: Wind and Seismic Applications*.

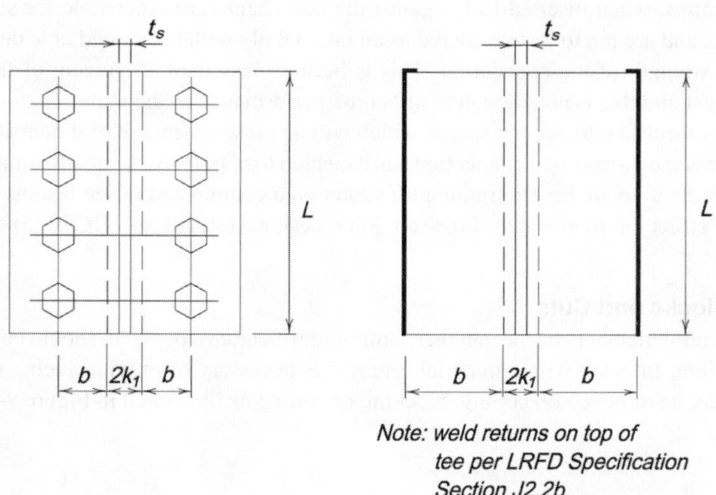

Fig. 9–5. Illustration of variables in shear connection ductility checks.

Local Web Yielding at Beam Ends

At beam ends, the design strength for local web yielding ϕR_n is determined per LRFD Specification Section K1.3 using Equation K1-3, which can be simplified using the constants ϕR_1 and ϕR_2 from Table 9-5 as follows:

$$\phi R_n = \phi R_1 + N(\phi R_2)$$

Note that, as a minimum, N must be equal to k, per LRFD Specification Section K1.3

Web Crippling at Beam Ends

At beam ends, the design strength for web crippling ϕR_n is determined per LRFD Specification Section K1.4 using Equation K1-5a or K1-5b, which can be simplified using constants ϕR_3, ϕR_4, ϕR_5 and ϕR_6 from Table 9-5 as follows:

For $N/d \leq 0.2$:

$$\phi R_n = \phi R_3 + N(\phi R_4)$$

For $N/d > 0.2$:

$$\phi R_n = \phi R_5 + N(\phi R_6)$$

Shims and Fillers

Shims are furnished to the erector for use in filling the spaces allowed for field clearance which might be present at connections such as simple shear connections, PR and FR moment connections, column base plates, and column splices. These shims, illustrated in Figure 9-6, may be either strip shims, with round punched holes, or finger shims, with slots cut through the edge. Whereas strip shims are less expensive to fabricate, finger shims may be laterally inserted and eliminate the need to remove erection bolts or pins already in place.

Finger shims, when inserted fully against the bolt shank, are acceptable for slip-critical connections and are not to be considered as an internal ply with the slotted hole determining the design strength of the connection. This is because less than 25 percent of the contact surface is lost and this is not enough to affect the performance of the joint.

A filler is furnished to occupy spaces which will be present because of dimensional separations between elements of a connection across which load transfer occurs. Examples where fillers might be used are beams framing off center on a column and raised beams.

For the effect of fillers and shims on joint design strength see RCSC Specification Section 5.1.

Copes, Blocks and Cuts

When structural members frame together, a minimum clearance of $^1/_2$ in. should be provided, when possible. In cases where material removal is necessary to provide such a clearance, material may be removed by coping, blocking or cutting as illustrated in Figure 9-7.

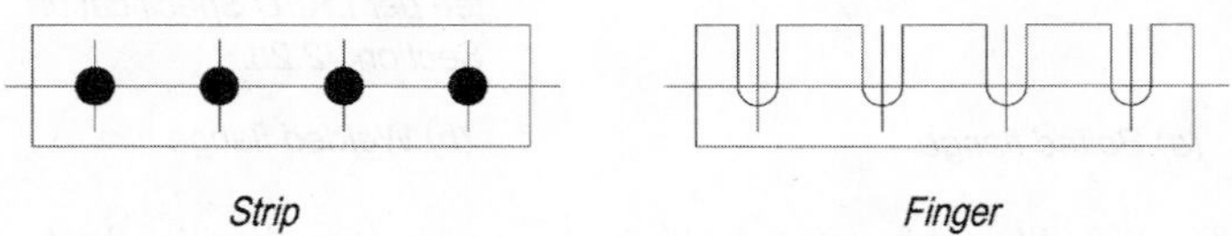

Fig. 9–6. Shims.

Material removal is costly and should be avoided when possible. In some cases, it may be possible to do so by setting the elevations of the tops of infill beams a sufficient distance below the tops of girders to clear the girder fillet radius. Alternatively, a connection such as that illustrated in Figure 9-8 could be used.

When material removal is necessary, coping is usually the most economical method to remove material. The recommended practices for coping are illustrated in Figure 9-9. The potential notch left by the first cut will occur in waste material and subsequently be removed by the second cut. All re-entrant corners must be shaped notch-free per AWS D1.1 to a radius. An approximate minimum radius to which this corner must be shaped is $^1/_2$ in.

Copes, blocks, and cuts can significantly reduce the design strengths of members and may require web reinforcement; it may be more economical to use a heavier member than to provide such reinforcement. The design strength of the unreinforced coped member is determined from the limit states of flexural yielding, local buckling, and lateral torsional buckling, if applicable.

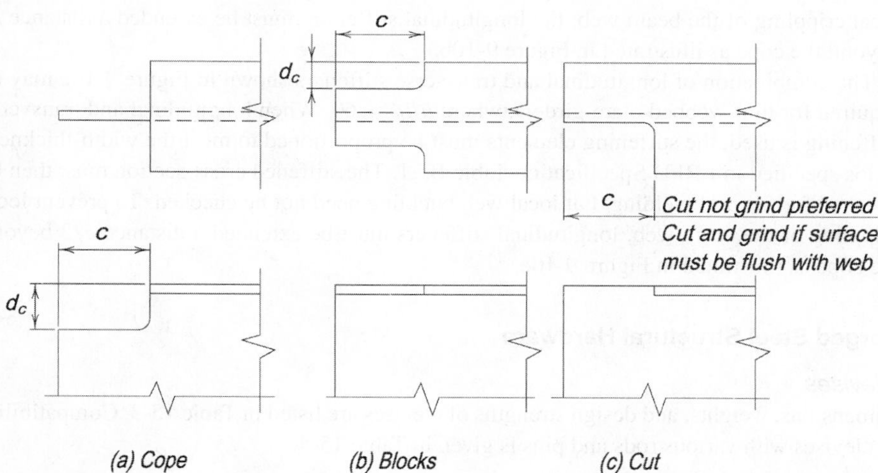

Fig. 9–7. Copes, blocks and cuts.

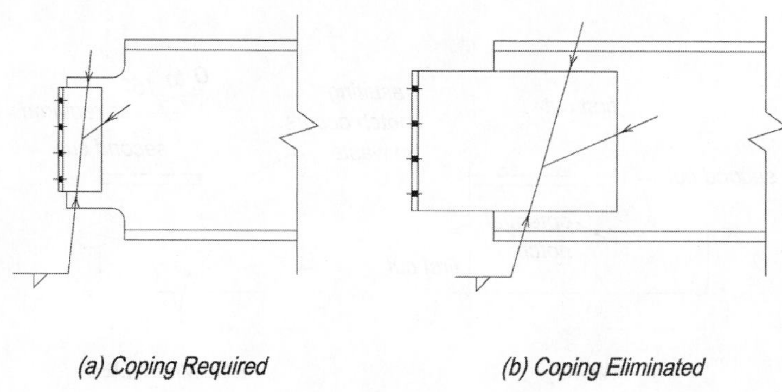

Fig. 9–8. Minimizing coping requirements.

Web Reinforcement of Coped Beams

When the strength of a coped beam is inadequate, either a different beam can be selected to eliminate the need for reinforcement, or reinforcement can be provided to increase the strength. In spite of the increase in material cost, the former solution may be the most economical option due to the appreciable labor cost associated with adding stiffeners and/or doubler plates. When the latter solution is required, some typical reinforcing details are illustrated in Figure 9-10.

The doubler plate illustrated in Figure 9-10a and the longitudinal stiffening illustrated in Figure 9-10b are ued with rolled sections where $h/t_w \leq 60$. When a doubler plate is used, the required doubler-plate thickness $t_{d\,req}$ is determined by substituting the quantity $(t_w + t_{d\,req})$ for t_w in the calculations of the design strength calculations for flexural yielding and local web buckling. To prevent local crippling of the beam web, the doubler plate must be extended at least a distance d_c (depth of cope) beyond the cope as illustrated in Figure 9-10a. When longitudinal stiffening is used, the stiffening elements must be proportioned to meet the width-thickness ratios specified in LRFD Specification Table B5.1. The stiffened cross-section must then be checked for flexural yielding, but local web buckling need not be checked. To prevent local crippling of the beam web, the longitudinal stiffening must be extended a distance d_c beyond the cope as illustrated in Figure 9-10b.

The combination of longitudinal and transverse stiffeners shown in Figure 9-10c may be required for thin-webbed plate-girders, where $h/t_w > 60$. When longitudinal and transverse stiffening is used, the stiffening elements must be proportioned to meet the width-thickness ratios specified in LRFD Specification Table B5.1. The stiffened cross-section must then be checked for flexural yielding, but local web buckling need not be checked. To prevent local crippling of the beam web, longitudinal stiffeners must be extended a distance $c/3$ beyond the cope as illustrated in Figure 9-10c.

Forged Steel Structural Hardware

Clevises

Dimensions, weights, and design strengths of clevises are listed in Table 15-3. Compatibility of clevises with various rods and pins is given in Table 15-4.

Turnbuckles

Dimensions, weights, and design strengths of turnbuckles are listed in Table 15-5.

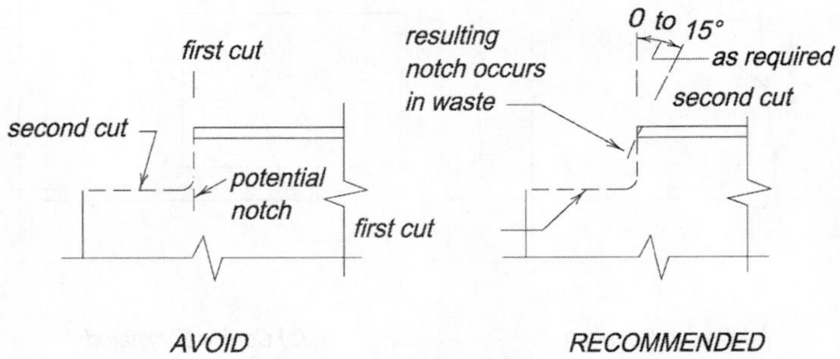

Fig. 9–9. Recommended coping practices.

Sleeve Nuts

Dimensions and weights of sleeve nuts are listed in Table 15-6.

Recessed-Pin Nuts

Dimensions and weights of recessed-pin nuts are listed in Table 15-7.

Cotter Pins

Dimensions and weights of cotter pins are listed in Table 15-8.

DESIGN EXAMPLES

EXAMPLE 9.1. Beam end coped at the top flange only.

Given: For a W21×62 (t_w = 0.400 in., d = 20.99 in., F_y = 50 ksi, F_u = 65 ksi) coped 8-in. deep by 9-in. long at the top flange only:

a. calculate the design strength of the beam end considering the limit states of flexural yielding and local buckling assuming $e = 9^1/_2$ in.

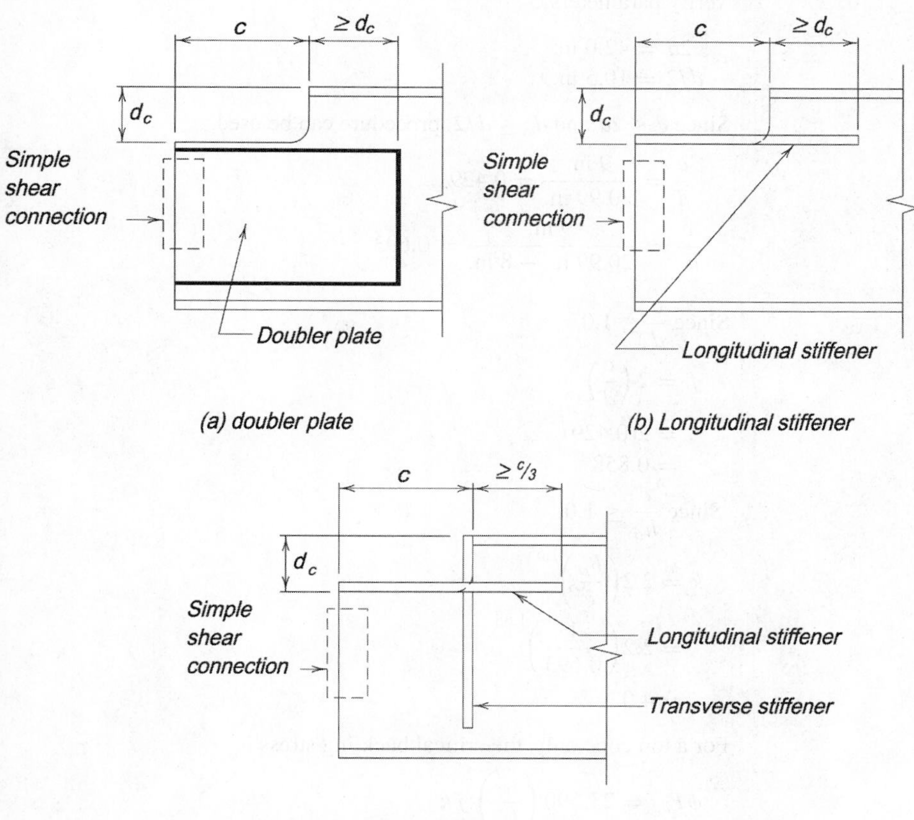

Fig. 9–10. Stiffening alternatives for coped beam ends.

b. determine the alternative W21 that would eliminate the need for stiffening for a required strength R_u of 95 kips

c. design a doubler plate to stiffen the W21×62 for a required strength R_u of 95 kips

d. design longitudinal stiffening for the W21×62 for a required strength R_u of 95 kips

Solution a: *Check flexural yielding*

From Table 9-2, S_{net} = 17.8 in.^{3}

$$\phi R_n = \frac{\phi F_y S_{net}}{e}$$
$$= \frac{0.9(50\text{ ksi})(17.8\text{ in.}^3)}{9^1/_2\text{ in.}}$$
$$= 84.3\text{ kips}$$

Check local buckling

Verify parameters

$$2d = 42.0\text{ in.}$$
$$d/2 = 10.5\text{ in.}$$

Since $c \leq 2d$ and $d_c \leq d/2$, procedure can be used.

$$\frac{c}{d} = \frac{9\text{ in.}}{20.99\text{ in.}} = 0.429$$
$$\frac{c}{h_o} = \frac{9\text{ in.}}{20.99\text{ in.} - 8\text{ in.}} = 0.693$$

Since $\frac{c}{d} \leq 1.0$,

$$f = 2\left(\frac{c}{d}\right)$$
$$= 2(0.429)$$
$$= 0.858$$

Since $\frac{c}{h_o} \leq 1.0$,

$$k = 2.2\left(\frac{h_o}{c}\right)^{1.65}$$
$$= 2.2\left(\frac{1}{0.693}\right)^{1.65}$$
$$= 4.03$$

For a top cope only, the critical buckling stress is

$$\phi F_{bc} = 23{,}590\left(\frac{t_w}{h_o}\right)^2 fk$$
$$= 23{,}590\left(\frac{0.400\text{ in.}}{20.99\text{ in} - 8\text{ in.}}\right)^2(0.858)(4.03)$$
$$= 77.3\text{ ksi}$$

and the design strength is

$$\phi R_n = \frac{\phi F_{bc} S_{net}}{e}$$
$$= \frac{(77.3\text{ ksi})(17.8\text{ in.}^3)}{(9^1/_2\text{ in.})}$$
$$= 145\text{ kips}$$

The design stength of the coped W21×62 is controlled by flexural yielding where

$$\phi R_n = 84.3\text{ kips}$$

Solution b:

If the required strength R_u were 95 kips, the W21×62 would be inadequate due to the limit state of flexural yielding. The required net elastic section modulus S_{req} would be

$$S_{req} = \frac{R_u e}{\phi F_y}$$
$$= \frac{95\text{ kips}(9^1/_2\text{ in.})}{0.9(50\text{ ksi})}$$
$$= 20.1\text{ in.}^3$$

From Table 9-2, a W21×73 with an 8-in. deep cope provides

$$S_{net} = 21.0\text{ in.}^3 > 20.1\text{ in.}^3 \quad \textbf{o.k.}$$

Check local buckling

Since the W21×62 provided $\phi R_n = 145$ kips > 95 kips for the limit state of local buckling, local buckling is not critical for the W21×73 with an 8-in. deep cope.

Solution c: *Design doubler plate*

From solutions a and b, the doubler plate must provide for 95 kips − 84.3 kips = 10.7 kips. Conservatively ignoring the effect of the Ad^2 term in computing the section modulus, the required section modulus for the doubler plate is

$$S_{req} = \frac{(R_u - \phi R_{n\ beam})e}{\phi F_y}$$
$$= \frac{(95\text{ kips} - 84.3\text{ kips})\ 9^1/_2\text{ in.}}{0.9(36\text{ ksi})}$$
$$= 3.14\text{ in.}^3$$

For an 8-in. deep plate,

$$t_{req} = \frac{6S_{req}}{d^2}$$
$$= \frac{6(3.14\text{ in.}^3)}{(8\text{ in.})^2}$$
$$= 0.294\text{ in.}$$

Thus, since the doubler plate must extend at least d_c beyond the cope, use PL $^5/_{16}$ in. × 8 in. × 1′-5

Solution d: *Design longitudinal stiffeners*

Try PL $^1/_4$ in. × 4 in. slotted to fit over beam web, $F_y = 50$ ksi. The neutral axis is located 4.40 in. from the bottom flange (8.84 in. from the top of the stiffener) and the elastic section modulus of the reinforced section is as follows:

	I_o (in.^{4})	Ad^2 (in.^{4})	$I_o + Ad^2$(in.^{4})
Stiffener	0.00521	76.0	76.0
W21×62 web	63.2	28.6	91.8
W21×62 bottom flange	0.160	84.9	85.1
			$I_x = 253$ in.^{4}

$$
\begin{aligned}
S_{net} &= \frac{I_x}{c} \\
&= \frac{253 \text{ in.}^4}{8.84 \text{ in.}} \\
&= 28.6 \text{ in.}^3
\end{aligned}
$$

and the design strength of the section is

$$
\begin{aligned}
\phi R_n &= \frac{\phi F_y S_{net}}{e} \\
&= \frac{0.9(50 \text{ ksi})(28.6 \text{ in.}^3)}{(9^1/_2\text{-in.})} \\
&= 136 \text{ kips} > 95 \text{ kips} \quad \textbf{o.k.}
\end{aligned}
$$

Thus, since the longitudinal stiffening must extend at least d_c beyond the cope,

Use PL $^1/_4$ in. × 4 in. × 1′-5.

EXAMPLE 9.2. Beam end coped at the top and bottom flanges.

Given: For a W21×62 ($t_w = 0.400$ in., $d = 20.99$ in., $F_y = 50$ ksi, $F_u = 65$ ksi) coped 3-in. deep by 7-in. long at the top flange and 4-in. deep by 7-in. long at the bottom flange. Calculate the design strength of the beam end considering the limit states of flexural yielding and local buckling assuming $e = 7^1/_2$ in.

Solution: *Check flexural yielding*

$$
\begin{aligned}
S_{net} &= \frac{t_w {h_o}^2}{6} \\
&= (0.400 \text{ in.})(20.99 \text{ in.} - 3 \text{ in.} - 4 \text{ in.})^2/6 \\
&= 13.1 \text{ in.}^3
\end{aligned}
$$

$$\phi R_n = \frac{\phi F_y S_{net}}{e}$$
$$= \frac{0.9(50\text{ ksi})(13.1\text{ in.}^3)}{(7^1/_2\text{ in.})}$$
$$= 78.6\text{ kips}$$

Check local buckling

Verify parameters

$$2d = 42.0\text{ in.}$$
$$0.2d = 4.20\text{ in.}$$

Since, for each cope, $c \leq 2d$ and $d_c \leq 0.2d$, procedure can be used.

$$f_d = 3.5 - 7.5\left(\frac{d_c}{d}\right)$$
$$= 3.5 - 7.5\left(\frac{4\text{ in.}}{20.99\text{ in.}}\right)$$
$$= 2.07$$

For the doubly coped beam, the critical stress is

$$\phi F_{bc} = 50{,}840\frac{t_w{}^2}{ch_o}f_d$$
$$= 50{,}840\left[\frac{(0.400\text{ in.})^2}{(7\text{ in.})(20.99\text{ in.} - 3\text{ in.} - 4\text{ in.})}\right](2.07)$$
$$= 172\text{ ksi}$$

and the design strength is

$$\phi R_n = \frac{\phi F_{bc} S_{net}}{e}$$
$$= \frac{(172\text{ ksi})(13.1\text{ in.}^3)}{(7^1/_2\text{ in})}$$
$$= 300\text{ kips}$$

The design strength of the coped W21×62 is controlled by flexural yielding where

$$\phi R_n = 78.6\text{ kips}$$

DESIGN TABLES

Table 9-1. Reduction in Area for Holes

Area reduction for standard, oversized, short-slotted, and long-slotted holes in material thicknesses from $^3/_{16}$ in. to 1 in. are given in Table 9-1. For material thicknesses not listed, the tabular value for 1-in. thickness can be multiplied by the actual thickness.

Table 9-2. Section Modulus of Coped W-Shapes

Values are given for the gross and net elastic section moduli for coped W-shapes as illustrated in the table header.

Tables 9-3 and 9-4. Block Shear Rupture

The terms in LRFD Specification Equations J4-3a and J4-3b are tabulated in Tables 9-3a, 9-3b, 9-4a and 9-4b. The indicated values are given per inch of material thickness.

Table 9-5. Beam End Bearing Constants

At beam ends, the design strength for local web yielding ϕR_n is determined per LRFD Specificaion Section K1.3 using Equation K1-3, which can be simplified using the constants ϕR_1 and ϕR_2 from Table 9-5 as follows:

$$\phi R_n = \phi R_1 + N(\phi R_2)$$

Note that, as a minimum, N must be equal to k, per LRFD Specification Section K1.3

Also, at beam ends, the design strength for web crippling ϕR_n is determined per LRFD Specification Section K1.4 using Equation K1-5a or K1-5b, which can be simplified using constants $\phi_r R_3$, $\phi_r R_4$, $\phi_r R_5$ and $\phi_r R_6$ from Table 9-5 as follows:

For $N/d \leq 0.2$:

$$\phi R_n = \phi_r R_3 + N(\phi_r R_4)$$

For $N/d > 0.2$:

$$\phi R_n = \phi_r R_5 + N(\phi_r R_6)$$

The design strength of the supported beam web ϕR (based on web crippling and web yielding, taking $N = 3^1/_4$) and the design shear strength of the beam web $\phi_v V_n$ are tabulated for convenience.

PART 9 REFERENCES

American Institute of Steel Construction, Inc., 1989, *Manual of Steel Construction, Allowable Stress Design and Plastic Design,* AISC, Chicago, IL.

Astaneh, A., 1985, "Procedure for Design Analysis of Hanger-Type Connections," *Engineering Journal*, Vol. 22, No. 2, (2nd Qtr.), pp. 63–66, AISC, Chicago, IL.

Cheng, J.J., J.A. Yura, and C.P. Johnston, 1984, "Design and Behavior of Coped Beams," Department of Civil Engineering, The University of Texas at Austin, Austin, TX.

Salmon, C.G. and J.E. Johnson, 1996, *Steel Structures: Design and Behavior*, 4th Edition, Harper Collins, New York, NY.

Thornton, W.A., 1985, "Prying Action—A General Treatment," *Engineering Journal*, Vol. 22, No. 2, (2nd Qtr.), pp. 67–75, AISC, Chicago, IL.

Whitmore, R.E., 1952, "Experimental Investigation of Stresses in Gusset Plates," *Bulletin No. 16*, Civil Engineering, The University of Tennessee Engineering Experiment Station, Knoxville, TN.

Table 9-1.
Reduction in Area for Holes, in.^{2}

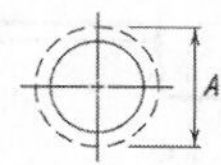

STD
Standard Hole

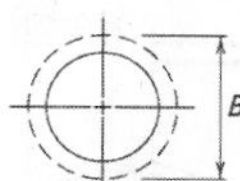

OVS
Oversized Hole

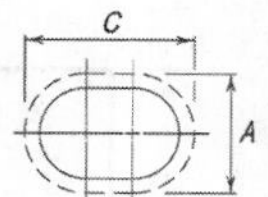

SSL
Short-Slotted Hole

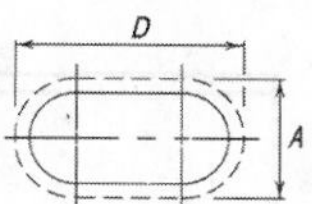

LSL
Long-Slotted Hole

Thickness t, in.	$A \times t$							$B \times t$						
	Bolt Diameter d_b, in.							Bolt Diameter d_b, in.						
	3/4	7/8	1	1 1/8	1 1/4	1 3/8	1 1/2	3/4	7/8	1	1 1/8	1 1/4	1 3/8	1 1/2
3/16	0.164	0.188	0.211	0.234	0.258	0.281	0.305	0.188	0.211	0.246	0.281	0.305	0.328	0.352
1/4	0.219	0.250	0.281	0.312	0.344	0.375	0.406	0.250	0.281	0.328	0.375	0.406	0.438	0.469
5/16	0.273	0.312	0.352	0.391	0.430	0.469	0.508	0.312	0.352	0.410	0.469	0.508	0.547	0.586
3/8	0.328	0.375	0.422	0.469	0.516	0.562	0.609	0.375	0.422	0.492	0.562	0.609	0.656	0.703
7/16	0.383	0.438	0.492	0.547	0.602	0.656	0.711	0.438	0.492	0.574	0.656	0.711	0.766	0.820
1/2	0.438	0.500	0.562	0.625	0.688	0.750	0.812	0.500	0.562	0.656	0.750	0.812	0.875	0.938
9/16	0.492	0.562	0.633	0.703	0.773	0.844	0.914	0.562	0.633	0.738	0.844	0.914	0.984	1.05
5/8	0.547	0.625	0.703	0.781	0.859	0.938	1.02	0.625	0.703	0.820	0.938	1.02	1.09	1.17
11/16	0.602	0.688	0.773	0.859	0.945	1.03	1.12	0.688	0.773	0.902	1.03	1.12	1.20	1.29
3/4	0.656	0.750	0.844	0.938	1.03	1.12	1.22	0.750	0.844	0.984	1.12	1.22	1.31	1.41
13/16	0.711	0.812	0.914	1.02	1.12	1.22	1.32	0.812	0.914	1.07	1.22	1.32	1.42	1.52
7/8	0.766	0.875	0.984	1.09	1.20	1.31	1.42	0.875	0.984	1.15	1.31	1.42	1.53	1.64
15/16	0.820	0.938	1.05	1.17	1.29	1.41	1.52	0.938	1.05	1.23	1.41	1.52	1.64	1.76
1	0.875	1.00	1.12	1.25	1.38	1.50	1.62	1.00	1.12	1.31	1.50	1.62	1.75	1.88

Thickness t, in.	$C \times t$							$D \times t$						
	Bolt Diameter d_b, in.							Bolt Diameter d_b, in.						
	3/4	7/8	1	1 1/8	1 1/4	1 3/8	1 1/2	3/4	7/8	1	1 1/8	1 1/4	1 3/8	1 1/2
3/16	0.199	0.223	0.258	0.293	0.316	0.340	0.363	0.363	0.422	0.480	0.539	0.598	0.656	0.715
1/4	0.266	0.297	0.344	0.391	0.422	0.453	0.484	0.484	0.562	0.641	0.719	0.797	0.875	0.953
5/16	0.332	0.371	0.430	0.488	0.527	0.566	0.605	0.605	0.703	0.801	0.898	0.996	1.09	1.19
3/8	0.398	0.445	0.516	0.586	0.633	0.680	0.727	0.727	0.844	0.961	1.08	1.20	1.31	1.43
7/16	0.465	0.520	0.602	0.684	0.738	0.793	0.848	0.848	0.984	1.12	1.26	1.39	1.53	1.67
1/2	0.531	0.594	0.688	0.781	0.844	0.906	0.969	0.969	1.12	1.28	1.44	1.59	1.75	1.91
9/16	0.598	0.668	0.773	0.879	0.949	1.02	1.09	1.09	1.27	1.44	1.62	1.79	1.97	2.14
5/8	0.664	0.742	0.859	0.977	1.05	1.13	1.21	1.21	1.41	1.60	1.80	1.99	2.19	2.38
11/16	0.730	0.816	0.945	1.07	1.16	1.25	1.33	1.33	1.55	1.76	1.98	2.19	2.41	2.62
3/4	0.797	0.891	1.03	1.17	1.27	1.36	1.45	1.45	1.69	1.92	2.16	2.39	2.62	2.86
13/16	0.863	0.965	1.12	1.27	1.37	1.47	1.57	1.57	1.83	2.08	2.34	2.59	2.84	3.10
7/8	0.930	1.04	1.20	1.37	1.48	1.59	1.70	1.70	1.97	2.24	2.52	2.79	3.06	3.34
15/16	0.996	1.11	1.29	1.46	1.58	1.70	1.82	1.82	2.11	2.40	2.70	2.99	3.28	3.57
1	1.06	1.19	1.38	1.56	1.69	1.81	1.94	1.94	2.25	2.56	2.88	3.19	3.50	3.81

Table 9-2.
Section Moduli for Coped W-Shapes

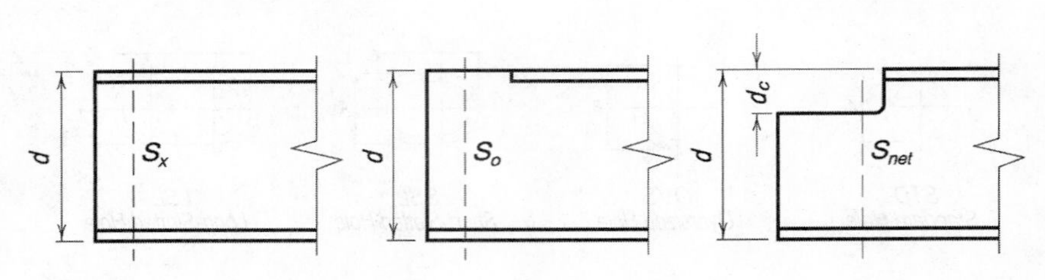

Shape	d, in.	t_f, in.	S_x, in.^{3}	S_o, in.^{3}	S_{net}, in.^{3}								
					d_c, in.								
					2	3	4	5	6	7	8	9	10
W44×335	44.0	1.77	1410	492	451	431	411	392	373	355	337	319	302
×290	43.6	1.58	1240	416	381	364	348	331	315	300	284	269	255
×262	43.3	1.42	1120	374	342	327	312	297	283	269	255	241	228
×230	42.9	1.22	971	330	301	288	274	261	249	236	224	212	200
W40×593	43.0	3.23	2340	810	–	–	672	639	607	576	545	516	487
×503	42.1	2.76	1980	674	–	585	557	529	502	476	450	425	401
×431	41.3	2.36	1690	568	–	492	468	444	421	399	377	355	335
×397	41.0	2.20	1560	513	–	445	423	401	380	360	340	320	301
×372	40.6	2.05	1460	479	–	415	394	374	354	335	316	298	280
×362	40.6	2.01	1420	464	–	401	381	362	343	324	306	288	271
×324	40.2	1.81	1280	409	371	353	335	318	301	284	268	253	237
×297	39.8	1.65	1170	373	338	322	305	289	274	259	244	229	215
×277	39.7	1.58	1100	336	304	289	275	260	246	232	219	206	193
×249	39.4	1.42	993	300	272	258	245	232	219	207	195	183	172
×215	39.0	1.22	859	256	232	220	209	198	187	176	166	156	146
×199	38.7	1.07	770	248	225	213	202	191	181	171	161	151	141
W40×392	41.6	2.52	1440	582	–	505	481	457	433	410	388	366	345
×331	40.8	2.13	1210	484	–	419	398	378	358	339	320	302	284
×327	40.8	2.13	1200	470	–	407	387	367	348	329	311	293	276
×278	40.2	1.81	1020	397	360	343	326	309	293	277	261	246	231
×264	40.0	1.73	971	371	337	321	305	289	273	258	244	230	216
×235	39.7	1.58	875	321	291	277	263	249	236	223	210	198	186
×211	39.4	1.42	786	286	260	247	234	222	210	198	187	176	165
×183	39.0	1.22	683	244	221	210	200	189	179	169	159	149	140
×167	38.6	1.02	600	234	212	201	191	181	171	161	152	142	134
×149	38.2	0.830	513	217	196	186	177	167	158	149	140	132	123
W36×798	42.0	4.29	2980	1020	–	–	–	795	754	713	674	635	598
×650	40.5	3.54	2420	795	–	–	650	616	583	551	519	488	458
×527	39.2	2.91	1950	618	–	530	503	475	449	423	398	373	350
×439	38.3	2.44	1620	504	–	431	408	385	363	342	321	301	281
×393	37.8	2.20	1450	443	–	378	358	338	318	299	281	263	245
×359	37.4	2.01	1320	400	–	341	322	304	286	269	252	236	220
×328	37.1	1.85	1210	360	324	307	290	274	257	242	227	212	198
×300	36.7	1.68	1110	328	295	279	263	248	233	219	205	192	179
×280	36.5	1.57	1030	304	274	259	244	230	216	203	190	177	165
×260	36.3	1.44	953	285	256	243	229	216	203	190	178	166	155
×245	36.1	1.35	895	269	241	228	215	203	191	179	167	156	145
×230	35.9	1.26	837	253	227	214	202	190	179	168	157	146	136

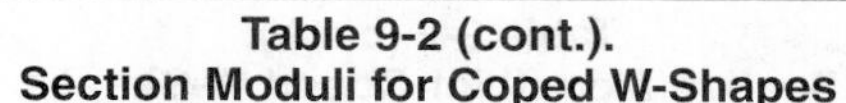

Table 9-2 (cont.). Section Moduli for Coped W-Shapes

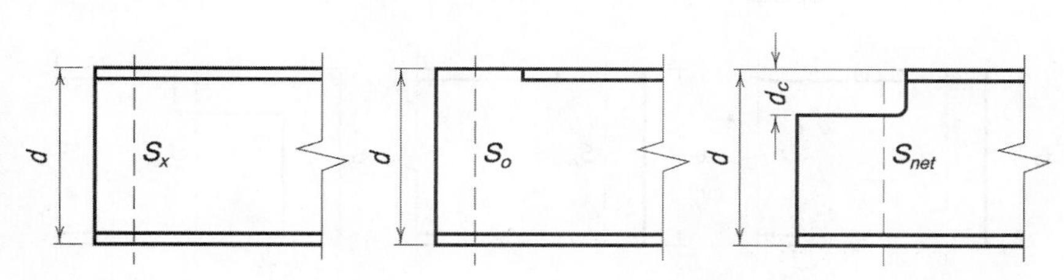

Shape	d, in.	t_f, in.	S_x, in.^{3}	S_o, in.^{3}	S_{net}, in.^{3}								
					d_c, in.								
					2	3	4	5	6	7	8	9	10
W36×256	37.4	1.73	895	329	296	281	265	251	236	222	208	195	182
×232	37.1	1.57	809	294	265	251	237	224	211	198	186	174	163
×210	36.7	1.36	719	273	245	232	220	207	195	183	172	161	150
×194	36.5	1.26	664	249	224	212	201	189	178	167	157	147	137
×182	36.3	1.18	623	234	210	199	188	177	167	157	147	137	128
×170	36.2	1.10	581	218	196	186	175	165	156	146	137	128	119
×160	36.0	1.02	542	206	185	175	165	156	147	138	129	120	112
×150	35.9	0.940	504	196	176	167	157	148	139	131	122	114	107
×135	35.6	0.790	439	182	163	154	146	137	129	121	113	106	98.5
W33×387	36.0	2.28	1350	414	–	350	330	311	292	273	255	238	221
×354	35.6	2.09	1240	374	–	316	298	280	263	246	230	214	198
×318	35.2	1.89	1110	330	296	279	262	247	231	216	202	188	174
×291	34.8	1.73	1020	299	267	252	237	223	208	195	181	169	156
×263	34.5	1.57	919	268	239	225	212	199	186	174	162	150	139
×241	34.2	1.40	831	250	223	210	198	185	173	162	151	140	129
×221	33.9	1.27	759	229	204	193	181	170	159	148	138	128	118
×201	33.7	1.15	686	209	186	175	165	154	144	135	125	116	107
W33×169	33.8	1.22	549	191	170	160	151	141	132	123	115	106	98.5
×152	33.5	1.06	487	177	157	148	139	130	122	114	106	98.1	90.6
×141	33.3	0.960	448	165	147	139	130	122	114	106	98.8	91.6	84.6
×130	33.1	0.855	406	155	138	130	122	114	107	99.6	92.6	85.8	79.2
×118	32.9	0.740	359	144	128	120	113	106	98.9	92.2	85.6	79.3	73.2
W30×391	33.2	2.44	1250	378	–	315	295	276	257	239	222	205	189
×357	32.8	2.24	1140	339	–	282	264	246	230	213	197	182	168
×326	32.4	2.05	1040	306	–	254	237	221	206	191	177	163	150
×292	32.0	1.85	930	269	237	223	208	194	180	167	154	142	130
×261	31.6	1.65	829	240	212	198	185	172	160	148	137	126	115
×235	31.3	1.50	748	211	186	174	163	152	141	130	120	110	101
×211	30.9	1.32	665	192	169	158	148	137	127	118	108	99.4	90.9
×191	30.7	1.19	600	174	153	143	133	124	115	106	97.9	89.8	82.0
×173	30.4	1.07	541	157	138	129	121	112	104	95.8	88.2	80.8	73.7
W30×148	30.7	1.18	436	152	134	126	117	109	101	93.6	86.2	79.1	72.3
×132	30.3	1.00	380	139	122	114	107	99.2	92.0	85.0	78.2	71.7	65.4
×124	30.2	0.930	355	131	116	108	101	93.6	86.7	80.1	73.7	67.6	61.7
×116	30.0	0.850	329	124	109	102	95.3	88.5	82.0	75.7	69.7	63.8	58.2
×108	29.8	0.760	299	117	103	96.4	89.8	83.4	77.2	71.3	65.5	60.0	54.7
×99	29.7	0.670	269	110	96.8	90.4	84.2	78.2	72.4	66.9	61.5	56.3	51.3
×90	29.5	0.610	245	98.5	86.5	80.8	75.2	69.8	64.6	59.6	54.7	50.1	45.6

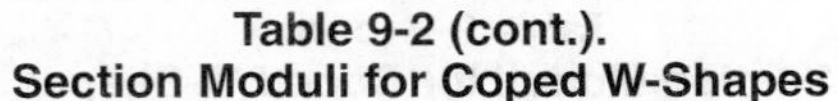

Table 9-2 (cont.). Section Moduli for Coped W-Shapes

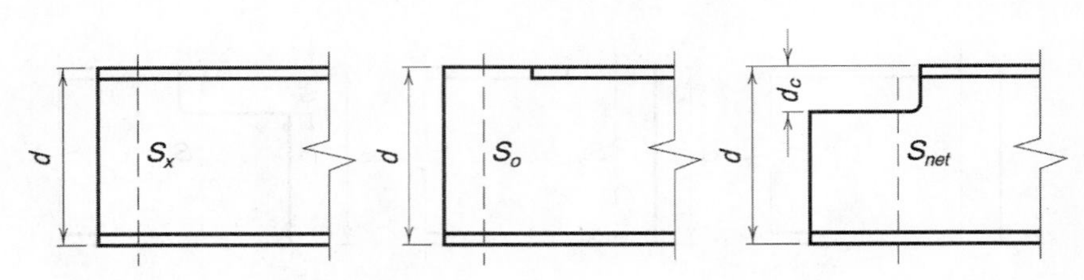

Shape	d, in.	t_f, in.	S_x, in.^{3}	S_o, in.^{3}	S_{net}, in.^{3}								
					d_c, in.								
					2	3	4	5	6	7	8	9	10
W27×539	32.5	3.54	1570	508	–	–	393	366	341	316	292	268	246
×368	30.4	2.48	1060	321	–	262	244	226	209	193	177	162	147
×336	30.0	2.28	972	287	–	234	218	202	186	172	157	144	130
×307	29.6	2.09	887	259	–	211	196	181	167	153	140	128	116
×281	29.3	1.93	814	233	204	190	176	163	150	138	126	115	104
×258	29.0	1.77	745	212	185	172	160	147	136	125	114	103	93.5
×235	28.7	1.61	677	193	168	157	145	134	123	113	103	93.6	84.5
×217	28.4	1.50	627	174	151	140	130	120	110	101	92.0	83.4	75.2
×194	28.1	1.34	559	154	134	125	115	106	97.5	89.2	81.2	73.5	66.2
×178	27.8	1.19	505	145	126	117	108	99.7	91.5	83.6	76.0	68.8	61.9
×161	27.6	1.08	458	131	113	105	97.2	89.5	82.1	75.0	68.1	61.6	55.3
×146	27.4	0.975	414	118	103	95.2	87.9	80.9	74.1	67.6	61.4	55.5	49.8
W27×129	27.6	1.10	345	116	101	93.8	86.7	79.9	73.3	67.0	60.9	55.1	49.5
×114	27.3	0.930	299	106	91.7	85.0	78.5	72.3	66.3	60.5	55.0	49.7	44.7
×102	27.1	0.830	267	94.2	81.6	75.7	69.9	64.3	58.9	53.8	48.8	44.1	39.6
×94	26.9	0.745	243	87.8	76.1	70.4	65.0	59.8	54.8	50.0	45.3	40.9	36.7
×84	26.7	0.640	213	80.4	69.6	64.4	59.5	54.7	50.0	45.6	41.4	37.3	33.4
W24×370	28.0	2.72	957	296	–	237	219	201	185	168	153	138	124
×335	27.5	2.48	864	261	–	209	192	177	161	147	133	120	107
×306	27.1	2.28	789	233	–	186	171	157	143	130	118	106	94.5
×279	26.7	2.09	718	210	–	167	153	140	128	116	105	93.9	83.7
×250	26.3	1.89	644	184	158	146	134	122	111	101	90.8	81.3	72.3
×229	26.0	1.73	588	167	143	132	121	110	100	90.8	81.6	73.0	64.7
×207	25.7	1.57	531	148	127	117	107	97.9	88.9	80.3	72.1	64.3	57.0
×192	25.5	1.46	491	137	117	108	98.5	89.8	81.5	73.6	66.0	58.8	52.0
×176	25.2	1.34	450	124	106	97.3	89.0	81.1	73.5	66.3	59.4	52.8	46.6
×162	25.0	1.22	414	115	98.1	90.1	82.4	75.0	67.9	61.2	54.7	48.6	42.9
×146	24.7	1.09	371	104	88.2	80.9	73.9	67.2	60.8	54.7	48.9	43.3	38.1
×131	24.5	0.960	329	94.6	80.5	73.8	67.4	61.3	55.4	49.8	44.5	39.4	34.6
×117	24.3	0.850	291	84.6	72.0	66.0	60.2	54.7	49.4	44.4	39.6	35.0	30.7
×104	24.1	0.750	258	75.7	64.3	58.9	53.7	48.8	44.0	39.5	35.2	31.1	27.3
W24×103	24.5	0.980	245	82.7	70.5	64.7	59.1	53.8	48.6	43.8	39.1	34.7	30.5
×94	24.3	0.875	222	76.1	64.8	59.4	54.3	49.3	44.6	40.1	35.8	31.7	27.8
×84	24.1	0.770	196	68.3	58.0	53.2	48.6	44.1	39.8	35.8	31.9	28.2	24.8
×76	23.9	0.680	176	62.5	53.1	48.7	44.4	40.3	36.4	32.6	29.1	25.7	22.5
×68	23.7	0.585	154	57.4	48.7	44.6	40.7	36.9	33.3	29.8	26.6	23.4	20.5
W24×62	23.7	0.590	132	56.7	48.2	44.1	40.3	36.5	33.0	29.6	26.4	23.3	20.4
×55	23.6	0.505	115	51.2	43.5	39.8	36.3	33.0	29.8	26.7	23.8	21.0	18.4

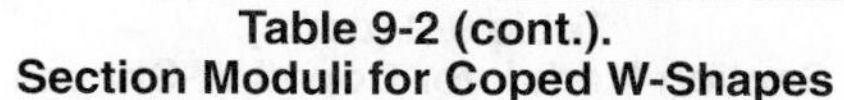

Table 9-2 (cont.). Section Moduli for Coped W-Shapes

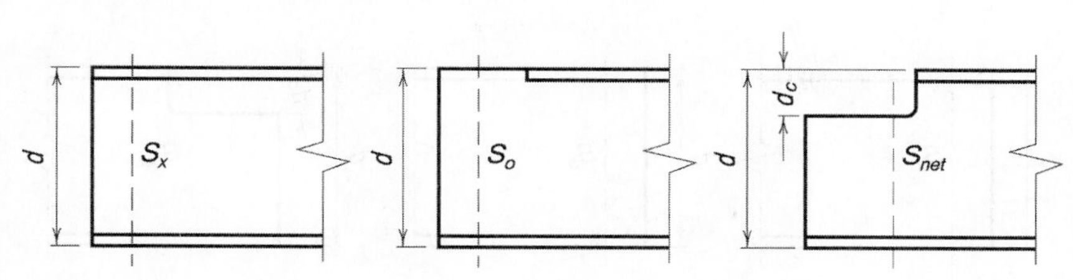

Shape	d, in.	t_f, in.	S_x, in.^{3}	S_o, in.^{3}	S_{net}, in.^{3}								
					d_c, in.								
					2	3	4	5	6	7	8	9	10
W21×201	23.0	1.63	461	124	104	95.0	86.0	77.3	69.1	61.4	54.0	47.1	40.6
×182	22.7	1.48	417	111	93.1	84.6	76.4	68.7	61.3	54.3	47.7	41.5	35.7
×166	22.5	1.36	381	99.4	83.1	75.5	68.1	61.1	54.5	48.2	42.3	36.7	31.5
×147	22.1	1.15	330	91.5	76.3	69.2	62.4	55.9	49.8	43.9	38.4	33.2	28.4
×132	21.8	1.03	296	80.7	67.2	60.8	54.8	49.0	43.5	38.3	33.4	28.9	24.6
×122	21.7	0.960	274	74.2	61.7	55.9	50.3	45.0	39.9	35.1	30.6	26.4	22.5
×111	21.5	0.875	249	67.0	55.6	50.3	45.2	40.4	35.8	31.5	27.4	23.6	20.0
×101	21.4	0.800	227	60.6	50.3	45.5	40.9	36.5	32.3	28.4	24.7	21.2	18.0
W21×93	21.6	0.930	192	67.1	55.9	50.6	45.6	40.8	36.2	31.9	27.9	24.0	20.5
×83	21.4	0.835	171	58.9	48.9	44.3	39.8	35.6	31.6	27.8	24.2	20.9	17.7
×73	21.2	0.740	151	51.3	42.6	38.5	34.6	30.9	27.4	24.0	20.9	18.0	15.2
×68	21.1	0.685	140	48.0	39.8	36.0	32.3	28.9	25.6	22.4	19.5	16.7	14.2
×62	21.0	0.615	127	44.1	36.6	33.1	29.7	26.5	23.4	20.6	17.9	15.3	13.0
×55	20.8	0.522	110	40.1	33.2	30.0	26.9	24.0	21.2	18.6	16.1	13.8	11.7
×48	20.6	0.430	93.0	36.1	29.9	27.0	24.2	21.5	19.0	16.7	14.4	12.4	10.4
W21×57	21.1	0.650	111	43.6	36.2	32.8	29.4	26.3	23.3	20.5	17.8	15.3	13.0
×50	20.8	0.535	94.9	39.1	32.4	29.3	26.3	23.5	20.8	18.2	15.8	13.6	11.5
×44	20.7	0.450	82.0	35.3	29.2	26.4	23.7	21.1	18.7	16.4	14.2	12.2	10.3
W18×175	20.0	1.59	344	91.8	74.8	66.9	59.4	52.3	45.7	39.4	33.6	28.2	
×158	19.7	1.44	310	81.6	66.3	59.2	52.5	46.1	40.1	34.5	29.3	24.5	
×143	19.5	1.32	282	72.6	58.9	52.5	46.5	40.8	35.4	30.4	25.8	21.5	
×130	19.3	1.20	256	65.6	53.1	47.3	41.8	36.6	31.8	27.2	23.0	19.1	
W18×119	19.0	1.06	231	61.9	50.0	44.5	39.2	34.3	29.7	25.4	21.4	17.7	
×106	18.7	0.940	204	54.2	43.6	38.8	34.1	29.8	25.7	21.9	18.4	15.1	
×97	18.6	0.870	188	48.9	39.3	34.9	30.7	26.8	23.1	19.6	16.5	13.5	
×86	18.4	0.770	166	43.1	34.6	30.7	27.0	23.5	20.2	17.1	14.3	11.7	
×76	18.2	0.680	146	37.5	30.1	26.6	23.4	20.3	17.4	14.7	12.3	10.0	
W18×71	18.5	0.810	127	42.5	34.3	30.4	26.8	23.4	20.2	17.2	14.4	11.8	
×65	18.4	0.750	117	38.5	31.0	27.5	24.2	21.1	18.2	15.5	12.9	10.6	
×60	18.2	0.695	108	34.9	28.0	24.8	21.8	19.0	16.3	13.8	11.5	9.45	
×55	18.1	0.630	98.3	32.3	25.9	23.0	20.2	17.6	15.1	12.8	10.7	8.71	
×50	18.0	0.570	88.9	29.2	23.4	20.7	18.2	15.8	13.6	11.5	9.56		
W18×46	18.1	0.605	78.8	29.0	23.3	20.7	18.2	15.8	13.6	11.5	9.63	7.88	
×40	17.9	0.525	68.4	24.9	20.0	17.7	15.5	13.5	11.6	9.80	8.16		
×35	17.7	0.425	57.6	22.7	18.2	16.1	14.1	12.3	10.5	8.88	7.37		
W16×100	17.0	0.985	176	44.6	35.0	30.6	26.5	22.7	19.1	15.8	12.8		
×89	16.8	0.875	156	39.3	30.8	26.9	23.2	19.8	16.7	13.8	11.1		
×77	16.5	0.760	135	33.1	25.8	22.5	19.4	16.5	13.8	11.3	9.09		
×67	16.3	0.665	118	28.2	22.0	19.1	16.4	13.9	11.6	9.51	7.60		

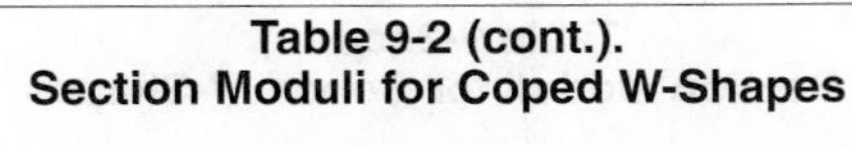

Table 9-2 (cont.). Section Moduli for Coped W-Shapes

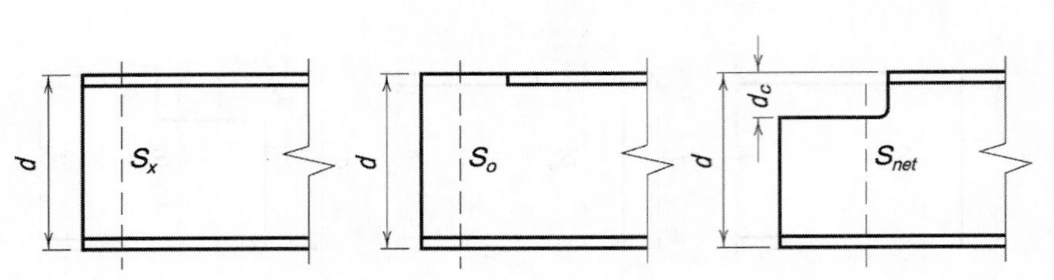

Shape	d, in.	t_f, in.	S_x, in.^{3}	S_o, in.^{3}	S_{net}, in.^{3}								
					d_c, in.								
					2	3	4	5	6	7	8	9	10
W16×57	16.4	0.715	92.2	29.3	22.9	20.0	17.2	14.7	12.3	10.1	8.11		
×50	16.3	0.630	81.0	25.7	20.1	17.5	15.1	12.8	10.7	8.81	7.06		
×45	16.1	0.565	72.7	22.9	17.8	15.5	13.3	11.3	9.42	7.70	6.14		
×40	16.0	0.505	64.7	20.1	15.6	13.6	11.6	9.87	8.22	6.71			
×36	15.9	0.430	56.5	18.9	14.7	12.8	11.0	9.28	7.73	6.30			
W16×31	15.9	0.440	47.2	17.2	13.4	11.6	9.99	8.47	7.05	5.76			
×26	15.7	0.345	38.4	15.0	11.6	10.1	8.66	7.32	6.09	4.96			
W14×808	22.8	5.12	1400	450	–	–	–	–	243	215	189	165	143
×730	22.4	4.91	1280	365	–	–	–	219	195	172	151	131	114
×665	21.6	4.52	1150	316	–	–	–	186	164	144	125	108	92.8
×605	20.9	4.16	1040	275	–	–	–	158	139	121	104	89.4	76.0
×550	20.2	3.82	931	237	–	–	152	134	117	101	86.4	73.3	61.7
×500	19.6	3.50	838	208	–	–	131	115	99.4	85.3	72.5	60.9	
×455	19.0	3.21	756	182	–	–	113	98.1	84.5	72.0	60.6	50.4	
W14×426	18.7	3.04	706	165	–	–	102	88.2	75.7	64.2	53.8	44.5	
×398	18.3	2.85	656	150	–	105	91.2	78.8	67.3	56.8	47.3	38.8	
×370	17.9	2.66	607	135	–	93.7	81.4	70.0	59.5	49.9	41.2		
×342	17.5	2.47	558	121	–	83.0	71.9	61.5	52.0	43.3	35.5		
×311	17.1	2.26	506	107	–	72.5	62.5	53.3	44.8	37.0	30.1		
×283	16.7	2.07	459	94.0	–	63.2	54.3	46.0	38.4	31.5	25.4		
×257	16.4	1.89	415	83.6	64.5	55.9	47.8	40.3	33.5	27.3	21.8		
×233	16.0	1.72	375	72.8	55.8	48.1	41.0	34.4	28.4	22.9			
×211	15.7	1.56	338	64.7	49.4	42.4	36.0	30.1	24.7	19.8			
×193	15.5	1.44	310	57.7	43.9	37.6	31.9	26.5	21.7	17.3			
×176	15.2	1.31	281	52.1	39.4	33.7	28.4	23.5	19.1	15.1			
×159	15.0	1.19	254	45.9	34.6	29.5	24.8	20.5	16.6	13.1			
×145	14.8	1.09	232	41.0	30.8	26.2	22.0	18.1	14.6	11.4			
W14×132	14.7	1.03	209	38.3	28.7	24.4	20.5	16.8	13.5	10.6			
×120	14.5	0.940	190	34.3	25.6	21.7	18.1	14.9	11.9	9.25			
×109	14.3	0.860	173	29.9	22.3	18.8	15.7	12.8	10.2	7.87			
×99	14.2	0.780	157	27.3	20.3	17.2	14.3	11.6	9.24	7.12			
×90	14.0	0.710	143	24.2	17.9	15.1	12.5	10.2	8.03				
W14×82	14.3	0.855	123	27.9	20.9	17.7	14.7	12.0	9.61	7.44			
×74	14.2	0.785	112	24.6	18.3	15.5	12.9	10.5	8.37	6.45			
×68	14.0	0.720	103	22.1	16.4	13.8	11.5	9.33	7.39				
×61	13.9	0.645	92.2	19.8	14.6	12.3	10.2	8.30	6.56				
W14×53	13.9	0.660	77.8	19.0	14.1	11.9	9.89	8.03	6.35				
×48	13.8	0.595	70.3	17.3	12.8	10.8	8.94	7.25	5.72				
×43	13.7	0.530	62.7	15.3	11.3	9.55	7.90	6.40	5.04				

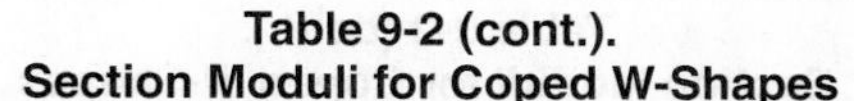

Table 9-2 (cont.). Section Moduli for Coped W-Shapes

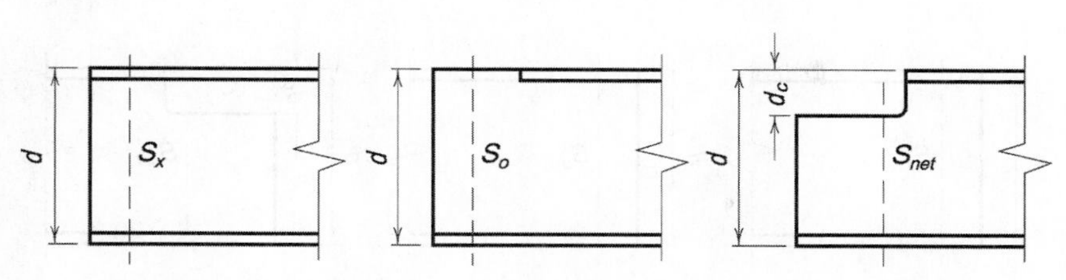

Shape	d, in.	t_f, in.	S_x, in.^{3}	S_o, in.^{3}	S_{net}, in.^{3}								
					d_c, in.								
					2	3	4	5	6	7	8	9	10
W14×38	14.1	0.515	54.3	16.0	12.0	10.2	8.48	6.94	5.54	4.28			
×34	14.0	0.455	48.2	14.5	10.8	9.17	7.65	6.25	4.98				
×30	13.8	0.385	41.6	13.2	9.82	8.31	6.91	5.63	4.47				
W14×26	13.9	0.420	34.9	12.3	9.18	7.79	6.49	5.30	4.22				
×22	13.7	0.335	28.6	10.6	7.92	6.70	5.57	4.54	3.60				
W12×336	16.8	2.96	483	123	–	83.1	71.4	60.6	50.7	41.9	34.0		
×305	16.3	2.71	435	107	–	71.4	61.0	51.4	42.7	34.8	28.0		
×279	15.9	2.47	393	96.7	–	63.5	54.0	45.2	37.3	30.1			
×252	15.4	2.25	353	83.8	–	54.3	45.8	38.1	31.0	24.8			
×230	15.1	2.07	321	74.9	–	48.1	40.4	33.4	27.1	21.4			
×210	14.7	1.90	292	65.6	49.0	41.5	34.7	28.4	22.8	17.9			
×190	14.4	1.74	263	57.1	42.4	35.8	29.8	24.3	19.4	15.0			
×170	14.0	1.56	235	49.4	36.3	30.5	25.2	20.4	16.1				
×152	13.7	1.40	209	43.2	31.6	26.4	21.7	17.4	13.6				
×136	13.4	1.25	186	37.8	27.5	22.9	18.7	14.9	11.5				
×120	13.1	1.11	163	32.7	23.6	19.6	15.9	12.6	9.64				
×106	12.9	0.990	145	27.6	19.8	16.4	13.2	10.4	7.93				
×96	12.7	0.900	131	24.3	17.4	14.3	11.5	9.01	6.81				
×87	12.5	0.810	118	22.1	15.7	12.9	10.3	8.05	6.04				
×79	12.4	0.735	107	20.0	14.1	11.6	9.27	7.20	5.38				
×72	12.3	0.670	97.4	18.0	12.8	10.4	8.33	6.45	4.80				
×65	12.1	0.605	87.9	15.9	11.2	9.12	7.25	5.58	4.12				
W12×58	12.2	0.640	78.0	14.8	10.5	8.54	6.80	5.25	3.90				
×53	12.1	0.575	70.6	13.9	9.82	8.01	6.37	4.91	3.63				
W12×50	12.2	0.640	64.2	14.8	10.5	8.56	6.83	5.29	3.93				
×45	12.1	0.575	57.7	13.2	9.34	7.63	6.08	4.69	3.47				
×40	11.9	0.515	51.5	11.4	7.97	6.48	5.14	3.93					
W12×35	12.5	0.520	45.6	12.3	8.85	7.30	5.89	4.61	3.48				
×30	12.3	0.440	38.6	10.4	7.41	6.10	4.90	3.82	2.86				
×26	12.2	0.380	33.4	9.06	6.45	5.30	4.25	3.30	2.47				
W12×22	12.3	0.425	25.4	9.59	6.88	5.68	4.58	3.58	2.70				
×19	12.2	0.350	21.3	8.45	6.05	4.99	4.02	3.14	2.36				
×16	12.0	0.265	17.1	7.44	5.31	4.37	3.50	2.73					
×14	11.9	0.225	14.9	6.60	4.70	3.86	3.09	2.40					

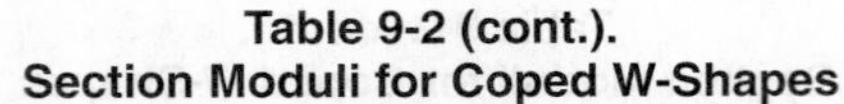

Table 9-2 (cont.). Section Moduli for Coped W-Shapes

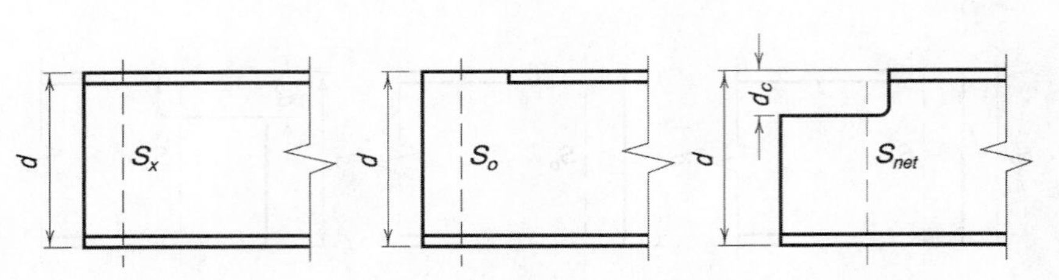

Shape	d, in.	t_f, in.	S_x, in.^{3}	S_o, in.^{3}	S_{net}, in.^{3}								
					d_c, in.								
					2	3	4	5	6	7	8	9	10
W10×112	11.4	1.25	126	25.9	17.6	15.8	14.1	12.4	10.9	9.45	8.12	6.90	
×100	11.1	1.12	112	22.3	15.0	13.4	11.9	10.5	9.12	7.87	6.72	5.66	
×88	10.8	0.990	98.5	19.0	12.6	11.2	9.93	8.69	7.53	6.46	5.47		
×77	10.6	0.870	85.9	16.2	10.7	9.49	8.36	7.29	6.29	5.37	4.52		
×68	10.4	0.770	75.7	13.9	9.13	8.08	7.09	6.17	5.30	4.50	3.77		
×60	10.2	0.680	66.7	12.0	7.84	6.92	6.06	5.25	4.49	3.79	3.15		
×54	10.1	0.615	60.0	10.5	6.80	6.00	5.24	4.53	3.87	3.26	2.70		
×49	9.98	0.560	54.6	9.46	6.10	5.37	4.68	4.04	3.44	2.89			
W10×45	10.1	0.620	49.1	9.75	6.33	5.59	4.88	4.22	3.61	3.04	2.52		
×39	9.92	0.530	42.1	8.49	5.48	4.82	4.20	3.62	3.08	2.59			
×33	9.73	0.435	35.0	7.49	4.80	4.22	3.67	3.15	2.67	2.23			
W10×30	10.5	0.510	32.4	8.69	5.79	5.15	4.55	3.97	3.44	2.94	2.47		
×26	10.3	0.440	27.9	7.29	4.83	4.28	3.77	3.28	2.83	2.40	2.01		
×22	10.2	0.360	23.2	6.54	4.32	3.83	3.37	2.93	2.52	2.14	1.79		
W10×19	10.2	0.395	18.8	6.47	4.29	3.81	3.35	2.92	2.52	2.14	1.79		
×17	10.1	0.330	16.2	6.00	3.97	3.52	3.10	2.69	2.32	1.97	1.65		
×15	9.99	0.270	13.8	5.52	3.64	3.22	2.83	2.46	2.12	1.79			
×12	9.87	0.210	10.9	4.43	2.91	2.58	2.26	1.96	1.68	1.42			

$\phi = 0.75$

Table 9-3a. Block Shear Rupture Tension Rupture Component per Inch of Thickness, $\phi[F_u A_{nt}]/t$, kips/in.

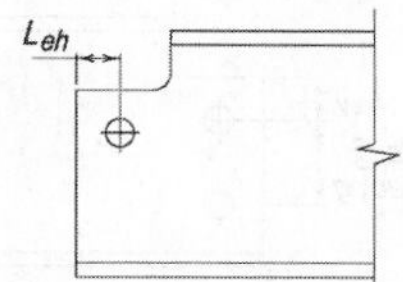

L_{eh}, in.	F_u, ksi								
	58			65			70		
	Bolt Diameter d_b, in.			Bolt Diameter d_b, in.			Bolt Diameter d_b, in.		
	3/4	7/8	1	3/4	7/8	1	3/4	7/8	1
1	24.5	21.8	19.0	27.4	24.4	21.3	29.5	26.3	23.0
1 1/8	29.9	27.2	24.5	33.5	30.5	27.4	36.1	32.8	29.5
1 1/4	35.3	32.6	29.9	39.6	36.6	33.5	42.7	39.4	36.1
1 3/8	40.8	38.1	35.3	45.7	42.7	39.6	49.2	45.9	42.7
1 1/2	46.2	43.5	40.8	51.8	48.8	45.7	55.8	52.5	49.2
1 5/8	51.7	48.9	46.2	57.9	54.8	51.8	62.3	59.1	55.8
1 3/4	57.1	54.4	51.7	64.0	60.9	57.9	68.9	65.6	62.3
1 7/8	62.5	59.8	57.1	70.1	67.0	64.0	75.5	72.2	68.9
2	68.0	65.3	62.5	76.2	73.1	70.1	82.0	78.8	75.5
2 1/4	78.8	76.1	73.4	88.4	85.3	82.3	95.2	91.9	88.6
2 1/2	89.7	87.0	84.3	101	97.5	94.5	108	105	102
2 3/4	101	97.9	95.2	113	110	107	121	118	115
3	111	109	106	125	122	119	135	131	128

$\phi = 0.75$

Table 9-3b.
Block Shear Rupture
Shear Yielding Component per Inch of Thickness, $\phi[0.6F_yA_{gv}]/t$, kips/in.

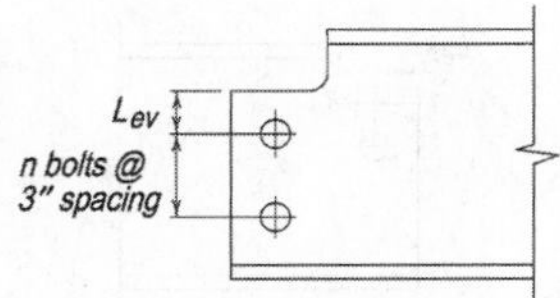

n	L_{ev}, in.	F_y, ksi 36	F_y, ksi 50	n	L_{ev}, in.	F_y, ksi 36	F_y, ksi 50
12	1 1/4	555	771	9	1 1/4	409	568
	1 3/8	557	773		1 3/8	411	571
	1 1/2	559	776		1 1/2	413	574
	1 5/8	561	779		1 5/8	415	577
	1 3/4	563	782		1 3/4	417	579
	1 7/8	565	785		1 7/8	419	582
	2	567	788		2	421	585
	2 1/4	571	793		2 1/4	425	591
	2 1/2	575	799		2 1/2	429	596
	2 3/4	579	804		2 3/4	433	602
	3	583	810		3	437	608
11	1 1/4	506	703	8	1 1/4	360	501
	1 3/8	508	706		1 3/8	362	503
	1 1/2	510	709		1 1/2	365	506
	1 5/8	512	712		1 5/8	367	509
	1 3/4	514	714		1 3/4	369	512
	1 7/8	516	717		1 7/8	371	515
	2	518	720		2	373	518
	2 1/4	522	726		2 1/4	377	523
	2 1/2	527	731		2 1/2	381	529
	2 3/4	531	737		2 3/4	385	534
	3	535	743		3	389	540
10	1 1/4	458	636	7	1 1/4	312	433
	1 3/8	460	638		1 3/8	314	436
	1 1/2	462	641		1 1/2	316	439
	1 5/8	464	644		1 5/8	318	442
	1 3/4	466	647		1 3/4	320	444
	1 7/8	468	650		1 7/8	322	447
	2	470	653		2	324	450
	2 1/4	474	658		2 1/4	328	456
	2 1/2	478	664		2 1/2	332	461
	2 3/4	482	669		2 3/4	336	467
	3	486	675		3	340	472

$\phi = 0.75$

Table 9-3b (cont.). Block Shear Rupture Shear Yielding Component per Inch of Thickness, $\phi[0.6F_yA_{gv}]/t$, kips/in.

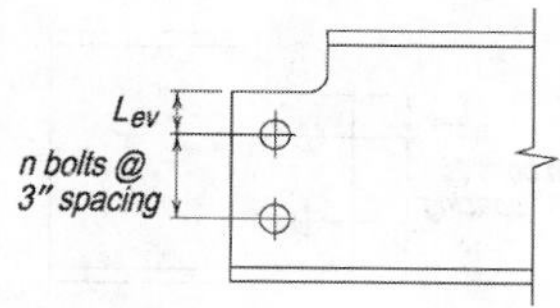

n	L_{ev}, in.	F_y, ksi 36	F_y, ksi 50
6	1 1/4	263	366
	1 3/8	265	368
	1 1/2	267	371
	1 5/8	269	374
	1 3/4	271	377
	1 7/8	273	380
	2	275	383
	2 1/4	279	388
	2 1/2	284	394
	2 3/4	288	399
	3	292	405
5	1 1/4	215	298
	1 3/8	217	301
	1 1/2	219	304
	1 5/8	221	307
	1 3/4	223	309
	1 7/8	225	312
	2	227	315
	2 1/4	231	321
	2 1/2	235	326
	2 3/4	239	332
	3	243	338
4	1 1/4	166	231
	1 3/8	168	233
	1 1/2	170	236
	1 5/8	172	239
	1 3/4	174	242
	1 7/8	176	245
	2	178	248
	2 1/4	182	253
	2 1/2	186	259
	2 3/4	190	264
	3	194	270
3	1 1/4	117	163
	1 3/8	119	166
	1 1/2	122	169
	1 5/8	124	172
	1 3/4	126	174
	1 7/8	128	177
	2	130	180
	2 1/4	134	186
	2 1/2	138	191
	2 3/4	142	197
	3	146	203
2	1 1/4	68.8	95.6
	1 3/8	70.9	98.4
	1 1/2	72.9	101
	1 5/8	74.9	104
	1 3/4	76.9	107
	1 7/8	79.0	110
	2	81.0	113
	2 1/4	85.0	118
	2 1/2	89.1	124
	2 3/4	93.1	129
	3	97.2	135

$\phi = 0.75$

Table 9-4a.
Block Shear Rupture
Shear Rupture Component per Inch of Thickness, $\phi[0.6 F_u A_{nv}]/t$, kips/in.

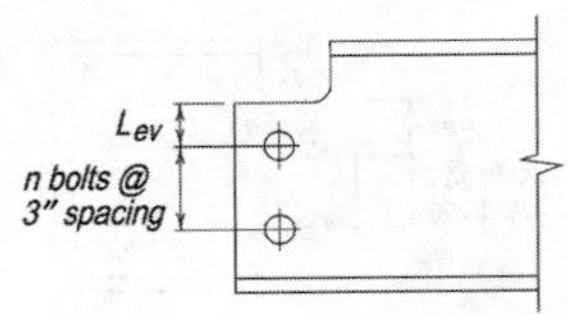

n	L_{ev}, in.	F_u, ksi								
		58			65			70		
		Bolt Diameter d_b, in.			Bolt Diameter d_b, in.			Bolt Diameter d_b, in.		
		3/4	7/8	1	3/4	7/8	1	3/4	7/8	1
12	1 1/4	631	594	556	707	665	623	762	717	671
	1 3/8	635	597	560	711	669	627	766	721	675
	1 1/2	638	600	563	715	673	631	770	725	679
	1 5/8	641	604	566	718	676	634	774	728	683
	1 3/4	644	607	569	722	680	638	778	732	687
	1 7/8	648	610	573	726	684	642	782	736	691
	2	651	613	576	729	687	645	786	740	695
	2 1/4	657	620	582	737	695	653	793	748	703
	2 1/2	664	626	589	744	702	660	801	756	711
	2 3/4	670	633	595	751	709	667	809	764	719
	3	677	639	602	759	717	675	817	772	726
11	1 1/4	576	542	507	645	607	569	695	654	612
	1 3/8	579	545	511	649	611	572	699	658	616
	1 1/2	582	548	514	653	614	576	703	662	620
	1 5/8	586	551	517	656	618	580	707	665	624
	1 3/4	589	555	520	660	622	583	711	669	628
	1 7/8	592	558	524	664	625	587	715	673	632
	2	595	561	527	667	629	590	719	677	636
	2 1/4	602	568	533	675	636	598	726	685	644
	2 1/2	608	574	540	682	643	605	734	693	652
	2 3/4	615	581	546	689	651	612	742	701	660
	3	622	587	553	697	658	620	750	709	667
10	1 1/4	520	489	458	583	548	514	628	591	553
	1 3/8	524	493	462	587	552	517	632	595	557
	1 1/2	527	496	465	590	556	521	636	599	561
	1 5/8	530	499	468	594	559	525	640	602	565
	1 3/4	533	502	471	598	563	528	644	606	569
	1 7/8	537	506	475	601	567	532	648	610	573
	2	540	509	478	605	570	536	652	614	577
	2 1/4	546	515	484	612	578	543	660	622	585
	2 1/2	553	522	491	620	585	550	667	630	593
	2 3/4	560	529	498	627	592	558	675	638	600
	3	566	535	504	634	600	565	683	646	608

$\phi = 0.75$

Table 9-4a (cont.). Block Shear Rupture
Shear Rupture Component per Inch of Thickness, $\phi[0.6 F_u A_{nv}]/t$, kips/in.

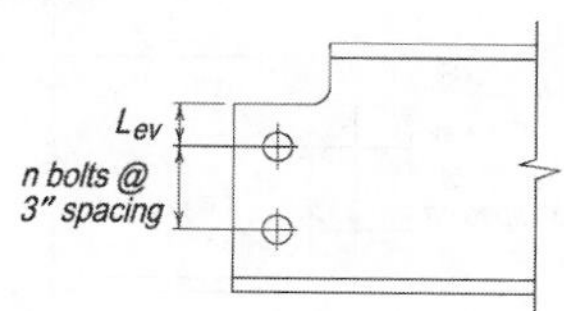

n	L_{ev}, in.	F_u, ksi								
		58			65			70		
		Bolt Diameter d_b, in.			Bolt Diameter d_b, in.			Bolt Diameter d_b, in.		
		$^3/_4$	$^7/_8$	1	$^3/_4$	$^7/_8$	1	$^3/_4$	$^7/_8$	1
9	$1^1/_4$	465	437	409	521	490	459	561	528	494
	$1^3/_8$	468	440	413	525	494	463	565	532	498
	$1^1/_2$	471	444	416	528	497	466	569	536	502
	$1^5/_8$	475	447	419	532	501	470	573	539	506
	$1^3/_4$	478	450	422	536	505	473	577	543	510
	$1^7/_8$	481	453	426	539	508	477	581	547	514
	2	484	457	429	543	512	481	585	551	518
	$2^1/_4$	491	463	436	550	519	488	593	559	526
	$2^1/_2$	498	470	442	558	527	495	600	567	534
	$2^3/_4$	504	476	449	565	534	503	608	575	541
	3	511	483	455	572	541	510	616	583	549
8	$1^1/_4$	409	385	361	459	431	404	494	465	435
	$1^3/_8$	413	388	364	463	435	408	498	469	439
	$1^1/_2$	416	392	367	466	439	411	502	473	443
	$1^5/_8$	419	395	370	470	442	415	506	476	447
	$1^3/_4$	422	398	374	473	446	419	510	480	451
	$1^7/_8$	426	401	377	477	450	422	514	484	455
	2	429	405	380	481	453	426	518	488	459
	$2^1/_4$	436	411	387	488	461	433	526	496	467
	$2^1/_2$	442	418	393	495	468	441	534	504	474
	$2^3/_4$	449	424	400	503	475	448	541	512	482
	3	455	431	406	510	483	455	549	520	490
7	$1^1/_4$	354	333	312	397	373	349	427	402	376
	$1^3/_8$	357	336	315	400	377	353	431	406	380
	$1^1/_2$	361	339	318	404	380	356	435	410	384
	$1^5/_8$	364	343	321	408	384	360	439	413	388
	$1^3/_4$	367	346	325	411	388	364	443	417	392
	$1^7/_8$	370	349	328	415	391	367	447	421	396
	2	374	352	331	419	395	371	451	425	400
	$2^1/_4$	380	359	338	426	402	378	459	433	408
	$2^1/_2$	387	365	344	433	410	386	467	441	415
	$2^3/_4$	393	372	351	441	417	393	474	449	423
	3	400	378	357	448	424	400	482	457	431

$\phi = 0.75$

Table 9-4a (cont.). Block Shear Rupture
Shear Rupture Component per Inch of Thickness, $\phi[0.6F_uA_{nv}]/t$, kips/in.

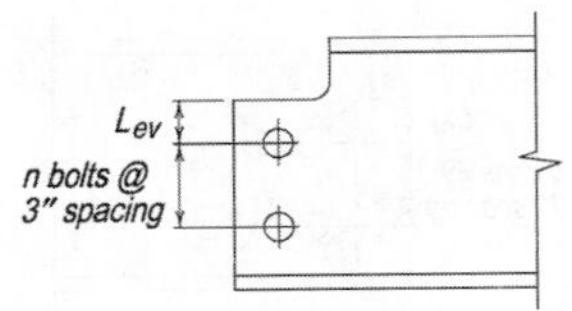

n	L_{ev}, in.	F_u, ksi								
		58			65			70		
		Bolt Diameter d_b, in.			Bolt Diameter d_b, in.			Bolt Diameter d_b, in.		
		3/4	7/8	1	3/4	7/8	1	3/4	7/8	1
6	1 1/4	299	281	263	335	314	294	360	339	317
	1 3/8	302	284	266	338	318	298	364	343	321
	1 1/2	305	287	269	342	322	302	368	346	325
	1 5/8	308	290	272	346	325	305	372	350	329
	1 3/4	312	294	276	349	329	309	376	354	333
	1 7/8	315	297	279	353	333	313	380	358	337
	2	318	300	282	356	336	316	384	362	341
	2 1/4	325	307	289	364	344	324	392	370	348
	2 1/2	331	313	295	371	351	331	400	378	356
	2 3/4	338	320	302	378	358	338	408	386	364
	3	344	326	308	386	366	346	415	394	372
5	1 1/4	243	228	214	272	256	239	293	276	258
	1 3/8	246	232	217	276	260	243	297	280	262
	1 1/2	250	235	220	280	263	247	301	284	266
	1 5/8	253	238	223	283	267	250	305	287	270
	1 3/4	256	241	227	287	271	254	309	291	274
	1 7/8	259	245	230	291	274	258	313	295	278
	2	263	248	233	294	278	261	317	299	282
	2 1/4	269	254	240	302	285	269	325	307	289
	2 1/2	276	261	246	309	293	276	333	315	297
	2 3/4	282	268	253	316	300	283	341	323	305
	3	289	274	259	324	307	291	348	331	313
4	1 1/4	188	176	165	210	197	185	226	213	199
	1 3/8	191	179	168	214	201	188	230	217	203
	1 1/2	194	183	171	218	205	192	234	221	207
	1 5/8	197	186	175	221	208	196	238	224	211
	1 3/4	201	189	178	225	212	199	242	228	215
	1 7/8	204	192	181	229	216	203	246	232	219
	2	207	196	184	232	219	207	250	236	222
	2 1/4	214	202	191	239	227	214	258	244	230
	2 1/2	220	209	197	247	234	221	266	252	238
	2 3/4	227	215	204	254	241	229	274	260	246
	3	233	222	210	261	249	236	282	268	254

$\phi = 0.75$

Table 9-4a (cont.). Block Shear Rupture Shear Rupture Component per Inch of Thickness, $\phi[0.6F_uA_{nv}]/t$, kips/in.

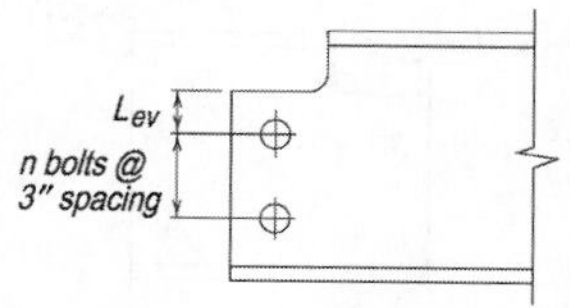

n	L_{ev}, in.	F_u, ksi								
		58			65			70		
		Bolt Diameter d_b, in.			Bolt Diameter d_b, in.			Bolt Diameter d_b, in.		
		3/4	7/8	1	3/4	7/8	1	3/4	7/8	1
3	1 1/4	132	124	116	148	139	130	159	150	140
	1 3/8	135	127	119	152	143	133	163	154	144
	1 1/2	139	131	122	155	146	137	167	158	148
	1 5/8	142	134	126	159	150	141	171	161	152
	1 3/4	145	137	129	163	154	144	175	165	156
	1 7/8	148	140	132	166	157	148	179	169	159
	2	152	144	135	170	161	152	183	173	163
	2 1/4	158	150	142	177	168	159	191	181	171
	2 1/2	165	157	148	185	175	166	199	189	179
	2 3/4	171	163	155	192	183	174	207	197	187
	3	178	170	161	199	190	181	215	205	195
2	1 1/4	76.7	71.8	66.9	85.9	80.4	75.0	92.5	86.6	80.7
	1 3/8	79.9	75.0	70.1	89.6	84.1	78.6	96.5	90.6	84.7
	1 1/2	83.2	78.3	73.4	93.2	87.7	82.3	100	94.5	88.6
	1 5/8	86.5	81.6	76.7	96.9	91.4	85.9	104	98.4	92.5
	1 3/4	89.7	84.8	79.9	101	95.1	89.6	108	102	96.5
	1 7/8	93.0	88.1	83.2	104	98.7	93.2	112	106	100
	2	96.2	91.4	86.5	108	102	96.9	116	110	104
	2 1/4	103	97.9	93.0	115	110	104	124	118	112
	2 1/2	109	104	99.5	122	117	112	132	126	120
	2 3/4	116	111	106	130	124	119	140	134	128
	3	122	117	113	137	132	126	148	142	136

$\phi = 0.75$

Table 9-4b. Block Shear Rupture Tension Yielding Component per Inch of Thickness, $\phi[F_y A_{gt}]/t$, kips/in.

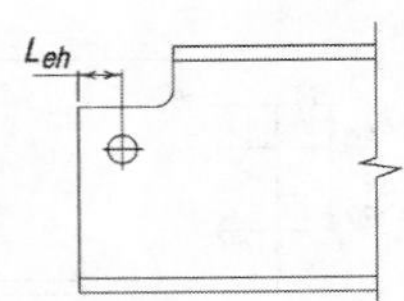

L_{eh}, in.	F_y, ksi	
	36	50
1	27.0	37.5
$1^1/_8$	30.4	42.2
$1^1/_4$	33.8	46.9
$1^3/_8$	37.1	51.6
$1^1/_2$	40.5	56.3
$1^5/_8$	43.9	60.9
$1^3/_4$	47.3	65.6
$1^7/_8$	50.6	70.3
2	54.0	75.0
$2^1/_4$	60.8	84.4
$2^1/_2$	67.5	93.8
$2^3/_4$	74.3	103
3	81.0	113

$F_y = 50$ ksi
$\phi = 1.00$
$\phi_r = 0.75$
$\phi_v = 0.90$

Table 9-5.
Beam End Bearing Constants

Shape	ϕR_1 kips	ϕR_2 kips/in.	$\phi_r R_3$ kips	$\phi_r R_4$ kips/in.	$\phi_r R_5$ kips	$\phi_r R_6$ kips/in.	$\phi R(N = 3^1/_4)$ kips	$\phi_v V_n$ kips
W44×335	335	51.0	495	14.8	452	19.7	500	1210
×290	265	43.5	368	10.4	338	13.8	402	1020
×262	222	39.5	302	8.69	277	11.6	331	924
×230	183	35.5	239	7.41	218	9.88	263	822
W40×593	1010	89.5	1550	44.8	1430	59.7	1300	2080
×503	770	77.0	1150	34.1	1050	45.4	1020	1750
×431	607	67.0	861	26.8	787	35.7	825	1490
×397	534	61.0	722	21.8	662	29.1	732	1350
×372	480	58.0	646	20.3	591	27.1	669	1270
×362	455	56.0	607	18.7	557	24.9	637	1230
×324	383	50.0	486	14.9	446	19.9	534	1090
×297	341	46.5	416	13.3	381	17.7	459	999
×277	298	41.5	343	9.88	317	13.2	375	890
×249	252	37.5	280	8.17	258	10.9	306	798
×215	203	32.5	209	6.26	193	8.34	229	684
×199	188	32.5	196	7.19	177	9.58	219	679
W40×392	677	71.0	970	29.6	888	39.5	907	1590
×331	515	61.0	710	22.6	649	30.2	713	1340
×327	498	59.0	676	20.5	620	27.3	690	1300
×278	390	51.0	501	15.8	458	21.1	552	1110
×264	360	48.0	447	13.9	410	18.5	492	1040
×235	298	41.5	343	9.88	317	13.2	375	890
×211	252	37.5	280	8.17	258	10.9	306	798
×183	203	32.5	209	6.26	193	8.34	229	684
×167	188	32.5	191	7.56	172	10.1	216	677
×149	167	31.5	165	8.55	143	11.4	192	650
W36×798	1650	119	2750	81.1	2520	108	2040	2700
×650	1190	98.5	1880	57.8	1720	77.1	1510	2150
×527	843	80.5	1260	39.6	1160	52.9	1100	1700
×439	627	68.0	895	29.2	820	38.9	848	1410
×393	524	61.0	722	23.7	662	31.6	722	1250
×359	455	56.0	607	20.3	557	27.0	637	1130
×328	398	51.0	506	16.8	465	22.3	561	1020
×300	347	47.3	430	14.8	394	19.8	478	936
×280	311	44.3	377	13.1	345	17.5	419	872
×260	282	42.0	334	12.3	304	16.4	374	823
×245	263	40.0	300	11.4	273	15.2	337	780
×230	238	38.0	269	10.5	243	14.0	303	737
W36×256	315	48.0	447	14.8	410	19.8	471	969
×232	272	43.5	367	12.3	337	16.3	407	871
×210	240	41.5	319	12.4	288	16.6	359	822
×194	209	38.3	271	10.5	246	14.1	306	754
×182	193	36.3	242	9.64	219	12.9	274	711
×170	170	34.0	212	8.56	192	11.4	240	665
×160	157	32.5	191	8.11	172	10.8	218	632
×150	146	31.3	173	7.84	154	10.5	199	606
×135	127	30.0	149	8.32	129	11.1	176	577
W33×387	502	63.0	771	26.4	708	35.2	707	1220
×354	426	58.0	652	22.7	599	30.3	614	1110
×318	358	52.0	527	18.3	484	24.4	527	988
×291	315	48.0	447	15.9	410	21.2	471	902
×263	265	43.5	367	13.2	337	17.6	406	810
×241	233	41.5	323	12.9	294	17.3	365	766
×221	206	38.8	278	11.7	251	15.6	316	709
×201	179	35.8	234	10.2	211	13.6	267	651

$F_y = 50$ ksi
$\phi = 1.00$
$\phi_r = 0.75$
$\phi_v = 0.90$

Table 9-5 (cont.). Beam End Bearing Constants

Shape	ϕR_1 kips	ϕR_2 kips/in.	$\phi_r R_3$ kips	$\phi_r R_4$ kips/in.	$\phi_r R_5$ kips	$\phi_r R_6$ kips/in.	$\phi R(N = 3^1/_4)$ kips	$\phi_v V_n$ kips
W33×169	178	33.5	219	7.90	201	10.5	245	611
×152	154	31.8	188	7.81	171	10.4	214	574
×141	137	30.3	167	7.51	150	10.0	191	544
×130	127	29.0	148	7.47	131	9.96	172	518
×118	112	27.5	127	7.41	111	9.87	151	489
W30×391	574	68.0	895	33.7	820	44.9	795	1220
×357	484	62.0	747	28.1	685	37.5	686	1100
×326	419	57.0	630	24.2	577	32.2	604	997
×292	351	51.0	506	19.4	465	25.9	516	881
×261	298	46.5	416	16.7	381	22.3	449	793
×235	246	41.5	335	13.2	307	17.6	377	701
×211	218	38.8	283	12.4	258	16.5	323	647
×191	183	35.5	236	10.6	214	14.2	270	589
×173	164	32.8	198	9.36	179	12.5	229	538
W30×148	168	32.5	206	8.22	189	11.0	232	539
×132	144	30.8	174	8.32	157	11.1	201	503
×124	133	29.3	156	7.73	140	10.3	181	477
×116	124	28.3	141	7.67	126	10.2	166	458
×108	115	27.3	127	7.75	111	10.3	152	439
×99	102	26.0	111	7.66	95.7	10.2	136	417
×90	88.1	23.5	90.9	6.25	78.6	8.34	111	374
W27×539	1090	98.5	1880	72.0	1720	96.0	1410	1730
×368	582	69.0	922	37.8	846	50.4	806	1130
×336	502	63.0	771	31.7	708	42.3	707	1020
×307	435	58.0	652	27.3	599	36.5	623	927
×281	373	53.0	548	22.8	503	30.4	545	839
×258	329	49.0	466	19.9	428	26.5	488	767
×235	284	45.5	398	17.7	364	23.6	432	705
×217	246	41.5	335	14.5	307	19.4	381	636
×194	211	37.5	272	12.1	249	16.2	311	569
×178	187	36.3	243	12.5	220	16.6	284	544
×161	165	33.0	201	10.5	182	13.9	235	492
×146	142	30.3	168	8.98	151	12.0	197	448
W27×129	153	30.5	181	8.10	166	10.8	207	455
×114	129	28.5	150	7.91	136	10.5	176	420
×102	113	25.8	122	6.58	110	8.77	143	377
×94	99.5	24.5	107	6.36	95.5	8.48	128	356
×84	89.8	23.0	90.2	6.17	79.2	8.23	110	332
W24×370	689	76.0	1120	50.0	1020	66.6	936	1150
×335	582	69.0	922	41.8	846	55.7	806	1020
×306	502	63.0	771	35.1	708	46.8	707	922
×279	435	58.0	652	30.3	599	40.4	623	836
×250	366	52.0	527	24.5	484	32.7	535	739
×229	315	48.0	447	21.3	410	28.4	471	674
×207	272	43.5	367	17.7	337	23.6	413	604
×192	240	40.5	318	15.5	292	20.6	368	558
×176	211	37.5	272	13.5	249	18.1	316	510
×162	187	35.3	236	12.5	215	16.6	277	476
×146	163	32.5	198	11.1	179	14.7	234	433
×131	142	30.3	167	10.2	150	13.6	200	400
×117	120	27.5	136	8.73	122	11.6	164	361
×104	102	25.0	111	7.49	98.6	9.99	135	325

$F_y = 50$ ksi
$\phi = 1.00$
$\phi_r = 0.75$
$\phi_v = 0.90$

Table 9-5 (cont.). Beam End Bearing Constants

Shape	ϕR_1 kips	ϕR_2 kips/in.	$\phi_r R_3$ kips	$\phi_r R_4$ kips/in.	$\phi_r R_5$ kips	$\phi_r R_6$ kips/in.	$\phi R (N = 3^1/_4)$ kips	$\phi_v V_n$ kips
W24×103	129	27.5	146	7.51	134	10.0	170	364
×94	113	25.8	125	6.96	114	9.28	148	338
×84	99.1	23.5	102	6.06	92.4	8.08	122	306
×76	85.9	22.0	86.9	5.68	77.9	7.57	105	284
×68	77.8	20.8	73.9	5.59	65.0	7.45	92.0	266
W24×62	80.6	21.5	78.2	6.16	68.5	8.22	98.3	275
×55	71.0	19.8	63.7	5.60	54.9	7.47	81.9	252
W21×201	284	45.5	400	21.8	367	29.0	432	565
×182	246	41.5	332	18.4	304	24.6	381	509
×166	211	37.5	274	14.9	251	19.9	322	456
×147	180	36.0	237	15.9	213	21.2	288	430
×132	157	32.5	192	13.3	173	17.7	235	383
×122	136	30.0	165	11.2	148	15.0	201	352
×111	120	27.5	138	9.58	124	12.8	169	319
×101	105	25.0	114	7.91	103	10.6	140	289
W21×93	118	29.0	154	10.5	139	14.0	188	338
×83	96.6	25.8	122	8.28	110	11.0	149	298
×73	81.8	22.8	95.4	6.51	86.2	8.68	117	260
×68	73.9	21.5	84.3	5.96	75.9	7.95	104	245
×62	65.6	20.0	71.7	5.37	64.2	7.16	89.1	227
×55	55.7	18.8	59.9	5.26	52.6	7.02	77.0	211
×48	49.2	17.5	49.1	5.25	41.8	6.99	66.1	195
W21×57	66.4	20.3	75.1	5.25	67.7	7.00	92.1	231
×50	59.4	19.0	61.9	5.34	54.5	7.13	79.3	213
×44	49.2	17.5	50.2	4.99	43.3	6.65	66.4	196
W18×175	271	44.5	382	24.0	350	32.0	416	481
×158	234	40.5	316	20.3	289	27.1	366	431
×143	200	36.5	259	16.4	238	21.8	312	384
×130	173	33.5	217	14.1	199	18.8	263	349
×119	159	32.8	197	15.1	178	20.2	246	336
×106	134	29.5	159	12.7	143	16.9	200	298
×97	117	26.8	132	10.3	119	13.7	165	269
×86	97.5	24.0	105	8.46	95.0	11.3	133	238
×76	83.0	21.3	82.5	6.72	74.4	8.96	104	209
W18×71	92.8	24.8	113	8.77	102	11.7	142	247
×65	80.9	22.5	94.4	7.16	85.7	9.54	118	224
×60	71.3	20.8	80.5	6.12	73.1	8.16	100	204
×55	64.0	19.5	69.8	5.64	63.0	7.52	88.2	191
×50	55.5	17.8	57.7	4.73	52.0	6.30	73.0	173
W18×46	56.3	18.0	60.7	4.62	55.1	6.16	75.7	176
×40	46.8	15.8	46.3	3.60	42.0	4.81	58.0	152
×35	42.2	15.0	38.7	3.89	34.1	5.19	51.3	143
W16×100	137	29.3	160	13.0	146	17.3	203	269
×89	115	26.3	129	10.7	117	14.2	163	238
×77	92.4	22.8	96.7	8.14	87.7	10.9	123	203
×67	77.1	19.8	73.1	6.16	66.4	8.22	93.2	174
W16×57	73.9	21.5	86.1	7.35	78.1	9.80	110	190
×50	62.3	19.0	67.2	5.79	60.9	7.72	86.0	167
×45	53.9	17.3	55.0	4.89	49.8	6.52	71.0	150
×40	45.3	15.3	43.2	3.81	39.2	5.07	55.7	132
×36	41.5	14.8	38.0	4.07	33.6	5.43	51.3	127

$F_y = 50$ ksi
$\phi = 1.00$
$\phi_r = 0.75$
$\phi_v = 0.90$

Table 9-5 (cont.). Beam End Bearing Constants

Shape	ϕR_1 kips	ϕR_2 kips/in.	$\phi_r R_3$ kips	$\phi_r R_4$ kips/in.	$\phi_r R_5$ kips	$\phi_r R_6$ kips/in.	$\phi R (N = 3^1/_4)$ kips	$\phi_v V_n$ kips
W16×31	38.7	13.8	34.6	3.22	31.1	4.30	45.1	118
×26	33.2	12.5	26.5	3.13	23.3	4.17	36.8	106
W14×808	3010	187	5910	486	5170	648	3620	2300
×730	2370	154	4310	285	3880	380	2870	1860
×665	2060	142	3660	252	3290	335	2520	1650
×605	1770	130	3090	219	2780	292	2190	1470
×550	1520	119	2590	189	2340	252	1910	1300
×500	1320	110	2190	166	1970	221	1670	1160
×455	1140	101	1860	146	1670	195	1460	1040
×426	1010	94.0	1620	127	1470	169	1320	949
×398	913	88.5	1440	115	1300	154	1200	875
×370	817	83.0	1260	104	1140	139	1090	802
×342	722	77.0	1090	91.6	978	122	972	728
×311	628	70.5	909	78.6	820	105	857	651
×283	544	64.5	762	67.3	687	89.7	754	582
×257	470	59.0	637	57.4	574	76.6	662	523
×233	401	53.5	524	48.2	473	64.3	575	462
×211	352	49.0	438	41.6	394	55.5	511	415
×193	306	44.5	364	34.2	329	45.6	451	372
×176	272	41.5	313	31.1	281	41.5	407	341
×159	233	37.3	253	25.1	228	33.5	337	302
×145	202	34.0	211	21.1	191	28.2	282	272
W14×132	186	32.3	190	19.2	171	25.6	254	256
×120	166	29.5	159	16.3	143	21.8	214	231
×109	144	26.3	127	12.8	115	17.0	171	203
×99	125	24.3	108	11.2	97.2	14.9	146	186
×90	110	22.0	88.8	9.29	80.2	12.4	120	166
W14×82	108	25.5	122	11.8	110	15.7	161	197
×74	91.4	22.5	96.6	8.86	88.2	11.8	127	173
×68	81.1	20.8	81.9	7.68	74.8	10.2	108	157
×61	70.3	18.8	66.6	6.37	60.7	8.50	88.3	141
W14×53	69.4	18.5	66.1	5.98	60.5	7.98	86.4	139
×48	61.1	17.0	55.2	5.19	50.5	6.92	73.0	127
×43	52.4	15.3	44.3	4.23	40.4	5.65	58.8	113
W14×38	41.2	15.5	44.7	4.45	40.6	5.93	59.8	118
×34	35.6	14.3	37.1	3.94	33.4	5.25	50.5	108
×30	31.6	13.5	31.4	4.01	27.8	5.35	45.2	101
W14×26	35.9	12.8	30.1	3.08	27.3	4.10	40.6	95.7
×22	30.5	11.5	23.1	2.87	20.4	3.83	32.9	85.1

$F_y = 50$ ksi
$\phi = 1.00$
$\phi_r = 0.75$
$\phi_v = 0.90$

Table 9-5 (cont.). Beam End Bearing Constants

Shape	ϕR_1 kips	ϕR_2 kips/in.	$\phi_r R_3$ kips	$\phi_r R_4$ kips/in.	$\phi_r R_5$ kips	$\phi_r R_6$ kips/in.	$\phi R(N = 3^1/_4)$ kips	$\phi_v V_n$ kips
W12×336	862	89.0	1480	123	1340	164	1150	807
×305	739	81.5	1240	106	1120	142	1000	717
×279	645	76.5	1070	98.8	970	132	894	657
×252	547	70.0	898	85.8	809	114	774	582
×230	474	64.5	762	74.4	687	99.2	683	526
×210	415	59.0	638	63.8	576	85.0	607	468
×190	348	53.0	520	51.5	471	68.7	520	412
×170	293	48.0	424	43.9	383	58.5	449	363
×152	251	43.5	347	37.2	313	49.6	393	322
×136	210	39.5	284	31.9	255	42.5	338	286
×120	178	35.5	228	26.7	204	35.6	293	251
×106	143	30.5	171	19.3	155	25.7	238	212
×96	125	27.5	140	15.8	126	21.0	195	189
×87	109	25.8	120	14.6	108	19.5	171	174
×79	95.5	23.5	99.8	12.3	89.6	16.5	143	157
×72	84.0	21.5	83.4	10.5	74.8	13.9	120	143
×65	73.1	19.5	68.4	8.78	61.4	11.7	99.4	127
W12×58	67.5	18.0	62.4	6.48	57.2	8.63	85.2	119
×53	59.3	17.3	55.5	6.40	50.3	8.53	78.1	113
W12×50	69.4	18.5	65.0	7.03	59.3	9.37	89.8	122
×45	57.6	16.8	53.1	5.86	48.4	7.81	73.8	109
×40	50.7	14.8	41.5	4.54	37.9	6.05	57.6	94.8
W12×35	44.5	15.0	42.8	4.50	39.1	6.00	58.6	101
×30	36.6	13.0	31.8	3.52	28.9	4.69	44.1	86.3
×26	30.5	11.5	24.6	2.84	22.3	3.79	34.6	75.8
W12×22	32.5	13.0	31.2	3.64	28.2	4.86	44.0	86.3
×19	27.5	11.8	24.3	3.29	21.7	4.39	35.9	77.4
×16	24.1	11.0	19.2	3.63	16.3	4.84	32.0	71.3
×14	20.3	10.0	15.3	3.24	12.8	4.32	26.8	64.3
W10×112	183	37.8	265	32.7	240	43.6	306	232
×100	154	34.0	214	27.4	194	36.5	265	204
×88	128	30.3	169	22.4	153	29.9	226	176
×77	104	26.5	130	17.5	118	23.3	190	152
×68	84.5	23.5	102	14.1	92.4	18.7	153	132
×60	72.2	21.0	81.1	11.6	73.2	15.4	123	116
×54	60.7	18.5	63.8	8.84	57.8	11.8	96.1	101
×49	53.1	17.0	53.6	7.62	48.5	10.2	81.6	91.6
W10×45	57.4	17.5	58.9	7.42	53.9	9.89	86.1	95.4
×39	46.8	15.8	46.5	6.44	42.2	8.59	70.2	84.4
×33	40.8	14.5	37.2	6.24	33.2	8.33	60.2	76.2
W10×30	42.2	15.0	42.4	5.46	38.6	7.29	62.2	85.0
×26	34.5	13.0	31.8	4.20	28.9	5.60	47.1	72.3
×22	28.1	12.0	25.5	4.08	22.7	5.44	40.4	66.1
W10×19	29.3	12.5	28.4	4.20	25.5	5.60	43.7	68.8
×17	26.3	12.0	24.4	4.49	21.4	5.99	40.9	65.4
×15	23.4	11.5	20.7	4.89	17.4	6.52	38.6	62.0
×12	17.8	9.50	13.7	3.59	11.4	4.78	26.9	50.6

SHEAR

PART 10

DESIGN OF SIMPLE SHEAR CONNECTIONS

SCOPE

The specification requirements and other design considerations summarized in this Part apply to the design of simple shear connections. For the design of flexible moment connections, see Part 11. For the design of fully restrained (FR) moment connections, see Part 12. For the design of simple shear connections for HSS and steel pipe, see the AISC *Hollow Structural Sections Connections Manual*.

FORCE TRANSFER

The required strength (factored end reaction) R_u is determined by analysis as indicated in LRFD Specification Section A5. Per LRFD Specification Sections A2 and J1.2, the ends of members with simple shear connections are normally assumed to be free to rotate under load as illustrated in Figure 10-1. While simple shear connections do actually possess some rotational restraint (see curve A in Figure 10-2), this small amount is usually neglected and the connection is idealized as completely flexible. The simple shear connections shown in this Manual are suitable to accommodate the end rotations required per LRFD Specification Section J1.2.

COMPARING CONNECTION ALTERNATIVES

Two-Sided Connections

Two-sided connections, such as double-angle and shear end-plate connections, offer the following advantages:

(1) suitability for use when the end reaction is large;
(2) compact connections (usually, the entire connection is contained within the flanges of the supported beam); and,
(3) eccentricity perpendicular to the beam axis need not be considered for workable gages (see Figure 10-6).

Note that two-sided connections may require consideration to ensure erectability. See "Constructability Considerations" below.

Seated Connections

Unstiffened and stiffened seated connections offer the following advantages:

(1) seats can be shop attached to the support, simplifying erection;
(2) ample erection clearance is provided;
(3) excellent safety during erection since double connections often can be eliminated; and,
(4) the bay length of the structure is easily maintained (seated connections may be preferable when maintaining bay length is a concern for repetitive bays of framing).

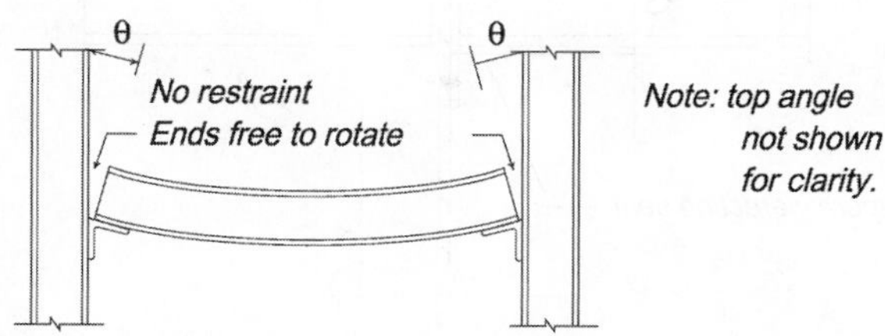

Fig. 10–1. Illustration of simple shear connection behavior.

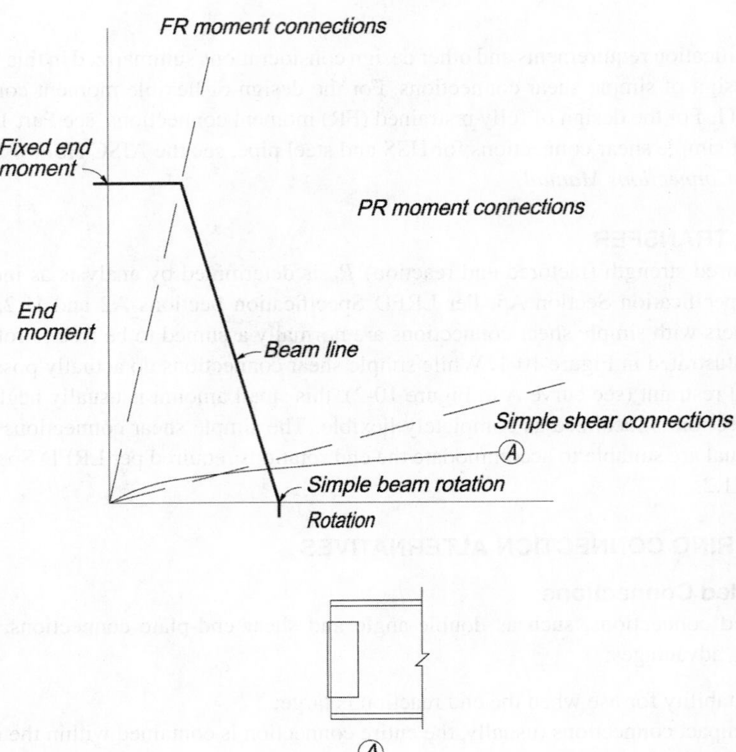

Fig. 10–2. Illustration of typical moment rotation curve for simple shear connection.

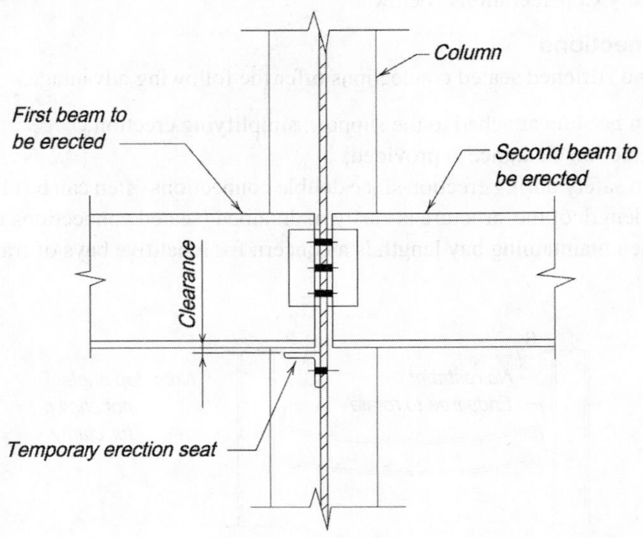

Fig. 10–3. Erection seat.

Note that seated connections can cause erection interference when floors are close, beams are deep, or seats protrude excessively from the column face. The practice of leaning or tilting the columns to erect a column-web connection is difficult, unsafe and should always be avoided.

One-Sided Connections

One-sided connections such as single-plate, single-angle and tee connections offer the following advantages:

(1) shop attachment of connection elements to the support, simplifying shop fabrication and erection;
(2) reduced material and shop labor requirements;
(3) ample erection clearance is provided; and,
(4) excellent safety during erection since double connections often can be eliminated.

CONSTRUCTABILITY CONSIDERATIONS

Double Connections

A double connection occurs in field-bolted construction when beams or girders frame opposite each other. Double connections are of concern to OSHA when they occur in the web of a column (see Figure 10-3) or the web of a beam that frames continuously over the top of a column[1] and all field bolts take the same open holes. A positive connection must be made and maintained for the first member to be erected while the second member to be erected is brought into its final position. Conditions requiring the connector to hang one beam temporarily on a partially inserted bolt or drift pin are not allowed by OSHA.

Framing details can be configured using staggered angles (see Figure 2-2a) or other similar details to provide a means to make a positive connection for the first member while the second member is brought into its final position. Alternatively, a temporary erection seat can be provided. The erection seat, usually an angle, is sized and attached to the column web to support the dead weight of the member, unless additional loading is indicated in the contract documents. It is located to clear the bottom flange of the supported member by approximately $^{3}/_{8}$ in. to accommodate mill, fabrication and erection tolerances.

The sequence of erection is most important in determining the need for erection seats. If the erection sequence is known, the erection seat is provided on the side needing the support. If the erection sequence is not known, a seat can be provided on both sides of the column web. Temporary erection seats may be reused at other locations after the connection(s) are made, but need not be removed unless they create an interference or removal is required in the contract documents.

See also the discussion under "Special Considerations for Simple Shear Connections."

Accessibility in Column Webs

Because of bolting and welding clearances, double-angle, shear end-plate, single-plate, single-angle and tee shear connections may not be suitable for connections to the webs of W-shape and similar columns, particularly for W8 columns, unless gages are reduced. Such connections may be impossible for W6, W5 and W4 columns.

[1] This requirement applies only at the location of the column, not at locations away from the column.

There is also an accessibility concern for entering and tightening the field bolts when the connection material is shop attached to the supporting column web and contained within the column flanges.

Field-Welded Connections

In field-welded connections, temporary erection bolts are usually provided to support the member until final welding is performed. A minimum of 2 bolts (one bolt in bracing members) must be placed for erection safety per OSHA requirements. Additional erection bolts may be required for loads during erection, to assist in pulling the connection angles up tightly against the web of the supporting beam prior to welding or for other reasons. Temporary erection bolts may be reused at other locations after final welding, but need not be removed unless they create an interference or removal is required in the contract documents.

Riding the Fillet

The detailed dimensions of connection elements must be compatible with the T-dimension of an uncoped beam and the remaining web depth, exclusive of fillets, of a coped beam. Note that the angle may encroach upon the fillet(s) as given in Figure 10-4.

DOUBLE-ANGLE CONNECTIONS

A double-angle connection is made with two angles, one on each side of the web of the beam to be supported, as illustrated in Figure 10-5. These angles may be bolted or welded to the supported beam as well as to the supporting member.

When the angles are welded to the support, adequate flexibility must be provided in the connection. As illustrated in Figure 10-5c, line welds are placed along the toes of the angles with a return at the top per LRFD Specification Section J2.2b. Note that welding across the entire top of the angles must be avoided as it would inhibit the flexibility and, therefore, the necessary end rotation of the connection. The performance of the resulting connection would not be as intended for double-angle connections.

Design Strength

The design strength of a double-angle connection is determined from the applicable limit states for the bolts (see Part 7), welds (see Part 8) and connection elements (see Part 9). In all cases, the design strength ϕR_n must equal or exceed the required strength R_u.

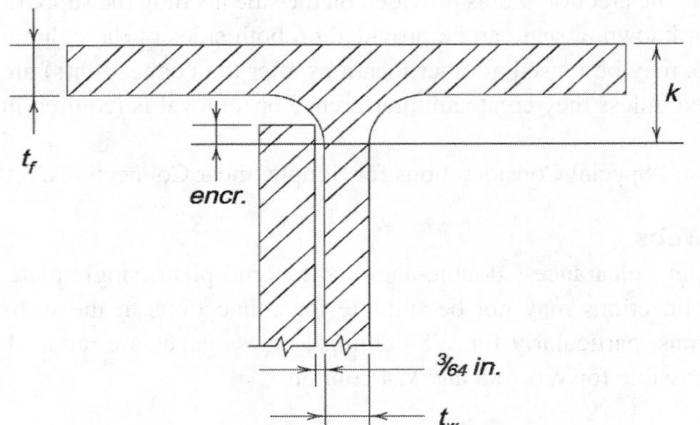

$k - t_f$ in.	Encr. in.
5/16	1/8
3/8 to 1/2	3/16
9/16 to 13/16	1/4
7/8 to 1 1/4	5/16
1 5/16 and 1 3/8	3/8

Fig. 10–4. Fillet encroachment (riding the fillet).

For the workable gages shown in Figure 10-6 and standard or short-slotted holes, eccentricity in double-angle connections may be neglected, except in the case of a double vertical row of bolts through the web of the supported beam. Eccentricity should always be considered in the design of welds for double-angle connections.

Recommended Angle Length and Thickness

To provide for stability during erection, it is recommended that the minimum angle length be one-half the T-dimension of the beam to be supported. The maximum length of the connection angles must be compatible with the T-dimension of an uncoped beam and the remaining web depth, exclusive of fillets, of a coped beam. Note that the angle may encroach upon the fillet(s) as given in Figure 10-4.

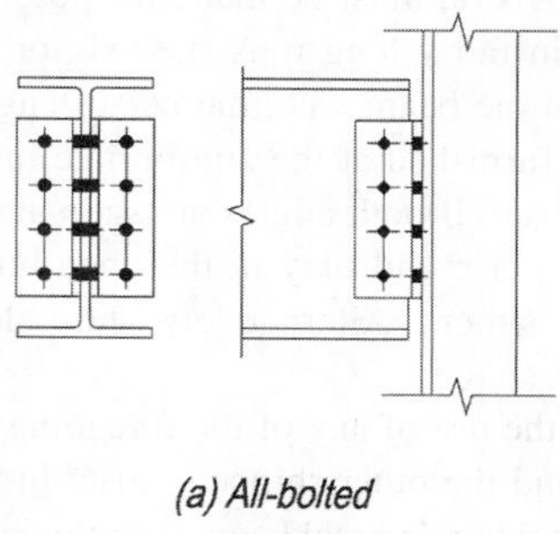

(a) All-bolted

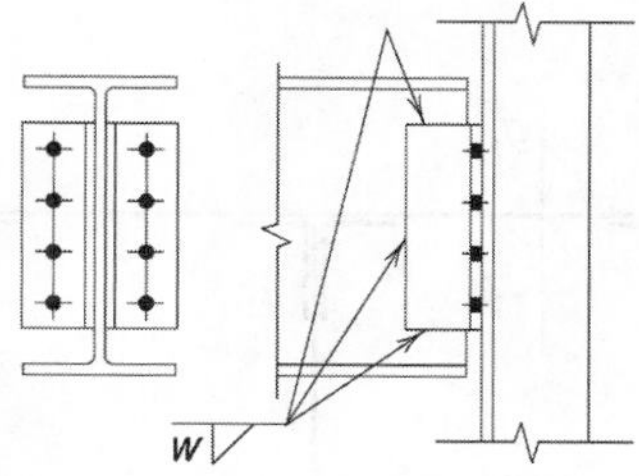

(b) Bolted/welded, angles welded to supported beam

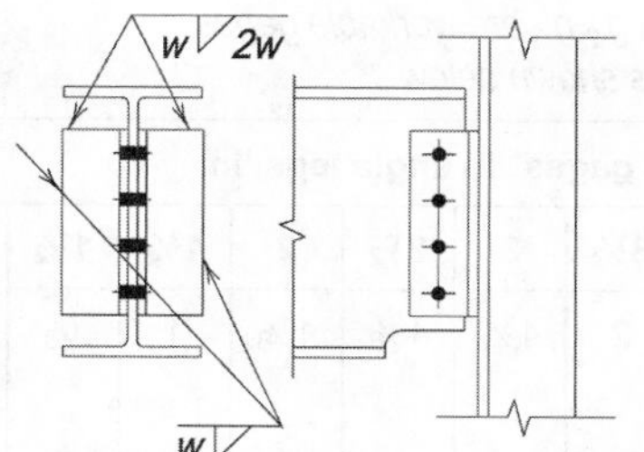

(c) Bolted/welded, angles welded to support

Fig. 10–5. Double-angle connections.

To provide for flexibility, the maximum angle thickness for use with workable gages should be limited to $^5/_8$ in. Alternatively, the shear-connection ductility checks illustrated in Part 9 can be used to justify other combinations of gage and angle thickness.

Shop and Field Practices

When framing to a girder web, both angles are usually shop attached to the web of the supported beam. When framing to a column web, both angles should be shop attached to the supported beam, when possible, and the associated constructability considerations should be addressed (see the preceding discussion under "Constructability Considerations").

When framing to a column flange, both angles can be shop attached to the column flange or the supported beam. In the former case, this is a knifed connection, as illustrated in Figure 10-5c, which requires an erection clearance as illustrated in Figure 10-7a and that the bottom flange be coped away. Also, provision must be made for possible mill variation in the depth of the columns, particularly in fairly long runs (i.e., six or more bays of framing). If both angles are shop attached to the beam web, the beam length can be shortened to provide for mill overrun with shims furnished at the appropriate intervals to fill the resulting gaps or to provide for mill underrun. If both angles are shop attached to the column flange, the erected beam is knifed into place and play in the open holes is normally sufficient to provide for the necessary adjustment. Alternatively, short-slotted holes can also be used.

When special requirements preclude the use of any of the foregoing practices, one angle could be shop attached to the support and the other shipped loose. In this case, the spread between the outstanding legs should equal the decimal beam web thickness plus a clearance

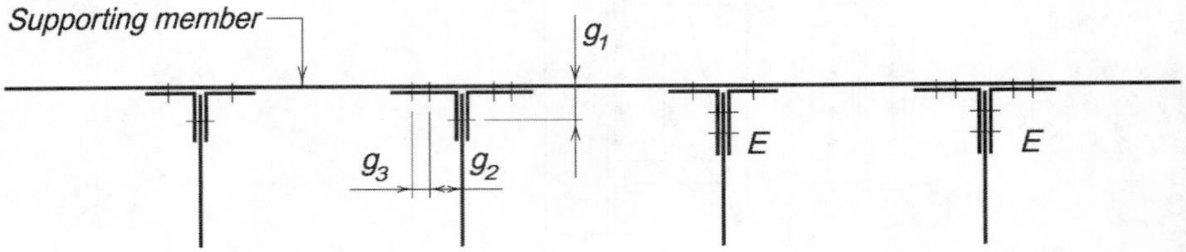

E indicates that eccentricity must be considered in this leg.
Gages g_1 g_2 g_3 are workable gages as shown below

Workable gages* in angle legs, in.														
Leg	**8**	**7**	**6**	**5**	**4**	**3½**	**3**	**2½**	**2**	**1¾**	**1½**	**1⅜**	**1¼**	**1**
g_1	4½	4	3½	3	2½	2	1¾	1⅜	1⅛	1	⅞	⅞	¾	⅝
g_2	3	2½	2¼	2										
g_3	3	3	2½	1¾										
*Other gages are permitted to suit specific requirements subject to clearances and edge distance limitations.														

Fig. 10–6. Eccentricity in double-angle connections.

that will produce an opening to the next higher $^1/_{16}$-in. increment, as illustrated in Figure 10-7b. Alternatively, short-slotted holes in the support-leg of the angle eliminate the need to provide for variations in web thickness. Note that the practice of shipping one angle loose is not desirable because it requires additional material handling as well as added erection costs and difficulty.

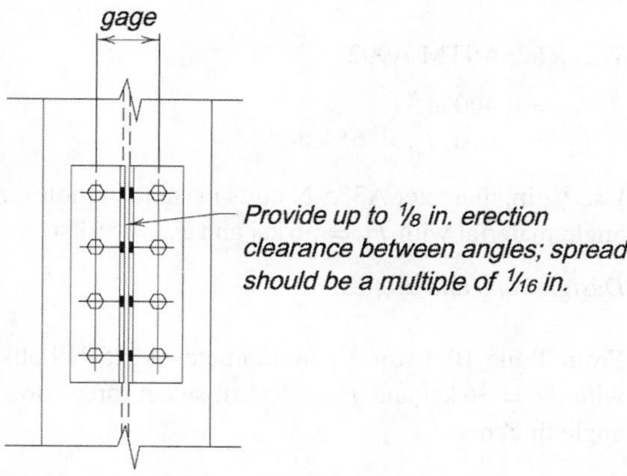

(a) Both angles shop attached to the column flange (beam knifed into place)

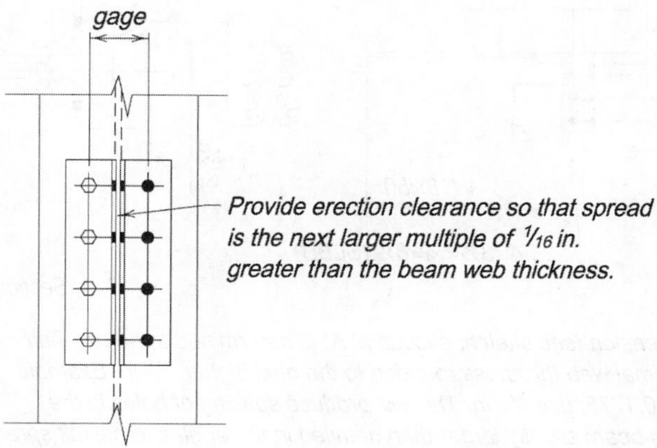

(b) One shop attached to the column flange, other shipped loose

Fig. 10–7. Erection clearances for double-angle connections.

EXAMPLE 10.1. All-bolted double-angle connection (beam-to-girder web).

Given: Refer to Figure 10-8. Use Table 10-1 to design an all-bolted double-angle connection for the W18×50 beam to W21×62 girder web connection.

$$R_u = 60\,\text{kips}$$

W18×50, ASTM A992

$t_w = 0.355$ in. $\qquad d = 18.0$ in.

$F_y = 50$ ksi, $F_u = 65$ ksi

top flange coped 2 in. deep by 4 in. long, $L_{ev} = 1^1/_4$ in., $L_{eh} = 1^3/_4$ in. (assumed to be $1^1/_2$ in. for calculation purposes to account for possible underrun in beam length)

W21×62, ASTM A992

$t_w = 0.400$ in.

$F_y = 50$ ksi, $F_u = 65$ ksi

Use $^3/_4$ in. diameter A325-N bolts in standard holes. Assume ASTM A36 angle material with $F_y = 36$ ksi and $F_u = 58$ ksi.

Solution: *Design bolts and angles*

From Table 10-1, for $^3/_4$-in. diameter A325-N bolts and angle material with $F_y = 36$ ksi and $F_u = 58$ ksi, select three rows of bolts and $^1/_4$-in. angle thickness.

$$\phi R_n = 76.7\,\text{kips} > 60\,\text{kips} \qquad \textbf{o.k.}$$

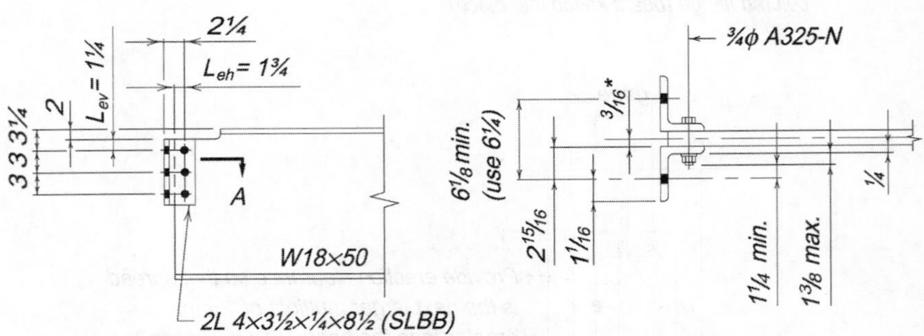

** This dimension (see sketch, section at A) is determined to be one-half of the decimal web thickness rounded to the next higher 1/16 in. Example: 0.355/2 = 0.1775; use 3/16 in. This will produce spacing of holes in the supporting beam slightly larger than detailed in the angles to permit spreading of angles (angles can be spread but not closed) at time of erection to supporting member. Alternatively, consider using horizontal slots in the support legs of the angles.*

Fig. 10–8. Illustration for Example 10.1.

Check supported beam web

From Table 10-1, for three rows of bolts, beam material with $F_y = 50$ ksi and $F_u = 65$ ksi, and $L_{ev} = 1^1/_4$ in. and $L_{eh} = 1^3/_4$ in. (assumed to be $1^1/_2$ in. for calculation purposes to account for possible underrun in beam length)

$$\phi R_n = (200\text{ kips/in.})(0.355\text{ in.})$$
$$= 71.0\text{ kips} > 60\text{ kips} \quad \textbf{o.k.}$$

Check flexural yielding on the coped section

From Table 9-2, $S_{net} = 23.4\text{ in.}^3$

$$\phi R_n = \frac{\phi F_y S_{net}}{e}$$
$$= \frac{0.9(50\text{ ksi})(23.4\text{ in.}^3)}{(4\text{ in.} + {}^1/_2\text{ in.})}$$
$$= 234\text{ kips} > 60\text{ kips} \quad \textbf{o.k.}$$

Check local web buckling at the cope

$$\frac{c}{d} = \frac{4\text{ in.}}{18.0\text{ in.}} = 0.222$$
$$\frac{c}{h_o} = \frac{4\text{ in.}}{(18.0\text{ in.} - 2\text{ in.})} = 0.250$$

Since $\frac{c}{d} \leq 1.0$,

$$f = 2\left(\frac{c}{d}\right)$$
$$= 2(0.222)$$
$$= 0.444$$

Since $\frac{c}{h_o} \leq 1.0$,

$$k = 2.2\left(\frac{h_o}{c}\right)^{1.65}$$
$$= 2.2\left(\frac{1}{0.250}\right)^{1.65}$$
$$= 21.7$$

$$\phi F_{bc} = 23{,}590\left(\frac{t_w}{h_o}\right)^2 fk$$
$$= 23{,}590\left(\frac{0.355\text{ in.}}{18.0\text{ in.} - 2\text{ in.}}\right)^2 (0.444)(21.7)$$
$$= 112\text{ ksi}$$

$$\phi R_n = \frac{\phi F_{bc} S_{net}}{e}$$
$$= \frac{(112\text{ ksi})(23.4\text{ in.}^3)}{(4\text{ in.} + {}^1/_2\text{ in.})}$$
$$= 582\text{ kips} > 60\text{ kips} \quad \textbf{o.k.}$$

Check supporting girder web

From Table 10-1, for three rows of bolts and girder material with $F_u =$ 65 ksi,

$$\begin{aligned}\phi R_n &= (526\,\text{kips/in.})(0.400\,\text{in.}) \\ &= 210\,\text{kips} > 60\,\text{kips} \quad \textbf{o.k.}\end{aligned}$$

The connection, as summarized in Figure 10-8, is adequate.

EXAMPLE 10.2. All-bolted double-angle connection (beam-to-column flange).

Given: Refer to Figure 10-9. Use Table 10-1 to design an all-bolted double-angle connection for the W36×230 beam to W14×90 column-flange connection.

$$R_u = 225\,\text{kips}$$

W36×230, ASTM A992

$$\begin{aligned}t_w &= 0.760\,\text{in.} \\ F_y &= 50\,\text{ksi},\ F_u = 65\,\text{ksi}\end{aligned}$$

W14×90, ASTM A992

$$\begin{aligned}t_f &= 0.710\,\text{in.} \\ F_y &= 50\,\text{ksi},\ F_u = 65\,\text{ksi}\end{aligned}$$

Use $^3/_4$-in. diameter A325-N bolts in standard holes. Assume ASTM A36 angle material with $F_y = 36$ ksi and $F_u = 58$ ksi.

Solution: *Design bolts and angles*

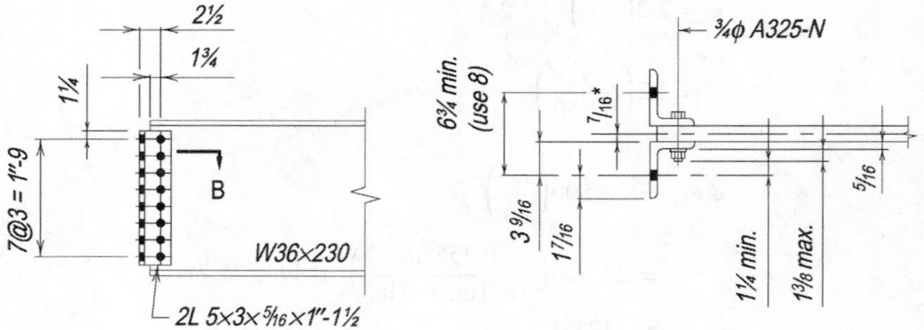

Fig. 10–9. Illustration for Example 10.2.

From Table 10-1, for $^3/_4$-in. diameter A325-N bolts and angle material with $F_y = 36$ ksi and $F_u = 58$ ksi, select eight rows of bolts and $^5/_{16}$-in. angle thickness.

$$\phi R_n = 254 \text{ kips} > 225 \text{ kips} \quad \textbf{o.k.}$$

Check supported beam web

From Table 10-1, for eight rows of bolts, beam material with $F_y = 50$ ksi and $F_u = 65$ ksi, and $L_{eh} = 1^3/_4$ in.,

$$\begin{aligned}\phi R_n &= (702 \text{ kips/in.})(0.760 \text{ in.}) \\ &= 534 \text{ kips} > 225 \text{ kips} \quad \textbf{o.k.}\end{aligned}$$

Check supporting column flange

From Table 10-1, for eight rows of bolts and column material with $F_y =$ 50 ksi and $F_u = 65$ ksi,

$$\begin{aligned}\phi R_n &= (1{,}400 \text{ kips/in.})(0.710 \text{ in.}) \\ &= 994 \text{ kips} > 225 \text{ kips} \quad \textbf{o.k.}\end{aligned}$$

EXAMPLE 10.3. Bolted/welded double-angle connection (beam-to-girder web).

Given: Refer to Example 10-1. Use Tables 10-2 to substitute welds for bolts in the supported-beam-web legs of the double-angle connection (welds A).

Solution: From Tables 10-2, for three rows of bolts (an angle length of $8^1/_2$ in.), a $^3/_{16}$ in. weld size provides $\phi R_n = 110$ kips. For beam web material with $F_y = 50$ ksi, the minimum web thickness is 0.286 in. Since $t_w =$ 0.355 in. > 0.286 in., no reduction in the tabulated value is required.

$$\phi R_n = 110 \text{ kips} > 60 \text{ kips} \quad \textbf{o.k.}$$

Check minimum angle thickness

The minimum angle thickness for Tables 10-2 is the weld size plus $^1/_{16}$ in., but not less than the thickness determined from Table 10-1.

$$\begin{aligned}t_{\min} &= {}^3/_{16} \text{ in.} + {}^1/_{16} \text{ in.} \\ &= {}^1/_4 \text{ in.}\end{aligned}$$

This thickness is equal to the thickness chosen previously from Table 10-1.

EXAMPLE 10.4. Bolted/welded double-angle connection (beam-to-column flange).

Given: Refer to Example 10-2. Use Tables 10-2 to substitute welds for bolts in the support legs of the double-angle connection (welds B).

Solution: From Tables 10-2, for eight rows of bolts (an angle length of $23^1/_2$ in.), a $^5/_{16}$-in. weld size provides $\phi R_n = 279$ kips. For beam web material with $F_y = 50$ ksi, the minimum column flange thickness is 0.238 in. Since $t_f = 0.710$ in. > 0.238 in., no reduction of the tabulated value is required.

$$\phi R_n = 279 \text{ kips} > 225 \text{ kips} \quad \textbf{o.k.}$$

Check minimum angle thickness

The minimum angle thickness for Tables 10-2 is the weld size plus $^1/_{16}$ in., but not less than the thickness determined from Table 10-1.

$$t_{min} = {}^5/_{16} \text{ in.} + {}^1/_{16} \text{ in.}$$
$$= {}^3/_8 \text{ in.}$$

Thus, the angle thickness must be increased to $^3/_8$ in. to accommodate the welded legs of the double-angle connection.

EXAMPLE 10.5. All-welded double-angle connection (beam-to-column flange).

Given: Refer to Example 10-2. Use Table 10-3 to design an all-welded double-angle connection for the W36×230 beam to W14×90 column-flange connection.

Solution: *Design supported-beam-web angle leg welds (welds A)*

From Table 10-3, for $L = 24$ in., a $^3/_{16}$-in. weld A size provides $\phi R_n =$ 258 kips. For beam web material with $F_y = 50$ ksi, the minimum supported beam web thickness is 0.286 in. Since $t_w = 0.760$ in. > 0.286 in., no reduction of the tabulated value is required.

$$\phi R_n = 258 \text{ kips} > 225 \text{ kips} \quad \textbf{o.k.}$$

Design support angle leg welds (welds B)

From Table 10-3, for $L = 24$ in., a $^1/_4$-in. weld B size provides $\phi R_n =$ 229 kips. For column flange material with $F_y = 50$ ksi, the minimum column flange thickness is 0.190 in. Since $t_f = 0.710$ in. > 0.190 in., no reduction of the tabulated value is required.

Check minimum angle thickness

The minimum angle thickness for Table 10-3 is the weld size plus $^1/_{16}$ in.

$$t_{min} = {}^1/_4 \text{ in.} + {}^1/_{16} \text{ in.}$$
$$= {}^5/_{16} \text{ in.}$$

Use 2L4×3×$^5/_{16}$

Table 10-1. All-Bolted Double-Angle Connections

Tables 10-1 are design aids for all-bolted double-angle connections. Design strengths are tabulated for supported and supporting member material with $F_y = 50$ ksi and $F_u = 65$ ksi and angle material with $F_y = 36$ ksi and $F_u = 58$ ksi. All values, including slip-critical bolt design strengths, are for comparison with factored loads.

Tabulated bolt and angle design strengths consider the limit states of bolt shear, bolt bearing on the angles, shear yielding of the angles, shear rupture of the angles and block shear rupture of the angles. Values are tabulated for 2 through 12 rows of $^3/_4$-in., $^7/_8$-in, and 1-in.-diameter

ASTM A325, F1852 and A490 bolts at 3-in. spacing. For calculation purposes, angle edge distances L_{ev} and L_{eh} are assumed to be $1^1/_4$ in.

Tabulated beam web design strengths, per in. of web thickness, consider the limit-state of bolt bearing on the beam web. For beams coped at the top flange only, the limit state of block shear rupture is also considered. Additionally, for beams coped at both the top and bottom flanges, the tabulated values consider the limit states of shear yielding and shear rupture of the beam web. Values are tabulated for beam web edge distances L_{ev} from $1^1/_4$ in. to 3 in. and for beam end distances L_{eh} of $1^1/_2$ in. and $1^3/_4$ in. For calculation purposes, these end distances have been reduced to $1^1/_4$ in. and $1^1/_2$ in., respectively, to account for possible underrun in beam length. For coped members, the limit states of flexural yielding and local buckling must be checked independently per Part 9. When required, web reinforcement of coped members is treated in Part 9.

Tabulated supporting member design strengths, per in. of flange or web thickness, consider the limit state of bolt bearing on the support.

Table 10-2. Bolted/Welded Double-Angle Connections

Table 10-2 is a design aid arranged to permit substitution of welds for bolts in connections designed with Tables 10-1. Electrode strength is assumed to be 70 ksi. All values are for comparison with factored loads. Holes for erection bolts may be placed as required in angle legs that are to be field welded.

Welds A may be used in place of bolts through the supported-beam-web legs of the double angles or welds B may be used in place of bolts through the support legs of the double angles. Although it is permissible to use welds A and B from Tables 10-2 in combination to obtain all-welded connections, it is recommended that such connections be selected from Table 10-3. This table will allow increased flexibility in the selection of angle lengths and connection strengths because Table 10-2 conforms to the bolt spacing and edge distance requirements for the all-bolted double-angle connections of Tables 10-1.

Weld design strengths are tabulated for the limit state of weld shear. Design strengths for welds A are determined by the instantaneous center of rotation method using Table 8-9 with $\theta = 0°$. Design strengths for welds B are determined by the elastic method. With the neutral axis assumed at one-sixth the depth of the angles measured downward and the tops of the angles in compression against each other through the beam web, the design strength of these welds is ϕR_n, where

$$\phi R_n = 2 \times \frac{1.392DL}{\sqrt{1 + \dfrac{12.96e^2}{L^2}}}$$

where

D = number of sixteenths-of-an-inch in the weld size
L = length of the connection angles, in.
e = width of the leg of the connection angle attached to the support, in.

The tabulated minimum thicknesses of the supported beam web for welds A and the support for welds B match the shear rupture strength of these elements with the strength of the weld metal. As derived in Part 9, the minimum supported beam web thickness for welds A (two lines of weld) is

$$t_{min} = \frac{6.19D}{F_u}$$

and the minimum supporting flange or web thickness for welds B (one line of weld) is

$$t_{min} = \frac{3.09D}{F_u}$$

When welds B line up on opposite sides of the support, the minimum thickness is the sum of the thicknesses required for each weld. In either case, when less than the minimum material thickness is present, the tabulated weld design strength must be reduced by the ratio of the thickness provided to the minimum thickness.

The minimum angle thickness when Table 10-2 is used is the weld size plus $^1/_{16}$ in. but not less than the angle thickness determined from Tables 10-1. The angle length L must be as tabulated in Table 10-2. In general, 2L4×3$^1/_2$ will accommodate workable gages, with the 4-in. leg attached to the supporting member. The width of web legs in Case I may be optionally reduced from 3$^1/_2$ in. to 3 in. The width of outstanding legs in Case II may be optionally reduced from 4 in. to 3 in. for values of L from 5$^1/_2$ through 17$^1/_2$ in.

Table 10-3. All-Welded Double-Angle Connections

Table 10-3 is a design aid for all-welded double-angle connections. Electrode strength is assumed to be 70 ksi. All values are for comparison with factored loads. Holes for erection bolts may be placed as required in angle legs that are to be field welded.

Weld design strengths are tabulated for the limit state of weld shear. Design strengths for welds A are determined by the instantaneous center of rotation method using Table 8-9 with $\theta = 0°$. Design strengths for welds B are determined by the elastic method as discussed previously for bolted/welded double-angle connections.

The tabulated minimum thicknesses of the supported beam web for welds A and the support for welds B match the shear rupture strength of these elements with the strength of the weld metal and are determined as discussed previously for Table 10-2. When welds B line up on opposite sides of the support, the minimum thickness is the sum of the thicknesses required for each weld. When less than the minimum material thickness is present, the tabulated weld design strength must be reduced by the ratio of the thickness provided to the minimum thickness.

The minimum angle thickness when Table 10-3 is used must be equal to the weld size plus $^1/_{16}$ in. The angle length L must be as tabulated in Table 10-3. 2L4×3 should be used for angle lengths equal to or greater than 18 in. 2L3×3 should be used otherwise.

Beam	$F_y = 50$ ksi, $F_u = 65$ ksi	Angle	$F_y = 36$ ksi, $F_u = 58$ ksi

Table 10-1. All-Bolted Double-Angle Connections

$\phi = 0.75$

3/4-in. Bolts — 12 Rows — W44

Bolt and Angle Design Strength, kips						
ASTM Desig.	Thread Cond.	Hole Type	Angle Thickness 1/4	5/16	3/8	1/2
A325/F1852	N	–	326	382	382	382
	X	–	326	408	477	477
	SC Class A	STD	251	251	251	251
		OVS	213	213	213	213
		SSLT	213	213	213	213
	SC Class B	STD	326	380	380	380
		OVS	307	323	323	323
		SSLT	323	323	323	323
A490	N	–	326	408	477	477
	X	–	326	408	489	596
	SC Class A	STD	313	313	313	313
		OVS	266	266	266	266
		SSLT	266	266	266	266
	SC Class B	STD	326	408	475	475
		OVS	307	383	403	403
		SSLT	326	403	403	403

Beam Web Design Strength per Inch Thickness, kips/in.

Hole Type	L_{eh},* in.	Un-coped	Coped at Top Flange Only, L_{ev}, in. 1 1/4	1 3/8	1 1/2	1 5/8	2	3	Coped at Both Flanges, L_{ev}, in. 1 1/4	1 3/8	1 1/2	1 5/8	2	3
STD	1 1/2	1050	747	751	754	758	769	798	731	739	746	753	769	798
	1 3/4	1050	759	763	767	770	781	810	731	739	746	753	775	810
OVS	1 1/2	1050	702	706	709	713	724	753	687	695	702	709	724	753
	1 3/4	1050	714	718	722	725	736	765	687	695	702	709	731	765
SSLT	1 1/2	1050	743	746	750	753	764	794	731	739	746	753	764	794
	1 3/4	1050	755	758	762	766	777	806	731	739	746	753	775	806

Support Design Strength per Inch Thickness, kips/in.
2110

Notes:
STD = Standard holes
OVS = Oversized holes
SSLT = Short-slotted holes oriented transverse to direction of load
N = Threads included
X = Threads excluded
SC = Slip critical

*Tabulated values include 1/4-in. reduction in end distance L_{eh} to account for possible underrun in beam length.

Angle	$F_y = 36$ ksi $F_u = 58$ ksi	$F_y = 50$ ksi $F_u = 65$ ksi	Beam

Table 10-1 (cont.). All-Bolted Double-Angle Connections

$\phi = 0.75$

3/4-in. Bolts
11 Rows
W44, 40

Varies
t
10@3 = 30
2 1/4
L_{eh}
L_{ev}
10@3 = 30
L_{ev}

Bolt and Angle Design Strength, kips

ASTM Desig.	Thread Cond.	Hole Type	Angle Thickness 1/4	5/16	3/8	1/2
A325/ F1852	N	–	299	350	350	350
	X	–	299	373	437	437
	SC Class A	STD	230	230	230	230
		OVS	195	195	195	195
		SSLT	195	195	195	195
	SC Class B	STD	299	348	348	348
		OVS	281	296	296	296
		SSLT	296	296	296	296
A490	N	–	299	373	437	437
	X	–	299	373	448	547
	SC Class A	STD	287	287	287	287
		OVS	244	244	244	244
		SSLT	244	244	244	244
	SC Class B	STD	299	373	435	435
		OVS	281	351	370	370
		SSLT	299	370	370	370

Beam Web Design Strength per Inch Thickness, kips/in.

Hole Type	L_{eh},* in.	Un-coped	Coped at Top Flange Only, L_{ev}, in. 1 1/4	1 3/8	1 1/2	1 5/8	2	3	Coped at Both Flanges, L_{ev}, in. 1 1/4	1 3/8	1 1/2	1 5/8	2	3
STD	1 1/2	965	685	689	692	696	707	736	669	676	684	691	707	736
	1 3/4	965	697	701	704	708	719	748	669	676	684	691	713	748
OVS	1 1/2	965	644	647	651	654	665	695	629	636	644	651	665	695
	1 3/4	965	656	659	663	667	678	707	629	636	644	651	673	707
SSLT	1 1/2	965	680	684	688	691	702	732	669	676	684	691	702	732
	1 3/4	965	693	696	700	704	714	744	669	676	684	691	713	744

Support Design Strength per Inch Thickness, kips/in.

1930

Notes:
STD = Standard holes
OVS = Oversized holes
SSLT = Short-slotted holes oriented transverse to direction of load
N = Threads included
X = Threads excluded
SC = Slip critical

*Tabulated values include 1/4-in. reduction in end distance L_{eh} to account for possible underrun in beam length.

Beam	
	$F_y = 50$ ksi
	$F_u = 65$ ksi

	Angle
$F_y = 36$ ksi	
$F_u = 58$ ksi	

Table 10-1 (cont.). All-Bolted Double-Angle Connections $\phi = 0.75$

3/4-in. Bolts, 10 Rows, W44, 40, 36

Bolt and Angle Design Strength, kips

ASTM Desig.	Thread Cond.	Hole Type	Angle Thickness 1/4	5/16	3/8	1/2
A325/ F1852	N	–	271	318	318	318
	X	–	271	338	398	398
	SC Class A	STD	209	209	209	209
		OVS	178	178	178	178
		SSLT	178	178	178	178
	SC Class B	STD	271	316	316	316
		OVS	254	269	269	269
		SSLT	269	269	269	269
A490	N	–	271	338	398	398
	X	–	271	338	406	497
	SC Class A	STD	261	261	261	261
		OVS	222	222	222	222
		SSLT	222	222	222	222
	SC Class B	STD	271	338	396	396
		OVS	254	318	336	336
		SSLT	271	336	336	336

Beam Web Design Strength per Inch Thickness, kips/in.

Hole Type	L_{eh},* in.	Un-coped	Coped at Top Flange Only, L_{ev}, in. 1 1/4	1 3/8	1 1/2	1 5/8	2	3	Coped at Both Flanges, L_{ev}, in. 1 1/4	1 3/8	1 1/2	1 5/8	2	3
STD	1 1/2	878	623	626	630	634	645	674	607	614	622	629	645	674
	1 3/4	878	635	639	642	646	657	686	607	614	622	629	651	686
OVS	1 1/2	878	585	589	592	596	607	636	570	578	585	592	607	636
	1 3/4	878	597	601	605	608	619	648	570	578	585	592	614	648
SSLT	1 1/2	878	618	622	626	629	640	669	607	614	622	629	640	669
	1 3/4	878	630	634	638	641	652	682	607	614	622	629	651	682

Support Design Strength per Inch Thickness, kips/in.

1760

Notes:
STD = Standard holes
OVS = Oversized holes
SSLT = Short-slotted holes oriented transverse to direction of load
N = Threads included
X = Threads excluded
SC = Slip critical

*Tabulated values include 1/4-in. reduction in end distance L_{eh} to account for possible underrun in beam length.

Angle	$F_y = 36$ ksi, $F_u = 58$ ksi

Beam	$F_y = 50$ ksi, $F_u = 65$ ksi

Table 10-1 (cont.). All-Bolted Double-Angle Connections
$\phi = 0.75$

3/4-in. Bolts
9 Rows
W44, 40, 36, 33

Varies
t
8@3 = 24
2 1/4
L_{eh}
L_{ev}
8@3 = 24
L_{ev}

Bolt and Angle Design Strength, kips

ASTM Desig.	Thread Cond.	Hole Type	Angle Thickness 1/4	5/16	3/8	1/2
A325/ F1852	N	–	243	286	286	286
	X	–	243	304	358	358
	SC Class A	STD	188	188	188	188
		OVS	160	160	160	160
		SSLT	160	160	160	160
	SC Class B	STD	243	285	285	285
		OVS	228	242	242	242
		SSLT	242	242	242	242
A490	N	–	243	304	358	358
	X	–	243	304	365	447
	SC Class A	STD	235	235	235	235
		OVS	200	200	200	200
		SSLT	200	200	200	200
	SC Class B	STD	243	304	356	356
		OVS	228	285	303	303
		SSLT	243	303	303	303

Beam Web Design Strength per Inch Thickness, kips/in.

Hole Type	L_{eh},* in.	Un-coped	Coped at Top Flange Only, L_{ev}, in. 1 1/4	1 3/8	1 1/2	1 5/8	2	3	Coped at Both Flanges, L_{ev}, in. 1 1/4	1 3/8	1 1/2	1 5/8	2	3
STD	1 1/2	790	561	564	568	572	583	612	545	552	559	567	583	612
	1 3/4	790	573	576	580	584	595	624	545	552	559	567	589	624
OVS	1 1/2	790	527	530	534	537	548	578	512	519	527	534	548	578
	1 3/4	790	539	542	546	550	561	590	512	519	527	534	556	590
SSLT	1 1/2	790	556	560	563	567	578	607	545	552	559	567	578	607
	1 3/4	790	568	572	576	579	590	619	545	552	559	567	589	619

Support Design Strength per Inch Thickness, kips/in.

1580

Notes:
STD = Standard holes
OVS = Oversized holes
SSLT = Short-slotted holes oriented transverse to direction of load
N = Threads included
X = Threads excluded
SC = Slip critical

*Tabulated values include 1/4-in. reduction in end distance L_{eh} to account for possible underrun in beam length.

Beam	$F_y = 50$ ksi $F_u = 65$ ksi

$F_y = 36$ ksi $F_u = 58$ ksi	Angle

Table 10-1 (cont.). All-Bolted Double-Angle Connections $\phi = 0.75$

3/4-in. Bolts
8 Rows
W44, 40, 36, 33, 30

Bolt and Angle Design Strength, kips						
ASTM Desig.	Thread Cond.	Hole Type	Angle Thickness 1/4	5/16	3/8	1/2
A325/F1852	N	–	215	254	254	254
	X	–	215	269	318	318
	SC Class A	STD	167	167	167	167
		OVS	142	142	142	142
		SSLT	142	142	142	142
	SC Class B	STD	215	253	253	253
		OVS	202	215	215	215
		SSLT	215	215	215	215
A490	N	–	215	269	318	318
	X	–	215	269	323	398
	SC Class A	STD	209	209	209	209
		OVS	178	178	178	178
		SSLT	178	178	178	178
	SC Class B	STD	215	269	316	316
		OVS	202	253	269	269
		SSLT	215	269	269	269

Beam Web Design Strength per Inch Thickness, kips/in.

Hole Type	L_{eh},* in.	Un-coped	Coped at Top Flange Only, L_{ev}, in. 1 1/4	1 3/8	1 1/2	1 5/8	2	3	Coped at Both Flanges, L_{ev}, in. 1 1/4	1 3/8	1 1/2	1 5/8	2	3
STD	1 1/2	702	498	502	506	509	520	550	483	490	497	505	520	550
	1 3/4	702	511	514	518	522	533	562	483	490	497	505	527	562
OVS	1 1/2	702	468	472	475	479	490	519	453	461	468	475	490	519
	1 3/4	702	480	484	488	491	502	531	453	461	468	475	497	531
SSLT	1 1/2	702	494	498	501	505	516	545	483	490	497	505	516	545
	1 3/4	702	506	510	513	517	528	557	483	490	497	505	527	557

Support Design Strength per Inch Thickness, kips/in.
1400

Notes:
STD = Standard holes
OVS = Oversized holes
SSLT = Short-slotted holes oriented transverse to direction of load
N = Threads included
X = Threads excluded
SC = Slip critical

*Tabulated values include 1/4-in. reduction in end distance L_{eh} to account for possible underrun in beam length.

Angle	$F_y = 36$ ksi $F_u = 58$ ksi	$F_y = 50$ ksi $F_u = 65$ ksi	Beam

Table 10-1 (cont.). All-Bolted Double-Angle Connections $\phi = 0.75$

$3/4$-in. Bolts
7 Rows
W44, 40, 36, 33, 30, 27, 24

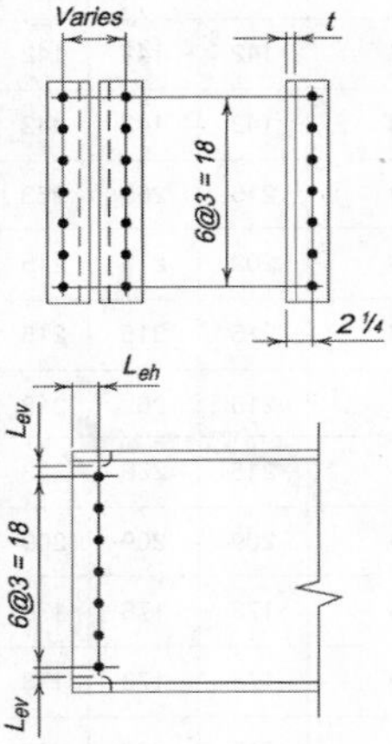

Bolt and Angle Design Strength, kips						
ASTM Desig.	Thread Cond.	Hole Type	Angle Thickness			
			$1/4$	$5/16$	$3/8$	$1/2$
A325/ F1852	N	–	188	223	223	223
	X	–	188	234	278	278
	SC Class A	STD	146	146	146	146
		OVS	124	124	124	124
		SSLT	124	124	124	124
	SC Class B	STD	188	221	221	221
		OVS	176	188	188	188
		SSLT	188	188	188	188
A490	N	–	188	234	278	278
	X	–	188	234	281	348
	SC Class A	STD	183	183	183	183
		OVS	155	155	155	155
		SSLT	155	155	155	155
	SC Class B	STD	188	234	277	277
		OVS	176	220	235	235
		SSLT	188	234	235	235

Beam Web Design Strength per Inch Thickness, kips/in.

Hole Type	L_{eh},* in.	Uncoped	Coped at Top Flange Only, L_{ev}, in.						Coped at Both Flanges, L_{ev}, in.					
			1 1/4	1 3/8	1 1/2	1 5/8	2	3	1 1/4	1 3/8	1 1/2	1 5/8	2	3
STD	1 1/2	614	436	440	444	447	458	488	420	428	435	442	458	488
	1 3/4	614	449	452	456	459	470	500	420	428	435	442	464	500
OVS	1 1/2	614	410	413	417	420	431	461	395	402	410	417	431	461
	1 3/4	614	422	425	429	433	444	473	395	402	410	417	439	473
SSLT	1 1/2	614	432	435	439	443	454	483	420	428	435	442	454	483
	1 3/4	614	444	448	451	455	466	495	420	428	435	442	464	495

Support Design Strength per Inch Thickness, kips/in.
1230

Notes:
STD = Standard holes
OVS = Oversized holes
SSLT = Short-slotted holes oriented transverse to direction of load
N = Threads included
X = Threads excluded
SC = Slip critical

*Tabulated values include $1/4$-in. reduction in end distance L_{eh} to account for possible underrun in beam length.

Beam	F_y = 50 ksi F_u = 65 ksi	F_y = 36 ksi F_u = 58 ksi	Angle

Table 10-1 (cont.). All-Bolted Double-Angle Connections $\phi = 0.75$

3/4-in. Bolts
6 Rows
W40, 36, 33, 30, 27, 24, 21

Varies
t
5@3 = 15
2 1/4
L_{eh}
L_{ev}
5@3 = 15
L_{ev}

Bolt and Angle Design Strength, kips						
ASTM Desig.	Thread Cond.	Hole Type	Angle Thickness			
			1/4	5/16	3/8	1/2
A325/ F1852	N	–	160	191	191	191
	X	–	160	200	239	239
	SC Class A	STD	125	125	125	125
		OVS	107	107	107	107
		SSLT	107	107	107	107
	SC Class B	STD	160	190	190	190
		OVS	150	161	161	161
		SSLT	160	161	161	161
A490	N	–	160	200	239	239
	X	–	160	200	240	298
	SC Class A	STD	157	157	157	157
		OVS	133	133	133	133
		SSLT	133	133	133	133
	SC Class B	STD	160	200	237	237
		OVS	150	188	202	202
		SSLT	160	200	202	202

Beam Web Design Strength per Inch Thickness, kips/in.

Hole Type	L_{eh},* in.	Un-coped	Coped at Top Flange Only						Coped at Both Flanges					
			L_{ev}, in.						L_{ev}, in.					
			1 1/4	1 3/8	1 1/2	1 5/8	2	3	1 1/4	1 3/8	1 1/2	1 5/8	2	3
STD	1 1/2	526	374	378	381	385	396	425	358	366	373	380	396	425
	1 3/4	526	386	390	394	397	408	438	358	366	373	380	402	438
OVS	1 1/2	526	351	355	358	362	373	402	336	344	351	358	373	402
	1 3/4	526	363	367	371	374	385	414	336	344	351	358	380	414
SSLT	1 1/2	526	370	373	377	381	392	421	358	366	373	380	392	421
	1 3/4	526	382	385	389	393	404	433	358	366	373	380	402	433

Support Design Strength per Inch Thickness, kips/in.
1050

Notes:
STD = Standard holes
OVS = Oversized holes
SSLT = Short-slotted holes oriented transverse to direction of load
N = Threads included
X = Threads excluded
SC = Slip critical

*Tabulated values include 1/4-in. reduction in end distance L_{eh} to account for possible underrun in beam length.

Angle	$F_y = 36$ ksi, $F_u = 58$ ksi		$F_y = 50$ ksi, $F_u = 65$ ksi	Beam

Table 10-1 (cont.). All-Bolted Double-Angle Connections $\phi = 0.75$

3/4-in. Bolts — 5 Rows — W30, 27, 24, 21, 18

Bolt and Angle Design Strength, kips

ASTM Desig.	Thread Cond.	Hole Type	Angle Thickness 1/4	5/16	3/8	1/2
A325/ F1852	N	–	132	159	159	159
	X	–	132	165	198	199
	SC Class A	STD	104	104	104	104
		OVS	88.8	88.8	88.8	88.8
		SSLT	88.8	88.8	88.8	88.8
	SC Class B	STD	132	158	158	158
		OVS	124	134	134	134
		SSLT	132	134	134	134
A490	N	–	132	165	198	199
	X	–	132	165	198	249
	SC Class A	STD	131	131	131	131
		OVS	111	111	111	111
		SSLT	111	111	111	111
	SC Class B	STD	132	165	198	198
		OVS	124	155	168	168
		SSLT	132	165	168	168

Beam Web Design Strength per Inch Thickness, kips/in.

Hole Type	L_{eh},* in.	Un-coped	Coped at Top Flange Only, L_{ev}, in. 1 1/4	1 3/8	1 1/2	1 5/8	2	3	Coped at Both Flanges, L_{ev}, in. 1 1/4	1 3/8	1 1/2	1 5/8	2	3
STD	1 1/2	439	312	316	319	323	334	363	296	303	311	318	334	363
	1 3/4	439	324	328	332	335	346	375	296	303	311	318	340	375
OVS	1 1/2	439	293	296	300	303	314	344	278	285	293	300	314	344
	1 3/4	439	305	308	312	316	327	356	278	285	293	300	322	356
SSLT	1 1/2	439	307	311	315	318	329	359	296	303	311	318	329	359
	1 3/4	439	320	323	327	331	342	371	296	303	311	318	340	371

Support Design Strength per Inch Thickness, kips/in.
877

Notes:
STD = Standard holes
OVS = Oversized holes
SSLT = Short-slotted holes oriented transverse to direction of load
N = Threads included
X = Threads excluded
SC = Slip critical

*Tabulated values include 1/4-in. reduction in end distance L_{eh} to account for possible underrun in beam length.

Beam	$F_y = 50$ ksi $F_u = 65$ ksi	$F_y = 36$ ksi $F_u = 58$ ksi	Angle

Table 10-1 (cont.). All-Bolted Double-Angle Connections

$\phi = 0.75$

3/4-in. Bolts
4 Rows
W24, 21, 18, 16

Bolt and Angle Design Strength, kips

ASTM Desig.	Thread Cond.	Hole Type	Angle Thickness 1/4	5/16	3/8	1/2
A325/ F1852	N	–	104	127	127	127
	X	–	104	131	157	159
	SC Class A	STD	83.5	83.5	83.5	83.5
		OVS	71.0	71.0	71.0	71.0
		SSLT	71.0	71.0	71.0	71.0
	SC Class B	STD	104	127	127	127
		OVS	97.9	108	108	108
		SSLT	104	108	108	108
A490	N	–	104	131	157	159
	X	–	104	131	157	199
	SC Class A	STD	104	104	104	104
		OVS	88.8	88.8	88.8	88.8
		SSLT	88.8	88.8	88.8	88.8
	SC Class B	STD	104	131	157	158
		OVS	97.9	122	134	134
		SSLT	104	131	134	134

Beam Web Design Strength per Inch Thickness, kips/in.

Hole Type	L_{eh},* in.	Un-coped	Coped at Top Flange Only, L_{ev}, in. 1 1/4	1 3/8	1 1/2	1 5/8	2	3	Coped at Both Flanges, L_{ev}, in. 1 1/4	1 3/8	1 1/2	1 5/8	2	3
STD	1 1/2	351	250	254	257	261	272	301	234	241	249	256	272	301
	1 3/4	351	262	266	269	273	284	313	234	241	249	256	278	313
OVS	1 1/2	351	234	238	241	245	256	285	219	227	234	241	256	285
	1 3/4	351	246	250	254	257	268	297	219	227	234	241	263	297
SSLT	1 1/2	351	245	249	253	256	267	296	234	241	249	256	267	296
	1 3/4	351	257	261	265	268	279	309	234	241	249	256	278	309

Support Design Strength per Inch Thickness, kips/in.
702

Notes:
STD = Standard holes
OVS = Oversized holes
SSLT = Short-slotted holes oriented transverse to direction of load
N = Threads included
X = Threads excluded
SC = Slip critical

*Tabulated values include 1/4-in. reduction in end distance L_{eh} to account for possible underrun in beam length.

Angle	$F_y = 36$ ksi, $F_u = 58$ ksi	Beam	$F_y = 50$ ksi, $F_u = 65$ ksi

Table 10-1 (cont.). All-Bolted Double-Angle Connections $\phi = 0.75$

3/4-in. Bolts
3 Rows
W18, 16, 14, 12, 10*

*Limited to W10×12, 15, 17, 19, 22, 26, 30.

Bolt and Angle Design Strength, kips						
ASTM Desig.	Thread Cond.	Hole Type	Angle Thickness 1/4	5/16	3/8	1/2
A325/F1852	N	–	76.7	95.4	95.4	95.4
	X	–	76.7	95.8	115	119
	SC Class A	STD	62.6	62.6	62.6	62.6
		OVS	53.3	53.3	53.3	53.3
		SSLT	53.3	53.3	53.3	53.3
	SC Class B	STD	76.7	94.9	94.9	94.9
		OVS	71.8	80.7	80.7	80.7
		SSLT	76.7	80.7	80.7	80.7
A490	N	–	76.7	95.8	115	119
	X	–	76.7	95.8	115	149
	SC Class A	STD	76.7	78.3	78.3	78.3
		OVS	66.6	66.6	66.6	66.6
		SSLT	66.6	66.6	66.6	66.6
	SC Class B	STD	76.7	95.8	115	119
		OVS	71.8	89.7	101	101
		SSLT	76.7	95.8	101	101

Beam Web Design Strength per Inch Thickness, kips/in.

Hole Type	L_{eh},* in.	Un-coped	Coped at Top Flange Only, L_{ev}, in. 1 1/4	1 3/8	1 1/2	1 5/8	2	3	Coped at Both Flanges, L_{ev}, in. 1 1/4	1 3/8	1 1/2	1 5/8	2	3
STD	1 1/2	263	188	191	195	199	210	239	172	179	186	194	210	239
	1 3/4	263	200	204	207	211	222	251	172	179	186	194	216	251
OVS	1 1/2	263	176	179	183	186	197	227	161	168	176	183	197	227
	1 3/4	263	188	191	195	199	210	239	161	168	176	183	205	239
SSLT	1 1/2	263	183	187	190	194	205	234	172	179	186	194	205	234
	1 3/4	263	195	199	203	206	217	246	172	179	186	194	216	246

Support Design Strength per Inch Thickness, kips/in.

526

Notes:
STD = Standard holes
OVS = Oversized holes
SSLT = Short-slotted holes oriented transverse to direction of load
N = Threads included
X = Threads excluded
SC = Slip critical

*Tabulated values include 1/4-in. reduction in end distance L_{eh} to account for possible underrun in beam length.

Beam	$F_y = 50$ ksi $F_u = 65$ ksi	$F_y = 36$ ksi $F_u = 58$ ksi	Angle

Table 10-1 (cont.). All-Bolted Double-Angle Connections $\phi = 0.75$

3/4-in. Bolts
2 Rows
W12, 10, 8

Bolt and Angle Design Strength, kips						
ASTM Desig.	**Thread Cond.**	**Hole Type**	**Angle Thickness**			
			$^1/_4$	$^5/_{16}$	$^3/_8$	$^1/_2$
A325/ F1852	N	–	48.9	61.2	63.6	63.6
	X	–	48.9	61.2	73.4	79.5
	SC Class A	STD	41.8	41.8	41.8	41.8
		OVS	35.5	35.5	35.5	35.5
		SSLT	35.5	35.5	35.5	35.5
	SC Class B	STD	48.9	61.2	63.3	63.3
		OVS	45.7	53.8	53.8	53.8
		SSLT	48.9	53.8	53.8	53.8
A490	N	–	48.9	61.2	73.4	79.5
	X	–	48.9	61.2	73.4	97.9
	SC Class A	STD	48.9	52.2	52.2	52.2
		OVS	44.4	44.4	44.4	44.4
		SSLT	44.4	44.4	44.4	44.4
	SC Class B	STD	48.9	61.2	73.4	79.1
		OVS	45.7	57.1	67.2	67.2
		SSLT	48.9	61.2	67.2	67.2

Beam Web Design Strength per Inch Thickness, kips/in.

Hole Type	L_{eh},* in.	Un-coped	Coped at Top Flange Only, L_{ev}, in.						Coped at Both Flanges, L_{ev}, in.					
			1 1/4	1 3/8	1 1/2	1 5/8	2	3	1 1/4	1 3/8	1 1/2	1 5/8	2	3
STD	1 1/2	175	126	129	133	137	147	175	110	117	124	132	147	175
	1 3/4	175	137	141	145	149	160	175	110	117	124	132	154	175
OVS	1 1/2	175	117	121	124	128	139	168	102	110	117	124	139	168
	1 3/4	175	129	133	137	140	151	175	102	110	117	124	146	175
SSLT	1 1/2	175	121	125	128	132	143	172	110	117	124	132	143	172
	1 3/4	175	133	137	140	144	155	175	110	117	124	132	154	175

Support Design Strength per Inch Thickness, kips/in.

351

Notes:
STD = Standard holes
OVS = Oversized holes
SSLT = Short-slotted holes oriented transverse to direction of load
N = Threads included
X = Threads excluded
SC = Slip critical

*Tabulated values include 1/4-in. reduction in end distance L_{eh} to account for possible underrun in beam length.

Angle	$F_y = 36$ ksi $F_u = 58$ ksi	$F_y = 50$ ksi $F_u = 65$ ksi	Beam

Table 10-1 (cont.). All-Bolted Double-Angle Connections

$\phi = 0.75$

7/8-in. Bolts — 12 Rows — W44

Varies; t; 11@3 = 33; 2 1/4; L_{eh}; L_{ev}; 11@3 = 33; L_{ev}

Bolt and Angle Design Strength, kips

ASTM Desig.	Thread Cond.	Hole Type	Angle Thickness 1/4	5/16	3/8	1/2
A325/F1852	N	–	307	383	460	520
	X	–	307	383	460	613
	SC Class A	STD	307	349	349	349
		OVS	287	297	297	297
		SSLT	297	297	297	297
	SC Class B	STD	307	383	460	520
		OVS	287	359	431	450
		SSLT	307	383	450	450
A490	N	–	307	383	460	613
	X	–	307	383	460	613
	SC Class A	STD	307	383	439	439
		OVS	287	359	373	373
		SSLT	307	373	373	373
	SC Class B	STD	307	383	460	613
		OVS	287	359	431	565
		SSLT	307	383	460	565

Beam Web Design Strength per Inch Thickness, kips/in.

Hole Type	L_{eh},* in.	Un-coped	Coped at Top Flange Only, L_{ev}, in. 1 1/4	1 3/8	1 1/2	1 5/8	2	3	Coped at Both Flanges, L_{ev}, in. 1 1/4	1 3/8	1 1/2	1 5/8	2	3
STD	1 1/2	1230	702	706	709	713	724	753	687	695	702	709	724	753
	1 3/4	1230	714	718	722	725	736	765	687	695	702	709	731	765
OVS	1 1/2	1230	657	661	664	668	679	708	644	651	658	665	679	708
	1 3/4	1230	669	673	676	680	691	720	644	651	658	665	687	720
SSLT	1 1/2	1230	697	701	705	708	719	749	687	695	702	708	719	749
	1 3/4	1230	710	713	717	721	732	761	687	695	702	709	731	761

Support Design Strength per Inch Thickness, kips/in.
2460

Notes:
STD = Standard holes
OVS = Oversized holes
SSLT = Short-slotted holes oriented transverse to direction of load
N = Threads included
X = Threads excluded
SC = Slip critical

*Tabulated values include 1/4-in. reduction in end distance L_{eh} to account for possible underrun in beam length.

Beam	$F_y = 50$ ksi, $F_u = 65$ ksi	$F_y = 36$ ksi, $F_u = 58$ ksi	Angle

Table 10-1 (cont.). All-Bolted Double-Angle Connections $\phi = 0.75$

7/8-in. Bolts
11 Rows
W44, 40

Varies, t, 10@3 = 30, 2 1/4, L_{eh}, L_{ev}, 10@3 = 30, L_{ev}

Bolt and Angle Design Strength, kips

ASTM Desig.	Thread Cond.	Hole Type	Angle Thickness 1/4	5/16	3/8	1/2
A325/ F1852	N	–	281	351	421	476
	X	–	281	351	421	561
	SC Class A	STD	281	320	320	320
		OVS	263	272	272	272
		SSLT	272	272	272	272
	SC Class B	STD	281	351	421	476
		OVS	263	328	394	412
		SSLT	281	351	412	412
A490	N	–	281	351	421	561
	X	–	281	351	421	561
	SC Class A	STD	281	351	402	402
		OVS	263	328	342	342
		SSLT	281	342	342	342
	SC Class B	STD	281	351	421	561
		OVS	263	328	394	518
		SSLT	281	351	421	518

Beam Web Design Strength per Inch Thickness, kips/in.

Hole Type	L_{eh},* in.	Un-coped	Coped at Top Flange Only, L_{ev}, in. 1 1/4	1 3/8	1 1/2	1 5/8	2	3	Coped at Both Flanges, L_{ev}, in. 1 1/4	1 3/8	1 1/2	1 5/8	2	3
STD	1 1/2	1130	644	647	651	654	665	695	629	636	644	651	665	695
	1 3/4	1130	656	659	663	667	678	707	629	636	644	651	673	707
OVS	1 1/2	1130	602	606	609	613	624	653	589	596	603	611	624	653
	1 3/4	1130	614	618	622	625	636	665	589	596	603	611	633	665
SSLT	1 1/2	1130	639	643	646	650	661	690	629	636	644	650	661	690
	1 3/4	1130	651	655	658	662	673	702	629	636	644	651	673	702

Support Design Strength per Inch Thickness, kips/in.

2250

Notes:
STD = Standard holes
OVS = Oversized holes
SSLT = Short-slotted holes oriented transverse to direction of load
N = Threads included
X = Threads excluded
SC = Slip critical

*Tabulated values include 1/4-in. reduction in end distance L_{eh} to account for possible underrun in beam length.

Angle	
$F_y = 36$ ksi	$F_u = 58$ ksi

Beam	
$F_y = 50$ ksi	$F_u = 65$ ksi

Table 10-1 (cont.). All-Bolted Double-Angle Connections $\phi = 0.75$

7/8-in. Bolts
10 Rows
W44, 40, 36

Bolt and Angle Design Strength, kips						
ASTM Desig.	Thread Cond.	Hole Type	Angle Thickness $^1/_4$	$^5/_{16}$	$^3/_8$	$^1/_2$
A325/ F1852	N	–	254	318	382	433
	X	–	254	318	382	509
	SC Class A	STD	254	291	291	291
		OVS	238	247	247	247
		SSLT	247	247	247	247
	SC Class B	STD	254	318	382	433
		OVS	238	298	357	375
		SSLT	254	318	375	375
A490	N	–	254	318	382	509
	X	–	254	318	382	509
	SC Class A	STD	254	318	365	365
		OVS	238	298	311	311
		SSLT	254	311	311	311
	SC Class B	STD	254	318	382	509
		OVS	238	298	357	471
		SSLT	254	318	382	471

Beam Web Design Strength per Inch Thickness, kips/in.

Hole Type	L_{eh},* in.	Un-coped	Coped at Top Flange Only, L_{ev}, in. 1 1/4	1 3/8	1 1/2	1 5/8	2	3	Coped at Both Flanges, L_{ev}, in. 1 1/4	1 3/8	1 1/2	1 5/8	2	3
STD	1 1/2	1020	585	589	592	596	607	636	570	578	585	592	607	636
	1 3/4	1020	597	601	605	608	619	648	570	578	585	592	614	648
OVS	1 1/2	1020	547	551	555	558	569	598	534	541	548	556	569	598
	1 3/4	1020	559	563	567	570	581	611	534	541	548	556	578	611
SSLT	1 1/2	1020	580	584	588	591	602	632	570	578	585	591	602	632
	1 3/4	1020	593	596	600	604	615	644	570	578	585	592	614	644

Support Design Strength per Inch Thickness, kips/in.
2050

Notes:
STD = Standard holes
OVS = Oversized holes
SSLT = Short-slotted holes oriented transverse to direction of load
N = Threads included
X = Threads excluded
SC = Slip critical

*Tabulated values include 1/4-in. reduction in end distance L_{eh} to account for possible underrun in beam length.

Beam	$F_y = 50$ ksi, $F_u = 65$ ksi		Angle	$F_y = 36$ ksi, $F_u = 58$ ksi

Table 10-1 (cont.). All-Bolted Double-Angle Connections

$\phi = 0.75$

7/8-in. Bolts
9 Rows
W44, 40, 36, 33

Varies
t
8@3 = 24
2 1/4
L_{eh}
L_{ev}
8@3 = 24
L_{ev}

Bolt and Angle Design Strength, kips

ASTM Desig.	Thread Cond.	Hole Type	Angle Thickness 1/4	5/16	3/8	1/2
A325/ F1852	N	–	228	285	343	390
	X	–	228	285	343	457
	SC Class A	STD	228	262	262	262
		OVS	214	223	223	223
		SSLT	223	223	223	223
	SC Class B	STD	228	285	343	390
		OVS	214	267	321	337
		SSLT	228	285	337	337
A490	N	–	228	285	343	457
	X	–	228	285	343	457
	SC Class A	STD	228	285	329	329
		OVS	214	267	280	280
		SSLT	228	280	280	280
	SC Class B	STD	228	285	343	457
		OVS	214	267	321	424
		SSLT	228	285	343	424

Beam Web Design Strength per Inch Thickness, kips/in.

Hole Type	L_{eh},* in.	Uncoped	Coped at Top Flange Only, L_{ev}, in. 1 1/4	1 3/8	1 1/2	1 5/8	2	3	Coped at Both Flanges, L_{ev}, in. 1 1/4	1 3/8	1 1/2	1 5/8	2	3
STD	1 1/2	921	527	530	534	537	548	578	512	519	527	534	548	578
	1 3/4	921	539	542	546	550	561	590	512	519	527	534	556	590
OVS	1 1/2	921	492	496	500	503	514	544	479	486	494	501	514	544
	1 3/4	921	505	508	512	516	527	556	479	486	494	501	523	556
SSLT	1 1/2	921	522	526	529	533	544	573	512	519	527	533	544	573
	1 3/4	921	534	538	541	545	556	585	512	519	527	534	556	585

Support Design Strength per Inch Thickness, kips/in.

1840

Notes:
STD = Standard holes
OVS = Oversized holes
SSLT = Short-slotted holes oriented transverse to direction of load
N = Threads included
X = Threads excluded
SC = Slip critical

*Tabulated values include 1/4-in. reduction in end distance L_{eh} to account for possible underrun in beam length.

Angle	$F_y = 36$ ksi, $F_u = 58$ ksi		$F_y = 50$ ksi, $F_u = 65$ ksi	Beam

Table 10-1 (cont.). All-Bolted Double-Angle Connections $\phi = 0.75$

7/8-in. Bolts
8 Rows
W44, 40, 36, 33, 30

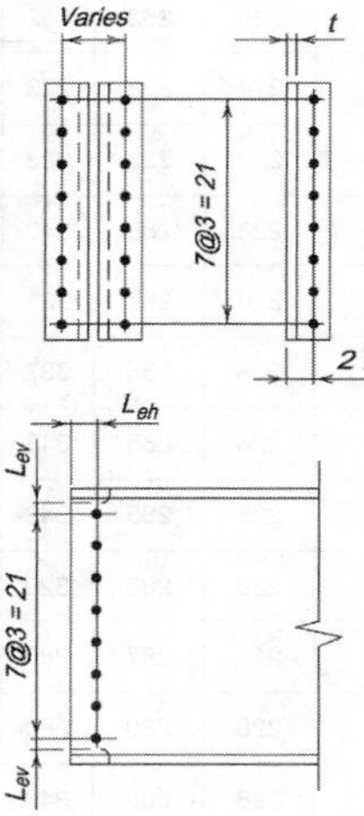

Varies
t
7@3 = 21
2 1/4
L_{eh}
L_{ev}
7@3 = 21
L_{ev}

Bolt and Angle Design Strength, kips

ASTM Desig.	Thread Cond.	Hole Type	Angle Thickness 1/4	5/16	3/8	1/2
A325/F1852	N	–	202	253	303	346
	X	–	202	253	303	405
	SC Class A	STD	202	233	233	233
		OVS	189	198	198	198
		SSLT	198	198	198	198
	SC Class B	STD	202	253	303	346
		OVS	189	237	284	300
		SSLT	202	253	300	300
A490	N	–	202	253	303	405
	X	–	202	253	303	405
	SC Class A	STD	202	253	292	292
		OVS	189	237	249	249
		SSLT	202	249	249	249
	SC Class B	STD	202	253	303	405
		OVS	189	237	284	377
		SSLT	202	253	303	377

Beam Web Design Strength per Inch Thickness, kips/in.

Hole Type	L_{eh},* in.	Un-coped	Coped at Top Flange Only, L_{ev}, in. 1 1/4	1 3/8	1 1/2	1 5/8	2	3	Coped at Both Flanges, L_{ev}, in. 1 1/4	1 3/8	1 1/2	1 5/8	2	3
STD	1 1/2	819	468	472	475	479	490	519	453	461	468	475	490	519
	1 3/4	819	480	484	488	491	502	531	453	461	468	475	497	531
OVS	1 1/2	819	438	441	445	449	459	489	424	431	439	446	459	489
	1 3/4	819	450	453	457	461	472	501	424	431	439	446	468	501
SSLT	1 1/2	819	463	467	471	474	485	515	453	461	468	474	485	515
	1 3/4	819	476	479	483	487	498	527	453	461	468	475	497	527

Support Design Strength per Inch Thickness, kips/in.
1640

Notes:
STD = Standard holes
OVS = Oversized holes
SSLT = Short-slotted holes oriented transverse to direction of load
N = Threads included
X = Threads excluded
SC = Slip critical

*Tabulated values include 1/4-in. reduction in end distance L_{eh} to account for possible underrun in beam length.

Beam	$F_y = 50$ ksi $F_u = 65$ ksi		$F_y = 36$ ksi $F_u = 58$ ksi	Angle

Table 10-1 (cont.). All-Bolted Double-Angle Connections

$\phi = 0.75$

7/8-in. Bolts, 7 Rows, W44, 40, 36, 33, 30, 27, 24

Bolt and Angle Design Strength, kips

ASTM Desig.	Thread Cond.	Hole Type	Angle Thickness 1/4	5/16	3/8	1/2
A325/ F1852	N	–	176	220	264	303
	X	–	176	220	264	352
	SC Class A	STD	176	204	204	204
		OVS	165	173	173	173
		SSLT	173	173	173	173
	SC Class B	STD	176	220	264	303
		OVS	165	206	247	262
		SSLT	176	220	262	262
A490	N	–	176	220	264	352
	X	–	176	220	264	352
	SC Class A	STD	176	220	256	256
		OVS	165	206	217	217
		SSLT	176	217	217	217
	SC Class B	STD	176	220	264	352
		OVS	165	206	247	329
		SSLT	176	220	264	329

Varies; t; 6@3 = 18; 2 1/4; L_{eh}; L_{ev}; 6@3 = 18; L_{ev}

Beam Web Design Strength per Inch Thickness, kips/in.

Hole Type	L_{eh},* in.	Un-coped	Coped at Top Flange Only, L_{ev}, in. 1 1/4	1 3/8	1 1/2	1 5/8	2	3	Coped at Both Flanges, L_{ev}, in. 1 1/4	1 3/8	1 1/2	1 5/8	2	3
STD	1 1/2	717	410	413	417	420	431	461	395	402	410	417	431	461
	1 3/4	717	422	425	429	433	444	473	395	402	410	417	439	473
OVS	1 1/2	717	383	386	390	394	405	434	369	377	384	391	405	434
	1 3/4	717	395	399	402	406	417	446	369	377	384	391	413	446
SSLT	1 1/2	717	405	409	412	416	427	456	395	402	410	416	427	456
	1 3/4	717	417	421	424	428	439	468	395	402	410	417	439	468

Support Design Strength per Inch Thickness, kips/in.	Notes:
1430	STD = Standard holes; OVS = Oversized holes; SSLT = Short-slotted holes oriented transverse to direction of load; N = Threads included; X = Threads excluded; SC = Slip critical. *Tabulated values include 1/4-in. reduction in end distance L_{eh} to account for possible underrun in beam length.

Angle	$F_y = 36$ ksi, $F_u = 58$ ksi

$F_y = 50$ ksi, $F_u = 65$ ksi	Beam

Table 10-1 (cont.). All-Bolted Double-Angle Connections

$\phi = 0.75$

7/8-in. Bolts
6 Rows
W40, 36, 33, 30, 27, 24, 21

Bolt and Angle Design Strength, kips

ASTM Desig.	Thread Cond.	Hole Type	Angle Thickness 1/4	5/16	3/8	1/2
A325/ F1852	N	–	150	188	225	260
	X	–	150	188	225	300
	SC Class A	STD	150	175	175	175
		OVS	140	148	148	148
		SSLT	148	148	148	148
	SC Class B	STD	150	188	225	260
		OVS	140	175	210	225
		SSLT	150	188	225	225
A490	N	–	150	188	225	300
	X	–	150	188	225	300
	SC Class A	STD	150	188	219	219
		OVS	140	175	186	186
		SSLT	150	186	186	186
	SC Class B	STD	150	188	225	300
		OVS	140	175	210	281
		SSLT	150	188	225	282

Beam Web Design Strength per Inch Thickness, kips/in.

Hole Type	L_{eh},* in.	Uncoped	Coped at Top Flange Only, L_{ev}, in. 1 1/4	1 3/8	1 1/2	1 5/8	2	3	Coped at Both Flanges, L_{ev}, in. 1 1/4	1 3/8	1 1/2	1 5/8	2	3
STD	1 1/2	614	351	355	358	362	373	402	336	344	351	358	373	402
	1 3/4	614	363	367	371	374	385	414	336	344	351	358	380	414
OVS	1 1/2	614	328	332	335	339	350	379	314	322	329	336	350	379
	1 3/4	614	340	344	347	351	362	391	314	322	329	336	358	391
SSLT	1 1/2	614	346	350	354	357	368	398	336	344	351	357	368	398
	1 3/4	614	359	362	366	370	381	410	336	344	351	358	380	410

Support Design Strength per Inch Thickness, kips/in.
1230

Notes:
STD = Standard holes
OVS = Oversized holes
SSLT = Short-slotted holes oriented transverse to direction of load
N = Threads included
X = Threads excluded
SC = Slip critical

*Tabulated values include 1/4-in. reduction in end distance L_{eh} to account for possible underrun in beam length.

Beam	$F_y = 50$ ksi $F_u = 65$ ksi	Angle	$F_y = 36$ ksi $F_u = 58$ ksi

Table 10-1 (cont.). All-Bolted Double-Angle Connections $\phi = 0.75$

$^7/_8$-in. Bolts
5 Rows
W30, 27, 24, 21, 18

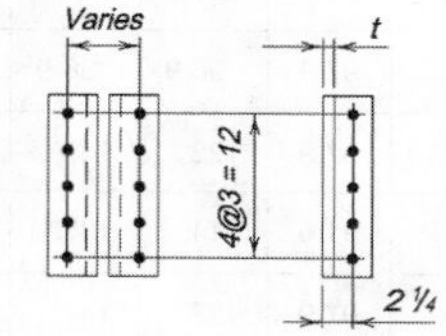

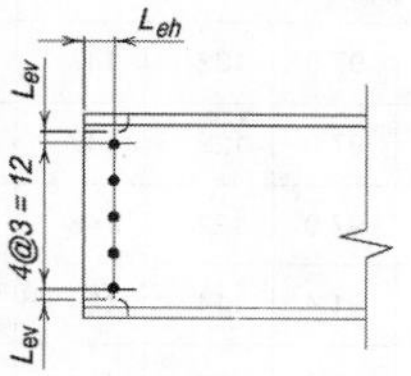

Bolt and Angle Design Strength, kips

ASTM Desig.	Thread Cond.	Hole Type	Angle Thickness $^1/_4$	$^5/_{16}$	$^3/_8$	$^1/_2$
A325/ F1852	N	–	124	155	186	216
	X	–	124	155	186	248
	SC Class A	STD	124	145	145	145
		OVS	116	124	124	124
		SSLT	124	124	124	124
	SC Class B	STD	124	155	186	216
		OVS	116	145	174	187
		SSLT	124	155	186	187
A490	N	–	124	155	186	248
	X	–	124	155	186	248
	SC Class A	STD	124	155	183	183
		OVS	116	145	155	155
		SSLT	124	155	155	155
	SC Class B	STD	124	155	186	248
		OVS	116	145	174	232
		SSLT	124	155	186	235

Beam Web Design Strength per Inch Thickness, kips/in.

Hole Type	L_{eh},* in.	Un-coped	Coped at Top Flange Only, L_{ev}, in. 1 1/4	1 3/8	1 1/2	1 5/8	2	3	Coped at Both Flanges, L_{ev}, in. 1 1/4	1 3/8	1 1/2	1 5/8	2	3
STD	1 1/2	512	293	296	300	303	314	344	278	285	293	300	314	344
	1 3/4	512	305	308	312	316	327	356	278	285	293	300	322	356
OVS	1 1/2	512	273	277	280	284	295	324	260	267	274	282	295	324
	1 3/4	512	285	289	293	296	307	336	260	267	274	282	303	336
SSLT	1 1/2	512	288	292	295	299	310	339	278	285	293	299	310	339
	1 3/4	512	300	304	307	311	322	351	278	285	293	300	322	351

Support Design Strength per Inch Thickness, kips/in.

1020

Notes:
STD = Standard holes
OVS = Oversized holes
SSLT = Short-slotted holes oriented transverse to direction of load
N = Threads included
X = Threads excluded
SC = Slip critical

*Tabulated values include $^1/_4$-in. reduction in end distance L_{eh} to account for possible underrun in beam length.

Angle	$F_y = 36$ ksi $F_u = 58$ ksi	$F_y = 50$ ksi $F_u = 65$ ksi	Beam

Table 10-1 (cont.). All-Bolted Double-Angle Connections

$\phi = 0.75$

7/8-in. Bolts, 4 Rows, W24, 21, 18, 16

Varies; t; 3@3 = 9; 2 1/4; L_{eh}; L_{ev}; 3@3 = 9; L_{ev}

Bolt and Angle Design Strength, kips

ASTM Desig.	Thread Cond.	Hole Type	Angle Thickness 1/4	5/16	3/8	1/2
A325/ F1852	N	–	97.9	122	147	173
	X	–	97.9	122	147	196
	SC Class A	STD	97.9	116	116	116
		OVS	91.4	98.9	98.9	98.9
		SSLT	97.9	98.9	98.9	98.9
	SC Class B	STD	97.9	122	147	173
		OVS	91.4	114	137	150
		SSLT	97.9	122	147	150
A490	N	–	97.9	122	147	196
	X	–	97.9	122	147	196
	SC Class A	STD	97.9	122	146	146
		OVS	91.4	114	124	124
		SSLT	97.9	122	124	124
	SC Class B	STD	97.9	122	147	196
		OVS	91.4	114	137	183
		SSLT	97.9	122	147	188

Beam Web Design Strength per Inch Thickness, kips/in.

Hole Type	L_{eh},* in.	Un-coped	Coped at Top Flange Only, L_{ev}, in. 1 1/4	1 3/8	1 1/2	1 5/8	2	3	Coped at Both Flanges, L_{ev}, in. 1 1/4	1 3/8	1 1/2	1 5/8	2	3
STD	1 1/2	409	234	238	241	245	256	285	219	227	234	241	256	285
	1 3/4	409	246	250	254	257	268	297	219	227	234	241	263	297
OVS	1 1/2	409	218	222	225	229	240	269	205	212	219	227	240	269
	1 3/4	409	230	234	238	241	252	282	205	212	219	227	249	282
SSLT	1 1/2	409	229	233	237	240	251	281	219	227	234	240	251	281
	1 3/4	409	242	245	249	253	264	293	219	227	234	241	263	293

Support Design Strength per Inch Thickness, kips/in.

819

Notes:
STD = Standard holes
OVS = Oversized holes
SSLT = Short-slotted holes oriented transverse to direction of load
N = Threads included
X = Threads excluded
SC = Slip critical

*Tabulated values include 1/4-in. reduction in end distance L_{eh} to account for possible underrun in beam length.

Beam	$F_y = 50$ ksi $F_u = 65$ ksi

$F_y = 36$ ksi $F_u = 58$ ksi	Angle

Table 10-1 (cont.). All-Bolted Double-Angle Connections $\phi = 0.75$

7/8-in. Bolts
3 Rows
W18, 16, 14, 12, 10*

*Limited to W10×12, 15, 17, 19, 22, 26, 30.

Bolt and Angle Design Strength, kips						
ASTM Desig.	Thread Cond.	Hole Type	Angle Thickness 1/4	5/16	3/8	1/2
A325/F1852	N	–	71.8	89.7	108	130
	X	–	71.8	89.7	108	144
	SC Class A	STD	71.8	87.3	87.3	87.3
		OVS	66.9	74.2	74.2	74.2
		SSLT	71.8	74.2	74.2	74.2
	SC Class B	STD	71.8	89.7	108	130
		OVS	66.9	83.6	100	112
		SSLT	71.8	89.7	108	112
A490	N	–	71.8	89.7	108	144
	X	–	71.8	89.7	108	144
	SC Class A	STD	71.8	89.7	108	110
		OVS	66.9	83.6	93.2	93.2
		SSLT	71.8	89.7	93.2	93.2
	SC Class B	STD	71.8	89.7	108	144
		OVS	66.9	83.6	100	134
		SSLT	71.8	89.7	108	141

Beam Web Design Strength per Inch Thickness, kips/in.

Hole Type	L_{eh},* in.	Uncoped	Coped at Top Flange Only, L_{ev}, in. 1 1/4	1 3/8	1 1/2	1 5/8	2	3	Coped at Both Flanges, L_{ev}, in. 1 1/4	1 3/8	1 1/2	1 5/8	2	3
STD	1 1/2	307	176	179	183	186	197	227	161	168	176	183	197	227
	1 3/4	307	188	191	195	199	210	239	161	168	176	183	205	239
OVS	1 1/2	307	163	167	171	174	185	215	150	157	165	172	185	215
	1 3/4	307	176	179	183	186	197	227	150	157	165	172	194	227
SSLT	1 1/2	307	171	175	178	182	193	222	161	168	176	182	193	222
	1 3/4	307	183	187	190	194	205	234	161	168	176	183	205	234

Support Design Strength per Inch Thickness, kips/in.

614

Notes:
STD = Standard holes
OVS = Oversized holes
SSLT = Short-slotted holes oriented transverse to direction of load
N = Threads included
X = Threads excluded
SC = Slip critical

*Tabulated values include 1/4-in. reduction in end distance L_{eh} to account for possible underrun in beam length.

Angle	$F_y = 36$ ksi, $F_u = 58$ ksi

Beam	$F_y = 50$ ksi, $F_u = 65$ ksi

Table 10-1 (cont.). All-Bolted Double-Angle Connections

$\phi = 0.75$

7/8-in. Bolts, 2 Rows, W12, 10, 8

Bolt and Angle Design Strength, kips

ASTM Desig.	Thread Cond.	Hole Type	Angle Thickness 1/4	5/16	3/8	1/2
A325/F1852	N	–	45.7	57.1	68.5	86.6
	X	–	45.7	57.1	68.5	91.4
	SC Class A	STD	45.7	57.1	58.2	58.2
		OVS	42.4	49.4	49.4	49.4
		SSLT	45.7	49.4	49.4	49.4
	SC Class B	STD	45.7	57.1	68.5	86.6
		OVS	42.4	53.0	63.6	74.9
		SSLT	45.7	57.1	68.5	74.9
A490	N	–	45.7	57.1	68.5	91.4
	X	–	45.7	57.1	68.5	91.4
	SC Class A	STD	45.7	57.1	68.5	73.1
		OVS	42.4	53.0	62.1	62.1
		SSLT	45.7	57.1	62.1	62.1
	SC Class B	STD	45.7	57.1	68.5	91.4
		OVS	42.4	53.0	63.6	84.8
		SSLT	45.7	57.1	68.5	91.4

Beam Web Design Strength per Inch Thickness, kips/in.

Hole Type	L_{eh},* in.	Un-coped	Coped at Top Flange Only, L_{ev}, in. 1 1/4	1 3/8	1 1/2	1 5/8	2	3	Coped at Both Flanges, L_{ev}, in. 1 1/4	1 3/8	1 1/2	1 5/8	2	3
STD	1 1/2	205	117	121	124	128	139	168	102	110	117	124	139	168
	1 3/4	205	129	133	137	140	151	180	102	110	117	124	146	180
OVS	1 1/2	205	108	112	116	119	130	160	95.1	102	110	117	130	160
	1 3/4	205	121	124	128	132	143	172	95.1	102	110	117	139	172
SSLT	1 1/2	205	112	116	120	123	134	164	102	110	117	123	134	164
	1 3/4	205	125	128	132	136	147	176	102	110	117	124	146	176

Support Design Strength per Inch Thickness, kips/in.

409

Notes:
STD = Standard holes
OVS = Oversized holes
SSLT = Short-slotted holes oriented transverse to direction of load
N = Threads included
X = Threads excluded
SC = Slip critical

*Tabulated values include 1/4-in. reduction in end distance L_{eh} to account for possible underrun in beam length.

Beam	$F_y = 50$ ksi $F_u = 65$ ksi		Angle	$F_y = 36$ ksi $F_u = 58$ ksi

Table 10-1 (cont.). All-Bolted Double-Angle Connections $\phi = 0.75$

1-in. Bolts, 12 Rows, W44

Varies

t

11@3 = 33

2 1/4

L_{eh}

L_{ev}

11@3 = 33

L_{ev}

Bolt and Angle Design Strength, kips

ASTM Desig.	Thread Cond.	Hole Type	Angle Thickness 1/4	5/16	3/8	1/2
A325/ F1852	N	–	287	359	431	574
	X	–	287	359	431	574
	SC Class A	STD	287	359	431	456
		OVS	258	322	387	388
		SSLT	287	359	388	388
	SC Class B	STD	287	359	431	574
		OVS	258	322	387	515
		SSLT	287	359	431	574
A490	N	–	287	359	431	574
	X	–	287	359	431	574
	SC Class A	STD	287	359	431	573
		OVS	258	322	387	487
		SSLT	287	359	431	487
	SC Class B	STD	287	359	431	574
		OVS	258	322	387	515
		SSLT	287	359	431	574

Beam Web Design Strength per Inch Thickness, kips/in.

Hole Type	L_{eh},* in.	Un-coped	Coped at Top Flange Only, L_{ev}, in. 1 1/4	1 3/8	1 1/2	1 5/8	2	3	Coped at Both Flanges, L_{ev}, in. 1 1/4	1 3/8	1 1/2	1 5/8	2	3
STD	1 1/2	1360	657	661	664	668	679	708	644	651	658	665	679	708
	1 3/4	1360	669	673	676	680	691	720	644	651	658	665	687	720
OVS	1 1/2	1240	589	593	597	600	611	640	578	585	592	600	611	640
	1 3/4	1240	601	605	609	612	623	653	578	585	592	600	622	653
SSLT	1 1/2	1360	651	654	658	662	673	702	644	651	658	662	673	702
	1 3/4	1360	663	667	670	674	685	714	644	651	658	665	685	714

Support Design Strength per Inch Thickness, kips/in.	
STD/SSLT	2730
OVS	2490

Notes:
STD = Standard holes
OVS = Oversized holes
SSLT = Short-slotted holes oriented transverse to direction of load
N = Threads included
X = Threads excluded
SC = Slip critical

*Tabulated values include 1/4-in. reduction in end distance L_{eh} to account for possible underrun in beam length.

Angle	$F_y = 36$ ksi $F_u = 58$ ksi

Beam	$F_y = 50$ ksi $F_u = 65$ ksi

Table 10-1 (cont.). All-Bolted Double-Angle Connections

$\phi = 0.75$

1-in. Bolts — 11 Rows — W44, 40

Bolt and Angle Design Strength, kips

ASTM Desig.	Thread Cond.	Hole Type	Angle Thickness $^1/_4$	$^5/_{16}$	$^3/_8$	$^1/_2$
A325/ F1852	N	–	263	328	394	525
	X	–	263	328	394	525
	SC Class A	STD	263	328	394	418
		OVS	236	295	354	356
		SSLT	263	328	356	356
	SC Class B	STD	263	328	394	525
		OVS	236	295	354	471
		SSLT	263	328	394	525
A490	N	–	263	328	394	525
	X	–	263	328	394	525
	SC Class A	STD	263	328	394	525
		OVS	236	295	354	446
		SSLT	263	328	394	446
	SC Class B	STD	263	328	394	525
		OVS	236	295	354	471
		SSLT	263	328	394	525

Beam Web Design Strength per Inch Thickness, kips/in.

Hole Type	L_{eh},* in.	Un-coped	Coped at Top Flange Only, L_{ev}, in. $1\,^1/_4$	$1\,^3/_8$	$1\,^1/_2$	$1\,^5/_8$	2	3	Coped at Both Flanges, L_{ev}, in. $1\,^1/_4$	$1\,^3/_8$	$1\,^1/_2$	$1\,^5/_8$	2	3
STD	$1\,^1/_2$	1250	602	606	609	613	624	653	589	596	603	611	624	653
	$1\,^3/_4$	1250	614	618	622	625	636	665	589	596	603	611	633	665
OVS	$1\,^1/_2$	1140	540	544	547	551	562	591	528	536	543	550	562	591
	$1\,^3/_4$	1140	552	556	559	563	574	603	528	536	543	550	572	603
SSLT	$1\,^1/_2$	1250	596	600	603	607	618	647	589	596	603	607	618	647
	$1\,^3/_4$	1250	608	612	615	619	630	659	589	596	603	611	630	659

Support Design Strength per Inch Thickness, kips/in.

STD/SSLT	2500
OVS	2280

Notes:
STD = Standard holes
OVS = Oversized holes
SSLT = Short-slotted holes oriented transverse to direction of load
N = Threads included
X = Threads excluded
SC = Slip critical

*Tabulated values include $^1/_4$-in. reduction in end distance L_{eh} to account for possible underrun in beam length.

Beam	$F_y = 50$ ksi $F_u = 65$ ksi

$F_y = 36$ ksi $F_u = 58$ ksi	Angle

Table 10-1 (cont.). All-Bolted Double-Angle Connections

$\phi = 0.75$

1-in. Bolts
10 Rows
W44, 40, 36

Varies, t, 9@3 = 27, 2 1/4, L_{eh}, L_{ev}, 9@3 = 27, L_{ev}

Bolt and Angle Design Strength, kips

ASTM Desig.	Thread Cond.	Hole Type	Angle Thickness 1/4	5/16	3/8	1/2
A325/ F1852	N	–	238	298	357	476
	X	–	238	298	357	476
	SC Class A	STD	238	298	357	380
		OVS	214	267	321	323
		SSLT	238	298	323	323
	SC Class B	STD	238	298	357	476
		OVS	214	267	321	427
		SSLT	238	298	357	476
A490	N	–	238	298	357	476
	X	–	238	298	357	476
	SC Class A	STD	238	298	357	476
		OVS	214	267	321	406
		SSLT	238	298	357	406
	SC Class B	STD	238	298	357	476
		OVS	214	267	321	427
		SSLT	238	298	357	476

Beam Web Design Strength per Inch Thickness, kips/in.

Hole Type	L_{eh},* in.	Un-coped	Coped at Top Flange Only, L_{ev}, in. 1 1/4	1 3/8	1 1/2	1 5/8	2	3	Coped at Both Flanges, L_{ev}, in. 1 1/4	1 3/8	1 1/2	1 5/8	2	3
STD	1 1/2	1140	547	551	555	558	569	598	534	541	548	556	569	598
	1 3/4	1140	559	563	567	570	581	611	534	541	548	556	578	611
OVS	1 1/2	1040	491	494	498	502	512	542	479	486	494	501	512	542
	1 3/4	1040	503	506	510	514	525	554	479	486	494	501	523	554
SSLT	1 1/2	1140	541	545	548	552	563	592	534	541	548	552	563	592
	1 3/4	1140	553	557	561	564	575	605	534	541	548	556	575	605

Support Design Strength per Inch Thickness, kips/in.

Hole Type	
STD/SSLT	2270
OVS	2080

Notes:
STD = Standard holes
OVS = Oversized holes
SSLT = Short-slotted holes oriented transverse to direction of load
N = Threads included
X = Threads excluded
SC = Slip critical

* Tabulated values include 1/4-in. reduction in end distance L_{eh} to account for possible underrun in beam length.

Angle	$F_y = 36$ ksi $F_u = 58$ ksi	$F_y = 50$ ksi $F_u = 65$ ksi	Beam

Table 10-1 (cont.). All-Bolted Double-Angle Connections $\phi = 0.75$

1-in. Bolts, 9 Rows, W44, 40, 36, 33

Bolt and Angle Design Strength, kips

ASTM Desig.	Thread Cond.	Hole Type	Angle Thickness 1/4	5/16	3/8	1/2
A325/ F1852	N	–	214	267	321	427
	X	–	214	267	321	427
	SC Class A	STD	214	267	321	342
		OVS	192	240	288	291
		SSLT	214	267	291	291
	SC Class B	STD	214	267	321	427
		OVS	192	240	288	383
		SSLT	214	267	321	427
A490	N	–	214	267	321	427
	X	–	214	267	321	427
	SC Class A	STD	214	267	321	427
		OVS	192	240	288	365
		SSLT	214	267	321	365
	SC Class B	STD	214	267	321	427
		OVS	192	240	288	383
		SSLT	214	267	321	427

Beam Web Design Strength per Inch Thickness, kips/in.

Hole Type	L_{eh},* in.	Un-coped	Coped at Top Flange Only, L_{ev}, in. 1 1/4	1 3/8	1 1/2	1 5/8	2	3	Coped at Both Flanges, L_{ev}, in. 1 1/4	1 3/8	1 1/2	1 5/8	2	3
STD	1 1/2	1020	492	496	500	503	514	544	479	486	494	501	514	544
	1 3/4	1020	505	508	512	516	527	556	479	486	494	501	523	556
OVS	1 1/2	936	441	445	449	452	463	492	430	437	444	452	463	492
	1 3/4	936	453	457	461	464	475	505	430	437	444	452	473	505
SSLT	1 1/2	1020	486	490	494	497	508	537	479	486	494	497	508	537
	1 3/4	1020	498	502	506	509	520	550	479	486	494	501	520	550

Support Design Strength per Inch Thickness, kips/in.

STD/SSLT	2050
OVS	1870

Notes:
STD = Standard holes
OVS = Oversized holes
SSLT = Short-slotted holes oriented transverse to direction of load
N = Threads included
X = Threads excluded
SC = Slip critical

*Tabulated values include 1/4-in. reduction in end distance L_{eh} to account for possible underrun in beam length.

Beam	
$F_y = 50$ ksi	
$F_u = 65$ ksi	

	Angle
$F_y = 36$ ksi	
$F_u = 58$ ksi	

Table 10-1 (cont.). All-Bolted Double-Angle Connections $\phi = 0.75$

1-in. Bolts
8 Rows
W44, 40, 36, 33, 30

Bolt and Angle Design Strength, kips						
ASTM Desig.	Thread Cond.	Hole Type	Angle Thickness $^1/_4$	$^5/_{16}$	$^3/_8$	$^1/_2$
A325/ F1852	N	–	189	237	284	378
	X	–	189	237	284	378
	SC Class A	STD	189	237	284	304
		OVS	170	212	254	259
		SSLT	189	237	259	259
	SC Class B	STD	189	237	284	378
		OVS	170	212	254	339
		SSLT	189	237	284	378
A490	N	–	189	237	284	378
	X	–	189	237	284	378
	SC Class A	STD	189	237	284	378
		OVS	170	212	254	325
		SSLT	189	237	284	325
	SC Class B	STD	189	237	284	378
		OVS	170	212	254	339
		SSLT	189	237	284	378

Beam Web Design Strength per Inch Thickness, kips/in.

Hole Type	L_{eh},* in.	Un-coped	Coped at Top Flange Only, L_{ev}, in. 1 1/4	1 3/8	1 1/2	1 5/8	2	3	Coped at Both Flanges, L_{ev}, in. 1 1/4	1 3/8	1 1/2	1 5/8	2	3
STD	1 1/2	910	438	441	445	449	459	489	424	431	439	446	459	489
	1 3/4	910	450	453	457	461	472	501	424	431	439	446	468	501
OVS	1 1/2	834	392	395	399	403	414	443	380	388	395	402	414	443
	1 3/4	834	404	408	411	415	426	455	380	388	395	402	424	455
SSLT	1 1/2	910	431	435	439	442	453	483	424	431	439	442	453	483
	1 3/4	910	444	447	451	455	466	495	424	431	439	446	466	495

Support Design Strength per Inch Thickness, kips/in.	
STD/SSLT	1820
OVS	1670

Notes:
STD = Standard holes
OVS = Oversized holes
SSLT = Short-slotted holes oriented transverse to direction of load
N = Threads included
X = Threads excluded
SC = Slip critical

*Tabulated values include $^1/_4$-in. reduction in end distance L_{eh} to account for possible underrun in beam length.

Angle	
$F_y = 36$ ksi	$F_u = 58$ ksi

Beam	
$F_y = 50$ ksi	$F_u = 65$ ksi

Table 10-1 (cont.). All-Bolted Double-Angle Connections

$\phi = 0.75$

1-in. Bolts
7 Rows
W44, 40, 36, 33, 27, 24

Varies
t
6@3 = 18
2 1/4
L_{eh}
L_{ev}
6@3 = 18
L_{ev}

Bolt and Angle Design Strength, kips

ASTM Desig.	Thread Cond.	Hole Type	Angle Thickness 1/4	5/16	3/8	1/2
A325/ F1852	N	–	165	206	247	330
	X	–	165	206	247	330
	SC Class A	STD	165	206	247	266
		OVS	148	185	221	226
		SSLT	165	206	226	226
	SC Class B	STD	165	206	247	330
		OVS	148	185	221	295
		SSLT	165	206	247	330
A490	N	–	165	206	247	330
	X	–	165	206	247	330
	SC Class A	STD	165	206	247	330
		OVS	148	185	221	284
		SSLT	165	206	247	284
	SC Class B	STD	165	206	247	330
		OVS	148	185	221	295
		SSLT	165	206	247	330

Beam Web Design Strength per Inch Thickness, kips/in.

Hole Type	L_{eh},* in.	Un-coped	Coped at Top Flange Only, L_{ev}, in. 1 1/4	1 3/8	1 1/2	1 5/8	2	3	Coped at Both Flanges, L_{ev}, in. 1 1/4	1 3/8	1 1/2	1 5/8	2	3
STD	1 1/2	797	383	386	390	394	405	434	369	377	384	391	405	434
	1 3/4	797	395	399	402	406	417	446	369	377	384	391	413	446
OVS	1 1/2	731	342	346	350	353	364	394	331	338	346	353	364	394
	1 3/4	731	355	358	362	366	377	406	331	338	346	353	375	406
SSLT	1 1/2	797	377	380	384	388	399	428	369	377	384	388	399	428
	1 3/4	797	389	392	396	400	411	440	369	377	384	391	411	440

Support Design Strength per Inch Thickness, kips/in.	
STD/SSLT	1590
OVS	1460

Notes:
STD = Standard holes
OVS = Oversized holes
SSLT = Short-slotted holes oriented transverse to direction of load
N = Threads included
X = Threads excluded
SC = Slip critical

*Tabulated values include 1/4-in. reduction in end distance L_{eh} to account for possible underrun in beam length.

Beam	
$F_y = 50$ ksi	$F_u = 65$ ksi

Angle	
$F_y = 36$ ksi	$F_u = 58$ ksi

Table 10-1 (cont.). All-Bolted Double-Angle Connections $\phi = 0.75$

1-in. Bolts
6 Rows
W40, 36, 33, 30, 27, 24, 21

Bolt and Angle Design Strength, kips						
ASTM Desig.	Thread Cond.	Hole Type	Angle Thickness, in.			
			1/4	5/16	3/8	1/2
A325/F1852	N	–	140	175	210	281
	X	–	140	175	210	281
	SC Class A	STD	140	175	210	228
		OVS	126	157	188	194
		SSLT	140	175	194	194
	SC Class B	STD	140	175	210	281
		OVS	126	157	188	251
		SSLT	140	175	210	281
A490	N	–	140	175	210	281
	X	–	140	175	210	281
	SC Class A	STD	140	175	210	281
		OVS	126	157	188	243
		SSLT	140	175	210	243
	SC Class B	STD	140	175	210	281
		OVS	126	157	188	251
		SSLT	140	175	210	281

Beam Web Design Strength per Inch Thickness, kips/in.

Hole Type	L_{eh},* in.	Uncoped	Coped at Top Flange Only, L_{ev}, in.						Coped at Both Flanges, L_{ev}, in.					
			1 1/4	1 3/8	1 1/2	1 5/8	2	3	1 1/4	1 3/8	1 1/2	1 5/8	2	3
STD	1 1/2	684	328	332	335	339	350	379	314	322	329	336	350	379
	1 3/4	684	340	344	347	351	362	391	314	322	329	336	358	391
OVS	1 1/2	629	293	297	300	304	315	344	282	289	296	303	315	344
	1 3/4	629	305	309	313	316	327	356	282	289	296	303	325	356
SSLT	1 1/2	684	322	325	329	333	344	373	314	322	329	333	344	373
	1 3/4	684	334	338	341	345	356	385	314	322	329	336	356	385

Support Design Strength per Inch Thickness, kips/in.	
STD/SSLT	1370
OVS	1260

Notes:
STD = Standard holes
OVS = Oversized holes
SSLT = Short-slotted holes oriented transverse to direction of load
N = Threads included
X = Threads excluded
SC = Slip critical

*Tabulated values include 1/4-in. reduction in end distance L_{eh} to account for possible underrun in beam length.

Angle	$F_y = 36$ ksi, $F_u = 58$ ksi

$F_y = 50$ ksi, $F_u = 65$ ksi	Beam

Table 10-1 (cont.). All-Bolted Double-Angle Connections

$\phi = 0.75$

1-in. Bolts — 5 Rows — W30, 27, 24, 21, 18

Bolt and Angle Design Strength, kips

ASTM Desig.	Thread Cond.	Hole Type	Angle Thickness $^1/_4$	$^5/_{16}$	$^3/_8$	$^1/_2$
A325/F1852	N	–	116	145	174	232
	X	–	116	145	174	232
	SC Class A	STD	116	145	174	190
		OVS	104	129	155	162
		SSLT	116	145	162	162
	SC Class B	STD	116	145	174	232
		OVS	104	129	155	207
		SSLT	116	145	174	232
A490	N	–	116	145	174	232
	X	–	116	145	174	232
	SC Class A	STD	116	145	174	232
		OVS	104	129	155	203
		SSLT	116	145	174	203
	SC Class B	STD	116	145	174	232
		OVS	104	129	155	207
		SSLT	116	145	174	232

Beam Web Design Strength per Inch Thickness, kips/in.

Hole Type	L_{eh},* in.	Uncoped	Coped at Top Flange Only, L_{ev}, in.: 1 $^1/_4$	1 $^3/_8$	1 $^1/_2$	1 $^5/_8$	2	3	Coped at Both Flanges, L_{ev}, in.: 1 $^1/_4$	1 $^3/_8$	1 $^1/_2$	1 $^5/_8$	2	3
STD	1 $^1/_2$	570	273	277	280	284	295	324	260	267	274	282	295	324
	1 $^3/_4$	570	285	289	293	296	307	336	260	267	274	282	303	336
OVS	1 $^1/_2$	526	244	247	251	255	266	295	232	239	247	254	266	295
	1 $^3/_4$	526	256	260	263	267	278	307	232	239	247	254	276	307
SSLT	1 $^1/_2$	570	267	271	274	278	289	318	260	267	274	278	289	318
	1 $^3/_4$	570	279	283	286	290	301	330	260	267	274	282	301	330

Support Design Strength per Inch Thickness, kips/in.

Hole Type	
STD/SSLT	1140
OVS	1050

Notes:
STD = Standard holes
OVS = Oversized holes
SSLT = Short-slotted holes oriented transverse to direction of load
N = Threads included
X = Threads excluded
SC = Slip critical

*Tabulated values include $^1/_4$-in. reduction in end distance L_{eh} to account for possible underrun in beam length.

Beam	$F_y = 50$ ksi $F_u = 65$ ksi

$F_y = 36$ ksi $F_u = 58$ ksi	Angle

Table 10-1 (cont.). All-Bolted Double-Angle Connections $\phi = 0.75$

1-in. Bolts
4 Rows
W24, 21, 18, 16

Bolt and Angle Design Strength, kips

ASTM Desig.	Thread Cond.	Hole Type	Angle Thickness 1/4	5/16	3/8	1/2
A325/ F1852	N	–	91.4	114	137	183
	X	–	91.4	114	137	183
	SC Class A	STD	91.4	114	137	152
		OVS	81.6	102	122	129
		SSLT	91.4	114	129	129
	SC Class B	STD	91.4	114	137	183
		OVS	81.6	102	122	163
		SSLT	91.4	114	137	183
A490	N	–	91.4	114	137	183
	X	–	91.4	114	137	183
	SC Class A	STD	91.4	114	137	183
		OVS	81.6	102	122	162
		SSLT	91.4	114	137	162
	SC Class B	STD	91.4	114	137	183
		OVS	81.6	102	122	163
		SSLT	91.4	114	137	183

Beam Web Design Strength per Inch Thickness, kips/in.

Hole Type	L_{eh},* in.	Un-coped	Coped at Top Flange Only, L_{ev}, in. 1 1/4	1 3/8	1 1/2	1 5/8	2	3	Coped at Both Flanges, L_{ev}, in. 1 1/4	1 3/8	1 1/2	1 5/8	2	3
STD	1 1/2	457	218	222	225	229	240	269	205	212	219	227	240	269
	1 3/4	457	230	234	238	241	252	282	205	212	219	227	249	282
OVS	1 1/2	424	194	198	202	205	216	246	183	190	197	205	216	246
	1 3/4	424	207	210	214	218	229	258	183	190	197	205	227	258
SSLT	1 1/2	457	212	216	219	223	234	263	205	212	219	223	234	263
	1 3/4	457	224	228	232	235	246	275	205	212	219	227	246	275

Support Design Strength per Inch Thickness, kips/in.

STD/SSLT	914
OVS	848

Notes:
STD = Standard holes
OVS = Oversized holes
SSLT = Short-slotted holes oriented transverse to direction of load
N = Threads included
X = Threads excluded
SC = Slip critical

*Tabulated values include 1/4-in. reduction in end distance L_{eh} to account for possible underrun in beam length.

Angle	$F_y = 36$ ksi, $F_u = 58$ ksi

Beam	$F_y = 50$ ksi, $F_u = 65$ ksi

Table 10-1 (cont.). All-Bolted Double-Angle Connections

$\phi = 0.75$

1-in. Bolts
3 Rows
W18, 16, 14, 12, 10*

*Limited to W10×12, 15, 17, 19, 22, 26, 30.

Bolt and Angle Design Strength, kips						
ASTM Desig.	Thread Cond.	Hole Type	Angle Thickness $^1/_4$	$^5/_{16}$	$^3/_8$	$^1/_2$
A325/ F1852	N	–	66.9	83.6	100	134
	X	–	66.9	83.6	100	134
	SC Class A	STD	66.9	83.6	100	114
		OVS	59.5	74.4	89.3	97.0
		SSLT	66.9	83.6	97.0	97.0
	SC Class B	STD	66.9	83.6	100	134
		OVS	59.5	74.4	89.3	119
		SSLT	66.9	83.6	100	134
A490	N	–	66.9	83.6	100	134
	X	–	66.9	83.6	100	134
	SC Class A	STD	66.9	83.6	100	134
		OVS	59.5	74.4	89.3	119
		SSLT	66.9	83.6	100	122
	SC Class B	STD	66.9	83.6	100	134
		OVS	59.5	74.4	89.3	119
		SSLT	66.9	83.6	100	134

Beam Web Design Strength per Inch Thickness, kips/in.

Hole Type	L_{eh},* in.	Un-coped	Coped at Top Flange Only, L_{ev}, in. 1 $^1/_4$	1 $^3/_8$	1 $^1/_2$	1 $^5/_8$	2	3	Coped at Both Flanges, L_{ev}, in. 1 $^1/_4$	1 $^3/_8$	1 $^1/_2$	1 $^5/_8$	2	3
STD	1 $^1/_2$	344	163	167	171	174	185	215	150	157	165	172	185	215
	1 $^3/_4$	344	176	179	183	186	197	227	150	157	165	172	194	227
OVS	1 $^1/_2$	322	145	149	152	156	167	196	133	141	148	155	167	196
	1 $^3/_4$	322	157	161	165	168	179	208	133	141	148	155	177	208
SSLT	1 $^1/_2$	344	157	161	165	168	179	208	150	157	165	168	179	208
	1 $^3/_4$	344	169	173	177	180	191	221	150	157	165	172	191	221

Support Design Strength per Inch Thickness, kips/in.	
STD/SSLT	687
OVS	644

Notes:
STD = Standard holes
OVS = Oversized holes
SSLT = Short-slotted holes oriented transverse to direction of load
N = Threads included
X = Threads excluded
SC = Slip critical

*Tabulated values include $^1/_4$-in. reduction in end distance L_{eh} to account for possible underrun in beam length.

Beam	
$F_y = 50$ ksi	$F_u = 65$ ksi

Angle	
$F_y = 36$ ksi	$F_u = 58$ ksi

Table 10-1 (cont.). All-Bolted Double-Angle Connections
$\phi = 0.75$

1-in. Bolts, 2 Rows, W12, 10, 8

Bolt and Angle Design Strength, kips						
ASTM Desig.	Thread Cond.	Hole Type	Angle Thickness 1/4	5/16	3/8	1/2
A325/F1852	N	–	42.4	53.0	63.6	84.8
	X	–	42.4	53.0	63.6	84.8
	SC Class A	STD	42.4	53.0	63.6	76.1
		OVS	37.5	46.9	56.3	64.7
		SSLT	42.4	53.0	63.6	64.7
	SC Class B	STD	42.4	53.0	63.6	84.8
		OVS	37.5	46.9	56.3	75.0
		SSLT	42.4	53.0	63.6	84.8
A490	N	–	42.4	53.0	63.6	84.8
	X	–	42.4	53.0	63.6	84.8
	SC Class A	STD	42.4	53.0	63.6	84.8
		OVS	37.5	46.9	56.3	75.0
		SSLT	42.4	53.0	63.6	81.1
	SC Class B	STD	42.4	53.0	63.6	84.8
		OVS	37.5	46.9	56.3	75.0
		SSLT	42.4	53.0	63.6	84.8

Beam Web Design Strength per Inch Thickness, kips/in.

Hole Type	L_{eh},* in.	Un-coped	Coped at Top Flange Only, L_{ev}, in. 1 1/4	1 3/8	1 1/2	1 5/8	2	3	Coped at Both Flanges, L_{ev}, in. 1 1/4	1 3/8	1 1/2	1 5/8	2	3
STD	1 1/2	230	108	112	116	119	130	160	95.1	102	110	117	130	160
	1 3/4	230	121	124	128	132	143	172	95.1	102	110	117	139	172
OVS	1 1/2	219	95.7	99.3	103	107	118	147	84.1	91.4	98.7	106	118	147
	1 3/4	219	108	112	115	119	130	159	84.1	91.4	98.7	106	128	159
SSLT	1 1/2	230	102	106	110	113	124	154	95.1	102	110	113	124	154
	1 3/4	230	115	118	122	126	137	166	95.1	102	110	117	137	166

Support Design Strength per Inch Thickness, kips/in.	
STD/SSLT	461
OVS	439

Notes:
STD = Standard holes
OVS = Oversized holes
SSLT = Short-slotted holes oriented transverse to direction of load
N = Threads included
X = Threads excluded
SC = Slip critical

*Tabulated values include 1/4-in. reduction in end distance L_{eh} to account for possible underrun in beam length.

$\phi = 0.75$ **Table 10-2.** $F_y = 50$ **ksi**

Combination Bolted/Welded Double-Angle Connections

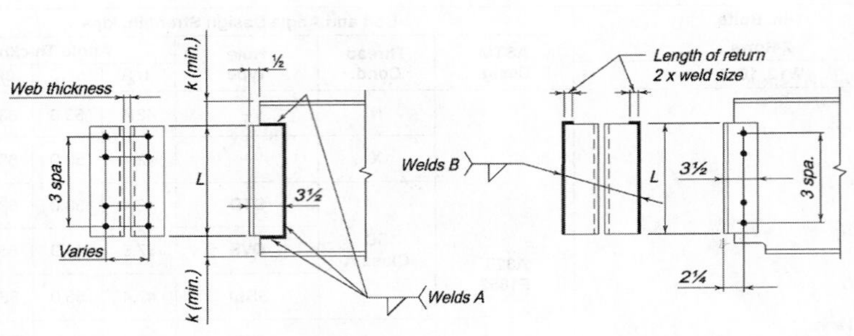

n	*L*	Welds A (70 ksi) Weld Size, in.	ϕR_n, kips	Min. Web Thickness, in.	Welds B (70 ksi) Weld Size, in.	ϕR_n, kips	Min. Support Thickness, in.
12	$35^1/_2$	$^5/_{16}$ $^1/_4$ $^3/_{16}$	592 473 355	0.476 0.381 0.286	$^3/_8$ $^5/_{16}$ $^1/_4$	550 458 366	0.285 0.238 0.190
11	$32^1/_2$	$^5/_{16}$ $^1/_4$ $^3/_{16}$	548 438 329	0.476 0.381 0.286	$^3/_8$ $^5/_{16}$ $^1/_4$	496 414 331	0.285 0.238 0.190
10	$29^1/_2$	$^5/_{16}$ $^1/_4$ $^3/_{16}$	505 404 303	0.476 0.381 0.286	$^3/_8$ $^5/_{16}$ $^1/_4$	443 369 295	0.285 0.238 0.190
9	$26^1/_2$	$^5/_{16}$ $^1/_4$ $^3/_{16}$	464 371 278	0.476 0.381 0.286	$^3/_8$ $^5/_{16}$ $^1/_4$	389 324 259	0.285 0.238 0.190
8	$23^1/_2$	$^5/_{16}$ $^1/_4$ $^3/_{16}$	422 338 253	0.476 0.381 0.286	$^3/_8$ $^5/_{16}$ $^1/_4$	335 279 223	0.285 0.238 0.190
7	$20^1/_2$	$^5/_{16}$ $^1/_4$ $^3/_{16}$	380 304 228	0.476 0.381 0.286	$^3/_8$ $^5/_{16}$ $^1/_4$	280 234 187	0.285 0.238 0.190
6	$17^1/_2$	$^5/_{16}$ $^1/_4$ $^3/_{16}$	334 267 200	0.476 0.381 0.286	$^3/_8$ $^5/_{16}$ $^1/_4$	226 188 150	0.285 0.238 0.190
5	$14^1/_2$	$^5/_{16}$ $^1/_4$ $^3/_{16}$	287 230 172	0.476 0.381 0.286	$^3/_8$ $^5/_{16}$ $^1/_4$	172 143 115	0.285 0.238 0.190
4	$11^1/_2$	$^5/_{16}$ $^1/_4$ $^3/_{16}$	237 190 142	0.476 0.381 0.286	$^3/_8$ $^5/_{16}$ $^1/_4$	120 99.9 79.9	0.285 0.238 0.190
3	$8^1/_2$	$^5/_{16}$ $^1/_4$ $^3/_{16}$	184 147 110	0.476 0.381 0.286	$^3/_8$ $^5/_{16}$ $^1/_4$	72.2 60.2 48.1	0.285 0.238 0.190
2	$5^1/_2$	$^5/_{16}$ $^1/_4$ $^3/_{16}$	125 100 75.3	0.476 0.381 0.286	$^3/_8$ $^5/_{16}$ $^1/_4$	32.8 27.3 21.9	0.285 0.238 0.190

$F_y = 50$ ksi

Table 10-3. All-Welded Double-Angle Connections

$\phi = 0.75$

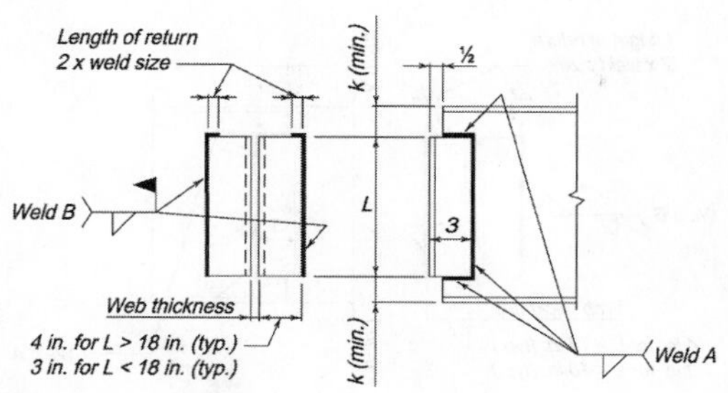

L	Welds A (70 ksi)			Welds B (70 ksi)		
	Weld Size, in.	ϕR_n, kips	Min. Web Thickness, in.	Weld Size, in.	ϕR_n, kips	Min. Support Thickness, in.
36	5/16	599	0.476	3/8	558	0.285
	1/4	479	0.381	5/16	465	0.238
	3/16	359	0.286	1/4	372	0.190
34	5/16	570	0.476	3/8	523	0.285
	1/4	456	0.381	5/16	436	0.238
	3/16	342	0.286	1/4	349	0.190
32	5/16	541	0.476	3/8	487	0.285
	1/4	432	0.381	5/16	406	0.238
	3/16	324	0.286	1/4	325	0.190
30	5/16	512	0.476	3/8	452	0.285
	1/4	410	0.381	5/16	377	0.238
	3/16	307	0.286	1/4	301	0.190
28	5/16	485	0.476	3/8	416	0.285
	1/4	388	0.381	5/16	347	0.238
	3/16	291	0.286	1/4	277	0.190
26	5/16	457	0.476	3/8	380	0.285
	1/4	366	0.381	5/16	317	0.238
	3/16	274	0.286	1/4	253	0.190
24	5/16	429	0.476	3/8	344	0.285
	1/4	343	0.381	5/16	286	0.238
	3/16	258	0.286	1/4	229	0.190
22	5/16	401	0.476	3/8	308	0.285
	1/4	321	0.381	5/16	256	0.238
	3/16	241	0.286	1/4	205	0.190
20	5/16	372	0.476	3/8	271	0.285
	1/4	298	0.381	5/16	226	0.238
	3/16	223	0.286	1/4	181	0.190
18	5/16	342	0.476	3/8	235	0.285
	1/4	273	0.381	5/16	196	0.238
	3/16	205	0.286	1/4	157	0.190

$\phi = 0.75$

Table 10-3 (cont.). All-Welded Double-Angle Connections

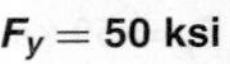

$F_y = 50$ ksi

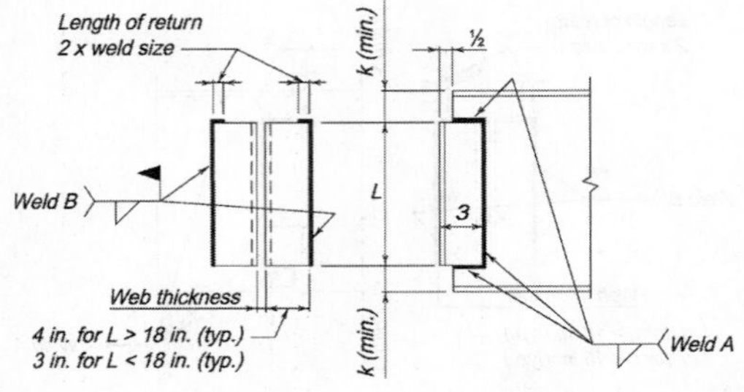

L	Welds A (70 ksi)			Welds B (70 ksi)		
	Weld Size, in.	ϕR_n, kips	Min. Web Thickness, in.	Weld Size, in.	ϕR_n, kips	Min. Support Thickness, in.
16	5/16	311	0.476	3/8	222	0.285
	1/4	249	0.381	5/16	185	0.238
	3/16	187	0.286	1/4	148	0.190
14	5/16	279	0.476	3/8	185	0.285
	1/4	223	0.381	5/16	154	0.238
	3/16	167	0.286	1/4	123	0.190
12	5/16	246	0.476	3/8	149	0.285
	1/4	197	0.381	5/16	124	0.238
	3/16	148	0.286	1/4	99.3	0.190
10	5/16	211	0.476	3/8	113	0.285
	1/4	168	0.381	5/16	94.6	0.238
	3/16	126	0.286	1/4	75.7	0.190
9	5/16	193	0.476	3/8	96.3	0.285
	1/4	154	0.381	5/16	80.2	0.238
	3/16	116	0.286	1/4	64.2	0.190
8	5/16	174	0.476	3/8	79.5	0.285
	1/4	139	0.381	5/16	66.3	0.238
	3/16	105	0.286	1/4	53.0	0.190
7	5/16	155	0.476	3/8	63.6	0.285
	1/4	124	0.381	5/16	53.0	0.238
	3/16	92.9	0.286	1/4	42.4	0.190
6	5/16	136	0.476	3/8	48.7	0.285
	1/4	109	0.381	5/16	40.6	0.238
	3/16	81.4	0.286	1/4	32.5	0.190
5	5/16	116	0.476	3/8	35.1	0.285
	1/4	92.5	0.381	5/16	29.2	0.238
	3/16	69.4	0.286	1/4	23.4	0.190
4	5/16	96.2	0.476	3/8	23.2	0.285
	1/4	77.0	0.381	5/16	19.3	0.238
	3/16	57.7	0.286	1/4	15.5	0.190

SHEAR END-PLATE CONNECTIONS

A shear end-plate connection is made with a plate length less than the supported beam depth, as illustrated in Figure 10-10. The end plate is always shop welded to the beam web with fillet welds on each side and usually field bolted to the supporting member. Welds connecting the end plate to the beam web should not be returned across the thickness of the beam web at the top or bottom of the end plate because of the danger of creating a notch in the beam web.

If the end plate is field welded to the support, adequate flexibility must be provided in the connection. Line welds are placed along the vertical edges of the plate with a return at the top per LRFD Specification Section J2.2b. Note that welding across the entire top of the plate must be avoided as it would inhibit the flexibility and, therefore, the necessary end rotation of the connection. The performance of the resulting connection would not be as intended for shear end-plate connections.

Design Checks

The design strength of a shear end-plate connection is determined from the applicable limit states for the bolts (see Part 7), welds (see Part 8) and connection elements (see Part 9). Note that the limit-state of shear rupture of the beam web must be checked along the length of weld connecting the end plate to the beam web. In all cases, the design strength ϕR_n must equal or exceed the required strength R_u.

Recommended End-Plate Dimensions and Thickness

To provide for stability during erection, it is recommended that the minimum end-plate length be one-half the T-dimension of the beam to be supported. The maximum length of the end-plate must be compatible with the clear distance between the flanges of an uncoped beam and the remaining clear distance of a coped beam.

To provide for flexibility, the combination of plate thickness and gage should be consistent with the recommendations given previously for a double-angle connection of similar thickness and gage.

Shop and Field Practices

When framing to a column web, the associated constructability considerations should be addressed (see the preceding discussion under "Constructability Considerations").

When framing to a column flange, provision must be made for possible mill variation in the depth of the columns, particularly in fairly long runs (i.e., six or more bays of framing). The beam length can be shortened to provide for mill overrun with shims furnished at the appropriate intervals to fill the resulting gaps or to provide for mill underrun. Shear end-plate

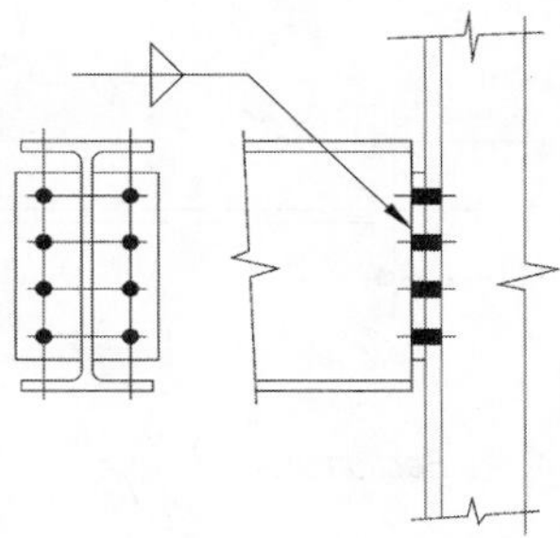

Fig. 10–10. Shear end-plate connections.

connections require close control in cutting the beam to the proper length and in squaring the beam ends such that both end plates are parallel, particularly when beams are cambered.

EXAMPLE 10.6. Shear end-plate connection (beam to girder web).

Given: Refer to Figure 10-11. Design a shear end-plate connection for the W18×50 beam to W21×62 girder web connection.

$R_u = 60\,\text{kips}$

W18×50, ASTM A992

$t_w = 0.355\,\text{in.}$ $\quad d = 18.0\,\text{in.}$
$F_y = 50\,\text{ksi},$ $\quad F_u = 65\,\text{ksi}$

top flange coped 2 in. deep by $4^1/_2$-in. long

W21×62, ASTM A992

$t_w = 0.400\,\text{in.}$
$F_y = 50\,\text{ksi},$ $\quad F_u = 65\,\text{ksi}$

Use $^3/_4$-in.-diameter A325-N bolts in standard holes and 70 ksi electrodes. Assume ASTM A36 plate material with $F_y = 36\,\text{ksi}$ and $F_u = 58\,\text{ksi}$.

Solution: *Design bolts and end plate*

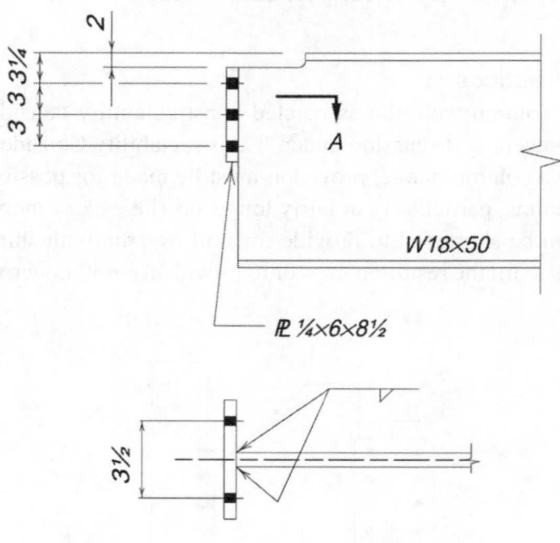

Fig. 10–11. Illustration for Example 10.6.

From Table 10-4, for $^3/_4$-in. diameter A325-N bolts and end-plate material with $F_y = 36\,\text{ksi}$ and $F_u = 58\,\text{ksi}$, select three rows of bolts and $^1/_4$-in. plate thickness

$$\phi R_n = 76.7\,\text{kips} > 60\,\text{kips} \qquad \textbf{o.k.}$$

Check weld and beam web

From Table 10-4, for a $^1/_4$-in. weld size and three rows of bolts (an end-plate length of $8^1/_2$ in.), a $^1/_4$-in. weld size provides $\phi R_n = 89.1$ kips. For beam web material with $F_y = 50\,\text{ksi}$, the minimum web thickness is 0.381 in. Since $t_w = 0.355\,\text{in.} < 0.381$ in. the tabular value must be reduced. Thus,

$$\begin{aligned}\phi R_n &= 89.1\,\text{kips}\left(\frac{0.355\,\text{in.}}{0.381\,\text{in.}}\right)\\ &= 83.0\,\text{kips} > 60\,\text{kips} \qquad \textbf{o.k.}\end{aligned}$$

Check flexural yielding on the coped section

From Table 9-2, $S_{net} = 23.4\,\text{in.}^3$

$$\begin{aligned}\phi R_n &= \frac{0.9F_yS_{net}}{e}\\ &= \frac{0.9(50\,\text{ksi})(23.4\,\text{in.}^4)}{(4^1/_2\,\text{in.} + {}^1/_4\,\text{in.})}\\ &= 222\,\text{kips} > 60\,\text{kips} \qquad \textbf{o.k.}\end{aligned}$$

Check local web buckling at the cope

$$\frac{c}{d} = \frac{4^1/_2\,\text{in.}}{18.0\,\text{in.}} = 0.250$$

$$\frac{c}{h_o} = \frac{4^1/_2\,\text{in.}}{(18.0\,\text{in.} - 2\,\text{in.})} = 0.281$$

Since $\dfrac{c}{d} \le 1.0$,

$$\begin{aligned}f &= 2\left(\frac{c}{d}\right)\\ &= 2(0.250)\\ &= 0.500\end{aligned}$$

Since $\dfrac{c}{h_o} \le 1.0$,

$$\begin{aligned}k &= 2.2\left(\frac{h_o}{c}\right)^{1.65}\\ &= 2.2\left(\frac{1}{0.281}\right)^{1.65}\\ &= 17.9\end{aligned}$$

$$\phi F_{bc} = 23{,}590\left(\frac{t_w}{h_o}\right)^2 fk$$
$$= 23{,}590\left(\frac{0.355\text{ in.}}{17.99\text{ in.} - 2\text{ in.}}\right)^2 (0.500)(17.9)$$
$$= 104\text{ ksi}$$
$$\phi R_n = \frac{\phi F_{bc} S_{net}}{e}$$
$$= \frac{(104\text{ ksi})(23.4\text{ in.}^3)}{(4^1/_2\text{ in.} + {}^1/_4\text{ in.})}$$
$$= 512\text{ kips} > 60\text{ kips} \qquad \textbf{o.k.}$$

Check supporting girder web:

From Table 10-4, for three rows of bolts and girder material with $F_u =$ 65 ksi,

$$\phi R_n = (526\text{ kips/in.})(0.400\text{ in.})$$
$$= 210\text{ kips} > 60\text{ kips} \qquad \textbf{o.k.}$$

The connection, as summarized in Figure 10-11, is adequate.

Table 10-4. Bolted/Welded Shear End-Plate Connections

Tables 10-3 are design aids for shear end-plate connections bolted to the supporting member and welded to the supported beam. Design strengths are tabulated for supported and supporting member material with $F_y = 50$ ksi and $F_u = 65$ ksi and end-plate material with $F_y = 36$, ksi and $F_u = 58$ ksi. Electrode strength is assumed to be 70 ksi. All values, including slip-critical bolt design strengths, are for comparison with factored loads.

Tabulated bolt and end-plate design strengths consider the limit states of bolt shear, bolt bearing on the end plate, shear yielding of the end plate, shear rupture of the end plate and block shear rupture of the end plate. Values are included for 2 through 12 rows of $^3/_4$-in., $^7/_8$-in., and 1-in.-diameter ASTM A325, F1852 and A490 bolts at 3-in. spacing. End-plate edge distances L_{ev} and L_{eh} are assumed to be $1^1/_4$ in.

Tabulated weld design strengths consider the limit state of weld shear assuming an effective weld length equal to the end-plate length minus twice the weld size. The tabulated minimum beam web thickness matches the shear rupture strength of the web material to the strength of the weld metal. As derived in Part 9, the minimum supported beam web thickness for two lines of weld is

$$t_{\min} = \frac{6.19D}{F_u}$$

where D is the number of sixteenths-of-an-inch in the weld size. When less than the minimum material thickness is present, the tabulated weld design strength must be reduced by the ratio of the thickness provided to the minimum thickness.

Tabulated supporting member design strengths, per in. of flange or web thickness, consider the limit-state of bolt bearing.

End Plate	Beam	$^3/_4$-in.-Diameter Bolts
$F_y = 36$ ksi	$F_y = 50$ ksi	12 Rows
$F_u = 58$ ksi	$F_u = 65$ ksi	W44

Table 10-4.
Bolted/Welded Shear End-Plate Connections

Bolt and End-Plate Design Strength, kips

ASTM Desig.	Thread Cond.	Hole Type	End-Plate Thickness		
			$^1/_4$	$^5/_{16}$	$^3/_8$
A325/ F1852	N	–	326	382	382
	X	–	326	408	477
	SC Class A	STD	251	251	251
		OVS	213	213	213
		SSLT	213	213	213
	SC Class B	STD	326	380	380
		OVS	307	323	323
		SSLT	323	323	323
A490	N	–	326	408	477
	X	–	326	408	489
	SC Class A	STD	313	313	313
		OVS	266	266	266
		SSLT	266	266	266
	SC Class B	STD	326	408	475
		OVS	307	383	403
		SSLT	326	403	403

Weld (70 ksi) and Beam Web Design Strength, kips

70 ksi Weld Size, in.	ϕR_n, kips	Minimum Beam Web Thickness, in.	Support Design Strength per Inch Thickness, kips/in.
$^3/_{16}$	293	0.286	2110
$^1/_4$	390	0.381	
$^5/_{16}$	486	0.476	
$^3/_8$	581	0.571	

STD = Standard holes
OVS = Oversized holes
SSLT = Short-slotted holes oriented transverse to direction of load

N = Threads included
X = Threads excluded
SC = Slip critical

$^3/_4$-in.-Diameter Bolts	End Plate	Beam
11 Rows	$F_y = 36$ ksi	$F_y = 50$ ksi
W44, 40	$F_u = 58$ ksi	$F_u = 65$ ksi

Table 10-4 (cont.). Bolted/Welded Shear End-Plate Connections

Bolt and End-Plate Design Strength, kips					
ASTM Desig.	Thread Cond.	Hole Type	End-Plate Thickness $^1/_4$	$^5/_{16}$	$^3/_8$
A325/ F1852	N	–	299	350	350
	X	–	299	373	437
	SC Class A	STD	230	230	230
		OVS	195	195	195
		SSLT	195	195	195
	SC Class B	STD	299	348	348
		OVS	281	296	296
		SSLT	296	296	296
A490	N	–	299	373	437
	X	–	299	373	448
	SC Class A	STD	287	287	287
		OVS	244	244	244
		SSLT	244	244	244
	SC Class B	STD	299	373	435
		OVS	281	351	370
		SSLT	299	370	370

70 ksi Weld Size, in.	ϕR_n, kips	Minimum Beam Web Thickness, in.	Support Design Strength per Inch Thickness, kips/in.
$^3/_{16}$	268	0.286	1930
$^1/_4$	356	0.381	
$^5/_{16}$	444	0.476	
$^3/_8$	530	0.571	

Weld (70 ksi) and Beam Web Design Strength, kips (first three columns)

STD = Standard holes
OVS = Oversized holes
SSLT = Short-slotted holes oriented transverse to direction of load
N = Threads included
X = Threads excluded
SC = Slip critical

End Plate	Beam	3/4-in.-Diameter Bolts
$F_y = 36$ ksi	$F_y = 50$ ksi	10 Rows
$F_u = 58$ ksi	$F_u = 65$ ksi	W44, 40, 36

Table 10-4 (cont.). Bolted/Welded Shear End-Plate Connections

Bolt and End-Plate Design Strength, kips

ASTM Desig.	Thread Cond.	Hole Type	End-Plate Thickness 1/4	5/16	3/8
A325/ F1852	N	–	271	318	318
	X	–	271	338	398
	SC Class A	STD	209	209	209
		OVS	178	178	178
		SSLT	178	178	178
	SC Class B	STD	271	316	316
		OVS	254	269	269
		SSLT	269	269	269
A490	N	–	271	338	398
	X	–	271	338	406
	SC Class A	STD	261	261	261
		OVS	222	222	222
		SSLT	222	222	222
	SC Class B	STD	271	338	396
		OVS	254	318	336
		SSLT	271	336	336

Weld (70 ksi) and Beam Web Design Strength, kips

70 ksi Weld Size, in.	ϕR_n, kips	Minimum Beam Web Thickness, in.	Support Design Strength per Inch Thickness, kips/in.
3/16	243	0.286	1760
1/4	323	0.381	
5/16	402	0.476	
3/8	480	0.571	

STD = Standard holes
OVS = Oversized holes
SSLT = Short-slotted holes oriented transverse to direction of load
N = Threads included
X = Threads excluded
SC = Slip critical

$^3/_4$-in.-Diameter Bolts	End Plate	Beam
9 Rows	$F_y = 36$ ksi	$F_y = 50$ ksi
W44, 40, 36, 33	$F_u = 58$ ksi	$F_u = 65$ ksi

Table 10-4 (cont.). Bolted/Welded Shear End-Plate Connections

Bolt and End-Plate Design Strength, kips

ASTM Desig.	Thread Cond.	Hole Type	End-Plate Thickness $^1/_4$	$^5/_{16}$	$^3/_8$
A325/ F1852	N	–	243	286	286
	X	–	243	304	358
	SC Class A	STD	188	188	188
		OVS	160	160	160
		SSLT	160	160	160
	SC Class B	STD	243	285	285
		OVS	228	242	242
		SSLT	242	242	242
A490	N	–	243	304	358
	X	–	243	304	365
	SC Class A	STD	235	235	235
		OVS	200	200	200
		SSLT	200	200	200
	SC Class B	STD	243	304	356
		OVS	228	285	303
		SSLT	243	303	303

Weld (70 ksi) and Beam Web Design Strength, kips

70 ksi Weld Size, in.	ϕR_n, kips	Minimum Beam Web Thickness, in.	Support Design Strength per Inch Thickness, kips/in.
$^3/_{16}$	218	0.286	1580
$^1/_4$	290	0.381	
$^5/_{16}$	360	0.476	
$^3/_8$	430	0.571	

STD = Standard holes
OVS = Oversized holes
SSLT = Short-slotted holes oriented transverse to direction of load
N = Threads included
X = Threads excluded
SC = Slip critical

End Plate	Beam	3/4-in.-Diameter Bolts
$F_y = 36$ ksi	$F_y = 50$ ksi	8 Rows
$F_u = 58$ ksi	$F_u = 65$ ksi	W44, 40, 36, 33, 30

Table 10-4 (cont.). Bolted/Welded Shear End-Plate Connections

Bolt and End-Plate Design Strength, kips

ASTM Desig.	Thread Cond.	Hole Type	End-Plate Thickness 1/4	5/16	3/8
A325/ F1852	N	–	215	254	254
	X	–	215	269	318
	SC Class A	STD	167	167	167
		OVS	142	142	142
		SSLT	142	142	142
	SC Class B	STD	215	253	253
		OVS	202	215	215
		SSLT	215	215	215
A490	N	–	215	269	318
	X	–	215	269	323
	SC Class A	STD	209	209	209
		OVS	178	178	178
		SSLT	178	178	178
	SC Class B	STD	215	269	316
		OVS	202	253	269
		SSLT	215	269	269

Weld (70 ksi) and Beam Web Design Strength, kips

70 ksi Weld Size, in.	ϕR_n, kips	Minimum Beam Web Thickness, in.	Support Design Strength per Inch Thickness, kips/in.
3/16	193	0.286	1400
1/4	256	0.381	
5/16	318	0.476	
3/8	380	0.571	

STD = Standard holes
OVS = Oversized holes
SSLT = Short-slotted holes oriented transverse to direction of load

N = Threads included
X = Threads excluded
SC = Slip critical

$^3/_4$-in.-Diameter Bolts	End Plate	Beam
7 Rows	$F_y = 36$ ksi	$F_y = 50$ ksi
W44, 40, 36, 33, 30, 27, 24	$F_u = 58$ ksi	$F_u = 65$ ksi

Table 10-4 (cont.).
Bolted/Welded Shear End-Plate Connections

Bolt and End-Plate Design Strength, kips

ASTM Desig.	Thread Cond.	Hole Type	End-Plate Thickness $^1/_4$	$^5/_{16}$	$^3/_8$
A325/ F1852	N	–	188	223	223
	X	–	188	234	278
	SC Class A	STD	146	146	146
		OVS	124	124	124
		SSLT	124	124	124
	SC Class B	STD	188	221	221
		OVS	176	188	188
		SSLT	188	188	188
A490	N	–	188	234	278
	X	–	188	234	281
	SC Class A	STD	183	183	183
		OVS	155	155	155
		SSLT	155	155	155
	SC Class B	STD	188	234	277
		OVS	176	220	235
		SSLT	188	234	235

Weld (70 ksi) and Beam Web Design Strength, kips

70 ksi Weld Size, in.	ϕR_n, kips	Minimum Beam Web Thickness, in.	Support Design Strength per Inch Thickness, kips/in.
$^3/_{16}$	168	0.286	1230
$^1/_4$	223	0.381	
$^5/_{16}$	277	0.476	
$^3/_8$	330	0.571	

STD = Standard holes
OVS = Oversized holes
SSLT = Short-slotted holes oriented transverse to direction of load
N = Threads included
X = Threads excluded
SC = Slip critical

End Plate	Beam	3/4-in.-Diameter Bolts
$F_y = 36$ ksi	$F_y = 50$ ksi	6 Rows
$F_u = 58$ ksi	$F_u = 65$ ksi	W40, 36, 33, 30, 27, 24, 21

Table 10-4 (cont.). Bolted/Welded Shear End-Plate Connections

Bolt and End-Plate Design Strength, kips

ASTM Desig.	Thread Cond.	Hole Type	End-Plate Thickness 1/4	5/16	3/8
A325/ F1852	N	–	160	191	191
	X	–	160	200	239
	SC Class A	STD	125	125	125
		OVS	107	107	107
		SSLT	107	107	107
	SC Class B	STD	160	190	190
		OVS	150	161	161
		SSLT	160	161	161
A490	N	–	160	200	239
	X	–	160	200	240
	SC Class A	STD	157	157	157
		OVS	133	133	133
		SSLT	133	133	133
	SC Class B	STD	160	200	237
		OVS	150	188	202
		SSLT	160	200	202

Weld (70 ksi) and Beam Web Design Strength, kips

70 ksi Weld Size, in.	ϕR_n, kips	Minimum Beam Web Thickness, in.	Support Design Strength per Inch Thickness, kips/in.
3/16	143	0.286	1050
1/4	189	0.381	
5/16	235	0.476	
3/8	280	0.571	

STD = Standard holes
OVS = Oversized holes
SSLT = Short-slotted holes oriented transverse to direction of load
N = Threads included
X = Threads excluded
SC = Slip critical

3/4-in.-Diameter Bolts	End Plate	Beam
5 Rows	$F_y = 36$ ksi	$F_y = 50$ ksi
W30, 27, 24, 21, 18	$F_u = 58$ ksi	$F_u = 65$ ksi

Table 10-4 (cont.). Bolted/Welded Shear End-Plate Connections

Bolt and End-Plate Design Strength, kips

ASTM Desig.	Thread Cond.	Hole Type	End-Plate Thickness 1/4	5/16	3/8
A325/ F1852	N	–	132	159	159
	X	–	132	165	198
	SC Class A	STD	104	104	104
		OVS	88.8	88.8	88.8
		SSLT	88.8	88.8	88.8
	SC Class B	STD	132	158	158
		OVS	124	134	134
		SSLT	132	134	134
A490	N	–	132	165	198
	X	–	132	165	198
	SC Class A	STD	131	131	131
		OVS	111	111	111
		SSLT	111	111	111
	SC Class B	STD	132	165	198
		OVS	124	155	168
		SSLT	132	165	168

Weld (70 ksi) and Beam Web Design Strength, kips

70 ksi Weld Size, in.	ϕR_n, kips	Minimum Beam Web Thickness, in.	Support Design Strength per Inch Thickness, kips/in.
3/16	118	0.286	878
1/4	156	0.381	
5/16	193	0.476	
3/8	230	0.571	

STD = Standard holes
OVS = Oversized holes
SSLT = Short-slotted holes oriented transverse to direction of load
N = Threads included
X = Threads excluded
SC = Slip critical

End Plate	Beam	3/4-in.-Diameter Bolts
$F_y = 36$ ksi	$F_y = 50$ ksi	4 Rows
$F_u = 58$ ksi	$F_u = 65$ ksi	W24, 21, 18, 16

Table 10-4 (cont.). Bolted/Welded Shear End-Plate Connections

Bolt and End-Plate Design Strength, kips

ASTM Desig.	Thread Cond.	Hole Type	End-Plate Thickness		
			1/4	5/16	3/8
A325/ F1852	N	–	104	127	127
	X	–	104	131	157
	SC Class A	STD	83.5	83.5	83.5
		OVS	71.0	71.0	71.0
		SSLT	71.0	71.0	71.0
	SC Class B	STD	104	127	127
		OVS	97.9	108	108
		SSLT	104	108	108
A490	N	–	104	131	157
	X	–	104	131	157
	SC Class A	STD	104	104	104
		OVS	88.8	88.8	88.8
		SSLT	88.8	88.8	88.8
	SC Class B	STD	104	131	157
		OVS	97.9	122	134
		SSLT	104	131	134

Weld (70 ksi) and Beam Web Design Strength, kips

70 ksi Weld Size, in.	ϕR_n, kips	Minimum Beam Web Thickness, in.	Support Design Strength per Inch Thickness, kips/in.
3/16	92.9	0.286	702
1/4	123	0.381	
5/16	151	0.476	
3/8	180	0.571	

STD = Standard holes
OVS = Oversized holes
SSLT = Short-slotted holes oriented transverse to direction of load
N = Threads included
X = Threads excluded
SC = Slip critical

$^3/_4$-in.-Diameter Bolts	End Plate	Beam
3 Rows	$F_y = 36$ ksi	$F_y = 50$ ksi
W18, 16, 14, 12, 10*	$F_u = 58$ ksi	$F_u = 65$ ksi

Table 10-4 (cont.). Bolted/Welded Shear End-Plate Connections

Bolt and End-Plate Design Strength, kips

ASTM Desig.	Thread Cond.	Hole Type	End-Plate Thickness $^1/_4$	$^5/_{16}$	$^3/_8$
A325/ F1852	N	–	76.7	95.4	95.4
	X	–	76.7	95.8	115
	SC Class A	STD	62.6	62.6	62.6
		OVS	53.3	53.3	53.3
		SSLT	53.3	53.3	53.3
	SC Class B	STD	76.7	94.9	94.9
		OVS	71.8	80.7	80.7
		SSLT	76.7	80.7	80.7
A490	N	–	76.7	95.8	115
	X	–	76.7	95.8	115
	SC Class A	STD	76.7	78.3	78.3
		OVS	66.6	66.6	66.6
		SSLT	66.6	66.6	66.6
	SC Class B	STD	76.7	95.8	115
		OVS	71.8	89.7	101
		SSLT	76.7	95.8	101

Weld (70 ksi) and Beam Web Design Strength, kips

70 ksi Weld Size, in.	ϕR_n, kips	Minimum Beam Web Thickness, in.	Support Design Strength per Inch Thickness, kips/in.
$^3/_{16}$	67.9	0.286	526
$^1/_4$	89.1	0.381	
$^5/_{16}$	110	0.476	
$^3/_8$	129	0.571	

STD = Standard holes
OVS = Oversized holes
SSLT = Short-slotted holes oriented transverse to direction of load
N = Threads included
X = Threads excluded
SC = Slip critical

*Limited to W10×12, 15, 17, 19, 22, 26, 30.

End Plate	Beam	$^{3}/_{4}$-in.-Diameter Bolts
$F_y = 36$ ksi	$F_y = 50$ ksi	2 Rows
$F_u = 58$ ksi	$F_u = 65$ ksi	W12, 10, 8

Table 10-4 (cont.). Bolted/Welded Shear End-Plate Connections

Bolt and End-Plate Design Strength, kips

ASTM Desig.	Thread Cond.	Hole Type	End-Plate Thickness $^{1}/_{4}$	End-Plate Thickness $^{5}/_{16}$	End-Plate Thickness $^{3}/_{8}$
A325/ F1852	N	–	48.9	61.2	63.6
	X	–	48.9	61.2	73.4
	SC Class A	STD	41.8	41.8	41.8
		OVS	35.5	35.5	35.5
		SSLT	35.5	35.5	35.5
	SC Class B	STD	48.9	61.2	63.3
		OVS	45.7	53.8	53.8
		SSLT	48.9	53.8	53.8
A490	N	–	48.9	61.2	73.4
	X	–	48.9	61.2	73.4
	SC Class A	STD	48.9	52.2	52.2
		OVS	44.4	44.4	44.4
		SSLT	44.4	44.4	44.4
	SC Class B	STD	48.9	61.2	73.4
		OVS	45.7	57.1	67.2
		SSLT	48.9	61.2	67.2

Weld (70 ksi) and Beam Web Design Strength, kips

70 ksi Weld Size, in.	ϕR_n, kips	Minimum Beam Web Thickness, in.	Support Design Strength per Inch Thickness, kips/in.
$^{3}/_{16}$	42.8	0.286	351
$^{1}/_{4}$	55.7	0.381	
$^{5}/_{16}$	67.9	0.476	
$^{3}/_{8}$	79.4	0.571	

STD = Standard holes
OVS = Oversized holes
SSLT = Short-slotted holes oriented transverse to direction of load
N = Threads included
X = Threads excluded
SC = Slip critical

$^7/_8$-in.-Diameter Bolts	End Plate	Beam
12 Rows	$F_y = 36$ ksi	$F_y = 50$ ksi
W44	$F_u = 58$ ksi	$F_u = 65$ ksi

Table 10-4 (cont.). Bolted/Welded Shear End-Plate Connections

Bolt and End-Plate Design Strength, kips

ASTM Desig.	Thread Cond.	Hole Type	End-Plate Thickness $^1/_4$	End-Plate Thickness $^5/_{16}$	End-Plate Thickness $^3/_8$
A325/ F1852	N	–	307	383	460
	X	–	307	383	460
	SC Class A	STD	307	349	349
		OVS	287	297	297
		SSLT	297	297	297
	SC Class B	STD	307	383	460
		OVS	287	359	431
		SSLT	307	383	450
A490	N	–	307	383	460
	X	–	307	383	460
	SC Class A	STD	307	383	439
		OVS	287	359	373
		SSLT	307	373	373
	SC Class B	STD	307	383	460
		OVS	287	359	431
		SSLT	307	383	460

Weld (70 ksi) and Beam Web Design Strength, kips

70 ksi Weld Size, in.	ϕR_n, kips	Minimum Beam Web Thickness, in.	Support Design Strength per Inch Thickness, kips/in.
$^3/_{16}$	293	0.286	2460
$^1/_4$	390	0.381	
$^5/_{16}$	486	0.476	
$^3/_8$	581	0.571	

STD = Standard holes
OVS = Oversized holes
SSLT = Short-slotted holes oriented transverse to direction of load
N = Threads included
X = Threads excluded
SC = Slip critical

End Plate	Beam	$^7/_8$-in.-Diameter Bolts
$F_y = 36$ ksi	$F_y = 50$ ksi	11 Rows
$F_u = 58$ ksi	$F_u = 65$ ksi	W44, 40

Table 10-4 (cont.). Bolted/Welded Shear End-Plate Connections

Bolt and End-Plate Design Strength, kips					
ASTM Desig.	**Thread Cond.**	**Hole Type**	**End-Plate Thickness**		
			$^1/_4$	$^5/_{16}$	$^3/_8$
A325/ F1852	N	–	281	351	421
	X	–	281	351	421
	SC Class A	STD	281	320	320
		OVS	263	272	272
		SSLT	272	272	272
	SC Class B	STD	281	351	421
		OVS	263	328	394
		SSLT	281	351	412
A490	N	–	281	351	421
	X	–	281	351	421
	SC Class A	STD	281	351	402
		OVS	263	328	342
		SSLT	281	342	342
	SC Class B	STD	281	351	421
		OVS	263	328	394
		SSLT	281	351	421

Weld (70 ksi) and Beam Web Design Strength, kips			Support Design Strength per Inch Thickness, kips/in.
70 ksi Weld Size, in.	**ϕR_n, kips**	**Minimum Beam Web Thickness, in.**	
$^3/_{16}$	268	0.286	2250
$^1/_4$	356	0.381	
$^5/_{16}$	444	0.476	
$^3/_8$	530	0.571	

STD = Standard holes
OVS = Oversized holes
SSLT = Short-slotted holes oriented transverse to direction of load
N = Threads included
X = Threads excluded
SC = Slip critical

7/8-in.-Diameter Bolts	End Plate	Beam
10 Rows	$F_y = 36$ ksi	$F_y = 50$ ksi
W44, 40, 36	$F_u = 58$ ksi	$F_u = 65$ ksi

Table 10-4 (cont.). Bolted/Welded Shear End-Plate Connections

Bolt and End-Plate Design Strength, kips

ASTM Desig.	Thread Cond.	Hole Type	End-Plate Thickness 1/4	5/16	3/8
A325/ F1852	N	–	254	318	382
	X	–	254	318	382
	SC Class A	STD	254	291	291
		OVS	238	247	247
		SSLT	247	247	247
	SC Class B	STD	254	318	382
		OVS	238	298	357
		SSLT	254	318	375
A490	N	–	254	318	382
	X	–	254	318	382
	SC Class A	STD	254	318	365
		OVS	238	298	311
		SSLT	254	311	311
	SC Class B	STD	254	318	382
		OVS	238	298	357
		SSLT	254	318	382

Weld (70 ksi) and Beam Web Design Strength, kips

70 ksi Weld Size, in.	ϕR_n, kips	Minimum Beam Web Thickness, in.	Support Design Strength per Inch Thickness, kips/in.
3/16	243	0.286	2050
1/4	323	0.381	
5/16	402	0.476	
3/8	480	0.571	

STD = Standard holes
OVS = Oversized holes
SSLT = Short-slotted holes oriented transverse to direction of load

N = Threads included
X = Threads excluded
SC = Slip critical

End Plate	Beam	$^7/_8$-in.-Diameter Bolts
$F_y = 36$ ksi	$F_y = 50$ ksi	9 Rows
$F_u = 58$ ksi	$F_u = 65$ ksi	W44, 40, 36, 33

Table 10-4 (cont.). Bolted/Welded Shear End-Plate Connections

Bolt and End-Plate Design Strength, kips

ASTM Desig.	Thread Cond.	Hole Type	End-Plate Thickness		
			$^1/_4$	$^5/_{16}$	$^3/_8$
A325/ F1852	N	–	228	285	343
	X	–	228	285	343
	SC Class A	STD	228	262	262
		OVS	214	223	223
		SSLT	223	223	223
	SC Class B	STD	228	285	343
		OVS	214	267	321
		SSLT	228	285	337
A490	N	–	228	285	343
	X	–	228	285	343
	SC Class A	STD	228	285	329
		OVS	214	267	280
		SSLT	228	280	280
	SC Class B	STD	228	285	343
		OVS	214	267	321
		SSLT	228	285	343

Weld (70 ksi) and Beam Web Design Strength, kips

70 ksi Weld Size, in.	ϕR_n, kips	Minimum Beam Web Thickness, in.	Support Design Strength per Inch Thickness, kips/in.
$^3/_{16}$	218	0.286	1840
$^1/_4$	290	0.381	
$^5/_{16}$	360	0.476	
$^3/_8$	430	0.571	

STD = Standard holes
OVS = Oversized holes
SSLT = Short-slotted holes oriented transverse to direction of load

N = Threads included
X = Threads excluded
SC = Slip critical

$^7/_8$-in.-Diameter Bolts	End Plate	Beam
8 Rows	$F_y = 36$ ksi	$F_y = 50$ ksi
W44, 40, 36, 33, 30	$F_u = 58$ ksi	$F_u = 65$ ksi

Table 10-4 (cont.). Bolted/Welded Shear End-Plate Connections

Bolt and End-Plate Design Strength, kips					
ASTM Desig.	Thread Cond.	Hole Type	End-Plate Thickness		
			$^1/_4$	$^5/_{16}$	$^3/_8$
A325/ F1852	N	–	202	253	303
	X	–	202	253	303
	SC Class A	STD	202	233	233
		OVS	189	198	198
		SSLT	198	198	198
	SC Class B	STD	202	253	303
		OVS	189	237	284
		SSLT	202	253	300
A490	N	–	202	253	303
	X	–	202	253	303
	SC Class A	STD	202	253	292
		OVS	189	237	249
		SSLT	202	249	249
	SC Class B	STD	202	253	303
		OVS	189	237	284
		SSLT	202	253	303

Weld (70 ksi) and Beam Web Design Strength, kips			Support Design Strength per Inch Thickness, kips/in.
70 ksi Weld Size, in.	ϕR_n, kips	Minimum Beam Web Thickness, in.	
$^3/_{16}$	193	0.286	1640
$^1/_4$	256	0.381	
$^5/_{16}$	318	0.476	
$^3/_8$	380	0.571	

STD = Standard holes
OVS = Oversized holes
SSLT = Short-slotted holes oriented transverse to direction of load
N = Threads included
X = Threads excluded
SC = Slip critical

End Plate	Beam
$F_y = 36$ ksi	$F_y = 50$ ksi
$F_u = 58$ ksi	$F_u = 65$ ksi

$^7/_8$-in.-Diameter Bolts
7 Rows
W44, 40, 36, 33, 30, 27, 24

Table 10-4 (cont.). Bolted/Welded Shear End-Plate Connections

Bolt and End-Plate Design Strength, kips

ASTM Desig.	Thread Cond.	Hole Type	End-Plate Thickness $^1/_4$	$^5/_{16}$	$^3/_8$
A325/ F1852	N	–	176	220	264
	X	–	176	220	264
	SC Class A	STD	176	204	204
		OVS	165	173	173
		SSLT	173	173	173
	SC Class B	STD	176	220	264
		OVS	165	206	247
		SSLT	176	220	262
A490	N	–	176	220	264
	X	–	176	220	264
	SC Class A	STD	176	220	256
		OVS	165	206	217
		SSLT	176	217	217
	SC Class B	STD	176	220	264
		OVS	165	206	247
		SSLT	176	220	264

Weld (70 ksi) and Beam Web Design Strength, kips

70 ksi Weld Size, in.	ϕR_n, kips	Minimum Beam Web Thickness, in.	Support Design Strength per Inch Thickness, kips/in.
$^3/_{16}$	168	0.286	1430
$^1/_4$	223	0.381	
$^5/_{16}$	277	0.476	
$^3/_8$	330	0.571	

STD = Standard holes
OVS = Oversized holes
SSLT = Short-slotted holes oriented transverse to direction of load
N = Threads included
X = Threads excluded
SC = Slip critical

7/8-in.-Diameter Bolts	End Plate	Beam
6 Rows	$F_y = 36$ ksi	$F_y = 50$ ksi
W40, 36, 33, 30, 27, 24, 21	$F_u = 58$ ksi	$F_u = 65$ ksi

Table 10-4 (cont.). Bolted/Welded Shear End-Plate Connections

Bolt and End-Plate Design Strength, kips

ASTM Desig.	Thread Cond.	Hole Type	End-Plate Thickness 1/4	5/16	3/8
A325/ F1852	N	–	150	188	225
	X	–	150	188	225
	SC Class A	STD	150	175	175
		OVS	140	148	148
		SSLT	148	148	148
	SC Class B	STD	150	188	225
		OVS	140	175	210
		SSLT	150	188	225
A490	N	–	150	188	225
	X	–	150	188	225
	SC Class A	STD	150	188	219
		OVS	140	175	186
		SSLT	150	186	186
	SC Class B	STD	150	188	225
		OVS	140	175	210
		SSLT	150	188	225

Weld (70 ksi) and Beam Web Design Strength, kips

70 ksi Weld Size, in.	ϕR_n, kips	Minimum Beam Web Thickness, in.	Support Design Strength per Inch Thickness, kips/in.
3/16	143	0.286	1230
1/4	189	0.381	
5/16	235	0.476	
3/8	280	0.571	

STD = Standard holes
OVS = Oversized holes
SSLT = Short-slotted holes oriented transverse to direction of load
N = Threads included
X = Threads excluded
SC = Slip critical

End Plate	Beam	$^7/_8$-in.-Diameter Bolts
$F_y = 36$ ksi	$F_y = 50$ ksi	5 Rows
$F_u = 58$ ksi	$F_u = 65$ ksi	W30, 27, 24, 21, 18

Table 10-4 (cont.). Bolted/Welded Shear End-Plate Connections

Bolt and End-Plate Design Strength, kips

ASTM Desig.	Thread Cond.	Hole Type	End-Plate Thickness $^1/_4$	$^5/_{16}$	$^3/_8$
A325/ F1852	N	–	124	155	186
	X	–	124	155	186
	SC Class A	STD	124	145	145
		OVS	116	124	124
		SSLT	124	124	124
	SC Class B	STD	124	155	186
		OVS	116	145	174
		SSLT	124	155	186
A490	N	–	124	155	186
	X	–	124	155	186
	SC Class A	STD	124	155	183
		OVS	116	145	155
		SSLT	124	155	155
	SC Class B	STD	124	155	186
		OVS	116	145	174
		SSLT	124	155	186

Weld (70 ksi) and Beam Web Design Strength, kips

70 ksi Weld Size, in.	ϕR_n, kips	Minimum Beam Web Thickness, in.	Support Design Strength per Inch Thickness, kips/in.
$^3/_{16}$	118	0.286	1020
$^1/_4$	156	0.381	
$^5/_{16}$	193	0.476	
$^3/_8$	230	0.571	

STD = Standard holes
OVS = Oversized holes
SSLT = Short-slotted holes oriented transverse to direction of load
N = Threads included
X = Threads excluded
SC = Slip critical

7/8-in.-Diameter Bolts	End Plate	Beam
4 Rows	$F_y = 36$ ksi	$F_y = 50$ ksi
W24, 21, 18, 16	$F_u = 58$ ksi	$F_u = 65$ ksi

Table 10-4 (cont.). Bolted/Welded Shear End-Plate Connections

Bolt and End-Plate Design Strength, kips

ASTM Desig.	Thread Cond.	Hole Type	End-Plate Thickness 1/4	5/16	3/8
A325/ F1852	N	–	97.9	122	147
	X	–	97.9	122	147
	SC Class A	STD	97.9	116	116
		OVS	91.4	98.9	98.9
		SSLT	97.9	98.9	98.9
	SC Class B	STD	97.9	122	147
		OVS	91.4	114	137
		SSLT	97.9	122	147
A490	N	–	97.9	122	147
	X	–	97.9	122	147
	SC Class A	STD	97.9	122	146
		OVS	91.4	114	124
		SSLT	97.9	122	124
	SC Class B	STD	97.9	122	147
		OVS	91.4	114	137
		SSLT	97.9	122	147

Weld (70 ksi) and Beam Web Design Strength, kips

70 ksi Weld Size, in.	ϕR_n, kips	Minimum Beam Web Thickness, in.	Support Design Strength per Inch Thickness, kips/in.
3/16	92.9	0.286	819
1/4	123	0.381	
5/16	151	0.476	
3/8	180	0.571	

STD = Standard holes
OVS = Oversized holes
SSLT = Short-slotted holes oriented transverse to direction of load
N = Threads included
X = Threads excluded
SC = Slip critical

End Plate	Beam	$^7/_8$-in.-Diameter Bolts
$F_y = 36$ ksi	$F_y = 50$ ksi	3 Rows
$F_u = 58$ ksi	$F_u = 65$ ksi	W18, 16, 14, 12, 10*

Table 10-4 (cont.). Bolted/Welded Shear End-Plate Connections

Bolt and End-Plate Design Strength, kips

ASTM Desig.	Thread Cond.	Hole Type	End-Plate Thickness		
			$^1/_4$	$^5/_{16}$	$^3/_8$
A325/ F1852	N	–	71.8	89.7	108
	X	–	71.8	89.7	108
	SC Class A	STD	71.8	87.3	87.3
		OVS	66.9	74.2	74.2
		SSLT	71.8	74.2	74.2
	SC Class B	STD	71.8	89.7	108
		OVS	66.9	83.6	100
		SSLT	71.8	89.7	108
A490	N	–	71.8	89.7	108
	X	–	71.8	89.7	108
	SC Class A	STD	71.8	89.7	108
		OVS	66.9	83.6	93.2
		SSLT	71.8	89.7	93.2
	SC Class B	STD	71.8	89.7	108
		OVS	66.9	83.6	100
		SSLT	71.8	89.7	108

Weld (70 ksi) and Beam Web Design Strength, kips			Support Design Strength per Inch Thickness, kips/in.
70 ksi Weld Size, in.	ϕR_n, kips	Minimum Beam Web Thickness, in.	
$^3/_{16}$	67.9	0.286	614
$^1/_4$	89.1	0.381	
$^5/_{16}$	110	0.476	
$^3/_8$	129	0.571	

STD = Standard holes
OVS = Oversized holes
SSLT = Short-slotted holes oriented transverse to direction of load
N = Threads included
X = Threads excluded
SC = Slip critical

* Limited to W10×12, 15, 17, 19, 22, 26, 30.

$^7/_8$-in.-Diameter Bolts	End Plate	Beam
2 Rows	$F_y = 36$ ksi	$F_y = 50$ ksi
W12, 10, 8	$F_u = 58$ ksi	$F_u = 65$ ksi

Table 10-4 (cont.). Bolted/Welded Shear End-Plate Connections

Bolt and End-Plate Design Strength, kips

ASTM Desig.	Thread Cond.	Hole Type	End-Plate Thickness $^1/_4$	$^5/_{16}$	$^3/_8$
A325/ F1852	N	–	45.7	57.1	68.5
	X	–	45.7	57.1	68.5
	SC Class A	STD	45.7	57.1	58.2
		OVS	42.4	49.4	49.4
		SSLT	45.7	49.4	49.4
	SC Class B	STD	45.7	57.1	68.5
		OVS	42.4	53.0	63.6
		SSLT	45.7	57.1	68.5
A490	N	–	45.7	57.1	68.5
	X	–	45.7	57.1	68.5
	SC Class A	STD	45.7	57.1	68.5
		OVS	42.4	53.0	62.1
		SSLT	45.7	57.1	62.1
	SC Class B	STD	45.7	57.1	68.5
		OVS	42.4	53.0	63.6
		SSLT	45.7	57.1	68.5

Weld (70 ksi) and Beam Web Design Strength, kips

70 ksi Weld Size, in.	ϕR_n, kips	Minimum Beam Web Thickness, in.	Support Design Strength per Inch Thickness, kips/in.
$^3/_{16}$	42.8	0.286	409
$^1/_4$	55.7	0.381	
$^5/_{16}$	67.9	0.476	
$^3/_8$	79.4	0.571	

STD = Standard holes
OVS = Oversized holes
SSLT = Short-slotted holes oriented transverse to direction of load
N = Threads included
X = Threads excluded
SC = Slip critical

End Plate	Beam	7/8-in.-Diameter Bolts
$F_y = 36$ ksi	$F_y = 50$ ksi	12 Rows
$F_u = 58$ ksi	$F_u = 65$ ksi	W44

Table 10-4 (cont.). Bolted/Welded Shear End-Plate Connections

Bolt and End-Plate Design Strength, kips

ASTM Desig.	Thread Cond.	Hole Type	End-Plate Thickness 1/4	5/16	3/8
A325/ F1852	N	–	287	359	431
	X	–	287	359	431
	SC Class A	STD	287	359	431
		OVS	258	322	387
		SSLT	287	359	388
	SC Class B	STD	287	359	431
		OVS	258	322	387
		SSLT	287	359	431
A490	N	–	287	359	431
	X	–	287	359	431
	SC Class A	STD	287	359	431
		OVS	258	322	387
		SSLT	287	359	431
	SC Class B	STD	287	359	431
		OVS	258	322	387
		SSLT	287	359	431

Weld (70 ksi) and Beam Web Design Strength, kips

70 ksi Weld Size, in.	ϕR_n, kips	Minimum Beam Web Thickness, in.
3/16	293	0.286
1/4	390	0.381
5/16	486	0.476
3/8	581	0.571

Support Design Strength per Inch Thickness, kips/in.

STD/SSLT	2730
OVS	2490

STD = Standard holes
OVS = Oversized holes
SSLT = Short-slotted holes oriented transverse to direction of load

N = Threads included
X = Threads excluded
SC = Slip critical

1-in.-Diameter Bolts	End Plate	Beam
11 Rows	$F_y = 36$ ksi	$F_y = 50$ ksi
W44, 40	$F_u = 58$ ksi	$F_u = 65$ ksi

Table 10-4 (cont.). Bolted/Welded Shear End-Plate Connections

Bolt and End-Plate Design Strength, kips

ASTM Desig.	Thread Cond.	Hole Type	End-Plate Thickness 1/4	5/16	3/8
A325/ F1852	N	–	263	328	394
	X	–	263	328	394
	SC Class A	STD	263	328	394
		OVS	236	295	354
		SSLT	263	328	356
	SC Class B	STD	263	328	394
		OVS	236	295	354
		SSLT	263	328	394
A490	N	–	263	328	394
	X	–	263	328	394
	SC Class A	STD	263	328	394
		OVS	236	295	354
		SSLT	263	328	394
	SC Class B	STD	263	328	394
		OVS	236	295	354
		SSLT	263	328	394

Weld (70 ksi) and Beam Web Design Strength, kips

70 ksi Weld Size, in.	ϕR_n, kips	Minimum Beam Web Thickness, in.
3/16	268	0.286
1/4	356	0.381
5/16	444	0.476
3/8	530	0.571

Support Design Strength per Inch Thickness, kips/in.

STD/SSLT	2500
OVS	2280

STD = Standard holes
OVS = Oversized holes
SSLT = Short-slotted holes oriented transverse to direction of load
N = Threads included
X = Threads excluded
SC = Slip critical

End Plate	Beam		$^7/_8$-in.-Diameter Bolts
$F_y = 36$ ksi	$F_y = 50$ ksi		10 Rows
$F_u = 58$ ksi	$F_u = 65$ ksi		W44,40, 36

Table 10-4 (cont.). Bolted/Welded Shear End-Plate Connections

Bolt and End-Plate Design Strength, kips

ASTM Desig.	Thread Cond.	Hole Type	End-Plate Thickness $^1/_4$	$^5/_{16}$	$^3/_8$
A325/ F1852	N	–	238	298	357
	X	–	238	298	357
	SC Class A	STD	238	298	357
		OVS	214	267	321
		SSLT	238	298	323
	SC Class B	STD	238	298	357
		OVS	214	267	321
		SSLT	238	298	357
A490	N	–	238	298	357
	X	–	238	298	357
	SC Class A	STD	238	298	357
		OVS	214	267	321
		SSLT	238	298	357
	SC Class B	STD	238	298	357
		OVS	214	267	321
		SSLT	238	298	357

Weld (70 ksi) and Beam Web Design Strength, kips

70 ksi Weld Size, in.	ϕR_n, kips	Minimum Beam Web Thickness, in.
$^3/_{16}$	243	0.286
$^1/_4$	323	0.381
$^5/_{16}$	402	0.476
$^3/_8$	480	0.571

Support Design Strength per Inch Thickness, kips/in.

STD/SSLT	2270
OVS	2080

STD = Standard holes
OVS = Oversized holes
SSLT = Short-slotted holes oriented transverse to direction of load

N = Threads included
X = Threads excluded
SC = Slip critical

1-in.-Diameter Bolts	End Plate	Beam
9 Rows	$F_y = 36$ ksi	$F_y = 50$ ksi
W44, 40, 36, 33	$F_u = 58$ ksi	$F_u = 65$ ksi

Table 10-4 (cont.). Bolted/Welded Shear End-Plate Connections

Bolt and End-Plate Design Strength, kips

ASTM Desig.	Thread Cond.	Hole Type	End-Plate Thickness 1/4	5/16	3/8
A325/ F1852	N	–	214	267	321
	X	–	214	267	321
	SC Class A	STD	214	267	321
		OVS	192	240	288
		SSLT	214	267	291
	SC Class B	STD	214	267	321
		OVS	192	240	288
		SSLT	214	267	321
A490	N	–	214	267	321
	X	–	214	267	321
	SC Class A	STD	214	267	321
		OVS	192	240	288
		SSLT	214	267	321
	SC Class B	STD	214	267	321
		OVS	192	240	288
		SSLT	214	267	321

Weld (70 ksi) and Beam Web Design Strength, kips

70 ksi Weld Size, in.	ϕR_n, kips	Minimum Beam Web Thickness, in.
3/16	218	0.286
1/4	290	0.381
5/16	360	0.476
3/8	430	0.571

Support Design Strength per Inch Thickness, kips/in.	
STD/SSLT	2050
OVS	1870

STD = Standard holes
OVS = Oversized holes
SSLT = Short-slotted holes oriented transverse to direction of load
N = Threads included
X = Threads excluded
SC = Slip critical

End Plate	Beam
$F_y = 36$ ksi	$F_y = 50$ ksi
$F_u = 58$ ksi	$F_u = 65$ ksi

1-in.-Diameter Bolts
8 Rows
W44, 40, 36, 33, 30

Table 10-4 (cont.). Bolted/Welded Shear End-Plate Connections

Bolt and End-Plate Design Strength, kips

ASTM Desig.	Thread Cond.	Hole Type	End-Plate Thickness 1/4	5/16	3/8
A325/ F1852	N	–	189	237	284
	X	–	189	237	284
	SC Class A	STD	189	237	284
		OVS	170	212	254
		SSLT	189	237	259
	SC Class B	STD	189	237	284
		OVS	170	212	254
		SSLT	189	237	284
A490	N	–	189	237	284
	X	–	189	237	284
	SC Class A	STD	189	237	284
		OVS	170	212	254
		SSLT	189	237	284
	SC Class B	STD	189	237	284
		OVS	170	212	254
		SSLT	189	237	284

Weld (70 ksi) and Beam Web Design Strength, kips

70 ksi Weld Size, in.	ϕR_n, kips	Minimum Beam Web Thickness, in.
3/16	193	0.286
1/4	256	0.381
5/16	318	0.476
3/8	380	0.571

Support Design Strength per Inch Thickness, kips/in.

Hole Type	kips/in.
STD/SSLT	1820
OVS	1670

STD = Standard holes
OVS = Oversized holes
SSLT = Short-slotted holes oriented transverse to direction of load
N = Threads included
X = Threads excluded
SC = Slip critical

1-in.-Diameter Bolts	End Plate	Beam
7 Rows	$F_y = 36$ ksi	$F_y = 50$ ksi
W44, 40, 36, 33, 30, 27, 24	$F_u = 58$ ksi	$F_u = 65$ ksi

Table 10-4 (cont.). Bolted/Welded Shear End-Plate Connections

Bolt and End-Plate Design Strength, kips					
ASTM Desig.	Thread Cond.	Hole Type	End-Plate Thickness		
			1/4	5/16	3/8
A325/ F1852	N	–	165	206	247
	X	–	165	206	247
	SC Class A	STD	165	206	247
		OVS	148	185	221
		SSLT	165	206	226
	SC Class B	STD	165	206	247
		OVS	148	185	221
		SSLT	165	206	247
A490	N	–	165	206	247
	X	–	165	206	247
	SC Class A	STD	165	206	247
		OVS	148	185	221
		SSLT	165	206	247
	SC Class B	STD	165	206	247
		OVS	148	185	221
		SSLT	165	206	247

Weld (70 ksi) and Beam Web Design Strength, kips			Support Design Strength per Inch Thickness, kips/in.	
70 ksi Weld Size, in.	ϕR_n, kips	Minimum Beam Web Thickness, in.		
3/16	168	0.286		
1/4	223	0.381		
5/16	277	0.476	STD/SSLT	1590
3/8	330	0.571	OVS	1460

STD = Standard holes
OVS = Oversized holes
SSLT = Short-slotted holes oriented transverse to direction of load
N = Threads included
X = Threads excluded
SC = Slip critical

End Plate	Beam	1-in.-Diameter Bolts
$F_y = 36$ ksi	$F_y = 50$ ksi	6 Rows
$F_u = 58$ ksi	$F_u = 65$ ksi	W40, 36, 33, 30, 27, 24, 21

Table 10-4 (cont.). Bolted/Welded Shear End-Plate Connections

Bolt and End-Plate Design Strength, kips

ASTM Desig.	Thread Cond.	Hole Type	End-Plate Thickness		
			1/4	5/16	3/8
A325/ F1852	N	–	140	175	210
	X	–	140	175	210
	SC Class A	STD	140	175	210
		OVS	126	157	188
		SSLT	140	175	194
	SC Class B	STD	140	175	210
		OVS	126	157	188
		SSLT	140	175	210
A490	N	–	140	175	210
	X	–	140	175	210
	SC Class A	STD	140	175	210
		OVS	126	157	188
		SSLT	140	175	210
	SC Class B	STD	140	175	210
		OVS	126	157	188
		SSLT	140	175	210

Weld (70 ksi) and Beam Web Design Strength, kips

70 ksi Weld Size, in.	ϕR_n, kips	Minimum Beam Web Thickness, in.
3/16	143	0.286
1/4	189	0.381
5/16	235	0.476
3/8	280	0.571

Support Design Strength per Inch Thickness, kips/in.

STD/SSLT	1370
OVS	1260

STD = Standard holes
OVS = Oversized holes
SSLT = Short-slotted holes oriented transverse to direction of load
N = Threads included
X = Threads excluded
SC = Slip critical

1-in.-Diameter Bolts		End Plate	Beam
5 Rows		$F_y = 36$ ksi	$F_y = 50$ ksi
W30, 27, 24, 21, 18		$F_u = 58$ ksi	$F_u = 65$ ksi

Table 10-4 (cont.). Bolted/Welded Shear End-Plate Connections

Bolt and End-Plate Design Strength, kips

ASTM Desig.	Thread Cond.	Hole Type	End-Plate Thickness $^1/_4$	$^5/_{16}$	$^3/_8$
A325/ F1852	N	–	116	145	174
	X	–	116	145	174
	SC Class A	STD	116	145	174
		OVS	104	129	155
		SSLT	116	145	162
	SC Class B	STD	116	145	174
		OVS	104	129	155
		SSLT	116	145	174
A490	N	–	116	145	174
	X	–	116	145	174
	SC Class A	STD	116	145	174
		OVS	104	129	155
		SSLT	116	145	174
	SC Class B	STD	116	145	174
		OVS	104	129	155
		SSLT	116	145	174

Weld (70 ksi) and Beam Web Design Strength, kips

70 ksi Weld Size, in.	ϕR_n, kips	Minimum Beam Web Thickness, in.
$^3/_{16}$	118	0.286
$^1/_4$	156	0.381
$^5/_{16}$	193	0.476
$^3/_8$	230	0.571

Support Design Strength per Inch Thickness, kips/in.

STD/SSLT	1140
OVS	1050

STD = Standard holes
OVS = Oversized holes
SSLT = Short-slotted holes oriented transverse to direction of load

N = Threads included
X = Threads excluded
SC = Slip critical

End Plate	Beam		1-in.-Diameter Bolts
$F_y = 36$ ksi	$F_y = 50$ ksi		4 Rows
$F_u = 58$ ksi	$F_u = 65$ ksi		W24, 21, 18, 16

Table 10-4 (cont.). Bolted/Welded Shear End-Plate Connections

Bolt and End-Plate Design Strength, kips					
ASTM Desig.	Thread Cond.	Hole Type	End-Plate Thickness		
			1/4	5/16	3/8
A325/ F1852	N	–	91.4	114	137
	X	–	91.4	114	137
	SC Class A	STD	91.4	114	137
		OVS	81.6	102	122
		SSLT	91.4	114	129
	SC Class B	STD	91.4	114	137
		OVS	81.6	102	122
		SSLT	91.4	114	137
A490	N	–	91.4	114	137
	X	–	91.4	114	137
	SC Class A	STD	91.4	114	137
		OVS	81.6	102	122
		SSLT	91.4	114	137
	SC Class B	STD	91.4	114	137
		OVS	81.6	102	122
		SSLT	91.4	114	137

Weld (70 ksi) and Beam Web Design Strength, kips			Support Design Strength per Inch Thickness, kips/in.	
70 ksi Weld Size, in.	ϕR_n, kips	Minimum Beam Web Thickness, in.		
3/16	92.9	0.286		
1/4	123	0.381		
5/16	151	0.476	STD/SSLT	914
3/8	180	0.571	OVS	848

STD = Standard holes
OVS = Oversized holes
SSLT = Short-slotted holes oriented transverse to direction of load

N = Threads included
X = Threads excluded
SC = Slip critical

1-in.-Diameter Bolts	End Plate	Beam
3 Rows	$F_y = 36$ ksi	$F_y = 50$ ksi
W18, 16, 14, 12, 10*	$F_u = 58$ ksi	$F_u = 65$ ksi

Table 10-4 (cont.). Bolted/Welded Shear End-Plate Connections

Bolt and End-Plate Design Strength, kips

ASTM Desig.	Thread Cond.	Hole Type	End-Plate Thickness 1/4	5/16	3/8
A325/ F1852	N	–	66.9	83.6	100
	X	–	66.9	83.6	100
	SC Class A	STD	66.9	83.6	100
		OVS	59.5	74.4	89.3
		SSLT	66.9	83.6	97.0
	SC Class B	STD	66.9	83.6	100
		OVS	59.5	74.4	89.3
		SSLT	66.9	83.6	100
A490	N	–	66.9	83.6	100
	X	–	66.9	83.6	100
	SC Class A	STD	66.9	83.6	100
		OVS	59.5	74.4	89.3
		SSLT	66.9	83.6	100
	SC Class B	STD	66.9	83.6	100
		OVS	59.5	74.4	89.3
		SSLT	66.9	83.6	100

Weld (70 ksi) and Beam Web Design Strength, kips

70 ksi Weld Size, in.	ϕR_n, kips	Minimum Beam Web Thickness, in.
3/16	67.9	0.286
1/4	89.1	0.381
5/16	110	0.476
3/8	129	0.571

Support Design Strength per Inch Thickness, kips/in.

Hole Type	
STD/SSLT	687
OVS	644

STD = Standard holes
OVS = Oversized holes
SSLT = Short-slotted holes oriented transverse to direction of load
N = Threads included
X = Threads excluded
SC = Slip critical

*Limited to W10×12, 15, 17, 19, 22, 26, 30.

End Plate	Beam
F_y = 36 ksi	F_y = 50 ksi
F_u = 58 ksi	F_u = 65 ksi

1-in.-Diameter Bolts

2 Rows

W12, 10, 8

Table 10-4 (cont.). Bolted/Welded Shear End-Plate Connections

Bolt and End-Plate Design Strength, kips					
ASTM Desig.	Thread Cond.	Hole Type	End-Plate Thickness		
			$^1/_4$	$^5/_{16}$	$^3/_8$
A325/ F1852	N	–	42.4	53.0	63.6
	X	–	42.4	53.0	63.6
	SC Class A	STD	42.4	53.0	63.6
		OVS	37.5	46.9	56.3
		SSLT	42.4	53.0	63.6
	SC Class B	STD	42.4	53.0	63.6
		OVS	37.5	46.9	56.3
		SSLT	42.4	53.0	63.6
A490	N	–	42.4	53.0	63.6
	X	–	42.4	53.0	63.6
	SC Class A	STD	42.4	53.0	63.6
		OVS	37.5	46.9	56.3
		SSLT	42.4	53.0	63.6
	SC Class B	STD	42.4	53.0	63.6
		OVS	37.5	46.9	56.3
		SSLT	42.4	53.0	63.6

Weld (70 ksi) and Beam Web Design Strength, kips		
70 ksi Weld Size, in.	ϕR_n, kips	Minimum Beam Web Thickness, in.
$^3/_{16}$	42.8	0.286
$^1/_4$	55.7	0.381
$^5/_{16}$	67.9	0.476
$^3/_8$	79.4	0.571

Support Design Strength per Inch Thickness, kips/in.	
STD/SSLT	461
OVS	439

STD = Standard holes
OVS = Oversized holes
SSLT = Short-slotted holes oriented transverse to direction of load

N = Threads included
X = Threads excluded
SC = Slip critical

UNSTIFFENED SEATED CONNECTIONS

An unstiffened seated connection is made with a seat angle and a top angle, as illustrated in Figure 10-12. These angles may be bolted or welded to the supported beam as well as to the supporting member.

While the seat angle is assumed to carry the entire end reaction of the supported beam, the top angle must be placed as shown or in the optional side location for satisfactory performance and stability (Roeder and Dailey, 1989). The top angle and its connections are not usually sized for any calculated strength requirement. A $^1/_4$-in.-thick angle with a 4-in. vertical leg dimension will generally be adequate. It may be bolted with two bolts through each leg or welded with minimum-size welds to either the supported or the supporting members.

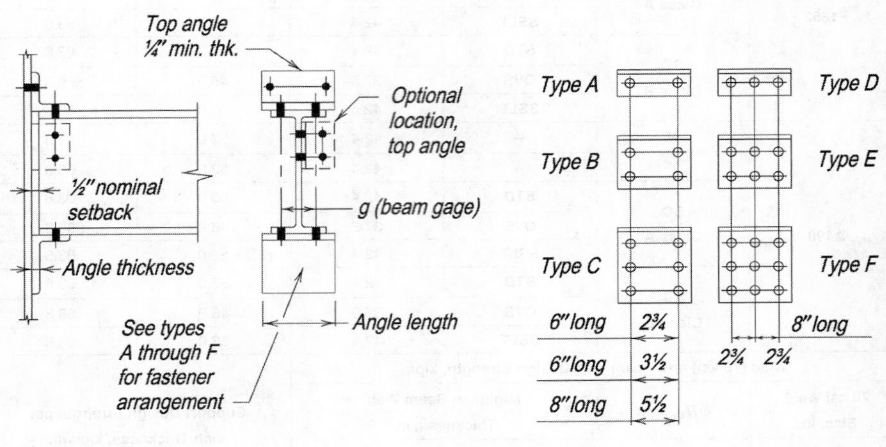

(a) All-bolted

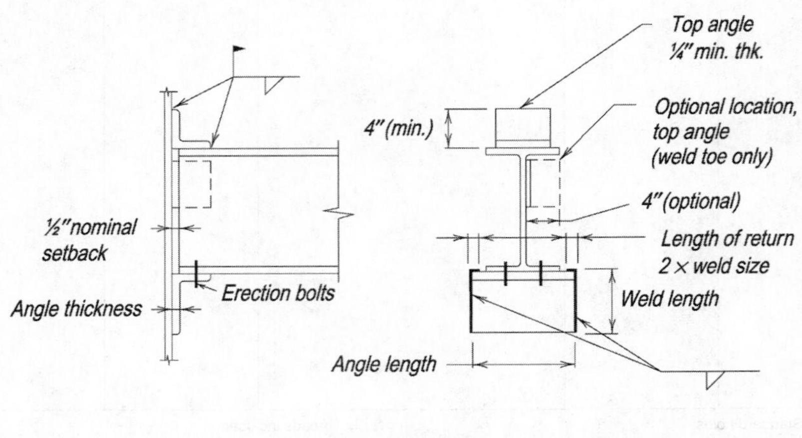

(b) All-welded

Fig. 10–12. Unstiffened seated connections.

When the top angle is welded to the support and/or the supported beam, adequate flexibility must be provided in the connection. As illustrated in Figure 10-12b, line welds are placed along the toe of each angle leg. Note that welding along the sides of the vertical angle leg must be avoided as it would inhibit the flexibility and, therefore, the necessary end rotation of the connection. The performance of such a connection would not be as intended for unstiffened seated connections.

Design Checks

The design strength of an unstiffened seated connection is determined from the applicable limit states for the bolts (see Part 7), welds (see Part 8) and connection elements (see Part 9). Additionally, the strength of the supported beam web must be checked for the limit states of local web yielding and web crippling. In all cases, the design strength ϕR_n must equal or exceed the required strength R_u.

The design strength for local web yielding ϕR_n is determined per LRFD Specification Section K1.3 using Equation K1-3, which is simplified using the constants in Table 9-6 as follows:

$$\phi R_n = \phi R_1 + N(\phi R_2)$$

Note that N must be equal to or greater than k, per LRFD Specification Section K1.3

The design strength for web crippling ϕR_n is determined per LRFD Specification Section K1.4 using Equation K1-5a or K1-5b, which are simplified using constants in Table 9-6 as follows:

For $N/d \leq 0.2$:

$$\phi R_n = \phi R_3 + N(\phi R_4)$$

For $N/d > 0.2$:

$$\phi R_n = \phi R_5 + N(\phi R_6)$$

For further information, see Carter et al. (1997).

Shop and Field Practices

Unstiffened seated connections may be made to the webs and flanges of supporting columns. If adequate clearance exists, unstiffened seated connections may also be made to the webs of supporting girders.

To provide for overrun in beam length, the nominal setback for the beam end is $^1/_2$ in. To provide for underrun in beam length, this setback is assumed to be $^3/_4$ in. for calculation purposes.

The seat angle is preferably shop attached to the support. Since the bottom flange typically establishes the plane of reference for seated connections, mill variation in beam depth may result in variation in the elevation of the top flange. Such variation is usually of no consequence with concrete slab and metal deck floors, but may be a concern when a grating or steel-plate floor is used. Unless special care is required, the usual mill tolerances for member depth of $^1/_8$ in. to $^1/_4$ in. are ignored. However, when the top angle is shop attached to the supported beam and field bolted to the support, mill variation in beam depth must be considered. Slotted holes, as illustrated in Figure 10-13a, will accommodate both overrun and underrun in the beam depth and are the preferred method for economy and convenience to both the fabricator and erector. Alternatively, the angle could be shipped loose with clearance provided as shown in Figure 10-13b. When the top angle is to be field welded to the support, no provision for mill variation in the beam depth is necessary.

When the top angle is shop attached to the support, an appropriate erection clearance is provided as required as illustrated in Figure 10-13c.

EXAMPLE 10.7. All-bolted unstiffened seated connection (beam-to-column web).

Given: Design an all-bolted unstiffened seated connection for a W16×50 beam to W14×90 column web connection

$$R_u = 55\,\text{kips}$$

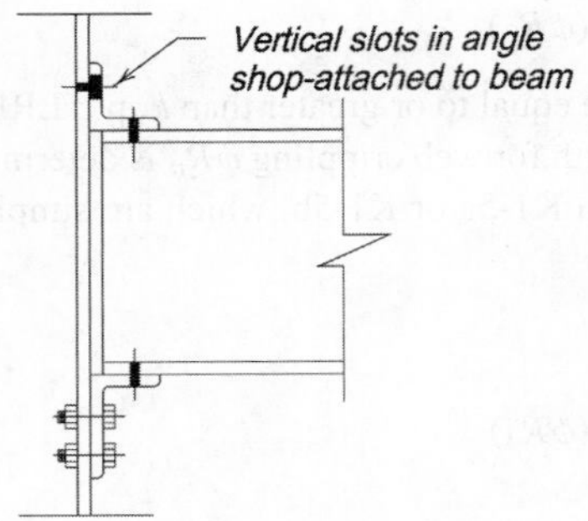

(a) Vertical slots

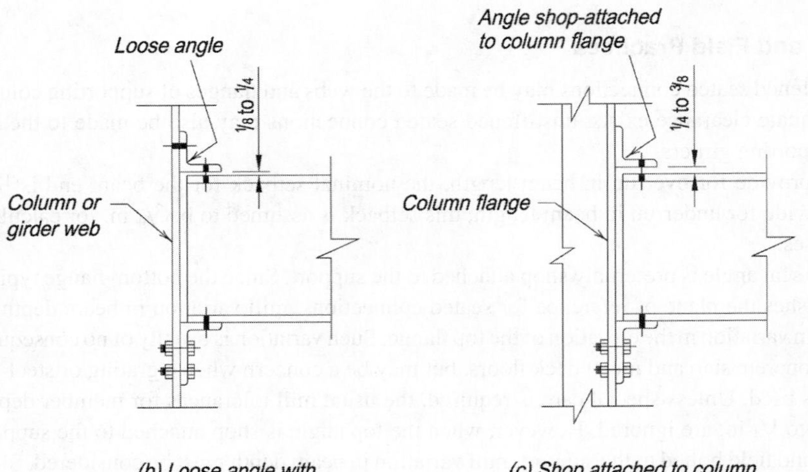

(b) Loose angle with clearance as shown

(c) Shop attached to column flange with clearance as shown

Fig. 10–13. Providing for variation in beam depth with seated connections.

W16×50, ASTM A992

$t_w = 0.380$ in. $\quad d = 16.3$ in. $\quad t_f = 0.630$ in. $\quad k = 1.03$ in.
$F_y = 50$ ksi, $\quad F_u = 65$ ksi

W14×90, ASTM A992

$t_w = 0.440$ in.
$F_y = 50$ ksi, $\quad F_u = 65$ ksi

Use $^7/_8$-in.-diameter A325-N bolts in standard holes. Assume ASTM A36 angle material with $F_y = 36$ ksi and $F_u = 58$ ksi.

Solution: *Design seat angle and bolts*

For local web yielding,

$$N_{min} = \frac{R_u - \phi R_1}{\phi R_2} \geq k$$
$$= \frac{55\text{ kips} - 44.9\text{ kips}}{13.7\text{ kips/in.}} \geq 1.03\text{ in.}$$
$$= 0.737\text{ in.} \geq 1.03\text{ in.}$$
$$= 1.03\text{ in.}$$

For web crippling,

When $\frac{N}{d} \leq 0.2$

$$N_{min} = \frac{R_u - \phi_r R_3}{\phi_r R_4}$$
$$= \frac{55\text{ kips} - 56.9\text{ kips}}{4.92\text{ kips/in.}}$$

which results in a negative quantity.

When $\frac{N}{d} > 0.2$

$$N_{min} = \frac{R_u - \phi_r R_5}{\phi_r R_6}$$
$$= \frac{55\text{ kips} - 51.6\text{ kips}}{6.56\text{ kips/in.}}$$
$$= 0.518\text{ in.}$$

Thus, $N_{min} = 1.03$ in.

From Table 10-5, an 8-in. angle length with a $^3/_4$-in. thickness and a $3^1/_2$-in. minimum outstanding leg will provide

$\phi R_n = 117\text{ kips} > 55\text{ kips}$ **o.k.**

Try L6×4×$^3/_4$, 8-in. long with $5^1/_2$-in. bolt gage.

For $^7/_8$-in.-diameter A325-N bolts, connection type B (four bolts) provides

$\phi R_n = 86.6\text{ kips} > 55\text{ kips}$ **o.k.**

The table indicates a 6×4×$^3/_4$ is available (4-in. OSL)

Check bolt bearing on the angle

The bearing strength per bolt from LRFD Specification Section J3.10 for the $^3/_4$-in. thick angle is

$$\begin{aligned}\phi R_n &= \phi(2.4dtF_u)\\ &= 0.75(2.4 \times {}^7/_8 \text{ in.} \times {}^3/_4 \text{ in.} \times 58 \text{ ksi})\\ &= 68.5 \text{ kips}\end{aligned}$$

Since this exceeds the strength of the bolts in single shear, bolt bearing is not critical.

Tabular values include checks of local yielding and web crippling strengths of beam web. **o.k.**

Use two $^7/_8$-in.-diameter A325-N bolts to connect the beam to the seat angle.

Select top angle and bolts

Use L4×4×$^1/_4$ with two $^7/_8$-in.-diameter A325-N bolts through each leg.

Check supporting column

The bearing strength per bolt from LRFD Specification Section J.3.10 for the W14×90 column web is

$$\begin{aligned}\phi R_n &= \phi(2.4dtF_u)\\ &= 0.75(2.4 \times {}^7/_8 \text{ in.} \times 0.440 \text{ in.} \times 65 \text{ ksi})\\ &= 45.0 \text{ kips}\end{aligned}$$

Since this exceeds the strength of the bolts in single shear, bolt bearing is not critical.

EXAMPLE 10.8. Bolted/welded unstiffened seated connection (beam-to-column flange).

Given: Design an unstiffened seated connection for a W21×62 beam to W14×61 column flange connection.

$$R_u = 55 \text{ kips}$$

W21×62, ASTM A992

$$t_w = 0.400 \text{ in.} \quad d = 21.0 \text{ in.} \quad t_f = 0.615 \text{ in.} \quad k = 1.12 \text{ in.}$$
$$F_y = 50 \text{ ksi}, \quad F_u = 65 \text{ ksi}$$

W14×61, ASTM A992

$$t_f = 0.645 \text{ in.}$$
$$F_y = 50 \text{ ksi}, F_u = 65 \text{ ksi}$$

Use $^3/_4$-in.-diameter A325-N bolts in standard holes to connect the supported beam to the seat and top angles. Use 70 ksi electrode welds to connect the seat and top angles to the column flange. Assume ASTM A36 angle material with $F_y = 36$ ksi and $F_u = 58$ ksi.

Solution: *Design seat angle and welds*

For local web yielding,

$$N_{\min} = \frac{R_u - \phi R_1}{\phi R_2} \geq k$$

$$= \frac{55\text{ kips} - 49.5\text{ kips}}{14.4\text{ kips/in.}} \geq 1.12\text{ in.}$$

$$= 0.382\text{ in.} \geq 1.12\text{ in.}$$

$$= 1.12\text{ in.}$$

For web crippling,

When $\frac{N}{d} \leq 0.2$

$$N_{\min} = \frac{R_u - \phi_r R_3}{\phi_r R_4}$$

$$= \frac{55\text{ kips} - 60.7\text{ kips}}{4.55\text{ kips/in.}}$$

which results in a negative quantity.

When $\frac{N}{d} > 0.2$

$$N_{\min} = \frac{R_u - \phi_r R_5}{\phi_r R_6}$$

$$= \frac{55\text{ kips} - 54.3\text{ kips}}{6.07\text{ kips/in.}}$$

$$= 0.115\text{ in.}$$

Thus, $N_{\min} = 1.12$ in.

From Table 10-5, an 8-in.-angle length with a $^3/_4$-in. thickness and a $3^1/_2$-in. minimum outstanding leg will provide

$\phi R_n = 117\text{ kips} > 55\text{ kips}$ **o.k.**

Tabular values include checks of local yielding and web crippling strengths of beam web. **o.k.**

Try L8×4×$^3/_4$, 8 in. long with $^5/_{16}$-in. fillet welds.

For weld strength, from Table 10-6,

$\phi R_n = 66.8\text{ kips} > 55\text{ kips}$ **o.k.**

Use two $^3/_4$-in.-diameter A325-N bolts to connect the beam to the seat angle.

Select top angle, bolts, and welds

Use L4×4×1/4 with two 3/4-in.-diameter A325-N bolts through the supported-beam leg of the angle. Use 3/16-in. fillet weld along the toe of the angle (minimum size from LRFD Specification Table J2.4).

Table 10-5. All-Bolted Unstiffened Seated Connections

Table 10-5 is a design aid for all-bolted unstiffened seats. Seat design strengths are tabulated, assuming a 4-in. outstanding leg, for angle material with $F_y = 36$ ksi and $F_u = 58$ ksi and beam material with $F_y = 50$ ksi and $F_u = 65$ ksi. All values are for comparison with factored loads.

Tabulated seat design strengths consider the limit states of shear yielding and flexural yielding of the outstanding angle leg, and local yielding and crippling of the beam web. A nominal beam setback of 1/2 in. is assumed in these tables. However, this setback is increased to 3/4 in. for calculation purposes in determining the tabulated values to account for the possibility of underrun in beam length.

Bolt design strengths are tabulated for the seat types illustrated in Figure 10-12a with 3/4-in., 7/8-in., and 1-in.-diameter ASTM A325, F1852 and A490 bolts. Vertical spacing of bolts and gages in seat angles may be arranged to suit conditions, provided the edge distance and spacing requirements in LRFD Specification Section J3 are met. Where thick angles are used, larger entering and tightening clearances may be required in the outstanding angle leg. The suitability of angle sizes and thicknesses for the seat types illustrated in Figure 10-12a is also listed in Table 10-5.

Bolted/Welded Unstiffened Seated Connections

Tables 10-5 and 10-6 may be used in combination to design unstiffened seated connections that are welded to the supporting member and bolted to the supported beam, or bolted to the supporting member and welded to the supported beam.

Table 10-6. All-Welded Unstiffened Seated Connections

Table 10-6 is a design aid for all-welded unstiffened seats (exception: the beam is bolted to the seat). Seat design strengths are tabulated, assuming either a 3 1/2-in. or 4-in. outstanding leg (as indicated in the table), for angle material with $F_y = 36$ ksi and $F_u = 58$ ksi and beam material with $F_y = 50$ ksi and $F_u = 65$ ksi. Electrode strength is assumed to be 70 ksi. All values are for comparison with factored loads.

Tabulated seat design strengths consider the limit states of shear yielding and flexural yielding of the outstanding angle leg, and local yielding and crippling of the beam web. A nominal beam setback of 1/2 in. is assumed in these tables. However, this setback is increased to 3/4 in. for calculation purposes in determining the tabulated values to account for the possibility of underrun in beam length.

Weld design strengths are tabulated using the elastic method. The minimum and maximum angle thickness for each case is also tabulated. While these tabular values are based upon 70 ksi electrodes, they may be used for other electrodes, provided the tabular values are adjusted for the electrodes used (e.g., for 60 ksi electrodes, multiply the tabular values by $60/70 = 0.866$, etc.) and the welds and base metal meet the required strength level provisions of LRFD Specification Section J2. Should combinations of material thickness and weld size selected from Table 10-6 exceed the limits in LRFD Specification Section J2.2, the weld size or material thickness should be increased as required.

As can be seen from the following, reduction of the tabulated weld strength is not normally required when unstiffened seats line up on opposite sides of the supporting web. From Salmon and Johnson (1996), the design strength of the welds to the support is ϕR_n, where

$$\phi R_n = 2 \times \frac{1.392DL}{\sqrt{1 + \frac{20.25e^2}{L^2}}}$$

where

D = number of sixteenths-of-an-inch in the weld size.
L = vertical leg dimension of the seat angle, in.
e = eccentricity of the beam end reaction with respect to the weld lines, in.

The term in the denominator that accounts for the eccentricity e increases the weld size far beyond what is required for shear alone, but with seats on both sides of the supporting member web, the forces due to eccentricity react against each other and have no effect on the web. Furthermore, as illustrated in Figure 10-14, there are actually two shear planes per weld, one at each weld toe and heel for a total of four shear planes. Thus, for an 8-in.-long 7×4×1 seat angle supporting factored reaction of 70 kips, the minimum support thickness would be

$$\frac{70 \text{ kips}}{0.75 \times 0.6 \times 65 \text{ ksi} \times 7 \text{ in.} \times 4 \text{ planes}} = 0.0855 \text{ in.}$$

For the identical connection on both sides of the support, the minimum support thickness would be less than $^3/_{16}$ in. Thus, supporting web thickness is generally not a concern.

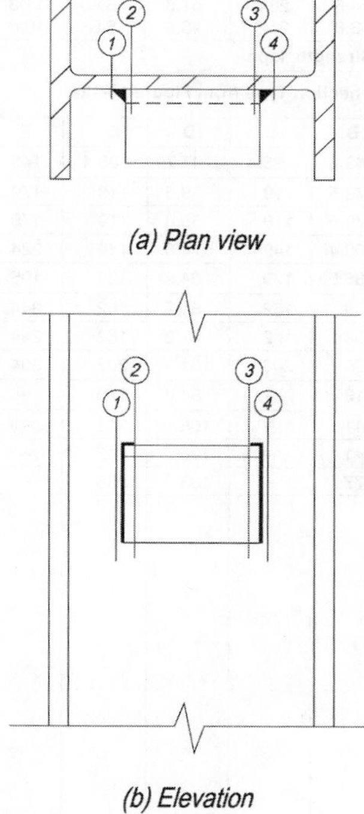

Fig. 10–14. Shear planes in column web for unstiffened seated connections.

Angle $F_y = 36$ ksi

Table 10-5.
All-Bolted Unstiffened Seated Connections

Outstanding Angle Leg Design Strength, kips

Required Bearing Length N_{req}, in.	Angle Length, in. 6					8					Min. Angle Leg, in.
	Angle Thickness, in.										
	3/8	1/2	5/8	3/4	1	3/8	1/2	5/8	3/4	1	
1/2	27.3					36.4					
9/16	24.3					32.4					
5/8	21.9	58.3				29.2	77.8				
11/16	19.9	55.5				26.5	74.1				
3/4	18.2	48.6				24.3	64.8				
13/16	16.8	43.2				22.4	57.6				
7/8	15.6	38.9				20.8	51.8				
15/16	14.6	35.3				19.4	47.1				
1	13.7	32.4	72.9			18.2	43.2	97.2			
1 1/16	12.9	29.9	67.5			17.2	39.9	90.0			
1 1/8	12.1	27.8	60.8			16.2	37.0	81.0			
1 3/16	11.5	25.9	55.2			15.3	34.6	73.6			
1 1/4	10.9	24.3	50.6			14.6	32.4	67.5			
1 5/16	10.4	22.9	46.7			13.9	30.5	62.3			3 1/2
1 3/8	9.94	21.6	43.4	87.5		13.3	28.8	57.9	117		
1 7/16	9.51	20.5	40.5	79.5		12.7	27.3	54.0	106		
1 1/2	9.11	19.4	38.0	72.9		12.1	25.9	50.6	97.2		
1 5/8	8.41	17.7	33.8	62.5		11.2	23.6	45.0	83.3		
1 3/4	7.81	16.2	30.4	54.7		10.4	21.6	40.5	72.9		
1 7/8	7.29	15.0	27.6	48.6		9.72	19.9	36.8	64.8		
2	6.83	13.9	25.3	43.7	117	9.11	18.5	33.8	58.3	156	
2 1/8	6.43	13.0	23.4	39.8	111	8.58	17.3	31.2	53.0	148	
2 1/4	6.07	12.1	21.7	36.4	97.2	8.10	16.2	28.9	48.6	130	
2 3/8	5.76	11.4	20.3	33.6	86.4	7.67	15.2	27.0	44.9	115	
2 1/2	5.47	10.8	19.0	31.2	77.8	7.29	14.4	25.3	41.7	104	
2 5/8	5.21	10.2	17.9	29.2	70.7	6.94	13.6	23.8	38.9	94.3	
2 3/4	4.97	9.72	16.9	27.3	64.8	6.63	13.0	22.5	36.4	86.4	
2 7/8	4.75	9.26	16.0	25.7	59.8	6.34	12.3	21.3	34.3	79.8	
3	4.56	8.84	15.2	24.3	55.5	6.07	11.8	20.3	32.4	74.1	
3 1/8	4.37	8.45	14.5	23.0	51.8	5.83	11.3	19.3	30.7	69.1	4
3 1/4	4.21	8.10	13.8	21.9	48.6	5.61	10.8	18.4	29.2	64.8	

Bolt Design Strength, kips

Bolt Dia., in.	ASTM Desig.	Thread Cond.	Connection Type from Figure 10-12a A	B	C	D	E	F
3/4	A325/F1852	N	31.8	63.6	95.4	47.7	95.4	143
		X	39.8	79.5	119	59.6	119	179
	A490	N	39.8	79.5	119	59.6	119	179
		X	49.7	99.4	149	74.6	149	224
7/8	A325/F1852	N	43.3	86.6	130	64.9	130	195
		X	54.1	108	162	81.2	162	244
	A490	N	54.1	108	162	81.2	162	244
		X	67.6	135	203	101	203	304
1	A325/F1852	N	56.5	113	–	84.8	170	–
		X	70.7	141	–	106	212	–
	A490	N	70.7	141	–	106	212	–
		X	88.4	177	–	133	265	–

Available Angles

Connection Type	Angle Size	t, in.
A, D	4×3	3/8 – 1/2
	4×3 1/2	3/8 – 1/2
	4×4	3/8 – 3/4
B, E	6×4	3/8 – 3/4
	7×4	3/8 – 3/4
	8×4	1/2 – 1
C, F[b]	8×4	1/2 – 1

[b] Not suitable for use with 1-in.-diameter bolts.

For tabulated values above the heavy line, shear yielding of the angle leg controls the design strength.

Table 10-6. All-Welded Unstiffened Seated Connections

Angle $F_y = 36$ ksi

Outstanding Angle Leg Design Strength, kips

Required Bearing Length N_{req}, in.	Angle Length 6 in. — Thickness 3/8	1/2	5/8	3/4	1	Angle Length 8 in. — Thickness 3/8	1/2	5/8	3/4	1	Min. Angle Leg, in.
1/2	27.3					36.4					
9/16	24.3					32.4					
5/8	21.9	58.3				29.2	77.8				
11/16	19.9	55.5				26.5	74.1				
3/4	18.2	48.6				24.3	64.8				
13/16	16.8	43.2				22.4	57.6				
7/8	15.6	38.9				20.8	51.8				
15/16	14.6	35.3				19.4	47.1				
1	13.7	32.4	72.9			18.2	43.2	97.2			
1 1/16	12.9	29.9	67.5			17.2	39.9	90.0			
1 1/8	12.1	27.8	60.8			16.2	37.0	81.0			
1 3/16	11.5	25.9	55.2			15.3	34.6	73.6			
1 1/4	10.9	24.3	50.6			14.6	32.4	67.5			
1 5/16	10.4	22.9	46.7			13.9	30.5	62.3			
1 3/8	9.94	21.6	43.4	87.5		13.3	28.8	57.9	117		3 1/2
1 7/16	9.51	20.5	40.5	79.5		12.7	27.3	54.0	106		
1 1/2	9.11	19.4	38.0	72.9		12.1	25.9	50.6	97.2		
1 5/8	8.41	17.7	33.8	62.5		11.2	23.6	45.0	83.3		
1 3/4	7.81	16.2	30.4	54.7		10.4	21.6	40.5	72.9		
1 7/8	7.29	15.0	27.6	48.6		9.72	19.9	36.8	64.8		
2	6.83	13.9	25.3	43.7	117	9.11	18.5	33.8	58.3	156	
2 1/8	6.43	13.0	23.4	39.8	111	8.58	17.3	31.2	53.0	148	
2 1/4	6.07	12.1	21.7	36.4	97.2	8.10	16.2	28.9	48.6	130	
2 3/8	5.76	11.4	20.3	33.6	86.4	7.67	15.2	27.0	44.9	115	
2 1/2	5.47	10.8	19.0	31.2	77.8	7.29	14.4	25.3	41.7	104	
2 5/8	5.21	10.2	17.9	29.2	70.7	6.94	13.6	23.8	38.9	94.3	
2 3/4	4.97	9.72	16.9	27.3	64.8	6.63	13.0	22.5	36.4	86.4	
2 7/8	4.75	9.26	16.0	25.7	59.8	6.34	12.3	21.3	34.3	79.8	
3	4.56	8.84	15.2	24.3	55.5	6.07	11.8	20.3	32.4	74.1	
3 1/8	4.37	8.45	14.5	23.0	51.8	5.83	11.3	19.3	30.7	69.1	4
3 1/4	4.21	8.10	13.8	21.9	48.6	5.61	10.8	18.4	29.2	64.8	

Weld (70 ksi) Design Strength, kips

70 ksi Weld Size, in.	Seat Angle Size (long leg vertical) 4 × 3 1/2	5 × 3 1/2	6 × 4	7 × 4	8 × 4
1/4	17.2	25.8	32.7	42.7	53.4
5/16	21.5	32.3	40.9	53.4	66.7
3/8	25.8	38.7	49.1	64.1	80.1
7/16	30.1	45.2	57.2	74.7	93.4
1/2	–	51.6	65.4	85.4	107
9/16	–	58.1	73.6	96.1	120
5/8	–	64.5	81.8	107	133
11/16	–	71.0	90.0	117	147
Available Angle Thickness, in.					
Minimum	3/8	3/8	3/8	3/8	1/2
Maximum	1/2	3/4	3/4	3/4	1

For tabulated values above the heavy line, shear yielding of the angle leg controls the design strength.

STIFFENED SEATED CONNECTIONS

A stiffened seated connection is made with a seat plate and stiffening element (e.g., a plate, structural tee or pair of angles) and a top angle, as illustrated in Figure 10-15. The top angle may be bolted or welded to the supported beam as well as to the supporting member and the

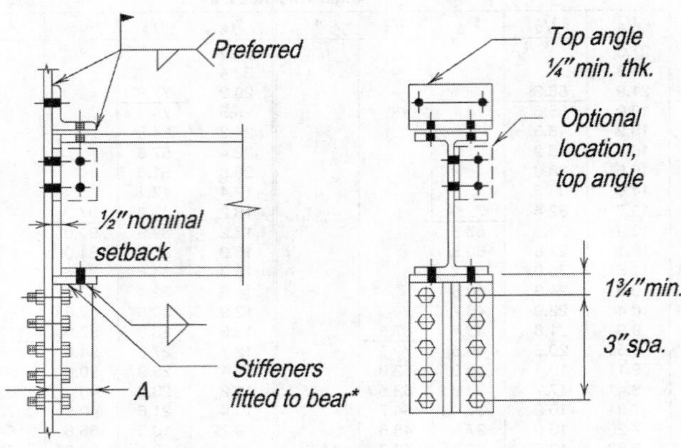

(a) All-bolted

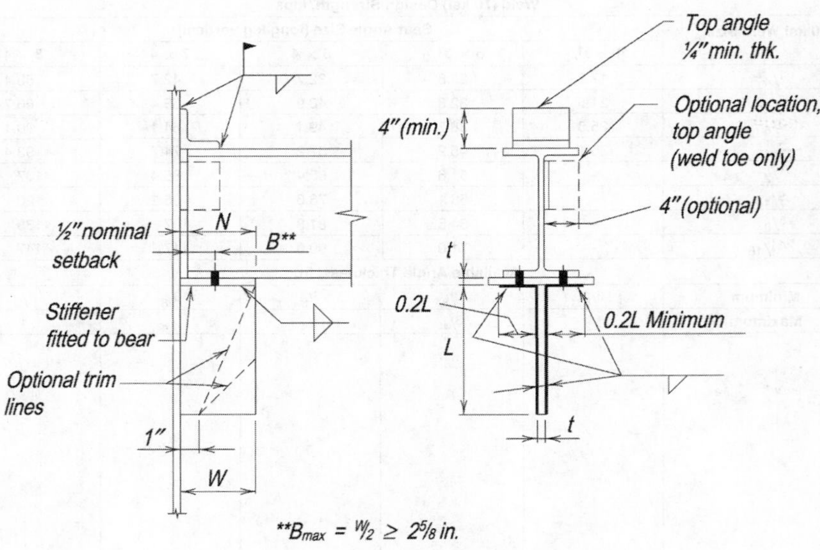

(b) Bolted/welded

Fig. 10–15. Stiffened seated connections.

stiffening element may be bolted or welded to the support. The seat plate should be bolted to the supported beam.

While the stiffening element is assumed to carry the entire end reaction of the supported beam, the top angle must be placed as shown or in the optional side location for satisfactory performance and stability (Roeder and Dailey, 1989). The top angle and its connections are not usually sized for any calculated strength requirement. A $^1/_4$-in.-thick angle with a 4-in. vertical leg dimension will generally be adequate. It may be bolted with two bolts through each leg or welded with minimum-size welds to either the supported or the supporting members.

When the top angle is welded to the support and/or the supported beam, adequate flexibility must be provided in the connection. As illustrated in Figure 10-12b, line welds are placed along the toe of each angle leg. Note that welding along the sides of the vertical angle leg must be avoided as it would inhibit the flexibility and, therefore, the necessary end rotation of the connection. The performance of such a connection would not be as intended for stiffened seated connections.

Design Checks

The design strength of a stiffened seated connection is determined from the applicable limit states for the bolts (see Part 7), welds (see Part 8) and connection elements (see Part 9). Additionally, the strength of the supported beam web must be checked for the limit states of local web yielding and web crippling. In all cases, the design strength ϕR_n must equal or exceed the required strength R_u.

The design strength for local web yielding ϕR_n is determined per LRFD Specification Section K1.3 using Equation K1-3, which is simplified using constants in Table 9-6 as follows:

$$\phi R_n = \phi R_1 + N(\phi R_2)$$

Note that N must be equal to or greater than k, per LRFD Specification Section K1.3.

The design strength for web crippling ϕR_n is determined per LRFD Specification Section K1.4 using Equation K1-5a or K1-5b, which are simplified using constants in Table 9-6 as follows:

For $N/d \leq 0.2$:

$$\phi R_n = \phi R_3 + N\,(\phi R_4)$$

For $N/d > 0.2$:

$$\phi R_n = \phi R_5 + N\,(\phi R_6)$$

When stiffened seated connections such as the one shown in Figure 10-15b are made to one side of a supporting column web, the column web may also need to be investigated for resistance to punching. In lieu of a more detailed analysis, Sputo and Ellifritt (1991) showed that punching will not be critical if the design parameters below and those summarized graphically in Figure 10-15b are met.

1. This simplified approach is applicable to the following column sections:

   W14×43-808 W12×40-336 W10×33-112
   W8×24-67 W6×20-25 W5×16-19

2. The supported beam must be bolted to the seat plate with high-strength bolts to account for the prying action caused by rotation of the connection. Welding the beam to the seat plate is not recommended because welds may lack the required strength and ductility. The centerline of the bolts should be located no more than the greater of $W/2$ or $2^5/_8$ in. from the column web face.
3. For seated connections where $W = 8$ in. or $W = 9$ in. and $3^1/_2$ in. $< B \leq W/2$, or where $W = 7$ in. and 3 in. $< B \leq W/2$ for a W14×43 column, refer to Sputo and Ellifritt (1991). These limitations are summarized at the bottom of Table 10-8.
4. The top angle may be bolted or welded, but must have a minimum $^1/_4$-in. thickness.
5. The seat plate should not be welded to the beam flange.

See also Ellifritt and Sputo (1999).

Shop and Field Practices

The comments for unstiffened seated connections are equally applicable to stiffened seated connections.

EXAMPLE 10.9. Stiffened seated connection (beam-to-column flange).

Given: Design a stiffened seated connection for a W21×68 beam to W14×90 column flange connection.

$$R_u = 125 \text{ kips}$$

W21×68, ASTM A992

$$t_w = 0.430 \text{ in.} \qquad d = 21.1 \text{ in.} \qquad t_f = 0.685 \text{ in.}$$
$$F_y = 50 \text{ ksi}, \qquad F_u = 65 \text{ ksi}$$

W14×90, ASTM A992

$$t_f = 0.710 \text{ in.}$$
$$F_y = 50 \text{ ksi}, \qquad F_u = 65 \text{ ksi}$$

Use $^3/_4$-in. diameter A325-N bolts in standard holes to connect the supported beam to the seat plate and top angle. Use 70 ksi electrode welds to connect the stiffener and top angle to the column flange. Assume ASTM A36 angle and plate material with $F_y = 36$ ksi and $F_u = 58$ ksi.

Solution: *Determine stiffener width W required for web crippling and local web yielding*

For web crippling, assume $N/d > 0.2$ and use constants ϕR_5 and ϕR_6 from Table 9-6.

$$W_{min} = \frac{R_u - \phi R_5}{\phi R_6} + \text{setback}$$
$$= \frac{125 \text{ kips} - 75.8 \text{ kips}}{7.92 \text{ kips/in.}} + {}^3/_4\text{-in.}$$
$$= 6.96 \text{ in.}$$

For local web yielding, use constants ϕR_1 and ϕR_2 from Table 9-6.

$$W_{min} = \frac{R_u - \phi R_1}{\phi R_2} + \text{setback}$$
$$= \frac{125\text{ kips} - 77.3\text{ kips}}{21.5\text{ kips/in.}} + {}^3\!/_4\text{-in.}$$
$$= 2.97\text{ in.}$$

The minimum stiffener width W for web crippling controls. Thus, use $W = 7$ in.

Check assumption

$$\frac{N}{d} = \frac{6.71\text{ in.} - {}^1\!/_2\text{ in.}}{21.1\text{ in.}}$$
$$= 0.294 > 0.2 \qquad \textbf{o.k.}$$

Determine stiffener length L and stiffener to column flange weld size

From Table 10-8, a stiffener with $L = 15$ in. and ${}^5\!/_{16}$-in. weld size provides

$$\phi R_n = 139\text{ kips} > 125\text{ kips} \qquad \textbf{o.k.}$$

Determine weld requirements for seat plate

Using ${}^5\!/_{16}$-in. fillet welds the minimum length of seat-plate-to-column-flange weld on each side of the stiffener is $0.2(L) = 3$ in. Use three inches of weld on each side of the stiffener. This also establishes the minimum weld between the seat plate and stiffener; use three inches of ${}^5\!/_{16}$-in. weld on both sides of the stiffener.

Determine seat plate dimensions

To accommodate two ${}^3\!/_4$ in. diameter A325-N bolts on a $5{}^1\!/_2$-in. gage connecting the beam flange to the seat plate, a width of eight inches is adequate. This is greater than the width required to accommodate the seat-plate-to-column-flange welds.

Use PL ${}^3\!/_8$ in.×7 in.×8 in. for the seat plate.

Determine stiffener plate thickness

To develop the stiffener-to-seat-plate welds, the minimum stiffener thickness is

$$t_{min} = 2({}^5\!/_{16}\text{ in.})$$
$$= {}^5\!/_8\text{ in.}$$

For a stiffener with $F_y = 36$ ksi and beam with $F_y = 50$ ksi, the minimum stiffener thickness is

$$t_{min} = 1.4t_w$$
$$= 1.4(0.430\text{ in.})$$
$$= 0.602\text{ in.}$$

The former controls; use PL$^5/_8$ in.×7 in.×15 in. for the stiffener.

Select top angle, bolts, and welds

Use L4×4×$^1/_4$ with two $^3/_4$-in.-diameter A325-N bolts through the supported-beam leg of the angle. Use $^1/_4$-in. fillet weld along the toe of the support leg of the angle (minimum size from LRFD Specification Table J2.4).

EXAMPLE 10.10. Stiffened seated connection (beam-to-column web).

Given:

Design a stiffened seated connection for a W21×68 beam to W14×90 column web connection.

$R_u = 125$ kips

W21×68, ASTM A992

$t_w = 0.430 \quad d = 21.13$ in. $\quad t_f = 0.685$ in.
$F_y = 50$ ksi, $\quad F_u = 65$ ksi

W14×90, ASTM A992

$t_w = 0.440$
$F_y = 50$ ksi, $F_u = 65$ ksi

Use $^3/_4$-in.-diameter A325-N bolts in standard holes to connect the supported beam to the seat plate and top angle. Use 70 ksi electrode welds to connect the stiffener and top angle to the column web. Assume ASTM A36 angle and plate material with $F_y = 36$ ksi and $F_u = 58$ ksi.

Solution:

Determine stiffener width W

As calculated previously in Example 10-9, use $W = 7$ in.

Determine stiffener length L and stiffener to column web weld size

As calculated previously in Example 10-9, use $L = 15$ in. and $^5/_{16}$-in. weld size.

Determine weld requirements for seat plate

As calculated previously in Example 10-9, use three inches of $^1/_4$-in. weld on both sides of the seat plate for the seat-plate-to-column-web welds and for the seat-plate-to-stiffener welds.

Determine seat plate dimensions

For a column-web support, from Table 10-8, the maximum distance from the face to the support to the line of bolts between the beam flange and seat plate is 3$^1/_2$ in. The PL$^3/_8$ in.×7 in.×8 in. chosen previously in Example 10-9 will accommodate these bolts.

Determine stiffener plate thickness

As calculated previously in Example 10-9, use PL$^5/_8$ in.×7 in.×15 in.

Select top angle, bolts, and welds

Use L4×4×$^1/_4$ with two $^3/_4$-in.-diameter A325-N bolts through the supported-beam leg of the angle. Use $^3/_{16}$ in. fillet weld along the toe of the support leg of the angle (minimum size from LRFD Specification Table J2.4).

Check column web

From Table 10-8, no limitation is placed on column web. Therefore, column web is **o.k.**

Table 10-7. All-Bolted Stiffened Seated Connections

Table 10-7 is a design aid for all-bolted stiffened seats. Stiffener design strengths are tabulated for stiffener material with $F_y = 36$ ksi and $F_u = 58$ ksi and with $F_y = 50$ ksi and $F_u = 65$ ksi. All values are for comparison with factored loads.

Tabulated values consider the limit state of bearing on the stiffening material. The designer must independently check the design strength of the beam web based upon the limit states of local web yielding and web crippling. A nominal beam setback of $^1/_2$ in. is assumed in these tables. However, this setback is increased to $^3/_4$ in. for calculation purposes in determining the tabulated values to account for the possibility of underrun in beam length.

Bolt design strengths are tabulated for two vertical rows of from three to seven $^3/_4$-in., $^7/_8$-in., and 1-in.-diameter ASTM A325, F1852 and A490 high-strength bolts based upon the limit state of bolt shear. Vertical spacing of bolts and gages in seat angles may be arranged to suit conditions, provided the edge distance and spacing requirements in LRFD Specification Section J3 are met.

Table 10-8. Bolted/Welded Stiffened Seated Connections

Table 10-8 is a design aid for stiffened seated connections welded to the support and bolted to the supported beam. Electrode strength is assumed to be 70 ksi. All values are for comparison with factored loads.

Weld design strengths are tabulated using the elastic method. While these tabular values are based upon 70 ksi electrodes, they may be used for other electrodes, provided the tabular values are adjusted for the electrodes used (e.g., for 60 ksi electrodes, multiply the tabular values by 60/70 = 0.866, etc.) and the weld and base metal meet the provisions of LRFD Specification Section J2.

The thickness of the horizontal seat plate or tee flange should not be less than $^3/_8$ in. If the seat and stiffener are built up from separate plates, the stiffener should be finished to bear under the seat. The welds connecting the two plates should have a strength equal to or greater than the horizontal welds to the support under the seat plate.

The designer must independently check the beam web for local web yielding and web crippling. The nominal beam setback of $^1/_2$ in. should be assumed to be $^3/_4$ in. for calculation purposes to account for possible underrun in beam length.

The stiffener thickness may be conservatively determined as follows. The minimum stiffener plate thickness t for supported beams with unstiffened webs should be the supported beam web thickness t_w multiplied by the ratio of F_y of the beam material to F_y of the stiffener material (e.g., F_y beam = 50 ksi, F_y stiffener = 36 ksi, $t = t_w \times 50/36$ minimum). Additionally, the minimum stiffener thickness t should be at least $2w$ for stiffener material with $F_y = 36$ ksi or $1.5w$ for stiffener material with $F_y = 50$ ksi, where w is the weld size for 70 ksi electrodes.

For 70 ksi electrodes, the minimum column web thickness is

$$t_{min} = \frac{3.09D}{F_u}$$

When welds line up on opposite sides of the support, the minimum thickness is the sum of the thicknesses required for each weld. In either case, when less than the minimum material thickness is present, the weld design strength must be reduced by the ratio of the thickness provided to the minimum thickness. As with unstiffened seated connections, the contribution of eccentricity to the required shear yielding strength is negligible. Should combinations of material thickness and weld size selected from Table 10-8 exceed the limits of LRFD Specification Section J2, increase the weld size or material thickness as required.

Table 10-7. All-Bolted Stiffened Seated Connections

Stiffener Angle Design Strength, kips[a]

Stiffener Material		$F_y = 36$ ksi $\phi R_n = 0.75\ (1.8 \times 36)\ A_{pb}$			$F_y = 50$ ksi $\phi R_n = 0.75\ (1.8 \times 50)\ A_{pb}$		
Stiffener Outstanding Leg A, in.[b]		3 1/2	4	5	3 1/2	4	5
Thickness of Stiffener Outstanding Legs, in.	5/16	83.5	98.7	129	116	137	179
	3/8	100	118	155	139	165	215
	1/2	134	158	207	186	219	287
	5/8	167	197	258	232	274	359
	3/4	200	237	310	278	329	430

Use minimum 3/8-in.-thick seat plate wide enough to extend beyond outstanding legs of stiffener.

[a] See LRFD Specification Sect. J8.

[b] Beam bearing length assumed 3/4 in. less for calculation purposes.

Bolt Design Strength, kips

Bolt Diameter, in.	ASTM Desig.	Thread Cond.	Number of Bolts in One Vertical Row				
			3	4	5	6	7
3/4	A325/F1852	N	95.4	127	159	191	223
		X	119	159	199	239	278
	A490	N	119	159	199	239	278
		X	149	199	249	298	348
7/8	A325/F1852	N	130	173	216	260	303
		X	162	216	271	325	379
	A490	N	162	216	271	325	379
		X	203	271	338	406	474
1	A325/F1852	N	170	226	283	339	396
		X	212	283	353	424	495
	A490	N	212	283	353	424	495
		X	265	353	442	530	619

$\phi = 0.75$

Table 10-8. Bolted/Welded Stiffened Seated Connections

L, in.	Width of Seat W, in.											
	4				5				6			
	70 ksi Weld Size, in.				70 ksi Weld Size, in.				70 ksi Weld Size, in.			
	1/4	5/16	3/8	7/16	5/16	3/8	7/16	1/2	5/16	3/8	7/16	1/2
6	34.0	42.5	51.1	59.6	35.2	42.2	49.3	56.3	29.9	35.9	41.9	47.8
7	44.9	56.1	67.3	78.6	46.9	56.2	65.6	75.0	40.1	48.1	56.1	64.1
8	56.7	70.8	85.0	99.2	59.8	71.7	83.7	95.6	51.4	61.7	72.0	82.2
9	69.2	86.5	104	121	73.7	88.5	103	118	63.8	76.6	89.3	102
10	82.3	103	123	144	88.5	106	124	142	77.2	92.6	108	123
11	95.8	120	144	168	104	125	146	167	91.3	110	128	146
12	110	137	165	192	120	144	168	192	106	127	149	170
13	124	155	186	217	137	164	192	219	122	146	170	195
14	138	173	207	242	154	185	216	246	138	165	193	220
15	152	191	229	267	171	206	240	274	154	185	216	247
16	167	209	250	292	189	227	265	302	171	205	240	274
17	181	227	272	318	207	248	290	331	188	226	264	301
18	196	245	294	343	225	270	315	360	206	247	288	329
19	211	263	316	369	243	291	340	388	223	268	313	357
20	225	281	338	394	261	313	365	417	241	289	337	386
21	240	300	359	419	279	335	391	446	259	311	362	414
22	254	318	381	445	297	357	416	476	277	332	388	443
23	269	336	403	470	315	378	442	505	295	354	413	472
24	283	354	425	495	334	400	467	534	313	376	438	501
25	297	372	446	520	352	422	492	563	331	397	464	530
26	312	390	468	546	370	444	518	592	349	419	489	559
27	326	408	489	571	388	466	543	621	368	441	515	588
Limitations for Connections to Column Webs												
B	2 5/8-in. max.				2 5/8-in. max.				3-in. max.			
	W12×40, W14×43 for L ≥ 9 in. limit weld ≤ 1/4 in.											

Notes:

1. Values shown assume 70 ksi electrodes. For 60 ksi electrodes, multiply tabular values by 0.857, or enter table with 1.17 times the required strength R_u. For 80 ksi electrodes, multiply tabular values by 1.14, or enter table with 0.875 times the required strength R_u.
2. Tabulated values are valid for stiffeners with minimum thickness of

$$t_{min} = \frac{F_{y\,beam}}{F_{y\,stiffener}} \times t_w$$

but not less than $2w$ for stiffeners with $F_y = 36$ ksi nor $1.5w$ for stiffeners with $F_y = 50$ ksi. In the above, t_w is the thickness of the unstiffened supported beam web and w is the nominal weld size.

3. Tabulated values may be limited by shear yielding of or bearing on the stiffener; refer to LRFD Specification Sections F2.2 and J8, respectively.

Table 10-8 (cont.). Bolted/Welded Stiffened Seated Connections

$\phi = 0.75$

L, in.	Width of Seat W, in.											
	7				8				9			
	70 ksi Weld Size, in.				70 ksi Weld Size, in.				70 ksi Weld Size, in.			
	5/16	3/8	7/16	1/2	5/16	3/8	1/2	5/8	5/16	3/8	1/2	5/8
11	81.0	97.2	113	130	72.5	87.1	116	145	65.6	78.7	105	131
12	94.7	114	133	151	85.1	102	136	170	77.1	92.5	123	154
13	109	131	153	174	98.3	118	157	197	89.3	107	143	179
14	124	149	174	198	112	135	180	224	102	123	164	204
15	139	167	195	223	127	152	203	253	116	139	185	232
16	155	186	217	249	142	170	227	283	130	156	208	260
17	172	206	240	275	157	189	251	314	144	173	231	289
18	188	226	264	301	173	208	277	346	159	191	255	319
19	205	246	287	329	189	227	303	378	175	210	280	350
20	223	267	312	356	206	247	329	411	191	229	305	381
21	240	288	336	384	222	267	356	445	207	248	331	413
22	258	309	361	412	240	287	383	479	223	268	357	446
23	275	330	385	440	257	308	411	514	240	288	384	480
24	293	352	410	469	274	329	439	548	257	308	411	513
25	311	373	435	498	292	350	467	584	274	329	438	548
26	329	395	461	526	309	371	495	619	291	349	466	582
27	347	417	486	555	327	393	524	655	308	370	494	617
28	365	438	511	584	345	414	552	690	326	391	522	652
29	383	460	537	613	363	436	581	726	344	412	550	687
30	402	482	562	643	381	457	610	762	362	434	578	723
31	420	504	588	672	399	479	639	799	379	455	607	759
32	438	526	613	701	417	501	668	835	397	477	636	795

Limitations for Connections to Column Webs

B	3 1/2-in. max.	3 1/2-in. max.	3 1/2-in. max.
	W14×43, limit B ≤ 3 in. See item 3 in preceding discussion "Design Checks"	See item 3 in preceding discussion "Design Checks"	See item 3 in preceding discussion "Design Checks"

Notes:

1. Values shown assume 70 ksi electrodes. For 60 ksi electrodes, multiply tabular values by 0.857, or enter table with 1.17 times the required strength R_u. For 80 ksi electrodes, multiply tabular values by 1.14, or enter table with 0.875 times the required
2. Tabulated values are valid for stiffeners with minimum thickness of

$$t_{min} = \frac{F_{y\,beam}}{F_{y\,stiffener}} \times t_w$$

but not less than $2w$ for stiffeners with $F_y = 36$ ksi nor $1.5w$ for stiffeners with $F_y = 50$ ksi. In the above, t_w is the thickness of the unstiffened supported beam web and w is the nominal weld size.

3. Tabulated values may be limited by shear yielding of or bearing on the stiffener; refer to LRFD Specification Sections F2.2 and J8, respectively.

SINGLE-PLATE CONNECTIONS

A single-plate connection is made with a plate as illustrated in Figure 10-16. The plate is always welded to the support on both sides of the plate and bolted to the supported member.

Design Checks

The design strength of a single-plate connection is determined from the applicable limit states for the bolts (see Part 7), welds (see Part 8) and connection elements (see Part 9). In all cases, the design strength ϕR_n must equal or exceed the required strength R_u.

Eccentricity must be considered in the design of the single-plate connection. That is, the bolts must be designed for the shear R_u and eccentric moment $R_u e_b$. The eccentricity on the bolts e_b depends upon the support condition present and whether standard or short-slotted holes are used in the plate (Astaneh et al., 1989).

A flexible support possesses relatively low rotational stiffness and permits the adjacent simply supported beam end rotation to be accommodated primarily through this supporting member's rotation. Such an end condition may exist with one-sided beam-to-girder-web connections or with deep beams connected to relatively light columns. For a flexible support with standard holes:

$$e_b = |(n-1) - a| \geq a$$

where

a = distance between the bolt line and weld line (see Figure 10-16), in.
n = number of bolts.

For a flexible support with short-slotted holes:

$$e_b = \left|\frac{2n}{3} - a\right| \geq a$$

In contrast, a rigid support possesses relatively high rotational stiffness, which constrains the adjacent simply-supported beam end rotation to occur primarily within the end connection, such as a beam-to-column-flange connection or two concurrent beam-to-girder-web

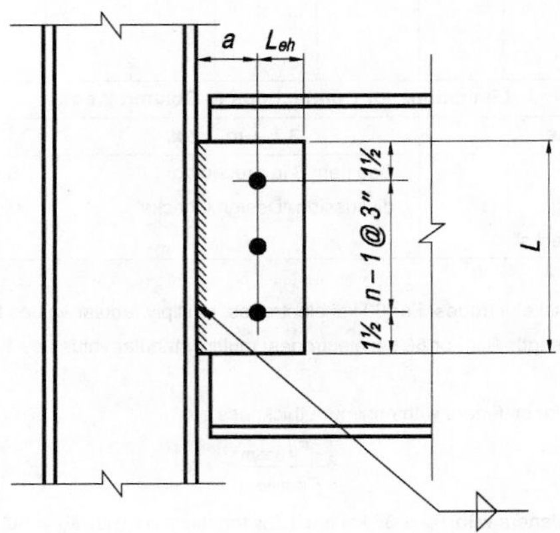

Fig. 10–16. Single-plate connections.

connections. For a rigid support with standard holes:

$$e_b = |(n-1) - a|$$

For a rigid support with short-slotted holes

$$e_b = \left|\frac{2n}{3} - a\right|$$

When the support condition is intermediate between flexible and rigid or cannot be readily classified as flexible or rigid, the larger value of e_b may conservatively be taken from the above equations.

For any combination of support condition and hole type, the 70 ksi electrode weld size should be equal to three-quarters of the plate thickness t_p for plate material with $F_y = 36$ ksi and $F_u = 58$ ksi. This weld size ensures that the plate will yield prior to weld fracture.

The foregoing procedure is valid for single-plate connections with $2^1/_2$ in. $\leq a \leq 3\ ^1/_2$ in. Single-plate connections with geometries and configurations other than those described above can be used based upon rational analysis.

Recommended Plate Length and Thickness

To provide for stability during erection, it is recommended that the minimum plate length be one-half the T-dimension of the beam to be supported. The maximum length of the plate must be compatible with the T-dimension of an uncoped beam and the remaining web depth, exclusive of fillets, of a coped beam. Note that the plate may encroach upon the fillet(s) as given in Figure 10-4.

To prevent local buckling of the plate, the minimum plate thickness should be such that,

$$t_{p\,min} \frac{L}{\sqrt{\frac{K}{F_y}}} \geq {}^1/_4 \text{ in.}$$

where K is the plate buckling coefficient tabulated in Part 9 for local buckling of beams coped at both the top and bottom flanges. To use the table in Part 9, calculate the plate aspect ratio as $L/2a$.

To provide for rotational ductility in the single plate, the maximum plate thickness should be such that

$$t_{p\,max} = \frac{d_b}{2} + {}^1/_{16} \text{ in.} \geq t_{p\,min}$$

where d_b is the bolt diameter, in. This ensures that bearing deformations will occur in the bolt holes prior to bolt shear.

Shop and Field Practices

Single-plate connections may be made to the webs of supporting girders and to the flanges of supporting columns. Because of bolting clearances, field-bolted single-plate connections may not be suitable for connections to the webs of supporting columns unless provision is made to extend the plate to locate the bolt line a sufficient distance beyond the column flanges. Such extension may require stiffening of the plate and the column web.

With the plate shop attached to the support, side erection of the beam is permitted. Play in the open holes usually compensates for mill variation in column flange supports and other field adjustments. Thus, slotted holes are not normally required.

EXAMPLE 10.11. Single-plate connection (beam-to-column flange).

Given:

Design a single-plate connection for a W16×50 beam to a W14×90 column flange.

$$R_u = 55\text{ kips}$$

W16×50, ASTM A992

$$t_w = 0.380\text{ in.}\qquad d = 16.3\text{ in.}\qquad t_f = 0.630\text{ in.}$$
$$F_y = 50\text{ ksi,}\qquad F_u = 65\text{ ksi}$$

W14×90, ASTM A992

$$t_f = 0.710\text{ in.}$$
$$F_y = 50\text{ ksi, } F_u = 65\text{ ksi}$$

Use $^3/_4$-in. diameter A325-N bolts in standard holes and 70 ksi electrode welds. Assume ASTM A36 single-plate material with $F_y = 36$ ksi and $F_u = 58$ ksi.

Solution:

Design bolts, single plate, and welds

From Table 10-9, assuming the column provides a rigid support, for $^3/_4$-in. diameter A325-N bolts and single-plate material with $F_y = 36$ ksi and $F_u = 58$ ksi, select four rows of bolts, $^1/_4$-in. single-plate thickness, and $^3/_{16}$-in. fillet weld size.

$$\phi R_n = 55.5\text{ kips} > 55\text{ kips}\qquad \textbf{o.k.}$$

Check supported beam web

From Table 10-1, for four rows of bolts, beam material with $F_y = 50$ ksi and $F_u = 65$ ksi, for an uncoped section,

$$\phi R_n = (351\text{ kips/in.})(0.380\text{ in.})$$
$$= 133\text{ kips} > 55\text{ kips}\qquad \textbf{o.k.}$$

EXAMPLE 10.12. Single-plate connection (beam-to-girder web).

Given:

Design a single-plate connection for a W18×35 beam to a W21×62 girder-web.

$$R_u = 40\text{ kips}$$

W18×35, ASTM A992

$$t_w = 0.300\text{ in.}\qquad d = 17.7\text{ in.}$$
$$F_y = 50\text{ ksi,}\qquad F_u = 65\text{ ksi}$$

top flange coped 2 in. deep by 3 in. long, $L_{ev} = 1^1/_2$ in., $L_{eh} = 1^1/_2$ in. (assumed to be $1^1/_4$ in. for calculation purposes to account for possible underrun in beam length),

W21×62, ASTM A992

$$t_w = 0.400\text{ in.}$$
$$F_y = 50\text{ ksi},\ F_u = 65\text{ ksi}$$

Use $^3/_4$-in.-diameter A325-N bolts in standard holes and 70 ksi electrode welds. Assume ASTM A36 single-plate material with $F_y = 36$ ksi and $F_u = 58$ ksi.

Solution:

Design bolts, single plate, and welds

From Table 10-9, assuming the girder provides a flexible support, for $^3/_4$-in.-diameter A325-N bolts and single-plate material with $F_y = 36$ ksi and $F_u = 58$ ksi, select four rows of bolts, $^1/_4$-in. single-plate thickness, and $^3/_{16}$-in. fillet weld size.

$$\phi R_n = 44.7\text{ kips} > 40\text{ kips} \quad \textbf{o.k.}$$

Check supported beam web

From Table 10-1, for four rows of bolts, beam material with $F_y = 50$ ksi and $F_u = 65$ ksi, and $L_{ev} = 1^1/_2$ in. and $L_{eh} = 1^1/_2$ in. (assumed to be $1^1/_4$ in. for calculation purposes to account for possible underrun in beam length),

$$\phi R_n = (257\text{ kips/in.})(0.300\text{ in.})$$
$$= 77.1\text{ kips} > 40\text{ kips} \quad \textbf{o.k.}$$

Check flexural yielding of the coped section

From Table 9-2, $S_{net} = 18.2\text{ in.}^3$

$$\phi R_n = \frac{0.9 F_y S_{net}}{e}$$
$$= \frac{0.9(50\text{ ksi})(18.2\text{ in}^3)}{3\text{ in.} + 1^3/_4\text{ in.}}$$
$$= 172\text{ kips} > 40\text{ kips} \quad \textbf{o.k.}$$

Check local web buckling at the cope

$$\frac{c}{d} = \frac{3\text{ in.}}{17.7\text{ in.}} = 0.170$$
$$\frac{c}{h_o} = \frac{3\text{ in.}}{17.7\text{ in.} - 2\text{ in.}} = 0.191$$

Since $\frac{c}{d} \leq 1.0$,

$$f = 2\left(\frac{c}{d}\right)$$
$$= 2(0.170)$$
$$= 0.340$$

Since $\dfrac{c}{h_o} \leq 1.0$,

$$k = 2.2\left(\frac{h_o}{c}\right)^{1.65}$$
$$= 2.2\left(\frac{1}{0.191}\right)^{1.65}$$
$$= 33.8$$
$$\phi F_{bc} = 23{,}590\left(\frac{t_w}{h_o}\right)^2 fk$$
$$= 23{,}590\left(\frac{0.300\text{ in.}}{17.7\text{ in.} - 2\text{ in.}}\right)^2 (0.340)(33.8)$$
$$= 99.0\text{ ksi}$$
$$\phi R_n = \frac{\phi F_{bc} S_{net}}{e}$$
$$= \frac{(99.0\text{ ksi})(18.2\text{ in.}^3)}{(3\text{ in.} + 1^3/_4\text{ in.})}$$
$$= 379\text{ kips} > 40\text{ kips} \quad \textbf{o.k.}$$

Table 10-9. Bolted/Welded Single-Plate Connections

Tables 10-9 are design aids for single-plate connections welded to the support and bolted to the supported beam. Design strengths are tabulated for plate material with $F_y = 36$ ksi and $F_u = 58$ ksi.

Tabulated bolt and plate design strengths consider the limit-states of bolt shear, bolt bearing on the plate, shear yielding of the plate, shear rupture of the plate, block shear rupture of the plate, and weld shear. Values are tabulated for two through twelve rows of $^3/_4$-in., $^7/_8$-in., 1-in. and $1^1/_8$-in.-diameter ASTM A325, F1852 and A490 bolts at 3-in. spacing. For calculation purposes, plate edge distances L_{ev} and L_{eh} are assumed to be $1^1/_2$ in. Weld sizes are tabulated equal to $^3/_4\, t_p$.

While the tabular values are based on $a = 3$ in., they may conservatively be used for values of a between $2^1/_2$ in. and 3 in. The tabulated values are valid for laterally supported beams, in steel and composite construction, all types of loading, snug-tightened and pretensioned bolts, and for supported and supporting members of all grades of steel.

Plate F_y = 36 ksi | 3/4-in. diameter bolts

Table 10-9. Single-Plate Connections

Bolt, Weld, and Single-Plate Design Strength, kips

n	ASTM Desig.	Thread Cond.	Support Cond.	Hole Type	Plate Thickness, in. 1/4	5/16	3/8	7/16	1/2	9/16
12 (L = 36)	A325 F1852	N	Flexible	STD	–	142	142	142	–	–
				SSLT	–	165	165	165	–	–
			Rigid	STD	–	142	142	142	–	–
				SSLT	–	165	165	165	–	–
		X	Flexible	STD	–	178	178	178	–	–
				SSLT	–	207	207	207	–	–
			Rigid	STD	–	178	178	178	–	–
				SSLT	–	207	207	207	–	–
	A490	N	Flexible	STD	–	178	178	178	–	–
				SSLT	–	207	207	207	–	–
			Rigid	STD	–	178	178	178	–	–
				SSLT	–	207	207	207	–	–
		X	Flexible	STD	–	208	222	222	–	–
				SSLT	–	208	250	258	–	–
			Rigid	STD	–	208	222	222	–	–
				SSLT	–	208	250	258	–	–
11 (L = 33)	A325 F1852	N	Flexible	STD	–	133	133	133	–	–
				SSLT	–	154	154	154	–	–
			Rigid	STD	–	133	133	133	–	–
				SSLT	–	154	154	154	–	–
		X	Flexible	STD	–	166	166	166	–	–
				SSLT	–	191	193	193	–	–
			Rigid	STD	–	166	166	166	–	–
				SSLT	–	191	193	193	–	–
	A490	N	Flexible	STD	–	166	166	166	–	–
				SSLT	–	191	193	193	–	–
			Rigid	STD	–	166	166	166	–	–
				SSLT	–	191	193	193	–	–
		X	Flexible	STD	–	191	208	208	–	–
				SSLT	–	191	229	241	–	–
			Rigid	STD	–	191	208	208	–	–
				SSLT	–	191	229	241	–	–
10 (L = 30)	A325 F1852	N	Flexible	STD	124	124	124	124	–	–
				SSLT	139	142	142	142	–	–
			Rigid	STD	124	124	124	124	–	–
				SSLT	139	142	142	142	–	–
		X	Flexible	STD	139	155	155	155	–	–
				SSLT	139	173	177	177	–	–
			Rigid	STD	139	155	155	155	–	–
				SSLT	139	173	177	177	–	–
	A490	N	Flexible	STD	139	155	155	155	–	–
				SSLT	139	173	177	177	–	–
			Rigid	STD	139	155	155	155	–	–
				SSLT	139	173	177	177	–	–
		X	Flexible	STD	139	173	194	194	–	–
				SSLT	139	173	208	222	–	–
			Rigid	STD	139	173	194	194	–	–
				SSLT	139	173	208	222	–	–
Weld Size					3/16	1/4	5/16	3/8	3/8	7/16

STD = Standard Holes
SSLT = Short-slotted holes transverse to direction of load
N = Threads included
X = Threads excluded

3/4-in. diameter bolts

Plate F_y = 36 ksi

Table 10-9 (cont.). Single-Plate Connections

Bolt, Weld, and Single-Plate Design Strength, kips

n	ASTM Desig.	Thread Cond.	Support Cond.	Hole Type	Plate Thickness, in. 1/4	5/16	3/8	7/16	1/2	9/16
9 (L = 27)	A325 F1852	N	Flexible	STD	115	115	115	115	–	–
				SSLT	125	130	130	130	–	–
			Rigid	STD	115	115	115	115	–	–
				SSLT	125	130	130	130	–	–
		X	Flexible	STD	125	144	144	144	–	–
				SSLT	125	156	162	162	–	–
			Rigid	STD	125	144	144	144	–	–
				SSLT	125	156	162	162	–	–
	A490	N	Flexible	STD	125	144	144	144	–	–
				SSLT	125	156	162	162	–	–
			Rigid	STD	125	144	144	144	–	–
				SSLT	125	156	162	162	–	–
		X	Flexible	STD	125	156	179	179	–	–
				SSLT	125	156	187	203	–	–
			Rigid	STD	125	156	179	179	–	–
				SSLT	125	156	187	203	–	–
8 (L = 24)	A325 F1852	N	Flexible	STD	106	106	106	106	–	–
				SSLT	111	113	113	113	–	–
			Rigid	STD	106	106	106	106	–	–
				SSLT	111	117	117	117	–	–
		X	Flexible	STD	111	132	132	132	–	–
				SSLT	111	139	142	142	–	–
			Rigid	STD	111	132	132	132	–	–
				SSLT	111	139	147	147	–	–
	A490	N	Flexible	STD	111	132	132	132	–	–
				SSLT	111	139	142	142	–	–
			Rigid	STD	111	132	132	132	–	–
				SSLT	111	139	147	147	–	–
		X	Flexible	STD	111	139	165	165	–	–
				SSLT	111	139	166	177	–	–
			Rigid	STD	111	139	165	165	–	–
				SSLT	111	139	166	183	–	–
7 (L = 21)	A325 F1852	N	Flexible	STD	96.4	96.4	96.4	96.4	–	–
				SSLT	96.4	96.4	96.4	96.4	–	–
			Rigid	STD	96.4	96.4	96.4	96.4	–	–
				SSLT	97.1	104	104	104	–	–
		X	Flexible	STD	97.1	120	120	120	–	–
				SSLT	97.1	120	120	120	–	–
			Rigid	STD	97.1	120	120	120	–	–
				SSLT	97.1	121	131	131	–	–
	A490	N	Flexible	STD	97.1	120	120	120	–	–
				SSLT	97.1	120	120	120	–	–
			Rigid	STD	97.1	120	120	120	–	–
				SSLT	97.1	121	131	131	–	–
		X	Flexible	STD	97.1	121	146	151	–	–
				SSLT	97.1	121	146	151	–	–
			Rigid	STD	97.1	121	146	151	–	–
				SSLT	97.1	121	146	163	–	–
Weld Size					3/16	1/4	5/16	3/8	3/8	7/16

STD = Standard Holes
SSLT = Short-slotted holes orientedtransverse to direction of load
N = Threads included
X = Threads excluded

Plate F_y = 36 ksi										3/4-in. diameter bolts

Table 10-9 (cont.). Single-Plate Connections

Bolt, Weld, and Single-Plate Design Strength, kips

n	ASTM Desig.	Thread Cond.	Support Cond.	Hole Type	Plate Thickness, in.					
					1/4	5/16	3/8	7/16	1/2	9/16
6 (L = 18)	A325 F1852	N	Flexible	STD	79.2	79.2	79.2	79.2	–	–
				SSLT	79.2	79.2	79.2	79.2	–	–
			Rigid	STD	83.2	86.7	86.7	86.7	–	–
				SSLT	83.2	91.1	91.1	91.1	–	–
		X	Flexible	STD	83.2	99.0	99.0	99.0	–	–
				SSLT	83.2	99.0	99.0	99.0	–	–
			Rigid	STD	83.2	104	108	108	–	–
				SSLT	83.2	104	114	114	–	–
	A490	N	Flexible	STD	83.2	99.0	99.0	99.0	–	–
				SSLT	83.2	99.0	99.0	99.0	–	–
			Rigid	STD	83.2	104	108	108	–	–
				SSLT	83.2	104	114	114	–	–
		X	Flexible	STD	83.2	104	124	124	–	–
				SSLT	83.2	104	124	124	–	–
			Rigid	STD	83.2	104	125	135	–	–
				SSLT	83.2	104	125	142	–	–
5 (L = 15)	A325 F1852	N	Flexible	STD	62.0	62.0	62.0	62.0	–	–
				SSLT	62.0	62.0	62.0	62.0	–	–
			Rigid	STD	69.3	74.7	74.7	74.7	–	–
				SSLT	69.3	77.9	77.9	77.9	–	–
		X	Flexible	STD	69.3	77.5	77.5	77.5	–	–
				SSLT	69.3	77.5	77.5	77.5	–	–
			Rigid	STD	69.3	86.7	93.3	93.3	–	–
				SSLT	69.3	86.7	97.4	97.4	–	–
	A490	N	Flexible	STD	69.3	77.5	77.5	77.5	–	–
				SSLT	69.3	77.5	77.5	77.5	–	–
			Rigid	STD	69.3	86.7	93.3	93.3	–	–
				SSLT	69.3	86.7	97.4	97.4	–	–
		X	Flexible	STD	69.3	86.7	96.9	96.9	–	–
				SSLT	69.3	86.7	96.9	96.9	–	–
			Rigid	STD	69.3	86.7	104	117	–	–
				SSLT	69.3	86.7	104	121	–	–
4 (L = 12)	A325 F1852	N	Flexible	STD	44.7	44.7	44.7	44.7	–	–
				SSLT	44.7	44.7	44.7	44.7	–	–
			Rigid	STD	55.5	63.6	63.6	63.6	–	–
				SSLT	55.5	61.8	61.8	61.8	–	–
		X	Flexible	STD	55.0	55.9	55.9	55.9	–	–
				SSLT	55.0	55.9	55.9	55.9	–	–
			Rigid	STD	55.5	69.3	79.5	79.5	–	–
				SSLT	55.5	69.3	77.3	77.3	–	–
	A490	N	Flexible	STD	55.0	55.9	55.9	55.9	–	–
				SSLT	55.0	55.9	55.9	55.9	–	–
			Rigid	STD	55.5	69.3	79.5	79.5	–	–
				SSLT	55.5	69.3	77.3	77.3	–	–
		X	Flexible	STD	55.0	68.8	69.8	69.8	–	–
				SSLT	55.0	68.8	69.8	69.8	–	–
			Rigid	STD	55.5	69.3	83.2	97.1	–	–
				SSLT	55.5	69.3	83.2	96.6	–	–
		Weld Size			3/16	1/4	5/16	3/8	3/8	7/16

STD = Standard Holes
SSLT = Short-slotted holes orientedtransverse to direction of load
N = Threads included
X = Threads excluded

3/4-in. diameter bolts | Plate F_y = 36 ksi

Table 10-9 (cont.). Single-Plate Connections

Bolt, Weld, and Single-Plate Design Strength, kips

n	ASTM Desig.	Thread Cond.	Support Cond.	Hole Type	Plate Thickness, in. 1/4	5/16	3/8	7/16	1/2	9/16
3 (L = 9)	A325 F1852	N	Flexible	STD	27.8	27.8	27.8	27.8	–	–
				SSLT	27.8	27.8	27.8	27.8	–	–
			Rigid	STD	41.6	41.6	41.6	41.6	–	–
				SSLT	41.6	41.6	41.6	41.6	–	–
		X	Flexible	STD	34.3	34.8	34.8	34.8	–	–
				SSLT	34.3	34.8	34.8	34.8	–	–
			Rigid	STD	41.6	52.0	52.0	52.0	–	–
				SSLT	41.6	52.0	52.0	52.0	–	–
	A490	N	Flexible	STD	34.3	34.8	34.8	34.8	–	–
				SSLT	34.3	34.8	34.8	34.8	–	–
			Rigid	STD	41.6	52.0	52.0	52.0	–	–
				SSLT	41.6	52.0	52.0	52.0	–	–
		X	Flexible	STD	34.3	42.8	43.5	43.5	–	–
				SSLT	34.3	42.8	43.5	43.5	–	–
			Rigid	STD	41.6	52.0	62.4	65	–	–
				SSLT	41.6	52.0	62.4	65	–	–
2 (L = 6)	A325 F1852	N	Flexible	STD	14.0	14.0	14.0	14.0	–	–
				SSLT	14.0	14.0	14.0	14.0	–	–
			Rigid	STD	18.8	18.8	18.8	18.8	–	–
				SSLT	20.9	20.9	20.9	20.9	–	–
		X	Flexible	STD	17.2	17.5	17.5	17.5	–	–
				SSLT	17.2	17.5	17.5	17.5	–	–
			Rigid	STD	23.1	23.5	23.5	23.5	–	–
				SSLT	25.8	26.2	26.2	26.2	–	–
	A490	N	Flexible	STD	17.2	17.5	17.5	17.5	–	–
				SSLT	17.2	17.5	17.5	17.5	–	–
			Rigid	STD	23.1	23.5	23.5	23.5	–	–
				SSLT	25.8	26.2	26.2	26.2	–	–
		X	Flexible	STD	17.2	21.5	21.9	21.9	–	–
				SSLT	17.2	21.5	21.9	21.9	–	–
			Rigid	STD	23.1	28.9	29.3	29.3	–	–
				SSLT	25.8	32.2	32.7	32.7	–	–
Weld Size					3/16	1/4	5/16	3/8	3/8	7/16

STD = Standard Holes
SSLT = Short-slotted holes orientedtransverse to direction of load
N = Threads included
X = Threads excluded

Plate F_y = 36 ksi

7/8-in. diameter bolts

Table 10-9 (cont.). Single-Plate Connections

Bolt, Weld, and Single-Plate Design Strength, kips

n	ASTM Desig.	Thread Cond.	Support Cond.	Hole Type	Plate Thickness, in. 1/4	5/16	3/8	7/16	1/2	9/16
12 (L = 36)	A325 F1852	N	Flexible	STD	–	193	193	193	193	–
				SSLT	–	196	225	225	225	–
			Rigid	STD	–	193	193	193	193	–
				SSLT	–	196	225	225	225	–
		X	Flexible	STD	–	196	235	242	242	–
				SSLT	–	196	235	274	281	–
			Rigid	STD	–	196	235	242	242	–
				SSLT	–	196	235	274	281	–
	A490	N	Flexible	STD	–	196	235	242	242	–
				SSLT	–	196	235	274	281	–
			Rigid	STD	–	196	235	242	242	–
				SSLT	–	196	235	274	281	–
		X	Flexible	STD	–	196	235	274	302	–
				SSLT	–	196	235	274	313	–
			Rigid	STD	–	196	235	274	302	–
				SSLT	–	196	235	274	313	–
11 (L = 33)	A325 F1852	N	Flexible	STD	–	179	181	181	181	–
				SSLT	–	179	210	210	210	–
			Rigid	STD	–	179	181	181	181	–
				SSLT	–	179	210	210	210	–
		X	Flexible	STD	–	179	215	226	226	–
				SSLT	–	179	215	251	262	–
			Rigid	STD	–	179	215	226	226	–
				SSLT	–	179	215	251	262	–
	A490	N	Flexible	STD	–	179	215	226	226	–
				SSLT	–	179	215	251	262	–
			Rigid	STD	–	179	215	226	226	–
				SSLT	–	179	215	251	262	–
		X	Flexible	STD	–	179	215	251	283	–
				SSLT	–	179	215	251	287	–
			Rigid	STD	–	179	215	251	283	–
				SSLT	–	179	215	251	287	–
10 (L = 30)	A325 F1852	N	Flexible	STD	131	163	169	169	169	–
				SSLT	131	163	193	193	193	–
			Rigid	STD	131	163	169	169	169	–
				SSLT	131	163	193	193	193	–
		X	Flexible	STD	131	163	196	211	211	–
				SSLT	131	163	196	228	241	–
			Rigid	STD	131	163	196	211	211	–
				SSLT	131	163	196	228	241	–
	A490	N	Flexible	STD	131	163	196	211	211	–
				SSLT	131	163	196	228	241	–
			Rigid	STD	131	163	196	211	211	–
				SSLT	131	163	196	228	241	–
		X	Flexible	STD	131	163	196	228	261	–
				SSLT	131	163	196	228	261	–
			Rigid	STD	131	163	196	228	261	–
				SSLT	131	163	196	228	261	–
Weld Size					3/16	1/4	5/16	3/8	3/8	7/16

STD = Standard Holes
SSLT = Short-slotted holes orientedtransverse to direction of load
N = Threads included
X = Threads excluded

7/8-in. diameter bolts

Plate F_y = 36 ksi

Table 10-9 (cont.). Single-Plate Connections

Bolt, Weld, and Single-Plate Design Strength, kips

n	ASTM Desig.	Thread Cond.	Support Cond.	Hole Type	Plate Thickness, in. 1/4	5/16	3/8	7/16	1/2	9/16
9 (L = 27)	A325 F1852	N	Flexible	STD	117	147	156	156	156	-
				SSLT	117	147	176	177	177	–
			Rigid	STD	117	147	156	156	156	–
				SSLT	117	147	176	177	177	–
		X	Flexible	STD	117	147	176	195	195	–
				SSLT	117	147	176	206	221	–
			Rigid	STD	117	147	176	195	195	–
				SSLT	117	147	176	206	221	–
	A490	N	Flexible	STD	117	147	176	195	195	–
				SSLT	117	147	176	206	221	–
			Rigid	STD	117	147	176	195	195	–
				SSLT	117	147	176	206	221	–
		X	Flexible	STD	117	147	176	206	235	–
				SSLT	117	147	176	206	235	–
			Rigid	STD	117	147	176	206	235	–
				SSLT	117	147	176	206	235	–
8 (L = 24)	A325 F1852	N	Flexible	STD	104	131	144	144	144	–
				SSLT	104	131	154	154	154	–
			Rigid	STD	104	131	144	144	144	–
				SSLT	104	131	157	160	160	–
		X	Flexible	STD	104	131	157	180	180	–
				SSLT	104	131	157	183	193	–
			Rigid	STD	104	131	157	180	180	–
				SSLT	104	131	157	183	200	–
	A490	N	Flexible	STD	104	131	157	180	180	–
				SSLT	104	131	157	183	193	–
			Rigid	STD	104	131	157	180	180	–
				SSLT	104	131	157	183	200	–
		X	Flexible	STD	104	131	157	183	209	–
				SSLT	104	131	157	183	209	–
			Rigid	STD	104	131	157	183	209	–
				SSLT	104	131	157	183	209	–
7 (L = 21)	A325 F1852	N	Flexible	STD	91.4	114	131	131	131	–
				SSLT	91.4	114	131	131	131	–
			Rigid	STD	91.4	114	131	131	131	–
				SSLT	91.4	114	137	142	142	–
		X	Flexible	STD	91.4	114	137	160	164	–
				SSLT	91.4	114	137	160	164	–
			Rigid	STD	91.4	114	137	160	164	–
				SSLT	91.4	114	137	160	178	–
	A490	N	Flexible	STD	91.4	114	137	160	164	–
				SSLT	91.4	114	137	160	164	–
			Rigid	STD	91.4	114	137	160	164	–
				SSLT	91.4	114	137	160	178	–
		X	Flexible	STD	91.4	114	137	160	183	–
				SSLT	91.4	114	137	160	183	–
			Rigid	STD	91.4	114	137	160	183	–
				SSLT	91.4	114	137	160	183	–
		Weld Size			3/16	1/4	5/16	3/8	3/8	7/16

STD = Standard Holes
SSLT = Short-slotted holes orientedtransverse to direction of load
N = Threads included
X = Threads excluded

Plate F_y = 36 ksi | 7/8-in. diameter bolts

Table 10-9 (cont.). Single-Plate Connections

Bolt, Weld, and Single-Plate Design Strength, kips

n	ASTM Desig.	Thread Cond.	Support Cond.	Hole Type	Plate Thickness, in. 1/4	5/16	3/8	7/16	1/2	9/16
6 (L = 18)	A325 F1852	N	Flexible	STD	78.3	97.9	108	108	108	–
				SSLT	78.3	97.9	108	108	108	–
			Rigid	STD	78.3	97.9	117	118	118	–
				SSLT	78.3	97.9	117	124	124	–
		X	Flexible	STD	78.3	97.9	117	135	135	–
				SSLT	78.3	97.9	117	135	135	–
			Rigid	STD	78.3	97.9	117	137	147	–
				SSLT	78.3	97.9	117	137	155	–
	A490	N	Flexible	STD	78.3	97.9	117	135	135	–
				SSLT	78.3	97.9	117	135	135	–
			Rigid	STD	78.3	97.9	117	137	147	–
				SSLT	78.3	97.9	117	137	155	–
		X	Flexible	STD	78.3	97.9	117	137	157	–
				SSLT	78.3	97.9	117	137	157	–
			Rigid	STD	78.3	97.9	117	137	157	–
				SSLT	78.3	97.9	117	137	157	–
5 (L = 15)	A325 F1852	N	Flexible	STD	65.3	81.6	84.4	84.4	84.4	–
				SSLT	65.3	81.6	84.4	84.4	84.4	–
			Rigid	STD	65.3	81.6	97.9	102	102	–
				SSLT	65.3	81.6	97.9	106	106	–
		X	Flexible	STD	65.3	81.6	97.9	106	106	–
				SSLT	65.3	81.6	97.9	106	106	–
			Rigid	STD	65.3	81.6	97.9	114	127	–
				SSLT	65.3	81.6	97.9	114	131	–
	A490	N	Flexible	STD	65.3	81.6	97.9	106	106	–
				SSLT	65.3	81.6	97.9	106	106	–
			Rigid	STD	65.3	81.6	97.9	114	127	–
				SSLT	65.3	81.6	97.9	114	131	–
		X	Flexible	STD	65.3	81.6	97.9	114	131	–
				SSLT	65.3	81.6	97.9	114	131	–
			Rigid	STD	65.3	81.6	97.9	114	131	–
				SSLT	65.3	81.6	97.9	114	131	–
4 (L = 12)	A325 F1852	N	Flexible	STD	52.2	60.8	60.8	60.8	60.8	–
				SSLT	52.2	60.8	60.8	60.8	60.8	–
			Rigid	STD	52.2	65.3	78.3	86.6	86.6	–
				SSLT	52.2	65.3	78.3	84.1	84.1	–
		X	Flexible	STD	52.2	65.3	76.0	76.0	76.0	–
				SSLT	52.2	65.3	76.0	76.0	76.0	–
			Rigid	STD	52.2	65.3	78.3	91.4	104	–
				SSLT	52.2	65.3	78.3	91.4	104	–
	A490	N	Flexible	STD	52.2	65.3	76.0	76.0	76.0	–
				SSLT	52.2	65.3	76.0	76.0	76.0	–
			Rigid	STD	52.2	65.3	78.3	91.4	104	–
				SSLT	52.2	65.3	78.3	91.4	104	–
		X	Flexible	STD	52.2	65.3	78.3	91.4	95.0	–
				SSLT	52.2	65.3	78.3	91.4	95.0	–
			Rigid	STD	52.2	65.3	78.3	91.4	104	–
				SSLT	52.2	65.3	78.3	91.4	104	–
Weld Size					3/16	1/4	5/16	3/8	3/8	7/16

STD = Standard Holes
SSLT = Short-slotted holes orientedtransverse to direction of load
N = Threads included
X = Threads excluded

$^{7}/_{8}$-in. diameter bolts | Plate F_y = 36 ksi

Table 10-9 (cont.). Single-Plate Connections

Bolt, Weld, and Single-Plate Design Strength, kips

n	ASTM Desig.	Thread Cond.	Support Cond.	Hole Type	Plate Thickness, in. 1/4	5/16	3/8	7/16	1/2	9/16
3 (L = 9)	A325 F1852	N	Flexible	STD	37.9	37.9	37.9	37.9	37.9	–
				SSLT	37.9	37.9	37.9	37.9	37.9	–
			Rigid	STD	39.1	48.9	56.6	56.6	56.6	–
				SSLT	39.1	48.9	56.6	56.6	56.6	–
		X	Flexible	STD	39.1	47.4	47.4	47.4	47.4	–
				SSLT	39.1	47.4	47.4	47.4	47.4	–
			Rigid	STD	39.1	48.9	58.7	68.5	70.8	–
				SSLT	39.1	48.9	58.7	68.5	70.8	–
	A490	N	Flexible	STD	39.1	47.4	47.4	47.4	47.4	–
				SSLT	39.1	47.4	47.4	47.4	47.4	–
			Rigid	STD	39.1	48.9	58.7	68.5	70.8	–
				SSLT	39.1	48.9	58.7	68.5	70.8	–
		X	Flexible	STD	39.1	48.9	58.7	59.2	59.2	–
				SSLT	39.1	48.9	58.7	59.2	59.2	–
			Rigid	STD	39.1	48.9	58.7	68.5	78.3	–
				SSLT	39.1	48.9	58.7	68.5	78.3	–
2 (L = 6)	A325 F1852	N	Flexible	STD	19.0	19.0	19.0	19.0	19.0	–
				SSLT	19.0	19.0	19.0	19.0	19.0	–
			Rigid	STD	25.5	25.5	25.5	25.5	25.5	–
				SSLT	26.1	28.5	28.5	28.5	28.5	–
		X	Flexible	STD	20.1	23.8	23.8	23.8	23.8	–
				SSLT	20.1	23.8	23.8	23.8	23.8	–
			Rigid	STD	26.1	31.9	31.9	31.9	31.9	–
				SSLT	26.1	32.6	35.6	35.6	35.6	–
	A490	N	Flexible	STD	20.1	23.8	23.8	23.8	23.8	–
				SSLT	20.1	23.8	23.8	23.8	23.8	–
			Rigid	STD	26.1	31.9	31.9	31.9	31.9	–
				SSLT	26.1	32.6	35.6	35.6	35.6	–
		X	Flexible	STD	20.1	25.1	29.8	29.8	29.8	–
				SSLT	20.1	25.1	29.8	29.8	29.8	–
			Rigid	STD	26.1	32.6	39.1	39.9	39.9	–
				SSLT	26.1	32.6	39.1	44.5	44.5	–
		Weld Size			3/16	1/4	5/16	3/8	3/8	7/16

STD = Standard Holes
SSLT = Short-slotted holes orientedtransverse to direction of load
N = Threads included
X = Threads excluded

Plate F_y = 36 ksi | **1-in. diameter bolts**

Table 10-9 (cont.). Single-Plate Connections

Bolt, Weld, and Single-Plate Design Strength, kips

n	ASTM Desig.	Thread Cond.	Support Cond.	Hole Type	Plate Thickness, in. 1/4	5/16	3/8	7/16	1/2	9/16
12 (L = 36)	A325 F1852	N	Flexible	STD	–	184	220	252	252	252
				SSLT	–	184	220	257	294	294
			Rigid	STD	–	184	220	252	252	252
				SSLT	–	184	220	257	294	294
		X	Flexible	STD	–	184	220	257	294	316
				SSLT	–	184	220	257	294	330
			Rigid	STD	–	184	220	257	294	316
				SSLT	–	184	220	257	294	330
	A490	N	Flexible	STD	–	184	220	257	294	316
				SSLT	–	184	220	257	294	330
			Rigid	STD	–	184	220	257	294	316
				SSLT	–	184	220	257	294	330
		X	Flexible	STD	–	184	220	257	294	330
				SSLT	–	184	220	257	294	330
			Rigid	STD	–	184	220	257	294	330
				SSLT	–	184	220	257	294	330
11 (L = 33)	A325 F1852	N	Flexible	STD	–	168	202	236	236	236
				SSLT	–	168	202	236	269	274
			Rigid	STD	–	168	202	236	236	236
				SSLT	–	168	202	236	269	274
		X	Flexible	STD	–	168	202	236	269	295
				SSLT	–	168	202	236	269	303
			Rigid	STD	–	168	202	236	269	295
				SSLT	–	168	202	236	269	303
	A490	N	Flexible	STD	–	168	202	236	269	295
				SSLT	–	168	202	236	269	303
			Rigid	STD	–	168	202	236	269	295
				SSLT	–	168	202	236	269	303
		X	Flexible	STD	–	168	202	236	269	303
				SSLT	–	168	202	236	269	303
			Rigid	STD	–	168	202	236	269	303
				SSLT	–	168	202	236	269	303
10 (L = 30)	A325 F1852	N	Flexible	STD	122	153	184	214	220	220
				SSLT	122	153	184	214	245	252
			Rigid	STD	122	153	184	214	220	220
				SSLT	122	153	184	214	245	252
		X	Flexible	STD	122	153	184	214	245	275
				SSLT	122	153	184	214	245	275
			Rigid	STD	122	153	184	214	245	275
				SSLT	122	153	184	214	245	275
	A490	N	Flexible	STD	122	153	184	214	245	275
				SSLT	122	153	184	214	245	275
			Rigid	STD	122	153	184	214	245	275
				SSLT	122	153	184	214	245	275
		X	Flexible	STD	122	153	184	214	245	275
				SSLT	122	153	184	214	245	275
			Rigid	STD	122	153	184	214	245	275
				SSLT	122	153	184	214	245	275
		Weld Size			3/16	1/4	5/16	3/8	3/8	7/16

STD = Standard Holes
SSLT = Short-slotted holes orientedtransverse to direction of load
N = Threads included
X = Threads excluded

1-in. diameter bolts									Plate F_y = 36 ksi	

Table 10-9 (cont.). Single-Plate Connections

Bolt, Weld, and Single-Plate Design Strength, kips

n	ASTM Desig.	Thread Cond.	Support Cond.	Hole Type	Plate Thickness, in. 1/4	5/16	3/8	7/16	1/2	9/16
9 (L = 27)	A325 F1852	N	Flexible	STD	110	138	165	193	204	204
				SSLT	110	138	165	193	220	231
			Rigid	STD	110	138	165	193	204	204
				SSLT	110	138	165	193	220	231
		X	Flexible	STD	110	138	165	193	220	248
				SSLT	110	138	165	193	220	248
			Rigid	STD	110	138	165	193	220	248
				SSLT	110	138	165	193	220	248
	A490	N	Flexible	STD	110	138	165	193	220	248
				SSLT	110	138	165	193	220	248
			Rigid	STD	110	138	165	193	220	248
				SSLT	110	138	165	193	220	248
		X	Flexible	STD	110	138	165	193	220	248
				SSLT	110	138	165	193	220	248
			Rigid	STD	110	138-	165	193	220	248
				SSLT	110	138	165	193	220	248
8 (L = 24)	A325 F1852	N	Flexible	STD	97.9	122	147	171	188	188
				SSLT	97.9	122	147	171	196	201
			Rigid	STD	97.9	122	147	171	188	188
				SSLT	97.9	122	147	171	196	209
		X	Flexible	STD	97.9	122	147	171	196	220
				SSLT	97.9	122	147	171	196	220
			Rigid	STD	97.9	122	147	171	196	220
				SSLT	97.9	122	147	171	196	220
	A490	N	Flexible	STD	97.9	122	147	171	196	220
				SSLT	97.9	122	147	171	196	220
			Rigid	STD	97.9	122	147	171	196	220
				SSLT	97.9	122	147	171	196	220
		X	Flexible	STD	97.9	122	147	171	196	220
				SSLT	97.9	122	147	171	196	220
			Rigid	STD	97.9	122	147	171	196	220
				SSLT	97.9	122	147	171	196	220
7 (L = 21)	A325 F1852	N	Flexible	STD	85.6	107	128	150	171	171
				SSLT	85.6	107	128	150	171	171
			Rigid	STD	85.6	107	128	150	171	171
				SSLT	85.6	107	128	150	171	186
		X	Flexible	STD	85.6	107	128	150	171	193
				SSLT	85.6	107	128	150	171	193
			Rigid	STD	85.6	107	128	150	171	193
				SSLT	85.6	107	128	150	171	193
	A490	N	Flexible	STD	85.6	107	128	150	171	193
				SSLT	85.6	107	128	150	171	193
			Rigid	STD	85.6	107	128	150	171	193
				SSLT	85.6	107	128	150	171	193
		X	Flexible	STD	85.6	107	128	150	171	193
				SSLT	85.6	107	128	150	171	193
			Rigid	STD	85.6	107	128	150	171	193
				SSLT	85.6	107	128	150	171	193
		Weld Size			3/16	1/4	5/16	3/8	3/8	7/16

STD = Standard Holes
SSLT = Short-slotted holes orientedtransverse to direction of load
N = Threads included
X = Threads excluded

Plate F_y = 36 ksi | 1-in. diameter bolts

Table 10-9 (cont.). Single-Plate Connections

Bolt, Weld, and Single-Plate Design Strength, kips										
n	ASTM Desig.	Thread Cond.	Support Cond.	Hole Type	Plate Thickness, in.					
					$^{1}/_{4}$	$^{5}/_{16}$	$^{3}/_{8}$	$^{7}/_{16}$	$^{1}/_{2}$	$^{9}/_{16}$
6 (L = 18)	A325 F1852	N	Flexible	STD	73.4	91.8	110	128	141	141
				SSLT	73.4	91.8	110	128	141	141
			Rigid	STD	73.4	91.8	110	128	147	154
				SSLT	73.4	91.8	110	128	147	162
		X	Flexible	STD	73.4	91.8	110	128	147	165
				SSLT	73.4	91.8	110	128	147	165
			Rigid	STD	73.4	91.8	110	128	147	165
				SSLT	73.4	91.8	110	128	147	165
	A490	N	Flexible	STD	73.4	91.8	110	128	147	165
				SSLT	73.4	91.8	110	128	147	165
			Rigid	STD	73.4	91.8	110	128	147	165
				SSLT	73.4	91.8	110	128	147	165
		X	Flexible	STD	73.4	91.8	110	128	147	165
				SSLT	73.4	91.8	110	128	147	165
			Rigid	STD	73.4	91.8	110	128	147	165
				SSLT	73.4	91.8	110	128	147	165
5 (L = 15)	A325 F1852	N	Flexible	STD	61.2	76.5	91.8	107	110	110
				SSLT	61.2	76.5	91.8	107	110	110
			Rigid	STD	61.2	76.5	91.8	107	122	133
				SSLT	61.2	76.5	91.8	107	122	138
		X	Flexible	STD	61.2	76.5	91.8	107	122	138
				SSLT	61.2	76.5	91.8	107	122	138
			Rigid	STD	61.2	76.5	91.8	107	122	138
				SSLT	61.2	76.5	91.8	107	122	138
	A490	N	Flexible	STD	61.2	76.5	91.8	107	122	138
				SSLT	61.2	76.5	91.8	107	122	138
			Rigid	STD	61.2	76.5	91.8	107	122	138
				SSLT	61.2	76.5	91.8	107	122	138
		X	Flexible	STD	61.2	76.5	91.8	107	122	138
				SSLT	61.2	76.5	91.8	107	122	138
			Rigid	STD	61.2	76.5	91.8	107	122	138
				SSLT	61.2	76.5	91.8	107	122	138
4 (L = 12)	A325 F1852	N	Flexible	STD	48.9	61.2	73.4	79.5	79.5	79.5
				SSLT	48.9	61.2	73.4	79.5	79.5	79.5
			Rigid	STD	48.9	61.2	73.4	85.6	97.9	110
				SSLT	48.9	61.2	73.4	85.6	97.9	110
		X	Flexible	STD	48.9	61.2	73.4	85.6	97.9	99.3
				SSLT	48.9	61.2	73.4	85.6	97.9	99.3
			Rigid	STD	48.9	61.2	73.4	85.6	97.9	110
				SSLT	48.9	61.2	73.4	85.6	97.9	110
	A490	N	Flexible	STD	48.9	61.2	73.4	85.6	97.9	99.3
				SSLT	48.9	61.2	73.4	85.6	97.9	99.3
			Rigid	STD	48.9	61.2	73.4	85.6	97.9	110
				SSLT	48.9	61.2	73.4	85.6	97.9	110
		X	Flexible	STD	48.9	61.2	73.4	85.6	97.9	110
				SSLT	48.9	61.2	73.4	85.6	97.9	110
			Rigid	STD	48.9	61.2	73.4	85.6	97.9	110
				SSLT	48.9	61.2	73.4	85.6	97.9	110
Weld Size					$^{3}/_{16}$	$^{1}/_{4}$	$^{5}/_{16}$	$^{3}/_{8}$	$^{3}/_{8}$	$^{7}/_{16}$

STD = Standard Holes
SSLT = Short-slotted holes orientedtransverse to direction of load
N = Threads included
X = Threads excluded

1-in. diameter bolts | Plate F_y = 36 ksi

Table 10-9 (cont.). Single-Plate Connections

Bolt, Weld, and Single-Plate Design Strength, kips

n	ASTM Desig.	Thread Cond.	Support Cond.	Hole Type	Plate Thickness, in.					
					1/4	5/16	3/8	7/16	1/2	9/16
3 (L = 9)	A325 F1852	N	Flexible	STD	36.7	45.9	49.5	49.5	49.5	49.5
				SSLT	36.7	45.9	49.5	49.5	49.5	49.5
			Rigid	STD	36.7	45.9	55.1	64.2	73.4	73.9
				SSLT	36.7	45.9	55.1	64.2	73.4	73.9
		X	Flexible	STD	36.7	45.9	55.1	61.9	61.9	61.9
				SSLT	36.7	45.9	55.1	61.9	61.9	61.9
			Rigid	STD	36.7	45.9	55.1	64.2	73.4	82.6
				SSLT	36.7	45.9	55.1	64.2	73.4	82.6
	A490	N	Flexible	STD	36.7	45.9	55.1	61.9	61.9	61.9
				SSLT	36.7	45.9	55.1	61.9	61.9	61.9
			Rigid	STD	36.7	45.9	55.1	64.2	73.4	82.6
				SSLT	36.7	45.9	55.1	64.2	73.4	82.6
		X	Flexible	STD	36.7	45.9	55.1	64.2	73.4	77.3
				SSLT	36.7	45.9	55.1	64.2	73.4	77.3
			Rigid	STD	36.7	45.9	55.1	64.2	73.4	82.6
				SSLT	36.7	45.9	55.1	64.2	73.4	82.6
2 (L = 6)	A325 F1852	N	Flexible	STD	23.0	24.9	24.9	24.9	24.9	24.9
				SSLT	23.0	24.9	24.9	24.9	24.9	24.9
			Rigid	STD	24.5	30.6	33.4	33.4	33.4	33.4
				SSLT	24.5	30.6	36.7	37.2	37.2	37.2
		X	Flexible	STD	23.0	28.7	31.1	31.1	31.1	31.1
				SSLT	23.0	28.7	31.1	31.1	31.1	31.1
			Rigid	STD	24.5	30.6	36.7	41.7	41.7	41.7
				SSLT	24.5	30.6	36.7	42.8	46.5	46.5
	A490	N	Flexible	STD	23.0	28.7	31.1	31.1	31.1	31.1
				SSLT	23.0	28.7	31.1	31.1	31.1	31.1
			Rigid	STD	24.5	30.6	36.7	41.7	41.7	41.7
				SSLT	24.5	30.6	36.7	42.8	46.5	46.5
		X	Flexible	STD	23.0	28.7	34.5	38.9	38.9	38.9
				SSLT	23.0	28.7	34.5	38.9	38.9	38.9
			Rigid	STD	24.5	30.6	36.7	42.8	48.9	52.1
				SSLT	24.5	30.6	36.7	42.8	48.9	55.1
Weld Size					3/16	1/4	5/16	3/8	3/8	7/16

STD = Standard Holes
SSLT = Short-slotted holes orientedtransverse to direction of load
N = Threads included
X = Threads excluded

Plate F_y = 36 ksi | **1 1/8-in. diameter bolts**

Table 10-9 (cont.). Single-Plate Connections

Bolt, Weld, and Single-Plate Design Strength, kips

n	ASTM Desig.	Thread Cond.	Support Cond.	Hole Type	Plate Thickness, in. 5/16	3/8	7/16	1/2	9/16	5/8
12 (L = 36)	A325 F1852	N	Flexible	STD	171	206	240	274	308	320
				SSLT	171	206	240	274	308	343
			Rigid	STD	171	206	240	274	308	320
				SSLT	171	206	240	274	308	343
		X	Flexible	STD	171	206	240	274	308	343
				SSLT	171	206	240	274	308	343
			Rigid	STD	171	206	240	274	308	343
				SSLT	171	206	240	274	308	343
	A490	N	Flexible	STD	171	206	240	274	308	343
				SSLT	171	206	240	274	308	343
			Rigid	STD	171	206	240	274	308	343
				SSLT	171	206	240	274	308	343
		X	Flexible	STD	171	206	240	274	308	343
				SSLT	171	206	240	274	308	343
			Rigid	STD	171	206	240	274	308	343
				SSLT	171	206	240	274	308	343
11 (L = 33)	A325 F1852	N	Flexible	STD	157	188	220	251	283	299
				SSLT	157	188	220	251	283	314
			Rigid	STD	157	188	220	251	283	299
				SSLT	157	188	220	251	283	314
		X	Flexible	STD	157	188	220	251	283	314
				SSLT	157	188	220	251	283	314
			Rigid	STD	157	188	220	251	283	314
				SSLT	157	188	220	251	283	314
	A490	N	Flexible	STD	157	188	220	251	283	314
				SSLT	157	188	220	251	283	314
			Rigid	STD	157	188	220	251	283	314
				SSLT	157	188	220	251	283	314
		X	Flexible	STD	157	188	220	251	283	314
				SSLT	157	188	220	251	283	314
			Rigid	STD	157	188	220	251	283	314
				SSLT	157	188	220	251	283	314
10 (L = 30)	A325 F1852	N	Flexible	STD	143	171	200	228	257	279
				SSLT	143	171	200	228	257	285
			Rigid	STD	143	171	200	228	257	279
				SSLT	143	171	200	228	257	285
		X	Flexible	STD	143	171	200	228	257	285
				SSLT	143	171	200	228	257	285
			Rigid	STD	143	171	200	228	257	285
				SSLT	143	171	200	228	257	285
	A490	N	Flexible	STD	143	171	200	228	257	285
				SSLT	143	171	200	228	257	285
			Rigid	STD	143	171	200	228	257	285
				SSLT	143	171	200	228	257	285
		X	Flexible	STD	143	171	200	228	257	285
				SSLT	143	171	200	228	257	285
			Rigid	STD	143	171	200	228	257	285
				SSLT	143	171	200	228	257	285
Weld Size					1/4	5/16	3/8	3/8	7/16	1/2

STD = Standard Holes
SSLT = Short-slotted holes orientedtransverse to direction of load
N = Threads included
X = Threads excluded

1 1/8-in. diameter bolts

Plate F_y = 36 ksi

Table 10-9 (cont.).
Single-Plate Connections

Bolt, Weld, and Single-Plate Design Strength, kips

n	ASTM Desig.	Thread Cond.	Support Cond.	Hole Type	Plate Thickness, in. 5/16	3/8	7/16	1/2	9/16	5/8
9 (L = 27)	A325 F1852	N	Flexible	STD	128	154	180	206	231	257
				SSLT	128	154	180	206	231	257
			Rigid	STD	128	154	180	206	231	257
				SSLT	128	154	180	206	231	257
		X	Flexible	STD	128	154	180	206	231	257
				SSLT	128	154	180	206	231	257
			Rigid	STD	128	154	180	206	231	257
				SSLT	128	154	180	206	231	257
	A490	N	Flexible	STD	128	154	180	206	231	257
				SSLT	128	154	180	206	231	257
			Rigid	STD	128	154	180	206	231	257
				SSLT	128	154	180	206	231	257
		X	Flexible	STD	128	154	180	206	231	257
				SSLT	128	154	180	206	231	257
			Rigid	STD	128	154	180	206	231	257
				SSLT	128	154	180	206	231	257
8 (L = 24)	A325 F1852	N	Flexible	STD	114	137	160	183	206	228
				SSLT	114	137	160	183	206	228
			Rigid	STD	114	137	160	183	206	228
				SSLT	114	137	160	183	206	228
		X	Flexible	STD	114	137	160	183	206	228
				SSLT	114	137	160	183	206	228
			Rigid	STD	114	137	160	183	206	228
				SSLT	114	137	160	183	206	228
	A490	N	Flexible	STD	114	137	160	183	206	228
				SSLT	114	137	160	183	206	228
			Rigid	STD	114	137	160	183	206	228
				SSLT	114	137	160	183	206	228
		X	Flexible	STD	114	137	160	183	206	228
				SSLT	114	137	160	183	206	228
			Rigid	STD	114	137	160	183	206	228
				SSLT	114	137	160	183	206	228
7 (L = 21)	A325 F1852	N	Flexible	STD	99.9	120	140	160	180	200
				SSLT	99.9	120	140	160	180	200
			Rigid	STD	99.9	120	140	160	180	200
				SSLT	99.9	120	140	160	180	200
		X	Flexible	STD	99.9	120	140	160	180	200
				SSLT	99.9	120	140	160	180	200
			Rigid	STD	99.9	120	140	160	180	200
				SSLT	99.9	120	140	160	180	200
	A490	N	Flexible	STD	99.9	120	140	160	180	200
				SSLT	99.9	120	140	160	180	200
			Rigid	STD	99.9	120	140	160	180	200
				SSLT	99.9	120	140	160	180	200
		X	Flexible	STD	99.9	120	140	160	180	200
				SSLT	99.9	120	140	160	180	200
			Rigid	STD	99.9	120	140	160	180	200
				SSLT	99.9	120	140	160	180	200
		Weld Size			1/4	5/16	3/8	3/8	7/16	1/2

STD = Standard Holes
SSLT = Short-slotted holes orientedtransverse to direction of load

N = Threads included
X = Threads excluded

Plate $F_y = 36$ ksi

$1\,{}^1/_8$-in. diameter bolts

Table 10-9 (cont.). Single-Plate Connections

Bolt, Weld, and Single-Plate Design Strength, kips

n	ASTM Desig.	Thread Cond.	Support Cond.	Hole Type	Plate Thickness, in.					
					$^5/_{16}$	$^3/_8$	$^7/_{16}$	$^1/_2$	$^9/_{16}$	$^5/_8$
6 (L = 18)	A325 F1852	N	Flexible	STD	85.6	103	120	137	154	171
				SSLT	85.6	103	120	137	154	171
			Rigid	STD	85.6	103	120	137	154	171
				SSLT	85.6	103	120	137	154	171
		X	Flexible	STD	85.6	103	120	137	154	171
				SSLT	85.6	103	120	137	154	171
			Rigid	STD	85.6	103	120	137	154	171
				SSLT	85.6	103	120	137	154	171
	A490	N	Flexible	STD	85.6	103	120	137	154	171
				SSLT	85.6	103	120	137	154	171
			Rigid	STD	85.6	103	120	137	154	171
				SSLT	85.6	103	120	137	154	171
		X	Flexible	STD	85.6	103	120	137	154	171
				SSLT	85.6	103	120	137	154	171
			Rigid	STD	85.6	103	120	137	154	171
				SSLT	85.6	103	120	137	154	171
5 (L = 15)	A325 F1852	N	Flexible	STD	71.4	85.6	99.9	114	128	140
				SSLT	71.4	85.6	99.9	114	128	140
			Rigid	STD	71.4	85.6	99.9	114	128	143
				SSLT	71.4	85.6	99.9	114	128	143
		X	Flexible	STD	71.4	85.6	99.9	114	128	143
				SSLT	71.4	85.6	99.9	114	128	143
			Rigid	STD	71.4	85.6	99.9	114	128	143
				SSLT	71.4	85.6	99.9	114	128	143
	A490	N	Flexible	STD	71.4	85.6	99.9	114	128	143
				SSLT	71.4	85.6	99.9	114	128	143
			Rigid	STD	71.4	85.6	99.9	114	128	143
				SSLT	71.4	85.6	99.9	114	128	143
		X	Flexible	STD	71.4	85.6	99.9	114	128	143
				SSLT	71.4	85.6	99.9	114	128	143
			Rigid	STD	71.4	85.6	99.9	114	128	143
				SSLT	71.4	85.6	99.9	114	128	143
4 (L = 12)	A325 F1852	N	Flexible	STD	57.1	68.5	79.9	91.4	101	101
				SSLT	57.1	68.5	79.9	91.4	101	101
			Rigid	STD	57.1	68.5	79.9	91.4	103	114
				SSLT	57.1	68.5	79.9	91.4	103	114
		X	Flexible	STD	57.1	68.5	79.9	91.4	103	114
				SSLT	57.1	68.5	79.9	91.4	103	114
			Rigid	STD	57.1	68.5	79.9	91.4	103	114
				SSLT	57.1	68.5	79.9	91.4	103	114
	A490	N	Flexible	STD	57.1	68.5	79.9	91.4	103	114
				SSLT	57.1	68.5	79.9	91.4	103	114
			Rigid	STD	57.1	68.5	79.9	91.4	103	114
				SSLT	57.1	68.5	79.9	91.4	103	114
		X	Flexible	STD	57.1	68.5	79.9	91.4	103	114
				SSLT	57.1	68.5	79.9	91.4	103	114
			Rigid	STD	57.1	68.5	79.9	91.4	103	114
				SSLT	57.1	68.5	79.9	91.4	103	114
Weld Size					$^1/_4$	$^5/_{16}$	$^3/_8$	$^3/_8$	$^7/_{16}$	$^1/_2$

STD = Standard Holes
SSLT = Short-slotted holes orientedtransverse to direction of load
N = Threads included
X = Threads excluded

1 1/8-in. diameter bolts | Plate F_y = 36 ksi

Table 10-9 (cont.). Single-Plate Connections

Bolt, Weld, and Single-Plate Design Strength, kips

n	ASTM Desig.	Thread Cond.	Support Cond.	Hole Type	Plate Thickness, in. 5/16	3/8	7/16	1/2	9/16	5/8
3 (L = 9)	A325 F1852	N	Flexible	STD	42.8	51.4	59.9	62.6	62.6	62.6
				SSLT	42.8	51.4	59.9	62.6	62.6	62.6
			Rigid	STD	42.8	51.4	59.9	68.5	77.1	85.6
				SSLT	42.8	51.4	59.9	68.5	77.1	85.6
		X	Flexible	STD	42.8	51.4	59.9	68.5	77.1	78.3
				SSLT	42.8	51.4	59.9	68.5	77.1	78.3
			Rigid	STD	42.8	51.4	59.9	68.5	77.1	85.6
				SSLT	42.8	51.4	59.9	68.5	77.1	85.6
	A490	N	Flexible	STD	42.8	51.4	59.9	68.5	77.1	78.3
				SSLT	42.8	51.4	59.9	68.5	77.1	78.3
			Rigid	STD	42.8	51.4	59.9	68.5	77.1	85.6
				SSLT	42.8	51.4	59.9	68.5	77.1	85.6
		X	Flexible	STD	42.8	51.4	59.9	68.5	77.1	85.6
				SSLT	42.8	51.4	59.9	68.5	77.1	85.6
			Rigid	STD	42.8	51.4	59.9	68.5	77.1	85.6
				SSLT	42.8	51.4	59.9	68.5	77.1	85.6
2 (L = 6)	A325 F1852	N	Flexible	STD	28.5	31.5	31.5	31.5	31.5	31.5
				SSLT	28.5	31.5	31.5	31.5	31.5	31.5
			Rigid	STD	28.5	34.3	40.0	42.2	42.2	42.2
				SSLT	28.5	34.3	40.0	45.7	47.1	47.1
		X	Flexible	STD	28.5	34.3	39.4	39.4	39.4	39.4
				SSLT	28.5	34.3	39.4	39.4	39.4	39.4
			Rigid	STD	28.5	34.3	40.0	45.7	51.4	52.8
				SSLT	28.5	34.3	40.0	45.7	51.4	57.1
	A490	N	Flexible	STD	28.5	34.3	39.4	39.4	39.4	39.4
				SSLT	28.5	34.3	39.4	39.4	39.4	39.4
			Rigid	STD	28.5	34.3	40.0	45.7	51.4	52.8
				SSLT	28.5	34.3	40.0	45.7	51.4	57.1
		X	Flexible	STD	28.5	34.3	40.0	45.7	49.2	49.2
				SSLT	28.5	34.3	40.0	45.7	49.2	49.2
			Rigid	STD	28.5	34.3	40.0	45.7	51.4	57.1
				SSLT	28.5	34.3	40.0	45.7	51.4	57.1
Weld Size					1/4	5/16	3/8	3/8	7/16	1/2

STD = Standard Holes
SSLT = Short-slotted holes orientedtransverse to direction of load
N = Threads included
X = Threads excluded

SINGLE-ANGLE CONNECTIONS

A single-angle connection is made with an angle on one side of the web of the beam to be supported, as illustrated in Figure 10-17. This angle is preferably shop bolted or welded to the supporting member and field bolted to the supported beam.

When the angle is welded to the support, adequate flexibility must be provided in the connection. As illustrated in Figure 10-17c, the weld is placed along the toe and across the bottom of the angle with a return at the top per LRFD Specification Section J2.2b. Note that welding across the entire top of the angle must be avoided as it would inhibit the flexibility and, therefore, the necessary end rotation of the connection. The performance of the resulting connection would not be as intended for single-angle connections.

Design Checks

The design strength of a double-angle connection is determined from the applicable limit states for the bolts (see Part 7), welds (see Part 8) and connection elements (see Part 9). In all cases, the design strength ϕR_n must equal or exceed the required strength R_u.

As illustrated in Figure 10-18, the effect of eccentricity should always be considered in the angle leg attached to the support. Additionally, eccentricity should be considered in the case of a double vertical row of bolts through the web of the supported beam or if the eccentricity exceeds 3 in. ($2^3/_4$-in gage plus $^1/_4$-in. half web). Eccentricity should always be considered in the design of welds for single-angle connections.

Recommended Angle Length and Thickness

To provide for stability during erection, it is recommended that the mimimum angle length be one-half the T-dimension of the beam to be supported. The maximum length of the connection angles must be compatible with the T-dimension of an uncoped beam and the remaining web depth, exclusive of fillets, of a coped beam. Note that the angle may encroach upon the fillet(s) as given in Figure 10-4.

A minimum angle thickness of $^3/_8$ in. for $^3/_4$-in. and $^7/_8$-in.-diameter bolts, and $^1/_2$ in. for 1-in.-diameter bolts should be used. A 4×3 angle is normally selected for a single angle welded to the support with the 3-in. leg being the welded leg.

Shop and Field Practices

Single-angle connections may be readily made to the webs of supporting girders and to the flanges of supporting columns. When framing to a column flange, provision must be made for possible mill variation in the depth of the columns. Since the angle is usually shop attached to the column flange, play in the open holes or horizontal slots in the angle leg may be used to provide the necessary adjustment to compensate for the mill variation. Attaching the angle to the column flange offers the advantage of side erection of the beam. The same is true for a girder web or truss support. Additionally, proper bay dimensions may be maintained without the need for shims. This advantage is lost in the case that the angle is shop attached to the supported beam web.

EXAMPLE 10.13. All-bolted single-angle connection (beam-to-girder web).

Given: Design an all-bolted single-angle connection (case I) for a W18×35 beam to W21×62 girder-web connection.

$$R_u = 40\,\text{kips}$$

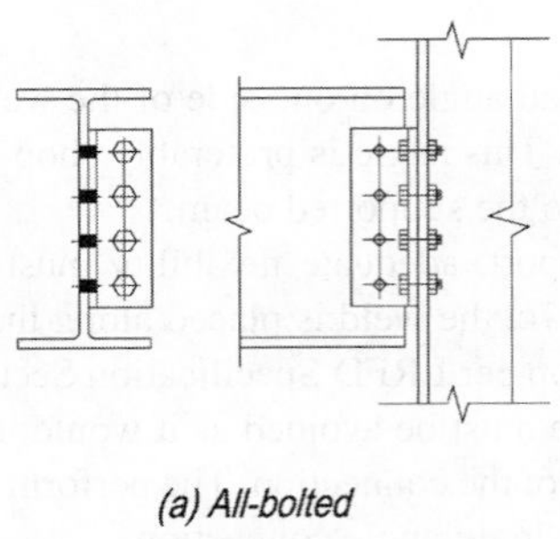

(a) All-bolted

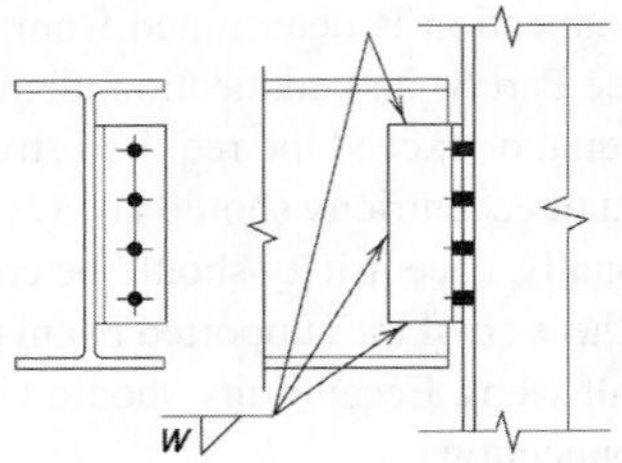

(b) Bolted/welded, angle welded to supported beam

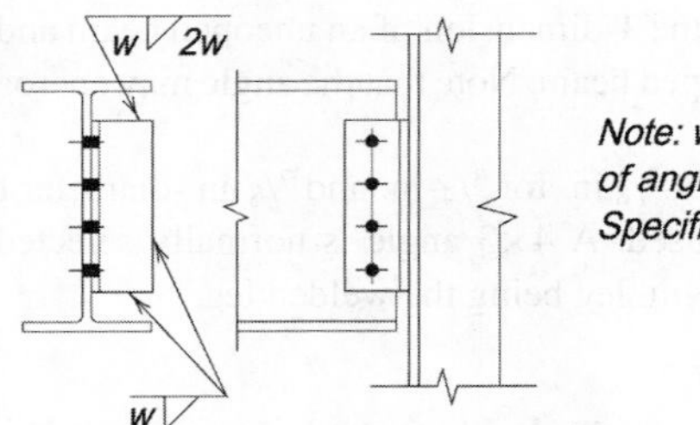

(c) Bolted/welded, angle welded to support

Fig. 10–17. Single-angle connections.

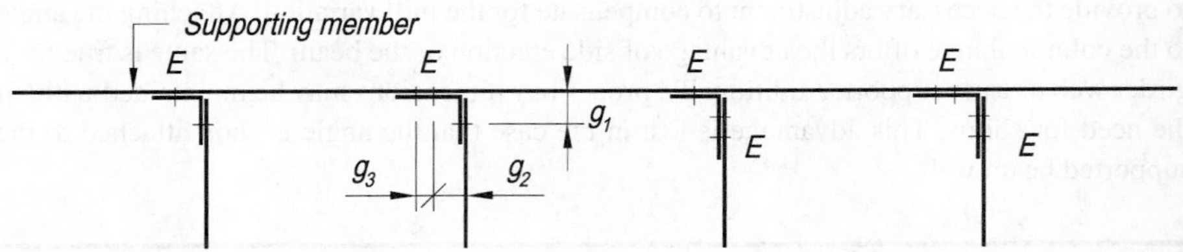

E indicates that eccentricity must be considered in this leg. Gages g_1, g_2, and g_3 are workable gages as shown in Figure 10-6.

Fig. 10–18. Eccentricity in single-angle connections.

W18×35, ASTM A992

$$t_w = 0.300 \text{ in.} \qquad d = 17.7$$
$$F_y = 50 \text{ ksi}, \qquad F_u = 65 \text{ ksi}$$

top flange coped 2 in. deep by 4 in. long

W21×62, ASTM A992

$$t_w = 0.400 \text{ in.}$$
$$F_y = 50 \text{ ksi}, \qquad F_u = 65 \text{ ksi}$$

Use $^3/_4$-in. diameter A325-N bolts in standard holes. Assume ASTM A36 angle material with $F_y = 36$ ksi and $F_u = 58$ ksi.

Solution: *Design bolts and single angle*

Since half-web dimension of W18×35 is less than $^1/_4$ in., tabular values in Table 10-10 may conservatively be used. Bolt shear is more critical than bolt bearing; thus, $\phi r_n = 15.9$ kips

$$\begin{aligned} C_{min} &= \frac{R_u}{\phi r_n} \\ &= \frac{40 \text{ kips}}{15.9 \text{ kips/bolt}} \\ &= 2.52 \end{aligned}$$

From Table 10-10, try a four-bolt connection with a $^3/_8$-in. thick angle (L4×3×$^3/_8$) $C = 3.07 > 2.52$ **o.k.**

The 3-in. leg will be shop bolted to the girder web and the 4-in. leg will be field bolted to the beam web.

Check shear yielding of the angle

$$\begin{aligned} \phi R_n &= 0.9(0.6F_yA_g) \\ &= 0.9[0.6 \times 36 \text{ ksi}(11^1/_2 \text{ in.} \times {}^3/_8 \text{ in.})] \\ &= 83.8 \text{ kips} > 40 \text{ kips} \quad \textbf{o.k.} \end{aligned}$$

Check shear rupture of the angle

$$\begin{aligned} \phi R_n &= 0.75(0.6F_uA_n) \\ &= 0.75[0.6 \times 58 \text{ ksi}(11^1/_2 \text{ in.} \\ &\quad \times {}^3/_8 \text{ in.} - 4 \times 0.875 \text{ in.} \times {}^3/_8 \text{ in.})] \\ &= 78.3 \text{ kips} > 40 \text{ kips} \quad \textbf{o.k.} \end{aligned}$$

Check block shear rupture of the angle

From Tables 9-3 and 9-4, with $L_{eh} = L_{ev} = 1^1/_4$ in., $0.6\,F_uA_{nv} > F_u\,A_{nt}$. Thus,

$$\phi R_n = \phi[0.6F_uA_{nv} + F_yA_{gt}]$$

From Tables 9-4a and 9-4b,

$$\begin{aligned}\phi R_n &= (188\,\text{kips/in.} + 33.8\,\text{kips/in.})^3/_8\,\text{in.}\\ &= 83.2\,\text{kips} > 40\,\text{kips} \quad \textbf{o.k.}\end{aligned}$$

Check flexure of the support-leg of the angle

The required strength M_u is

$$\begin{aligned}M_u &= R_u e\\ &= 40\,\text{kips} \times (2^1/_4\,\text{in.} + 0.300\,\text{in.}/2)\\ &= 96.0\,\text{in.-kips}\end{aligned}$$

For flexural yielding

$$\begin{aligned}\phi M_n &= \phi F_y S_x\\ &= 0.9 \times (36\,\text{ksi})\left[\frac{(^3/_8\,\text{in.})(11^1/_2\,\text{in.})^2}{6}\right]\\ &= 268\,\text{in.-kips} > 96.0\,\text{in.-kips} \quad \textbf{o.k.}\end{aligned}$$

For flexural rupture using general equation from Table 15-2 (bracket plates),

$$\begin{aligned}\phi M_n &= \phi F_u S_{net}\\ &= 0.75(58\,\text{ksi})\left[\frac{^3/_8\,\text{in.}}{6}\left((11^1/_2\,\text{in.})^2 - \frac{(3\,\text{in.})^2(4)(4^2-1)(0.875\,\text{in.})}{11^1/_2\,\text{in.}}\right)\right]\\ &= 248\,\text{in.-kips} > 96.0\,\text{in.-kips} \quad \textbf{o.k.}\end{aligned}$$

Check the supported beam web

From Table 10-1, for four rows of bolts, beam material with $F_y = 50\,\text{ksi}$ and $F_u = 65\,\text{ksi}$, and $L_{ev} = 1^1/_4$ in. and $L_{eh} = 1^1/_2$ in. (Assumed to be $1^1/_4$ in. for calculation purposes to provide for possible underrun in beam length),

$$\begin{aligned}\phi R_n &= (250\,\text{kips/in.})(0.300\,\text{in.})\\ &= 75.0\,\text{kips} > 40\,\text{kips} \quad \textbf{o.k.}\end{aligned}$$

Check flexural yielding on the coped section

From Table 9-2, $S_{net} = 18.2\,\text{in.}^3$

$$\begin{aligned}\phi R_n &= \frac{0.9\,F_y S_{net}}{e}\\ &= \frac{0.9(50\,\text{ksi})(18.2\,\text{in.}^3)}{(4\,\text{in.} + ^1/_2\,\text{in.})}\\ &= 182\,\text{kips} > 40\,\text{kips} \quad \textbf{o.k.}\end{aligned}$$

Check local web buckling at the cope

$$\frac{c}{d} = \frac{4\,\text{in.}}{17.7\,\text{in.}} = 0.226$$

$$\frac{c}{h_o} = \frac{4\,\text{in.}}{17.7\,\text{in.} - 2\,\text{in.}} = 0.255$$

Since $\dfrac{c}{d} \leq 1.0$,

$$f = 2\left(\frac{c}{d}\right)$$
$$= 2(0.226)$$
$$= 0.452$$

Since $\dfrac{c}{h_o} \leq 10$,

$$k = 2.2\left(\frac{h_o}{c}\right)^{1.65}$$
$$= 2.2\left(\frac{1}{0.255}\right)^{1.65}$$
$$= 21.0$$

$$\phi F_{bc} = 23{,}590\left(\frac{t_w}{h_o}\right)^2 fk$$
$$= 23{,}590\left(\frac{0.300\text{ in.}}{17.7\text{ in.} - 2\text{ in.}}\right)^2 (0.452)(21.0)$$
$$= 81.8\text{ ksi}$$

$$\phi R_n = \frac{\phi F_{bc} S_{net}}{e}$$
$$= \frac{(81.8\text{ ksi})(18.2\text{ in.}^3)}{(4\text{ in.} + 1/2\text{ in.})}$$
$$= 331\text{ kips} > 40\text{ kips} \quad \textbf{o.k.}$$

Check supporting girder web

From Table 10-1 for four rows of bolts and girder material with F_u = 65 ksi. Taking half the tabulated value,

$$\phi R_n = 1/2(702\text{ kips/in.})(0.400\text{ in.})$$
$$= 140\text{ kips} > 40\text{ kips} \quad \textbf{o.k.}$$

EXAMPLE 10.14. Bolted/welded single-angle connection (beam-to-column flange).

Given: Design a single-angle connection for a W16×50 beam to W14×90 column flange connection.

$R_u = 55$ kips

W16×50, ASTM A992

$t_w = 0.380$ in. $\quad d = 16.3$ in. $\quad t_f = 0.630$ in.
$F_y = 50$ ksi, $\quad F_u = 65$ ksi

W14×90, ASTM A992

$t_f = 0.710$
$F_y = 50$ ksi, $\quad F_u = 65$ ksi

Use $^3/_4$-in. diameter A325-N bolts to connect the supported beams to the single angle. Use 70 ksi electrode welds to connect the single angle to the column flange. Assume ASTM A36 angle material with $F_y = 36$ ksi and $F_u = 58$ ksi.

Solution: *Design single angle, bolts, and welds*

Since half-web dimension of W16×50 is less than $^1/_4$ in., tabular values in Table 10-11 may conservatively be used.

From Table 10-11, try a four-bolt single angle (L4×3×$^3/_8$).

$$\phi R_n = 63.6\,\text{kips} > 55\,\text{kips} \qquad \textbf{o.k.}$$

Also from Table 10-11, with a $^3/_{16}$-in. fillet weld size

$$\phi R_n = 56.4\,\text{kips} > 55\,\text{kips} \qquad \textbf{o.k.}$$

Use four-bolt single-angle (L4×3×$^3/_8$). The 3-in. leg will be shop welded to the column flange and the 4-in. leg will be field bolted to the beam web.

Check supported beam web

The bearing strength of the beam web per bolt is

$$\begin{aligned} \phi r_n &= \phi(2.4dtF_u) \\ &= 0.75(2.4\times{}^3/_4\text{ in.} \times 0.380\text{ in.} \times 65\text{ ksi}) \\ &= 33.3\text{ kips} \end{aligned}$$

Since this exceeds the single shear strength per bolt, bolt bearing on the beam web is not critical.

Check support

From Table 10-11, the minimum support thickness for the $^3/_{16}$-in. welds is 0.286 in.

$$t_w = 0.710 > 0.286 \qquad \textbf{o.k.}$$

Table 10-10. All-Bolted Single-Angle Connections

Table 10-10 is a design aid for all-bolted single-angle connections. The tabulated eccentrically loaded bolt group coefficients C are useful in determining the design strength ϕR_n, where

$$\phi R_n = C \times \phi r_n$$

In the above equation,

C = coefficient from Table 10-10
ϕr_n = the lesser of the design strength of one bolt in shear or bearing, kips

Table 10-11. Bolted/Welded Single-Angle Connections

Table 10-11 is a design aid for bolted/welded single angle connections. Electrode strength is assumed to be 70 ksi. All values are for comparison with factored loads. In the rare case where a single-angle connection must be field welded, erection bolts may be placed in the leg to be field welded.

Weld design strengths are determined by the instantaneous center of rotation method using Table 8-11 with $\theta = 0°$. The tabulated values assume a half-web thickness of $^1/_4$ in. and may be used conservatively for lesser half-web thicknesses. For half-web thicknesses greater than $^1/_4$ in., the tabulated values should be reduced proportionally to eight percent at a half-web thickness of $^1/_2$ in. The tabulated minimum supporting flange or web thickness is the thickness that matches the strength of the support material to the strength of the weld material. In a manner similar to that illustrated previously for Tables 10-2, the minimum material thickness (for one line of weld) may be calculated as:

$$t = \frac{3.09D}{F_u}$$

where D is the number of sixteenths in the weld size. When welds line up on opposite sides of the support, the minimum thickness is the sum of the thicknesses required for each weld. In either case, when less than the minimum material thickness is present, the tabulated weld design strength should be multiplied by the ratio of the thickness provided to the minimum thickness.

Table 10-10.
All-Bolted Single-Angle Connections

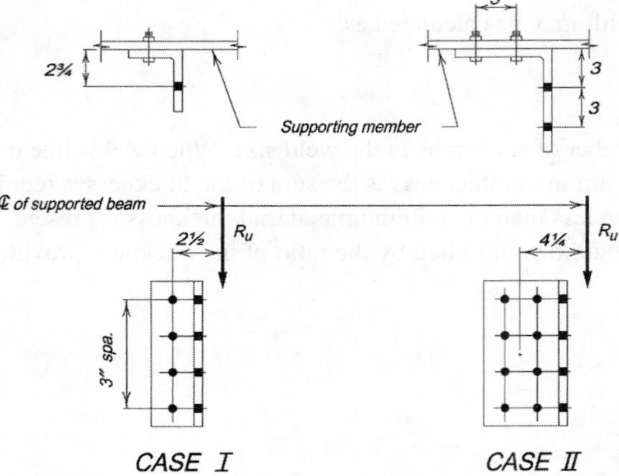

Note: standard holes in support leg of angle

Eccentrically Loaded Bolt Group Coefficients, *C*

Number of Bolts in One Vertical Row, *n*	Case I	Case II
12	11.4	21.5
11	10.4	19.4
10	9.37	17.3
9	8.34	15.2
8	7.31	13.0
7	6.27	10.9
6	5.22	8.70
5	4.15	6.63
4	3.07	4.70
3	1.99	2.94
2	1.03	1.61
1	–	0.518

$\phi R_n = C \times \phi r_n$

where

C = coefficient from Table above
ϕr_n = design strength of one bolt in shear or bearing, kips/bolt

Notes:
For eccentricities less than or equal to those shown above, tabulated values may be conservatively used.
For greater eccentricities, coefficient C should be recalculated from Table 7-17 or Table 7-18.
Connection may be bearing-type or slip-critical.

Table 10-11. Bolted/Welded Single-Angle Connections

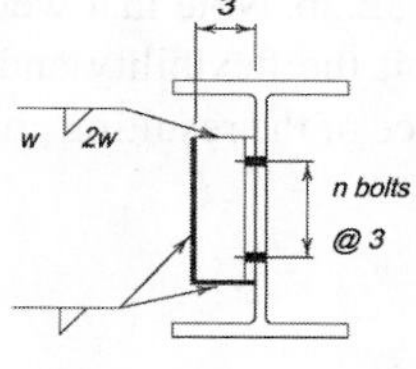

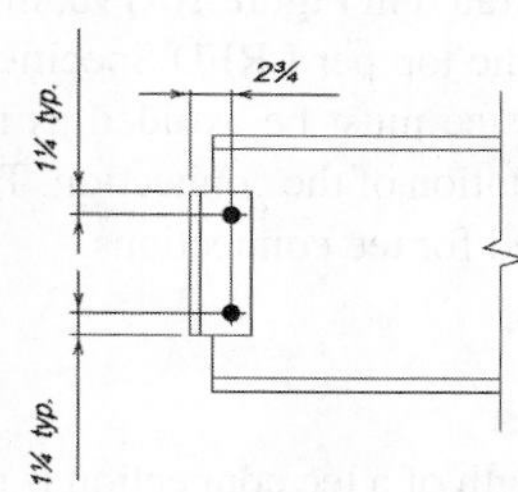

Number of Bolts in One Vertical Row	A325-N Bolt Shear Strength, kips		Angle Size ($F_y = 36$ ksi)	Angle Length, in.	Weld (70 ksi)		
	3/4 in.	7/8 in.			Design Strength, kips	Size, in.	Minimum t_w of Supporting Member with Angles Both Sides of Web
12	191	260		35 1/2	269 215 161	5/16 1/4 3/16	0.476 0.381 0.286
11	175	238		32 1/2	247 198 148	5/16 1/4 3/16	0.476 0.381 0.286
10	159	216		29 1/2	226 181 136	5/16 1/4 3/16	0.476 0.381 0.286
9	143	195		26 1/2	205 164 123	5/16 1/4 3/16	0.476 0.381 0.286
8	127	173	L4×3×3/8	23 1/2	185 148 111	5/16 1/4 3/16	0.476 0.381 0.286
7	111	152		20 1/2	164 131 98.3	5/16 1/4 3/16	0.476 0.381 0.286
6	95.4	130		17 1/2	141 113 84.8	5/16 1/4 3/16	0.476 0.381 0.286
5	79.5	108		14 1/2	118 94.6 71.0	5/16 1/4 3/16	0.476 0.381 0.286
4	63.6	86.6		11 1/2	94.1 75.3 56.4	5/16 1/4 3/16	0.476 0.381 0.286
3	47.7	64.9		8 1/2	68.5 54.8 41.1	5/16 1/4 3/16	0.476 0.381 0.286
2	31.8	43.3		5 1/2	42.2 33.8 25.3	5/16 1/4 3/16	0.476 0.381 0.286

Notes:
Gage in angle leg attached to beam web as well as leg width may be decreased. 3-in. welded leg may not be increased or decreased.
Tabulated weld design strengths are based on a 1/4-in. half web for the supported member. Smaller half webs will result in these values being conservative. For half webs over 1/4 in., weld values must be reduced proportionally to 8% for a 1/2-in. half web or recalculated.
When the beam web thickness of the supporting member is less than the minimum and single-angle connections are back to back, either stagger the angles or multiply the weld design strength by the ratio of the actual web thickness to the tabulated minimum thickness to determine the reduced weld design strength.

TEE CONNECTIONS

A tee connection is made with a structural tee as illustrated in Figure 10-19. The tee is preferably shop bolted or welded to the supporting member and field bolted to the supported beam.

When the tee is welded to the support, adequate flexibility must be provided in the connection. As illustrated in Figure 10-19b, line welds are placed along the toes of the tee flange with a return at the top per LRFD Specification Section J2.2b. Note that welding across the entire top of the tee must be avoided as it would inhibit the flexibility and, therefore, the necessary end rotation of the connection. The performance of the resulting connection would not be as intended for tee connections.

Design Checks

The design strength of a tee connection is determined from the applicable limit states for the bolts (see Part 7), welds (see Part 8) and connection elements (see Part 9). In all cases, the design strength ϕR_n must equal or exceed the required strength R_u.

Eccentricity must be considered in the design of tee connections. For a flexible support, the bolts or welds attaching the tee flange to the support must be designed for the shear R_u. Also, the bolts through the tee stem must be designed for the shear R_u and the eccentric moment $R_u a$ where a is the distance from the face of the support to the centroid of the bolt group through the tee stem.

For a rigid support, the bolts or welds attaching the tee flange to the support must be designed for the shear R_u and the eccentric moment $R_u a$; the bolts through the tee stem must be designed for the shear R_u.

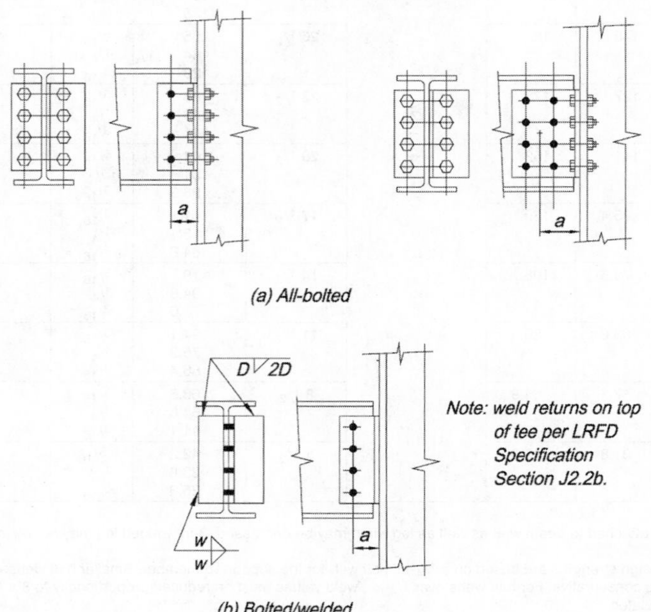

Fig. 10–19. Tee connections.

Recommended Tee Length and Flange and Web Thicknesses

To provide for stability during erection, it is recommended that the mimimum tee length be one-half the T-dimension of the beam to be supported. The maximum length of the tee must be compatible with the T-dimension of an uncoped beam and the remaining web depth, exclusive of fillets, of a coped beam. Note that the tee may encroach upon the fillet(s) as given in Figure 10-4.

To provide for flexibility, the tee selected should meet the ductility checks illustrated in Part 9. The flange thickness of tees used in simple shear connections should be held to a minimum to permit the flexure necessary to accommodate the end rotation of the beam, unless the tee connection is proportioned to meet the geometric requirements for single-plate connections.

Shop and Field Practices

When framing to a column flange, provision must be made for possible mill variation in the depth of the columns. If the tee is shop attached to the column flange, play in the open holes usually furnishes the necessary adjustment to compensate for the mill variation. This approach offers the advantage of side erection of the beam. Alternatively, if the tee is shop attached to the supported beam web, the beam length could be shortened to provide for mill overrun and shims could be furnished at the appropriate intervals to fill the resulting gaps or to provide for mill underrun.

When a single vertical row of bolts is used in a tee stem, a 4-in. or 5-in. stem is required to accommodate the end distance of the supported beam and possible overrun/underrun in beam length. A double vertical row of bolts will require a 7-in. or 8-in. tee stem. There is no maximum limit on l_h for the tee stem.

EXAMPLE 10.15. All-bolted tee connection (beam-to-column flange).

Given:

Design an all-bolted tee connection for a W16×50 beam to a W14×90 column flange.

$$R_u = 55\text{ kips}$$

W16×50, ASTM A992

$$t_w = 0.380\text{ in.} \qquad d = 16.3\text{ in.} \qquad t_f = 0.630\text{ in.}$$
$$F_y = 50\text{ ksi}, \qquad F_u = 65\text{ ksi}$$

W14×90, ASTM A992

$$t_f = 0.710\text{ in.}$$
$$F_y = 50\text{ ksi}, \qquad F_u = 65\text{ ksi}$$

Use $3/4$-in. diameter A325-N bolts in standard holes. Assume the ASTM A992 tee has $F_y = 50$ ksi and $F_u = 65$ ksi.

Solution:

Try WT5×22.5($d = 5.05$ in., $b_f = 8.02$ in., $t_f = 0.620$ in., $t_s = 0.350$ in., $k_1 = 13/16$ in.) with a four-bolt connection ($L = 11\,1/2$ in.) and $L_{eh} = 1\,1/4$ in.

Check limitation on tee stem thickness

$$t_{s\,max} = \frac{d_b}{2} + {}^1\!/_{16}\text{ in.}$$
$$= \frac{{}^3\!/_4\text{ in.}}{2} + {}^1\!/_{16}\text{ in.}$$
$$= 0.438\text{ in.} > 0.350\text{ in.} \qquad \textbf{o.k.}$$

Check limitation on bolt diameter for bolts through tee flange

Assuming a 5½-in. gage,

$$b = \frac{g - 2k_1}{2}$$
$$= \frac{5{}^1\!/_2\text{ in.} - 2({}^{13}\!/_{16}\text{ in.})}{2}$$
$$= 1.94\text{ in.}$$
$$d_{b\,min} = 0.613t_f\sqrt{\frac{F_y}{b}\left(\frac{b^2}{L^2}+2\right)} \le 0.69\sqrt{t_s}$$
$$= 0.613(0.620\text{ in.})\sqrt{\frac{50\text{ ksi}}{1.94\text{ in.}}\left(\frac{(1.94\text{ in.})^2}{(11{}^1\!/_2\text{ in.})^2}+2\right)}$$
$$\le 0.69\sqrt{0.350\text{ in.}}$$
$$= 2.75\text{ in.} \le 0.408\text{ in.}$$
$$= 0.408\text{ in.}$$

Since $d_b = {}^3\!/_4$ in. $> d_{b\,min} = 0.408$ in., **o.k.**

Check bolt group through beam web for shear and bearing

$$a = d - L_{eh}$$
$$= 5.05\text{ in.} - 1{}^1\!/_4\text{ in.}$$
$$= 3.80\text{ in.}$$

Assuming the column provides a rigid support,

$$e_b = 0$$

Since bolt shear is more critical than bolt bearing, $\phi r_n = 15.9$ kips, Thus,

$$\phi R_n = n \times \phi r_n$$
$$= 4\text{ bolts} \times 15.9\text{ kips}$$
$$= 63.6\text{ kips} > 55\text{ kips} \qquad \textbf{o.k.}$$

Check shear yielding of the tee stem

$$\phi R_n = 0.9(0.6F_yA_g)$$
$$= 0.9[0.6 \times 50\text{ ksi}(11{}^1\!/_2\text{ in.} \times 0.350\text{ in.})]$$
$$= 109\text{ kips} > 55\text{ kips} \qquad \textbf{o.k.}$$

Check shear rupture of the tee stem

$$\phi R_n = 0.75(0.6F_uA_n)$$
$$= 0.75[0.6 \times 65\text{ ksi}(11{}^1\!/_2\text{ in.} - 4 \times 0.875\text{ in.})(0.350\text{ in.})]$$
$$= 81.9\text{ kips} > 55\text{ kips} \qquad \textbf{o.k.}$$

Check block shear rupture of the tee stem

From Tables 9-3 and 9-4, with $L_{eh} = L_{ev} = 1^1/_4$ in., $0.6\,F_u A_{nv} > F_u A_{nt}$. Thus,

$$\phi R_n = \phi[0.6 F_u A_{nv} + F_y A_{gt}]$$

From Tables 9-4a and 9-4b,

$$\begin{aligned}\phi R_n &= (210\,\text{kips/in.} + 46.9\,\text{kips/in.})(0.350\,\text{in.})\\ &= 89.9\,\text{kips} > 55\,\text{kips} \quad \textbf{o.k.}\end{aligned}$$

Check bolt group through support for shear and bearing combined with tension due to eccentricity

Calculate tensile force per bolt r_{ut}.

$$2r_{ut}\,[2 \times (1.5\text{ in.} + 4.5\text{ in.})] = R_u e$$

$$\begin{aligned}r_{ut} &= \frac{55\,\text{kips}\,(5.05\,\text{in.} - 1^1/_4\,\text{in.})}{2\,\text{bolts}(12\,\text{in.})}2\,\text{bolts}(12\,\text{in.})\\ &= 8.71\,\text{kips/bolts}\end{aligned}$$

Check design strength of bolts for tension-shear interaction

$$\begin{aligned}r_{uv} &= \frac{55\,\text{kips}}{8\,\text{bolts}}\\ &= 6.88\,\text{kips/bolt} < 15.9\,\text{kips/bolt} \qquad \textbf{o.k.}\end{aligned}$$

$$\begin{aligned}F_t &= 117\,\text{ksi} - 2.5 f_v \le 90\,\text{ksi}\\ &= 117\,\text{ksi} - 2.5\left(\frac{6.88\,\text{kips/bolts}}{0.442\,\text{in.}^2}\right) \le 90\,\text{ksi}\\ &= 78.1\,\text{ksi}\end{aligned}$$

$$\begin{aligned}\phi r_n &= \phi F_t A_b\\ &= 0.75(78.1\,\text{ksi})(0.442\,\text{in.}^2)\\ &= 25.9\,\text{kips/bolt} > 8.71\,\text{kips/bolt} \qquad \textbf{o.k.}\end{aligned}$$

Check bearing strength at bolt holes

With $L_e = 1^1/_4$ in. and $s = 3$ in., the bearing strength of the tee flange exceeds the single shear strength of the bolts. Therefore, bearing strength is **o.k.**

Check prying action

$$\begin{aligned}b &= \frac{g - t_s}{2}\\ &= \frac{5^1/_2\,\text{in.} - 0.350\,\text{in.}}{2}\\ &= 2.58\,\text{in.}\end{aligned}$$

$$\begin{aligned}a &= \frac{b_f - g}{2}\\ &= \frac{8.02\,\text{in.} - 5^1/_2\,\text{in.}}{2}\\ &= 1.26\,\text{in.}\end{aligned}$$

Since $a = 1.26$ in. is less than $1.25b = 3.23$ in., use $a = 1.26$ in. for calculation purposes.

$$\begin{aligned} b' &= b - d/2 \\ &= 2.58\text{ in.} - (^3\!/_4\text{ in.}/2) \\ &= 2.21\text{ in.} \end{aligned}$$

$$\begin{aligned} a' &= a + d/2 \\ &= 1.26\text{ in.} + (^3\!/_4\text{ in.}/2) \\ &= 1.64\text{ in.} \end{aligned}$$

$$\begin{aligned} \rho &= \frac{b'}{a'} \\ &= \frac{2.21\text{ in.}}{1.64\text{ in.}} \\ &= 1.35 \end{aligned}$$

$$\begin{aligned} \beta &= \frac{1}{\rho}\left(\frac{\phi r_n}{r_{ut}} - 1\right) \\ &= \frac{1}{1.35}\left(\frac{29.0\text{ kips/bolts}}{8.71\text{ kips/bolts}} - 1\right) \\ &= 1.72 \end{aligned}$$

Since $\beta \geq 1$, set $\alpha' = 1.0$

$$\begin{aligned} p &= \frac{11^1\!/_2\text{ in.}}{4\text{ bolts}} \\ &= 2.88\text{ in./bolt} \end{aligned}$$

$$\begin{aligned} \delta &= 1 - \frac{d'}{p} \\ &= 1 - \frac{^{13}\!/_{16}\text{ in.}}{2.88\text{ in.}} \\ &= 0.718 \end{aligned}$$

$$\begin{aligned} t_{req} &= \sqrt{\frac{4.44 r_{ut} b'}{p F_y (1 + \delta\alpha')}} \\ &= \sqrt{\frac{4.44(8.71\text{ kips/bolt})(2.21\text{ in.})}{(2.88\text{ in./bolt})(50\text{ ksi})[1 + (0.718)(1.0)]}} \\ &= 0.588\text{ in.} < 0.620\text{ in.} \qquad \textbf{o.k.} \end{aligned}$$

Similarly, checks of the tee flange for shear yielding, shear rupture, and block shear will show that the tee flange is **o.k.**

Check the supported beam web

From Table 10-1, for four rows of $^3\!/_4$-in. diameter bolts and an uncoped beam with $F_y = 50$ ksi and $F_u = 65$ ksi,

$$\phi R_n = (351\text{ kips/in.})(0.380\text{ in.})$$
$$= 133\text{ kips} > 55\text{ kips} \qquad \textbf{o.k.}$$

Check the supporting column flange

From Table 10-1, for four rows of $^3/_4$-in. diameter bolts with $F_y = 50$ ksi and $F_u = 65$ ksi,

$$\phi R_n = (702\text{ kips/in.})(0.710\text{ in.})$$
$$= 498\text{ kips} > 55\text{ kips} \qquad \textbf{o.k.}$$

EXAMPLE 10.16. Bolted/welded tee connection (beam-to-column flange).

Given: Redesign the tee connection of Example 10-15 to be bolted to the supported beam and welded to the support for a factored end reaction

$$R_u = 37\text{ kips}.$$

Solution: Try WT5×22.5 ($d = 5.05$ in., $b_f = 8.02$ in., $t_f = 0.620$ in., $t_s = 0.350$ in., $k_1 = {}^{13}/_{16}$-in.) with a four-bolt connection ($L = 11^1/_2$ in.) and $L_{eh} = 1^1/_4$ in.

Check limitation on tee stem thickness

$$t_{s\,\max} = \frac{d_b}{2} + {}^1/_{16}\text{ in.}$$
$$= \frac{^3/_4\text{ in.}}{2} + {}^1/_{16}\text{ in.}$$
$$= 0.438\text{ in.} > 0.350\text{ in.} \qquad \textbf{o.k.}$$

Design the welds connecting the tee flange to the column flange

This connection is inherently flexible because the welds are at the toes of the WT flanges. This is true independent of the rigidity of the support. Therefore, it is recommended that this connection be designed with a flexible support condition; any rigidity-induced weld forces are accounted for by the minimum required weld size $w_{\min}$. Thus,

$$b = \frac{b_f - 2k_1}{2}$$
$$= \frac{8.02\text{-in.} - 2(^{13}/_{16}\text{ in.})}{2}$$
$$= 3.20\text{ in.}$$

$$w_{\min} = 0.0158\frac{F_y {t_f}^2}{b}\left(\frac{b^2}{L^2} + 2\right) \le {}^3/_4\, t_s$$
$$= 0.0158\frac{(50\text{ ksi})(0.620\text{ in.})^2}{(3.20\text{ in.})}\left(\frac{(3.20\text{ in.})^2}{(11^1/_2\text{ in.})^2} + 2\right) \le {}^3/_4(0.350\text{ in.})$$
$$= 0.197\text{ in.} \le 0.263\text{ in.}$$
$$= 0.197\text{ in.}$$

Try $^1/_4$-in. fillet welds.

$$\begin{aligned}\phi R_n &= 1.392Dl\\ &= 1.392(4\text{ sixteenths})(2\times 11^1/_2\text{ in.})\\ &= 128\text{ kips} > 37\text{ kips} \qquad \textbf{o.k.}\end{aligned}$$

Use $^1/_4$-in. fillet welds.

Check stem side of connection

Since the connection is flexible, the tee stem and bolts must be designed for both the shear and the eccentric moment e_b where

$$\begin{aligned}e_b &= a\\ &= 3.80\text{ in.}\end{aligned}$$

Thus the tee stem and bolts must be designed for $R_u = 37$ kips and $R_u e_b = 141$ in.-kips.

Check bolt group through beam web for shear and bearing

From Table 7-17 for $\theta = 0°$ with $s = 3$ in., $e_x = e_b = 3.80$ in., and $n =$ 4 bolts,

$$C = 2.45$$

and, since bolt shear is more critical than bolt bearing,

$$\begin{aligned}\phi R_n &= C \times \phi r_n\\ &= 2.45(15.9\text{ kips/bolt})\\ &= 39.0\text{ kips} > 37\text{ kips} \qquad \textbf{o.k.}\end{aligned}$$

Check flexure on the tee stem

For flexural yielding,

$$\begin{aligned}\phi M_n &= \phi F_y S_x\\ &= 0.9(50\text{ ksi})\frac{(0.350\text{ in.})(11^1/_2\text{ in.})^2}{6}\\ &= 347\text{ in.-kips} > 141\text{ in.-kips} \qquad \textbf{o.k.}\end{aligned}$$

For flexural rupture (see Table 15-2),

$$\begin{aligned}S_{net} &= \frac{t}{6}\left[d^2 - \frac{s^2 n(n^2-1)(d_b + 0.125\text{ in.})}{d}\right]\\ &= \frac{0.350\text{ in.}}{6}\left[(11^1/_2\text{ in.})^2 - \frac{(3\text{ in.})^2 4(4^2-1)(0.875\text{ in.})}{11^1/_2\text{ in.}}\right]\\ &= 5.32\text{ in.}^3\end{aligned}$$

$$\begin{aligned}\phi M_n &= \phi F_u S_{net}\\ &= 0.75(65\text{ ksi})(5.32\text{ in.}^3)\\ &= 259\text{ in.-kips} > 141\text{ in.-kips} \qquad \textbf{o.k.}\end{aligned}$$

Check shear yielding of the tee stem

$$\begin{aligned}\phi R_n &= 0.9(0.6F_yA_g)\\ &= 0.9[0.6 \times 50\,\text{ksi}(11^1/_2\text{ in.} \times 0.350\,\text{in.})]\\ &= 109\,\text{kips} > 37\,\text{kips} \qquad \textbf{o.k.}\end{aligned}$$

Check shear rupture of the tee stem

$$\begin{aligned}\phi R_n &= 0.75(0.6F_uA_n)\\ &= 0.75[0.6 \times 65\,\text{ksi}(11^1/_2\text{ in.} - 4 \times 0.875\,\text{in.})(0.350\,\text{in.})]\\ &= 81.9\,\text{kips} > 37\,\text{kips} \qquad \textbf{o.k.}\end{aligned}$$

Check block shear rupture of the tee stem

From Tables 9-3 and 9-4, with $L_{eh} = L_{ev} = 1^1/_4$ in., $0.6\,F_uA_{nv} > F_u\,A_{nt}$. Thus,

$$\phi R_n = \phi[0.6FuA_{nv} + F_yA_{gt}]$$

From Tables 9-4a and 9-4b,

$$\begin{aligned}\phi R_n &= (210\,\text{kips/in.} + 46.9\,\text{kips/in.})(0.350\,\text{in.})\\ &= 89.9\,\text{kips} > 37\,\text{kips} \qquad \textbf{o.k.}\end{aligned}$$

Check supported beam web

From Tables 10-1, for four rows of $^3/_4$-in. diameter bolts and an uncoped beam with $F_y = 50$ ksi and $F_u = 65$ ksi,

$$\begin{aligned}\phi R_n &= (351\,\text{kips/in.})(0.380\,\text{in.})\\ &= 133\,\text{kips} > 37\,\text{kips} \qquad \textbf{o.k.}\end{aligned}$$

Check supporting column flange

From Tables 10-2, for beam web material with $F_y = 50$ ksi, the minimum support thickness is 0.238 in. for $^1/_4$-in. fillet welds

$$t_f = 0.710\,\text{in.} > 0.238\,\text{in.} \qquad \textbf{o.k.}$$

SHEAR SPLICES

Shear splices are usually made with a single plate, as shown in Figure 10-20a, or two plates, as shown in Figures 10-20b and 10-20c. Although the rotational flexibility required at a shear splice is usually much less than that required at the end of a simple-span beam, when a highly flexible splice is desired, the splice utilizing four framing angles, shown in Figure 10-21, is especially useful. These shear splices may be bolted and/or welded.

The design strength of a shear splice is determined from the applicable limit states for the bolts (see Part 7), welds (see Part 8) and connection elements (see Part 9). In all cases, the design strength ϕR_n must equal or exceed the required strength R_u.

Eccentricity must be considered in the design of shear splices except all-bolted shear splices utilizing four framing angles as illustrated in Figure 10-21. When the splice is symmetrical, as shown for the bolted splice in Figure 10-20a, each side of the splice is equally restrained regardless of the relative flexibility of the spliced members. Accordingly, as illustrated in Figure 10-22, the eccentricity of the shear to the center of gravity of either bolt group is equal to half the distance between the centroids of the bolt groups. Therefore, each bolt group can be designed for the shear R_u and one-half the eccentric moment $R_u e$ (Kulak and Green, 1990). This approach is also applicable to symmetrical welded splices.

When the splice is not symmetrical, as shown in Figures 10-20b and 10-20c, one side of the splice will possess a higher degree of rigidity. For the splice shown in Figure 10-20b, the right side is more rigid because the stiffness of the weld group exceeds the stiffness of the bolt group, even if the bolts are pretensioned or slip-critical. Also, for the splice shown in Figure 10-20c, the right side is more rigid since there are two vertical rows of bolts while

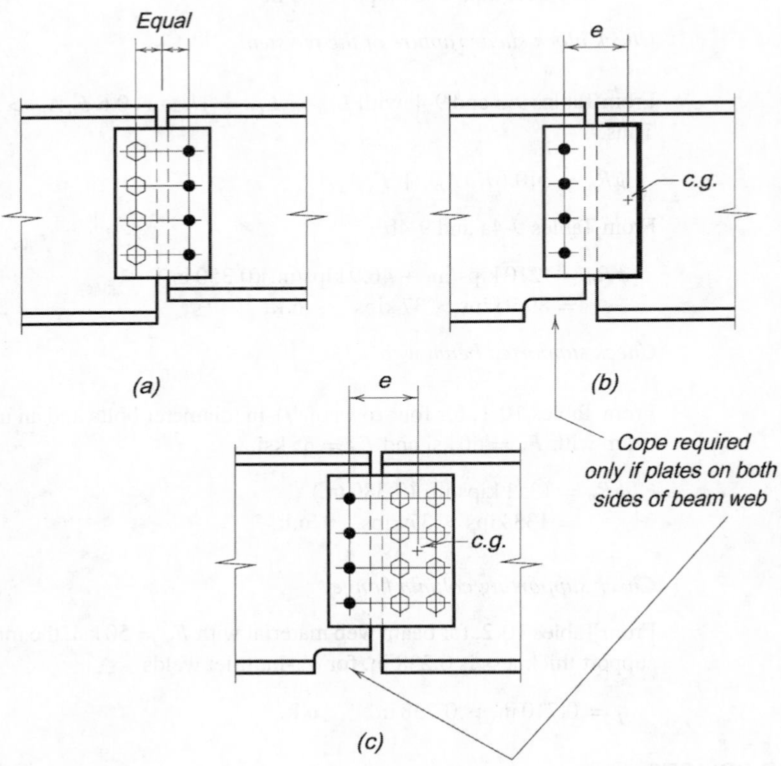

Fig. 10–20. Plate-type shear splices.

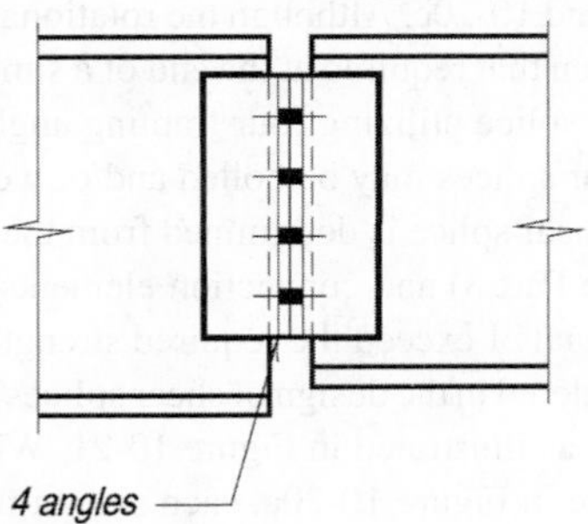

Fig. 10–21. Angle-type shear splice.

the left side has only one. In these cases, it is conservative to design the side with the higher rigidity for the shear R_u and the full eccentric moment $R_u e$. The side with the lower rigidity can then be designed for the shear R_u only. This approach is applicable regardless of the relative flexibility of the spliced members.

Some splices, such as those that occur at expansion joints, require special attention and are beyond the scope of this Manual.

EXAMPLE 10.17. All-bolted single-plate shear splice.

Given:

Design an all-bolted single-plate shear splice between a W24×55 beam and W24×68 beam.

$$R_u = 60\,\text{kips}$$

W24×55, ASTM A992

$$t_w = 0.395\,\text{in.}$$
$$F_y = 50\,\text{ksi},\ F_u = 65\,\text{ksi}$$

W24×68, ASTM A992

$$t_w = 0.415\,\text{in.}$$
$$F_y = 50\,\text{ksi},\ F_u = 65\,\text{ksi}$$

Use $^7/_8$-in. diameter A325-N bolts with five inches between vertical bolt rows. Assume ASTM A36 plate material with $F_y = 36\,\text{ksi}$ and $F_u = 58\,\text{ksi}$.

Solution:

Design bolt groups

Using a symmetrical splice, each bolt group will carry one-half the eccentric moment. Thus, the eccentricity on each bolt group $e = 2^1/_2$ in. For bolt shear, $\phi r_n = 21.6$ kips/bolt. For bearing on the web of the W24×55,

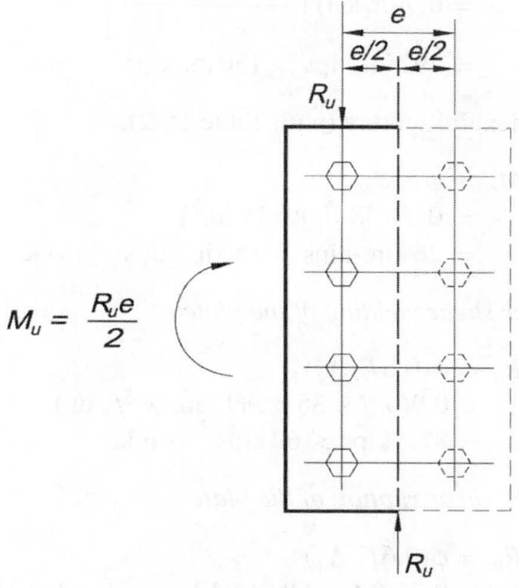

Fig. 10–22. Eccentricity in a symmetrical shear splice.

$\phi r_n = 40.4$ kips/bolt. Since bolt shear is more critical,

$$C_{\min} = \frac{R_u}{\phi R_n} = \frac{60 \text{ kips}}{21.6 \text{ kips/bolt}} = 2.78$$

From Table 7-17 with $\theta = 0°$ and $e_x = 2^1/_2$ in., a four-bolt connection provides

$C = 3.07 > 2.78$ **o.k.**

Design splice plate

Try PL $^3/_8$ in.×8 in.×1′-0.

Check bolt bearing on plate

$$\begin{aligned}\phi R_n &= C(2.4dtF_u)\\ &= 3.07(2.4\times{}^7/_8 \text{ in.} \times {}^3/_8 \text{ in.} \times 58 \text{ ksi})\\ &= 140 \text{ kips} > 60 \text{ kips} \quad \textbf{o.k.}\end{aligned}$$

Check flexure of the plate

$$\begin{aligned}M_u &= \frac{R_u e}{2}\\ &= \frac{60 \text{ kips} \times 5 \text{ in.}}{2}\\ &= 150 \text{ in.-kips}\end{aligned}$$

For flexural yielding,

$$\begin{aligned}\phi M_n &= \phi F_y S_x\\ &= 0.9(36 \text{ ksi})\left[\frac{{}^3/_8 \text{ in.}(12 \text{ in.})^2}{6}\right]\\ &= 292 \text{ in.-kips} > 150 \text{ in.-kips} \quad \textbf{o.k.}\end{aligned}$$

For flexural rupture (with Table 15-2),

$$\begin{aligned}\phi M_n &= \phi F_u S_{net}\\ &= 0.75(58 \text{ ksi})(6.19 \text{ in.}^3)\\ &= 269 \text{ in.-kips} > 150 \text{ in.-kips} \quad \textbf{o.k.}\end{aligned}$$

Check shear yielding of the plate

$$\begin{aligned}\phi R_n &= \phi(0.6F_y A_g)\\ &= 0.9(0.6 \times 36 \text{ ksi})(12 \text{ in.} \times {}^3/_8 \text{ in.})\\ &= 87.5 \text{ kips} > 60 \text{ kips} \quad \textbf{o.k.}\end{aligned}$$

Check shear rupture of the plate

$$\begin{aligned}\phi R_n &= \phi(0.6F_u A_n)\\ &= 0.75(0.6 \times 58 \text{ ksi})(12 \text{ in.} - 4 \times 1 \text{ in.}){}^3/_8 \text{ in.}\\ &= 78.3 \text{ kips} > 60 \text{ kips} \quad \textbf{o.k.}\end{aligned}$$

Check block shear rupture of the plate

From Tables 9-3 and 9-4, with four $^7/_8$-in. diameter bolts and $L_{ev} = L_{eh} = 1^1/_2$ in., $0.6F_u A_{nv} > F_y A_{nt}$. Thus,

$$\begin{aligned}\phi R_n &= \phi[0.6F_u A_{nv} + F_y A_{gt}] \\ &= (183\,\text{kips/in.} + 40.5\,\text{kips/in.})^3/_8\,\text{in.} \\ &= 83.8\,\text{kips} > 60\,\text{kips} \quad \textbf{o.k.}\end{aligned}$$

Use PL$^3/_8$ in.×8 in.×1′-0.

EXAMPLE 10.18. Bolted/welded single-plate shear splice.

Given: Refer to Figure 10-23. Design a single-plate shear splice between a W16×31 beam and W16×50 beam (not illustrated)

$R_u = 50$ kips

W16×31, ASTM A992

$t_w = 0.275$ in.
$F_y = 50$ ksi, $F_u = 65$ ksi

W16×50, ASTM A992

$t_w = 0.380$
$F_y = 50$ ksi, $F_u = 65$ ksi

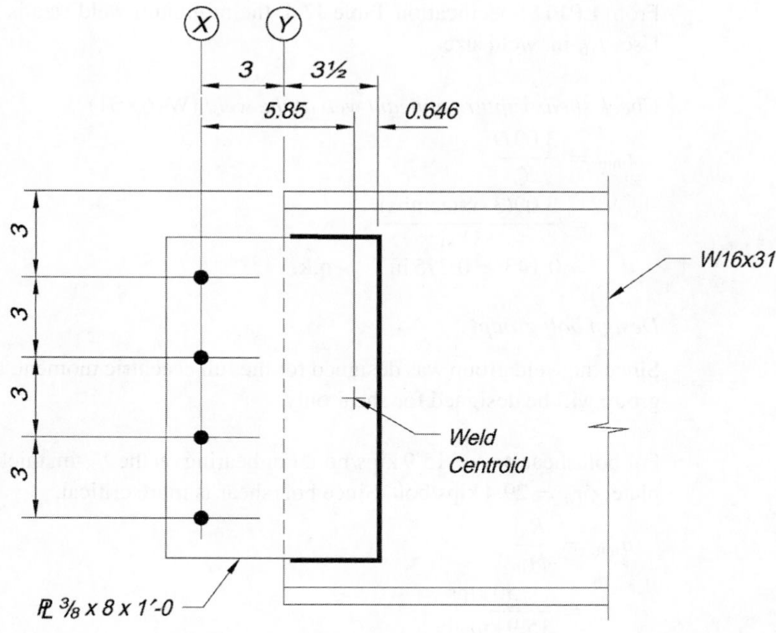

Fig. 10–23. Illustration for Example 10.18.

Use $^3/_4$-in. diameter A325-N bolts through the web of the W16×50 and 70 ksi electrode welds to the web of the W16×31. Assume ASTM A36 plate material with $F_y = 36$ ksi and $F_u = 58$ ksi.

Solution:

Design weld group

Since splice is unsymmetrical and the weld group is more rigid, it will be designed for the full eccentric moment.

Assume PL $^3/_8$ in.×8 in.×1′-0 as illustrated in Figure 10-23.

$$k = \frac{kl}{l} = \frac{3^1/_2 \text{ in.}}{12 \text{ in.}} = 0.292$$

By interpolation from Table 8-9 with $\theta = 0°$, $x = 0.0538$ and $xl = 0.646$ in.

$$\begin{aligned} al &= 6^1/_2 \text{ in.} - 0.646 \text{ in.} \\ &= 5.85 \text{ in.} \end{aligned}$$

$$\text{Thus } a = \frac{al}{l} = \frac{5.85 \text{ in.}}{12 \text{ in.}} = 0.488$$

By interpolation, $C = 1.61$ and the required weld size is

$$\begin{aligned} D_{req} &= \frac{R_u}{CC_1 l} \\ &= \frac{50 \text{ kips}}{(1.61)(1.0)(12 \text{ in.})} \\ &= 2.59 \rightarrow 3 \text{ sixteenths} \end{aligned}$$

From LRFD Specification Table J2.4, the minimum weld size is $^3/_{16}$ in. Use $^3/_{16}$-in. weld size.

Check shear rupture of beam web at the weld (W16×31)

$$\begin{aligned} t_{\min} &= \frac{3.09D}{F_u} \\ &= \frac{3.09(3 \text{ sixteenths})}{65 \text{ ksi}} \\ &= 0.143 < 0.275 \text{ in.} \qquad \textbf{o.k.} \end{aligned}$$

Design bolt group

Since the weld group was designed for the full eccentric moment, the bolt group will be designed for shear only.

For bolt shear $\phi r_n = 15.9$ kips/bolt. For bearing on the $^3/_8$-in. thick single plate, $\phi r_n = 29.4$ kips/bolt. Since bolt shear is more critical,

$$\begin{aligned} n_{\min} &= \frac{R_u}{\phi r_n} \\ &= \frac{50 \text{ kips}}{15.9 \text{ kips/bolt}} \\ &= 3.14 \rightarrow 4 \text{ bolts} \end{aligned}$$

Design single plate

As before, try PL$^3/_8$ in.×8 in.×1′-0.

Check flexure of the plate

$$
\begin{aligned}
M_u &= R_u e \\
&= 50\,\text{kips}\,(3\,\text{in.}) \\
&= 150\,\text{in.-kips}
\end{aligned}
$$

For flexural yielding

$$
\begin{aligned}
\phi M_n &= \phi F_y S_x \\
&= 0.9(36\,\text{ksi})\left[\frac{{}^3/_8\,\text{in.}(12\,\text{in.})^2}{6}\right] \\
&= 292\,\text{in.-kips} > 150\,\text{in.-kips} \qquad \textbf{o.k.}
\end{aligned}
$$

For flexural rupture (with Table 15-2),

$$
\begin{aligned}
\phi M_n &= \phi F_u S_{net} \\
&= 0.75(58\,\text{ksi})(6.54\,\text{in.}^3) \\
&= 285\,\text{in.-kips} > 150\,\text{in.-kips} \qquad \textbf{o.k.}
\end{aligned}
$$

Check shear yielding of the plate

$$
\begin{aligned}
\phi R_n &= \phi(0.6 F_y A_g) \\
&= 0.9(0.6 \times 36\,\text{ksi})(12\,\text{in.} \times {}^3/_8\,\text{in.}) \\
&= 87.5\,\text{kips} > 50\,\text{kips} \qquad \textbf{o.k.}
\end{aligned}
$$

Check shear rupture of the plate

$$
\begin{aligned}
\phi R_n &= \phi(0.6 F_u A_n) \\
&= 0.75(0.6 \times 58\,\text{ksi})(12\,\text{in.} - 4 \times 1\,\text{in.})^3/_8\,\text{in.} \\
&= 78.3\,\text{kips} > 50\,\text{kips} \qquad \textbf{o.k.}
\end{aligned}
$$

Check block shear rupture of the plate

From Tables 9-3 and 9-4, with four $^3/_4$-in. diameter bolts and $L_{ev} = L_{eh} = 1^1/_2$ in., $0.6F_u A_{nv} > F_y A_{nt}$. Thus,

$$
\begin{aligned}
\phi R_n &= \phi[0.6 F_u A_{nv} + F_y A_{gt}] \\
&= (194\,\text{kips/in.} + 40.5\,\text{kips/in.})^3/_8\,\text{in.} \\
&= 87.9\,\text{kips} > 50\,\text{kips} \qquad \textbf{o.k.}
\end{aligned}
$$

Use PL$^3/_8$ in.×8 in.×1′-0

SPECIAL CONSIDERATIONS FOR SIMPLE SHEAR CONNECTIONS

Simple Shear Connections Subject to Axial Forces

When simple shear connections are subjected to axial load in addition to the shear, the important limit states are angle leg bending and prying action. These tend to require that the angle, plate or flange thickness increase or the gage decrease, or both, and these requirements may compromise the connection's ability to remain flexible enough to accommodate the simple beam end rotation. The shear connection ductility checks derived in Part 9 can be used to ensure that adequate ductility exists.

Simple Shear Connections at Stiffened Column-Web Locations

Stiffeners are obstacles to direct connections to column web. Figure 10-24a illustrates a seat angle welded to the toes of the column flanges; Figure 10-24d shows a vertical plate extended beyond the column flanges. Figures 10-24b and 10-24c offer two additional options for framing at locations of diagonal stiffeners; these should be examined carefully as they may create erection problems. Additionally, the deep cope of Figure 10-24c may significantly reduce the design strength of the beam at the end connection. Alternatively, the bottom transverse stiffener could be extended to serve as a seat plate with a bearing stiffener provided to distribute the beam reaction.

Eccentric Effect of Extended Gages

Consider a simple shear connection to the web of a column that requires transverse stiffeners for two concurrent beam-to-column-flange moment connections. If it were not possible to eliminate the stiffeners by selection of a heavier column section, the field connection would have to be located clear of the column flanges, as shown in Figure 10-25, to provide for access and erectability.

The extension of the connection beyond normal gage lines results in an eccentric moment. While this eccentric moment is usually neglected in a connection framing to a column flange, the resistance of the column to weak-axis bending is typically only 20 to 50 percent of that in the strong axis. Thus the eccentric moment should be considered in this column-web connection, especially if the eccentricity e is large. Similarly, eccentricities larger than normal gages may also be a concern in connections to girder webs.

Column-Web Supports

There are two components contributing to the total eccentric moment: (1) $R_u e$ the eccentricity of the beam end reaction; and (2) M_{pr} the partial restraint of the connection. To determine what eccentric moment must be considered in the design, first assume that the column is part of a braced frame for weak-axis bending, is pinned-ended with $K = 1$, and will be concentrically loaded, as illustrated in Figure 10-26. The beam is loaded before the column and will deflect under load as shown in Figure 10-26. Because of the partial restraint of the connection, a couple M_{pr} develops between the beam and column and adds to the eccentric couple $R_u e$. Thus,

$$M_{con} = R_u e + M_{pr}$$

As the loading of the column begins, the assembly will deflect further in the same direction under load, as indicated in Figure 10-27, until the column load reaches some magnitude P_{sbr} when the rotation of the column will equal the simply supported beam end rotation. At this load, the rotation of the column negates M'_{pr} since it also relieves the partial restraint effect of the connection and,

$$M_{con} = R_u e$$

As the column load is increased above P_{sbr}, the column rotation exceeds the simply supported beam end rotation and a moment M'_{pr} results such that

$$M_{con} = R_u e - M'_{pr}$$

Note that the partial restraint of the connection now actually stabilizes the column and reduces its effective length factor K below the originally assumed value of 1. Thus, since M'_{pr} must be greater than zero, it must also be true that $R_u e > M_{con}$. It is therefore conservative to design the connection for the shear R_u and the eccentric moment $R_u e$.

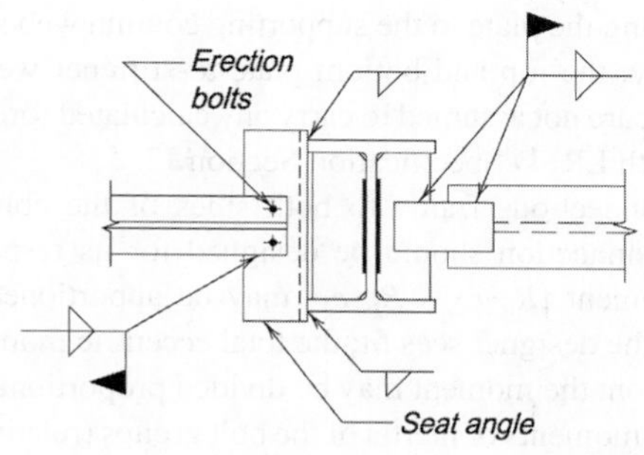

SECTION A-A

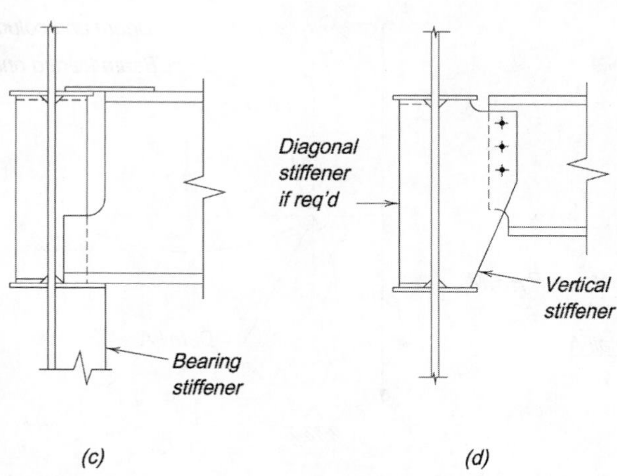

Fig. 10–24. Simple shear connections at stiffened column-web locations.

The welds connecting the plate to the supporting column web should be designed to resist the full shear R_u only; the top and bottom plate-to-stiffener welds have minimal strength normal to their length, are not assumed to carry any calculated force, and may be of minimum size in accordance with LRFD Specification Section J2.

If simple shear connections frame to both sides of the column web as illustrated in Figure 10-28, each connection should be designed for its respective shear R_{u1} and R_{u2}, and the eccentric moment $|R_{u2}e_2 - R_{u1}e_1|$ may be apportioned between the two simple shear connections as the designer sees fit; the total eccentric moment may be assumed to act on the larger connection, the moment may be divided proportionally among the connections according to the polar moments of inertia of the bolt groups (relative stiffness), or the moment may be divided proportionally between the connections according to the section moduli of the bolt groups (relative moment strength). If provision is made for ductility and stability,

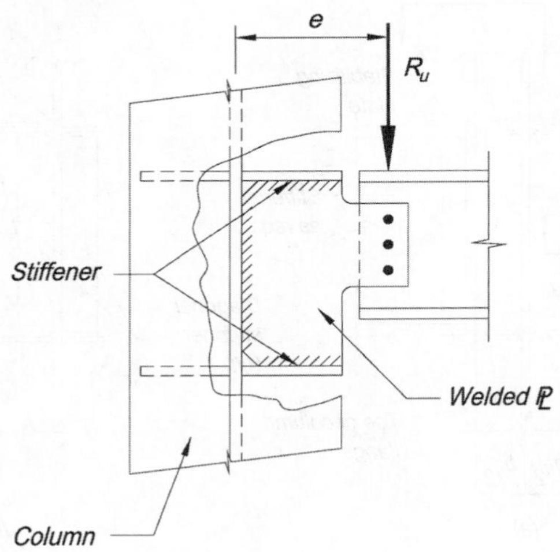

Fig. 10–25. Eccentric effect of extended gages.

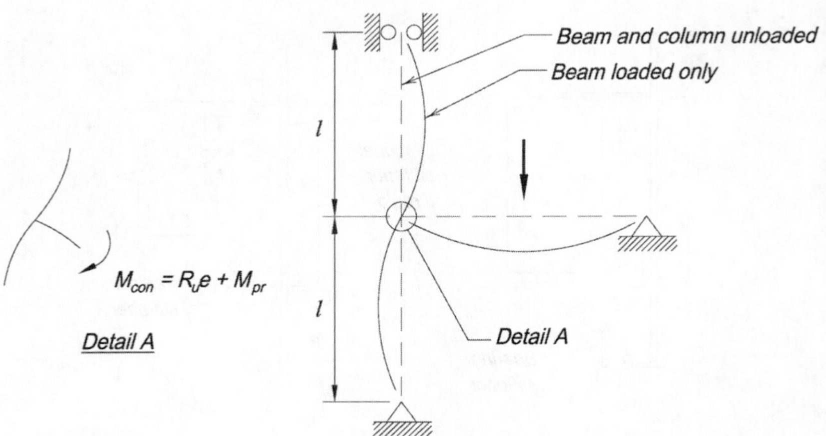

Fig. 10–26. Illustration of beam, column and connection behavior under loading of beam only.

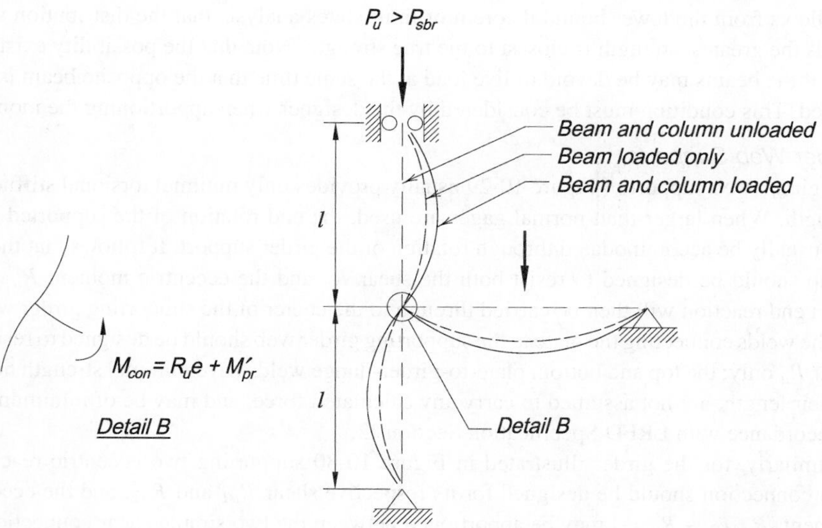

Fig. 10–27. Illustration of beam, column and connection behavior under loading of beam and column.

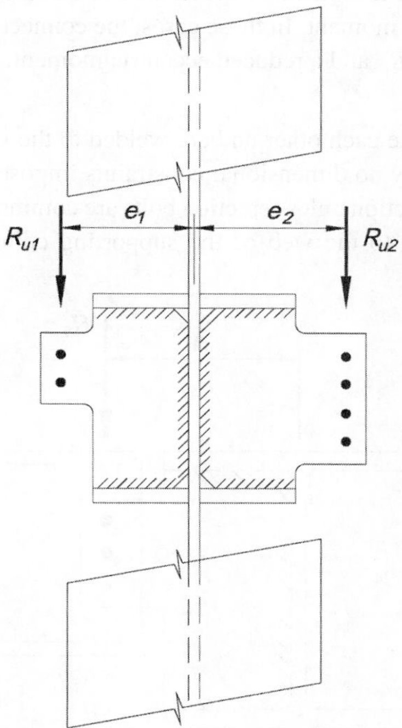

Fig. 10–28. Column subject to dual eccentric moments.

it follows from the lower bound theorem of limit-states analysis that the distribution which yields the greatest strength is closest to the true strength. Note that the possibility exists that one of the beams may be devoid of live load at the same time that the opposite beam is fully loaded. This condition must be considered by the designer when apportioning the moment.

Girder-Web Supports

The girder-web support of Figure 10-29 usually provides only minimal torsional stiffness or strength. When larger-than-normal gages are used, the end rotation of the supported beam will usually be accommodated through rotation of the girder support. It follows that the bolt group should be designed to resist both the shear R_u and the eccentric moment $R_u e$. The beam end reaction will then be carried through to the center of the supporting girder web.

The welds connecting the plate to the supporting girder web should be designed to resist the shear R_u only; the top and bottom plate-to-girder-flange welds have minimal strength normal to their length, are not assumed to carry any calculated force, and may be of minimum size in accordance with LRFD Specification Section J2.

Similarly, for the girder illustrated in Figure 10-30 supporting two eccentric reactions, each connection should be designed for its respective shear R_{u1} and R_{u2}, and the eccentric moment $|R_{u2}e_2 - R_{u1}e_1|$ may be apportioned between the two simple shear connections as the designer sees fit.

Alternative Treatment of Eccentric Moment

In the foregoing treatment of eccentric moments with column- and girder-web supports, it is possible to design the support (instead of the connection) for the eccentric moment $R_u e$. Additionally, when metal deck is used with puddle welds or self-tapping screws, the metal deck tends to reduce relative movement between the two members and thus will tend to carry all or some of the eccentric moment. In these cases, the connection may be designed for the shear R_u only or the shear R_u and a reduced eccentric moment.

Double Connections

When beams frame opposite each other and are welded to the web of the supporting girder or column, there are usually no dimensional constraints imposed on one connection by the presence of the other connection unless erection bolts are common to each connection. When the connections are bolted to the web of the supporting column or girder, however, the

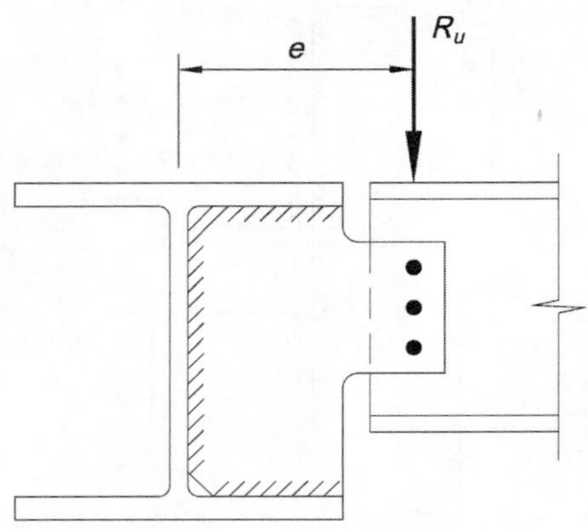

Fig. 10–29. Eccentric moment on girder-web support.

close proximity of the connections requires that some or all fasteners be common to both connections. This is known as a double connection. See also the discussion under Erectability Considerations.

Supported Beams of Different Nominal Depths

When beams of different nominal depths frame into a double connection, care must be taken to avoid interference from the bottom flange of the shallower beam with the entering and tightening clearances for the bolts of the connection for the deeper beam. Access to the bolts that will support the deeper beam may be provided by coping or blocking the bottom flange of the shallower beam. Alternatively, stagger may be used to favorably position the bolts around the bottom flange of the shallower beam.

Supported Beams Offset Laterally

Frequently, beams do not frame exactly opposite each other, but are offset slightly as illustrated in Figure 10-31. Several connection configurations are possible, depending on the offset dimension.

If the offset were equal to the gage on the support, the connection could be designed with all bolts on the same gage lines as shown in Figure 10-31b and the angles arranged as shown in Figure 10-31d. If the offset were less than the gage on the support, staggering the bolts as shown in Figure 10-31c would reduce the required gage and the angles could be arranged as shown in Figure 10-31c. In any case, each bolt transmits an equal share of its beam reaction(s) to the supporting member. Once the geometry of the connection has been determined, the distribution of the forces is patterned after that in the design of a typical connection. For normal gages, eccentricity may be ignored in this type of connection.

Beams Offset from Column Centerline

Framing to the Column Flange from the Strong Axis

As illustrated in Figure 10-32, beam-to-column-flange connections offset from the column centerline may be supported on a typical welded seat, stiffened or unstiffened, provided the welds for the seat can be spaced approximately equally on either side of the beam centerline. Two such seats offset from the W12×65 column centerline by $2^1/_4$ in. and $3^1/_2$ in. are shown in Figures 10-32a and 10-32b, respectively. While not shown, top angles should be used with this connection.

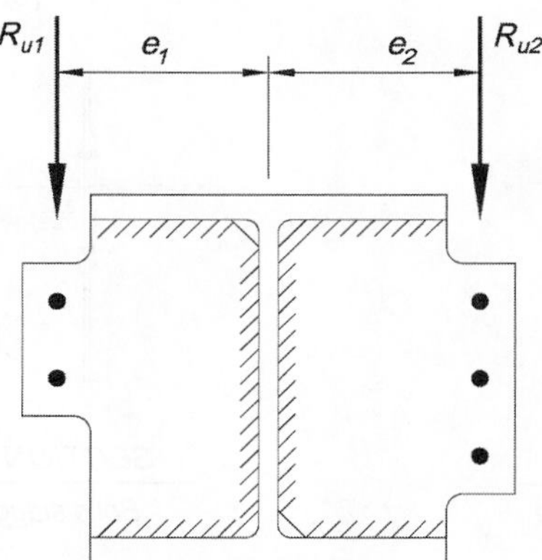

Fig. 10–30. Girder-web support subject to dual eccentric moments.

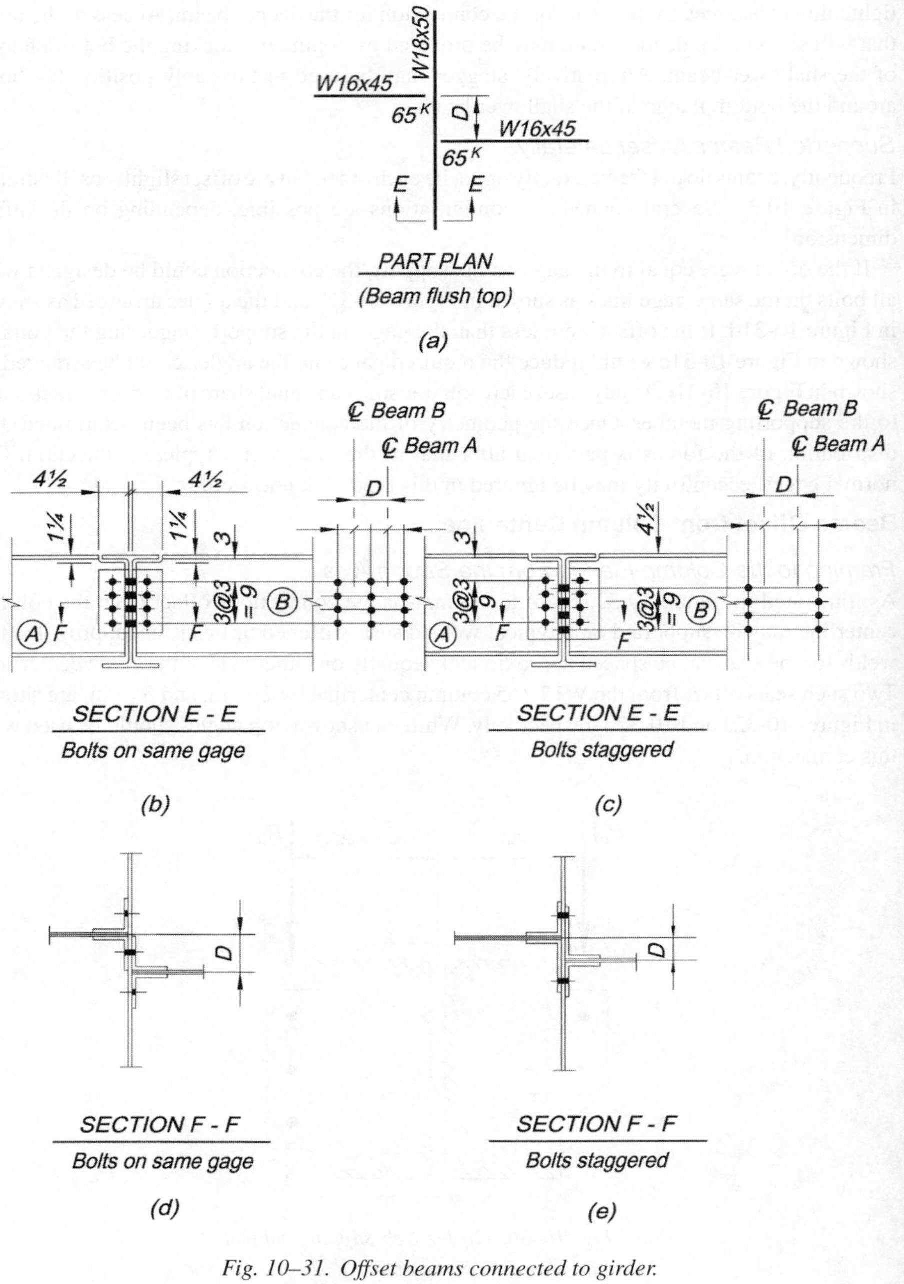

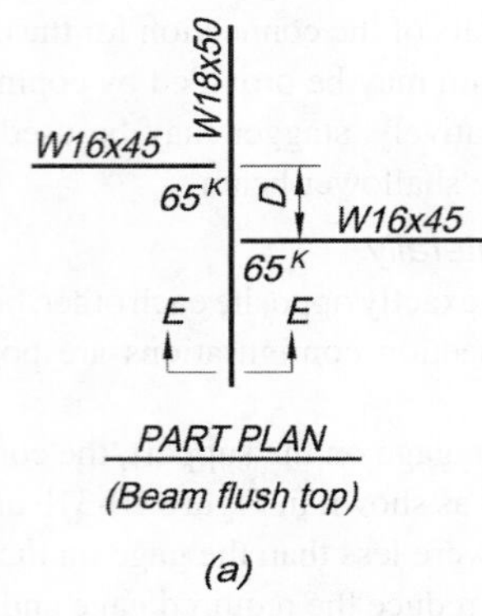

W18x50
W16x45
65K
D
W16x45
65K
E
E
PART PLAN
(Beam flush top)
(a)
℄ Beam B
℄ Beam A
4½
4½
D
1¼
1¼
3
3@3 = 9
A
F
F
B
SECTION E - E
Bolts on same gage
(b)
℄ Beam B
℄ Beam A
D
3
4½
3@3 = 9
A
F
3@3 = 9
B
SECTION E - E
Bolts staggered
(c)
D
SECTION F - F
Bolts on same gage
(d)
D
SECTION F - F
Bolts staggered
(e)

Fig. 10–31. Offset beams connected to girder.

Since the entire seat fits within the flange width of the column, the connection of Figure 10-32a is readily selected from the design aids presented previously. However, the larger beam offsets in Figures 10-32b and 10-32c require that one of the welds be made along the edge of the column flange against the back side of the seat angle. Note that the end return is omitted because weld returns should not be carried around such a corner.

For the beam offset of $5^{1}/_{2}$ in. shown in Figure 10-32c, the seat angle overhangs the edge of the beam and the horizontal distance between the vertical welds is reduced to $3^{1}/_{2}$ in.;

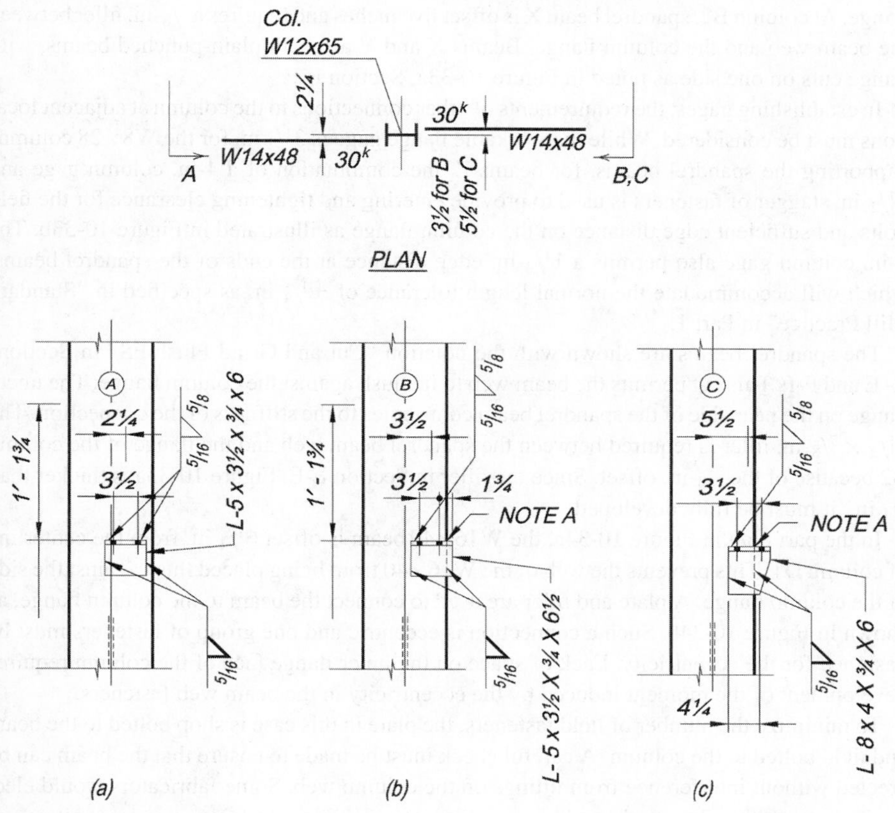

NOTE A

End return is omitted because the AWS Code does not permit weld returns to be carried around the corner formed by the column flange toe and seat angle heel.

NOTE B

Beam and top angle not shown for clarity.

Fig. 10–32. Offset beams connected to column flanges.

the center of gravity of the weld group is located $1^{1}/_{4}$ in. to the left of the beam centerline. The force on each weld may be determined by statics. In this case, the larger force is in the right-hand weld and may be determined by summing moments about the left hand weld. Once the larger force has been determined, the seat should conservatively be designed to carry twice the force in the more highly loaded weld as illustrated in Example 10-23.

Framing to Column Flange from the Weak Axis

Spandrel beams X and Y in the part plan shown in Figure 10-33 are offset $4^{1}/_{8}$ in. from the centerline of column C1, permitting the beam web to be connected directly to the column flange. At column B2, spandrel beam X is offset five inches and requires a $^{7}/_{8}$-in. filler between the beam web and the column flange. Beams X and Y are both plain-punched beams, with flange cuts on one side as noted in Figure 10-33a, Section F-F.

In establishing gages, the requirements of other connections to the column at adjacent locations must be considered. While the workable flange gage is $3^{1}/_{2}$ in. for the W8×28 columns supporting the spandrel beams, for beams Z, the combination of a 4-in. column gage and $1^{1}/_{2}$-in. stagger of fasteners is used to provide entering and tightening clearance for the field bolts and sufficient edge distance on the column flange as illustrated in Figure 10-33b. The 4-in. column gage also permits a $1^{1}/_{2}$-in. edge distance at the ends of the spandrel beams, which will accommodate the normal length tolerance of $\pm^{1}/_{4}$ in. as specified in "Standard Mill Practice" in Part 1.

The spandrel beams are shown with the notation "Cut and Grind Flush FS" in Sections E- E and F-F. This cut permits the beam web to lie flush against the column flange. The uncut flange on the near side of the spandrel beam contributes to the stiffness of the connection. The $2^{1}/_{2} \times {}^{7}/_{8}$-in. filler is required between the spandrel beam web and the flange of the column B2 because of the $^{7}/_{8}$-in. offset. Since the filler in Section E-E, Figure 10-33a is thicker than $^{3}/_{4}$ in., it must be fully developed.

In the part plan in Figure 10-34a, the W16×40 beam is offset $6^{1}/_{4}$ in. from the centerline of column D1. This prevents the web of the W16×40 from being placed flush against the side of the column flange. A plate and filler are used to connect the beam to the column flange, as shown in Figure 10-34b. Such a connection is eccentric and one group of fasteners must be designed for the eccentricity. Lack of space on the inner flange face of the column requires development of the moment induced by the eccentricity in the beam web fasteners.

To minimize the number of field fasteners, the plate in this case is shop bolted to the beam and field bolted to the column. A careful check must be made to ensure that the beam can be erected without interference from fittings on the column web. Some fabricators would elect to shop attach the plate to the column to eliminate possible interference and permit use of plain-punched beams. Additionally, if the column were a heavy section, the fabricator may elect to shop weld the plate to the column to avoid drilling the thick flanges. The welding of this plate to the column creates a much stiffer connection and the design should be modified to recognize the increased rigidity.

If the centerline of the W16 were offset $6^{1}/_{16}$ in. from line 1, it would be possible to cope or cut the flanges flush top and bottom and frame the web directly to the column flange with details similar to those shown in Figure 10-33. This type of framing also provides a connection with more rigidity than normally contemplated in simple construction. A coped connection of this type would create a bending moment at the root of the cope that might require reinforcement of the beam web.

One method frequently adopted to avoid moment transfer to the column because of beam connection rigidity is to use slotted holes and a bearing connection to provide some flexibility.

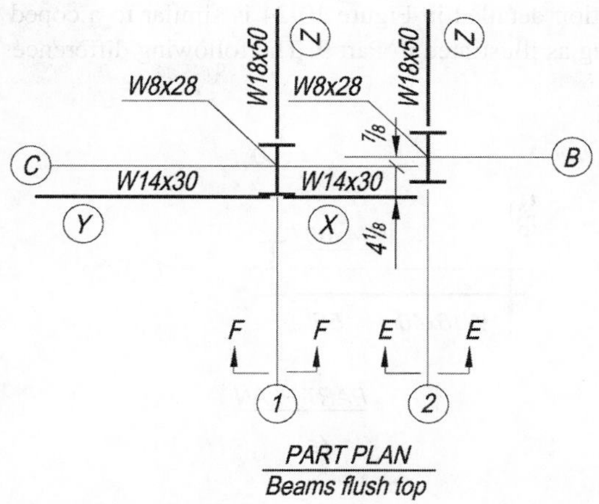

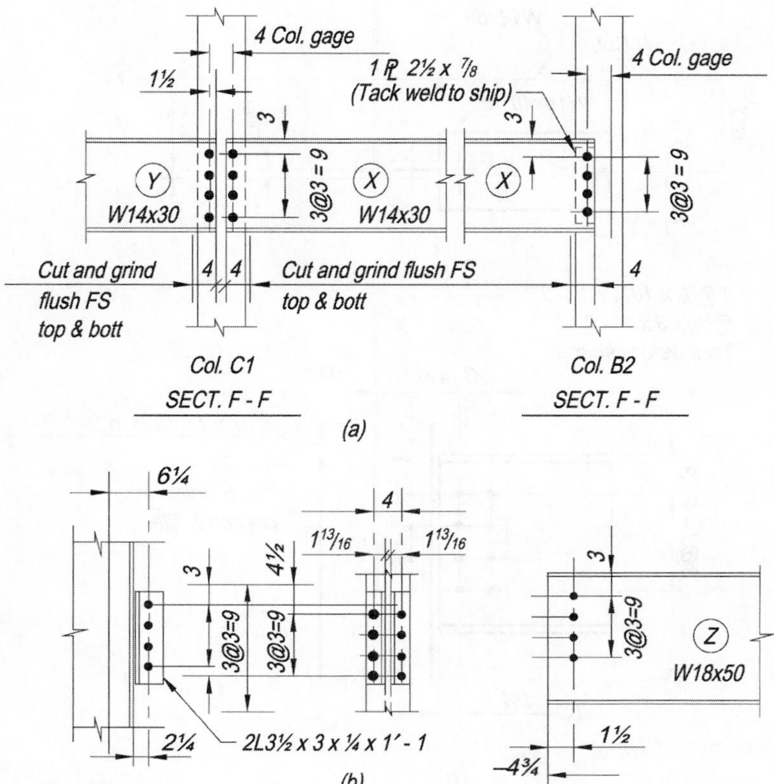

Fig. 10–33. Offset beams connected to column.

The slotted holes would be provided in the connection plate only and would be in the field connection only. These slotted connections also would accommodate fabrication and erection tolerances.

The type of connection detailed in Figure 10-34 is similar to a coped beam and should be checked for buckling as illustrated in Part 9. The following differences are apparent and

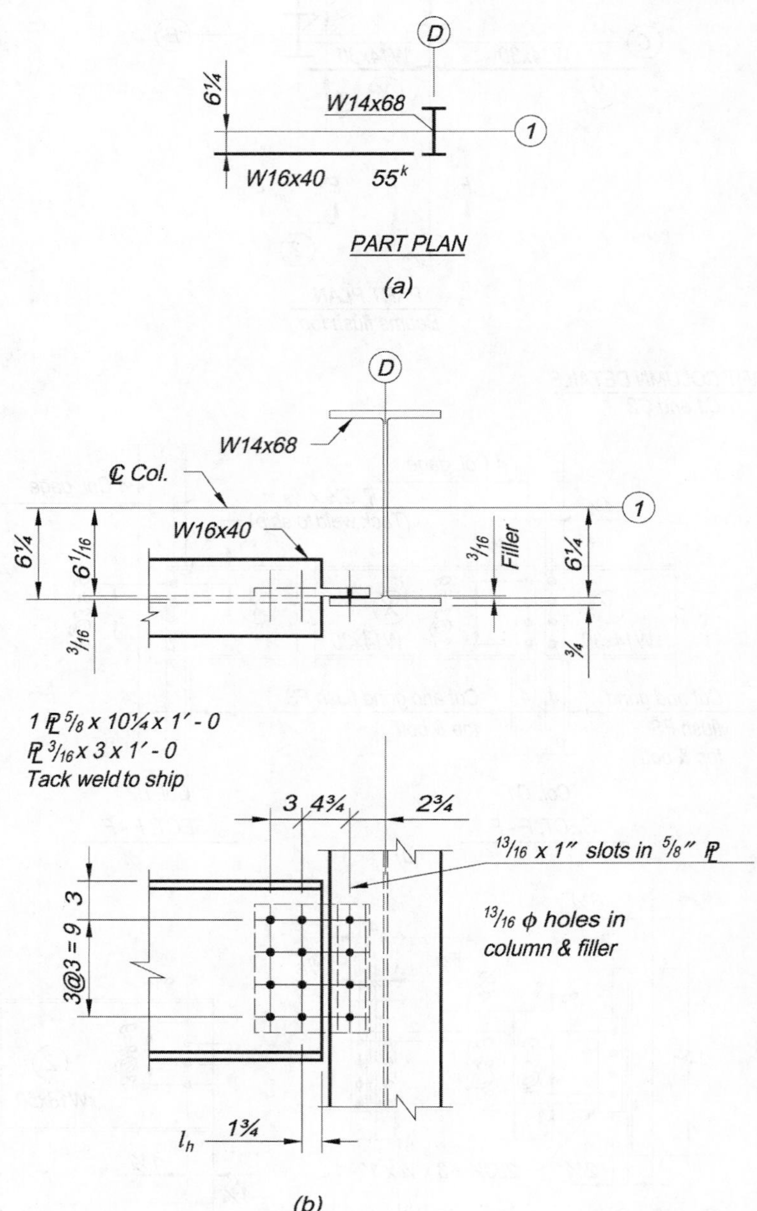

Fig. 10–34. Offset beam connected to column.

should be recognized in the analysis:

1. The effective length of equivalent "cope" is longer by the amount of end distance to the first bolt gage line.
2. There is an inherent eccentricity due to the beam web and plate thickness. The ordinary web and plate thicknesses normally will not require an analysis for this condition, since the inelastic rotation allowed by the LRFD Specification will relieve this secondary moment effect. Two plates may sometimes be required to counter this eccentricity when dimensions are significant.
3. The connection plate can be made of sufficient thickness as required for bending or buckling stresses with a minimum thickness of $^3/_8$ in.

Framing to the Column Web

If the offset of the beam from the centerline of the column web is small enough that the connection may still be centered on or under the supported beam, no special considerations need be made. However, when the offset of the beam is too large to permit the centering of the connection under the beam as in Figure 10-35, it may be necessary to consider the effect of eccentricity in the fastener group.

The offset of the beam in Figure 10-35 requires that the top and bottom flanges be blocked to provide erection clearance at the column flange. Since only half of each flange, then, remains in which to punch holes, a 6-in. outstanding leg is used for both the seat and top angles of these connections; this permits the use of two field bolts to each of the seat and top angles, which are required by OSHA.

Connections for Raised Beams

When raised beams are connected to column flanges or webs, there is usually no special consideration required. However, when the support is a girder, the differing tops of steel may preclude the use of typical connections. Figure 10-36 shows several typical details commonly used for such cases in bolted construction. Figure 10-37 shows several typical details commonly used in welded construction.

In Figure 10-36a, since the top of the W12×35 is located somewhat less than 12 inches above the top of the W18 supporting beam, a double-angle connection is used. This connection would be designed for the beam reaction and the shop bolts would be governed by double shear or bearing, just as if they were located in a vertical position. However, the field bolts are not required to carry any calculated force under gravity loading.

The maximum permissible distance m depends on the beam reaction, since the web remaining after the bottom cope must provide sufficient area to resist the vertical shear as well as the bending moment which would be critical at the end of the cope. The beam can be reinforced by extending the angles beyond the cope and adding additional shop bolts for development. The angle size and/or thickness can be increased to gain shear area or section modulus, if required. The effect of any eccentricity would be a matter of judgment, but could be neglected for small dimensions.

When this connection is used for flexure or for dynamic or cyclical loading, the web is subjected to high stress concentrations at the end of the cope, and it is good practice to extend the angles as shown in Figure 10-36a by the dashed lines to add at least two additional web fasteners.

Figure 10-36b covers the case where the bottom flange of the W12×35 is located a few inches above the top of the W18. The beam bears directly upon fillers and is connected to the W18 by four field bolts which are not required to transmit a calculated gravity load. If the distance m exceeds the thickest plate which can be punched, two or more plates may be used. Even though the fillers in this case need only be $6^1/_2$ in. square, the amount of material required increases rapidly as m increases. If m exceeds 2 or 3 in., another type of detail may be more economical.

The detail shown in Figure 10-36c is used frequently when m is up to 6 or 7 in. The load on the shop bolts in this case is no greater than that in Figure 10-36a. However, to provide more lateral stiffness, the fittings are cut from a 15-in. channel and are detailed to overlap the beam web sufficiently to permit four shop bolts on two gage lines.

A stool or pedestal, cut from a rolled shape, can be used with or without fillers to provide for the necessary m-distance as in Figure 10-36d. A pair of connection angles and a tee will also serve a similar purpose, as shown in Figure 10-36e. To provide adequate strength to carry the beam end reaction and to provide lateral stiffness, the web thickness of the pedestal in each of these cases should be at least as thick as the member being supported.

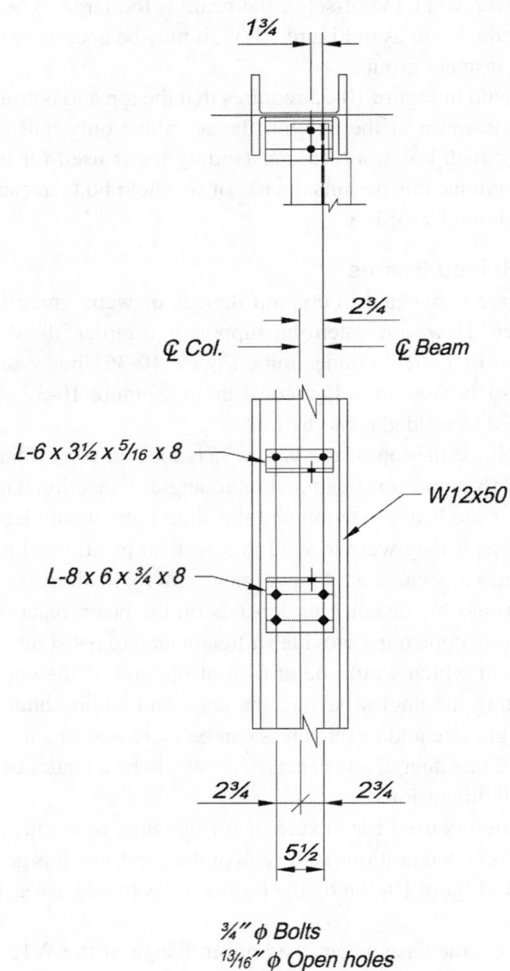

Fig. 10–35. Offset beam connected to column web.

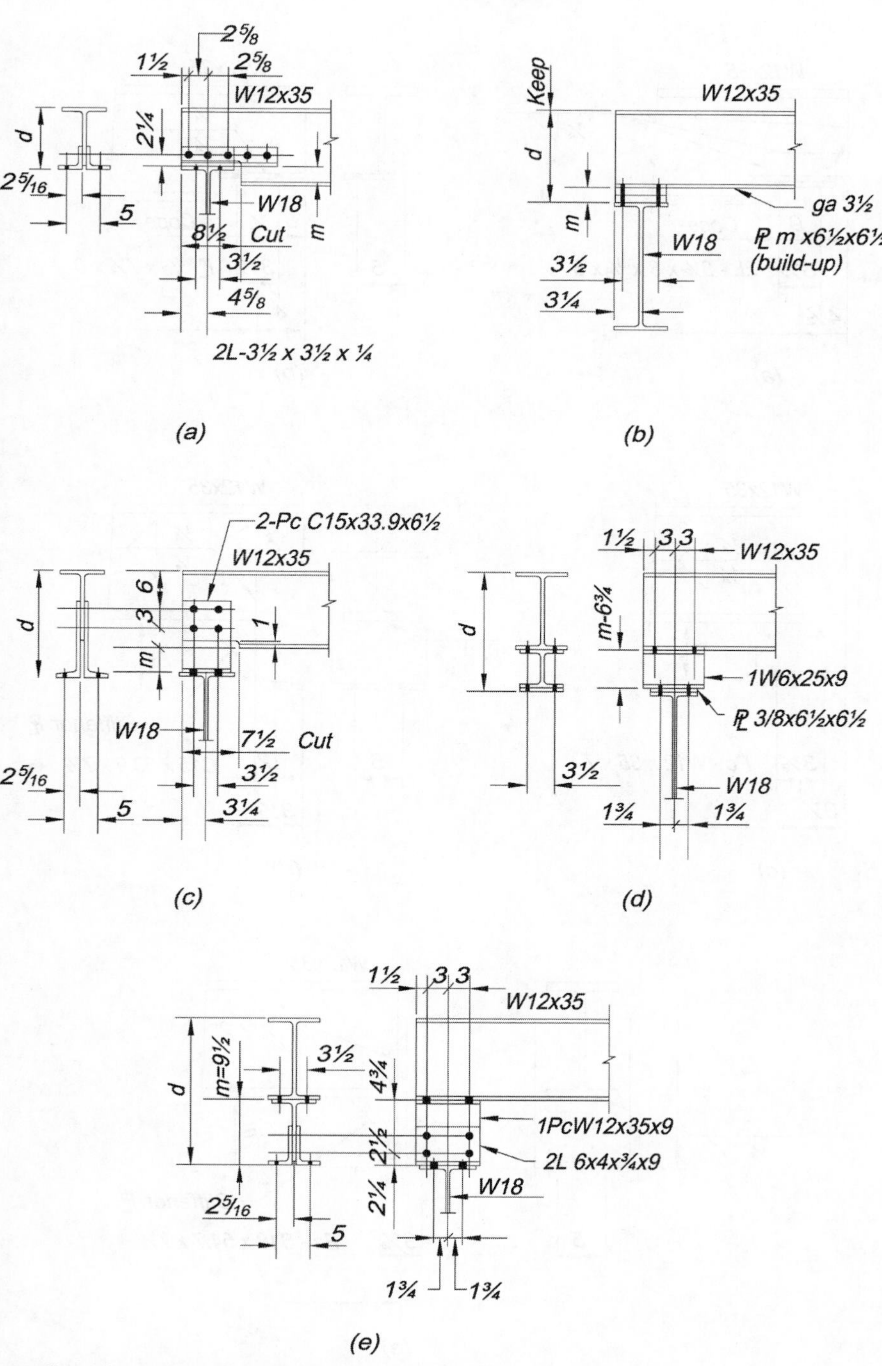

Note: See discussion "Stability Bracing" in Part 2.

Fig. 10–36. Bolted raised beam connections.

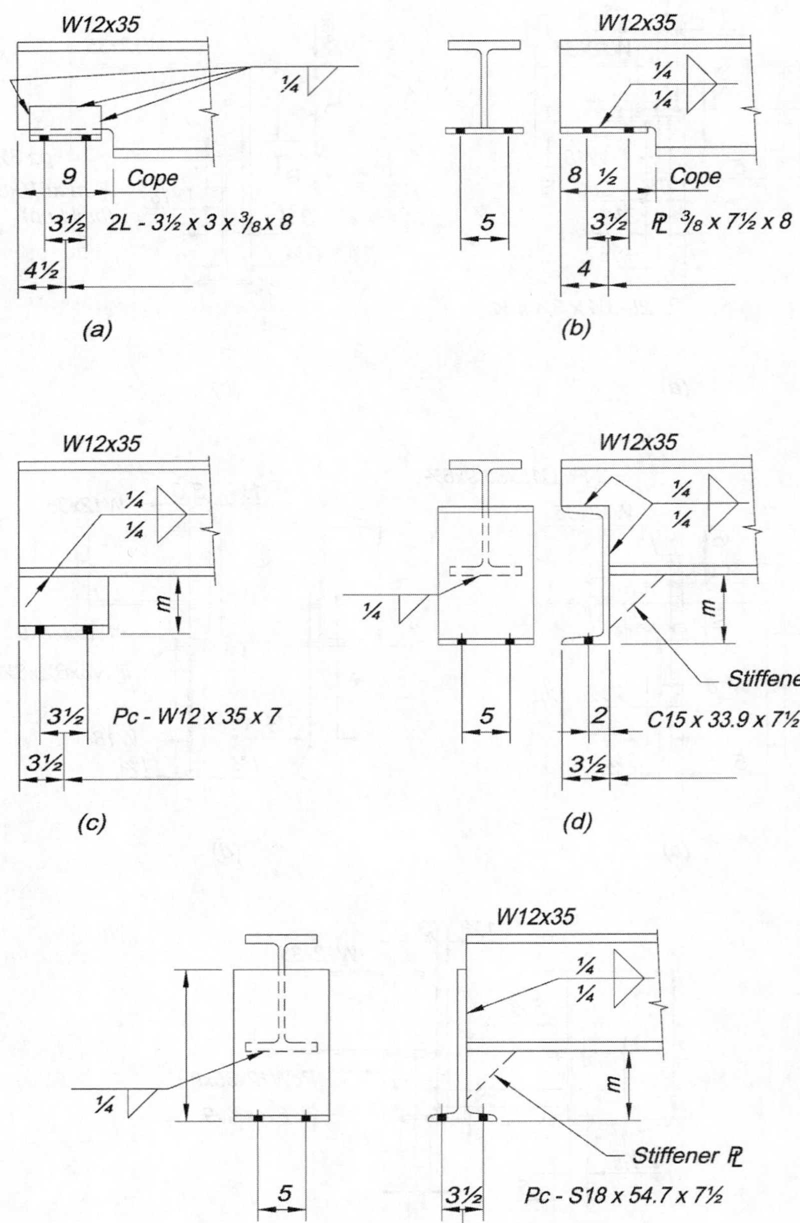

Note: See discussion "Stability Bracing" in Part 2.

Fig. 10–37. Welded raised beam connections.

In Figure 10-37a, welded framing angles are substituted for the bolted angles of Figure 10-36a. In Figure 10-37b, a single horizontal plate is shown replacing the pair of framing angles; this results in a savings in material and the amount of shop welding. In this case, particular care must be taken in cutting the beam web and positioning the plate at right angles to the beam web. For this reason, if only a few connections of this type are to be made, some fabricators prefer to use the angles as in Figure 10-37a. If sufficient duplication were available to warrant making a simple jig to position the plate during welding, the solution of Figure 10-37b may be economical.

Figure 10-37c shows a tee centered on the beam web and welded to the bottom flange of the beam. The tee stem thickness should not be less than the beam web thickness. The welded solutions shown in Figures 10-37d and 10-37e are capable of providing good lateral stiffness. The latter two types also permit end rotation as the beam deflects under load. However, if the m distance exceeds 3 or 4 in., it is advisable to shop weld a triangular bracket plate at one end of the beam, as indicated by the dashed lines, to prevent the beam from deflecting along its longitudinal axis.

Other equally satisfactory details may be devised to meet the needs of connections for raised beams. They will vary depending on the size of the supported beam and the distance m. When using this type of connection where the load is transmitted through bearing, the provisions of LRFD Specification Sections K1.3 and K1.4 must be satisfied for both the supported and supporting members. For the detail of Figure 10-37b, since the rolled fillet has been removed by the cut, the value of k would be taken as the thickness of the plate plus the fillet weld size.

LRFD Specification Section B6 requires stability and restraint against rotation about the beam's longitudinal axis. This provision is most easily accomplished with a floor on top of the supported beam. In the absence of a floor, the top flange may be supported by a strut or bracket attached to the supporting member. When the beam is encased in a wall, this stability may also be provided with wall anchors; refer to "Wall Anchors" in Part 15.

This discussion has considered that the field bolts which attach the beam to the pedestal or support beam, are subject to no calculated load. It is important, however, to recognize that when the beam deflects about its neutral axis, a tensile force can be exerted on the outside bolts. The intensity of this tensile force is a function of the dimension d indicated in Figure 10-36, the span length of the supported member, and the beam stiffness. If these forces are large, high-strength bolts should be used and the connection analyzed for the effects of prying action.

Raised beam connections such as these are used frequently as equipment or machinery supports where it is important to maintain a true and level surface or elevation. When this tolerance becomes important, the dimension d should be noted "keep" to advise the fabricator of this importance, as shown in Figure 10-36b. Since the supporting beam is subject to certain camber/deflection tolerances, it also may be appropriate to furnish shim packs between the connection and the supporting member.

Non-Rectangular Simple Shear Connections

It is often necessary to design connections for beams that do not frame into a support orthogonally. Such a beam may be inclined with respect to the supporting member in various directions. Depending upon the relative angular position which a beam assumes, the connection may be classified among three categories: skewed, sloped, or canted. These conditions are illustrated in Figure 10-38 for beam-to-girder web connections; the same descriptions apply to beam-to-column flange and web connections. Additionally, beams may be oriented in a combination of any or all of these conditions. For any condition of skewed, sloped, or canted framing, the single-plate connection is generally the simplest and most economical of those illustrated in this text.

Skewed Connections

A beam is said to be skewed when its flanges lie in a plane perpendicular to the plane of the face of the supporting member, but its web inclined to the face of the supporting member.

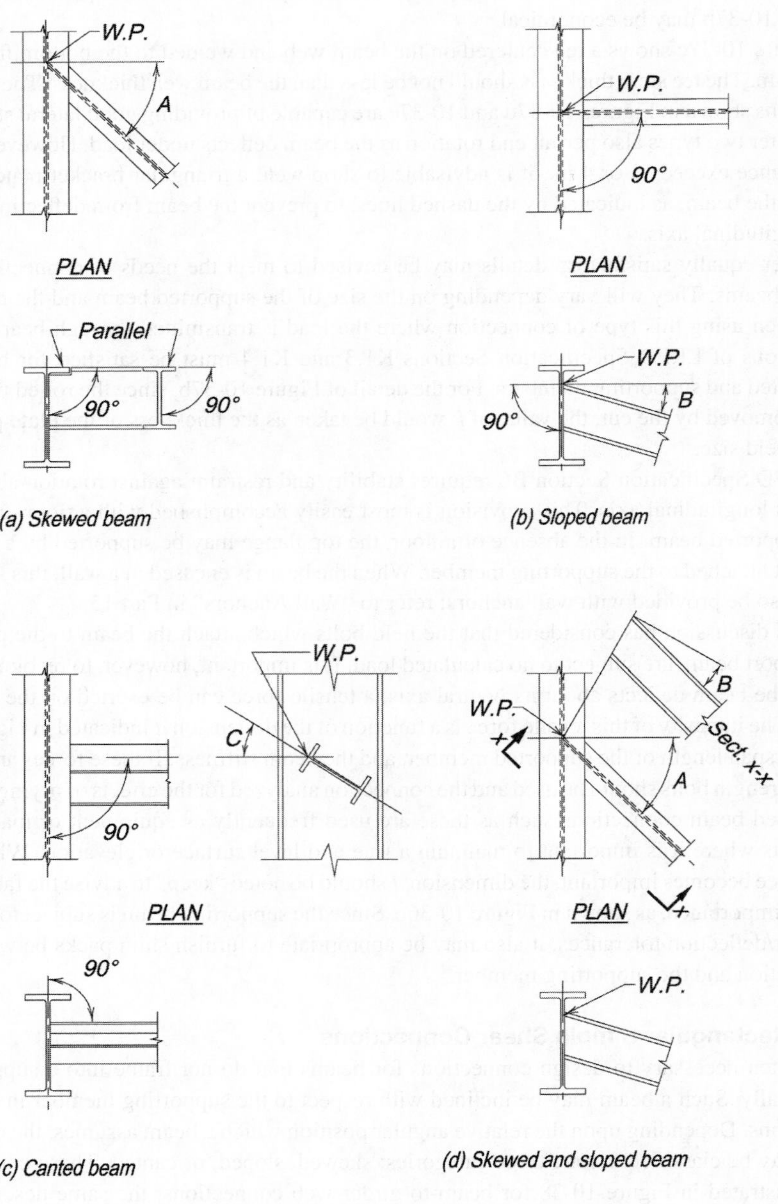

Fig. 10–38. Non-rectangular connections.

The angle of skew A appears in Figure 10-38a and represents the horizontal bevel to which the fittings must be bent or set, or the direction of gage lines on a seated connection.

When the skew angle is less than 15° (3-in-12 slope), a pair of double angles can be bent inward or outward to make the connection as shown in Figure 10-39. While bent angle sections are usually drawn as bending in a straight line from the heel, rolled angles will tend to bend about the root of the fillet (dimension *k* in Manual Part 1). This produces a significant jog in the leg alignment, which is magnified by the amount of bend. Above this angle of skew, it becomes impractical to bend rolled angles.

For skews approximately greater than 5° (1-in.-12 slope), a pair of bent plates, shown in Figure 10-40, may be a more practical solution. Bent plates are not subject to the deformation problem described for bent angles, but the radius and direction of the bend must be considered to avoid cracking during the cold-bending operation.

Bent plates exhibit better ductility when bent perpendicular to the rolling direction and are, therefore, less likely to crack. Whenever possible, bent connection plates should be billed with the width dimension parallel to the bend line. The length of the plate is measured on its mid-thickness, without regard to the radius of the bend. While this will provide a plate that is slightly longer than necessary, this will be corrected when the bend is laid out to the proper radius prior to fabrication.

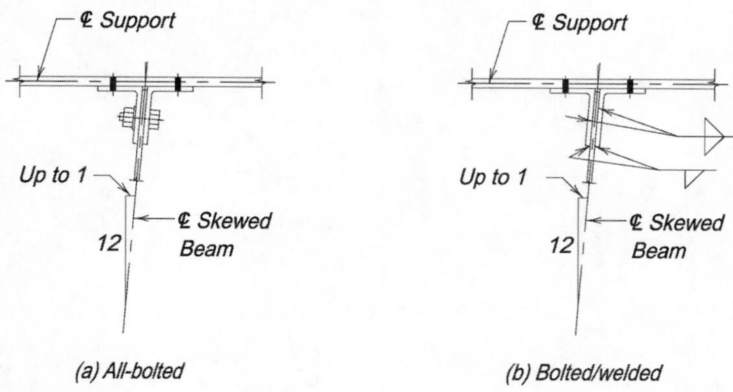

Fig. 10–39. Skewed beam connections with bent double angles.

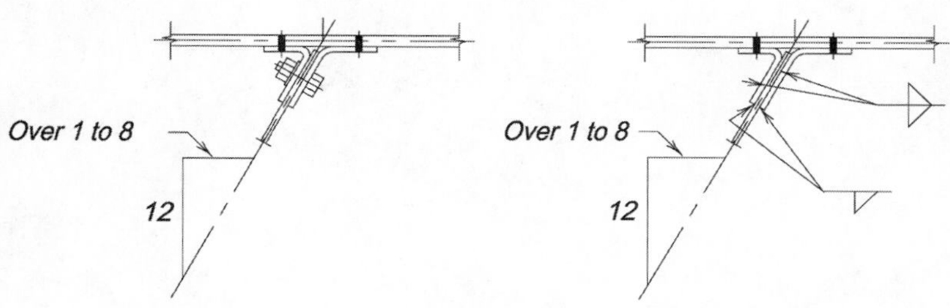

Fig. 10–40. Skewed beam connections with double bent plates.

Table 10-12 gives the generally accepted minimum inside-bending radius for plate thickness t for various grades of steel. Values are for bend lines transverse to the direction of final rolling (Brockenbrough, 1998). When bend lines are parallel to the direction of final rolling, the tabular values may have to be approximately doubled. When bend lines are longer than 36 inches, all radii may have to be increased if problems in bending are encountered.

Before bending, special attention should be given to the condition of plate edges transverse to the bend lines. Flame-cut edges of hardenable steels should be machined or softened by heat treatment. Nicks should be ground out and sharp corners should be rounded.

Table 10-12.
Minimum Inside Radius for Cold Bending[1]

ASTM Designation[2]	Thickness t, in.			
	Up to $3/4$	Over $3/4$ to 1	Over 1 to 2	Over 2
A36, A572-42	$1^1/_2\ t$	$1^1/_2\ t$	$1^1/_2\ t$	$2\ t$
A242, A529-50, A529-55, A572-50, A588, A992	$1^1/_2\ t$	$1^1/_2\ t$	$2\ t$	$2^1/_2\ t$
A572-55, A852	$1^1/_2\ t$	$1^1/_2\ t$	$2^1/_2\ t$	$3\ t$
A572-60, A572-65	$1^1/_2\ t$	$1^1/_2\ t$	$3\ t$	$3^1/_2\ t$
A514	$1^3/_4\ t$	$2^1/_4\ t$	$4^1/_2\ t$	$5^1/_2\ t$

[1] Values are for bend lines perpendicular to direction of final rolling. If bend lines are parallel to final rolling direction, multiply values by 1.5.
[2] The grade designation follows the dash; where no grade is shown, all grades and/or classes are included.

The strength of bent angles and bent plate connections may be calculated in the same manner as for square framed beams, making due allowances for eccentricity. The load is assumed to be applied at the point where the skewed beam center line intersects the face of the supporting member.

As the angle of skew increases, entering and tightening clearances on the acutely angled side of the connection will require a larger gage on the support. If the gage were to become objectionable, a single bent plate, illustrated in Figure 10-41, may provide a better solution. Note that the single bent plate may be of the conventional type, or a more compact connection may be developed by "wrapping" the single bent plate as illustrated in Figure 10-41c.

In all-bolted construction, both the shop and field bolts should be designed for shear and the eccentric moment. A C-shaped weld is preferable to avoid turning the beam during shop fabrication. Single bent plates should be checked for flexural strength.

Table 10-13 gives clearance dimensions for bent double-angle connections and double and single bent plate connections, and specifies beam set-backs and gages. Since these dimensions are based on the maximum material thicknesses and fastener sizes indicated, it is suggested that in cases where many duplicate connections with less than maximum material or fasteners are required, savings can be effected if these dimensions are developed from specific bevels, beam sizes, and fitting thicknesses.

Skewed single plate and skewed end plate connections, shown in Figures 10-42 and 10-43, provide a simple, direct connection with a minimum of fittings and multiple punching requirements. When fillet welded, these connections may be used for skews up to 30° (or a slope of 6 $^{5}/_{16}$ in 12) provided the root opening formed does not exceed $^{3}/_{16}$ in. as specified in AWS D1.1 paragraph 3.3.1. For skew angles greater than 30°, see AWS D1.1, Section 2.11.

The maximum beam web thickness which may be supported is a function of the maximum root opening and the angle of skew. If the thickness of the beam web were such that a larger root opening were encountered, the skewed single plate or the web connecting to the skewed end

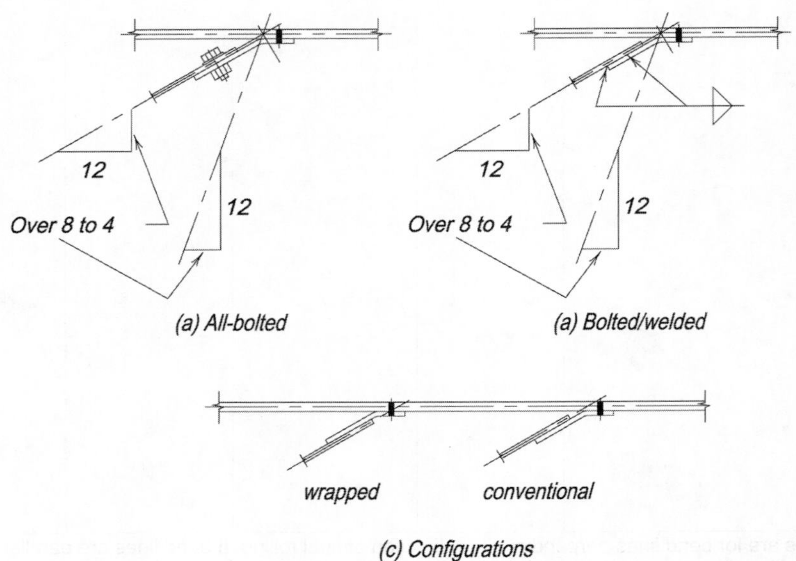

Fig. 10–41. Skewed beam connections with single bent plates.

plate may be beveled, as shown in Figures 10-42b and 10-43b. Since no root opening occurs with the bevel, there is no limitation on the thickness of the beam web. However, beveling, especially of the beam web, requires careful finishing and is an expensive procedure which may outweigh its advantages.

The design of skewed end plate connections is similar to that discussed previously in "Shear End-Plate Connections" in this Part. However, when the gage of the bolts is not centered on the beam web, this eccentric loading should be considered. The design of skewed single-plate connections is similar to that discussed previously in "Single-Plate Connections" in this Part.

Table 10-13 specifies gages and the dimension *A* which is added to the fillet weld size to compensate for the root opening for skewed end-plate connections. This table is based conservatively on a gap of $^1/_8$ in. For beam webs beveled to the appropriate skew, $A = 0$ and the tabulated values do not apply. Table 10-13 also provides similar information for skewed single-plate connections. Additionally, this table provides clearances and dimensions for groove welded single-plate connections with backing bars for skews greater than 30°; refer to AWS D1.1 for prequalified welds for both types of joints.

When skewed stiffened seated connections are used, the stiffening element should be located so as to cross the skewed beam centerline well out on the seat. This can be accomplished by shifting the stiffener to the left or right of center to support beams which skew to the left or to the right, respectively. Alternatively, it may be possible to skew the stiffening element.

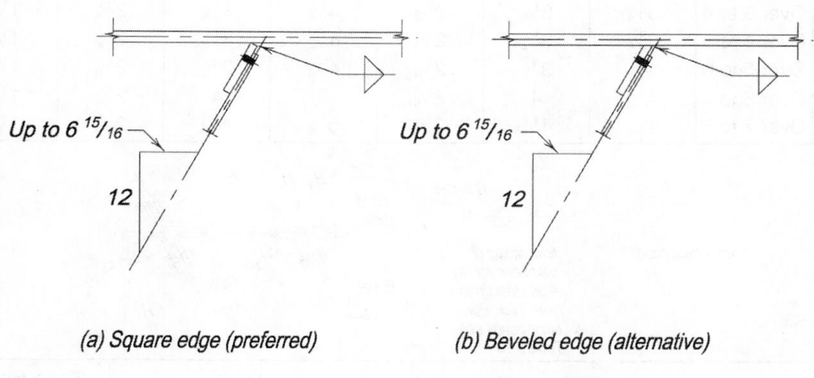

Fig. 10–42. Skewed single-plate connections.

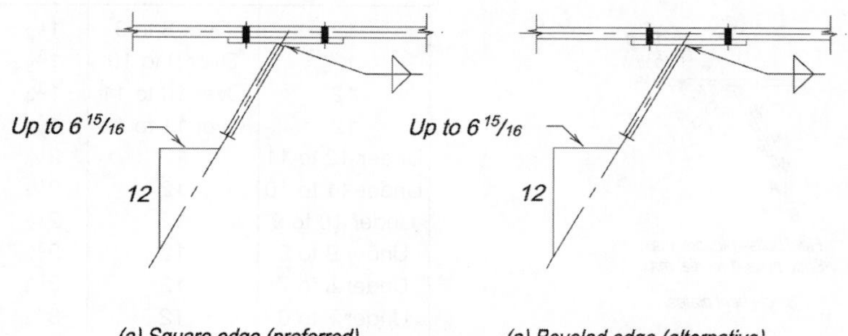

Fig. 10–43. Skewed shear end-plate connections.

Table 10-13.
Clearance Dimensions for Skewed Connections
All-Bolted

Values given are for webs up to ¾ in. thick, angles up to ⅝ in. thick, and bent plates up to ½ in. thick. Bolts are either ⅞-in. diameter or 1-in. diameter, as noted. Values will be conservative for material thinner than the maximums listed, or for work with smaller bolts, and may be reduced to suit conditions by calculation or layout. For thicker material or larger bolts, check entering, driving, and tightening clearances and increase D and bolt gages as necessary. All dimensions are in inches. Enter bolts as shown.

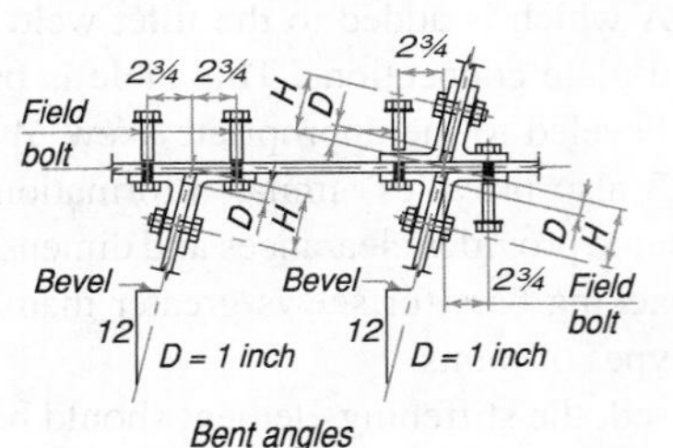

Bent angles

Values of *H* for Various Fastener Combinations

	Field Bolts	⅞	1
	Shop Bolts	⅞	1
Bevel	Up to 1	4*	4¼*
	Over 1 to 2	4⅛	4⅜
	Over 2 to 3	4⅜	4¾

*For back-to-back connections, stagger shop and field bolts or increase the 2¾-in. field bolt dimension to 3¼.

Values of H, H_1, H_2, and D for Various Bolt Combinations

	Field Fastener	⅞			1			
	Shop Fastener	⅞			1			
	Dimension	*H*	*H*₁	*H*₂	*H*	*H*₁	*H*₂	*D*
Bevel	Over 3 to 4	3¾	3¼	2½	4¼	3¼	2¾	1¼
	Over 4 to 5	3¾	3½	2¼	4½	3½	2½	1¼
	Over 5 to 6	4	3¾	2¼	4¾	3¾	2¼	1½
	Over 6 to 7	4½	4	2¼	5	4	2¼	1½
	Over 7 to 8	4¾	4¼	2¼	5¼	4¼	2¼	1½

Double bent plates

t
R = 1½t

Min. radius of cold bend for A 36 steel up to ½ in. thick. For other bends see Table 9-13

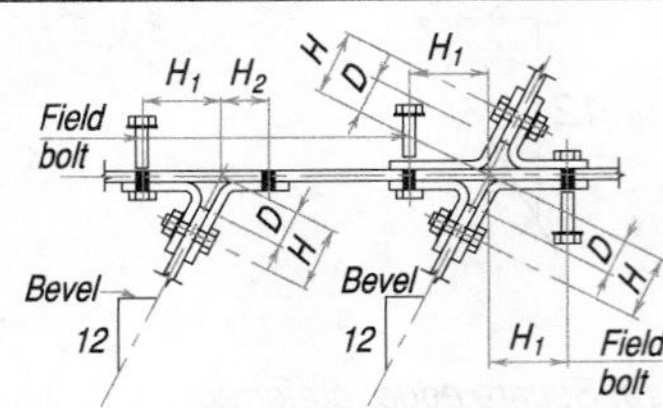

2¼
Field bolt
D
H
A
B

Field bolts–1 in. dia. max.
Shop bolts–1 in. dia. max.

Single bent plates

A	B	Shop Bolts D	Shop Bolts H
12	Over 8 to 9	1½	3
12	Over 9 to 10	1⅝	3⅛
12	Over 10 to 11	1¾	3¼
12	Over 11 to 12	1⅞	3⅜
Under 12 to 11	12	2⅛	3⅝
Under 11 to 10	12	2¼	3¾
Under 10 to 9	12	2½	4
Under 9 to 8	12	2¾	4¼
Under 8 to 7	12	3¼	4¾
Under 7 to 6	12	3¾	5¼
Under 6 to 5	12	4½	6
Under 5 to 4	12	5⅝	7⅛

Table 10-13 (cont.). Clearance Dimensions for Skewed Connections Bolted/Welded

Values given are for webs up to ¾ in. thick, angles up to ⅝ in. thick, and bent plates up to ½ in. thick, with bolts 1-in. diameter maximum. Values will be conservative for thinner material and for work with smaller bolts, and may be reduced to suit conditions by calculation or layout. For thicker material or larger bolts check entering and tightening clearances and increase beam set-back D and bolt gages as necessary. Enter bolts as shown. All dimensions are in inches.

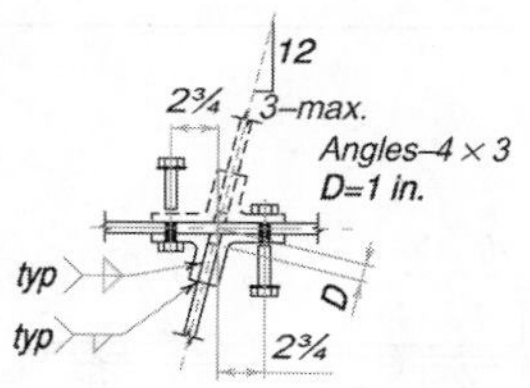

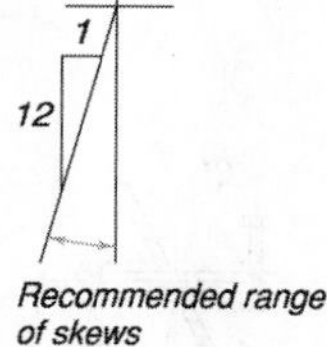

Recommended range of skews

Bent angles

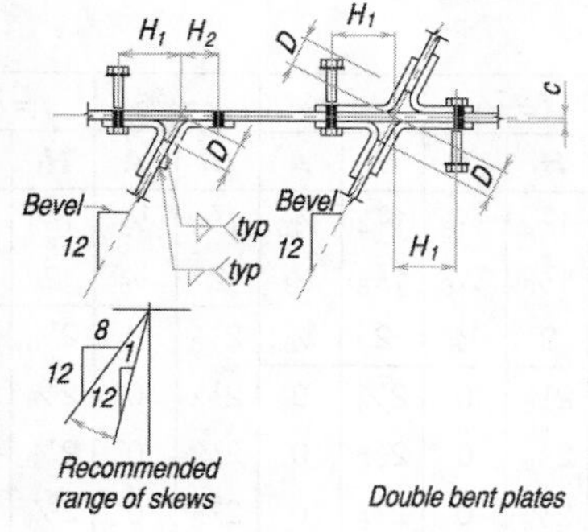

Recommended range of skews

Double bent plates

R = 1½t

Min. radius of cold bend for A 36 steel up to ½ in. thick. For other bends see Table 9-13

Bevel	D	H_1	H_2
Over 3 to 4	c + ⅝	3¼	2¾
Over 4 to 5	c + 11⁄16	3½	2½
Over 5 to 6	c + ¾	3¾	2¼
Over 6 to 7	c + 13⁄16	4	2¼
Over 7 to 8	c + ⅞	4¼	2¼

$$C = \frac{t_w}{2} + 1/16''$$

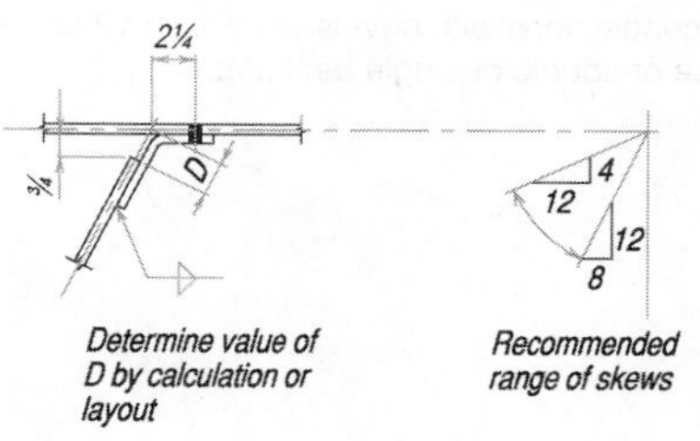

Determine value of D by calculation or layout

Recommended range of skews

Single bent plates

Table 10-13 (cont.). Clearance Dimensions for Skewed Connections Bolted/Welded

Values given are for material and bolt sizes noted below. See "Shear End-Plate Connections" in Part 10 for proportioning these connections. *S* indicates weld size required for strength, or a size suitable to the thickness of material. When the beam web is cut square, only that portion of the table above the heavy lines is applicable. Dimension *A* is added to the weld size to compensate for the root opening caused by the skew. When the beam web is beveled to the required skew, values of H_1 for the entire table are valid, and A = 0. In either case, where weld strength is critical, increase the weld size to obtain the required throat dimension. Enter bolts as shown. All dimensions are in inches.

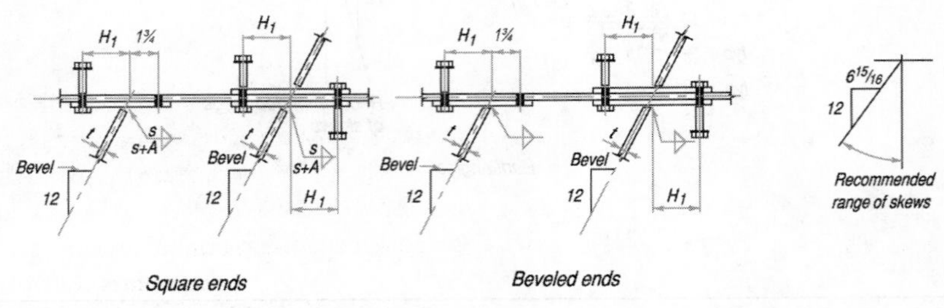

End plates

Bevel	t = 1/4 H1	t = 1/4 A	t = 5/16 H1	t = 5/16 A	t = 3/8 H1	t = 3/8 A	t = 7/16 H1	t = 7/16 A	t = 1/2 H1	t = 1/2 A	t = 5/8 H1	t = 5/8 A	t = 3/4 H1	t = 3/4 A
Up to 1 5/8	1 3/4	0	1 3/4	0	1 3/4	1/16	1 3/4	1/16	1 3/4	1/16	1 7/8	1/8	1 7/8	1/8
Over 1 5/8 to 2 1/8	1 3/4	0	1 3/4	1/16	1 7/8	1/16	1 7/8	1/16	1 7/8	1/8	2	1/8	2	1/8
Over 2 1/8 to 3 1/4	1 7/8	1/16	1 7/8	1/8	2	1/8	2	1/8	2	1/8	2 1/8	0	2 1/8	0
Over 3 1/4 to 4 3/8	2 1/8	1/8	2 1/8	1/8	2 1/8	1/8	2 1/8	0	2 1/4	0	2 1/4	0	2 3/8	0
Over 4 3/8 to 5 5/8	2 1/4	1/8	2 1/4	1/8	2 3/8	0	2 3/8	0	2 3/8	0	2 1/2	0	2 1/2	0
Over 5 5/8 to 6 15/16	2 1/2	1/8	2 1/2	0	2 1/2	0	2 1/2	0	2 5/8	0	2 5/8	0	2 3/4	0

Bolts: 7/8-in. diameter maximum
End Plate thickness: 3/8 in. maximum
Supporting web thickness: 3/4 in. maximum

Use of fillet welds is limited to connections with bevels of 6 15/16 in 12 and less.
For greater bevels consider use of double or single bent plates.

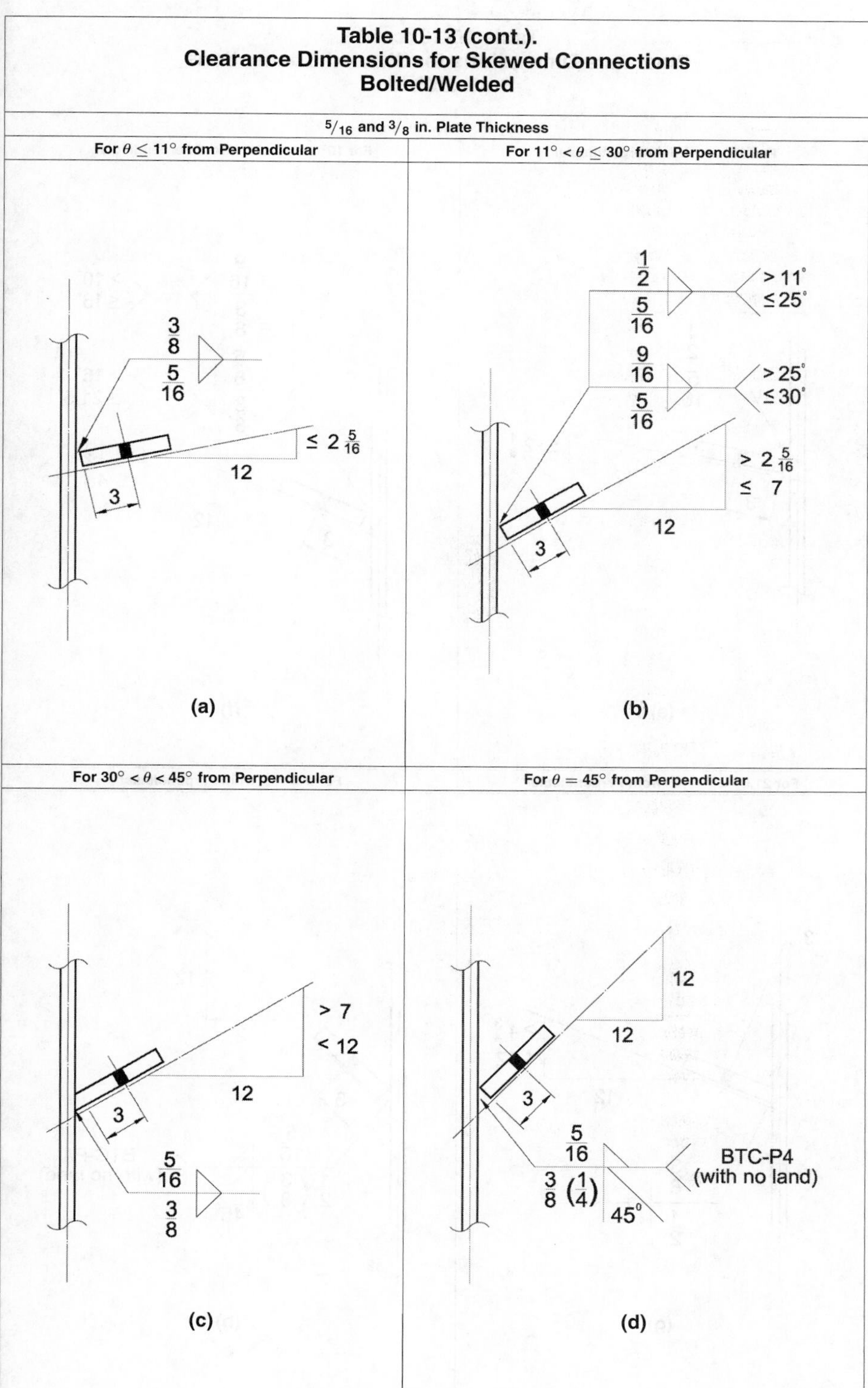
Table 10-13 (cont.).
Clearance Dimensions for Skewed Connections
Bolted/Welded
5/16 and 3/8 in. Plate Thickness
For θ ≤ 11° from Perpendicular
3/8
5/16
≤ 2 5/16
12
3
(a)
For 11° < θ ≤ 30° from Perpendicular
1/2
5/16
> 11°
≤ 25°
9/16
5/16
> 25°
≤ 30°
> 2 5/16
≤ 7
12
3
(b)
For 30° < θ < 45° from Perpendicular
> 7
< 12
12
3
5/16
3/8
(c)
For θ = 45° from Perpendicular
12
12
3
5/16
3/8 (1/4)
45°
BTC-P4
(with no land)
(d)

Table 10-13 (cont.). Clearance Dimensions for Skewed Connections Bolted/Welded

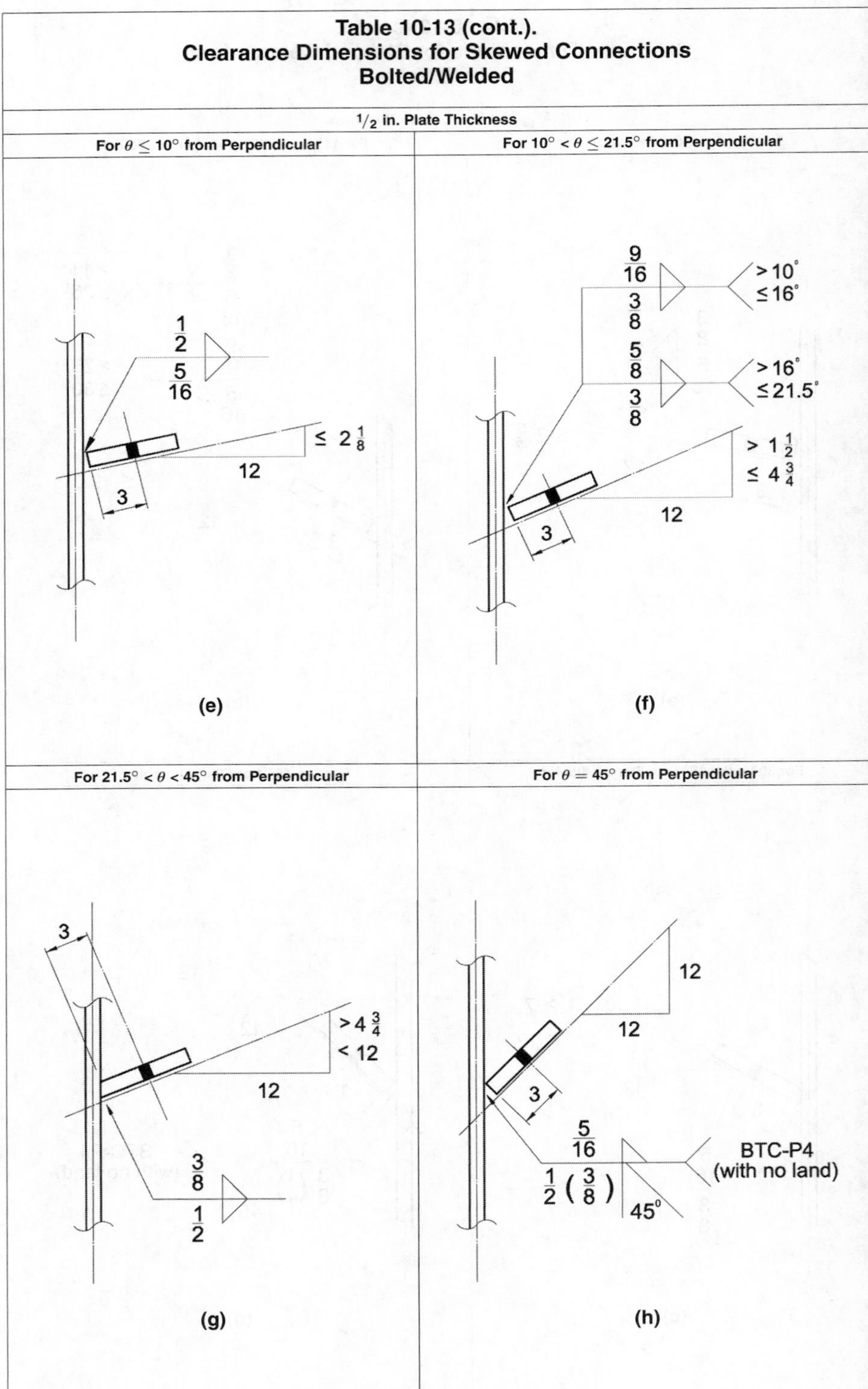

Sloped Connections

A beam is said to be sloped if the plane of its web is perpendicular to the plane of the face of the supporting member, but its flanges are not perpendicular to this face. The angle of slope B is shown in Figure 10-38b and represents the vertical angle to which the fittings must be set to the web of the sloped beam, or the amount that seat and top angles must be bent.

The design of sloped connections usually can be adapted directly from the rectangular connections covered earlier in this Part, with consideration of the geometry of the connection to establish the location of fittings and fasteners. Note that sloped beams often require copes to clear supporting girders, as illustrated in Figure 10-44.

Figure 10-45 shows a sloped beam with double-angle connections, welded to the beam and bolted to the support. The design of this connection is essentially similar to that for rectangular double-angle connections. Alternatively, shear end-plate, tee, single-angle, single-plate, or seated connections could be used. Selection of a particular connection type may be influenced by fabrication economy, erectability, and/or by the types of connections used elsewhere in the structure.

Sloped seated beam connections may utilize either bent angles or plates, depending on the angle of slope. Dimensioning and entering and clearance requirements for sloped seated connections are generally similar to those for skewed connections. The bent seat and top plate shown in Figure 10-46 may be used for smaller bevels.

When the angle of slope is small, it is economical to place transverse holes in the beam web on lines perpendicular to the beam flange; this requires only one stroke of a multiple punch per line. Since non-standard hole arrangements, then, usually occur in the connecting materials (which are single punched), this requires that sufficient dimensions be provided for

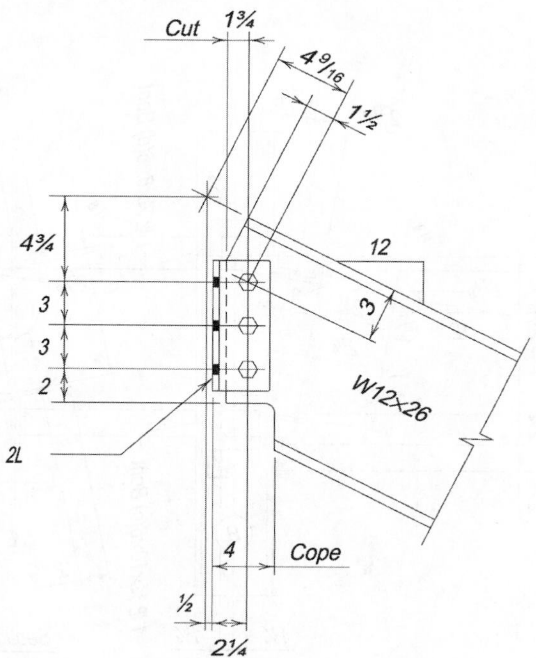

Fig. 10–44. Sloped double-angle connection.

the connecting material to contain fasteners with adequate edges and gages, and at the same time fit the angle to the web without encroaching on the flange fillets of the beam. For the end connection of the beam, this was accomplished by using a 6-in. angle leg; a 4-in. or even a 5-in. leg would not have furnished sufficient edge distance at the extreme fastener.

As the angle of slope increases, however, bolts for the end connections cannot conveniently be lined up to permit simultaneous punching of all holes in a transverse row. In this case, the fabricator may choose to disregard beam gage lines and arrange the hole punching so that ordinary square framed connection material can be used throughout, as shown in Figure 10-47.

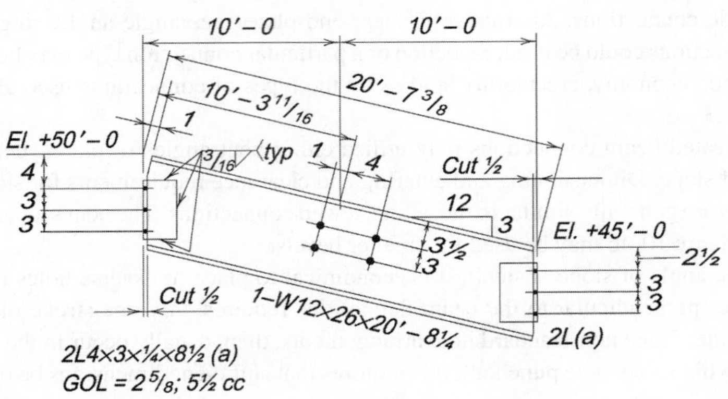

Fig. 10–45. Sloped double-angle connection.

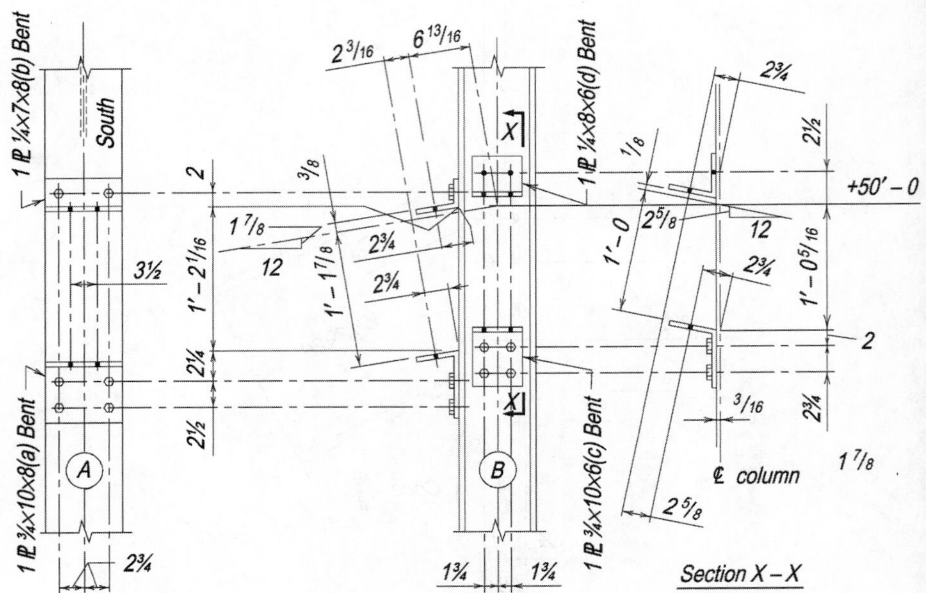

Fig. 10–46. Sloped seated connections.

Canted Connections

A beam perpendicular to the face of a supporting member, but rotated so that its flanges are tilted with respect to those of the support, is said to be canted. The angle of cant C is shown in Figure 10-38c.

The design of canted connections usually can be adapted directly from the rectangular connections covered earlier in this Part. In Figure 10-48, a double-angle connection is used. Alternatively, shear end-plate, seated, single angle, single-plate, and tee connections may also be used.

For channel B2, which is supported by a sloping member B1 (not shown), to match the hole pattern in supporting member B1, the holes in the connecting materials must be canted. As shown in Figure 10-49, the top flange of the channel and the connection angles d^R and d^L are cut to clear the flanges of beam B1. In this detail, with a 3-in-12 angle of cant, 4-in. legs were wide enough to contain the pattern of hole punching.

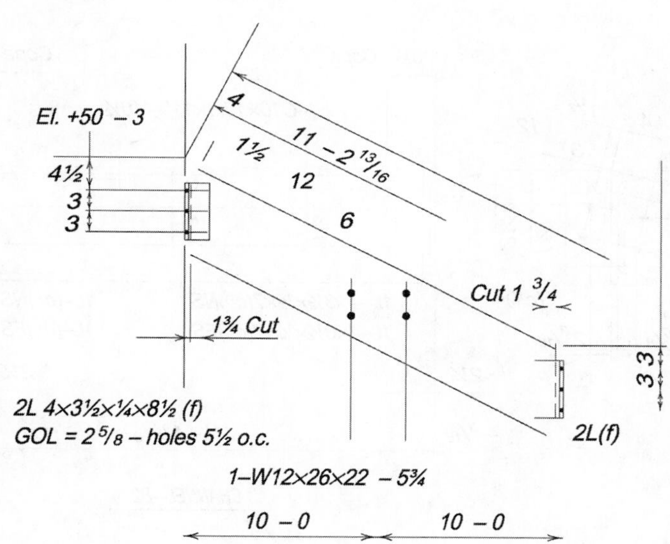

Fig. 10–47. Sloped beam with rectangular connections.

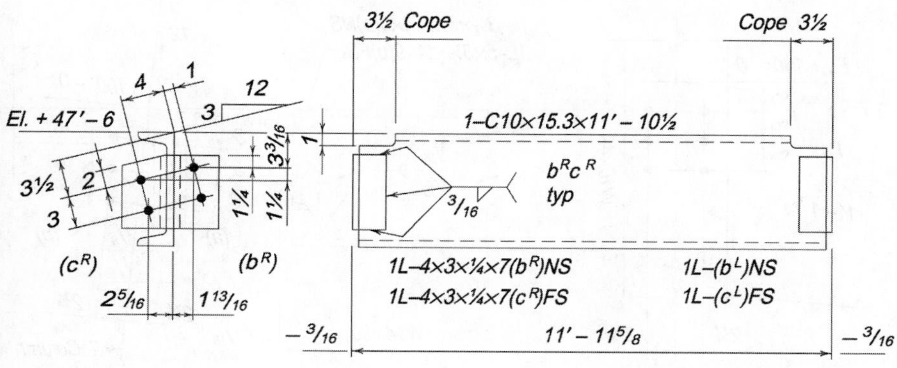

Fig. 10–48. Canted double-angle connections.

Since the multiple punching or drilling of column flanges requires strict adherence to column gage lines, punching is generally skewed in the fittings. When, for some reason, this is not possible, as in Figure 10-50, skewed reference lines are shown on the column to aid in matching connections.

When canted connecting materials are assembled on the beam, particular care must be used in determining the direction of skew for punching the connection angles. An error reversing this skew may permit matching of holes in both members, but the beam will be canted opposite to the intended direction.

Note the connection angles in Figure 10-50 are shown shop welded to the beam. This was done to provide tightening clearance for $^{3}/_{4}$-in. high-strength field bolts in the opposite leg. Had the shop fasteners been bolts, it would have been necessary to stagger the field and shop fasteners and provide longer angles for the increased spacing.

Canted seated beams, shown in Figure 10-51, present few problems other than those in ordinary square-end seated beams. Sufficient width and length of angle leg must be provided

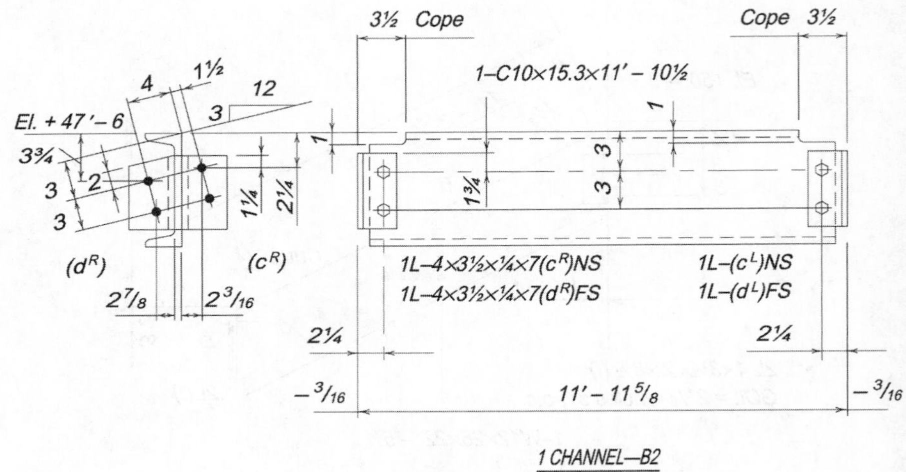

Fig. 10–49. Canted connections to a sloping support.

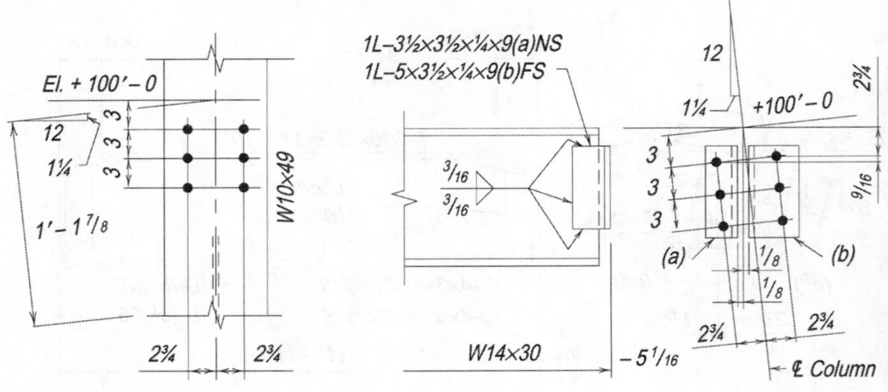

Fig. 10–50. Canted connection to column flange.

to contain the gage line punching or drilling in the column face, as well as the off-center location of the holes matching the punching in the beam flange. The elevation of the top flange centerline and the bevel of the beam flange may be given for reference on the beam detail, although the bevel shown will not affect the fabrication.

Inclines in Two or More Directions (Hip and Valley Framing)

When a beam inclines in two or more directions with respect to the axis of its supporting member, it can be classified as a combination of those inclination directions. For example, the beam of Figure 10-38d is both skewed and sloped. Angle A shows the skew and angle B shows the slope. Note that, since the inclined beam is foreshortened in the elevation, the true angle B appears only in the auxiliary projection, Section X-X. The development of these details is quite complicated and graphical solutions to this compound angle work can be found in any textbook on descriptive geometry. Accurate dimensions may then be determined with basic trigonometry.

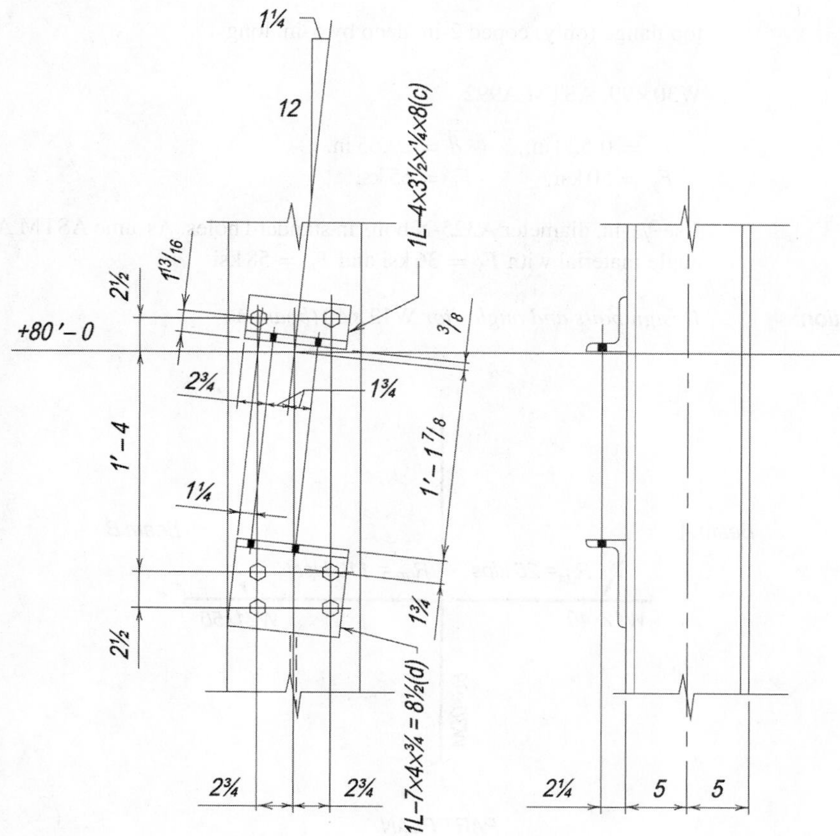

Fig. 10–51. Canted seated connections.

EXAMPLE 10.19. All-bolted double-angle connections (beams-to-girder web).

Given: Refer to Figure 10-52. Design all-bolted double-angle connections for the W12×40 beam (A) and W21×50 beam (B) to W30×99 girder-web connection.

$$R_{uA} = 25\,\text{kips}$$
$$R_{uB} = 110\,\text{kips}$$

W12×40, ASTM A992

$t_w = 0.295$ in., $d = 11.9$ in.
$F_y = 50$ ksi, $F_u = 65$ ksi

top and bottom flanges coped 2-in. deep by 5-in. long (bottom cope allows for entering and tightening bolts through support)

W21×50, ASTM A992

$t_w = 0.380$ in., $d = 20.8$ in.
$F_y = 50$ ksi, $F_u = 65$ ksi

top flange (only) coped 2-in. deep by 5-in. long

W30×99, ASTM A992

$t_w = 0.520$ in., $d = 29.65$ in.
$F_y = 50$ ksi, $F_u = 65$ ksi

Use 3/4-in. diameter A325-N bolts in standard holes. Assume ASTM A36 angle material with $F_y = 36$ ksi and $F_u = 58$ ksi

Solution: *Design bolts and angles for* W12×40 *(beam A)*

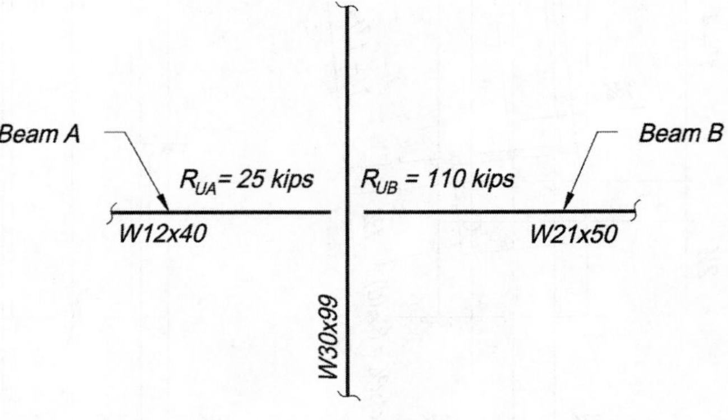

Fig. 10–52. Illustration for Example 10.19.

From Table 10-1, for $^3/_4$-in. diameter A325-N bolts and angle material with $F_y = 36\,\text{ksi}$ and $F_u = 58\,\text{ksi}$, select two rows of bolts and $^1/_4$-in. angle thickness

$$\phi R_n = 48.9 > 25\,\text{kips} \qquad \textbf{o.k.}$$

Check supported beam web (beam A)

From Table 10-1, for two rows of bolts and beam material with $F_y = 50\,\text{ksi}$ and $F_u = 65\,\text{ksi}$, and $L_{ev} = 1^1/_4$ in. and $L_{eh} = 1^1/_2$ in. (assumed to be $1^1/_4$ in. for calculation purposes to account for possible underrun in beam length)

$$\begin{aligned}\phi R_n &= (110\,\text{kips/in.})(0.295\,\text{in.})\\ &= 32.5\,\text{kips} > 25\,\text{kips} \qquad \textbf{o.k.}\end{aligned}$$

Check flexural yielding of the coped sections (beam A)

$$\begin{aligned}S_{net} &= \frac{t_w h_o{}^2}{6}\\ &= (0.295\,\text{in.})(11.94\,\text{in.} - 2\,\text{in.} - 2\,\text{in.})^2/6\\ &= 3.10\,\text{in.}^3\\ \phi R_n &= \frac{\phi F_y S_{net}}{e}\\ &= \frac{0.9(50\,\text{ksi})(3.10\,\text{in.}^3)}{(5^1/_2\,\text{in.})}\\ &= 25.4\,\text{kips} > 25\,\text{kips} \qquad \textbf{o.k.}\end{aligned}$$

Check local buckling at the cope (beam A)

$$\begin{aligned}f_d &= 3.5 - 7.5\left(\frac{d_c}{d}\right)\\ &= 3.5 - 7.5\left(\frac{2\,\text{in.}}{11.9\,\text{in.}}\right)\\ &= 2.24\end{aligned}$$

$$\begin{aligned}\phi F_{bc} &= 50{,}840\frac{t_w{}^2}{c h_o} f_d\\ &= 50{,}840\left[\frac{(0.295\,\text{in.})^2}{(5\,\text{in.})(11.9\,\text{in.} - 2\,\text{in.} - 2\,\text{in.})}\right](2.24)\\ &= 250\,\text{ksi}\\ \phi R_n &= \frac{\phi F_{bc} S_{net}}{e}\\ &= \frac{(250\,\text{ksi})(3.10\,\text{in.}^3)}{(5^1/_2\,\text{in.})}\\ &= 141\,\text{kips} > 25\,\text{kips} \qquad \textbf{o.k.}\end{aligned}$$

Design bolts and angles for W21×50 *(beam B)*

From Table 10-1, for $^3/_4$-in. diameter A325-N bolts and angle material with $F_y = 36\,\text{ksi}$ and $F_u = 58\,\text{ksi}$, select five rows of bolts and $^1/_4$-in.

angle thickness.

$$\phi R_n = 132\,\text{kips} > 110\,\text{kips} \qquad \textbf{o.k.}$$

Check supported beam web (beam B)

From Table 10-1, for five rows of bolts and beam material with $F_y = 50$ ksi and $F_u = 65$ ksi, and $L_{ev} = 1^1/_4$ in. and $L_{eh} = 1^1/_2$ in. (assumed to be $1^1/_4$ in. for calculation purposes to account for possible underrun in beam length)

$$\begin{aligned}\phi R_n &= (312\,\text{kips/in.})(0.380\,\text{in.})\\ &= 119\,\text{kips} > 110\,\text{kips} \qquad \textbf{o.k.}\end{aligned}$$

Check flexural yielding of the coped section (beam B)

From Table 9-2, $S_{net} = 32.5$ in.^{3}

$$\begin{aligned}\phi R_n &= \frac{\phi F_y S_{net}}{e}\\ &= \frac{0.9(50\,\text{ksi})(32.5\,\text{in.}^3)}{(6^1/_2\,\text{in.})}\\ &= 266\,\text{kips} > 110\,\text{kips}\end{aligned}$$

Check local web buckling at the cope (beam B)

$$\frac{c}{d} = \frac{5\,\text{in.}}{20.8\,\text{in.}} = 0.240$$

$$\frac{c}{h_o} = \frac{5\,\text{in.}}{20.8\,\text{in.} - 2\,\text{in.}} = 0.266$$

Since $\dfrac{c}{d} \le 1.0$,

$$\begin{aligned}f &= 2\left(\frac{c}{d}\right)\\ &= 2(0.240)\\ &= 0.480\end{aligned}$$

Since $\dfrac{c}{h_o} \le 1.0$,

$$\begin{aligned}k &= 2.2\left(\frac{h_o}{c}\right)^{1.65}\\ &= 2.2\left(\frac{1}{0.266}\right)^{1.65}\\ &= 19.6\end{aligned}$$

$$\begin{aligned}\phi F_{bc} &= 23{,}590\left(\frac{t_w}{h_o}\right)^2 fk\\ &= 23{,}590\left(\frac{0.380\,\text{in.}}{20.8\,\text{in.} - 2\,\text{in.}}\right)^2 (0.480)(19.6)\\ &= 90.4\,\text{ksi}\end{aligned}$$

$$\phi R_n = \frac{\phi F_{bc} S_{net}}{e}$$

$$= \frac{(90.4\,\text{ksi})(32.5\,\text{in.}^3)}{(6^1/_2\,\text{in.})}$$

$$= 534\,\text{kips} > 110\,\text{kips} \quad \textbf{o.k.}$$

Check supporting girder web

The required bearing strength per bolt is maximum for the bolts that are common to both connections. From beam A, each bolt transmits one-fourth of 25 kips or 6.25 kips/bolt. From beam B, each bolt transmits one-tenth of 110 kips or 11.0 kips. Thus,

$$r_u = 6.25\,\text{kips/bolt} + 11.0\,\text{kips/bolt}$$
$$= 17.3\,\text{kips/bolt}$$

From LRFD Specification Section J3.10, the design bearing strength per bolt is

$$\phi r_n = \phi(2.4dtF_u)$$
$$= 0.75(2.4 \times {}^3/_4\,\text{in.} \times 0.520\,\text{in.} \times 65\,\text{ksi})$$
$$= 45.6\,\text{kips/bolt} > 17.3\,\text{kips} \quad \textbf{o.k.}$$

EXAMPLE 10.20. Offset all-bolted double-angle connections (beam-to-girder web).

Given:

For the all-bolted double-angle connection design of Example 10-1, suppose that two such connections were made back to back for beams with an offset. Determine the design changes necessary to accommodate an offset of 6 in.

Solution:

Since the offset dimension (6 in.) is approximately equal to the gage on the support from Example 10-1 ($6^1/_4$ in.), use a connection configuration similar to that illustrated in Figure 10-31d. All aspects of these connections than are unchanged with the exception of the middle vertical row of bolts (through both connections) which now carry their proportional share of the reaction of both connections.

Check supporting girder web

The required bearing strength per bolt is

$$r_u = \frac{2\,\text{connections} \times 60\,\text{kips/connection}}{6\,\text{bolts}}$$
$$= 20\,\text{kips/bolt}$$

From LRFD Specification Section J3.10, the design strength per bolt is

$$\phi r_n = \phi(2.4dtF_u)$$
$$= 0.75(2.4 \times {}^3/_4\,\text{in.} \times 0.400\,\text{in.} \times 65\,\text{ksi})$$
$$= 35.1\,\text{kips/bolt} > 20\,\text{kips/bolt} \quad \textbf{o.k.}$$

EXAMPLE 10.21. Offset unstiffened seated connection (beam-to-column flange).

Given:

Refer to Figure 10-32c. Determine the seat angle and weld size required for the unstiffened seated connection for the W14×48 beam to W12×65 column-flange connection with an offset of 5½ in.

$R_u = 30\,\text{kips}$

W14×48, ASTM A992

$t_w = 0.340\,\text{in.}, \quad d = 13.8\,\text{in.}, \quad t_f = 0.595\,\text{in.}$
$F_y = 50\,\text{ksi}, \quad F_u = 65\,\text{ksi}$

W12×65, ASTM A992

$t_w = 0.390$
$F_y = 50\,\text{ksi}, \quad F_u = 65\,\text{ksi}$

Use 70 ksi electrode welds to connect the seat angle to the column flange. Assume a 4-in. outstanding angle leg is adequate and the ASTM A36 angle material has $F_y = 36\,\text{ksi}$ and $F_u = 58\,\text{ksi}$.

Solution:

Design seat angle and welds

The required strength for the right-hand weld can be determined by summing moments about the left-hand weld.

$$R_{uR} = \frac{(30\,\text{kips})(3\,\text{in.})}{3^1/_2\,\text{in.}} = 25.7\,\text{kips}$$

Selecting the welds on both sides of the seat to resist this force, the total required strength would be 51.4 kips.

For local web yielding,

$$N_{\min} = \frac{R_u - \phi R_1}{\phi R_2} \geq k$$
$$= \frac{51.4\,\text{kips} - 61.1\,\text{kips}}{17.0\,\text{kips/in.}} \geq 1.19\,\text{in.}$$
$$= (\text{a negative result}) \geq 1.19\,\text{in.}$$
$$= 1.19\,\text{in.}$$

For web crippling,

When $\dfrac{N}{d} \leq 0.2$

$$N_{\min} = \frac{R_u - \phi_r R_3}{\phi_r R_4}$$
$$= \frac{51.4\,\text{kips} - 55.2\,\text{kips}}{5.19\,\text{kips/in.}}$$

which results in a negative quantity.

When $\dfrac{N}{d} > 0.2$

$$N_{min} = \frac{R_u - \phi_r R_5}{\phi_r R_6}$$
$$= \frac{51.4\text{ kips} - 50.5\text{ kips}}{6.92\text{ kips/in.}}$$
$$= 0.130\text{ in.}$$

Thus, $N_{min} = 1.19$ in.

From Table 10-6, a 6-in. angle length with a 3/4-in. thickness and a 3 1/2-in. minimum outstanding leg will provide

$$\phi R_n = 87.5\text{ kips} > 51.4\text{ kips} \qquad \textbf{o.k.}$$

Tabular values include checks of local yielding and web crippling strengths of beam web. **o.k.**

Try L8×4×3/4, 6 in. long with 5/16-in. fillet welds.

For weld strength, from Table 10-6,

$$\phi R_n = 66.7\text{ kips} > 51.4\text{ kips} \qquad \textbf{o.k.}$$

Use L8×4×3/4×6 in. for the seat angle. Use two 3/4-in. diameter A325-N bolts to connect the beam to the seat angle.

Select top angle, bolts, and welds

Use L4×4×1/4 with two 3/4-in.-diameter A325-N bolts through the supported-beam leg of the angle. Use 1/4-in. fillet weld along the toe of the angle (minimum size from LRFD Specification Table J2.4).

EXAMPLE 10.22. Offset single-plate connection (beam-to-column web).

Given:

Refer to Figure 10-34. Design the connection between the W16×40 beam and W14×68 column web (flange).

$$R_u = 55\text{ kips}$$

W16×40, ASTM A992

$$t_w = 0.305\text{ in.}, \qquad d = 16.0\text{ in.}$$
$$F_y = 50\text{ ksi}, \qquad F_u = 65\text{ ksi}$$

W14×68, ASTM A992

$$t_f = 0.720\text{ in.}$$
$$F_y = 50\text{ ksi}, \qquad F_u = 65\text{ ksi}$$

Use 3/4-in. diameter A325-N bolts in standard holes except use short-slotted holes in plate for bolts through the column flange. Use ASTM A36 plate material with $F_y = 36$ ksi and $F_u = 58$ ksi.

Solution: *Design bolts connecting beam web to plate*

For bolt shear, $\phi r_n = 15.9$ kips/bolt. For bolt bearing on the beam web, $\phi r_n = 26.8$ kips/bolt. Since bolt shear is more critical,

$$\begin{aligned} C_{\min} &= \frac{R_u}{\phi R_n} \\ &= \frac{55\,\text{kips}}{15.9\,\text{kips/bolt}} \\ &= 3.45 \end{aligned}$$

From Table 7-18 with $\theta = 0°$ and an eccentricity of $6^1/_4$ in. as shown in Figure 10-34, a four-row by two-vertical-row bolt group provides

$C = 3.59 > 3.45$ **o.k.**

Design bolts connecting plate to column flange

Try one vertical row of four $^3/_4$-in. diameter A325-N bolts.

Check bolt shear

$$\begin{aligned} \phi R_n &= \phi(F_v A_v) \times n \\ &= \left[0.75 \times 48\,\text{ksi} \times \frac{\pi}{4}(^3/_4\,\text{in.})^2\right] \times 4\,\text{bolts} \\ &= 63.6\,\text{kips} > 55\,\text{kips} \qquad \textbf{o.k.} \end{aligned}$$

Check bolt bearing on column flange

$$\begin{aligned} \phi R_n &= \phi(2.4dtF_u) \times n \\ &= 0.75(2.4 \times {}^3/_4\,\text{in.} \times 0.720\,\text{in.} \times 65\,\text{ksi}) \times 4\,\text{bolts} \\ &= 253\,\text{kips} > 55\,\text{kips} \qquad \textbf{o.k.} \end{aligned}$$

Design connection plate

Try PL $^5/_8$ in.×12 in.

Check flexural strength of the plate

The required strength is

$$\begin{aligned} M_u &= R_u e \\ &= (55\,\text{kips})(4^3/_4\,\text{in.}) \\ &= 261\,\text{in.-kips} \end{aligned}$$

For flexural yielding

$$\begin{aligned} \phi M_n &= \phi F_y S_x \\ &= 0.9(36\,\text{ksi})\left[\frac{^5/_8\,\text{in.}(12\,\text{in.})^2}{6}\right] \\ &= 486\,\text{in.-kips} > 261\,\text{in.-kips} \qquad \textbf{o.k.} \end{aligned}$$

For flexural rupture (from Table 15-2),

$$\begin{aligned} \phi M_n &= \phi F_u S_{net} \\ &= 0.75(58\,\text{ksi})(11.0\,\text{in.}^3) \\ &= 479\,\text{in.-kips} > 261\,\text{in.-kips} \qquad \textbf{o.k.} \end{aligned}$$

Check shear yielding of the plate

$$\begin{aligned}\phi R_n &= \phi(0.6F_y)A_g \\ &= 0.9(0.6 \times 36\,\text{ksi})(12\,\text{in.} \times {}^5\!/_8\,\text{in.}) \\ &= 146\,\text{kips} > 55\,\text{kips} \quad \textbf{o.k.}\end{aligned}$$

Check shear rupture of the plate

$$\begin{aligned}\phi R_n &= \phi(0.6F_u)A_n \\ &= 0.75(0.6 \times 58\,\text{ksi})(12\,\text{in.} - 4 \times 0.875\,\text{in.}){}^5\!/_8\,\text{in.} \\ &= 139\,\text{kips} > 55\,\text{kips} \quad \textbf{o.k.}\end{aligned}$$

Check block shear rupture of the plate

From Table 9-3 and 9-4 with $n = 4$, $L_{ev} = 1^1\!/_2$ in., $L_{eh} = 1^1\!/_4$ in. $0.6F_uA_{nv} > F_uA_{nt}$. Thus,

$$\phi R_n = \phi[0.6F_uA_{nv} + F_yA_{gt}]$$

From Table 9-4a and 9-4b,

$$\begin{aligned}\phi R_n &= (194\,\text{kips/in.} + 33.8\,\text{kips/in.}){}^5\!/_8\,\text{in.} \\ &= 142\,\text{kips} > 55\,\text{kips} \quad \textbf{o.k.}\end{aligned}$$

Check local buckling of the plate

This check is analogous to the local buckling check for doubly coped beams as illustrated previously in Part 9 where $c = 6$ in. and $d_c = 1^1\!/_2$-in. at both the top and bottom flanges.

$$\begin{aligned}f_d &= 3.5 - 7.5\left(\frac{d_c}{d}\right) \\ &= 3.5 - 7.5\left(\frac{1^1\!/_2\,\text{in.}}{16.0\,\text{in.}}\right) \\ &= 2.80 \\ \phi F_{bc} &= 50{,}840\frac{t_w{}^2}{ch_o}f_d \\ &= 50{,}840\left[\frac{({}^5\!/_8\,\text{in.})^2}{(6\,\text{in.})(16.0\,\text{in.} - 1.5\,\text{in.} - 1.5\,\text{in.})}\right](2.80) \\ &= 712\,\text{ksi} \\ \phi R_n &= \frac{\phi F_{bc}S_{net}}{e} \\ &= \frac{(712\,\text{ksi})(11.0\,\text{in.}^3)}{(4^3\!/_4\,\text{in.})} \\ &= 1{,}650\,\text{kips} > 55\,\text{kips} \quad \textbf{o.k.}\end{aligned}$$

EXAMPLE 10.23. Offset unstiffened seated connection (beam-to-column web).

Given: Refer to Figure 10-53. Design the seat angle and weld size required for the unstiffened seated connection for the W16×45 beam to W12×53 column-web connection.

$$R_u = 30\,\text{kips}$$

W16×45, ASTM A992

$$t_w = 0.345\text{ in.}$$
$$F_y = 50\text{ ksi}, \qquad F_u = 65\text{ ksi}$$

W12×53, ASTM A992

$$t_f = 0.575\text{ in.}, \qquad d = 12.1\text{ in.}$$
$$F_y = 50\text{ ksi}, \qquad F_u = 65\text{ ksi}$$

Use 70 ksi electrode welds to connect the seat angle to the column-flange toes. Assume a 4-in. outstanding angle leg is adequate and the ASTM A36 angle material has $F_y = 36$ ksi and $F_u = 58$ ksi.

Solution: *Design seat angle and welds*

The required strength for the left-hand weld can be determined by summing moments about the right-hand weld.

$$R_{uL} = \frac{(30\text{ kips})(10^1/_2\text{-in.})}{12.06\text{ in.}}$$
$$= 26.1\text{ kips}$$

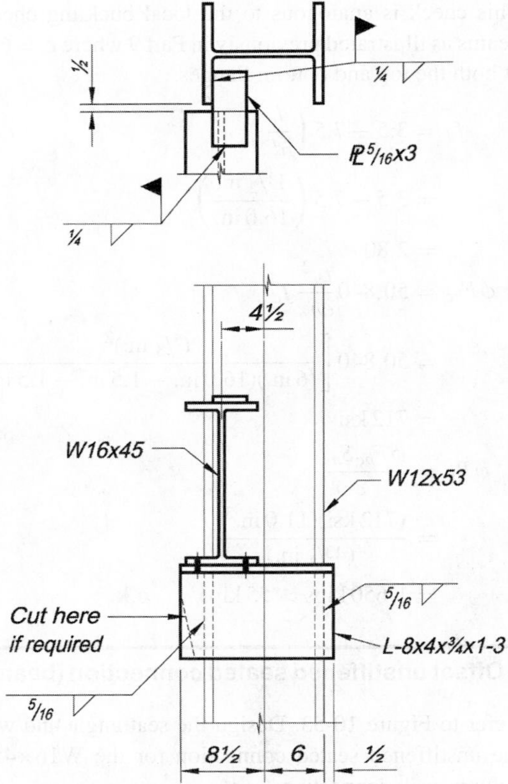

Fig. 10–53. Illustration for Example 10.23.

Selecting the welds on both sides of the seat to resist this force, the total required strength would be 52.2 kips.

For local web yielding,

$$N_{min} = \frac{R_u - \phi R_1}{\phi R_2} \geq k$$

$$= \frac{52.2\,\text{kips} - 53.9\,\text{kips}}{17.3\,\text{kips/in.}} \geq 0.967\,\text{in.}$$

$$= \text{(a negative result)} \geq 0.967\,\text{in.}$$

$$= 0.967\,\text{in.}$$

For web crippling,

When $\frac{N}{d} \leq 0.2$

$$N_{min} = \frac{R_u - \phi_r R_3}{\phi_r R_4}$$

$$= \frac{52.2\,\text{kips} - 55.0\,\text{kips}}{4.89\,\text{kips/in.}}$$

which results in a negative quantity.

When $\frac{N}{d} > 0.2$

$$N_{min} = \frac{R_u - \phi_r R_5}{\phi_r R_6}$$

$$= \frac{52.2\,\text{kips} - 49.8\,\text{kips}}{6.52\,\text{kips/in.}}$$

$$= 0.368\,\text{in.}$$

Thus, $N_{min} = 0.967$ in.

From Table 10-6, a 6-in. angle length with a $^3/_4$-in. thickness and a $3^1/_2$-in. minimum outstanding leg will provide

$$\phi R_n = 87.5\,\text{kips} > 52.2\,\text{kips} \qquad \textbf{o.k.}$$

Tabular values include checks of local yielding and web crippling strengths of beam web. **o.k.**

Try L8×4×$^3/_4$, 6 in. long with $^5/_{16}$-in. fillet welds.

For weld strength, from Table 10-6,

$$\phi R_n = 66.7\,\text{kips} > 52.2\,\text{kips} \qquad \textbf{o.k.}$$

Use L8×4×$^3/_4$×6 in. for the seat angle. Use two $^3/_4$-in. diameter A325-N bolts to connect the beam to the seat angle.

Select top angle, bolts, and welds

Use L4×4×$^1/_4$ with two $^3/_4$-in.-diameter A325-N bolts through the supported-beam leg of the angle. Use $^1/_4$-in. fillet weld along the toe of the angle (minimum size from LRFD Specification Table J2.4).

EXAMPLE 10.24. Skewed double-bent-plate connection (beam-to-girder web).

Given:

Refer to Figure 10-54. Design the skewed double-bent-plate connection for the W16×77 beam to W27×94 girder-web connection.

$$R_u = 80\,\text{kips}$$

W16×77, ASTM A992

$$t_w = 0.455\text{ in.}, \qquad d = 16.5\text{ in.}$$
$$F_y = 50\,\text{ksi}, \qquad F_u = 65\,\text{ksi}$$

W27×94, ASTM A992

$$t_w = 0.490$$
$$F_y = 50\,\text{ksi}, \qquad F_u = 65\,\text{ksi}$$

Use $^7/_8$-in. diameter A325-N bolts in standard holes through the support. Use 70 ksi electrode welds to the supported beam. Assume ASTM A36 plate material with $F_y = 36$ ksi and $F_u = 58$ ksi.

Solution:

From the scaled layout of Figure 10-54c, assuming the welds across the top and bottom of the plates will be $2^1/_2$ in. long, the load is assumed to act at the intersection of the beam centerline and the support face. While the welds do not coincide on opposite faces of the beam web and the weld groups are offset, the locations of the weld groups will be averaged and considered identical.

Design welds

Assume plate length of $8^1/_2$ in.

$$k = \frac{kl}{l}$$
$$= \frac{2^1/_2\text{ in.}}{8^1/_2\text{ in.}}$$
$$= 0.294$$

From Table 8-9, with $\theta = 0°$ and $k = 0.294$

$x = 0.054$ by interpolation

Thus,

$$a = \frac{3^5/_8\text{ in.} - xl}{l}$$
$$a = \frac{3^5/_8\text{ in.} - 0.054(8^1/_2\text{ in.})}{8^1/_2\text{ in.}}$$
$$= 0.372$$

Interpolation from Table 8-9 with $\theta = 0°$, $a = 0.372$, and $k = 0.294$,

$$C = 1.89$$

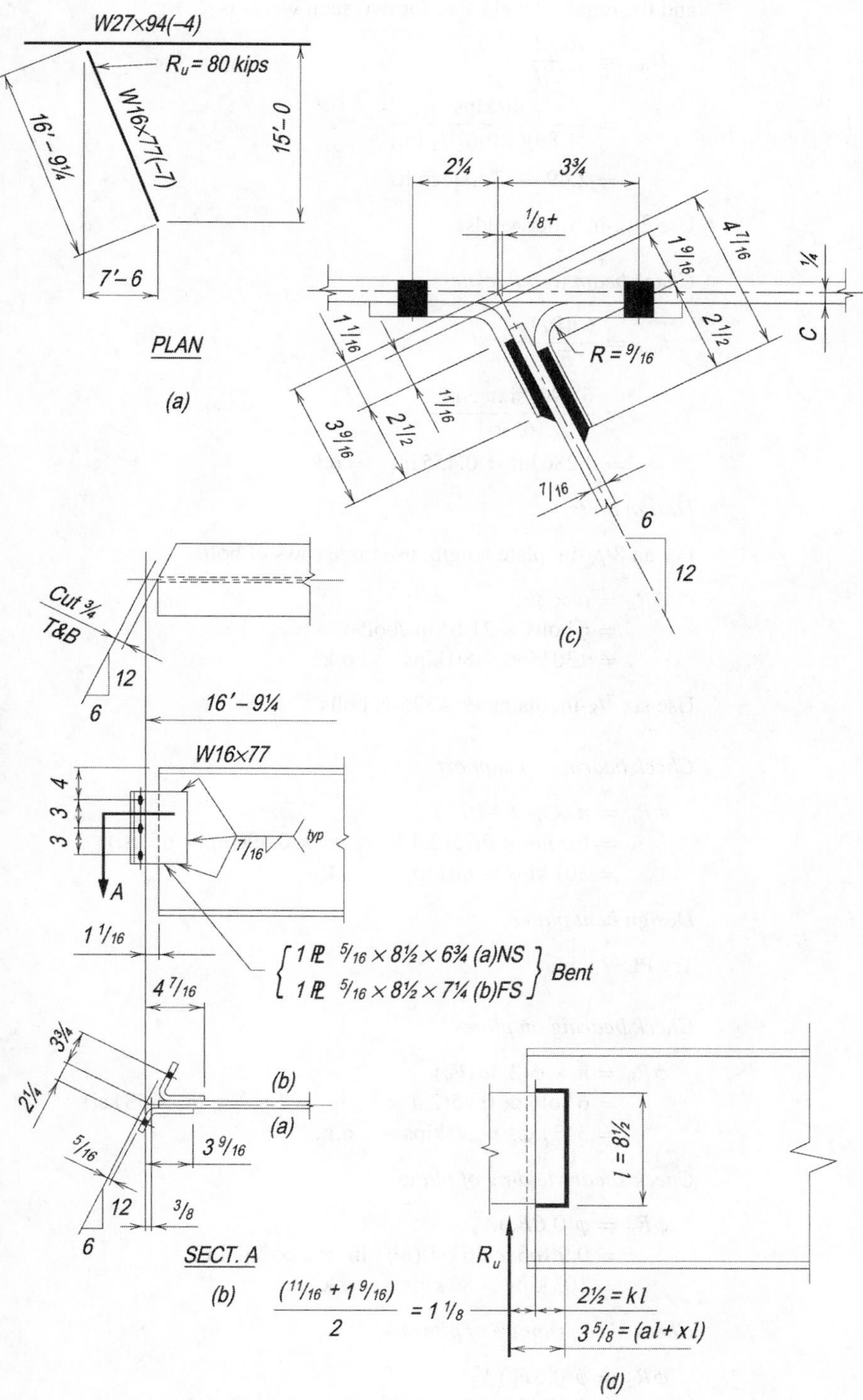

Fig. 10–54. Illustration for Example 10.24.

and the required weld size for two such welds is

$$D_{req} = \frac{R_u/2}{CC_1 l}$$
$$= \frac{40\text{ kips}}{(1.89)(1.0)(8^1/_2\text{ in.})}$$
$$= 2.49 \rightarrow 3\text{ sixteenths}$$

Use $^3/_{16}$-in. fillet welds.

Check beam web thickness

$$t_{\min} = \frac{6.19D}{F_u}$$
$$= \frac{6.19(3\text{ sixteenths})}{65\text{ ksi}}$$
$$= 0.286\text{ in.} < 0.455\text{ in.} \qquad \textbf{o.k.}$$

Design bolts

For an $8^1/_2$-in. plate length, use three rows of bolts.

$$\phi R_n = n \times \phi r_n$$
$$= 6\text{ bolts} \times 21.6\text{ kips/bolt}$$
$$= 130\text{ kips} > 80\text{ kips} \qquad \textbf{o.k.}$$

Use six $^7/_8$-in. diameter A325-N bolts.

Check bearing on support

$$\phi R_n = n \times \phi(2.4dtF_u)$$
$$= 6\text{ bolts} \times 0.75(2.4 \times {}^7/_8\text{ in.} \times 0.490\text{ in.} \times 65\text{ ksi})$$
$$= 301\text{ kips} > 80\text{ kips} \qquad \textbf{o.k.}$$

Design bent plates

Try PL $^5/_{16}$ in.

Check bearing on plates

$$\phi R_n = n \times \phi(2.4dtF_u)$$
$$= 6\text{ bolts} \times 0.75(2.4 \times {}^7/_8\text{ in.} \times 2 \times {}^5/_{16}\text{ in.} \times 58\text{ ksi})$$
$$= 343\text{ kips} > 80\text{ kips} \qquad \textbf{o.k.}$$

Check shear yielding of plates

$$\phi R_n = \phi(0.6F_y)A_g$$
$$= 0.9(0.6 \times 36\text{ ksi})(8^1/_2\text{ in.} \times 2 \times {}^5/_{16}\text{ in.})$$
$$= 103\text{ kips} > 80\text{ kips} \qquad \textbf{o.k.}$$

Check shear rupture of plates

$$\phi R_n = \phi(0.6F_u)A_n$$
$$= 0.75(0.6 \times 58\text{ ksi})(8^1/_2\text{ in.} - 3 \times 1\text{ in.})(2 \times {}^5/_{16}\text{ in.})$$
$$= 90.0\text{ kips} > 80\text{ kips} \qquad \textbf{o.k.}$$

Check block shear rupture of the plates

From Tables 9-3 and 9-4, $0.6F_u A_{nt} > F_u A_{nt}$. Thus,

$$\phi R_n = \phi[0.6F_u A_{nv} + F_y A_{gt}]$$

From Tables 9-4a and 9-4b, with $n = 3$ and $L_{ev} = L_{eh} = 1^1/_4$,

$$\begin{aligned}\phi R_n &= (124\,\text{kips/in.} + 33.8\,\text{kips/in.})(2 \times {}^5/_{16}\,\text{in.})\\ &= 98.6\,\text{kips} > 80\,\text{kips} \quad \textbf{o.k.}\end{aligned}$$

PART 10 REFERENCES

Astaneh, A., S.M. Call, and K.M. McMullin, 1989, “Design of Single-Plate Shear Connections,” *Engineering Journal*, Vol. 26, No. 1, (1st Qtr.), pp. 21–32, AISC, Chicago, IL.

Brockenbrough, R.L., 1998, *Fabrication Guidelines for Cold Bending*, R.L. Brockenbrough and Associates, Pittsburgh, PA.

Carter, C.J., W.A. Thornton, and T.M. Murray, 1997, “Discussion – The Behavior and Load-Carrying Capacity of Unstiffened Seated Beam Connections,” *Engineering Journal*, Vol. 34, No. 4, (4th Qtr.), pp. 151–156, AISC, Chicago, IL.

Roeder, C.W., and R.H. Dailey, 1989, “The Results of Experiments on Seated Beam Connections,” *Engineering Journal*, Vol. 26, No. 3, (3rd Qtr.), pp. 90–95, AISC, Chicago, IL.

Ellifritt, D.S., and T. Sputo, 1999, “Design Criteria for Stiffened Seated Connections to Column Webs,” *Engineering Journal*, Vol. 36, No. 4, (4th Qtr.), pp. 160–167, AISC, Chicago, IL.

Kulak, G.L., and D.L. Green, 1990, “Design of Connectors in Web-Flange Beam or Girder Splices,” *Engineering Journal*, Vol. 27, No. 2, (2nd Qtr.), pp. 41–48, AISC, Chicago, IL.

Salmon, C.G. and J.E. Johnson, 1996, *Steel Structures: Design and Behavior*, 4th Edition, Harper Collins, New York, NY.

Sputo, T., and D.S. Ellifritt, 1991, “Proposed Design Criteria for Stiffened Seated Connections to Column Webs,” *Proceedings of the 1991 National Steel Construction Conference*, pp. 8.1–8.26, AISC, Chicago, IL.

PART 11

DESIGN OF FLEXIBLE MOMENT CONNECTIONS

SCOPE

The specification requirements and other design considerations summarized in this Part apply to the design of flexible moment connections. For the design of simple shear connections, see Part 10. For the design of fully restrained moment connections, see Part 12. For the design of connections for HSS and steel pipe, see the AISC *Hollow Structural Sections Connections Manual*. For connections that are part of a seismic-force-resisting-system in which the seismic response modification factor R is taken greater than 3, see the Seismic Provisions, which are available from AISC at www.aisc.org.

LOAD DETERMINATION

The behavior of PR moment connections, as illustrated in Figure 11-1, is intermediate in degree between the flexibility of simple shear connections and the full rigidity of FR moment connections. Per LRFD Specification Section A2, PR moment connections are permitted upon evidence that the connections to be used are capable of furnishing, as a minimum, a predictable percentage of full end restraint. For further information on the use of PR moment connections, see Geschwindner (1991), Nethercot and Chen (1988), Gerstle and Ackroyd (1989), Deierlein et al. (1990), Goverdhan (1984) and Kishi and Chen (1986).

As an alternative, flexible moment connections can be used as a simplified approach (Disque, 1964; Disque, 1975; Gerstle and Ackroyd, 1989). The flexible moment connection is designed in two stages. First, the beam and its web connections are designed as simple considering only the gravity loads. Second, the lateral loads only are arbitrarily distributed to selected beam flange-to-column connections to form the lateral frames. The resulting flexible moment connections are then designed as "fully restrained" for the calculated required strength due to lateral loads only.

While flexible moment connections (see Figure 11-2a) are not true PR moment connections, they do provide a simple, reliable and economical alternative in the design of connections that must resist lateral-load-induced moments. Flexible moment connections usually result in heavier beams, lighter columns, and reduced stiffening requirements. Additionally, there are several advantages to their use: (1) simplified analysis and calculations; (2) the beams and girders may be designed as simply connected members for gravity loads; and (3) the columns may be designed as axially loaded members with applied moments due to lateral load. Certain provisions, however, must be met when using this type of moment connection:

1. The lateral frames must resist the lateral moments throughout the entire structure from top to bottom.
2. The beams, columns, and their connections must resist the applied lateral moments.

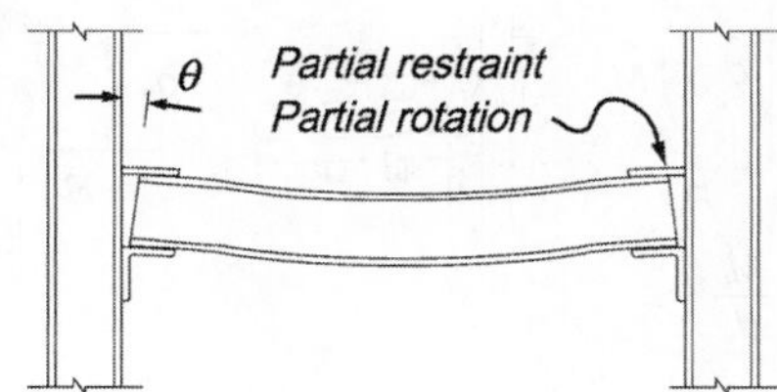

Fig. 11–1. Flexible moment connection behavior.

3. The girders must be capable of carrying the full gravity load as simply supported beams.
4. The connection material must have sufficient inelastic rotation capacity to prevent the welds and/or fasteners from failing due to combined gravity and lateral loading.

The loading and unloading sequence that occurs in the flexible moment connections is described in detail by Disque (1964) and Disque (1975). The assumed distribution of this loading, the assumed connection deformation, and the locations of the points of inflection for use in calculating the bending moments are illustrated in Figure 11-2. Reasonably proportioned connections will result despite these apparently arbitrary assumptions which are required to overcome the complexities of a more exact PR moment frame analysis. An in-depth investigation of the analysis and design of flexibly connected lateral frames is given in Ackroyd (1987). Ackroyd reports that the flexible wind frame approach is valid for frames less than 10 stories in height.

Flexible moment connections deliver concentrated forces to the flanges of columns that must be accounted for in the design of the column and column panel-zone per LRFD Specification Section K1. Either the column size can be selected with adequate flange and web thicknesses to eliminate the need for column stiffening, or transverse stiffeners and/or web doubler plates can be provided. For further information, refer to AISC Design Guide No. 13 *Stiffening of Wide-Flange Columns at Moment Connections: Wind and Seismic Applications* (Carter, 1999).

FLANGE-ANGLE FLEXIBLE MOMENT CONNECTIONS

Flange-angle flexible moment connections are made with top and bottom angles and a simple shear connection.

The design strength of a flange-angle flexible moment connection is determined from the applicable limit states for the bolts (see Part 7), welds (see Part 8) and connection elements (see Part 9). In all cases, the design strength ϕR_n must equal or exceed the required strength R_u.

The tensile force is carried to the angle by the flange bolts, with the angle assumed to deform as illustrated in Figure 11-2. A point of inflection is assumed between the bolt gage line and the top face of the connection angle, for use in calculating the local bending moment and the corresponding required angle thickness. The effect of prying action must also be considered.

The strength of this type of connection is often limited by the available angle thickness and the maximum number of fasteners that can be placed on a single gage line of the vertical

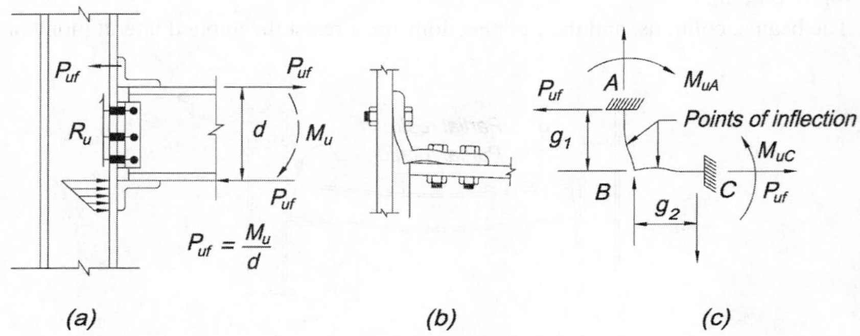

Fig. 11–2. Flange-angle flexible moment connection behavior.

leg of the connection angle at the tension flange. Figure 11-3 illustrates the column flange deformation and shows that only the fasteners closest to the column web are fully effective in transferring forces.

FLANGE-PLATED FLEXIBLE MOMENT CONNECTIONS

Originally proposed by Blodgett (1966), as illustrated in Figure 11-4, a flange-plated flexible moment connection consists of a simple shear connection and top and bottom flange plates that connect the flanges of the supported beam to the supporting column. These flange plates are welded to the supporting column and may be bolted or welded to the flanges of the supporting beam. An unwelded length of $1^1/_2$ times the flange-plate width b_A is normally

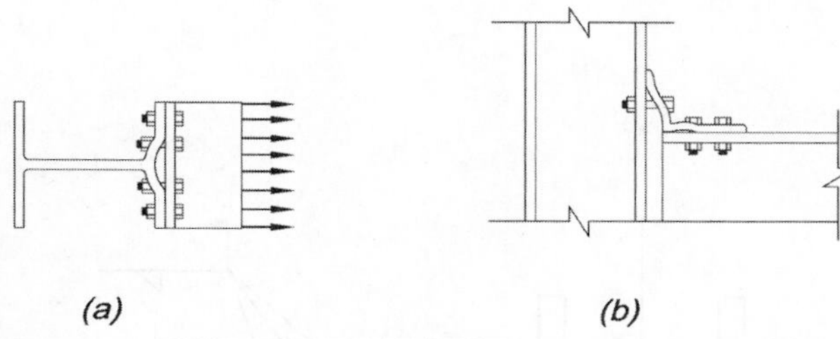

Fig. 11–3. Illustration of deformations in flange-angle flexible moment connections.

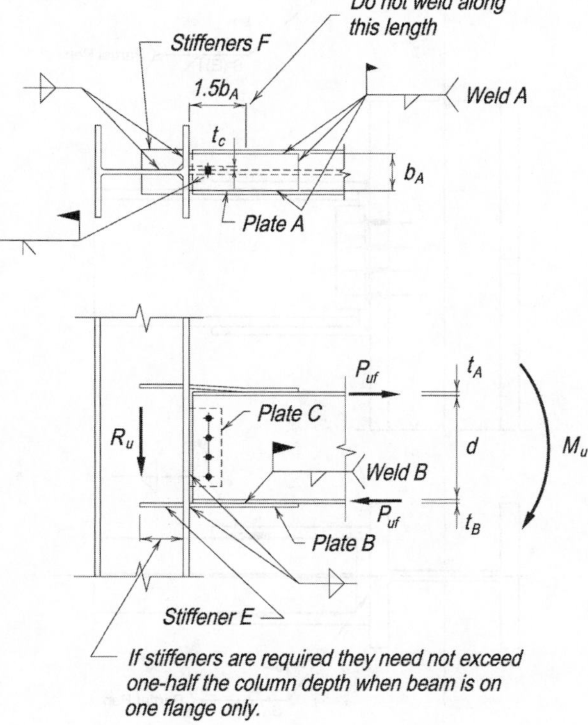

Fig. 11–4. Flange-plated flexible moment connection.

assumed to permit the elongation of the plate necessary for PR behavior. Other flange-plated details are illustrated in Figures 11-5 and 11-6.

The design strength of a flange-plated flexible moment connection is determined from the applicable limit states for the bolts (see Part 7), welds (see Part 8) and connection elements (see Part 9). In all cases, the design strength ϕR_n must equal or exceed the required strength R_u.

The shop and field practices for flange-plated FR moment connections (see Part 12) are equally applicable to flange-plated flexible moment connections.

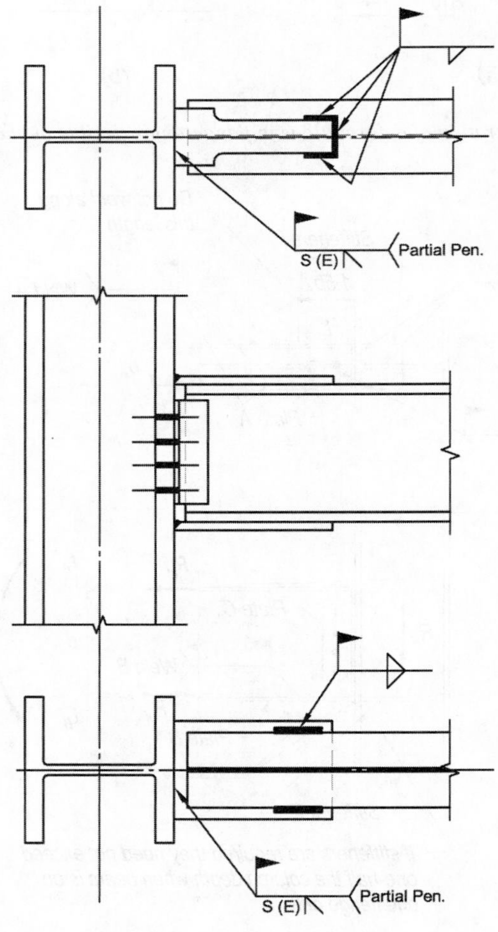

Fig. 11–5. Typical strong-axis flange-plated flexible moment connections.

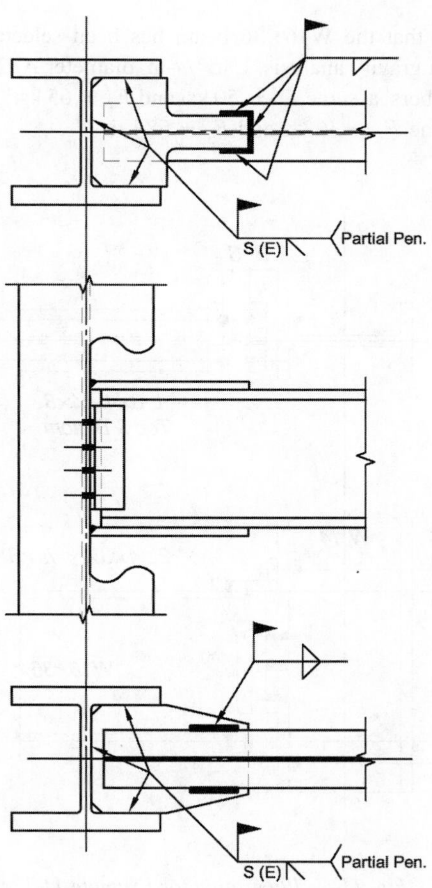

Fig. 11–6. Typical weak-axis flange-plated flexible moment connections.

DESIGN EXAMPLES

EXAMPLE 11.1. Flange-angle flexible moment connections.

Given:

Refer to Figure 11-7. Design the flexible wind connection shown for the W16×36 beam to W14 column flange connection. From the simple beam gravity analysis with 3.4 kips/ft on a 20-ft span,

$$R_u = 34.0 \text{ kips}$$
$$M_{ug} = 170 \text{ ft-kips (at beam centerline)}$$

From the portal analysis shown in Figure 11-8 the wind moment is

$$M_{uw} = 56.0 \text{ ft-kips (at connection)}$$

W16×36, ASTM A992

$d = 15.9$ in. $\quad b_f = 6.99$ in. $\quad Z_x = 64.0$ in.^{3}
$t_w = 0.295$ in. $\quad t_f = 0.430$ in.

Note that the W16×36 beam has been selected based upon a simple beam gravity analysis. Use 3/4-in. diameter A325-N bolts. For structural members, assume $F_y = 50$ ksi and $F_u = 65$ ksi; for connecting materials, assume $F_y = 36$ ksi and $F_u = 58$ ksi.

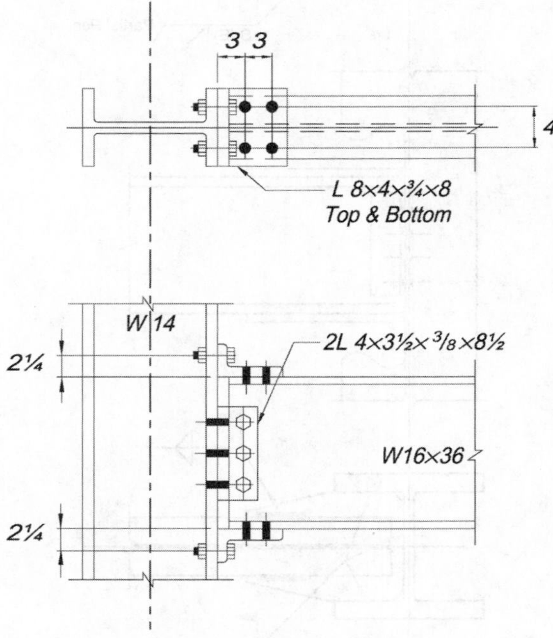

Fig. 11–7. Illustration for Example 11.1.

Solution: *Check beam design flexural strength (at connection)*

$$Z_{req} = \frac{M_{uw} \times 12 \text{ in./ft}}{0.9F_y}$$
$$= \frac{(56.0 \text{ ft-kips})(12 \text{ in./ft})}{0.9(50 \text{ ksi})}$$
$$= 14.9 \text{ in.}^3$$

Assuming two rows of 3/4-in. diameter A325-N bolts in standard holes, from LRFD Specification Section B10:

$$A_{fg} = b_f \times t_f$$
$$= 6.99 \text{ in.} \times 0.430 \text{ in.}$$
$$= 3.00 \text{ in.}^2$$
$$A_{fn} = A_{fg} - 2\,(d_b + 1/8 \text{ in.})t_f$$
$$= 3.00 \text{ in.}^2 - 2(3/4 \text{ in.} + 1/8 \text{ in.})(0.430 \text{ in.})$$
$$= 2.25 \text{ in.}^2$$

Since $0.75F_uA_{fn}(= 110 \text{ kips})$ is less than $0.9F_yA_{fg}(= 135 \text{ kips})$, the effective tension flange area A_{fe} is

$$A_{fe} = \frac{5F_u}{6F_y}A_{fn}$$
$$= \frac{5}{6}\left(\frac{65 \text{ ksi}}{50 \text{ ksi}}\right)2.25 \text{ in.}^2$$
$$= 2.44 \text{ in.}^2$$

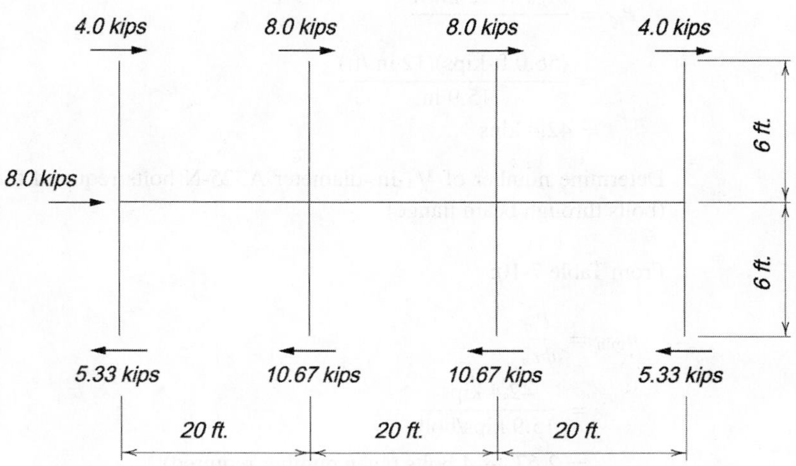

Fig. 11–8. Illustration of portal analysis for Example 11.1.

This is an 18.7 percent reduction from the gross flange area A_{fg} and the effective plastic section modulus Z_e is

$$
\begin{aligned}
Z_e &\approx Z_x - 2\left(0.187 A_{fg} \frac{d}{2}\right) \\
&\approx 63.0 \text{ in.}^3 - 2\left(0.187 \times 3.00 \text{ in.}^2 \times \frac{15.9 \text{ in.}}{2}\right) \\
&\approx 55.1 \text{ in.}^3
\end{aligned}
$$

Since $Z_e > Z_{req}$, the beam design flexural strength is **o.k.**

Design the double-angle web connection

From Table 10-2, for 3/4-in. diameter A325-N bolts and angle material with $F_y = 36$ ksi and $F_u = 58$ ksi, select three rows of bolts and 1/4-in. angle thickness.

$$\phi R_n = 76.7 \text{ kips} > 34.0 \text{ kips} \quad \textbf{o.k.}$$

From Table 10-2, for three rows of bolts and an uncoped beam with $F_y = 50$ ksi and $F_u = 65$ ksi

$$
\begin{aligned}
\phi R_n &= (263 \text{ kips/in.})(0.295 \text{ in.}) \\
&= 77.6 \text{ kips} > 34.0 \text{ kips} \quad \textbf{o.k.}
\end{aligned}
$$

Note: If the column section were given, it could also be checked using Table 10-2

Design the tension flange angle and connection

Calculate the flange force P_{uf}

$$
\begin{aligned}
P_{uf} &= \frac{M_{uw} \times 12 \text{ in./ft}}{d} \\
&= \frac{(56.0 \text{ ft-kips})(12 \text{ in./ft})}{15.9 \text{ in.}} \\
&= 42.4 \text{ kips}
\end{aligned}
$$

Determine number of 3/4-in. diameter A325-N bolts required for shear (bolts through beam flange)

From Table 7-10:

$$
\begin{aligned}
n_{\min} &= \frac{P_{uf}}{\phi r_n} \\
&= \frac{42.4 \text{ kips}}{15.9 \text{ kips/bolt}} \\
&= 2.67 \rightarrow 4 \text{ bolts (even number required)}
\end{aligned}
$$

Determine number of 3/4-in. diameter A325-N bolts required for tension (bolts through column flange)

From Table 7-14:

$$n_{\min} = \frac{P_{uf}}{\phi r_n}$$
$$= \frac{42.4 \text{ kips}}{29.8 \text{ kips/bolt}}$$
$$= 1.42 \rightarrow 2 \text{ bolts}$$

Determine flange angle thickness for flexure

Try L8×4 8-in. long. The tributary load in bending is then 42.4 kips/ 8 in. = 5.3 kips/in. The preliminary angle thickness may now be selected from Table 9-2. Since this table is based upon a symmetrical connection, enter table with twice the tributary load or 10.6 kips/in. and $b = 1^1/_2$ in. For $F_y = 36$ ksi angle material, a $^3/_4$-in. thickness provides for 12.2 kips/in.

Try L8×4×$^3/_4$×8 in.

Check angle thickness for prying action assuming a 4-in. gage

$$r_{ut} = \frac{P_{uf}}{2 \text{ bolts}}$$
$$= \frac{42.4 \text{ kips}}{2 \text{ bolts}}$$
$$= 21.2 \text{ kips/bolt}$$

$$b = 1^1/_2 \text{ in.}$$

$$a = 4 \text{ in.} - b - t$$
$$= 4 \text{ in.} - 1^1/_2 \text{ in.} - {}^3/_4 \text{ in.}$$
$$= 1^3/_4 \text{ in.}$$

Since $a = 1^3/_4$ in. is less than $1.25b$, use $a = 1\,^3/_4$ in. in calculations

$$b' = b - d/2$$
$$= 1^1/_2 \text{ in.} - {}^3/_4 \text{ in.}/2$$
$$= 1^1/_8 \text{ in.}$$

$$a' = a + d/2$$
$$= 1^3/_4 \text{ in.} + {}^3/_4 \text{ in.}/2$$
$$= 2^1/_8 \text{ in.}$$

$$\rho = \frac{b'}{a'}$$
$$= \frac{1^1/_8 \text{ in.}}{2^1/_8 \text{ in.}}$$
$$= 0.529$$

$$\beta = \frac{1}{\rho}\left(\frac{\phi r_n}{r_{ut}} - 1\right)1$$
$$= \frac{1}{0.529}\left(\frac{29.8 \text{ kips/bolt}}{21.2 \text{ kips/bolt}} - 1\right)$$
$$= 0.767$$
$$\delta = 1 - d'/p$$
$$= 1 - \frac{13/16 \text{ in.}}{4 \text{ in.}}$$
$$= 0.797$$

Since $b < 1$, α' is equal to the lesser of 1.0 and

$$\frac{1}{\delta}\left(\frac{\beta}{1-\beta}\right) = \frac{1}{0.797}\left(\frac{0.767}{1-0.767}\right)$$

Thus, set $\alpha' = 1.0$ and

$$t_{req} = \sqrt{\frac{4.44 r_{ut} b'}{p F_y (1 + \delta\alpha')}}$$
$$= \sqrt{\frac{4.44(21.2 \text{ kips/bolt})(1^1/_8 \text{ in.})}{(4 \text{ in.})(36 \text{ ksi})[1 + (0.797)(1.0)]}}$$
$$= 0.640 \text{ in.} < 3/4 \text{ in.} \qquad \textbf{o.k.}$$

Check tension yielding of the angle

$$\phi R_n = \phi F_y A_g$$
$$= 0.9(36 \text{ ksi})(8 \text{ in.} \times 3/4 \text{ in.})$$
$$= 194 \text{ kips} > 42.4 \text{ kips} \qquad \textbf{o.k.}$$

Check tension rupture of the angle

$$\phi R_n = \phi F_u A_n$$
$$= 0.75\,(58 \text{ ksi})(8 \text{ in.} - 2 \times 0.875 \text{ in.})(3/4 \text{ in.})$$
$$= 204 \text{ kips} > 42.4 \text{ kips} \qquad \textbf{o.k.}$$

Check shear yielding of the angle

$$\phi R_n = \phi\,(0.6 F_y) A_g$$
$$= 0.9\,(0.6 \times 36 \text{ ksi})(8 \text{ in.} \times 3/4 \text{ in.})$$
$$= 116 \text{ kips} > 42.4 \text{ kips} \qquad \textbf{o.k.}$$

Check shear rupture of the angle

$$\phi R_n = \phi\,(0.6 F_u) A_n$$
$$= 0.75\,(0.6 \times 58 \text{ ksi})(8 \text{ in.} - 2 \times 0.875 \text{ in.})(3/4 \text{ in.})$$
$$= 122 \text{ kips} > 42.4 \text{ kips} \qquad \textbf{o.k.}$$

Check block shear rupture of the angle

From Tables 9-3 and 9-4, with $L_{ev} = 2$ in., $L_{eh} = 2$ in., and $n = 2$, $0.6\,F_u A_{nv} > F_u A_{nt}$. Thus, from Tables 9-4a and 9-4b,

$$\begin{aligned}\phi R_n &= \phi(0.6 F_u A_{nv} + F_y A_{gt})\\ &= (96.0 \text{ kips/in.} + 54.0 \text{ kips/in.})(3/4 \text{ in.})\\ &= 113 \text{ kips} > 42.4 \text{ kips} \qquad \textbf{o.k.}\end{aligned}$$

Design the compression flange angle and connection

For symmetry, try L8×4×3/4×8 in. with four 3/4-in. diameter A325-N bolts through beam flange and two 3/4-in. diameter A325-N bolts through column flange.

Check design compressive strength of angle assuming $K = 0.65$ and $l = 3$ in. (normal gage).

$$\begin{aligned}\frac{Kl}{r} &= \frac{0.65(3 \text{ in.})}{\sqrt{\dfrac{(8 \text{ in.})(3/4 \text{ in.})^3/12}{(8 \text{ in.})(3/4 \text{ in.})}}}\\ &= 9.01\end{aligned}$$

From LRFD Specification Table 3-36 with $\dfrac{Kl}{r} = 9.01$,

$$\phi_c F_{cr} = 30.47 \text{ ksi}$$

and the design compressive strength of the angle is

$$\begin{aligned}\phi R_n &= \phi_c F_{cr} A\\ &= (30.47 \text{ ksi})(8 \text{ in.} \times 3/4 \text{ in.})\\ &= 183 \text{ kips} > 42.4 \text{ kips} \qquad \textbf{o.k.}\end{aligned}$$

Comment: In all situations where flexible connections are used, the impact of connection rotation on drift of the wind frame must be checked. In addition, the column design must account for the reduced beam stiffness due to connection rotation.

EXAMPLE 11.2. Flange-plated flexible moment connection.

Given: Design a welded flange-plated PR moment connection for a 20-ft-long W18×50 beam to W14×109 column flange connection. The beam supports a 7.29 kip/ft uniform load. For structural members, $F_y = 50$ ksi and $F_u = 65$ ksi; for connecting materials, $F_y = 36$ ksi and $F_u = 58$ ksi.

$$\begin{aligned}R_u &= 73.0 \text{ kips}\\ M_u &= 225 \text{ ft-kips}\end{aligned}$$

W18×50

$d = 17.99$ in. $\quad b_f = 7.495$ in. $\quad Z_x = 101$ in.^{3}
$t_w = 0.355$ in. $\quad t_f = 0.570$ $\quad I_x = 800$ in.^{4}

W14×109

$d = 14.32$ in. $b_f = 14.605$ in. $k = 1\frac{9}{16}$ in.
$t_w = 0.525$ in. $t_f = 0.860$ in. $T = 11\frac{1}{4}$ in.

Use 70-ksi electrodes and 7/8-in. diameter A325-N bolts.

Solution: *Check beam design flexural strength*

$$Z_{req} = \frac{M_u \times 12 \text{ in./ft}}{0.9F_y} = \frac{(225 \text{ ft-kips})(12 \text{ in./ft})}{0.9(50 \text{ ksi})} = 60 \text{ in.}^3$$

$$Z_x = 101 \text{ in.}^3$$

Since $Z_x > Z_{req}$, the beam design flexural strength is **o.k.**

Design the single-plate web connection

Determine number of 7/8-in. diameter A325-N bolts required for shear. From Table 7-10

$$n_{\min} = \frac{R_u}{\phi r_n} = \frac{73.0 \text{ kips}}{21.6 \text{ kips/bolt}} = 3.38 \rightarrow 4 \text{ bolts}$$

Try PL 3/8

Determine number of 7/8-in. diameter A325-N bolts required for material bearing, assuming $L_e = 1\frac{1}{2}$ in. and $s = 3$ in. The 3/8-in. thick plate ($F_u = 58$ ksi) is more critical than the 0.355-in.-thick beam web ($F_u = 65$ ksi). From Table 7-12,

$$n_{\min} = \frac{R_u}{\phi r_n} = \frac{73.0 \text{ kips}}{34.3 \text{ kips/bolt}} = 2.13 \rightarrow 3 \text{ bolts} < 4 \text{ bolts}$$

Therefore, bolt shear is more critical. Try a four-bolt single-plate connection.

Check shear yielding of the plate

$$\phi R_n = 0.9\,(0.6F_y A_g) = 0.9[0.6 \times 36 \text{ ksi } (12 \text{ in.} \times 3/8 \text{ in.})] = 87.5 \text{ kips} > 73.0 \text{ kips} \quad \textbf{o.k.}$$

Check shear rupture of the plate

$$\begin{aligned}\phi R_n &= 0.75\,(0.6 F_u A_n)\\ &= 0.75[0.6 \times 58 \text{ ksi } (12 \text{ in.} - 4(7/8 \text{ in.} + 1/8 \text{ in.}))3/8 \text{ in.}]\\ &= 78.3 \text{ kips} > 73.0 \text{ kips} \quad \textbf{o.k.}\end{aligned}$$

Check block shear rupture of the plate

With $L_{eh} = 1\frac{1}{2}$ in. and $L_{ev} = 1\frac{1}{2}$ in., from Tables 9-3a and 9-4a, $0.6\ F_u A_{nv} > F_u A_{nt}$. Thus,

$$\phi R_n = \phi[0.6 F_u A_{nv} + F_y A_{gt}]$$

From Tables 9-4a and 9-4b,

$$\begin{aligned}\phi R_n &= (183 \text{ kips/in.} + 40.5 \text{ kips/in.})3/8 \text{ in.}\\ &= 83.8 \text{ kips} > 73.0 \text{ kips} \quad \textbf{o.k.}\end{aligned}$$

Determine required weld size for fillet welds to supporting column flange:

$$\begin{aligned}D_{\min} &= \frac{R_u}{2 \times 1.392\, l}\\ &= \frac{73.0 \text{ kips}}{2 \times 1.392(12 \text{ in.})}\\ &= 2.19 \rightarrow 3 \text{ sixteenths}\end{aligned}$$

From LRFD Specification Table J2.4, since the column flange thickness is over 3/4 in., the minimum fillet weld size is 5/16 in., use two 5/16-in. fillet welds.

Design the tension flange plate and connection

Calculate the flange force P_{uf}

$$\begin{aligned}P_{uf} &= \frac{M_u \times 12 \text{ in./ft}}{d + t_{p\ avg}}\\ &= \frac{(225 \text{ ft-kips})(12 \text{ in./ft})}{18.0 \text{ in.} + \dfrac{0.75 \text{ in.} + 0.625 \text{ in.}}{2}}\\ &= 144 \text{ kips}\end{aligned}$$

Determine tension flange-plate dimensions

From Figure 8-12, assume a shelf dimension of 5/8 in. on both sides of the plate. The plate width, then, is 7.495 in. – 2(5/8 in.) = 6.245. Try a 3/4 in. × 6 1/4 in. flange plate.

Check tension yielding of the flange plate:

$$\begin{aligned}\phi R_n &= \phi F_y A_g\\ &= 0.9(36 \text{ ksi})(6\,1/4 \text{ in.})(3/4 \text{ in.})\\ &= 152 \text{ kips} > 144 \text{ kips} \quad \textbf{o.k.}\end{aligned}$$

Determine required weld size and length for fillet welds to beam flange.

Try a $^5/_{16}$-in. fillet weld. The minimum length of weld l_{min} is:

$$\begin{aligned} l_{min} &= \frac{P_{uf}}{1.392D} \\ &= \frac{150 \text{ kips}}{1.392(5 \text{ sixteenths})} \\ &= 21.6 \text{ in.} \end{aligned}$$

Use 8 in. of weld along each side and $6\,^1/_4$ in. of weld along the end of the flange plate.

Select tension flange plate dimensions

To provide for an 8-in. weld length and an unwelded length of $1\,^1/_2$ times the plate width, use PL $^3/_4$ in.$\times 6\,^1/_4$ in.$\times 17\,^1/_2$ in.

Determine required weld size for fillet welds to supporting column flange.

$$\begin{aligned} D_{min} &= \frac{P_{uf}}{2 \times 1.392\,l} \\ &= \frac{144 \text{ kips}}{2 \times 1.392(6\,^1/_4 \text{ in.})} \\ &= 8.28 \rightarrow 9 \text{ sixteenths} \end{aligned}$$

Use $^9/_{16}$-in. fillet welds.

Since these fillet welds are large, groove welds may be more economical.

Design the compression flange plate and connection

The compression flange plate should have approximately the same area as the tension flange plate (4.69 in.^{2}). Assume a shelf dimension of $^5/_8$ in. The plate width, then, is 7.495 in. + 2($^5/_8$ in.) = 8.745 in. To approximately balance the flange-plate areas, try a $^5/_8$ in.$\times 8\,^3/_4$ in. compression flange plate.

Check design compressive strength of flange plate

Assuming $K = 0.65$ and $l = {}^3/_4$ in. ($^1/_2$-in. setback plus $^1/_4$-in. tolerance).

$$\begin{aligned} \frac{Kl}{r} &= \frac{0.65(^3/_4 \text{ in.})}{\sqrt{\dfrac{(8\,^3/_4 \text{ in.})(^5/_8 \text{ in.})^3/12}{(8\,^3/_4 \text{ in.})(^5/_8 \text{ in.})}}} \\ &= 2.70 \end{aligned}$$

From LRFD Specification Table 3-36 with $\dfrac{Kl}{r} = 2.70$,

$$\phi_c F_{cr} = 30.59 \text{ ksi}$$

and the design compressive strength of the flange-plate is

$$\begin{aligned}\phi R_n &= \phi_c F_{cr} A \\ &= (30.59\text{ ksi})(8\,{}^{3}\!/_{4}\text{ in.} \times {}^{5}\!/_{8}\text{ in.}) \\ &= 167\text{ kips} > 144\text{ kips} \qquad \textbf{o.k.}\end{aligned}$$

Determine required weld size and length for fillet welds to beam flange

As before for the tension flange plate, with $^{5}/_{16}$-in. fillet welds, use 8 in. along each side and 6 $^{1}/_{4}$ in. along the end of the compression flange plate.

Select compression flange plate dimensions

Use PL $^{5}/_{8}$ in. × 8 $^{3}/_{4}$ in. × 1′-0.

Determine required weld size for fillet welds to supporting column flange

$$\begin{aligned}D_{\text{min}} &= \frac{P_{uf}}{2 \times 1.392\, l} \\ &= \frac{144\text{ kips}}{2 \times 1.392(8\,{}^{3}\!/_{4}\text{ in.})} \\ &= 5.91 \rightarrow 6\text{ sixteenths}\end{aligned}$$

Use $^{3}/_{8}$-in. fillet welds.

Since these fillet welds are large, groove welds may be more economical.

Comment: The column section should be checked for stiffening requirements. A check of the applicable limit states from LRFD Specification Section K1 (refer to "Column Stiffening at Moment Connections in this Part) will show the W14×109 column in the above example is adequate without stiffening.

PART 11 REFERENCES

Ackroyd, M.H., 1987, "Simplified Frame Design of Type PR Construction," *Engineering Journal*, Vol. 24, No. 4, (4th Qtr.), pp. 141–146, AISC, Chicago, IL.

Blodgett, O.W., 1966, *Design of Welded Structures*, James F. Lincoln Arc Welding Foundation, Cleveland, OH.

Carter, C.J., 1999, AISC Design Guide No. 13 *Wide-Flange Column Stiffening at Moment Connections: Wind and Seismic Applications*, AISC, Chicago, IL.

Deierlein, G.G., S.H. Hsieh, and Y.J. Shen, 1990, "Computer-Aided Design of Steel Structures with Flexible Connections," *Proceedings of the 1990 National Steel Construction Conference*, pp. 9.1–9.21, AISC, Chicago, IL.

Disque, R.O., 1975, "Directional Moment Connections—A Proposed Design Method for Unbraced Steel Frames," *Engineering Journal*, Vol. 12, No. 1, (1st Qtr.), pp. 14–18, AISC, Chicago, IL.

Disque, R.O., 1964, "Wind Connections with Simple Framing," *Engineering Journal*, Vol. 1, No. 3, (July), pp. 101–103, AISC, Chicago, IL.

Gerstle, K.H., and M.H. Ackroyd, 1989, "Behavior and Design of Flexibly Connected Building Frames," *Proceedings of the 1989 National Steel Construction Conference*, pp. 1.1–1.28, AISC, Chicago, IL.

Geschwindner, L.F., 1991, "A Simplified Look at Partially Restrained Connections," *Engineering Journal*, Vol. 28, No. 2, (2nd Qtr.), pp. 73–78, AISC, Chicago, IL.

Goverdhan, A.V., 1984, "A Collection of Experimental Moment Rotation Curves and Evaluation of Prediction Equations for Semi-Rigid Connections," Master of Science Thesis, Vanderbilt University, Nashville, TN.

Kishi, N., and W.F. Chen, 1986, "Database of Steel Beam-to-Column Connections," CE-STR-86-26, Purdue University, School of Engineering, West Lafayette, IN.

Nethercot, D.A., and W.F. Chen, 1988, "Effects of Connections on Columns," *Journal of Constructional Steel Research*, pp. 201–239, Elsevier Applied Science Publishers, Essex, England.

PART 12

DESIGN OF FULLY RESTRAINED (FR) MOMENT CONNECTIONS

SCOPE

The specification requirements and other design considerations summarized in this Part apply to the design of fully restrained (FR) moment connections. For the design of simple shear connections, see Part 10. For the design of flexible moment connections, see Part 11. For the design of FR moment connections for HSS and steel pipe, see the AISC *Hollow Structural Sections Connections Manual*. For FR moment connections that are part of a seismic-force-resisting system in which the seismic response modification factor R is taken greater than 3, see the Seismic Provisions, which are available from AISC at www.aisc.org.

FR MOMENT CONNECTIONS

Load Determination

As defined in LRFD Specification Section A2.2, FR moment connections possess sufficient rigidity to maintain the angles between intersecting members as illustrated in Figure 12-1. While connections considered to be fully restrained seldom provide for zero rotation between members, the small amount of flexibility present is usually neglected and the connection, such as those illustrated in Figure 12-2, is idealized to prevent relative rotation.

As stated in LRFD Specification Section B9, end connections in FR construction shall be designed to carry the factored forces and moments, except that some inelastic but self-limiting deformation of a part of the connection is permitted. Huang, et al. (1973) showed that the moment can be resolved into an effective tension-compression couple acting as axial forces at the beam flanges. The flange force P_{uf} may be calculated as:

$$P_{uf} = \frac{M_u}{d_m}$$

where

P_{uf} = factored beam flange force, tensile or compressive, kips
M_u = beam end moment, kip-in.
d_m = moment arm between the flange forces, in. (varies for all FR connections and for stiffener design)

Shear is transferred through the beam web shear connection. Since, by definition, the angle between the beam and column in an FR moment connection remains unchanged under loading, eccentricity can be neglected entirely in the shear connection. Additionally, it is permissible to use bolts in bearing in either standard or slotted holes perpendicular to the line of force. Axial forces, if present, are normally assumed to be distributed uniformly across the beam flange cross-sectional area. However, if the beam web connection has sufficient stiffness, it can also be assumed to participate in the transfer of beam axial force.

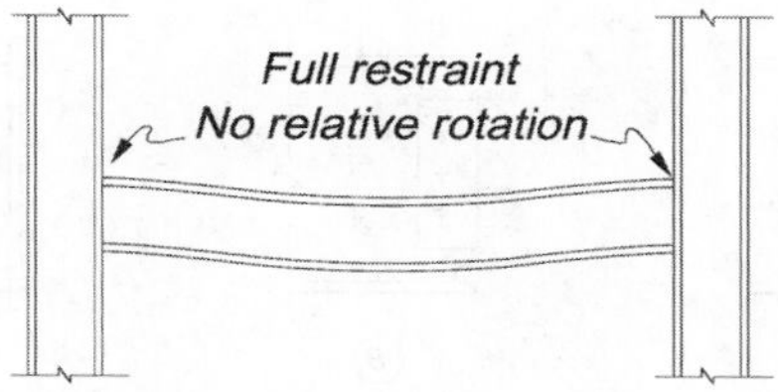

Fig. 12–1. FR moment connection behavior.

Moment connections deliver concentrated forces to the flanges of columns that must be accounted for in the design of the column and column panel-zone per LRFD Specification Section K1. Either the column size can be selected with adequate flange and web thickness to eliminate the need for column stiffening, or transverse stiffeners and/or web doubler plates can be provided. For further information, refer to AISC Design Guide No. 13 *Stiffening of Wide-Flange Columns at Moment Connections: Wind and Seismic Applications* (Carter, 1999).

Design Checks

The design strength of an FR moment connection is determined from the applicable limit states for the bolts (see Part 7), welds (see Part 8) and connection elements (see Part 9). The effect of eccentricity in the shear connection can be neglected. Additionally, the strength of the supporting column (and thus the need for stiffening) must be checked; refer to AISC Design Guide No. 13 *Wide-Flange Column Stiffening at Moment Connections—Wind and Seismic Applications* (Carter, 1999). In all cases, the design strength ϕR_n must equal or exceed the required strength R_u.

Temporary Support During Erection

Bolted construction provides a ready means to erect and temporarily connect members by use of the bolt holes. In contrast, FR moment connections in welded construction must be given special attention so that all pieces affecting the alignment of the welded joint may be erected,

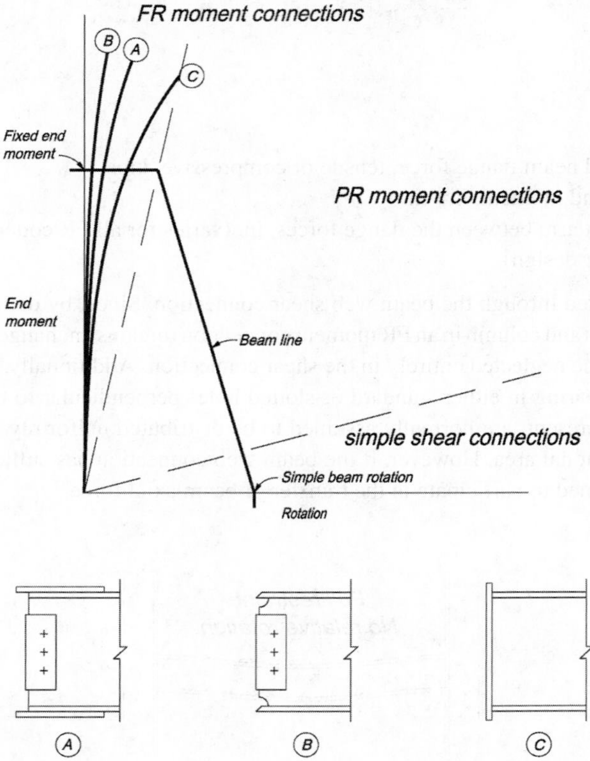

Fig. 12–2. FR moment connection behavior.

fitted and supported until the necessary welds are made. Temporary support can be provided in welded construction by furnishing holes for erection bolts, temporary seats, special lugs or by other means.

The effects of temporary erection aids on the finished structure should be considered, particularly on members subjected to tension loading or fatigue. They should be permitted to remain in place whenever possible since they seldom are reusable and the cost to remove them can be significant. If left in place, erection aids should be located so as not to cause a stress concentration. If, however, erection aids are to be removed, care should be taken so that the base metal is not damaged.

Temporary supports should be sufficient to carry any loads imposed by the erection process, such as the dead weight of the member, additional construction equipment, or material storage. Additionally, they must be flexible enough to allow plumbing of the structure, particularly in tier buildings.

Welding Considerations for Fully Restrained Moment Connections

Field welding should be arranged for welding in the flat or horizontal position and preference should be given to fillet welds over groove welds, whenever possible. Additionally, the joint detail and welding procedure should be constructed to minimize distortion and the possibility of lamellar tearing.

The typical complete-joint-penetration groove weld in a directly welded flange connection for a rolled beam can be expected to shrink about $\frac{1}{16}$ in. in the length dimension of the beam when it cools and contracts. Thicker welds, such as for welded plate-girder flanges will shrink even more—up to $\frac{1}{8}$ in. or $\frac{3}{16}$ in. This amount of shrinkage can cause erection problems in locating and plumbing the columns along lines of continuous beams. A method of calculating weld shrinkage can be found in Lincoln Electric Co. (1973). Unnecessarily thick stiffeners with complete-joint-penetration groove welds should be avoided since the accompanying weld shrinkage may contribute to lamellar tearing.

Weld shrinkage can best be controlled by fabricating the beam longer than required by the amount of the anticipated weld shrinkage. Alternatively, the weld-joint root opening can be increased. For further information, refer to AWS D1.1.

Flange-Plated FR Moment Connections

As illustrated in Figure 12-3, a flange-plated FR moment connection consists of a shear connection and top and bottom flange plates that connect the flanges of the supported beam to the supporting column. These flange plates are welded to the supporting column and may be bolted or welded to the flanges of the supported beam.

In a column flange connection, the flange plates are usually located with respect to the column web centerline. Because of the column-flange mill tolerance on out-of-squareness with the web, it is desirable to shop-fit long flange plates from the theoretical column-web centerline to assure good field fit-up with the beam. Misalignment on short connections, as illustrated in Figure 12-4, can be accommodated by providing oversized holes in the plates. Since mill tolerances in both the beam and the column may cause significant shop and/or field assembly problems, it may be desirable to ship the flange plates loose for field attachment to the column.

Directly Welded Flange FR Moment Connections

As illustrated in Figure 12-5, a directly welded flange FR moment connection consists of a shear connection and complete-joint-penetration groove welds, which directly connect the top and bottom flanges of the supported beam to the supporting column. Note, in

Figure 12-5b, the stiffener extends beyond the toe of the column flange to eliminate the effects of triaxial stresses.

The plastic moment of the supported beam $\phi_b M_p$ can be developed with sufficient inelastic rotation and deformation capacity through such a connection. This apparent increase in beam strength above the prediction of elastic theory occurs because of strain hardening in the flanges. See Huang, et al. (1973), Krawinkler and Popov (1982) and Beedle, et al. (1973).

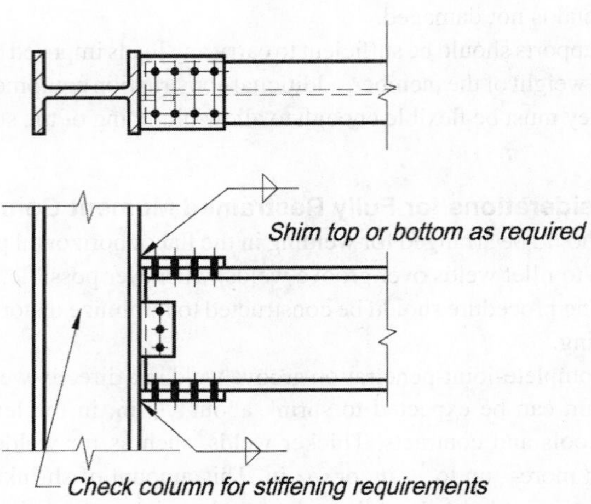

(a) Column flange support, bolted flange plates

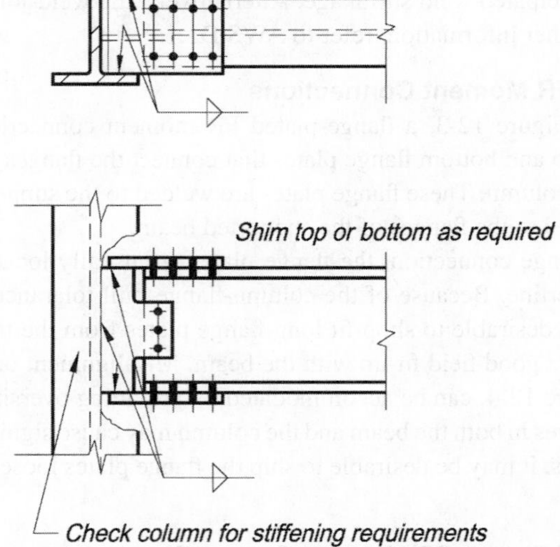

(b) Column web support, bolted flange plates

Fig. 12–3. Flange-plated FR moment connections.

Extended End-Plate FR Moment Connections

As illustrated in Figure 12-6, an extended end-plate connection consists of a plate of length greater than the beam depth, perpendicular to the longitudinal axis of the supported beam. The end-plate is always welded to the web and flanges of the supported beam on each side and bolted to the supporting member.

As illustrated in Figure 12-7, extended end-plate connections are classified by the number of bolts at the tension flange and may be used with or without end-plate stiffeners. The

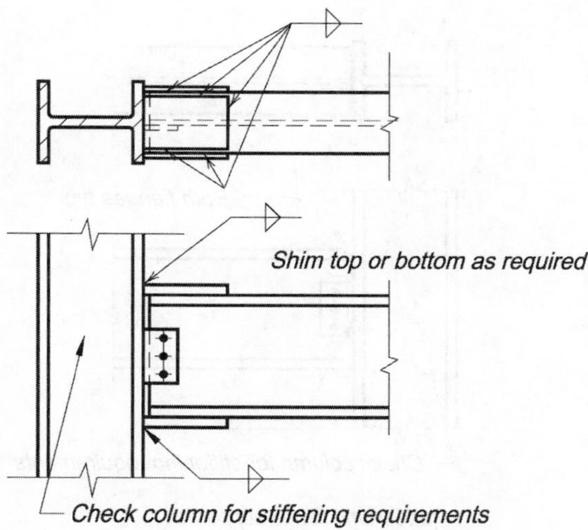

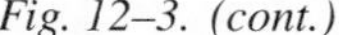

(c) Column flange support, welded flange plates

Fig. 12–3. (cont.)

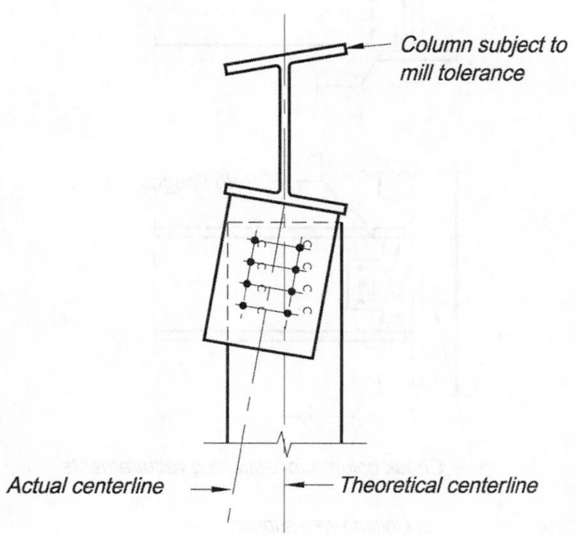

Fig. 12–4. Effect of mill tolerances on flange-plated connections.

four-bolt unstiffened extended end-plate connection of Figure 12-7a is generally limited by bolt strength to use with less than one-half of the available beam sections. An alternative is the eight-bolt stiffened extended end-plate connection shown in Figure 12-7b.

Design assumptions and basic procedures for the four-bolt unstiffened and eight-bolt stiffened configurations follow. For a more detailed discussion of the corresponding design procedures, refer to the AISC Design Guide No. 4 *Extended End-Plate Moment Connections* (Murray, 1990). For alternative design procedures, see Murray et al. (1992), which covers

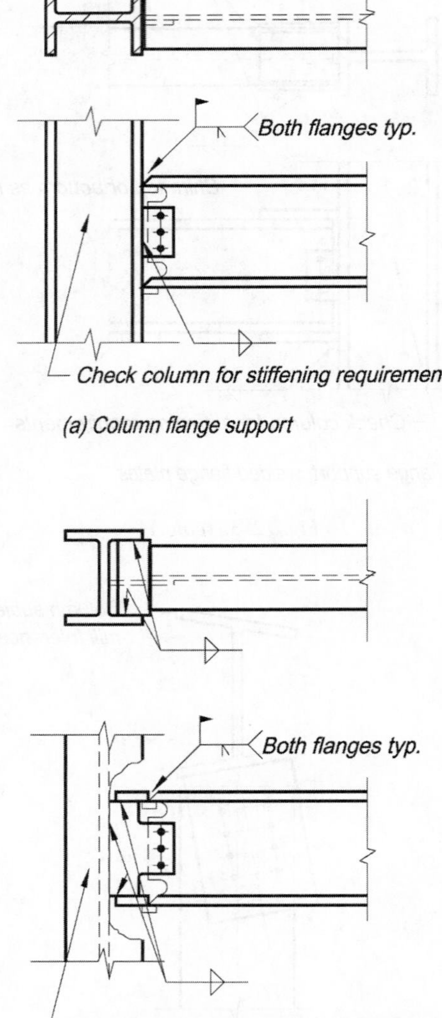

Fig. 12–5. Directly welded flange FR moment connections.

moment end plates with snug-tightened bolts, and FEMA (2000), which covers seismic applications with R taken greater than 3.

When pretensioned bearing bolts (N or X) are used, they must be designed using the shear-tension interaction equation of LRFD Specification Table J3.5. If bolts are to be slip-critical, all bolts may be designed for shear only and the shear-tension interaction equation may be ignored. Although some of the bolts lose a part of their clamping force due to applied tension, the overall connection suffers no loss of frictional resistance because the bolt tension produced by the moment is coupled with a compensating compressive force on the other side of the axis in bending. Thus, the net clamping force is maintained in the connection.

Shop and Field Practices

This type of connection requires extra care in shop fabrication and field erection. The fit-up of extended end-plate connections is sensitive to the column flanges and may be affected by column flange-to-web squareness, beam camber, or squareness of the beam end. The beam is frequently fabricated short to accommodate the column overrun tolerances with shims furnished to fill any gaps which might result.

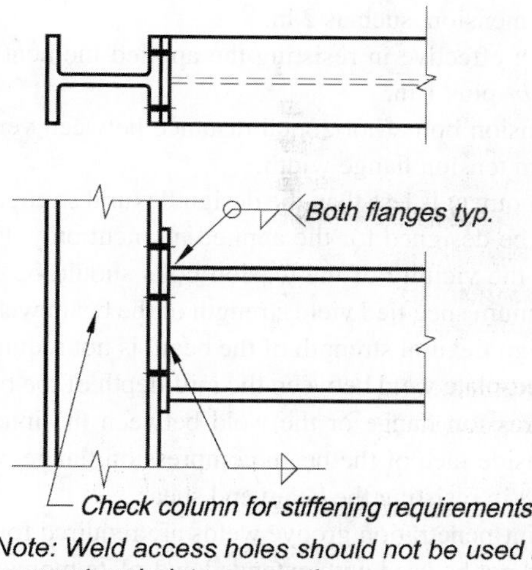

Fig. 12–6. Extended end-plate FR moment connection.

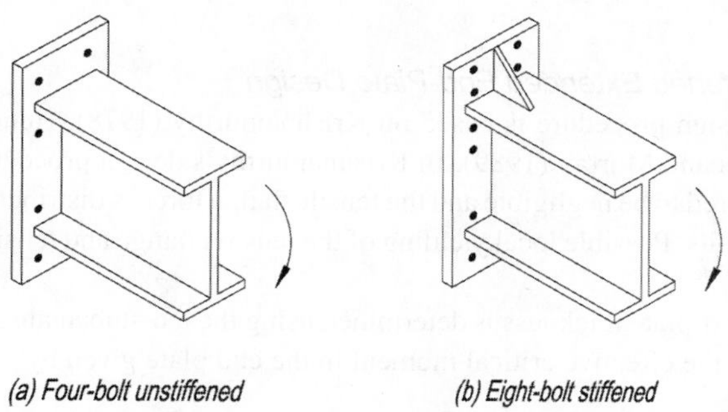

Fig. 12–7. Configurations of extended end-plate FR moment connections.

Design Assumptions

Several assumptions have been made in the design procedures that follow for four-bolt unstiffened and eight-bolt stiffened extended end-plate connections. These assumptions are as follows:

1. Pretensioned ASTM A325 or A490 high-strength bolts in diameters not greater than $1\frac{1}{2}$ in. must be used, except that ASTM A490 bolts should not be used in the eight-bolt stiffened configuration. If it is desired to use FR moment end-plate connections with snug-tightened bolts, see Murray et al. (1992).
2. End-plate material should preferably be ASTM A36.
3. Only static loading is permitted (wind, seismic with the seismic response modification factor R taken equal to or less than 3, snow, and temperature loads are considered static loads). If it is desired to use FR moment end plate connections in seismic applications with R taken greater than 3, see FEMA (2000).
4. The recommended minimum distance from the face of the beam flange to the nearest bolt centerline is the bolt diameter d_b plus $\frac{1}{2}$ in. Note that, although the smallest possible distance will generally result in the most economical connection, many fabricators prefer to use a standard dimension, such as 2 in.
5. The end-plate width effective in resisting the applied moment is not greater than the beam flange width b_f plus 1 in.
6. The gage of the tension bolts (horizontal distance between vertical bolt lines) should not exceed the beam tension flange width.
7. When the applied moment is less than the design flexural strength of the beam, the bolts and end plate may be designed for the applied moment only. However, beam-web-to-end-plate welds in the vicinity of the tension bolts should be designed to develop 60 percent of the minimum specified yield strength of the beam web. This is recommended even if the full design flexural strength of the beam is not required for frame strength.
8. Only the web-to-end-plate weld between the mid-depth of the beam and the inside face of the beam compression flange or the weld between the inner row of tension bolts plus $2d_b$ and the inside face of the beam compression flange, whichever is smaller, is considered effective in resisting the beam end shear.
9. When complete-joint-penetration groove welds are required for the flange welds, weld access holes should not be used with extended end-plate moment connections (FEMA, 2000). Instead, the backing bars should be placed inside the beam flange at both the top and bottom flanges, terminating at the flange-to-web fillets. The weld should be treated as a partial-joint-penetration groove weld between the toes of the web-to-flange fillets.

Four-Bolt Unstiffened Extended End-Plate Design

The following design procedure is based on Krishnamurthy (1978), Hendrick and Murray (1984), and Curtis and Murray (1989). In Krishnamurthy's design procedure, prying action forces are considered to be negligible and the tensile flange force is distributed equally among the four tension bolts. Possible local yielding of the tension flange and tensile area of the web is neglected.

The required end-plate thickness is determined using the tee-stub analogy, as illustrated in Figure 12-8, with the effective critical moment in the end plate given by

$$M_{eu} = \frac{\alpha_m P_{uf} p_e}{4}$$

where

P_{uf} = factored beam flange force, kips
$\alpha_m = C_a C_b (A_f/A_w)^{1/3} (p_e/d_b)^{1/4}$
C_a = constant from Table 12-1
$C_b = (b_f/b_p)^{1/2}$
b_f = beam flange width, in.
b_p = effective end-plate width, in., not to exceed $b_f + 1$ in.
A_f = area of beam tension flange, in.^{2}
A_w = area of beam web, clear of flanges, in.^{2}
p_e = effective pitch, in.
$= p_f - (d_b/4) - w_t$
p_f = distance from centerline of bolt to nearer surface of the tension flange, in. Generally, $d_b + 1/2$ in. is enough to provide entering and tightening clearance; 2 in. is a common standard.
w_t = fillet weld throat size or size of reinforcement for groove weld, in.
d_b = nominal bolt diameter, in.

Values of C_a are tabulated for various combinations of beam and end-plate material grades and ASTM A325 or A490 bolts in Table 12-1. Values of A_f/A_w for the W-shapes listed in Part 1 are found in Table 12-2. The required end-plate thickness $t_{p\ req}$ is then determined as

$$t_{p\ req} = \sqrt{\frac{4M_{eu}}{\phi F_y b_p}}$$

where F_y is the specified minimum yield stress of the end-plate material, ksi, and $\phi = 0.90$.

The strength of the column should then be investigated for stiffening requirements; refer to AISC Design Guide No. 13 *Stiffening of Wide-Flange Columns at Moment Connections: Wind and Seismic Applications* (Carter, 1999). Note that, since column web stiffeners add considerable fabrication expense and may interfere with weak-axis framing, it is often advantageous to eliminate the need for stiffening. The designer should therefore consider increasing the column size to a section with adequate strength. Alternatively, if the column were inadequate

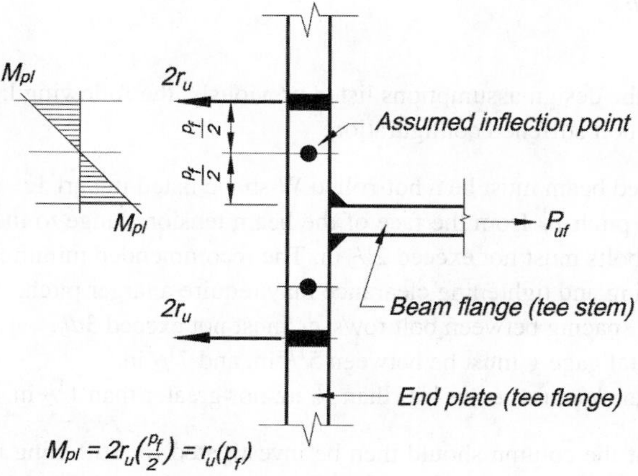

Fig. 12–8. Tee-stub analogy for extended end-plate FR moment connections.

due to local flange bending strength, increasing the tension-bolt pitch p_f or switching to an eight-bolt stiffened extended end-plate configuration may increase the length of column flange effective in flange bending and thereby eliminate the need for stiffening.

Eight-Bolt Stiffened Extended End-Plate Design

The following design procedure is based on Murray and Kukreti (1988), Hendrick and Murray (1984), and Curtis and Murray (1989). Murray and Kukreti (1988) present two methods for determining the required end-plate thickness and bolt diameter; both methods are limited to the use of ASTM A36 end-plate material with ASTM A325 bolts and include the effects of prying action.

The first method was developed from a regression analysis of finite-element-analysis data including second-order geometric effects and inelastic plate and bolt material properties. The resulting equations are elaborate and beyond the scope of this Manual; refer to the AISC Design Guide No. 4 *Extended End-Plate Moment Connections* (Murray, 1990) and Murray and Kukreti (1988).

The second method offers a simplified approach which was developed with the first method by generating end-plate thicknesses and bolt diameters for all W-shapes listed in Part 1 assuming ASTM A36 steel and beam sections at various moment levels. The number of bolts effective in resisting the tensile flange force was then determined for each connection; a conservative lower bound of six effective bolts was established. Next, it was assumed that the plate thickness could be established from tee-stub analogy bending, as illustrated in Figure 12-8, where

$$M_{eu} = 2r_u\left(\frac{p_{eff}}{2}\right) = r_u p_{eff}$$

where r_u is the force per bolt based on six effective bolts, kips.

From the generated designs, it was determined the effective pitch p_{eff} is

$$p_{eff} = \frac{p_f}{4.17}\sqrt{g^2 + p_f^{\,2}}$$

The required end-plate thickness is then determined from

$$t_{preq} = \sqrt{\frac{4M_{eu}}{\phi F_y b_p}}$$

where $\phi = 0.9$

In addition to the design assumptions listed previously, the following limitations must be met for the eight-bolt stiffened configuration:

1. The supported beam must be a hot-rolled W-shape listed in Part 1.
2. The vertical pitch p_f from the face of the beam tension flange to the centerline of the first row of bolts must not exceed $2\frac{1}{2}$ in. The recommended minimum pitch is d_b plus $\frac{1}{2}$ in.; entering and tightening clearance may require a larger pitch.
3. The vertical spacing between bolt rows p_b must not exceed $3d_b$.
4. The horizontal gage g must be between $5\frac{1}{2}$ in. and $7\frac{1}{2}$ in.
5. Bolt diameter d_b must be not less than $\frac{3}{4}$ in. nor greater than $1\frac{1}{2}$ in.

The strength of the column should then be investigated for stiffening requirements. The recommendations of Hendrick and Murray (1984) can be used to check column web strengths in local yielding, buckling, and panel zone shear; refer to AISC Design Guide No. 13 *Stiffening of Wide-Flange Columns at Moment Connections: Wind and Seismic Applications* (Carter, 1999).

Unless the column flange is 1.5 to 2 times thicker than the end plate, transverse stiffening is required. If effective-flange-length effects are neglected, the behavior of the column flange is identical to that of the end plate. Therefore, the column flange must be at least as thick as the end plate and the transverse stiffeners must be at least as thick as the beam flange. Additionally, the weld connecting the transverse stiffener to the flange must be sufficient to develop the strength of the full thickness of the stiffener plate.

A column flange that is 1.5 to 2 times thicker than the end plate may not require transverse stiffening. From Curtis and Murray (1989), an unstiffened flange may be evaluated according to the flange bending equation presented for extended end-plate connections in AISC Design Guide No. 13 *Stiffening of Wide-Flange Columns at Moment Connections: Wind and Seismic Applications* (Carter, 1999) with $b_s = 3.5p_b + c$. Because this reference considered only ASTM A36 steel, it is recommended that column material with greater yield strength be checked as if ASTM A36 material were used.

Note that, since column web stiffeners add considerable fabrication expense and may interfere with weak-axis framing, it is often advantageous to eliminate the need for stiffening. The designer should therefore consider increasing the column size to a section with adequate strength.

FR MOMENT SPLICES

Beams and girders sometimes are spliced in locations where both shear and moment must be transferred across the splice. Per LRFD Specification Section J7, the full strength of the smaller section being spliced must be developed in groove-welded butt splices. Other types of beam or girder splices must develop the strength required by the actual forces at the point of the splice.

Location of Moment Splices

A careful analysis is particularly important in continuous structures where a splice may be located at or near the point of inflection. Since this inflection point can and does migrate under service loading, actual forces and moments may differ significantly from those assumed. Furthermore, since loading application and frequency can change in the lifetime of the structure, it is prudent for the designer to specify some minimum strength requirement at the splice. Hart and Milek (1965) propose that splices in fixed-ended beams be located at the one-sixth point of the span and be adequate to resist a moment equal to one-sixth of the flexural strength of the member, as a minimum.

Force Transfer in Moment Splices

Force transfer in moment splices can be assumed to occur in a manner similar to that developed for FR moment connections. That is, the shear R_u is primarily transferred through the beam web connection and the moment can be resolved into an effective tension-compression couple where the force at each flange is P_{uf} where:

$$P_{uf} = \frac{M_u}{d_m}$$

where

P_{uf} = actored beam flange force, tensile or compressive, kips
M_u = moment in the beam at the splice, kip-in.
d_m = moment arm, in.

Axial forces, if present, are normally assumed to be distributed uniformly across the beam flange cross-sectional area. However, if the beam web connection has sufficient stiffness, it can also be assumed to participate in the transfer of beam axial force.

Flange-Plated FR Moment Splices

Moment splices can be designed as shown in Figure 12-9, to utilize flange plates and a web connection. The flange plates and web connection may be bolted or welded.

The splice and spliced beams should be checked in a manner similar to that described previously under "Flange-Plated FR Moment Connections," except that the web connection should be designed as illustrated previously for shear splices in Part 10.

Figure 12-9 is a composite detail illustrating two types of splices, bolted and welded. The left side of the splice in Figure 12-9 illustrates the detail of a bolted flange-plated moment splice. For this case, the flange plates are normally made approximately the same width as the beam flange as shown in Section A-A.

Alternatively, the right side of the splice in Figure 12-9 illustrates the detail of a welded splice. As shown in Section B-B, the top plate is narrower and the bottom plate is wider than the beam flange, permitting the deposition of weld metal in the downhand or horizontal position without inverting the beam. While this is a benefit in shop fabrication (the beam does not have to be turned over), it is of extreme importance in the field where the weld can be

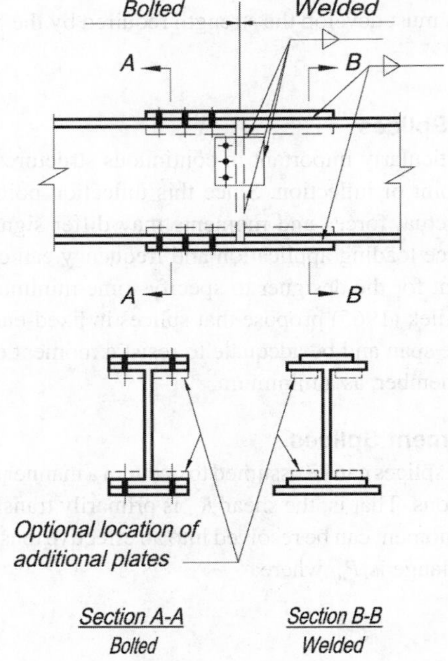

Fig. 12–9. Flange-plated moment splice.

made in the horizontal instead of the overhead position since the beam cannot be turned over. This detail also provides tolerance for field alignment, since the joint gap can be opened or closed. When splices are field welded, some means for temporary support must be provided as discussed previously in "Temporary Support During Erection."

If the beam or girder flange is thick and the flange forces are large, it may be desirable to place additional plates on the insides of the flanges. In a bolted splice (Section A-A), the bolts are then loaded in double shear and a more compact joint may result. Note that these additional plates must have sufficient area to develop their share of the double-shear bolt load.

In a welded splice (Section B-B), these additional plates must have sufficient area to match the strength of the welds that connect them. Additionally, these plates must be set away from the beam web a distance sufficient to permit deposition of weld metal as shown in Figure 12-10a. This distance is a function of the beam depth and flange width, as well as the welding equipment to be used. A distance of 2 to 2 ½ in. or more may be required for this access. One alternative is to bevel the bottom edge of the plate to clear the beam fillet and place the plate tight to the beam web with a fillet weld as illustrated in Figure 12-10a. The effects of this bevel on the area of the plate must be considered in determining the required plate width and thickness. Another alternative would be to use unbeveled inclined plates as shown in Figure 12-10b.

Directly Welded Flange FR Moment Splices

Moment splices can be designed, as shown in Figure 12-11, to utilize a complete-joint-penetration groove weld connecting the flanges of the members being spliced. The web connection may then be bolted or welded.

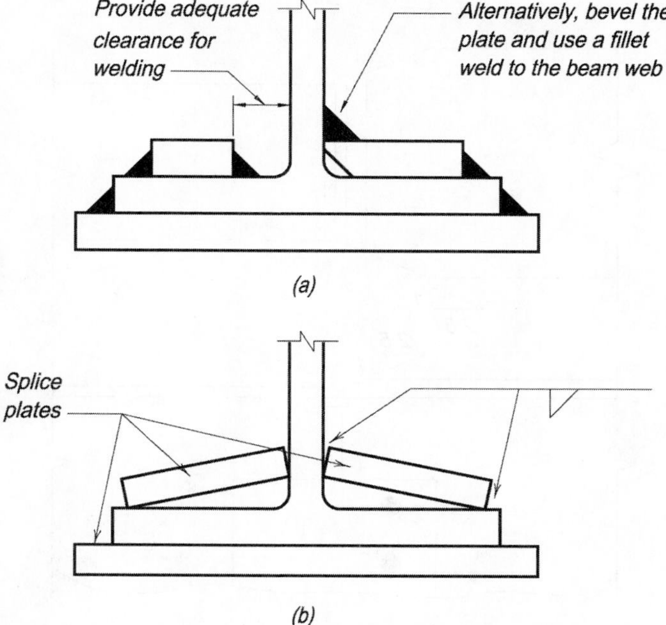

Fig. 12–10. Welding clearances for flange-plated moment splices.

The splice and spliced beams should be checked in a manner similar to that described previously under "Directly Welded Flange FR Moment Connections," except that the web connection should be designed as illustrated previously for shear splices in Part 10.

When the flange thickness or width varies across the splice and the calculated stress is greater than one-third of the specified tensile stress F_u, Figure 12-12 shows the detail required at the tension butt joint per AWS D1.1 Section 2.20. A transition slope of not less than 1 in 2½ must be provided by chamfering the thicker part, tapering the wider part, sloping the weld metal, or by any combination of these. When the calculated stress does not exceed one-third of the specified tensile stress F_u, no transition is required in statically loaded structures. Compression butt joints do not require transitional tapering.

Although rare in occurrence, some spliced members must be level on top. Where the depths of these spliced members differ, consideration should be given to the use of a flange plate

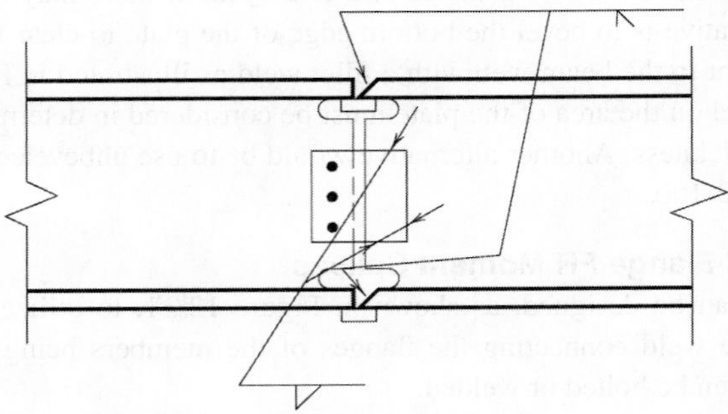

Fig. 12–11. Directly welded flange moment splice.

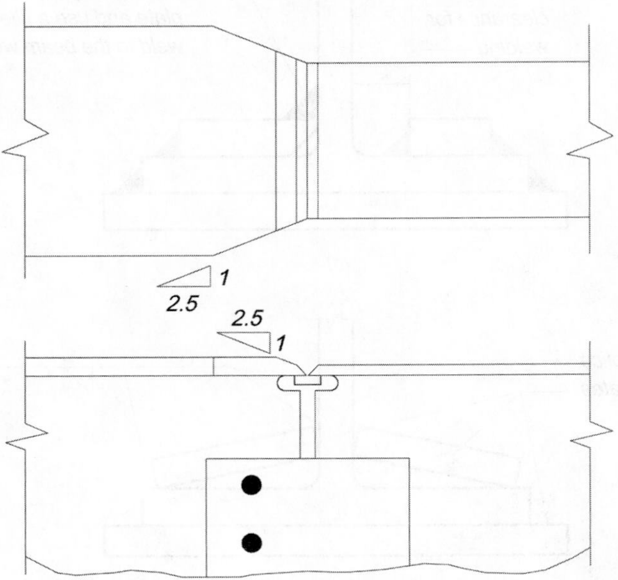

Fig. 12–12. Transitions at tension flange for directly welded flange moment splices, when required.

of uniform thickness for the full length of the shallower member. This avoids the fabrication problems created by an inverted transition.

In Figure 12-12, the web depth is kept constant (this is always the case with rolled shapes of the same nominal depth). This avoids an offset cut with a transition in the web for a built-up girder. Eccentricity resulting from differing flange thicknesses is usually ignored in the design. The web plates normally are aligned to their center lines and the 1 in $2\frac{1}{2}$ slope is chamfered into the plate or the weld is sloped, depending upon the relative thicknesses.

The groove (butt) welded splice preparation shown in Figure 12-11 may be used for either shop or field welding. Alternatively, for shop welding where the beam may be turned over, the joint preparation of the bottom flange could be inverted.

In splices subjected to dynamic or fatigue loading, the backing bar should be removed and the weld should be ground flush when it is normal to the applied stress (AISC, 1977). The access holes should be free of notches and should provide a smooth transition at the juncture of the web and flange.

Extended End-Plate FR Moment Splices

Moment splices can be designed as shown in Figure 12-13, to utilize four-bolt unstiffened extended end plates connecting the members being spliced. If the end-plate and the bolts are designed properly, it is possible to load this type of connection to reach the full plastic moment capacity of the beam, $\phi_b M_p$.

The splice and spliced beams should be checked in a manner similar to that described previously under "Extended End-Plate FR Moment Connections."

The comments for "Extended End-Plate Connections" are equally applicable to extended end-plate moment splices.

SPECIAL CONSIDERATIONS

FR Moment Connections to Column-Web Supports

It is frequently required that FR moment connections be made to column web supports. While the mechanics of analysis and design do not differ from FR moment connection to column flange supports, the details of the connection design as well as the ductility considerations required are significantly different.

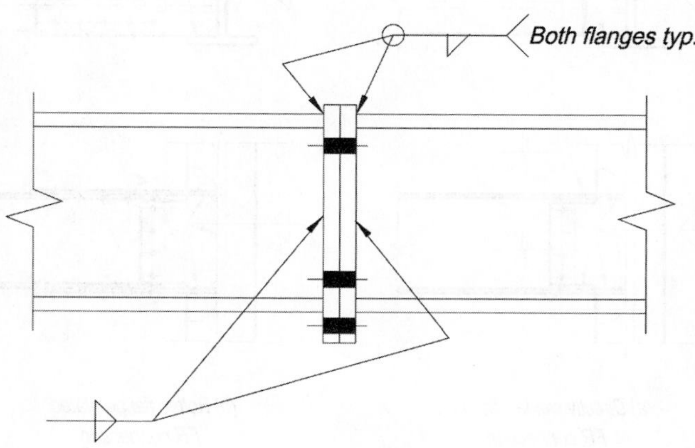

Fig. 12–13. Extended end-plate moment splice.

Recommended Details

When an FR moment connection is made to a column web, it is normal practice to stop the beam short and locate all bolts outside of the column flanges as illustrated in Figures 12-3b and 12-5b. This simplifies the erection of the beam and permits the use of an impact wrench to tighten all bolts. It is also preferable to locate welds outside the column flanges to provide adequate clearance.

Ductility Considerations

Driscoll and Beedle (1982) discuss the testing and failure of two FR moment connections to column-web supports: a directly welded flange connection and a bolted flange-plated connection, shown respectively in Figures 12-14a and 12-14b. Although the connections in these tests were proportioned to be "critical," they were expected to provide inelastic rotations at full plastic load. Failure occurred unexpectedly, however, on the first cycle of loading; brittle fracture occurred in the tension connection plate at the load corresponding to the plastic moment before significant inelastic rotation had occurred.

Examination and testing after the unexpected failure revealed that the welds were of proper size and quality and that the plate had normal strength and ductility. The following is quoted, with minor editorial changes relative to figure numbers, directly from Driscoll and Beedle (1982).

> "Calculations indicate that the failures occurred due to high strain concentrations. These concentrations are: (1) at the junction of the connection plate and the column flange tip and (2) at the edge of the butt weld joining the beam flange and the connection plate."
>
> Figure 12-15 illustrates the distribution of longitudinal stress across the width of the connection plate and the concentration of stress in the plate at the column flange tips. It also illustrates the uniform longitudinal stress distribution in the connection plate at some distance away from the connection. The stress distribution shown represents schematically the values measured during the load tests and those obtained from finite element analysis. (σ_o is a nominal stress in the elastic range.) The results of the analyses are valid up to the loading that causes the combined stress to equal the yield point. Furthermore, the analyses

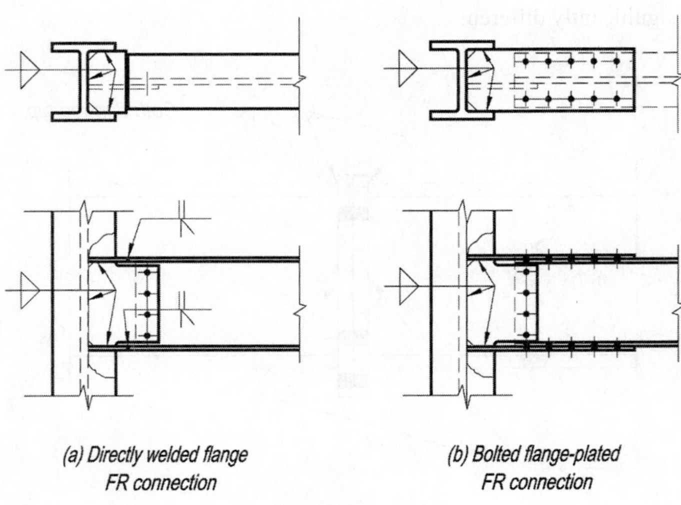

Fig. 12–14. Test specimens used by Driscoll and Beedle (1982).

indicate that localized yielding could begin when the applied uniform stress is less than one-third of the yield point. Another contribution of the non-uniformity is the fact that there is no back-up stiffener. This means that the welds to the web near its center are not fully effective.

"The longitudinal stresses in the moment connection plate introduce strains in the transverse and the through-thickness directions (the Poisson effect). Because of the attachment of the connection plate to the column flanges, restraint is introduced; this causes tensile stresses in the transverse and the through-thickness directions. Thus, referring to Figure 12-15, tri-axial tensile stresses are present along Section A-A, and they are at their maximum values at the intersections of Sections A-A and C-C. In such a situation, and when the magnitudes of the stresses are sufficiently high, materials that are otherwise ductile may fail by premature brittle fracture."

The results of nine simulated weak-axis FR moment connection tests performed by Driscoll et al. (1983) are summarized in Figure 12-16. In these tests, the beam flange was simulated by a plate measuring either 1 in.×10 in. or $1\frac{1}{8}$ in.×9 in. The fracture strength exceeds the yield strength in every case, and sufficient ductility is provided in all cases except for that of Specimen D. Also, if the rolling direction in the first five specimens (A, B, C, D, and E) were parallel to the loading direction, which would more closely approximate an actual beam flange, the ductility ratios for these would be higher. The connections with extended connection plates (i.e., projection of three inches), with extensions either rectangular or tapered, appeared equally suitable for the static loads of the tests.

Based on the tests, Driscoll, et. al. (1983) report that those specimens with extended connection plates have better toughness and ductility and are preferred in design for seismic loads, even though the other connection types (except D) may be deemed adequate to meet the requirements of many design situations.

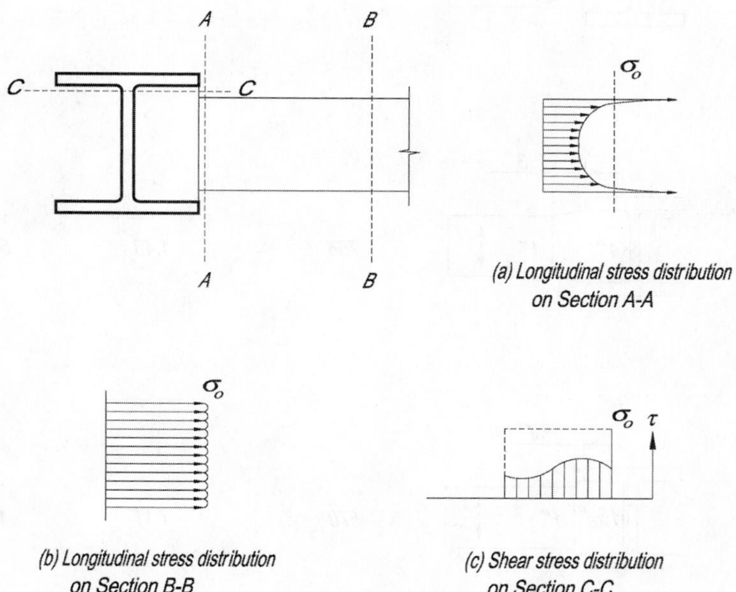

Fig. 12–15. Stress distributions in test specimens used by Driscoll and Beedle (1982).

In accordance with the preceding discussion, the following suggestions are made regarding the design of this type of connection:

1. For directly welded (butt) flange-to-plate connections, the connection plate should be thicker than the beam flange. This greater area accounts for shear lag and also provides for misalignment tolerances.

   AWS D1.1, Section 3.3.3 restricts the misalignment of abutting parts such as this to 10 percent of the thickness, with $^1/_8$ in. maximum for a part restrained against bending due to eccentricity of alignment. Considering the various tolerances in mill rolling ($\pm^1/_8$ in. for W-shapes), fabrication, and erection, it is prudent design to call for the

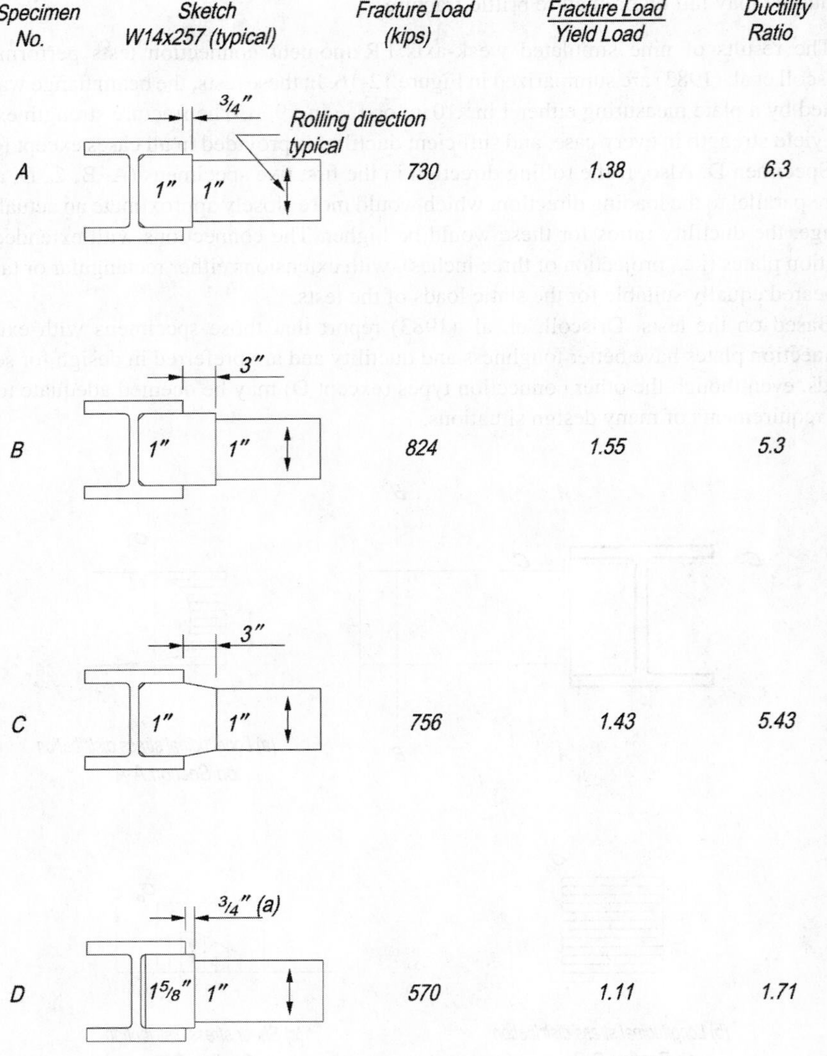

Fig. 12–16. Results of weak-axis FR moment connection ductility tests performed by Driscoll et al. (1983).

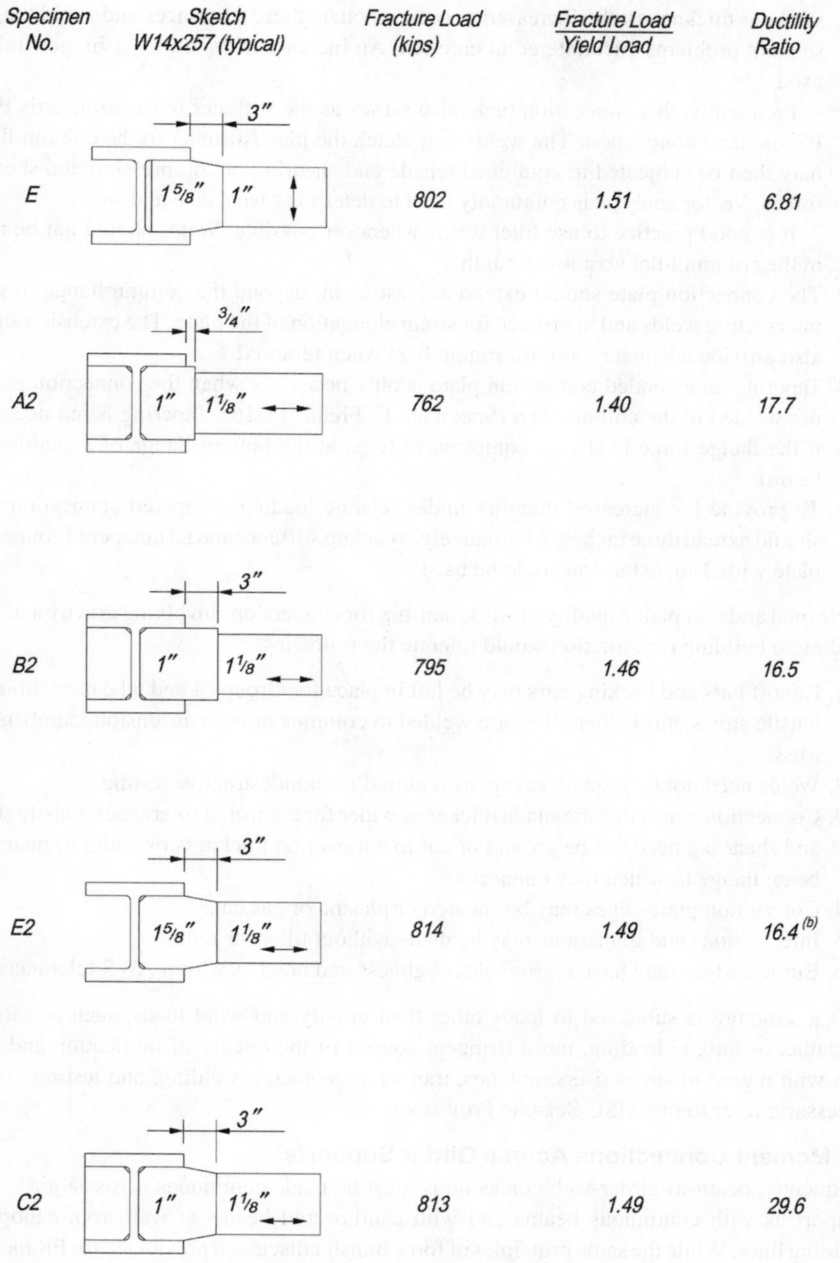

Notes: (a) 3/4″ dimension is estimated—no dimension given.

(b) Ductility ratio estimated. Actual value not known due to malfunction in deflection gage.

Fig. 12–16. (cont.)

stiffener thickness to be increased to accommodate these tolerances and avoid the subsequent problems encountered at erection. An increase of 1/8 in. to 1/4 in. generally is used.

Frequently, this connection plate also serves as the stiffener for a strong axis FR or PR moment connection. The welds that attach the plate/stiffener to the column flange may then be subjected to combined tensile and shearing or compression and shearing forces. Vector analysis is commonly used to determine weld size and stress.

It is good practice to use fillet welds whenever possible. Welds should not be made in the column fillet area for strength.

2. The connection plate should extend at least 3/4 in. beyond the column flange to avoid intersecting welds and to provide for strain elongation of the plate. The extension should also provide adequate room for runout bars when required.
3. Tapering an extended connection plate is only necessary when the connection plate is not welded to the column web (Specimen E, Figure 12-16). Tapering is not necessary if the flange force is always compressive (e.g., at the bottom flange of a cantilevered beam).
4. To provide for increased ductility under seismic loading, a tapered connection plate should extend three inches. Alternatively, a backup stiffener and an untapered connection plate with 3-in. extension could be used.

Normal and acceptable quality of workmanship for connections involving gravity and wind loading in building construction would tolerate the following:

1. Runoff bars and backing bars may be left in place for Groups 4 and 5 beams (subject to tensile stress only) where they are welded to columns or used as tension members in a truss.
2. Welds need not be ground, except as required for nondestructive testing.
3. Connection plates that are made thicker or wider for control of tolerances, tensile stress, and shear lag need not be ground or cut to a transition thickness or width to match the beam flange to which they connect.
4. Connection plate edges may be sheared or plasma or gas cut.
5. Intersections and transitions may be made without fillets or radii.
6. Burned edges may have reasonable roughness and notches within AWS tolerances.

If a structure is subjected to loads other than gravity and wind loads, such as seismic, dynamic, or fatigue loading, more stringent control of the quality of fabrication and erection with regard to stress risers, notches, transition geometry, welding, and testing may be necessary; refer to the AISC Seismic Provisions.

FR Moment Connections Across Girder Supports

Frequently, beam-to-girder-web connections must be made continuous across a girder-web support as with continuous beams and with cantilevered beams at wall, roof-canopy, or building lines. While the same principles of force transfer discussed previously for FR moment connections may be applied, the designer must carefully investigate the relative stiffness of the assembled members being subjected to moment or torsion and provide the fabricator and erector with reliable camber ordinates.

Additionally, the design should still provide some means for final field adjustment to accommodate the accumulated tolerances of mill production, fabrication, and erection; it is very desirable that the details of field connections provide for some adjustment during erection. Figure 12-17 illustrates several details that have been used in this type of connection and the designer may select the desirable components of one or more of the sketches to suit a particular application. Therefore, these components are discussed here as a top flange, bottom flange, and web connection.

Top Flange Connection

As shown in Figure 12-17a, the top flange connection may be directly welded to the top flange of the supporting girder. Figures 12-17b and 12-17c illustrate an independent splice plate that ties the two beams together by use of a longitudinal fillet weld or bolts. This tie plate does not require attachment to the girder flange, although it is sometimes so connected to control noise if the connection is subjected to vibration.

Bottom Flange Connection

When the bottom flanges deliver a compressive force only, the flange forces are frequently developed by directly welding these flanges to the girder web as illustrated in Figure 12-17a. Figure 12-17b illustrates the use of an angle or channel extending beyond the beam flange to provide for a horizontal fillet weld; Figure 12-17c is similar, but uses bolts instead of welds to develop the flange force.

Web Connection

While a single-plate connection is shown in Figure 12-17a and unstiffened seated connections are shown in Figures 12-17b and 12-17c, any of the shear connections in Part 10 may be used. Note that the effect of eccentricity in the shear connection may be neglected.

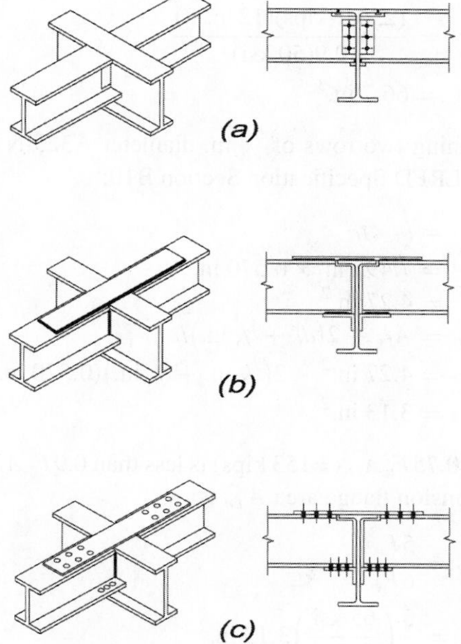

Fig. 12–17. FR moment connections across girder-web supports.

DESIGN EXAMPLES

EXAMPLE 12.1. Bolted flange-plated FR moment connection (beam-to-column flange).

Given: Design a bolted flange-plated FR moment connection for a W18×50 beam to W14×99 column-flange connection. For structural members, assume $F_y = 50$ ksi and $F_u = 65$ ksi; for connecting material, assume $F_y = 36$ ksi and $F_u = 58$ ksi. Use ⅞ in. diameter ASTM A325-N bolts and 70 ksi electrodes.

$$R_u = 45.0 \text{ kips}$$
$$M_u = 250 \text{ ft-kips}$$

W18×50, ASTM A992

$d = 18.0$ in. $b_f = 7.50$ in. $Z_x = 101$ in.^{3}
$t_w = 0.355$ in. $t_f = 0.570$ in.

W14×99, ASTM A992

$d = 14.2$ in. $b_f = 14.6$ in. $k = 2\,{}^1\!/_{16}$ in.
$t_w = 0.485$ in. $t_f = 0.780$ in. $T = 10$ in.

Solution: *Check beam design flexural strength:*

$$\begin{aligned} Z_{req} &= \frac{M_u \times 12 \text{ in./ft}}{0.9F_y} \\ &= \frac{(250 \text{ ft-kips})(12 \text{ in./ft})}{0.9(50 \text{ ksi})} \\ &= 66.7 \text{ in.}^3 \end{aligned}$$

Assuming two rows of ⅞ in. diameter A325-N bolts in standard holes, from LRFD Specification Section B10:

$$\begin{aligned} A_{fg} &= b_f \times t_f \\ &= 7.495 \text{ in.} \times 0.570 \text{ in.} \\ &= 4.27 \text{ in.}^2 \\ A_{fn} &= A_{fg} - 2(d_b + \tfrac{1}{8} \text{ in.})t_f \\ &= 4.27 \text{ in.}^2 - 2(\tfrac{7}{8} \text{ in.} + \tfrac{1}{8} \text{ in.})(0.570 \text{ in.}) \\ &= 3.13 \text{ in.}^2 \end{aligned}$$

since $0.75F_uA_{fn}(=153 \text{ kips})$ is less than $0.9F_yA_{fg}(=192 \text{ kips})$, the effective tension flange area A_{fe} is:

$$\begin{aligned} A_{fe} &= \frac{5F_u}{6F_y}A_{fn} \\ &= \frac{5}{6}\left(\frac{65 \text{ ksi}}{50 \text{ ksi}}\right)3.13 \text{ in.}^2 \\ &= 3.39 \text{ in.}^2 \end{aligned}$$

This is a 20.6 percent reduction from the gross flange area A_{fg}, and the effective plastic section modulus Z_e is:

$$Z_e \approx Z_x - 2\left(0.206A_{fg}\frac{d}{2}\right)$$
$$\approx 101 \text{ in.}^3 - 2\left(0.206 \times 4.27 \text{ in.}^2 \times \frac{17.99 \text{ in.}}{2}\right)$$
$$\approx 85.2 \text{ in.}^3$$

Since $Z_e > Z_{req}$, the beam design flexural strength is **o.k.**

Design the single-plate web connection.

Determine number of $^7\!/_8$ in. diameter A325-N bolts required for shear.

From Table 7-10:

$$n_{\min} = \frac{R_u}{\phi r_n}$$
$$= \frac{45.0 \text{ kips}}{21.6 \text{ kips/bolt}}$$
$$= 2.08 \rightarrow 3 \text{ bolts}$$

Try PL $^5\!/_{16}\times 9$

Determine number of $^7\!/_8$ in. diameter A325-N bolts required for material bearing, assuming $L_e = 1\,^1\!/_2$ in. and $s = 3$ in. The $^5\!/_{16}$ in. plate is more critical than the 0.355-in.-thick beam web. From Table 7-12:

$$n_{\min} = \frac{R_u}{\phi r_n}$$
$$= \frac{45.0 \text{ kips}}{28.6 \text{ kips/bolt}}$$
$$= 1.57 \rightarrow 2 \text{ bolts} < 3 \text{ bolts}$$

Therefore, bolt shear is more critical. Try a three-bolt single-plate connection.

Check shear yielding of the plate:

$$\phi R_n = 0.9(0.6F_yA_g)$$
$$= 0.9[0.6 \times 36 \text{ ksi}(9 \text{ in.} \times {}^5\!/_{16} \text{ in.})]$$
$$= 54.7 \text{ kips} > 45.0 \text{ kips} \quad \textbf{o.k.}$$

Check shear rupture of the plate:

$$\phi R_n = 0.75(0.6F_uA_n)$$
$$= 0.75\left[0.6 \times 58 \text{ ksi}\left(9 \text{ in.} - 3({}^7\!/_8 \text{ in.} + {}^1\!/_8 \text{ in.})\right){}^5\!/_{16} \text{ in.}\right]$$
$$= 48.9 \text{ kips} > 45.0 \text{ kips} \quad \textbf{o.k.}$$

Check block shear rupture of the plate:

With $L_{eh} = 1\frac{1}{2}$ in. and $L_{ev} = 1\frac{1}{2}$ in., from Tables 9-3a and 9-4a, $0.6\ F_u A_{nv} > F_u A_{nt}$. Thus,

$$\phi R_n = \phi[0.6F_u A_{nv} + F_y A_{gt}] \leq \phi[0.6F_u A_{nv} + F_u A_{nt}]$$

From Tables 9-3a, 9-4a and 9-4b,

$$\begin{aligned}
\phi R_n &= (131 \text{ kips/in.} + 40.5 \text{ kips/in.})\tfrac{5}{16} \text{ in.} \\
&\leq (131 \text{ kips/in.} + 43.5 \text{ kips/in.})\tfrac{5}{16} \text{ in.} \\
&= 53.6 \text{ kips} \leq 54.5 \text{ kips} \\
&= 53.6 \text{ kips} > 45.0 \text{ kips} \qquad \textbf{o.k.}
\end{aligned}$$

Determine required weld size for fillet welds to supporting column flange:

$$\begin{aligned}
D_{\min} &= \frac{R_u}{2\times 1.392\,l} \\
&= \frac{45.0 \text{ kips}}{2\times 1.392(9 \text{ in.})} \\
&= 1.80 \rightarrow 2 \text{ sixteenths}
\end{aligned}$$

From LRFD Specification Table J2.4, since the column flange thickness is over $\frac{3}{4}$ in., the minimum fillet weld size is $\frac{5}{16}$ in., use two $\frac{5}{16}$ in. fillet welds.

Design the tension flange plate and connection.

Calculate the flange force P_{uf}:

$$\begin{aligned}
P_{uf} &= \frac{M_u \times 12 \text{ in./ft}}{d} \\
&= \frac{(250 \text{ ft-kips})(12 \text{ in./ft})}{18.0 \text{ in.}} \\
&= 167 \text{ kips}
\end{aligned}$$

Determine number of $\frac{7}{8}$ in. diameter A325-N bolts required for shear. From Table 7-10:

$$\begin{aligned}
n_{\min} &= \frac{P_{uf}}{\phi r_n} \\
&= \frac{167 \text{ kips}}{21.6 \text{ kips/bolt}} \\
&= 7.73 \rightarrow 8 \text{ bolts}
\end{aligned}$$

Try PL $\frac{3}{4}$ in.$\times$7 in.

Determine number of $\frac{7}{8}$ in. diameter A325-N bolts required for material bearing on beam flange (more critical than flange plate), from Table 7-12:

$$n_{\min} = \frac{P_{uf}}{\phi r_n} = \frac{167 \text{ kips}}{58.1 \text{ kips/bolt}} = 2.9 \rightarrow 4 \text{ bolts (even number required)}$$

Bolt shear is more critical. Try two rows of four bolts on a 4-in. gage.

Check tension yielding of flange plate:

$$\begin{aligned}\phi R_n &= \phi F_y A_g \\ &= 0.9 \times 36 \text{ ksi} \times 7 \text{ in.} \times {}^3\!/\!_4 \text{ in.} \\ &= 170 \text{ kips} > 167 \text{ kips} \qquad \textbf{o.k.}\end{aligned}$$

Check tension rupture of flange plate:

$$\begin{aligned}\phi R_n &= \phi F_u A_n \\ &= 0.75 \times 58 \text{ ksi}\left[7 \text{ in.} - 2 \times \left({}^7\!/\!_8 \text{ in.} + {}^1\!/\!_8 \text{ in.}\right)\right]{}^3\!/\!_4 \text{ in.} \\ &= 163 \text{ kips} < 167 \text{ kips required} \qquad \textbf{n.g.}\end{aligned}$$

Try PL ${}^3\!/\!_4$ in.×$7{}^1\!/\!_4$ in.:

$$\begin{aligned}\phi R_n &= 0.75 \times 58 \text{ ksi}\left[7{}^1\!/\!_4 \text{ in.} - 2 \times \left({}^7\!/\!_8 \text{ in.} + {}^1\!/\!_8 \text{ in.}\right)\right]{}^3\!/\!_4 \text{ in.} \\ &= 171 \text{ kips} > 167 \text{ kips} \qquad \textbf{o.k.}\end{aligned}$$

Check block shear rupture of flange plate:

There are two cases for which block shear must be checked. The first case involves the tearout of the two blocks outside the two rows of bolt holes in the flange plate; for this case $L_{eh} = 1{}^5\!/\!_8$ in. and $L_{ev} = 1{}^1\!/\!_2$ in. The second case involves the tearout of the block between the two rows of holes in the flange plate. Tables 9-3 and 9-4 may be adapted for this calculation by considering the 4-in. width to be comprised of two 2-in.-wide blocks where $L_{eh} = 2$ in. and $L_{ev} = 1{}^1\!/\!_2$ in. Thus, the former case is more critical. From Tables 9-3a and 9-4a, $0.6F_u A_{nv} > F_u A_{nt}$. Thus,

$$\phi R_n = \phi[0.6F_u A_{nv} + F_y A_{gt}] \leq \phi[0.6F_u A_{nv} + F_u A_{nt}]$$

From Tables 9-3a, 9-4a and 9-4b,

$$\begin{aligned}\phi R_n &= 2(183 \text{ kips/in.} + 43.9 \text{ kips/in.}){}^3\!/\!_4 \text{ in.} \\ &\leq 2(183 \text{ kips/in} + 48.9 \text{ kips/in.}){}^3\!/\!_4 \text{ in.} \\ &= 340 \text{ kips} \leq 348 \text{ kips} \\ &= 340 \text{ kips} > 167 \text{ kips} \qquad \textbf{o.k.}\end{aligned}$$

Determine required weld size for fillet welds to supporting column flange:

$$\begin{aligned}D_{\min} &= \frac{P_{uf}}{2 \times 1.5 \times 1.392\, l} \\ &= \frac{167 \text{ kips}}{2 \times 1.5 \times 1.392(7{}^1\!/\!_4 \text{ in.})} \\ &= 5.52 \rightarrow 6 \text{ sixteenths}\end{aligned}$$

Note that the 1.5 factor is from LRFD Specification Appendix J2.4.

Use $^3\!/_8$ in. fillet welds.

Design the compression flange plate and connection.

Check design compressive strength of flange plate assuming $K = 0.65$ and $l = 2$ in. ($1^1\!/_2$ in. edge distance plus $^1\!/_2$ in. setback)

$$\frac{Kl}{r} = \frac{0.65(2\text{ in.})}{\sqrt{\dfrac{(7^1\!/_4\text{ in.})(^3\!/_4\text{ in.})^3/12}{(7\,^1\!/_4\text{ in.})(^3\!/_4\text{ in.})}}}$$
$$= 6.00$$

From LRFD Specification Table 3-36 with $\dfrac{Kl}{r} = 6.00$,

$$\phi F_{cr} = 30.54\text{ ksi}$$

and the design compressive strength of the flange plate is

$$\begin{aligned}\phi R_n &= \phi_c F_{cr} A \\ &= (30.54\text{ ksi})(7^1\!/_4\text{ in.} \times {}^3\!/_4\text{ in.}) \\ &= 167\text{ kips}\end{aligned}$$

Since the design strength equals the required strength, the flange plate is adequate.

The compression flange plate will be identical to the tension flange plate: a $^3\!/_4$ in.×$7\,^1\!/_4$ in. plate with eight bolts in two rows of four bolts on a 4 in. gage and $^3\!/_8$ in. fillet welds to the supporting column flange.

Comment: The column must be checked for stiffening requirements. For further information, see AISC Design Guide No. 13 *Wide-Flange Column Stiffening at Moment Connections—Wind and Seismic Applications* (Carter, 1999).

EXAMPLE 12.2. Welded flange-plated FR moment connection (beam-to-column flange).

Given: Design a welded flange-plated FR moment connection for a W18×50 beam to W14×99 column flange connection. For structural members, $F_y = 50$ ksi; for connecting material $F_y = 36$ ksi. Use 70 ksi electrodes and ASTM A325-N bolts.

$$\begin{aligned}R_u &= 45.0\text{ kips} \\ M_u &= 250\text{ kips}\end{aligned}$$

W18×50, ASTM A992

$d = 18.0$ in. $\quad b_f = 7.50$ in. $\quad Z_x = 101$ in.^{3}
$t_w = 0.355$ in. $\quad t_f = 0.570$ in.

W14×99, ASTM A992

$d = 14.2$ in. $\quad b_f = 14.6$ in. $\quad k = 2^1\!/_{16}$ in.
$t_w = 0.485$ in. $\quad t_f = 0.780$ in. $\quad T = 10$ in.

Solution:

Check beam design flexural strength:

$$Z_{req} = \frac{M_u \times 12 \text{ in./ft}}{0.9F_y}$$
$$= \frac{(250 \text{ ft-kips})(12 \text{ in./ft})}{0.9(50 \text{ ksi})}$$
$$= 66.7 \text{ in.}^3$$
$$Z_x = 101 \text{ in.}^3$$

Since $Z_x > Z_{req}$, the beam design flexural strength is **o.k.**

Design the single-plate web connection.

From Example 12-1, a three-bolt, $^5\!/_{16}$ in.-thick single plate with two $^5\!/_{16}$ in. fillet welds will be adequate.

Design the tension flange plate and connection.

Calculate the flange force P_{uf}.

$$P_{uf} = \frac{M_u \times 12 \text{ in./ft}}{d}$$
$$= \frac{(250 \text{ ft-kips})(12 \text{ in./ft})}{18.0 \text{ in.}}$$
$$= 167 \text{ kips}$$

Determine tension flange-plate dimensions.

From Figure 8-14, assume a shelf dimension of $^5\!/_8$ in. on both sides of the plate. The plate width, then, is 7.495 in.$-2(^5\!/_8$ in.$) = 6.245$. Try a 1 in.×6¼-in. flange plate.

Check tension yielding of the flange plate:

$$\phi R_n = \phi F_y A_g$$
$$= 0.9 \times 36 \text{ ksi} \times 6^1\!/_4 \text{ in.} \times 1 \text{ in.}$$
$$= 202.5 \text{ kips} \quad \textbf{o.k.}$$

Determine required weld size and length for fillet welds to beam flange. Try a $^5\!/_{16}$ in. fillet weld. The minimum length of weld $l_{\min}$ is:

$$l_{\min} = \frac{P_{uf}}{1.392D}$$
$$= \frac{167 \text{ kips}}{1.392(5 \text{ sixteenths})}$$
$$= 12.0 \text{ in.}$$

Use 3 in. of weld along each side and 6 ¼ in. of weld along the end of the flange plate.

Determine required weld size for fillet welds to supporting column flange:

$$D_{\min} = \frac{P_{uf}}{2 \times 1.5 \times 1.392\, l}$$
$$= \frac{167 \text{ kips}}{2 \times 1.5 \times 1.392(6\tfrac{1}{4} \text{ in.})}$$
$$= 6.40 \rightarrow 7 \text{ sixteenths}$$

Use $\frac{7}{16}$ in. fillet welds.

Design the compression flange plate and connection:

The compression flange plate should have approximately the same area as the tension flange plate (6.25 in.^{2}). Assume a shelf dimension of ⅝ in. The plate width, then, is 7.495 in. + 2(⅝ in.) = 8.745. To approximately balance the flange-plate areas, try a ¾ in. × 8 ¾ in. compression flange plate.

Check design compressive strength of flange plate assuming $K = 0.65$ and $l = 2$ in. (1 ½ in. edge distance plus ½ in. setback).

$$\frac{Kl}{r} = \frac{0.65(2 \text{ in.})}{\sqrt{\dfrac{(8\tfrac{3}{4} \text{ in.})(\tfrac{3}{4} \text{ in.})^3/12}{(8\tfrac{3}{4} \text{ in.})(\tfrac{3}{4} \text{ in.})}}}$$
$$= 6.00$$

From LRFD Specification Table 3-36 with $\dfrac{Kl}{r} = 6.00$,

$$\phi F_{cr} = 30.54 \text{ ksi}$$

and the design compressive strength of the flange plate is

$$\phi R_n = \phi_c F_{cr} A$$
$$= (30.54 \text{ ksi})(8\tfrac{3}{4} \text{ in.} \times \tfrac{3}{4} \text{ in.})$$
$$= 200 \text{ kips} > 167 \text{ kips} \quad \textbf{o.k.}$$

Determine required weld size and length for fillet welds to beam flange.

Based upon the weld length required for the tension flange plate, use a $\frac{5}{16}$ in. fillet weld and 6 in. of weld along each side of the beam flange.

Determine required weld size for fillet welds to supporting column flange.

As before for the tension flange plate, use $\frac{7}{16}$ in. fillet welds.

Comment: The column must be checked for stiffening requirements. For further information, see AISC Design Guide No. 13 *Wide-Flange Column Stiffening at Moment Connections–Wind and Seismic Applications* (Carter, 1999).

EXAMPLE 12.3. Directly welded flange FR moment connection (beam-to-column flange).

Given:

Design a directly welded flange FR moment connection for a W18×50 beam to W14×99 column-flange connection. For structural members, assume $F_y = 50$ ksi and $F_u = 65$ ksi; for connecting material, assume $F_y = 36$ ksi and $F_u = 50$ ksi. Use 70 ksi electrodes and ASTM A325-N bolts.

$R_u = 45.0$ kips
$M_u = 250$ kips

W18×50, ASTM A992

$d = 18.0$ in. $b_f = 7.50$ in. $Z_x = 101$ in.^{3}
$t_w = 0.355$ in. $t_f = 0.570$ in.

W14×99, ASTM A992

$d = 14.2$ in. $b_f = 14.6$ in. $k = 2\,{}^1\!/_{16}$ in.
$t_w = 0.485$ in. $t_f = 0.780$ in. $T = 10$ in.

Solution:

Check beam design flexural strength.

From Example 12.2, the beam design flexural strength is **o.k.**

Design the single-plate connection.

From Example 12.1, three ${}^7\!/_8$ in. diameter A325-N bolts, ${}^5\!/_{16}$ in.-thick single plate with two ${}^5\!/_{16}$ in. fillet welds will be adequate.

A complete-joint-penetration groove weld will transfer the entire flange force in tension and compression.

Comment:

The column must be checked for stiffening requirements. For further information, see AISC Design Guide No. 13 *Wide-Flange Column Stiffening at Moment Connections—Wind and Seismic Applications* (Carter, 1999).

EXAMPLE 12.4. Four-bolt unstiffened extended end-plate FR moment connection (beam-to-column flange).

Given:

Design a four-bolt unstiffened extended end-plate FR moment connection for a W18×50 beam to W14×99 column-flange connection. For structural members, assume $F_y = 50$ ksi and $F_u = 65$ ksi; for connecting material, assume $F_y = 36$ ksi and $F_u = 58$ ksi. Use ASTM A325-SC bolts (Class A surfaces) and 70 ksi electrodes.

$R_u = 45.0$ kips
$M_u = 250$ ft-kips

W18×50, ASTM A992

$d = 18.0$ in. $b_f = 7.50$ in. $Z_x = 101$ in.^{3}
$t_w = 0.355$ in. $t_f = 0.570$ in. $A_f / A_w = 0.714$

W14×99, ASTM A992

$d = 14.2$ in. $b_f = 14.6$ in. $k = 2^1\!/_{16}$ in.
$k_1 = 1^7\!/_{16}$ in. $t_w = 0.485$ in. $t_f = 0.780$ in.
$T = 10$ in.

Solution: *Check beam design flexural strength.*

From Example 12.2, the beam design flexural strength is **o.k.**

Design the bolts (a minimum of four bolts is required at the tension flange; a minimum of two bolts is required at the compression flange).

Calculate the flange force P_{uf}.

$$
\begin{aligned}
P_{uf} &= \frac{M_u \times 12 \text{ in./ft}}{(d - t_f)} \\
&= \frac{(250 \text{ ft-kips})(12 \text{ in./ft})}{18.0 \text{ in.} - 0.570 \text{ in.}} \\
&= 172 \text{ kips}
\end{aligned}
$$

Determine number of 1-in. diameter A325-SC bolts required for tension (Note that pretensioned bearing-type bolts would also be acceptable). From Table 7-14

$$
\begin{aligned}
n_{\min} &= \frac{P_{uf}}{\phi r_n} \\
&= \frac{172 \text{ kips}}{53.0 \text{ kips/bolt}} \\
&= 3.25 \rightarrow 4 \text{ bolts}
\end{aligned}
$$

Determine number of 1-in. diameter A325-SC bolts required for slip resistance. From Table 7-16

$$
\begin{aligned}
n_{\min} &= \frac{R_u}{\phi r_n} \\
&= \frac{45 \text{ kips}}{19.0 \text{ kips/bolt}} \\
&= 2.37 \rightarrow 3 \text{ bolts}
\end{aligned}
$$

Minimum of four bolts at tension flange and two bolts at compression flange controls. Try six 1-in. diameter A325-SC bolts (N for bolt shear check).

Check bolt shear:

From Table 7-10 for six 1-in. diameter A325-N bolts:

$$
\begin{aligned}
\phi R_n &= 6 \times 28.3 \text{ kips/bolt} \\
&= 170 \text{ kips} > 45.0 \text{ kips}
\end{aligned}
$$
o.k.

Try $^3\!/_4$ in.-thick end plate.

Check material bearing.

The thickness of the end plate is more critical than the column flange. From Table 7-12, with the conservative assumption that only the bolts at the compression flange are in bearing,

$$\begin{aligned}\phi R_n &= 2 \text{ bolts} \times 104 \text{ kips/bolt/in.} \times {}^3\!/_4 \text{ in.}\\ &= 156 \text{ kips} > 45.0 \text{ kips} \qquad \textbf{o.k.}\end{aligned}$$

Design the end plate and its connection to beam.

Calculate the effective end-plate width b_p.

Try an end plate with $L_e = 1^1\!/_2$ in., $g = 5^1\!/_2$ in., and $p_f = d_b + {}^1\!/_2$ in. $= 1^1\!/_2$ in.

$$\begin{aligned}b_p &= 2L_e + g\\ &= 2(1^1\!/_2 \text{ in.}) + 5^1\!/_2 \text{ in.}\\ &= 8^1\!/_2 \text{ in.}\end{aligned}$$

Since $b_p \approx b_f + 1$ in., the full width of the end plate may be considered effective.

Determine the required end-plate thickness.

$$M_{eu} = \frac{\alpha_m P_{uf} p_e}{4}$$

$$\alpha_m = C_a C_b \left(\frac{A_f}{A_w}\right)^{1/3} \left(\frac{p_e}{d_b}\right)^{1/4}$$

$$C_a = 1.45 \text{ from Table 12-1.}$$

$$C_b = \sqrt{\frac{b_f}{b_p}} = \sqrt{\frac{7.50}{8^1\!/_2}} = 0.939$$

$$\begin{aligned}p_e &= p_f - \frac{d_b}{4} - w_t \text{ (assuming } {}^1\!/_2 \text{ in. fillet weld)}\\ &= 1^1\!/_2 \text{ in.} - \frac{1 \text{ in.}}{4} - {}^1\!/_2 \text{ in.}\\ &= 0.75 \text{ in.}\end{aligned}$$

$$\begin{aligned}\alpha_m &= 1.45 \times 0.939 \times (0.714)^{1/3} \left(\frac{0.75 \text{ in.}}{1 \text{ in.}}\right)^{1/4} 0.75 \text{ in.}\\ &= 1.13\end{aligned}$$

$$\begin{aligned}M_{eu} &= 1.13 \times 172 \text{ kips} \times \frac{0.75 \text{ in.}}{4}\\ &= 36.4 \text{ in.-kips}\end{aligned}$$

$$\begin{aligned}t_{p\ \min} &= \sqrt{\frac{4M_{eu}}{\phi F_y b_p}}\\ &= \sqrt{\frac{4(36.4 \text{ in.-kips})}{0.9 \times 36 \text{ ksi} \times 8^1\!/_2 \text{ in.}}}\\ &= 0.727 \rightarrow {}^3\!/_4 \text{ in.}\end{aligned}$$

Try a $\frac{3}{4}$ in.×$8\frac{1}{2}$ in. end plate.

Check shear yielding of the end plate.

From LRFD Specification Section J5.3:

$$\begin{aligned}\phi R_n &= 2 \times \phi(0.60 F_y A_g)\\ &= 2 \times 0.9(0.6 \times 36 \text{ ksi} \times 8\tfrac{1}{2} \text{ in.} \times \tfrac{3}{4} \text{ in.})\\ &= 248 \text{ kips} > 172 \text{ kips} \qquad \textbf{o.k.}\end{aligned}$$

Determine required fillet weld for beam-web-to-end-plate connection.

From LRFD Specification Table J2.4, the minimum size is $\frac{5}{16}$ in. Determine size required to develop web flexural strength near tension bolts:

$$\begin{aligned}D_{\min} &= \frac{0.9 F_y t_w}{2 \times 1.392}\\ &= \frac{0.9 \times 36 \text{ ksi} \times 0.355 \text{ in.}}{2 \times 1.392}\\ &= 4.13 \rightarrow 5 \text{ sixteenths}\end{aligned}$$

Use $\frac{5}{16}$ in. fillet weld on both sides of the beam web from the inside face of the beam flange to the centerline of the inside bolt holes plus two bolt diameters.

Determine size required for the factored shear R_u. R_u is resisted by weld between the mid-depth of the beam and the inside face of the compression flange or between the inner row of tension bolts plus two bolt diameters, whichever is smaller. By inspection the former governs for this example.

$$\begin{aligned}l &= \frac{d}{2} - t_f\\ &= \frac{18.0 \text{ in.}}{2} - 0.570 \text{ in.}\\ &= 8.43 \text{ in.}\\ D_{\min} &= \frac{R_u}{2 \times 1.392 l}\\ &= \frac{172 \text{ kips}}{2 \times 1.392(8.43 \text{ in.})}\\ &= 1.92 \rightarrow 5 \text{ sixteenths (minimum size)}\end{aligned}$$

Use $\frac{5}{16}$ in. fillet weld on both sides of the beam web below the tension-bolt region.

Determine required fillet weld size for beam flange to end-plate connection.

$$\begin{aligned}l &= 2(b_f + t_f) - t_w\\ &= 2(7.50 \text{ in.} + 0.570 \text{ in.}) - 0.355 \text{ in.}\\ &= 15.8 \text{ in.}\end{aligned}$$

$$D_{\min} = \frac{P_{uf}}{1.5 \times 1.392\,l}$$
$$= \frac{172 \text{ kips}}{1.5 \times 1.392 \times 15.8 \text{ in.}}$$
$$= 5.21 \rightarrow 6 \text{ sixteenths}$$

Note that the 1.5 factor is from LRFD Specification Appendix J2.4.

Use 3⁄8 in. fillet welds at beam tension flange. Welds at compression flange may be 5⁄16-in. fillet welds (minimum size from LRFD Specification Table J2.4)

Comment: The column must be checked for stiffening requirements. For further information, see AISC Design Guide No. 13 *Wide-Flange Column Stiffening at Moment Connections—Wind and Seismic Applications* (Carter, 1999).

EXAMPLE 12.5. Eight-bolt stiffened extended end-plate FR moment connection (beam-to-column flange).

Given: Design an eight-bolt stiffened extended end-plate FR moment connection for a W33×118 beam to W14×311 column-flange connection. For structural members, assume $F_y = 50$ ksi and $F_u = 65$ ksi; for connecting material, assume $F_y = 36$ ksi and $F_u = 58$ ksi. Use ASTM A325-SC bolts (Class A surfaces) and 70 ksi electrodes.

$$R_u = 135 \text{ kips}$$
$$M_u = 1{,}050 \text{ ft-kips}$$

W33×118, ASTM A992

$d = 32.9$ in. $\quad b_f = 11.5$ in. $\quad Z_x = 415$ in.^{3}
$t_w = 0.550$ in. $\quad t_f = 0.740$ in.

W14×311, ASTM A992

$d = 17.1$ in. $\quad b_f = 16.2$ in. $\quad k = 3\tfrac{9}{16}$ in.
$k_1 = 1\tfrac{15}{16}$ in. $\quad t_w = 1.41$ in. $\quad t_f = 2.26$ in.
$T = 10$ in.

Solution: *Check beam design flexural strength:*

$$Z_{req} = \frac{M_u \times 12 \text{ in./ft}}{0.9F_y}$$
$$= \frac{(1{,}050 \text{ ft-kips})(12 \text{ in./ft})}{0.9(50 \text{ ksi})}$$
$$= 280 \text{ in.}^3$$
$$Z_x = 415 \text{ in.}^3$$

Since $Z_x > Z_{req}$, the beam design flexural strength is **o.k.**

Design the bolts (a minimum of eight bolts is required at the tension flange; a minimum of two bolts is required at the compression flange).

Calculate the flange force P_{uf}:

$$P_{uf} = \frac{M_u \times 12 \text{ in./ft}}{(d - t_f)}$$
$$= \frac{(1{,}050 \text{ ft-kips})(12 \text{ in./ft})}{(32.9 \text{ in.} - 0.740 \text{ in.})}$$
$$= 392 \text{ kips}$$

Try eight $1\frac{1}{8}$ in. diameter A325-SC bolts (six effective).

$$\phi r_{n\ req} = \frac{P_{uf}}{6 \text{ bolts}}$$
$$= \frac{392 \text{ kips}}{6 \text{ bolts}}$$
$$= 65.3 \text{ kips/bolt} < 67.1 \text{ kips/bolt} \qquad \textbf{o.k. for tension}$$

Check slip resistance with eight bolts at tension flange and two bolts at compression flange.

From Table 7-16:

$$\phi R_n = 10 \text{ bolts} \times 20.9 \text{ kips/bolt}$$
$$= 209 \text{ kips} > 135 \text{ kips} \qquad \textbf{o.k.}$$

Try eight $1\frac{1}{8}$ in. diameter A325-SC bolts (N for bolt shear check).

Check bolt shear.

From Table 7-10 for ten $1\frac{1}{8}$ in. diameter A325-N bolts:

$$\phi R_n = 10 \text{ bolts} \times 35.8 \text{ kips/bolt}$$
$$= 358 \text{ kips} > 135 \text{ kips} \qquad \textbf{o.k.}$$

Try PL $1\frac{1}{4}$.

Check material bearing.

From Table 7-12, the design bearing strength of one bolt is

$$\phi r_n = 147 \text{ kips/bolt}$$

Since this exceeds the design shear strength of the bolts, bearing is not critical.

Design the end plate and its connection to the beam.

Calculate the effective end-plate width b_p.

Try an end plate with $L_e = 1\frac{3}{4}$ in., $g = 6$ in., $p_f = d_b + \frac{1}{2}$ in. $= 1\frac{5}{8}$ in., $p_b = 3d_b$, and stiffener thickness $t_s = \frac{5}{8}$ in. ($t_{w\ beam} = 0.550$ in.). Note that all of the specified limitations for this simplified method have been met.

$$b_p = 2L_e + g = 2\left(1\tfrac{3}{4} \text{ in.}\right) + 6 \text{ in.} = 9.5 \text{ in.}$$

This dimension is less than the flange width of the beam b_f. Thus, use a plate with $b_p = b_f + 1$ in. ≈ 12½ in. This allows for runoff.

Determine the required end-plate thickness:

$$p_{eff} = \frac{p_f}{4.17}\sqrt{g^2 + {p_f}^2}$$
$$= \frac{1^5/_8 \text{ in.}}{4.17}\sqrt{(6 \text{ in.})^2 + (1^5/_8 \text{ in.})^2}$$
$$= 2.42 \text{ in.}$$

$$M_{eu} = r_u\, p_{eff}$$
$$= \frac{392 \text{ kips}}{6 \text{ bolts effective}} \times 2.42 \text{ in.}$$
$$= 158 \text{ in.-kips}$$

$$t_{p\ req} = \sqrt{\frac{4M_{eu}}{\phi F_y b_p}}$$
$$= \sqrt{\frac{4 \times 158 \text{ in.-kips}}{0.9 \times 36 \text{ ksi} \times 12^1/_2 \text{ in.}}}$$
$$= 1.25 \rightarrow 1^1/_4 \text{ in.}$$

Try PL 1¼ in.×12½ in.

Check shear yielding of the end plate.

From LRFD Specification Section J5.3.

$$\phi R_n = 2 \times \phi(0.60 F_y A_g)$$
$$= 2 \times 0.9(0.60 \times 36 \text{ ksi} \times 12^1/_2 \text{ in.} \times 1^1/_4 \text{ in.})$$
$$= 607.5 \text{ kips} > 392 \text{ kips} \qquad \textbf{o.k.}$$

Determine required fillet weld size for beam web to end-plate connection.

From LRFD Specification Table J2.4, the minimum size is $^5/_{16}$ in.

Determine size required to develop web flexural strength near tension bolts:

$$D_{min} = \frac{0.9 F_y t_w}{2 \times 1.5 \times 1.392}$$
$$= \frac{0.9 \times 50 \text{ ksi} \times 0.550 \text{ in.}}{2 \times 1.5 \times 1.392}$$
$$= 5.93 \rightarrow 6 \text{ sixteenths}$$

Use ⅜ in. fillet welds on both sides of the beam web from the inside face of the beam flange to the centerline of the inside bolt holes plus two bolt diameters.

Determine size required for the factored shear R_u.

R_u is resisted by weld between the mid-depth of the beam and the inside face of the compression flange or between the inner row of tension bolts plus two bolt diameters, whichever is smaller. By inspection, the former governs for this example.

$$
\begin{aligned}
l &= \frac{d}{2} - t_f \\
&= \frac{32.9 \text{ in.}}{2} - 0.740 \text{ in.} \\
&= 15.7 \text{ in.} \\
D_{\min} &= \frac{R_u}{2 \times 1.392\, l} \\
&= \frac{135 \text{ kips}}{2 \times 1.392(15.7 \text{ in.})} \\
&= 3.09 \rightarrow 5 \text{ sixteenths (minimum size)}
\end{aligned}
$$

Use 5/16 in. fillet welds on both sides of the beam web below the tension-bolt region.

Determine required weld for beam flange to end-plate connection.

Use 3/8 in. fillet welds at the tension flange. Welds at the compression flange may be 5/16 in. fillet welds (minimum size from LRFD Specification Table J2.4).

Comment: The column must be checked for stiffening requirements. For further information, see AISC Design Guide No. 13 *Wide-Flange Column Stiffening at Moment Connections—Wind and Seismic Applications* (Carter, 1999).

PART 12 REFERENCES

American Institute of Steel Construction, 1977, *Bridge Fatigue Guide Design and Details*, AISC, Chicago, IL.

Beedle, L.S., L.W. Lu, and E. Ozer, 1973, "Recent Developments in Steel Building Design," *Engineering Journal*, Vol. 10, No. 4, (4th Qtr.), pp. 98–111, AISC, Chicago, IL.

Carter, C.J., 1999, AISC Design Guide No. 13 *Wide-Flange Column Stiffening at Moment Connections: Wind and Seismic Applications*, AISC, Chicago, IL.

Curtis, L.E. and T.M. Murray, 1989, "Column Flange Strength at Moment End-Plate Connections," *Engineering Journal*, Vol. 26, No. 2, (2nd Qtr), pp. 41–50, AISC, Chicago, IL.

Driscoll, G.C., A. Pourbohloul, and X. Wang, 1983, "Fracture of Moment Connections—Tests on Simulated Beam-to-Column Web Moment Connection Details," *Fritz Engineering Laboratory Report No. 469.7*, Lehigh University, Bethlehem, PA.

Driscoll, G.C. and L.S. Beedle, 1982, "Suggestions for Avoiding Beam-to-Column Web Connection Failures," *Engineering Journal*, Vol. 19, No. 1, (1st Qtr.), pp. 16–19, AISC, Chicago, IL.

Federal Emergency Management Agency, 2000, FEMA 350 *Recommended Seismic Design Criteria for New Steel Moment-Frame Buildings*, FEMA, Washington, D.C.

Hart, W.H. and W.A. Milek, 1965, "Splices in Plastically Designed Continuous Structures," *Engineering Journal*, Vol. 2, No. 2, (April), pp. 33–37, AISC, Chicago, IL.

Hendrick, R.A. and T.M. Murray, 1984, "Column Web Compression Strength at End-Plate Connections," *Engineering Journal*, Vol. 21, No. 3, (3rd Qtr.), pp. 161–169, AISC, Chicago, IL.

Huang, J.S., W.F. Chen, and L.S. Beedle, 1973, "Behavior and Design of Steel Beam-to-Column Moment Connections," *Bulletin 188*, October, Welding Research Council, New York, NY.

Krawinkler, H. and E.P. Popov, 1982, "Seismic Behavior of Moment Connections and Joints," *Journal of the Structural Division*, Vol. 108, No. ST2, (February), pp. 373–391, ASCE, New York, NY.

Krishnamurthy, N., 1978, "A Fresh Look at Bolted End-Plate Behavior and Design," *Engineering Journal*, Vol. 15, No. 2, (2nd Qtr.), pp. 39–49, AISC, Chicago, IL.

Lincoln Electric Company, 1973, *The Procedure Handbook of Arc Welding*, Lincoln Electric Company, Cleveland, OH.

Murray, T.M., D.P. Kline, and K.B. Rojani, 1992, "Use of Snug-Tightened Bolts in End-Plate Connections," *Connections in Steel Structures II*, R. Bjorhovde, A. Colson, G. Haaijer and J.W.B. Stark, Editors, AISC, Chicago, IL.

Murray, T.M., 1990, AISC Design Guide No. 4 *Extended End-Plate Moment Connections*, AISC, Chicago, IL.

Murray, T.M. and A. Kukreti, 1988, "Design of Eight-Bolt Stiffened Moment End-Plates," *Engineering Journal*, Vol. 25, No. 2, (2nd Qtr.), pp. 45–52, AISC, Chicago, IL.

Table 12-1.
Values of C_a for Extended End-Plate Design

ASTM Bolt Designation	End-Plate F_y, ksi	C_a
A325/F1852	36	1.45
	50	1.31
A490	36	1.48
	50	1.33

Notes:

Tabulated values apply to beams with $F_y = 50$ ksi.

$$C_a = 1.2\left[1.29\left(\frac{F_{avg}}{F_{bu}}\right)^{0.4}\left(\frac{F_{bt}}{F_b}\right)^{0.5}\right]$$

$F_{bu} = 93.0$ ksi for A325 bolts; 115 ksi for A490 bolts.
$F_{bt} = 44.0$ ksi for A325 bolts; 54.0 ksi for A490 bolts.

Table 12-2. Values of A_f / A_w for Extended End-Plate Design

Shape	A_f / A_w	Shape	A_f / A_w	Shape	A_f / A_w	Shape	A_f / A_w	Shape	A_f / A_w
W44×335	0.686	W33×169	0.668	W24×103	0.711	W16×57	0.791	W12×336	2.05
×290	0.710	×152	0.617	×94	0.683	×50	0.779	×305	2.02
×262	0.702	×141	0.582	×84	0.655	×45	0.770	×279	1.93
×230	0.671	×130	0.540	×76	0.616	×40	0.773	×252	1.92
		×118	0.492	×68	0.561	×36	0.677	×230	1.89
W40×593	0.825							×210	1.89
×503	0.804	W30×391	0.988	W24×62	0.429	W16×31	0.589	×190	1.91
×431	0.780	×357	0.989	×55	0.397	×26	0.506	×170	1.88
×397	0.793	×326	0.979					×152	1.85
×372	0.780	×292	0.981	W21×201	1.14	W14×808	2.03	×136	1.80
×362	0.785	×261	0.953	×182	1.13	×730	2.28	×120	1.77
×324	0.787	×235	0.964	×166	1.14	×665	2.25	×106	1.81
×297	0.768	×211	0.910	×147	1.01	×605	2.21	×96	1.83
×277	0.823	×191	0.888	×132	0.995	×550	2.20	×87	1.75
×249	0.818	×173	0.867	×122	1.00	×500	2.16	×79	1.73
×215	0.811			×111	0.991	×455	2.12	×72	1.71
×199	0.711	W30×148	0.673	×101	0.994	×426	2.14	×65	1.71
		×132	0.603			×398	2.12		
W40×392	0.602	×124	0.589	W21×93	0.684	×370	2.10	W12×58	1.63
×331	0.583	×116	0.558	×83	0.687	×342	2.09	×53	1.52
×327	0.598	×108	0.518	×73	0.685	×311	2.06		
×278	0.582	×99	0.477	×68	0.668	×283	2.06	W12×50	1.28
×264	0.587	×90	0.477	×62	0.641	×257	2.03	×45	1.26
×235	0.620			×55	0.579	×233	2.03	×40	1.29
×211	0.611	W27×539	1.08	×48	0.507	×211	2.00		
×183	0.606	×368	1.04			×193	2.01	W12×35	0.992
×167	0.506	×336	1.04	W21×57	0.532	×176	1.97	×30	0.966
×149	0.425	×307	1.02	×50	0.466	×159	1.97	×26	0.937
		×281	1.03	×44	0.422	×145	1.97		
W36×798	0.971	×258	1.01					W12×22	0.575
×650	0.946	×235	0.986	W18×175	1.21	W14×132	1.86	×19	0.519
×527	0.931	×217	1.00	×158	1.19	×120	1.86	×16	0.419
×439	0.913	×194	0.984	×143	1.20	×109	1.90	×14	0.390
×393	0.907	×178	0.910	×130	1.19	×99	1.86		
×359	0.898	×161	0.901	×119	1.08	×90	1.86	W10×112	1.93
×328	0.901	×146	0.887	×106	1.06			×100	1.91
×300	0.890			×97	1.07	W14×82	1.34	×88	1.91
×280	0.883	W27×129	0.710	×86	1.06	×74	1.40	×77	1.89
×260	0.852	×114	0.648	×76	1.05	×68	1.38	×68	1.87
×245	0.834	×102	0.634			×61	1.36	×60	1.85
×230	0.820	×94	0.598	W18×71	0.741			×54	1.87
0		×84	0.545	×65	0.749	W14×53	1.14	×49	1.86
W36×256	0.648			×60	0.753	×48	1.11		
×232	0.643	W24×370	1.09	×55	0.722	×43	1.10	W10×45	1.60
×210	0.588	×335	1.08	×50	0.714			×39	1.52
×194	0.587	×306	1.08			W14×38	0.861	×33	1.35
×182	0.580	×279	1.06	W18×46	0.603	×34	0.823		
×170	0.571	×250	1.07	×40	0.595	×30	0.736	W10×30	1.04
×160	0.554	×229	1.05	×35	0.504			×26	1.04
×150	0.531	×207	1.04			W14×26	0.634	×22	0.910
×135	0.464	×192	1.04	W16×100	1.17	×22	0.559		
		×176	1.02	×89	1.15			W10×19	0.675
W33×387	0.932	×162	0.997	×77	1.15			×17	0.584
×354	0.923	×146	0.961	×67	1.15			×15	0.497
×318	0.925	×131	0.907					×12	0.463
×291	0.914	×117	0.875						
×263	0.909	×104	0.850						
×241	0.854								
×221	0.826								
×201	0.804								

PART 13

DESIGN OF BRACING CONNECTIONS AND TRUSS CONNECTIONS

BRACE

SCOPE

The specification requirements and other design considerations summarized in this Part apply to the design of concentric bracing connections and truss connections. For the design of bracing connections and truss connections for HSS and steel pipe, see the AISC *Hollow Structural Sections Connections Manual*. For bracing connections and truss connections that are part of a seismic-force-resisting system in which the seismic response modification factor R is taken greater than 3, see the Seismic Provisions, which are available from AISC at www.aisc.org. For the design of eccentric bracing connections, see Lindsay and Goverdahn (1989) and the Seismic Provisions.

BRACING CONNECTIONS

Diagonal Bracing Members

Diagonal bracing members can be rods, single angles or channels, double angles or channels, tees, W-shapes, or HSS as required by the lateral loads. Slender diagonal bracing members are relatively flexible and, thus, vibration and sag may be considerations. In slender tension-only bracing, these problems can be minimized with "draw" or pretension created by shortening the fabricated length of the diagonal brace from the theoretical length L between member working points. In general, the following deductions will be sufficient: no deduction for $L \leq 10$ ft; deduct $^1/_{16}$ in. for $10 \text{ ft} < L \leq 20$ ft; deduct $^1/_8$ in. for $20 < L \leq 35$ ft; and, deduct $^3/_{16}$ in. for $L > 35$ ft. This principle is not applicable to diagonal bracing members other than light angles since it is difficult to stretch heavier members; vibration and sag are not usually design considerations in heavier diagonal bracing members. In any diagonal bracing member, however, it is permissible to deduct an additional $^1/_{32}$ in. when necessary to avoid dimensioning to thirty-seconds of an inch.

When double-angle diagonal bracing members are separated, as at "sandwiched" end connections to gussets, intermittent connections must be provided if the unsupported length of the diagonal brace exceeds the limits specified in LRFD Specification Section D2 for tension members or LRFD Specification Section E4 for compression members; note that a minimum of two stitch-fillers is required. These may be made with either bolted or welded stitch-fillers. Many fabricators prefer ring or rectangular bolted stitch-fillers when the angles require other punching, as at the end connections. In welded construction, a stitch-filler with protruding ends, as shown in Figure 13-1a is preferred because it is easy to fit and weld. The short stitch-filler shown in Figure 13-1b is used if a smooth appearance is desired.

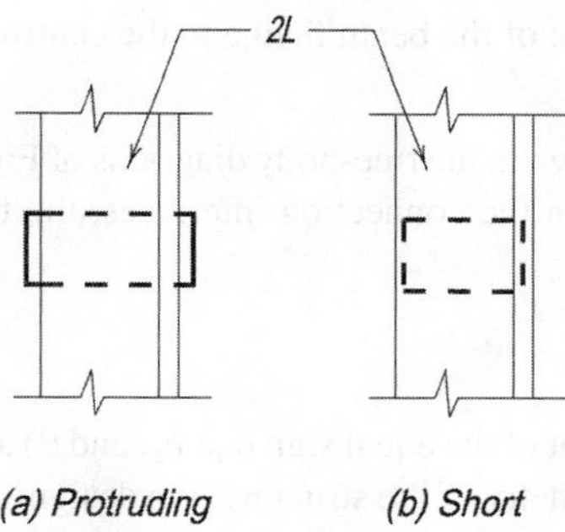

Fig. 13–1. Welded stitch fillers.

When a full-length filler is provided, as in corrosive environments, the maximum spacing of stitch bolts should be as specified in LRFD Specification Section J3.5. Alternatively, the edges of the filler may be seal welded.

Force Transfer in Diagonal Bracing Connections

There has been some controversy as to which of several available analysis methods provides the best means for the safe and economical design and analysis of diagonal bracing connections. To resolve this situation, starting in 1981, AISC sponsored extensive computer studies of this connection by Richard (1986). Associated with Richard's work, full scale tests were performed by Bjorhovde and Chakrabarti (1985), Gross and Cheok (1988), and Gross (1990). Also, AISC and ASCE formed a task group to recommend a design method for this connection. In 1990, this task group recommended three methods for further study; refer to Appendix A of Thornton (1991).

Using the results of the aforementioned full scale tests, Thornton (1991) showed that these three methods yield safe designs, and that of the three methods, the Uniform Force Method (see Model 3 of Thornton, 1991) best predicts both the design strength and critical limit state of the connection. Furthermore, Thornton (1992) showed that the Uniform Force Method yields the most economical design through comparison of actual designs by the different methods and through consideration of the efficiency of force transmission. For the above reasons, and also because it is the most versatile method, the Uniform Force Method has been adopted for use in this book.

The Uniform Force Method

The essence of the Uniform Force Method is to select the geometry of the connection so that moments do not exist on the three connection interfaces; i.e., gusset-to-beam, gusset-to-column, and beam-to-column. In the absence of moment, these connections may then be designed for shear and/or tension only, hence the origin of the name Uniform Force Method.

With the control points as illustrated in Figure 13-2 and the working point chosen at the intersection of the centerlines of the beam, column, and diagonal brace as shown in Figure 13-2a, four geometric parameters e_b, e_c, α, and β can be identified,

where

e_b = one-half the depth of the beam, in.

e_c = one-half the depth of the column, in. Note that, for a column web support, $e_c \approx 0$

α = distance from the face of the column flange or web to the centroid of the gusset-to-beam connection, in.

β = distance from the face of the beam flange to the centroid of the gusset-to-column connection, in.

For the force distribution shown in the free-body diagrams of Figures 13-2b, 13-2c, and 13-2d to remain free of moments on the connection interfaces, the following expression must be satisfied.

$$\alpha - \beta \tan\theta = e_b \tan\theta - e_c \qquad (11\text{-}1)$$

Since the variables on the right of the equal sign (e_b, e_c, and θ) are all defined by the members being connected and the geometry of the structure, the designer may select values of α and β for which the equation is true, thereby locating the centroids of the gusset-to-beam and gusset-to-column connections.

Once α and β have been determined, the factored axial and shear forces for which these connections must be designed can be determined from the following equations.

$$V_{uc} = \frac{\beta}{r} P_u \qquad H_{uc} = \frac{e_c}{r} P_u$$

$$H_{ub} = \frac{\alpha}{r} P_u \qquad V_{ub} = \frac{e_b}{r} P_u$$

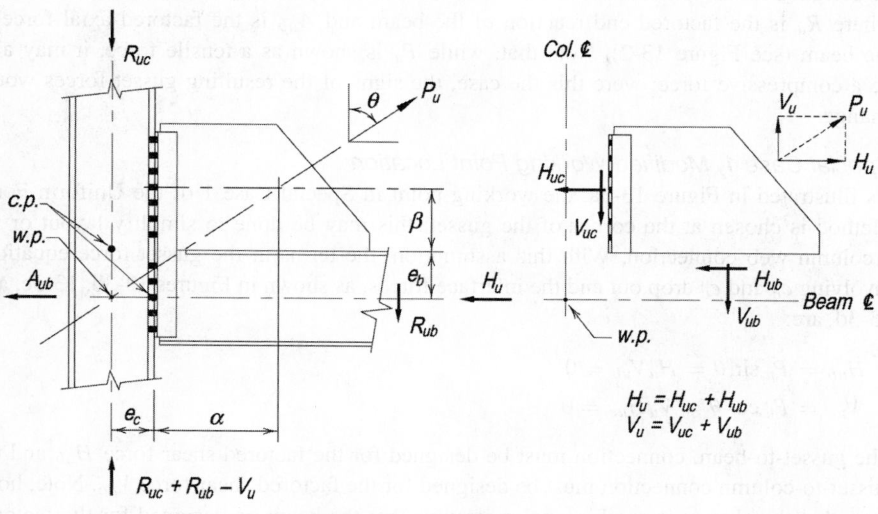

(a) Diagonal bracing connection and external forces

(b) Gusset free-body diagram

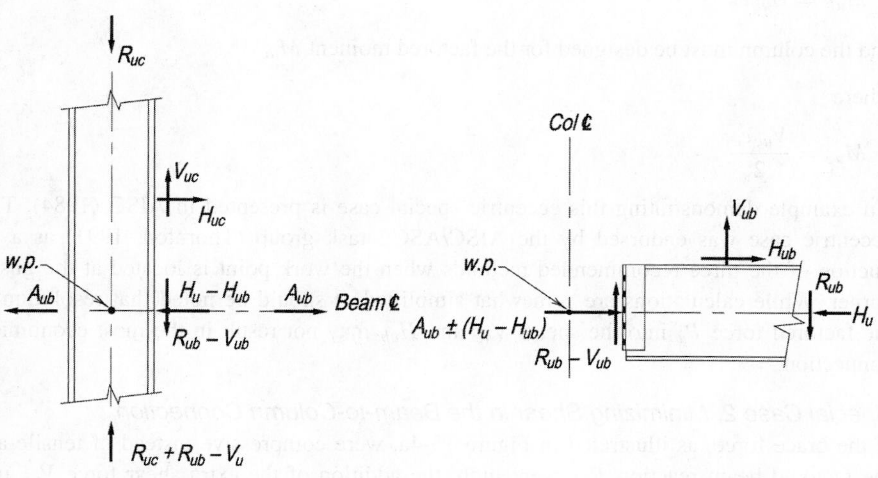

(c) Column free-body diagram

(d) Beam free-body diagram

Fig. 13–2. Force transfer by the uniform force (UF) method, work point (w.p.) and control points (c.p.) as indicated.

where

$$r = \sqrt{(\alpha + e_c)^2 + (\beta + e_b)^2}$$

The gusset-to-beam connection must be designed for the factored shear force H_{ub} and the factored axial force V_{ub}, the gusset-to-column connection must be designed for the factored shear force V_{uc} and the factored axial force H_{uc}, and the beam-to-column connection must be designed for the factored shear $R_u - V_{ub}$ and the factored axial force $A_{ub} \pm (H_u - H_{ub})$, where R_u is the factored end reaction of the beam and A_{ub} is the factored axial force in the beam (see Figure 13-2). Note that, while P_u is shown as a tensile force, it may also be a compressive force; were this the case, the signs of the resulting gusset forces would change.

Special Case 1, Modified Working Point Location

As illustrated in Figure 13-3a, the working point in Special Case 1 of the Uniform Force Method is chosen at the corner of the gusset; this may be done to simplify layout or for a column web connection. With this assumption, the terms in the gusset force equations involving e_b and e_c drop out and the interface forces, as shown in Figures 13-3b, 13-3c, and 13-3d, are:

$$H_{ub} = P_u \sin\theta = H_u V_{ub} = 0$$

$$V_{uc} = P_u \cos\theta = V_u H_{uc} = 0$$

The gusset-to-beam connection must be designed for the factored shear force H_{ub} and the gusset-to-column connection must be designed for the factored shear force V_{uc}. Note, however, that the change in working point requires that the beam be designed for the factored moment M_{ub},

where

$$M_{ub} = H_{ub} e_b$$

and the column must be designed for the factored moment M_{uc}

where

$$M_{uc} = \frac{V_{uc} e_c}{2}$$

An example demonstrating this eccentric special case is presented in AISC (1984). This eccentric case was endorsed by the AISC/ASCE task group (Thornton, 1991) as a reduction of the three recommended methods when the work point is located at the gusset corner. While calculations are somewhat simplified, it should be noted that resolution of the factored force P_u into the shears V_{uc} and H_{ub} may not result in the most economical connection.

Special Case 2, Minimizing Shear in the Beam-to-Column Connection

If the brace force, as illustrated in Figure 13-4a, were compressive instead of tensile and the factored beam reaction R_{ub} were high, the addition of the extra shear force V_{ub} into the beam might exceed the design strength of the beam and require doubler plates or a haunched connection. Alternatively, the vertical force in the gusset-to-beam connection V_{ub} can be limited in a manner which is somewhat analogous to using the gusset itself as a haunch.

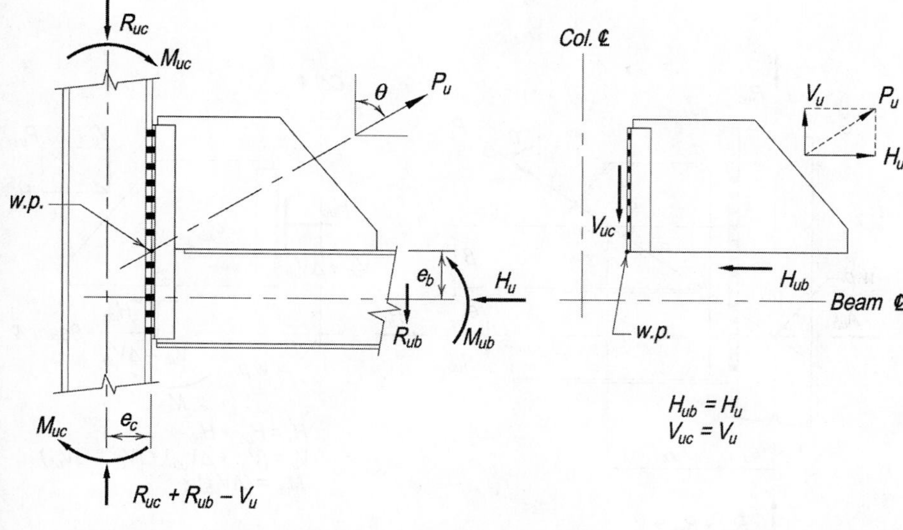

(a) Diagonal bracing connection (b) Gusset free-body diagram

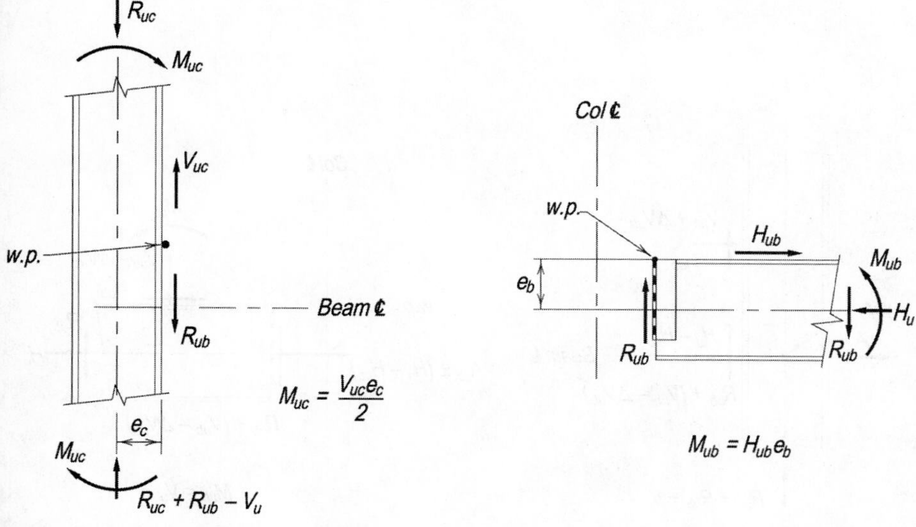

(c) Column free-body diagram (d) Beam free-body diagram

Fig. 13–3. Force transfer, UF method special case 1.

As illustrated in Figure 13-4b, assume that V_{ub} is reduced by an arbitrary amount ΔV_{ub}. By statics, the vertical force at the gusset-to-column interface will be increased to $V_{uc} + \Delta V_{ub}$, and a moment M_{ub} will result on the gusset-to-beam connection,

where

$$M_{ub} = (\Delta V_{ub})\alpha$$

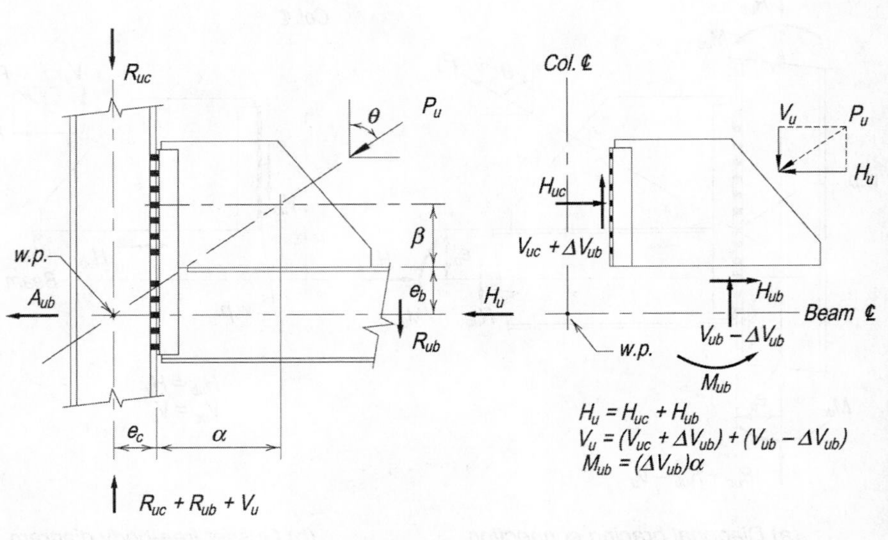

(a) Diagonal bracing connection

(b) Gusset free-body diagram

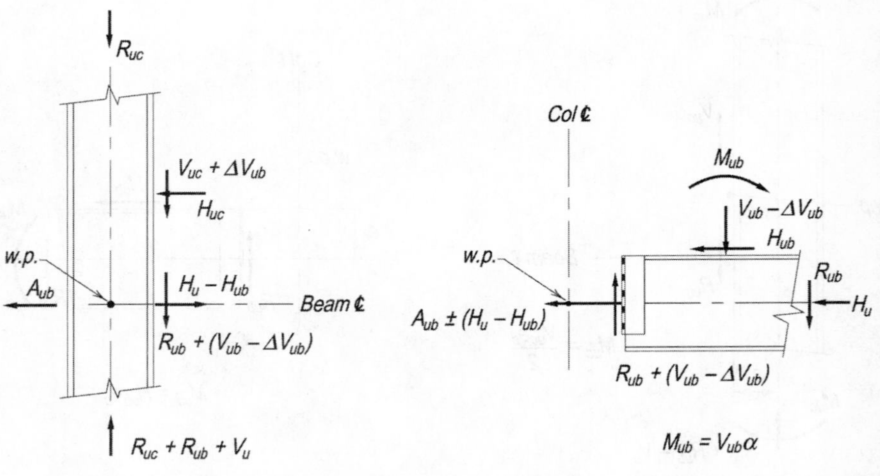

(c) Column free-body diagram

(d) Beam free-body diagram

Fig. 13–4. Force transfer, UF method special case 2.

If ΔV_{ub} is taken equal to V_{ub}, none of the vertical component of the brace force is transmitted to the beam; the resulting procedure is that presented by AISC (1984) for concentric gravity axes, extended to connections to column flanges. This method was also recommended by the AISC/ASCE task group.

Design by this method may be uneconomical. It is very punishing to the gusset and beam because of the moment M_{ub} induced on the gusset-to-beam connection. This moment will require a larger connection and a thicker gusset. Additionally, the limit state of local web yielding may limit the strength of the beam. This special case interrupts the natural flow of forces assumed in the Uniform Force Method and thus is best used when the beam-to-column interface is already highly loaded, independently of the brace, by a high shear R_u in the beam-to-column connection.

Special Case 3, No Gusset-to-Column Web Connection

When the connection is to a column web and the brace is shallow (as for large θ) or the beam is deep, it may be more economical to eliminate the gusset-to-column connection entirely and connect the gusset to the beam only. The Uniform Force Method can be applied to this situation by setting β and e_c equal to zero as illustrated in Figure 13-5. Since there is to be no gusset-to-column connection, V_{uc} and H_{uc} also equal zero. Thus, $V_{ub} = V_u$ and $H_{ub} = H_u$.

If $\bar{\alpha} = \alpha = e_b \tan\theta$, there is no moment on the gusset-to-beam interface and the gusset-to-beam connection can be designed for the factored shear force H_{ub} and the factored axial force V_{ub}. If $\bar{\alpha} \neq \alpha = e_B \tan\theta$, the gusset-to-beam interface must be designed for the moment M_{ub} in addition to H_{ub} and V_{ub},

where

$$M_{ub} = V_{ub}(\alpha - \bar{\alpha})$$

The beam-to-column connection must be designed for the factored shear force $R_u + V_{ub}$.

Note that, since the connection is to a column web, e_c is zero and hence H_c is also zero. For a connection to a column flange, if the gusset-to-column-flange connection is eliminated, the beam-to-column connection must be a moment connection designed for the moment $V_u e_c$ in addition to the shear V_u. Thus, uniform forces on all interfaces are no longer possible.

Analysis of Existing Diagonal Bracing Connections

A combination of α and β which provides for no moments on the three interfaces can usually be achieved when a connection is being designed. However, when analyzing an existing connection or when other constraints exist on gusset dimensions, the values of α and β may not satisfy Equation 11-1. When this happens, uniform interface forces will not satisfy equilibrium and moments will exist on one or both gusset edges or at the beam-to-column interface.

To illustrate this point, consider an existing design where the actual centroids of the gusset-to-beam and gusset-to-column connections are at $\bar{\alpha}$ and $\bar{\beta}$, respectively. If the connection at one edge of the gusset is more rigid than the other, it is logical to assume that the more rigid edge takes all of the moment necessary for equilibrium. For instance, the gusset of Figure 13-2 is shown welded to the beam and bolted with double angles to the column. For this configuration, the gusset-to-beam connection will be much more rigid than the gusset-to-column connection.

Take α and β as the ideal centroids of the gusset-to-beam and gusset-to-column connections, respectively. Setting $\beta = \bar{\beta}$, the α required for no moment on the gusset-to-beam connection may be calculated as:

$$\alpha = K + \bar{\beta} \tan\theta$$

where

$$K = e_b \tan\theta - e_c$$

If $\alpha \neq \bar{\alpha}$, a moment M_{ub} will exist on the gusset-to-beam connection, where

$$M_{ub} = V_{ub}(\alpha - \bar{\alpha})$$

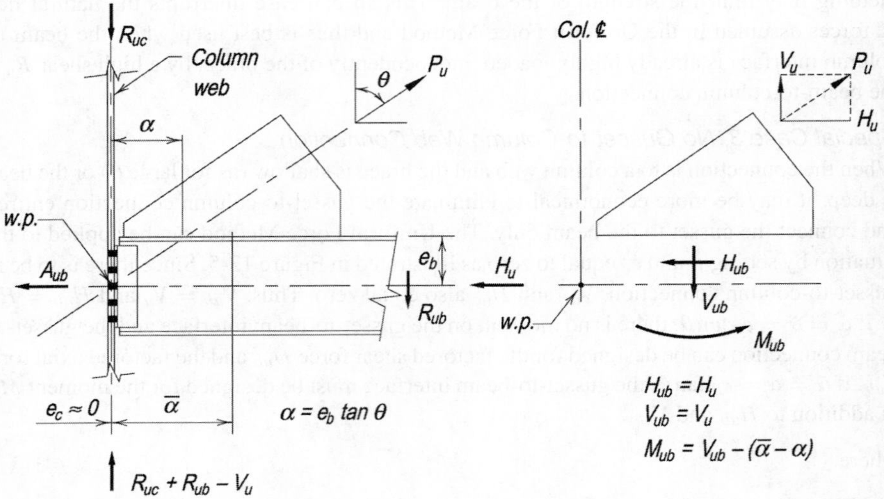

(a) Diagonal bracing connection (b) Gusset free-body diagram

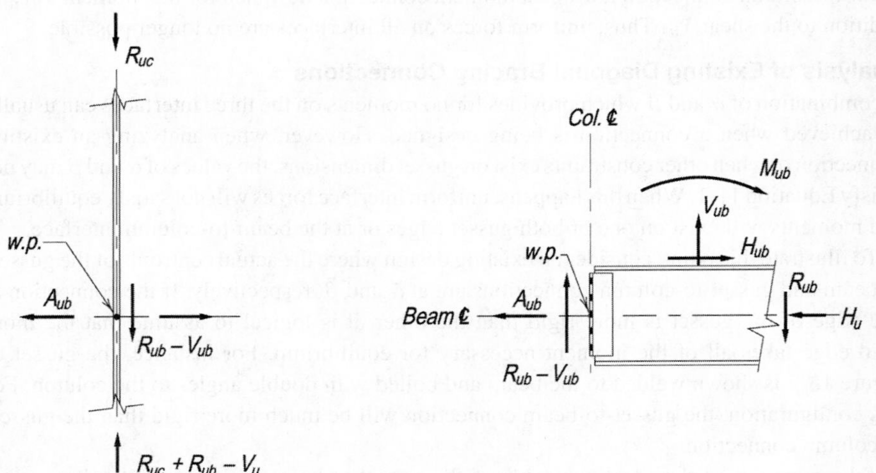

(c) Column free-body diagram (d) Beam free-body diagram

Fig. 13–5. Force transfer, UF method special case 3.

Conversely, suppose the gusset-to-column connection were judged to be more rigid. Setting $\alpha = \bar{\alpha}$, the β required for no moment on the gusset-to-column connection may be calculated as:

$$\beta = \frac{\bar{\alpha} - K}{\tan\theta}$$

If $\beta \neq \bar{\beta}$, a moment M_{uc} will exist on the gusset-to-column connection, where

$$M_{uc} = H_{uc}(\beta - \bar{\beta})$$

If both connections were equally rigid and no obvious allocation of moment could be made, the moment could be distributed based on minimized eccentricities $\alpha - \bar{\alpha}$ and $\beta - \bar{\beta}$ by minimizing the objective function ϕ, where

$$\phi = \left(\frac{\alpha - \bar{\alpha}}{\bar{\alpha}}\right)^2 + \left(\frac{\beta - \bar{\beta}}{\bar{\beta}}\right)^2 - \lambda(\alpha - \beta\tan\theta - K)$$

In the above equation, λ is a Lagrange multiplier.

The values of α and β which minimize ϕ are:

$$\alpha = \frac{K'\tan\theta + K\left(\dfrac{\bar{\alpha}}{\bar{\beta}}\right)^2}{D}$$

and

$$\beta = \frac{K' - K\tan\theta}{D}$$

where

$$K' = \bar{\alpha}\left(\tan\theta + \frac{\bar{\alpha}}{\bar{\beta}}\right)$$

$$D = \tan^2\theta + \left(\frac{\bar{\alpha}}{\bar{\beta}}\right)^2$$

Design Strength

The design strength of a diagonal bracing connection is determined from the applicable limit states for the bolts (see Part 7), welds (see Part 8) and connection elements (see Part 9). In all cases, the design strength ϕR_n must exceed the required strength R_u. Note that when the gusset is directly welded to the beam or column, the connection must be designed for the larger of the peak stress and 1.4 times the average stress, but the weld size need not be larger than that required to develop the strength of the gusset. This 40 percent increase is recommended to provide ductility to allow adequate force redistribution in the weld group. However, the weld size need not exceed that required to develop the strength of the gusset plate.

TRUSS CONNECTIONS

Members in Trusses

For light loads, trusses are commonly composed of tees for the top and bottom chords with single-angle or double-angle web members. In welded construction, the single-angle and double-angle web members may, in many cases, be welded to the stem of the tee, thus, eliminating the need for gussets. When single-angle web members are used, all web members should be placed on the same side of the chord; staggering the web members causes a torque on the chord, as illustrated in Figure 13-6.

Double-angle truss members are designed to act as a single composite unit. When unequal-leg angles are used, long legs are normally assembled back to back. A simple notation for this is LLBB (long legs back-to-back) and SLBB (short legs back-to-back). Alternatively, the notation might be graphical in nature as ⅃L and ⸏⅃L⸏.

For large loads, W-shapes may be used with the web vertical and gussets welded to the flange for the truss connections. Web members may be single angles or double angles, although W-shapes are sometimes used for both chord and web members as shown in Figure 13-7. Heavy shapes in trusses must meet the design and fabrication restrictions and special requirements in LRFD Specification Section A3.1c. With member orientation as shown for the field-welded truss joint in Figure 13-7a, connections usually are made by groove welding flanges to flanges and fillet welding webs directly or indirectly by the use of gussets. Fit-up of joints in this type of construction are very sensitive to dimensional variations in the rolled shapes; fabricators sometimes prefer to use built-up shapes in these cases.

The web connection plate in Figure 13-7a is a typical detail. While the diagonal member could theoretically be cut so that the diagonal web would be extended into the web of the chord for a direct connection, such a detail is difficult to fabricate. Additionally, welding access becomes very limited; note the obvious difficulty of welding the gusset or diagonal directly to the chord web. As illustrated, this weld is usually omitted.

When stiffeners and doubler plates are required for concentrated flange forces, the designer should consider selecting a heavier section to eliminate the need for stiffening. Although this will increase the material cost of the member, the heavier section will likely provide a more economical solution due to the reduction in labor cost associated with the elimination of stiffening (Ricker, 1992 and Thornton, 1992).

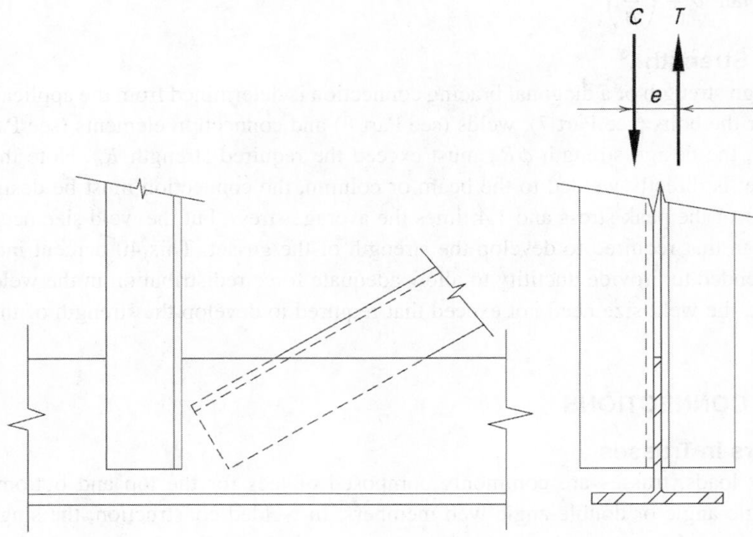

Fig. 13–6. Staggered truss web members result in a torque on the truss chord.

Minimum Connection Strength

From LRFD Specification Section J1.7, truss connections must be designed for a minimum factored load of 10 kips. Additionally, when trusses are shop assembled or field assembled on the ground for subsequent erection, consideration should be given to loads induced during handling, shipping, and erection.

Panel-Point Connections

A panel-point connection connects diagonal and/or vertical web members to the chord member of a truss. These web members deliver axial forces, tensile or compressive, to the truss chord. In bolted construction, a gusset is usually required because of bolt spacing and edge distance requirements. In welded construction, it is sometimes possible to eliminate the need for a gusset.

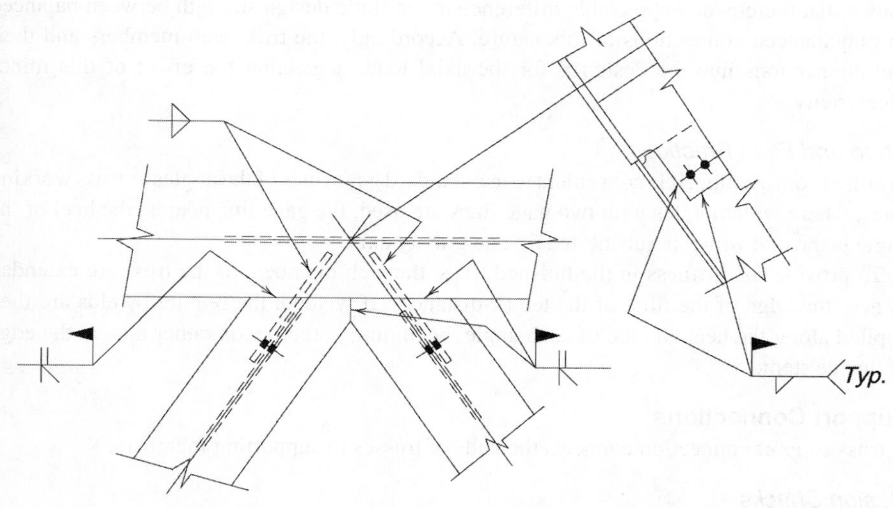

(a) Shop and field welding

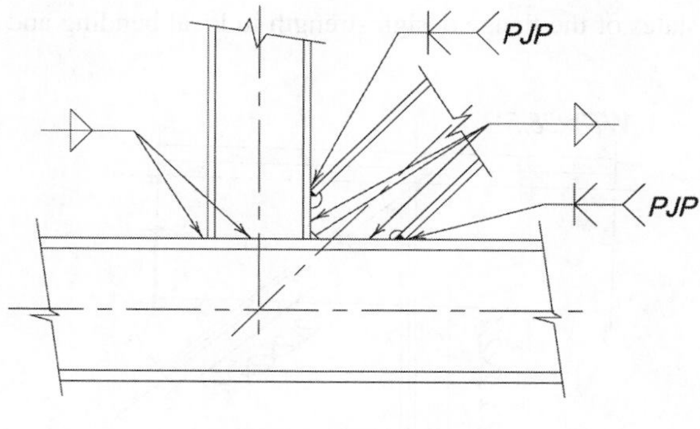

(b) Shop welding

Fig. 13–7. Truss panel-point connections for W-shape truss members.

Design Checks

The design strength of a panel-point connection is determined from the applicable limit states for the bolts (see Part 7), welds (see Part 8) and connection elements (see Part 9). In all cases, the design strength ϕR_n must exceed the required strength R_u.

In the panel-point connection of Figure 13-8, the neutral axes of the vertical and diagonal truss members intersect on the neutral axis of the truss chord. As a result, the forces in all members of the truss are axial. It is common practice, however, to modify working lines slightly from the gravity axes to establish repetitive panels and avoid fractional dimensions less than $^1/_8$ in. or to accommodate a larger panel-point connection or a connection for bottom-chord lateral bracing, a purlin, or a sway-frame. This eccentricity and the resulting moment must be considered in the design of the truss chord.

In contrast, for the design of the truss web members, LRFD Specification Section J1.8 states that the center of gravity of the end connection of a statically loaded truss member need not coincide with the gravity axis of the connected member. This is because tests have shown that there is no appreciable difference in the static design strength between balanced and unbalanced connections of this nature. Accordingly, the truss web members and their end connections may be designed for the axial load, neglecting the effect of this minor eccentricity.

Shop and Field Practices

In bolted construction, it is convenient to use standard gage lines of the angles as truss working lines; where wider angles with two gage lines are used, the gage line nearest the heel of the angle is the one which is substituted for the gravity axis.

To provide for stiffness in the finished truss, the web members of the truss are extended to near the edge of the fillet of the tee (k-distance). If welded, the required welds are then applied along the heel and toe of each angle, beginning at their ends rather than at the edge of the tee stem.

Support Connections

A truss support connection connects the ends of trusses to supporting members.

Design Checks

The design strength of a support connection is determined from the applicable limit states for the bolts (see Part 7), welds (see Part 8) and connection elements (see Part 9). Additionally, truss support connections produce tensile or compressive single concentrated forces at the beam end; the limit states of the flange design strength in local bending and the limit states

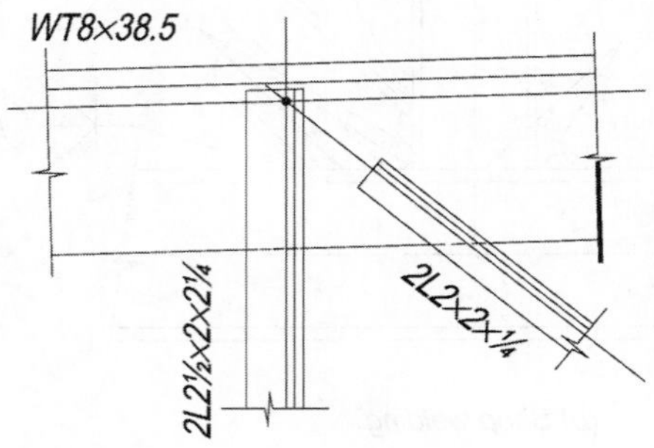

Fig. 13–8. Truss panel-point connection.

of the web design strength in local yielding, crippling, and compression buckling may have to be checked. In all cases, the design strength ϕR_n must exceed the required strength R_u.

At the end of a truss supported by a column, all member axes may not intersect at a common point. When this is the case, an eccentricity results. Typically, it is the neutral axis of the column that does not meet at the working point.

If trusses with similar reactions line up on opposite sides of the column, consideration of eccentricity would not be required since any moment would be transferred through the column and into the other truss. However, if there is little or no load on the opposite side of the column, the resulting eccentricity must be considered.

In Figure 13-9, the truss chord and diagonal intersect at a common working point on the face of the column flange. In this detail, there is no eccentricity in the gusset, gusset-to-column connection, truss chord, or diagonal. However, the column must be designed for the moment due to the eccentricity of the truss reaction from the neutral axis of the column.

For the truss support connection illustrated in Figure 13-10, this eccentricity results in a moment. Assuming the connection between the members is adequate, joint rotation is resisted by the combined flexural strength of the column, the truss top chord, and the truss diagonal. However, the distribution of moment between these members will be proportional to the stiffness of the members. Thus, when the stiffness of the column is much greater than the stiffness of the other elements of the truss support connection, it is good practice to design the column and gusset-to-column connection for the full eccentricity.

Due to its importance, the truss support connection is frequently shown in detail on the design drawing.

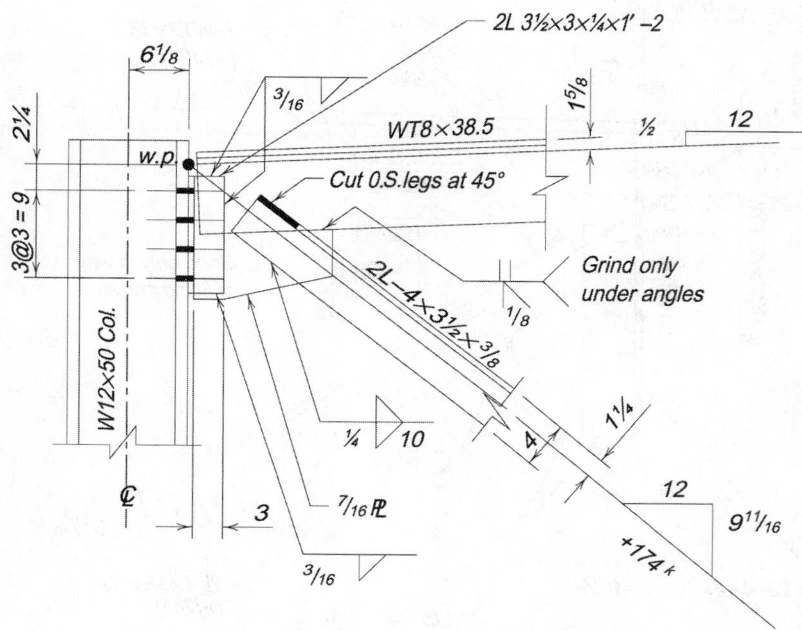

Fig. 13–9. Truss support connection, working point (w.p.) on column face.

Shop and Field Practices

When a truss is erected in place and loaded, truss members in tension will lengthen and truss members in compression will shorten. At the support connection, this may cause the tension chord of a "square-ended" truss to encroach on its connection to the supporting column. When the connection is shop-attached to the truss, erection clearance must be provided with shims to fill out whatever space remains after the truss is erected and loaded. In field-erected connections, however, provision must be made for the necessary adjustment in the connection.

When the tension chord delivers no calculated force to the connection, adjustment can usually be provided with slotted holes. For short spans with relatively light loads, the comparatively small deflections can be absorbed by the normal hole clearances provided for bolted construction. Slightly greater misalignment can be corrected in the field by reaming the holes. If appreciable deflection is expected, the connection may be welded. Alternatively, bolt holes may be field-drilled, but this is an expensive operation which should be avoided if at all possible.

An approximation of the elongation which may be expected can be determined from the relationship between stress *s* and strain *e*, where *E*, the modulus of elasticity, is

$$E = \frac{\sigma}{\varepsilon}$$

With $\sigma = P/A$, $\varepsilon = \Delta/l$, and $E = 29{,}000$ ksi, Δ, the elongation in in. will be:

$$\Delta = \frac{Pl}{29{,}000A}$$

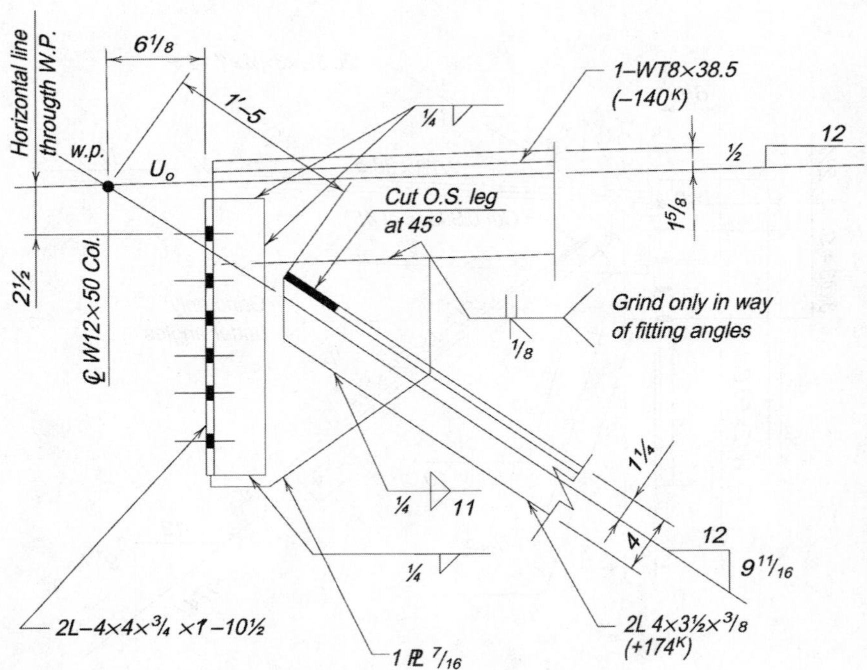

Fig. 13–10. Truss-support connection, working point (w.p.) at column centerline.

In the above equation,

P = unfactored axial force, kips
A = gross area of the truss chord, in.^{2}
l = length, in.

The total change in length of the truss chord is $\Sigma\Delta_i$, the sum of the changes in the lengths of the individual panel segments of the truss chord. The misalignment at each support connection of the tension chord is one-half the total elongation.

Truss Chord Splices

Truss chord splices are expensive to fabricate and should be avoided whenever possible. In general, chord splices in ordinary building trusses are confined to cases where: (1) the finished truss is too large to be shipped in one piece; (2) the truss chord exceeds the available material length; (3) the reduction in member size of the chord justifies the added cost of a splice; or, (4) a sharp change in direction occurs in working line of the chord and bending does not provide a satisfactory alternative.

DESIGN EXAMPLES

EXAMPLE 13.1. Bracing connection design.

Given: Refer to Figure 13-11. Design the diagonal bracing connection between the W12×87 brace and the W18×106 beam and the W14×605 column. Use $^7/_8$ in. diameter A325-N bolts in standard holes and 70 ksi electrodes. For structural members, assume $F_y = 50$ ksi and $F_u = 65$ ksi; for connecting material, assume $F_y = 36$ ksi and $F_u = 58$ ksi.

W12×87, ASTM A992

$d = 12.5$ in. $\quad b_f = 12.1$ in. $\quad A = 25.6$ in.^{2}
$t_w = 0.515$ in. $\quad t_f = 0.810$ in.

W18×106, ASTM A992

$d = 18.7$ in. $\quad b_f = 11.2$ in. $\quad k = 1.34$ in.
$t_w = 0.590$ in. $\quad t_f = 0.940$ in.

W14×605, ASTM A992

$d = 20.9$ in. $\quad b_f = 17.4$ in.
$t_w = 2.60$ in. $\quad t_f = 4.16$ in.

Solution: *Brace-to-gusset connection*

Distribute brace force in proportion to web and flange areas.

Force in flange

$$P_{uf} = \frac{P_u(b_f t_f)}{A} = \frac{675\text{ kips}(12.1\text{ in.} \times 0.810\text{ in.})}{25.6\text{ in}^2} = 259\text{ kips}$$

Force in web

$$\begin{aligned} P_{uw} &= P_u - 2P_{uf} \\ &= 675 \text{ kips} - 2(259 \text{ kips}) \\ &= 157 \text{ kips} \end{aligned}$$

Design brace-flange-to-gusset connection.

Determine number of $^7/_8$ in. diameter A325-N bolts required on the brace side (single shear) for shear.

$$\begin{aligned} n_{\min} &= \frac{P_{uf}}{\phi r_n} \\ &= \frac{259 \text{ kips}}{21.6 \text{ kips/bolt}} \\ &= 11.99 \rightarrow 12 \text{ bolts} \end{aligned}$$

On the gusset side, since these bolts are in double shear, half as many bolts will be required. Try six rows of two bolts each through the flange, six bolts through the gusset, and 2L4×4×$^3/_4$ angles (A = 10.9 in.^{2}, $\bar{x}$ = 1.27 in.).

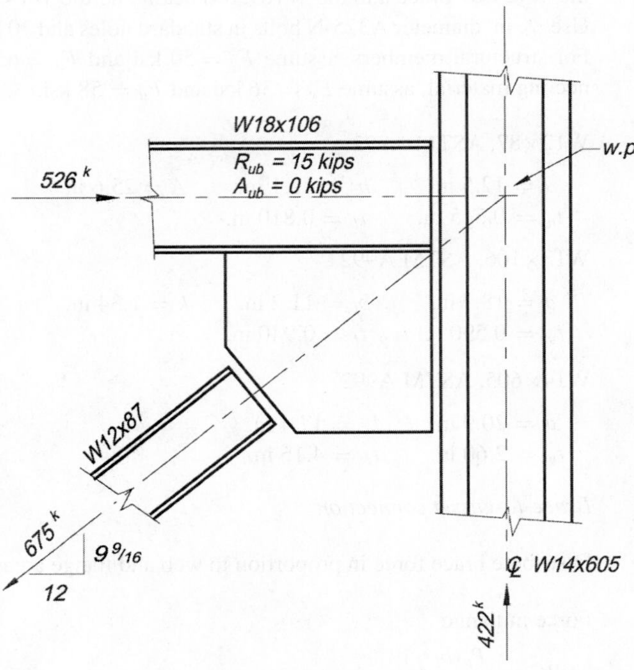

Fig. 13–11. Illustration for Example 13.1.

Check tension yielding of the angles

$$\begin{aligned}\phi R_n &= \phi F_y A_g \\ &= 0.90(36\text{ ksi})(10.9\text{ in.}^2) \\ &= 353\text{ kips} > 259\text{ kips} \quad \textbf{o.k.}\end{aligned}$$

Check tension rupture of the angles.

From LRFD Specification Sections B3.2,

$$\begin{aligned}U &= \frac{1-\bar{x}}{l} \le 0.9 \\ &= \frac{1-1.27\text{ in.}}{15\text{ in.}} \le 0.9 \\ &= 0.92 \rightarrow 0.9 \\ UA_n &= 0.9(10.9\text{ in.}^2 - 2 \times 0.75\text{ in.} \times 1\text{ in.}) \\ &= 8.46\text{ in.}^2\end{aligned}$$

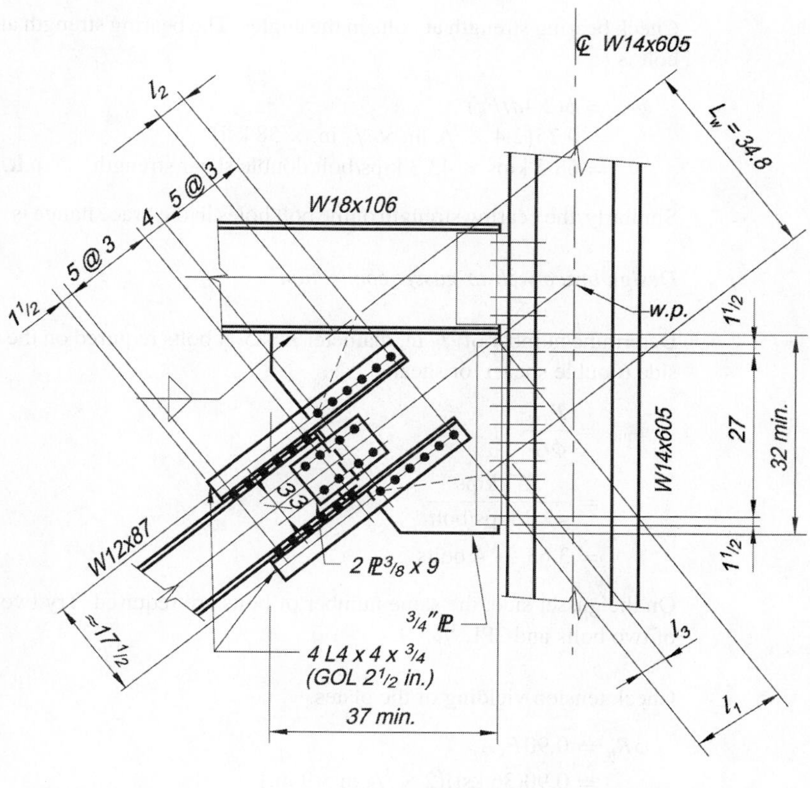

(b) connection

Fig. 13–11. (Continued)

Thus $A_e = UA_n = 8.46$ in.^{2}

$$\begin{aligned}\phi R_n &= \phi F_u A_e \\ &= 0.75(58 \text{ ksi})8.46 \text{ in.}^2 \\ &= 368 \text{ kips} > 259 \text{ kips} \qquad \textbf{o.k.}\end{aligned}$$

Check block shear rupture of angles.

From Tables 9-3 and 9-4 with $n = 6$, $L_{ev} = 1^1/_2$ in., and $L_{eh} = 1^1/_2$ in., $0.6F_u A_{nv} > F_u A_{nt}$. Thus,

$$\begin{aligned}\phi R_n &= \phi[0.6F_u A_{nv} + F_y A_{gt}] \times 2 \text{ blocks} \\ &\leq \phi[0.6F_u A_{nv} + F_u A_{nt}] \times 2 \text{ blocks} \\ &= (287 \text{ kips/in.} + 40.5 \text{ kips/in.})(^3/_4 \text{ in.}) \times 2 \\ &\leq (287 \text{ kips/in.} + 43.5 \text{ kips/in.})(^3/_4 \text{ in.}) \times 2 \\ &= 491 \text{ kips} \leq 496 \text{ kips} \\ &= 491 \text{ kips} > 259 \text{ kips} \qquad \textbf{o.k.}\end{aligned}$$

Similarly, the block shear rupture strength of the brace flange is **o.k.**

Check bearing strength at bolts in the angles. The bearing strength at each bolt is

$$\begin{aligned}\phi r_n &= \phi(2.4dtF_u) \\ &= 0.75(2.4 \times {^7/_8} \text{ in.} \times {^3/_4} \text{ in.} \times 58 \text{ ksi}) \\ &= 68.5 \text{ kips} > 43.3 \text{ kips/bolt double shear strength} \qquad \textbf{o.k.}\end{aligned}$$

Similarly, the bearing strength of the bolt holes in the brace flange is **o.k.**

Design brace-web-to-gusset connection

Determine number of $^7/_8$ in. diameter A325-N bolts required on the brace side (double shear) for shear.

$$\begin{aligned}n_{\min} &= \frac{P_{uw}}{\phi r_n} \\ &= \frac{157 \text{ kips}}{43.3 \text{ kips/bolt}} \\ &= 3.63 \rightarrow 4 \text{ bolts}\end{aligned}$$

On the gusset side, the same number of bolts are required. Try two rows of two bolts and 2PL $^3/_8 \times 9$.

Check tension yielding of the plates.

$$\begin{aligned}\phi R_n &= 0.90F_y A_g \\ &= 0.90(36 \text{ ksi})(2 \times {^3/_8} \text{ in.} \times 9 \text{ in.}) \\ &= 219 \text{ kips} > 157 \text{ kips} \qquad \textbf{o.k.}\end{aligned}$$

Check tension rupture of the plates (LRFD Specification Section J5.2).

Taking A_e as the lesser of A_n and $0.85A_g$,

$$\begin{aligned} A_n &= 2 \times {}^3\!/_8 \text{ in.} \times 9 \text{ in.} - 4 \times {}^3\!/_8 \text{ in.} \times 1 \text{ in.} \\ &= 5.25 \text{ in.}^2 \\ 0.85A_g &= 0.85(2 \times {}^3\!/_8 \text{ in.} \times 9 \text{ in.}) \\ &= 5.74 \text{ in.}^2 \end{aligned}$$

Thus, $A_e = A_n = 5.25 \text{ in.}^2$

$$\begin{aligned} \phi R_n &= 0.75F_uA_e \\ &= 0.75(58 \text{ ksi})5.25 \text{ in.}^2 \\ &= 228 \text{ kips} > 157 \text{ kips} \qquad \textbf{o.k.} \end{aligned}$$

Check block shear rupture of the plates (outer blocks) from Tables 9-3 and 9-4 with $n = 2$, $L_{ev} = 1^1\!/_2$ in., and $L_{eh} = 1^1\!/_2$ in., $0.6F_uA_{nv} > F_uA_{nt}$. Thus,

$$\begin{aligned} \phi R_n &= \phi[0.6F_uA_{nv} + F_yA_{gt}] \times 2 \text{ blocks} \times 2 \text{ plates} \\ &\leq \phi[0.6F_uA_{nv} + F_uA_{nt}] \times 2 \text{ blocks} \times 2 \text{ plates} \\ &= (78.3 \text{ kips/in.} + 40.5 \text{ kips/in.})({}^3\!/_8 \text{ in.}) \times 4 \\ &\leq (78.3 \text{ kips/in.} + 43.5 \text{ kips/in.})({}^3\!/_8 \text{ in.}) \times 4 \\ &= 178 \text{ kips} \leq 183 \text{ kips} \\ &= 178 \text{ kips} > 157 \text{ kips} \qquad \textbf{o.k.} \end{aligned}$$

Similarly, the block shear rupture strength of the interior blocks of the brace-web plates and the brace web are **o.k.**

Check bearing strength of bolt holes in the plates. The bearing strength at each bolt hole is 68.5 kips which exceeds the double shear strength of the bolt and is **o.k.**

Check tension yielding of the brace.

$$\begin{aligned} \phi R_n &= \phi F_yA_g \\ &= 0.90(50 \text{ ksi})(25.6 \text{ in.}^2) \\ &= 1{,}150 \text{ kips} > 675 \text{ kips} \qquad \textbf{o.k.} \end{aligned}$$

Check tension rupture of the brace.

Taking A_e as A_n,

$$\begin{aligned} A_n &= 25.6 \text{ in.}^2 - (4 \times 0.810 \text{ in.} + 2 \times 0.515 \text{ in.})(1 \text{ in.}) \\ &= 21.3 \text{ in.}^2 \end{aligned}$$

Thus, $A_e = A_n = 21.3 \text{ in.}^2$

$$\begin{aligned} \phi R_n &= \phi F_uA_n \\ &= 0.75(65 \text{ ksi})21.3 \text{ in.}^2 \\ &= 1{,}040 \text{ kips} > 675 \text{ kips} \qquad \textbf{o.k.} \end{aligned}$$

Design gusset

From edge distance, spacing, and clearance requirements, try PL $^3/_4$ in. Check bearing strength at bolt holes. The bearing strength at each bolt hole is

$$\begin{aligned}\phi r_n &= \phi(2.4\,dt\,F_u)\\ &= 0.75\left(2.4 \times {}^7/_8 \text{ in.} \times {}^3/_4 \text{ in.} \times 58 \text{ ksi}\right)\\ &= 68.5 \text{ kips} > 43.3 \text{ kips/bolt double shear strength} \qquad \textbf{o.k.}\end{aligned}$$

Check block shear rupture for force transmitted through web. From Tables 9-3 and 9-4 with $n = 2$, $L_{ev} = 1^1/_2$ in., $L_{eh} = 3$ in., $F_u A_{nt} > 0.6 F_u A_{nv}$. Thus

$$\begin{aligned}\phi R_n &= \phi[0.6 F_y A_{gv} + F_u A_{nt}] \times 2 \text{ blocks}\\ &\le \phi[0.6 F_u A_{nv} + F_u A_{nt}] \times 2 \text{ blocks}\\ &= (72.9 \text{ kips/in.} + 109 \text{ kips/in.})\left({}^3/_4 \text{ in.}\right) \times 2 \text{ blocks}\\ &\le (78.3 \text{ kips/in.} + 109 \text{ kips/in.})\left({}^3/_4 \text{ in.}\right) \times 2 \text{ blocks}\\ &= 273 \text{ kips} \le 281 \text{ kips}\\ &= 273 \text{ kips} > 157 \text{ kips} \qquad \textbf{o.k.}\end{aligned}$$

Check block shear rupture for total brace force.

With $A_{gv} = 24.8 \text{ in.}^2$, $A_{gt} = 13.2 \text{ in.}^2$, $A_{nv} = 15.8 \text{ in.}^2$, and $A_{nt} = 12.4 \text{ in.}^2$, $F_u A_{nt} > 0.6 F_u A_{nv}$. Thus

$$\begin{aligned}\phi R_n &= \phi[0.6 F_y A_{gv} + F_u A_{nt}] \le \phi[0.6 F_u A_{nv} + F_u A_{nt}]\\ &= 0.75[0.6(36 \text{ ksi})(24.8 \text{ in.}^2) + (58 \text{ ksi})(12.4 \text{ in.}^2)]\\ &\le 0.75[0.6(58 \text{ ksi})(16.5 \text{ in.}^2) + (58 \text{ ksi})(12.4 \text{ in.}^2)]\\ &= 941 \text{ kips} \le 970 \text{ kips}\\ &= 941 \text{ kips} > 675 \text{ kips} \qquad \textbf{o.k.}\end{aligned}$$

Check tension yielding on the Whitmore section of the gusset. The Whitmore section, as illustrated with dashed lines in Figure 13-11b, is 34.8 in. long; 30.9 in. occurs in the gusset and 3.90 in. occurs in the beam web. Thus

$$\begin{aligned}\phi R_n &= \phi F_y A_w\\ &= 0.90\left[(36 \text{ ksi})\left(30.9 \times {}^3/_4 \text{ in.}\right) + (50 \text{ ksi})(3.90 \text{ in.} \times 0.590 \text{ in.})\right]\\ &= 854 \text{ kips} > 675 \text{ kips} \qquad \textbf{o.k.}\end{aligned}$$

The beam web thickness is used, conservatively ignoring the larger thickness in the beam-flange and flange-to-web-fillet area.

Note that, were this a compressive force, gusset buckling would have to be checked; refer to the comments at the end of this example.

Distribution of brace force to beam and column

From the members and frame geometry

$$e_b = \frac{d_b}{2} = \frac{18.7 \text{ in.}}{2} = 9.35 \text{ in.}$$

$$e_c = \frac{d_c}{2} = \frac{20.9 \text{ in.}}{2} = 10.5 \text{ in.}$$

$$\tan\theta = \frac{12}{9\,{}^{9}\!/_{16}} = 1.25$$

and

$$\begin{aligned} e_b \tan\theta - e_c &= 9.35 \text{ in.}(1.25) - 10.5 \text{ in.} \\ &= 1.19 \text{ in.} \end{aligned}$$

Try gusset PL $^3\!/_4$ in. × 42 in. horizontally ×33 in. vertically (Several intermediate gusset dimensions were inadequate). With connection centroids at the midpoint of the gusset edges

$$\begin{aligned} \alpha &= \frac{42 \text{ in.}}{2} + {}^1\!/_2 \text{ in.} \\ &= 21.5 \text{ in.} \end{aligned}$$

where $^1\!/_2$ in. is allowed for the setback between the gusset and the column, and

$$\begin{aligned} \bar{\beta} &= \frac{33 \text{ in.}}{2} \\ &= 16.5 \text{ in.} \end{aligned}$$

Choosing $\beta = \bar{\beta}$, the $\bar{\alpha}$ required for uniform forces is

$$\begin{aligned} \bar{\alpha} &= e_b \tan\theta - e_c + \beta \tan\theta \\ &= 1.19 \text{ in.} + (16.5 \text{ in.})(1.25) \\ &= 21.8 \text{ in.} \end{aligned}$$

The resulting eccentricity is $\alpha - \bar{\alpha}$, where

$$\begin{aligned} \alpha - \bar{\alpha} &= 21.5 \text{ in.} - 21.8 \text{ in.} \\ &= -0.3 \text{ in.} \end{aligned}$$

This slight eccentricity is negligible. Use $\alpha = 21.8$ in. and $\beta = 16.5$ in.

Calculate gusset interface forces

$$\begin{aligned} r &= \sqrt{(\alpha + e_c)^2 + (\beta + e_b)^2} \\ &= \sqrt{(21.8 \text{ in.} + 10.5 \text{ in.})^2 + (16.5 \text{ in.} + 9.35 \text{ in.})^2} \\ &= 41.4 \text{ in.} \end{aligned}$$

On the gusset-to-column connection

$$\begin{aligned} H_{uc} &= \frac{e_c}{r} P_u \\ &= \frac{10.5 \text{ in.}}{41.4 \text{ in.}}(675 \text{ kips}) \\ &= 171 \text{ kips} \end{aligned}$$

$$
\begin{aligned}
V_{uc} &= \frac{\beta}{r} P_u \\
&= \frac{16.5 \text{ in.}}{41.4 \text{ in.}}(675 \text{ kips}) \\
&= 269 \text{ kips}
\end{aligned}
$$

On the gusset-to-beam connection

$$
\begin{aligned}
H_{ub} &= \frac{\alpha}{r} P_u \\
&= \frac{21.8 \text{ in.}}{41.4 \text{ in.}}(675 \text{ kips}) \\
&= 355 \text{ kips}
\end{aligned}
$$

$$
\begin{aligned}
V_{ub} &= \frac{e_b}{r} P_u \\
&= \frac{9.35 \text{ in.}}{41.4 \text{ in.}}(675 \text{ kips}) \\
&= 153 \text{ kips}
\end{aligned}
$$

Design gusset-to-column connection

Try 2L4×4×$^5/_8$×2′-6 welded to the gusset and bolted with 10 rows of $^7/_8$ in. diameter A325-N bolts in standard holes to the column flange.

Calculate tensile force per bolt r_{ut}.

$$
\begin{aligned}
r_{ut} &= \frac{H_{uc}}{n} \\
&= \frac{171 \text{ kips}}{20 \text{ bolts}} \\
&= 8.55 \text{ kips/bolt}
\end{aligned}
$$

Check design strength of bolts for tension-shear interaction.

$$
\begin{aligned}
r_{uv} &= \frac{V_{uc}}{n} \\
&= \frac{269 \text{ kips}}{20 \text{ bolts}} \\
&= 13.5 \text{ kips/bolt} < 21.6 \text{ kips/bolt} \qquad \textbf{o.k.}
\end{aligned}
$$

$$
\begin{aligned}
F_t &= 117 \text{ ksi} - 2.5 f_v \le 90 \text{ ksi} \\
&= 117 \text{ ksi} - 2.5 \left(\frac{13.5 \text{ kips/bolt}}{\frac{\pi}{4}(^7/_8 \text{ in.})^2} \right) \\
&= 60.9 \text{ ksi}
\end{aligned}
$$

$$
\begin{aligned}
\phi r_n &= \phi F_t A_b \\
&= 0.75(60.9 \text{ ksi}) \left[\frac{\pi}{4}(^7/_8 \text{ in.})^2 \right] \\
&= 27.5 \text{ kips} > 8.55 \text{ kips/bolt} \qquad \textbf{o.k.}
\end{aligned}
$$

Check bearing strength at bolt holes.

The bearing strength per bolt is

$$\begin{aligned}\phi r_n &= \phi\,(2.4\,dt\,F_u)\\ &= 0.75\,(2.4 \times {}^7\!/_8\text{ in.} \times {}^5\!/_8\text{ in.} \times 58\text{ ksi})\\ &= 57.1\text{ kips/bolt}\end{aligned}$$

Since this exceeds the single-shear strength of the bolts, bearing strength is **o.k.**

Check prying action.

$$\begin{aligned}b &= g - t/2\\ &= 2^1\!/_2\text{ in.} - {}^5\!/_8\text{ in.}/2\\ &= 2.19\text{ in.}\end{aligned}$$

Note: $1^1\!/_4$ in. entering and tightening clearance accommodated, **o.k.**

$$\begin{aligned}a &= 4\text{ in.} - g\\ &= 4\text{ in.} - 2^1\!/_2\text{ in.}\\ &= 1.5\text{ in.}\end{aligned}$$

Since $a = 1.5$ in. is less than $1.25b = 2.73$ in., use $a = 1.5$ in.

$$\begin{aligned}b' &= b - d/2\\ &= 2.19\text{ in.} - ({}^7\!/_8\text{ in.}/2)\\ &= 1.75\text{ in.}\end{aligned}$$

$$\begin{aligned}a' &= a + d/2\\ &= 1.5\text{ in.} + ({}^7\!/_8\text{ in.}/2)\\ &= 1.94\text{ in.}\end{aligned}$$

$$\begin{aligned}\rho &= \frac{b'}{a'}\\ &= \frac{1.75\text{ in.}}{1.94\text{ in.}}\\ &= 0.902\end{aligned}$$

$$\begin{aligned}\beta &= \frac{1}{\rho}\left(\frac{\phi r_n}{r_{ut}} - 1\right)\\ &= \frac{1}{0.902}\left(\frac{27.5\text{ kips/bolt}}{8.55\text{ kips/bolt}} - 1\right)\\ &= 2.45\end{aligned}$$

Since $\beta \geq 1$, set $\alpha' = 1.0$

$$\begin{aligned}\delta &= 1 - \frac{d'}{p}\\ &= 1 - \frac{{}^{15}\!/_{16}\text{ in.}}{3\text{ in.}}\\ &= 0.688\end{aligned}$$

$$t_{req} = \sqrt{\frac{4.44 r_{ut} b'}{p F_y (1 + \delta\alpha')}}$$

$$= \sqrt{\frac{4.44\ (8.55\ \text{kips/bolt})(1.75\ \text{in.})}{(3\ \text{in.})(36\ \text{ksi})[1 + (0.688)(1.0)]}}$$

$$= 0.604\ \text{in.}$$

Since $t = {}^5\!/_8$ in. > 0.604 in., angles are **o.k.**

Design welds

Try fillet welds around perimeter (3 sides) of both angles.

$$P_{uc} = \sqrt{{H_{uc}}^2 + {V_{uc}}^2}$$

$$= \sqrt{(171\ \text{kips})^2 + (269\ \text{kips})^2}$$

$$= 319\ \text{kips}$$

$$\theta = \tan^{-1}\left(\frac{H_{uc}}{V_{uc}}\right)$$

$$= \tan^{-1}\left(\frac{171\ \text{kips}}{269\ \text{kips}}\right)$$

$$= 32.4°$$

From Table 8-7 with $\theta = 30°$,

$$l = 30\ \text{in.}$$
$$kl = 3^1\!/_2\ \text{in.}$$
$$k = 0.117$$

By interpolation

$$x = 0.011$$
$$xl = 0.011(30\ \text{in.})$$
$$= 0.33\ \text{in.}$$
$$al = 4\ \text{in.} - xl$$
$$= 4\ \text{in.} - 0.33\ \text{in.}$$
$$= 3.67\ \text{in.}$$
$$a = 0.122$$

By interpolation

$$C = 1.95$$

and

$$D_{req} = \frac{P_{uc}}{C C_1 l}$$

$$= \frac{319\ \text{kips}}{1.95 \times 1.0 \times (2\ \text{welds} \times 30\ \text{in.})}$$

$$= 2.73 \rightarrow 3\ \text{sixteenths required for strength}$$

From LRFD Specification Table J2.4, minimum weld size is 1/4 in. Use 1/4 in. fillet welds.

Check gusset thickness (against weld size required for strength)

For two fillet welds

$$t_{min} = \frac{6.19D}{F_u}$$
$$= \frac{6.19(2.73 \text{ sixteenths})}{58 \text{ ksi}}$$
$$= 0.291 \text{ in.} < 3/4 \text{ in.} \quad \textbf{o.k.}$$

Check strength of angles.

Shear yielding (due to V_{uc})

$$\phi R_n = \phi\,(0.60 F_y A_g)$$
$$= 0.90\big[0.60(36 \text{ ksi})\big(2 \times 30 \text{ in.} \times 5/8 \text{ in.}\big)\big]$$
$$= 729 \text{ kips} > 269 \text{ kips} \quad \textbf{o.k.}$$

Similarly, shear yielding of the angles due to H_{uc} is not critical.

Shear rupture

$$\phi R_n = \phi\,(0.60\, F_u\, A_{nv})$$
$$= 0.75\big[0.60(58 \text{ ksi})\big(2 \times 5/8 \text{ in.} \times 30 \text{ in.} - 20 \times 5/8 \text{ in.} \times 1 \text{ in.}\big)\big]$$
$$= 653 \text{ kips} > 269 \text{ kips} \quad \textbf{o.k.}$$

Block shear rupture

From Tables 9-3 and 9-4, with $n = 10$, $L_{ev} = 1^1/_2$ in., and $L_{eh} = 1^1/_2$ in., $0.6F_u\, A_{nv} > F_u\, A_{nt}$. Thus,

$$\phi R_n = \phi\,[0.6F_u\, A_{nv} + F_y\, A_{gt}] \times 2 \text{ blocks}$$
$$\leq \phi\,[0.6F_u\, A_{nv} + F_u\, A_{nt}] \times 2 \text{ blocks}$$
$$= (496 \text{ kips/in.} + 40.5 \text{ kips/in.})\big(5/8 \text{ in.}\big) \times 2 \text{ blocks}$$
$$\leq (496 \text{ kips/in.} + 43.5 \text{ kips/in.})\big(5/8 \text{ in.}\big) \times 2 \text{ blocks}$$
$$= 671 \text{ kips} \leq 674 \text{ kips}$$
$$= 671 \text{ kips} > 269 \text{ kips} \quad \textbf{o.k.}$$

Check column flange.

By inspection, the 4.16 in.-thick column flange has adequate flexural strength, stiffness, and bearing strength.

Design gusset-to-beam connection

$$P_{ub} = \sqrt{H_{ub}{}^2 + V_{ub}{}^2}$$
$$= \sqrt{(355 \text{ kips})^2 + (153 \text{ kips})^2}$$
$$= 387 \text{ kips}$$

From Richard (1986) it is recommended that the design factored force be increased by 40 percent to ensure adequate force redistribution in the weld group and the validity of the Uniform Force Method. Thus,

$$\begin{aligned} D_{req} &= \frac{1.4P_{ub}}{1.392\,l} \\ &= \frac{1.4(387\text{ kips})}{1.392(2 \times 42\text{ in.})} \\ &= 4.63 \rightarrow 5\text{ sixteenths} \end{aligned}$$

(Note that, if a moment existed on this interface, the connection would be designed for the larger of the peak stress and 1.4 times the average stress.)

This is equal to the minimum weld size from LRFD Specification Table J2.4.

Check gusset thickness (against weld size required for strength)

For two fillet welds

$$\begin{aligned} t_{\min} &= \frac{6.19D}{F_u} \\ &= \frac{6.19(2.73\text{ sixteenths})}{58\text{ ksi}} \\ &= 0.291\text{ in.} < 3/4\text{ in.} \qquad \textbf{o.k.} \end{aligned}$$

Check local web yielding of the beam.

$$\begin{aligned} \phi R_n &= \phi\,(N + 2.5\,k)F_{yw}\,t_w \\ &= 1.0\,[2.5\,(1.34\text{ in.}) + 42\text{ in.}](50\text{ ksi})(0.590\text{ in.}) \\ &= 1{,}340\text{ kips} > 153\text{ kips} \qquad \textbf{o.k.} \end{aligned}$$

Design beam-to-column connection

Since the brace may be in tension or compression, the required strength of the beam-to-column connection is as follows. The required shear strength is

$$\begin{aligned} R_{ub} \pm V_{ub} &= 15\text{ kips} \pm 153\text{ kips} \\ &= 168\text{ kips} \end{aligned}$$

and the required axial strength is

$$\begin{aligned} A_{ub} \pm (H_u - H_{ub}) &= 0\text{ kips} \pm (171\text{ kips}) \\ &= 171\text{ kips} \end{aligned}$$

Try 2L8×6×1×1′-2½ (Leg gage = 2¾ in.) welded to the beam web and bolted with five rows of ⅞ in. diameter A325-N bolts in standard holes to the column flange.

Calculate tensile force per bolt r_{ut}.

$$\begin{aligned} r_{ut} &= \frac{171\text{ kips}}{10\text{ bolts}} \\ &= 17.1\text{ kips/bolt} \end{aligned}$$

Check design strength of bolts for tension-shear interaction.

$$r_{uv} = \frac{168 \text{ kips}}{10 \text{ bolts}}$$
$$= 16.8 \text{ kips/bolt} < 21.6 \text{ kips/bolt} \quad \textbf{o.k.}$$
$$F_t = 117 \text{ ksi} - 2.5 f_v \le 90 \text{ ksi}$$
$$= 117 \text{ ksi} - 2.5\left(\frac{16.8 \text{ kips/bolt}}{\frac{\pi}{4}(^7/_8 \text{ in.})^2}\right)$$
$$= 47.2 \text{ ksi}$$
$$\phi r_n = \phi F_t A_b$$
$$= 0.75(47.2 \text{ ksi})\left[\frac{\pi}{4}(^7/_8 \text{ in.})^2\right]$$
$$= 21.3 \text{ kips} > 17.1 \text{ kips/bolt} \quad \textbf{o.k.}$$

Check bearing strength at bolt holes.

Since the strength of each bolt exceeds the single shear strength of the bolts, bearing strength is **o.k.**

Check prying action

$$b = g - t/2$$
$$= 2^3/_4 \text{ in.} - 1 \text{ in.}/2$$
$$= 2^1/_4 \text{ in.}$$

Note: $1^1/_4$ in. entering and tightening clearance is accommodated, **o.k.**

$$a = 6 \text{ in.} - g$$
$$= 6 \text{ in.} - 2^3/_4 \text{ in.}$$
$$= 3^1/_4 \text{ in.}$$

Since $a = 3^1/_4$ in. exceeds $1.25b = 2.81$ in., use $a = 2.81$ in. for calculation purposes.

$$b' = b - d/2$$
$$= 2^1/_4 \text{ in.} - (^7/_8 \text{ in.}/2)$$
$$= 1.81 \text{ in.}$$
$$a' = a + d/2$$
$$= 2.81 \text{ in.} + (^7/_8 \text{ in.}/2)$$
$$= 3.24 \text{ in.}$$
$$\rho = \frac{b'}{a'}$$
$$= \frac{1.81 \text{ in.}}{3.24 \text{ in.}}$$
$$= 0.559$$

$$\begin{aligned}
\beta &= \frac{1}{\rho}\left(\frac{\phi r_n}{r_{ut}} - 1\right) \\
&= \frac{1}{0.559}\left(\frac{21.3 \text{ kips/bolt}}{17.1 \text{ kips/bolt}} - 1\right) \\
&= 0.439
\end{aligned}$$

$$\begin{aligned}
\delta &= 1 - \frac{d'}{p} \\
&= 1 - \frac{{}^{15}/_{16} \text{ in.}}{2.90 \text{ in.}} \\
&= 0.677
\end{aligned}$$

$$\begin{aligned}
\alpha' &= \frac{1}{\delta}\left(\frac{\beta}{1-\beta}\right) \leq 1.0 \\
&= \frac{1}{0.677}\left(\frac{0.439}{1-0.439}\right) \leq 1.0 \\
&= 1.16 \leq 1.0 \\
&= 1.0
\end{aligned}$$

$$\begin{aligned}
p &= \frac{14^1/_2 \text{ in.}}{5 \text{ bolts}} \\
&= 2.90 \text{ in./bolt}
\end{aligned}$$

$$\begin{aligned}
t_{req} &= \sqrt{\frac{4.44 r_{ut} b'}{p F_y (1 + \delta\alpha')}} \\
&= \sqrt{\frac{4.44(17.1 \text{ kips/bolt})(1.81 \text{ in.})}{(2.90 \text{ in.})(36 \text{ ksi})[1 + (0.677)(1.0)]}} \\
&= 0.886 \text{ in.}
\end{aligned}$$

Since $t = 1$ in. > 0.886 in., angles are **o.k.**

Design welds

Try fillet welds around perimeter (three sides) of both angles.

$$\begin{aligned}
P_{uc} &= \sqrt{(171 \text{ kips})^2 + (168 \text{ kips})^2} \\
&= 240 \text{ kips}
\end{aligned}$$

$$\begin{aligned}
\theta &= \tan^{-1}\left(\frac{171 \text{ kips}}{168 \text{ kips}}\right) \\
&= 45.5°
\end{aligned}$$

From Table 8-7 with $\theta = 45°$

$$\begin{aligned}
l &= 14^1/_2 \text{ in.} \\
kl &= 7^1/_2 \text{ in.} \\
k &= 0.517
\end{aligned}$$

By interpolation

$$x = 0.132$$
$$xl = 0.132(14^1/_2 \text{ in.})$$
$$= 1.91 \text{ in.}$$
$$al = 8 \text{ in.} - xl$$
$$= 8 \text{ in.} - 1.91 \text{ in.}$$
$$= 6.09 \text{ in.}$$

$$a = 0.420$$

By interpolation

$$C = 2.66$$

and

$$D_{req} = \frac{P_{uc}}{CC_1 l}$$
$$= \frac{240 \text{ kips}}{2.66 \times 1.0 \times (2 \text{ welds} \times 14^1/_2 \text{ in.})}$$
$$= 3.11 \rightarrow 4 \text{ sixteenths required for strength}$$

From LRFD Specification Table J2.4, minimum weld size is $^5/_{16}$ in. Use $^5/_{16}$in. fillet welds.

Check beam web thickness (against weld size required for strength)

For two fillet welds,

$$t_{\min} = \frac{6.19D}{F_u}$$
$$= \frac{6.19(3.11 \text{ sixteenths})}{65 \text{ ksi}}$$
$$= 0.296 \text{ in.} < 0.590 \text{ in.} \quad \textbf{o.k.}$$

Check the strength of angles

Shear yielding

$$\phi R_n = \phi(0.60 F_y A_g)$$
$$= 0.9[0.60(36 \text{ ksi})(2 \times 14^1/_2 \text{ in.} \times 1 \text{ in.})]$$
$$= 564 \text{ kips} > 168 \text{ kips} \quad \textbf{o.k.}$$

Similarly, shear yielding of the angles due to H_{uc} is not critical.

Shear rupture

$$\phi R_n = \phi(0.60 F_u A_{nv})$$
$$= 0.75[0.60(58 \text{ ksi})(2 \times 1 \text{ in.} \times 14^1/_2 \text{ in.} - 10 \times 1 \text{ in.} \times 1 \text{ in.})]$$
$$= 496 \text{ kips} > 168 \text{ kips} \quad \textbf{o.k.}$$

Block shear rupture

With $n = 5$, $L_{ev} = 1^1/_4$ in., $L_{eh} = 3^1/_4$ in., $0.6F_uA_{nv} > F_uA_{nt}$. Thus

$$\begin{aligned}\phi R_n &= \phi[0.6F_uA_{nv} + F_yA_{gt}] \times 2 \text{ blocks}\\ &\leq \phi[0.6F_uA_{nv} + F_uA_{nt}] \times 2 \text{ blocks}\end{aligned}$$

By inspection, the first part of this equation governs, and

$$\begin{aligned}\phi R_n &= 0.75\big[0.6(58 \text{ ksi})\big(13^1/_4 \text{ in.} - 5 \times 1 \text{ in.}\big)(1 \text{ in.})\\ &\quad + (36 \text{ ksi})\big(3^1/_4 \text{ in.}\big)(1 \text{ in.})\big](2)\\ &= 606 \text{ kips} > 168 \text{ kips} \qquad \textbf{o.k.}\end{aligned}$$

Check column flange.

By inspection, the 4.16 in.-thick column flange has adequate flexural strength, stiffeners, and bearing strength.

Comments: Were the brace in compression, the buckling strength of the gusset would have to be checked, where

$$\phi R_n = \phi_c F_{cr} A_w$$

In the above equation, $\phi_c F_{cr}$ may be determined from $\frac{kl_1}{r}$ with LRFD Specification Table C-36, where l_1 is the perpendicular distance from the Whitmore section to the interior corner of the gusset. Alternatively, the average value of

$$\frac{l_1 + l_2 + l_3}{3}$$

may be substituted (AISC, 1984), where these quantities are illustrated in Figure 13-11. Note that, for this example, l_2 is negative since part of the Whitmore section is in the beam web.

The effective length factor K has been established as 0.5 by full scale tests on bracing connections (Gross, 1990). It assumes that the gusset is supported on both edges as is the case in Figure 13-11. In cases where the gusset is supported on one edge only, such as that illustrated in Figure 13-12d (and possibly Figure 13-12a), the brace can more readily move out-of-plane and a sidesway mode of buckling can occur in the gusset. For this case, K should be taken as 1.2.

EXAMPLE 13.2. Bracing connection design.

Given: Refer to Figure 13-12. Each of the four designs shown for the diagonal bracing connection between the W14×68 brace, W24×55 beam, and W14×211 column web have been developed using the Uniform Force Method (the General Case and Special Cases 1, 2, and 3) for the load case of $1.2D + 1.3W$. Refer the AISC (1992) for the unfactored loads and complete designs. For the given values of α and β, determine the interface forces on the gusset-to-column and gusset-to-beam connections for

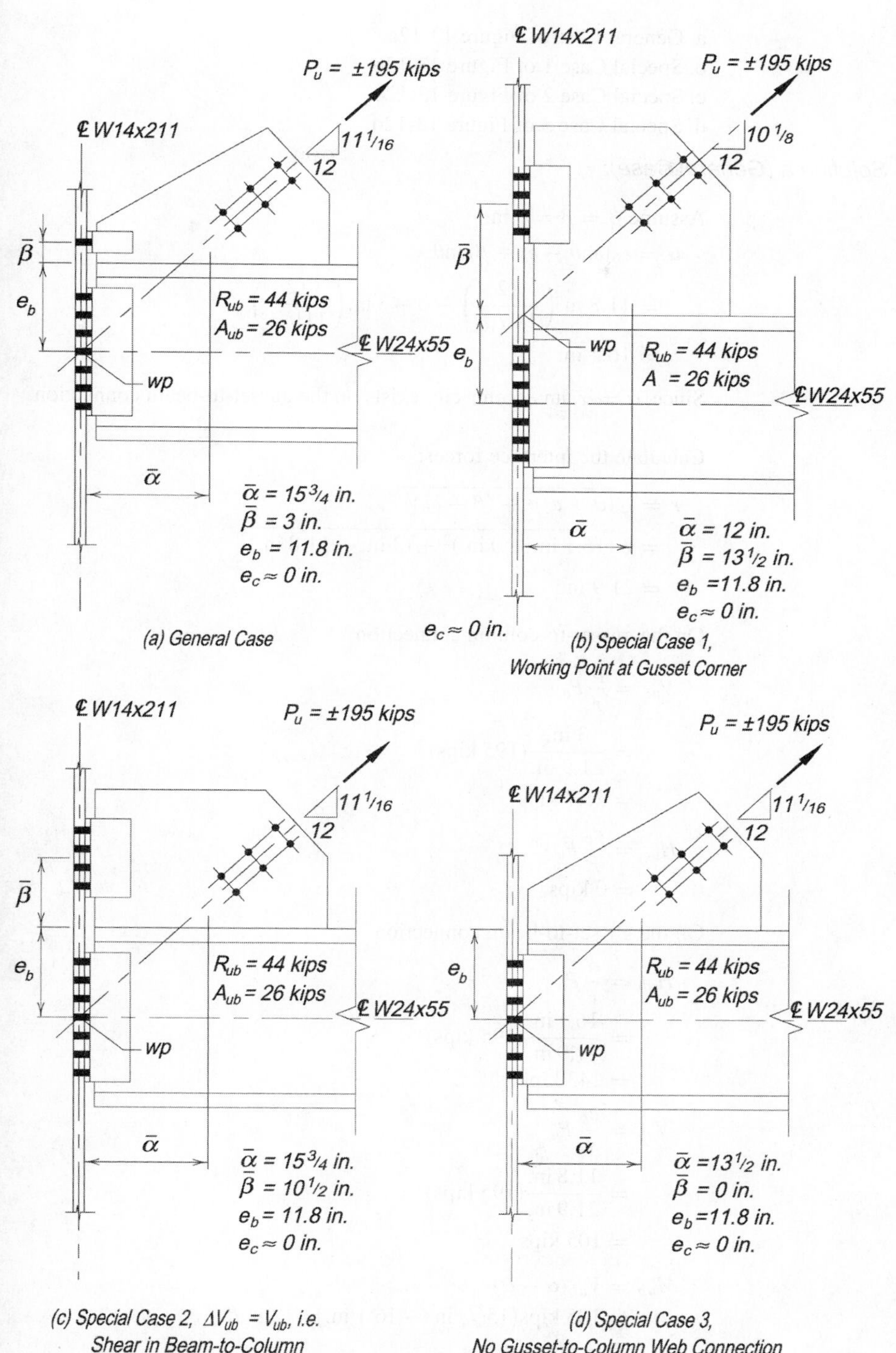

Fig. 13–12. Illustration for Example 13.2.

a. General Case of Figure 13-12a.
b. Special Case 1 of Figure 13-12b.
c. Special Case 2 of Figure 13-12c.
d. Special Case 3 of Figure 13-12d.

Solution a (General Case):

Assume $\beta = \bar{\beta} = 3$ in.

$$\begin{aligned}\alpha &= e_b \tan\theta - e_c + \beta\tan\theta \\ &= 11.8\text{ in.}\left(\frac{12}{11^1/_{16}}\right) - 0 + 3\text{ in.}\left(\frac{12}{11^1/_{16}}\right) \\ &= 16.1\text{ in.}\end{aligned}$$

Since, $\alpha \neq \overline{\alpha}$, an eccentricity exists on the gusset-to-beam connection.

Calculate the interface forces:

$$\begin{aligned}r &= \sqrt{(\alpha + e_c)^2 + (\beta + e_b)^2} \\ &= \sqrt{(16.1\text{ in.} + 0\text{ in.})^2 + (3\text{ in.} + 11.8\text{ in.})^2} \\ &= 21.9\text{ in.}\end{aligned}$$

On the gusset-to-column connection

$$\begin{aligned}V_{uc} &= \frac{\beta}{r}P_u \\ &= \frac{3\text{ in.}}{21.9\text{ in.}}(195\text{ kips}) \\ &= 26.7\text{ kips}\end{aligned}$$

$$\begin{aligned}H_{uc} &= \frac{e_c}{r}P_u \\ &= 0\text{ kips}\end{aligned}$$

On the gusset-to-beam connection

$$\begin{aligned}H_{ub} &= \frac{\alpha}{r}P_u \\ &= \frac{16.1\text{ in.}}{21.9\text{ in.}}(195\text{ kips}) \\ &= 143\text{ kips}\end{aligned}$$

$$\begin{aligned}V_{ub} &= \frac{e_b}{r}P_u \\ &= \frac{11.8\text{ in.}}{21.9\text{ in.}}(195\text{ kips}) \\ &= 105\text{ kips}\end{aligned}$$

$$\begin{aligned}M_{ub} &= V_{ub}(\alpha - \bar{\alpha}) \\ &= 105\text{ kips}\big(15^3/_4\text{ in.} - 16.1\text{ in.}\big)/(12\text{ in./ft}) \\ &= -3.06\text{ kip-ft}\end{aligned}$$

In this case, this small moment is negligible.

On the beam-to-column connection, the factored shear is

$$\begin{aligned} R_{ub} + V_{ub} &= 44 \text{ kips} + 105 \text{ kips} \\ &= 149 \text{ kips} \end{aligned}$$

and the factored axial force is

$$\begin{aligned} A_{ub} \pm H_{uc} &= 26 \text{ kips} \pm 0 \text{ kips} \\ &= 26 \text{ kips} \end{aligned}$$

For a discussion of the sign to use between A_{ub} and H_{uc}, refer to AISC (1992).

Solution b (Special Case 1):

In this case, the centroidal positions of the gusset-edge connections are irrelevant; $\bar{\alpha}$ and $\bar{\beta}$ are given to define the geometry of the connection, but are not needed to determine the gusset edge forces.

The angle of the brace from the vertical is

$$\begin{aligned} \theta &= \tan^{-1}\left(\frac{12}{10^1/_8}\right) \\ &= 49.8^\circ \end{aligned}$$

The horizontal component of the brace force is

$$\begin{aligned} H_u &= P_u \sin\theta \\ &= 195 \text{ kips} \times \sin(49.8^\circ) \\ &= 149 \text{ kips} \end{aligned}$$

and the vertical component of the brace force is

$$\begin{aligned} V_u &= P_u \cos\theta \\ &= 195 \text{ kips} \times \sin(49.8^\circ) \\ &= 126 \text{ kips} \end{aligned}$$

On the gusset-to-column connection

$$\begin{aligned} V_{uc} &= V_u = 126 \text{ kips} \\ H_{uc} &= 0 \text{ kips} \end{aligned}$$

On the gusset-to-beam connection

$$\begin{aligned} V_{ub} &= 0 \text{ kips} \\ H_{ub} &= H_u = 149 \text{ kips} \end{aligned}$$

On the beam-to-column connection

$$\begin{aligned} R_{ub} &= 44 \text{ kips (shear)} \\ A_{ub} &= 26 \text{ kips (axial transfer force)} \end{aligned}$$

In addition to the forces on the connection interfaces, the beam is subjected to a moment M_{ub} (see Figure 13-3d), where

$$\begin{aligned} M_{ub} &= H_{ub}e_b \\ &= 149 \text{ kips} \times 11.8 \text{ in./12 in./ft} \\ &= 147 \text{ kips-ft} \end{aligned}$$

This moment, as well as the beam axial load $H_u = 149$ kips and the moment and shear in the beam associated with the end reaction R_{ub}, must be considered in the design of the beam.

Solution c (Special Case 2):

Assume $\beta = \bar{\beta} = 10^1/_2$ in.

$$\begin{aligned}\alpha &= e_b \tan\theta - e_c + \beta\tan\theta \\ &= 11.8 \text{ in.}\left(\frac{12}{11^1/_{16}}\right) - 0 + 10^1/_2 \text{ in.}\left(\frac{12}{11^1/_{16}}\right) \\ &= 24.2 \text{ in.}\end{aligned}$$

Calculate the interface forces for the general case before applying Special Case 2.

$$\begin{aligned}r &= \sqrt{(\alpha + e_c)^2 + (\beta + e_b)^2} \\ &= \sqrt{(24.2 \text{ in.} + 0 \text{ in.})^2 + (10^1/_2 \text{ in.} + 11.8 \text{ in.})^2} \\ &= 32.9 \text{ in.}\end{aligned}$$

On the gusset-to-beam connection

$$\begin{aligned}H_{ub} &= \frac{\alpha}{r} P_u \\ &= \frac{24.2 \text{ in.}}{32.9 \text{ in.}}(195 \text{ kips}) \\ &= 143 \text{ kips}\end{aligned}$$

$$\begin{aligned}V_{ub} &= \frac{e_b}{r} P_u \\ &= \frac{11.8 \text{ in.}}{32.9 \text{ in.}}(195 \text{ kips}) \\ &= 69.9 \text{ kips}\end{aligned}$$

On the gusset-to-column connection

$$\begin{aligned}H_{uc} &= \frac{e_c}{r} P_u \\ &= 0 \text{ kips}\end{aligned}$$

$$\begin{aligned}V_{uc} &= \frac{\beta}{r} P_u \\ &= \frac{10.5 \text{ in.}}{32.9 \text{ in.}}(195 \text{ kips}) \\ &= 62.2 \text{ kips}\end{aligned}$$

On the beam-to-column connection, the factored shear is

$$\begin{aligned}R_{ub} + V_{ub} &= 44.0 \text{ kips} + 66.9 \text{ kips} \\ &= 111 \text{ kips}\end{aligned}$$

and the factored axial force is

$$\begin{aligned}A_{ub} \pm H_{uc} &= 26.0 \text{ kips} \pm 0 \text{ kips} \\ &= 26.0 \text{ kips}\end{aligned}$$

Next, applying Special Case 2 with $\Delta V_{ub} = V_{ub} = 69.9$ kips, calculate the interface forces.

On the gusset-to-beam connection (where V_{ub} is replaced by $V_{ub} - \Delta V_{ub}$)

$$\begin{aligned} H_{ub} &= 143 \text{ kips (unchanged)} \\ V_{ub} &= 69.9 \text{ kips} - 69.9 \text{ kips} \\ &= 0 \text{ kips} \\ M_{ub} &= (\Delta V_{ub})\alpha \\ &= (69.9 \text{ kips})(24.2 \text{ in.})/(12 \text{ in./ft}) \\ &= 141 \text{ kips-ft} \end{aligned}$$

On the gusset-to-column connection (where V_{uc} is replaced by $V_{uc} + \Delta V_{ub}$)

$$\begin{aligned} H_{uc} &= 0 \text{ kips (unchanged)} \\ V_{uc} &= 62.2 \text{ kips} + 69.9 \text{ kips} \\ &= 132 \text{ kips} \end{aligned}$$

On the beam-to-column connection, the factored shear is

$$\begin{aligned} R_{ub} + \Delta V_{ub} - \Delta V_{ub} &= 44 \text{ kips} + 69.9 \text{ kips} - 69.9 \text{ kips} \\ &= 44 \text{ kips} \end{aligned}$$

and the factored axial force is

$$\begin{aligned} A_{ub} \pm H_{uc} &= 26 \text{ kips} \pm 0 \text{ kips} \\ &= 26 \text{ kips} \end{aligned}$$

Solution d (Special Case 3):

Assume $\beta = \bar{\beta} = 0$ in.

$$\begin{aligned} \alpha &= e_b \tan\theta \\ &= 11.8 \text{ in.}\left(\frac{12}{11^1/_{16}}\right) \\ &= 12.8 \text{ in.} \end{aligned}$$

Since, $\alpha \neq \bar{\alpha}$, an eccentricity exists on the gusset-to-beam connection.

Calculate the interface forces.

$$\begin{aligned} r &= \sqrt{\alpha^2 + e_b{}^2} \\ &= \sqrt{(12.8 \text{ in.})^2 + (11.8 \text{ in.})^2} \\ &= 17.4 \text{ in.} \end{aligned}$$

On the gusset-to-beam connection

$$\begin{aligned} H_{ub} &= \frac{\alpha}{r} P_u \\ &= \frac{12.8 \text{ in.}}{17.4 \text{ in.}}(195 \text{ kips}) \\ &= 143 \text{ kips} \end{aligned}$$

$$\begin{aligned} V_{ub} &= \frac{e_b}{r} P_u \\ &= \frac{11.8 \text{ in.}}{17.4 \text{ in.}}(195 \text{ kips}) \\ &= 132 \text{ kips} \end{aligned}$$

$$\begin{aligned} M_{ub} &= V_{ub}(\alpha - \bar{\alpha}) \\ &= 132 \text{ kips}\left(12.8 \text{ in.} - 13^1/_2 \text{ in.}\right) \\ &= -7.70 \text{ kip-ft} \end{aligned}$$

In this case, this small moment is negligible.

On the beam-to-column connection, the factored shear is

$$\begin{aligned} R_{ub} + V_{ub} &= 44 \text{ kips} + 132 \text{ kips} \\ &= 176 \text{ kips} \end{aligned}$$

and the factored axial force is

$$\begin{aligned} A_{ub} \pm H_{uc} &= 26 \text{ kips} \pm 0 \text{ kips} \\ &= 26 \text{ kips} \end{aligned}$$

Comments: From the foregoing results, designs by Special Case 3 and the General Case of the Uniform Force Method provide more economical designs. Additionally, note that designs by Special Case 1 and Special Case 2 result in moments on the beam and/or column which must be considered.

EXAMPLE 13.3. Truss panel-point connection design

Given: Refer to Figure 13-13. Determine the requirements for the following cases:

a. joint L_1
b. joint U_1

Assume 70 ksi electrodes. For the WT truss chord, assume $F_y = 50$ ksi and $F_u = 65$ ksi. For angle and splice material, assume $F_y = 36$ ksi and $F_u = 58$ ksi.

Solution a: *Check shear yield of the tee stem (on Section A-A)*

$$\begin{aligned} \phi R_n &= \phi(0.6 F_y A_w) \\ &= 0.90(0.6 \times 50 \text{ ksi} \times 8.215 \text{ in.} \times 0.430 \text{ in.}) \\ &= 95.4 \text{ kips} < 104 \text{ kips} \qquad \textbf{n.g.} \end{aligned}$$

Additional shear area must be provided.

Try PL$^7/_{16}$ in.×4 in. complete-joint-penetration groove welded to the stem of the WT.

$$\begin{aligned} \phi R_n &= 95.4 \text{ kips} + 0.9\left(0.6 \times 36 \text{ ksi} \times 4 \text{ in.} \times {}^7/_{16} \text{ in.}\right) \\ &= 129 \text{ kips} > 104 \text{ kips} \qquad \textbf{o.k.} \end{aligned}$$

Design welds for member U_1L_1

The minimum weld size from LRFD Specification Table J2.4 is $^3/_{16}$ in. The maximum weld size cannot exceed $^5/_{16}$ in. Calculate the minimum length of $^3/_{16}$ in. fillet weld:

$$L_{min} = \frac{R_n}{1.392D}$$

$$= \frac{104\text{ kips}}{1.392(3\text{ sixteenths})}$$

$$= 24.9\text{ in.}$$

Use $6^1/_2$ in. of $^3/_{16}$ in. weld at the heel and toe of both angles for a total of 26 inches.

Design welds for member U_0L_1

The minimum weld size from LRFD Specification Table J2.4 is $^3/_{16}$ in. The maximum weld size cannot exceed $^1/_4$ in. Calculate the minimum length of $^3/_{16}$ in. fillet weld:

$$L_{min} = \frac{R_u}{1.392D}$$

$$= \frac{166\text{ kips}}{1.392(3\text{ sixteenths})}$$

$$= 39.8\text{ in.}$$

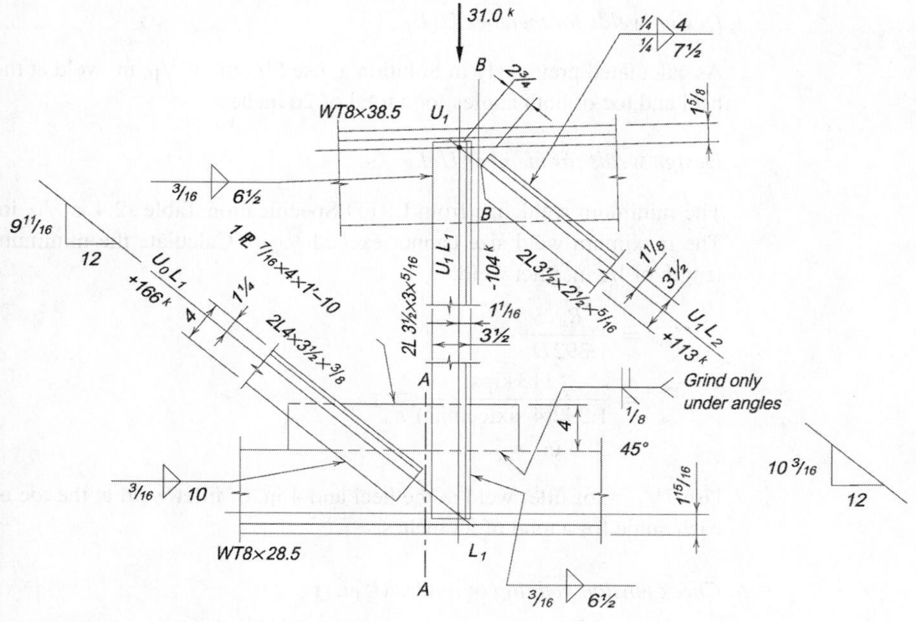

Fig. 13–13. Illustration for Example 13.3.

Use 10 in. of $^3/_{16}$ in. weld at the heel and toe of both angles for a total of 40 inches.

Check tension yielding of angles (U_0L_1)

$$\begin{aligned}\phi R_n &= \phi F_y A_g \\ &= 0.9(36\text{ ksi})(2 \times 2.67\text{ in.}^2) \\ &= 173\text{ kips} > 166\text{ kips} \qquad \textbf{o.k.}\end{aligned}$$

Check tension rupture of angles (U_0L_1)

From LRFD Specification Section B3,

$$\begin{aligned}U &= 1 - \frac{\bar{x}}{L} \le 0.9 \\ &= 1 - \frac{1.21\text{ in.}}{10\text{ in.}} \\ &= 0.879\end{aligned}$$

$$\begin{aligned}\phi R_n &= \phi F_u A_n \\ &= 0.75(58\text{ ksi})(0.879 \times 2 \times 2.67\text{ in.}^2) \\ &= 204\text{ kips} > 166\text{ kips} \qquad \textbf{o.k.}\end{aligned}$$

Solution b: *Check shear yielding of the tee stem (on Section B-B)*

$$\begin{aligned}\phi R_n &= \phi(0.6 F_y A_w) \\ &= 0.90(0.6 \times 50\text{ ksi} \times 8.26\text{ in.} \times 0.455) \\ &= 101\text{ kips} > 73.1\text{ kips} \qquad \textbf{o.k.}\end{aligned}$$

Design welds for member U_1L_1

As calculated previously in Solution a, use $6^1/_2$ in. of $^3/_{16}$ in. weld at the heel and toe of both angles for a total of 26 inches.

Design welds for member U_1L_2

The minimum weld size from LRFD Specification Table J2.4 is $^3/_{16}$ in. The maximum weld size cannot exceed $^1/_4$ in. Calculate the minimum length of $^1/_4$ in. fillet weld:

$$\begin{aligned}L_{\min} &= \frac{R_n}{1.392D} \\ &= \frac{113\text{ kips}}{1.392(4\text{ sixteenths})} \\ &= 20.3\text{ in.}\end{aligned}$$

Use $7^1/_2$ in. of fillet weld at the heel and 4 in. of fillet weld at the toe of each angle for a total of 23 inches.

Check tension yielding of angles (U_1L_2)

$$\begin{aligned}\phi R_n &= \phi F_y A_g \\ &= 0.9(36\text{ ksi})(2 \times 1.78\text{ in.}^2) \\ &= 115\text{ kips} > 113\text{ kips} \qquad \textbf{o.k.}\end{aligned}$$

Check tension rupture of angles (U_1L_2)

From LRFD Specification Section B3,

$$U = 1 - \frac{\bar{x}}{L} \le 0.9$$
$$= 0.802$$

Thus,

$$A_e = UA_g$$
$$= 0.802(2 \times 1.78 \text{ in.}^2)$$
$$= 2.86 \text{ in.}^2$$

From LRFD Specification Section J5.2,

$$\phi R_n = \phi F_u A_e$$
$$= 0.75(58 \text{ ksi})(2.86 \text{ in.}^2)$$
$$= 124 \text{ kips} > 113 \text{ kips} \quad \textbf{o.k.}$$

Check block shear rupture

Because of the cut end of the angle, the block shear rupture model presented in Part 9 does not directly apply. Conservatively, the block shear rupture strength will be based on the shear rupture strength of the WT stem along the length of the welds. Thus, the design strength is:

$$\phi R_n = \phi(0.6F_u A_w)$$
$$= 0.90[0.6 \times 50 \text{ ksi} \times (7.5 \text{ in.} + 4 \text{ in.}) \times 0.455 \text{ in.}]$$
$$= 141 \text{ kips} > 113 \text{ kips} \quad \textbf{o.k.}$$

EXAMPLE 13.4. Truss support connection design.

Given: Refer to Figure 13-14. Determine:
a. the connection requirements between the gusset and column,
b. the required gusset size and the weld requirements for member U_0L_1 at the gusset.

Solution a: *Design bolts connecting angles to column (shear and tension)*

From Table 7-10, the number of 7/8 in. diameter A325-N bolts required for shear only is

$$n_{\min} = \frac{R_u}{\phi r_n}$$
$$= \frac{111 \text{ kips}}{21.6 \text{ kips/bolt}}$$
$$= 5.14 \rightarrow 6 \text{ bolts}$$

Assuming an angle thickness of 5/8 in., bearing is not critical. For a trial calculation, the number of bolts was increased to 12 in pairs at 3 in. spacing; the flexural strength of the angles was found to be insufficient.

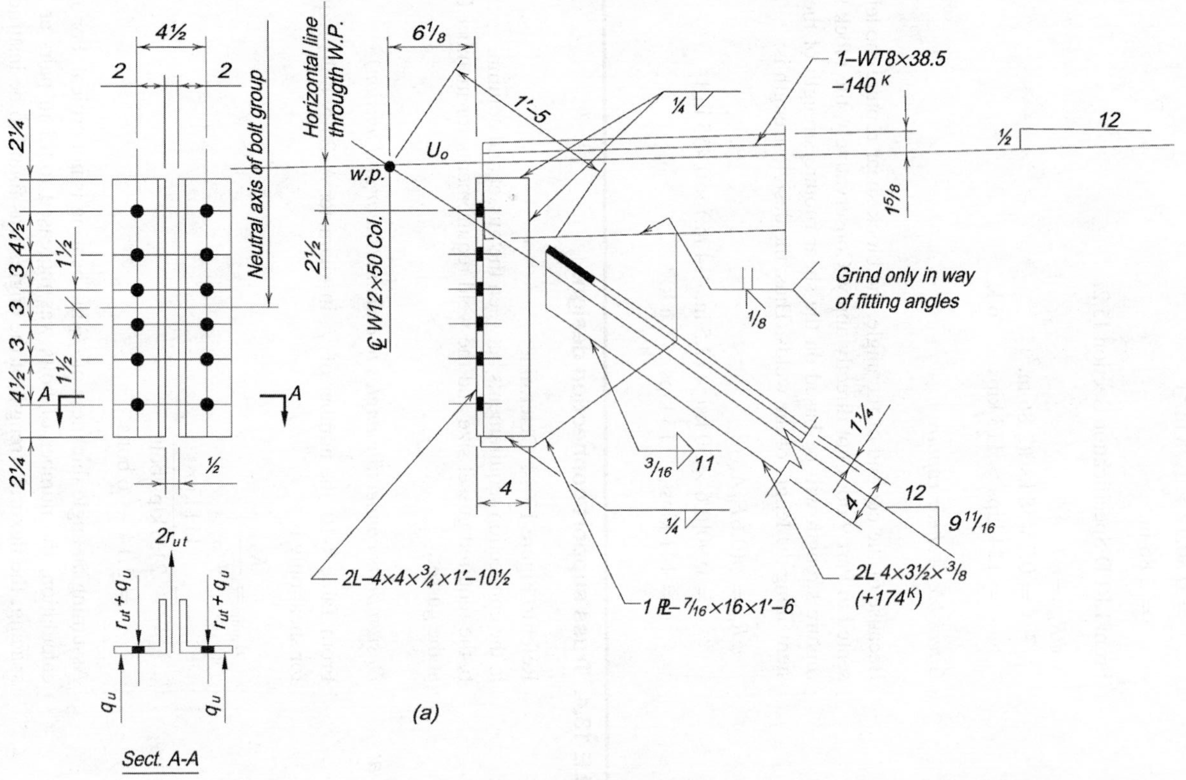

Fig. 13–14. Illustration for Example 13.4.

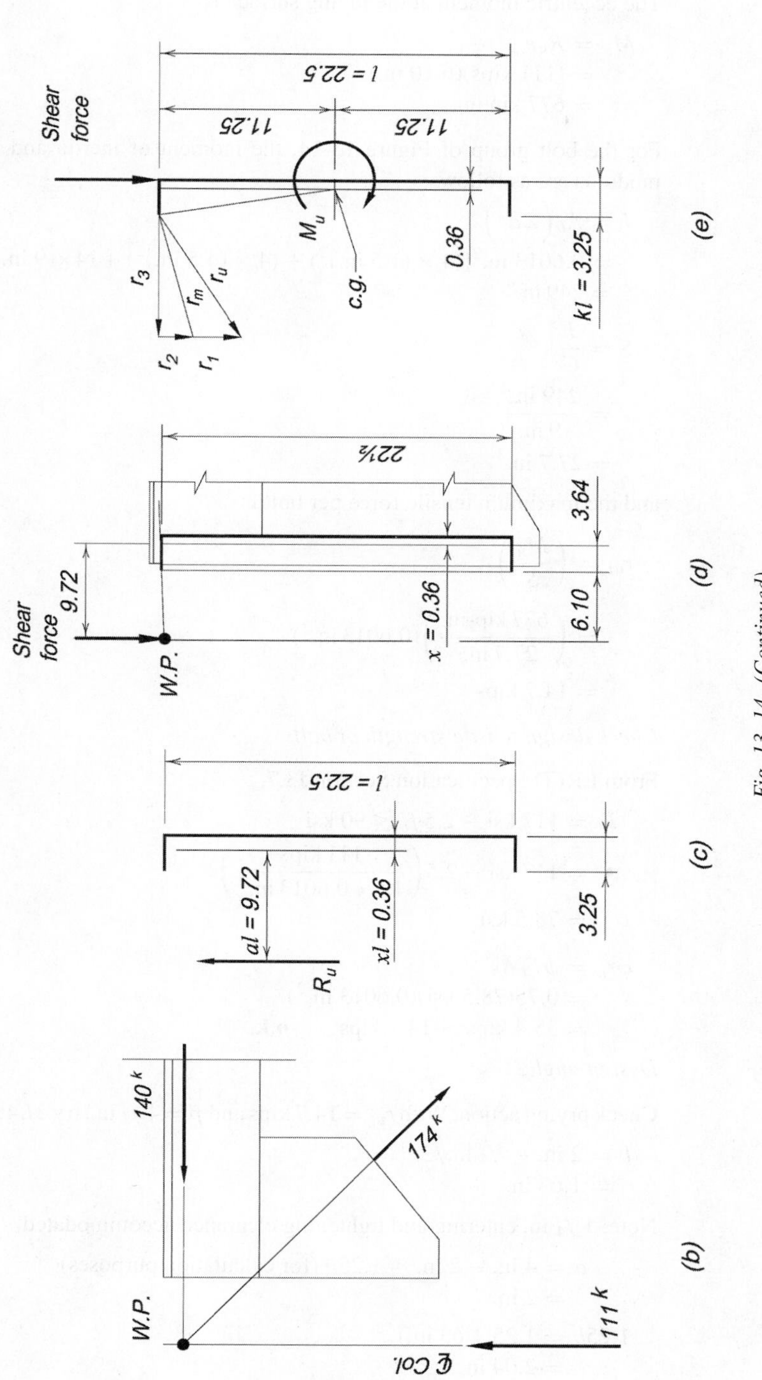

Fig. 13–14. (Continued)

Subsequently, the spacing was revised to $4^1/_2$ in. between the two rows of bolts at the top and bottom of the connection as illustrated in Figure 13-14.

The eccentric moment at the faying surface is

$$\begin{aligned} M_u &= R_u e \\ &= (111 \text{ kips})(6.10 \text{ in.}) \\ &= 677 \text{ kip in.} \end{aligned}$$

For the bolt group of Figure 13-14, the moment of inertia and section modulus are as follows.

$$\begin{aligned} I &= A_b\left(\Sigma d^2\right) \\ &= 0.6013 \text{ in.}^2[(4 \times (1.5 \text{ in.})^2) + (4 \times (4.5 \text{ in.})^2) + (4 \times (9 \text{ in.})^2)] \\ &= 249 \text{ in.}^4 \end{aligned}$$

$$\begin{aligned} S &= \frac{I}{C} \\ &= \frac{249 \text{ in.}^4}{9 \text{ in.}} \\ &= 27.7 \text{ in.}^3 \end{aligned}$$

and the maximum tensile force per bolt is

$$\begin{aligned} r_{ut} &= \left(\frac{M_u}{S}\right)A \\ &= \left(\frac{677 \text{ kip-in.}}{27.7 \text{ in.}^3}\right)(0.6013 \text{ in.}^2) \\ &= 14.7 \text{ kips} \end{aligned}$$

Check design tensile strength of bolts

From LRFD Specification Section J3.7,

$$\begin{aligned} F_t &= 117 \text{ ksi} - 2.5 f_v \leq 90 \text{ ksi} \\ &= 117 \text{ ksi} - 2.5\left(\frac{111 \text{ kips}}{12 \times 0.6013 \text{ in.}^2}\right) \\ &= 78.5 \text{ ksi} \end{aligned}$$

$$\begin{aligned} \phi r_n &= \phi F_t A_b \\ &= 0.75(78.5 \text{ ksi})(0.6013 \text{ in.}^2) \\ &= 35.4 \text{ kips} > 14.7 \text{ kips} \qquad \textbf{o.k.} \end{aligned}$$

Design angles

Check prying action. With $r_{ut} = 14.7$ kips and $p = 4^1/_2$ in., try $2L4\times4\times{}^3/_4$

$$\begin{aligned} b &= 2 \text{ in.} - {}^3/_4 \text{ in.}/2 \\ &= 1.63 \text{ in.} \end{aligned}$$

Note: $1^1/_4$ in. entering and tightening clearance accommodated, **o.k.**

$$\begin{aligned} a &= 4 \text{ in.} - 2 \text{ in.} \leq 1.25b \text{ (for calculation purposes)} \\ &= 2 \text{ in.} \end{aligned}$$

$$\begin{aligned} 1.25b &= 1.25(1.63 \text{ in.}) \\ &= 2.04 \text{ in.} \end{aligned}$$

Thus, use $a = 2$ in.

$$\begin{aligned} b' &= b - d/2 \\ &= 1.63\text{ in.} - (^7/_8\text{ in.}/2) \\ &= 1.19\text{ in.} \end{aligned}$$

$$\begin{aligned} a' &= a + d/2 \\ &= 2\text{ in.} + (^7/_8\text{ in.}/2) \\ &= 2.44\text{ in.} \end{aligned}$$

$$\begin{aligned} \rho &= \frac{b'}{a'} \\ &= \frac{1.19\text{ in.}}{2.44\text{ in.}} \\ &= 0.488 \end{aligned}$$

$$\begin{aligned} \beta &= \frac{1}{\rho}\left(\frac{\phi r_n}{r_{ut}} - 1\right)1 \\ &= \frac{1}{0.488}\left(\frac{35.4\text{ kips/bolt}}{14.7\text{ kips/bolt}} - 1\right) \\ &= 2.89 \end{aligned}$$

Since $\beta \geq 1$, set $\alpha' = 1.0$

$$\begin{aligned} \delta &= 1 - \frac{d'}{p} \\ &= 1 - \frac{^{15}/_{16}\text{ in.}}{3\text{ in.}} \\ &= 0.688 \end{aligned}$$

$$\begin{aligned} t_{req} &= \sqrt{\frac{4.44 r_{ut} b'}{p F_y (1 + \delta\alpha')}} \\ &= \sqrt{\frac{4.44(14.7\text{ kips/bolt})(1.19\text{ in.})}{(3\text{ in.})(36\text{ ksi})[1 + (0.688)(1.0)]}} \\ &= 0.652\text{ in.} \end{aligned}$$

Since $t = {}^3/_4$ in. > 0.652 in., angles are **o.k.**

Check shear yielding of the angles.

$$\begin{aligned} \phi R_n &= \phi(0.6F_y)A_g \\ &= 0.9(0.6 \times 36\text{ ksi})(2 \times 22.5\text{ in.} \times {}^3/_4\text{ in.}) \\ &= 656\text{ kips} > 111\text{ kips} \quad \textbf{o.k.} \end{aligned}$$

Check shear rupture of the angles.

$$\begin{aligned} \phi R_n &= \phi(0.6F_u)A_n \\ &= 0.75(0.6 \times 58\text{ ksi})[2 \times (22.5\text{ in.} - 6 \times 1\text{ in.}) \times {}^3/_4\text{ in.}] \\ &= 646\text{ kips} > 111\text{ kips} \quad \textbf{o.k.} \end{aligned}$$

Check block shear rupture of the angles.

$$A_{gv} = 2 \times \left(22^1/_2 \text{ in.} - 2^1/_4 \text{ in.}\right)\left(^3/_4 \text{ in.}\right) = 30.4 \text{ in.}^2$$

$$A_{gt} = 2 \times \left(2 \text{ in.} \times {}^3/_4 \text{ in.}\right) = 3.00 \text{ in.}^2$$

$$A_{nv} = 30.4 \text{ in.}^2 - 2 \times \left[5.5(1 \text{ in.})(^3/_4 \text{ in.})\right] = 22.2 \text{ in.}^2$$

$$A_{nt} = 3.00 \text{ in.}^2 - 2 \times \left[0.5(1 \text{ in.})(^3/_4 \text{ in.})\right] = 2.25 \text{ in.}^2$$

Since $0.6F_uA_{nv} > F_uA_{nt}$,

$$\phi R_n = \phi[0.6F_uA_{nv} + F_yA_{gt}] = 0.75[0.6(58 \text{ ksi})(22.2 \text{ in.}^2) + (36 \text{ ksi})(3.00 \text{ in.}^2)] = 660 \text{ kips} > 111 \text{ kips} \quad \textbf{o.k.}$$

Use $2L4 \times 4 \times {}^3/_4$

Design angle-to-gusset connection

From LRFD Specification Table J2.4, the minimum weld size is $^1/_4$ in.

From Table 8-7 with $\theta = 0°$

$$kl = 3.25 \text{ in.}$$

$$l = 22.5 \text{ in.}$$

$$k = \frac{3.25 \text{ in.}}{22.5 \text{ in.}} = 0.144$$

by interpolation, $x = 0.017$ and

$$al + xl = 10.1 \text{ in.}$$

$$a = [10.1 \text{ in.} - 0.017(22.5 \text{ in.})]/22.5 \text{ in.} = 0.432$$

By interpolation, $C = 1.33$ and

$$D_{req} = \frac{R_u}{2 \times CC_1l} = \frac{111 \text{ kips}}{2 \times (1.33)(1.0)(22.5 \text{ in.})} = 1.86 \rightarrow 2 \text{ sixteenths}$$

Use $^1/_4$ in. fillet welds.

Solution b: *Design chord stem extension plate*

The chord stem extension plate thickness must match that of the tee stem; approximately $^7/_{16}$ in.

Check tension yielding on the Whitmore section.

$$\begin{aligned} L_w &= 4 \text{ in.} + 2 \times (11 \text{ in.} \times \tan 30^\circ) \\ &= 16.7 \text{ in.} \end{aligned}$$

$$\begin{aligned} \phi R_n &= \phi F_y A_g \\ &= 0.9(36 \text{ ksi})(16.7 \text{ in.} \times {}^{7}\!/_{16} \text{ in.}) \\ &= 237 \text{ kips} > 174 \text{ kips} \quad \textbf{o.k.} \end{aligned}$$

Check block shear rupture of the gusset.

From LRFD Specification Section J4.3,

$$\begin{aligned} 0.6F_u A_{nv} &= 0.6(58 \text{ ksi})(2 \times 11 \text{ in.} \times {}^{7}\!/_{16} \text{ in.}) \\ &= 335 \text{ kips} \\ F_u A_{nt} &= (58 \text{ ksi})(4 \text{ in.} \times {}^{7}\!/_{16} \text{ in.}) \\ &= 102 \text{ kips} \end{aligned}$$

Since $0.6F_u A_{nv} > F_u A_{nt}$,

$$\begin{aligned} \phi R_n &= \phi[0.6F_u A_{nv} + F_y A_{gt}] \\ &= 0.75[335 \text{ kips} + (36 \text{ ksi})(4 \text{ in.} \times {}^{7}\!/_{16} \text{ in.})] \\ &= 299 \text{ kips} > 174 \text{ kips} \quad \textbf{o.k.} \end{aligned}$$

The gusset width must be such that the groove weld connecting it to the stem of the tee can transfer the 140 kip force between the gusset and the top chord (note that the slight slope of the top chord has been ignored). The required length is

$$\begin{aligned} L_{req} &= \frac{R_u}{\phi(0.6F_u)t} \\ &= \frac{140 \text{ kips}}{0.75(0.6 \times 58 \text{ ksi})({}^{7}\!/_{16} \text{ in.})} \\ &= 12.3 \text{ in.} \end{aligned}$$

Use $L = 16$ in. to allow for weld runout and offset between the gusset and tee stem at the end of the chord.

The gusset length depends upon the connection angles. From a scaled layout, the gusset must extend 1′–6 below the tee stem.

Use PL$^{7}/_{16}$ in. × 16 in. × 1′–6. Note that fabricators may prefer to use $^{1}/_{2}$ in. plate from stock instead of ordering $^{7}/_{16}$ in. plate. Were this the case, the weld joining the angles to the tee stem could be increased slightly to accommodate the resulting small gap.

Design weld connecting diagonal to gusset

From LRFD Specification Table J2.4, the minimum weld size is $^{3}/_{16}$ in.

Try $^3/_{16}$ in. fillet weld.

$$L_{req} = \frac{R_u}{2 \times D \times 1.392}$$
$$= \frac{174\text{kips}}{2(3 \text{ sixteenths})(1.392)}$$
$$= 20.8 \text{ in.}$$

Use 11 in. at the heel and 11 in. at the toe.

Check gusset thickness

For two $^3/_{16}$ in. fillet welds

$$t_{\min} = \frac{6.19D}{F_u}$$
$$= \frac{6.19(3 \text{ sixteenths})}{58 \text{ ksi}}$$
$$= 0.320 \text{ in.} < {}^7/_{16} \text{ in.} \quad \textbf{o.k.}$$

PART 13 REFERENCES

American Institute of Steel Construction, Inc., 1992, *Manual of Steel Construction, Volume II—Connections*, ASD 9th Ed./LRFD 1st Ed., AISC, Chicago, IL.

American Institute of Steel Construction, Inc., 1984, *Engineering for Steel Construction*, pp. 7.55–7.62, AISC, Chicago, IL.

Bjorhovde, R. and S.K. Chakrabarti, 1985, "Tests of Full-Size Gusset Plate Connections," *Journal of Structural Engineering*, Vol. 111, No. 3, (March), pp. 667–684, ASCE, New York, NY.

Gross, J.L., 1990, "Experimental Study of Gusseted Connections," *Engineering Journal*, Vol. 27, No. 3, (3rd Qtr.), pp. 89–97, AISC, Chicago, IL.

Gross, J.L. and G. Cheok, 1988, *Experimental Study of Gusseted Connections for Laterally Braced Steel Buildings*, National Institute of Standards and Technology Report NISTIR 88-3849, NIST, Gaithersburg, MD.

Lindsay, S.D. and A.V. Goverdahn, 1989, "Eccentrically Braced Frames: Suggested Design Procedures for Wind and Low Seismic Forces," *National Steel Construction Conference Proceedings*, pp. 17.1–17.25, AISC, Chicago, IL.

Richard, R.M., 1986, "Analysis of Large Bracing Connection Designs for Heavy Construction," *National Steel Construction Conference Proceedings*, pp. 31.1–31.24, AISC, Chicago, IL.

Ricker, D.T., 1992, "Value Engineering and Steel Economy," *Modern Steel Construction*, Vol. 32, No. 2, (February), AISC, Chicago, IL.

Thornton, W.A., 1992, "Designing for Cost Efficient Fabrication and Construction," *Constructional Steel Design—An International Guide*, (Chapter 7), pp. 845–854, Elsevier, London, UK.

Thornton, W.A., 1991, "On the Analysis and Design of Bracing Connections," *National Steel Construction Conference Proceedings*, pp. 26.1–26.33, AISC, Chicago, IL.

PART 14

DESIGN OF BEAM BEARING PLATES, COLUMN BASE PLATES, ANCHOR RODS AND COLUMN SPLICES

SCOPE

The specification requirements and other design considerations summarized in this Part apply to the design of beam bearing plates, column base plates, anchor rods and column splices. For the design of similar connections for HSS and steel pipe, see the AISC *Hollow Structural Sections Connections Manual*. For column splices that are part of a seismic-force-resisting system in which the seismic response modification factor R is taken greater than 3, see the Seismic Provisions, which are available from AISC at www.aisc.org.

BEAM BEARING PLATES

A beam bearing plate is made with a plate as illustrated in Figure 14-1.

Force Transfer

As illustrated in Figure 14-1, the required strength (beam end reaction) R_u is distributed from the beam bottom flange to the bearing plate over an area equal to $2Nk$, where N is the bearing length (length of contact between the beam bottom flange and the bearing plate), in. The bearing plate is then assumed to distribute the beam end reaction to the concrete or masonry as a uniform bearing pressure by cantilevered bending of the plate. The bearing-plate cantilever dimension is taken as

$$n = \frac{B}{2} - k$$

where B is the bearing plate width, in.

In the rare case where a bearing plate is not required, the beam end reaction R_u is assumed to be uniformly distributed from the beam bottom flange to the concrete or masonry as a uniform bearing pressure by cantilevered bending of the beam flanges. The beam-flange cantilever dimension is calculated as for a bearing-plate, but using the beam flange width b_f in place of B.

Recommended Bearing-Plate Dimensions and Thickness

The length of bearing N may be established by available wall thickness, clearance requirements, or by the minimum requirements based on local web yielding or web crippling. The selected dimensions of the bearing plate B and N should preferably be in full inches. Bearing-plate thickness should be specified in multiples of $1/8$ in. up to $1\,1/4$-in. thickness and in multiples of $1/4$ in. thereafter.

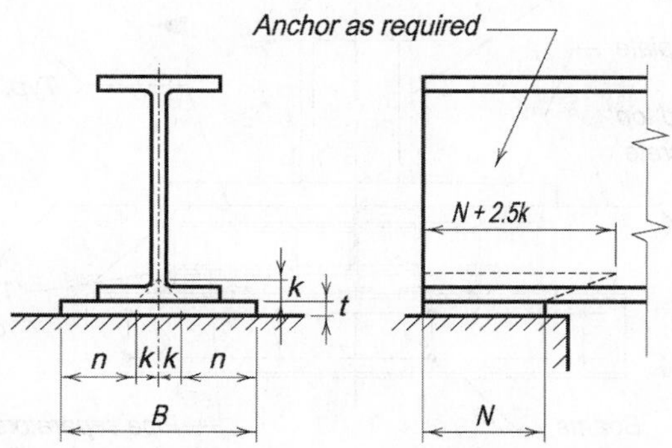

Fig. 14–1. Beam bearing plate variables.

Design Strength

The design strength of a beam bearing plate is determined from the applicable limit states from Part 9 (connection elements), as illustrated in Example 14-1. In all cases, the design strength ϕR_n must exceed the required strength R_u. The stability of the beam end must also be addressed as discussed in "Design Basis, Stability Bracing, Beam Ends Supported on Bearing Plates" in Part 2.

COLUMN BASE PLATES FOR AXIAL COMPRESSION

A column base plate is made with a plate and a minimum of four anchor rods as illustrated in Figure 14-2. The base plate is often attached to the bottom of the column in the shop. For extremely heavy loads, or where subsoil conditions are poor, a grillage as shown in Figure 14-3 may be required. This grillage consists of one or more layers of closely spaced beams (usually S-shapes because of the thicker webs) encased in the concrete foundation.

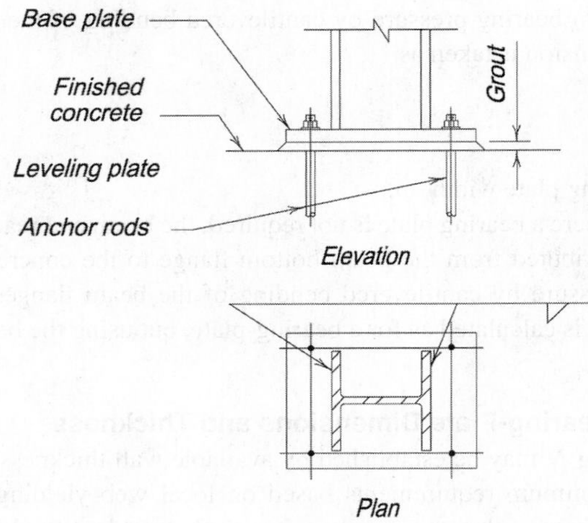

Fig. 14–2. Typical column base for axial compressive loads.

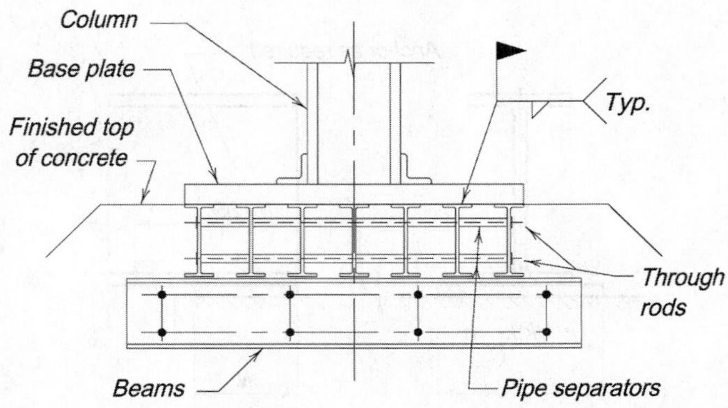

Fig. 14–3. Grillage foundation.

Force Transfer

As illustrated in Figure 14-4, the required strength (column axial force) P_u is distributed from the column end to the column base plate in direct bearing. The column base plate is then assumed to distribute the column axial force to the concrete or masonry as a uniform bearing pressure by cantilevered bending of the plate. The critical base-plate cantilever dimension l is determined as the larger of m, n, and $\lambda n'$, where

$$m = \frac{N - 0.95d}{2}$$

$$n = \frac{B - 0.8b_f}{2}$$

$$n' = \frac{\sqrt{db_f}}{4}$$

$$\lambda = \frac{2\sqrt{X}}{1 + \sqrt{1 - X}} \leq 1$$

$$X = \left(\frac{4db_f}{(d + b_f)^2}\right)\frac{P_u}{\phi_c P_p}$$

Note that, because both the term in parentheses and the ratio of P_u to $\phi_c P_p$ are always less than or equal to one, the value of X will always be less than or equal to one. Note also that λ can always be taken conservatively as 1. For further information, see Thornton (1990) and AISC Design Guide No. 1 *Column Base Plates* (DeWolf and Ricker, 1990).

Recommended Base-Plate Dimensions and Thickness

The selected dimensions of the base plate B and N should preferably be in full inches. Base-plate thickness should be specified in multiples of $^1/_8$ in. up to $1^1/_4$-in. thickness and in multiples of $^1/_4$ in. thereafter.

Design Strength

The design strength of an axially loaded column base plate is determined from the applicable limit states in Part 9 (connection elements), as illustrated in Example 14-2. From

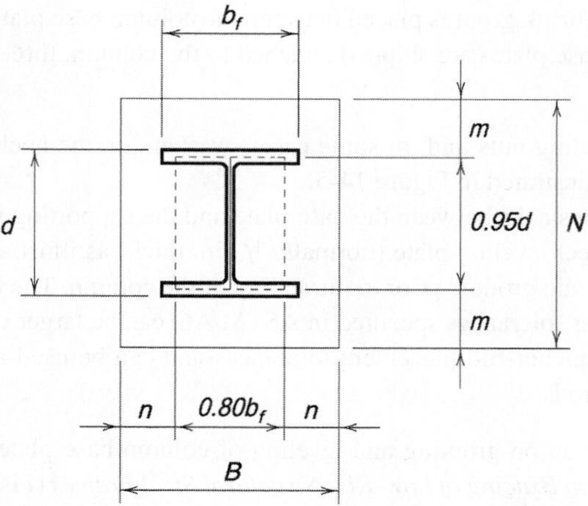

Fig. 14–4. Column base-plate design variables.

Thornton (1990), the minimum base-plate thickness can be calculated as:

$$t_{min} = l\sqrt{\frac{2P_u}{0.9F_yBN}}$$

In all cases, the design strength ϕR_n must exceed the required strength R_u.

Finishing Requirements

Base-plate finishing requirements are given in LRFD Specification Section M2.8. When finishing is required, the plate material must be ordered thicker than the specified base-plate thickness to allow for the material removed in finishing. Finishing allowances are given in Table 14-1 per ASTM A6 flatness tolerances for steel base plates with F_u equal to or less than 60 ksi based upon the width, thickness, and whether one or both sides are to be finished. Finishing allowances for steel base plates with F_u greater than 60 ksi should be increased by 50 percent.

The criteria for fit-up of column splices are also applicable to column base plates.

Holes for Anchor Rods and Grouting

Recommended maximum anchor-rod hole sizes are given in Table 14-2. These hole sizes will accommodate reasonable misalignments in the setting of the anchor rods and allow better adjustment of the column base to the correct centerlines. It is normally unnecessary to deduct the area of holes when determining the required base-plate area. An adequate washer should be provided for each anchor rod.

When base plates with large areas are used, at least one grout hole should be provided near the center of the base plate through which grout may be poured. This will provide for a more even distribution of the grout and also prevent air pockets. Note that a grout hole may not be required when the grout is dry-packed. Grout holes do not require the same accuracy for size and location as anchor-rod holes.

Holes in base plates for anchor rods and grouting often must be flame-cut, because drill sizes and punching capabilities may be limited to smaller diameters. Flame-cut holes may have a slight taper and should be inspected to assure proper clearances for anchor rods.

Grouting and Leveling

High-strength, non-shrink grout is placed between the column base plate and the supporting foundation. When base plates are shipped attached to the column, three methods of column support are:

1. The use of leveling nuts and, in some cases, washers on the anchor rods beneath the base plate, as illustrated in Figure 14-5.
2. The use of shim stacks between the base plate and the supporting foundation.
3. The use of a steel leveling plate (normally $^1/_4$-in. thick) as illustrated in Figure 14-2a, set to elevation and grouted prior to the setting of the column. The leveling plate should meet the flatness tolerances specified in ASTM A6, can be larger than the base plate to accommodate anchor-rod placement tolerances and can be used as a setting template for the anchor rods.

For further information on grouting and leveling of column base plates, see AISC Design Guide No. 10 *Erection Bracing of Low-Rise Structural Steel Frames* (Fisher and West, 1997).

When base plates are shipped loose, the base plates are usually grouted after the base plate has been aligned and leveled with one of the preceding methods. For heavy loose base plates, three-point leveling bolts, illustrated in Figure 14-6, are commonly used. These threaded attachments may consist of a nut or an angle and nut welded to the base plate. Leveling bolts must be of sufficient length to compensate for the space provided for grouting. Rounding the point of the leveling bolt will prevent it from "walking" or moving laterally as it is turned. Additionally, a small steel pad under the point reduces friction and prevents damage to the concrete.

Heavy loose base plates should be provided with some means of handling at the erection site. Lifting holes can be provided in the vertical legs of shop-attached connection angles. Lifting lugs can also be used and can remain in place after erection, unless they create an interference or removal is required in the contract documents.

Leveling bolts or nuts should not be used to support the column during erection. If grouting is delayed until after steel erection, the base plate must be shimmed to properly distribute loads

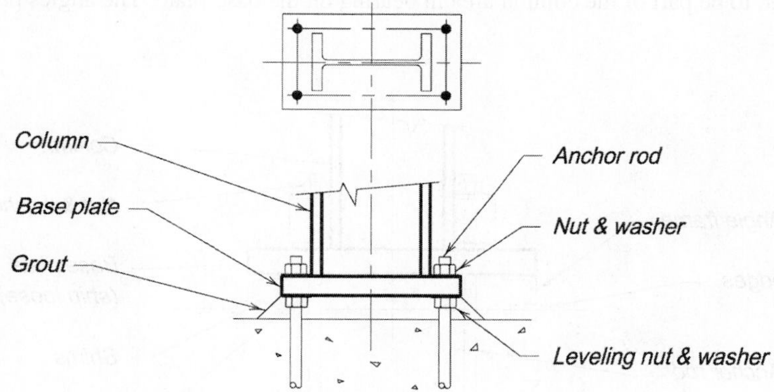

Fig. 14–5. Leveling nuts and washers.

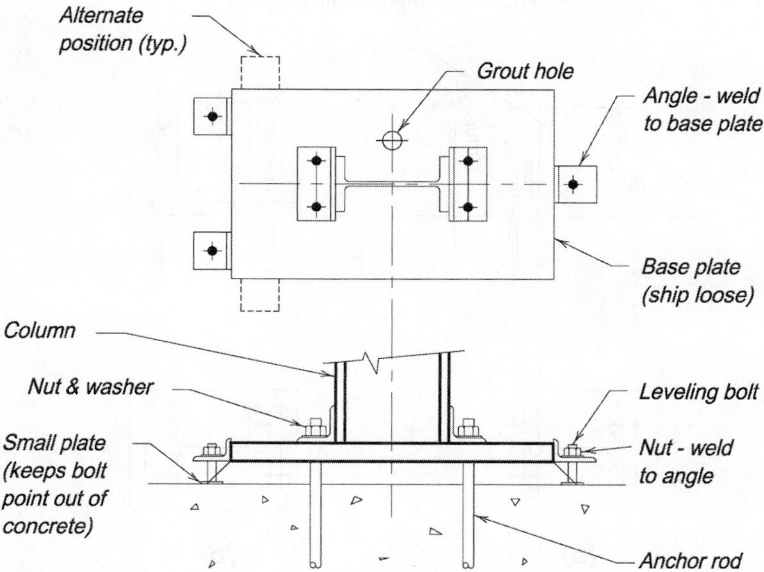

Fig. 14–6. Three-point leveling.

to the foundation without overstressing either the base plate or the concrete. This difficulty of supporting columns while leveling and grouting their bases makes it advisable that footings be finished to near the proper elevation (Ricker, 1989). The top of the rough footing should be set approximately 1 to 2 in. below the bottom of the base plate to provide for adjustment. Alternatively, an angle frame as illustrated in Figure 14-7 could be constructed to the proper elevation and filled with grout prior to erection.

COLUMN BASE PLATES FOR AXIAL TENSION, SHEAR OR MOMENT

For anchor rod diameters not greater than 1¼ in., angles bolted or welded to the column as shown in Figure 14-8a are generally adequate to transfer uplift forces resulting from axial loads and moments. When greater resistance is required, stiffeners may be used with horizontal plates or angles as illustrated in Figure 14-8b. These stiffeners are not usually considered to be part of the column area in bearing on the base-plate. The angles preferably

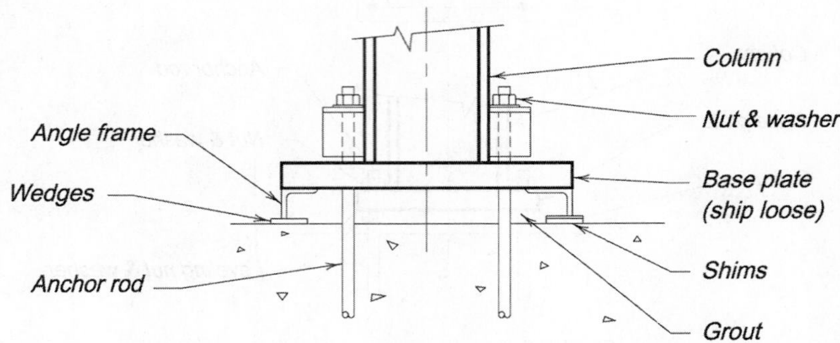

Fig. 14–7. Angle-frame leveling.

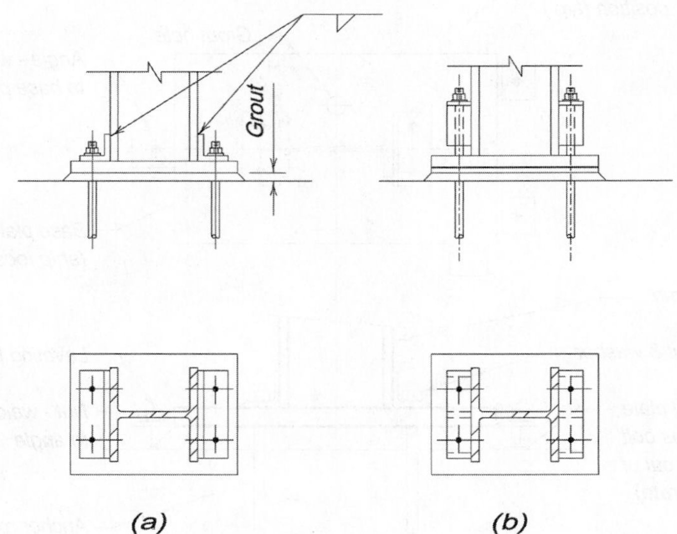

Fig. 14–8. Typical column bases for uplift.

should be set back from the column end about 1/8 in. Stiffeners preferably should be set back about 1 in. from the base plate to eliminate a pocket that might prevent drainage and thus protect the column and column base plate from corrosion.

For further information, see AISC Design Guide No. 1 *Column Base Plates* (DeWolf and Ricker, 1990).

ANCHOR RODS

Cast-in-place anchor rods, illustrated in Figure 14-9, are generally made from unheaded rod material or headed bolt material. Drilled-in (post-set) anchors can be used for corrective work or in new work as determined by the owner's designated representative for design and as permitted in the applicable building code. The design of post-set anchors is governed by manufacturer's specifications; see also ACI 349 Appendix B. Post-set anchors that rely upon torque or tension to develop anchorage by wedging action should not be used unless the stability of the column during erection is provided by means other than the post-set anchors.

Minimum Edge Distance and Embedment Length

Minimum edge distances, embedment lengths and the design of anchorages into concrete, in general, are now provided in the International Building Code (ICC, 2000) and ASCE 7. Both references contain the same material, which was originally intended for including in the 1999 edition of ACI 318, but was not finished by the deadline. The strength design of steel-to-concrete anchorage is covered in IBC Section 1913 and ASCE7-98 Section A.9.9.6. Working-stress design provisions are also contained in IBC Section 1912. These provisions include methods to account for edge distance and group action, as does ACI 349. AISC Design Guides 1, 7 and 10 provide additional material on the design of anchor rods in concrete.

In addition to providing the recommended minimum embedment length, anchor rods must extend a distance above the foundation that is sufficient to permit adequate thread engagement of the nut. Adequate thread engagement for anchor rods is identical to the condition described in the RCSC Specification as adequate for steel-to-steel structural joints using high-strength bolts: having the end of the [anchor rod] flush with or outside the face of the nut.

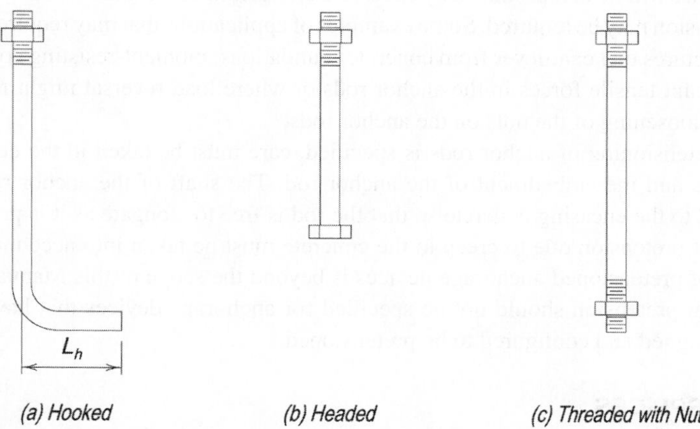

Fig. 14–9. Cast-in-place anchor rods.

Washer Requirements

Because base plates typically have holes larger than oversized holes to allow for tolerances on the location of the anchor rod, washers are usually furnished from ASTM A36 steel plate. They may be round, square, or rectangular, and generally have holes that are $^{1}/_{16}$-in. larger than the anchor rod diameter. The thickness must be suitable for the forces to be transferred. Minimum washer sizes are given in Table 14-2.

Hooked Anchor Rods

Hooked anchor rods should be used only for axially loaded compression members subject to compression only to locate and prevent the displacement or overturning of columns due to erection loads or accidental collisions during erection. Additionally, high-strength steels are not recommended for use in hooked rods since bending with heat may materially affect their strength.

Headed or Threaded and Nutted Anchor Rods

When anchor rods are required for a calculated tensile force T_u, a more positive anchorage is formed when headed anchor rods, illustrated in Figure 14-9b, are used. With adequate embedment and edge distance, the limit state is either a tensile failure of the anchor rod or the pull-out of a cone of concrete radiating outward from the head (Marsh and Burdette, 1985) as illustrated in Figure 14-10. Marsh and Burdette (1985) showed that the head of the anchor rod usually provides sufficient anchorage and the use of an additional washer or plate does not add significantly to the anchorage. The nut and threading shown in Figure 14-9c is acceptable in lieu of a bolt head. The nut should be welded to the rod to prevent the rod from turning out when the top nut is tightened.

Anchor-Rod Nut Installation

The majority of anchorage applications in buildings do not require special anchor-rod nut installation procedures or pretension in the anchor rod, and the anchor-rod nuts should be "drawn down tight" as columns and bases are erected, per ANSI A10.13 Section 9.6. This condition can be achieved by following the same practices as recommended for snug-tightened installation in steel-to-steel bolted joints in the RCSC Specification. That is, most anchor-rod nuts can be installed using the full effort of an ironworker with an ordinary spud wrench.

When in the judgment of the owner's designated representative for design, the performance of the structure will be compromised by excessive elongation of the anchor rods under tensile loads, pretension may be required. Some examples of applications that may require pretension include structures that cantilever from concrete foundations, moment-resisting column bases with significant tensile forces in the anchor rods or where load reversal might result in the progressive loosening of the nuts on the anchor rods.

When pretensioning of anchor rods is specified, care must be taken in the design of the column base and the embedment of the anchor rod. The shaft of the anchor rod must be free of bond to the encasing concrete so that the rod is free to elongate as it is pretensioned. Also, loss of pretension due to creep in the concrete must be taken into account. Although the design of pretensioned anchorage devices is beyond the scope of this Manual, it should be noted that pretension should not be specified for anchorage devices that have not been properly designed and configured to be pretensioned.

COLUMN SPLICES

When the height of a building exceeds the available length of column sections, or when it is economically advantageous to change the column size at a given floor level, it becomes necessary to splice two columns together. Column splices at the final exterior and interior perimeter and at interior openings must be located a minimum of 48 in. above the finished floor to accommodate the attachment of safety cables, except when constructability does not allow. For simplicity and uniformity, other column splices should be located at the same height. Note that column splices placed significantly higher than this are impractical in terms of field assembly.

Fit-Up of Column Splices

From LRFD Specification Section M2.6, the ends of columns in a column splice which depend upon contact bearing for the transfer of axial forces must be finished to a common plane by milling, sawing, or other suitable means. In theory, if this were done and the pieces were erected truly plumb, there would be full-contact bearing across the entire surface; this is true in most cases. However, LRFD Specification Section M4.4 recognizes that a perfect fit on the entire available surface will not exist in all cases.

A $^{1}/_{16}$-in. gap is permissible with no requirements for repair or shimming. During erection, at the time of tightening the bolts or depositing the welds, columns will usually be subjected to loads which are significantly less than the design loads. Full scale tests (Popov and

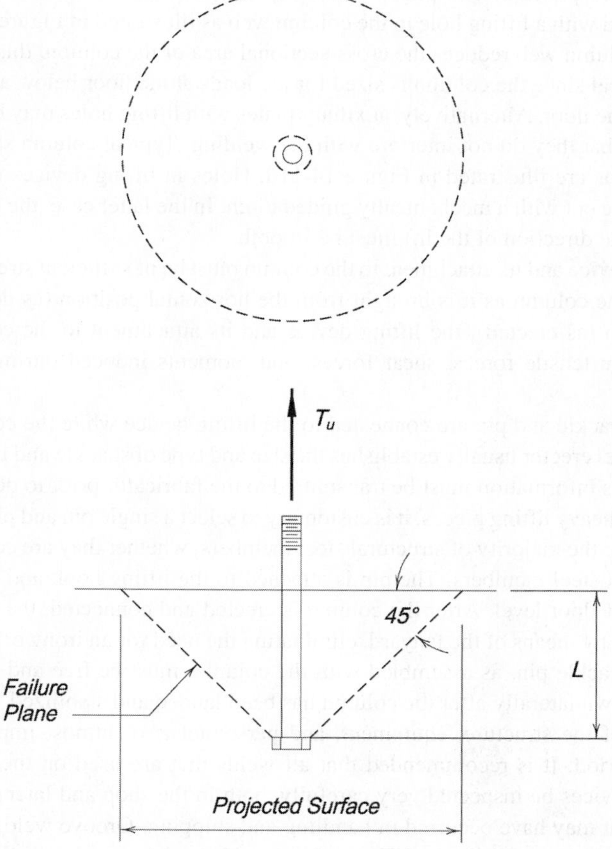

Fig. 14–10. Concrete cone subject to pull-out.

Steven, 1977) which progressed to column failure have demonstrated that subsequent loading to the design loads does not result in distress in the bolts or welds of the splice.

If the gap exceeds $^1/_{16}$-in., but is less than $^1/_4$-in., non-tapered steel shims are required if sufficient contact area does not exist. Mild steel shims are acceptable regardless of the steel grade of the column or bearing material. If required, these shims must be contained, usually with a tack weld, so that they cannot be worked out of the joint.

There is no provision in the LRFD Specification for gaps larger than $^1/_4$-in. When such a gap exists, an engineering evaluation should be made of this condition based upon the type of loading transferred by the column splice. Tightly driven tapered shims may be required or the required strength may be developed through flange and web splice plates. Alternatively, the gap may be ground or gouged to a suitable profile and filled with weld metal.

Lifting Devices

As illustrated in Figure 14-11, lifting devices are typically used to facilitate the handling and erection of columns. When flange-plated or web-plated column splices are used for W-shape columns, it is convenient to place lifting holes in these flange plates as illustrated in Figure 14-11a. When butt-plated column splices are used, additional temporary plates with lifting holes may be required as illustrated in Figure 14-11b. W-shape column splices which do not utilize web-plated or butt-plated column splices, i.e., groove-welded column splices, may be provided with a lifting hole in the column web as illustrated in Figure 14-11c. While a hole in the column web reduces the cross-sectional area of the column, this reduction will seldom be critical since the column is sized for the loads at the floor below and the splice is located above the floor. Alternatively, auxiliary plates with lifting holes may be connected to the column so that they do not interfere with the welding. Typical column splices for tubes and box-columns are illustrated in Figure 14-11d. Holes in lifting devices may be drilled, reamed, or flame cut with a mechanically guided torch. In the latter case, the bearing surface of the hole in the direction of the lift must be smooth.

The lifting device and its attachment to the column must be of sufficient strength to support the weight of the column as it is brought from the horizontal position (as delivered) to the vertical position (as erected); the lifting device and its attachment to the column must be adequate for the tensile forces, shear forces, and moments induced during handling and erection.

A suitable shackle and pin are connected to the lifting device while the column is on the ground. The steel erector usually establishes the size and type of shackle and pin to be used in erection and this information must be transmitted to the fabricator prior to detailing. Except for excessively heavy lifting pieces, it is customary to select a single pin and pinhole diameter to accommodate the majority of structural steel members, whether they are columns or other heavy structural steel members. The pin is attached to the lifting hook and a lanyard trails to the ground or floor level. After the column is erected and connected, the pin is removed from the device by means of the lanyard, eliminating the need for an ironworker to climb the column. The shackle pin, as assembled with the column, must be free and clear, so that it may be withdrawn laterally after the column has been landed and stabilized.

The safety of the structure, equipment, and personnel is of utmost importance during the erection period. It is recommended that all welds that are used on the lifting devices and stability devices be inspected very carefully, both in the shop and later in the field, for any damage that may have occurred in handling and shipping. Groove welds frequently are inspected with ultrasonic methods (UT) and fillet welds are inspected with magnetic particle (MT) or liquid dye penetrant (PT) methods.

Column Alignment and Stability During Erection

Column splices should provide for safety and stability during erection when the columns might be subjected to wind, construction, and/or accidental loading prior to the placing of the floor system. The nominal flange-plated, web-plated, and butt-plated column splices developed here consider this type of loading.

In other splices, column alignment and stability during erection are achieved by the addition of temporary lugs for field bolting as illustrated in Figure 14-12. The material thickness, weld size, and bolt diameter required are a function of the loading. A conservative resisting moment arm is normally taken as the distance from the compressive toe or flange face to the gage line of the temporary lug. The overturning moment should be checked about both axes of the column. The recommended minimum plate or angle thickness is $^1/_2$ in.; the recommended minimum weld size is $^5/_{16}$ in.; additionally, high-strength bolts are normally used for stability devices.

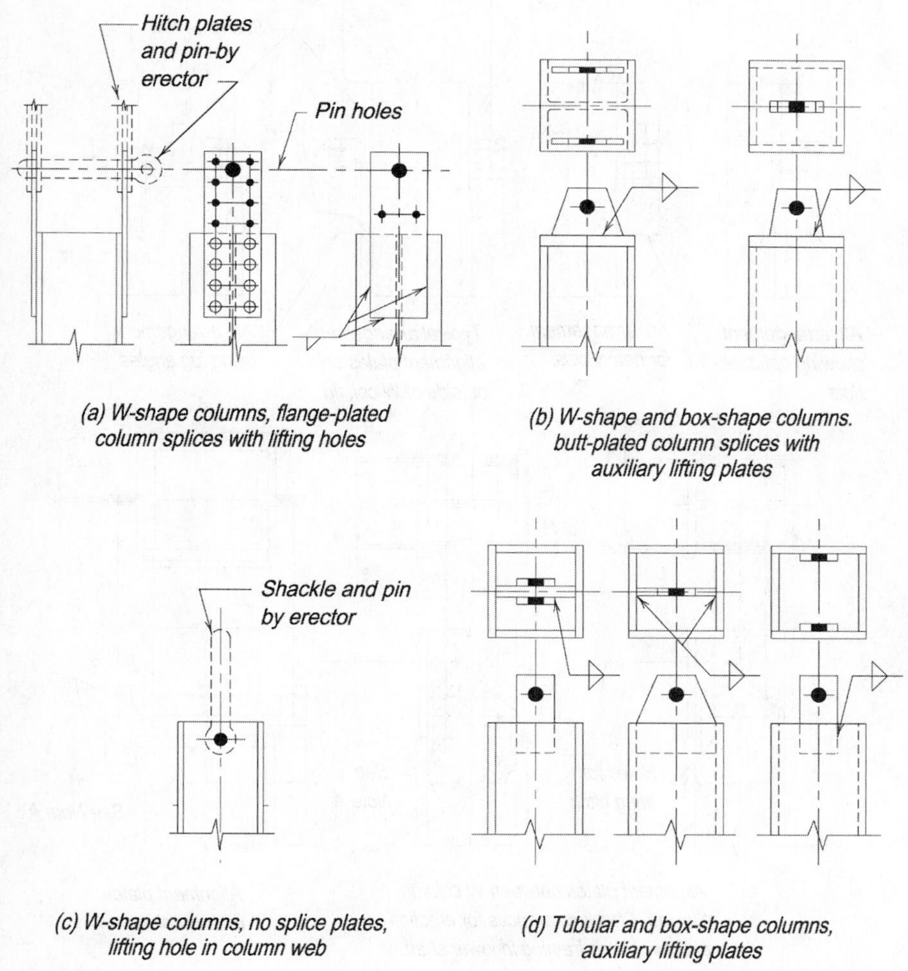

Fig. 14–11. Lifting devices for columns.

Temporary lugs are not normally used as lifting devices. Unless required to be removed in the contract documents, these temporary lugs may remain.

Column alignment is provided with centerpunch marks that are useful in centering the columns in two directions.

Force Transfer in Column Splices

As illustrated in Figure 14-13, for the W-shapes most frequently used as columns, the distance between the inner faces of the flanges is constant throughout any given nominal depth; as the

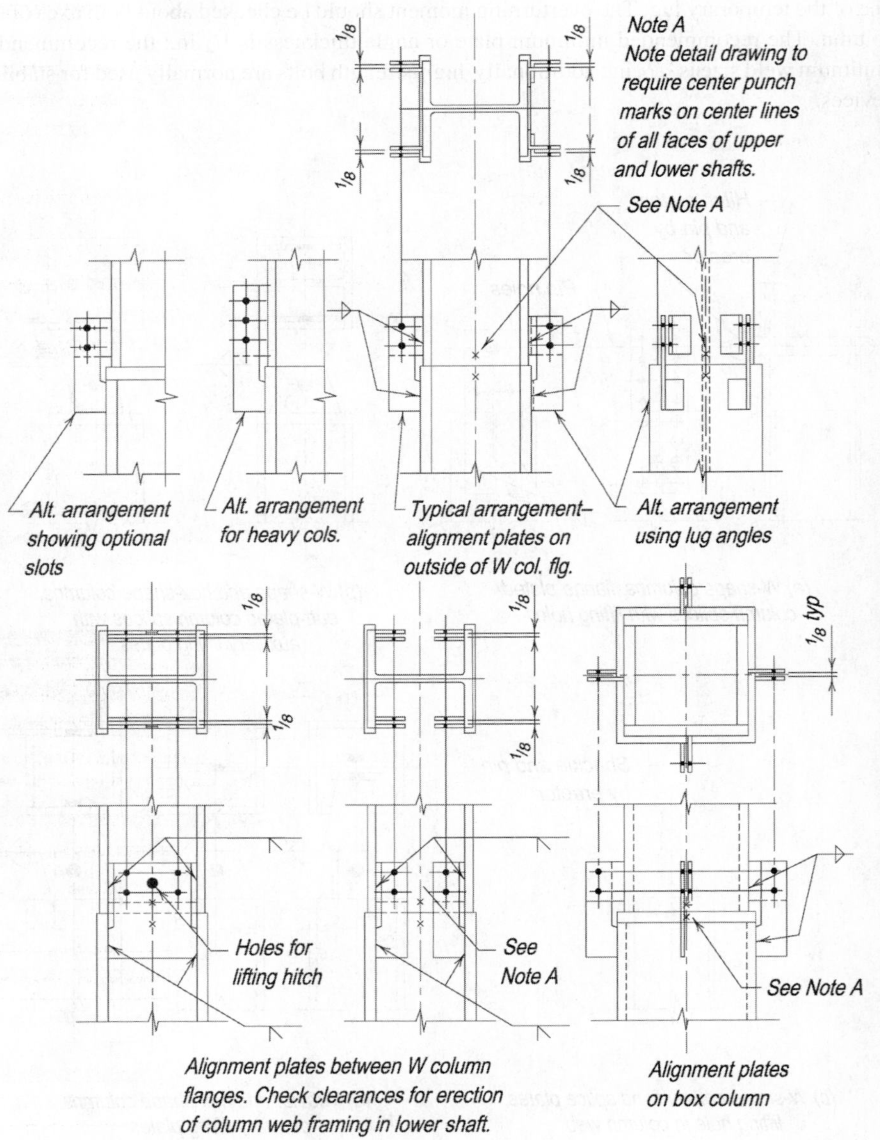

Fig. 14–12. Column stability and alignment devices.

nominal weight per foot increases for each nominal depth, the flange and web thicknesses increase. From LRFD Specification Section J8, the design bearing strength of the contact area of a finished surface is

$$\phi R_n = 0.75(1.8F_y A_{pb})$$

This bearing strength is much greater than the axial strength of the column and will seldom prove critical in the member design. In column splices transferring only axial forces, then, complete axial force transfer may be achieved through bearing on finished surfaces; bolts or welds are required by LRFD Specification Section J1.4 to be sufficient to hold all parts securely in place.

In addition to axial forces, from LRFD Specification Section J1.4, column splices must be proportioned to resist tension developed by the factored load combination $0.9D \pm (1.6W$ or $1.0E)$. Note that it is not permissible to use forces due to live load to offset the tensile forces from wind or seismic loads.

For dead and wind loads, the required strength is $0.9D - 1.6W$, where D is the compressive force due to the dead load and W is the tensile force due to wind load. If $0.9D \geq 1.6W$, the splice is not subjected to tension and a nominal splice may be selected from those in Tables 14-3. When $0.9D < 1.6W$, the splice will be subjected to tension and the nominal splices from Tables 14-3 are acceptable if the design tensile strength of the splice $\phi_t P_n$ is greater than or equal to the required strength. Otherwise, a splice must be designed with sufficient area and attachment.

When shear from lateral loads is divided among several columns, the force on any single column is relatively small and can usually be resisted by friction on the contact bearing surfaces and/or by the flange plates, web plates, or butt plates. If the required shear strength exceeds the design shear strength of the column splice selected from Tables 14-3, a column splice must be designed with sufficient area and attachment.

The column splices shown in Tables 14-3 meet the OSHA requirement for 300 lbs located 18 in. from the column face.

Flange-Plated Column Splices

Tables 14-3 give typical flange-plated column splice details for W-shape columns. These details are not splice standards, but rather, typical column splices in accordance with LRFD Specification provisions and typical erection requirements. Other splice designs may also be developed. It is assumed in all cases that the lower shaft will be the heavier, although not necessarily the deeper, section.

Full-contact bearing is always achieved when lighter sections are centered over heavier sections of the same nominal depth. If the upper column is not centered on the lower column,

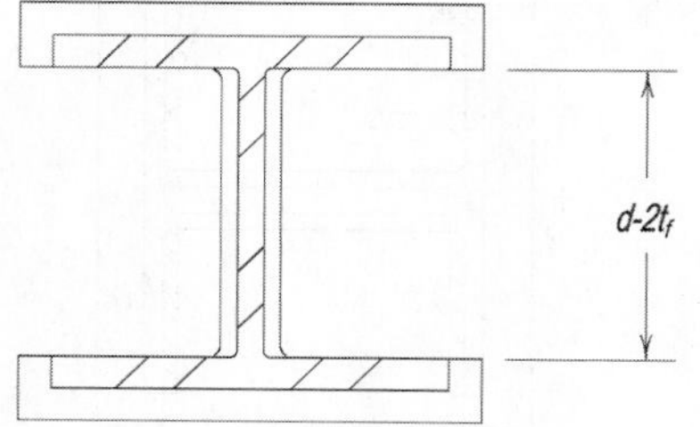

Column Size	d-$2t_f$ (in.)
W8×24–67	7.13
W10×33–112	8.86
W12×40–336	10.91
W14×43–730	12.60

Fig. 14–13. Distance between flanges for typical W-shape columns.

or if columns of different nominal depths must bear on each other, some areas of the upper column will not be in contact with the lower column. These areas are hatched in Figure 14-14.

When additional bearing area is not required, unfinished fillers may be used. These fillers are intended for "pack-out" of thickness and are usually set back ¼ in. or more from the finished column end. Since no force is transferred by these fillers, only nominal attachment to the column is required.

When additional bearing area is required, fillers finished to bear on the larger column may be provided. Such fillers are proportioned to carry bearing loads at the bearing strength calculated from LRFD Specification Section J8 and must be connected to the column to transfer this calculated force.

Although flange plates are shown shop assembled to the lower column, it is equally acceptable to invert this arrangement and place them on the upper column. This will usually require fills of increased thickness to maintain erection clearances.

In Tables 14-3, Cases I and II are for all-bolted flange-plated column splices for W-shape columns. Bolts in column splices are usually the same size and type as for other bolts on the column. Bolt spacing, end distance, and edge distances resulting from the billed plate sizes permit the use of ¾-in. and ⅞-in. bolts in the splice details shown. Larger diameter bolts may require an increase in edge or end distances. Refer to LRFD Specification Chapter J. The use of high-strength bolts in bearing-type connections is assumed in all field and shop splices. However, when slotted or oversized holes are utilized, or in splices employing underdeveloped fillers over ¼-in. thick, slip-critical connections may be required; refer to LRFD Specification Section J6. For ease of erection, field clearances for lap splices fastened by bolts range from ⅛ in. to 3/16 in. under each plate.

Cases IV and V are for all-welded flange-plated column splices for W-shape columns. Splice welds are assumed to be made with E70XX electrodes and are proportioned as required by the LRFD Specification provisions. The SAW, GMAW, and FCAW equivalents to E70XX electrodes may be substituted if desired. Field clearance for welded splices are limited to 1/16 in. to control the expense of building up welds to close openings. Note that the fillet weld lengths Y as compared to the lengths $L/2$, provide 2-in. unwelded distance below and above the column shaft finish line. This provides a degree of flexibility in the splice plates to assist the erector.

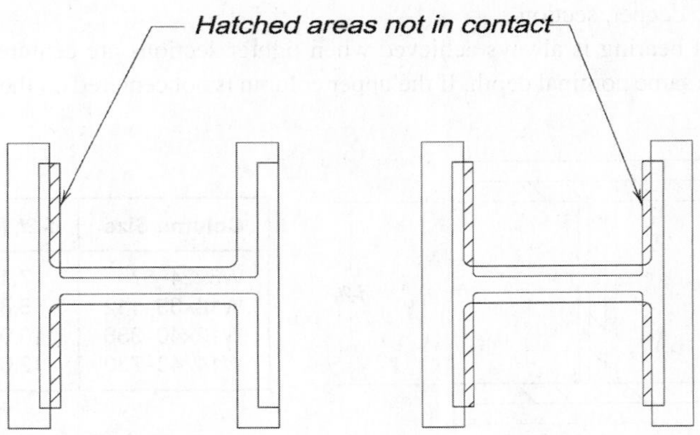

Fig. 14–14. Columns not centered or of different nominal depth.

Cases VI and VII are for combination bolted and welded column splices. Since the design strength of the welds will, in most cases, exceed the strength of the bolts, the weld and splice lengths shown may be reduced, if desired, to balance the strength of the fasteners to the upper or lower column, provided that the design strength of the splice is still greater than the required strength of the splice, including erection loading.

Directly Welded Flange Column Splices

Tables 14-3 also include typical directly welded flange column splice details for W-shape and tubular or box-shaped columns. These details are not splice standards, but rather, typical column splices in accordance with LRFD Specification provisions and typical erection requirements. Other splice designs may also be developed. It is assumed in all cases that the lower shaft will be the heavier, although not necessarily the deeper, section.

Case VIII is for W-shape columns spliced with either partial-joint-penetration or complete-joint-penetration groove welds. Case X is for tubular or box-shaped columns spliced with partial-joint-penetration or complete-joint-penetration groove welds.

Butt-Plated Column Splices

Tables 14-3 further include typical butt-plated column splice details for W-shape and tubular or box-shaped columns. These details are not splice standards, but rather, present typical column splices in accordance with LRFD Specification provisions and typical erection requirements. Other splice designs may also be developed. It is assumed in all cases that the lower shaft will be the heavier, although not necessarily the deeper, section.

Butt plates are used frequently on welded splices where the upper and lower columns are of different nominal depths, but may not be economical for bolted splices since fillers cannot be eliminated. Typical butt plates are $1^1/_2$-in. thick for a W8 over W10 splice, and 2-in. thick for other W-shape combinations such as W10 over W12 and W12 over W14. Butt plates which are subjected to substantial bending stresses, such as required on boxed columns, will require a more careful review and analysis. One method of extensive experience is to assume forces are transferred through the butt plate on a 45° angle and check the thickness obtained for shear and bearing strength. Finishing requirements for butt plates are specified in LRFD Specification Section M2.8.

Case III is a combination flange-plated and butt-plated column splice for W-shape columns. Case IX is for welded butt-plated column splices for W-shape columns. Case XI is for welded butt-plated column splices for tubular or box-shaped columns. Case XII is for welded butt-plated column splices between W-shape and tubular or box-shaped columns.

DESIGN EXAMPLES

EXAMPLE 14.1. Beam bearing plate design.

Given: A W18×50 beam with a factored end reaction of 85 kips is supported by a 10-in.-thick concrete wall. If the beam has $F_y = 50$ ksi, the concrete has $f'_c = 3$ ksi, and the bearing plate has $F_y = 36$ ksi, determine:

a. if a bearing plate is required if the beam is supported by the full wall thickness,
b. the bearing plate required if $N = 10$ in. (the full wall thickness),
c. the bearing plate required if $N = 6$ in. and the bearing plate is centered on the thickness of the wall.

W18×50, ASTM A992

$d = 18.0$ in. $b_f = 7.50$ in. $k = 0.972$ in.
$t_w = 0.355$ in. $t_f = 0.570$ in. $k_1 = {}^{13}/_{16}$ in.

Solution a: $N = 10$ in.

Check local web yielding

From Table 9-6,

$$N_{req} = \frac{R_u - \phi R_1}{\phi R_2}$$
$$= \frac{85 \text{ kips} - 55.5 \text{ kips}}{17.8 \text{ kips/in.}}$$
$$= 1.66 \text{ in.} < 10 \text{ in.} \quad \textbf{o.k.}$$

Check web crippling

$$\frac{N}{d} = \frac{10 \text{ in.}}{18.0 \text{ in.}}$$
$$= 0.556$$

Since $\frac{N}{d} > 0.2$, from Table 9-6,

$$N_{req} = \frac{R_u - \phi R_5}{\phi R_6}$$
$$= \frac{85 \text{ kips} - 51.9 \text{ kips}}{6.29 \text{ kips/in.}}$$
$$= 5.26 \text{ in.} < 10 \text{ in.} \quad \textbf{o.k.}$$

Check bearing strength of concrete

$$\phi_c P_p = \phi_c (0.85 f_c') A_1$$
$$= 0.60\,(0.85 \times 3 \text{ ksi})(7.50 \text{ in.} \times 10 \text{ in.})$$
$$= 115 \text{ kips} > 85 \text{ kips} \quad \textbf{o.k.}$$

Check beam flange thickness

$$n = \frac{b_f}{2} - k$$
$$= \frac{7.50}{2} - 0.972 \text{ in.}$$
$$= 2.78 \text{ in.}$$
$$t_{req} = \sqrt{\frac{2.22 R_u n^2}{A_1 F_y}}$$
$$= \sqrt{\frac{2.22(85 \text{ kips})(2.78 \text{ in.})^2}{(7.50 \text{ in.} \times 10 \text{ in.})(50 \text{ ksi})}}$$
$$= 0.624 \text{ in.} > 0.570 \text{ in.} \quad \textbf{n.g.}$$

A bearing plate is required.

Solution b: $N = 10$ in.

From Solution a, local web yielding and web crippling are not critical.

Calculate required bearing-plate width.

$$A_{1\,req} = \frac{R_u}{\phi_c(0.85 f_c')}$$
$$= \frac{85 \text{ kips}}{0.60(0.85 \times 3 \text{ ksi})}$$
$$= 55.6 \text{ in.}^2$$

$$B_{req} = \frac{A_{1\,req}}{N}$$
$$= \frac{55.6 \text{ in.}^2}{10 \text{ in.}}$$
$$= 5.56 \text{ in.}$$

Use $B = 8$ in. (least whole-inch dimension that exceeds b_f)

Calculate required bearing-plate thickness.

$$n = \frac{B}{2} - k$$
$$= \frac{8 \text{ in.}}{2} - 0.972 \text{ in.}$$
$$= 3.03 \text{ in.}$$

$$t_{\min} = \sqrt{\frac{2.22 R_u n^2}{A_1 F_y}}$$
$$= \sqrt{\frac{2.22(85 \text{ kips})(3.03 \text{ in.})^2}{(10 \text{ in.} \times 8 \text{ in.})(36 \text{ ksi})}}$$
$$= 0.776 \text{ in.}$$

Use PL $1 \times 10 \times 0'$-8

Solution c: $N = 6$ in.

From Solution a, local web yielding and web crippling are not critical.

Try $B = 8$ in.

$$A_1 = B \times N$$
$$= (8 \text{ in.})(6 \text{ in.})$$
$$= 48 \text{ in.}^2$$

To determine the dimensions of the area A_2, the load is spread into the concrete at 2 : 1 until an edge is met. The 6-in. dimension spreads 2 in. to each side to meet the concrete edge. Thus, the 8-in. dimension can also be spread 2 in. to each side. Therefore,

$$\begin{aligned} N_1 &= 6 \text{ in.} + 2(2 \text{ in.}) \\ &= 10 \text{ in.} \end{aligned}$$

$$\begin{aligned} B_1 &= 8 \text{ in.} + 2(2 \text{ in.}) \\ &= 12 \text{ in.} \end{aligned}$$

$$\begin{aligned} A_2 &= B_1 \times N_1 \\ &= 12 \text{ in.} (10 \text{ in.}) \\ &= 120 \text{ in.}^2 \end{aligned}$$

Check $\sqrt{A_2/A_1} = 1.58 \leq 2$ **o.k.**

$$\begin{aligned} A_{1\ req} &= \frac{1}{A_2}\left(\frac{R_u}{\phi_c(0.85 f'_c)}\right)^2 \\ &= \frac{1}{120 \text{ in.}^2}\left(\frac{85 \text{ kips}}{0.6(0.85 \times 3 \text{ ksi})}\right)^2 \\ &= 25.7 \text{ in.}^2 < 48 \text{ in.}^2 \qquad \textbf{o.k.} \end{aligned}$$

Calculate required bearing-plate thickness

$$\begin{aligned} n &= \frac{B}{2} - k \\ &= \frac{8 \text{ in.}}{2} - 0.972 \text{ in.} \\ &= 3.03 \text{ in.} \end{aligned}$$

$$\begin{aligned} t_{min} &= \sqrt{\frac{2.22 R_u n^2}{A_1 F_y}} \\ &= \sqrt{\frac{2.22(85 \text{ kips})(3.03 \text{ in.})^2}{(6 \text{ in.} \times 8 \text{ in.})(36 \text{ ksi})}} \\ &= 1.00 \text{ in.} \end{aligned}$$

Use PL1×6×0′-8.

EXAMPLE 14.2. Column base plate design for axial compression.

Given: A W12×170 column with a factored axial load of 1,100 kips bears on a concrete pedestal. If the column has $F_y = 50$ ksi, the concrete has $f'_c =$ 3 ksi, and the base-plate has $F_y = 36$ ksi, determine:

a. the base plate and pedestal dimensions required if the base plate is to cover the full pedestal area,
b. the base plate dimensions required for a 32-in. × 32-in. concrete pedestal.

W12×170, ASTM A992

$d = 14.0$ in. $\quad b_f = 12.6$ in.
$t_w = 0.960$ in. $\quad t_f = 1.56$ in.

Solution a: *Calculate required base plate area.*

$$A_{1\ req} = \frac{P_u}{\phi_c(0.85 f'_c)}$$
$$= \frac{1{,}100 \text{ kips}}{0.60(0.85 \times 3 \text{ ksi})}$$
$$= 719 \text{ in.}^2$$

Optimize base plate dimensions.

$$\Delta = \frac{0.95d - 0.8b_f}{2}$$
$$= \frac{0.95(14.0 \text{ in.}) - 0.8(12.6 \text{ in.})}{2}$$
$$= 1.64 \text{ in.}$$
$$N \approx \sqrt{A_1} + \Delta$$
$$\approx \sqrt{719 \text{ in.}^2} + 1.64 \text{ in.}$$
$$\approx 28.5 \text{ in.}$$

Try $N = 28$ in. and $B = 26$ in. (pedestal dimensions same)

Calculate required base-plate thickness

$$m = \frac{N - 0.95d}{2}$$
$$= \frac{28 \text{ in.} - 0.95(14.0 \text{ in.})}{2}$$
$$= 7.34 \text{ in.}$$
$$n = \frac{B - 0.8b_f}{2}$$
$$= \frac{26 \text{ in.} - 0.8(12.6 \text{ in.})}{2}$$
$$= 7.97 \text{ in.}$$
$$\phi_c P_p = 0.6(0.85 f'_c A_1)$$
$$= 0.6(0.85 \times 3 \text{ ksi} \times 26 \text{ in.} \times 28 \text{ in.})$$
$$= 1{,}110 \text{ kips}$$
$$X = \left(\frac{4db_f}{(d+b_f)^2}\right)\frac{P_u}{\phi_c P_p}$$
$$= \left(\frac{4(14.0 \text{ in.})(12.6 \text{ in.})}{(14.0 \text{ in.} + 12.6 \text{ in.})^2}\right)\left(\frac{1{,}100 \text{ kips}}{1{,}110 \text{ kips}}\right)$$
$$= 0.988$$

$$\begin{aligned}\lambda &= \frac{2\sqrt{X}}{1+\sqrt{1-X}} \le 1 \\ &= \frac{2\sqrt{0.988}}{1+\sqrt{1-0.988}} \\ &= 1.79 \rightarrow 1\end{aligned}$$

$$\begin{aligned}\lambda n' &= \lambda\frac{\sqrt{db_f}}{4} \\ &= \frac{(1)\sqrt{14.0\text{ in.} \times 12.6\text{ in.}}}{4} \\ &= 3.32\text{ in.}\end{aligned}$$

From this

$$\begin{aligned}l &= \max(m, n, \lambda n') \\ &= \max(7.34\text{ in.}, 7.97\text{ in.}, 3.32\text{ in.}) \\ &= 7.97\text{ in.}\end{aligned}$$

and

$$\begin{aligned}t_{req} &= l\sqrt{\frac{2P_u}{0.9F_yBN}} \\ &= (7.97\text{ in.})\sqrt{\frac{2(1{,}100\text{ kips})}{0.9(36\text{ ksi})(26\text{ in.})(28\text{ in.})}} \\ &= 2.43\text{ in.}\end{aligned}$$

Use PL2$^1/_2$×26×2′-4.

Solution b: Assume from pedestal dimensions (32 in. × 32 in.) that $A_2 = 900$ in.^{2}

Calculate required base-plate area.

$$\begin{aligned}A_{1\ req} &= \frac{1}{A_2}\left(\frac{R_u}{\phi_c(0.85f'_c)}\right)^2 \\ &= \frac{1}{900\text{ in.}^2}\left(\frac{1{,}100\text{ kips}}{0.6(0.85 \times 3\text{ ksi})}\right)^2 \\ &= 575\text{ in.}^2\end{aligned}$$

Optimize base-plate dimensions.

From Solution a, $\Delta = 1.64$ in.

$$\begin{aligned}N &\approx \sqrt{A_1} + \Delta \\ &\approx \sqrt{575\text{ in.}^2} + 1.64\text{ in.} \\ &\approx 25.6\text{ in.}^2\end{aligned}$$

Try $N = 25$ in. and $B = 23$ in.

Spreading the load at 2 : 1 to determine A_2, the 25-in. dimension spreads $3^1/_2$ in. to each side. The 23-in. dimension will also spread $3^1/_2$ in. to each side. Therefore,

$$N_1 = 25 \text{ in.} + 2(3^1/_2 \text{ in.})$$
$$= 32 \text{ in.}$$
$$B_1 = 23 \text{ in.} + 2(3^1/_2 \text{ in.})$$
$$= 30 \text{ in.}$$
$$A_2 = B_1 \times N_1$$
$$= 30 \text{ in.}(32 \text{ in.})$$
$$= 960 \text{ in.}^2$$

Check $\sqrt{A_2/A_1} = 1.29 \leq 2$ **o.k.**

Calculate required base-plate thickness.

$$m = \frac{N - 0.95\,d}{2}$$
$$= \frac{25 \text{ in.} - 0.95(14.03 \text{ in.})}{2}$$
$$= 5.84 \text{ in.}$$
$$n = \frac{B - 0.8b_f}{2}$$
$$= \frac{23 \text{ in.} - 0.8(12.570 \text{ in.})}{2}$$
$$= 6.47 \text{ in.}$$

$$\phi_c P_p = \phi_c \left(0.85 f_c' A_1 \sqrt{\frac{A_2}{A_1}}\right)$$
$$= 0.6 \left(0.85 \times 3 \text{ ksi} \times 575 \text{ in.}^2 \sqrt{\frac{960 \text{ in.}^2}{575 \text{ in.}^2}}\right)$$
$$= 1{,}140 \text{ kips}$$
$$X = \left(\frac{4db_f}{(d + b_f)^2}\right)\frac{P_u}{\phi_c P_p}$$
$$= \left(\frac{4(14.0 \text{ in.})(12.6 \text{ in.})}{(14.0 \text{ in.} + 12.6 \text{ in.})^2}\right)\left(\frac{1{,}100 \text{ kips}}{1{,}140 \text{ kips}}\right)$$
$$= 0.962$$

$$\lambda = \frac{2\sqrt{X}}{1 + \sqrt{1 - X}} \leq 1$$
$$= \frac{2\sqrt{0.962}}{1 + \sqrt{1 - 0.962}}$$
$$= 1.64 \rightarrow 1$$

$$\lambda n' = \lambda \frac{\sqrt{d b_f}}{4}$$

$$= \frac{(1)\sqrt{14.0 \text{ in.} \times 12.6 \text{ in.}}}{4}$$

$$= 3.32 \text{ in.}$$

From this

$$l = \max(m, n, \lambda n')$$
$$= \max(5.84 \text{ in.}, 6.47 \text{ in.}, 3.32 \text{ in.})$$
$$= 6.47 \text{ in.}$$

and

$$t_{req} = l\sqrt{\frac{2P_u}{0.9F_y BN}}$$

$$= (6.47 \text{ in.})\sqrt{\frac{2(1{,}100 \text{ kips})}{0.9(36 \text{ ksi})(23 \text{ in.})(25 \text{ in.})}}$$

$$= 2.22$$

Use PL2$^1/_2$×23×2′-1.

PART 14 REFERENCES

DeWolf, J.T. and D.T. Ricker, 1990, AISC Design Guide No. 1, *Column Base Plates*, AISC, Chicago, IL.

Fisher, J.M. and M.A. West, 1997, AISC Design Guide No. 10 *Erection Bracing of Low-Rise Structural Steel Frames*, AISC, Chicago, IL.

Fling, R.S., 1970, "Design of Steel Bearing Plates," *Engineering Journal*, Vol. 7, No. 2, (April), pp. 37–39, AISC, Chicago, IL.

International Code Council, 2000, *International Building Code*, ICC, Falls Church, VA.

Marsh, M.L., and E.G. Burdette, 1985, "Anchorage of Steel Building Components to Concrete," *Engineering Journal*, Vol. 15, No. 4, (4th Qtr.), pp. 33–39, AISC, Chicago, IL.

Marsh, M.L., and E.G. Burdette, 1985, "Multiple Bolt Anchorages: Method for Determining the Effective Projected Area of Overlapping Stress Cones," *Engineering Journal*, Vol. 15, No. 4, (4th Qtr.), pp. 29–32, AISC, Chicago, IL.

Murray, T.M., 1983, "Design of Lightly Loaded Column Base Plates," *Engineering Journal*, Vol. 20, No. 4, (4th Qtr.), pp. 143–152, AISC, Chicago, IL.

Popov, E.P. and R.M. Stephen, 1977, "Capacity of Columns with Splice Imperfections," *Engineering Journal*, Vol. 14, No. 1, (1st Qtr.), pp. 16–23, AISC, Chicago, IL.

Ricker, D.T., 1989, "Some Practical Aspects of Column Base Selection," *Engineering Journal*, Vol. 26, No. 3, (3rd Qtr.), AISC, Chicago, IL.

Shipp, J.G. and E.R. Haninger, 1983, "Design of Headed Anchor Bolts," *Engineering Journal*, Vol. 20, No. 2, (2nd Qtr.), pp. 58–69, AISC, Chicago. IL.

Thornton, W.A., 1990a, "Design of Small Base Plates for Wide-Flange Columns," *Engineering Journal*, Vol. 27, No. 3, (3rd Qtr.), pp. 108–110, AISC, Chicago, IL.

Thornton, W.A., 1990b, "Design of Small Base Plates for Wide-Flange Columns—A Concatenation of Methods," *Engineering Journal*, Vol. 27, No. 4, (4th Qtr.), pp. 173–174, AISC, Chicago, IL.

Table 14-1.
Finish Allowances

Size	Thickness (in.)	Add to Fin. One Side (in.)	Add to Fin. Two Sides (in.)
Maximum dimension 24 in. or less	1¼ or less over 1¼ to 2, incl.	1⁄16 ⅛	⅛ ¼
Maximum dimension over 24 in.	1¼ or less over 1¼ to 2, incl.	⅛ 3⁄16	¼ ⅜
56 in. wide or less	over 2 to 7½, incl. over 7½ to 10, incl. over 10 to 15, incl.	¼ ½ ¾	⅜ ⅝ ⅞
Over 56 in. wide to 72 in. wide	over 2 to 6, incl. over 6 to 10, incl. over 10 to 15, incl.	¼ ½ ¾	⅜ ⅝ ⅞

Table 14-2.
Sizes for Anchor-Rod Holes in Base Plates

Anchor-Rod Diameter, in.	Hole Diameter, in.	Min. Washer Diameter, in.	Anchor-Rod Diameter, in.	Hole Diameter, in.	Min. Washer Diameter, in.
3/4	1 5/16	1 7/8	1 1/2	2 5/16	3 1/8
7/8	1 9/16	2 1/4	1 3/4	2 3/4	3 3/4
1	1 13/16	2 5/8	2	3 1/4	4 1/2
1 1/4	2 1/16	2 7/8	2 1/2	3 3/4	5

Note:
The Min. Washer Diameter is calculated assuming the anchor rod is located at the hole edge and is the minimum diameter required to completely cover the hole. Actual washer diameter (or width for rectangular plate) washer may vary. Washer thickness and size must be selected based upon forces to be transferred by the anchor rods.

Table 14-3. Typical Column Splices

Case I:
All-bolted flange-plated column splices between columns with depth d_u and d_l nominally the same.

Column Size	Gage g_u or g_l	Flange Plates			
		Type	Width	Thk.	Length
W14×455 to 730	13½	1	16	¾	1′ 6½
257 to 426	11½	1	14	⅝	1′ 6½
145 to 233	11½	1	14	½	1′ 6½
90 to 132	11½	2	14	⅜	1′ 0½
43 to 82	5½	2	8	⅜	1′ 0½
W12×120 to 336	5½	2	8	⅝	1′ 0½
40 to 106	5½	2	8	⅜	1′ 0½
W10×33 to 112	5½	2	8	⅜	1′ 0½
W8×31 to 67	5½	2	8	⅜	1′ 0½
24 & 28	3½	2	6	⅜	1′ 0½

Gages shown may be modified if necessary to accommodate fittings elsewhere on the column.

Case I-A: $d_l = (d_u + 1/4$ in.) to $(d_u + 5/8$ in.)	Flange plates: Select g_u for upper column; select g_l and flange plate dimensions for lower columns (see table above). Fillers: None. Shims: Furnish sufficient strip shims 2½×1⅛ to provide 0 to $^1/_{16}$-in. clearance each side.
Case I-B: $d_l = (d_u - 1/4$ in.) to $(d_u + 1/8$ in.)	Flange plates: Same as Case I-A. Fillers (shop bolted under flange plates): Select thickness as ⅛-in. for $d_l = d_u$ and $d_l = (d_u + 1/8$ in.) or as ¼-in. for $d_l = (d_u - 1/8$ in.) and $d_l = (d_u - 1/4$ in.) Select width to match flange plate and length as 0′ 9 for Type 1 or 0′ 6 for Type 2. Shims: Same as Case I-A.
Case I-C: $d_l = (d_u + 3/4$ in.) and over.	Flange plates: Same as Case I-A. Fillers (shop bolted to upper column): Select thickness as $(d_l - d_u) / 2$ minus ⅛-in., whichever results in ⅛-in. multiples of filler thickness. Select width to match flange plate, but not greater than upper column flange width. Select length as 1′ 0 for Type 1 or 0′ 9 for Type 2. Shims: Same as Case I-A.

For lifting devices, see Figure 11-21.

Table 14-3 (cont.).
Typical Column Splices

Case I:
All-bolted flange-plated column splices between columns with depth d_u and d_l nominally the same.

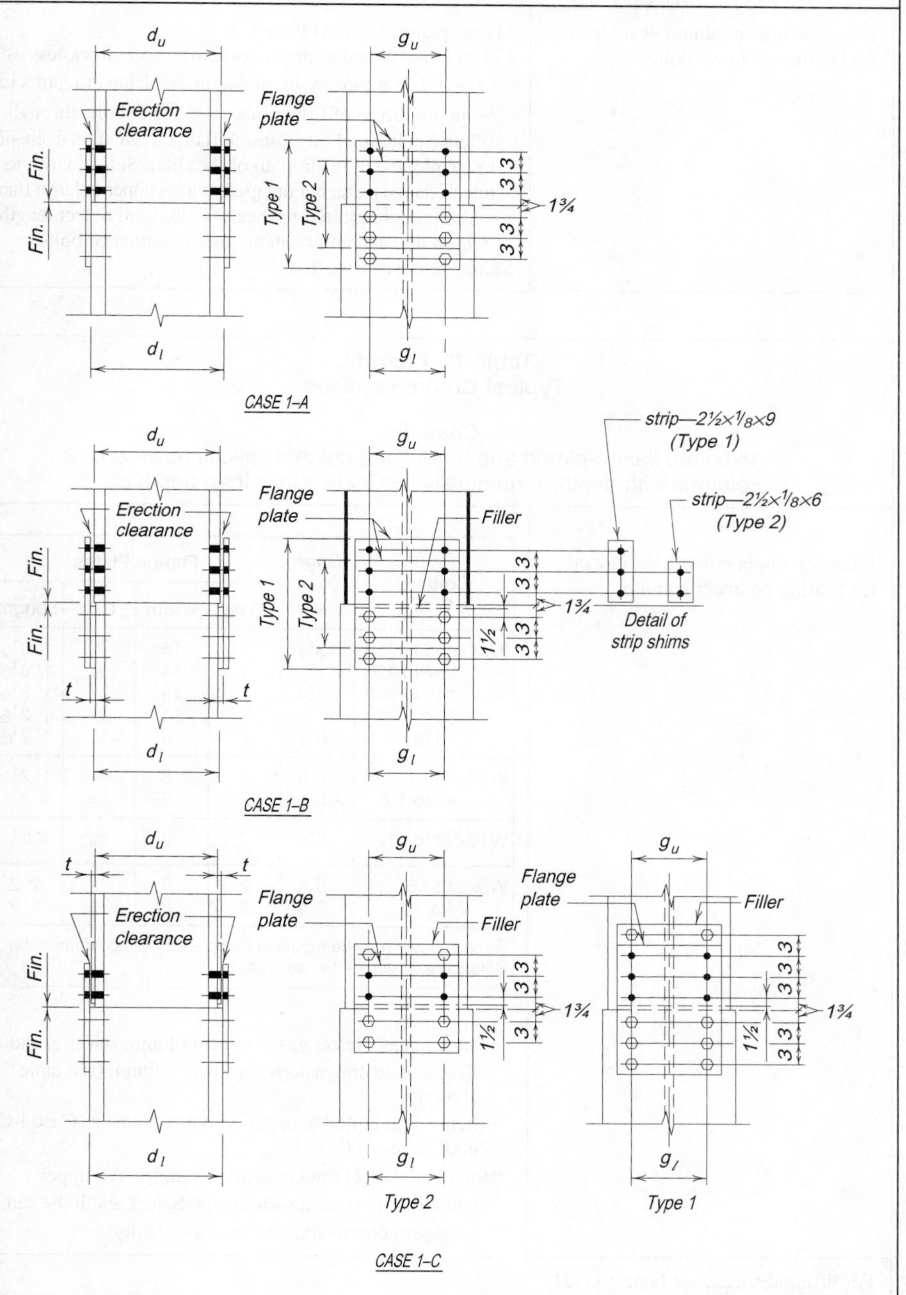

Table 14-3 (cont.). Typical Column Splices

Case II:

All-bolted flange-plated column splices between columns with depth d_u nominally two inches less than depth d_l.

Fillers on upper column developed for bearing on lower column.	Flange plates: Same as Case I-A. Fillers (shop bolted to upper column): Select thickness as $(d_l - d_u) / 2$ minus ⅛-in. or $^3\!/_{16}$-in., whichever results in ⅛-in. multiples of filler thickness. Select bolts through fillers (including bolts through flange plates) on each side to develop bearing stregnth of the filler. Select width to match flange plate, but not greater than upper column flange width unless required for bearing strength. Select length as required to accommodate required number of bolts. Shims: Same as Case I-A.

Table 11-4 (cont.). Typical Column splices

Case III:

All-bolted flange-plated and butt-plated column splices between columns with depth d_u nominally two inches less than depth d_l.

Fillers on upper column developed for bearing on lower column.

Column Size	Gage g_u or g_l	Flange Plates			
		Type	Width	Thk.	Length
W14×455 to 730	13½	1	16	¾	1′ 8½
257 to 426	11½	1	14	⅝	1′ 8½
145 to 233	11½	1	14	½	1′ 8½
90 to 132	11½	2	14	⅜	1′ 2½
43 to 82	5½	2	8	⅜	1′ 2½
W12×120 to 336	5½	2	8	⅝	1′ 2½
40 to 106	5½	2	8	⅜	1′ 2½
W10×33 to 112	5½	2	8	⅜	1′ 2½
W8×31 to 67	5½	2	8	⅜	1′ 2
24 & 28	3½	2	8	⅜	1′ 2

Gages shown may be modified if necessary to accommodate fittings elsewhere on the column.

Flange plates: Select g_u for upper column, select g_l and flange plate dimensions for lower column (see table above).

Fillers (shop bolted to upper column): Same as Case I-C.

Shims: Same as Case I-A.

Butt plate: Select thickness as 1½-in. for W8 upper column or two inches for others. Select width the same as upper column and length as $d_l - \frac{1}{4}$ in.

For lifting devices, see Figure 11-21.

Table 14-3 (cont.).
Typical Column Splices

Case II:
All-bolted flange-plated column splices between columns with depth d_u nominally two inches less than depth d_l.

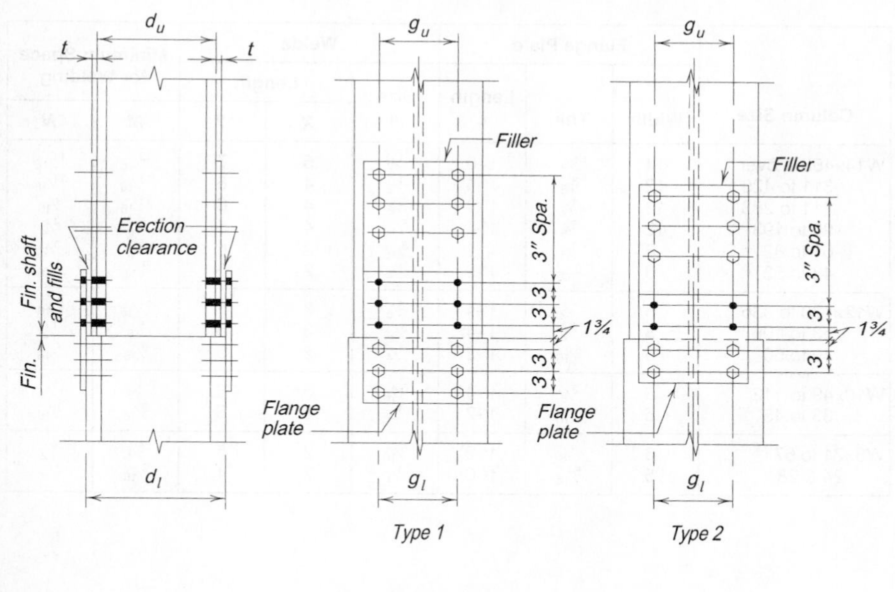

CASE II

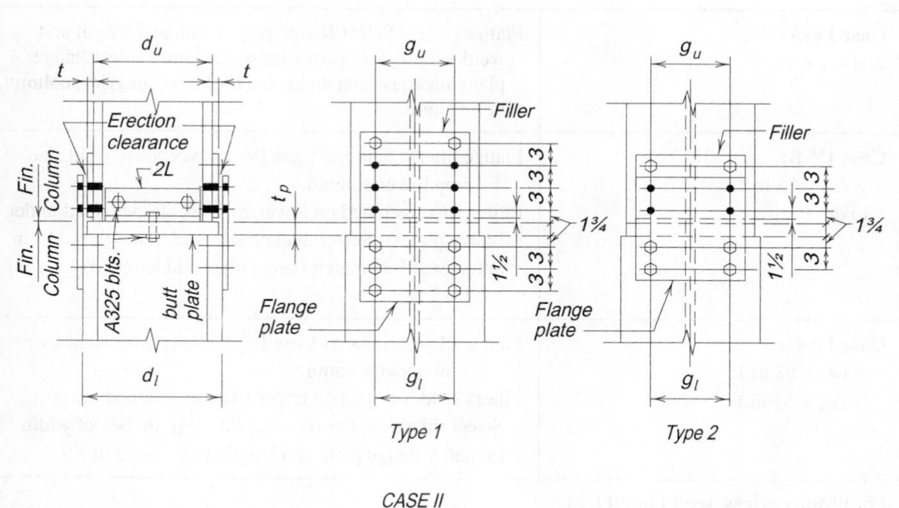

CASE II

Table 14-3 (cont.). Typical Column Splices

Case IV:

All-welded flange-plated column splices between columns with depths d_u and d_l nominally the same.

Column Size	Flange Plate			Welds			Minimum Space for Welding	
	Width	Thk.	Length L	Size A	Length X	Length Y	M	N
W14×455 & over	14	5/8	1′-6	1/2	5	7	13/16	11/16
311 to 426	12	5/8	1′-4	1/2	4	6	13/16	11/16
211 to 283	12	1/2	1′-4	3/8	4	6	11/16	9/16
90 to 193	12	3/8	1′-4	5/16	4	6	5/8	1/2
61 to 82	8	3/8	1′-4	5/16	3	6	5/8	1/2
43 to 53	6	5/16	1′-2	1/4	2	5	9/16	7/16
W12×120 to 336	8	1/2	1′-4	3/8	3	6	11/16	9/16
53 to 106	8	3/8	1′-4	5/16	3	6	5/8	1/2
40 to 50	6	5/16	1′-2	1/4	2	5	9/16	7/16
W10×49 to 112	8	3/8	1′-4	5/16	3	6	5/8	1/2
33 to 45	6	5/16	1′-2	1/4	2	5	9/16	7/16
W8×31 to 67	6	3/8	1′-2	5/16	2	5	5/8	1/2
24 & 28	5	5/16	1′-0	1/4	2	4	9/16	7/16

Case IV-A: $d_l = (d_u + 1/8)$	Flange plates: Select flange-plate width and length and weld lengths for upper (lighter) column; select flange-plate thickness and weld size for lower (heavier) column. Fillers: None.
Case IV-B: $d_l = (d_u - 1/4$ in.) to d_u	Flange plates: Same as Case IV-A, except use weld size $A + t$ on lower column. Fillers (undeveloped on lower column, shop welded under flange plates): Select thickness t as $(d_l - d_u) / 2 + 1/16$ in. Select width to match flange plate and length as $L / 2 - 2$ in.
Case IV-C: $d_l = (d_u + 1/4$ in.) to $(d_u + 1/2$ in.)	Flange plates: Same as Case IV-A, except use weld size $A + t$ on upper column. Fillers (undeveloped on upper column, shipped loose): Select thickness t as $(d_l - d_u) / 2 - 1/16$ in. Select width to match flange plate and length as $L / 2 - 2$ in.

For lifting devices, see Figure 11-21.

Table 14-3 (cont.). Typical Column Splices

Case IV:
All-welded flange-plated column splices between columns with depths d_u and d_l nominally the same.

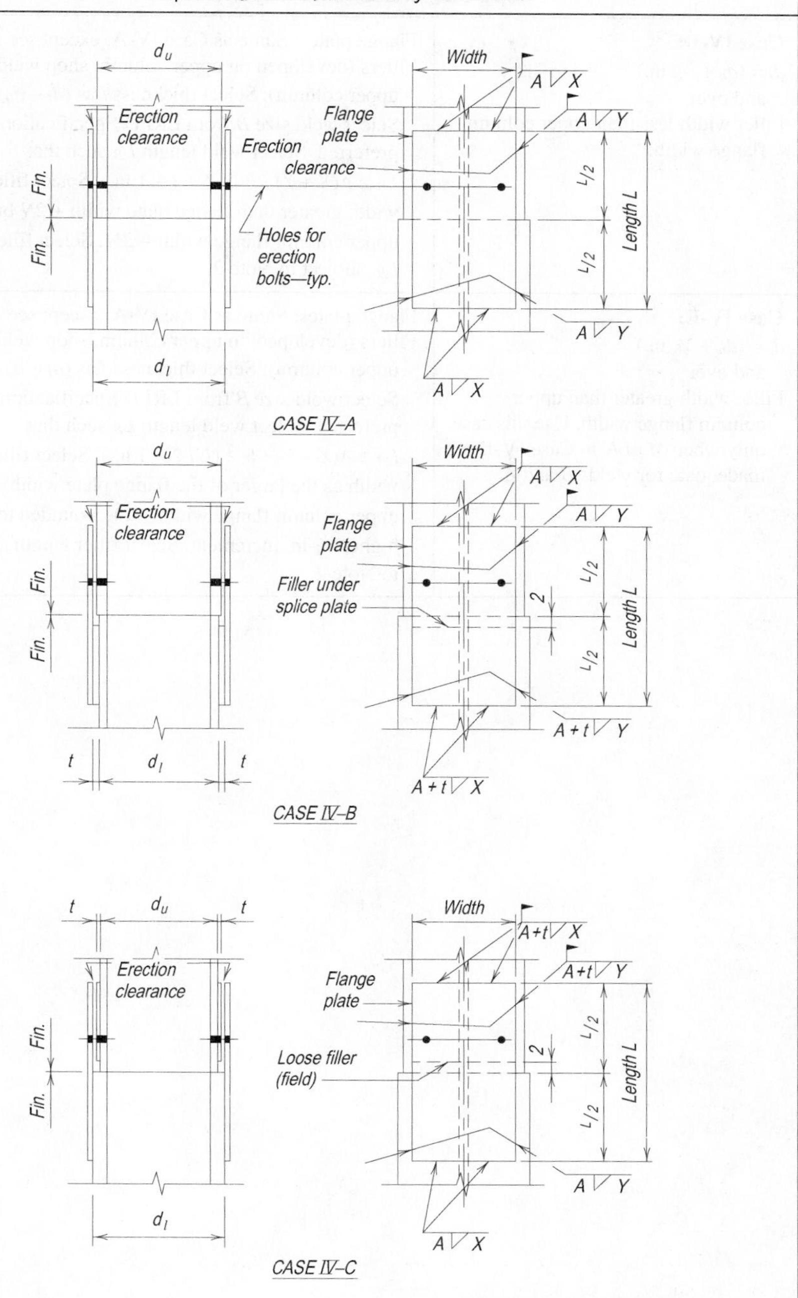

Table 14-3 (cont.). Typical Column Splices **Case IV:** All-welded flange-plated column splices between columns with depths d_u and d_l nominally the same	
Case IV-D: $d_l = (d_u + \frac{5}{8}$ in.) and over Filler width less than upper column flange width.	Flange plates: Same as Case IV-A, except see Note 1. Fillers (developed on upper column, shop welded to upper column): Select thickness t as $(d_l - d_u) / 2 - \frac{1}{16}$ in. Select weld size B from LRFD Specification; ≤5⁄16-in. preferred. Select weld length L_B such that $L_B \geq A(X + Y) / B \geq (L / 2 + 1$ in.). Select filler width greater than flange plate width + $2N$ but less than upper column flange width – $2M$. Select filler length as L_B, subject to Note 2.
Case IV-E: $d_l = (d_u + \frac{5}{8}$ in.) and over Filler width greater than upper column flange width. Use this case only when M or N in Case IV-D are inadequate for welds B and A.	Flange plates: Same as Case IV-A, except see Note 1. Fillers (developed on upper column, shop welded to upper column): Select thickness t as $(d_l - d_u) / 2 - \frac{1}{16}$ in. Select weld size B from LRFD Specification; ≤5⁄16-in. preferred. Select weld length L_B such that $L_B \geq A(X + Y) / B \geq (L / 2 + 1$ in.). Select filler width as the larger of the flange plate width + $2N$ and the upper column flange width + $2M$, rounded to the next higher 1⁄4-in. increment. Select filler length as L_B subject to Note 2.

Table 14-3 (cont.). Typical Column Splices

Case IV:

All-welded flange-plated column splices between columns with depths d_u and d_l nominally the same

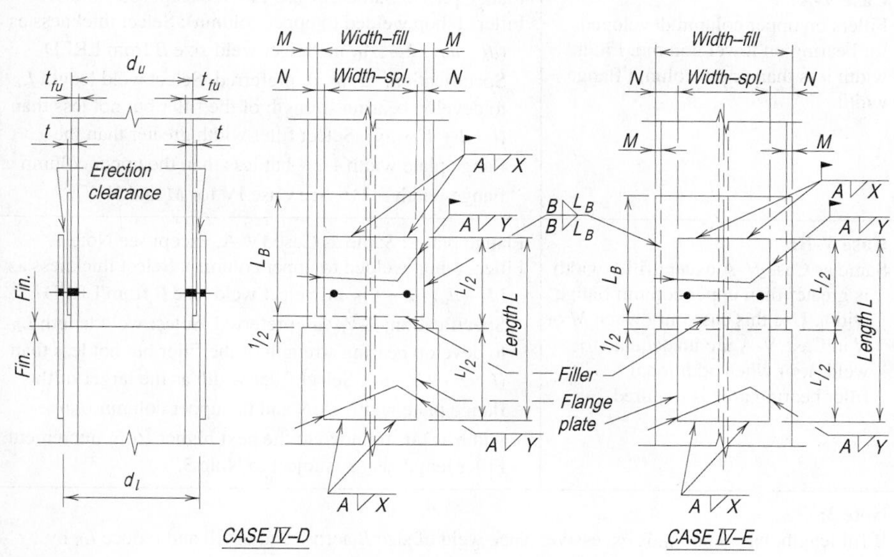

Note 1:
Where welds fasten flange plates to developed fillers, or developed fillers to column flanges (Cases IV-E and V-B), use the table to the right to check minimum fill thickness for balanced fill and weld shear strength.

Weld A E70XX	Minimum Fill Thickness for Balanced Weld and Plate Shear	
	F_y	
	36	50
¼	0.26	0.19
5⁄16	0.32	0.23
3⁄8	0.38	0.28
7⁄16	0.45	0.33
½	0.51	0.37

Assume that an E70XX weld with $A = ½$, $X = 4$, and $Y = 6$ is to be used at full strength on an A36 fill ¼-in. thick. Since this table shows that the minimum fill thickness to develop this ½-in. weld is 0.51 in., the ¼-in. fill will be overstressed. A balanced condition is obtained by multiplying the length $(X + Y)$ by the ratio of the minimum to the actual thickness of fill, thus:

$$(4 + 6) \times \frac{0.51}{0.25} = 20.4$$

use $(X + Y) = 20½$-in.

Placing this additional increment of $(X + Y)$ can be done by making weld lengths X continuous across the end of the splice plate and by increasing lengths Y (and therefore the plate length) if required.

Note 2:
If fill length, based on L_B, is excessive, place weld of size B across one or both ends of fill and reduce L_B accordingly, but not to less than $(L / 2 + 1)$. Omit return welds in Cases IV-E and V-B.

<table>
<tr><td colspan="2">

Table 14-3 (cont.).
Typical Column Splices

Case V:
All-welded flange-plated column splices between columns with depth d_u nominally two inches less than depth d_l.

</td></tr>
<tr><td>

Case V-A:
Fillers on upper column developed for bearing on lower column. Filler width less than upper column flange width.

</td><td>

Flange plates: Same as Case IV-A, except see Note 1.
Fillers (shop welded to upper column): Select thickness as $(d_l - d_u) / 2 - 1/16$ in. Select weld size *B* from LRFD Specification; $\leq 5/16$ in. preferred. Select weld length L_B to develop bearing strength of the filler but not less than $(L / 2 + 1\frac{1}{2}$ in.). Select filler width greater than the flange plate width + 2*N* but less than the upper column flange width – 2*M*. See Case IV for *M* and *N*.

</td></tr>
<tr><td>

Case V-B:
Same as Case V-A except filler width is greater than upper column flange width. Use this case only when *M* or *N* in Case V-A are inadequate for weld *A*, or when additional filler bearing area is required.

</td><td>

Flange plates: Same as Case IV-A, except see Note 1.
Fillers (shop welded to upper column): Select thickness as $(d_l - d_u) / 2 - 1/16$ in. Select weld size *B* from LRFD Specification; $\leq 5/16$ in. preferred. Select weld length L_B to develop bearing strength of the filler but not less than $(L / 2 + 1\frac{1}{2}$ in.). Select filler width as the larger of the flange plate width + 2*N* and the upper column flange width + 2*M*, rounded to the next higher 1/4 in. increment. Filler length as L_B, subject to Note 3.

</td></tr>
<tr><td colspan="2">

Note 3:
If fill length, based on L_B, is excessive, place weld of size *B* across end of fill and reduce L_B by one-half of such additional weld length, but not to less than $(L / 2 + 1\frac{1}{2})$. Omit return welds in Case V-B.

</td></tr>
</table>

Table 14-3 (cont.).
Typical Column Splices

Case V:
All-welded flange-plated column splices between columns with depth d_u nominally two inches less than depth d_l.

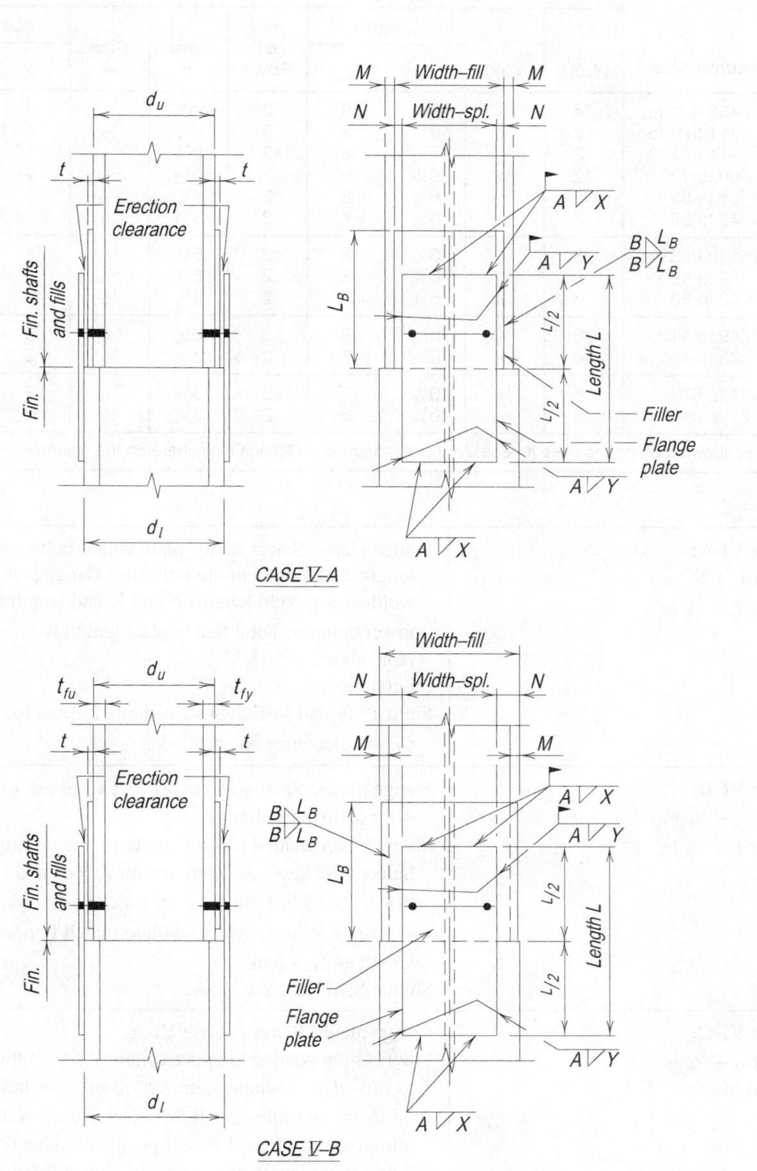

Table 14-3 (cont.). Typical Column Splices

Case VI:
Combination bolted and welded column splices between columns with depths d_n and d_l nominally the same.

Column Size	Flange Plate				Bolts		Welds		
	Width	Thk.	Length		No. of Rows	Gage g	Size A	Length	
			L_U	L_L				X	Y
W14×455 & over	14	5/8	9¼	9	3	11½	½	5	7
311 to 426	12	5/8	9¼	8	3	9½	½	4	6
211 to 283	12	½	9¼	8	3	9½	3/8	4	6
90 to 193	12	3/8	6¼	8	2	9½	5/16	4	6
61 to 82	8	3/8	6¼	8	2	5½	5/16	3	6
43 to 53	6	5/16	6¼	7	2	3½	¼	2	5
W12×120 to 336	8	½	6¼	8	2	5½	3/8	3	6
53 to 106	8	3/8	6¼	8	2	5½	5/16	3	6
40 to 50	6	5/16	6¼	7	2	3½	¼	2	5
W10×49 to 112	8	3/8	6¼	8	2	5½	5/16	3	6
33 to 45	6	5/16	6¼	7	2	3½	¼	2	5
W8×31 to 67	6	3/8	6¼	7	2	3½	5/16	2	5
24 & 28	5	5/16	6¼	6	2	3½	¼	2	4

Gages shown may be modified if necessary to accommodate fittings elsewhere on the columns.

Case VI-A: $d_l = (d_u + 1/4$ in.) to $(d_u + 5/8$ in.)	Flange plates: Select flange plate width, bolts, gage and length L_U for upper column; select flange plate thickness, weld size A, weld lengths X and Y, and length L_L for lower column. Total flange plate length is $L_U + L_L$ (see table above). Fillers: None. Shims: Furnish sufficient strip shims 2½×⅛ to obtain 0 to 1/16-in. clearance on each side.
Case VI-B: $d_l = (d_u - 1/4$ in.) to $(d_u + 1/8$ in.)	Flange plates: Same as Case VI-A, except use weld size $A + t$ on lower column. Fillers (shop welded to lower column under flange plate): Select thickness t as ⅛-in. for for $d_l = d_u$ and $d_l = (d_u + 1/8$ in.) or as 3/16-in. for $d_l = (d_u - 1/8$ in.) and $d_l = (d_u - 1/4$ in.). Select width to match flange plate and length as L_L – 2 in. Shims: Same as Case VI-A.
Case VI-C: $d_l = (d_u + 3/4$ in.) and over	Flange plates: Same as Case VI-A. Fillers (shop welded to upper column): Select thickness t as $(d_l - d_u) / 2$ minus ⅛-in. or 3/16-in., whichever results in ⅛-in. multiples of fill thickness. Select weld size B as minimum size from LRFD Specification Section J2. Select weld length as L_U – ¼ in. Select filler width as flange plate width and filler length as L_U – ¼-in. Shims: Same as Case VI-A.

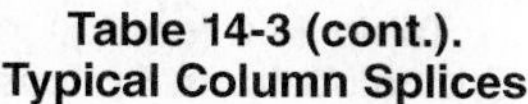

Table 14-3 (cont.). Typical Column Splices

Case VI:

Combination bolted and welded column splices between columns with depths d_n and d_l nominally the same.

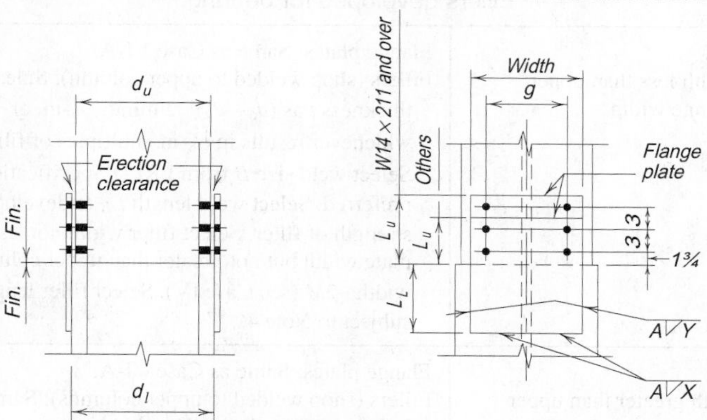

CASE VI–A

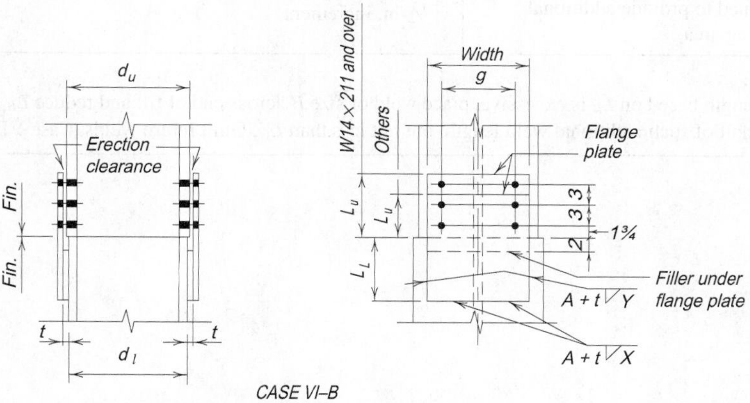

CASE VI–B

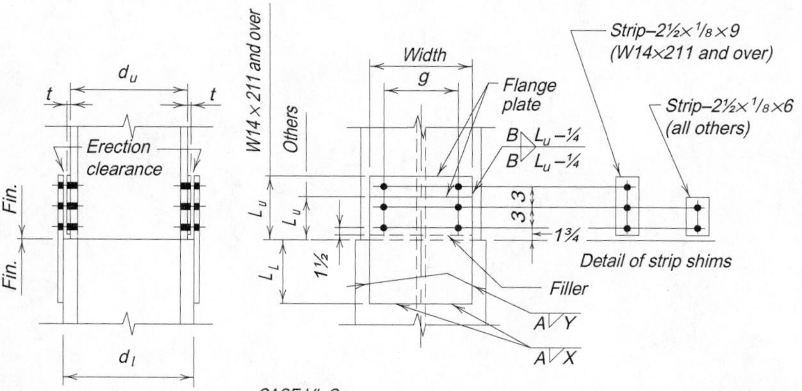

CASE VI–C

Table 14-3 (cont.). Typical Column Splices

Case VII:

Combination bolted and welded flange-plated column splices between columns with depth d_u nominally two inches less than depth d_l
Fillers developed for bearing.

Case VII-A: Fillers of width less than upper column flange width.	Flange plates: Same as Case VI-A. Fillers (shop welded to upper column): Select filler thickness t as $(d_l - d_u)$ / 2 minus ⅛-in. or $^3\!/_{16}$-in., whichever results in ⅛-in. multiples of filler thickness. Select weld size B from LRFD Specification; $\leq ^5\!/_{16}$-in. preferred. Select weld length L_B to develop bearing strength of filler. Select filler width not less than flange plate width but not greater than upper column flange width $-2M$ (see Case IV). Select filler length as L_B, subject to Note 4.
Case VII-B: Filler of width greater than upper column flange width. Use Case VII-B only when fillers must be widened to provide additional bearing area.	Flange plates: Same as Case VI-A. Fillers (shop welded to upper columns): Same as Case VII-A except select filler width as upper column flange width + $2M$ (see Case IV) rounded to the next larger ½-in. increment.

Note 4:
If fill length based on L_B is excessive, place weld of size B across end of fill and reduce L_B by one-half of such additional weld length, but not less than L_U. Omit return welds, Case VII-B.

Table 14-3 (cont.). Typical Column Splices

Case VII:

Combination bolted and welded flange-plated column splices between columns with depth d_u nominally two inches less than depth d_l
Fillers developed for bearing.

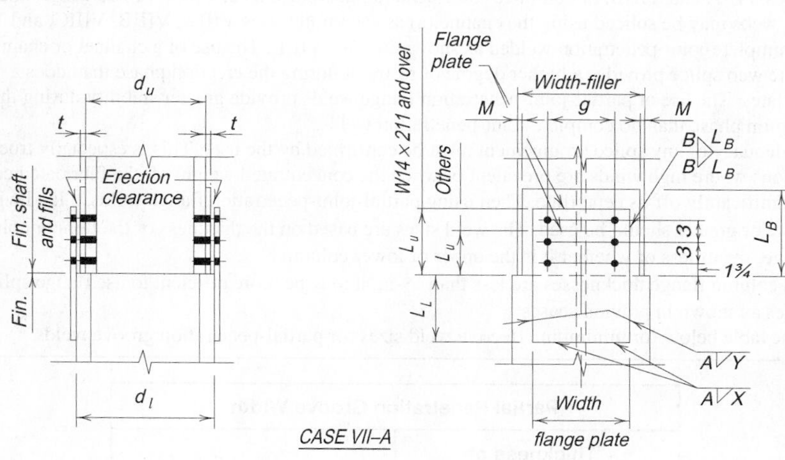

CASE VII–A

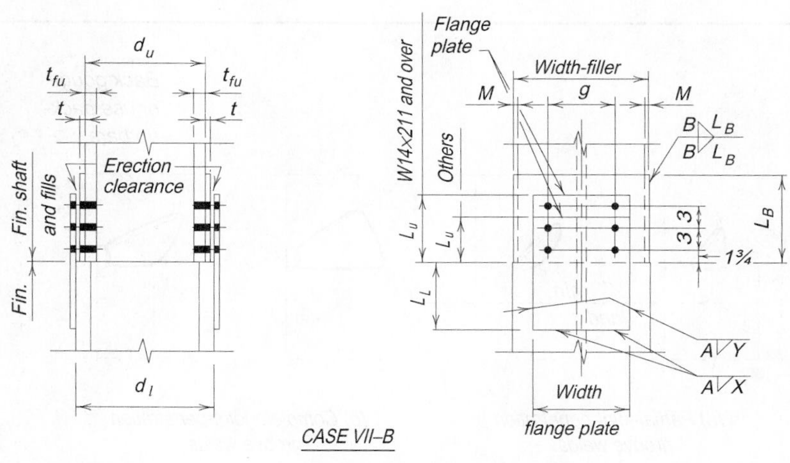

CASE VII–B

Table 14-3 (cont.). Typical Column Splices

Case VIII: Directly welded flange column splices between columns with depths d_u and d_l nominally the same.

These types of splices exhibit versatility. The flanges may be partial-joint-penetration welded as in Cases VIIIA and VIIIB, or complete-joint-penetration welded as in Cases VIIIC, VIIID, and VIIIE. The webs may be spliced using the channel(s) as shown in Cases VIIIA, VIIIB, VIIIC, and VIIID, or complete-joint-penetration welded as shown in Case VIIIE. The use of a channel or channels at the web splice provides a higher degree of restraint during the erection phase than does a plate or plates. The use of partial-joint-penetration flange welds provide greater stability during the erection phase than do complete-joint-penetration welds.

The adequacy of any splice arrangement must be confirmed by the user. This is especially true in regions where high winds are prevalent or when the concentrated weight of the fabricated column is significantly off its centerline. Then using partial-joint-penetration flange welds, a land width of ¼-in. or greater should be used. The weld sizes are based on the thickness of the thinner column flange, regardless of whether it is the upper or lower column.

When column flange thicknesses are less than ½-in. it may be more efficient to use flange splice plates as shown in previous cases.

See the table below for minimum effective weld sizes for partial-penetration groove welds.

Partial Penetration Groove Width

[a]Thickness of Column Material T_u	Minimum Effective Welds Size E
[b]Over ½ to ¾, incl.	¼
Over ¾ to 1½, incl.	5⁄16
Over 1½ to 2¼, incl.	⅜
Over 2¼ to 6, incl.	½
Over 6	⅝

[a]Thickness of thicker part joined.
[b]For less than ½, use splice plates.

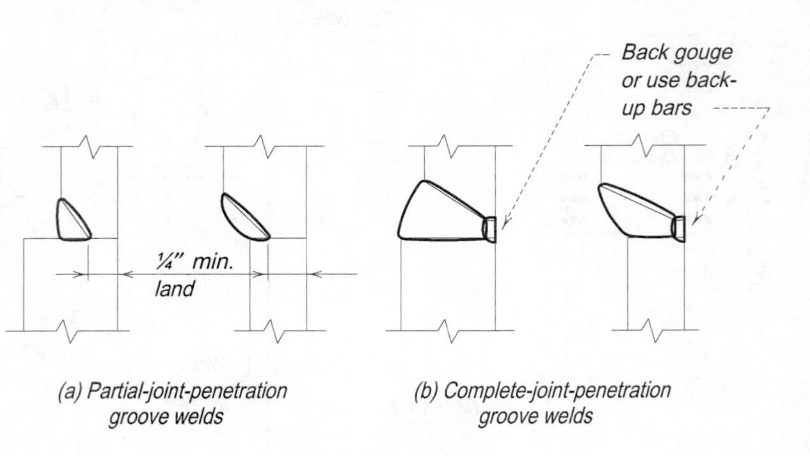

(a) Partial-joint-penetration groove welds

(b) Complete-joint-penetration groove welds

Table 14-3 (cont.).
Typical Column Splices

Directly welded flange column splices between columns with depths d_u and d_l nominally the same.

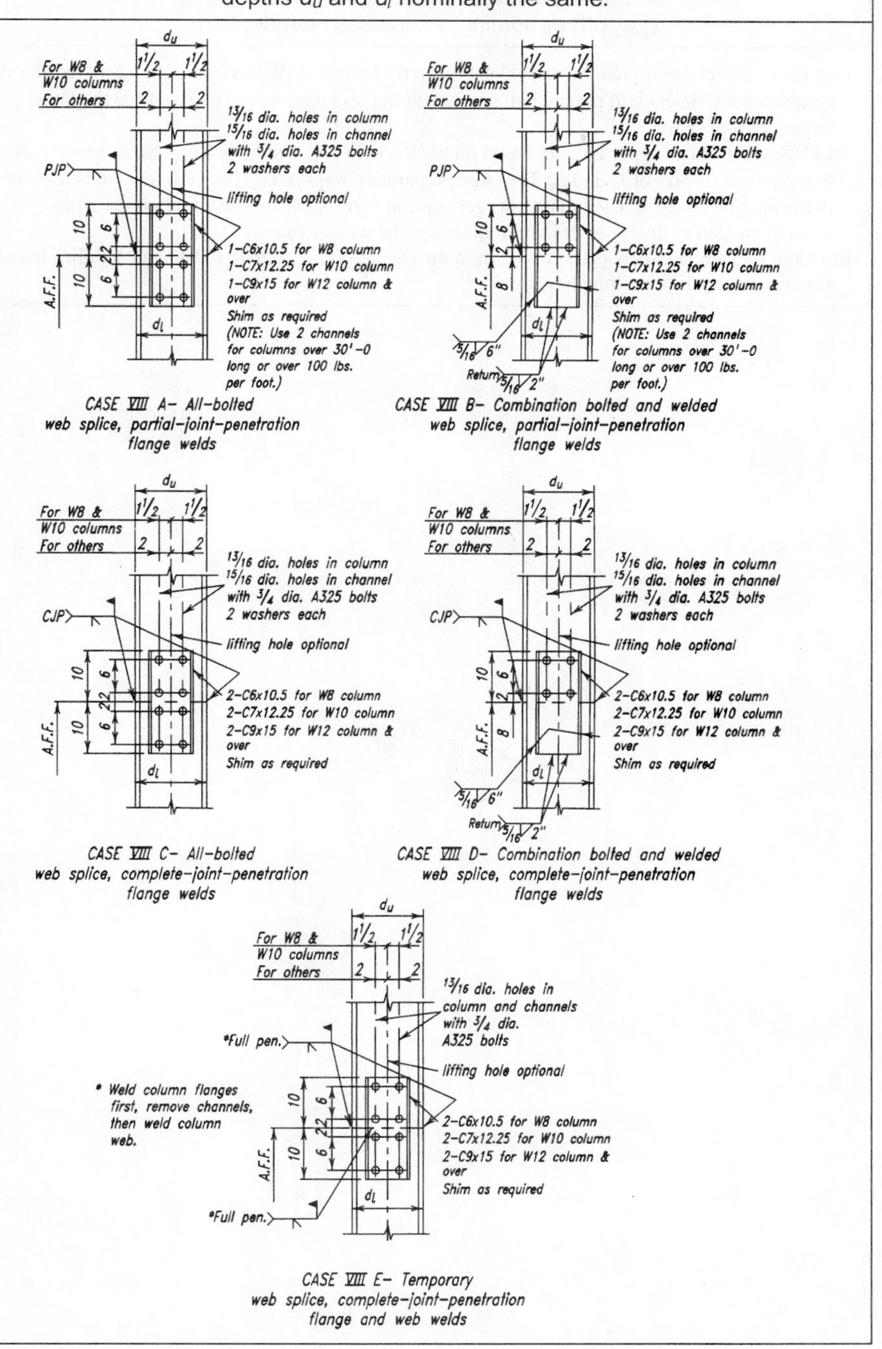

CASE VIII A- All-bolted web splice, partial-joint-penetration flange welds

CASE VIII B- Combination bolted and welded web splice, partial-joint-penetration flange welds

CASE VIII C- All-bolted web splice, complete-joint-penetration flange welds

CASE VIII D- Combination bolted and welded web splice, complete-joint-penetration flange welds

CASE VIII E- Temporary web splice, complete-joint-penetration flange and web welds

Table 14-3 (cont.). Typical Column Splices

Case IX:
Butt-plated column splices between columns with depth d_u nominally 2 in. less than depth d_l.

Butt plate: Select a butt plate thickness of 1½-in. for W8 over W10 columns and 2 in. for all other combinations. Select butt plate width and length not less than w_l and d_l assuming the lower is the larger column shaft.

Weld: Select weld to upper column based on the thicker of t_{fu} and t_p. Select weld to lower column based on the thicker of t_{fl} and t_p. The edge preparation required by the groove weld is usually performed on the column shafts. However, special cases such as when the butt plate must be field welded to the lower column require special consideration.

Erection: clip angles, such as those shown in the sketch below, help to locate and stabilize the upper column during the erection phase.

Table 14-3 (cont.).
Typical Column Splices

Case IX:
Butt-plated column splices between columns with depth d_u nominally 2 in. less than depth d_l.

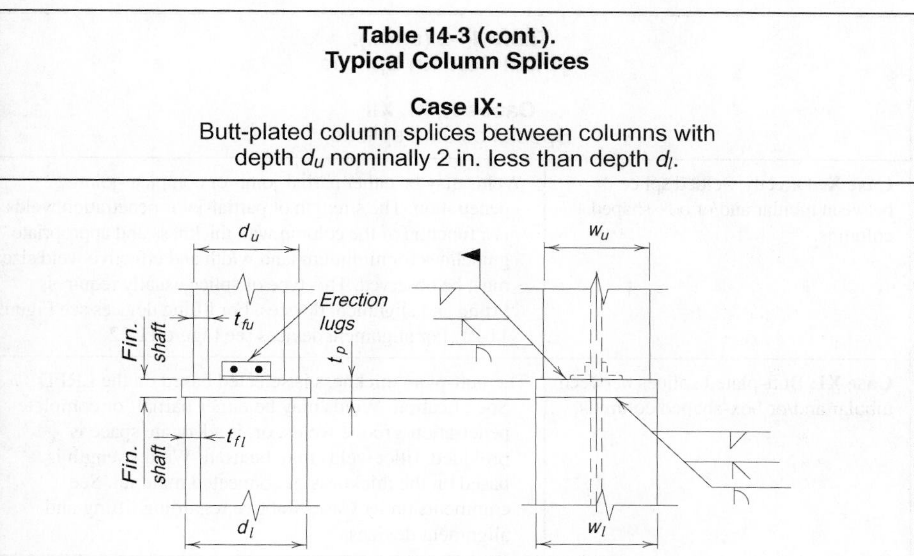

CASE IX

Table 14-3 (cont.). Typical Column Splices **Cases X, XI, XII** Special column splices	
Case X: Directly welded splice between tubular and/or box-shaped columns.	Welds may be either partial-joint- or complete-joint-penetration. The strength of partial-joint-penetration welds is a function of the column wall thickness and appropriate guidelines for minimum land width and effective weld size must be observed. This type of splice usually requires lifting and alignment devices. For lifting devices see Figure 11-21. For alignment devices see Figure 11-22.
Case XI: Butt-plated splices between tubular and/or box-shaped columns.	The butt-plate thickness is selected based on the LRFD Specification. Welds may be either partial- or complete-penetration-groove welds, or, if adequate space is provided, fillet welds may be used. Weld strength is based on the thickness of connected material. See comments under Case X above regarding lifting and alignment devices.
Case XII: Butt-plated column splices between W-shape columns and tubular or box-shaped columns.	See comments under Case XI above.

Table 14-3 (cont.). Typical Column Splices

Cases X, XI, XII
Special column splices

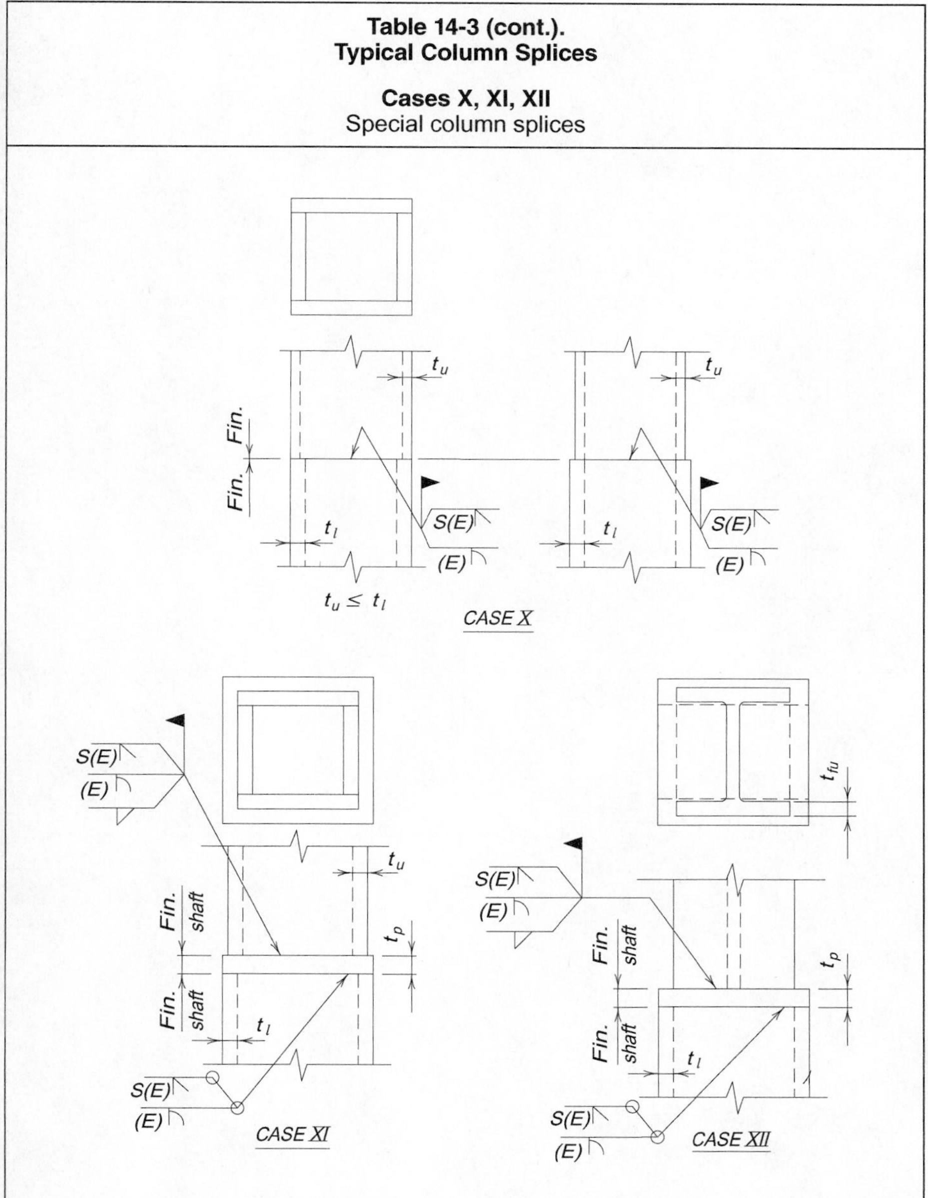

PART 15

DESIGN OF HANGER CONNECTIONS, BRACKET PLATES AND CRANE-RAIL CONNECTIONS

OTHER

SCOPE

The specification requirements and other design considerations summarized in this Part apply to the design of hanger connections, bracket plates and crane-rail connections. For the design of similar connections for HSS and steel pipe, see the AISC *Hollow Structural Sections Connections Manual*.

HANGER CONNECTIONS

Hanger connections, illustrated in Figure 15-1, are usually made with a plate, tee, angle, or pair of angles. The design strength of a hanger connection is determined from the applicable limit states for the bolts (see Part 7), welds (see Part 8) and connection elements (see Part 9), as illustrated in Example 15-1. In all cases, the design strength ϕR_n must exceed the required strength R_u.

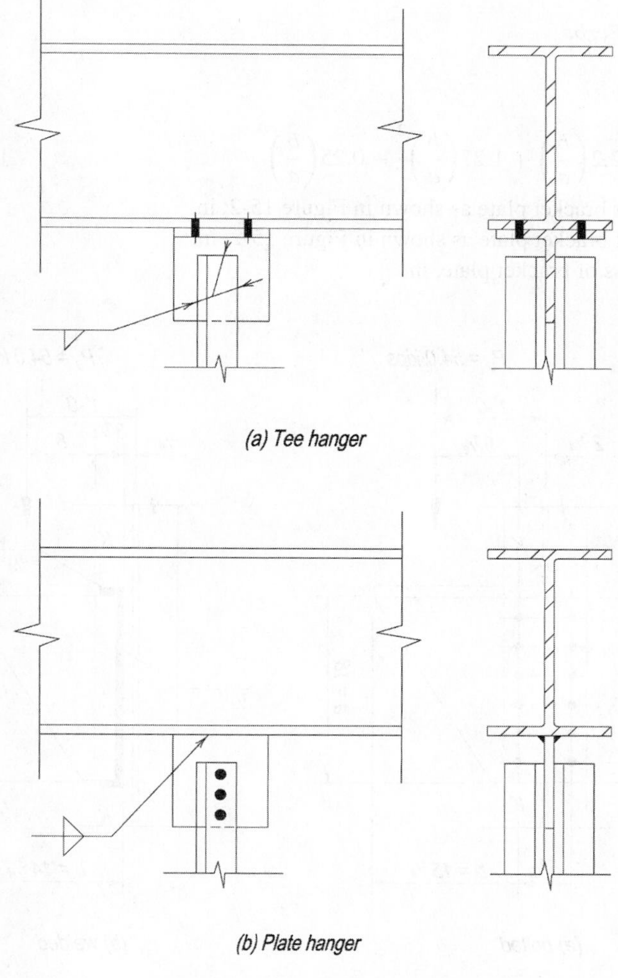

Fig. 15–1. Typical hanger connections.

BRACKET PLATES

A bracket plate, illustrated in Figure 15-2, acts as a cantilevered beam. The design strength of a bracket plate is determined from the applicable limit states for the bolts (see Part 7), welds (see Part 8) and connection elements (see Part 9), as illustrated in Examples 15-2 and 15-3. Additionally, flexural yielding, flexural rupture and local buckling must be checked.

For flexural yielding, the design strength of the bracket plate is ϕM_n, where

$$\phi = 0.90$$
$$M_n = F_y S_x$$

where S_x is the gross elastic section modulus of the bracket plate. Additionally, triangular-shaped bracket plates should be checked for flexural yielding on the free edge (Salmon and Johnson, 1996). In lieu of a more detailed analysis, the load on the bracket plate can be limited by the following design strength:

$$\phi P_n = 0.85 F_y z b t$$

where

$$z = 1.39 - 2.2\left(\frac{b}{a}\right) + 1.27\left(\frac{b}{a}\right)^2 - 0.25\left(\frac{b}{a}\right)^3$$

b = width of bracket plate as shown in Figure 15-2, in.
a = depth of bracket plate as shown in Figure 15-2, in.
t = thickness of bracket plate, in.

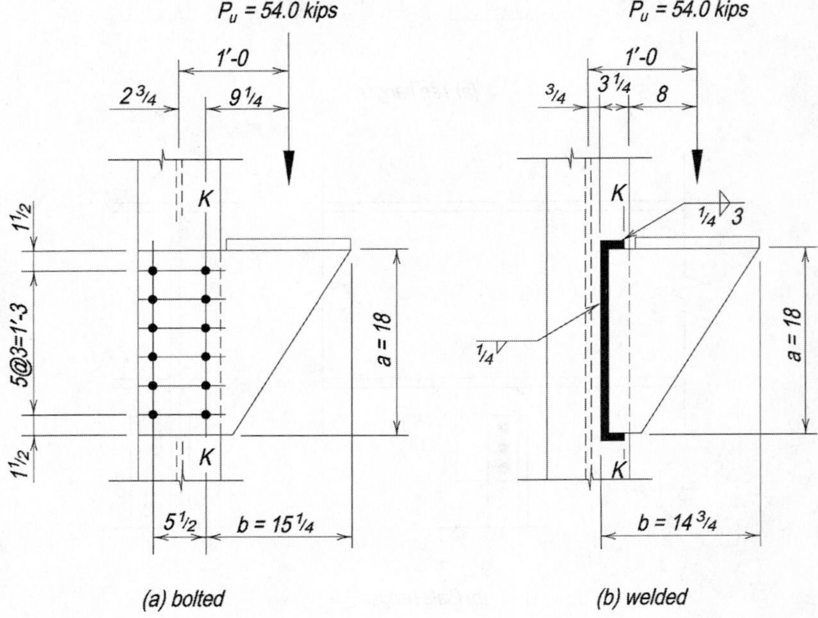

Fig. 15–2. Bracket-plate connections.

For flexural rupture, the design strength of the bracket plate is ϕM_n, where

$$\phi = 0.75$$
$$M_n = F_u S_{net}$$

where S_{net} is the net elastic section modulus of the bracket plate. Values of S_{net} are given in Table 15-2 for various bolt hole diameters and numbers of fasteners at 3-in. spacing.

Local buckling can be prevented in bracket plates if the following width-thickness ratios are satisfied, provided the centroid of the applied load is approximately $0.6b$ from the line of support (line K in Figure 15-2a) and lateral movement of the outstanding portion of the bracket plate is prevented (Salmon and Johnson, 1996):

$$\frac{b}{t} \leq \frac{250}{\sqrt{F_y}} \quad \text{for} \quad 0.5 < b/a \leq 1.0$$

$$\frac{b}{t} \leq \frac{250}{\sqrt{F_y}}\left(\frac{b}{a}\right) \quad \text{for} \quad 1.0 < b/a \leq 2.0$$

CRANE-RAIL CONNECTIONS

Bolted splices, properly maintained, provide a suitable means of connecting rail joints for light to medium duty cycle cranes. For high duty cycle cranes welded rail splices are preferred. Welded splice bars are not recommended.

Bolted Splices

Standard rail drilling and joint-bar punching, as furnished by manufacturers of light standard rails for track work, include round holes in rail ends and slotted holes in joint bars to receive standard oval-neck track bolts. Holes in rails are oversized and punching in joint bars is spaced to allow 1/16-in. to 1/8-in. clearance between rail ends (see manufacturers' catalogs for spacing and dimensions of holes and slots). Although this construction is satisfactory for track and light crane service, its use in general crane service may lead to high maintenance and joint failure. Welded splices are therefore preferable.

For best service in bolted splices, it is recommended that tight joints be required for all rails for crane service. This will require rail ends to be finished, and the special rail drilling and joint-bar punching tabulated below. Special rail drilling is accepted by some mills, or rails may be ordered blank for shop drilling. End finishing of standard rails can be done at the mill. However, light rails often must be end-finished in the shop or ground at the site prior to erection. In the crane rail range from 104 to 175 lbs per yard, rails and joint bars are manufactured to obtain a tight fit and no further special end finishing, drilling, or punching is required. Because of cumulative tolerance variations in holes, bolt diameters, and rail ends, a slight gap may sometimes occur. It may sometimes be necessary to ream holes through joined bar and rail to permit entry of bolts.

Joint bars for crane service are provided in various sections to match the rails. Joint bars for light and standard rails can be purchased blank for special shop punching to obtain tight joints. See Bethlehem Steel Corp. Booklet 3351 for dimensions, material specifications, and the identification necessary to match the crane rail section.

Joint-bar bolts, as distinguished from oval-neck track bolts, have straight shanks to the head and are manufactured to ASTM A449 specifications. Nuts are manufactured to ASTM A563 grade B specifications. Alternatively, ASTM A325 bolts and compatible ASTM A563 nuts can be used. Bolt assembly includes an alloy steel spring washer, furnished to AREMA specifications. After installation, bolts should be retightened within 30 days and every three months thereafter.

Welded Splices

When welded splices are specified, consult the manufacturer for recommended rail-end preparation, welding procedure, and method of ordering. Although the joint continuity made possible by this method of splicing is desirable, the careful control required in all stages of the welding operation may be difficult to meet during crane rail installation. Rails should not be attached to structural supports by welding. Rails with holes for joint-bar bolts should not be used in making welded splices.

Hook Bolt Fastenings

Hook bolts (Figure 15-3) are used primarily with light rails when attached to beams that are too narrow for clamps. Rail adjustment to $\pm 1/2$ in. is inherent in the threaded shank. Hook bolts are paired alternately 3 to 4 in. apart, spaced at about 24 in. on center. The special rail drilling required must be done in the fabricator's shop. Hook bolts are not recommended for use with heavy-duty cycle cranes (CMAA Classes, D, E, and F). It is generally recommended that hook bolts should not be used in runway systems that are longer than 500 ft because the bolts do not allow for longitudinal movement of the rail.

Rail Clip Fastenings

Rail clips are forged or cast devices that are shaped to match specific rail profiles. They are usually bolted to the runway girder flange with one bolt or are sometimes welded. Rail clips have been used satisfactorily with all classes of cranes. However, one drawback is that when a single bolt is used, the clip can rotate in response to rail longitudinal movement. This clip rotation can cause cam action that might force the rail out of alignment. Because of this limitation, rail clips should only be used in crane systems subject to infrequent use, and for runways less than 500 ft in length.

Rail Clamp Fastenings

Rail clamps are a common method of attachment for heavy-duty cycle cranes. Rail clamps are detailed to provide two types: tight and floating (see Figure 15-4). Each clamp consists of two plates: an upper clamp plate and a lower filler plate. Dimensions shown are suggested. See manufacturers' catalogs for recommended gages, bolt sizes, and detail dimensions not shown.

The lower plate is flat and nominally matches the height of the toe of the rail flange. The upper plate covers the lower plate and extends over the top of the lower rail flange. In the tight clamp, the upper plate is detailed to fit tightly to the lower tail flange top, thus "clamping" it tightly in place when the fasteners are tightened. In the past, the tight clamp had been illustrated with the filler plates fitted tightly against the rail flange toe. This tight fit-up was

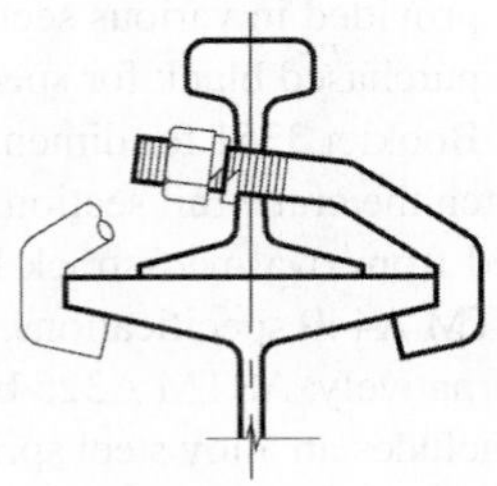

Fig. 15–3. Hook bolts.

rarely achieved in practice and is not considered to be necessary to achieve a tight type clamp. In the floating type clamp, the pieces are detailed to provide a clearance both alongside the rail flange toe and below the upper plate. The floating type does not, in reality, clamp the rail but merely holds the rail within the limits of the clamp clearances. High-strength bolts are recommended for both clamp types. Both types should be spaced 3 ft or less apart.

Patented Rail Clip Fastenings

Each manufacturer's literature presents in detail the desirable aspects of the various designs. In general, patented rail clips are easy to install due to their range of adjustment and provide both limitation of lateral movement and allowance for longitudinal movement. Patented rail clips should be considered as a viable alternative to conventional hook bolts, clips, or clamps. Because of their desirable characteristics, patented rail clips can be used without restriction except as limited by the specific manufacturer's recommendations. Installations using patented rail clips sometimes incorporate pads beneath the rail. When this is done the lateral float of the rail should be limited as in the case of the tight rail clamps.

DESIGN EXAMPLES

EXAMPLE 15.1. Hanger connection design.

Given: Refer to Figure 15-5. Design a WT tension-hanger connection for a 2L3×3×5/16 tension member to W24×94 beam connection. For the beam and WT, $F_y = 50$ ksi and $F_u = 65$ ksi; for the double angles, $F_y = 36$ ksi

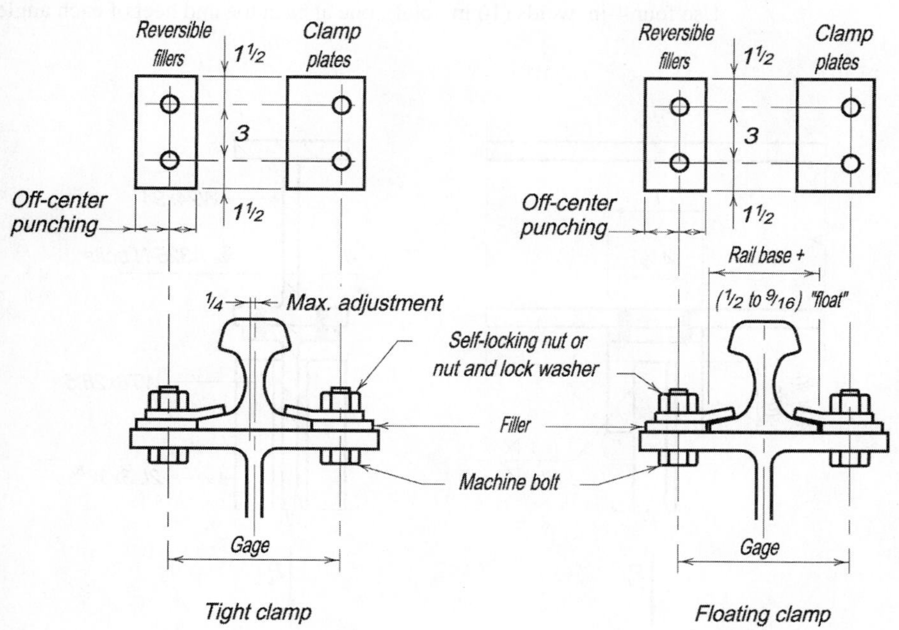

Fig. 15–4. Rail clamps.

and $F_u = 58$ ksi. Use 3/4-in. diameter ASTM A325-N bolts and 70 ksi electrodes.

$$P_u = 80 \text{ kips}$$

W24×94, ASTM A992

$$d = 24.3 \text{ in.} \qquad b_f = 9.07 \text{ in.}$$
$$t_w = 0.515 \text{ in.} \qquad t_f = 0.875 \text{ in.}$$

2L3×3×5/16, ASTM A36

$$A = 3.55 \text{ in.}^2 \qquad \bar{x} = 0.865 \text{ in.}$$

Solution: *Check tension yielding of angles*

$$\phi R_n = \phi F_y A_g$$
$$= 0.9(36 \text{ ksi})(3.55 \text{ in.}^2)$$
$$= 115 \text{ kips} > 80 \text{ kips} \quad \textbf{o.k.}$$

Check tension rupture of angles (Design welds to find length of connection and U)

Try 1/4-in. fillet welds

$$L_{\min} = \frac{P_u}{1.392D}$$
$$= \frac{80 \text{ kips}}{1.392(4 \text{ sixteenths})}$$
$$= 14.4 \text{ in.}$$

Use four 4-in. welds (16 in. total), one at each toe and heel of each angle.

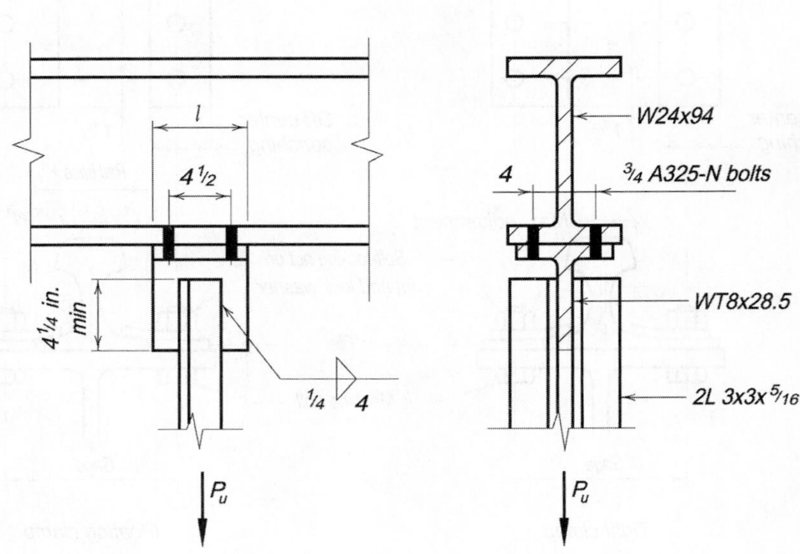

Fig. 15–5. Illustration for Example 15.1.

Calculate effective net area

From LRFD Specification Section B3

$$U = 1 - \frac{\bar{x}}{L} \le 0.9$$
$$= 1 - \frac{0.865 \text{ in.}}{4 \text{ in.}}$$
$$= 0.784$$

(Note: in lieu of the calculation shown above, U may be taken as 0.75 since $1.5w > l \ge w$ per LRFD Specification Section B3.2d)

$$A_e = UA_n$$
$$= 0.784(3.55 \text{ in.}^2)$$
$$= 2.78 \text{ in.}^2$$
$$\phi R_n = \phi F_u A_e$$
$$= 0.75(58 \text{ ksi})(2.78 \text{ in.}^2)$$
$$= 121 \text{ kips} > 80 \text{ kips} \quad \textbf{o.k.}$$

Select preliminary WT using beam gage $g = 4$ in.

With four 3/4-in. diameter A325-N bolts,

$$r_{ut} = \frac{P_u}{n}$$
$$= \frac{80 \text{ kips}}{4 \text{ bolts}}$$
$$= 20 \text{ kips/bolt}$$

Since for 3/4-in. diameter A325N bolts $\phi r_n = 29.8$ kips(>20 kips), the bolts are **o.k.**

With four bolts, the maximum effective length is $2g = 8$ in. Thus, there are 4 in. of tee length tributary to each pair of bolts and

$$\frac{2 \text{ bolts } (20 \text{ kips/bolt})}{4 \text{ in.}} = 10.0 \text{ kips/in.}$$

The minimum depth of WT that can be used is equal to the sum of the weld length plus the weld size plus the k-dimension for the selected section. From Table 9-2 with an assumed $b = 4 \text{ in.}/2 = 2$ in., to $t_o \approx$ 11/16 in., and $d_{\text{min}} = 4 \text{ in.} + 1/4 \text{ in.} + k \approx 6$ in., appropriate selections include:

WT6×39.5	WT8×28.5
WT7×34	WT9×30

Try WT8×28.5; $b_f = 7.12$ in., $t_f = 0.715$ in., $t_w = 0.430$ in.

Check prying action with WT8×28.5×0′-8

$$b = \frac{g - t_w}{2}$$
$$= (4 \text{ in.} - 0.430 \text{ in.})/2$$
$$= 1.79 \text{ in.} > 1^1/_4\text{-in. entering and tightening clearance,} \quad \textbf{o.k.}$$

$$a = \frac{b_f - g}{2}$$
$$= (7.12 \text{ in.} - 4 \text{ in.})/2$$
$$= 1.56 \text{ in.}$$

Since $a = 1.56$ in. is less than $1.25b = 2.24$ in., use $a = 1.56$ in.

$$b' = b - \frac{d}{2}$$
$$= 1.79 \text{ in.} - (^3/_4 \text{ in.}/2)$$
$$= 1.42 \text{ in.}$$

$$a' = a + \frac{d}{2}$$
$$= 1.56 \text{ in.} + (^3/_4 \text{ in.}/2)$$
$$= 1.94 \text{ in.}$$

$$\rho = \frac{b'}{a'}$$
$$= 1.42 \text{ in.}/1.94 \text{ in.}$$
$$= 0.732$$

$$\beta = \frac{1}{\rho}\left(\frac{\phi r_n}{r_{ut}} - 1\right)$$
$$= \frac{1}{0.732}\left(\frac{29.8 \text{ kips/bolt}}{20 \text{ kips/bolt}} - 1\right)$$
$$= 0.669$$

Since $\beta < 1.0$,

$$\delta = 1 - \frac{d'}{p}$$
$$= 1 - (^{13}/_{16} \text{ in.}/4 \text{ in.})$$
$$= 0.797$$

$$\alpha' = \frac{1}{\delta}\left(\frac{\beta}{1-\beta}\right) \leq 1.0$$
$$= \frac{1}{0.797}\left(\frac{0.669}{1 - 0.669}\right)$$
$$= 2.54 \rightarrow 1.0$$

$$t_{req} = \sqrt{\frac{4.44 r_{ut} b'}{p F_y (1 + \delta\alpha')}}$$
$$= \sqrt{\frac{4.44(20 \text{ kips/bolt})(1.42 \text{ in.})}{(4 \text{ in.})(50 \text{ ksi})[1 + (0.797)(1.0)]}}$$
$$= 0.592 \text{ in.} < t_f = 0.715 \text{ in.} \quad \textbf{o.k.}$$

Check design tensile strength of bolts.

(Note this calculation is optional; the required thickness t_{req}, calculated above, will keep the total bolt tensile force $r_{ut} + q_u$ less than the design strength ϕr_n. It is included for information only.)

Calculate q_u

$$t_c = \sqrt{\frac{4.44\phi r_n b'}{pF_y}}$$
$$= \sqrt{\frac{4.44(29.8 \text{ kips/bolt})(1.42 \text{ in.})}{(4 \text{ in.})(50 \text{ ksi})}}$$
$$= 0.969 \text{ in.}$$

$$\alpha = \frac{1}{\delta}\left[\frac{r_{ut}}{\phi r_n}\left(\frac{t_c}{t}\right)^2 - 1\right] \geq 0$$
$$= \frac{1}{0.797}\left(\frac{20 \text{ kips/bolt}}{29.8 \text{ kips/bolt}}\left(\frac{0.969 \text{ in.}}{0.715 \text{ in.}}\right)^2 - 1\right)$$
$$= 0.292$$

$$q_u = \phi r_n\left[\delta\alpha\rho\left(\frac{t}{t_c}\right)^2\right]$$
$$= 29.8 \text{ kips/bolt}\left[0.797(0.292)(0.732)\left(\frac{0.715 \text{ in.}}{0.969 \text{ in.}}\right)^2\right]$$
$$= 2.76 \text{ kips/bolt}$$

Total tension on bolt

$$r_{ut} + q_u = 20 \text{ kips/bolt} + 2.76 \text{ kips/bolt}$$
$$= 22.8 \text{ kips/bolt} < 29.8 \text{ kips/bolt} \qquad \textbf{o.k.}$$

Check the WT hanger as follows:

Check tension yielding of the tee stem on the Whitmore section (see sketch below)

The effective width of the tee stem (which cannot exceed the actual width of 8 in.) is

$$L_w = 3 \text{ in.} + 2(4 \text{ in.} \times \tan 30°) \leq 8 \text{ in.}$$
$$= 7.62 \text{ in.}$$

and the design strength is

$$\phi R_n = \phi F_y A_{geff}$$
$$= 0.9(50 \text{ ksi})(7.62 \text{ in.} \times 0.430 \text{ in.})$$
$$= 147 \text{ kips} > 80 \text{ kips} \qquad \textbf{o.k.}$$

Check shear yielding of the base metal along the toe and heel of each weld line.

$$\phi R_n = \phi(0.6F_y)A_g$$
$$= 0.9(0.6 \times 50 \text{ ksi})(4 \times 4 \text{ in.} \times 0.430 \text{ in.})$$
$$= 186 \text{ kips} > 80 \text{ kips} \qquad \textbf{o.k.}$$

Check shear rupture of the base metal along the toe and heel of each weld line.

$$\begin{aligned}\phi R_n &= \phi(0.6F_u)A_n \\ &= 0.75(0.6 \times 65 \text{ ksi})(4 \times 4 \text{ in.} \times 0.430 \text{ in.}) \\ &= 201 \text{ kips} > 80 \text{ kips} \quad \textbf{o.k.}\end{aligned}$$

Check shear rupture of the flanges.

$$\begin{aligned}\phi R_n &= \phi[0.6F_u A_n] \\ &= 0.75[0.6(65 \text{ ksi})(2 \times 8 \text{ in.} - 4 \times 0.875 \text{ in.})(0.715 \text{ in.})] \\ &= 261 \text{ kips} > 80 \text{ kips} \quad \textbf{o.k.}\end{aligned}$$

Check shear yielding of the flanges.

$$\begin{aligned}\phi R_n &= \phi[0.6F_y A_g] \\ &= 0.9[0.6(50 \text{ ksi})(2 \times 8 \text{ in.} \times 0.715 \text{ in.})] \\ &= 309 \text{ kips} > 80 \text{ kips} \quad \textbf{o.k.}\end{aligned}$$

Check block shear rupture of the tee stem.

From LRFD Specification Section J4.3

$$\begin{aligned}0.6F_u A_{nv} &= 0.6(65 \text{ ksi})(2 \times 4 \text{ in.} \times 0.430 \text{ in.}) \\ &= 134 \text{ kips} \\ F_u A_{nt} &= (65 \text{ ksi})(3 \text{ in.} \times 0.430 \text{ in.}) \\ &= 83.9 \text{ kips}\end{aligned}$$

Since $0.6F_u A_{nv} > F_u A_{nt}$,

$$\begin{aligned}\phi R_n &= \phi[0.6F_u A_{nv} + F_y A_{gt}] \leq \phi[0.6F_u A_{nv} + F_u A_{nt}] \\ &= 0.75[134 \text{ kips} + (50 \text{ ksi})(3 \text{ in.} \times 0.430 \text{ in.})] \\ &\leq 0.75[134 \text{ kips} + 83.9 \text{ kips}] \\ &= 149 \text{ kips} \leq 163 \text{ kips} \\ &= 149 \text{ kips} > 80 \text{ kips} \quad \textbf{o.k.}\end{aligned}$$

Comments: Alternatively, a WT tension hanger could be selected with a flange thickness to reduce the effect of prying action to an insignificant amount, i.e., $q_u \approx 0$. Using $b' = 1.42$ in. as an assumption,

$$\begin{aligned}t_{req} &= \sqrt{\frac{4.44 r_{ut} b'}{pF_y}} \\ &= \sqrt{\frac{4.44(20 \text{ kips/bolt})(1.42 \text{ in.})}{(4 \text{ in./bolt})(50 \text{ ksi})}} \\ &= 0.794 \text{ in.}\end{aligned}$$

A WT9×35.5 with $t_f = 0.810$ in., $t_w = 0.495$ in. (>0.430 in.), and $b_f = 7.635$ in. is adequate.

EXAMPLE 15.2. Bolted bracket plate design.

Given:

Refer to Figure 15-2a. Design a bracket plate to support the factored force indicated. Use $^7/_8$-in. diameter A325-N bolts. For the plate, assume $F_y = 36$ ksi and $F_u = 58$ ksi.

Solution:

Select bolt group.

For $^7/_8$-in. diameter A325-N bolts in single shear, $\phi r_n = 21.6$ kips (Table 7-10). Thus,

$$\begin{aligned} C_{\min} &= \frac{R_u}{\phi r_n} \\ &= 54.0 \text{ kips}/21.6 \text{ kips} \\ &= 2.50 \end{aligned}$$

From Table 7-19 with $\theta = 0°$, a $5^1/_2$-in. gage with $s = 3$ in., $e_x = 12$ in., and $n = 6$ provides

$$C = 4.53 > 2.50 \qquad \textbf{o.k.}$$

Try PL$^3/_8$-in. × 18 in.

Check bolt bearing.

From Table 7-12, $\phi r_n = 91.4$ kips/bolt. Since this is greater than the single-shear strength of one bolt, bolt bearing is not critical.

Check flexure in the bracket plate.

On line K, the required strength M_u is

$$\begin{aligned} M_u &= P_u e_b \\ &= 54.0 \text{ kips}(12 \text{ in.} - 2^3/_4 \text{ in.}) \\ &= 500 \text{ in.-kips} \end{aligned}$$

For flexural yielding on line K,

$$\begin{aligned} \phi M_n &= \phi F_y S_x \\ &= 0.9(36 \text{ ksi})\frac{(^3/_8 \text{ in.})(18 \text{ in.})^2}{6} \\ &= 656 \text{ in.-kips} > 500 \text{ in.-kips} \qquad \textbf{o.k.} \end{aligned}$$

For flexural rupture on line K, with $S_{net} = 13.7$ in.^{3} from Table 15-1,

$$\begin{aligned} \phi M_n &= \phi F_u S_{net} \\ &= 0.75(58 \text{ ksi})(13.7 \text{ in.}^3) \\ &= 596 \text{ in.-kips} > 500 \text{ in.-kips} \qquad \textbf{o.k.} \end{aligned}$$

For flexural yielding on the free edge of the triangular plate,

$$
\begin{aligned}
z &= 1.39 - 2.2\left(\frac{b}{a}\right) + 1.27\left(\frac{b}{a}\right)^2 - 0.25\left(\frac{b}{a}\right)^3 \\
&= 1.39 - 2.2\left(\frac{15^1/_4 \text{ in.}}{18 \text{ in.}}\right) + 1.27\left(\frac{15^1/_4 \text{ in.}}{18 \text{ in.}}\right)^2 \\
&\quad - 0.25\left(\frac{15^1/_4 \text{ in.}}{18 \text{ in.}}\right)^3 \\
&= 0.286 \\
\phi P_n &= 0.85 F_y z b t \\
&= 0.85(36 \text{ ksi})(0.286)(15^1/_4 \text{ in.})(^3/_8 \text{ in.}) \\
&= 50.0 \text{ kips} < 54.0 \text{ kips} \qquad \textbf{n.g.}
\end{aligned}
$$

Try PL$^1/_2$ in.$\times$18 in.

$$
\begin{aligned}
\phi P_n &= 0.85(36 \text{ ksi})(0.286)(15^1/_4 \text{ in})(^1/_2 \text{ in.}) \\
&= 66.7 \text{ kips} > 54.0 \text{ kips} \qquad \textbf{o.k.}
\end{aligned}
$$

Check local buckling of the bracket plate.

$$\frac{b}{a} = \frac{15^1/_4 \text{ in.}}{18 \text{ in.}} = 0.847$$

Since $0.5 \leq \dfrac{b}{a} < 1.0$

$$
\begin{aligned}
t_{min} &= b\left(\frac{\sqrt{F_y}}{250}\right) \\
&= 15\,^1/_4 \text{ in.}\left(\frac{\sqrt{36 \text{ ksi}}}{250}\right) \\
&= 0.366 \text{ in.} < {}^1/_2 \text{ in.} \qquad \textbf{o.k.}
\end{aligned}
$$

Check shear yielding of the bracket plate.

$$
\begin{aligned}
\phi R_n &= \phi(0.6F_y)A_g \\
&= 0.9(0.6 \times 36 \text{ ksi})(18 \text{ in.} \times {}^1/_2 \text{ in.}) \\
&= 175 \text{ kips} > 54.0 \text{ kips} \qquad \textbf{o.k.}
\end{aligned}
$$

Check shear rupture of the bracket plate.

$$
\begin{aligned}
\phi R_n &= \phi(0.6F_u)A_n \\
&= 0.75(0.6 \times 58 \text{ ksi})[18 - (6 \times 1 \text{ in.})](^1/_2 \text{ in.}) \\
&= 157 \text{ kips} > 54.0 \text{ kips} \qquad \textbf{o.k.}
\end{aligned}
$$

Check block shear rupture of the bracket plate (shear plane on line K, tension plane across bottom two bolts).

Since the eccentricity reduces the 12 bolts to 4.53 effective, the equivalent block shear rupture force applied concentric to the bolt group would be 54 kips times 12/4.53, or 143 kips.

$$
\begin{aligned}
A_{gv} &= (18 \text{ in.} - 1^1/_2 \text{ in.})(^1/_2 \text{ in.}) \\
&= 8.25 \text{ in.}^2 \\
A_{nv} &= 8.25 \text{ in.}^2 - 5.5(1 \text{ in.})(^1/_2 \text{ in.}) \\
&= 5.50 \text{ in.}^2
\end{aligned}
$$

$$A_{gt} = (5\tfrac{1}{2}\text{ in.} + 1\tfrac{1}{2}\text{ in.})(\tfrac{1}{2}\text{ in.})$$
$$= 3.50\text{ in.}^2$$
$$A_{nt} = 3.50\text{ in.}^2 - 1.5(1\text{ in.})(\tfrac{1}{2}\text{ in.})$$
$$= 2.75\text{ in.}^2$$

Since $0.6F_uA_{nv} > F_uA_{nt}$,

$$\phi R_n = \phi[0.6F_uA_{nv} + F_yA_{gt}] \leq \phi[0.6F_uA_{nv} + F_uA_{nt}]$$
$$= 0.75[0.6(58\text{ ksi})(5.50\text{ in.}^2) + (36\text{ ksi})(3.50\text{ in.}^2)]$$
$$\leq 0.75[0.6(58\text{ ksi})(5.50\text{ in.}^2) + (58\text{ ksi})(2.75\text{ in.}^2)]$$
$$= 238\text{ kips} \leq 263\text{ kips}$$
$$= 238\text{ kips} > 143\text{ kips} \qquad \textbf{o.k.}$$

EXAMPLE 15.3. Welded bracket plate design.

Given: Refer to Figure 15-2b. Design a bracket plate to support the factored force indicated. Use 70 ksi electrodes. For the plate, assume $F_y = 36$ ksi and $F_u = 58$ ksi.

Solution: *Select weld group assuming PL* $^1/_2$ *in.×18 in.*

Try "C"-shaped weld with $kl = 3$ in. and $l = 18$ in.

Interpolating from Table 8-7 with $\theta = 0°$,

$$x = 0.0221$$

and

$$al + xl = 11\tfrac{1}{4}\text{ in.}$$
$$a(18\text{ in.}) + (0.0221 \times 18\text{ in.}) = 11\tfrac{1}{4}\text{ in.}$$
$$a = 0.603$$

Interpolating from Table 8-7 with $\theta = 0°$, $k = 0.167$, and $a = 0.647$,

$$C = 1.09$$

Thus, the weld size required for strength is

$$D_{req} = \frac{P_u}{CC_1l}$$
$$= \frac{54.0\text{ kips}}{(1.09)(1.0)(18\text{ in.})}$$
$$= 2.75 \rightarrow 3\text{ sixteenths}$$

Minimum weld size from LRFD Specification Table J2.4 is $^3/_{16}$ in.

Therefore, use $^3/_{16}$ in. fillet weld.

Check flexure on the bracket plate.

Conservatively taking the moment in the plate equal to the moment on the weld group,

$$\begin{aligned} M_u &= P_u(al) \\ &= 54.0 \text{ kips}(10.85 \text{ in.}) \\ &= 586 \text{ in.-kips} \end{aligned}$$

For flexural yielding of the plate,

$$\begin{aligned} \phi M_n &= \phi F_y S_x \\ &= 0.9(36 \text{ ksi})\frac{(^1/_2 \text{ in.})(18 \text{ in.})^2}{6} \\ &= 875 \text{ in.-kips} > 586 \text{ in.-kips} \qquad \textbf{o.k.} \end{aligned}$$

For yielding on the free edge of the triangular plate,

$$\begin{aligned} z &= 1.39 - 2.2\left(\frac{b}{a}\right) + 1.27\left(\frac{b}{a}\right)^2 - 0.25\left(\frac{b}{a}\right)^3 \\ &= 1.39 - 2.2\left(\frac{10.85 \text{ in.}}{18 \text{ in.}}\right) + 1.27\left(\frac{10.85 \text{ in.}}{18 \text{ in.}}\right)^2 \\ &\quad - 0.25\left(\frac{10.85 \text{ in.}}{18 \text{ in.}}\right)^3 \\ &= 0.381 \end{aligned}$$

$$\begin{aligned} \phi P_n &= 0.85 F_y z b t \\ &= 0.85(36 \text{ ksi})(0.381)(10.85 \text{ in.})(^1/_2 \text{ in.}) \\ &= 63.2 \text{ kips} > 54.0 \text{ kips} \qquad \textbf{o.k.} \end{aligned}$$

Check local buckling of the bracket plate.

$$\frac{b}{a} = \frac{10.85 \text{ in.}}{18 \text{ in.}} = 0.603$$

Since $0.5 \le \dfrac{b}{a} < 1.0$

$$\begin{aligned} t_{min} &= b\left(\frac{\sqrt{F_y}}{250}\right) \\ &= 10.85 \text{ in.}\left(\frac{\sqrt{36 \text{ ksi}}}{250}\right) \\ &= 0.260 \text{ in.} < {}^1/_2 \text{ in.} \qquad \textbf{o.k.} \end{aligned}$$

Check shear yielding of the bracket plate.

$$\begin{aligned} \phi R_n &= \phi(0.6F_y)A_g \\ &= 0.9(0.6 \times 36 \text{ ksi})(18 \text{ in.} \times {}^1/_2 \text{ in.}) \\ &= 175 \text{ kips} > 54.0 \text{ kips} \qquad \textbf{o.k.} \end{aligned}$$

DESIGN TABLES

Table 15-1. Preliminary Hanger Connection Selection Table

Values are given for the tensile force per in. of fitting length design strength in flexural yielding of a tee fitting flange with $F_y = 36$ ksi and $F_y = 50$ ksi. Once the number and size of bolts required such that the design tensile strength of one bolt ϕr_n exceeds the factored tensile force per bolt r_{ut} (exclusive of pretension) is known, Table 15-1 can be used to select a trial fitting.

In this table, it is assumed that equal critical moments exist at the face of the tee stem (M_{u1}) and at the bolt line (M_{u2}). From LRFD Specification Section F1.1, the design flexural yielding strength of the tee flange is $\phi_b M_n$, where $\phi_b = 0.90$ and

$$M_n = M_p = F_y Z_x$$

In the above equation, the plastic section modulus Z_x per unit length of the tee flange is

$$Z_x = \frac{t^2}{4}$$

where t is the thickness of the angle or tee flange, in. Thus, for a unit length of the tee flange

$$\phi_b M_n = \frac{0.90 F_y t^2}{4}$$

and the factored tensile force on the fitting $2r_{ut}$ must be such that

$$2r_{ut} \leq \frac{0.9 F_y t^2}{b}$$

where b is the distance from bolt centerline to face of the angle leg or tee stem, in. For $F_y = 36$ ksi, the above equation may be simplified as

$$2r_{ut} \leq \frac{32.4t^2}{b}$$

and for $F_y = 50$ ksi, the above equation may be simplified as

$$2r_{ut} \leq \frac{45.0t^2}{b}$$

Table 15-2. Net Elastic Section Modulus S_{net}

Values of the net elastic section modulus S_{net} are given in Table 15-2 for various hole diameters and numbers of fasteners spaced three inches on center, the usual spacing for these connections.

FORGED STEEL STRUCTURAL HARDWARE

Table 15-3. Dimensions and Weights of Clevises

Dimensions, weights, and design strengths of clevises are listed in Table 15-3.

Table 15-4. Clevis Numbers Compatible with Various Rods and Pins

Compatibility of clevises with various rods and pins is given in Table 15-4.

Table 15-5. Dimensions and Weights of Turnbuckles

Dimensions, weights, and design strengths of turnbuckles are listed in Table 15-5.

Table 15-6. Dimensions and Weights of Sleeve Nuts

Dimensions and weights of sleeve nuts are listed in Table 15-6.

Table 15-7. Dimensions and Weights of Recessed-Pin Nuts

Dimensions and weights of recessed-pin nuts are listed in Table 15-7.

Table 15-8. Dimensions and Weights of Cotter Pins

Dimensions and weights of cotter pins are listed in Table 15-8.

PART 15 REFERENCE

Salmon, C.G. and J.E. Johnson, 1996, *Steel Structures: Design and Behavior*, 4th Edition, Harper Collins, New York, NY.

$F_y = 36$ ksi

Table 15-1.
Preliminary Hanger Connection Selection Table
Design Tensile Strength, kips per linear in.,
Limited by Flexural Yielding of the Flange

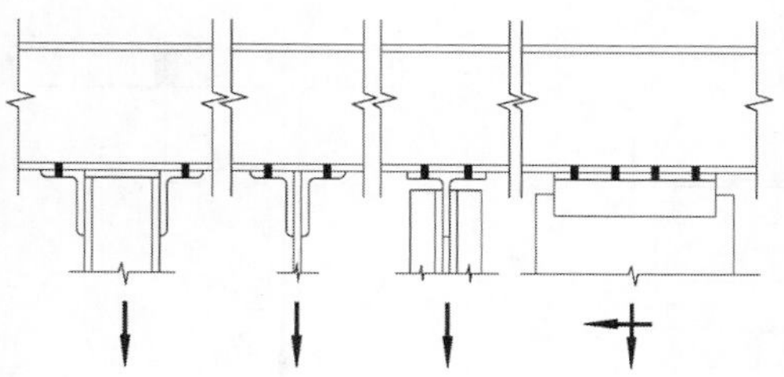

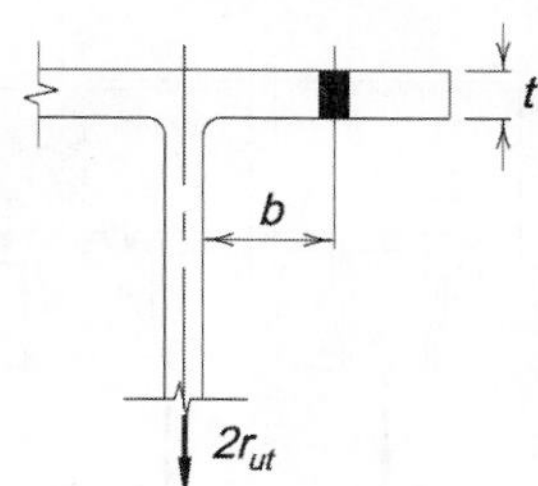

t, in.	*b*, in.									
	1	1 1/4	1 1/2	1 3/4	2	2 1/4	2 1/2	2 3/4	3	3 1/4
5/16	3.16	2.53	2.11	1.81	1.58	1.41	1.27	1.15	1.05	0.974
3/8	4.56	3.64	3.04	2.60	2.28	2.02	1.82	1.66	1.52	1.40
7/16	6.20	4.96	4.13	3.54	3.10	2.76	2.48	2.26	2.07	1.91
1/2	8.10	6.48	5.40	4.63	4.05	3.60	3.24	2.95	2.70	2.49
9/16	10.3	8.20	6.83	5.86	5.13	4.56	4.10	3.73	3.42	3.15
5/8	12.7	10.1	8.44	7.23	6.33	5.63	5.06	4.60	4.22	3.89
11/16	15.3	12.3	10.2	8.75	7.66	6.81	6.13	5.57	5.10	4.71
3/4	18.2	14.6	12.1	10.4	9.11	8.10	7.29	6.63	6.07	5.61
13/16	21.4	17.1	14.3	12.2	10.7	9.51	8.56	7.78	7.13	6.58
7/8	24.8	19.8	16.5	14.2	12.4	11.0	9.92	9.02	8.27	7.63
15/16	28.5	22.8	19.0	16.3	14.2	12.7	11.4	10.4	9.49	8.76
1	32.4	25.9	21.6	18.5	16.2	14.4	13.0	11.8	10.8	10.0
1 1/16	36.6	29.3	24.4	20.9	18.3	16.3	14.6	13.3	12.2	11.3
1 1/8	41.0	32.8	27.3	23.4	20.5	18.2	16.4	14.9	13.7	12.6
1 3/16	45.7	36.6	30.5	26.1	22.8	20.3	18.3	16.6	15.2	14.1
1 1/4	50.6	40.5	33.8	28.9	25.3	22.5	20.3	18.4	16.9	15.6

$F_y = 50$ ksi

Table 15-1 (cont.).
Preliminary Hanger Connection Selection Table
Design Tensile Strength, kips per linear in.,
Limited by Flexural Yielding of the Flange

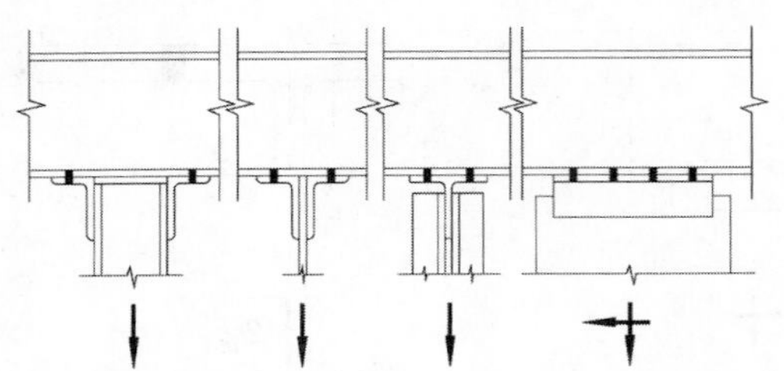

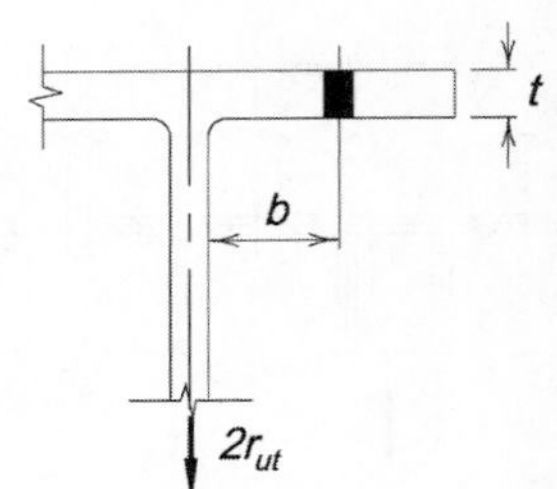

t, in.	*b*, in.									
	1	1 1/4	1 1/2	1 3/4	2	2 1/4	2 1/2	2 3/4	3	3 1/4
5/16	4.39	3.52	2.93	2.51	2.20	1.95	1.76	1.60	1.46	1.35
3/8	6.33	5.06	4.22	3.62	3.16	2.81	2.53	2.30	2.11	1.95
7/16	8.61	6.89	5.74	4.92	4.31	3.83	3.45	3.13	2.87	2.65
1/2	11.3	9.00	7.50	6.43	5.63	5.00	4.50	4.09	3.75	3.46
9/16	14.2	11.4	9.49	8.14	7.12	6.33	5.70	5.18	4.75	4.38
5/8	17.6	14.1	11.7	10.0	8.79	7.81	7.03	6.39	5.86	5.41
11/16	21.3	17.0	14.2	12.2	10.6	9.45	8.51	7.73	7.09	6.54
3/4	25.3	20.3	16.9	14.5	12.7	11.3	10.1	9.20	8.44	7.79
13/16	29.7	23.8	19.8	17.0	14.9	13.2	11.9	10.8	9.90	9.14
7/8	34.5	27.6	23.0	19.7	17.2	15.3	13.8	12.5	11.5	10.6
15/16	39.6	31.6	26.4	22.6	19.8	17.6	15.8	14.4	13.2	12.2
1	45.0	36.0	30.0	25.7	22.5	20.0	18.0	16.4	15.0	13.8
1 1/16	50.8	40.6	33.9	29.0	25.4	22.6	20.3	18.5	16.9	15.6
1 1/8	57.0	45.6	38.0	32.5	28.5	25.3	22.8	20.7	19.0	17.5
1 3/16	63.5	50.8	42.3	36.3	31.7	28.2	25.4	23.1	21.2	19.5
1 1/4	70.3	56.3	46.9	40.2	35.2	31.3	28.1	25.6	23.4	21.6

Table 15-2.
Net Elastic Section Modulus S_{net}, in.^{3}

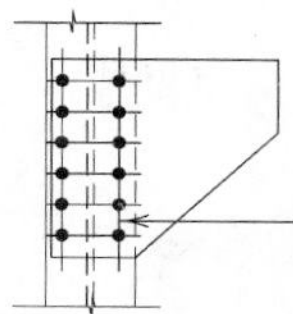

Net elastic section modulus taken along this line

# Bolts in One Vertical Row n	Bracket Plate Depth d, in.	Nominal Bolt Diameter d_b, in.														
		3/4					7/8					1				
		Bracket Plate Thickness t, in.														
		1/4	3/8	1/2	5/8	3/4	3/8	1/2	5/8	3/4	7/8	1/2	5/8	3/4	7/8	1
2	6	1.17	1.76	2.34	2.93	3.52	1.69	2.25	2.81	3.38	3.94	2.16	2.70	3.23	3.77	4.31
3	9	2.50	3.75	5.0	6.25	7.50	3.56	4.75	5.94	7.12	8.31	4.50	5.62	6.75	7.88	9.00
4	12	4.36	6.54	8.72	10.9	13.1	6.19	8.25	10.3	12.4	14.4	7.78	9.73	11.7	13.6	15.6
5	15	6.75	10.1	13.5	16.9	20.2	9.56	12.7	15.9	19.1	22.3	12.0	15.0	18.0	21.0	24.0
6	18	9.67	14.5	19.3	24.2	29.0	13.7	18.2	22.8	27.4	31.9	17.2	21.4	25.7	30.0	34.3
7	21	13.1	19.7	26.2	32.8	39.4	18.6	24.8	30.9	37.1	43.3	23.3	29.1	34.9	40.7	46.5
8	24	17.1	25.7	34.2	42.8	51.3	24.2	32.2	40.3	48.4	56.4	30.3	37.9	45.4	53.0	60.6
9	27	21.6	32.4	43.2	54.1	64.9	30.6	40.8	50.9	61.1	71.3	38.2	47.8	57.4	66.9	76.5
10	30	26.7	40.0	53.3	66.7	80.0	37.7	50.3	62.8	75.4	87.9	47.2	58.9	70.7	82.5	94.3
12	36	38.4	57.5	76.7	95.9	115	54.2	72.2	90.3	108	126	67.8	84.7	102	119	136
14	42	52.2	78.3	104	130	157	73.7	98.2	123	147	172	92.2	115	138	161	184
16	48	68.1	102	136	170	204	96.2	128	160	192	224	120	150	180	210	241
18	54	86.2	129	172	215	259	122	162	203	243	284	152	190	228	266	304
20	60	106	160	213	266	319	150	200	250	300	350	188	235	282	329	376
22	66	129	193	257	322	386	182	242	303	363	424	227	284	341	398	454
24	72	153	230	306	383	459	216	288	360	432	504	270	338	405	473	541
26	78	180	270	359	449	539	254	338	423	507	592	317	396	476	555	634
28	84	208	313	417	521	625	294	392	490	588	686	368	460	552	644	736
30	90	239	359	478	598	718	338	450	563	675	788	422	528	633	739	844
32	96	272	408	544	680	816	384	512	640	768	896	480	600	720	840	961
34	102	307	461	614	768	922	434	578	723	867	1010	542	678	813	949	1080
36	108	344	517	689	861	1030	486	648	810	972	1130	608	760	912	1060	1220

Notes:
Diameter of holes is assumed to be 1/16 in. larger than a nominal diameter of hole.
Bolts spaced 3 in. vertically with 1 1/2-in. edge distance at top and bottom.
Interpolate for intermediate plate thicknesses. Otherwise use the general equation,

$$S_{net} = \frac{t}{6}\left[d^2 - \frac{s^2 n(n^2-1)(d_h+1/16)}{d}\right]$$

where
d = bracket plate depth, in.
d_h = nominal diameter of hole, in.
n = number of bolts in one vertical row, in.
s = bolt spacing, in.
t = bracket plate thickness, in.

Table 15-3.
Dimensions and Weights of Clevises

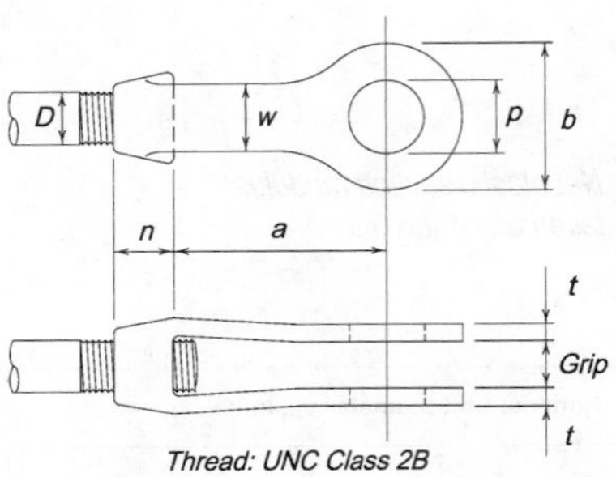

Grip = plate thickness + ¼ in.

Clevis Number	Dimensions, in.							Weight, pounds	Design Strength ϕR_n*, kips
	Max. *D*	Max. *p*	*b*	*n*	*a*	*w*	*t*		
2	5/8	3/4	17/16	5/8	39/16	11/16	5/16 (+1/32, −0)	1	8.75
21/2	7/8	11/2	21/2	11/8	4	11/4	5/16 (+1/32, −0)	2	18.8
3	13/8	13/4	3	11/4	51/16	11/2	1/2 (+1/16, −1/32)	4	37.5
31/2	11/2	2	31/2	11/2	6	13/4	1/2 (+1/16, −1/16)	6	45.0
4	13/4	21/4	4	13/4	515/16	2	1/2 (+1/16, −1/16)	8	52.5
5	21/8	21/2	5	21/4	7	21/2	5/8 (+3/32, −0)	16	93.8
6	21/2	3	6	23/4	8	3	3/4 (+3/32, −0)	26	135
7	3	33/4	7	3	9	31/2	7/8 (+1/8, −1/16)	36	171
8	4	41/4	8	4	101/8	4	11/2 (+1/8, −1/16)	90	338

Notes:
Weights and Dimensions of clevises are typical; products of all suppliers are essentially similar. User shall verify with the manufacturer that product meets design-strength specifications above.
* Tabulated design strengths for comparison with factored loads are based on $\phi = 0.5$. To determine safe working load (kips) for comparison with service loads, divide tabular design strength by 1.5. Safe working load, then, corresponds to a 3:1 factor of safety using maximum pin diameter.

Table 15-4. Clevis Numbers Compatible with Various Rods and Pins

Dia. Of Tap, in.	Diameter of pin, in.																	
	1/2	5/8	3/4	7/8	1	1 1/4	1 1/2	1 3/4	2	2 1/4	2 1/2	2 3/4	3	3 1/4	3 1/2	3 3/4	4	4 1/4
3/8	2	2	2															
1/2	2	2	2															
5/8	2	2	2	2 1/2	2 1/2	2 1/2	2 1/2											
3/4			2 1/2	2 1/2	2 1/2	2 1/2	2 1/2											
7/8				2 1/2	2 1/2	2 1/2	2 1/2	3										
1					3	3	3	3										
1 1/8					3	3	3	3	3 1/2									
1 1/4					3	3	3	3	3 1/2									
1 3/8						3	3	3 1/2	3 1/2	4								
1 1/2						3 1/2	3 1/2	4	4	5								
1 5/8						4	4	4	5	5	5							
1 3/4							4	5	5	5	5							
1 7/8							5	5	5	5	5							
2							5	5	5	5	5	6	6					
2 1/8								5	5	6	6	6	6					
2 1/4									6	6	6	6	6	7	7			
2 3/8									6	6	6	6	7	7	7	7		
2 1/2									6	6	6	7	7	7	7	7		
2 5/8											7	7	7	7	7	8		
2 3/4											7	7	7	7	8	8		
2 7/8											7	8	8	8	8	8	8	8
3											7	8	8	8	8	8	8	8
3 1/8												8	8	8	8	8	8	8
3 1/4												8	8	8	8	8	8	8
3 3/8												8	8	8	8	8	8	8
3 1/2													8	8	8	8	8	8
3 5/8													8	8	8	8	8	
3 3/4													8	8	8	8	8	
3 7/8														8	8	8		
4														8	8			

Notes:
Tabular values assume that the net area of the clevis through the pin hole is greater than or equal to 125 percent of the net area of the rod, and is applicable to round rods without upset ends. For other net area ratios, the required clevis size may be calculated by reference to the dimensions tabulated in Table 7-4 and 15-3.

Table 15-5. Dimensions and Weights of Turnbuckles

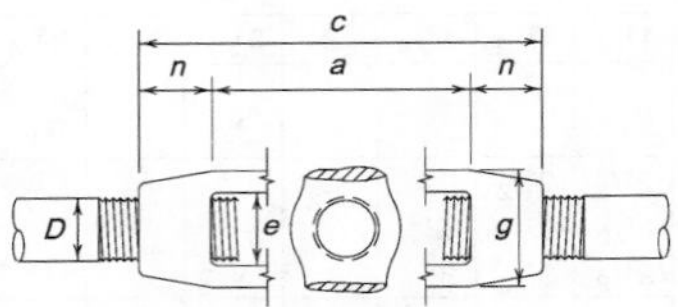

Threads: UNC and 4UN Class 2B

Diameter D, in.	Dimensions, in.					Weight (pounds) for Length *a*, in.						Design Strength, φRn*, kips
	a	*n*	*c*	*e*	*g*	6	9	12	18	24	26	
$\frac{3}{8}$	6	$\frac{9}{16}$	$7\frac{1}{8}$	$\frac{9}{16}$	$1\frac{1}{32}$	0.42						3.00
$\frac{1}{2}$	6	$\frac{25}{32}$	$7\frac{9}{16}$	$\frac{11}{16}$	$1\frac{5}{16}$	0.65	0.90	1.20				5.50
$\frac{5}{8}$	6	$\frac{15}{16}$	$7\frac{7}{8}$	$\frac{13}{16}$	$1\frac{1}{2}$	0.98	1.35	1.58	2.43			8.75
$\frac{3}{4}$	6	$1\frac{1}{16}$	$8\frac{1}{8}$	$\frac{15}{16}$	$1\frac{23}{32}$	1.45	1.84	2.35	3.06	4.25		13.0
$\frac{7}{8}$	6	$1\frac{5}{16}$	$8\frac{5}{8}$	$1\frac{3}{32}$	$1\frac{7}{8}$	1.85		3.02	4.20	5.43		18.0
1	6	$1\frac{7}{16}$	$8\frac{7}{8}$	$1\frac{9}{32}$	$2\frac{1}{32}$	2.60		4.02	4.40	6.85	10.0	23.3
$1\frac{1}{8}$	6	$1\frac{9}{16}$	$9\frac{1}{8}$	$1\frac{13}{32}$	$2\frac{9}{32}$	4.06		4.70	6.10			29.0
$1\frac{1}{4}$	6	$1\frac{9}{16}$	$9\frac{1}{8}$	$1\frac{9}{16}$	$2\frac{17}{32}$	4.00		6.49	7.13	11.3	13.1	38.0
$1\frac{3}{8}$	6	$1\frac{13}{16}$	$9\frac{5}{8}$	$1\frac{11}{16}$	$2\frac{3}{4}$	6.15						43.5
$1\frac{1}{2}$	6	$1\frac{7}{8}$	$9\frac{3}{4}$	$1\frac{27}{32}$	$3\frac{1}{32}$	6.15		9.70	9.13	16.8	19.4	52.5
$1\frac{5}{8}$	6	$2\frac{1}{2}$	11	$1\frac{31}{32}$	$3\frac{9}{32}$	9.80						61.3
$1\frac{3}{4}$	6	$2\frac{1}{2}$	11	$2\frac{1}{8}$	$3\frac{9}{16}$	9.80		15.3	16.0	19.5		70.8
$1\frac{7}{8}$	6	$2\frac{13}{16}$	$11\frac{5}{8}$	$2\frac{3}{8}$	4	14.0		15.3				93.0
2	6	$2\frac{13}{16}$	$11\frac{5}{8}$	$2\frac{3}{8}$	4	14.0		15.3		27.5		93.0
$2\frac{1}{4}$	6	$3\frac{5}{16}$	$12\frac{5}{8}$	$2\frac{11}{16}$	$4\frac{5}{8}$	19.6		30.9		43.5		120
$2\frac{1}{2}$	6	$3\frac{3}{4}$	$13\frac{1}{2}$	3	5	23.3		30.9		42.4		150
$2\frac{3}{4}$	6	$4\frac{3}{16}$	$14\frac{3}{8}$	$3\frac{1}{4}$	$5\frac{5}{8}$	31.5				54.0		188
3	6	$4\frac{5}{16}$	$14\frac{5}{8}$	$3\frac{5}{8}$	$6\frac{1}{8}$	39.5						242
$3\frac{1}{4}$	6	$5\frac{7}{16}$	$16\frac{7}{8}$	$3\frac{7}{8}$	$6\frac{3}{4}$	60.5		79.5				305
$3\frac{1}{2}$	6	$5\frac{7}{16}$	$16\frac{7}{8}$	$3\frac{7}{8}$	$6\frac{3}{4}$	60.5	70.0	79.5				305
$3\frac{3}{4}$	6	6	18	$4\frac{5}{8}$	$8\frac{1}{2}$	95.0						420
4	6	6	18	$4\frac{5}{8}$	$8\frac{1}{2}$	95.0						420
$4\frac{1}{4}$	9	$6\frac{3}{4}$	$22\frac{1}{2}$	$5\frac{1}{4}$	$9\frac{3}{4}$		152					585
$4\frac{1}{2}$	9	$6\frac{3}{4}$	$22\frac{1}{2}$	$5\frac{1}{4}$	$9\frac{3}{4}$		152					585
$4\frac{3}{4}$	9	$6\frac{3}{4}$	$22\frac{1}{2}$	$5\frac{1}{4}$	$9\frac{3}{4}$		152					585
5	9	$7\frac{1}{2}$	24	6	10		200					737

Notes:

Weights and dimensions of turnbuckles are typical; products of all suppliers are essentially similar. Users shall verify with the manufacturer that product meets design strength specifications above.

*Tabulated design strengths for comparison with factored loads are based on $\phi = 0.5$. To determine safe working load (kips) for comparison with service loads, divide tabular design strength by 1.5. Safe working load, then, corresponds to a 3:1 factor of safety using maximum pin diameter.

Table 15-6. Dimensions and Weights of Sleeve Nuts

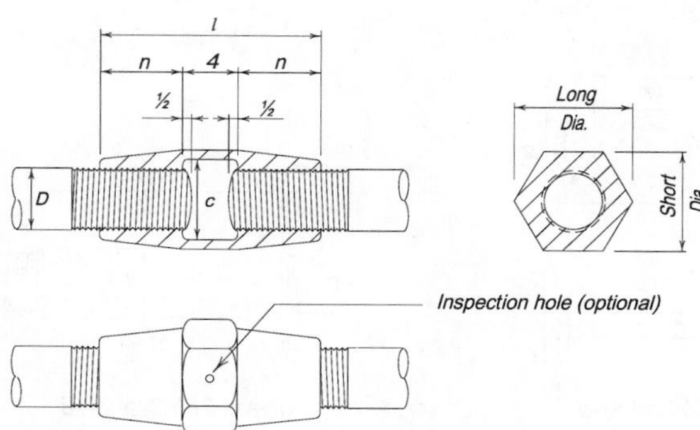

Thread: UNC and 4 UN Class 2B

Screw Dia. D, in.	Dimensions, in.					Weight, pounds
	Short Dia.	Long Dia.	Length l	Nut n	Clear c	
3/8	11/16	25/32	4	–	–	0.27
7/16	25/32	7/8	4	–	–	0.34
1/2	7/8	1	4	–	–	0.43
9/16	15/16	1 1/16	5	–	–	0.64
5/8	1 1/16	1 7/32	5	–	–	0.93
3/4	1 1/4	1 7/16	5	–	–	1.12
7/8	1 7/16	1 5/8	7	1 7/16	1	1.75
1	1 5/8	1 13/16	7	1 7/16	1 1/8	2.46
1 1/8	1 13/16	2 1/16	7 1/2	1 5/8	1 1/4	3.10
1 1/4	2	2 1/4	7 1/2	1 5/8	1 3/8	4.04
1 3/8	2 3/16	2 1/2	8	1 7/8	1 1/2	4.97
1 1/2	2 3/8	2 11/16	8	1 7/8	1 5/8	6.16
1 5/8	2 9/16	2 15/16	8 1/2	2 1/16	1 3/4	7.36
1 3/4	2 3/4	3 1/8	8 1/2	2 1/16	1 7/8	8.87
1 7/8	2 15/16	3 5/16	9	2 5/16	2	10.4
2	3 1/8	3 1/2	9	2 5/16	2 1/8	12.2
2 1/4	3 1/2	3 15/16	9 1/2	2 1/2	2 3/8	16.2
2 1/2	3 7/8	4 3/8	10	2 3/4	2 5/8	21.1
2 3/4	4 1/4	4 13/16	10 1/2	2 15/16	2 7/8	26.7
3	4 5/8	5 1/4	11	3 3/16	3 1/8	33.2
3 1/4	5	5 5/8	11 1/2	3 3/8	3 3/8	40.6
3 1/2	5 3/8	6	12	3 5/8	3 5/8	49.1
3 3/4	5 3/4	6 3/8	12 1/2	3 13/16	3 7/8	58.6
4	6 1/8	6 7/8	13	4 1/16	4 1/8	69.2
4 1/4	6 1/2	7 1/2	13 1/2	4 3/4	4 3/8	75.0
4 1/2	6 7/8	7 15/16	14	5	4 3/4	90.0
4 3/4	7 1/4	8 3/8	14 1/2	5 1/4	5	98.0
5	7 5/8	8 7/8	15	5 1/2	5 1/4	110
5 1/4	8	9 1/4	15 1/2	5 3/4	5 1/2	122
5 1/2	8 3/8	9 3/4	16	6	5 3/4	142
5 3/4	8 3/4	10 1/8	16 1/2	6 1/4	6	157
6	9 1/8	10 5/8	17	6 1/2	6 1/4	176

Notes:
Weights and dimensions of sleeve nuts are typical; products of all suppliers are essentially similar. User shall verify with the manufacturer that strengths of sleeve nut are greater than the corresponding connecting rod when the same material is used.

Table 15-7. Dimensions and Weights of Recessed-Pin Nuts

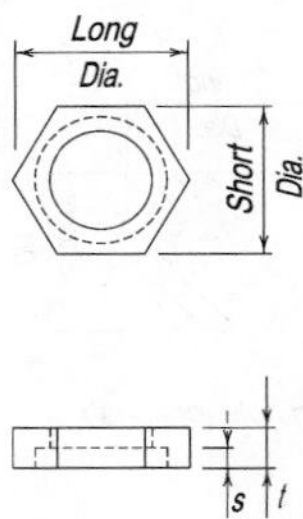

Material: Steel

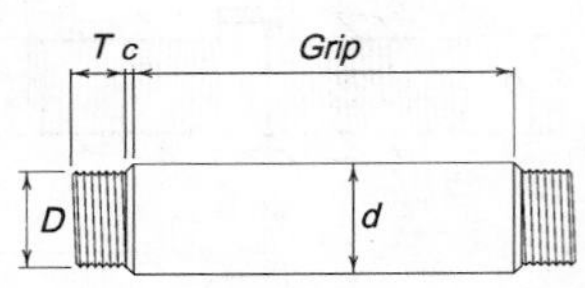

Thread: 6 UN Class 2A/2B

Pin Dia. d, in.	Pin Dimensions, in.			Nut Dimensions, in.					Weight, pounds
	Thread		c	Thickness t	Diameter		Recess		
	D	T			Short Dia.	Long Dia.	Rough Dia.	s	
2, 2 1/4	1 1/2	1	1/8	7/8	3	3 3/8	2 5/8	1/4	1
2 1/2, 2 3/4	2	1 1/8	1/8	1	3 5/8	4 1/8	3 1/8	1/4	2
3, 3 1/4, 3 1/2	2 1/2	1 1/4	1/8	1 1/8	4 3/8	5	3 7/8	3/8	3
3 3/4, 4	3	1 3/8	1/4	1 1/4	4 7/8	5 5/8	4 3/8	3/8	4
4 1/4, 4 1/2, 4 3/4	3 1/2	1 1/2	1/4	1 3/8	5 3/4	6 5/8	5 1/4	1/2	5
5, 5 1/4	4	1 5/8	1/4	1 1/2	6 1/4	7 1/4	5 3/4	1/2	6
5 1/2, 5 3/4, 6	4 1/2	1 3/4	1/4	1 5/8	7	8 1/8	6 1/2	5/8	8
6 1/4, 6 1/2	5	1 7/8	3/8	1 3/4	7 5/8	8 7/8	7	5/8	10
6 3/4, 7	5 1/2	2	3/8	1 7/8	8 1/8	9 3/8	7 1/2	3/4	12
7 1/4, 7 1/2	5 1/2	2	3/8	1 7/8	8 5/8	10	8	3/4	14
7 3/4, 8, 8 1/4	6	2 1/4	3/8	2 1/8	9 3/8	10 7/8	8 3/4	3/4	19
8 1/2, 8 3/4, 9	6	2 1/4	3/8	2 1/8	10 1/4	11 7/8	9 5/8	3/4	24
9 1/4, 9 1/2	6	2 3/8	3/8	2 1/4	11 1/4	13	10 5/8	3/4	32
9 3/4, 10	6	2 3/8	3/8	2 1/4	11 1/4	13	10 5/8	3/4	32

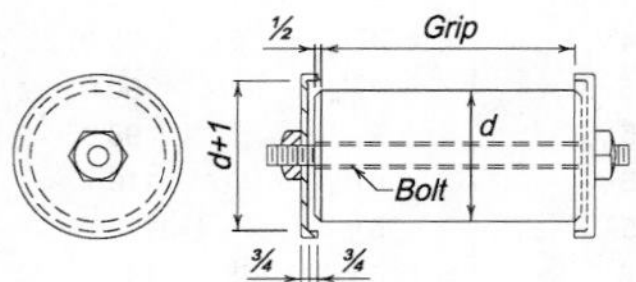

Typical Pin Cap Detail for Pins over 10 in. in dia.
Dimensions shown are approximate

Notes:
Although nuts may be used on all sizes of pins as shown above, a detail similar to that shown at the left is preferable for pin diameters over 10 inches. In this detail, the pin is held in place by a recessed cap at each end and secured by a bolt passing completely through the caps and pin. Suitable provisions must be made for attaching pilots and driving nuts.

Table 15-8. Dimensions and Weights of Cotter Pins

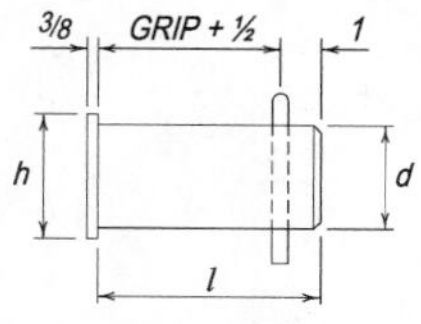

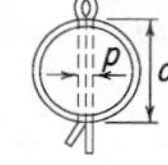

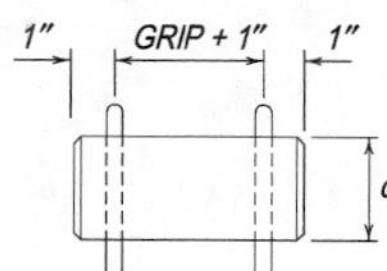

l = Length of pin, in.

Pin Diameter d, in.	Pins with Heads		Cotter		
	Head Diameter *h*, in.	Weight of One, pounds	Length *c*, in.	Diameter *p*, in.	Weight per 100, pounds
$1^{1}/_{4}$	$1^{1}/_{2}$	$0.19 + 0.35l$	$^{7}/_{8}$	$^{1}/_{4}$	2.64
$1^{1}/_{2}$	$1^{3}/_{4}$	$0.26 + 0.50l$	1	$^{1}/_{4}$	3.10
$1^{3}/_{4}$	2	$0.33 + 0.68l$	$1^{1}/_{8}$	$^{1}/_{4}$	3.50
2	$2^{3}/_{8}$	$0.47 + 0.89l$	$1^{1}/_{4}$	$^{3}/_{8}$	9.00
$2^{1}/_{4}$	$2^{5}/_{8}$	$0.58 + 1.13l$	$1^{3}/_{8}$	$^{3}/_{8}$	9.40
$2^{1}/_{2}$	$2^{7}/_{8}$	$0.70 + 1.39l$	$1^{1}/_{2}$	$^{3}/_{8}$	10.9
$2^{3}/_{4}$	$3^{1}/_{8}$	$0.82 + 1.68l$	$1^{5}/_{8}$	$^{3}/_{8}$	11.4
3	$3^{1}/_{2}$	$1.02 + 2.00l$	$1^{3}/_{4}$	$^{1}/_{2}$	28.5
$3^{1}/_{4}$	$3^{3}/_{4}$	$1.17 + 2.35l$	$1^{7}/_{8}$	$^{1}/_{2}$	28.5
$3^{1}/_{2}$	4	$1.34 + 2.73l$	$1^{7}/_{8}$	$^{1}/_{2}$	33.8
$3^{3}/_{4}$	$4^{1}/_{4}$	$1.51 + 3.13l$	$2^{1}/_{4}$	$^{1}/_{2}$	33.8

PART 16

SPECIFICATIONS AND CODES

Notes:
The above documents are available for free download in *.pdf format at www.aisc.org. While not included in this Manual, the AISC *Seismic Provisions for Structural Steel Buildings*, April 15, 1997 and *Seismic Provisions Supplement No. 2*, November 10, 2000 are available for free download in *.pdf format at www.aisc.org. Information about AISC Certification of steel fabricators and steel erectors is available at www.aisc.org/quality.html.

Load and Resistance Factor Design Specification for Structural Steel Buildings

December 27, 1999
with errata incorporated as of September 4, 2001

Supersedes the *Load and Resistance Factor Design Specification for Structural Steel Buildings* dated December 1, 1993 and all previous versions.

Prepared by the
American Institute of Steel Construction, Inc.
Under the Direction of the
AISC Committee on Specifications and approved by
the AISC Board of Directors.

AMERICAN INSTITUTE OF STEEL CONSTRUCTION, INC.
One East Wacker Drive, Suite 3100
Chicago, Illinois 60601-2001

The information presented in this publication has been prepared in accordance with recognized engineering principles and is for general information only. While it is believed to be accurate, this information should not be used or relied upon for any specific application without competent professional examination and verification of its accuracy, suitability, and applicability by a licensed engineer, architect or other professional. The publication of the material contained herein is not intended as a representation or warranty on the part of the American Institute of Steel Construction, Inc. or of any other person named herein, that this information is suitable for any general or particular use or of freedom from infringement of any patent or patents. Anyone making use of this information assumes all liability arising from such use.

Caution must be exercised when relying upon other specifications and codes developed by other bodies and incorporated by reference herein since such material may be modified or amended from time to time subsequent to the printing of this edition. The American Institute of Steel Construction, Inc. bears no responsibility for such material other than to refer to it and incorporate it by reference at the time of the initial publication of this edition.

Printed in the United States of America

First Revision, September 4, 2001

PREFACE

The AISC *Load and Resistance Factor Design (LRFD) Specification for Structural Steel Buildings* is based on reliability theory. As have all AISC Specifications, this Specification has been based upon past successful usage, advances in the state of knowledge, and changes in design practice. This Specification has been developed as a consensus document to provide a uniform practice in the design of steel-framed buildings. The intention is to provide design criteria for routine use and not to provide specific criteria for infrequently encountered problems, which occur in the full range of structural design.

This Specification is the result of the consensus deliberations of a committee of structural engineers with wide experience and high professional standing, representing a wide geographical distribution throughout the U.S. The committee includes approximately equal numbers of engineers in private practice and code agencies, engineers involved in research and teaching, and engineers employed by steel fabricating and producing companies. The contributions and assistance of more than 50 additional professional volunteers working in 15 task committees are also hereby acknowledged.

The Symbols, Glossary, and Appendices to this Specification are an integral part of the Specification. A non-mandatory Commentary has been prepared to provide background for the Specification provisions and the user is encouraged to consult it.

The principal changes incorporated in this edition of the Specification include:

- Dual units format. Values and equations are given in both U.S. customary and metric units. The metric conversions (given in parentheses following the U.S. units) are based on ASTM E380, *Standard Practice for Use of the International System of Units (SI)*. The equations are non-dimensionalized where possible by factoring out material constants, such as E and G.
- Inclusion of new structural steels ASTM A913 and A992.
- Additional notch toughness requirements for complete-joint-penetration groove welds with tension applied normal to the effective area.
- New provisions for stability bracing of beams, columns, and frames.
- New Chapter N for evaluation of existing structures.
- Revised provisions for member design under fatigue loading in Appendix K.

- Reorganization of material on pin-connected members and eyebars.
- Revised provisions for concrete-encased beams.
- New limitation on the stud reduction factor when a single stud is used in a rib.
- Revised bolt bearing strength criteria.

The reader is cautioned that professional judgment must be exercised when data or recommendations in the Specification are applied, as described more fully in the disclaimer notice preceding this Preface.

By the Committee,

Stanley D. Lindsey, Chairman
Roger E. Ferch, Vice Chairman
Hansraj G. Ashar
William F. Baker
John M. Barsom
Reidar Bjorhovde
Roger L. Brockenbrough
Wai-Fah Chen
Gregory G. Deierlein
Robert O. Disque
Duane S. Ellifritt
Bruce R. Ellingwood
Shu-Jin Fang
Steven J. Fenves
James M. Fisher
John W. Fisher
Theodore V. Galambos
Lawrence G. Griffis
James R. Harris
Tony C. Hazel
Mark V. Holland
Nestor R. Iwankiw
Lawrence A. Kloiber
Roberto T. Leon
H.S. Lew
James O. Malley
Richard W. Marshall
Harry W. Martin
William A. Milek
Duane K. Miller
Thomas M. Murray
R. Shankar Nair
Clarkson W. Pinkham
Douglas D. Rees-Evans
Thomas Z. Scarangello
Donald R. Sherman
W. Lee Shoemaker
Frank F. Sowokinos
William A. Thornton
Raymond H. R. Tide
Ivan M. Viest
Joseph A. Yura
Cynthia J. Lanz, Secretary

TABLE OF CONTENTS

SPECIFICATION (Cont'd)

SPECIFICATION (Cont'd)

SPECIFICATION (Cont'd)

SPECIFICATION (Cont'd)

SPECIFICATION (Cont'd)

COMMENTARY

COMMENTARY (Cont'd)

COMMENTARY (Cont'd)

COMMENTARY (Cont'd)

COMMENTARY (Cont'd)

SYMBOLS

The section number in the right hand column refers to the section where the symbol is first used.

Symbol	Definition	Section
A	Area of directly connected elements	B3
A_B	Loaded area of concrete, in.^{2} (mm^2)	I2.4
A_b	Nominal unthreaded body area of bolt or threaded part, in.^{2} (mm^2)	J3.6
A_c	Area of concrete, in.^{2} (mm^2)	I2.2
A_c	Area of concrete slab within effective width, in.^{2} (mm^2)	I5.2
A_D	Area of an upset rod based on the major thread diameter, in.^{2} (mm^2)	J3.6
A_e	Effective area, in.^{2} (mm^2)	B3
A_f	Area of the compression flange, in.^{2} (mm^2)	App. F3
A_{fe}	Effective tension flange area, in.^{2} (mm^2)	B10
A_{fg}	Gross area of flange, in.^{2} (mm^2)	B10
A_{fn}	Net area of flange, in.^{2} (mm^2)	B10
A_g	Gross area, in.^{2} (mm^2)	A5
A_{gt}	Gross area subject to tension, in.^{2} (mm^2)	J4.3
A_{gv}	Gross area subject to shear, in.^{2} (mm^2)	J4.3
A_n	Net area, in.^{2} (mm^2)	B2
A_{nt}	Net area subject to tension, in.^{2} (mm^2)	J4.2
A_{nv}	Net area subject to shear, in.^{2} (mm^2)	J4.1
A_{pb}	Projected bearing area, in.^{2} (mm^2)	J8.1
A_r	Area of reinforcing bars, in.^{2} (mm^2)	I2.2
A_s	Area of steel cross section, in.^{2} (mm^2)	I2.2
A_{sc}	Cross-sectional area of stud shear connector, in.^{2} (mm^2)	I5.3
A_{sf}	Shear area on the failure path, in.^{2} (mm^2)	D3
A_t	Net tensile area, in.^{2} (mm^2)	App. K3
A_w	Web area, in.^{2} (mm^2)	F2.1
A_1	Area of steel concentrically bearing on a concrete support, in.^{2} (mm^2)	J9
A_2	Total cross-sectional area of a concrete support, in.^{2} (mm^2)	J9
B	Factor for bending stress in tees and double angles	F1.2
B	Factor for bending stress in web-tapered members, in. (mm), defined by Equations A-F3-8 through A-F3-11	App. F3
B_1, B_2	Factors used in determining M_u for combined bending and axial forces when first-order analysis is employed	C1
C_{PG}	Plate-girder coefficient	App. G2
C_b	Bending coefficient dependent on moment gradient	F1.2a
C_f	Constant based on stress category, given in Table A-K3.1	App. K3.3
C_m	Coefficient applied to bending term in interaction formula for prismatic members and dependent on column curvature caused by applied moments	C1
C_m'	Coefficient applied to bending term in interaction formula for tapered members and dependent on axial stress at the	

	small end of the member	App. F3
C_p	Ponding flexibility coefficient for primary member in a flat roof	K2
C_s	Ponding flexibility coefficient for secondary member in a flat roof	K2
C_v	Ratio of "critical" web stress, according to linear buckling theory, to the shear yield stress of web material	App. G3
C_w	Warping constant, in.^{6} (mm^6)	F1.2
D	Outside diameter of circular hollow section, in. (mm)	App. B5.3
D	Dead load due to the weight of the structural elements and permanent features on the structure; nominal dead load	App. K2
D	Factor used in Equation A-G4-2, dependent on the type of transverse stiffeners used in a plate girder	App. G4
E	Modulus of elasticity of steel, $E = 29{,}000$ ksi (200 000 MPa)	E2
E	Earthquake load	
E_c	Modulus of elasticity of concrete, ksi (MPa)	I2.2
E_m	Modified modulus of elasticity, ksi (MPa)	I2.2
F_{BM}	Nominal strength of the base material to be welded, ksi (MPa)	J2.4
F_{EXX}	Classification number of weld metal (minimum specified strength), ksi (MPa)	J2.4
F_L	Smaller of $(F_{yf} - F_r)$ or F_{yw}, ksi (MPa)	F1.2
F_{SR}	Design stress range, ksi (MPa)	App. K3.3
F_{TH}	Threshold fatigue stress range, maximum stress range for indefinite design life, ksi (MPa)	App. K3.3
$F_{b\gamma}$	Flexural stress for tapered members defined by Equations A-F3-4 and A-F3-5	App. F3
F_{cr}	Critical stress, ksi (MPa)	E2
F_{crft}, F_{cry}, F_{crz}	Flexural-torsional buckling stresses for double-angle and tee-shaped compression members, ksi (MPa)	E3
F_e	Elastic buckling stress, ksi (MPa)	App. E3
F_{ex}	Elastic flexural buckling stress about the major axis, ksi (MPa)	App. E3
F_{ey}	Elastic flexural buckling stress about the minor axis, ksi (MPa)	App. E3
F_{ez}	Elastic torsional buckling stress, ksi (MPa)	App. E3
F_{my}	Modified yield stress for composite columns, ksi (MPa)	I2.2
F_n	Nominal shear rupture strength, ksi (MPa)	J4
F_r	Compressive residual stress in flange [10 ksi (69 MPa) for rolled shapes; 16.5 ksi (114 MPa) for welded built-up shapes]	B5.1
$F_{s\gamma}$	Stress for tapered members defined by Equation A-F3-6, ksi (MPa)	App. F3
F_u	Specified minimum tensile strength of the type of steel being used, ksi (MPa)	B10
F_w	Nominal strength of the weld electrode material, ksi (MPa)	J2.4
$F_{w\gamma}$	Stress for tapered members defined by Equation A-F3-7, ksi (MPa)	App. F3
F_y	Specified minimum yield stress of the type of steel being used, ksi (MPa). As used in this Specification, "yield stress" denotes either the specified minimum yield point (for those steels that have a yield point) or specified yield	

strength (for those steels that do not have a yield point) A5

F_{yf} Specified minimum yield stress of the flange, ksi (MPa). B5.1

F_{yr} Specified minimum yield stress of reinforcing bars, ksi (MPa) I2.2

F_{yst} Specified minimum yield stress of the stiffener material, ksi (MPa) . . . . . App. G4

F_{yw} Specified minimum yield stress of the web, ksi (MPa) B5.1

G Shear modulus of elasticity of steel, $G = 11{,}200$ ksi (77 200 MPa) F1.2

H Horizontal force, kips (N) C1

H Flexural constant E3

H_s Length of stud connector after welding, in. (mm) I3.5

I Moment of inertia, in.^{4} (mm^4) F1.2

I_d Moment of inertia of the steel deck supported on secondary members, in.^{4} (mm^4) K2

I_p Moment of inertia of primary members, in.^{4} (mm^4) K2

I_s Moment of inertia of secondary members, in.^{4} (mm^4) K2

I_{st} Moment of inertia of a transverse stiffener, in.^{4} (mm^4) . . . . . App. G4

I_{yc} Moment of inertia about y-axis referred to compression flange, or if reverse curvature bending referred to smaller flange, in.^{4} (mm^4) . . App. F1

J Torsional constant for a section, in.^{4} (mm^4) F1.2

K Effective length factor for prismatic member B7

K_z Effective length factor for torsional buckling . . . . . App. E3

K_γ Effective length factor for a tapered member . . . . . App. F3

L Story height or panel spacing, in. (mm) C1

L Length of connection in the direction of loading, in. (mm) B3

L Live load due to occupancy and moveable equipment N4

L_b Laterally unbraced length; length between points which are either braced against lateral displacement of compression flange or braced against twist of the cross section, in. (mm) F1.2

L_c Length of channel shear connector, in. (mm) I5.4

L_c Edge distance, in. (mm) J3.10

L_p Limiting laterally unbraced length for full plastic bending capacity, uniform moment case ($C_b = 1.0$), in. (mm) F1.2

L_p Column spacing in direction of girder, ft (m) K2

L_{pd} Limiting laterally unbraced length for plastic analysis, in. (m) F1.2

L_q Maximum unbraced length for the required column force with K equal to one, in. (mm) C3

L_r Limiting laterally unbraced length for inelastic lateral-torsional buckling, in. (mm) F1.2

L_r Roof live load N4

L_s Column spacing perpendicular to direction of girder, ft (m) K2

M_A Absolute value of moment at quarter point of the unbraced beam segment, kip-in. (N-mm) F1.2

M_B Absolute value of moment at centerline of the unbraced beam segment, kip-in. (N-mm) F1.2

M_C Absolute value of moment at three-quarter point of the unbraced beam segment, kip-in. (N-mm) F1.2

M_{cr} Elastic buckling moment, kip-in. (N-mm) F1.2

M_{lt}	Required flexural strength in member due to lateral frame translation only, kip-in. (N-mm)	C1
M_{max}	Absolute value of maximum moment in the unbraced beam segment, kip-in. (N-mm)	F1.2
M_n	Nominal flexural strength, kip-in. (N-mm)	F1.1
M'_{nx}, M'_{ny}	Flexural strength defined in Equations A-H3-7 and A-H3-8 for use in alternate interaction equations for combined bending and axial force, kip-in. (N-mm)	App. H3
M_{nt}	Required flexural strength in member assuming there is no lateral translation of the frame, kip-in. (N-mm)	C1
M_p	Plastic bending moment, kip-in. (N-mm)	F1.1
M_p	Moment defined in Equations A-H3-5 and A-H3-6, for use in alternate interaction equations for combined bending and axial force, kip-in. (N-mm)	App. H3
M_r	Limiting buckling moment, M_{cr}, when $\lambda = \lambda_r$ and $C_b = 1.0$, kip-in. (N-mm)	F1.2
M_u	Required flexural strength, kip-in. (N-mm)	C1
M_y	Moment corresponding to onset of yielding at the extreme fiber from an elastic stress distribution (= $F_y S$ for homogeneous sections), kip-in. (N-mm)	F1.1
M_1	Smaller moment at end of unbraced length of beam or beam-column, kip-in. (N-mm)	F1.2
M_2	Larger moment at end of unbraced length of beam or beam-column, kip-in. (N-mm)	F1.2
N	Length of bearing, in. (mm)	K1.3
N	Number of stress range fluctuations in design life	App. K3.3
N_r	Number of stud connectors in one rib at a beam intersection	I3.5
P_{br}	Required story or panel bracing shear force, kip (N)	C3
P_{e1}, P_{e2}	Elastic Euler buckling load for braced and unbraced frame, respectively, kips (N)	C1
P_n	Nominal axial strength (tension or compression), kips (N)	D1
P_p	Bearing load on concrete, kips (N)	J9
P_u	Required axial strength (tension or compression), kips (N)	B5.1
P_y	Yield strength, kips (N)	B5.1
Q	Full reduction factor for slender compression elements	App. E3
Q_a	Reduction factor for slender stiffened compression elements	App. B5
Q_n	Nominal strength of one stud shear connector, kips (N)	I5
Q_s	Reduction factor for slender unstiffened compression elements	App. B5.3
R	Nominal load due to initial rainwater or ice exclusive of the ponding contribution	N4
R_{PG}	Plate girder bending strength reduction factor	App. G
R_e	Hybrid girder factor	App. F1
R_n	Nominal strength	A5.3
R_v	Web shear strength, kips (N)	K1.7
S	Elastic section modulus, in.^{3} (mm^3)	F1.2
S	Spacing of secondary members, ft (m)	K2
S	Snow load	N4
S_x'	Elastic section modulus of larger end of tapered member about its major axis, in.^{3} (mm^3)	App. F3

Symbol	Definition	Section
S_{eff}	Effective section modulus about major axis, in.^{3} (mm^3)	App. F1
S_{xt} , S_{xc}	Elastic section modulus referred to tension and compression flanges, respectively, in.^{3} (mm^3)	App. F1
T	Tension force due to service loads, kips (N)	J3.9
T_b	Specified pretension load in high-strength bolt, kips (N)	J3.9
T_u	Required tensile strength due to factored loads, kips (N)	J3.9
U	Reduction coefficient, used in calculating effective net area	B3
V_n	Nominal shear strength, kips (N)	F2.2
V_u	Required shear strength, kips (N)	App. G4
W	Wind load	Comm. A4
X_1	Beam buckling factor defined by Equation F1-8	F1.2
X_2	Beam buckling factor defined by Equation F1-9	F1.2
Z	Plastic section modulus, in.^{3} (mm^3)	F1.1
a	Clear distance between transverse stiffeners, in. (mm)	App. F2.2
a	Distance between connectors in a built-up member, in. (mm)	E4
a	Shortest distance from edge of pin hole to edge of member measured parallel to direction of force, in. (mm)	D3
a_r	Ratio of web area to compression flange area	App. G2
a'	Weld length, in. (mm)	B10
b	Compression element width, in. (mm)	B5.1
b_e	Reduced effective width for slender compression elements, in. (mm)	App. B5.3
b_{eff}	Effective edge distance, in. (mm)	D3
b_f	Flange width, in. (mm)	B5.1
b_s	Stiffener width for one-sided stiffeners, in. (mm)	C3.4
c_1, c_2, c_3	Numerical coefficients	I2.2
d	Nominal fastener diameter, in. (mm)	J3.3
d	Overall depth of member, in. (mm)	B5.1
d	Pin diameter, in. (mm)	D3
d	Roller diameter, in. (mm)	J8.2
d_L	Depth at larger end of unbraced tapered segment, in. (mm)	App. F3
d_b	Beam depth, in. (mm)	K1.7
d_b	Nominal diameter (body or shank diameter), in. (mm)	App. K3.3
d_c	Column depth, in. (mm)	K1.7
d_o	Depth at smaller end of unbraced tapered segment, in. (mm)	App. F3
e	Base of natural logarithm = 2.71828	Comm. E2
f	Computed compressive stress in the stiffened element, ksi (MPa)	App. B5.3
f_{b1}	Smallest computed bending stress at one end of a tapered segment, ksi (MPa)	App. F3
f_{b2}	Largest computed bending stress at one end of a tapered segment, ksi (MPa)	App. F3
f_c'	Specified compressive strength of concrete, ksi (MPa)	I2.2
f_o	Stress due to 1.2D + 1.2R, ksi (MPa)	App. K2
f_{un}	Required normal stress, ksi (MPa)	H2
f_{uv}	Required shear stress, ksi (MPa)	H2
f_v	Required shear stress due to factored loads in bolts or rivets, ksi (MPa)	J3.7
g	Transverse center-to-center spacing (gage) between fastener gage lines, in. (mm)	B2
h	Clear distance between flanges less the fillet or corner radius for	

	rolled shapes; and for built-up sections, the distance between adjacent lines of fasteners or the clear distance between flanges when welds are used, in. (mm)	B5.1
h	Distance between centroids of individual components perpendicular to the member axis of buckling, in. (mm)	E4
h_c	Twice the distance from the centroid to the following: the inside face of the compression flange less the fillet or corner radius, for rolled shapes; the nearest line of fasteners at the compression flange or the inside faces of the compression flange when welds are used, for built-up sections, in. (mm)	B5.1
h_o	Distance between flange centroids, in. (mm)	C3
h_r	Nominal rib height, in. (mm)	I3.5
h_s	Factor used in Equation A-F3-6 for web-tapered members	App. F3
h_w	Factor used in Equation A-F3-7 for web-tapered members	App. F3
j	Factor defined by Equation A-F2-4 for minimum moment of inertia for a transverse stiffener	App. F2.3
k	Distance from outer face of flange to web toe of fillet, in. (mm)	K1.3
k_v	Web plate buckling coefficient	App. F2.2
l	Laterally unbraced length of member at the point of load, in. (mm)	B7
l	Length of bearing, in. (mm)	J8.2
l	Length of connection in the direction of loading, in. (mm)	B3
l	Length of weld, in. (mm)	B3
m	Ratio of web to flange yield stress or critical stress in hybrid beams	App. G2
n	Number of nodal braced points within the span	C3
n	Threads per inch (per mm)	App. K3.4
r	Governing radius of gyration, in. (mm)	B7
r_{To}	For the smaller end of a tapered member, the radius of gyration, considering only the compression flange plus one-third of the compression web area, taken about an axis in the plane of the web, in. (mm)	App. F3.4
r_i	Minimum radius of gyration of individual component in a built-up member, in. (mm)	E4
r_{ib}	Radius of gyration of individual component relative to centroidal axis parallel to member axis of buckling, in. (mm)	E4
r_m	Radius of gyration of the steel shape, pipe, or tubing in composite columns. For steel shapes it may not be less than 0.3 times the overall thickness of the composite section, in. (mm)	I2
$\bar{r}_o$	Polar radius of gyration about the shear center, in. (mm)	E3
r_{ox}, r_{oy}	Radius of gyration about x and y axes at the smaller end of a tapered member, respectively, in. (mm)	App. F3.3
r_x, r_y	Radius of gyration about x and y axes, respectively, in. (mm)	E3
r_{yc}	Radius of gyration about y axis referred to compression flange, or if reverse curvature bending, referred to smaller flange, in. (mm)	App. F1
s	Longitudinal center-to-center spacing (pitch) of any two consecutive holes, in. (mm)	B2
t	Thickness of element, in. (mm)	B5.1
t	HSS design wall thickness, in. (mm)	B5.1
t_f	Flange thickness, in. (mm)	B5.1
t_f	Flange thickness of channel shear connector, in. (mm)	I5.4

t_s	Web stiffener thickness, in. (mm)	C3.4
t_w	Web thickness of channel shear connector, in. (mm)	I5.4
t_w	Web thickness, in. (mm)	B5.3
w	Leg size of the fillet weld, in. (mm)	J2.2
w	Plate width; distance between welds, in. (mm)	B3
w	Unit weight of concrete, lbs/cu ft. (kg/m^3)	I2
w_r	Average width of concrete rib or haunch, in. (mm)	I3.5
x	Subscript relating symbol to member strong axis	
x_o, y_o	Coordinates of the shear center with respect to the centroid, in. (mm)	E3
$\bar{x}$	Connection eccentricity, in. (mm)	B3
y	Subscript relating symbol to member weak axis	
z	Distance from the smaller end of tapered member used in Equation A-F3-1 for the variation in depth, in. (mm)	App. F3
z	Subscript relating symbol to member longitudinal axis	
α	Separation ratio for built-up compression members = $h / 2r_{ib}$	E4
β	Reduction factor given by Equation J2-1	J2
β_T	Brace stiffness requirement when there is no web distortion	C3.4
β_{Tb}	Required nodal torsional bracing stiffness	C3.4
β_{br}	Required story or panel shear stiffness	C3
β_{sec}	Web distortional stiffness, including the effect of web transverse stiffeners, if any	C3.4
Δ_{oh}	Translation deflection of the story under consideration, in. (mm)	C1
γ	Depth tapering ratio	App. F3
γ	Subscript for tapered members	App. F3
ζ	Exponent for alternate beam-column interaction equation	App. H3
η	Exponent for alternate beam-column interaction equation	App. H3
λ_c	Column slenderness parameter	C1
λ_e	Equivalent slenderness parameter	App. E3
λ_{eff}	Effective slenderness ratio defined by Equation A-F3-2	App. F3
λ_p	Limiting slenderness parameter for compact element	B5.1
λ_r	Limiting slenderness parameter for noncompact element	B5.1
ϕ	Resistance factor	A5.3
ϕ_b	Resistance factor for flexure	F1
ϕ_c	Resistance factor for compression	A5.1
ϕ_c	Resistance factor for axially loaded composite columns	I2.2
ϕ_{sf}	Resistance factor for shear on the failure path	D3
ϕ_t	Resistance factor for tension	D1
ϕ_v	Resistance factor for shear	F2.2

GLOSSARY

Alignment chart for columns. A nomograph for determining the effective length factor K for some types of columns

Amplification factor. A multiplier of the value of moment or deflection in the unbraced length of an axially loaded member to reflect the secondary values generated by the eccentricity of the applied axial load within the member

Aspect ratio. In any rectangular configuration, the ratio of the lengths of the sides

Batten plate. A plate element used to join two parallel components of a built-up column, girder, or strut rigidly connected to the parallel components and designed to transmit shear between them

Beam. A structural member whose primary function is to carry loads transverse to its longitudinal axis

Beam-column. A structural member whose primary function is to carry loads both transverse and parallel to its longitudinal axis

Bent. A plane framework of beam or truss members which support loads and the columns which support these members

Biaxial bending. Simultaneous bending of a member about two perpendicular axes

Bifurcation. The phenomenon whereby a perfectly straight member under compression may either assume a deflected position or may remain undeflected, or a beam under flexure may either deflect and twist out of plane or remain in its in-plane deflected position

Braced frame. A frame in which the resistance to lateral load or frame instability is primarily provided by a diagonal, a K brace, or other auxiliary system of bracing

Brittle fracture. Abrupt cleavage with little or no prior ductile deformation

Buckling load. The load at which a perfectly straight member under compression assumes a deflected position

Built-up member. A member made of structural metal elements that are welded, bolted, or riveted together

Charpy V-notch impact test. A standard dynamic test in which a notched specimen is struck and broken by a single blow in a specially designed testing machine. The measured test values may be the energy absorbed, the percentage shear fracture, the lateral expansion opposite the notch, or a combination thereof

Cladding. The exterior covering of the structural components of a building

Cold-formed members. Structural members formed from steel without the application of heat

Column. A structural member whose primary function is to carry loads parallel to its longitudinal axis

Column curve. A curve expressing the relationship between axial column strength and slenderness ratio

Combined mechanism. A mechanism determined by plastic analysis procedure which combines elementary beam, panel, and joint mechanisms

Compact section. Compact sections are capable of developing a fully plastic stress distribution and possess rotation capacity of approximately three before the onset of local buckling

Composite beam. A steel beam structurally connected to a concrete slab so that the beam and slab respond to loads as a unit

Concrete-encased beam. A beam totally encased in concrete cast integrally with the slab

Connection. Combination of joints used to transmit forces between two or more members. Categorized by the type and amount of force transferred (moment, shear, end reaction). See also *Splices*

Critical load. The load at which bifurcation occurs as determined by a theoretical stability analysis

Curvature. Rotation per unit length due to bending

Design documents. Documents prepared by the designer (design drawings, design details, and job specifications)

Design strength. Resistance (force, moment, stress, as appropriate) provided by element or connection; the product of the nominal strength and the resistance factor

Diagonal bracing. Inclined structural members carrying primarily axial load enabling a structural frame to act as a truss to resist horizontal loads

Diaphragm. Floor slab, metal wall, or roof panel possessing a large in-plane shear stiffness and strength adequate to transmit horizontal forces to resisting systems

Diaphragm action. The in-plane action of a floor system (also roofs and walls) such that all columns framing into the floor from above and below are maintained in the same position relative to each other

Double concentrated forces. Two equal and opposite forces which form a couple on the same side of the loaded member

Double curvature. A bending condition in which end moments on a member cause the member to assume an S shape

Drift. Lateral deflection of a building

Drift index. The ratio of lateral deflection to the height of the building

Ductility factor. The ratio of the total deformation at maximum load to the elastic-limit deformation

Effective length. The equivalent length *KL* used in compression formulas and determined by a bifurcation analysis

Effective length factor K. The ratio between the effective length and the unbraced length of the member measured between the centers of gravity of the bracing members

Effective moment of inertia. The moment of inertia of the cross section of a member that remains elastic when partial plastification of the cross section takes place, usually under the combination of residual stress and applied stress. Also, the moment of inertia based on effective widths of elements that buckle locally. Also, the moment of inertia used in the design of partially composite members

Effective stiffness. The stiffness of a member computed using the effective moment of inertia of its cross section

Effective width. The reduced width of a plate or slab which, with an assumed uniform stress distribution, produces the same effect on the behavior of a structural member as the actual plate width with its nonuniform stress distribution

Elastic analysis. Determination of load effects (force, moment, stress, as appropriate) on members and connections based on the assumption that material deformation disappears on removal of the force that produced it

Elastic-perfectly plastic. A material which has an idealized stress-strain curve that varies linearly from the point of zero strain and zero stress up to the yield point of the material, and then increases in strain at the value of the yield stress without any further increases in stress

Embedment. A steel component cast in a concrete structure which is used to transmit externally applied loads to the concrete structure by means of bearing, shear, bond, friction, or any combination thereof. The embedment may be fabricated of structural-steel plates, shapes, bars, bolts, pipe, studs, concrete reinforcing bars, shear connectors, or any combination thereof

Encased steel structure. A steel-framed structure in which all of the individual frame members are completely encased in cast-in-place concrete

Euler formula. The mathematical relationship expressing the value of the Euler load in terms of the modulus of elasticity, the moment of inertia of the cross section, and the length of a column

Euler load. The critical load of a perfectly straight, centrally loaded pin-ended column

Eyebar. A particular type of pin-connected tension member of uniform thickness with forged or flame-cut head of greater width than the body proportioned to provide approximately equal strength in the head and body

Factored load. The product of the nominal load and a load factor

Fastener. Generic term for welds, bolts, rivets, or other connecting device

Fatigue. A fracture phenomenon resulting from a fluctuating stress cycle

First-order analysis. Analysis based on first-order deformations in which equilibrium conditions are formulated on the undeformed structure

Flame-cut plate. A plate in which the longitudinal edges have been prepared by oxygen cutting from a larger plate

Flat width. For a rectangular HSS, the nominal width minus twice the outside corner radius. In absence of knowledge of the corner radius, the flat width may be taken as the total section width minus three times the thickness

Flexible connection. A connection permitting a portion, but not all, of the simple beam rotation of a member end

Floor system. The system of structural components separating the stories of a building

Force. Resultant of distribution of stress over a prescribed area. A reaction that develops in a member as a result of load (formerly called total stress or stress). Generic term signifying axial loads, bending moment, torques, and shears

Fracture toughness. Measure of the ability to absorb energy without fracture. Generally determined by impact loading of specimens containing a notch having a prescribed geometry

Frame buckling. A condition under which bifurcation may occur in a frame

Frame instability. A condition under which a frame deforms with increasing lateral deflection under a system of increasing applied monotonic loads until a maximum value of the load called the stability limit is reached, after which the frame will continue to deflect without further increase in load

Fully composite beam. A composite beam with sufficient shear connectors to develop the full flexural strength of the composite section

High-cycle fatigue. Failure resulting from more than 20,000 applications of cyclic stress

HSS. Hollow structural sections that are prismatic square, rectangular or round products of a pipe or tubing mill and meet the geometric tolerance, tensile strength and chemical composition requirements of a standard specification

Hybrid beam. A fabricated steel beam composed of flanges with a greater yield strength than that of the web. Whenever the maximum flange stress is less than or equal to the web yield stress the girder is considered homogeneous

Hysteresis loop. A plot of force versus displacement of a structure or member subjected to reversed, repeated load into the inelastic range, in which the path followed during release and removal of load is different from the path for the addition of load over the same range of displacement

Inclusions. Nonmetallic material entrapped in otherwise sound metal

Incomplete fusion. Lack of union by melting of filler and base metal over entire prescribed area

Inelastic action. Material deformation that does not disappear on removal of the force that produced it

Instability. A condition reached in the loading of an element or structure in which continued deformation results in a decrease of load-resisting capacity

Joint. Area where two or more ends, surfaces, or edges are attached. Categorized by type of fastener or weld used and method of force transfer

K bracing. A system of struts used in a braced frame in which the pattern of the struts resembles the letter K, either normal or on its side

Lamellar tearing. Separation in highly restrained base metal caused by through-thickness strains induced by shrinkage of adjacent filler metal

Lateral bracing member. A member utilized individually or as a component of a lateral bracing system to prevent buckling of members or elements and/or to resist lateral loads

Lateral (or lateral-torsional) buckling. Buckling of a member involving lateral deflection and twist

Leaning column. Gravity-loaded column where connections to the frame (simple connections) do not provide resistance to lateral loads

Limit state. A condition in which a structure or component becomes unfit for service and is judged either to be no longer useful for its intended function (*serviceability limit state*) or to be unsafe (*strength limit state*)

Limit states. Limits of structural usefulness, such as brittle fracture, plastic collapse, excessive deformation, durability, fatigue, instability, and serviceability

Load factor. A factor that accounts for unavoidable deviations of the actual load from the nominal value and for uncertainties in the analysis that transforms the load into a load effect

Loads. Forces or other actions that arise on structural systems from the weight of all permanent construction, occupants and their possessions, environmental effects, differential settlement, and restrained dimensional changes. *Permanent* loads are those loads in which variations in time are rare or of small magnitude. All other loads are *variable* loads. See *Nominal loads*

LRFD (Load and Resistance Factor Design). A method of proportioning structural components (members, connectors, connecting elements, and assemblages) such that no applicable limit state is exceeded when the structure is subjected to all appropriate load combinations

Local buckling. The buckling of a compression element which may precipitate the failure of the whole member

Low-cycle fatigue. Fracture resulting from a relatively high-stress range resulting in a relatively small number of cycles to failure

Lower bound load. A load computed on the basis of an assumed equilibrium moment diagram in which the moments are not greater than M_p that is less than or at best equal to the true ultimate load

Mechanism. An articulated system able to deform without an increase in load, used in the special sense that the linkage may include real hinges or plastic hinges, or both

Mechanism method. A method of plastic analysis in which equilibrium between external forces and internal plastic hinges is calculated on the basis of an assumed mechanism. The failure load so determined is an upper bound

Nodal Brace. A brace that prevents the lateral movement or twist at the particular brace location along the length of the beam or column without any direct attachment to other braces at adjacent brace points. (See relative brace)

Nominal loads. The magnitudes of the loads specified by the applicable code

Nominal strength. The capacity of a structure or component to resist the effects of loads, as determined by computations using specified material strengths and dimensions and formulas derived from accepted principles of structural mechanics or by field tests or laboratory tests of scaled models, allowing for modeling effects and differences between laboratory and field conditions

Noncompact section. Noncompact sections can develop the yield stress in compression elements before local buckling occurs, but will not resist inelastic local buckling at strain levels required for a fully plastic stress distribution

P-Delta effect. Secondary effect of column axial loads and lateral deflection on the moments in members

Panel zone. The zone in a beam-to-column connection that transmits moment by a shear panel

Partially composite beam. A composite beam for which the shear strength of shear connectors governs the flexural strength

Plane frame. A structural system assumed for the purpose of analysis and design to be two-dimensional

Plastic analysis. Determination of load effects (force, moment, stress, as appropriate) on members and connections based on the assumption of rigid-plastic behavior, i.e., that equilibrium is satisfied throughout the structure and yield is not exceeded anywhere. Second order effects may need to be considered

Plastic design section. The cross section of a member which can maintain a full plastic moment through large rotations so that a mechanism can develop; the section suitable for design by plastic analysis

Plastic hinge. A yielded zone which forms in a structural member when the plastic moment is attained. The beam is assumed to rotate as if hinged, except that it is restrained by the plastic moment M_p

Plastic-limit load. The maximum load that is attained when a sufficient number of yield zones have formed to permit the structure to deform plastically without further increase in load. It is the largest load a structure will support, when perfect plasticity is assumed and when such factors as instability, second-order effects, strain hardening, and fracture are neglected

Plastic mechanism. See *Mechanism*

Plastic modulus. The section modulus of resistance to bending of a completely yielded cross section. It is the combined static moment about the plastic neutral axis of the cross-sectional areas above and below that axis

Plastic moment. The resisting moment of a fully-yielded cross section

Plastic strain. The difference between total strain and elastic strain

Plastic zone. The yielded region of a member

Plastification. The process of successive yielding of fibers in the cross section of a member as bending moment is increased

Plate girder. A built-up structural beam

Post-buckling strength. The load that can be carried by an element, member, or frame after buckling

Prying Action. Lever action that exists in connections in which the line of application of the applied load is eccentric to the axis of the bolt, causing deformation of the fitting and an amplification of the axial force in the bolt

Redistribution of moment. A process which results in the successive formation of plastic hinges so that less highly stressed portions of a structure may carry increased moments

Relative Brace. A brace that controls the relative movement of two adjacent brace points along the length of a beam or column or the relative lateral displacement of two stories in a frame. (See nodal brace)

Required strength. Load effect (force, moment, stress, as appropriate) acting on element or connection determined either by structural analysis from the factored loads (using appropriate critical load combinations) or explicitly specified

Residual stress. The stresses that remain in an unloaded member after it has been formed into a finished product. (Examples of such stresses include, but are not limited to, those induced by cold bending, cooling after rolling, or welding)

Resistance. The capacity of a structure or component to resist the effects of loads. It is determined by computations using specified material strengths, dimensions and formulas derived from accepted principles of structural mechanics, or by field tests or laboratory tests of scaled models, allowing for modeling effects and differences between laboratory and field conditions. Resistance is a generic term that includes both strength and serviceability limit states

Resistance factor. A factor that accounts for unavoidable deviations of the actual strength from the nominal value and the manner and consequences of failure

Rigid frame. A structure in which connections maintain the angular relationship between beam and column members under load

Root of the flange. Location on the web of the corner radius termination point or the toe of the flange-to-web weld. Measured as the k distance from the far side of the flange

Rotation capacity. The incremental angular rotation that a given shape can accept prior to local failure defined as $R = (\theta_u / \theta_p)$ where θ_u is the overall rotation attained at the factored load state and θ_p is the idealized rotation corresponding to elastic theory applied to the case of $M = M_p$

St. Venant torsion. That portion of the torsion in a member that induces only shear stresses in the member

Second-order analysis. Analysis based on second-order deformations, in which equilibrium conditions are formulated on the deformed structure

Service load. Load expected to be supported by the structure under normal usage; often taken as the nominal load

Serviceability limit state. Limiting condition affecting the ability of a structure to preserve its appearance, maintainability, durability, or the comfort of its occupants or function of machinery under normal usage

Shape factor. The ratio of the plastic moment to the yield moment, or the ratio of the plastic modulus to the section modulus for a cross section

Shear friction. Friction between the embedment and the concrete that transmits shear loads. The relative displacement in the plane of the shear load is considered to be resisted by shear-friction anchors located perpendicular to the plane of the shear load

Shear lugs. Plates, welded studs, bolts, and other steel shapes that are embedded in the concrete and located transverse to the direction of the shear force and that transmit shear loads, introduced into the concrete by local bearing at the shear lug-concrete interface

Shear wall. A wall that resists, in its own plane, shear forces resulting from applied wind, earthquake, or other transverse loads or provides frame stability. Also called a structural wall

Sidesway. The lateral movement of a structure under the action of lateral loads, unsymmetrical vertical loads, or unsymmetrical properties of the structure

Sidesway buckling. The buckling mode of a multistory frame precipitated by the relative lateral displacements of joints, leading to failure by sidesway of the frame

Simple plastic theory. See *Plastic design*

Single curvature. A deformed shape of a member having one smooth continuous arc, as opposed to double curvature which contains a reversal

Slender-element section. The cross section of a member which will experience local buckling in the elastic range

Slenderness ratio. The ratio of the effective length of a column to the radius of gyration of the column, both with respect to the same axis of bending

Slip-critical joint. A bolted joint in which the slip resistance of the connection is required

Space frame. A three-dimensional structural framework (as contrasted to a plane frame)

Splice. The connection between two structural elements joined at their ends to form a single, longer element

Stability-limit load. Maximum (theoretical) load a structure can support when second-order instability effects are included

Stepped column. A column with changes from one cross section to another occurring at abrupt points within the length of the column

Stiffener. A member, usually an angle or plate, attached to a plate or web of a beam or girder to distribute load, to transfer shear, or to prevent buckling of the member to which it is attached

Stiffness. The resistance to deformation of a member or structure measured by the ratio of the applied force to the corresponding displacement

Story drift. The difference in horizontal deflection at the top and bottom of a story

Strain hardening. Phenomenon wherein ductile steel, after undergoing considerable deformation at or just above yield point, exhibits the capacity to resist substantially higher loading than that which caused initial yielding

Strain-hardening strain. For structural steels that have a flat (plastic) region in the stress-strain relationship, the value of the strain at the onset of strain hardening

Strength design. A method of proportioning structural members using load factors and resistance factors such that no applicable limit state is exceeded (also called load and resistance factor design)

Strength limit state. Limiting condition affecting the safety of the structure, in which the ultimate load-carrying capacity is reached

Stress. Force per unit area

Stress concentration. Localized stress considerably higher than average (even in uniformly loaded cross sections of uniform thickness) due to abrupt changes in geometry or localized loading

Strong axis. The major principal axis of a cross section

Structural system. An assemblage of load-carrying components which are joined together to provide regular interaction or interdependence

Stub column. A short compression-test specimen, long enough for use in measuring the stress-strain relationship for the complete cross section, but short enough to avoid buckling as a column in the elastic and plastic ranges

Subassemblage. A truncated portion of a structural frame

Supported frame. A frame which depends upon adjacent braced or unbraced frames for resistance to lateral load or frame instability. (This transfer of load is frequently provided by the floor or roof system through diaphragm action or by horizontal cross bracing in the roof)

Tangent modulus. At any given stress level, the slope of the stress-strain curve of a material in the inelastic range as determined by the compression test of a small specimen under controlled conditions

Temporary structure. A general term for anything that is built or constructed (usually to carry construction loads) that will eventually be removed before or after completion of construction and does not become part of the permanent structural system

Tensile strength. The maximum tensile stress that a material is capable of sustaining

Tension field action. The behavior of a plate girder panel under shear force in which diagonal tensile stresses develop in the web and compressive forces develop in the transverse stiffeners in a manner analogous to a Pratt truss

Toe of the fillet. Termination point of fillet weld or of rolled section fillet

Torque-tension relationship. Term applied to the wrench torque required to produce specified pre-tension in high-strength bolts

Turn-of-nut method. Procedure whereby the specified pre-tension in high-strength bolts is controlled by rotation of the wrench a predetermined amount after the nut has been tightened to a snug fit

Unbraced frame. A frame in which the resistance to lateral load is provided by the bending resistance of frame members and their connections

Unbraced length. The distance between braced points of a member, measured between the centers of gravity of the bracing members

Undercut. A notch resulting from the melting and removal of base metal at the edge of a weld

Universal-mill plate. A plate in which the longitudinal edges have been formed by a rolling process during manufacture. Often abbreviated as UM plate

Upper bound load. A load computed on the basis of an assumed mechanism which will always be at best equal to or greater than the true ultimate load

Vertical bracing system. A system of shear walls, braced frames, or both, extending through one or more floors of a building

Von Mises yield criterion. A theory which states that inelastic action at any point in a body under any combination of stresses begins only when the strain energy of distortion per unit volume absorbed at the point is equal to the strain energy of distortion absorbed per unit volume at any point in a simple tensile bar stressed to the elastic limit under a state of uniaxial stress. It is often called the maximum strain-energy-of-distortion theory. Accordingly, shear yield occurs at 0.58 times the yield strength

Warping torsion. That portion of the total resistance to torsion that is provided by resistance to warping of the cross section

Weak axis. The minor principal axis of a cross section

Weathering steel. A type of high-strength, low-alloy steel which can be used in normal environments (not marine) and outdoor exposures without protective paint covering. This steel develops a tight adherent rust at a decreasing rate with respect to time

Web buckling. The buckling of a web plate

Web crippling. The local failure of a web plate in the immediate vicinity of a concentrated load or reaction

Working load. Also called service load. The actual load assumed to be acting on the structure

Yield moment. In a member subjected to bending, the moment at which an outer fiber first attains the yield stress

Yield plateau. The portion of the stress-strain curve for uniaxial tension or compression in which the stress remains essentially constant during a period of substantially increased strain

Yield point. The first stress in a material at which an increase in strain occurs without an increase in stress

Yield strength. The stress at which a material exhibits a specified limiting deviation from the proportionality of stress to strain. Deviation expressed in terms of strain

Yield stress. Yield point, yield strength, or yield stress level as defined

Yield-stress level. The average stress during yielding in the plastic range, the stress determined in a tension test when the strain reaches 0.005 in. per in. (0.005 mm per mm)

CHAPTER A

GENERAL PROVISIONS

A1. SCOPE

The *Load and Resistance Factor Design Specification for Structural Steel Buildings* shall govern the design, fabrication, and erection of steel-framed buildings. As an alternative, the AISC *Specification for Structural Steel Buildings, Allowable Stress Design and Plastic Design* is permitted.

This Specification includes the list of symbols, the glossary, and the appendices. The tables of numerical values are provided for design convenience.

Seismic design of buildings shall comply with the AISC *Seismic Provisions for Structural Steel Buildings*, *Seismic Provisions Supplement No. 1*, and with this Specification.

Single angle members shall comply with the *Specification for Load and Resistance Factor Design of Single-Angle Members* and with this Specification.

Hollow structural sections (HSS) shall comply with the *Specification for the Design of Steel Hollow Structural Sections* and with this Specification.

Design of nuclear structures shall comply with the *Specification for the Design, Fabrication and Erection of Steel Safety Related Structures for Nuclear Facilities* and with this Specification.

As used in this Specification, the term *structural steel* refers to the steel elements of the structural steel frame essential to the support of the required loads. Such elements are enumerated in Section 2.1 of the AISC *Code of Standard Practice for Steel Buildings and Bridges*. For the design of cold-formed steel structural members, whose profiles contain rounded corners and slender flat elements, the provisions of the American Iron and Steel Institute *Specification for the Design of Cold-Formed Steel Structural Members* are recommended.

A2. TYPES OF CONSTRUCTION

Two basic types of construction and associated design assumptions shall be permitted under the conditions stated herein, and each will govern in a specific manner the strength of members and the types and strength of their connections.

Type FR (fully restrained), commonly designated as "rigid-frame" (continuous frame), assumes that connections have sufficient stiffness to maintain the angles between intersecting members.

Type PR (partially restrained) assumes that connections have insufficient stiffness to maintain the angles between intersecting members. When connection restraint is considered, use of Type PR construction under this Specification requires that the strength, stiffness and ductility characteristics of the connections be incorporated in the analysis and design. These characteristics shall be documented in the technical literature or established by analytical or experimental means.

When connection restraint is ignored, commonly designated "simple framing," it is assumed that for the transmission of gravity loads the ends of the beams and girders are connected for shear only and are free to rotate. For "simple framing" the following requirements apply:

(1) The connections and connected members shall be adequate to resist the factored gravity loads as "simple beams."

(2) The connections and connected members shall be adequate to resist the factored lateral loads.

(3) The connections shall have sufficient inelastic rotation capacity to avoid overload of fasteners or welds under combined factored gravity and lateral loading.

The type of construction assumed in the design shall be indicated on the design documents. The design of all connections shall be consistent with the assumption.

A3. MATERIAL

1. Structural Steel

1a. ASTM Designations

Material conforming to one of the following standard specifications is approved for use under this Specification:

Carbon Structural Steel, ASTM A36/A36M

Pipe, Steel, Black and Hot-Dipped, Zinc-Coated Welded and Seamless ASTM A53/A53M, Gr. B

High-Strength Low-Alloy Structural Steel, ASTM A242/A242M

Cold-Formed Welded and Seamless Carbon Steel Structural Tubing in Rounds and Shapes, ASTM A500

Hot-Formed Welded and Seamless Carbon Steel Structural Tubing, ASTM 501

High-Yield-Strength, Quenched and Tempered Alloy Steel Plate, Suitable for Welding, ASTM A514/A514M

High-Strength Carbon-Manganese Steel of Structural Quality, ASTM A529/A529M

Steel, Sheet and Strip, Carbon, Hot-Rolled, Structural Quality, ASTM A570/A570M, Gr. 40 (275), 45 (310), and 50 (345)

High-Strength Low-Alloy Columbium-Vanadium Steels of Structural Quality, ASTM A572/A572M

High-Strength Low-Alloy Structural Steel with 50 ksi (345 MPa) Minimum Yield Point to 4-in. (100 mm) Thick, ASTM A588/A588M

Steel, Sheet and Strip, High-Strength, Low-Alloy, Hot-Rolled and Cold-Rolled, with Improved Atmospheric Corrosion Resistance, ASTM A606

Steel, Sheet and Strip, High-Strength, Low-Alloy, Columbium or Vanadium, or Both, Hot-Rolled and Cold-Rolled, ASTM A607

Hot-Formed Welded and Seamless High-Strength Low-Alloy Structural Tubing, ASTM A618

Carbon and High-Strength Low-Alloy Structural Steel Shapes, Plates and Bars

Cold-Formed Welded and Seamless High-Strength, Low-Alloy Structural Tubing with Improved Atmospheric Corrosion Resistance, ASTM A847 Errata 9/4/01

and Quenched-and-Tempered Alloy Structural Steel Plates for Bridges, ASTM A709/A709M

Quenched and Tempered Low-Alloy Structural Steel Plate with 70 ksi (485 MPa) Minimum Yield Strength to 4 in. (100 mm) Thick, ASTM A852/A852M

High-Strength Low-Alloy Steel Shapes of Structural Quality, Produced by Quenching and Self-Tempering Process (QST), ASTM A913/A913M

Steel for Structural Shapes for Use in Building Framing, ASTM A992/A992M

Certified mill test reports or certified reports of tests made by the fabricator or a testing laboratory in accordance with ASTM A6/A6M, Standard Specification for General Requirements for Rolled Structural Steel Bars, Plates, Shapes, and Sheet Piling or A568/A568M, Standard Specification for Steel, Sheet, Carbon, and High-Strength, Low-Alloy, Hot-Rolled and Cold-Rolled, General Requirements for, as applicable, shall constitute sufficient evidence of conformity with one of the above ASTM standards. If requested, the fabricator shall provide an affidavit stating that the structural steel furnished meets the requirements of the grade specified.

1b. Unidentified Steel

Unidentified steel, if surface conditions are acceptable according to criteria contained in ASTM A6/A6M, is permitted to be used for unimportant members or details, where the precise physical properties and weldability of the steel would not affect the strength of the structure.

1c. Heavy Shapes

For ASTM A6/A6M Group 4 and 5 rolled shapes to be used as members subject to primary tensile stresses due to tension or flexure, toughness need not be specified if splices are made by bolting. If such members are spliced using complete-joint-penetration groove welds, the steel shall be specified in the contract documents to be supplied with Charpy V-notch (CVN) impact testing in accordance with ASTM A6/A6M, Supplementary Requirement S5. The impact test shall meet a minimum average value of 20 ft-lbs. (27 J) absorbed energy at +70°F (+21°C) and shall be conducted in accordance with ASTM A673/A673M, with the following exceptions:

(1) The center longitudinal axis of the specimens shall be located as near as practical to midway between the inner flange surface and the center of the flange thickness at the intersection with the web mid-thickness.

(2) Tests shall be conducted by the producer on material selected from a location representing the top of each ingot or part of an ingot used to produce the product represented by these tests.

For plates exceeding two-in. (50 mm) thick used for built-up cross-sections with bolted splices and subject to primary tensile stresses due to tension or flexure, material toughness need not be specified. If such cross-sections are spliced using complete-joint-penetration welds, the steel shall be specified in the contract documents to be supplied with Charpy V-notch testing in accordance with ASTM A6/A6M, Supplementary Requirement S5. The impact test shall be conducted by the producer in accordance with ASTM A673/A673M, Frequency P, and shall meet a minimum average value of 20 ft-lbs. (27 J) absorbed energy at +70°F (+21°C).

The above supplementary requirements also apply when complete-joint-penetraion welded joints through the thickness of ASTM A6/A6M Group 4 and 5 shapes and built-up cross sections with thickness exceeding two in. (50 mm) are used in connections subjected to primary tensile stress due to tension or flexure of such members. The requirements need not apply to ASTM A6/A6M Group 4 and 5 shapes and built-up members with thickness exceeding two in. (50 mm) to which members other than ASTM A6/A6M Group 4 and 5 shapes and built-up members are connected by complete-joint- penetration welded joints through the thickness of the thinner material to the face of the heavy material.

Additional requirements for joints in heavy rolled and built-up members are given in Sections J1.5, J1.6, J2.6, J2.8 and M2.2.

2. Steel Castings and Forgings

Cast steel shall conform to one of the following standard specifications:

Steel Castings, Carbon, for General Application, ASTM A27/A27M, Gr. 65-35 (450-240)

Steel Castings, High Strength, for Structural Purposes, ASTM A148/148M Gr. 80-50 (550-345)

Steel forgings shall conform to the following standard specification:

Steel Forgings Carbon and Alloy, for General Industrial Use, ASTM A668/ A668M

Certified test reports shall constitute sufficient evidence of conformity with standards.

3. Bolts, Washers, and Nuts

Steel bolts, washers, and nuts shall conform to one of the following standard specifications:

Carbon and Alloy Steel Nuts for Bolts for High-Pressure or High-Temperature Service, or Both, ASTM A194/A194M

Carbon Steel Bolts and Studs, 60,000 PSI Tensile Strength, ASTM A307

Structural Bolts, Steel, Heat Treated, 120/105 ksi Minimum Tensile Strength, ASTM A325

High-Strength Bolts for Structural Steel Joints [Metric], ASTM A325M

Quenched and Tempered Steel Bolts and Studs, ASTM A449

Heat-Treated Steel Structural Bolts, 150 ksi Minimum Tensile Strength, ASTM A490

High-Strength Steel Bolts, Classes 10.9 and 10.9.3, for Structural Steel Joints [Metric], ASTM A490M

Carbon and Alloy Steel Nuts, ASTM A563

Carbon and Alloy Steel Nuts [Metric], ASTM A563M

Hardened Steel Washers, ASTM F436

Hardened Steel Washers [Metric], ASTM F436M

Compressible-Washer-Type Direct Tension Indicators for Use with Structural Fasteners, ASTM F959

Compressible-Washer-Type Direct Tension Indicators for Use with Structural Fasteners [Metric], ASTM F959M

"Twist Off" Type Tension Control Structural Bolt/Nut/Washer Assemblies, Steel, Heat Treated, 120/105 ksi Minimum Tensile Strength, ASTM F1852

ASTM A449 bolts are permitted to be used only in connections requiring bolt diameters greater than $1\frac{1}{2}$-in. (38 mm) and shall not be used in slip-critical connections.

Manufacturer's certification shall constitute sufficient evidence of conformity with the standards.

4. Anchor Rods and Threaded Rods

Anchor rods and threaded rod steel shall conform to one of the following standard specifications:

Carbon Structural Steel, ASTM A36/A36M
Alloy Steel and Stainless Steel Bolting Materials for High-Temperature Service, ASTM A193/A193M
Quenched and Tempered Alloy Steel Bolts, Studs and Other Externally Threaded Fasteners, ASTM A354
High-Strength Low-Alloy Columbium-Vanadium Structural Steel, ASTM A572/A572M
High-Strength Low-Alloy Structural Steel with 50 ksi [345 MPa] Minimum Yield Point to 4-in. [100 mm] Thick, ASTM A588/A588M
Anchor Bolts, Steel, 36, 55, and 105-ksi Yield Strength, ASTM F1554

Threads on anchor rods and threaded rods shall conform to the Unified Standard Series of ASME B18.2.6 and shall have Class 2A tolerances.

Steel bolts conforming to other provisions of Section A3.3 are permitted as anchor rods. A449 material is acceptable for high-strength anchor rods and threaded rods of any diameter.

Manufacturer's certification shall constitute sufficient evidence of conformity with the standards.

5. Filler Metal and Flux for Welding

Filler metals and fluxes shall conform to one of the following specifications of the American Welding Society:

Specification for Carbon Steel Electrodes for Shielded Metal Arc Welding, AWS A5.1
Specification for Low-Alloy Steel Electrodes for Shielded Metal Arc Welding, AWS A5.5
Specification for Carbon Steel Electrodes and Fluxes for Submerged Arc Welding, AWS A5.17/A5.17M
Specification for Carbon Steel Electrodes and Rods for Gas Shielded Arc Welding, AWS A5.18
Specification for Carbon Steel Electrodes for Flux Cored Arc Welding, AWS A5.20
Specification for Low-Alloy Steel Electrodes and Fluxes for Submerged Arc Welding, AWS A5.23/A5.23M
Specification for Carbon and Low-Alloy Steel Electrodes and Fluxes for Electroslag Welding, AWS A5.25/A5.25M
Specification for Carbon and Low-Alloy Steel Electrodes for Electrogas Welding, AWS A5.26/A5.26M
Specification for Low-Alloy Steel Electrodes and Rods for Gas Shielded Arc Welding, AWS A5.28

Specification for Low-Alloy Steel Electrodes for Flux Cored Arc Welding, AWS A5.29

Specification for Welding Shielding Gases, AWS A5.32/A5.32M

Manufacturer's certification shall constitute sufficient evidence of conformity with the standards. Filler metals and fluxes that are suitable for the intended application shall be selected.

6. Stud Shear Connectors

Steel stud shear connectors shall conform to the requirements of *Structural Welding Code—Steel*, AWS D1.1.

Manufacturer's certification shall constitute sufficient evidence of conformity with the code.

A4. LOADS AND LOAD COMBINATIONS

The nominal loads and factored load combinations shall be as stipulated by the applicable code under which the structure is designed or dictated by the conditions involved. In the absence of a code, the loads and factored load combinations, including impact and crane loads, shall be those stipulated in ASCE 7. For design purposes, the loads stipulated by the applicable code or ASCE 7 shall be taken as nominal loads.

A5. DESIGN BASIS

1. Required Strength at Factored Loads

The required strength of structural members and connections shall be determined by structural analysis for the appropriate factored load combinations as stipulated in Section A4.

Design by either elastic or plastic analysis is permitted, except that design by plastic analysis is permitted only for steels with specified minimum yield stresses not exceeding 65 ksi (450 MPa) and is subject to provisions of Sections B5.2, C1.1, C2.1a, C2.2a, E1.2, F1.3, H1, and I1.

Beams and girders composed of compact sections, as defined in Section B5.1, and satisfying the unbraced length requirements of Section F1.3 (including composite members) which are continuous over supports or are rigidly framed to columns may be proportioned for nine-tenths of the negative moments produced by the factored gravity loading at points of support, provided that the maximum positive moment is increased by one-tenth of the average negative moments. This reduction is not permitted for hybrid beams, members of A514/A514M steel, or moments produced by loading on cantilevers. If the negative moment is resisted by a column rigidly framed to the beam or girder, the one-tenth reduction may be used in proportioning the column for combined axial force and flexure, provided that the axial force does not exceed ϕ_c times $0.15A_gF_y$,

where

A_g = gross area, in.^{2} (mm^2)
F_y = specified minimum yield stress, ksi (MPa)
ϕ_c = resistance factor for compression

2. Limit States

LRFD is a method of proportioning structures so that no applicable limit state is exceeded when the structure is subjected to all appropriate factored load combinations.

Strength limit states are related to safety and concern maximum load carrying capacity. Serviceability limit states are related to performance under normal service conditions. The term "resistance" includes both strength limit states and serviceability limit states.

3. Design for Strength

The required strength shall be determined for each applicable load combination as stipulated in Section A4.

The design strength of each structural component or assemblage shall equal or exceed the required strength based on the factored loads. The design strength ϕR_n for each applicable limit state is calculated as the nominal strength R_n multiplied by a resistance factor ϕ. Nominal strengths R_n and resistance factors ϕ are given in Chapters D through K.

4. Design for Serviceability and Other Considerations

The overall structure and the individual members, connections, and connectors shall be checked for serviceability. Provisions for design for serviceability are given in Chapter L.

A6. REFERENCED CODES AND STANDARDS

The following documents are referenced in this Specification:

ACI International (ACI)

Building Code Requirements for Structural Concrete and Commentary, ACI 318-99

Metric Building Code Requirements for Structural Concrete and Commentary, ACI 318M-99

American Institute of Steel Construction, Inc. (AISC)

Code of Standard Practice for Steel Buildings and Bridges, 2000

Seismic Provisions for Structural Steel Buildings, 1997

Seismic Provisions for Structural Steel Buildings Supplement No. 1, 1999

Specification for Structural Steel Buildings-Allowable Stress Design and Plastic Design, 1989

Specification for Load and Resistance Factor Design of Single-Angle Members, 1993

Specification for the Design of Steel Hollow Structural Sections, 1997

Specification for the Design, Fabrication and Erection of Steel Safety Related Structures for Nuclear Facilities, 1994

American Iron and Steel Institute (AISI)

Specification for the Design of Cold-Formed Steel Structural Members, 1996

Specification for the Design of Cold-Formed Steel Structural Members, Supplement No. 1, 1999

American Society of Civil Engineers (ASCE)
Minimum Design Loads for Buildings and Other Structures, ASCE 7-98

American Society of Mechanical Engineers (ASME)
Fasteners for Use in Structural Applications, ASME B18.2.6-96
Surface Texture, Surface Roughness, Waviness, and Lay, ASME B46.1-85

American Society for Testing and Materials (ASTM)

ASTM A6/A6M-2000	ASTM A27/A27M-95
ASTM A36/A36M-97a	ASTM A53/A53M-99b
ASTM A148/A148M-93b	ASTM A193/A193M-99
ASTM A194/A194M-98b	ASTM A242/A242M-98
ASTM A307-97	ASTM A325-97
ASTM A325M-97	ASTM A354-98
ASTM A370-97a	ASTM A449-93
ASTM A490-97	ASTM A490M-93
ASTM A500-99	ASTM A501-99
ASTM A502-93	ASTM A514/A514M-94a
ASTM A529/A529M-96	ASTM A563-97
ASTM A563M-97	ASTM A568/568M-98
ASTM A570/A570M-98	ASTM A572/A572M-99
ASTM A588/A588M-97a	ASTM A606-98
ASTM A607-96	ASTM A618-99
ASTM A668/A668M-96	ASTM A673/A673M-95
ASTM A709/A709M-97b	ASTM A751-96
ASTM A847-99a	ASTM A852/A852M-97
ASTM A913/A913M-97	ASTM A992/A992M-98
ASTM C33-97	ASTM C330-97
ASTM F436-93	ASTM F436M-93
ASTM F606-98	ASTM F606M-95b
ASTM F959-96	ASTM F959M-97
ASTM F1554-99	ASTM F1852-98

American Welding Society (AWS)

AWS D1.1:2000	AWS A5.25/A5.25M-97
AWS A5.17/A5.17M-98	AWS A5.29-98
AWS A5.23/A5.23M-97	AWS A5.5-96
AWS A5.28-96	AWS A5.20-95
AWS A5.1-91	AWS A5.26/A5.26M-97
AWS A5.18-93	AWS A5.32/A5.32M-97

Research Council on Structural Connections (RCSC)
Load and Resistance Factor Design Specification for Structural Joints Using ASTM A325 or A490 Bolts, 1994

A7. DESIGN DOCUMENTS

The design drawings shall show a complete design with sizes, sections, and relative locations of all members. Floor levels, column centers and offsets shall be dimensioned. Drawings shall be drawn to a scale large enough to show the information clearly.

Design documents shall indicate the type or types of construction as defined in Section A2 and include the required strengths (moments and forces) if necessary for preparation of shop drawings.

Where joints are to be assembled with high-strength bolts, the design documents shall indicate the connection type (i.e., snug-tightened, pretensioned, or slip-critical).

Camber of trusses, beams, and girders, if required, shall be specified in the design documents.

The requirements for stiffeners and bracing shall be shown in the design documents.

Welding and inspection symbols used on design and shop drawings shall be the American Welding Society symbols. Welding symbols for special requirements not covered by AWS are permitted to be used provided complete explanations thereof are shown in the design documents.

Weld lengths called for in the design documents and on the shop drawings shall be the net effective lengths.

CHAPTER B

DESIGN REQUIREMENTS

This chapter contains provisions which are common to the Specification as a whole.

B1. GROSS AREA

The gross area A_g of a member at any point is the sum of the products of the thickness and the gross width of each element measured normal to the axis of the member. For angles, the gross width is the sum of the widths of the legs less the thickness.

B2. NET AREA

The net area A_n of a member is the sum of the products of the thickness and the net width of each element computed as follows:

In computing net area for tension and shear, the width of a bolt hole shall be taken as $\frac{1}{16}$-in. (2 mm) greater than the nominal dimension of the hole.

For a chain of holes extending across a part in any diagonal or zigzag line, the net width of the part shall be obtained by deducting from the gross width the sum of the diameters or slot dimensions as provided in Section J3.2, of all holes in the chain, and adding, for each gage space in the chain, the quantity $s^2/4g$

where

s = longitudinal center-to-center spacing (pitch) of any two consecutive holes, in. (mm)

g = transverse center-to-center spacing (gage) between fastener gage lines, in. (mm)

For angles, the gage for holes in opposite adjacent legs shall be the sum of the gages from the back of the angles less the thickness.

In determining the net area across plug or slot welds, the filler metal shall not be considered as adding to the net area.

B3. EFFECTIVE AREA OF TENSION MEMBERS

The effective area of tension members shall be determined as follows:

(1) When tension load is transmitted directly to each of the cross-sectional elements by fasteners or welds, the effective area A_e is equal to the net area A_n.

(2) When the tension load is transmitted by fasteners or welds through some but not all of the cross-sectional elements of the member, the effective area A_e shall be computed as follows:

(a) When the tension load is transmitted only by fasteners

$$A_e = A_n U \tag{B3-1}$$

where

U = reduction coefficient
$= 1-(\bar{x}/l) \leq 0.9$
$\bar{x}$ = connection eccentricity, in. (mm)
l = length of the connection in the direction of loading, in. (mm)

(b) When the tension load is transmitted only by longitudinal welds to other than a plate member or by longitudinal welds in combination with transverse welds

$$A_e = A_g U \tag{B3-2}$$

where

$U = 1-(\bar{x}/l) \leq 0.9$
A_g = gross area of member, in.^{2} (mm^2)

(c) When the tension load is transmitted only by transverse welds

$$A_e = AU \tag{B3-3}$$

where

A = area of directly connected elements, in.^{2} (mm^2)
$U = 1.0$

(d) When the tension load is transmitted to a plate only by longitudinal welds along both edges at the end of the plate

$$A_e = A_g U \tag{B3-4}$$

where

For $l \geq 2w$ . . . . $U = 1.00$
For $2w > l \geq 1.5w$ . . . . $U = 0.87$
For $1.5w > l \geq w$ . . . . $U = 0.75$

where

l = length of weld, in. (mm)
w = plate width (distance between welds), in. (mm)

Larger values of U are permitted to be used when justified by tests or other rational criteria.

For effective area of connecting elements, see Section J5.2.

B4. STABILITY

General stability shall be provided for the structure as a whole and for each of its elements.

Consideration shall be given to the significant effects of the loads on the deflected shape of the structure and its individual elements.

B5. LOCAL BUCKLING

1. Classification of Steel Sections

Steel sections are classified as compact, noncompact, or slender-element sections. For a section to qualify as compact, its flanges must be continuously connected to the web or webs and the width-thickness ratios of its compression elements must not exceed the limiting width-thickness ratios λ_p from Table B5.1. If the width-thickness ratio of one or more compression elements exceeds λ_p, but does not exceed λ_r, the section is noncompact. If the width-thickness ratio of any element exceeds λ_r from Table B5.1, the section is referred to as a slender-element compression section.

For unstiffened elements which are supported along only one edge parallel to the direction of the compression force, the width shall be taken as follows:

(a) For flanges of I-shaped members and tees, the width b is half the full-flange width, b_f.

(b) For legs of angles and flanges of channels and zees, the width b is the full nominal dimension.

(c) For plates, the width b is the distance from the free edge to the first row of fasteners or line of welds.

(d) For stems of tees, d is taken as the full nominal depth.

For stiffened elements which are supported along two edges parallel to the direction of the compression force, the width shall be taken as follows:

(a) For webs of rolled or formed sections, h is the clear distance between flanges less the fillet or corner radius at each flange; h_c is twice the distance from the centroid to the inside face of the compression flange less the fillet or corner radius.

(b) For webs of built-up sections, h is the distance between adjacent lines of fasteners or the clear distance between flanges when welds are used, and h_c is twice the distance from the centroid to the nearest line of fasteners at the compression flange or the inside face of the compression flange when welds are used; h_p is twice the distance from the plastic neutral axis to the nearest line of fasteners at the compression flange or the inside face of the compression flange when welds are used.

(c) For flange or diaphragm plates in built-up sections, the width b is the distance between adjacent lines of fasteners or lines of welds.

(d) For flanges of rectangular hollow structural sections, the width b is the clear distance between webs less the inside corner radius on each side. If the corner radius is not known, the width may be taken as the total section width minus three times the thickness. The thickness t shall be taken as the design wall thickness. When the design wall thickness is not known, it is permitted to be taken as 0.93 times the nominal wall thickness.

For tapered flanges of rolled sections, the thickness is the nominal value halfway between the free edge and the corresponding face of the web.

2. Design by Plastic Analysis

Design by plastic analysis is permitted, as limited in Section A5.1, when flanges subject to compression involving hinge rotation and all webs have a width-thickness ratio less than or equal to the limiting λ_p from Table B5.1. For circular hollow sections see Footnote d of Table B5.1.

3. Slender-Element Compression Sections

For the flexural design of I-shaped sections, channels and rectangular or circular sections with slender flange elements, see Appendix F1. For other shapes in flexure or members in axial compression that have slender compression elements, see Appendix B5.3. For plate girders with slender web elements, see Appendix G.

B6. BRACING AT SUPPORTS

At points of support for beams, girders and trusses, restraint against rotation about their longitudinal axis shall be provided.

B7. LIMITING SLENDERNESS RATIOS

For members in which the design is based on compression, the slenderness ratio Kl/r preferably should not exceed 200.

For members in which the design is based on tension, the slenderness ratio l/r preferably should not exceed 300. The above limitation does not apply to rods in tension. Members in which the design is dictated by tension loading, but which may be subject to some compression under other load conditions, need not satisfy the compression slenderness limit.

B8. SIMPLE SPANS

Beams, girders and trusses designed on the basis of simple spans shall have an effective length equal to the distance between centers of gravity of the members to which they deliver their end reactions.

B9. END RESTRAINT

Beams, girders, and trusses designed on the assumptions of full or partial end restraint, as well as the sections of the members to which they connect, shall have design strengths, as prescribed in Chapters D through K, equal to or exceeding the effect of factored forces and moments except that some inelastic but self-limiting deformation of a part of the connection is permitted.

B10. PROPORTIONS OF BEAMS AND GIRDERS

When rolled or welded shapes, plate girders and cover-plated beams are proportioned on the basis of flexural strength of the gross section:

(a) If

$$0.75F_u A_{fn} \geq 0.9F_y A_{fg} \tag{B10-1}$$

no deduction shall be made for bolt or rivet holes in either flange,

where

TABLE B5.1
Limiting Width-Thickness Ratios for Compression Elements

	Description of Element	Width Thickness Ratio	Limiting Width-Thickness Ratios λ_p (compact)	λ_r (noncompact)
Unstiffened Elements	Flanges of I-shaped rolled beams and channels in flexure	b/t	$0.38\sqrt{E/F_y}$ [c]	$0.83\sqrt{E/F_L}$ [e]
	Flanges of I-shaped hybrid or welded beams in flexure	b/t	$0.38\sqrt{E/F_{yf}}$	$0.95\sqrt{E/(F_L/k_c)}$ [e], [f]
	Flanges projecting from built-up compression members	b/t	NA	$0.64\sqrt{E/(F_y/k_c)}$ [f]
	Flanges of I-shaped sections in pure compression, plates projecting from compression elements; outstanding legs of pairs of angles in continuous contact; flanges of channels in pure compression	b/t	NA	$0.56\sqrt{E/F_y}$
	Legs of single angle struts; legs of double angle struts with separators; unstiffened elements, i.e., supported along one edge	b/t	NA	$0.45\sqrt{E/F_y}$
	Stems of tees	d/t	NA	$0.75\sqrt{E/F_y}$

A_{fg} = gross flange area, in.^{2} (mm^2)
A_{fn} = net tension flange area calculated in accordance with the provisions of Sections B1 and B2, in.^{2} (mm^2)
F_u = specified minimum tensile strength, ksi (MPa)

(b) If

$$0.75F_u A_{fn} < 0.9F_y A_{fg} \tag{B10-2}$$

the member flexural properties shall be based on an effective tension flange area A_{fe}

$$A_{fe} = \frac{5}{6}\frac{F_u}{F_y}A_{fn} \tag{B10-3}$$

and the maximum flexural strength shall be based on the elastic section modulus.

Hybrid girders shall be proportioned by the flexural strength of their gross section,

TABLE B5.1 (cont.)
Limiting Width-Thickness Ratios for Compression Elements

	Description of Element	Width Thick-ness Ratio	Limiting Width-Thickness Ratios: λ_p (compact)	Limiting Width-Thickness Ratios: λ_r (noncompact)
Stiffened Elements	Flanges of rectangular box and hollow structural sections of uniform thickness subject to bending or compression; flange cover plates and diaphragm plates between lines of fasteners or welds	b/t		
	for uniform compression		$1.12\sqrt{E/F_y}$	$1.40\sqrt{E/F_y}$
	for plastic analysis		$0.939\sqrt{E/F_y}$	-
	Unsupported width of cover plates perforated with a succession of access holes [b]	b/t	NA	$1.86\sqrt{E/F_y}$
	Webs in flexural compression [a]	h/t_w	$3.76\sqrt{E/F_y}$ [c], [g]	$5.70\sqrt{E/F_y}$ [h]
	Webs in combined flexural and axial compression	h/t_w	for $P_u/\phi_b P_y \le 0.125$ [c],[g] $3.76\sqrt{\dfrac{E}{F_y}}\left(1-\dfrac{2.75P_u}{\phi_b P_y}\right)$ for $P_u/\phi_b P_y > 0.125$ [c][g] $1.12\sqrt{\dfrac{E}{F_y}}\left(2.33-\dfrac{P_u}{\phi_b P_y}\right) \ge 1.49\sqrt{\dfrac{E}{F_y}}$	[h] $5.70\sqrt{\dfrac{E}{F_y}}\left(1-0.74\dfrac{P_u}{\boxed{\phi_b P_y}}\right)$ Errata 9/4/01
	All other uniformly compressed stiffened elements, i.e., supported along two edges	b/t h/t_w	NA	$1.49\sqrt{E/F_y}$
	Circular hollow sections In axial compression In flexure	D/t	[d] NA $0.07E/F_y$	 $0.11E/F_y$ $0.31E/F_y$

[a] For hybrid beams, use the yield strength of the flange F_{yf} instead of F_y.

[b] Assumes net area of plate at widest hole.

[c] ~~Assumes an inelastic rotation capacity of 3 radians. For structures in zones of high seismicity, a greater rotation capacity may be required.~~

Assumes an inelastic ductility ratio (ratio of strain at fracture to strain at yield) of 3. When the seismic response modification factor ***R*** is taken greater than 3, a greater rotation capacity may be required.

[d] For plastic design use $0.045E/F_y$.

[e] F_L = smaller of $(F_{yf} - F_r)$ or F_{yw}, ksi (MPa)
F_r = compressive residual stress in flange
= 10 ksi (69 MPa) for rolled shapes
= 16.5 ksi (114 MPa) for welded shapes

[f] $k_c = \dfrac{4}{\sqrt{h/t_w}}$ and $0.35 \le k_c \le 0.763$

[g] For members with unequal flanges, use h_p instead of h when comparing to λ_p.

[h] For members with unequal flanges, see Appendix B5.1.

Errata 9/4/01

subject to the applicable provisions in Appendix G1, provided they are not required to resist an axial force greater than ϕ_b times $0.15F_{yf}A_g$, where F_{yf} is the specified minimum yield stress of the flange material and A_g is the gross area. No limit is placed on the web stresses produced by the applied bending moment for which a hybrid girder is designed, except as provided in Section K3 and Appendix K3. To qualify as hybrid girders, the flanges at any given section shall have the same cross-sectional area and be made of the same grade of steel.

Flanges of welded beams or girders may be varied in thickness or width by splicing a series of plates or by the use of cover plates.

The total cross-sectional area of cover plates of bolted or riveted girders shall not exceed 70 percent of the total flange area.

High-strength bolts, rivets, or welds connecting flange to web, or cover plate to flange, shall be proportioned to resist the total horizontal shear resulting from the bending forces on the girder. The longitudinal distribution of these bolts, rivets, or intermittent welds shall be in proportion to the intensity of the shear. However, the longitudinal spacing shall not exceed the maximum permitted for compression or tension members in Section E4 or D2, respectively. Bolts, rivets, or welds connecting flange to web shall also be proportioned to transmit to the web any loads applied directly to the flange, unless provision is made to transmit such loads by direct bearing.

Partial length cover plates shall be extended beyond the theoretical cutoff point and the extended portion shall be attached to the beam or girder by high-strength bolts in a slip-critical connection, rivets, or fillet welds. The attachment shall be adequate, at the applicable design strength given in Sections J2.2, J3.8, or K3 to develop the cover plate's portion of the flexural design strength in the beam or girder at the theoretical cutoff point.

For welded cover plates, the welds connecting the cover plate termination to the beam or girder shall have continuous welds along both edges of the cover plate in the length a', defined below, and shall be adequate, at the applicable design strength, to develop the cover plate's portion of the design strength in the beam or girder at the distance a' from the end of the cover plate.

(a) When there is a continuous weld equal to or larger than three-fourths of the plate thickness across the end of the plate

$$a' = w \tag{B10-4}$$

where

w = width of cover plate, in. (mm)

(b) When there is a continuous weld smaller than three-fourths of the plate thickness across the end of the plate

$$a' = 1.5w \tag{B10-5}$$

(c) When there is no weld across the end of the plate

$$a' = 2w \tag{B10-6}$$

CHAPTER C

FRAMES AND OTHER STRUCTURES

This chapter contains general requirements for stability of the structure as a whole.

C1. SECOND ORDER EFFECTS

Second order ($P\Delta$) effects shall be considered in the design of frames.

1. Design by Plastic Analysis

In structures designed on the basis of plastic analysis, as limited in Section A5.1, the required flexural strength M_u shall be determined from a second-order plastic analysis that satisfies the requirements of Section C2.

2. Design by Elastic Analysis

In structures designed on the basis of elastic analysis, M_u for beam-columns, connections, and connected members shall be determined from a second-order elastic analysis or from the following approximate second-order analysis procedure:

$$M_u = B_1 M_{nt} + B_2 M_{lt} \tag{C1-1}$$

where

M_{nt} = required flexural strength in member assuming there is no lateral translation of the frame, kip-in. (N-mm)

M_{lt} = required flexural strength in member as a result of lateral translation of the frame only, kip-in. (N-mm)

$$B_1 = \frac{C_m}{(1 - P_u / P_{e1})} \geq 1 \tag{C1-2}$$

$$P_{e1} = \frac{\pi^2 EI}{(KL)^2}$$

where I is the moment of inertia in the plane of bending and K is the effective length factor in the plane of bending determined in accordance with Section C2.1, for the braced frame.

P_u = required axial compressive strength for the member under consideration, kips (N)

C_m = a coefficient based on elastic first-order analysis assuming no lateral translation of the frame whose value shall be taken as follows:

(a) For compression members not subject to transverse loading between their supports in the plane of bending,

$$C_m = 0.6 - 0.4(M_1 / M_2) \tag{C1-3}$$

where M_1/M_2 is the ratio of the smaller to larger moments at the ends of that portion of the member unbraced in the plane of bending under consideration. M_1/M_2 is positive when the member is bent in reverse curvature, negative when bent in single curvature.

(b) For compression members subjected to transverse loading between their supports, the value of C_m shall be determined either by rational analysis or by the use of the following values:

For members whose ends are restrained. . . . . $C_m = 0.85$
For members whose ends are unrestrained . . . . $C_m = 1.00$

$$B_2 = \frac{1}{1-\Sigma P_u\left(\dfrac{\Delta_{oh}}{\Sigma HL}\right)} \qquad \text{(C1-4)}$$

or

$$B_2 = \frac{1}{1-\left(\dfrac{\Sigma P_u}{\Sigma P_{e2}}\right)} \qquad \text{(C1-5)}$$

ΣP_u = required axial strength of all columns in a story, kips (N)
Δ_{oh} = lateral inter-story deflection, in. (mm)
ΣH = sum of all story horizontal forces producing Δ_{oh}, kips (N)
L = story height, in. (mm)

$$P_{e2} = \frac{\pi^2 EI}{(KL)^2}, \text{ kips (N)}$$

where I is the moment of inertia in the plane of bending and K is the effective length factor in the plane of bending determined in accordance with Section C2.2, for the unbraced frame.

C2. FRAME STABILITY

1. Braced Frames

In trusses and frames where lateral stability is provided by diagonal bracing, shear walls, or equivalent means, the effective length factor K for compression members shall be taken as unity, unless structural analysis shows that a smaller value may be used.

The vertical bracing system for a braced multistory frame shall be determined by structural analysis to be adequate to prevent buckling of the structure and to maintain the lateral stability of the structure, including the overturning effects of drift, under the factored load combinations stipulated in Section A4.

The vertical bracing system for a braced multistory frame may be considered to function together with in-plane shear-resisting exterior and interior walls, floor slabs, and roof decks, which are properly secured to the structural frames. The columns, girders, beams, and diagonal members, when used as the vertical bracing system, may be considered to comprise a vertically cantilevered simply connected truss in the analyses for frame buckling and lateral stability. Axial deformation of all members in the vertical bracing system shall be included in the lateral stability analysis.

1a. Design by Plastic Analysis

In braced frames designed on the basis of plastic analysis, as limited in Section A5.1, the axial force in these members caused by factored gravity plus factored horizontal loads shall not exceed $0.85\phi_c$ times $A_g F_y$.

2. Unbraced Frames

In frames where lateral stability depends upon the bending stiffness of rigidly connected beams and columns, the effective length factor K of compression members shall be determined by structural analysis. The destabilizing effects of gravity loaded columns whose simple connections to the frame do not provide resistance to lateral loads shall be included in the design of the moment-frame columns. Stiffness reduction adjustment due to column inelasticity is permitted.

Analysis of the required strength of unbraced multistory frames shall include the effects of frame instability and column axial deformation under the factored load combinations stipulated in Section A4.

2a. Design by Plastic Analysis

In unbraced frames designed on the basis of plastic analysis, as limited in Section A5.1, the axial force in the columns caused by factored gravity plus factored horizontal loads shall not exceed $0.75\phi_c$ times $A_g F_y$.

C3. STABILITY BRACING

1. Scope

These requirements address the minimum brace strength and stiffness necessary to ensure member design strengths based on the unbraced length between braces with an effective length factor K equal to unity. Bracing is assumed to be perpendicular to the member(s) to be braced; for inclined or diagonal bracing, the brace strength (force or moment) and stiffness (force per unit displacement or moment per unit rotation) must be adjusted for the angle of inclination. The evaluation of the stiffness furnished by a brace shall include its member and geometric properties, as well as the effects of connections and anchoring details.

Two general types of bracing systems are considered, relative and nodal. A relative brace controls the movement of the brace point with respect to adjacent braced points. A nodal brace controls the movement at the braced point without direct interaction with adjacent braced points. The strength and stiffness furnished by the stability bracing shall not be less than the required limits. A second order analysis that includes an initial out-of-plumbness of the structure or out-of-straightness of the member to obtain brace strength and stiffness can be used in lieu of the requirements of this section.

2. Frames

In braced frames where lateral stability is provided by diagonal bracing, shear walls, or other equivalent means, the required story or panel bracing shear force is:

$$P_{br} = 0.004 \Sigma P_u \quad \text{(C3-1)}$$

The required story or panel shear stiffness is:

$$\beta_{br} = \frac{2\Sigma P_u}{\phi L} \qquad \text{(C3-2)}$$

where

ϕ = 0.75
ΣP_u = summation of the factored column axial loads in the story or panel supported by the bracing, kips (N)
L = story height or panel spacing, in. (mm)

These story stability requirements shall be combined with the lateral forces and drift requirements from other sources, such as wind or seismic loading.

3. Columns

An individual column can be braced at intermediate points along its length by relative or nodal bracing systems. It is assumed that nodal braces are equally spaced along the column.

(a) *Relative Bracing*

The required brace strength is:

$$P_{br} = 0.004 P_u \qquad \text{(C3-3)}$$

The required brace stiffness is:

$$\beta_{br} = \frac{2P_u}{\phi L_b} \qquad \text{(C3-4)}$$

where

ϕ = 0.75
P_u = required compressive strength, kips (N)
L_b = distance between braces, in. (mm)

(b) *Nodal Bracing*

The required brace strength is:

$$P_{br} = 0.01 P_u \qquad \text{(C3-5)}$$

The required brace stiffness is:

$$\beta_{br} = \frac{8P_u}{\phi L_b} \qquad \text{(C3-6)}$$

where

ϕ = 0.75

When the actual spacing of braced points is less than L_q, where L_q is the maximum unbraced length for the required column force with K equal to one, then L_b in Equations C3-4 and C3-6 is permitted to be taken equal to L_q.

4. Beams

Beam bracing must prevent the relative displacement of the top and bottom flanges, i.e. twist of the section. Lateral stability of beams shall be provided by lateral bracing, torsional bracing, or a combination of the two. In members subjected to double curvature bending, the inflection point shall not be considered a brace point.

4a. Lateral Bracing

Bracing shall be attached near the compression flange, except for a cantilevered member, where an end brace shall be attached near the top (tension) flange. Lateral bracing shall be attached to both flanges at the brace point near the inflection point for beams subjected to double curvature bending along the length to be braced.

(a) *Relative Bracing*

The required brace strength is:

$$P_{br} = 0.008 M_u C_d / h_o \tag{C3-7}$$

The required brace stiffness is:

$$\beta_{br} = \frac{4 M_u C_d}{\phi L_b h_o} \tag{C3-8}$$

where

$\phi = 0.75$
M_u = required flexural strength, kip-in. (N-mm)
h_o = distance between flange centroids, in. (mm)
C_d = 1.0 for bending in single curvature; 2.0 for double curvature; $C_d = 2.0$ only applies to the brace closest to the inflection point.
L_b = distance between braces, in. (mm)

(b) *Nodal Bracing*

The required brace strength is:

$$P_{br} = 0.02 M_u C_d / h_o \tag{C3-9}$$

The required brace stiffness is:

$$\beta_{br} = \frac{10 M_u C_d}{\phi L_b h_o} \tag{C3-10}$$

where

$\phi = 0.75$

When the actual spacing of braced points is less than L_q, the maximum unbraced length for M_u, then L_b in Equations C3-8 and C3-10 shall be permitted to be taken equal to L_q.

4b. Torsional Bracing

Torsional bracing can be nodal or continuous along the beam length. The bracing can be attached at any cross-sectional location and need not be attached near the compression flange. The connection between a torsional brace and the beam must be able to support the required moment given below.

(a) *Nodal Bracing*

The required bracing moment is:

$$M_{br} = \frac{0.024 M_u L}{n C_b L_b} \tag{C3-11}$$

The required cross-frame or diaphragm bracing stiffness is:

$$\beta_{Tb} = \frac{\beta_T}{\left(1 - \frac{\beta_T}{\beta_{\text{sec}}}\right)} \tag{C3-12}$$

where

$$\beta_T = \frac{2.4 L M_u^2}{\phi n E I_y C_b^2} \tag{C3-13}$$

$$\beta_{\text{sec}} = \frac{3.3E}{h_o}\left(\frac{1.5 h_o t_w^3}{12} + \frac{t_s b_s^3}{12}\right) \tag{C3-14}$$

ϕ = 0.75
L = span length, in. (mm)
n = number of nodal braced points within the span
E = 29,000 ksi (200 000 MPa)
I_y = out-of-plane moment of inertia, in.^{4} (mm^4)
C_b = is a modification factor defined in Chapter F
t_w = beam web thickness, in. (mm)
t_s = web stiffener thickness, in. (mm)
b_s = stiffener width for one-sided stiffeners (use twice the individual stiffener width for pairs of stiffeners), in. (mm)
β_T = brace stiffness excluding web distortion, kip-in/radian (N-mm/radian)
β_{sec} = web distortional stiffness, including the effect of web transverse stiffeners, if any, kip-in/radian (N-mm/radian)

If $\beta_{sec} < \beta_T$, Equation C3-12 is negative, which indicates that torsional beam bracing will not be effective due to inadequate web distortional stiffness.

When required, the web stiffener shall extend the full depth of the braced member and shall be attached to the flange if the torsional brace is also attached to the flange. Alternatively, it shall be permissible to stop the stiffener short by a distance equal to $4t_w$ from any beam flange that is not directly attached to the torsional brace. When the actual spacing of braced points is less than L_q, then L_b in Equation C3-11 shall be permitted to be taken equal to L_q.

(b) *Continuous Torsional Bracing*

For continuous bracing, use Equations C3-11, C3-12 and C3-13 with L/n taken as

1.0; the bracing moment and stiffness are given per unit span length. The distortional stiffness for an unstiffened web is

$$\beta_{sec} = \frac{3.3Et_w^3}{12h_o} \tag{C3-15}$$

CHAPTER D

TENSION MEMBERS

This chapter applies to prismatic members subject to axial tension caused by static forces acting through the centroidal axis. For members subject to combined axial tension and flexure, see Section H1.1. For threaded rods, see Section J3. For block shear rupture strength at end connections of tension members, see Section J4.3. For the design tensile strength of connecting elements, see Section J5.2. For members subject to fatigue, see Section K3.

D1. DESIGN TENSILE STRENGTH

The design strength of tension members, $\phi_t P_n$, shall be the lower value obtained according to the limit states of yielding in the gross section and fracture in the net section.

(a) For yielding in the gross section:

$$\phi_t = 0.90$$

$$P_n = F_y A_g \tag{D1-1}$$

(b) For fracture in the net section:

$$\phi_t = 0.75$$

$$P_n = F_u A_e \tag{D1-2}$$

where

A_e = effective net area, in.^{2} (mm^2)
A_g = gross area of member, in.^{2} (mm^2)
F_y = specified minimum yield stress, ksi (MPa)
F_u = specified minimum tensile strength, ksi (MPa)

When members without holes are fully connected by welds, the effective net section used in Equation D1-2 shall be as defined in Section B3. When holes are present in a member with welded-end connections, or at the welded connection in the case of plug or slot welds, the net section through the holes shall be used in Equation D1-2.

D2. BUILT-UP MEMBERS

For limitations on the longitudinal spacing of connectors between elements in continuous contact consisting of a plate and a shape or two plates, see Section J3.5.

The longitudinal spacing of connectors between components should preferably limit the slenderness ratio in any component between the connectors to 300.

Either perforated cover plates or tie plates without lacing are permitted to be used on the open sides of built-up tension members. Tie plates shall have a length not less than two-thirds the distance between the lines of welds or fasteners connecting them to the components of the member. The thickness of such tie plates shall not be less than one-fiftieth of the distance between these lines. The longitudinal spacing of intermittent welds or fasteners at tie plates shall not exceed six in. (150 mm). The spacing of tie plates shall be such that the slenderness ratio of any component in the length between tie plates should preferably not exceed 300.

D3. PIN-CONNECTED MEMBERS AND EYEBARS

1. Pin-Connected Members

1a. Design Strength

The design strength of a pin-connected member, ϕP_n, shall be the lowest value of the following limit states:

(a) Tension on the net effective area:

$$\phi = \phi_t = 0.75 \qquad \text{(D3-1)}$$

$$P_n = 2tb_{eff} F_u$$

(b) Shear on the effective area:

$$\phi = \phi_{sf} = 0.75 \qquad \text{(D3-2)}$$

$$P_n = 0.6 A_{sf} F_u$$

(c) For bearing on the projected area of the pin, see Section J8.

(d) For yielding in the gross section, use Equation D1-1.

where

$A_{sf} = 2t(a + d/2)$, in.^{2} (mm^2)

a = shortest distance from edge of the pin hole to the edge of the member measured parallel to the direction of the force, in. (mm)

$b_{eff} = 2t + 0.63$, in. ($= 2t + 16$, mm) but not more than the actual distance from the edge of the hole to the edge of the part measured in the direction normal to the applied force

d = pin diameter, in. (mm)

t = thickness of plate, in. (mm)

1b. Detailing Requirements

The pin hole shall be located midway between the edges of the member in the direction normal to the applied force. When the pin is expected to provide for relative movement between connected parts while under full load, the diameter of the pin hole shall not be more than 1/32-in. (1 mm) greater than the diameter of the pin.

The width of the plate beyond the pin hole shall not be less than $2b_{eff} + d$ and the minimum extension, a, beyond the bearing end of the pin hole, parallel to the axis of the member, shall not be less than 1.33 x b_{eff}.

The corners beyond the pin hole are permitted to be cut at 45° to the axis of the member, provided the net area beyond the pin hole, on a plane perpendicular to the cut, is not less than that required beyond the pin hole parallel to the axis of the member.

2. Eyebars

2a. Design Strength

The design strength of eyebars shall be determined in accordance with Section D1, with A_g taken as the cross-sectional area of the body.

For calculation purposes, the width of the body of the eyebars shall not exceed eight times its thickness.

2b. Detailing Requirements

Eyebars shall be of uniform thickness, without reinforcement at the pin holes, and have circular heads whose periphery is concentric with the pin hole.

The radius of transition between the circular head and the eyebar body shall not be less than the head diameter.

The pin diameter shall not be less than seven-eighths times the eyebar body width, and the pin hole diameter shall not be more than $\frac{1}{32}$-in (1 mm) greater than the pin diameter.

For steels having F_y greater than 70 ksi (485 MPa), the hole diameter shall not exceed five times the plate thickness and the width of the eyebar body shall be reduced accordingly.

A thickness of less than $\frac{1}{2}$-in. (13 mm) is permissible only if external nuts are provided to tighten pin plates and filler plates into snug contact. The width from the hole edge to the plate edge perpendicular to the direction of applied load shall be greater than two-thirds and, for the purpose of calculation, not more than three-fourths times the eyebar body width.

CHAPTER E

COLUMNS AND OTHER COMPRESSION MEMBERS

This chapter applies to compact and non-compact prismatic members subject to axial compression through the centroidal axis. For members subject to combined axial compression and flexure, see Section H1.2. For members with slender compression elements, see Appendix B5.3. For tapered members, see Appendix F3.

E1. EFFECTIVE LENGTH AND SLENDERNESS LIMITATIONS

1. Effective Length

The effective length factor K shall be determined in accordance with Section C2.

2. Design by Plastic Analysis

Design by plastic analysis, as limited in Section A5.1, is permitted if the column slenderness parameter λ_c does not exceed 1.5K.

E2. DESIGN COMPRESSIVE STRENGTH FOR FLEXURAL BUCKLING

The design strength for flexural buckling of compression members whose elements have width-thickness ratios less than λ_r from Section B5.1 is $\phi_c P_n$:

$$\phi_c = 0.85$$

$$P_n = A_g F_{cr} \qquad \text{(E2-1)}$$

(a) For $\lambda_c \leq 1.5$

$$F_{cr} = \left(0.658^{\lambda_c^2}\right) F_y \qquad \text{(E2-2)}$$

(b) For $\lambda_c > 1.5$

$$F_{cr} = \left[\frac{0.877}{\lambda_c^2}\right] F_y \qquad \text{(E2-3)}$$

where

$$\lambda_c = \frac{Kl}{r\pi}\sqrt{\frac{F_y}{E}} \qquad \text{(E2-4)}$$

A_g = gross area of member, in.^{2} (mm^2)
F_y = specified minimum yield stress, ksi (MPa)
E = modulus of elasticity, ksi (MPa)
K = effective length factor
l = laterally unbraced length of member, in. (mm)
r = governing radius of gyration about the axis of buckling, in. (mm)

For members whose elements do not meet the requirements of Section B5.1, see Appendix B5.3.

E3. DESIGN COMPRESSIVE STRENGTH FOR FLEXURAL-TORSIONAL BUCKLING

The design strength for flexural-torsional buckling of double-angle and tee-shaped compression members whose elements have width-thickness ratios less than λ_r from Section B5.1 is $\phi_c P_n$:

where

$$\phi_c = 0.85$$

$$P_n = A_g F_{crft} \quad \text{(E3-1)}$$

$$F_{crft} = \left(\frac{F_{cry} + F_{crz}}{2H}\right)\left[1 - \sqrt{1 - \frac{4F_{cry}F_{crz}H}{(F_{cry} + F_{crz})^2}}\right] \quad \text{(E3-2)}$$

$F_{crz} = \dfrac{GJ}{A\bar{r}_o^2}$

$\bar{r}_o$ = polar radius of gyration about shear center, in. (mm) (see Equation A-E3-8)

$H = 1 - \dfrac{y_o^2}{\bar{r}_o^2}$

y_o = distance between shear center and centroid, in. (mm)

F_{cry} is determined according to Section E2 for flexural buckling about the y-axis of symmetry for $\lambda_c = \dfrac{Kl}{r_y \pi}\sqrt{\dfrac{F_y}{E}}$.

For double-angle and tee-shaped members whose elements do not meet the requirements of Section B5.1, see Appendix B5.3 to determine F_{cry} for use in Equation E3-1.

Other singly symmetric and unsymmetric columns, and doubly symmetric columns, such as cruciform or built-up columns, with very thin walls shall be designed for the limit states of flexural-torsional and torsional buckling in accordance with Appendix E3.

E4. BUILT-UP MEMBERS

1. Design Strength

The design strength of built-up members composed of two or more shapes shall be determined in accordance with Section E2 and Section E3 subject to the following modification. If the buckling mode involves relative deformations that produce shear forces in the connectors between individual shapes, Kl/r is replaced by $(Kl/r)_m$ determined as follows:

(a) For intermediate connectors that are snug-tight bolted:

$$\left(\frac{Kl}{r}\right)_m = \sqrt{\left(\frac{Kl}{r}\right)_o^2 + \left(\frac{a}{r_i}\right)^2} \qquad \text{(E4-1)}$$

(b) For intermediate connectors that are welded or fully tensioned bolted:

$$\left(\frac{Kl}{r}\right)_m = \sqrt{\left(\frac{Kl}{r}\right)_o^2 + 0.82\frac{\alpha^2}{(1+\alpha^2)}\left(\frac{a}{r_{ib}}\right)^2} \qquad \text{(E4-2)}$$

where

$\left(\frac{Kl}{r}\right)_o$ = column slenderness of built-up member acting as a unit

$\left(\frac{Kl}{r}\right)_m$ = modified column slenderness of built-up member

a = distance between connectors, in. (mm)

r_i = minimum radius of gyration of individual component, in. (mm)

r_{ib} = radius of gyration of individual component relative to its centroidal axis parallel to member axis of buckling, in. (mm)

α = separation ratio = $h / 2r_{ib}$

h = distance between centroids of individual components perpendicular to the member axis of buckling, in. (mm)

2. Detailing Requirements

At the ends of built-up compression members bearing on base plates or milled surfaces, all components in contact with one another shall be connected by a weld having a length not less than the maximum width of the member or by bolts spaced longitudinally not more than four diameters apart for a distance equal to $1\frac{1}{2}$ times the maximum width of the member.

Along the length of built-up compression members between the end connections required above, longitudinal spacing for intermittent welds, bolts, or rivets shall be adequate to provide for the transfer of the required forces. For limitations on the longitudinal spacing of connectors between elements in continuous contact consisting of a plate and a shape or two plates, see Section J3.5. Where a component of a built-up compression member consists of an outside plate, the maximum spacing shall not exceed the thickness of the thinner outside plate times $0.75\sqrt{E / F_y}$, nor 12 in. (305 mm), when intermittent welds are provided along the edges of the components or when fasteners are provided on all gage lines at each section. When fasteners are staggered, the maximum spacing on each gage line shall not exceed the thickness of the thinner outside plate times $1.12\sqrt{E / F_y}$ nor 18 in. (460 mm).

Individual components of compression members composed of two or more shapes shall be connected to one another at intervals, a, such that the effective slenderness ratio Ka / r_i of each of the component shapes, between the connectors, does not exceed three-fourths times the governing slenderness ratio of the built-up member. The least radius of gyration r_i shall be used in computing the slenderness ratio of each component part. The end connection shall be welded or fully tensioned bolted with clean mill scale or blast-cleaned faying surfaces with Class A coatings.

Open sides of compression members built up from plates or shapes shall be provided with continuous cover plates perforated with a succession of access holes. The unsupported width of such plates at access holes, as defined in Section B5.1, is assumed to contribute to the design strength provided that:

(1) The width-thickness ratio conforms to the limitations of Section B5.1.

(2) The ratio of length (in direction of stress) to width of hole shall not exceed two.

(3) The clear distance between holes in the direction of stress shall be not less than the transverse distance between nearest lines of connecting fasteners or welds.

(4) The periphery of the holes at all points shall have a minimum radius of 1½-in. (38 mm).

As an alternative to perforated cover plates, lacing with tie plates is permitted at each end and at intermediate points if the lacing is interrupted. Tie plates shall be as near the ends as practicable. In main members providing design strength, the end tie plates shall have a length of not less than the distance between the lines of fasteners or welds connecting them to the components of the member. Intermediate tie plates shall have a length not less than one-half of this distance. The thickness of tie plates shall be not less than one-fiftieth of the distance between lines of welds or fasteners connecting them to the segments of the members. In welded construction, the welding on each line connecting a tie plate shall aggregate not less than one-third the length of the plate. In bolted and riveted construction, the spacing in the direction of stress in tie plates shall be not more than six diameters and the tie plates shall be connected to each segment by at least three fasteners.

Lacing, including flat bars, angles, channels, or other shapes employed as lacing, shall be so spaced that l/r of the flange included between their connections shall not exceed the governing slenderness ratio for the member as a whole. Lacing shall be proportioned to provide a shearing strength normal to the axis of the member equal to two percent of the compressive design strength of the member. The l/r ratio for lacing bars arranged in single systems shall not exceed 140. For double lacing this ratio shall not exceed 200. Double lacing bars shall be joined at the intersections. For lacing bars in compression, l is permitted to be taken as the unsupported length of the lacing bar between welds or fasteners connecting it to the components of the built-up member for single lacing, and 70 percent of that distance for double lacing. The inclination of lacing bars to the axis of the member shall preferably be not less than 60° for single lacing and 45° for double lacing. When the distance between the lines of welds or fasteners in the flanges is more than 15 in. (380 mm), the lacing shall preferably be double or be made of angles.

For additional spacing requirements, see Section J3.

E5. CONNECTIONS FOR PIN-CONNECTED COMPRESSION MEMBERS

Pin connections of pin-connected compression members shall conform to the requirements of Sections D3.1 and D3.2, except Equations D3-1 and D3-2 do not apply.

CHAPTER F

BEAMS AND OTHER FLEXURAL MEMBERS

This chapter applies to compact and noncompact prismatic members subject to flexure and shear. For members subject to combined flexure and axial force, see Section H1. For members subject to fatigue, see Section K3. For members with slender compression elements, see Appendix B5. For web-tapered members, see Appendix F3. For members with slender web elements (plate girders), see Appendix G.

F1. DESIGN FOR FLEXURE

The nominal flexural strength M_n is the lowest value obtained according to the limit states of: (a) yielding; (b) lateral-torsional buckling; (c) flange local buckling; and (d) web local buckling. For laterally braced compact beams with $L_b \leq L_p$, only the limit state of yielding is applicable. For unbraced compact beams and noncompact tees and double angles, only the limit states of yielding and lateral-torsional buckling are applicable. The lateral-torsional buckling limit state is not applicable to members subject to bending about the minor axis, or to square or circular shapes.

This section applies to homogeneous and hybrid shapes with at least one axis of symmetry and which are subject to simple bending about one principal axis. For simple bending, the beam is loaded in a plane parallel to a principal axis that passes through the shear center or the beam is restrained against twisting at load points and supports. Only the limit states of yielding and lateral-torsional buckling are considered in this section. The lateral-torsional buckling provisions are limited to doubly symmetric shapes, channels, double angles, and tees. For lateral-torsional buckling of other singly symmetric shapes and for the limit states of flange local buckling and web local buckling of noncompact or slender-element sections, see Appendix F1. For unsymmetric shapes and beams subject to torsion combined with flexure, see Section H2. For biaxial bending, see Section H1.

1. Yielding

The flexural design strength of beams, determined by the limit state of yielding, is $\phi_b M_n$:

$$\phi_b = 0.90$$

$$M_n = M_p \tag{F1-1}$$

where

M_p = plastic moment ($= F_y Z \leq 1.5 M_y$ for homogeneous sections), kip-in. (N-mm)
M_y = moment corresponding to onset of yielding at the extreme fiber from an elastic stress distribution ($= F_y S$ for homogeneous section and $F_{yf} S$ for hybrid sections), kip-in. (N-mm)

See Section B10 for further limitations on M_n where there are holes in the tension flange.

2. Lateral-Torsional Buckling

This limit state is only applicable to members subject to major axis bending. The flexural design strength, determined by the limit state of lateral-torsional buckling, is $\phi_b M_n$:

ϕ_b = 0.90
M_n = nominal flexural strength determined as follows

2a. Doubly Symmetric Shapes and Channels with $L_b \le L_r$

The nominal flexural strength is:

$$M_n = C_b \left[M_p - \left(M_p - M_r\right)\left(\frac{L_b - L_p}{L_r - L_p}\right)\right] \le M_p \qquad \text{(F1-2)}$$

where

L_b = distance between points braced against lateral displacement of the compression flange, or between points braced to prevent twist of the cross section, in. (mm)
L_p = limiting laterally unbraced length as defined below, in. (mm)
L_r = limiting laterally unbraced length as defined below, in. (mm)
M_r = limiting buckling moment as defined below, kip-in. (N-mm)

In the above equation, C_b is a modification factor for non-uniform moment diagrams where, when both ends of the beam segment are braced:

$$C_b = \frac{12.5 M_{max}}{2.5 M_{max} + 3M_A + 4M_B + 3M_C} \qquad \text{(F1-3)}$$

where

$M_{\max}$ = absolute value of maximum moment in the unbraced segment, kip-in. (N-mm)
M_A = absolute value of moment at quarter point of the unbraced segment, kip-in. (N-mm)
M_B = absolute value of moment at centerline of the unbraced beam segment, kip-in. (N-mm)
M_C = absolute value of moment at three-quarter point of the unbraced beam segment, kip-in. (N-mm)

C_b is permitted to be conservatively taken as 1.0 for all cases. Equations F1-4 and F1-6 are conservatively based on $C_b = 1.0$. For cantilevers or overhangs where the free end is unbraced, $C_b = 1.0$.

The limiting unbraced length, L_p, shall be determined as follows.

(a) For I-shaped members including hybrid sections and channels:

$$L_p = 1.76 r_y \sqrt{\frac{E}{F_{yf}}} \tag{F1-4}$$

(b) For solid rectangular bars and box sections:

$$L_p = \frac{0.13 r_y E}{M_p} \sqrt{JA} \tag{F1-5}$$

where

A = cross-sectional area, in.^{2} (mm^2)
J = torsional constant, in.^{4} (mm^4)

The limiting laterally unbraced length L_r and the corresponding buckling moment M_r shall be determined as follows.

(a) For doubly symmetric I-shaped members and channels:

$$L_r = \frac{r_y X_1}{F_L} \sqrt{1 + \sqrt{1 + X_2 F_L^2}} \tag{F1-6}$$

$$M_r = F_L S_x \tag{F1-7}$$

where

$$X_1 = \frac{\pi}{S_x} \sqrt{\frac{EGJA}{2}} \tag{F1-8}$$

$$X_2 = 4 \frac{C_w}{I_y} \left(\frac{S_x}{GJ} \right)^2 \tag{F1-9}$$

S_x = section modulus about major axis, in.^{3} (mm^3)
E = modulus of elasticity of steel, 29,000 ksi (200 000 MPa)
G = shear modulus of elasticity of steel, 11,200 ksi (77 200 MPa)
F_L = smaller of $(F_{yf} - F_r)$ or F_{yw}, ksi (MPa)
F_r = compressive residual stress in flange; 10 ksi (69 MPa) for rolled shapes, 16.5 ksi (114 MPa) for welded built-up shapes
F_{yf} = yield stress of flange, ksi (MPa)
F_{yw} = yield stress of web, ksi (MPa)
I_y = moment of inertia about y-axis, in.^{4} (mm^4)
C_w = warping constant, in.^{6} (mm^6)

(b) For solid rectangular bars and box sections:

$$L_r = \frac{2 r_y E \sqrt{JA}}{M_r} \tag{F1-10}$$

$$M_r = F_{yf} S_x \tag{F1-11}$$

2b. Doubly Symmetric Shapes and Channels with $L_b > L_r$

The nominal flexural strength is:

$$M_n = M_{cr} \leq M_p \tag{F1-12}$$

where M_{cr} is the critical elastic moment, determined as follows.

(a) For doubly symmetric I-shaped members and channels:

$$M_{cr} = C_b \frac{\pi}{L_b} \sqrt{EI_y GJ + \left(\frac{\pi E}{L_b}\right)^2 I_y C_w} \tag{F1-13}$$

$$= \frac{C_b S_x X_1 \sqrt{2}}{L_b / r_y} \sqrt{1 + \frac{X_1^2 X_2}{2(L_b / r_y)^2}}$$

(b) For solid rectangular bars and symmetric box sections:

$$M_{cr} = \frac{57\,000 C_b \sqrt{JA}}{L_b / r_y} \tag{F1-14}$$

2c. Tees and Double Angles

For tees and double-angle beams loaded in the plane of symmetry:

$$M_n = M_{cr} = \frac{\pi \sqrt{EI_y GJ}}{L_b} \left[B + \sqrt{1 + B^2}\right] \tag{F1-15}$$

where

$M_n \leq 1.5 M_y$ for stems in tension
$M_n \leq 1.0 M_y$ for stems in compression
$B = \pm 2.3 (d / L_b) \sqrt{I_y / J}$ (F1-16)

The plus sign for B applies when the stem is in tension and the minus sign applies when the stem is in compression. If the tip of the stem is in compression anywhere along the unbraced length, use the negative value of B.

3. Design by Plastic Analysis

Design by plastic analysis, as limited in Section A5.1, is permitted for a compact section member bent about the major axis when the laterally unbraced length L_b of the compression flange adjacent to plastic hinge locations associated with the failure mechanism does not exceed L_{pd}, determined as follows.

(a) For doubly symmetric and singly symmetric I-shaped members with the compression flange equal to or larger than the tension flange (including hybrid members) loaded in the plane of the web:

$$L_{pd} = \left[0.12 + 0.076\left(\frac{M_1}{M_2}\right)\right]\left(\frac{E}{F_y}\right)r_y \quad \text{(F1-17)}$$

where

F_y = specified minimum yield stress of the compression flange, ksi (MPa)
M_1= smaller moment at end of unbraced length of beam, kip-in. (N-mm)
M_2= larger moment at end of unbraced length of beam, kip-in. (N-mm)
r_y = radius of gyration about minor axis, in. (mm)

(M_1 / M_2) is positive when moments cause reverse curvature and negative for single curvature

(b) For solid rectangular bars and symmetric box beams:

$$L_{pd} = \left[0.17 + 0.10\left(\frac{M_1}{M_2}\right)\right]\left(\frac{E}{F_y}\right)r_y \geq 0.10\left(\frac{E}{F_y}\right)r_y \quad \text{(F1-18)}$$

There is no limit on L_b for members with circular or square cross sections nor for any beam bent about its minor axis.

In the region of the last hinge to form, and in regions not adjacent to a plastic hinge, the flexural design strength shall be determined in accordance with Section F1.2.

F2. DESIGN FOR SHEAR

This section applies to unstiffened webs of singly or doubly symmetric beams, including hybrid beams, and channels subject to shear in the plane of the web. For the design shear strength of webs with stiffeners, see Appendix F2 or Appendix G3. For shear in the weak direction of the shapes above, pipes, and unsymmetric sections, see Section H2. For web panels subject to high shear, see Section K1.7. For shear strength at connections, see Sections J4 and J5.

1. Web Area Determination

The web area A_w shall be taken as the overall depth d times the web thickness t_w.

2. Design Shear Strength

The design shear strength of unstiffened webs, with $h / t_w \leq 260$, is $\phi_v V_n$,

where

$\phi_v = 0.90$
V_n= nominal shear strength defined as follows.

(a) For $h / t_w \leq 2.45\sqrt{E / F_{yw}}$

$$V_n = 0.6 F_{yw} A_w \quad \text{(F2-1)}$$

(b) For $2.45\sqrt{E / F_{yw}} < h / t_w \leq 3.07\sqrt{E / F_{yw}}$

$$V_n = 0.6F_{yw}A_w\left(\frac{2.45\sqrt{E/F_{yw}}}{h/t_w}\right) \tag{F2-2}$$

(c) For $3.07\sqrt{E/F_{yw}} < h/t_w \le 260$

$$V_n = A_w\left[\frac{4.52E}{(h/t_w)^2}\right] \tag{F2-3}$$

The general design shear strength of webs with or without stiffeners is given in Appendix F2.2 and an alternative method utilizing tension field action is given in Appendix G3.

3. Transverse Stiffeners

See Appendix F2.3.

F3. WEB-TAPERED MEMBERS

See Appendix F3.

F4. BEAMS AND GIRDERS WITH WEB OPENINGS

The effect of all web openings on the design strength of steel and composite beams shall be determined. Adequate reinforcement shall be provided when the required strength exceeds the design strength of the member at the opening.

CHAPTER G

PLATE GIRDERS

I-shaped plate girders shall be distinguished from I-shaped beams on the basis of the web slenderness ratio h/t_w. When this value is greater than λ_r the provisions of Appendices G1 and G2 shall apply for design flexural strength. For $h/t_w \le \lambda_r$, the provisions of Chapter F or Appendix F shall apply for design flexural strength. For girders with unequal flanges, see Appendix B5.1.

The design shear strength and transverse stiffener design shall be based on either Section F2 (without tension-field action) or Appendix G3 (with tension-field action). For girders with unequal flanges, see Appendix B5.1.

CHAPTER H

MEMBERS UNDER COMBINED FORCES AND TORSION

This chapter applies to prismatic members subject to axial force and flexure about one or both axes of symmetry, with or without torsion, and torsion only. For web-tapered members, see Appendix F3.

H1. SYMMETRIC MEMBERS SUBJECT TO BENDING AND AXIAL FORCE

1. Doubly and Singly Symmetric Members in Flexure and Tension

The interaction of flexure and tension in symmetric shapes shall be limited by Equations H1-1a and H1-1b.

(a) For $\dfrac{P_u}{\phi P_n} \geq 0.2$

$$\frac{P_u}{\phi P_n} + \frac{8}{9}\left(\frac{M_{ux}}{\phi_b M_{nx}} + \frac{M_{uy}}{\phi_b M_{ny}}\right) \leq 1.0 \quad \text{(H1-1a)}$$

(b) For $\dfrac{P_u}{\phi P_n} < 0.2$

$$\frac{P_u}{2\phi P_n} + \left(\frac{M_{ux}}{\phi_b M_{nx}} + \frac{M_{uy}}{\phi_b M_{ny}}\right) \leq 1.0 \quad \text{(H1-1b)}$$

where

P_u = required tensile strength, kips (N)
P_n = nominal tensile strength determined in accordance with Section D1, kips (N)
M_u = required flexural strength determined in accordance with Section C1, kip-in. (N-mm)
M_n = nominal flexural strength determined in accordance with Section F1, kip-in. (N-mm)
x = subscript relating symbol to strong axis bending
y = subscript relating symbol to weak axis bending
ϕ = ϕ_t = resistance factor for tension (see Section D1)
ϕ_b = resistance factor for flexure = 0.90

A more detailed analysis of the interaction of flexure and tension is permitted in lieu of Equations H1-1a and H1-1b.

2. Doubly and Singly Symmetric Members in Flexure and Compression

The interaction of flexure and compression in symmetric shapes shall be limited by Equations H1-1a and H1-1b

where

P_u = required compressive strength, kips (N)
P_n = nominal compressive strength determined in accordance with Section E2, kips (N)
ϕ = ϕ_c = resistance factor for compression = 0.85 (see Section E2)
ϕ_b = resistance factor for flexure = 0.90

H2. UNSYMMETRIC MEMBERS AND MEMBERS UNDER TORSION AND COMBINED TORSION, FLEXURE, SHEAR, AND/OR AXIAL FORCE

The design strength, ϕF_n, of the member shall equal or exceed the required strength expressed in terms of the normal stress f_{un} or the shear stress f_{uv}, determined by elastic analysis for the factored loads:

(a) For the limit state of yielding under normal stress:

$$f_{un} \leq \phi F_n \tag{H2-1}$$

$$\phi = 0.90$$
$$F_n = F_y$$

(b) For the limit state of yielding under shear stress:

$$f_{uv} \leq 0.6\phi F_n \tag{H2-2}$$

Errata 9/4/01

$$\phi = 0.90$$
$$F_n = F_y$$

(c) For the limit state of buckling:

$$f_{un} \text{ or } f_{uv} \leq \phi_c F_n, \text{ as applicable} \tag{H2-3}$$

$$\phi_c = 0.85$$
$$F_n = F_{cr}$$

Some constrained local yielding is permitted adjacent to areas which remain elastic.

H3. ALTERNATIVE INTERACTION EQUATIONS FOR MEMBERS UNDER COMBINED STRESS

See Appendix H3.

CHAPTER I

COMPOSITE MEMBERS

This chapter applies to composite columns composed of rolled or built-up structural steel shapes, pipe or HSS, and structural concrete acting together and to steel beams supporting a reinforced concrete slab so interconnected that the beams and the slab act together to resist bending. Simple and continuous composite beams with shear connectors and concrete-encased beams, constructed with or without temporary shores, are included.

I1. DESIGN ASSUMPTIONS AND DEFINITIONS

Force Determination. In determining forces in members and connections of a structure that includes composite beams, consideration shall be given to the effective sections at the time each increment of load is applied.

Elastic Analysis. For an elastic analysis of continuous composite beams without haunched ends, it is permissible to assume that the stiffness of a beam is uniform throughout the beam length. The stiffness is permitted to be computed using the weighted average of the moments of inertia in the positive moment region and the negative moment region.

Plastic Analysis. When plastic analysis is used, as limited in Section A5.1, the strength of flexural composite members shall be determined from plastic stress distributions.

Plastic Stress Distribution for Positive Moment. If the slab in the positive moment region is connected to the steel beam with shear connectors, a concrete stress of $0.85f_c'$ is permitted to be assumed uniformly distributed throughout the effective compression zone, where f_c' is the specified compressive strength of the concrete. Concrete tensile strength shall be neglected. A uniformly distributed steel stress of F_y shall be assumed throughout the tension zone and throughout the compression zone in the structural steel section. The net tensile force in the steel section shall be equal to the compressive force in the concrete slab.

Plastic Stress Distribution for Negative Moment. If the slab in the negative moment region is connected to the steel beam with shear connectors, a tensile stress of F_{yr} shall be assumed in all adequately developed longitudinal reinforcing bars within the effective width of the concrete slab. Concrete tensile strength shall be neglected. A uniformly distributed steel stress of F_y shall be assumed throughout the tension zone and throughout the compression zone in the structural steel section. The net compressive force in the steel section shall be equal to the total tensile force in the reinforcing steel.

Elastic Stress Distribution. When a determination of elastic stress distribution is required, strains in steel and concrete shall be assumed directly proportional to the distance from the neutral axis. The stress shall equal strain times modulus of elasticity for steel, E, or modulus of elasticity for concrete, E_c. Concrete tensile strength shall be neglected. Maximum stress in the steel shall not exceed F_y. Maximum compressive stress in the concrete shall not exceed $0.85f_c'$. In composite hybrid beams, the maximum stress in the steel flange shall not exceed F_{yf} but the strain in the web may exceed the yield strain; the stress shall be taken as F_{yw} at such locations.

Fully Composite Beam. Shear connectors are provided in sufficient numbers to develop the maximum flexural strength of the composite beam. For elastic stress distribution it shall be assumed that no slip occurs.

Partially Composite Beam. The shear strength of shear connectors governs the flexural strength of the partially composite beam. Elastic computations such as those for deflections, fatigue, and vibrations shall include the effect of slip.

Concrete-Encased Beam. A beam totally encased in concrete cast integrally with the slab may be assumed to be interconnected to the concrete by natural bond, without additional anchorage, provided that: (1) concrete cover over the beam sides and soffit is at least two in. (50 mm); (2) the top of the beam is at least 1½-in. (38 mm) below the top and two in. (50 mm) above the bottom of the slab; and (3) concrete encasement contains adequate mesh or other reinforcing steel to prevent spalling of concrete.

Composite Column. A steel column fabricated from rolled or built-up steel shapes and encased in structural concrete or fabricated from steel pipe or HSS and filled with structural concrete.

Encased Composite Column. A steel column fabricated from rolled or built-up shapes and encased in structural concrete.

Filled Composite Column. Structural steel HSS or pipes that are filled with structural concrete.

I2. COMPRESSION MEMBERS

1. Limitations

To qualify as a composite column, the following limitations shall be met:

(1) The cross-sectional area of the steel shape, pipe, or HSS shall comprise at least four percent of the total composite cross section.

(2) Concrete encasement of a steel core shall be reinforced with longitudinal load-carrying bars, longitudinal bars to restrain concrete, and lateral ties. Longitudinal load-carrying bars shall be continuous at framed levels; longitudinal restraining bars may be interrupted at framed levels. The spacing of ties shall be not greater than two-thirds of the least dimension of the composite cross section. The cross-sectional area of the transverse and longitudinal reinforcement shall be at least 0.007 sq. in. per in. (180 mm^2 per m) of bar spacing. The encasement shall provide at least 1½-in. (38 mm) of clear cover outside of both transverse and longitudinal reinforcement.

(3) Concrete shall have a specified compressive strength f_c' of not less than 3 ksi (21 MPa) nor more than 8 ksi (55 MPa) for normal weight concrete and not less than 4 ksi (28 MPa) for light weight concrete.

(4) The specified minimum yield stress of structural steel and reinforcing bars

used in calculating the strength of a composite column shall not exceed 60 ksi (415 MPa).

(5) The minimum wall thickness of structural steel pipe or HSS filled with concrete shall be equal to $b\sqrt{F_y/3E}$ for each face of width b in rectangular sections and $D\sqrt{F_y/8E}$ for circular sections of outside diameter D.

2. Design Strength

The design strength of axially loaded composite columns is $\phi_c P_n$,

where

$\phi_c = 0.85$

P_n = nominal axial compressive strength determined from Equations E2-1 through E2-4 with the following modifications:

(1) A_g is replaced by A_s, the gross area of steel shape, pipe, or HSS, in.^{2} (mm^2)

(2) r is replaced by r_m, the radius of gyration of the steel shape, pipe, or HSS except that for steel shapes it shall not be less than 0.3 times the overall thickness of the composite cross section in the plane of buckling, in. (mm)

(3) F_y is replaced by F_{my}, the modified yield stress from Equation I2-1

$$F_{my} = F_y + c_1 F_{yr}(A_r/A_s) + c_2 f_c'(A_c/A_s) \quad \text{(I2-1)}$$

(4) E is replaced by E_m, the modified modulus of elasticity from Equation I2-2.

$$E_m = E + c_3 E_c (A_c/A_s) \quad \text{(I2-2)}$$

where

A_c = area of concrete, in.^{2} (mm^2)
A_r = area of longitudinal reinforcing bars, in.^{2} (mm^2)
A_s = area of steel, in.^{2} (mm^2)
E = modulus of elasticity of steel, ksi (MPa)
E_c = modulus of elasticity of concrete. E_c is permitted to be computed from $E_c = w^{1.5}\sqrt{f_c'}$ (metric: $E_c = 0.041 w^{1.5}\sqrt{f_c'}$) where w, the unit weight of concrete, is expressed in lbs./cu. ft (kg/m^3) and f_c' is expressed in ksi (MPa).
F_y = specified minimum yield stress of steel shape, pipe, or HSS, ksi (MPa)
F_{yr} = specified minimum yield stress of longitudinal reinforcing bars, ksi (MPa)
f_c' = specified compressive strength of concrete, ksi (MPa)
c_1, c_2, c_3 = numerical coefficients. For concrete-filled pipe and HSS: $c_1 = 1.0$, $c_2 = 0.85$, and $c_3 = 0.4$; for concrete-encased shapes $c_1 = 0.7$, $c_2 = 0.6$, and $c_3 = 0.2$

3. Columns with Multiple Steel Shapes

If the composite cross section includes two or more steel shapes, the shapes shall be interconnected with lacing, tie plates, or batten plates to prevent buckling of individual shapes before hardening of concrete.

4. Load Transfer

Loads applied to axially loaded encased composite columns shall be transferred between the steel and concrete in accordance with the following requirements:

(a) When the external force is applied directly to the steel section, shear connectors shall be provided to transfer the force $V_u{}'$ as follows:

$$V_u{}' = V_u(1 - A_s F_y / P_n) \qquad \text{(I2-3)}$$

where

V_u = force introduced to column, kips (N)
A_s = area of steel section, in.^{2} (mm^2)
F_y = yield strength of the steel section, ksi (MPa)
P_n = nominal compressive strength of the composite column without consideration of slenderness effects, kips (N)

(b) When the external force is applied directly to the concrete encasement, shear connectors shall be provided to transfer the force $V_u{}'$ as follows:

$$V_u{}' = V_u(A_s F_y / P_n) \qquad \text{(I2-4)}$$

Shear connectors transferring the force $V_u{}'$ shall be distributed along the length of the member. The maximum connector spacing shall be 16 in. (405 mm) and connectors shall be placed on at least two faces of the steel shape in a configuration symmetrical about the steel shape axes.

Where the supporting concrete area in direct bearing is wider than the loaded area on one or more sides and otherwise restrained laterally on the remaining sides, the maximum design strength shall be:

$$\phi_B \, 1.7 f_c{}' A_B \qquad \text{(I2-5)}$$

where

$\phi_B = 0.65$
A_B = loaded area, in.^{2} (mm^2)

I3. FLEXURAL MEMBERS

1. Effective Width

The effective width of the concrete slab is the sum of the effective widths for each side of the beam center-line, each of which shall not exceed:

(1) one-eighth of the beam span, center-to-center of supports;

(2) one-half the distance to the center-line of the adjacent beam; or

(3) the distance to the edge of the slab.

2. Design Strength of Beams with Shear Connectors

The positive design flexural strength $\phi_b M_n$ shall be determined as follows:

(a) For $h / t_w \leq 3.76\sqrt{E / F_{yf}}$:

$\phi_b = 0.85$; M_n shall be determined from the plastic stress distribution on the composite section.

(b) For $h / t_w > 3.76\sqrt{E / F_{yf}}$:

$\phi_b = 0.90$; M_n shall be determined from the superposition of elastic stresses, considering the effects of shoring.

The negative design flexural strength $\phi_b M_n$ shall be determined for the steel section alone, in accordance with the requirements of Chapter F.

Alternatively, the negative design flexural strength $\phi_b M_n$ shall be computed with: $\phi_b = 0.85$ and M_n determined from the plastic stress distribution on the composite section, provided that:

(1) Steel beam is an adequately braced compact section, as defined in Section B5.

(2) Shear connectors connect the slab to the steel beam in the negative moment region.

(3) Slab reinforcement parallel to the steel beam, within the effective width of the slab, is properly developed.

3. Design Strength of Concrete-Encased Beams

The design flexural strength $\phi_b M_n$ shall be computed with $\phi_b = 0.90$ and M_n determined from the superposition of elastic stresses, considering the effects of shoring.

Alternatively, the design flexural strength $\phi_b M_n$ shall be computed with $\phi_b = 0.90$ and M_n determined from the plastic stress distribution on the steel section alone.

If shear connectors are provided and the concrete meets the requirements of Section I2.1(2), the design flexural strength $\phi_b M_n$ shall be computed based upon the plastic stress distribution on the composite section with $\phi_b = 0.85$.

4. Strength During Construction

When temporary shores are not used during construction, the steel section alone shall have adequate strength to support all loads applied prior to the concrete attaining 75 percent of its specified strength f_c'. The design flexural strength of the steel section shall be determined in accordance with the requirements of Section F1.

5. Formed Steel Deck

5a. General

The design flexural strength, $\phi_b M_n$, of composite construction consisting of concrete slabs on formed steel deck connected to steel beams shall be determined by the applicable portions of Section I3.2, with the following modifications:

(1) This section is applicable to decks with nominal rib height not greater than three in. (75 mm). The average width of concrete rib or haunch w_r shall be not less than two in. (50 mm), but shall not be taken in calculations as more than the minimum clear width near the top of the steel deck. See Section I3.5c for additional restrictions.

(2) The concrete slab shall be connected to the steel beam with welded stud shear connectors ¾-in. (19 mm) or less in diameter (AWS D1.1). Studs shall be

welded either through the deck or directly to the steel beam. Stud shear connectors, after installation, shall extend not less than 1½-in. (38 mm) above the top of the steel deck.

The slab thickness above the steel deck shall be not less than two in. (50 mm).

5b. Deck Ribs Oriented Perpendicular to Steel Beam

Concrete below the top of the steel deck shall be neglected in determining section properties and in calculating A_c for deck ribs oriented perpendicular to the steel beams.

The spacing of stud shear connectors along the length of a supporting beam shall not exceed 36 in. (915 mm).

The nominal strength of a stud shear connector shall be the value stipulated in Section I5 multiplied by the following reduction factor:

$$\frac{0.85}{\sqrt{N_r}}(w_r / h_r)[(H_s / h_r) - 1.0] \leq 1.0 \qquad \text{(I3-1)}$$

where

h_r = nominal rib height, in. (mm)

H_s = length of stud connector after welding, in. (mm), not to exceed the value $h_r + 3$ in. (75 mm) in computations, although actual length may be greater

N_r = number of stud connectors in one rib at a beam intersection, not to exceed three in computations, although more than three studs may be installed

w_r = average width of concrete rib or haunch (as defined in Section I3.5a), in. (mm)

Where there is only a single stud placed in a rib oriented perpendicular to the steel beam, the reduction factor of Equation I3-1 shall not exceed 0.75.

To resist uplift, steel deck shall be anchored to all supporting members at a spacing not to exceed 18 in. (460 mm). Such anchorage shall be provided by stud connectors, a combination of stud connectors and arc spot (puddle) welds, or other devices specified by the designer.

5c. Deck Ribs Oriented Parallel to Steel Beam

Concrete below the top of the steel deck may be included in determining section properties and shall be included in calculating A_c in Section I5.

Steel deck ribs over supporting beams may be split longitudinally and separated to form a concrete haunch.

When the nominal depth of steel deck is 1½-in. (38 mm) or greater, the average width w_r of the supported haunch or rib shall be not less than two in. (50 mm) for the first stud in the transverse row plus four stud diameters for each additional stud.

The nominal strength of a stud shear connector shall be the value stipulated in Section I5, except that when w_r / h_r is less than 1.5, the value from Section I5 shall be multiplied by the following reduction factor:

$$0.6(w_r / h_r)[(H_s / h_r) - 1.0] \leq 1.0 \qquad \text{(I3-2)}$$

where h_r and H_s are as defined in Section I3.5b and w_r is the average width of concrete rib or haunch as defined in Section I3.5a.

6. Design Shear Strength

The design shear strength of composite beams shall be determined by the shear strength of the steel web, in accordance with Section F2.

I4. COMBINED COMPRESSION AND FLEXURE

The interaction of axial compression and flexure in the plane of symmetry on composite members shall be limited by Section H1.2 with the following modifications:

M_n = nominal flexural strength determined from plastic stress distribution on the composite cross section except as provided below, kip-in. (N-mm)
P_{e1}, P_{e2} = A_sF_{my} / λ_c^2 elastic buckling load, kips (N)
F_{my} = modified yield stress, ksi (MPa), see Section I2
ϕ_b = resistance factor for flexure from Section I3
ϕ_c = resistance factor for compression = 0.85
λ_c = column slenderness parameter defined by Equation E2-4 as modified in Section I2.2

When the axial term in Equations H1-1a and H1-1b is less than 0.3, the nominal flexural strength M_n shall be determined by straight line transition between the nominal flexural strength determined from the plastic distribution on the composite cross sections at $(P_u/\phi_cP_n) = 0.3$ and the flexural strength at $P_u = 0$ as determined in Section I3. If shear connectors are used at $P_u = 0$, they shall be provided whenever P_u/ϕ_cP_n is less than 0.3.

I5. SHEAR CONNECTORS

This section applies to the design of stud and channel shear connectors. For connectors of other types, see Section I6.

1. Materials

Shear connectors shall be headed steel studs not less than four stud diameters in length after installation, or hot rolled steel channels. The stud connectors shall conform to the requirements of Section A3.6. The channel connectors shall conform to the requirements of Section A3. Shear connectors shall be embedded in concrete slabs made with ASTM C33 aggregate or with rotary kiln produced aggregates conforming to ASTM C330, with concrete unit weight not less than 90 pcf (1 440 kg/m^3).

2. Horizontal Shear Force

The entire horizontal shear at the interface between the steel beam and the concrete slab shall be assumed to be transferred by shear connectors, except for concrete-encased beams as defined in Section I1. For composite action with concrete subject to flexural compression, the total horizontal shear force between the point of maximum positive moment and the point of zero moment shall be taken as the smallest of the following: (a) $0.85f_c'A_c$; (b) A_sF_y; and (c) ΣQ_n;

where

A_c = area of concrete slab within effective width, in.^{2} (mm^2)
A_s = area of steel cross section, in.^{2} (mm^2)
ΣQ_n = sum of nominal strengths of shear connectors between the point of maximum positive moment and the point of zero moment, kips (N)

For hybrid beams, the yield force shall be computed separately for each component of the cross section; A_sF_y of the entire cross section is the sum of the component yield forces.

In continuous composite beams where longitudinal reinforcing steel in the negative moment regions is considered to act compositely with the steel beam, the total horizontal shear force between the point of maximum negative moment and the point of zero moment shall be taken as the smaller of A_rF_{yr} and ΣQ_n;

where

A_r = area of adequately developed longitudinal reinforcing steel within the effective width of the concrete slab, in.^{2} (mm^2)
F_{yr} = minimum specified yield stress of the reinforcing steel, ksi (MPa)

3. Strength of Stud Shear Connectors

The nominal strength of one stud shear connector embedded in a solid concrete slab is

$$Q_n = 0.5A_{sc}\sqrt{f_c'E_c} \leq A_{sc}F_u \tag{I5-1}$$

where

A_{sc} = cross-sectional area of stud shear connector, in.^{2} (mm^2)
F_u = specified minimum tensile strength of a stud shear connector, ksi (MPa)
E_c = modulus of elasticity of concrete, ksi (MPa)

For a stud shear connector embedded in a slab on a formed steel deck, refer to Section I3 for reduction factors given by Equations I3-1 and I3-2 as applicable. The reduction factors apply only to the $0.5A_{sc}\sqrt{f_c'E_c}$ term in Equation I5-1.

4. Strength of Channel Shear Connectors

The nominal strength of one channel shear connector embedded in a solid concrete slab is

$$Q_n = 0.3(t_f + 0.5t_w)L_c\sqrt{f_c'E_c} \tag{I5-2}$$

where

t_f = flange thickness of channel shear connector, in. (mm)
t_w = web thickness of channel shear connector, in. (mm)
L_c = length of channel shear connector, in. (mm)

5. Required Number of Shear Connectors

The number of shear connectors required between the section of maximum bending moment, positive or negative, and the adjacent section of zero moment shall be equal to the horizontal shear force as determined in Section I5.2 divided by the nominal strength of one shear connector as determined from Section I5.3 or Section I5.4.

6. Shear Connector Placement and Spacing

Shear connectors required on each side of the point of maximum bending moment, positive or negative, shall be distributed uniformly between that point and the adjacent points of zero moment, unless otherwise specified. However, the number of shear connectors placed between any concentrated load and the nearest point of zero moment shall be sufficient to develop the maximum moment required at the concentrated load point.

Shear connectors shall have at least one in. (25 mm) of lateral concrete cover, except for connectors installed in the ribs of formed steel decks. The diameter of studs shall not be greater than 2.5 times the thickness of the flange to which they are welded, unless located over the web. The minimum center-to-center spacing of stud connectors shall be six diameters along the longitudinal axis of the supporting composite beam and four diameters transverse to the longitudinal axis of the supporting composite beam, except that within the ribs of formed steel decks oriented perpendicular to the steel beam the minimum center-to-center spacing shall be four diameters in any direction. The maximum center-to-center spacing of shear connectors shall not exceed eight times the total slab thickness. Also see Section I3.5b.

I6. SPECIAL CASES

When composite construction does not conform to the requirements of Section I1 through Section I5, the strength of shear connectors and details of construction shall be established by a suitable test program.

CHAPTER J

CONNECTIONS, JOINTS, AND FASTENERS

This chapter applies to connecting elements, connectors, and the affected elements of the connected members subject to static loads. For connections subject to fatigue, see Appendix K3.

J1. GENERAL PROVISIONS

1. Design Basis

Connections consist of affected elements of connected members (e.g., beam webs), connecting elements (e.g., gussets, angles, brackets), and connectors (e.g., welds, bolts, rivets). These components shall be proportioned so that their design strength equals or exceeds the required strength determined by structural analysis for factored loads acting on the structure or a specified proportion of the strength of the connected members, whichever is appropriate.

2. Simple Connections

Connections of beams, girders, or trusses shall be designed as flexible and are permitted to be proportioned for the reaction shears only, except as otherwise indicated in the design documents. Flexible beam connections shall accommodate end rotations of unrestrained (simple) beams. To accomplish this, some inelastic but self-limiting deformation in the connection is permitted.

3. Moment Connections

End connections of restrained beams, girders, and trusses shall be designed for the combined effect of forces resulting from moment and shear induced by the rigidity of the connections.

4. Compression Members with Bearing Joints

When columns bear on bearing plates or are finished to bear at splices, there shall be sufficient connectors to hold all parts securely in place.

When compression members other than columns are finished to bear, the splice material and its connectors shall be arranged to hold all parts in line and shall be proportioned for 50 percent of the required strength of the member.

All compression joints shall be proportioned to resist any tension developed by the factored load combinations stipulated in Section A4.

5. Splices in Heavy Sections

This paragraph applies to ASTM A6/A6M Group 4 and 5 rolled shapes, or shapes built-up by welding plates more than two in. (50 mm) thick together to form the cross section, and where the cross section is to be spliced and subject to primary tensile stresses due to tension or flexure. When the individual elements of the cross section are spliced prior to being joined to form the cross section in accordance with AWS D1.1, Article 5.21.6, the applicable provisions of AWS D1.1 apply in lieu of the requirements of this section. When tensile forces in these sections are to be transmitted through splices by complete-joint-penetration groove welds, material notch-toughness requirements as given in Section A3.1c, weld access hole details as given in Section J1.6, welding preheat requirements as given in Section J2.8, and thermal-cut surface preparation and inspection requirements as given in Section M2.2 apply.

At tension splices in ASTM A6/A6M Group 4 and 5 shapes and built-up members of material more than two in. (50 mm) thick, weld tabs and backing shall be removed and the surfaces ground smooth.

When splicing ASTM A6/A6M Group 4 and 5 rolled shapes or shapes built-up by welding plates more than two in. (50 mm) thick to form a cross section, and where the section is to be used as a primary compression member, all weld access holes required to facilitate groove welding operations shall satisfy the provisions of Section J1.6.

Alternatively, splicing of such members subject to compression, including members which are subject to tension due to wind or seismic loads, shall be accomplished using splice details which do not induce large weld shrinkage strains; for example partial-joint-penetration flange groove welds with fillet-welded surface lap plate splices on the web, bolted lap plate splices, or combination bolted/fillet-welded lap plate splices.

6. Beam Copes and Weld Access Holes

All weld access holes required to facilitate welding operations shall have a length from the toe of the weld preparation not less than $1\frac{1}{2}$ times the thickness of the material in which the hole is made. The height of the access hole shall be adequate for deposition of sound weld metal in the adjacent plates and provide clearance for weld tabs for the weld in the material in which the hole is made, but not less than the thickness of the material. In hot-rolled shapes and built-up shapes, all beam copes and weld access holes shall be shaped free of notches and sharp re-entrant corners, except that when fillet web-to-flange welds are used in built-up shapes, access holes are permitted to terminate perpendicular to the flange.

For ASTM A6/A6M Group 4 and 5 shapes and built-up shapes of material more than two in. (50 mm) thick, the thermally cut surfaces of beam copes and weld access holes shall be ground to bright metal and inspected by either magnetic particle or dye penetrant methods prior to deposition of splice welds. If the curved transition portion of weld access holes and beam copes are formed by predrilled or sawed holes, that portion of the access hole or cope need not be ground. Weld access holes and beam copes in other shapes need not be ground nor inspected by dye penetrant or magnetic particle methods.

7. Minimum Strength of Connections

Connections providing design strength shall be designed to support a factored load not less than 10 kips (44 kN), except for lacing, sag rods, or girts.

8. Placement of Welds and Bolts

Groups of welds or bolts at the ends of any member which transmit axial force into that member shall be sized so that the center of gravity of the group coincides with the center of gravity of the member, unless provision is made for the eccentricity. The foregoing provision is not applicable to end connections of statically-loaded single angle, double angle, and similar members.

9. Bolts in Combination with Welds

In new work, A307 bolts or high-strength bolts proportioned as bearing-type connections shall not be considered as sharing the load in combination with welds. Welds, if used, shall be proportioned for the entire force in the connection. In slip-critical connections, high-strength bolts are permitted to be considered as sharing the load with the welds. These calculations shall be made at factored loads.

In making welded alterations to structures, existing rivets and high-strength bolts tightened to the requirements for slip-critical connections are permitted to be utilized for carrying loads present at the time of alteration and the welding need only provide the additional design strength required.

10. High-Strength Bolts in Combination with Rivets

In both new work and alterations, in connections designed as slip-critical connections in accordance with the provisions of Section J3, high-strength bolts are permitted to be considered as sharing the load with rivets.

11. Limitations on Bolted and Welded Connections

Fully pretensioned high-strength bolts (see Table J3.1 or J3.1M) or welds shall be used for the following connections:

Column splices in all tier structures 200 ft (60 m) or more in height.

Column splices in tier structures 100 (30 m) to 200 ft (60 m) in height, if the least horizontal dimension is less than 40 percent of the height.

Column splices in tier structures less than 100 ft (30 m) in height, if the least horizontal dimension is less than 25 percent of the height.

Connections of all beams and girders to columns and of any other beams and girders on which the bracing of columns is dependent, in structures over 125 ft (38 m) in height.

In all structures carrying cranes of over five-ton (50 kN) capacity: roof-truss splices and connections of trusses to columns, column splices, column bracing, knee braces, and crane supports.

Connections for supports of running machinery, or of other live loads which produce impact or reversal of stress.

Any other connections stipulated on the design drawings.

In all other cases connections are permitted to be made with A307 bolts or snug-tight high-strength bolts.

For the purpose of this section, the height of a tier structure shall be taken as the vertical distance from the curb level to the highest point of the roof beams in the case of flat roofs, or to the mean height of the gable in the case of roofs having a slope of more than 25 percent. Where the curb level has not been established, or where the structure does not adjoin a street, the mean level of the adjoining land shall be used instead of curb level. It is permissible to exclude penthouses in computing the height of the structure.

J2. WELDS

All provisions of AWS D1.1, apply under this specification, except the provisions applicable to Tubular Structures, which are outside the scope of this specification, and except that the provisions of the listed AISC LRFD Specification Sections apply under this Specification in lieu of the cited AWS Code provisions as follows:

AISC Specification Section J1.5 and J1.6 in lieu of AWS D1.1 Section 5.17
AISC Specification Section J2.2 in lieu of AWS D1.1 Section 2.4.1.1
AISC Specification Table J2.5 in lieu of AWS D1.1 Table 2.3
AISC Specification Table A-K3.1 in lieu of AWS D1.1 Section 2.27.1
AISC Specification Section K3 and Appendix K3 in lieu of AWS Section 2, Part C
AISC Specification Section M2.2 in lieu of AWS Sections 5.15.1.2, 5.15.4.3 and 5.15.4.4

The length and disposition of welds, including end returns shall be indicated on the design and shop drawings.

1. Groove Welds

1a. Effective Area

The effective area of groove welds shall be considered as the effective length of the welds times the effective throat thickness.

The effective length of a groove weld shall be the width of the part joined.

The effective throat thickness of a complete-joint-penetration groove weld shall be the thickness of the thinner part joined.

The effective throat thickness of a partial-joint-penetration groove weld shall be as shown in Table J2.1.

The effective throat thickness of a flare groove weld when flush to the surface of a bar or 90° bend in formed section shall be as shown in Table J2.2. Random sections of production welds for each welding procedure, or such test sections as may be required by design documents, shall be used to verify that the effective throat is consistently obtained.

Larger effective throat thicknesses than those in Table J2.2 are permitted, provided the fabricator can establish by qualification the consistent production of such larger effective throat thicknesses. Qualification shall consist of sectioning the weld normal to its axis, at mid-length and terminal ends. Such sectioning shall be made on a number of combinations of material sizes representative of the range to be used in the fabrication or as required by the designer.

1b. Limitations

The minimum effective throat thickness of a partial-joint-penetration groove weld shall be as shown in Table J2.3. Weld size is determined by the thicker of the two parts joined, except that the weld size need not exceed the thickness of the thinnest part joined, even when a larger size is required by calculated strength. For this

TABLE J2.1
Effective Throat Thickness of Partial-Joint-Penetration Groove Welds

Welding Process	Welding Position	Included Angle at Root of Groove	Effective Throat Thickness
Shielded metal arc Submerged arc	All	J or U joint	Depth of chamfer
Gas metal arc		Bevel or V joint ≥ 60°	
Flux-cored arc		Bevel or V joint < 60° but ≥ 45°	Depth of chamfer Minus ⅛-in. (3 mm)

TABLE J2.2
Effective Throat Thickness of Flare Groove Welds

Type of Weld	Radius (*R*) of Bar or Bend	Effective Throat Thickness
Flare bevel groove	All	$\frac{5}{16}R$
Flare V-groove	All	$\frac{1}{2}R$ [a]
[a] Use $\frac{3}{8}R$ for Gas Metal Arc Welding (except short circuiting transfer process) when $R \geq 1$ in. (25 mm)		

Errata 9/4/01

TABLE J2.3
Minimum Effective Throat Thickness of Partial-Joint-Penetration Groove Welds

Material Thickness of Thicker Part Joined, in. (mm)	Minimum Effective Throat Thickness [a], in. (mm)
To ¼ (6) inclusive	⅛ (3)
Over ¼ (6) to ½ (13)	3/16 (5)
Over ½ (13) to ¾ (19)	¼ (6)
Over ¾ (19) to 1½ (38)	5/16 (8)
Over 1½ (38) to 2¼ (57)	⅜ (10)
Over 2¼ (57) to 6 (150)	½ (13)
Over 6 (150)	⅝ (16)
[a] See Table J2.1	

exception, particular care shall be taken to provide sufficient preheat for soundness of the weld.

2. Fillet Welds

2a. Effective Area

The effective area of fillet welds shall be as defined in AWS D1.1 Section 2.4.3 and 2.11. The effective throat thickness of a fillet weld shall be the shortest distance from the root of the joint to the face of the diagrammatic weld, except that for fillet welds made by the submerged arc process, the effective throat thickness shall be

TABLE J2.4
Minimum Size of Fillet Welds

Material Thickness of Thicker Part Joined, in. (mm)	Minimum Size of Fillet Weld[a] in. (mm)
To 1/4 (6) inclusive	1/8 (3)
Over 1/4 (6) to 1/2 (13)	3/16 (5)
Over 1/2 (13) to 3/4 (19)	1/4 (6)
Over 3/4 (19)	5/16 (8)

[a] Leg dimension of fillet welds. Single pass welds must be used.
[b] See Section J2.2b for maximum size of fillet welds.

taken equal to the leg size for 3/8-in. (10 mm) and smaller fillet welds, and equal to the theoretical throat plus 0.11-in. (3 mm) for fillet welds over 3/8-in.(10 mm).

For fillet welds in holes and slots, the effective length shall be the length of the centerline of the weld along the center of the plane through the throat. In the case of overlapping fillets, the effective area shall not exceed the nominal cross-sectional area of the hole or slot, in the plane of the faying surface.

2b. Limitations

The *minimum size of fillet welds* shall be not less than the size required to transmit calculated forces nor the size as shown in Table J2.4 which is based upon experiences and provides some margin for uncalculated stress encountered during fabrication, handling, transportation, and erection. These provisions do not apply to fillet weld reinforcements of partial- or complete-joint-penetration welds.

The *maximum size of fillet welds* of connected parts shall be:

(a) Along edges of material less than 1/4-in. (6 mm) thick, not greater than the thickness of the material.

(b) Along edges of material 1/4-in. (6 mm) or more in thickness, not greater than the thickness of the material minus 1/16-in. (2 mm), unless the weld is especially designated on the drawings to be built out to obtain full-throat thickness. In the as-welded condition, the distance between the edge of the base metal and the toe of the weld is permitted to be less than 1/16-in. (2 mm) provided the weld size is clearly verifiable.

For flange-web welds and similar connections, the actual weld size need not be larger than that required to develop the web capacity, and the requirements of Table J2.4 need not apply.

The *minimum effective length of fillet welds* designed on the basis of strength shall be not less than four times the nominal size, or else the size of the weld shall be considered not to exceed 1/4 of its effective length. If longitudinal fillet welds are used alone in end connections of flat-bar tension members, the length of each fillet weld shall be not less than the perpendicular distance between them. For the effect of longitudinal fillet weld length in end connections upon the effective area of the connected member, see Section B3.

For end-loaded fillet welds with a length up to 100 times the leg dimension, it is permitted to take the effective length equal to the actual length. When the length of the end-loaded fillet weld exceeds 100 times the weld size, the effective length shall be determined by multiplying the actual length by the reduction factor, β,

where

$$\beta = 1.2 - 0.002(L/w) \leq 1.0 \qquad \text{(J2-1)}$$

L = actual length of end-loaded weld, in. (mm)
w = weld leg size, in. (mm)

When the length of the weld exceeds 300 times the leg size, the value of β shall be taken as 0.60.

Intermittent fillet welds may be used to transfer calculated stress across a joint or faying surfaces when the strength required is less than that developed by a continuous fillet weld of the smallest permitted size, and to join components of built-up members. The effective length of any segment of intermittent fillet welding shall be not less than four times the weld size, with a minimum of 1½ -in. (38 mm).

In lap joints, the minimum amount of lap shall be five times the thickness of the thinner part joined, but not less than one in. (25 mm). Lap joints joining plates or bars subjected to axial stress that utilize transverse fillet welds only shall be fillet welded along the end of both lapped parts, except where the deflection of the lapped parts is sufficiently restrained to prevent opening of the joint under maximum loading.

Fillet weld terminations are permitted to extend to the ends or sides of parts or be stopped short or boxed except as limited by the following:

(1) For lap joints in which one part extends beyond an edge subject to calculated tensile stress, fillet welds shall terminate not less than the size of the weld from that edge.

(2) For connections and structural elements with cyclic forces, normal to outstanding legs, of frequency and magnitude that would tend to cause a progressive fatigue failure initiating from a point of maximum stress at the end of the weld, fillet welds shall be returned around the corner for a distance not less than the smaller of two times the weld size or the width of the part.

(3) For connections whose design requires flexibility of the outstanding legs, if end returns are used, their length shall not exceed four times the nominal size of the weld.

(4) Fillet welds joining transverse stiffeners to plate girder webs shall end not less than four times nor more than six times the thickness of the web from the web toe of the web-to-flange welds, except where the ends of stiffeners are welded to the flange.

(5) Fillet welds, which occur on opposite sides of a common plane, shall be interrupted at the corner common to both welds.

Fillet welds in holes or slots may be used to transmit shear in lap joints or to prevent the buckling or separation of lapped parts and to join components of built-up members. Such fillet welds may overlap, subject to the provisions of Section J2. Fillet welds in holes or slots are not to be considered plug or slot welds.

3. Plug and Slot Welds

3a. Effective Area

The effective shearing area of plug and slot welds shall be considered as the nominal cross-sectional area of the hole or slot in the plane of the faying surface.

3b. Limitations

Plug or slot welds are permitted to be used to transmit shear in lap joints or to prevent buckling of lapped parts and to join component parts of built-up members.

The diameter of the holes for a plug weld shall not be less than the thickness of the part containing it plus $\frac{5}{16}$-in. (8 mm), rounded to the next larger odd $\frac{1}{16}$-in. (even mm), nor greater than the minimum diameter plus $\frac{1}{8}$-in.(3 mm) or $2\frac{1}{4}$ times the thickness of the weld.

The minimum center-to-center spacing of plug welds shall be four times the diameter of the hole.

The length of slot for a slot weld shall not exceed 10 times the thickness of the weld. The width of the slot shall be not less than the thickness of the part containing it plus $\frac{5}{16}$-in. (8 mm) rounded to the next larger odd $\frac{1}{16}$-in. (even mm), nor shall it be larger than $2\frac{1}{4}$ times the thickness of the weld. The ends of the slot shall be semicircular or shall have the corners rounded to a radius of not less than the thickness of the part containing it, except those ends which extend to the edge of the part.

The minimum spacing of lines of slot welds in a direction transverse to their length shall be four times the width of the slot. The minimum center-to-center spacing in a longitudinal direction on any line shall be two times the length of the slot.

The thickness of plug or slot welds in material $\frac{5}{8}$-in. (16 mm) or less in thickness shall be equal to the thickness of the material. In material over $\frac{5}{8}$-in. (16 mm) thick, the thickness of the weld shall be at least one-half the thickness of the material but not less than $\frac{5}{8}$-in. (16 mm).

4. Design Strength

The design strength of welds shall be the lower value of (a) $\phi F_{BM} A_{BM}$ and (b) $\phi F_w A_w$, when applicable. The values of ϕ, F_{BM}, and F_w and limitations thereon are given in Table J2.5,

where

F_{BM} = nominal strength of the base material, ksi (MPa)
F_w = nominal strength of the weld electrode, ksi (MPa)
A_{BM} = cross-sectional area of the base material, in.^{2} (mm^2)
A_w = effective cross-sectional area of the weld, in.^{2} (mm^2)
ϕ = resistance factor

Alternatively, fillet welds loaded in-plane are permitted to be designed in accordance with Appendix J2.4.

TABLE J2.5
Design Strength of Welds

Types of Weld and Stress [a]	Material	Resistance Factor ϕ	Nominal Strength F_{BM} or F_w	Filler Metal Requirements [b, c]
Complete-Joint-Penetration Groove Weld				
Tension normal to effective area	Base	0.90	F_y	Matching filler metal shall be used. For CVN requirements see footnote [d].
Compression normal to effective area	Base	0.90	F_y	Filler metal with a strength level equal to or less than matching filler metal is permitted to be used.
Tension or compression parallel to axis of weld				
Shear on effective area	Base Weld	0.90 0.80	$0.60F_y$ $0.60F_{EXX}$	
Partial-Joint-Penetration Groove Weld				
Compression normal to effective area	Base	0.90	F_y	Filler metal with a strength level equal to or less than matching filler metal is permitted to be used.
Tension or compression parallel to axis of weld [e]				
Shear parallel to axis of weld	Base Weld	[f] 0.75	[f] $0.60F_{EXX}$	
Tension normal to effective area	Base Weld	0.90 0.80	F_y $0.60F_{EXX}$	
Fillet Welds				
Shear on effective area	Base Weld	[f] 0.75	[f] $0.60F_{EXX}$ [g]	Filler metal with a strength level equal to or less than matching filler metal is permitted to be used.
Tension or compression parallel to axis of weld [e]	Base	0.90	F_y	
Plug or Slot Welds				
Shear parallel to faying surfaces (on effective area)	Base Weld	[f] 0.75	[f] $0.60F_{EXX}$	Filler metal with a strength level equal to or less than matching filler metal is permitted to be used.

[a] For definition of effective area, see Section J2.
[b] For matching filler metal, see Table 3.1, AWS D1.1.
[c] Filler metal one strength level stronger than matching filler metal is permitted.
[d] For T and corner joints with the backing bar left in place during service, filler metal with a classification requiring a minimum Charpy V-notch (CVN) toughness of 20 ft-lbs. (27 J) @ +40°F (4°C) shall be used. If filler metal without the required toughness is used and the backing bar is left in place, the joint shall be sized using the resistance factor and nominal strength for a partial-joint-penetration weld.
[e] Fillet welds and partial-joint-penetration groove welds joining component elements of built-up members, such as flange-to-web connections, are not required to be designed with regard to the tensile or compressive stress in these elements parallel to the axis of the welds.
[f] The design of connected material is governed by Sections J4 and J5.
[g] For alternative design strength, see Appendix J2.4.

5. Combination of Welds

If two or more of the general types of welds (groove, fillet, plug, slot) are combined in a single joint, the design strength of each shall be separately computed with reference to the axis of the group in order to determine the design strength of the combination.

6. Weld Metal Requirements

The choice of electrode for use with complete-joint-penetration groove welds subject to tension normal to the effective area shall comply with the requirements for matching weld metals given in AWS D1.1.

Weld metal with a specified Charpy V-notch (CVN) toughness of 20 ft-lbs (27 J) at 40°F (4°C) shall be used in the following joints:

(a) Complete-joint-penetration groove welded T and corner joints with steel backing left in place, subject to tension normal to the effective area, unless the joints are designed as noted in Table J2.5 (see footnote d).

(b) Complete-joint-penetration groove welded splices subject to tension normal to the effective area in Group 4 and Group 5 shapes and shapes built up by welding plates more than two in. (50 mm) thick.

The manufacturer's Certificate of Conformance shall be sufficient evidence of compliance.

7. Mixed Weld Metal

When notch-toughness is specified, the process consumables for all weld metal, tack welds, root pass, and subsequent passes deposited in a joint shall be compatible to assure notch-tough composite weld metal.

8. Preheat for Heavy Shapes

For ASTM A6/A6M Group 4 and 5 shapes and welded built-up members made of plates more than two in. (50 mm) thick, a preheat equal to or greater than 350°F (175°C) shall be used when making groove-weld splices.

J3. BOLTS AND THREADED PARTS

1. High-Strength Bolts

Use of high-strength bolts shall conform to the provisions of the *Load and Resistance Factor Design Specification for Structural Joints Using ASTM A325 or A490 Bolts*, as approved by the Research Council on Structural Connections, except as otherwise provided in this Specification.

If required to be tightened to more than 50 percent of their specified minimum tensile strength, A449 bolts in tension and bearing-type shear connections shall have an ASTM F436 hardened washer installed under the bolt head, and the nuts shall meet the requirements of ASTM A563. When assembled, all joint surfaces, including those adjacent to the washers, shall be free of scale, except tight mill scale. All A325 or A325M and A490 or A490M bolts shall be tightened to a bolt tension not less than that given in Table J3.1 or J3.1M, except as noted below. Tightening shall be done by any of the following methods: turn-of-nut method, a direct tension indicator, calibrated wrench, or alternative design bolt.

Bolts need only be tightened to the snug-tight condition when in: (a) bearing-type connections where slip is permitted, or (b) tension or combined shear and tension applications, for ASTM A325 or A325M bolts only, where loosening or fatigue due to vibration or load fluctuations are not design considerations. The snug-tight condition is defined as the tightness attained by either a few impacts of an impact wrench or the full effort of a worker with an ordinary spud wrench that brings the connected plies into firm contact. The nominal strength value given in Table J3.2 and Table J3.5 shall be used for bolts tightened to the snug-tight condition. Bolts tightened only to the snug-tight condition shall be clearly identified on the design and erection drawings.

When A490 or A490M bolts over one in. (25 mm) in diameter are used in slotted or oversize holes in external plies, a single hardened washer conforming to ASTM F436, except with 5/16-in. (8 mm) minimum thickness, shall be used in lieu of the standard washer.

In slip-critical connections in which the direction of loading is toward an edge of a connected part, adequate design bearing strength shall be provided based upon the applicable requirements of Section J3.10.

2. Size and Use of Holes

The *maximum sizes* of holes for rivets and bolts are given in Table J3.3 or J3.3M, except that larger holes, required for tolerance on location of anchor rods in concrete foundations, are allowed in column base details.

Standard holes shall be provided in member-to-member connections, unless oversized, short-slotted, or long-slotted holes in bolted connections are approved by the designer. Finger shims up to 1/4-in. (6 mm) are permitted in slip-critical connections designed on the basis of standard holes without reducing the nominal shear strength of the fastener to that specified for slotted holes.

Oversized holes are allowed in any or all plies of slip-critical connections, but they shall not be used in bearing-type connections. Hardened washers shall be installed over oversized holes in an outer ply.

Short-slotted holes are allowed in any or all plies of slip-critical or bearing-type connections. The slots are permitted to be used without regard to direction of loading in slip-critical connections, but the length shall be normal to the direction of the load in bearing-type connections. Washers shall be installed over short-slotted holes in an outer ply; when high-strength bolts are used, such washers shall be hardened.

Long-slotted holes are allowed in only one of the connected parts of either a slip-critical or bearing-type connection at an individual faying surface. Long-slotted holes are permitted to be used without regard to direction of loading in slip-critical connections, but shall be normal to the direction of load in bearing-type connections. Where long-slotted holes are used in an outer ply, plate washers, or a continuous bar with standard holes, having a size sufficient to completely cover the slot after installation, shall be provided. In high-strength bolted connections, such plate washers or continuous bars shall be not less than 5/16-in. (8 mm) thick and shall be of structural grade material, but need not be hardened. If hardened washers are required for use of high-strength bolts, the hardened washers shall be placed over the outer surface of the plate washer or bar.

TABLE J3.1
Minimum Bolt Pretension, kips*

Bolt Size, in.	A325 Bolts	A490 Bolts
½	12	15
⅝	19	24
¾	28	35
⅞	39	49
1	51	64
1⅛	56	80
1¼	71	102
1⅜	85	121
1½	103	148

*Equal to 0.70 of minimum tensile strength of bolts, rounded off to nearest kip, as specified in ASTM specifications for A325 and A490 bolts with UNC threads.

TABLE J3.1M
Minimum Bolt Pretension, kN*

Bolt Size, mm	A325M Bolts	A490M Bolts
M16	91	114
M20	142	179
M22	176	221
M24	205	257
M27	267	334
M30	326	408
M36	475	595

*Equal to 0.70 of minimum tensile strength of bolts, rounded off to nearest kN, as specified in ASTM specifications for A325M and A490M bolts with UNC threads.

3. Minimum Spacing

The distance between centers of standard, oversized, or slotted holes, shall not be less than 2⅔ times the nominal diameter of the fastener; a distance of $3d$ is preferred. Refer to Section J3.10 for bearing strength requirements.

4. Minimum Edge Distance

The distance from the center of a standard hole to an edge of a connected part shall not be less than either the applicable value from Table J3.4 or J3.4M, or as required in Section J3.10. The distance from the center of an oversized or slotted hole to an edge of a connected part shall be not less than that required for a standard hole to an edge of a connected part plus the applicable increment C_2 from Table J3.6 or J3.6M. Refer to Section J3.10 for bearing strength requirements.

5. Maximum Spacing and Edge Distance

The maximum distance from the center of any bolt or rivet to the nearest edge of parts in contact shall be 12 times the thickness of the connected part under consideration, but shall not exceed six in. (150 mm). The longitudinal spacing of connectors between elements in continuous contact consisting of a plate and a shape or two plates shall be as follows:

TABLE J3.2
Design Strength of Fasteners

Description of Fasteners	Tensile Strength: Resistance Factor φ	Tensile Strength: Nominal Strength, ksi (MPa)	Shear Strength in Bearing-type Connections: Resistance Factor φ	Shear Strength in Bearing-type Connections: Nominal Strength, ksi (MPa)
A307 bolts	0.75	45 (310) [a]	0.75	24 (165) [b,e]
A325 or A325M bolts, when threads are not excluded from shear planes		90 (620) [d]		48 (330) [e]
A325 or A325M bolts, when threads are excluded from shear planes		90 (620) [d]		60 (414) [e]
A490 or A490M bolts, when threads are not excluded from shear planes		113 (780) [d]		60 (414) [e]
A490 or A490M bolts, when threads are excluded from shear planes		113 (780) [d]		75 (520) [e]
Threaded parts meeting the requirements of Section A3, when threads are not excluded from shear planes		$0.75F_u$ [a,c]		$0.40F_u$
Threaded parts meeting the requirements of Section A3, when threads are excluded from shear planes		$0.75F_u$ [a,c]		$0.50F_u$ [a,c]
A502, Gr. 1, hot-driven Rivets		45 (310) [a]		25 (172) [e]
A502, Gr. 2 & 3, hot-driven Rivets		60 (414) [a]		33 (228) [e]

[a] Static loading only.
[b] Threads permitted in shear planes.
[c] The nominal tensile strength of the threaded portion of an upset rod, based upon the cross-sectional area at its major thread diameter, A_D shall be larger than the nominal body area of the rod before upsetting times F_y.
[d] For A325 or A325M and A490 or A490M bolts subject to tensile fatigue loading, see Appendix K3.
[e] When bearing-type connections used to splice tension members have a fastener pattern whose length, measured parallel to the line of force, exceeds 50 in. (1 270 mm), tabulated values shall be reduced by 20 percent.

(a) For painted members or unpainted members not subject to corrosion, the spacing shall not exceed 24 times the thickness of the thinner plate or 12 in. (305 mm).

(b) For unpainted members of weathering steel subject to atmospheric corrosion, the spacing shall not exceed 14 times the thickness of the thinner plate or seven-in. (180 mm).

TABLE J3.3
Nominal Hole Dimensions, in.

Bolt Diameter	Hole Dimensions			
	Standard (Dia.)	Oversize (Dia.)	Short-slot (Width × Length)	Long-slot (Width × Length)
$\frac{1}{2}$	$\frac{9}{16}$	$\frac{5}{8}$	$\frac{9}{16} \times \frac{11}{16}$	$\frac{9}{16} \times 1\frac{1}{4}$
$\frac{5}{8}$	$\frac{11}{16}$	$\frac{13}{16}$	$\frac{11}{16} \times \frac{7}{8}$	$\frac{11}{16} \times 1\frac{9}{16}$
$\frac{3}{4}$	$\frac{13}{16}$	$\frac{15}{16}$	$\frac{13}{16} \times 1$	$\frac{13}{16} \times 1\frac{7}{8}$
$\frac{7}{8}$	$\frac{15}{16}$	$1\frac{1}{16}$	$\frac{15}{16} \times 1\frac{1}{8}$	$\frac{15}{16} \times 2\frac{3}{16}$
1	$1\frac{1}{16}$	$1\frac{1}{4}$	$1\frac{1}{16} \times 1\frac{5}{16}$	$1\frac{1}{16} \times 2\frac{1}{2}$
$\geq 1\frac{1}{8}$	$d + \frac{1}{16}$	$d + \frac{5}{16}$	$(d + \frac{1}{16}) \times (d + \frac{3}{8})$	$(d + \frac{1}{16}) \times (2.5 \times d)$

TABLE J3.3M
Nominal Hole Dimensions, mm

Bolt Diameter	Hole Dimensions			
	Standard (Dia.)	Oversize (Dia.)	Short-slot (Width × Length)	Long-slot (Width × Length)
M16	18	20	18 × 22	18 × 40
M20	22	24	22 × 26	22 × 50
M22	24	28	24 × 30	24 × 55
M24	27 [a]	30	27 × 32	27 × 60
M27	30	35	30 × 37	30 × 67
M30	33	38	33 × 40	33 × 75
≥ M36	$d + 3$	$d + 8$	$(d + 3) \times (d + 10)$	$(d + 3) \times 2.5d$

[a] Clearance provided allows the use of a 1 in. bolt if desirable.

6. Design Tension or Shear Strength

The design tension or shear strength of a high-strength bolt or threaded part is $\phi F_n A_b$,

where

ϕ = resistance factor tabulated in Table J3.2

F_n = nominal tensile strength F_t, or shear strength, F_v, tabulated in Table J3.2, ksi (MPa)

A_b = nominal unthreaded body area of bolt or threaded part (for upset rods, see Footnote c, Table J3.2), in.^{2} (mm^2)

The applied load shall be the sum of the factored loads and any tension resulting from prying action produced by deformation of the connected parts.

TABLE J3.4
Minimum Edge Distance,[a] in., From Center of Standard Hole[b] to Edge of Connected Part

Nominal Rivet or Bolt Diameter (in.)	At Sheared Edges	At Rolled Edges of Plates, Shapes or Bars, or Gas Cut Edges [c]
½	⅞	¾
⅝	1⅛	⅞
¾	1¼	1
⅞	1½ [d]	1⅛
1	1¾ [d]	1¼
1⅛	2	1½
1¼	2¼	1⅝
Over 1¼	1¾ × Diameter	1¼ × Diameter

[a] Lesser edge distances are permitted to be used provided Equations from Section J3.10, as appropriate, are satisfied.
[b] For oversized or slotted holes, see Table J3.6.
[c] All edge distances in this column are permitted to be reduced ⅛-in. when the hole is at a point where stress does not exceed 25 percent of the maximum design strength in the element.
[d] These are permitted to be 1¼-in. at the ends of beam connection angles and shear end plates.

TABLE J3.4M
Minimum Edge Distance,[a] mm, From Center of Standard Hole[b] to Edge of Connected Part

Nominal Rivet or Bolt Diameter (mm)	At Sheared Edges	At Rolled Edges of Plates, Shapes or Bars, or Gas Cut Edges [c]
16	28	22
20	34	26
22	38 [d]	28
24	42 [d]	30
27	48	34
30	52	38
36	64	46
Over 36	1.75d	1.25d

[a] Lesser edge distances are permitted to be used provided Equations from Section J3.10, as appropriate, are satisfied.
[b] For oversized or slotted holes, see Table J3.6M.
[c] All edge distances in this column are permitted to be reduced 3 mm when the hole is at a point where stress does not exceed 25 percent of the maximum design strength in the element.
[d] These are permitted to be 32 mm at the ends of beam connection angles and shear end plates.

7. Combined Tension and Shear in Bearing-Type Connections

The design strength of a bolt or rivet subject to combined tension and shear is $\phi F_t A_b$,

where

$\phi = 0.75$

F_t = nominal tension stress computed from the equations in Table J3.5 as a function of f_v, the required shear stress produced by the factored loads. Alternately, the use of the equations in Table A-J3.1 in Appendix J is permitted. The design shear strength ϕF_v, tabulated in Table J3.2, shall equal or exceed the shear stress, f_v.

8. High-Strength Bolts in Slip-Critical Connections

The design for shear of high-strength bolts in slip-critical connections shall be in accordance with either Section J3.8a or J3.8b and checked for shear in accordance with Sections J3.6 and J3.7 and bearing in accordance with Sections J3.1 and J3.10.

8a. Slip-Critical Connections Designed at Factored Loads

The design slip resistance per bolt, ϕr_{str}, shall equal or exceed the required force per bolt due to factored loads,

where

$$r_{str} = 1.13\mu T_b N_s \quad \text{(J3-1)}$$

T_b = minimum fastener tension given in Table J3.1 or J3.1M, kips (kN)

N_s = number of slip planes

μ = mean slip coefficient for Class A, B, or C surfaces, as applicable, or as established by tests

(a) For Class A surfaces (unpainted clean mill scale steel surfaces or surfaces with Class A coatings on blast-cleaned steel),

$$\mu = 0.33$$

(b) For Class B surfaces (unpainted blast-cleaned steel surfaces or surfaces with Class B coatings on blast-cleaned steel),

$$\mu = 0.50$$

(c) For Class C surfaces (hot-dip galvanized and roughened surfaces),

$$\mu = 0.35$$

ϕ = resistance factor

(a) For standard holes,

$$\phi = 1.0$$

(b) For oversized and short-slotted holes,

$$\phi = 0.85$$

(c) For long-slotted holes transverse to the direction of load,

$$\phi = 0.70$$

(d) For long-slotted holes parallel to the direction of load,

$$\phi = 0.60$$

Finger shims up to ¼-in. (6 mm) are permitted to be introduced into slip-critical

TABLE J3.5
Nominal Tension Stress (F_t), ksi (MPa)
Fasteners in Bearing-type Connections

Description of Fasteners	Threads Included in the Shear Plane	Threads Excluded from the Shear Plane
A307 bolts	$59 - 2.5f_v \leq 45$ $(171 - 2.5f_v \leq 310)$	
A325 bolts A325M bolts	$117 - 2.5f_v \leq 90$ $(807 - 2.5f_v \leq 621)$	$117 - 2.0f_v \leq 90$ $(807 - 2.0f_v \leq 621)$
A490 bolts A490M bolts	$147 - 2.5f_v \leq 113$ $(1010 - 2.5f_v \leq 779)$	$147 - 2.0f_v \leq 113$ $(1010 - 2.0f_v \leq 779)$
Threaded parts A449 bolts over 1½ diameter	$0.98F_u - 2.5f_v \leq 0.75F_u$	$0.98F_u - 2.0f_v \leq 0.75F_u$
A502 Gr. 1 rivets	$59 - 2.4f_v \leq 45$ $(407 - 2.4f_v \leq 310)$	
A502 Gr. 2 rivets	$78 - 2.4f_v \leq 60$ $(538 - 2.4f_v \leq 414)$	

TABLE J3.6
Values of Edge Distance Increment C_2, in.

Nominal Diameter of Fastener (in.)	Oversized Holes	Slotted Holes		
		Long Axis Perpendicular to Edge		Long Axis Parallel to Edge
		Short Slots	Long Slots [a]	
≤ 7/8	1/16	1/8	¾d	0
1	1/8	1/8		
≥ 1⅛	1/8	3/16		

[a] When length of slot is less than maximum allowable (see Table J3.5), C_2 are permitted to be reduced by one-half the difference between the maximum and actual slot lengths.

TABLE J3.6M
Values of Edge Distance Increment C_2, mm

Nominal Diameter of Fastener (mm)	Oversized Holes	Slotted Holes		
		Long Axis Perpendicular to Edge		Long Axis Parallel to Edge
		Short Slots	Long Slots [a]	
≤ 22	2	3	0.75d	0
24	3	3		
≥ 27	3	5		

[a] When length of slot is less than maximum allowable (see Table J3.5), C_2 are permitted to be reduced by one-half the difference between the maximum and actual slot lengths.

connections designed on the basis of standard holes without reducing the design shear stress of the fastener to that specified for slotted holes.

8b. Slip-Critical Connections Designed at Service Loads

See Appendix J3.8b.

9. Combined Tension and Shear in Slip-Critical Connections

The design of slip-critical connections subject to tensile forces shall be in accordance with either Sections J3.9a and J3.8a or Sections J3.9b and J3.8b.

9a. Slip-Critical Connections Designed at Factored Loads

When a slip-critical connection is subjected to an applied tension T_u that reduces the net clamping force, the slip resistance ϕr_{str} according to Section J3.8a, shall be multiplied by the following factor:

$$1-\frac{T_u}{(1.13T_bN_b)}$$

where

T_b= minimum bolt pre-tension from Table J3.1or J3.1M, kips (kN)
N_b= number of bolts carrying factored-load tension T_u

9b. Slip-Critical Connections Designed at Service Loads

See Appendix J3.9b.

10. Bearing Strength at Bolt Holes

Bearing strength shall be checked for both bearing-type and slip-critical connections. The use of oversized holes and short- and long-slotted holes parallel to the line of force is restricted to slip-critical connections per Section J3.2.

The design bearing strength at bolt holes is ϕR_n,

where

ϕ = 0.75

and R_n is determined as follows:

(a) For a bolt in a connection with standard, oversized, and short-slotted holes independent of the direction of loading, or a long-slotted hole with the slot parallel to the direction of the bearing force:

when deformation at the bolt hole at service load is a design consideration:

$$R_n=1.2L_ctF_u \le 2.4dtF_u \tag{J3-2a}$$

when deformation at the bolt hole at service load is not a design consideration:

$$R_n = 1.5L_ctF_u \le 3.0dtF_u \tag{J3-2b}$$

(b) For a bolt in a connection with long-slotted holes with the slot perpendicular to the direction of force:

$$R_n=1.0L_ctF_u \le 2.0dtF_u \tag{J3-2c}$$

In the foregoing,

R_n = nominal bearing strength of the connected material, kips (N)
F_u = specified minimum tensile strength of the connected material, ksi (MPa)
L_c = clear distance, in the direction of the force, between the edge of the hole and the edge of the adjacent hole or edge of the material, in. (mm)
d = nominal bolt diameter, in. (mm)
t = thickness of connected material, in. (mm)

For connections, the bearing resistance shall be taken as the sum of the bearing resistances of the individual bolts.

11. Long Grips

A307 bolts providing design strength, and for which the grip exceeds five diameters, shall have their number increased one percent for each additional $\frac{1}{16}$-in. (2 mm) in the grip.

J4. DESIGN RUPTURE STRENGTH

1. Shear Rupture Strength

The design strength for the limit state of rupture along a shear failure path in the affected elements of connected members shall be taken as ϕR_n

where

$\phi = 0.75$
$$R_n = 0.6 F_u A_{nv} \tag{J4-1}$$
A_{nv} = net area subject to shear, in.^{2} (mm^2)

2. Tension Rupture Strength

The design strength for the limit state of rupture along a tension path in the affected elements of connected members shall be taken as ϕR_n

where

$\phi = 0.75$
$$R_n = F_u A_{nt} \tag{J4-2}$$
A_{nt} = net area subject to tension, in.^{2} (mm^2)

3. Block Shear Rupture Strength

Block shear is a limit state in which the resistance is determined by the sum of the shear strength on a failure path(s) and the tensile strength on a perpendicular segment. It shall be checked at beam end connections where the top flange is coped and in similar situations, such as tension members and gusset plates. When ultimate rupture strength on the net section is used to determine the resistance on one segment, yielding on the gross section shall be used on the perpendicular segment. The block shear rupture design strength, ϕR_n, shall be determined as follows:

(a) When $F_u A_{nt} \geq 0.6 F_u A_{nv}$:

$$\phi R_n = \phi[0.6 F_y A_{gv} + F_u A_{nt}] \leq \phi[0.6 F_u A_{nv} + F_u A_{nt}] \tag{J4-3a}$$

(b) When $F_u A_{nt} < 0.6 F_u A_{nv}$:

$$\phi R_n = \phi[0.6F_u A_{nv} + F_y A_{gt}] \leq \phi[0.6F_u A_{nv} + F_u A_{nt}] \tag{J4-3b}$$

where

ϕ = 0.75
A_{gv} = gross area subject to shear, in.^{2} (mm^2)
A_{gt} = gross area subject to tension, in.^{2} (mm^2)
A_{nv} = net area subject to shear, in.^{2} (mm^2)
A_{nt} = net area subject to tension, in.^{2} (mm^2)

J5. CONNECTING ELEMENTS

This section applies to the design of connecting elements, such as plates, gussets, angles, brackets, and the panel zones of beam-to-column connections.

1. Eccentric Connections

Intersecting axially stressed members shall have their gravity axis intersect at one point, if practicable; if not, provision shall be made for bending and shearing stresses due to the eccentricity. Also see Section J1.8.

2. Design Strength of Connecting Elements in Tension

The design strength, ϕR_n, of welded, bolted, and riveted connecting elements statically loaded in tension (e.g., splice and gusset plates) shall be the lower value obtained according to limit states of yielding, rupture of the connecting element, and block shear rupture.

(a) For tension yielding of the connecting element:

$$\phi = 0.90$$

$$R_n = A_g F_y \tag{J5-1}$$

(b) For tension rupture of the connecting element:

$$\phi = 0.75$$

$$R_n = A_n F_u \tag{J5-2}$$

where A_n is the net area, not to exceed $0.85A_g$.

(c) For block shear rupture of connecting elements, see Section J4.3.

3. Other Connecting Elements

For all other connecting elements, the design strength, ϕR_n, shall be determined for the applicable limit state to ensure that the design strength is equal to or greater than the required strength, where R_n is the nominal strength appropriate to the geometry and type of loading on the connecting element. For shear yielding of the connecting element:

$$\phi = 0.90$$

$$R_n = 0.60 A_g F_y \tag{J5-3}$$

If the connecting element is in compression an appropriate limit state analysis shall be made.

J6. FILLERS

In welded construction, any filler ¼-in. (6 mm) or more in thickness shall extend beyond the edges of the splice plate and shall be welded to the part on which it is fitted with sufficient weld to transmit the splice plate load, applied at the surface of the filler. The welds joining the splice plate to the filler shall be sufficient to transmit the splice plate load and shall be long enough to avoid overloading the filler along the toe of the weld. Any filler less than ¼-in. (6 mm) thick shall have its edges made flush with the edges of the splice plate and the weld size shall be the sum of the size necessary to carry the splice plus the thickness of the filler plate.

When a bolt that carries load passes through fillers that are equal to or less than ¼-in. (6 mm) thick, the design shear strength shall be used without reduction. When a bolt that carries load passes through fillers that are greater than ¼-in. (6 mm) thick, one of the following requirements shall apply:

(1) For fillers that are equal to or less than ¾-in. (19 mm) thick, the design shear strength of the bolts shall be multiplied by the factor $[1 - 0.4(t - 0.25)]$ [Metric: $[1 - 0.0154(t - 6)]$], where t is the total thickness of the fillers up to ¾-in. (19 mm).

(2) The fillers shall be extended beyond the joint and the filler extension shall be secured with enough bolts to uniformly distribute the total force in the connected element over the combined cross-section of the connected element and the fillers;

(3) The size of the joint shall be increased to accommodate a number of bolts that is equivalent to the total number required in (2) above; or

(4) The joint shall be designed as a slip-critical joint.

J7. SPLICES

Groove-welded splices in plate girders and beams shall develop the full strength of the smaller spliced section. Other types of splices in cross sections of plate girders and beams shall develop the strength required by the forces at the point of the splice.

J8. BEARING STRENGTH

The strength of surfaces in bearing is ϕR_n,

where

$\phi = 0.75$
R_n is defined below for the various types of bearing

(a) For milled surfaces, pins in reamed, drilled, or bored holes, and ends of fitted bearing stiffeners,

$$R_n = 1.8 F_y A_{pb} \tag{J8-1}$$

where

F_y = specified minimum yield stress, ksi (MPa)
A_{pb} = projected bearing area, in.^{2} (mm^2)

(b) For expansion rollers and rockers,

If $d \leq 25$ in. (635 mm),

$$R_n = 1.2(F_y - 13)ld / 20 \quad \text{(J8-2)}$$

$$(\text{Metric: } R_n = 1.2(F_y - 90)ld / 20) \quad \text{(J8-2M)}$$

If $d > 25$ in. (635 mm),

$$R_n = 6.0(F_y - 13)l\sqrt{d} / 20 \quad \text{(J8-3)}$$

$$(\text{Metric: } R_n = 6.0(F_y - 90)l\sqrt{d} / 20) \quad \text{(J8-3M)}$$

where

d = diameter, in. (mm)
l = length of bearing, in. (mm)

J9. COLUMN BASES AND BEARING ON CONCRETE

Proper provision shall be made to transfer the column loads and moments to the footings and foundations.

In the absence of code regulations, design bearing loads on concrete may be taken as $\phi_c P_p$:

(a) On the full area of a concrete support

$$P_p = 0.85 f_c' A_1 \quad \text{(J9-1)}$$

(b) On less than the full area of a concrete support

$$P_p = 0.85 f_c' A_1 \sqrt{A_2 / A_1} \quad \text{(J9-2)}$$

where

ϕ_c = 0.60
A_1 = area of steel concentrically bearing on a concrete support, in.^{2} (mm^2)
A_2 = maximum area of the portion of the supporting surface that is geometrically similar to and concentric with the loaded area, in.^{2} (mm^2)
$\sqrt{A_2 / A_1} \leq 2$

J10. ANCHOR RODS AND EMBEDMENTS

Steel anchor rods and embedments shall be proportioned to develop the factored load combinations stipulated in Section A4. If the load factors and combinations stipulated in Section A4 are used to design concrete structural elements, the provisions of ACI 318 shall be used with appropriate ϕ factors as given in ACI 318, Appendix C.

CHAPTER K

CONCENTRATED FORCES, PONDING, AND FATIGUE

This chapter covers member strength design considerations pertaining to concentrated forces, ponding, and fatigue.

K1. FLANGES AND WEBS WITH CONCENTRATED FORCES

1. Design Basis

Sections K1.2 through K1.7 apply to single and double concentrated forces as indicated in each Section. A single concentrated force is tensile or compressive. Double concentrated forces, one tensile and one compressive, form a couple on the same side of the loaded member.

Transverse stiffeners are required at locations of concentrated tensile forces in accordance with Section K1.2 for the limit state of flange local bending, and at unframed ends of beams and girders in accordance with Section K1.8. Transverse stiffeners or doubler plates are required at locations of concentrated forces in accordance with Sections K1.3 through K1.6 for the limit states of web local yielding, crippling, sidesway buckling, and compression buckling. Doubler plates or diagonal stiffeners are required in accordance with Section K1.7 for the limit state of web panel-zone shear.

Transverse stiffeners and diagonal stiffeners required by Sections K1.2 through K1.8 shall also meet the requirements of Section K1.9. Doubler plates required by Sections K1.3 through K1.6 shall also meet the requirements of Section K1.10.

2. Flange Local Bending

This Section applies to both tensile single-concentrated forces and the tensile component of double-concentrated forces.

A pair of transverse stiffeners extending at least one-half the depth of the web shall be provided adjacent to a concentrated tensile force centrally applied across the flange when the required strength of the flange exceeds ϕR_n,

where

$$\phi = 0.90$$

$$R_n = 6.25 t_f^{\,2} F_{yf} \qquad \text{(K1-1)}$$

F_{yf} = specified minimum yield stress of the flange, ksi (MPa)
t_f = thickness of the loaded flange, in. (mm)

If the length of loading across the member flange is less than $0.15b$, where b is the member flange width, Equation K1-1 need not be checked.

When the concentrated force to be resisted is applied at a distance from the member end that is less than $10t_f$, R_n shall be reduced by 50 percent.

When transverse stiffeners are required, they shall be welded to the loaded flange to develop the welded portion of the stiffener. The weld connecting transverse stiffeners to the web shall be sized to transmit the unbalanced force in the stiffener to the web. Also, see Section K1.9.

3. Web Local Yielding

This Section applies to single-concentrated forces and both components of double-concentrated forces.

Either a pair of transverse stiffeners or a doubler plate, extending at least one-half the depth of the web, shall be provided adjacent to a concentrated tensile or compressive force when the required strength of the web at the toe of the fillet exceeds ϕR_n,

where

$\phi = 1.0$
and R_n is determined as follows:

(a) When the concentrated force to be resisted is applied at a distance from the member end that is greater than the depth of the member d,

$$R_n = (5k + N)F_{yw}\, t_w \tag{K1-2}$$

(b) When the concentrated force to be resisted is applied at a distance from the member end that is less than or equal to the depth of the member d,

$$R_n = (2.5k + N)F_{yw}\, t_w \tag{K1-3}$$

In Equations K1-2 and K1-3, the following definitions apply:

F_{yw} = specified minimum yield stress of the web, ksi (MPa)
N = length of bearing (not less than k for end beam reactions), in. (mm)
k = distance from outer face of the flange to the web toe of the fillet, in. (mm)
t_w = web thickness, in. (mm)

When required, for a tensile force normal to the flange, transverse stiffeners shall be welded to the loaded flange to develop the connected portion of the stiffener. When required for a compressive force normal to the flange, transverse stiffeners shall either bear on or be welded to the loaded flange to develop the force transmitted to the stiffener. The weld connecting transverse stiffeners to the web shall be sized to transmit the unbalanced force in the stiffener to the web. Also, see Section K1.9.

Alternatively, when doubler plates are required, see Section K1.10.

4. Web Crippling

This Section applies to both compressive single-concentrated forces and the compressive component of double-concentrated forces.

Either a transverse stiffener, a pair of transverse stiffeners, or a doubler plate, extending at least one-half the depth of the web, shall be provided adjacent to a concentrated compressive force when the required strength of the web exceeds ϕR_n,

where

$\phi = 0.75$
and R_n is determined as follows:

(a) When the concentrated compressive force to be resisted is applied at a distance from the member end that is greater than or equal to $d/2$,

$$R_n = 0.80 t_w^2 \left[1 + 3\left(\frac{N}{d}\right)\left(\frac{t_w}{t_f}\right)^{1.5} \right] \sqrt{\frac{EF_{yw} t_f}{t_w}} \tag{K1-4}$$

(b) When the concentrated compressive force to be resisted is applied at a distance from the member end that is less than $d/2$,

For $N/d \le 0.2$,

$$R_n = 0.40 t_w^2 \left[1 + 3\left(\frac{N}{d}\right)\left(\frac{t_w}{t_f}\right)^{1.5} \right] \sqrt{\frac{EF_{yw} t_f}{t_w}} \tag{K1-5a}$$

For $N/d > 0.2$,

$$R_n = 0.40 t_w^2 \left[1 + \left(\frac{4N}{d} - 0.2\right)\left(\frac{t_w}{t_f}\right)^{1.5} \right] \sqrt{\frac{EF_{yw} t_f}{t_w}} \, . \tag{K1-5b}$$

In Equations K1-4 and K1-5, the following definitions apply:

d = overall depth of the member, in. (mm)
t_f = flange thickness, in. (mm)

When transverse stiffeners are required, they shall either bear on or be welded to the loaded flange to develop the force transmitted to the stiffener. The weld connecting transverse stiffeners to the web shall be sized to transmit the unbalanced force in the stiffener to the web. Also, see Section K1.9.

Alternatively, when doubler plates are required, see Section K1.10.

5. Web Sidesway Buckling

This Section applies only to compressive single-concentrated forces applied to members where relative lateral movement between the loaded compression flange and the tension flange is not restrained at the point of application of the concentrated force.

The design strength of the web is ϕR_n,

where

$\phi = 0.85$
and R_n is determined as follows:

(a) If the compression flange is restrained against rotation:

For $(h/t_w)/(l/b_f) \le 2.3$,

$$R_n = \frac{C_r t_w^3 t_f}{h^2}\left[1+0.4\left(\frac{h/t_w}{l/b_f}\right)^3\right] \tag{K1-6}$$

for $(h / t_w) / (l / b_f) > 2.3$, the limit state of sidesway web buckling does not apply.

When the required strength of the web exceeds ϕR_n, local lateral bracing shall be provided at the tension flange or either a pair of transverse stiffeners or a doubler plate, extending at least one-half the depth of the web, shall be provided adjacent to the concentrated compressive force.

When transverse stiffeners are required, they shall either bear on or be welded to the loaded flange to develop the full-applied force. The weld connecting transverse stiffeners to the web shall be sized to transmit the force in the stiffener to the web. Also, see Section K1.9.

Alternatively, when doubler plates are required, they shall be sized to develop the full-applied force. Also, see Section K1.10.

(b) If the compression flange is *not* restrained against rotation:

For $(h / t_w) / (l / b_f) \leq 1.7$,

$$R_n = \frac{C_r t_w^3 t_f}{h^2}\left[0.4\left(\frac{h/t_w}{l/b_f}\right)^3\right] \tag{K1-7}$$

for $(h / t_w) / (l / b_f) > 1.7$, the limit state of sidesway web buckling does not apply.

When the required strength of the web exceeds ϕR_n, local lateral bracing shall be provided at both flanges at the point of application of the concentrated forces.

In Equations K1-6 and K1-7, the following definitions apply:

l = largest laterally unbraced length along either flange at the point of load, in. (mm)
b_f = flange width, in. (mm)
t_f = flange thickness, in. (mm)
t_w = web thickness, in. (mm)
h = clear distance between flanges less the fillet or corner radius for rolled shapes; distance between adjacent lines of fasteners or the clear distance between flanges when welds are used for built-up shapes, in. (mm)
C_r = 960,000 ksi (6.62x10^6 MPa) when $M_u < M_y$ at the location of the force
= 480,000 ksi (3.31x10^6 MPa) when $M_u \geq M_y$ at the location of the force

6. Web Compression Buckling

This Section applies to a pair of compressive single-concentrated forces or the compressive components in a pair of double-concentrated forces, applied at both flanges of a member at the same location.

Either a single transverse stiffener, or pair of transverse stiffeners, or a doubler plate, extending the full depth of the web, shall be provided adjacent to concentrated compressive forces at both flanges when the required strength of the web exceeds ϕR_n,

where

$$\phi = 0.90$$

and

$$R_n = \frac{24t_w^3\sqrt{EF_{yw}}}{h} \tag{K1-8}$$

When the pair of concentrated compressive forces to be resisted is applied at a distance from the member end that is less than $d/2$, R_n shall be reduced by 50 percent.

When transverse stiffeners are required, they shall either bear on or be welded to the loaded flange to develop the force transmitted to the stiffener. The weld connecting transverse stiffeners to the web shall be sized to transmit the unbalanced force in the stiffener to the web. Also, see Section K1.9.

Alternatively, when doubler plates are required, see Section K1.10.

7. Web Panel-Zone Shear

Either doubler plates or diagonal stiffeners shall be provided within the boundaries of the rigid connection of members whose webs lie in a common plane when the required strength exceeds ϕR_v,

where

$\phi = 0.90$
and R_v is determined as follows:

(a) When the effect of panel-zone deformation on frame stability is *not* considered in the analysis,

For $P_u \leq 0.4P_y$

$$R_v = 0.60F_y\, d_c\, t_w \tag{K1-9}$$

For $P_u > 0.4P_y$

$$R_v = 0.60F_y d_c t_w \left(1.4 - \frac{P_u}{P_y}\right) \tag{K1-10}$$

(b) When frame stability, including plastic panel-zone deformation, is considered in the analysis:

For $P_u \leq 0.75P_y$

$$R_v = 0.60F_y d_c t_w \left(1 + \frac{3b_{cf}t_{cf}^{\;2}}{d_b d_c t_w}\right) \tag{K1-11}$$

For $P_u > 0.75P_y$

$$R_v = 0.60F_y d_c t_w \left(1 + \frac{3b_{cf}t_{cf}^2}{d_b d_c t_w}\right)\left(1.9 - \frac{1.2P_u}{P_y}\right) \tag{K1-12}$$

In Equations K1-9 through K1-12, the following definitions apply:

t_w = column web thickness, in. (mm)
b_{cf}= width of column flange, in. (mm)
t_{cf} = thickness of the column flange, in. (mm)
d_b = beam depth, in. (mm)
d_c = column depth, in. (mm)
F_y = yield strength of the column web, ksi (MPa)
P_y = F_yA, axial yield strength of the column, kips (N)
A = column cross-sectional area, in.^{2} (mm^2)

When doubler plates are required, they shall meet the criteria of Section F2 and shall be welded to develop the proportion of the total shear force which is to be carried.

Alternatively, when diagonal stiffeners are required, the weld connecting diagonal stiffeners to the web shall be sized to transmit the stiffener force caused by unbalanced moments to the web. Also, see Section K1.9.

8. Unframed Ends of Beams and Girders

At unframed ends of beams and girders not otherwise restrained against rotation about their longitudinal axes, a pair of transverse stiffeners, extending the full depth of the web, shall be provided. Also, see Section K1.9.

9. Additional Stiffener Requirements for Concentrated Forces

Transverse and diagonal stiffeners shall also comply with the following criteria:

(1) The width of each stiffener plus one-half the thickness of the column web shall not be less than one-third of the width of the flange or moment connection plate delivering the concentrated force.

(2) The thickness of a stiffener shall not be less than one-half the thickness of the flange or moment connection plate delivering the concentrated load, and not less than its width times $1.79\sqrt{F_y / E}$.

Full depth transverse stiffeners for compressive forces applied to a beam or plate girder flange shall be designed as axially compressed members (columns) in accordance with the requirements of Section E2, with an effective length of $0.75h$, a cross section composed of two stiffeners and a strip of the web having a width of $25t_w$ at interior stiffeners and $12t_w$ at the ends of members.

The weld connecting bearing stiffeners to the web shall be sized to transmit the excess web shear force to the stiffener. For fitted bearing stiffeners, see Section J8.

10. Additional Doubler Plate Requirements for Concentrated Forces

Doubler plates required by Sections K1.3 through K1.6 shall also comply with the following criteria:

(1) The thickness and extent of the doubler plate shall provide the additional material necessary to equal or exceed the strength requirements.

(2) The doubler plate shall be welded to develop the proportion of the total force transmitted to the doubler plate.

K2. PONDING

The roof system shall be investigated by structural analysis to assure adequate strength and stability under ponding conditions, unless the roof surface is provided with sufficient slope toward points of free drainage or adequate individual drains to prevent the accumulation of rainwater.

The roof system shall be considered stable and no further investigation is needed if:

$$C_p + 0.9C_s \leq 0.25 \tag{K2-1}$$

$$I_d \geq 25(S^4)10^{-6} \tag{K2-2}$$

$$(\text{Metric: } I_d \geq 3\,940\, S^4) \tag{K2-2M}$$

where

$$C_p = \frac{32 L_s L_p^4}{10^7 I_p}$$

$$\left(\text{Metric: } C_p = \frac{504 L_s L_p^4}{I_p}\right)$$

$$C_s = \frac{32 S L_s^4}{10^7 I_s}$$

$$\left(\text{Metric: } C_s = \frac{504 S L_s^4}{I_s}\right)$$

L_p = column spacing in direction of girder (length of primary members), ft (m)
L_s = column spacing perpendicular to direction of girder (length of secondary members), ft (m)
S = spacing of secondary members, ft (m)
I_p = moment of inertia of primary members, in.^{4} (mm^4)
I_s = moment of inertia of secondary members, in.^{4} (mm^4)
I_d = moment of inertia of the steel deck supported on secondary members, in.^{4} per ft (mm^4 per m)

For trusses and steel joists, the moment of inertia I_s shall be decreased 15 percent when used in the above equation. A steel deck shall be considered a secondary member when it is directly supported by the primary members.

See Appendix K2 for an alternate determination of flat roof framing stiffness.

K3. DESIGN FOR CYCLIC LOADING (FATIGUE)

Few members or connections in conventional buildings need to be designed for fatigue, since most load changes in such structures occur only a small number of times or produce only minor stress fluctuations. The occurrence of full design wind or earthquake loads is too infrequent to warrant consideration in fatigue design. However, crane runways and supporting structures for machinery and equipment are often subject to fatigue loading conditions.

Members and their connections subject to fatigue loading shall be proportioned in accordance with the provisions of Appendix K3 for service loads.

CHAPTER L

SERVICEABILITY DESIGN CONSIDERATIONS

This chapter is intended to provide design guidance for serviceability considerations.

Serviceability is a state in which the function of a building, its appearance, maintainability, durability, and comfort of its occupants are preserved under normal usage. The general design requirement for serviceability is given in Section A5.4. Limiting values of structural behavior to ensure serviceability (e.g., maximum deflections, accelerations, etc.) shall be chosen with due regard to the intended function of the structure. Where necessary, serviceability shall be checked using realistic loads for the appropriate serviceability limit state.

L1. CAMBER

If any special camber requirements are necessary to bring a loaded member into proper relation with the work of other trades, as for the attachment of runs of sash, the requirements shall be set forth in the design documents. If camber involves the erection of any member under a preload, this shall be noted in the design documents.

Beams and trusses detailed without specified camber shall be fabricated so that after erection any camber due to rolling or shop assembly shall be upward.

L2. EXPANSION AND CONTRACTION

Adequate provision shall be made for expansion and contraction appropriate to the service conditions of the structure.

L3. DEFLECTIONS, VIBRATION, AND DRIFT

1. Deflections

Deformations in structural members and structural systems due to service loads shall not impair the serviceability of the structure.

2. Floor Vibration

Vibration shall be considered in designing beams and girders supporting large areas free of partitions or other sources of damping where excessive vibration due to pedestrian traffic or other sources within the building is not acceptable.

3. Drift

Lateral deflection or drift of structures due to code-specified wind or seismic loads shall not cause collision with adjacent structures nor exceed the limiting values of such drifts which may be specified or appropriate.

L4. CONNECTION SLIP

For the design of slip-critical connections see Sections J3.8 and J3.9.

L5. CORROSION

When appropriate, structural components shall be designed to tolerate corrosion or shall be protected against corrosion that may impair the strength or serviceability of the structure.

CHAPTER M

FABRICATION, ERECTION, AND QUALITY CONTROL

This chapter provides requirements for shop drawings, fabrication, shop painting, erection, and quality control.

M1. SHOP DRAWINGS

Shop drawings giving complete information necessary for the fabrication of the component parts of the structure, including the location, type, and size of all welds, bolts, and rivets, shall be prepared in advance of the actual fabrication. These drawings shall clearly distinguish between shop and field welds and bolts and shall clearly identify pretensioned and slip-critical high-strength bolted connections.

Shop drawings shall be made in conformity with good practice and with due regard to speed and economy in fabrication and erection.

M2. FABRICATION

1. Cambering, Curving, and Straightening

Local application of heat or mechanical means is permitted to be used to introduce or correct camber, curvature, and straightness. The temperature of heated areas, as measured by approved methods, shall not exceed 1,100°F (593°C) for A514/A514M and A852/A852M steel nor 1,200°F (649°C) for other steels.

2. Thermal Cutting

Thermally cut edges shall meet the requirements of AWS 5.15.1.2, 5.15.4.3 and 5.15.4.4 with the exception that thermally cut free edges which will be subject to calculated static tensile stress shall be free of round bottom gouges greater than 3/16-in. (5 mm) deep and sharp V-shaped notches. Gouges greater than 3/16-in. (5 mm) deep and notches shall be removed by grinding or repaired by welding.

Re-entrant corners, except re-entrant corners of beam copes and weld access holes, shall meet the requirements of AWS 5.16. If another specified contour is required it must be shown on the contract documents.

Beam copes and weld access holes shall meet the geometrical requirements of Section J1.6. For beam copes and weld access holes in ASTM A6/A6M Group 4 and 5 shapes and welded built-up shapes with material thickness greater than two in. (50 mm), a preheat temperature of not less than 150°F (66°C) shall be applied prior to thermal cutting.

3. Planing of Edges

Planing or finishing of sheared or thermally cut edges of plates or shapes is not required unless specifically called for in the design documents or included in a stipulated edge preparation for welding.

4. Welded Construction

The technique of welding, the workmanship, appearance, and quality of welds and the methods used in correcting nonconforming work shall be in accordance with AWS D1.1 except as modified in Section J2.

5. Bolted Construction

All parts of bolted members shall be pinned or bolted and rigidly held together during assembly. Use of a drift pin in bolt holes during assembly shall not distort the metal or enlarge the holes. Poor matching of holes shall be cause for rejection.

If the thickness of the material is not greater than the nominal diameter of the bolt plus ⅛-in. (3 mm), the holes are permitted to be punched. If the thickness of the material is greater than the nominal diameter of the bolt plus ⅛-in. (3 mm), the holes shall be either drilled or sub-punched and reamed. The die for all sub-punched holes, and the drill for all sub-drilled holes, shall be at least $\frac{1}{16}$-in. (2 mm) smaller than the nominal diameter of the bolt. Holes in ASTM A514/A514M steel plates over ½-in. (13 mm) thick shall be drilled.

Fully-inserted finger shims, with a total thickness of not more than ¼-in. (6 mm) within a joint, are permitted in joints without changing the design strength (based upon hole type) for the design of connections. The orientation of such shims is independent of the direction of application of the load.

The use of high-strength bolts shall conform to the requirements of the RCSC *Load and Resistance Factor Design Specification for Structural Joints Using ASTM A325 or A490 Bolts*.

6. Compression Joints

Compression joints which depend on contact bearing as part of the splice strength shall have the bearing surfaces of individual fabricated pieces prepared by milling, sawing, or other suitable means.

7. Dimensional Tolerances

Dimensional tolerances shall be in accordance with the AISC *Code of Standard Practice*.

8. Finish of Column Bases

Column bases and base plates shall be finished in accordance with the following requirements:

(1) Steel bearing plates two in. (50 mm) or less in thickness are permitted without milling, provided a satisfactory contact bearing is obtained. Steel bearing plates over two in. (50 mm) but not over four in. (100 mm) in thickness are permitted to be straightened by pressing or, if presses are not available, by milling for all bearing surfaces (except as noted in subparagraphs 2 and 3 of this section), to obtain a satisfactory contact bearing. Steel bearing plates over four in. (100 mm) in thickness shall be milled for all bearing surfaces (except as noted in subparagraphs 2 and 3 of this section).

(2) Bottom surfaces of bearing plates and column bases which are grouted to ensure full bearing contact on foundations need not be milled.

(3) Top surfaces of bearing plates need not be milled when complete-joint-penetration groove welds are provided between the column and the bearing plate.

M3. SHOP PAINTING

1. General Requirements

Shop painting and surface preparation shall be in accordance with the provisions of the AISC *Code of Standard Practice*.

Shop paint is not required unless specified by the contract documents.

2. Inaccessible Surfaces

Except for contact surfaces, surfaces inaccessible after shop assembly shall be cleaned and painted prior to assembly, if required by the design documents.

3. Contact Surfaces

Paint is permitted unconditionally in bearing-type connections. For slip-critical connections, the faying surface requirements shall be in accordance with the RCSC *Specification for Structural Joints Using ASTM A325 or A490 Bolts*, paragraph 3(b).

4. Finished Surfaces

Machine-finished surfaces shall be protected against corrosion by a rust inhibitive coating that can be removed prior to erection, or which has characteristics that make removal prior to erection unnecessary.

5. Surfaces Adjacent to Field Welds

Unless otherwise specified in the design documents, surfaces within two in. (50 mm) of any field weld location shall be free of materials that would prevent proper welding or produce objectionable fumes during welding.

M4. ERECTION

1. Alignment of Column Bases

Column bases shall be set level and to correct elevation with full bearing on concrete or masonry.

2. Bracing

The frame of steel skeleton buildings shall be carried up true and plumb within the limits defined in the AISC *Code of Standard Practice*. Temporary bracing shall be provided, in accordance with the requirements of the *Code of Standard Practice*, wherever necessary to support all loads to which the structure may be subjected, including equipment and the operation of same. Such bracing shall be left in place as long as required for safety.

3. Alignment

No permanent bolting or welding shall be performed until the adjacent affected portions of the structure have been properly aligned.

4. Fit of Column Compression Joints and Base Plates

Lack of contact bearing not exceeding a gap of $\frac{1}{16}$-in. (2 mm), regardless of the type of splice used (partial-joint-penetration groove welded, or bolted), is permitted. If the gap exceeds $\frac{1}{16}$-in. (2 mm), but is less than $\frac{1}{4}$-in. (6 mm), and if an engineering investigation shows that sufficient contact area does not exist, the gap shall be packed out with non-tapered steel shims. Shims need not be other than mild steel, regardless of the grade of the main material.

5. Field Welding

Shop paint on surfaces adjacent to joints to be field welded shall be wire brushed if necessary to assure weld quality.

Field welding of attachments to installed embedments in contact with concrete shall be done in such a manner as to avoid excessive thermal expansion of the embedment which could result in spalling or cracking of the concrete or excessive stress in the embedment anchors.

6. Field Painting

Responsibility for touch-up painting, cleaning, and field painting shall be allocated in accordance with accepted local practices, and this allocation shall be set forth explicitly in the design documents.

7. Field Connections

As erection progresses, the structure shall be securely bolted or welded to support all dead, wind, and erection loads.

M5. QUALITY CONTROL

The fabricator shall provide quality control procedures to the extent that the fabricator deems necessary to assure that all work is performed in accordance with this Specification. In addition to the fabricator's quality control procedures, material and workmanship at all times may be subject to inspection by qualified inspectors representing the purchaser. If such inspection by representatives of the purchaser will be required, it shall be so stated in the design documents.

1. Cooperation

As far as possible, all inspection by representatives of the purchaser shall be made at the fabricator's plant. The fabricator shall cooperate with the inspector, permitting access for inspection to all places where work is being done. The purchaser's inspector shall schedule this work for minimum interruption to the work of the fabricator.

2. Rejections

Material or workmanship not in reasonable conformance with the provisions of this Specification may be rejected at any time during the progress of the work.

The fabricator shall receive copies of all reports furnished to the purchaser by the inspection agency.

3. Inspection of Welding

The inspection of welding shall be performed in accordance with the provisions of AWS D1.1 except as modified in Section J2.

When visual inspection is required to be performed by AWS certified welding inspectors, it shall be so specified in the design documents.

When nondestructive testing is required, the process, extent, and standards of acceptance shall be clearly defined in the design documents.

4. Inspection of Slip-Critical High-Strength Bolted Connections

The inspection of slip-critical high-strength bolted connections shall be in accordance with the provisions of the RCSC *Load and Resistance Factor Design Specification for Structural Joints Using ASTM A325 or A490 Bolts*.

5. Identification of Steel

The fabricator shall be able to demonstrate by a written procedure and by actual practice a method of material application and identification, visible at least through the "fit-up" operation, of the main structural elements of a shipping piece.

The identification method shall be capable of verifying proper material application as it relates to:

(1) Material specification designation

(2) Heat number, if required

(3) Material test reports for special requirements

CHAPTER N

EVALUATION OF EXISTING STRUCTURES

This chapter applies to the evaluation of the strength and stiffness under static vertical (gravity) loads of existing structures by structural analysis, by load tests, or by a combination of structural analysis and load tests when specified by the Engineer of Record or in the contract documents. For such evaluation, the steel grades are not limited to those listed in Section A3.1. This chapter does not address load testing for the effects of seismic loads or moving loads (vibrations).

N1. GENERAL PROVISIONS

These provisions shall be applicable when the evaluation of an existing steel structure is specified for (a) verification of a specific set of design loadings or (b) determination of the design strength of a load resisting member or system. The evaluation shall be performed by structural analysis (Section N3), by load tests (Section N4), or by a combination of structural analysis and load tests, as specified in the contract documents. Where load tests are used, the Engineer of Record shall first analyze the structure, prepare a testing plan, and develop a written procedure to prevent excessive permanent deformation or catastrophic collapse during testing.

N2. MATERIAL PROPERTIES

1. Determination of Required Tests

The Engineer of Record shall determine the specific tests that are required from Section N2.2 through N2.6 and specify the locations where they are required. Where available, the use of applicable project records shall be permitted to reduce or eliminate the need for testing.

2. Tensile Properties

Tensile properties of members shall be considered in evaluation by structural analysis (Section N3) or load tests (Section N4). Such properties shall include the yield stress, tensile strength, and percent elongation. Where available, certified mill test reports or certified reports of tests made by the fabricator or a testing laboratory in accordance with ASTM A6/A6M or A568/A568M, as applicable, shall be permitted for this purpose. Otherwise, tensile tests shall be conducted in accordance with ASTM A370 from samples cut from components of the structure.

3. Chemical Composition

Where welding is anticipated for repair or modification of existing structures, the chemical composition of the steel shall be determined for use in preparing a welding procedure specification (WPS). Where available, results from certified mill test reports or certified reports of tests made by the fabricator or a testing laboratory in accordance with ASTM procedures shall be permitted for this purpose. Otherwise, analyses shall be conducted in accordance with ASTM A751 from the samples used to determine tensile properties, or from samples taken from the same locations.

4. Base Metal Notch Toughness

Where welded tension splices in heavy shapes and plates as defined in Section A3.1c are critical to the performance of the structure, the Charpy V-notch toughness shall be determined in accordance with the provisions of Section A3.1c. If the notch toughness so determined does not meet the provisions of Section A3.1c, the Engineer of Record shall determine if remedial actions are required.

5. Weld Metal

Where structural performance is dependent on existing welded connections, representative samples of weld metal shall be obtained. Chemical analysis and mechanical tests shall be made to characterize the weld metal. A determination shall be made of the magnitude and consequences of imperfections. If the requirements of AWS D1.1 are not met, the Engineer of Record shall determine if remedial actions are required.

6. Bolts and Rivets

Representative samples of bolts shall be inspected to determine markings and classifications. Where bolts can not be properly identified visually, representative samples shall be removed and tested to determine tensile strength in accordance with ASTM F606 or ASTM F606M and the bolt classified accordingly. Alternatively, the assumption that the bolts are A307 shall be permitted. Rivets shall be assumed to be A502, Grade 1, unless a higher grade is established through documentation or testing.

N3. EVALUATION BY STRUCTURAL ANALYSIS

1. Dimensional Data

All dimensions used in the evaluation, such as spans, column heights, member spacings, bracing locations, cross section dimensions, thicknesses, and connection details, shall be determined from a field survey. Alternatively, when available, it shall be permitted to determine such dimensions from applicable project design or shop drawings with field verification of critical values.

2. Strength Evaluation

Forces (load effects) in members and connections shall be determined by structural analysis applicable to the type of structure evaluated. The load effects shall be determined for the loads and factored load combinations stipulated in Section A4.

The design strength of members and connections shall be determined from applicable provisions of Chapters B through K of this Specification.

3. Serviceability Evaluation

Where required, the deformations at service loads shall be calculated and reported.

N4. EVALUATION BY LOAD TESTS

1. Determination of Live Load Rating by Testing

To determine the live load rating of an existing floor or roof structure by testing, a test load shall be applied incrementally in accordance with the Engineer of Record's plan. The structure shall be visually inspected for signs of distress or imminent failure at each load level. Appropriate measures shall be taken if these or any other unusual conditions are encountered.

The tested design strength of the structure shall be taken as the maximum applied test load plus the in-situ dead load. The live load rating of a floor structure shall be determined by setting the tested design strength equal to $1.2D + 1.6L$, where D is the nominal dead load and L is the nominal live load rating for the structure. The nominal live load rating of the floor structure shall not exceed that which can be calculated using applicable provisions of the specification. For roof structures, L_r, S, or R as defined in the Symbols, shall be substituted for L. More severe load combinations shall be used where required by applicable building codes.

Periodic unloading shall be considered once the service load level is attained and after the onset of inelastic structural behavior is identified to document the amount of permanent set and the magnitude of the inelastic deformations. Deformations of the structure, such as member deflections, shall be monitored at critical locations during the test, referenced to the initial position before loading. It shall be demonstrated, while maintaining maximum test load for one hour, that the deformation of the structure does not increase by more than 10 percent above that at the beginning of the holding period. It is permissible to repeat the sequence if necessary to demonstrate compliance.

Deformations of the structure shall also be recorded 24 hours after the test loading is removed to determine the amount of permanent set. Because the amount of acceptable permanent deformation depends on the specific structure, no limit is specified for permanent deformation at maximum loading. Where it is not feasible to load test the entire structure, a segment or zone of not less than one complete bay, representative of the most critical conditions, shall be selected.

2. Serviceability Evaluation

When load tests are prescribed, the structure shall be loaded incrementally to the service load level. Deformations shall be monitored for a period of one hour. The structure shall then be unloaded and the deformation recorded.

N5. EVALUATION REPORT

After the evaluation of an existing structure has been completed, the Engineer of Record shall prepare a report documenting the evaluation. The report shall indicate whether the evaluation was performed by structural analysis, by load testing or by a combination of structural analysis and load testing. Furthermore, when testing is performed, the report shall include the loads and load combination used and the load-deformation and time-deformation relationships observed. All relevant information obtained from design drawing, mill test reports, and auxiliary material testing shall also be reported. Finally, the report shall indicate whether the design strength of the structure, including all members and connections, is adequate to withstand the load effects.

APPENDIX B

DESIGN REQUIREMENTS

Appendix B5.1 provides an expanded definition of limiting width-thickness ratio for webs in combined flexure and axial compression. Appendix B5.3 applies to the design of members containing slender compression elements.

B5. LOCAL BUCKLING

1. Classification of Steel Sections

For members with unequal flanges and with webs in combined flexural and axial compression, λ_r for the limit state of web local buckling is

$$\lambda_r = 1.49\sqrt{\frac{E}{F_y}}\left[1 + 2.83\left(\frac{h}{h_c}\right)\left(1 - \frac{P_u}{\phi_b P_y}\right)\right] \quad \text{(A-B5-1)}$$

$$\frac{3}{4} \le \frac{h}{h_c} \le \frac{3}{2}$$

For members with unequal flanges with webs subjected to flexure only, λ_r for the limit state of web local buckling is

$$\lambda_r = 1.49\sqrt{\frac{E}{F_y}}\left[1 + 2.83\left(\frac{h}{h_c}\right)\right] \quad \text{(A-B5-2)}$$

$$\frac{3}{4} \le \frac{h}{h_c} \le \frac{3}{2}$$

where λ_r, h, and h_c are as defined in Section B5.1.

These substitutions shall be made in Appendices F and G when applied to members with unequal flanges. If the compression flange is larger than the tension flange, λ_r shall be determined using Equation A-B5-1, A-B5-2, or Table B5.1.

3. Slender-Element Compression Sections

Axially loaded members containing elements subject to compression which have a width-thickness ratio in excess of the applicable λ_r as stipulated in Section B5.1 shall be proportioned according to this Appendix. Flexural members with slender compression elements shall be designed in accordance with Appendices F and G. Flexural members with proportions not covered by Appendix F1 shall be designed in accordance with this Appendix.

3a. Unstiffened Compression Elements

The design strength of unstiffened compression elements whose width-thickness ratio exceeds the applicable limit λ_r as stipulated in Section B5.1 shall be subject to a reduction factor Q_s. The value of Q_s shall be determined by Equations A-B5-3 through A-B5-10, as applicable. When such elements comprise the compression flange of a flexural member, the design flexural strength, in ksi, shall be computed using $\phi_b F_y Q_s$, where $\phi_b = 0.90$. The design strength of axially loaded compression members shall be modified by the appropriate reduction factor Q, as provided in Appendix B5.3d.

(a) For single angles:

when $0.45\sqrt{E/F_y} < b/t < 0.91\sqrt{E/F_y}$:

$$Q_s = 1.340 - 0.76(b/t)\sqrt{F_y/E} \qquad \text{(A-B5-3)}$$

Errata 9/4/01 when $b/t \geq 0.91\sqrt{E/F_y}$:

$$Q_s = 0.53E/\left[F_y (b/t)^2\right] \qquad \text{(A-B5-4)}$$

(b) For flanges, angles, and plates projecting from rolled beams or columns or other compression members:

when $0.56\sqrt{E/F_y} < b/t < 1.03\sqrt{E/F_y}$:

$$Q_s = 1.415 - 0.74(b/t)\sqrt{F_y/E} \qquad \text{(A-B5-5)}$$

when $b/t \geq 1.03\sqrt{E/F_y}$:

$$Q_s = 0.69E/\left[F_y (b/t)^2\right] \qquad \text{(A-B5-6)}$$

(c) For flanges, angles and plates projecting from built-up columns or other compression members:

when $0.64\sqrt{E/(F_y/k_c)} < b/t < 1.17\sqrt{E/(F_y/k_c)}$:

$$Q_s = 1.415 - 0.65(b/t)\sqrt{(F_y/k_c E)} \qquad \text{(A-B5-7)}$$

when $b/t \geq 1.17\sqrt{E/(F_y/k_c)}$:

$$Q_s = 0.90Ek_c/\left[F_y (b/t)^2\right] \qquad \text{(A-B5-8)}$$

The coefficient, k_c, shall be computed as follows:

(a) For I-shaped sections:

$$k_c = \frac{4}{\sqrt{h/t_w}},\ 0.35 \le k_c \le 0.763$$

where

h = depth of web, in. (mm)
t_w = thickness of web, in. (mm)

(b) For other sections:

$k_c = 0.763$

(d) For stems of tees:

when $0.75\sqrt{E/F_y} < \boxed{d/t} < 1.03\sqrt{E/F_y}$:

$$Q_s = 1.908 - 1.22\boxed{(d/t)}\sqrt{F_y/E} \tag{A-B5-9}$$

when $\boxed{d/t} \ge 1.03\sqrt{E/F_y}$: Errata 9/4/01

$$Q_s = 0.69 \quad / \left[F_y \boxed{(d/t)}^2 \right] \tag{A-B5-10}$$

where

$\boxed{d}$ = width of unstiffened compression element as defined in Section B5.1, in. (mm)
t = thickness of unstiffened element, in. (mm)

3b. Stiffened Compression Elements

When the width-thickness ratio of uniformly compressed stiffened elements (except perforated cover plates) exceeds the limit λ_r stipulated in Section B5.1, a reduced effective width b_e shall be used in computing the design properties of the section containing the element.

(a) For flanges of square and rectangular sections of uniform thickness:

when $\frac{b}{t} \ge 1.40\sqrt{\frac{E}{f}}$:

$$b_e = 1.91t\sqrt{\frac{E}{f}}\left[1 - \frac{0.38}{(b/t)}\sqrt{\frac{E}{f}}\right] \tag{A-B5-11}$$

otherwise $b_e = b$.

(b) For other uniformly compressed elements:

when $\frac{b}{t} \ge 1.49\sqrt{\frac{E}{f}}$:

$$b_e = 1.91t\sqrt{\frac{E}{f}}\left[1 - \frac{0.34}{(b/t)}\sqrt{\frac{E}{f}}\right] \tag{A-B5-12}$$

otherwise $b_e = b$.

where

b = actual width of a stiffened compression element, as defined in Section B5.1, in. (mm)
b_e = reduced effective width, in. (mm)
t = element thickness, in. (mm)
f = computed elastic compressive stress in the stiffened elements, based on the design properties as specified in Appendix B5.3c, ksi (MPa). If unstiffened elements are included in the total cross section, f for the stiffened element must be such that the maximum compressive stress in the unstiffened element does not exceed $\phi_c F_{cr}$ as defined in Appendix B5.3d with $Q = Q_s$ and $\phi_c = 0.85$, or $\phi_b F_y Q_s$ with $\phi_b = 0.90$, as applicable.

(c) For axially loaded circular sections with diameter-to-thickness ratio D / t greater than $0.11E/F_y$ but less than $0.45E / F_y$

$$Q = Q_a = \frac{0.038E}{F_y(D/t)} + \frac{2}{3} \tag{A-B5-13}$$

where

D = outside diameter, in. (mm)
t = wall thickness, in. (mm)

3c. Design Properties

Properties of sections shall be determined using the full cross section, except as follows:

In computing the moment of inertia and elastic section modulus of flexural members, the effective width of uniformly compressed stiffened elements b_e, as determined in Appendix B5.3b, shall be used in determining effective cross-sectional properties.

For unstiffened elements of the cross section, Q_s is determined from Appendix B5.3a. For stiffened elements of the cross section

$$Q_a = \frac{\text{effective area}}{\text{actual area}} \tag{A-B5-14}$$

where the effective area is equal to the summation of the effective areas of the cross section.

3d. Design Strength

For axially loaded compression members the gross cross-sectional area and the radius of gyration r shall be computed on the basis of the actual cross section. The critical stress F_{cr} shall be determined as follows:

(a) For $\lambda_c \sqrt{Q} \le 1.5$:

$$F_{cr} = Q(0.658^{Q\lambda_c^2})F_y \tag{A-B5-15}$$

(b) For $\lambda_c \sqrt{Q} > 1.5$:

$$F_{cr} = \left[\frac{0.877}{\lambda_c^2} \right] F_y \quad \text{(A-B5-16)}$$

where

$$Q = Q_s Q_a \quad \text{(A-B5-17)}$$

Cross sections comprised of only unstiffened elements, $Q = Q_s$, ($Q_a = 1.0$)

Cross sections comprised of only stiffened elements, $Q = Q_a$, ($Q_s = 1.0$)

Cross sections comprised of both stiffened and unstiffened elements, $Q = Q_s Q_a$

APPENDIX E

COLUMNS AND OTHER COMPRESSION MEMBERS

This Appendix applies to the strength of doubly symmetric columns with thin plate elements, and singly symmetric and unsymmetric columns for the limit states of flexural-torsional and torsional buckling.

E3. DESIGN COMPRESSIVE STRENGTH FOR FLEXURAL-TORSIONAL BUCKLING

The design strength of compression members determined by the limit states of torsional and flexural-torsional buckling is $\phi_c P_n$,

where

$\phi_c = 0.85$
P_n = nominal resistance in compression, kips (N)
$= A_g F_{cr}$ (A-E3-1)
A_g = gross area of cross section, in.^{2} (mm^2)

The nominal critical stress F_{cr} is determined as follows:

(a) For $\lambda_e \sqrt{Q} \leq 1.5$:

$$F_{cr} = Q(0.658^{Q\lambda_e^2})F_y \quad \text{(A-E3-2)}$$

(b) For $\lambda_e \sqrt{Q} > 1.5$:

$$F_{cr} = \left[\frac{0.877}{\lambda_e^2}\right]F_y \quad \text{(A-E3-3)}$$

where

$$\lambda_e = \sqrt{F_y / F_e} \quad \text{(A-E3-4)}$$

Q = 1.0 for elements meeting the width-thickness ratios λ_r of Section B5.1
= $Q_s Q_a$ for elements not meeting the width-thickness ratios λ_r of Section B5.1 and determined in accordance with the provisions of Appendix B5.3

The critical torsional or flexural-torsional elastic buckling stress F_e is determined as follows:

(a) For doubly symmetric shapes:

$$F_e = \left[\frac{\pi^2 E C_w}{(K_z l)^2} + GJ \right] \frac{1}{I_x + I_y} \tag{A-E3-5}$$

(b) For singly symmetric shapes where y is the axis of symmetry:

$$F_e = \frac{F_{ey} + F_{ez}}{2H} \left(1 - \sqrt{1 - \frac{4 F_{ey} F_{ez} H}{(F_{ey} + F_{ez})^2}} \right) \tag{A-E3-6}$$

(c) For unsymmetric shapes, the critical flexural-torsional elastic buckling stress F_e is the lowest root of the cubic equation

$$(F_e - F_{ex})(F_e - F_{ey})(F_e - F_{ez}) - F_e^2 (F_e - F_{ey}) \left(\frac{x_o}{\bar{r}_o} \right)^2 - F_e^2 (F_e - F_{ex}) \left(\frac{y_o}{\bar{r}_o} \right)^2 = 0 \tag{A-E3-7}$$

where

K_z = effective length factor for torsional buckling
G = shear modulus, ksi (MPa)
C_w = warping constant, in.^{6} (mm^6)
J = torsional constant, in.^{4} (mm^4)
I_x, I_y = moment of inertia about the principal axes, in.^{4} (mm^4)
x_o, y_o = coordinates of shear center with respect to the centroid, in. (mm)

$$\bar{r}_o^2 = x_o^2 + y_o^2 + \frac{I_x + I_y}{A} \tag{A-E3-8}$$

$$H = 1 - \left(\frac{x_o^2 + y_o^2}{\bar{r}_o^2} \right) \tag{A-E3-9}$$

$$F_{ex} = \frac{\pi^2 E}{(K_x l / r_x)^2} \tag{A-E3-10}$$

$$F_{ey} = \frac{\pi^2 E}{(K_y l / r_y)^2} \tag{A-E3-11}$$

$$F_{ez} = \left(\frac{\pi^2 E C_w}{(K_z l)^2} + GJ \right) \frac{1}{A \bar{r}_o^2} \tag{A-E3-12}$$

A = cross-sectional area of member, in.^{2} (mm^2)
l = unbraced length, in. (mm)
K_x, K_y = effective length factors in x and y directions
r_x, r_y = radii of gyration about the principal axes, in. (mm)
$\bar{r}_o$ = polar radius of gyration about the shear center, in. (mm)

APPENDIX F

BEAMS AND OTHER FLEXURAL MEMBERS

Appendix F1 provides the design flexural strength of beams and other flexural members. Appendix F2 provides the design shear strength of webs with and without stiffeners and requirements on transverse stiffeners. Appendix F3 applies to web-tapered members.

F1. DESIGN FOR FLEXURE

The design strength for flexural members is $\phi_b M_n$ where $\phi_b = 0.90$ and M_n is the nominal strength.

Table A-F1.1 provides a tabular summary of Equations F1-1 through F1-15 for determining the nominal flexural strength of beams and other flexural members. For slenderness parameters of cross sections not included in Table A-F1.1, see Appendix B5.3. For flexural members with unequal flanges see Appendix B5.1 for the determination of λ_r for the limit state of web local buckling.

The nominal flexural strength M_n is the lowest value obtained according to the limit states of yielding: lateral-torsional buckling (LTB); flange local buckling (FLB); and web local buckling (WLB).

The nominal flexural strength M_n shall be determined as follows for each limit state:

(a) For $\lambda \le \lambda_p$:

$$M_n = M_p \qquad \text{(A-F1-1)}$$

(b) For $\lambda_p < \lambda \le \lambda_r$:

For the limit state of lateral-torsional buckling:

$$M_n = C_b\left[M_p - (M_p - M_r)\left(\frac{\lambda - \lambda_p}{\lambda_r - \lambda_p}\right)\right] \le M_p \qquad \text{(A-F1-2)}$$

For the limit states of flange and web local buckling:

$$M_n = M_p - (M_p - M_r)\left(\frac{\lambda - \lambda_p}{\lambda_r - \lambda_p}\right) \qquad \text{(A-F1-3)}$$

(c) For $\lambda > \lambda_r$:

For the limit state of lateral-torsional buckling and flange local buckling:

$$M_n = M_{cr} = SF_{cr} \le M_p \qquad \text{(A-F1-4)}$$

For the design of other flexural members with slender webs, the limit state of web local buckling is not applicable. See Appendix G2.

For λ of the flange > λ_r in shapes not included in Table A-F1.1, see Appendix B5.3.

For λ of the web > λ_r, see Appendix G.

The terms used in the above equations are:

M_n = nominal flexural strength, kip-in. (N-mm)
M_p = $F_y Z$, plastic moment ≤ 1.5 $F_y S$, kip-in. (N-mm)
M_{cr} = buckling moment, kip-in. (N-mm)
M_r = limiting buckling moment (equal to M_{cr} when $\lambda = \lambda_r$), kip-in. (N-mm)
λ = controlling slenderness parameter
= minor axis slenderness ratio L_b / r_y for lateral-torsional buckling
= flange width-thickness ratio b/t for flange local buckling as defined in Section B5.1
= web depth-thickness ratio h/t_w for web local buckling as defined in Section B5.1
λ_p = largest value of λ for which $M_n = M_p$
λ_r = largest value of λ for which buckling is inelastic
F_{cr} = critical stress, ksi (MPa)
C_b = bending coefficient dependent on moment gradient, see Section F1.2a, Equation F1-3
S = section modulus, in.^{3} (mm^3)
L_b = laterally unbraced length, in. (mm)
r_y = radius of gyration about minor axis, in. (mm)

The applicable limit states and equations for M_p, M_r, F_{cr}, λ, λ_p, and λ_r are given in the Table A-F1.1 for shapes covered in this Appendix. The terms used in the table are:

A = cross-sectional area, in.^{2} (mm^2)
F_L = smaller of ($F_{yf} - F_r$) or F_{yw}, ksi (MPa)
F_r = compressive residual stress in flange
= 10 ksi (69 N/mm^2) for rolled shapes
= 16.5 ksi (114 N/mm^2) for welded shapes
F_y = specified minimum yield strength, ksi (MPa)
F_{yf} = yield strength of the flange, ksi (MPa)
F_{yw} = yield strength of the web, ksi (MPa)
I_{yc} = moment of inertia of compression flange about y axis or if reverse curvature bending, moment of inertia of smaller flange, in.^{4} (mm^4)
J = torsional constant, in.^{4} (mm^4)
R_e = see Appendix G2
S_{eff} = effective section modulus about major axis, in.^{3} (mm^3)
S_{xc} = section modulus of the outside fiber of the compression flange, in.^{3} (mm^3)
S_{xt} = section modulus of the outside fiber of the tension flange, in.^{3} (mm^3)
Z = plastic section modulus, in.^{3} (mm^3)
b = flange width, in. (mm)
d = overall depth, in. (mm)
f = computed compressive stress in the stiffened element, ksi (MPa)

TABLE A-F1.1
Nominal Strength Parameters

Shape	Plastic Moment M_p	Limit State of Buckling	Limiting Buckling Moment M_r
Channels and doubly and singly symmetric I-shaped beams (including hybrid beams) bent about major axis [a]	$F_y Z_x$ [b]	LTB doubly symmetric members and channels	$F_L S_x$
		LTB singly symmetric members	$F_L S_{xc} \le F_{yt} S_{xt}$
		FLB	$F_L S_x$
		WLB	$R_e F_{yf} S_x$
Channels and doubly and singly symmetric I-shaped members bent about minor axis [a]	$F_y Z_y$	FLB	$F_y S_y$

NOTE: LTB applies only for strong axis bending.
[a] Excluding double angles and tees.
[b] Computed from fully plastic stress distribution for hybrid sections.

Errata 9/4/01

[c] $X_1 = \dfrac{\pi}{S_x}\sqrt{\dfrac{EGJA}{2}}$ $\quad X_2 = 4\dfrac{C_w}{I_y}\left(\dfrac{S_x}{GJ}\right)^2$

[d] $\lambda_r = \dfrac{X_1}{F_L}\sqrt{1+\sqrt{1+X_2 F_L^2}}$

[e] $F_{cr} = \dfrac{M_{cr}}{S_{xc}}$, where $M_{cr} = \dfrac{2EC_b}{L_b}\sqrt{I_y J}\left[B_1+\sqrt{\left(1+B_2+B_1^2\right)}\right] \le M_p$

where

$$B_1 = 2.25\left[2(I_{yc}/I_y)-1\right](h/L_b)\sqrt{(I_y/J)}$$

$$B_2 = 25(1-I_{yc}/I_y)(I_{yc}/J)(h/L_b)^2$$

$$C_b = 1.0 \text{ if } I_{yc}/I_y < 0.1 \text{ or } I_{yc}/I_y > 0.9$$

TABLE A-F1.1 (cont'd)
Nominal Strength Parameters

Critical Stress F_{cr}	Slenderness Parameters λ	λ_p	λ_r	Limitations
$\frac{C_b X_1 \sqrt{2}}{\lambda}\sqrt{1+\frac{X_1^2 X_2}{2\lambda^2}}$	$\frac{L_b}{r_y}$	$1.76\sqrt{\frac{E}{F_{yf}}}$	[c,d]	Applicable for I-shaped members if $h/t_w \le \lambda_r$. When $h/t_w > \lambda_r$ see Appendix G.
[e]	$\frac{L_b}{r_{yc}}$	$1.76\sqrt{\frac{E}{F_{yf}}}$	Value of λ for which $M_{cr}(C_b = 1) = M_r$	
[f]	$\frac{b}{t}$	$0.38\sqrt{\frac{E}{F_{yf}}}$	[g]	
Not applicable	$\frac{h}{t_w}$	$3.76\sqrt{\frac{E}{F_{yf}}}$	λ_r as defined in Section B5.1	
$\frac{0.69E}{\lambda^2}$	$\frac{b}{t}$	$0.38\sqrt{\frac{E}{F_{yf}}}$	$0.83\sqrt{\frac{E}{F_L}}$	None

Notes (cont'd):

[f] $F_{cr} = \frac{0.69E}{\lambda^2}$ for rolled shapes

$F_{cr} = \frac{0.90Ek_c}{\lambda^2}$ for welded shapes

where
$k_c = 4 / \sqrt{h / t_w}$ and $0.35 \le k_c \le 0.763$

[g] $\lambda_r = 0.83\sqrt{\frac{E}{F_L}}$ for rolled shapes

$\lambda_r = 0.95\sqrt{\frac{E}{(F_L / k_c)}}$ for welded shapes

TABLE A-F1.1 (cont'd) Nominal Strength Parameters

Shape	Plastic Moment M_p	Limit State of Buckling	Limiting Buckling Moment M_r
Solid symmetric shapes, except rectangular bars, bent about major axis	$F_y Z_x$	Not Applicable	
Solid rectangular bars bent about major axis	$F_y Z_x$	LTB	$F_y S_x$
Symmetric box sections loaded in a plane of symmetry	$F_y Z$	LTB	$F_{yf} S_{eff}$
		FLB	$F_y S_{eff}$
		WLB	Same as for I-shape
Round HSS	$F_y Z$	LTB	Not applicable
		FLB	$M_n = \left(\frac{0.021E}{D/t} + F_y\right) S$ [h]
		WLB	Not applicable

Notes (cont'd):
[h] This equation is to be used in place of Equation A-F1-3.

TABLE A-F1.1 (cont'd)
Nominal Strength Parameters

Critical Stress F_{cr}	Slenderness Parameters λ	λ_p	λ_r	Limitations
Not applicable				
$\frac{2EC_b\sqrt{JA}}{\lambda S_x}$	$\frac{L_b}{r_y}$	$\frac{0.13E\sqrt{JA}}{M_p}$	$\frac{2E\sqrt{JA}}{M_r}$	None
$\frac{2EC_b\sqrt{JA}}{\lambda S_x}$	$\frac{L_b}{r_y}$	$\frac{0.13E\sqrt{JA}}{M_p}$	$\frac{2E\sqrt{JA}}{M_r}$	Applicable if $h/t_w \le 5.70\sqrt{E/F_{yf}}$
$\frac{S_{eff}}{S}F_y$ [i]	$\frac{b}{t}$	$1.12\sqrt{\frac{E}{F_y}}$	$1.40\sqrt{\frac{E}{F_y}}$	None
Same as for I-Shape				
Not applicable				
$\frac{0.33E}{D/t}$	D/t	$\frac{0.071E}{F_y}$	$\frac{0.31E}{F_y}$	$D/t < \frac{0.45E}{F_y}$
Not applicable				

Notes (cont'd):

[i] S_{eff} is the effective section modulus for the section with a compression flange b_e defined in Appendix B5.3b.

h = clear distance between flanges less the fillet or corner radius at each flange, in. (mm)
r_{yc} = radius of gyration of compression flange about y axis or if reverse curvature bending, smaller flange, in. (mm)
t_f = flange thickness, in. (mm)
t_w = web thickness, in. (mm)

F2. DESIGN FOR SHEAR

2. Design Shear Strength

The design shear strength of stiffened or unstiffened webs is ϕV_n,

where
$\phi_v = 0.90$
V_n = nominal shear strength defined as follows:

(a) For $h/t_w \leq 1.10\sqrt{k_v E/F_{yw}}$:

$$V_n = 0.6F_{yw}A_w \tag{A-F2-1}$$

(b) For $1.10\sqrt{k_v E/F_{yw}} < h/t_w \leq 1.37\sqrt{k_v E/F_{yw}}$:

$$V_n = 0.6F_{yw}A_w(1.10\sqrt{k_v E/F_{yw}})/(h/t_w) \tag{A-F2-2}$$

(c) For $h/t_w > 1.37\sqrt{k_v E/F_{yw}}$:

$$V_n = A_w(0.91Ek_v)/(h/t_w)^2 \tag{A-F2-3}$$

where

$k_v = 5 + 5/(a/h)^2$
= 5 when $a/h > 3$ or $a/h > [260/(h/t)]^2$
a = distance between transverse stiffeners, in. (mm)
h = for rolled shapes, the clear distance between flanges less the fillet or corner radius, in. (mm)
= for built-up welded sections, the clear distance between flanges, in. (mm)
= for built-up bolted or riveted sections, the distance between fastener lines, in. (mm)

3. Transverse Stiffeners

Transverse stiffeners are not required in plate girders where $h/t_w \leq 2.45\sqrt{E/F_{yw}}$ or where the required shear, V_u, as determined by structural analysis for the factored loads, is less than or equal to $0.6\phi_v A_w F_{yw} C_v$, where $\phi_v = 0.90$ and the shear coefficient C_v defined in Appendix G3 is determined for $k_v = 5$.

Transverse stiffeners used to develop the web design shear strength as provided in Appendix F2.2 shall have a moment of inertia about an axis in the web center for stiffener pairs, or about the face in contact with the web plate for single stiffeners, which shall not be less than $at_w^3 j$, where

$$j = 2.5 / (a/h)^2 - 2 \geq 0.5 \qquad \text{(A-F2-4)}$$

Intermediate stiffeners are permitted to be stopped short of the tension flange, provided bearing is not needed to transmit a concentrated load or reaction. The weld by which intermediate stiffeners are attached to the web shall be terminated not less than four times nor more than six times the web thickness from the near toe of the web-to-flange weld. When single stiffeners are used, they shall be attached to the compression flange, if it consists of a rectangular plate, to resist any uplift tendency due to torsion in the flange. When lateral bracing is attached to a stiffener, or a pair of stiffeners, these, in turn, shall be connected to the compression flange to transmit one percent of the total flange stress, unless the flange is composed only of angles.

Bolts connecting stiffeners to the girder web shall be spaced not more than 12 in. (300 mm) on center. If intermittent fillet welds are used, the clear distance between welds shall not be more than 16 times the web thickness nor more than 10 in. (250 mm).

F3. WEB-TAPERED MEMBERS

The design of tapered members meeting the requirements of this section shall be governed by the provisions of Chapters D through H, except as modified by this Appendix.

1. General Requirements

In order to qualify under this Specification, a tapered member shall meet the following requirements:

(1) It shall possess at least one axis of symmetry, which shall be perpendicular to the plane of bending if moments are present.

(2) The flanges shall be of equal and constant area.

(3) The depth shall vary linearly as

$$d = d_o\left(1 + \gamma \frac{z}{L}\right) \qquad \text{(A-F3-1)}$$

where

$\gamma = (d_L - d_o) / d_o \leq$ the smaller of $0.268(L / d_o)$ or 6.0
d_o = depth at smaller end of member, in. (mm)
d_L = depth at larger end of member, in. (mm)
z = distance from the smaller end of member, in. (mm)
L = unbraced length of member measured between the center of gravity of the bracing members, in. (mm)

2. Design Tensile Strength

The design strength of tapered tension members shall be determined in accordance with Section D1.

3. Design Compressive Strength

The design strength of tapered compression members shall be determined in accordance with Appendix E3, using an effective slenderness parameter λ_{eff} computed as follows:

$$\lambda_{eff} = \frac{S}{\pi}\sqrt{\frac{QF_y}{E}} \tag{A-F3-2}$$

where

$S = KL/r_{oy}$ for weak axis buckling and $K_\gamma L/r_{ox}$ for strong axis buckling
K = effective length factor for a prismatic member
K_γ = effective length factor for a tapered member as determined by a rational analysis
r_{ox} = strong axis radius of gyration at the smaller end of a tapered member, in. (mm)
r_{oy} = weak axis radius of gyration at the smaller end of a tapered member, in. (mm)
F_y = specified minimum yield stress, ksi (MPa)
Q = reduction factor
= 1.0 if all elements meet the limiting width-thickness ratios λ_r of Section B5.1
= $Q_s Q_a$, determined in accordance with Appendix B5.3, if any stiffened and/or unstiffened elements exceed the ratios λ_r of Section B5.1
E = modulus of elasticity for steel, ksi (MPa)

The smallest area of the tapered member shall be used for A_g in Equation E2-1.

4. Design Flexural Strength

The design flexural strength of tapered flexural members for the limit state of lateral-torsional buckling is $\phi_b M_n$, where $\phi_b = 0.90$ and the nominal strength is

$$M_n = (5/3)S'_x F_{b\gamma} \tag{A-F3-3}$$

where

S'_x = the section modulus of the critical section of the unbraced beam length under consideration

$$F_{b\gamma} = \frac{2}{3}\left[1.0 - \frac{F_y}{6B\sqrt{F_{s\gamma}^2 + F_{w\gamma}^2}}\right]F_y \le 0.60F_y \tag{A-F3-4}$$

unless $F_{b\gamma} \le F_y/3$, in which case

$$F_{b\gamma} = B\sqrt{F_{s\gamma}^2 + F_{w\gamma}^2} \tag{A-F3-5}$$

In the preceding equations,

$$F_{s\gamma} = \frac{0.41E}{h_s L d_o / A_f} \tag{A-F3-6}$$

$$F_{w\gamma} = \frac{5.9E}{(h_w L / r_{To})^2} \tag{A-F3-7}$$

where

h_s = factor equal to $1.0 + 0.230\gamma\sqrt{Ld_o / A_f}$

h_w = factor equal to $1.0 + 0.00385\gamma\sqrt{L / r_{To}}$

r_{To} = radius of gyration of a section at the smaller end, considering only the compression flange plus one-third of the compression web area, taken about an axis in the plane of the web, in. (mm)

A_f = area of the compression flange, in.^{2} (mm^2)

and where B is determined as follows:

(a) When the maximum moment M_2 in three adjacent segments of approximately equal unbraced length is located within the central segment and M_1 is the larger moment at one end of the three-segment portion of a member:

$$B = 1.0 + 0.37\left(1.0 + \frac{M_1}{M_2}\right) + 0.50\gamma\left(1.0 + \frac{M_1}{M_2}\right) \geq 1.0 \tag{A-F3-8}$$

(b) When the largest computed bending stress f_{b2} occurs at the larger end of two adjacent segments of approximately equal unbraced lengths and f_{b1} is the computed bending stress at the smaller end of the two-segment portion of a member:

$$B = 1.0 + 0.58\left(1.0 + \frac{f_{b1}}{f_{b2}}\right) - 0.70\gamma\left(1.0 + \frac{f_{b1}}{f_{b2}}\right) \geq 1.0 \tag{A-F3-9}$$

(c) When the largest computed bending stress f_{b2} occurs at the smaller end of two adjacent segments of approximately equal unbraced length and f_{b1} is the computed bending stress at the larger end of the two-segment portion of a member:

$$B = 1.0 + 0.55\left(1.0 + \frac{f_{b1}}{f_{b2}}\right) + 2.20\gamma\left(1.0 + \frac{f_{b1}}{f_{b2}}\right) \geq 1.0 \tag{A-F3-10}$$

In the foregoing, $\gamma = (d_L - d_o) / d_o$ is calculated for the unbraced length that contains the maximum computed bending stress. M_1 / M_2 is considered as negative when producing single curvature. In the rare case where M_1 / M_2 is positive, it is recommended that it be taken as zero. f_{b1} / f_{b2} is considered as negative when producing single curvature. If a point of contraflexure occurs in one of two adjacent unbraced segments, f_{b1} / f_{b2} is considered as positive. The ratio $f_{b1} / f_{b2} \neq 0$.

(d) When the computed bending stress at the smaller end of a tapered member or segment thereof is equal to zero:

$$B = \frac{1.75}{1.0 + 0.25\sqrt{\gamma}} \qquad \text{(A-F3-11)}$$

where $\gamma = (d_L - d_o) / d_o$ is calculated for the unbraced length adjacent to the point of zero bending stress.

5. Design Shear Strength

The design shear strength of tapered flexural members shall be determined in accordance with Section F2.

6. Combined Flexure and Axial Force

For tapered members with a single web taper subject to compression and bending about the major axis, Equation H1-1 applies, with the following modifications. P_n and P_{ex} shall be determined for the properties of the smaller end, using appropriate effective length factors. M_{nx}, M_u, and M_{px} shall be determined for the larger end; $M_{nx} = (5/3)S'_x F_{b\gamma}$, where S'_x is the elastic section modulus of the larger end, and $F_{b\gamma}$ is the design flexural stress of tapered members. C_{mx} is replaced by C'_m, determined as follows:

(a) When the member is subjected to end moments which cause single curvature bending and approximately equal computed moments at the ends:

$$C'_m = 1.0 + 0.1\left(\frac{P_u}{\phi_b P_{ex}}\right) + 0.3\left(\frac{P_u}{\phi_b P_{ex}}\right)^2 \qquad \text{(A-F3-12)}$$

(b) When the computed bending moment at the smaller end of the unbraced length is equal to zero:

$$C'_m = 1.0 - 0.9\left(\frac{P_u}{\phi_b P_{ex}}\right) + 0.6\left(\frac{P_u}{\phi_b P_{ex}}\right)^2 \qquad \text{(A-F3-13)}$$

When the effective slenderness parameter $\lambda_{eff} \geq 1.5$ and combined stress is checked incrementally along the length, the actual area and the actual section modulus at the section under investigation is permitted to be used.

APPENDIX G

PLATE GIRDERS

This appendix applies to I-shaped plate girders with slender webs.

G1. LIMITATIONS

Doubly and singly symmetric single-web non-hybrid and hybrid plate girders loaded in the plane of the web shall be proportioned according to the provisions of this Appendix or Section F2, provided that the following limits are satisfied:

(a) For $\dfrac{a}{h} \le 1.5$:

$$\frac{h}{t_w} \le 11.7\sqrt{\frac{E}{F_{yf}}} \quad \text{(A-G1-1)}$$

(b) For $\dfrac{a}{h} > 1.5$:

$$\frac{h}{t_w} \le \frac{0.48E}{\sqrt{F_{yf}(F_{yf}+16.5)}} \quad \text{(A-G1-2)}$$

$$\left(\text{Metric: } \frac{h}{t_w} \le \frac{0.48E}{\sqrt{F_{yf}(F_{yf}+114)}}\right) \quad \text{(A-G1-2M)}$$

where

a = clear distance between transverse stiffeners, in. (mm)
h = clear distance between flanges less the fillet or corner radius for rolled shapes; and for built-up sections, the distance between adjacent lines of fasteners or the clear distance between flanges when welds are used, in. (mm)
t_w = web thickness, in. (mm)
F_{yf} = specified minimum yield stress of a flange, ksi (MPa)

In unstiffened girders h / t_w shall not exceed 260.

G2. DESIGN FLEXURAL STRENGTH

The design flexural strength for plate girders with slender webs shall be $\phi_b M_n$, where $\phi_b = 0.90$ and M_n is the lower value obtained according to the limit states of tension-flange yield and compression-flange buckling. For girders with unequal flanges, see Appendix B5.1 for the determination of λ_r for the limit state of web local buckling.

(a) For tension-flange yield:

$$M_n = S_{xt} R_e F_{yt} \tag{A-G2-1}$$

(b) For compression-flange buckling:

$$M_n = S_{xc} R_{PG} R_e F_{cr} \tag{A-G2-2}$$

where

$$R_{PG} = 1 - \frac{a_r}{1{,}200 + 300a_r}\left(\frac{h_c}{t_w} - 5.70\sqrt{\frac{E}{F_{cr}}}\right) \le 1.0 \tag{A-G2-3}$$

R_e = hybrid girder factor

$= \dfrac{12 + a_r(3m - m^3)}{12 + 2a_r} \le 1.0$ (for non-hybrid girders, $R_e = 1.0$)

a_r = ratio of web area to compression flange area (≤ 10)
m = ratio of web yield stress to flange yield stress or to F_{cr}
F_{cr} = critical compression flange stress, ksi (MPa)
F_{yt} = yield stress of tension flange, ksi (MPa)
S_{xc} = section modulus referred to compression flange, in.^{3} (mm^3)
S_{xt} = section modulus referred to tension flange, in.^{3} (mm^3)
h_c = twice the distance from the centroid to the nearest line of fasteners at the compression flange or the inside of the face of the compression flange when welds are used, in. (mm)

The critical stress F_{cr} to be used is dependent upon the slenderness parameters λ, λ_p, λ_r, and C_{PG} as follows:

(a) For $\lambda \le \lambda_p$:

$$F_{cr} = F_{yf} \tag{A-G2-4}$$

(b) For $\lambda_p < \lambda \le \lambda_r$:

$$F_{cr} = C_b F_{yf}\left[1 - \frac{1}{2}\left(\frac{\lambda - \lambda_p}{\lambda_r - \lambda_p}\right)\right] \le F_{yf} \tag{A-G2-5}$$

(c) For $\lambda > \lambda_r$:

$$F_{cr} = \frac{C_{PG}}{\lambda^2} \tag{A-G2-6}$$

In the foregoing, the slenderness parameter shall be determined for both the limit state of lateral-torsional buckling and the limit state of flange local buckling; the slenderness parameter which results in the lowest value of F_{cr} governs.

(a) For the limit state of lateral-torsional buckling:

$$\lambda = \frac{L_b}{r_T} \qquad \text{(A-G2-7)}$$

$$\lambda_p = 1.76\sqrt{\frac{E}{F_{yf}}} \qquad \text{(A-G2-8)}$$

$$\lambda_r = 4.44\sqrt{\frac{E}{F_{yf}}} \qquad \text{(A-G2-9)}$$

$$C_{PG} = 286{,}000 C_b \qquad \text{(A-G2-10)}$$

$$(\text{Metric: } C_{PG} = 1\,970\,000 C_b) \qquad \text{(A-G2-10M)}$$

where

C_b= see Section F1.2, Equation F1-3
r_T = radius of gyration of compression flange plus one-third of the compression portion of the web, in. (mm)

(b) For the limit state of flange local buckling:

$$\lambda = \frac{b_f}{2t_f} \qquad \text{(A-G2-11)}$$

$$\lambda_p = 0.38\sqrt{\frac{E}{F_{yf}}} \qquad \text{(A-G2-12)}$$

$$\lambda_r = 1.35\sqrt{\frac{E}{F_{yf}/k_c}} \qquad \text{(A-G2-13)}$$

$$C_{PG} = 26{,}200 k_c \qquad \text{(A-G2-14)}$$

$$(\text{Metric: } C_{PG} = 180\,650 K_c) \qquad \text{(A-G2-14M)}$$

$$C_b = 1.0$$

where $k_c = 4/\sqrt{h/t_w}$ and $0.35 \le k_c \le 0.763$.

The limit state of flexural web local buckling is not applicable.

G3. DESIGN SHEAR STRENGTH

The design shear strength with tension field action shall be $\phi_v V_n$, kips (kN), where

$\phi_v = 0.90$ and V_n is determined as follows:

(a) For $h/t_w \le 1.10\sqrt{k_v E/F_{yw}}$:

$$V_n = 0.6 F_{yw} A_w \qquad \text{(A-G3-1)}$$

(b) For $h/t_w > 1.10\sqrt{k_v E/F_{yw}}$:

$$V_n = 0.6F_{yw}A_w\left(C_v + \frac{1-C_v}{1.15\sqrt{1+(a/h)^2}}\right) \tag{A-G3-2}$$

Also see Appendix G4 and G5.

Tension field action is not permitted for end-panels in non-hybrid plate girders, all panels in hybrid and web-tapered plate girders, and when a/h exceeds 3.0 or $[260/(h/t_w)]$. For these cases, the nominal strength is:

$$V_n = 0.6F_{yw}A_wC_v \tag{A-G3-3}$$

The web plate buckling coefficient k_v is given as

$$k_v = 5 + \frac{5}{(a/h)^2} \tag{A-G3-4}$$

except that k_v shall be taken as 5.0 if a/h exceeds 3.0 or $[260/(h/t_w)]^2$.

The shear coefficient C_v is determined as follows:

(a) For $1.10\sqrt{\dfrac{k_vE}{F_{yw}}} \le \dfrac{h}{t_w} \le 1.37\sqrt{\dfrac{k_vE}{F_{yw}}}$:

$$C_v = \frac{1.10\sqrt{k_vE/F_{yw}}}{h/t_w} \tag{A-G3-5}$$

(b) For $\dfrac{h}{t_w} > 1.37\sqrt{\dfrac{k_vE}{F_{yw}}}$:

$$C_v = \frac{1.51k_vE}{(h/t_w)^2F_{yw}} \tag{A-G3-6}$$

G4. TRANSVERSE STIFFENERS

Transverse stiffeners are not required in plate girders where $h/t_w \le 2.45\sqrt{E/F_{yw}}$, or where the required shear V_u, as determined by structural analysis for the factored loads, is less than or equal to $0.6\phi_v F_{yw} A_w C_v$, where C_v is determined for $k_v = 5$ and $\phi_v = 0.90$. Stiffeners may be required in certain portions of a plate girder to develop the required shear or to satisfy the limitations given in Appendix G1. Transverse stiffeners shall satisfy the requirements of Appendix F2.3.

When designing for tension field action, the stiffener area A_{st} shall not be less than

$$\frac{F_{yw}}{F_{yst}}\left[0.15Dht_w(1-C_v)\frac{V_u}{\phi_v V_n}-18t_w^2\right]\geq 0 \tag{A-G4-1}$$

where

F_{yst} = specified yield stress of the stiffener material, ksi (MPa)
D = 1 for stiffeners in pairs
= 1.8 for single angle stiffeners
= 2.4 for single plate stiffeners

C_v and V_n are defined in Appendix G3, and V_u is the required shear at the location of the stiffener.

G5. FLEXURE-SHEAR INTERACTION

For $0.6\phi V_n \leq V_u \leq \phi V_n$ and $0.75\phi M_n \leq M_u \leq \phi M_n$, plate girders with webs designed for tension field action shall satisfy the additional flexure-shear interaction criterion:

$$\frac{M_u}{\phi M_n}+0.625\frac{V_u}{\phi V_n}\leq 1.375 \tag{A-G5-1}$$

where

M_n = nominal flexural strength of plate girder from Appendix G2 or Section F1
ϕ = 0.90
V_n = nominal shear strength from Appendix G3

APPENDIX H

MEMBERS UNDER COMBINED FORCES AND TORSION

This appendix provides alternative interaction equations for braced frames with I-shaped members with $b_f / d \leq 1.0$ and box-shaped members.

H3. ALTERNATIVE INTERACTION EQUATIONS FOR MEMBERS UNDER COMBINED STRESS

For I-shaped members with $b_f / d \leq 1.0$ and box-shaped members, the use of the following interaction equations in lieu of Equations H1-1a and H1-1b is permitted for braced frames only. Both Equations A-H3-1 and A-H3-2 shall be satisfied.

$$\left(\frac{M_{ux}}{\phi_b M'_{px}}\right)^{\zeta} + \left(\frac{M_{uy}}{\phi_b M'_{py}}\right)^{\zeta} \leq 1.0 \qquad \text{(A-H3-1)}$$

$$\left(\frac{C_{mx} M_{ux}}{\phi_b M'_{nx}}\right)^{\eta} + \left(\frac{C_{my} M_{uy}}{\phi_b M'_{ny}}\right)^{\eta} \leq 1.0 \qquad \text{(A-H3-2)}$$

The terms in Equations A-H3-1 and A-H3-2 are determined as follows:

(a) For I-shaped members:

For $b_f / d < 0.5$:

$$\zeta = 1.0$$

For $0.5 \leq b_f / d \leq 1.0$:

$$\zeta = 1.6 - \frac{P_u / P_y}{2\left[\ln(P_u / P_y)\right]} \qquad \text{(A-H3-3)}$$

For $b_f / d < 0.3$:

$$\eta = 1.0$$

For $0.3 \leq b_f / d \leq 1.0$:

$$\eta = 0.4 + \frac{P_u}{P_y} + \frac{b_f}{d} \geq 1.0 \qquad \text{(A-H3-4)}$$

where

b_f = flange width, in. (mm)
d = member depth, in. (mm)

C_m = coefficient applied to the bending term in the interaction equation for prismatic members and dependent on column curvature caused by applied moments, see Section C1.

$$M'_{px} = 1.2M_{px}\left[1-(P_u/P_y)\right] \le M_{px} \quad \text{(A-H3-5)}$$

$$M'_{py} = 1.2M_{py}\left[1-(P_u/P_y)^2\right] \le M_{py} \quad \text{(A-H3-6)}$$

$$M'_{nx} = M_{nx}\left(1-\frac{P_u}{\phi_c P_n}\right)\left(1-\frac{P_u}{P_{ex}}\right) \quad \text{(A-H3-7)}$$

$$M'_{ny} = M_{ny}\left(1-\frac{P_u}{\phi_c P_n}\right)\left(1-\frac{P_u}{P_{ey}}\right) \quad \text{(A-H3-8)}$$

(b) For box-section members:

$$\zeta = 1.7 - \frac{P_u/P_y}{\ln(P_u/P_y)} \quad \text{(A-H3-9)}$$

$$\eta = 1.7 - \frac{P_u/P_y}{\ln(P_u/P_y)} - a\lambda_x\left(\frac{P_u}{P_y}\right)^b > 1.1 \quad \text{(A-H3-10)}$$

For $P_u/P_y \le 4.0$, $a = 0.06$, and $b = 1.0$;

For $P_u/P_y > 4.0$, $a = 0.15$, and $b = 2.0$:

$$M'_{px} = 1.2M_{px}\left[1-P_u/P_y\right] \le M_{px} \quad \text{(A-H3-11a)}$$

$$M'_{py} = 1.2M_{py}\left[1-P_u/P_y\right] \le M_{py} \quad \text{(A-H3-11b)}$$

$$M'_{nx} = M_{nx}\left(1-\frac{P_u}{\phi_c P_n}\right)\left(1-\frac{P_u}{P_{ex}}\frac{1.25}{(B/H)^{1/3}}\right) \quad \text{(A-H3-12)}$$

$$M'_{ny} = M_{ny}\left(1-\frac{P_u}{\phi_c P_n}\right)\left(1-\frac{P_u}{P_{ey}}\frac{1.25}{(B/H)^{1/2}}\right) \quad \text{(A-H3-13)}$$

where

P_n = nominal compressive strength determined in accordance with Section E2, kips (N)
P_u = required axial strength, kips (N)
P_y = compressive yield strength A_gF_y, kips (N)
ϕ_b = resistance factor for flexure = 0.90
ϕ_c = resistance factor for compression = 0.85
P_e = Euler buckling strength A_gF_y/λ_c^2, where λ_c is the column slenderness parameter defined by Equation E2-4, kips (N)
M_u = required flexural strength, kip-in. (N-mm)

M_n = nominal flexural strength, determined in accordance with Section F1, kip-in. (N-mm)
M_p = plastic moment $\leq 1.5 F_y S$, kip-in. (N-mm)
λ_x = column slenderness parameter with respect to the strong axis
B = outside width of box section parallel to major principal axis x, in. (mm)
H = outside depth of box section perpendicular to major principal axis x, in. (mm)

APPENDIX J

CONNECTIONS, JOINTS, AND FASTENERS

Appendix J2.4 provides the alternative design strength for fillet welds. Appendices J3.8 and J3.9 pertain to the design of slip-critical connections using service loads.

J2. WELDS

4. Design Strength

In lieu of the constant design strength for fillet welds given in Table J2.5, the following procedure is permitted.

(a) For a linear weld group loaded in-plane through the center of gravity, the design strength is $\phi F_w A_w$,

where

$$F_w = 0.60 F_{EXX}\left(1.0 + 0.50 \sin^{1.5}\theta\right) \qquad \text{(A-J2-1)}$$

$\phi = 0.75$

F_{EXX} = electrode classification number, i.e., minimum specified strength, ksi (MPa)

θ = angle of loading measured from the weld longitudinal axis, degrees

A_w = effective area of weld throat, in.^{2} (mm^2)

(b) For weld elements within a weld group that are loaded in-plane and analyzed using an instantaneous center of rotation method, the components of the design strength are $\phi F_{wx} A_w$ and $\phi F_{wy} A_w$,

where

$F_{wx} = \Sigma F_{wix}$

$F_{wy} = \Sigma F_{wiy}$

$F_{wi} = 0.60 F_{EXX} (1.0 + 0.50 \sin^{1.5} \theta) f(p)$

$f(p) = [p(1.9 - 0.9p)]^{0.3}$

$\phi = 0.75$

F_{wi} = nominal stress in any ith weld element, ksi (MPa)

F_{wix} = x component of stress F_{wi}

F_{wiy} = y component of stress F_{wi}

p = Δ_i / Δ_m, ratio of element i deformation to its deformation at maximum stress

Δ_m = $0.209(\theta + 2)^{-0.32} w$, deformation of weld element at maximum stress, in. (mm)

Δ_i = deformation of weld elements at intermediate stress levels, linearly

TABLE A-J3.1
Nominal Tension Stress (F_t), ksi (MPa)
Fasteners in Bearing-type Connections

Description of Fasteners	Threads Included in the Shear Plane	Threads Excluded from the Shear Plane
A307 bolts	$\sqrt{45^2 - 6.25f_v^2}$	
(Metric)	$\left(\sqrt{310^2 - 6.25f_v^2}\right)$	
A325 bolts	$\sqrt{90^2 - 6.25f_v^2}$	$\sqrt{90^2 - 4.00f_v^2}$
(A325M bolts)	$\left(\sqrt{621^2 - 6.25f_v^2}\right)$	$\left(\sqrt{621^2 - 4.00f_v^2}\right)$
A490 bolts	$\sqrt{113^2 - 6.31f_v^2}$	$\sqrt{113^2 - 4.04f_v^2}$
(A490M bolts)	$\left(\sqrt{779^2 - 6.31f_v^2}\right)$	$\left(\sqrt{779^2 - 4.04f_v^2}\right)$
Threaded parts A449 bolts over 1½ in. (38 mm)	$\sqrt{(0.75F_u)^2 - 6.25f_v^2}$	$\sqrt{(0.75F_u)^2 - 4.00f_v^2}$
A502 Gr. 1 rivets	$\sqrt{45^2 - 5.76f_v^2}$	
(Metric)	$\left(\sqrt{310^2 - 5.76f_v^2}\right)$	
A502 Gr. 2 rivets	$\sqrt{60^2 - 5.86f_v^2}$	
(Metric)	$\left(\sqrt{414^2 - 5.86f_v^2}\right)$	

Errata 9/4/01

TABLE A-J3.2
Slip-Critical Resistance to Shear at Service Loads, F_v, ksi (MPa), of High-Strength Bolts[a]

Errata 9/4/01

Type of Bolt	Resistance to Shear at Service Loads, ksi (MPa)			
	Standard Size Holes	Oversized and Short-slotted Holes	Long-slotted Holes: Perpendicular to Line of Force	Long-slotted Holes: Parallel to Line of Force
A325 (A325M)	17 (117)	15 (103)	12 (83)	10 (69)
A490 (A490M)	21 (145)	18 (124)	15 (103)	13 (90)

[a] For each shear plane.

proportioned to the critical deformation based on distance from the instantaneous center of rotation, r_i, in. (mm)
$= r_i \Delta_u / r_{crit}$

Δ_u $= 1.087(\theta + 6)^{-0.65} w \leq 0.17w$, deformation of weld element at ultimate stress (fracture), usually in element furthest from instantaneous center of rotation, in. (mm)

w = leg size of the fillet weld, in. (mm)

r_{crit} = distance from instantaneous center of rotation to weld element with minimum Δ_u / r_i ratio, in. (mm)

J3. BOLTS AND THREADED PARTS

7. Combined Tension and Shear in Bearing-Type Connections

As an alternative to the use of the equations in Table J3.5, the use of the equations in Table A-J3.1 is permitted.

8. High-Strength Bolts in Slip-Critical Connections

8b. Slip-Critical Connections Designed at Service Loads

The design resistance to shear per bolt $\phi F_v A_b$ for use at service loads shall equal or exceed the shear per bolt due to service loads,

where

ϕ = 1.0 for standard, oversized, and short-slotted holes and long-slotted holes when the long slot is perpendicular or parallel to the line of force

F_v = ~~nominal~~ slip-critical shear resistance tabulated in Table A-J3.2, ksi (MPa). The values for F_v in Table A-J3.2 are based on Class A surfaces with slip coefficient $\mu = 0.33$. When specified by the designer, the ~~nominal~~ slip resistance for connections having special faying surface conditions is permitted to be adjusted to the applicable values in the RCSC Load and Resistance Factor Design Specification. Errata 9/4/01

When the loading combination includes wind loads in addition to dead and live loads, the total shear on the bolt due to combined load effects, at service load, may be multiplied by 0.75.

9. Combined Tension and Shear in Slip-Critical Connections

9b. Slip-Critical Connections Designed at Service Loads

When a slip-critical connection is subjected to an applied tension T that reduces the net clamping force, the slip resistance per bolt, $\phi F_v A_b$, according to Appendix J3.8b shall be multiplied by the following factor:

$$1 - \frac{T}{0.8 T_b N_b}$$

where

T_b = minimum fastener tension from Table J3.1, kips (N)
N_b = number of bolts carrying service-load tension T

APPENDIX K

CONCENTRATED FORCES, PONDING, AND FATIGUE

Appendix K2 provides an alternative determination of roof stiffness. Appendix K3 pertains to the design of members and connections subject to high cyclic loading (fatigue).

K2. PONDING

The provisions of this Appendix are permitted to be used when a more exact determination of flat roof framing stiffness is needed than that given by the provision of Section K2 that $C_p + 0.9C_s \leq 0.25$.

For any combination of primary and secondary framing, the stress index is computed as

$$U_p = \left[\frac{F_y - f_o}{f_o} \right]_p \quad \text{for the primary member} \qquad \text{(A-K2-1)}$$

Errata 9/4/01

$$U_s = \left[\frac{F_y - f_o}{f_o} \right]_s \quad \text{for the secondary member} \qquad \text{(A-K2-2)}$$

where

f_o = the stress due to $1.2D + 1.2R$ (D = nominal dead load, R = nominal load due to rain water or ice exclusive of the ponding contribution),* ksi (MPa)

Enter Figure A-K2.1 at the level of the computed stress index U_p determined for the primary beam; move horizontally to the computed C_s value of the secondary beams and then downward to the abscissa scale. The combined stiffness of the primary and secondary framing is sufficient to prevent ponding if the flexibility constant read from this latter scale is more than the value of C_p computed for the given primary member; if not, a stiffer primary or secondary beam, or combination of both, is required. In the above,

$$C_p = \frac{32 L_s L_p^4}{10^7 I_p}$$

$$\left(\text{Metric: } C_p = \frac{504 L_s L_p^4}{I_p} \right)$$

$$C_s = \frac{32 S L_s^4}{10^7 I_s}$$

*Depending upon geographic location, this loading should include such amount of snow as might also be present, although ponding failures have occurred more frequently during torrential summer rains when the rate of precipitation exceeded the rate of drainage runoff and the resulting hydraulic gradient over large roof areas caused substantial accumulation of water some distance from the eaves. A load factor of 1.2 shall be used for loads resulting from these phenomena.

$$\left(\text{Metric: } C_s = \frac{504\,S L_s^4}{I_s}\right)$$

where

L_p = column spacing in direction of girder (length of primary members), ft (m)
L_s = column spacing perpendicular to direction of girder (length of secondary members), ft (m)
S = spacing of secondary members, ft (m)
I_p = moment of inertia of primary members, in.^{4} (mm^4)
I_s = moment of inertia of secondary members, in.^{4} (mm^4)

A similar procedure must be followed using Figure A-K2.2.

Roof framing consisting of a series of equally spaced wall-bearing beams is consid-

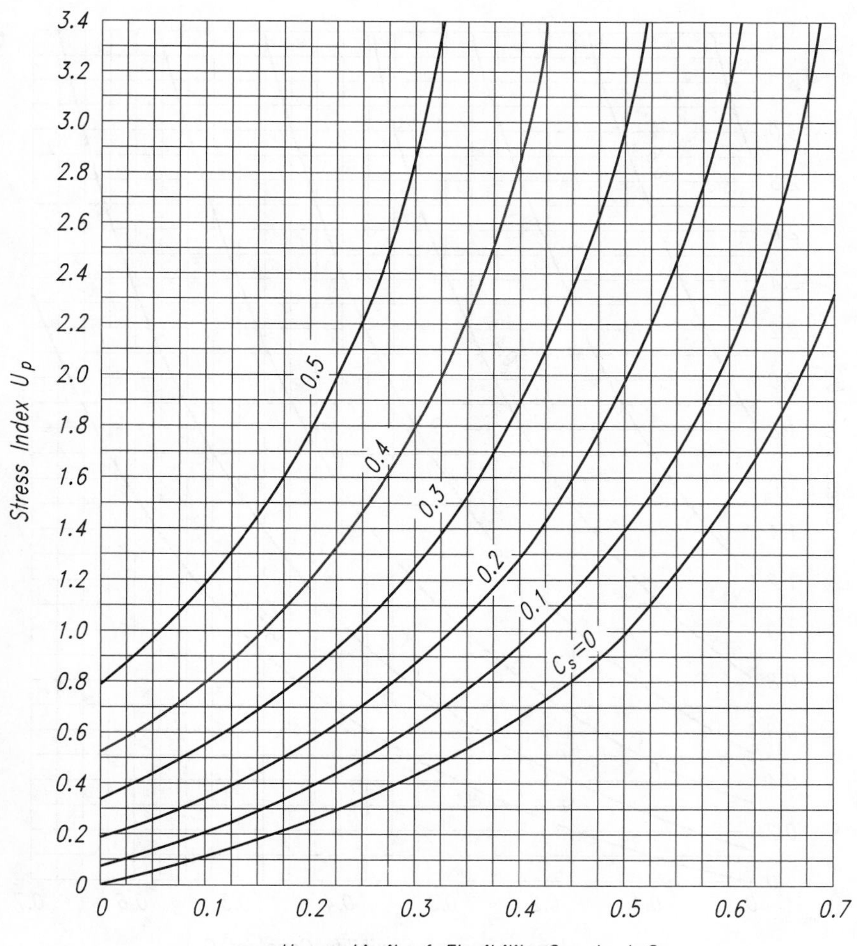

Fig. A-K2.1. Limiting flexibility coefficient for the primary systems.

ered as consisting of secondary members supported on an infinitely stiff primary member. For this case, enter Figure A-K2.2 with the computed stress index U_s. The limiting value of C_s is determined by the intercept of a horizontal line representing the U_s value and the curve for $C_p = 0$.

The ponding deflection contributed by a metal deck is usually such a small part of the total ponding deflection of a roof panel that it is sufficient merely to limit its moment of inertia (per foot (meter) of width normal to its span) to 0.000025 (3 940) times the fourth power of its span length. However, the stability against ponding of a roof consisting of a metal roof deck of relatively slender depth-span ratio, spanning between beams supported directly on columns, may need to be checked. This can be done using Figure A-K2.1 or A-K2.2 using as C_s the flexibility constant for a one-foot (one-meter) width of the roof deck ($S = 1.0$).

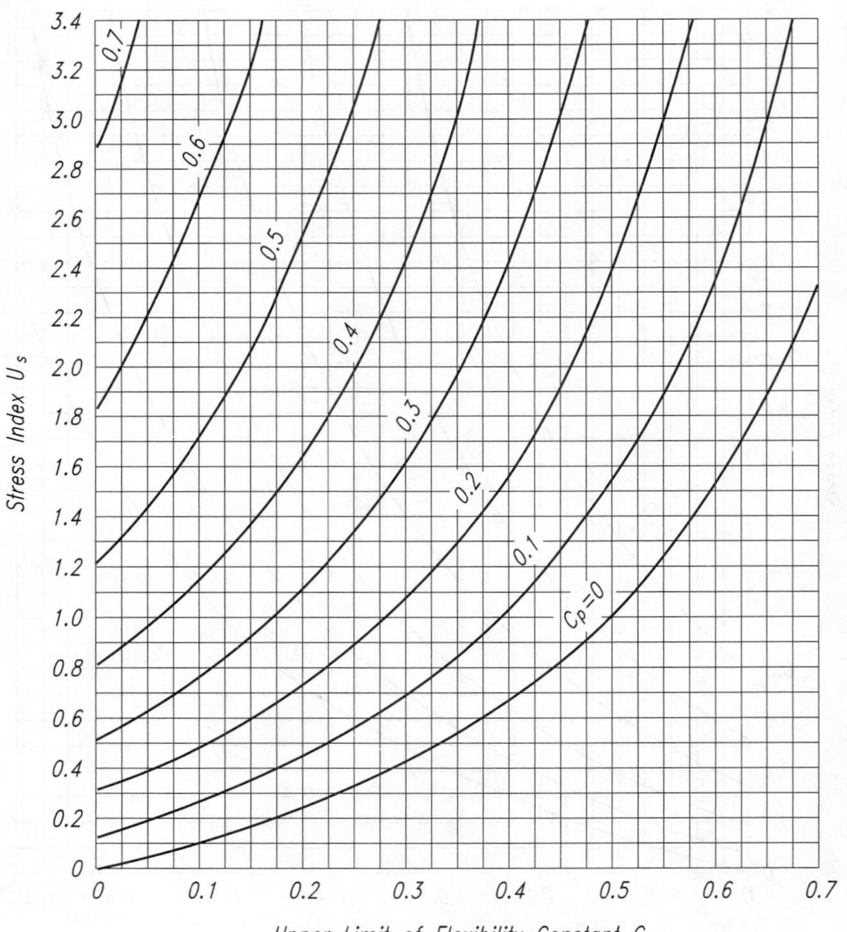

Fig. A-K2.2. Limiting flexibility coefficient for the secondary systems.

Since the shear rigidity of the web system of steel joists and trusses is less than that of a solid plate, their moment of inertia shall be taken as 85 percent of their chords.

K3. DESIGN FOR CYCLIC LOADING (FATIGUE)

This appendix applies to members and connections subject to high cycle loading within the elastic range of stresses of frequency and magnitude sufficient to initiate cracking and progressive failure (fatigue).

1. General

The provisions of this section apply to stresses calculated on the basis of unfactored loads. The maximum permitted stress due to unfactored loads is $0.66F_y$.

Stress range is defined as the magnitude of the change in stress due to the application or removal of the unfactored live load. In the case of a stress reversal, the stress range shall be computed as the numerical sum of maximum repeated tensile and compressive stresses or the numerical sum of maximum shearing stresses of opposite direction at the point of probable crack initiation.

In the case of complete-joint-penetration butt welds, the maximum design stress range calculated by Equation A-K3.1 applies only to welds with internal soundness meeting the acceptance requirements of Section 6.12.2 or 6.13.2 of AWS D1.1.

No evaluation of fatigue resistance is required if the live load stress range is less than the threshold stress range, F_{TH}. See Table A-K3.1.

No evaluation of fatigue resistance is required if the number of cycles of application of live load is less than 2×10^4.

The cyclic load resistance determined by the provisions of this appendix is applicable to structures with suitable corrosion protection or subject only to mildly corrosive atmospheres, such as normal atmospheric conditions.

The cyclic load resistance determined by the provisions of this appendix is applicable only to structures subject to temperatures not exceeding 300°F (150°C).

The Engineer of Record shall provide either complete details including weld sizes or shall specify the planned cycle life and the maximum range of moments, shears and reactions for the connections.

2. Calculation of Maximum Stresses and Stress Ranges

Calculated stresses shall be based upon elastic analysis. Stresses shall not be amplified by stress concentration factors for geometrical discontinuities.

For bolts and threaded rods subject to axial tension, the calculated stresses shall include the effects of prying action, if any.

In the case of axial stress combined with bending, the maximum stresses, of each kind, shall be those determined for concurrent arrangements of the applied load.

For members having symmetric cross sections, the fasteners and welds shall be arranged symmetrically about the axis of the member, or the total stresses including those due to eccentricity shall be included in the calculation of the stress range.

For axially loaded angle members where the center of gravity of the connecting welds lies between the line of the center of gravity of the angle cross section and the center of the connected leg, the effects of eccentricity shall be ignored. If the center of gravity of the connecting welds lies outside this zone, the total stresses, including those due to joint eccentricity, shall be included in the calculation of stress range.

3. Design Stress Range

The range of stress at service loads shall not exceed the stress range computed as follows.

(a) For stress categories A, B, B′, C, D, E and E′ the design stress range, F_{SR}, shall be determined by Equation A-K3.1 or A-K3.1M.

$$F_{SR} = \left(\frac{C_f}{N}\right)^{0.333} \geq F_{TH} \quad \text{(A-K3.1)}$$

$$\left(\text{Metric: } F_{SR} = \left(\frac{C_f \times 327}{N}\right)^{0.333} \geq F_{TH}\right) \quad \text{(A-K3.1M)}$$

where

F_{SR} = Design stress range, ksi (MPa)
C_f = Constant from Table A-K3.1 for the category
N = Number of stress range fluctuations in design life
= Number of stress range fluctuations per day × 365 × years of design life
F_{TH} = Threshold fatigue stress range, maximum stress range for indefinite design life from Table A-K3.1, ksi (MPa)

(b) For stress category F, the design stress range, F_{SR}, shall be determined by Equation A-K3.2 or A-K3.2M.

$$F_{SR} = \left(\frac{C_f}{N}\right)^{0.167} \geq F_{TH} \quad \text{(A-K3.2)}$$

$$\left(\text{Metric: } F_{SR} = \left(\frac{C_f \times 11 \times 10^4}{N}\right)^{0.167} \geq F_{TH}\right) \quad \text{(A-K3.2M)}$$

(c) For tension-loaded plate elements connected at their end by cruciform, T- or corner details with complete-joint-penetration groove welds or partial-joint-penetration groove welds, fillet welds, or combinations of the preceding, transverse to the direction of stress, the design stress range on the cross section of the tension-loaded plate element at the toe of the weld shall be determined as follows:

Based upon crack initiation from the toe of the weld on the tension loaded plate element the design stress range, F_{SR}, shall be determined by Equation A-K3.1 or A-K3.1M, for Category C which is equal to

$$F_{SR} = \left(\frac{44 \times 10^8}{N}\right)^{0.333} \geq 10$$

$$\left(\text{Metric: } F_{SR} = \left(\frac{14.4 \times 10^{11}}{N} \right)^{0.333} \geq 68.9 \right)$$

Based upon crack initiation from the root of the weld the design stress range, F_{SR}, on the tension loaded plate element using transverse partial-joint-penetration groove welds, with or without reinforcing or contouring fillet welds, the design stress range on the cross section at the toe of the weld shall be determined by Equation A-K3.3 or A-K3.3M, Category C′ as follows:

$$F_{SR} = R_{PJP} \left(\frac{44 \times 10^8}{N} \right)^{0.333} \tag{A-K3.3}$$

$$\left(\text{Metric: } F_{SR} = 1.72 R_{PJP} \left(\frac{14.4 \times 10^{11}}{N} \right)^{0.333} \right) \tag{A-K3.3M}$$

where:

R_{PJP} = reduction factor for reinforced or non-reinforced transverse partial-joint-penetration (PJP) joints. Use Category C if $R_{PJP} = 1.0$.

$$= \left(\frac{0.65 - 0.59\left(\frac{2a}{t_p} \right) + 0.72\left(\frac{w}{t_p} \right)}{t_p^{0.167}} \right) \leq 1.0$$

$2a$ = the length of the non-welded root face in the direction of the thickness of the tension-loaded plate, in. (mm)

w = the leg size of the reinforcing or contouring fillet, if any, in the direction of the thickness of the tension-loaded plate, in. (mm)

t_p = thickness of tension loaded plate, in. (mm)

Based upon crack initiation from the roots of a pair of transverse fillet welds on opposite sides of the tension loaded plate element the design stress range, F_{SR}, on the cross section at the toe of the welds shall be determined by Equation A-K3.4 or A-K3.4M, Category C″ as follows:

$$F_{SR} = R_{FIL} \left(\frac{44 \times 10^8}{N} \right)^{0.333} \tag{A-K3.4}$$

$$\left(\text{Metric: } F_{SR} = 1.72 R_{FIL} \left(\frac{14.4 \times 10^{11}}{N} \right)^{0.333} \right) \tag{A-K3.4M}$$

where

R_{FIL} = reduction factor for joints using a pair of transverse fillet welds only. Use Category C if $R_{FIL} = 1.0$.

$$= \left(\frac{0.06 + 0.72(w/t_p)}{t_p^{0.167}} \right) \leq 1.0$$

4. Bolts and Threaded Parts

The range of stress at service loads shall not exceed the stress range computed as follows.

(a) For mechanically fastened connections loaded in shear, the maximum range of stress in the connected material at service loads shall not exceed the design stress range computed using Equation A-K3.1 where C_f and F_{TH} are taken from Section 2 of Table A-K3.1.

(b) For high-strength bolts, common bolts, and threaded anchor rods with cut, ground or rolled threads, the maximum range of tensile stress on the net tensile area from applied axial load and moment plus load due to prying action shall not exceed the design stress range computed using Equation A-K3.1 or A-K3.1M. The factor C_f shall be taken as 3.9×10^8 (as for category E′). The threshold stress, F_{TH} shall be taken as 7 ksi (48 MPa) (as for category D). The net tensile area is given by Equation A-K3.5 and A-K3.6M.

$$A_t = \frac{\pi}{4}\left(d_b - \frac{0.9743}{n}\right)^2 \quad \text{(A-K3.5)}$$

$$\left(\text{Metric bolts: } A_t = \frac{\pi}{4}(d_b - 0.9382P)^2\right) \quad \text{(A-K3.6M)}$$

where

P = pitch, mm per thread
d_b = the nominal diameter (body or shank diameter), in. (mm)
n = threads per in.

For joints in which the material within the grip is not limited to steel or joints which are not tensioned to the requirements of Table J3.1 or J3.1M, all axial load and moment applied to the joint plus effects of prying action (if any) shall be assumed to be carried exclusively by the bolts or rods.

For joints in which the material within the grip is limited to steel and which are tensioned to the requirements of Table J3.1 or J3.1M, an analysis of the relative stiffness of the connected parts and bolts shall be permitted to be used to determine the tensile stress range in the pretensioned bolts due to the total service live load and moment plus effects of prying action. Alternatively, the stress range in the bolts shall be assumed to be equal to the stress on the net tensile area due to 20 percent of the absolute value of the service load axial load and moment from dead, live and other loads.

5. Special Fabrication and Erection Requirements

Longitudinal backing bars are permitted to remain in place, and if used, shall be continuous. If splicing is necessary for long joints, the bar shall be joined with complete penetration butt joints and the reinforcement ground prior to assembly in the joint.

In transverse joints subject to tension, backing bars, if used, shall be removed and the joint back gouged and welded.

In transverse complete-joint-penetration T- and corner joints, a reinforcing fillet weld, not less than ¼ in. (6 mm) in size shall be added at re-entrant corners.

The surface roughness of flame cut edges subject to significant cyclic tensile stress ranges shall not exceed 1,000 μin. (25 μm), where ASME B46.1 is the reference standard.

Re-entrant corners at cuts, copes and weld access holes shall form a radius of not less than ⅜ in. (10 mm) by pre-drilling or sub-punching and reaming a hole, or by thermal cutting to form the radius of the cut. If the radius portion is formed by thermal cutting, the cut surface shall be ground to a bright metal surface.

For transverse butt joints in regions of high tensile stress, run-off tabs shall be used to provide for cascading the weld termination outside the finished joint. End dams shall not be used. Run-off tabs shall be removed and the end of the weld finished flush with the edge of the member.

See Section J2.2b *Fillet Weld Terminations* for requirements for end returns on certain fillet welds subject to cyclic service loading.

TABLE A-K3.1
Fatigue Design Parameters

Description	Stress Category	Constant C_f	Threshold F_{TH} ksi (MPa)	Potential Crack Initiation Point
SECTION 1 – PLAIN MATERIAL AWAY FROM ANY WELDING				
1.1 Base metal, except non-coated weathering steel, with rolled or cleaned surface. Flame-cut edges with surface roughness value of 1,000 μin. (25 μm) or less, but without re-entrant corners.	A	250×10^8	24 (165)	Away from all welds or structural connections
1.2 Non-coated weathering steel base metal with rolled or cleaned surface. Flame-cut edges with surface roughness value of 1,000 μin. (25 μm) or less, but without re-entrant corners.	B	120×10^8	16 (110)	Away from all welds or structural connections
1.3 Member with drilled or reamed holes. Member with re-entrant corners at copes, cuts, block-outs or other geometrical discontinuities made to requirements of Appendix K3.5, except weld access holes.	B	120×10^8	16 (110)	At any external edge or at hole perimeter
1.4 Rolled cross sections with weld access holes made to requirements of Section J1.6 and Appendix K3.5. Members with drilled or reamed holes containing bolts for attachment of light bracing where there is a small longitudinal component of brace force.	C	44×10^8	10 (69)	At re-entrant corner of weld access hole or at any small hole (may contain bolt for minor connections)
SECTION 2 – CONNECTED MATERIAL IN MECHANICALLY FASTENED JOINTS				
2.1 Gross area of base metal in lap joints connected by high-strength bolts in joints satisfying all requirements for slip-critical connections.	B	120×10^8	16 (110)	Through gross section near hole
2.2 Base metal at net section of high-strength bolted joints, designed on the basis of bearing resistance, but fabricated and installed to all requirements for slip-critical connections.	B	120×10^8	16 (110)	In net section originating at side of hole
2.3 Base metal at the net section of other mechanically fastened joints except eye bars and pin plates.	D	22×10^8	7 (48)	In net section originating at side of hole
2.4 Base metal at net section of eyebar head or pin plate.	E	11×10^8	4.5 (31)	In net section originating at side of hole

TABLE A-K3.1 (Cont'd) Fatigue Design Parameters

Illustrative Typical Examples
SECTION 1 - PLAIN MATERIAL AWAY FROM ANY WELDING
1.1 and 1.2 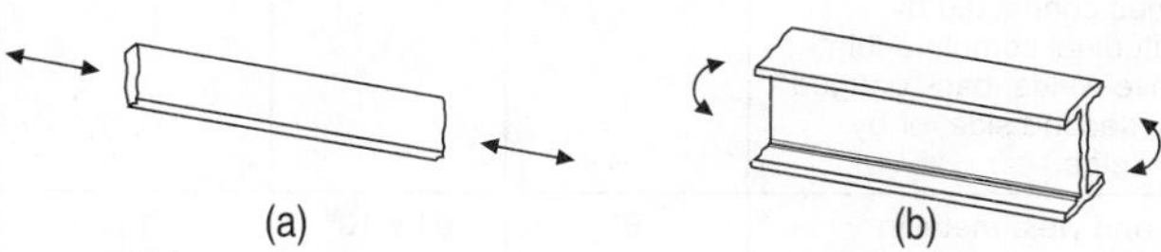
1.3 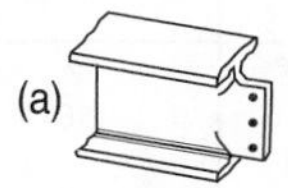 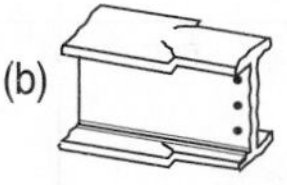 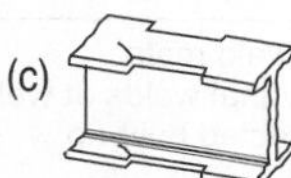
1.4 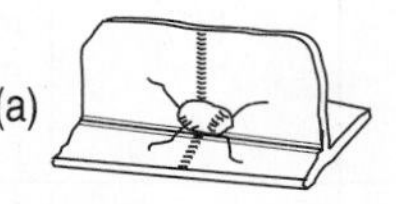 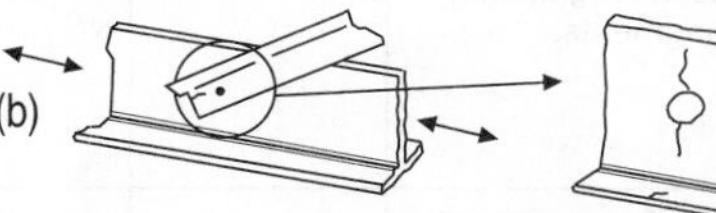
SECTION 2 - CONNECTED MATERIAL IN MECHANICALLY FASTENED JOINTS
2.1 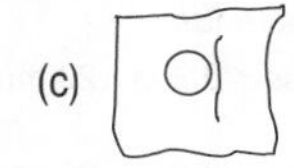
2.2 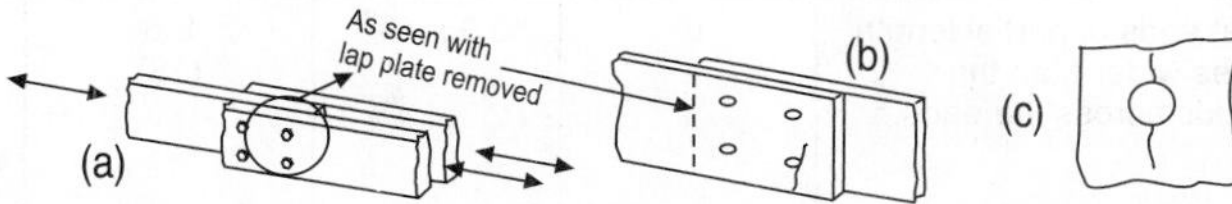
2.3 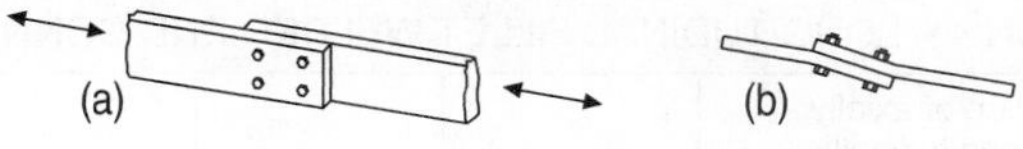
2.4 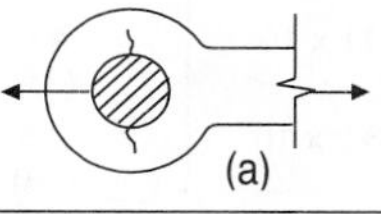 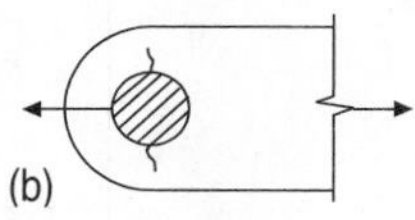

TABLE A-K3.1 (Cont'd)
Fatigue Design Parameters

Description	Stress Category	Constant C_f	Threshold F_{TH} ksi (MPa)	Potential Crack Initiation Point
SECTION 3 – WELDED JOINTS JOINING COMPONENTS OF BUILT-UP MEMBERS				
3.1 Base metal and weld metal in members without attachments built-up of plates or shapes connected by continuous longitudinal complete-joint-penetration groove welds, back gouged and welded from second side, or by continuous fillet welds.	B	120 x 10^8	16 (110)	From surface or internal discontinuities in weld away from end of weld
3.2 Base metal and weld metal in members without attachments built-up of plates or shapes, connected by continuous longitudinal complete penetration groove welds with backing bars not removed, or by continuous partial-joint-penetration groove welds.	B′	61 x 10^8	12 (83)	From surface or internal discontinuities in weld, including weld attaching backing bars
3.3 Base metal and weld metal termination of longitudinal welds at weld access holes in connected built-up members.	D	22 x 10^8	7 (48)	From the weld termination into the web or flange
3.4 Base metal at ends of longitudinal intermittent fillet weld segments.	E	11 x 10^8	4.5 (31)	In connected material at start and stop locations of any weld deposit
3.5 Base metal at ends of partial length welded coverplates narrower than the flange having square or tapered ends, with or without welds across the ends of coverplates wider than the flange with welds across the ends.				In flange at toe of end weld or in flange at termination of longitudinal weld or in edge of flange with wide coverplates
Flange thickness ≤ 0.8 in. (20 mm)	E	11 x 10^8	4.5 (31)	
Flange thickness > 0.8 in. (20 mm)	E′	3.9 x 10^8	2.6 (18)	
3.6 Base metal at ends of partial length welded coverplates wider than the flange without welds across the ends.	E′	3.9 x 10^8	2.6 (18)	In edge of flange at end of coverplate weld
SECTION 4 – LONGITUDINAL FILLET WELDED END CONNECTIONS				
4.1 Base metal at junction of axially loaded members with longitudinally welded end connections. Welds shall be on each side of the axis of the member to balance weld stresses.				Initiating from end of any weld termination extending into the base metal
$t \leq$ ½-in. (13 mm)	E	11 x 10^8	4.5 (31)	
$t >$ ½-in. (13 mm)	E′	3.9 x 10^8	2.6 (18)	

TABLE A-K3.1 (Cont'd) Fatigue Design Parameters

Illustrative Typical Examples
SECTION 3 - WELDED JOINTS JOINING COMPONENTS OF BUILT-UP MEMBERS
3.1 (a) or CJP (b) or (c)
3.2 (a) CJP (b)
3.3 (a) (b)
3.4 2-6 (a) (b) (c)
3.5 (a) (b) (c)
3.6 No Weld (a) Typ (b)
SECTION 4 - LONGITUDINAL FILLET WELDED END CONNECTIONS
4.1 t = thickness (a) t = thickness (b)

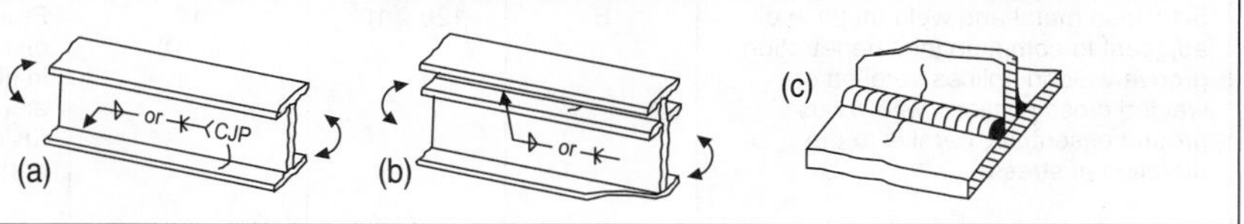

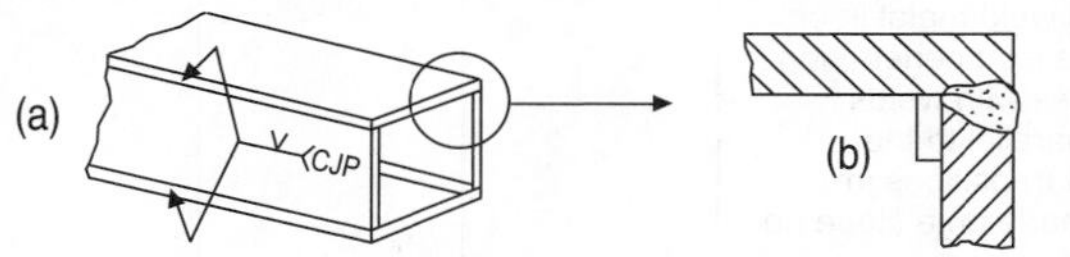

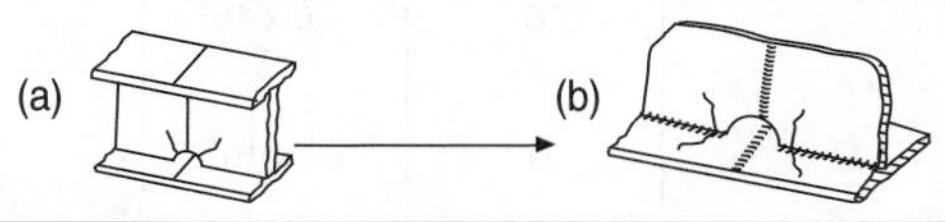

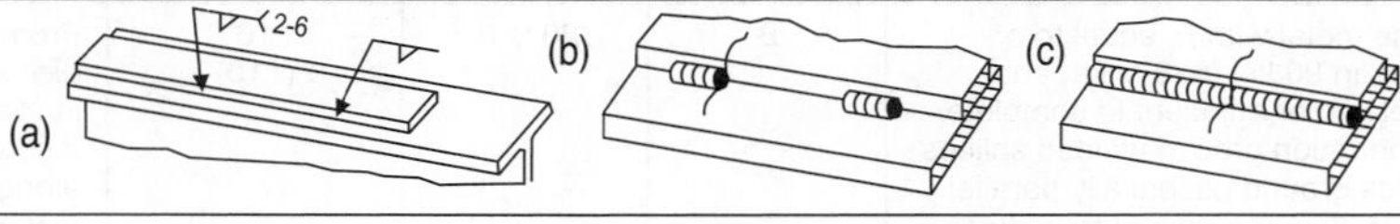

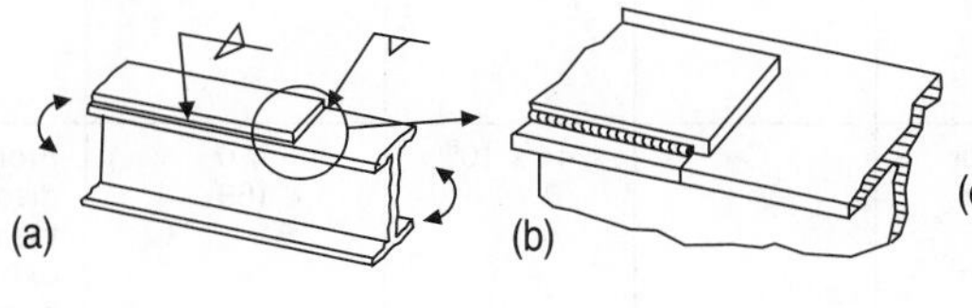

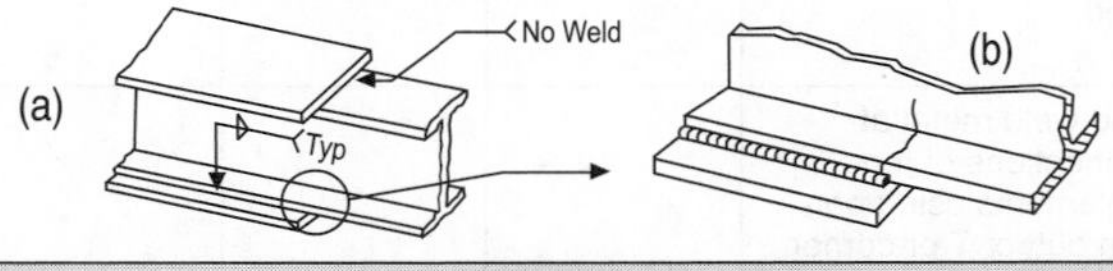

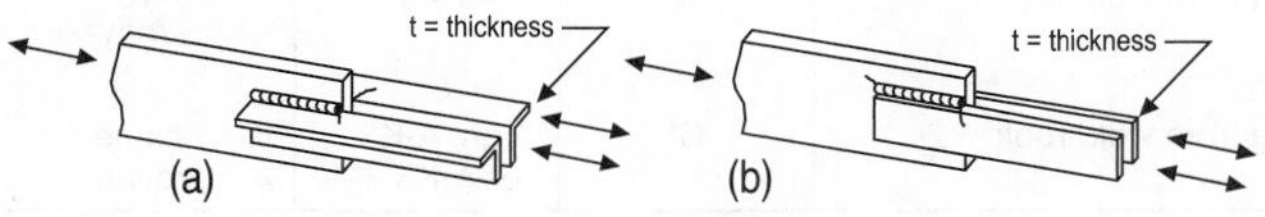

TABLE A-K3.1 (Cont'd) Fatigue Design Parameters

Description	Stress Category	Constant C_f	Threshold F_{TH} ksi (MPa)	Potential Crack Initiation Point
SECTION 5 – WELDED JOINTS TRANSVERSE TO DIRECTION OF STRESS				
5.1 Base metal and weld metal in or adjacent to complete joint penetration groove welded splices in rolled or welded cross sections with welds ground essentially parallel to the direction of stress.	B	120 x 10^8	16 (110)	From internal discontinuities in filler metal or along the fusion boundary
5.2 Base metal and weld metal in or adjacent to complete joint penetration groove welded splices with welds ground essentially parallel to the direction of stress at transitions in thickness or width made on a slope no greater than 8 to 20%.				From internal discontinuities in filler metal or along fusion boundary or at start of transition when $F_y \geq$ 90 ksi (620 MPa)
F_y < 90 ksi (620 MPa)	B	120 x 10^8	16 (110)	
$F_y \geq$ 90 ksi (620 MPa)	B′	61 x 10^8	12 (83)	
5.3 Base metal with F_y equal to or greater than 90 ksi (620 MPa) and weld metal in or adjacent to complete joint penetration groove welded splices with welds ground essentially parallel to the direction of stress at transitions in width made on a radius of not less than 2 ft. (600 mm) with the point of tangency at the end of the groove weld.	B	120 x 10^8	16 (110)	From internal discontinuities in filler metal or discontinuities along the fusion boundary
5.4 Base metal and weld metal in or adjacent to the toe of complete joint penetration T or corner joints or splices, with or without transitions in thickness having slopes no greater than 8 to 20%, when weld reinforcement is not removed.	C	44 x 10^8	10 (69)	From surface discontinuity at toe of weld extending into base metal or along fusion boundary.
5.5 Base metal and weld metal at transverse end connections of tension-loaded plate elements using partial joint penetration butt or T or corner joints, with reinforcing or contouring fillets, F_{SR} shall be the smaller of the toe crack or root crack stress range.				Initiating from geometrical discontinuity at toe of weld extending into base metal or, initiating at weld root subject to tension extending up and then out through weld
Crack initiating from weld toe:	C	44 x 10^8	10 (69)	
Crack initiating from weld root:	C′	Eqn. A-K3.3 or A-K3.3M	None provided	

TABLE A-K3.1 (Cont'd) Fatigue Design Parameters

Illustrative Typical Examples

SECTION 5 – WELDED JOINTS TRANSVERSE TO DIRECTION OF STRESS

5.1

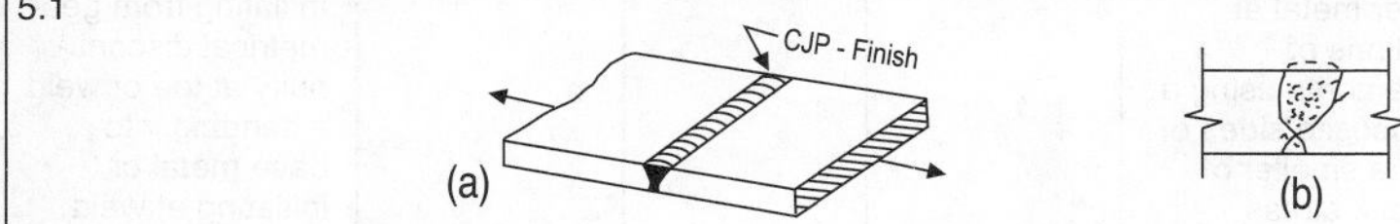

5.2

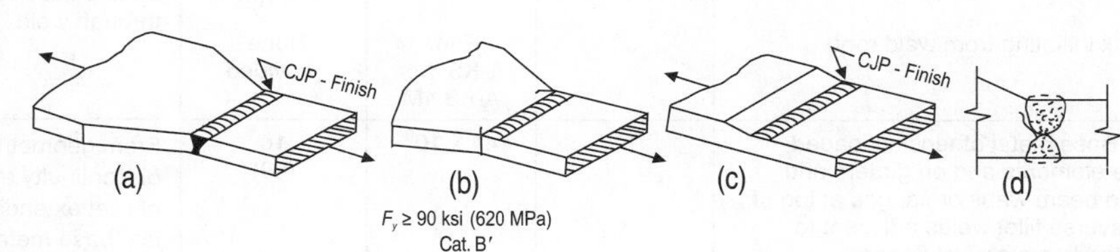

5.3

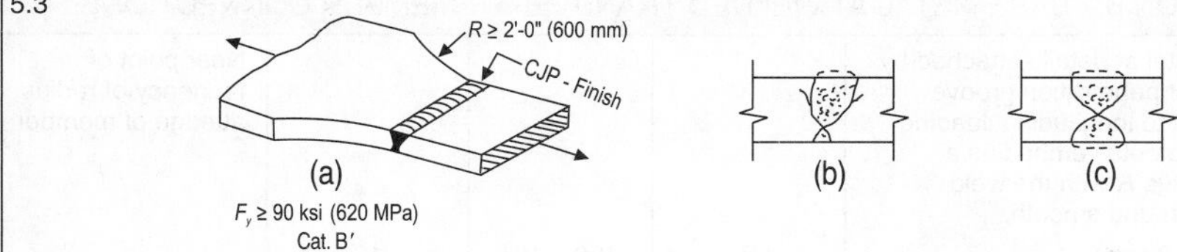

5.4

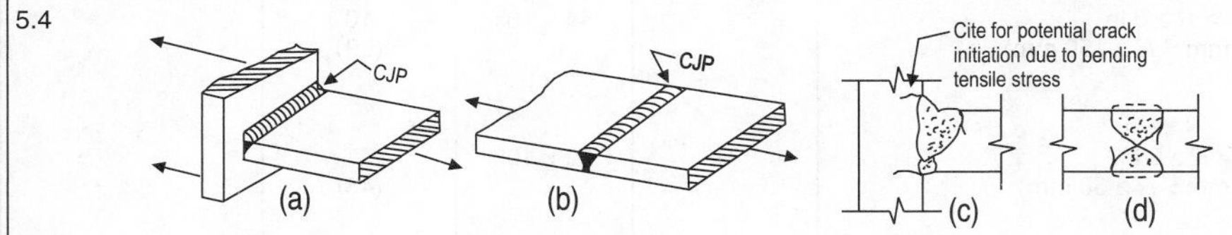

5.5

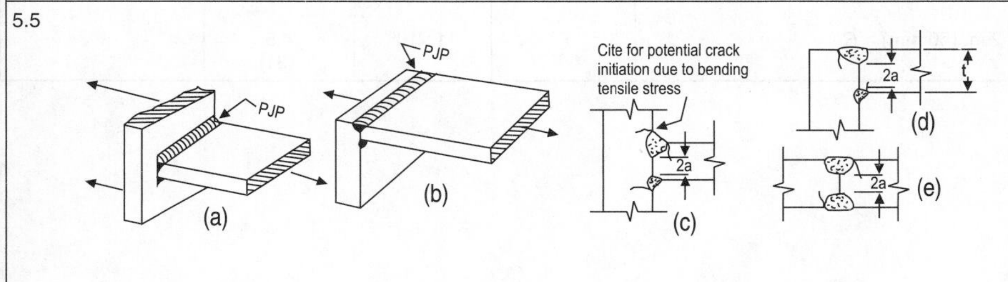

TABLE A-K3.1 (Cont'd)
Fatigue Design Parameters

Description	Stress Category	Constant C_f	Threshold F_{TH} ksi (MPa)	Potential Crack Initiation Point
SECTION 5 – WELDED JOINTS TRANSVERSE TO DIRECTION OF STRESS (cont'd)				
5.6 Base metal and filler metal at transverse end connections of tension-loaded plate elements using a pair of fillet welds on opposite sides of the plate. F_{SR} shall be the smaller of the toe crack or root crack stress range.				Initiating from geometrical discontinuity at toe of weld extending into base metal or, initiating at weld root subject to tension extending up and then out through weld
Crack initiating from weld toe:	C	44 x 10^8	10 (69)	
Crack initiating from weld root:	C″	Eqn. A-K3.4 or A-K3.4M	None provided	
5.7 Base metal of tension loaded plate elements and on girders and rolled beam webs or flanges at toe of transverse fillet welds adjacent to welded transverse stiffeners.	C	44 x 10^8	10 (69)	From geometrical discontinuity at toe of fillet extending into base metal
SECTION 6 – BASE METAL AT WELDED TRANSVERSE MEMBER CONNECTIONS				
6.1 Base metal at details attached by complete joint penetration groove welds subject to longitudinal loading only when the detail embodies a transition radius R with the weld termination ground smooth.				Near point of tangency of radius at edge of member
$R \geq 24$ in. (600 mm)	B	120×10^8	16 (110)	
24 in. > $R \geq 6$ in. (600 mm > $R \geq 150$ mm)	C	44×10^8	10 (69)	
6 in. > $R \geq 2$ in. (150 mm > $R \geq 50$ mm)	D	22×10^8	7 (48)	
2 in. (50 mm) > R	E	11×10^8	4.5 (31)	

TABLE A-K3.1 (Cont'd) Fatigue Design Parameters

Illustrative Typical Examples

SECTION 5 – WELDED JOINTS TRANSVERSE TO DIRECTION OF STRESS (cont'd)

5.6

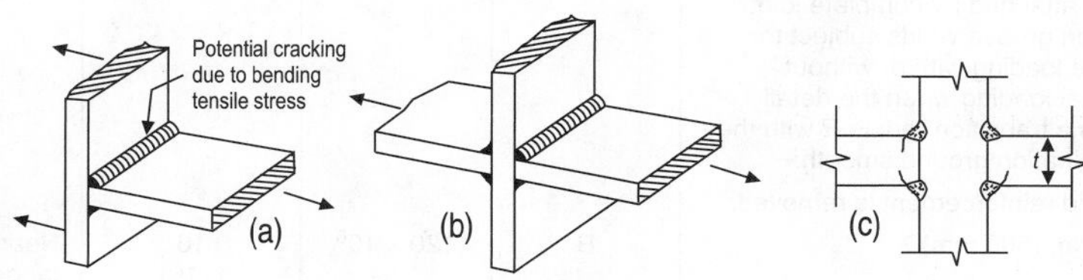

5.7

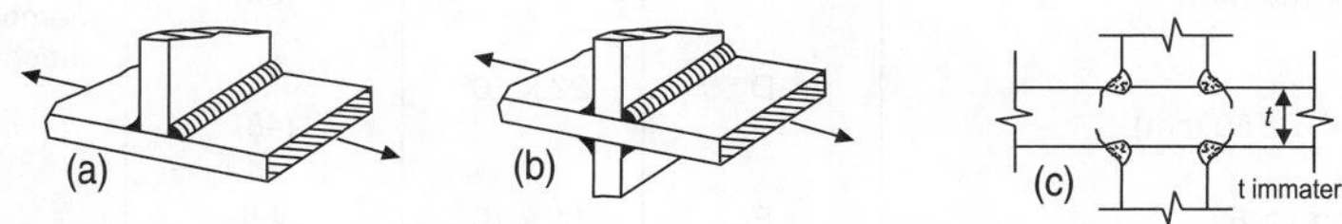

SECTION 6 – BASE METAL AT WELDED TRANSVERSE MEMBER CONNECTIONS

6.1

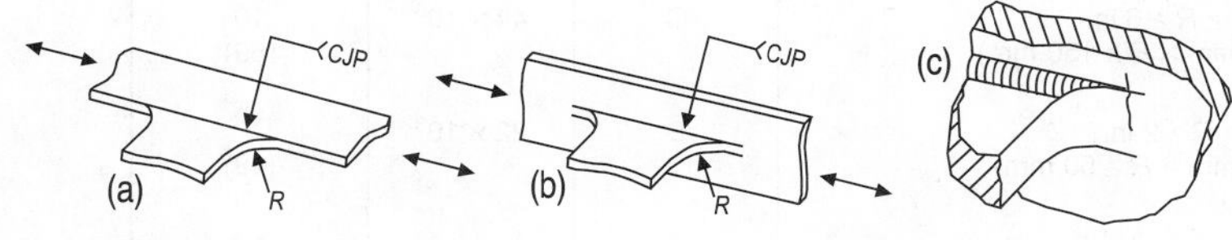

TABLE A-K3.1 (Cont'd) Fatigue Design Parameters

Description	Stress Category	Constant C_f	Threshold F_{TH} ksi (MPa)	Potential Crack Initiation Point
SECTION 6 – BASE METAL AT WELDED TRANSVERSE MEMBER CONNECTIONS (cont'd)				
6.2 Base metal at details of equal thickness attached by complete joint penetration groove welds subject to transverse loading with or without longitudinal loading when the detail embodies a transition radius R with the weld termination ground smooth:				
When weld reinforcement is removed:				
$R \geq 24$ in. (600 mm)	B	120×10^8	16 (110)	Near points of tangency of radius or in the weld or at fusion boundary or member or attachment
24 in. > $R \geq 6$ in. (600 mm > 150 mm)	C	44×10^8	10 (69)	
6 in. > $R \geq 2$ in. (150 mm > $R \geq 50$ mm)	D	22×10^8	7 (48)	
2 in. (50 mm) > R	E	11×10^8	4.5 (31)	
When weld reinforcement is not removed:				At toe of the weld either along edge of member or the attachment
$R \geq 24$ in. (600 mm)	C	44×10^8	10 (69)	
24 in. > $R \geq 6$ in. (600 mm > $R \geq 150$ mm)	C	44×10^8	10 (69)	
6 in. > $R \geq 2$ in. (150 mm > $R \geq 50$ mm)	D	22×10^8	7 (48)	
2 in. (50 mm) > R	E	11×10^8	4.5 (31)	
6.3 Base metal at details of unequal thickness attached by complete joint penetration groove welds subject to transverse loading with or without longitudinal loading when the detail embodies a transition radius R with the weld termination ground smooth.				
When weld reinforcement is removed:				
$R > 2$ in. (50 mm)	D	22×10^8	7 (48)	At toe of weld along edge of thinner material
$R \leq 2$ in. (50 mm)	E	11×10^8	4.5 (31)	In weld termination in small radius
When reinforcement is not removed:				At toe of weld along edge of thinner material
Any radius	E	11×10^8	4.5 (31)	

TABLE A-K3.1 (Cont'd)
Fatigue Design Parameters

Illustrative Typical Examples

SECTION 6 – BASE METAL AT WELDED TRANSVERSE MEMBER CONNECTIONS (cont'd)

6.2

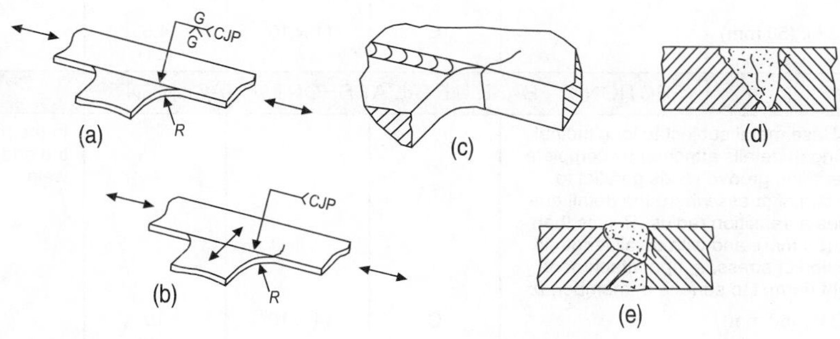

6.3

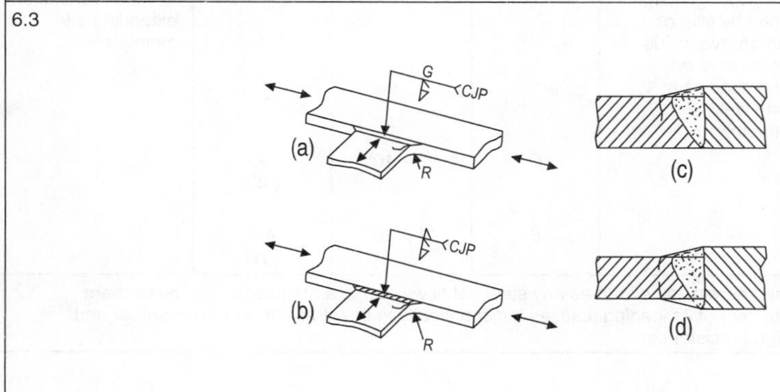

TABLE A-K3.1 (Cont'd) Fatigue Design Parameters

Description	Stress Category	Constant C_f	Threshold F_{TH} Ksi (MPa)	Potential Crack Initiation Point
SECTION 6 – BASE METAL AT WELDED TRANSVERSE MEMBER CONNECTIONS (cont'd)				
6.4 Base metal subject to longitudinal stress at transverse members, with or without transverse stress, attached by fillet or partial penetration groove welds parallel to direction of stress when the detail embodies a transition radius, *R*, with weld termination ground smooth:				In weld termination or from the toe of the weld extending into member
$R > 2$ in. (50 mm)	D	22×10^8	7 (48)	
$R \leq 2$ in. (50 mm)	E	11×10^8	4.5 (31)	
SECTION 7 – BASE METAL AT SHORT ATTACHMENTS[1]				
7.1 Base metal subject to longitudinal loading at details attached by complete penetration groove welds parallel to direction of stress where the detail embodies a transition radius, *R*, less than 2 in. (50 mm), and with detail length in direction of stress, *a*, and attachment height normal to surface of member, *b*:				In the member at the end of the weld
a < 2 in. (50 mm)	C	44×10^8	10 (69)	
2 in. (50 mm) $\leq a \leq 12b$ or 4 in (100 mm)	D	22×10^8	7 (48)	
a > 12*b* or 4 in. (100 mm) when *b* is $\leq$ 1 in. (25 mm)	E	11×10^8	4.5 (31)	
a > 12*b* or 4 in. (100 mm) when *b* is > 1 in. (25 mm)	E′	3.9×10^8	2.6 (18)	
7.2 Base metal subject to longitudinal stress at details attached by fillet or partial joint penetration groove welds, with or without transverse load on detail, when the detail embodies a transition radius, *R*, with weld termination ground smooth:				In weld termination extending into member
R > 2 in. (50 mm)	D	22×10^8	7 (48)	
R $\leq$ 2 in. (50 mm)	E	11×10^8	4.5 (31)	

[1] "Attachment" as used herein, is defined as any steel detail welded to a member which, by its mere presence and independent of its loading, causes a discontinuity in the stress flow in the member and thus reduces the fatigue resistance.

TABLE A-K3.1 (Cont'd) Fatigue Design Parameters

Illustrative Typical Examples

SECTION 6 – BASE METAL AT WELDED TRANSVERSE MEMBER CONNECTIONS (cont'd)

6.4

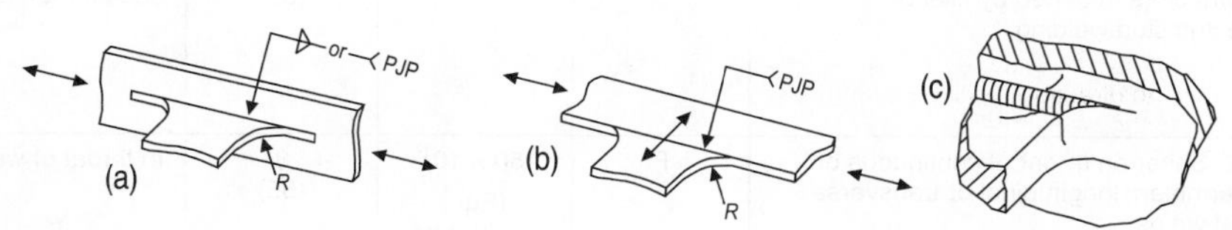

SECTION 7 – BASE METAL AT SHORT ATTACHMENTS[1]

7.1

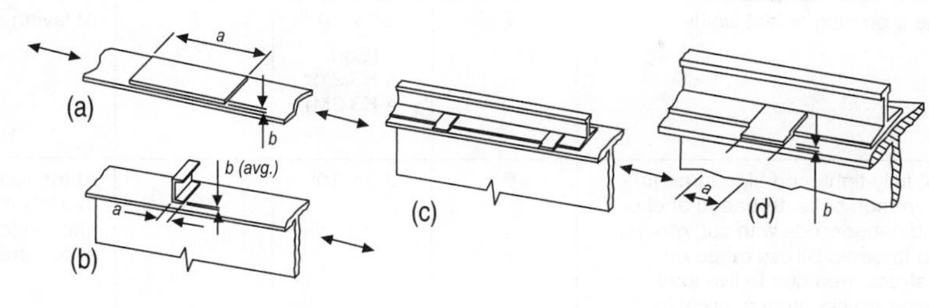

7.2

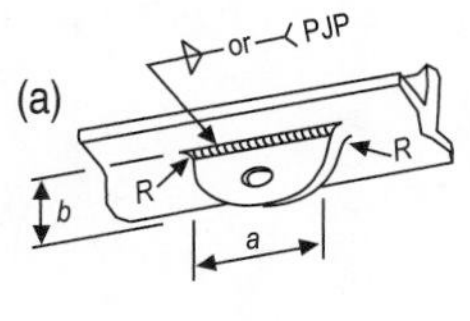

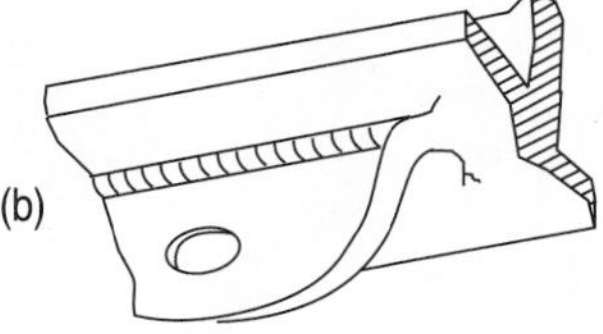

TABLE A-K3.1 (Cont'd) Fatigue Design Parameters

Description	Stress Category	Constant C_f	Threshold F_{TH} Ksi (MPa)	Potential Crack Initiation Point
SECTION 8 - MISCELLANEOUS				
8.1 Base metal at stud-type shear connectors attached by fillet or electric stud welding.	C	44×10^8	10 (69)	At toe of weld in base metal
8.2 Shear on throat of continuous or intermittent longitudinal or transverse fillet welds.	F	150×10^{10} (Eqn. A-K3.2 or A-K3.2M)	8 (55)	In throat of weld
8.3 Base metal at plug or slot welds.	E	11×10^8	4.5 (31)	At end of weld in base metal
8.4 Shear on plug or slot welds.	F	150×10^{10} (Eqn. A-K3.2 or A-K3.2M)	8 (55)	At faying surface
8.5 Not fully-tightened high-strength bolts, common bolts, threaded anchor rods and hanger rods with cut, ground or rolled threads. Stress range on tensile stress area due to live load plus prying action when applicable.	E′	3.9×10^8	7 (48)	At the root of the threads extending into the tensile stress area

TABLE A-K3.1 (Cont'd)
Fatigue Design Parameters

Illustrative Typical Examples

SECTION 8 - MISCELLANEOUS

8.1

(a) (b)

8.2

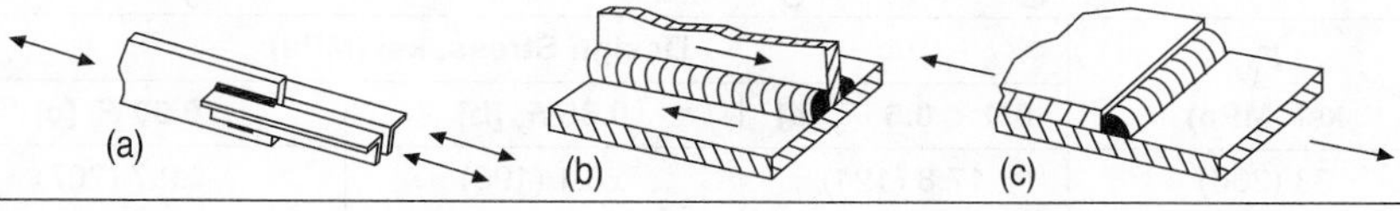

(a) (b) (c)

8.3

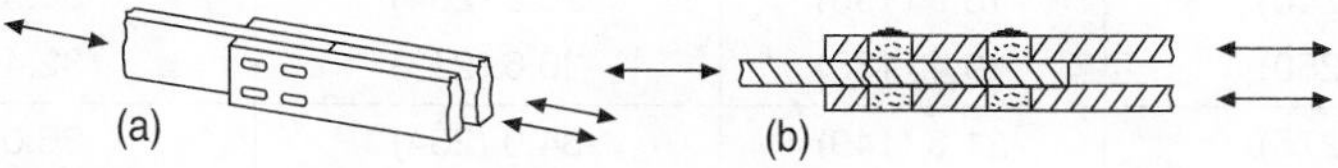

(a) (b)

8.4

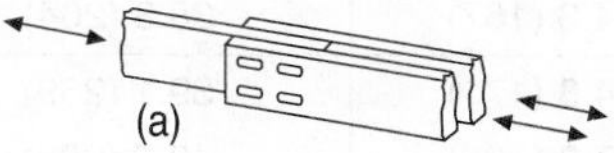

(a)

8.5

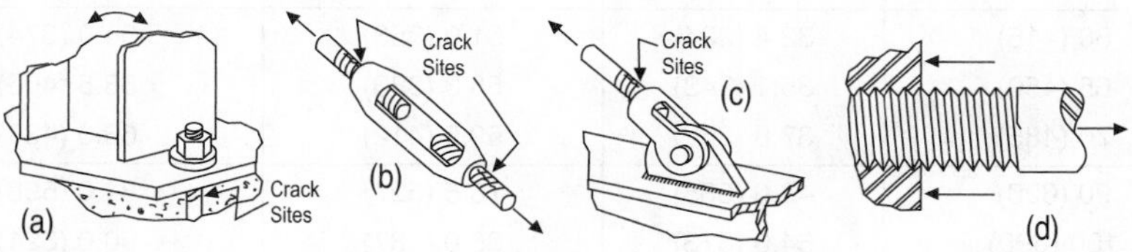

(a) Crack Sites

(b) Crack Sites

(c) Crack Sites

(d)

NUMERICAL VALUES

TABLE 1
Design Strength as a Function of F_y

F_y ksi (MPa)	Design Stress, ksi (MPa)		
	0.9 × 0.6 F_y [a]	**0.85 F_y [b]**	**0.90 F_y [c]**
33 (230)	17.8 (124)	28.1 (196)	29.7 (207)
35 (240)	18.9 (130)	29.8 (204)	31.5 (216)
36 (250)	19.4 (135)	30.6 (213)	32.4 (225)
40 (275)	21.6 (149)	34.0 (234)	36.0 (248)
42 (290)	22.7 (157)	35.7 (247)	37.8 (261)
45 (310)	24.3 (167)	38.3 (264)	40.5 (279)
46 (317)	24.8 (171)	39.1 (269)	41.4 (285)
50 (345)	27.0 (186)	42.5 (293)	45.0 (311)
55 (380)	29.7 (205)	46.8 (323)	49.5 (342)
60 (415)	32.4 (224)	51.0 (353)	54.0 (374)
65 (450)	35.1 (243)	55.3 (383)	58.5 (405)
70 (485)	37.8 (262)	59.5 (412)	63.0 (437)
90 (620)	48.6 (335)	76.5 (527)	81.0 (558)
100 (690)	54.0 (373)	85.0 (587)	90.0 (621)

[a] See Section F2, Equation F2-1
[b] See Section E2, Equation E2-1
[c] See Section D1, Equation D1-1

TABLE 2
Design Strength as a Function of F_u

Item	ASTM Designation	Grade	F_y Ksi (MPa)	F_u Ksi (MPa)	Design Strength, ksi (MPa): Connected Part of Designated Steel: Tension $0.75 \times F_u$ [a]	Bearing $0.75 \times 2.4 F_u$ [b]	Bolt or Threaded Part of Designated Steel: Tension $0.75 \times 0.75 F_u$ [c]	Shear $0.75 \times 0.40 F_u$ [d]	Shear $0.75 \times 0.50 F_u$ [e]
Shapes, Plates, Bars, Sheet and Tubing or Threaded parts	A36/A36M	-	36 (250)	58-80 (400-500)	43.5 (300)	104 (720)	32.6 (225)	17.4 (120)	21.8 (150)
	A53	-	35 (240)	60 (415)	45.0 (311)	108 (747)	-	-	-
	A242/A242M, A588/A588M	-	50 (345)	70 (485)	52.5 (364)	126 (873)	39.4 (273)	21.0 (146)	26.3 (182)
			42 (290)	63 (435)	47.3 (326)	113 (783)	35.4 (245)	18.9 (131)	23.2 (163)
			20 (275)	60 (415)	45.0 (311)	108 (747)	33.8 (233)	18.0 (125)	22.5 (156)
	A500	-	33/39 [f] (230/269)	45 (310)	33.8 (233)	81.0 (558)	-	-	-
			42/46 [f] (290/315)	58 (400)	43.5 (300)	104 (720)	-	-	-
			46/50 [f] (315/345)	62 (427)	46.5 (320)	112 (769)	-	-	-
	A501	-	36 (250)	58 (400)	43.5 (300)	104 (720)	-	-	-
	A514/A514M	-	100 (690)	110-130 (760-895)	82.5 (570)	198 (1370)	61.9 (428)	33.0 (228)	41.3 (285)
			90 (620)	100-130 (690-895)	75.0 (518)	180 (1240)	56.3 (388)	30.0 (207)	37.5 (259)
	A529/A529M	-	50 (345)	70-100 (485-690)	52.5 (364)	126 (873)	39.4 (273)	21.0 (146)	26.3 (182)
			55 (380)	70-100 (485-690)	52.5 (364)	126 (873)	39.4 (273)	21.0 (146)	26.3 (182)
	A570/A570M	-	40 (275)	55 (380)	41.3 (285)	99.0 (684)	-	-	-
			42 (290)	58 (400)	43.5 (300)	104 (720)	-	-	-
	A572/A572M	42 (290)	42 (290)	60 (415)	45.0 (311)	108 (747)	33.8 (233)	18.0 (125)	22.5 (156)
		50 (345)	50 (345)	65 (450)	48.8 (338)	117 (810)	36.6 (253)	19.5 (135)	24.2 (169)
		55 (380)	55 (380)	70 (485)	52.5 (364)	126 (873)	39.4 (273)	21.0 (146)	26.3 (182)
		60 (415)	60 (415)	75 (520)	56.3 (390)	135 (936)	42.2 (293)	22.5 (156)	28.1 (195)
		65 (450)	65 (450)	80 (550)	60.0 (413)	144 (990)	45.0 (309)	24.0 (165)	30.0 (206)

TABLE 2 (cont'd)
Design Strength as a Function of F_u

Item	ASTM Designation	Grade	F_y Ksi (MPa)	F_u Ksi (MPa)	Design Strength, ksi (MPa)				
					Connected Part of Designated Steel		Bolt or Threaded Part of Designated Steel		
					Tension $0.75 \times F_u$ [a]	Bearing $0.75 \times 2.4\ F_u$ [b]	Tension $0.75 \times 0.75\ F_u$ [c]	Shear $0.75 \times 0.40\ F_u$ [d]	Shear $0.75 \times 0.50\ F_u$ [e]
Shapes, Plates, Bars, Sheet and Tubing or Threaded parts	A606	-	45 (310)	65 (450)	48.8 (338)	117 (810)	-	-	-
			50 (345)	70 (485)	52.5 (364)	126 (873)	-	-	-
	A607	-	45 (310)	60 (415)	45.0 (311)	108 (747)	-	-	-
			50 (345)	65 (450)	48.8 (338)	117 (810)	-	-	-
			55 (380)	70 (485)	52.5 (364)	126 (873)	-	-	-
			60 (415)	75 (520)	56.3 (390)	135 (936)	-	-	-
			65 (450)	80 (550)	60.0 (413)	144 (990)	-	-	-
			70 (485)	85 (585)	63.8 (439)	153 (1050)	-	-	-
	A618	-	50 (345)	70 (485)	52.5 (364)	126 (873)	-	-	-
			50 (345)	65 (450)	48.8 (388)	117 (810)	-	-	-
	A913/A913M	50 (345)	50 (345)	65 (450)	48.8 (338)	117 (810)	36.6 (253)	19.5 (135)	24.4 (169)
		60 (415)	60 (415)	75 (520)	56.3 (390)	135 (936)	42.2 (293)	22.5 (156)	28.1 (195)
		65 (450)	65 (450)	80 (550)	60.0 (413)	144 (990)	45.0 (309)	24.0 (165)	30.0 (206)
		70 (485)	70 (485)	90 (620)	67.5 (465)	162 (1120)	50.6 (349)	27.0 (186)	33.8 (233)
	A992/A992M	-	50 to 65 (345 to 450)	65 (450)	48.8 (338)	117 (810)	36.6 (253)	19.5 (135)	24.4 (169)
Bolts	A449	-	92 (635)	120 (825)	-	-	67.5 (464)	36.0 (248)	45.0 (309)
			81 (560)	105 (725)	-	-	59.1 (408)	31.5 (218)	39.4 (272)
			58 (400)	90 (620)	-	-	50.6 (349)	27.0 (186)	33.8 (233)

[a] On effective net area, see Sections D1, J5.2.
[b] Produced by fastener in shear, see Section J3.10. Note that smaller maximum design bearing stresses, as a function of hole type spacing, are given.
[c] On nominal body area, see Table J3.2.
[d] Threads not excluded from shear plane, see Table J3.2.
[e] Threads excluded from shear plane, see Table J3.2.
[f] Smaller value for circular shapes, larger for square or rectangular shapes.
Note: For dimensional and size limitations, see the appropriate ASTM Specification.

TABLE 3-36

Design Stress for Compression Members of 36 ksi Specified Yield Stress Steel, ϕ_c = 0.85[a]

$\frac{Kl}{r}$	$\phi_c F_{cr}$ ksi	$\frac{Kl}{r}$	$\phi_c F_{cr}$ ksi	$\frac{Kl}{r}$	$\phi_c F_{cr}$ ksi	$\frac{Kl}{r}$	$\phi_c F_{cr}$ ksi	$\frac{Kl}{r}$	$\phi_c F_{cr}$ ksi
1	30.6	41	28.0	81	21.7	121	14.2	161	8.23
2	30.6	42	27.9	82	21.5	122	14.0	162	8.13
3	30.6	43	27.8	83	21.3	123	13.8	163	8.03
4	30.6	44	27.6	84	21.1	124	13.6	164	7.93
5	30.6	45	27.5	85	20.9	125	13.4	165	7.84
6	30.5	46	27.4	86	20.7	126	13.3	166	7.74
7	30.5	47	27.2	87	20.5	127	13.1	167	7.65
8	30.5	48	27.1	88	20.4	128	12.9	168	7.56
9	30.5	49	27.0	89	20.2	129	12.7	169	7.47
10	30.4	50	26.8	90	20.0	130	12.6	170	7.38
11	30.4	51	26.7	91	19.8	131	12.4	171	7.30
12	30.4	52	26.5	92	19.6	132	12.2	172	7.21
13	30.3	53	26.4	93	19.4	133	12.1	173	7.13
14	30.3	54	26.3	94	19.2	134	11.9	174	7.05
15	30.2	55	26.1	95	19.0	135	11.7	175	6.97
16	30.2	56	25.9	96	18.8	136	11.5	176	6.89
17	30.1	57	25.8	97	18.7	137	11.4	177	6.81
18	30.1	58	25.6	98	18.5	138	11.2	178	6.73
19	30.0	59	25.5	99	18.3	139	11.0	179	6.66
20	30.0	60	25.3	100	18.1	140	10.9	180	6.59
21	29.9	61	25.2	101	17.9	141	10.7	181	6.51
22	29.8	62	25.0	102	17.7	142	10.6	182	6.44
23	29.8	63	24.8	103	17.5	143	10.4	183	6.37
24	26.7	64	24.7	104	17.3	144	10.3	184	6.30
25	29.6	65	24.5	105	17.1	145	10.2	185	6.23
26	29.5	66	24.3	106	16.9	146	10.0	186	6.17
27	29.5	67	24.2	107	16.8	147	9.87	187	6.10
28	29.4	68	24.0	108	16.6	148	9.74	188	6.04
29	29.3	69	23.8	109	16.4	149	9.61	189	5.97
30	29.2	70	23.6	110	16.2	150	9.48	190	5.91
31	29.1	71	23.5	111	16.0	151	9.36	191	5.85
32	29.0	72	23.3	112	15.8	152	9.23	192	5.79
33	28.9	73	23.1	113	15.6	153	9.11	193	5.73
34	28.8	74	22.9	114	15.4	154	9.00	194	5.67
35	28.7	75	22.8	115	15.3	155	8.88	195	5.61
36	28.6	76	22.6	116	15.1	156	8.77	196	5.55
37	28.5	77	22.4	117	14.9	157	8.66	197	5.50
38	28.4	78	22.2	118	14.7	158	8.55	198	5.44
39	28.3	79	22.0	119	14.5	159	8.44	199	5.39
40	28.1	80	21.9	120	14.3	160	8.33	200	5.33

[a] When element width-to-thickness ratio exceeds λ_r, see Appendix B5.3.

TABLE 3-36M
Design Stress for Compression Members of 250 MPa Specified Yield Stress Steel, ϕ_c = 0.85[a]

$\frac{Kl}{r}$	$\phi_c F_{cr}$ MPa	$\frac{Kl}{r}$	$\phi_c F_{cr}$ MPa	$\frac{Kl}{r}$	$\phi_c F_{cr}$ MPa	$\frac{Kl}{r}$	$\phi_c F_{cr}$ MPa	$\frac{Kl}{r}$	$\phi_c F_{cr}$ MPa
1	212	41	194	81	150	121	97.8	161	56.8
2	212	42	194	82	149	122	96.5	162	56.1
3	212	43	193	83	147	123	95.3	163	55.4
4	212	44	192	84	146	124	94.1	164	54.7
5	212	45	191	85	145	125	92.8	165	54.1
6	212	46	190	86	144	126	91.6	166	53.4
7	212	47	189	87	142	127	90.4	167	52.8
8	212	48	188	88	141	128	89.2	168	52.1
9	212	49	187	89	140	129	88.0	169	51.5
10	211	50	186	90	138	130	86.8	170	50.9
11	211	51	185	91	137	131	85.6	171	50.3
12	211	52	184	92	136	132	84.4	172	49.7
13	211	53	183	93	134	133	83.2	173	49.2
14	210	54	182	94	133	134	82.0	174	48.6
15	210	55	181	95	132	135	80.7	175	48.1
16	210	56	180	96	130	136	79.6	176	47.5
17	209	57	179	97	129	137	78.4	177	47.0
18	209	58	178	98	128	138	77.3	178	46.4
19	208	59	177	99	126	139	76.2	179	45.9
20	208	60	176	100	125	140	75.1	180	45.4
21	208	61	174	101	124	141	74.0	181	44.9
22	207	62	173	102	122	142	73.0	182	44.4
23	207	63	172	103	121	143	72.0	183	43.9
24	206	64	171	104	120	144	71.0	184	43.5
25	206	65	170	105	118	145	70.0	185	43.0
26	205	66	169	106	117	146	69.0	186	42.5
27	204	67	168	107	116	147	68.1	187	42.1
28	204	68	166	108	115	148	67.2	188	41.6
29	203	69	165	109	113	149	66.3	189	41.2
30	203	70	164	110	112	150	65.4	190	40.8
31	202	71	163	111	111	151	64.5	191	40.3
32	201	72	161	112	109	152	63.7	192	39.9
33	201	73	160	113	108	153	62.9	193	39.5
34	200	74	159	114	107	154	62.1	194	39.1
35	199	75	158	115	105	155	61.3	195	38.7
36	198	76	156	116	104	156	60.5	196	38.3
37	198	77	155	117	103	157	59.7	197	37.9
38	197	78	154	118	102	158	58.9	198	37.5
39	196	79	153	119	100	159	58.2	199	37.2
40	195	80	151	120	99.1	160	57.5	200	36.8

[a] When element width-to-thickness ratio exceeds λ_r, see Appendix B5.3.

TABLE 3-50
Design Stress for Compression Members of 50 ksi Specified Yield Stress Steel, $\phi_c = 0.85$[a]

$\frac{Kl}{r}$	$\phi_c F_{cr}$ ksi	$\frac{Kl}{r}$	$\phi_c F_{cr}$ ksi	$\frac{Kl}{r}$	$\phi_c F_{cr}$ ksi	$\frac{Kl}{r}$	$\phi_c F_{cr}$ ksi	$\frac{Kl}{r}$	$\phi_c F_{cr}$ ksi
1	42.5	41	37.6	81	26.3	121	14.6	161	8.23
2	42.5	42	37.4	82	26.0	122	14.3	162	8.13
3	42.5	43	37.1	83	25.7	123	14.1	163	8.03
4	42.5	44	36.9	84	25.4	124	13.9	164	7.93
5	42.4	45	36.7	85	25.1	125	13.7	165	7.84
6	42.4	46	36.4	86	24.8	126	13.4	166	7.74
7	42.4	47	36.2	87	24.4	127	13.2	167	7.65
8	42.3	48	35.9	88	24.1	128	13.0	168	7.56
9	42.3	49	35.7	89	23.8	129	12.8	169	7.47
10	42.2	50	35.4	90	23.5	130	12.6	170	7.38
11	42.1	51	35.1	91	23.2	131	12.4	171	7.30
12	42.1	52	34.9	92	22.9	132	12.3	172	7.21
13	42.0	53	34.6	93	22.6	133	12.1	173	7.13
14	41.9	54	34.3	94	22.3	134	11.9	174	7.05
15	41.8	55	34.1	95	22.0	135	11.7	175	6.97
16	41.7	56	33.8	96	21.7	136	11.5	176	6.89
17	41.6	57	33.5	97	21.4	137	11.4	177	6.81
18	41.5	58	33.2	98	21.1	138	11.2	178	6.73
19	41.4	59	33.0	99	20.8	139	11.0	179	6.66
20	41.3	60	32.7	100	20.5	140	10.9	180	6.59
21	41.2	61	32.4	101	20.2	141	10.7	181	6.51
22	41.0	62	32.1	102	19.9	142	10.6	182	6.44
23	40.9	63	31.8	103	19.6	143	10.4	183	6.37
24	40.8	64	31.5	104	19.3	144	10.3	184	6.30
25	40.6	65	31.2	105	19.0	145	10.2	185	6.23
26	40.5	66	30.9	106	18.7	146	10.0	186	6.17
27	40.3	67	30.6	107	18.4	147	9.87	187	6.10
28	40.1	68	30.3	108	18.1	148	9.74	188	6.04
29	40.0	69	30.0	109	17.8	149	9.61	189	5.97
30	39.8	70	29.7	110	17.6	150	9.48	190	5.91
31	39.6	71	29.4	111	17.3	151	9.36	191	5.85
32	39.4	72	29.1	112	17.0	152	9.23	192	5.79
33	39.3	73	28.8	113	16.7	153	9.11	193	5.73
34	39.1	74	28.5	114	16.4	154	9.00	194	5.67
35	38.9	75	28.2	115	16.1	155	8.88	195	5.61
36	38.7	76	27.9	116	15.9	156	8.77	196	5.55
37	38.5	77	27.6	117	15.6	157	8.66	197	5.50
38	38.2	78	27.2	118	15.3	158	8.55	198	5.44
39	38.0	79	26.9	119	15.1	159	8.44	199	5.39
40	37.8	80	26.6	120	14.8	160	8.33	200	5.33

[a] When element width-to-thickness ratio exceeds λ_r, see Appendix B5.3.

TABLE 3-50M

Design Stress for Compression Members of 345 MPa Specified Yield Stress Steel, $\phi_c = 0.85$[a]

$\frac{Kl}{r}$	$\phi_c F_{cr}$ MPa	$\frac{Kl}{r}$	$\phi_c F_{cr}$ MPa	$\frac{Kl}{r}$	$\phi_c F_{cr}$ MPa	$\frac{Kl}{r}$	$\phi_c F_{cr}$ MPa	$\frac{Kl}{r}$	$\phi_c F_{cr}$ MPa
1	293	41	259	81	181	121	101	161	56.8
2	293	42	258	82	179	122	98.9	162	56.1
3	293	43	256	83	177	123	97.3	163	55.4
4	293	44	255	84	175	124	95.7	164	54.7
5	293	45	253	85	173	125	94.2	165	54.1
6	292	46	251	86	171	126	92.7	166	53.4
7	292	47	249	87	169	127	91.2	167	52.8
8	292	48	248	88	166	128	89.8	168	52.1
9	292	49	246	89	164	129	88.4	169	51.5
10	291	50	244	90	162	130	87.1	170	50.9
11	291	51	242	91	160	131	85.7	171	50.3
12	290	52	241	92	158	132	84.5	172	49.7
13	290	53	239	93	156	133	83.2	173	49.2
14	289	54	237	94	154	134	82.0	174	48.6
15	288	55	235	95	152	135	80.7	175	48.1
16	288	56	233	96	149	136	79.6	176	47.5
17	287	57	231	97	147	137	78.4	177	47.0
18	286	58	229	98	145	138	77.3	178	46.4
19	286	59	227	99	143	139	76.2	179	45.9
20	285	60	225	100	141	140	75.1	180	45.4
21	284	61	223	101	139	141	74.0	181	44.9
22	283	62	221	102	137	142	73.0	182	44.4
23	282	63	219	103	135	143	72.0	183	43.9
24	281	64	217	104	133	144	71.0	184	43.5
25	280	65	215	105	131	145	70.0	185	43.0
26	279	66	213	106	129	146	69.0	186	42.5
27	278	67	211	107	127	147	68.1	187	42.1
28	277	68	209	108	125	148	67.2	188	41.6
29	276	69	207	109	123	149	66.3	189	41.2
30	275	70	205	110	121	150	65.4	190	40.8
31	273	71	203	111	119	151	64.5	191	40.3
32	272	72	201	112	117	152	63.7	192	39.9
33	271	73	199	113	115	153	62.9	193	39.5
34	269	74	196	114	113	154	62.1	194	39.1
35	268	75	194	115	111	155	61.3	195	38.7
36	267	76	192	116	109	156	60.5	196	38.3
37	265	77	190	117	107	157	59.7	197	37.9
38	264	78	188	118	106	158	58.9	198	37.5
39	262	79	186	119	104	159	58.2	199	37.2
40	261	80	184	120	102	160	57.5	200	36.8

[a] When element width-to-thickness ratio exceeds λ_r, see Appendix B5.3.

TABLE 4

Values of $\phi_c F_{cr} / F_y$, $\phi_c = 0.85$ for Determining Design Stress for Compression Members for Steel of Any Yield Stress[a]

λ_c	$\phi_c F_{cr}/F_y$	λ_c	$\phi_c F_{cr}/F_y$	λ_c	$\phi_c F_{cr}/F_y$	λ_c	$\phi_c F_{cr}/F_y$
0.02	0.850	0.82	0.641	1.62	0.284	2.42	0.127
0.04	0.849	0.84	0.632	1.64	0.277	2.44	0.125
0.06	0.849	0.86	0.623	1.66	0.271	2.46	0.123
0.08	0.848	0.88	0.614	1.68	0.264	2.48	0.121
0.10	0.846	0.90	0.605	1.70	0.258	2.50	0.119
0.12	0.845	0.92	0.596	1.72	0.252	2.52	0.117
1.04	0.843	0.94	0.587	1.74	0.246	2.54	0.116
0.16	0.841	0.96	0.578	1.76	0.241	2.56	0.114
0.18	0.839	0.98	0.568	1.78	0.235	2.58	0.112
0.20	0.836	1.00	0.559	1.80	0.230	2.60	0.110
0.22	0.833	1.02	0.550	1.82	0.225	2.62	0.109
0.24	0.830	1.04	0.540	1.84	0.220	2.64	0.107
0.26	0.826	1.06	0.531	1.86	0.215	2.66	0.105
0.28	0.823	1.08	0.521	1.88	0.211	2.68	0.104
0.30	0.819	1.10	0.512	1.90	0.206	2.70	0.102
0.32	0.814	1.12	0.503	1.92	0.202	2.72	0.101
0.34	0.810	1.14	0.493	1.94	0.198	2.74	0.099
0.36	0.805	1.16	0.484	1.96	0.194	2.76	0.098
0.38	0.800	1.18	0.474	1.98	0.190	2.78	0.096
0.40	0.795	1.20	0.465	2.00	0.186	2.80	0.095
0.42	0.789	1.22	0.456	2.02	0.183	2.82	0.094
0.44	0.784	1.24	0.446	2.04	0.179	2.84	0.092
0.46	0.778	1.26	0.437	2.06	0.176	2.86	0.091
0.48	0.772	1.28	0.428	2.08	0.172	2.88	0.090
0.50	0.765	1.30	0.419	2.10	0.169	2.90	0.089
0.52	0.759	1.32	0.410	2.12	0.166	2.92	0.087
0.54	0.752	1.34	0.401	2.14	0.163	2.94	0.086
0.56	0.745	1.36	0.392	2.16	0.160	2.96	0.085
0.58	0.738	1.38	0.383	2.18	0.157	2.98	0.084
0.60	0.731	1.40	0.374	2.20	0.154	3.00	0.083
0.62	0.724	1.42	0.365	2.22	0.151	3.02	0.082
0.64	0.716	1.44	0.357	2.24	0.149	3.04	0.081
0.66	0.708	1.46	0.348	2.26	0.146	3.06	0.080
0.68	0.700	1.48	0.339	2.28	0.143	3.08	0.079
0.70	0.692	1.50	0.331	2.30	0.141	3.10	0.078
0.72	0.684	1.52	0.323	2.32	0.138	3.12	0.077
0.74	0.676	1.54	0.340	2.34	0.136	3.14	0.076
0.76	0.667	1.56	0.306	2.36	0.134	3.16	0.075
0.78	0.659	1.58	0.299	2.38	0.132	3.18	0.074
0.80	0.650	1.60	0.291	2.40	0.129	3.20	0.073

[a] When element width-to-thickness ratios exceed λ_r, see Appendix B5.3
Values of $\lambda_c > 2.24$ exceed Kl/r of 200 for F_y = 36 ksi (250 MPa)
Values of $\lambda_c > 2.64$ exceed Kl/r of 200 for F_y = 50 ksi (345 MPa)

TABLE 5
Values of Kl/r for $F_y = 36$ and 50 ksi*

λ_c	Kl/r		λ_c	Kl/r	
	$F_y = 36$	$F_y = 50$		$F_y = 36$	$F_y = 50$
0.02	1.8	1.5	0.82	73.1	62.0
0.04	3.6	3.0	0.84	74.9	63.6
0.06	5.3	4.5	0.86	76.7	65.1
0.08	7.1	6.1	0.88	78.5	66.6
0.10	8.9	7.6	0.90	80.2	68.1
0.12	10.7	9.1	0.92	82.0	69.6
0.14	12.5	10.6	0.94	83.8	71.1
0.16	14.3	12.1	0.96	85.6	72.6
0.18	16.0	13.6	0.98	87.4	74.1
0.20	17.8	15.1	1.00	89.2	75.7
0.22	19.6	16.6	1.02	90.9	77.2
0.24	21.4	18.2	1.04	92.7	78.7
0.26	23.2	19.7	1.06	94.5	80.2
0.28	25.0	21.2	1.08	96.3	81.7
0.30	26.7	22.7	1.10	98.1	83.2
0.32	28.5	24.2	1.12	99.9	84.7
0.34	30.3	25.7	1.14	102	86.3
0.36	32.1	27.2	1.16	103	87.8
0.38	33.9	28.8	1.18	106	89.3
0.40	35.7	30.3	1.20	107	90.8
0.42	37.4	31.8	1.22	109	92.3
0.44	39.2	33.3	1.24	111	93.8
0.46	41.0	34.8	1.26	112	95.3
0.48	42.8	36.3	1.28	114	96.8
0.50	44.6	37.8	1.30	116	98.4
0.52	46.4	39.3	1.32	118	99.9
0.54	48.1	40.9	1.34	120	101
0.56	49.9	42.4	1.36	121	103
0.58	51.7	43.9	13.8	123	104
0.60	53.5	45.4	1.40	125	106
0.62	55.3	46.9	1.42	127	107
0.64	57.1	48.4	1.44	128	109
0.66	58.8	49.9	1.46	130	111
0.68	60.6	51.4	1.48	132	112
0.70	62.4	53.0	1.50	134	114
0.72	64.2	54.5	1.52	136	115
0.74	66.0	56.0	1.54	137	117
0.76	67.8	57.5	1.56	139	118
0.78	69.5	59.0	1.58	141	120
0.80	71.3	60.5	1.60	143	121

* Note: Calculated values are based on U.S. customary units. Metric units give values within 1 percent of those listed.

TABLE 5 (cont'd)
Values of Kl/r for $F_y = 36$ and 50 ksi*

λ_c	Kl/r		λ_c	Kl/r
	$F_y = 36$	$F_y = 50$		$F_y = 50$
1.62	144	122	2.42	183
1.64	146	124	2.44	185
1.66	148	126	2.46	186
1.68	150	127	2.48	188
1.70	152	129	2.50	189
1.72	153	130	2.52	191
1.74	155	132	2.54	192
1.76	157	133	2.56	194
1.78	159	135	2.58	195
1.80	161	136	2.60	197
1.82	162	138	2.62	198
1.84	164	139	2.64	200
1.86	166	141		
1.88	168	142		
1.90	169	144		
1.92	171	145		
1.94	173	147		
1.96	175	148		
1.98	177	150		
2.00	178	151		
2.02	180	153		
2.04	182	154		
2.06	184	156		
2.08	186	157		
2.10	187	159		
2.12	189	160		
2.14	191	162		
2.16	193	163		
2.18	194	165		
2.20	196	167		
2.22	198	168		
2.24	200	170		
2.26		171		
2.28		173		
2.30		174		
2.32		176		
2.34		177		
2.36		179		
2.38		180		
2.40		182		

Heavy line indicates Kl/r of 200.
* Note: Calculated values are based on U.S. customary units. Metric units give values within 1 percent of those listed.

TABLE 6
Slenderness Ratios of Elements as a Function of F_y From Table B5.1*

Ratio	F_y, ksi (MPa)					
	36 (250)	**42 (290)**	**46 (317)**	**50 (345)**	**60 (415)**	**65 (450)**
$0.38\sqrt{E/F_y}$	10.8	9.99	9.54	9.15	8.35	8.03
$0.45\sqrt{E/F_y}$	12.8	11.8	11.3	10.8	9.89	9.51
$0.56\sqrt{E/F_y}$	15.9	14.7	14.1	13.5	12.3	11.8
$0.75\sqrt{E/F_y}$	21.3	19.7	18.8	18.1	16.5	15.8
$1.12\sqrt{E/F_y}$	31.8	29.4	28.1	27.0	24.6	23.7
$1.40\sqrt{E/F_y}$	39.7	36.8	35.2	33.7	30.8	29.6
$1.49\sqrt{E/F_y}$	42.3	39.2	37.4	35.9	32.8	31.5
$1.86\sqrt{E/F_y}$	52.8	48.9	46.7	44.8	40.9	39.3
$3.76\sqrt{E/F_y}$	107	98.8	94.4	90.6	82.7	79.4
$5.70\sqrt{E/F_y}$	162	150	143	137	125	120
$0.045E/F_y$	36.3	31.0	28.4	26.1	21.8	20.0
$0.07E/F_y$	56.4	48.3	44.1	40.6	33.8	31.2
$0.11E/F_y$	88.6	76.0	69.3	63.8	53.2	49.1
$0.31E/F_y$	250	214	195	180	150	138

* Note: Calculated values are based on U.S. Customary Units. Metric units give values within 1 percent of those listed.

E = 29,000 ksi (200 000 MPa)

TABLE 7
Values of C_m for Use in Section C1

$\frac{M_1}{M_2}$	C_m	$\frac{M_1}{M_2}$	C_m	$\frac{M_1}{M_2}$	C_m
-1.00	1.00	-0.45	0.78	0.10	0.56
-0.95	0.98	-0.40	0.76	0.15	0.54
-0.90	0.96	-0.35	0.74	0.20	0.52
-0.85	0.94	-0.30	0.72	0.25	0.50
-0.80	0.92	-0.25	0.70	0.30	0.48
-0.75	0.90	-0.20	0.68	0.35	0.46
-0.70	0.88	-0.15	0.66	0.40	0.44
-0.65	0.86	-0.10	0.64	0.45	0.42
-0.60	0.84	-0.05	0.62	0.50	0.40
				0.60	0.36
-0.55	0.82	0	0.60	0.80	0.28
-0.50	0.80	0.05	0.58	1.00	0.20

Note 1: $C_m = 0.6 - 0.4\,(M_1 / M_2)$

Note 2: M_1 / M_2 is positive for reverse curvature and negative for single curvature. $|M_1| \leq |M_2|$

TABLE 8-36

$\dfrac{\phi_v V_n}{A_w}$ (ksi) for Plate Girders by Appendix F2 for 36 ksi Yield Stress Steel, Tension Field Action Not Included

$\dfrac{h}{t_w}$	Aspect ratio *a / h:* Stiffener Spacing to Web Depth													
	0.5	**0.6**	**0.7**	**0.8**	**0.9**	**1.0**	**1.2**	**1.4**	**1.6**	**1.8**	**2.0**	**2.5**	**3.0**	**Over 3.0**
60	19.4	19.4	19.4	19.4	19.4	19.4	19.4	19.4	19.4	19.4	19.4	19.4	19.4	19.4
70	19.4	19.4	19.4	19.4	19.4	19.4	19.4	19.4	19.4	19.4	19.4	19.4	19.4	19.4
80	19.4	19.4	19.4	19.4	19.4	19.4	19.4	19.4	19.4	19.4	18.9	18.2	17.9	16.9
90	19.4	19.4	19.4	19.4	19.4	19.4	19.4	18.5	17.8	17.2	16.8	16.2	15.9	14.7
100	19.4	19.4	19.4	19.4	19.4	19.2	17.6	16.6	16.0	15.5	14.9	13.8	13.2	11.9
110	19.4	19.4	19.4	19.4	18.4	17.4	16.0	14.8	13.7	12.8	12.3	11.4	10.9	9.8
120	19.4	19.4	19.4	18.1	16.9	16.0	14.0	12.5	11.5	10.8	10.8	10.3	9.6	9.2
130	19.4	19.4	18.2	16.7	15.6	14.1	11.9	10.6	9.8	9.2	8.8	8.2	7.8	7.0
140	19.4	18.8	16.9	15.5	13.5	12.1	10.3	9.2	8.4	7.9	7.6	7.0	6.7	6.1
150	19.4	17.6	15.7	13.5	11.8	10.6	8.9	8.0	7.3	6.9	6.6	6.1	5.9	5.3
160	18.9	16.5	14.1	11.9	10.4	9.3	7.9	7.0	6.5	6.1	5.8	5.4		4.6
170	17.8	15.5	12.5	10.5	9.2	8.2	7.0	6.2	5.7	5.4	5.1			4.1
180	16.8	13.9	11.1	9.4	8.2	7.3	6.2	5.5	5.1	4.8	4.6			3.7
200	14.9	11.2	9.0	7.6	6.6	5.9	5.0	4.5	4.1					3.0
220	12.3	9.3	7.5	6.3	5.5	4.9	4.2							2.5
240	10.3	7.8	6.3	5.3	4.6	4.1								2.1
260	8.8	6.6	5.3	4.5	3.9	3.5								1.8
280	7.6	5.7	4.6	3.9										
300	6.6	5.0	4.0											
320	5.8	4.4												

TABLE 8-36 M

$\frac{\phi_v V_n}{A_w}$ (MPa) for Plate Girders by Appendix F2 for 250 MPa Yield Stress Steel, Tension Field Action Not Included

$\frac{h}{t_w}$	Aspect ratio *a* / *h*: Stiffener Spacing to Web Depth													
	0.5	**0.6**	**0.7**	**0.8**	**0.9**	**1.0**	**1.2**	**1.4**	**1.6**	**1.8**	**2.0**	**2.5**	**3.0**	**Over 3.0**
60	135	135	135	135	135	135	135	135	135	135	135	135	135	135
70	135	135	135	135	135	135	135	135	135	135	135	135	135	134
80	135	135	135	135	135	135	135	135	135	134	131	126	124	117
90	135	135	135	135	135	135	135	128	123	119	117	112	110	101
100	135	135	135	135	135	133	122	115	111	107	102	95.0	91.0	81.9
110	135	135	135	135	128	121	111	102	94.1	88.6	84.6	78.5	75.2	67.7
120	135	135	135	125	117	111	96.4	85.9	79.1	74.4	71.1	66.0	63.2	56.9
130	135	135	126	116	108	96.9	82.1	73.2	67.4	63.4	60.6	56.2	53.8	48.5
140	135	130	117	107	93.4	83.6	70.8	63.1	58.1	54.7	52.2	48.5	46.4	41.8
150	135	122	109	93.3	81.3	72.8	61.7	55.0	50.6	47.6	45.5	42.2	40.4	36.4
160	131	114	97.3	82.0	71.5	64.0	54.2	48.3	44.5	41.9	40.0	37.1		32.0
170	124	107	86.2	72.6	63.3	56.7	48.0	42.8	39.4	37.1	35.4			28.3
180	117	95.5	76.9	64.8	56.5	50.6	42.8	38.2	35.2	33.1	31.6			25.3
200	102	77.3	62.3	52.5	45.8	41.0	34.7	30.9	28.5					20.5
220	84.6	63.9	51.5	43.4	37.8	33.8	28.7							16.9
240	71.1	53.7	43.2	36.4	31.8	28.4								14.2
260	60.6	45.8	36.8	31.0	27.1	24.2								12.1
280	52.2	39.5	31.8	26.8										
300	45.5	34.4	27.7											
320	40.0	30.2												

TABLE 8-50

$\frac{\phi_v V_n}{A_w}$ (ksi) for Plate Girders by Appendix F2 for 50 ksi Yield Stress Steel, Tension Field Action Not Included

$\frac{h}{t_w}$	**Aspect ratio *a*/ *h*: Stiffener Spacing to Web Depth**													
	0.5	**0.6**	**0.7**	**0.8**	**0.9**	**1.0**	**1.2**	**1.4**	**1.6**	**1.8**	**2.0**	**2.5**	**3.0**	**Over 3.0**
60	27.0	27.0	27.0	27.0	27.0	27.0	27.0	27.0	27.0	27.0	27.0	27.0	27.0	26.6
70	27.0	27.0	27.0	27.0	27.0	27.0	27.0	27.0	26.9	26.1	25.5	24.6	24.0	22.8
80	27.0	27.0	27.0	27.0	27.0	27.0	26.0	24.5	23.5	22.8	22.3	21.5	20.6	18.6
90	27.0	27.0	27.0	27.0	26.5	25.1	23.1	21.8	20.4	19.2	18.3	17.0	16.3	14.7
100	27.0	27.0	27.0	25.6	23.9	22.6	20.1	17.9	16.5	15.5	14.9	13.8	13.2	11.9
110	27.0	27.0	25.3	23.2	21.7	19.6	16.6	14.8	13.7	12.8	12.3	11.4	10.9	9.8
120	27.0	25.9	23.2	21.1	18.4	16.5	14.0	12.5	11.5	10.8	10.3	9.6	9.2	8.3
130	27.0	23.9	21.4	18.0	15.7	14.1	11.9	10.6	9.8	9.2	8.8	8.2	7.8	7.0
140	25.5	22.2	18.4	15.5	13.5	12.1	10.3	9.2	8.4	7.9	7.6	7.0	6.7	6.1
150	23.8	19.9	16.1	13.5	11.8	10.6	8.9	8.0	7.3	6.9	6.6	6.1	5.9	5.3
160	22.3	17.5	14.1	11.9	10.4	9.3	7.9	7.0	6.5	6.1	5.8	5.4		4.6
170	20.6	15.5	12.5	10.5	9.2	8.2	7.0	6.2	5.7	5.4	5.1			4.1
180	18.3	13.9	11.1	9.4	8.2	7.3	6.2	5.5	5.1	4.8	4.6			3.7
200	14.9	11.2	9.0	7.6	6.6	5.9	5.0	4.5	4.1					3.0
220	12.3	9.3	7.5	6.3	5.5	4.9	4.2							2.5
240	10.3	7.8	6.3	5.3	4.6	4.1								2.1
260	8.8	6.6	5.3	4.5	3.9	3.5								1.8
280	7.6	5.7	4.6	3.9										

TABLE 8-50 M
$\frac{\phi_v V_n}{A_w}$ (MPa) for Plate Girders by Appendix F2 for 345 MPa Yield Stress Steel, Tension Field Action Not Included

$\frac{h}{t_w}$	Aspect ratio *a* / *h*: Stiffener Spacing to Web Depth													
	0.5	**0.6**	**0.7**	**0.8**	**0.9**	**1.0**	**1.2**	**1.4**	**1.6**	**1.8**	**2.0**	**2.5**	**3.0**	**Over 3.0**
60	186	186	186	186	186	186	186	186	186	186	186	186	186	184
70	186	186	186	186	186	186	186	186	186	180	176	170	166	158
80	186	186	186	186	186	186	180	169	163	158	154	148	142	128
90	186	186	186	186	183	173	160	151	141	132	126	117	112	101
100	186	186	186	177	165	156	139	124	114	107	102	95.0	91.0	81.9
110	186	186	175	161	150	135	115	102	94.1	88.6	84.6	78.5	75.2	67.7
120	186	179	160	146	127	114	96.4	85.9	79.1	74.4	71.1	66.0	63.2	56.9
130	186	165	147	124	108	96.9	82.1	73.2	67.4	63.4	60.6	56.2	53.8	48.5
140	176	153	127	107	93.4	83.6	70.8	63.1	58.1	54.7	52.2	48.5	46.4	41.8
150	164	138	111	93.3	81.3	72.8	61.7	55.0	50.6	47.6	45.5	42.2	40.4	36.4
160	154	121	97.3	82.0	71.5	64.0	54.2	48.3	44.5	41.9	40.0	37.1		32.0
170	142	107	86.2	72.6	63.3	56.7	48.0	42.8	39.4	37.1	35.4			28.3
180	126	95.5	76.9	64.8	56.5	50.6	42.8	38.2	35.2	33.1	31.6			25.3
200	102	77.3	62.3	52.5	45.8	41.0	34.7	30.9	28.5					20.5
220	84.6	63.9	51.5	43.4	37.8	33.8	28.7							16.9
240	71.1	53.7	43.2	36.4	31.8	28.4								14.2
260	60.6	45.8	36.8	31.0	27.1	24.2								
280	52.2	39.5	31.8	26.8										

TABLE 9-36
$\frac{\phi_v V_n}{A_w}$ (ksi) for Plate Girders by Appendix G for 36 ksi Yield Stress Steel, Tension Field Action Included[b]

(Italic values indicate gross area, as percent of $(h \times t_w)$ required for pairs of intermediate stiffeners of 36 ksi yield stress steel with $V_u / \phi V_n = 1.0$)[a]

$\frac{h}{t_w}$	Aspect ratio *a / h*: Stiffener Spacing to Web Depth													
	0.5	0.6	0.7	0.8	0.9	1.0	1.2	1.4	1.6	1.8	2.0	2.5	3.0	Over 3.0 [c]
60	19.4	19.4	19.4	19.4	19.4	19.4	19.4	19.4	19.4	19.4	19.4	19.4	19.4	19.4
70	19.4	19.4	19.4	19.4	19.4	19.4	19.4	19.4	19.4	19.4	19.4	19.4	19.4	19.4
80	19.4	19.4	19.4	19.4	19.4	19.4	19.4	19.4	19.4	19.4	19.1	18.6	18.3	16.9
90	19.4	19.4	19.4	19.4	19.4	19.4	19.4	19.0	18.5	18.2	17.8	17.3	16.8	14.7
100	19.4	19.4	19.4	19.4	19.4	19.3	18.6	18.1	17.6	17.2	16.6	15.6	14.9	11.9
110	19.4	19.4	19.4	19.4	19.1	18.7	17.9	17.2	16.3	15.6	15.1	14.0	13.3	9.8
120	19.4	19.4	19.4	19.0	18.5	18.1	17.0	16.0	15.1	14.4	13.9	12.8	12.0	8.3
130	19.4	19.4	19.1	18.6	18.1	17.4	16.1	15.1	14.2	13.5	12.9	11.8	11.0	7.0
140	19.4	19.3	18.7	18.2	17.4	16.6	15.4	14.4	13.5	12.8	12.2	11.0	10.2	6.1
150	19.4	19.0	18.4	17.5	16.7	16.0	14.8	13.8	12.9	12.2	11.6	10.4	9.6	5.3
160	19.3	18.7	17.9	17.0	16.2	15.5	14.3	13.3	12.4	11.7	11.1	9.9		4.6
170	19.1	18.4	17.4	16.6	15.8	15.1	13.9	12.9	12.0	11.3 *0.3*	10.7 *0.4*			4.1
180	18.9	18.0	17.1	16.2	15.5	14.8	13.6 *0.2*	12.6 *0.7*	11.7 *1.1*	11.0 *1.3*	10.4 *1.5*			3.7
200	18.4	17.3	16.4	15.6 *0.1*	14.9 *0.9*	14.2 *1.4*	13.1 *2.1*	12.0 *2.5*	11.2 *2.8*					3.0
220	17.8	16.9	16.0 *1.1*	15.2 *2.0*	14.5 *2.6*	13.8 *3.0*	12.7 *3.6*							2.5
240	17.4	16.5 *1.5*	15.7 *2.7*	14.9 *3.4*	14.2 *3.9*	13.5 *4.3*								2.1
260	17.1 *1.3*	16.2 *3.0*	15.4 *4.0*	14.6 *4.6*	14.0 *5.0*	13.3 *5.4*								1.8
280	16.8 *2.7*	16.0 *4.2*	15.2 *5.0*	14.4 *5.6*										

TABLE 9-36 (cont'd)
$\frac{\phi_v V_n}{A_w}$ (ksi) for Plate Girders by Appendix G for 36 ksi Yield Stress Steel, Tension Field Action Included[b]

(Italic values indicate gross area, as percent of $(h \times t_w)$ required for pairs of intermediate stiffeners of 36 ksi yield stress steel with $V_u / \phi V_n = 1.0$)[a]

$\frac{h}{t_w}$	Aspect ratio *a* / *h* : Stiffener Spacing to Web Depth													
	0.5	0.6	0.7	0.8	0.9	1.0	1.2	1.4	1.6	1.8	2.0	2.5	3.0	Over 3.0 [c]
300	16.6	15.8	15.0											
	3.9	*5.2*	*5.9*											
320	16.4	15.6												
	4.9	*6.0*												

[a] For area of single-angle and single-plate stiffeners, or when $V_u / \phi V_n < 1.0$, see Equation A-G4-1.
[b] For end-panels and all panels in hybrid and web-tapered plate girders, use Table 9-36.
[c] Same as for Table 9-36.
Note: Girders so proportioned that the computed shear is less than that given in the right-hand column do not require intermediate stiffeners.

TABLE 9-36 M

$\frac{\phi_v V_n}{A_w}$ (MPa) for Plate Girders by Appendix G for 250 MPa Yield Stress Steel, Tension Field Action Included[b]

(Italic values indicate gross area, as percent of ($h \times t_w$) required for pairs of intermediate stiffeners of 250 MPa yield stress steel with $V_u / \phi V_n = 1.0$)[a]

	Aspect ratio *a* / *h*: Stiffener Spacing to Web Depth													
$\frac{h}{t_w}$	**0.5**	**0.6**	**0.7**	**0.8**	**0.9**	**1.0**	**1.2**	**1.4**	**1.6**	**1.8**	**2.0**	**2.5**	**3.0**	**Over 3.0 [c]**
60	135	135	135	135	135	135	135	135	135	135	135	135	135	135
70	135	135	135	135	135	135	135	135	135	135	135	135	135	134
80	135	135	135	135	135	135	135	135	135	135	133	129	127	117
90	135	135	135	135	135	135	135	132	129	126	124	120	117	101
100	135	135	135	135	135	134	129	125	122	119	115	108	103	81.9
110	135	135	135	135	132	130	124	119	113	108	104	96.9	91.8	67.7
120	135	135	135	132	129	126	118	111	105	100	96.0	88.4	83.0	56.9
130	135	135	132	129	126	120	112	105	98.6	93.7	89.6	81.7	76.2	48.5
140	135	134	130	126	120	115	107	99.5	93.6	88.7	84.5	76.5	70.9	41.8
150	135	132	128	122	116	111	103	95.5	89.6	84.6	80.4	72.2	66.5	36.4
160	134	130	124	118	113	108	99.2	92.2	86.3	81.2	77.0	68.8		32.0
170	132	128	121	115	110	105	96.5	89.4	83.5	78.5 *0.3*	74.2 *0.4*			28.3
180	131	125	118	113	107	103	94.2 *0.2*	87.2 *0.7*	81.2 *1.1*	76.2 *1.3*	71.9 *1.5*			25.3
200	128	120	114	109 *0.1*	103 *0.9*	98.8 *1.4*	90.6 *2.1*	83.6 *2.5*	77.6 *2.8*					20.5
220	124	117	111 *1.1*	106 *2.0*	101 *2.6*	96.1 *3.0*	87.9 *3.6*							16.9
240	121	114 *1.5*	109 *2.7*	103 *3.4*	98.5 *3.9*	94.0 *4.3*								14.2
260	118 *1.3*	112 *3.0*	107 *4.0*	102 *4.6*	96.9 *5.0*	92.4 *5.4*								12.1
280	117 *2.7*	111 *4.2*	105 *5.0*	100 *5.6*										

TABLE 9-36 M (cont'd)

$\frac{\phi_v V_n}{A_w}$ (MPa) for Plate Girders by Appendix G for 250 MPa Yield Stress Steel, Tension Field Action Included[b]

(Italic values indicate gross area, as percent of ($h \times t_w$) required for pairs of intermediate stiffeners of 250 MPa yield stress steel with $V_u / \phi V_n = 1.0$)[a]

$\frac{h}{t_w}$	Aspect ratio *a / h:* Stiffener Spacing to Web Depth													
	0.5	**0.6**	**0.7**	**0.8**	**0.9**	**1.0**	**1.2**	**1.4**	**1.6**	**1.8**	**2.0**	**2.5**	**3.0**	**Over 3.0 [c]**
300	115 *3.9*	109 *5.2*	104 *5.9*											
320	114 *4.9*	108 *6.0*												

[a] For area of single-angle and single-plate stiffeners, or when $V_u / \phi V_n < 10$, see Equation A-G4-1.
[b] For end-panels and all panels in hybrid and web-tapered plate girders, use Table 9-36M.
[c] Same as for Table 9-36M.
Note: Girders so proportioned that the computed shear is less than that given in the right-hand column do not require intermediate stiffeners.

TABLE 9-50

$\frac{\phi_v V_n}{A_w}$ (ksi) for Plate Girders by Appendix G for 50 ksi Yield Stress Steel, Tension Field Action Included [b]

(Italic values indicate gross area, as percent of $(h \times t_w)$ required for pairs of intermediate stiffeners of 50 ksi yield stress steel with $V_u / \phi V_n = 1.0$) [a]

$\frac{h}{t_w}$	**Aspect ratio *a* / *h* : Stiffener Spacing to Web Depth**													
	0.5	**0.6**	**0.7**	**0.8**	**0.9**	**1.0**	**1.2**	**1.4**	**1.6**	**1.8**	**2.0**	**2.5**	**3.0**	**Over 3.0 [c]**
60	27.0	27.0	27.0	27.0	27.0	27.0	27.0	27.0	27.0	27.0	27.0	27.0	27.0	26.6
70	27.0	27.0	27.0	27.0	27.0	27.0	27.0	27.0	26.9	26.5	26.1	25.4	24.9	22.8
80	27.0	27.0	27.0	27.0	27.0	27.0	26.5	25.8	25.1	24.6	24.1	23.3	22.4	18.6
90	27.0	27.0	27.0	27.0	26.8	26.3	25.3	24.4	23.4	22.5	21.7	20.2	19.2	14.7
100	27.0	27.0	27.0	26.5	25.9	25.3	24.0	22.5	21.4	20.4	19.6	18.0	17.0	11.9
110	27.0	27.0	26.5	25.8	25.1	24.2	22.4	21.0	19.8	18.8	18.0	16.4	15.3	9.8
120	27.0	26.7	25.9	25.1	24.0	23.0	21.2	19.8	18.6	17.6	16.8	15.2	14.1	8.3
130	27.0	26.2	25.4	24.1	23.0	22.0	20.3	18.9	17.7	16.7	15.9	14.2	13.1	7.0
140	26.7	25.8	24.5	23.3	22.2	21.3	19.6	18.2	17.0	16.0	15.1	13.5	12.3	6.1
150	26.3	25.2	23.9	22.7	21.6	20.7	19.0	17.6	16.4	15.4	14.5	12.9	11.7	5.3
160	26.0	24.6	23.3	22.2	21.1	20.2	18.5	17.1	15.9 *0.2*	14.9 *0.4*	14.0 *0.5*	12.4 *0.8*		4.6
170	25.6	24.1	22.8	21.7	20.7	19.8	18.1 *0.5*	16.7 *1.0*	15.2 *1.2*	14.5 *1.4*	13.6 *1.6*			4.1
180	25.1	23.7	22.4	21.3	20.3 *0.4*	19.4 *0.9*	17.8 *1.5*	16.4 *1.9*	15.2 *2.2*	14.2 *2.3*	13.3 *2.5*			3.7
200	24.3	23.0	21.8 *1.0*	20.8 *1.8*	19.8 *2.3*	18.9 *2.7*	17.3 *3.2*	15.9 *3.5*	14.7 *3.7*					3.0
220	23.7	22.5 *1.7*	24.1 *2.7*	20.4 *3.3*	19.4 *3.8*	18.5 *4.1*	16.9 *4.5*							2.5
240	23.2 *1.8*	22.1 *3.2*	21.0 *4.0*	20.0 *4.6*	19.1 *4.9*	18.2 *5.2*								2.1
260	23.0 *3.2*	21.8 *4.4*	20.8 *5.1*	19.8 *5.6*	18.8 *5.9*	18.0 *6.1*								
280	22.7 *4.4*	21.6 *5.4*	20.6 *6.0*	19.6 *6.4*										

[a] For area of single-angle and single-plate stiffeners, or when $V_u / \phi V_n < 1.0$, see Equation A-G4-1.

[b] For end-panels and all panels in hybrid and web-tapered plate girders, use Table 9-50.

[c] Same as for Table 9-50.

Note: Girders so proportioned that the computed shear is less than that given in the right-hand column do not require intermediate stiffeners.

TABLE 9-50 M

$\frac{\phi_v V_n}{A_w}$ (MPa) for Plate Girders by Appendix G for 345 MPa Yield Stress Steel, Tension Field Action Included[b]

(Italic values indicate gross area, as percent of $(h \times t_w)$ required for pairs of intermediate stiffeners of 345 MPa yield stress steel with $V_u / \phi V_n = 1.0$)[a]

$\frac{h}{t_w}$	**Aspect ratio *a* / *h* : Stiffener Spacing to Web Depth**													
	0.5	**0.6**	**0.7**	**0.8**	**0.9**	**1.0**	**1.2**	**1.4**	**1.6**	**1.8**	**2.0**	**2.5**	**3.0**	**Over 3.0 [c]**
60	186	186	186	186	186	186	186	186	186	186	186	186	186	184
70	186	186	186	186	186	186	186	186	186	183	180	175	172	158
80	186	186	186	186	186	186	183	178	174	170	167	161	155	128
90	186	186	186	186	185	181	174	169	162	155	150	140	133	101
100	186	186	186	183	179	175	165	155	147	141	135	125	117	81.9
110	186	186	183	178	173	167	155	145	137	130	124	113	106	67.7
120	186	184	179	173	165	158	147	137	129	122	116	105	97.1	56.9
130	186	181	175	166	159	152	140	130	122	115	110	98.3	90.4	48.5
140	184	178	169	161	154	147	135	125	117	110	104	93.1	85.0	41.8
150	181	174	165	157	149	143	131	121	113	106	100	88.8	80.6	36.4
160	179	170	161	153	146	139	128	118	110 *0.2*	103 *0.4*	96.9 *0.5*	85.3 *0.8*		32.0
170	176	166	158	150	143	136	125 *0.5*	115 *1.0*	107 *1.2*	100 *1.4*	94.1 *1.6*			28.3
180	173	163	155	147	140 *0.4*	134 *0.9*	123 *1.5*	113 *1.9*	105 *2.2*	97.8 *2.3*	91.8 *2.5*			25.3
200	168	159	151 *1.0*	143 *1.8*	137 *2.3*	130 *2.7*	119 *3.2*	109 *3.5*	101 *3.7*					20.5
220	164	155 *1.7*	148 *2.7*	140 *3.3*	134 *3.8*	128 *4.1*	116 *4.5*							16.9
240	161 *1.8*	153 *3.2*	145 *4.0*	138 *4.6*	132 *4.9*	126 *5.2*								14.2
260	158 *3.2*	151 *4.4*	143 *5.1*	136 *5.6*	130 *5.9*	124 *6.1*								
280	157 *4.4*	149 *5.4*	142 *6.0*	135 *6.4*										

[a] For area of single-angle and single-plate stiffeners, or when $V_u / \phi V_n < 1.0$, see Appendix A-G4-1.
[b] For end-panels and all panels in hybrid and web-tapered plate girders, use Table 8-50M.
[c] Same as for Table 8-50M.
Note: Girders so proportioned that the computed shear is less than that given in the right-hand column do not require intermediate stiffeners.

TABLE 10
Nominal Horizontal Shear Load for One Connector Q_n, kips [a] From Equations I5-1 and I5-2

Connector [b]	Specified Compressive Strength of concrete f_c', ksi [d]		
	3.0	3.5	4.0
½-in. dia. × 2-in. hooked or headed stud	9.4	10.5	11.6
⅝-in. dia. × 2½ in. hooked or headed stud	14.6	16.4	18.1
¾-in. dia. × 3-in. hooked or headed stud	21.0	23.6	26.1
⅞-in. dia. × 3½-in. hooked or headed stud	28.6	32.1	35.5
Channel C3 × 4.1	10.2 L_c [c]	11.5 L_c [c]	12.7 L_c [c]
Channel C4 × 5.4	11.1 L_c [c]	12.4 L_c [c]	13.8 L_c [c]
Channel C5 × 6.7	11.9 L_c [c]	13.3 L_c [c]	14.7 L_c [c]

[a] Applicable only to concrete made with ASTM C33 aggregates.
[b] The nominal horizontal loads tabulated may also be used for studs longer than shown.
[c] L_c = length of channel, inches.
[d] $F_u > 0.5 \ (f_c' w)^{0.75}$, w = 145 lbs./cu. ft.

TABLE 10 M
Nominal Horizontal Shear Load for One Connector Q_n, kN[a] From Equations I5-1 and I5-2

Connector [b]	Specified Compressive Strength of Concrete, f_c', MPa [d]		
	20	25	30
12.7 mm dia.x 50 mm hooked or headed stud	42	48	55
15.9 mm dia.x 63 mm hooked or headed stud	65	75	86
19.0 mm dia.x 75 mm hooked or headed stud	93	108	123
22.2 mm dia.x 88 mm hooked or headed stud	127	145	167
Channel C75 x 6.1	1.8 L_c [c]	2.1 L_c [c]	2.4 L_c [c]
Channel C100 x 8	1.9L_c [c]	2.2 L_c [c]	2.6 L_c [c]
Channel C130 x 10.4	2.1 L_c [c]	2.4 L_c [c]	2.7 L_c [c]

[a] Applicable only to concrete made with ASTM C33 aggregates.
[b] The nominal horizontal loads tabulated may also be used for studs longer than shown.
[c] L_c = length of channel, mm.
[d] $F_u > 0.1 \ (f_c' w)^{0.75}$, w = 2325 kg/m^3

COMMENTARY on the Load and Resistance Factor Design Specification for Structural Steel Buildings

December 27, 1999

INTRODUCTION

The Specification is intended to be complete for normal design usage.

The Commentary furnishes background information and references for the benefit of the engineer seeking further understanding of the basis, derivations and limits of the specification.

The Specification and Commentary are intended for use by design professionals with demonstrated engineering competence.

CHAPTER A

GENERAL PROVISIONS

A1. SCOPE

Load and Resistance Factor Design (LRFD) is an improved approach to the design of structural steel for buildings. It involves explicit consideration of limit states, multiple load factors, and resistance factors, and implicit probabilistic determination of reliability. The designation LRFD reflects the concept of factoring both loads and resistance. This type of factoring differs from the AISC allowable stress design (ASD) Specification (AISC, 1989), where only the resistance is divided by a factor of safety (to obtain allowable stress) and from the plastic design portion of that Specification, where only the loads are multiplied by a common load factor. The LRFD method was devised to offer the designer greater flexibility, more rationality, and possible overall economy.

The format of using resistance factors and multiple load factors is not new, as several such design codes are in effect [the ACI-318M Building Code Requirements for Structural Concrete (ACI, 1999) and the AASHTO Load and Resistance Factor Design for Bridges (AASHTO, 1996)]. Nor should the new LRFD method give designs radically different from the older methods, since it was tuned, or "calibrated," to typical representative designs of the earlier methods. The principal new ingredient is the use of a probabilistic mathematical model in the development of the load and resistance factors, which made it possible to give proper weight to the accuracy with which the various loads and resistances can be determined. Also, it provides a rational methodology for transference of test results into design provisions. A more rational design procedure leading to more uniform reliability is the practical result.

A2. TYPES OF CONSTRUCTION

Connection Classification. In the first edition of the LRFD Specification (AISC, 1986), the types of construction were changed from Type 1 through Type 3 (AISC, 1978) to the more general terms of fully-restrained (FR) and partially restrained (PR), to provide appropriate recognition of connection stiffness. The third edition of the LRFD Specification emphasizes the combined importance of stiffness, strength and ductility in connection design.

Examples of connection classification schemes include those described by Ackroyd and Gerstle (1982), Bjorhovde, Colson, and Brozzetti (1990), and Eurocode 3 (1992) (Leon, 1994). The basic assumption made in classifying connections is that their most important behavioral characteristics can be modeled by a moment-rotation (M-θ) curve such as shown in Figure C-A2.1. Implicit in the moment-rotation curve is the definition of the connection as a region of the connected members, along with the connecting elements. The connection is defined in this way because the rotation is measured over a gage length that incorporates the contributions of both the connecting elements and the members being connected.

Connection Strength. Referring to Figure C-A2.1, it is presumed that the nominal connection strength M_n can be determined on the basis of an ultimate limit state model of the connection or from test data. Further, many PR connections do not exhibit a plateau in their moment-rotation relationship, even at large rotations. In determining their strength based on tests, it is necessary to assume a rotation at which to define the nominal strength. For this purpose, the connection strength can be defined at a rotation of approximately θ_n=0.02 radians (Hsieh and Deierlein, 1991, and Leon, Hoffman, and Staeger, 1996).

An important aspect of the nominal strength of a connection, M_n, is its relationship to the strength of the connected beam $M_{p,beam}$. A connection is *full strength* if $M_n > M_{p,beam}$, otherwise the connection is *partial strength.*

A partial strength PR connection must be designed with sufficient ductility to permit the connection components to deform and to avoid any brittle failure modes.

It is also useful to define a lower limit for the strength, below which the connection can be treated as simple. Connections that transmit less than $0.2M_{p,beam}$ at a rotation of 0.02 radians can be considered to have no flexural strength for design. It should be recognized, however, that the aggregate strength of many weak partial strength connections (e.g. those with a capacity less than $0.2M_{p,beam}$) can be significant when compared to that of a few strong connections (FEMA, 1997).

Connection ductility. Connection ductility is a key parameter when the deformations are concentrated in the connection elements, as is the typical case in partial strength PR connections. The ductility required will depend on the flexibility of the connections and the particular application. For example, the ductility requirement for a braced frame in a non-seismic area will generally be less than for an unbraced frame in a high seismic area. Referring again to Figure C-A2.1, the rotation capacity, θ_u, can be defined as the value of the connection rotation at the point where either (a) the resisting strength of the connection has dropped to $0.8M_n$, i.e. to 80 percent of the nominal connection strength, or (b) the connection has deformed beyond a reasonable limit, e.g. 0.03 radians. This second criterion for determining θ_u is intended to apply to connections where there is no loss in strength until very large rotations occur. For example, tests of double-angle web connections show that some details will deform in a ductile manner beyond the point where the beam

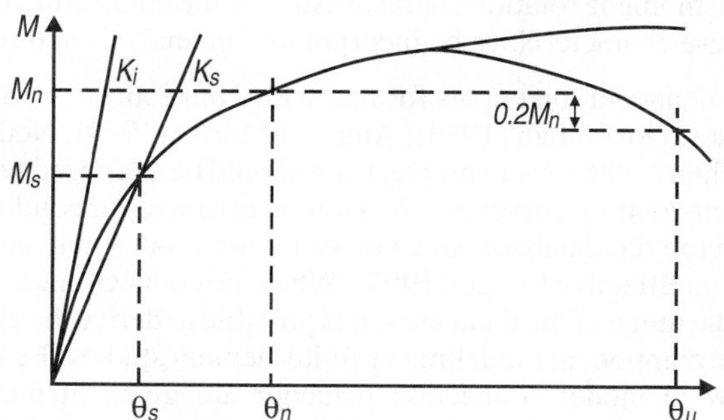

Fig. C-A2.1. Typical moment-rotation response of a partially-restrained connection.

comes into contact with the supporting column. However, it would not be appropriate to rely on these extremely large rotations (typically more than 0.1 radians) for design.

The available ductility, θ_u, should be compared with the required rotational ductility under the full factored loads, as determined by an analysis that takes into account the nonlinear behavior of the connection. In the absence of accurate analyses of the required rotation capacity, the connection ductility may be considered adequate when the available ductility is greater than 0.03 radians. This rotation is equal to the minimum beam-to-column connection ductility as specified in the AISC seismic provisions for special moment frames (AISC, 1997 and 1999). Many types of partial strength PR connections, such as top and seat-angle details, meet this criterion.

Connection Stiffness. Because many PR connections manifest nonlinear behavior even at low force levels, the initial stiffness of the connection, K_i, does not characterize the connection response adequately. Short of modeling the nonlinear response, a better measure of behavior is the secant stiffness, K_s (see Figure C-A2.1). The secant stiffness is defined on the basis of either the moment, M_s, or the rotation, θ_s, that would occur under the applied loads. Generally, two distinct values of secant stiffness should be considered in design, with one corresponding to the behavior under service loads and the other to the behavior under factored loads.

The ratio of connection stiffness to beam stiffness can be defined as $\alpha = K_s L/EI$, where L and EI are the length and bending rigidity, respectively, of the connected beam. Limiting values of α are approximate ways of categorizing connection stiffness in order to simplify the analysis. The limits are not exact values, and generally depend on the structural geometry and the limit state used to establish the criterion. For continuous beams in braced frames, for example, limits based on achieving a certain percentage of the fixed-end moment or reaching a deflection limit can be used to establish stiffness criteria (Leon, 1994).

Following such an approach, where α is defined using the secant stiffness for the serviceability limit state, it is reasonable to classify connections as fully restrained if $\alpha > 20$. On the other hand, connections with $\alpha < 2$ may be approximated as simple.

Structural Analysis and Design. When the secant stiffness falls below the fully restrained limit, engineers should account for the PR behavior in determining member and connection forces, displacements, and frame stability effects. This requires, first, that the moment-rotation characteristics of the connection be known, and second, that these characteristics be incorporated in analysis and member design.

Typical moment-rotation curves for many PR connections are available from several databases: Goverdhan (1983); Ang and Morris (1984); Nethercot (1985); and Kishi and Chen (1986), for example. Care should be exercised when utilizing tabulated moment-rotation curves not to extrapolate to sizes or conditions beyond those used to develop the database, since other failure modes may control (ASCE Task Committee on Effective Length, 1997). When the connections to be modeled do not fall within the range of the databases, it is possible to derive the characteristics from tests, simple component modeling, or finite element studies (FEMA, 1995). Examples of how to model connection behavior are given in numerous references (Bjorhovde, Brozzetti, and Colson, 1988; Bjorhovde, Colson, Haaijer, and Stark, 1992; Bjorhovde, Colson, and Zandonini, 1996; Chen and Lui, 1991; Lorenz, Kato, and Chen, 1993; Chen and Toma, 1994; Chen, Goto, and Liew, 1995; and Leon et al., 1996).

The degree of sophistication of the analysis depends on the problem at hand. Usually design for PR construction requires separate analyses for the serviceability and ultimate limit states. For serviceability, an analysis using linear springs with a secant stiffness K_s is often sufficient. Under factored loads, a more careful procedure is needed to ensure that the characteristics assumed in the analysis are consistent with those of the connection response. As shown in Figure C-A2.1, the response is especially nonlinear as the applied moment approaches the nominal connection strength, M_n. In particular, the effect of the connection nonlinearity on second-order moments and other stability checks need to be considered (ASCE Task Committee on Effective Length, 1997). The preferable method of incorporating these effects in the ultimate limit state check is through a second-order analysis that models the nonlinear moment-rotation response of the connections explicitly. Alternatively, for regular structures in regions of low to moderate seismicity, properly calibrated second-order elastic analysis methods or plastic hinge methods (e.g., Leon et al., 1996) can be used. Elastic procedures may, for example, use linear springs with reduced secant-stiffness values determined to be consistent with the maximum rotations calculated under the factored loads.

Prior to this Specification, an analysis procedure was used that ignored the restraining effect of connections for gravity loads. However, the effect was taken into account for lateral load resistance to wind loads. This method was referred to as "simple framing", and required that three conditions be satisfied (Disque, 1964). Today's approach of using a predictable degree of connection restraint is a more accurate representation of the structural behavior.

A3. MATERIAL

1. Structural Steel

1a. ASTM Designations

The grades of structural steel approved for use under the LRFD Specification, covered by ASTM standard specifications, extend to a yield stress of 100 ksi (690 MPa). Some of these ASTM standards specify a minimum yield point, while others specify a minimum yield strength. The term "yield stress" is used in the Specification as a generic term to denote either the yield point or the yield strength.

It is important to be aware of limitations of availability that may exist for some combinations of strength and size. Not all structural section sizes are included in the various material specifications. For example, the 60 ksi (415 MPa) yield strength steel in the A572/A572M specification includes plate only up to 1¼-in. (32 mm) in thickness. Another limitation on availability is that even when a product is included in the specifications, it may be infrequently produced by the mills. Specifying these products may result in procurement delays or require ordering large quantities directly from the producing mills. Consequently, it is prudent to check availability before completing the details of a design.

Properties in the direction of rolling are of principal interest in the design of steel structures. Hence, yield stress as determined by the standard tensile test is the principal mechanical property recognized in the selection of the steels approved for use under the Specification. It must be recognized that other mechanical and physical properties of rolled steel, such as anisotropy, ductility, notch toughness, formability, corrosion resistance, etc., may also be important to the satisfactory performance of a structure.

It is not possible to incorporate in the Commentary adequate information to impart full understanding of all factors which might merit consideration in the selection and specification of materials for unique or especially demanding applications. In such a situation the user of the Specification is advised to make use of reference material contained in the literature on the specific properties of concern and to specify supplementary material production or quality requirements as provided for in ASTM material specifications. One such case is the design of highly restrained welded connections (AISC, 1973). Rolled steel is anisotropic, especially insofar as ductility is concerned; therefore, weld contraction strains in the region of highly restrained welded connections may exceed the capabilities of the material if special attention is not given to material selection, details, workmanship, and inspection.

Another special situation is that of fracture control design for certain types of service conditions (AASHTO, 1996). The relatively warm temperatures of steel in buildings, the essentially static strain rates, the stress intensity, and the number of cycles of full design stress make the probability of fracture in building structures extremely remote. Good workmanship and good design details incorporating joint geometry that avoids severe stress concentrations are generally the most effective means of providing fracture-resistant construction. However, for especially demanding service conditions such as low temperatures with impact loading, the specification of steels with superior notch toughness may be warranted.

1c. Heavy Shapes

The web-to-flange intersection and the web center of heavy hot-rolled shapes as well as the interior portions of heavy plates may contain a coarser grain structure and/or lower toughness material than other areas of these products. This is probably caused by ingot segregation, as well as somewhat less deformation during hot rolling, higher finishing temperature, and a slower cooling rate after rolling for these heavy sections. This characteristic is not detrimental to suitability for service for compression members, or for non-welded members.

However, when heavy cross sections are joined by splices or connections using complete-joint-penetration welds which extend through the coarser and/or lower notch-tough interior portions, tensile strains induced by weld shrinkage may result in cracking, for example in a complete-joint-penetration welded connection of a heavy cross section beam to any column section. When members of lesser thickness are joined by complete-joint-penetration welds, which induce smaller weld shrinkage strains, to the finer grained and/or more notch-tough surface material of ASTM A6/A6M Group 4 and 5 shapes and heavy built-up cross sections, the potential for cracking is significantly lower, for example in a complete-joint-penetration groove welded connection of a non-heavy cross-section beam to a heavy cross-section column.

For critical applications such as primary tension members, material should be specified to provide adequate toughness at service temperatures. Because of differences in the strain rate between the Charpy V-notch (CVN) impact test and the strain rate experienced in actual structures, the CVN test is conducted at a temperature higher than the anticipated service temperature for the structure. The location of the CVN test is shown in Figure C-A3.1.

The toughness requirements of Section A3.1c are intended only to provide material of reasonable toughness for ordinary service applications. For unusual applications and/or low temperature service, more restrictive requirements and/or toughness requirements for other section sizes and thicknesses may be appropriate. To minimize the potential for fracture, the notch toughness requirements of Section A3.1c must be used in conjunction with good design and fabrication procedures. Specific requirements are given in Sections J1.5, J1.6, J2.6, J2.8, and M2.2.

For rotary-straightened W-shapes, an area of reduced notch toughness has been documented in a limited region of the web immediately adjacent to the flange as illustrated in Figure C-A3.2. Preliminary recommendations have been issued (AISC, 1997) and AISC is currently exploring the associated implications for design and construction. It is anticipated that recommendations will be forthcoming, albeit after the publication of this document. For this reason, the reader is encouraged to maintain an awareness of AISC recommendations as they become available.

3. Bolts, Washers, and Nuts

The ASTM standard for A307 bolts covers two grades of fasteners. Either grade may be used under the LRFD Specification; however, it should be noted that Gr. B

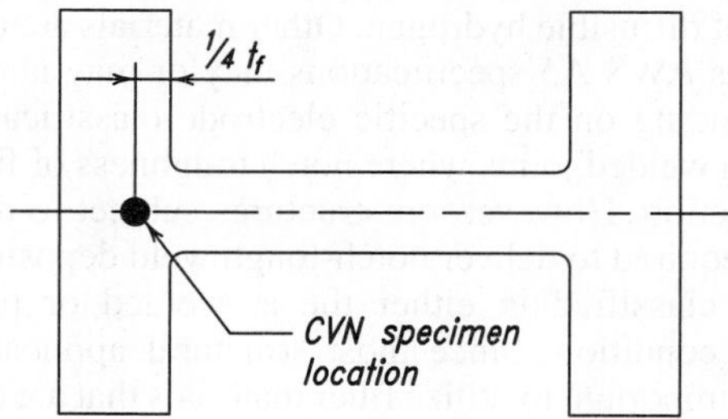

Fig. C-A3.1. Location from which Charpy impact specimen shall be taken.

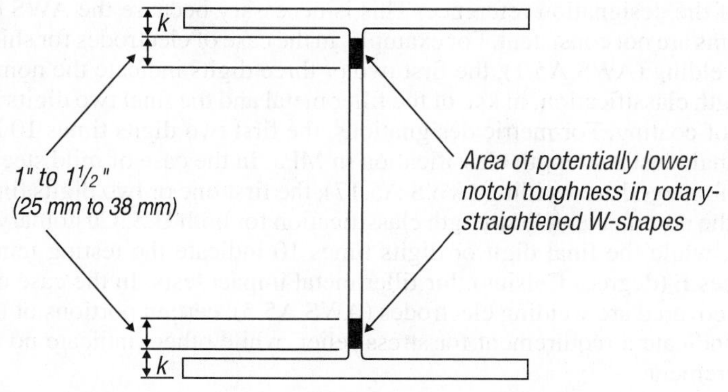

Fig. C-A3.2. "k-area."

is intended for pipe flange bolting and Gr. A is the grade long in use for structural applications.

4. Anchor Rods and Threaded Rods

Since there is a limit on the maximum available length of A325 or A325M and A490 or A490M bolts, the attempted use of these bolts for anchor rods with design lengths longer than the maximum available lengths has presented problems in the past. The inclusion of A449 and A354 materials in this Specification allows the use of higher strength material for bolts longer than A325 or A325M and A490 or A490M bolts. The designer should be aware that pretensioning of anchor rods is not recommended due to relaxation and the potential for stress corrosion after pretensioning.

The designer should specify the appropriate thread and SAE fit for threaded rods used as load-carrying members.

5. Filler Metal and Flux for Welding

The filler metal specifications issued by the American Welding Society (AWS) are general specifications which include filler metals suitable for building construction, as well as consumables that would not be suitable for building construction. For example, some electrodes covered by the specifications are specifically limited to single pass applications, while others are restricted to sheet metal applications. Many of the filler metals listed are "low hydrogen," that is, they deposit filler metal with low levels of diffusible hydrogen. Other materials are not. Filler metals listed under the various AWS A5 specifications may or may not have required impact toughness, depending on the specific electrode classification. Section J2.6 has identified certain welded joints where notch toughness of filler metal is needed in building construction. However, on structures subject to dynamic loading, filler metals may be required to deliver notch-tough weld deposits in other joints. Filler metals may be classified in either the as-welded or post weld heat-treated (stress-relieved) condition. Since most structural applications will not involve stress relief, it is important to utilize filler materials that are classified in conditions similar to those experienced by the actual structure.

When specifying filler metal and/or flux by AWS designation, the applicable standard specifications should be carefully reviewed to assure a complete understanding of the designation reference. This is necessary because the AWS designation systems are not consistent. For example, in the case of electrodes for shielded metal arc welding (AWS A5.1), the first two or three digits indicate the nominal tensile strength classification, in ksi, of the filler metal and the final two digits indicate the type of coating. For metric designations, the first two digits times 10 indicate the nominal tensile strength classification in MPa. In the case of mild steel electrodes for submerged arc welding (AWS A5.17), the first one or two digits times 10 indicate the nominal tensile strength classification for both U.S. Customary and metric units, while the final digit or digits times 10 indicate the testing temperature in degrees F (degrees Celsius), for filler metal impact tests. In the case of low-alloy steel covered arc welding electrodes (AWS A5.5), certain portions of the designation indicate a requirement for stress relief, while others indicate no stress relief requirement.

Engineers do not, in general, specify the exact filler metal to be employed on a particular structure. Rather, the decision as to which welding process and which filler metal is to be utilized is usually left with the fabricator or erector. To ensure that the proper filler metals are used, codes restrict the usage of certain filler materials, or impose qualification testing to prove the suitability of the specific electrode.

A4. LOADS AND LOAD COMBINATIONS

The load factors and load combinations are developed in Ellingwood, MacGregor, Galambos, and Cornell (1982) based on the recommended minimum loads given in ASCE 7 (ASCE, 1998).

The load factors and load combinations recognize that when several loads act in combination with the dead load (e.g., dead plus live plus wind), only one of these takes on its maximum lifetime value, while the other load is at its "arbitrary point-in-time value" (i.e., at a value which can be expected to be on the structure at any time). For example, under dead, live, and wind loads the following combinations are appropriate:

$$\gamma_D D + \gamma_L L \qquad \text{(C-A4-1)}$$

$$\gamma_D D + \gamma_{L_a} L_a + \gamma_W W \qquad \text{(C-A4-2)}$$

$$\gamma_D D + \gamma_L L + \gamma_{W_a} W_a \qquad \text{(C-A4-3)}$$

where γ is the appropriate load factor as designated by the subscript symbol.Subscript a refers to an "arbitrary point-in-time" value.

The mean value of arbitrary point-in-time live load L_a is on the order of 0.24 to 0.4 times the mean maximum lifetime live load L for many occupancies, but its dispersion is far greater. The arbitrary point-in-time wind load W_a, acting in conjunction with the maximum lifetime live load, is the maximum daily wind. It turns out that $\gamma_{W_a} W_a$ is a negligible quantity so only two load combinations remain:

$$1.2D + 1.6L \qquad \text{(C-A4-4)}$$

$$1.2D + 0.5L + \boxed{1.6}W \qquad \text{(C-A4-5)}$$

Errata 9/4/01

The load factor 0.5 assigned to L in the second formula reflects the statistical properties of L_a, but to avoid having to calculate yet another load, it is reduced so it can be combined with the maximum lifetime wind load.

The nominal loads *D, L, W, E,* and *S* are the code loads or the loads given in ASCE 7. The latest edition of the ASCE 7 Standard on structural loads released in 1998 has adopted, in most aspects, the seismic design provisions from NEHRP (1997), as has the AISC *Seismic Provisions for Structural Steel Buildings* (AISC, 1997 and 1999). The reader is referred to the commentaries to these documents for an expanded discussion on seismic loads, load factors, and seismic design of steel buildings.

A5. DESIGN BASIS

1. Required Strength at Factored Loads

LRFD permits the use of both elastic and plastic structural analyses. LRFD provisions result in essentially the same methodology for, and end product of, plastic design as included in the AISC ASD Specification (AISC, 1989) except that the LRFD provisions tend to be slightly more liberal, reflecting added experience and the results of further research. The 10 percent redistribution permitted is consistent with that in the AISC ASD Specification (AISC, 1989).

In some circumstances, as in the proportioning of the bracing members that carry no calculated forces (see Section C3) and of connection components (see Section J1.7), the required strength is explicitly stated in the Specification.

2. Limit States

A limit state is a condition which represents the limit of structural usefulness. Limit states may be dictated by functional requirements, such as maximum deflections or drift; they may be conceptual, such as plastic hinge or mechanism formation; or they may represent the actual collapse of the whole or part of the structure, such as fracture or instability. Design criteria ensure that a limit state is violated only with an acceptably small probability by selecting the combination of load and resistance factors and nominal load and resistance values which will never be exceeded under the design assumptions.

Two kinds of limit states apply for structures: limit states of strength which define safety against extreme loads during the intended life of the structure, and limit states of serviceability which define functional requirements. The LRFD Specification, like other structural codes, focuses on the limit states of strength because of overriding considerations of public safety for the life, limb, and property of human beings. This does not mean that limit states of serviceability are not important to the designer, who must equally ensure functional performance and economy of design. However, these latter considerations permit more exercise of judgment on the part of designers. Minimum considerations of public safety, on the other hand, are not matters of individual judgment and, therefore, specifications dwell more on the limit states of strength than on the limit states of serviceability.

Limit states of strength vary from member to member, and several limit states may apply to a given member. The following limit states of strength are the most common: onset of yielding, formation of a plastic hinge, formation of a plastic mechanism, overall frame or member instability, lateral-torsional buckling, local buckling, tensile fracture, development of fatigue cracks, deflection instability, alternating plasticity, and excessive deformation. The most common serviceability limit states include unacceptable elastic deflections and drift, unacceptable vibrations, and permanent deformations.

3. Design for Strength

The general format of the LRFD Specification is given by the formula:

$$\Sigma\gamma_i Q_i \leq \phi R_n \qquad \text{(C-A5-1)}$$

where

Σ = summation
i = type of load, i.e., dead load, live load, wind, etc.
Q_i = nominal load effect
γ_i = load factor corresponding to Q_i
$\Sigma\gamma_i Q_i$ = required strength
R_n = nominal strength
ϕ = resistance factor corresponding to R_n
ϕR_n = design strength

The left side of Equation C-A5-1 represents the required resistance computed by structural analysis based upon assumed loads, and the right side of Equation C-A5-1 represents a limiting structural capacity provided by the selected members. In LRFD, the designer compares the effect of factored loads to the strength actually provided. The term design strength refers to the resistance or strength ϕR_n that must be provided by the selected member. The load factors γ and the resistance factors ϕ reflect the fact that loads, load effects (the computed forces and moments in the structural elements), and the resistances can be determined only to imperfect degrees of accuracy. The resistance factor ϕ is equal to or less than 1.0 because there is always a chance for the actual resistance to be less than the nominal value R_n computed by the equations given in Chapters D through K. Similarly, the load factors γ reflect the fact that the actual load effects may deviate from the nominal values of Q_i computed from the specified nominal loads. These factors account for unavoidable inaccuracies in the theory, variations in the material properties and dimensions, and uncertainties in the determination of loads. They provide a margin of reliability to account for unexpected loads. They do not account for gross error or negligence.

The LRFD Specification is based on (1) probabilistic models of loads and resistance, (2) a calibration of the LRFD criteria to the 1978 edition of the AISC ASD Specification for selected members, and (3) the evaluation of the resulting criteria by judgment and past experience aided by comparative design office studies of representative structures.

The following is a brief summary of the probabilistic basis for LRFD (Ravindra and Galambos, 1978, and Ellingwood, MacGregor, Galambos, and Cornell, 1982). The load effects Q and the resistance factors R are assumed to be statistically independent random variables. In Figure C-A5.1, frequency distributions for Q and R are portrayed as separate curves on a common plot for a hypothetical case. As long as the resistance R is greater than (to the right of) the effects of the loads Q, a margin of safety for the particular limit state exists. However, because Q and R are random variables, there is some small probability that R may be less than Q, ($R<Q$). The probability of this limit state is related to the degree of overlap of the frequency distributions in Figure C-A5.1, which depends on their relative positioning (R_m *vs.* Q_m) and their dispersions.

An equivalent situation may be represented as in Figure C-A5.2. If the expression $R < Q$ is divided by Q and the result expressed logarithmically, the result will be a single frequency distribution curve combining the uncertainties of both R and Q. The probability of attaining a limit state ($R < Q$) is equal to the probability that ln $(R / Q) < 0$ and is represented by the shaded area in the diagram.

The shaded area may be reduced and thus reliability increased in either of two ways: (1) by moving the mean of ln (R/Q) to the right, or (2) by reducing the spread of the curve for a given position of the mean relative to the origin. A convenient way of combining these two approaches is by defining the position of the mean using the standard deviation of ln (R/Q) as the unit of measure. Thus, the distance from the origin to the mean is measured as the number of standard deviations of the function ln (R/Q). As shown in Figure C-A5.2, this is stated as β times $\sigma_{\ln(R/Q)}$, the standard deviation of ln (R/Q). The factor β therefore is called the "reliability index."

If the actual shape of the distribution of ln (R/Q) were known, and if an acceptable value of the probability of reaching the limit state could be agreed upon, one could establish a completely probability-based set of design criteria. Unfortunately, this much information frequently is not known. The distribution shape of each of the many variables (material, loads, etc.) has an influence on the shape of the distribution of ln (R/Q). Often only the means and the standard deviations of the many variables involved in the makeup of the resistance and the load effect can be estimated. However, this information is enough to build an approximate design criterion which is independent of the knowledge of the distribution, by stipulating the following design condition:

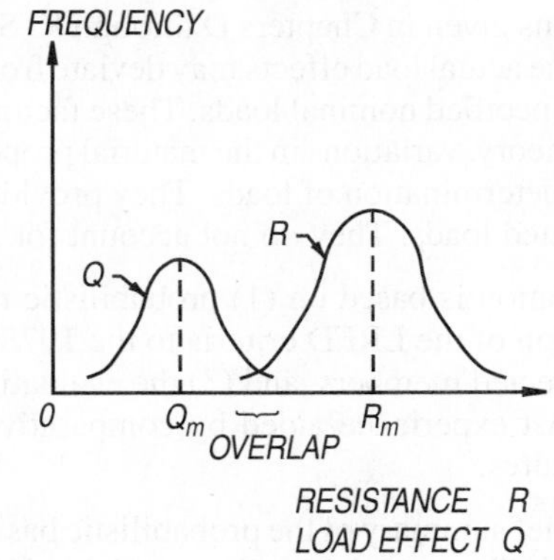

Fig. C-A5.1. Frequency distribution of load effect Q and resistance R.

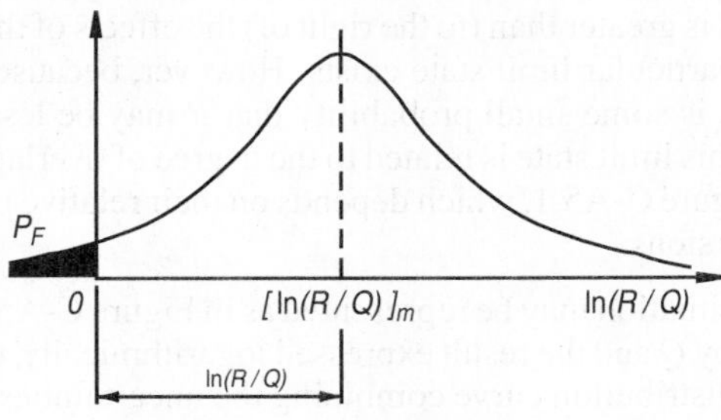

Fig. C-A5.2. Definition of Reliability Index.

$$\beta\sigma_{\ln(R/Q)} \approx \beta\sqrt{V_R^2 + V_Q^2} \leq \ln\left(R_m / Q_m\right) \tag{C-A5-2}$$

In this equation, the standard deviation has been replaced by the approximation $\sqrt{V_R^2 + V_Q^2}$, where $V_R = \sigma_R / R_m$ and $V_Q = \sigma_Q / Q_m$ (σ_R and σ_Q are the standard deviations, R_m and Q_m are the mean values, V_R and V_Q are the coefficients of variation, respectively, of the resistance R and the load effect Q). For structural elements and the usual loading, R_m, Q_m, and the coefficients of variation, V_R and V_Q, can be estimated, so a calculation of

$$\beta = \frac{\ln\left(R_m / Q_m\right)}{\sqrt{V_R^2 + V_Q^2}} \tag{C-A5-3}$$

will give a comparative value of the measure of reliability of a structure or component.

The description of the determination of β as given above is a simple way of defining the probabilistic method used in the development of LRFD. A more refined method, which can accommodate more complex design situations (such as the beam-column interaction equation) and include probabilistic distributions other than the lognormal distribution used to derive Equation C-A5-3, has been developed since the publication of Ravindra and Galambos (1978), and is fully described in Galambos, Ellingwood, MacGregor, and Cornell (1982). This latter method has been used in the development of the recommended load combinations in ASCE 7. The two methods give essentially the same β values for most steel structural members and connections.

Statistical properties (mean values and coefficients of variations) are presented for the basic material properties and for steel beams, columns, composite beams, plate girders, beam-columns, and connection elements in a series of eight articles in the September 1978 issue of the *Journal of the Structural Division of ASCE* (Vol. 104, ST9). The corresponding load statistics are given in Galambos, et al. (1982). Based on these statistics, the values of β inherent in the 1978 edition of the AISC ASD Specification were evaluated under different load combinations (live/dead, wind/dead, etc.), and for various tributary areas for typical members (beams, columns, beam-columns, structural components, etc.). As might be expected, there was a considerable variation in the range of β values. Examination of the many β values associated with ASD revealed certain trends. For example, compact rolled beams (flexure) and tension members (yielding) had β values that decreased from about 3.1 at $L/D = 0.50$ to 2.4 at $L/D = 4$. This decrease is a result of ASD applying the same factor to dead load, which is relatively predictable, and live load, which is more variable. For bolted or welded connections, β was on the order of 4 to 5. Reliability indices for load combinations involving wind and earthquake loads tended to be lower. Based on a thorough assessment of implied reliabilities in existing acceptable design practice, common load factors for various structural materials (steel, reinforced concrete, etc.) were developed in Ellingwood et al. (1982).

One of the features of the probability-based method used in the development of LRFD is that the variations of β values can be reduced by specifying several "target" β values and selecting multiple load and resistance factors to meet these targets. The Committee on Specifications set the point at which LRFD is calibrated to ASD at $L/D = 3.0$ for braced compact beams in flexure and tension members at yield. The resistance factor, ϕ, for these limit states is 0.90, and the implied β is approximately 2.6 for members and 4.0 for connections; this larger β value for connections reflects the fact that connections are expected to be stronger than the members that they connect. Limit states for other members are handled consistently.

Computer methods as well as charts are given in Ellingwood et al. (1982) for the use of specification writers to determine the resistance factors ϕ. These factors can also be approximately determined by the following:

$$\phi = \left(R_m / R_n\right) \exp\left(-0.55\beta V_r\right) \quad \text{(C-A5-4)*}$$

where

R_m = mean resistance
R_n = nominal resistance according to the equations in Chapters D through K
V_r = coefficient of variation of the resistance

4. Design for Serviceability and Other Considerations

Nominally, serviceability should be checked at the unfactored loads. For combinations of gravity and wind or seismic loads some additional reduction factor may be warranted.

*Note that exp(x) is identical to the more familiar e^x

CHAPTER B

DESIGN REQUIREMENTS

B2. NET AREA

Critical net area is based on net width and load transfer at a particular chain.

B3. EFFECTIVE AREA OF TENSION MEMBERS

Section B3 deals with the effect of shear lag, which is applicable to both welded and bolted tension members. The reduction coefficient U is applied to the net area A_n of bolted members and to the gross area A_g of welded members. As the length of connection l is increased, the shear lag effect is diminished. This concept is expressed empirically by the equation for U. Munse and Chesson (1963) have shown that using this expression to compute an effective area, with few exceptions, the estimated strength of some 1,000 bolted and riveted connection test specimens correlated with observed test results within a scatterband of ±10 percent. Newer research (Easterling and Gonzales, 1993) provides further justification for current provisions.

For any given profile and connected elements, $\bar{x}$ is a fixed geometric property. It is illustrated as the distance from the connection plane, or face of the member, to the centroid of the member section resisting the connection force. See Figure C-B3.1. The length l is dependent upon the number of fasteners or equivalent length of weld required to develop the given tensile force, and this in turn is dependent upon the mechanical properties of the member and the capacity of the fasteners or weld used. The length l is illustrated as the distance, parallel to the line of force, between the first and last fasteners in a line for bolted connections. The number of bolts in a line, for the purpose of the determination of l, is determined by the line with the maximum number of bolts in the connection. For staggered bolts, the out-to-out dimension is used for l. See Figure C-B3.2. There is insufficient data to establish a value of U if all lines have only one bolt, but it is probably conservative to use A_e equal to the net area of the connected element. For welded connections, l is the length of the weld parallel to the line of force. For combinations of longitudinal and transverse welds (see Figure C-B3.3), l is the length of longitudinal weld because the transverse weld has little or no effect on the shear lag problem, i.e., it does little to get the load into the unattached portions of the member.

Previous issues of this Specification have presented values for U for bolted or riveted connections of W, M, and S shapes, tees cut from these shapes, and other shapes. These values are acceptable for use in lieu of calculated values and are retained here for the convenience of designers.

For bolted or riveted connections the following values of U may be used:

(a) W, M, or S shapes with flange widths not less than two-thirds the depth, and structural tees cut from these shapes, provided the connection is to the flanges and has no fewer than three fasteners per line in the direction of stress, $U = 0.90$.

(b) W, M, or S shapes not meeting the conditions of subparagraph a, structural tees cut from these shapes, and all other shapes including built-up cross sections, provided the connection has no fewer than three fasteners per line in the direction of stress, $U = 0.85$.

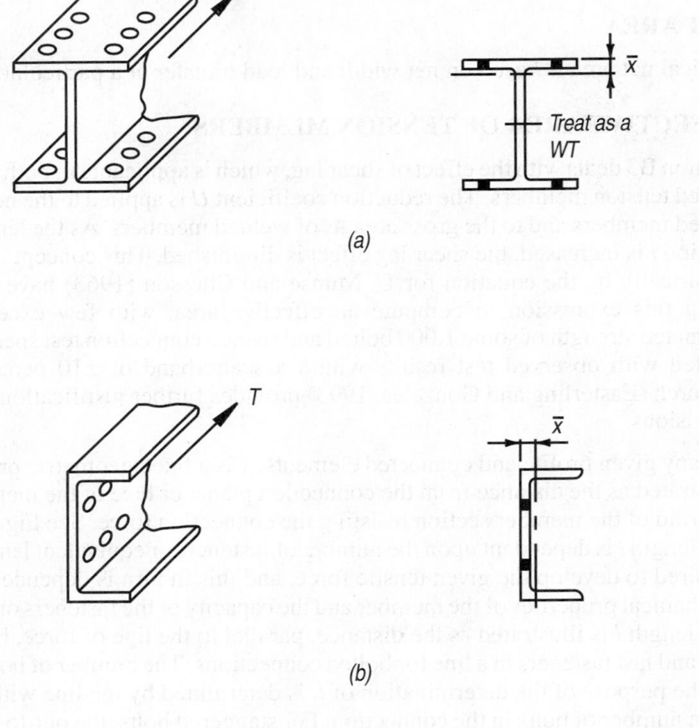

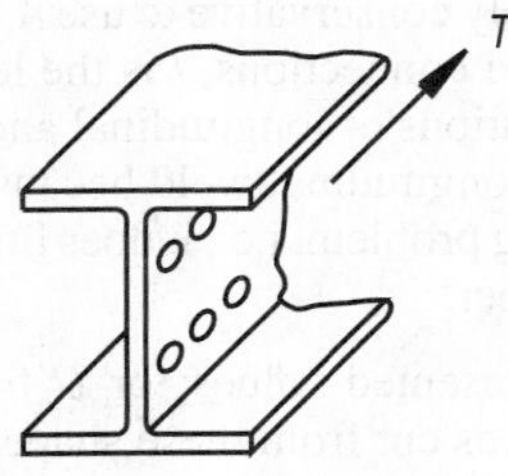

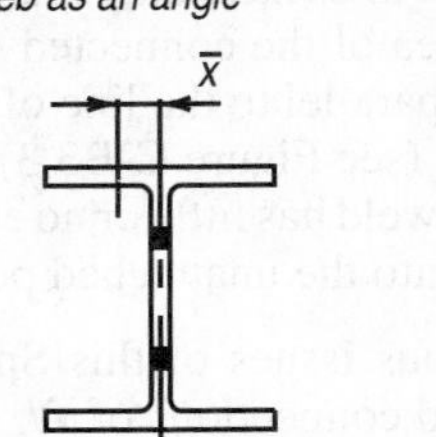

(c)

Fig. C-B3.1. Determination of $\bar{x}$ for U.

(c) All members having only two fasteners per line in the direction of stress, U = 0.75.

When a tension load is transmitted by fillet welds to some but not all elements of a cross section, the weld strength will control.

B5. LOCAL BUCKLING

For the purposes of this Specification, steel sections are divided into compact sections, noncompact sections, and sections with slender compression elements. Compact sections are capable of developing a *fully plastic* stress distribution and they possess a rotational capacity of approximately 3 before the onset of local buckling (Yura, Galambos, and Ravindra, 1978). Noncompact sections can develop the yield stress in compression elements before local buckling occurs, but will not resist inelastic local buckling at the strain levels required for a fully plastic stress distribution. Slender compression elements buckle elastically before the yield stress is achieved.

The dividing line between compact and noncompact sections is the limiting width-thickness ratio λ_p. For a section to be compact, all of its compression elements must have width-thickness ratios equal to or smaller than the limiting λ_p.

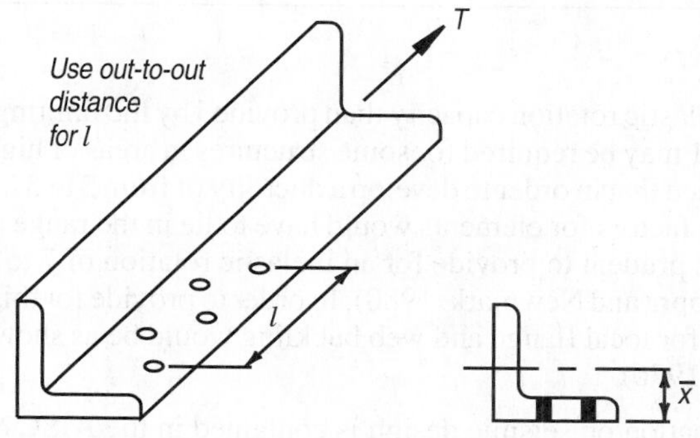

Fig. C-B3.2. Staggered holes.

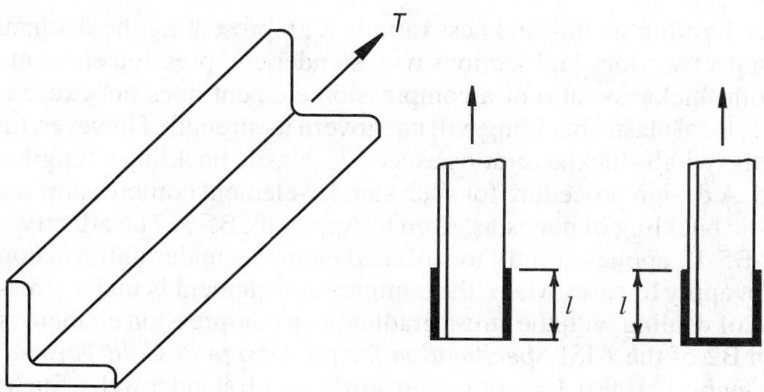

Fig. C-B3.3. Longitudinal and transverse welds.

TABLE C-B5.1
Limiting Width-Thickness Ratios for Compression Elements

Description of Element	Width-Thickness Ratio	Limiting Width-thickness Ratios λ_p	
		Non-seismic	Seismic
Flanges of I-shaped sections (including hybrid sections) and channels in flexure [a]	b/t	$0.38\sqrt{E/F_y}$	$0.31\sqrt{E/F_y}$
Webs in combined flexural and axial compression	h/t_w	For $P_u / \phi_b P_y \leq 0.125$	
		$3.76\sqrt{\dfrac{E}{F_y}}\left(1-\dfrac{2.75P_u}{\phi_b P_y}\right)$	$3.05\sqrt{\dfrac{E}{F_y}}\left(1-\dfrac{1.54P_u}{\phi_b P_y}\right)$
		For $P_u / \phi_b P_y > 0.125$	
		$1.12\sqrt{\dfrac{E}{F_y}}\left(2.33-\dfrac{P_u}{\phi_b P_y}\right) \geq 1.49\sqrt{\dfrac{E}{F_y}}$	
[a] For hybrid beams use F_{yf} in place of F_y			

A greater inelastic rotation capacity than provided by the limiting values λ_p given in Table C-B5.1 may be required for some structures in areas of high seismicity. It has been suggested that in order to develop a ductility of from 3 to 5 in a structural member, ductility factors for elements would have to lie in the range of 5 to 15. Thus, in this case it is prudent to provide for an inelastic rotation of 7 to 9 times the elastic rotation (Chopra and Newmark, 1980). In order to provide for this rotation capacity, the limits λ_p for local flange and web buckling would be as shown in Table C-B5.1 (Galambos, 1976).

More information on seismic design is contained in the AISC *Seismic Provisions for Structural Steel Buildings* (AISC, 1997) and the *Seismic Provisions for Structural Steel Buildings Supplement No. 1* (AISC, 1999).

Another limiting width-thickness ratio is λ_r, representing the distinction between noncompact sections and sections with slender compression elements. As long as the width-thickness ratio of a compression element does not exceed the limiting value λ_r, local elastic buckling will not govern its strength. However, for those cases where the width-thickness ratios exceed λ_r, elastic buckling strength must be considered. A design procedure for such slender-element compression sections, based on elastic buckling of plates, is given in Appendix B5.3. The effective width Equation A-B5-12 applies strictly to stiffened elements under uniform compression. It does not apply to cases where the compression element is under stress gradient. A method of dealing with the stress gradient in a compression element is provided in Section B2 of the AISI *Specification for the Design of Cold-Formed Steel Structural Members* (1996). Exceptions are girders with slender webs. Such plate girders are capable of developing postbuckling strength in excess of the elastic buckling load. A design procedure for plate girders including tension field action is given in Appendix G.

The values of the limiting ratios λ_p and λ_r specified in Table B5.1 are similar to those in AISC (1989) and Table 2.3.3.3 of Galambos (1976), except that: (1) $\lambda_p = 0.38\sqrt{E/F_y}$, limited in Galambos (1976) to indeterminate beams when moments are determined by elastic analysis and to determinate beams, was adopted for all conditions on the basis of Yura et al. (1978); and (2) $\lambda_p = 0.045E/F_y$ for plastic design of circular hollow sections was obtained from Sherman (1976).

The high shape factor for circular hollow sections makes it impractical to use the same slenderness limits to define the regions of behavior for different types of loading. In Table B5.1, the values of λ_p for a compact shape that can achieve the plastic moment, and λ_r for bending, are based on an analysis of test data from several projects involving the bending of pipes in a region of constant moment (Sherman and Tanavde, 1984, and Galambos, 1998). The same analysis produced the equation for the inelastic moment capacity in Table A-F1.1 in Appendix F1. However, a more restrictive value of λ_p is required to prevent inelastic local buckling from limiting the plastic hinge rotation capacity needed to develop a mechanism in a circular hollow beam section (Sherman, 1976).

The values of λ_r for axial compression and for bending are both based on test data. The former value has been used in building specifications since 1968 (Winter, 1970). Appendices B5 and F1 also limit the diameter-to-thickness ratio for any circular section to $0.45E/F_y$. Beyond this, the local buckling strength decreases rapidly, making it impractical to use these sections in building construction.

Following the SSRC recommendations (Galambos, 1998) and the approach used for other shapes with slender compression elements, a Q factor is used for circular sections to account for interaction between local and column buckling. The Q factor is the ratio between the local buckling stress and the yield stress. The local buckling stress for the circular section is taken from the inelastic AISI criteria (Winter, 1970) and is based on tests conducted on fabricated and manufactured cylinders. Subsequent tests on fabricated cylinders (Galambos, 1998) confirm that this equation is conservative.

The definitions of the width and thickness of compression elements agree with the 1978 AISC ASD Specification with minor modifications. Their applicability extends to sections formed by bending and to unsymmetrical and hybrid sections.

For built-up I-shaped sections under axial compression, modifications have been made to the flange local buckling criterion to include web-flange interaction. The k_c in the λ_r limit, in Equations A-B5-7 and A-B5-8 and the elastic buckling Equation A-B5-8 are the same that are used for flexural members. Theory indicates that the web-flange interaction in axial compression is at least as severe as in flexure. Rolled shapes are excluded from this criterion because there are no standard sections with proportions where the interaction would occur. In built-up sections where the interaction causes a reduction in the flange local buckling strength, it is likely that the web is also a thin stiffened element.

The k_c factor accounts for the interaction of flange and web local buckling demonstrated in experiments conducted by Johnson (1985). The maximum limit of 0.763 corresponds to $F_{cr} = 0.69E / \lambda^2$ which was used as the local buckling strength in earlier editions of both the ASD and LRFD Specifications. An $h / t_w = 27.5$ is required to reach $k_c = 0.763$. Fully fixed restraint for an unstiffened compression element corresponds to $k_c = 1.3$ while zero restraint gives $k_c = 0.42$. Because of web-flange interactions it is possible to get $k_c < 0.42$ from the new k_c formula. If $h / t_w > 5.70\sqrt{E / F_y}$ use $h / t_w = 5.70\sqrt{E / F_y}$ in the k_c equation, which corresponds to the 0.35 limit.

Illustrations of some of the requirements of Table B5.1 are shown in Figure C-B5.1.

B7. LIMITING SLENDERNESS RATIOS

Chapters D and E provide reliable criteria for resistance of axially loaded members based on theory and confirmed by tests for all significant parameters including slenderness. The advisory upper limits on slenderness contained in Section B7 are based on professional judgment and practical considerations of economics, ease of handling, and care required to minimize inadvertent damage during fabrication, transport, and erection. Out-of-straightness within reasonable tolerances does not affect the strength of tension members, and the effect of out-of-straightness within specified tolerances on the strength of compression members is accounted for in formulas for resistance. Applied tension tends to reduce, whereas compression tends to amplify, out-of-straightness. Therefore, more liberal criteria are suggested for tension members, including those subject to small compressive forces resulting from transient loads such as earthquake and wind. For members with slenderness ratios greater than 200, these compressive forces correspond to $\phi_c F_{cr}$ less than 5.33 ksi (18 MPa).

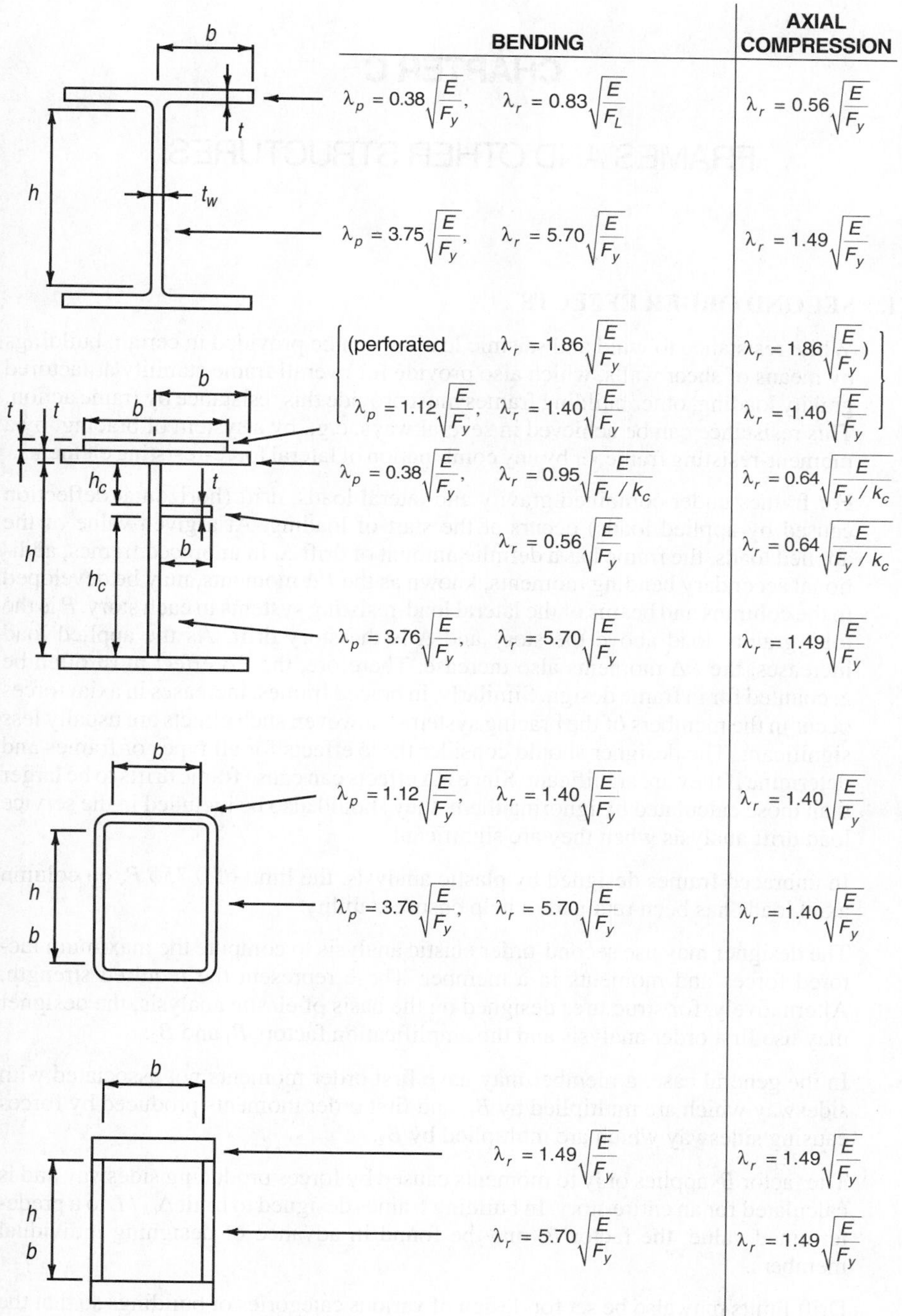

Fig. C-B5.1 Selected examples of Table B5.1 requirements.

CHAPTER C

FRAMES AND OTHER STRUCTURES

C1. SECOND ORDER EFFECTS

While resistance to wind and seismic loading can be provided in certain buildings by means of shear walls, which also provide for overall frame stability at factored gravity loading, other building frames must provide this resistance by frame action. This resistance can be achieved in several ways, e.g., by a system of bracing, by a moment-resisting frame, or by any combination of lateral force-resisting elements.

For frames under combined gravity and lateral loads, drift (horizontal deflection caused by applied loads) occurs at the start of loading. At a given value of the applied loads, the frame has a definite amount of drift Δ. In unbraced frames, additional secondary bending moments, known as the $P\Delta$ moments, may be developed in the columns and beams of the lateral load-resisting systems in each story. P is the total gravity load above the story and Δ is the story drift. As the applied load increases, the $P\Delta$ moments also increase. Therefore, the $P\Delta$ effect must often be accounted for in frame design. Similarly, in braced frames, increases in axial forces occur in the members of the bracing systems; however, such effects are usually less significant. The designer should consider these effects for all types of frames and determine if they are significant. Since $P\Delta$ effects can cause frame drifts to be larger than those calculated by ignoring them, they should also be included in the service load drift analysis when they are significant.

In unbraced frames designed by plastic analysis, the limit of $0.75\phi_c P_y$ on column axial loads has been retained to help ensure stability.

The designer may use second-order elastic analysis to compute the maximum factored forces and moments in a member. These represent the required strength. Alternatively, for structures designed on the basis of elastic analysis, the designer may use first order analysis and the amplification factors B_1 and B_2.

In the general case, a member may have first order moments not associated with sidesway which are multiplied by B_1, and first order moments produced by forces causing sidesway which are multiplied by B_2.

The factor B_2 applies only to moments caused by forces producing sidesway and is calculated for an entire story. In building frames designed to limit Δ_{oh} / L to a predetermined value, the factor B_2 may be found in advance of designing individual members.

Drift limits may also be set for design of various categories of buildings so that the effect of secondary bending can be insignificant (Kanchanalai and Lu, 1979, and ATC, 1978). It is conservative to use the B_2 factor with the sum of the sway and the no-sway moments, i.e., with $M_{lt} + M_{nt}$.

The two kinds of first order moment M_{nt} and M_{lt} may both occur in sidesway frames from gravity loads. M_{nt} is defined as a moment developed in a member with frame sidesway prevented. If a significant restraining force is necessary to prevent sidesway of an unsymmetrical structure (or an unsymmetrically loaded symmetrical structure), the moments induced by releasing the restraining force will be M_{lt} moments, to be multiplied by B_2. In most reasonably symmetric frames, this effect will be small. If such a moment B_2M_{lt} is added algebraically to the B_1M_{nt} moment developed with sidesway prevented, a fairly accurate value of M_u will result. End moments produced in sidesway frames by lateral loads from wind or earthquake will always be M_{lt} moments to be multiplied by B_2.

When first order end moments in members subjected to axial compression are magnified by B_1 and B_2 factors, equilibrium requires that they be balanced by moments in connected members (Figure C-C1.1). This can generally be accomplished satisfactorily by distributing the difference between the magnified moment and the first order moment to any other moment-resisting members attached to the compressed member (or members) in proportion to the relative stiffness of the uncompressed members. Minor imbalances may be neglected in the judgment of the engineer. However, complex conditions, such as occur when there is significant magnification in several members meeting at a joint, may require a second order elastic analysis. Connections shall also be designed to resist the magnified end moments.

The center-to-center member length is usually used in the structural analysis. In braced and unbraced frames, P_n is governed by the maximum slenderness ratio regardless of the plane of bending. However, P_{e1} and P_{e2} are always calculated using the slenderness ratio in the plane of bending. Thus, when flexure is about the strong

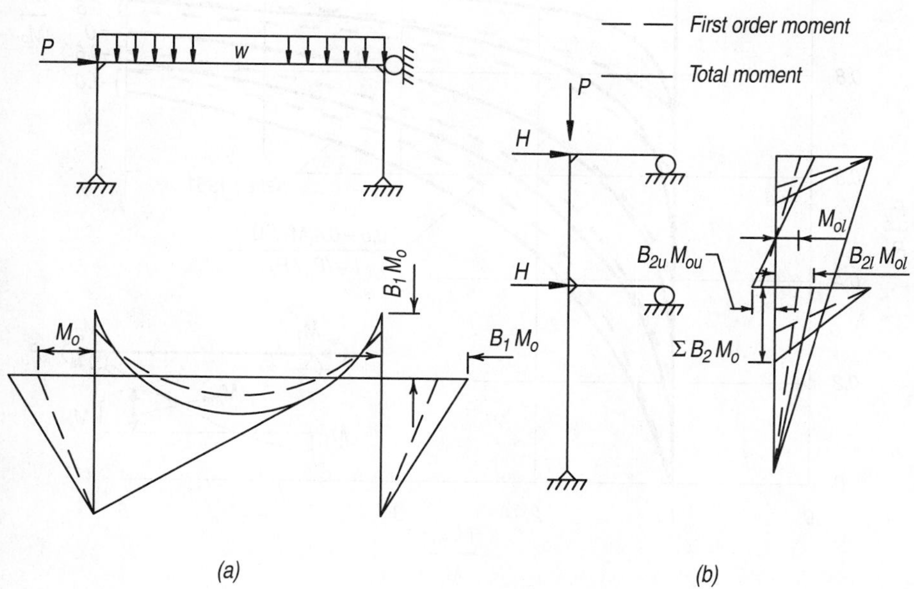

Fig. C-C1.1. Moment amplification.

axis only, two different values of slenderness ratio may be involved in solving a given problem.

When second order analysis is used, it must account for the interaction of the factored load effects, that is, combinations of factored loads must be used in analysis. Superposition of forces obtained from separate analyses is not adequate.

When bending occurs about both the x and the y axes, the required flexural strength calculated about each axis is adjusted by the value of C_m and P_{e1} or P_{e2} corresponding to the distribution of moment and the slenderness ratio in its plane of bending, and is then taken as a fraction of the design bending strength, $\phi_b M_n$, about that axis, with due regard to the unbraced length of the compression flange where this is a factor.

Equations C1-2 and C1-3 approximate the maximum second order moments in compression members with no relative joint translation and no transverse loads between the ends of the member. This approximation is compared to an exact solution (Ketter, 1961) in Figure C-C1.2. For single curvature, Equation C1-3 is slightly unconservative, for a zero end moment it is almost exact, and for double curvature it is conservative. The 1978 AISC ASD Specification imposed the limit $C_m \geq 0.4$ which corresponds to a M_1 / M_2 ratio of 0.5. However, Figure C-C1.2 shows that if, for example, $M_1 / M_2 = 0.8$, the $C_m = 0.28$ is already very conservative, so the limit has been removed. The limit was originally adopted from Austin (1961), which was intended to apply to lateral-torsional buckling, not second-order in-plane bending strength. The AISC Specifications, both in the 1989 ASD and LRFD, use a modification factor C_b as given in Equation F1-3 for lateral-torsional buckling. C_b is approximately the inverse of C_m as presented in Austin (1961) with a 0.4 limit. In

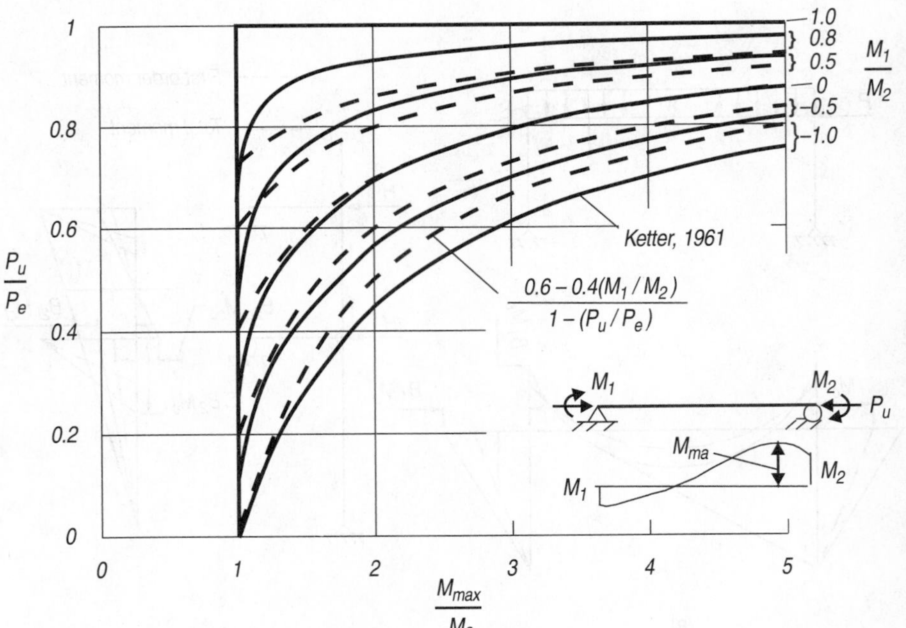

Fig, C-C1.2. Second-order moments for beam-columns in braced frames.

TABLE C-C1.1
Amplification Factors ψ and C_m

Case	Ψ	C_m
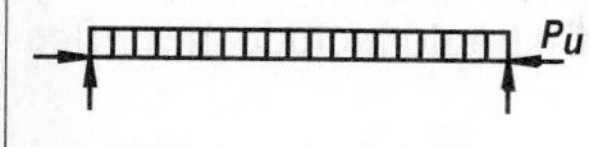 Simply supported, uniform load, P_u	0	1.0
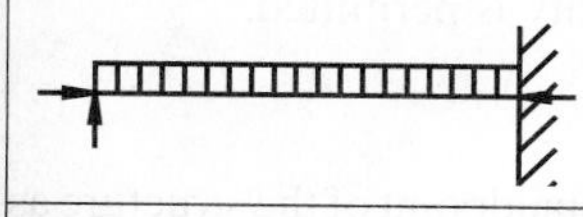 Pinned–fixed, uniform load	−0.4	$1-0.4\dfrac{P_u}{P_{e1}}$
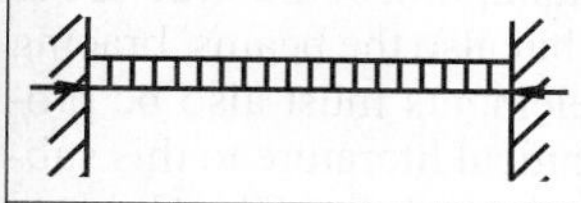 Fixed–fixed, uniform load	−0.4	$1-0.4\dfrac{P_u}{P_{e1}}$
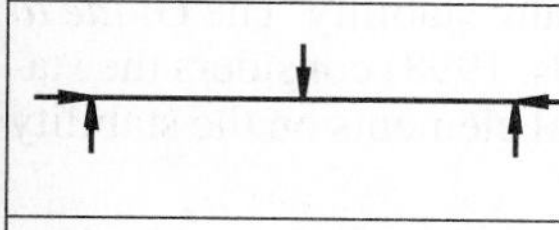 Simply supported, concentrated load at midspan	−0.2	$1-0.2\dfrac{P_u}{P_{e1}}$
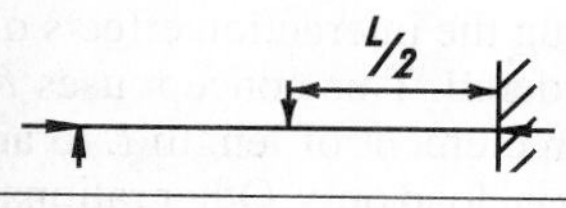 Pinned–fixed, concentrated load at $L/2$	−0.3	$1-0.3\dfrac{P_u}{P_{e1}}$
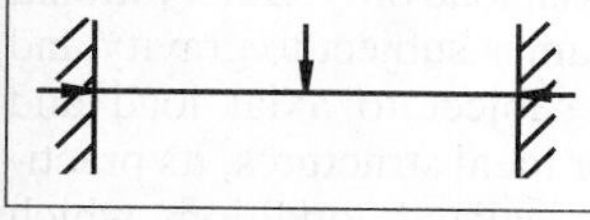 Fixed–fixed, concentrated load at midspan	−0.2	$1-0.2\dfrac{P_u}{P_{e1}}$

Galambos (1961) it was pointed out that Equation C1-3 could be used for in-plane second order moments if the 0.4 limit was eliminated. Unfortunately, Austin (1961) was misinterpreted and a lateral-torsional buckling solution was used for an in-plane second-order analysis. This oversight has now been corrected.

For beam columns with transverse loadings, the second-order moment can be approximated by using the following equation

$$C_m = 1 + \psi P_u / P_{e1}$$

for simply supported members

where

$$\psi = \frac{\pi^2 \delta_o EI}{M_o L^2} - 1$$

δ_o = maximum deflection due to transverse loading, in. (mm)

M_o = maximum factored design moment between supports due to transverse loading, kip-in. (N-mm)

For restrained ends, some limiting cases are given in Table C-C1.1 together with two cases of simply supported beam-columns (Iwankiw, 1984). These values of C_m are always used with the maximum moment in the member. For the restrained-end cases, the values of B_1 will be most accurate if values of $K < 1.0$ corresponding to the end boundary conditions are used in calculating P_{e1}. In lieu of using the equations above, $C_m = 1.0$ can be used conservatively for transversely loaded members with unrestrained ends and 0.85 for restrained ends.

If, as in the case of a derrick boom, a beam-column is subject to transverse (gravity) load and a calculable amount of end moment, the value δ_0 should include the deflection between supports produced by this moment.

Stiffness reduction adjustment due to column inelasticity is permitted.

C2. FRAME STABILITY

The stability of structures must be considered from the standpoint of the structure as a whole, including not only the compression members, but also the beams, bracing system, and connections. The stability of individual elements must also be provided. Considerable attention has been given in the technical literature to this subject, and various methods of analysis are available to assure stability. The *Guide to Stability Design Criteria for Metal Structures* (Galambos, 1998) considers the stability of individual elements, and the effects of individual elements on the stability of the structure as a whole.

The effective length concept is one method of estimating the interaction effects of the total frame on a compression element being considered. This concept uses K factors to equate the strength of a framed compression element of length L to an equivalent pin-ended member of length KL subject to axial load only. Other rational methods are available for evaluating the stability of frames subject to gravity and side loading and individual compression members subject to axial load and moments. Although the concept is completely valid for ideal structures, its practical implementation involves several assumptions of idealized conditions which will be mentioned later.

Two conditions, opposite in their effect upon column strength under axial loading, must be considered. If enough axial load is applied to the columns in an unbraced frame dependent entirely on their own bending stiffness for resistance to lateral deflection of the tops of the columns with respect to their bases (see Figure C-C2.1), the effective length of these columns will exceed the actual length. On the other hand, if the same frame were braced to resist such lateral movement, the effective length would be less than the actual length, due to the restraint (resistance to joint translation) provided by the bracing or other lateral support. The ratio K, effective column length to actual unbraced length, may be greater or less than 1.0.

The theoretical K values for six idealized conditions in which joint rotation and translation are either fully realized or nonexistent are tabulated in Table C-C2.1.

Also shown are suggested design values recommended by the Structural Stability Research Council (SSRC) for use when these conditions are approximated in actual design. In general, these suggested values are slightly higher than their theoretical equivalents, since joint fixity is seldom fully realized.

If the column base in Case (f) of Table C-C2.1 were truly pinned, K would actually exceed 2.0 for a frame such as that pictured in Figure C-C2.1, because the flexibility of the horizontal member would prevent realization of full fixity at the top of the column. On the other hand, it has been shown (Galambos, 1960) that the restraining

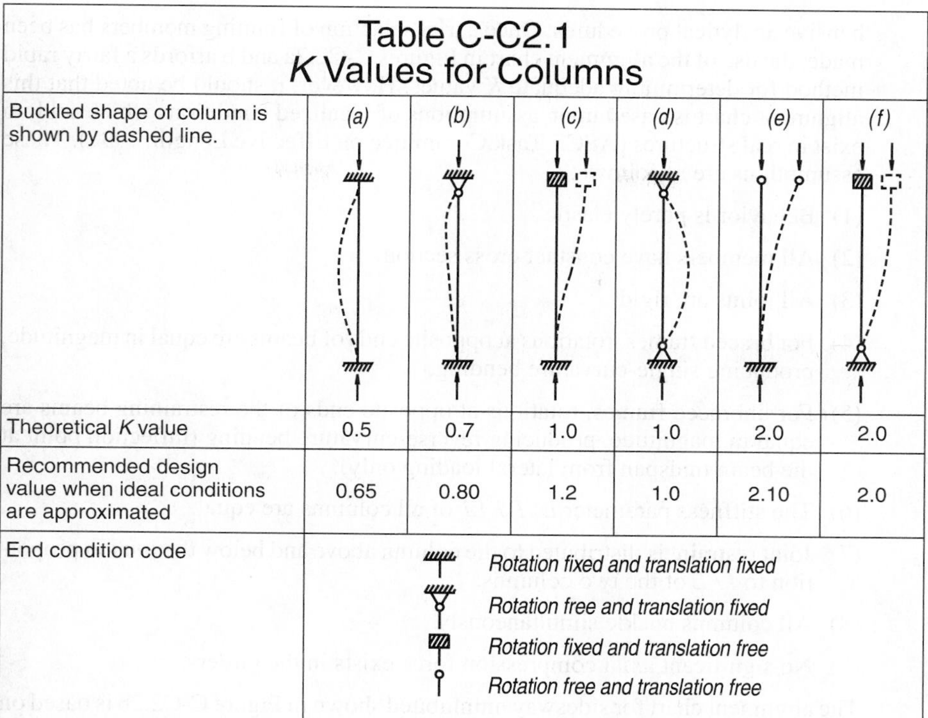

Table C-C2.1
***K* Values for Columns**

Buckled shape of column is shown by dashed line.	(a)	(b)	(c)	(d)	(e)	(f)
Theoretical *K* value	0.5	0.7	1.0	1.0	2.0	2.0
Recommended design value when ideal conditions are approximated	0.65	0.80	1.2	1.0	2.10	2.0
End condition code	Rotation fixed and translation fixed Rotation free and translation fixed Rotation fixed and translation free Rotation free and translation free					

influence of foundations, even where these footings are designed only for vertical load, can be very substantial in the case of flat-ended column base details with ordinary anchorage. For this condition, a design *K* value of 1.5 would generally be conservative in Case (f).

While in some cases masonry walls provide enough lateral support for building frames to control lateral deflection, light curtain wall construction and wide column spacing can create a situation where only the bending stiffness of the frame provides this support. In this case the effective length factor *K* for an unbraced length of column *L* is dependent upon the bending stiffness provided by the other in-plane members entering the joint at each end of the unbraced segment. If the combined stiffness provided by the beams is sufficiently small, relative to that of the unbraced column segments, *KL* could exceed two or more story heights (Bleich, 1952).

Translation of the joints in the plane of a truss is inhibited and, due to end restraint, the effective length of compression members might be assumed to be less than the distance between panel points. However, it is usual practice to take *K* as equal to 1.0 (ASCE Task Committee on Effective Length, 1997). If all members of the truss reached their ultimate load capacity simultaneously, the restraints at the ends of the compression members would be greatly reduced.

Several rational methods are available to estimate the effective length of the columns in an unbraced frame with sufficient accuracy for design. These range from simple interpolation between the idealized cases shown in Table C-C2.1 to comprehensive analytical procedures. Once a trial selection of framing members has been made, the use of the alignment chart in Figures C-C2.2a and b affords a fairly rapid method for determining adequate K values. However, it should be noted that this alignment chart is based upon assumptions of idealized conditions which seldom exist in real structures (ASCE Task Committee on Effective Length, 1997). These assumptions are as follows:

(1) Behavior is purely elastic.

(2) All members have constant cross section.

(3) All joints are rigid.

(4) For braced frames, rotations at opposite ends of beams are equal in magnitude, producing single-curvature bending.

(5) For unbraced frames, rotations at opposite ends of the restraining beams are equal in magnitude, producing reverse-curvature bending (inflection point at the beam midspan from lateral loading only).

(6) The stiffness parameter $L\sqrt{P/EI}$ of all columns are equal.

(7) Joint restraint is distributed to the column above and below the joint in proportion to I/L of the two columns.

(8) All columns buckle simultaneously.

(9) No significant axial compression force exists in the girders.

The alignment chart for sidesway uninhibited shown in Figure C-C2.2b is based on the following equation:

$$\frac{G_A G_B (\pi/K)^2 - 36}{6(G_A + G_B)} - \frac{(\pi/K)}{\tan(\pi/K)} = 0 \qquad \text{(C-C2-1)}$$

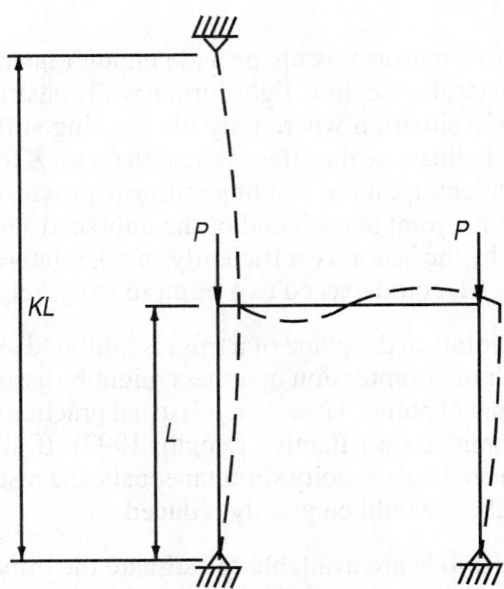

Fig. C-C2.1. Column effective length.

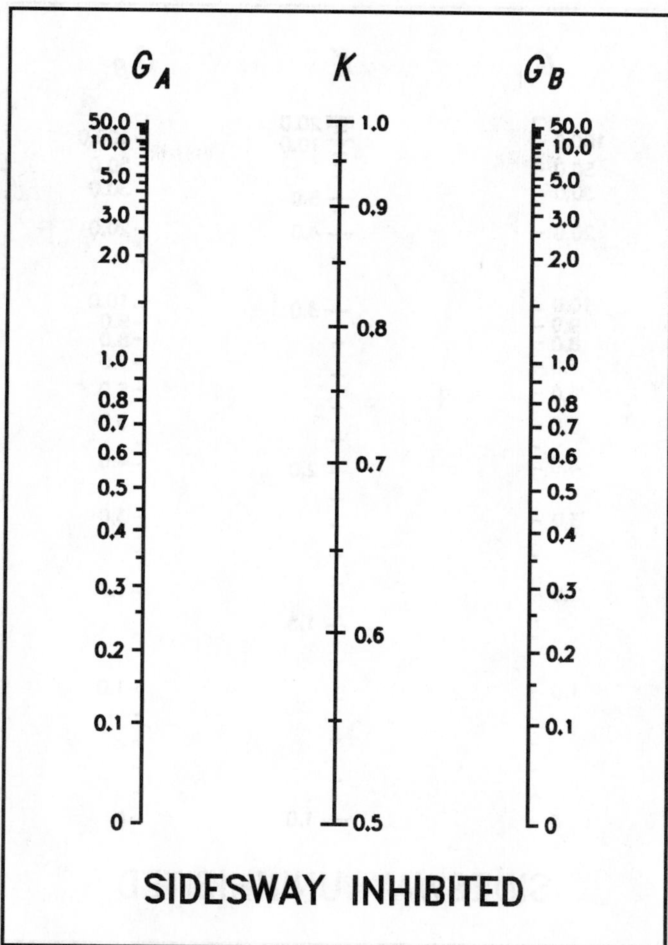

Notes for Fig. C-C2.2a and b: The subscripts A and B refer to the joints at the two ends of the column section being considered. G is defined as

$$G = \frac{\Sigma(I_c / L_c)}{\Sigma(I_g / L_g)}$$

in which Σ indicates a summation of all members rigidly connected to that joint and lying on the plane in which buckling of the column is being considered. I_c *is the moment of inertia and* L_c *the unsupported length of a column section, and* I_g *is the moment of inertia and* L_g *the unsupported length of a girder or other restraining member.* I_c *and* I_g *are taken about axes perpendicular to the plane of buckling being considered.*

For column ends supported by but not rigidly connected to a footing or foundation, G is theoretically infinity, but, unless actually designed as a true friction-free pin, may be taken as "10" for practical designs. If the column end is rigidly attached to a properly designed footing, G may be taken as 1.0. Smaller values may be used if justified by analysis. Errata 9/4/01

Fig. C-C2.2a. Alignment chart for effective length of columns in continuous frames – Sidesway Inhibited.

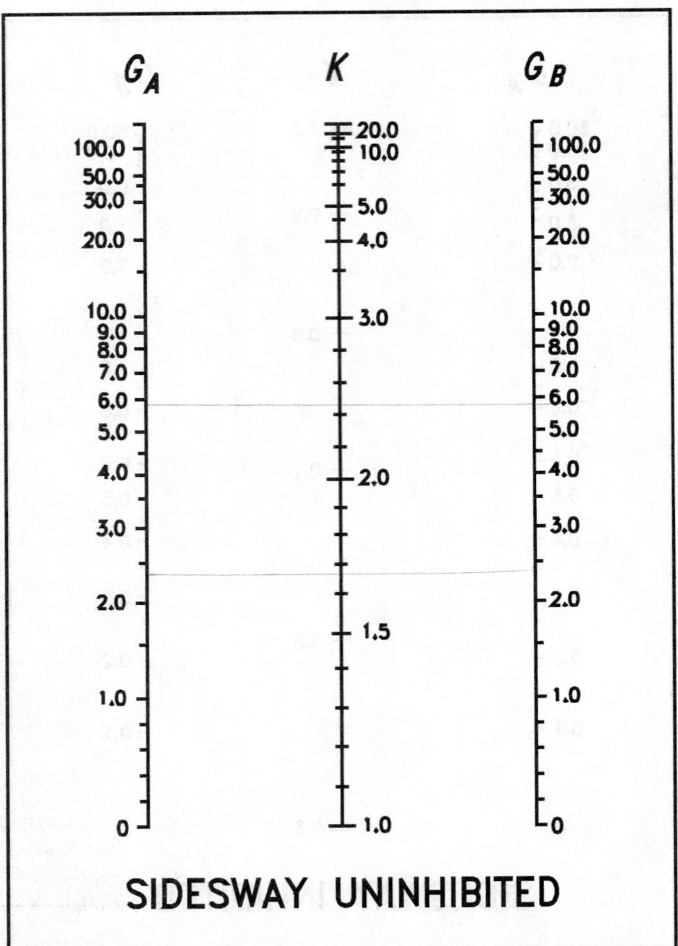

See Notes in Fig. C-C2.2a.

Fig. C-C2.2b. Alignment chart for effective length of columns in continuous frames – Sidesway Uninhibited.

with G defined as

$$G = \frac{\Sigma(EI/L)_c}{\Sigma(EI/L)_g} \tag{C-C2-2}$$

The expression for G given in the footnote of the alignment chart has assumed that E of the beams and columns are the same. However the alignment chart is valid for different materials if Equation C-C2-2 is used. An equation for the sidesway-inhibited chart can be found in ASCE Task Committee on Effective Length (1997).

Where the actual conditions differ from the assumptions above, unrealistic K factors may result. There are modifications available that may be used with Figure C-C2.2b or Equation C-C2-1 to give buckling loads that better reflect the conditions in real structures (ASCE Task Committee on Effective Length, 1997 and Chen and Lui, 1987). Some of the modifications are summarized below.

Columns loaded into the inelastic range of column behavior can be viewed as having a tangent modulus E_T that is smaller than E. For such columns, $E_c/E_g = E_T/E$ in Equation C-C2-2, which gives smaller G values, and therefore, smaller K factors than those based on elastic behavior (assumption 1). It is conservative to base the column design on elastic K factors. For less conservative solutions, inelastic K factors can be determined by using τE for E_c in Equation C-C2-2 where $\tau = E_T/E$ is a stiffness reduction factor (SRF). Yura (1971) and Disque (1973) showed that the SRF could be determined from the ratio of the inelastic column design strength to the elastic column design strength. Using the column design strengths ϕP_n from Equation E2-2 (inelastic) and E2-3 (elastic) gives

(a) For $(P_u / P_y) \le \frac{1}{3}$ (elastic); $\tau = 1.0$

(b) For $(P_u / P_y) > \frac{1}{3}$ (inelastic)

$$\tau = -7.38(P_u / P_y)\log\left(\frac{(P_u / P_y)}{0.85}\right) \qquad \text{(C-C2-3)}$$

where P_y is the column squash load, $(F_y A_g)$, and P_u is the required column strength. P_u must not exceed $\phi_c P_y$.

When a beam connection at the column under consideration is a shear connection (no moment), then that beam cannot be considered in the $\Sigma(EI/L)_g$ term of Equation C-C2-2. Only FR connections can be used directly in the determination of G (assumption 3). PR connections with a documented moment-rotation response can be utilized, but the $(EI/L)_g$ of each beam must be adjusted to account for the connection flexibility. ASCE Task Committee on Effective Length (1997) provides a detailed discussion of frame stability with PR connections. PR connections cannot be considered as rigid (FR) connections when assessing frame stability. Section A2 contains additional information on PR connections.

A beam stiffness of $6EI/L$ was used in the development of Equation C-C2-1 (assumption 5). For other values of beam stiffness, use $m(EI/L)_g$ in determining G where m = (actual girder stiffness coefficient)/6. When the far end of a girder has a shear connection instead of a FR connection, $m = 0.5$. A general expression for m when the inflection point from a lateral load analysis is located anywhere along the girder span is available (ASCE Task Committee on Effective Length, 1997).

Compressive axial load in a girder reduces its stiffness, which will have an adverse effect on K of the column (see assumption 9). To account for any compressive axial load in a girder, the girder stiffness parameter $(EI/L)_g$ in Equation C-C2-2 should be modified by the factor

$$\left[1 - \frac{Q}{Q_{cr}}\right] \qquad \text{(C-C2-4)}$$

where Q is the axial load in the girder and Q_{cr} is the in plane buckling load of the girder based on $K = 1.0$. Tensile axial load in the girders can be ignored when determining G.

Sidesway instability of an unbraced frame is a story phenomenon involving the sum of the sway resistances of each column in the story and the sum of the factored gravity loads in the columns in that story. If each column in a story of an unbraced frame is designed to support its own P and $P\Delta$ moment, then all the columns will buckle simultaneously (assumption 8). Under this condition, there is no interaction among the columns in the story; column sway instability and frame instability occur at the same time. Framing systems can be used that redistribute the story $\Sigma P\Delta$ to the columns in that story in proportion to their individual stiffnesses. In an unbraced frame that contains columns that contribute little or nothing to the sway stiffness of the story, such columns can be designed using K = 1.0 (leaning columns), but the other columns in the story must be designed to support the destabilizing $P\Delta$ moments developed from the loads on such columns. Similarly, more highly loaded columns in a story will redistribute some of their $P\Delta$ moments to more lightly loaded columns.

Two methods for evaluating story frame stability are recognized, the story stiffness method (LeMessurier, 1976 and 1977) and the story buckling method (Yura, 1971) as reflected in Equations C1-4 and C1-5, respectively. For an individual column in the sway-resisting system,

$$\lambda_c^2 = \frac{(KL)^2}{\pi^2 EI} A_g F_y = \frac{P_y}{P_{e2}}$$

To account for the redistribution of the $P\Delta$ moments within a story, determine P_n for an individual column in the sway-resisting system by substituting λ'_c for λ_c, where

$$\lambda'_c = \frac{K'L}{\pi r}\sqrt{\frac{F_y}{E}}$$

and K' is given below.

For the story stiffness method

$$K' = \sqrt{\frac{P_e}{0.822 P_u} \Sigma P_u \left(\frac{\Delta_{oh}}{\Sigma HL}\right)} \qquad \text{(C-C2-5)}$$

where

$$P_e = \frac{\pi^2 EI}{L^2}.$$

The 0.822 factor is the ratio of the lateral column shear force per radian of drift to the buckling load of a sway permitted column with large end restraint, $G = 0$. This factor will approach 1.0 for more flexible systems or systems with a large percentage of leaner columns. Less conservative methods are given in ASCE Task Committee on Effective Length (1997). The term,

$$\Sigma P_u\left(\frac{\Delta_{oh}}{\Sigma HL}\right)$$

is a constant for all rigidly connected columns in a story and is the same term used in Equation C1-4.

For the story buckling approach

$$K' = \sqrt{\frac{P_e}{P_u}\left(\frac{\Sigma P_u}{\Sigma P_{e2}}\right)} \qquad \text{(C-C2-6)}$$

where $(\Sigma P_u \,/\, \Sigma P_{e2})$ is the same term found in Equation C1-5 and is a constant for all rigidly-connected columns in a story.

The value of P_n calculated using K' by either method cannot be taken greater than P_n based on sidesway inhibited buckling. Additional simplified methods were given in the previous edition of this commentary. Although they are not repeated here, they are equally valid within the limitations placed on them in that edition. A comparison of the influence of those methods may be found in Geschwindner (1994).

The theoretical K-factors that are less than 1.0 (Cases (a) and (b) in Table C-C2.1 and the sidesway inhibited alignment chart in Figure C-C2.2a), are based on the assumption that there is no relative lateral movement of the ends of the column. When bracing is proportioned by the requirements of Section C3, K equal to 1.0 should be used, not values less than 1.0, because a small relative movement of the brace points is anticipated.

C3. STABILITY BRACING

1. Scope

The design requirements consider two general types of bracing systems, relative and nodal, as shown in Fig. C-C3.1. A relative column brace system (such as diagonal bracing or shear walls) is attached to two locations along the length of the column that defines the unbraced length. The relative brace system shown consists of the diagonal and the strut that controls the movement at one end of the unbraced length, A, with respect to the other end of the unbraced length, B. The diagonal and the strut both contribute to the strength and stiffness of the relative brace system. However, when the strut is a floor beam, its stiffness is large compared to the diagonal so the diagonal controls the strength and stiffness of the relative brace. A nodal brace controls the movement only at the particular brace point, without direct interaction with adjacent braced points. Therefore to define an unbraced length there must be additional adjacent brace points as shown in Figure C-C3.1. The two nodal column braces at C and D that are attached to the rigid abutment define the unbraced length for which $K = 1.0$ can be used. For beams a cross frame between two adjacent beams at midspan is a nodal brace because it prevents twist of the beams only at the particular cross frame location. The unbraced length is half the span length. The twist at the ends of the two beams is prevented by the beam-to-column connections at the end supports. Similarly, a nodal lateral brace attached at midspan to the top flange of the beams and a rigid support assumes that there is no lateral movement at the column locations.

The brace requirements will enable a member to potentially reach a maximum load based on the unbraced length between the brace points and $K = 1.0$. This is not the same as the no-sway buckling load as illustrated in Figure C-C3.2 for the braced cantilever. The critical stiffness is $1.0\,P_e/L$, corresponding to $K = 1.0$. A brace with five times this stiffness is necessary to reach 95 percent of the $K = 0.7$ limit. Theoretically, an infinitely stiff brace is required to reach the no-sway limit. Bracing required to reach specified rotation capacities or ductility limits is beyond the scope of these recommendations. Member inelasticity has no significant effect on the brace requirements (Yura, 1995).

Winter (1958 and 1960) developed the concept of dual criterion for bracing design, strength and stiffness. The brace force is a function of the initial column out-of-straightness, Δ_o, and the brace stiffness, β. For a relative brace system, the relationship between column load, brace stiffness and sway displacement is shown in Figure C-C3.3. If $\beta = \beta_i$, the critical brace stiffness for a perfectly plumb member, then $P = P_e$ only if the sway deflection gets very large. Unfortunately, such large dis-

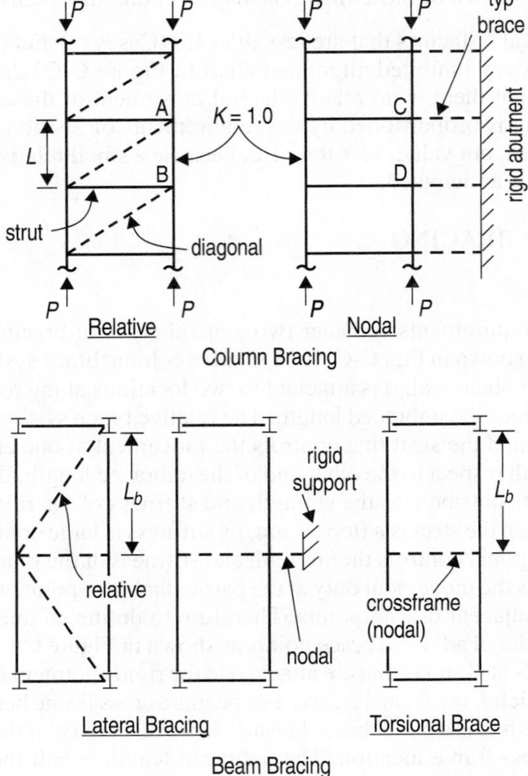

Fig. C-C3.1. Types of bracing.

placements produce large brace forces. For practical design, Δ must be kept small at the factored load level.

The brace stiffness requirements, β_{br}, for frames, columns, and beams were chosen as twice the critical stiffness. The $\phi = 0.75$ specified for all brace stiffness requirements is consistent with the implied resistance factor for elastic Euler column buckling, i.e. $0.877 \times \phi_c = 0.75$. For the relative brace system shown in Figure C-C3.3, $\beta_{br} = 2\beta_i$ gives $P_{br} = 0.4\%\ P_e$ for $\Delta_o = 0.002L$. If the brace stiffness provided, β_{act}, is different from the requirement, then the brace force or brace moment can be multiplied by the following factor:

$$\frac{1}{2 - \dfrac{\beta_{br}}{\beta_{act}}} \qquad \text{(C-C3-1)}$$

No ϕ is specified in the brace strength requirements since ϕ is included in the component design strength provisions in other chapters of this Specification.

The initial displacement, Δ_o, for relative and nodal braces is defined with respect to the distance between adjacent braces, as shown in Figure C-C3.4. The initial Δ_o is a displacement from the straight position at the brace points caused by sources other than brace elongations from gravity loads or compressive forces, such as displacements caused by wind or other lateral forces, erection tolerances, column shorten-

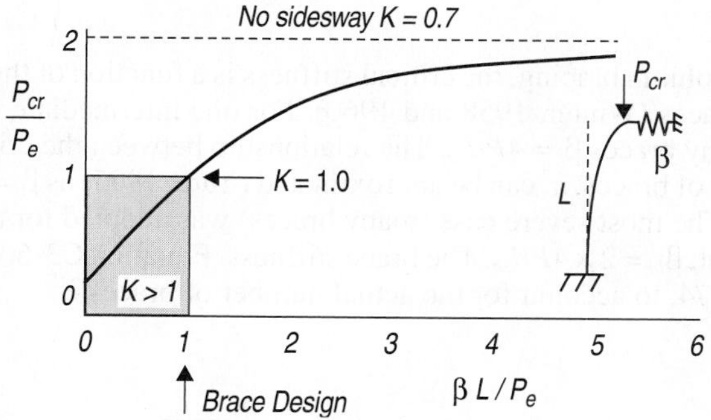

Fig. C-C3.2. Braced cantilever.

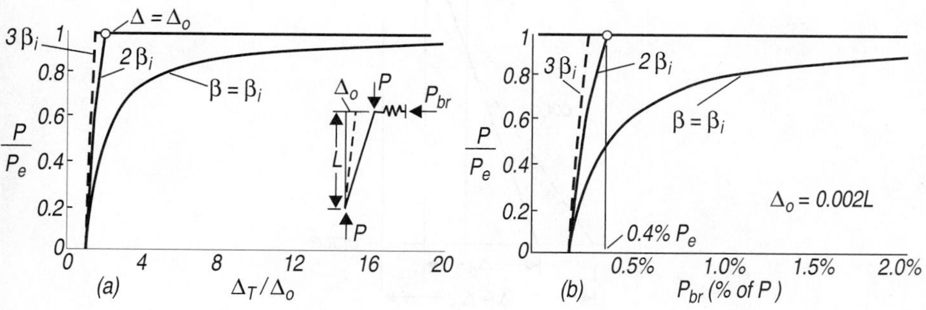

Fig. C-C3.3. Effect of initial out-of-plumbness.

ing, etc. The brace force recommendations for frames, columns and beam lateral bracing are based on an assumed $\Delta_o = 0.002L$, where L is the distance between adjacent brace points. For torsional bracing of beams, an initial twist angle, θ_o, is assumed where $\theta_o = 0.002L/h_o$, and h_o is the distance between flange centroids. For other Δ_o and θ_o values, use direct proportion to modify the brace strength requirements, P_{br} and M_{br}. For cases where it is unlikely that all columns in a story are out-of-plumb in the same direction, Chen and Tong (1994) recommend an average $\Delta_o = 0.002L/\sqrt{n_o}$ where n_o columns, each with a random Δ_o, are to be stabilized by the brace system. This reduced Δ_o would be appropriate when combining the stability brace forces with wind and seismic forces.

Brace connections, if they are flexible or can slip, should be considered in the evaluation of the bracing stiffness as follows:

$$\frac{1}{\beta_{act}} = \frac{1}{\beta_{conn}} + \frac{1}{\beta_{brace}} \qquad \text{(C-C3-2)}$$

The brace system stiffness, β_{act}, is less than the smaller of the connection stiffness, β_{conn}, or the stiffness of the brace, β_{brace}. Slip in connections with standard holes need not be considered except when only a few bolts are used. When evaluating the bracing of rows of columns or beams, consideration must be given to the accumulation of the brace forces along the length of the brace that results in a different displacement at each beam or column location. In general, brace forces can be minimized by increasing the number of braced bays and using stiff braces.

3. Columns

For nodal column bracing, the critical stiffness is a function of the number of intermediate braces (Winter, 1958 and 1960). For one intermediate brace, $\beta_i = 2P/L_b$, and for many braces $\beta_i = 4P/L_b$. The relationship between the critical stiffness and the number of braces, n, can be approximated (Yura, 1995) as $\beta_i = N_iP/L_b$, where $N_i = 4 - 2/n$. The most severe case (many braces) was adopted for the brace stiffness requirement, $\beta_{br} = 2 \times 4P/L_b$. The brace stiffness, Equation C3-6, can be reduced by the ratio, $N_i/4$, to account for the actual number of braces.

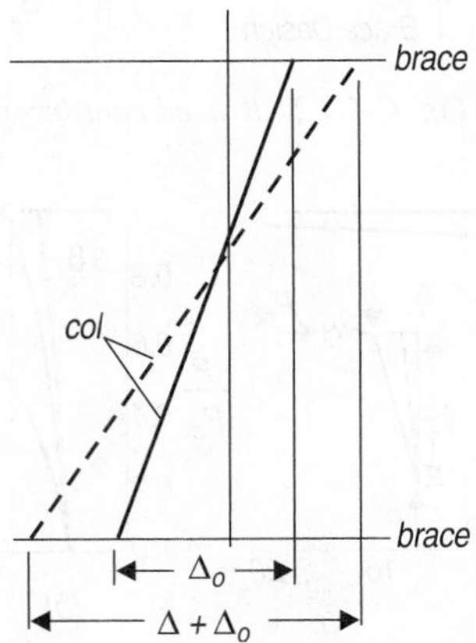

Fig. C-C3.4. Definitions.

The unbraced length, L_b, in Equations C3-4 and C3-6 is assumed to be equal to the length L_q that enables the column to reach P_u. When the actual bracing spacing is less than L_q, the calculated required stiffness may become quite conservative since the stiffness equations are inversely proportional to L_b. In such cases, L_q can be substituted for L_b. For example, a W12×53 (W310×79) with P_u= 400 kips (1 780 kN) can have a maximum unbraced length of 14 ft (4.3 m) for A36 (A36M) steel. If the actual bracing spacing is 8 ft (2.4 m), then 14 ft (4.3 m) may be used in Equations C3-4 and C3-6 to determine the required stiffness.

Winter's rigid model would derive a brace force of 0.8 percent P_u which accounts only for lateral displacement force effects. To account for the additional force due to member curvature, this theoretical force has been increased to one percent P_u.

4. Beams

Beam bracing must prevent twist of the section, not lateral displacement. Both lateral bracing (for example, joists attached to the compression flange of a simply supported beam) and torsional bracing (for example, a cross frame or diaphragm between adjacent girders) can effectively control twist. Lateral bracing systems that are attached near the beam centroid are ineffective. For beams with double curvature, the inflection point can not be considered a brace point because twist occurs at that point (Galambos, 1998). A lateral brace on one flange near the inflection point also is ineffective. In double curvature cases the lateral brace near the inflection point must be attached to both flanges to prevent twist, or torsional bracing must be used. The beam brace requirements are based on the recommendations by Yura (1993).

4a. Lateral Bracing

For lateral bracing, the following stiffness requirement was derived following Winter's approach:

$$\beta_{br} = 2N_i\,(C_bP_f)\,C_tC_d\,/\,\phi L_b \qquad \text{(C-C3-3)}$$

where

N_i = 1.0 for relative bracing
= (4-2/n) for discrete bracing
n = number of intermediate braces
P_f = beam compressive flange force
= $\pi^2 EI_{yc}/L_b^2$
I_{yc} = out-of-plane moment of inertia of the compression flange
C_b = moment modifier from Chapter F
C_t = accounts for top flange loading (use C_t=1.0 for centroidal loading)
= 1 + (1.2/n)
C_d = double curvature factor (compression in both flanges)
=1+ $(M_S/M_L)^2$
M_S = smallest moment causing compression in each flange
M_L = largest moment causing compression in each flange

The C_d factor varies between 1.0 and 2.0 and is applied only to the brace closest to the inflection point. The term ($2N_iC_t$) can be conservatively approximated as 10 for any number of nodal braces and 4 for relative bracing and (C_bP_f) can be approximated by M_u / h which simplifies Equation C-C3-3 to the stiffness requirements given by Equations C3-8 and C3-10. Equation C-C3-3 can be used in lieu of Equations C3-8 and C3-10.

The brace strength requirement for relative bracing is

$$P_{br} = 0.004\, M_u C_t C_d / h_o \tag{C-C3-4a}$$

and for nodal bracing

$$P_{br} = 0.01 M_u C_t C_d / h_o \tag{C-C3-4b}$$

They are based on an assumed initial lateral displacement of the compression flange of $0.002L_b$. The brace strength requirements of Equations C3-7 and C3-9 are derived from Equations C-C3-4a and C-C3-4b assuming top flange loading (C_t = 2). Equations C-C3-4a and C-C3-4b can be used in lieu of Equations C3-7 and C3-9 respectively.

4b. Torsional Bracing

Torsional bracing can either be attached continuously along the length of the beam (for example, metal deck or slabs) or be located at discrete points along the length of the member (for example, cross frames). Torsional bracing attached to the tension flange is just as effective as a brace attached at mid depth or the compression flange. Partially restrained connections can be used if their stiffness is considered in evaluating the torsional brace stiffness.

The torsional brace requirements are based on the buckling strength of a beam with a continuous torsional brace along its length developed by Taylor and Ojalvo (1966) and modified for cross-section distortion by Yura (1993).

$$M_u \le M_{cr} = \sqrt{(C_{bu} M_o)^2 + \frac{C_b^2 E I_y \bar{\beta}_T}{2C_{tt}}} \tag{C-C3-5}$$

The term ($C_{bu}M_o$) is the buckling strength of the beam without torsional bracing. C_{tt} = 1.2 when there is top flange loading and C_{tt} = 1.0 for centroidal loading. $\bar{\beta}_T = n\beta_T / L$ is the continuous torsional brace stiffness per unit length or its equivalent when n nodal braces, each with a stiffness β_T, are used along the span L and the 2 accounts for initial out-of-straightness. Neglecting the unbraced beam buckling term gives a conservative estimate of the torsional brace stiffness requirement (Equation C3-13). A more accurate estimate of the brace requirements can be obtained by replacing M_u with ($M_u - C_{bu}M_o$) in Equations C3-11 and C3-13. The β_{sec} term in Equations C3-12, C3-14 and C3-15 accounts for cross-section distortion. A web stiffener at the brace point reduces cross-sectional distortion and improves the effectiveness of a torsional brace. When a cross frame is attached near both flanges or a diaphragm is approximately the same depth as the girder, then web distortion will be insignificant so β_{sec} equals infinity. The required bracing stiffness, β_{Tb}, given by Equation C3-12 was obtained by solving the following expression that represents the brace system stiffness including distortion effects:

$$\frac{1}{\beta_T} = \frac{1}{\beta_{Tb}} + \frac{1}{\beta_{\text{sec}}} \tag{C-C3-6}$$

The brace moment requirements are based on an assumed initial twist of $0.002L_b / h_o$.

Parallel chord trusses with both chords extended to the end of the span and attached to supports can be treated like beams. In Equations C3-7 through C3-11, M_u may be taken as the maximum compressive chord force times the depth of the truss to determine the brace strength and stiffness requirements. Cross-section distortion effects, $\beta_{sec,}$ need not be considered when full-depth cross frames are used for bracing. When either chord does not extend to the end of the span, consideration should be given to control twist near the ends of the span by the use of cross frames or ties.

CHAPTER D

TENSION MEMBERS

D1. DESIGN TENSILE STRENGTH

Due to strain hardening, a ductile steel bar loaded in axial tension can resist, without fracture, a force greater than the product of its gross area and its coupon yield stress. However, excessive elongation of a tension member due to uncontrolled yielding of its gross area not only marks the limit of its usefulness, but can precipitate failure of the structural system of which it is a part. On the other hand, depending upon the reduction of area and other mechanical properties of the steel, the member can fail by fracture of the net area at a load smaller than required to yield the gross area. Hence, general yielding of the gross area and fracture of the net area both constitute failure limit states. The relative values of ϕ_t given for yielding and fracture reflect the same basic difference in factor of safety as between design of members and design of connections in the AISC ASD Specification.

The length of the member in the net area is negligible relative to the total length of the member. As a result, the strain hardening condition is quickly reached and yielding of the net area at fastener holes does not constitute a limit state of practical significance.

D2. BUILT-UP MEMBERS

The slenderness ratio L / r of tension members other than rods, HSS, or straps should preferably not exceed the limiting value of 300. This slenderness limit recommended for tension members is not essential to the structural integrity of such members; it merely assures a degree of stiffness such that undesirable lateral movement ("slapping" or vibration) will be unlikely.

See Section B7 and Commentary Section E4.

D3. PIN-CONNECTED MEMBERS AND EYEBARS

Forged eyebars have generally been replaced by pin-connected plates or eyebars thermally cut from plates. Provisions for the proportioning of eyebars contained in the LRFD Specification are based upon standards evolved from long experience with forged eyebars. Through extensive destructive testing, eyebars have been found to provide balanced designs when they are thermally cut instead of forged. The somewhat more conservative rules for pin-connected members of nonuniform cross section and those not having enlarged "circular" heads are likewise based on the results of experimental research (Johnston, 1939).

Somewhat stockier proportions are provided for eyebars and pin-connected members fabricated from steel having a yield stress greater than 70 ksi (485 MPa), in order to eliminate any possibility of their "dishing" under the higher design stress.

CHAPTER E

COLUMNS AND OTHER COMPRESSION MEMBERS

E1. EFFECTIVE LENGTH AND SLENDERNESS LIMITATIONS

1. Effective Length

The Commentary on Section C2 regarding frame stability and effective length factors applies here. Further analytic methods, formulas, charts, and references for the determination of effective length are provided in Chapter 15 of the SSRC Guide (Galambos, 1998).

2. Design by Plastic Analysis

The limitation on λ_c is essentially the same as that for l/r in Chapter N of the 1989 AISC Specification—Allowable Stress Design and Plastic Design.

E2. DESIGN COMPRESSIVE STRENGTH FOR FLEXURAL BUCKLING*

Equations E2-2 and E2-3 are based on a reasonable conversion of research data into design equations. Conversion of the allowable stress design (ASD) equations which was based on the CRC—Column Research Council—curve (Galambos, 1998) was found to be cumbersome for two reasons. The first was the nature of the ASD variable safety factor. Secondly, the difference in philosophical origins of the two design procedures requires an assumption of a live load-to-dead load ratio (L/D).

Since all L/D ratios could not be considered, a value of approximately 1.1 at λ equal to 1.0 was used to calibrate the exponential equation for columns with the lower range of λ against the appropriate ASD provision. The coefficient with the Euler equation was obtained by equating the ASD and LRFD expressions at λ of 1.5.

Equations E2-2 and E2-3 are essentially the same curve as column-strength curve 2P of the Structural Stability Research Council which is based on an initial out-of-straightness curve of $l/1500$ (Bjorhovde, 1972 and 1988; Galambos, 1998; Tide, 1985).

It should be noted that this set of column equations has a range of reliability (β) values. At low- and high-column slenderness, β values exceeding 3.0 and 3.3 respectively are obtained compared to β of 2.60 at L/D of 1.1. This is considered satisfactory, since the limits of out-of-straightness combined with residual stress have not been clearly established. Furthermore, there has been no history of unacceptable

*For tapered members see Commentary Appendix F3.

behavior of columns designed using the ASD procedure. This includes cases with L/D ratios greater than 1.1.

Equations E2-2 and E2-3 can be restated in terms of the more familiar slenderness ratio Kl/r. First, Equation E2-2 is expressed in exponential form,

$$F_{cr} = \left[\exp\left(-0.419\lambda_c^2\right)\right]F_y \tag{C-E2-1}$$

Note that exp(x) is identical to e^x. Substitution of λ_c according to definition of λ_c in Section E2 gives,

For $\dfrac{Kl}{r} \le 4.71\sqrt{\dfrac{E}{F_y}}$

$$F_{cr} = \left\{\exp\left[-0.0424\frac{F_y}{E}\left(\frac{Kl}{r}\right)^2\right]\right\}F_y \tag{C-E2-2}$$

For $\dfrac{Kl}{r} > 4.71\sqrt{\dfrac{E}{F_y}}$

$$F_{cr} = \frac{0.877\pi^2 E}{\left(\dfrac{Kl}{r}\right)^2} \tag{C-E2-3}$$

E3. DESIGN COMPRESSIVE STRENGTH FOR FLEXURAL-TORSIONAL BUCKLING

Torsional buckling of symmetric shapes and flexural-torsional buckling of unsymmetric shapes are failure modes usually not considered in the design of hot-rolled columns. They generally do not govern, or the critical load differs very little from the weak axis planar buckling load. Such buckling loads may, however, control the capacity of symmetric columns made from relatively thin plate elements and unsymmetric columns. Design equations for determining the strength of such columns are given in Appendix E3. The AISC Design Guide, *Torsional Analysis of Structural Steel Members* (Seaburg and Carter, 1997) provides an overview of the fundamentals and basic theory of torsional loading for structural steel members. Design examples are also included.

Tees that conform to the limits in Table C-E3.1 need not be checked for flexural-torsional buckling.

A simpler and more accurate design strength for the special case of tees and double-angles is based on Galambos (1991) wherein the y-axis of symmetry flexural-buckling strength component is determined directly from the column formulas.

The separate AISC *Specification for Load and Resistance Factor Design of Single-Angle Members* contains detailed provisions not only for the limit state of compression, but also for tension, shear, flexure, and combined forces.

TABLE C-E3.1 Limiting Proportions for Tees		
Shape	Ratio of Full Flange Width to Profile Depth	Ratio of Flange Thickness to Web or Stem Thickness
Built-up tees	≥ 0.50	≥ 1.25
Rolled tees	≥ 0.50	≥ 1.10

E4. BUILT-UP MEMBERS

Requirements for detailing and design of built-up members, which cannot be stated in terms of calculated stress, are based upon judgment and experience.

The longitudinal spacing of connectors connecting components of built-up compression members must be such that the slenderness ratio l/r of individual shapes does not exceed three-fourths of the slenderness ratio of the entire member. Additional requirements are imposed for built-up members consisting of angles. However, these minimum requirements do not necessarily ensure that the effective slenderness ratio of the built-up member is equal to that for the built-up member acting as a single unit. Section E4 gives formulas for modified slenderness ratios that are based on research and take into account the effect of shear deformation in the connectors (Zandonini, 1985). Equation E4-1 for snug-tight intermediate connectors is empirically based on test results (Zandonini, 1985). Equation E4-2 is derived from theory and verified by test data. In both cases the end connection must be welded or slip-critical bolted (Aslani and Goel, 1991). The connectors must be designed to resist the shear forces which develop in the buckled member. The shear stresses are highest where the slope of the buckled shape is maximum (Bleich, 1952).

Maximum fastener spacing less than that required for strength may be needed to ensure a close fit over the entire faying surface of components in continuous contact. Specific requirements are given for weathering steel members exposed to atmospheric corrosion (Brockenbrough, 1983).

The provisions governing the proportioning of perforated cover plates are based upon extensive experimental research (Stang and Jaffe, 1948).

CHAPTER F

BEAMS AND OTHER FLEXURAL MEMBERS

F1. DESIGN FOR FLEXURE

1. Yielding

The bending strength of a laterally braced compact section is the plastic moment M_p. If the shape has a large shape factor (ratio of plastic moment to the moment corresponding to the onset of yielding at the extreme fiber), significant inelastic deformation may occur at service load if the section is permitted to reach M_p at factored load. The limit of $1.5M_y$ at factored load will control the amount of inelastic deformation for sections with shape factors greater than 1.5. This provision is not intended to limit the plastic moment of a hybrid section with a web yield stress lower than the flange yield stress. Yielding in the web does not result in significant inelastic deformations. In hybrid sections, $M_y = F_{yf} S$.

Lateral-torsional buckling cannot occur if the moment of inertia about the bending axis is equal to or less than the moment of inertia out of plane. Thus, for shapes bent about the minor axis and shapes with $I_x = I_y$, such as square or circular shapes, the limit state of lateral-torsional buckling is not applicable and yielding controls if the section is compact.

2. Lateral-Torsional Buckling

2a. Doubly Symmetric Shapes and Channels with $L_b \le L_r$

The basic relationship between nominal moment M_n and unbraced length L_b is shown in Figure C-F1.1 for a compact section with $C_b = 1.0$. There are four principal zones defined on the basic curve by L_{pd}, L_p, and L_r. Equation F1-4 defines the maximum unbraced length L_p to reach M_p with uniform moment. Elastic lateral-torsional buckling will occur when the unbraced length is greater than L_r given by Equation F1-6. Equation F1-2 defines the inelastic lateral-torsional buckling as a straight line between the defined limits L_p and L_r. Buckling strength in the elastic region $L_b > L_r$ is given by Equation F1-14 for I-shaped members.

For other moment diagrams, the lateral buckling strength is obtained by multiplying the basic strength by C_b as shown in Figure C-F1.1. The maximum M_n, however, is limited to M_p. Note that L_p given by Equation F1-4 is merely a definition which has physical meaning when $C_b = 1.0$. For C_b greater than 1.0, larger unbraced lengths are permitted to reach M_p as shown by the curve for $C_b > 1.0$. For design, this length could be calculated by setting Equation F1-2 equal to M_p and solving this equation for L_b using the desired C_b value.

The equation

$$C_b = 1.75 + 1.05(M_1 / M_2) + 0.3(M_1 / M_2)^2 \le 2.3 \qquad \text{(C-F1-1)}$$

has been used since 1961 to adjust the flexural-torsional buckling equation for variations in the moment diagram within the unbraced length. This equation is applicable only to moment diagrams that are straight lines between braced points. The equation provides a lower bound fit to the solutions developed by Salvadori (1956) which are shown in Figure C-F1.2. Another equation

$$C_b = \frac{1}{0.6 - 0.4\dfrac{M_1}{M_2}} \leq 2.5 \qquad \text{(C-F1-2)}$$

fits the average value theoretical solutions when the beams are bent in reverse curvature and also provides a reasonable fit to the theory. If the maximum moment

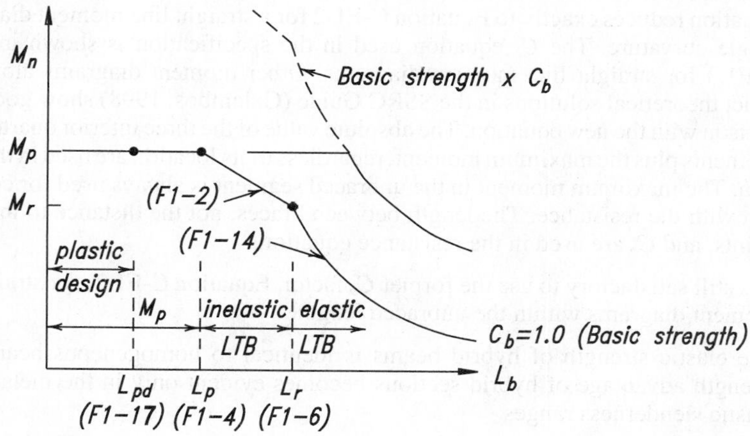

Fig. C-F1.1. Nominal moment as a function of unbraced length and moment gradient.

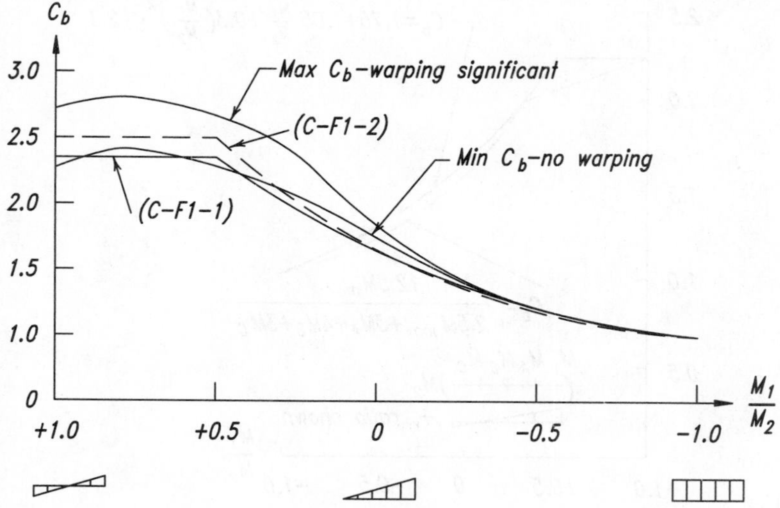

Fig. C-F1.2. Moment modifier C_b for beams.

within the unbraced segment is equal to or larger than the end moment, $C_b = 1.0$ is used.

The equations above can be easily misinterpreted and misapplied to moment diagrams that are not straight within the unbraced segment. Kirby and Nethercot (1979) presented an equation which applies to various shapes of moment diagrams within the unbraced segment. Their equation has been adjusted slightly to the following

$$C_b = \frac{12.5M_{\max}}{2.5M_{\max} + 3M_A + 4M_B + 3M_C} \qquad \text{(C-F1-3)}$$

This equation gives more accurate solutions for fixed-end beams, and the adjusted equation reduces exactly to Equation C-F1-2 for a straight line moment diagram in single curvature. The C_b equation used in the specification is shown in Figure C-F1.3 for straight line moment diagrams. Other moment diagrams along with exact theoretical solutions in the SSRC Guide (Galambos, 1998) show good comparison with the new equation. The absolute value of the three interior quarter-point moments plus the maximum moment, regardless of its location are used in the equation. The maximum moment in the unbraced segment is always used for comparison with the resistance. The length between braces, not the distance to inflection points, and C_b are used in the resistance equation.

It is still satisfactory to use the former C_b factor, Equation C-F1-1, for straight line moment diagrams within the unbraced length.

The elastic strength of hybrid beams is identical to homogeneous beams. The strength advantage of hybrid sections becomes evident only in the inelastic and plastic slenderness ranges.

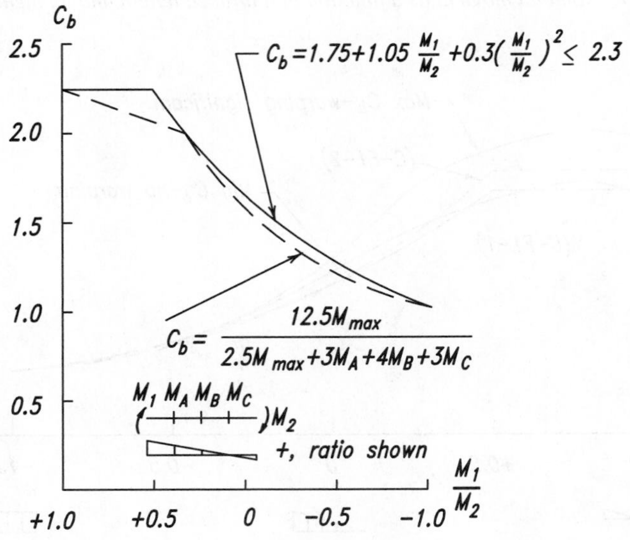

C-F1.3. C_b for a straight line moment diagram–prismatic beam.

2b. Doubly Symmetric Shapes and Channels with $L_b > L_r$

The equation given in the Specification assumes that the loading is applied along the beam centroidal axis. If the load is placed on the top flange and the flange is not braced, there is a tipping effect that reduces the critical moment; conversely, if the load is suspended from the bottom flange and is not braced, there is a stabilizing effect which increases the critical moment (Galambos, 1998). For unbraced top flange loading, the reduced critical moment may be conservatively approximated by setting the warping buckling factor X_2 to zero.

An effective length factor of unity is implied in these critical moment equations to represent a worst case pinned-pinned unbraced segment. Including consideration of any end restraint of the adjacent segments on the critical segment can increase its buckling capacity. The effects of beam continuity on lateral-torsional buckling have been studied and a simple and conservative design method, based on the analogy of end-restrained nonsway columns with an effective length factor less than one, has been proposed (Galambos, 1998).

2c. Tees and Double-Angles

The lateral-torsional buckling strength (LTB) of singly symmetric tee beams is given by a fairly complex formula (Galambos, 1998). Equation F1-15 is a simplified formulation based on Kitipornchai and Trahair (1980). See also Ellifritt, Wine, Sputo, and Samuel (1992).

The C_b used for I-shaped beams is unconservative for tee beams with the stem in compression. For such cases $C_b = 1.0$ is appropriate. When beams are bent in reverse curvature, the portion with the stem in compression may control the LTB resistance even though the moments may be small relative to other portions of the unbraced length with $C_b \approx 1.0$. This is because the LTB strength of a tee with the stem in compression may be only about one-fourth of the capacity for the stem in tension. Since the buckling strength is sensitive to the moment diagram, C_b has been conservatively taken as 1.0. In cases where the stem is in tension, connection details should be designed to minimize any end restraining moments which might cause the stem to be in compression.

3. Design by Plastic Analysis

Equation F1-17 sets a limit on unbraced length adjacent to a plastic hinge for plastic analysis. There is a substantial increase in unbraced length for positive moment ratios (reverse curvature) because the yielding is confined to zones close to the brace points (Yura et al., 1978).

Equation F1-18 is an equation in similar form for solid rectangular bars and symmetric box beams. Equations F1-17 and F1-18 assume that the moment diagram within the unbraced length next to plastic hinge locations is reasonably linear. For nonlinear diagrams between braces, judgment should be used in choosing a representative ratio.

Equations F1-17 and F1-18 were developed to provide rotation capacities of at least 3.0, which are sufficient for most applications (Yura et al., 1978). When inelastic rotations of 7 to 9 are deemed appropriate in areas of high seismicity, as discussed in Commentary Section B5, Equation F1-17 would become:

$$L_{pd} = 0.086\left(\frac{E}{F_y}\right)r_y \qquad \text{(C-F1-4)}$$

F2. DESIGN FOR SHEAR

For unstiffened webs $k_v = 5.0$, therefore

$$1.10\sqrt{Ek_v / F_{yw}} = 2.45\sqrt{E / F_{yw}}, \text{ and } 1.37\sqrt{Ek_v / F_{yw}} = 3.07\sqrt{E / F_{yw}}$$

For webs with $h / t_w \leq 1.10\sqrt{Ek_v / F_{yw}}$, the nominal shear strength V_n is based on shear yielding of the web, Equation F2-1 and Equation A-F2-1. This h / t_w limit was determined by setting the critical stress causing shear buckling F_{cr} equal to the yield stress of the web F_{yw} in Equation 35 of Cooper, Galambos, and Ravindra (1978) and Timoshenko and Gere (1961). When $h / t_w > 1.10\sqrt{Ek_v / F_{yw}}$, the web shear strength is based on buckling. Basler (1961) suggested taking the proportional limit as 80 percent of the yield stress of the web. This corresponds to $h / t_w = (1.10 / 0.8)\sqrt{Ek_v / F_{yw}}$. Thus, when $h / t_w > 1.37\sqrt{Ek_v / F_{yw}}$, the web strength is determined from the elastic buckling stress given by Equation 6 of Cooper et al. (1978) and Timoshenko and Gere (1961):

$$F_{cr} = \frac{\pi^2 E k_v}{12\left(1 - v^2\right)\left(h / t_w\right)^2} \qquad \text{(C-F2-1)}$$

The nominal shear strength, given by Equation F2-3 and A-F2-3, was obtained by multiplying F_{cr} by the web area and using $E = 29{,}000$ ksi (200 000 MPa) and $v = 0.3$. A straight line transition, Equation F2-2 and A-F2-2, is used between the limits $1.10\sqrt{Ek_v / F_{yw}}$ and $1.37\sqrt{Ek_v / F_{yw}}$.

The shear strength of flexural members follows the approach used in the AISC ASD Specification, except for two simplifications. First, the expression for the plate buckling coefficient k_v has been simplified; it corresponds to that given by AASHTO *Standard Specification for Highway Bridges* (1996). The earlier expression for k_v was a curve fit to the exact expression; the new expression is just as accurate. Second, the alternate method (tension field action) for web shear strength is placed in Appendix G because it was desired that only one method appear in the main body of the Specification with alternate methods given in the Appendix. When designing plate girders, thicker unstiffened webs will frequently be less costly than lighter stiffened web designs because of the additional fabrication. If a stiffened girder design has economic advantages, the tension field method in Appendix G will require fewer stiffeners.

The equations in this section were established assuming monotonically increasing loads. If a flexural member is subjected to load reversals causing cyclic yielding over large portions of a web, such as may occur during a major earthquake, special design considerations may apply (Popov, 1980).

F4. BEAMS AND GIRDERS WITH WEB OPENINGS

Web openings in structural floor members may be necessary to accommodate various mechanical, electrical, and other systems. Strength limit states, including local buckling of the compression flange, web, and tee-shaped compression zone above or below the opening, lateral buckling and moment-shear interaction, or serviceability may control the design of a flexural member with web openings. The location, size, and number of openings are important and empirical limits for them have been identified. One general procedure for assessing these effects and the design of any needed reinforcement for both steel and composite beams is given in Darwin (1990) and in ASCE Task Committee on Design Criteria for Composite Structures in Steel and Concrete (1992 and 1992a).

CHAPTER H

MEMBERS UNDER COMBINED FORCES AND TORSION

H1. SYMMETRIC MEMBERS SUBJECT TO BENDING AND AXIAL FORCE

Equations H1-1a and H1-1b are simplifications and clarifications of similar equations used in the AISC ASD Specification since 1961. Previously, both equations had to be checked. In the new formulation the applicable equation is governed by the value of the first term, $P_u / \phi P_n$. For bending about one axis only, the equations have the form shown in Figure C-H1.1.

The first term $P_u / \phi P_n$ has the same significance as the axial load term f_a / F_a in Equations H1-1 of the AISC ASD Specification. This means that for members in compression P_n must be based on the largest effective slenderness ratio Kl / r. In the development of Equations H1-1a and H1-1b, a number of alternative formulations were compared to the exact inelastic solutions of 82 sidesway cases reported in Kanchanalai (1977). In particular, the possibility of using Kl / r as the actual column length ($K = 1$) in determining P_n, combined with an elastic second order moment M_u, was studied. In those cases where the true P_n based on Kl / r, with $K = 1.0$, was in the inelastic range, the errors proved to be unacceptably large without the additional check that $P_u \leq \phi_c P_n$, P_n being based on effective length. Although deviations from exact solutions were reduced, they still remained high.

In summary, it is not possible to formulate a safe general interaction equation for compression without considering effective length directly (or indirectly by a sec-

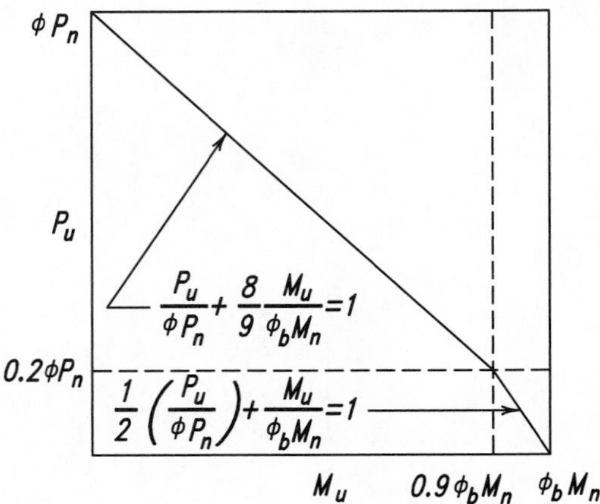

Fig. C-H1.1. Beam-column interaction equations.

ond equation). Therefore, the requirement that the nominal compressive strength P_n be based on the effective length KL in the general equation is continued in the LRFD Specification as it has been in the AISC ASD Specification since 1961. It is not intended that these provisions be applicable to limit nonlinear secondary flexure that might be encountered in large amplitude earthquake stability design (ATC, 1978).

The defined term M_u is the maximum moment in a member. In the calculation of this moment, inclusion of beneficial second order effects of tension is optional. But consideration of detrimental second order effects of axial compression and translation of gravity loads is required. Provisions for calculation of these effects are given in Chapter C.

The interaction equations in Appendix H3 have been recommended for biaxially loaded H and wide flange shapes in Galambos (1998) and Springfield (1975). These equations which can be used only in braced frames represent a considerable liberalization over the provisions given in Section H1; it is, therefore, also necessary to check yielding under service loads, using the appropriate load and resistance factors for the serviceability limit state in Equation H1-1a or H1-1b with $M_{ux} = S_x F_y$ and $M_{uy} = S_y F_y$. Appendix H3 also provides interaction equations for rectangular box-shaped beam-columns. These equations are taken from Zhou and Chen (1985).

H2. UNSYMMETRIC MEMBERS AND MEMBERS UNDER TORSION AND COMBINED TORSION, FLEXURE, SHEAR, AND/OR AXIAL FORCE

This section deals with types of cross sections and loadings not covered in Section H1, especially where torsion is a consideration. For such cases it is recommended to perform an elastic analysis based on the theoretical numerical methods available from the literature for the determination of the maximum normal and shear stresses, or for the elastic buckling stresses. In the buckling calculations an equivalent slenderness parameter is determined for use in Equation E2-2 or E2-3, as follows:

$$\lambda_e = \sqrt{F_y / F_e}$$

where F_e is the elastic buckling stress determined from a stability analysis. This procedure is similar to that of Appendix E3.

For the analysis of members with open sections under torsion refer to AISC (1997).

CHAPTER I

COMPOSITE MEMBERS

I1. DESIGN ASSUMPTIONS AND DEFINITIONS

Force Determination. Loads applied to an unshored beam before the concrete has hardened are resisted by the steel section alone, and only loads applied after the concrete has hardened are considered as resisted by the composite section. It is usually assumed for design purposes that concrete has hardened when it attains 75 percent of its design strength. In beams properly shored during construction, all loads may be assumed as resisted by the composite cross section. Loads applied to a continuous composite beam with shear connectors throughout its length, after the slab is cracked in the negative moment region, are resisted in that region by the steel section and by properly anchored longitudinal slab reinforcement.

For purposes of plastic analysis all loads are considered resisted by the composite cross section, since a fully plastic strength is reached only after considerable yielding at the locations of plastic hinges.

Elastic Analysis. The use of constant stiffness in elastic analyses of continuous beams is analogous to the practice in reinforced concrete design. The stiffness calculated using a weighted average of moments of inertia in the positive moment region and negative moment regions may take the following form:

$$I_t = aI_{pos} + bI_{neg}$$

where

I_{pos} = effective moment of inertia for positive moment, in.^{4} (mm^4)
I_{neg} = effective moment of inertia for negative moment, in.^{4} (mm^4)

The effective moment of inertia shall be based on the cracked transformed section considering degree of composite actions. For continuous beams subjected to gravity loads only, the value of a may be taken as 0.6 and the value of b may be taken as 0.4. For the case of composite beams in moment resisting frames, the value of a and b may be taken as 0.5.

Plastic Analysis. For composite beams with shear connectors, plastic analysis may be used only when the steel section in the positive moment region has a compact web, i.e., $h / t_w \leq 3.76\sqrt{E / F_{yf}}$, and when the steel section in the negative moment region is compact, as required for steel beams alone. No compactness limitations are placed on encased beams, but plastic analysis is permitted only if the direct contribution of concrete to the strength of sections is neglected; the concrete is relied upon only to prevent buckling.

Plastic Stress Distribution for Positive Moment. Plastic stress distributions are described in Commentary Section I3, and a discussion of the composite participation of slab reinforcement is presented.

Plastic Stress Distribution for Negative Moment. Plastic stress distributions are described in Commentary Section I3.

Elastic Stress Distribution. The strain distribution at any cross section of a composite beam is related to slip between the structural steel and concrete elements. Prior to slip, strain in both steel and concrete is proportional to the distance from the neutral axis for the elastic transformed section. After slip, the strain distribution is discontinuous, with a jump at the top of the steel shape. The strains in steel and concrete are proportional to distances from separate neutral axes, one for steel and the other for concrete.

Fully Composite Beam. Either the tensile yield strength of the steel section or the compressive stress of the concrete slab governs the maximum flexural strength of a fully composite beam subjected to a positive moment. The tensile yield strength of the longitudinal reinforcing bars in the slab governs the maximum flexural strength of a fully composite beam subjected to a negative moment. When shear connectors are provided in sufficient numbers to fully develop this maximum flexural strength, any slip that occurs prior to yielding is minor and has negligible influence both on stresses and stiffness.

Partially Composite Beam. The effects of slip on elastic properties of a partially composite beam can be significant and should be accounted for in calculations of deflections and stresses at service loads. Approximate elastic properties of partially composite beams are given in Commentary Section I3. For simplified design methods, see Hansell, Galambos, Ravindra, and Viest (1978).

Concrete-Encased Beam. When the dimensions of a concrete slab supported on steel beams are such that the slab can effectively serve as the flange of a composite T-beam, and the concrete and steel are adequately tied together so as to act as a unit, the beam can be proportioned on the assumption of composite action.

Two cases are recognized: fully encased steel beams, which depend upon natural bond for interaction with the concrete, and those with mechanical anchorage to the slab (shear connectors), which do not have to be encased.

I2. COMPRESSION MEMBERS

1. Limitations

(1) The lower limit of four percent on the cross-sectional area of structural steel differentiates between composite and reinforced concrete columns. If the area is less than four percent, a column with a structural steel core should be designed as a reinforced concrete column.

(2) The specified minimum quantity of transverse and longitudinal reinforcement in the encasement should be adequate to prevent severe spalling of the surface concrete during fires.

(3) Very little of the supporting test data involved concrete strengths in excess of 6 ksi (41 MPa), even though the cylinder strength for one group of four columns was 9.6 ksi (66 MPa). Normal weight concrete is believed to have been used in all tests. Thus, the upper limit of concrete strength is specified as 8 ksi (55

MPa) for normal weight concrete. A lower limit of 3 ksi (21 MPa) is specified for normal weight concrete and 4 ksi (28 MPa) for lightweight concrete to encourage the use of good quality, yet readily available, grades of structural concrete.

(4) In addition to the work of Bridge and Roderick (1978), SSRC Task Group 20 (1979), and Galambos and Chapuis (1980), recent work by Kenny, Bruce, and Bjorhovde (1994) has shown that due to concrete confinement effects, the previous limitation of 55 ksi (380 MPa) for the maximum steel yield stress is highly restrictive. Further, the most commonly used reinforcing steel grade has a yield stress of 60 ksi (415 MPa). The increase is therefore a rational recognition of material properties and structural behavior.

The 60 ksi (415 MPa) limitation for the yield stress is very conservative for tubular composite columns, where the concrete confinement provided by the tube walls is very significant. Kenny et al. have proposed raising the value of F_y for such columns to whatever the yield stress is for the steel grade used, but not higher than 80 ksi (550 MPa).

(5) The specified minimum wall thicknesses are identical to those in the 1995 ACI Building Code (1995). The purpose of this provision is to prevent buckling of the steel pipe or HSS before yielding.

2. Design Strength

The procedure adopted for the design of axially loaded composite columns is described in detail in Galambos and Chapuis (1980). It is based on the equation for the strength of a short column derived in Galambos and Chapuis (1980), and the same reductions for slenderness as those specified for steel columns in Section E2. The design follows the same path as the design of steel columns, except that the yield stress of structural steel, the modulus of elasticity of steel, and the radius of gyration of the steel section are modified to account for the effect of concrete and longitudinal reinforcing bars. A detailed explanation of the origin of these modifications may be found in SSRC Task Group 20 (1979). Galambos and Chapuis (1980) includes comparisons of the design procedure with 48 tests of axially loaded stub columns, 96 tests of concrete-filled pipes or tubing (HSS), and 26 tests of concrete-encased steel shapes. The mean ratio of the test failure loads to the predicted strengths was 1.18 for all 170 tests, and the corresponding coefficient of variation was 0.19.

3. Columns with Multiple Steel Shapes

This limitation is based on Australian research reported in Bridge and Roderick (1978), which demonstrated that after hardening of the concrete the composite column will respond to loading as a unit even without lacing, tie plates, or batten plates connecting the individual steel sections.

4. Load Transfer

To avoid overstressing either the structural steel section or the concrete at connections, a transfer of load by direct bearing, shear connectors, or a combination of both is required. When shear connectors are used, a uniform spacing is appropriate in most situations, but when large forces are applied, other connector arrangements may be needed to avoid overloading the component (steel section or concrete encasement) to which the load is applied directly.

Although it is recognized that force transfer also occurs by bond between the steel and concrete, this is disregarded for encased sections (Griffis, 1992). Force transfer by bond is commonly used in concrete-filled hollow structural sections (API, 1993) as long as the connections are detailed to limit local deformations, but no guidelines are available for structures other than fixed offshore platforms.

When a supporting concrete area is wider on all sides than the loaded area, the nominal bearing strength for concrete can be taken as

$$0.85\phi_B f'_c \sqrt{A_2 / A_1}$$

where A_1 is the loaded area and A_2 is the base of a frustum extending 45° in plan and at a 50 percent slope in elevation from the loaded area (ACI, 1995, 1995a). The value of $\sqrt{A_2 / A_1}$ must be less than or equal to 2. In most practical cases, this limit will be reached and thus the Specification uses a nominal bearing strength of $1.7\phi_B f'_c A_B$. The resistance factor for bearing, ϕ_B, is 0.65 in accordance with Appendix C in ACI 318 and ACI 318M.

I3. FLEXURAL MEMBERS

1. Effective Width

LRFD provisions for effective width omit any limit based on slab thickness, in accordance with both theoretical and experimental studies, as well as composite beam codes in other countries (ASCE, 1979). The same effective width rules apply to composite beams with a slab on either one side or both sides of the beam. To simplify design, effective width is based on the full span, center-to-center of supports, for both simple and continuous beams.

2. Design Strength of Beams with Shear Connectors

This section applies to simple and continuous composite beams with shear connectors, constructed with or without temporary shores.

Positive Flexural Design Strength. Flexural strength of a composite beam in the positive moment region may be limited by the plastic strength of the steel section, the concrete slab, or shear connectors. In addition, web buckling may limit flexural strength if the web is slender and a significantly large portion of the web is in compression.

According to Table B5.1, local web buckling does not reduce the plastic strength of a bare steel beam if the beam depth-to-web thickness ratio is not larger than $3.76\sqrt{E / F_y}$. In the absence of web buckling research on composite beams, the same ratio is conservatively applied to composite beams. Furthermore, for more slender webs, the LRFD Specification conservatively adopts first yield as the flexural strength limit. In this case, stresses on the steel section from permanent loads applied to unshored beams before the concrete has hardened must be superimposed on stresses on the composite section from loads applied to the beams after hardening of concrete. In this superposition, all permanent loads should be multiplied by the dead load factor and all live loads should be multiplied by the live load factor. For shored beams, all loads may be assumed as resisted by the composite section.

When first yield is the flexural strength limit, the elastic transformed section is used to calculate stresses on the composite section. The modular ratio $n = E/E_c$ used to determine the transformed section depends on the specified unit weight and strength of concrete. Note that this procedure for compact beams differs from the requirements of Section I2 of the 1989 AISC ASD Specification.

Plastic Stress Distribution for Positive Moment. When flexural strength is determined from the plastic stress distribution shown in Figure C-I3.1, the compression force C in the concrete slab is the smallest of:

$$C = A_{sw} F_{yw} + 2 A_{sf} F_{yf} \tag{C-I3-1}$$

$$C = 0.85 f_c' A_c \tag{C-I3-2}$$

$$C = \Sigma Q_n \tag{C-I3-3}$$

For a non-hybrid steel section, Equation C-I3-1 becomes $C = A_s F_y$

where

f_c' = specified compressive strength of concrete, ksi (MPa)
A_c = area of concrete slab within effective width, in.^{2} (mm^2)
A_s = area of steel cross section, in.^{2} (mm^2)
A_{sw} = area of steel web, in.^{2} (mm^2)
A_{sf} = area of steel flange, in.^{2} (mm^2)
F_y = minimum specified yield stress of steel, ksi (MPa)
F_{yw} = minimum specified yield stress of web steel, ksi (MPa)
F_{yf} = minimum specified yield stress of flange steel, ksi (MPa)
ΣQ_n = sum of nominal strengths of shear connectors between the point of maximum positive moment and the point of zero moment to either side, kips (N)

Longitudinal slab reinforcement makes a negligible contribution to the compression force, except when Equation C-I3-2 governs. In this case, the area of longitudinal reinforcement within the effective width of the concrete slab times the yield stress of the reinforcement may be added in determining C.

The depth of the compression block is

$$a = \frac{C}{0.85 f_c' b} \tag{C-I3-4}$$

where

b = effective width of concrete slab, in. (mm)

A fully composite beam corresponds to the case of C governed by the yield strength of the steel beam or the compressive strength of the concrete slab, as in Equation C-I3-1 or C-I3-2. The number and strength of shear connectors govern C for a partially composite beam as in Equation C-I3-3.

The plastic stress distribution may have the plastic neutral axis (PNA) in the web, in the top flange of the steel section or in the slab, depending on the value of C.

The nominal plastic moment resistance of a composite section in positive bending is given by the following equation and Figure C-I3.1:

$$M_n = C(d_1 + d_2) + P_y\,(d_3 - d_2) \tag{C-I3-5}$$

where

P_y = tensile strength of the steel section; for a non-hybrid steel section, $P_y = A_s F_y$, kips (N)
d_1 = distance from the centroid of the compression force C in concrete to the top of the steel section, in. (mm)
d_2 = distance from the centroid of the compression force in the steel section to the top of the steel section, in. (mm). For the case of no compression in the steel section $d_2 = 0$.
d_3 = distance from P_y to the top of the steel section, in. (mm)

Equation C-I3-5 is generally applicable including both non-hybrid and hybrid steel sections symmetrical about one or two axes.

Approximate Elastic Properties of Partially Composite Beams. Elastic calculations for stress and deflection of partially composite beams should include the effects of slip.

The effective moment of inertia I_{eff} for a partially composite beam is approximated by

$$I_{eff} = I_s + \sqrt{\left(\Sigma Q_n / C_f\right)}\left(I_{tr} - I_s\right) \tag{C-I3-6}$$

where

I_s = moment of inertia for the structural steel section, in.^{4} (mm^4)
I_{tr} = moment of inertia for the fully composite uncracked transformed section, in.^{4} (mm^4)
ΣQ_n = strength of shear connectors between the point of maximum positive moment and the point of zero moment to either side, kips (N)
C_f = compression force in concrete slab for fully composite beam; smaller of Equations C-I3-1 and C-I3-2, kips (N)

The effective section modulus S_{eff}, referred to the tension flange of the steel section for a partially composite beam, is approximated by

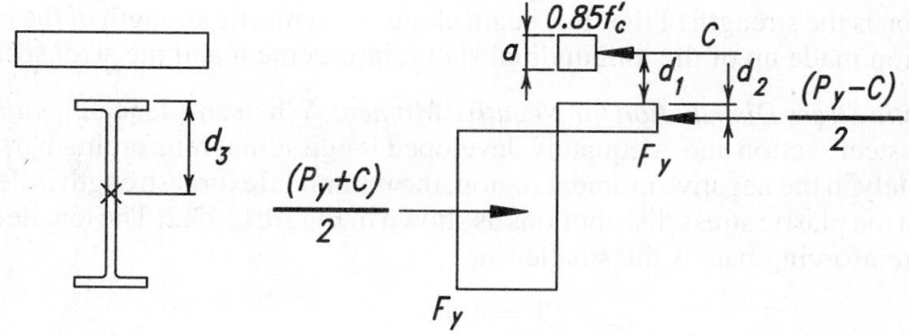

Fig. C-I3.1. Plastic stress distribution for positive moment in composite beams.

$$S_{eff} = S_s + \sqrt{(\Sigma Q_n / C_f)}(S_{tr} - S_s) \tag{C-I3-7}$$

where

S_s = section modulus for the structural steel section, referred to the tension flange, in.^{3} (mm^3)

S_{tr} = section modulus for the fully composite uncracked transformed section, referred to the tension flange of the steel section, in.^{3} (mm^3)

Equations C-I3-6 and C-I3-7 should not be used for ratios $\Sigma Q_n / C_f$ less than 0.25. This restriction is to prevent excessive slip, as well as substantial loss in beam stiffness. Studies indicate that Equations C-I3-6 and C-I3-7 adequately reflect the reduction in beam stiffness and strength, respectively, when fewer connectors are used than required for full composite action (Grant, Fisher, and Slutter, 1977).

It is not practical to make accurate deflection calculations of composite flexural sections in the design office. Careful comparisons to short-term deflection tests indicate that the effective moment of inertia, I_{eff}, is 15 to 30 percent lower than that calculated based on linear elastic theory. Therefore, for realistic deflection calculations, I_{eff} should be taken as 0.80 I_{eff} or 0.75 I_{eff}. As an alternative, it has been shown that one may use lower bound moment of inertia, I_{lb}, as defined below:

$$I_{lb} = I_x + A_s(Y_{ENA} - d_3)^2 + (\Sigma Q_n / F_y)(2d_3 + d_1 - Y_{ENA})^2 \tag{C-I3-8}$$

where

d_1 = distance from the centroid of the longitudinal slab reinforcement to the top of the steel section, in. (mm)

d_3 = distance from P_{yc} to the top of the steel section, in. (mm)

I_{lb} = lower bound moment of inertia, in.^{3} (mm^3)

$Y_{ENA} = [A_3 d_3 + (\Sigma Q_n/F_y)(2d_3 + d_1)/(A_s + (\Sigma Q_n/F_y)]$

Calculations for long-term deformations due to creep and shrinkage may also be carried out. Because the basic properties of the concrete are not known to the designer, simplified models such as those proposed by Viest, Fountain, and Singleton (1958), Branson (1964), Chien and Ritchie (1984), and Viest, Colaco, Furlong, Griffis, Leon, and Wyllie (1997) can be used.

Negative Flexural Design Strength. The flexural strength in the negative moment region is the strength of the steel beam alone or the plastic strength of the composite section made up of the longitudinal slab reinforcement and the steel section.

Plastic Stress Distribution for Negative Moment. When an adequately braced compact steel section and adequately developed longitudinal reinforcing bars act compositely in the negative moment region, the nominal flexural strength is determined from the plastic stress distributions as shown in Figure C-I3.2. The tensile force T in the reinforcing bars is the smaller of:

$$T = A_r F_{yr} \tag{C-I3-9}$$

$$T = \Sigma Q_n \tag{C-I3-10}$$

where

A_r = area of properly developed slab reinforcement parallel to the steel beam and within the effective width of the slab, in.^{2} (mm^2)
F_{yr} = specified yield stress of the slab reinforcement, ksi (MPa)
ΣQ_n = sum of the nominal strengths of shear connectors between the point of maximum negative moment and the point of zero moment to either side, kips (N)

A third theoretical limit on T is the product of the area and yield stress of the steel section. However, this limit is redundant in view of practical limitations on slab reinforcement.

The nominal plastic moment resistance of a composite section in negative bending is given by the following equation:

$$M_n = T\left(d_1 + d_2\right) + P_{yc}\left(d_3 - d_2\right) \quad \text{(C-I3-11)}$$

where

P_{yc} = the compressive strength of the steel section; for a non-hybrid section $P_{yc} = A_s F_y$, kips (N)
d_1 = distance from the centroid of the longitudinal slab reinforcement to the top of the steel section, in. (mm)
d_2 = distance from the centroid of the tension force in the steel section to the top of the steel section, in. (mm)
d_3 = distance from P_{yc} to the top of the steel section, in. (mm)

Transverse Reinforcement for the Slab. Where experience has shown that longitudinal cracking detrimental to serviceability is likely to occur, the slab should be reinforced in the direction transverse to the supporting steel section. It is recommended that the area of such reinforcement should be at least 0.002 times the concrete area in the longitudinal direction of the beam and should be uniformly distributed.

3. Design Strength of Concrete-Encased Beams

Tests of concrete-encased beams demonstrated that (1) the encasement drastically reduces the possibility of lateral-torsional instability and prevents local buckling of the encased steel, (2) the restrictions imposed on the encasement practically prevent bond failure prior to first yielding of the steel section, and (3) bond failure does not necessarily limit the moment capacity of an encased steel beam (ASCE, 1979).

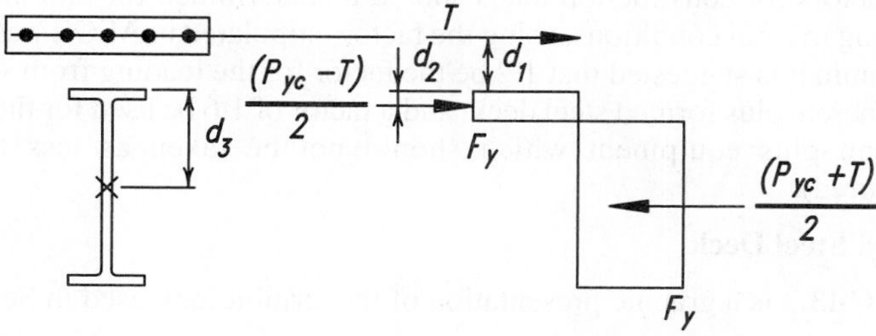

Fig. C-I3.2. Plastic stress distribution for negative moment.

Accordingly, the LRFD Specification permits three alternate design methods: one based on the first yield in the tension flange of the composite section; one based on the plastic moment capacity of the steel beam alone; and a third method based upon the plastic moment capacity of the composite section applicable only when shear connectors are provided along the steel section and reinforcement of the concrete encasement meets the specified detailing requirements. No limitations are placed on the slenderness of either the composite beam or the elements of the steel section, since the encasement effectively inhibits both local and lateral buckling.

In the method based on first yield, stresses on the steel section from permanent loads applied to unshored beams before the concrete has hardened must be superimposed on stresses on the composite section from loads applied to the beams after hardening of the concrete. In this superposition, all permanent loads should be multiplied by the dead load factor and all live loads should be multiplied by the live load factor. For shored beams, all loads may be assumed as resisted by the composite section. Complete interaction (no slip) between the concrete and steel is assumed.

The contribution of concrete to the strength of the composite section is ordinarily larger in positive moment regions than in negative moment regions. Accordingly, design based on the composite section is more advantageous in the regions of positive moments.

4. Strength During Construction

When temporary shores are not used during construction, the steel beam alone must resist all loads applied before the concrete has hardened enough to provide composite action. Unshored beam deflection caused by wet concrete tends to increase slab thickness and dead load. For longer spans this may lead to instability analogous to roof ponding. An excessive increase of slab thickness may be avoided by beam camber.

When forms are not attached to the top flange, lateral bracing of the steel beam during construction may not be continuous and the unbraced length may control flexural strength, as defined in Section F1.

The LRFD Specification does not include special requirements for a margin against yield during construction. According to Section F1, maximum factored moment during construction is $0.90F_yZ$ where F_yZ is the plastic moment ($0.90F_y Z \approx 0.90 \times 1.1F_y S$). This is equivalent to approximately the yield moment, F_yS. Hence, required flexural strength during construction prevents moment in excess of the yield moment.

Load factors for construction loads should be determined for individual projects according to local conditions, using the factors stipulated in ASCE 7 as a guide. As a minimum it is suggested that 1.2 be the factor for the loading from steel framing plus concrete plus formed steel deck, and a factor of 1.6 be used for the live load of workmen plus equipment which should not be taken as less than 20 psf (unfactored).

5. Formed Steel Deck

Figure C-I3.3 is a graphic presentation of the terminology used in Section I3.5.

When studs are used on beams with formed steel deck, they may be welded directly through the deck or through prepunched or cut-in-place holes in the deck. The usual procedure is to install studs by welding directly through the deck; however, when the deck thickness is greater than 16 gage (1.5 mm) for single thickness, or 18 (1.2 mm) gage for each sheet of double thickness, or when the total thickness of galvanized coating is greater than 1.25 ounces/sq. ft (0.38 kg/m^2), special precautions and procedures recommended by the stud manufacturer should be followed.

The design rules for composite construction with formed steel deck are based upon a study (Grant et al., 1977) of the then available test results. The limiting parameters listed in Section I3.5 were established to keep composite construction with formed steel deck within the available research data.

Seventeen full size composite beams with concrete slab on formed steel deck were

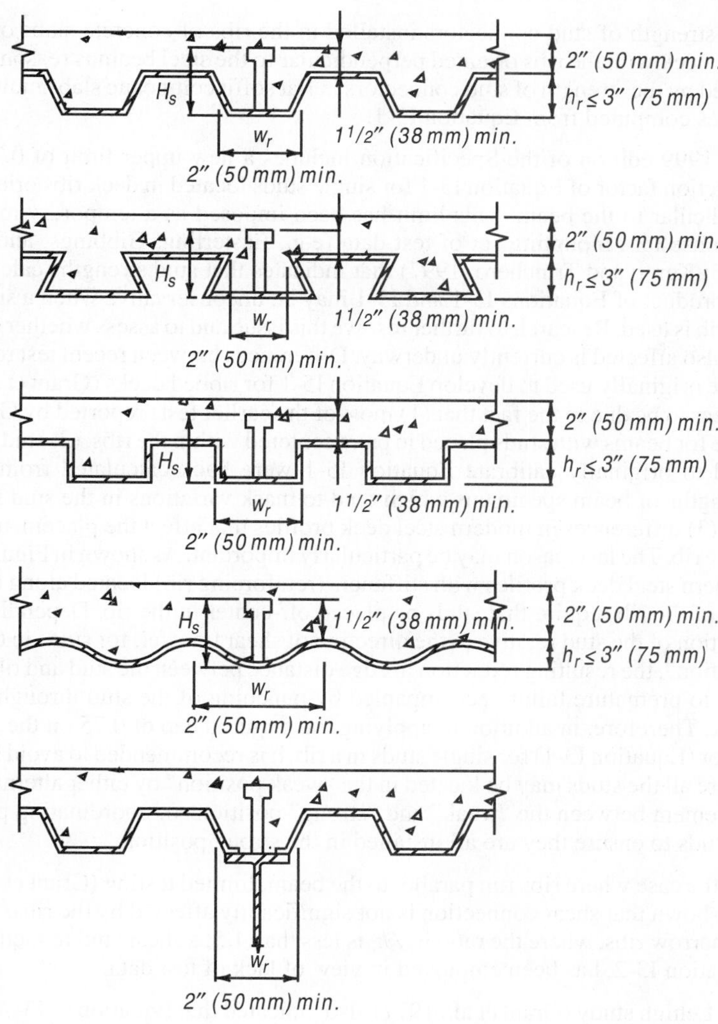

Fig. C-I3.3. Steel deck limits.

tested at Lehigh University and the results supplemented by the results of 58 tests performed elsewhere. The range of stud and steel deck dimensions encompassed by the 75 tests were limited to:

(1) Stud dimensions: ¾-in. dia. × 3.00 to 7.00 in.

(2) Rib width: 1.94 in. to 7.25 in.

(3) Rib height: 0.88 in. to 3.00 in.

(4) Ratio w_r / h_r: 1.30 to 3.33

(5) Ratio H_s / h_r: 1.50 to 3.41

(6) Number of studs in any one rib: 1, 2, or 3

The strength of stud connectors installed in the ribs of concrete slabs on formed steel deck with the ribs oriented perpendicular to the steel beam is reasonably estimated by the strength of stud connectors in flat soffit composite slabs multiplied by values computed from Equation I3-1.

The 1999 edition of the Specification includes a new upper limit of 0.75 on the reduction factor of Equation I3-1 for single studs located in deck ribs oriented perpendicular to the beam. This limit has been imposed as a temporary measure in response to a mounting set of test data (e.g., Easterling, Gibbings, and Murray, 1993; Kemp and Trinchero, 1997) that indicates that stud strengths calculated by the product of Equations I3-1 and I5-1 may be unconservative when a single stud per rib is used. Research to further resolve this issue and to assess whether stud pairs are also affected is currently underway. Differences between recent test results and those originally used to develop Equation I5-1 for ribbed decks (Grant et al., 1977) appear to be due to the fact that (1) most of the earlier tests reported by Grant et al. were for beams with studs placed in pairs centered within the ribs, (2) stud strengths used to originally calibrate Equation I5-1 were back calculated from moment strengths of beam specimens which tend to mask variations in the stud strengths, and (3) differences in modern steel deck profiles that affect the placement of studs in the rib. The last reason may be particularly important. As shown in Figure C-I3.4, modern steel deck profiles with stiffeners (reinforcing rib) located along the center line of the rib require that studs be placed off-center in the rib. Depending on the location of the stud relative to the direction of shear transfer, for studs in the "weak position", the resulting reduction in edge distance between the stud and rib wall can lead to premature failure accompanied by punching of the stud through the steel deck. Therefore, in addition to applying the required cap of 0.75 on the reduction factor (Equation I3-1) for single studs in a rib, it is recommended to avoid situations where all the studs may be located in the "weak position" by either alternating stud placement between the "weak" and "strong" positions or coordinating placement of studs to ensure they are all installed in the strong position.

For the case where ribs run parallel to the beam, limited testing (Grant et al., 1977) has shown that shear connection is not significantly affected by the ribs. However, for narrow ribs, where the ratio w_r / h_r is less than 1.5, a shear stud reduction factor, Equation I3-2, has been employed in view of lack of test data.

The Lehigh study (Grant et al., 1977) also indicated that Equation C-I3-7 for effective section modulus and Equation C-I3-6 for effective moment of inertia were valid for composite construction with formed steel deck.

Based on the Lehigh test data (Grant et al., 1977), the maximum spacing of steel deck anchorage to resist uplift was increased from 16 (405 mm) to 18 in. (460 mm) in order to accommodate current production profiles.

When metal deck includes units for carrying electrical wiring, crossover headers are commonly installed over the cellular deck perpendicular to the ribs. They create trenches which completely or partially replace sections of the concrete slab above the deck. These trenches, running parallel to or transverse to a composite beam, may reduce the effectiveness of the concrete flange. Without special provisions to replace the concrete displaced by the trench, the trench should be considered as a complete structural discontinuity in the concrete flange.

When trenches are parallel to the composite beam, the effective flange width should be determined from the known position of the trench.

Trenches oriented transverse to composite beams should, if possible, be located in areas of low bending moment and the full required number of studs should be placed between the trench and the point of maximum positive moment. Where the trench cannot be located in an area of low moment, the beam should be designed as non-composite.

6. Design Shear Strength

A conservative approach to vertical shear provisions for composite beams is adopted by assigning all shear to the steel section web. This neglects any concrete slab contribution and serves to simplify the design.

I4. COMBINED COMPRESSION AND FLEXURE

The procedure adopted for the design of beam-columns is described and supported by comparisons with test data in Galambos and Chapuis (1980). The basic approach is identical to that specified for steel columns in Section H1.

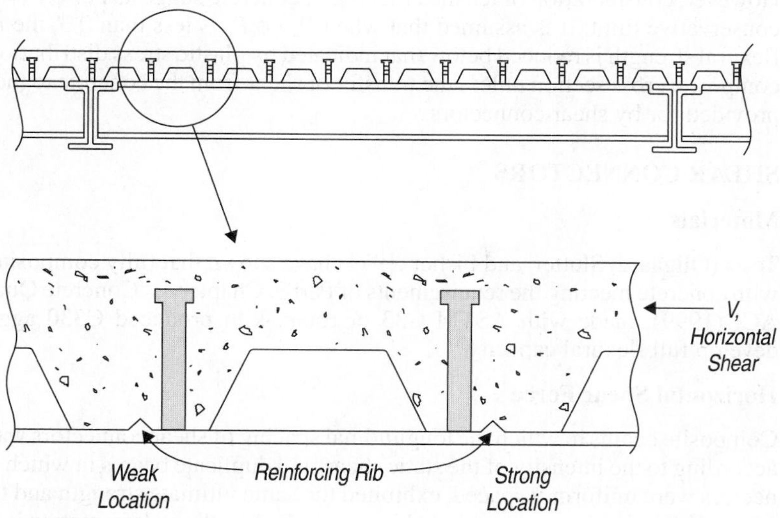

Fig. C-I3.4. Alternative shear stud positions in rib decked profiles.

The nominal axial strength of a beam-column is obtained from Section I2.2, while the nominal flexural strength is determined from the plastic stress distribution on the composite section. An approximate formula for this plastic moment resistance of a composite column is given in Galambos and Chapuis (1980).

$$M_n = M_p = ZF_y + \frac{(h_2 - 2c_r)}{3} A_r F_{yr} + \left(\frac{h_2}{2} - \frac{A_w F_y}{1.7 f_c' h_1} \right) A_w F_y \qquad \text{(C-I4-1)}$$

where

A_w = web area of encased steel shape; for concrete-filled HSS, $A_w = 0$, in.^{2} (mm^2)
Z = plastic section modulus of the steel section, in.^{3} (mm^3)
c_r = average of distance from compression face to longitudinal reinforcement in that face and distance from tension face to longitudinal reinforcement in that face, in. (mm)
h_1 = width of composite cross section perpendicular to the plane of bending, in. (mm)
h_2 = width of composite cross section parallel to the plane of bending, in. (mm)

The supporting comparisons with beam-column tests included 48 concrete filled pipes or tubing and 44 concrete-encased steel shapes (Galambos and Chapuis, 1980). The overall mean test-to-prediction ratio was 1.23 and the coefficient of variation 0.21.

The last paragraph in Section I4 provides a transition from beam-columns to beams. It involves bond between the steel section and concrete. Section I3 for beams requires either shear connectors or full, properly reinforced encasement of the steel section. Furthermore, even with full encasement, it is assumed that bond is capable of developing only the moment at first yielding in the steel of the composite section. No test data are available on the loss of bond in composite beam-columns. However, consideration of tensile cracking of concrete suggests $P_u / \phi_c P_n = 0.3$ as a conservative limit. It is assumed that when $P_u / \phi_c P_n$ is less than 0.3, the nominal flexural strength is reduced below that indicated by plastic stress distribution on the composite cross section unless the transfer of shear from the concrete to the steel is provided for by shear connectors.

I5. SHEAR CONNECTORS

1. Materials

Tests (Ollgaard, Slutter, and Fisher, 1971) have shown that fully composite beams with concrete meeting the requirements of Part 3, Chapter 4, "Concrete Quality," of ACI (1999), made with ASTM C33 or rotary-kiln produced C330 aggregates, develop full flexural capacity.

2. Horizontal Shear Force

Composite beams in which the longitudinal spacing of shear connectors was varied according to the intensity of the static shear, and duplicate beams in which the connectors were uniformly spaced, exhibited the same ultimate strength and the same amount of deflection at normal working loads. Only a slight deformation in the concrete and the more heavily stressed connectors is needed to redistribute the horizontal shear to other less heavily stressed connectors. The important consideration is that the total number of connectors be sufficient to develop the shear V_h on either side of the point of maximum moment. The provisions of the LRFD Specification are based upon this concept of composite action.

In computing the design flexural strength at points of maximum negative bending, reinforcement parallel to the steel beam within the effective width of the slab may be included, provided such reinforcement is properly anchored beyond the region of negative moment. However, enough shear connectors are required to transfer the ultimate tensile force in the reinforcement, from the slab to the steel beam.

3. Strength of Stud Shear Connectors

Studies have defined stud shear connector strength in terms of normal weight and lightweight aggregate concretes as a function of both concrete modulus of elasticity and concrete strength as given by Equation I5-1.

Equation I5-1, obtained from Ollgaard et al. (1971), corresponds to Tables I4.1 and I4.2 in Section I4 of the 1989 AISC ASD Specification. Note that an upper bound on stud shear strength is the product of the cross-sectional area of the stud times its ultimate tensile strength.

The LRFD Specification does not specify a resistance factor for shear connector strength. The resistance factor for the flexural strength of a composite beam accounts for all sources of variability, including those associated with the shear connectors.

4. Strength of Channel Shear Connectors

Equation I5-2 is a modified form of the formula for the strength of channel connectors developed by Slutter and Driscoll (1965). The modification has extended its use to lightweight concrete.

6. Shear Connector Placement and Spacing

Uniform spacing of shear connectors is permitted except in the presence of heavy concentrated loads.

Studs not located directly over the web of a beam tend to tear out of a thin flange before attaining full shear-resisting capacity. To guard against this contingency, the size of a stud not located over the beam web is limited to $2\frac{1}{2}$ times the flange thickness (Goble, 1968).

The minimum spacing of connectors along the length of the beam, in both flat soffit concrete slabs and in formed steel deck with ribs parallel to the beam, is six diameters; this spacing reflects development of shear planes in the concrete slab (Ollgaard et al., 1971). Since most test data are based on the minimum transverse spacing of four diameters, this transverse spacing was set as the minimum permitted. If the steel beam flange is narrow, this spacing requirement may be achieved by staggering the studs with a minimum transverse spacing of three diameters between the staggered row of studs. The reduction in connector capacity in the ribs of formed steel decks is provided by the factor $0.85/\sqrt{N_r}$, which accounts for the reduced capacity of multiple connectors, including the effect of spacing. When deck ribs are parallel to the beam and the design requires more studs than can be placed in the rib, the deck may be split so that adequate spacing is available for stud installation. Figure C-I5.1 shows possible connector arrangements.

I6. SPECIAL CASES

Tests are required for construction that falls outside the limits given in the Specification. Different types of shear connectors may require different spacing and other detailing than stud and channel connectors.

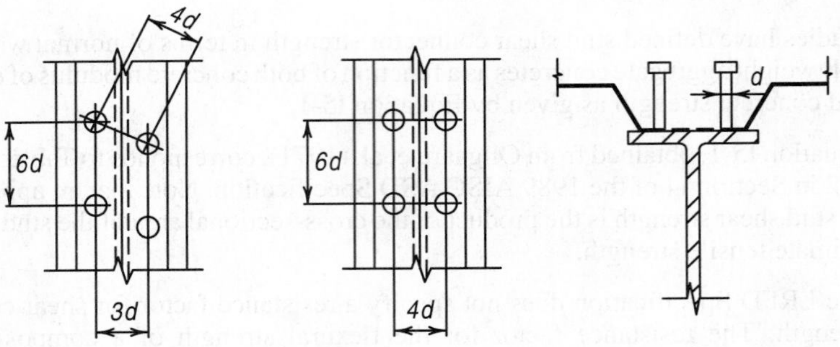

Fig. C-I5.1. Shear connector arrangements.

CHAPTER J

CONNECTIONS, JOINTS, AND FASTENERS

J1. GENERAL PROVISIONS

5. Splices in Heavy Sections

Solidified but still-hot filler metal contracts significantly as it cools to ambient temperature. Shrinkage of large welds between elements which are not free to move to accommodate the shrinkage causes strains in the material adjacent to the weld that can exceed the yield point strain. In thick material the weld shrinkage is restrained in the thickness direction, as well as in the width and length directions, causing triaxial stresses to develop that may inhibit the ability of ductile steel to deform in a ductile manner. Under these conditions, the possibility of brittle fracture increases.

When splicing ASTM A6/A6M Group 4 and 5 rolled sections or heavy welded built-up members, the potentially harmful weld shrinkage strains can be avoided by using bolted splices or fillet-welded lap splices or splices that combine a welded and bolted detail (see Figure C-J1.1). Details and techniques that perform well for materials of modest thickness usually must be changed or supplemented by more demanding requirements when welding thick material. Also, the provisions of the *Structural Welding Code,* AWS D1.1, are minimum requirements that apply to most structural welding situations; however, when designing and fabricating welded splices of ASTM A6/A6M Group 4 and 5 shapes and similar built-up cross sections, special consideration must be given to all aspects of the welded splice detail.

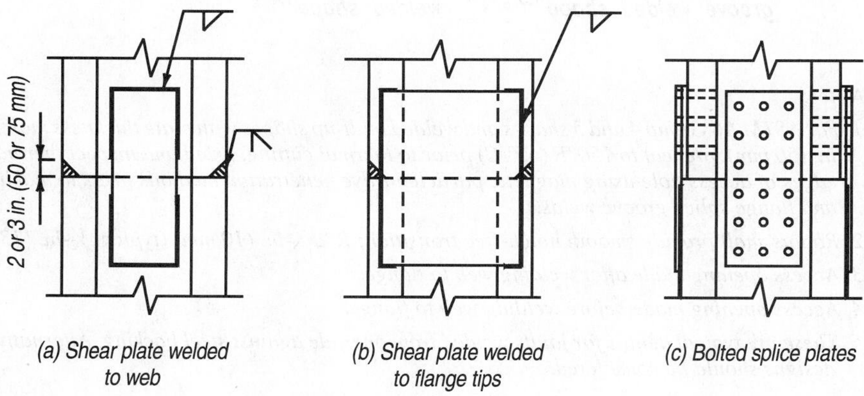

Fig. C-J1.1. Alternative splices that minimize weld restraint tensile stresses.

- Notch-toughness requirements should be specified for tension members. See Commentary Section A3.
- Generously sized weld access holes, Figure C-J1.2, are required to provide increased relief from concentrated weld shrinkage strains, to avoid close juncture of welds in orthogonal directions, and to provide adequate clearance for the exercise of high quality workmanship in hole preparation, welding, and ease of inspection.

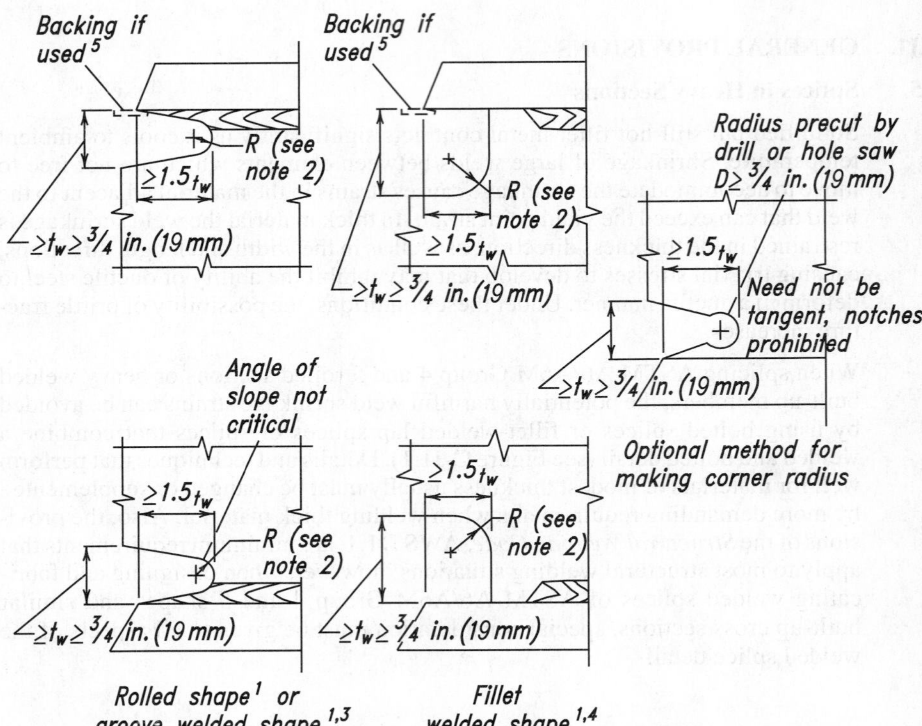

Notes:

1. *For ASTM A6 Group 4 and 5 shapes and welded built-up shapes with plate thickness more than 2 in. (50 mm), preheat to 150°F (65° C) prior to thermal cutting, grind and inspect thermally cut edges of access hole using magnetic particle or dye penetration methods prior to making web and flange splice groove welds.*
2. *Radius shall provide smooth notch-free transition; R ≥ 3/8-in. (10 mm) (typical 1/2-in. (13 mm))*
3. *Access opening made after welding web to flange.*
4. *Access opening made before welding web to flange.*
5. *These are typical details for joints welded from one side against steel backing. Alternative joint designs should be considered.*

Fig. C-J1.2. Weld access hole geometry.

- Preheating for thermal cutting is required to minimize the formation of a hard surface layer.
- Grinding to bright metal and inspection using magnetic particle or dye-penetrant methods is required to remove the hard surface layer and to assure smooth transitions free of notches or cracks.

In addition to tension splices of truss chord members and tension flanges of flexural members, other joints fabricated of heavy sections subject to tension should be given special consideration during design and fabrication.

8. Placement of Welds and Bolts

Slight eccentricities between the gravity axis of single and double angle members and the center of gravity of connecting rivets or bolts have long been ignored as having negligible effect on the static strength of such members. Tests (Gibson and Wake, 1942) have shown that similar practice is warranted in the case of welded members in statically loaded structures.

However, the fatigue life of eccentrically loaded welded angles has been shown to be very short (Kloppel and Seeger, 1964). Notches at the roots of fillet welds are harmful when alternating tensile stresses are normal to the axis of the weld, as could occur due to bending when axial cyclic loading is applied to angles with end welds not balanced about the neutral axis. Accordingly, balanced welds are indicated when such members are subjected to cyclic loading (see Figure C-J1.3).

9. Bolts in Combination with Welds

Welds will not share the load equally with mechanical fasteners in bearing-type connections. Before ultimate loading occurs, the fastener will slip and the weld will carry an indeterminately larger share of the load.

Accordingly, the sharing of load between welds and A307 bolts or high-strength bolts in a bearing-type connection is not recommended. For similar reasons, A307 bolts and rivets should not be assumed to share loads in a single group of fasteners.

For high-strength bolts in slip-critical connections to share the load with welds it is

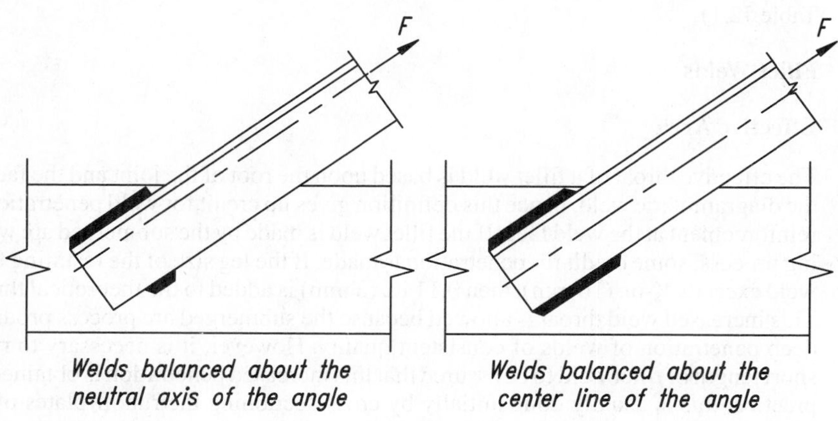

Figure C-J1.3. Balanced welds.

advisable to fully tension the bolts before the weld is made. If the weld is placed first, angular distortion from the heat of the weld might prevent the faying action required for development of the slip-critical force. When the bolts are fully tensioned before the weld is made, the slip-critical bolts and the weld may be assumed to share the load on a common shear plane (Kulak, Fisher, and Struik, 1987). The heat of welding near bolts will not alter the mechanical properties of the bolts.

In making alterations to existing structures, it is assumed that whatever slip is likely to occur in high-strength bolted bearing-type connections or riveted connections will have already taken place. Hence, in such cases the use of welding to resist all stresses, other than those produced by existing dead load present at the time of making the alteration, is permitted.

It should be noted that combinations of fasteners as defined herein does not refer to connections such as shear plates for beam-to-column connections which are welded to the column and bolted to the beam flange or web (Kulak et al., 1987) and other comparable connections.

10. High-Strength Bolts in Combination with Rivets

When high-strength bolts are used in combination with rivets, the ductility of the rivets permits the direct addition of the strengths of both fastener types.

J2. WELDS

1. Groove Welds

The engineer preparing contract design drawings cannot specify the depth of groove without knowing the welding process and the position of welding. Accordingly, only the effective throat for partial-joint-penetration groove welds should be specified on design drawings, allowing the fabricator to produce this effective throat with his own choice of welding process and position. The weld reinforcement is not used in determining the effective throat thickness of a groove weld (see Table J2.1).

2. Fillet Welds

2a. Effective Area

The effective throat of a fillet weld is based upon the root of the joint and the face of the diagrammatic weld, hence this definition gives no credit for weld penetration or reinforcement at the weld face. If the fillet weld is made by the submerged arc welding process, some credit for penetration is made. If the leg size of the resulting fillet weld exceeds ⅜-in. (10 mm), then 0.11 in. (3 mm) is added to the theoretical throat. This increased weld throat is allowed because the submerged arc process produces deep penetration of welds of consistent quality. However, it is necessary to run a short length of fillet weld to be assured that this increased penetration is obtained. In practice, this is usually done initially by cross-sectioning the runoff plates of the joint. Once this is done, no further testing is required, as long as the welding procedure is not changed.

2b. Limitations

Table J2.4 provides a minimum size of fillet weld for a given thickness of the thicker part joined.

The requirements are not based upon strength considerations, but upon the quench effect of thick material on small welds. Very rapid cooling of weld metal may result in a loss of ductility. Further, the restraint to weld metal shrinkage provided by thick material may result in weld cracking. Because a 5/16-in. (8 mm) fillet weld is the largest that can be deposited in a single pass by SMAW process, 5/16-in. (8 mm) applies to all material 3/4-in. (19 mm) and greater in thickness, but minimum preheat and interpass temperature are required by AWS D1.1.* Both the design engineer and the shop welder must be governed by the requirements.

Table J2.3 gives the minimum effective throat of a partial-joint-penetration groove weld. Notice that Table J2.3 for partial-joint-penetration groove welds goes up to a plate thickness of over 6 in. (150 mm) and a minimum weld throat of 5/8-in. (16 mm), whereas, for fillet welds Table J2.4 goes up to a plate thickness of over 3/4-in. (19 mm) and a minimum leg size of fillet weld of only 5/16-in. (8 mm). The additional thickness for partial-joint-penetration groove welds is to provide for reasonable proportionality between weld and material thickness.

For plates of 1/4-in. (6 mm) or more in thickness, it is necessary that the inspector be able to identify the edge of the plate to position the weld gage. This is assured if the weld is kept back at least 1/16-in. (2 mm) from the edge, as shown in Figure C-J2.1.

Where longitudinal fillet welds are used alone in a connection (see Figure C-J2.2), Section J2.2b requires the length of each weld to be at least equal to the width of the connecting material because of shear lag (Freeman, 1930).

By providing a minimum lap of five times the thickness of the thinner part of a lap

* See Table J2.4.

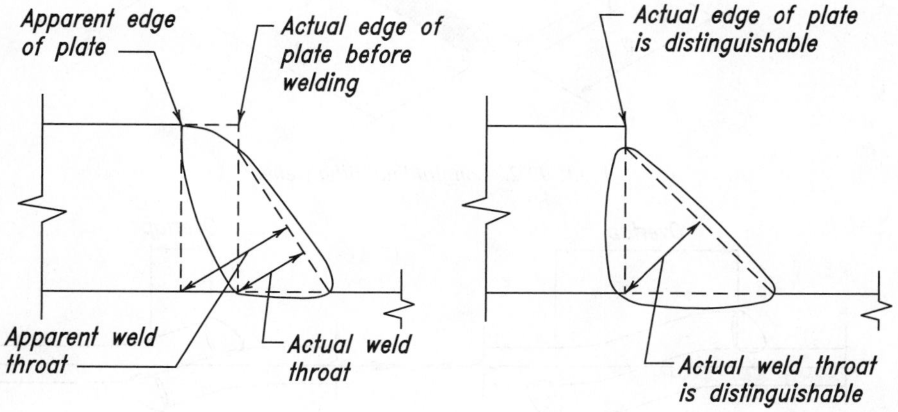

Fig. C-J2.1. Identification of plate edge.

joint, the resulting rotation of the joint when pulled will not be excessive, as shown in Figure C-J2.3. Fillet welded lap joints under tension tend to open and apply a tearing action at the root of the weld as shown in Figure C-J2.4b, unless restrained by a force *F* as shown in Figure C-J2.4a.

End returns are not essential for developing the capacity of fillet welded connections and have a negligible effect on their strength. Their use has been encouraged to insure that the weld size is maintained over the length of the weld, to enhance the fatigue resistance of cyclically loaded flexible end connections, and to increase the plastic deformation capability of such connections.

The weld capacity database on which the specifications were developed had no end returns. This includes the study by Higgins and Preece (1968), seat angle tests by Lyse and Schreiner (1935), the seat and top angle tests by Lyse and Gibson (1937), beam webs welded directly to column or girder by fillet welds by Johnston and Deits (1942), and the eccentrically loaded welded connections reported by Butler, Pal, and Kulak (1972). Hence, the current design-resistance values and joint-capacity models do not require end returns when the required weld size is provided. Johnston and Green (1940) noted that movement consistent with the design assumption of no end restraint (i.e., joint flexibility) was enhanced without end returns. They also verified that greater plastic deformation of the connection was achieved when end returns existed, although the strength was not significantly different.

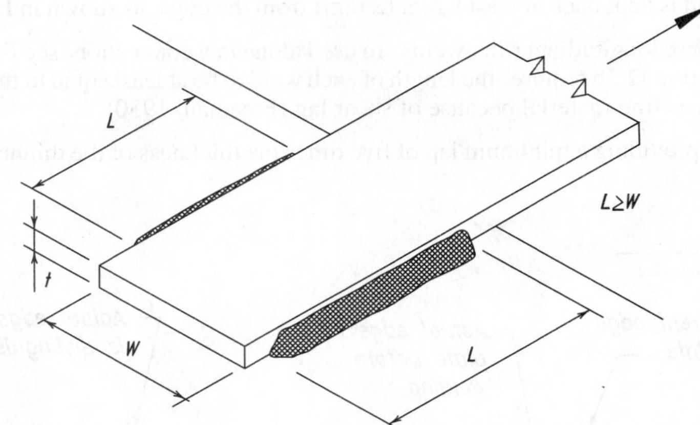

Fig. C-J2.2. Longitudinal fillet welds.

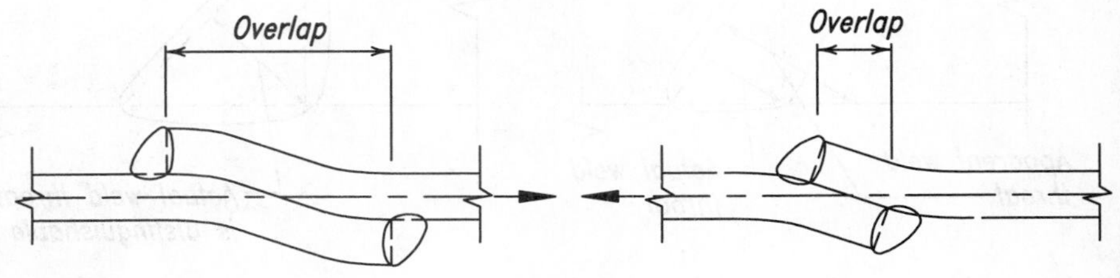

Fig. C-J2.3. Minimum lap.

When longitudinal fillet welds parallel to the stress are used to transmit the load to the end of an axially loaded member, the welds are termed "end loaded". Typical examples of such welds would include, but are not necessarily limited to, longitudinally welded lap joints at the end of axially loaded members, welds attaching bearing stiffeners, and similar cases. Typical examples of longitudinally loaded fillet welds which are not considered end loaded include, but are not limited to, welds that connect plates or shapes to form built-up cross sections in which the shear force is applied to each increment of length of weld stress depending upon the distribution of shear load along the length of the member, welds attaching beam web connection angles and shear plates because the flow of shear force from the beam or girder web to the weld is essentially uniform throughout the weld length, that is, the weld is not end-loaded despite the fact that it is loaded parallel to the weld axis. Neither does the reduction factor apply to welds attaching stiffeners to webs because the stiffeners and welds are not subject to calculated axial stress but merely serve to keep the web flat.

The distribution of stress along the length of end loaded fillet welds is far from uniform and is dependent upon complex relationships between the stiffness of the longitudinal fillet weld relative to the stiffness of the connected materials. Beyond some length, it is non-conservative to assume that the average stress over the total length of the weld may be taken as equal to the full design strength. Experience has shown that when the length of the weld is equal to approximately 100 times the weld size or less, it is reasonable to assume the effective length is equal to the actual length. For weld lengths greater than 100 times the weld size, the effective length should be taken less than the actual length. The reduction coefficient, β, provided in Section J2.2b is the equivalent of Eurocode 3, which is a simplified approximation to exponential formulas developed by finite element studies and tests preformed in Europe over many years. The criterion is based upon combined consideration of nominal strength for fillet welds with leg size less than $\frac{1}{4}$ in. (6 mm) and upon a judgement based serviceability limit of slightly less than $\frac{1}{32}$ in. (1 mm) displacement at the end of the weld for welds with leg size $\frac{1}{4}$ in. (6 mm) and larger. Mathematically, the application of the β factor implies that the minimum strength of an end-loaded weld is achieved when the length is approximately 300 times the leg size. Because it is illogical to conclude that the total strength of a weld longer than 300 times the weld size would be less than a shorter weld, the length reduction coefficient is taken as 0.6 when the weld length is greater than 300 times the leg size.

Fillet weld terminations do not affect the strength or serviceability of connections

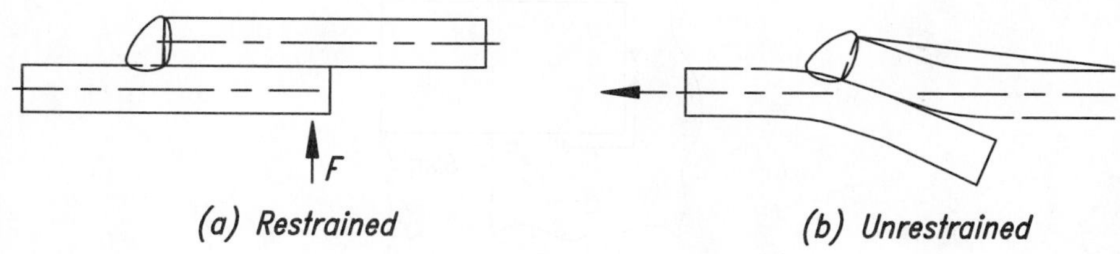

Fig. C-J2.4. Restraint of lap joints.

in most cases. However, in certain cases, the disposition of welds affect the planned function of connections, and notches may affect the static strength and/or the resistance to crack initiation if cyclic loads of sufficient magnitude and frequency occur. For these cases, limitations are specified to assure desired performance.

(a) At lapped joints where one part extends beyond the end or edge of the part to which it is welded and if the parts are subject to calculated tensile stress at the start of the overlap, it is important that the weld terminate a short distance from the stressed edge. For one typical example, the lap joint between the tee chord and the web members of a truss, the weld should not extend to the edge of the tee stem. See Figure C-J2.5. The best technique to avoid inadvertent notches at this critical location is to strike the welding arc at a point slightly back from the edge and proceed with welding in the direction away from the edge. See Figure C-J2.6. On the other hand, where framing angles extend beyond the end of the beam web to which they are welded, the free end of the beam web is subject to zero stress; thus, it is permissible for the fillet weld to extend continuously across the top end, along the side and along the bottom end of the angle to the extreme end of the beam. See Figure C-J2.7.

(b) For connections which are subject to maximum stress at the weld termination

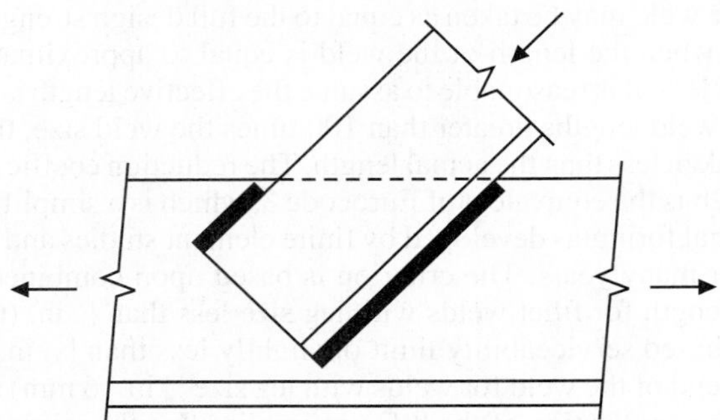

Fig. C-J2.5. Fillet welds near tension edges.

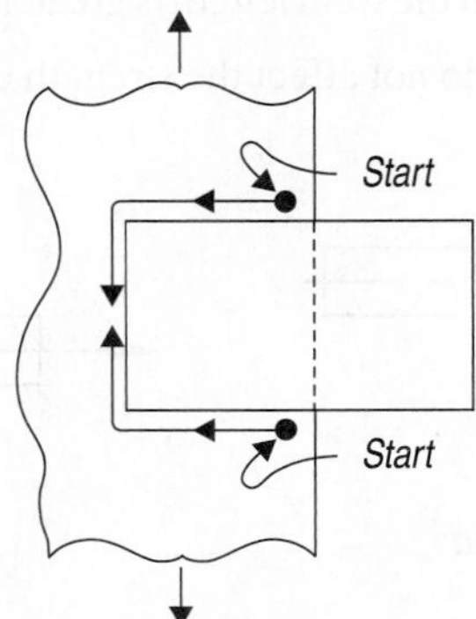

Fig. C-J2.6. Suggested direction of welding travel to avoid notches.

due to cyclic forces and/or moments of sufficient magnitude and frequency to initiate cracks emanating from unfilled start or stop craters or other discontinuities, at the end of the weld must be protected by boxing or returns. If the bracket is a plate projecting from the face of a support, extra care must be exercised in the deposition of the boxing weld across the thickness of the plate to assure that a fillet free of notches is provided.

(c) For connections such as framing angles and simple end plates which are assumed in design of the structure to be flexible connections, the top and bottom edges of the outstanding legs must be left unwelded over a substantial portion of their length in order to assure flexibility of the connection. Research tests (Johnston and Green, 1940) have shown that the static strength of the connection is the same with or without end returns; therefore the use of returns is optional, but if used, their length must be restricted to not more than four times the weld size, see Figure C-J2.8.

(d) Experience has shown that when ends of intermediate transverse stiffeners on the webs of plate girders are not welded to the flanges (the usual practice), small torsional distortions of the flange which occur near shipping bearing points in the normal course of shipping by rail or truck may cause high out-of-plane bending stresses (yield point) and fatigue cracking at the toe of the web-to-flange welds. This has been observed even with closely fitted stiff-

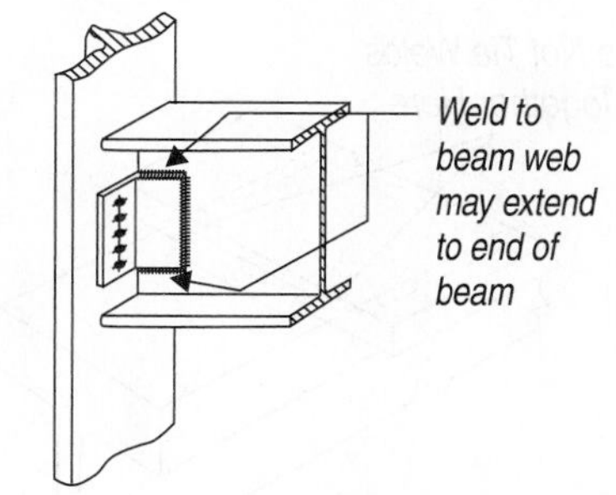

Fig.C-J2.7. Fillet weld details on framing angles.

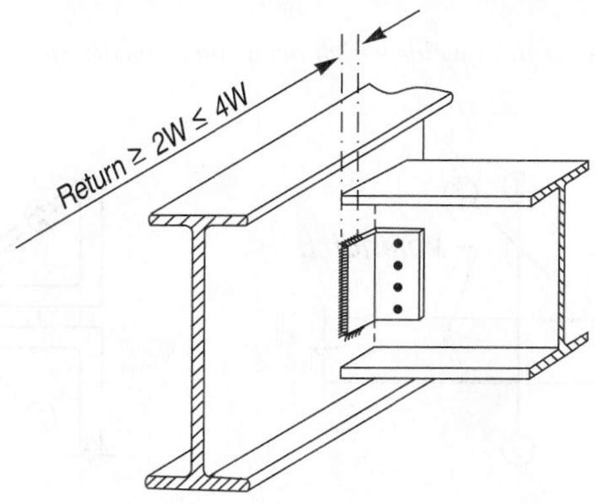

Fig.C-J2.8. Flexible connection returns optional unless subject to fatigue.

eners. The intensity of these out-of-plane stresses may be effectively limited and cracking prevented if "breathing room" is provided by terminating web-to-flange welds. The unwelded distance should not exceed six times the web thickness to assure that column buckling of the web within the unwelded length does not occur.

(e) For fillet welds which occur on opposite sides of a common plane, it is not possible to deposit a weld continuously around the corner from one side to the other without causing a gouge in the corner of the parts joined; therefore the welds must be interrupted at the corner. See Figure C-J2.9.

4. Design Strength

The strength of welds is governed by the strength of either the base material or the deposited weld metal. Table J2.5 contains the resistance factors and nominal weld strengths, as well as a number of limitations.

It should be noted that in Table J2.5 the nominal strength of fillet welds is determined from the effective throat area, whereas the strength of the connected parts is governed by their respective thicknesses. Figure C-J2.10 illustrates the shear planes for fillet welds and base material:

(a) Plane 1-1, in which the resistance is governed by the shear strength for material A.

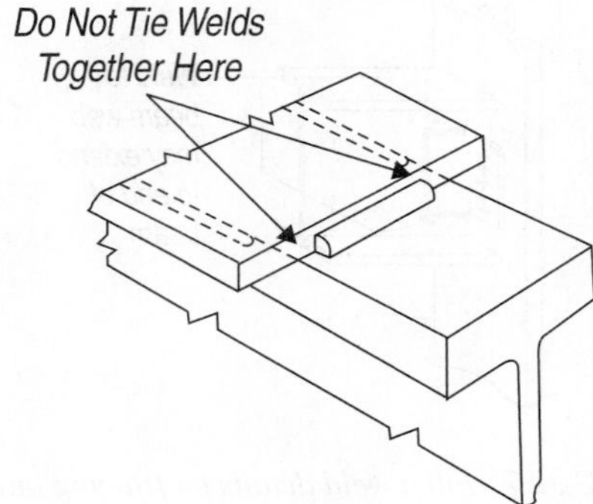

Fig. C-J2.9. Details for fillet welds which occur on opposite sides of a common plane.

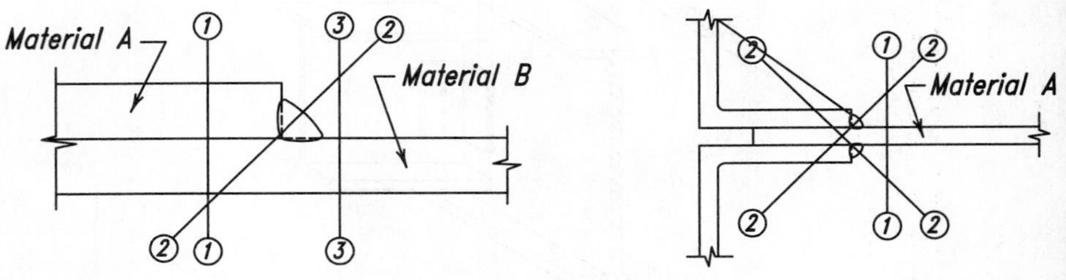

Fig. C-J2.10. Shear planes for fillet welds loaded in longitudinal shear.

(b) Plane 2-2, in which the resistance is governed by the shear strength of the weld metal.

(c) Plane 3-3, in which the resistance is governed by the shear strength of the material B.

The resistance of the welded joint is the lowest of the resistance calculated in each plane of shear transfer. Note that planes 1-1 and 3-3 are positioned away from the fusion areas between the weld and the base material. Tests have demonstrated that the stress on this fusion area is not critical in determining the shear strength of fillet welds (Preece, 1968).

The shear planes for plug and partial-joint-penetration groove welds are shown in Figure C-J2.11 for the weld and base metal. Generally the base metal will govern the shear strength.

5. Combination of Welds

This method of adding weld strengths does not apply to a welded joint using a partial-joint-penetration single bevel groove weld with a superimposed fillet weld. In this case, the effective throat of the combined joint must be determined and the design strength based upon this throat area.

6. Weld Metal Requirements

Applied and residual stresses and geometrical discontinuities from back-up bars with associated notch effects contribute to sensitivity to fracture. Some weld metals in combination with certain procedures result in welds with low notch toughness. The Specification requires a minimum specified toughness for weld metals in those joints that are subject to more significant applied stresses and toughness demands.

The level of toughness required was selected as one level more conservative than

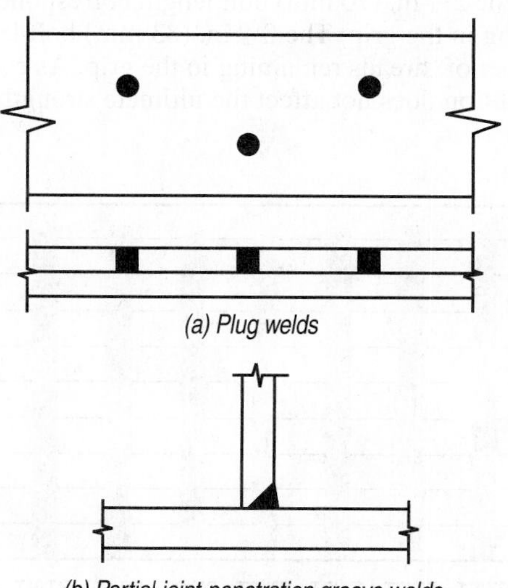

Fig. C-J2.11. Shear planes for plug and partial-joint-penetration groove welds.

the base metal requirement for Group 4 and 5 shapes. Research continues on this subject.

7. Mixed Weld Metal

Problems can occur when incompatible weld metals are used in combination and notch-tough composite weld metal is required. For instance, tack welds deposited using a self-shielded process with aluminum deoxidizers in the electrodes and subsequently covered by SAW weld passes can result in composite weld metal with low notch-toughness, despite the fact that each process by itself could provide notch-tough weld metal.

J3. BOLTS AND THREADED PARTS

1. High-Strength Bolts

In general, the use of high-strength bolts is required to conform to the provisions of the *Load and Resistance Factor Design Specification for Structural Joints Using ASTM A325 or A490 Bolts* (RCSC, 1994) as approved by the Research Council on Structural Connections.

Occasionally the need arises for the use of high-strength bolts of diameters and lengths in excess of those available for A325 or A325M and A490 or A490M bolts, as for example, anchor rods for fastening machine bases. For this situation Section A3.3 permits the use of A449 bolts and A354 threaded rods.

With this edition of the Specification snug-tightened installation is permitted for static applications involving ASTM A325 or A325M bolts (only) in tension or combined shear and tension. Figures C-J3.1 and C-J3.2 illustrate the results of testing by Johnson (1996) on tee stubs bolted flange-to-flange with two ASTM A325 bolts. This testing involved two lengths (2¾ in. and 3¼ in.) (70 mm and 83 mm) and several combinations of installed pretension (finger-tight, snug-tight, and pretensioned). The 2¾-in. (70 mm) bolt length corresponds to the least number of threads remaining in the grip. The 3¼-in. (83 mm) bolt length corresponds to the maximum number of threads remaining in the grip. As evidenced by the data, the installation condition does not affect the ultimate strength.

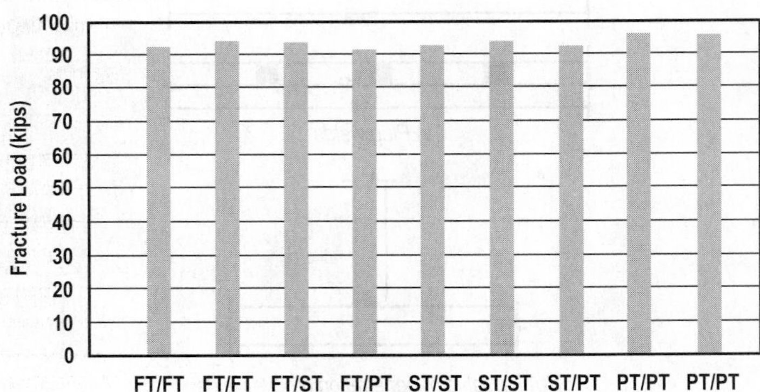

Fig. C-J3.1. Johnson (1996) tests, 2¾-in.-long, ¾-in.-diameter ASTM A325 bolts.

There are practical cases in the design of structures where slip of the connection is desirable in order to allow for expansion and contraction of a joint in a controlled manner. Regardless of whether force transfer is required in the directions normal to the slip direction, the nuts should be hand-tightened with a spud wrench and then backed off one-quarter turn. Furthermore, it is advisable to deform the bolt threads or use a locking nut or jamb nut to insure that the nut does not back off under service conditions. Thread deformation is commonly accomplished with a cold chisel and hammer applied at one location. Note that tack-welding of the nut to the bolt threads is discouraged.

2. Size and Use of Holes

To provide some latitude for adjustment in plumbing up a frame during erection, three types of enlarged holes are permitted, subject to the approval of the designer. The nominal maximum sizes of these holes are given in Table J3.3 or J3.3M. The use of these enlarged holes is restricted to connections assembled with bolts and is subject to the provisions of Sections J3.3 and J3.4.

3. Minimum Spacing

The *maximum* factored strength R_n at a bolt or rivet hole in bearing requires that the distance between the centerline of the first fastener and the edge of a plate toward which the force is directed should not be less than $1\frac{1}{2}d$ where d is the fastener diameter (Kulak et al., 1987). By similar reasoning the distance measured in the line of force, from the centerline of any fastener to the nearest edge of an adjacent hole, should not be less than $3d$, to ensure maximum design strength in bearing. Plotting of numerous test results indicates that the critical bearing strength is directly proportional to the above defined distances up to a maximum value of $3d$, above which no additional bearing strength is achieved (Kulak et al., 1987). ~~Table J3.7 lists the increments that must be added to adjust the spacing upward to compensate for an increase in hole dimension parallel to the line of force.~~ Section J3.10 gives the bearing strength criteria as a function of spacing. Errata 9/4/01

4. Minimum Edge Distance

Critical bearing stress is a function of the material tensile strength, the spacing of

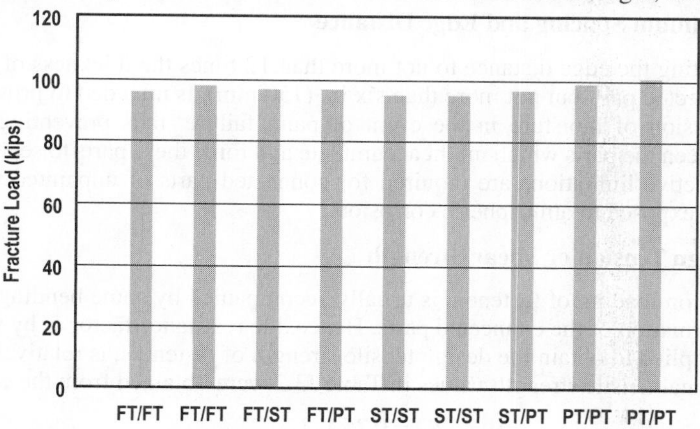

Fig. C-J3.2. Johnson (1996) tests, $3\frac{1}{4}$-in.-long, $\frac{3}{4}$-in.-diameter ASTM A325 bolts.

fasteners, and the distance from the edge of the part to the center line of the nearest fastener. Tests have shown (Kulak et al., 1987) that a linear relationship exists between the ratio of critical bearing stress to tensile strength (of the connected material) and the ratio of fastener spacing (in the line of force) to fastener diameter. The following equation affords a good lower bound to published test data for single-fastener connections with standard holes, and is conservative for adequately spaced multi-fastener connections:

$$\frac{F_{pcr}}{F_u} = \frac{l_e}{d} \tag{C-J3-1}$$

where

F_{pcr} = critical bearing stress, ksi (MPa)
F_u = tensile strength of the connected material, ksi (MPa)
l_e = distance, along a line of transmitted force, from the center of a fastener to the nearest edge of an adjacent fastener or to the free edge of a connected part (in the direction of stress), in. (mm)
d = diameter of fastener, in. (mm)

The provisions of Section J3.3 are concerned with l_e as hole spacing, whereas Section J3.4 is concerned with l_e as edge distance in the direction of stress.

Section J3.10 establishes a maximum bearing strength. Spacing and/or edge distance may be increased to provide for a required bearing strength, or bearing force may be reduced to satisfy a spacing and/or edge distance limitation.

It has long been known that the critical bearing stress of a single fastener connection is more dependent upon a given edge distance than multi-fastener connections (Jones, 1940). For this reason, longer edge distances (in the direction of force) are required for connections with one fastener in the line of transmitted force than required for those having two or more. The recommended minimum distance transverse to the direction of load is primarily a workmanship tolerance. It has little, if any, effect on the strength of the member.

5. Maximum Spacing and Edge Distance

Limiting the edge distance to not more than 12 times the thickness of an outside connected part, but not more than six in. (150 mm), is intended to provide for the exclusion of moisture in the event of paint failure, thus preventing corrosion between the parts which might accumulate and force these parts to separate. More restrictive limitations are required for connected parts of unpainted weathering steel exposed to atmospheric corrosion.

6. Design Tension or Shear Strength

Tension loading of fasteners is usually accompanied by some bending due to the deformation of the connected parts. Hence, the resistance factor ϕ, by which R_n is multiplied to obtain the design tensile strength of fasteners, is relatively low. The nominal tensile strength values in Table J3.2 were obtained from the equation

$$R_n = 0.75 A_b F_u \tag{C-J3-2}$$

This tensile strength given by Equation C-J3-2 is independent of whether the bolt was initially installed pretensioned or snug-tightened. Recent tests (Johnson, 1996 and Murray, Kline, and Rojani, 1992) confirm that the performance of ASTM A325 and A325M bolts in tension not subjected to fatigue are unaffected by the original installation condition. While the equation was developed for bolted connections (Kulak et al., 1987), it was also conservatively applied to threaded parts and to rivets.

In connections consisting of only a few fasteners, the effects of strain on the shear in bearing fasteners is negligible (Kulak et al., 1987 and Fisher et al., 1978). In longer joints, the differential strain produces an uneven distribution between fasteners (those near the end taking a disproportionate part of the total load), so that the maximum strength per fastener is reduced. The AISC ASD Specification permits connections up to 50 in. (1 270 mm) in length without a reduction in maximum shear stress. With this in mind the resistance factor ϕ for shear in bearing-type connections has been selected to accommodate the same range of connections.

The values of nominal shear strength in Table J3.2 were obtained from the equation

$$R_n / mA_b = 0.50 F_u \qquad \text{(C-J3-3)}$$

when threads are excluded from the shear planes and

$$R_n / mA_b = 0.40 F_u \qquad \text{(C-J3-4)}$$

when threads are not excluded from the shear plane, where m is the number of shear planes (Kulak et al., 1987). While developed for bolted connections, the equations were also conservatively applied to threaded parts and rivets. The value given for A307 bolts was obtained from Equation C-J3-4 but is specified for all cases regardless of the position of threads. For A325 or A325M bolts, no distinction is made between small and large diameters, even though the minimum tensile strength F_u is lower for bolts with diameters in excess of one inch. It was felt that such a refinement of design was not justified, particularly in view of the low resistance factor ϕ, the increasing ratio of tensile area to gross area, and other compensating factors.

7. Combined Tension and Shear in Bearing-Type Connections

Tests have shown that the strength of bearing fasteners subject to combined shear and tension resulting from externally applied forces can be closely defined by an ellipse (Kulak et al., 1987). Such a curve can be replaced, with only minor deviations, by three straight lines as shown in Figure C-J3.3. This latter representation offers the advantage that no modification of either type stress is required in the presence of fairly large magnitudes of the other type. This linear representation was adopted for Table J3.5, giving a limiting tensile stress F_t as a function of the shearing stress f_v for bearing-type connections. Following a change in the 1994 RCSC *LRFD Specification for Structural Joints Using ASTM A325 or A490 Bolts,* the coefficients in the equations in Table J3.5 have been modified for consistency (Carter, Tide, and Yura, 1997).

8. High-Strength Bolts in Slip-Critical Connections

Connections classified as slip-critical include those cases where slip could theoretically exceed an amount deemed by the Engineer of Record to affect the suitability for service of the structure by excessive distortion or reduction in strength or stability, even though the nominal strength of the connection may be adequate. Also included are those cases where slip of any magnitude must be prevented, for example, joints subject to fatigue, connectors between elements of built-up members at their ends (Sections D2 and E4), and bolts in combination with welds (Section J1.9).

The onset of slipping in a high-strength bolted, slip-critical connection is not an indication that the maximum strength of the connection has been reached. Its occurrence may be only a serviceability limit state. The design check for slip resistance can be made at two different load levels, factored loads (Sections J3.8a and J3.9a) and service loads (Appendices J3.8b and J3.9b). The nominal slip resistances r_{str} and $F_v A_b$ to be used with factored loads and service loads, respectively, are based on two different design concepts. The slip resistance r_{str} with factored loads is the mean resistance per bolt, which is a function of the mean slip coefficient and the clamping force. The 1.13 factor in (Equation J3-1) accounts for the expected 13 percent increase above the minimum specified preload provided by calibrated wrench tightening procedures. This was used to represent typical installations. The factored load resistance r_{str} uses the β reliability index approach that is used for the other design checks such as tension and bearing. The service load approach uses a probability of slip concept that implies a 90 percent reliability that slip will not occur if the calibrated wrench method of bolt installation is used.

The Engineer of Record must make the determination to use factored loads, service loads, or both in checking the slip resistance of a slip-critical connection. The following commentary is provided as guidance and an indication of the intent of the Specification.

In the case of slip-critical connections with three or more bolts in holes with only a small clearance, such as standard holes and slotted holes loaded transversely to the axis of the slot, the freedom to slip does not generally exist because one or more bolts are in bearing even before load is applied due to normal fabrication tolerances and erection procedures. If connections with standard holes have only one or two bolts in the direction of the applied force, a small slip may occur. In this case, slip-critical connections subjected to vibration or wind should be checked for slip at service-load levels. In built-up compression members, such as double-angle struts

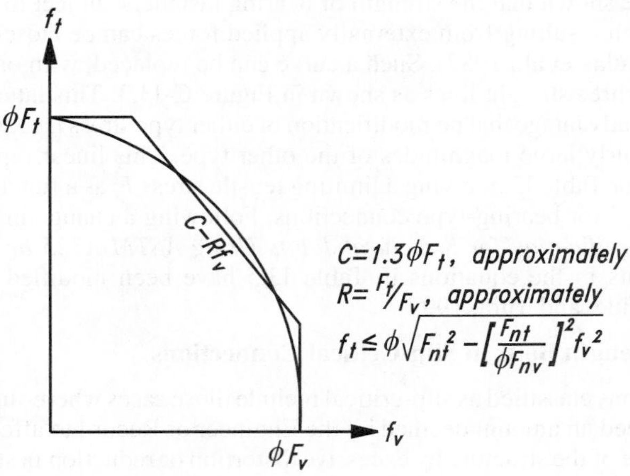

Fig. C-J3.3. Three straight line approximation.

in trusses, a small slip in the end connections can significantly reduce the strength of the compression member so the slip-critical end connection should be checked for slip at the factored-load level, whether or not a slip-critical connection is required by a serviceability requirement.

In connections with long slots that are parallel to the direction of the applied load, slip of the connection prior to attainment of the factored load might be large enough to alter the usual assumption of analysis that the undeformed structure can be used to obtain the internal forces. The Specification allows the designer two alternatives in this case. If the connection is designed so that it will not slip under the effects of service loads, then the effect of the factored loads acting on the deformed structure (deformed by the maximum amount of slip in the long slots at all locations) must be included in the structural analysis. Alternatively, the connection can be designed so that it will not slip at loads up to the factored load level.

Joints subjected to full reverse cyclical loading are clearly slip-critical joints since slip would permit back and forth movement of the joint and early fatigue. However, for joints subjected to pulsating load that does not involve reversal of direction, proper fatigue design could be provided either as a slip-critical joint on the basis of stress on the gross section, or as a non-slip-critical joint on the basis of stress on the net section. Because fatigue results from repeated application of the service load rather than the overload load, design should be based upon service-load criteria.

For high-strength bolts in combination with welds in statically loaded conditions and considering new work only, the nominal strength may be taken as the sum of the slip resistances provided by the bolts and the shear resistance of the welds. Section J1.9 requires that the slip resistance be determined at factored load levels. If one type of connector is already loaded when the second type of connector is introduced, the nominal strength cannot by obtained by adding the two resistances. The Guide (Kulak et al., 1987) should be consulted in these cases.

Slip of slip-critical connections is likely to occur at approximately 1.4 to 1.5 times the service loads. For standard holes, oversized holes, and short-slotted holes the connection can be designed either at service loads (Appendix J3.8b) or at factored loads (Section J3.8a). The nominal loads and ϕ factors have been adjusted accordingly. The number of connectors will be essentially the same for the two procedures because they have been calibrated to give similar results. Slight differences will occur because of variation in the ratio of live load to dead load.

In connections containing long slots that are parallel to the direction of the applied load, slip of the connection prior to attainment of the factored load might be large enough to alter the usual assumption of analysis that the undeformed structure can be used to obtain the internal forces. To guard against this occurring, the design slip resistance is further reduced by 0.8 when designing at service load (Appendix J3.8b) and by setting ϕ to 0.60 in conjunction with factored loads (Section J3.8a).

While the possibility of a slip-critical connection slipping into bearing under anticipated service conditions is small, such connections must comply with the provisions of Section J3.10 in order to prevent connection failure at the maximum load condition.

10. Bearing Strength at Bolt Holes

Provisions for bearing strength of pins differ from those for bearing strength of bolts; refer to Section J8.

Bearing values are provided as a measure of the strength of the material upon which a bolt bears, not as a protection to the fastener, which needs no such protection. Accordingly, the same bearing value applies to all joints assembled by bolts, regardless of fastener shear strength or the presence or absence of threads in the bearing area.

Material bearing strength may be limited either by bearing deformation of the hole or by block shear rupture of the material upon which the bolt bears. Recent testing by Kim and Yura (1996) and Lewis and Zwerneman (1996) has confirmed the bearing strength provisions for the former case wherein the nominal bearing strength R_n is equal to $CdtF_u$ and C is 2.4, 3.0, or 2.0 depending upon hole type and/or acceptability of hole ovalization at ultimate load as indicated in LRFD Specification Section J3.10. However, this same research indicated the need for more accurate bearing strength provisions when block-shear-rupture-type failure would control. Appropriate equations for bearing strength as a function of clear distance L_c are therefore provided and this formulation is consistent with that adopted by RCSC in the *Load and Resistance Factor Design Specification for Structural Joints Using ASTM A325 or A490 Bolts* (RCSC, 1994).

Frank and Yura (1981) demonstrated that hole elongation greater than ¼ in. (6 mm) will generally begin to develop as the bearing force is increased beyond $\phi(2.4dtF_u)$, especially if it is combined with high tensile stress on the net section, even though rupture does not occur; for a long-slotted hole with the slot perpendicular to the direction of force, the same is true for a bearing force greater than $\phi(2.0dtF_u)$. An upper bound of $\phi(3.0dtF_u)$ anticipates hole ovalization (deformation greater than ¼ in. (6 mm)) at maximum load.

Additionally, to simplify and generalize such bearing strength calculations, the current provisions have been based upon a clear-distance formulation. Previous provisions utilized edge distances and bolt spacings measured to hole centerlines with adjustment factors to account for varying hole type and orientation, as well as minimum edge distance requirements.

11. Long Grips

Provisions requiring a decrease in calculated stress for A307 bolts having long grips (by arbitrarily increasing the required number in proportion to the grip length) are not required for high-strength bolts. Tests (Bendigo, Hansen, and Rumpf, 1963) have demonstrated that the ultimate shearing strength of high-strength bolts having a grip of eight or nine diameters is no less than that of similar bolts with much shorter grips.

J4. DESIGN RUPTURE STRENGTH

Tests (Birkemoe and Gilmor, 1978) on coped beams indicated that a tearing failure mode (rupture) can occur along the perimeter of the bolt holes as shown in Figure C-J4.1. This block shear mode combines tensile strength on one plane and shear strength on a perpendicular plane. The failure path is defined by the center lines of the bolt holes. The block shear failure mode is not limited to the coped ends of beams. Other examples are shown in Figure C-J4.1 and C-J4.2.

The block shear failure mode should also be checked around the periphery of welded connections. Welded connection block shear is determined using $\phi = 0.75$ in conjunction with the area of both the fracture and yielding planes (Yura, 1988).

The LRFD Specification has adopted a conservative model to predict block shear strength. Test results suggest that it is reasonable to add the yield strength on one plane to the rupture strength of the perpendicular plane (Ricles and Yura, 1983, and Hardash and Bjorhovde, 1985). Therefore, two possible block shear strengths can be calculated; rupture strength F_u on the net tensile section along with shear yielding $0.6F_y$ on the gross section on the shear plane(s), or rupture $0.6F_u$ on the net shear area(s) combined with yielding F_y on the gross tensile area. This is the basis of Equations J4-3a and J4-3b.

These equations are consistent with the philosophy in Chapter D for tension members, where gross area is used for the limit state of yielding and net area is used for rupture. The controlling equation is the one that produces the *larger* rupture force.

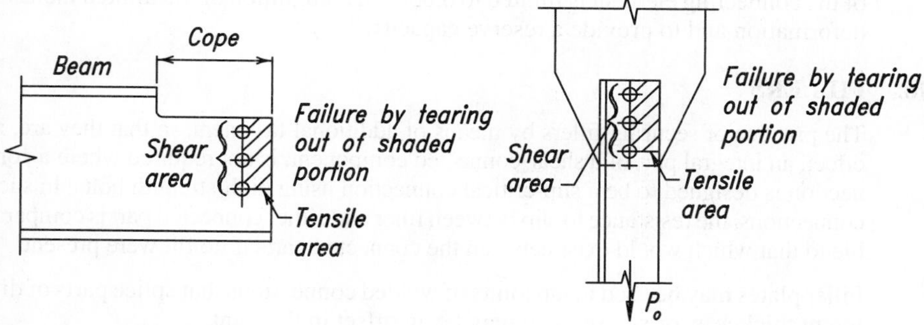

Fig. C-J4.1. Failure for block shear rupture limit state.

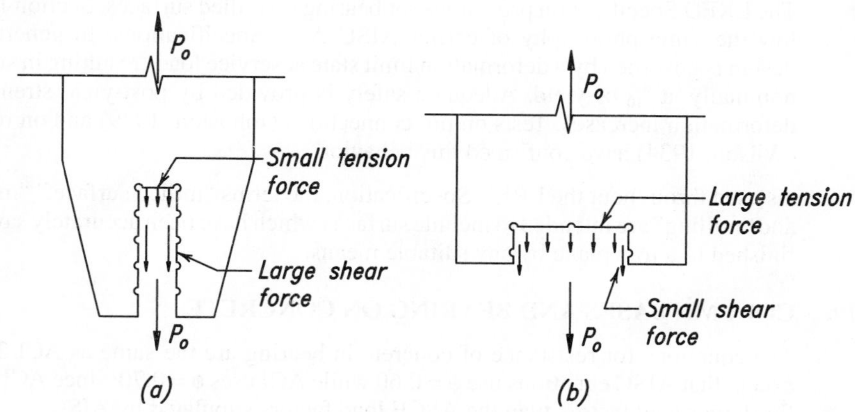

Fig. C-J4.2. Block shear rupture in tension.

This can be explained by the two extreme examples given in Figure C-J4.2. In Case (a), the total force is resisted primarily by shear, so shear rupture, not shear yielding, should control the block shear tearing mode; therefore, use Equation J4-3b. For Case (b), block shear cannot occur until the tension area ruptures as given by Equation J4-3a. If Equation J4-3b (shear rupture on the small area and yielding on the large tension area) is checked for Case (b), a smaller P_o will result. In fact, as the shear area gets smaller and approaches zero, the use of Equation J4-3b for Case (b) would give a block shear strength based totally on *yielding* of the gross tensile area. Block shear is a rupture or tearing phenomenon not a yielding limit state. Therefore, the proper equation to use is the one with the larger rupture term.

J5. CONNECTING ELEMENTS

2. Design Strength of Connecting Elements in Tension

Tests have shown that yield will occur on the gross section area before the tensile capacity of the net section is reached, if the ratio $A_n / A_g \leq 0.85$ (Kulak et al., 1987). Since the length of connecting elements is small compared to the member length, inelastic deformation of the gross section is limited. Hence, the effective net area A_n of the connecting element is limited to $0.85A_g$ in recognition of the limited inelastic deformation and to provide a reserve capacity.

J6. FILLERS

The practice of securing fillers by means of additional fasteners, so that they are, in effect, an integral part of a shear-connected component, is not required where a connection is designed to be a slip-critical connection using high-strength bolts. In such connections, the resistance to slip between filler and either connected part is comparable to that which would exist between the connected parts if no fill were present.

Filler plates may be used in lap joints of welded connections that splice parts of different thickness, or where there may be an offset in the joint.

J8. BEARING STRENGTH

The LRFD Specification provisions for bearing on milled surfaces, Section J8, follow the same philosophy of earlier AISC ASD Specifications. In general, the design is governed by a deformation limit state at service loads resulting in stresses nominally at $\frac{9}{10}$ of yield. Adequate safety is provided by post-yield strength as deformation increases. Tests on pin connections (Johnston, 1939) and on rockers (Wilson, 1934) have confirmed this behavior.

As used throughout the LRFD Specification, the terms "milled surface," "milled," and "milling" are intended to include surfaces which have been accurately sawed or finished to a true plane by any suitable means.

J9. COLUMN BASES AND BEARING ON CONCRETE

The equations for resistance of concrete in bearing are the same as ACI 318-99 except that AISC equations use $\phi = 0.60$ while ACI uses $\phi = 0.70$, since ACI specifies larger load factors than the ASCE load factors stipulated by AISC.

See DeWolf and Ricker (1990) for guidelines on the design of column base plates.

CHAPTER K

CONCENTRATED FORCES, PONDING, AND FATIGUE

K1. FLANGES AND WEBS WITH CONCENTRATED FORCES

1. Design Basis

The LRFD Specification separates flange and web strength requirements into distinct categories representing different limit state criteria, i.e., flange local bending (Section K1.2), web local yielding (Section K1.3), web crippling (Section K1.4), web sidesway buckling (Section K1.5), web compression buckling (Section K1.6), and web panel-zone shear (Section K1.7).

These criteria are applied to two distinct types of concentrated forces which act on member flanges. *Single concentrated forces* may be tensile, such as those delivered by tension hangers, or compressive, such as those delivered by bearing plates at beam interior positions, reactions at beam ends, and other *bearing* connections. *Double concentrated forces,* one tensile and one compressive, form a couple on the same side of the loaded member, such as that delivered to column flanges through welded and bolted *moment* connections. See Carter (1999) for guidelines on column stiffener design.

2. Flange Local Bending

Where a tensile force is applied through a plate welded across a flange, that flange must be sufficiently rigid to prevent deformation of the flange and the corresponding high-stress concentration in the weld in line with the web.

The effective column flange length for local flange bending is $12t_f$ (Graham, et al., 1959). Thus, it is assumed that yield lines form in the flange at $6t_f$ in each direction from the point of the applied concentrated force. To develop the fixed edge consistent with the assumptions of this model, an additional $4t_f$ and therefore a total of $10t_f$, is required for the full flange-bending strength given by Equation K1-1. In the absence of applicable research, a 50 percent reduction has been introduced for cases wherein the applied concentrated force is less than $10t_f$ from the member end.

This criterion given by Equation K1-1 was originally developed for *moment* connections, but it also applies to *single concentrated forces* such as tension hangers consisting of a plate welded to the bottom flange of a beam and transverse to the beam web.

3. Web Local Yielding

The web strength criteria have been established to limit the stress in the web of a member into which a force is being transmitted. It should matter little whether the member receiving the force is a beam or a column; however, Galambos (1976) and AISC (1978), references upon which the LRFD Specification is based, did make such a distinction. For beams, a 2:1 stress gradient through the flange was used, whereas the gradient through column flanges was 2½:1. In Section K1.3, the 2½ :1 gradient is used for both cases.

This criterion applies to both *bearing* and *moment* connections.

4. Web Crippling

The expression for resistance to web crippling at a concentrated force is a departure from earlier specifications (IABSE, 1968; Bergfelt, 1971; Hoglund, 1971; and Elgaaly, 1983). Equations K1-4 and K1-5 are based on research by Roberts (1981). The increase in Equation K1-5b for $N/d > 0.2$ was developed after additional testing (Elgaaly and Salkar, 1991) to better represent the effect of longer bearing lengths at ends of members. All tests were conducted on bare steel beams without the expected beneficial contributions of any connection or floor attachments. Thus, the resulting criteria are considered conservative for such applications.

These equations were developed for *bearing* connections, but are also generally applicable to *moment* connections. However, for the rolled shapes listed in Part 1 of the LRFD Manual with F_y not greater than 50 ksi (345 MPa), the web crippling criterion will never control the design in a *moment* connection except for a W12×50 (W310×74) or W10×33 (W250×49.1) column.

The web crippling phenomenon has been observed to occur in the web adjacent to the loaded flange. For this reason, a half-depth stiffener (or stiffeners) or a half-depth doubler plate is expected to eliminate this limit state.

5. Web Sidesway Buckling

The web sidesway buckling criterion was developed after observing several unexpected failures in tested beams (Summers and Yura, 1982). In those tests the compression flanges were braced at the concentrated load, the web was squeezed into compression, and the tension flange buckled (see Figure C-K1.1).

Web sidesway buckling will not occur in the following cases. For flanges restrained against rotation:

$$\frac{h/t_w}{l/b_f} > 2.3 \tag{C-K1-1}$$

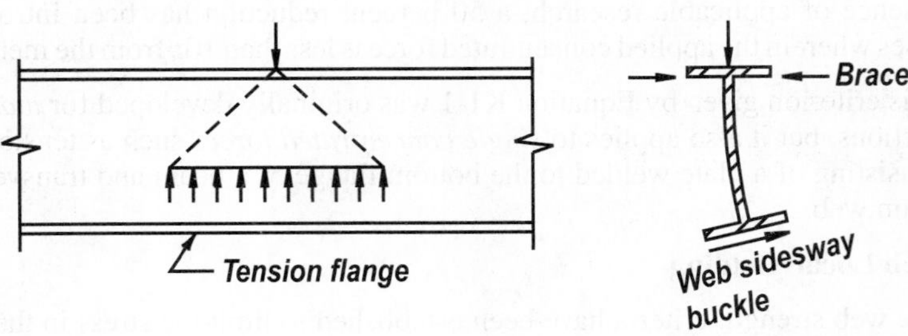

Fig. C-K1.1. Web sidesway buckling.

For flanges *not* restrained against rotation:

$$\frac{h / t_w}{l / b_f} > 1.7 \qquad \text{(C-K1-2)}$$

where l is as shown in Figure C-K1.2.

Web sidesway buckling can also be prevented by the proper design of lateral bracing or stiffeners at the load point. It is suggested that local bracing at both flanges be designed for one percent of the concentrated force applied at that point. Stiffeners must extend from the load point through at least one-half the beam or girder depth. In addition, the pair of stiffeners should be designed to carry the full load. If flange rotation is permitted at the loaded flange, neither stiffeners nor doubler plates will be effective.

In the 1st Edition LRFD Manual, the web sidesway buckling equations were based on the assumption that $h / t_f = 40$, a convenient assumption which is generally true for economy beams. This assumption has been removed so that the equations will be applicable to all sections.

This criterion was developed only for *bearing* connections and does not apply to *moment* connections.

6. Web Compression Buckling

When compressive forces are applied to both flanges of a member at the same loca-

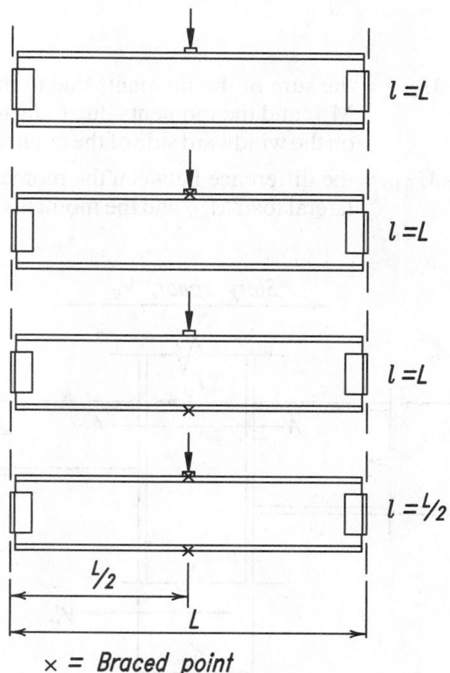

Fig. C-K1.2. Unbraced flange length.

tion, as by *moment* connections at both flanges of a column, the member web must have its slenderness ratio limited to avoid the possibility of buckling. This is done in the LRFD Specification with Equation K1-8, which is a modified form of a similar equation used in the ASD Specification. This equation is applicable to a pair of *moment* connections, and to other pairs of compressive forces applied at both flanges of a member, for which N/d is small (<1). When N/d is not small, the member web should be designed as a compression member in accordance with Chapter E.

Equation K1-8 is predicated on an interior member loading condition. In the absence of applicable research, a 50 percent reduction has been introduced for cases wherein the compressive forces are close to the member end.

Equation K1-8 has also traditionally been applied when there is a *moment* connection to only one flange of the column and compressive force is applied to only one flange. Its use in this case is conservative.

7. Web Panel-Zone Shear

The column web shear stresses may be high within the boundaries of the rigid connection of two or more members whose webs lie in a common plane. Such webs should be reinforced when the calculated factored force ΣF_u along plane A-A in Figure C-K1.3 exceeds the column web design strength ϕR_v, where

$$\Sigma F_u = \frac{M_{u1}}{d_{m1}} + \frac{M_{u2}}{d_{m2}} - V_u \qquad \text{(C-K1-3)}$$

and

$M_{u1} = M_{u1L} + M_{u1G}$ = the sum of the moments due to the factored lateral load M_{u1L} and the moments due to factored gravity load M_{u1G} on the windward side of the connection, kip-in. (N-mm)

$M_{u2} = M_{u2L} + M_{u2G}$ = the difference between the moments due to the factored lateral load M_{u2L} and the moments due to factored gravity

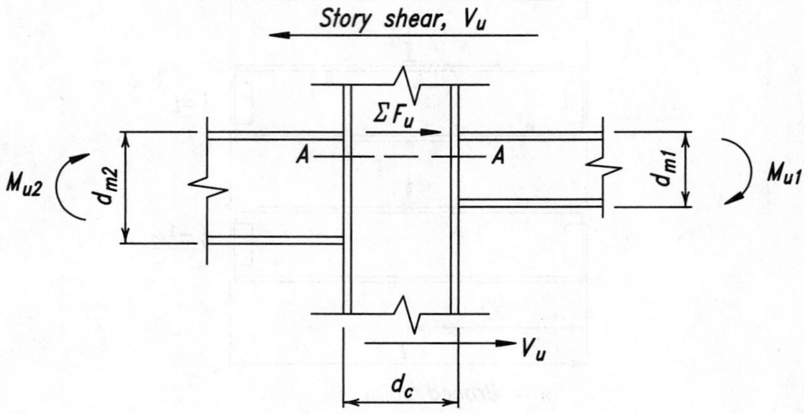

Fig. C-K1.3. Forces in panel zone.

load M_{u2G} on the windward side of the connection, kip-in. (N-mm)

d_{m1}, d_{m2} = distance between flange forces in a moment connection, in. (mm)

Conservatively, 0.95 times the beam depth has been used for d_m in the past. If $\Sigma F_u \leq \phi R_v$, no reinforcement is necessary, i.e., $t_{req} \leq t_w$, where t_w is the column web thickness.

Consistent with elastic first order analysis, Equations K1-9 and K1-10 limit panel-zone behavior to the elastic range. While such connection panels possess large reserve capacity beyond initial general shear yielding, the corresponding inelastic joint deformations may adversely affect the strength and stability of the frame or story (Fielding and Huang, 1971, and Fielding and Chen, 1973). Panel-zone shear yielding affects the overall frame stiffness and, therefore, the ultimate-strength second-order effects may be significant. The shear/axial interaction expression of Equation K1-10, as shown in Figure C-K1.4, is chosen to ensure elastic panel behavior.

If adequate connection ductility is provided and the frame analysis considers the inelastic panel-zone deformations, then the additional inelastic shear strength is recognized in Equations K1-11 and K1-12 by the factor

$$\left(1+\frac{3b_{cf}t_{cf}^2}{d_b d_c t_w}\right)$$

This inelastic shear strength has been most often utilized for design of frames in high seismic zones and should be used when the panel zone is to be designed to match the strength of the members from which it is formed.

The shear/axial interaction expression incorporated in Equation K1-12 (see Figure C-K1.5) is similar to that contained in the previous editions of this Specification

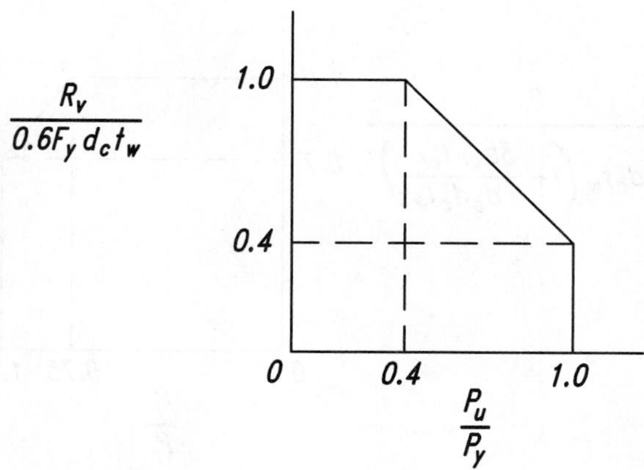

Fig. C-K1.4. Interaction of shear and axial force—elastic.

and recognizes the observed fact that when the panel-zone web has completely yielded in shear, the axial column load is carried in the flanges.

K2. PONDING

As used in the LRFD Specification, *ponding* refers to the retention of water due solely to the deflection of flat roof framing. The amount of this water is dependent upon the flexibility of the framing. Lacking sufficient framing stiffness, its accumulated weight can result in collapse of the roof if a strength evaluation is not made.

Representing the deflected shape of the primary and critical secondary member as a half-sine wave, the weight and distribution of the ponded water can be estimated and, from this, the contribution that the deflection each of these members makes to the total ponding deflection can be expressed (Marino, 1966):

For the primary member:

$$\Delta_w = \frac{\alpha_p \Delta_o [1 + 0.25\pi\alpha_s + 0.25\pi\rho(1+\alpha_s)]}{1 - 0.25\pi\alpha_p\alpha_s}$$

For the secondary member:

$$\delta_w = \frac{\alpha_s \delta_o \left[1 + \frac{\pi^2}{32}\alpha_p + \frac{\pi^2}{8\rho}(1+\alpha_p) + 0.185\alpha_s\alpha_p\right]}{1 - 0.25\pi\alpha_p\alpha_s}$$

In these expressions Δ_o and δ_o are, respectively, the primary and secondary beam deflections due to loading present at the initiation of ponding,

$$\alpha_p = C_p / (1 - C_p),\ \alpha_s = C_s / (1 - C_s),\ \text{and}\ \rho = \delta_o / \Delta_o = C_s / C_p$$

Using the above expressions for Δ_w and δ_w, the ratios Δ_w / Δ_o and δ_w / δ_o can be computed for any given combination of primary and secondary beam framing using, respectively, the computed value of parameters C_p and C_s defined in the LRFD Specification.

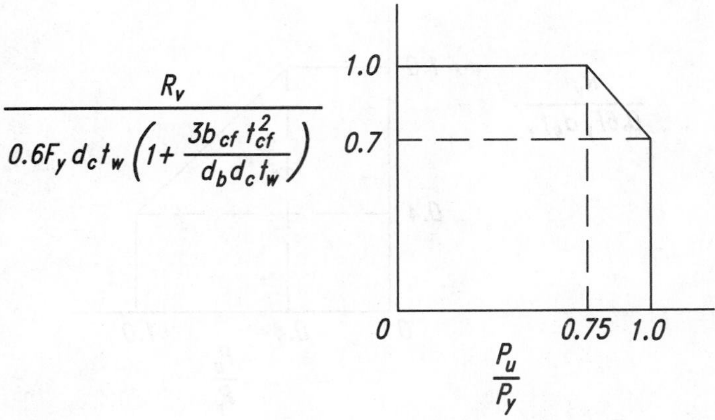

Fig. C-K1.5. Interaction of shear and axial force—inelastic.

Even on the basis of unlimited elastic behavior, it is seen that the ponding deflections would become infinitely large unless

$$\left(\frac{C_p}{1-C_p}\right)\left(\frac{C_s}{1-C_s}\right) < \frac{4}{\pi}$$

Since elastic behavior is not unlimited, the effective bending strength available in each member to resist the stress caused by ponding action is restricted to the difference between the yield stress of the member and the stress f_o produced by the total load supported by it before consideration of ponding is included.

Note that elastic deflection is directly proportional to stress. The admissible amount of ponding in either the primary or critical (midspan) secondary member, in terms of the applicable ratio Δ_w / Δ_o and δ_w / δ_o, can be represented as $(F_y - f_o)/f_o$. Substituting this expression for Δ_w / Δ_o and δ_w / δ_o, and combining with the foregoing expressions for Δ_w and δ_w, the relationship between critical values for C_p and C_s and the available elastic bending strength to resist ponding is obtained. The curves presented in Figures A-K2.1 and A-K2.2 are based upon this relationship. They constitute a design aid for use when a more exact determination of required flat roof framing stiffness is needed than given by the LRFD Specification provision that $C_p + 0.9C_s \leq 0.25$.

Given any combination of primary and secondary framing, the stress index is computed as

$$U_p = \left(\frac{F_y - f_o}{f_o}\right)_p \quad \text{for the primary member}$$

$$U_s = \left(\frac{F_y - f_o}{f_o}\right)_s \quad \text{for the secondary member}$$

where f_o, in each case, is the computed bending stress, ksi (MPa), in the member due to the supported loading, neglecting ponding effect. Depending upon geographic location, this loading should include such amount of snow as might also be present, although ponding failures have occurred more frequently during torrential summer rains when the rate of precipitation exceeded the rate of drainage runoff and the resulting hydraulic gradient over large roof areas caused substantial accumulation of water some distance from the eaves.

Given the size, spacing, and span of a tentatively selected combination of primary and secondary beams, for example, one may enter Figure A-K2.1 at the level of the computed stress index U_p, determined for the primary beam; move horizontally to the computed C_s value of the secondary beams; then move downward to the abscissa scale. The combined stiffness of the primary and secondary framing is sufficient to prevent ponding if the flexibility constant read from this latter scale is more than the value of C_p computed for the given primary member; if not, a stiffer primary or secondary beam, or combination of both, is required.

If the roof framing consists of a series of equally-spaced wall-bearing beams, they would be considered as secondary members, supported on an infinitely stiff primary member. For this case, one would use Figure A-K2.2. The limiting value of C_s

would be determined by the intercept of a horizontal line representing the U_s value and the curve for $C_p = 0$.

The ponding deflection contributed by a metal deck is usually such a small part of the total ponding deflection of a roof panel that it is sufficient merely to limit its moment of inertia [in.^{4} per foot (mm^4 per meter) of width normal to its span] to 0.000025 (3 940) times the fourth power of its span length, as provided in the LRFD Specification. However, the stability against ponding of a roof consisting of a metal roof deck of relatively slender depth-span ratio, spanning between beams supported directly on columns, may need to be checked. This can be done using Figure A-K2.1 or A-K2.2 with the following computed values:

U_p = stress index for the supporting beam
U_s = stress index for the roof deck
C_p = flexibility constant for the supporting beams
C_s = flexibility constant for one foot width of the roof deck ($S = 1.0$)

Since the shear rigidity of the web system is less than that of a solid plate, the moment of inertia of steel joists and trusses should be taken as somewhat less than that of their chords.

CHAPTER L

SERVICEABILITY DESIGN CONSIDERATIONS

Serviceability criteria are formulated to prevent disruptions of the functional use and damage to the structure during its normal everyday use. While malfunctions may not result in the collapse of a structure or in loss of life or injury, they can seriously impair the usefulness of the structure and lead to costly repairs. Neglect of serviceability may result in unacceptably flexible structures.

There are essentially three types of structural behavior which may impair serviceability:

(1) Excessive local damage (local yielding, buckling, slip, or cracking) that may require excessive maintenance or lead to corrosion.

(2) Excessive deflection or rotation that may affect the appearance, function, or drainage of the structure, or may cause damage to nonstructural components and their attachments.

(3) Excessive vibrations induced by wind or transient live loads which affect the comfort of occupants of the structure or the operation of mechanical equipment.

In allowable stress design, the AISC Specification accounts for possible local damage with factors of safety included in the allowable stresses, while deflection and vibration are controlled, directly or indirectly, by limiting deflections and span-depth ratios. In the past, these rules have led to satisfactory performance of structures, with perhaps the exception of large open floor areas without partitions. In LRFD the serviceability checks should consider the appropriate loads, the response of the structure, and the reaction of the occupants to the structural response.

Examples of loads that may require consideration of serviceability include permanent live loads, wind, and earthquake; effects of human activities such as walking, dancing, etc.; temperature fluctuations; and vibrations induced by traffic near the building or by the operation of mechanical equipment within the building.

Serviceability checks are concerned with adequate performance under the appropriate load conditions. Elastic behavior can usually be assumed. However, some structural elements may have to be examined with respect to their long-term behavior under load.

It is difficult to specify limiting values of structural performance based on serviceability considerations because these depend to a great extent on the type of structure, its intended use, and subjective physiological reaction. For example, acceptable structural motion in a hospital clearly would be much less than in an ordinary industrial building. It should be noted that humans perceive levels of structural motion that are far less than motions that would cause any structural damage. Serviceability limits must be determined through careful consideration by the designer and client.

Serviceability guidelines for low-rise buildings are given in Fisher and West (1990).

L1. CAMBER

The engineer should consider specifying camber when deflections at the appropriate load level present a serviceability problem.

L2. EXPANSION AND CONTRACTION

As in the case of deflections, the satisfactory control of expansion cannot be reduced to a few simple rules, but must depend largely upon the good judgment of qualified engineers.

The problem is more serious in buildings with masonry walls than with prefabricated units. Complete separation of the framing, at widely spaced expansion joints, is generally more satisfactory than more frequently located devices dependent upon the sliding of parts in bearing, and usually less expensive than rocker or roller expansion bearings.

Creep and shrinkage of concrete and yielding of steel are among the causes, other than temperature, for dimensional changes.

L3. DEFLECTIONS, VIBRATION, AND DRIFT

1. Deflections

Excessive transverse deflections or lateral drift may lead to permanent damage to building elements, separation of cladding, or loss of weathertightness, damaging transfer of load to non-load-supporting elements, disruption of operation of building service systems, objectionable changes in appearance of portions of the buildings, and discomfort of occupants.

The LRFD Specification does not provide specific limiting deflections for individual members or structural assemblies. Such limits would depend on the function of the structure (ASCE, 1979; CSA, 1989; and Ad Hoc Committee on Servicability Research, 1986). Provisions that limit deflections to a percentage of span may not be adequate for certain long-span floor systems; a limit on maximum deflection that is independent of span length may also be necessary to minimize the possibility of damage to adjoining or connecting nonstructural elements.

Deflection calculations for composite beams should include an allowance for slip for short-term deflection calculations, and for creep and shrinkage for long-term deflection calculations (see Commentary Section I3.2).

2. Floor Vibration

The increasing use of high-strength materials and efficient structural schemes leads to longer spans and more flexible floor systems. Even though the use of a deflection limit related to span length generally precluded vibration problems in the past, some floor systems may require explicit consideration of the dynamic, as well as the static, characteristics of the floor system.

The dynamic response of structures or structural assemblies may be difficult to analyze because of difficulties in defining the actual mass, stiffness, and damping characteristics. Moreover, different load sources cause varying responses. For example, a steel beam-concrete slab floor system may respond to live loading as a non-composite system, but to transient excitation from human activity as an orthotropic composite plate. Nonstructural partitions, cladding, and built-in furniture significantly increase the stiffness and damping of the structure and frequently eliminate potential vibration problems. The damping can also depend on the amplitude of excitation.

The general objective in minimizing problems associated with excessive structural motion is to limit accelerations, velocities, and displacements to levels that would not be disturbing to the building occupants. Generally, occupants of a building find sustained vibrations more objectionable than transient vibrations.

The levels of peak acceleration that people find annoying depend on frequency of response. Thresholds of annoyance for transient vibrations are somewhat higher and depend on the amount of damping in the floor system. These levels depend on the individual and the activity at the time of excitation (ASCE, 1979; ISO, 1974; CSA, 1989; Murray, Allen, and Ungar, 1997; and Ad Hoc Committee on Serviceability Research, 1986).

The most effective way to reduce effects of continuous vibrations is through vibration isolation devices. Care should be taken to avoid resonance, where the frequency of steady-state excitation is close to the fundamental frequency of the system. Transient vibrations are reduced most effectively by increasing the damping in the structural assembly. Mechanical equipment which can produce objectionable vibrations in any portion of a structure should be adequately isolated to reduce the transmission of such vibrations to critical elements of the structure.

3. Drift

The LRFD Specification does not provide specific limiting values for lateral drift. If a drift analysis is desired, the stiffening effect of non-load-supporting elements such as partitions and infilled walls may be included in the analysis of drift.

Some irrecoverable inelastic deformations may occur at given load levels in certain types of construction. The effect of such deformations may be negligible or serious, depending on the function of the structure, and should be considered by the designer on a case by case basis.

The deformation limits should apply to structural assemblies as a whole. Reasonable tolerance should also be provided for creep. Where load cycling occurs, consideration should be given to the possibility of increases in residual deformation that may lead to incremental failure.

L5. CORROSION

Steel members may deteriorate in particular service environments. This deterioration may appear either in external corrosion, which would be visible upon inspection, or in undetected changes that would reduce its strength. The designer should recognize these problems by either factoring a specific amount of damage tolerance into the design or providing adequate protection systems (e.g., coatings, cathodic protection) and/or planned maintenance programs so that such problems do not occur.

CHAPTER M

FABRICATION, ERECTION, AND QUALITY CONTROL

M2. FABRICATION

1. Cambering, Curving, and Straightening

The use of heat for straightening or cambering members is permitted for A514/A514M and A852/A852M steel, as it is for other steels. However, the maximum temperature permitted is 1,100°F (593°C) compared to 1,200°F (649°C) for other steels.

Cambering of flexural members, when required by the contract documents, may be accomplished in various ways. In the case of trusses and girders, the desired curvature can be built in during assembly of the component parts. Within limits, rolled beams can be cold-cambered at the producing mills.

Local application of heat has come into common use as a means of straightening or cambering beams and girders. The method depends upon an ultimate shortening of the heat-affected zones. A number of such zones, on the side of the member that would be subject to compression during cold-cambering or "gagging," are heated enough to be "upset" by the restraint provided by surrounding unheated areas. Shortening takes place upon cooling.

While the final curvature or camber can be controlled by these methods, it must be realized that some deviation, due to workmanship error and permanent change due to handling, is inevitable.

2. Thermal Cutting

Preferably thermal cutting shall be done by machine. The requirement for a positive preheat of 150°F (66°C) minimum when thermal cutting beam copes and weld access holes in ASTM A6/A6M Group 4 and 5 shapes, and in built-up shapes made of material more than two in. (50 mm) thick, tends to minimize the hard surface layer and the initiation of cracks.

5. Bolted Construction

In the past, it has been required to tighten all ASTM A325 or A325M and A490 or A490M bolts in both slip-critical and bearing-type connections to a specified tension. The requirement was changed in 1985 to permit most bearing-type connections to be tightened to a snug-tight condition.

In a snug-tight bearing connection, the bolts cannot be subjected to tension loads, slip can be permitted, and loosening or fatigue due to vibration or load fluctuations are not design considerations.

It is suggested that snug-tight bearing-type connections be used in applications where A307 bolts would be permitted.

This section provides rules for the use of oversized and slotted holes paralleling the provisions which have been in the RCSC Specification (RCSC, 1994) since 1972, extended to include A307 bolts which are outside the scope of the high-strength bolt specifications.

M3. SHOP PAINTING

The surface condition of steel framing disclosed by the demolition of long-standing buildings has been found to be unchanged from the time of its erection, except at isolated spots where leakage may have occurred. Even in the presence of leakage, the shop coat is of minor influence (Bigos, Smith, Ball, and Foehl, 1954).

The LRFD Specification does not define the type of paint to be used when a shop coat is required. Conditions of exposure and individual preference with regard to finish paint are factors which bear on the selection of the proper primer. Hence, a single formulation would not suffice. For a comprehensive treatment of the subject, see SSPC (1989).

5. Surfaces Adjacent to Field Welds

The Specification allows for welding through surface materials, including appropriate shop coatings, that do not adversely affect weld quality nor create objectionable fumes.

M4. ERECTION

2. Bracing

For information on the design of temporary lateral support systems and components for low-rise building, see Fisher and West (1997).

4. Fit of Column Compression Joints and Base Plates

Tests at the University of California-Berkeley (Popov and Stephen, 1977) on spliced full-size columns with joints that had been intentionally milled out-of-square, relative to either strong or weak axis, demonstrated that the load-carrying capacity was the same as that for a similar unspliced column. In the tests, gaps of $\frac{1}{16}$-in. (2 mm) were not shimmed; gaps of $\frac{1}{4}$-in. (6 mm) were shimmed with non-tapered mild steel shims. Minimum size partial-joint-penetration welds were used in all tests. No tests were performed on specimens with gaps greater than $\frac{1}{4}$-in. (6 mm).

5. Field Welding

The purpose of wire brushing shop paint on surfaces adjacent to joints to be field welded is to reduce the possibility of porosity and cracking and also to reduce any environmental hazard. Although there are limited tests which indicate that painted surfaces result in sound welds without wire brushing, other studies have resulted in excessive porosity and/or cracking when welding coated surfaces. Wire brushing to reduce the paint film thickness minimizes rejectable welds. Grinding or other procedures beyond wire brushing is not necessary.

CHAPTER N

EVALUATION OF EXISTING STRUCTURES

N1. GENERAL PROVISIONS

The load combinations referred to in this chapter reflect gravity loading because it is the most prevalent condition encountered. If other loading conditions are a consideration, such as lateral loads, the appropriate load combination from ASCE 7 (ASCE, 1998) or from the applicable building code should be used. Guidelines for seismic evaluation are available in other publications (FEMA, 1997a and 1997b). The Engineer of Record for a project is generally established by the owner.

N2. MATERIAL PROPERTIES

1. Determination of Required Tests

The extent of tests required depends on the nature of the project, the criticality of the structural system or member evaluated, and the availability of records pertinent to the project. Thus, the Engineer of Record is required to determine the specific tests required and the locations from which specimens are to be obtained.

2. Tensile Properties

Samples required for tensile tests should be removed from regions of reduced stress, such as at flange tips at beam ends and external plate edges, to minimize the effects of the reduced area. The number of tests required will depend on whether they are conducted to merely confirm the strength of a known material or to establish the strength of some other steel. Guidance on the appropriate minimum number of tests is available (FEMA, 1997a).

It should be recognized that the yield stress determined by standard ASTM methods and reported by mills and testing laboratories is somewhat greater than the *static* yield stress because of dynamic effects of testing. Also, the test specimen location may have an effect. These effects have already been accounted for in the nominal strength equations in the Specification. However, when strength evaluation is done by load testing, this effect should be accounted for in test planning because yielding will tend to occur earlier than otherwise anticipated. The static yield stress, F_{ys}, can be estimated from that determined by routine application of ASTM methods, F_y, by the following equation (Galambos, 1978 and 1998):

$$F_{ys} = R(F_y - 4) \quad \text{(C-N2-1)}$$

$$(\text{Metric: } F_{ys} = R(F_y - 27)) \quad \text{(C-N2-1M)}$$

where

F_{ys} = static yield stress, ksi (MPa)
F_y = reported yield stress, ksi (MPa)

R = 0.95 for tests taken from web specimens
R = 1.00 for tests taken from flange specimens

The R factor in Equation C-N2-1 accounts for the effect of the coupon location on the reported yield stress. Prior to 1997, certified mill test reports for structural shapes were based on specimens removed from the web, in accordance with ASTM A6/A6M. Subsequently the specified coupon location was changed to the flange. During 1997-1998, there was a transition from web specimens to flange specimens as the new provisions of ASTM A6/A6M were adopted.

4. Base Metal Notch Toughness

The Engineer of Record shall specify the location of samples. Samples shall be cored, flame cut, or saw cut. The Engineer of Record will determine if remedial actions are required, such as the possible use of bolted splice plates.

5. Weld Metal

Because connections typically have a greater reliability index than structural members, strength testing of weld metal is not usually necessary. However, field investigations have sometimes indicated that complete-joint-penetration welds, such as at beam-to-column connections, were not made in accordance with AWS D1.1 (AWS, 1998). The specified provisions in Section N2.4 provide a means for judging the quality of such a weld. Where feasible, any samples removed should be obtained from compression splices rather than tension splices, because the effects of repairs to restore the sampled area are less critical.

6. Bolts and Rivets

Because connections typically have a greater reliability index than structural members, removal and strength testing of fasteners is not usually necessary. However, strength testing of bolts is required where they can not be properly identified otherwise. Because removal and testing of rivets is difficult, assuming the lowest rivet strength grade simplifies the investigation.

N3. EVALUATION BY STRUCTURAL ANALYSIS

2. Strength Evaluation

Resistance factors reflect variations in determining strength of members and connections, such as uncertainty in theory and variations in material properties and dimensions. If an investigation of an existing structure indicates that there are variations in material properties or dimensions significantly greater than those anticipated in new construction, the Engineer of Record should consider the use of more conservative values.

N4. EVALUATION BY LOAD TESTS

1. Determination of Live Load Rating by Testing

Generally, structures that can be designed according to the provisions of this Specification need no confirmation of calculated results by test. However, special situations may arise when it is desirable to confirm by tests the results of calculations. Minimal test procedures are provided to determine the live load rating of a structure. However, in no case is the live load rating determined by test to exceed that which can be calculated using the provisions of the Specification. This is not intended to preclude testing to evaluate special conditions or configurations that are not adequately covered by this Specification.

It is essential that the Engineer of Record take all necessary precautions to ensure that the structure does not fail catastrophically during testing. A careful assessment of structural conditions before testing is a fundamental requirement. This includes accurate measurement and characterization of the size and strength of members, connections, and details. All safety regulations of OSHA and other pertinent bodies must be strictly adhered to. Shoring and scaffolding should be used as required in the proximity of the test area to mitigate against unexpected circumstances. Deformations must be carefully monitored and structural conditions must be continually evaluated. In some cases it may be desirable to monitor strains as well.

The Engineer of Record must use judgement to determine when deflections are becoming excessive and terminate the tests at a safe level even if the desired loading has not been achieved. Incremental loading is specified so that deformations can be accurately monitored and the performance of the structure carefully observed. Load increments should be small enough initially so that the onset of significant yielding can be determined. The increment can be reduced as the level of inelastic behavior increases, and the behavior at this level carefully evaluated to determine when to safely terminate the test. Periodic unloading after the onset of inelastic behavior will help the Engineer of Record determine when to terminate the test to avoid excessive permanent deformation or catastrophic failure.

It must be recognized that the margin of safety at the maximum load level used in the test may be very small, depending on such factors as the original design, the purpose of the tests, and the condition of the structure. Thus, it is imperative that all appropriate safety measures be adopted. It is recommended that the maximum live load used for load tests be selected conservatively. It should be noted that experience in testing more than one bay of a structure is limited.

Criteria limiting increases in deformations for a period of one hour have been given to ensure that the structure is stable at the loads evaluated.

A detailed discussion of reliability-based condition assessment of existing structures has been provided by Ellingwood (1996).

2. Serviceability Evaluation

In certain cases serviceability criteria must be determined by load testing. It should be recognized that complete recovery (i.e., return to initial deflected shape) after removal of maximum load is unlikely because of phenomena such as local yielding, slip at the slab interface in composite construction, creep in concrete slabs, localized crushing or deformation at shear connections in slabs, slip in bolted connections, and effects of continuity. Because most structures exhibit some slack when load is first applied, it is appropriate to project the load-deformation curve back to zero load to determine the slack and exclude it from the recorded deformations. Where desirable, the applied load sequence can be repeated to demonstrate that the structure is essentially elastic under service loads and that the permanent set is not detrimental.

N5. EVALUATION REPORT

Extensive evaluation and load testing of existing structures is often performed when appropriate documentation no longer exists or when there is considerable disagreement about the condition of a structure. The resulting evaluation is only effective if well documented, particularly when load testing is involved. Furthermore, as time passes, various interpretations of the results can arise unless all parameters of the structural performance, including material properties, strength, and stiffness, are well documented.

APPENDIX B

DESIGN REQUIREMENTS

B5. LOCAL BUCKLING

1. Classification of Steel Sections

The limiting width-thickness λ_p and λ_r ratios for webs in pure flexure ($P_u / \phi_b P_y = 0$) and with axial compression have been revised in terms of (h / t) rather than (h_c / t). The simplified formulation in Table B5.1 for λ_r based on double symmetry with equal flanges ($h / h_c = 1$) is unconservative when the compression flange is smaller than the tension flange, and conservative if the reverse is true. The more accurate limit is given in Appendix B5.1 as a function of h_c. Figure C-A-B5.1 illustrates the λ_r variation for axial compression and flange asymmetry effects.

The $\frac{3}{4}$ minimum and $\frac{3}{2}$ maximum restrictions on h / h_c in Equations A-B5-1 and A-B5-2 approximately correspond to the 0.1 and 0.9 range of I_{yc} / I_y for a member to be considered a singly symmetric I shape. Otherwise, when the flange areas differ by more than a factor of two, the member should be conservatively designed as a tee section.

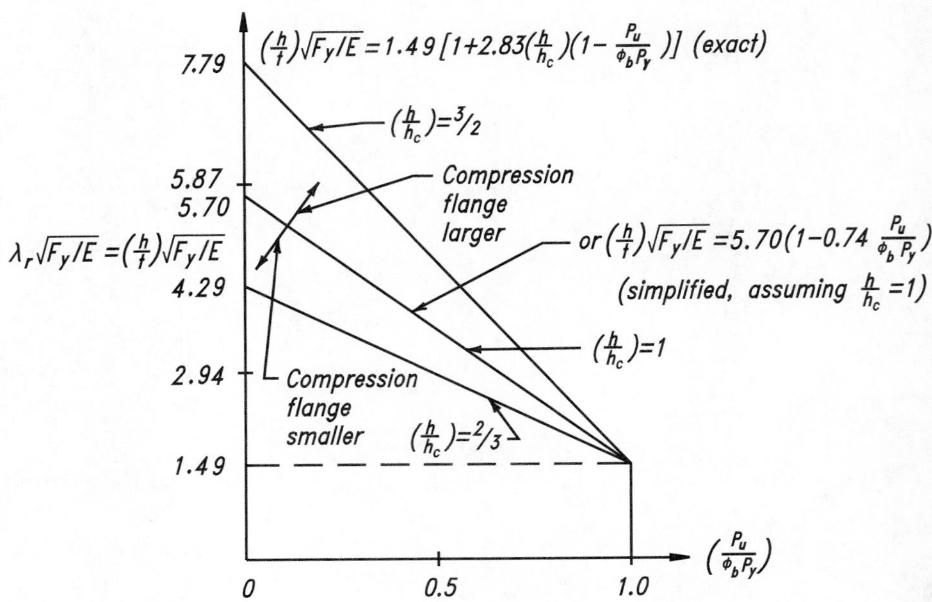

Fig. C-A-B5.1. Web local buckling for I-shaped members.

APPENDIX E

COLUMNS AND OTHER COMPRESSION MEMBERS

E3. DESIGN COMPRESSIVE STRENGTH FOR FLEXURAL-TORSIONAL BUCKLING

The equations in Appendix E3 for determining the flexural-torsional elastic buckling loads of columns are derived in texts on structural stability (Timoshenko and Gere, 1961; Bleich, 1952; Galambos, 1968; and Chen and Atsuta, 1977, for example). Since these equations for flexural-torsional buckling apply only to elastic buckling, they must be modified for inelastic buckling when $F_{cr} > 0.5F_y$. This is accomplished through the use of the equivalent slenderness factor $\lambda_e = \sqrt{F_y / F_e}$.

APPENDIX F

BEAMS AND OTHER FLEXURAL MEMBERS

F1. DESIGN FOR FLEXURE

Three limit states must be investigated to determine the moment capacity of flexural members: lateral-torsional buckling (LTB), local buckling of the compression flange (FLB), and local buckling of the web (WLB). These limit states depend, respectively, on the beam slenderness ratio L_b / r_y, the width-thickness ratio b / t of the compression flange and the width-thickness ratio h / t_w of the web. For convenience, all three measures of slenderness are denoted by λ.

Variations in M_n with L_b are shown in Figure C-F1.1. The discussion of plastic, inelastic, and elastic buckling in Commentary Section F1 with reference to lateral-torsional buckling applies here except for an important difference in the significance of λ_p for lateral-torsional buckling and local buckling. Values of λ_p for FLB and WLB produce a compact section with a rotation capacity of about three (after reaching M_p) before the onset of local buckling, and therefore meet the requirements for plastic analysis of load effects (Commentary Section B5). On the other hand, values of λ_p for LTB do not allow plastic analysis because they do not provide rotation capacity beyond that needed to develop M_p. Instead $L_b \leq L_{pd}$ (Section F1.3) must be satisfied.

Analyses to include restraint effects of adjoining elements are discussed in Galambos (1998). Analysis of the lateral stability of members with shapes not covered in this appendix must be performed according to the available literature (Galambos, 1998).

See the Commentary for Section B5 for the discussion of the equation regarding the bending capacity of circular sections.

F3. WEB-TAPERED MEMBERS

1. General Requirements

The provision contained in Appendix F3 covers only those aspects of the design of tapered members that are unique to tapered members. For other criteria of design not specifically covered in Appendix F3, see the appropriate portions of this Specification and Commentary.

The design of wide-flange columns with a single web taper and constant flanges follows the same procedure as for uniform columns according to Section E2, except the column slenderness parameter λ_c for major axis buckling is determined for a slenderness ratio $K_\gamma L / r_{ox}$, and for minor axis buckling for KL / r_{oy}, where K_γ is an effective length factor for tapered members, K is the effective length factor for prismatic members, and r_{ox} and r_{oy} are the radii of gyration about the x and the y axes, respectively, taken at the smaller end of the tapered member.

For stepped columns or columns with other than a single web taper, the elastic critical stress is determined by analysis or from data in reference texts or research reports (Chapters 11 and 13 in Timoshenko and Gere (1961), Bleich (1952), and Kitipornchai and Trahair (1980)), and then the same procedure of using λ_{eff} is utilized in calculating the factored resistance.

This same approach is recommended for open section built-up columns (columns with perforated cover plates, lacing, and battens) where the elastic critical buckling stress determination must include a reduction for the effect of shear. Methods for calculating the elastic buckling strength of such columns are given in Chapter 12 of the SSRC Guide (Galambos, 1998) and in Timoshenko and Gere (1961) and Bleich (1952).

3. Design Compressive Strength

The approach in formulating $F_{a\gamma}$ of tapered columns is based on the concept that the critical stress for an axially loaded tapered column is equal to that of a prismatic column of different length, but of the same cross section as the smaller end of the tapered column. This has resulted in an equivalent effective length factor K_γ for a tapered member subjected to axial compression (Lee, Morrell, and Ketter, 1972). This factor, which is used to determine the value of S in Equations A-F3-2 and λ_c in Equation E2-3, can be determined accurately for a symmetrical rectangular rigid frame comprised of prismatic beams and tapered columns.

With modifying assumptions, such a frame can be used as a mathematical model to determine with sufficient accuracy the influence of the stiffness $\Sigma(I/b)_g$ of beams and rafters which afford restraint at the ends of a tapered column in other cases such as those shown in Figure C-A-F3.3. From Equations A-F3-2 and E2-3, the critical load P_{cr} can be expressed as $\pi^2 EI_o/(K_\gamma l)^2$. The value of K_γ can be obtained by interpolation, using the appropriate chart from Lee et al. (1972) and restraint modifiers G_T and G_B. In each of these modifiers the tapered column, treated as a prismatic member having a moment of inertia I_o, computed at the smaller end, and its actual length

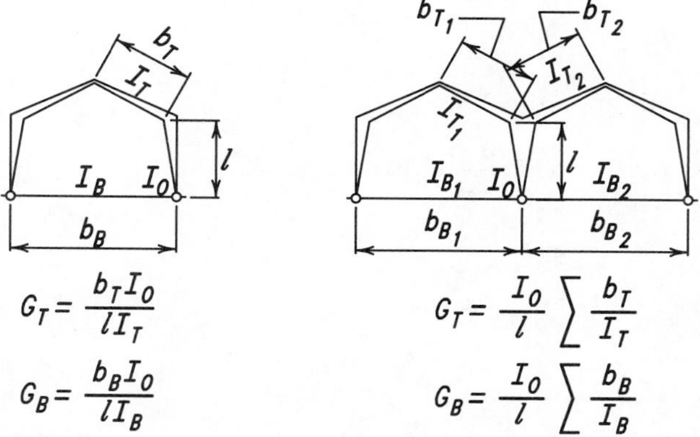

Fig. C-A-F3.3. Restraint modifiers for tapered columns.

l, is assigned the stiffness I_o/l, which is then divided by the stiffness of the restraining members at the end of the tapered column under consideration.

4. Design Flexural Strength

The development of the design bending stress for tapered beams follows closely with that for prismatic beams. The basic concept is to replace a tapered beam by an equivalent prismatic beam with a different length, but with a cross section identical to that of the smaller end of the tapered beam (Lee et al., 1972). This has led to the modified length factors h_s and h_w in Equations A-F3-6 and A-F3-7.

Equations A-F3-6 and A-F3-7 are based on total resistance to lateral buckling, using both St. Venant and warping resistance. The factor B modifies the basic $F_{b\gamma}$ to members which are continuous past lateral supports. Categories a, b, and c of Appendix F3.4 usually apply; however, it is to be noted that they apply only when the axial force is small and adjacent unbraced segments are approximately equal in length. For a single member, or segments which do not fall into category a, b, c, or d, the recommended value of B is unity. The value of B should also be taken as unity when computing the value of $F_{b\gamma}$ to obtain M_n to be used in Equations H1-1 and C1-1, since the effect of moment gradient is provided for by the factor C_m. The background material is given in WRC Bulletin No. 192 (Morrell and Lee, 1974).

APPENDIX G

PLATE GIRDERS

Appendix G is taken from AISI Bulletin 27 (Galambos, 1978). Comparable provisions are included in the AISC ASD Specification. The provisions are presented in an appendix as they are seldom used and produce designs which are often less economical than plate girders designed without tension-field action.

The web slenderness ratio $h / t_w = 5.70\sqrt{E / F_{yf}}$ that distinguishes plate girders from beams is written in terms of the flange yield stress, because for hybrid girders inelastic buckling of the web due to bending depends on the flange strain.

The equation for R_e used in the 1986 LRFD Specification was the same as that used in the AASHTO *Standard Specification for Highway Bridges*. In this edition, the equation for R_e, used in the AISC ASD Specification since 1969, is used because its derivation is published (Gaylord, Gaylord, and Stallmeyer, 1992 and Joint ASCE-AASHTO Committee on Flexural Members, 1968) and it is more accurate than the AASHTO equation.

G2. DESIGN FLEXURAL STRENGTH

In previous versions of the AISC Specification a coefficient of $0.0005a_r$ was used in R_{PG} based on the work of Basler (1961). This value is valid for $a_r \leq 2$. In that same paper, Basler developed a more general coefficient, applicable to all ratios of A_w / A_f which has been adopted because application of the previous equation to sections with large a_r values gives unreasonable results. An arbitrary limit of $a_r \leq 10$ is imposed so that the R_{PG} expression is not applied to sections approaching a tee shape.

APPENDIX H

MEMBERS UNDER COMBINED FORCES AND TORSION

H3. ALTERNATIVE INTERACTION EQUATIONS FOR MEMBERS UNDER COMBINED STRESS

In the case of members not subject to flexural buckling, i.e., $L_b < L_{pd}$, the use of somewhat more liberal interaction Equations A-H3-5 and A-H3-6 is acceptable as an alternative when the flexure is about one axis only.

The alternative interaction Equations A-H3-1 and A-H3-2 for H and wide-flange column shapes were taken from Galambos (1998), Springfield (1975), and Tebedge and Chen (1974).

For I-shaped members with $b_f / d > 1.0$, use of Section H1 is recommended, because no additional research is available for this case.

APPENDIX J

CONNECTIONS, JOINTS, AND FASTENERS

J2. WELDS

4. Design Strength

When weld groups are loaded in shear by an external load that does not act through the center of gravity of the group, the load is eccentric and will tend to cause a relative rotation and translation between the parts connected by the weld. The point about which rotation tends to take place is called the instantaneous center of rotation. Its location is dependent upon the load eccentricity, geometry of the weld group, and deformation of the weld at different angles of the resultant elemental force relative to the weld axis.

The individual resistance force of each unit weld element can be assumed to act on a line perpendicular to a ray passing through the instantaneous center and that element's location (see Figure C-A-J2.1).

The ultimate shear strength of weld groups can be obtained from the load deformation relationship of a single-unit weld element. This relationship was originally given by Butler et al. (1972) for E60 (E43) electrodes. Curves for E70 (E48) electrodes used in the Appendix were obtained by Lesik and Kennedy (1990).

Unlike the load-deformation relationship for bolts, strength and deformation performance in welds are dependent on the angle that the resultant elemental force makes with the axis of the weld element (see Figure C-A-J2.1). The actual load deformation relationship for welds is given in Figure C-A-J2.2, taken from Lesik

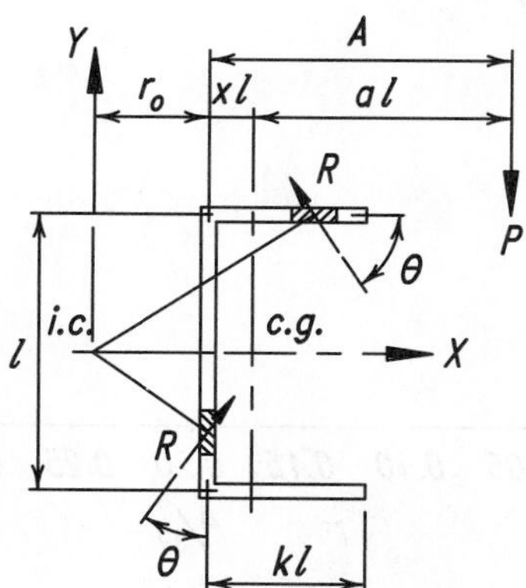

Fig. C-A-J2.1. Weld element nomenclature.

and Kennedy (1990). Conversion of the SI equation to foot-pound units results in the following weld strength equation for R_n:

$$R_n = 0.852(1.0 + 0.50 \sin^{1.5} \theta) F_{EXX} A_w$$

Because the maximum strength is limited to $0.60F_{EXX}$ for longitudinally loaded welds ($\theta = 0°$), the LRFD Specification provision provides, in the reduced equation coefficient, a reasonable margin for any variation in welding techniques and procedures. To eliminate possible computational difficulties, the maximum deformation in the weld elements is limited to $0.17w$. For design convenience, a simple elliptical formula is used for $f(p)$ to closely approximate the empirically derived polynomial in Lesik and Kennedy (1990).

The total resistance of all the weld elements combine to resist the eccentric ultimate load, and when the correct location of the instantaneous center has been selected, the three in-plane equations of statics (ΣF_x, ΣF_y, ΣM) will be satisfied. Numerical techniques, such as those given by Brandt (1982), have been developed to locate the instantaneous center of rotation subject to convergent tolerances. Earlier editions of the AISC *Manual of Steel Construction* (AISC, 1980, 1986a, 1989) took advantage of the inelastic redistribution of stresses that is inherent in the Appendix J2.4 procedure. However, in each of the utilized computational techniques the resulting coefficients were factored down so that the maximum stress, at any point in the weld group, did not exceed the limiting value specified by either the Allowable Stress Design or LRFD Specifications, $0.3F_u$ or $0.6F_u$, respectively. As a result, the tabulated weld-capacity data shown in the appropriate referenced manual tables will be

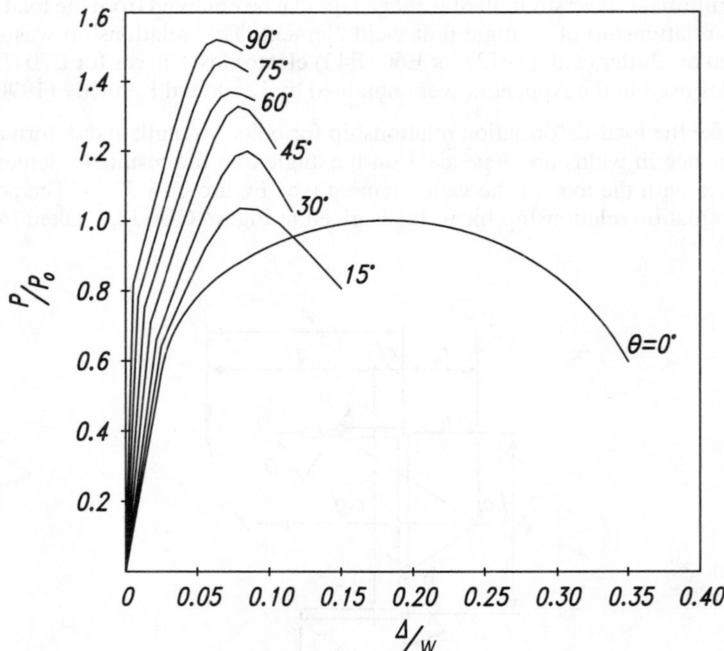

Fig. C-A-J2.2. Load deformation relationship.

found to be conservative relative to the data obtained using the computational procedure presented in Appendix J2.4.

APPENDIX K

CONCENTRATED FORCES, PONDING, AND FATIGUE

K3. DESIGN FOR CYCLIC LOADING (FATIGUE)

Because most members in building frames are not subject to a large enough number of cycles of full design stress application to require design for fatigue, the provisions covering such designs have been placed in Appendix K3.

When fatigue is a design consideration, its severity is most significantly affected by the number of load applications, the magnitude of the stress range, and the severity of the stress concentrations associated with the particular details. These factors are not encountered in normal building designs; however, when encountered and when fatigue is of concern, all provisions of Appendix K3 must be satisfied.

In general, members or connections subject to less than a few thousand cycles of loading will not constitute a fatigue condition except possibly for cases involving full reversal of loading and particularly sensitive categories of details. This is because the admissible static design stress range will be limited by the admissible static design stress. At low levels of cyclic tensile stress, a point is reached where the stress range is so low that fatigue cracking will not initiate regardless of the number of cycles of loading. This level of stress is defined as the fatigue threshold, F_{TH}.

Fluctuation in stress which does not involve tensile stress does not cause crack propagation and is not considered to be a fatigue situation. On the other hand, in elements of members subject solely to calculated compressive stress, fatigue cracks may initiate in regions of high tensile residual stress. In such situations, the cracks generally do not propagate beyond the region of the residual tensile stress, because the residual stress is relieved by the crack. For this reason, stress ranges that are completely in compression need not be investigated for fatigue. For cases involving cyclic reversal of stress the calculated stress range must be taken as the sum of the compressive stress and the tensile stress caused by different direction or pattern of applied live load.

When fabrication details involving more than one category occur at the same location in a member, the stress range at that location must be limited to that of the most restrictive category. By locating notch-producing fabrication details in regions subject to a small range of stress, the need for a member larger than required by static loading will often be eliminated.

Extensive test programs (Fisher, Frank, Hirt, and McNamee, 1970, and Fisher, Albrecht, Yen, and Klingerman, 1974) using full size specimens, substantiated by theoretical stress analysis, have confirmed the following general conclusions:

(1) Stress range and notch severity are the dominant stress variables for welded details and beams.

(2) Other variables such as minimum stress, mean stress, and maximum stress are not significant for design purposes.

(3) Structural steels with yield points of 36 (250) to 100 ksi (690 MPa) do not exhibit significantly different fatigue strength for given welded details fabricated in the same manner.

Because the design stress range may be readily calculated from the equation for the mean curve minus two standard deviations of the actual test data using modern hand calculators or computers, the past method which relied upon multiple tables of cycles of loading, stress categories, design stress ranges, and illustrative examples has been replaced by a single table (Table A-K3.1). In the new format, the situation description, the stress category, the ingredients for the applicable equation, instructive information and pertinent illustrative examples are presented in separate cells across individual table rows. The sites of concern for potential crack initiation are present in text and in example sketches. Similar format and consistent criteria is being developed for the AWS Code and other Specifications.

A detail not covered by earlier editions of the Specification has been added (Frank and Fisher, 1979) to cover tension-loaded plate elements connected at their end by transverse groove or fillet welds in which there is more than a single site for the initiation of fatigue cracking, one of which will be more critical than the others depending upon welded joint type and size and material thickness. Regardless of the site within the joint at which potential crack initiation is considered, the design stress range provided is applicable to connected material at the toe of the weld.

The fatigue resistance of bolts subject to tension is predictable in the absence of pretension and prying action; and in this edition of the specification criteria are provided for such non-pretensioned details as hanger rods and anchor rods. In the case of pretensioned bolts, deformation of the connected parts through which pretension is applied introduces prying action, the magnitude of which is not completely predictable (Kulak et al., 1987). The effects of prying are not limited to a change in the average axial tension on the bolt but includes bending in the threaded area under the nut. Because of the uncertainties, definitive criteria for calculating prying effects and definitive criteria for design stress range are not included in the specification. To limit the uncertainties regarding prying action on the fatigue of pretensioned bolts in details which introduce prying, the design stress range provided in Table A-K3.1 is appropriate for extended cyclic loading only if the prying included in the applied load is small.

Non-pretensioned fasteners are not permitted under the Specification for joints subject to cyclic shear forces. Bolts installed in joints meeting all the requirements for slip-critical connections survive unharmed when subject to cyclic shear stresses sufficient to fracture the connected parts, for which criteria are provided in Section 2 of Table A-K3.1.

References

Ackroyd, M. H., and Gerstle, K. H. (1982), "Behavior of Type 2 Steel Frames," *Journal of the Structural Division*, ASCE, Vol. 108, No. ST7, July, pp. 1541-1556.

Ad Hoc Committee on Serviceability Research (1986), "Structural Serviceability: A Critical Appraisal and Research Needs," *Journal of Structural Engineering*, ASCE,Vol. 112, No. 12, pp. 2646-2664.

American Association of State Highway and Transportation Officials (AASHTO) (1996), *Load and Resistance Factor Design Specification for Highway Bridges*, Washington, D.C.

American Concrete Institute (ACI) (1999), *Building Code Requirements for Structural Concrete,* ACI 318-99, Detroit, MI.

American Concrete Institute (ACI) (1999a), *Building Code Requirements for Structural Concrete,* ACI 318M-99, Detroit, MI.

American Institute of Steel Construction, Inc. (AISC) (1973), "Commentary on Highly Restrained Welded Connections," *Engineering Journal,* AISC, Vol. 10, No. 3, 3rd Quarter, pp. 61-73.

American Institute of Steel Construction, Inc. (AISC) (1978), *Specification for the Design,Fabrication, and Erection of Structural Steel for Buildings,* Chicago, IL.

American Institute of Steel Construction, Inc. (AISC) (1980), *Manual of Steel Construction*, Chicago, IL.

American Institute of Steel Construction, Inc. (AISC) (1986), *Load and Resistance FactorDesign Specification for Structural Steel Buildings,* Chicago, IL.

American Institute of Steel Construction, Inc. (AISC) (1986a), *Load and Resistance Factor Design Manual of Steel Construction*, Chicago, IL.

American Institute of Steel Construction, Inc. (AISC) (1989), *Specification for Structural Steel Buildings—Allowable Stress Design and Plastic Design,* Chicago, IL.

American Institute of Steel Construction, Inc. (AISC) (1989a), *Allowable Stress Design Manual of Steel Construction*, Chicago, IL.

American Institute of Steel Construction, Inc. (AISC) (1997), *Seismic Provisions for Structural Steel Buildings,* Chicago, IL.

American Institute of Steel Construction, Inc. (AISC) (1997a), "k-area Advisory Statement," *Modern Steel Construction,* February.

American Institute of Steel Construction, Inc. (AISC) (1999), *Seismic Provisions for Structural Steel Buildings Supplement No. 1*, Chicago, IL.

American Petroleum Institute (API) (1993), *Recommended Practice for Planning, Designing and Constructing Fixed Offshore Platforms—Load and Resistance Factor Design,* First Edition, July 1, 1993, Washington, D.C.

American Society of Civil Engineers (ASCE) (1979), *Structural Design of Tall Steel Buildings,* New York, NY.

American Society of Civil Engineers (ASCE) (1998), *Minimum Design Loads for Buildings and Other Structures,* ASCE7-98, New York, NY.

Ang, K. M., and Morris, G. A. (1984), "Analysis of Three-Dimensional Frames with Flexible Beam-Column Connections," *Canadian Journal of Civil Engineering*, Vol. 11, No. 2, pp. 245-254.

ASCE Task Committee on Design Criteria for Composite Structures in Steel and Concrete (1992), "Proposed Specification for Structural Steel Beams with Web Openings," Darwin, D., Chmn., *Journal of Structural Engineering,* ASCE, Vol. 118, No. ST12, December, pp. 3315-3324.

ASCE Task Committee on Design Criteria for Composite Structures in Steel and Concrete (1992a), "Commentary on Proposed Specification for Structural Steel Beams with Web Openings," Darwin, D., Chmn., *Journal of Structural Engineering,* ASCE, Vol. 118, No. ST12, December, pp. 3325-3349.

ASCE Task Committee on Effective Length (1997), *Effective Length and Notional Load Approaches for Assessing Frame Stability: Implications for American Steel Design,* ASCE, Reston, VA.

Joint ASCE-AASHTO Committee on Flexural Members (1968), "Design of Hybrid Steel Beams," Report of Subcommittee 1, *Journal of the Structural Division,* ASCE, Vol. 94, No. ST6, June, pp. 1397-1426.

Aslani, F., and Goel, S. C. (1991), "An Analytical Criteria for Buckling Strength of Built-Up Compression Members," *Engineering Journal,* AISC, Vol. 28, No. 4, 4th Quarter, pp. 159-168.

ATC (1978), *Tentative Provisions for the Development of Seismic Regulations for Buildings,* ATC Publication 3-06, June.

Austin, W. J. (1961), "Strength and Design of Metal Beam-Columns," *Journal of the Structural Division,* ASCE, Vol. 87, No. ST4, April, pp. 1-32.

Basler, K. (1961), "Strength of Plate Girders in Shear," *Journal of the Structural Division,* ASCE, Vol. 104, No. ST9, October, pp. 151-180.

Bendigo, R. A., Hansen, R. M., and Rumpf, J. L. (1963), "Long Bolted Joints," *Journal of the Structural Division,* ASCE, Vol. 89, No. ST6, December, pp. 187-213.

Bergfelt, A. (1971), *Studies and Tests on Slender Plate Girders Without Stiffeners,* March.

Bigos, J., Smith, G. W., Ball, E. F., and Foehl, P. J. (1954), "Shop Paint and Painting Practice," *Proceedings of AISC National Engineering Conference*, Milwaukee, WI.

Birkemoe, P. C., and Gilmor, M. I. (1978), "Behavior of Bearing-Critical Double-Angle Beam Connections," *Engineering Journal,* AISC, Vol. 15, No. 4, 4th Quarter, pp. 109-115.

Bjorhovde, R. (1972), *Deterministic and Probabilistic Approaches to the Strength of Steel Columns,* Ph.D. Dissertation, Lehigh University, Bethlehem, PA, May.

Bjorhovde, R. (1988), "Columns: From Theory to Practice," *Engineering Journal,* AISC, Vol. 25, No. 1, 1st Quarter, pp. 21-34.

Bjorhovde, R., Brozzetti, J., and Colson, A. (Eds.) (1988), *Connections in Steel Structures: Behaviour, Strength and Design*, Elsevier Applied Science, London, England.

Bjorhovde, R., Colson, A., and Brozzetti, J. (1990), "Classification System for Beam-to-Column Connections," *Journal of Structural Engineering*, ASCE, Vol. 116, No. 11, pp. 3059-3076.

Bjorhovde, R., Colson, A., Haaijer, G., and Stark, J. W. B. (Eds.) (1992), *Connections in Steel Structures II: Behavior, Strength and Design*, AISC, Chicago, IL.

Bjorhovde, R., Colson, A., and Zandonini, R. (Eds.) (1996), *Connections in Steel Structures III: Behaviour, Strength and Design*, Pergamon Press, London, England.

Bleich, F. (1952), *Buckling Strength of Metal Structures,* McGraw-Hill Book Co., New York, NY.

Brandt, G. D. (1982), "A General Solution for Eccentric Loads on Weld Groups," *Engineering Journal,* AISC, Vol. 19, No. 3, 3rd Quarter, pp.150-159.

Branson, D. E. (1964), "Time-Dependent Effects in Composite Concrete Beams," *Journal of the American Concrete Institute*, Vol. 61, February.

Bridge, P. Q., and Roderick, J. W. (1978), "Behavior of Built-Up Composite Columns," *Journal of the Structural Division,* ASCE, Vol. 104, No. ST7, July, pp. 1141– 1165.

Brockenbrough, R. L. (1983), "Considerations in the Design of Bolted Joints for Weathering Steel," *Engineering Journal,* AISC, Vol. 20, No. 1, 1st Quarter, pp. 40-45.

Butler, L. J., Pal, S., and Kulak, G. L. (1972), "Eccentrically Loaded Welded Connections," *Journal of the Structural Division,* ASCE, Vol. 98, No. ST5, May, pp. 989-1005.

Canadian Standards Association (1989), *Limit States Design of Steel Structures,* Appendices G, H, and I, CSA S16.1-M89, Rexdale, Ontario, Canada.

Carter, C. J., Tide, R. H. R., and Yura, J. A. (1997), "A Summary of Changes and Derivation of LRFD Bolt Design Provisions," *Engineering Journal*, AISC, Vol. 34, No. 3, 3rd Quarter, pp.75-81.

Carter, C. J. (1999), *Stiffening of Wide-Flange Columns at Moment Connections: Wind and Seismic Applications*, Steel Design Guide Series No. 13, AISC, Chicago, IL.

Chen, S., and Tong, G. (1994), "Design for Stability: Correct Use of Braces," *Steel Structures, J. Singapore Struct. Steel Soc.*, Vol. 5, No.1, Dec., pp. 15-23.

Chen, W. F., and Atsuta, T. (1977), *Theory of Beam Columns, Volume II: Space Behavior and Design,* McGraw-Hill, New York, NY.

Chen, W. F., Goto, Y., and Liew, J. Y. R. (1995), *Stability Design of Semi-Rigid Frames*, John Wiley and Sons, New York, NY.

Chen, W. F., and Lui, E. M. (1991), *Stability Design of Steel Frames*, CRC Press, Boca Raton, FL.

Chen, W. F., and Toma, S. (1994), *Advanced Analysis of Steel Frames*, CRC Press, Boca Raton, FL.

Chien, E. Y. L., and Ritchie, J. K. (1984), *Composite Floor Systems*, Canadian Institute of Steel Construction, Willowdale, Ontario, Canada.

Chopra, A. K., and Newmark, N. M. (1980), *Design of Earthquake Resistant Structures,* John Wiley and Sons, Inc., New York, NY.

Cooper, P. B., Galambos, T. V., and Ravindra, M. K. (1978), "LRFD Criteria for Plate Girders," *Journal of the Structural Division,* ASCE, Vol. 104, No. ST9, September, pp. 1389-1407.

Darwin, D. (1990), *Steel and Composite Beams with Web Openings,* AISC Steel Design Guide Series No. 2, AISC, Chicago, IL.

DeWolf, J. T. and Ricker, D. T. (1990), *Column Base Plates*, Steel Design Guide Series No. 1, AISC, Chicago, IL.

Disque, R. O. (1964), "Wind Connections with Simple Framing," *Engineering Journal,* AISC, Vol. 1, No. 3, July, pp. 101-103.

Disque, R. O. (1973), "Inelastic K-Factor in Design," *Engineering Journal,* AISC, Vol. 10, No. 2, 2nd Quarter, pp. 33-35.

Easterling, W. S., Gibbings, D. R., and Murray, T. M. (1993), "Strength of Shear Studs in Steel Deck on Composite Beams and Joists," *Engineering Journal*, AISC, Vol. 30, No. 2, 2nd Quarter, pp. 44-55.

Easterling, W. S., and Gonzales, L. (1993), "Shear Lag Effects in Steel Tension Members," *Engineering Journal,* AISC, Vol. 30, No. 3, 3rd Quarter, pp. 77-89.

Elgaaly, M. (1983), "Web Design Under Compressive Edge Loads," *Engineering Journal,* AISC, Vol. 20, No. 4, 4th Quarter, pp. 153-171.

Elgaaly, M., and Salkar, R. (1991), "Web Crippling Under Edge Loading," *Proceedings of AISC National Steel Construction Conference,* Washington, D.C.

Ellifritt, D. S., Wine, G., Sputo, T., and Samuel, S. (1992), "Flexural Strength of WT Sections," *Engineering Journal,* AISC, Vol. 29, No. 2, 2nd Quarter, pp. 67-74.

Ellingwood, B. E., MacGregor, J. G., Galambos, T. V., and Cornell, C. A. (1982), "Probability-Based Load Criteria: Load Factors and Load Combinations," *Journal of the Structural Division,* ASCE, Vol. 108, No. 5, pp. 978-997.

Eurocode 3 (1992), *Design of Steel Structures, Part 1: General Rules and Rules for Buildings*, ENV 1993-1-1:1992, Comite European de Normalisation (CEN), Brussels, Belgium.

Federal Emergency Management Agency (FEMA) (1995), *Interim Guidelines: Evaluation, Repair, Modification and Design of Welded Steel Moment Frame Structures*, Bulletin No. 267, FEMA, Washington, D.C.

Federal Emergency Management Agency (FEMA) (1997), *Seismic Performance of Bolted and Riveted Connections" in "Background Reports; Metallurgy, Fracture Mechanics, Welding, Moment Connections and Frame Systems Behavior*, Bulletin No. 288, FEMA, Washington, D.C.

Fielding, D. J., and Huang, J. S. (1971), "Shear in Steel Beam-to-Column Connections," *Welding Journal,* AWS, Vol. 50, No. 7, Research Supplement, pp. 313–326.

Fielding, D. J., and Chen, W. F. (1973), "Steel Frame Analysis and Connection Shear Deformation," *Journal of the Structural Division,* ASCE, Vol. 99, No. ST1, January, pp. 1-18.

Fisher, J. M., and West, M. A. (1990), *Serviceability Design Considerations for Low-Rise Buildings*, Steel Design Guide Series No. 3, AISC, Chicago, IL.

Fisher, J. M., and West, M. A. (1997), *Erection Bracing of Low-Rise Structural Steel Buildings,* Steel Design Guide Series No. 10, AISC, Chicago, IL.

Fisher, J. W., Albrecht, P. A., Yen, B. T., Klingerman, D. J., and McNamee, B. M. (1974), *Fatigue Strength of Steel Beams With Welded Stiffeners and Attachments,* National Cooperative Highway Research Program, Report 147, Washington, D.C.

Fisher, J. W., Frank, K. H., Hirt, M. A., and McNamee, B. M. (1970), *Effect of Weldments on the Fatigue Strength of Beams,* National Cooperative Highway Research Program, Report 102, Washington, D.C.

Fisher, J. W., Galambos, T. V., Kulak, G. L., and Ravindra, M. K. (1978), "Load and Resistance Factor Design Criteria for Connectors," *Journal of the Structural Division,* ASCE, Vol. 104, No. ST9, September, pp. 1427-1441.

Frank, K. H., and Fisher, J. W. (1979), "Fatigue Strength of Fillet Welded Cruciform Joints," *Journal of the Structural Division*, ASCE, Vol. 105, No. ST9, September.

Frank, K. H., and Yura, J. A. (1981), *An Experimental Study of Bolted Shear Connections,* FHWA/RD-81/148, Washington, D.C., December.

Freeman, F. R. (1930), *The Strength of Arc-Welded Joints,* Proc. Inst. Civil Engineers, Vol. 231, London, England.

Galambos, T. V. (1960), "Influence of Partial Base Fixity on Frame Stability," *Journal of the Structural Division,* ASCE, Vol. 86, No. ST5, May.

Galambos, T. V. (1961), Discussion of "Strength and Design of Metal Beam-Columns" by W. J. Austin, *Journal of the Structural Division*, ASCE, Vol 87, No. ST8, p. 247.

Galambos, T. V. (1968), *Structural Members and Frames,* Prentice-Hall, Englewood Cliffs, NJ.

Galambos, T. V. (1976), *Proposed Criteria for Load Resistance Factor Design of Steel Building Structures,* Research Report No. 45, Civil Engineering Dept., Washington University, St. Louis, MO, May.

Galambos, T. V. (1978), *Bulletin No. 27,* American Iron and Steel Institute, Washington, D.C., January.

Galambos, T. V. (1991), "Design of Axially Loaded Compressed Angles," *Proceedings 1991 Annual Technical Session*, Structural Stability Research Council, Chicago, IL, pp. 353-367.

Galambos, T. V. (ed.) (1998), *Guide to Stability Design Criteria for Metal Structures,* Structural Stability Research Council, 5th Edition, John Wiley & Sons, New York, NY.

Galambos, T. V., and Chapuis, J. (1980), *LRFD Criteria for Composite Columns and Beam-Columns,* Revised Draft, Washington University, Dept. of Civil Engineering, St. Louis, MO, December.

Galambos, T. V., Ellingwood, B., MacGregor, J. G., and Cornell, C. A. (1982), "Probability-Based Load Criteria: Assessment of Current Design Practice," *Journal of the Structural Division,* ASCE, Vol. 108, No. ST5, May, pp. 959-977.

Galambos, T. V., and Ravindra, M. K. (1978), "Properties of Steel for Use in LRFD," *Journal of the Structural Division,* ASCE, Vol. 104, No. ST9, September, pp. 1459-1468.

Gaylord, E. H. Jr., Gaylord, C. N., and Stallmeyer, J. E. (1992), *Design of Steel Structures,* 3rd Edition, McGraw-Hill Book Co., New York, NY.

Geschwindner, L. F. (1994) "A Practical Approach to the Leaning Column," *Engineering Journal*, AISC, Vol. 31, No. 4, 4th Quarter, pp. 141-149.

Gibson, G. T., and Wake, B. T. (1942), "An Investigation of Welded Connections for Angle Tension Members," *The Welding Journal,* American Welding Society, January, p. 44.

Goble, G. G. (1968), "Shear Strength of Thin Flange Composite Specimens," *Engineering Journal,* AISC, Vol. 5., No. 2, 2nd Quarter, pp. 62-65.

Goverdhan, A. V. (1983), *A Collection of Experimental Moment Rotation Curves: Evaluation of Predicting Equations for Semi-Rigid Connections*, M.S. Thesis, Vanderbilt University, Nashville, TN.

Grant, J. A., Fisher, J. W., and Slutter, R. G. (1977), "Composite Beams with Formed Steel Deck," *Engineering Journal,* AISC, Vol. 14, No. 1, 1st Quarter, pp. 24-43.

Griffis, L. G. (1992), *Load and Resistance Factor Design of W-Shapes Encased in Concrete*, Steel Design Guide Series No. 6, AISC, Chicago, IL.

Hansell, W. C., Galambos, T. V., Ravindra, M. K., and Viest, I. M. (1978), "Composite Beam Criteria in LRFD," *Journal of the Structural Division,* ASCE, Vol. 104, No. ST9, September, pp. 1409-1426.

Hardash, S. G., and Bjorhovde, R. (1985), "New Design Criteria for Gusset Plates in Tension," *Engineering Journal,* AISC, Vol. 22, No. 2, 2nd Quarter, pp. 77-94.

Hoglund, T. (1971), *Simply Supported Long Thin Plate I-Girders Without Web Stiffeners, Subjected to Distributed Transverse Load,* Dept. of Building Statics and Structural Engineering of the Royal Institute of Technology, Stockholm, Sweden.

Hsieh, S. H., and Deierlein, G. G. (1991), "Nonlinear Analysis of Three-Dimensional Steel Frames with Semi-Rigid Connections," *Computers and Structures*, Vol. 41. No 5, pp. 995-1009.

International Association of Bridge and Structural Engineering (IABSE) (1968), *Final Report of the Eighth Congress,* Zurich, Switzerland, September.

International Organization for Standardization (ISO) (1974), *Guide for the Evaluation of Human Exposure to Whole-Body Vibration,* Document ISO 2631, September.

Iwankiw, N. (1984), "Note on Beam-Column Moment Amplification Factor," *Engineering Journal,* AISC, Vol. 21, No. 1, 1st Quarter, pp. 21-23.

Johnson, D. L. (1985), "An Investigation into the Interaction of Flanges and Webs in Wide-Flange Shapes," *Proceedings 1985 Annual Technical Session*, Cleveland, OH, Structural Stability Research Council, Lehigh University, Bethlehem, PA.

Johnson, D. L. (1996), *Final Report on Tee Stub Tests*, Butler Research Report, Grandview, MO, May.

Johnston, B. G. (1939), *Pin-Connected Plate Links,* 1939 ASCE Transactions.

Johnston, B. G., and Deits, G. R. (1942) "Tests of Miscellaneous Welded Building Connections," *Welding Journal,* American Welding Society, November, p. 5.

Johnston, B. G., and Green, L. F. (1940), "Flexible Welded Angle Connections," *The Welding Journal,* American Welding Society, October.

Jones, J. (1940), "Static Tests on Riveted Joints," *Civil Engineering,* ASCE, May.

Kanchanalai, T. (1977), *The Design and Behavior of Beam-Columns in Unbraced Steel Frames,* AISI Project No. 189, Report No. 2, Civil Engineering/Structures Research Lab, University of Texas-Austin, TX, October.

Kanchanalai, T., and Lu, L. W. (1979), "Analysis and Design of Framed Columns Under Minor Axis Bending," *Engineering Journal,* AISC, Vol. 16, No. 2, 2nd Quarter, pp. 29-41.

Kemp, A. R., and Trinchero, P. E. (1997), "Horizontal Shear Failures Around Connectors Used With Steel Decking," *Composite Construction in Steel and Concrete III*, ASCE, pp. 104-118.

Kenny, J. R., Bruce, D. A., and Bjorhovde, R. (1994), "Removal of Yield Stress Limitation for Composite Tubular Columns," *Engineering Journal*, AISC, Vol. 31, No. 1, pp. 1-11.

Ketter, R. L. (1961), "Further Studies of the Strength of Beam-Columns," *Journal of the Structural Division,* ASCE, Vol. 87, No. ST6, August, pp. 135-152.

Kim, H. J., and Yura, J. A. (1996), *The Effect of End Distance on the Bearing Strength of Bolted Connections,* PMFSEL Report No. 96-1, University of Texas-Austin.

Kirby, P. A., and Nethercot, D. A. (1979), *Design for Structural Stability,* John Wiley and Sons, Inc., New York, NY.

Kishi, N., and Chen, W. F. (1986), *Data Base of Steel Beam-to-Column Connections*, Vol. 1 & 2, Structural Engineering Report No. CE-STR-86-26, School of Civil Engineering, Purdue University, West Lafayette, IN.

Kitipornchai, S., and Trahair, N. S. (1980), "Buckling Properties of Monosymmetric I-Beams," *Journal of the Structural Division,* ASCE, Vol. 109, No. ST5, May, pp. 941-957.

Kloppel, K., and Seeger, T. (1964), "Dauerversuche Mit Einsohnittigen Hv-Verbindurgen Aus ST37," *Der Stahlbau,* Vol. 33, No. 8, August 1964, pp. 225–245 and Vol. 33, No. 11, November, pp. 335–346.

Kulak, G. L., Fisher, J. W., and Struik, J. H. A. (1987), *Guide to Design Criteria for Bolted and Riveted Joints,* 2nd Edition, John Wiley & Sons, New York, NY.

Lee, G. D., Morrell, M. L., and Ketter, R. L. (1972), "Design of Tapered Members," *WRC Bulletin,* No. 173, June.

LeMessurier, W. J. (1976), "A Practical Method of Second Order Analysis, Part 1—Pin-Jointed Frames," *Engineering Journal,* AISC, Vol. 13, No. 4, 4th Quarter, pp. 89-96.

LeMessurier, W. J. (1977), "A Practical Method of Second Order Analysis, Part 2—Rigid Frames," *Engineering Journal,* AISC, Vol. 14, No. 2, 2nd Quarter, pp. 49-67.

Leon, R. T. (1994), "Composite Semi-Rigid Construction," *Engineering Journal*, AISC, Vol. 31. No. 2, pp. 57-67.

Leon, R. T., Hoffman, J., and Staeger, T. (1996), *Design of Partially-Restrained Composite Connections*, Steel Design Guide Series No. 9, AISC, Chicago, IL.

Lewis, B. E., and Zwerneman, F. J. (1996), *Edge Distance, Spacing, and Bearing in Bolted Connections*, Oklahoma State University, July.

Lesik, D. F., and Kennedy, D. J. L. (1990), "Ultimate Strength of Fillet Welded Connections Loaded in Plane," *Canadian Journal of Civil Engineering,* National Research Council of Canada, Ottawa, Canada, Vol. 17, No. 1.

Lorenz, R. F., Kato, B., and Chen, W. F. (Eds.) (1993), *Semi-Rigid Connections in Steel Frames*, Council for Tall Buildings and Urban Habitat (CTBUH), Lehigh University, Bethlehem, PA.

Lyse, I., and Schreiner (1935), "An Investigation of Welded Seat Angle Connections," *Welding Journal,* February, p. 1.

Lyse, I., and Gibson, G. J. (1937), "Effect of Welded Top Angles on Beam-Column Connections," *Welding Journal,* October.

Marino, F. J. (1966), "Ponding of Two-Way Roof Systems," *Engineering Journal,* AISC, Vol. 3, No. 3, 3rd Quarter, pp. 93-100.

Morrell, M. L., and Lee, G. C. (1974), *Allowable Stress for Web-Tapered Members,* WRC Bulletin 192, Welding Research Council, New York, NY, February.

Munse, W. H., and Chesson, Jr., E., (1963), "Riveted and Bolted Joints: Net Section Design," *Journal of the Structural Division,* ASCE, Vol. 89, No. ST1, February, pp. 49-106.

Murray, T. M., Allen, D. E., and Ungar, E. E. (1997), *Floor Vibrations Due to Human Activity*, Steel Design Guide Series No. 11, AISC, Chicago, IL.

Murray, T. M., Kline, D. P., and Rojani, K. B. (1992), "Use of Snug-Tightened Bolts in End-Plate Connections," in *Connections in Steel Structures II,* R. Bjorhovde, A. Colson, G. Haaijer, and J. W. B. Stark, (eds.), AISC, Chicago, IL.

NEHRP (1997), *NEHRP Recommended Provisions for the Development of Seismic Regulations for New Buildings,* Federal Emergency Management Agency Report, FEMA 302, Washington, D. C.

Nethercot, D. A. (1985), *Steel Beam to Column Connections—A Review of Test Data and Their Applicability to the Evaluation of the Joint Behaviour of the Performance of Steel Frames*, CIRIA, London, England.

Ollgaard, J. G., Slutter, R. G., and Fisher, J. W. (1971), "Shear Strength of Stud Shear Connections in Lightweight and Normal Weight Concrete," *Engineering Journal,* AISC, Vol. 8, No. 2, 2nd Quarter, pp. 55-64.

Popov, E. P. (1980), "An Update on Eccentric Seismic Bracing," *Engineering Journal,* AISC, Vol. 17, No. 3, 3rd Quarter, pp. 70-71.

Popov, E. P., and Stephen, R. M. (1977), "Capacity of Columns with Splice Imperfections," *Engineering Journal,* AISC, Vol. 14, No. 1, 1st Quarter, pp. 16-23.

Preece, F. R. (1968), *AWS-AISC Fillet Weld Study—Longitudinal and Transverse Shear Tests,* Testing Engineers, Inc., Los Angeles, CA, May.

Ravindra, M. K., and Galambos, T. V. (1978), "Load and Resistance Factor Design for Steel," *Journal of the Structural Division,* ASCE, Vol. 104, No. ST9, September, pp. 1337-1353.

Research Council on Structural Connections (RCSC) (1994), *Load and Resistance Factor Design Specification for Structural Joints Using ASTM A325 or A490 Bolts,* AISC, Chicago, IL.

Ricles, J. M., and Yura, J. A. (1983), "Strength of Double-Row Bolted Web Connections," *Journal of the Structural Division,* ASCE, Vol. 109, No. ST1, January, pp. 126-142.

Roberts, T. M. (1981), "Slender Plate Girders Subjected to Edge Loading," *Proceedings, Institution of Civil Engineers,* Part 2, 71, September, London, England.

Salvadori, M. (1956), "Lateral Buckling of Eccentrically Loaded I-Columns," *1956 ASCE Transactions,* Vol. 122-1.

Seaburg, P. A., and Carter, C. J. (1997), *Torsional Analysis of Structural Steel Members*, Steel Design Guide Series No. 9, AISC, Chicago, IL.

Sherman, D. R. (1976), *Tentative Criteria for Structural Applications of Steel Tubing and Pipe,* American Iron and Steel Institute, Washington, D.C., August.

Sherman, D. R., and Tanavde, A. S. (1984), *Comparative Study of Flexural Capacity of Pipes,* Civil Engineering Department Report, University of Wisconsin-Milwaukee, WI, March.

Slutter, R. G., and Driscoll, G. C. (1965), "Flexural Strength of Steel-Concrete Composite Beams," *Journal of the Structural Division,* ASCE, Vol. 91, No. ST2, April.

Springfield, J. (1975), "Design of Column Subject to Biaxial Bending," *Engineering Journal,* AISC, Vol. 12, No. 3, 3rd Quarter, pp. 73-81.

Stang, A. H., and Jaffe, B. S. (1948), *Perforated Cover Plates for Steel Columns,* Research Paper RP1861, National Bureau of Standards, Washington, D.C.

Steel Structures Painting Council (SSPC) (1989), *Steel Structures Painting Manual, Vol. 2, Systems and Specifications,* SSPC, Pittsburgh, PA.

Structural Stability Research Council Task Group 20 (1979), "A Specification for the Design of Steel-Concrete Composite Columns," *Engineering Journal,* AISC, Vol. 16, No. 4, 4th Quarter, pp. 101-115.

Summers, P. A., and Yura, J. A. (1982), *The Behavior of Beams Subjected to Concentrated Loads,* Phil M. Ferguson Structural Engineering Laboratory Report No. 82-5, University of Texas, Austin, TX, August.

Taylor, A. C., and Ojalvo, M. (1966), "Torsional Restraint of Lateral Buckling," *Journal of the Structural Division,* ASCE, Vol. 92, No. ST2, pp. 115-129.

Tebedge, N., and Chen, W. F. (1974), "Design Criteria for H-Columns Under Biaxial Loading," *Journal of the Structural Division,* ASCE, Vol. 100, ST3, pp. 579-598.

Tide, R. H. R. (1985), "Reasonable Column Design Equations," *Proceedings 1985 Annual Technical Session*, Structural Stability Research Council, April, Cleveland, OH, Lehigh University, Bethlehem, PA.

Timoshenko, S. P., and Gere, J. M. (1961), *Theory of Elastic Stability,* McGraw-Hill Book Company, New York, NY.

Viest, I. M., Colaco, J. P., Furlong, R. W., Griffis, L. G., Leon, R. T., and Wyllie, Jr., L. A. (1997), *Composite Construction Design for Buildings*, ASCE and McGraw-Hill Book Company, New York, NY.

Viest, I. M., Fountain, R. S., and Singleton, R. C. (1958), *Composite Construction in Steel and Concrete for Bridges and Buildings*, McGraw-Hill Book Company, New York, NY.

Wilson, W. M. (1934), *The Bearing Value of Rollers,* Bulletin No. 263, University of Illinois Engineering Experiment Station, Urbana, IL.

Winter, G. (1958), "Lateral Bracing of Columns and Beams," *Journal of the Structural Division,* ASCE, Vol. 84, No. ST2, March, pp. 1561-1 – 1561-22.

Winter, G. (1960), "Lateral Bracing of Columns and Beams," *Transactions, ASCE*, Vol. 125, Part 1, pp. 809-825.

Winter, G. (1970), *Commentary on the 1968 Edition of Light Gage Cold-Formed Steel Design Manual,* American Iron and Steel Institute, Washington, D.C.

Yura, J. A. (1971), "The Effective Length of Columns in Unbraced Frames," *Engineering Journal,* AISC, Vol. 8, No. 2, April, pp. 37-42.

Yura, J. A., Galambos, T. V., and Ravindra, K. (1978), "The Bending Resistance of Steel Beams," *Journal of the Structural Division,* ASCE, Vol. 104, No. ST9, September, pp. 1355-1370.

Yura, J. A. (1988), *Elements for Teaching Load & Resistance Factor Design,* AISC, Chicago, IL, April.

Yura, J. A. (1993), "Fundamentals of Beam Bracing," *Proc., SSRC Conf., "Is Your Structure Suitably Braced?,"* Milwaukee, WI, April.

Yura, J. A. (1995),"Bracing for Stability-State-of-the-Art," *Proceedings, Structures Congress XIII, ASCE*, Boston, MA, April, pp. 88-103.

Zandonini, R. (1985), "Stability of Compact Built-Up Struts: Experimental Investigation and Numerical Simulation," (in Italian) *Construzioni Metalliche,* No. 4.

Zhou, S. P., and Chen, W. F. (1985), "Design Criteria for Box Columns Under Biaxial Loading," *Journal of Structural Engineering,* ASCE, Vol. 111, No. ST12, December, pp. 2643-2658.

Supplementary Bibliography

American Society of Civil Engineers (ASCE) (1971), "Plastic Design in Steel," *ASCE Manual of Engineering Practice,* No. 41, 2nd Edition, New York, NY.

American Welding Bureau (1931), *Report of Structural Welding Committee.*

Basler, K., and Thurlimann, B. (1963), "Strength of Plate Girders in Bending," *Journal of the Structural Division,* ASCE, Vol. 89, No. ST4, August.

Barakat, M., and Chen, W. F. (1991), "Design Analysis of Semi-Rigid Frames: Evaluation and Implementation," *Engineering Journal,* AISC, Vol. 28, No. 2, 2nd Quarter, pp. 55-64.

Beedle, L. S., and Tall, L. (1960), "Basic Column Strength," *Journal of the Structural Division,* ASCE, Vol. 86, No. ST7, July, pp. 139-173.

Bjorhovde, R., Galambos, T. V., and Ravindra, M. K. (1978), "LRFD Criteria for Steel Beam-Columns," *Journal of the Structural Division,* ASCE, Vol. 104, No. ST9, September, pp. 1371-1387.

Chen, W. F., and Atsuta, T. (1976), *Theory of Beam Columns, Volume I: In-Plane Behavior and Design,* McGraw-Hill, New York, NY.

Cheong-Siat Moy, F., Ozer, E., and Lu, L. W. (1977), "Strength of Steel Frames Under Gravity Loads," *Journal of the Structural Division*, ASCE, Vol. 103, No. ST6, June, pp. 1223-1235.

Daniels, J. H., and Lu, L. W. (1972), "Plastic Subassemblage Analysis for Unbraced Frames," *Journal of the Structural Division,* ASCE, Vol. 98, No. ST8, August, pp. 1769-1788.

Galambos, T. V. (1980), *Reliability of Axially Loaded Columns,* Washington University, Department of Civil Engineering, St. Louis, MO, December.

Galambos, T. V., and Ravindra, M. K. (1973), *Tentative Load and Resistance Factor Design Criteria for Steel Buildings,* Research Report No. 18, Washington University, Dept. of Civil Engineering, St. Louis, MO, September.

Galambos, T. V., and Ravindra, M. K. (1976), *Load and Resistance Factor Design Criteria for Steel Beams,* Research Report No. 27, Washington University, Dept. of Civil Engineering, St. Louis, MO, February.

Hall, D. H. (1981), "Proposed Steel Column Strength Criteria," *Journal of the Structural Division,* ASCE, Vol. 107, No. ST4, April, pp. 649-670.

Keating, P. B., and Fisher, J. W. (1985), *Review of Fatigue Tests and Design Criteria on Welded Details,* NCHRP Project 12-15(50), October 1985, Washington, D.C.

Kotecki, D. S., and Moll, R. A. (1970), "A Toughness Study of Steel Weld Metal from Self-Shielded, Flux-Cored Electrodes, Part 1," *Welding Journal,* AWS, Vol. 49, April.

Kotecki, D. S., and Moll, R. A. (1972), "A Toughness Study of Steel Weld Metal from Self-Shielded, Flux-Cored Electrodes, Part 2," *Welding Journal,* AWS, Vol. 51, March, p. 138.

LeMessurier, W. J., McNamara, R. J., and Scrivener, J. C. (1974), "Approximate Analytical Model for Multi-Story Frames," *Engineering Journal,* AISC, Vol. 11, No. 4, 4th Quarter, pp. 92-98.

Liapunov, S. (1974), "Ultimate Load Studies of Plane Multi-Story Steel Rigid Frames," *Journal of the Structural Division,* ASCE, Vol. 100, No. ST8, Proc. Paper 10750, August.

Lim, L. C., and Lu, L. W. (1970), *The Strength and Behavior of Laterally Unsupported Columns*, Fritz Engineering Laboratory Report No. 329.5, Lehigh University, Bethlehem, PA, June.

Lu, L. W. (1967), "Design of Braced Multi-Story Frames by the Plastic Method," *Engineering Journal,* AISC, Vol. 4, No. 1, 1st Quarter, pp. 1-9.

Lu, L. W., Ozer, E., Daniels, J. H., Okten, O. S., and Morino, S. (1977), "Strength and Drift Characteristics of Steel Frames," *Journal of the Structural Division,* ASCE, Vol. 103, No. ST11, November, pp. 2225-2241.

Rao, N. R. N., Lohrmann, M., and Tall, L. (1966), "Effect of Strain Rate on the Yield Stress of Structural Steels," *Journal of Materials,* Vol. 1, No. 1, ASTM, March.

Ross, D. A., and Chen, W. F. (1976), "Design Criteria for Steel I-Columns Under Axial Load and Biaxial Bending," *Canadian Journal of Civil Engineering,* Vol. 3, No. 3.

Springfield, J., and Adams, P. F. (1972), "Aspects of Column Design in Tall Steel Buildings," *Journal of the Structural Division,* ASCE, Vol. 9, No. ST5, May, pp. 1069-1083.

Terashima, H., and Hart, P. H. M. (1984), "Effect of Aluminum on Carbon, Manganese, Niobium Steel Submerged Arc Weld Metal Properties," *Welding Journal,* Vol. 63, June, p. 173.

Wood, B. R., Beaulieu, D., and Adams, P. F. (1976), "Column Design by P-Delta Method," *Journal of the Structural Division,* ASCE, Vol. 102, No. ST2, February, pp. 411-427.

Metric Conversion Factors for Common Steel Design Units Used in the LRFD Specification

Unit	Multiply	by	to obtain
length	inch (in.)	25.4	millimeters (mm)
length	foot (ft)	0.3048	meters (m)
mass	pound-mass (lbm)	0.4536	kilogram (kg)
stress	ksi	6.895	megapascals (MPa), N/mm^2
moment	kip-in	113,000	N-mm
energy	ft-lbf	1.356	joule (J)
force	kip (1,000 lbf)	4,448	newton (N)
force	psf	47.88	pascal (Pa), N/m^2
force	plf	14.59	N/m
temperature	To convert °F to °C: $t_c^\circ = (t_f^\circ - 32)/1.8$		

force in lbf or N = mass × g
where g, acceleration due to gravity = 32.2 ft/sec^2 = 9.81 m/sec^2

AMERICAN INSTITUTE OF STEEL CONSTRUCTION, INC.
One East Wasker Drive, Suite 3100, Chicago, Illinois 60601-2001

Pub. No. S350L

Load and Resistance Factor Design Specification for Steel Hollow Structural Sections

November 10, 2000

Supersedes *Specification for the Design of Steel Hollow Structural Sections* dated April 15, 1997 and all previous versions.

Prepared by the American Institute of Steel Construction, Inc. under the direction of the AISC Committee on Specifications and approved by the AISC Board of Directors.

AMERICAN INSTITUTE OF STEEL CONSTRUCTION, INC.
One East Wacker Drive, Suite 3100 Chicago, Illinois 60601-2001

The information presented in this publication has been prepared in accordance with recognized engineering principles and is for general information only. While it is believed to be accurate, this information should not be used or relied upon for any specific application without competent professional examination and verification of its accuracy, suitability, and applicability by a licensed engineer, architect, or other professional. The publication of the material contained herein is not intended as a representation or warranty on the part of the American Institute of Steel Construction, Inc., or of any other person named herein, that this information is suitable for any general or particular use or of freedom from infringement of any patent or patents. Anyone making use of this information assumes all liability arising from such use.

Caution must be exercised when relying upon other specifications and codes developed by other bodies and incorporated by reference herein since such material may be modified or amended from time to time subsequent to the printing of this edition. The American Institute of Steel Construction, Inc. bears no responsibility for such material other than to refer to it and incorporate it by reference at the time of the initial publication of this edition.

Printed in the United States of America

PREFACE

The AISC *Load and Resistance Factor Design (LRFD) Specification for Structural Steel Buildings* is intended to cover the common design criteria in routine office practice. Accordingly, it is not feasible to also cover the many special and unique problems encountered within the full range of structural design practice. This AISC *Load and Resistance Factor Design Specification for Steel Hollow Structural Sections* is a separate document that addresses one such topic: the design and construction of building systems that utilize steel hollow structural sections (HSS). A list of Symbols and a non-mandatory Commentary with background information are provided.

The AISC Committee on Specifications, Task Committee 13—Hollow Structural Sections is responsible for its ongoing development. Additionally, the AISC Committee on Specifications has enhanced these provisions through careful scrutiny, discussion, suggestion for improvements, and endorsement.

The reader is cautioned that professional judgment must be exercised when data or recommendations in this Specification are applied. The publication of the material contained herein is not intended as a representation or warranty on the part of the American Institute of Steel Construction, Inc.—or any other person named herein—that this information is suitable for general or particular use, or freedom from infringement of any patent or patents. Anyone making use of this information assumes all liability arising from such use. The design of structures is within the scope of expertise of a competent licensed structural engineer, architect, or other licensed professional for the application of principles to a particular structure.

By the AISC Committee on Specifications, Task Committee 13—Hollow Structural Sections,

D.R. Sherman, Chairman
R. Kaehler
L.A. Kloiber
J.A. Packer
F.J. Palmer
C. Pierson
J.M. Ricles
S.A. Ashton, Secretary

R. Kaz[illegible] [illegible]
L.A. Kloiber [illegible] Ricles
[illegible] [illegible], Secretary

CROSS REFERENCE TO THE AISC LRFD SPECIFICATION

This table provides a cross reference from Sections in this Specification to the relevant Sections and related Appendices in the *LRFD Specification for Structural Steel Buildings* dated December 27, 1999.

	2000 HSS Specification	1999 LRFD Specification
1.1	Scope	A1
1.2	Material	–
	1. Structural Steel	A3.1
	2. Design Wall Thickness	–
1.3	Loads and Load Combinations	A4
2.1	Effective Area of Tension Members	B3
2.2	Local Buckling	–
	1. Classification of Steel Sections	B5.1
	2. Design by Plastic Analysis	B5.2
	3. Design in Seismic Applications	–
2.3	Limiting Slenderness Ratios	B7
3.1	Design Tensile Strength	D1
4.1	Effective Length and Slenderness Limitations	E1
4.2	Design Compressive Strength	App. B5
5.1	Design Flexural Strength	App. F1.1
5.2	Design Shear Strength	F2
5.3	Design by Plastic Analysis	F1.3
5.4	Design in Seismic Applications	–
6.	Torsion Members	H2
7.1	Design for Combined Flexure and Axial Force	H1
7.2	Design for Combined Torsion, Shear, Flexure, and/or Axial Force	H2
8.	Concentrated Forces on HSS	K1
9.1	General Provisions for Connections and Fasteners	J1, J3
9.2	Welds	J2
9.3	Other Connection Requirements	–
	1. Shear Rupture Strength	J4.1
	2. Tension Rupture Strength	J4.2
	3. Punching Shear Rupture Strength	–
	4. Eccentric Connections	J5.1
9.4	HSS-to-HSS Truss Connections	–
10.	General Requirements for HSS Fabrication	M

TABLE OF CONTENTS

SYMBOLS

The section number in the right hand column refers to the section where the symbol is first used.

Symbol	Definition	Section
A	Area used to calculate A_e, in.^{2} (mm^2)	2.1
A_g	Gross area of cross-section, in.^{2} (mm^2)	2.1
	Chord gross area, in.^{2} (mm^2)	9.4
A_e	Effective net area for tension member, in.^{2} (mm^2)	2.1
A_n	Net area, in.^{2} (mm^2)	2.1
A_w	Web area, in.^{2} (mm^2)	5.3
B	Overall width of rectangular HSS, in. (mm)	1.3
B_b	Overall width of rectangular HSS branch member in a truss connection, in. (mm)	9.2
C	HSS torsional constant	6.
D	Outside diameter of round HSS, in. (mm)	2.1
D_b	Outside diameter of round HSS branch member in a truss connection, in. (mm)	9.4
E	Modulus of elasticity, ksi (MPa)	2.2.1
F_{cr}	Critical stress for column buckling, ksi (MPa)	4.2
F_n	Nominal stress for rectangular HSS shear resistance, ksi (MPa)	5.2
F_u	Specified minimum tensile strength of the HSS, ksi (MPa)	3.1
F_y	Specified minimum yield strength of the HSS, ksi (MPa)	2.2.1
F_{y1}	Specified minimum yield strength of plate or connecting element that is welded to an HSS, ksi (MPa)	8.1
F_{yb}	Specified minimum yield strength of HSS branch member in a truss connection, ksi (MPa)	9.4
H	Overall height of rectangular HSS, in. (mm)	1.3
H_b	Overall height of rectangular HSS branch member in a truss connection, in. (mm)	9.2
K	Compression member effective length factor	4.1
L_b	Unbraced length, in. (mm)	5.1
L_{pd}	Maximum unbraced length for plastic moment M_p in plastic analysis, in. (mm)	5.3
M_n	Nominal flexural strength, kip-in. (N-mm)	5.1
M_p	Plastic moment of section, kip-in. (N-mm)	5.1
M_r	Yield moment of section, kip-in. (N-mm)	5.1
M_u	Required flexural strength, kip-in. (N-mm)	7.1
M_{ur}	Resultant required flexural strength for round HSS, kip-in. (N-mm)	7.1
N	Bearing length of concentrated load along length of HSS, in. (mm)	8.1
P_n	Nominal axial strength, kips (N)	2.2.1
P_u	Required axial strength, kips (N)	4.2
P_y	Axial yield load, kips (N)	2.2.1
Q	Effective area factor	4.2

Q_f	Connection resistance reduction factor for compression in HSS, parameter used for truss connections	9.4
Q_q	Parameter used for truss connections	9.4
Q_β	Parameter used for truss connections	9.4
R_f	Reduction factor for wind forces on exposed HSS	1.3
R_n	Nominal resistance of connections to HSS, kips (N)	8.
S	Elastic section modulus, in.^{3} (mm^3)	5.1
S_{eff}	Effective elastic section modulus for thin-walled rectangular HSS, in.^{3} (mm^3)	5.1
T_n	Nominal torsional strength, kips (N)	6.1
T_u	Required torsional strength, kips (N)	7.2
U	Shear lag factor, parameter used for truss connections	2.1
V_n	Nominal shear strength, kips (N)	5.2
V_u	Required shear strength, kips (N)	7.2
Z	Plastic section modulus, in.^{3} (mm^3)	5.1
a	Length of essentially constant shear in a beam, in. (mm)	5.2
b	Flat width of rectangular HSS flange or side, which is permitted to be taken as $B - 3t$, in. (mm)	2.2.1
b_1	Width of plate or connecting element that is welded to an HSS, in. (mm)	8.1
b_{eoi}	Parameter used for truss connections	9.4
b_{gap}	Parameter used for truss connections	9.4
c	Constant for bending in rectangular HSS branches of truss connections	9.4
d	Bolt diameter, in. (mm)	9.1
f	Stress, ksi (MPa)	4.2
g	Gap between branch members in a gapped K-connection, in. (mm)	9.4
h	Flat width of rectangular HSS web or side, which is permitted to be taken as $H - 3t$, in. (mm)	2.2.1
k	Distance from point of application of concentrated force to critical section of HSS, in. (mm)	8.1
l	Member length, in. (mm)	2.3
l	Connection length, in. (mm)	2.1
r	Radius of gyration, in. (mm)	2.3
r_y	Radius of gyration about the y-axis, in. (mm)	5.1
t	Design HSS wall thickness as given in Section 1.2, in. (mm)	2.2.1
t_1	Thickness of plate or connecting element that is welded to an HSS, in. (mm)	8.1
t_b	Thickness of branch member in an HSS truss connection, in. (mm)	9.4
$\bar{x}$	Eccentricity for shear lag, in. (mm)	2.1
α	Parameter used for truss connections	9.4
β	Parameter used for truss connections	9.4
β_{eff}	Parameter used for truss connections	9.4
β_{eop}	Parameter used for truss connections	9.4
β_{gap}	Parameter used for truss connections	9.4
γ	Parameter used for truss connections	9.4
ϕ	Resistance factor	
η	Parameter used for truss connections	9.4
λ	Wall slenderness	2.2.1

λ_c	Column slenderness	4.2
λ_p	Maximum wall slenderness for compact section	2.2.1
λ_r	Maximum wall slenderness for non-compact section	2.2.1
ζ	Parameter used for truss connections	9.4

Load and Resistance Factor Design Specification for Steel Hollow Structural Sections

November 10, 2000

1. GENERAL PROVISIONS

1.1. Scope

This Specification is intended for the design of round and rectangular hollow structural sections (HSS) that are used as structural members in buildings and for the design of connections to HSS. HSS are: (1) prismatic structural shapes; and, (2) products of a pipe or tubing mill that meet the geometric tolerances, tensile requirements, and chemical requirements of a standard specification. Rectangular HSS include square and rectangular cross-sections that have rounded corners within the tolerances of an appropriate product specification. Only unstiffened non-composite HSS in non-fatigue applications are considered in this Specification.

This Specification includes the list of symbols.

This Specification is based on the AISC *Load and Resistance Factor Design Specification for Structural Steel Buildings* (AISC, 1999), hereinafter referred to as the LRFD Specification. In some cases, criteria taken from the LRFD Specification have been modified to appear in non-dimensional form and to apply directly to rectangular HSS, which have two webs. For situations that are not covered in this Specification, the criteria in the LRFD Specification shall apply. In seismic applications, HSS shall be designed to meet the requirements of the AISC *Seismic Provisions for Structural Steel Buildings* (AISC, 1997) and the *Seismic Provisions Supplement No. 2* (AISC, 2000).

1.2. Material

1. Structural Steel

HSS material that meets the requirements in one of the following ASTM specifications is approved for use under this Specification:

Standard Specification for Pipe, Steel, Black and Hot-Dipped, Zinc Coated, Welded and Seamless, ASTM A53/A53M-99b Gr. B

Standard Specification for Cold-Formed Welded and Seamless Carbon Steel Structural Tubing in Rounds and Shapes, ASTM A500-99

Standard Specification for Hot-Formed Welded and Seamless Carbon Steel Structural Tubing, ASTM A501-99

Standard Specification for Hot-Formed Welded and Seamless High-Strength Low-Alloy Structural Tubing, ASTM A618-99

Standard Specification for Cold-Formed Welded and Seamless High-Strength, Low-Alloy Structural Tubing with Improved Atmospheric Corrosion Resistance, ASTM A847-99a

Certified mill test reports or certified reports of tests made by the fabricator or a qualified testing laboratory that meet the requirements in ASTM A370, Test Methods and Definitions for Mechanical Testing of Steel Products, and the governing specification shall constitute sufficient evidence of conformity with one of the above ASTM standards. If requested, the fabricator shall provide an affidavit stating that the structural steel furnished meets the requirements of the grade specified.

2. Design Wall Thickness

The design wall thickness t shall be used in calculations involving the HSS wall thickness. When the design wall thickness is not known, it is permitted to be taken as 0.93 times the nominal wall thickness.

1.3. Loads and Load Combinations

The nominal loads and factored load combinations shall be as stipulated by the applicable code under which the structure is designed or dictated by the condition involved. In the absence of a code, the loads and factored load combinations, including impact and crane loads, shall be those stipulated in ASCE 7. For design purposes, the loads stipulated by the applicable code or ASCE 7 shall be taken as nominal loads.

If permitted by the applicable building code, wind forces on the projected areas of exposed HSS are permitted to be reduced by the factor R_f from the forces on frameworks with similar configurations but using sections or shapes with flat elements. R_f shall be taken as follows:

(a) For round HSS, $R_f = 2/3$.

(b) For rectangular HSS with outside corner radii that are greater than or equal to 0.05 times the width B and wind force acting on the short side (B), $R_f = 0.4 + 0.6\,B/H \le 2/3$, where H is the depth of the HSS. For rectangular HSS under other conditions, $R_f = 1.0$.

2. DESIGN REQUIREMENTS

2.1. Effective Area of Tension Members

The effective area A_e of tension members shall be determined as follows:

$$A_e = AU \qquad (2.1\text{-}1)$$

(a) For a welded connection that is continuous around the perimeter, $A = A_g$, where A_g is the gross area and $U = 1$.

(b) For connections with concentric gusset plates and slotted HSS, $A = A_n$, where the net area A_n at the end of the gusset plate is the gross area minus the product of the thickness and total width of material that is removed to form the slots and

$$U = 1 - (\bar{x}/l) \le 0.9 \qquad (2.1\text{-}2)$$

In the above equation, $\bar{x}$ is the perpendicular distance from the weld to the centroid of the cross-sectional area that is tributary to the weld.

For round HSS with a single concentric gusset plate

$$\bar{x} = \frac{D}{\pi} \tag{2.1-3}$$

For rectangular HSS with a single concentric gusset plate

$$\bar{x} = \frac{B^2 + 2BH}{4(B + H)} \tag{2.1-4}$$

(c) For connections with rectangular HSS and a pair of side gusset plates, $A = A_g$, where A_g is the gross area and U shall be calculated using Equation 2.1-2 with

$$\bar{x} = \frac{B^2}{4(B + H)} \tag{2.1-5}$$

where

l = length of the connection in the direction of loading, in. (mm)
D = outside diameter of round HSS, in. (mm)
B = overall width of rectangular HSS, in. (mm)
H = overall height of rectangular HSS, in. (mm)

Larger values of U are permitted to be used in the foregoing cases when justified by tests or other rational criteria. For other end-connection configurations, U shall be determined by tests or other rational criteria.

2.2. Local Buckling

1. Classification of Steel Sections

HSS are classified for local buckling of the wall in compression as compact, noncompact, or slender-element cross-sections according to the limiting wall slenderness ratios λ_p and λ_r in Table 2.2-1. For an HSS to qualify as compact, the wall slenderness ratio λ must be less than or equal to λ_p. If λ exceeds λ_p but is less than or equal to λ_r, the HSS is noncompact. If λ exceeds λ_r, the HSS is a slender-element cross-section. The wall slenderness ratio λ shall be calculated as follows:

(a) For round HSS, $\lambda = D/t$, where D is the outside diameter and t is the wall thickness. This Specification is applicable only to round HSS with λ less than or equal to $0.448E/F_y$, where E is the modulus of elasticity and F_y is the specified minimum yield stress.

(b) For flanges of rectangular HSS, $\lambda = b/t$, where b is the clear distance between webs less the inside corner radius at each web and t is the wall thickness. If the corner radius is not known, b is permitted to be taken as the overall flange width B minus three times the wall thickness t.

(c) For webs of rectangular HSS, $\lambda = h/t$, where h is the clear distance between flanges less the inside corner radius at each flange and t is the wall thickness. If the corner radius is not known, h is permitted to be taken as the overall web depth H minus three times the wall thickness t.

2. Design by Plastic Analysis

Design by plastic analysis is permitted when λ is less than or equal to λ_p for plastic analysis in Table 2.2-1.

TABLE 2.2-1
Limiting Wall Slenderness for Compression Elements

Element	Wall Slenderness Ratio, λ	Limiting Wall Slenderness	
		λ_p (compact)	λ_r (noncompact)
Round HSS for axial compression for flexure for plastic analysis	D/t [a]	 n.a. $0.0714E/F_y$ $0.0448E/F_y$	 $0.114E/F_y$ $0.309E/F_y$ n.a.
Rectangular HSS wall for uniform compression for plastic analysis	b/t or h/t	 $1.12\sqrt{E/F_y}$ $0.939\sqrt{E/F_y}$	 $1.40\sqrt{E/F_y}$ n.a.
Rectangular HSS wall as a web in flexural compression	h/t	$3.76\sqrt{E/F_y}$	$5.70\sqrt{E/F_y}$
Rectangular HSS wall as a web in combined flexure and axial compression	h/t	[b]	$5.70\sqrt{E/F_y}\left(1-\dfrac{0.74P_u}{\phi_b P_y}\right)$

[a] D/t must be less than or equal to $0.448E/F_y$

[b] For $P_u/\phi_b P_y \le 0.125$

$$3.76\sqrt{E/F_y}\left(1-\frac{2.75P_u}{\phi_b P_y}\right)$$

For $P_u/\phi_b P_y > 0.125$

$$1.12\sqrt{E/F_y}\left(2.33-\frac{P_u}{\phi_b P_y}\right) \ge 1.49\sqrt{E/F_y}$$

3. Design in Seismic Applications

In seismic applications, λ shall also meet the requirements in the AISC *Seismic Provisions for Structural Steel Buildings* (AISC, 1997) and the *Seismic Provisions Supplement No. 2* (AISC, 2000).

2.3. Limiting Slenderness Ratios

For compression members, the slenderness ratio Kl/r preferably should not exceed 200.

For tension members, the slenderness ratio l/r preferably should not exceed 300. Members that are primarily tension members but that are subject to some compression under other load conditions need not satisfy the compression slenderness limit.

For bracing members in seismic applications, l/r shall meet the requirements in AISC *Seismic Provisions for Structural Steel Buildings* (AISC, 1997) and the *Seismic Provisions Supplement No. 2* (AISC, 2000).

3. TENSION MEMBERS

3.1. Design Tensile Strength

The design tensile strength $\phi_t P_n$ shall be the lower value obtained according to the limit states of yielding in the gross section and fracture in the net section.

(a) For yielding on the gross area:

$$\phi_t = 0.9$$
$$P_n = F_y A_g \tag{3.1-1}$$

(b) For rupture on the net effective area:

$$\phi_t = 0.75$$
$$P_n = F_u A_e \tag{3.1-2}$$

where

A_e = effective net area, in.^{2} (mm^2)
A_g = gross area of HSS, in.^{2} (mm^2)
F_y = specified minimum yield strength, ksi (MPa)
F_u = specified minimum tensile strength, ksi (MPa)
P_n = nominal axial strength, ksi (MPa)

4. COLUMNS AND OTHER COMPRESSION MEMBERS

4.1. Effective Length and Slenderness Limitations

1. Effective Length

The effective length factor K for compression members shall be taken as follows or as determined by rational analysis:

(a) In trusses that are made with HSS branch (web) members that are welded around their full perimeter to continuous HSS chord members, the effective length factor K that is used to modify the length between panel points for in-plane buckling, or between locations of lateral bracing for out-of-plane buckling, shall be not less than:

$$K = 0.75 \text{ for branch members}$$
$$K = 0.9 \text{ for chord members}$$

(b) In trusses that are made with HSS branch members that do not meet the requirements in Section 4.1.1(a) or with non-HSS branch members connected to continuous HSS chord members, the effective length factor K that is used to modify the length between panel points for in-plane buckling shall be not less than:

$$K = 1.0 \text{ for branch members}$$
$$K = 0.9 \text{ for chord members}$$

(c) In frames for which lateral stability is provided by diagonal bracing, shear walls or equivalent means, K shall be taken as unity, unless a lesser value can be justified by rational analysis.

(d) In frames for which lateral stability is dependent upon the flexural stiffness of rigidly connected beams and columns, K shall be determined by rational analysis.

2. Design by Plastic Analysis

Design by plastic analysis is permitted if the column slenderness parameter λ_c is less than or equal to $1.5K$ and the axial force in columns of unbraced frames due to factored gravity loads plus factored lateral loads does not exceed ϕ_c times $0.75F_yA_g$.

4.2. Design Compressive Strength

The design strength for flexural buckling of compression members is $\phi_c P_n$.

$$\phi_c = 0.85$$

$$P_n = F_{cr} A_g \tag{4.2-1}$$

F_{cr} shall be determined as follows:

(a) For $\lambda_c \sqrt{Q} \leq 1.5$,

$$F_{cr} = Q\left(0.658^{Q\lambda_c^2}\right) F_y \tag{4.2-2}$$

(b) For $\lambda_c \sqrt{Q} > 1.5$,

$$F_{cr} = \left[\frac{0.877}{\lambda_c^2}\right] F_y \tag{4.2-3}$$

where

$$\lambda_c = \frac{Kl}{r\pi}\sqrt{\frac{F_y}{E}} \tag{4.2-4}$$

Q shall be determined as follows:

(a) For $\lambda \leq \lambda_r$ in Section 2.2, $Q = 1$

(b) For $\lambda > \lambda_r$ in Section 2.2,

(i) For round HSS with $\lambda < 0.448E/F_y$,

$$Q = \frac{0.0379E}{F_y(D/t)} + \frac{2}{3} \tag{4.2-5}$$

(ii) For rectangular HSS,

$$Q = \frac{\text{effective area}}{A_g} \tag{4.2-6}$$

where the effective area is equal to the summation of the effective areas of the sides using

$$b_e = 1.91t\sqrt{\frac{E}{f}}\left[1 - \frac{0.381}{(b/t)}\sqrt{\frac{E}{f}}\right] \leq b \tag{4.2-7}$$

with $f = P_u/A_g$

5. BEAMS AND OTHER FLEXURAL MEMBERS

5.1. Design Flexural Strength

The design flexural strength $\phi_b M_n$ shall be determined as follows:

$$\phi_b = 0.90$$

(a) For round HSS, for $\lambda \leq \lambda_p$ in Section 2.2,

$$M_n = M_p = F_y Z \tag{5.1-1}$$

(i) For $\lambda_p < \lambda \leq \lambda_r$,

$$M_n = \left(\frac{0.0207}{D/t}\frac{E}{F_y} + 1\right) F_y S \tag{5.1-2}$$

(ii) For $\lambda_r < \lambda \leq 0.448E/F_y$,

$$M_n = \frac{0.330E}{D/t} S \tag{5.1-3}$$

(b) For rectangular HSS, for $\lambda \leq \lambda_p$ in Section 2.2,

$$M_n = M_p = F_y Z \tag{5.1-4}$$

(i) For $\lambda_p < \lambda \leq \lambda_r$,

$$M_n = \left[M_p - (M_p - M_r) \left(\frac{\lambda - \lambda_p}{\lambda_r - \lambda_p} \right) \right] \tag{5.1-5}$$

where

$$M_r = F_y S$$

(ii) For $\lambda > \lambda_r$,

$$M_n = F_y S_{eff} \tag{5.1-6}$$

where S_{eff} is the effective section modulus with the effective width of the compression flange taken as

$$b_e = 1.91t \sqrt{\frac{E}{F_y}} \left[1 - \frac{0.381}{(b/t)} \sqrt{\frac{E}{F_y}} \right] \leq b \tag{5.1-7}$$

L_b is not limited for HSS structures designed by elastic analysis.

5.2. Design Shear Strength

The design shear strength of unstiffened HSS $\phi_v V_n$ shall be determined as follows:

$$\phi_v = 0.9$$

(a) For round HSS,

$$V_n = F_{cr} A_g / 2 \tag{5.2-1}$$

where F_{cr} shall be the larger of

$$\frac{1.60E}{\sqrt{a/D}(D/t)^{5/4}} \quad \text{and} \quad \frac{0.78E}{(D/t)^{3/2}} \tag{5.2-2}$$

but shall not exceed $0.6F_y$ and a is the distance from maximum to zero shear force.

(b) For rectangular HSS,

$$V_n = F_n A_w \tag{5.2-3}$$

where

$$A_w = 2Ht \tag{5.2-4}$$

F_n shall be determined as follows:

(i) For $h/t \leq 2.45\sqrt{E/F_y}$,

$$F_n = 0.6F_y \tag{5.2-5}$$

(ii) For $2.45\sqrt{E/F_y} < h/t \leq 3.07\sqrt{E/F_y}$,

$$0.6F_y(2.45\sqrt{E/F_y})/(h/t) \quad (5.2\text{-}6)$$

(iii) For $3.07\sqrt{E/F_y} < h/t \leq 260$,

$$F_n = 0.458\pi^2 E/(h/t)^2 \quad (5.2\text{-}7)$$

5.3. Design by Plastic Analysis

Design by plastic analysis is permitted for compact round HSS with λ less than or equal to $0.0448E/F_y$ and for rectangular HSS with λ less than or equal to λ_p in Section 2.2.

For rectangular HSS bent about the major axis, the laterally unbraced length L_b of the compression flange adjacent to plastic hinge locations that are associated with the failure mechanism shall not exceed L_{pd}, where

$$L_{pd} = \frac{5000 + 3000(M_1/M_2)}{F_y} r_y \geq 3000r_y/F_y \quad (5.3\text{-}1)$$

and

F_y = specified minimum yield stress, ksi (MPa)
M_1 = smaller moment at the end of the unbraced length, kip-in. (N-mm)
M_2 = larger moment at the end of the unbraced length, kip-in. (N-mm)
r_y = radius of gyration about the minor axis, in. (mm)
M_1/M_2 is positive when moments cause reverse curvature and negative for single curvature

5.4. Design in Seismic Applications

For seismic applications, refer to AISC *Seismic Provisions for Structural Steel Buildings* (AISC, 1997) and *Seismic Provisions Supplement No. 2* (AISC, 2000).

6. TORSION MEMBERS

6.1. Design Torsional Strength

The design torsional strength $\phi_T T_n$ shall be determined as follows:

$$\phi_T = 0.90$$
$$T_n = F_{cr}C \quad (6.1\text{-}1)$$

where C is the HSS torsional constant.

F_{cr} shall be determined as follows:

(a) For round HSS, F_{cr} shall be the larger of

$$\frac{1.23E}{\sqrt{L/D}(D/t)^{5/4}} \quad \text{and} \quad \frac{0.6E}{(D/t)^{3/2}} \quad (6.1\text{-}2)$$

but shall not exceed $0.6F_y$.

(b) For rectangular HSS,

(i) For $h/t \leq 2.45\sqrt{E/F_y}$,

$$F_{cr} = 0.6F_y \quad (6.1\text{-}3)$$

(ii) For $2.45\sqrt{E/F_y} < h/t \leq 3.07\sqrt{E/F_y}$,

$$F_{cr} = 0.6F_y(2.45\sqrt{E/F_y})/(h/t) \tag{6.1-4}$$

(iii) For $3.07\sqrt{E/F_y} < h/t \leq 260$,

$$F_{cr} = 0.458\pi^2 E/(h/t)^2 \tag{6.1-5}$$

7. MEMBERS UNDER COMBINED FORCES

7.1. Design for Combined Flexure and Axial Force

The interaction of flexure and axial force shall be limited by Equations 7.1-1 and 7.1-2.

(a) For $P_u/\phi P_n \geq 0.2$,

$$\frac{P_u}{\phi P_n} + \frac{8}{9}\left(\frac{M_{ux}}{\phi_b M_{nx}} + \frac{M_{uy}}{\phi_b M_{ny}}\right) \leq 1.0 \tag{7.1-1}$$

(b) For $P_u/\phi P_n < 0.2$,

$$\frac{P_u}{2\phi P_n} + \left(\frac{M_{ux}}{\phi_b M_{nx}} + \frac{M_{uy}}{\phi_b M_{ny}}\right) \leq 1.0 \tag{7.1-2}$$

where

P_u = required axial tensile or compressive strength, kips (N)
P_n = nominal tensile or compressive strength determined in accordance with Sections 3.1 or 4.2, kips (N)
M_u = required flexural strength determined in accordance with LRFD Specification Section C1, kip-in. (N-mm)
M_n = nominal flexural strength determined in accordance with Section 5.1, kip-in. (N-mm)
x = subscript relating symbol to strong-axis bending
y = subscript relating symbol to weak-axis bending
ϕ = ϕ_t from Section 3.1 for tension
= 0.85 for compression
ϕ_b = 0.90

For biaxial flexure of round HSS that are laterally unbraced along their length and with end conditions such that the effective length factor K is the same for any direction of bending, the design is permitted to be based upon a single resultant moment M_{ur}, where:

$$M_{ur} = \sqrt{M_{ux}^2 + M_{uy}^2} \tag{7.1-3}$$

Alternatively, use of the provisions in LRFD Specification Appendix H3 (b) is permitted.

7.2. Design for Combined Torsion, Shear, Flexure, and/or Axial Force

When the required torsional strength is significant, the interaction of torsion, shear, flexure, and/or axial force shall be limited by Equation 7.2-1:

$$\left(\frac{P_u}{\phi P_n} + \frac{M_u}{\phi_b M_n}\right) + \left(\frac{V_u}{\phi_v V_n} + \frac{T_u}{\phi_T T_n}\right)^2 \leq 1.0 \qquad (7.2\text{-}1)$$

where

P_u = required axial tensile or compressive strength, kips (N)
P_n = nominal tensile or compressive strength determined in accordance with Sections 3.1 or 4.2, kips (N)
M_u = required flexural strength determined in accordance with LRFD Specification Section C1, kip-in. (N-mm)
M_n = lesser of $F_y S$ and M_n determined in accordance with Section 5.1, kip-in. (N-mm)
S = elastic section modulus
V_u = required shear strength at the section corresponding to M_u, kips (N)
V_n = nominal shear strength determined in accordance with Section 5.2, kips (N)
T_u = required torsional strength, kip-in. (N-mm)
T_n = nominal torsional strength determined in accordance with Section 6, kip-in. (N-mm)
ϕ = ϕ_t from Section 3.1 for tension
= 0.85 for compression
$\phi_b = \phi_v = \phi_T = 0.90$

8. CONCENTRATED FORCES ON HSS

The design strength ϕR_n at locations of concentrated forces on unstiffened HSS shall be determined from the applicable criteria in Sections 8.1 through 8.3.

8.1. Concentrated Force Distributed Transversely

When a concentrated force is distributed transversely to the axis of the HSS, the design strength ϕR_n shall be determined as follows:

(a) For round HSS,

$$\phi = 1.0$$

$$R_n = \frac{5F_y t^2}{1 - 0.81 b_1/D} Q_f \qquad (8.1\text{-}1)$$

where

b_1 = the width of the load, in. (mm)
Q_f = 1 for tension in the HSS
= $1 - 0.3f/F_y - 0.3(f/F_y)^2 \leq 1$ for compression in the HSS
f = the magnitude of the maximum compressive stress in the HSS due to axial force and bending at the location of the concentrated force, ksi (MPa)

(b) For rectangular HSS,

$$\phi = 1.0$$

$$R_n = \frac{10F_y t}{B/t} b_1 \leq F_{y1} t_1 b_1 \quad (8.1\text{-}2)$$

where

b_1 = the width of the loaded plate, in. (mm)
t_1 = the thickness of the loaded plate, in. (mm)
F_{y1} = specified minimum yield strength of the loaded plate, ksi (MPa)

(i) When the force is distributed across the full width of the rectangular HSS, the limit state of local web yielding shall be checked for both tensile and compressive forces and the limit state of web crippling shall be checked for compressive forces.

For local web yielding,

$$\phi = 1.0$$

$$R_n = 2F_y t(5k + N) \quad (8.1\text{-}3)$$

For web crippling,

$$\phi = 0.75$$

$$R_n = 1.6t^2[1 + 3N/h]\sqrt{EF_y} \quad (8.1\text{-}4)$$

where

k = outside corner radius of the HSS, which if not known is permitted to be taken as $1.5t$, in. (mm)
N = bearing length of the load along the length of the HSS, in. (mm)
h = flat width of side wall of the HSS as defined in Section 2.2.1, in. (mm)

(ii) When the force is distributed across a width of the rectangular HSS that is greater than $0.85B$ but less than $B - 2t$, the design strength shall not exceed ϕR_n, where

$$\phi = 1.0$$

$$R_n = 0.6F_y t(2t_1 + 2b_{ep}) \quad (8.1\text{-}5)$$

where $b_{ep} = 10b_1/(B/t) \leq b_1$

(iii) When compressive forces coincide on opposite faces of the rectangular HSS, the limit state of compression buckling of the webs shall be checked and the design strength shall not exceed ϕR_n, where

$$\phi = 0.90$$

$$R_n = \frac{48t^3\sqrt{EF_y}}{h} \quad (8.1\text{-}6)$$

8.2. Concentrated Force Distributed Longitudinally at the Center of the HSS Face

When a concentrated force is distributed longitudinally along the axis of the HSS at the center of the HSS face, the design strength ϕR_n shall be determined as follows:

(a) For round HSS,

$$\phi = 1.0$$

$$R_n = 5F_y t^2(1 + 0.25N/D)Q_f \tag{8.2-1}$$

(b) For rectangular HSS,

$$\phi = 1.0$$

$$R_n = \frac{F_y t^2}{1 - t_1/B}\left[\frac{2N}{B} + 4\sqrt{1 - t_1/B}\right] Q_f \tag{8.2-2}$$

where t_1 is the thickness of the loaded plate.

8.3. Concentrated Axial Force on the End of a Rectangular HSS with a Cap Plate

When a concentrated force acts on the end of an HSS with a cap plate and along the axis of the HSS, the design strength ϕR_n shall be determined for each loaded wall as follows. The limit state of local wall yielding shall be checked for both tensile and compressive forces and the limit state of wall crippling shall be checked for compressive forces.

For local wall yielding,

$$\phi = 1.0$$

$$R_n = (5t_1 + N)F_y t \leq BF_y t \tag{8.3-1}$$

For wall crippling,

$$\phi = 0.75$$

$$R_n = 0.80t^2\left[1 + 3\left(\frac{N}{B/2}\right)\left(\frac{t}{t_1}\right)^{1.5}\right]\sqrt{EF_y(t_1/t)} \tag{8.3-2}$$

where

t_1 = thickness of cap plate, in. (mm)
N = bearing length of the load across the width of the HSS, in. (mm)

9. CONNECTIONS AND FASTENERS

9.1. General Provisions for Connections and Fasteners

The provisions of LRFD Specification Section J1.1 through J1.11 and the provisions for bolts and threaded parts in LRFD Specification Sections J3.1 through J3.11 shall apply with the following additions and modifications.

1. Through Bolts

When connections are made using bolts that pass completely through an unstiffened HSS, the bolts shall be installed to only the snug-tight condition and the connection shall be considered to be a bearing-type connection. The bearing strength per loaded wall is ϕR_n, where

$$\phi = 0.75$$

$$R_n = 1.8 F_y d t \quad (9.1\text{-}1)$$

where

F_y = specified minimum yield strength of the HSS, ksi (MPa)
d = bolt diameter, in. (mm)
t = HSS wall thickness, in. (mm)

2. Special Connectors

The design strength of special connectors other than the bolts considered in LRFD Specification Table J3.2 shall be verified by tests.

3. Tension Connectors

When bolts or other connectors in tension are attached to an HSS wall, the strength of the HSS wall shall be determined by rational analysis.

9.2. Welds

The non-uniformity of load transfer along the line of weld due to differences in relative flexibility of HSS walls in HSS-to-HSS and similar connections shall be considered in proportioning such connections. In such cases, the strength of fillet welds shall be determined from LRFD Specification Section J2.4, excluding the alternative in Appendix J2.4, and the effective weld length L_e of groove and fillet welds shall be limited as follows:

(a) In T-, Y-, and Cross-connections with rectangular HSS as defined in Section 9.4,

$$L_e = 2H_b + B_b \text{ for } \theta \leq 50 \text{ degrees} \quad (9.2\text{-}1)$$

$$L_e = 2H_b \text{ for } \theta \geq 60 \text{ degrees} \quad (9.2\text{-}2)$$

Linear interpolation shall be used to determine L_e for values of θ between 50 and 60 degrees.

(b) In gapped K-connections with rectangular HSS as defined in Section 9.4,

$$L_e = 2H_b + 2B_b \text{ for } \theta \leq 50 \text{ degrees} \quad (9.2\text{-}3)$$

$$L_e = 2H_b + B_b \text{ for } \theta \geq 60 \text{ degrees} \quad (9.2\text{-}4)$$

Linear interpolation shall be used to determine L_e for values of θ between 50 and 60 degrees.

(c) When a transverse plate is welded to the face of an HSS member,

$$L_e = 2 \frac{10}{B/t} \frac{F_y t}{F_{y1} t_1} b_1 \leq 2b_1 \quad (9.2\text{-}5)$$

where

H_b = width of branch member wall that is parallel to the axis of the chord member, in. (mm)
B_b = width of branch member wall that is transverse to the axis of the chord member, in. (mm)
θ = least angle between branch member and chord member
B = width of chord member wall to which plate is attached, in. (mm)
b_1 = width of attached plate, in. (mm)
t = thickness of chord member wall, in. (mm)
t_1 = thickness of attached plate, in. (mm)
F_y = yield strength of HSS, ksi (MPa)
F_{y1} = yield strength of plate, ksi (MPa)

In lieu of the above, other rational criteria are permitted.

9.3. Other Connection Requirements

The provisions of LRFD Specification Sections J4 through J10 shall apply with the following additions and modifications.

1. Shear Rupture Strength

The design shear rupture strength along a path adjacent to a fillet weld on the HSS wall shall be taken as ϕR_n, where

$$\phi = 0.75$$
$$R_n = 0.6 F_u t L \quad (9.3\text{-}1)$$

where

t = HSS wall thickness, in. (mm)
L = length of weld, in. (mm)

2. Tension Rupture Strength

The design tension rupture strength along a path adjacent to a fillet weld on the HSS wall shall be taken as ϕR_n, where

$$\phi = 0.75$$
$$R_n = F_u t L \quad (9.3\text{-}2)$$

where

t = HSS wall thickness, in. (mm)
L = length of weld, in. (mm)

3. Punching Shear Rupture Strength

When a plate that is parallel to the longitudinal axis of an HSS and projects from the wall is subjected to a load that is parallel but eccentric or has a component perpendicular to the HSS wall,

$$\phi_t f t_p \leq 1.2 \phi_v F_u t \quad (9.3\text{-}3)$$

where

$\phi_v = 0.75$
$\phi_t = 0.90$
f = maximum stress in the plate perpendicular to the HSS wall, ksi (MPa)
t_p = plate thickness, in. (mm)
F_u = specified minimum tensile strength of the HSS, ksi (MPa)
t = HSS wall thickness, in. (mm)

4. Eccentric Connections

For trusses that are made with HSS that are connected by welding branch members to chord members, eccentricities within the limits of applicability in Section 9.4 are permitted without consideration of the resulting moments for the design of the connection, except in fatigue applications. In fatigue applications, refer to AWS D1.1.

9.4. HSS-to-HSS Truss Connections

HSS-to-HSS truss connections are defined as connections that consist of one or more branch members that are directly welded to a continuous chord that passes through the connection and shall be classified as follows:

(a) When the punching load in a branch member is equilibrated by beam shear in the chord member, the connection shall be classified as a T-connection when the branch is perpendicular to the chord and a Y-connection otherwise.
(b) When the punching load in a branch member is essentially equilibrated by loads in other branch member(s) on the same side of the joint, the connection shall be classified as a K-connection.
(c) When the punching load is transmitted through the chord member and is equilibrated by branch member(s) on the opposite side, the connection shall be classified as a Cross-connection.

When branch members transmit part of their load as K-connections and part of their load as T-, Y-, or Cross-connections, the design strength shall be determined by interpolation on the proportion of each in total.

For the purposes of this Specification, the centerlines of the branch member(s) and the chord members shall lie in a single plane and K-connections shall be used in the gapped configuration. For other configurations such as a multi-planar connection, a connection with a branch member that is offset so that its centerline does not intersect with the centerline of the chord, or when an overlapped K-connection is used, the provisions of AWS D1.1, other verified design procedures, tests, or rational analysis shall be used.

1. Definitions of Parameters

β = the width ratio; the ratio of branch diameter to chord diameter $= D_b/D$ for round HSS; the ratio of overall branch width to chord width $= B_b/B$ for rectangular HSS.

β_{eff} = the effective width ratio; the sum of the perimeters of the two branch members in a K-connection divided by eight times the chord width.

γ = the chord slenderness ratio; the ratio of one-half the diameter to the wall thickness $= D/2t$ for round HSS; the ratio of one-half the width to wall thickness $= B/2t$ for rectangular HSS.

η = the load length parameter, applicable only to rectangular HSS; the ratio of the length of contact of the branch with the chord in the plane of the connection to the chord width $= N/B$, where $N = H_b/\sin\theta$ and θ is the angle between the branch and chord.

ζ = the gap ratio; the ratio of the gap between the branches of a gapped K-connection to the width of the chord $= g/D$ for round HSS $= g/B$ for rectangular HSS.

2. Criteria for Round HSS

The design strength of the branch ϕP_n and/or ϕM_n for axial loads in the branch and for flexure in the branch, respectively, shall be determined from the limit states of chord wall plastification, punching shear rupture, and general collapse as applicable below.

The interaction of stress due to chord member forces and local branch connection forces shall be considered. The chord-stress interaction parameter Q_f shall be determined as

$$Q_f = 1.0 - \lambda\gamma U^2 \tag{9.4-1}$$

where U is the utilization ratio given by

$$U^2 = \left(\frac{P_u}{A_g F_y}\right)^2 + \left(\frac{M_u}{S F_y}\right)^2 \tag{9.4-2}$$

and

λ = 0.030 for axial load in the branch
= 0.044 for in-plane bending in the branch
= 0.018 for out-of-plane bending in the branch
P_u = required axial strength in chord, kips (N)
A_g = chord gross area, in.^{2} (mm^2)
F_y = chord yield strength, ksi (MPa)
M_u = larger required flexural strength in chord and connection, kip-ft (N-mm)
S = chord elastic section modulus, in.^{3} (mm^3)

2a. Limits of Applicability

The criteria herein are applicable only when the connection configuration is within the following limits of applicability:

(1) joint eccentricity: $-0.55D \le e \le 0.25D$, where D is the chord diameter and e is positive away from the branches
(2) branch angle: $\theta \ge 30°$
(3) wall stiffness: ratio of diameter to wall thickness less than or equal to 50 for chords and branches in T-, Y- and K-connections and less than or equal to 40 for chords of Cross-connections
(4) width ratio: $0.2 < D_b/D \le 1.0$
(5) gap: g greater than or equal to the sum of the branch wall thicknesses

2b. Branches with Axial Loads

For T-, Y-, and gapped K-connections, the design strength of the branch ϕP_n shall be the lower value obtained according to the limit states of chord wall plastification and punching shear rupture. For Cross-connections, the design strength of the branch

ϕP_n shall be the lowest value obtained according to the limit states of chord wall plastification, punching shear rupture, and general collapse.

(a) For the limit state of chord wall plastification,

$$\phi = 0.80$$
$$P_n \sin\theta = t^2 F_y [6\pi \beta Q_q] Q_f \quad (9.4\text{-}3)$$

where

$$Q_q = \left(\frac{1.7}{\alpha} + \frac{0.18}{\beta}\right) Q_\beta^{0.7(\alpha-1)}$$

(i) For $\beta \leq 0.6$, $Q_\beta = 1.0$

(ii) For $\beta > 0.6$, $Q_\beta = \dfrac{0.3}{\beta(1-0.833\beta)}$

α = chord ovalization parameter
= 1.7 for T- and Y-connections
= $1.0 + 0.7g/D_b$, $1 \leq \alpha < 1.7$ for gapped K-connections
= 2.4 for Cross-connections

(b) For the limit state of punching shear rupture,

$$\phi = 0.95$$
$$P_n \sin\theta = \pi D_b t (0.6F_y) \quad (9.4\text{-}4)$$

(c) For the limit state of general collapse,

$$\phi = 0.80$$
$$P_n \sin\theta = 1.8 t^2 F_y (1.9 + 7.2\beta) Q_\beta Q_f \quad (9.4\text{-}5)$$

2c. Branches with Flexure

For T-, Y-, gapped K-, and Cross-connections, the design strength of the branch ϕM_n shall be the lower value obtained according to the limit states of chord wall plastification and punching shear rupture.

(a) For the limit state of chord wall plastification,

$$\phi = 0.80$$
$$M_n \sin\theta = t^2 F_y [D_b/4][6\pi \beta Q_q] Q_f \quad (9.4\text{-}6)$$

where

$$Q_q = \left(\frac{2.1}{\alpha} + \frac{0.6}{\beta}\right) Q_\beta^{1.2(\alpha-0.67)}$$

(i) For $\beta \leq 0.6$, $Q_\beta = 1.0$

(ii) For $\beta > 0.6$, $Q_\beta = \dfrac{0.3}{\beta(1-0.833\beta)}$

α = chord ovalization parameter
= 0.67 for in-plane bending
= 1.5 for out-of-plane bending

For combinations of in-plane and out-of-plane bending, α shall be determined by interpolation and Q_f shall be determined with interpolated values of λ.

(b) For the limit state of punching shear rupture,

$$\phi = 0.95$$

$$M_n \sin\theta = D_b^2 t(0.6F_y) \quad (9.4\text{-}7)$$

2d. Branches with Combined Axial Loads and Flexure

The interaction of combined axial loads and flexure in HSS-to-HSS truss connections shall meet the following requirement:

$$\left(\frac{P_u}{\phi P_n}\right)^{1.75} + \frac{M_u}{\phi M_n} \le 1.0 \quad (9.4\text{-}8)$$

where

P_u = required axial strength of the branch, kips (N)
M_u = required flexural strength of the branch, kip-ft (N-mm)

3. Criteria For Rectangular HSS

The design axial strength of the branch ϕP_n and the design flexural strength ϕM_n of the branch shall be the lowest value obtained according to the limit states of chord wall plastification, punching shear rupture, sidewall strength, and uneven load distribution as applicable below.

The interaction of stress due to chord member forces and local branch connection forces shall be considered with the chord-stress interaction parameter Q_f, where

Q_f = 1 when the chord is in tension
$= 1.3 - 0.4U/\beta \le 1$ (9.4-9a)
in T-, Y-, and Cross-connections when the chord is in compression
$= 1.3 - 0.4U/\beta_{eff} \le 1$ (9.4-9b)
in K-connections when the chord is in compression

where

$$U = \left|\frac{P_u}{A_g F_y}\right| + \left|\frac{M_u}{SF_y}\right| \quad (9.4\text{-}10)$$

where

P_u = required axial strength of the chord, kips (N)
M_u = required flexural strength of the chord, kip-ft (N-mm)

3a. Limits of Applicability

The criteria herein are applicable only when the connection configuration is within the following limits:

(1) joint eccentricity: $-0.55H \le e \le 0.25H$, where H is the chord depth and e is positive away from the branches
(2) branch angle: $\theta \ge 30°$
(3) wall stiffness: ratio of wall width to wall thickness less than or equal to 35 for chords and branches; also less than or equal to $1.25\sqrt{E/F_{yb}}$ for branches in compression

(4) strength: F_y less than or equal to 52 ksi (360 MPa) for chord and branches
(5) chord and branch aspect ratio: $0.5 \leq$ ratio of depth to width ≤ 2.0
(6) ductility: $F_y/F_u \leq 0.8$
(7) other limits apply for specific criteria

3b. Branches with Axial Loads in T-, Y- and Cross-connections

For T-, Y-, and Cross-connections, the design strength of the branch ϕP_n shall be the lowest value obtained according to the limit states of chord wall plastification, punching shear rupture, sidewall strength, and uneven load distribution. In addition to the limits of applicability in Section 9.4.3a, β shall not be less than 0.25.

(a) For the limit state of chord wall plastification,

$$\phi = 1.0$$

$$P_n \sin\theta = F_y t^2 \left[\frac{2\eta}{1-\beta} + \frac{4}{\sqrt{(1-\beta)}} \right] Q_f \qquad (9.4\text{-}11)$$

This limit state need not be checked when $\beta > 0.85$.

(b) For the limit state of punching shear rupture,

$$\phi = 0.95$$

$$P_n \sin\theta = 0.6 F_y t B [2\eta + 2\beta_{eop}] \qquad (9.4\text{-}12)$$

In the above equation, the effective outside punching parameter $\beta_{eop} = 5\beta/\gamma$ shall not exceed β.

This limit state need not be checked when $\beta > 1 - 1/\gamma$ nor when $\beta < 0.85$.

(c) For the limit state of sidewall strength, the design strength for branches in tension shall be taken as the design strength for local sidewall yielding. For the limit state of sidewall strength, the design strength for branches in compression shall be taken as the lesser of the design strengths for local sidewall yielding and sidewall crippling.

This limit state need not be checked unless the chord member and branch member have the same width ($\beta = 1.0$).

(i) For the limit state of local yielding,

$$\phi = 1.0 \text{ for a branch in tension}$$
$$= 0.8 \text{ for a branch in compression}$$

$$P_n \sin\theta = 2t F_y (5k + N) \qquad (9.4\text{-}13)$$

where

k = outside corner radius of the HSS, which is permitted to be taken as $1.5t$ if unknown, in. (mm)
N = bearing length of the load along the length of the HSS, $H_b/\sin\theta$, in. (mm)
H_b = height of the branch, in. (mm)

(ii) For the limit state of sidewall crippling, in T- and Y-connections,

$$\phi = 0.75$$

$$P_n \sin\theta = 1.6t^2[1 + 3N/H]\sqrt{EF_yQ_f} \tag{9.4-14}$$

(iii) For the limit state of sidewall crippling in Cross-connections,

$$\phi = 0.80$$

$$P_n \sin\theta = \left[\frac{48t^3}{H - 4t}\right]\sqrt{EF_yQ_f} \tag{9.4-15}$$

(d) For the limit state of uneven load distribution,

$$\phi = 0.95$$

$$P_n = F_{yb}t_b[2H_b + 2b_{eoi} - 4_{tb}] \tag{9.4-16}$$

where

$$b_{eoi} = \frac{10}{B/t}\frac{F_y t}{F_{yb}t_b}B_b \leq B_b \tag{9.4-17}$$

F_{yb} = branch yield strength, ksi (MPa)
t_b = branch thickness, in. (mm)

This limit state need not be checked when $\beta < 0.85$.

3c. Branches with Axial Loads in Gapped K-connections

For gapped K-connections, the design strength of each branch ϕP_n shall be the lowest value obtained according to the limit states of chord wall plastification, punching shear rupture, shear yielding, and uneven load distribution. In addition to the limits of applicability in Section 9.4.3a, the following limits shall apply:

(1) $B_b/B \geq 0.1 + \gamma/50$
(2) $\beta_{eff} \geq 0.35$
(3) $\xi \geq 0.5(1 - \beta_{eff})$
(4) the smaller $B_b > 0.63$ times the larger B_b

(a) For the limit state of chord wall plastification,

$$\phi = 0.90$$

$$P_n \sin\theta = F_y t^2[9.8\beta_{eff}\sqrt{\gamma}\,]Q_f \tag{9.4-18}$$

(b) For the limit state of punching shear rupture,

$$\phi = 0.95$$

$$P_n \sin\theta = (0.6F_y)tB[2\eta + \beta + \beta_{eop}] \tag{9.4-19}$$

This limit state need only be checked if $B_b < B - 2t$ or the branch is not square.

(c) For the limit state of shear yielding of the chord in the gap, the design strength shal[l] be checked in accordance with Section 5.2.

(d) For the limit state of uneven load distribution,

$$\phi = 0.95$$

$$P_n = F_{yb}t_b[2H_b + B_b + b_{eoi} - 4t_b] \tag{9.4-20}$$

where

$$b_{eoi} = \frac{10}{B/t}\frac{F_y t}{F_{yb}t_b}B_b \leq B_b \tag{9.4-21}$$

This limit need only be checked if the branch is not square.

3d. Branches with Bending

Primary bending moments M_u due to applied loads, cantilevered beams, sidesway of unbraced frames, and other sources, shall be considered in the design as an additional axial tension or compression load

$$P_{ua} = \frac{M_u}{c \sin\theta} \tag{9.4-22}$$

where

$c = N/4$ for in-plane bending
$c = B_b/4$ for out-of-plane bending

10. GENERAL REQUIREMENTS FOR HSS FABRICATION

The following requirements shall be met in addition to the requirements of LRFD Specification Chapter M.

(1) When water can collect inside an HSS, either during construction or during service, HSS shall be sealed, provided with a drain hole at the base, or protected by other suitable means.
(2) HSS shall be cleaned with a suitable solvent if paint is specified per LRFD Specification Section M3.1.
(3) HSS shall be cleaned with a suitable solvent at locations of welding.

$$\phi R_n = F_{yb} t_b (2H_b + b_{eoi} + b_{eoi} - 4t_b) \qquad (9.4\text{-}20)$$

where

$$b_{eoi} = \frac{10}{B/t} \frac{F_y t}{F_{yb} t_b} B_b \le B_b \qquad (9.4\text{-}21)$$

$$P_u = \frac{M_u}{c \sin\theta} \qquad (9.4\text{-}22)$$

COMMENTARY
on the Load and Resistance Factor Design Specification for Steel Hollow Structural Sections

November 10, 2000

1. GENERAL PROVISIONS

1.1. Scope

For the purposes of this Specification, HSS are defined as hollow structural sections with constant wall thickness and a round, square, or rectangular cross-section that is constant along the length of the member. HSS are manufactured by forming skelp (strip or plate) to the desired shape and joining the edges with a continuously welded seam. Although the term pipe is commonly associated with round members that are used for fluid transmission, only steel pipe products that are used for structural purposes are included in the HSS definition. Published information is available describing the details of the various methods used to manufacture HSS (STI, 1996; Graham, 1965).

Because the design requirements for guaranteed pressure containment systems are more stringent than those for structural members, this Specification does not apply to members for which pressure containment is essential. Several other potential applications are also excluded from the scope of this Specification: (1) buried cylindrical shapes for which soil interaction is an important factor in the required strength; (2) stiffened HSS; (3) composite HSS; and, (4) HSS in fatigue applications. However, it is not intended that HSS with connection elements that also stiffen the cross-section be excluded from the scope.

Non-HSS products such as fabricated pipes and stiffened shells are excluded from the scope of this Specification. These are defined as members that are formed by shaping plates and joining them with one or more longitudinal seam welds, but neither in a tubing mill nor in accordance with a product specification. Although it is certainly possible to fabricate large pipes and shells to the same quality level as is commonly obtained with manufactured HSS, such quality is not universally assured by standard product specifications. Because the buckling strength of cylindrical sections is greatly influenced by geometric imperfections, there is good justification for excluding such products from the scope of this Specification. Accordingly, it is left to the Structural Engineer of Record (SER) to determine the suitability of such products for use with this Specification.

HSS are efficient structural members for the resistance of compressive and torsional forces. Consequently, they are increasingly selected in structural applications such as columns and members in plane trusses or space frames. HSS generally have a lower ratio of exposed surface area to volume when compared with other shapes, which results in reduced painting, fireproofing, and maintenance expense. Additionally, their low resistance to external fluid flow provides a distinct advantage for frameworks that are exposed to wind or water currents. The use of HSS has been limited in the past by difficulty in joining, but modern fabricating technology has overcome this disadvantage.

This Specification combines design guidelines from several sources. The primary basis for the recommendations is the design philosophy and criteria contained in the LRFD Specification. Because much of the LRFD Specification reflects the behavior of wide flange members, modifications have been made where HSS have been shown to behave differently or when interpretation of LRFD criteria to HSS applications can be clarified or simplified. Such modifications are explained in the Commentary. The criteria have also been modified to appear in a non-dimensional form.

In areas where the LRFD Specification contains little direct guidance for the design of buildings with HSS, such as connections, basic research, and criteria from other sources have been used. Much of the basic research is taken from the Comite International pour le Developpement et l'Etude de la Construction Tubulaire (CIDECT) programs, which have sponsored many projects in Europe and Canada concerning HSS construction. These have been incorporated into this Specification with modifications to provide design strengths that are comparable with those in the LRFD Specification. Consequently, this Specification is intended for the design of structural members in building structures that are normally encountered in structural engineering practice and experience. When the general uncertainty in loading or quality of control is substantially different, this Specification may not apply.

When HSS are used in the seismic force resisting system of buildings in high-seismic regions, the requirements in the AISC *Seismic Provisions for Structural Steel Buildings* (AISC, 1997) and Supplement No. 2 to the AISC *Seismic Provisions* (AISC, 2000) are applicable.

1.2. Material

ASTM A53 Grade B is included as an approved HSS material specification because it is the most readily available round HSS product in the United States. Other North American HSS products that have properties and characteristics that are similar to the approved ASTM products are produced in Canada under CAN/CSA-G40.21-M, "Structural Quality Steels." In addition, steel pipe is produced to other specifications that meet the strength, ductility, and weldability requirements of the materials in Section 1.2, but may have additional requirements for notch toughness or pressure testing.

Specified minimum yield and tensile strengths are summarized in Table C1.2-1 for various HSS material specifications and grades.

Round HSS can be readily obtained in most of the material specifications and grades in Table C1.2-1, although atmospheric-corrosion-resistant material (ASTM A618 and A847) may require a special order. For rectangular HSS, ASTM A500 Grade B is the most commonly available material, and a special order would be required for any other

TABLE C1.2-1
Minimum Tensile Properties of HSS Steels

Specification	Grade	F_y, ksi (MPa)	F_u, ksi (MPa)
ASTM A53	B	35 (240)	60 (415)
ASTM A500 (round)	A B C	33 (228) 42 (290) 46 (317)	45 (311) 58 (400) 62 (428)
ASTM A500 (rectangular)	A B C	39 (269) 46 (317) 50 (345)	45 (311) 58 (400) 62 (428)
ASTM A501	–	36 (248)	58 (400)
ASTM A618 (round)	I and II III	50 (345) 50 (345)	70 (483) 65 (448)
ASTM A847	–	50 (345)	70 (483)
CAN/CSA G40.21	350W	51 (350)	65 (450)

material. Depending upon size, either welded or seamless round HSS can be obtained. In North America, however, all ASTM A500 rectangular HSS for structural purposes are welded. Rectangular HSS differ from box sections in that they have uniform thickness except for some thickening in the rounded corners.

ASTM A500 Grade A material does not meet the ductility "limit of applicability" for direct connections in Section 9.4.3a. This limit requires that $F_y/F_u \leq 0.8$. In determining that other materials meet the ductility limit, it is important to note that ASTM A500 permits the yield to be determined by either the 0.2% offset method or at 0.5% elongation under load (EUL). Since ASTM A500 materials are cold-formed and have rounded stress-strain curves with no yield plateau, the latter method indicates yield strengths greater than the 0.2% offset. The ductility limit is intended to apply to yield strengths determined by 0.2% offset. However, mill reports may indicate the EUL yield, raising concerns that the material does not have adequate ductility. Supplemental tension tests may be required to determine the 0.2% offset yield.

Even though ASTM A501 includes rectangular HSS, hot-formed rectangular HSS are not currently produced in the United States. CAN/CSA G40.21 includes hot-formed Class H and cold-formed Class C. However, Class H rectangular HSS are produced by hot-finishing HSS that were manufactured by cold-forming. Hot-formed HSS have relatively low levels of residual stress, which enhances their performance in compression and may provide better ductility in the corners of rectangular HSS.

ASTM A500 tolerances allow for a wall thickness that is not greater than plus/minus 10 percent of the nominal value. Because the plate and strip from which electric-resistance-welded (ERW) HSS are made are produced to a much smaller thickness tolerance, manufacturers in the United States consistently produce ERW HSS with a wall thickness that is near the lower-bound wall thickness limit. Consequently, AISC and the Steel Tube Institute of North America (STI) recommend that 0.93 times the nominal wall thickness should be used for calculations involving engineering design properties of ERW HSS. This results in a mass variation that is similar to that found in other structural shapes. Submerged-arc-welded (SAW) HSS are produced with a wall thickness that is near the nominal thickness and require no such reduction. The design wall thickness and section properties based upon this thickness have been tabulated in AISC and STI publications since 1997.

1.3. Loads and Load Combinations

In many instances, the members in a framing system have no influence on the type or magnitude of the loads that must be considered in design. This is certainly true for dead loads, live loads, and impact loads. Horizontal crane forces, when they are present, and wind forces on enclosed structures are also not influenced by the type of members used in the framing system. Consequently, reference is made in this Specification to the applicable building code or ASCE 7. There are, however, two situations in which the use of HSS may allow a reduction in the design forces that must be considered: wind forces on exposed frameworks and pressures created by the enclosed nature of the HSS.

Wind forces on exposed frameworks can occur either in the final structural configuration or during construction. The shape of a round HSS has a lower resistance to fluid flow than shapes with flat elements (e.g., W shapes) and therefore reduces the wind forces. The general determination of wind pressures is given in the applicable building code or ASCE 7 when building codes do not apply. The determination of wind forces on exposed frameworks is a complex problem involving the solidarity ratio, shielding, and wind angle. In the absence of other wind-force reduction provisions that consider member shape, the provisions in this Specification can be used.

Wind forces on an exposed profile are proportional to a drag coefficient C, which varies with the type of profile. The value of C for a square shape with sharp corners is 2.03. Research (Hayus, 1968) indicates that the rounded corners of a square HSS reduce this drag coefficient. As shown in Figure C1.3-1, further reductions occur when rectangular HSS are oriented with the short side perpendicular to the wind. However,

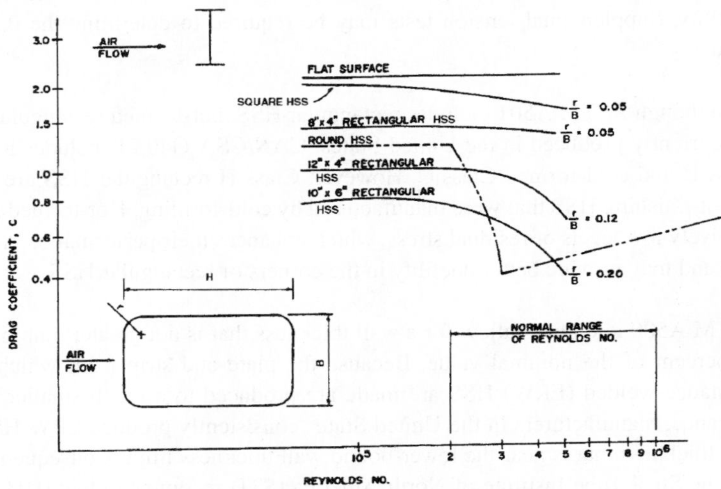

Fig. C1.3-1. Variation in drag coefficient.

TABLE C1.3-1
Drag Coefficients

Section	Corner Radius	C	$C/2.03$
Square HSS	$0.05B$	2.03	1.00
Round HSS	–	1.25	0.62
8-in. × 4-in. HSS	$0.05B$	1.4	0.69
12-in. × 4-in. HSS	$0.12B$	1.0	0.49
10-in. × 6-in. HSS	$0.20B$	0.8	0.39

C varies considerably with the orientation of the section relative to the wind and with the Reynolds number. The maximum values of C as indicated by the flat portions of the curves in Figure C1.3-1 provide conservative values for use in design. Table C1.3-1 lists these values along with their relative magnitude normalized by 2.03, which is the value of C for the square HSS. The Figure and Table also include the corresponding C for a round HSS. In this case, a one-third reduction (two-thirds factor) from the force on a flat surface is justified.

A similar reduction concept can be applied to wind acting on the short side of an HSS. Using the two data points for HSS with the sharpest corners (square and 8-in.× 4-in. rectangular HSS,) a linear reduction factor on the wind force is approximated by $0.4 + 0.6(B/H)$, in which the aspect ratio of the HSS is used as the variable. Conservatively, the reduction has been cut off at the one-third reduction for round HSS, even though the data indicates that with a larger corner radius, C may be less than that for a round HSS. There is no reduction for the wind force on the long side of a rectangular HSS.

2. DESIGN REQUIREMENTS

2.1. Effective Area of Tension Members

End connections for HSS in tension are commonly made by welding around the perimeter of the HSS. Alternatively, an end connection with gusset plates can be used. Single gusset plates are welded in longitudinal slots that are located at a centerline of the cross-section. Because welding around the end of the gusset plate is not recommended, the net area at the end of the slot in the HSS will be less than the gross area, as illustrated in Figure C2.1-1. A pair of gusset plates can be welded to opposite sides of a rectangular HSS with flare bevel groove welds.

For end connections of these three types, the general provisions of LRFD Specification Section B3 are simplified and the connection eccentricity $\bar{x}$ can be explicitly defined. These types of gusset-plate connections and the definitions of $\bar{x}$ and L are illustrated in Figure C2.1-2.

2.2. Local Buckling

The wall slenderness parameters and the slenderness limits λ_p and λ_r in Table 2.2-1 are taken from LRFD Specification Section B5, but have been presented in a non-dimensional form. The design wall thickness as defined in Section 1.2.2 is used to determine slenderness.

The limits for rectangular HSS walls in uniform compression have been used in AISC Specifications since 1969. They are based upon the work of Winter (1947) where adjacent stiffened compression elements in box sections of uniform thickness were observed to provide negligible torsional restraint for one another along their corner edges. The λ_p limit for plastic analysis is adopted from CSA (1994). The web slenderness limits are the same as those used for webs in wide-flange shapes.

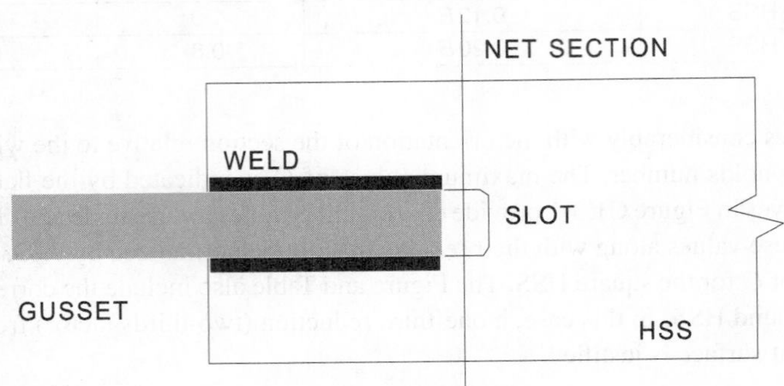

Fig. C2.1-1. Net area through slot for single gusset plate.

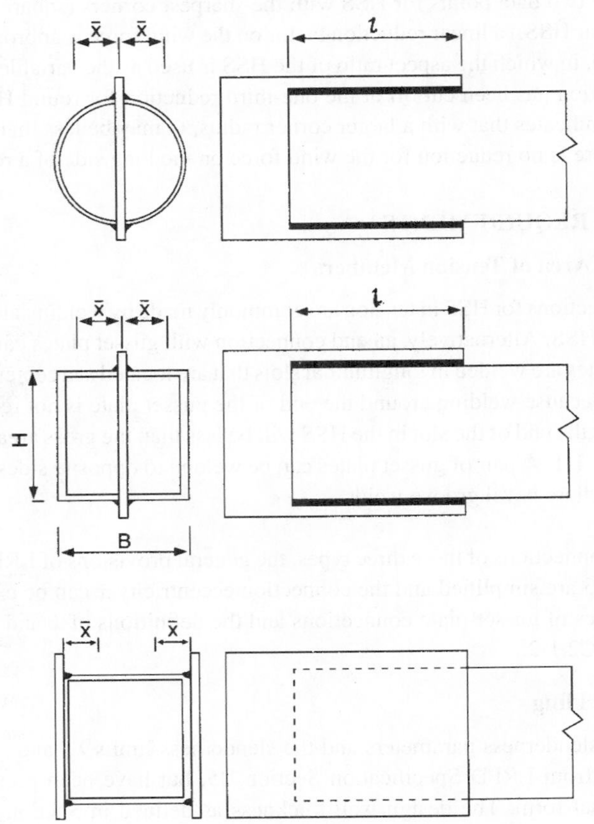

Fig. C2.1-2. Determination of $\bar{x}$.

The λ_r limit for round HSS in compression was first used by AISC in the 1978 AISC ASD Specification. It was recommended by Schilling (1965) based upon research at Cornell University that produced provisions in the 1968 AISI Cold-Formed Specification (Winter, 1968). The same limit was also used to define a compact shape in bending in the 1978 AISC ASD Specification. However, the limits for λ_p and λ_r were changed in the 1986 AISC LRFD Specification based upon experimental research on round HSS in bending (Sherman, 1985; Galambos, 1988). Excluding the use of round HSS with $D/t > 0.448E/F_y$ was also recommended by Schilling (1965).

Lower values of λ_r are specified for high-seismic design in the AISC *Seismic Provisions for Structural Steel Buildings* (AISC, 1997; AISC, 2000) based upon tests (Lui & Goel, 1987) that have shown that rectangular HSS braces subjected to reversed axial load fracture catastrophically under relatively few cycles if a local buckle forms. This was confirmed more recently in tests (Sherman, 1995) where rectangular HSS braces sustained over 500 cycles when a local buckle did not form, even though general column buckling had occurred, but failed in less than 40 cycles when a local buckle developed. The seismic λ_r is based upon tests (Lui & Goel, 1987) of HSS that had a small enough b/t so that braces performed satisfactorily for members with reasonable column slenderness. Filling the rectangular HSS with lean concrete (concrete mixed with a low proportion of cement) has been shown to effectively stiffen the HSS walls and improve cyclic performance. Because fracture at a low number of cycles is also possible with cold-formed round HSS braces that form local buckles, the limiting λ_r in compression in the 1992 AISC *Seismic Provisions* was chosen to be the same as for plastic analysis.

3. TENSION MEMBERS

3.1. Design Tensile Strength

Except for HSS that are subjected to cyclic load reversals, there is no information that the factors governing the strength of HSS in tension differ from those for other structural shapes. Therefore, the criteria in Section 3.1 are identical to those in LRFD Specification Section D1. However, because the number of different end connection types that are practical for HSS is limited, the determination of the net effective area A_e can be simplified as it has been in Section 2.1.

4. COLUMNS AND OTHER COMPRESSION MEMBERS

4.1. Effective Length and Slenderness Limitations

The high torsional strength and stiffness of an HSS provides increased restraint for members that frame into it when compared to that provided by other structural shapes. For example, the connection between an HSS branch member and a continuous HSS chord member is commonly made with a continuous weld around the perimeter of the branch member. In such a connection, the chord then provides considerable end restraint both in-plane and out-of-plane of the truss; the HSS branch member also provides a degree of lateral restraint against rotation of the chord. In both cases, advantage can be taken of the end restraint by using the effective lengths in Section 4.1.1(a). The use of K equal to 0.75 for the branch members and 0.9 for the chord between bracing points is based upon the recommendations of CIDECT research (Rondal, 1992).

It is important to note that even though end restraint is present, it is still reasonable to assume that the truss joints are pinned. Secondary moments due to end fixity may be neglected unless the joint eccentricity exceeds the limits of applicability in Section 9.4 or fatigue is a design consideration. For fatigue applications, refer to AWS D1.1.

The provisions for unbraced frames and braced frames are taken from the LRFD Specification. Values of K for compression members in frames can be determined from the alignment charts and equations in the corresponding LRFD Specification Commentary.

4.2. Design Compressive Strength

The axial compressive strength of an HSS is influenced by its method of production, shape, and dimensions and is further complicated by large differences between theoretical predictions and experimental results for local buckling, especially for round HSS. Rather than repeat the excellent discussions that can be found elsewhere concerning the behavior of various hollow cross-sections (Schilling, 1965; McGuire, 1968; Galambos, 1988; and Sherman, 1992), the results and basis for the design equations is explained herein. Some of the major considerations that must be included in comprehensive criteria for HSS are as follows:

1. As with any thin-walled member of constant cross-section, either overall flexural buckling or local buckling can be the controlling limit state. Flexural buckling strength under axial load is governed by the slenderness ratio KL/r, the yield strength F_y, residual stresses, and initial out-of-straightness. Local buckling of rectangular HSS is based upon the principles of plate buckling theory and is governed by the square of the width-thickness ratio. In very short round sections, local buckling is similar to that for an infinitely wide plate and the ratio of the length to the thickness is of prime importance. For longer round sections, the local buckling configuration consists of approximately square waves along the length and around the circumference and the strength is a function of the ratio of the diameter to the thickness. The local buckling strength of thin round sections is extremely sensitive to initial distortion from the perfect cylindrical surface. Because manufactured HSS generally have less initial distortion than a comparable fabricated cylinder, local buckling generally occurs at a higher load.

2. The strength of short compression members is governed by local buckling, whereas the strength of long compression members is governed by flexural buckling. In compression members of intermediate length, there is an interaction between local and flexural buckling.

3. Rectangular HSS that are cold-formed from round HSS have a rounded stress-strain curve due to through-thickness residual stresses that vary from about $0.4F_y$ to $0.8F_y$. This reduces the flexural buckling strength of intermediate-length members to below that of comparable hot-formed HSS with similar yield strength. Residual stresses are not as large in cold-formed round HSS and are negligible in hot-formed HSS.

4. Manufactured HSS tend to have small initial out-of-straightness, which increases the strength of intermediate-length HSS relative to other shapes.

Because many HSS have a wall slenderness ratio that exceeds λ_r, the equations for flexural buckling include the interaction with local buckling as given in LRFD Specification Appendix B. In effect, the local buckling reduction factor Q reduces the yield stress in both the inelastic flexural buckling equation and in the slenderness λ_c. Its influence is illustrated in Figure C4.2-1. Of course, if λ is less than or equal to λ_r, the local buckling reduction factor Q is equal to unity and the equations reduce to the flexural buckling equations of LRFD Specification Chapter E.

For round HSS, Q has its origins in the critical stress from local buckling test data with conservative adjustments in an early AISI Specification (Winter, 1968). The constants have been further adjusted for a non-dimensional form and a design strength format rather than an allowable stress format.

The effective width equation that is used to obtain Q for rectangular HSS is also from an early AISI Specification (Winter, 1968). The constants in this equation were established for closed cross-sections with uniform thickness. The flexural buckling equations and ϕ_c equal to 0.85 are the same as those used in the LRFD Specification and can be used conservatively for all HSS. However, for hot-formed HSS or ASTM A500 Grade D HSS, which have lower residual stresses but are generally unavailable in the United States, a higher resistance factor could be justified and would give factored resistances that are comparable to those in other specifications that use multiple column curves. This is based upon the extensive CIDECT column test program in Europe and several less extensive studies in North America, as summarized in Table C4.2-1. A large amount of CIDECT data for seamless HSS is not included. The test data for HSS in axial compression are illustrated in Figure C4.2-2 for round HSS and Figure C4.2-3 for rectangular HSS. The curve for the LRFD nominal column strength ($\phi = 1.0$) is superimposed on the Figures. The yield load used to non-dimensionalize the ordinates is the cross-sectional area times the tensile-test yield strength, which was determined by

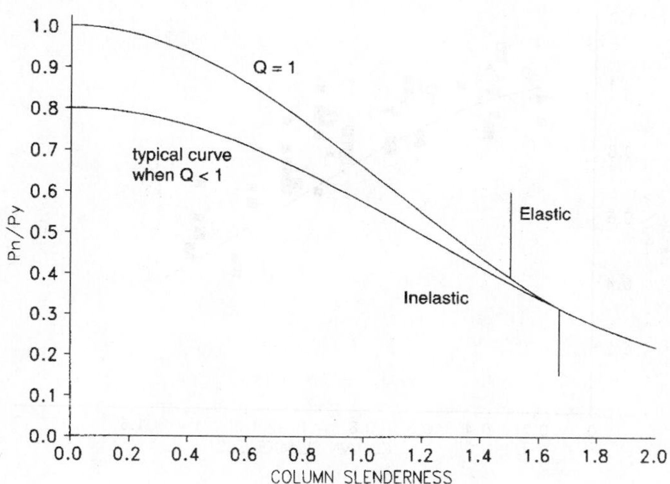

Fig. C4.2-1. Influence of Q on column strength.

Table C4.2-1
Summary of Test Programs on HSS Columns

Round HSS			
Type	**Symbol in Figure C4.2-2**	**Reference**	**No. of Tests**
Hot-Formed	□	CIDECT #	10
Cold-Formed	■	CIDECT #	65
		Sherman, 1980	4
Fabricated Pipe	✱	Chen, 1977	10
		Yang, 1987	6
Rectangular HSS			
Type	**Symbol in Figure C4.2-2**	**Reference**	**No. of Tests**
Hot-Formed	□	CIDECT #	88
		Estuar, 1965	10
Cold-Formed	■	CIDECT #	132
		Bjorhovde, 1979	1
		Key, 1985	11
Cold-Formed Stress Relieved	✱	Bjorhovde, 1979	19
		Sherman, 1969	2

No reference cited for CIDECT data.

the 0.2 percent offset method for the cold-formed HSS that exhibit rounded stress-strain curves. In the figures, solid squares represent data from tests on cold-formed HSS while open squares represent those on hot-formed HSS. Asterisks represent special cases of fabricated pipe in Figure C4.2-2 and stress-relieved rectangular HSS in Figure C4.2-3. Note that cold-formed ASTM A500 rectangular HSS that are stress relieved to 840°F (450°C) have strengths that are similar to those for hot-formed HSS.

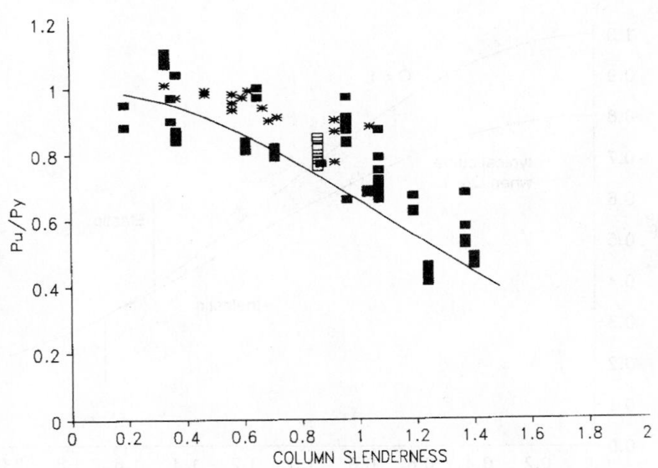

Fig. C4.2-2. Test Data for Round HSS.

5. BEAMS AND OTHER FLEXURAL MEMBERS

5.1. Design Flexural Strength

The provisions for the nominal flexural strength of HSS include the limit states of yielding, inelastic local buckling, and elastic local buckling. Round HSS, square HSS, and rectangular HSS bent about the minor axis are not subject to lateral-torsional buckling. Furthermore, for rectangular HSS bent about the major axis, the limit state of lateral-torsional buckling is not included in this Specification in spite of LRFD Specification provisions that reduce the flexural strength when the unbraced length exceeds limiting unbraced length values.

Because of the high torsional resistance of the closed cross-section, the critical unbraced lengths L_p and L_r that correspond to the development of the plastic moment and the yield moment, respectively, are very large. For example, as shown in Figure C5.1-1, an HSS 20×4×$^5/_{16}$, which has one of the largest depth-width ratios among standard HSS, has L_p of 8.7 ft and L_r of 137 ft as determined in accordance with the LRFD Specification. An extreme deflection limit might correspond to a length-to-depth ratio of 24 or a length of 40 ft for this member. Using the specified linear reduction between the plastic moment and the yield moment for lateral-torsional buckling, the plastic moment is reduced by only 7 percent for the 40-ft length. In most practical designs where the moment gradient C_b is also a factor, the reduction will be nonexistent or insignificant.

The failure modes and post-buckling behavior of round HSS can be grouped into three categories (Galambos, 1998; Sherman, 1992).

(a) For low D/t, a long plastic plateau occurs in the moment-rotation curve. The cross-section gradually ovalizes, local wave buckles eventually form, and the moment

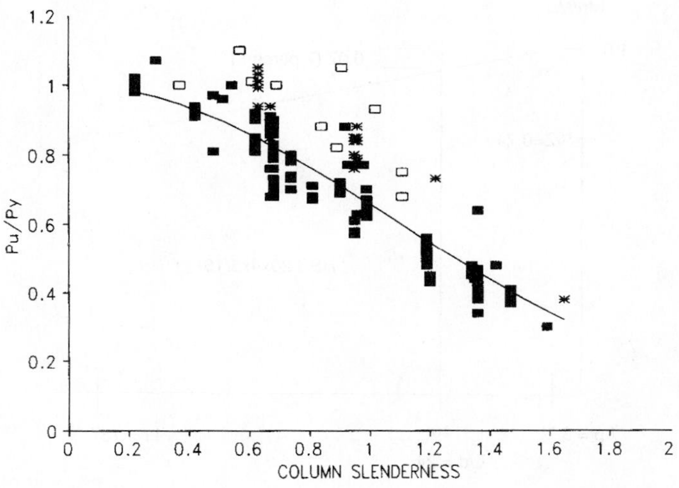

Fig. C4.2-3. Test Data for Rectangular HSS.

resistance subsequently decays slowly. Flexural strength may exceed the theoretical plastic moment due to strain hardening.

(b) For intermediate D/t, the plastic moment is nearly achieved but a single local buckle develops and the moment resistance decays slowly with little or no plastic plateau region.

(c) For high D/t HSS, multiple buckles form suddenly with very little ovalization and the flexural strength drops quickly.

The flexural strength criteria for round HSS reflect these three regions of behavior and are based upon five experimental programs involving hot-formed seamless pipe, electric-resistance-welded pipe, and fabricated tubing (Galambos, 1998). The criteria are the same as those in LRFD Specification Appendix F.

The criteria for local buckling of rectangular HSS are also the same as those in LRFD Specification Appendix F. The equation for the effective width of the compression flange when b/t exceeds λ_r is the same as that used for rectangular HSS in axial compression except that the stress is taken as the yield stress. This implies that the stress in the corners of the compression flange is at yield when the ultimate post-buckling strength of the flange is reached. When using the effective width, the nominal flexural strength is determined from the effective section modulus to the compression flange using the distance from the shifted neutral axis. A slightly conservative estimate of the nominal flexural strength can be obtained by using the effective width for both the compression and tension flange, thereby maintaining the symmetry of the cross-section and simplifying the calculation.

The shape factor (Z/S) for HSS is generally between 1.15 and 1.4. Hence, the maximum limit on M_p of $1.5M_y$ in LRFD Specification Chapter F is satisfied for HSS and need not be explicitly checked.

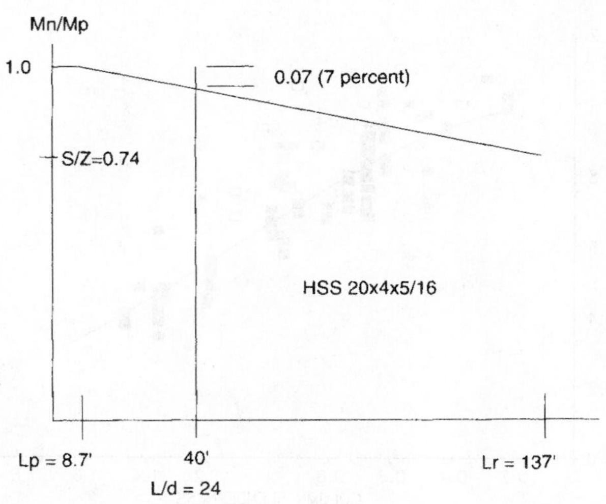

Fig. C5.1-1. Lateral-torsional buckling of rectangular HSS.

In order to use plastic analysis for design, the HSS must have sufficient rotational capacity at the plastic moment to develop the hinge mechanism. This requires a compact cross-section to prevent premature local buckling and lateral bracing for rectangular HSS that are bent about the major axis to prevent lateral-torsional buckling in the vicinity of hinges. The requirement for L_{pd} is taken from the LRFD Specification. The provisions for the wall slenderness defining the compact section are more restrictive than λ_p, which defines when M_p can be achieved without consideration of rotational capacity. This is due to the higher shape factors for HSS relative to wide-flange shapes where λ_p also applies to plastic analysis. The more restrictive D/t for round HSS is also used in the LRFD Specification, and the b/t limit for rectangular HSS is based upon test results for cold-formed sections (Korol, 1972), which are most frequently used in North America.

5.2. Design Shear Strength

Little information is available on round HSS subjected to transverse shear, and recommendations are based upon criteria for local buckling of cylinders subjected to torsion.

However, since torsion is generally constant along the member length and transverse shear usually has a gradient, it is recommended to take the critical stress for transverse shear as 1.3 times the critical stress for torsion (Galambos, 1998; Brockenbrough and Johnston, 1981). The torsion equations apply over the full length of the member, but for transverse shear it is reasonable to use a length to the location of zero shear. Only thin HSS may require a reduction in the shear strength based upon first shear yield. Even in this case, shear will only govern the design of round HSS beams for the case of thin sections with short spans.

In the equation for the nominal shear strength V_n of round HSS, it is assumed that the shear stress is at the neutral axis, which is calculated as VQ/Ib, is at F_{cr}. For a thin round section with radius R and thickness t, I is $\pi R^3 t$, the first moment of area Q is $2R^2 t$, and $b = 2t$. This gives the stress at the centroid as $V/(\pi Rt)$, in which the denominator is recognized as half the area of the round HSS.

The provisions for the nominal shear strength of rectangular HSS are the same as those for unstiffened webs in the LRFD Specification. For rectangular HSS, the nominal shear strength considers the two webs in the section.

6. TORSION MEMBERS

HSS are frequently used in space-frame construction and in other situations wherein significant torsional moments must be resisted by the members. Because of its closed cross-section, an HSS is far more efficient in resisting torsion than an open cross-section such as a W-shape or channel. While normal and shear stresses due to warping of the cross-section are usually significant in shapes of open cross-section, they are insignificant for closed cross-sections, and the total torsional moment can be assumed to be resisted by pure torsional shear stresses, which are sometimes called St. Venant torsional stresses.

In HSS, the pure torsional shear stress is assumed to be uniformly distributed and is equal to the torsional moment T_u divided by a torsional shear constant for the cross-section C. In a limit state format, the nominal torsional resisting moment is the shear constant times the critical shear stress F_{cr}.

For a round HSS, the torsional shear constant is equal to the polar moment of inertia divided by the radius:

$$C = \frac{\pi\left(D^4 - D_i^4\right)}{32D/2} = \frac{\pi(D-t)^2 t}{2} \tag{C6-1}$$

where D_i is the inside diameter.

For a rectangular HSS, the torsional shear constant is obtained as $2tA_o$ using the membrane analogy (Timoshenko, 1956), where A_o is the area bounded by the midline of the section. Conservatively assuming an outside corner radius that is equal to $2t$, the midline corner radius is $1.5t$, and

$$A_o = (B-t)(H-t) - \frac{9(4-\pi)}{4}t^2 \tag{C6-2}$$

which yields

$$C = 2(B-t)(H-t)t - 4.5(4-\pi)t^3 \tag{C6-3}$$

The resistance factor and nominal strengths used for torsion are the same as those used for flexural shear.

When considering local buckling in round HSS subjected to torsion, most structural members will either be long or of moderate length and the criteria for short cylinders will not apply. The elastic local buckling strength of long cylinders is unaffected by end conditions, and the critical stress is as given by Galambos (1998) as:

$$F_{cr} = \frac{K_t E}{(D/t)^{3/2}} \tag{C6-4}$$

The theoretical value of K_t is 0.73, but a value of 0.6 is recommended to account for initial imperfections. An equation for the critical elastic local stress for round HSS of moderate length ($L > 5.1D^2/t$) where the edges at the ends are not fixed against rotation is given by Schilling (1965) and Galambos (1988) as:

$$F_{cr} = \frac{1.23E}{\sqrt{L/D}(D/t)^{5/4}} \tag{C6-5}$$

This equation includes a 15 percent reduction to account for initial imperfections.

The local buckling equations are plotted in Figure C6-1 with modulus of elasticity E equal to 29,000 ksi. Although there is some inconsistency concerning the division between long and moderately long HSS, it appears from Figure C6-1 that the expression for sections of moderate length is valid for most practical lengths. It is also evident that it would be uneconomical to neglect the increase in torsional strength for members of moderate length. Therefore, in this Specification, the length effect is included for the simple end conditions and the approximately 10 percent increase in buckling strength is neglected for edges fixed at the ends. A limitation is provided so that the shear yield strength is not exceeded.

The critical stress criteria for rectangular HSS are identical to the flexural shear criteria of Section 5.2. The shear distribution due to torsion is uniform in the longest sides of a rectangular HSS, which is the same distribution that is assumed to exist in a beam web. Therefore, it is reasonable that the criteria for buckling is the same in both cases.

7. MEMBERS UNDER COMBINED FORCES

7.1. Design for Combined Flexure and Axial Force

The provisions for interaction between flexure and axial force are taken directly from LRFD Specification Section H1. As stated in LRFD Specification Section C1, for structures designed on the basis of elastic analysis, the required flexural strength M_u shall be determined from a second-order analysis. Second-order effects must be considered not only for the beam-column member, but also for the other framing members that connect to it and the associated connections. In lieu of a second-order analysis, an approximate second-order procedure is permitted to determine M_u as detailed in LRFD Specification Section C1.

In the case of biaxial flexure on round HSS, which have the same section modulus about any axis, the LRFD Specification interaction equations lead to inconsistent results when compared with design for a resultant moment (Pillai and Ellis, 1971). The interaction equation adds the effects of M_{ux} and M_{uy} which can be conservative by as much as 40 percent when compared to combining them vectorially. Therefore, provision has been included in this Specification for the direct design based upon the resultant moment M_{ur}.

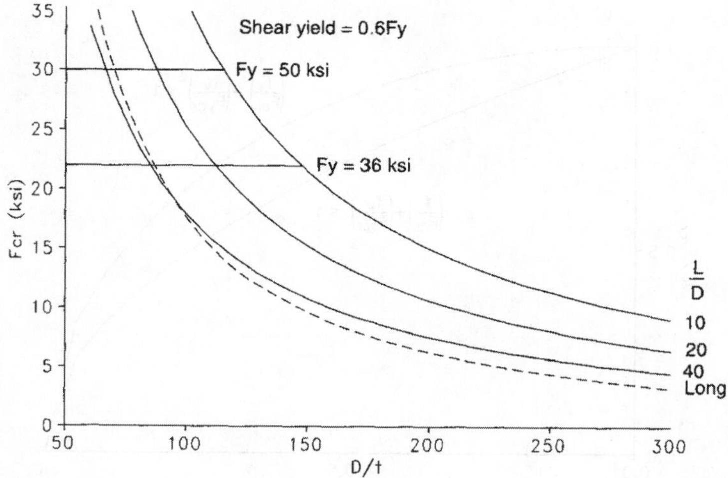

Fig. C6-1. Torsion F_{cr} for round HSS.

7.2. Design for Combined Torsion, Shear, Flexure, and/or Axial Force

Several interaction equation forms have been proposed for load combinations that produce both normal and shear stresses. In one common form, the normal and shear stresses are combined elliptically with the sum of the squares (Felton and Dobbs, 1967).

$$\left(\frac{f}{F_{cr}}\right)^2 + \left(\frac{f_v}{F_{vcr}}\right)^2 \leq 1 \tag{C7.2-1}$$

In a second form, the first power of the normal stresses is used.

$$\left(\frac{f}{F_{cr}}\right) + \left(\frac{f_v}{F_{vcr}}\right)^2 \leq 1 \tag{C7.2-2}$$

These equations are plotted in Figure C7.2-1. The latter form is more conservative but not overly conservative (Schilling, 1965) and is the basis for the interaction equation used in this Specification in a limit-states format. Flexure and axial force effects are combined linearly and then combined with the square of a linear combination of flexural shear and torsion effects. When an axial load is present, the required flexural strength M_u is obtained by second-order analysis. Because the interaction is based upon an elastic combination of stresses, the nominal flexural strength M_n is limited to the yield moment.

8. CONCENTRATED FORCES ON HSS

Concentrated forces result from line loads that are applied to the HSS through a plate, a connecting element that has a flange, or a similar element. The line load can be distributed either transversely or longitudinally as shown in Figure C8.1-1.

Round HSS with either transverse or longitudinal line loads fail by local plastic distortion of the section. The resistances for these loadings are given by Packer and Henderson (1997). Because the resistance equations given by Packer and Henderson

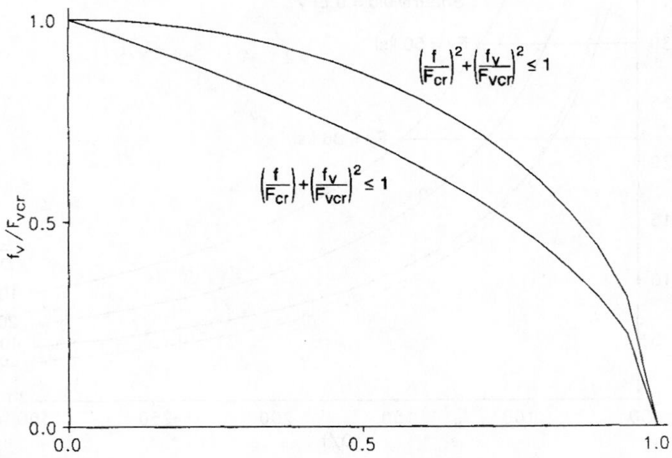

Fig. C7.2-1. Stress interaction of shear and normal effects.

account for variations between experimental strengths and theory, the appropriate resistance factor ϕ is 1.0. When there are compressive normal stresses in the HSS in the vicinity of the line load, the resistance to local plastic distortion is reduced and a reduction factor Q_f is applied. In Packer and Henderson, Q_f is determined as a function of f, which is the maximum compressive stress in the chord (a negative number). In this Specification, the Packer and Henderson equation for Q_f has been modified so that the magnitude of f is used (an absolute value).

The limit state for a rectangular HSS depends upon the width of the load relative to the width of the loaded face of the HSS. For transverse loads, the variable stiffness across the face of the HSS causes a non-uniform distribution of the force in the connecting element delivering the load. Failure may occur due to an excessive concentration of force on an effective width of the connecting element at the resistance given by

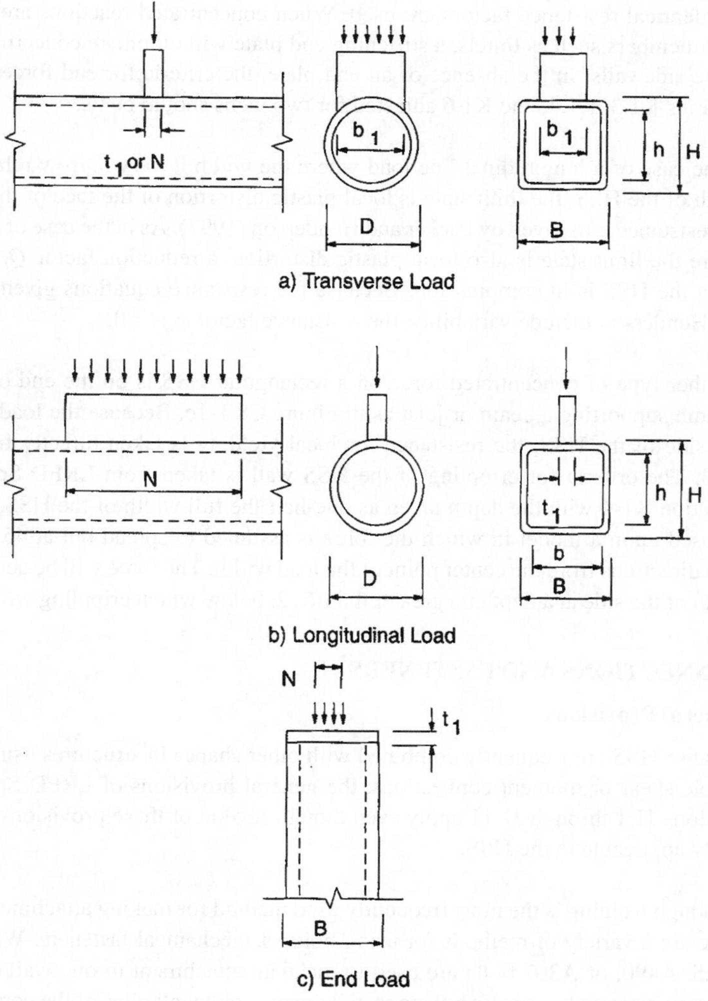

Fig. C8.1-1. Concentrated force configurations on HSS.

Packer and Henderson (1997). The limit of $F_{y1}t_1b_1$ does not represent the limit state for the axial strength of the loaded plate, but limits the effective width given by Packer and Henderson (1997) from exceeding the width of the HSS. When the force is across nearly the full width of the HSS, the concentration of force at the end of the connecting element can cause a punching shear rupture through the wall of the HSS. Again, the resistance is as given by Packer and Henderson (1997).

When the force is across the full width, failure of the sidewalls of the HSS is possible. The resistance for local sidewall yielding is taken from LRFD Specification Section K1.3, adjusted for the presence of two sidewalls acting as webs. The resistance for sidewall crippling is consistent with LRFD Specification Section K1.4 in a non-dimensional form and accounting for two sidewalls. The same is true for the compression buckling criteria from LRFD Specification Section K1.6 when compressive forces are applied on opposite faces of the HSS. For the limit states taken from the LRFD Specification, the identical resistance factors are used. When concentrated reactions are present in HSS members such as lintels, a stiffening end plate will eliminate concerns of failure of the sidewalls. In the absence of an end plate, the criteria for end forces in LRFD Sections K1.3, K1.4, and K1.6 adjusted for two webs should be used.

In the case of a longitudinal line load where the width is very narrow relative to the width of the HSS, the limit state is local plastic distortion of the face of the HSS and the resistance is as given by Packer and Henderson (1997). As in the case of round HSS where the limit state is also local plastic distortion, a reduction factor Q_f is applied when the HSS is in compression. Because the resistance equations given by Packer and Henderson include variability, the resistance factor ϕ is 1.0.

Another type of concentrated force on a rectangular HSS is on the end of a capped column supporting a beam or joist as in Figure C8.1-1c. Because the load is on only one side of the HSS, the resistance for local yielding is taken directly from LRFD K1.3. The criteria for crippling of the HSS wall is taken from LRFD Specification Equation K1-4 with the depth taken as one-half the full width of the HSS, $B/2$. This is based upon a model in which the force is assumed to spread out at 45 degrees in both directions from the center point of the load width. The force will be across the full width of the side at a depth no greater than $B/2$, below which crippling will not occur.

9. CONNECTIONS AND FASTENERS

9.1. General Provisions

Because HSS are frequently combined with other shapes in structures using standard simple shear or moment connections, the general provisions of LRFD Specification Sections J1.1 through J1.11 apply even though several of these provisions are not directly applicable to the HSS.

Although welding is the most frequently used method for making attachments to HSS, there are a variety of methods for using bolts or mechanical fasteners. When ASTM A325, A490, or A307 bolts are used to make an attachment to one wall of an HSS, using an open end or access hole to install the nut so that all plies of the connection are in contact within the grip of the bolt, the provision of LRFD Specification Section J3 apply. These provisions also apply to other portions of a structure that includes HSS. However, if a bolt passes completely through the HSS, an attempt to fully tension the bolt would distort the cross-section of the HSS wall and the specified minimum installed tension could not be achieved without damaging the member. Therefore, through-bolts should only be installed in the snug-tight condition. The bearing strength of such connections as given in Section 9.1.1 is consistent with that in LRFD Specification Section J8. The designer should note that some connection details do not distribute the load to both walls equally. Additionally, some details may induce bending in the through-bolt.

There are a variety of blind bolts, unique installation methods, welded studs, and special connectors that could be used with HSS. Many of these alternatives are still in a developmental stage and general criteria for their use are not available. In such cases, test data must be used to justify their suitability for a particular application.

With any connection configuration where the fasteners transmit a tensile force to the HSS wall, a rational analysis must be used to determine the appropriate limit states. These may include a yield-line mechanism in the HSS wall and/or pull-out through the HSS wall in addition to applicable limit states for the fasteners subject to tension.

9.2. Welds

These provisions are based upon similar provisions in AWS D1.1. The design strength of a connection, for a member welded to a wall of an HSS, is a function of the geometric parameters of the connected members. It is often less than the strength of the member and in many cases cannot be increased by increasing the weld strength. The weld, however, must be sized to provide for the uneven distribution of load along the weld line at the required strength.

The effective length provisions included in this Section are intended to provide for ductile joint behavior through prevention of progressive failure or "unzipping" of the weld. The variables are as illustrated for the Y-connection in Figure C9.4-1.

Other rational approaches are available in AWS D1.1, such as the use of fillet welds with an effective throat thickness of at least 1.1 times the thickness of the branch member or the use of prequalified fillet and partial-joint-penetration groove welded details that are sized to ensure ductile behavior.

9.3. Other Connection Requirements

The requirements of LRFD Specification Section J4 through J10 are applicable to connections to HSS as well as to other portions of a structure. The notation for shear rupture strength and tension rupture strength in LRFD Specification Section J4 has been modified when applied specifically to the HSS. An additional limit state for punching shear has been added based upon failures that were observed in tests of single plate connections to HSS (Sherman, 1996).

The tension force in a unit length of the plate is ft_p. Limiting this force to the through-thickness shear strength of the HSS for the two welded planes on the sides of the plate,

$$ft_p \leq 2(0.6F_u)t \tag{C9.3-1}$$

The punching shear rupture criterion includes the standard resistance factors of 0.9 for yielding and 0.75 for rupture. For a single plate connection, the stress f is a bending stress that results from the factored shear acting at an eccentricity (Sherman and Ales, 1991). A simple and conservative criterion is obtained by setting f equal to the yield strength of the plate F_{yp}, which is the maximum possible value. This reduces the above equation (with resistance factors) to

$$t_p < \frac{F_u}{F_{yp}}t \tag{C9.3-2}$$

An exception to LRFD Specification Section J5.1 on eccentric connections is that bending moments caused by eccentricities in HSS-to-HSS truss connections need not be considered for connection design as long as the eccentricities are within the limits of applicability in Section 9.4. These eccentricities have been included in the database upon which the connection strength criteria in Section 9.4 have been based.

9.4. HSS-to-HSS Truss Connections

A wide variety of connection configurations is possible where branch members are directly welded to a chord that is continuous through the connection. The criteria in this specification are limited to a few of the configurations that are typically used for planar HSS trusses in building applications: T-, Y-, gapped K-, and Cross-connections, as illustrated in Figure C9.4-1.

Overlapped K-connections are not covered in this Specification because they are more expensive to fabricate than connections with a gap, especially for round HSS. For this and other connection configurations that are not included in Figure C9.4-1, the design must be based upon the requirements in AWS D1.1, experimental test results, or other verified design procedures. Design procedures such as those contained in the recommendations of the International Institute of Welding (IIW, 1989) would generally be acceptable.

The criterion that is used for the design strength of a branch of a gapped K- connection is based upon the mechanism of load transfer rather than the configuration of the connection. The gapped-K-connection criteria apply only when the branch axial force is equilibrated by the forces in other branch members on the same side of the chord. If the force is equilibrated by shear in the chord, the criteria for a T- or Y-connection apply. If the load is transferred through the chord, the criteria for a Cross-connection must be used. Examples of these principles are shown in Figure C9.4-2. When the load is equilibrated by a combination of two of the mechanisms, the design strength is determined by interpolation between the two criteria based upon the percentage of the load transfer by each mechanism.

The symbols for key dimensions that are used in HSS-to-HSS truss connection design criteria are also defined in Figure C9.4-1. Parameters derived from ratios of these dimensions that are used in AWS D1.1 and other specifications are defined in Section 9.4.1.

The criteria in Section 9.4.2 for round HSS and Section 9.4.3 for rectangular HSS have been taken from AWS D1.1. The parameter Q_f, which is a strength reduction factor that accounts for the level of axial and bending stress in the chord at the connection, is defined in both sections. The criteria are partly empirical and are based upon numerous worldwide test programs involving hundreds of tests. Since the criteria are in some cases the result of empirical curve fitting, they may not be accurate outside the range of tested parameters. AWS D1.1 does not contain a set of limits of applicability for round HSS except for an upper limit on yield strength of 60 ksi. The HSS material specifications listed in Section 3.1 meet this limitation. However, since the database used by various international specifications and design recommendations is essentially the same, the limits of applicability given by Packer and Henderson (1997) for round HSS have been used in Section 9.4.2a.

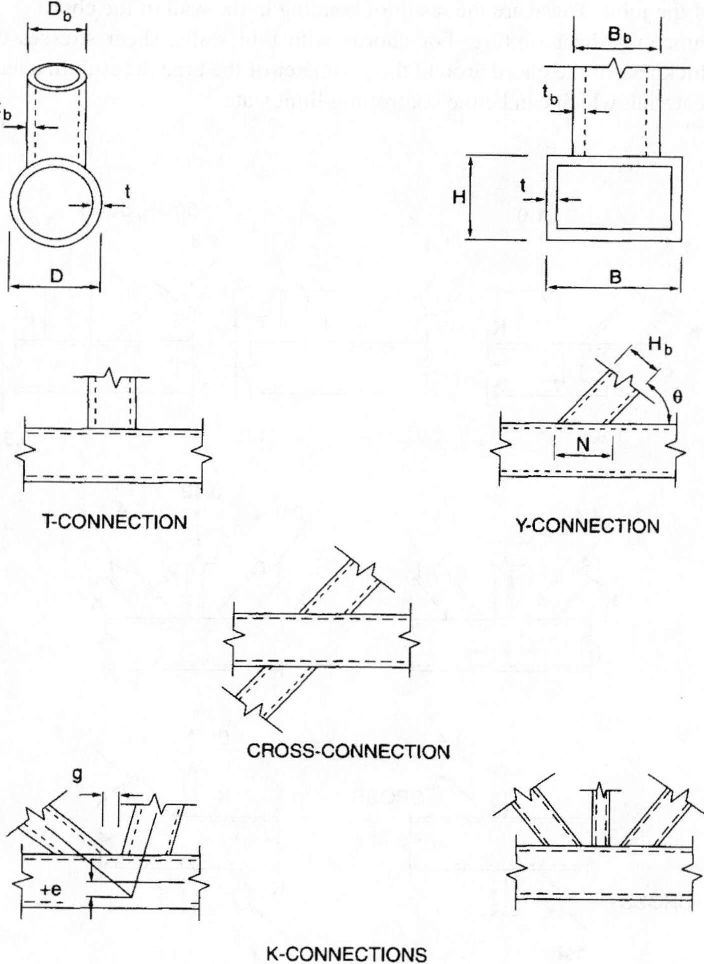

Fig. C9.4-1. HSS-to-HSS truss connection configurations.

The criteria in Section 9.4.3 for rectangular HSS are similar to those in other established HSS design specifications and recommendations throughout the world. They include the limits of applicability for rectangular HSS connections in Section 9.4.3a. One other limit has been specified for T-, Y-, and Cross-connections in Section 9.4.3b and several more for K-connections in Section 9.4.3c. In addition to the lower limit for the gap, the gap must have sufficient size to accommodate the welds to both branch members in the gap region. If the upper limit for the gap size is exceeded, as controlled by the limit on eccentricity, the branches should be considered separately as T- or Y-connections.

The criteria place limits on the magnitude of the component of the branch force that is perpendicular to the chord for various limit states. These limit states include:

(1) Chord wall plastification: High local stresses and distortions occur in the vicinity of the joint. These are the result of bending in the wall of the chord.
(2) Punching shear rupture: For chords with thin walls, shear stresses through the thickness of the chord around the perimeter of the branch result in a tearing out of material, which can be the controlling limit state.

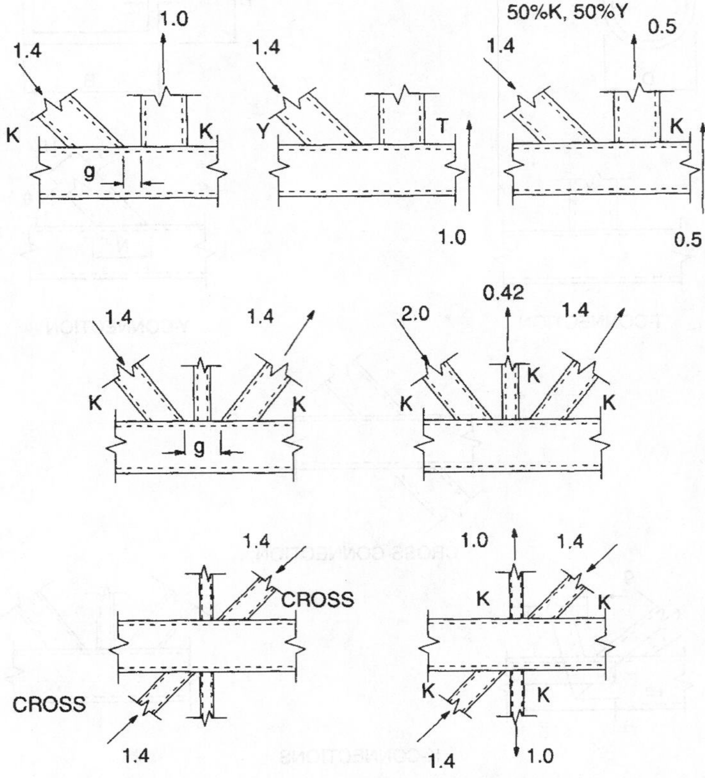

Fig. C9.4-2. Examples of connection criteria classifications (all inclined members at 45 degrees).

(3) General collapse: This is most common in Cross-connections where the branches are in compression or in other situations when the load is transferred through the chord. It results in a squashing of a round HSS chord or a buckling of the side wall of a rectangular HSS chord. The sidewall strength criteria in Section 9.4.3b, are based upon similar provisions in LRFD Specification Sections K1.3, K1.4, and K1.6. They have been modified to account for the presence of two sidewalls in the HSS and the buckling criteria have been expressed in a non-dimensional form.

(4) Uneven load distribution: In connections where the branch is equal or nearly equal in width to the chord, the increase in the stiffness of the chord wall from its center toward its sides concentrates the transfer of the branch force toward the sides of the chord and might cause a premature failure of the branch.

These limit states are illustrated in Figure C9.4-3. In some situations, it can be determined beforehand that a particular limit state will not apply. For example, chord wall plastification will not occur in rectangular HSS T-connections when the branch has the same or nearly the same width as the chord. Hence, ranges of parameters are indicated in Section 9.4.3b and Section 9.4.3c when a limit state need not be checked. In Section

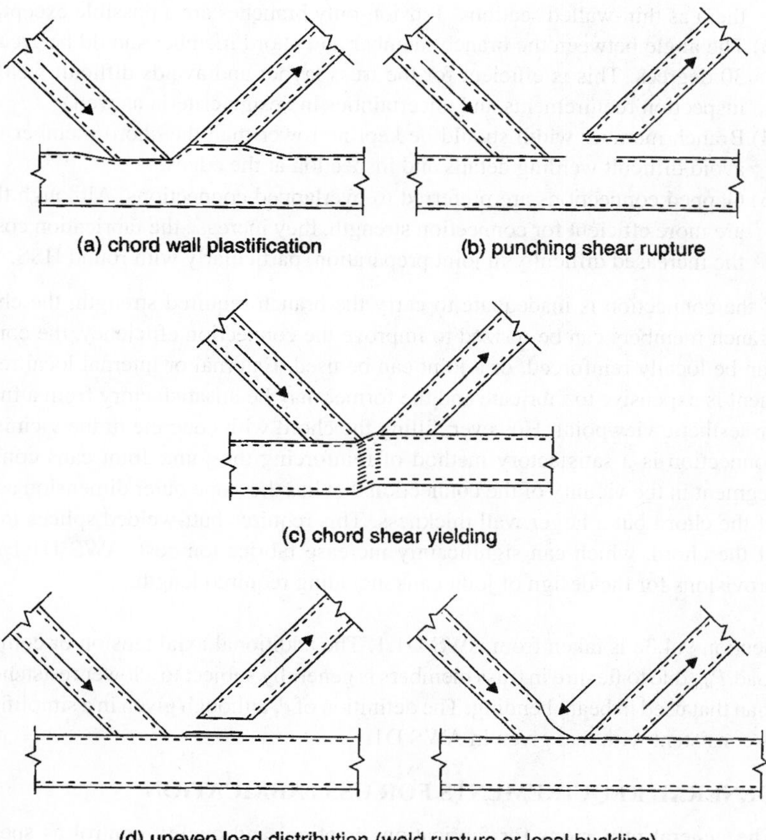

Fig. C9.4-3. Limit states in gapped K-connections.

9.4.3c, one such limit is given as $B_b/B \geq 0.1 + \gamma/50$, which is based upon Packer and Henderson (1997). This is a departure from AWS D1.1, which specifies a limit of $\beta_{eff} \geq 0.1 + \gamma/50$.

Since the criteria place a limit on the force that the branch can carry, it is not always possible to develop the full strength of the branch without reinforcing the chord with stiffeners. It should be noted that such reinforcement involves expensive fabrication. To minimize cost, it may be more desirable to maximize the efficiency of the connection with respect to the force that can be developed in the branch. Consider the following suggestions:

(1) Chord members should be relatively thick and branch members should be relatively thin. This is efficient for the strength of the connection, and thinner branches reduce the required weld size, if the branch member is joined to develop its wall strength. There may be some compromise required in the design of compression chord members because larger sizes with thinner walls make efficient compression members. This, however, may be detrimental to the strength of the connection.

(2) All members should have D/t or b/t values below the limit that would classify them as thin-walled sections. Tension-only branches are a possible exception.

(3) The angle between the branch member and chord member should be greater than 30 degrees. This is efficient for the truss layout and avoids difficult welding and inspection requirements and uncertainties in design criteria as well.

(4) Branch member width should be kept narrower than the chord member width to avoid difficult welding details and inspection at the edges.

(5) Gapped connections are preferred to overlapped connections. Although the latter are more efficient for connection strength, they increase the fabrication cost due to the increased difficulty in joint preparation, particularly with round HSS.

If the connection is inadequate to carry the branch required strength, the chord and branch members can be resized to improve the connection efficiency, the connection can be locally reinforced, or a joint can be used. External or internal local reinforcement is expensive to fabricate and the former may be unsatisfactory from a functional or aesthetic viewpoint. However, filling the chord with concrete in the vicinity of the connection is a satisfactory method of reinforcing the joint. Joint cans consist of a segment in the vicinity of the connection that has the same outer dimension as the rest of the chord but a larger wall thickness. This requires butt-welded splices to the rest of the chord, which can significantly increase fabrication cost. AWS D1.1 contains provisions for the design of joint cans including required length.

Section 9.4.3c is taken from AWS D1.1. The additional axial tension or compression load P_{ua} due to flexure in truss members is generally subject to a lower resistance factor than that used in beam bending. The definition of c, although given in a simplified form, is consistent with that given in AWS D1.1.

10. GENERAL REQUIREMENTS FOR HSS FABRICATION

The general provisions for fabrication, erection, and quality control as specified in LRFD Specification Chapter M are also applicable to HSS. In addition, the following HSS-specific concerns are addressed:

(1) Because the interior of an HSS is difficult to inspect, some concern has been expressed regarding internal corrosion. However, good design practice can eliminate the concern and the need for expensive protection.

Corrosion occurs in the presence of oxygen and water. In an enclosed building, it is improbable that there would be sufficient reintroduction of moisture to cause severe corrosion. Therefore, internal corrosion protection is a consideration only in HSS that are exposed to weather.

In a sealed HSS, internal corrosion cannot progress beyond the point where the oxygen or moisture necessary for chemical oxidation is consumed (AISI, 1970). The oxidation depth is insignificant when the corrosion process must stop, even when a corrosive atmosphere exists at the time of sealing. If fine openings exist at connections, moisture and air can enter the HSS through capillary action or by aspiration due to the partial vacuum that is created if the HSS is cooled rapidly (Blodgett, 1967). This can be prevented by providing pressure-equalizing holes in locations that make it impossible for water to flow into the HSS by gravity.

Situations where conservative practice would recommend an internal protective coating include: (1) open HSS where changes in the air volume by ventilation or direct flow of water is possible; and (2) open HSS subject to a temperature gradient that would cause condensation. In such instances it may also be prudent to use a minimum $^5/_{16}$-in. wall thickness.

HSS that are filled or partially filled with concrete should not be sealed. In the event of fire, water in the concrete will vaporize and may create pressure sufficient to burst a sealed HSS. Care should be taken to ensure that water does not remain in the HSS during or after construction, since the expansion caused by freezing can create pressure that is sufficient to burst an HSS.

Galvanized HSS assemblies should not be completely sealed because rapid pressure changes during the galvanizing process tend to burst sealed assemblies.

(2) As a result of manufacturing, a light oil coating is generally present on the outer surface of the HSS. If paint is specified, HSS must be cleaned of this oil coating with a suitable solvent; see the *Steel Structures Painting Manual*. (SSPC, 1991)

(3) To avoid weld contamination, the light oil coating that is generally present after manufacturing an HSS should be removed with a suitable solvent in locations where welding will be performed. In cases where an external coating has been applied at the mill, the coating should be removed at the location of welding or the manufacturer should be consulted regarding the suitability of welding in the presence of the coating.

REFERENCE

American Institute of Steel Construction (1997), *Seismic Provisions for Structural Steel Buildings*, AISC, Chicago, IL.

American Institute of Steel Construction (1999), *Load and Resistance Design Specification for Structural Steel Buildings*, AISC, Chicago, IL.

American Institute of Steel Construction (2000), *Seismic Provisions for Structural Steel Buildings Supplement No. 2*, November 10, 2000, AISC, Chicago, IL.

American Society of Civil Engineers (1995), ASCE 7-95 *Minimum Design Loads for Buildings and Other Structures*, ASCE, Reston, VA.

American Iron and Steel Institute (1970), *Interior Corrosion of Structural Steel Closed Sections*, Bulletin 18, February, Washington, DC.

Bjorhovde R. and Birkemoe, P. C. (1979), "Limit State Design of HSS Columns," *Canadian Journal of Civil Engineering*, Vol. 6, No. 2, June (pp. 276–291).

Blodgett, O. W. (1967), "The Question of Corrosion in Hollow Steel Sections," *Welding Design Studies in Steel Structures*, Lincoln Electric, D610.163, August, Cleveland, OH.

Brockenbrough, R. B. and Johnston, B. G. (1981), *USS Steel Design Manual*, United States Steel Corporation, Pittsburgh, PA.

Canadian Standards Association (1994), CSA Standard S16.1 *Limit States Design of Steel Structures*, CSA, Rexdale, Ontario, Canada.

Chen, W. F. and Ross, D. A. (1977), "Tests of Fabricated Tubular Columns," *Journal of the Structural Division*, Vol. 103, ST3, paper 12809, ASCE, Reston, VA.

Estuar, F. R. and Tall, L. (1965), "The Column Strength of Hot-Rolled Tubular Shapes – An Experimental Evaluation," *Fritz Engineering Laboratory Report No. 296.1*, Lehigh University, Bethlehem, PA.

Felton, L. P. and Dobbs, M. W. (1967), "Optimum Design of Tubes for Bending and Torsion," *Journal of the Structural Division*, Vol. 93, ST4, paper 5397, ASCE, Reston, VA.

Galambos, T. V. (ed.) (1998), *Guide to Design Criteria for Metal Compression Members*, Structural Stability Research Council, 5th Ed., John Wiley & Sons, New York, NY.

Graham, R. R. (1965), "Manufacture and Use of Structural Tubing," *Journal of Metals*, September.

Hayus, F. (1968), *Drag Measurements on One Square Section and Two Rectangular Sections with Different Corner Radii*, English Translation of CIDECT Report 1-NK1-68-41, September.

IIW (1989), *Design Recommendations for Hollow Structural Section Joints – Predominantly Statically Loaded*, 2nd Ed., International Institute of Welding Subcommission XV-E, Doc. XV-582-85, IIW Annual Assembly, Helsinki, Finland.

Key, P. W. and Hancock, G. J. (1985), *An Experimental Investigation of the Column Behaviour of Cold-Formed Square Hollow Sections*, Research Report R493, University of Sydney, Sydney, Australia.

Korol, R. M. (1972), *The Plastic Behaviour of Hollow Structural Sections with Implication for Design*, Canadian Structural Engineering Conference.

Lui, Z. and Goel, S. C. (1987), *Investigation of Concrete Filled Steel Tubes under Cyclic Bending and Buckling*, UMCE Report 87-3, University of Michigan, Ann Arbor, MI.

Packer, J. A. and Henderson, J. E. (1997), *Design Guide for Hollow Structural Section Connections*, Canadian Institute of Steel Construction, Willowdale, Ontario, Canada.

Rondal, J. (1992), *Structural Stability of Hollow Sections*, Comite International pour le Developpement et l'Etude de la Construction Tubulaire, Verlag TÜV Rheinland GmbH, Köln, Germany.

Schilling, C. G. (1965), "Buckling Strength of Circular Tubes," *Journal of the Structural Division*, Vol. 91, ST5, paper 4520, ASCE, Reston, VA.

Sherman, D. R. and Lukas, D. E. (1969), "Torsionally Stiff Columns Under Eccentric Loads," *Journal of the Structural Division*, Vol. 96, ST2, paper 7090, ASCE, Reston, VA.

Sherman, D. R. (1980), *Local Buckling Behavior of Tubular Strut Type Beam-Columns*, Civil Engineering Department Report, University of Wisconsin-Milwaukee, Milwaukee, WI.

Sherman, D. R. (1985), "Bending Equations for Circular Tubes," *Proceedings Annual Technical Session*, Structural Stability Research Council, Lehigh University, Bethlehem, PA.

Sherman, D. R. and Ales, J. M. (1991), "The Design of Shear Tabs with Tubular Columns," *Proceedings of the 1991 AISC National Steel Construction Conference*, AISC, Chicago, IL.

Sherman, D. R. (1992), "Tubular Members," *Constructional Steel Design – An International Guide*, edited by P. J. Dowling, J. H. Harding, and R. Bjorhovde, Chapter 2.4, pp. 91–104, Elsevier Applied Science, London, England.

Sherman, D. R. (1995), *Stability Related Deterioration of Structures*, 1995 Theme Conference, Structural Stability Research Council, Lehigh University, Bethlehem, PA.

SSPC (1991), *Steel Structures Painting Manual*, Steel Structures Painting Council, Pittsburgh, PA.

Steel Tube Institute (1996), *Principle Producers and Capabilities*, STI, Mentor, OH.

Timoshenko, S. (1956), *Strength of Materials, Part II*, 3rd Ed., D. Van Nostrand Company, Inc.

Winter, G. (1968), *Commentary on the Specification for the Design of Cold-Formed Steel Members*, American Iron and Steel Institute, Washington, D.C.

Yang, X. M. (1987), "Study of Carrying Capacity of Fabricated Tubular Columns Under Axial Compression," *Proceedings Annual Technical Session*, Structural Stability Research Council, Lehigh University, Bethlehem, PA.

Load and Resistance Factor Design Specification for Single-Angle Members

November 10, 2000

Supersedes the *Specification for Load and Resistance Factor Design of Single-Angle Members* dated December 1, 1993

Prepared by the
American Institute of Steel Construction, Inc.
Under the Direction of the
AISC Committee on Specifications and approved by
the AISC Board of Directors

AMERICAN INSTITUTE OF STEEL CONSTRUCTION, INC.
One East Wacker Drive, Suite 3100
Chicago, IL 60601-2001

PREFACE

The intention of the AISC Specification is to cover the common everyday design criteria in routine design office usage. It is not feasible to also cover the many special and unique problems encountered within the full range of structural design practice. This separate Specification and Commentary addresses one such topic—single-angle members—to provide needed design guidance for this more complex structural shape under various load and support conditions.

The revised single-angle design criteria were developed through a consensus process by the AISC Task Committee 12 on Single Angles:

James M. Fisher, Chairman
Leroy A. Lutz, Vice-Chairman
Mohamed Elgaaly
Shu-Jin Fang
Theodore V. Galambos
Subhash Goel
Charlotte S. Harman
Todd Helwig
Donald W. White
Sergio Zoruba, Secretary

The full AISC Committee on Specifications has reviewed and approved this Specification.

A non-mandatory Commentary provides background for the Specification provisions and the user is encouraged to consult it.

The principal changes in this edition include:

Revisions to flexural design strength criteria

a. For the limit state of local buckling when the angle leg is in compression
b. For the limit state of yielding when the tip of an angle leg is in tension
c. For the limit state of lateral-torsional buckling
d. For bending about geometric axes

The reader is cautioned that professional judgment must be exercised when data or recommendations in this Specification are applied. The publication of the material contained herein is not intended as a representation or warranty on the part of the American Institute of Steel Construction, Inc.—or any other person named herein—that this information is suitable for general or particular use, or freedom from infringement of any patent or patents. Anyone making use of this information assumes all liability arising from such use. The design of structures is within the scope of expertise of a competent licensed structural engineer, architect, or other licensed professional for the application of principles to a particular structure.

Load and Resistance Factor Design Specification for Single-Angle Members

November 10, 2000

1. SCOPE

This document contains Load and Resistance Factor Design (LRFD) criteria for hot-rolled, single-angle members with equal and unequal legs in tension, shear, compression, flexure, and for combined forces. It is intended to be compatible with, and a supplement to, the 1999 AISC *Load and Resistance Factor Design Specification for Structural Steel Buildings* and repeats some common criteria for ease of reference. For design purposes, the conservative simplifications and approximations in the Specification provisions for single angles are permitted to be refined through a more precise analysis. As an alternative to this Specification, the 1989 AISC *Specification for Allowable Stress Design of Single-Angle Members* is permitted.

The Specification for single-angle design supersedes any comparable but more general requirements of the AISC LRFD. All other design, fabrication, and erection provisions not directly covered by this document shall be in compliance with the AISC LRFD. For design of slender, cold-formed steel angles, the AISI *LRFD Specification for the Design of Cold-Formed Steel Structural Members* referenced in Section A6 of the AISC LRFD is applicable.

2. TENSION

The tensile design strength $\phi_t P_n$ shall be the lower value obtained according to the limit states of yielding, $\phi_t = 0.9$, $P_n = F_y A_g$, and fracture, $\phi_t = 0.75$, $P_n = F_u A_e$.

a. For members connected by bolting, the net area and effective net area shall be determined from AISC LRFD Specification Sections B1 to B3 inclusive.

b. When the load is transmitted by longitudinal welds only or a combination of longitudinal and transverse welds through just one leg of the angle, the effective net area A_e shall be:

$$A_e = A_g U \tag{2-1}$$

where

A_g = gross area of member
$U = \left(1 - \frac{\bar{x}}{l}\right) \leq 0.9$
$\bar{x}$ = connection eccentricity
l = length of connection in the direction of loading

c. When a load is transmitted by transverse weld through just one leg of the angle, A_e is the area of the connected leg and $U = 1$.

For members whose design is based on tension, the slenderness ratio l/r preferably should not exceed 300. Members in which the design is dictated by tension loading, but which may be subject to some compression under other load conditions, need not satisfy the compression slenderness limits.

3. SHEAR

For the limit state of yielding in shear, the shear stress, f_{uv}, due to flexure and torsion shall not exceed:

$$f_{uv} \leq \phi_v 0.6 F_y$$
$$\phi_v = 0.9 \tag{3-1}$$

4. COMPRESSION

The design strength of compression members shall be $\phi_c P_n$

where

$\phi_c = 0.90$
$P_n = A_g F_{cr}$

a. For $\lambda_c \sqrt{Q} \leq 1.5$

$$F_{cr} = Q(0.658^{Q\lambda_c^2})F_y \tag{4-1}$$

b. For $\lambda_c \sqrt{Q} > 1.5$

$$F_{cr} = \left[\frac{0.877}{\lambda_c^2}\right] F_y \tag{4-2}$$

where

$\lambda_c = \frac{Kl}{r\pi}\sqrt{\frac{F_y}{E}}$
F_y = specified minimum yield stress of steel
Q = reduction factor for local buckling

The reduction factor Q shall be:

when $\dfrac{b}{t} \leq 0.446\sqrt{\dfrac{E}{F_y}}$:

$$Q = 1.0 \tag{4-3a}$$

when $0.446\sqrt{\dfrac{E}{F_y}} < \dfrac{b}{t} < 0.910\sqrt{\dfrac{E}{F_y}}$:

$$Q = 1.34 - 0.761\frac{b}{t}\sqrt{\frac{F_y}{E}} \tag{4-3b}$$

when $\dfrac{b}{t} \geq 0.910\sqrt{\dfrac{E}{F_y}}$:

$$Q = \frac{0.534E}{F_y\left(\frac{b}{t}\right)^2} \tag{4-3c}$$

where

b = full width of longest angle leg
t = thickness of angle

For members whose design is based on compressive force, the largest effective slenderness ratio preferably should not exceed 200.

5. FLEXURE

The flexure design strengths of Section 5.1 shall be used as indicated in Sections 5.2 and 5.3.

5.1. Flexural Design Strength

The flexural design strength shall be limited to the minimum value $\phi_b M_n$ determined from Sections 5.1.1, 5.1.2, and 5.1.3, as applicable, with $\phi_b = 0.9$.

5.1.1. For the limit state of local buckling when the tip of an angle leg is in compression:

when $\dfrac{b}{t} \leq 0.54\sqrt{\dfrac{E}{F_y}}$:

$$M_n = 1.5\,F_y S_c \tag{5-1a}$$

when $0.54\sqrt{\frac{E}{F_y}} < \frac{b}{t} \leq 0.91\sqrt{\frac{E}{F_y}}$:

$$M_n = F_y S_c \left[1.5 - 0.93 \left(\frac{b/t}{0.54\sqrt{\frac{E}{F_y}}} - 1 \right) \right] \quad \text{(5-1b)}$$

when $\frac{b}{t} > 0.91\sqrt{\frac{E}{F_y}}$:

$$M_n = 1.34 Q F_y S_c \quad \text{(5-1c)}$$

where

b = full width of angle leg with tip in compression
Q = reduction factor per Equation 4-3c
S_c = elastic section modulus to the tip in compression relative to axis of bending
E = modulus of elasticity

5.1.2. For the limit state of yielding when the tip of an angle leg is in tension

$$M_n = 1.5 M_y \quad \text{(5-2)}$$

where

M_y = yield moment about the axis of bending

5.1.3. For the limit state of lateral-torsional buckling:

when $M_{ob} \leq M_y$:

$$M_n = [0.92 - 0.17 M_{ob}/M_y] M_{ob} \quad \text{(5-3a)}$$

when $M_{ob} > M_y$:

$$M_n = [1.92 - 1.17\sqrt{M_y/M_{ob}}] M_y \leq 1.5 M_y \quad \text{(5-3b)}$$

where

M_{ob} = elastic lateral-torsional buckling moment, from Section 5.2 or 5.3 as applicable

5.2. Bending about Geometric Axes

5.2.1. a. Angle bending members with lateral-torsional restraint along the length shall be designed on the basis of geometric axis bending with the nominal flexural strength M_n limited to the provisions of Sections 5.1.1 and 5.1.2.

b. For equal-leg angles if the lateral-torsional restraint is only at the point of maximum moment, the required moment shall be limited to $\phi_b M_n$ per Section 5.1. M_y shall be computed using the geometric axis section modulus and M_{ob} shall be substituted by using 1.25 times M_{ob} computed from Equation 5-4.

5.2.2. Equal-leg angle members without lateral-torsional restraint subjected to flexure applied about one of the geometric axes are permitted to be designed considering only geometric axis bending provided:

a. The yield moment shall be based on use of 0.80 of the geometric axis section modulus.

b. With maximum compression of the angle-leg tips, the nominal flexural strength M_n shall be determined by the provisions in Section 5.1.1 and in Section 5.1.3,

where

$$M_{ob} = \frac{0.66Eb^4tC_b}{l^2}\left[\sqrt{1 + 0.78(lt/b^2)^2} - 1\right] \tag{5-4}$$

l = unbraced length

$$C_b = \frac{12.5M_{\max}}{2.5M_{\max} + 3M_A + 4M_B + 3M_C} \leq 1.5$$

where

M_{max} = absolute value of maximum moment in the unbraced beam segment
M_A = absolute value of moment at quarter point of the unbraced beam segment
M_B = absolute value of moment at centerline of the unbraced beam segment
M_C = absolute value of moment at three-quarter point of the unbraced beam segment

c. With maximum tension at the angle-leg tips, the nominal flexural strength shall be determined according to Section 5.1.2 and in Section 5.1.3 using M_{ob} in Equation 5-4 with -1 being replaced by $+1$.

5.2.3. Unequal-leg angle members without lateral-torsional restraint subjected to bending about one of the geometric axes shall be designed using Section 5.3.

5.3. Bending about Principal Axes

Angles without lateral-torsional restraint shall be designed considering principal-axis bending, except for the alternative of Section 5.2.2, if appropriate. Bending about both of the principal axes shall be evaluated as required in Section 6.

5.3.1. Equal-leg angles:

a. Major-axis bending:

The nominal flexural strength M_n about the major principal axis shall be determined by the provisions in Section 5.1.1 and in Section 5.1.3,

where

$$M_{ob} = C_b \frac{0.46Eb^2t^2}{l} \tag{5-5}$$

b. Minor-axis bending:

The nominal design strength M_n about the minor principal axis shall be determined by Section 5.1.1 when the leg tips are in compression, and by Section 5.1.2 when the leg tips are in tension.

5.3.2. Unequal-leg angles:

a. Major-axis bending:

The nominal flexural strength M_n about the major principal axis shall be determined by the provisions in Section 5.1.1 for the compression leg and in Section 5.1.3,

where

$$M_{ob} = 4.9E \frac{I_z}{l^2} C_b \left[\sqrt{\beta_w^2 + 0.052(lt/r_z)^2} + \beta_w\right] \tag{5-6}$$

I_z = minor principal axis moment of inertia

r_z = radius of gyration for minor principal axis

$\beta_w = \left[\frac{1}{I_w}\int_A z_o(w^2 + z^2)dA\right] - 2z_o$, special section property for unequal-leg angles, positive for short leg in compression and negative for long leg in compression (see Commentary for values for common angle sizes). If the long leg is in compression anywhere along the unbraced length of the member, the negative value of β_w shall be used.

z_o = coordinate along z axis of the shear center with respect to centroid

I_w = moment of inertia for major principal axis

b. Minor-axis bending:

The nominal design strength M_n about the minor principal axis shall be determined by Section 5.1.1 when leg tips are in compression and by Section 5.1.2 when the leg tips are in tension.

6. COMBINED FORCES

The interaction equation shall be evaluated for the principal bending axes either by addition of all the maximum axial and flexural terms, or by considering the sense of the associated flexural stresses at the critical points of the cross section, the flexural terms are either added to or subtracted from the axial load term.

6.1. Members in Flexure and Axial Compression

6.1.1. The interaction of flexure and axial compression applicable to specific locations on the cross section shall be limited by Equations 6-1a and 6-1b:

For $\dfrac{P_u}{\phi P_n} \geq 0.2$

$$\left| \frac{P_u}{\phi P_n} + \frac{8}{9}\left(\frac{M_{uw}}{\phi_b M_{nw}} + \frac{M_{uz}}{\phi_b M_{nz}} \right) \right| \leq 1.0 \qquad \text{(6-1a)}$$

For $\dfrac{P_u}{\phi P_n} \leq 0.2$

$$\left| \frac{P_u}{2\phi P_n} + \left(\frac{M_{uw}}{\phi_b M_{nw}} + \frac{M_{uz}}{\phi_b M_{nz}} \right) \right| \leq 1.0 \qquad \text{(6-1b)}$$

where

P_u = required compressive strength

P_n = nominal compressive strength determined in accordance with Section 4

M_u = required flexural strength

M_n = nominal flexural strength for tension or compression in accordance with Section 5, as appropriate. Use section modulus for specific location in the cross section and consider the type of stress.

ϕ = ϕ_c = resistance factor for compression = 0.90

ϕ_b = resistance factor for flexure = 0.90

w = subscript relating symbol to major-axis bending

z = subscript relating symbol to minor-axis bending

In Equations 6-1a and 6-1b when M_n represents the flexural strength of the compression side, the corresponding M_u shall be multiplied by B_1.

$$B_1 = \frac{C_m}{1 - \frac{P_u}{P_{e1}}} \geq 1.0 \tag{6-2}$$

where

C_m = bending coefficient defined in AISC LRFD
P_{e1} = elastic buckling load for the braced frame defined in AISC LRFD

6.1.2. For members constrained to bend about a geometric axis with nominal flexural strength determined per Section 5.2.1, the radius of gyration r for P_{e1} shall be taken as the geometric axis value. The bending terms for the principal axes in Equations 6-1a and 6-1b shall be replaced by a single geometric axis term.

6.1.3. Alternatively, for equal-leg angles without lateral-torsional restraint along the length and with bending applied about one of the geometric axes, the provisions of Section 5.2.2 are permitted for the required and design bending strength. If Section 5.2.2 is used for M_n, the radius of gyration about the axis of bending r for P_{e1} shall be taken as the geometric axis value of r divided by 1.35 in the absence of a more detailed analysis. The bending terms for the principal axes in Equations 6-1a and 6-1b shall be replaced by a single geometric axis term.

6.2. Members in Flexure and Axial Tension

The interaction of flexure and axial tension shall be limited by Equations 6-1a and 6-1b where

P_u = required tensile strength
P_n = nominal tensile strength determined in accordance with Section 2
M_u = required flexural strength
M_n = nominal flexural strength for tension or compression in accordance with Section 5, as appropriate. Use section modulus for specific location in the cross section and consider the type of stress.
ϕ = ϕ_t = resistance factor for tension = 0.90
ϕ_b = resistance factor for flexure = 0.90

For members subject to bending about a geometric axis, the required bending strength evaluation shall be in accordance with Sections 6.1.2 and 6.1.3. Second-order effects due to axial tension and bending interaction are permitted to be considered in the determination of M_u for use in Formulas 6-1a and 6-1b. In lieu of using Formulas 6-1a and 6-1b, a more detailed analysis of the interaction of flexure and tension is permitted.

COMMENTARY on the Load and Resistance Factor Design Specification for Single-Angle Members

November 10, 2000

INTRODUCTION

This Specification is intended to be complete for normal design usage in conjunction with the main 1999 AISC LRFD Specification and Commentary.

This Commentary furnishes background information and references for the benefit of the engineer seeking further understanding of the derivation and limits of the specification.

The Specification and Commentary are intended for use by design professionals with demonstrated engineering competence.

C2. TENSION

The criteria for the design of tension members in AISC LRFD Specification Section D1 have been adopted for angles with bolted connections. However, recognizing the effect of shear lag when the connection is welded, the criteria in Section B3 of the AISC LRFD Specification have been applied.

The advisory upper slenderness limits are not due to strength considerations but are based on professional judgment and practical considerations of economics, ease of handling, and transportability. The radius of gyration about the z axis will produce the maximum l/r and, except for very unusual support conditions, the maximum Kl/r. Since the advisory slenderness limit for compression members is less than for tension members, an accommodation has been made for members with $Kl/r > 200$ that are always in tension, except for unusual load conditions which produce a small compression force.

C3. SHEAR

Shear stress due to factored loads in a single-angle member are the result of the gradient in the bending moment along the length (flexural shear) and the torsional moment.

The maximum elastic stress due to flexural shear may be computed by

$$f_v = \frac{1.5V_b}{bt} \tag{C3-1}$$

where

V_b = component of the shear force parallel to the angle leg with length b and thickness t, kips

The stress, which is constant through the thickness, should be determined for both legs to determine the maximum.

The 1.5 factor is the calculated elastic value for equal-leg angles loaded along one of the principal axes. For equal-leg angles loaded along one of the geometric axes (laterally braced or unbraced) the factor is 1.35. Constants between these limits may be calculated conservatively from V_bQ/It to determine the maximum stress at the neutral axis.

Alternatively, if only flexural shear is considered, a uniform flexural shear stress in the leg of V_b/bt may be used due to inelastic material behavior and stress redistribution.

If the angle is not laterally braced against twist, a torsional moment is produced equal to the applied transverse load times the perpendicular distance e to the shear center, which is at the heel of the angle cross section. Torsional moments are resisted by two types of shear behavior: pure torsion (St. Venant) and warping torsion (Seaburg and Carter, 1997). If the boundary conditions are such that the cross section is free to warp, the applied torsional moment M_T is resisted by pure shear stresses as shown in Figure C3.1a. Except near the ends of the legs, these stresses are constant along the length of the leg, and the maximum value can be approximated by

$$f_v = M_T t/J = \frac{3M_t}{At} \tag{C3-2}$$

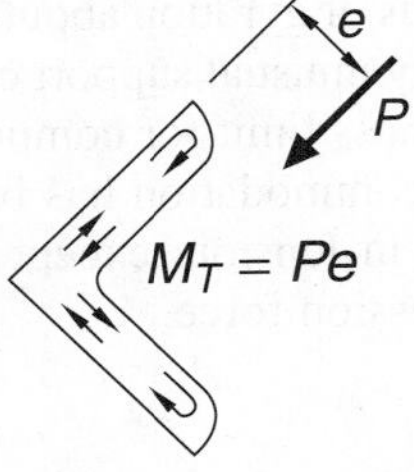

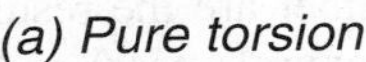

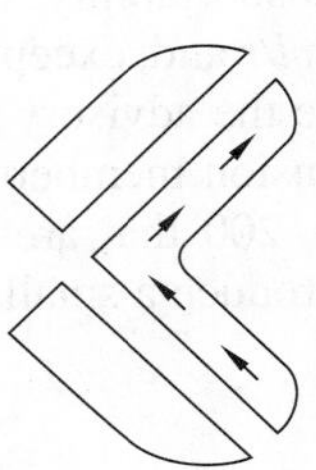

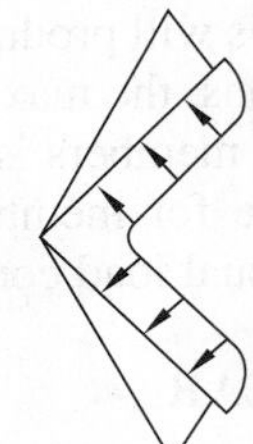

(b) In-plane warping (c) Across-thickness warping

Fig. C.3.1. Shear stresses due to torsion.

where

J = torsional constant (approximated by $\Sigma bt^3/3$ when precomputed value is unavailable)
A = angle cross-sectional area

At a section where warping is restrained, the torsional moment is resisted by warping shear stresses of two types (Gjelsvik, 1981). One type is in-plane (contour) as shown in Figure C3.1b, which varies from zero at the toe to a maximum at the heel of the angle. The other type is across the thickness and is sometimes referred to as secondary warping shear. As indicated in Figure C3.1c, it varies from zero at the heel to a maximum at the toe.

In an angle with typical boundary conditions and an unrestrained load point, the torsional moment produces all three types of shear stresses (pure, in-plane warping, and secondary warping) in varying proportions along its length. The total applied moment is resisted by a combination of three types of internal moments that differ in relative proportions according to the distance from the boundary condition. Using typical angle dimensions, it can be shown that the two warping shears are approximately the same order of magnitude and are less than 20 percent of the pure shear stress for the same torsional moment. Therefore, it is conservative to compute the torsional shear stress using the pure shear equation and total applied torsional moment M_T as if no warping restraint were present. This stress is added directly to the flexural shear stress to produce a maximum surface shear stress near the mid-length of a leg. Since this sum is a local maximum that does not extend through the thickness, applying the limit of $\phi_v 0.6F_y$ adds another degree of conservatism relative to the design of other structural shapes.

In general, torsional moments from laterally unrestrained transverse loads also produce warping normal stresses that are superimposed on bending stresses. However, since the warping strength for a single angle is relatively small, this additional bending effect is negligible and often ignored in design practice.

C4. COMPRESSION

The provisions for the critical compression stress account for the three possible limit states that may occur in an angle column depending on its proportions: general column flexural buckling, local buckling of thin legs, and flexural-torsional buckling of the member. The Q-factor in the equation for critical stress accounts for the local buckling, and the expressions for Q are nondimensionalized from AISC LRFD Specification (AISC, 1999) Appendix B5. Flexural-torsional buckling is covered in Appendix E of the AISC LRFD Specification (AISC, 1999). This strength limit state is approximated by the Q-factor reduction for slender-angle legs. For non-slender sections where $Q = 1$, flexural-torsional buckling is relevant for relatively short columns, but it was shown by Galambos (1991) that the error of neglecting this effect is not significant. For this reason no explicit consideration of this effect is required in these single-angle specifications. The provisions of Appendix E of AISC LRFD may be conservatively used to directly consider flexural-torsional buckling for single-angle members.

The effective length factors for angle columns may be determined by consulting the paper by Lutz (1992).

The resistance factor ϕ was increased from 0.85 in AISC LRFD for all cross sections to 0.90 for single angles only because it was shown that a ϕ of 0.90 provides an equivalent degree of reliability (Galambos, 1992).

C5. FLEXURE

Flexural strength limits are established for yielding, local buckling, and lateral-torsional buckling. In addition to addressing the general case of unequal-leg single angles, the equal-leg angle is treated as a special case. Furthermore, bending of equal-leg angles about a geometric axis, an axis parallel to one of the legs, is addressed separately as it is a very common situation.

The tips of an angle refer to the free edges of the two legs. In most cases of unrestrained bending, the flexural stresses at the two tips will have the same sign (tension or compression). For constrained bending about a geometric axis, the tip stresses will differ in sign. Criteria for both tension and compression at the tip should be checked as appropriate, but in most cases it will be evident which controls.

Appropriate serviceability limits for single-angle beams need also to be considered. In particular, for longer members subjected to unrestrained bending, deflections are likely to control rather than lateral-torsional or local buckling strength.

C5.1.1. These provisions follow the LRFD format for nominal flexural resistance. There is a region of full yielding, a linear transition to the yield moment, and a region of local buckling. The strength at full yielding is limited to a shape factor of 1.50 applied to the yield moment. This leads to a lower bound plastic moment for an angle that could be bent about any axis, inasmuch as these provisions are applicable to all flexural conditions. The 1.25 factor originally used was known to be a conservative value. Recent research work (Earls and Galambos, 1997) has indicated that the 1.50 factor represents a better lower bound value.

The *b/t* limits have been modified to be more representative of flexural limits rather than using those for single angles under uniform compression. Typically the flexural stresses will vary along the leg length permitting the use of the stress limits given. Even for the geometric axis flexure case which produces uniform compression along one leg, use of these limits will provide a conservative value when compared to the results obtained by Earls and Galambos, 1997.

C5.1.2. Since the shape factor for angles is in excess of 1.50, the nominal design strength $M_n = 1.5M_y$ for compact members is justified provided that instability does not control.

C5.1.3. Lateral-torsional instability may limit the flexural strength of an unbraced single-angle beam. As illustrated in Figure C5.1, Equation 5-3a represents the elastic buckling portion with the nominal flexural strength, M_n, varying from 75 percent to 92 percent of the theoretical buckling moment, M_{ob}. Equation 5-3b represents the inelastic buckling transition expression between $0.75M_y$ and $1.5M_y$. Equation 5-3b has been modified to better reflect its use with the increased upper limit of 1.5 M_y. The maximum beam flexural strength $M_n = 1.5$ M_y will occur when the theoretical buckling moment M_{ob} reaches or exceeds $7.7M_y$ as illustrated in Figure C5.1. These equations are modifications of those developed from the results of Australian research on single angles in flexure and on an analytical model consisting of two rectangular elements of length equal to the actual angle leg width minus one-half the thickness (Leigh and Lay, 1984; Australian Institute of Steel Construction, 1975; Leigh and Lay, 1978; Madugula and Kennedy, 1985).

A more general C_b moment gradient formula consistent with the 1999 AISC LRFD Specification is used to correct lateral-torsional stability equations from the assumed most severe case of uniform moment throughout the unbraced length ($C_b = 1.0$). The equation for

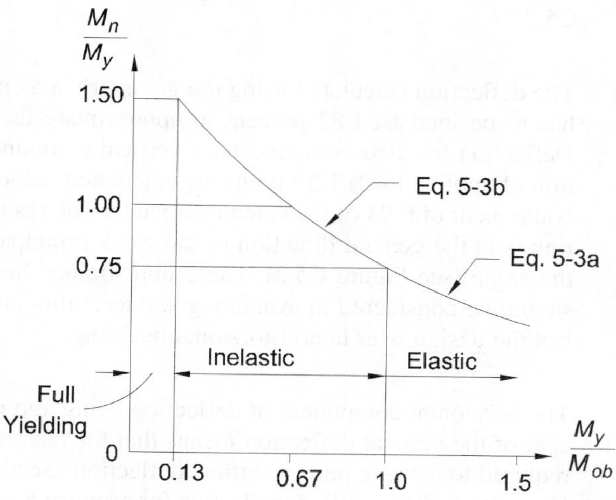

Fig. C5.1. Lateral-torsional buckling limits of a single-angle beam.

C_b used in the ASD version is applicable only to moment diagrams that are straight lines between brace points. In lieu of a more detailed analysis, the reduced maximum limit of 1.5 is imposed for single-angle beams to represent conservatively the lower envelope of this cross section's non-uniform bending response.

C5.2.1. An angle beam loaded parallel to one leg will deflect and bend about that leg only if the angle is restrained laterally along the length. In this case simple bending occurs without any torsional rotation or lateral deflection and the geometric axis section properties should be used in the evaluation of the flexural design strength and deflection. If only the point of maximum moment is laterally braced, lateral-torsional buckling of the unbraced length under simple bending must also be checked, as outlined in Section 5.2.1b.

C5.2.2. When bending is applied about one leg of a laterally unrestrained single angle, it will deflect laterally as well as in the bending direction. Its behavior can be evaluated by resolving the load and/or moments into principal axis components and determining the sum of these principal axis flexural effects. Section 5.2.2 is provided to simplify and expedite the design calculations for this common situation with equal-leg angles.

For such unrestrained bending of an equal-leg angle, the resulting maximum normal stress at the angle tip (in the direction of bending) will be approximately 25 percent greater than calculated using the geometric axis section modulus. The value of M_{ob} in Equation 5-4 and the evaluation of M_y using 0.80 of the geometric axis section modulus reflect bending about the inclined axis shown in Figure C5.2.

The deflection calculated using the geometric axis moment of inertia has to be increased 82 percent to approximate the total deflection. Deflection has two components, a vertical component (in the direction of applied load) 1.56 times the calculated value and a horizontal component of 0.94 of the calculated value. The resultant total deflection is in the general direction of the weak principal axis bending of the angle (see Figure C5.2). These unrestrained bending deflections should be considered in evaluating serviceability and will often control the design over lateral-torsional buckling.

The horizontal component of deflection being approximately 60 percent of the vertical deflection means that the lateral restraining force required to achieve purely vertical deflection (Section 5.2.1) must be 60 percent of the applied load value (or produce a moment 60 percent of the applied value) which is very significant.

Lateral-torsional buckling is limited by M_{ob} (Leigh and Lay, 1984 and 1978) in Equation 5-4, which is based on

$$M_{cr} = \frac{2.33Eb^4t}{(1 + 3\cos^2\theta)(Kl)^2} \times \left[\sqrt{\sin^2\theta + \frac{0.156(1 + 3\cos^2\theta)(Kl)^2t^2}{b^4}} + \sin\theta\right] \quad \text{(C5-1)}$$

(the general expression for the critical moment of an equal-leg angle) with $\theta = -45°$ for the condition where the angle tip stress is compression (see Figure C5.3). Lateral-torsional buckling can also limit the moment capacity of the cross section when the maximum angle tip stress is tension from geometric axis flexure, especially with use of the new flexural capacity limits in Section 5.1. Using $\theta = 45°$ in Equation C5-1, the resulting expression is Equation 5-4 with a $+1$ instead of -1 as the last term.

Stress at the tip of the angle leg parallel to the applied bending axis is of the same sign as the maximum stress at the tip of the other leg when the single angle is unrestrained. For an equal-leg angle this stress is about one-third of the maximum stress. It is only necessary to check the nominal bending strength based on the tip of the angle leg with the maximum stress when evaluating such an angle. Since this maximum moment per Section 5.2.2 represents combined principal axis moments and Equation 5-4 represents the design limit for these combined flexural moments, only a single flexural term needs to be considered when evaluating combined flexural and axial effects.

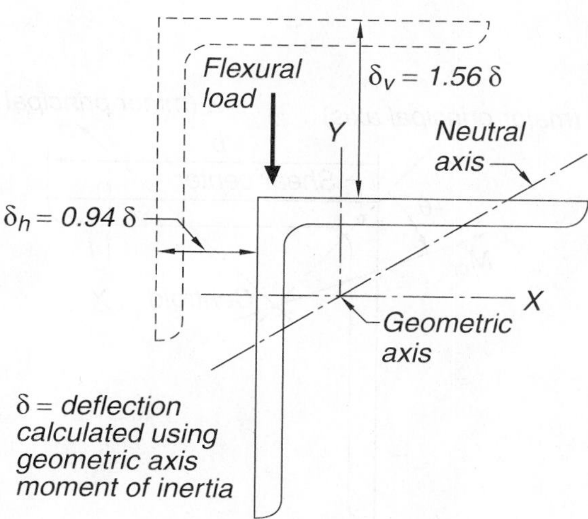

Fig. C5.2. Geometric axis bending of laterally unrestrained equal-leg angles.

C5.2.3. For unequal-leg angles without lateral-torsional restraint the applied load or moment must be resolved into components along the two principal axes in all cases and designed for biaxial bending using the interaction equation.

C5.3.1. Under major axis bending of equal-leg angles Equation 5-5 in combination with 5-3a or 5-3b controls the nominal design moment against overall lateral-torsional buckling of the angle. This is based on M_{cr}, given earlier with $\theta = 0$.

Lateral-torsional buckling for this case will reduce the stress below $1.5M_y$ only for $l/t \geq 7350C_b/F_y$ ($M_{ob} = 7.7M_y$). If the lt/b^2 parameter is small (less than approximately $0.87C_b$ for this case), local buckling will control the nominal design moment and M_n based on lateral-torsional buckling need not be evaluated. Local buckling must be checked using Section 5.1.1.

C5.3.2. Lateral-torsional buckling about the major principal W axis of an unequal-leg angle is controlled by M_{ob} in Equation 5-6. Section property β_w reflects the location of the shear center relative to the principal axis of the section and the bending direction under uniform bending. Positive β_w and maximum M_{ob} occurs when the shear center is in flexural compression while negative β_w and minimum M_{ob} occurs when the shear center is in flexural tension (see Figure C5.4). This β_w effect is consistent with behavior of singly symmetric I-shaped beams which are more stable when the compression flange is larger than the tension flange. For principal W-axis bending of equal-leg angles, β_w is equal to zero due to symmetry and Equation 5-6 reduces to Equation 5-5 for this special case.

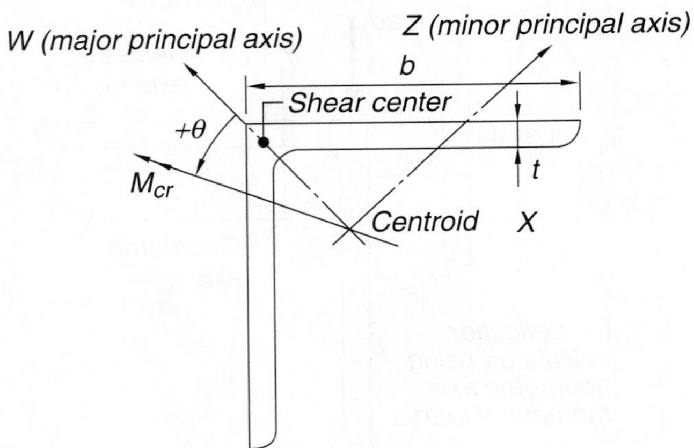

Fig. C5.3. Equal-leg angle with general moment loading.

TABLE C5.1
β_w Values for Angles

Angle Size (in.)	β_w (in.)*
9 × 4	6.54
8 × 6 8 × 4	3.31 5.48
7 × 4	4.37
6 × 4 6 × 3.5	3.14 3.69
5 × 3.5 5 × 3	2.40 2.99
4 × 3.5 4 × 3	0.87 1.65
3.5 × 3 3.5 × 2.5	0.87 1.62
3 x 2.5 3 x 2	0.86 1.56
2.5 x 2	0.85
Equal legs	0.00

* Has positive or negative value depending on direction of bending (see Figure C5.4).

For reverse curvature bending, part of the unbraced length has positive β_w, while the remainder has negative β_w and conservatively, the negative value is assigned for that entire unbraced segment.

β_w is essentially independent of angle thickness (less than one percent variation from mean value) and is primarily a function of the leg widths. The average values shown in Table C5.1 may be used for design.

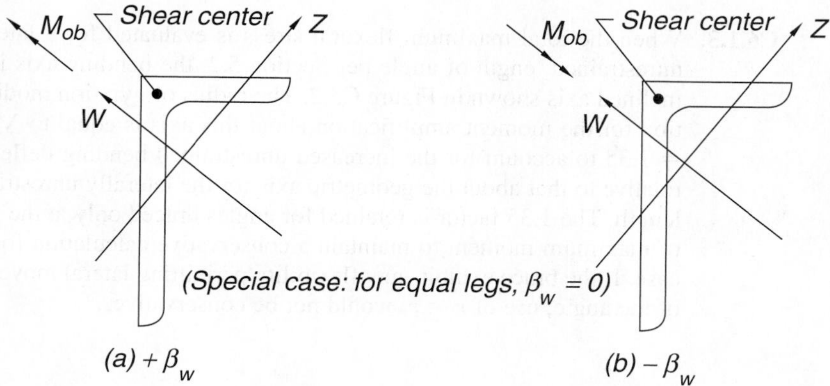

Fig. C5.4. Unequal-leg angle in bending.

C6. COMBINED STRESSES

The stability and strength interaction equations of AISC LRFD Specification Chapter H have been adopted with modifications to account for various conditions of bending that may be encountered. Bending will usually accompany axial loading in a single-angle member since the axial load and connection along the legs are eccentric to the centroid of the cross section. Unless the situation conforms to Section 5.2.1 or 5.2.2 in that Section 6.1.2 or 6.1.3 may be used, the applied moment should be resolved about the principal axes for the interaction check.

For the non-symmetric and singly symmetric single angles, the interaction expression related to stresses at a particular location on the cross section is the most accurate due to lack of double symmetry. At a particular location, it is possible to have stresses of different sign from the various components such that a combination of tensile and compressive stress will represent a critical condition. The absolute value of the combined terms must be checked at the angle-leg tips and heel and compared with 1.0.

When using the combined force expressions for single angles, M_{uw} and M_{uz} are positive as customary. The evaluation of M_n in Section 5.1 is dependent on the location on the cross section being examined by using the appropriate value of section modulus, S. Since the sign of the stress is important in using Equations 6-1a and 6-1b, M_n is considered either positive or negative by assigning a sign to S to reflect the stress condition as adding to, or subtracting from, the axial load effect. A designer may choose to use any consistent sign convention.

It is conservative to ignore this refinement and simply use positive critical M_n values in the bending terms and add the absolute values of all terms (Elgaaly, Davids, and Dagher, 1992 and Adluri and Madugula, 1992).

Alternative special interaction equations for single angles have been published (Adluri and Madugula, 1992).

C6.1.3. When the total maximum flexural stress is evaluated for a laterally unrestrained length of angle per Section 5.2, the bending axis is the inclined axis shown in Figure C5.2. The radius of gyration modification for the moment amplification about this axis is equal to $\sqrt{1.82} = 1.35$ to account for the increased unrestrained bending deflection relative to that about the geometric axis for the laterally unrestrained length. The 1.35 factor is retained for angles braced only at the point of maximum moment to maintain a conservative calculation for this case. If the brace exhibits any flexibility permitting lateral movement of the angle, use of $r = r_x$ would not be conservative.

List of References

Alduri, S. M. and Madugula, M. K. S. (1992), "Eccentrically Loaded Steel Single-Angle Struts," *Engineering Journal,* AISC, 2nd Quarter.

American Institute of Steel Construction, Inc. (1999), *Load and Resistance Factor Design Specification for Structural Steel Buildings,* Chicago, IL.

American Institute of Steel Construction, Inc. (1989), *Specification for Allowable Stress Design of Single-Angle Members,* Chicago, IL.

Australian Institute of Steel Construction (1975), *Australian Standard AS1250,* 1975.

Earls, C. J., and Galambos, T.V. (1997), "Design Recommendations for Equal Leg Single Angle Flexural Members," *Journal of Constructional Steel Research,* Vol. 43, Nos. 1-3, pp. 65–85.

Elgaaly, M., Davids, W. and Dagher, H. (1992), "Non-Slender Single-Angle Struts," *Engineering Journal,* AISC, 2nd Quarter.

Galambos, T. V. (1991), "Stability of Axially Loaded Compressed Angles," Structural Stability Research Council, *Annual Technical Session Proceedings,* Apr. 15–17, Chicago, IL.

Gjelsvik, A. (1981), *The Theory of Thin-walled Bars,* John Wiley and Sons, New York.

Leigh, J. M. and Lay, M. G. (1978), "Laterally Unsupported Angles with Equal and Unequal Legs," Report MRL 22/2 July 1978, Melbourne Research Laboratories, Clayton.

Leigh, J. M. and Lay, M. G. (1984), "The Design of Laterally Unsupported Angles," in *Steel Design Current Practice,* Section 2, Bending Members, AISC, January.

Lutz, L. A. (1992), "Critical Slenderness of Compression Members with Effective Lengths About Nonprincipal Axes," Structural Stability Research Council, Annual Technical Session Proceedings, Apr. 6–7, Pittsburgh, PA.

Madugula, M. K. S. and Kennedy, J. B. (1985), *Single and Compound Angle Members,* Elsevier Applied Science, New York.

Seaburg, P. A., and Carter, C. J. (1997), *Torsional Analysis of Structural Steel Members,* Steel Design Guide Series No. 9, AISC, Chicago, IL.

NOTES

American Institute of Steel Construction, Inc.
One East Wacker Drive, Suite 3100
Chicago, IL 60601-2001

Pub No. S351L (3M201)

Specification for Structural Joints Using ASTM A325 or A490 Bolts

June 23, 2000

Supersedes the June 3, 1994 *LRFD Specification for Structural Joints Using ASTM A325 or A490 Bolts* and the June 3, 1994 *ASD Specification for Structural Joints Using ASTM A325 or A490 Bolts.*

Prepared by RCSC Committee 15—Specifications and approved by the Research Council on Structural Connections.

RESEARCH COUNCIL ON STRUCTURAL CONNECTIONS
www.boltcouncil.org

c/o American Institute of Steel Construction, Inc.
One East Wacker Drive, Suite 3100, Chicago, Illinois 60601-2001

The information presented in this publication has been prepared in accordance with recognized engineering principles and is for general information only. While it is believed to be accurate, this information should not be used or relied upon for any specific application without competent professional examination and verification of its accuracy, suitability, and applicability by a licensed engineer, architect or other design professional. The publication of the material contained herein is not intended as a representation or warranty on the part of the Research Council on Structural Connections or of any other person named herein, that this information is suitable for any general or particular use or of freedom from infringement of any patent or patents. Anyone making use of this information assumes all liability arising from such use.

Caution must be exercised when relying upon other specifications and codes developed by other bodies and incorporated by reference herein since such material may be modified or amended from time to time subsequent to the printing of this edition. The Research Council on Structural Connections bears no responsibility for such material other than to refer to it and incorporate it by reference at the time of the initial publication of this edition.

Printed in the United States of America

PREFACE

The purpose of the Research Council on Structural Connections (RCSC) is:

(1) To stimulate and support such investigation as may be deemed necessary and valuable to determine the suitability, strength and behavior of various types of structural connections;

(2) To promote the knowledge of economical and efficient practices relating to such structural connections; and,

(3) To prepare and publish related standards and such other documents as necessary to achieving its purpose.

The Council membership consists of qualified structural engineers from academic and research institutions, practicing design engineers, suppliers and manufacturers of fastener components, fabricators, erectors and code-writing authorities.

The first Specification approved by the Council, called the *Specification for Assembly of Structural Joints Using High Tensile Steel Bolts*, was published in January 1951. Since that time the Council has published fourteen successive editions. Each was developed through the deliberations and approval of the full Council membership and based upon past successful usage, advances in the state of knowledge and changes in engineering design practice. This edition of the Council's *Specification for Structural Joints Using ASTM A325 or A490 Bolts* is significantly reorganized and revised from earlier editions. The major changes are:

- The previously separate LRFD and ASD versions of this Specification have been unified, with LRFD as the basis in the main body and ASD as an alternative in Appendix B.
- A Symbols (nomenclature) section has been added.
- A Glossary section has been added.
- An Index section has been added.
- Commentary information has been placed immediately following its corresponding Specification provision to provide convenient reference to background and further explanation.
- A summary of the drawing information and approvals required from the Engineer of Record has been added in Section 1.4.
- Explicit coverage of material and geometric requirements for washer-type indicating devices, twist-off-type tension-control bolt assemblies and alternative design fasteners has been added in Sections 2.6, 2.7 and 2.8, respectively.
- Provisions allowing the thermal cutting of bolt holes with the approval of the Engineer of Record have been added in Section 3.3.
- Requirements for the treatment of burrs have been clarified and expanded in Section 3.4.
- New information has been added in Section 4 to address the applicability and suitability of the various joint types: snug-tightened joints, pretensioned joints and slip-critical joints. The requirements in this Section also serve to identify the applicable design, installation and inspection requirements for each of the joint types.

- Requirements have been added to Section 5.1 to address the design implications of the presence of fillers or shims.
- In Section 5.5, the limitation on prying force for applications that involve tensile fatigue has been reduced from 60 percent of the total applied load to 30 percent.
- Washer requirements have been simplified and clarified in Section 6.
- Provisions for pre-installation verification of fastener assemblies have been clarified in Section 7, as invoked in Section 8.2 for pretensioned joints and slip-critical joints.
- Installation requirements have been simplified and clarified significantly in Section 8.
- Inspection requirements in Section 9 have been significantly expanded to clarify the intent and applicability.
- Arbitration provisions to be used in the case of a dispute have been clarified in Section 10.

In addition, many other changes have been made throughout this Specification.

By the Research Council on Structural Connections,

James M. Doyle Chairman	Peter C. Birkemoe Vice Chairman	Benjamin J. Wallace Secretary/Treasurer
Abolhassan Astaneh-Asl	Peter F. Kasper	Robert E. Shaw, Jr.
Joseph Bahadrian	Lawrence A. Kloiber	Firas I. Sheikh-Ibrahim
Frederick R. Beckmann	Richard F. Knoblock	W. Lee Shoemaker
David W. Bogaty	Geoffrey L. Kulak	Arun Syam
Charles J. Carter	Thomas J. Langill	Thomas S. Tarpy, Jr.
Michael Chu	Kenneth B. Lohr	William A. Thornton
Nick E. Deal	Richard W. Marshall	Anthony Thyen
Robert J. Dexter	William A. Milek, Jr.	Raymond H. R. Tide
Robert O. Disque	Eugene R. Mitchell	Emile W. J. Troup
John E. Dutton	Heath E. Mitchell	Richard B. Vincent
Edward R. Estes	William H. Munse	I. Wayne Wallace
John W. Fisher	Thomas M. Murray	Charles J. Wilson
Karl H. Frank	Rex V. Owen	Ted W. Winneberger
Rodney D. Gibble	Frederick J. Palmer	Wei-Wen Yu
Michael I. Gilmor	Frank A. Petrigliano	Joseph A. Yura
John Hoffman	Suresh C. Satsangi	Farrel Zwerneman
Mark V. Holland	Gerald E. Schroeder	
Donald L. Johnson	David F. Sharp	

This roster is current as of the date of approval of this Specification.

TABLE OF CONTENTS

SYMBOLS

The following symbols are used in this Specification.

A_b Cross-sectional area based upon the nominal diameter of bolt, in.^{2}

D Slip probability factor as described in Section 5.4.2

D_u Multiplier that reflects the ratio of the mean installed bolt pretension to the specified minimum bolt pretension T_m as described in Section 5.4.1

F_n Nominal strength (per unit area), ksi

F_u Specified minimum tensile strength (per unit area), ksi

I Moment of inertia of the built-up member about the axis of buckling (see the Commentary to Section 5.4), in.^{4}

L Total length of the built-up member (see the Commentary to Section 5.4), in.

L_c Clear distance, in the direction of load, between the edge of the hole and the edge of the adjacent hole or the edge of the material, in.

N_b Number of bolts in the joint

P_u Required strength in compression, kips; Axial compressive force in the built-up member (see the Commentary to Section 5.4), kips

Q First moment of area of one component about the axis of buckling of the built-up member (see the Commentary to Section 5.4), in.^{3}

R_n Nominal strength, kips

R_s Service-load slip resistance, kips

T Applied service load in tension, kips

T_m Specified minimum bolt pretension (for pretensioned joints as specified in Table 8.1), kips

T_u Required strength in tension (factored tensile load), kips

V_u Required strength in shear (factored shear load), kips

d_b Nominal diameter of bolt, in.

t Thickness of the connected material, in.

t' Total thickness of fillers or shims (see Section 5.1), in.

k_s Slip coefficient for an individual specimen determined in accordance with Appendix A

ϕ Resistance factor

ϕR_n Design strength, kips

μ Mean slip coefficient

GLOSSARY

The following terms are used in this Specification. Where used, they are italicized to alert the user that the term is defined in this Glossary.

Coated Faying Surface. A *faying surface* that has been primed, primed and painted or protected against corrosion, except by hot-dip galvanizing.

Connection. An assembly of one or more *joints* that is used to transmit forces between two or more members.

Contractor. The party or parties responsible to provide, prepare and assemble the fastener components and connected parts described in this Specification.

Design Strength. ϕR_n, the resistance provided by an element or *connection*; the product of the *nominal strength* R_n and the resistance factor ϕ.

Engineer of Record. The party responsible for the design of the structure and for the approvals that are required in this Specification (see Section 1.4 and the corresponding Commentary).

Fastener Assembly. An assembly of fastener components that is supplied, tested and installed as a unit.

Faying Surface. The plane of contact between two plies of a *joint*.

Firm Contact. The condition that exists on a *faying surface* when the plies are solidly seated against each other, but not necessarily in continuous contact.

Galvanized Faying Surface. A *faying surface* that has been hot-dip galvanized.

Grip. The total thickness of the plies of a *joint* through which the bolt passes, exclusive of washers or direct-tension indicators.

Guide. The *Guide to Design Criteria for Bolted and Riveted Joints*, 2nd Edition (Kulak et al., 1987).

High-Strength Bolt. An ASTM A325 or A490 bolt, an ASTM F1852 twist-off-type tension-control bolt or an alternative-design fastener that meets the requirements in Section 2.8.

Inspector. The party responsible to ensure that the *contractor* has satisfied the provisions of this Specification in the work.

Joint. A bolted assembly with or without collateral materials that is used to join two structural elements.

Lot. In this Specification, the term *lot* shall be taken as that given in the ASTM Standard as follows:

Product	ASTM Standard	See Lot Definition in Section
Bolts	A325	9.4
	A490	11.3.2 or 11.4.2
Twist-off-type tension control bolt assemblies	F1852	13.4
Nuts	A563	9.2
Washers	F436	9.2
Compressible-washer-type direct tension indicators	F959	10.2.2

Manufacturer. The party or parties that produce the components of the *fastener assembly*.

Mean Slip Coefficient. μ, the ratio of the frictional shear load at the *faying surface* to the total normal force when slip occurs.

Nominal Strength. The capacity of a structure or component to resist the effects of loads, as determined by computations using the specified material strengths and dimensions and equations derived from accepted principles of structural mechanics or by field tests or laboratory tests of scaled models, allowing for modeling effects and differences between laboratory and field conditions.

Pretensioned Joint. A *joint* that transmits shear and/or tensile loads in which the bolts have been installed in accordance with Section 8.2 to provide a pretension in the installed bolt.

Protected Storage. The continuous protection of fastener components in closed containers in a protected shelter as described in the Commentary to Section 2.2.

Prying Action. Lever action that exists in *connections* in which the line of application of the applied load is eccentric to the axis of the bolt, causing deformation of the fitting and an amplification of the axial tension in the bolt.

Required Strength. The load effect acting on an element or *connection* determined by structural analysis from the factored loads using the most appropriate critical load combination.

Routine Observation. Periodic monitoring of the work in progress.

Shear/Bearing Joint. A *snug-tightened joint* or *pretensioned joint* with bolts that transmit shear loads and for which the design criteria are based upon the shear strength of the bolts and the bearing strength of the connected materials.

Slip-Critical Joint. A *joint* that transmits shear loads or shear loads in combination with tensile loads in which the bolts have been installed in accordance with Section 8.2 to provide a pretension in the installed bolt (clamping force on the *faying surfaces*), and with *faying surfaces* that have been prepared to provide a calculable resistance against slip.

Snug-Tightened Joint. A *joint* in which the bolts have been installed in accordance with Section 8.1. The snug-tightened condition is the tightness that is attained with a few impacts of an impact wrench or the full effort of an ironworker using an ordinary spud wrench to bring the plies into *firm contact*.

Start of Work. Any time prior to the installation of *high-strength bolts* in structural *connections* in accordance with Section 8.

Sufficient Thread Engagement. Having the end of the bolt extending beyond or at least flush with the outer face of the nut; a condition that develops the strength of the bolt.

Supplier. The party that sells the fastener components to the party that will install them in the work.

Tension Calibrator. A calibrated tension-indicating device that is used to verify the acceptability of the pretensioning method when a *pretensioned joint* or *slip-critical joint* is specified.

Uncoated Faying Surface. A *faying surface* that has neither been primed, painted, nor galvanized and is free of loose scale, dirt and other foreign material.

NOTES

SPECIFICATION FOR STRUCTURAL JOINTS USING ASTM A325 OR A490 BOLTS
June 23, 2000

SECTION 1. GENERAL REQUIREMENTS

1.1. Scope

This Specification covers the design of bolted *joints* and the installation and inspection of the assemblies of fastener components listed in Section 1.3, the use of alternative-design fasteners as permitted in Section 2.8 and alternative washer-type indicating devices as permitted in Section 2.6.2, in structural steel *joints*. This Specification relates only to those aspects of the connected materials that bear upon the performance of the fastener components. The Symbols, Glossary and Appendices are a part of this Specification.

Commentary:

This Specification deals principally with two strength grades of *high-strength bolts*, ASTM A325 and A490, and with their design, installation and inspection in structural steel *joints*. Equivalent fasteners, however, such as ASTM F1852 (equivalent to ASTM A325) twist-off-type tension-control bolt assemblies, are also covered. These provisions may not be relied upon for high-strength fasteners of other chemical composition, mechanical properties, or size. These provisions do not apply when material other than steel is included in the *grip*; nor are they applicable to anchor rods.

This Specification relates only to the performance of fasteners in structural steel *joints* and those few aspects of the connected material that affect this performance. Many other aspects of *connection* design and fabrication are of equal importance and must not be overlooked. For more general information on design and issues relating to *high-strength bolting* and the connected material, refer to current steel design textbooks and the *Guide to Design Criteria for Bolted and Riveted Joints*, 2nd Edition (Kulak et al., 1987).

1.2. Loads, Load Factors and Load Combinations

The design and construction of the structure shall conform to an applicable load and resistance factor design specification for steel structures. Because factored load combinations account for the reduced probabilities of maximum loads acting concurrently, the *design strengths* given in this Specification shall not be increased. Appendix B is included as an alternative approach.

Commentary:

This Specification is written in the load and resistance factor design (LRFD) format, which provides a method of proportioning structural components such that no applicable limit state is exceeded when the structure is subject to all appropriate load combinations. When a structure or structural component ceases to fulfill the intended purpose in some way, it is said to have exceeded a limit state. Strength limit states concern maximum load-carrying capability, and are related to safety. Serviceability limit states are usually related to performance under normal service conditions, and usually are not related to strength or safety. The term "resistance" includes both strength limit states and serviceability limit states.

The *design strength* ϕR_n is the *nominal strength* R_n multiplied by the resistance factor ϕ. The factored load is the sum of the nominal loads multiplied by load factors, with due recognition of load combinations that account for the improbability of simultaneous occurrence of multiple transient load effects at their respective maximum values. The *design strength* ϕR_n of each structural component or assemblage must equal or exceed the *required strength* (V_u, T_u, etc.).

Although loads, load factors and load combinations are not explicitly specified in this Specification, the resistance factors herein are based upon those specified in ASCE 7. When the design is governed by other load criteria, the resistance factors specified herein should be adjusted as appropriate.

1.3. Referenced Standards and Specifications

The following standards and specifications are referenced herein:

American Institute of Steel Construction
Load and Resistance Factor Design Specification for Structural Steel Buildings, December 27,1999

American National Standards Institute
ANSI/ASME B18.2.6-96 *Fasteners for Use in Structural Applications*

American Society for Testing and Materials
ASTM A123-97a *Standard Specification for Zinc (Hot-Dip Galvanized) Coatings on Iron and Steel Products*

ASTM A153-98 *Standard Specification for Zinc Coating (Hot-Dip) on Iron and Steel Hardware*

ASTM A194-98b *Specification for Carbon and Alloy Steel Nuts for Bolts for High Pressure or High-Temperature Service, or Both*

ASTM A325-97 *Standard Specification for Structural Bolts, Steel, Heat Treated, 120/105 ksi Minimum Tensile Strength*

ASTM A490-97 *Standard Specification for Heat-Treated Steel Structural Bolts, 150 ksi Minimum Tensile Strength*

ASTM A563-97 *Standard Specification for Carbon and Alloy Steel Nuts*

ASTM B695-91[1] *Standard Specification for Coatings of Zinc Mechanically Deposited on Iron and Steel*

---

[1] Reapproved 1997.

ASTM F436-93 *Standard Specification for Hardened Steel Washers*

ASTM F959-99a *Standard Specification for Compressible-Washer-Type Direct Tension Indicators for Use with Structural Fasteners*

ASTM F1852-98 *"Twist off" Type Tension Control Structural Bolt/Nut/Washer Assemblies, Steel, Heat Treated, 120/105 ksi Minimum Tensile Strength*

American Society of Civil Engineers
ASCE 7-98 *Minimum Design Loads for Buildings and Other Structures*

SSPC: The Society for Protective Coatings
SSPC-PA2-96 *Measurement of Dry Coating Thickness With Magnetic Gages*

Commentary:
Familiarity with the referenced AISC, ASCE, ASME, ASTM and SSPC specification requirements is necessary for the proper application of this Specification. The discussion of referenced specifications in this Commentary is limited to only a few frequently overlooked or misunderstood items.

1.4. Drawing Information

The *Engineer of Record* shall specify the following information in the contract documents

(1) The ASTM designation and type (Section 2) of bolt to be used;
(2) The *joint* type (Section 4);
(3) The required class of slip resistance if *slip-critical joints* are specified (Section 4); and,
(4) Whether slip is checked at the factored-load level or the service-load level, if *slip-critical joints* are specified (Section 5).

Commentary:
A summary of the information that the *Engineer of Record* is required to provide in the contract documents is provided in this Section. The parenthetical reference after each listed item indicates the location of the actual requirement in this Specification. In addition, the approval of the *Engineer of Record* is required in this Specification in the following cases:

(1) For the reuse of non-galvanized ASTM A325 bolts (Section 2.3.3);
(2) For the use of alternative washer-type indicating devices that differ from those that meet the requirements of ASTM F959, including the corresponding installation and inspection requirements that are provided by the *manufacturer* (Section 2.6.2);
(3) For the use of alternative-design fasteners, including the corresponding installation and inspection requirements that are provided by the *manufacturer* (Section 2.8);

(4) For the use of faying-surface coatings in *slip-critical joints* that provide a *mean slip coefficient* determined per Appendix A, but differing from Class A or Class B (Section 3.2.2(b));
(5) For the use of thermal cutting in the production of bolt holes (Section 3.3);
(6) For the use of oversized (Section 3.3.2), short-slotted (Section 3.3.3) or long slotted holes (Section 3.3.4) in lieu of standard holes;
(7) For the use of a value of D_u other than 1.13 (Section 5.4.1); and,
(8) For the use of a value of D other than 0.80 (Section 5.4.2).

SECTION 2. FASTENER COMPONENTS

2.1. Manufacturer Certification of Fastener Components

Manufacturer certifications documenting conformance to the applicable specifications required in Sections 2.3 through 2.8 for all fastener components used in the *fastener assemblies* shall be available to the *Engineer of Record* and *inspector* prior to assembly or erection of structural steel.

Commentary:
Certification by the *manufacturer* or *supplier* of *high-strength bolts*, nuts, washers and other components of the *fastener assembly* is required to ensure that the components to be used are identifiable and meet the requirements of the applicable ASTM Specifications.

2.2. Storage of Fastener Components

Fastener components shall be protected from dirt and moisture in closed containers at the site of installation. Only as many fastener components as are anticipated to be installed during the work shift shall be taken from *protected storage*. Fastener components that are not incorporated into the work shall be returned to *protected storage* at the end of the work shift. Fastener components shall not be cleaned or modified from the as-delivered condition.

Fastener components that accumulate rust or dirt shall not be incorporated into the work unless they are requalified as specified in Section 7. ASTM F1852 twist-off-type tension-control bolt assemblies and alternative-design fasteners that meet the requirements in Section 2.8 shall not be relubricated, except by the *manufacturer*.

Commentary:
Protected storage requirements are specified for *high-strength bolts*, nuts, washers and other fastener components with the intent that the condition of the components be maintained as nearly as possible to the as-manufactured condition until they are installed in the work. This involves:

(1) The storage of the fastener components in closed containers to protect from dirt and corrosion;
(2) The storage of the closed containers in a protected shelter;
(3) The removal of fastener components from *protected storage* only as necessary; and,
(4) The prompt return of unused fastener components to *protected storage*.

To facilitate manufacture, prevent corrosion and facilitate installation, the *manufacturer* may apply various coatings and oils that are present in the as manufactured condition. As such, the condition of supplied fastener components or the *fastener assembly* should not be altered to make them unsuitable for pre-tensioned installation.

If fastener components become dirty, rusty, or otherwise have their as received condition altered, they may be unsuitable for pre-tensioned installation.

It is also possible that a *fastener assembly* may not pass the pre-installation verification requirements of Section 7. Except for ASTM F1852 twist-off-type tension-control bolt assemblies (Section 2.7) and some alternative-design fasteners (Section 2.8), fastener components can be cleaned and lubricated by the fabricator or the erector. Because the acceptability of their installation is dependent upon specific lubrication, ASTM F1852 twist-off-type tension-control bolt assemblies and some alternative-design fasteners are suitable only if the *manufacturer* lubricates them.

2.3. Heavy-Hex Structural Bolts

2.3.1. Specifications: Heavy-hex structural bolts shall meet the requirements of ASTM A325 or ASTM A490. The *Engineer of Record* shall specify the ASTM designation and type of bolt (see Table 2.1) to be used.

2.3.2. Geometry: Heavy-hex structural bolt dimensions shall meet the requirements of ANSI/ASME B18.2.6. The bolt length used shall be such that the end of the bolt extends beyond or is at least flush with the outer face of the nut when properly installed.

Table 2.1. Acceptable ASTM A563 Nut Grade and Finish and ASTM F436 Washer Type and Finish

ASTM Desig.	Bolt Type	Bolt Finish[d]	ASTM A563 nut grade and finish[d]	ASTM F436 washer type and finish[a,d]
A325	1	Plain (uncoated)	C, C3, D, DH[c] and DH3; plain	1; plain
		Galvanized	DH[c]; galvanized And lubricated	1; galvanized
	3	Plain	C3 and DH3; plain	3; plain
F1852	1	Plain (uncoated)	C, C3, DH[c] and DH3; plain	1; plain[b]
		Mechanically Galvanized	DH[c]; mechanically galvanized and lubricated	1; mechanically galvanized[b]
	3	Plain	C3 and DH3; plain	3; plain[b]
A490	1	Plain	DH[c] and DH3; plain	1; plain
	3	Plain	DH3; plain	3; plain

[a] Applicable only if washer is required in Section 6.
[b] Required in all cases under nut per Section 6.
[c] The substitution of ASTM A194 grade 2H nuts in place of ASTM A563 grade DH nuts is permitted.
[d] "Galvanized" as used in this table refers to hot-dip galvanizing in accordance with ASTM A153 or mechanical galvanizing in accordance with ASTM B695.

2.3.3. Reuse: ASTM A490 bolts and galvanized ASTM A325 bolts shall not be reused. When approved by the *Engineer of Record*, black ASTM A325 bolts are permitted to be reused. Touching up or re-tightening bolts that may have been loosened by the installation of adjacent bolts shall not be considered to be a reuse.

Commentary:
ASTM A325 and ASTM A490 currently provide for two types (according to metallurgical classification) of *high-strength bolts*, supplied in diameters from ½ in. to 1½ in. inclusive. Type 1 covers medium carbon steel for ASTM A325 bolts and alloy steel for ASTM A490 bolts. Type 3 covers *high-strength bolts* that have improved atmospheric corrosion resistance and weathering characteristics. (Reference to Type 2 ASTM A325 and Type 2 A490 bolts, which appeared in previous editions of this Specification, has been removed following the removal of similar reference within the ASTM A325 and A490 Specifications). When the bolt type is not specified, either Type 1 or Type 3 may be supplied at the option of the *manufacturer*. Note that ASTM F1852 twist-off-type tension-control bolt assemblies may be manufactured with a button head or hexagonal head; other requirements for these *fastener assemblies* are found in Section 2.7.

Regular heavy-hex structural bolts and twist-off-type tension-control bolt assemblies are required by ASTM Specifications to be distinctively marked. Certain markings are mandatory. In addition to the mandatory markings, the *manufacturer* may apply additional distinguishing markings. The mandatory and sample optional markings are illustrated in Figure C-2.1.

ASTM Specifications permit the galvanizing of ASTM A325 bolts but not ASTM A490 bolts. Similarly, the application of zinc to ASTM A490 bolts by metallizing or mechanical coating is not permitted because the effect of mechanical galvanizing on embrittlement and delayed cracking of ASTM A490 bolts has not been fully investigated to date.

Galvanized *high-strength bolts* and nuts must be considered as a manufactured *fastener assembly*. Insofar as the hot-dip galvanized bolt and nut assembly is concerned, four principal factors must be considered so that the provisions of this Specification are understood and properly applied. These are:

(1) The effect of the hot-dip galvanizing process on the mechanical properties of high-strength steels;
(2) The effect of over-tapping for hot-dip galvanized coatings on the nut stripping strength;
(3) The effect of galvanizing and lubrication on the torque required for pretensioning; and,
(4) Shipping requirements.

Birkemoe and Herrschaft (1970) showed that, in the as-galvanized condition, galvanizing increases the friction between the bolt and nut threads as well as the variability of the torque-induced pretension. A lower required torque and more consistent results are obtained if the nuts are lubricated. Thus, it is required in ASTM A325 that a galvanized bolt and lubricated galvanized nut be assembled in a steel *joint* with a galvanized washer and tested by the *supplier* prior to shipment. This testing must show that the galvanized nut with the lubricant provided may be rotated from the snug-tight condition well in excess of the rotation required for pretensioned installation without stripping. This requirement applies to both hot-dip and mechanically galvanized fasteners. The above requirements clearly indicate that:

(1) Galvanized *high-strength bolts* and nuts must be treated as a *fastener assembly*;

(2) The *supplier* must supply nuts that have been lubricated and tested with the supplied *high-strength bolts*;

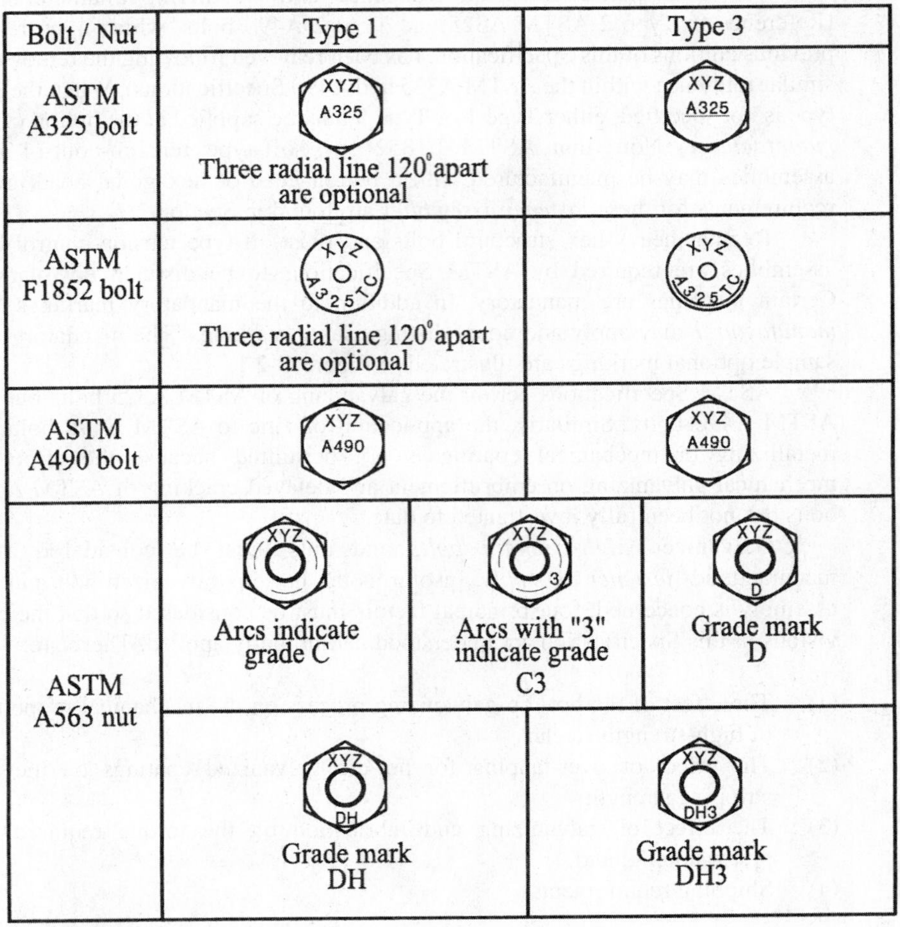

Bolt / Nut	Type 1		Type 3
ASTM A325 bolt	XYZ A325 Three radial line 120° apart are optional		XYZ A325
ASTM F1852 bolt	XYZ A325TC Three radial line 120° apart are optional		XYZ A325TC
ASTM A490 bolt	XYZ A490		XYZ A490
ASTM A563 nut	XYZ Arcs indicate grade C	XYZ 3 Arcs with "3" indicate grade C3	XYZ D Grade mark D
	XYZ DH Grade mark DH		XYZ DH3 Grade mark DH3

Notes:
1. XYZ represents the manufacturer's identification mark.
2. ASTM F1852 twist-off-type tension-control bolt assemblies are also produced with heavy-hex head that has similar markings.

Figure C-2.1. Required marks for acceptable bolt and nut assemblies.

(3) Nuts and *high-strength bolts* must be shipped together in the same shipping container; and,

(4) The purchase of galvanized *high-strength bolts* and galvanized nuts from separate *suppliers* is not in accordance with the intent of the ASTM Specifications because the control of over-tapping, the testing and application of lubricant and the *supplier* responsibility for the performance of the assembly would clearly not have been provided as required.

Because some of the lubricants used to meet the requirements of ASTM Specifications are water soluble, it is advisable that galvanized *high-strength bolts* and nuts be shipped and stored in plastic bags or in sealed wood or metal containers. Containers of fasteners with hot-wax-type lubricants should not be subjected to heat that would cause depletion or change in the properties of the lubricant.

Both the hot-dip galvanizing process (ASTM A153) and the mechanical galvanizing process (ASTM B695) are recognized in ASTM A325. The effects of the two processes upon the performance characteristics and requirements for proper installation are distinctly different. Therefore, distinction between the two must be noted in the comments that follow. In accordance with ASTM A325, all threaded components of the *fastener assembly* must be galvanized by the same process and the *supplier's* option is limited to one process per item with no mixed processes in a *lot*. Mixing *high-strength bolts* that are galvanized by one process with nuts that are galvanized by the other may result in an unworkable assembly.

Steels in the 200 ksi and higher tensile-strength range are subject to embrittlement if hydrogen is permitted to remain in the steel and the steel is subjected to high tensile stress. The minimum tensile strength of ASTM A325 bolts is 105 ksi or 120 ksi, depending upon the diameter, and maximum hardness limits result in production tensile strengths well below the critical range. The maximum tensile strength for ASTM A490 bolts was set at 170 ksi to provide a little more than a ten-percent margin below 200 ksi. However, because *manufacturers* must target their production slightly higher than the required minimum, ASTM A490 bolts close to the critical range of tensile strength must be anticipated. For black *high-strength bolts*, this is not a cause for concern. However, if the bolt is hot-dip galvanized, delayed brittle fracture in service is a concern because of the possibility of the introduction of hydrogen during the pickling operation of the hot-dip galvanizing process and the subsequent "sealing-in" of the hydrogen by the zinc coating. There also exists the possibility of cathodic hydrogen absorption arising from the corrosion process in certain aggressive environments.

ASTM A325 and A490 bolts are manufactured to dimensions as specified in ANSI/ASME B18.2.6. The basic dimensions, as defined in Figure C-2.2, are shown in Table C-2.1.

The principal geometric features of heavy-hex structural bolts that distinguish them from bolts for general application are the size of the head and the unthreaded body length. The head of the heavy-hex structural bolt is specified to be the same size as a heavy-hex nut of the same nominal diameter so that the ironworker may use the same wrench or socket either on the bolt head and/or on the nut. With the specific exception of fully threaded ASTM A325T bolts as discussed below, heavy-hex structural bolts have shorter threaded lengths than bolts for general applications. By making the body length of the bolt the control dimension, it has been possible to exclude the thread from all shear planes when desirable, except for the case of thin outside parts adjacent to the nut.

The shorter threaded lengths provided with heavy-hex structural bolts tend to minimize the threaded portion of the bolt within the *grip*. Accordingly, care must also be exercised to provide adequate threaded length between the nut and the bolt head to enable appropriate installation without jamming the nut on the thread run-out.

Depending upon the increments of supplied bolt lengths, the full thread may extend into the *grip* for an assembly without washers by as much as ⅜ in. for ½, ⅝, ¾, ⅞, 1¼, and 1½ in. diameter *high-strength bolts* and as much as ½ in. for 1,

Table C-2.1. Bolt and Nut Dimensions

Nominal Bolt Diameter d_b, in.	Heavy Hex Structural Bolt Dimensions			Heavy Hex Nut Dimensions	
	Width across flats F, in.	Height H_1, in.	Thread Length T, in.	Width across flats W, in.	Height H_2, in.
1/2	7/8	5/16	1	7/8	31/64
5/8	1 1/16	25/64	1 1/4	1 1/16	39/64
3/4	1 1/4	15/32	1 3/8	1 1/4	47/64
7/8	1 7/16	35/64	1 1/2	1 7/16	55/64
1	1 5/8	39/64	1 3/4	1 5/8	63/64
1 1/8	1 13/16	11/16	2	1 13/16	1 7/64
1 1/4	2	25/32	2	2	1 7/32
1 3/8	2 3/16	27/32	2 1/4	2 3/16	1 11/32
1 1/2	2 3/8	15/16	2 1/4	2 3/8	1 15/32

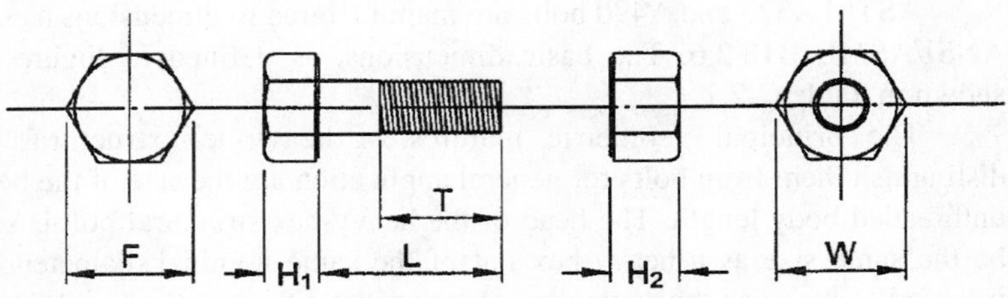

Figure C-2.2. Heavy-hex structural bolt and heavy-hex nut.

1⅛, and 1⅜ in. diameter *high-strength bolts*. When the thickness of the ply closest to the nut is less than the ⅜ in. or ½ in. dimensions given above, it may still be possible to exclude the threads from the shear plane, when required, depending upon the specific combination of bolt length, *grip* and number of washers used under the nut (Carter, 1996). If necessary, the next increment of bolt length can be specified with ASTM F436 washers in sufficient number to both exclude the threads from the shear plane and ensure that the assembly can be installed with adequate threads included in the *grip* for proper installation.

At maximum accumulation of tolerances from all components in the *fastener assembly*, the thread run-out will cross the shear plane for the critical combination of bolt length and *grip* used to select the foregoing rules of thumb for ply thickness required to exclude the threads. This condition is not of concern, however, for two reasons. First, it is too unlikely that all component tolerances will accumulate at their maximum values to warrant consideration. Second, even if the maximum accumulation were to occur, the small reduction in shear strength due to the presence of the thread run-out (not a full thread) would be negligible.

There is an exception to the foregoing thread length requirements for ASTM A325 bolts but not for ASTM A490 bolts nor ASTM F1852 twist-off-type tension-control bolt assemblies. Supplementary requirements in ASTM A325 permit the purchaser to specify a bolt that is threaded for the full length of the shank, when the bolt length is equal to or less than four times the nominal diameter. This exception is provided to increase economy through simplified ordering and inventory control in the fabrication and erection of some structures. It is particularly useful in those structures in which the strength of the *connection* is dependent upon the bearing strength of relatively thin connected material rather than the shear strength of the bolt, whether with threads in the shear plane or not. As required in ASTM A325, *high-strength bolts* ordered to such supplementary requirements must be marked with the symbol A325T.

To determine the required bolt length, the value shown in Table C-2.2 should be added to the *grip* (i.e., the total thickness of all connected material, exclusive of washers). For each ASTM F436 washer that is used, add 5⁄32 in.; for each beveled washer, add 5⁄16 in. The tabulated values provide appropriate allowances for manufacturing tolerances and also provide *sufficient thread engagement* with an installed heavy-hex nut. The length determined by the use of Table C-2.2 should be adjusted to the next longer ¼-in. length increment (½-in. length increment for lengths exceeding 6 in.). A more extensive table for bolt length selection based upon these rules is available (Carter, 1996).

Pretensioned installation involves the inelastic elongation of the portion of the threaded length between the nut and the thread run-out. ASTM A490 bolts and galvanized ASTM A325 bolts possess sufficient ductility to undergo one pretensioned installation, but are not consistently ductile enough to undergo a second pretensioned installation. Black ASTM A325 bolts, however, possess sufficient ductility to undergo more than one pretensioned installation as suggested in the *Guide* (Kulak et al., 1987). As a simple rule of thumb, a black ASTM A325 bolt is suitable for reuse if the nut can be run up the threads by hand.

Table C- 2.2. Bolt Length Selection Increment

Nominal Bolt Diameter d_b, in.	To Determine the Required Bolt Length, Add to Grip, in.
$\frac{1}{2}$	$\frac{11}{16}$
$\frac{5}{8}$	$\frac{7}{8}$
$\frac{3}{4}$	1
$\frac{7}{8}$	$1\frac{1}{8}$
1	$1\frac{1}{4}$
$1\frac{1}{8}$	$1\frac{1}{2}$
$1\frac{1}{4}$	$1\frac{5}{8}$
$1\frac{3}{8}$	$1\frac{3}{4}$
$1\frac{1}{2}$	$1\frac{7}{8}$

2.4. Heavy-Hex Nuts

2.4.1. Specifications: Heavy-hex nuts shall meet the requirements of ASTM A563. The grade and finish of such nuts shall be as given in Table 2.1.

2.4.2. Geometry: Heavy-hex nut dimensions shall meet the requirements of ANSI/ASME B18.2.6.

Commentary:
Heavy-hex nuts are required by ASTM Specifications to be distinctively marked. Certain markings are mandatory. In addition to the mandatory markings, the *manufacturer* may apply additional distinguishing markings. The mandatory markings and sample optional markings are illustrated in Figure C-2.1.

Hot-dip galvanizing affects the stripping strength of the bolt-nut assembly because, to accommodate the relatively thick zinc coatings of non-uniform thickness on bolt threads, it is usual practice to hot-dip galvanize the blank nut and then to tap the nut over-size. This results in a reduction of thread engagement with a consequent reduction of the stripping strength. Only the stronger hardened nuts have adequate strength to meet ASTM thread strength requirements after over-tapping. Therefore, as specified in ASTM A325, only ASTM A563 grade DH are suitable for use as galvanized nuts. This requirement should not be overlooked if non-galvanized nuts are purchased and then sent to a local galvanizer for hot-dip galvanizing. Because the mechanical galvanizing process results in a more uniformly distributed and smooth zinc coating, nuts may be tapped over-size before galvanizing by an amount that is less than that required for the hot-dip process before galvanizing.

In earlier editions, this Specification permitted the use of ASTM A194 grade 2H nuts in the same finish as that permitted for ASTM A563 nuts in the following cases: with ASTM A325 Type 1 plain, Type 1 galvanized and Type 3 plain bolts and with ASTM A490 Type 1 plain bolts. Reference to ASTM A194 grade 2H nuts has been removed following the removal of similar reference within the ASTM A325 and A490 Specifications. However, it should be noted that ASTM A194 grade 2H nuts remain acceptable in these applications as indicated by footnote in Table 2.1, should they be available.

ASTM A563 nuts are manufactured to dimensions as specified in ANSI/ASME B18.2.6. The basic dimensions, as defined in Figure C-2.2, are shown in Table C-2.1.

2.5. Washers

Flat circular washers and square or rectangular beveled washers shall meet the requirements of ASTM F436, except as provided in Table 6.1. The type and finish of such washers shall be as given in Table 2.1.

2.6. Washer-Type Indicating Devices

The use of washer-type indicating devices is permitted as described in Sections 2.6.1 and 2.6.2.

2.6.1. Compressible-Washer-Type Direct Tension Indicators: Compressible-washer-type direct tension indicators shall meet the requirements of ASTM F959.

2.6.2. Alternative Washer-Type Indicating Devices: When approved by the *Engineer of Record*, the use of alternative washer-type indicating devices that differ from those that meet the requirements of ASTM F959 is permitted.

Detailed installation instructions shall be prepared by the *manufacturer* in a supplemental specification that is approved by the *Engineer of Record* and shall provide for:

(1) The required character and frequency of pre-installation verification;
(2) The alignment of bolt holes to permit insertion of the bolt without undue damage to the threads;
(3) The placement of *fastener assemblies* in all types and sizes of holes, including placement and orientation of the alternative and regular washers;
(4) The systematic assembly of the *joint*, progressing from the most rigid part of the *joint* until the connected plies are in *firm contact*; and;
(5) The subsequent systematic pretensioning of all bolts in the *joint*, progressing from the most rigid part of the *joint* in a manner that will minimize relaxation of previously pretensioned bolts.

Detailed inspection instructions shall be prepared by the *manufacturer* in a supplemental specification that is approved by the *Engineer of Record* and shall provide for:

(1) Observation of the required pre-installation verification testing; and,
(2) Subsequent *routine observation* to ensure the proper use of the alternative washer-type indicating device.

2.7. Twist-Off-Type Tension-Control Bolt Assemblies

2.7.1. Specifications: Twist-off-type tension-control bolt assemblies shall meet the requirements of ASTM F1852. The *Engineer of Record* shall specify the type of bolt (Table 2.1) to be used.

2.7.2. Geometry: Twist-off-type tension-control bolt assembly dimensions shall meet the requirements of ASTM F1852. The bolt length used shall be such that the end of the bolt extends beyond or is at least flush with the outer face of the nut when properly installed.

Commentary:

It is the policy of the Research Council on Structural Connections to directly recognize only those fastener components that are manufactured to meet the requirements in an approved ASTM specification. Prior to this edition, the RCSC Specification provided for the use of ASTM A325 and A490 bolts directly and alternative-design fasteners meeting detailed requirements similar to those in Section 2.8 when approved by the *Engineer of Record*. With this edition, ASTM F1852 twist-off-type tension-control bolt assemblies are now recognized directly. Essentially, ASTM F1852 relates an ASTM A325-equivalent product to a specific method of installation that is suitable for use in all *joint* types as described in Section 8. Provision has also been retained for approval by the *Engineer of Record* of other alternative-design fasteners that meet the detailed requirements in 2.8. As an example of one such approval, the use of twist-off-type tension-control bolt assemblies with ASTM A490 mechanical properties is usually deemed acceptable.

If galvanized, ASTM F1852 twist-off-type tension-control bolt assemblies are required in ASTM F1852 to be mechanically galvanized.

While specific provisions for reuse of ASTM F1852 twist-off-type tension control bolts have not been included in this Specification, those given in Section 2.3.3 for reuse of heavy-hex structural bolts are equally applicable if the use of an alternative pretensioning method, such as the turn-of-nut pretensioning method, is practical. It is assumed that rotation of the non-turned element can be restrained.

2.8. Alternative-Design Fasteners

When approved by the *Engineer of Record*, the use of alternative-design fasteners is permitted if they:

(1) Meet the materials, manufacturing and chemical composition requirements of ASTM A325 or ASTM A490, as applicable;
(2) Meet the mechanical property requirements of ASTM A325 or ASTM A490 in full-size tests;
(3) Have a body diameter and bearing area under the bolt head and nut that is equal to or greater than those provided by a bolt and nut of the same nomi-

nal dimensions specified in Sections 2.3 and 2.4; and,
(4) Are supplied and used in the work as a *fastener assembly*.

Such alternative-design fasteners are permitted to differ in other dimensions from those of the specified *high-strength bolts* and nuts.

Detailed installation instructions shall be prepared by the *manufacturer* in a supplemental specification that is approved by the *Engineer of Record* and shall provide for:

(1) The required character and frequency of pre-installation verification;
(2) The alignment of bolt holes to permit insertion of the alternative-design fastener without undue damage;
(3) The placement of *fastener assemblies* in all holes, including any washer requirements as appropriate;
(4) The systematic assembly of the *joint*, progressing from the most rigid part of the *joint* until the connected plies are in *firm contact*; and,
(5) The subsequent systematic pretensioning of all *fastener assemblies* in the *joint*, progressing from the most rigid part of the *joint* in a manner that will minimize relaxation of previously pretensioned bolts.

Detailed inspection instructions shall be prepared by the *manufacturer* in a supplemental specification that is approved by the *Engineer of Record* and shall provide for:

(1) Observation of the required pre-installation verification testing; and,
(2) Subsequent *routine observation* to ensure the proper use of the alternative-design fastener.

SECTION 3. BOLTED PARTS

3.1. Connected Plies

All connected plies that are within the *grip* of the bolt and any materials that are used under the head or nut shall be steel (uncoated, coated or galvanized) as defined in Section 3.2. Compressible materials shall not be placed within the *grip* of the bolt. The slope of the surfaces of parts in contact with the bolt head and nut shall be equal to or less than 1:20 with respect to a plane that is normal to the bolt axis.

Commentary:

The presence of gaskets, insulation or any compressible materials other than the specified coatings within the *grip* would preclude the development and/or retention of the installed pretensions in the bolts, when required.

ASTM A325, F1852 and A490 bolt assemblies are ductile enough to deform to a surface with a slope that is less than or equal to 1:20 with respect to a plane normal to the bolt axis. Greater slopes are undesirable because the resultant localized bending decreases both the strength and the ductility of the bolt.

3.2. Faying Surfaces

Faying surfaces and surfaces adjacent to the bolt head and nut shall be free of dirt and other foreign material. Additionally, *faying surfaces* shall meet the requirements in Sections 3.2.1 or 3.2.2.

3.2.1. *Snug-Tightened Joints* and *Pretensioned Joints*: The *faying surfaces* of *snug-tightened joints* and *pretensioned joints* as defined in Sections 4.1 and 4.2 are permitted to be uncoated, coated with coatings of any formulation or galvanized.

Commentary:

In both *snug-tightened joints* and *pretensioned joints*, the ultimate strength is dependent upon shear transmitted by the bolts and bearing of the bolts against the connected material. It is independent of any frictional resistance that may exist on the *faying surfaces*. Consequently, since slip resistance is not an issue, the *faying surfaces* are permitted to be uncoated, coated, or galvanized without regard to the resulting slip coefficient obtained.

3.2.2. *Slip-Critical Joints*: The *faying surfaces* of *slip-critical joints* as defined in Section 4.3, including those of filler plates and finger shims, shall meet the following requirements:

(a) *Uncoated Faying Surfaces*: *Uncoated faying surfaces* shall be free of scale, except tight mill scale, and free of coatings, including inadvertent overspray, in areas closer than one bolt diameter but not less than 1 in. from the edge of any hole and in all areas within the bolt pattern.

(b) *Coated Faying Surfaces*: *Coated faying surfaces* shall first be blast cleaned and subsequently coated with a coating that is qualified in accordance with the requirements in Appendix A as a Class A or Class B coating as defined in Section 5.4. Alternatively, when approved by the *Engineer of Record*, coatings that provide a *mean slip coefficient* that differs from Class A or Class B are permitted when:

(1) The *mean slip coefficient* μ is established by testing in accordance with the requirements in Appendix A; and,
(2) The design slip resistance is determined in accordance with Section 5.4 using this coefficient, except that, for design purposes, a value of μ greater than 0.50 shall not be used.

The plies of *slip-critical joints* with *coated faying surfaces* shall not be assembled before the coating has cured for the minimum time that was used in the qualifying tests.

(c) *Galvanized Faying Surfaces*: *Galvanized faying surfaces* shall first be hot-dip galvanized in accordance with the requirements of ASTM A123 and subsequently roughened by means of hand wire brushing. Power wire brushing is not permitted. When prepared by roughening, the *galvanized faying surface* is designated as Class C for design.

Commentary:
Slip-critical joints are those *joints* that have specified *faying surface* conditions that, in the presence of the clamping force provided by pretensioned fasteners, resist a design load solely by friction and without displacement at the *faying surfaces*. Consequently, it is necessary to prepare the *faying surfaces* in a manner so that the desired slip performance is achieved.

Clean mill scale steel surfaces (Class A, see Section 5.4.1) and blast-cleaned steel surfaces (Class B, see Section 5.4.1) can be used within *slip-critical joints*. When used, it is necessary to keep the *faying surfaces* free of coatings, including inadvertent overspray.

Corrosion often occurs on uncoated blast-cleaned steel surfaces (Class B, see Section 5.4.1) due to exposure between the time of fabrication and subsequent erection. In normal atmospheric exposures, this corrosion is not detrimental and may actually increase the slip resistance of the *joint*. Yura et al. (1981) found that the Class B slip coefficient could be maintained for up to one year prior to *joint* assembly.

Polyzois and Frank (1986) demonstrated that, for plate material with thickness in the range of ⅜ in. to ¾ in., the contact pressure caused by bolt pretension is concentrated on the *faying surfaces* in annular rings around and close to the bolts. In this study, unqualified paint on the *faying surfaces* away from the edge of the bolt hole by not less than 1 in. nor the bolt diameter did not reduce the slip resistance. However, this would not likely be the case for *joints* involving thicker material, particularly those with a large number of bolts on multiple gage lines; the Table 8.1 minimum bolt pretension might not be adequate to completely flatten and pull thicker material into tight contact around every bolt. Instead, the bolt pretension would be balanced by contact pressure on the regions of the *faying surfaces* that are in contact. To account for both possibilities, it is required in this Specification that all areas between the bolts be free of coatings, including overspray, as illustrated in Figure C-3.1.

As a practical matter, the smaller coating-free area can be laid out and protected more easily using masking located relative to the bolt-hole pattern than relative to the limits of the complete area of *faying surface* contact with varying and uncertain edge distance. Furthermore, the narrow coating strip around the perimeter of the *faying surface* minimizes the required field touch-up of uncoated material outside of the *joint*.

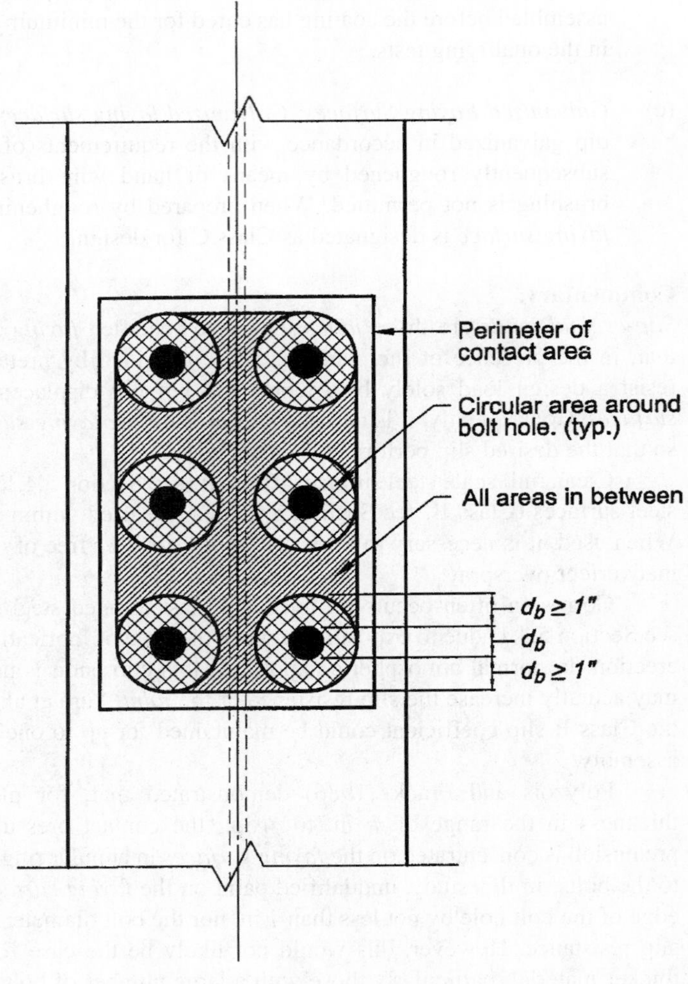

Figure C-3.1. Faying surfaces of slip-critical connections painted with unqualified paints.

Polyzois and Frank (1986) also investigated the effect of various degrees of inadvertent overspray on slip resistance. It was found that even a small amount of overspray of unqualified paint (that is, not qualified as a Class A or Class B coating) within the specified coating-free area on clean mill scale can reduce the slip resistance significantly. On blast-cleaned surfaces, however, the presence of a small amount of overspray was not as detrimental. For simplicity, this Specification requires that all overspray be prohibited from areas that are required to be free of coatings in *slip-critical joints* regardless of whether the surface is clean mill scale steel or blast-cleaned steel.

In the 1980 edition of this Specification, generic names for coatings applied to *faying surfaces* were the basis for categories of allowable working stresses in *slip-critical* (friction) *joints*. Frank and Yura (1981) demonstrated that the slip coefficients for coatings described by a generic type are not unique values for a given generic coating description or product, but rather depend also upon the type of vehicle used. Small differences in formulation from *manufacturer* to *manufacturer* or from *lot* to *lot* with a single *manufacturer* can significantly affect slip coefficients if certain essential variables within a generic type are changed. Consequently, it is unrealistic to assign coatings to categories with relatively small incremental differences between categories based solely upon a generic description.

When the *faying surfaces* of a *slip-critical joint* are to be protected against corrosion, a qualified coating must be used. A qualified coating is one that has been tested in accordance with Appendix A, the sole basis for qualification of any coating to be used in conjunction with this Specification. Coatings can be qualified as follows:

(1) As a Class A coating as defined in Section 5.4.1;
(2) As a Class B coating as defined in Section 5.4.1; or,
(3) As a coating with a *mean slip coefficient* μ other than 0.33 (Class A) but not greater than 0.50 (Class B).

Requalification is required if any essential variable associated with surface preparation, paint manufacture, application method or curing requirements is changed. See Appendix A.

Frank and Yura (1981) also investigated the effect of varying the time between coating the *faying surfaces* and assembly of the *joint* and pretensioning the bolts in order to ascertain if partially cured paint continued to cure within the assembled *joint* over a period of time. The results indicated that all curing effectively ceased at the time the *joint* was assembled and paint that was not fully cured at that time acted as a lubricant. The slip resistance of a *joint* that was assembled after a time less than the curing time used in the qualifying tests was severely reduced. Thus, the curing time prior to mating the *faying surfaces* is an essential parameter to be specified and controlled during construction.

The *mean slip coefficient* for clean hot-dip galvanized surfaces is on the order of 0.19 as compared with a factor of about 0.33 for clean mill scale. Birkemoe and Herrschaft (1970) showed that this *mean slip coefficient* can be significantly improved by treatments such as hand wire brushing or light "brush-off" grit blasting. In either case, the treatment must be controlled to achieve visible roughening or scoring. Power wire brushing is unsatisfactory because it may polish rather than roughen the surface, or remove the coating.

Field experience and test results have indicated that galvanized assemblies may continue to slip under sustained loading (Kulak et al., 1987; pp. 198-208). Tests of hot-dip galvanized *joints* subjected to sustained loading show a creep-type behavior that was not observed in short-duration or fatigue-type load application. See also the Commentary to Appendix A.

3.3. Bolt Holes

The nominal dimensions of standard, oversized, short-slotted and long-slotted holes for *high-strength bolts* shall be equal to or less than those shown in Table 3.1. Thermally cut bolt holes shall be permitted if approved by the *Engineer of Record*. For statically loaded *joints*, thermally cut surfaces need not be ground. For cyclically loaded *joints*, thermally cut surfaces shall be ground smooth.

Commentary:

The footnotes in Table 3.1 provide for slight variations in the dimensions of bolt holes from the nominal dimensions. When the dimensions of bolt holes are such that they exceed these permitted variations, the bolt hole must be treated as the next larger type.

3.3.1. Standard Holes: In the absence of approval by the *Engineer of Record* for the use of other hole types, standard holes shall be used in all plies of bolted *joints*.

Table 3.1. Nominal Bolt Hole Dimensions

Nominal Bolt Diameter, d_b, in.	Nominal Bolt Hole Dimensions[a,b], in.			
	Standard (diameter)	Oversized (diameter)	Short-slotted (width × length)	Long-slotted (width × length)
$^1/_2$	$^9/_{16}$	$^5/_8$	$^9/_{16} \times {}^{11}/_{16}$	$^9/_{16} \times 1^1/_4$
$^5/_8$	$^{11}/_{16}$	$^{13}/_{16}$	$^{11}/_{16} \times {}^7/_8$	$^{11}/_{16} \times 1^9/_{16}$
$^3/_4$	$^{13}/_{16}$	$^{15}/_{16}$	$^{13}/_{16} \times 1$	$^{13}/_{16} \times 1^7/_8$
$^7/_8$	$^{15}/_{16}$	$1^1/_{16}$	$^{15}/_{16} \times 1^1/_8$	$^{15}/_{16} \times 2^3/_{16}$
1	$1^1/_{16}$	$1^1/_4$	$1^1/_{16} \times 1^5/_{16}$	$1^1/_{16} \times 2^1/_2$
$\geq 1^1/_8$	$d_b + {}^1/_{16}$	$d_b + {}^5/_{16}$	$(d_b + {}^1/_{16}) \times (d_b + {}^3/_8)$	$(d_b + {}^1/_{16}) \times (2.5d_b)$

[a] The upper tolerance on the tabulated nominal dimensions shall not exceed $^1/_{32}$-in. Exception: In the width of slotted holes, gouges not more than $^1/_{16}$-in. deep are permitted.

[b] The slightly conical hole that naturally results from punching operations with properly matched punches and dies is acceptable.

Commentary:
The use of bolt holes 1/16 in. larger than the bolt installed in them has been permitted since the first publication of this Specification. Allen and Fisher (1968) showed that larger holes could be permitted for *high-strength bolts* without adversely affecting the bolt shear or member bearing strength. However, the slip resistance can be reduced by the failure to achieve adequate pretension initially or by the relaxation of the bolt pretension as the highly compressed material yields at the edge of the hole or slot. The provisions for oversized and slotted holes in this Specification are based upon these findings and the additional concern for the consequences of a slip of significant magnitude if it should occur in the direction of the slot. Because an increase in hole size generally reduces the net area of a connected part, the use of oversized holes or of slotted holes is subject to approval by the *Engineer of Record*.

3.3.2. Oversized Holes: When approved by the *Engineer of Record*, oversized holes are permitted in any or all plies of *slip-critical joints* as defined in Section 4.3.

Commentary:
See the Commentary to Section 3.3.1.

3.3.3. Short-Slotted Holes: When approved by the *Engineer of Record*, short-slotted holes are permitted in any or all plies of *snug-tightened joints* as defined in Section 4.1, and *pretensioned joints* as defined in Section 4.2, provided the applied load is approximately perpendicular (between 80 and 100 degrees) to the axis of the slot. When approved by the *Engineer of Record*, short-slotted holes are permitted in any or all plies of *slip-critical joints* as defined in Section 4.3 without regard for the direction of the applied load.

Commentary:
See the Commentary to Section 3.3.1.

3.3.4. Long-Slotted Holes: When approved by the *Engineer of Record*, long-slotted holes are permitted in only one ply at any individual *faying surface* of *snug-tightened joints* as defined in Section 4.1, and *pretensioned joints* as defined in Section 4.2, provided the applied load is approximately perpendicular (between 80 and 100 degrees) to the axis of the slot. When approved by the *Engineer of Record*, long-slotted holes are permitted in one ply only at any individual *faying surface* of *slip-critical joints* as defined in Section 4.3 without regard for the direction of the applied load. Fully inserted finger shims between the *faying surfaces* of load-transmitting elements of bolted *joints* are not considered a long-slotted element of a *joint*; nor are they considered to be a ply at any individual *faying surface*.

Commentary:
See the Commentary to Section 3.3.1.

Finger shims are devices that are often used to permit the alignment and plumbing of structures. When these devices are fully and properly inserted, they do not have the same effect on bolt pretension relaxation or the *connection* performance, as do long-slotted holes in an outer ply. When fully inserted, the shim provides support around approximately 75 percent of the perimeter of the bolt in contrast to the greatly reduced area that exists with a bolt that is centered in a long slot. Furthermore, finger shims are always enclosed on both sides by the connected material, which should be effective in bridging the space between the fingers.

3.4. Burrs

Burrs that extend 1/16 in. or less above the surface are permitted to remain on the *faying surfaces* of *snug-tightened joints* as defined in Section 4.1 and *pretensioned joints* as defined in Section 4.2. Burrs that extend over 1/16 in. above the surface shall be removed from all *joints*. Burrs that would prevent solid seating of the connected plies prior to the pretensioning of *slip-critical joints* as defined in Section 4.3 shall be removed.

Commentary:
Polyzois and Yura (1985) and McKinney and Zwerneman (1993) demonstrated that the slip resistance of *joints* was either unchanged or slightly improved by the presence of burrs. Therefore, small (1/16 in. or less) burrs that do not prevent solid seating of the connected parts need not be removed. On the other hand, parallel tests in the same program demonstrated that large burrs (over 1/16 in.) could cause a small increase in the required nut rotation from the snug-tight condition to achieve the specified pretension with the turn-of—nut pretensioning method. In the interest of simplicity, this Specification requires that all large burrs be removed.

SECTION 4. JOINT TYPE

For *joints* with fasteners that are loaded in shear or combined shear and tension, the *Engineer of Record* shall specify the *joint* type in the contract documents as snug-tightened, pretensioned or slip-critical. For *slip-critical joints*, the required class of slip resistance in accordance with Section 5.4 shall also be specified. For *joints* with fasteners that are loaded in tension only, the *Engineer of Record* shall specify the *joint* type in the contract documents as snug-tightened or pretensioned. Table 4.1 summarizes the applications and requirements of the three *joint* types.

Table 4.1. Summary of Applications and Requirements for Bolted Joints

Load Transfer	Application	Joint Type[a,b]	Faying Surface Prep.?	Install per Section	Inspect per Section	Arbitrate per Section 10?
Shear only	Resistance to shear load by shear/bearing	ST	No	8.1	9.1	No
	Resistance to shear by shear/bearing. Bolt pretension is required, but for reasons other than slip resistance.	PT	No	8.2	9.2	No
	Shear-load resistance by friction on faying surfaces is required.	SC	Yes[d]	8.2	9.3	If required to resolve dispute
Combined shear and tension	Resistance to shear load by shear/bearing. Tension load is static only.[c]	ST	No	8.1	9.1	No
	Resistance to shear by shear/bearing. Bolt pretension is required, but for reasons other than slip resistance.	PT	No	8.2	9.2	If required to resolve dispute
	Shear-load resistance by friction on faying surfaces is required.	SC	Yes[d]	8.2	9.3	If required to resolve dispute
Tension only	Static loading only.[c]	ST	No	8.1	9.1	No
	All other conditions of tension-only loading.	PT	No	8.2	9.2	If required to resolve dispute

[a] Under Joint Type: ST = snug-tightened, PT = pretensioned and SC = slip-critical; See Section 4.
[b] See Sections 4 and 5 for the design requirements for each joint type.
[c] Per Section 4.2, the use of ASTM A490 bolts in snug-tightened joints with tensile loads is not permitted.
[d] See Section 3.2.2.

Commentary:

When first approved by the Research Council on Structural Connections, in January, 1951, the "Specification for Assembly of Structural Joints Using High-Strength Bolts" merely permitted the substitution of a like number of ASTM A325 bolts for hot driven ASTM A141[2] steel rivets of the same nominal diameter. Additionally, it was required that all bolts be pretensioned and that all *faying surfaces* be free of paint; hence, satisfying the requirements for a *slip-critical joint* by the present-day definition. As revised in 1954, the omission of paint was required to apply only to "*joints* subject to stress reversal, impact or vibration, or to cases where stress redistribution due to *joint* slippage would be undesirable." This relaxation of the earlier provision recognized the fact that, in many applications, movement of the connected parts that brings the bolts into bearing against the sides of their holes is in no way detrimental. Bolted *joints* were then designated as "bearing type", "friction type" or "direct tension". With the 1985 edition of this Specification, these designations were changed to "shear/bearing", "slip-critical" and "direct tension", respectively, and snug-tightened installation was permitted for many *shear/bearing joints*. With this edition of this Specification, *snug-tightened joints* are also permitted for qualified applications involving ASTM A325 bolts in direct tension.

If non-pretensioned bolts are used in the type of *joint* that places the bolts in shear, load is transferred by shear in the bolts and bearing stress in the connected material. At the ultimate limit state, failure will occur by shear failure of the bolts, by bearing failure of the connected material or by failure of the member itself. On the other hand, if pretensioned bolts are used in such a *joint*, the frictional force that develops between the connected plies will initially transfer the load. Until the frictional force is exceeded, there is no shear in the bolts and no bearing stress in the connected components. Further increase of load places the bolts into shear and against the connected material in bearing, just as was the case when non-pretensioned bolts were used. Since it is known that the pretension in bolts will have been dissipated by the time bolt shear failure takes place (Kulak et al., 1987; p. 49), the ultimate limit state of a pretensioned bolted *joint* is the same as an otherwise identical *joint* that uses non-pretensioned bolts.

Because the consequences of slip into bearing vary from application to application, the determination of whether a *joint* can be designated as snug-tightened or as pre-tensioned or rather must be designated as slip-critical is best left to judgment and a decision on the part of the *Engineer of Record*. In the case of *joints* with three or more bolts in holes with only a small clearance, the freedom to slip generally does not exist. It is probable that normal fabrication tolerances and erection procedures are such that one or more bolts are in bearing even before additional load is applied. Such is the case for standard holes and for slotted holes loaded transverse to the axis of the slot.

Joints that are required to be *slip-critical joints* include:

(1) Those cases where slip movement could theoretically exceed an amount deemed by the *Engineer of Record* to affect the serviceability of the structure or through excessive distortion to cause a reduction in strength or stability, even though the resistance to fracture of the *connection* and yielding of the member may be adequate; and,

[2] ASTM A141 (discontinued in 1967) became identified as A502 Grade 1 (discontinued 1999).

(2) Those cases where slip of any magnitude must be prevented, such as in *joints* subject to significant load reversal and *joints* between elements of built-up compression members in which any slip could cause a reduction of the flexural stiffness required for the stability of the built-up member.

In this Specification, the provisions for the design, installation and inspection of bolted *joints* are dependent upon the type of *joint* that is specified by the *Engineer of Record*. Consequently, it is required that the *Engineer of Record* identify the *joint* type in the contract documents.

4.1. Snug-Tightened Joints

Except as required in Sections 4.2 and 4.3, *snug-tightened joints* are permitted.

Bolts in *snug-tightened joints* shall be designed in accordance with the applicable provisions of Sections 5.1, 5.2 and 5.3, installed in accordance with Section 8.1 and inspected in accordance with Section 9.1. As indicated in Section 4 and Table 4.1, requirements for *faying surface* condition shall not apply to *snug-tightened joints*.

Commentary:

Recognizing that the ultimate strength of a *connection* is independent of the bolt pretension and slip movement, there are numerous practical cases in the design of structures where, if slip occurs, it will not be detrimental to the serviceability of the structure. Additionally, there are cases where slip of the *joint* is desirable to permit rotation in a *joint* or to minimize the transfer of moment. To provide for these cases while at the same time making use of the shear strength of *high-strength bolts*, *snug-tightened joints* are permitted.

The maximum amount of slip that can occur in a *joint* is, theoretically, equal to twice the hole clearance. In practical terms, it is observed in laboratory and field experience to be much less; usually, about one-half the hole clearance. Acceptable inaccuracies in the location of holes within a pattern of bolts usually cause one or more bolts to be in bearing in the initial, unloaded condition. Furthermore, even with perfectly positioned holes, the usual method of erection causes the weight of the connected elements to put some of the bolts into direct bearing at the time the member is supported on loose bolts and the lifting crane is unhooked. Additional loading in the same direction would not cause additional *joint* slip of any significance.

With this edition of this Specification, *snug-tightened joints* are also permitted for statically loaded applications involving ASTM A325 bolts and ASTM F1852 twist-off-type tension-control bolt assemblies in direct tension. However, snug-tightened installation is not permitted for these fasteners in applications involving non-static loading, nor for applications involving ASTM A490 bolts.

4.2. Pretensioned Joints

Pretensioned joints are only required in the following applications:

(1) *Joints* in which fastener pretension is required in the specification or code that invokes this Specification;
(2) *Joints* that are subject to significant load reversal;
(3) *Joints* that are subject to fatigue load with no reversal of the loading direction;
(4) *Joints* with ASTM A325 or F1852 bolts that are subject to tensile fatigue; and,
(5) *Joints* with ASTM A490 bolts that are subject to tension or combined shear and tension, with or without fatigue.

Bolts in *pretensioned joints* subject to shear shall be designed in accordance with the applicable provisions of Sections 5.1 and 5.3, installed in accordance with Section 8.2 and inspected in accordance with Section 9.2. Bolts in *pretensioned joints* subject to tension or combined shear and tension shall be designed in accordance with the applicable provisions of Sections 5.1, 5.2, 5.3 and 5.5, installed in accordance with Section 8.2 and inspected in accordance with Section 9.2. As indicated in Section 4 and Table 4.1, requirements for *faying surface* condition shall not apply to *pretensioned joints*.

Commentary:
Under the provisions of some other specifications, certain shear *connections* are required to be pretensioned, but are not required to be slip-critical. Several cases are given, for example, in AISC LRFD Specification Section J1.11 (AISC, 1999) wherein certain bolted *joints* in bearing *connections* are to be pretensioned regardless of whether or not the potential for slip is a concern. The AISC Specification requires that *joints* be pretensioned in the following circumstances:

(1) Column splices in buildings with high ratios of height to width;
(2) *Connections* of members that provide bracing to columns in tall buildings;
(3) Various *connections* in buildings with cranes over 5-ton capacity; and,
(4) *Connections* for supports of running machinery and other sources of impact or stress reversal.

When pretension is desired for reasons other than the necessity to prevent slip, a *pretensioned joint* should be specified in the contract documents.

4.3. Slip-Critical Joints

Slip-critical joints are only required in the following applications involving shear or combined shear and tension:

(1) *Joints* that are subject to fatigue load with reversal of the loading direction;
(2) *Joints* that utilize oversized holes;
(3) *Joints* that utilize slotted holes, except those with applied load approximately normal (within 80 to 100 degrees) to the direction of the long dimension of the slot; and,
(4) *Joints* in which slip at the *faying surfaces* would be detrimental to the performance of the structure.

Bolts in *slip-critical joints* shall be designed in accordance with the applicable provisions of Sections 5.1, 5.2, 5.3, 5.4 and 5.5, installed in accordance with Section 8.2 and inspected in accordance with Section 9.3.

Commentary:
In certain cases, slip of a bolted *joint* in shear under service loads would be undesirable or must be precluded. Clearly, *joints* that are subject to reversed fatigue load must be slip-critical since slip may result in back-and-forth movement of the *joint* and the potential for accelerated fatigue failure. Unless slip is intended, as desired in a sliding expansion *joint*, slip in *joints* with long-slotted holes that are parallel to the direction of the applied load might be large enough to invalidate structural analyses that are based upon the assumption of small displacements.

For *joints* subject to fatigue load with respect to shear of the bolts that does not involve a reversal of load direction, there are two alternatives for fatigue design. The designer can provide either a *slip-critical joint* that is proportioned on the basis of the applied stress range on the gross section, or a *pretensioned joint* that is proportioned on the basis of applied stress range on the net section.

SECTION 5. LIMIT STATES IN BOLTED JOINTS

The design shear strength and design tensile strength of bolts shall be determined in accordance with Section 5.1. The interaction of combined shear and tension on bolts shall be limited in accordance with Section 5.2. The design bearing strength of the connected parts at bolt holes shall be determined in accordance with Section 5.3. Each of these *design strengths* shall be equal to or greater than the *required strength*. The axial load in bolts that are subject to tension or combined shear and tension shall be calculated with consideration of the effects of the externally applied tensile load and any additional tension resulting from *prying action* produced by deformation of the connected parts.

When slip resistance is required at the *faying surfaces* subject to shear or combined shear and tension, slip resistance shall be checked at either the factored-load level or service-load level, at the option of the *Engineer of Record*. When slip of the *joint* under factored loads would affect the ability of the structure to support the factored loads, the *design strength* determined in accordance with Section 5.4.1 shall be equal to or greater than the *required strength*. When slip resistance under service loads is the design criterion, the strength determined in accordance with Section 5.4.2 shall be equal to or greater than the effect of the service loads. In addition, the strength requirements in Sections 5.1, 5.2 and 5.3 shall also be met.

When bolts are subject to cyclic application of axial tension, the stress determined in accordance with Section 5.5 shall be equal to or greater than the stress due to the effect of the service loads, including any additional tension resulting from *prying action* produced by deformation of the connected parts. In addition, the strength requirements in Sections 5.1, 5.2 and 5.3 shall also be met.

Commentary:
This section of the Specification provides the design requirements for *high-strength bolts* in bolted *joints*. However, this information is not intended to provide comprehensive coverage of the design of *high-strength bolted connections*. Other design considerations of importance to the satisfactory performance of the connected material, such as block shear rupture, shear lag, *prying action* and *connection* stiffness and its effect on the performance of the structure, are beyond the scope of this Specification and Commentary.

The design of bolted *joints* that transmit shear requires consideration of the shear strength of the bolts and the bearing strength of the connected material. If such *joints* are designated as *slip-critical joints*, the slip resistance must also be checked. This serviceability check can be made at the factored-load level (Section 5.4.1) or at the service-load level (Section 5.4.2). Regardless of which load level is selected for the check of slip resistance, the prevention of slip in the service-load range is the design criterion.

Parameters that influence the shear strength of bolted *joints* include:

(1) Geometric parameters—the ratio of the net area to the gross area of the connected parts, the ratio of the net area of the connected parts to the total shear-resisting area of the bolts and the length of the *joint*; and,

(2) Material parameter—the ratio of the yield strength to the tensile strength of the connected parts.

Using both mathematical models and physical testing, it was possible to study the influences of these parameters (Kulak et al., 1987; pp. 89-116 and 126-132). These showed that, under the rules that existed at that time the longest (and often the most important) *joints* had the lowest factor of safety, about 2.0 based on ultimate strength.

In general, bolted *joints* that are designed in accordance with the provisions of this Specification will have a higher reliability than will the members they connect. This occurs primarily because the resistance factors used in limit states for the design of bolted *joints* were chosen to provide a reliability higher than that used for member design. Additionally, the controlling strength limit state in the structural member, such as yielding or deflection, is usually reached well before the strength limit state in the *connection*, such as bolt shear strength or bearing strength of the connected material. The installation requirements vary with *joint* type and influence the behavior of the *joints* within the service-load range, however, this influence is ignored in all strength calculations. Secondary tensile stresses that may be produced in bolts in *shear/bearing joints*, such as through the flexing of double-angle *connections* to accommodate the simple-beam end rotation, need not be considered.

It is sometimes necessary to use *high-strength bolts* and fillet welds in the same *connection*, particularly as the result of remedial work. When these fastening elements act in the same shear plane, the combined strength is a function of whether the bolts are snug-tightened or pretensioned, the location of the bolts relative to the holes in which they are located and the orientation of the fillet welds. The fillet welds can be parallel or transverse to the direction of load. Recent work (Manuel and Kulak, 1999) can be used to calculate the *design strength* of such *joints*.

5.1. Design Shear and Tensile Strengths

Shear and tensile strengths shall not be reduced by the installed bolt pretension. For *joints*, the design shear and tensile strengths shall be taken as the sum of the strengths of the individual bolts.

The *design strength* in shear or the *design strength* in tension for an ASTM A325, A490 or F1852 bolt is ϕR_n, where $\phi = 0.75$ and:

$$R_n = F_n A_b \qquad \text{(Equation 5.1)}$$

where

R_n = *nominal strength* (shear strength per shear plane or tensile strength) of a bolt, kips;

F_n = *nominal strength* per unit area from Table 5.1 for the appropriate applied load conditions, ksi, adjusted for the presence of fillers as required below; and,

A_b = cross-sectional area based upon the nominal diameter of bolt, in.^{2}

When a bolt that carries load passes through fillers or shims in a shear plane that are equal to or less than ¼-in. thick, F_n from Table 5.1 shall be used without reduction. When a bolt that carries load passes through fillers or shims that are greater than ¼-in. thick, they shall be designed in accordance with one of the following procedures:

Table 5.1. Nominal Strength per Unit Area of Bolts

<table>
<tr><th colspan="2" rowspan="2">Applied Load Condition</th><th colspan="2">Nominal Strength per Unit Area F_n, ksi</th></tr>
<tr><th>ASTM A325 or F1852 Bolt</th><th>ASTM A490 Bolt</th></tr>
<tr><td rowspan="2">Tension[a]</td><td>Static</td><td>90</td><td>113</td></tr>
<tr><td>Fatigue</td><td colspan="2">See Section 5.5</td></tr>
<tr><td rowspan="2">Shear[a,b]</td><td>Threads included in shear plane</td><td>48</td><td>60</td></tr>
<tr><td>Threads excluded from shear plane</td><td>60</td><td>75</td></tr>
<tr><td colspan="4">[a] Except as required in Section 5.2.
[b] In shear connections that transmit axial force and have length between extreme bolts measured parallel to the line of force exceeds 50 in., tabulated values shall be reduced by 20 percent.</td></tr>
</table>

(1) For fillers or shims that are equal to or less than ¾ in. thick, F_n from Table 5.1 shall be multiplied by the factor $[1 - 0.4(t' - 0.25)]$, where t' is the total thickness of fillers or shims, in., up to ¾ in.;

(2) The fillers or shims shall be extended beyond the *joint* and the filler or shim extension shall be secured with enough bolts to uniformly distribute the total force in the connected element over the combined cross-section of the connected element and the fillers or shims;

(3) The size of the *joint* shall be increased to accommodate a number of bolts that is equivalent to the total number required in (2) above; or

(4) The *joint* shall be designed as a *slip-critical joint*. The slip resistance of the *joint* shall not be reduced for the presence of fillers or shims.

Commentary:

The nominal shear and tensile strengths of ASTM A325, F1852 and A490 bolts are given in Table 5.1. These values are based upon the work of a large number of researchers throughout the world, as reported in the *Guide* (Kulak et al., 1987). The *design strength* equals the *nominal strength* multiplied by a resistance factor ϕ. On average, the *design strengths* result in bolted *joint* designs that are approximately equivalent to those provided under the allowable stress rules given in the 1980 edition of this Specification.

The nominal shear strength is based upon the observation that the shear strength of a single *high-strength bolt* is about 0.62 times the tensile strength of that bolt (Kulak et al., 1987; pp. 44-50). However, in lap splices transmitting axial force between members with more than two bolts in the line of force, non-uniform deformation of the connected material between fasteners causes a non-uniform distribution of the shear force to the bolts. Consequently, the strength of the *joint* decreases in terms of the average strength of all the bolts in the *joint* as the *joint* length increases (Kulak et al., 1987; pp. 99-104). Rather than provide a decreasing function that reflects this decrease in average fastener strength with *joint* length, a single reduction factor of 0.80 is applied to the 0.62 multiplier. This accommodates bolts in all *joints* up to 50 in. in length without seriously affecting the economy of very short *joints*. As noted in Footnote b in Table 5.1, the average shear strength of bolts in *joints* longer than 50 in. in length must be further reduced by 20 percent. Note that this reduction does not apply in cases when the distribution of force is essentially uniform along the *joint*, such as the bolted *joints* in a shear *connection* at the end of a deep plate girder.

The average ratio of nominal shear strength for bolts with threads included in the shear plane to the nominal shear strength for bolts with threads excluded from the shear plane is 0.83 with a standard deviation of 0.03 (Frank and Yura, 1981). Conservatively, a reduction factor of 0.80 is used to account for the reduction in shear strength for a bolt with threads included in the shear plane but calculated with the area corresponding to the nominal bolt diameter. The case of a bolt in double shear with a non-threaded section in one shear plane and a threaded section in the other shear plane is not covered in this Specification for two reasons. First, the manner in which load is shared between these two dissimilar shear areas is uncertain. Second, the detailer's lack of certainty as to the orientation of the bolt placement might leave both shear planes in the threaded section. Thus, if threads are included in one shear plane, the conservative assumption is made that threads are included in all shear planes.

The tensile strength of a *high-strength bolt* is the product of its ultimate tensile strength (per unit area) and some area through the threaded portion. This area, called the tensile stress area, is a derived quantity that is a function of the relative thread size and pitch. For the usual sizes of structural bolts, it is about 75 percent of the nominal cross-sectional area of the bolt. Hence, the nominal tensile strengths per unit area given in Table 5.1 are 0.75 times the tensile strength of the bolt material. According to Equation 5.1, the nominal area of the bolt is then used to calculate the *design strength* in tension. The *nominal strengths* so-calculated are intended to form the basis for comparison with the externally applied bolt tension plus any additional tension that results from *prying action* that is produced by deformation of the connected elements.

If pretensioned bolts are used in a *joint* that loads the bolts in tension, the question arises as to whether the pretension and the applied tension are additive. Because the compressed parts are being unloaded during the application of the external tensile force, the increase in bolt tension is minimal until the parts separate (Kulak et al., 1987; pp. 263-266). Thus, there will be little increase in bolt force above the pretension load under service loads. After the parts separate, the bolt acts as a tension member, as expected, and its *design strength* is that given in Equation 5.1 multiplied by the resistance factor ϕ.

Pretensioned bolts have torsion present during the installation process. Once the installation is completed, any residual torsion is quite small and will disappear entirely when the fastener is loaded to the point of plate separation. Hence, there is no question of torsion—tension interaction when considering the ultimate tensile strength of a *high-strength bolt* (Kulak et al., 1987; pp. 41-47).

When required, pretension is induced in a bolt by imposing a small axial elongation during installation, as described in the Commentary to Section 8. When the *joint* is subsequently loaded in shear, tension or combined shear and tension, the bolts will undergo significant deformations prior to failure that have the effect of overriding the small axial elongation that was introduced during installation and, thereby, removing the pretension. Measurements taken in laboratory tests confirm that the pretension that would be sustained if the applied load were removed is essentially zero before the bolt fails in shear (Kulak et al., 1987; pp. 93-94). Thus, the shear and tensile strengths of a bolt are not affected by the presence of an initial pretension in the bolt.

See also the Commentary to Section 5.5.

5.2. Combined Shear and Tension

When combined shear and tension loads are transmitted by an ASTM A325, A490 or F1852 bolt, the ultimate limit-state interaction shall be:

$$\left[\frac{T_u}{(\phi R_n)_t}\right]^2 + \left[\frac{V_u}{(\phi R_n)_v}\right]^2 \leq 1 \qquad \text{(Equation 5.2)}$$

where

T_u = *required strength* in tension (factored tensile load) per bolt, kips;

V_u = *required strength* in shear (factored shear load) per bolt, kips;

$(\phi R_n)_t$ = *design strength* in tension determined in accordance with Section 5.1, kips; and,

$(\phi R_n)_v$ = *design strength* in shear determined in accordance with Section 5.1, kips

Commentary:

When both shear forces and tensile forces act on a bolt, the interaction can be conveniently expressed as an elliptical solution (Chesson et al., 1965) that includes the elements of the bolt acting in shear alone and the bolt acting in tension alone. Although the elliptical solution provides the best estimate of the strength of bolts subject to combined shear and tension and is thus used in this Specification, the nature of the elliptical solution is such that it can be approximated conveniently using three straight lines (Carter et al., 1997). Earlier editions of this specification have used such linear representations for the convenience of design calculations. The elliptical interaction equation in effect shows that, for design purposes, significant interaction does not occur until either force component exceeds 20 percent of the limiting strength for that component.

5.3. Design Bearing Strength at Bolt Holes

For *joints*, the design bearing strength shall be taken as the sum of the strengths of the connected material at the individual bolt holes.

The design bearing strength of the connected material at a standard bolt hole, oversized bolt hole, short-slotted bolt hole independent of the direction of loading or long-slotted bolt hole with the slot parallel to the direction of the bearing load is ϕR_n, where $\phi = 0.75$ and:

(1) when deformation of the bolt hole at service load is a design consideration;

$$R_n = 1.2 L_c t F_u \leq 2.4 d_b t F_u \qquad \text{(Equation 5.3)}$$

(2) when deformation of the bolt hole at service load is not a design consideration;

$$R_n = 1.5L_c tF_u \leq 3d_b tF_u \qquad \text{(Equation 5.4)}$$

The design bearing strength of the connected material at a long-slotted bolt hole with the slot perpendicular to the direction of the bearing load is ϕR_n, where ϕ = 0.75 and:

$$R_n = L_c tF_u \leq 2d_b tF_u \qquad \text{(Equation 5.5)}$$

In Equations 5.3, 5.4 and 5.5,

R_n = *nominal strength* (bearing strength of the connected material), kips;
F_u = specified minimum tensile strength (per unit area) of the connected material, ksi;
L_c = clear distance, in the direction of load, between the edge of the hole and the edge of the adjacent hole or the edge of the material, in.;
d_b = nominal diameter of bolt, in.; and,
t = thickness of the connected material, in.

Commentary:
The contact pressure at the interface between a bolt and the connected material can be expressed as a bearing stress on the bolt or on the connected material. The connected material is always critical. For simplicity, the bearing area is expressed as the bolt diameter times the thickness of the connected material in bearing. The governing value of the bearing stress has been determined from extensive experimental research and a further limitation on strength was derived from the case of a bolt at the end of a tension member or near another fastener.

The design equations are based upon the models presented in the *Guide* (Kulak et al., 1987; pp. 141-143), except that the clear distance to another hole or edge is used in the Specification formulation rather than the bolt spacing or end distance as used in the *Guide* (see Figure C-5.1). Equation 5.3 is derived from tests (Kulak et al., 1987; pp. 112-116) that showed that the total elongation, including local bearing deformation, of a standard hole that is loaded to obtain the ultimate strength equal to $3dtF_u$ in Equation 5.4 was on the order of the diameter of the bolt. This apparent hole elongation results largely from bearing deformation of the material that is immediately adjacent to the bolt. The lower value of $2.4dtF_u$ in Equation 5.3 provides a bearing strength limit-state that is attainable at reasonable deformation (¼ in.). Strength and deformation limits were thus used to jointly evaluate bearing strength test results for design.

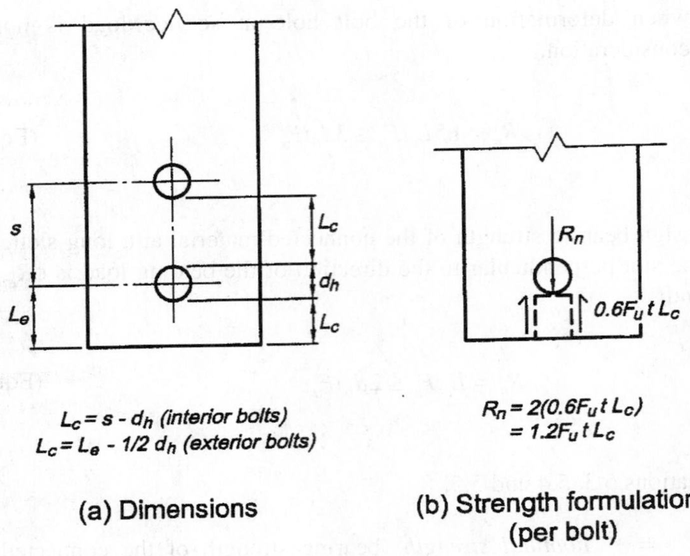

Figure. C-5.1. Bearing strength formulation.

When long-slotted holes are oriented with the long dimension perpendicular to the direction of load, the bending component of the deformation in the material between adjacent holes or between the hole and the edge of the plate is increased. The nominal bearing strength is limited to $2dtF_u$, which again provides a bearing strength limit-state that is attainable at reasonable deformation.

The design bearing strength has been expressed as that of a single bolt, although it is really that of the connected material that is immediately adjacent to the bolt. In calculating the design bearing strength of a connected part, the total bearing strength of the connected part can be taken as the sum of the bearing strengths of the individual bolts.

5.4. Design Slip Resistance

5.4.1. At the Factored-Load Level: The design slip resistance is ϕR_n, where ϕ is as defined below and:

$$R_n = \mu D_u T_m N_b \left(1 - \frac{T_u}{D_u T_m N_b}\right) \quad \text{(Equation 5.6)}$$

where

ϕ = 1.0 for standard holes
= 0.85 for oversized and short-slotted holes
= 0.70 for long-slotted holes perpendicular to the direction of load
= 0.60 for long-slotted holes parallel to the direction of load;

R_n = *nominal strength* (slip resistance) of a slip plane, kips;

μ = *mean slip coefficient* for Class A, B or C *faying surfaces*, as applicable, or as established by testing in accordance with Appendix A (see Section 3.2.2(b))

= 0.33 for Class A *faying surfaces* (uncoated clean mill scale steel surfaces or surfaces with Class A coatings on blast-cleaned steel)

= 0.50 for Class B surfaces (uncoated blast-cleaned steel surfaces or surfaces with Class B coatings on blast-cleaned steel

= 0.35 for Class C surfaces (roughened hot-dip galvanized surfaces);

D_u = 1.13, a multiplier that reflects the ratio of the mean installed bolt pretension to the specified minimum bolt pretension T_m; the use of other values of D_u shall be approved by the *Engineer of Record;*

T_m = specified minimum bolt pretension (for *pretensioned joints* as specified in Table 8.1), kips;

N_b = number of bolts in the *joint*; and

T_u = *required strength* in tension (tensile component of applied factored load for combined shear and tension loading), kips

= zero if the *joint* is subject to shear only

5.4.2. At the Service-Load Level: The service-load slip resistance is ϕR_s, where ϕ is as defined in Section 5.4.1 and:

$$R_s = \mu D T_m N_b \left(1 - \frac{T}{D T_m N_b}\right) \qquad \text{(Equation 5.7)}$$

where

D = 0.80, a slip probability factor that reflects the distribution of actual slip coefficient values about the mean, the ratio of mean installed bolt pretension to the specified minimum bolt pretension, T_m, and a slip probability level; the use of other values of D must be approved by the *Engineer of Record*

T = applied service load in tension (tensile component of applied service load for combined shear and tension loading), kips

= zero if the *joint* is subject to shear only

and all other variables are as defined for Equation 5.6.

Commentary:

The design check for slip resistance can be made either at the factored-load level (Section 5.4.1) or at the service-load level (Section 5.4.2). These alternatives are based upon different design philosophies, which are discussed below. They have been calibrated to produce results that are essentially the same. The factored-load level approach is provided for the expedience of only working with factored loads. Irrespective of the approach, the limit state is based upon the prevention of slip at service-load levels.

If the factored-load provision is used, the *nominal strength* R_n represents the mean resistance, which is a function of the *mean slip coefficient* μ and the specified minimum bolt pretension (clamping force) T_m. The 1.13 multiplier in Equation 5.6 accounts for the expected 13 percent higher mean value of the installed bolt pretension provided by the calibrated wrench pretensioning method compared to the specified minimum bolt pretension T_m used in the calculation. In the absence of other field test data, this value is used for all methods.

If the service-load approach is used, a probability of slip is identified. It implies that there is 90 percent reliability that slip will not occur at the calculated slip load if the calibrated wrench pretensioning method is used, or that there is 95 percent reliability that slip will not occur at the calculated slip load if the turn-of-nut pretensioning method is used. The probability of loading occurrence was not considered in developing these slip probabilities (Kulak et al., 1987; pg. 135).

For most applications, the assumption that the slip resistance at each fastener is equal and additive with that at the other fasteners is based on the fact that all locations must develop the slip force before a total *joint* slip can occur at that plane. Similarly the forces developed at various slip planes do not necessarily develop simultaneously, but one can assume that the full slip resistances must be mobilized at each plane before full *joint* slip can occur. Equations 5.6 and 5.7 are formulated for the general case of a single slip plane. The total slip resistance of a *joint* with multiple slip planes can be calculated as that for a single slip plane multiplied by the number of slip planes.

Only the *Engineer of Record* can determine whether the potential slippage of a *joint* is critical at the service-load level as a serviceability consideration only or whether slippage could result in distortions of the frame such that the ability of the frame to resist the factored loads would be reduced. The following comments reflect the collective thinking of the Council and are provided as guidance and an indication of the intent of the Specification. See also the Commentary to Sections 4.2 and 4.3.

(1) If *joints* with standard holes have only one or two bolts in the direction of the applied load, a small slip may occur. In this case, *joints* subject to vibration should be proportioned to resist slip at the service-load level.

(2) In built-up compression members, such as double-angle struts in trusses, a small relative slip between the elements especially at the end *connections* can increase the effective length of the combined cross-section to that of the individual components and significantly reduce the compressive strength of the strut. Therefore, the *connection* between the elements at the ends of built-up members should be checked at the factored-load level, whether or not a *slip-critical joint* is required for serviceability. As given by Sherman and Yura (1998), the required slip resistance is $0.008P_uLQ/I$, where P_u is the axial compressive force in the built-up member, kips, L is the total length of the built-up member, in., Q is the first moment of area of one component about the axis of buckling of the built-up member, in.^{3}, and I is the moment of inertia of the built-up member about the axis of buckling, in.^{4}

(3) In *joints* with long-slotted holes that are parallel to the direction of the applied load, the designer has two alternatives. The *joint* can be designed to prevent slip in the service-load range using either the factored-load-level provision in Section 5.4.1 or the service-load-level provision in Section 5.4.2. In either case, however, the effect of the factored loads acting on the deformed structure (deformed by the maximum amount of slip in the long slots at all locations) must be included in the structural analysis.

(4) In *joints* subject to fatigue, design should be based upon service-load criteria and the design slip resistance of Section 5.4.2 because fatigue is a function of the service load performance rather than that of the factored load.

Extensive data developed through research sponsored by the Council and others during the past twenty years has been statistically analyzed to provide improved information on slip probability of *joints* in which the bolts have been pretensioned to the requirements of Table 8.1. Two variables, the *mean slip coefficient* of the *faying surfaces* and the bolt pretension, were found to affect the slip resistance of *joints*. Field studies (Kulak and Birkemoe, 1993) of installed bolts in various structural applications indicate that the Table 8.1 pretensions have been achieved as anticipated in the laboratory research.

An examination of the slip-coefficient data for a wide range of surface conditions indicates that the data are distributed normally and the standard deviation is essentially the same for each surface condition class. This means that different reduction factors should be applied to classes of surfaces with different *mean slip coefficients*—the smaller the mean value of the coefficient of friction, the smaller (more severe) the appropriate reduction factor—to provide equivalent reliability of slip resistance.

The bolt clamping force data indicate that bolt pretensions are distributed normally for each pretensioning method. However, the data also indicate that the mean value of the bolt pretension is different for each method. As noted previously, if the calibrated wrench method is used to pretension ASTM A325 bolts, the mean value of bolt pretension is about 1.13 times the specified minimum pretension in Table 8.1. If the turn-of-nut pretensioning method is used, the mean pretension is about 1.35 times the specified minimum pretension for ASTM A325 bolts and about 1.26 for ASTM A490 bolts.

The combined effects of the variability of the *mean slip coefficient* and bolt pretension have been accounted for approximately in the single value of the slip probability factor D in the equation for nominal slip resistance in Section 5.4.2. This implies 90 percent reliability that slip will not occur if the calibrated wrench pretensioning method is used and 95 percent reliability if the turn-of-nut pretensioning method is used. For values of D that are appropriate for other *mean slip coefficients* and slip probabilities, refer to the *Guide* (Kulak et al., 1987; pg. 135). The values given therein are suitable for direct substitution into the formula for slip resistance in Section 5.4.2.

The calibrated wrench installation method targets a specific bolt pretension, which is 5 percent greater than the specified minimum value given in Table 8.1. Thus, regardless of the actual strength of production bolts, this target value is unique for a given fastener grade. On the other hand, the turn-of-nut installation method imposes an elongation on the fastener. Consequently, the inherent strength of the bolts being installed will be reflected in the resulting pretension because this elongation will bring the fastener to its proportional limit under combined torsion and tension. As a result of these differences, the mean value and nature of the frequency distribution of pretensions for the two installation methods differ. Turn-of-nut installations result in higher mean levels of pretension than do calibrated wrench installations. These differences were taken into account when the design criteria for *slip-critical joints* were developed.

Statistical information on the pretension characteristics of bolts installed in the field using direct tension indicators and twist-off-type tension-control bolts is limited.

In any of the foregoing installation methods, it can be expected that a portion of the bolt assembly (the threaded portion of the bolt within the *grip* length and/or the engaged threads of the nut and bolt) will reach the inelastic region of behavior. This permanent distortion has no undesirable effect on the subsequent performance of the bolt.

Because of the greater likelihood that significant deformation can occur in *joints* with oversized or slotted holes, lower values of design slip resistance are provided for *joints* with these hole types through a modification of the resistance factor ϕ. For the case of long-slotted holes, even though the slip load is the same for loading transverse or parallel to the axis of the slot, the value for loading parallel to the axis has been further reduced, based upon judgment, in recognition of the greater consequences of slip.

Although the design philosophy for *slip-critical joints* presumes that they do not slip into bearing when subject to loads in the service range, it is mandatory that *slip-critical joints* also meet the requirements of Sections 5.1, 5.2 and 5.3. Thus, they must meet the strength requirements to resist the factored loads as *shear/bearing joints*.

Section 3.2.2(b) permits the *Engineer of Record* to authorize the use of *faying surfaces* with a *mean slip coefficient* μ that is less than 0.50 (Class B) and other than 0.33 (Class A) This authorization requires that the following restrictions are met:

(1) The *mean slip coefficient* μ must be determined in accordance with Appendix A; and,
(2) The appropriate slip probability factor *D* must be selected from the *Guide* (Kulak et al., 1987) for design at the service-load level.

Prior to the 1994 edition of this Specification, μ for Class C surfaces was taken as 0.40. This value was reduced to 0.35 in the 1994 edition for better agreement with the available research (Kulak et al., 1987; pp. 78-82).

5.5. Tensile Fatigue

The tensile stress in the bolt that results from the cyclic application of externally applied service loads and the prying force, if any, but not the pretension, shall not exceed the stress in Table 5.2. The nominal diameter of the bolt shall be used in

Table 5.2. Maximum Tensile Stress for Fatigue Loading

Number of Cycles	Maximum Bolt Stress for Design at Service Loads[a], ksi	
	ASTM A325 or F1852 Bolt	ASTM A490 Bolt
Not more than 20,000	44	54
From 20,000 to 500,000	40	49
More than 500,000	31	38

[a] Including the effects of *prying action*, if any, but excluding the pretension.

calculating the bolt stress. The connected parts shall be proportioned so that the calculated prying force does not exceed 30 percent of the externally applied load. *Joints* that are subject to tensile fatigue loading shall be specified as pretensioned in accordance with Section 4.2 or slip-critical in accordance with Section 4.3.

Commentary:
As described in the Commentary to Section 5.1, *high-strength bolts* in *pretensioned joints* that are nominally loaded in tension will experience little, if any, increase in axial stress under service loads. For this reason, pretensioned bolts are not adversely affected by repeated application of service-load tensile stress. However, care must be taken to ensure that the calculated prying force is a relatively small part of the total applied bolt tension (Kulak et al., 1987; p. 272). The provisions that cover bolt fatigue in tension are based upon research results where various single-bolt assemblies and *joints* with bolts in tension were subjected to repeated external loads that produced fatigue failure of the pretensioned fasteners. A limited range of prying effects was investigated in this research. As a matter of judgment, in this edition of the Specification the limit on prying forces as a percentage of the total externally applied tensile force has been reduced from 60 percent to 30 percent.

SECTION 6. USE OF WASHERS

6.1. Snug-Tightened Joints

Washers are not required in snug-tightened joints, except as required in Sections 6.1.1 and 6.1.2.

6.1.1. Sloping Surfaces: When the outer face of the *joint* has a slope that is greater than 1:20 with respect to a plane that is normal to the bolt axis, an ASTM F436 beveled washer shall be used to compensate for the lack of parallelism.

6.1.2. Slotted Hole: When a slotted hole occurs in an outer ply, an ASTM F436 washer or $^5/_{16}$ in. thick common plate washer shall be used to cover the hole.

6.2. Pretensioned Joints and Slip-Critical Joints

Washers are not required in *pretensioned joints* and *slip-critical joints*, except as required in Sections 6.1.1, 6.1.2, 6.2.1, 6.2.2, 6.2.3, 6.2.4 and 6.2.5.

6.2.1. Specified Minimum Yield Strength of Connected Material Less Than 40 ksi: When ASTM A490 bolts are pretensioned in connected material of specified minimum yield strength less than 40 ksi, ASTM F436 washers shall be used under both the bolt head and nut.

6.2.2. Calibrated Wrench Pretensioning: When the calibrated wrench pretensioning method is used, an ASTM F436 washer shall be used under the turned element.

6.2.3. Twist-Off-Type Tension-Control Bolt Pretensioning: When the twist-off-type tension-control bolt pretensioning method is used, an ASTM F436 washer shall be used under the nut as part of the *fastener assembly*.

6.2.4. Direct-Tension-Indicator Pretensioning: When the direct-tension-indicator pretensioning method is used, an ASTM F436 washer shall be used as follows:

(1) When the nut is turned and the direct tension indicator is located under the bolt head, an ASTM F436 washer shall be used under the nut;

(2) When the nut is turned and the direct tension indicator is located under the nut, an ASTM F436 washer shall be used between the nut and the direct tension indicator;

(3) When the bolt head is turned and the direct tension indicator is located under the nut, an ASTM F436 washer shall be used under the bolt head; and,

(4) When the bolt head is turned and the direct tension indicator is located under the bolt head, an ASTM F436 washer shall be used between the bolt head and the direct tension indicator.

6.2.5. Oversized or Slotted Hole: When an oversized or slotted hole occurs in an outer ply, the washer requirements shall be as given in Table 6.1. The washer used shall be of sufficient size to completely cover the hole.

Table 6.1. Washer Requirements for Bolted Joints with Oversized and Slotted Holes in the Outer Ply

<table>
<tr><th rowspan="2">ASTM Designation</th><th rowspan="2">Nominal Bolt Diameter d_b, in.</th><th colspan="3">Hole Type in Outer Ply</th></tr>
<tr><th>Oversized</th><th>Short-Slotted</th><th>Long-Slotted</th></tr>
<tr><td>A325 or F1852</td><td>$\frac{1}{2}$—$1\frac{1}{2}$</td><td colspan="2" rowspan="2">ASTM F436 [a]</td><td rowspan="2">$\frac{5}{16}$-in.-thick plate washer or continuous bar[b,c]</td></tr>
<tr><td rowspan="2">A490</td><td>≤ 1</td></tr>
<tr><td>> 1</td><td colspan="2">ASTM F436 with $\frac{5}{16}$-in.-thickness[b]</td><td>$\frac{5}{16}$-in.-thick hardened plate washer or hardened continuous bar[b]</td></tr>
</table>

[a] This requirement shall not apply to heads of tension-control bolt assemblies that meet the requirements in Section 2.7 and provide a bearing circle with a diameter that is equal to or greater than the diameter of the corresponding ASTM F436 washer.

[b] Multiple washers with a combined thickness of $\frac{5}{16}$ in. or larger do not satisfy this requirement.

[c] The plate washer or bar shall be of structural-grade steel material, but need not be hardened.

Commentary:

It is important that shop drawings and *connection* details clearly reflect the number and disposition of washers when they are required, especially the thick hardened washers or plate washers that are required for some slotted hole applications. The total thickness of washers in the *grip* affects the length of bolt that must be supplied and used.

The primary function of washers is to provide a hardened non-galling surface under the turned element, particularly for torque-based pretensioning methods such as the calibrated wrench pretensioning method and twist-off-type tension-control bolt pretensioning method. Circular flat washers that meet the requirements of ASTM F436 provide both a hardened non-galling surface and an increase in bearing area that is approximately 50 percent larger than that provided by a heavy-hex bolt head or nut. However, tests have shown that washers of the standard $\frac{5}{32}$ in. thickness have a minor influence on the pressure distribution of the induced bolt pretension. Furthermore, they showed that a larger thickness is required when ASTM A490 bolts are used with material that has a minimum specified yield strength that is less than 40 ksi. This is necessary to mitigate the effects of local yielding of the material in the vicinity of the contact area of the head and nut. The requirement for standard thickness hardened washers, when such washers are specified, is waived for alternative design fasteners that incorporate a bearing surface under the head of the same diameter as the hardened washer.

Heat-treated washers not less than $\frac{5}{16}$ in. thick are required to cover oversized and short-slotted holes in external plies, when ASTM A490 bolts of diameter larger than 1 in. are used. This was found necessary to distribute the high clamping pressure so as to prevent collapse of the hole perimeter and enable the development of the desired clamping force. Preliminary investigation has shown that a similar but less severe deformation occurs when oversized or slotted holes are in the interior plies. The reduction in clamping force may be offset by "keying," which tends to increase the resistance to slip. These effects are accentuated in *joints* of thin plies.

When long-slotted holes occur in an outer ply, 5⁄16 in. thick plate washers or continuous bars are required in Table 6.1. This requirement can be satisfied with material of any structural grade, provided such material is hardened when ASTM A490 bolts of diameter greater than 1 in. are used. Alternatively, either of the following options can be used:

(1) The use of material with F_y greater than 40 ksi will eliminate the need to also provide ASTM F436 washers in accordance with the requirements in Section 6.2.1 for ASTM A490 bolts of any diameter.

(2) Material with F_y equal to or less than 40 ksi can be used with ASTM F436 washers in accordance with the requirements in Section 6.2.1

SECTION 7. PRE-INSTALLATION VERIFICATION

The requirements in this Section shall apply only as indicated in Sections 8.2 to verify that the *fastener assemblies* and pretensioned installation procedures perform as required prior to installation.

7.1. Tension Calibrator

A *tension calibrator* shall be used where bolts are to be installed in *pretensioned joints* and *slip-critical joints* to:

(1) Confirm the suitability of the complete *fastener assembly*, including lubrication, for pretensioned installation; and,
(2) Confirm the procedure and proper use by the bolting crew of the pretensioning method to be used.

The accuracy of the *tension calibrator* shall be confirmed through calibration at least annually.

Commentary:
A *tension calibrator* is a hydraulic device that indicates the pretension that is developed in a bolt that is installed in it. Such a device is an economical and valuable tool and it must be readily available whenever *high-strength bolts* are to be pretensioned. A bolt *tension calibrator* is essential for:

(1) The pre-installation verification of the suitability of the *fastener assembly*, including the lubrication that is applied by the *manufacturer* or specially applied, to develop the specified minimum pretension;
(2) Verifying the adequacy and proper use of the specified pretensioning method to be used;
(3) Determining the installation torque for the calibrated wrench pretensioning method; and,
(4) Determining an arbitration torque as specified in Section 10, if required to resolve dispute.

It is the only economically available tool for the described essential uses in the shop and field.

Hydraulic *tension calibrators* undergo a slight deformation during bolt pretensioning. Hence, when bolts are pretensioned according to Section 8.2.1, the nut rotation corresponding to a given pretension reading may be somewhat larger than it would be if the same bolt were pretensioned in a solid steel assembly. Stated differently, the reading of an hydraulic *tension calibrator* tends to underestimate the pretension that a given rotation of the turned element would induce in a bolt in a *pretensioned joint*.

7.2. Required Testing

A representative sample of not fewer than three complete *fastener assemblies* of each combination of diameter, length, grade and *lot* to be used in the work shall be checked at the site of installation in a *tension calibrator* to verify that the pretensioning method develops a pretension that is equal to or greater than 1.05 times that specified for installation and inspection in Table 8.1. Washers shall be used in the pre-installation verification assemblies as required in the work in accordance with the requirements in Section 6.2.

If the actual pretension developed in any of the *fastener assemblies* is less than 1.05 times that specified for installation and inspection in Table 8.1, the cause(s) shall be determined and resolved before the *fastener assemblies* are used in the work. Cleaning, lubrication and retesting of these *fastener assemblies*, except ASTM F1852 twist-off-type tension-control bolt assemblies, (see Section 2.2) are permitted, provided that all assemblies are treated in the same manner.

Commentary:
The fastener components listed in Section 1.3 are manufactured under separate ASTM specifications, each of which includes tolerances that are appropriate for the individual component covered. While these tolerances are intended to provide for a reasonable and workable fit between the components when used in an assembly, the cumulative effect of the individual tolerances permits a significant variation in the installation characteristics of the complete *fastener assembly*. It is the intent in this Specification that the responsibility rests with the *supplier* for proper performance of the *fastener assembly*, the components of which may have been produced by more than one *manufacturer*.

When pretensioned installation is required, it is essential that the effects of the accumulation of tolerances, surface condition and lubrication be taken into account. Hence, pre-installation verification testing of the complete *fastener assembly* is required as indicated in Section 8 to ensure that the *fastener assemblies* and installation method to be used in the work will provide a pretension that exceeds those specified in Table 8.1. It is not, however, intended simply to verify conformance with the individual ASTM specifications.

It is recognized in this Specification that a natural scatter is found in the results of the pre-installation verification testing that is required in Section 8. Furthermore, it is recognized that the pretensions developed in tests of a representative sample of the fastener components that will be installed in the work must be slightly higher to provide confidence that the majority of *fastener assemblies* will achieve the minimum required pretension as given in Table 8.1. Accordingly, the minimum pretension to be used in pre-installation verification is 1.05 times that required for installation and inspection.

Pre-installation verification testing of as-received bolts and nuts is also a requirement in this Specification because of instances of under-strength and counterfeit bolts and nuts. Pre-installation verification testing provides a practical means for ensuring that non-conforming *fastener assemblies* are not incorporated into the work. Experience on many projects has shown that bolts and/or nuts not meeting the requirements of the applicable ASTM Specification would have been identified prior to installation if they had been tested as an assembly in a *tension calibrator*. The expense of replacing bolts installed in the structure when the non-conforming bolts were discovered at a later date would have been avoided.

Additionally, pre-installation verification testing clarifies for the bolting crew and the *inspector* the proper implementation of the selected pretensioning method and the adequacy of the installation equipment. It will also identify potential sources of problems, such as the need for lubrication to prevent failure of bolts by combined high torque with tension, under-strength assemblies resulting from excessive over-tapping of hot-dip galvanized nuts or other failures to meet strength or geometry requirements of applicable ASTM specifications.

The pre-installation verification requirements in this Section presume that *fastener assemblies* so verified will be pretensioned before the condition of the *fastener assemblies*, the equipment and the steelwork have changed significantly. Research by Kulak and Undershute (1998) on twist-off-type tension-control bolt assemblies from various *manufacturers* showed that installed pretensions could be a function of the time and environmental conditions of storage and exposure. The reduced performance of these bolts was caused by a deterioration of the lubricity of the assemblies. Furthermore, all bolt pre-tensioning that is achieved through rotation of the nut (or the head) is affected by the presence of torque, the excess of which has been demonstrated to adversely affect the development of the desired pretension. Thus, it is required that the condition of the *fastener assemblies* must be replicated in pre-installation verification. When time of exposure between the placement of *fastener assemblies* in the field work and the subsequent pretensioning of those *fastener assemblies* is of concern, pre-installation verification can be performed on *fastener assemblies* removed from the work or on extra *fastener assemblies* that, at the time of placement, were set aside to experience the same degree of exposure.

SECTION 8. INSTALLATION

Prior to installation, the fastener components shall be stored in accordance with Section 2.2. For *joints* that are designated in the contract documents as *snug-tightened joints*, the bolts shall be installed in accordance with Section 8.1. For *joints* that are designated in the contract documents as pretensioned or slip-critical, the bolts shall be installed in accordance with Section 8.2.

8.1. Snug-Tightened Joints

All bolt holes shall be aligned to permit insertion of the bolts without undue damage to the threads. Bolts shall be placed in all holes with washers positioned as required in Section 6.1 and nuts threaded to complete the assembly. Compacting the *joint* to the snug-tight condition shall progress systematically from the most rigid part of the *joint*. The snug-tightened condition is the tightness that is attained with a few impacts of an impact wrench or the full effort of an ironworker using an ordinary spud wrench to bring the connected plies into *firm contact*.

Commentary:

As discussed in the Commentary to Section 4, the bolted *joints* in most shear *connections* and in many tension *connections* can be specified as *snug-tightened joints*. The snug-tightened condition is the tightness that is attained with a few impacts of an impact wrench or the full effort of an ironworker using an ordinary spud wrench to bring the plies into *firm contact*. More than one cycle through the bolt pattern may be required to achieve the *snug-tightened joint*.

The actual pretensions that result in individual fasteners in *snug-tightened joints* will vary from *joint* to *joint* depending upon the thickness, flatness, and degree of parallelism of the connected plies as well as the effort applied. In most *joints*, plies of *joints* involving material of ordinary thickness and flatness can be drawn into complete contact at relatively low levels of pretension. However, in some *joints* in thick material, or in material with large burrs, it may not be possible to reach continuous contact throughout the *faying surface* area as is commonly achieved in *joints* of thinner plates. This is generally not detrimental to the performance of the *joint*.

As used in Section 8.1, the term "undue damage" is intended to mean damage that would be sufficient to render the product unfit for its intended use.

8.2. Pretensioned Joints

One of the pretensioning methods in Sections 8.2.1 through 8.2.4 shall be used, except when alternative-design fasteners that meet the requirements of Section 2.8 or alternative washer-type indicating devices that meet the requirements of Section 2.6.2 are used, in which case, installation instructions provided by the *manufacturer* and approved by the *Engineer of Record* shall be followed. When it is impractical to turn the nut, pretensioning by turning the bolt head is permitted while rotation of the nut is prevented, provided that the washer requirements in Section 6.2 are met. A pretension that is equal to or greater than the value in Table 8.1 shall be provided. The pre-installation verification procedures specified in Section 7 shall be performed using *fastener assemblies* that are representative of the condition of those that will be pretensioned in the work.

Table 8.1. Minimum Bolt Pretension for Pretensioned and Slip-Critical Joints

Nominal Bolt Diameter d_b, in.	Specified Minimum Bolt Pretension T_m, kips[a]	
	ASTM A325 and F1852 Bolts	ASTM A490 Bolts
1/2	12	15
5/8	19	24
3/4	28	35
7/8	39	49
1	51	64
1 1/8	56	80
1 1/4	71	102
1 3/8	85	121
1 1/2	103	148

[a] Equal to 70 percent of the specified minimum tensile strength of bolts as specified in ASTM Specifications for tests of full-size ASTM A325 and A490 bolts with UNC threads loaded in axial tension, rounded to the nearest kip.

Commentary:
The minimum pretension for ASTM A325 and A490 bolts is equal to 70 percent of the specified minimum tensile strength. As tabulated in Table 8.1, the values have been rounded to the nearest kip.

Four pretensioning methods are provided without preference in this Specification. Each method may be relied upon to provide satisfactory results when conscientiously implemented with the specified *fastener assembly* components in good condition. However, it must be recognized that misuse or abuse is possible with any method. With all methods, it is important to first install bolts in all holes of the *joint* and to compact the *joint* until the connected plies are in *firm contact*. Only after completion of this operation can the *joint* be reliably pretensioned. Both the initial phase of compacting the *joint* and the subsequent phase of pretensioning should begin at the most rigidly fixed or stiffest point.

In some *joints* in thick material, it may not be possible to reach continuous contact throughout the *faying surface* area, as is commonly achieved in *joints* of thinner plates. This is not detrimental to the performance of the *joint*. If the specified pretension is present in all bolts of the completed *joint*, the clamping force, which is equal to the total of the pretensions in all bolts, will be transferred at the locations that are in contact and the *joint* will be fully effective in resisting slip through friction.

If individual bolts are pretensioned in a single continuous operation in a *joint* that has not first been properly compacted or fitted up, the pretension in the bolts that are pretensioned first may be relaxed or removed by the pretensioning of adjacent bolts. The resulting reduction in total clamping force will reduce the slip resistance.

In the case of hot-dip galvanized coatings, especially if the *joint* consists of many plies of thickly coated material, relaxation of bolt pretension may be significant and re-pretensioning of the bolts may be required subsequent to the initial pretensioning. Munse (1967) showed that a loss of pretension of approximately 6.5 percent occurred for galvanized plates and bolts due to relaxation as compared with 2.5 percent for uncoated *joints*. This loss of bolt pretension occurred in five days; loss recorded thereafter was negligible. Either this loss can be allowed for in design or pretension may be brought back to the prescribed level by re-pretensioning the bolts after an initial period of "settling-in".

As stated in the *Guide*, Kulak et al (1987; p. 61), "…it seems reasonable to expect an increase in bolt force relaxation as the *grip* length is decreased. Similarly, increasing the number of plies for a constant *grip* length might also lead to an increase in bolt relaxation."

8.2.1. Turn-of-Nut Pretensioning: All bolts shall be installed in accordance with the requirements in Section 8.1, with washers positioned as required in Section 6.2. Subsequently, the nut or head rotation specified in Table 8.2 shall be applied to all *fastener assemblies* in the *joint*, progressing systematically from the most rigid part of the *joint* in a manner that will minimize relaxation of previously pretensioned bolts. The part not turned by the wrench shall be prevented from rotating during this operation.

Table 8.2. Nut Rotation from Snug-Tight Condition for Turn-of-Nut Pretensioning [a,b]

Bolt Length[c]	Disposition of Outer Face of Bolted Parts		
	Both faces normal to bolt axis	One face normal to bolt axis, other sloped not more than 1:20[d]	Both faces sloped not more than 1:20 from normal to bolt axis[d]
Not more than $4d_b$	1/3 turn	1/2 turn	2/3 turn
More than $4d_b$ but not more than $8d_b$	1/2 turn	2/3 turn	5/6 turn
More than $8d_b$ but not more than $12d_b$	2/3 turn	5/6 turn	1 turn

[a] Nut rotation is relative to bolt regardless of the element (nut or bolt) being turned. For required nut rotations of ½ turn and less, the tolerance is plus or minus 30 degrees; for required nut rotations of ⅔ turn and more, the tolerance is plus or minus 45 degrees.

[b] Applicable only to joints in which all material within the grip is steel.

[c] When the bolt length exceeds $12d_b$, the required nut rotation shall be determined by actual testing in a suitable tension calibrator that simulates the conditions of solidly fitting steel.

[d] Beveled washer not used.

Commentary:
The turn-of-nut pretensioning method results in more uniform bolt pretensions than is generally provided with torque-controlled pretensioning methods. Strain-control that reaches the inelastic region of bolt behavior is inherently more reliable than a method that is dependent upon torque control. However, proper implementation is dependent upon ensuring that the *joint* is properly compacted prior to application of the required partial turn and that the bolt head (or nut) is securely held when the nut (or bolt head) is being turned.

Match-marking of the nut and protruding end of the bolt after snug-tightening can be helpful in the subsequent installation process, and is certainly an aid to inspection.

As indicated in Table 8.2, there is no available research that establishes the required nut rotation for bolt lengths exceeding $12d_b$. The required turn for such bolts can be established on a case-by-case basis using a *tension calibrator*.

8.2.2. Calibrated Wrench Pretensioning: The pre-installation verification procedures specified in Section 7 shall be performed daily for the calibration of the installation wrench. Torque values determined from tables or from equations that claim to relate torque to pretension without verification shall not be used.

All bolts shall be installed in accordance with the requirements in Section 8.1, with washers positioned as required in Section 6.2. Subsequently, the installation torque determined in the pre-installation verification of the *fastener assembly* (Section 7) shall be applied to all bolts in the *joint*, progressing systematically from the most rigid part of the *joint* in a manner that will minimize relaxation of previously pretensioned bolts. The part not turned by the wrench shall be prevented from rotating during this operation. Application of the installation torque need not produce a relative rotation between the bolt and nut that is greater than the rotation specified in Table 8.2.

Commentary:
The scatter in installed pretension can be significant when torque-controlled methods of installation are used. The variables that affect the relationship between torque and pretension include:

(1) The finish and tolerance on the bolt and nut threads;
(2) The uniformity, degree and condition of lubrication;
(3) The shop or job-site conditions that contribute to dust and dirt or corrosion on the threads;
(4) The friction that exists to a varying degree between the turned element (the nut face or bearing area of the bolt head) and the supporting surface;
(5) The variability of the air supply parameters on impact wrenches that results from the length of air lines or number of wrenches operating from the same source;
(6) The condition, lubrication and power supply for the torque wrench, which may change within a work shift; and,
(7) The repeatability of the performance of any wrench that senses or responds to the level of the applied torque.

In the first edition of this Specification, which was published in 1951, a table of torque-to-pretension relationships for bolts of various diameters was included. It was soon demonstrated in research that a variation in the torque-to-pretension of as high as ±40 percent must be anticipated unless the relationship is established individually for each bolt *lot*, diameter, and fastener condition. Hence, in the 1954 edition of this Specification, recognition of relationships between torque and pretension in the form of tabulated values or equations was withdrawn. Recognition of the calibrated wrench pretensioning method was retained however until 1980, but with the requirement that the torque required for installation be determined specifically for the bolts being installed on a daily basis. Recognition of the method was withdrawn in 1980 because of the continuing controversy that resulted from the failure of users to adhere to the requirements for the valid use of the method during both installation and inspection.

In the 1985 edition of this Specification, the calibrated wrench pretensioning method was reinstated, but with more emphasis on detailed requirements that must be carefully followed. For calibrated wrench pretensioning, wrenches must be calibrated:

(1) Daily;
(2) When the *lot* of any component of the *fastener assembly* is changed;
(3) When the *lot* of any component of the *fastener assembly* is relubricated;
(4) When significant differences are noted in the surface condition of the bolt threads, nuts or washers; or,
(5) When any major component of the wrench including lubrication, hose and air supply are altered.

It is also important that:

(1) Fastener components be protected from dirt and moisture at the shop or jobsite as required in Section 2;
(2) Washers be used as specified in Section 6; and
(3) The time between removal from *protected storage* and wrench calibration and final pretensioning be minimal.

8.2.3. Twist-Off-Type Tension-Control Bolt Pretensioning: Twist-off-type tension-control bolt assemblies that meet the requirements of ASTM F1852 shall be used.

All *fastener assemblies* shall be installed in accordance with the requirements in Section 8.1 without severing the splined end and with washers positioned as required in Section 6.2. If a splined end is severed during this operation, the *fastener assembly* shall be removed and replaced. Subsequently, all bolts in the *joint* shall be pretensioned with the twist-off-type tension-control bolt installation wrench, progressing systematically from the most rigid part of the *joint* in a manner that will minimize relaxation of previously pretensioned bolts.

Commentary:
ASTM F1852 twist-off-type tension-control bolt assemblies have a splined end that extends beyond the threaded portion of the bolt. During installation, this

splined end is gripped by a specially designed wrench chuck and provides a means for turning the nut relative to the bolt. This product is, in fact, based upon a torque-controlled installation method to which the *fastener assembly* variables affecting torque that were discussed in the Commentary to Section 8.2.2 apply, except for wrench calibration, because torque is controlled within the *fastener assembly*.

Twist-off-type tension-control bolt assemblies must be used in the as-delivered, clean, lubricated condition as specified in Section 2. Adherence to the requirements in this Specification, especially those for storage, cleanliness and verification, is necessary for their proper use.

8.2.4. Direct-Tension-Indicator Pretensioning: Direct tension indicators that meet the requirements of ASTM F959 shall be used. The pre-installation verification procedures specified in Section 7 shall demonstrate that, when the pretension in the bolt reaches 1.05 times that specified for installation and inspection in Table 8.1, the gap is not less than the job inspection gap in accordance with ASTM F959.

All bolts shall be installed in accordance with the requirements in Section 8.1, with washers positioned as required in Section 6.2. The installer shall verify that the direct-tension-indicator protrusions have not been compressed to a gap that is less than the job inspection gap during this operation, and if this has occurred, the direct tension indicator shall be removed and replaced. Subsequently, all bolts in the *joint* shall be pretensioned, progressing systematically from the most rigid part of the *joint* in a manner that will minimize relaxation of previously pretensioned bolts. The installer shall verify that the direct tension indicator protrusions have been compressed to a gap that is less than the job inspection gap.

Commentary:

ASTM F959 direct tension indicators are recognized in this Specification as a bolt-tension-indicating device. Direct tension indicators are hardened, washer-shaped devices incorporating small arch-like protrusions on the bearing surface that are designed to deform in a controlled manner when subjected to compressive load.

During installation, care must be taken to ensure that the direct-tension-indicator arches are oriented to bear against the hardened bearing surface of the bolt head or nut or against a hardened flat washer if used under turned element whether that turned element is the nut or the bolt. Proper use and orientation is illustrated in Figure C-8.1.

In some cases, more than a single cycle of systematic partial pretensioning may be required to deform the direct-tension-indicator protrusions to the gap that is specified by the *manufacturer*. If the gaps fail to close or when the washer *lot* is changed, another verification procedure using the *tension calibrator* must be performed.

Provided the connected plies are in *firm contact*, partial compression of the direct tension indicator protrusions is commonly taken as an indication that the snug-tight condition has been achieved.

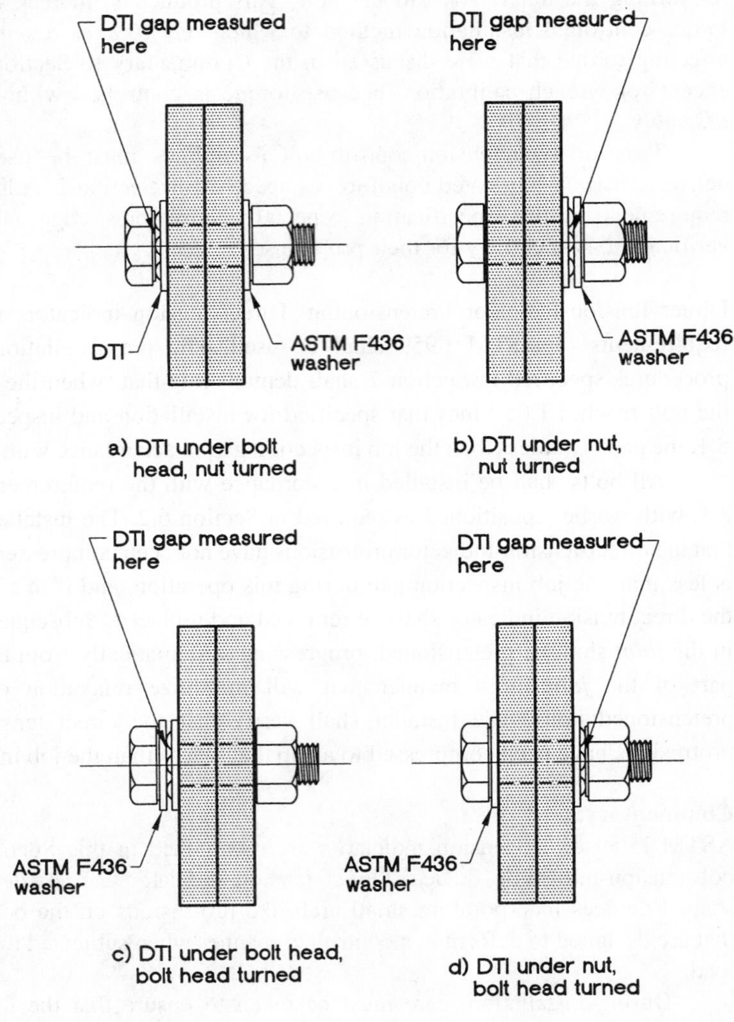

Figure. C-8.1. Proper use and orientation of ASTM F959 direct-tension indicator.

SECTION 9. INSPECTION

When inspection is required in the contract documents, the *inspector* shall ensure while the work is in progress that the requirements in this Specification are met. When inspection is not required in the contract documents, the *contractor* shall ensure while the work is in progress that the requirements in this Specification are met.

For *joints* that are designated in the contract documents as *snug-tightened joints*, the inspection shall be in accordance with Section 9.1. For *joints* that are designated in the contract documents as pretensioned, the inspection shall be in accordance with Section 9.2. For *joints* that are designated in the contract documents as slip-critical, the inspection shall be in accordance with Section 9.3.

9.1. Snug-Tightened Joints

Prior to the *start of work*, it shall be ensured that all fastener components to be used in the work meet the requirements in Section 2. Subsequently, it shall be ensured that all connected plies meet the requirements in Section 3.1 and all bolt holes meet the requirements in Sections 3.3 and 3.4. After the *connections* have been assembled, it shall be visually ensured that the plies of the connected elements have been brought into *firm contact* and that washers have been used as required in Section 6. No further evidence of conformity is required for *snug-tightened joints*. The magnitude of the clamping force that exists in a *snug-tightened joint* is not a consideration.

Commentary:
Inspection requirements for *snug-tightened joints* consist of verification that the proper fastener components were used, the connected elements were fabricated properly, and the bolted *joint* was drawn into firm contact. Because pretension is not required for the proper performance of a *snug-tightened joint*, the installed bolts should not be inspected to determine the actual installed pretension. Likewise, the arbitration procedures described in Section 10 are not appropriate.

9.2. Pretensioned Joints

For *pretensioned joints*, the following inspection shall be performed in addition to that required in Section 9.1:

(1) When the turn-of-nut pretensioning method is used for installation, the inspection shall be in accordance with Section 9.2.1.
(2) When the calibrated wrench pretensioning method is used for installation, the inspection shall be in accordance with Section 9.2.2.
(3) When the twist-off-type tension-control bolt pretensioning method is used for installation, the inspection shall be in accordance with Section 9.2.3.
(4) When the direct-tension-indicator pretensioning method is used for installation, the inspection shall be in accordance with Section 9.2.4.
(5) When alternative-design fasteners that meet the requirements of Section 2.8 or alternative washer-type indicating devices that meet the requirements of Section 2.6.2 are used, the inspection shall be in accordance with inspection instructions provided by the manufacturer and approved by the Engineer of Record.

Commentary:
When *joints* are designated as pretensioned, they are not subject to the same faying-surface-treatment inspection requirements as is specified for *slip-critical joints* in Section 9.3.

9.2.1. Turn-of-Nut Pretensioning: The *inspector* shall observe the pre-installation verification testing required in Section 8.2.1. Subsequently, it shall be ensured by *routine observation* that the bolting crew properly rotates the turned element relative to the unturned element by the amount specified in Table 8.2. Alternatively, when *fastener assemblies* are match-marked after the initial fit-up of the *joint* but prior to pretensioning, visual inspection after pretensioning is permitted in lieu of routine observation. No further evidence of conformity is required. A pretension that is greater than the value specified in Table 8.1 shall not be cause for rejection.

Commentary:
Match-marking of the assembly during installation as discussed in the Commentary to Section 8.2.1 improves the ability to inspect bolts that have been pretensioned with the turn-of-nut pretensioning method. The sides of nuts and bolt heads that have been impacted sufficiently to induce the Table 8.1 minimum pretension will appear slightly peened.

The turn-of-nut pretensioning method, when properly applied and verified during the construction, provides more reliable installed pretensions than after-the-fact *inspection* testing. Therefore, proper inspection of the method is for the inspector to observe the required pre-installation verification testing of the *fastener assemblies* and the method to be used, followed by monitoring of the work in progress to ensure that the method is routinely and properly applied, or visual inspection of match-marked assemblies.

Some problems with the turn-of-nut pretensioning method have been encountered with hot-dip galvanized bolts. In some cases, the problems have been attributed to an especially effective lubricant applied by the *manufacturer* to ensure that bolts and nuts from stock will meet the ASTM Specification requirements for minimum turns testing of galvanized fasteners. Job-site testing in the *tension calibrator* demonstrated that the lubricant reduced the coefficient of friction between the bolt and nut to the degree that "the full effort of an ironworker using an ordinary spud wrench" to snug-tighten the *joint* actually induced the full required pretension. Also, because the nuts could be removed with an ordinary spud wrench, they were erroneously judged by the *inspector* to be improperly pretensioned. Excessively lubricated *high-strength bolts* may require significantly less torque to induce the specified pretension. The required pre-installation verification will reveal this potential problem.

Conversely, the absence of lubrication or lack of proper over-tapping can cause seizing of the nut and bolt threads, which will result in a twist failure of the bolt at less than the specified pretension. For such situations, the use of a *tension calibrator* to check the bolt assemblies to be installed will be helpful in establishing the need for lubrication.

9.2.2. Calibrated Wrench Pretensioning: The *inspector* shall observe the pre-installation verification testing required in Section 8.2.2. Subsequently, it shall be ensured by *routine observation* that the bolting crew properly applies the calibrated wrench to the turned element. No further evidence of conformity is required. A pretension that is greater than the value specified in Table 8.1 shall not be cause for rejection.

Commentary:
For proper inspection of the method, it is necessary for the *inspector* to observe the required pre-installation verification testing of the *fastener assemblies* and the method to be used, followed by monitoring of the work in progress to ensure that the method is routinely and properly applied within the limits on time between removal from *protected storage* and final pretensioning.

9.2.3. Twist-Off-Type Tension-Control Bolt Pretensioning: The *inspector* shall observe the pre-installation verification testing required in Section 8.2.3. Subsequently, it shall be ensured by *routine observation* that the splined ends are properly severed during installation by the bolting crew. No further evidence of conformity is required. A pretension that is greater than the value specified in Table 8.1 shall not be cause for rejection.

Commentary:
The sheared-off splined end of an installed ASTM F1852 twist-off-type tension-control bolt assembly merely signifies that at some time the bolt was subjected to a torque that was adequate to cause the shearing. If in fact all fasteners are individually pretensioned in a single continuous operation without first properly snug-tightening all fasteners, they may give a misleading indication that the bolts have been properly pretensioned. Therefore it is necessary that the *inspector* observe the required pre-installation verification testing of the *fastener assemblies*, and the ability to apply partial tension prior to twist-off is demonstrated. This is followed by monitoring of the work in progress to ensure that the method is routinely and properly applied within the limits on time between removal from *protected storage* and final twist-off of the splined end.

9.2.4. Direct-Tension-Indicator Pretensioning: The *inspector* shall observe the pre-installation verification testing required in Section 8.2.4. Subsequently, but prior to pretensioning, it shall be ensured by *routine observation* that the appropriate feeler gage is accepted in at least half of the spaces between the protrusions of the direct tension indicator and that the protrusions are properly oriented away from the work. If the appropriate feeler gage is accepted in fewer than half of the spaces, the direct tension indicator shall be removed and replaced. After pretensioning, it shall be ensured by *routine observation* that the appropriate feeler gage is refused entry into at least half of the spaces between the protrusions. No further evidence of conformity is required. A pretension that is greater than that specified in Table 8.1 shall not be cause for rejection.

Commentary:
When the *joint* is initially snug tightened, the direct tension indicator arch-like protrusions will generally compress partially. Whenever the snug-tightening operation causes one-half or more of the gaps between these arch-like protrusions to close to 0.015 in. or less (0.005 in. or less for coated direct tension indicators), the direct tension indicator should be replaced. Only after this initial operation should the bolts be pretensioned in a systematic manner. If the bolts are installed and pretensioned in a single continuous operation, direct tension indicators may give the *inspector* a misleading indication that the bolts have been properly pretensioned. Therefore, it is necessary that the *inspector* observe the required pre-installation verification testing of the *fastener assemblies* with the direct-tension indicators properly located and the method to be used. Following this operation, the *inspector* should monitor the work in progress to ensure that the method is routinely and properly applied.

9.3. Slip-Critical Joints

Prior to assembly, it shall be visually verified that the *faying surfaces* of *slip-critical joints* meet the requirements in Section 3.2.2. Subsequently, the inspection required in Section 9.2 shall be performed.

Commentary:
When *joints* are specified as slip-critical, it is necessary to verify that the *faying surface* condition meets the requirements as specified in the contract documents prior to assembly of the *joint* and that the bolts are properly pretensioned after they have been installed. Accordingly, the inspection requirements for *slip-critical joints* are identical to those specified in Section 9.2, with additional *faying surface* condition inspection requirements.

SECTION 10. ARBITRATION

When it is suspected after inspection in accordance with Section 9.2 or Section 9.3 that bolts in pretensioned or *slip-critical joints* do not have the proper pretension, the following arbitration procedure is permitted. If verification of bolt pretension is required after the passage of a period of time and exposure of the completed *joints*, an alternative arbitration procedure that is appropriate to the specific situation shall be used.

(1) A representative sample of five bolt and nut assemblies of each combination of diameter, length, grade and *lot* in question shall be installed in a *tension calibrator*. The material under the turned element shall be the same as in the actual installation; that is, structural steel or hardened washer. The bolt shall be partially pretensioned to approximately 15 percent of the pretension specified in Table 8.1. Subsequently, the bolt shall be pretensioned to the minimum value specified in Table 8.1.

(2) A manual torque wrench that indicates torque by means of a dial, or one that may be adjusted to give an indication that a defined torque has been reached, shall be applied to the pretensioned bolt. The torque that is necessary to rotate the nut or bolt head five degrees (approximately 1 in. at 12-in. radius) relative to its mating component in the tightening direction shall be determined. The arbitration torque shall be determined by rejecting the high and low values and averaging the remaining three.

(3) Bolts represented by the above sample shall be tested by applying, in the tightening direction, the arbitration torque to 10 percent of the bolts, but no fewer than two bolts, selected at random in each *joint* in question. If no nut or bolt head is turned relative to its mating component by application of the arbitration torque, the *joint* shall be accepted as properly pretensioned.

If any nut or bolt is turned relative to its mating component by an attempted application of the arbitration torque, all bolts in the *joint* shall be tested. Those bolts whose nut or head is turned relative to its mating component by application of the arbitration torque shall be re-pretensioned by the Fabricator or Erector and reinspected. Alternatively, the Fabricator or Erector, at their option, is permitted to re-pretension all of the bolts in the *joint* and subsequently resubmit the *joint* for inspection.

Commentary:

When bolt pretension is arbitrated using torque wrenches after pretensioning, such arbitration is subject to all of the uncertainties of torque-controlled calibrated wrench installation that are discussed in the Commentary to Section 8.2.2. Additionally, the reliability of after-the-fact torque wrench arbitration is reduced by the absence of many of the controls that are necessary to minimize the variability of the torque-to-pretension relationship, such as:

(1) The use of hardened washers[3];

(2) Careful attention to lubrication; and,

(3) The uncertainty of the effect of passage of time and exposure in the installed condition.

Furthermore, in many cases such arbitration may have to be based upon an arbitration torque that is determined either using bolts that can only be assumed to be representative of the bolts used in the actual job or using bolts that are removed from completed *joints*. Ultimately, such arbitration may wrongly reject bolts that were subjected to a properly implemented installation procedure. The arbitration procedure contained in this Specification is provided, in spite of its limitations, as the most feasible available at this time.

Arbitration using an ultrasonic extensometer or a mechanical one capable of measuring changes in bolt length can be performed on a sample of bolts that is representative of those that have been installed in the work. Several *manufacturers* produce equipment specifically for this application. The use of appropriate techniques, which includes calibration, can produce a very accurate measurement of the actual pretension. The method involves measurement of the change in bolt length during the release of the nut, combined with either a load calibration of the removed *fastener assembly* or a theoretical calculation of the force corresponding to the measured elastic release or "stretch". Reinstallation of the released bolt or installation of a replacement bolt is required.

The required release suggests that the direct use of extensometers as an inspection tool be used in only the most critical cases. The problem of reinstallation may require bolt replacement unless torque can be applied slowly using a manual or hydraulic wrench, which will permit the restoration of the original elongation.

---

[3] For example, because the reliability of the turn-of-nut pretensioning method is not dependent upon the presence or absence of washers under the turned element, washers are not generally required, except for other reasons as indicated in Section 6. Thus, in the absence of washers,after-the-fact, torque-based arbitration is particularly unreliable when the turn-of-nut pretensioning method has been used for installation.

APPENDIX A. TESTING METHOD TO DETERMINE THE SLIP COEFFICIENT FOR COATINGS USED IN BOLTED JOINTS

SECTION A1. GENERAL PROVISIONS

A1.1. Purpose and Scope

The purpose of this testing procedure is to determine the *mean slip coefficient* of a coating for use in the design of *slip-critical joints*. Adherence to this testing method provides that the creep deformation of the coating due to both the clamping force of the bolt and the service-load *joint* shear are such that the coating will provide satisfactory performance under sustained loading.

Commentary:

The Research Council on Structural Connections on June 14, 1984, first approved the testing method developed by Yura and Frank (1985). It has since been revised to incorporate changes resulting from the intervening years of experience with the testing method, and is now included as an appendix to this Specification.

The slip coefficient under short-term static loading has been found to be independent of the magnitude of the clamping force, variations in coating thickness and bolt hole diameter.

The proposed test methods are designed to provide the necessary information to evaluate the suitability of a coating for *slip-critical joints* and to determine the *mean slip coefficient* to be used in the design of the *joints*. The initial testing of the compression specimens provides a measure of the scatter of the slip coefficient.

The creep tests are designed to measure the creep behavior of the coating under the service loads, determined by the slip coefficient of the coating based upon the compression test results. The slip test conducted at the conclusion of the creep test is to ensure that the loss of clamping force in the bolt does not reduce the slip load below that associated with the design slip coefficient. ASTM A490 bolts are specified, since the loss of clamping force is larger for these bolts than that for ASTM A325 bolts. Qualification of the coating for use in a structure at an average thickness of 2 mils less than that to be used for the test specimen is to ensure that a casual buildup of the coating due to overspray and other causes does not jeopardize the coating's performance.

A1.2. Definition of Essential Variables

Essential variables are those that, if changed, will require retesting of the coating to determine its *mean slip coefficient*. The essential variables and the relationship of these variables to the limitations of application of the coating for structural *joints* are given below. The slip coefficient testing shall be repeated if there is any change in these essential variables.

A1.2.1. Time Interval: The time interval between application of the coating and the time of testing is an essential variable. The time interval must be recorded in hours and any special curing procedures detailed. Curing according to published *manufacturer's* recommendations would not be considered a special curing procedure. The coatings are qualified for use in structural *connections* that are assembled after coating for a time equal to or greater than the interval used in the test specimens. Special curing conditions used in the test specimens will also apply to the use of the coating in the structural *connections*.

A1.2.2. Coating Thickness: The coating thickness is an essential variable. The maximum average coating thickness, as per SSPC PA2 (SSPC 1993; SSPC 1991), allowed on the faying surfaces is 2 mils less than the average thickness, rounded to the nearest whole mil, of the coating that is used on the test specimens.

A1.2.3. Coating Composition and Method of Manufacture: The composition of the coating, including the thinners used, and its method of manufacture are essential variables.

A1.3. Retesting

A coating that fails to meet the creep or the post-creep slip test requirements in Section A4 may be retested in accordance with methods in Section A4 at a lower slip coefficient without repeating the static short-term tests specified in Section A3. Essential variables shall remain unchanged in the retest.

SECTION A2. TEST PLATES AND COATING OF THE SPECIMENS

A2.1. Test Plates

The test specimen plates for the short-term static tests are shown in Figure A1. The plates are 4 in. × 4 in. × ⅝ in. thick, with a 1 in. diameter hole drilled 1½ in. ± 1/16 in. from one edge. The test specimen plates for the creep tests are shown in Figure A2. The plates are 4 in. × 7 in. × ⅝ in. thick with two 1 in. diameter holes drilled 1½ in. ± 1/16 in. from each end. The edges of the plates may be milled, as-rolled or saw-cut; thermally cut edges are not permitted. The plates shall be flat enough to ensure that they will be in reasonably full contact over the *faying surface*. All burrs, lips or rough edges shall be removed. The arrangement of the specimen plates for the testing is shown in Figure A2. The plates shall be fabricated from a steel with a specified minimum yield strength that is between 36 and 50 ksi.

If specimens with more than one bolt are desired, the contact surface per bolt shall be 4 in. × 3 in. as shown for the single-bolt specimen in Figure A1.

Commentary:

The use of 1-in.-diameter bolt holes in the specimens is to ensure that adequate clearance is available for slip. Fabrication tolerances, coating buildup on the holes, and assembly tolerances tend to reduce the apparent clearances.

A2.2. Specimen Coating

Coatings are to be applied to the specimens in a manner that is consistent with that to be used in the actual intended structural application. The method of applying the coating and the surface preparation shall be given in the test report. The specimens are to be coated to an average thickness that is 2 mils greater than the maximum thickness to be used in the structure on both of the plate surfaces (the faying and outer surfaces). The thickness of the total coating and the primer, if used, shall be measured on the contact surface of the specimens. The thickness shall be measured in accordance with SSPC-PA2 (SSPC, 1993; SSPC, 1991). Two spot readings (six gage readings) shall be made for each contact surface. The overall average thickness from the three plates comprising a specimen is the average thickness for the specimen. This value shall be reported for each specimen. The average coating thickness of the creep specimens shall be calculated and reported.

The time between application of the coating and specimen assembly shall be the same for all specimens within ±4 hours. The average time shall be calculated and reported.

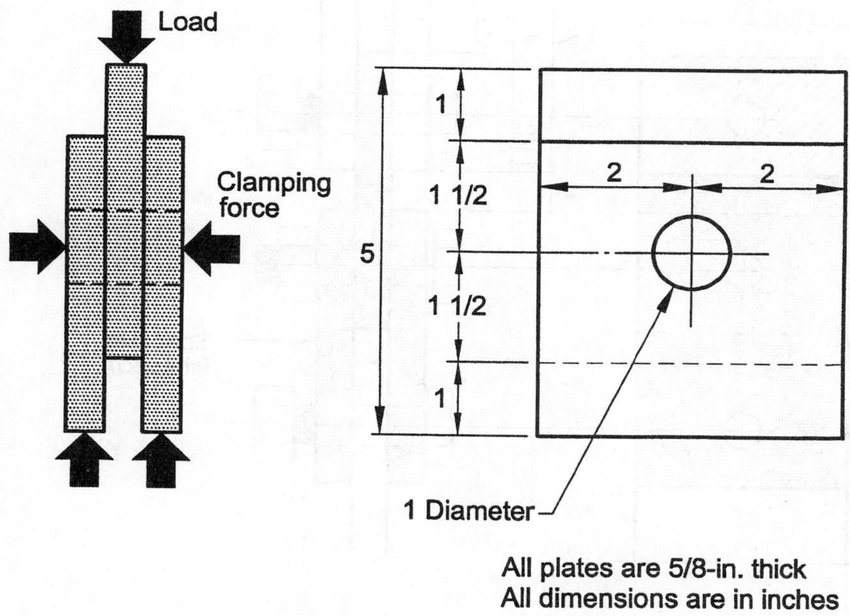

Figure A-1. Compression slip test specimen.

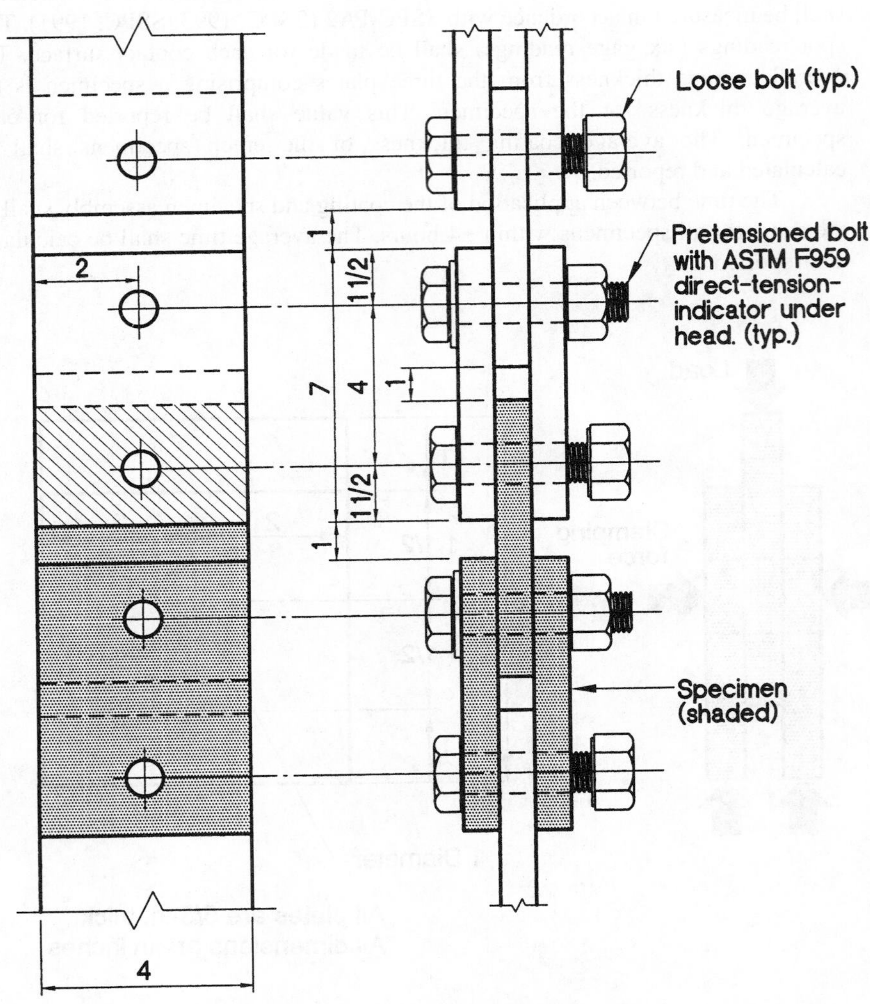

Figure A-2. Creep test specimen assembly.

SECTION A3. SLIP TESTS

The methods and procedures described herein are used to experimentally determine the *mean slip coefficient* under short-term static loading for *high-strength bolted joints*. The *mean slip coefficient* shall be determined by testing one set of five specimens.

Commentary:
The slip load measured in this setup yields the slip coefficient directly since the clamping force is controlled and measured directly. The resulting slip coefficient has been found to correlate with both tension and compression tests of bolted specimens. However, tests of bolted specimens revealed that the clamping force may not be constant but decreases with time due to the compressive creep of the coating on the *faying surfaces* and under the nut and bolt head. The reduction in clamping force can be considerable for *joints* with high clamping force and thick coatings (as much as a 20 percent loss). This reduction in clamping force causes a corresponding reduction in the slip load. The resulting reduction in slip load must be considered in the procedure used to determine the design allowable slip loads for the coating.

The loss in clamping force is a characteristic of the coating. Consequently, it cannot be accounted for by an increase in the factor of safety or a reduction in the clamping force used for design without unduly penalizing coatings that do not exhibit this behavior.

A3.1. Compression Test Setup

The test setup shown in Figure A3 has two major loading components, one to apply a clamping force to the specimen plates and another to apply a compressive load to the specimen so that the load is transferred across the *faying surfaces* by friction.

A3.1.1. Clamping Force System: The clamping force system consists of a ⅞ in. diameter threaded rod that passes through the specimen and a centerhole compression ram. An ASTM A563 grade DH nut is used at both ends of the rod and a hardened washer is used at each side of the test specimen. Between the ram and the specimen is a specially modified ⅞ in. diameter ASTM A563 grade DH nut in which the threads have been drilled out so that it will slide with little resistance along the rod. When oil is pumped into the centerhole ram, the piston rod extends, thus forcing the special nut against one of the outside plates of the specimen. This action puts tension in the threaded rod and applies a clamping force to the specimen, thereby simulating the effect of a pretensioned bolt. If the diameter of the centerhole ram is greater than 1 in., additional plate washers will be necessary at the ends of the ram. The clamping force system shall have a capability to apply a load of at least 49 kips and shall maintain this load during the test with an accuracy of 0.5 kips.

Commentary:
The slip coefficient can be easily determined using the hydraulic bolt test setup included in this Specification. The clamping force system simulates the clamping action of a pretensioned *high-strength bolt*. The centerhole ram applies a clamping force to the specimen, simulating that due to a pretensioned bolt.

A3.1.2. Compressive Load System: A compressive load shall be applied to the specimen until slip occurs. This compressive load shall be applied with a compression test machine or a reaction frame using a hydraulic loading device. The loading device and the necessary supporting elements shall be able to support a force of 120 kips. The compression loading system shall have a minimum accuracy of 1 percent of the slip load.

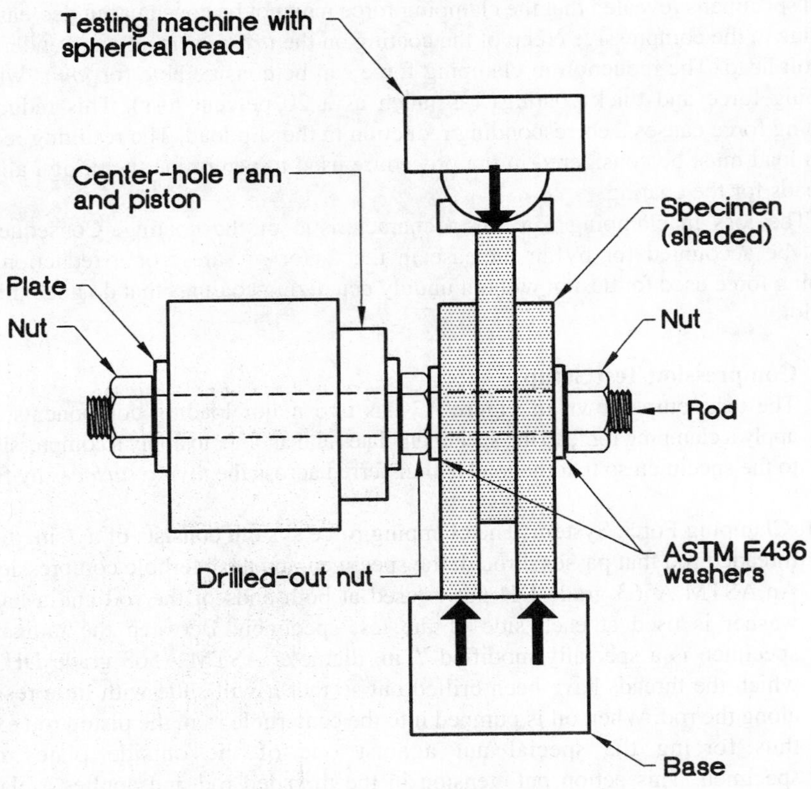

Figure A-3. Compression slip test setup.

A3.2. Instrumentation

A3.2.1. Clamping Force: The clamping force shall be measured within 0.5 kips. This is accomplished by measuring the pressure in the calibrated ram or placing a load cell in series with the ram.

A3.2.2. Compression Load: The compression load shall be measured during the test by direct reading from a compression testing machine, a load cell in series with the specimen and the compression loading device or pressure readings on a calibrated compression ram.

A3.2.3. Slip Deformation: The displacement of the center plate relative to the two outside plates shall be measured. This displacement, called "slip" for simplicity, shall be the average or that which occurs at the centerline of the specimen. This can be accomplished by using the average of two gages placed on the two exposed edges of the specimen or by monitoring the movement of the loading head relative to the base. If the latter method is used, due regard shall be taken for any slack that may be present in the loading system prior to application of the load. Deflections shall be measured by dial gages or any other calibrated device that has an accuracy of at least 0.001 in.

A3.3. Test Procedure

The specimen shall be installed in the test setup as shown in Figure A3. Before the hydraulic clamping force is applied, the individual plates shall be positioned so that they are in, or close to, full bearing contact with the ⅞ in. threaded rod in a direction that is opposite to the planned compressive loading to ensure obvious slip deformation. Care shall be taken in positioning the two outside plates so that the specimen is perpendicular to the base with both plates in contact with the base. After the plates are positioned, the centerhole ram shall be engaged to produce a clamping force of 49 kips. The applied clamping force shall be maintained within ±0.5 kips during the test until slip occurs.

The spherical head of the compression loading machine shall be brought into contact with the center plate of the specimen after the clamping force is applied. The spherical head or other appropriate device ensures concentric loading. When 1 kip or less of compressive load is applied, the slip gages shall be engaged or attached. The purpose of engaging the deflection gage(s), after a slight load is applied, is to eliminate initial specimen settling deformation from the slip reading.

When the slip gages are in place, the compression load shall be applied at a rate that does not exceed 25 kips per minute nor 0.003 in. of slip displacement per minute until the slip load is reached. The test should be terminated when a slip of 0.05 in. or greater is recorded. The load-slip relationship should preferably be monitored continuously on an X-Y plotter throughout the test, but in lieu of continuous data, sufficient load-slip data shall be recorded to evaluate the slip load defined below.

A3.4. Slip Load

Typical load-slip response is shown in Figure A4. Three types of curves are usually observed and the slip load associated with each type is defined as follows:

Curve (a)	Slip load is the maximum load, provided this maximum occurs before a slip of 0.02 in. is recorded.
Curve (b)	Slip load is the load at which the slip rate increases suddenly.
Curve (c)	Slip load is the load corresponding to a deformation of 0.02 in. This definition applies when the load vs. slip curves show a gradual change in response.

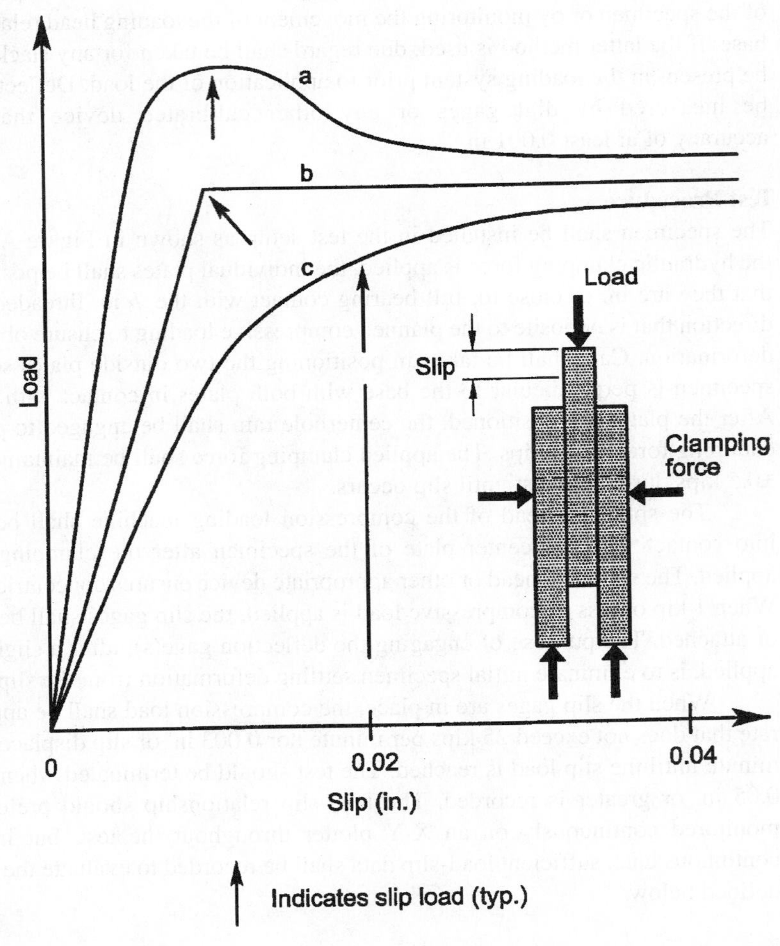

Figure A-4. Definition of slip load.

A3.5. Slip Coefficient

The slip coefficient for an individual specimen k_s shall be calculated as follows: The *mean slip coefficient* μ for one set of five specimens shall be reported.

$$k_s = \frac{slip\ load}{2 \times\ clamping\ force} \qquad \text{(Equation A3.1)}$$

A3.6. Alternative Test Methods

Alternative test methods to determine slip are permitted, provided the accuracy of load measurement and clamping satisfies the conditions presented in the previous sections. For example, the slip load may be determined from a tension-type test setup rather than the compression-type test setup as long as the contact surface area per bolt of the test specimen is the same as that shown in Figure A1. The clamping force of at least 49 kips may be applied by any means, provided the force can be established within ± 1 percent.

Commentary:

Alternative test procedures and specimens may be used as long as the accuracy of load measurement and specimen geometry are maintained as prescribed. For example, strain-gaged bolts can usually provide the desired accuracy. However, bolts that are pretensioned by the turn-of-nut, calibrated wrench, alternative-design fastener, or direct-tension-indicator pretensioning method usually show too much variation to meet the ± 1 percent requirement of the slip test.

SECTION A4. TENSION CREEP TEST

The test method outlined is intended to ensure that the coating will not undergo significant creep deformation under sustained service loading. The test also indicates the loss in clamping force in the bolt due to the compression or creep of the coating. Three replicate specimens are to be tested.

Commentary:
The creep deformation of the bolted *joint* under the applied shear loading is also an important characteristic and a function of the coating applied. Thicker coatings tend to creep more than thinner coatings. Rate of creep deformation increases as the applied load approaches the slip load. Extensive testing has shown that the rate of creep is not constant with time, rather it decreases with time. After about 1,000 hours of loading, the additional creep deformation is negligible.

A4.1. Test Setup

Tension-type specimens, as shown in Figure A2, are to be used. The replicate specimens are to be linked together in a single chain-like arrangement, using loose pin bolts, so the same load is applied to all specimens. The specimens shall be assembled so the specimen plates are bearing against the bolt in a direction opposite to the applied tension loading. Care shall be taken in the assembly of the specimens to ensure the centerline of the holes used to accept the pin bolts is in line with the bolts used to assemble the *joint*. The load level, specified in Section A4.2, shall be maintained constant within ±1 percent by springs, load maintainers, servo controllers, dead weight or other suitable equipment. The bolts used to clamp the specimens together shall be ⅞ in. diameter ASTM A490 bolts. All bolts shall come from the same *lot*.

The clamping force in the bolts shall be a minimum of 49 kips. The clamping force shall be determined by calibrating the bolt force with bolt elongation, if standard bolts are used. Alternatively, special *fastener assemblies* that control the clamping force by other means, such as calibrated bolt torque or strain gages, are permitted. A minimum of three bolt calibrations shall be performed using the technique selected for bolt force determination. The average of the three-bolt calibration shall be calculated and reported. The method of measuring bolt force shall ensure the clamping force is within ±2 kips of the average value.

The relative slip between the outside plates and the center plates shall be measured to an accuracy of 0.001 in. These slips are to be measured on both sides of each specimen.

A4.2. Test Procedure

The load to be placed on the creep specimens is the service load permitted for ⅞ in. diameter ASTM A490 bolts in *slip-critical joints* in Section 5 for the particular slip coefficient category under consideration. The load shall be placed on the specimen and held for 1,000 hours. The creep deformation of a specimen is calculated using the average reading of the two displacements on either side of the specimen. The difference between the average after 1,000 hours and the initial average reading taken within one-half hour after loading the specimens is defined as the creep deformation of the specimen. This value shall be reported for each specimen. If the creep deformation of any specimen exceeds 0.005 in., the coating has failed the test for the slip coefficient used. The coating may be retested using new specimens in accordance with this Section at a load corresponding to a lower value of slip coefficient.

If the value of creep deformation is less than 0.005 in. for all specimens, the specimens shall be loaded in tension to a load that is equal to the average clamping force times the design slip coefficient times 2, since there are two slip planes. The average slip deformation that occurs at this load shall be less than 0.015 in. for the three specimens. If the deformation is greater than this value, the coating is considered to have failed to meet the requirements for the particular *mean slip coefficient* used. The value of deformation for each specimen shall be reported.

Commentary:
See Commentary in Section A1.1

APPENDIX B. ALLOWABLE STRESS DESIGN (ASD) ALTERNATIVE

As an alternative to the load and resistance factor design provisions given in Sections 1 through 10, the following allowable stress design provisions are permitted. The provisions in Sections 1 through 10 in this Specification shall apply to ASD, except as follows:

B1.2 Loads, Load Factors and Load Combinations

The design and construction of the structure shall conform to an applicable allowable stress design specification for steel structures. When permitted in the applicable building code or specification, the allowable stresses in Section B5 are permitted to be increased to account for the effects of multiple transient loads in combination. When a load reduction factor is used to account for the effects of multiple transient loads in combination, the allowable stresses in Section B5 shall not be increased.

Commentary:
Although loads, load factors and load combinations are not explicitly specified in this Specification, the allowable stresses herein are based upon those specified in ASCE 7. When the design is governed by other load criteria, the allowable stresses specified herein shall be adjusted as appropriate.

SECTION B5. LIMIT STATES IN BOLTED JOINTS

The allowable shear strength and the allowable tensile strength of bolts shall be determined in accordance with Section B5.1. The interaction of combined shear and tension on bolts shall be limited in accordance with Section B5.2. The allowable bearing strength of the connected parts at bolt holes shall be determined in accordance with Section B5.3. Each of these allowable strengths shall be equal to or greater than the effect of the service loads. The axial load in bolts that are subject to tension or combined shear and tension shall be calculated with consideration of the externally applied tensile load and any additional tension resulting from *prying action* produced by deformation of the connected parts.

When slip resistance is required at the *faying surfaces* subject to shear or combined shear and tension, the slip resistance determined in accordance with Section B5.4 shall be equal to or greater than the effect of the service loads. In addition, the strength requirements in Sections B5.1, B5.2 and B5.3 shall also be met.

When bolts are subject to cyclic application of axial tension, the allowable stress determined in accordance with Section B5.5 shall be equal to or greater than the stress due to the effect of the service loads, including any additional tension resulting from *prying action* produced by deformation of the connected parts. In addition, the strength requirements in Sections B5.1, B5.2 and B5.3 shall also be met.

B5.1. Allowable Shear and Tensile Stresses

Shear and tensile strengths shall not be reduced by the installed bolt pretension. For *joints*, the allowable strength shall be based upon the allowable shear and tensile stresses of the individual bolts and shall be taken as the sum of the allowable strengths of the individual bolts.

The allowable shear strength or allowable tensile strength for an ASTM A325, A490 or F1852 bolt is R_a, where:

$$R_a = F_a A_b \qquad \text{(Equation B5.1)}$$

where

R_a = allowable shear strength per shear plane or allowable tensile strength of a bolt, kips;

F_a = allowable stress from Table B5.1 for the appropriate applied load conditions, ksi, adjusted for the presence of fillers or shims as required below; and,

A_b = cross-sectional area based upon the nominal diameter of bolt, in.^{2}

When a bolt that carries load passes through fillers or shims in a shear plane that are equal to or less than ¼ in. thick, F_a from Table B5.1 shall be used without reduction. When a bolt that carries load passes through fillers or shims that are greater than ¼ in. thick, one of the following requirements shall apply:

(1) For fillers or shims that are equal to or less than ¾ in. thick, F_a from Table B5.1 shall be multiplied by the factor $[1 - 0.4(t' - 0.25)]$, where t' is the total thickness of fillers or shims, in., up to ¾ in.;

(2) The fillers or shims shall be extended beyond the *joint* and the filler extension shall be secured with enough bolts to uniformly distribute the total force in the connected element over the combined cross-section of the connected element and the fillers or shims;

(3) The size of the *joint* shall be increased to accommodate a number of bolts that is equivalent to the total number required in (2) above; or

(4) The *joint* shall be designed as a *slip-critical joint*. The slip resistance of the *joint* shall not be reduced for the presence of fillers or shims.

B5.2. Combined Shear and Tension Stress

When combined shear and tension loads are transmitted by an ASTM A325, A490 or F1852 bolt, the bolt shall be proportioned so that the tensile stress F_t, ksi, on the cross-sectional area based upon the nominal diameter of bolt A_b produced by forces applied to the connected parts, shall not exceed the values computed from the equations in Table B5.2, where f_v, the shear stress produced by the same forces, shall not exceed the value for shear determined in accordance with the requirements in Section B5.1.

Table B5.1. Allowable Stress in Bolts

Applied Load Condition		Allowable Stress F_a, ksi	
		ASTM A325 or F1852 Bolt	ASTM A490 Bolt
Tension[a]	Static	44	54
	Fatigue	See Section B5.5	
Shear[a,b]	Threads included in shear plane	21	28
	Threads excluded from shear plane	30	40

[a] Except as required in Section B5.2.

[b] In shear *connections* that transmit axial force and have length between extreme bolts measured parallel to the line of force exceeds 50 in., tabulated values shall be reduced by 20 percent.

Table B5.2. Allowable Tensile Stress F_t for Bolts Subject to Combined Shear and Tension

Thread Condition	Allowable Tensile Stress F_t, ksi	
	ASTM A325 or F1852 Bolt	ASTM A490 Bolt
Threads included in Shear plane	$\sqrt{(44)^2 - 4.39 f_v^2}$	$\sqrt{(54)^2 - 3.71 f_v^2}$
Threads excluded From shear plane	$\sqrt{(44)^2 - 2.15 f_v^2}$	$\sqrt{(54)^2 - 1.82 f_v^2}$

B5.3. Allowable Bearing at Bolt Holes

For *joints*, the allowable bearing strength shall be taken as the sum of the strengths of the connected material at the individual bolt holes.

The allowable bearing strength of the connected material at a standard bolt hole, oversized bolt hole, short-slotted bolt hole independent of the direction of loading or long-slotted bolt hole with the slot parallel to the direction of the bearing load is R_a, where:

(1) when deformation of the bolt hole at service load is a design consideration;

$$R_a = 0.6 L_c t F_u \leq 1.2 d_b t F_u \qquad \text{(Equation B5.2)}$$

(2) when deformation of the bolt hole at service load is not a design consideration;

$$R_a = 0.75 L_c t F_u \leq 1.5 d_b t F_u \qquad \text{(Equation B5.3)}$$

The allowable bearing strength of the connected material at a long-slotted bolt hole with the slot perpendicular to the direction of the bearing load is R_a, where:

$$R_a = 0.5L_c tF_u \leq d_b tF_u \qquad \text{(Equation B5.4)}$$

In Equations B5.2, B5.3 and B5.4,

R_a = allowable bearing strength of the connected material, kips;
F_u = specified minimum tensile strength (per unit area) of the connected material, ksi;
L_c = clear distance, in the direction of load, between the edge of the hole and the edge of the adjacent hole or the edge of the material, in.;
d_b = nominal bolt diameter, in.; and,
t = thickness of the connected material, in.

B5.4. Allowable Slip Resistance

The allowable slip resistance is R_a, where:

$$R_a = H\mu DT_m N_b \left(1 - \frac{T}{DT_m N_b}\right) \qquad \text{(Equation B5.5)}$$

where

H = 1.0 for standard holes
= 0.85 for oversized and short-slotted holes
= 0.70 for long-slotted holes perpendicular to the direction of load
= 0.60 for long-slotted holes parallel to the direction of load;
μ = *mean slip coefficient* for Class A, B or C faying surfaces, as applicable, or as established by testing in accordance with Appendix A (see Section 3.2.2(b))

Table B5.3. Allowable Stress for Fatigue Loading

Number of Cycles	Maximum Bolt Stress for Design at Service Loads[a], ksi	
	ASTM A325 or F1852 Bolt	ASTM A490 Bolt
Not more than 20,000	44	54
From 20,000 to 500,000	40	49
More than 500,000	31	38

[a] Including the effects of *prying action*, if any, but excluding the pretension.

= 0.33 for Class A *faying surfaces* (uncoated clean mill scale steel sur faces or surfaces with Class A coatings on blast cleaned steel)

= 0.50 for Class B surfaces (uncoated blast-cleaned steel surfaces or surfaces with Class B coatings on blast-cleaned steel)

= 0.35 for Class C surfaces (roughened hot-dip galvanized surfaces);

D = 0.80, a slip probability factor that reflects the distribution of actual slip coefficient values about the mean, the ratio of measured bolt tensile strength to the specified minimum values, and a slip probability level; the use of other values of D shall be approved by the *Engineer of Record*;

T_m = specified minimum bolt pretension (for pretensioned joints as specified in Table 8.1), kips;

N_b = number of bolts in the joint; and,

T = applied service load in tension (tensile component of applied service load for combined shear and tension loading), kips

= zero if the joint is subject to shear only

B5.5. Tensile Fatigue

The tensile stress in the bolt that results from the cyclic application of externally applied service loads and the prying force, if any, but not the pretension, shall not exceed the stress in Table B5.3. The nominal diameter of the bolt shall be used in calculating the bolt stress. The connected parts shall be proportioned so that the calculated prying force does not exceed 30 percent of the externally applied load. *Joints* that are subject to tensile fatigue loading shall be pretensioned in accordance with Section 4.2 or slip-critical in accordance with Section 4.

REFERENCES

Allen, R.N. and J.W. Fisher, 1968, "Bolted Joints With Oversize or Slotted Holes," *Journal of the Structural Division*, Vol. 94, No. ST9, September, ASCE, Reston, VA.

American Institute of Steel Construction, 1999, *Load and Resistance Factor Design (LRFD) Specification for Structural Steel Buildings*, AISC, Chicago, IL.

Birkemoe, P.C. and D.C. Herrschaft, 1970, "Bolted Galvanized Bridges—Engineering Acceptance Near," *Civil Engineering*, April, ASCE, Reston, VA.

Carter, C.J., R.H.R. Tide and J.A. Yura, 1997, "A Summary of Changes and Derivation of LRFD Bolt Design Provisions," *Engineering Journal*, Vol. 34, No. 3, (3rd Qtr.), AISC, Chicago, IL.

Carter, C.J., 1996, "Specifying Bolt Length for High-Strength Bolts," *Engineering Journal*, Vol. 33, No. 2, (2nd Qtr.), AISC, Chicago, IL.

Chesson, Jr., E, N.L. Faustino and W.H. Munse, 1965, "High-Strength Bolts Subjected to Tension and Shear," *Journal of the Structural Division*, Vol. 91, No. ST5, October, ASCE, Reston, VA.

Fisher, J.W. and J.L. Rumpf, 1965, "Analysis of Bolted Butt Joints," *Journal of the Structural Division*, Vol. 91, No. ST5, October, ASCE, Reston, VA.

Frank, K.H. and J.A. Yura, 1981, "An Experimental Study of Bolted Shear Connections," FHWA/RD-81/148, December, Federal Highway Administration, Washington, D.C.

Kulak, G.L., J.W. Fisher and J.H.A. Struik, 1987, *Guide to Design Criteria for Bolted and Riveted Joints*, Second Edition, John Wiley & Sons, New York, NY.

Kulak, G.L. and P.C. Birkemoe, 1993, "Field Studies of Bolt Pretension," *Journal of Constructional Steel Research*, No. 25, pp. 95-106.

Kulak, G.L. and S.T. Undershute, 1998, "Tension Control Bolts: Strength and Installation," *Journal of Bridge Engineering*, Vol. 3 No. 1, February, ASCE, Reston, VA.

Manuel, T.J. and G.L. Kulak, 2000, "Strength of Joints that Combine Bolts and Welds," *Journal of Structural Engineering*, Vol. 126, No. 3, March, ASCE, Reston, VA.

McKinney, M. and F.J. Zwerneman, 1993, "The Effect of Burrs on the Slip Capacity in

Multiple Bolt Connections," *Final Report to the Research Council on Structural Connections*, August.

Munse, W. H., 1967, "Structural Behavior of Hot Galvanized Bolted Connections," *Proceedings of the 8th International Conference on Hot-dip Galvanizing*, June, London, England.

Polyzois, D. and K.H. Frank, 1986, "Effect of Overspray and Incomplete Masking of Faying Surfaces on the Slip Resistance of Bolted Connections," *Engineering Journal*, Vol. 23, No. 2, (2nd Qtr), AISC, Chicago, IL.

Polyzois, D. and J.A. Yura, 1985, "Effect of Burrs on Bolted Friction Connections," *Engineering Journal*, Vol.22, No. 3, (3rd Qtr), AISC, Chicago, IL.

Sherman, D.R. and J.A. Yura, 1998, "Bolted Double-Angle Compression Members," *Journal of Constructional Steel Research*, 46:1-3, Paper No. 197, Elsevier Science Ltd., Kidlington, Oxford, UK.

SSPC, 1993, Steel Structures Painting Manual, Vol. 1, Third Edition, SSPC: The Society for Protective Coatings, Pittsburgh, PA.

SSPC, 1991, *Steel Structures Painting Manua*l, Vol. 2, Sixth Edition, SSPC: The Society for Protective Coatings, Pittsburgh, PA.

Yura, J.A. and K.H. Frank, 1985, "Testing Method to Determine Slip Coefficient for Coatings Used in Bolted Joints," *Engineering Journal*, Vol. 22, No. 3, (3rd Qtr.), AISC, Chicago, IL.

Yura, J.A., K.H. Frank and L. Cayes, 1981, "Bolted Friction Connections with Weathering Steel," *Journal of the Structural Division*, Vol. 107, No. ST11, November, ASCE, Reston, VA.

INDEX

RESEARCH COUNCIL ON STRUCTURAL CONNECTIONS
www.boltcouncil.org

c/o American Institute of Steel Construction, Inc.
One East Wacker Drive, Suite 3100, Chicago, Illinois 60601-2001

Pub. No. S348 (30M1200)

Code of Standard Practice for Steel Buildings and Bridges

March 7, 2000

Supersedes the June 10, 1992 AISC *Code of Standard Practice for Steel Buildings and Bridges* and all previous versions.

Prepared by the American Institute of Steel Construction, Inc. under the direction of the AISC Committee on the Code of Standard Practice and approved by the AISC Board of Directors.

AMERICAN INSTITUTE OF STEEL CONSTRUCTION, INC.
One East Wacker Drive, Suite 3100, Chicago, Illinois 60601-2001

The information presented in this publication has been prepared in accordance with recognized engineering principles and is for general information only. While it is believed to be accurate, this information should not be used or relied upon for any specific application without competent professional examination and verification of its accuracy, suitability and applicability by a licensed engineer, architect or other professional. The publication of the material contained herein is not intended as a representation or warranty on the part of the American Institute of Steel Construction, Inc. or of any other person named herein, that this information is suitable for any general or particular use or of freedom from infringement of any patent or patents. Anyone making use of this information assumes all liability arising from such use.

Caution must be exercised when relying upon other specifications and codes developed by other bodies and incorporated by reference herein since such material may be modified or amended from time to time subsequent to the printing of this edition. The American Institute of Steel Construction, Inc. bears no responsibility for such material other than to refer to it and incorporate it by reference at the time of the initial publication of this edition.

Printed in the United States of America

PREFACE

As in any industry, trade practices have developed among those that are involved in the design, purchase, fabrication and erection of structural steel. This Code provides a useful framework for a common understanding of the acceptable standards when contracting for structural steel. As such, it is useful for owners, architects, engineers, general contractors, construction managers, fabricators, steel detailers, erectors and others that are associated with construction in structural steel. Unless specific provisions to the contrary are contained in the contract documents, the existing trade practices that are contained herein are considered to be the standard custom and usage of the industry and are thereby incorporated into the relationships between the parties to a contract.

The Symbols and Glossary are an integral part of this Code. In many sections of this Code, a non-mandatory Commentary has been prepared to provide background and further explanation for the corresponding Code provisions. The user is encouraged to consult it.

Since the first edition of this Code was published in 1924, AISC has continuously surveyed the structural steel design community and construction industry to determine standard trade practices. Since then, this Code has been periodically updated to reflect new and changing technology and industry practices.

This edition is the fifth complete revision of this Code since it was first published. It is the result of the deliberations of a fair and balanced Committee, the membership of which included six structural engineers, two architects, one general contractor, seven fabricators, one steel detailer, three erectors and one attorney. The following major changes have been made in this revision:

- Commentary information, when available, has been placed immediately following its corresponding Code provisions.
- The use of the term "Owner" throughout this Code has been generally (but not completely) eliminated, where appropriate. Instead, one or both of the terms "Owner's Designated Representative for Design" and "Owner's Designated Representative for Construction" has been used.
- Both U.S. customary units and metric units have been provided. See Section 1.3.

- Requirements for existing structures have been added to cover demolition and shoring, protection against damage, surveying or field dimensioning and hazardous materials.
- The classifications of materials in Section 2 have been editorially revised and expanded.
- Provisions for the resolution of discrepancies have been added in Section 3.3.
- Also in Section 3.3, the order of precedence of contract documents has been changed for simplicity and to reflect current practices.
- Provisions for fast-track project delivery have been added in Section 3.6.
- The responsibilities of the various entities involved in the shop and erection drawing approval process have been simplified and clarified in Section 4.
- Issues regarding the use of design drawings by the fabricator and/or the erector are now covered in Section 4.3.
- The permissible variation from theoretical curvature for a curved member is now covered in Section 6.4.2.
- Provisions have been added in Section 6.4.5 to cover permissible variations in camber for fabricated trusses.
- Section 6.5 has been editorially restructured and substantively modified to recognize that the majority of steel in building structures need not be primed or painted.
- Coverage of bearing devices has been revised: installation of bearing devices is now covered in Section 7.6 and grouting is covered in Section 7.7.
- Use of the terms self-supporting and non-self-supporting has been eliminated and replaced with the provisions for temporary support in Section 7.10.
- Provisions in Section 7.10.3 for the loads that must be considered during erection have been revised.
- The intent of the provisions that address the accumulation of mill tolerances and fabrication tolerances and their relationship to the erection tolerances has been clarified in Section 7.12.
- Quality-assurance provisions in Section 8 have been revised to recognize both the AISC Quality Certification program for fabricators and the AISC Erector Certification program.

- AESS requirements for welds have been clarified in Sections 10.2.5.
- AESS requirements for HSS weld seams have been added in Section 10.2.8.

In addition, many other changes have been made throughout this Code.

By the AISC Committee on the Code of Standard Practice,

Frank B. Wylie, III, Chairman
Barry L. Barger, Vice Chairman
Paul M. Brosnahan
James R. Burdette, Jr.
Richard B. Cook
William B. Cooper
William R. Davidson
Joseph A. Free, Jr.
Lawrence G. Griffis
D. Kirk Harman
James L. Larson
William F. McEleney
Leonard R. Middleton
James Mirgliotta
Donald G. Moore
Homer R. Peterson, II
David B. Ratterman
Rex D. Smith
James A. Stori
Thomas S. Tarpy, Jr.
Michael J. Tylk
Michael A. West
Charles J. Carter, Secretary

TABLE OF CONTENTS

GLOSSARY

The following terms are used in this Code. Where used, they are capitalized to alert the user that the term is defined in this Glossary.

AASHTO. American Association of State Highway and Transportation Officials.

Adjustable Items. See Section 7.13.1.3.

AESS. See Architecturally Exposed Structural Steel.

AISC. American Institute of Steel Construction, Inc.

Anchor Bolt. See Anchor Rod.

Anchor Rod. A mechanical device that is either cast or drilled and chemically adhered, grouted or wedged into concrete and/or masonry for the purpose of the subsequent attachment of Structural Steel.

Anchor-Rod Group. A set of Anchor Rods that receives a single fabricated Structural Steel shipping piece.

ANSI. American National Standards Institute.

Architect. The entity that is professionally qualified and duly licensed to perform architectural services.

Architecturally Exposed Structural Steel. See Section 10.

AREMA. American Railway Engineering and Maintenance of Way Association.

ASME. American Society of Mechanical Engineers.

ASTM. American Society for Testing and Materials.

that is responsible to the Owner for the overall structural design of the project, including the Structural Steel frame. This is usually the Structural Engineer of Record.

Plans. See Design Drawings.

RCSC. Research Council on Structural Connections.

Released for Construction. The term that describes the status of Contract Documents that are in such a condition that the Fabricator and the Erector can rely upon them for the performance of their work, including the ordering of material and the preparation of Shop and Erection Drawings.

SER. See Structural Engineer of Record.

Shop Drawings. Drawings of the individual Structural Steel shipping pieces that are to be produced in the fabrication shop.

SJI. Steel Joist Institute.

Specifications. The portion of the Contract Documents that consists of the written requirements for materials, standards and workmanship.

SSPC. SSPC: The Society for Protective Coatings, which was formerly known as the Steel Structures Painting Council.

Standard Structural Shapes. Hot-rolled W-, S-, M- and HP-shapes, channels and angles listed in ASTM A6/A6M; structural tees split from the hot-rolled W-, S- and M- shapes listed in ASTM A6/A6M; hollow structural sections produced to ASTM A500, A501, A618 or A847; and, steel pipe produced to ASTM A53/A53M.

Steel Detailer. The entity that produces the Shop and Erection Drawings.

Structural Engineer of Record. The licensed professional who is responsible for sealing the Contract Documents, which indicates that he or she has performed or supervised the analysis, design and document preparation for the structure and has knowledge of the load-carrying structural system.

Structural Steel. The elements of the structural frame as given in Section 2.1.

Tier. The Structural Steel framing defined by a column shipping piece.

Weld Show-Through. In Architecturally Exposed Structural Steel, visual indication of the presence of a weld or welds on the side of the member opposite the weld.

NOTES

CODE OF STANDARD PRACTICE FOR STEEL BUILDINGS AND BRIDGES

March 7, 2000

SECTION 1. GENERAL PROVISIONS

1.1. Scope

In the absence of specific instructions to the contrary in the Contract Documents, the trade practices that are defined in this Code shall govern the fabrication and erection of Structural Steel.

Commentary:
The practices defined in this Code are the commonly accepted standards of custom and usage for Structural Steel fabrication and erection, which generally represent the most efficient approach. This Code is not applicable to steel joists or metal building systems, which are addressed by SJI and MBMA, respectively.

1.2. Referenced Specifications, Codes and Standards

The following documents are referenced in this Code:

AASHTO Specification—The 1998 AASHTO *LRFD Bridge Design Specifications*, 2nd Edition, with interims up to and including 1999, or the 1996 AASHTO *Standard Specifications for Highway Bridges*, 16th Edition with interims up to and including 1999.

AISC Manual of Steel Construction—The AISC *Manual of Steel Construction, Volumes I and II, 2nd Edition LRFD or 9th Edition ASD.*

AISC Seismic Provisions—The AISC *Seismic Provisions for Structural Steel Buildings*, April 15, 1997 with *Seismic Provisions for Structural Steel Buildings (1997) Supplement No. 1*, February 15, 1999.

AISC Specification—The AISC *Specification for Structural Steel Buildings, 1999 LRFD or 1989 ASD*, as adopted by the American Institute of Steel Construction, Inc.

ANSI/ASME B46.1—ANSI/ASME B46.1-95, Surface Texture (Surface Roughness, Waviness and Lay).

AREMA Specification—The 1999 AREMA *Manual for Railway Engineering, Volume II—Structures, Chapter 15.*

ASTM A6/A6M—98, *Specification for General Requirements for Rolled Structural Steel Bars, Plates, Shapes, and Sheet Piling.*

ASTM A53/A53M—99b, *Specification for Pipe, Steel, Black and Hot-Dipped, Zinc-Coated, Welded and Seamless.*

ASTM A325—97, *Specification for Structural Bolts, Steel, Heat Treated, 120/105 ksi Minimum Tensile Strength.*

ASTM A325M—97, *Specification for High-Strength Bolts for Structural Steel Joints (Metric).*

ASTM A490—97, *Specification for Heat-Treated Steel Structural Bolts, 150 ksi Minimum Tensile Strength.*

ASTMA490M—93, *Specification for High-Strength Steel Bolts, Classes 10.9 and 10.9.3, for Structural Steel Joints (Metric).*

ASTM A500—99, *Specification for Cold-Formed Welded and Seamless Carbon Steel Structural Tubing in Rounds and Shapes*. No metric equivalent exists.

ASTM A501—99, *Specification for Hot-Formed Welded and Seamless Carbon Steel Structural Tubing*. No metric equivalent exists.

ASTM A618—99, *Specification for Hot-Formed Welded and Seamless High-Strength Low-Alloy Structural Tubing*. No metric equivalent exists.

ASTM A847—99a, *Specification for Cold-Formed Welded and Seamless High-Strength, Low-Alloy Structural Tubing with Improved Atmospheric Corrosion Resistance*. No metric equivalent exists.

ASTM F1852/F1852M—98, *Specification for "Twist-Off" Type Tension Control Structural Bolt/Nut/Washer Assemblies, Steel, Heat Treated, 120/105 ksi Minimum Tensile Strength.*

AWS D1.1—The AWS D1.1 *Structural Welding Code—Steel*, 1998.

CASE Document 11—*An Agreement Between Structural Engineer of Record and Contractor for Transfer of Computer Aided Drafting (CAD) files on Electronic Media*, 1996

CASE Document 962—The *National Practice Guidelines for the Structural Engineer of Record*, Third Edition, 1997.

RCSC Specification—The *Specification for Structural Joints Using ASTM A325 or A490 Bolts, 1994 LRFD or 1994 ASD.*

SSPC SP2—SSPC *Surface Preparation Specification No. 2, Hand Tool Cleaning*, July 5, 1995.
SSPC SP6—SSPC *Surface Preparation Specification No. 6, Commercial Blast Cleaning*, September 15, 1994.

1.3. Units

In this Code, the values stated in either U.S. customary units or metric units shall be used. Each system shall be used independently of the other.

Commentary:
In this Code, dimensions, weights and other measures are given in U.S. customary units with rounded or rationalized metric-unit equivalents in brackets. Because the values stated in each system are not exact equivalents, the selective combination of values from each of the two systems is not permitted.

1.4. Design Criteria

For buildings, in the absence of other design criteria, the provisions in the AISC Specification shall govern the design of the Structural Steel. For bridges, in the absence of other design criteria, the provisions in the AASHTO Specification and AREMA Specification shall govern the design of the Structural Steel, as applicable.

1.5. Responsibility for Design

1.5.1. When the Owner's Designated Representative for Design provides the design, Design Drawings and Specifications, the Fabricator and the Erector are not responsible for the suitability, adequacy or building-code conformance of the design.

1.5.2. When the Owner enters into a direct contract with the Fabricator to both design and fabricate an entire, completed steel structure, the Fabricator shall be responsible for the suitability, adequacy and building-code conformance of the Structural Steel design. The Owner shall be responsible for the suitability, adequacy and building-code conformance of the non-Structural Steel arrangement and the performance criteria for the Structural Steel frame.

1.6. Patents and Copyrights

The entity or entities that are responsible for the specification and/or selection of proprietary structural designs shall secure all intellectual property rights necessary for the use of those designs.

1.7. Existing Structures

1.7.1. Demolition and shoring of any part of an existing structure are not within the scope of work that is provided by either the Fabricator or the Erector. Such demolition and shoring shall be performed in a timely manner so as not to interfere with or delay the work of the Fabricator and the Erector.

1.7.2. Protection of an existing structure and its contents and equipment, so as to prevent damage from normal erection processes, is not within the scope of work that is provided by either the Fabricator or the Erector. Such protection shall be performed in a timely manner so as not to interfere with or delay the work of the Fabricator or the Erector.

1.7.3. Surveying or field dimensioning of an existing structure is not within the scope of work that is provided by either the Fabricator or the Erector. Such surveying or field dimensioning, which is necessary for the completion of Shop and Erection Drawings and fabrication, shall be performed and furnished to the Fabricator in a timely manner so as not to interfere with or delay the work of the Fabricator or the Erector.

1.7.4. Abatement or removal of Hazardous Materials is not within the scope of work that is provided by either the Fabricator or the Erector. Such abatement or removal shall be performed in a timely manner so as not to interfere with or delay the work of the Fabricator and the Erector.

1.8. Means, Methods and Safety of Erection

1.8.1. The Erector shall be responsible for the means, methods and safety of erection of the Structural Steel frame.

1.8.2. The Structural Engineer of Record shall be responsible for the structural adequacy of the structure in the completed project. The Structural Engineer of Record shall not be responsible for the means, methods and safety of erection of the Structural Steel frame. See also Sections 3.1.4 and 7.10.

SECTION 2. CLASSIFICATION OF MATERIALS

2.1. Definition of Structural Steel

Structural Steel shall consist of the elements of the structural frame that are shown and sized in the structural Design Drawings, essential to support the design loads and described as:

Anchor Rods that will receive Structural Steel.
Base plates.
Beams, including built-up beams, if made from Standard Structural Shapes and/or plates.
Bearing plates.
Bearings of steel for girders, trusses or bridges.
Bracing, if permanent.
Canopy framing, if made from Standard Structural Shapes and/or plates.
Columns, including built-up columns, if made from Standard Structural Shapes and/or plates.
Connection materials for framing Structural Steel to Structural Steel.
Crane stops, if made from Standard Structural Shapes and/or plates.
Door frames, if made from Standard Structural Shapes and/or plates and if part of the Structural Steel frame.
Edge angles and plates, if attached to the Structural Steel frame or steel (open-web) joists.
Embedded Structural Steel parts, other than bearing plates, that will receive Structural Steel.
Expansion joints, if attached to the Structural Steel frame.
Fasteners for connecting Structural Steel items: permanent shop bolts, nuts and washers; shop bolts, nuts and washers for shipment; field bolts, nuts and washers for permanent Connections; and, permanent pins.
Floor-opening frames, if made from Standard Structural Shapes and/or plates and attached to the Structural Steel frame or steel (open-web) joists.
Floor plates (checkered or plain), if attached to the Structural Steel frame.

Girders, including built-up girders, if made from Standard Structural Shapes and/or plates.
Girts, if made from Standard Structural Shapes.
Grillage beams and girders.
Hangers, if made from Standard Structural Shapes, plates and/or rods and framing Structural Steel to Structural Steel.
Leveling nuts and washers.
Leveling plates.
Leveling screws.
Lintels, if attached to the Structural Steel frame.
Marquee framing, if made from Standard Structural Shapes and/or plates.
Machinery supports, if made from Standard Structural Shapes and/or plates and attached to the Structural Steel frame.
Monorail elements, if made from Standard Structural Shapes and/or plates and attached to the Structural Steel frame.
Posts, if part of the Structural Steel frame.
Purlins, if made from Standard Structural Shapes.
Relieving angles, if attached to the Structural Steel frame.
Roof-opening frames, if made from Standard Structural Shapes and/or plates and attached to the Structural Steel frame or steel (open-web) joists.
Roof-screen support frames, if made from Standard Structural Shapes.
Sag rods, if part of the Structural Steel frame and connecting Structural Steel to Structural Steel.
Shear stud connectors, if specified to be shop attached.
Shims, if permanent.
Struts, if permanent and part of the Structural Steel frame.
Tie rods, if part of the Structural Steel frame.
Trusses, if made from Standard Structural Shapes and/or built-up members.
Wall-opening frames, if made from Standard Structural Shapes and/or plates and attached to the Structural Steel frame.
Wedges, if permanent.

Commentary:
The Fabricator normally fabricates the items listed in Section 2.1.

Such items must be shown, sized and described in the structural Design Drawings. Bracing includes vertical bracing for resistance to wind and seismic load and structural stability, horizontal bracing for floor and roof systems and permanent stability bracing for components of the Structural Steel frame.

2.2. Other Steel, Iron or Metal Items

Structural Steel shall not include other steel, iron or metal items that are not generally described in Section 2.1, even where such items are shown in the structural Design Drawings or are attached to the Structural Steel frame. Other steel, iron or metal items include but are not limited to:

Bearings, if non-steel.
Cables for permanent bracing or suspension systems.
Castings.
Catwalks.
Chutes.
Cold-formed steel products.
Cold-rolled steel products, except those that are specifically covered in the AISC Specification.
Corner guards.
Crane rails, splices, bolts and clamps.
Crane stops, if not made from Standard Structural Shapes or plates.
Door guards.
Embedded steel parts, other than bearing plates, that do not receive Structural Steel or that are embedded in precast concrete.
Expansion joints, if not attached to the Structural Steel frame.
Flagpole support steel.
Floor plates (checkered or plain), if not attached to the Structural Steel frame.
Forgings.
Gage-metal products.
Grating.
Handrail.
Hangers, if not made from Standard Structural Shapes, plates

and/or rods or not framing Structural Steel to Structural Steel.

Hoppers.

Items that are required for the assembly or erection of materials that are furnished by trades other than the Fabricator or Erector.

Ladders.

Lintels, if not attached to the Structural Steel frame.

Masonry anchors.

Miscellaneous metal.

Ornamental metal framing.

Pressure vessels.

Reinforcing steel for concrete or masonry.

Relieving angles, if not attached to the Structural Steel frame.

Roof screen support frames, if not made from Standard Structural Shapes.

Safety cages.

Shear stud connectors, if specified to be field installed.

Stacks.

Stairs.

Steel deck.

Steel (open-web) joists.

Steel joist girders.

Tanks.

Toe plates.

Trench or pit covers.

Commentary:
Section 2.2 includes many items that may be furnished by the Fabricator if contracted to do so by specific notation and detail in the Contract Documents. When such items are contracted to be provided by the Fabricator, coordination will normally be required between the Fabricator and other material suppliers and trades. The provisions in this Code are not intended to apply to items in Section 2.2.

In previous editions of this Code, provisions regarding who should normally furnish field-installed shear stud connectors and cold-formed steel deck support angles were included in Section 7.8. These provisions have been eliminated since field-installed shear stud connectors and steel deck support angles are not defined as Structural Steel in this Code.

SECTION 3. DESIGN DRAWINGS AND SPECIFICATIONS

3.1. Structural Design Drawings and Specifications

Unless otherwise indicated in the Contract Documents, the structural Design Drawings shall be based upon consideration of the design loads and forces to be resisted by the Structural Steel frame in the completed project.

The structural Design Drawings shall clearly show the work that is to be performed and shall give the following information with sufficient dimensions to accurately convey the quantity and nature of the Structural Steel to be fabricated:

(a) The size, section, material grade and location of all members;
(b) All geometry and working points necessary for layout;
(c) Floor elevations;
(d) Column centers and offsets;
(e) The camber requirements for members; and,
(f) The information that is required in Sections 3.1.1 through 3.1.6.

The Structural Steel Specification shall include any special requirements for the fabrication and erection of the Structural Steel. The structural Design Drawings, Specifications and addenda shall be numbered and dated for the purposes of identification.

Commentary:
Contract Documents vary greatly in complexity and completeness. Nonetheless, the Fabricator and the Erector must be able to rely upon the accuracy and completeness of the Contract Documents. This allows the Fabricator and the Erector to provide the Owner with bids that are adequate and complete. It also enables the preparation of the Shop and Erection Drawings, the ordering of materials and the timely fabrication and erection of shipping pieces.

In some cases, the Owner can benefit when reasonable latitude is allowed in the Contract Documents for alternatives that can reduce cost without compromising quality. However, critical requirements that are necessary to protect the Owner's interest, that affect the integrity of the structure or that are necessary for the Fabricator and the Erector to proceed with their work must be included in the Contract Documents. Some examples of critical information include:

Standard specifications and codes that govern Structural Steel design and construction, including bolting and welding.
Material specifications.
Special material requirements to be reported on the certified mill test reports.
Welded-joint configuration.
Weld-procedure qualification.
Special requirements for work of other trades.
Final disposition of backing bars and runoff tabs.
Lateral bracing.
Stability bracing.
Connections or data for Connection selection and/or completion.
Restrictions on Connection types.
Column stiffeners (also known as continuity plates).
Column web doubler plates.
Bearing stiffeners on beams and girders.
Web reinforcement.
Openings for other trades.
Surface preparation and shop painting requirements.
Shop and field inspection requirements.
Non-destructive testing requirements, including acceptance criteria.
Special requirements on delivery.
Special erection limitations.
Identification of non-Structural Steel elements that interact with the Structural Steel frame to provide for the lateral stability of the Structural Steel frame (see Section 3.1.4).
Column differential shortening information.
Special fabrication and erection tolerances for AESS.
Special pay-weight provisions.

3.1.1. Permanent bracing, column stiffeners, column web doubler plates, bearing stiffeners in beams and girders, web reinforcement, openings for other trades and other special details, where required, shall be shown in sufficient detail in the structural Design Drawings so that the quantity, detailing and fabrication requirements for these items can be readily understood.

3.1.2. The Owner's Designated Representative for Design shall either show the complete design of the Connections in the structural Design Drawings or allow the Fabricator to select or complete the Connection details while preparing the Shop and Erection Drawings. When the Fabricator is allowed to select or complete the Connection details, the following information shall be provided in the structural Design Drawings:

(a) Any restrictions on the types of Connections that are permitted;
(b) Data concerning the loads, including shears, moments, axial forces and transfer forces, that are to be resisted by the individual members and their Connections, sufficient to allow the Fabricator to select or complete the Connection details while preparing the Shop and Erection Drawings;
(c) Whether the data required in (b) is given at the service-load level or the factored-load level; and,
(d) Whether LRFD or ASD is to be used in the selection or completion of Connection details.

When the Fabricator selects or completes the Connection details, the Fabricator shall utilize the requirements in the AISC Specification and the Contract Documents and submit the Connection details to the Owner's Designated Representative for Design for approval.

Commentary:
When the Owner's Designated Representative for Design shows the complete design of the Connections in the structural Design Drawings, the following information is included:

(a) All weld sizes and lengths;
(b) All bolt sizes, locations, quantities and grades;

(c) All plate and angle sizes, thicknesses and dimensions; and,
(d) All work point locations and related information.

The intent of this approach is that complete information necessary for Connection detailing, fabrication and erection is shown in the structural Design Drawings. The Steel Detailer will then be able to transfer this information to the Shop and Erection Drawings, applying it to the individual pieces being detailed.

When the Owner's Designated Representative for Design allows the Fabricator to select or complete the Connections, this is commonly done by referring to tables in the Contract Documents or in the AISC Manual of Steel Construction, or by schematically showing the types of Connections required in the structural Design Drawings. The Steel Detailer will then configure the Connections based upon the design loads and other information given in the structural Design Drawings. If the desired Connection is not covered in those tables, a detail of the "special" Connection should be contained in the structural Design Drawings. This detail should provide such information as weld sizes, plate thicknesses and quantities of bolts. However, there may be some geometry and dimensional information that the Steel Detailer must develop. The intent of this method is that the Steel Detailer will select the Connection materials and configuration from the referenced tables or complete the specific Connection configuration (i.e. dimensions, edge distances and bolt spacing) based upon the Connection details that are shown in the structural Design Drawings.

This method will require the skill of an experienced Steel Detailer, who is familiar with the AISC requirements for Connection configurations, capable and experienced in the use of the Connection tables in the AISC Manual of Steel Construction and capable of calculating dimensions and adapting a typical Connection detail to similar situations. Notations of loadings in the structural Design Drawings are only to facilitate selection of the Connections from the referenced tables. It is not the intent of this method that the Steel Detailer practice engineering.

If there are any restrictions as to the types of Connections to be used, particularly as it relates to simple shear Connections, it is required that these limitations be set forth in the structural Design

Drawings and Specifications. There are a variety of Connections available in the AISC Manual of Steel Construction for a given situation. Preference for a particular type will vary between Fabricators and Erectors. Stating these limitations, if any, in the structural Design Drawings and Specifications will help to avoid repeated changes to the Shop and Erection Drawings due to the selection of a Connection that is not acceptable to the Owner's Designated Representative for Design, thereby avoiding additional cost and/or delay for the redrawing of the Shop and Erection Drawings.

The structural Design Drawings must indicate the method of design used as LRFD or ASD. In order to conform to the spirit of the AISC Specification, the Connections must be selected using the same method and the corresponding references.

3.1.3. When leveling plates are to be furnished as part of the contract requirements, their locations and required thickness and sizes shall be specified in the Contract Documents.

3.1.4. When the Structural Steel frame, in the completely erected and fully connected state, requires interaction with non-Structural Steel elements (see Section 2) for strength and/or stability, those non-Structural Steel elements shall be identified in the Contract Documents as required in Section 7.10.

Commentary:
Examples of non-Structural Steel elements include diaphragms made of steel deck, diaphragms made of concrete on steel deck and masonry and/or concrete shear walls.

3.1.5. When camber is required, the magnitude, direction and location of camber shall be specified in the structural Design Drawings.

Commentary:
For cantilevers, the specified camber may be up or down, depending upon the framing and loading.

3.1.6. Specific members or portions thereof that are to be left unpainted shall be identified in the Contract Documents. When shop painting is required, the painting requirements shall be specified in the Contract Documents, including the following information:

(a) The identification of specific members or portions thereof to be painted;
(b) The surface preparation that is required for these members;
(c) The paint specifications and manufacturer's product identification that are required for these members; and,
(d) The minimum dry-film shop-coat thickness that is required for these members.

Commentary:
Some members or portions thereof may be required to be left unpainted, such as those that will be in contact and acting compositely with concrete, or those that will receive spray-applied fire protection materials.

3.2. Architectural, Electrical and Mechanical Design Drawings and Specifications

All requirements for the quantities, sizes and locations of Structural Steel shall be shown or noted in the structural Design Drawings. The use of architectural, electrical and/or mechanical Design Drawings as a supplement to the structural Design Drawings is permitted for the purposes of defining detail configurations and construction information.

3.3. Discrepancies

When a discrepancy is discovered in the Contract Documents in the course of the Fabricator's work, the Fabricator shall promptly notify the Owner's Designated Representative for Construction so that the discrepancy can be resolved by the Owner's Designated Representative for Design. Such resolution shall be timely so as not to delay the Fabricator's work.

When discrepancies exist between the Design Drawings and Specifications, the Design Drawings shall govern. When discrepancies exist between scale dimensions in the Design Drawings and the figures written in them, the figures shall govern. When discrepancies exist between the structural Design Drawings and the architectural, electrical or mechanical Design Drawings or Design Drawings for other trades, the structural Design Drawings shall govern.

Commentary:
While it is the Fabricator's responsibility to report any discrepancies that are discovered in the Contract Documents, it is not the Fabricator's responsibility to discover discrepancies, including those that are associated with the coordination of the various design disciplines. The quality of the Contract Documents is the responsibility of the entities that produce those documents.

3.4. Legibility of Design Drawings

Design Drawings shall be clearly legible and drawn to a scale that is not less than 1/8 in. to the foot [10 mm per 1 000 mm]. More complex information shall be drawn to a scale that is adequate to clearly convey the information.

3.5. Revisions to the Design Drawings and Specifications

Revisions to the Design Drawings and Specifications shall be made either by issuing new Design Drawings and Specifications or by reissuing the existing Design Drawings and Specifications. In either case, all revisions, including revisions that are communicated through the annotation of Shop and/or Erection Drawings (see Section 4.4.2), shall be clearly and individually indicated in the Contract Documents. The Contract Documents shall be dated and identified by revision number. Each Design Drawing shall be identified by the same drawing number throughout the duration of the project, regardless of the revision. See also Section 9.3.

Commentary:
Revisions to the Design Drawings and Specifications can be made by issuing sketches and supplemental information separate from the Design Drawings and Specifications. These sketches and supplemental information become amendments to the Design Drawings and Specifications and are considered new Contract Documents. All sketches and supplemental information must be

uniquely identified with a number and date as the latest instructions until such time as they may be superseded by new information.

When revisions are made by revising and re-issuing the existing structural Design Drawings and/or Specifications, a unique revision number and date must be added to those documents to identify that information as the latest instructions until such time as they may be superseded by new information. The same unique drawing number must identify each Design Drawing throughout the duration of the project so that revisions can be properly tracked, thus avoiding confusion and miscommunication among the various entities involved in the project.

When revisions are communicated through the annotation of Shop or Erection Drawings or contractor submissions, such changes must be confirmed in writing by one of the aforementioned methods. This written confirmation is imperative to maintain control of the cost and schedule of a project and to avoid potential errors in fabrication.

3.6. Fast-Track Project Delivery

When the fast-track project delivery system is selected, release of the structural Design Drawings and Specifications shall constitute a Release for Construction, regardless of the status of the architectural, electrical, mechanical and other interfacing designs and Contract Documents. Subsequent revisions, if any, shall be the responsibility of the Owner and shall be made in accordance with Sections 3.5 and 9.3.

Commentary:
The fast-track project delivery system generally provides for a condensed schedule for the design and construction of a project. Under this delivery system, the Owner elects to Release for Construction the structural Design Drawings and Specifications, which may be partially complete, at a time that may precede the completion of and coordination with architectural, mechanical, electrical and other design work and Contract Documents. The release of these structural Design Drawings and Specifications may also precede the release of the General Conditions and Division 1 Specifications.

Release of the structural Design Drawings and

Specifications to the Fabricator for ordering of material constitutes a Release for Construction. Accordingly, the Fabricator and the Erector may begin their work based upon those partially complete documents. As the architectural, mechanical, electrical and other design elements of the project are completed, revisions may be required in design and/or construction. Thus, when considering the fast-track project delivery system, the Owner should balance the potential benefits to the project schedule with the project cost contingency that may be required to allow for these subsequent revisions.

SECTION 4. SHOP AND ERECTION DRAWINGS

4.1. Owner Responsibility

The Owner shall furnish, in a timely manner and in accordance with the Contract Documents, complete structural Design Drawings and Specifications that have been Released for Construction. Unless otherwise noted, Design Drawings that are provided as part of a contract bid package shall constitute authorization by the Owner that the Design Drawings are Released for Construction

Commentary:
When the Owner issues Released-for-Construction Design Drawings and Specifications, the Fabricator and the Erector rely on the fact that these are the Owner's requirements for the project. This release is required by the Fabricator prior to the ordering of material and the preparation and completion of Shop and Erection Drawings.

To ensure the orderly flow of material procurement, detailing, fabrication and erection activities, on phased construction projects, it is essential that designs are not continuously revised after they have been Released for Construction. In essence, once a portion of a design is Released for Construction, the essential elements of that design should be "frozen" to ensure adherence to the contract price and construction schedule. Alternatively, all parties should reach a common understanding of the effects of future changes, if any, as they affect scheduled deliveries and added costs.

4.2. Fabricator Responsibility

Except as provided in Section 4.5, the Fabricator shall produce Shop and Erection Drawings for the fabrication and erection of the Structural Steel and is responsible for the following:

(a) The transfer of information from the Contract Documents into accurate and complete Shop and Erection Drawings; and,
(b) The development of accurate, detailed dimensional information to provide for the fit-up of parts in the field.

When the Fabricator submits a request to change Connection details that are described in the Contract Documents, the Fabricator shall notify the Owner's Designated Representatives for Design and Construction in writing in advance of the submission of the Shop and Erection Drawings. The Owner's Designated Representative for Design shall review and approve or reject the request in a timely manner.

When requested to do so by the Owner's Designated Representative for Design, the Fabricator shall advise the Owner's Designated Representatives for Design and Construction of its schedule for the submittal of Shop and Erection Drawings so as to facilitate the timely flow of information between all parties.

Commentary:
As the Fabricator develops the detailed dimensional information for production of the Shop and Erection Drawings, there may be discrepancies, missing information or conflicts discovered in the Contract Documents. See Section 3.3.

When the Fabricator intends to make a submission of alternative Connection details to those shown in the Contract Documents, the Fabricator must notify the Owner's Designated Representatives for Design and Construction in advance. This will allow the parties involved to plan for the increased effort that may be required to review the alternative Connection details. In addition, the Owner will be able to evaluate the potential for cost savings and/or schedule improvements against the additional design cost for review of the alternative Connection details by the Owner's Designated Representative for Design. This evaluation by the Owner may result in the rejection of the alternative Connection details or acceptance of the submission for review based upon cost savings, schedule improvements and/or job efficiencies.

When the Fabricator provides a schedule for the submission of the Shop and Erection Drawings, it must be recognized that this schedule may be affected by revisions and the response time to requests for missing information or the resolution of discrepancies.

4.3. Use of CAD Files and/or Copies of Design Drawings

The Fabricator shall neither use nor reproduce any part of the

Design Drawings as part of the Shop or Erection Drawings without the written permission of the Owner's Designated Representative for Design. When CAD files or copies of the Design Drawings are made available for the Fabricator's use, the Fabricator shall accept this information under the following conditions:

(a) All information contained in the CAD files or copies of the Design Drawings shall be considered instruments of service of the Owner's Designated Representative for Design and shall not be used for other projects, additions to the project or the completion of the project by others. CAD files and copies of the Design Drawings shall remain the property of the Owner's Designated Representative for Design and in no case shall the transfer of these CAD files or copies of the Design Drawings be considered a sale.
(b) The CAD files or copies of the Design Drawings shall not be considered to be Contract Documents. In the event of a conflict between the Design Drawings and the CAD files or copies thereof, the Design Drawings shall govern;
(c) The use of CAD files or copies of the Design Drawings shall not in any way obviate the Fabricator's responsibility for proper checking and coordination of dimensions, details, member sizes and fit-up and quantities of materials as required to facilitate the preparation of Shop and Erection Drawings that are complete and accurate as required in Section 4.2; and,
(d) The Fabricator shall remove information that is not required for the fabrication or erection of the Structural Steel from the CAD files or copies of the Design Drawings.

Commentary:
With the advent of electronic media and the internet, electronic copies of Design Drawings are becoming readily available to the Fabricator. As a result, the Owner's Designated Representative for Design may have reduced control over the unauthorized use of the Design Drawings. There are many copyright and other legal issues to be considered.

The Owner's Designated Representative for Design may

choose to make CAD files or copies of the Design Drawings available to the Fabricator, and may charge a service or licensing fee for this convenience. In doing so, a carefully negotiated agreement should be established to set out the specific responsibilities of both parties in view of the liabilities involved for both parties. For a sample contract, see CASE Document 11.

The CAD files and/or copies of the Design Drawings are provided to the Fabricator for convenience only. The information therein should be adapted for use only in reference to the placement of Structural Steel members during erection. The Fabricator should treat this information as if it were fully produced by the Fabricator and undertake the same level of checking and quality assurance. When amendments or revisions are made to the Contract Documents, the Fabricator must update this reference material.

When CAD files or copies of the Design Drawings are provided to the Fabricator, they often contain other information, such as architectural backgrounds or references to other Contract Documents. This additional material should be removed when producing Shop and Erection Drawings to avoid the potential for confusion.

4.4. Approval

Except as provided in Section 4.5, the Shop and Erection Drawings shall be submitted to the Owner's Designated Representatives for Design and Construction for review and approval. These drawings shall be returned to the Fabricator within 14 calendar days. Approved Shop and Erection Drawings shall be individually annotated by the Owner's Designated Representatives for Design and Construction as either approved or approved subject to corrections noted. When so required, the Fabricator shall subsequently make the corrections noted and furnish corrected Shop and Erection Drawings to the Owner's Designated Representatives for Design and Construction.

Commentary:
As used in this Code, the 14-day allotment for the return of Shop and Erection Drawings is intended to represent the Fabricator's portal-to-portal time. The intent in this Code is that, in the absence

of information to the contrary in the Contract Documents, 14 days may be assumed for the purposes of bidding, contracting and scheduling. A submittal schedule is commonly used to facilitate the approval process.

4.4.1. Approval of the Shop and Erection Drawings, approval subject to corrections noted and similar approvals shall constitute the following:

(a) Confirmation that the Fabricator has correctly interpreted the Contract Documents in the preparation of those submittals;
(b) Confirmation that the Owner's Designated Representative for Design has reviewed and approved the Connection details shown on the Shop and Erection Drawings and submitted in accordance with Section 3.1.2, if applicable; and,
(c) Release by the Owner's Designated Representatives for Design and Construction for the Fabricator to begin fabrication using the approved submittals.

Such approval shall not relieve the Fabricator of the responsibility for either the accuracy of the detailed dimensions in the Shop and Erection Drawings or the general fit-up of parts that are to be assembled in the field.

The Fabricator shall determine the fabrication schedule that is necessary to meet the requirements of the contract.

Commentary:
When considering the current language in this Section, the Committee sought language that would parallel the practices of CASE. In CASE Document 962, CASE indicates that when the design of some element of the primary structural system is left to someone other than the Structural Engineer of Record, "…such elements, including connections designed by others, should be reviewed by the Structural Engineer of Record. He [or she] should review such designs and details, accept or reject them and be responsible for their effects on the primary structural system." Historically, this Code has embraced this same concept.

From the inception of this Code, AISC and the industry in

general have recognized that only the Owner's Designated Representative for Design has all the information necessary to evaluate the total impact of Connection details on the overall structural design of the project. This authority has traditionally been exercised during the approval process for Shop and Erection Drawings. The Owner's Designated Representative for Design has thus retained responsibility for the adequacy and safety of the entire structure since at least the 1927 edition of this Code.

4.4.2. Unless otherwise noted, any additions, deletions or revisions that are indicated on the approved Shop and Erection Drawings shall constitute authorization by the Owner that the additions, deletions or revisions are Released for Construction. The Fabricator and the Erector shall promptly notify the Owner's Designated Representative for Construction when any direction or notation on the Shop or Erection Drawings or other information will result in an additional cost and/or a delay. See Sections 3.5 and 9.3.

Commentary:
When the Fabricator notifies the Owner's Designated Representative for Construction that a direction or notation on the Shop or Erection Drawings will result in an additional cost or a delay, it is then normally the responsibility of the Owner's Designated Representative for Construction to subsequently notify the Owner's Designated Representative for Design.

4.5. Shop and/or Erection Drawings Not Furnished by the Fabricator

When the Shop and Erection Drawings are not prepared by the Fabricator, but are furnished by others, they shall be delivered to the Fabricator in a timely manner. These Shop and Erection Drawings shall be prepared, insofar as is practical, in accordance with the shop fabrication and detailing standards of the Fabricator. The Fabricator shall neither be responsible for the completeness or accuracy of Shop and Erection Drawings so furnished, nor for the general fit-up of the members that are fabricated from them.

SECTION 5. MATERIALS

5.1. Mill Materials

Unless otherwise noted in the Contract Documents, the Fabricator is permitted to order the materials that are necessary for fabrication when the Fabricator receives Contract Documents that have been Released for Construction.

Commentary:
The Fabricator may purchase materials in stock lengths, exact lengths or multiples of exact lengths to suit the dimensions shown in the structural Design Drawings. Such purchases will normally be job-specific in nature and may not suitable for use on other projects or returned for full credit if subsequent design changes make these materials unsuitable for their originally intended use. The Fabricator should be paid for these materials upon delivery from the mill, subject to appropriate additional payment or credit if subsequent unanticipated modification or reorder is required. Purchasing materials to exact lengths is not considered fabrication.

5.1.1. Unless otherwise specified by means of special testing requirements in the Contract Documents, mill testing shall be limited to those tests that are required for the material in the ASTM specifications indicated in the Contract Documents. Certified mill test reports shall be furnished by the Fabricator if requested to do so by the Owner's Designated Representative for Design, either in the Contract Documents or in separate written instructions given to the Fabricator prior to ordering Mill Materials.

Commentary:
Mill tests are performed to demonstrate material conformance to ASTM specifications in accordance with the contract requirements.

5.1.2. When Mill Material does not satisfy ASTM A6/A6M tolerances for camber, profile, flatness or sweep, the Fabricator shall be permitted to perform corrective procedures, including the use of controlled heating and/or mechanical straightening, subject to the limitations in the AISC Specification.

Commentary:
Mill dimensional tolerances are completely set forth in ASTM A6/A6M. Normal variations in the cross-sectional geometry of Standard Structural Shapes must be recognized by the designer, the Fabricator, the Steel Detailer and the Erector (for example, see Figure C–5.1). Such tolerances are mandatory because roll wear, thermal distortions of the hot cross-section immediately after leaving the forming rolls and differential cooling distortions that take place on the cooling beds are all unavoidable. Geometric perfection of the cross-section is not necessary for either structural or architectural reasons, if the tolerances are recognized and provided for.

ASTM A6/A6M also stipulates tolerances for straightness that are adequate for typical construction. However, these characteristics may be controlled or corrected to closer tolerances during the fabrication process when the added cost is justified by the special requirements for an atypical project.

5.1.3. When variations that exceed ASTM A6/A6M tolerances are discovered or occur after the receipt of Mill Material the Fabricator shall, at the Fabricator's option, be permitted to perform the ASTM A6/A6M corrective procedures for mill reconditioning of the surface of Structural Steel shapes and plates.

5.1.4. When special tolerances that are more restrictive than those in ASTM A6/A6M are required for Mill Materials, such special tolerances shall be specified in the Contract Documents. The Fabricator shall, at the Fabricator's option, be permitted to order material to ASTM A6/A6M tolerances and subsequently perform the corrective procedures described in Sections 5.1.2 and 5.1.3.

5.2. Stock Materials

5.2.1. If used for structural purposes, materials that are taken from stock by the Fabricator shall be of a quality that is at least equal to that required in the ASTM specifications indicated in the Contract Documents.

5.2.2. Certified mill test reports shall be accepted as sufficient record of the quality of materials taken from stock by the Fabricator. The

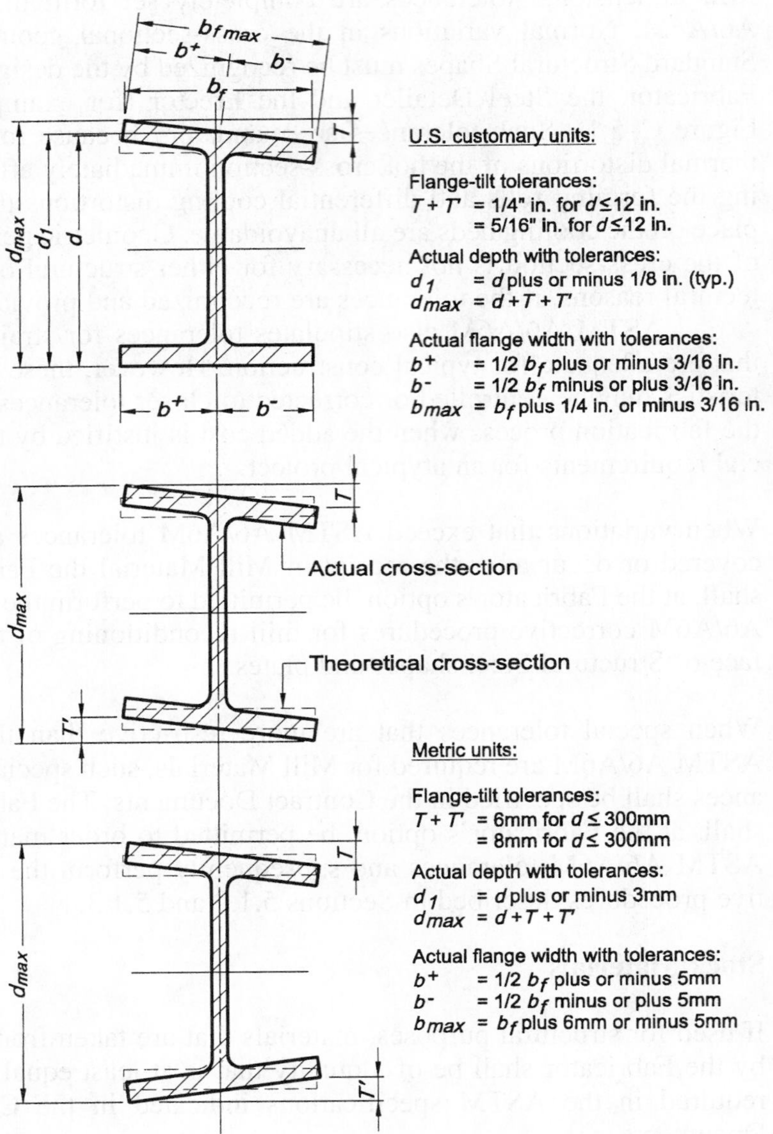

Figure C-5.1. Mill tolerances on the cross-section of a W-shape.

Fabricator shall review and retain the certified mill test reports that cover such stock materials. However, the Fabricator need not maintain records that identify individual pieces of stock material against individual certified mill test reports, provided the Fabricator purchases stock materials that meet the requirements for material grade and quality in the applicable ASTM specifications.

5.2.3. Stock materials that are purchased under no particular specification, under a specification that is less rigorous than the applicable ASTM specifications or without certified mill test reports or other recognized test reports shall not be used without the approval of the Owner's Designated Representative for Design.

SECTION 6. SHOP FABRICATION AND DELIVERY

6.1. Identification of Material

6.1.1. Material that is ordered to special requirements shall be marked by the supplier as specified in ASTM A6/A6M Section 12 prior to delivery to the Fabricator's shop or other point of use. Material that is ordered to special requirements, but not so marked by the supplier, shall not be used until:

(a) Its identification is established by means of testing in accordance with the applicable ASTM specifications; and,
(b) A Fabricator's identification mark, as described in Section 6.1.2 and 6.1.3, has been applied.

6.1.2. During fabrication, up to the point of assembling members, each piece of material that is ordered to special requirements shall carry a Fabricator's identification mark or an original supplier's identification mark. The Fabricator's identification mark shall be in accordance with the Fabricator's established identification system, which shall be on record and available prior to the start of fabrication for the information of the Owner's Designated Representative for Construction, the building-code authority and the Inspector.

6.1.3. Members that are made of material that is ordered to special requirements shall not be given the same assembling or erection mark as members made of other material, even if they are of identical dimensions and detail.

6.2. Preparation of Material

6.2.1. The thermal cutting of Structural Steel by hand-guided or mechanically guided means is permitted.

6.2.2. Surfaces that are specified as "finished" in the Contract Documents shall have a roughness height value measured in accordance with ANSI/ASME B46.1 that is equal to or less than 500. The use of any fabricating technique that produces such a finish is permitted.

Commentary:
Most cutting processes, including friction sawing and cold sawing, and milling processes meet a surface roughness limitation of 500 per ANSI/ASME B46.1.

6.3. Fitting and Fastening

6.3.1. Projecting elements of Connection materials need not be straightened in the connecting plane, subject to the limitations in the AISC Specification.

6.3.2. Backing bars and runoff tabs shall be used in accordance with AWS D1.1 as required to produce sound welds. The Fabricator or Erector need not remove backing bars or runoff tabs unless such removal is specified in the Contract Documents. When the removal of backing bars is specified in the Contract Documents, such removal shall meet the requirements in AWS D1.1. When the removal of runoff tabs is specified in the Contract Documents, hand flame-cutting close to the edge of the finished member with no further finishing is permitted, unless other finishing is specified in the Contract Documents.

Commentary:
In most cases, the treatment of backing bars and runoff tabs is left to the discretion of the Owner's Designated Representative for Design. In some cases, treatment beyond the basic cases described in this Section may be required. As one example, special treatment is required for backing bars and runoff tabs in beam-to-column moment Connections when the requirements in the AISC Seismic Provisions must be met. In all cases, the Owner's Designated Representative for Design should specify the required treatments in the Contract Documents.

6.3.3. Unless otherwise noted in the Shop Drawings, high-strength bolts for shop-attached Connection material shall be installed in the shop in accordance with the requirements in the AISC Specification.

6.4. Fabrication Tolerances

The tolerances on Structural Steel fabrication shall be in accordance with the requirements in Section 6.4.1 through 6.4.6.

Commentary:
Fabrication tolerances are stipulated in several specifications and codes, each applicable to a specialized area of construction. Basic fabrication tolerances are stipulated in this Section. For Architecturally Exposed Structural Steel, see Section 10. Other specifications and codes are also commonly incorporated by reference in the Contract Documents, such as the AISC Specification, the RCSC Specification, AWS D1.1 and the AASHTO Specification.

6.4.1. For members that have both ends finished (see Section 6.2.2) for contact bearing, the variation in the overall length shall be equal to or less than 1/32 in. [1 mm]. For other members that frame to other Structural Steel elements, the variation in the detailed length shall be as follows:

(a) For members that are equal to or less than 30 ft [9 000 mm] in length, the variation shall be equal to or less than 1/16 in. [2 mm].

(b) For members that are greater than 30 ft [9 000 mm] in length, the variation shall be equal to or less than 1/8 in. [3 mm].

6.4.2. For straight structural members other than compression members, whether of a single Standard Structural Shape or built-up, the variation in straightness shall be equal to or less than that specified for wide-flange shapes in ASTM A6/A6M, except when a smaller variation in straightness is specified in the Contract Documents. For straight compression members, whether of a Standard Structural Shape or built-up, the variation in straightness shall be equal to or less than 1/1000 of the axial length between points that are to be laterally supported. For curved structural members, the variation from the theoretical curvature shall be equal to or less than the variation in sweep that is specified for an equivalent straight member of the same straight length in ASTM A6/A6M.

In all cases, completed members shall be free of twists, bends and open joints. Sharp kinks or bends shall be cause for rejection.

6.4.3. For beams and trusses that are detailed without specified camber, the member shall be fabricated so that, after erection, any incidental camber due to rolling or shop fabrication is upward.

6.4.4. For beams that are specified in the Contract Documents with camber, beams received by the Fabricator with 75% of the specified camber shall require no further cambering. Otherwise, the variation in camber shall be as follows:

(a) For beams that are equal to or less than 50 ft [15 000 mm] in length, the variation shall be equal to or less than minus zero / plus 1/2 in. [13 mm].
(b) For beams that are greater than 50 ft [15 000 mm] in length, the variation shall be equal to or less than minus zero / plus 1/2 in. plus 1/8 in. for each 10 ft or fraction thereof [13 mm plus 3 mm for each 3 000 mm or fraction thereof] in excess of 50 ft [15 000 mm] in length.

For the purpose of inspection, camber shall be measured in the Fabricator's shop in the unstressed condition.

Commentary:
There is no known way to inspect beam camber after the beam is received in the field because of factors that include:

(a) The release of stresses in members over time and in varying applications;
(b) The effects of the dead weight of the member;
(c) The restraint caused by the end Connections in the erected state; and,
(d) The effects of additional dead load that may ultimately be intended to be applied, if any.

Therefore, inspection of the Fabricator's work on beam camber must be done in the fabrication shop in the unstressed condition.

6.4.5. For fabricated trusses that are specified in the Contract Documents with camber, the variation in camber at each specified camber point shall be equal to or less than plus or minus 1/800 of the distance to that point from the nearest point of support. For the purpose of inspection, camber shall be measured in the Fabricator's shop in the unstressed condition.

Commentary:
There is no known way to inspect truss camber after the truss is received in the field because of factors that include:

(a) The effects of the dead weight of the member;
(b) The restraint caused by the truss Connections in the erected state; and,
(c) The effects of additional dead load that may ultimately be intended to be applied, if any.

Therefore, inspection of the Fabricator's work on truss camber must be done in the fabrication shop in the unstressed condition. See Figure C–6.1.

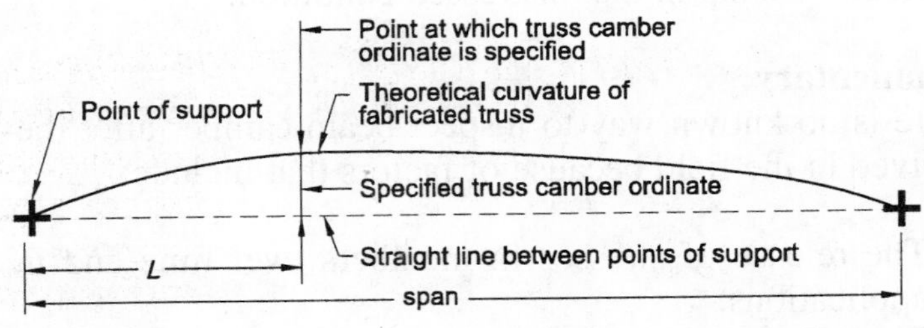

Taking L as the distance from the point at which truss camber is specified to the closer point of support, in. [mm], the tolerance on truss camber at that point is calculated as $L/800$. L must be equal to or less than one-half the span.

Figure C-6.1. Illustration of the tolerance on camber for fabricated trusses with specified camber.

6.4.6. When permissible variations in the depths of beams and girders result in abrupt changes in depth at splices, such deviations shall be accounted for as follows:

(a) For splices with bolted joints, the variations in depth shall be taken up with filler plates; and,
(b) For splices with welded joints, the weld profile shall be adjusted to conform to the variations in depth, the required cross-section of weld shall be provided and the slope of the weld surface shall meet the requirements in AWS D1.1.

6.5. Shop Cleaning and Painting (see also Section 3.1.6)
Structural Steel that does not require shop paint shall be cleaned of oil and grease with solvent cleaners, and of dirt and other foreign material by sweeping with a fiber brush or other suitable means. For Structural Steel that is required to be shop painted, the requirements in Sections 6.5.1 through 6.5.4 shall apply.

Commentary:
Extended exposure of unpainted Structural Steel that has been cleaned for the subsequent application of fire protection materials can be detrimental to the fabricated product. Most levels of cleaning require the removal of all loose mill scale, but permit some amount of tightly adhering mill scale. When a piece of Structural Steel that has been cleaned to an acceptable level is left exposed to a normal environment, moisture can penetrate behind the scale, and some "lifting" of the scale by the oxidation process is to be expected. Cleanup of "lifted" mill scale is not the responsibility of the Fabricator, but is to be assigned by contract requirement to an appropriate contractor.

Section 6.5.4 of this Code is not applicable to weathering steel, for which special cleaning specifications are always required in the Contract Documents.

6.5.1. The Fabricator is not responsible for deterioration of the shop coat that may result from exposure to ordinary atmospheric conditions or corrosive conditions that are more severe than ordinary atmospheric conditions.

Commentary:
The shop coat of paint is the prime coat of the protective system. It is intended as protection for only a short period of exposure in ordinary atmospheric conditions, and is considered a temporary and provisional coating.

6.5.2. Unless otherwise specified in the Contract Documents, the Fabricator shall, as a minimum, hand clean the Structural Steel of loose rust, loose mill scale, dirt and other foreign matter, prior to painting, by means of wire brushing or by other methods elected by the Fabricator, to meet the requirements of SSPC-SP2. If the Fabricator's workmanship on surface preparation is to be inspected by the Inspector, such inspection shall be performed in a timely manner prior to the application of the shop coat.

Commentary:
The selection of a paint system is a design decision involving many factors including:

(a) The Owner's preference;
(b) The service life of the structure;
(c) The severity of environmental exposure;
(d) The cost of both initial application and future renewals; and,
(e) The compatibility of the various components that comprise the paint system (surface preparation, shop coat and subsequent coats).

Because the inspection of shop painting must be concerned with workmanship at each stage of the operation, the Fabricator provides notice of the schedule of operations and affords the Inspector access to the work site. Inspection must then be coordinated with that schedule so as to avoid delay of the scheduled operations.

Acceptance of the prepared surface must be made prior to the application of the shop coat because the degree of surface preparation cannot be readily verified after painting. Time delay between surface preparation and the application of the shop coat can result in unacceptable deterioration of a properly prepared surface, necessitating a repetition of surface preparation. This is especially true with blast-cleaned surfaces. Therefore, to avoid potential deterioration of the surface, it is assumed that surface preparation is accepted unless it is inspected and rejected prior to the scheduled application of the shop coat.

The shop coat in any paint system is designed to maximize the wetting and adherence characteristics of the paint, usually at the expense of its weathering capabilities. Deterioration of the shop coat normally begins immediately after exposure to the elements and worsens as the duration of exposure is extended. Consequently, extended exposure of the shop coat will likely lead to its deterioration and may necessitate repair, possibly including the repetition of surface preparation and shop coat application in limited areas. With the introduction of high-performance paint systems, avoiding delay in the application of the shop coat has become more critical. High-performance paint systems generally require a greater degree of surface preparation, as well as early application of weathering protection for the shop coat.

Since the Fabricator does not control the selection of the paint system, the compatibility of the various components of the total paint system, or the length of exposure of the shop coat, the Fabricator cannot guarantee the performance of the shop coat or any other part of the system. Instead, the Fabricator is responsible only for accomplishing the specified surface preparation and for applying the shop coat (or coats) in accordance with the Contract Documents.

This Section stipulates that the Structural Steel is to be cleaned to meet the requirements in SSPC-SP2. This stipulation is not intended to represent an exclusive cleaning level, but rather the level of surface preparation that will be furnished unless otherwise specified in the Contract Documents if the Structural Steel is to be painted.

Further information regarding shop painting is available in *A Guide to Shop Painting of Structural Steel*, published jointly by SSPC and AISC.

6.5.3. Unless otherwise specified in the Contract Documents, paint shall be applied by brushing, spraying, rolling, flow coating, dipping or other suitable means, at the election of the Fabricator. When the term "shop coat", "shop paint" or other equivalent term is used with no paint system specified, the Fabricator's standard shop paint shall be applied to a minimum dry-film thickness of one mil [25 μm].

6.5.4. Touch-up of abrasions caused by handling after painting shall be the responsibility of the contractor that performs touch-up in the field or field painting.

Commentary:
Touch-up in the field and field painting are not normally part of the Fabricator's or the Erector's contract.

6.6. Marking and Shipping of Materials

6.6.1. Unless otherwise specified in the Contract Documents, erection marks shall be applied to the Structural Steel members by painting or other suitable means.

6.6.2. Bolt assemblies and loose bolts, nuts and washers shall be shipped in separate closed containers according to length and diameter, as applicable. Pins and other small parts and packages of bolts, nuts and washers shall be shipped in boxes, crates, kegs or barrels. A list and description of the material shall appear on the outside of each closed container.

Commentary:
In most cases bolts, nuts and other components in a fastener assembly can be shipped loose in separate containers. However, ASTM F1852/F1852M twist-off-type tension-control bolt assemblies and galvanized ASTM A325, A325M and F1852/F1852M bolt assemblies must be assembled and shipped in the same container according to length and diameter.

6.7. Delivery of Materials

6.7.1. Fabricated Structural Steel shall be delivered in a sequence that will permit efficient and economical fabrication and erection, and that is consistent with requirements in the Contract Documents. If the Owner or Owner's Designated Representative for Construction wishes to prescribe or control the sequence of delivery of materials, that entity shall specify the required sequence in the Contract Documents. If the Owner's Designated Representative for Construction contracts separately for delivery and for erection, the Owner's Designated Representative for Construction shall coordinate planning between contractors.

6.7.2. Anchor Rods, washers, nuts and other anchorage or grillage materials that are to be built into concrete or masonry shall be shipped so that they will be available when needed. The Owner's Designated Representative for Construction shall allow the Fabricator sufficient time to fabricate and ship such materials before they are needed.

6.7.3. If any shortage is claimed relative to the quantities of materials that are shown in the shipping statements, the Owner's Designated Representative for Construction or the Erector shall promptly notify the Fabricator so that the claim can be investigated.

Commentary:
The quantities of material that are shown in the shipping statement are customarily accepted as correct by the Owner's Designated Representative for Construction, the Fabricator and the Erector.

6.7.4. Unless otherwise specified in the Contract Documents, and subject to the approved Shop and Erection Drawings, the Fabricator shall limit the number of field splices to that consistent with minimum project cost.

Commentary:
This Section recognizes that the size and weight of Structural Steel assemblies may be limited by shop capabilities, the permissible

weight and clearance dimensions of available transportation or job-site conditions.

6.7.5. If material arrives at its destination in damaged condition, the receiving entity shall promptly notify the Fabricator and carrier prior to unloading the material, or promptly upon discovery prior to erection.

SECTION 7. ERECTION

7.1. Method of Erection

Fabricated Structural Steel shall be erected using methods and a sequence that will permit efficient and economical performance of erection, and that is consistent with the requirements in the Contract Documents. If the Owner or Owner's Designated Representative for Construction wishes to prescribe or control the method and/or sequence of erection, or specifies that certain members cannot be erected in their normal sequence, that entity shall specify the required method and sequence in the Contract Documents. If the Owner's Designated Representative for Construction contracts separately for fabrication services and for erection services, the Owner's Designated Representative for Construction shall coordinate planning between contractors.

Commentary:
Design modifications are sometimes requested by the Erector to allow or facilitate the erection of the Structural Steel frame. When this is the case, the Erector should notify the Fabricator prior to the preparation of Shop and Erection Drawings so that the Fabricator may refer the Erector's request to the Owner's Designated Representatives for Design and Construction for resolution.

7.2. Job-Site Conditions

The Owner's Designated Representative for Construction shall provide and maintain the following for the Fabricator and the Erector:

(a) Adequate access roads into and through the job site for the safe delivery and movement of the material to be erected and of derricks, cranes, trucks and other necessary equipment under their own power;
(b) A firm, properly graded, drained, convenient and adequate space at the job site for the operation of the Erector's equipment, free from overhead obstructions, such as power lines, telephone lines or similar conditions; and,
(c) Adequate storage space, when the structure does not occupy the full available job site, to enable the Fabricator and the Erector to operate at maximum practical speed.

Otherwise, the Owner's Designated Representative for Construction shall inform the Fabricator and the Erector of the actual job-site conditions and/or special delivery requirements prior to bidding.

7.3. Foundations, Piers and Abutments

The accurate location, strength and suitability of, and access to, all foundations, piers and abutments shall be the responsibility of the Owner's Designated Representative for Construction.

7.4. Building Lines and Bench Marks

The Owner's Designated Representative for Construction shall be responsible for the accurate location of building lines and benchmarks at the job site and shall furnish the Erector with a plan that contains all such information. The Owner's Designated Representative for Construction shall establish offset building lines and reference elevations at each level for the Erector's use in the positioning of Adjustable Items (see Section 7.13.1.3), if any.

7.5. Installation of Anchor Rods, Foundation Bolts and Other Embedded Items

7.5.1. Anchor Rods, foundation bolts and other embedded items shall be set by the Owner's Designated Representative for Construction in accordance with an approved Embedment Drawing. The variation in location of these items from the dimensions shown in the Embedment Drawings shall be as follows:

(a) The variation in dimension between the centers of any two Anchor Rods within an Anchor-Rod Group shall be equal to or less than 1/8 in. [3 mm].

(b) The variation in dimension between the centers of adjacent Anchor-Rod Groups shall be equal to or less than 1/4 in. [6 mm].

(c) The variation in elevation of the tops of Anchor Rods shall be equal to or less than plus or minus 1/2 in. [13 mm].

(d) The accumulated variation in dimension between centers of Anchor-Rod Groups along the Established Column Line through multiple Anchor-Rod Groups shall be equal to or less than 1/4 in. per 100 ft [2 mm per 10 000 mm], but not to exceed a total of 1 in. [25 mm].

(e) The variation in dimension from the center of any Anchor-Rod Group to the Established Column Line through that group shall be equal to or less than 1/4 in. [6 mm].

The tolerances that are specified in (b), (c) and (d) shall apply to offset dimensions shown in the structural Design Drawings, measured parallel and perpendicular to the nearest Established Column Line, for individual columns that are shown in the structural Design Drawings as offset from Established Column Lines.

Commentary:
The tolerances established in this Section have been selected for compatibility with the holes sizes that are recommended for base plates in the AISC Manual of Steel Construction. If special conditions require more restrictive tolerances, the contractor responsible for setting the Anchor Rods should be so informed in the Contract Documents. When the Anchor Rods are set in sleeves, the adjustment provided may be used to satisfy the required Anchor-Rod setting tolerances.

7.5.2. Unless otherwise specified in the Contract Documents, Anchor Rods shall be set with their longitudinal axis perpendicular to the theoretical bearing surface.

7.5.3. Embedded items and Connection materials that are part of the work of other trades, but that will receive Structural Steel, shall be located and set by the Owner's Designated Representative for Construction in accordance with an approved Embedment Drawing. The variation in location of these items shall be limited to a magnitude that is consistent with the tolerances that are specified in Section 7.13 for the erection of the Structural Steel.

7.5.4. All work that is performed by the Owner's Designated Representative for Construction shall be completed so as not to delay or interfere with the work of the Fabricator and the Erector.

The Owner's Designated Representative for Construction shall conduct a survey of the as-built locations of Anchor Rods, foundation bolts and other embedded items, and shall verify that all items covered in Section 7.5 meet the corresponding tolerances. When corrective action is necessary, the Owner's Designated Representative for Construction shall obtain the guidance and approval of the Owner's Designated Representative for Design.

Commentary:
Few Fabricators or Erectors have the capability to provide this survey. Under standard practice, it is the responsibility of others.

7.6. Installation of Bearing Devices

All leveling plates, leveling nuts and washers and loose base and bearing plates that can be handled without a derrick or crane are set to line and grade by the Owner's Designated Representative for Construction. Loose base and bearing plates that require handling with a derrick or crane shall be set by the Erector to lines and grades established by the Owner's Designated Representative for Construction. The Fabricator shall clearly scribe loose base and bearing plates with lines or other suitable marks to facilitate proper alignment.

Promptly after the setting of Bearing Devices, the Owner's Designated Representative for Construction shall check them for line and grade. The variation in elevation relative to the established grade for all Bearing Devices shall be equal to or less than plus or minus 1/8 in. [3 mm]. The final location of Bearing Devices shall be the responsibility of the Owner's Designated Representative for Construction.

Commentary:
The 1/8 in. [3 mm] tolerance on elevation of Bearing Devices relative to established grades is provided to permit some variation in setting Bearing Devices, and to account for the accuracy that is attainable with standard surveying instruments. The use of leveling plates larger than 22 in. by 22 in. [550 mm by 550 mm] is discouraged and grouting is recommended with larger sizes. For the purposes of erection stability, the use of leveling nuts and washers is discouraged when base plates have less than four Anchor Rods.

7.7. Grouting

Grouting shall be the responsibility of the Owner's Designated Representative for Construction. Leveling plates and loose base and bearing plates shall be promptly grouted after they are set and checked for line and grade. Columns with attached base plates, beams with attached bearing plates and other similar members with attached Bearing Devices that are temporarily supported on leveling nuts and washers, shims or other similar leveling devices, shall be promptly grouted after the Structural Steel frame or portion thereof has been plumbed.

Commentary:
In the majority of structures the vertical load from the column bases is transmitted to the foundations through structural grout. In general, there are three methods by which support is provided for column bases during erection:

(a) Pre-grouted leveling plates or loose base plates;
(b) Shims; and,
(c) Leveling nuts and washers on the Anchor Rods beneath the column base.

Standard practice provides that loose base plates and leveling plates are to be grouted as they are set. Bearing Devices that are set on shims or leveling nuts are grouted after plumbing, which means that the weight of the erected Structural Steel frame is supported on the shims or washers, nuts and Anchor Rods. The Erector must take care to ensure that the load that is transmitted in this temporary condition does not exceed the strength of the shims or washers, nuts and Anchor Rods. These considerations are presented in greater detail in AISC Design Guides No. 1 and 10.

7.8. Field Connection Material

7.8.1. The Fabricator shall provide field Connection details that are consistent with the requirements in the Contract Documents and that will, in the Fabricator's opinion, result in economical fabrication and erection.

7.8.2. When the Fabricator is responsible for erecting the Structural Steel, the Fabricator shall furnish all materials that are required for both temporary and permanent Connection of the component parts of the Structural Steel frame.

7.8.3. When the erection of the Structural Steel is not performed by the Fabricator, the Fabricator shall furnish the following field Connection material:

(a) Bolts, nuts and washers of the required grade, type and size and in sufficient quantity for all Structural Steel-to-Structural Steel field Connections that are to be permanently bolted, including an extra 2 percent of each bolt size (diameter and length);
(b) Shims that are shown as necessary for make-up of permanent Structural Steel-to-Structural Steel Connections; and,
(c) Backing bars and run-off tabs that are required for field welding.

7.8.4. The Erector shall furnish all welding electrodes, fit-up bolts and drift pins used for the erection of the Structural Steel.

Commentary:
See the commentary for Section 2.2.

7.9. Loose Material

Unless otherwise specified in the Contract Documents, loose Structural Steel items that are not connected to the Structural Steel frame shall be set by the Owner's Designated Representative for Construction without assistance from the Erector.

7.10. Temporary Support of Structural Steel Frames

7.10.1. The Owner's Designated Representative for Design shall identify the following in the Contract Documents:

(a) The lateral-load-resisting system and connecting diaphragm elements that provide for lateral strength and stability in the completed structure; and,

(b) Any special erection conditions or other considerations that are required by the design concept, such as the use of shores, jacks or loads that must be adjusted as erection progresses to set or maintain camber, position within specified tolerances or prestress.

Commentary:
See Commentary Section 7.10.3.

7.10.2. The Owner's Designated Representative for Construction shall indicate to the Erector prior to bidding, the installation schedule for non-Structural Steel elements of the lateral-load-resisting system and connecting diaphragm elements identified by the Owner's Designated Representative for Design in the Contract Documents.

Commentary:
See Commentary Section 7.10.3.

7.10.3. Based upon the information provided in accordance with Sections 7.10.1 and 7.10.2, the Erector shall determine, furnish and install all temporary supports, such as temporary guys, beams, falsework, cribbing or other elements required for the erection operation. These temporary supports shall be sufficient to secure the bare Structural Steel framing or any portion thereof against loads that are likely to be encountered during erection, including those due to wind and those that result from erection operations.

The Erector need not consider loads during erection that result from the performance of work by, or the acts of, others, except as specifically identified by the Owner's Designated Representatives for Design and Construction, nor those that are unpredictable, such as loads due to hurricane, tornado, earthquake, explosion or collision.

Temporary supports that are required during or after the erection of the Structural Steel frame for the support of loads caused by non-Structural Steel elements, including cladding, interior partitions and other such elements that will induce or transmit loads to the Structural Steel frame during or after erection, shall be the responsibility of others.

Commentary:
Many Structural Steel frames have lateral-load-resisting systems that are activated during the erection process. Such lateral-load-resisting systems may consist of welded moment frames, braced frames or, in some instances, columns that cantilever from fixed-base foundations. Such frames are normally braced with temporary guys that, together with the steel deck floor and roof diaphragms, or other diaphragm bracing that may be included as part of the design, provide stability during the erection process. The guy cables are also commonly used to plumb the Structural Steel frame. The Erector normally furnishes and installs the required temporary supports and bracing to secure the bare Structural Steel frame, or portion thereof, during the erection process.

If the Owner's Designated Representative for Construction determines that steel decking is not installed by the Erector, temporary diaphragm bracing may be required if a horizontal diaphragm is not available to distribute loads to the vertical and lateral load resisting system. If the steel deck will not be available as a diaphragm during Structural Steel erection, the Owner's Designated Representative for Construction must communicate this condition to the Erector prior to bidding. If such diaphragm bracing is required, it must be furnished and installed by the Erector.

Sometimes structural systems that are employed by the Owner's Designated Representative for Design rely upon other elements besides the Structural Steel frame for lateral-load resistance. For instance, concrete or masonry shear walls or precast spandrels may be used to provide resistance to vertical and lateral loads in the completed structure. Because these situations may not be obvious to the contractor or the Erector, it is required in this Code that the Owner's Designated Representative for Design identify such situations in the Contract Documents. Similarly, if a structure is designed so that special erection techniques are required, such as jacking to impose certain loads or position during erection, it is required in this Code that such requirements be specifically identified in the Contract Documents.

In some instances, the Owner's Designated Representative for Design may elect to show erection bracing in the Design Drawings. When this is the case, the Owner's Designated Representative for Design should then confirm that the bracing requirements were understood by review and approval of the Erection Drawings during the submittal process.

Sometimes during construction of a building, collateral building elements, such as exterior cladding, may be required to be installed on the bare Structural Steel frame prior to completion of the lateral-load-resisting system. These elements may increase the potential for lateral loads on the temporary supports. Such temporary supports may also be required to be left in place after the Structural Steel frame has been erected. Special provisions should be made by the Owner's Designated Representative for Construction for these conditions.

7.10.4. All temporary supports that are required for the erection operation and furnished and installed by the Erector shall remain the property of the Erector and shall not be modified, moved or removed without the consent of the Erector. Temporary supports provided by the Erector shall remain in place until the portion of the Structural Steel frame that they brace is complete and the lateral-load-resisting system and connecting diaphragm elements identified by the Owner's Designated Representative for Design in accordance with Section 7.10.1 are installed. Temporary supports that are required to be left in place after the completion of Structural Steel erection shall be removed when no longer needed by the Owner's Designated Representative for Construction and returned to the Erector in good condition.

7.11. Safety Protection

7.11.1. The Erector shall provide floor coverings, handrails, walkways and other safety protection for the Erector's personnel as required by law and the applicable safety regulations. Unless otherwise specified in the Contract Documents, the Erector is permitted to remove such safety protection from areas where the erection operations are completed.

7.11.2 When safety protection provided by the Erector is left in an area for the use of other trades after the Structural Steel erection activity is completed, the Owner's Designated Representative for Construction shall:

(a) Accept responsibility for and maintain this protection;
(b) Indemnify the Fabricator and the Erector from damages that may be incurred from the use of this protection by other trades;
(c) Ensure that this protection is adequate for use by other affected trades;
(d) Ensure that this protection complies with applicable safety regulations when being used by other trades; and,
(e) Remove this protection when it is no longer required and return it to the Erector in the same condition as it was received.

7.11.3. Safety protection for other trades that are not under the direct employment of the Erector shall be the responsibility of the Owner's Designated Representative for Construction.

7.11.4. When permanent steel decking is used for protective flooring and is installed by the Owner's Designated Representative for Construction, all such work shall be scheduled and performed in a timely manner so as not to interfere with or delay the work of the Fabricator or the Erector. The sequence of installation that is used shall meet all safety regulations.

7.11.5. Unless the interaction and safety of activities of others, such as construction by others or the storage of materials that belong to others, are coordinated with the work of the Erector by the Owner's Designated Representative for Construction, such activities shall not be permitted until the erection of the Structural Steel frame or portion thereof is completed by the Erector and accepted by the Owner's Designated Representative for Construction.

7.12. Structural Steel Frame Tolerances

The accumulation of the mill tolerances and fabrication tolerances shall not cause the erection tolerances to be exceeded.

Commentary:
In previous editions of this Code, it was stated that "…variations are deemed to be within the limits of good practice when they do not exceed the cumulative effect of rolling tolerances, fabricating tolerances and erection tolerances." It is recognized in the current provision in this Section that accumulations of mill tolerances and fabrication tolerances generally occur between the locations at which erection tolerances are applied, and not at the same locations.

7.13. Erection Tolerances

Erection tolerances shall be defined relative to member working points and working lines, which shall be defined as follows:

(a) For members other than horizontal members, the member work point shall be the actual center of the member at each end of the shipping piece.
(b) For horizontal members, the working point shall be the actual centerline of the top flange or top surface at each end.
(c) The member working line shall be the straight line that connects the member working points.

The substitution of other working points is permitted for ease of reference, provided they are based upon the above definitions.

The tolerances on Structural Steel erection shall be in accordance with the requirements in Sections 7.13.1 through 7.13.3.

Commentary:
The erection tolerances defined in this Section have been developed through long-standing usage as practical criteria for the erection of Structural Steel. Erection tolerances were first defined in the 1924 edition of this Code in Section 7(f), "Plumbing Up." With the changes that took place in the types and use of materials in building construction after World War II, and the increasing demand by

Architects and Owners for more specific tolerances, AISC adopted new standards for erection tolerances in Section 7(h) of the March 15, 1959 edition of this Code. Experience has proven that those tolerances can be economically obtained.

Differential column shortening may be a consideration in design and construction. In some cases, it may occur due to variability in the accumulation of dead load among different columns (see Figure C–7.1). In other cases, it may be characteristic of the structural system that is employed in the design. Consideration of the effects of differential column shortening may be very important, such as when the slab thickness is reduced, when electrical and other similar fittings mounted on the Structural Steel are intended to be flush with the finished floor and when there is little clearance between bottoms of beams and the tops of door frames or ductwork.

Expansion and contraction in a Structural Steel frame may also be a consideration in the design and construction. Steel will expand or contract approximately 1/8 in. per 100 ft for each change of 15°F [2 mm per 10 000 mm for each change of 15°C] in temperature. This change in length can be assumed to act about the center of rigidity. When anchored to their foundations, end columns will be plumb only when the steel is at normal temperature (see Figure C–7.2). It is therefore necessary to correct field measurements of offsets to the structure from established baselines for the expansion or contraction of the exposed Structural Steel frame. For example, a 200-ft-long [60 000-m-long] building that is plumbed up at 100°F [38°C] should have working points at the tops of the end columns positioned 1/2 in. [14 mm] further apart than the working points at the corresponding bases in order for the columns to be plumb at 70°F [21°C]. Differential temperature effects on column length should also be taken into account in plumbing surveys when tall Structural Steel frames are subjected to sun exposure on one side.

The alignment of lintels, spandrels, wall supports and similar members that are used to connect other building construction units to the Structural Steel frame should have an adjustment of sufficient magnitude to allow for the accumulation of mill tolerances and fabrication tolerances, as well as the erection tolerances. See Figure C–7.3.

7.13.1. The tolerances on position and alignment of member working points and working lines shall be as described in Sections 7.13.1.1 through 7.13.1.3.

7.13.1.1. For an individual column shipping piece, the angular variation of the working line from a plumb line shall be equal to or less than

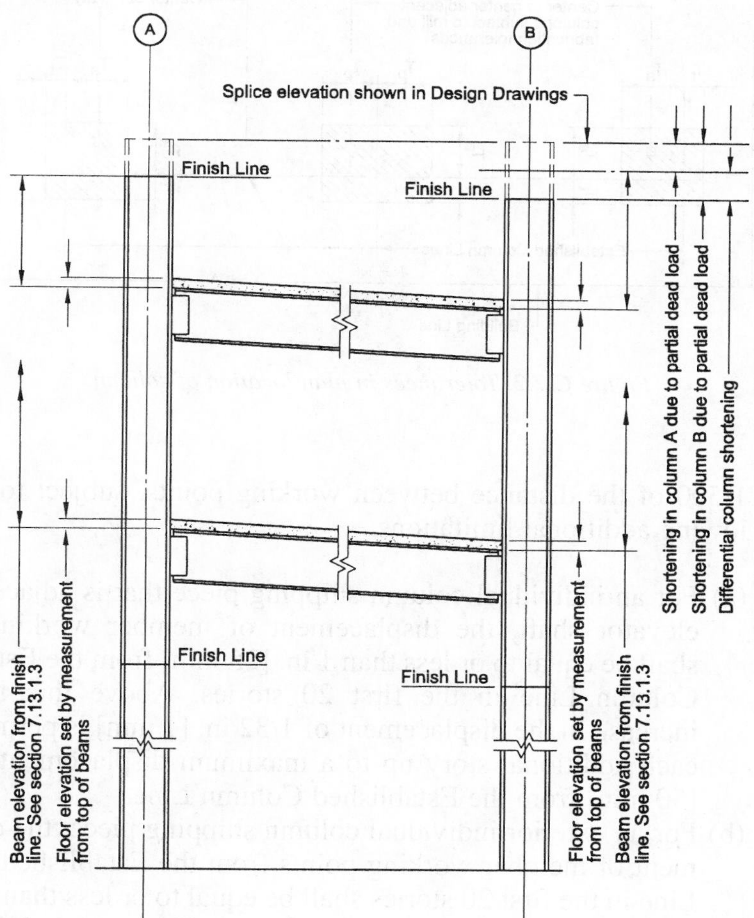

Figure C-7.1. Effects of differential column shortening.

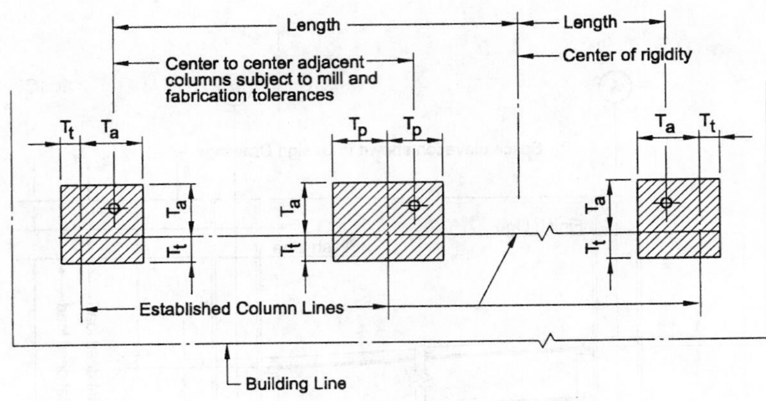

Figure C-7.2. Tolerances in plan location of column.

1/500 of the distance between working points, subject to the following additional limitations:

(a) For an individual column shipping piece that is adjacent to an elevator shaft, the displacement of member working points shall be equal to or less than 1 in. [25 mm] from the Established Column Line in the first 20 stories. Above this level, an increase in the displacement of 1/32 in. [1 mm] is permitted for each additional story up to a maximum displacement of 2 in. [50 mm] from the Established Column Line.

(b) For an exterior individual column shipping piece, the displacement of member working points from the Established Column Line in the first 20 stories shall be equal to or less than 1 in. [25 mm] toward and 2 in. [50 mm] away from the building line. Above this level, an increase in the displacement of 1/16 in. [2 mm] is permitted for each additional story up to a maximum displacement of 2 in. [50 mm] toward and 3 in. [75 mm] away from the building line.

Commentary:
The limitations that are described in this Section and illustrated in Figures C–7.4 and C–7.5 make it possible to maintain built-in-place or prefabricated facades in a true vertical plane up to the 20th story, if Connections that provide for 3 in. [75 mm] of adjustment are used. Above the 20th story, the facade may be maintained within 1/16 in. [2 mm] per story with a maximum total deviation of 1 in. [25 mm] from a true vertical plane, if Connections that provide for 3 in. [75 mm] of adjustment are used. Connections that permit adjustments of plus 2 in. [50 mm] to minus 3 in. [75 mm] (5 in. [125 mm] total) will be necessary in cases where it is desired to construct the facade to a true vertical plane above the 20th story.

(c) For an exterior individual column shipping piece, the member working points at any splice level for multi-Tier buildings and at the tops of columns for single-Tier buildings shall fall within a horizontal envelope, parallel to the building line, that is equal to or less than 1 1/2 in. [38 mm] wide for buildings up to 300 ft [90 000 mm] in length. An increase in the width of this horizontal envelope of 1/2 in. [13 mm] is permitted for each additional 100 ft [30 000 m] in length up to a maximum width of 3 in. [75 mm].

Commentary:
This Section limits the position of exterior column working points at any given splice elevation to a narrow horizontal envelope parallel to the building line (see Figure C–7.6). This envelope is limited to a width of 1 1/2 in. [38 mm], normal to the building line, in up to 300 ft [90 000 mm] of building length. The horizontal location of this envelope is not necessarily directly above or below the corresponding envelope at the adjacent splice elevations, but should be within the limitation of the 1 in 500 plumbness tolerance specified for the controlling columns (see Figure C–7.5).

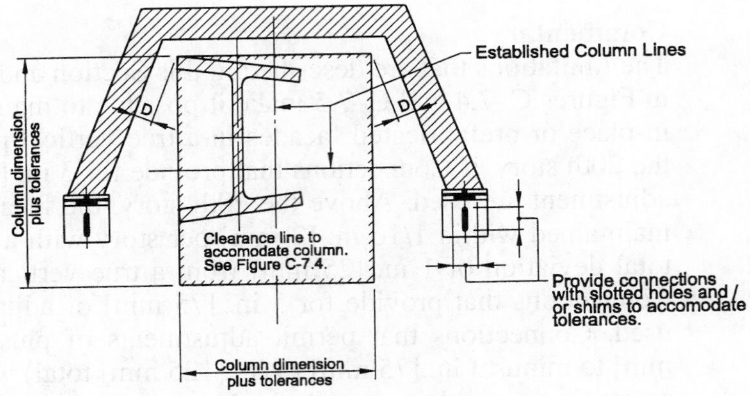

Figure C-7.3. Clearance required to accommodate fascia.

(d) For an exterior column shipping piece, the displacement of member working points from the Established Column Line, parallel to the building line, shall be equal to or less than 2 in. [50 mm] in the first 20 stories. Above this level, an increase in the displacement of 1/16 in. [2 mm] is permitted for each additional story up to a maximum displacement of 3 in. [75 mm] parallel to the building line.

7.13.1.2. For members other than column shipping pieces, the following limitations shall apply:

(a) For a member that consists of an individual, straight shipping piece without field splices, other than a cantilevered member,

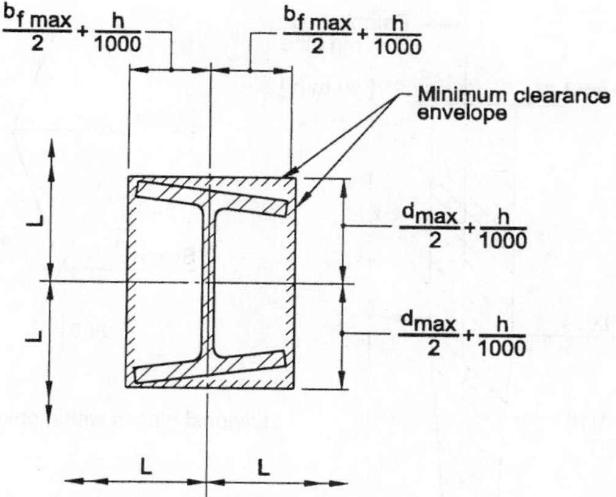

For enclosures or attachments that may follow column alignment.

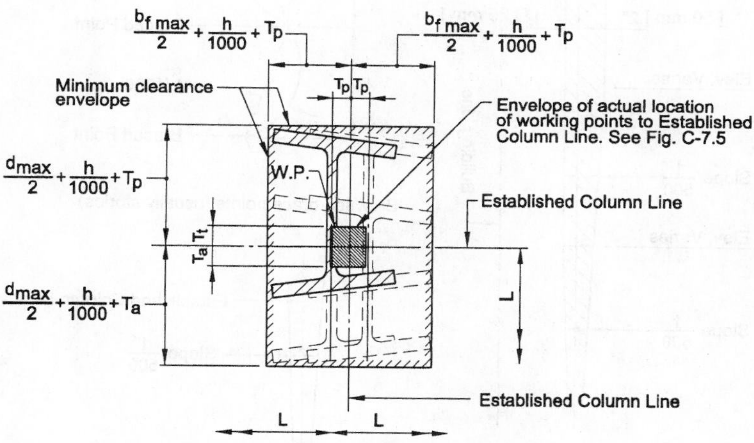

For enclosures or attachments that must be held to precise plan location.

L = Actual center to center of columns = plan dimensions ± column cross section tolerance of columns ± beam length tolerance.

T_a = Plumbness tolerance away from building line (varies, see Fig. C-7.5)

T_t = Plumbness tolerance toward building line (varies, see Fig. C-7.5)

T_p = Plumbness tolerance parallel to building line (=T_a)

Figure C-7.4. Clearance required to accommodate accumulated column tolerances.

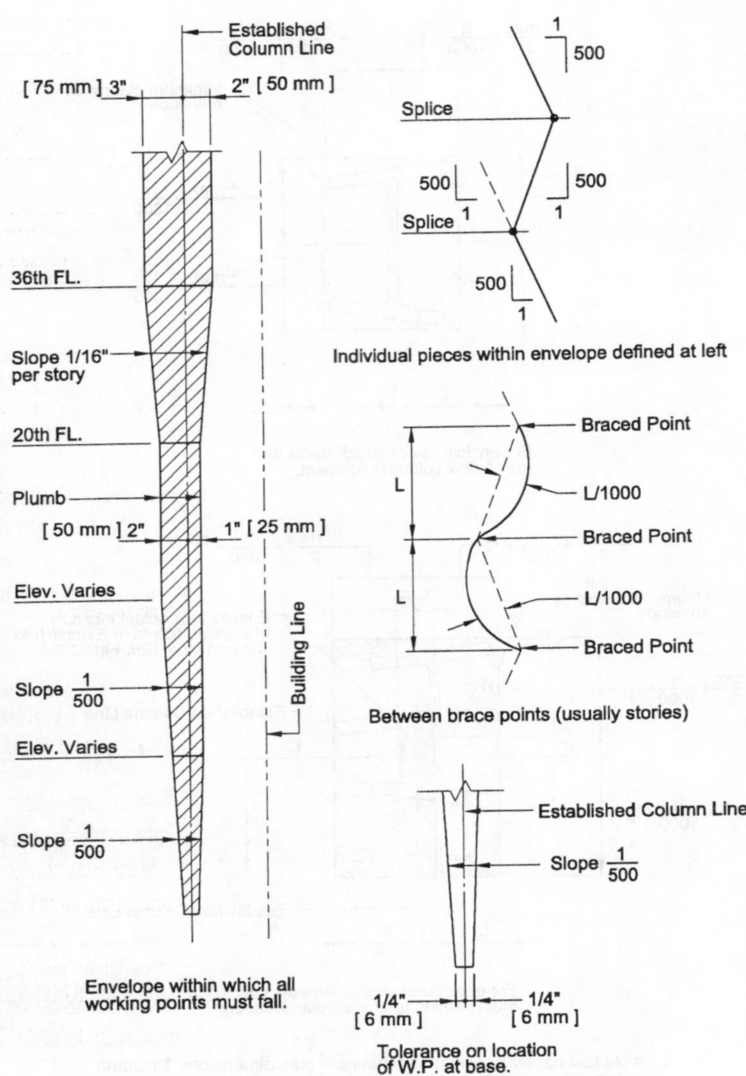

Figure C-7.5. Exterior column plumbness tolerances normal to building line.

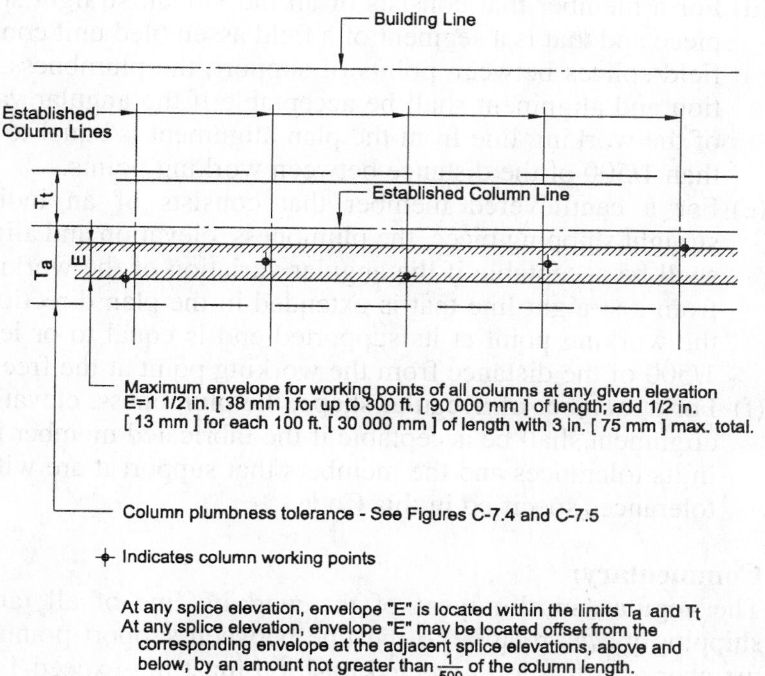

Figure C-7.6. Tolerances in plan at any splice elevation of exterior columns.

the variation in alignment shall be acceptable if it is caused solely by variations in column alignment and/or primary supporting member alignment that are within the permissible variations for the fabrication and erection of such members.

(b) For a member that consists of an individual, straight shipping piece that connects to a column, the variation in the distance from the member working point to the upper finished splice line of the column shall be equal to or less than plus 3/16 in. [5 mm] and minus 5/16 in. [8 mm].

(c) For a member that consists of an individual shipping piece that does not connect to a column, the variation in elevation shall be acceptable if it is caused solely by the variations in the elevations of the supporting members within the permissible variations for the fabrication and erection of those members.

(d) For a member that consists of an individual, straight shipping piece and that is a segment of a field assembled unit containing field splices between points of support, the plumbness, elevation and alignment shall be acceptable if the angular variation of the working line from the plan alignment is equal to or less than 1/500 of the distance between working points.

(e) For a cantilevered member that consists of an individual, straight shipping piece, the plumbness, elevation and alignment shall be acceptable if the angular variation of the working line from a straight line that is extended in the plan direction from the working point at its supported end is equal to or less than 1/500 of the distance from the working point at the free end.

(f) For a member of irregular shape, the plumbness, elevation and alignment shall be acceptable if the fabricated member is within its tolerances and the members that support it are within the tolerances specified in this Code.

Commentary:

The angular misalignment of the working line of all fabricated shipping pieces relative to the line between support points of the member as a whole in erected position must not exceed 1 in 500. Note that the tolerance is not stated in terms of a linear displacement at any point and is not to be taken as the overall length between supports divided by 500. Typical examples are shown in Figure C–7.7. Numerous conditions within tolerance for these and other cases are possible. This condition applies to both plan and elevation tolerances.

7.13.1.3. For members that are identified as Adjustable Items by the Owner's Designated Representative for Design in the Contract Documents, the Fabricator shall provide adjustable Connections for these members to the supporting Structural Steel frame. Otherwise, the Fabricator is permitted to provide non-adjustable Connections. When Adjustable Items are specified, the Owner's Designated Representative for Design shall indicate the total adjustability that is required for the proper alignment of these supports for other trades. The variation in the position and alignment of Adjustable Items shall be as follows:

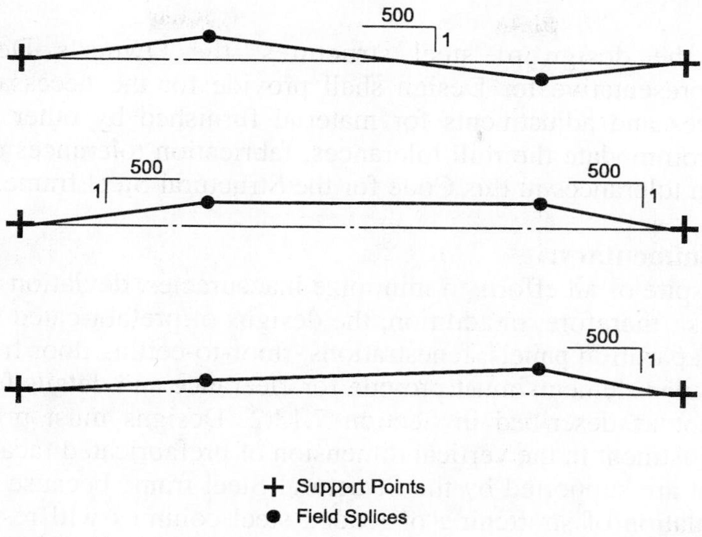

Figure C-7.7. Alignment tolerances for members with field splices.

(a) The variation in the vertical distance from the upper finished splice line of the nearest column to the support location specified in the structural Design Drawings shall be equal to or less than plus or minus 3/8 in. [10 mm].
(b) The variation in the horizontal distance from the established finish line at the particular floor shall be equal to or less than plus or minus 3/8 in. [10 mm].
(c) The variation in vertical and horizontal alignment at the abutting ends of Adjustable Items shall be equal to or less than plus or minus 3/16 in. [5 mm].

Commentary:
When the alignment of lintels, wall supports, curb angles, mullions and similar supporting members for the use of other trades is required to be closer than that permitted by the foregoing tolerances for Structural Steel, the Owner's Designated Representative for

Design must identify such items in the Contract Documents as Adjustable Items.

7.13.2. In the design of steel structures, the Owner's Designated Representative for Design shall provide for the necessary clearances and adjustments for material furnished by other trades to accommodate the mill tolerances, fabrication tolerances and erection tolerances in this Code for the Structural Steel frame.

Commentary:
In spite of all efforts to minimize inaccuracies, deviations will still exist; therefore, in addition, the designs of prefabricated wall panels, partition panels, fenestrations, floor-to-ceiling door frames and similar elements must provide for clearance and details for adjustment as described in Section 7.13.2. Designs must provide for adjustment in the vertical dimension of prefabricated facade panels that are supported by the Structural Steel frame because the accumulation of shortening of loaded steel columns will result in the unstressed facade supported at each floor level being higher than the Structural Steel framing to which it must be attached. Observations in the field have shown that where a heavy facade is erected to a greater height on one side of a multistory building than on the other, the Structural Steel framing will be pulled out of alignment. Facades should be erected at a relatively uniform rate around the perimeter of the structure.

7.13.3. Prior to placing or applying any other materials, the Owner's Designated Representative for Construction shall determine that the location of the Structural Steel is acceptable for plumbness, elevation and alignment. The Erector shall be given either timely notice of acceptance by the Owner's Designated Representative for Construction, or a listing of specific items that are to be corrected in order to obtain acceptance. Such notice shall be rendered promptly upon completion of any part of the work and prior to the start of work by other trades that may be supported, attached or applied to the Structural Steel frame.

7.14. Correction of Errors

The correction of minor misfits by moderate amounts of reaming, grinding, welding or cutting, and the drawing of elements into line with drift pins, shall be considered to be normal erection operations. Errors that cannot be corrected using the foregoing means, or that require major changes in member or Connection configuration, shall be promptly reported to the Owner's Designated Representatives for Design and Construction and the Fabricator by the Erector, to enable the responsible entity to either correct the error or approve the most efficient and economical method of correction to be used by others.

Commentary:
As used in this Section, the term "moderate" refers to the amount of reaming, grinding, welding or cutting that must be done on the project as a whole, not the amount that is required at an individual location. It is not intended to address limitations on the amount of material that is removed by reaming at an individual bolt hole, for example, which is limited by the bolt-hole size and tolerance requirements in the AISC and RCSC Specifications.

7.15. Cuts, Alterations and Holes for Other Trades

Neither the Fabricator nor the Erector shall cut, drill or otherwise alter their work, nor the work of other trades, to accommodate other trades, unless such work is clearly specified in the Contract Documents. When such work is so specified, the Owner's Designated Representatives for Design and Construction shall furnish complete information as to materials, size, location and number of alterations in a timely manner so as not to delay the preparation of Shop and Erection Drawings.

7.16. Handling and Storage

The Erector shall take reasonable care in the proper handling and storage of the Structural Steel during erection operations to avoid the accumulation of excess dirt and foreign matter. The Erector shall not be responsible for the removal from the Structural Steel of dust, dirt or other foreign matter that may accumulate during erection as the result of job-site conditions or exposure to the elements. The Erector shall handle and store all bolts, nuts, washers and related fastening products in accordance with the requirements of the RCSC Specification.

Commentary:
During storage, loading, transport, unloading and erection, blemish marks caused by slings, chains, blocking, tie-downs, etc., occur in varying degrees. Abrasions caused by handling or cartage after painting are to be expected. It must be recognized that any shop-applied coating, no matter how carefully protected, will require touching-up in the field. Touching-up of these blemished areas is the responsibility of the contractor performing the field touch-up or field painting.

The Erector is responsible for the proper storage and handling of fabricated Structural Steel at the job site during erection. Shop-painted Structural Steel that is stored in the field pending erection should be kept free of the ground and positioned so as to minimize the potential for water retention. The Owner or Owner's Designated Representative for Construction is responsible for providing suitable job-site conditions and proper access so that the Fabricator/Erector may perform its work.

Job-site conditions are frequently muddy, sandy, dusty or a combination thereof during the erection period. Under such conditions it may be impossible to store and handle the Structural Steel in such a way as to completely avoid any accumulation of mud, dirt or sand on the surface of the Structural Steel, even though the Fabricator and the Erector manages to proceed with their work.

Repairs of damage to painted surfaces and/or removal of foreign materials due to adverse job-site conditions are outside the scope of responsibility of the Fabricator and the Erector when reasonable attempts at proper handling and storage have been made.

7.17. Field Painting

Neither the Fabricator nor the Erector is responsible to paint field bolt heads and nuts or field welds, nor to touch up abrasions of the shop coat, nor to perform any other field painting.

7.18. Final Cleaning Up

Upon the completion of erection and before final acceptance, the Erector shall remove all of the Erector's falsework, rubbish and temporary buildings.

SECTION 8. QUALITY ASSURANCE

8.1. General

8.1.1. The Fabricator shall maintain a quality assurance program to ensure that the work is performed in accordance with the requirements in this Code, the AISC Specification and the Contract Documents. The Fabricator shall have the option to use the AISC Quality Certification Program to establish and administer the quality assurance program.

Commentary:
The AISC Quality Certification Program confirms to the construction industry that a certified Structural Steel fabrication shop has the capability by reason of commitment, personnel, organization, experience, procedures, knowledge and equipment to produce fabricated Structural Steel of the required quality for a given category of work. The AISC Quality Certification Program is not intended to involve inspection and/or judgment of product quality on individual projects. Neither is it intended to guarantee the quality of specific fabricated Structural Steel products.

8.1.2. The Erector shall maintain a quality assurance program to ensure that the work is performed in accordance with the requirements in this Code, the AISC Specification and the Contract Documents. The Erector shall be capable of performing the erection of the Structural Steel, and shall provide the equipment, personnel and management for the scope, magnitude and required quality of each project. The Erector shall have the option to use the AISC Erector Certification Program to establish and administer the quality assurance program.

Commentary:
The AISC Erector Certification Program confirms to the construction industry that a certified Structural Steel Erector has the capability by reason of commitment, personnel, organization, experience, procedures, knowledge and equipment to erect fabricated Structural Steel to the required quality for a given category of work.

The AISC Erector Certification Program is not intended to involve inspection and/or judgment of product quality on individual projects. Neither is it intended to guarantee the quality of specific erected Structural Steel products.

8.1.3. When the Owner requires more extensive quality assurance or independent inspection by qualified personnel, or requires that the Fabricator be certified under the AISC Quality Certification Program and/or requires that the Erector be certified under the AISC Erector Certification Program, this shall be clearly stated in the Contract Documents, including a definition of the scope of such inspection.

8.2. Inspection of Mill Material

Certified mill test reports shall constitute sufficient evidence that the mill product satisfies material order requirements. The Fabricator shall make a visual inspection of material that is received from the mill, but need not perform any material tests unless the Owner's Designated Representative for Design specifies in the Contract Documents that additional testing is to be performed at the Owner's expense.

8.3. Non-Destructive Testing

When non-destructive testing is required, the process, extent, technique and standards of acceptance shall be clearly specified in the Contract Documents.

8.4. Surface Preparation and Shop Painting Inspection

Inspection of surface preparation and shop painting shall be planned for the acceptance of each operation as the Fabricator completes it. Inspection of the paint system, including material and thickness, shall be made promptly upon completion of the paint application. When wet-film thickness is to be inspected, it shall be measured during the application.

8.5. Independent Inspection

When inspection by personnel other than those of the Fabricator and/or Erector is specified in the Contract Documents, the requirements in Sections 8.5.1 through 8.5.6 shall be met.

8.5.1. The Fabricator and the Erector shall provide the Inspector with access to all places where the work is being performed. A minimum of 24 hours notification shall be given prior to the commencement of work.

8.5.2. Inspection of shop work by the Inspector shall be performed in the Fabricator's shop to the fullest extent possible. Such inspections shall be timely, in-sequence and performed in such a manner as will not disrupt fabrication operations and will permit the repair of non-conforming work prior to any required painting while the material is still in-process in the fabrication shop.

8.5.3. Inspection of field work shall be promptly completed without delaying the progress or correction of the work.

8.5.4. Rejection of material or workmanship that is not in conformance with the Contract Documents shall be permitted at any time during the progress of the work. However, this provision shall not relieve the Owner or the Inspector of the obligation for timely, in-sequence inspections.

8.5.5. The Fabricator and the Erector shall be informed of deficiencies that are noted by the Inspector promptly after the inspection. Copies of all reports prepared by the Inspector shall be promptly given to the Fabricator and the Erector. The necessary corrective work shall be performed in a timely manner.

8.5.6. The Inspector shall not suggest, direct, or approve the Fabricator or Erector to deviate from the Contract Documents or the approved Shop and Erection Drawings, or approve such deviation, without the written approval of the Owner's Designated Representatives for Design and Construction.

SECTION 9. CONTRACTS

9.1. Types of Contracts

9.1.1. For contracts that stipulate a lump sum price, the work that is required to be performed by the Fabricator and the Erector shall be completely defined in the Contract Documents.

9.1.2. For contracts that stipulate a price per pound, the scope of work that is required to be performed by the Fabricator and the Erector, the type of materials, the character of fabrication and the conditions of erection shall be based upon the Contract Documents, which shall be representative of the work to be performed.

9.1.3. For contracts that stipulate a price per item, the work that is required to be performed by the Fabricator and the Erector shall be based upon the quantity and the character of the items that are described in the Contract Documents.

9.1.4. For contracts that stipulate unit prices for various categories of Structural Steel, the scope of work that is required to be performed by the Fabricator and the Erector shall be based upon the quantity, character and complexity of the items in each category as described in the Contract Documents, and shall also be representative of the work to be performed in each category.

9.2. Calculation of Weights

Unless otherwise specified in the contract, for contracts stipulating a price per pound for fabricated Structural Steel that is delivered and/or erected, the quantities of materials for payment shall be determined by the calculation of the gross weight of materials as shown in the Shop Drawings.

Commentary:
The standard procedure for calculation of weights that is described in this Code meets the need for a universally acceptable system for defining "pay weights" in contracts based upon the weight of delivered and/or erected materials. These procedures permits the Owner

to easily and accurately evaluate price-per-pound proposals from potential suppliers and enables all parties to a contract to have a clear and common understanding of the basis for payment.

The procedure in this Code affords a simple, readily understood method of calculation that will produce pay weights that are consistent throughout the industry and that may be easily verified by the Owner. While this procedure does not produce actual weights, it can be used by purchasers and suppliers to define a widely accepted basis for bidding and contracting for Structural Steel. However, any other system can be used as the basis for a contractual agreement. When other systems are used, both the supplier and the purchaser should clearly understand how the alternative procedure is handled.

9.2.1. The unit weight of steel shall be taken as 490 lb/ft^3 [7 850 kg/m^3]. The unit weight of other materials shall be in accordance with the manufacturer's published data for the specific product.

9.2.2. The weights of Standard Structural Shapes, plates and bars shall be calculated on the basis of Shop Drawings that show the actual quantities and dimensions of material to be fabricated, as follows:

(a) The weights of all Standard Structural Shapes shall be calculated using the nominal weight per ft [mass per m] and the detailed overall length.

(b) The weights of plates and bars shall be calculated using the detailed overall rectangular dimensions.

(c) When parts can be economically cut in multiples from material of larger dimensions, the weight shall be calculated on the basis of the theoretical rectangular dimensions of the material from which the parts are cut.

(d) When parts are cut from Standard Structural Shapes, leaving a non-standard section that is not useable on the same contract, the weight shall be calculated using the nominal weight per ft [mass per m] and the detailed overall length of the Standard Structural Shapes from which the parts are cut.

(e) Deductions shall not be made for material that is removed for cuts, copes, clips, blocks, drilling, punching, boring, slot milling, planing or weld joint preparation.

9.2.3. The items for which weights are shown in tables in the AISC Manual of Steel Construction shall be calculated on the basis of the tabulated weights shown therein.

9.2.4. The weights of items that are not shown in tables in the AISC Manual of Steel Construction shall be taken from the manufacturer's catalog and the manufacturer's shipping weight shall be used.

Commentary:
Many items that are weighed for payment purposes are not tabulated with weights in the AISC Manual of Steel Construction. These include, but are not limited to, Anchor Rods, clevises, turnbuckles, sleeve nuts, recessed-pin nuts, cotter pins and similar devices.

9.2.5. The weights of shop or field weld metal and protective coatings shall not be included in the calculated weight for the purposes of payment.

9.3. Revisions to the Contract Documents

Revisions to the Contract Documents shall be confirmed by change order or extra work order. Unless otherwise noted, the issuance of a revision to the Contract Documents shall constitute authorization by the Owner that the revision is Released for Construction. The contract price and schedule shall be adjusted in accordance with Sections 9.4 and 9.5.

9.4. Contract Price Adjustment

9.4.1. When the scope of work and responsibilities of the Fabricator and the Erector are changed from those previously established in the Contract Documents, an appropriate modification of the contract price shall be made. In computing the contract price adjustment, the Fabricator and the Erector shall consider the quantity of work that is added or deleted, the modifications in the character of the work and the timeliness of the change with respect to the status of material ordering, detailing, fabrication and erection operations.

Commentary:
The fabrication and erection of Structural Steel is a dynamic process. Typically, material is being acquired at the same time that the Shop and Erection Drawings are being prepared. Additionally, the fabrication shop will normally fabricate pieces in the order that the Structural Steel is being shipped and erected.

Items that are revised or placed on hold generally upset these relationships and can be very disruptive to the detailing, fabricating and erecting processes. The provisions in Sections 3.5, 4.4.2 and 9.3 are intended to minimize these disruptions so as to allow work to continue. Accordingly, it is required in this Code that the reviewer of requests for contract price adjustments recognize this and allow compensation to the Fabricator and the Erector for these inefficiencies and for the materials that are purchased and the detailing, fabrication and erection that has been performed, when affected by the change.

9.4.2. Requests for contract price adjustments shall be presented by the Fabricator and/or the Erector in a timely manner and shall be accompanied by a description of the change that is sufficient to permit evaluation and timely approval by the Owner.

9.4.3. Price-per-pound and price-per-item contracts shall provide for additions or deletions to the quantity of work that are made prior to the time the work is Released for Construction. When changes are made to the character of the work at any time, or when additions and/or deletions are made to the quantity of the work after it is released for detailing, fabrication or erection, the contract price shall be equitably adjusted.

9.5. Scheduling

9.5.1. The contract schedule shall state when the Design Drawings will be Released for Construction, if the Design Drawings are not available at the time of bidding, and when the job site, foundations, piers and abutments will be ready, free from obstructions and accessible to the Erector, so that erection can start at the designated time and continue without interference or delay caused by the Owner's Designated Representative for Construction or other trades.

9.5.2. The Fabricator and the Erector shall advise the Owner's Designated Representatives for Design and Construction, in a timely manner, of the effect any revision has on the contract schedule.

9.5.3. If the fabrication or erection is significantly delayed due to revisions to the requirements of the contract, or for other reasons that are the responsibility of others, the Fabricator and/or Erector shall be compensated for the additional costs incurred.

9.6. Terms of Payment

The Fabricator shall be paid for Mill Materials and fabricated product that is stored off the job site. Other terms of payment for the contract shall be outlined in the Contract Documents.

Commentary:

These terms include such items as progress payments for material, fabrication, erection, retainage, performance and payment bonds and final payment. If a performance or payment bond, paid for by the Owner, is required by contract, no retainage shall be required.

SECTION 10. ARCHITECTURALLY EXPOSED STRUCTURAL STEEL

10.1. General Requirements

When members are specifically designated as "Architecturally Exposed Structural Steel" or "AESS" in the Contract Documents, the requirements in Sections 1 through 9 shall apply as modified in Section 10. AESS members or components shall be fabricated and erected with the care and dimensional tolerances that are stipulated in Sections 10.2 through 10.4. The following additional information shall be provided in the Contract Documents when AESS is specified:

(a) Specific identification of members or components that are AESS;
(b) Fabrication and/or erection tolerances that are to be more restrictive than provided for in this Section, if any; and,
(c) Requirements, if any, of a mock-up panel or components for inspection and acceptance standards prior to the start of fabrication.

Commentary:
This Section of this Code defines additional requirements that apply only to members that are specifically designated by the Contract Documents as "Architecturally Exposed Structural Steel" (AESS). The rapidly increasing use of exposed Structural Steel as a medium of architectural expression has given rise to a demand for closer dimensional tolerances and smoother finished surfaces than required for ordinary Structural Steel framing.

This Section of this Code establishes standards for these requirements that take into account both the desired finished appearance and the abilities of the fabrication shop to produce the desired product. It should be pointed out that the term "Architecturally Exposed Structural Steel" (AESS), as covered in this Section, must be specified in the Contract Documents if the Fabricator is required to meet the fabricating standards in this Section, and applies only to that portion of the Structural Steel so identified.

AESS requirements usually involve significant cost in excess of that for Structural Steel that is fabricated in the absence of an AESS requirement. Therefore, the designation AESS should be applied rationally, with visual acceptance criteria that are appropriate for the distance at which the exposed element will be viewed in the completed structure. In order to avoid misunderstandings and to hold costs to a minimum, only those Structural Steel surfaces and Connections that will remain exposed and subject to normal view by pedestrians or occupants of the completed structure should be designated as AESS.

10.2. Fabrication

10.2.1. The permissible tolerances for out-of-square or out-of-parallel, depth, width and symmetry of rolled shapes shall be as specified in ASTM A6/A6M. Unless otherwise specified in the Contract Documents, the exact matching of abutting cross-sectional configurations shall not be necessary. The as-fabricated straightness tolerances of members shall be one-half of the standard camber and sweep tolerances in ASTM A6/A6M.

10.2.2. The tolerances on overall profile dimensions of members that are built-up from a series of Standard Structural Shapes, plates and/or bars by welding shall be taken as the accumulation of the variations that are permitted for the component parts in ASTM A6/A6M. The as-fabricated straightness tolerances for the member as a whole shall be one-half the standard camber and sweep tolerances for rolled shapes in ASTM A6/A6M.

10.2.3. Unless specific visual acceptance criteria for Weld Show-Through are specified in the Contract Documents, the members or components shall be acceptable as produced.

Commentary:
Weld Show-Through is generally a function of weld size and material thickness.

10.2.4. All copes, miters and cuts in surfaces that are exposed to view shall be made with uniform gaps of 1/8 in. [3 mm] if shown as open joints, or in reasonable contact if shown without gap.

10.2.5. All welds that are exposed to view shall be visually acceptable if they meet the requirements in AWS D1.1, except all groove and plug welds that are exposed to view shall not project more than 1/16 in. [2 mm] above the exposed surface. Finishing or grinding of welds shall not be necessary, unless such treatment is required to provide for clearances or fit of other components.

10.2.6. Erection marks or other painted marks shall not be made on those surfaces of weathering steel AESS members that are to be exposed in the completed structure. Unless otherwise specified in the Contract Documents, the Fabricator shall clean weathering steel AESS members to meet the requirements of SSPC-SP6.

10.2.7. Stamped or raised manufacturer's identification marks shall not be filled, ground or otherwise removed.

10.2.8. Seams of hollow structural sections shall be acceptable as produced. Seams shall be oriented away from view or as directed in the Contract Documents.

10.3. Delivery of Materials

The Fabricator shall use special care to avoid bending, twisting or otherwise distorting the Structural Steel.

10.4. Erection

10.4.1. The Erector shall use special care in unloading, handling and erecting the Structural Steel to avoid marking or distorting the Structural Steel. Care shall also be taken to minimize damage to any shop paint. If temporary braces or erection clips are used, care shall be taken to avoid the creation of unsightly surfaces upon removal. Tack welds shall be ground smooth and holes shall be filled with weld metal or body solder and smoothed by grinding or filing. The Erector shall plan and execute all operations in such a manner that the close fit and neat appearance of the structure will not be impaired.

10.4.2. Unless otherwise specified in the Contract Documents, AESS members and components shall be plumbed, leveled and aligned to a tolerance that is one-half that permitted for non-AESS members. To accommodate these erection tolerances for AESS, the Owner's Designated Representative for Design shall specify Connections between AESS members and non-AESS members, masonry, concrete and other supports as Adjustable Items, in order to provide the Erector with means for adjustment.

When AESS is backed with concrete, the Owner's Designated Representative for Construction shall provide sufficient shores, ties and strongbacks to prevent sagging, bulging or similar deformation of the AESS members due to the weight and pressure of the wet concrete.

AMERICAN INSTITUTE OF STEEL CONSTRUCTION, INC.
One East Wacker Drive, Suite 3100, Chicago, Illinois 60601-2001

Pub. No. S303 (20M500)

PART 17

MISCELLANEOUS DATA AND MATHEMATICAL INFORMATION

Table 17-1.
SI Equivalents of Standard U.S. Shape Profiles
W-Shapes

Shape	SI Equivalent	Shape	SI Equivalent	Shape	SI Equivalent
in. × lb/ft	mm × kg/m	in. × lb/ft	mm × kg/m	in. × lb/ft	mm × kg/m
W44×335	W1100×499	W33×387	W840×576	W24×370	W610×551
×290	×433	×354	×527	×335	×498
×262	×390	×318	×473	×306	×455
×230	×343	×291	×433	×279	×415
		×263	×392	×250	×372
W40×593	W1000×883	×241	×359	×229	×341
×503	×748	×221	×329	×207	×307
×431	×642	×201	×299	×192	×285
×397	×591			×176	×262
×372	×554	W33×169	W840×251	×162	×241
×362	×539	×152	×226	×146	×217
×324	×483	×141	×210	×131	×195
×297	×443	×130	×193	×117	×174
×277	×412	×118	×176	×104	×155
×249	×371				
×215	×321	W30×391	W760×582	W24×103	W610×153
×199	×296	×357	×531	×94	×140
		×326	×484	×84	×125
W40×392	W1000×584	×292	×434	×76	×113
×331	×494	×261	×389	×68	×101
×327	×486	×235	×350		
×278	×415	×211	×314	W24×62	W610×92
×264	×393	×191	×284	×55	×82
×235	×350	×173	×257		
×211	×314			W21×201	W530×300
×183	×272	W30×148	W760×220	×182	×272
×167	×249	×132	×196	×166	×248
×149	×222	×124	×185	×147	×219
		×116	×173	×132	×196
W36×798	W920×1188	×108	×161	×122	×182
×650	×967	×99	×147	×111	×165
×527	×784	×90	×134	×101	×150
×439	×653				
×393	×585	W27×539	W690×802	W21×93	W530×138
×359	×534	×368	×548	×83	×123
×328	×488	×336	×500	×73	×109
×300	×446	×307	×457	×68	×101
×280	×417	×281	×419	×62	×92
×260	×387	×258	×384	×55	×82
×245	×365	×235	×350	×48	×72
×230	×342	×217	×323		
		×194	×289	W21×57	W530×85
W36×256	W920×381	×178	×265	×50	×74
×232	×345	×161	×240	×44	×66
×210	×313	×146	×217		
×194	×289			W18×175	W460×260
×182	×271	W27×129	W690×192	×158	×235
×170	×253	×114	×170	×143	×213
×160	×238	×102	×152	×130	×193
×150	×223	×94	×140	×119	×177
×135	×201	×84	×125	×106	×158
				×97	×144
				×86	×128
				×76	×113

Table 17-1 (cont.). SI Equivalents of Standard U.S. Shape Profiles W-Shapes

Shape	SI Equivalent	Shape	SI Equivalent	Shape	SI Equivalent
in. × lb/ft	mm × kg/m	in. × lb/ft	mm × kg/m	in. × lb/ft	mm × kg/m
W18×71	W460×106	W14×53	W360×79	W10×45	W250×67
×65	×97	×48	×72	×39	×58
×60	×89	×43	×64	×33	×49.1
×55	×82	W14×38	W360×57.8	W10×30	W250×44.8
×50	×74	×34	×51	×26	×38.5
W18×46	W460×68	×30	×44	×22	×32.7
×40	×60	W14×26	W360×39	W10×19	W250×28.4
×35	×52	×22	×32.9	×17	×25.3
W16×100	W410×149	W12×336	W310×500	×15	×22.3
×89	×132	×305	×454	×12	×17.9
×77	×114	×279	×415	W8×67	W200×100
×67	×100	×252	×375	×58	×86
W16×57	W410×85	×230	×342	×48	×71
×50	×75	×210	×313	×40	×59
×45	×67	×190	×283	×35	×52
×40	×60	×170	×253	×31	×46.1
×36	×53	×152	×226	W8×28	W200×41.7
W16×31	W410×46.1	×136	×202	×24	×35.9
×26	×38.8	×120	×179	W8×21	W200×31.3
W14×808	W360×1202	×106	×158	×18	×26.6
×730	×1086	×96	×143	W8×15	W200×22.5
×665	×990	×87	×129	×13	×19.3
×605	×900	×79	×117	×10	×15
×550	×818	×72	×107	W6×25	W150×37.1
×500	×744	×65	×97	×20	×29.8
×455	×677	W12×58	W310×86	×15	×22.5
×426	×634	×53	×79	W6×16	W150×24
×398	×592	W12×50	W310×74	×12	×18
×370	×551	×45	×67	×9	×13.5
×342	×509	×40	×60	×8.5	×13
×311	×463	W12×35	W310×52	W5×19	W130×28.1
×283	×421	×30	×44.5	×16	×23.8
×257	×382	×26	×38.7	W4×13	W100×19.3
×233	×347	W12×22	W310×32.7		
×211	×314	×19	×28.3		
×193	×287	×16	×23.8		
×176	×262	×14	×21		
×159	×237	W10×112	W250×167		
×145	×216	×100	×149		
W14×132	W360×196	×88	×131		
×120	×179	×77	×115		
×109	×162	×68	×101		
×99	×147	×60	×89		
×90	×134	×54	×80		
W14×82	W360×122	×49	×73		
×74	×110				
×68	×101				
×61	×91				

Table 17-2.
SI Equivalents of Standard U.S. Shape Profiles
M-, S- and HP-Shapes

Shape	SI Equivalent	Shape	SI Equivalent	Shape	SI Equivalent
in. × lb/ft	mm × kg/m	in. × lb/ft	mm × kg/m	in. × lb/ft	mm × kg/m
M12×11.8	M310×17.6	S24×121	S610×180	HP14×117	HP360×174
×10.8	×16.1	×106	×158	×102	×152
×10	×14.9	S24×100	S610×149	×89	×132
M10×9	M250×13.4	×90	×134	×73	×108
×8	×11.9	×80	×119	HP12×84	HP310×125
×7.5	×11.2	S20×96	S510×143	×74	×110
M8×6.5	M200×9.7	×86	×128	×63	×93
×6.2	×9.2	S20×75	S510×112	×53	×79
M6×4.4	M150×6.6	×66	×98.2	HP10×57	HP250×85
×3.7	×5.5	S18×70	S460×104	×42	×62
M5×18.9	M130×28.1	×54.7	×81.4	HP8×36	HP200×53
M4×6	M100×8.9	S15×50	S380×74		
		×42.9	×64		
		S12×50	S310×74		
		×40.8	×60.7		
		S12×35	S310×52		
		×31.8	×47.3		
		S10×35	S250×52		
		×25.4	×37.8		
		S8×23	S200×34		
		×18.4	×27.4		
		S6×17.25	S150×25.7		
		×12.5	×18.6		
		S5×10	S130×15		
		S4×9.5	S100×14.1		
		×7.7	×11.5		
		S3×7.5	S75×11.2		
		×5.7	×8.5		

Table 17-3. SI Equivalents of Standard U.S. Shape Profiles Channels

Shape	SI Equivalent
in. × lb/ft	mm × kg/m
C15×50	C380×74
×40	×60
×33.9	×50.4
C12×30	C310×45
×25	×37
×20.7	×30.8
C10×30	C250×45
×25	×37
×20	×30
×15.3	×22.8
C9×20	C230×30
×15	×22
×13.4	×19.9
C8×18.75	C200×27.9
×13.75	×20.5
×11.5	×17.1
C7×14.75	C180×22
×12.25	×18.2
×9.8	×14.6
C6×13	C150×19.3
×10.5	×15.6
×8.2	×12.2
C5×9	C130×13
×6.7	×10.4
C4×7.25	C100×10.8
×5.4	×8
×4.5	×6.7
C3×6	C75×8.9
×5	×7.4
×4.1	×6.1
×3.5	×5.2

Shape	SI Equivalent
in. × lb/ft	mm × kg/m
MC18×58	MC460×86
×51.9	×77.2
×45.8	×68.2
×42.7	×63.5
MC13×50	MC330×74
×40	×60
×35	×52
×31.8	×47.3
MC12×50	MC310×74
×45	×67
×40	×60
×35	×52
×31	×46
MC12×10.6	MC310×15.8
MC10×41.1	MC250×61.2
×33.6	×50
×28.5	×42.4
MC10×25	MC250×37
×22	×33
MC10×8.4	MC250×12.5
MC9×25.4	MC230×37.8
×23.9	×35.6
MC8×22.8	MC200×33.9
×21.4	×31.8
MC8×20	MC200×29.8
×18.7	×27.8
MC8×8.5	MC200×12.6
MC7×22.7	MC180×33.8
×19.1	×28.4
MC6×18	MC150×26.8
×15.3	×22.8
MC6×16.3	MC150×24.3
×15.1	×22.5
MC6×12	MC150×17.9

Table 17-4.
SI Equivalents of Standard U.S. Shape Profiles
Angles

Shape in. × in. × in.	SI Equivalent mm × mm × mm
L8×8×1 1/8	L203×203×28.6
×1	×25.4
×7/8	×22.2
×3/4	×19
×5/8	×15.9
×9/16	×14.3
×1/2	×12.7
L8×6×1	L203×152×25.4
×7/8	×22.2
×3/4	×19
×5/8	×15.9
×9/16	×14.3
×1/2	×12.7
×7/16	×11.1
L8×4×1	L203×102×25.4
×7/8	×22.2
×3/4	×19
×5/8	×15.9
×9/16	×14.3
×1/2	×12.7
×7/16	×11.1
L7×4×3/4	L178×102×19
×5/8	×15.9
×1/2	×12.7
×7/16	×11.1
×3/8	×9.5
L6×6×1	L152×152×25.4
×7/8	×22.2
×3/4	×19
×5/8	×15.9
×9/16	×14.3
×1/2	×12.7
×7/16	×11.1
×3/8	×9.5
×5/16	×7.9
L6×4×7/8	L152×102×22.2
×3/4	×19
×5/8	×15.9
×9/16	×14.3
×1/2	×12.7
×7/16	×11.1
×3/8	×9.5
×5/16	×7.9
L6×3 1/2×1/2	L152×89×12.7
×3/8	×9.5
×5/16	×7.9
L5×5×7/8	L127×127×22.2
×3/4	×19
×5/8	×15.9
×1/2	×12.7
×7/16	×11.1
×3/8	×9.5
×5/16	×7.9
L5×3 1/2×3/4	L127×89×19
×5/8	×15.9
×1/2	×12.7
×3/8	×9.5
×5/16	×7.9
×1/4	×6.4
L5×3×1/2	L127×76×12.7
×7/16	×11.1
×3/8	×9.5
×5/16	×7.9
×1/4	×6.4
L4×4×3/4	L102×102×19
×5/8	×15.9
×1/2	×12.7
×7/16	×11.1
×3/8	×9.5
×5/16	×7.9
×1/4	×6.4
L4×3 1/2×1/2	L102×89×12.7
×3/8	×9.5
×5/16	×7.9
×1/4	×6.4
L4×3×5/8	L102×76×15.9
×1/2	×12.7
×3/8	×9.5
×5/16	×7.9
×1/4	×6.4
L3 1/2×3 1/2×1/2	L89×89×12.7
×7/16	×11.1
×3/8	×9.5
×5/16	×7.9
×1/4	×6.4
L3 1/2×3×1/2	L89×76×12.7
×7/16	×11.1
×3/8	×9.5
×5/16	×7.9
×1/4	×6.4
L3 1/2×2 1/2×1/2	L89×64×12.7
×3/8	×9.5
×5/16	×7.9
×1/4	×6.4
L3×3×1/2	L76×76×12.7
×7/16	×11.1
×3/8	×9.5
×5/16	×7.9
×1/4	×6.4
×3/16	×4.8
L3×2 1/2×1/2	L76×64×12.7
×7/16	×11.1
×3/8	×9.5
×5/16	×7.9
×1/4	×6.4
×3/16	×4.8
L3×2×1/2	L76×51×12.7
×3/8	×9.5
×5/16	×7.9
×1/4	×6.4
×3/16	×4.8
L2 1/2×2 1/2×1/2	L64×64×12.7
×3/8	×9.5
×5/16	×7.9
×1/4	×6.4
×3/16	×4.8
L2 1/2×2×3/8	L64×51×9.5
×5/16	×7.9
×1/4	×6.4
×3/16	×4.8
L2×2×3/8	L51×51×9.5
×5/16	×7.9
×1/4	×6.4
×3/16	×4.8
×1/8	×3.2

Table 17-5.
SI Equivalents of Standard U.S. Shape Profiles
WT-Shapes

Shape	SI Equivalent	Shape	SI Equivalent	Shape	SI Equivalent
in. × lb/ft	**mm × kg/m**	**in. × lb/ft**	**mm × kg/m**	**in. × lb/ft**	**mm × kg/m**
WT22×167.5	WT550×249.5	WT16.5×193.5	WT420×288	WT12×185	WT305×275.5
×145	×216.5	×177	×263.5	×167.5	×249
×131	×195	×159	×236.5	×153	×227.5
×115	×171.5	×145.5	×216.5	×139.5	×207.5
		×131.5	×196	×125	×186
WT20×296.5	WT500×441.5	×120.5	×179.5	×114.5	×170.5
×251.5	×374	×110.5	×164.5	×103.5	×153.5
×215.5	×321	×100.5	×149.5	×96	×142.5
×198.5	×295.5			×88	×131
×186	×277	WT16.5×84.5	WT420×125.5	×81	×120.5
×181	×269.5	×76	×113	×73	×108.5
×162	×241.5	×70.5	×105	×65.5	×97.5
×148.5	×221.5	×65	×96.5	×58.5	×87
×138.5	×206	×59	×88	×52	×77.5
×124.5	×185.5				
×107.5	×160.5	WT15×195.5	WT380×291	WT12×51.5	WT305×76.5
×99.5	×148	×178.5	×265.5	×47	×70
		×163	×242	×42	×62.5
WT20×196	WT500×292	×146	×217	×38	×56.5
×165.5	×247	×130.5	×194.5	×34	×50.5
×163.5	×243	×117.5	×175		
×139	×207.5	×105.5	×157	WT12×31	WT305×46
×132	×196.5	×95.5	×142	×27.5	×41
×117.5	×175	×86.5	×128.5		
×105.5	×157			WT10.5×100.5	WT265×150
×91.5	×136	WT15×74	WT380×110	×91	×136
×83.5	×124.5	×66	×98	×83	×124
×74.5	×111	×62	×92.5	×73.5	×109.5
		×58	×86.5	×66	×98
WT18×399	WT460×594	×54	×80.5	×61	×91
×325	×483.5	×49.5	×73.5	×55.5	×82.5
×263.5	×392	×45	×67	×50.5	×75
×219.5	×326.5				
×196.5	×292.5	WT13.5×269.5	WT345×401	WT10.5×46.5	WT265×69
×179.5	×267	×184	×274	×41.5	×61.5
×164	×244	×168	×250	×36.5	×54.5
×150	×223	×153.5	×228.5	×34	×50.5
×140	×208.5	×140.5	×209.5	×31	×46
×130	×193.5	×129	×192	×27.5	×41
×122.5	×182.5	×117.5	×175	×24	×36
×115	×171	×108.5	×161.5		
		×97	×144.5	WT10.5×28.5	WT265×42.5
WT18×128	WT460×190.5	×89	×132.5	×25	×37
×116	×172.5	×80.5	×120	×22	×33
×105	×156.5	×73	×108.5		
×97	×144.5			WT9×87.5	WT230×130
×91	×135.5	WT13.5×64.5	WT345×96	×79	×117.5
×85	×126.5	×57	×85	×71.5	×106.5
×80	×119	×51	×76	×65	×96.5
×75	×111.5	×47	×70	×59.5	×88.5
×67.5	×100.5	×42	×62.5	×53	×79
				×48.5	×72
				×43	×64
				×38	×56.5

Table 17-5 (cont.).
SI Equivalents of Standard U.S. Shape Profiles
WT-Shapes

Shape	SI Equivalent
in. × lb/ft	mm × kg/m
WT9×35.5	WT230×53
×32.5	×48.5
×30	×44.5
×27.5	×41
×25	×37
WT9×23	WT230×34
×20	×30
×17.5	×26
WT8×50	WT205×74.5
×44.5	×66
×38.5	×57
×33.5	×50
WT8×28.5	WT205×42.5
×25	×37.5
×22.5	×33.5
×20	×30
×18	×26.5
WT8×15.5	WT205×23.05
×13	×19.4
WT7×404	WT180×601
×365	×543
×332.5	×495
×302.5	×450
×275	×409
×250	×372
×227.5	×338.5
×213	×317
×199	×296
×185	×275.5
×171	×254.5
×155.5	×231.5
×141.5	×210.5
×128.5	×191
×116.5	×173.5
×105.5	×157
×96.5	×143.5
×88	×131
×79.5	×118.5
×72.5	×108
WT7×66	WT180×98
×60	×89.5
×54.5	×81
×49.5	×73.5
×45	×67
WT7×41	WT180×61
×37	×55
×34	×50.5
×30.5	×45.5

Shape	SI Equivalent
in. × lb/ft	mm × kg/m
WT7×26.5	WT180×39.5
×24	×36
×21.5	×32
WT7×19	WT180×28.9
×17	×25.5
×15	×22
WT7×13	WT180×19.5
×11	×16.45
WT6×168	WT155×250
×152.5	×227
×139.5	×207.5
×126	×187.5
×115	×171
×105	×156.5
×95	×141.5
×85	×126.5
×76	×113
×68	×101
×60	×89.5
×53	×79
×48	×71.5
×43.5	×64.5
×39.5	×58.5
×36	×53.5
×32.5	×48.5
WT6×29	WT155×43
×26.5	×39.5
WT6×25	WT155×37
×22.5	×33.5
×20	×30
WT6×17.5	WT155×26
×15	×22.25
×13	×19.35
WT6×11	WT155×16.35
×9.5	×14.15
×8	×11.9
×7	×10.5
WT5×56	WT125×83.5
×50	×74.5
×44	×65.5
×38.5	×57.5
×34	×50.5
×30	×44.5
×27	×40
×24.5	×36.5

Shape	SI Equivalent
in. × lb/ft	mm × kg/m
WT5×22.5	WT125×33.5
×19.5	×29
×16.5	×24.55
WT5×15	WT125×22.4
×13	×19.25
×11	×16.35
WT5×9.5	WT125×14.2
×8.5	×12.65
×7.5	×11.15
×6	×8.95
WT4×33.5	WT100×50
×29	×43
×24	×35.5
×20	×29.5
×17.5	×26
×15.5	×23.05
WT4×14	WT100×20.85
×12	×17.95
WT4×10.5	WT100×15.65
×9	×13.3
WT4×7.5	WT100×11.25
×6.5	×9.65
×5	×7.5
WT3×12.5	WT75×18.55
×10	×14.9
×7.5	×11.25
WT3×8	WT75×12
×6	×9
×4.5	×6.75
×4.25	×6.5
WT2.5×9.5	WT65×14.05
×8	×11.9
WT2×6.5	WT50×9.65

Table 17-6.
SI Equivalents of Standard U.S. Shape Profiles
MT- and ST-Shapes

Shape in. × lb/ft	SI Equivalent mm × kg/m	Shape in. × lb/ft	SI Equivalent mm × kg/m
MT6×5.9	MT155×8.8	ST12×60.5	ST305×90
×5.4	×8.05	×53	×79
×5	×7.45		
		ST12×50	ST305×74.5
MT5×4.5	MT125×6.7	×45	×67
×4	×5.95	×40	×59.5
×3.75	×5.6		
		ST10×48	ST255×71.5
MT4×3.25	MT100×4.85	×43	×64
×3.1	×4.6		
		ST10×37.5	ST255×56
MT3×2.2	MT75×3.3	×33	×49.1
×1.85	×2.75		
		ST9×35	ST230×52
MT2.5×9.45	MT65×14.05	×27.35	×40.7
MT2×3	MT50×4.45	ST7.5×25	ST190×37
		×21.45	×32
		ST6×25	ST155×37
		×20.4	×30.35
		ST6×17.5	ST155×26
		×15.9	×23.65
		ST5×17.5	ST125×26
		×12.7	×18.9
		ST4×11.5	ST100×17
		×9.2	×13.7
		ST3×8.625	ST75×12.85
		×6.25	×9.3
		ST2.5×5	ST65×7.5
		ST2×4.75	ST50×7.05
		×3.85	×5.75
		ST1.5×3.75	ST37.5×5.6
		×2.85	×4.25

Table 17-7. SI Equivalents of Standard U.S. Shape Profiles Rectangular HSS

Shape	SI Equivalent
in. × in. × in.	mm × mm × mm
HSS20×12×5/8	HSS508×304.8×15.9
×1/2	×12.7
×3/8	×9.5
×5/16	×7.9
HSS20×8×5/8	HSS508×203.2×15.9
×1/2	×12.7
×3/8	×9.5
×5/16	×7.9
HSS20×4×1/2	HSS508×101.6×12.7
×3/8	×9.5
×5/16	×7.9
HSS18×12×5/8	HSS457.2×304.8×15.9
×1/2	×12.7
×3/8	×9.5
HSS18×6×5/8	HSS457.2×152.4×15.9
×1/2	×12.7
×3/8	×9.5
×5/16	×7.9
×1/4	×6.4
HSS16×16×5/8	HSS406.4×406.4×15.9
×1/2	×12.7
×3/8	×9.5
×5/16	×7.9
HSS16×12×5/8	HSS406.4×304.8×15.9
×1/2	×12.7
×3/8	×9.5
×5/16	×7.9
HSS16×8×5/8	HSS406.4×203.2×15.9
×1/2	×12.7
×3/8	×9.5
×5/16	×7.9
HSS16×4×1/2	HSS406.4×101.6×12.7
×3/8	×9.5
×5/16	×7.9
HSS14×14×5/8	HSS355.6×355.6×15.9
×1/2	×12.7
×3/8	×9.5
×5/16	×7.9
HSS14×12×1/2	HSS355.6×304.8×12.7
×3/8	×9.5
HSS14×10×5/8	HSS355.6×254×15.9
×1/2	×12.7
×3/8	×9.5
×5/16	×7.9
×1/4	×6.4
HSS14×6×5/8	HSS355.6×152.4×15.9
×1/2	×12.7
×3/8	×9.5
×5/16	×7.9
×1/4	×6.4
×3/16	×4.8
HSS14×4×5/8	HSS355.6×101.6×15.9
×1/2	×12.7
×3/8	×9.5
×5/16	×7.9
×1/4	×6.4
×3/16	×4.8
HSS12×12×5/8	HSS304.8×304.8×15.9
×1/2	×12.7
×3/8	×9.5
×5/16	×7.9
×1/4	×6.4
HSS12×10×1/2	HSS304.8×254×12.7
×3/8	×9.5
×5/16	×7.9
×1/4	×6.4
HSS12×8×5/8	HSS304.8×203.2×15.9
×1/2	×12.7
×3/8	×9.5
×5/16	×7.9
×1/4	×6.4
×3/16	×4.8
HSS12×6×5/8	HSS304.8×152.4×15.9
×1/2	×12.7
×3/8	×9.5
×5/16	×7.9
×1/4	×6.4
×3/16	×4.8
HSS12×4×5/8	HSS304.8×101.6×15.9
×1/2	×12.7
×3/8	×9.5
×5/16	×7.9
×1/4	×6.4
×3/16	×4.8

Table 17-7 (cont.). SI Equivalents of Standard U.S. Shape Profiles Rectangular HSS

Shape in. × in. × in.	SI Equivalent mm × mm × mm
HSS12×3½×3/8	HSS304.8×88.9×9.5
×5/16	×7.9
HSS12×3×5/16	HSS304.8×76.2×7.9
×1/4	×6.4
×3/16	×4.8
HSS12×2×1/4	HSS304.8×50.8×6.4
×3/16	×4.8
HSS10×10×5/8	HSS254×254×15.9
×1/2	×12.7
×3/8	×9.5
×5/16	×7.9
×1/4	×6.4
×3/16	×4.8
HSS10×8×1/2	HSS254×203.2×12.7
×3/8	×9.5
×5/16	×7.9
×1/4	×6.4
×3/16	×4.8
HSS10×6×5/8	HSS254×152.4×15.9
×1/2	×12.7
×3/8	×9.5
×5/16	×7.9
×1/4	×6.4
×3/16	×4.8
HSS10×5×3/8	HSS254×127×9.5
×5/16	×7.9
×1/4	×6.4
×3/16	×4.8
HSS10×4×5/8	HSS254×101.6×15.9
×1/2	×12.7
×3/8	×9.5
×5/16	×7.9
×1/4	×6.4
×3/16	×4.8
HSS10×3½×3/16	HSS254×88.9×4.8
HSS10×3×3/8	HSS254×76.2×9.5
×5/16	×7.9
×1/4	×6.4
×3/16	×4.8
×1/8	×3.2
HSS10×2×3/8	HSS254×50.8×9.5
×5/16	×7.9
×1/4	×6.4
×3/16	×4.8

Shape in. × in. × in.	SI Equivalent mm × mm × mm
HSS9×7×5/8	HSS228.6×177.8×15.9
×1/2	×12.7
×3/8	×9.5
×5/16	×7.9
×1/4	×6.4
×3/16	×4.8
HSS9×5×5/8	HSS228.6×127×15.9
×1/2	×12.7
×3/8	×9.5
×5/16	×7.9
×1/4	×6.4
×3/16	×4.8
HSS9×3×1/2	HSS228.6×76.2×12.7
×3/8	×9.5
×5/16	×7.9
×1/4	×6.4
×3/16	×4.8
HSS8×8×5/8	HSS203.2×203.2×15.9
×1/2	×12.7
×3/8	×9.5
×5/16	×7.9
×1/4	×6.4
×3/16	×4.8
HSS8×6×5/8	HSS203.2×152.4×15.9
×1/2	×12.7
×3/8	×9.5
×5/16	×7.9
×1/4	×6.4
×3/16	×4.8
HSS8×4×5/8	HSS203.2×101.6×15.9
×1/2	×12.7
×3/8	×9.5
×5/16	×7.9
×1/4	×6.4
×3/16	×4.8
×1/8	×3.2
HSS8×3×1/2	HSS203.2×76.2×12.7
×3/8	×9.5
×5/16	×7.9
×1/4	×6.4
×3/16	×4.8
×1/8	×3.2
HSS8×2×3/8	HSS203.2×50.8×9.5
×5/16	×7.9
×1/4	×6.4
×3/16	×4.8
×1/8	×3.2

Table 17-7 (cont.). SI Equivalents of Standard U.S. Shape Profiles Rectangular HSS

Shape	SI Equivalent
in. × in. × in.	mm × mm × mm
HSS7×7×5/8	HSS177.8×177.8×15.9
×1/2	×12.7
×3/8	×9.5
×5/16	×7.9
×1/4	×6.4
×3/16	×4.8
HSS7×5×5/8	HSS177.8×127×15.9
×1/2	×12.7
×3/8	×9.5
×5/16	×7.9
×1/4	×6.4
×3/16	×4.8
×1/8	×3.2
HSS7×4×1/2	HSS177.8×101.6×12.7
×3/8	×9.5
×5/16	×7.9
×1/4	×6.4
×3/16	×4.8
×1/8	×3.2
HSS7×3×1/2	HSS177.8×76.2×12.7
×3/8	×9.5
×5/16	×7.9
×1/4	×6.4
×3/16	×4.8
×1/8	×3.2
HSS6×6×5/8	HSS152.4×152.4×15.9
×1/2	×12.7
×3/8	×9.5
×5/16	×7.9
×1/4	×6.4
×3/16	×4.8
×1/8	×3.2
HSS6×5×3/8	HSS152.4×127×9.5
×5/16	×7.9
×1/4	×6.4
×3/16	×4.8
HSS6×4×1/2	HSS152.4×101.6×12.7
×3/8	×9.5
×5/16	×7.9
×1/4	×6.4
×3/16	×4.8
×1/8	×3.2
HSS6×3×1/2	HSS152.4×76.2×12.7
×3/8	×9.5
×5/16	×7.9
×1/4	×6.4
×3/16	×4.8
×1/8	×3.2
HSS6×2×3/8	HSS152.4×50.8×9.5
×5/16	×7.9
×1/4	×6.4
×3/16	×4.8
×1/8	×3.2
HSS5½×5½×3/8	HSS139.7×139.7×9.5
×5/16	×7.9
×1/4	×6.4
×3/16	×4.8
×1/8	×3.2
HSS5×5×1/2	HSS127×127×12.7
×3/8	×9.5
×5/16	×7.9
×1/4	×6.4
×3/16	×4.8
×1/8	×3.2
HSS5×4×1/2	HSS127×101.6×12.7
×3/8	×9.5
×5/16	×7.9
×1/4	×6.4
×3/16	×4.8
HSS5×3×1/2	HSS127×76.2×12.7
×3/8	×9.5
×5/16	×7.9
×1/4	×6.4
×3/16	×4.8
×1/8	×3.2
HSS5×2½×1/4	HSS127×63.5×6.4
×3/16	×4.8
×1/8	×3.2
HSS5×2×3/8	HSS127×50.8×9.5
×5/16	×7.9
×1/4	×6.4
×3/16	×4.8
×1/8	×3.2
HSS4½×4½×1/2	HSS114.3×114.3×12.7
×3/8	×9.5
×5/16	×7.9
×1/4	×6.4
×3/16	×4.8
×1/8	×3.2
HSS4×4×1/2	HSS101.6×101.6×12.7
×3/8	×9.5
×5/16	×7.9
×1/4	×6.4
×3/16	×4.8
×1/8	×3.2

Table 17-7 (cont.). SI Equivalents of Standard U.S. Shape Profiles Rectangular HSS

Shape	SI Equivalent
in. × in. × in.	mm × mm × mm
HSS4×3×3/8	HSS101.6×76.2×9.5
×5/16	×7.9
×1/4	×6.4
×3/16	×4.8
×1/8	×3.2
HSS4×2½×5/16	HSS101.6×63.5×7.9
×1/4	×6.4
×3/16	×4.8
HSS4×2×3/8	HSS101.6×50.8×9.5
×5/16	×7.9
×1/4	×6.4
×3/16	×4.8
×1/8	×3.2
HSS3½×3½×3/8	HSS88.9×88.9×9.5
×5/16	×7.9
×1/4	×6.4
×3/16	×4.8
×1/8	×3.2
HSS3½×2½×3/8	HSS88.9×63.5×9.5
×5/16	×7.9
×1/4	×6.4
×3/16	×4.8
×1/8	×3.2
HSS3×3×3/8	HSS76.2×76.2×9.5
×5/16	×7.9
×1/4	×6.4
×3/16	×4.8
×1/8	×3.2
HSS3×2½×5/16	HSS76.2×63.5×7.9
×1/4	×6.4
×3/16	×4.8
×1/8	×3.2
HSS3×2×5/16	HSS76.2×50.8×7.9
×1/4	×6.4
×3/16	×4.8
×1/8	×3.2
HSS3×1½×1/4	HSS76.2×38.1×6.4
×3/16	×4.8
×1/8	×3.2
HSS3×1×1/8	HSS76.2×25.4×3.2
HSS2½×2½×5/16	HSS63.5×63.5×7.9
×1/4	×6.4
×3/16	×4.8
×1/8	×3.2
HSS2½×1½×1/4	HSS63.5×38.1×6.4
×3/16	×4.8
×1/8	×3.2
HSS2¼×2¼×1/4	HSS57.2×57.2×6.4
×3/16	×4.8
×1/8	×3.2
HSS2×2×1/4	HSS50.8×50.8×6.4
×3/16	×4.8
×1/8	×3.2
HSS2×1½×3/16	HSS50.8×38.1×4.8
HSS2×1×3/16	HSS50.8×25.4×4.8
×1/8	×3.2
HSS1¾×1¾×3/16	HSS44.5×44.5×4.8
HSS1⅝×1⅝×3/16	HSS41.3×41.3×4.8
×1/8	×3.2
HSS1½×1½×3/16	HSS38.1×38.1×4.8
×1/8	×3.2
HSS1¼×1¼×3/16	HSS31.8×31.8×4.8
×1/8	×3.2

Table 17-8.
SI Equivalents of Standard U.S. Shape Profiles
Round HSS

Shape in. × in.	SI Equivalent mm × mm
HSS20.000×0.500	HSS508×12.7
×0.375	×9.5
HSS18.000×0.500	HSS457.2×12.7
×0.375	×9.5
HSS16.000×0.500	HSS406.4×12.7
×0.438	×11.1
×0.375	×9.5
×0.312	×7.9
HSS14.000×0.500	HSS355.6×12.7
×0.375	×9.5
×0.312	×7.9
HSS12.750×0.500	HSS323.9×12.7
×0.375	×9.5
×0.250	×6.4
HSS12.500×0.625	HSS317.5×15.9
×0.500	×12.7
×0.375	×9.5
×0.312	×7.9
×0.250	×6.4
×0.188	×4.8
HSS11.250×0.625	HSS285.8×15.9
×0.500	×12.7
×0.375	×9.5
×0.312	×7.9
×0.250	×6.4
×0.188	×4.8
HSS10.750×0.500	HSS273.1×12.7
×0.250	×6.4
HSS10.000×0.625	HSS254×15.9
×0.500	×12.7
×0.375	×9.5
×0.312	×7.9
×0.250	×6.4
×0.188	×4.8
HSS9.625×0.500	HSS244.5×12.7
×0.375	×9.5
×0.312	×7.9
×0.250	×6.4
×0.188	×4.8
HSS8.750×0.500	HSS222.3×12.7
×0.375	×9.5
×0.312	×7.9
×0.250	×6.4
×0.188	×4.8
HSS8.625×0.500	HSS219.1×12.7
×0.375	×9.5
×0.322	×8.2
×0.250	×6.4
×0.188	×4.8
HSS7.625×0.125	HSS193.7×3.2
HSS7.500×0.500	HSS190.5×12.7
×0.375	×9.5
×0.312	×7.9
×0.250	×6.4
×0.188	×4.8
HSS7.000×0.500	HSS177.8×12.7
×0.375	×9.5
×0.312	×7.9
×0.250	×6.4
×0.188	×4.8
×0.125	×3.2
HSS6.875×0.500	HSS174.6×12.7
×0.375	×9.5
×0.312	×7.9
×0.250	×6.4
×0.188	×4.8
HSS6.625×0.500	HSS168.3×12.7
×0.432	×11
×0.375	×9.5
×0.312	×7.9
×0.280	×7.1
×0.250	×6.4
×0.188	×4.8
×0.125	×3.2
HSS6.125×0.500	HSS155.6×12.7
×0.375	×9.5
×0.312	×7.9
×0.250	×6.4
×0.188	×4.8
HSS6.000×0.500	HSS152.4×12.7
×0.375	×9.5
×0.312	×7.9
×0.280	×7.1
×0.250	×6.4
×0.188	×4.8
×0.125	×3.2
HSS5.563×0.375	HSS141.3×9.5
×0.258	×6.6
×0.188	×4.8
×0.134	×3.4

Table 17-8 (cont.). SI Equivalents of Standard U.S. Shape Profiles Round HSS

Shape	SI Equivalent	Shape	SI Equivalent
in. × in.	mm × mm	in. × in.	mm × mm
HSS5.500×0.500	HSS139.7×12.7	HSS2.375×0.250	HSS60.3×6.4
×0.375	×9.5	×0.218	×5.5
×0.258	×6.6	×0.188	×4.8
HSS5.000×0.500	HSS127×12.7	×0.154	×3.9
×0.375	×9.5	×0.125	×3.2
×0.312	×7.9	HSS1.900×0.145	HSS48.3×3.7
×0.258	×6.6		
×0.250	×6.4	HSS1.660×0.140	HSS42.2×3.6
×0.188	×4.8		
×0.125	×3.2		
HSS4.500×0.337	HSS114.3×8.6		
×0.237	×6		
×0.188	×4.8		
×0.125	×3.2		
HSS4.000×0.337	HSS101.6×8.6		
×0.313	×8		
×0.250	×6.4		
×0.237	×6		
×0.226	×5.7		
×0.220	×5.6		
×0.188	×4.8		
×0.125	×3.2		
HSS3.500×0.313	HSS88.9×8		
×0.300	×7.6		
×0.250	×6.4		
×0.216	×5.5		
×0.203	×5.2		
×0.188	×4.8		
×0.125	×3.2		
HSS3.000×0.300	HSS76.2×7.6		
×0.250	×6.4		
×0.216	×5.5		
×0.203	×5.2		
×0.188	×4.8		
×0.152	×3.9		
×0.134	×3.4		
×0.120	×3		
HSS2.875×0.250	HSS73×6.4		
×0.203	×5.2		
×0.188	×4.8		
×0.125	×3.2		
HSS2.500×0.250	HSS63.5×6.4		
×0.188	×4.8		
×0.125	×3.2		

Table 17-9.
SI Equivalents of Standard U.S. Shape Profiles
Steel Pipe

Standard Weight		Extra Strong		Double-Extra Strong	
Shape	SI Equivalent	Shape	SI Equivalent	Shape	SI Equivalent
PIPE 1/2 STD	PIPE 13 STD	PIPE 1/2 XS	PIPE 13 XS	PIPE 2 XXS	PIPE 51 XXS
PIPE 3/4 STD	PIPE 19 STD	PIPE 3/4 XS	PIPE 19 XS	PIPE 21/2 XXS	PIPE 64 XXS
PIPE 1 STD	PIPE 25 STD	PIPE 1 XS	PIPE 25 XS	PIPE 3 XXS	PIPE 75 XXS
PIPE 11/4 STD	PIPE 32 STD	PIPE 11/4 XS	PIPE 32 XS	PIPE 4 XXS	PIPE 102 XXS
PIPE 11/2 STD	PIPE 38 STD	PIPE 11/2 XS	PIPE 38 XS	PIPE 5 XXS	PIPE 127 XXS
PIPE 2 STD	PIPE 51 STD	PIPE 2 XS	PIPE 51 XS	PIPE 6 XXS	PIPE 152 XXS
PIPE 21/2 STD	PIPE 64 STD	PIPE 21/2 XS	PIPE 64 XS	PIPE 8 XXS	PIPE 203 XXS
PIPE 3 STD	PIPE 75 STD	PIPE 3 XS	PIPE 75 XS		
PIPE 31/2 STD	PIPE 89 STD	PIPE 31/2 XS	PIPE 89 XS		
PIPE 4 STD	PIPE 102 STD	PIPE 4 XS	PIPE 102 XS		
PIPE 5 STD	PIPE 127 STD	PIPE 5 XS	PIPE 127 XS		
PIPE 6 STD	PIPE 152 STD	PIPE 6 XS	PIPE 152 XS		
PIPE 8 STD	PIPE 203 STD	PIPE 8 XS	PIPE 203 XS		
PIPE 10 STD	PIPE 254 STD	PIPE 10 XS	PIPE 254 XS		
PIPE 12 STD	PIPE 310 STD	PIPE 12 XS	PIPE 310 XS		

Table 17-10.
Wire and Sheet Metal Gages
Equivalent thickness in decimals of an inch

Gage No.	U.S. Standard Gage for Uncoated Hot & Cold-Rolled Sheets[b]	Galvanized Sheet Gage for Hot-Dipped Zinc Coated Sheets[b]	USA Steel Wire Gage	Gage No.	U.S. Standard Gage for Uncoated Hot & Cold-Rolled Sheets[b]	Galvanized Sheet Gage for Hot-Dipped Zinc Coated Sheets[b]	USA Steel Wire Gage
7/0	–	–	.490	13	.0897	.0934	.092[a]
6/0	–	–	.462[a]	14	.0747	.0785	.080
5/0	–	–	.430[a]	15	.0673	.0710	.072
4/0	–	–	.394[a]	16	.0598	.0635	.062[a]
3/0	–	–	.362[a]	17	.0538	.0575	.054
2/0	–	–	.331	18	.0478	.0516	.048[a]
1/0	–	–	.306	19	.0418	.0456	.041
1	–	–	.283	20	.0359	.0396	.035[a]
2	–	–	.262[a]	21	.0329	.0366	–
3	.2391	–	.244[a]	22	.0299	.0336	–
4	.2242	–	.225[a]	23	.0269	.0306	–
5	.2092	–	.207	24	.0239	.0276	–
6	.1943	–	.192	25	.0209	.0247	–
7	.1793	–	.177	26	.0179	.0217	–
8	.1644	.1681	.162	27	.0164	.0202	–
9	.1495	.1532	.148[a]	28	.0149	.0187	–
10	.1345	.1382	.135	29	–	.0172	–
11	.1196	.1233	.120[a]	30	–	.0157	–
12	.1046	.1084	.106[a]				

[a] Rounded value. The steel wire gage has been taken from ASTM A510 "General Requirements for Wire Rods and Coarse Round Wire, Carbon Steel." Sizes originally quoted to four decimal equivalent places have been rounded to three decimal places in accordance with rounding procedures of ASTM "Recommended Practice" E29.

[b] The equivalent thicknesses are for information only. The product is commonly specified to decimal thickness, not to gage number.

Table 17-11. Coefficients of Expansion

The coefficient of linear expansion (ε) is the change in length, per unit, for a change of one degree of temperature. The coefficient of surface expansion is approximately two times the linear coefficient, and the coefficient of volume expansion, for solids, is approximately three times the linear coefficient.

A bar, free to move, will increase in length with an increase in temperature and will decrease in length with a decrease in temperature. The change in length will be $\varepsilon t l$, where ε is the coefficient of linear expansion, t the change in temperature and l the length. If the ends of a bar are fixed, a change in temperature (t) will cause a change in the unit stress of $E\varepsilon t$, and in force of $AE\varepsilon t$, where A is the cross-sectional area of the bar and E the modulus of elasticity.

The following table gives the coefficient of linear expansion for 100°, or 100 times the value indicated above.

Example: A piece of medium steel is exactly 40 ft long at 60°F. Find the length at 90°F assuming the ends are free to move.

$$\text{change of length} = \varepsilon t l = \frac{.00065 \times 30 \times 40}{100} = .0078 \text{ ft}$$

The length at 90° is 40.0078 ft

Example: A piece of medium carbon steel is exactly 40 ft long and the ends are fixed. If the temperature increases 30°F, what is the resulting change in the unit stress?

$$\text{change in unit stress} = E\varepsilon t = \frac{29{,}000 \times .00065 \times 30}{100} = 5.7 \text{ ksi}$$

COEFFICIENTS OF EXPANSION FOR 100 DEGREES = 100ε

Materials	Linear Expansion		Materials	Linear Expansion	
	Celsius	Fahrenheit		Celsius	Fahrenheit
METALS AND ALLOYS			**STONE AND MASONRY**		
Aluminum, wrought	.00231	.00128	Ashlar masonry	.00063	.00035
Brass	.00188	.00104	Brick masonry	.00061	.00034
Bronze	.00181	.00101	Cement, portland	.00126	.00070
Copper	.00168	.00093	Concrete	.00099	.00055
Iron, cast, gray	.00106	.00059	Granite	.00080	.00044
Iron, wrought	.00120	.00067	Limestone	.00076	.00042
Iron, wire	.00124	.00069	Marble	.00081	.00045
Lead	.00286	.00159	Plaster	.00166	.00092
Magnesium, various alloys	.0029	.0016	Rubble masonry	.00063	.00035
Nickel	.00126	.00070	Sandstone	.00097	.00054
Steel, mild	.00117	.00065	Slate	.00080	.00044
Steel, stainless, 18-8	.00178	.00099			
Zinc, rolled	.00311	.00173			
TIMBER			**TIMBER**		
Fir (parallel to fiber)	.00037	.00021	Fir (perpendicular to fiber)	.0058	.0032
Maple (parallel to fiber)	.00064	.00036	Maple (perpendicular to fiber)	.0048	.0027
Oak (parallel to fiber)	.00049	.00027	Oak (perpendicular to fiber)	.0054	.0030
Pine (parallel to fiber)	.00054	.00030	Pine (perpendicular to fiber)	.0034	.0019

EXPANSION OF WATER
Maximum Density = 1

°C	Volume	°C	Volume	°C	Volume	°C	Volume	°C	Volume	°C	Volume
0	1.000126	10	1.000257	30	1.004234	50	1.011877	70	1.022384	90	1.035829
4	1.000000	20	1.001732	40	1.007627	60	1.016954	80	1.029003	100	1.043116

Table 17-12. Weights and Specific Gravities

Substance	Weight lb per cu ft	Specific Gravity	Substance	Weight lb per cu ft	Specific Gravity
ASHLAR, MASONRY			**MINERALS**		
Granite, syenite, gneiss	165	2.3–3.0	Asbestos	153	2.1–2.8
Limestone, marble	160	2.3–2.8	Barytes	281	4.50
Sandstone, bluestone	140	2.1–2.4	Basalt	184	2.7–3.2
			Bauxite	159	2.55
MORTAR RUBBLE MASONRY			Borax	109	1.7–1.8
Granite, syenite, gneiss	155	2.2–2.8	Chalk	137	1.8–2.6
Limestone, marble	150	2.2–2.6	Clay, marl	137	1.8–2.6
Sandstone, bluestone	130	2.0–2.2	Dolomite	181	2.9
			Feldspar, orthoclase	159	2.5–2.6
DRY RUBBLE MASONRY			Gneiss, serpentine	159	2.4–2.7
Granite, syenite, gneiss	130	1.9–2.3	Granite, syenite	175	2.5–3.1
Limestone, marble	125	1.9–2.1	Greenstone, trap	187	2.8–3.2
Sandstone, bluestone	110	1.8–1.9	Gypsum, alabaster	159	2.3–2.8
			Hornblende	187	3.0
BRICK MASONRY			Limestone, marble	165	2.5–2.8
Pressed brick	140	2.2–2.3	Magnesite	187	3.0
Common brick	120	1.8–2.0	Phosphate rock, apatite	200	3.2
Soft brick	100	1.5–1.7	Porphyry	172	2.6–2.9
			Pumice, natural	40	0.37–0.90
CONCRETE MASONRY			Quartz, flint	165	2.5–2.8
Cement, stone, sand	144	2.2–2.4	Sandstone, bluestone	147	2.2–2.5
Cement, slag, etc.	130	1.9–2.3	Shale, slate	175	2.7–2.9
Cement, cinder, etc.	100	1.5–1.7	Soapstone, talc	169	2.6–2.8
VARIOUS BUILDING MATERIALS			**STONE, QUARRIED, PILED**		
Ashes, cinders	40–45	–	Basalt, granite, gneiss	96	–
Cement, portland, loose	90	–	Limestone, marble, quartz	95	–
Cement, portland, set	183	2.7–3.2	Sandstone	82	–
Lime, gypsum, loose	53–64	–	Shale	92	–
Mortar, set	103	1.4–1.9	Greenstone, hornblende	107	–
Slags, bank slag	67–72	–			
Slags, bank screenings	98–117	–			
Slags, machine slag	96	–	**BITUMINOUS SUBSTANCES**		
Slags, slag sand	49–55	–	Asphaltum	81	1.1–1.5
EARTH, ETC., EXCAVATED			Coal, anthracite	97	1.4–1.7
Clay, dry	63	–	Coal, bituminous	84	1.2–1.5
Clay, damp, plastic	110	–	Coal, lignite	78	1.1–1.4
Clay and gravel, dry	100	–	Coal, peat, turf, dry	47	0.65–0.85
Earth, dry, loose	76	–	Coal, charcoal, pine	23	0.28–0.44
Earth, dry, packed	95	–	Coal, charcoal, oak	33	0.47–0.57
Earth, moist, loose	78	–	Coal, coke	75	1.0–1.4
Earth, moist, packed	96	–	Graphite	131	1.9–2.3
Earth, mud, flowing	108	–	Paraffin	56	0.87–0.91
Earth, mud, packed	115	–	Petroleum	54	0.87
Riprap, limestone	80–85	–	Petroleum, refined	50	0.79–0.82
Riprap, sandstone	90	–	Petroleum, benzine	46	0.73–0.75
Riprap, shale	105	–	Petroleum, gasoline	42	0.66–0.69
Sand, gravel, dry, loose	90–105	–	Pitch	69	1.07–1.15
Sand, gravel, dry, packed	100–120	–	Tar, bituminous	75	1.20
Sand, gravel, wet	118–120	–			
EXCAVATIONS IN WATER			**COAL AND COKE, PILED**		
Sand or gravel	60	–	Coal, anthracite	47–58	–
Sand or gravel and clay	65	–	Coal, bituminous, lignite	40–54	–
Clay	80	–	Coal, peat, turf	20–26	–
River mud	90	–	Coal charcoal	10–14	–
Soil	70	–	Coal coke	23–32	–
Stone riprap	65	–			

The specific gravities of solids and liquids refer to water at 4°C, those of gases to air at 0°C and 760 mm pressure. The weights per cubic foot are derived from average specific gravities, except where stated that weights are for bulk, heaped, or loose material, etc.

Table 17-12 (cont.). Weights and Specific Gravities

Substance	Weight lb per cu ft	Specific Gravity
METALS, ALLOYS, ORES		
Aluminum, cast, hammered	165	2.55–2.75
Brass, cast, rolled	534	8.4–8.7
Bronze, 7.9 to 14% Sn	509	7.4–8.9
Bronze, aluminum	481	7.7
Copper, cast, rolled	556	8.8–9.0
Copper ore, pyrites	262	4.1–4.3
Gold, cast, hammered	1205	19.25–19.3
Iron, cast, pig	450	7.2
Iron, wrought	485	7.6–7.9
Iron, speigel-eisen	468	7.5
Iron, ferro-silicon	437	6.7–7.3
Iron ore, hematite	325	5.2
Iron ore, hematite in bank	160–180	–
Iron ore, hematite loose	130–160	–
Iron ore, limonite	237	3.6–4.0
Iron ore, magnetite	315	4.9–5.2
Iron slag	172	2.5–3.0
Lead	710	11.37
Lead ore, galena	465	7.3–7.6
Magnesium, alloys	112	1.74–1.83
Manganese	475	7.2–8.0
Manganese ore, pyrolusite	259	3.7–4.6
Mercury	849	13.6
Monel Metal	556	8.8–9.0
Nickel	565	8.9–9.2
Platinum, cast, hammered	1330	21.1–21.5
Silver, cast, hammered	656	10.4–10.6
Steel, rolled	490	7.85
Tin, cast, hammered	459	7.2–7.5
Tin ore, cassiterite	418	6.4–7.0
Zinc, cast, rolled	440	6.9–7.2
Zinc ore, blende	253	3.9–4.2
VARIOUS SOLIDS		
Cereals, oats ... bulk	32	–
Cereals, barley ... bulk	39	–
Cereals, corn, rye ... bulk	48	–
Cereals, wheat ... bulk	48	–
Hay and Straw ... bales	20	–
Cotton, Flax, Hemp	93	1.47–1.50
Fats	58	0.90–0.97
Flour, loose	28	0.40–0.50
Flour, pressed	47	0.70–0.80
Glass, common	156	2.40–2.60
Glass, plate or crown	161	2.45–2.72
Glass, crystal	184	2.90–3.00
Leather	59	0.86–1.02
Paper	58	0.70–1.15
Potatoes, piled	42	–
Rubber, caoutchouc	59	0.92–0.96
Rubber goods	94	1.0–2.0
Salt, granulated, piled	48	–
Saltpeter	67	–
Starch	96	1.53
Sulphur	125	1.93–2.07
Wool	82	1.32
TIMBER, U.S. SEASONED		
Moisture content by weight:		
Seasoned timber 15 to 20%		
Green timber up to 50%		
Ash, white, red	40	0.62–0.65
Cedar, white, red	22	0.32–.038
Chestnut	41	0.66
Cypress	30	0.48
Fir, Douglas spruce	32	0.51
Fir, eastern	25	0.40
Elm, white	45	0.72
Hemlock	29	0.42–0.52
Hickory	49	0.74–0.84
Locust	46	0.73
Maple, hard	43	0.68
Maple, white	33	0.53
Oak, chestnut	54	0.86
Oak, live	59	0.95
Oak, red, black	41	0.65
Oak, white	46	0.74
Pine, Oregon	32	0.51
Pine, red	30	0.48
Pine, white	26	0.41
Pine, yellow, long-leaf	44	0.70
Pine, yellow, short-leaf	38	0.61
Poplar	30	0.48
Redwood, California	26	0.42
Spruce, white, black	27	0.40–0.46
Walnut, black	38	0.61
Walnut, white	26	0.41
VARIOUS LIQUIDS		
Alcohol, 100%	49	0.79
Acids, muriatic 40%	75	1.20
Acids, nitric 91%	94	1.50
Acids, sulphuric 87%	112	1.80
Lye, soda 66%	106	1.70
Oils, vegetable	58	0.91–0.94
Oils, mineral, lubricants	57	0.90–0.93
Water, 4°C max. density	62.428	1.0
Water, 100°C	59.830	0.9584
Water, ice	56	0.88–0.92
Water, snow, fresh fallen	8	.125
Water, sea water	64	1.02–1.03
GASES		
Air, 0°C 760 mm	.08071	1.0
Ammonia	.0478	0.5920
Carbon dioxide	.1234	1.5291
Carbon monoxide	.0781	0.9673
Gas, illuminating	.028–.036	0.35–0.45
Gas, natural	.038–.039	0.47–0.48
Hydrogen	.00559	0.0693
Nitrogen	.0784	0.9714
Oxygen	.0892	1.1056

The specific gravities of solids and liquids refer to water at 4°C, those of gases to air at 0°C and 760 mm pressure. The weights per cubic foot are derived from average specific gravities, except where stated that weights are for bulk, heaped, or loose material, etc.

Table 17-13. Weights of Building Materials

Materials	Weight lb per sq ft	Materials	Weight lb per sq ft
CEILINGS		**PARTITIONS**	
Channel suspended system	1	Clay tile	
Lathing and plastering	See Partitions	3 in.	17
Acoustical fiber tile	1	4 in.	18
		6 in.	28
		8 in.	34
		10 in.	40
FLOORS		Gypsum block	
Steel deck	See Manufacturer	2 in.	$9^1/_2$
		3 in.	$10^1/_2$
Concrete-Reinforced 1 in.		4 in.	$12^1/_2$
Stone	$12^1/_2$	5 in.	14
Slag	$11^1/_2$	6 in.	$18^1/_2$
Lightweight	6 to 10	Wood studs 2×4	
		12–16 in. o.c.	2
Concrete-Plain 1 in.		Steel partitions	4
Stone	12	Plaster 1 in.	
Slag	11	Cement	10
Lightweight	3 to 9	Gypsum	5
		Lathing	
Fills 1 inch		Metal	$^1/_2$
Gypsum	6	Gypsum board $^1/_2$ in.	2
Sand	8		
Cinders	4		
Finishes			
Terrazzo 1 in.	13		
Ceramic or Quarry Tile $^3/_4$-in.	10	**WALLS**	
Linoleum $^1/_4$-in.	1	Brick	
Mastic $^3/_4$-in.	9	4 in.	40
Hardwood $^7/_8$-in.	4	8 in.	80
Softwood $^3/_4$-in.	$2^1/_2$	12 in.	120
		Hollow concrete block	
		(Heavy aggregate)	
ROOFS		4 in.	30
Copper or tin	1	6 in.	43
Corrugated steel	See Manufactuer	8 in.	55
3-ply ready roofing	1	$12^1/_2$ in.	80
3-ply felt and gravel	$5^1/_2$	Hollow concrete block	
5-ply felt and gravel	6	(Light aggregate)	
		4 in.	21
Shingles		6 in.	30
Wood	2	8 in.	38
Asphalt	3	12 in.	55
Clay tile	9 to 14	Clay tile (Load bearing)	
Slate $^1/_4$ in.	10	4 in.	25
		6 in.	30
Sheathing		8 in.	33
Wood $^3/_4$ in.	3	12 in.	45
Gypsum 1 in.	4	Stone 4 in.	55
		Glass block 4 in.	18
Insulation 1 in.		Window, Glass, Frame, & Sash	8
Loose	$^1/_2$	Curtain walls	See Manufacturer
Poured	2	Structural glass 1 in.	15
Rigid	$1^1/_2$	Corrugated Cement Asbestos $^1/_4$ in.	3

For weights of other materials used in building construction, see Table 17-12.

Table 17-14. Weights and Measures United States System

LINEAR MEASURE

Inches	*Feet*	*Yards*	*Rods*	*Furlongs*	*Miles*
1.0	.08333	.02778	.0050505	.00012626	.00001578
12.0	1.0	.33333	.0606061	.00151515	.00018939
36.0	3.0	1.0	.1818182	.00454545	.00056818
198.0	16.5	5.5	1.0	.025	.003125
7,920.0	660.0	220.0	40.0	1.0	.125
63,360.0	5,280.0	1,760.0	320.0	8.0	1.0

SQUARE AND LAND MEASURE

Sq. Inches	*Square Feet*	*Square Yards*	*Square Rods*	*Acres*	*Sq. Miles*
1.0	.006944	.000772			
144.0	1.0	.111111			
1,296.0	9.0	1.0	.03306	.000207	
39,204.0	272.25	30.25	1.0	.00625	.0000098
	43,560.0	4,840.0	160.0	1.0	.0015625
		3,097,600.0	102,400.0	640.0	1.0

AVOIRDUPOIS WEIGHTS

Grains	*Drams*	*Ounces*	*Pounds*	*Tons*
1.0	.03657	.002286	.000143	.0000000714
27.34375	1.0	.0625	.003906	.00000195
437.5	16.0	1.0	.0625	.00003125
7,000.0	256.0	16.0	1.0	.0005
14,000,000.0	512,000.0	32,000.0	2,000.0	1.0

DRY MEASURE

Pints	*Quarts*	*Pecks*	*Cubic Feet*	*Bushels*
1.0	.5	.0625	.01945	.01563
2.0	1.0	.125	.03891	.03125
16.0	8.0	1.0	.31112	.25
51.42627	25.71314	3.21414	1.0	.80354
64.0	32.0	4.0	1.2445	1.0

LIQUID MEASURE

Gills	*Pints*	*Quarts*	*U.S. Gallons*	*Cubic Feet*
1.0	.25	.125	.03125	.00418
4.0	1.0	.5	.125	.01671
8.0	2.0	1.0	.250	.03342
32.0	8.0	4.0	1.0	.1337
			7.48052	1.0

SI UNITS FOR STRUCTURAL STEEL DESIGN

Although there are seven metric base units in the SI system, only four are currently used by AISC in structural steel design. These base units are listed in Table 17-15.

Table 17-15. Base SI Units for Steel Design

Quantity	Unit	Symbol
length	meter	m
mass	kilogram	kg
time	second	s
temperature	celsius	°C

Similarly, of the numerous decimal prefixes included in the SI system, only three are used in steel design; see Table 17-16.

Table 17-16. SI Prefixes for Steel Design

Prefix	Symbol	Order of Magnitude	Expression
mega	M	10^6	1,000,000 (one million)
kilo	k	10^3	1,000 (one thousand)
milli	m	10^{-3}	0.001 (one thousandth)

In addition, three derived units are applicable to the present conversion. They are shown in Table 17-17.

Table 17-17. Derived SI Units for Steel Design

Quantity	Name	Symbol	Expression
force	newton	N	$N = kg \times m/s^2$
stress	pascal	Pa	$Pa = N/m^2$
energy	joule	J	$J = N \times m$

Although specified in SI, the pascal is not universally accepted as the unit of stress. Because section properties are expressed in millimeters, it is more convenient to express stress in newtons per square millimeter ($1\ N/mm^2 = 1\ MPa$). This is the practice followed in recent international structural design standards. It should be noted that the joule, as the unit of energy, is used to express energy absorption requirements for impact tests. Moments are expressed in terms of $N \times m$.

A summary of the conversion factors relating traditional U.S. units of measurement to the corresponding SI units is given in Table 17-18.

Table 17-18. Summary of SI Conversion Factors

Multiply	By:	To obtain:
inch (in.)	25.4	millimeters (mm)
foot (ft)	305	millimeters (mm)
pound-mass (lb)	0.454	kilogram (kg)
pound-force (lbf)	4.448	newton (N)
ksi	6.895	N/mm^2
ft-lbf	1.356	joule (J)
psf	47.88	N/m^2
plf	14.59	N/m

Note that fractions resulting from metric conversion should be rounded to whole millimeters. Common fractions of inches and their metric equivalent are in Table 17-19.

Table 17-19. SI Equivalents of Fractions of an Inch

Fraction, in.	Exact conversion, mm	Rounded to: (mm)
$^1/_{16}$	1.5875	2
$^1/_8$	3.175	3
$^3/_{16}$	4.7625	5
$^1/_4$	6.35	6
$^5/_{16}$	7.9375	8
$^3/_8$	9.525	10
$^7/_{16}$	11.1125	11
$^1/_2$	12.7	13
$^5/_8$	15.875	16
$^3/_4$	19.05	19
$^7/_8$	22.225	22
1	25.4	25

Bolt diameters are taken directly from the ASTM Specifications A325M and A490M rather than converting the diameters of bolts dimensioned in inches. The metric bolt designations are in Table 17-20.

Table 17-20. SI Bolt Designation

Designation	Diameter, mm	Diameter, in.
M16	16	0.63
M20	20	0.79
M22	22	0.87
M24	24	0.94
M27	27	1.06
M30	30	1.18
M36	36	1.42

The yield strengths of structural steels are taken from the metric ASTM Specifications. It should be noted that the yield points are slightly different from the traditional values. See Table 17-21. The modulus of elasticity of steel E is taken as 200,000 N/mm^2. The shear modulus of elasticity of steel G is 77,000 N/mm^2.

Table 17-21. SI Steel Yield Stresses

ASTM Designation	Yield stress, N/mm^2	Yield stress, ksi
A36M	250	36.26
A572M Gr. 345 A588M	345	50.04
A852M	485	70.34
A514M	690	100.07

Table 17-22.
Weights and Measures
International System of Units (SI)[a]
(Metric practice)

BASE UNITS

Quantity	Unit	Symbol
length	meter	m
mass	kilogram	kg
time	second	s
electric current	ampere	A
thermodynamic temperature	kelvin	K
amount of substance	mole	mol
luminous intensity	candela	cd

SUPPLEMENTARY UNITS

Symbol	Unit	Symbol
plane angle	radian	rad
solid angle	steradian	sr

DERIVED UNITS (WITH SPECIAL NAMES)

Quantity	Unit	Symbol	Formula
force	newton	N	kg-m/s^2
pressure, stress	pascal	Pa	N/m^2
energy, work, quantity of heat	joule	J	N-m
power	watt	W	J/s

DERIVED UNITS (WITHOUT SPECIAL NAMES)

Quantity	Unit	Formula
area	square meter	m^2
volume	cubic meter	m^3
velocity	meter per second	m/s
acceleration	meter per second squared	m/s^2
specific volume	cubic meter per kilogram	m^3/kg
density	kilogram per cubic meter	kg/m^3

SI PREFIXES

Multiplication Factor	Prefix	Symbol
1 000 000 000 000 000 000 = 10^{18}	exa	E
1 000 000 000 000 000 = 10^{15}	peta	P
1 000 000 000 000 = 10^{12}	tera	T
1 000 000 000 = 10^{9}	giga	G
1 000 000 = 10^{6}	mega	M
1 000 = 10^{3}	kilo	k
100 = 10^{2}	hecto[b]	h
10 = 10^{1}	deka[b]	da
0.1 = 10^{-1}	deci[b]	d
0.01 = 10^{-2}	centi[b]	c
0.001 = 10^{-3}	milli	m
0.000 001 = 10^{-6}	micro	μ
0.000 000 001 = 10^{-9}	nano	n
0.000 000 000 001 = 10^{-12}	pico	p
0.000 000 000 000 001 = 10^{-15}	femto	f
0.000 000 000 000 000 001 = 10^{-18}	atto	a

[a] Refer to ASTM E380 for more complete information on SI.
[b] Use is not recommended.

Table 17-23. SI Conversion Factors[a]

Quantity	Multiply	by	to obtain	
Length	inch	25.400	millimeter	mm
	foot	0.305	meter	m
	yard	0.914	meter	m
	mile (U.S. Statute)	1.609	kilometer	km
	millimeter	39.370×10^{-3}	inch	in
	meter	3.281	foot	ft
	meter	1.094	yard	yd
	kilometer	0.621	mile	mi
Area	square inch	0.645×10^{3}	square millimeter	mm^2
	square foot	0.093	square meter	m^2
	square yard	0.836	square meter	m^2
	square mile (U.S. Statute)	2.590	square kilometer	km^2
	acre	4.047×10^{3}	square meter	m^2
	acre	0.405	hectare	
	square millimeter	1.550×10^{-3}	square inch	in^2
	square meter	10.764	square foot	ft^2
	square meter	1.196	square yard	yd^2
	square kilometer	0.386	square mile	mi^2
	square meter	0.247×10^{-3}	acre	
	hectare	2.471	acre	
Volume	cubic inch	16.387×10^{3}	cubic millimeter	mm^3
	cubic foot	28.317×10^{-3}	cubic meter	m^3
	cubic yard	0.765	cubic meter	m^3
	gallon (U.S. liquid)	3.785	liter	l
	quart (U.S. liquid)	0.946	liter	l
	cubic millimeter	61.024×10^{-6}	cubic inch	in^3
	cubic meter	35.315	cubic foot	ft^3
	cubic meter	1.308	cubic yard	yd^3
	liter	0.264	gallon (U.S. liquid)	gal
	liter	1.057	quart (U.S. liquid)	qt
Mass	ounce (avoirdupois)	28.35	gram	g
	pound (avoirdupois)	0.454	kilogram	kg
	short ton	0.907×10^{3}	kilogram	kg
	gram	35.274×10^{-3}	ounce (avoirdupois)	oz av
	kilogram	2.205	pound (avoirdupois)	lb av
	kilogram	1.102×10^{-3}	short ton	

[a] Refer to ASTM E380 for more complete information on SI.
The conversion factors tabulated herein have been rounded.

Table 17-23 (cont.). SI Conversion Factors[a]

Quantity	Multiply	by	to obtain	
Force	ounce-force	0.278	newton	N
	pound-force	4.448	newton	N
	newton	3.597	ounce-force	
	newton	0.225	pound-force	lbf
Bending Moment	pound-force-inch	0.113	newton-meter	N-m
	pound-force-foot	1.356	newton-meter	N-m
	newton-meter	8.851	pound-force-inch	lbf-in
	newton-meter	0.738	pound-force-foot	lbf-ft
Pressure, Stress	pound-force per square inch	6.895	kilopascal	kPa
	foot of water (39.2 F)	2.989	kilopascal	kPa
	inch of mercury (32 F)	3.386	kilopascal	kPa
	kilopascal	0.145	pound-force per square inch	lbf/in^2
	kilopascal	0.335	foot of water (39.2 F)	
	kilopascal	0.295	inch of mercury (32 F)	
Energy, Work, Heat	foot-pound-force	1.356	joule	J
	[b]British thermal unit	1.055×10^3	joule	J
	[b]calorie	4.187	joule	J
	kilowatt hour	3.600×10^6	joule	J
	joule	0.738	foot-pound-force	ft-lbf
	joule	0.948×10^{-3}	[b]British thermal unit	Btu
	joule	0.239	[b]calorie	
	joule	0.278×10^{-6}	kilowatt hour	kW-h
Power	foot-pound-force/second	1.356	watt	W
	[b]British thermal unit per hour	0.293	watt	W
	horsepower (550 ft lbf/s)	0.746	kilowatt	kW
	watt	0.738	foot-pound-force/ second	ft-lbf/s
	watt	3.412	[b]British thermal unit per hour	Btu/h
	kilowatt	1.341	horsepower (550 ft-lbf/s)	hp
Angle	degree	17.453×10^{-3}	radian	rad
	radian	57.296	degree	
Temperature	degree Fahrenheit	$t°C = (t°F - 32)/1.8$	degree Celsius	
	degree Celsius	$t°F = 1.8 \times t°C + 32$	degree Fahrenheit	

[a]Refer to ASTM E380 for more complete information on SI.
[b]International Table.
The conversion factors tabulated herein have been rounded.

Bracing Formulas

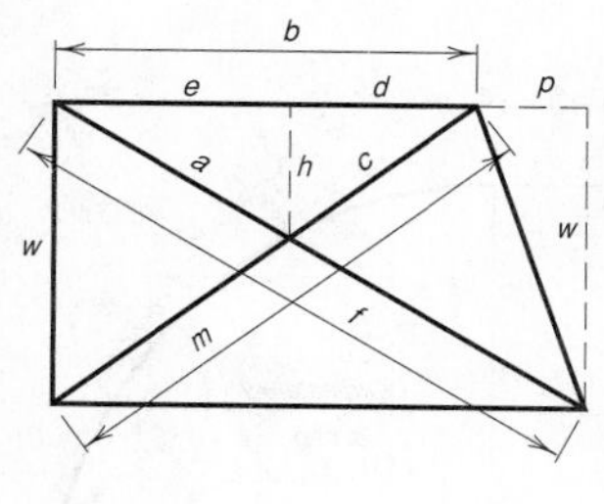

Given	To Find	Formula
bpw	f	$\sqrt{(b+p)^2+w^2}$
bw	m	$\sqrt{b^2+w^2}$
bp	d	$b^2 \div (2b+p)$
bp	e	$b(b+p) \div (2b+p)$
bfp	a	$bf \div (2b+p)$
bmp	c	$bm \div (2b+p)$
bpw	h	$bw \div (2b+p)$
afw	h	$aw \div f$
cmw	h	$cw \div m$

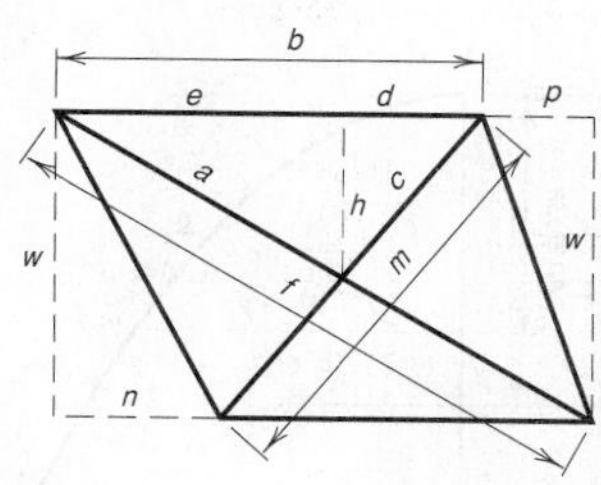

Given	To Find	Formula
bpw	f	$\sqrt{(b+p)^2+w^2}$
bnw	m	$\sqrt{(b-n)^2+w^2}$
bnp	d	$b(b-n) \div (2b+p-n)$
bnp	e	$b(b+p) \div (2b+p-n)$
bfnp	a	$bf \div (2b+p-n)$
bmnp	c	$bm \div (2b+p-n)$
bnpw	h	$bw \div (2b+p-n)$
afw	h	$aw \div f$
cmw	h	$cw \div m$

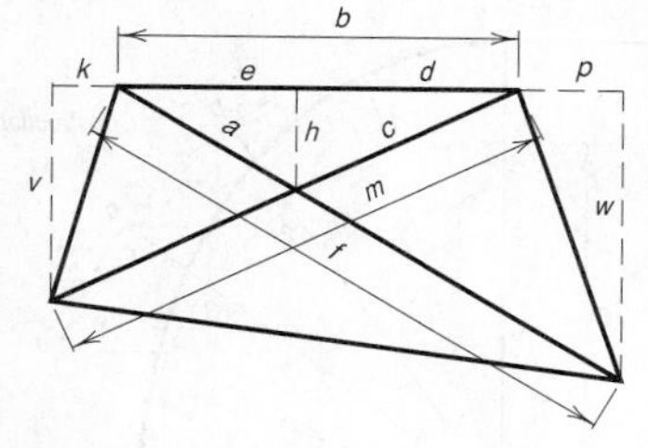

Given	To Find	Formula
bpw	f	$\sqrt{(b+p)^2+w^2}$
bkv	m	$\sqrt{(b+k)^2+v^2}$
bkpvw	d	$bw(b+k) \div [v(b+p)+w(b+k)]$
bkpvw	e	$bv(b+p) \div [v(b+p)+w(b+k)]$
bfkpvw	a	$f\,bv \div [v(b+p)+w(b+k)]$
bkmpvw	c	$bmw \div [v(b+p)+w(b+k)]$
bkpvw	h	$bvw \div [v(b+p)+w(b+k)]$
afw	h	$aw \div f$
cmv	h	$cv \div m$

PARALLEL BRACING

$k = (\log B - \log T) \div$ no. of panels. Constant k plus the logarithm of any line equals the log of the corresponding line in the next panel below.

$a = TH \div (T+e+p)$

$b = Th \div (T+e+p)$

$c = \sqrt{(^1/_2 T + {}^1/_2 e)^2 + a^2}$

$d = ce \div (T+e)$

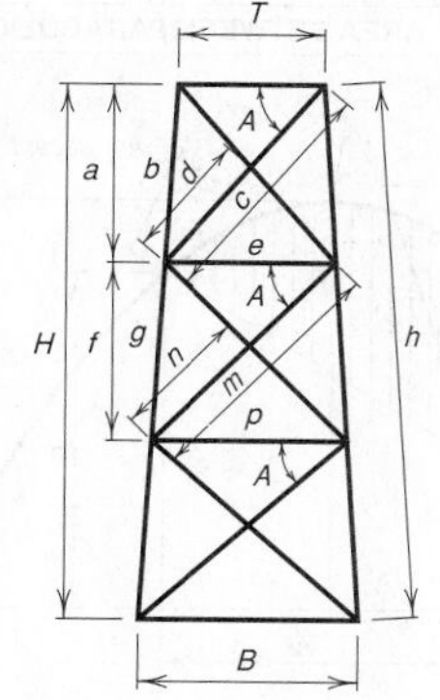

$\log e = k + \log T$

$\log f = k + \log a$

$\log g = k + \log b$

$\log m = k + \log c$

$\log n = k + \log d$

$\log p = k + \log e$

The above method can be used for any number of panels. In the formulas for “a” and “b” the sum in parenthesis, which in the case shown is $(T+e+p)$, is always composed of all the horizontal distances except the base.

Properties of Parabola and Ellipse

PARABOLA

if $\frac{H}{B} \leq 0.1$, Approx. $1/2$ perimeter $= \sqrt{B^2 + 4/3\,H^2}$

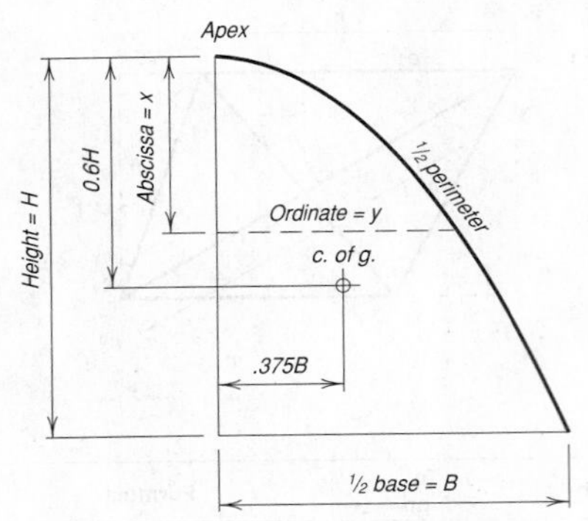

Parameter $P = \frac{B^2}{H}$ Area $= 2/3\,HB$

$x = \frac{Y^2}{P}$

$y = \sqrt{xP}$

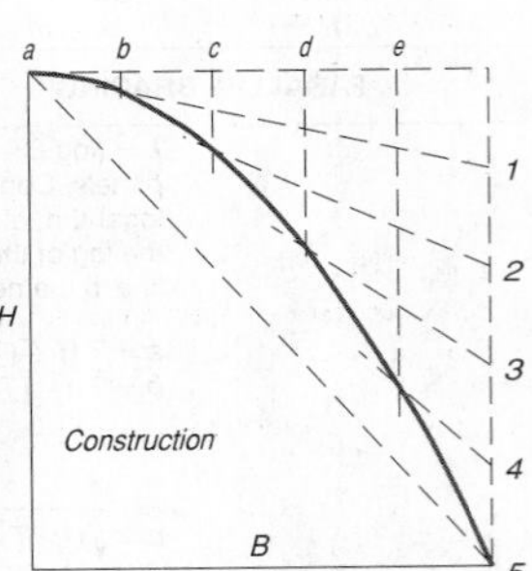

ELLIPSE

Approx. $1/4$ perimeter $= \frac{\pi}{4}\sqrt{2(H^2 + B^2)}$

$(x^2 \div H^2) + (y^2 \div B^2) = 1$

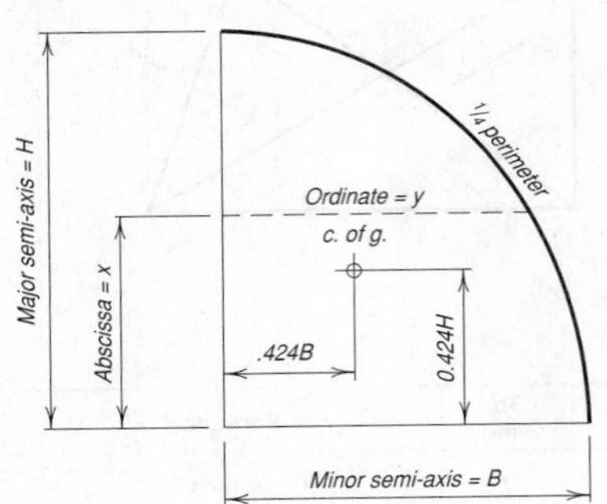

Area = .7854Dd

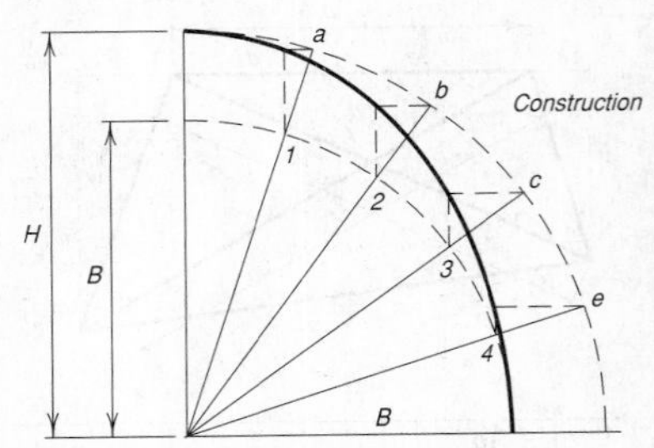

AREA BETWEEN PARABOLIC CURVE AND SECANT

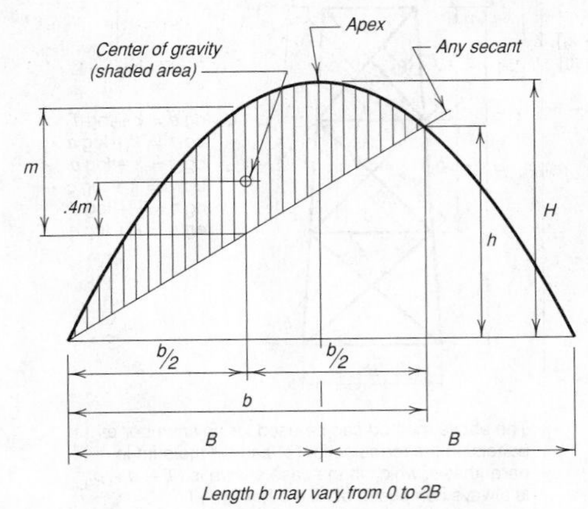

$h = Hb\left(\frac{2B - b}{B^2}\right)$

$m = \frac{Hb^2}{4B^2}$

Shaded area

$= 2/3\,bm$

$= \frac{Hb^3}{6B^2}$

Properties of the Circle

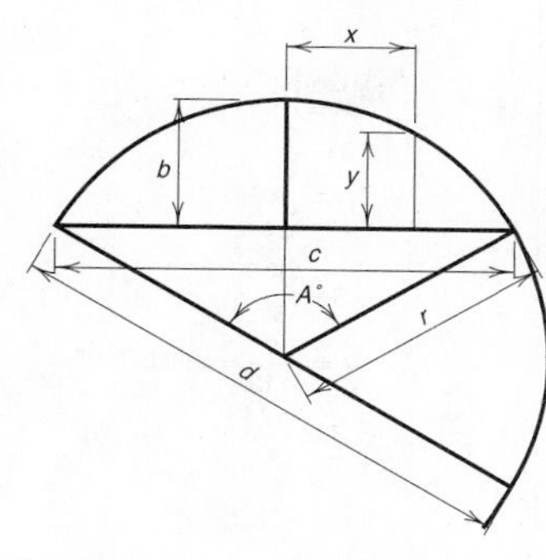

Circumference = 6.28318 r = 3.14159 d

Diameter = 0.31831 circumference

Area = $3.14159r^2$

$$\text{Arc } a = \frac{\pi r A^\circ}{180^\circ} = 0.017453 r A^\circ$$

$$\text{Angle } A^\circ = \frac{180^\circ a}{\pi r} = 57.29578 \frac{a}{r}$$

$$\text{Radius } r = \frac{4b^2 + c^2}{8b}$$

$$\text{Chord } c = 2\sqrt{2br - b^2} = 2r \sin \frac{A}{2}$$

$$\text{Rise } b = r - \tfrac{1}{2}\sqrt{4r^2 - c^2} = \frac{c}{2} \tan \frac{A}{4}$$

$$= 2r \sin^2 \frac{A}{4} = r + y - \sqrt{r^2 - x^2}$$

$$y = b - r + \sqrt{r^2 - x^2}$$

$$x = \sqrt{r^2 - (r + y - b)^2}$$

Diameter of circle of equal periphery as square = 1.27324 side of square
Side of square of equal periphery as circle = 0.78540 diameter of circle
Diameter of circle circumscribed about square = 1.41421 side of square
Side of square inscribed in circle = 0.70711 diameter of circle

CIRCULAR SECTOR

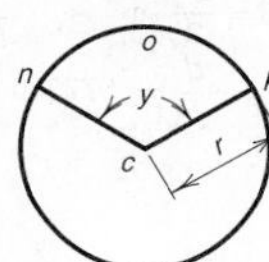

r = radius of circle y = angle ncp in degrees

Area of Sector $ncpo$ = ½ (length of arc $nop \times r$)

$$= \text{Area of Circle} \times \frac{y}{360}$$

$$= 0.0087266 \times r^2 \times y$$

CIRCULAR SEGMENT

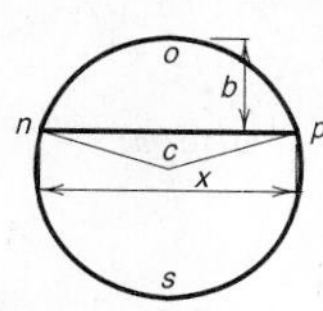

r = radius of circle x = chord b = rise

Area of Segment nop = Area of Sector $ncpo$ – Area of triangle ncp

$$= \frac{(\text{Length of arc } nop \times r) - x(r - b)}{2}$$

Area of Segment nsp = Area of Circle – Area of Segment nop

VALUES FOR FUNCTIONS OF π

$\pi = 3.14159265359$, log = 0.4971499

$\pi^2 = 9.8696044$, log = 0.9942997

$\frac{1}{\pi} = 0.3183099$, log = $\bar{1}.5028501$

$\sqrt{\frac{1}{\pi}} = 05641896$, log = $\bar{1}.7514251$

$\pi^3 = 31.0062767$, log = 1.4914496

$\frac{1}{\pi^2} = 0.1013212$, log = $\bar{1}.0057003$

$\frac{\pi}{180} = 0.0174533$, log = $\bar{2}.2418774$

$\sqrt{\pi} = 1.7724539$, log = 0.2485749

$\frac{1}{\pi^3} = 0.0322515$, log = $\bar{2}.5085504$

$\frac{180}{\pi} = 57.2957795$, log = 1.7581226

Note: Logs of fractions such as $\bar{1}.5028501$ and $\bar{2}.5085500$ may also be written 9.5028501–10 and 8.5085500–10 respectively.

Properties of Geometric Sections

SQUARE *Axis of moments through center* 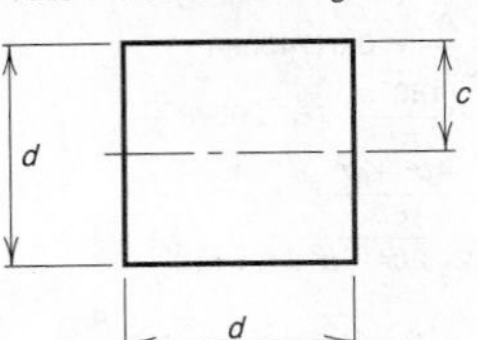	$A = d^2$ $c = \frac{d}{2}$ $I = \frac{d^4}{12}$ $S = \frac{d^3}{6}$ $r = \frac{d}{\sqrt{12}} = .288675\, d$ $Z = \frac{d^3}{4}$
SQUARE *Axis of moments on base* 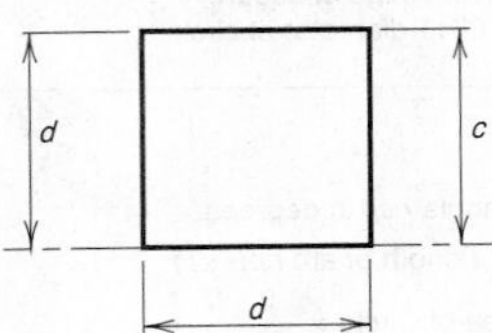	$A = d^2$ $c = d$ $I = \frac{d^4}{3}$ $S = \frac{d^3}{3}$ $r = \frac{d}{\sqrt{3}} = .577350\, d$
SQUARE *Axis of moments on diagonal* 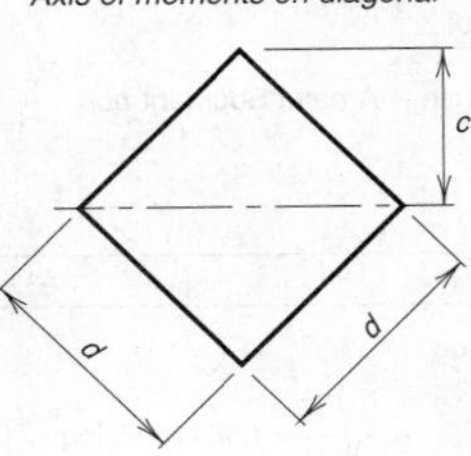	$A = d^2$ $c = \frac{d}{\sqrt{2}} = .707107\, d$ $I = \frac{d^4}{12}$ $S = \frac{d^3}{6\sqrt{2}} = .117851\, d^3$ $r = \frac{d}{\sqrt{12}} = .288675\, d$ $Z = \frac{2c^3}{3} = \frac{d^3}{3\sqrt{2}} = .235702\, d^3$
RECTANGLE *Axis of moments through center* 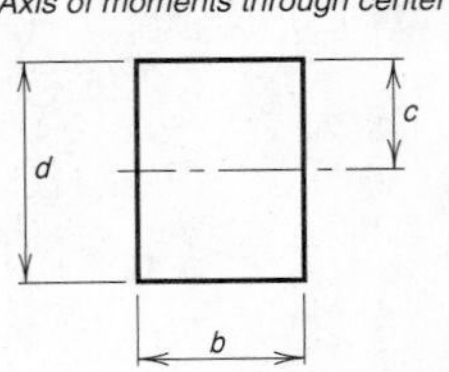	$A = bd$ $c = \frac{d}{2}$ $I = \frac{bd^3}{12}$ $S = \frac{bd^2}{6}$ $r = \frac{d}{\sqrt{12}} = .288675\, d$ $Z = \frac{bd^2}{4}$

Properties of Geometric Sections (cont.)

RECTANGLE
Axis of moments on base

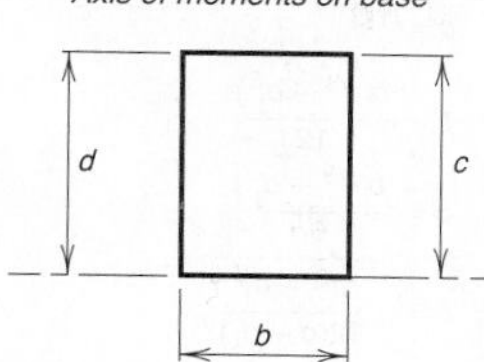

$$A = bd$$
$$c = d$$
$$I = \frac{bd^3}{3}$$
$$S = \frac{bd^2}{3}$$
$$r = \frac{d}{\sqrt{3}} = .577350\ d$$

RECTANGLE
Axis of moments on diagonal

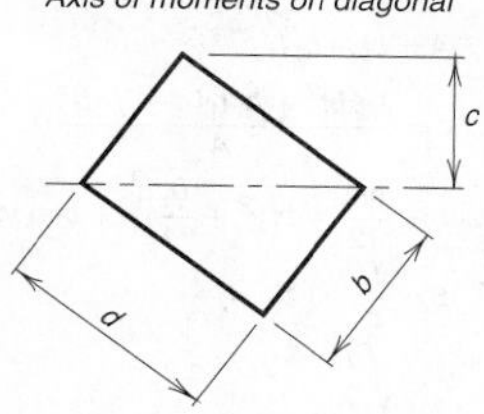

$$A = bd$$
$$c = \frac{bd}{\sqrt{b^2 + d^2}}$$
$$I = \frac{b^3 d^3}{6(b^2 + d^2)}$$
$$S = \frac{b^2 d^2}{6\sqrt{b^2 + d^2}}$$
$$r = \frac{bd}{\sqrt{6(b^2 + d^2)}}$$

RECTANGLE
Axis of moments any line through center of gravity

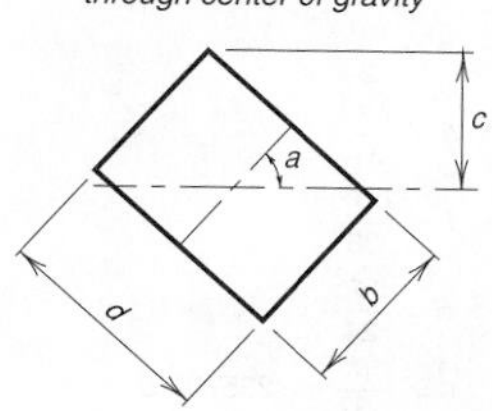

$$A = bd$$
$$c = \frac{b \sin a + d \cos a}{2}$$
$$I = \frac{bd(b^2 \sin^2 a + d^2 \cos^2 a)}{12}$$
$$S = \frac{bd(b^2 \sin^2 a + d^2 \cos^2 a)}{6(b \sin a + d \cos a)}$$
$$r = \sqrt{\frac{b^2 \sin^2 a + d^2 \cos^2 a}{12}}$$

HOLLOW RECTANGLE
Axis of moments through center

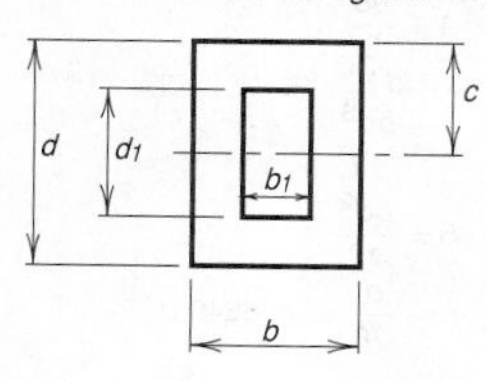

$$A = bd - b_1 d_1$$
$$c = \frac{d}{2}$$
$$I = \frac{bd^3 - b_1 d_1^3}{12}$$
$$S = \frac{bd^3 - b_1 d_1^3}{6d}$$
$$r = \sqrt{\frac{bd^3 - b_1 d_1^3}{12A}}$$
$$Z = \frac{bd^2}{4} - \frac{b_1 d_1^2}{4}$$

Properties of Geometric Sections (cont.)

EQUAL RECTANGLES Axis of moments through center of gravity 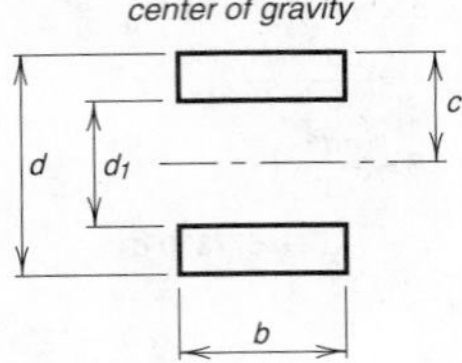 	$A = b(d - d_1)$ $c = \frac{d}{2}$ $I = \frac{b(d^3 - d_1^3)}{12}$ $S = \frac{b(d^3 - d_1^3)}{6d}$ $r = \sqrt{\frac{d^3 - d_1^3}{12(d - d_1)}}$ $Z = \frac{b}{4}(d^2 - d_1^2)$
UNEQUAL RECTANGLES Axis of moments through center of gravity 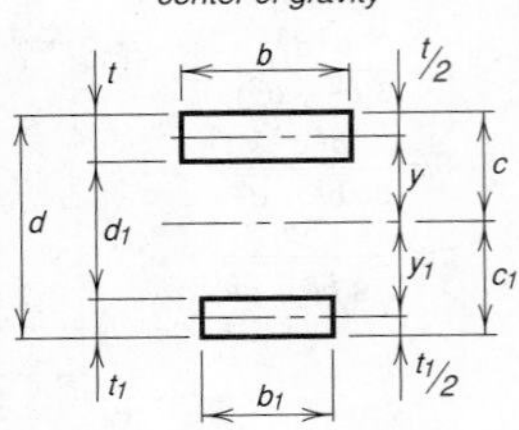 	$A = bt + b_1 t_1$ $c = \frac{1/2\, bt^2 + b_1 t_1 (d - 1/2\, t_1)}{A}$ $I = \frac{bt^3}{12} + bty^2 + \frac{b_1 t_1^3}{12} + b_1 t_1 y_1^2$ $S = \frac{I}{c} \quad S_1 = \frac{I}{c_1}$ $r = \sqrt{\frac{I}{A}}$ $Z = bty + b_1 t_1 y_1$
TRIANGLE Axis of moments through center of gravity 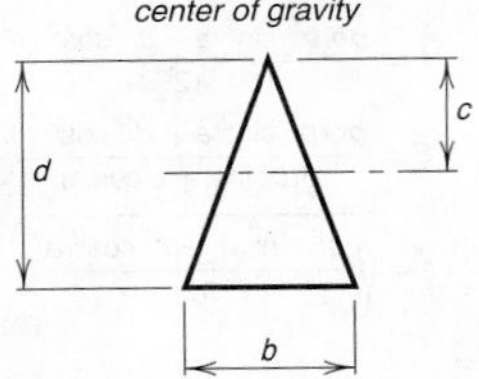 	$A = \frac{bd}{2}$ $c = \frac{2d}{3}$ $I = \frac{bd^3}{36}$ $S = \frac{bd^2}{24}$ $r = \frac{d}{\sqrt{18}} = .235702\, d$
TRIANGLE Axis of moments on base 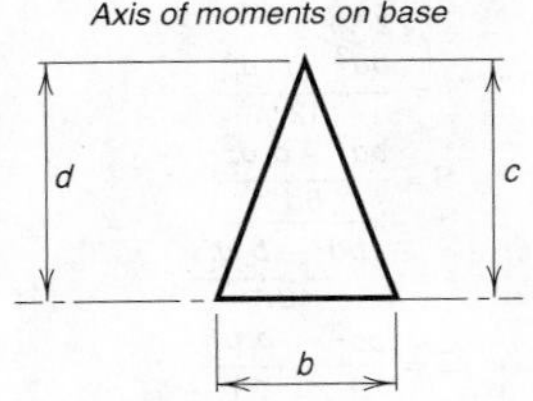 	$A = \frac{bd}{2}$ $c = d$ $I = \frac{bd^3}{12}$ $S = \frac{bd^2}{12}$ $r = \frac{d}{\sqrt{6}} = .408248\, d$

Properties of Geometric Sections (cont.)

TRAPEZOID *Axis of moments through center of gravity* 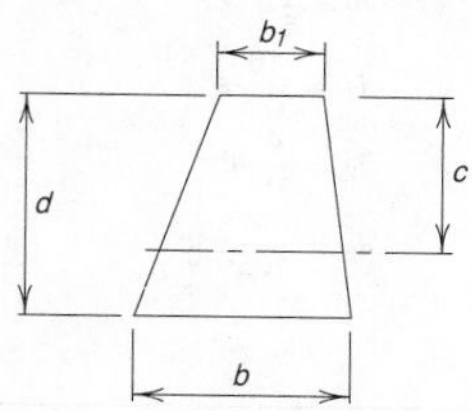	$A = \dfrac{d(b+b_1)}{2}$ $c = \dfrac{d(2b+b_1)}{3(b+b_1)}$ $I = \dfrac{d^3(b^2+4bb_1+b_1^2)}{36(b+b_1)}$ $S = \dfrac{d^2(b^2+4bb_1+b_1^2)}{12(2b+b_1)}$ $r = \dfrac{d}{6(b+b_1)}\sqrt{2(b^2+4bb_1+b_1^2)}$
CIRCLE *Axis of moments through center* 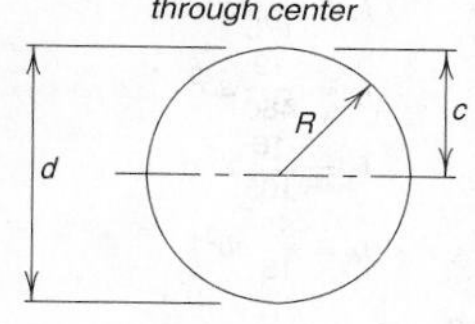	$A = \dfrac{\pi d^2}{4} = \pi R^2 = .785398 \quad d^2 = 3.141593 R^2$ $c = \dfrac{d}{2} = R$ $I = \dfrac{\pi d^4}{64} = \dfrac{\pi R^4}{4} = .049087 \quad d^4 = .785398 R^4$ $S = \dfrac{\pi d^3}{32} = \dfrac{\pi R^3}{4} = .098175 \quad d^3 = .785398 R^3$ $r = \dfrac{d}{4} = \dfrac{R}{2}$ $Z = \dfrac{d^3}{6}$
HOLLOW CIRCLE *Axis of moments through center* 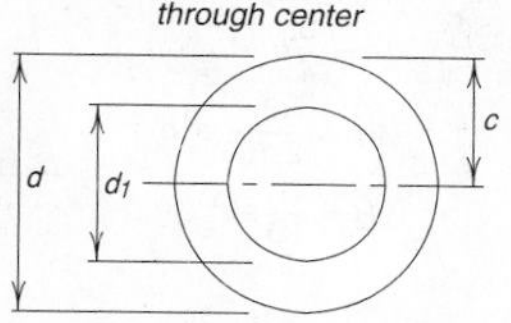	$A = \dfrac{\pi(d^2-d_1^2)}{4} = .785398\,(d^2-d_1^2)$ $c = \dfrac{d}{2}$ $I = \dfrac{\pi(d^4-d_1^4)}{64} = .049087\,(d^4-d_1^4)$ $S = \dfrac{\pi(d^4-d_1^4)}{32d} = .098175\,\dfrac{d^4-d_1^4}{d}$ $r = \dfrac{\sqrt{d^2+d_1^2}}{4}$ $Z = \dfrac{d^3}{6} - \dfrac{d_1^3}{6}$
HALF CIRCLE *Axis of moments through center of gravity* 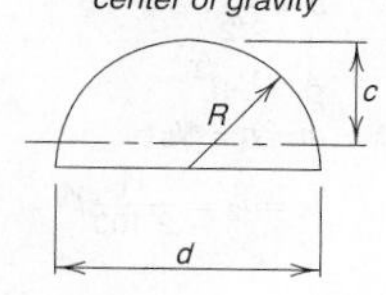	$A = \dfrac{\pi R^2}{2} = 1.570796\ R^2$ $c = R\left(1-\dfrac{4}{3\pi}\right) = .575587\ R$ $I = R^4\left(\dfrac{\pi}{8}-\dfrac{8}{9\pi}\right) = .109757\ R^4$ $S = \dfrac{R^3}{24}\,\dfrac{(9\pi^2-64)}{(3\pi-4)} = .190687\ R^3$ $r = R\dfrac{\sqrt{9\pi^2-64}}{6\pi} = .264336\ R$

Properties of Geometric Sections (cont.)

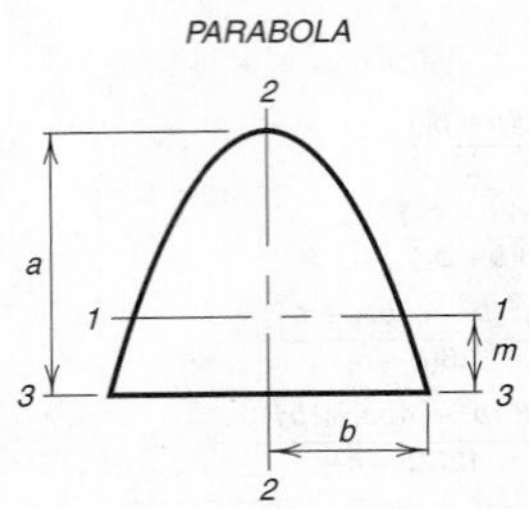

$$A = \frac{4}{3}ab$$

$$m = \frac{2}{5}a$$

$$I_1 = \frac{16}{175}a^3b$$

$$I_2 = \frac{4}{15}ab^3$$

$$I_3 = \frac{32}{105}a^3b$$

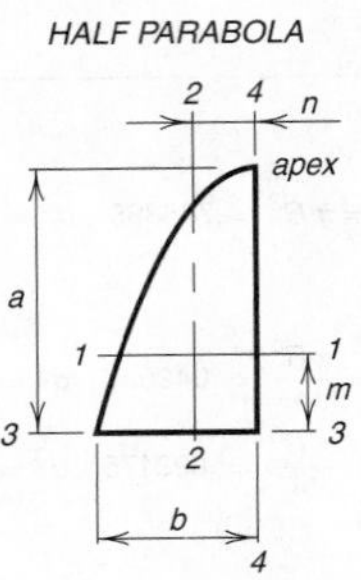

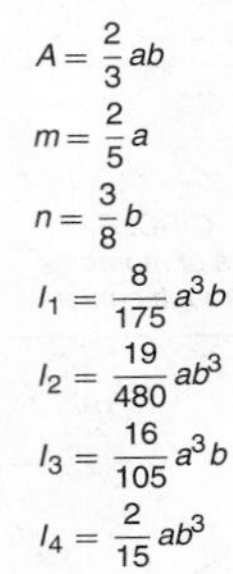

$$A = \frac{2}{3}ab$$

$$m = \frac{2}{5}a$$

$$n = \frac{3}{8}b$$

$$I_1 = \frac{8}{175}a^3b$$

$$I_2 = \frac{19}{480}ab^3$$

$$I_3 = \frac{16}{105}a^3b$$

$$I_4 = \frac{2}{15}ab^3$$

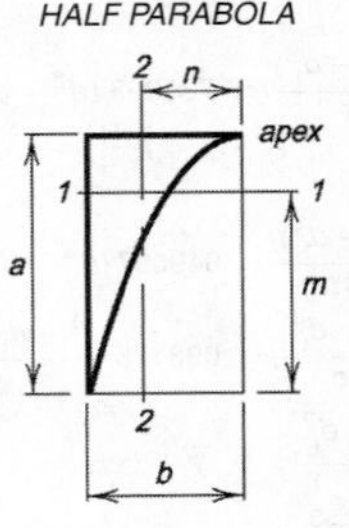

$$A = \frac{1}{3}ab$$

$$m = \frac{7}{10}a$$

$$n = \frac{3}{4}b$$

$$I_1 = \frac{37}{2{,}100}a^3b$$

$$I_2 = \frac{1}{80}ab^3$$

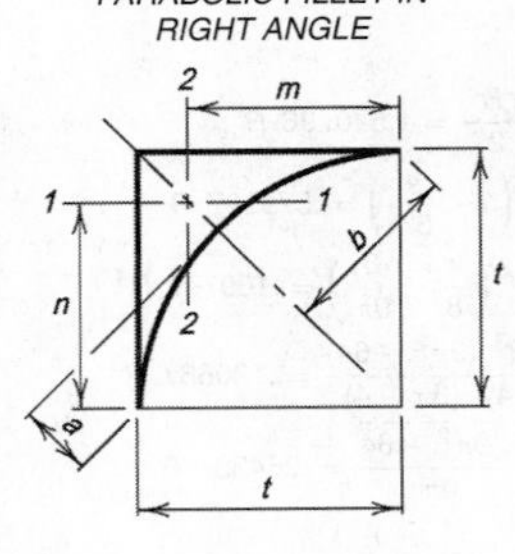

$$a = \frac{t}{2\sqrt{2}}$$

$$b = \frac{t}{\sqrt{2}}$$

$$A = {}^1/_6\,t^2$$

$$m = n = {}^4/_5\,t$$

$$I_1 = I_2 = \frac{11}{2{,}100}t^4$$

Properties of Geometric Sections (cont.)

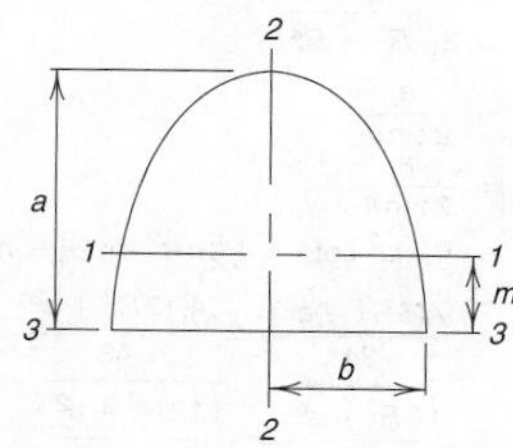

$$A = {}^1\!/_2 \pi ab$$

$$m = \frac{4a}{3\pi}$$

$$I_1 = a^3 b \left(\frac{\pi}{8} - \frac{8}{9\pi} \right)$$

$$I_2 = {}^1\!/_8 \pi a b^3$$

$$I_3 = {}^1\!/_8 \pi a^3 b$$

*QUARTER ELLIPSE

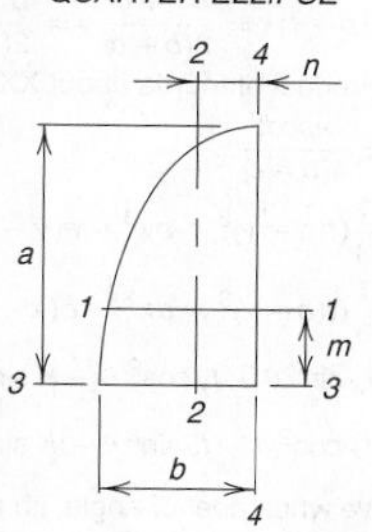

$$A = \frac{1}{4} \pi ab$$

$$m = \frac{4a}{3\pi}$$

$$n = \frac{4b}{3\pi}$$

$$I_1 = a^3 b \left(\frac{\pi}{16} - \frac{4}{9\pi} \right)$$

$$I_2 = a b^3 \left(\frac{\pi}{16} - \frac{4}{9\pi} \right)$$

$$I_3 = {}^1\!/_{16} \pi a^3 b$$

$$I_4 = {}^1\!/_{16} \pi a b^3$$

*ELLIPTIC COMPLEMENT

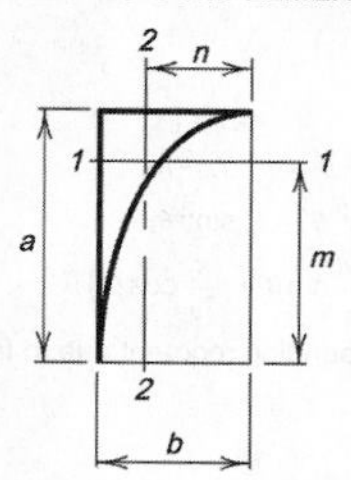

$$A = ab \left(1 - \frac{\pi}{4} \right)$$

$$m = \frac{a}{6 \left(1 - \frac{\pi}{4} \right)}$$

$$n = \frac{b}{6 \left(1 - \frac{\pi}{4} \right)}$$

$$I_1 = a^3 b \left(\frac{1}{3} - \frac{\pi}{16} - \frac{1}{36 \left(1 - \frac{\pi}{4} \right)} \right)$$

$$I_2 = a b^3 \left(\frac{1}{3} - \frac{\pi}{16} - \frac{1}{36 \left(1 - \frac{\pi}{4} \right)} \right)$$

*To obtain properties of half circle, quarter circle, and circular complement, substitute $a = b = R$.

Properties of Geometric Sections (cont.)

REGULAR POLYGON

Axis of moments through center

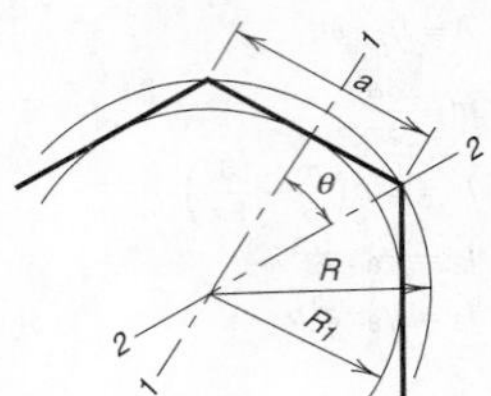

n = Number of sides

$$\theta = \frac{180°}{n}$$

$$a = 2\sqrt{R^2 - R_1^2}$$

$$R = \frac{a}{2\sin\theta}$$

$$R_1 = \frac{a}{2\tan\theta}$$

$$A = {}^1\!/_4 na^2 \cot\theta = {}^1\!/_2 nR^2 \sin 2\theta = nR_1^2 \tan\theta$$

$$I_1 = I_2 = \frac{A(6R^2 - a^2)}{24} = \frac{A(12R_1^2 + a^2)}{48}$$

$$r_1 = r_2 = \sqrt{\frac{6R^2 - a^2}{24}} = \sqrt{\frac{12R_1^2 + a^2}{48}}$$

ANGLE

Axis of moments through center of gravity

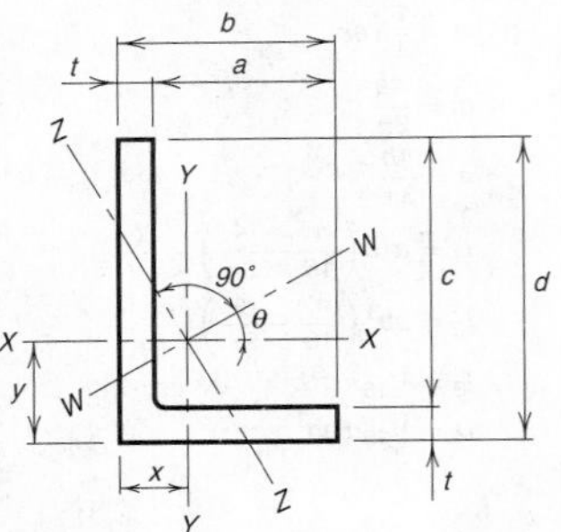

$$\tan 2\theta = \frac{2K}{I_y - I_x}$$

$$A = t(b+c) \quad x = \frac{b^2 + ct}{2(b+c)} \quad y = \frac{d^2 + at}{2(b+c)}$$

K = Product of Inertia about XX and YY

$$= \pm\frac{abcdt}{4(b+c)}$$

$$I_x = \frac{1}{3}(t(d-y)^3 + by^3 - a(y-t)^3)$$

$$I_y = \frac{1}{3}(t(b-x)^3 + dx^3 - c(x-t)^3)$$

$$I_z = I_x \sin^2\theta + I_y \cos^2\theta + K \sin 2\theta$$

$$I_w = I_x \cos^2\theta + I_y \sin^2\theta - K \sin 2\theta$$

K is negative when heel of angle, ith respect to center of gravity, is in 1st or 3rd quadrant, postitive when in 2nd or 4th quadrant.

BEAMS AND CHANNELS

Transverse force oblique through center of gravity

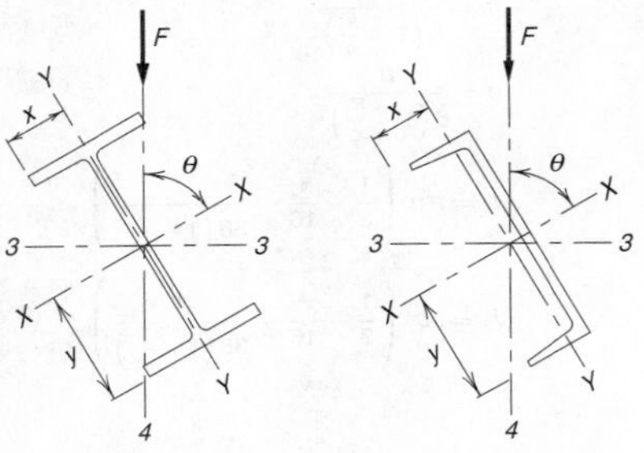

$$I_3 = I_x \sin^2\theta + I_y \cos^2\theta$$

$$I_4 = I_x \cos^2\theta + I_y \sin^2\theta$$

$$f_b = M\left(\frac{y}{I_x}\sin\theta + \frac{x}{I_y}\cos\theta\right)$$

where M is bending moment due to force F.

Trigonometric Formulas

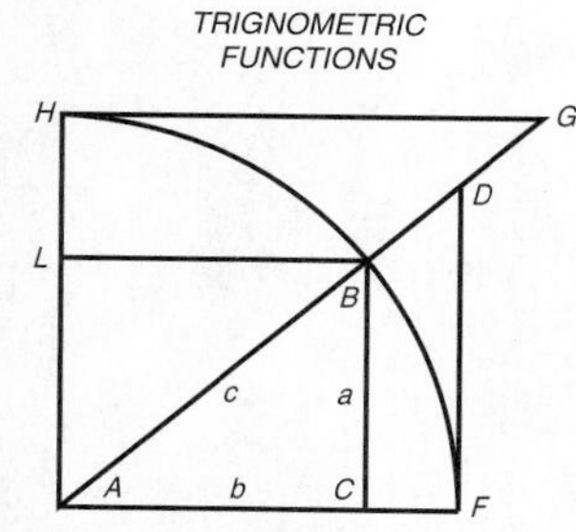

$$\text{Radius } AF = 1 = \sin^2 A + \cos^2 A = \sin A \operatorname{cosec} A = \cos A \sec A = \tan A \cot A$$

$$\sin A = \frac{\cos A}{\cot A} = \frac{1}{\operatorname{cosec} A} = \cos A \tan A = \sqrt{1-\cos^2 A} = BC$$

$$\cos A = \frac{\sin A}{\tan A} = \frac{1}{\sec A} = \sin A \cot A = \sqrt{1-\sin^2 A} = AC$$

$$\tan A = \frac{\sin A}{\cos A} = \frac{1}{\cot A} = \sin A \sec A = FD$$

$$\cot A = \frac{\cos A}{\sin A} = \frac{1}{\tan A} = \cos A \operatorname{cosec} A = HG$$

$$\sec A = \frac{\tan A}{\sin A} = \frac{1}{\cos A} = AD$$

$$\operatorname{cosec} A = \frac{\cot A}{\cos A} = \frac{1}{\sin A} = AG$$

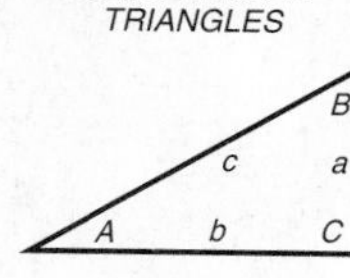

$$a^2 = c^2 - b^2$$
$$b^2 = c^2 - a^2$$
$$c^2 = a^2 + b^2$$

Known	Required					
	A	B	a	b	c	Area
a, b	$\tan A = \frac{a}{b}$	$\tan B = \frac{b}{a}$			$\sqrt{a^2+b^2}$	$\frac{ab}{2}$
a, c	$\sin A = \frac{a}{c}$	$\cos B = \frac{a}{c}$		$\sqrt{c^2-a^2}$		$\frac{a\sqrt{c^2-a^2}}{2}$
A, a		$90° - A$		$a \cot A$	$\frac{a}{\sin A}$	$\frac{a^2 \cot A}{2}$
A, b		$90° - A$	$b \tan A$		$\frac{b}{\cos A}$	$\frac{b^2 \tan A}{2}$
A, c		$90° - A$	$c \sin A$	$c \cos A$		$\frac{c^2 \sin 2A}{4}$

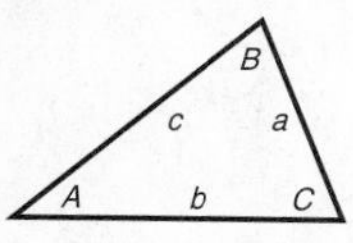

$$s = \frac{a+b+c}{2}$$
$$K = \sqrt{\frac{(s-a)(s-b)(s-c)}{s}}$$

$$a^2 = b^2 + c^2 - 2bc \cos A$$
$$b^2 = a^2 + c^2 - 2ac \cos B$$
$$c^2 = a^2 + b^2 - 2ab \cos C$$

Known	Required					
	A	B	C	b	c	Area
a, b, c	$\tan \frac{1}{2} A = \frac{K}{s-a}$	$\tan \frac{1}{2} B = \frac{K}{s-b}$	$\tan \frac{1}{2} C = \frac{K}{s-c}$			$\sqrt{s(s-a)(s-b)(s-c)}$
a, A, B			$180° - (A+B)$	$\frac{a \sin B}{\sin A}$	$\frac{a \sin C}{\sin A}$	
a, b, A		$\sin B = \frac{b \sin A}{a}$			$\frac{b \sin C}{\sin B}$	
a, b, C	$\tan A = \frac{a \sin C}{b - a \cos C}$				$\sqrt{a^2+b^2-2ab \cos C}$	$\frac{ab \sin C}{2}$

NOMENCLATURE

A	Area of directly connected elements
A	Cross-sectional area, in.^{2}
A	Horizontal distance from end panel point to mid-span of a truss, ft.
A	Minimum side dimension for square or rectangular beveled washer, in.
A_B	Loaded area of concrete, in.^{2}
A_{BM}	Cross-sectional area of base metal for a welded joint, in.^{2}
A_b	Nominal unthreaded body area of bolt or threaded part, in.^{2}
A_c	Area of concrete, in.^{2}
A_c	Area of concrete in a composite column, in.^{2}
A_c	Area of concrete slab within effective width, in.^{2}
A_{cp}	Projected surface area of concrete cone surrounding headed anchor rods, in.^{2}
A_D	Area of an upset rod based on the major thread diameter, in.^{2}
A_e	Effective area, in.^{2}
A_f	Area of flange, in.^{2}
A_{fe}	Effective tension flange area, in.^{2}
A_{fg}	Gross area of flange, in.^{2}
A_{fn}	Net area of flange, in.^{2}
A_g	Gross area, in.^{2}
A_{gt}	Gross area subject to tension, in.^{2}
A_{gv}	Gross area subject to shear, in.^{2}
A_n	Net area, in.^{2}
A_{nt}	Net area subject to tension, in.^{2}
A_{nv}	Net area subject to shear, in.^{2}
A_{pb}	Projected bearing area, in.^{2}
A_r	Area of reinforcing bars, in.^{2}
A_s	Area of steel cross section, in.^{2}
A_{sc}	Cross-sectional area of shear stud connector, in.^{2}
A_{sf}	Shear area on the failure path, in.^{2}
A_{st}	Cross-sectional area of stiffener or pair of stiffeners, in.^{2}
A_t	Net tensile area, in.^{2}
A_v	Seismic coefficient representing the effective peak velocity-related acceleration
A_w	Area of web clear of flanges, in.^{2}
A_w	Effective area of weld, in.^{2}
A_1	Area of steel bearing concentrically on a concrete support (See LRFD Specification Section J9), in.^{2}
A_2	Total cross-sectional area of a concrete support (See LRFD Specification Section J9), in.^{2}
B	Factor for bending stress in tees and double angles
B	Factor for bending stress in web-tapered members, in., defined by Equations A-F3-8 through A-F3-11
B	Horizontal distance from mid-span of a truss to a given panel point, ft.
B	Base plate width, in.
B_1, B_2	Factors used in determining M_u for combined bending and axial forces when first-order analysis is employed
BF	A factor that can be used to calculate the flexural strength for unbraced length L_b between L_p and L_r, defined in Part 5
C	Required mid-span camber, in.
C	Width across points of square or hex bolt head or nut, or maximum diameter of countersunk bolt head, in.
C	Coefficient for eccentrically loaded bolt and weld groups

C_{PG} Plate girder coefficient

C_{Tot} Sum of compressive forces in a composite beam, kips

C_a, C_b Coefficients used in extended end-plate connection design

C_b Bending coefficient dependent upon moment gradient

C_c Beam reaction coefficient (Part 5)

C_{con} Effective concrete flange force for a composite beam, kips

C_m Coefficient applied to bending term in interaction formula for prismatic members and dependent upon column curvature caused by applied moments

C_m' Coefficient applied to bending term in interaction formula for tapered members and dependent upon axial stress at the small end of the member

C_p Ponding flexibility coefficient for primary member in a flat roof

C_s Ponding flexibility coefficient for secondary member in a flat roof

C_s Seismic response factor related to the fundamental period of the building

C_{stl} Compressive force in steel in a composite beam, kips

C_v Ratio of ìcriticalî web stress, according to linear buckling theory, to the shear yield stress of web material

C_w Warping constant, $in.^6$

C_1 Loading constant used in deflection calculations (Part 5)

C_1 Clearance for tightening, in. (see Tables 7-3)

C_1 Electrode coefficient for relative strength of electrodes where, for E70 electrodes, C1 = 1.00 (see Table 8-4)

C_2 Clearance for entering, in. (see Tables 7-3)

C_3 Clearance for fillet based on one standard hardened washer, in. (see Tables 7-3)

CG Center of gravity

D Outside diameter of circular hollow section, in.

D Dead load, due to the weight of the structural elements and permanent features on the structure

D Factor used in LRFD Specification Equation A-G4-2, dependent on the type of transverse stiffeners used in a plate girder

D Offset from the base line at a panel point of a truss, in.

D Damping in percent of critical

D Slip probability factor for bolts

D Number of sixteenths-of-an-inch in the weld size

DLF Dynamic load factor

E Modulus of elasticity of steel (E = 29,000 ksi)

E Earthquake load

E Minimum edge distance for clipped washer, in.

E Minimum effective throat thickness for partial-joint-penetration groove weld, in.

E_c Modulus of elasticity of concrete, ksi

E_m Modified modulus of elasticity for the design of composite columns, ksi

ENA Elastic neutral axis

F Width across flats of bolt head, in.

F Clearance for tightening staggered bolts, in. (see Tables 7-3)

F_{BM} Nominal strength of the base material to be welded, ksi

F_{EXX} Classification number of weld metal (minimum specified strength), ksi

F_L Smaller of $(F_{yf} - F_r)$ or F_{yw}, ksi

F_{SR} Design stress range, ksi

F_{TH} Threshold fatigue stress range, maximum stress range for indefinite design life, ksi

$F_{b\gamma}$ Flexural stress for tapered members defined by Equations A-F3-4 and A-F3-5

F_{cr} Critical stress, ksi

F_{crft}, F_{cry}, F_{crz}	Flexural-torsional buckling stresses for double-angle and tee-shaped compression members, ksi
F_e	Elastic buckling stress, ksi
F_{ex}	Elastic flexural buckling stress about the major axis, ksi
F_{ey}	Elastic flexural buckling stress about the minor axis, ksi
F_{ez}	Elastic torsional buckling stress, ksi
F_{my}	Modified yield stress for the design of composite columns, ksi
F_n	Nominal shear rupture strength, ksi
F_n, F_{nt}	Nominal strength of bolt, ksi
F_p	Nominal bearing stress on fastener, ksi
F_r	Compressive residual stress in flange [10 ksi for rolled shapes; 16.5 ksi for welded built-up shapes]
$F_{s\gamma}$	Stress for tapered members defined by LRFD Specification Equation A-F3-6, ksi
F_t	Nominal tensile strength of bolt from LRFD Specification Table J3.2, ksi
F_u	Specified minimum tensile strength of the type of steel being used, ksi
F_v	Nominal shear strength of bolt from LRFD Specification Table J3.2, ksi
F_w	Nominal strength of the weld electrode material, ksi
$F_{w\gamma}$	Stress for tapered members defined by Equation A-F3-7, ksi
F_y	Specified minimum yield stress of the type of steel being used, ksi. As used in the LRFD Specification, ìyield stressî denotes either the specified minimum yield point (for steels that have a yield point) or specified yield strength (for steels that do not have a yield point)
F_y'''	The theoretical maximum yield stress (ksi) based on the web depth-thickness ratio (h / t_w) above which the web of a column is considered a ìslender elementî (See LRFD Specification Table B5.1) $= \left(\frac{253}{h/t_w}\right)^2$ Note: In the tables, ó indicates $F_y''' > 65$ ksi.
F_{yb}	F_y of a beam, ksi
F_{yc}	F_y of a column, ksi
F_{yf}	Specified minimum yield stress of the flange, ksi
F_{yr}	Specified minimum yield stress of reinforcing bars, ksi
$F_{y\,st}$	Specified minimum yield stress of the stiffener material, ksi
F_{yw}	Specified minimum yield stress of the web, ksi
G	Shear modulus of elasticity of steel (G = 11,200 ksi)
G	Ratio of the total column stiffness framing into a joint to that of the stiffening members framing into the same joint
H	Horizontal force, kips
H	Flexural constant
H	Average story height
H	Height of bolt head or nut, in.
H	Theoretical thread height, in. (see Table 7-4)
H_s	Length of shear stud connector after welding, in.
H_1	Height of bolt head, in. (see Tables 7-3)
H_2	Maximum bolt shank extension based on one standard hardened washer, in. (see Tables 7-3)
I	Moment of inertia, in.^{4}
I_{LB}	Lower bound moment of inertia for composite section, in.^{4}
I_c	Moment of inertia of column section about axis perpendicular to plane of buckling, in.^{4}
I_d	Moment of inertia of the steel deck supported on secondary members, in.^{4}
I_g	Moment of inertia of girder about axis perpendicular to plane of buckling, in.^{4}
I_p	Moment of inertia of primary member, in.^{4}

I_p Polar moment of inertia of bolt and weld groups ($I_p = I_x + I_y$), in.^{4} per in.^{2}
I_s Moment of inertia of secondary member, in.^{4}
I_{st} Moment of inertia of a transverse stiffener, in.^{4}
I_t Transformed moment of inertia of the composite section, in.^{4}
I_x Moment of inertia of bolt and weld groups about X-axis, in.4 per in.^{2}
I_y Moment of inertia of bolt and weld groups about Y-axis, in.4 per in.^{2}
I_{yc} Moment of inertia about y-axis referred to compression flange, or if reverse curvature bending referred to smaller flange, in.^{4}
IC Instantaneous center of rotation
ID Nominal inside diameter of flat circular washer, in.
J Torsional constant for a section, in.^{4}
K Effective length factor for a prismatic member
K Minimum root diameter of threaded fastener, in. (see Table 7-4)
K_{dep} Fillet depth, $(k - t_f)$, in.
K_i' Modified effective length factor of a column
K_z Effective length factor for torsional buckling
K_γ Effective length factor for a tapered member
L Unbraced length of member measured between the centers of gravity of the bracing members, in. or ft, as indicated
L Span length, ft
L Length of connection in the direction of loading, in.
L Story height or panel spacing, in.
L Live load due to occupancy and moveable equipment
L Edge distance or center-to-center distance for holes, in.
L′ Total live load
L_b Laterally unbraced length; length between points that are either braced against lateral displacement of the compression flange or braced against twist of the cross section, in. or ft, as indicated
L_c Length of channel shear connector, in.
L_c Unsupported length of a column section, ft
L_e Edge distance, in.
L_{eh} Horizontal edge distance, in.
L_{ev} Vertical edge distance, in.
L_g Unsupported length of a girder or other restraining member, ft
L_h Hook length for hooked anchor rods, in.
L_m Limiting laterally unbraced length for full plastic flexural strength ($C_b > 1.0$), in. or ft, as indicated
L_m' Limiting laterally unbraced length for the maximum design flexural strength for noncompact shapes ($C_b > 1.0$), in. or ft, as indicated
L_p Column spacing in direction of girder, ft
L_p Limiting laterally unbraced length for full plastic flexural strength, uniform moment case ($C_b = 1.0$), in. or ft, as indicated
L_p' Limiting laterally unbraced length for the maximum design flexural strength for noncompact shapes, uniform moment case ($C_b = 1.0$), in. or ft, as indicated
L_{pd} Limiting laterally unbraced length for plastic analysis, in. or ft, as indicated
L_q Maximum unbraced length for the required column force with *K* equal to one, in.
L_r Limiting laterally unbraced length for inelastic lateral-torsional buckling, in. or ft, as indicated
L_r Roof live load
L_s Column spacing perpendicular to direction of girder, ft
M Beam bending moment, kip-in. or kip-ft, as indicated
M_A Absolute value of moment at quarter point of the unbraced beam segment, kip-in.

M_B	Absolute value of moment at centerline of the unbraced beam segment, kip-in.
M_C	Absolute value of moment at three-quarter point of the unbraced beam segment, kip-in.
M_{LL}	Beam moment due to live load, kip-in. or kip-ft, as indicated
M_T	Applied torsional moment, kip-in.
M_{cr}	Elastic buckling moment, kip-in. or kip-ft, as indicated
M_{eu}	Required flexural strength for extended end-plate connections, kip-in.
M_{lt}	Required flexural strength in member due to lateral frame translation, kip-in.
M_{max}	Maximum bending moment, kip-in. or kip-ft, as indicated
M_{max}	Absolute value of maximum moment in the unbraced beam segment, kip-in.
M_n	Nominal flexural strength, kip-in. or kip-ft, as indicated
M_p'	Maximum design flexural strength for noncompact shapes, when $L_b \leq L_p'$, kip-in. or kip-ft, as indicated
M_{nt}	Required flexural strength in member assuming there is no lateral translation of the frame, kip-in.
M_{nx}', M_{ny}'	Flexural strength defined in LRFD Specification Equations A-H3-7 and A-H3-8, for use in the alternate interaction equations for combined bending and axial force, kip-in. or kip-ft, as indicated
M_p	Plastic bending moment, kip-in. or kip-ft, as indicated
M_p'	Moment defined in LRFD Specification Equations A-H3-5 and A-H3-6, for use in the alternate interaction formulas for combined bending and axial force, kip-in. or kip-ft, as indicated
M_{pa}	Plastic bending moment modified by axial load ratio, kip-in.
M_r	Limiting buckling moment, M_{cr}, when $\lambda = \lambda_r$ and $C_b = 1.0$, kip-in. or kip-ft, as indicated
M_u	Required flexural strength, kip-in. or kip-ft, as indicated
M_y	Moment corresponding to onset of yielding at the extreme fiber from an elastic stress distribution (= F_yS for homogeneous sections), kip-in.
M_{ob}	Elastic lateral-torsional buckling moment, kip-in. or kip-ft, as indicated
M_1	Smaller moment at end of unbraced length of beam or beam-column, kip-in.
M_2	Larger moment at end of unbraced length of beam or beam-column, kip-in.
N	Length of bearing, in.
N	Ratio of the factored gravity load supported by all columns in a story to that supported by the columns in the rigid frame
N	Length of base plate, in.
N	Number of stress range fluctuations in design life
N_b	Number of bolts in a joint
N_r	Number of shear stud connectors in one rib at a beam intersection.
N_s	Number of slip planes
OD	Nominal outside diameter of flat circular washer, in.
P	Concentrated load, kips
P	Bolt stagger, in.
P	Thread pitch, in. (see Table 8-7)
P_D	Unfactored dead load, kips
P_E	Unfactored earthquake load, kips
P_L	Unfactored live load, kips
P_S	Unfactored snow load, kips
P_{bf}	Applied factored beam flange force in moment connections, kips
P_{br}	Required story or panel bracing shear force, kip
P_{e1}, P_{e2}	Elastic Euler buckling load for braced and unbraced frame, respectively, kips
P_{fb}	Resistance to flange local bending per LRFD Specification Equation K1-1 (used to check need for column web stiffeners), kips
P_n	Nominal axial strength (tension or compression), kips

P_p	Bearing load on concrete, kips
P_u	Factored concentrated beam load, kips
P_u	Required axial strength (tension or compression), kips
$P_u e$	Induced moment due to eccentricity e in an eccentrically loaded bolt or weld group, kip-in.
P_{uf}	Factored beam flange force, tensile or compressive, kips
P_{wb}	Resistance to web compression buckling per LRFD Specification Equation K1-8 (used to check need for column web stiffening), kips
P_{wi}	A factor consisting of terms from the second portion of LRFD Specification Equation K1-2 (used in a column web stiffener check for web local yielding), kips/in.
P_{wo}	A factor consisting of the first portion of LRFD Specification Equation K1-2 (used in a column web stiffener check for web local yielding), kips
P_y	Yield strength, kips
PNA	Plastic neutral axis
Q	Full reduction factor for slender compression elements
Q_a	Reduction factor for slender stiffened compression elements
Q_f	Statical moment for a point in the flange directly above the vertical edge of the web, in.^{3}
Q_i	Load effects
Q_n	Nominal strength of one stud shear connector, kips
Q_s	Reduction factor for slender unstiffened compression elements
Q_w	Statical moment at mid-depth of the section, in.^{3}
R	Nominal load due to initial rainwater or ice exclusive of the ponding contribution
R	Nominal reaction, kips
R	Earthquake response modification coefficient
R_{PG}	Plate girder bending strength reduction factor
R_e	Hybrid girder factor
R_n	Nominal strength, kips
R_s	Nominal slip resistance of a bolt, kips
R_u	Required strength determined from factored loads; must be less than or equal to design strength ϕR_n
$R_{u\,st}$	Required strength for transverse stiffener (factored force delivered to stiffener), kips
R_v	Web shear strength, kips
R_1	An expression consisting of the first portion of LRFD Specification Equation K1-3, kips
R_2	An expression consisting of terms from the second portion of LRFD Specification Equation K1-3, kips/in.
R_3	An expression consisting of the first portion of LRFD Specification Equation K1-5a, kips
R_4	An expression consisting of terms from the second portion of LRFD Specification Equation K1-5a, kips/in.
R_5	An expression consisting of terms from LRFD Specification Equation K1-5b, kips
R_6	An expression consisting of terms from LRFD Specification Equation K1-5b, kips/in.
S	Elastic section modulus, in.^{3}
S	Spacing, in. or ft, as indicated
S	Snow load
S	Groove depth for partial-joint-penetration groove welds, in.
S'	Additional elastic section modulus corresponding to $^1/_{16}$-in. increase in web thickness for built-up wide flange sections, in.^{3}
S_c	Elastic section modulus to the tip of the angle in compression, in.^{3}

S_{eff}	Effective section modulus about major axis, in.^{3}
S_{net}	Net elastic section modulus, in.^{3}
S_w	Warping statical moment at a point on the cross section, in.^{4}
S_x	Elastic section modulus about major axis, in.^{3}
S_x'	Elastic section modulus of larger end of tapered member about its major axis, in.^{3}
S_{xt}, S_{xc}	Elastic section modulus referred to tension and compression flanges, respectively, in.^{3}
SRF	Stiffness reduction factors (Table 4-1), for use with the alignment charts (LRFD Specification Figure C-C2.2) in the determination of effective length factors *K* for columns
T	Distance between web toes of fillets at top and at bottom of web, in. = d - $2k$
T	Tension force due to service loads, kips
T	Thickness of flat circular washer or mean thickness of square or rectangular beveled washer, in.
T	Unfactored tensile force on slip-critical connections designed at service loads, kips
T_b, T_m	Specified pretension load in high-strength bolt (LRFD Specification Table J3.1), kips
T_{stl}	Tensile force in steel in a composite beam, kips
T_{Tot}	Sum of tensile forces in a composite beam, kips
T_u	Required tensile strength due to factored loads, kips
U	Reduction coefficient, used in calculating effective net area
V	Shear force, kips
V_b	Shear force component, kips
V_h	Total horizontal force transferred by the shear connections, kips
V_n	Nominal shear strength, kips
V_u	Required shear strength, kips
W	Wind load
W	Uniformly distributed load, kips
W	Weight, lbs or kips, as indicated
W	Width across flats of nut, in.
W_c	Uniform load constant for beams, kip-ft
W_{no}	Normalized warping function at a point at the flange edge, in.^{2}
W_u	Total factored uniformly distributed load, kips
Workable Gage	Gage for fasteners in flange (Part 1) that provides for entering and tightening clearances and edge distance and spacing requirements, in. When the listed value is shaded, the actual size, combination and orientation of fastener components should be compared with the geometry of the cross-section to ensure compatibility. Other gages that provide for entering and tightening clearances and edge distance and spacing requirements can also be used.
X_1	Beam buckling factor defined by LRFD Specification Equation F1-8
X_2	Beam buckling factor defined by LRFD Specification Equation F1-9
Y_{ENA}	Distance from bottom of steel beam to elastic neutral axis, in.
Y_{con}	Distance from top of steel beam to top of concrete, in.
Y_1	Distance from top of steel beam to the plastic neutral axis, in.
Y_2	Distance from top of steel beam to the concrete flange force in a composite beam, in.
Z	Plastic section modulus, in.^{3}
Z′	Additional plastic section modulus corresponding to $^1/_{16}$-inch increase in web thickness for built-up wide flange section, in.^{3}
Z_e	Effective plastic section modulus, in.^{3}
a	Clear distance between transverse stiffeners, in.
a	Distance between connectors in a built-up member, in.

a	Effective concrete flange thickness of a composite beam, in.
a	Shortest distance from edge of pinhole to edge of member measured parallel to direction of force, in.
a	Coefficient for eccentrically loaded weld group
a	Distance from bolt centerline to edge of fitting subjected to prying action, but not greater than 1.25*b*, in.
a	Depth of bracket plate, in.
a_r	Ratio of web area to compression flange area
a'	Weld length, in.
b	Compression element width, in.
b	Effective concrete flange width in a composite beam, in.
b	Width of composite column section, in.
b	Minimum shelf dimension for deposition of fillet weld, in.
b	Width of bracket plate, in.
b	Distance from bolt centerline to face of fitting subjected to prying action, in.
b	Coefficient related to axial compressive strength for beam-column design (Part 6), $(\text{kips})^{-1}$
b_e	Reduced effective width for slender compression elements, in.
b_{eff}	Effective edge distance, in.
b_f	Flange width of rolled beam or plate girder, in.
b_s	Stiffener width for one-sided stiffeners, in.
b_s	Width of transverse stiffener, in.
b_s	Width of extended end-plate, in.
c	Distance from the neutral axis to the extreme fiber of the cross section, in.
c	Cope length, in.
c_1, c_2, c_3	Numerical coefficients
d	Nominal fastener diameter, in.
d	Overall depth of member, in.
d	Pin diameter, in.
d	Roller diameter, in.
d_L	Depth at larger end of unbraced tapered segment, in.
d_b	Nominal bolt diameter, in.
d_c	Column depth, in.
d_c	Cope depth, in.
d_{ct}	Top-flange cope depth, in.
d_{cb}	Bottom-flange cope depth, in.
d_h	Hole diameter, in.
d_m	Moment arm between resultant tensile and compressive forces due to a moment or eccentric force, in.
d_z	Overall panel-zone depth, in.
d_0	Depth at smaller end of unbraced tapered segment, in.
e	Eccentricity, in.
e	Base of natural logarithms = 2.71828...
e_o	Horizontal distance from the outer edge of a channel web to its shear center, in.
f	Computed compressive stress in the stiffened element, ksi
f	Plate buckling model adjustment factor for beams coped at top flange only
f_b	Maximum bending stress, ksi
f_{b1}	Smallest computed bending stress at one end of a tapered segment, ksi
f_{b2}	Largest computed bending stress at one end of a tapered segment, ksi
f_c'	Specified compressive strength of concrete, ksi
f_d	Adjustment factor for beams coped at both flanges
f_{un}	Required normal stress, ksi
f_{uv}	Required shear stress, ksi
f_v	Computed shear stress, ksi

f_o	Stress due to $1.2D + 1.2R$, ksi
g	Transverse center-to-center spacing (gage) between fastener gage lines, in.
g	Acceleration due to gravity = 32.2 ft/sec^2 = 386 in./sec^2
h	Clear distance between flanges less the fillet or corner radius for rolled shapes; and for built-up sections, the distance between adjacent lines of fasteners or the clear distance between flanges when welds are used, in.
h	Depth of composite column section, in.
h	Distance between centroids of individual components perpendicular to the member axis of buckling, in.
h_c	Twice the distance from the centroid to the following: the inside face of the compression flange less the fillet or corner radius, for rolled shapes; the nearest line of fasteners at the compression flange or the inside faces of the compression flange when welds are used, for built-up sections, in.
h_o	Distance between flange centroids, in.
h_r	Nominal rib height, in.
h_s	Factor used in LRFD Specification Equation A-F3-6 for web-tapered members
h_w	Factor used in LRFD Specification Equation A-F3-7 for web-tapered members
h_o	Remaining web depth of coped beam, in.
j	Factor defined by LRFD Specification Equations A-F2-4 for minimum moment of inertia for a transverse stiffener
k	Distance from outer face of flange to web toe of fillet, in.
k	Slenderness parameter
k	Plate buckling coefficient for beams coped at top flange only
k_s	Bolt slip coefficient
k_v	Web plate buckling coefficient
k_1	Distance from web center line to flange toe of fillet, in.
l	Laterally unbraced length of member at the point of load, in.
l	Span length, in.
l	Length of bearing, in.
l	Length of connection in the direction of loading, in.
l	Length of weld, in.
l	Characteristic length of weld group (see Tables 8-5 through 8-12), in.
l_o	Distance from center of gravity (CG) to instantaneous center of rotation (IC) of bolt or weld group, in.
m	Ratio of web to flange yield stress or critical stress in hybrid beams
m	Coefficient related to strong-axis flexural strength for beam-column design (Part 6), (kip-ft)$^{-1}$
m	Cantilever dimension for base plate (see Part 14), in.
n	Number of shear connectors between point of maximum positive moment and the point of zero moment to each side
n	Number of bolts in a vertical row
n	Number of threads per inch on threaded fasteners
n	Coefficient related to weak-axis flexural strength for beam-column design (Part 6), (kip-ft)$^{-1}$
n	Cantilever dimension for base plate (see Part 14), in.
n'	Number of bolts above the neutral axis (in tension)
p	Length of supporting flange parallel to stem or leg of hanger tributary to each bolt in determining prying action, in.
p_e	Effective span used to compute M_{eu} for extended end-plate connections, in.
p_f	Distance from centerline of bolt to nearer surface of tension flange in extended end-plate connections, in.
q_u	Additional tension per bolt resulting from prying action produced by deformation of the connected parts, kips/bolt

r	Governing radius of gyration, in.
r_T	Radius of gyration of compression flange plus one third of the compression portion of the web taken about an axis in the plane of the web, in.
r_{To}	For the smaller end of a tapered member, the radius of gyration, considering only the compression flange plus one-third of the compression web area, taken about an axis in the plane of the web, in.
r_i	Minimum radius of gyration of individual component in a built-up member, in.
r_{ib}	Radius of gyration of individual component relative to centroidal axis parallel to member axis of buckling, in.
r_m	Radius of gyration of steel shape, pipe, or tubing in composite columns. For steel shapes it may not be less than 0.3 times the overall thickness of the composite section, in.
r_n	Nominal strength per bolt from LRFD Specification
r_x, r_y	Radius of gyration about *x* and *y* axes respectively, in.
r_{ut}	Required tensile strength per bolt or per inch of weld (factored force per bolt or per inch of weld due to a tensile force), kips/bolt
r_{ut}	Required shear strength per bolt or per inch of weld (factored force per bolt or per inch of weld due to a shear force), kips/bolt
r_{yc}	Radius of gyration about *y* axis referred to compression flange, or if reverse curvature bending, referred to smaller flange, in.
$\bar{r}_o$	Polar radius of gyration about the shear center, in.
r_{ox}, r_{oy}	Radius of gyration about *x* and *y* axes at the smaller end of a tapered member respectively, in.
s	Longitudinal center-to-center spacing (pitch) of any two consecutive holes, in.
s	Bolt spacing, in.
t	Thickness of element, in.
t	HSS design wall thickness, in.
t	Change in temperature, degrees Fahrenheit or Celsius, as indicated
t_b	Thickness of beam flange or connection plate delivering concentrated force, in.
t_c	Flange or angle thickness required to develop design tensile strength of bolts with no prying action, in.
t_e	Total required effective thickness of column web with doubler plate, in.
t_f	Flange thickness, in.
t_f	Flange thickness of channel shear connector, in.
t_p	Thickness of base plate, in.
t_p	Panel zone thickness including doubler plates, in.
$t_{p\,req}$	Required doubler plate thickness, in.
t_s	Extended end-plate thickness, in.
t_s	Web stiffener thickness, in.
t_w	Web thickness, in.
t_w	Web thickness of channel shear connector, in.
t_{wb}	Beam web thickness, in.
t_{wc}	Column web thickness, in.
t_z	Panel zone thickness, in.
w	Uniformly distributed load per unit of length, kips/in.
w	Leg size of the fillet weld, in.
w	Plate width; distance between welds, in.
w	Subscript relating symbol to strong principal axis of angle
w	Unit weight of concrete, lbs/ft^3
w_r	Average width of concrete rib or haunch, in.
w_z	Panel zone width, in.
x	Subscript relating symbol to strong axis

x	Horizontal distance, in.
$\bar{x}$	Horizontal distance from the outer edge of a channel web to its centroid, in.
$\bar{x}$	Connection eccentricity, in.
x_p	Horizontal distance from the designated edge of member to its plastic neutral axis, in.
x_o	Horizontal distance, in.
x_o, y_o	Coordinates of the shear center with respect to the centroid, in.
y	Moment arm between centroid of tensile forces and compressive forces, in.
y	Subscript relating symbol to weak axis
y_p	Vertical distance from the designated edge of member to its plastic neutral axis, in.
y_1, y_2	Vertical distance from designated edge of member to center of gravity, in.
z	Distance from the smaller end of tapered member used in LRFD Specification Equation A-F3-1 for the variation in depth, in.
z	Subscript relating symbol to member longitudinal axis
z	Subscript relating symbol to weak principal axis of angle
z	Coefficient for buckling of triangular-shaped bracket plate
Δ	Deflection, in.
Δ_{LL}	Live load deflection, in.
Δ_{oh}	Translation deflection of the story under consideration, in.
α	Separation ratio for built-up compression members, LRFD Specification Equation E4
α	Fraction of member force transferred across a particular net section
α	Ratio of moment at bolt line to moment at stem line for determining prying action in hanger connections
α	Ideal distance from face of column flange or web to centroid of gusset-to-beam connection for bracing connections and uniform force method, in.
$\bar{\alpha}$	Actual distance from face of column flange or web to centroid of gusset-to-beam connection for bracing connections and uniform force method, in.
α_m	Coefficient for calculating M_{eu} for extended end-plate connections
β	Ideal distance from face of beam flange to centroid of gusset-to-column connection for bracing connections and uniform force method, in.
β	Reduction factor given by Equation J2-1
$\bar{\beta}$	Actual distance from face of beam flange to centroid of gusset-to-column connection for bracing connections and uniform force method, in.
β_T	Brace stiffness requirement when there is no web distortion
β_{Tb}	Required nodal torsional bracing stiffness
β_{br}	Required story or panel shear stiffness
β_{sec}	Web distortional stiffness, including the effect of web transverse stiffeners, if any
β_w	Special section property for unequal-leg angles (Single-Angle Specification)
γ	Depth tapering ratio
γ	Subscript for tapered members
γ_i	Load factor
δ	Deflection, in.
δ	Ratio of net area at bolt line to gross area at face of stem or angle leg used to determine prying action for hanger connections
ε	Coefficient of linear expansion, with units as indicated
ζ	Exponent for alternate beam-column interaction equation
η	Exponent for alternate beam-column interaction equation
λ	Slenderness parameter
λ_c	Column slenderness parameter
λ_e	Equivalent slenderness parameter

λ_{eff}	Effective slenderness ratio defined by LRFD Specification Equation A-F3-2
λ_p	Limiting slenderness parameter for compact element
λ_r	Limiting slenderness parameter for noncompact element
μ	Coefficient of friction; mean slip coefficient for bolts
ϕ	Resistance factor
ϕ_b	Resistance factor for flexure
ϕ_c	Resistance factor for compression
ϕ_c	Resistance factor for axially loaded composite columns
ϕ_r	Resistance factor for compression, used in web crippling equations
ϕ_{sf}	Resistance factor for shear on the failure path
ϕ_t	Resistance factor for tension
ϕ_v	Resistance factor for shear
ϕ_w	Resistance factor for welds
ϕF_{bc}	Design buckling stress for coped beams, ksi
ϕR_n	Design strength from LRFD Specification; must equal or exceed required strength R_u
ϕr_n	Design strength per bolt or per inch of weld from LRFD Specification; must equal or exceed required strength per bolt or per inch of weld r_u
kip	1,000 pounds
ksi	Stress, kips/in.^{2}

INDEX

NOTES

NOTES

NOTES

NOTES

NOTES

NOTES

NOTES

NOTES

NOTES

NOTES

NOTES